PROF. JOHN SAVILL
DEPT. OF MEDICINE
UNIVERSITY HOSPITAL
NOTTINGHAM
NG7 2UH

OXFORD MEDICAL PUBLICATIONS

OXFORD TEXTBOOK OF CLINICAL NEPHROLOGY

EDITORS

STEWART CAMERON
Professor of Renal Medicine,
Guy's Hospital, London, UK

ALEX M. DAVISON
Consultant Renal Physician,
St James's University Hospital, Leeds, UK

JEAN-PIERRE GRÜNFELD
Professor of Nephrology,
Hôpital Necker, Paris, France

DAVID KERR
Professor of Renal Medicine,
Royal Postgraduate Medical School, London, UK

EBERHARD RITZ
Head of Division of Nephrology,
University of Heidelberg, Germany

ASSOCIATE EDITORS

A. J. REES
Professor of Nephrology,
Royal Postgraduate Medical School, London, UK

C. G. WINEARLS
Consultant Nephrologist,
Churchill Hospital, Oxford, UK

L. HERNANDO
Professor, Servicio de Nefrología,
University of Madrid, Spain

C. VAN YPERSELE DE STRIHOU
Professor of Medicine,
University of Louvain Medical School, Belgium

CLAUDIO PONTICELLI
Director, Division of Nephrology and Dialysis,
Ospedale Maggiore, Milan, Italy

TOPIC EDITORS

T. M. BARRATT
Professor of Paediatric Nephrology,
Institute for Child Health, London, UK

R. GREGER
Professor, Physiologisches Institut,
Albert Ludwig University, Freiburg, Germany

J. RITTER
Professor of Pharmacology,
Guy's Campus, United Medical and Dental Schools, London, UK

OXFORD TEXTBOOK OF CLINICAL NEPHROLOGY

Edited by

**STEWART CAMERON, ALEX M. DAVISON,
JEAN-PIERRE GRÜNFELD, DAVID KERR,
and EBERHARD RITZ**

with the assistance of

**A. J. REES, C. G. WINEARLS,
LUIS HERNANDO, CHARLES VAN YPERSELE DE STRIHOU,
CLAUDIO PONTICELLI, MARTIN BARRATT, RAINER GREGER,
and JIM RITTER**

VOLUME 1

Oxford New York Tokyo
OXFORD UNIVERSITY PRESS

Oxford University Press, Walton Street, Oxford OX2 6DP
Oxford   New York   Toronto
Delhi   Bombay   Calcutta   Madras   Karachi
Petaling Jaya   Singapore   Hong Kong   Tokyo
Nairobi   Dar es Salaam   Cape Town
Melbourne   Auckland
and associated companies in
Berlin   Ibadan

Oxford is a trade mark of Oxford University Press

Published in the United States
by Oxford University Press, New York

© The editors listed on p. ii, 1992

First published 1992
Reprinted (with corrections) 1992

All rights reserved. No part of this publication may be reproduced,
stored in a retrieval system, or transmitted, in any form or by any means,
electronic, mechanical, photocopying, recording, or otherwise, without
the prior permission of Oxford University Press

A catalogue record for this book is
available from the British Library

Library of Congress Cataloging in Publication Data
(Cataloging data available)

ISBN 0–19–261825–3 (vol 1)
ISBN 0–19–262093–2 (3 vol set)
(Available only as a three volume set)

Printed in Great Britain by
William Clowes Ltd, Beccles, Suffolk

Preface

Why another large text on nephrology? Because this one is *different*. It begins not with the anatomy of the nephron, but with the approach to the renal patient. It is intended as a text on *clinical nephrology*, of primary use to those caring for patients with renal disease. Not that we do not value the science that underlies our clinical practice—far from it. In each section the basic science relevant to the problem under discussion will be found incorporated at the appropriate point in the text for the clinician. In this area we have had the assistance of one of the foremost renal physiologists.

We deal in this book with many of the rarer renal problems and renal manifestations of systemic disease that are not dealt with in other texts—as a glance at the index and that of other similar volumes will show. A unique feature of the book is that at the end we have provided a guide to the book from the point of view of other specialist physicians—gastroenterologists, rheumatologists, neurologists, and so on—so that both they and generalists can enter the complex world of nephrology more easily. We have paid special attention to the handling of drugs by the kidney, and to the effects of drugs upon the kidney and renal tract. In this we have been assisted by our distinguished editor in clinical pharmacology.

We have tried to look at nephrology in a global context, remembering that the great majority of patients with renal diseases live in the developing world. Several chapters deal specifically with nephrology as it is seen in the tropics. We have also included chapters which deal with renal disease at the extremes of life. Paediatric nephrology has been blended into the text throughout with the assistance of our able paediatric editor, and several chapters deal with the special problems of the growing number of elderly patients with renal disease.

Finally, we hope that these volumes will be, as well as for day-to-day use when needed, a useful and pleasurable source for browsing when the pressure is off. Above all we hope that these volumes will be a literate as well as a comprehensive guide to diseases of the kidney, and diseases affecting the kidney. We thank our Associate Editors Luis Hernando, Claudio Ponticelli, Andy Rees, Charles van Ypersele de Strihou, and C. J. Winearls without whom these volumes could never have been completed.

Stewart Cameron
Alex M. Davison
Jean-Pierre Grünfeld
David Kerr
Eberhard Ritz

Dose schedules are being continually revised and new side effects recognized. Oxford University Press makes no representation, express or implied, that the drug dosages in this book are correct. For these reasons the reader is strongly urged to consult the drug company's printed instructions before administering any of the drugs recommended in this book.

Contents

Volume 1
Colour plate section follows prelims

Contributors xiii

SECTION 1 Assessment of the patient with renal disease 1

1.1 History and clinical examination
A. M. Davison and Jean-Pierre Grünfeld 3

1.2 Urinalysis and microscopy Giovanni Battista Fogazzi 16

1.3 Renal function testing J. Stewart Cameron 24

1.4 Renal function in the neonate Henrik Ekblad and Anita Aperia 50

1.5 Renal function in the elderly Juan F. Macias-Nuñez and J. Stewart Cameron 56

1.6 Visualizing the kidney

   1.6.1 Radiological tactics: rationale Thomas Sherwood 70

   1.6.2 Conventional uroradiology and contrast media Jean-François Moreau, Olivier Helenon, Joël Chabriais, and Mourad Souissi 76

   1.6.3 Renal ultrasonography (including Doppler echography) Paul A. Dubbins 86

   1.6.4 Computed tomography and magnetic resonance imaging of the kidney and upper urinary tract A. N. Dardenne and G. C. Dooms 99

   1.6.5 Renal angiography F. Joffre 111

   1.6.6 Nuclear imaging in nephrology Michael Maisey and K.E. Britton 116

1.7 Renal structure/ultrastructure

   1.7.1 The renal glomerulus—the structural basis of ultrafiltration Marlies Elger and Wilhelm Kriz 129

   1.7.2 Renal biopsy: performance and interpretation Claudio Ponticelli, M. J. Mihatsch, and Enrico Imbasciati 141

SECTION 2 Pharmacology and drug use in kidney patients 157

2.1 Drug-induced nephropathies J. P. Fillastre and M. Godin 159

2.2 Handling of drugs in kidney disease D. J. S. Carmichael 175

2.3 Action and clinical use of diuretics R. Greger and A. Heidland 197

SECTION 3 The patient with glomerular disease 225

3.1 The clinical approach to haematuria and proteinuria N. P. Mallick and Colin D. Short 227

3.2 Immune mechanisms of glomerular damage Philippe Druet and Denis Glotz 240

3.3 Fluid retention in renal disease: the genesis of renal oedema E. J. Dorhout Mees 262

3.4 Clinical consequences of the nephrotic syndrome J. Stewart Cameron 276

3.5 Minimal changes and focal segmental glomerular sclerosis Michel Broyer, Alain Meyrier, and Patrick Niaudet 298

3.6 IgA nephropathies F. Paolo Schena 339

3.7 Membranous nephropathy Pietro Zucchelli and Sonia Pasquali 370

3.8 Mesangiocapillary glomerulonephritis D. Gwyn Williams 388

3.9 Acute endocapillary glomerulonephritis Bernardo Rodriguez-Iturbe 405

3.10 Crescentic glomerulonephritis A. J. Rees and J. Stewart Cameron 418

3.11 Antiglomerular basement membrane disease Neil Turner and A. J. Rees 438

3.12 Infection-associated glomerulonephritis A. M. Davison 456

3.13 Malignancy-associated glomerular disease A. M. Davison and D. Thomson 475

3.14 Glomerular disease in the tropics K. S. Chugh and Vinay Sakhuja 486

SECTION 4 The patient with systemic disease affecting the kidney 503

4.1 The patient with diabetes mellitus José Reimão Pinto and GianCarlo Viberti 505

4.2 Amyloidosis G. K. Van der Hem and M. H. van Rijswijk 545

4.3 Kidney involvement in plasma cell dyscrasias Luigi Minetti 562

4.4 Sarcoidosis Sabine Kenouch and Jean-Philippe Mery 576

4.5 Vasculitis

   4.5.1 Pathogenesis of angiitis F. J. van der Woude 583

   4.5.2 The nephritis of Henoch-Schönlein purpura George B. Haycock 595

- 4.5.3 Systemic vasculitis *Gillian Gaskin and Charles D. Pusey* 612
- 4.6 **Systemic lupus erythematosus**
    - 4.6.1 Recent advances in the pathogenesis of systemic lupus erythematosus *Laurent Jacob, Jean-Paul Viard, and Jean-François Bach* 636
    - 4.6.2 Systemic lupus erythematosus (clinical) *Claudio Ponticelli and Giovanni Banfi* 646
- 4.7 **Scleroderma—systemic sclerosis** *Carol M. Black* 667
- 4.8 **Rheumatoid arthritis, mixed connective tissue disease, and polymyositis** *Paul Emery and D. Adu* 678
- 4.9 **Essential mixed cryoglobulinaemia** *Giuseppe D'Amico* 686
- 4.10 **Sjögren's syndrome and overlap syndromes** *Patrick J. W. Venables* 693
- 4.11 **Sickle-cell disease and the kidney** *L. W. Statius van Eps* 700

SECTION 5   The patient with tubular disease   721

- 5.1 **Fanconi syndrome** *J. Brodehl* 723
- 5.2 **Isolated defects of tubular function** *George B. Haycock* 741
- 5.3 **Renal tubular acidosis** *J. Rodriguez-Soriano* 763
- 5.4 **Bartter's syndrome** *R. O. B. Gans and S. J. Hoorntje* 782
- 5.5 **Nephrogenic diabetes insipidus** *Daniel G. Bichet* 789

INDEX

Volume 2

Contributors   xiii

SECTION 6   The patient with chronic interstitial disease   801

- 6.1 Analgesic nephropathy   *M. Molzahn and Wolfgang Pommer*   803
- 6.2 Non-steroidal anti-inflammatory drugs and the kidney   *David H. Adams and D. Adu*   819
- 6.3 Uric acid and the kidney   *R. Verberckmoes*   825
- 6.4 Heavy metals and the kidney   *Richard P. Wedeen*   837
- 6.5 Radiation   *Jerry M. Bergstein*   849
- 6.6 Lithium   *Rowan G. Walker and Priscilla Kincaid-Smith*   853
- 6.7 Balkan nephropathy   *Momir H. Polenakovic and Vladisav Stefanovic*   857

SECTION 7   The patient with water, electrolyte, or acid/base disorders   867

- 7.1 Hypo-hypernatraemia   *Patricia A. Gabow*   869
- 7.2 Hypo-hyperkalaemia   *Richard L. Tannen*   895
- 7.3 Clinical acid/base disorders   *Serafino Garella*   917

SECTION 8   The patient with acute renal insufficiency   967

- 8.1 The investigative approach to the patient with acute renal failure   *David J. Rainford and Paul E. Stevens*   969
- 8.2 Pathophysiology of acute renal failure   *Peter J. Ratcliffe*   982
- 8.3 Acute renal failure from tubular injury   *David J. Rainford and Paul E. Stevens*   1006
- 8.4 Management of acute renal failure   *Dieter Kleinknecht*   1015
- 8.5 Continuous renal replacement therapy in acute renal failure (haemofiltration and dialysis)   *Heinz-Gunter Sieberth*   1026
- 8.6 Acute renal failure complicating nephrotic syndrome   *H. A. Koomans, R. J. Hene, and E. J. Dorhout Mees*   1034
- 8.7 Haemolytic-uraemic syndrome   *Guy Neild*   1041
- 8.8 Acute renal failure due to vasculitis and glomerulonephritis   *Charles D. Pusey and A. J. Rees*   1060
- 8.9 Ischaemic renal disease   *Harry R. Jacobson and Julia A. Breyer*   1077
- 8.10 Acute renal failure from interstitial disease   *Dieter Kleinknecht and Dominique Droz*   1084
- 8.11 Acute renal failure in liver disease: clinical, diagnostic, and therapeutical aspects   *Jose M. Lopez-Novoa*   1098
- 8.12 Acute renal failure in the neonate and older child   *Alan M. Robson and J. Stewart Cameron*   1110
- 8.13 Acute renal failure in the tropics and Hantavirus disease   *R. Suvanapha, V. Sitprija, and C. van Ypersele de Strihou*   1124

SECTION 9   The patient with chronic renal failure   1147

- 9.1 The assessment of the patient with chronic renal insufficiency   *Michael J. D. Cassidy and David N. S. Kerr*   1149
- 9.2 Initial management of chronic renal failure   *P. M. ter Wee*   1173
- 9.3 Mechanisms of progression and consequences of nephron reduction   *A. M. El Nahas*   1191
- 9.4 Causes of end-stage renal failure   *Antony J. Wing*   1227
- 9.5 The uraemic syndrome   *Raymond C. Vanholder and Severin M. G. Ringoir*   1236
- 9.6 Organ involvement in uraemia
  - 9.6.1 Metabolic disorders in chronic renal failure   *Franz Schaefer and Eberhard Ritz*   1251
  - 9.6.2 Cardiological problems in uraemic patients   *W. Kramer, V. Wizemann, Alexander P. Mandelbaum, and Eberhard Ritz*   1264
  - 9.6.3 Gastrointestinal effects of chronic renal failure   *Ciaran C. Doherty*   1278
  - 9.6.4 Systemic consequences of uraemia; growth   *Cyril Chantler*   1294
  - 9.6.5 Effects of chronic renal failure on the immune response   *Lucienne Chatenoud*   1312
  - 9.6.6 Endocrine disorders in chronic renal failure   *Eberhard Ritz and Jurgen Bommer*   1317
  - 9.6.7 Sexual dysfunction in chronic renal failure   *Jurgen Bommer*   1329
  - 9.6.8 Anaemia in chronic renal failure   *Judith M. Stevens, Armin Kurtz, Kai-Uwe Eckardt, and C. G. Winearls*   1344
  - 9.6.9 Coagulation abnormalities of patients with chronic renal failure   *GianLuigi Vigano and Giuseppi Remuzzi*   1361
  - 9.6.10 Bony complications in chronic renal failure   *Helmut Reichel, T. Drueke, and Eberhard Ritz*   1365

Contents

9.6.11 The skin in chronic renal failure
Claudio Ponticelli and Pier Luca Bencini — 1390

9.6.12 Neuropsychiatric alterations in uraemia
Raymond C. Vanholder — 1396

SECTION 10 The dialysis patient — 1403

10.1 Vascular access *Kazuo Ota* — 1405

10.2 Haemodialysis, complications during haemodialysis, and adequacy of haemodialysis *Christoph J. Olbricht, U. Frei, and K. M. Koch* — 1417

10.3 Medical complications of the long-term dialysis patient *Jurgen Bommer* — 1436

10.4 Control of blood pressure in patients on haemodialysis *Pietro Zucchelli and Alessandro Zuccala* — 1458

10.5 Psychological aspects of treatment of renal failure *A. Peter Lundin and Richard B. Weiner* — 1467

10.6 Peritoneal dialysis *Ram Gokal* — 1477

SECTION 11 The transplant patient — 1507

11.1 Preparation of the recipient *Robert A. P. Koene* — 1509

11.2 Management of the transplant recipient *Henri Kreis and Christophe Legendre* — 1520

11.3 Transplant immunology *Peter J. Morris* — 1543

11.4 Graft rejection and immunosuppression in kidney transplantation *Terry B. Strom* — 1551

11.5 Cyclosporin nephrotoxicity *Hugh S. Cairns and Guy Neild* — 1560

11.6 Long-term complications and results in the transplant patient *J. Douglas Briggs and Brian J. R. Junor* — 1570

11.7 Hypertension in the transplant patient *C. van Ypersele de Strihou and J.-M. Pochet* — 1594

INDEX

Volume 3

Contributors — xiii

SECTION 12 Specific problems in chronic renal failure — 1603

12.1 Chronic renal failure in children *Otto Mehls and Richard N. Fine* — 1605

12.2 Treatment of end-stage renal disease in the elderly *Juan F. Macias-Nuñez and J. Stewart Cameron* — 1621

12.3 Renal failure in diabetic nephropathy *Eberhard Ritz and Michael Koch* — 1635

SECTION 13 Urinary tract infection — 1657

13.1 Microbiology and defences of the urinary tract *Max Sussman* — 1659

13.2 Lower and upper urinary tract infection in the adult *W. R. Cattell* — 1676

13.3 Lower and upper urinary tract infection in the child *Kate Verrier Jones* — 1699

13.4 Renal tuberculosis *L. Hernando and R. Vela-Navarrete* — 1719

13.5 Schistosomiasis *Rashad S. Barsoum* — 1729

13.6 Fungal infections and the kidney *Rosalinde Hurley* — 1742

SECTION 14 Disorders of mineral metabolism and renal stone disease — 1751

14.1 Hypercalcaemia and hypocalcaemia *John A. Kanis, Neveen A. T. Hamdy, and Eugene V. McCloskey* — 1753

14.2 Hypo- and hyperphosphataemia *Claude Amiel, Claire Bailly, Brigitte Escoubet, and Gerard Friedlander* — 1782

14.3 Hypo- and hypermagnesaemia *I. M. Shafik and John H. Dirks* — 1802

14.4 Aetiological factors in stone formation *William G. Robertson* — 1822

14.5 The medical management of stone disease *G. Alan Rose* — 1847

14.6 The surgical management of renal stones *Hugh N. Whitfield* — 1869

14.7 Nephrocalcinosis *Oliver Wrong* — 1882

SECTION 15 The pregnant patient — 1907

15.1 The normal physiological changes which occur during pregnancy *Chris Baylis and John M. Davison* — 1909

15.2 Renal complications which can occur in pregnancy *John M. Davison* — 1928

15.3 Pregnancy in patients with underlying renal disease *John M. Davison and Chris Baylis* — 1936

15.4 Pregnancy-induced hypertension *Ian A. Greer* — 1956

SECTION 16 Structural abnormalities — 1981

16.1 Vesicoureteric reflux and reflux nephropathy *Ross R. Bailey* — 1983

16.2 The patient with urinary tract obstruction *L. R. I. Baker and Hugh N. Whitfield* — 2002

16.3 Congenital anomalies of the urinary tract *W. Rascher, Martin Meyer-Schwickerath, and H. Olbing* — 2023

SECTION 17 The patient with renal hypertension — 2045

- 17.1 Clinical approach to hypertension *Robert Wilkinson* — 2047
- 17.2 The kidney and the control of blood pressure *J. D. Swales* — 2058
- 17.3 Clinical investigation of the renin-angiotensin-aldosterone system *Pierre Corvol, T. T. Guyene, J. Menard, and P. F. Plouin* — 2069
- 17.4 Effects of hypertension on renal vasculature and structure *Udo Helmchen* — 2075
- 17.5 Hypertension and progression of renal disease *Giuseppe Maschio, Lamberto Oldrizzi, and Carlo Rugiu* — 2083
- 17.6 Hypertension and unilateral renal parenchymal disease *A. Ramming and A. Heidland* — 2088
- 17.7 Renovascular hypertension *J. Mann, J.-R. Allenberg, Christine Reisch, R. Dietz, M. Weber, and F. C. Luft* — 2096
- 17.8 Hypertension in advanced renal failure *Pietro Zucchelli and Alessandro Zuccala* — 2117
- 17.9 Accelerated hypertension *Anthony E. G. Raine* — 2124
- 17.10 The hypertensive child *W. Rascher* — 2139

SECTION 18 The patient with inherited disease — 2153

- 18.1 Molecular genetics of renal disorders *Stephen T. Reeders* — 2155
- 18.2 Cystic diseases
  - 18.2.1 Polycystic kidney disease in children *Michel Broyer and Marie-France Gagnadoux* — 2163
  - 18.2.2 Autosomal-dominant polycystic kidney disease *Yves Pirson and Jean-Pierre Grünfeld* — 2171
- 18.3 Nephronophthisis *C. Kleinknecht and Renée Habib* — 2188
- 18.4 Inherited glomerular diseases
  - 18.4.1 Alport's syndrome *Jean-Pierre Grünfeld* — 2197
  - 18.4.2 Anderson-Fabry disease and other inherited metabolic disorders with significant renal involvement *Stephen H. Morgan* — 2206
  - 18.4.3 Nail-patella syndrome and other rare inherited disorders with glomerular involvement *Guillaume Bobrie and Jean-Pierre Grünfeld* — 2214
  - 18.4.4 Congenital nephrotic syndrome *T. Martin Barratt* — 2218
- 18.5 Inherited metabolic diseases of the kidney
  - 18.5.1 Cystinosis *J. Brodehl* — 2220
  - 18.5.2 The primary hyperoxalurias *Stephen H. Morgan* — 2226
- 18.6 Rare inherited syndromes with renal involvement *J. Stewart Cameron* — 2231

SECTION 19 The patient with cancer of the kidney or upper tract — 2235

- 19.1 Renal carcinoma and other tumours *Manuel Urrutia Avisrror* — 2237
- 19.2 Wilms' tumour *Christopher D. Mitchell* — 2253
- 19.3 Tumours of the renal pelvis and ureter *Peter Whelan* — 2266

SECTION 20 Miscellaneous renal conditions *J. Stewart Cameron* — 2275

SECTION 21 Other organ systems in relation to kidney diseases — 2291

- 21.1 Rheumatic diseases and the kidney *J. Stewart Cameron* — 2293
- 21.2 Diseases of skin and kidney *Anthony C. Chu and A. J. Rees* — 2299
- 21.3 Gastrointestinal problems *Eberhard Ritz* — 2310
- 21.4 Liver diseases and the kidney *Jean-Pierre Grünfeld* — 2312
- 21.5 Oncology and the kidney *Claudio Ponticelli* — 2316
- 21.6 The nervous system and the kidney *Michael Johnson and A. M. Davison* — 2323
- 21.7 The cardiorespiratory system and the kidney *M. Jadoul and C. van Ypersele de Strihou* — 2334
- 21.8 Haematology and renal diseases *C. G. Winearls* — 2340

INDEX

Contributors

DAVID H. ADAMS
Liver Unit, Queen Elizabeth Hospital, Birmingham, UK.

D. ADU
Consultant Physician, Queen Elizabeth Hospital, Birmingham, UK.

J.-R. ALLENBERG
Professor, Department of Surgery, Section of Vascular Surgery, University of Heidelberg, Germany.

CLAUDE AMIEL
Professor of Physiology, INSERM Unit 251 and Department of Physiology, Faculté Xavier Bichat, Université Paris 7, France.

ANITA APERIA
Professor of Pediatrics and Chairman of Department, Department of Pediatrics, St Goran's Children's Hospital, Karolinska Institute, Stockholm, Sweden.

JEAN-FRANÇOIS BACH
Professor of Immunology, INSERM U25, Hôpital Necker, Paris, France.

ROSS R. BAILEY
Head of Department of Nephrology, Christchurch Hospital, New Zealand.

CLAIRE BAILLY
INSERM U251 and Department of Physiology, Faculté Xavier Bichat, Université Paris 7, France.

L. R. I. BAKER
Consultant Physician and Nephrologist, St Bartholomew's Hospital, London, UK.

GIOVANNI BANFI
Senior Assistant, Division of Nephrology and Dialysis, Istituto Scientifico Ospedale Maggiore, Milan, Italy.

T. MARTIN BARRATT
Professor of Paediatric Nephrology, Medical Unit, Institute of Child Health, London, UK.

RASHAD S. BARSOUM
Professor of Medicine and Chief of Nephrology, Kasr-El-Aini Medical School, Cairo University; Chairman, Cairo Kidney Center, Egypt.

CHRIS BAYLIS
Associate Professor of Physiology, West Virginia University, Morgantown, USA.

PIER LUCA BENCINI
Consultant Dermatologist, Istituto Scientifico San Raffaele, Department of Dermatology, University of Milan, Italy.

JERRY M. BERGSTEIN
Professor and Head, Section of Nephrology, Department of Pediatrics, Indiana University School of Medicine, Indianapolis, USA.

DANIEL G. BICHET
Associate Professor, Université de Montreal; Director, Clinical Research Unit, Research Center and Nephrology Division, Hôpital du Sacré-Coeur de Montreal, Canada.

CAROL M. BLACK
Rheumatology Unit and School of Medicine, Royal Free Hospital, London, UK.

GUILLAUME BOBRIE
Department of Nephrology, Hôpital Necker, Paris, France.

JURGEN BOMMER
Professor of Medicine, I. Medizinische Universitätsklinik Heidelberg, Germany.

JULIA A. BREYER
Assistant Professor of Medicine; Director, Outpatient Renal Clinic, Vanderbilt University Medical Center, Nashville, Tennessee, USA.

J. DOUGLAS BRIGGS
Consultant Nephrologist, Western Infirmary, Glasgow, UK.

K. E. BRITTON
Guy's Hospital, London, UK.

J. BRODEHL
Professor, Department of Paediatric Nephrology and Metabolic Diseases, Medical School Hannover, Germany.

MICHEL BROYER
Professor of Paediatrics; Head of Paediatric Nephrology Department, Hôpital Necker Enfants Malades, Paris, France.

HUGH S. CAIRNS
Lecturer and Honorary Senior Registrar, Department of Nephrology and Institute of Urology, University College and Middlesex School of Medicine, University College London, UK.

J. STEWART CAMERON
Professor of Renal Medicine, Guy's Hospital, London, UK.

Contributors

D. J. S. CARMICHAEL
Consultant Renal Physician, Southend General Hospital, UK.

MICHAEL J. D. CASSIDY
Consultant Physician, Groote Schuur Hospital; Senior Lecturer in Medicine, Department of Medicine, University of Cape Town, South Africa.

W. R. CATTELL
Physician and Senior Consultant Nephrologist, St Bartholomew's Hospital, London, UK.

JOËL CHABRIAIS
Consultant Radiologist, Hôpital Necker, Paris, France.

CYRIL CHANTLER
Nationwide Professor of Paediatric Nephrology, United Medical and Dental Schools of Guy's and St Thomas's Hospitals, London, UK.

LUCIENNE CHATENOUD
INSERM U25—CNRS UA 122, Hôpital Necker, Paris, France.

ANTHONY C. CHU
Senior Lecturer and Consultant Dermatologist, Royal Postgraduate Medical School, Hammersmith Hospital, London, UK.

K. S. CHUGH
Professor, Department of Nephrology, Postgraduate Institute of Medical Education and Research, Chandigarh, India.

PIERRE CORVOL
Professor of Experimental Medicine, INSERM U36, College de France, Paris, France.

GIUSEPPE D'AMICO
Professor of Medicine, Division of Nephrology, San Carlo Hospital, Milan, Italy.

A. N. DARDENNE
Associate Professor of Radiology, Université Catholique de Louvain; Chief, Genitourinary and Ultrasound Section, Department of Radiology and Diagnostic Imaging, St Luc University Hospital, Brussels, Belgium.

A. M. DAVISON
Consultant Renal Physician, St James's University Hospital, Leeds, UK.

JOHN M. DAVISON
Scientific Staff and Consultant Obstetrician and Gynaecologist, Medical Research Council Human Reproduction Group, Princess Mary Maternity Hospital, Newcastle upon Tyne, UK.

FRANÇOISE DEGOS
Service d'Hepatologie and INSERM U24, Hôpital Beaujon, Clichy, France.

R. DIETZ
Professor, Department of Cardiology, University of Heidelberg, Germany.

JAVIER DIEZ
Professor of Medicine, University of Zaragoza; Associate Professor of Medicine, Center for Biomedical Research, University of Navarro, Spain.

JOHN H. DIRKS
Dean of the Faculty of Medicine, University of Toronto, Canada.

CIARAN C. DOHERTY
Consultant Nephrologist, Regional Nephrology Unit, Belfast City Hospital, Northern Ireland.

G. C. DOOMS
Department of Radiology, University Hospital St-Luc, Brussels, Belgium.

E. J. DORHOUT MEES
Professor, Department of Nephrology, Ege University, Izmir, Turkey.

DOMINIQUE DROZ
Laboratoire d'Anatomie Pathologique et Clinique Nephrologique, Hôpital Necker, Paris, France.

T. DRUEKE
Associate Professor, Department of Nephrology, Hôpital Necker, Paris, France.

PHILIPPE DRUET
Professor, INSERM U28, Hôpital Broussais, Paris, France.

PAUL A. DUBBINS
Consultant Radiologist, Ultrasound Department, Plymouth General Hospital, Plymouth, UK.

KAI-UWE ECKARDT
Physiologisches Institut, Universität Zurich, Switzerland.

HENRIK EKBLAD
Department of Paediatrics, Children's Hospital, University of Turku, Finland.

A. M. EL NAHAS
Sheffield Kidney Institute, Northern General Hospital, Sheffield, UK.

MARLIES ELGER
Institut für Anatomie und Zellbiologie, Universität Heidelberg, Germany.

PAUL EMERY
Department of Rheumatology, University of Birmingham, UK.

Contributors

BRIGITTE ESCOUBET
Department of Physiology, Faculté de Médecine Xavier Bichat, Paris, France.

J. P. FILLASTRE
Professor, CHU de Rouen, France.

RICHARD N. FINE
Professor and Chairman, Department of Pediatrics, Children's Medical Center at Stony Brook, State University of New York at Stony Brook, USA.

GIOVANNI BATTISTA FOGAZZI
Division of Nephrology and Dialysis, Ospedale Maggiore, Milan, Italy.

U. FREI
Medizinische Hochschule Hannover, Germany.

GERARD FRIEDLANDER
Department of Physiology and INSERM U251, Faculté Xavier Bichat, Université Paris 7, Paris, France.

PATRICIA A. GABOW
Denver General Hospital and University of Colorado Health Sciences Center, USA.

MARIE-FRANCE GAGNADOUX
Pediatric Nephrology Department, Hôpital Necker Enfants Malades, Paris, France.

R.O.B. GANS
Department of Medicine, Free University Hospital, Amsterdam, The Netherlands.

SERAFINO GARELLA
Chairman, Department of Medicine, St Joseph Hospital; Professor of Clinical Medicine, Northwestern University Medical School, Chicago, USA.

GILLIAN GASKIN
MRC Training Fellow and Honorary Senior Registrar, Renal Unit, Department of Medicine, Royal Postgraduate Medical School, Hammersmith Hospital, London, UK.

DENIS GLOTZ
Service du Professeur Bariety, Hôpital Broussais, Paris, France.

M. GODIN
Professor, CHU de Rouen, France.

RAM GOKAL
Royal Infirmary, Manchester, UK.

IAN A. GREER
Clinical Consultant, MRC Reproductive Biology Unit; Honorary Senior Lecturer, Department of Obstetrics and Gynaecology, University of Edinburgh; Honorary Consultant Obstetrician and Gynaecologist, Simpson Memorial Maternity Pavilion and Royal Infirmary, Edinburgh, UK.

R. GREGER
Professor, Physiologisches Institut der Universität, Freiburg, Germany.

JEAN-PIERRE GRÜNFELD
Professor of Nephrology, Hôpital Necker, Paris, France.

T. T. GUYENE
INSERM U36, College de France, Paris, France.

RENÉE HABIB
Director of Research (INSERM); Director of the Unit of Paediatric Nephrology, Hôpital Necker Enfants Malades, Paris, France.

NEVEEN A. T. HAMDY
Chef de Clinique, Metabolic Unit, Department of Endocrinology and Metabolic Diseases, University Hospital Leiden, The Netherlands.

GEORGE B. HAYCOCK
Ferdinand James De Rothschild Professor of Paediatrics, United Medical and Dental Schools of Guy's and St Thomas's Hospitals, University of London, UK.

A. HEIDLAND
Professor of Nephrology, Department of Nephrology, Medical University, Wurzburg, Germany.

OLIVIER HELENON
Assistant Professor of Radiology, Hôpital Necker, Paris, France.

UDO HELMCHEN
Institute of Pathology, University Hospital Eppendorf, Hamburg, Germany.

R. J. HENE
Department of Nephrology, University Hospital Utrecht, The Netherlands.

L. HERNANDO
Professor of Medicine, Fundación Jimenez Diaz, Universidad Autónoma, Madrid, Spain.

S. J. HOORNTJE
Department of Medicine, Catharina Hospital, Eindoven, The Netherlands.

ROSALINDE HURLEY
Professor of Microbiology, the Royal Postgraduate Medical School's Institute of Obstetrics and Gynaecology, Queen Charlotte's and Chelsea Hospital, London, UK.

ENRICO IMBASCIATI
Director, Servizio Nefrologia e Dialisi, Ospedale Civile Sondrio, Italy.

Contributors

LAURENT JACOB
INSERM U25, Hôpital Necker, Paris, France.

HARRY R. JACOBSON
Harry R. Johnson Professor of Medicine; Director, Division of Nephrology, Vanderbilt University Medical Center, Nashville, Tennessee, USA.

M. JADOUL
Renal Unit, University of Louvain Medical School, Clinic Universitaires St-Luc, Brussels, Belgium.

F. JOFFRE
Professor of Radiology, University Paul Sabatier; Head, Service Central de Radiologie au Centre Hospitalier et Universitaire de Toulouse-Rangueil, Toulouse, France.

MICHAEL JOHNSON
Consultant Neurologist, St James's University Hospital, Leeds, UK.

BRIAN J. R. JUNOR
Consultant Nephrologist, Western Infirmary, Glasgow, UK.

JOHN A. KANIS
Professor in Human Metabolism and Clinical Biochemistry, WHO Collaborating Centre for Metabolic Bone Disease, University of Sheffield Medical School, UK.

SABINE KENOUCH
Senior Staff Nephrologist, Department of Nephrology, Université Xavier Bichat, Paris, France.

DAVID N. S. KERR
Professor of Renal Medicine, Royal Postgraduate Medical School, Hammersmith Hospital, London, UK.

PRISCILLA KINCAID-SMITH
Director of Nephrology, Royal Melbourne Hospital, Victoria, Australia.

C. KLEINKNECHT
INSERM U192, Hôpital Necker Enfants Malades, Paris, France.

DIETER KLEINKNECHT
Head of Department of Nephrology, Centre Hospitalier, Montreuil, France.

MICHAEL KOCH
Medizinische Einrichtungen der Heinrich Heine Universität, Dusseldorf, Germany.

K. M. KOCH
Department of Nephrology, Centre of Internal Medicine, Medizinisch Hochschule Hannover, Germany.

ROBERT A. P. KOENE
Department of Medicine, Division of Nephrology, University Hospital, Nijmegen, The Netherlands.

H. A. KOOMANS
Professor of Nephrology and Hypertension, University Hospital, Utrecht, The Netherlands.

W. KRAMER
University of Heidelberg, Germany.

HENRI KREIS
Service de Transplantation, Hôpital Necker, Paris, France.

WILHELM KRIZ
Professor, Institut für Anatomie und Zellbiologie, Universität Heidelberg, Germany.

ARMIN KURTZ
Physiologisches Institut, Universität Zurich, Switzerland.

CHRISTOPHE LEGENDRE
Service de Transplantation, Hôpital Necker, Paris, France.

JOSE M. LOPEZ-NOVOA
Professor of Medicine, University of Salamanca, Spain.

F. C. LUFT
Professor of Medicine, University of Erlanger-Nurnberg; Adjunct Professor of Medicine and Pharmacology, Indiana University, Indianapolis, USA.

A. PETER LUNDIN
Associate Professor of Medicine, SUNY Health Science Center at Brooklyn, New York, USA.

EUGENE V. McCLOSKEY
University of Sheffield Medical School, UK.

JUAN F. MACIAS-NUÑEZ
Titular Professor of Medicine, University of Salamanca, Spain.

MICHAEL MAISEY
Professor, Division of Radiological Sciences, United Medical and Dental Schools of Guy's and St Thomas's Hospitals, London, UK.

N. P. MALLICK
Department of Renal Medicine, Royal Infirmary, Manchester, UK.

ALEXANDER P. MANDELBAUM
Department of Internal Medicine, Ruperto Carola University, Heidelberg, Germany.

J. MANN
Professor, Universität Erlanger-Nurnberg, 4 Medizinische Klinik, Nurnberg, Germany.

GIUSEPPE MASCHIO
Professor of Nephrology and Chief of Division of Nephrology, University Hospital, Verona, Italy.

Contributors

OTTO MEHLS
Professor of Paediatrics, University Children's Hospital, Heidelberg, Germany.

J. MENARD
INSERM U36, Paris, France.

JEAN-PHILIPPE MERY
Professor of Nephrology, Department of Nephrology, Université Xavier Bichat, Paris, France.

MARTIN MEYER-SCHWICKERATH
Associate Professor of Urology, Department of Urology, University of Essen, Germany.

ALAIN MEYRIER
Professor of Nephrology, Bobigny Medical School, Paris-North University, Bobigny, France.

M. J. MIHATSCH
Institut für Pathologie de Universität Basel, Switzerland.

LUIGI MINETTI
Head of Division of Nephrology and Dialysis, Ospedale Ca' Granda Niguarda, Milan, Italy.

CHRISTOPHER D. MITCHELL
Lecturer, Department of Haematology and Oncology, Institute of Child Health, London, UK.

M. MOLZAHN
Professor, Department of Internal Medicine III (Nephrology/Hypertension), Humboldt-Krankenhaus Berlin, Germany.

JEAN-FRANÇOIS MOREAU
Professor of Radiology and Chairman, Department of Radiology, Hôpital Necker, Paris, France.

STEPHEN H. MORGAN
Senior Registrar in Renal Medicine, South West Thames Regional Health Authority, St Helier Hospital, Carshalton and St George's Hospitals, London, UK.

PETER J. MORRIS
Professor of Surgery, Nuffield Department of Surgery, University of Oxford, John Radcliffe Hospital, Oxford, UK.

GUY NEILD
Professor, Department of Nephrology and Institute of Urology, University College and Middlesex School of Medicine, London, UK.

PATRICK NIAUDET
Assistant Professor, Pediatric Nephrology Department and INSERM U192, Hôpital Necker Enfants Malades, Paris, France.

H. OLBING
Direktor der Abteilung für Kindernephrologie, Universitäts-Kinderklinik, Essen, Germany.

CHRISTOPH J. OLBRICHT
Professor of Medicine, Medizinische Hochschule Hannover, Germany.

LAMBERTO OLDRIZZI
Assistant Professor of Nephrology, University Hospital, Verona, Italy.

KAZUO OTA
Professor of Surgery, Kidney Center, Tokyo Women's Medical College, Tokyo, Japan.

SONIA PASQUALI
Department of Nephrology and Dialysis-Malpighi, Policlinico S. Orsola Malpighi, Bologna, Italy.

JOSÉ REIMÃO PINTO
Clinical Assistant, Nephrology Department, Santa Maria University Hospital, Lisbon, Portugal.

YVES PIRSON
Charge de Cours; Associate Chief of Renal Unit, University of Louvain Medical School, Cliniques Universitaires St-Luc, Brussels, Belgium.

P. F. PLOUIN
Service d'Hypertension Arterielle, Hôpital Broussais, Paris, France.

J. M. POCHET
Cliniques Universitaires St-Luc, Brussels and Cliniques Universitaires U.C.L. de Mont-Godinne, Yvoir, Belgium.

MOMIR H. POLENAKOVIC
Professor of Medicine, Department of Nephrology, Faculty of Medicine, University 'Cyril and Methodius'. Skopje, Yugoslavia.

WOLFGANG POMMER
Department of Internal Medicine III (Nephrology/Hypertension), Humboldt-Krankenhaus Berlin, Germany.

CLAUDIO PONTICELLI
Director, Division of Nephrology and Dialysis, Istituto Scientifico Ospedale Maggiore, Milan, Italy.

CHARLES D. PUSEY
Wellcome Senior Research Fellow in Clinical Science; Senior Lecturer and Honorary Consultant Physician, Renal Unit, Department of Medicine, Royal Postgraduate Medical School, Hammersmith Hospital, London, UK.

ANTHONY E. G. RAINE
Department of Nephrology, St Bartholomew's Hospital, London, UK.

DAVID J. RAINFORD
Consultant in Renal Medicine, Princess Mary's Royal Air Force Hospital, Halton, Aylesbury, UK.

Contributors

A. RAMMING
Department of Nephrology, Medical University, Wurzburg, Germany.

W. RASCHER
Professor, Universitäts-Kinderklinik, Essen, Germany.

PETER J. RATCLIFFE
Clinical Lecturer in Medicine, Nuffield Department of Medicine, John Radcliffe Hospital, Oxford, UK.

STEPHEN REEDERS
Assistant Professor of Internal Medicine and Human Genetics, Yale University School of Medicine, USA.

A. J. REES
Professor of Nephrology, Hammersmith Hospital, London, UK.

HELMUT REICHEL
Department of Internal Medicine, Division of Nephrology, University of Heidelberg, Germany.

CHRISTINE REISCH
Medizinische Klinik, Universität Erlangen-Nurnberg, Nurnberg, Germany.

GIUSEPPE REMUZZI
Mario Negri Institute for Pharmacological Research, and Division of Nephrology, Ospedali Riuniti di Bergamo, Italy.

SEVERIN M. G. RINGOIR
Professor of Medicine, University Hospital, Ghent, Belgium.

EBERHARD RITZ
Professor of Medicine, Ruperto Carola University, Heidelberg, Germany.

WILLIAM G. ROBERTSON
Chairman of Biological and Medical Research, King Faisal Specialist Hospital and Research Centre, Riyadh, Saudi Arabia.

ALAN M. ROBSON
Medical Director, Children's Hospital, New Orleans, Louisiana, USA.

BERNARDO RODRIGUEZ-ITURBE
Professor of Medicine, Renal Service and Laboratory, Hospital Universitario and Instituto de Investigaciones Biomedicas (INBIOMED), Maracaibo, Venezuela.

J. RODRIGUEZ-SORIANO
Professor of Paediatrics, Basque University School of Medicine; Head, Department of Paediatrics, Hospital Infantíl de Cruces, Bilbao, Spain.

G. ALAN ROSE
Consultant Clinical Pathologist, St Peter's Hospitals, London, and Royal National Orthopaedic Hospital, Stanmore, UK.

CARLO RUGIU
Assistant Professor of Nephrology, University Hospital, Verona, Italy.

VINAY SAKHUJA
Additional Professor, Department of Nephrology, Postgraduate Institute of Medical Education and Research, Chandigarh, India.

FRANZ SCHAEFER
Division of Paediatric Nephrology, University Children's Hospital, Heidelberg, Germany.

F. PAOLO SCHENA
Professor, Department of Nephrology, University of Bari, Italy.

I. M. SHAFIK
Lecturer, Department of Medicine, University of Toronto, Canada.

THOMAS SHERWOOD
Professor of Radiology, University of Cambridge, UK.

COLIN D. SHORT
Department of Renal Medicine, Royal Infirmary, Manchester, UK.

HEINZ-GUNTER SIEBERTH
Director, Medizinische Klinik II, Medical Faculty, Technical University of Northrhine Westphalia (RWTH), Aachen, Germany.

VISITH SITPRIJA
Professor of Medicine and Chief of Division of Nephrology and Co-ordinating Kidney Center, Faculty of Medicine, Chulalongkorn University, Bangkok, Thailand.

MOURAD SOUISSI
Consultant Radiologist, Hôpital Necker, Paris, France.

L. W. STATIUS VAN EPS
Head, Department of Internal Medicine, Slotervaart Hospital, Amsterdam, The Netherlands.

VLADISAV STEFANOVIC
Professor of Medicine, Institute of Nephrology and Haemodialysis, University School of Medicine, Nis, Yugoslavia.

PAUL E. STEVENS
Department of Renal Medicine, Royal Air Force Hospital, Halton, Aylesbury, Buckinghamshire, UK.

JUDITH M. STEVENS
Research Fellow, Royal Postgraduate Medical School, London, UK.

TERRY B. STROM
Professor of Medicine, Harvard Medical School and Beth Israel Hospital, Boston, Massachusetts, USA.

Contributors

MAX SUSSMAN
Professor, Department of Microbiology, Medical School, University of Newcastle upon Tyne, UK.

R. SUVANAPHA
Associate Professor of Medicine, Division of Nephrology, Department of Medicine, Chulalongkorn University, Bangkok, Thailand.

J. D. SWALES
Professor of Medicine, University of Leicester, UK.

RICHARD L. TANNEN
Professor and Chairman, Department of Medicine, University of Southern California, USA.

P. M. TER WEE
Department of Internal Medicine, Free University Hospital, Amsterdam, The Netherlands.

D. THOMSON
Senior Lecturer, Department of Pathology, University of Edinburgh, UK.

NEIL TURNER
Renal Unit, Department of Medicine, Royal Postgraduate Medical School, Hammersmith Hospital, London, UK.

MANUEL URRUTIA AVISRROR
Department of Urology, University of Salamanca, Spain.

G. K. VAN DER HEM
Professor of Nephrology, Division of Nephrology, Department of Internal Medicine, University Hospital, Groningen, The Netherlands.

F. J. VAN DER WOUDE
Department of Nephrology, University Hospital of Leiden, The Netherlands.

L. A. VAN ES
Professor, Department of Nephrology, University Hospital, Leiden, The Netherlands.

M. H. VAN RIJSWIJK
Professor, Division of Rheumatology, Department of Internal Medicine, University Hospital, Groningen, The Netherlands.

C. VAN YPERSELE DE STRIHOU
Professor of Medicine, Clinic Universitaires St-Luc, Brussels, Belgium.

RAYMOND C. VANHOLDER
Department of Nephrology, University Hospital, Ghent, Belgium.

R. VELA-NAVARRETE
Professor and Chairman, Fundación Jimenez Diaz, Universidad Autonoma, Madrid, Spain.

PATRICK J. W. VENABLES
Senior Lecturer and Consultant in Rheumatology, Kennedy Institute and Charing Cross Hospital, London, UK.

R. VERBERCKMOES
Professor, Renal Unit, Department of Internal Medicine, Universitair Ziekenhuis Gasthuisberg, Belgium.

KATE VERRIER JONES
Laura Ashley Senior Lecturer in Paediatric Nephrology, Department of Child Health, Royal Infirmary, Cardiff, UK.

JEAN-PAUL VIARD
Service d'Immunologie Clinique, INSERM U25, Hôpital Necker, Paris, France.

GIANCARLO VIBERTI
Professor of Diabetic Medicine; Director, Unit for Metabolic Medicine, Division of Medicine, United Medical and Dental Schools, Guy's Hospital, London, UK.

GIANLUIGI VIGANO'
Mario Negri Institute for Pharmacological Research, Bergamo, Italy.

ROWAN G. WALKER
Department of Nephrology, Royal Melbourne Hospital, Victoria, Australia.

M. WEBER
Professor, Medizinische Klinik IV, University Erlangen-Nurnberg, Erlangen, Germany.

RICHARD P. WEDEEN
VA Medical Center, East Orange, and UMDNJ, New Jersey Medical School, Newark, USA.

RICHARD B. WEINER
Clinical Assistant Professor of Pyschiatry S.U.N.Y., Health Science Center at Brooklyn and Unit Chief, Children's Inpatient Psychiatric Unit, Brooklyn, New York, USA.

PETER WHELAN
Consultant Urologist, Department of Urology, St James' University Hospital, Leeds, UK.

HUGH N. WHITFIELD
St Bartholomew's Hospital, London, UK.

ROBERT WILKINSON
Reader in Medicine, Freeman Hospital, University of Newcastle upon Tyne, UK.

D. GWYN WILLIAMS
Senior Lecturer in Medicine and Consultant Nephrologist, United Medical and Dental Schools of Guy's and St Thomas's Hospitals, Guy's Hospital, London, UK.

C. G. WINEARLS
Consultant Nephrologist, Renal Unit, Churchill Hospital, Oxford, UK.

ANTONY J. WING
United Medical and Dental Schools of Guy's and St Thomas's Hospitals, St Thomas's Hospital, London, UK.

V. WIZEMANN
Georg Haas Dialysezentrum, Giessen, Germany.

OLIVER WRONG
Emeritus Professor of Medicine, University College, London, UK.

ALESSANDRO ZUCCALA
Divisione di Nefrologia e Dialisi, Malpighi, Policlinico S. Orsola-Malpighi, Bologna, Italy.

PIETRO ZUCCHELLI
Professor, Department of Nephrology and Dialysis-Malpighi, Policlinico S. Orsola-Malpighi, Bologna, Italy.

Chapter 3.8

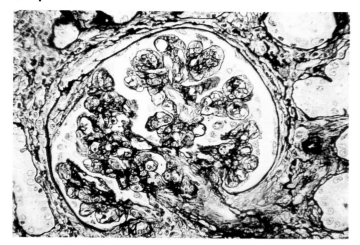

Plate 1 Mesangiocapillary Type I. The appearances on staining with silver methenamine are shown. The mesangial matrix is diffusely increased, whilst the capillary walls show the 'double contour' appearance with large silver deficient subendothelial areas corresponding to immune aggregates and mesangial cytoplasm. A small sclerosing crescent is seen at top left (silver methenamine counterstained with haematoxylin-eosin; original magnification × 200). (By courtesy of Dr Barrie Hartley.)

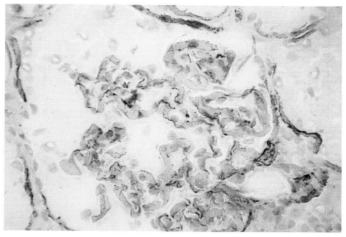

Plate 2 Mesangiocapillary Type II. Frozen section stained with immunoperoxidase-conjugated anti-C1q antibody to show the staining of the inner and outer basement membrane (laminae rarae interna and externa) with complement components. (original magnification × 200). (By courtesy of Dr Barrie Hartley.)

Chapter 4.2

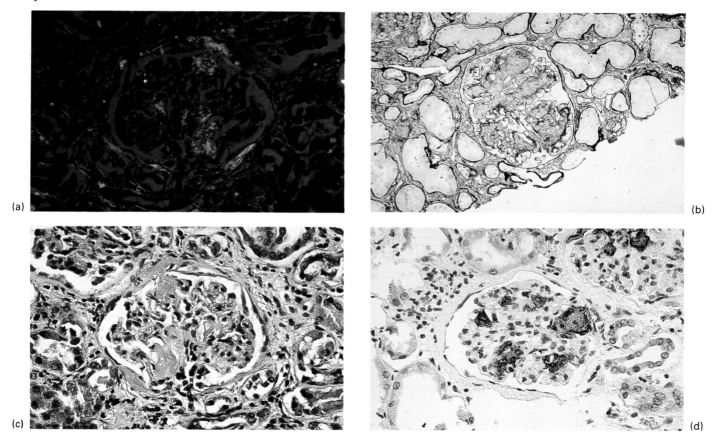

Plate 1 (a) Glomerulus showing amyloid deposits (magnification × 240; staining with alkaline Congo red). (b) Glomerulus shown in (a) viewed with polarized light, showing the characteristic apple-green birefringence of the Congo red-stained amyloid deposits. (c) Glomerulus with amyloid deposits showing irregular staining of the basement membranes (methenamine-silver staining; magnification × 240. (d) Glomerulus with amyloid deposits. Immunoperoxidase staining with anti-AA antibodies, counterstaining with haematoxylin (magnification × 240).

Chapter 9.6.3

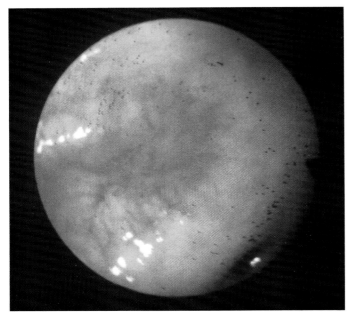

Plate 1 Angiodysplasia of the caecum seen at colonoscopy in a 70-year-old female patient (by courtesy of Dr K. Porter).

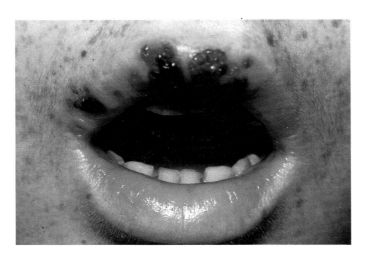

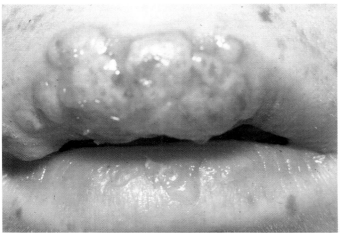

Plate 2 Effect of topical acyclovir on orolabial herpes in a patient with SLE on immunosuppression.

Chapter 18.4.2

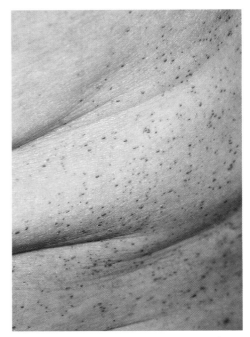

Plate 1 Angiokeratomata, the characteristic rash of α-galactosidase A deficiency (Anderson–Fabry disease).

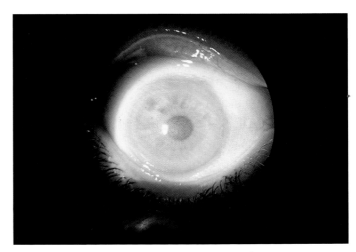

Plate 2 Corneal changes in lecithin:cholesterol acyltransferase deficiency (by courtesy of E. Gjone).

SECTION 1
Assessment of the patient with renal disease

1.1 History and clinical examination

1.2 Urinalysis and microscopy

1.3 Renal function testing

1.4 Renal function in the neonate

1.5 Renal function in the elderly

1.6 Visualizing the kidney

    1.6.1 Radiological tactics: rationale

    1.6.2 Conventional uroradiology and contrast media

    1.6.3 Renal ultrasonography (including Doppler echography)

    1.6.4 Computed tomography and magnetic resonance imaging of the kidney and upper urinary tract

    1.6.5 Renal angiography

    1.6.6 Nuclear imaging in nephrology

1.7 Renal structure/ultrastructure

    1.7.1 The renal glomerulus—the structural basis of ultrafiltration

    1.7.2 Renal biopsy: performance and interpretation

1.1 History and clinical examination of the patient with renal disease

A.M. DAVISON AND JEAN-PIERRE GRÜNFELD

Introduction

Effective management of the patient with renal disease is dependent upon establishing an accurate diagnosis. To achieve this the clinician must be aware of the possible presentations of renal diseases and the constellation of symptoms and signs which form recognized diseases and syndromes. It is essential to obtain a full and detailed history relating to the presenting symptoms, but it is also important to obtain an accurate past history, family history, and social history. The need for a specific, detailed documentation of exposure to environmental factors, such as chemicals which may be encountered at work or during recreation, is being increasingly recognized and full details of all recent medications must also be obtained. Following this, a careful and thorough examination should be undertaken, paying particular attention to those clinical signs which are known to reflect underlying renal disease. Finally, a properly structured investigative approach should be formulated to ensure that the maximum information is obtained with the minimum inconvenience to the patient. In this way unnecessary investigations are avoided and the most cost-effective use is made of available resources in achieving an accurate diagnosis. Thereafter a plan can be formulated for effective management and follow-up.

Patient management extends beyond achieving an accurate diagnosis and progresses into the area of effective treatment. In this respect, the nephrologist is particularly privileged because, unlike other organs, the kidney can be effectively replaced by dialysis and transplantation. This, however, poses particular and unique problems for the clinical nephrologist as it generates social and ethical consequences which have to be carefully considered in the integrated management of the patient. This chapter, although devoted to the history and clinical examination of the patient with renal disease, is not intended to be comprehensive, but rather an overview, and the reader is referred throughout to the relevant chapters of this book for a detailed discussion of individual syndromes.

Clinical presentation of renal disease

The patient with renal disease may come to the attention of the clinician for one of several reasons.

1. The patient is asymptomatic, but an abnormality has been detected on clinical or laboratory examination which indicates an underlying renal abnormality.
2. The patient complains of a symptom or has a physical sign which directly or indirectly indicates underlying renal disease.
3. The patient has a systemic disease which is known to be associated with renal involvement.

Asymptomatic patients

Asymptomatic patients are most commonly detected following routine investigations such as urine analysis, blood pressure measurement, or blood chemistry analysis following hospital admission, or undertaken as part of a health screening programme (see Chapter 3.1). In a number of patients, renal disease is detected during clinical examination for either health insurance, occupational purposes, or during pregnancy. In a small number of instances there is regular screening in view of a known association between an employment and development of renal disease, e.g. aniline dye workers have a higher than normal incidence of urothelial tumours. Asymptomatic patients are often detected as a result of the investigation of family members following the diagnosis of a familial renal disease. Finally, as a recent development, a considerable number of neonates are diagnosed as having urological disorders because of routine ultrasound screening of the mother during pregnancy. It will be appreciated therefore that the detection of asymptomatic patients depends on many factors, including the policy with respect to regular medical screening in particular age groups, such as at school entry and/or school leaving. It will also depend to a certain extent on chance, as not all occupations have pre-employment medical examinations and not all patients undergo medical examinations for insurance purposes. To some extent this may well explain purported differences in the incidence of various forms of renal disease reported from different countries.

Symptomatic patients

Symptomatic renal disease most commonly presents as a disorder of micturition, urine volume, or urine composition, or as pain, oedema, or symptoms related to impairment of renal function.

Disorders of micturition

The most common disorder of micturition is frequency, a term used to indicate that the bladder is emptied more often than normal. This may be associated with an increased urine volume (polyuria; see below) or with a normal urine volume. The latter may result from irritation of the bladder by inflammation, stone, or tumour; from a reduction in the bladder capacity as a consequence of fibrotic contraction; or external pressure from a pelvic mass or gravid uterus. In many conditions the frequency will be accompanied by nocturnal micturition. In patients with frequency it is important to determine whether a normal volume of urine at each bladder emptying, or whether only small quantities are passed. The former indicates increased urine formation; the latter indicates diminished bladder capacity. Nocturia may arise from sleep disturbances, since it is normally sleep that stimulates

ADH secretion, with a consequent diminution in urine volume. Patients who lie awake at night do not have increased ADH production and, in the recumbent position, have an increased renal blood flow with consequent increased urine volume and subsequent nocturia.

Men past middle age frequently have prostatic enlargement which characteristically causes the stream of urine produced on voiding to be poor. There may also be some difficulty in initiating micturition (hesitancy) or in stopping (terminal dribbling). Eventually the enlargement may cause complete urethral obstruction resulting in urinary retention and acute bladder distension. In a number of patients with prostatic hypertrophy, there is retention and back pressure, diminishing filtrate flow in the nephrons and interfering with the ability of the medulla to maintain a concentration gradient. This results in impairment of urinary concentration and, consequently, increased urine volume. Paradoxically, therefore, some patients with developing obstruction may present with increased urine output, and this may falsely reassure the patient and clinician, resulting in late diagnosis (see Chapter 16.2).

Dysuria is most commonly used to describe pain or discomfort during micturition. This is usually described as a burning or tingling sensation felt at the urethral meatus or in the suprapubic area during or immediately after micturition. It usually arises as a consequence of bladder, prostatic, or urethral inflammation (see Section 13). When associated with frequency and urgency of micturition it usually indicates cystitis, which is particularly common in young women and is often related to sexual activity. In older women or in men there is usually an underlying condition, such as a structural abnormality of the bladder or prostate. In men perineal or rectal pain suggest inflammation of the prostate.

In young children, cystitis should be suspected if the child cries on micturition or has unexplained fever; investigations for structural abnormalities of the urinary tract should be undertaken in these patients.

Disorders of urine volume

Disorders of urinary volume consist of polyuria, where the volume is increased, oliguria, where the urine volume is diminished, and anuria, where there is an absence of urine.

Polyuria may be due to: (1) an excess intake of fluid such as in compulsive water drinking; (2) an increase in tubular solute load, as in chronic renal failure, glycosuria due to diabetes mellitus, hypercalcaemia, or myeloma; (3) a diminution in ADH production, which may occur following trauma to the head or tumours of the hypothalamus or pituitary; (4) disordered medullary concentration gradient as a consequence of medullary disease such as nephrocalcinosis, renal papillary necrosis, medullary cystic disease, and other conditions which mainly affect the medulla; (5) conditions which impair the tubular response to ADH, such as hypercalcaemia, potassium depletion, and a rare inherited form of tubular insensitivity to ADH called congenital nephrogenic diabetes insipidus (see Chapter 5.5)

Oliguria describes a reduction in urine volume to less than that required for the excretion of the residues from normal daily metabolic functions. In normal adults, under extreme conditions, homeostasis can usually be maintained with a daily urine output of 500 ml. Thus in the adult a urinary output of less than 400 ml per day would constitute oliguria. In young children an approximately equivalent figure is 1.0 ml/kg body weight/h. This usually indicates an underlying acute renal failure whether from prerenal causes, acute vasculitis, acute glomerular lesions, tubular necrosis from toxins or sepsis or acute interstitial nephritis.

Anuria, the absence of any urine output, usually indicates obstruction of the urinary tract, although rarely it may arise from renal infarction or cortical necrosis. In a patient with anuria, there must be careful examination of the lower abdomen; rectal examination and abdominal ultrasound are mandatory investigations. Not uncommonly, patients with pelvic malignancies such as carcinoma of the cervix or rectum present with anuria due to the tumour spreading laterally to the pelvic wall, causing external compression of the lower ureter.

Alteration in urinary composition (see also Chapters 1.2 and 3.1)

Visible haematuria (see Chapter 3.1) is the clinical symptom which most commonly causes patients to seek medical advice as a consequence of an altered urine composition. Haematuria can arise from any part of the renal tract from the glomerulus to the urethra and may be due to conditions such as glomerulonephritis, infection, calculi, or tumours. However, it may indicate an underlying disorder of coagulation such as thrombocytopenia, a deficiency in coagulation factors, or may occur as a result of anticoagulant drugs given for other conditions. Not all discoloured urine is due to the presence of blood (Table 1), and factitious causes must be considered.

In some patients haematuria is detectable only on microscopic examination or by chemical testing; in others it is visible to the naked eye (macroscopic haematuria) and may be even sufficient to produce clots. Blood arising from the glomerulus frequently gives rise to a red-brown discoloration of the urine which is sometimes described as a 'smoky' appearance or likened to tea or coca cola. In some patients macroscopic haematuria is only intermittent; this recurrent haematuria is commonly associated with IgA nephropathy. In such patients the macroscopic episodes are usually closely accompanied by mucosal inflammation (frequently of the respiratory tract), and may last for 1 to 3 days. Between episodes, microscopic haematuria is usually present, although in children the urine may clear completely. Blood arising from the urethra is dislodged by the urinary flow and therefore appears only in the initial urinary stream. Bleeding from the bladder and prostatic bed is however, more commonly noticed at the end of micturition and this is described as terminal haema-

Table 1 Differential diagnosis of red/brown urine

Haematuria
Haemoglobinuria
Myoglobinuria
Urates
Porphyria
Alkaptonuria (homogentisic acid)
Drugs*
    Analgesics: phenacetin, antipyrine
    Antibiotics: rifampicin, metronidazole, nitrofurantoin
    Anticoagulants: phenindione, warfarin
    Anticonvulsants: phenytoin
Vegetable dyes
    Beetroot and some berries (anthocyanins)
    Paprika
    Food colouring materials

* Many more have been described in occasional case reports.

turia. Some patients present with haemospermia and this is usually indicative of a prostatic problem or a bleeding diathesis. Overall, however, the most common cause of haematuria, both micro- and macroscopic, is infection in the urinary tract (see Section 13).

Proteinuria is usually determined chemically, although some patients will have noticed that their urine appears particularly frothy. A small amount of protein is usually present (200 mg/24 h) and more than 50 per cent of this is tubular in origin (see Chapter 3.1). Urine screening for protein is usually undertaken by a 'stix' test but it must be remembered that in myeloma this test is negative as it does not detect light chains. In such patients this negative test in the presence of a positive sulphosalicylic acid test is a useful diagnostic aid. In some patients minor degrees of proteinuria may be present (<1 g/24 h), without being of any pathological significance. In others proteinuria is related to posture, being absent in the urine passed immediately on rising in the morning, but being present after the subject has been up and about for some time. This is described as orthostatic or postural proteinuria and is usually considered benign. In some subjects proteinuria is only present following exercise.

Pathological proteinuria may indicate glomerular or interstitial disease. Generally, that arising from interstitial disease is mild, up to 2 g daily, whereas that from glomerular disease is variable, up to 10 g or more daily. In some instances the type of proteinuria is diagnostic, e.g. 'selective' proteinuria in minimal lesion glomerulonephritis (see Chapter 3.5), or Bence–Jones proteinuria in myeloma (see Chapter 4.3).

Bacteriuria is the presence of bacteria in the urine and may be symptomatic or asymptomatic. Bladder urine is normally sterile, but the urethra, and particularly the urethral meatus is not sterile, and urine may become contaminated during micturition. For this reason bacteriuria is usually only considered significant if the numbers of bacteria in a midstream specimen of properly collected urine exceed 10^5 organisms per ml (see Section 13).

Leucocyturia—the presence of white blood cells in the urine—occurs in a number of conditions such as nephrocalcinosis, papillary necrosis, and analgesic nephropathy. Eosinophils may be detected in the urine in tubulointerstitial nephritis of immunoallergic origin.

Pyuria is always considered significant and indicative of infection. Sterile pyuria should always raise the possibility of underlying tuberculosis, but this also occurs in chlamydial infections.

Calculi may be passed spontaneously, usually accompanied by symptoms of severe colic but, surprisingly, sometimes without any symptoms.

Occasionally, small pieces of tissue may be passed in the urine and these may arise from papillary necrosis or tumours. Other patients may pass small pieces of material which have entered the bladder as a consequence of a vesicocolonic fistula.

Pneumaturia is an uncommon presentation and indicates a communication between the urinary and alimentary tracts, most commonly as a consequence of vesicocolonic fistula resulting from diverticular disease or carcinoma of the colon.

Pain

Pain is an inconsistent symptom of renal disease but, when present, most commonly represents inflammation or obstruction. Inflammation of the kidney, in the form of pyelonephritis, usually causes pain localized to the renal angle on the side of the affected kidney. It develops gradually and is variable in severity but is usually constant in nature. A perirenal abscess can give symptoms related to diaphragmatic irritation if the abscess tracks upwards and to psoas muscle irritation if tracking downwards. Glomerular inflammation is usually asymptomatic but may be associated with a dull lumbar ache, particularly in acute glomerulonephritis and IgA-associated nephropathy.

Some patients present with an intermittent dull aching loin pain, which may be variable in severity, associated with visible haematuria; this has been termed the loin pain haematuria syndrome. Renal biopsy yields unremarkable findings apart from complement (C3) deposition in the walls of afferent arterioles (Burden *et al.* 1979), the significance of which is not clear. In addition, renal angiography may reveal focal or generalized vascular lesions of tortuosity, beading, stenosis or occlusion, sometimes associated with cortical infarcts (Burden *et al.* 1975).

Pain arising from an acute obstruction is usually sudden in onset, severe and colicky in nature, and radiates to the groin or scrotum. Chronic obstruction may, however, be remarkably asymptomatic. It is important to emphasize that significant renal destruction can occur without any pain or discomfort and in many patients the first symptoms relate to the metabolic consequences of severely impaired kidney function.

Oedema

Oedema may arise due to the hypoproteinaemia which is a consequence of significant proteinuria (in adults the prolonged daily excretion of >3.5 g). This complex is termed the nephrotic syndrome (see Chapters 3.3 and 3.4). The oedema is usually most noticeable around the eyes in the morning, and in the feet and ankles in the evening; with increasing severity this diurnal change is lost and the patient notices more generalized swelling throughout the day. An increase in abdominal girth may be noticed if ascites develops.

Oedema may also occur as a consequence of salt and water retention in patients with chronic renal failure, congestive cardiac failure, chronic hepatic disease such as cirrhosis, or following the administration of certain drugs, such as non-steroid anti-inflammatory drugs or calcium-channel antagonists.

Impairment of renal function

Some patients present with uraemia as a consequence of impaired renal function. This is usually of a chronic nature (see Section 8 and Section 9).

Associated with systemic disease

Many patients with systemic disease (see Section 4) have renal disease, and this may become manifest before, coincidentally with or subsequent to the diagnosis of the systemic condition. As a general rule, however, the occurrence or development of renal disease in a patient with systemic disease is indicative of a worse prognosis.

Specific clinical syndromes

Asymptomatic proteinuria

This is the detection of proteinuria during a coincidental medical examination such as, for example, at school, for life insurance purposes, or for employment. Protein excretion is usually less than 3.5 g daily and may represent 'physiological' proteinuria, which is either orthostatic or exercise-induced, or pathological proteinuria due to glomerular, tubular or interstitial disease.

The nephrotic syndrome

This is a consequence of hypoproteinaemia and develops as a direct result of prolonged heavy proteinuria (see Chapters 3.2, 3.3, and 3.4).

Acute nephritic syndrome

Haematuria, proteinuria, oliguria, and hypertension are diagnostic of this syndrome. Poststreptococcal glomerulonephritis is one example and occurs some 10 to 20 days following the introduction of an antigen. It may also be due to an acute exacerbation of a chronic glomerular disease such as IgA nephropathy. The syndrome is considered to be due to immune complex formation with consequent activation of the complement and coagulation systems, giving rise to an inflammatory response.

Microscopic haematuria

This is usually detected at routine clinical examination and may be due to bleeding from any part of the renal tract. Red cells originating from the glomerulus usually have a dysmorphic appearance and/or are present as red cell casts. The more normal the microscopic appearance of red cells the more likely it is that bleeding originates from the lower urinary tract.

Recurrent haematuria

Intermittent episodes of macroscopic haematuria are referred to as recurrent haematuria. Most commonly this occurs at the time of a mucosal inflammation in patients with IgA nephropathy or more rarely in children with Alport's syndrome, but it can, of course, be the result of intermittent bleeding from a structural lesion in the urinary tract, such as a tumour, or in adult polycystic kidney disease.

Hypertension

Patients with glomerular disease are frequently hypertensive; raised blood pressure is seen less frequently in tubular or interstitial conditions. It is particularly common, but not invariable, in the late stages of chronic renal failure (see Chapter 17.8). Sustained severe hypertension in children is usually due to chronic renal disease.

Acute renal failure

Such an acute reduction in renal function may be prerenal, renal, or postrenal in origin (see Section 8).

Chronic renal failure

A host of non-specific symptoms is usually present, none of which individually indicates an underlying renal problem. However, when taken together clearly, the symptoms point to a widespread metabolic abnormality (see Section 9).

Clinical history

History of presenting complaint

The first step in a clinical interview should be to elicit carefully the patient's symptoms in a clear, chronological order. Many patients, particularly when being interviewed for the first time, are anxious and tense and it is worth spending some time making the patient feel at ease prior to starting the interview. Patients should be encouraged to describe their symptoms in their own words and if they use diagnostic terms, they should be asked to explain in detail what they mean; for example, a patient may use the term 'cystitis' to indicate dysuria, frequency, nocturia, or foul-smelling urine. In some instances the patient will previously have sought medical opinion and will reiterate the previously postulated diagnosis. It is sometimes only with the most careful probing that the exact nature of the underlying symptom is determined. Wherever possible precise dating should be obtained, but many patients have a poor recall for chronological events.

Enquiry must be made relating to associated symptoms, but frequently patients will omit what they consider to be insignificant facts, e.g. the presence of upper respiratory tract symptoms at the time of macroscopic haematuria unless asked directly. In this respect it is important to ask patients about any activity being undertaken at the time of or immediately before the development of symptoms, to establish relationships between, for example, exercise and haematuria, or sexual activity and dysuria. It should be remembered, however, that patients may falsely attribute symptoms with an associated event, and caution must be undertaken when interpreting the information gained.

Pain is possibly the most common reason for a patient seeking medical opinion. It is important to determine the site of the pain and whether it is localized to a well-described area, or is diffuse. The mode of onset is important; a gradually increasing pain is suggestive of inflammation whereas sudden severe onset suggests obstruction or rupture. In the renal tract sudden onset of pain may indicate ureteric obstruction. In addition, the clinician must determine whether the pain radiates and, if so, whether this follows any recognized pathway. The character of the pain should be determined, although many patients find this difficult to describe. A pain which waxes and wanes may be indicative of a colic and may cause the patient to be restless, whereas a steady pain such as that from pyelonephritis may cause the patient to remain still in an attempt to avoid aggravation by movement. Factors which are known to precipitate or aggravate the pain should be determined, as well as those measures which cause relief. Although enquiry is frequently made regarding severity of pain, it is usually very difficult for the patient to describe this. Very careful and sympathetic questioning is needed to identify the exact character and pattern of the pain.

In patients with haematuria and/or proteinuria, it is particularly important to establish whether previous urine analyses have been undertaken and the results of such tests. In some instances the patients will know the results; if not, a careful review of the records of previous hospital attendances may well provide much useful information. In any event it is important to obtain the results of any previously arranged investigations whenever possible, as these can be of value in dating the onset of a particular illness.

A history of hypertension may be closely related to renal disease and it is important to document the date of its appearance and subsequent complications, as well as any efficacy, tolerance, and patient compliance associated with prescribed therapy.

The rate of progression of various forms of renal disease is variable and can frequently only be assessed by retrospective analysis. Slowly progressive renal failure may extend over years or decades, in contrast to the rapid deterioration in renal function which may be due to an acute condition such as crescentic glomerulonephritis, or an exacerbation superimposed on some other chronic renal disease. Patients with chronic renal failure may have an acute deterioration related to dehydration, urinary tract obstruction, drug toxicity, the development of an accelerated phase of hypertension or to hypotension induced by excessive antihypertensive therapy.

Thus in obtaining the history of the present complaint, it is important to determine all the factors surrounding the particular illness and care should be taken to ensure that each symptom is fully described and documented in a proper chronological order.

Past history

In many renal diseases, and particularly in chronic renal failure, the past history is of particular relevance. Wherever possible this should be documented in as much detail as can be obtained and should include childhood as well as adolescence and all major illnesses and hospital admissions.

In childhood, a history of protracted nocturnal enuresis may indicate an abnormality of the urinary tract, or a disordered concentrating mechanism. The age of becoming 'dry' at night, if available, may be useful. Unexplained febrile episodes in childhood are often manifestations of upper urinary tract infections, and suggest structural abnormalities such as vesicoureteric reflux. Some adults who present with proteinuria and haematuria give a history of repeated upper respiratory tract infections or scarlet fever in childhood; such episodes may have been the initiating events in a chronic glomerulonephritis leading to impaired renal function in adult life.

A history of hypertension, particularly in children and young adults, seldom leads to a diagnosis of idiopathic or 'essential' hypertension and usually indicates an underlying disease, most commonly of the kidney.

A large number of conditions may be associated with renal involvement, either directly, indirectly, or as a complication of treatment. In this introductory text it is impossible to cover all of them, and a few examples will be used for illustrative purposes. A long-standing condition such as lupus erythematosus may initially be confined to skin or joints and only later affect the kidneys. Other conditions, such as chronic sepsis from either bronchiectasis or osteomyelitis, may indirectly by amyloid deposition result in glomerular lesions and the development of a nephrotic syndrome. Finally, patients with rheumatoid arthritis may be treated with a variety of agents such as analgesics, gold, or penicillamine, all of which may result in renal damage.

A number of metabolic diseases will eventually cause renal problems. For instanced, hyperparathyroidism may manifest as unexplained abdominal pain and bone problems; these may significantly precede the development of renal calculi. Similarly, diabetes mellitus of either type may be associated with the development of diabetic glomerulosclerosis. This most commonly develops after 10 years of known diabetes mellitus, but can occur surprisingly early following diagnosis, particularly of the maturity-onset type (type II) (see Chapter 4.1).

When constructing the past history, it is not uncommon to find that the patient's memory is unreliable, and every effort should be made to resort to previous medical notes, whether from the family physician or from hospital attendances. The past history is not only important with respect to achieving a diagnosis, but is also of particular value when drug treatment is being considered. For instance, in hypertensive patients it is important to avoid the use of β-blockers in patients with a history of asthma. Similarly, centrally-acting antihypertensive drug should be avoided in patients with a history of depression. In other circumstances, therapy can only be introduced with the concomitant use of prophylactic drugs: for example H_2-receptor blocking drugs such as ranitidine should be given to patients with a history of peptic ulceration who require steroid therapy.

Gynaecological and obstetric history (see Section 15)

In women it is important to obtain full details of the menstrual history, contraception, and pregnancy. The menarche may be delayed in patients with impaired renal function. Some patients with renal failure may develop amenorrhoea, although in some, particularly older patients, menorrhagia may occur due to abnormalities of coagulation and/or platelet function. Heavy and prolonged menstrual bleeding, a common aggravating factor of anaemia in uraemic women, may be considerably improved by progesterone treatment. Underlying renal disease significantly increases the risk of hypertension in women taking the combined oestrogen-progesterone contraceptive pill; this may not be reversible. There may also be exacerbation of symptoms in women with systemic lupus erythematosus.

Pregnancy may be associated with aggravation or initiation of renal disease. Hypertension may develop during pregnancy, while pre-existing hypertension may be exacerbated. In obtaining a history, therefore, the course of blood pressure throughout pregnancy should be ascertained, together with details of drug therapy and whether or not hypertension was complicated by proteinuria and whether the pregnancy went, or was allowed to proceed to term. Commonly, patients with proteinuria have increased urinary protein excretion during pregnancy and this may be sufficient to produce a nephrotic syndrome, precipitating symptomatic disease. In such patients, it is difficult to differentiate between pre-eclampsia and previously asymptomatic glomerular disease.

Pregnancy predisposes to urinary tract infection, particularly of the upper tract, in patients with polycystic disease or reflux nephropathy. Bacteriuria is common, but symptomatic upper urinary tract infections usually indicate a structural abnormality requiring detailed investigations once the pregnancy is completed.

Recurrent fetal loss may indicate the presence of antiphospholipid antibodies (lupus anticoagulant) and should raise the possibility of underlying systemic lupus erythematosus. Details of parturition should be obtained, particularly with respect to the term and circumstances of delivery and the weight and Apgar score of the neonate. Acute renal failure developing shortly after delivery may indicate a form of haemolytic–uraemic syndrome, which is frequently irreversible.

Drug history

Recent drug intake should be carefully noted with respect to the first day of administration, dosage, duration of administration, and clinical manifestations reported by the patient. A single tablet or dose of a large variety of drugs may be sufficient to induce an acute allergic interstitial nephritis in a very small number of hypersensitive subjects. In contrast, interstitial nephritis induced by ingestion of non-steroidal anti-inflammatory agents may develop after several months of administration. Elderly patients, who often use non-steroidal anti-inflammatory drugs to relieve articular pain, are highly susceptible to their deleterious effect on the kidney. Furthermore, these patients are particularly at risk of renal drug toxicity for many reasons, including the normal reduction of renal function with age, changes in pharmaco-

Table 2 Drug-induced hypertension

1. Steroid-mediated hypertension
    Corticosteroids (mainly in patients with pre-existent renal failure)
    Oestroprogestative pills or oestrogens (oral administration)
    Mineralocorticoids (9α-fluoroprednisolone by nasal spray or antihaemorrhoid cream)
    Androgen derivatives (danazol)
    Liquorice intoxication
2. Hypertension mediated by sympathomimetic drugs
    Isoproterenol
    Various sympathomimetic drugs in nasal or eye drops or in anorectic preparations (such as phenylpropanolamine)
    Monoamineoxidase inhibitor associated with sympathomimetic drug or tyramine-rich food
    Sulpiride, metoclopramide, haloperidol, methylergometrine, tricyclic derivatives (such as imipramine) in very rare cases
3. Cyclosporin-associated hypertension (increased incidence in patients also receiving steroids)
4. Erythropoietin
    In dialysis patients, particularly those with a history of previous hypertension

kinetics with ageing, the common association of other organ impairment (such as borderline heart failure), and the frequent use of several drugs which may interact.

Hypoproteinaemia may reduce drug binding to plasma proteins resulting in an increased availability of 'free' drug and increasing the potential for toxicity. It must also be remembered that corticosteroids, particularly in pulse form, may affect drug binding to plasma proteins resulting in an increase in the free or unbound portion.

Hypotensive drugs may cause a deterioration in renal function, particularly in patients with long-standing disease or when the blood pressure is lowered precipitously. In addition angiotensin-converting enzyme inhibitors may produce profound renal failure in patients with undiagnosed bilateral renovascular disease or vascular disease, such as severe atheroma, in a solitary kidney.

Information on chronic drug intake is important. Long-term renal and tubulointerstitial toxicity of analgesics has been clearly demonstrated, and similar nephrotoxicity has been more recently discussed with regard to lithium. Other drugs (such as gold salts or D-penicillamine) may induce glomerular changes which reverse on ceasing administration.

In addition to exerting direct toxic effects on the kidney, some drugs elevate blood pressure (Table 2). The possibility of drug-induced or aggravated hypertension should be considered in all hypertensive patients (Grünfeld and Bellet 1982; Stokes 1976). Hypertension is most frequently associated with oestrogen/progesterone compounds (hypertension may be moderate, accelerated, or associated with the haemolytic–uraemic syndrome), corticosteroids, cyclosporin, and liquorice. With regard to the last, hypertension is probably not related to the mineralcorticoid activity of liquorice *per se*, but to its inhibition of 11-hydroxysteroid dehydrogenase activity, which leads to high tissue concentrations of cortisol which are not metabolized to cortisone (Stewart *et al.* 1987).

Finally it should be emphasized that the extrarenal side effects of many drugs are increased in patients with renal diseases (the nephrotic syndrome and/or renal failure), and great caution should be exercised in the use of any recent drug for which medical experience is limited. In some cases, a toxic effect may be predicted on a pathophysiological basis. For example lovastatin, an HMG-CoA reductase inhibitor may induce rhabdomyolysis, particularly in patients treated with cyclosporin or gemfibrozil. Lovastatin-induced myositis may severely aggravate hyperkalaemia in diabetic patients with slight renal failure receiving an angiotensin converting enzyme inhibitor. (Edelman and Witztum 1989). Many patients do not relate medicines that they can buy without prescription, commonly referred to as 'over-the-counter drugs', to drugs which are prescribed by medical practitioners. Careful enquiry about domestic use of non-prescription drugs, particularly analgesics and non-steroidal anti-inflammatory drugs is always useful.

Nephrologists should also be aware of covert drug intake, especially of laxatives or diuretics, or surreptitious vomiting in emotionally labile women. These patients experience severe potassium depletion or recurrent oedema on drug withdrawal, leading to the incorrect belief that diuretics are required for the so-called 'idiopathic oedema syndrome' and resulting in a perpetuation of drug intake (MacGregor *et al.* 1975; MacGregor and de Wardener 1988).

Dietary history

A dietetic history can be of diagnostic value as well as being of importance in patient management, particularly in the presence of renal failure. The help and advice of a skilled dietitian is frequently required.

Excessive sodium intake may account for resistance to antihypertensive therapy or for recurrent pulmonary oedema in patients with advanced renal failure. In contrast, sudden and excessive sodium restriction or increased sodium loss through the skin or gastrointestinal tract may precipitate severe hypovolaemia in a patient with salt-losing nephropathy.

It is important to estimate protein intake in patients with impaired renal function. In children an estimate of calorie intake is important; an inadequate energy intake is a common cause of growth failure in children with chronic renal failure.

Idiopathic stone formers ingest more animal protein than normal subjects; such a diet is associated with increased urinary excretion of calcium, oxalate, and uric acid, all of which are known risk factors for calcium stone formation. The prevalence of stones in vegetarians is less than half of that in the general population. Inadequate fluid intake may explain the frequent recurrence of urinary stones in certain patients. In patients with absorptive hypercalciuria, stone recurrence may also be favoured by the ingestion of calcium-rich water. Some patients have a high intake of milk and alkali, which can lead to both nephrolithiasis and nephrocalcinosis. Patients with a history of food allergy may develop a nephrotic syndrome (Lagrue 1989).

Alcohol intake should also be estimated: excessive consumption induces an increase in blood pressure and is responsible for a substantial percentage of cases of essential hypertension (Saunders *et al.* 1981; Lang *et al.* 1987). It may also account for poor compliance with therapy.

Some patients have a particular desire for acidic foods such as fruit juices and rhubarb, many of which have a high oxalate content and may, following chronic excessive ingestion, result in calcium oxalate precipitation in the kidney. Other patients may consume large quantities of strong tea or coffee every day, leading both to extrasystoles and polyuria, depending upon the methyl xanthine content.

Vegetarian patients may be prone to vitamin B_{12} deficiency, depending upon the strictness of their diet. Ethnic diets contain either very large amounts of sodium (Japanese, Indian) or potassium (Indian). The use of unleavened bread, such as chapatis, results in a high phosphate intake, calcium binding, and acceleration of osteomalacia or the rachitic component of renal osteodystrophy.

Obtaining a detailed dietary history may, therefor, help establish a diagnosis, but, in the majority of patients, is of particular help in establishing a baseline value for subsequent dietary manipulations.

Factitious history

Self-induced or factitious disease may also be encountered in nephrology. This commonly involves the addition of material, such as blood, protein, or sugar, to a normal urine specimen. In some instances a parent will provide false symptoms and signs or will tamper with the child's urine or other laboratory samples, resulting in excessive and unnecessary investigations (frequently invasive) being undertaken: this has been called 'Munchausen by proxy' (Meadow 1977). The lengths to which adult patients will go in falsifying information are surprising: intravenous injection of metaraminol or isoprenaline has been used to produce hypertension (Lurvey et al. 1973; Portioli and Valcavi 1981) the intravenous injection of milk, tetanus toxoid, or faeces produces immune complex nephritis; 'stones' which clearly could not originate in the renal tract (e.g. date stones) may be produced, and physical signs, such as temperature, may be altered to mislead the attending physician. In the majority of these patients a clue may be obtained from the willingness of the patients to go along with prolonged, frequently repetitive investigations, the bizarre nature of the symptoms/sign complex or the nature of the laboratory findings e.g. *Mycobacterium smegmatis* from sputum or Lactobacillus from blood. On confrontation most patients will deny any possibility of factitious illness and they will frequently seek medical attention and investigation elsewhere.

Social history

Socioeconomic and educational status influence the incidence and mode of presentation of many renal diseases and may also give an indication as to the required degree of follow-up supervision. Acute poststreptococcal glomerulonephritis following skin infection is more prevalent in children with poor hygiene. The frequency of bacteriuria is much greater in multiparous patients and in pregnant women of low socioeconomic status (Turck et al. 1962). Poor compliance with therapy is common in hypertensive patients of low socioeconomic class, and this may explain in part the relatively high incidence of accelerated hypertension in this group of patients. Socioeconomic factors are also important in idiopathic calcium stone formation, which is more common in men of higher social class. This may of course, relate to the increased protein intake associated with increased affluence.

A knowledge of the social environment is of great importance in planning the management of patients with end-stage renal disease. The choice of a particular dialysis mode is influenced by the family support available to the patient and the willingness of the family to provide care and attention.

Details of tobacco consumption should be obtained, since this represents a significant vascular risk factor. The deleterious effect of smoking contributes to the development of artherosclerosis in dialysis patients, and it is also a risk factor for renovascular hypertension, (MacKay et al. (1979), accelerated hypertension (Isles et al. 1979), Goodpasture's syndrome in young males, and for the development of diabetic nephropathy (Stegmayr and Lithner 1987; Telmer et al. 1984). Drug addiction should also be sought, since it exposes the patient to a number of renal complications, such as acute renal failure due to rhabdomyolysis, amyloidosis, vasculitis, proliferative glomerulonephritis, and septicaemia. In addition, these patients are at risk from a number of infections, including HIV, which may be associated with a focal and segmental glomerulosclerosis (see Chapter 3.12).

Occupational history

A number of occupational factors are important in the development of renal disease. Working in hot atmospheres with increased passive fluid loss increases the incidence of urinary stone formation. Certain workers are exposed to toxins and chemicals, and hydrocarbon inhalation may play a role in triggering glomerulonephritis, although convincing evidence has been reported only for Goodpasture's syndrome. It may be that solvent exposure does not cause glomerulonephritis, but rather delays its resolution. Certain chemicals are also associated with the induction of renal disease; for example, aniline dye workers have an increased incidence of urothelial tumours.

Certain infections with a predilection for the kidney may also be contracted by certain occupational groups. Acute renal failure due to leptospirosis is more common in miners, sewage workers, and farm labourers; Hantavirus infection may arise in laboratory workers handling animals such as rats, or in farmers from endemic areas. Occupational details should be recorded, particularly in patients who handle chemicals or are exposed to toxic vapours. Exposure to lead, particularly in a vaporized form, such as occurs during welding lead pipes or in using oxyacetylene to cut metal previously covered with a lead-based paint, may result in lead nephropathy and subsequent chronic renal failure.

Ethnic and geographical factors

Ethnic factors may be important in various renal diseases. Amyloidosis complicates familial Mediterranean fever in Arabs from the Mediterranean area, Turks, and Sephardic Jews, whereas Sephardic Jews from Baghdad, Southern Russia, and the Balkans, and Ashkenazic Jews and Armenians are rarely affected. The incidence of IgA nephropathy is greater in white populations and in some Asian countries (Japan, Singapore, and China) than in black populations in North America and Africa or in North African patients. Mesangioproliferative glomerulonephritis is common in Navajo and Zuni Indians in the United States, whereas the incidence of renal disease due to non-insulin dependent diabetes mellitus is increased in Zuni and Pima Indians. Systemic lupus erythematosus is several times more common in oriental and black populations than amongst Caucasians, and within the latter group it is more common in the Middle East and India than in Europe. Glomerular disease related to tropical infections such as malaria obviously has a geographical factor. Tubulointerstitial disease such as Balkan nephropathy may also have a geographical basis (see Chapter 6.7).

Sickle-cell disease is widespread amongst African, Middle Eastern, and Indian populations and both severe hypertension and diabetes mellitus with renal failure are more common in black patients. Black and Indian patients may develop tuberculosis during immunosuppressive treatment for systemic disease or following transplantation: many have advocated prophylactic antituberculous chemotherapy in the latter situation. Patients from the tropics may have positive treponemal serology due to previous exposure to yaws, and the incidence of a positive antinuclear factor is much greater. In addition, the 'normal' values and rate of growth, size, and weight derived from an exclusively Caucasian population may not apply to other ethnic groups, e.g. the lower peripheral white cell count in black populations. Finally, it must be remembered that there are ethnic differences in enzymes involved in the handling of drugs, such as the reduced oxidoreductase activity of oriental patients.

Ethnic and geographical factors also relate to bladder stone disease, hypertension and tuberculosis.

Travel abroad exposes many patients to diseases such as malaria which may affect the kidneys. It is not uncommon for patients to neglect proper prophylactic antimalarial therapy, and details of foreign travel, even years previously, may be important. For instance, fulminant and disseminated strongyloidiasis may develop in immunosuppressed transplant patients who visited endemic tropical areas even up to 30 years previously. In such patients a prophylactic course of thiabendazole or mebendazole is justified before surgery, and this should be repeated 2 to 3 weeks after transplantation.

Family history

It is important to obtain medical information on the spouse, children, parents, siblings, and other relatives. The family pedigree should be established in as much detail as possible; this may involve obtaining medical notes and interviewing and examining relatives. This is obviously essential in hereditary renal diseases (the main groups of which are listed in Table 3 and described in Section 18), although it may be time-consuming to collect precise data on unaffected as well as affected family members. Some families show psychological reluctance to undergo genetic investigation and considerable care and tact are required. Genetic predispostion is observed in other renal diseases in addition to the well-established inherited disorders. The familial predisposition to systemic lupus erythematosus and other autoimmune diseases is well known. Vesicoureteric reflux is clearly familial in some kindreds. This is not only of theoretical interest, but may also be clinically relevant; there should be adequate diagnosis and treatment of recurrent febrile episodes in the child of a woman with reflux nephropathy. Familial occurrence has been documented in primary immune glomerulonephritis and in IgA nephropathy (Berger's disease and/or Henoch–Schönlein nephritis). The urine of first-degree relatives of children presenting with haematuria should be examined for blood.

Genetic heterogeneity should also be kept in mind: renal diseases that are similar on clinical grounds (on the basis of our current knowledge) may have different modes of inheritance in different families. This is well substantiated in Alport's syndrome and the nephronophthisis-medullary cystic disease complex. Familial predisposition to some renal complications of other diseases may be more complex. Genetic susceptibility to diabetic nephropathy has been demonstrated, and it has also been suggested that patients with insulin-dependent diabetes mellitus and a family history of hypertension are more prone to diabetic nephropathy than those without such a family history (see Chapter 4.1). This may apply to other renal diseases. For instance, genetic susceptibility to essential hypertension may contribute to an increase in blood pressure; even in patients with renal disease, a family history of hypertension should be sought in these patients. On occasions it is difficult to differentiate between genetic and environmental factors.

Table 3 Classification of the main groups of inherited kidney diseases

1. Cystic kidney disease (Chapter 18.2)
2. Alport's syndrome and variants (Chapter 18.4.1)
3. Inherited metabolic diseases with kidney involvement
    With non-glomerular involvement—such as cystinosis primary hyperoxaluria (Chapter 18.5) or inherited urate nephropathy (Chapter 6.3)
    With glomerular involvement—such as Fabry's disease or lecithin-cholesterol acyltransferase deficiency (Chapter 18.2), and also associated with diabetes mellitus, defects of the complement system, genetic amyloidosis etc.
4. Other inherited diseases with glomerular or non-glomerular involvement
    With glomerular involvement, such as the congenital nephrotic syndrome (Chapter 18.4.4) or the nail-patella syndrome and other syndromes with extrarenal defects (Chapter 18.4.3)
    With non-glomerular involvement, such as nephronophthisis (Chapter 18.3), tuberous sclerosis with renal angiomyolipoma (Chapter 19.1)
    With cystic kidney disease (Chapter 18.2), neurofibromatosis with renal artery changes (Chapter 17.7), and von Hippel-Lindau's disease with renal cell carcinoma (Chapters 18.1, 18.2, and 19.1)
5. Primary immune glomerulonephritis, occasionally familial such as in IgA nephropathy, or more rarely in other primary glomerular diseases (Section 3)
6. Inherited tubular disorders such as cystinuria (Section 14) and various inherited tubular defects described in Section 5
7. Various renal diseases with 'genetic influence' such as calcium nephrolithiasis, vesicoureteric reflux and related nephropathy, haemolytic uraemic syndrome
8. Unclassified cases

Systematic enquiry

Even when the patient interview has proceeded along the previously described lines, there may still be areas which have not been covered, or a reluctance of the patient to reveal details. There is, therefore, a need for a systematic enquiry which should cover all aspects of general health and include an enquiry as to changes in weight and appetite. Enquiries should always be made to determine whether the patient has at any time suffered from dysuria, frequency, nocturia, or haematuria. In patients with chronic renal failure, it is important to document whether the patient has suffered from pruritus, breathlessness, peripheral oedema, sleep disturbance, restless legs, visual changes, or gritty eyes. The duration of nocturia may give a valuable clue to the duration and rate of progression of the disease. The majority of patients are reluctant to discuss, particularly at a first interview, problems related to sexual function but it is important to enquire about libido, and of impotence in males. These factors can cause considerable anxiety in a patient and are commonly only discussed once the examining clinician has introduced the subject.

Clinical examination in adults

Examination of the kidneys and urinary tract

Clinical examination of the urinary tract follows the standard pattern of inspection, palpation, percussion, and auscultation. On abdominal inspection, it may be possible to see one or both kidneys, particularly if advanced polycystic kidney disease is present. Occasionally, chronic obstructive uropathy may cause such distension of the ureters that they become visible. Obstruction to the bladder outflow track results in bladder distension which may be readily visible in patients, particularly those who are thin.

Palpation of the kidneys is best carried out with the patient in the recumbent position with the head slightly raised on a pillow and the arms resting at the side of the body. The right kidney is palpated by placing the left hand posteriorly in the loin and the right hand horizontally on the anterior abdominal wall to the right of the umbilicus. By pushing forward with the left hand and asking the patient to take a deep breath, the lower pole of the kidney is commonly palpable in thin patients by pressing the right hand inwards and upwards. To palpate the left kidney, the left hand should be placed posteriorly in the left loin and the right hand on the anterior abdominal wall to the left of the umbilicus. The left kidney is not as readily palpable as the right and care should be taken to distinguish kidney from spleen. On palpation, an estimate of the size and shape of the kidney should be made if possible, although this may be difficult except in those who are very thin. In normal patients the surface of the kidney is smooth and relatively hard, but in those with cystic disease an irregular surface may be detected. Any tenderness on palpation should be recorded.

Abdominal percussion may be of value if there is difficulty distinguishing between an enlarged left kidney and splenomegaly or in patients with hepatomegaly. Percussion is also of value in determining the presence and degree of ascites.

Auscultation of the abdomen is essential in all patients with hypertension; the stethoscope should be placed posteriorly in the loin, laterally in the flank and anteriorly and in each area the examiner should listen carefully for the presence of a bruit. In addition, auscultation is mandatory in any patient who has had a renal biopsy and who subsequently develops hypertension, in view of the possibility of the development of a postbiopsy arteriovenous fistula.

Examination of the urinary tract is incomplete without a carefully conducted rectal examination and, where indicated, a vaginal examination. Prostatic hypertrophy is common and, when present, may aggravate impaired renal function. In females, carcinoma of the cervix may extend laterally and result in a 'frozen' pelvis. On occasions an unexpected finding, such as a villous papilloma giving rise to profound hypokalaemia, will be detected.

Measurement of blood pressure

The measurement of blood pressure is an important step in the clinical examination of a renal patient. It is worth recalling the main precautions to prevent artefacts in these measurements (Frohlich *et al.* 1988). The patient should be supine or sitting, then standing, in a quiet environment with an arm resting at heart level. An appropriately sized cuff should be selected: the width should be at least 40 per cent of arm circumference, and length at least 80 per cent of arm circumference. For example, for an arm circumference of 30 cm and a bladder width of 12 cm, blood pressure readings will be correct; in contrast, for an arm circumference of 40 cm, using the same cuff, systolic blood pressure will be overestimated by 10 mmHg. Cuff errors are of great importance in obese individuals with thick arms and lead to overestimation of blood pressure. A cuff that is too wide causes less error than one that is too narrow. The use of a wide bladder (15 cm) is recommended for all adults except those with thin arms that are out of the cuff range. The use of mercury manometers should be encouraged; aneroid apparatus should be calibrated at least once every 6 months with a mercury gravity manometer. In adults, cessation of sound (Phase V) rather than muffling (Phase IV) corresponds to diastolic blood pressure. However, in certain subjects, including those with aortic valvular insufficiency or high cardiac output (as in anaemia, thyrotoxicosis, or pregnancy) the Korotkoff sounds do not disappear. In such conditions, Phase IV is the only reliable index of diastolic blood pressure. In patients with cardiac dysrhythmias, such as atrial fibrillation, multiple readings are needed and blood pressure values should be considered as only approximations.

Overestimation of systolic blood pressure is common in elderly patients with hard, calcified vessels. This may be detected clinically when the radial artery is palpable even when the cuff is inflated above systolic blood pressure (positive Osler's maneouvre, suggesting pseudohypertension). Bias in reading on the part of the observer may be overcome by devices that 'blind' to the actual pressure values (e.g. the Hawksley Random-Zero Device and the London School of Hygiene sphygmomanometer).

Measurement of blood pressure is essential in all children with renal disease, indeed in all sick children, yet is frequently omitted, occasionally with disastrous results. Patience and practice is required; the correct cuff is the largest that can be applied to the upper arm, and results must be related to published normal data.

In addition to casual blood pressure measurements, two other modes of monitoring have recently been developed but have so far not been extensively tested in renal patients. Home blood pressure measurements (Hunt *et al.* 1985) can be obtained with either the usual manometers or with electronic manometers, whose accuracy is difficult to assess. This technique may be useful in some patients who are willing to participate in their own management. However, in those who have a tendency to become alarmed by minor changes in blood pressure it can have deleterious effects. Ambulatory blood pressure measurements can be performed with various devices (Winslow *et al.* 1986). As stated by Frohlich *et al.* (1988) 'they are currently not practical for widespread and general clinical application' and data are not yet available concerning 'normal' ambulatory 24 h pressure. This method, however, is valuable in research programmes involving essential hypertension, hypertension associated with renal disease or in end-stage renal failure patients.

General examination

In any patient with renal disease it is important to undertake a full clinical examination and carefully document all detected abnormalities. It is important to remember that the absence of certain findings may be as important as the presence of a diagnostic sign. Each clinician develops their own particular examination technique and therefore, there is no standardized 'correct' method. The most important fact to remember is that the patient must be examined from head to toe.

The skin of patients with renal disease may reveal a number of interesting abnormalities. Patients with uraemia frequently have pallor and pigmentation and the skin is dry and flaky. There may be evidence of purpura and frequently scratch marks will indicate pruritus. Palpable subcutaneous nodules may be present, indicating dystrophic calcification. In terminal renal failure uraemic frost may be visible, particularly on the face. In Fabry's disease, small dark red hyperkeratotic papules are visible, most commonly in the periumbilical area, and are more marked in males than females. In transplant recipients, manifestations of steroid side-effects, such as mooning of the face, a central redistribution of body fat, purpura, and acne, are common. Long-term immunosuppression is also associated with hyperkeratotic lesions which can progress to keratoacanthoma and undergo malignant change (see Chapter 21.2)

The general appearance of the patient may also be diagnostic. In some patients with type II mesangiocapillary glomerulonephritis there is a partial lipodystrophy, where there is a loss of subcutaneous fat in the upper part of the body with a normal or increased distribution in the lower part (see Chapter 3.8). A number of inherited conditions are associated with typical facial appearances. Skin turgor is also an important sign: lax and slowly collapsing skin tension is indicative of salt and water depletion whereas a plastic consistency may be indicative of water intoxication.

Examination of the eyes may reveal perilimbal calcification in patients with long-standing uraemia, subconjunctival haemorrhages in patients with vasculitis, and lenticonus in Marfan's syndrome. Fundoscopic examination may reveal the changes typical of hypertension, diabetes mellitus, or vasculitis. In some patients, the fundus is particularly difficult to visualize, due to the presence of cataracts, particularly common in patients who have received high dose-corticosteroids.

The facial appearance may be suggestive of an underlying disease, for instance, the thickening and rigidity of the skin in systemic sclerosis, which may be associated with multiple telangiectasia. Hearing may be abnormal in patients with Alport's syndrome and also in patients with impaired renal function who have been exposed to excessive doses of aminoglycosides. In examination of the mouth it is important to document the state of the teeth, the condition of the palate and gums, and the presence of any fungal infections.

On examination of the praecordium, particular attention should be paid to murmurs and the presence of a pericardial rub. Any murmurs should be carefully documented—as a changing murmur is particularly suggestive of infective endocarditis. Flow murmurs are particularly common in patients with uraemic anaemia, but although pericarditis will most commonly be due to uraemia it may also arise from infection and vasculitis. Additional heart sounds may indicate fluid overload. In examining the chest particular attention should be paid to the presence of crepitations, pleural fluid, and pleuritic rubs.

Abdominal examination mainly involves an estimation of kidney size, but the presence of any other organomegaly or tumours should be determined. Lymphadenopathy may indicate an underlying lymphoproliferative disease. Testicular examination should be undertaken to determine whether atrophy or tumour is present. In patients in chronic renal failure, who may subsequently require chronic ambulatory peritoneal dialysis, it is necessary to exclude inguinal or umbilical hernias which may not be evident with the patient lying down.

It is important to assess whether there is growth retardation in children and adolescents and whether there is any abnormality of bones indicative of an underlying renal osteodystrophy. Height and weight should be accurately measured using a stadiometer and plotted on appropriate centile charts. Secondary sexual characteristics should be documented as puberty and is commonly delayed in uraemia.

In examination of the limbs attention should be paid to the hands, with specific reference to the nails. Transverse ridges (Beau's lines) may indicate serious preceding illness. In a nephrotic syndrome, the nails are pale and opaque, whereas splinter haemorrhages may be visible in patients with vasculitis and endocarditis. In the lower limbs, oedema may be obvious and (particularly in older patients) examination should be made of the vascular supply, as atherosclerosis is particularly frequent in long-standing renal failure. The presence of proximal myopathy can be assessed by asking the patient to rise from a chair without using the upper limbs. The patella should be examined for the abnormalities of the nail-patella syndrome.

Nervous examination should include the central nervous system, the peripheral nervous system, and the autonomic nervous system. Motor and sensory function should be tested and the integrity of the autonomic nervous system can be tested by measuring the increase in blood pressure with sustained hand grip or the changes in the R-R interval during a Valsalva manoeuvre. Clinical neuropathy is particularly likely to be manifest in patients with diabetes mellitus.

A full examination in patients with renal disease is mandatory and while particular attention should be paid to the urinary tract and the clinical signs indicative of underlying renal disease, all systems should be examined thoroughly, so that the degree of systemic involvement is carefully documented.

Particular aspects of the neonate and child with renal disease

In these cases the informant will usually be a parent rather than the patient himself. A careful family history is important (including evidence of consanguinity) because of the high frequency with which renal disease in childhood has a genetic basis. The pregnancy should be reviewed: enquiries should be made about the results of antenatal ultrasound examination, which may be a pointer to congenital obstructive uropathy or cystic disease. Oligohydramnios suggests severe renal dysplasia or agenesis. Perinatal asphyxia, particularly associated with macroscopic haematuria, may cause ischaemic medullary necrosis and result in a renal scarring, difficult to differentiate from reflux nephropathy. Fever and failure to thrive in early infancy suggests urinary tract infection and may be a pointer to underlying urological malformation, particularly obstructive uropathy or vesicoureteric reflux. A poor urinary stream in a male infant suggests a posterior urethral valve, but may also be due to neuropathic bladder. Failure to thrive, vomiting, and episodes of dehydration may be a pointer to renal dysplasia with chronic renal failure or to a renal tubular disorder. A review of the growth records are often illuminating as they may give an indication of the duration of severe renal disease.

The physical examination of young children requires a certain expertise. Height (length), weight, and blood pressure measurements are mandatory at all ages, as is a brief developmental assessment. The state of hydration should be reviewed: the jugular venous pulse is not easily observed in babies, but intravascu-

Table 4 Main renal disorders having relatively high incidence in elderly patients*

Glomerular
  Membranous nephropathy (with nephrotic syndrome)
  Crescentic glomerulonephritis (with rapidly progressive renal failure)
  Anti-GBM nephritis (in women)
  Amyloidosis
Tubulointerstitial
  Analgesic nephropathy
  Obstructive uropathy (e.g. prostate obstruction, retroperitoneal fibrosis, pelvic cancer)
  Multiple myeloma and related diseases
Vascular
Nephrosclerosis
  Diabetic nephropathy (mainly type II diabetes, with vascular and glomerular involvement)
  Renal atheromatous disease (including renal artery stenosis or thrombosis, associated or not with abdominal aorta aneurysm, and multiple cholesterol emboli disease)
  Vasculitis (Wegener's granulomatosis; microscopic polyarteritis)
Drug-induced
Cystic
  Solitary or multiple renal cysts
  Autosomal dominant polycystic disease
Malignant
  Renal adenocarcinoma
  Transitional cell carcinoma of urinary tract
Water and electrolyte disorder

* Adapted from Macias Nunez and Cameron (1987).

lar volume overload is usually reflected in hepatomegaly. The kidneys are often palpable in a normal neonate. Congenital malformations such as polydactyly, external ear abnormalities, or meningomyelocele may be a pointer to renal problems. Absent abdominal wall musculature ('prune-belly') is associated with cryptorchidism and urological abnormality, and is probably the consequence of transient urethral obstruction in fetal life. The genitalia should always be examined, particularly in girls with recurrent urinary tract infections, because of the possibility of sexual abuse.

Renal disease in elderly patients

Elderly patients have an increased incidence of certain conditions (Table 4). The changes in renal function which occur with age (see Chapter 1.5) must be taken into account when assessing the results of investigations. Care must also be exercised in the use of therapeutic agents and appropriate adjustments of dosage must be made (see Chapter 2.2).

Emergencies in patients with renal diseases

A number of renal diseases and complications of renal disease present as acute medical emergencies requiring urgent action. Some, such as hyperkalaemia, require immediate treatment whereas in others there is time to undertake certain specific investigations to confirm the diagnosis, e.g. renal biopsy and serological testing for anti-GBM antibodies in Goodpasture's syndrome.

In patients with primary renal disease the major acute emergencies involve rapidly progressing glomerulonephritis due to crescentic glomerulonephritis, Goodpasture's syndrome, vasculitis (systemic lupus erythematosus and polyarteritis), mesangiocapillary glomerulonephritis, and, rarely, IgA nephropathy. In such situations urgent investigations, including renal biopsy, should be undertaken so that appropriate therapy may be undertaken at the earliest opportunity. In many instances, if the glomerular disease has progressed to the stage of oliguria the prognosis for recovery of renal function is very poor, in spite of all therapeutic interventions. If there is a clinical suspicion that any of these conditions underly an acute deterioration then referral to a specialist centre is required.

In patients with acute renal failure the most common medical emergencies involve disorders of fluid and electrolytes and sepsis. Pulmonary oedema, particularly of acute onset, may have few physical signs other than breathlessness and trachypnoea. Hyperkalaemia is asymptomatic and can be life-threatening; children seem to tolerate hyperkalaemia better than adults. In patients with prolonged acute renal failure, particularly if associated with hypotension, there is always the risk of gastrointestinal haemorrhage from multiple small areas of mucosal infarction.

Fluid and electrolyte disorders are also common in chronic renal failure but they tend to be more insidious in progression. Pericarditis is common and may give rise to cardiac tamponade. Patients who have progressed to an end-stage renal failure programme may have acute medical emergencies relating to vascular access (fistula thrombosis) dialysis technique (such as allergy to dialyser sterilizing material, ethylene dioxide), dialysis-associated medications (heparin and bleeding; erythropoietin and convulsions and/or hypertension), dietary management (hyperkalaemia, pulmonary oedema) and infection. Transplant recipients may present with acute infections (Herpes simplex encephalitis, Pneumocystis pneumonia) and rejection episodes.

The nephrologist may therefore be faced with many diverse acute medical emergencies, and must be aware of their presentations and management.

Table 5 Emergencies in the renal patient

Malignant or severe hypertension
Acute renal failure with oligoanuria
Acute glomerulonephritis (in childhood)
Severe water and electrolyte disorders, particularly hyperkalaemia
Renal colic and any acute urinary tract obstruction
Cardiovascular and respiratory emergencies: not only myocardial infarction, acute pulmonary oedema, aortic dissection, etc., but also uraemic pericarditis, pulmonary embolism in the nephrotic patient, severe haemoptysis in a patient with the Goodpasture syndrome etc.
Acute thrombosis of arteriovenous fistula
Severe haemorrhage such as gastrointestinal bleeding in patients with acute renal failure, severe bleeding after renal biopsy or from arteriovenous fistula, acute rupture of a transplanted kidney, intracranial haemorrhage due to ruptured aneurysm in polycystic kidney disease.
Fever such as that due to upper urinary tract infection or to septicaemia of any other origin; or fever in immunodepressed patients

Structured investigation

Having obtained a full history and undertaken a careful clinical examination, the clinician will be in a position to formulate a differential diagnosis with respect to the underlying disease. The investigations which are subsequently undertaken should be structured in such a way as to confirm or refute the postulated diagnosis.

A number of basic investigations are essential in any patient who is known to have or suspected of having an underlying renal disease. Urine analysis should be performed at each clinic visit and the results carefully recorded in the patient's notes. As a minimum the urine pH and the presence of blood, protein, and glucose should be documented. Urine should also be obtained for microscopy; particular attention should be paid to the presence and type of casts and, if haematuria is detected, the morphology of red cells. Urine culture should also be undertaken; where appropriate, this should include investigation for anaerobic organisms and *Mycobacterium tuberculosis*. All patients should also have blood urea, serum creatinine, and electrolytes determined and a full blood count should be performed.

If a 'stix' examination reveals proteinuria then further investigations to determine its type and extent are indicated. Since it is difficult to obtain an accurate 24-h urine collection for protein quantitation in outpatients, it has been suggested that determination of protein:creatinine ratio in a single sample is adequate; this is particularly helpful in children, and is probably sufficient for patient follow-up. However, it should be remembered that there is a diurnal variation in urinary protein and creatinine excretion, and samples should be collected at approximately the same time of day so that reasonable comparisons can be made. The type of protein being excreted in the urine can be determined by electrophoresis.

In the presence of urinary abnormalities, particularly proteinuria and haematuria, consideration should be given to more detailed investigation of the urinary tract and kidney histology. In most circumstances the first investigation should be a renal ultrasound examination. If investigation of the urine reveals the presence of red cell casts, or dysmorphic red cells, then it is more appropriate to proceed to renal biopsy than to undertake an intravenous pyelogram. Before biopsy, it is important to determine that the patient has two kidneys of a reasonable size, and this can best be undertaken by ultrasound examination just prior to the biopsy procedure, which may be ultrasound guided. Material obtained on renal biopsy should be placed in appropriate fixative so that subsequent examination can include light microscopy, immunological investigation, and electron microscopy.

In hypertensive patients renal perfusion should be investigated particularly if renovascular disease is suspected; abnormal results should be followed by either digital vascular imaging or arteriography. In patients with vascular abnormalities, it may be necessary to proceed to the estimation of renal vein renin concentrations.

In patients suspected of having renal tubular defects, specific investigation can be undertaken to determine urinary concentration, acidification, and urinary amino-acid excretion.

In a properly structured investigative plan, test are undertaken in a sequential fashion, commencing with simple screening tests and proceeding to specific diagnostic tests. There is always a tendency to undertake investigations because the test is available, rather than because it is believed to be of particular value in reaching a diagnosis in a particular patient. The urge to do this should be resisted unless the investigation is being undertaken as part of a specific study to which the patient consents.

Conclusion of initial interview

After obtaining the relevant clinical history and undertaking a full clinical examination it is possible to formulate a differential diagnosis from cognitive associations. In some instances this diagnosis will be obvious at this stage, as in patients with gross adult type polycystic kidney disease, but in many it will only be obtained following the results of specific investigations such as renal biopsy. The nature of the investigations should be explained in full to the patient and, where appropriate, to the relatives. This is particularly important for the so-called invasive investigations such as biopsy and arteriography. Many investigations are not diagnostic, but add to the overall basic data base and increase the ability to predict prognosis. Failure to explain this to the patient may subsequently result in considerable confusion. Some conditions, such as Goodpasture's syndrome, require rapid invasive investigations, but many can be treated in a more considered way and patients should be given the opportunity of considering the advice offered before committing themselves to potentially hazardous investigations. At all times the patient should be encouraged to raise questions, although many, particularly the elderly, are reluctant to trouble the 'busy' physician. Finally, the patient should be advised how long the investigation will take and also how long it will take for laboratory studies to be completed and for the results to become available. It is sometimes worth documenting this advice in the patient's notes so that important information is not 'lost'.

It is worth remembering that the most important points to emerge may only come at the conclusion of the interview, when a patient may remark 'while I am here may I ask you about . . . '. Sometimes it takes this time for the patient to pluck up courage to ask about something which is a major cause for concern. Commonly patients believe they have an illness of a more serious nature than reality and much anxiety may develop. To avoid mentioning certain diagnoses, such as malignancy, may only serve to reinforce the patient's mistaken belief, and time must be spent in explaining the nature of investigations and the results obtained together with the details of treatment and prognosis.

Planning follow-up

The aim of patient follow-up is:

1. To measure the effectiveness of therapy
2. To detect any change in the clinical condition particularly with respect to deteriorating renal function and more rarely
3. To gain experience of a particular clinical condition or the effects of newly introduced therapeutic measures.

Such follow-up should inconvenience the patient as little as possible, while ensuring that these aims are properly achieved.

Planning of a follow-up programme should consider who is to undertake the observations, the intervals at which the patient should be seen and the measurements which should be taken.

In many conditions, follow-up can be undertaken adequately by the family physician, with advice as and when required from the specialist centre. In some patients, particularly those who live some distance from the specialist centre, follow-up is best achieved by a local physician with a specific interest in renal disease. For patients with conditions which require treatment with potentially hazardous medications, such as the combination of corticosteroids and immunosuppressive drugs, and for patients whose clinical status is likely to change rapidly observation by a nephrologist is advisable.

If follow-up is to be undertaken outside a specialist centre, the clinician must be provided with adequate information. This should include the diagnosis and details regarding the natural history of the condition, outlining what could be expected to happen to a particular patient. Full details relating to specific therapy should be clearly documented, along with specific advice relating to when the patient should be referred back to the specialist centre.

The frequency of follow-up visits should be carefully reviewed at regular intervals. Patient review frequently needs to be made at short intervals to determine the adequacy of therapy and to detect deterioration in renal function. Even in apparently homogeneous conditions such as membranous glomerulonephritis the rate of progression of renal functional impairment is very variable; the only means of determining this at present is through regular follow-up visits at which renal function is estimated. Fortunately, for many conditions the rate of progression in an individual is relatively constant, and it is possible to lengthen the time between clinic visits for those patients with stable or only slowly deteriorating disease, whilst increasing the frequency in other patients in whom function is deteriorating more rapidly. In patients with deteriorating renal function it is important to plan ahead, introducing dietary manipulations at an appropriate time and preparing for dialysis by creating a fistula early, so that it can mature before being used. Such planning also affords the opportunity of introducing to the patient the concept of end-stage renal failure and management by dialysis and/or transplantation, allowing time for the patient to come to terms with his illness and for an introduction to the staff of the dialysis unit. Experience shows that patients who enter a dialysis programme following a period of management in a clinic progress very much more favourably than those patients whose first contact with a dialysis unit is at a time when they are in terminal renal failure.

In planning a follow-up programme it is also important to give thought to the clinical investigations which are required at each visit. Urine analysis, blood pressure measurement, and estimation of urea and electrolytes would appear to be mandatory for all patients, but other more detailed investigations should only be undertaken as indicated. Effective planning of follow-up appointments allow for the most cost-effective use of both investigative resource and clinical time.

References

Burden, R.P., Booth, L.J., Ockenden, B.G., Boyd, W.N., Higgins, P. McR., and Aber, G.M. (1975). Intrarenal vascular changes in adult patients with recurrent haematuria and loin pain—a clinical, histological and angiographic study. *Quarterly Journal of Medicine*, **44**, 433–474.

Burden, R.P., Dathan, J.R., Etherington, M.D., Guyer, P.P., and MacIver, A.G. (1979). The loin-pain/haematuria syndrome. *Lancet*, **i**, 897–900.

Edelman, S. and Witztum, J.L. (1989). Hyperkalaemia during treatment with HMG-CoA reductase inhibitor. *New England Journal of Medicine*, **320**, 1219–20.

Frohlich, E.D., Grim, C., Labarthe, D.R., Maxwell, M.H., Perloff, D., and Weidman, W.H. (1988). Recommendations for human blood pressure determination by sphygmomanometers; report of a special task force appointed by the Steering Committee. *American Heart Association Circulation*, **77**, 502A–514A.

Grünfeld, J-P. and Bellet, M. (1982). A propos des hypertensions artérielles induites par les medicaments. *Néphrologie*, **3**, 167–170.

Hunt, J.C., Frohlich, E.D., Moser, M., Roccella, E.J., and Keighley, E.A. (1985). Devices used for self-measurement of blood pressure. Revised statement of the National High Blood Pressure Education Programme. *Archives of Internal Medicine*, **145**, 2231–2234.

Isles, C., *et al.* (1979). Excess smoking in malignant phase hypertension. *British Medical Journal*, **1**, 579–581.

Lagrue, G., Laurent, J., and Rostoker, G. (1989). Food allergy and idiopathic nephrotic syndrome. *Kidney International*, **Suppl. 27**, S147–51.

Lang, T., Degoulet, P., Aime, F., Devries, C., Jacquinet-Salord, M.C., and Fouriaud, C. (1987). Relationship between alcohol consumption and hypertension prevalence and control in a French population. *Journal of Chronic Diseases*, **40**, 713–720.

Lurvey, A., Ysin, A., and Dequattro, V. (1973). Pseudopheochromocytoma after self-administered isoproterenol. *Journal of Clinical Endocrinology and Metabolism*, **36**, 766–769.

MacGregor, G.A. and De Wardener, H.E. (1988). Idiopathic oedema. In *Diseases of the Kidney*. 4th Ed., R.W., Schrier and C.W., Gottsshalk eds., Little, Brown and Co., Boston, p. 2743–2753.

MacGregor, G.A., Tasker, P.R.W., and De Wardener, H.E. (1975). Diuretic induced oedema. *Lancet*, **i**, 489.

MacKay, A., Brown, J.J., Cumming, A.M.M., Isles, C., Lever, A.F., and Robertson, J.I.S. (1979). Smoking and renal artery stenosis. *British Medical Journal*, **2**, 770.

Meadow, R. (1977). Munchhausen syndrome by proxy. The hinterland of child abuse. *Lancet*, **ii**, 343–345.

Portioli, I. and Valcavi, R. (1981). Factitious phaeochromocytoma: a case for Sherlock Holmes. *British Medical Journal*, **283**, 1660–1661.

Saunders, J.B., Beevers, D.G., and Paton, A. (1981). Alcohol-induced hypertension. *Lancet*, **ii**, 653–656.

Stegmayr, B. and Lithner, F. (1987). Tobacco and end stage diabetic nephropathy. *British Medical Journal*, **295**, 581–582.

Stewart, P.M., Valentino, R., Wallace, A.M., Burt, D., Shackleton, C.H.L., and Edwards, C.R.W. (1987). Mineralocorticoid activity of licorice: 11-β-hydroxysteroid dehydrogenase deficiency comes of age. *Lancet*, **ii**, 821–823.

Stokes, G.S. (1976). Drug-induced hypertension: pathogenesis and management. *Drugs*, **121**, 222–230.

Telmer, S., Sandahl Christiansen, J.M., Andersen, A.R., Nerup, J., and Deckert, T. (1984). Smoking habits and prevalence of clinical diabetic microangiography in insulin-dependent diabetics. *Acta Medica Scandinavica*, **215**, 64–68.

Turck, M., Goffe, B.S., and Petersdorf, R.G. (1962). Bacteriuria of pregnancy. Relation to socioeconomic factors. *New England Journal of Medicine*, **266**, 857–860.

Winslow, C.M., *et al.* (1986). Automated ambulatory blood pressure monitoring. *Annals of Internal Medicine*, **104**, 275–278.

1.2 Urinalysis and microscopy

GIOVANNI BATTISTA FOGAZZI

While it is still debated whether it is worthwhile to perform urinalysis in unselected patients (Lancet 1988), there is no doubt that urinalysis is mandatory when the patient's history or the findings on examination suggest the presence of a urinary tract disease. In such patients urinalysis is of cardinal importance in determining the type and severity of the disease and in evaluating its course over time. Urinalysis includes both physical and chemical examination of the urine, as well as a microscopic examination of the urine sediment.

Physical examination of the urine

Colour

The urine colour must be evaluated in optimal conditions, i.e., under good lighting and with the urine in a glass container and seen against a white background. The colour of normal urine ranges from pale to dark yellow and amber, which depends on the urochrome concentration. Gross haematuria is the most important and frequent cause of alteration of urine colour. In this condition the urine is pink to black. The factors that influence the final colour are the amount of erythrocytes or of haemoglobin contained, the pH, and the duration of the contact between the haemoglobin and the urine. The lower the pH and the longer the contact, the darker the colour (Berman 1977). Urine is of a variable red colour also in haemoglobinuria, myoglobinuria, after eating beets by some genetically susceptible people, and after rifampin. All these conditions can easily be distinguished from haematuria by the absence of erythrocytes under the microscope. Another frequent cause of hyperpigmented urine (yellow to brown) is jaundice and all other states associated with high conjugated bilirubin levels. In other conditions, urine may be normal in colour when fresh, but darkens upon standing. This occurs in porphyria because of urinary porphobilinogen, in melanoma because of melanogen, and in alkaptonuria because of homogentisic acid. Therefore, when such diseases are suspected the urine must be exposed to light for some time. Drugs may also influence colour of urine. Urine containing nitrofurantoin is dark yellow to orange and that containing levodopa brown to black; amitryptyline and methylene blue colour urine green or blue-green.

Appearance

Urine may be cloudy whenever one or more of the formed elements within it is present in high concentration. These are more frequently leucocytes, erythrocytes, epithelial cells, bacteria, or amorphous material (Schumann and Greenberg 1979). Moreover, a clear fresh urine may cloud after standing, especially when kept in a refrigerator, because of the precipitation of phosphates or urates.

Odour

A pungent smell is typical of urine contaminated with bacteria, while a sweet or fruity odour is due to ketones. Some rare diseases confer a characteristic smell to the urine. These are maple syrup urine disease (maple syrup odour), phenylketonuria ('musty' or 'mousy' odour), isovaleric acidaemia ('sweaty feet' odour), hypermethioninaemia ('rancid butter' or 'fishy' odour) (Graff 1983).

Specific gravity

Urine specific gravity is the ratio of the weight of a volume of urine to the weight of the same volume of distilled water at a constant temperature. Specific gravity evaluates the concentrating and diluting ability of the kidney and depends on both the number and size of the particles dissolved in the urine. It is commonly measured with the urinometer, which is a weighed float marked with a scale for specific gravity values from 1.000 to 1.060. The use of the urinometer is very simple and fast, but has some drawbacks since it requires at least 30 ml of urine and correction of the readings if the temperature of the urine is different from that for which the urinometer is calibrated (usually 20°C) or if the urine contains large amounts of proteins or of glucose (Strasinger 1985). The concentration of the urine may also be measured by a refractometer, which actually measures the refractive index of the urine; by a dipstick, which changes its colour according to the ion concentration of urine; by an osmometer, which measures the number of particles in milliosmoles/kg of water. These methods have the advantages of requiring only small volumes of urine (1 drop to 2 ml) and of being little or not at all influenced by the size of the particles. In random samples from a single subject specific gravity ranges from 1.003 to 1.035. When it is very low, urine is not suitable for analysis since protein concentration may be underestimated and formed elements may lyse, dissolve, or become deformed.

Chemical examination of the urine

pH

Although in the normal subject urinary pH may range from 4.5 to 8.0, it usually averages 5.0 to 6.0, with variations caused mainly by the food intake. The pH is commonly evaluated by dipsticks containing methyl red and bromothymol blue, which turn from yellow or orange to blue as it increases. These sticks are reliable although misclassifications may occur; these are due mainly to the reader's attitude and motivation, as shown in a study performed by James et al. (1978). When a more accurate determination of the pH is needed, a pH meter with a glass electrode is used. Urinary pH is useful in detecting acid–base balance disorders and in conditions that require the urine be maintained at a specific pH. This occurs in patients with urolithiasis, in patients with urinary tract infection and when a certain pH is desired to increase the efficacy of some antimicrobial agents (aminoglycosides and erythromycin are more active in alkaline urine; nitrofurantoin and tetracycline are more active in acid urine). Knowledge of the pH is mandatory in urine sediment analysis for identification of crystals and to assess whether the sample is reliable for the study of casts, which dissolve as the pH increases.

Glucose

In the normal subject, only 2 to 20 mg/dl of glucose are found in the urine. Pathological amounts are found in diabetes mellitus and in renal glycosuria, conditions in which the evaluation of urinary glucose is of paramount importance.

Glucose can be detected by the 'glucose oxidase test' or by the 'copper reduction test'. The glucose oxidase test is carried out using dipsticks which are impregnated with a mixture of glucose oxidase, peroxidase, chromogen, and buffer. In the presence of glucose, a sequential enzyme reaction occurs which causes a colour change of the reagent pad, on the basis of which a semiquantitative evaluation can be made. This test detects only glucose and is sensitive to concentrations of 50 to 100 mg/dl. False-positive results are possible when hypochlorite or peroxide contaminate the urine container; false-negative results may be seen if urine contains large amounts of ascorbic acid.

The copper reduction test is based on the property of glucose and other sugars to reduce copper sulphate to cuprous oxide in the presence of alkali or heat. It can be performed by either Benedict's test or Clinitest (Ames). Clinitest is simpler and faster and consists of tablets which contain copper sulphate and other reagents necessary to its reduction. When the tablet is added to urine which contains reducing substances, the colour changes from negative blue to green, yellow, orange, and brick-red. Clinitest can measure as little as 250 mg/dl of reducing substances, while Benedict's test can reveal as little as 20 mg/dl. For this reason Benedict's test may yield mildly positive results in normal subjects. Because copper reduction tests detect other sugars besides glucose, they are of value in diagnosing diseases such as galactosuria and fructosuria. False-positive results may be seen after various drugs and urine preservatives. False-negative results are usually absent.

Haemoglobin

This is detected by dipsticks based on the pseudoperoxidase capacity of haemoglobin to catalyse a reaction between hydrogen peroxide and the chromogen o-tolidine. In the presence of haemoglobin, o-tolidine turns blue with two possible patterns, spotted or uniform. Spotted positivity indicates intact erythrocytes, while the uniform pattern indicates free haemoglobin. Free haemoglobin may occur in the urine as the result of haematuria associated with red cell lysis in the urinary tract, favoured by alkaline pH or specific gravity below 1.007, or as the result of intravascular haemolysis. Urine microscopy is useful in distinguishing these two conditions, since erythrocytes are seen in haematuria, while they are absent in intravascular haemolysis. Dipsticks give a uniform positive pattern also for myoglobin, which appears in the urine after muscle damage. Haemoglobin can be distinguished from myoglobin by adding ammonium sulphate to the urine. Haemoglobin precipitates and leaves a clear supernatant, while myoglobin stays in solution and gives a pigmented supernatant. The two pigments can also be distinguished by electrophoresis, immunodiffusion, haemagglutination inhibition, or immunoelectrophoresis.

Dipsticks for haemoglobin may give false-positive results in urine containing oxidizing agents or large numbers of bacteria. False-negative results occur with high concentrations of urinary ascorbic acid.

Protein

Under physiological conditions urine protein excretion does not exceed 150 mg/day for adults and 140 mg/m^2 body surface for children. The main proteins excreted in the normal subject are albumin, immunoglobulins, immunoglobulin light chains, and Tamm-Horsfall protein. Proteinuria is a typical finding in renal disease, and the underlying mechanism serves for classification of the proteinurias. These are classified as 'glomerular', 'tubular', 'overload', and 'benign' (Abuelo 1983). Glomerular proteinuria is due to increased glomerular permeability to proteins and occurs in primary and secondary glomerulopathies. Tubular proteinuria is due to decreased tubular reabsorption of proteins contained in the glomerular filtrate and is seen in tubular and interstitial diseases. Overload proteinuria is secondary to increased production of low-molecular-weight proteins, such as immunoglobulin light chains (which are increased in monoclonal gammopathies) or lysozyme (which increases in some leukaemias). Benign proteinuria includes functional proteinuria, as seen in fever or after exercise, idiopathic transient proteinuria, and orthostatic proteinuria; its pathogenetic mechanism is still poorly understood.

Detection

Use of dipsticks is the method most used to detect proteinuria. These contain a pH indicator which undergoes a colour change from yellow to green after binding to proteins, according to the protein concentration. This method is very simple and fast but allows only a rough quantitation of the proteinuria, which is expressed on a semiquantitative scale from 0 to 4+. Dipsticks are more sensitive to albumin than to other proteins, with the result that false-negative results can occur in serious conditions such as those associated with Bence-Jones proteinuria. They should therefore be used with reservation and for the first evaluation of a patient rather than for a patient with certain renal disease.

Quantitative evaluation

A semiquantitative evaluation of urine proteins can also be obtained with 'precipitation methods', which evaluate the turbidity occurring after protein precipitation by sulphosalicylic, or trichloroacetic or nitric acids, or by heat and acetic acid. These methods are more sensitive than the dipstick method and detect all proteins, including albumin, globulins, glycoprotein, and Bence-Jones protein. Turbidity can be evaluated by photometry or nephelometry and this allows a reliable quantitative measurement of proteinuria (Pesce and Roy First 1979). For this reason the precipitation methods are those most used in nephrological laboratories. The false-negative and false-positive results which may occur with both dipstick and precipitation methods are shown in Table 1.

Table 1 Causes of false-negative and false-positive results with the dipstick and the precipitation methods

	Dipstick method	Precipitation methods
False-negative results	Dilute urine Light chains	Dilute urine Urine pH >8
False-positive results	Highly concentrated urine Gross haematuria Urine pH >8 Urine containing quaternary ammonium compounds or phenazopyridine Too long immersion of the dipstick in the urine	Highly concentrated urine Gross haematuria Radiographic contrast media Large amounts of urinary tolbutamide or sulphonamide metabolites, or cephalosporin or penicillin analogues

The protein/creatinine ratio

For a precise quantitation of proteinuria 24-h urine samples are needed. However, because the 24-h collection is both time-consuming and subject to error, the evaluation of protein/creatinine ratio in a single voided urine sample has been proposed for quantitating proteinuria (Ginsberg et al. 1983). The measurements obtained with this method correlated significantly with those obtained in the classical way; all patients with proteinuria more than 3.5 g/24 h had a protein/creatinine ratio higher than 3.5 in the single voided samples, and all patients who had a proteinuria of less than 0.2 g/24 h had a protein/creatinine ratio of less than 0.2. This approach is especially good for non-compliant patients. When the measurement of very low urinary albumin concentrations is needed, as in the initial phases of the renal disease in diabetes mellitus, a radioimmunoassay method is used (Keen and Chlouverakis 1963).

Qualitative analysis

After quantitation, a qualitative analysis of the proteinuria is carried out to evaluate whether the proteinuria is glomerular, tubular, or due to overload. Several methods are available for this purpose (Pesce and Roy First 1979), but electrophoresis on cellulose acetate or agarose after urine concentration is one of the most widely used. To distinguish glomerular from tubular proteinuria, measurement of urinary β_2-microglobulin by radial immunodiffusion method is also used. In tubular diseases the excretion of this microglobulin is higher than that observed in patients with glomerular diseases (Hall and Vasiljevic 1973). For Bence-Jones protein, the heat test is no longer recommended because of frequent false-negative and false-positive results. In addition to electrophoresis, which reveals a dense band in the β- and/or γ- regions, immunoelectrophoresis is used to identify the monoclonal component. However, since this method may give false-positive results immunofixation is now considered more sensitive and more reliable than immunoelectrophoresis (Kyle 1988).

Finally, the 'selectivity' of proteinuria can be evaluated. This is better done by the ratio of the clearance of IgG (molecular weight 160 000) to the clearance of transferrin (molecular weight 88 000). When the ratio is lower than 0.1 the proteinuria is defined as selective (Cameron and Blandford 1966).

Enzymuria and brush border antigens in the urine

Urinary excretion of small amounts of enzymes located in the cells of the renal tubules, and in particular in the brush border of the proximal tubules, is found in normal individuals (see Pesce and First 1979, for a detailed listing). Increased excretion of these enzymes has been shown to be a sensitive index of renal damage (Price 1982; Maruhn 1982; Mondorf 1982; Mondorf et al. 1984). These measurements have the advantage that they can be made easily and with great sensitivity, but unfortunately are non-specific so far as disease processes are concerned. Thus, they are not very useful for diagnostic purposes, and their main value lies on the one hand in screening populations at risk for renal damage, and on the other the examining the fine nephrotoxicity of agents used clinically or commercially; or for examining consecutively the activity of processes already identified. Most urinary enzyme excretion measurements indicate tubular damage with some specificity, since many of these enzymes are molecules too large to be capable of passing the glomerular filter under normal circumstances.

A number of brush border enzymes have been studied in normal and pathological urines, including alkaline phosphatase (EC.3.1.3.1) (Hartmann et al. 1985), leucine aminopeptidase (EC.3.4.11.2.), (Hartmann et al. 1985; Cavaliere et al. 1987), γ-glutamyltransferase (EC.2.3.2), (Harmann et al., 1985; Cavaliere et al., 1987) α-glucosidase (EC.3.2.1.20), (Hartmann et al. 1985) and trehalase (EC.3.2.1.28) (Hartmann et al. 1985), all located in the brush border of the proximal tubule.

Most data, however, relate to N-acetyl-β-D-glucosaminidase, a hydrolytic enzyme present in lysosomes, of molecular weight 130 kDa (Wellwood et al. 1975; Price 1979; Sherman et al. 1983). Excretion of brush border enzymes or N-acetyl-β-D-glucosaminidase has been shown to increase after a variety of nephrotoxic insults, including aminoglycoside antibiotics (Mondorf 1982; Mondorf et al. 1984), cisplatin (Litterst et al. 1985–6), some contrast media (Cavaliere et al. 1987), heavy metals and organic solvents (Meyer et al. 1984), and ingestion of large doses of aspirin (Lockwood and Bosmann 1979), as well as during transplant rejection (Wellwood et al. 1978), urinary tract infections involving the kidney, (Sherman et al. 1983), glomerulonephritis (Sherman et al. 1983), and following renal surgery (Price et al. 1970). Interestingly, theraputic doses of cyclosporin A are not associated with increased excretion of N-acetyl-β-D-glucosaminidase in man (Finn et al. 1985). In addition the excretion of N-acetyl-β-D-glucosaminidase excretion by patients with both renal artery stenosis (Mansell et al. 1978a), and with essential hypertension (Mansell et al. 1978b), is raised, and returns to normal after treatment of the hypertension (Alderman et al. 1983).

Thus assay of urinary N-acetyl-β-D-glucosaminidase or brush border enzymes seems capable of detecting minimal degrees of renal dysfunction, which is at the same time their advantage and their disadvantage for clincial use. The problem in using these assays diagnostically is that (for example) modest doses of aspirin, intercurrent urinary tract infections, and theraputic doses of aminoglycosides will all increase urinary enzyme excretion. Thus, the clinical value of urinary N-acetyl-β-D-glucosaminidase estimations in ill patients is limited, and initial hopes that the assay would provide an early and sensitive index of renal transplant rejection have not been sustained. The main value seems to be in the study of defined nephrotoxins and in screening populations at risk of renal damage.

Another parallel approach has been to avoid the many problems of enzyme assay in urines of variable chemical composition, by measuring the excretion of brush border proteins using radioimmunoassays, employing either specific polyclonal, or more recently monoclonal antibodies. Two examples of this approach are the assay of the brush border antigen BB-50, a 50 kDa protein whose function is not yet identified (Mutti et al. 1985), and of the adenosine deaminase binding protein, a 120 kDa brush border protein (Tolkoff-Rubin et al. 1986, 1987) whose excretion increases in transplant rejection and tubular injury, and also in acute bacterial pyelonephritis involving the kidney (Tolkoff-Rubin et al. 1987).

Urine microscopy

The microscopic analysis of urine sediment is a simple diagnostic tool which, when properly performed, may be of great value in the evaluation of the urinary tract diseases.

Methods

The first urine of the morning is preferred, since it is the most concentrated and acidic, and preserves the formed elements better. Meals can cause an 'alkaline tide' in the urine, which precipitates phosphates or causes cast dissolution. Strenuous physical activity may cause variable changes in the urinary sediment (Fasset et al. 1982a). Secretions and leakage from the genital tract may devalue the urinary sediment findings. Urinary tract catheterization may cause variable haematuria, leucocyturia, and bacteriuria. Therefore, all these should be avoided or the observer should be aware of them. Urine should be handled and analysed while fresh to prevent the lysis of the formed elements, the changes caused by bacterial growth, and airborne contamination. It should not be kept in the refrigerator since low temperatures may precipitate phosphates or urates which can mask all the other constituents. Specific gravity and pH should be known before analysis of the urinary sediment since these factors influence both size and shape of erythrocytes and leucocytes and cast formation is hindered at high pH (Burton et al. 1975). Finally, each laboratory should have data for urinary sediment of normal subjects according to the methods used within it.

Routinely, a fixed volume (10 ml) of midstream urine is centrifuged at 2000 to 2500 r.p.m. for 5 to 10 min; the supernatant is poured off and the sediment, after a gentle but thorough resuspension, is poured on to a slide, placed under a coverslip and examined. This method, which is simple and fast, provides only a qualitative analysis of the urinary sediment. It is quite rough and lacks the reproducibility needed for quantitative evaluation (Gadeholt 1964). Hence, the content of formed elements is usually scored semiquantitatively from 0 to 3+ or given in terms such as 'rare', 'some', 'moderate amount', or as mean number per high-power field. More reliable counts are done on timed urine, which after centrifugation and resuspension is placed in a counting chamber. The elements actually seen are multiplied by correction factors to arrive at the excretion rate, and the number of elements is given per unit of time (Addis 1925) or per volume of urine (Kesson et al. 1978). However, even with these methods errors are possible, because the multiplication may greatly amplify even small errors in counting and the counting chambers are inadequate for good morphological analysis.

A bright-field microscope is commonly used for analysis of urinary sediment. With appropriate light and condenser adjustment it is sufficient for routine work, and when polarized light is applied anisotropic particles can also be identified. However, the phase-contrast microscope permits better vision of all elements, in particular those with a low refractive index, as well as cell membranes (Brody et al. 1968). For this reason and because it is a practical and inexpensive device, it is preferred to the traditional microscope. Immunofluorescence microscopy was used for studies of the nature and composition of urinary casts (Rutecki et al. 1971), but is no longer employed in routine work. Electron microscopy has been used for identification of amyloid fibrils in urine (Derosena et al. 1975), for studying urinary casts (Lindner et al. 1980) and for studying the urinary sediment of acute tubular necrosis (Mandal et al. 1985), but this technique is restricted to research work.

Urinary sediment is commonly viewed without staining. This allows rapid preparation and provides reliable identification of most elements, although leucocyte subtypes, renal tubular cells and cellular changes can be recognized only poorly or not at all. When better cytology or more details are needed, stains such as Papanicolaou's, Sternheimer-Malbin's, or Wright's are used.

Two magnifications are required, the lower ($100\times$ or $160\times$) for a general overview and quantification of casts, the higher ($400\times$) for details and cell evaluation. The larger the number of microscopic fields examined, the more reliable the results.

Formed elements of urinary sediment
Erythrocytes (Fig. 1(a,b))

In the normal subject there are fewer than one erythrocytes per high power field (one every three to four high-power fields in our experience) (Fogazzi et al. 1989). In pathological conditions, erythrocyturia can originate in any point of the urinary tract. Pathological casts or other elements of renal origin suggest renal bleeding, while epithelial transitional cells or microscopical clots suggests that the erythrocytes are from the extrarenal urinary tract. It is also possible to distinguish glomerular from non-glomerular bleeding by erythrocyte morphology. In phase-contrast microscopy erythrocytes that originate from the glomeruli have several abnormalities in size, shape and membrane appearance (the so-called dysmorphic erythrocytes), while red cells from a non-glomerular source have normal and regular morphology (the so-called isomorphic erythrocytes) (Fairley and Birch 1982). This simple method is now widely used and when more than 80 per cent of red cells are dysmorphic the haematuria is considered to be 'glomerular', while when more than 80 per cent of red cells are isomorphic it is classified as 'non-glomerular' (Fasset et al. 1982b). However, apart from the possibility of 'mixed' haematurias, erythrocyte morphology is not always easy to define and isomomorphic red cells have been found in glomerular diseases during gross haematuria (Van Iseghem et al. 1983), after water or frusemide-induced diuresis (Schuetz et al. 1985), in renal insufficiency (Fogazzi and Moroni 1984), and in Berger's disease (Fairley and Birch 1982). Finally, the method was shown to have poor inter observer reproducibility in a recent study (Raman et al. 1986). Thus the study of red cell morphology is useful for

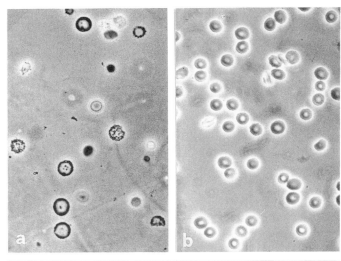

Fig. 1 Blood cells in the urine: (a) dysmorphic erythrocytes, typical of glomerular bleeding; (b) isomorphic erythrocytes, found in haematurias of non-glomerular origin; (c) cluster of polymorphonuclear leucocytes (and bacteria) as seen in urinary tract infections (phase-contrast microscope).

many but not all conditions. The cause of erythrocyte dysmorphism is not yet known. It has been suggested that it occurs during passage through the glomerular basement membrane (Kubota et al. 1988) or during passage along the distal tubules (Schuetz et al. 1985). More recently, automated blood-cell volume analysers have been proposed to distinguish glomerular from non-glomerular haematurias, a method which could avoid the current controversial aspects of microscopic study of erythrocyte morphology (Sichiri et al. 1988).

Leucocytes (Fig. 1(c))

There is no agreement about the normal number of leucocytes in urinary sediment. In our own experience they are rare (one every five to seven) high-power fields (Fogazzi et al. 1989). White cells can enter the urinary tract anywhere from the glomerulus to the urethra, and in glomerular diseases the underlying mechanism is thought to escape through glomerular basement gaps (Burkolder 1969). Infection, inflammation, haemorrhage, or contamination cause leucocyturia and only the formed elements which are found together with leucocytes can help to identify its source. In most conditions, urinary leucocytes are polymorphonuclear; however, in drug-induced interstitial nephritides eosinophils are also present and are considered strongly suggestive of these diseases. However, eosinophils can be identified only after staining. Wright's stain is commonly employed but recently Hansel's stain has been found to be more sensitive and with it eosinophils have also been found in both rapidly progressive and exudative glomerulonephritides and in urinary tract infections (Nolan et al. 1986). Lymphocytes too can be identified only in stained urinary sediment. Some investigators consider them a reliable and early marker of acute cellular allograft rejection (Sandoz et al. 1986).

Renal tubular cells (Fig. 2(a))

These are very rare in the urinary sediment of normal subjects and are hard to identify in unstained urine. Tubular cells in the urine are a marker for tubular damage and are found in several conditions, such as acute tubular necrosis (Mandal et al. 1985), cortical necrosis, acute pyelonephritis, acute glomerulonephritides, or acute cellular allograft rejection. They are also found in viral infections, after renal toxin administration and after some chemotherapy (Haber 1981). When filled with lipids, which occurs mainly in the nephrotic syndrome, these cells are seen as 'oval fat bodies'.

Transitional epithelial cells (Fig. 2(b))

These line the urinary tract from renal calyces to the proximal urethra. Rare in normal urine, they increase in all conditions affecting the urothelium. Thus, they are found in inflammation, neoplasias and after urinary tract catheterization.

Squamous epithelial cells (Fig. 2(c))

These line the bladder trigone, almost the entire female urethra, the distal third of the male urethra, and the vagina. Their presence in urine is never pathological.

Lipids

These are seen as 'oval fat bodies', free fat droplets and fatty casts. In bright-field and phase-contrast microscopy lipids are easily identifiable, but they are seen better under polarized light as 'Maltese crosses'. However, only cholesterol esters and free cholesterol appear as Maltese crosses, while neutral fat, free fatty acids, and phospholipids are identified with certitude only after Sudan III, Sudan IV, or oil red O staining. Lipids appear in the urine mainly in patients with the nephrotic syndrome, but in a recent study they were found also in patients with mild proteinuria and non-glomerular diseases (Braden et al. 1988). Lipids may also be found in urinary sediment after prostate expression and as contaminants from unclean utensils and lubricants.

Casts (Fig. 3)

These are cylindrical elements which form in the distal tubules and in the collecting ducts after aggregation of Tamm-Horsfall protein, which is the matrix for all types of cast. Various factors favour the aggregation of this protein, produced in the thick segment of Henle's loop, and hence cast formation: these are increasing intratubular concentration of hydrogen ions, electrolytes, ultrafiltered proteins, and dehydration (Hoyer and Seiler 1979). Casts are among the most important elements in urinary sediment since their presence correlates well with renal damage (Györy et al. 1984).

There are several types of cast. Those such as hyaline and

1.2 Urinalysis and microscopy

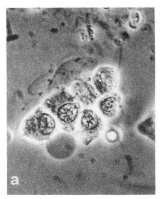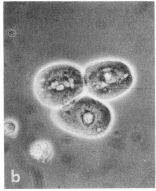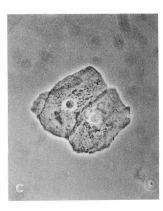

Fig. 2 Epithelial cells: (a) renal tubular cells; (b) transitional cells; (c) squamous cells (phase-contrast microscope).

Waxy casts are typical of severe renal insufficiency and they too are thought to derive from cellular degeneration. Fatty casts contain variable amounts of lipids as oval fat bodies or fat droplets and are mainly seen in the nephrotic syndrome. Pigmented casts include bilirubin casts, which are typical of the conditions with direct hyperbilirubinaemia, and haemoglobin casts, which derive from both red cell cast degeneration and haemolytic conditions. Bacterial casts, which would be pathognomonic for acute pyelonephritis, have recently been described (Lindner *et al.* 1980).

Pseudocasts are elements which resemble casts, but are mucus threads or aggregations of cells, crystals, organisms, or contaminants. They do not form in the tubules nor do they have a Tamm-Horsfall matrix. Cylindroids also resemble casts, but have one extremity which ends like a mucous strand. They are considered as true casts by the majority of workers.

Crystals (Fig. 4)

A large variety of crystals may be found in the urine, but only the crystals of cystine, leucine, tyrosine, and cholesterol have pathological significance. Very frequently crystals of uric acid, calcium oxalate, and triple phosphate are found, as well as amorphous phosphates or urates. None of these has clinical relevance, since their formation is pH- and temperature-dependent or is related to the ingestion of some type of food. In recurrent stone-forming patients, larger crystals and crystal aggregates may be seen. However, study of crystalluria in these patients is useful only when the urine is collected, handled, and read at a temperature of 37°C while fresh.

Organisms

Bacteria are frequently seen in urinary sediment. Since urine is usually not collected and handled under sterile conditions, bacteria may be due to contamination rather than infection. Infection is suspected when polymorphonuclear leucocytes are also present. Fungi are also a frequent finding, especially in women. On most occasions they are due to contamination from the genitalia. *Trichomonas vaginalis* is a common protozoon and can often be found in the urine, where it may come from either genitalia or the lower urinary tract. Of the parasites, *Enterobius vermicularis* is rare in urinary sediment and ordinarily is a contaminant from genitalia or anus. On the other hand, *Schistosoma haematobium* can actually be present in the urinary tract, since the adult worms live in the veins of the pelvis and deposit ova in the walls of ureters, bladder, or urethra. *Schistosoma haematobium* is very common in Africa and the Middle East (Chapter 13.5).

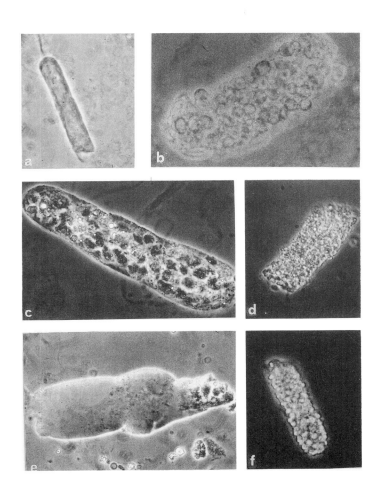

Fig. 3 Casts: (a) hyaline; (b) erythrocytic; (c) cellular (containing renal tubular cells); (d) granular; (e) waxy; (f) lipidic (phase-contrast microscope).

finely granular casts may be found in the normal subject also. Pathological casts are cellular, granular, waxy, fatty, or pigmented. Cellular casts contain erythrocytes or leucocytes or renal tubular cells and are an unequivocal marker of the renal origin of the cells included. The cells within the cast may degenerate upon standing in the tubules; granular casts derive from this process. However the granules in casts may also be due to various serum proteins filtered through the glomerulus, as has been shown in immunofluorescence studies (Rutecki *et al.* 1971).

Table 2 Typical findings of urinary sediment in the main groups of diseases of the urinary tract. The conditions in which the element is more prominent are in parentheses

Glomerular diseases	Dysmorphic erythrocytes*	(Proliferative glomerulonephritis)
	Renal tubular cells	
	Leucocytes	
	Cellular and granular casts	
	Lipids	(Non-proliferative glomerulonephritis with nephrotic syndrome)
Interstitial renal diseases	Renal tubular cells	
	Cellular and granular casts	
	Leucocytes	
	Eosinophils†	(Drug-induced acute interstitial nephritis)
	Lymphocytes†	(Acute allograft rejection)
	Bacteria ± bacterial casts	(Acute pyelonephritis)
Non-renal diseases	Isomorphic erythrocytes	
	Epithelial transitional cells	
	Leucocytes	
	Bacteria	(Urinary tract infection)
	Crystals‡	(Urolithiasis)

* Isomorphic erythrocytes in some circumstances (see text).
† Stain is needed.
‡ Fresh urine at 37°C is needed for a reliable study of crystals.

Contaminants

Many contaminants are possible: these include fibres, starch particles, hair, glass, oil, or pollens. They get into the urine through unclean laboratory tools, genitalia, or air. Contamination should be avoided but it is necessary for the observer to be able to recognize them.

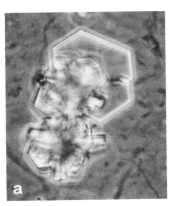
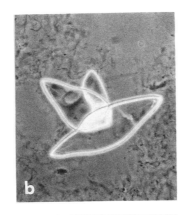

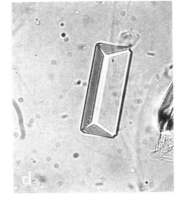

Fig. 4 Crystals: (a) cystine; (b) uric acid; (c) calcium oxalate dihydrated; (d) triple phosphate (phase-contrast microscope).

Urine sediment and drugs

Some drugs may produce changes in urinary sediment. Large numbers of granular casts may be observed in patients treated with high doses of ammonium exchange resins (Friedman et al. 1951). These casts are merely the result of increased acidity and solute concentration by the resin and have no clinical significance. Intravenous injection of frusemide or etacrinic acid causes the appearance of hyaline casts which disappear after 24 h. These casts have no clinical relevance either (Imhof et al. 1972). In acid urine, triamterene causes the appearance of granular birefringent casts and of brown birefringent crystals (Fairley et al. 1986). It is thought that these crystals might favour renal stone formation. Crystals of various sizes and shapes may also appear after sulphonamides, and these crystals are also said to form renal stones (Haber 1981). Ampicillin, when given in large doses, may cause the appearance of masses of long needles in acid urine (Graff 1983).

Typical findings in the urinary sediment in the main categories of urinary tract diseases are shown in Table 2.

References

Abuelo, J.G. (1983). Proteinuria: diagnostic principles and procedures. *Annals of Internal Medicine*, **98**, 186–91.

Addis, T. (1925). A clinical classification of Bright's diseases. *Journal of the American Medical Association*, **5**, 163–7

Alderman, M.H., Melcher, L., Drayer, D.E., and Reidenberg, M.M. (1983). Increased excretion of urinary N-acetyl-β-glucosaminidase in essential hypertension and its decline with antihypertensive therapy. *New England Journal of Medicine*, **309**, 1213–17.

Berman, L.B. (1977). When the urine is red. *Journal of the American Medical Association*, **237**, 2753–4.

Braden, L., Sanchez, P.G., Fitzgibbons, J.P., Stupak, W.J., and Germain, M.J. (1988). Urinary doubly refractile lipid bodies in nonglomerular diseases. *American Journal of Kidney Diseases*, **11**, 332–7.

Brody, L., Webster, M.C., and Kark, R.M. (1968). Identification of elements of urinary sediment with phase-contrast microscopy. *Journal of American Medical Association*, **206**, 1777–81.

Burkolder, P.M. (1969). Ultrastructural demonstration of injury and perforation of glomerular capillary basement membrane in acute proliferative glomerulonephritis. *American Journal of Pathology*, **56**, 251–65.

Burton, J.R., Rowe, J.W., and Hill, R.N. (1975). Quantitation of casts in urine sediment. *Annals of Internal Medicine*, **83**, 518–9.

Cameron, J.S. and Blandford, G. (1966). The simple assessment of selectivity in heavy proteinuria. *Lancet*, **ii**, 242–7.

Cavaliere, G. *et al.* (1987). Tubular nephrotoxicity after intravenous urography with ionic high-osmolal and nonionic low-osmolal contrast media in patients with chronic renal insufficiency. *Nephron*, **46**, 128–33.

Derosena, R., Koss, M.N., and Pirani, C.L. (1975). Demonstration of amyloid fibrils in urinary sediment. *New England Journal of Medicine*, **293**, 1131–3.

Fairley, K.F. and Birch, D.F. (1982). Hematuria: a simple method for identifying glomerular bleeding. *Kidney International*, **21**, 105–8.

Fairley, K.F., Woo, K.T., Birch, D.F., Leaker, B.R., and Ratnaike, S. (1986). Triamterene-induced crystalluria and cylindruria: clinical and experimental studies. *Clinical Nephrology*, **26**, 169–73.

Fasset, R.G., Owen, J.E., Fairley, J., Birch, D.F., and Fairley, K.F. (1982*a*). Urinary red-cell morphology during exercise. *British Medical Journal*, **285**, 1455–7.

Fasset, R.G., Horgan, B.A., and Mathew, T.H. (1982*b*). Detection of glomerular bleeding by phase-contrast microscopy. *Lancet*, **i**, 1432–4.

Fogazzi, G.B. and Moroni, G. (1984). Ematuria glomerulare e non glomerulare: studio della morfologia delle emazie urinarie in pazienti portatori di malattie di vario tipo e con diverso grado della funzione renale. *Giornale Italiano di Nefrologia*, **1**, 45–9.

Fogazzi, G.B., Passerini, P., Bazzi, M., Bogetic, L., and Barletta, L. (1989). Use of high power field in the evaluation of formed elements of urine. *Journal of Nephrology*, **2**, 107–12.

Friedman, I.S., Zuckerman, S., and Cohn, T.D. (1951). The production of urinary casts during the use of cation exchange resins. *American Journal of Medical Sciences*, **221**, 672–7.

Gadeholt, H. (1964). Quantitative estimation of urinary sediment, with special regard to sources of error. *British Medical Journal*, **1**, 1547–9.

Ginsberg, J.M., Chang, B.S., Matarese, R.A., and Garella, S. (1983). Use of single voided urine samples to estimate quantitative proteinuria. *New England Journal of Medicine*, **309**, 1543–6.

Graff, L. (1983). *A Handbook of Routine Urinalysis*. Lippincot, Philadelphia.

Györy, A.Z., Hadfield, C., and Lauer, C.S. (1984). Value of urine microscopy in predicting histological changes in the kidney: double blind comparison. *British Medical Journal*, **288**, 819–22.

Haber, M.H. (1981). *Urinary Sediment: A Textbook Atlas*. American Society of Clinical Pathologists, Chicago.

Hall, P.W. and Vasiljevic, M. (1973). Beta$_2$-microglobulin excretion as an index of renal tubular disorders with special reference to endemic Balkan nephropathy. *Journal of Laboratory and Clinical Medicine*, **81**, 897–904.

Hartmann, H.G., Braedel, H.E., and Jutzler, G.A. (1985). Detection of renal tubular lesions after abdominal aortography and selective arteriography by quantitative measurements of brush-border enzymes in the urine. *Nephron*, **39**, 95–101.

Hoyer, J.R. and Seiler, M.W. (1979). Pathophysiology of Tamm-Horsfall protein. *Kidney International*, **16**, 279–89.

Imhof, P.R., Hushak, J., Schumann, G., Dukor, P., Wagner, J., and Keller, H.M. (1972). Excretion of urinary casts after the administration of diuretics. *British Medical Journal*, **2**, 199–202.

James, G.P., Bee, D.E., and Fuller, J.B. (1978). Accuracy and precision of urinary pH determinations using two commercially available dipsticks. *American Journal of Clinical Pathology*, **70**, 368–74.

Keen, H. and Chlouverakis, C. (1963). An immunoassay method for urinary albumin at low concentrations. *Lancet*, **ii**, 913–4.

Kesson, A.M., Talbott, J.M., and Györy, A.Z. (1978). Microscopic examination of urine. *Lancet*, **ii**, 809–12.

Kubota, H., *et al.* (1988). Mechanism of urinary erythrocyte deformity in patients with glomerular disease. *Nephron*, **48**, 338–9.

Kyle, R.A. (1988). Kidney in multiple myeloma and related disorders. In *The Kidney in Dyscrasias* (eds. L. Minetti, G. D'Amico, and C. Ponticelli), pp. 141–51, Kluwer Academic Publishers, Dordrecht.

Lancet (1988). Is routine urinalysis worthwhile? *Lancet*, **i**, 747.

Lindner, L.E., Jones, R.N., and Haber, M.H. (1980). A specific urinary cast in acute pyelonephritis. *American Journal of Clinical Pathology*, **73**, 809–11.

Litterst, C., Smith, J.H., Smith, N.A., Vozonir, J., and Copley, M. (1985–6). Sensitivity of urinary enzymes as indicators of renal toxicity of the anti-cancer agent cis-platin. *Uremia Investigation*, **9**, 111–18.

Lockwood, T.D. and Bosmann, H.B. (1979). The use of urinary N-acetyl-beta-glucosaminidase in renal toxicology. II. Elevation in human excretion after aspirin and sodium salicylate. *Toxicology and Applied Pharmacology*, **49**, 337–45.

Mandal, A.K., Sklar, A.H., and Hudson, J.B. (1985). Transmission electron microscopy of urinary sediment in human acute renal failure. *Kidney International*, **28**, 58–63.

Mansell, M.A., Jones, N.F., Ziroyannis, P.N., Tucker, S.M., and Marson, W.S. (1978*a*). Detection of renal artery stenosis by measuring urinary N-acetyl-beta-D-glucosaminidase. *British Medical Journal*, **1**, 414–15.

Mansell, M.A., Jones, N.F., Ziroyannis, P.N., and Marson, W.S. (1978*b*). N-acetyl-β-D-glucosaminidase: a new approach to the screening of hypertensive patients for renal disease. *Lancet*, **ii**, 803–5.

Maruhn, D. (1979). Evaluation of urinary enzyme patterns in patients with kidney diseases and primary' benign hypertension. *Current Problems in Clinical Biochemistry*, **9**, 135–9.

Meyer, B.R., Fischbein, A., Rosenmann, K., Lerman, Y., Drayer, D.E., and Reidenberg, M.M. (1984). Increased urinary enzyme excretion in workers exposed to nephrotoxic chemicals. *American Journal of Medicine*, **76**, 989–96.

Mondorf, A.W. (1982). Urinary enzymatic markers of renal damage. In *The Aminoglycosides* (eds. A. Whelton, and H. Neu), pp. 283–301. Marcel Dekker, New York.

Mondorf, A.W., *et al.* (1984). Brush border enzymes and drug nephrotoxicity. In *Acute Renal Failure* (eds. K. Solez and A. Whelton), pp. 281–98. Marcel Dekker, New York.

Mutti, A., *et al.* (1985). Urinary excretion of brush-border antigen revealed by monoclonal antibody: early indicator of toxic nephropathy. *Lancet*, **ii**, 914–17.

Nolan, C.R. III, Anger, M.S., and Kelleher, S.P. (1986). Eosinophiluria, a new method of detection and definition of the clinical spectrum. *New England Journal of Medicine*, **315**, 1516–9.

Pesce, A.J. and Roy First, M. (1979). *Proteinuria. An Integrated Review*. Dekker, New York.

Price, R.G. (1979). Urinary N-acetyl-beta-D-glucosaminidase as an indicator of renal disease. *Current Problems in Clinical Biochemistry*, **9**, 150–3.

Price, R.G. (1982). Urinary enzymes, nephrotoxicity and renal disease. *Toxicology*, **26**, 99–134.

Price, R.G., Dance, N., Richards, B., and Cattell, W.R. (1970). The excretion of N-acetyl-β-glucosaminidase and beta-galactosidase following surgery to the kidney. *Clinica Chimica Acta*, **27**, 65–72.

Raman, V.G., Pead, L., Lee, H.A., and Maskell, R. (1986). A blind controlled trial of phase-contrast microscopy by two observers for evaluating the source of haematuria. *Nephron*, **44**, 304–8.

Rutecki, G.J., Goldsmith, C., and Schreiner, G.E. (1971). Characterization of proteins in urinary casts. *New England Journal of Medicine*, **284**, 1049–52.

Sandoz, P.F., Bielmann, D., Mihatsch, M.J., and Thiel, G. (1986). Value of urinary sediment in the diagnosis of interstitial rejection in renal transplant. *Transplantation*, **41**, 343–8.

Schuetz, E., Schaefer, R.M., Heidbreder, E., and Heidland, A. (1985). Effect of diuresis on urinary erythrocyte morphology in glomerulonephritis. *Klinische Wochenschrift*, **63**, 575–7.

Schumann, G.B. and Greenberg, N.F. (1979). Usefulness of macroscopic urinalysis as a screening procedure. *American Journal of Clinical Pathology*, **71**, 452–6.

Sherman, R.L., Drayer, D.E., Leyland-Jones, B.R., and Reidenberg, M.M. (1983). N-acetyl-β-glucosaminidase and β₂-microglobulin: their urinary excretion in patients with parenchymatous renal disease. *Archives of Internal Medicine*, **143**, 1183–5.

Sichiri, M., *et al.* (1988). Red-cell volume distribution curves in diagnosis of glomerular and non-glomerular haematuria. *Lancet*, **i**, 908–11.

Strasinger, S.K. (1985). *Urinalysis and Body Fluids*. Davis, Philadelphia.

Tolkoff-Rubin, N.E., *et al.* (1986). Diagnosis of tubular injury in renal transplant patients by a urinary assay for a proximal tubular antigen, the adenosine-deaminase-binding protein. *Transplantation*, **41**, 593–7.

Tolkoff-Rubin, N., *et al.* (1987). Diagnosis of renal proximal tubular injury by urinary immunoassay for a proximal tubular antigen, the adenosine deaminase binding protein. *Nephrology, Dialysis, Transplantation*, **2**, 143–8.

Van Iseghem, Ph., Ouglastaine, D., Bollens, W., and Michielsen, P. (1983). Urinary erythrocyte morphology in acute glomerulonephritis. *British Medical Journal*, **287**, 1183.

Wellwood, J.M., Ellis, B.G., Price, R.G., Hammond, K., and Thompson, A.E. (1975). Urinary N-acetyl-beta-D-glucosaminidase activities in patients with renal disease. *British Medical Journal*, **3**, 408–11.

Wellwood, J.M., Davies, D., Leighton, M., and Thompson, A.E. (1978). Urinary N-acetyl-beta-D-glucosaminidase assay in renal transplant recipients. *Transplantation*, **26**, 396–400.

1.3 Renal function testing

J. STEWART CAMERON

This chapter deals with the testing of renal function in clinical practice. Principally this concerns the physiological basis for, and the measurement of the glomerular filtration rate, renal blood flow, the concentrating capacity of the kidney, and its ability to excrete hydrogen ions. The handling of electrolytes is discussed further in Chapters 2.1, 7.1 and 7.2, and proximal tubular functions and their assessment (with the exception of bicarbonate reabsorption, which is dealt with a little here) in Chapters 5.1 to 5.3.

Renal function

In general terms, excretion of soluble waste products through the kidney is achieved by the process of glomerular filtration, using energy supplied by the heart; only a small number of important metabolites and exogenous compounds are actually secreted by the renal tubules. In contrast, regulation of the composition of the body fluids is achieved almost entirely by variations in the tubular absorption or secretion of individual components, using energy generated locally. The energy requirement for the reabsorption of solute accounts for the great proportion of the kidney's consumption of oxygen, whereas reabsorption of water is entirely a passive process, in pursuit of solute. Much of this occurs in the proximal tubule, but the main final concentrating and diluting mechanism depends upon the 'counter-current' arrangement of tubules and blood vessels in the renal medulla, so as to form an exchanger and multiplier of chemical gradients.

The kidney is also an important site of production and secretion for several hormones in particular renin, erythropoietin, and the active form of vitamin D, 1,25-di hydroxycholecalciferol, which circulate and act distantly from the kidney as well as locally, and autocoids such as prostaglandins, which act as local hormones in the regulation of renal function. Extensive accounts of the kidney's functioning in normal circumstances can be found in Seldin and Giebisch (1985), Brenner and Rector (1986), and Schrier and Gottschalk (1988).

The physiology of tubular function is dealt with in other chapters of this book; here, we will consider mainly the physiology of renal blood flow and glomerular filtration rate, paying particular attention to points of importance for pathophysiology and drug action.

Renal blood flow; the renal circulation(s) (Ofstad 1985; Brenner *et al.* 1986; Steinhausen *et al.* 1988)

One-fifth of the cardiac output flows through the two kidneys which weigh only about 250 g, i.e. 1000 to 1200 ml for an adult male weighing 70 kg, or 4 ml/g.min (about 50–70 nl/min per glomerulus). Although the energy requirements of the kidney are high, most of this huge blood flow is related to the passive process of filtration, and not to the energy-requiring reabsorption of solute. Indeed, the parts of the kidney most in need of oxygen have a very low Pa_{O_2} in health, with a low margin for electron transport and ATP generation should systemic hypoxaemia occur (see Chapter 8.2).

Blood flow, at any level of cardiac output and perfusion pressure, is determined by the resistance within the organ, and this forms a complex gradient through the kidney (Fig. 1). The main resistance is found at the level of the afferent and efferent arterioles, with little fall in pressure across the capillary beds of the glomeruli and the peritubular plexus; nevertheless, the small pressure fall within the glomeruli is of considerable importance to filtration (see below).

Distribution of blood flow (Aukland 1976; Knox *et al.* 1986; Arendshorst and Navar 1988)

Renal blood flow is not distributed equally throughout the kidney substance: the outer cortex receives perhaps 50 per cent of the total, the inner cortex 33 per cent, and the medulla only about 15 to 17 per cent; the smallest portion goes to the deep medulla and papilla. Moreover, intrarenal blood flow is not distributed according to oxygen consumption, and this mismatch

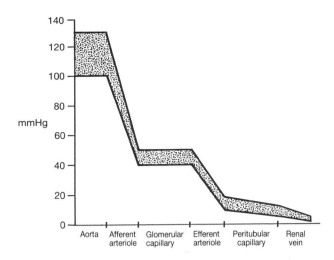

Fig. 1 The profile of vascular pressures along the intrarenal vasculature in the rat, in which the pressures have been measured by direct manometry. The major resistance is found at the afferent arteriole, with a smaller contribution from the efferent arteriole. A similar situation probably applies in the human kidney, but can only be inferred.

renders some areas, such as the thick ascending limbs, with their very high oxygen consumption devoted to reabsorption of sodium and chloride, on the verge of hypoxia even in normal circumstances (see Chapter 8.2). The distinctive and important anatomy of the renal medullary vasculature has been reviewed by Bankir *et al.* (1984), Brenner *et al.* (1986), and Kriz (1981).

Regulation of vascular tone (Steinhausen *et al.* 1988)

The resistance at the afferent and efferent arteriolar level is regulated by the tone in the myocytes of the vessel wall. These are subject to influence by a number of known vasoactive substances, listed in Table 1; this list is almost certainly incomplete, and the complexity of regulation is immediately evident. Most of these substances are released and destroyed locally (see below), and this is probably the main regulatory mechanism, but some (such as angiotensin II, atrial natriuretic peptide (ANP), and arginine vasopressin (AVP)) circulate more widely or arrive in the plasma from distant sites of synthesis to influence local renal events. To make things even more complicated, there are many interactions known amongst these various individual agents (Fig. 2). Most of these agents act also upon mesangial cells to alter K_f (see below) and thus regulate glomerular filtration rate (GFR) independent of blood flow or arteriolar tone.

Each myocyte bears receptors specific for these various vasoactive agents, and each ligand, on binding to its receptor, activates (by signal transduction through the cell membrane) intracellular events which lead to phosphorylation of the actomyosin within the cell, which in turn affects its tone. The transduction mechanisms of the agents are listed in Table 1; the key role for cytosolic Ca^{2+} in contraction is evident. Clearly, manipulation by pharmacological agents of these interactions is a powerful way of influencing both renal blood flow and GFR.

Recently, it has become clear that an important source of substances regulating smooth muscle tone in vessel walls is the endothelial cell lining of the vessels themselves. Again in Table 1, those substances known to be produced by endothelial cells are listed. This list will almost certainly grow in coming years, and both vasodilator and vasoconstrictor substances are represented. The myocyte and the endothelial cells bear the same receptors, but often respond to the same agent in opposite ways, so that the differential action of agents on the two types of cell provides a mechanism for balancing vascular tone. Two new endothelially-derived vasoactive agents, a dilator endothelial-derived relaxing factor (EDRF) (Vanhoutte 1988) and a constrictor endothelin (Yanigasawa *et al.* 1988) have been described, and their role in pathophysiology is being explored.

Regulation of the renal blood flow (Navar 1978; Ofstad and Aukland 1985; Brenner *et al.* 1986)

A striking characteristic of the normal renal circulation is autoregulation in the face of wide variations in perfusion pressure; this persists even in the denervated or isolated perfused kidney. Thus, the renal blood flow is essentially constant at mean arterial pressures from 70 to 200 mmHg, and therefore so is the GFR. The advantages for maintaining a constant renal function are obvious, although why renal function should need to be constant is less clear. Obviously also, renal vascular resistance must vary inversely with perfusion pressure, and this is mainly regulated at the afferent arteriolar level. Initially it was supposed that autoregulation proceeded from some intrinsic activity of vessel wall myocytes in response to stretch. This 'myogenic' theory has been shown to be incomplete, although it survives in a modified form, in that a number of important vasoactive substances can be released from renal endothelium in response to stretch. Autoregulation derives, primarily, from the cardinal role of these locally derived effectors—mostly released from endothelium (Table 1) or local renal nerves, acting upon the local myocytes of the small resistance vessels.

Clearly also, any self-regulatory system such as this requires a sensing system which assesses some function of renal perfusion pressure. It appears that this is mainly undertaken by the juxtaglomerular apparatus, both by cells sensitive to stretch within the afferent arteriolar wall, and also by signals from the adjacent macula densa cells of the distal tubule. These signals respond to changes in distal delivery of sodium and chloride, and provide a second autoregulatory loop based upon the net effects of both glomerular filtration and tubular reabsorption—both flow-dependent phenomena in part. This tubuloglomerular feedback (Bell and Navar 1984) provides a powerful mechanism for the fine tuning of renal perfusion, through alterations in afferent arteriolar tone, and perhaps explains why the main regulatory resistance step is found at this site. Despite the fact that autoregulation can occur in isolated kidneys, in the intact subject the system is modulated by the sympathetic outflow and catecholamine secretion. However, the importance of this mechanism is the subject of much controversy.

The various humoral mediators are listed in Table 1, and their interactions depicted diagrammatically in Fig. 2. A central role for the renin–angiotensin system, for which there is much evidence, is described. Understanding the role of angiotensin II is complicated by the fact that, in addition to being a locally produced hormone, it has a long enough half-life to circulate to the kidney from distant sources, either as angiotensin I which is cleaved by the angiotensin-converting enzyme in the endothelial cells, or as angiotensin II itself. Angiotensin, together with eicosanoids (prostaglandins and leukotrienes) and the kinin system, which releases EDRF from endothelium, all interact, predominantly to influence afferent arteriolar tone, but also directly, or

Table 1 Some vasoactive substances acting on arteriolar myocytes in afferent arteriole, efferent arteriole, and mesangium

Agent	Action	Signal	Immediate effect	Derived from endothelium
Angiotensin II	Constrictor	Inositol triphosphate	Direct Ca^{2+} diacylglycerol	Yes, through angiotensin converting enzyme
Thromboxane A_2	Constrictor	Inositol triphosphate	Direct Ca^{2+} diacylglycerol	Yes
Leukotrienes	Constrictor	Inositol triphosphate	Direct Ca^{2+} diacylglycerol	?
Arginine vasopressin	Constrictor	Inositol triphosphate	Direct Ca^{2+} diacylglycerol	No (from outside kidney)
(Nor)adrenaline (alpha)	Constrictor	Inositol triphosphate	Direct Ca^{2+} diacylglycerol	No (vessel nerves)
AGCP (PAF)§	Constrictor	Inositol triphosphate	Direct Ca^{2+} diacylglycerol	Yes
Endothelin	Constrictor	Inositol triphosphate	Direct Ca^{2+}	Yes
(Nor)adrenaline (beta)	Constrictor†	cAMP	P kinase A	No (vessel nerves)
Adenosine	Constrictor‡	?Inositol triphosphate?A	Direct Ca^{2+}	From local ATP
Prostaglandin I_2	Dilator	cAMP	P kinase A	Yes
Prostaglandin E_2	Dilator	cAMP	P kinase A	Yes
EDRF (?nitric oxide)*	Dilator	cGMP	P kinase G	Yes
ANP	Dilator	cGMP	P kinase G	No (from outside kidney)
Dopamine	Dilator	cAMP	P kinase A	No (vessel nerves)

* Bradykinin, histamine, and acetylcholine act through endothelial release of endothelial derived relaxing factor (EDRF), which may be nitric oxide, and endothelin, and so are not listed separately.
† Probably vasoconstrictor through renin release.
‡ The exact mode(s) of the complex actions of adenosine are not known; it influences angiotensin release (see text).
§ PAF = platelet activating factor, AGCP (1-alkyl-2-acetyl-sn-glycero-3-phosphorylcholine).
Direct Ca^{2+} = principal action to raise cytosolic Ca^{2+}, which binds to calmodulin and hence activates myosin kinase and leads to increased tone.
P kinase = principal initial action to raise cAMP (A) or cGMP (G), leading to phosphorylation of myosin light chain and its inactivation.

indirectly through peritubular capillary flow, to influence the handling of sodium by the renal tubules.

Other substances have a profound effect on renal perfusion and are listed in Table 1. Adenosine (Osswald et al. 1978; Spielman and Thomson 1982; Schnermann 1988; Cosmes and Macias 1988) is of course produced during degradation of adenine nucleotides such as ATP, and unusually (in other vascular beds it is a dilator) in the kidney is a net constrictor, predominantly of the afferent arteriole, with dilatation of the efferent arteriole and hence reduction in GFR. It inhibits renin release and may also modulate tubuloglomerular feedback.

Arginine vasopressin (AVP) can bind to arteriolar V1 receptors to cause vasoconstriction, as well as to tubular V2 receptors to influence permeability and water transport. It, like angiotensin II, circulates and when released in response to systemic stimuli, both osmotic and non-osmotic (mainly volume-related), will enter the kidney and influence local vascular tone. It will also, by acting elsewhere within the circulation, tend to raise the perfusion pressure, and finally leads to release of both renin and prostaglandins within the kidney. Its effects are therefore very complex. The role of vasopressin, like that of adenosine, has not been defined clearly as yet, but it probably plays a crucial role in regulating medullary perfusion (Zimmerhackl 1988).

Equally the role of histamine as a vasodilator after binding to vascular H1 receptors is clear in experimental systems, and has been demonstrated to depend upon release of EDRF, but its role in whole systems is still being explored. Dopamine is a vasodilator at low concentrations, but a constrictor agonist at higher concentrations, a fact of clinical importance since it has a long half-life and therefore can be infused. Acetyl glyceryl phosphorylcholine (AGPC) (platelet activating factor or PAF) (Schlondorff and Neuwirth 1986) is a lipid-derived vasoconstrictor whose role in normal and pathological states is not clear, but which is produced within endothelial cells and influences vascular tone profoundly.

Other influences on renal blood flow

Appropriately, the water and sodium balance of the subject will influence renal blood flow, but the autoregulatory systems just described limit the changes observed; however, the reserve of compensation will be used up at extremes of salt and water balance, especially in water or sodium depletion. Great interest has been expressed lately in the effects of protein intake in modulating renal perfusion, for reasons detailed below. Acute intravenous protein or amino-acid loading, or chronic high protein diets, lead to glomerular vasodilatation, at least in part mediated by vasodilator prostaglandins, since the response is blockable by indomethacin. A prior role for glucagon release blockable by somatostatin now seems certain (Hirschberg et al. 1988), and this topic is discussed further below. Conversely, a low protein diet is associated with renal vasoconstriction. As well as an effect on resistance vessels, protein intake also modulates tubuloglomerular feedback (Seney and Wright 1985).

Glomerular filtration (Brenner et al. 1977; Deen and Satrat 1981; Brenner et al. 1986)

As noted above, each minute 1 l of blood is distributed to about 2 million glomerular tufts, whose combined filtering area is about 1 m². The glomerular filters retain within the circulation all cells and almost all the protein of the plasma, but allow free passage of water and solutes, so that 180 to 200 l of filtrate is formed each 24 h (130 ml/min, or about 2 ml/s), that is 100 μl per glomerulus, or 50 to 70 nl/min. Exactly how this exquisite discrimination between protein and solute is achieved at such high rates of solvent flow is still not clear: the normal microanatomy of the glomerulus is described in detail in Chapter 1.7.1. The glomerular capillary wall is highly specialized, in that the endothelial cells are reduced to a thin fenestrated sheet of cytoplasm, whilst on the external surface of the capillary is a unique cell, the podocyte or epithelial cell, whose surface directed towards the capillary basement membrane forms an

1.3 Renal function testing

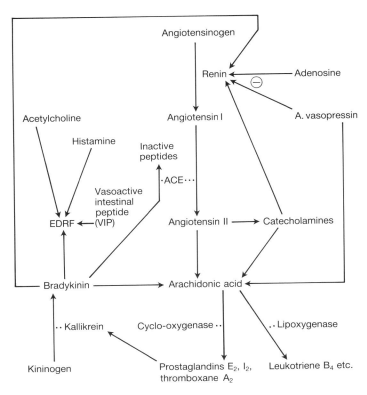

Fig. 2 Some inter-relationships between some of the vasoactive substances important in determining intrarenal vascular tone. The total position is almost certainly more complicated than shown, and most of the relationships have been demonstrated or inferred from *in-vitro*, or partially isolated, systems. All the arrows except that of arginine vasopressin and renin indicates stimulation or release of the agent at the end of the arrow. Note that the regulatory and inhibitory mechanisms of the substances present in all these systems are not shown. How such relationships integrate or apply *in vivo* can only be speculative at the moment.

intricate series of interlocking foot processes, applied to its outer surface. It is known also that there is a high density of negative charge distributed over the filtration barrier in the form of negatively charged glycosaminoglycans, particularly on the podocyte surface, discontinuously in the lamina rara externa of the basement membrane, and over the subendothelial layer on the inner surface of the capillary wall.

Determinants of glomerular filtration

As in all capillary beds, as described by Starling almost 100 years ago, the determinants of glomerular ultrafiltration are first, the net ultrafiltration pressure (P_{UF}), second the hydraulic permeability of the capillary wall (k), and third the area of available filtering surface (S) (Brenner *et al.* 1986)

$$\text{GFR} = P_{UF} \times k \times S$$

The hydraulic permeability, k, of glomerular capillaries is much higher than that found in other capillary beds, emphasizing the specialized function of the glomerulus as a filter. The ultrafiltration pressure, P_{UF}, depends in turn upon the hydrostatic pressure operating across the glomerular capillary wall (P) and the osmotic pressure of the plasma proteins (π), such that

$$P_{UF} = (P - \pi)$$

Now

$$P = P_{gc} - P_{tf} \text{ and } \pi = \pi_{gc} - \pi_{tf}$$

where gc = glomerular capillary, and tf = tubular fluid. Therefore

$$P_{UF} = (P_{gc} - P_{tf}) - (\pi_{gc} - \pi_{tf})$$

In practice, in species without surface glomeruli and in man, the effective area of filtering surface (S) is impossible to determine, so that the term K_f, which is the product of the area of filtration S and the permeability k, ($K_f = kS$) is used, so that

$$\text{GFR} = K_f (P_{gc} - P_{tf}) - (\pi_{gc} - \pi_{tf})$$

In rats where glomeruli are exposed on the surface of their dystrophic kidneys (the Munich-Wistar rat) it is possible using micropuncture techniques to measure these quantities directly (Fig. 3) at least in outer cortical nephrons; in man, and in juxtamedullary glomeruli in rats, they can only be inferred. There is some evidence that the larger juxtamedullary glomeruli have higher filtration rates than outer cortical nephrons, and the local dynamics may be different in detail, although of course similar in principle.

Regulation of GFR (Schnermann 1981; Schnermann and Briggs 1985; Romero and Knox 1988; Davis *et al.* 1988)

One of the other crucial determinants of the GFR is of course the renal blood flow and hence plasma flow, although the precise

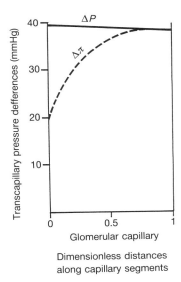

Fig. 3 The relationships of the transmembrane pressure ΔP, and the net oncotic pressure $\Delta \pi$, along glomerular and peritubular capillaries, as measured directly in the rat. $\Delta \pi$ rises sharply along the glomerular capillary (shown as dimensionless units of 0 to 1.0) as a result of a rise in protein concentration when plasma water is removed as glomerular filtrate; meanwhile, P falls by only a small amount. Filtration equilibrium ($\Delta P = \Delta \pi$) is reached before the end of the capillary, but throughout the major part of the capillary the pressures favour filtration. In the peritubular capillary, in contrast, the net pressure favours reabsorption throughout, since the oncotic pressure in the postglomerular capillaries is so high, and the perfusion pressure low. (Modified from data of Deen, Brenner, and others.)

relationship between them is very complex (see below). The GFR is thus dependent upon changes in overall renal blood flow discussed above, and all of the substances listed in Table 1 are capable of modifying the GFR. The main mechanisms are through differential effects on afferent and efferent glomerular arteriolar tone, and also through specific receptors for these agents present also upon the mesangial cells (Schlondorff 1987; Mené et al. 1989).

Obviously, dilatation of the efferent arteriole will (other things remaining constant) increase glomerular blood flow, decrease P and hence reduce GFR. Similarly, efferent arteriolar constriction will raise ΔP and increase GFR. Exactly opposite effects will be seen from afferent ateriolar dilatation or constriction, the former leading to an increase in GFR and the latter to a reduction. The interplay between these pre- and postcapillary sphincters of the glomerular capillary bed—an arrangement unique in the body—can permit exquisite regulation of glomerular blood flow and GFR. Steinhausen et al. (1987) describe the directly observed effects of several vasoactive agents on the two arterioles, but in a very non-physiological system; other data come of course, indirectly from micropuncture studies of glomerular pressure and flows. Other authors (Heller et al. 1988) have suggested that the efferent, rather than the afferent, arteriole is the main target for several vasoactive substances of physiological importance, including angiotensin II.

Further, as discussed in Chapter 1.7.1, mesangial cells are modified arteriolar smooth muscle cells to the point where the glomerular 'capillaries' can be considered as hemiarterioles, with their inner surface invested by cells capable of contraction and relaxation, should any one of the ligands listed in Table 1 meet receptors on the mesangial cells. At least in the rat, the mesangial cell has receptors for almost all the substances listed in Table 1, as well as some agents not discussed here, including PTH, insulin, and serotonin, whose significance is not yet clear. Most of these substances can be shown to influence mesangial cells in vitro, and putatively in vivo, and therefore to be capable—independent of, or in addition to, afferent and efferent arteriolar changes—of modifying glomerular blood flow, and hence GFR. In addition, mesangial contraction is capable of diminishing the area of filtration surface available (S) and thus K_f (kS) and hence GFR.

The manner in which these various changes will modify the GFR can be derived from the equations given above, and Fig. 3. This Figure indicates that in the latter third of each glomerular capillary, $\Delta \pi$ rises to equal ΔP, so that there is no net filtration. i.e. filtration equilibrium is found. In the case of an increase in capillary plasma flow, $\Delta \pi$ rises more slowly so that ($\Delta P - \Delta \pi$) remains above unity for longer, filtration proceeds over a greater length of glomerular capillary, and GFR rises. At high capillary plasma flows, the possibility of filtration disequilibrium arises, i.e. $\Delta \pi$ never equals ΔP even at the end of the capillary, and therefore net filtration occurs right along the loop. Conversely a fall in renal plasma flow will cause $\Delta \pi$ to rise more rapidly and a greater length of capillary will permit filtration, provided that ΔP does not decrease also. Similarly an increase in plasma flow will increase K_f through an increase in S, and increase GFR, and vice versa. The importance of the pre- and postcapillary sphincters of the glomerulus is immediately evident; whether renal plasma flow and pressure is modified by precapillary or by postcapillary constriction (or dilatation) predicts a variety of effects on GFR.

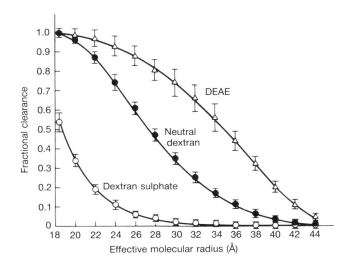

Fig. 4 Fractional clearances (versus $C_{\text{inulin}} = 1.0$) of neutral dextran (●), anionic dextran sulphate (○), and cationic DEAE dextran (△), plotted as a function of effective molecular radius (1 Å = 0.1 nm). Data for normal rats from studies of Deen et al. (1979). Relative to the neutral molecules, anionic molecules are excluded, and cationic molecules facilitated in their passage through the capillary wall, as a result of the high negative charge on this structure.

The glomerulus as charge- and size-selective filter (Brenner et al. 1978; Deen and Satrat 1981)

Another important property of the glomerular filter is that it functions both as a charge-selective and a size-selective barrier, i.e., smaller molecules and those with a small negative or even a net positive charge at physiological pH will pass more easily through the glomerulus. The charge of the molecule only becomes critical above the limit at which the size-selective barrier function becomes important. Up to an Einstein-Stokes radius of 2nm (molecular weight approximately 15 000 Da for a protein) molecules are filtered as freely as water, i.e. their clearance (see below) is equal to the glomerular filtration rate (Fig. 4). These data were obtained in rats but a similar situation exists in humans (Deen et al. 1985). Thus, charge selectivity is most important in determining the retention of plasma protein molecules, since at the pH of plasma most plasma proteins carry a net negative charge; also, as mentioned above there is a negative charge of the glomerular capillary wall. Thus, the glomerular filtrate is almost devoid of protein (about 1 mg/l in the rat, compared with 6000 mg/l in the plasma). Again in the rat, the penetration of negatively charged macromolecules is hindered, and that of positively charged molecules facilitated (Fig. 4); similar data can be inferred for humans from studies of proteinuria, given the disadvantage that tubular protein reabsorption intervenes to modify data derived from urine rather than glomerular filtrate. Exactly how this barrier is disturbed in disease is not clear; both charge- and size-selective defects seem to be present in many instances.

Variations in glomerular filtration rate

In most early reports on the GFR, the constancy of the measurement on repeated estimations was emphasized (Smith 1951), even though it was well known that it could vary extensively in laboratory animals. However, it now appears that much of the apparent constancy in the GFR has been merely an artefact of the stable conditions and long periods required to measure it. The GFR is very sensitive to renal blood flow, which is known to vary with varying conditions such as degree of activity, or posture, as noted above. There are also large diurnal variations; GFR is 25 to 30 per cent higher in the afternoon (Toor *et al.* 1965; Lang *et al.* 1980).

These variations are of little consequence, since GFR is usually assessed under basal conditions; but the effect of diet on the GFR is by no means negligible. Pullman *et al.* (1954) showed that changes in protein intake would induce changes in GFR in man, as had been well known from studies in animals (Smith 1951), although the changes in humans are less than in many animal species when expressed in relation to GFR before protein ingestion. For example, in the original studies of Pullman *et al.* (1954) the inulin clearance rose from 95.2 ± 3.1 to 117.3 ± 4.3 ml/min when protein intake was raised from between 0.1 and 0.4 to between 2.3 and 3.0 g/kg.24 h.

This observation has since been frequently confirmed in normal individuals (Bosch *et al.* 1984; Hostetter 1986; Rodriguez-Iturbe *et al.* 1985; Viberti *et al* 1987) and in patients with renal disease (Bosch *et al.* 1984; Rodriguez-Iturbe *et al.* 1985), as well as after kidney donation (Rodriguez-Iturbe *et al.* 1985; Tapson *et al.* 1986). The same rise in GFR is seen after amino-acid infusions (Graf *et al.* 1983; ter Wee *et al.* 1985). These observations have given rise to the concept of a 'renal reserve' of GFR, normally under-utilized. This 'renal reserve' is present at least down to a GFR of 50 ml/min, beyond which the ability of the kidneys to augment renal blood flow and GFR seems to have been exploited to the maximum in the face of the disease, as judged by the failure of the GFR to rise in these patients.

The exact mechanisms by which the GFR is elevated by protein or amino-acid loading are still unclear; it seems most likely at the time of writing that glucagon release is a major early event, since the reaction can be blocked by somatostatin. Glucagon release leads in turn to the release of renal vasodilators, including vasodilator prostaglandins such as prostaglandin I_2 and prostaglandin E_2 (Hirschberg *et al.* 1988), perhaps through the prior release of a poorly characterized hepatic vasodilator, glomerulopressin (Alvestrand and Bergstrom 1984). In addition, Seney and Wright (1985) have evidence that protein diminishes tubuloglomerular feedback.

The converse applies also; diminution in the GFR can be expected in states of low protein intake or protein-calorie malnutrition, such as anorexia nervosa, even when corrected for the diminished body surface area (Klahr and Alleyne 1973). Vegetarians have lower creatinine clearances than meat-eaters (Bosch *et al.* 1984), and this is also true of inulin clearances (Viberti *et al.* unpublished).

Pregnancy (see Chapter 15.1) is another state in which increases in the GFR are found (Davison and Dunlop 1980; Lindheimer and Katz 1986); at maximum this rises to some 150 per cent of baseline levels; correspondingly the plasma creatinine concentration falls. These changes in GFR parallel and follow increases in extracellular volume and renal blood flow (Lindheimer and Katz 1986). Even in diseased or single kidneys (e.g. transplanted kidneys) the creatinine clearance normally rises, at a lower level but parallel to the normal changes (Davison 1985). In acromegaly also (Ikkos *et al.* 1956), and in the early hyperglycaemic phases of untreated diabetes mellitus, the GFR is raised well above normal limits (Ditzel and Junker 1972; Mogensen 1972).

Tests used clinically in the investigation of renal function

The amount of a substance excreted through the kidneys is the product of the urine flow rate (V), and the concentration of that substance in the urine (U). The quantity (UV), the excretion per unit time, is of itself important in the context of metabolic balance and reflects production rates, but by itself does not define renal performance in respect of that substance. For that purpose, it is necessary also to know the plasma concentration, P. Gradually during the 1920s (see Harvey (1980) for the history) the concept of renal clearance (C) was introduced as a way of expressing the relationship between the excretion per unit time and the concentration in the plasma, which is obviously an index of the kidney's ability to 'clear' the blood of any substance. This is given by the formula

$$C_{\text{substance}} = \frac{(U_{\text{substance}} \times V)}{P_{\text{substance}}}$$

This represents the notional volume which would, in a given unit of time, be completely 'cleared' of a substance. It is usually expressed in ml/min, less often ml/s or l/24 h. The power of this concept is that it can be used to express the relative ability of the kidney to excrete any substance in comparison with any other.

Excretory (glomerular) function

Obviously the most important substance with which to compare the filtration of any substance is water, i.e. the glomerular filtration rate. Of itself this cannot be measured, but a moment's consideration of the clearance formula just given suggests that if a substance is neither reabsorbed nor secreted by the renal tubules, then the amount UV excreted in the urine will equal that filtered at the glomeruli in the same unit of time. Equally, if the same substance is not metabolized, and has no extrarenal route of excretion, then the clearance of the substance UV/P will equal the filtration of water, the glomerular filtration rate.

Unfortunately, no normally-occurring substance yet described exactly satisfies these conditions, and the major problem of clinical measurement of the GFR concerns the way in which creatinine falls short of this ideal (see below). The alternative is to infuse or inject a foreign substance which does satisfy these conditions completely, but this complicates the measurement of GFR considerably.

The inulin clearance as a measure of GFR

The clearance of inulin (Shannon and Smith 1935; Smith 1951), or synthetic inulin-like polyfructosans (Inutest®) (Berglund 1965), performed using a continuous intravenous infusion to maintain constant plasma concentrations, and an indwelling bladder catheter to ensure complete urine collections, has long been the benchmark for the measurement of GFR.

Inulin, prepared from the tubers of dahlia or Jerusalem artichoke, is a disperse polymer of fructose with an average molecular weight of 5000 (Einstein-Stokes radius 1.5 nm). Inevitably, commercial preparations contain some free fructose, which may be increased by the heating necessary to get the substance into solution at a concentration of 100 g/l. One advantage of the polyfructosans (e.g. Inutest®) is that they are easily soluble in water at room temperature. Since all methods for measuring inulin (see below) depend upon hydrolysis and measurement of the resultant free fructose, plasma and urine blank measurements are of particular importance. The high plasma and urine blank measurements characteristic of the chemical methods for inulin determination have been reduced by more specific enzymatic methods capable of automation, measuring finally sorbitol (Gutmann and Bergmeyer 1974; Dalton and Turner 1987), or an inulinase to hydrolyse the inulin to fructose (Day and Workman, 1984; Degenaar et al. 1987).

Nowadays, the total coefficient of variation on an infusion inulin clearance using automated methods for measuring the inulin is probably about 5 per cent. This suggests that two clearance measurements differing by 14 per cent have a 95 per cent chance of representing a real difference in GFR. Newer methods of inulin measurement based on high performance liquid chromatography (HPLC) techniques and refractometry are being investigated at the moment.

Although rigorously validated in normal man, it is still uncertain to some extent whether inulin clearances accurately reflect the GFR in diseased kidneys. It must be assumed that inulin can penetrate the diseased glomerulus like water, as in health; and that the diseased tubule, like the normal tubule, does not permit back diffusion of the inulin. It has been suggested (Arturson et al. 1971) that the glomerulus of newborn and especially preterm babies does not permit the free passage of inulin, but the work of Coulthard and Ruddock (1983) using chromatography of plasma and urinary inulin in preterm babies during and after infusions, reassures us that this is not the case.

Also, Rosenbaum et al. (1979) have suggested that, on the basis that creatinine and iothalamate clearances (see sections below) greatly exceed inulin clearances in transplant recipients, in some way filtration of inulin might be retarded in these subjects. Again, work from Myers and colleagues (Carrie et al. 1980; Shemesh et al. 1985) is reassuring in that neutral dextrans of Einstein-Stokes radii of 1.5 to 2.5 nm are cleared as freely as water, even in diseased glomeruli. Thus the measurement of inulin clearance during constant infusion still remains the 'gold standard' for the measurement of the GFR.

Problems with the clearance concept in practice

Constancy of plasma concentration of substance

If the substance is exogenous, such as inulin, an infusion using a very accurate pump is required to produce constant levels; filtration-equilibrium techniques take too long in practice. Diurnal and other variations occur in the data from endogenous substances (e.g. creatinine) which reduce their value, and are usually ignored. Although plasma creatinine concentrations are relatively constant and independent of diet including protein intake, they are higher in the afternoon (Sirota et al. 1950).

Urine collections

By far the greatest disadvantage of clearance measurements—inulin, creatinine, p-aminohippurate, or other—is that a value for UV is required, and hence a timed urine volume measurement must be made. The best way to obtain this is to pass a catheter into the bladder and expel timed volumes of urine by air displacement. Today, in humans timed urine specimens after a diuresis has been induced are substituted even in research protocols. No comparisons appear to have been done either in adults or in children comparing the two techniques of urine collection, and such studies are now unlikely to be performed because of reluctance to catheterize.

If clearances are done on ambulant inpatients or outpatients, the problem of achieving complete and accurately timed urine collections becomes crucial. The exact period of the collection is unimportant provided that it is known, and the bladder emptied before the start of the collection and again at the end. The difficulties are well known: patients 'lose' specimens and are afraid to mention it; the timing may be inaccurate so that the specimen supplied may correspond to less or more than 24 h without the laboratory knowing this; urine already in the bladder at the start of the period may be included, and so on. Undoubtedly the urine collection is the weakest factor in the assessment of the creatinine clearance.

In patients with urinary tract anomalies, obstruction, or reflux, accurate collections can never be obtained. In the elderly, collections are rarely accurate (Goldberg and Finkelstein 1987). In children the problems of timed urinary collections mean that 24-h urine measurements are not appropriate; the protocol of the Pediatric Nephrology Department of the Albert Einstein College of Medicine, New York may be used. The child is given an oral water load of 600 to 800 ml/m^2, and over the next hour or two under direct supervision he or she produces three or four timed specimens. A single blood specimen is collected, and this may be used to calculate three or four clearances, which are then averaged to give the final result. Alternatively, a timed voiding at bedtime and an early morning urine and blood collection immediately on rising can be used.

Correction of glomerular filtration rate for size

One major problem which still exists—although it is rarely discussed today—is exactly how the glomerular filtration rates of individuals of different ages and sizes should be expressed and compared. In one of the original papers by Møller, Mackintosh, and van Slyke in which the clearance concept was first named and used (Harvey 1980), it was suggested that renal functional data be corrected according to surface area, derived from tables of height and weight, using the American adult male surface area of 1.73 m^2 as the standard. This practice is now almost universal, sometimes with the substitution of 'ideal weight for height' to allow for wasting or oedema. This in essence means that the GFR is being expressed for height only, an approach which has been used to predict the GFR of children from the plasma creatinine (see below). With the use of creatinine clearances, there is an additional problem in that creatinine excretion, and hence clearance, varies widely with body composition, and in particular muscle mass, not to mention the analytical difficulties of measuring creatinine.

Masarei (1975) showed that correcting creatinine clearances for surface area did not improve on a simple correction for body weight, but again this becomes inaccurate with oedema. The commonly displayed nomograms for deriving surface area from height and weight measurements do not apply well to the extremes of age; the subject is well reviewed by Haycock et al. (1978) who, from available data, derive the empirical formula

Surface area (m^2) = weight (kg)$^{0.54}$ × height (cm)$^{0.40}$ × 0.024

and show that it holds from 0.2 to 2.0 m^2.

Reference ranges for the GFR in normal individuals given by Smith (1951) are 88 to 174 ml/min.1.73 m^2 for males and 87 to 147 ml/min.1.73 m^2 for females (95 per cent confidence limits). However, the effect of age on GFR is important (see Fig. 3, Chapter 1.5). Up to the age of 2 years the GFR rises, even when corrected for body surface area, and from the age of 35 or so steadily falls. Although these changes are not surprising, some of this fall-off may be attributed to the correction for body surface area just mentioned, which is made without any consideration of body composition.

Creatinine clearance as a measure of GFR

Ever since the suggestion of Popper and Mandel (1937) that the clearance of endogenous creatinine approximates to the GFR, this test has been popular in clinical medicine in one form or another. However, its performance and interpretation presents formidable difficulties (Payne 1986):

1. variations in the generation rate of creatinine
2. the accurate measurement of creatinine, especially in plasma
3. the secretion of creatinine by the renal tubules
4. the difficulty of obtaining complete, accurately timed urine collections.

The generation of creatinine

Creatinine is formed by the non-enzymatic dehydration of muscle creatine, itself synthesized in the liver and transported to muscle (see Levey *et al.* 1988 for a review). The main determinant (98 per cent) of the creatine pool (approximately 100 g) therefore is muscle mass. The only other source of creatine is, of course, meat in the diet, supplying 600 to 800 mg creatine/24 h. About 1.6 per cent of this creatine pool is converted to creatinine each day. Thus wasting diseases, paralysis, and meat intake will all influence the production—and therefore the plasma concentration and excretion—of creatinine. In uraemia, urinary creatinine excretion is reduced to two-thirds or less of normal. Why is this?

There are two major possible explanations. The first is that the actual production of creatinine goes down in uraemia (Goldman and Moss 1959). This might be expected, since appetite is suppressed in uraemia and wasting is common. However, the careful studies of Jones and Burnett (1974), using [^{14}C-methyl] creatinine, have shown that creatinine, like urea and uric acid, is secreted into the gut and broken down there by bacterial action. Jones and Burnett found creatinine 'deficits' (extrarenal excretion) of from 16 to 66 per cent in the five subjects studied. Mitch and coworkers (Mitch and Walser 1978; Mitch *et al.* 1980) have studied this important point further, and suggest an average figure for extrarenal clearance of creatinine in advanced renal failure to be 400 ml/kg. 24 h, i.e. about 2 ml/min for a 70 kg subject.

The measurement of plasma creatinine

Because of problems with the accurate measurement of plasma creatinine, and the wide use of this measurement in nephrology, it is essential that an nephrologist knows precisely how creatinine is being measured in the laboratory he or she uses, the advantages and disadvantages of the methodology used, the interfering substances, and the accuracy from day to day of the method. Spencer (1986) gives an extensive review of the measurement of creatinine in plasma.

Creatinine is usually measured by the Jaffé reaction, now known for more than a century, using the reaction of creatinine with alkaline picrate (Bonsnes and Taussky 1945; Husdan and Rapaport 1968) to form an orange-red coloured Janovsky complex (Spencer 1986) (Fig.5). Unfortunately, other chromogens which react to give a similar or identical colour are present in plasma, but not in urine (Spencer 1986). Thus the creatinine concentration is variably overestimated in plasma, and consequently

(a) $(Picrate) + (OH^-) \xrightarrow{K_0} (Picrate-OH^-)^*$

(b) $(Picrate-OH^-)^* + (Creatinine) \underset{K_2}{\overset{K_1}{\rightleftharpoons}} (Picrate-creatinine\ complex) + (OH^-)$
   $\downarrow$
   Yellow bis complex

(c) [Pielic acid]   [Creatinine]

[Janovsky complex]

Fig. 5 the reaction of creatinine with picrate under alkaline conditions in the Jaffé reaction to form a Janovsky complex (after Spencer 1986). Only recently, after 100 years, has the chemistry of this reaction become clear!

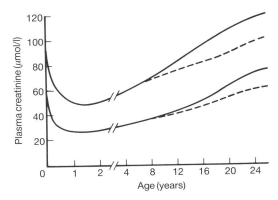

Fig. 6 Data on plasma creatinine concentrations measured by rate-dependant methods in children and young adults. (Modified from unpublished data of T.M. Barratt and J.M. Round, by permission.)

the calculated creatinine clearance is underestimated. This effect is especially important at low concentrations of creatinine, and so in adults is most notable around normal values. A particular problem exists in paediatrics, since the plasma creatinine in infants and children is much lower than normal adults (Fig. 6, and Chapter 1.4).

Table 2 gives a list of some substances which are known to interfere with the analysis of creatinine; these vary somewhat with the method used. Cook (1975), Rosano and Brown (1982) and especially Spencer (1986) review in detail the factors which influence the assay of creatinine by different methods, and conclude that chemical alternatives to the Jaffé reaction are not more convenient or specific. Enzyme-based methods, available as kits from Boehringer and Wako are becoming widely used, especially in Europe, despite being more expensive. Spencer (1986) suggests that they have greater precision at concen-

Table 2 (a) Substances which interfere with the measurement of plasma creatinine or the creatinine clearance

Endogenous

Protein	Deproteination essential; N.B. influence of hyperproteinaemia in some kinetic methods
Ketones and ketoacids	Actetoacetate and acetone elevate levels (red complex); depends on timing of rate analysis, temperature etc.; N.B. also ascorbic acid
Glucose and other sugars	Picrate to picramate (red); variable with temperature, rate, and method
Bilirubin	Reduces kinetic levels by increasing blank (biliverdin)
Fatty acids	Especially palmitic, raise levels
Urate	Jaffe-reactive on heating; elevates levels
Urea	Weakly Jaffé-reactive; elevates levels

Exogenous

Cephalosphorins	Strongly raise levels (up to 500 μmol/l); varies with drug and method of analysis. Cefoxitin, cephalothin, cefamandole, cephaloridine, cefotaxime all react, but rate of reaction varies
5-Fluorocytosine	Only with methods using creatinine deaminase—chromogen strongly raises levels
Phenylacetylurea	Elevates levels—picrate reactive
Acetoheximide	Elevates in some methods (not SMA II)
Methanol metabolites	Not identified, raised levels

(See Cook (1976) and Spencer (1986) for further discussion and references.)

(b) Substances which block tubular secretion of creatinine and induce true elevation of plasma creatinine (reducing creatinine clearance)

Triamterene	Probenecid
Spironolactone	Cimetidine (not ranitidine)
Amiloride	Trimethoprim

trations below 100 μmol/l, but this is really dependent upon the automation and the photometer used rather than an inherent trait of the method. However, enzymatic methods are usually more accurate than their Jaffé counterparts because there is less non-specific blank. HPLC techniques can be used also (Ambrose *et al.* 1983) and with greater availability of the apparatus may well become more common in clinical practice. These are highly reproducible, and not susceptible to interference.

If a Jaffé-based method is used, several approaches have been used to deal with the problem of non-creatinine Jaffé-reacting substances in the plasma (Spencer 1986): originally these were removed by absorption using Fuller's earth or resins, and currently they are dialysed off; alternatively, rate-dependent reactions can be used, which rely on the fact that the reaction of the creatinine with the reagent is in general faster than that of the non-creatinine chromogens. Reaction rate methods give greater accuracy as well as specificity, provided that the reaction conditions and period are carefully controlled. Finally the role of interfering pharmacological substances which may react like creatinine must be considered (Young *et al.* 1975), and this may vary from method to method (Table 2). Some interfering substances can be oxidized, or extracted by solvents, but these methods are little used.

The measurement of plasma creatinine is discussed further below in the section on prediction of GFR from plasma creatinine.

Renal handling of creatinine

Varying degrees of creatinine secretion in the tubule are found in different species, and the kidneys of many do not appear to transport it in either direction. In the adult rat and dog there is even a difference between sexes, but this is not true of humans (Harvey *et al.* 1966). In man, infused creatinine is secreted by the tubule (Shannon 1935) and usually the creatinine clearance equals or exceeds the GFR (as measured by the clearance of inulin) by a factor of 1.1 to 1.2 at clearances above 80 to 90 ml/min (Doolan *et al.* 1962; Bauer *et al.* 1982*b*; Shemesh *et al.* 1985). This is also true of transplanted kidneys (Ross *et al.* 1987). In children, creatinine clearances showed a $C_{creatinine}/C_{inulin}$ ratio of 0.98:1 at values for the GFR greater than 100 ml/min, and 1.1:1 at figures less than this (Arant *et al.* 1971; Mak *et al.* 1983).

A most important point, however, is that this ratio steadily increases as the GFR falls and the plasma creatinine rises, to the point where the creatinine clearance may be more than twice the inulin clearance (Manz *et al.* 1977; Jagenburg *et al.* 1978; Bauer *et al.* 1982*a*; Shemesh *et al.* 1985). Overestimation of the GFR by creatinine clearance increases proportionally as the GFR declines. The observation that the creatinine clearance may be reduced to values comparable to the GFR by the administration of cimetidine, which inhibits creatinine secretion in the renal tubule (Burgess *et al.* 1982; Shemes *et al.* 1985) suggests that this is the result of creatinine secretion, analogous to the effect when creatinine is infused.

Creatinine reabsorption is unusual in man, but is known to occur at very low urine flow rates (Brod and Sirota 1948; Levinsky and Berliner 1959) such as in congestive heart failure (Baldwin *et al.* 1950) either by tubular reabsorption or diffusion of creatinine through intact urinary epithelia.

The effect of proteinuria on the creatinine clearance is more controversial. Brod and Sirota (1948) first noted that in the presence of massive proteinuria the creatinine clearance can be double the GFR. This observation was confirmed by Berlyne *et al.* (1965), and more recently by Shemesh *et al.* (1985), but a number of authors have failed to find any relationship between proteinuria and creatinine clearance (Rapoport and Husdan 1968; Hilton *et al.* 1969; Maisey *et al.* 1969). Rosenbaum (1970) noted no difference with proteinuria, but oedema was associated with a high creatinin: inulin clearance ratio. Brod and Sirota

(1948) were able to inhibit the creatinine secretion with carinamide, but no further experiments with other agents such as cimetidine appear to have been performed since.

Finally the effect on the calculated creatinine clearance of using the more specific enzymatic methods or rate-specific Jaffé reactions with sodium tetraborate or lithium hydroxide buffers for measuring plasma creatinine requires consideration. It is purely coincidental that in normal subjects, using older nonspecific methods to measure plasma creatinine, the false elevation in the plasma creatinine result approximately cancels out the amount by which the creatinine clearance exceeds the GFR. With more specific methods, the true plasma creatinine is lower, with a corresponding increase in the calculated creatinine clearance, and a further overestimation of the GFR. No study of a large normal population has yet been done with the latest methodology to determine what the 'new' normal ranges may be.

Accurate urine collections
This has been mentioned above. Some check on the completeness of the urine collections can be obtained by measuring the creatinine output in the specimen (Wibell and Bjørsell-Östling 1973); overnight urine samples can be used in children. this depends upon the assumption that the excretion of creatinine is relatively constant from day to day, and is related in a simple fashion to body weight. The data of Kampmann *et al.* (1974) show clearly that the change in body composition with age is an important factor: a fit muscular 25-year-old weighing 60 kg may excrete more than twice as much creatinine as an obese 60-year-old with the same weight.

Also, creatinine excretion is not as constant from day to day as has been assumed from the original work of Folin (Scott and Hurley 1968; Edwards *et al.* 1969; Gowans and Fraser 1988). The coefficient of variation in creatinine excretion in a single individual is never less than 6 per cent, and may be more than 20 per cent. In some studies of nitrogen excretion, including Folin's original work, the variance in the excretion of urea was less than the variance in the excretion of creatinine!

The creatinine clearance in practice
Despite its many disadvantages and problems (Payne 1986; Spencer 1986), creatinine clearance remains the most widely used test for estimation of the GFR in clinical practice. Table 3 gives reference values from the literature for children and for adults using widely employed methodology. The Table gives the analytical method used in each study since this will influence the plasma creatinine value and hence the clearance, as noted above.

The profound effect of age upon creatinine clearance must not be forgotten, and this topic is dealt with in detail in Chapter 1.5. Other factors leading to biological variation in the GFR as

Table 3 GFR and creatinine clearance in young adults

(a) Inulin clearances (ml/min. 1.73 m^2)

	Mean	± 2 SD	Source	Method†
Children				
Newborn	38	26–60	Chantler and Barratt (1975)*	Various, mainly resorcinol
3 months	58	30–86	Chantler and Barratt (1975)*	
6 months	77	41–103	Chantler and Barratt (1975)*	
1 year	103	49–157	Chantler and Barratt (1975)*	
2–12 years	127	89–165	Chantler and Barratt (1975)*	
Males aged 20–29 years	132	90–174	Wesson (1969)*	
Females aged 20–29 years	119	84–156		

(b) Creatinine clearances (ml/min. 1.73 m^2)

	Mean	+ 2 SD	Source	Method†
Children aged 3–13 years	113	94–142	Chantler and Barratt (1975)*	True
Males and females aged 20–29 years				
Males	108	76–140	Edwards and Whyte (1959)	True
Females	104	86–132‡		
Males	123	84–162	Doolan *et al.* (1962)	True
Females	114	84–146‡		
Males	110	64–156	Kampmann *et al.* (1974)	Auto Analyzer method
Females	95	57–133		
Males	117	49–185	Wibell *et al.* (1973)	Auto Analyzer method
Females	114	74–154§		(mean of duplicate clearances)
Males	112	78–146ø	Rowe *et al.* (1976)	True
Females	—	—		

* Collected data from the world literature.
‡ Adults aged 17 to 59 years.
§ Adults aged 17 to 36 years.
ø Adults aged 17 to 34 years.
† True = plasma creatinine measured after extraction/adsorption of chromogens. Using the total chromogen methods, and therefore a falsely high plasma creatinine, Doolan *et al.* (1962) obtained figures of 103 (71–135) and 97 (77–117) for males and females respectively.
For GFR and creatinine clearances at different ages and in the elderly, see Chapter 1.5.

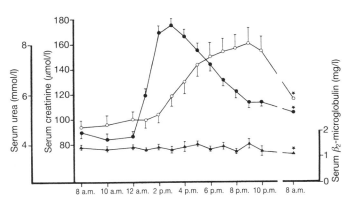

Fig. 7 The effect of a meal of stewed meat given at 1200 h on serum urea, serum creatinine and serum β_2-microglobulin concentrations. (From Jacobsen *et al.* (1980). *Lancet*, **i**, 319, with permission.)

Plasma creatinine concentrations as an indicator of GFR
(Fig. 8) (Cockroft and Gault 1976; Lott and Hayton 1978; Levey *et al.* 1988)

Because of the complexity of measuring inulin clearances, and the difficulties of obtaining accurately timed urine collections for creatinine clearances, the possibility of predicting the GFR from measurements of the plasma creatinine alone has received attention for many years. Creatinine excretion is only relatively constant for each individual, as already discussed, and it also varies with muscle mass. However, if as noted above

$$C_{cr} = \frac{(U_{cr} \times V)}{P_{cr}} \text{ and } (U_{cr} \times V) = k$$

then

$$C_{cr} \text{ varies as } \frac{k}{P_{cr}}$$

the relationship of plasma creatinine clearance and GFR itself has the general characteristics of a hyperbolic function. Unfortunately this means that plasma creatinine is a very poor index of GFR at modest reductions in GFR—just where precision is most needed. Also, since

$$P_{cr} = \frac{\text{Generation}_{cr}}{(UV_{cr} + \text{Extrarenal}_{cr})}$$

measured by the inulin clearance have been discussed in the section above. Naturally, all of these will apply equally to the clearance of creatinine.

In addition, however, there are several factors which apply only to creatinine clearances. If muscle mass changes, then changes in $UV_{creatinine}$ may be interpreted as changes of GFR. Mitch and Walser (1978) studied the rate at which urinary creatinine excretion equilibrates after a change in muscle mass, and found that this takes about 3 weeks. Thus urinary creatinine is an unreliable comparison for patients on different diets, or after changing protein intake (El Nahas and Coles 1986).

There is also one circumstance during which plasma creatinine may suddenly double. This is after the ingestion of meat subjected to prolonged stewing—the so-called 'goulash' effect (Fig. 7) (Camara *et al.* 1951; Jacobsen *et al.* 1979, 1980; Tapson *et al.* 1986). This results from conversion, by the stewing, of creatine in the meat to creatinine, so that in effect a creatinine meal is being fed. This conversion does not take place with grilling, frying, or short-term boiling, or with non-meat protein. In women, there is a slight increase premenstrually in the creatinine clearance, with a fall in plasma creatinine concentration (Davison and Noble 1981), presumably a minor version of changes which occur in pregnancy (Moniz *et al.* 1985).

Consecutive 24-h collections may increase the accuracy of the creatinine clearance (Wibell and Bjørsell-Östling 1973) and give some indication of how good a patient may be at collecting specimens. Chantler and Barratt (1972) found that in well-motivated adults the coefficient of variation in replicate measurements from the same individual was 11 per cent. This implies that two estimates of the creatinine clearance must differ by as much as 31 per cent for there to be a 95 per cent chance that a real change in GFR has taken place. Morgan *et al.* (1978) similarly arrived at a figure of 33 per cent, and Gowans and Fraser (1988) gave 35 to 45 per cent. These figures are, of course, much worse than variations in measurement of the plasma creatinine.

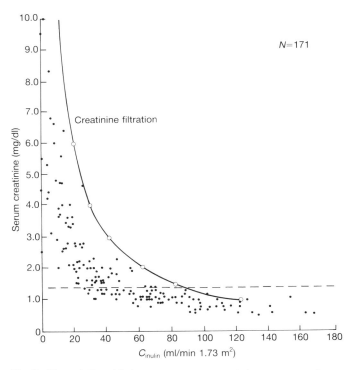

Fig. 8 The relationship between serum creatinine concentration and the GFR, measured as the clearance of inulin, in 171 patients with glomerular disease. The hypothetical relationship between GFR and serum creatinine is shown as the continuous line, assuming that only filtration of creatinine takes place. The broken horizontal line represents the upper limit of a normal serum creatinine (1.4 mg/dl). It can be seen that because of creatinine secretion and/or creatinine deficit through gut excretion, the serum creatinine consistently overestimates the GFR. (Reproduced from Shemesh *et al.* (1985). *Kidney International*, **28**, 830–8 with permission.)

plasma creatinine is subject to all the variations in creatinine generation and extrarenal excretion discussed above.

The measurement of 'true' plasma creatinine concentration usually has a coefficient of variation of about 6 to 7 per cent for replicate measurements of the same sample around a level of 120 μmol/l (Morgan *et al.* 1978); more recently, Spencer (1986) quotes 4 per cent. Thus a change of about 17 per cent (approximately 14 μmol/l) in successive samples gives a 95 per cent chance that there is a real difference between the two observations, and hence in the GFR. There is considerable variation in the performance of different laboratories in measuring plasma creatinine, some being as good as Wibell and Bjørsell-Östling (1973) who achieved a coefficient of variation as low as 2 per cent; the important point is to assess what degree of precision is available in the local laboratory. Plasma creatinine concentrations vary little throughout the day, although there is a tendency for a slight rise in the afternoon (Sirota *et al.* 1950); they are relatively independent of protein intake (Addis *et al.* 1951), except for the effect of stewed meat which is discussed above.

Effect of age on plasma creatinine

Although the GFR and the creatinine clearance fall steadily with increasing age over the age of 25 years or so, there is no corresponding rise in plasma creatinine with age, the levels being substantially the same at 80 years of age as at 20 (see Chapters 1.4 and 1.5 for detailed discussion). By contrast, in children there is a steady rise in plasma creatinine as muscle mass increases throughout childhood, and up to puberty there is only a small difference between girls and boys (Chantler and Barratt 1975; Schwartz *et al.* 1976a; Marris *et al* 1982), as shown in Fig. 6. Other data on plasma creatinine for both children and adults are summarized in the report by Morgan *et al* (1978).

Prediction of GFR from plasma creatinine

Given the relative imprecision of the plasma creatinine measurements and variations in creatinine generation and in extrarenal secretion, obviously the prediction of GFR from plasma creatinine is limited in its accuracy. Further, there is a systematic error in the prediction, in that plasma creatinine always overestimates the GFR at all levels of renal function (Fig. 7), i.e., the rise in plasma creatinine with decreasing GFR falls short of that predicted from constant creatinine production, even allowing for the somewhat enhanced secretion of creatinine at low values for GFR discussed above (Shemesh *et al.* 1985). This holds true for the transplanted kidney also (Ross *et al.* 1987).

Thus, measurement of plasma creatinine in patients with reduced renal function will substantially overestimate the GFR because of renal tubular creatinine secretion and extrarenal losses in the gut. Nevertheless, it continues to be a popular test of renal function because of its simplicity. Rather than use graphs to derive a creatinine clearance or GFR from the plasma creatinine measurement, a number of workers have tried to develop nomograms or formulae which will allow immediate prediction of GFR (or more usually, creatinine clearance) from the plasma creatinine. One of these nomograms is that produced by Kampmann *et al.* (1974). Obviously allowance must be made for both age and body size in their construction, and Rowe *et al.* (1976b) give a similar nomogram.

Alternatively one can use an empirically-derived formula. Perhaps the most widely employed and best validated for use in adults is that of Cockroft and Gault (1976)

$$C_{cr} = \frac{140 - (\text{age (years)} \times \text{weight (kg)})}{(72 \times P_{cr} \text{ (mg/dl)})} \text{ (ml/min)}$$

This formula assumes that the 'true' plasma creatinine is measured by an AutoAnalyzer, and the authors suggest that in females, the result should be factored further by 0.85 to allow for the smaller creatinine production in women. This empirical correction demonstrates how imprecise such estimates are.

Mitch and Walser (1978) suggest that the extrarenal excretion of creatinine should be taken into consideration in constructing any formula to predict creatinine clearance from plasma creatinine in renal failure; they give the following formulae, based on the data of Cockroft and Gault, and on their own measurements of extrarenal degradation of creatinine, for use in patients with advanced renal failure (plasma creatinine greater than 800 μmol/l. For males

$$C_{cr} \text{ (l/kg. 24 h)} = \frac{(28 - 0.2)}{P_{cr} \text{ (mg/dl)}} - 0.04$$

For females

$$C_{cr} \text{ (l/kg. 24 h)} = \frac{(23.8 - 0.17)}{P_{cr} \text{ (mg/dl)}} - 0.04$$

Obviously to convert these figures to the more usual ml/min, they should be multiplied by the weight of the patient in kg $\times$ 0.69.

Very obese patients present particular problems, and Salazar and Corcoran (1988) have derived formulae which better apply to this group. For males

$$C_{cr} = \frac{(137 - \text{age (years)}) \times (0.285 \times \text{weight (kg)})}{(51 \times P_{cr} \text{ (mg/dl)})}$$
$$+ \frac{(12.1 \times \text{height}^2 \text{ (cm)})}{(51 \times P_{cr} \text{ (mg/dl)})}$$

For females

$$C_{cr} = \frac{(146 - \text{age (years)}) \times (0.287 \times \text{weight (kg)})}{60 \times P_{cr} \text{ (mg/dl)}}$$
$$+ \frac{(9.74 \times \text{height}^2 \text{ (cm)})}{60 \times P_{cr} \text{ (mg/dl)}}$$

Barrett and Chantler (1975) developed formulae for use in children, in whom muscle mass can be predicted with some accuracy from height and weight (Graystone 1968). In fact length (or height in older children) must relate the plasma creatinine and the GFR, provided that (a) creatinine production relates directly to body mass (in turn proportional to body volume), and (b) that the GFR bears a constant relation to body surface area (itself proportional to the square of length). Barratt and Chantler (1975) derived the following formula

$$P_{cr} (\mu\text{mol/l}) = (\text{height (cm)} \times 0.331)$$

(This formula applies to the 'true' creatinine, measured manually; for AutoAnalyzer results, add 12.4 μmol/l). Thus the GFR can be predicted (Morris *et al.* 1982)

$$\text{GFR} = 40 \times \left(\frac{\text{height (cm)}}{P_{cr} (\mu\text{mol/l})}\right)$$

(the constant is 52.5 if AutoAnalyzer results are used).

However widely accepted this type of approach may be, and despite the fact that Morgan *et al.* (1978) and Payne (1986) concluded that the plasma creatinine level was preferable to the creatinine clearance, it must never be forgotten

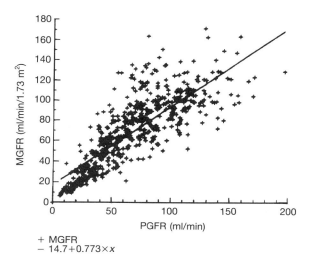

Fig. 9 Correlation between measured glomerular filtration rate using the iothalamate infusion method and predicted GFR from plasma creatinine measurements using the formula of Cockroft and Gault (1976) in 500 patients with various degrees of renal function. It can be seen that the prediction of GFR from plasma creatinine measurements by the Cockroft and Gault formula is far from precise. The correlation was improved by including a term for the patient's surface area, even though weight is included already in the formula. (Reproduced from Rolin et al. (1984) American Journal of Kidney Diseases, **4**, 58–54, with permission.)

that prediction of the glomerular filtration rate from the plasma creatinine is fraught with uncertainties, and whilst useful in clinical practice has no place in research studies. For example, Rolin et al. (1984) studied 500 pairs of estimates of plasma creatinine and GFR, measured by the clearance of iodothalamate (see below) at all levels of renal function (Fig. 9). They examined the effect of predicting the GFR using the formula of Cockroft and Gault (1976) and as can be seen from Fig. 9, the scatter was considerable, remaining an approximately constant proportion of measured GFR in the range studied. Prediction was particularly inaccurate in the very old (as in the study of Goldberg and Finkelstein 1987), the very young, and the very heavy. They found that prediction was improved by correcting the final result for body surface area, even though the weight of the patient is included in the formula. Similarly, in children Davies et al. (1982) have pointed out the limitations of the height/creatinine approach to the estimation of GFR in children.

Whilst the clinical usefulness of measurement of the plasma creatinine is beyond doubt, particularly if repeated several times as in the construction of slopes of reciprocal or logarithmic creatinines, the limitations of this method of assessing the GFR have in general been forgotten (Shemesh et al. 1985). Certainly the plasma creatinine, or one of its manipulations, whether by formula as discussed here, or as reciprocal or logarithm as discussed in the next section, provides an insecure basis for research projects or the critical evaluation of new therapies (Shemesh et al. 1985; El Nahas and Coles 1986). Mitch and Walser (1978) point out that on average, creatinine excretion and hence creatinine clearance do not stabilize until 41 days from a reduction in nitrogen intake; hence changes can only be evaluated over periods of 4 months or more. Mitch himself (1986) presents the best defence of the use of the slope method employing reciprocal creatinines, but the present writer is left with the feeling that whilst perhaps of use in the clinic, the are certainly inadequate when dealing with critical evaluation of dietary manipulations in renal failure.

Reciprocal and logarithm of plasma creatinine

The clearance formula given above, and the derivation which shows that the creatinine clearance will vary directly as the reciprocal of the plasma creatinine, has awakened interest in expressing serial plasma creatinines as their reciprocal, in the hope that thereby a straight line plot of fall-off in glomerular function might be obtained (Mitch et al. 1976; Rutherford et al. 1977) (Fig. 10). Specifically it was hoped that predictions of the time of onset of terminal renal failure could thus be made relatively easily, and so planning for substitution therapy could be made in advance. A logarithmic transformation of the raw plasma creatinine data, which assumes a constant fractional loss of renal function rather than the constant rate of loss implied in the analysis of reciprocals, gives a similar plot (Rutherford et al. 1977).

Obviously, the accuracy of any transformation of the plasma creatinine concentration is no better than the plasma concentration measurement from which it is derived, so that all the remarks made above about the plasma creatinine apply to reciprocally- or logarithmically-transformed data. This simple

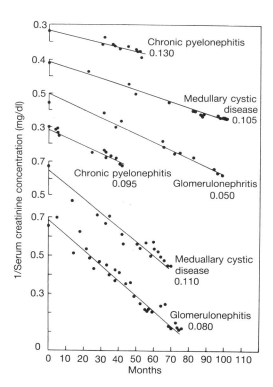

Fig. 10 Composite plot of reciprocal serum creatinine concentrations (in mg/dl) on the vertical axis versus time (months of observation) on the horizontal axis, in six patients with progressive renal failure. The number at the right of each line is the final reciprocal creatinine measurement for each patient. Diagnoses are shown also. (Reproduced from Mitch et al. (1976). Lancet, **ii**, 1326–8, with permission.)

but important point seems to have been forgotten in much of the recent discussion of the use of transformed creatinine results, but of course the power of the method rests with the fact that the slope is based upon many serial measurements of the plasma creatinine.

Nevertheless, the rate of fall-off in GFR (and hence in reciprocal or log. creatinine) is essentially linear in many patients (Mitch *et al.* 1976; Rutherford *et al.* 1977; Oksa *et al.* 1983), although the rate of fall-off varies greatly between individuals (Fig. 8), and 30 to 50 per cent of patients may show a very non-linear pattern. In general, patients with tubulointerstitial disease show slow rates of decline, while those with glomerular disease and diabetes show a more rapid fall-off in renal function. (Maschio *et al.* 1987; Jones *et al.* 1979). A critical assessment of this method of analysis to predict time of terminal renal failure is that of Gretz *et al.* (1983), and of the whole concept by Levey *et al.* (1988). These papers should be consulted for more details.

Levey *et al.* (1988) also discuss in detail the problem of assessing whether the slope of $1/P_{cr}$ inflects, either spontaneously or in response to treatment, so that two lines fit the data better than a single slope (Jones and Molitoris 1984; Kirschbaum 1986).

Assessment of the GFR by measurement of plasma β_2-microglobulin

During the past three decades it has been realized that the kidney provides an important arena for the disposal of peptides and proteins, particularly those of small molecular weight (see Chapters 9.5 and 10.3). This breakdown of microproteins occurs principally by unrestrained filtration at the glomerulus, followed by pinocytotic reabsorption in the tubule, with catabolism to constituent amino acids. Thus a large number of peptides (including those with hormonal action such as PTH and insulin) accumulate in the plasma with declining renal function. A number of small proteins, including α_1-microglobulin (Kusano *et al.* 1985) have been studied in this respect but the microglobulin assessed in the greatest detail as a measure of GFR has been β_2-microglobulin (Schardijn and Statius van Eps 1987).

This microglobulin has a molecular weight of 11 815, a single amino-acid chain of 99 amino acids and an Einstein-Stokes radius of 2.1 nm. Thus it is filtered at the glomerulus like water (Wibell *et al.* 1973; Karlsson *et al.* 1980). Subsequently not less than 99.9 per cent is reabsorbed and subsequently degraded in the renal tubule. Because it is filtered so readily, its plasma concentration in health is low (less than 2 mg/l and averaging about 1.5 mg/l) (Wibell *et al.* 1973). The plasma concentration rises as the GFR falls, reaching about 40 mg/l in terminal uraemia (Wibell *et al.* 1973; Karlsson *et al.* 1980), and a similar level in patients on haemodialysis (Vincent *et al.* 1978; Shea *et al.* 1981).

The logarithm of the plasma concentration is linearly related to the logarithm of GFR throughout the whole range (Wibell *et al* 1973) (Fig.11), so that it provides an excellent marker for renal dysfunction (Trollfors and Norrby 1981; Shea *et al.* 1981), particularly at modest reductions of GFR; at this point it provides the best plasma indication of GFR, at least in older subjects with reduced muscle mass and low creatinine output. The plasma concentration of β_2-microglobulin is not affected by muscle mass, nor by the sex of the subject.

There are several main reasons why β_2-microglobulin has not become more useful in clinical practice. First, until recently its measurement involved an expensive radioimmunoassay. As cheaper methods become available, for example those based on ELISA techniques and laser nephelometry, the demand for the

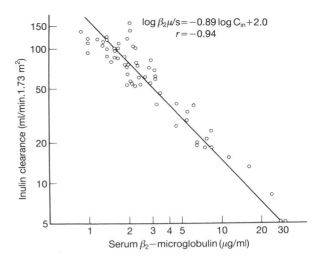

Fig. 11 The relationship between serum β_2-microglobulin concentrations in serum and GFR measured by inulin clearance in 60 patients with varying degrees of renal dysfunction. Both scales are logarithmic, and the relationship is effectively linear. Note that serum concentration of β_2-microglobulin are raised even at a GFR of 80 to 100 ml/min. (Reproduced from Karlsson *et al.* (1980). *Scandinavian Journal of Clinical and Laboratory Medicine*, **40**, (suppl. 154), 27–37, with permission.)

measurement of β_2-microglobulin will probably increase. Second, the main biological disadvantage of β_2-microglobulin as an index of GFR is that although in most patients its turnover is rather constant, in patients with some tumours—especially those involving cells of the lymphoid series—and with some inflammatory diseases such as systemic lupus erythematosus, Sjögren's syndrome, and rheumatoid arthritis, there may be elevations of the plasma concentration which represent increased production rather than reduced clearance (Karlsson *et al.* 1980).

Measurement of GFR using radionuclides

Several radiolabelled chelates which are believed to be handled entirely, or almost entirely by glomerular filtration have been used to assess the GFR in man. Chief amongst them are ^{51}Cr-edetic acid (Chantler *et al.* 1969) (available for human use in Europe but not in the United States), ^{125}I-iothalamate (Sigman *et al.* 1966; Maher *et al.* 1971) and ^{99}Tc^m-DTPA (Hilson *et al.* 1976; Barbour *et al.* 1976; Johansson and Falch 1978). They can of course simply be used as substitutes for inulin in conventional infusion techniques for assessing the GFR, but more interest has centred on alternative techniques which avoid urine collections and/or infusions. Iodothalamate can be used also as the compound labelled with stable iodine, measurement being made by fluorescent excitation analysis of the iodine moiety (Mak *et al.* 1983; Al-Dahhan *et al.* unpublished work).

Following injection of a bolus of a non-metabolized substance not handled by the tubules such as inulin, it should be possible to derive the GFR by compartmental analysis from some function of the disappearance curve of the substance from the plasma. This approach was investigated quite early by Homer Smith himself (Smith 1951) using inulin itself (see above), but the theoretical difficulties led to its abandonment. However, the introduction of easily-assayed radiolabelled compounds reawakened interest in the area.

Edetate

^{51}Cr-edetate has been the compound most studied (Chantler *et al.* 1969; Chantler and Barratt 1975; Ditzel *et al.* 1972; Brøchner-Mortensen 1972; Jagenberg *et al.* 1978; Trollfors and Norrby 1981). Standard infusion clearances using this agent are virtually identical to those obtained with inulin, even down to a GFR of 3 to 15 ml/min (Chantler *et al.* 1969; Brøchner-Mortensen 1972; Manz *et al.* 1977; Jagenburg *et al.* 1978), although there is evidence of a minor degree of tubular reabsorption whereas in some other species (for example the rat) it is secreted.

The principles of the single-injection technique are the same whatever chelate is used, and the description here applies to the other materials mentioned below. In the single-injection technique the material is injected intravenously and it is important to know the exact amount given (which is usually estimated by weighing the syringe before and after the injection). Also all the material must enter the vein; this may be ensured by setting up an infusion for a few minutes and injecting into the free-flowing stream. However, this complicates the technique for outpatient use and increases the cost considerably. Following injection, the plasma is sampled from the opposite arm and concentration plotted against time. Accurate timing of the samples is important. After 90 to 120 min in non-oedematous patients the fall-off usually can be described by a single exponential, and samples may be taken after this time to assess the GFR.

The number of samples may be quite small; equally accurate results have been obtained with from two to 12 samples (Chantler *et al.* 1969; Ditzel *et al.* 1972). To obtain the best results, it may be necessary to prolong the sampling times in patients with severe renal functional impairment (Maisey *et al.* 1969; Jagenburg *et al.* 1978). Radioactivity is plotted against time on semilogarithmic paper (log. radioactivity), and the slope of fall-off in counts is extrapolated back to zero time to give a notional volume of distribution at time zero (V_0). The GFR is then derived from the equation

$$\text{GFR} = V_o \left(\frac{\log_e 2}{t_{1/2}} \right)$$

where V_o = notional volume at zero time by extrapolation and $t_{1/2}$ = half-time for fall-off in radioactivity.

This simple monoexponential analysis has inherent disadvantages, although in practice the 'clearance' calculated is highly reproducible. The calculated result must be multiplied by 0.87 to correct for:

1. an overestimate of 22 per cent of V_o which arises from oversimplification of the monoexponential approach
2. a discrepancy of 5 per cent from inulin and ^{51}Cr-edetate clearances caused by tubular reabsorption of the chelate
3. a higher concentration (7 per cent) of chelate in venous blood compared with arterial blood.

To avoid these corrections in the simple approach, other authors have returned to the original exponential analysis, taking additional samples during the initial equilibration period (Ditzel *et al.* 1972; Silkalns *et al.* 1973) and using a two-compartment model, or basing their calculations on the whole area under the disappearance curve (Harries *et al.* 1972; Hagstam *et al.* 1974). Both of these methods have theoretical advantages, but in practice have the disadvantage that a greater number of samples must be taken (usually five or six at least) to obtain satisfactory analysis.

The simple monoexponential analysis gives confidence limits such that there is a 95 per cent probability that a change in ^{51}Cr-edetate clearance of ± 11 per cent represents a real difference in GFR (Chantler and Barratt 1972); it is thus more reproducible than even the infusion inulin clearance, and much superior to the creatinine clearance, as discussed above. Fisher and Veall (1975) have described the use of a single blood sample to determine the GFR after injection of ^{51}Cr-edetate, but this is not valid in oedematous subjects. The correlation between the single injection method using ^{51}Cr-edetate and inulin infusion clearances is excellent at normal and modest reduction of GFR (Brøchner-Mortensen 1972), but at low GFR (<15 ml/min) the two-compartment analysis overestimates GFR considerably (Manz *et al.* 1977; Jagenburg *et al.* 1978).

Naturally, the radiation dose imposed by a measurement of GFR using a radioisotopic technique must be given careful consideration. Lui and Ell (1986) quote the radiation dose from 2 MBq of ^{51}Cr-edetate to be 0.08 mGy, an effective dose-equivalent of 0.004 mSv, that is about 4 per cent of that of a routine posteroanterior chest radiograph.

Iothalamate

Labelled contrast media were first investigated using ^{125}I-labelled diatrizoate (Donaldson 1968) but this turned out to be secreted by the renal tubules and was abandoned in favour of ^{125}I-iothalamate (Sigman *et al.* 1966). Iothalamate has principally been used to measure GFR after a subcutaneous injection of a bolus of the marker, which is followed by slow release of the compound into the circulation, and provides stable plasma and urine levels for some hours (Israelit *et al.* 1973; Adefuin *et al.* 1976; Shea *et al.* 1981; Rolin *et al.* 1984), but it has also been used for infusion clearances (Sigman *et al.* 1966; Griep and Nelp 1969; Maher *et al.* 1971) as well as for the single-injection technique (Hall *et al.* 1976). The radiation dose is higher with this compound than with the other chelates discussed here, and thus the non-radioactive measurement of iothalamate by fluorescent activation of the iodine moiety has attractions (Moss *et al.* 1972), especially in paediatrics (Mak *et al.* 1983; Al-Dahhan *et al.*, unpublished work). The non-ionic contrast medium Iohexol has been used similarly using single-injection and fluorescent activation (O'Reilly *et al.* 1986).

Most observers have found that using the infusion techniques, iothalamate clearances are sensibly identical to inulin clearances, both in adults (Sigman *et al.* 1966; Maher *et al.* 1971; Eberstat 1984) and children (Mak *et al.* 1983; Al-Dahhan *et al.* unpublished work), although a few workers have found clearances exceeding that of inulin. Rosenbaum *et al.* (1979) and Odlin *et al.* (1985) have pointed out that it is an unsuitable agent for measurement of the GFR in some non-human species, because of extensive tubular handling; however, the effect is not present in man. Since the subcutaneous injection technique is essentially equivalent to the infusion technique, it also gives very good agreement with inulin clearances (Israelit *et al.* 1973; Adefuin *et al.* 1976). The single injection method using iothalamate (Hall *et al.* 1976) is of course subject to all the problems of compartmental analysis discussed above with reference to ^{51}Cr-edetate.

Diethylene triamine penta-acetic acid tin (DTPA-Sn or simply DTPA)

This chelate, labelled either with ^{113}In^m or with ^{99}Tc^m has been less used as an agent for the measurement of GFR (Hilson *et al.* 1976; Barbour *et al.* 1976; Johansson and Falch 1978; Shemesh *et al.* 1985) than other chelates, despite the fact that it is widely available and used for renal perfusion studies. Infusion clearances of this chelate are equal to those using ^{51}Cr-EDTA or ^{125}I-iothalamate (Hilson *et al.* 1976; Johansson and Falch 1978), and also correspond very closely to those of inulin over a wide range of renal function in diseased kidneys (Shemesh *et al.* 1985). However the latter authors point out that (hardly surprisingly) the correlation of the clearance calculated from the slope method is much less close, the DTPA clearance being less than the inulin clearance using the two-compartment model discussed above for analysis (Shemesh *et al.* 1985).

Gamma-camera images of the kidneys themselves can be obtained using ^{99}Tc^m-DTPA and GFR obtained from these sequences. Rehling and colleagues (1985, 1986) have described a technique 60 1-s wide-field frames and 120 10-s frames after intravenous injection of a bolus of ^{99}Tc^m-DTPA Sn, correcting the renal handling by background data obtained from the left ventricle. Unfortunately the method was compared with an inulin clearance based on the single injection technique (Rehling *et al.* 1984), but the correlation of 0.97 is nevertheless impressive. This method appears to be more precise than other similar analyses which have been published (Gates 1981; Fawdry *et al.* 1985; Chacati *et al.* 1985) and deserves more exploration.

Renal blood flow

Renal blood flow is not often estimated in clinical practice, although with the advent of isotopic methods it is becoming a practicable technique. Clearly the reference method is to place a flowmeter directly around the renal artery, assuming that it is single. Obviously this is not often feasible. The total renal blood flow can of course be calculated from the plasma flow and the haematocrit, but the haematocrit varies in different parts of the circulation. The classical method of determining the renal plasma flow is to employ the clearance of *p*-aminohippurate (PAH). At low plasma concentrations, PAH is substantially removed by filtration and secretion in a single circulation through the kidney.

PAH clearance

Even though the chemical estimation of PAH in plasma and urine (Smith *et al.* 1945) is simple compared to that of inulin, the snag is that the renal extraction of the compound in a single passage through the kidney, is incomplete (about 85 per cent), so that without a catheter in the renal vein to monitor the extraction of PAH, the estimate of plasma flow remains just that—an estimate only. The extraction of PAH may well be disturbed to an unknown extent in disease, but this has not been studied in detail for most conditions. In our own unpublished studies (Fluck, D.C. Cameron, J.S. Ogg, C.S., and Brown, C.B.) of patients with severe valvular heart disease, even prior to operation the extraction of PAH was a low as 30 per cent in some patients (Fig. 12), which correlated with renal arteriovenous oxygen extraction. Thus a standard PAH extraction cannot be assumed, although this is almost always done.

Data for renal blood flow in normal adults and children have been summarized by Wesson (1969). At the age of 20 years, renal plasma flow is at its peak of 650 ± 100 ml/min.1.73 m^2 (mean ± SD), which gives a calculated renal blood flow of 1050 ± 150 ml/min.1.73 m^2. The apparently very low plasma flow in infants is mainly the result of much lower extraction of PAH by

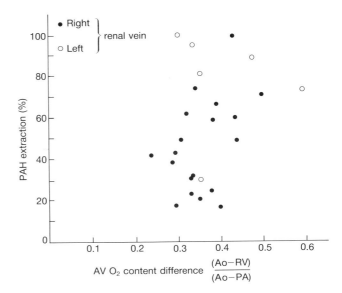

Fig. 12 Data from patients undergoing cardiopulmonary bypass operations who had renal venous vatheters inserted for direct measurement of extraction and clearance of *p*-aminohippuric acid. Note the very low extractions present in many patients even before operation, and that there is some correlation with arteriovenous oxygen differences across the kidney (Ao = aorta).

the neonatal kidney (see Chapter 1.4), averaging about 60 per cent (Calcagno and Rubin 1963).

Even more than GFR, renal blood flow varies with pain, stress, activity, and pregnancy (Sims and Kranz 1958; Lindheimer and Katz 1986), protein content of the diet (Pullman *et al.* 1954; Hostetter 1986; Viberti *et al.* 1987; Rodriguez-Iturbe *et al.* 1988), and fever, apart from any changes with disease states. In general, in chronic renal diseases the GFR declines at the same rate as the renal blood flow; the exception is, interestingly, in the minimal change nephrotic syndrome, in which the renal plasma flow in both adults and children is relatively well preserved, with a marked fall in the GFR (Berg and Bohlin 1982). In this condition there is a fall in calculated filtration fraction, presumably as a result of changes in K_f (see Chapter 3.3). In acute conditions such as shock, the two may deviate substantially, most notably in acute renal failure from tubular necrosis. Using an extraction method employing ^{85}Kr, Brun *et al.* (1955) showed in clinical acute renal failure in humans that despite almost zero GFR, plasma flow remained about 50 per cent of normal values. These data are in agreement with a large amount of data from experimental animals.

An old problem in the use of PAH is that in the presence of glucose, PAH clearance is depressed (Smith 1951). Competition between glucose and PAH for proximal tubular transport sites was suggested as the explanation, but Greene *et al.* (1987) and Dalton *et al.* (1988) pointed out recently that in glycosuric urine stored at $-20°$ Schiff-base formation between the PAH and the glucose takes place, with an apparent reduction in U_{PAH}. This problem can be overcome by collecting the urine into alkali, or by acid hydrolysis before measurement of the PAH.

^{125}I-*o*-iodohippurate clearance

All the criticisms of the use of PAH to measure renal plasma flow and blood flow mentioned above apply to the use of radiolabelled *o*-iodohippurate used in the classical infusion-equilibrium, urine-collection technique. The extraction of *o*-iodohippurate is also lower than that of PAH (B.D. Myers, personal communication). In addition, as with the GFR, attempts have been made to use the single-injection technique to estimate renal blood flow (Farmer *et al.* 1967; Cohen *et al.* 1971; Silkalns *et al.* 1973) using multicomponent analysis of the fall-off in plasma radioactivity. The agreement with the classical technique may well be fortuitous, since *o*-iodohippurate is secreted to some extent by the liver, arteriovenous differences may be as large as 30 per cent, and at no time after injection can the curve be described by a single exponential. The value of such clearances as an estimate of renal plasma flow is dubious.

Imaging techniques using ^{131}I-hippuran

Again as with GFR measurement, attempts have been made to measure renal plasma flow non-invasively, using gamma-camera images after injection of ^{131}I-labelled hippuran (Schlegel *et al.* 1979). This involves the following relationship

$$\text{relative uptake (1–2 min)} = \left(\frac{\text{(kidney-background)} \times y^2}{\text{1 min count of injected dose}}\right) \times 100$$

In the study of Chacati *et al.* (1985), this method correlated well ($r = 0.86$) with conventionally measured PAH clearances in 22 patients and nine additional single kidneys.

The curve of disappearance of ^{85}Kr from the kidney may also be analysed into four components; it has been suggested that these correspond to regions of differential blood flow through anatomical compartments of the kidney (Barger and Herd 1971). Although these deconvolution techniques (see Butt *et al.* (1983) for references) appear to be applicable to dogs, their validity in man is not established, although in acute renal failure they appeared to confirm the preferential ischaemia of the outer cortex already predicted from histological studies.

5-Hydroxyindoleacetic acid clearance

Until recently it was believed that no endogenous substance provided an index of renal blood flow. However, Hannedouche *et al.* (1989) have described an extraction of 78 per cent for 5-hydroxyindoleacetic acid (5-HIAA), a major metabolite of serotonin, and showed that its clearance corresponded closely with, and competed for, that of PAH. Since this substance is now readily measurable by chromatographic techniques, it may represent a major advance in the measurement of renal plasma flow.

Tests of tubular function

Unlike glomerular function, 'tubular function' presents a bewildering variety of different aspects. Also, in clinical terms, tests of tubular function are far less commonly required than is some assessment of glomerular filtration rate, so that in this section only a summary of tests is given.

For convenience, tests of 'tubular function' can be divided into those mainly concerned with proximal tubular function, and those mainly concerned with distal tubular function. However, many functions such as sodium reabsorption are carried out in almost all postglomerular segments of the nephron, and so this division is to some extent artificial. Tests of urinary concentrating ability depend upon the ability of the whole nephron to establish a medullary concentration of solute, as well as the delivery of solute to the distal nephron from the proximal tubule. Similarly, tests of acidifying capacity depend upon the ability of the proximal nephron to reabsorb bicarbonate, as well as the capacity of the distal nephron to establish and maintain a gradient of hydrogen ion between the tubular fluid and the plasma.

Tests of mainly proximal tubular function are dealt with in Chapters 5.1 to 5.5, in which glycosuria, phosphaturia, aminoaciduria, uricosuria, and the Fanconi syndrome are discussed. Here we will consider sodium handling in the nephron, and tests of mainly distal nephron function. However, for convenience in discussing tests of the net excretory capacity of the kidney for hydrogen ion at this point here, we will include the proximal reabsorption of bicarbonate (see also Chapters 5.3 and 7.3).

Sodium handling in the nephron

This topic is discussed *in extenso* in Chapter 2.3 in relation to diuretic therapy, and again in Chapter 7.1; only a brief summary is given here.

Between 90 and 99 per cent of the filtered water (150–200 l/24 h) and corresponding amounts of sodium chloride (1.5–1.7 kg/24 h) are reabsorbed in the nephrons of the kidney. The reabsorption of water is entirely passive, following osmotically the active absorption of NaCl. These mechanisms of NaCl absorption are heterogenous, differing in the various nephron segments. In all segments however, the primary active mechanism is the same: pumping of Na^+ out of the cell across the basal cell pole via the (Na^+,K^+-ATPase. The sodium uptake mechanisms across the luminal membrane are, however, different in the various segments.

In the proximal tubular luminal absorptive process Na^+ is taken up mainly in exchange for protons, hence coupling the reabsorption of Na^+ and HCO_3^-: this accounts for some 60 per cent of the filtered load. In the thick ascending limb, luminal uptake of Na^+ occurs via a Na^+2Cl^- K^+ cotransport system. Reabsorption at this site accounts for another 25 to 30 per cent of the filtered load of NaCl, so that only 10 to 15 per cent leaves the ascending limb. Because the thick ascending limb has a very poor permeability for water, the tubular fluid is diluted and the interstitial fluid concentrated at this site in the nephron. This diluting mechanism is responsible for the ability of the kidney sometimes to produce diluted, and at other times (e.g. in the presence of AVP), concentrated urine. The luminal NaCl absorptive process of the thick ascending limb is inhibited by so-called 'loop' diuretics (see Chapter 2.3). In the early distal tubule, luminal NaCl most probably proceeds via a NaCl cotransport system. This portion of the nephron is capable of reabsorbing some 10 per cent of filtered NaCl, and its luminal uptake system is inhibited by thiazides (see Chapter 2.3) In the collecting duct, only a small percentage of filtered NaCl and less than 10 per cent of water is reabsorbed. Sodium transport across the luminal membrane occurs via sodium channels. Water transport is controlled via the density of AVP-dependent water channels, and the Na^+ absorption is under the control of aldosterone. Potassium secretion at the nephron site is proportional to sodium reabsorption, and hence potassium loss will be augmented whenever Na^+ absorption at this site is increased. Sodium reabsorption at this site is inhibited by amiloride and by triamterene. Therefore, both these drugs spare potassium excretion (see Chapter 2.3).

Estimation of distal tubular sodium delivery; the lithium clearance

Both in research, and in a growing number of clinical situations, it is of interest to know what proportion of sodium is being reabsorbed in the proximal nephron, and how much is being delivered to a more distal site. In animals this can be measured directly by micropuncture techniques, but in man only indirect estimates are available.

The lithium clearance

One of these estimates is the lithium clearance C_{Li} (Thomsen et al. 1988). In the proximal tubule lithium is absorbed at the same rate as sodium, but normally it is not handled at all in the distal nephron, i.e. beyond the thick ascending limb (Thomsen et al. 1981). Thus lithium clearance gives some estimate of proximal tubular reabsorption of sodium. In normal individuals, the ratio C_{Li} to $C_{creatinine}$ is about 0.25 in women and 0.23 in men (Schou et al. 1986).

Although C_{Li} may provide a valid measure of proximal sodium reabsorption under normal circumstances, this is not so under extreme conditions of either salt depletion in rats or humans (Kirchner 1987; Atherton et al. 1987; Thomsen et al. 1988) or under osmotic diuresis (Skott et al. 1987) during which proximal clearance of sodium is respectively over- and underestimated. Vincent et al. (1987) give a diagram which allows one to correct for sodium excretion and creatinine in single measurements of C_{Li}.

The reasons for these shortcomings of the lithium clearance technique are probably twofold: first, some Li^+ is reabsorbed paracellularly in the thick ascending limb, driven by the lumen-positive voltage. This process is increased during volume contraction because of the increase in lumen-positive voltage, and is unmasked by frusemide. Second, a small amount of Li^+ is absorbed in the collecting duct, and this process is amiloride-sensitive (Koomans et al. 1989).

Calculation of distal sodium delivery in absence of AVP

Another approach is to estimate the proximal abstraction of sodium from final clearances of water (calculated as usual from P_{osm}, U_{osm}, and V, see below), and that of sodium. Work in animals (Danovitch 1978) has established that during a water diuresis (i.e under conditions of zero plasma AVP), the distal delivery of sodium and water may be approximated by the expression

$$C_{Na^+} + C_{H_2O}$$

These approximations must be interpreted with caution, because the entire concept is empirical and probably too simplistic. Many unknown factors will bear upon the results, and even the assumption that AVP secretion is completely suppressed may be invalid because of non-osmotic release of AVP, e.g. by pain, nausea, drugs, or diminished plasma volume. Also in many situations it is difficult or impossible to induce a maximum diuresis because of inability to tolerate a water load. Finally, even in the complete absence of AVP, some water is reabsorbed in the collecting duct.

Tests mainly of distal nephron function

Tests of concentrating ability

One of the earliest tests of renal function used in clinical medicine was the estimation of the concentrating ability of the kidney on the basis of the specific gravity of the urine, since it was recognized very early that this was diminished as renal excretory capacity fell. Tests of concentrating ability have been displaced from this central position by better measurements of excretory capacity, such as plasma creatinine and urea concentrations, and measurements of GFR itself. It is of interest, however, that when glomerular changes are scored semiquantitatively in glomerulonephritis, the correlation of the histology with the maximal concentration of the urine is better than that between the GFR itself and the glomerular changes (Schainuck et al. 1970).

Failure to concentrate the urine adequately may depend upon either a failure (or partial failure) of the central release of antidiuretic hormone ADH, now identified as arginine vasopressin (AVP), or upon a peripheral insensitivity of the distal nephrons of the kidney to the hormone, which again may be partial or complete, arising from congenital defects in (rare), or damage to (common), the countercurrent multiplier system. Thus diabetes insipidus may be divided into central, or pituitary diabetes insipidus, and renal or nephrogenic diabetes insipidus.

The most common cause of a failure of urine concentration is, of course, a fall in the glomerular filtration rate in renal failure, but the exact mechanisms by which this limitation is brought about are obscure. Usually the fall-off in concentrating ability parallels the fall in GFR, and begins before a detectable rise in plasma creatinine concentration. However, in some diseases with disproportionate medullary damage (such as reflux nephropathy or polycystic kidneys) concentrating ability is profoundly reduced even when glomerular filtration is relatively normal. This is also seen in sickle-cell disease (see Chapter 4.11), and in the recovery phase of tubular necrosis (see Chapter 8.2) in both autologous and in transplanted kidneys.

Less common, in contrast, are acquired forms of nephrogenic diabetes insipidus from exogenous agents, such as lithium (see Chapter 6.6), demeclocycline, amphotericin B, methoxyflurane, cisplatin, and a number of miscellaneous drugs on rare occasions (see Chapters 2.1 and 5.5). Prolonged potassium depletion, and also hypercalcaemia, will induce an AVP-resistant diabetes insipidus. Finally, a rare primary inherited familial diabetes insipidus also exists (Braden et al. 1985). Males are almost exclusively affected, but the inheritance overall falls short of a classic X-linked condition, probably because some heterozygotes have clinical manifestations. This is discussed in more detail in Chapter 5.5

Urine osmolality

The measurement of urinary specific gravity, with all its attendant difficulties and inaccuracies, is not discussed here since this measurement has now been entirely superseded by measurement of urinary osmolality. The relevant measurement to assess urine concentrating capacity is the osmolality of the urine (Dormandy 1967; Wolf and Pillay 1969). This, usually measured by the depression of the freezing point, gives a measure of the number of particles in solution in the urine. Ideally, to test tubular concentrating ability, the absorption of solute-free water under conditions of maximal antidiuresis should be measured. This is usually referred to as TcM_{H_2O}, since the reabsorption of water is usually denoted by Tc_{H_2O}. This may be measured by augmenting distal sodium delivery by mannitol or saline infusion, and then subtracting the urine volume, V, from the total osmolar clearance, which like any other clearance, may be calculated from

$$O_{osm} = \frac{U_{osm} V}{P_{osm}}$$

and then

$$Tc_{H_2O} = C_{osm} - V$$

In situations of water loading, when the osmolality of the plasma is greater than that of the urine and the urine volume is high, there is net excretion of solute-free water, i.e. $V > C_{osm}$, usually called simply 'free water'; conversely, in situations of antidiuresis there is reabsorption of free water. The measurement of Tcm_{H_2O} requires timed samples of urine and simultaneous measurement of plasma and urine osmolality. It should also be noted that the ability to reabsorb the 'solute-free' water depends upon the delivery of solute to the distal nephron, and this is also true of simpler concentrating tests done clinically. The normal value for Tcm_{H_2O} is about 5 ml/min. 1.73 m² (Boyarsky and Smith 1957) but may exceed this by a factor of 2 in situations of high solute intake and excretion, so that its definition is useless in the absence of data on solute excretion.

In practice, the osmolality of the urine (U_{osm}) can give much useful information, especially when the plasma osmolality (P_{osm}) is measured as well; in most situations, the plasma osmolality remains relatively constant (but not in uraemia or hyperglycaemia, of course, or after alcohol ingestion) and so U_{osm} alone suffices as an indicator of the U/P osmolality. In conditions of antidiuresis also, V becomes small (e.g. 0.5 ml/min or less) so that the effect on the equation for Tc_{H_2O} given above is small.

Early-morning urine osmolality and fluid deprivation tests

The simplest test—and a very useful one in clinical practice—is to measure the osmolality of a random specimen of early-morning urine. This will often exceed 800 mosm/l, and thus immediately exclude a major concentrating defect.

If it is wished to perform a more standardized test, then dehydration will be needed. After a normal lunch, no further fluid is given, a dry supper is taken and the early morning specimen of urine is collected the next day. Miles *et al.* (1954) showed that from 18 to 26 h of fluid deprivation were necessary to achieve maximum concentration in adults. A check on the efficiency of the dehydration may be made by weighing the patient; a loss of 3 to 5 per cent should be achieved, and surreptitious water ingestion detected. Conversely in very polyuric patients, a check on the weight may be made to avoid excessive or even dangerous dehydration, and the test stopped when this weight loss has been achieved. An alternative and much better test is to monitor P_{osm} and demonstrate a rise. A paediatric version of the test has been described (Edelmann *et al.* 1967a; Frasier *et al.* 1967).

Isaacson (1960) found a mean maximum U_{osm} of 1057 mosm/l for normal adult subjects using the concentrating test, with 95 per cent confidence limits of 807 to 1407 mosm/l. Concentrating ability is poorer at the extremes of life. In infants, Edelmann *et al.* (1967a) found that although older children could concentrate their urine to an average of 1089 mosm/l (95 per cent confidence limits 869–1309 mosm/l), in infants concentrating ability depended very much upon the nitrogen intake to allow 'priming' of the medullary concentration gradient by the urea generated (Edelmann *et al.* 1966). In the elderly, Rowe *et al.* (1976a) found maximum osmolalities of 1109 ± 22 mmosm/l (means ± SD) at 20 to 39 years of age, 1051 ± 19 at 40 to 49, and 882 ± 49 at 60 to 79 years of age after a 12-h thirst (see Chapter 1.5). Moreover, the 'washout' effect of a prolonged diuresis, however induced, must not be forgotten. Thus, compulsive water drinkers may show a concentrating defect which mimics that of true diabetes insipidus to a certain extent, even in the presence of endogenous or exogenous antidiuretic hormone (de Wardener 1985). Also in primary polydipsia, U_{osm} did not rise above a mean of 738 mosm/l (Braden *et al.* 1985).

Tests using exogenous AVP

The first part of the test just outlined examines the ability, not only of the kidney to respond to AVP, but also of the hypothalamic–pituitary axis to release it. If there is failure of concentration in response to dehydration, or if it is preferred to go to the second phase immediately, then exogenous AVP may be give in the form of 5 i.u. aqueous vasopressin subcutaneously, or usually DDAVP (desamino-*cis*-1,8-D-arginine vasopressin) (Curtis and Donovan 1979). A similar trap arising from long-standing polyuria exists here, since patients with central deficiency of AVP may fail to respond adequately to the exogenous hormone, again because of the 'washout' effect. Patients with true nephrogenic diabetes insipidus will fail to respond to exogenous vasopressin even when the system has already been 'primed' with a maximal dehydration. Osmolalities after injection of vasopressin are usually slightly lower than after maximum dehydration: de Wardener (1985) found 95 per cent confidence limits of 750 to 1150 mosm/l in adults, and Winberg (1959) 813 to 1327 mosm/l in children, with a mean of 1069 mosm/l.

Measurements of plasma AVP

In recent years it has become possible to measure the very small amounts of AVP present in the plasma by a specific radioimmunoassay, and to relate these concentrations to plasma osmolality in normal and diseased states (Robertson *et al.* 1973) Figure 13 shows this relationship, and demonstrates where the different types of diabetes insipidus fall. To bring out the differences more clearly, an intravenous infusion of hypertonic saline may be given, and the plasma AVP concentration and P_{osm} followed during the infusion (Zerbe and Robertson 1981).

Urinary dilution

Diluting capacity is difficult to assess under routine conditions. Normal individuals are capable of passing urine of only 40 to 80 mosm/l under a water load of 1 or 2 l taken slowly (Schoen 1957) provided that they have a normal solute intake. ADH is not required for urine dilution, so that in states of ADH resistance or absence, dilution is relatively well preserved without concentrating ability being present. This is true of potassium depletion, sickle-cell disease, and hypercalcaemia. Conversely, concentrating capacity may be preserved with loss of diluting capacity in cardiac failure, liver disease, and adrenal insufficiency, probably because of non-osmotic release of ADH. Dilution is rarely assessed except as a part of research projects.

Tests of hydrogen ion generation

The ability of the kidney to excrete hydrogen ion is reflected in several functions. These include the pH of the urine, the excretion of titrable acid buffered by phosphate and other urinary buffers, and of ammonium. In some situations, the amount of bicarbonate in the urine is also included—on a normal mixed diet there is none (see Chapter 5.3). In general, inability of the kidney to excrete enough hydrogen ion results from either a failure to secrete hydrogen or ammonium ions into the tubular fluid, or on the other hand by failure to reabsorb filtered bicarbonate.

The 'classical' form of renal tubular acidosis is distal, type 1, or gradient acidosis. In this condition, the fundamental fault is an inability to establish a gradient of hydrogen ion across the secret-

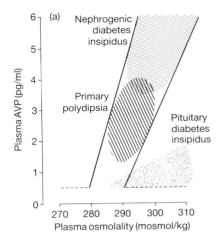

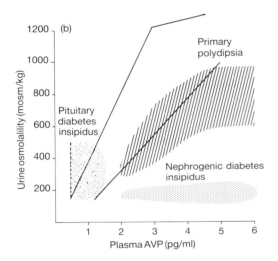

Fig. 13 The relationships between the concentration of plasma immunoreactive arginine vasopressin (AVP), and the plasma osmolality (a) and urine osmolality (b) in normal individuals and disorders of urinary concentration. The normal limits are shown by the continuous lines, and the limit of detection of plasma AVP by the broken line (0.4 pg /ml). Typical findings, either spontaneously or in response to hypertonic saline infusion, are shown for the three main causes of diabetes insipidus: pituitary diabetes insipidus, nephrogenic diabetes insipidus, and primary polydipsia. Note that in the last condition, although the kidney is structurally intact, the high water intake results in a 'wash-out' of medullary solute, and hence the upper limit of urinary concentration is reduced in relation to plasma AVP, even though the relationship between plasma tonicity and plasma AVP concentration is normal. (Redrawn from data of Robertson *et al.* (1973).)

ing epithelia of the distal tubule and collecting duct, so as to maintain a concentration of hydrogen ions in the tubular fluid (pH 4.8 at minimum) up to 600 times greater than that of the blood (pH 7.37). This can arise most commonly, from inability of the nephron to secrete hydrogen ions, but also from immediate leak back of secreted ions. A third form is that dependent upon the ionic gradient (voltage) generated by reabsorption of sodium in the distal nephron; if this is inadequate or absent, then hydrogen ion secretion will be defective; hyperkalaemia is usually present. Finally, renal tubular acidosis of distal type may arise in association with either aldosterone deficiency (sometimes called type 4 renal tubular acidosis) or aldosterone resistance (Sebastian *et al.* 1985), both of which are again associated with hyperkalaemia. All these conditions are discussed in detail in Chapter 5.4.

Plasma bicarbonate and early morning urine pH

At least to begin with, the presence of an acidosis is not accompanied by a fall in the pH of the blood from the usual value of 7.37 (a hydrogen ion activity of 40 nmol/l). There is however a consumption of buffer which is reflected in a diminution of the plasma bicarbonate concentration. Thus the first step in the examination of any patient with acidosis is to measure as accurately as possible the actual plasma bicarbonate concentration. This is best done using the Astrup technique on capillary, or better, arterial blood, since manual or AutoAnalyzer measurements on venous blood may give misleadingly low values. If the arterial concentration of bicarbonate at rest is below 20 mmol/l (normal 22 mmol/l), then the second crucial measurement is the pH of the urine. If plasma bicarbonate concentration is diminished and the urine pH is more than 5.5, then some defect of renal acidification must be present. Conversely, a pH of less than 5.3 in an early morning casual specimen of urine excludes a defect of hydrogen ion secretion. It is best that the pH be measured immediately after voiding, since the urine pH may change on standing for a variety of reasons; if this is impossible, then urine should be collected under mineral oil.

Acid-loading tests

In milder forms of defect, it will be necessary to stress the physiological system to make the diagnosis. The usual way to perform this is to give a 'short' acid load of ammonium chloride 0.1 g/kg orally, as described by Wrong and Davies (1959). For an average adult this is 7 g of ammonium chloride (84 mmol), while for children, 75 mmol/m^2 has been suggested, increasing the dose if the urine is not acidified (Edelmann *et al.* 1967b). Ammonium chloride is very nauseating, and should be given 0.5 g or 1.0 g at a time during a meal—above all not on an empty stomach first thing in the morning! Usually the whole dose can be taken over about 30 min in gelatin capsules or as a mixture. Normal individuals should pass at least one specimen of urine with a pH of 5.3 or less during the next 6 h; urine is best collected at 2, 4 and 6 h and the pH measured promptly, or collections taken under oil. At 4 h, capillary or venous blood should be taken to check that the plasma bicarbonate has been lowered to 18 mmol/l or less, otherwise the test is invalid since the system has not been stressed sufficiently.

In infants, although a urine pH of 5.0 or lower is attained, this usually corresponds to a lower plasma bicarbonate concentration than is found in adults at a similar urine pH (Edelmann *et al.* 1967b,c).

A 'long' test is sometimes used, but it is tedious, expensive, and unpleasant for the patient; in some hands, however, it gives more reliable results (Batlle and Kurtzman 1985). In this test, the dose of ammonium chloride is repeated for 3 to 5 days, with 24-h collections of urine. Usually the urine pH will become less than 5.0 after 3 days or so, and this version has the advantage that the ammonium excretion can be measured also, since this will be maximally raised after 4 to 5 days. Calcium chloride (2 mmol/kg) can be used if subjects find ammonium chloride too nauseating, or if it is contraindicated as in the presence of liver disease. The infusion of arginine hydrochloride or dilute hydrochloric acid has been used also.

Urine: plasma P_{CO_2} gradient; bicarbonate loading and threshold

An alternative, but still experimental approach is to use the P_{CO_2} of the urine as an indicator of distal hydrogen ion secretion (see Batlle and Kurtzman (1985) for discussion). Normally, the P_{CO_2} of the urine is considerably higher than that of the blood during bicarbonate loading. Secreted hydrogen ion reacts with filtered bicarbonate to form carbonic acid. In urine, the dehydration of carbonic acid to CO_2 and water proceeds only slowly, since carbonic anhydrase is not present, and infusion of the enzyme abolishes the urine: blood P_{CO_2} gradient, which has been interpreted to indicate that the urine: blood gradient depends upon distal secretion of hydrogen ions. Subjects with distal secretory defects of hydrogen ion fail to raise their urinary P_{CO_2} above that of blood, provided that the urine pH is raised above 7.8, which indicates that there is sufficient bicarbonate present. The crucial point is to ensure that the urinary pH exceeds that of the plasma whilst making the measurement.

The other main group of patients with renal tubular acidosis have proximal type 2 acidosis in which there is bicarbonate wasting into the urine because of proximal tubular deficiency. (Soriano *et al.* 1967; Rodriguez-Soriano and Edelmann 1969; Morris 1969; Edelmann 1985). This can occur as part of a generalized proximal tubular deficiency in the Fanconi syndrome, discussed in Chapter 5.1, or as an isolated defect. The former is easily distinguished by the additional presence of glycosuria, aminoaciduria and hypouricaemia. Patients with type 2 renal tubular acidosis, although they may have a resting acidosis and alkaline urine, can generate an acid urine with a normally low urine pH in response to acid loading as described above; the difference is that they will achieve this low urine pH at a markedly lower level of plasma bicarbonate than normal individuals (Fig. 14).

In adults with bicarbonate wasting, the plasma bicarbonate concentration at which bicarbonate becomes detectable in the urine is 16 to 20 mmol/l, compared with the normal 20 to 26 mmol/l (rather lower in infants (Edelmann *et al.* 1967c)). Bicarbonate infusion (which is feasible but tedious) demonstrates that the basic defect is in bicarbonate reabsorption; generation of a distal hydrogen ion gradient is normal. Correspondingly, patients with a proximal tubular acidosis require much larger doses of alkali to correct their acidosis than those with distal defects. Neither the Fanconi-associated or the primary form is common, and the latter is usually seen in paediatric clinics.

Chronic renal failure and ammonium excretion

Much more common than any of the types of renal tubular acidosis discussed above is the inability of patients with chronic renal failure to excrete hydrogen ion adequately. The excretion of ammonium ion is particularly reduced (Wrong 1965) but in relation to the number of surviving nephrons, i.e. the [NH_4/GFR] secretion is actually increased. If careful balance studies are performed, most patients in advanced renal failure are in positive hydrogen ion balance (Goodman *et al.* 1965), the excess hydrogen ion probably being buffered in bone at the expense of bone mineral, as happens in untreated renal tubular acidosis. As phosphate excretion falls in renal failure, the titrable acid also decreases. Again, hydrogen ion secretion appears reduced, but factored by GFR, each nephron is revealed as secreting many times the normal amount of total acid. The urine pH is usually low (about 5.0) in renal failure.

A failure of ammonium ion excretion has also been reported to be a specific feature of patients with gout, so that they pass

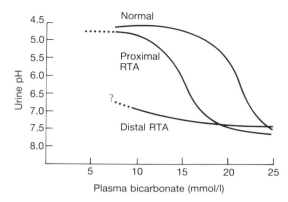

Fig. 14 The relationship between urinary pH and plasma bicarbonate (Astrup) in normal individuals and in patients with distal (gradient, type 1) and proximal (threshold, type 2) renal tubular acidosis. In normal individuals, bicarbonate only spills into the urine if the plasma concentration exceeds 22 mmol/l, and thus urine only becomes alkaline if the plasma bicarbonate concentration exceeds this level. Correspondingly, if the plasma bicarbonate concentration falls below 18 mmol/l, urine of maximum acidity will be passed.

In patients with a proximal 'leak' of bicarbonate, the threshold for appearance of bicarbonate in the urine may be only 15 or 16 mmol/l, and the urine will become alkaline at plasma bicarbonate concentrations as low as 18 mmol/l. If the plasma bicarbonate concentration is lowered still further by an acid load, the kidney is capable of elaborating urine with a normally low pH, but only at plasma bicarbonate concentrations below 15 mmol/l

In classical distal type 1 tubular acidosis, the threshold for bicarbonate is normal, but since the distal tubule is not capable of generating or maintaining the gradient of hydrogen ion between blood and urine, the urine remains alkaline even when the plasma bicarbonate is lowered to 10 to 15 mmol/l or even less. (Modified from data of Soriano *et al.* (1967).)

urine of maximal acidity over 24 h without the usual diurnal variation (Pak Poy 1965; Cameron and Simmonds 1981).

The excretion of ammonium ion is rarely measured in general clinical practice, because of technical difficulties with the measurement. However, a number of papers suggest that the urinary anion gap ($Na^+ + K^+ - Cl^-$) gives a close approximation of ammonium ion generation (Goldstein *et al.* 1986; Batlle *et al.* 1988; Halperin *et al.* 1988), and may be valuable in the diagnosis of patients with hyperchloraemic acidosis, although this is controversial.

References

Addis, T., Barratt, E., Poo, L.J., Ureen, H.J., and Lippman, R.W. (1951). The relation between protein consumption and diurnal variations of the endogenous creatinine clearance in normal individuals. *Journal of Clinical Investigation*, **30**, 206–9.

Adefuin, P.Y., Gur, A., Siegel, N.J., Spencer, R.P., and Hayslett, J.P. (1976). Single subcutaneous injection of iothalamate sodium I–125 to measure a glomerular filtration rate. *Journal of the American Medical Association*, **235**, 1467–9.

Alvestrand, A. and Bergstrom, J. (1984). Glomerular hyperfiltration after protein ingestion, during glucagon infusion, and insulin-dependent diabetes is induced by a liver hormone: deficient production of this hormone in hepatic failure causes hepatorenal syndrome. *Lancet*, **i**, 195–7.

Ambrose, R.T., Ketchum, D.F., and Smith, J.W. (1983). Creatinine determined by 'high performance liquid chromatography'. *Clinical Chemistry*, **29**, 256–59.

Arant, B.S., Edelmann, C.M., and Spitzer, A. (1971). The congruence of creatinine and inulin clearances in children: use of the Technicon AutoAnalyzer. *Journal of Pediatrics*, **81**, 559–61.

Arendshorst, W.J. and Navar, G.B. (1988). Renal circulation and glomerular hemodynamics. In *Strauss and Welt's diseases of the kidney*, (Eds. R.W. Schrier and C.W. Gottschalk), pp. 65–118. Little, Brown, Boston.

Arturson, G., Groth, T., and Grotte, G. (1971). Human glomerular membrane porosity and filtration pressure: dextran clearance analysed by theoretical models. *Clinical Science*, **40**, 137–58.

Atherton, J.C., et al. (1987). Lithium clearance in man: effects of dietary salt intake, acute changes of extracellular fluid volume, amiloride and frusemide. *Clinical Science*, **73**, 645–51.

Aukland, K. (1976). Renal blood flow. In *International Revenue of Physiology, Kidney and Urinary Tract Physiology. II.* (ed. K. Thurau), pp. 23–80. University Park Press, Baltimore.

Baldwin, D.S., Sirota, J.H., and Villareal, H. (1950). Diurnal variations of renal function in congestive heart failure. *Proceedings of the Society for Experimental Biology and Medicine*, **74**, 578–81.

Bankir, L., Bouby, N., and Trinh-Trang-Tan, M-M. (1984). Organization of the medullary circulation: functional implications. In *Nephrology* (Ed. R.R. Robinson), pp. 84–106. Springer, New York.

Barbour, G.L., Crumb, C.K., Boyd, C.M., Reenes, R.D., Rastogi, S.P., and Patterson, R.M. (1976). Comparison of inulin, iothlamate and 99mTc-DTPA for measurement of glomerular filtration rate. *Journal of Nuclear Medicine*, **17**, 317–20.

Barger, A.C. and Herd, J.A. (1971). The renal circulation. *New England Journal of Medicine*, **284**, 482–90.

Battle, D.C. and Kurtzman, N.A. (1985). The defect in distal (type 1) renal tubular acidosis. In *Renal Tubular Disorders* (eds. H.C. Gonick and V.M. Buckalew), pp. 281–306. Marcel Dekker, New York.

Batlle, D.C., Hizon, M., Cohen, E., Gutterman, C., and Gupta, R. (1988). The use of the urinary anion gap in the diagnosis of hyperchloremic metabolic acidosis. *New England Journal of Medicine*, **318**, 594–9.

Bauer, J.H., Brooks, C.S., and Burch, R.N. (1982a). Renal function studies in man with advanced renal insufficiency. *American Journal of Kidney Diseases*, **2**, 30–5.

Bauer, J.H., Brooks, C.S., and Burch, R.N. (1982b). Critical appraisal of creatinine clearance as a measurement of glomerular filtration rate. *American Journal of Kidney Diseases*, **2**, 337–46.

Bell, P.D. and Navar, L.G. (1984). Tubuloglomerular feedback. In *Nephrology* (Ed. R.R. Robinson) pp. 130–42. Springer, New York.

Berg, U. and Bohlin, A-B. (1982). Renal hemodynamics in minimal change nephrotic syndrome of childhood. *International Journal of Pediatric Nephrology*, **3**, 187–92.

Berglund, F. (1965). Renal clearances of inulin, polyfructosan-S and a polyetheylene glycol (PEG 1000) in the rat. *Acta Physiologica Scandinavica*, **64**, 238–44.

Berlyne, G.M., Varley, H., Nilwarangkur, S., and Hoerni, M. (1965). Endogenous creatinine clearance and glomerular filtration rate. *Lancet*, **ii**, 874–6.

Bonsnes, R.W. and Taussky, K.K. (1945). On the colorimetric determination of creatinine by the Jaffe reaction. *Journal of Biological Chemistry*, **158**, 581–91.

Bosch, J.P., Lauer, A., and Glabman, S. (1984). Short-term protein loading in assessment of patients with renal disease. *American Journal of Medicine*, **77**, 873–79.

Boyarsky, S. and Smith, H.W. (1957). Renal concentrating operation at low urine flows. *Journal of Urology*, **78**, 511–24.

Braden, G.L., Singer, I., and Cox, M. (1985). Nephrogenic diabetes insipidus. In *Renal tubular disorders* (Eds. H.C. Gonick, and V.M. Buckalew), pp. 387–456. Marcel Dekker, New York.

Brenner, B.M., Bohrer, M.P., Baylis, C., and Deen, W.M. (1977). Determinants of glomerular permelectivity: insights derived from observations *in vivo*. *Kidney International*, **12**, 229–37.

Brenner, B.M., Dworkin, L.D., and Ichikawa, I. (1986). Glomerular filtration. In *The Kidney* (eds. B.M. Brenner, and F.C. Rector). 3rd edn., pp. 124–44. Saunders, Philadelphia.

Brenner, B.M. and Rector, F.C. (eds.). (1986). *The Kidney*. 3rd edn. Saunders, Philadelphia.

Brochner-Mortensen, J. (1972). A simple method for the determination of glomerular filtration rate. *Scandinavian Journal of Clinical and Laboratory Investigation*, **30**, 271–4.

Brochner-Mortensen, J. (1978). *Routine Methods and their Reliability for Assessment of Glomerular Filtration Rate in Adults, with Special Reference to Total ^{51}Cr-EDTA Plasma Clearance*. pp. 1–76. Copenhagen, Ladeforenings Forlag.

Brod, J. and Sirota, J.H. (1948). Renal clearance of endogenous creatinine in man. *Journal of Clinical Investigation*, **27**, 645–45.

Brun, C., et al. (1955). Renal blood flow in anuric humans determined by the use of radioactive krypton 85. *Proceedings of the Society for Experimental Biology and Medicine*, **89**, 687–90.

Burgess, E., Blair, A., Krichman, K., and Cutler, R.E. (1982). Inhibition of renal creatinine secretion by cimetidine. *Renal Physiology*, **5**, 27–30.

Butt, T.J., Jones, D.R., Wallis, A.T., and Simpson, F.O. (1983). Comparison of published values from renal blood flow. *New Zealand Medical Journal*, **96**, 861–4.

Calcagno, P.L. and Rubin, M.I. (1963). Renal extraction of PAH in infants and children. *Journal of Clinical Investigation*, **42**, 1632–9.

Camara, A.A., Arn, K.D., and Reimer, A. (1951). The 24 hourly endogenous creatinine clearance as a clinical measure of the functional state of the kidneys. *Journal of Laboratory and Clinical Medicine*, **37**, 743–63.

Cameron, J.S. and Simmonds, H.A. (1981). Uric acid and the kidney. *Journal of Clinical Pathology*, **34**, 1245–54.

Carrie, B.J., Golbetz, H., Michaels, A.B., and Myers, B.D. (1980). Creatinine: an inadequate filtration marker in glomerular diseases. *American Journal of Medicine*, **69**, 177–82.

Chacati, A., Meyers, A., Rigo, P., and Godon, J.P. (1985). Validation of a simple isotopic technique for the measurement of global and separated renal function. *Uremia Investigation*, **9**, 177–9.

Chantler, C. and Barratt, T.M. (1972). Estimation of the glomerular filtration rate from the plasma clearance of 51–Chromium edetic acid. *Archives of Disease in Childhood*, **47**, 613–7.

Chantler, C., Garnett, E.S., Parsons, V., and Veall, N. (1969). Glomerular filtration rate measurement in man by the single injection method using ^{51}Cr-EDTA. *Clinical Science*, **37**, 169–80.

Cockroft, D. and Gault, M.K. (1976). Prediction of creatinine clearance from serum creatinine. *Nephron*, **16**, 31–41.

Cohen, M.L., Patel, J.K., and Baxter, D.L. (1971). External monitoring and plasma disappearance for the determination of renal function: comparison of effective renal plasma flow and glomerular filtration rate. *Pediatrics*, **48**, 377–92.

Cook, J.G.H. (1975). Technical bulletin no. 36. Factors influencing the assay of creatinine. *Annals of Clinical Biochemistry*, **12**, 219–32.

Cosmes, P.G. and Macias, J.F. (1988). Adenosina y autorregulacion renal. *Nefrologia*, **8**, 89–93.

Coulthard, M.G. and Ruddock, V. (1983). Validation of inulin as a marker for glomerular filtration in preterm babies. *Kidney International*, **23**, 407–9.

Curtis, J.R. and Donovan, B.A. (1979). Assessment of renal concentrating ability. *British Medical Journal*, **1**, 301–5.

Dalton, R.N. and Turner, C. (1987). A sensitive and specific method for the measurement of inulin. *Annals of Clinical Biochemistry*, **24**, (supplement 1) sl–s31.

Dalton, R.N., Wiseman, M.J., Turner, C., and Viberti, G. (1988). Measurement of urinary para-aminohippuric acid in glycosuric diabetics: problems and solutions. *Kidney International*, **34**, 117–20.

Danovitch, G.M. (1978). Clearance methodology in the study of the function of the distal tubule. *Renal Physiology*, **1**, 56–60.

Davies, J.G., Taylor, C.M., White R.H.R, and Marshall, T. (1982). Clinical limitations of the estimation of glomerular filtration rate from the height/plasma creatinine ratio: a comparison with simultaneous ^{51}Cr-edetic acid slope clearances. *Archives of Disease in Childhood*, **57**, 607–10.

Davis, J.M., Haberle, D.A., and Kawata, T. (1988). Function of the juxtaglomerular apparatus: local control of glomerular hemodynamics. *Journal of Physiology*, **402**, 473–95.

Davison, J.M. (1985). The effect of pregnancy on kidney function in renal allograft recipients. *Kidney International*, **27**, 74–9.

Davison, J.M. and Dunlop, W. (1980). Renal hemodynamics and tubular function in normal human pregnancy. *Kidney International*, **18**, 152–61.

Davison, J.M. and Noble, M.C.B. (1981). Serial changes in 24 hour creatinine clearance during normal menstrual cycles and the first trimester of pregnancy. *British Journal of Obstetrics and Gynaecology*, **81**, 10–17.

Day, D.F. and Workman, W.E. (1984). A simple assay for renal clearance determination using immobilized B-furanosidase. *Annals of the New York Academy of Science*, **434**, 504–7.

Deen, W.M., Bohrer, M.P., and Brenner, B.M. (1979). Macromolecule transport across glomerular capillaries: application of pore theory. *Kidney International*, **16**, 353–65.

Deen, W.M. and Satrat, B. (1981). Determinants of the glomerular filtration of proteins. *Amercian Journal of Physiology*, **241**: F162–F170.

Deen, W.M., Bridges, C.E., Brenner, B.M., and Myers, B.D. (1985). Heteroporous model of glomerular size selectivity: application to normal and nephrotic humans. *American Journal of Physiology*, **249**, F374–F389.

Degenaar, C.P., Frenken, L.A.M. and van Hooff, J.P. (1987). Enzymatic method for determination of inulin. *Clinical Chemistry*, **33**, 1070–1.

Ditzel, J. and Junker, K. (1972) Abnormal glomerular filtration rate, renal plasma flow, and protein excretion in recent and short-term diabetics. *British Medical Journal*, **1**, 13–19.

Ditzel, J., Westergaard, P., and Bronklov, M. (1972). Glomerular filtration rate determined by ^{51}Cr EDTA complex. *Scandinavian Journal of Urology and Nephrology*, **6**, 166–70.

Donaldson, I.M.C. (1968). Comparison of the renal clearance of inulin and radioactive diatrizoate ('Hypaque') as measures of the glomerular filtration rate in man. *Clinical Science*, **35**, 513–24.

Doolan, P.D., Alben, E.L., and Theil, G.B. (1962). A clinical appraisal of the plasma concentration and endogenous clearance of creatinine. *American Journal of Medicine*, **32**, 65–79.

Dormandy, T. (1967). Osmometry. *Lancet*, i, 267–70.

Eberstadt, P.L. (1984). Comparative study of glomerular filtration rate with diverse labelled agents. *International Urology and Nephrology*, **16**, 3–11.

Edelmann, C.M. Jr. (1985). Isolated proximal (type 2) renal tubular acidosis. In *Renal Tubular Disorders. Pathoghysiology, diagnosis, management*. (Eds. H.C. Gonick and V.M. Buckalew), pp. 261–80. Marcel Dekker, New York.

Edelmann, C.M., Barnett, H.L., Stark, H., Boichis, H. and Soriano, J.R. (1967a). A standardized test of renal concentrating capacity in children. *American Journal of Diseases of Children*, **114**, 639–44.

Edelmann, C.M., Boichis, H., Soriano, J.R., and Stark, H. (1967b). The renal response of children to acute ammonium chloride acidosis. *Pediatric Research*, **1**, 452–60.

Edelmann, C.M., Soriano, J.R., Boichis, H., Gruskin, A., and Aosta, M. (1967c). Renal bicarbonate reabsorption and hydrogen ion excretion in normal infants. *Journal of Clinical Investigation*, **46**, 1309–17.

Edwards, K.D.G. and Whyte, H.W. (1969). Plasma creatinine level and creatinine clearance as tests of renal function. *Australasian Annals of Medicine*, **8**, 218–24.

Edwards, O.M., Bayliss, R.I.S., and Millen, S. (1969). Urinary creatinine excretion as an index of the completeness of 24 hour urine collections. *Lancet*, ii, 1165–6.

El Nahas, A.M. and Coles, G.A. (1986). Dietary treatment of chronic renal failure: ten unanswered questions. *Lancet*, i, 597–600.

Farmer, C.D., Tauxe, W.N., Maher, F.T., and Hunt, J.C. (1967). Measurement of renal function with radio-iodinated diatrizoate and *o*-iodohippurate. *American Journal of Clinical Pathology*, **47**, 9–16.

Fawdry, R.M., Gruenewald, S.M., Collins, L.T., and Roberts, A.J. (1985). Comparative assessment of techniques for estimation of glomerular filtration rate with ^{99m}Tc-DTPA. *European Journal of Nuclear Medicine*, **11**, 7–12.

Fisher, M. and Veall, N. (1975). Glomerular filtration rate estimation on a single blood sample. *British Medical Journal*, **2**, 542.

Gates, G.F. (1981/2). Glomerular filtration rate: estimation from fractional accumulation of ^{99m}Tc–DTPA. *American Journal of Roentgenology*, **138**, 565–70.

Goldman, R. and Moss, J. (1959). Synthesis of creatinine in nephrectomized rats. *Amercian Journal of Physiology*, **197**, 865–8.

Goodman, A.D., Lemann, J. Jr., Lennon, E.J., and Relman, A.S. (1965). Production, excretion and net balance of acid in patients with renal acidosis. *Journal of Clinical Investigation*, **44**, 495–506.

Goldberg, T.H. and Finkelstein, M.S. (1987). Difficulties in estimating glomerular filtration rate in the elderly. *Archives of Internal Medicine*, **147**, 1430–3.

Goldstein, M.B., Bear, R., Richardson, R.M.A., Marsden, P.A., and Halperin, M.L. (1986). The urine anion gap: a clinically useful index of ammonium excretion. *American Journal of Medical Science*, **292**, 198–202.

Gowans, E.M.S. and Fraser, C.G. (1988). Biological variation of serum and urine creatinine and creatinine clearance: ramifications for interpretation of results and patient care. *Annals of Clinical Chemistry and Biochemistry*, **25**, 259–63.

Graf, H., Stummvoll, H.K., Luger, A., and Prager, R. (1983). Effect of amino acid infusion on glomerular filtration rate. *New England Journal of Medicine*, **308**, 159–60.

Graystone, J.E. (1968). Creatinine excretion during growth. In *Human Growth* (Ed. D.B. Chek), pp. 182–97. Lea and Febiger, Philadelphia.

Greene, S.A., Dalton, R.N., Turner, C., Haycock, G.B., and Chantler, C. (1987). Effect of hyperglycemia with and without glycosuria on inulin and para-aminohippurate clearance. *Kidney International*, **32**, 896–99.

Griep, R.J. and Nelp, W.B. (1969). Mechanism of excretion of radio iodinated sodium iothalamate. *Radiology*, **93**, 807–11.

Gretz, N., Manz, F., and Strauch, M. (1983). Predictability of the progression of chronic renal failure. *Kidney International*, **24**, (suppl 15), S-2–S-5.

Gutmann, I. and Bergmeyer, H.U. (1974). Inulin. In *Methods of Enzymatic Anaylsis* (Ed. H.U. Bergmeyer). 2nd edn., Vol. III pp. 1149–50. Verlag Chemie, Weinheim, Academic Press, New York.

Hagstam, K.E., Nordenfelt, I., Svensson, L., and Svensson, S.E. (1974). Comparison of different methods for determination of glomerular filtration rate in disease. *Scandinavian Journal of Clinical and Laboratory Investigation*, **34**, 31–6.

Hall, J.E., Guyton, A.C., and Farr, B.M. (1976). A single injection method for measuring glomerular filtration rate. *American Journal of Physiology*, **232**, F72–F76.

Halperin, M.L., Richardson, R.M.A., Bear, R.A., Magner, P.O., Kamel, K., and Ethier, J. (1988). Urine ammonium: the key to the diagnosis of distal renal tubular acidosis. *Nephron*, **50**, 1–4.

Hannedouche, T., Laude, D., Dechaux, M., Grunfeld, J-P., and Elghozi, J-L. (1989). Plasma 5–hydroxyindoleacetic acid as an endogenous index of renal plasma flow. *Kidney International*, **35**, 95–8.

Harvey, A.M., Malvin, R.L., and Vander, A.J. (1966). Comparison of creatinine secretion in men and women. *Nephron*, **3**, 201–5.

Harvey, A. McG. (1980). Classics in clinical science: the concept of renal clearance. *American Journal of Medicine*, **68**, 6–8.

Haycock, G.B., Schwartz, G.J., and Wisotsky, D.H. (1978). Geometric method for measuring body surface area: a height–weight formula validated in infants, children and adults. *Journal of Pediatrics*, **93**, 62–6.

Heller, J., Horacek, V., and Kamaradova, S. (1988). Efferent arteriole: the main target vessel for some vasoconstrictors? In *Nephrology*, (Ed. A.M. Davison), pp. 75–86. Ballière Tindall, London.

Hilson, A.J.W., Mistry, R.D., and Maisey, N.M. (1976). Technetium–99 metastable diethylene-triamine-penta-acetic acid for the measurement of glomerular filtration rate. *British Journal of Radiology*, **49**, 794–6.

Hilton, P.J., Lavender, S., Roth, Z., and Jones, N.F. (1969). Creatinine clearance in patients with proteinuria. *Lancet*, **ii**, 1215–6.

Hirschberg, R.R., Zipser, R.D., Slomowitz, L.A., and Kopple, J.D. (1988). Glucagon and prostaglandins are mediators of amino acid-induced rise in renal hemodynamics. *Kidney International*, **33**, 1147–55.

Hostetter, T. (1986). Human renal response to a meat meal. *Amercian Journal of Physiology*, **250**, F613–F618.

Husdan, H. and Rapoport, A. (1968). Estimation of creatinine by the Jaffe reaction. *Clinical Chemistry*, **14**, 222–38.

Ikkos, D., Ljunggren, R., and Luft, R. (1956). Glomerular filtration rate and renal plasma flow in acromegaly. *Acta Endocrinologica*, **21**, 226–36.

Isaacson, L.C. (1960). Urinary osmolarity in thirsting subjects. *Lancet*, **i**, 467–8.

Israelit, A.H., Long, D.L., White, M.G., and Hull, A.R. (1973). Measurement of glomerular filtration rate utilizing a single subcutaneous injection of ^{125}I-iothalamate. *Kidney International*, **4**, 346–9.

Jacobsen, F.K., Christensen, C.K., Mogensen, C.E., Andreasen, F., and Heilskov, N.S.C. (1979). Pronounced increase in serum creatinine concentration after eating cooked meat. *British Medical Journal*, **1**, 1049–50.

Jacobsen, F.K., Christensen, C.K., Mogensen, C.E., and Heilskov, N.S.C. (1980). Evaluation of kidney function after meals. *Lancet*, **i**, 319.

Jagenburg, R., Attman, P-O., Aurell, M., and Bucht, H. (1978). Determination of glomerular filtration rate in advanced renal insufficiency. *Scandinavian Journal of Urology and Nephrology*, **12**, 133–7.

Johansson, R. and Falch, D. (1978). ^{113m}In-DTPA, a useful compound for the determination of glomerular filtration rate (G.F.R.). The binding of ^{113m}In to DTPA and comparison between G.F.R. estimated with ^{113m}In-DTPA and ^{125}I-iothalamate. *European Journal of Nuclear Medicine*, **3**, 179–81.

Jones, J.D. and Burnett, P.C. (1974). Creatinine metabolism in humans with decreased renal function: creatinine deficit. *Clinical Chemistry*, **20**, 1204–12.

Jones, R.H., Hayakawa, H., McKay, J.D., Parsons, V., and Watkins, P.F. (1979). The progression of diabetic nephropathy. *Lancet*, **i**, 1105–7.

Jones, R.H. and Molitoris, B.A. (1984). A statistical method for determining the breakpoint of two lines. *Analytical Biochemistry*, **141**, 287–90.

Kampmann, J., Siersbaek-Nielsen, K., Kristensen, M., and Hansen, J.M. (1974). Rapid evaluation of creatinine clearance. *Acta Medica Scandinavica*, **196**, 517–20.

Karlsson, F.A., Wibell, L., and Evrin, P.E. (1980). B2-microglobulin in clinical medicine. *Scandinavian Journal of Clinical and Laboratory Investigation*, **40**, (suppl. 154), 27–37.

Kirschbaum, B.B. (1986). Analysis of reciprocal creatinine plots in renal failure. *American Journal of Medical Science*, **29**, 401–4.

Klahr, S. and Alleyne, G.A.O. (1973). Effect of chronic protein-calorie malnutrition on the kidney. *Kidney International*, **3**, 129–41.

Knox, F.G., *et al.* (1984). Intrarenal distribution of blood flow: evolution of a new approach to measurement. *Kidney International*, **25**, 473–9.

Koomans, H.A., Boer, W.H., and Dorhout Mees, E.J. (1989). Evaluation of lithium clearance as a marker of proximal tubule sodium handling. *Kidney International*, **36**, 2–12.

Kriz, W. (1981). Structural organization of the renal medulla: comparative and functional aspects. *American Journal of Physiology*, **241**, R3–R16.

Kusano, E., Suzuki, M., Asano, Y., Itoh, Y., Tagaki, K., and Kawai, T. (1985). Human o 1-microglobulin and its relationship to renal function. *Nephron*, **41**, 320–4.

Lang, F., Greger, R., Oberleithner, H., Griss, E., Lang, K., and Pastner, D. (1980). Renal handling of urate in healthy man in hyperuricaemia and renal insufficiency: circadian fluctuation, effect of water diuresis and of uricosuric agents. *European Journal of Clinical Investigation*, **10**, 285–92.

Levey, A.S., Perrone, R.D., and Madias, N.E. (1988). Serum creatinine and renal function. *Annual Review of Medicine*, **39**, 465–90.

Levinsky, N.G. and Berliner, R.W. (1959). Changes in the composition of the urine in ureter and bladder at low urine flow. *American Journal of Physiology*, **196**, 549–53.

Lindheimer, M.H. and Katz, A.S. (1986). The kidney in pregnancy. In *The Kidney*, (Eds. B.M. Brenner and F.C. Rector), 3rd. edn., pp. 1253–95. Saunders, Philadelphia.

Lott, R.S. and Hayton, W.L. (1978). Estimation of creatinine clearance from serum creatinine concentration—a review. *Drug Intelligence and Clinical Pharmacology*, **12**, 140–50.

Lui, D. and Ell, P.J. (1986). Time to scrap creatinine clearance? *British Medical Journal*, **293**, 1371.

Maher, F.T., Nolan N.G., and Elveback, L.R. (1971). Comparison of simultaneous clearance of ^{125}I labeled sodium iothalamate (Glofil) and of inulin. *Mayo Clinic Proceedings*, **46**, 690–1.

Maisey, M.N., Ogg, C.S., and Cameron, J.S. (1969). Measuring glomerular filtration rate. *Lancet*, **i**, 733.

Mak, R.H., Dahhan, J.A., Azzopardi, D., Bosque, M., Chantler, C., and Haycock, G.H. (1983). Measurement of glomerular filtration rate in children after renal transplantation. *Kidney International*, **23**, 410–3.

Manz, F., Alatas, H., Kochen, W., Lutz, P., Rebien, W., and Scharer, K. (1977). Determination of glomerular function in advanced renal failure. *Archives of Disease in Childhood*, **52**, 721–4.

Maschio, G., *et al.* (1987). Factors affecting progression of renal failure in patients on long-term dietary protein restriction. *Kidney International*, **32**, (suppl. 22): S–49–S–52.

Mene, P., Simonsen, M.S., and Dunn, M.J. (1989). Phospholipids in signal transduction of mesangial cells. *American Journal of Physiology*, **256**, F375–F386.

Miles, B.E., Paton, A., and de Wardener, H.E. (1954). Maximum urinary concentration, *British Medical Journal*, **2**, 901–5.

Mitch, W.E. (1986). Measuring the rate of progression of renal insufficiency. In Mitch, W.E. (Ed) *The Progressive Nature of Renal Disease*, pp. 167–87. Churchill-Livingstone, New York.

Mitch, W.E., Collier, V.U., and Walser, M. (1980). Creatinine metabolism in chronic renal failure. *Clinical Science*, **58**, 327–35.

Mitch, W.E. and Walser, M. (1978). A proposed mechanism for reduced creatinine excretion in severe chronic renal failure. *Nephron*, **21**, 248–54.

Mitch, W.E., Walser, M., Buffington, G.A., and Lemann, J. Jr. (1976). A simple method of estimating progression of renal failure. *Lancet*, **ii**, 1326–8.

Mogensen, C.E. (1972). Kidney function and glomerular permeability to macromolecules in juvenile diabetes. *Danish Medical Bulletin*, **19**, (supplement 3), 1–40.

Moniz, C.F., Nicolaides, K.H., Bamforth, F.J., and Rodeck, C.H. (1985). Normal reference ranges for biochemical substances relating to renal, hepatic and bone function in fetal and maternal plasma throughout pregnancy. *Journal of Clinical Pathology*, **38**, 468–72.

Morgan, D.B., Carver, M.E., and Payne, R.B. (1977). Plasma creatinine and urea: creatinine ratio in patients with raised blood urea. *British Medical Journal*, **2**, 929–32.

Morgan D.B., Dillon, S., and Payne, R.B. (1978). The assessment of glomerular function: creatinine clearance or plasma creatinine? *Postgraduate Medical Journal*, **54**, 302–410.

Morris, M.C., Allanby, C.W., Toseland, P., Haycock, G.B., and Chantler, C. (1982). Evaluation of a height/plasma creatinine formula for the measurement of glomerular filtration rate. *Archives of Disease in Childhood*, 57, 611–6.

Morris, R.C. (1969). Renal tubular acidosis. Mechanisms, classification and implications. *New England Journal of Medicine*, 281, 1405–13.

Moss, A., Kauffman, L., and Nelson, J.A. (1972). Fluorescent excitation analysis: a method of iodine determination in vitro. *Investigative Radiology*, 7, 335–8.

Navar, L.G. (1978). Renal autoregulation: perspectives from whole kidney and single nephron studies. *American Journal of Physiology*, 234, F357–F435.

Odlind, B., Hallgren, R., Sohtell, M., and Lindstrom, B. (1985). Is ^{125}I-iothalamate an ideal marker for glomerular filtration? *Kidney International*, 27, 9–16.

Ofstad, J. and Aukland, K. (1985). Renal circulation. In *The Kidney. Physiology and Pathophysiology*. (Eds. D.W. Seldin and G. Giebisch), pp. 471–96. Raven Press, New York.

Oken, D.E. (1989). Does the utrafiltration coefficent play a key role in regulating glomerular filtration in the rat? *Amercian Journal of Physiology*, 256, F505–F515.

Oksa, H., Pasternack, A., Luomala, M., and Sirvio, M. (1983). progression of chronic renal failure. *Nephron*, 35, 31–4.

O'Reilly, P.H., Brooman, P.J.C., Martin, P.J., Pollard, A.J., Farah, N.B., and Mason, G.C. (1986). Accuracy and reproducibility of a new contrast clearance method for the determination of glomerular filtration rate. *British Medical Journal*, 293, 234–6.

Osswald, H., Schmitz, H.J., and Kemper, R. (1978). Renal action of adenosine: effect on renin secretion in the rat. *Naunyn Schmiederberg's Archives of Pharmacology*, 303, 95–9.

Pak Poy, R.K. (1965). Urinary pH in gout. *Australian Annals of Medicine*, 14, 35–9.

Payne, R.B. (1986). Creatinine clearance; an redundant clinical investigation. *Annals of Clinical Biochemistry*, 23, 243–50.

Popper, H. and Mandel, E. (1937). Filtrations and Reabsorptionsleitung in der Nierenpathologie. *Ergebnisse der Inneren Medizin und Kinderheilkunde*, 53, 685–94.

Pullman, T.N., Alving, A.S., Dein, R.J., and Lansdowne, M. (1954). The influence of dietary protein intake on specific renal function in normal man. *Journal of Laboratory and Clinical Medicine*, 44, 320–32.

Rapoport, A. and Husdan, H. (1968). Endogenous creatinine clearance and serum creatinine in the clinical assessment of kidney function. *Canadian Medical Association Journal*, 98, 149–56.

Rehling, M., Moller, M.L., Thamdrup, B., Lund, J.O., and Trap–Jensen, J. (1986) Reliability of a ^{99m}Tc-DTPA gamma camera technique for determination of single kidney glomerular filtration rate. *Scandinavian Journal of Urology and Nephrology*, 20, 57–62.

Rehling, M., Moller, M.L., Lund, L.O., Jensen, K.B., Thamdrup, B., and Trap-Jensen, J. (1985). ^{99m}Tc-DTPA gamma-camera renography: normal values and rapid determination of single-kidney glomerular filtration rate. *European Journal of Nuclear Medicine*, 11, 1–6.

Rehling, M., Moeller, M.L., Thram-drup, B., Lund, J.O., and Trap-Jensen, J. (1984). Simultaneous measurement of renal clearance and plasma clearance of ^{99m}Tc-labelled diethylenetriaminepentaacetate, ^{51}Cr-labelled ethylenediaminetetraacetate and inulin in man. *Clinical Science*, 66, 613–9.

Robertson, G.L., Mahr, E.A., Athar, S., and Sinha, T. (1973). Development and clinical application of a new method for the radioimmunoassay of arginine vasopressin in human plasma. *Journal of Clincial Investigation*, 52, 2340–52.

Rodriguez-Iturbe, B., Herrera, J., and Garcia, R. (1985). Response to acute protein load in kidney donors and in apparently normal post acute glomerulonephritis patients: evidence for glomerular hyperfiltration. *Lancet*, ii, 461–3.

Rodriguez-Iturbe, B., Herrera, J., and Garcia, R. (1988). Relationship between glomerular filtration rate and renal blood flow at different levels of protein-induced hyperfiltration in man. *Clinical Science*, 74, 11–15.

Rodriguez-Soriano, J. and Edelmann, C.M. (1969). Renal tubular acidosis. *Annual Review of Medicine*, 20, 363–82.

Rolin, H.A., Hall, P.M., and Wei, R. (1984). Inaccuracy of estimated creatinine clearance for prediction of iothalamate glomerular filtration rate. *American Journal of Kidney Diseases*, 4, 48–54.

Romero, J.C. and Knox, G.G. (1988). Mechanisms underlying pressure-related natriuresis: the role of the renin–angiotensin and prostaglandin systems. *Hypertension*, 11, 724–38.

Rosano, T.G. and Brown, H.H. (1982). Analytical and biological variability of serum creatinine and creatinine clearance: implications for clincial interpretation. *Clinical Chemistry*, 28, 2330–1.

Rosenbaum, J.L. (1970). Evaluation of clearance studies in chronic kidney disease. *Journal of Chronic Disease*, 22, 507–14.

Rosenbaum, R.W., Hruska, K.A., Anderson, C., Robson, A.M., Slatopolsky, E., and Klahr, S. (1979). Inulin: an inadequate marker of glomerular filtration rate in kidney donors and transplant recipients? *Kidney International*, 16, 179–86.

Ross, E.A., Wilkinson, A., Hawkins, R.A., and Danovitch, G.M. (1987). The plasma creatinine concentration is not an accurate reflection of the glomerular filtration rate in stable transplant patients receiving cyclosporine. *Amercian Journal of Kidney Disease*, 10, 113–7.

Rowe, J.W., Shock, N.W., and de Fronzo, R.A. (1976a). The influence of age on the renal response to water deprivation in man. *Nephron*, 17, 277–8.

Rowe, J.W., Andres, R., Tobin, J.D., Norris, A.H., and Shock, N.W. (1976b). Age-adjusted standards for creatinine clearance. *Annals of Internal Medicine*, 84, 567–8.

Rutherford, W.E., Blodin, J., Miller, J.P., Grenwalt, A.S., and Vavra, J.D. (1977). Chronic progressive renal disease: rate of change of serum creatinine concentration. *Kidney International*, 11, 62–70.

Salazar, D.E. and Corcoran G.B. (1988). Predicting creatinine clearance and renal drug clearance in obese patients from estimated fat-free body mass. *American Journal of Medicine*, 84, 1053–60.

Schainuck, L.I., Striker, G.E., Cutler, R.E., and Bendit, E.P. (1970). Structural–functional correlations in renal disease II: the correlations. *Human Pathology*, 1, 631–41.

Schardijn, G.H.C. and Statius van Eps, L.W. (1987). B2-microglobulin: its significance in the evaluation of renal function. *Kidney International*, 32, 635–41.

Schlegel, J.U., Halikiopoulos, H.L., and Prima, R. (1979). Determination of filtration fraction using the gamma camera. *Journal of Urology*, 122, 447–50.

Schlondorff, D. (1987). The glomerular cell: an expanding role for a specialized pericyte. *FASEB Journal*, 1, 272–81.

Schlondorff, D. and Neuwirth, R. (1986). Platelet-activating factor and the kidney. *American Journal of Physiology*, 251, F1-F11.

Schnermann, J. (1981). Localization, mediators and function of the glomerular vascular response to alterations of distal fluid delivery. *Federation Proceedings*, 40, 109–15.

Schnermann, J. (1988). Effect of adenosine analogues on tubuloglomerular feedback responses. *American Journal of Physiology*, 255, F33–F42.

Schnermann, J. and Briggs, J. (1985). Function of the juxtaglomerular apparatus: local control of glomerular haemodynamics. In *The Kidney*, (eds. D.W. Seldin and G. Giebisch), pp. 669–98. Raven Press, New York.

Schoen, E.J. (1957). Minimum urine total solute concentration in response to water loading in normal men. *Journal of Applied Physiology*, 10, 267–70.

Schou, M., Thomsen, K., and Vestergaard, P. (1986). The renal lithium clearance and its correlations with other biological variables: observations in a large group of physically healthy persons. *Clincial Nephrology*, 25, 207–11.

Schrier, R.W. and Gottschalk, C.W. (1988). *Strauss and Welt's Diseases of the Kidney*. Little, Brown, Boston.

Schwartz, G.J., Haycock, G.B., Chir, B., and Spitzer, A. (1976a). Plasma creatinine and urea concentration in children: normal values for age and sex. *Journal of Pediatrics*, **88**, 828–30.

Schwartz, G.J., Haycock, G.B., and Edelmann, C.M. (1976b). A simple estimate of glomerular filtration rate in children derived from body length and plasma creatinine. *Pediatrics*, **58**, 259–63.

Scott, P.J. and Hurley, P.J. (1968). Demonstration of individual variation in constancy of 24-hour urinary creatinine excretion. *Clinica Chimica Acta*, **21**, 411–4.

Sebastian, A., *et al.* (1985). Hyperkalaemic renal tubular acidosis. In *Renal Tubular Disorders. Pathophysiology, Diagnosis and Management*, (Eds. H.C. Gonick and V.M. Buckalew), pp. 307–56. Marcel Dekker, New York.

Seldin, D.W. and Giebisch, G. (1985). *The Kidney. Physiology and Pathophysiology*. Raven Press, New York.

Seney, F.D. and Wright, F.S. (1985). Dietary protein suppresses feedback control of glomerular filtration rate in rats. *Journal of Clinical Investigation*, **75**, 558–68.

Shannon, J.A. (1935). The renal excretion of creatinine in man. *Journal of Clinical Investigation*, **14**, 403–10.

Shannon, J.A. and Smith, H.W. (1935). The excretion of inulin, xylose, and urea in normal and phlorizinized man. *Journal of Clinical Investigation*, **14**, 393–401.

Shea, P.H., Maher, J.F., and Horak, E. (1981). Prediction of glomerular filtration rate by serum creatinine and B2-microglobulin. *Nephron*, **29**, 30–5.

Shemesh, O., Golbetz., H., Kriss, J.P., and Myers, B.D. (1985). Limitations of creatinine as a filtration marker in glomerulopathic patients. *Kidney International*, **28**, 830–8.

Sigman, E.M., Elwood, C.M., and Knox, F. (1966). The measurement of glomerular filtration rate in man with iothalamate ^{131}I (Conray). *Journal of Nuclear Medicine*, **7**, 60–8.

Silkalns, G.L., *et al.* (1973). Simultaneous measurement of glomerular filtration rate and renal plasma flow using plasma disappearance curves. *Journal of Pediatrics*. **83**, 749–57.

Sims, E.A.H. and Krantz, K.E. (1958). Serial studies of renal function during pregnancy and the puerperium in normal women. *Journal of Clinicial Investigations*, **37**, 1966–74.

Sirota, J.H., Baldwin, D.S., and Villareal, H. (1950). Diurnal variations in renal function in man. *Journal of Clinical Investigation*, **29**, 187–92.

Skott, O., Bruun, N.E., Giese, J., Holstein-Rathlou, N-H., and Leyssac, P.P. (1987). What does lithium clearance measure during osmotic diuresis? *Clinical Science*, **73**, 126–7.

Smith, H.W. (1951). *The Kidney: Structure and Function in Health and Disease*. Oxford University Press, Oxford.

Smith, H.W., Goldring, W., and Chasis, H. (1938). The measurement of the tubular excretory mass, effective blood flow and filtration rate in the normal human kidney. *Journal of Clinical Investigation*, **17**, 263–78.

Smith, H.W., Goldring, W., Chasis, H., Ranger, H.A., and Bradley, S.E. (1943). The application of saturation methods to the study of glomerular and tubular function in the human kidney. *Journal of the Mount Sinai Hospital*, **10**, 59–108.

Smith, H.W., Finkelstein, N., Aliminosa, L., Crawford, B., and Graber, M. (1945). Renal clearance of substituted hippuric acid derviatives and other aromatic acids in dog and man. *Journal of Clincial Investigation*, **24**, 388–404.

Soriano, J.R., Boichis, H., Stark, H., and Edelmann, C.M. (1967). Proximal renal tubular acidosis. A defect in bicarbonate reabsorption with normal urinary acidifcation. *Pediatric Research*, **1**, 81–98.

Spencer, K. (1986). Analytical reviews in clinical chemistry: the estimation of creatinine. *Annals of Clinical Biochemistry*, **23**, 1–25.

Spielman, W.S. and Thomson, C.I. (1982). A proposed role for adenosine in the regulation of renal hemodynamics and renin release. *American Journal of Physiology*, **242**, F423–F435.

Steinhausen, M., Holz, F.G., and Parekh, N. (1988). Regulation of pre- and post-glomerular resistances visualized in the split hydronephrotic kidney. In *Nephrology* (ed. A.M. Davison), pp. 37–45. Ballière Tindall, London.

Tapson, J.S., Mansy, H., Marshall, S.M., and Tisdall, S.R. (1986). Renal functional reserve in kidney donors. *Quarterly Journal of Medicine*, **60**, 725–32.

Thomsen, K., Holstein-Rathlou, N.H., and Leyssac, P.P. (1981). Comparison of three measures of proximal tubular reabsorption: lithium clearance, occlusion time, and micropuncture. *American Journal of Physiology*, **241**, F348-F355.

Thomsen, K., Schou, M., and Westergaard, P. (1988). Distinction between proximal and distal regulations of sodium and potassium excretion in humans. *Clinical Nephrology*, **29**, 12–18.

Toor, M., Massry, S., Katz, A.I., and Agmon, J. (1965). Diurinal variations in the composition of blood and urine in man living in a hot climate. *Nephron*, **2**, 334–54.

Trollfors, B. and Norrby, R. (1981). Estimation of glomerular filtration rate by serum creatinine and serum β2-microglobulin. *Nephron*, **28**, 196–9.

Vanhoutte, P.M. (1988). The endothelium-modulator of smooth vascular tone. *New England Journal of Medicine*, **319**, 512–3.

Viberti, G., Bognetti, E., Wiseman, M.J., Dodds, R., Gross, J.L., and Keen, H. (1987). Effect of protein-restricted diet on renal response to a meat meal in humans. *American Journal of Physiology*, **253**, F388–F393.

Vincent, C., Revillard, J.P., Galland, M., and Traeger, J. (1978). Serum β2-microglobulin in hemodialyzed patients. *Nephron*, **21**, 260–8.

Vincent, H.H., Boer, W.H., Wenting, G.J., Schalekamp, M.A.D.H., and Weimar, W. (1987). Simplified method for evaluating proximal tubular fluid reabsorption. *Lancet*, **ii**, 1406–7.

de Wardener, H.E. (1985). *The Kidney*, 4th edn. Churchill Livingstone, London.

ter Wee, P.M., Geerlings, W., Rosman, J.B., Sluter, W.J., Van der Geest, S., and Donker, A.J.M. (1985). Testing renal reserve filtration capacity with an amino acid solution. *Nephron*, **41**, 193–9.

Wesson, L.G. (1969). *Physiology of the Human Kidney*. Grune and Stratton, New York.

Wibell, L. and Bjørsell-Östling, E. (1973). Endogenous creatinine clearance in apparently healthy individuals as determined by 24 hour ambulatory urine collection. *Uppsala Journal of Medical Science*, **78**, 45–7.

Wibell, L., Evrin, P.E., and Berggård, I. (1973) Serum β_2-microglobulin in renal disease. *Nephron*, **10**, 320–31.

Winberg, J. (1959). Determination of renal concentrating capacity in infants and children without renal disease. *Acta Paediatrica (Stockholm)*, **48**, 318–28.

Wolf, A.V. and Pillay, V.K.G. (1969). Renal concentration tests. Osmotic pressure, specific gravity, refraction and electrical conductivity compared. *American Journal of Medicine*, **46**, 837–43.

Wrong, O. (1965). Urinary hydrogen ion excretion. *Journal of Clinical Pathology*, **18**, 520–6.

Wrong, O. and Davies, H.E.F. (1959). The excretion of acid in renal disease. *Quarterly Journal of Medicine*, **28**, 259–313.

Yanigasawa, M., *et al.* (1988). A novel potent vasoconstrictor peptide produced by vascular endothelial cells. *Nature*, **332**, 411–15.

Young, D.S., Pestaner, L.C., and Gibberman, V. (1975). Effects of drugs on clinical laboratory tests. *Clinical Chemistry*, **21** (Suppl A), 1D–432D.

Zerbe, R.L. and Robertson, G.L. (1981). A comparison of plasma vasopressin measurements with a standard indirect test in the differential diagnosis of polyuria. *New England Journal of Medicine*, **305**, 1539–46.

Zimmerhackl, L.B. (1988). Renal medullary microcirculation and renal water and solute handling. In *Nephrology* (Ed. A.M. Davision), pp. 69–74. Ballière Tindall, London.

1.4 Renal function in the neonate

HENRIK EKBLAD AND ANITA APERIA

Introduction

This chapter will present the concept that the renal functional capacity is low in the newborn infant and that consequently the ability to regulate electrolyte and water balance is less good in infants than in adults.

Many parameters can be used to assess renal function. Glomerular filtration rate is a stable function which is generally related to the size of the organism. The renal tubular cells are able to adapt their transporting function to the physiological needs of the body. Renal tubular function will, therefore, vary widely under normal physiological conditions and will have to be tested under rigorously standardized conditions or following a known challenge to normal homeostasis. Regardless of the parameter used to estimate renal function, the capacity of the kidney is generally more limited in the infant than in the adult.

Differences in renal function between infants and adults may be due to the fact that we are applying the wrong standards of reference to infants; body surface area might not reflect the same needs of the infant as of the adult. During the last decade, however, numerous experimental studies have shown that the renal cells are not fully differentiated at birth and that many of the differences in renal function seen between infants and adults can probably be attributed to immaturity. Although this chapter will focus on renal function in the human neonate, experimental data will be used to show that the kidney undergoes terminal differentiation during postnatal life.

Assessment of renal function

Glomerular filtration rate in newborn infants is generally determined as the clearance of endogenous creatinine (Aperia et al. 1981a; Brion et al. 1986). Timed urine samples are collected during a 6 to 8 h period. Since voiding is frequent in the infant and bladder emptying is fairly complete, the urine sampling is generally quite accurate. In the infant, as in the adult, some creatinine may be secreted from tubular cells. During the period of urine collection, one or more capillary blood samples are taken for the determination of serum creatinine. Serum creatinine is much lower in the infant than in the adult in the range 10 to 50 μmol/l and the method used for its determination should have a high sensitivity. Although there are obviously possibilities for methodological error, creatinine clearance shows a surprisingly good correlation with the inulin clearance in newborn infants (Aperia et al. 1981b; Brion et al. 1986). Normal values for clearance of creatinine for preterm and full term infants are shown in Fig 1.

A rough estimate of the glomerular filtration rate in children can be made by determining serum creatinine only (Schwartz et al. 1976). In neonates, however, serum creatinine during the first

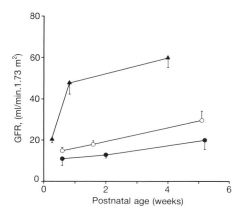

Fig. 1 Glomerular filtration rate in infants with a gestational age of 25–30 weeks (●), 31–34 weeks (o) and full term infants (▲) in relation to postnatal age. The bars represent ±SEM. (From Vanpeé et al. 1988.)

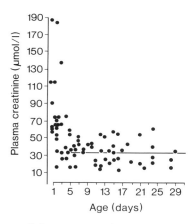

Fig. 2 Plasma creatinine concentrations during the first month of life of 34 neonates receiving no drug therapy. (From Feldman and Guignard 1982.)

few days of life represents the creatinine level in the mother since creatinine passes freely through the placental barrier (Feldman and Guignard 1982), and this must be taken into account (Fig. 2).

Studies of endogenous creatinine clearance can also be used for evaluation of some tubular parameters by determining the fractional excretion of the solute (FE), i.e. the percentage of the filtered solute that is excreted into the urine. FE_{Na} is calculated as

$$\frac{U_{Na} \times S_{Cr}}{S_{Na} \times U_{Cr}} \times 100$$

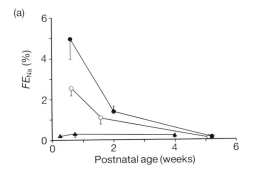

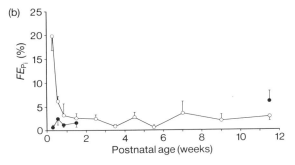

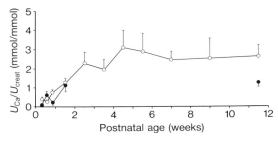

Fig. 4 Urinary calcium/creatinine ratio in preterm (o) and full term (●) infants. The bars represent ± SEM. (From Karlén et al. 1985.)

Fig. 3(a) Fractional urinary sodium excretion in infants with a gestational age of 25–30 weeks (●), 31–34 weeks (o) and full term infants (▲) in relation to postnatal age. The bars represent ± SEM. (From Vanpeé et al. 1988). (b) Fractional excretion of phosphate in preterm (o) and full term (●) infants. The bars represent ± SEM. (From Karlén et al. 1985.)

U_{Na} and U_{Cr} are urine concentrations of Na and creatinine respectively and S_{Na} and S_{Cr} are serum concentrations of Na and creatinine.

The normal values for the fractional excretion of sodium and phosphate in preterm and full term infants are shown in Fig. 3(a and b).

The urinary K:Na ratio is a useful index of the renal response to aldosterone. Aldosterone enhances tubular K^+ secretion and Na^+ reabsorption, and if the kidney is responding to high levels of aldosterone, the urinary K:Na ratio usually exceeds 1. In newborn infants serum aldosterone is usually higher than in adults and $U_K:U_{Na}$ ratios of more than 1 are common. The urinary K:Na ratio is generally lower in newborn preterm than in full term infants (Aperia et al. 1979).

Urinary calcium excretion can be evaluated as the calcium:creatinine ratio in spot urine samples (Ghazali and Barratt 1974). The calcium creatinine ratio is generally higher in infants than in older children and adults (Karlén et al. 1985). The normal values for the urinary calcium creatinine ratio in preterm and full term infants are shown in Fig. 4.

Urinary concentrating capacity of infants can be assessed in the same way as in older children. Urine osmolality is measured in spot urine samples following the intranasal administration of synthetic antidiuretic hormone (Svenningsen and Aronson 1974). Normal values for urine osmolality in preterm and full term infants that have been exposed to antidiuretic hormone are shown in Fig. 5.

The neonate's capacity for urinary acidification can be assessed by measuring urine pH and plasma bicarbonate concentration (Guignard 1987). Although normal values for urine pH have not been established, this is usually about 6 during the first

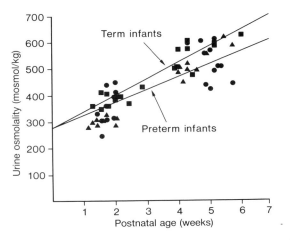

Fig. 5 Postnatal development of renal concentration capacity, i.e. maximum response after DDAVP test, during the first 6 postnatal weeks of life. ■ = term infants (group I); ● = preterm infants (group II); ▲ = asphyxiated infants (group III). (From Svenningsen and Aronson 1974.)

few days of life. The plasma bicarbonate concentrations are low in the full term neonate (21–23 mmol/l) and even lower in the preterm neonate (19–22 mmol/l) (Kildeberg 1964; Schwartz et al. 1979). Urine pH changes in response to plasma bicarbonate concentration in neonates under stable conditions (Torrado et al. 1974).

Glomerular filtration rate

Glomerular filtration rate is low relative to body surface area (BSA) during the first two to three postnatal days in all newborn infants, but increases rapidly during the first week of life in healthy full term infants (from approximately 25 to 50 ml/min.1.73 m^2 BSA). Thereafter the glomerular filtration rate increases more slowly, reaching adult values (100–140 ml/min. 1.73 m^2 BSA) between 1 and 2 years of age. The postnatal increase in glomerular filtration rate is much slower in preterm infants: at 1 month of age glomerular filtration rate is approximately 30 ml/min.1.73 m^2 in infants with a gestational age less than 30 weeks. Experimental studies have demonstrated that high renal vascular resistance and relatively low hydrostatic pressure in the glomerular capillaries are important determinants of the low glomerular filtration rate in newborn infants (Gruskin et al. 1970; Horster and Valtin 1971; Spitzer and

Brandis 1974; Aperia and Herin 1975). So far there is no evidence to suggest that the water permeability of the glomerular capillary is lower in the neonate than in the adult (Savin *et al.* 1985).

Tubular transporting capacity

Sodium transport

Most of the energy that is consumed by the kidney is used for active sodium transport mediated by Na^+, K^+-ATPase against a gradient along the basolateral membrane (Katz 1982). This active transport creates a driving force for sodium entry into all tubule cells, generally in co- or countertransport with other solutes. Amino acids and glucose participate in cotransport while ions are involved in countertransport mechanisms. The activity of Na^+, K^+-ATPase per unit length increases postnatally in all nephron segments (Schmidt and Horster 1977; Aperia *et al.* 1981*c*; Schwartz and Evan 1984; Rane and Aperia 1985). Consequently the transporting capacity of each tubular segment will also increase postnatally (Aperia and Larson 1979).

Preterm infants are generally considered to be in a sodium losing state during their first 2 to 3 weeks of life: even if the sodium intake is very low, its fractional excretion is higher in preterm than in full term infants. Those only given the amount of sodium that breast milk would provide will generally develop a negative sodium balance during the first few weeks of life. Circulating levels of angiotensin and aldosterone are increased as a result.

Fractional excretion of sodium decreases fairly rapidly in most preterm infants and generally reaches the low level of that observed in full term infants at 2 to 3 weeks of age (Sulyok *et al.* 1979*a*; Aperia *et al.* 1981*a*; Al-Dahhan *et al.* 1983). Infants that have severe respiratory problems or are in poor condition will maintain a high fractional sodium excretion for longer.

Breast milk provides the infant with a fairly low daily sodium intake. Since the newborn infant has a restricted capacity to excrete excess sodium (Aperia *et al.* 1975), intake should not be increased, provided that the urinary losses are not very large. The capacity to respond to a positive sodium balance by increasing urinary Na excretion is not fully developed until the age of 1 to 2 years.

Bicarbonate reabsorption and hydrogen excretion

The kidney regulates acid–base balance by hydrogen ion secretion via Na^+-H^+ exchange, by reabsorption of filtered bicarbonate and excretion of hydrogen ion as titrable acid and ammonia.

The neonatal kidney has a low capacity to reabsorb filtered bicarbonate (Schwartz and Evan 1983), a low level of Na^+-H^+ exchange activity in the proximal tubule (Ekblad *et al.* 1990), and also a low level of H^+-ATPase activity in the more distal parts of the nephron (Brion *et al.* 1989). As a consequence, neonates have a somewhat lower plasma bicarbonate concentration than the adult and also have a lower capacity to excrete titrable acids, which further restricts their ability to respond to an increased acid load (Svenningsen 1974).

Kildeberg (1964) introduced the term 'late metabolic acidosis' to describe apparently healthy preterm infants with metabolic acidosis and growth failure at 1 to 3 weeks of age. Schwartz *et al.* (1979) found no growth failure in preterm infants with 'late metabolic acidosis' and suggested that these infants in fact represent one extreme of the normal distribution that describes this population.

Some clinical studies indicate that the low renal threshold for bicarbonate is linked to the high urinary sodium excretion in preterm infants (Sulyok *et al.* 1972). This same group showed in a later study that sodium supplementation in preterm infants prevents the development of metabolic acidosis (Sulyok *et al.* 1981).

Renal tubular bicarbonate reabsorption is further decreased in hypoxia and sepsis, and metabolic acidosis is common (Torrado *et al.* 1974). In neonates with renal disease bicarbonate reabsorption is frequently reduced to the extent that metabolic acidosis develops.

Urinary concentrating capacity

Urinary concentrating capacity is low in the neonate and increases exponentially during the first 18 months of life. Of the intrarenal processes that contribute to the concentration of urine, active sodium reabsorption in the water-impermeable, thick, ascending loop of Henle is probably the primary event in this cascade of events. This active sodium reabsorption takes place against a high sodium gradient and is a very energy demanding process. Recirculation of urea between different nephron segments also contributes to the hypertonicity in the medullary interstitium. Finally, the collecting duct cells are made permeable to water by the action of antidiuretic hormone. Observations from experimental studies suggest that all of those intrarenal processes undergo postnatal maturation (Schlondorff *et al.* 1978; Rane and Aperia 1985; Rane *et al.* 1985).

Calcium and phosphorus transport

Results from clinical studies suggest that the capacity of the neonatal kidney to reabsorb filtered phosphate and calcium is generally high (Karlén *et al.* 1985). Neiberger *et al.* (1989) showed that brush-border membrane vesicles from newborn guinea pigs have a high capacity to reabsorb phosphate. It is therefore unlikely that kidney immaturity plays a major role for the development of calcium and phosphate deficits, which are common in breast milk and standard formula fed preterm infants that are not given phosphate and calcium supplements.

Transport of peptides

The capacity to reabsorb filtered peptides in the proximal tubule is less good in the neonate than in the adult. β_2-Microglobulin reabsorption in the proximal tubule is decreased in the newborn infant as compared to the adult, and in the preterm as compared to the full term neonate (Aperia and Broberger 1979). The biological significance of this has not yet been fully evaluated, but it may well influence protein metabolism.

Hormonal regulation of renal function

Salt and water excretion by the kidney is regulated by a number of endocrine and autocrine factors, whose production and release are generally developed early in fetal life. There is, however, ample evidence that the end-organ response to many of these factors is blunted, especially in the kidney of the preterm baby.

Renin–angiotensin–aldosterone system

Preterm infants generally have very high urinary aldosterone excretion rates and high plasma renin activity, suggesting an increased activity of the renin–angiotensin–aldosterone system (Beitins et al. 1972; Dillon et al. 1976). This has been attributed to a salt deficit since oral supplementation with sodium suppresses the hormone levels (Sulyok et al. 1981). Even in the presence of high serum levels of aldosterone, preterm babies may still have high urinary sodium excretion and a urinary K:Na ratio below 1 suggesting renal tubular unresponsiveness to aldosterone (Aperia et al. 1979; Sulyok et al. 1979b). Experimental studies support this concept: Stephenson et al. (1984) showed that young rats (7–9 days) were essentially insensitive to aldosterone, in contrast to animals aged 13 to 15 days and older.

Antidiuretic hormone

Secretion of antidiuretic hormone may be very high during fetal life. In the lamb fetus antidiuretic hormone is released in response to hypovolaemia (Robillard et al. 1979) and hypoxaemia (Weismann and Robillard 1988). The renal response to antidiuretic hormone is blunted in the neonate: although plasma levels are often high, urine osmolality is generally below that of plasma in newborn infants (Svenningsen and Aronson 1974), although in certain stressful situations it can be as high as 550 mosm/kg H_2O (Rees et al. 1984a).

The effects of high serum antidiuretic hormone levels on water and salt homeostasis in the neonate are not clear. It is important to distinguish between inappropriate secretion and non-osmolar release of antidiuretic hormone: according to Schwartz et al. (1957) high rates of secretion should not be described as inappropriate unless volume contraction has been excluded. Sulyok et al. (1985) have shown a progressive increase in antidiuretic hormone secretion in preterm infants with late hyponatraemia. They postulated that chronic high urinary sodium excretion leads to a contraction of the extracellular volume, which may stimulate antidiuretic hormone secretion. In view of the hyponatraemia and hypo-osmolality, this rise in secretion seems to be inappropriate, but in view of the postulated contraction of the extracellular volume, it can be regarded as appropriate in helping to restore the normal volume of body fluids.

Like many other hormones, antidiuretic hormone has several effects. It interacts not only with renal tubule receptors but also with vascular receptors, and thus influences peripheral vascular resistance. During hypovolaemia antidiuretic hormone is of great importance for the maintenance of the arterial blood pressure, and experimental evidence suggests that this effect is at least as pronounced in the infant as in the adult (Herin et al. 1984). It is therefore possible that the high antidiuretic levels seen in the perinatal period are of major importance for the control of blood pressure and regional blood flow and of minor importance for water balance.

Atrial natriuretic factor

Atrial natriuretic factor increases glomerular filtration and inhibits tubular sodium transport in the adult kidney. Its net natriuretic effect depends on several complex interactions with other hormonal systems, such as the renin–angiotensin–aldosterone system (Raine et al. 1989).

The role of atrial natriuretic factor in the regulation of fluid and electrolyte homeostasis in the neonate is not clear. The capacity to release this factor in response to volume overload appears to be well developed both in the fetus and the newborn baby. Robillard and Weiner (1988) showed that the human fetus is able to release atrial natriuretic factor in response to volume expansion when receiving in utero blood transfusion for treatment of immune haemolytic anaemia. High plasma concentrations of atrial natriuretic factor have been shown during the first few days of life; preterm babies may have especially high concentrations, in the range of 1000 to 2000 pg/ml (Shaffer et al. 1986). The end-organ response to atrial natriuretic factor in infancy is not clear: it may not be involved in the natriuretic response to increased sodium intake (Shaffer and Meade 1989). Experimental studies in rats have given evidence for a blunted renal response to atrial natriuretic factor in infancy (Karlén et al. 1990), although other studies have disagreed on this point (Shine et al. 1987). Inscho et al. (1987) showed significant age-dependent differences in levels and, interestingly, also of the biological and immunological activity of rat atrial natriuretic factor. Recently Ito and Yamashita (1989) showed two different molecular forms of the factor in the plasma and urine of neonates; their biological activities have not yet been clarified.

Common electrolyte disturbances

Hyper- and hyponatraemia

Hyper- and hyponatraemia are relatively common in preterm infants in intensive care. Hypernatraemia is usually seen during the first few days of life, when large transepidermal water losses may cause dehydration (Sedin et al. 1985). Such water losses exceed sodium losses, and plasma sodium levels above 150 mmol/l are not uncommon in these situations. Hypernatraemia is best avoided by manoeuvres aimed at reducing extrarenal water loss.

After the first week of life the rate of passive water loss declines, and hyponatraemia becomes a more common problem. This late hyponatraemia has been attributed to excessive renal sodium loss (Al-Dahhan et al. 1983), but other factors may also be involved: plasma might be diluted from absorption of interstitial oedema fluid, which is common in critically ill preterm neonates (Rees et al. 1984a,b). Hyponatraemia in very low birth weight infants might also be attributed to a redistribution of sodium between the intra- and extracellular spaces, due to immaturity of the Na^+,K^+-ATPase in the cell membrane (Aperia et al. 1981c).

In neonatal hyponatraemia the plasma sodium concentration usually is between 120 and 130 mmol/l and sodium supplementation may be necessary. Avoidance of tissue hypoxia and control of infection also play a role in its prevention, since they adversely affect renal tubular function, resulting in even higher sodium excretion than that due to immaturity alone. They may also lead to fluid retention and development of tissue oedema with consequent dilutional hyponatraemia.

Metabolic acidosis

Preterm infants may develop moderate metabolic acidosis during the second week of life even if the postnatal course is uncomplicated (see above). This condition is transient and does not generally require treatment, unlike the situation in neonates with a more complicated clinical course. In critically ill neonates (e.g. those with septicaemia, severe respiratory distress, extreme prematurity) hypoventilation leads to hypercapnia and acidosis, which might not always be fully corrected by appropriate ventilatory support. Severe metabolic acidosis may develop due to circulatory failure and decreased tissue perfusion. Net acid input exceeds net acid excretion, as the neonate is unable to respond adequately to an increased acid load. Metabolic acidosis may also develop when bicarbonate is lost in the urine or through the gastrointestinal tract.

The treatment of metabolic acidosis should concentrate on establishing adequate circulation, tissue perfusion, and renal output, controlling infection, and providing appropriate nutrition. Rapid infusion of sodium bicarbonate should be avoided, since rapid changes in osmolar concentration may be deleterious to the central nervous system (Finberg 1977), and may be related to the development of intraventricular haemorrhage (Dykes et al. 1980). Sodium acetate, given as a slow and continuous infusion, has been used in the prevention of metabolic acidosis in stressed neonates (Ekblad et al. 1985).

Prerenal failure

In the adult, glomerular filtration rate can fall from 100 to 50 ml/min.1.73 m^2 without any symptoms of renal insufficiency: dietary restriction and electrolyte supplementation are not generally considered until it falls below 25 ml/min.1.73 m^2. In most newborn infants, however, glomerular filtration rate is in the range of 20 to 30 ml/min.1.73 m^2, and all sick neonates are, therefore, at risk of developing prerenal failure. This condition is fairly common in infants with severe respiratory problems, patent ductus arteriosus or other circulatory disturbances, or septicaemia (Anand 1982). Typical laboratory manifestations are increased serum creatinine, acidosis, and hyponatraemia (Reimold et al. 1977). The acidosis is probably due to renal bicarbonate losses and the low capacity of the kidney to excrete hydrogen ions; urinary sodium losses, water retention and redistribution of sodium from the extracellular to the intracellular space may contribute to the hyponatraemia. Fractional sodium excretion is generally very high. Electrolyte and water balance should be carefully monitored in infants with prerenal failure; low doses of dopamine and thiazide diuretics are generally beneficial (see below). If renal parenchymal damage occurs, medical therapy may fail and peritoneal dialysis should be considered. The long-term prognosis of infants with prerenal failure is generally good.

Drugs influencing renal function

Dopamine

Dopamine is a natriuretic hormone in the infant as well as in the adult; its renal effects are probably mediated both by increased glomerular filtration and by inhibition of tubular sodium reabsorption (Goldberg 1972). Exogenous dopamine is commonly used to stabilize renal function and increase sodium excretion in sick infants. In order to trigger dopamine receptors only, doses of 0.5 to 2.0 μg/kg.min should be used. Such low doses significantly increases urine flow, creatinine clearance, and sodium excretion in preterm infants with the respiratory distress syndrome (Tulassay et al. 1983). If higher doses are given, other catecholamine receptors might also be triggered and the effect on the kidney might be the opposite.

Dopamine has been used to prevent the renal side effects of indomethacin in preterm infants with patent ductus arteriosus (Seri et al. 1984), and has also been combined with frusemide in the treatment of severe functional renal failure of preterm infants with the respiratory distress syndrome (Tulassay and Seri 1986).

Frusemide

Frusemide increases natriuresis and diuresis by inhibiting sodium chloride reabsorption in the ascending loop of Henle. In neonates the onset of action is slower and the diuretic effect is less intense but of longer duration than in adults (Woo et al. 1978). There is also a great variability in the response to frusemide, which has been attributed to low and unstable renal function (Guignard 1987).

Frusemide therapy is indicated in neonates with severe pulmonary oedema and severe functional renal failure (Ross et al. 1978), but should be used with caution in critically ill preterm neonates, since it may have side-effects, the most important of which are hypovolaemia and lowering of arterial blood pressure, which further impair renal function (Green et al. 1988). Frusemide has also been implicated in failure of ductal closure (Green et al. 1983), but there is no general agreement on this point. Calcium excretion is promoted by frusemide and prolonged treatment has been associated with renal calcifications (Hufnagle et al. 1982; Ezzedeen et al. 1988). Secondary hyperparathyroidism and bone disease have also been described after long-term administration of frusemide to preterm infants (Venkataraman et al. 1983).

Indomethacin

Indomethacin, an inhibitor of prostaglandin synthesis, is used to induce the closure of the ductus arteriosus in preterm infants. In neonates indomethacin decreases urine flow rate, glomerular filtration rate, free water clearance, and electrolyte excretion rate (Cifuentes et al. 1979). These changes are reversible, and renal function returns to pretreatment values within a few days after the drug is discontinued.

Aminoglycosides

Aminoglycosides are commonly used in neonates in the treatment of suspected or confirmed sepsis. Gentamicin-associated nephrotoxicity is generally considered to be infrequent in neonates (McCracken and Gay Jones 1970). More recent studies have shown that plasma creatinine concentrations may increase during treatment of these patients (Elinder and Aperia 1983). The low perfusion of the cortex is thought to protect against gentamicin accumulation in preterm infants.

References

Al-Dahhan, J., Haycock, G.B., Chantler, C., and Stimmler, L. (1983). Sodium homeostasis in term and preterm neonates. I. Renal aspects. *Archives of Disease in Childhood*, **58**, 335–42.

Anand, S.K. (1982). Acute renal failure in the neonate. *Pediatric Clinics of North America*, **29**, 791–800.

Aperia, A. and Herin, P. (1975). Development of glomerular perfusion rate and filtration rate in rats 17–60 days old. *American Journal of Physiology*, **228**, 1319–25.

Aperia, A. and Broberger, U. (1979). β2-Microglobulin, an indicator of renal tubular maturation and dysfunction in the newborn. *Acta Paediatrica Scandinavica*, **68**, 669–76.

Aperia, A. and Larsson, L. (1979). Correlation between fluid reabsorption and proximal tubule ultrastructure during development of the rat kidney. *Acta Physiologica Scandinavica*, **105**, 11–22.

Aperia, A., Broberger, O., Thodenius, K., and Zetterström, R. (1975). Development of renal control of salt and fluid homeostasis during the first year of life. *Acta Paediatrica Scandinavica*, **64**, 393–8.

Aperia, A., Broberger, O., Herin, P., and Zetterström, R. (1979). Sodium excretion in relation to sodium intake and aldosterone excretion in newborn pre-term and full-term infants. *Acta Paediatrica Scandinavica*, **68**, 813–7.

Aperia, A., Broberger, O., Elinder, G., Herin, P., and Zetterström, R. (1981a). Postnatal development of renal function in preterm and fullterm infants. *Acta Paediatrica Scandinavica*, **70**, 183–7.

Aperia, A., Broberger, O., Elinder, G., Herin, P., Thodenius, K., and Zetterström, R. (1981b). Postnatal changes of glomerular filtration rate in preterm and fullterm infants. In *The kidney during development* (ed. A. Spitzer), pp. 133–7. Masson Publ. Inc., New York.

Aperia, A., Larsson, L., and Zetterström, R. (1981c). Hormonal induction of NaKATPase in developing proximal tubular cells. *American Journal of Physiology*, **241**, F356–60.

Beitins, I.Z., Bayard, F., Levitsky, L., Ances, I.G., Kowarski, A., and Migeon, C.J. (1972). Plasma aldosterone concentration at delivery and during the newborn period. *Journal of Clinical Investigation*, **51**, 386–94.

Brion, L.P., Fleischman, A.R., McCarton, C., and Schwartz, G.J. (1986). A simple estimate of glomerular filtration rate in low birth weight infants during the first year of life: Noninvasive assessment of body composition and growth. *Journal of Pediatrics*, **109**, 698–707.

Brion, L.P., Schwartz, J.H., Lachman, H.M., Zavilowitz, B.J., and Schwartz, G.J. (1989). Development of H^+ secretion by cultured renal inner medullary collecting duct cells. *American Journal of Physiology*, **257**, F486–F501.

Cifuentes, R.F., Olley, P.M., Balfe, J.W., Radde, I.C., and Soldin, S.J. (1979). Indomethacin and renal function in premature infants with persistent patent ductus arteriosus. *Journal of Pediatrics*, **95**, 583–7.

Dillon, M.J., Gillin, M.E.A., Ryness, J.M., and de Swiet, M. (1976). Plasma renin activity and aldosterone concentration in the human newborn. *Archives of Disease in Childhood*, **51**, 537–40.

Dykes, F.D., Lazzara, A., Ahmann, P., Blumenstein, B., Schwartz, J., and Brann, A.W. (1980). Intraventricular hemorrhage: A prospective evaluation of etiopathogenesis. *Pediatrics*, **66**, 42–9.

Ekblad, H., Kero, P., and Takala, J. (1985). Slow sodium acetate infusion in the correction of metabolic acidosis in premature infants. *American Journal of Diseases of Children*, **139**, 708–10.

Ekblad, H., Larsson, S.H., and Aperia, A. (1990). The capacity to recover from intracellular acidosis is lower in infant than in adolescent renal cells. *Pediatric Research*, 27, 327A.

Elinder, G. and Aperia, A. (1983). Development of glomerular filtration rate and excretion of β_2-microglobulin in neonates during gentamicin treatment. *Acta Paediatrica Scandinavica*, **72**, 219–24.

Ezzedeen, F., Adelman, R.D., and Ahlfors, C.E. (1988). Renal calcification in preterm infants: pathophysiology and long-term sequelae. *Journal of Pediatrics*, **113**, 532–9.

Feldman, H. and Guignard, J.-P. (1982). Plasma creatinine in the first month of life. *Archives of Disease in Childhood*, **57**, 123–6.

Finberg, L. (1977). The relationship of intravenous infusions and intracranial hemorrhage–a commentary. *Journal of Pediatrics*, **91**, 777–8.

Ghazali, S. and Barratt, T.M. (1974). Urinary excretion of calcium and magnesium in children. *Archives of Disease in Childhood*, **49**, 97–101.

Goldberg, L.I. (1972). Cardiovascular and renal actions of dopamine: potential clinical applications. *Pharmacological Reviews*, **24**, 1–29.

Green, T.P., Thompson, T.R., Johnson, D.E., and Lock, J.E. (1983). Furosemide promotes patent ductus arteriosus in premature infants with the respiratory-distress syndrome. *New England Journal of Medicine*, **308**, 743–8.

Green, T.P., Johnson, D.E., Bass, J.L., Landrum, B.G., Ferrara, T.B., and Thompson, T.R. (1988) Prophylactic furosemide in severe respiratory distress syndrome: Blinded prospective study. *Journal of Pediatrics*, **112**, 605–12.

Gruskin, A.B., Edelmann, C.M., Jr., and Yuan, S. (1970). Maturational changes in renal blood flow in piglets. *Pediatric Research*, **4**, 7–13.

Guignard, J.-P. (1987). Neonatal nephrology. In *Pediatric nephrology*, (2nd edn), (eds. M. A. Holliday, T. M. Barratt, and R. L. Vernier), pp. 921–44. Williams & Wilkins, Baltimore.

Herin, P., Eklöf, A.-C., and Aperia, A. (1984). Role of arginine-vasopressin in blood pressure control in young and adult rats. *Pediatric Research*, **18**, 701–4.

Horster, M. and Valtin, H. (1971). Postnatal development of renal function: micropuncture and clearance studies in the dog. *Journal of Clinical Investigation*, **50**, 779–95.

Hufnagle, K.G., Khan, S.N., Penn, D., Cacciarelli, A., and Williams, P. (1982). Renal calcifications: A complication of long-term furosemide therapy in preterm infants. *Pediatrics*, **70**, 360–3.

Inscho, E.W., Wilfinger, W.W., and Banks, R.O. (1987). Age-related differences in the natriuretic and hypotensive properties of rat atrial extracts. *Endocrinology*, **121**, 1662–70.

Ito, Y. and Yamashita, F. (1989). Physiological role and characteristic molecular forms of atrial natriuretic peptide in plasma and urine of neonates. *Pediatric Nephrology*, **3**, C142.

Karlén, J., Aperia, A., and Zetterström, R. (1985). Renal excretion of calcium and phosphate in preterm and term infants. *Journal of Pediatrics*, **106**, 814–19.

Karlén, J., Rane, S., and Aperia, A. (1990). Tubular response to hormones is blunted in weanling rats. *Acta Physiologica Scandinavica*, **138**, 443–9.

Katz, A.I. (1982). Renal Na-K-ATPase: its role in tubular sodium and potassium transport. *American Journal of Physiology*, **242**, F207–19.

Kildeberg, P. (1964). Disturbances of hydrogen ion balance occurring in premature infants. II. Late metabolic acidosis. *Acta Paediatrica Scandinavica*, **53**, 517–26.

McCracken, G.H., Jr. and Gay Jones, L. (1970). Gentamicin in the neonatal period. *American Journal of Diseases of Children*, **120**, 524–33.

Neiberger, R.E., Barac-Nieto, M., and Spitzer, A. (1989). Renal reabsorption of phosphate during development: transport kinetics in BBMV. *American Journal of Physiology*, **257**, F268–74.

Raine, A.E.G., Firth, J.G., and Ledingham, J.G.G. (1989). Renal actions of atrial natriuretic factor. *Clinical Science*, **76**, 1–8.

Rane, S. and Aperia, A. (1985). Ontogeny of Na-K-ATPase activity in thick ascending limb and of concentrating capacity. *American Journal of Physiology*, **249**, F723–8.

Rane, S., Aperia, A., Eneroth, P., and Lundin, S. (1985). Development of urinary concentrating capacity in weanling rats. *Pediatric Research*, **19**, 472–5.

Rees, L., Brook, C.G.D., Shaw, J.C.L., and Forsling, M.L. (1984a). Hyponatraemia in the first week of life in preterm infants. Part I Arginine vasopressin secretion. *Archives of Disease in Childhood*, **59**, 414–22.

Rees, L., Shaw, J.C.L., Brook, C.G.D., and Forsling, M.L. (1984b). Hyponatraemia in the first week of life in preterm infants. Part II Sodium and water balance. *Archives of Disease in Childhood*, **59**, 423–9.

Reimold, E.W., Dinh Don, T., and Worthen, H.G. (1977). Renal failure during the first year of life. *Pediatrics*, **59**, 987–94.

Robillard, J.E. and Weiner, C. (1988). Atrial natriuretic factor in the human fetus: Effect of volume expansion. *Journal of Pediatrics*, **113**, 552–5.

Robillard, J.E., Weitzman, R.E., Fisher, D.A., and Smith, F.G., Jr. (1979). The dynamics of vasopressin release and blood volume regulation during fetal hemorrhage in the lamb fetus. *Pediatric Research*, **13**, 606–10.

Ross, B.S., Pollak, A., and Oh, W. (1978). The pharmacologic effects of furosemide therapy in the low-birthweight infant. *Journal of Pediatrics*, **92**, 149–53.

Savin, V.J., Beason-Griffin, C., and Richardson, W.P. (1985). Ultrafiltration coefficient of isolated glomeruli of rats aged 4 days to maturation. *Kidney International*, **28**, 926–31.

Schlondorff, D., Weber, H., Trizna, W., and Fine, L.G. (1978). Vasopressin responsiveness of renal adenylate cyclase in newborn rats and rabbits. *American Journal of Physiology*, **234**, F16–F21.

Schmidt, U. and Horster, M. (1977). Na-K-activated ATPase: activity maturation in rabbit nephron segments dissected in vitro. *American Journal of Physiology*, **233**, F55–F60.

Schwartz, G.J. and Evan, A.P. (1983). Development of solute transport in rabbit proximal tubule. I. HCO_3^- and glucose absorption. *American Journal of Physiology*, **245**, F382–90.

Schwartz, G.J. and Evan, A.P. (1984). Development of solute transport in rabbit proximal tubule. III. Na-K-ATPase activity. *American Journal of Physiology*, **246**, F845–52.

Schwartz, G.J., Haycock, G.B., and Spitzer, A. (1976). Plasma creatinine and urea concentration in children: Normal values for age and sex. *Journal of Pediatrics*, **88**, 828–30.

Schwartz, G.J., Haycock, G.B., Edelmann, C.M., Jr., and Spitzer, A. (1979). Late metabolic acidosis: A reassessment of the definition. *Journal of Pediatrics*, **95**, 102–7.

Schwartz, W.B., Bennett, W., Curelop, S., and Bartter, F.C. (1957). A syndrome of renal sodium loss and hyponatremia probably resulting from inappropriate secretion of antidiuretic hormone. *American Journal of Medicine*, **23**, 529–42.

Sedin, G., Hammarlund, K., Nilsson, G.E., Strömberg, B., and Öberg, P.Å. (1985). Measurement of transepidermal water loss in newborn infants. *Clinics in Perinatology*, **12**, 79–99.

Seri, I., Tulassay, T., Kiszel, J., Machay, T., and Csömör, S. (1984). Cardiovascular response to dopamine in hypotensive preterm neonates with severe hyaline membrane disease. *European Journal of Pediatrics*, **142**, 3–9.

Shaffer, S.G. and Meade, V.M. (1989). Sodium balance and extracellular volume regulation in very low birth weight infants. *Journal of Pediatrics*, **115**, 285–90.

Shaffer, S.G., Geer, P.G., and Goetz, K.L. (1986). Elevated atrial natriuretic factor in neonates with respiratory distress syndrome. *Journal of Pediatrics*, **109**, 1028–33.

Shine, P., McDougall, J.G., Towstoless, M.K., and Wintour, E.M. (1987). Action of atrial natriuretic peptide in the immature ovine kidney. *Pediatric Research*, **22**, 11–15.

Spitzer, A. and Brandis, M. (1974). Functional and morphologic maturation of the superficial nephrons. Relationship to total kidney function. *Journal of Clinical Investigation*, **53**, 279–87.

Stephenson, G., Hammet, M., Hadaway, G., and Funder, J.W. (1984). Ontogeny of renal mineralocorticoid receptors and urinary electrolyte responses in the rat. *American Journal of Physiology*, **247**, F665–71.

Sulyok, E., Heim, T., Soltész, G., and Jászai, V. (1972). The influence of maturity on renal control of acidosis in newborn infants. *Biology of the Neonate*, **21**, 418–35.

Sulyok, E., Varga, F., Györy, E., Jobst, K., and Csaba, I.F. (1979a). Postnatal development of renal sodium handling in premature infants. *Journal of Pediatrics*, **95**, 787–92.

Sulyok, E., *et al.* (1979b). Postnatal development of renin–angiotensin–aldosterone system, RAAS, in relation to electrolyte balance in premature infants. *Pediatric Research*, **13**, 817–20.

Sulyok, E., Németh, M., Tényi, I., Csaba, I.F., Varga, L., and Varga, F. (1981). Relationship between the postnatal development of the renin–angiotensin–aldosterone system and the electrolyte and acid–base status in the sodium chloride supplemented premature infant. *Acta Paediatrica Academiae Scientiarum Hungaricae*, **22**, 109–21.

Sulyok, E., *et al.* (1985). Late hyponatremia in premature infants: Role of aldosterone and arginine vasopressin. *Journal of Pediatrics*, **106**, 990–4.

Svenningsen, N.W. (1974). Renal acid–base titration studies in infants with and without metabolic acidosis in the postneonatal period. *Pediatric Research*, **8**, 659–72.

Svenningsen, N.W. and Aronson, A.S. (1974). Postnatal development of renal concentrating capacity as estimated by DDAVP-test in normal and asphyxiated neonates. *Biology of the Neonate*, **25**, 230–41.

Torrado, A., Guignard, J.-P., Prod'hom, L.S., and Gautier, E. (1974). Hypoxemia and renal function in newborns with respiratory distress syndrome. *Helvetica Paediatrica Acta*, **29**, 399–403.

Tulassay, T. and Seri, I. (1986). Acute oliguria in preterm infants with hyaline membrane disease: Interaction of dopamine and furosemide. *Acta Paediatrica Scandinavica*, **75**, 420-4.

Tulassay, T., Seri, I., Machay, T., Kiszel, J., Varga, J., and Csömör, S. (1983). Effects of dopamine on renal functions in premature neonates with respiratory distress syndrome. *International Journal of Pediatric Nephrology*, **4**, 19–23.

Vanpeé, M., Herin, P., Zetterström, R., and Aperia, A. (1988). Postnatal development of renal function in very low birth-weight infants. *Acta Paediatrica Scandinavica*, **77**, 191–7.

Venkataraman, P.S., Han, B.K., Tsang, R.C., and Daugherty, C.C. (1983). Secondary hyperparathyroidism and bone disease in infants receiving long-term furosemide therapy. *American Journal of Diseases of Children*, **137**, 1157–61.

Weismann, D.N. and Robillard, J.E. (1988). Renal hemodynamic responses to hypoxemia during development: relationships to circulating vasoactive substances. *Pediatric Research*, **23**, 155–62.

Woo, W.C.R., Dupont, C., Collinge, J., and Aranda, J.V. (1978). Effects of furosemide in the newborn. *Clinical Pharmacology and Therapeutics*, **23**, 266–69.

1.5 Renal function in the elderly

JUAN F. MACIAS-NUNEZ AND J. STEWART CAMERON

'Years steal
Fire from the mind as vigour from the limb'
    Lord Byron : *Childe Harold's Pilgrimage*, 1812–18.

'Old age isn't so bad when you consider the alternative'
    Maurice Chevalier, 1960

Introduction

Profound changes take place with age in the structure and function of the kidneys (Macias-Nuñez and Cameron 1987). Some of these are of great clinical importance, others of physiological interest. In addition, handling of drugs by the aged individual, and by the ageing kidney, changes, As the proportion of elderly citizens in Western populations rises, so a greater and greater proportion of patients seen in nephrological services can be described as 'elderly' or 'very old'. Numerous studies have addressed the physiology and function of the ageing kidney (Davies and Shock 1950; Papper 1973; Rowe *et al.* 1976; Brown *et al.* 1986) but it should be noted that almost without exception these studies have been cross-sectional and not longitudinal studies, which to date are generally lacking.

Anatomy of the ageing kidney (McLachlan 1987)

Macroscopic changes

The normal aged kidney has a smooth or a fine granularity of the surface. Renal weight decreases 20 to 30 per cent between the age of 30 and 90 years from between 200 and 270 g, to between 180 and 200 g (Tauchi *et al.* 1971; Brown *et al.* 1986). Only 12 to 14 per cent of kidneys in elderly subjects are coarsely scarred at autopsy. This can be differentiated from pyelonephritic scarring, which is more evident at the poles of the kidney; usually the scarring is associated with caliectasis. The scars of the ageing kidney occur at all sites without any deformity of the calyces. Fat in the renal sinus increases with age. Renal length diminishes by 2 cm between the age of 50 and 80 years, which represents a loss of volume of around 40 per cent (McLachlan 1987). The loss of substance affects the cortex more than the medulla. Glomeruli may become more crowded together as the cortex thins, but this crowding may be masked in routine histological preparations because there is in addition a fourfold increase of cortical volume occupied by tissue other than glomeruli and tubules, which is independent of age (Goyal 1982).

Glomeruli

Although the total number of glomeruli diminishes from the usual figure of about 1.3 million/kidney to half or two-thirds this number in the seventh decade as the kidney ages (McLachlan *et al.* 1977) (Fig. 1(a)), the scatter of data at all ages is very large, and many aged individuals are within the normal range for younger adults. Some remaining glomeruli appear partially or totally hyalinized, which is the origin of the glomerular sclerosis which accompanies ageing (Fig. 1(b)). Glomerular sclerosis begins at approximately 30 years of age. The percentage of obsolete glomeruli varies between 1 and 30 per cent in persons aged 50 years or more (Kaplan *et al.* 1975; McLachlan *et al.* 1977). Glomerular basement membranes are patchily reduplicated and, in general, thicker than in the young (but see Steffes *et al.* 1983). On microangiographic examination, there is an obliteration particularly of juxtamedullary nephrons, but not of those sited more peripherally, with the formation of a direct channel between afferent and efferent arterioles in this area of the kidney (Takazakura *et al.* 1972) (Fig. 2). This loss of juxtamedullary glomeruli may play some role in the impairment of the countercurrent multiplier in the aged (see below). The mesangium which accounts for 8 per cent of glomerular volume at 45 years of age, increases to nearly 12 per cent at the age of 70 (Sorensen 1977). At the same time, however, glomerular cross-sectional area decreases (McLachlan *et al.* 1977) which may account, at least in part, for these date. The lobularity and complexity of the tuft decreases. In any case the area of filtering surface per surviving glomerulus falls with age.

The reasons for the progressive glomerulosclerosis with age are not clear. Obviously, the marked vascular changes with age

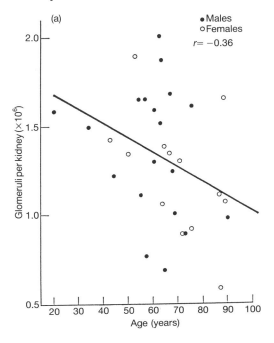

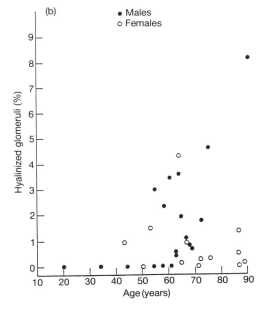

Fig. 1(a) The number of glomeruli in human kidneys. Reproduced with permission from McLachlan *et al.* (1977). There is very wide scatter, but the number of surviving glomeruli tends to fall with age. (b) The proportion of sclerosed glomeruli in the kidney as a function of age. Reproduced with permission from McLachlan *et al.* (1977). See the text for comment.

could play a part (see below), but a primary involution of the glomeruli, as suggested by Jean Oliver (McLachlan 1987) may also take place. It is notable that there is only a poor correlation between vascular changes and glomerulosclerosis. These important points are discussed in detail by McLachlan (1987).

Tubular changes

Tubular cells undergo fatty degeneration with age, showing an irregular thickening of their basal membrane. By microdissec-

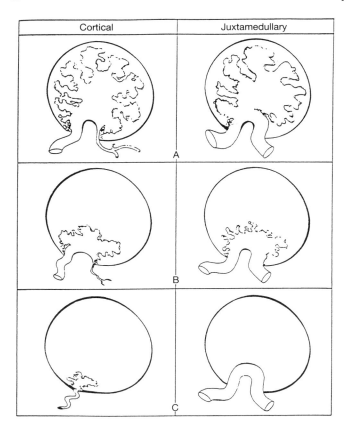

Fig. 2 Summary of the changes seen on microangiography in juxtamedullary glomeruli (left panels) and more superficial nephrons (right panels). Changes are sequential and more severe from top to bottom in each case (A to C). In healthy peripheral nephrons, the afferent arteriole forms glomerular capillaries which then join to become efferent arterioles. In juxtamedullary glomeruli, the glomerular capillaries more closely resemble side branches of a continuous afferent and efferent arteriolar system. When peripheral glomeruli degenerate, afferent arterioles end blindly; when juxtamedullary glomeruli degenerate, direct communications between afferent and efferent arterioles are formed. From Mclachlan et al. (1987) with permission. See also Takazakura et al. (1972).

tion, the existence of diverticula arising from the distal and convoluted tubules, becoming more frequent with age, has been demonstrated. It seems that these result from progressive weakening of the tubular basal membrane; the significance of this observation remains speculative. Mean proximal tubular volume decreases with age, from 0.136 mm^3 to 0.061 mm^3 at the age of 84 (Darmady et al. 1973). In general, as the kidney ages the reduction in proximal tubular volume parallels that of the glomerular volume, so that anatomical glomerulotubular balance is preserved into old age. As will be seen below this is not true of renal function.

The interstitium

Interstitial changes consist mainly of increasing zones of tubular atrophy and fibrosis, which may relate to the defects in concentration and dilution observed as part of the normal renal ageing process (see below). There is a variable interstitial infiltrate of mononuclear cells; whether these are damaging agents or a reparative mechanism (or both) is unknown. Interstitial volume, as noted above, increases as a proportion of renal volume.

Renal vessels

The age-related changes in renal vessels have been well documented (Darmady et al. 1973) but are little understood. In the literature there is some confusion in distinguishing what exactly may be arterial changes of normal ageing, and what are pathological changes of 'arteriosclerosis'. In addition, the possible role of increasing blood pressure with age is unclear. There is disagreement over definitions of 'arteriosclerosis': some authors describe intimal proliferation, hyalinization, hypertrophy, and fibrosis, either alone or in combination, with this term.

In apparently normal aged individuals (McLachlan et al. 1977), prearterioles, from which afferent arterioles arise, show subendothelial deposition of hyaline and collagen fibres resulting in an intimal thickening. Changes in arterioles are similar except that fibrous material is less common. Small arteries exhibit a thickening of the intima due to proliferation of the elastic tissue. In vessels larger than arterioles, there is general agreement that progressive reduplication of elastic tissue and thickening of the intima are predominantly age related. In radiographic studies (Davidson et al. 1969) there is increasing tortuosity of the interlobular arteries and abnormalities in the arcuate arteries. Finally, as many as one-third of individuals with arteriosclerotic disease over the age of 50 develop clinical hypertension. Postmortem studies show significant atherosclerotic main renal artery stenosis, often without associated hypertension in 49 per cent of individuals (Holley et al. 1964).

Renal plasma flow

Davies and Shock (1950) demonstrated an age-dependent fall-off in both renal plasma flow using Diodrast clearances, and a lesser fall-off in GFR as measured by inulin clearance with a rise in the filtration fraction. Diodrast clearance, as a measure of renal plasma flow, fell from 613 to 290 ml/min in the ninth decade of life (Davies and Shock 1950). This implies a renal blood flow of about 1150 ml/min falling to 650 ml/min. Wesson (1969) analysed all the available literature, and showed a fall of approximately 10 per cent per decade in renal plasma flow. This is associated with an increase in renal vascular resistance with age (McDonald et al. 1951) and although the site of this resistance is not known precisely in man, it is well known that the calculated filtration fraction increases with age (Davies and Shock 1950). To what extent the fall-off in renal perfusion is fixed, as a result of anatomical or functional changes, has been examined by McDonald et al. (1951) and Hollenberg et al. (1974): response to vasoconstrictor stimulus (angiotensin) was unimpaired in the aged, whereas vasodilatation in response to pyrogen or acetylcholine (an endothelium-dependent vasodilator) was blunted.

Hollenberg et al. (1974), in their study of prospective kidney donors aged up to 76 years, confirmed the fall-off in renal plasma flow using p-aminohippurate (PAH) clearances, assuming normal PAH extractions; it should noted, however, that there are no published data on renal extraction of PAH in the aged. They showed also a fall in cortical blood flow, analysed by washout curves of radioactive xenon. Medullary flow, in contrast, was relatively well preserved. This decline in renal perfusion could, of course, result from a decrease in cardiac output with age, but data on this point are controversial: some studies show, and

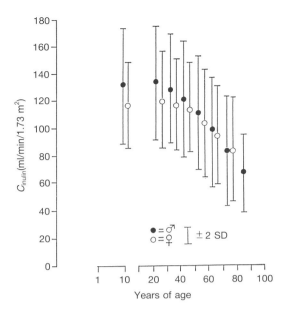

Fig. 3 The inulin clearance in healthy individuals at various ages. Graphical representation of data collected from the literature by Wesson (1969). Note that in older individuals not only does the inulin clearance fall, but the difference between the sexes seen in young subjects disappears. Modified from Macias and Cameron (1987) with permission.

some do not, a fall off in cardiac output in the aged (Brandfonbrener et al. 1955; Rodeheffer et al. 1984; Lakatta 1987).

Glomerular filtration rate

Inulin clearance

In the data from Davies and Shock (1950), the drop in inulin clearance with ageing was from a mean of 122 to a mean of 65 ml/min between 30 and 90 years respectively. Wesson (1969) summarized the available literature, from which Fig. 3 is derived, demonstrating the fall-off in renal function from age 30 onwards. Note that the lower limit if inulin clearance (mean-2 SD) for normal 80-year-olds is only 40 ml/min.1.73 m², and that in the elderly, as in prepubertal children, there is no sex difference in glomerular filtration rate.

This fall-off in GFR parallels the anatomical changes described in the previous section. Remaining nephrons will contain a group of hyperperfused and hyperfiltering nephrons, and Anderson et al. (1987) have speculated as to the role of this in progressive glomerulosclerosis.

Creatinine clearance and excretion

Using the creatinine clearance as an index of GFR, Kampmann et al. (1974) and Rowe et al. (1976a) among others (Laine et al. 1977) made the same observation that GFR declines with age (Table 1) One problem which arises with these data is that similar populations were not studied. Rowe et al. (1976a) studied healthy aged individuals, whilst Kampmann et al. studied a hospital population, but excluded any patient with a raised plasma creatinine. Cockroft and Gault (1976) in deriving their formula (see below) used hospital patients, but included all subjects, whatever their renal function. A second point is that all these data were obtained with creatinine methodology which is now rarely used; age-corrected standards for true plasma creatinine and creatinine clearance using rate-dependent chemical methods and modern buffers were not then available (see Chapter 1.4 for further discussion).

Finally all these studies are cross-sectional (except for some data in the study of Rowe et al. 1976a), and not longitudinal. Lindeman et al. (1985) have since published an important extended longitudinal study of old people in Baltimore, some of whom were followed with repeat measurements for more than 30 years. It is interesting to note that despite a mean calculated fall-off in creatinine clearance of 0.75 ml/min.year, 92 of 254 individuals studied showed no fall-off in creatinine clearance, and a few even increased their clearances! Larsson et al. (1986) also found no decline in GFR in individuals between the ages of 70 and 79, although serum creatinine increased from 91 to 96 μmol/l in women and from 100 to 107 mmol/l in men.

A very important practical point is that, in spite of the usual decline in creatinine clearance, there is, in general, no corresponding increase in plasma creatinine with age in otherwise healthy individuals (Rowe et al. 1976a), so that the normal plasma creatinine in an aged subject conceals the physiological fall-off in GFR displayed in Fig. 3. The normal levels of plasma creatinine are substantially the same at the ages of 20 and 80 years (Fig. 4) (Kampmann et al. 1974; Rowe et al. 1976a). This apparent paradox arises because the production of creatinine, and hence the urinary creatinine output, falls steadily with age in relation to the decreasing muscle mass and total body weight (Fig. 4). The significance of a modestly raised plasma creatinine in an aged subject is thus even greater than in a younger patient.

Bearing these data in mind, it is clear that plasma creatinine is not an adequate method to estimate GFR in the elderly, since it will systematically overestimate it in addition to all the other factors discussed in detail in Chapter 1.4. Because of this, it is necessary to estimate the GFR using not only plasma creatinine, but also age and body weight, and in the middle-aged, sex, or without the performance of a formal creatinine clearance. Rowe et al. (1976a), Kampmann et al. (1974) and Cockroft and Gault (1976) have constructed nomograms which allow a better estimate of the GFR in clinical practice from plasma creatinine concentrations; that of the last authors is the most reliable and most used. This is discussed in more detail in Chapter 1.4. Keller (1987) using the data of Kampmann et al. (1974), pointed out that the simplest formula to estimate the expected normal mean creatinine clearance in ml/min from 25 to 100 years of age is

Table 1 The mean creatinine clearance at different ages (ml/min.1.73 m²)

Age (years)	Denmark (n = 249) (x ± 2 SD)	France (n = 1797) (x ± 2 SD)	Spain (n = 23) (x)
20–29	110	107	
30–39	97	104	126
40–49	88	100	
50–59	81	95	
60–69	72	86	91
70–79	64	77	
80–89	47	73	
90–99	34		

Data of Kampmann et al. (1974), Laine et al. (1977), and Macias et al. (1981).

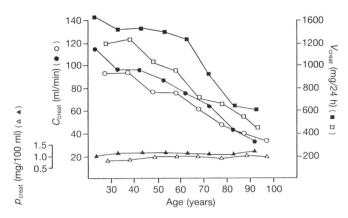

Fig. 4 Changes in excretion of creatinine, plasma creatinine, and creatinine clearance in healthy individuals at different ages. Male patients are represented by closed, and female by open symbols. In older subjects, creatinine clearance falls considerably, but plasma creatinine remains the same because creatinine excretion (and by inference creatinine production) decreases *pari passu*. Graphical representation of data from Kampmann *et al.* (1974), reproduced from Macias and Cameron (1987) with permission.

[130−age (years)]. Although this is true in general for older persons, individual variation obtained with any of these formulae may be considerable (Durakovic 1986). If those caring for elderly individuals are not aware of all these problems, excessive doses of drugs may be prescribed, and if these are excreted mainly or only by the kidney, will accumulate, with a high risk of nephrotoxicity, renal failure, or other drug side-effects (Goldberg *et al.* 1987) (see below).

Despite the fall-off in creatinine clearance and GFR in the elderly, the ability to respond with an increase to a protein load ('renal reserve') is undiminished with age (Böhler *et al.* 1989).

β_2-microglobulin

Unlike creatinine, the plasma concentration of β_2-microglobulin does rise as renal function declines with age (Evrin and Wibell 1972). In a study of 39 individuals aged 57 to 76 years, they found mean (± SD) plasma concentrations of β_2-microglobulin as shown in Table 2. Even so, there is a great deal of overlap and this method, despite the increased availability of β_2-microglobulin assays, has not been used in clinical practice in the aged to any extent.

Radioisotopic methods

The only radioisotopic method to be used in the aged has been the ^{51}Cr-edetate method. Granerus and Aurell (1981) give reference values for the single injection method using this chelate for young and elderly individuals, which show the expected fall-off

Table 2 Plasma concentrations of β_2-microglobulin (mg/l)

Age (years)	Males	Females
17–36	1.37 ± 0.20	1.40 ± 0.30
37–56	1.56 ± 0.31	1.65 ± 0.36
57–76	1.83 ± 0.41	1.75 ± 0.30

Mean ± SD.

with age. This fall has been confirmed by Macias-Nuñez *et al.* (1981), but Larsson *et al.* (1986) were unable to show any decline in GFR in elderly subjects between 70 and 79 years of age with this measure of GFR. The mean (± 2 SD) for GFR in 79-year-olds in this study was 46 to 94 ml/min.1.73m².

Tubular function

Acidification

As early as 1948, Shock and Yiengst observed that although resting pH and pCO_2 showed no differences in the elderly with respect to young controls in basal conditions, in response to a challenge with acid the fall in blood pH and pCO_2 was greater and longer lasting than in young individuals. Adler *et al.* (1968) noted that elderly individuals excreted only 19 per cent of the acid load given within a mean of 5 days (variable collection periods were used), compared with 35 per cent in younger subjects. It has since become a generally accepted idea that old people have a diminished capacity for renal acidification, attributed principally to a decrease in the urinary elimination of ammonium.

Ammonium ion

The effect of ageing on ammonium ion production is, however, controversial. First, Adler *et al.* (1968) noted that ammonium ion excretion formed a lower proportion of the excreted acid in the elderly, supposing this to be a result of the decrease in GFR, since they found a correlation between the two, even though much lower levels of GFR than those found in the aged would be necessary for diminution of ammonium ion excretion to the level found. Also, data on other factors influencing urinary acidification were not available: the fall in blood bicarbonate after acid loading was only 4.6 mmol/l, the subjects were allowed to walk or lie down with consequent variable aldosterone secretion, and finally some subjects needed bladder catheterization; the duration of collections varied from subject to subject.

Agarwal and Cabebe (1980) agreed that elderly individuals excrete less ammonium ion than the young in response to a short oral ammonium chloride load but since they found no difference in GFR between their elderly subjects, they concluded that a primary tubular defect accounted for the diminution in urinary output of ammonium ion. There are problems with these data also, since again the fall in plasma bicarbonate averaged only 2.5 mmol/l, the collection periods varied from subject to subject, and no data on nutritional status were available.

Our own group (Macias-Nuñez *et al.* 1983) found no differences in the capacity for renal ammonium elimination with age in a group of carefully studied and controlled elderly individuals (mean age 72 years) in whom the study was prolonged to 8 h—an important point because in the elderly the majority of the excretion of ammonium ion was during the last 3 h—and a fall of 7 mmol/l in plasma bicarbonate was obtained. The elderly subjects, however, took longer to reach peak excretion and the same dose of ammonium chloride induced a greater fall in blood bicarbonate in the elderly. These findings suggest that in healthy elderly subjects renal ammonium ion secretion in response to acid load is not different, provided that an adequate stimulus is provided in the absence of electrolyte and nutritional disturbances, and extra time is available for slower adaptation.

Schück and Nádvorníková (1987) studied the influence of ageing on ammonium excretion, titrable acid and urinary pH under basal conditions; there were no data concerning blood electrolytes nor nutritional status. GFR was reduced in the aged group, but there was no difference in the excretion of titrable acid or ammonium ion. Thus when titrable acid and ammonium ion were corrected per 100 ml of GFR, both ratios increased with age. Following an oral acid load, the elderly group was unable to increase their urinary NH_4^+ and titrable acid, but their excretion per 100 ml of GFR was consistently higher than the controls throughout the test. One interpretation of these data is that, as in chronic renal failure, in the elderly with reduced renal function the nephron is in the resting state secreting more ammonium ion than normal, and hence has a blunted response to the acid load.

Thus the more recent and comprehensive studies suggest merely slower adaptation, rather than a true deficiency in function.

Titrable acid

There is almost a complete consensus that elderly subjects are able to eliminate the same amount of titrable acid in response to an acute acid load as younger individuals (Agarwal and Cabebe 1980; Macias-Nuñez et al. 1983). Schück and Nádvorníková (1987) found that titrable acid per 100 ml of GFR increases with age, but that also the ageing kidney was unable significantly to increase it after a short acute acid load. This aspect of acid excretion thus resembled that found also for ammonium ion in this study, as noted above.

Urinary pH

In the available literature, there is no disagreement that urine pH can be lowered as effectively in age as in youth (Adler et al. 1986; Macias-Nuñez et al. 1983; Agarwal and Cabebe, 1980; Schück and Nádvorníková 1987). Adler (1986) even found a more intense fall in urinary pH in the elderly, to a mean 4.85 versus 4.96 in the young, but in the study of Agarwal and Cabebe (1980), the young performed slightly better.

Bicarbonate

The tubular capacity for bicarbonate reabsorption, as far as renal threshold for bicarbonate is concerned, is not impaired by age in the only study which has examined this point (Macias-Nuñez et al. 1983). There are no data concerning bicarbonate 'T_m' in the aged.

Acid excretion in the elderly: summing up

It seems that there are no or few differences between young and elderly persons regarding resting blood pH and bicarbonate under basal conditions (Shock and Yiengst 1948; Hilton et al. 1955; Adler et al. 1968; Agarwal and Cabebe 1980; Macias-Nuñez et al. 1983). However, when a healthy elderly person receives an acute load of strong acid, buffering is less effective than in the young, with a lower trough blood bicarbonate. Also, there is some delay in eliminating the acid load compared with younger persons (Macias-Nuñez et al. 1983). Under basal conditions, ammonium ion elimination is similar in the aged, and when corrected per 100 ml of GFR, ammonium excretion may even increase above levels observed in the young (Schück and Nádvorníková 1987).

Glucose

There are few publications regarding the renal handling of other substances by the ageing kidney. Among them, glucose has been shown to have a normal threshold in the elderly (Tabernero Romo 1987)—although Butterfield et al. (1967) had earlier suggested a rise in threshold with age—but a reduced apparent T_{mG} has been demonstrated in parallel with the fall in GFR (Miller et al. 1952). The decline in the apparent glucose T_m averaged 359 mg/min.1.73 m^2 in the third decade, and 219 mg/min.1.73 m^2 in the ninth. This peculiarity of the ageing kidney has some interest in everyday clinical practice, given the frequency of diabetes mellitus in the elderly population. With a discrete rise in glycaemia, elderly persons are more prone to develop glycosuria, which does not mean that they are diabetic. In this case blood sugar must be tested before starting antidiabetic care.

Calcium and phosphate

In general, dietary and environmental factors such as exposure to sunlight are much greater determinants of the frequent and important calcium, phosphate, and bone problems in the elderly than primary alterations in ageing renal function. It is a usual clinical finding to observe relatively low blood calcium and a high phosphate level in old age. In the urine, hypocalciuria and hyperphosphaturia are often found (Galinsky et al. 1987). Phosphate T_m is lower in the aged than in the young.

These changes in the metabolism of calcium and phosphate, may be related to low levels of 1,25-dihydroxy-vitamin D found in the elderly, probably as a result of a deficit of renal 1α-hydroxylase as an expression of the normal ageing process, which is known to be present in aged rats. This also could explain the high levels of PTH which accompany ageing, although retention of PTH fragments with decreasing GFR may play a part. Serum levels of 24, 25-dihydroxy-vitamin D are also low in the elderly (Galinsky et al 1987). The response of the 1α,1-hydroxylase to PTH and forskolin *in vitro* has been shown to be reduced in ageing rats, and presumably humans also.

Sodium, potassium, and water handling

In general, the aged kidney is quite capable of maintaining normal plasma electrolyte content as well as pH under normal circumstances. The functional deficits appearing with age usually relate to a reduction in the capacity of the kidney and its capability with respect to rate of change in response to stresses, especially when they are extreme.

Sodium

In the elderly, the renal response to both sodium loading and especially sodium depletion is blunted.

Sodium loading

There are minor discrepancies regarding the ability of the ageing kidney to excrete extra sodium, but most researchers suggest that its capacity to excrete a sodium load, either under basal conditions or during volume expansion, diminishes with age (Yamada et al. 1979; Hiraide 1981; Bengele et al. 1981; Hackbarth and Harrison 1982; Myers et al. 1982). One study did not show a difference in regard to age in the renal handling of sodium (Karlberg and Tolagen 1977). The majority of observations are in agreement with the high incidence of volume overload occurring in elderly patients given intravenous saline in geriatric units.

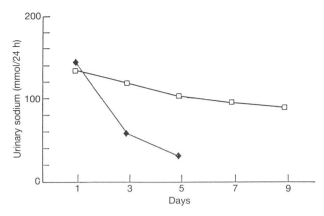

Fig. 5 Effect of a 50 mmol/24 h sodium diet on urinary excretion of sodium of young (closed diamonds) and elderly (open diamonds) subjects. The younger subjects adapted quickly and reached sodium balance within 5 days; the elderly failed to reach balance within the 9 days of the study, continuing to excrete more than 100 mmol of sodium per 24 h (see text). Data of Macias et al. (1983), reproduced from Macias et al. (1987) with permission.

Table 3 Sodium excretion and plasma sodium on a low sodium diet

	Young ($n = 6$)	Elderly ($n = 6$)	p value
Plasma Na (mmol/l)	139.5 ± 2.0	139.9 ± 2.2	NS
$U_{Na} V$ (mmol/24-h)	66.4 ± 42.0	165.5 ± 97.6	< 0.05
FE_{Na} (%)	0.93 ± 0.33	1.58 ± 0.71	< 0.05

Data of Macias Nuñez et al. (1980).
Data after 9 days on a 50 mmol Na^+ diet (mean ± SD).

Tubular handling of sodium

All observers agree that there is a diminished capacity to conserve sodium with increasing age, which is again consonant with clinical observations that large sodium excretion rates and sodium depletion are frequently found in patients on geriatric wards (Epstein and Hollenberg 1976; Macias-Nuñez et al. 1978; Macias-Nuñez et al. 1980). The renal and hormonal response to mild acute sodium restriction differs substantially between young and elderly well-nourished healthy people (Fig. 5.) Under basal conditions, plasma sodium levels were similar in both populations (Table 3). As the GFR was lower in the aged, the amount of filtered sodium would also be lower than in young subjects (Davies and Shock 1950); Sourander 1983). In spite of the lower sodium tubular load, 24-h urinary sodium output and fractional excretion of sodium were significantly greater in the elderly (Table 3). This suggests that the renal tubule of the elderly subject is unable to retain sodium adequately, either in absolute terms or when compared to glomerular filtration.

The capacity of the ageing kidney to adapt chronically to a low salt intake (50 mmol/24 h) is clearly blunted also (Epstein and Hollenberg 1976; Macias-Nuñez et al. 1978). In the latter study, young people reached sodium balance in 5 days, whereas the elderly were unable to reach sodium balance, in spite of a mean loss of 1.3 kg of body weight during the 9 days of 50 mmol/24 h sodium diet (Fig. 5, Table 4).

In spite of this, Kirkland et al. (1983) found that daily sodium and potassium excretions were reduced in healthy aged subjects. It is important to notice that in this study the elderly subjects were taking their usual diet, with no instructions to restrict salt intake. But as the authors pointed out, the lower urinary sodium output found may be more an expression of a lower salt content in diet than a renal 'avidity' for salt by the ageing kidney. It is known that the majority of aged individuals have a tendency to reduce their salt intake spontaneously. This study also noted a greater urinary elimination of water and electrolytes during the night in the elderly, which can at least in part explain the nocturia observed in 70 per cent of elderly persons.

Table 4 Change in body weight (kg) in young and elderly subjects after 9 days on a 50 mmol/24 h sodium diet (mean ± SD)

	Young subjects ($n = 6$)	Elderly subjects ($n = 6$)
Normal diet	66.4 ± 14.9	70.7 ± 11.3
Low sodium diet	65.5 ± 14.7	69.3 ± 10.7
Change	−0.9 ± 0.3	−1.3 ± 0.3, $p = 0.05$

Data of Macias et al. (1980).

Site of sodium loss within the nephron

From studies on lithium clearance (which is mainly reabsorbed in the proximal tubule to a comparable degree as sodium and water (see Chapter 1.4), it is probable that fractional delivery of sodium from the proximal tubule does not alter significantly with age (Schou et al. 1986); in this study however the ages of the individuals studied are not given, and how many were very elderly is not clear. With this in mind and in an attempt to identify the site in the nephron of the ageing kidney responsible for the defect, we (Macias-Nuñez et al. 1978) measured sodium excretion when patients were challenged with a saline load. This test is believed to be able to discriminate the functional capacity of the 'proximal nephron' from that of the 'distal nephron' (thick ascending limb of Henle's loop) (see Chapter 1.4 for discussion). It is clear from the results of this study (Table 5, Fig. 6) that, in agreement with the date of Schou et al. the 'proximal nephron' behaved similarly in the young and the elderly whereas in the 'distal nephron' a clear-cut difference in the handling of sodium in the elderly was evident, present in 85 per cent of tested population of healthy elderly persons.

This diminished capacity to reabsorb sodium by the ascending limb of Henle's loop of healthy elderly persons has two direct important consequences: first, the amount of sodium arriving at more distal segments of the nephron (distal convoluted and collecting tubules) increases, and second, the capacity to concen-

Table 5 Study of the segmental handling of sodium in different parts of the nephron in young and elderly subjects

	C_{Na} (ml/min)	'Proximal' nephron (ml/min)	'Distal' nephron (ml/min)
Young ($n = 10$)	3.20 ± 2.12	18.3 ± 7.2	83.0 ± 6.3
	$p < 0.05$	NS	$p < 0.05$
Elderly ($n = 12$)	6.52 ± 2.35	18.9 ± 5.9	59.1 ± 7.9

Data of Macias et al. (1978). Subjects were studied during zero ADH secretion during sodium loading.
All figures mean ± SD.

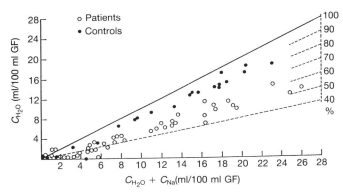

Fig. 6 Graphical representation of the calculated segmental capacity to reabsorb sodium by the 'proximal' and 'distal' nephron, studied under conditions of volume expansion from sodium loading. The normal elderly subjects are indicated by open circles, and the younger subjects by closed circles. Throughout the study, the data from the elderly subjects showed a lower percentage of sodium reabsorption attributable to the 'distal' nephron when compared with the younger subjects (see text). Reproduced from Macias *et al.* (1978) with permission.

Table 7 Plasma renin concentrations in young and elderly subjects on a normal and sodium restricted diet, and at rest and walking

Rest		Walking	
Young	Elderly	Young	Elderly
Normal diet			
4.55 ± 0.36*	0.79 ± 0.60†	7.06 ± 4.00*	0.98 ± 0.85†
Sodium restriction			
6.20 ± 4.70*	1.75 ± 1.80†	8.10 ± 4.20*	3.77 ± 3.77†

All data mean ± SD expressed in ng angiotensin I/ml.h.
* $p < 0.05$ between resting and walking. † $p < 0.05$ between young and elderly. (Data of Macias *et al.* 1978.)

trate the medullary interstitium is also diminished. As a consequence, elderly subjects exhibit both increased sodium excretion and inability maximally to concentrate the urine.

Causes of salt wasting in the aged

Several frankly pathological factors found rather frequently in the elderly population could account for renal sodium losses in the aged: obstructive uropathy, urinary infections, surreptitious diuretic intake, and pyelonephritis (Brocklehurst and Hanley 1979). Nevertheless, these conditions are not present in most elderly subjects who exhibit a defect in renal handling. Thus there also seems to be sodium losses not due to any identifiable pathological processes (Narins 1970; Sunderam and Mankikar 1983).

Plasma and urinary aldosterone concentrations were significantly reduced in the elderly under basal conditions, and the difference is enhanced after salt restriction (Table 6) or the assumption of an upright posture, or frusemide administration. These results are in agreement with the data of other workers (Flood *et al.* 1967; Sambhi *et al.* 1973; Crane and Harris 1976; Epstein and Hollenberg 1976). The tubular response to the administration of exogenous aldosterone, as assessed from the fall in sodium excretion, is attenuated with age also (Ceruso *et al.*

Table 6 Blood and urinary aldosterone concentrations in young and elderly subjects on a normal diet and during sodium restriction

Blood		Urine	
Young ($n = 6$)	Elderly ($n = 6$)	Young ($n = 6$)	Elderly ($n = 6$)
Normal diet			
19.6 ± 9.6	5.6 ± 4.2†	10.95 ± 8.8	4.2 ± 2.8†
Sodium restriction			
30.6 ± 4.8	8.6 ± 3.9†	29.3 ± 19.6*	9.6 ± 3.8†

All data expressed as mean ± SD in pg/ml (blood) and ng/24 h (urine).
* $p < 0.05$ normal diet versus salt restriction; † $p < 0.05$ young versus elderly. (Data of Macias *et al.* 1978.)

1970). Plasma renin concentrations are, in addition, lower in the elderly than in the young population (Table 7) despite normal concentrations of renin substrate, which have been demonstrated previously in other studies (Weidmann *et al.* 1977). Thus, a dual effect of low aldosterone secretion and a relative insensitivity of the distal nephron to the hormone could account for diminished sodium reabsorption at this site, and hence for the renal sodium losses in the healthy aged population.

It is unlikely that the inability to conserve sodium normally is related to excess prostaglandin production in the elderly. These substances act on the distal segments of the nephron (Strandhoy *et al.* 1974) presumably to inhibit sodium reabsorption (Lee *et al.* 1976). In fact, indomethacin, an inhibitor of cyclo-oxygenase and hence prostaglandin synthesis, has no effect on the reabsorption of sodium in the ascending limb of Henle's loop of elderly subjects (Macias-Nuñez *et al.* 1980), but reabsorption at this site is reduced already, as just noted.

Despite the tendency to an exaggerated natriuresis in the elderly, total body sodium is not significantly decreased with age (Weidmann *et al.* 1977; Cox and Orme 1973; Fulöp *et al.* 1985). In contrast to the tendency to lose sodium through the kidney, the red cell sodium content is increased in the aged population (Naylor 1970; Nagaki and Teraoka 1976; Cumberbatch and Morgan 1981). In other tissues, such as skeletal muscle, intracellular sodium seems to vary with age, although the data obtained were from a population no older than 57 years (Campana *et al.* 1971).

The high red cell sodium content in the aged has been attributed to an alteration in ion transport by the red cell membrane (Naylor 1970; Cox and Shalaby 1981; Cumberbatch and Morgan 1981). This alteration could account for the high sodium and low potassium content (see below) of the red cells (Cox and Shalaby 1981). The cause of this alteration has been attributed in turn to an effect of ageing itself, in altering membrane synthesis or metabolic membrane support (Naylor 1970).

Others claim that a functional defect of the Na^+K^+-ATPase dependent pump is present (Ibrahim *et al.* 1978). However, in the brain of old rats there is no decrease in Na^+K^+-ATPase activity (La Manna *et al.* 1983) and others have shown that with age, the number of units able to extract sodium from the cell diminishes (Cumberbatch and Morgan 1981). The Na^+K^+-ATPase activity of the renal medulla of old rats has been said to be reduced (Bengele *et al.* 1983), but details on tubule length and weight are lacking to interpret these data fully. While a Na^+K^+-ATPase deficit may contribute to a loss of sodium in various models, it is not clear that it is important in the salt-losing defect in the elderly human.

Other factors, such as interstitial fibrosis, peritubular forces or the action of sex hormones, could contribute to the defect. However, definitive evidence is lacking for any of these (Epstein and Hollenberg 1979). In summary, the ascending limb of Henle's loop and other tubular segments where aldosterone exerts its effect (Ceruso *et al.* 1970) are primarily responsible for the impairment in tubular renal sodium handling by the ageing kidney.

Potassium

Plasma potassium

Plasma potassium concentrations in the aged do not differ from those of the younger population (Videbaek and Ackermann 1953), a finding that has since been confirmed by numerous studies (Dall *et al.* 1971; Burini *et al.* 1973). Under normal conditions, plasma potassium is normal in the elderly, but when diuretics are taken, they develop hypokalaemia more rapidly than do the young (Dall *et al.*; Sunderam and Mankikar 1983).

Red cell and total body potassium

It is generally accepted that red cell potassium is a reasonable reflection of intracellular potassium content (Videbaek and Ackermann 1953), and a significant decrease of the red cell potassium content in the elderly has been confirmed (Kromhout *et al.* 1977; Hiraide 1981; Young 1983) despite a normal plasma potassium, while there is only one study in which red cell potassium was found to be elevated compared with levels in the young (Nagaki and Teraoka 1976).

Total exchangeable body potassium measurements using isotopic dilution methods employing radiolabelled potassium-42-alpha, potassium-43 or whole-body potassium detection of potassium-40, all have been in agreement that there is a low total body potassium in elderly subjects compared to the young (Allen *et al.* 1960; Forbes and Reina 1970; Molaschi and Molaschi 1974; Cohn *et al.* 1976; Rajagopalan *et al.* 1980; Hiraide 1981; Lye 1981; Cox and Shalaby 1981; Young 1983). The mean total body potassium of 2500 mmol in the elderly is approximately 20 per cent less than in the young (3000 mmol).

Why are the elderly potassium depleted?

Since 85 per cent of potassium is in muscle, efforts to explain the different content of potassium have focused on this tissue, principally in relation to the diminution of muscle mass that occurs with age (Allen *et al.* 1960; Forbes and Reina 1970; Kromhout *et al.* 1977; Rajagopalan *et al.* 1980; Lye 1981). However, when correction of total body potassium content is made using the lean body mass as a point of reference, the difference between young and elderly is even more pronounced (Lye 1981). When potassium supplements are given to the healthy elderly, the red cell deficit of potassium vanishes, but total body potassium deficit remains unchanged (Kromhout *et al.* 1977). This finding could mean: (1) that the red cell potassium content is not an accurate method to assess total body potassium content, at least in the elderly; (2) that the muscle cells of the elderly are relatively impermeable to potassium; or (3) that the elderly possess a reduced number of muscle cells, which are already potassium replete (Kromhout *et al.* 1977). This last observation is supported by the fact that the tissue potassium found in muscle biopsies remains constant throughout life (Campana *et al.* 1971; Moller *et al.* 1983).

Other theories have been proposed to explain potassium

Table 8 Renal handling of potassium in young and elderly subjects

	Plasma K^+ (mmol/l)	$U_{K^+} V$ (mmol/24 h)	FE_{K^+} (%)
Young	3.60 ± 0.24	43.2 ± 9.1	8.60 ± 3.10
Elderly	3.60 ± 0.16	30.5 ± 9.70	9.30 ± 3.40
p	NS	< 0.05	NS

All data expressed as mean $\pm$ SD (Macias *et al.* 1980).
Patients were taking their usual diet throughout the study.

depletion in the aged. One claims that an alteration in the cell membrane may be responsible for a potassium loss from, and sodium gain by, (see above) the cell. Support for this comes from studies performed in the red cell of the elderly (Naylor 1970; Nagaki and Teraoka 1976; Cumberbatch and Morgan 1981) demonstrating a high content of sodium in these cells. The demonstration that muscle cell potassium may be normal despite ageing militates, however, against this view. Conceivably, an alteration of the erythrocyte membrane could be specific and not be present in any other cells, but proof for either idea is lacking.

It has also been suggested that the causes of both low calcium and potassium levels are the nutritional deficiencies, the hormonal alteration, and/or reduction in physical activity found in the aged (Cohn *et al.* 1976). Daily potassium intake ranges from 60 to 150 mmol/24 h in the healthy young population. In the United Kingdom and the United States of America, potassium intake is much lower in the old than in young populations, being on the average less than 60 mmol/24 h (Dall *et al.* 1971; Davies *et al.* 1973; Abdulla *et al.* 1979). This low potassium ingestion is attributed to the inclination that the elderly have to eat more carbohydrates (sweets, cakes) instead of fresh fruit, meat, or fish which offer a higher potassium content (Dall and Gardiner 1971; Davies *et al.* 1973). This low potassium intake may be responsible also for the low red cell potassium content (Kromhout *et al.* 1977; Barbagallo *et al.* 1979; Hiraide 1981) and could explain why there is correction of the defect after potassium administration (Kromhout *et al.* 1977).

The final possible explanation for the tendency to potassium deficiency is an inability of the kidney to conserve potassium. In fact, we have observed that the total renal excretion of potassium is significantly lower in the aged population than in the young (Table 8) which is an accordance with the data of Kirkland *et al.* (1983). However, when corrected for the reduced GFR, the fractional excretion of potassium tends to be greater in the elderly than in young (Table 8). Possibly, as the number of functioning nephrons is decreased with age, the elimination of potassium per single nephron becomes greater. The increased distal delivery of sodium discussed above would lead to greater reabsorption of sodium in the distal nephron and hence to potassium losses. Whether this potassium loss is in part responsible for the low erythrocyte potassium in the aged is unknown. The question of diuretics and potassium losses in the aged is discussed below.

Water

Water equilibrium is achieved through a balance between water intake and water disposal. This balance is controlled by the regulation of thirst, neurohypophyseal function, and the renal capacity for water excretion (Shannon *et al.* 1984). Thirst and the intake of liquids diminish with age (Goldman 1981). When healthy active elderly volunteers are water restricted for 24 h, the threshold for thirst is found to be increased and water intake reduced with respect to a control group of younger subjects. Despite the reduction in water intake and thirst, a considerable increase in blood osmolality, plasma sodium concentration, and in circulating vasopressin occurred during water deprivation (Phillips *et al.* 1984). Osmotic release of AVP in response to intravenous hypertonic saline in the elderly is in fact greater than in the young; however the AVP response to volume depletion is blunted (Rowe *et al.* 1982), as in the aged rat (Bengele *et al.* 1983), in which insensitivity of the cortical collecting duct to AVP rather than a reduced medullary hyperosmolality appeared to be responsible, with a reduced generation of cAMP.

This lack of a sensation of thirst in the elderly, despite an increase in plasma tonicity, remains unexplained. Dryness of the mouth and a decrease of taste with age, may contribute to the diminution of the thirst observed in the aged (Hugonot *et al.* 1978). Another mechanism for diminished thirst in the elderly may be an alteration in mental capacity (Seymour *et al.* 1980) or cortical cerebral dysfunction (Miller *et al.* 1982). It has also been suggested that a reduction in the sensitivity of the osmoreceptors responsible for thirst regulation may play a role in the water handling alterations in elderly persons. A contrary view has arisen from the studies of Helderman *et al.* (1978). In these studies an increase in the sensitivity of the osmoreceptors that regulate vasopresssin release was observed.

Thirst diminution may also result from an inappropriate response to hypovolaemia. Age diminishes the sensitivity of the baroreceptors, and the release of vasopressin mediated by the baroreceptors (Rowe *et al.* 1982). Finally angiotensin concentrations—a powerful generator of thirst—are lower in the elderly. Therefore stimulation of thirst requires a severe hypovolaemia and/or hypotension (Robertson 1984). There is, as noted above, no difference in the plasma concentration of vasopressin between the young and the elderly before or following water deprivation (Shannon *et al.* 1984).

Urinary concentration and dilution

Urinary concentration

Ageing reduces the capacity of the kidney to concentrate the urine (Dontas *et al.* 1972; Rowe *et al.* 1976b), the maximum capacity diminishing by 5 per cent with every 10 years of age (Lewis and Alving 1938) from a maximum urinary specific gravity of 1030 at 40 years of age, to only 1023 at 89 years of age. Rowe *et al.* (1976b) found a mean maximum osmolality of 1109 mosm/kg in subjects aged 20 to 39 years, with a minimum urine flow rate of 0.49±0.03 ml/min, 1051 mosm/kg in 40 to 59-year-olds, and only 882 mosm/kg in those aged 60 to 79 years, in whom minimum urinary flow rate was more than double that of the young at 1.03 ml/min. The data of Lindeman *et al.* (1960) are similar. Kirkland *et al.* (1983) noted that even healthy individuals aged 60 to 80 years showed a higher excretion of water, sodium, and potassium during the night than younger controls, with complaints of nocturia in a high proportion. This alteration in circadian rhythm does not depend upon inability to concentrate the urine alone, but upon the defects in sodium handling and aldosterone secretion and sensitivity already mentioned.

The origin of the diminution in concentration is complex also; the small differences between the amount of liquid taken by young and elderly subjects cannot completely explain the differences in solutes and urinary elimination observed during water deprivation (Rowe *et al.* 1976b). The diminution of the concentrating ability has been related to the decrement in glomerular filtration rate that occurs with age. It is assumed that as the number of functioning nephrons decreases with age, the remainder are subjected to osmotic diuresis which impairs the ability to concentrate urine (Lindeman *et al.* 1960; Kleeman 1972). Despite this, some data do not reveal a close relationship between the reduction of glomerular filtration rate and the capacity to concentrate urine in the elderly (Rowe *et al.* 1976b). The relative increase in medullary blood flow noted above could contribute to the impairment of renal concentration capacity (Takazakura *et al.* 1972; Hollenberg *et al.* 1974).

Nevertheless, data supporting this suggestion are not available. Inappropriately low ADH is not, as already discussed, a factor in the genesis of the defect. The defect in sodium chloride reabsorption in the ascending limb of the loop of Henle, which is the basic mechanism for the operation of the countercurrent concentration mechanism, may be an important factor for the decrease in the capacity to concentrate urine seen in the aged; this is also discussed above.

Urinary dilution

There are only a few reports dealing with the capacity of the ageing kidney to dilute urine, but it has been found to be decreased (Lindeman *et al.* 1960; Dontas *et al.* 1972; Phillips *et al.* 1984; Editorial 1984). Dontas *et al.* (1972) found a minimum urine concentration of only 92 mosm/kg in the elderly compared with 52 mosm/kg in the young. Maximum free water clearance (C_{H_2O}) was also reduced in the elderly from 16.2 ml/min to 5.9 ml/min. This is probably dependent in the main upon the lower GFR in the elderly, but the C_{H_2O}/GFR was even so somewhat reduced in the older subjects (9.1 vs.10.2 per cent). Again, the functional impairment of the diluting segment of the thick ascending limb described above seems to account for the remainder of the diminution in the capacity to dilute urine observed in the aged (Macias-Núñez *et al.* 1978).

Regulation of plasma volume

Total body water is slightly diminished with age, so that it comprises only 54 per cent of total body weight (Edelman and Leibman 1959), probably because old people have a greater proportion of their body weight as fat than the young. The diminution seems to be predominantly intracellular (Edmonds *et al.* 1975). Cross-sectional and longitudinal studies have revealed that plasma and blood volume do not alter as a result of age alone in healthy adults (Cohn and Shock 1949; Chien *et al.* 1966), although Wörum *et al.* (1984) found that elderly women have significantly higher plasma volume than young women. We have found that plasma and blood volume measurements using radiolabelled albumin do not differ between young and elderly healthy volunteers. Males have a higher volume than females, regardless of age (Table 9). Thus a low plasma volume in an elderly subject is almost always the result of disease.

Psychiatric disturbances and electrolytes in the elderly

In elderly untreated patients with psychiatric pathological changes, an alteration of the intracellular, extracellular, and total body water has been found (Cox and Orme 1973). Deterioration of verbal learning is associated with an increase in body water, intracellular, and extracellular fluid and exchangeable

Table 9 Plasma volume in young and elderly subjects

Young		Elderly	
Male	Female	Male	Female
1516 ± 184	1369 ± 85*	1513 ± 155	1398 ± 191*

* $p < 0.05$ males versus females; there was no difference between young and elderly subjects.
Data in ml/m², mean ± SD (from Macias et al. 1980).

sodium and potassium in relation to dry body weight. The diminution in verbal ability was associated with a shift of water from the extracellular to the intracellular compartments and diminution of the interchangeable sodium in relation to lean body weight. In addition to this, there may be effects of lithium or other drugs on water and electrolyte handling (see below and Chapter 6.6). The importance of checking plasma electrolytes and urea in elderly confused individuals scarcely needs mention.

Renal pharmacokinetics in the elderly
(Cohen 1986; Brown et al. 1986; Dominguez-Gil et al. 1987; Montamat et al. 1989)

The ageing process is accompanied not only by modifications in renal function, but also in the function of other pharmacokinetically important organs such as the intestine and liver, and by lower serum albumin concentrations. Although prescribing for the elderly is one of the most common tasks for physicians, guidelines to avoid drug misuse are often forgotten.

It is difficult to draw conclusions from studies directed at clarifying the effects of ageing on pharmacokinetics because, as in studies of other aspects of the ageing process, most data are cross-sectional and not longitudinal. It is difficult to control for every factor that can modify pharmacokinetics, such as concomitant diseases and their drug therapy, multiple primary diagnoses, multiple drug ingestion and even poorer compliance with prescriptions in the elderly (Zimmer 1985).

Absorption and distribution of drugs in the elderly

Although the kidney plays a key role in the elimination process, it cannot be blamed for all disturbances caused by drug therapy in the elderly, since the changes in pharmacokinetics with ageing are extremely complex. Drugs, once (and if!) they have been taken are subject to gastrointestinal absorption and hepatic action (first-pass hepatic metabolism). Since age delays gastric emptying time, the speed and quantity of absorption will vary. Diminished gastric acid secretion and consequent higher gastric pH also occur with age. This may affect drug handling either to enhance or diminish blood levels of the active drug. Splanchnic blood flow is reduced by 30 to 40 per cent in the aged. As a consequence, there is an increase in the systemic availability of some drugs that are normally extracted in the liver. Atrophy of intestinal macro- and microvilli, with an increase in connective tissue and slower bowel motility, are frequent in the elderly.

The proportion of adipose tissue is increased in the elderly, more so in females, and lean body mass, total body water, and especially intracellular water are diminished. Therefore lipid-soluble drugs show an increased volume of distribution. In contrast, water-soluble drugs have a diminished volume of distribution. Plasma proteins, in particular albumin, are diminished—from 40 to 35 g/l at the age of 80, which accounts for the decrement in protein binding of drugs in the elderly, with a corresponding increase of the free fraction of the drug.

Elimination

As has been discussed earlier in this chapter, both GFR and renal blood flow diminish as the individual ages. Consequently drugs excreted by the kidney exclusively or predominantly by filtration have slower rates of elimination, increased half-lives in the plasma, and many accumulate leading to undesirable side-effects, including nephrotoxicity. In addition to excretion by filtration, many drugs are secreted by the renal tubules, which in some cases may be the dominant mode of excretion. These secretory mechanisms are complex, and many drugs and their metabolites compete for them. For some drugs, alterations in urinary pH are important in permitting or impeding tubular secretion.

Because of all these factors, it is necessary to modify the intervals of administration, and sometimes also the priming dosage of such drugs. On top of this, states of dehydration, cardiac failure, and occult renal disease, such as prostatic or renal vascular obstruction in the elderly, may reduce renal drug elimination even further, temporarily or in the long term.

Bearing all this in mind, it is essential to calculate the approximate GFR, using the actual, or calculated creatinine clearance derived from the plasma creatinine measurement. Despite its inaccuracies and disadvantages, as discussed in Chapter 1.4., the formula of Cockroft and Gault (1976) is the best validated. Drugs such as digoxin, aminoglycosides, lithium, practolol, quinidine, tetracycline, procainamide, and methotrexate are well known examples of those with reduced rates of renal elimination in the elderly (Dominguez-Gil et al. 1987).

However, drug elimination is a very complicated process of which renal excretion is only one component. Metabolism, with the potential for accumulation of metabolites as well as (or instead of) the parent drug, hepatic (biliary) as well as renal excretion, represent a further complication. For more information the reader can consult Dominguez-Gil et al. (1987) and Cohen (1986), and Chapter 2.1 of this textbook.

Common drugs that are frequently misused in the aged

There are a few drugs in common use whose elimination is predominantly renal which are frequently misused in the elderly.

Digoxin

Digoxin is frequently used in an inappropriate fashion in clinical practice. As the muscular mass of elderly subjects is reduced in relation to body weight and digoxin distributes in water rather than in fat, its volume of distribution is reduced. Consequently the loading dose of digoxin per kg body weight also needs to be reduced. Furthermore, since clearance is by renal elimination which is impaired, accumulation may occur and the maintenance dose may also need to be reduced. If, in addition, the patient is receiving diuretic therapy or has hypokalaemia for other reasons (which is very common in the aged—see Chapter 1.4), cardiotoxicity is commonly present, since the dose-response curve to digoxin is shifted to the left by hypokalaemia.

Aminoglycosides

Aminoglycosides are excreted exclusively by the kidney. When they are used in a similar dose to that recommended for the young, renal tubular damage may occur with consequent renal failure and further drug accumulation. The acute renal failure is usually polyuric (see Chapter 2.1), which often delays the diagnosis of nephrotoxicity. Tobramycin is slightly less nepohrotoxic than gentamicin, but only slightly so, and netilmycin also has some nephrotoxicity; its principal advantage in the elderly patient, whose balance is often already impaired, is that it possesses relatively little ototoxicity (Kahlmeter and Dahlager 1984). To use aminoglycosides safely in the elderly one must monitor plasma levels; nomograms and formulae are inadequate. In hospitals where there are no facilities to measure plasma levels, the drugs are best avoided in the elderly where possible; if they must be used, no more than 160 mg/24 h should be used as a starting dose, and the drug prescribed for no more than 5 days.

Non-steroidal anti-inflammatory drugs (NSAIDs) (Clive and Stoff 1984)

NSAIDs are taken by a very large number of elderly persons for osteoarticular and muscular complaints. In persons with normal renal function and well-hydrated individuals they cause little problem, but in situations of renal hyperfusion with low effective circulating volume, such as in dehydration or desalination, both frequent in the elderly, or in cardiac failure they can induce an acute renal failure that is often accompanied by oedema, hypertension, and hyperkalaemia (see Chapter 6.2). Aspirin, ibuprofen, fenoprofen, naproxen, piroxicam, and indomethacin are all able to produce this syndrome. Some of these drugs produce an acute interstitial nephritis in addition, particularly fenoprofen, which may be accompanied by a nephrotic syndrome and leads to an acute renal failure, usually polyuric. Vasculitis may occasionally also be seen.

Antidepressants

Antidepressants such as imipramine and to a lesser degree tricyclic and tetracyclic agents, may induce acute retention of urine in elderly males with prostatic hypertrophy and subclinical obstruction.

Chlorpropamide

Chlorpropamide is an oral antidiabetic compound, excreted by the kidney and with a very long half-life; it should be avoided in the treatment of aged diabetic individuals. Tolbutamide, glibenclamide, gliquidone, or glipazide are all safer for this purpose in the elderly.

Lithium

Lithium, extensively used in the management of manic depressive illness at a plasma concentration of 1 mmol/l, is eliminated exclusively by the kidney, and also may produce nephrotoxicity, especially at plasma levels of 1.5 mmol/l and higher (see Chapter 6.7). As with the aminoglycosides, the first effect is impairment of renal concentrating ability. Renal tubular acidosis is also frequently produced by lithium. Acute renal failure, when present, is more often polyuric than oliguric and as usual in polyuric acute renal failure its diagnosis is often late or even missed altogether (see Chapter 6.6).

Summary: the clinical importance of alterations in renal function in the elderly subject

Under normal circumstances, without stresses, homeostasis dependent on renal function is adequate in the elderly. When, however, stress is applied, the reduced maximal effort of which the ageing kidney is capable may be insufficient, and its response too slow, to cope with acute changes.

The normal elderly individual tends to take too little sodium in his or her diet, to waste sodium, adapts poorly to salt restriction and to salt loading, has hyporeninaemic hypoaldosteronism, and is often chronically potassium depleted because of reduced intake over long periods or because of conventional diuretics prescribed unthinkingly. Paradoxically, because of their hyporeninaemia and hypoaldosteronaemia, the elderly are at the same time vulnerable to hyperkalaemia with potassium-sparing diuretics, NSAIDs, and ACE inhibitors.

They often take inappropriate diets, and easily become confused and stop eating or drinking appropriately or at all. They cannot concentrate their urine as well as the young despite adequate release of AVP, and when their plasma tonicity rises they have blunted appreciation of thirst. Again, their sodium balance and ability to concentrate and dilute urine may be further compromised by diuretics.

They are thus at much greater risk of developing hyponatraemia or (less commonly) hypernatraemia than younger people, whether at home because of spontaneous failure of intake or gastrointestinal losses, or in the hospital as a result of inappropriate intravenous infusions. Drugs such as sedatives and tranquillizers which interfere with AVP release, or with its action on the tubule such as lithium and demeclocycline, are best avoided. Urinary electrolytes and osmolality are not reliable as indicators of renal ischaemia or function in the elderly, and measurements of renin and aldosterone in elderly hypertensives are of little value.

The glomerular filtration rate in the elderly population is lower, so that retention of any endogenous or exogenous substance is likely if its excretion depends substantially upon renal clearance. Because of the failure of plasma creatinine to rise with age, those caring for the aged may easily forget just how low renal function may be even in healthy elderly subjects. In addition, further renal impairment from what can be regarded as pathological processes even in the aged, such as renovascular atheroma, urinary tract obstruction, urinary infection, stones, glomerulonephritis, compounded by ingestion of drugs such as ACE inhibitors or non-steroidal anti-inflammatory drugs, is frequently present. The reduced renal plasma flow and GFR of the elderly kidney makes it more vulnerable to nephrotoxic substances, or to ischaemic insults, which are the more likely because of local vascular disease and/or reduced cardiovascular responses.

As well as having osteoporosis, many old people are vitamin-D deficient because of low dietary intake, plus poor 1-hydroxylation of the vitamin into the active hormone. On top of this their low plasma phosphate may lead to mild hyperparathyroidism, with further loss of bone.

Geriatric nephrology is a growing part of the specialty, and will continue to grow into the 21st century; we owe it to our elderly patients to be aware of their special vulnerability and special needs.

References

Abdulla, M., Jagerstad, M., Norden, A., Thulin, T., and Svensson, S. (1979). Nutrition and old age. Sodium. *Scandinavian Journal of Gastroenterology*, **14**, 138–42.

Adler, S., Lindeman, R.D., Yiengst, M.J., Beard, E., and Shock, N.W. (1968). Effect of acute acid loading on urinary acid excretion by the aging human kidney. *Journal of Laboratory and Clinical Medicine*, **72**, 278–89.

Agarwal, B.N. and Cabebe, F.G. (1980). Renal acidification in elderly subjects. *Nephron*, **26**, 291–5.

Allen, T.H., Anderson, E.C., and Langham, W.H. (1960). Total body potassium and gross body composition in relation to age. *Journal of Gerontology*, **15**, 348–57.

Anderson, S., Meyer, T.W., and Brenner, B.M. (1987). Mechanisms of age-associated glomerular sclerosis. In *Renal Function and Disease in the Elderly* (eds. J.F. Macias-Nuñez, and J.S. Cameron), pp. 49–66. Butterworths, London.

Barbagallo, G., Disciacca, A., and Pardo, A. (1979). Potassium depletion in aged patients: an evaluation through red-blood cell potassium determination. *Age and Ageing*, **8**, 190–5.

Bengele, H.H., Mathias, R.S., and Alexander E.A. (1981). Impaired natriuresis after volume expansion in the aged rat. *Renal Physiology*, **4**, 22–9.

Bengele, H.H., Mathias, R., Perkins, J.H., McNamara, E.R., and Alexander, E.A. (1983). Impaired renal and extrarenal adaption in old rats. *Kidney International*, **23**, 684–90.

Böhler, J., Gloer, D., Keller, E., and Schollmeyer, P.J. (1989). Renal functional reserve in the elderly. *Nephrology Dialysis Transplantation*, **4**, 416–7.

Brandfonbrener, M., Lansdowne, M., and Shock, N.W. (1955). Changes in cardiac output with age. *Circulation*, **12**, 557–66.

Brown, W.W., et al. (1986). Aging and the kidney. *Archives of Internal Medicine*, **146**, 1790–6.

Brocklehurst, J.C. and Hanely, T. (1979). *Geriatria Fundamental*, 1st edn, pp. 195–214. Toray S.A., Barcelona.

Burini, R., Da Silva, C.A., Ribeiro, M.A.C., and Campana, A.D. (1973). Concentracão de sodio e de potassio no soro e plasma de individuos normais. Influencia da idade, do sexo e do sistema de colheita do sangue sobre os resultados. *Revista de Hospitale Clinico da Facultad do Medicina de São Paulo*, **28**, 9–14.

Butterfield, W.J.H., Keen, H., and Whichelow, M. (1967). Renal glucose threshold variations with age. *British Medical Journal*, **4**, 505–7.

Campana, A.O., De Andrade, D.R., Gazoni, E., Burini, R.C., and Neves, D.P. (1971). Water, fat, sodium, potassium and chloride content of skeletal muscle in 'normal' subjects. *Revista Brasileira de Pesquisas Medicas e Biológicas*, **4**, 409–15.

Ceruso, D., Squadrito, G., Quartarone, M., and Parisi, M. (1970). Comportamento della funzionalita renale e degli elettroliti ematici ed urinari dopo aldosterone in soggetti anziani. *Giornale di Gerontologia*, **18**, 1–6.

Chien, S., Usami, S., and Simmons, R.L. (1966). Blood volume and age: repeated measurements on normal men. *Journal of Applied Physiology*, **21**, 583–8.

Clive, D.M. and Stoff, J.S. (1984). Renal syndromes associated with nonsteroidal antiinflammatory drugs. *New England Journal of Medicine*, **310**, 563–72.

Cockroft, D.W. and Gault, M.H. (1976). Prediction of creatinine clearance from serum creatinine. *Nephron*, **16**, 31–41.

Cohen, J.L. (1986). Pharamacokinetic changes in aging. *American Journal of Medicine*, **80**, (suppl 5A), 31–8.

Cohn, J.E. and Shock, N.W. (1949). Blood volume studies in middle-aged and elderly males. *American Journal of Medical Sciences*, **217**, 388–91.

Cohn, S.H., Vaswani, A., Zanzi, I., Aloia, J.F., Roginsky, M.S., and Ellis, K.J., (1976). Changes in body chemical composition with age measured by total body neutron activation. *Metabolism*, **25**, 85–96.

Cox, J.R. and Orme, J.E. (1973). Body water, electrolytes and psychological test performance in elderly patients. *Gerontologica Clinica*, **15**, 203–8.

Cox, J.R. and Shalaby, W.A. (1981). Potassium changes with age. *Gerontolgoy*, **27**, 340–4.

Crane, M. and Harris, J.J. (1976). Effect of aging on renin activity and aldosterone excretion. *Journal of Laboratory and Clinical Medicine*, **87**, 947–59.

Cumberbatch, M. and Morgan, D.B. (1981). Relations between sodium transport and sodium concentration in human erythrocytes in health and disease. *Clinical Science*, **60**, 555–64.

Dall, J.L.C. and Gardiner, H.S. (1971). Dietary intake of potassium by geriatric patients. *Gerontologia Clinics*, **13**, 119–24.

Dall, J.L.C., Paulose, S., and Fergusson, J.A. (1971). Potassium intake of elderly patients in hospital. *Gerontologia Clinica*, **13**, 114–18.

Darmady, E.M., Offer, J., and Woodhouse, M.A. (1973). The parameters of the ageing kidney. *Journal of Pathology*, **109**, 195–207.

Davidson, A.J., Talner, D.B., and Downs, M. (1969). A study of the angiographic appearances of the kidney in an aging normal population. *Radiology*, **92**, 975–83.

Davies, D.F. and Shock, N.W. (1950). Age changes in glomerular filtration rate, effective renal plasma flow, and tubular excretory capacity in adult males. *Journal of Clinical Investigation*, **29**, 496–507.

Davies, L., Hastrop, K., and Bender, A.E. (1973) Potassium intake of the elderly. *Modern Geriatrics*, **4**, 482–8.

Dominguez-Gil, A., Garcia, M.J., and Navarro, A.S. (1987). Pharmacokinetics in the aged. In *Renal Function and Renal Disease in the Elderly* (eds. J.F. Macias-Nuñez and J.S. Cameron), pp. 184–203. Butterworths, London.

Dontas, A.S., Marketos, S., and Papanayioutou, P. (1972). Mechanisms of renal tubular defects in old age. *Postgraduate Medical Journal*, **48**, 295–303.

Durakovic, Z. (1986). Creatinine clearance in the elderly: a comparison of direct measurement and calculation from serum creatinine. *Nephron*, **44**, 66–9.

Edelman, I.S. and Leibman, J. (1959). Anatomy of body water and electrolytes. *American Journal of Medicine*, **27**, 256–60.

Editorial (1984). Thirst and osmoregulation in the elderly. *Lancet*, **ii**, 1017–18.

Edmonds, C.J., Jasani, B.M., and Smith, T. (1975). Total body potassium and body fat estimation in relationship to height, sex, age, malnutrition and obesity. *Clinical Science and Molecular Medicine*, **48**, 431–40.

Epstein, M. and Hollenberg, N.K. (1976). Age as a determinant of renal sodium conservation in normal man. *Journal of Laboratory and Clinical Medicine*, **87**, 411–17.

Epstein, M. and Hollenberg, N.K. (1979). Renal 'salt wasting' despite apparently normal renal, adrenal and central nervous system function. *Nephron*, **24**, 121–6.

Evrin, P-E. and Wibell, L. (1972). The serum levels and urinary excretion of β_2-microglobulin in apparently health subjects. *Scandinavian Journal of Clinical and Laboratory Medicine*, **29**, 69–74.

Flood, C., Gerondache, C., Pincus, G., Tait, J.F., Tait, S.A.S, and Willoughby, S. (1967). The metabolism and secretion of aldosterone in elderly subjects. *Journal of Clinical Investigation*, **46**, 960–6.

Forbes, G.B. and Reina, J.C. (1970). Adult lean body mass declines with age: some longitudinal observations. *Metabolism*, **19**, 653–663.

Fulöp, T., Wórum, I., Csöngor, J., Foris, G., and Leóvey, A. (1985). Body composition in elderly people. *Gerontology*, **31**, 6–14.

Galinsky, D., Meller, Y., and Shany, S. (1987). The aging kidney and calcium-regulating hormones: vitamin D metabolites, parathyroid hormone and calcitonin. In *Renal Function and Disease in the elderly* (eds. J.F. Macias-Nuñez and J.S. Cameron, pp. 121–42. Butterworths, London.

Goldberg, T.H., Martin, S., and Finkelstein, S. (1987). Difficulties in estimating glomerular filtration rate in the elderly. *Archives of Internal Medicine*, **147**, 1430–3.

Goldman, R. (1981). Modern ideas about the renal function in the elderly. In *Geriatrics for the Practitioner* (eds. A.N.J Reinders Folmer and J. Schouten), pp. 157–66. Excerpta Medica, Amsterdam.

Goyal, V.K. (1982). Changes with age in the human kidney. *Experimental Gerontology*, **17**, 321–31.

Granerus, G. and Aurell, M. (1981). Reference values for ^{51}Cr-EDTA clearance as a measure of glomerular filtration rate. *Scandinavian Journal of Clinical and Laboratory Investigation*, **41**, 611–6.

Hackbarth, H. and Harrison, D.E. (1982). Changes with age in renal function and morphology in C57BL/6, CBA/HT6, and B6CBAF1 mice. *Journal of Gerontology*, **37**, 540–7.

Helderman, J.H., Vestal, R.E., Rowe, J.W., Tobin, J.D., Andres, R., and Robertson, G.L. (1978). The response of arginine vasopressin to intravenous ethanol and hypertonic saline in man: the impact of aging. *Journal of Gerontology*, **33**, 39–47.

Hilton, G.F., Goodbody, M.F., and Kreusi, O.R. (1955). The effect of prolonged administration of ammonium chloride on the blood acid-base equilibrium of geriatric subjects. *Journal of the American Geriatrics Society*, **3**, 697–703.

Hiraide, K. (1981). Alterations in electrolytes with aging. *Nihon University Journal of Medicine*, **23**, 21–31.

Hollenberg, N.K., Adams, D.F., Solomon, H.S., Rashid, A., Abrams, H.L., and Merrill, J.P. (1974). Senescence and the renal vasculature in normal man. *Circulation Research*, **34**, 309–16.

Holley, K.E., Hunt, J.C., Brown, A.L., Kincaid, O.W., and Sheps, S.G. (1964). Renal artery stenosis. A clinical pathological study of normotensive and hypertensive patients. *American Journal of Medicine*, **37**, 14–22.

Hugonot, R., Dubos, G., and Mathes, G. (1978). Étude experimental des troubles de la soif du viellard. *La Revue de Gériatrie*, **4 September**, 179–91.

Ibrahim, I.K., Ritch, A.E.S., McLennan, W.J., and May T. (1978). Are potassium supplements for the elderly necessary? *Age and Ageing*, **7**, 165–70.

Kahlmeter, G. and Dahlager, J.I. (1984). Aminoglycoside toxicity—a review of studies published between 1975 and 1982. *Journal of Antimicrobial Chemotherapy*, **13**, **(Suppl A)**, 9–22.

Kampmann, J., Siersbaek-Nielsen., Kristensen, K., and Molholm Hansen, J. (1974). Rapid evaluation of creatinine clearance. *Acta Medica Scandinavica*, **196**, 517–20.

Kaplan, C., Pasternack, B., Shah, H., and Gallo, G. (1975). Age-related incidence of sclerotic glomeruli in human kidneys. *American Journal of Pathology*, **80**, 227–34.

Karlberg, B.E. and Tolagen, K. (1977). Relationships between blood pressure, age, plasma renin activity and electrolyte excretion in normotensive subjects. *Scandinavian Journal of Clinical and Laboratory Investigation*, **37**, 521–8.

Keller, F. (1987). Kidney function and age. *Nephrology Dialysis Transplantation*, **2**, 382 (letter).

Kirkland, J.L., Lye, M., Levy, D.W., and Banerjee, A.K. (1983). Patterns of urine flow and electrolyte secretion in healthy elderly people. *British Medical Journal*, **285**, 1665–7.

Kleeman, C.R. (1972). Water metabolism. In *Clinical Disorders of Fluid and Electrolyte Balance* (eds. M.H. Maxwell and C.R. Kleeman), p. 697. McGraw-Hill, New York.

Kromhout, D., Broberg, U., Carlmark, B., Karlsson, S., Nisell, O., and Reizenstein, P. (1977). Potassium depletion and ageing. *Comprehensive Therapeutics*, **3**, 32–7.

Laine, G., Goulle, J.P., Houlbreque, P., Gruchy, D., and Leblanc, J. (1977). Clairance de la créatinine. Valeurs de référence en fonction de l'age et du sexe. *Nouvelle Presse Médicale*, **30**, 2690–1.

Lakatta, E.G. (1987). Cardiac muscle changes in senescense. *Annual Review of Physiology*, **49**, 519–31.

La Manna, J.C., Doull, G., McCracken, K., and Harik, S.I. (1983). (Na$^+$–K$^+$)–ATPase activity and oubain-binding sites in the cerebral cortex of young and aged Fischer-344 rats. *Gerontology*, **29**, 242–7.

Larsson, M., Jagenburg, R., and Landahl, S. (1986). Renal function in an elderly population. *Scandinavian Journal of Clinical Laboratory Investigation*, **46**, 593–8.

Lee, J.B., Patak, R.V., and Mookerjee, B.K. (1976). Renal prostaglandins and the regulation of blood pressure and sodium and water homeostasis. *American Journal of Medicine*, **60**, 781–98.

Lewis, W.H. and Alving, A.S. (1938). Changes with age in the renal function of adult men. Clearance of urea, amount of urea nitrogen in the blood, concentrating ability of kidneys. *American Journal of Physiology*, **123**, 505–15.

Lindeman, R.D., Van Buren, H.C., and Raisz, L.G. (1960). Osmolar renal concentrating ability in healthy young men and hospitalized patients without renal disease. *New England Journal of Medicine*, **262**, 1396–409.

Lindeman, R.D., Tobin, J., and Shock, N.W. (1985). Longitudinal studies on the rate of decline in renal function with age. *Journal of the American Geriatrics Society*, **33**, 278–85.

Lye, M. (1981). Distribution of body potassium in healthy elderly subjects. *Gerontology*, **27**, 286.

McDonald, R.K., Solomon, D.H., and Shock, N.W. (1951). Aging as a factor in the hemodynamic changes induced by a standard pyrogen. *Journal of Clinical Investigation*, **30**, 457–62.

Macias-Nuñez, J.F. and Cameron, J.S. (1987). Renal function and disease in the elderly. Butterworths, London.

Macias-Nuñez, J.F., *et al.* (1978). Renal handling of sodium in old people: a functional study. *Age and Ageing*, **7**, 178–81.

Macias-Nuñez, J.F., Garcia-Iglesias, C., Tabernero-Romo, J.M., Rodriguez-Commes, J.L., Corbacho-Becerra, L., and Sanchez-Tomero, J.A. (1980). Renal management of sodium under indometacin and aldosterone in the elderly. *Age and Ageing*, **9**, 165–72.

Macias-Nuñez, J.F., *et al.* (1981). Estudio del filtrado glomerular en viejos sanos. *Revista Española de Geriatria y Gerontologia*, **16**, 113–24.

Macias-Nuñez, J.F. *et al.* (1983). Comportamiento del riñon del viejo en la sobrecarga de acidos. *Nefrologia*, **3**, 11–16.

Macias-Nuñez, J.F., Bondia Roman, A., and Rodriguez-Commes, J.L. (1987). Physiology and disorders of water balance and electrolytes in the elderly. In *Renal Function and Disease in the Elderly* (eds. J.F. Macias-Nuñez and J.S. Cameron), pp. 67–93. Butterworths, London.

McLachlan, M. (1987). Anatomic structural and vascular changes in the aging kidney. In *Renal Function and Disease in the Elderly* (eds. J.F. Macias-Nuñez and J.S. Cameron), pp. 3–26. Butterworths, London.

McLachlan M.S.F., Guthrie, J.C., Anderson, C.K., and Fulker, M.J. (1977). Vascular and glomerular changes in the ageing kidney. *Journal of Pathology*, **121**, 65–77.

Miller, J.H., McDonald R.K., and Shock, N.W. (1952). Age changes in the maximal rate of renal tubular reabsorption of glucose. *Journal of Gerontology*, **7**, 196–200.

Miller, P.D., Krebs, R.A., Neal, B.J., and McIntyre, D.O. (1982). Hypodipsia in geriatric patients. *American Journal of Medicine*, **73**, 354–56.

Molaschi, M. and Molaschi, E.S. (1974). Correlazione tra metabolismo del potassio, funzionalita renale ed invecchiamento nell'uomo. *Giornale di Gerontologia*, **22**, 241–50.

Moller, P., Alvestrand, A., Bergstrom, J., and Fürst, P. (1983). Electrolytes and free amino acids in leg skeletal muscle of young and elderly women. *Gerontology*, **29**, 1–8.

Montamat, S.C., Cusack, B.J., and Vestal, R.E. (1989). Management of drug therapy in the elderly. *New England Journal of Medicine*, **321**, 303–8.

Myers, J., Morgan, T., Waga, S., and Manley, K. (1982). The effect of sodium intake on blood pressure related to the age of the patients. *Clinical and Experimental Pharmacology and Physiology*, **9**, 287–9.

Nagaki, J. and Teraoka, M. (1976). Age and sex differences of sodium and potassium concentration in red blood cells. *Clinica Chimica Acta*, **66**, 453–5.

Narins, R.G. (1970). Post-obstructive diuresis: a review. *Journal of the American Geriatric Society*, **18**, 925–36.

Naylor, G.J. (1970). The relationship between age and sodium metabolism in human erythrocytes. *Gerontologia*, **16**, 217–22.

Papper, S. (1973). The effects of age in reducing renal function. *Geriatrics*, **28**, 83–7.

Phillips, P.A., et al. (1984). Reduced thirst after water deprivation in healthy elderly men. *New England Journal of Medicine*, **311**, 753–9.

Rajagopalan, B., Thomas, G.W., Beilin, L.J., and Ledingham J.G.G. (1980). Total body potassium falls with age. *Clinical Science*, **59**, 427s–9s.

Robertson, G.L. (1984). Abnormalities of thirst regulation. *Kidney International*, **25**, 460–9.

Rodeheffer, R.J., Gerstenblith, G., Becker, L.C., Fleg, J.L., Wesifeldt, M.L., and Lakatta, E.G. (1984). Exercise cardiac output is maintained with advancing age in healthy human subjects: cardiac dilatation and increased stroke volume compensate for a diminished heart rate. *Circulation*, **69**, 203–13.

Rowe, J.W., Andres, R., Tobin, J.D. Nomis A.H., and Shock, N.W. (1976a), The effect of age on creatinine clearance in men: a cross sectional and longitudinal study. *Journal of Gerontology*, **31**, 155–63.

Rowe, J.W., Shock, N.W., and De Fronzo, R.A. (1976b). The influence of age on the renal response to water deprivation in man. *Nephron*, **17**, 270–8.

Rowe, J.W., Minaker, K.L., Sparrow, D., and Robertson, G.L. (1982). Age-related failure of volume–pressure-mediated vasopressin release. *Journal of Clinical Endocrinology and Metabolism*, **54**, 661–4.

Sambhi, M.P., Crane, M.G., and Genest, J. (1973). Essential hypertension: new concepts about mechanisms. *Annals of Internal Medicine*, **79**, 411–24.

Schou, M., Thomson, K., and Vestergaard, P. (1986). The renal lithium clearance and its correlation with other biological variables: observation in a large group of physically healthy persons. *Clinical Nephrology*, **25**, 207–11.

Schück, O. and Nádvorníková, H. (1987). Short acidification test and its interpretation with respect to age. *Nephron*, **46**, 215–16.

Seymour, D.G., Henschke, P.J., Cape, R.D.T., and Campbell, A.J. (1980). Acute confusional states and dementia in the elderly: the role of dehydration/volume depletion, physical illness and age. *Age and Ageing*, **9**, 137–46.

Shannon, R.P., Minaker, K.L., and Rowe, J.W. (1984), Aging and water balance in humans. *Seminars in Nephrology*, **4**, 346–53.

Shock, N.W. and Yiengst, M.J. (1948). Experimental displacement of the acid-base equilibrium of the blood in aged males. *Federation Proceedings*, **7**, 114–19.

Sorensen, F.H. (1977). Quantitative studies of the renal corpuscle IV. *Acta Microbiologica et Pathologica Scandinavica*, **85**, 356–66.

Sourander, L. (1983). The kidney. In *Geriatrics* (ed. D. Platt), pp. 202–21. Springer, Berlin.

Strandhoy, J.W., et al. (1974). Effects of protaglandins E_1 and E_2 on renal sodium reabsorption and Starling forces. *American Journal of Physiology*, **53**, 389–92.

Steffes, M.W., Barbosa, J., Bagsen, J.M., Matas, A.J., and Mauer, M.W. (1983). Quantitative glomerular morphology of the normal human kidney. *Laboratory Investigation*, **49**, 82–6.

Sunderam, S.G. and Mankikar, G.D. (1983). Hyponatremia in the elderly. *Age and Aging*, **12**, 77–80.

Tabernero Romo, J.M. (1987). Proximal tubular function and renal acidification in the aged. In *Renal Function and Disease in the Elderly* (eds J.F. Macias-Nuñez and J.S. Cameron), pp. 143–61. Butterworths, London.

Takazakura, E., Sawabu, N., Handa, A. Takada, A., Shindda, A., and Takeuchi, J. (1972). Intrarenal vascular change with age and disease. *Kidney International*, **2**, 224–30.

Tauchi, H., Tsuboi, K., and Okutani, J. (1971). Age changes in the human kidney of the different ages. *Gerontologia*, **17**, 87–97.

Videbaek, A. and Ackermann, P.G. (1953). The potassium content of plasma red cells in various age groups. *Journal of Gerontology*, **8**, 63–4.

Weidmann, P., et al. (1977). Interrelations between age and plasma renin, aldosterone and cortisol, urinary cathecholamines, and the sodium/volume state in normal man. *Klinische Wochenschrift*, **55**, 725–33.

Wesson, L.G. (1969). Renal hemodynamics in physiological states. In *Physiology of the Human Kidney* (ed. L.G. Wesson), p. 96. Grune and Stratton, New York.

Worum, I., Fulop, T., Csongor, J., Foris, G., and Leovey, A. (1984). Interrelation between body composition and endocrine system in healthy elderly people. *Mechanisms of Aging and Development*, **28**, 315–24.

Yamada, T., Endo, T., Ito, K., Nagata, H., and Izumiyama, T. (1979). Age-related changes in endocrine and renal function in patients with essential hypertension. *Journal of the American Geriatrics Society*, **27**, 286–92.

Young, V.R. (1983). Protein and aminoacid catabolism and nutrition during human aging. In *Geriatrics* (ed. D. Platt), pp. 393–416. Springer, New York.

Zimmer, A.W., Calkins, E., Hadley, E., Ostfeld, A.M., Kaye, K.M., and Kaye, K. (1985). Conducting clinical research in geriatric populations. *Annals of Internal Medicine*, **103**, 276–83.

1.6.1 Radiological tactics: rationale

THOMAS SHERWOOD

This chapter points to the main lines of radiological attack on particular clinical problems. It is dominated by the idea that nephrologists, like all good doctors, do best for their patients if they act as humane clinical scientists. That means having informed ideas about what is likely to be wrong, and using investigations as critical experiments on these hypotheses. Great scientists impress by their elegant, direct experiments on good hypotheses. This approach stands at the far end of the process known as the diagnostic work-up, where all available data are gathered first, in more or less indiscriminate fashion, in the hope that the findings will produce ideas. Ticking a shopping list in the high technology supermarket is sometimes seen as the proper diagnostic endeavour of twentieth century clinical science. However, such a strategy is unkind to the patient, uncritical for the doctor, and has little to do with scientific method.

Unilateral kidney failure

The clinical prompts leading to this finding are usually very helpful, i.e. pain or haematuria. Alternatively this could be a chance

1.6.1 Radiological tactics: rationale

Table 1 Unilateral kidney failure

	First imaging tests	Definitive test
Absent kidney	Ultrasound/DMSA scan	Cystoscopy
Ectopic kidney	Ultrasound/DMSA scan	IVU
Arterial stenosis/occlusion	Ultrasound/DMSA scan	Arteriography
Venous obstruction	Ultrasound/DMSA scan	Venography
Urinary tract obstruction		
Acute	IVU	IVU
Chronic	Ultrasound/DTPA scan/IVU	Whitaker test if indicated
Infection		
Acute (suppurative pyelonephritis)	Ultrasound/white cell (indium) scan	CT
Chronic (atrophic pyelonephritis)	Ultrasound/DMSA scan	IVU and cystography if indicated
Tumour	Ultrasound	CT
Trauma	IVU/DMSA scan	CT ± arteriography

discovery, on an imaging test done for some quite different reason.

When unilateral renal impairment is such a silent finding, a first question may be whether there is a kidney at all in the correct position on this side. The only wholly reliable sign of a congenitally absent kidney is the asymmetric trigone found at cystoscopy. Once this diagnosis is suspected in a patient with symptoms, there is therefore little point in pursuing elaborate, costly procedures such as computed tomography (CT) or magnetic resonance imaging (MRI): a urologist with a cystoscope is needed. These tests are however quite apt for the patient without symptoms in whom an absent kidney is an unimportant chance finding which simply needs confirmation and no further action. Ultrasound (US) and intravenous urography (IVU) certainly have important error rates in distinguishing between small and absent kidneys on one side.

For the ectopic kidney, commonly pelvic, IVU is the most helpful investigation, since this will clearly demonstrate any urinary drainage problems as well as identifying the ectopic site. Ultrasound may well locate the oddly placed kidney, but is clearly less effective for the additional questions asked of such a finding.

Radionuclide dimercaptosuccinic acid (DMSA) scans can be extremely helpful, allowing functional mapping in the hunt for an absent or ectopic kidney. They do not replace the definitive tests (Table 1).

Ultrasound has gathered a powerful new ingredient in Doppler flow studies of the renal vessels, of obvious relevance in the diagnosis of arterial or venous obstruction. In the present state of knowledge, arteriography and venography remain as important arbiters of diagnostic decisions here. There is little point in subjecting a patient to these invasive procedures, however, if ultrasound studies have shown patent renal vessels with normal flow patterns.

Obstruction of the urinary tract is a common cause of unilateral kidney failure. If ureteric colic is the obvious clinical pointer to a likely ureteric stone, IVU is the test of choice for making a complete diagnosis, except in such settings as pregnancy. More often, chronic obstruction needs to be considered by the nephrologist. Where ultrasound and DTPA (diethylamide triamine penta-acetic acid) scans raise the likelihood of this diagnosis, further delineation of the cause is usually required. An IVU (for intrinsic urinary tract lesions) or CT (for extrinsic causes) may follow. A pressure/flow study or Whitaker test is useful in selected patients for deciding whether a dilated urinary

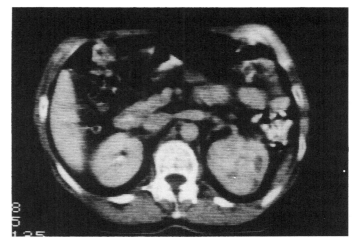

Fig. 1 A diabetic patient with severe loin symptoms and fever: note multiple abscesses in the left kidney which resolved on antibiotic therapy.

tract is really obstructed or not: can urine be transported at normal pressure and flow from kidney to ureter (Whitaker 1979)?

A feverish diabetic patient with unilateral loin pain and infected urine is an obvious candidate for the diagnosis of suppurative pyelonephritis. Following abnormal ultrasound and indium-labelled white cell scans, CT is very effective here (Fig. 1). The late effects of childhood reflux nephropathy can be a big problem, and are considered in Chapter 16.1. Ultrasound and DMSA scans are a good first screen for examining this state.

The kidney infiltrated by tumour is an important consideration for the nephrologist: at least 20 per cent of patients present to physicians, with systemic, constitutional symptoms (Chisholm 1982). The 'classical triad' of haematuria, loin pain and mass is a nineteenth century hangover. CT is definitive.

Trauma as an acute event is unlikely to present to nephrologists, but the late finding of a previously damaged kidney may prove a difficult diagnostic problem. Arterial compromise or renal compression by haematoma may result in a small kidney (Table 2).

The small kidney

Radiological tests have much to offer in unravelling the main causes of this problem (Table 2). When both kidneys are found

Table 2 Small kidneys

	First imaging tests	Definitive test
1. Small kidney with normal papillary architecture, e.g.		
Vascular lesion	Ultrasonography	Arteriography
Glomerulonephritis	Ultrasonography	Renal biopsy
Atypical postobstructive atrophy (rare)	Ultrasound/DMSA scan	IVU
2. Small kidney with uniform papillary deformities and even parenchymal thinning, e.g.		
Postobstructive atrophy	Ultrasound/DMSA scan	IVU
Renal papillary necrosis (late)	Ultrasonography	IVU
3. Small kidney with irregular papillary deformities and uneven parenchymal thinning, e.g.		
Chronic pyelonephritis	Ultrasound/DMSA scan	IVU and cystography if indicated
Renal papillary necrosis	Ultrasonography	IVU
Renal stones	Plain film	IVU
Tuberculosis	IVU	Urine culture

Table 3 Large kidneys

	First imaging tests	Definitive test
Urinary tract obstruction	Ultrasound/DTPA scan/IVU	Whitaker test if indicated
Glomerulonephritis	Ultrasonography	Renal biopsy
Polycystic disease	Ultrasonography	CT
Renal infarction (acute)	Ultrasound/DMSA scan	CT
Renal vein occlusion	Ultrasound/DMSA scan	Venography
Renal infiltration e.g. amyloid	Ultrasonography	CT and biopsy

to be small in the patient presenting with renal failure, there may be no point in such detailed radiological investigation for successful management. Obstruction must be excluded of course (see section on renal failure below).

The large kidney

Table 3 outlines examples of the major causes under this heading. The timetable of the pathological processes involved here is particularly important for renal size. Renal infarction gives rise to an enlarged kidney first, which atrophies in the weeks following the insult. Amyloid infiltration is often quoted as a renal enlarger, but this only holds true in the early stages. Most late amyloid kidneys are small (Davidson 1985).

Renal failure

The radiological questions posed in renal failure are:

1. How large (and where) are the kidneys?
2. Are the kidneys obstructed?
3. Are there any clues as to diagnosis?

Ultrasound examination has taken over from IVU as the standard first test for tackling these questions. It involves no appreciable risk, and is not dependent on renal function. It is worth special emphasis, however, that ultrasound examination in this setting should always be accompanied by a plain abdominal radiograph. Important findings such as renal stones may otherwise be overlooked.

Ultrasonography is usually sufficient to answer question 1. A difficulty about question 2 is that, unlike most other diagnostic problems in medicine, what is required is an answer with 100 per cent accuracy. It will not do to miss any patient in renal failure with obstruction. Table 4 is a decision tree based on the work of Webb *et al.* (1984), showing how a reliable diagnostic pathway can be constructed. It reaches for CT (without contrast medium, avoiding this hazard in the patient with renal impairment) where ultrasonography delivers a doubtful answer. To reinforce the point about plain films already made, these authors advocate additional abdominal tomography if the clinical background or inadequate abdominal film raises any query on possible renal stones.

In 51 per cent of patients renal dilatation is reliably ruled out by ultrasonography (i.e. no obstruction). The demonstration of small kidneys points to end-stage renal failure in just over half this group. Ultrasonography shows dilated calyces in 37 per cent of all the patients. Many of these will be found obstructed, but a complete diagnosis by ultrasonography as to level and cause is unusual. This leaves only a small doubtful group of patients (12 per cent) in whom obstruction cannot be excluded at this stage because of indeterminate findings. Staghorn calculi, polycystic disease, and other multiple renal cysts are particular culprits here.

In summary, plain films and ultrasonography can deliver a complete radiological diagnosis in about 60 per cent of patients with renal impairment. CT without contrast medium as the next step increases the successful diagnostic harvest to 84 per cent. Intravenous urography, or retrograde studies (involving ureteric catheterization at cystoscopy), are required only for the remaining 16 per cent of patients. Antegrade (percutaneous) pyelography is a useful stand-by as an alternative to retrograde studies in a sick patient, if cystoscopy with its attendant risks and general

1.6.1 Radiological tactics: rationale

anaesthesia is best avoided. In experienced hands it is of course perfectly possible to puncture undilated kidneys by this means. If necessary, percutaneous drainage can follow at once.

Some anxiety has been cast on the possibility of error in this pathway (Evans 1987). Ultrasound examination relies on finding dilated upper urinary tracts for diagnosing obstruction, and there are two circumstances in particular where this may be missed. These are:

1. Very recent obstruction, i.e. days old only, where there has not been time for notable dilatation to occur. The well-known fact that in acute obstruction secondary to a ureteric stone, as in unilateral renal colic, there is only mild distension, should serve as a reminder of this timescale problem.
2. A kidney made stiff by infiltration, e.g. from extensive tumour, will not distend when obstructed.

Where anuria in renal failure raises strong suspicion that obstruction is the cause, it is always worth thinking twice over a negative ultrasound scan. Outside these very rare instances of difficulty, ultrasonography is an excellent and reliable way of excluding upper urinary tract obstruction.

Urinary infection

This is of course common in both children and adults, but the need for radiological investigation is sharply different between the two age groups. It can be reasonably argued that every child with a proved urinary tract infection should have a plain abdominal radiograph (to exclude stones) and an ultrasound scan (to discover drainage anomalies) (Whitaker and Sherwood 1984). Adults by contrast need to be protected from much useless radiological investigation, even plain films and ultrasound scans if inappropriate.

Recurrent cystitis in women

This can be such a troublesome problem that in despair all available diagnostic help is sought. Unhappily the pay-off from routine imaging in this condition is very poor indeed.

The occasional find of an important lesion will be used to justify the practice. Renal stones may be discovered, apparently unsuspected, or a pelviureteric junction obstruction with a hydronephrotic kidney may be seen for the first time.

It is probably better to ask how such near-chance findings could be anticipated on the clinical presentation of recurrent urinary infection. The identity of the infecting organism can help. Proteus infections should always raise the alarm about a possible stone, and plain film tomography may be needed to look for a poorly calcified stone. Loin symptoms may point to a hydronephrosis. Renal papillary necrosis is worth remembering as a cause of unexplained infection having an upper urinary tract flavour. Women with recurrent infection need to be questioned about possible analgesic abuse.

A small kidney with reflux may occasionally be found in adults presenting with cystitis and loin pain. Whilst the renal damage was clearly done long ago, persisting troublesome pain, reflux, and infection on the side of the abnormality may add up to a good reason for surgical intervention.

Cystitis in men

Lower urinary tract symptoms in young men as part of a venereal infection often raise the question of urethral stricture if persistent. The pay-off from routine urethrography is poor, but the investigation may be appropriate in selected, recalcitrant cases.

Much more commonly, in older men bladder outflow obstruction secondary to an enlarged prostate will be suspected following a urinary infection. Time was when IVU was a routine requirement with this diagnosis; plain film and ultrasound scanning have now taken over. The traditional questions 'does the

Table 4 Ultrasonography and CT in renal failure (reconstructed from Webb *et al.* 1984)

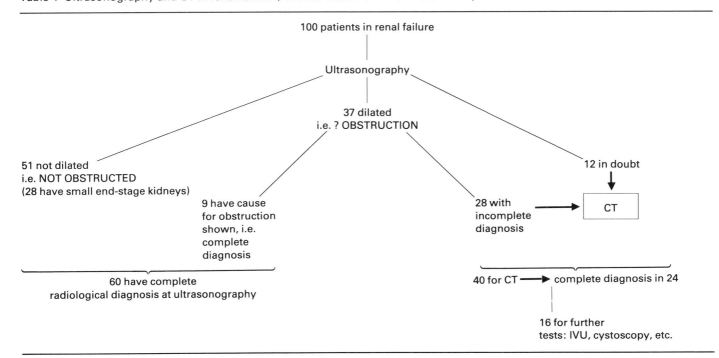

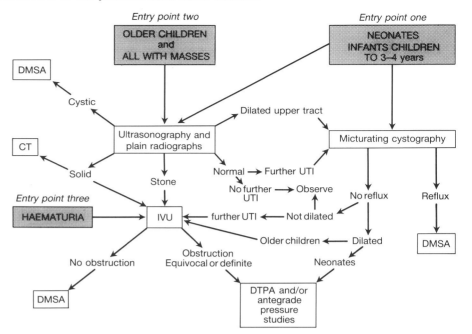

Fig. 2 A diagnostic map for planning children's investigations, adapted from Whitaker and Sherwood (1987). It allows for several entry points, depending on presenting symptoms and signs.

bladder empty and are the kidneys all right' can certainly be answered in this way, but there is no good evidence that the replies alter the management of these patients (de Lacey et al. 1988). Clinical analysis of symptoms and signs, together with a voiding flow rate, probably determine the correct course of action for most patients. There is a small group of patients with high-pressure bladder outflow obstruction who develop dilated upper tracts and move toward renal failure (George et al. 1983). They need early surgical attention, but they make up only a very small proportion of all the men suspected of 'prostatism'. Routine ultrasound scanning can be defended as a mechanism for finding this rare subgroup, but its cost-effectiveness is in doubt—a familiar dilemma in health services research.

Children with urinary infection

Correctable surgical causes are uncommon but important, and make up the easy rationale for investigating all children with a proved urinary tract infection (Whitaker and Sherwood 1984). The more common problem of vesicoureteric reflux is a vexed question, made particularly difficult by the fact that the standard screen of plain film and ultrasound scan cannot detect reflux itself. Natural history and the outcome of treatment are further troublesome though fascinating hurdles for devising a sensible diagnostic strategy. The subject is dealt with in greater detail later in Chapters 13.3 and 16.1. A first diagnostic map can be devised as guide for tackling the problem of the child whose urinary symptoms demand radiological investigation (Fig. 2). Such mapping of diagnostic tactics may illustrate a useful elaboration of the more usual algorithmic decision tree with its rigidly prescribed paths. In doubtful, subtle territories, maps are of obvious attraction for explorers; in addition they probably represent a more apt account of real-life diagnostic method (Campbell 1987).

Hydronephrosis

This is a useful if rather vague term, indicating a dilated pelvicalyceal system. It may be used to imply a narrow pelviureteric junction as cause. The investigative approach to excluding obstruction in renal failure has already been considered. This section deals with finding a dilated upper urinary tract, either on the basis of symptoms, or as a chance discovery. The latter event is increasingly common, especially in children.

Table 5 presents a summary of the main lines of attack on the problem. There is of course a wide range of severity in hydronephrosis. At one end is the chance finding in utero of a mildly distended upper urinary tract which will turn out to revert to normal in the first weeks or months of life. In a fit, growing infant with clear urine, no more than serial ultrasound scans are needed to monitor this progress. At the other end are children or adults with unilateral loin symptoms, or infection/haematuria or stones, coupled with an obvious hydronephrosis demanding surgical attention. In this group IVU will usually be required to assess the level of obstruction and anatomical detail, but operation may well follow in clear-cut cases without further ado. For children an additional cystogram is important in order to exclude ureteric reflux as a cause/complication.

Most patients, however, fall into a grey zone between these extremes. Dilatation may well be intermittent, the tie-up with symptoms doubtful, and appropriate management therefore uncertain.

A distinction must then be made between stasis of urine, say in a mildly distended upper urinary tract on the basis of poor material (e.g. congenital hydronephrosis or megaureter), and obstruction. Surgery may be quite inappropriate in the first group, and essential for preserving nephrons in the second. Once reflux has been excluded in children, further tests are needed. The quantitative assessment of renal function and urine transport offered by DTPA scanning (including diuretic challenge on renography curves as well as imaging) is generally extremely helpful, and should resolve the question in most patients, especially on serial studies. In cases of doubt, a Whitaker test is the gold standard on which pressure/flow relations can be judged. Debate continues on the issue (Lancet 1987), and the test is obviously unattractive as a universal prescription since it is invasive. We have found it an excellent court of appeal, much to be preferred to unnecessary surgery.

Table 5 Hydronephrosis

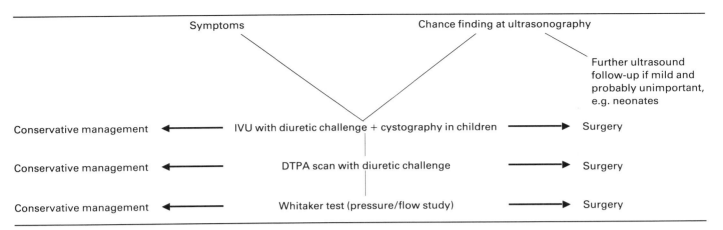

Haematuria

There are obvious differences in diagnostic import between this finding in children and adults. In childhood, drainage anomalies, or causes more usually presenting with infection, are to the fore; investigative pathways for such conditions have already been considered.

For adults, there is no need to elaborate further here the investigation of painful haematuria linked with infection (e.g. acute cystitis), renal trauma, or ureteric colic. Appropriate tests and action are clear. Painless haematuria is a more difficult problem for the radiologist. Nephrologists will not need reminding that other urinary findings are of great importance, for instance the presence of casts leading, in the relevant clinical setting, to the diagnosis of glomerulonephritis.

'Painless haematuria is due to tumour unless proved to the contrary' is a convenient adage, providing it is not driven to an absurd scenario of never-ending investigations. The next dictum must be that 'bladder tumour is a cystoscopic diagnosis'. For this reason all adults with painless haematuria, not obviously part of the presentation of glomerulonephritis, require referral to a urologist. This rule is in danger of being subverted by the glitter of new imaging advances. The danger that report of 'IVU normal' would be misinterpreted in the sense of 'bladder tumour excluded' was well understood in past decades. It is important that the same reservation extends to ultrasonography and CT, and probably also magnetic resonance imaging (MRI). 'Let's just have a high-tech image of the urinary tract to make sure it's not serious' will not do for the patient with haematuria.

Once a bladder tumour has been found or excluded on cystoscopy, imaging will of course be needed to look further. It is worth reiterating that at least a fifth of renal carcinomas present with symptoms that lead to the physician rather than the surgeon (Chisholm 1982). Because other lesions of the upper urinary tracts must also be considered (for instance ureteric tumours), our own preference is to keep IVU as the radiological investigation of first choice in painless haematuria. This stance is not universally favoured since renal tumours may occasionally be missed by IVU. Renal masses in 'IVU unfriendly' sites (directly anterior or posterior on the kidney) can be elusive, even on tomography. For this reason ultrasonography is preferred in some centres as first test.

Ultrasonography is in any event the investigation that will follow at once if a mass is found, since the differentiation of carcinoma from the much more common renal cyst is highly reliable by this means. No further renal imaging is needed for the patient shown to have a characteristic renal cyst on ultrasonography. This conclusion naturally assumes that a plain film has been done: without it the exclusion of an additional stone is never secure.

Polycystic disease is a condition of obvious nephrological interest, and ultrasonography is also well placed to make this diagnosis in adults. Below the age of 20 years cysts may be too small to make the exclusion of this disease reliable in members of an affected family.

If ultrasonography confirms a solid renal mass, i.e. likely carcinoma, further information is of obvious surgical value. The questions to be answered concern staging of the tumour, particularly extension into the renal vein and inferior vena cava, normality of the contralateral kidney, and of course possible lung metastases (chest radiography). CT had been taken as the natural test of first choice for solving the local queries on the tumour, and is certainly as good as angiography. However, it now seems likely that ultrasonography itself is as reliable for appropriate surgical management, and that there is no need to reach for CT as the inevitable preoperative investigation (Webb *et al.* 1987).

CT comes into its own for assessing the doubtful renal mass, i.e. where IVU and ultrasonography do not give characteristic findings. Renal cysts that have bled, infarcts (Fig. 3), or large septa of Bertin ('cortical islands') are well-known simulators of renal tumours, and are generally sorted out satisfactorily on CT. MRI may also have worth while results for the difficult renal mass.

A brief note on the loin pain–haematuria syndrome appears appropriate for a nephrological textbook. The initial documentation of unique arteriographic abnormalities in women with this otherwise unexplained symptom complex has not been substantiated by controlled studies. This is not to deny the undoubted value of demonstrating signs of renal arteritis on arteriograms in well-defined states such as polyarteritis nodosa and drug-abuse angiitis.

Simplicity, elegance and economy—these watchwords of good radiological tactics are also those ensuring the kindest diagnostic path for the patient, and a scientific one for the nephrologist.

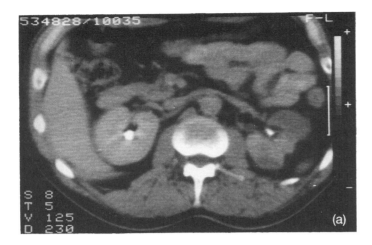

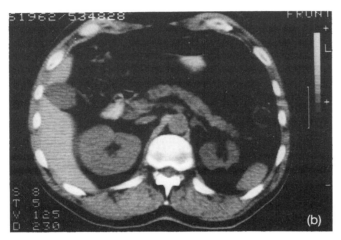

Fig. 3 CT scans in renal infarction. (a) On presentation with left loin pain and haematuria: note the swollen, avascular segments in the left kidney (contrast medium enhancement). (b) Five years later, there is characteristic shrinkage of the affected segments.

References

Campbell, E.J.M. (1987). The diagnosing mind. *Lancet*, **i**, 849–51.
Chisholm, G.D. (1982). Systemic effects—special oncology: kidney. In *Scientific Foundations of Urology* (2nd edn) (eds. G.D. Chisholm and D.I. Williams), pp. 677–80. Heinemann, London.
Davidson, A.J. (1985). Amyloidosis. In *Radiology of the kidney*. pp. 229–31. Saunders, Philadelphia.
de Lacey, G., Johnson, S., and Mee, D. (1988). Prostatism: how useful is routine imaging of the urinary tract? *British Medical Journal*, **296**, 965–7.
Evans, C. (1987). Renal failure radiology—1987. *Clinical Radiology*, **38**, 457–62.
George, N.J.R., O'Reilly, P.H., Barnard, R.J., and Blacklock, N.J. (1983). High pressure chronic retention. *British Medical Journal*, **286**, 1780–3.
Lancet. (1987). Hydronephrosis, renal obstruction, and renography. **i**, 1301–2.
Webb, J.A.W., Reznek, R.H., White, F.E., Cattell, W.R., Fry, I.K., and Baker, L.R.I. (1984). Can ultrasound and computed tomography replace high-dose urography in patients with impaired renal function? *Quarterly Journal of Medicine*, **53**, 411–25.
Webb, J.A.W., Murray, A., Bary, P.R., and Hendry, W.F. (1987). The accuracy and limitations of ultrasound in the assessment of venous extension in renal carcinoma. *British Journal of Urology*, **60**, 14–17.
Whitaker, R.H. (1979). The Whitaker test. *Urologic Clinics of North America*, **6**, 529–39.
Whitaker, R.H. and Sherwood, T. (1984). Another look at diagnostic pathways in children with urinary tract infection. *British Medical Journal*, **288**, 839–41.
Whitaker, R.H. and Sherwood, T. (1987). Diagnostic pathways. *Lancet*, **i**, 1266.

1.6.2 Conventional uroradiology and contrast media

JEAN-FRANÇOIS MOREAU, OLIVIER HELENON, JOËL CHABRIAIS, AND MOURAD SOUISSI

Radiocontrast media

Although uroradiology started with Röentgen's discovery of X-rays 100 years ago, for more than 25 years, it was restricted to plain films and retrograde vizualization of the urinary tract using opaque catheters or injections of gas. The first revolution was brought about by the synthesis of an organic iodinated molecule (Uroselectan) by von Lichtemberg and Swick in 1929: intravenous urography was born (Strain 1971). At almost the same time, Dos Santos introduced translumbar arteriography, enabling opacification of the renal vasculature using thorium dioxide before its replacement by iodinated molecules when its carcinogenic effects were demonstrated (Strain 1971).

The second revolution occurred in the mid-1950s, when new tri-iodinated benzoate molecules (diatrizoate, iothalamate), fully eliminated by the glomeruli, were synthesized and which, combined with Seldinger's technique of retrograde vascular catheterization, enabled selective arteriograms to be obtained (Strain 1971; Abrams 1971). The third revolution started during the 1970s, when the new low osmolar molecules (ioxaglate, non-ionics) became available, along with computerized imaging technology, based not only on radiographs but also on ultrasound waves and on magnetic fields.

Contrast media available in 1990

Members of the first generation of water soluble iodinated contrast media (Uroselectan, Iodopyracet, Acetrizoate) are no longer available. The radiologist can now choose the most suitable molecule for an examination from the hyperosmolar family and the low osmolar family of contrast media (Sovak 1984). All are water soluble, stable in solution, eliminated through glomerular filtration and have clearances similar to that of inulin. For conventional uroradiological techniques, the iodine concentration range of the solutions is 30 to 38 g/100 ml. Lower concentrations are available for computerized radiology. The dose to be injected should be calculated as grams of iodine per kilogram body weight (g/kg body weight).

The hyperosmolar molecules available in Europe are sodium and/or methylglucamine salts of diatrizoic, iothalamic, metrizoic and ioxitalamic acids. In the American literature they are often termed 'ionics', because they are ionizable. The osmolality of these solutions is about five times that of the plasma, and they have an LD_{50} in mice near 6 g/kg body weight. They induce a strong osmotic diuresis. Their patents are old; their costs of synthesis, and thus their retail prices, are low.

The low-osmolality molecules can be divided into ionics and non-ionics. The only representative of the ionics is the French dimer ioxaglate, introduced in 1976. At present, most of the non-ionics, introduced later, are monomers: iopamidol, iohexol, ioversol; non-ionic dimers are not yet available, except in Germany, but should emerge in the near future. The osmolality of these solutions is about twice that of plasma, and their LD_{50} is two to three times that of the hyperosmolar ionics. Their osmotic diuresis rate is very low, especially when ioxaglate is injected. Unfortunately, the cost of their synthesis, especially that of non-ionics, is high; in France, their retail price is at least 2.5 times that of the hyperosmolar ionics; in North America, the ratio may be 1 to 10, but the latter group is much cheaper there than in Western Europe (McClennan 1987; Jacobson and Rosenquist 1988).

General toxicity

The general toxicity of the iodinated molecules available at the beginning of the decade, whether ionic or non-ionic, hyperosmolar or not, monomer or dimer, is low. Iodine itself is not believed to play a part in toxic reactions.

Lethal accidents

The incidence of death is about 1:75 000 to 1:100 000 intravenous injections (Derobert *et al.* 1964; Hartman *et al.* 1982; Pinet *et al.* 1982). The number of lethal accidents induced by the molecules themselves is probably overestimated in some papers (Shehadi and Toniolo 1980), where the potential role of the underlying disease and associated treatments is not taken into account.

Death is due to anaphylactoid shock, and is unpredictable, although it is supposed to occur more frequently in truly allergic patients (see below). A recent Japanese survey (Katayama *et al.* 1988; Yamagushi *et al.* 1989) showed the risk in young subjects is the same as that in the elderly, although in other series the risk of severe reactions increased with age (Lasser *et al.* 1987). Intravenous injections (for computed tomography and digitalized angiography as well as for urography) are more often responsible for such deaths than intra-arterial injections. A lower risk might be expected from the use of new molecules, but fatal complications have not been completely eliminated (Jacobson and Rosenquist 1988; McClennan 1987). The possibility of lethal complications underlines the need of facilities for rapid treatment of collapse and for resuscitation in the radiology ward.

Minor complications

These complications occur mainly after bolus intravenous injections, and include fever, nausea or vomiting, metallic taste in the mouth, headache, thirst, and proximal pain. Their incidence is reduced by the use of low osmolality molecules, which improve the comfort of the patient. Intra-arterial injections are practically painless, especially with ioxaglate.

Allergic accidents

This group of complications is manifested by cutaneous and respiratory manifestations: only patients exhibiting such symptoms should be labelled 'allergic'. Erythemas and urticaria are common, even with the new molecules; they are weakly pruritic, benign, and transient, even without the administration of antihistamine drugs. Quincke's oedema is rare but may be fatal when it affects the larynx. Asthma is rarely observed but subclinical bronchospasm may be more common (Errfmeyer *et al.* 1986).

Cardiovascular complications

Vagal syncope (Lalli 1980) should be differentiated from anaphylactoid shock. In our experience, it is more often related to local analgesia for arteriography than to the injection of radiocontrast medium; atropine is active as premedication or therapy. In the absence of acute myocardial insufficiency, the best therapy for cardiovascular collapse induced by contrast medium is a large, rapid intravenous infusion of macromolecular fluids. Although electrocardiographic changes have been observed during intravenous injections, myocardial toxicity of the iodinated molecules plays only a minor role in patients submitted to uroradiological examinations.

Superficial thrombophlebitis is a common complication of intravenous injections, especially when a hyperosmolar medium is injected into a dependent vein of the cephalic network of the forearm. Local anti-inflammatory drugs are indicated, but anticoagulant therapy is not usually required.

Deep thrombophlebitis (phlegmatia alba dolens) is rare. Several cases of severe phlegmatia coerulea or of catastrophic arterial thrombosis have recently been described following radiocontrast media injection into patients with lupus anticoagulant and/or anticardiolipin antibodies (Collins 1989; J.F. Moreau, unpublished observations). These patients are usually sensitive to the thrombotic effect of radiocontrast media and are also predisposed to recurrent spontaneous thrombosis.

Hyperosmolar ionics tend to have anticoagulant properties; however, the procoagulant action of non-ionics has been recently debated (Robertson 1987, 1989).

Pathophysiology and preventive therapy

Lasser (1987) proposed that trauma induced by catheterization of the vein and by injection of the contrast medium activates Factor XII; this initiates cascade reactions, involving the kallikrein–kinin system, the complement system, the contact-system activators, and the coagulation system (Belleville et al. 1985; Dawson et al. 1986; Moreau et al. 1988a; Neoh et al. 1981; Rice et al 1983; Westaby et al. 1985). Lalli (1980) has emphasized the protective effect of relaxation and sedative drugs such as diazepam; anxiety and stress may contribute by activating thrombin, itself a powerful activator of the C1 complement fraction. This cascade is normally inhibited in vivo by the C1-esterase inhibitor. Such protection is enhanced by steroid pretreatment.

Various drugs (steroids, histamine H_1-H_2 receptor antagonists, or caproic acid) have been used for preventing contrast medium general toxicity in presumed high-risk groups; contradictory results have emerged from these uncontrolled studies. Lasser et al. (1987) showed in a randomized prospective study that two doses of oral corticosteroids (methylprednisolone, 32 mg), administered approximately 12 and 2 h before challenge with ionic contrast material, significantly reduced the incidence of reactions of all types, including the most severe.

Nephrotoxicity

Epidemiology and clinical course

Contrast-associated nephropathy is one of the most common causes of drug-induced oliguric acute renal failure, and may follow intravenous urography as well as computerized tomography and angiography (Moreau et al. 1988b).

The incidence of contrast-associated renal failure is less than 1 per cent in retrospective studies; however, it reaches approximately 10 per cent in prospective studies. This difference is explained by various factors, such as selection of patients, definition of renal failure, state of hydration, and dose of contrast agent (Berns 1989; Brezis and Epstein 1989; Deray et al. 1990).

Contrast-associated acute renal failure is associated with oliguria in approximately two-thirds of the reported cases, and is usually resistant to loop diuretics. Renal failure develops within 2 days of the injection, rarely later. Serum creatinine concentration usually begins to increase within the first 24 h and peaks within 2 to 4 days, with a spontaneous return to baseline values within 1 week. Temporary dialysis may, however, be needed, particularly in patients with a baseline serum creatinine concentration above 300 μmol/l. Renal sequelae are unusual, except in patients with advanced renal failure, (mainly diabetics), in whom end-stage renal failure may be precipitated. Low urine sodium concentration and fractional excretion of sodium have been reported, but this finding has been challenged as has the value of persistent visualization of the kidneys on plain films (Berns 1989).

Most cases of contrast-associated renal failure are unrecognized because serum creatinine concentration is not systematically monitored following radiocontrast administration.

Predisposing factors

A pure contrast medium-induced nephropathy almost never involves patients with normal baseline renal function. If renal failure appears in such patients after contrast administration, other factors must be looked for: for example, myeloma cast nephropathy (Vix 1966) or acute renal ischaemia complicating anaphylactoid or cardiogenic shock. When the radiological procedure has required intra-arterial catheterization and/or selective injection or interventional radiology procedure, anuria may result from renal artery dissection or embolization. Renal failure may also be due to multiple intrarenal cholesterol microemboli. This is characterized by insidious onset of progressive renal failure following an intra-arterial contrast study, associated with peripheral extrarenal microembolization, particularly involving the extremities and retina (Chapter 8.1).

Pre-existing renal failure (serum creatinine concentration >140 μmol/l) is, therefore, the most common predisposing factor. The risk is very low (<2 to 4 per cent) at normal serum creatinine levels, and rises exponentially as basal serum creatinine concentrations increase (Berns 1989).

The second most important risk factor appears to be diabetes mellitus but only if it is associated with nephropathy and chronic renal failure. The incidence and severity of contrast-induced renal failure are higher in diabetic than in non-diabetic patients (Harkonen and Kjellstrand 1981; Shieh et al. 1982; Parfrey et al. 1989).

Other clinical risk factors which have been identified include dehydration, age over 60 years, concurrent administration of nephrotoxic drugs (Cochran et al. 1983), solitary kidney and multiple myeloma (Berns 1989) (see also Chapter 4.3 for myeloma). The radiological risk factors are the dose of iodine and repeated injections. Overdosage is not clearly defined: severe accidents may occur with low doses and vice versa. The risk is unpredictable in a given patient with slight renal failure. When the dose of iodine injected is more than 1 g/kg body weight in patients with advanced renal failure, the need for a preventive dialysis session following the radiological procedure has been advocated, but there is no study to substantiate this approach (Berns 1989). In contrast, repeated radiocontrast injections over a short period of time seem to be deleterious. At least 5 days, or a return to baseline creatinine, should be allowed between two injections of radiocontrast agents in high-risk patients.

Pathophysiology

The precise pathogenesis of radiocontrast-induced nephropathy remains unclear. Renal biopsies performed in azotaemic patients after injection of contrast medium, whatever its chemical structure, often showed intense cytoplasmic vacuolization of proximal tubular cells (called 'osmotic nephrosis') (Moreau et al. 1980). It is now definitely established that the effects of these lesions on renal function are minimal or absent, even though enzymuria is evidence of tubular cell injury. Neither acute tubular necrosis nor glomerular changes has been demonstrated in man.

Contrast medium injected into the renal artery induces renal vasoconstriction, following a transient vasodilatory phase. Angiotensin II antagonist and calcium-channel blockers blunt the vasoconstrictor response in animals (Berns 1989).

The absence of a reproducible animal model resembling the clinical syndrome has limited our understanding of pathophysiology. New models have recently been developed based on the assumption that combined renal insults are often involved in triggering radiocontrast nephropathy. Low sodium diet, indomethacin and meglumine iothalamate produced reversible renal failure in rabbits; this was ascribed to a decrease in glomerular ultrafiltration coefficient and was prevented by 1 week of saline and desoxycorticosterone (Vari et al. 1988). An acute infusion of indomethacin and sodium iothalamate induced acute renal insufficiency in salt-depleted uninephrectomized rats. The authors stressed the role of selective hypoxic damage to the thick ascending limbs of Henle's loops in the outer medulla, and pointed to the beneficial role of frusemide in preventing medullary damage (Heyman et al. 1988, 1990).

Other hypotheses which have been put forward to account for radiocontrast nephropathy include intratubular obstruction and

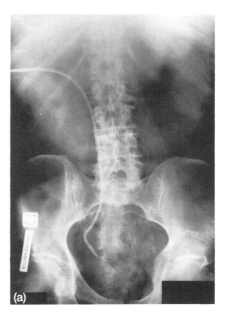
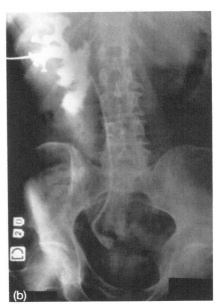

Fig. 1 Right retroperitoneal fluid collection resulting from a leakage of urine through the pelviureteric junction after percutaneous nephrostomy and JJ ureteric catheterization during treatment of carcinomatous stenosis of the pelvic ureter. (a) Plain film of the abdomen: the external margin of the right psoas muscle and the homolateral renal outline are not visible, unlike the ipsilateral ones. (b) Antegrade pyelography: opacification of the fistula and of the fluid collection.

microcirculatory erythrocyte sloughing, but these are not supported by strong clinical or experimental evidence. The development of antibodies to contrast medium has very rarely been related to the occurrence of acute renal failure (Berns 1989).

Prevention

Recommendations

The following recommendations can be proposed.

1. Identify high-risk patients (pre-existing azotaemia, diabetic renal failure, severe cardiovascular disease, jaundice, or multiple myeloma).
2. Discuss the rationale and tactics of investigation with the radiologist to weigh risks and benefits for each patient.
3. Minimize contrast volume and space procedures (see above).
4. Avoid volume depletion and ensure adequate hydration and slight volume expansion with half-normal saline or sodium bicarbonate solution, with or without the addition of mannitol or frusemide, before and for several hours after the administration of contrast material.
5. Discontinue the concomitant use of other potentially nephrotoxic medications, such as non-steroidal anti-inflammatory drugs.

These recommendations have not completely prevented radiocontrast renal failure, but have probably contributed to its decreased incidence, as shown in the recent study by Parfrey *et al.* (1989) in diabetic patients with pre-existing renal insufficiency in whom 0.45 per cent saline solution was infused for 2 days after the imaging procedure.

New radiocontrast agents

The nephrotoxicity of new radiocontrast agents (ionic dimers and non-ionic monomers) has been amply demonstrated (McClennan 1987; Schwab *et al.* 1989; Deray *et al.* 1990). Prospective studies in high-risk patients have shown a similar incidence of increase in serum creatinine with iopamidol (a non-ionic contrast medium) and with an ionic radiocontrast agent (Schwab *et al.* 1989; Berns 1989).

Techniques in conventional uroradiology

The only diagnostic imaging technique which can provide anatomical and physiological information on the whole urinary tract is intravenous urography, starting with the plain films of the abdomen and the pelvis (Moreau and Affre 1979; Moreau and Mazzara 1983). The radiologist should be aware that he will often not find what the physician expects but rather will find unexpected abnormalities.

Plain films

The plain film should be set on the umbilicus; the upper margin should be over the adrenal glands, and the lower margin should be under the intraperineal part of the urethra in males. Since this field is rarely covered by the largest films in adults complementary exposures should be taken, before the contrast medium injection. When body weight is high enough, the external margin of the psoas muscle and the renal outlines are spontaneously visible and renal tomography can be performed successfully; when there is no retroperitoneal fat or when there is a retroperitoneal fluid collection, the renal outline and the psoas muscle cannot be shown inside the hydric shade of grey given by the surrounding solid organs, even on the best tomograms (Fig. 1(a,b)). Gaseous structures should be carefully examined: it is more difficult to detect the radiolucency of fistulas or of emphysematous lesions than it is to detect radio-opacities.

Radio-opaque urolithiasis is only detectable on plain films and is easily seen when the calcium stone concentration is high, and/or when it is located on non-skeletal structures, and/or when it is not masked by intestinal gas. The homolateral posterior oblique film (true profile of the kidney) is better than a profile of the abdomen; on the latter, the kidneys are superimposed on the spine and on the pelvic bones; on the true profile of the kidney, the psoas external margin is visible and the ureter is localized on the vertical line drawn from the middle of the renal sinus. Low-opaque renal calculi are shown better by tomography. Spontaneously opaque calculi vanish on the urogram if the radio-opacity of the iodinated urine is nearly similar (Felson's silhouette sign): the injection of contrast medium should,

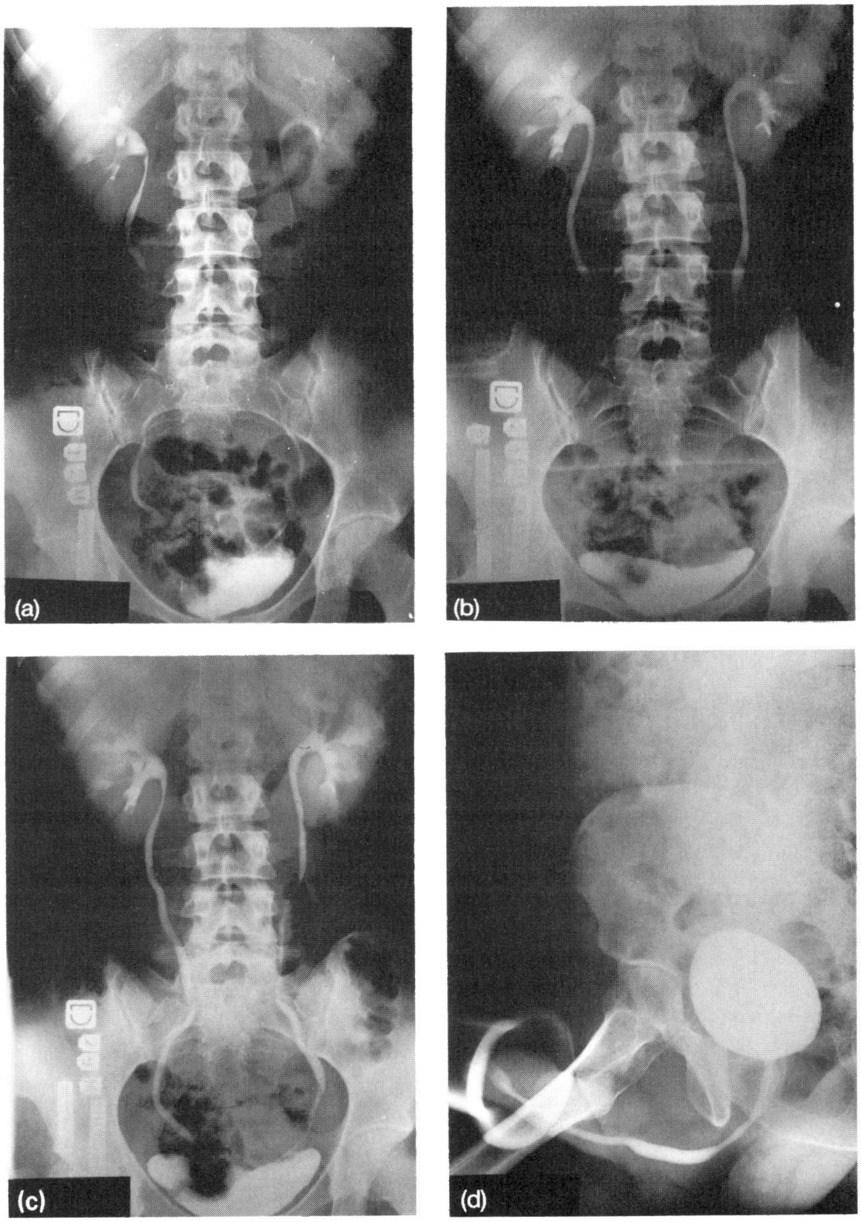

Fig. 2 Normal intravenous urogram in a male patient. (a) Early film (12th minute following the beginning of IV injection) exhibiting the lack of distension of the upper urinary tract and a normal posterior face of the bladder.
(b) Distension of the upper tract during efficient placement of a compression device on the iliac ureters. (c) Excellent visibility of both ureters immediately after decompression. (d) The last stage is the micturating urethrocystogram showing a normal urethra.

therefore, not be performed before the plain film(s) has been interpreted definitely.

Intravenous urography

Contrast medium can be injected in to all kinds of veins; urograms can be obtained after arteriography or computerized tomography, and timing starts with the beginning of the injection. The calibre of the vessel and the flow rate of the contrast medium determines whether bolus injection, slower injection, or drip infusion is best. Bolus administration allows (i) opacification of the vein; when retroperitoneal fibrosis is suspected it is useful to opacify the inferior vena cava through the iliac vein; (ii) opacification of the abdominal aorta and its renal branches, the top of which is obtained by digital subtraction; (iii) a sophisticated nephrographic study; the arterial stage is characterized by well defined corticomedullary differentiation, the tubular stage by a homogeneous nephrogram; (iv) a well set minute sequence pyelogram, which normally appears simultaneously on both sides approximately 2.5 min after the beginning of the injection; (v) a denser total body opacification. The radiologist is free to determine how to manage the sequence, according to his interpretation of the earliest exposures. These should normally be taken without ureteric compression (Fig 2(a)); they enable evaluation of the synchronization of tubular excretion, ureteric dynamics, and the posterior face of the bladder. Delayed and non-aggressive ureteric compression is used for the morphological study of the pelvicalyceal system (Fig. 2(b)); an exposure taken immediately after the release of the compression device often provides an excellent filling of both ureters (Fig. 2(c)). If the amount of iodine is large enough, the last stage should be a voiding urethrocystogram, especially in males (Michel 1989) (Fig. 2(d)). According to the earliest findings and to the clinical pattern, complementary injections of contrast media, oblique and profile views, nephrotomographies, wash-out test with frusemide (Fig. 3), or late films can be ordered.

Fluoroscopic study is rarely useful during the procedure.

1.6.2 Conventional uroradiology and contrast media

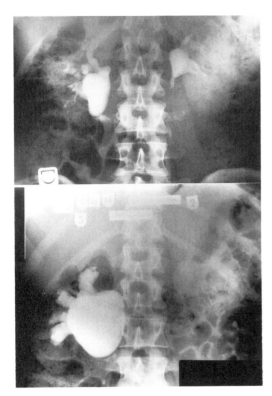

Fig. 3 Intermittent hydronephrosis inducing right acute renal colics. Upper panel: conventional intravenous urogram after injection of 60 ml of contrast medium after a 6-h dehydration. Lower panel: after re-injection of 60 ml of contrast medium and intravenous injection of frusemide, normal wash-out of the left pyelogram and acute distension of the right pelvicalyceal system coexisting with acute pain.

Improvements in film and screen technology have decreased the irradiation of the gonads; however, in children the testicles should be covered with lead sheet. Intravenous urography is contraindicated in pregnant women, if the investigation can be delayed to the postpartum stage; the prohibition of pelvic radiography during the second half of the menstrual cycle has also recently been reviewed (Harvey *et al.* 1985; Hopton *et al.* 1985).

Retrograde and antegrade ureteropyelography

Improvement in intravenous urography techniques and the development of new technologies have decreased the indications for retrograde ureteropyelography, which requires cystoscopy. Its major application is in the preoperative visualization of the distal part of the ureter beyond a stenosis. Under cystoscopy it is easy in the female but often difficult in the male, in whom it may be associated with traumatic and infectious complications (prostatitis, pyonephrosis, septicaemia). Antegrade ureteropyelography can be performed by percutaneous puncture of the pelvis or through the catheter used for percutaneous nephrostomy (Fig. 1(b)); it shows the proximal part of the ureter over the stenosis.

Retrograde cystography

This the most convenient means of exploring the lower urinary tract in females. After catheterization of the urethra, the bladder is filled under pressure; in our unit, exposures are taken sequentially to avoid the risk of misdiagnosing intermittent passive reflux. A voiding urethrocystogram, aimed at detecting active reflux and/or urethral abnormalities, should be performed systematically (Fig. 4(a, b)); the postmicturition film should be performed immediately after the end of the micturition, because many active reflexes develop during its latest stages (Fig. 4(c)). Video recording of micturition may be fruitful, provided that the diameter of the screen is large enough to show a large part of the upper urinary tract; this is more easily obtained in children than in adults. In female patients investigated for upper urinary tract infection, we recommend performing retrograde and micturating cystography immediately followed by a short intravenous urogram; this aims to show reflux and renal damage in the same procedure. Complications due to excessive pressure in the urinary tract, superinfection, or urethral trauma are less rare in males than in females.

Retrograde urethrocystography

In females, this is useful only for the detection of fistulae or urethral diverticuli; we recommend catheterization of the urethral meatus by a virgo hysterographic device. In males, it is rarely indicated if other techniques of micturating urethrocystography are available very often the prostate is studied by ultrasound (Chapter 1.3). The major risk is infection and leakage of fluid through the mucosa.

Suprapubic urethrocystography

Suprapubic puncture of the bladder followed by opacification through a flexible thin catheter is our favorite technique in males (Fig. 5). Failures are very rare, provided that the bladder is full at puncture, which can be assessed by ultrasound; the risks of infection are minimal, even when there is a leak of urine into the peritoneal cavity or the Retzius space; the urethra and the bladder neck are not injured and urodynamics can be studied without artefacts. Repeated fillings can be made for different views of the lower urinary tract.

X-ray exposure

Gonads, and the embryo of a pregnant woman, are exposed to the potentially deleterious effects of X-rays. In males gonad protection can be afforded, at least partially, by sheet lead covering the testicles, as recommended by paediatric radiologists, and by precise setting of the X-ray beam. Ovarian exposure in females cannot be prevented by such methods, except when the study is restricted to the kidneys. The genetic risk, which is not clearly defined, even in children, is today rather limited by the improvement in films and in screens which enable a reduced dose to be delivered during examination. The total dose delivered during an average intravenous urogrant is around 10 to 15 mGy; moreover, most uroradiological procedures do not require prolonged fluoroscopy. One should not forget that X-ray doses are cumulative; thus repeated examinations are more dangerous when they are started in infants and children.

The risk of fetal exposure *in utero* is of clinical concern in sexually active women. Recent studies, although based on small numbers, have suggested that low dose prenatal irradiation may increase the risk of childhood cancer (Harvey *et al.* 1985; Hopton *et al.* 1985). Embryos *in utero* can be killed with high doses, above 3.5 Gy. There is a risk of radiation-induced congenital malformation when the dose is up to 1 cGy. In practice, pelvic

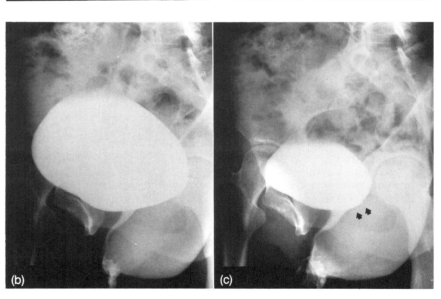

Fig. 4 Repeated episodes of left acute pyelonephritis in a young adult female related to a left active vesicoureteric reflux. Diagnosis by retrograde cystography followed by voiding urethrocystography. (a) No passive reflux is found on the premicturating film. (b) The reflux is not visible on the early film. (c) Opacification of the last few centimetres of the ectopic left ureter from the initial urethra, on the late film (double arrows). (d) The ureteric reflux is visible on the postvoiding film.

exposure should be limited to the strict minimum in the known pregnant woman. The main risks involve the earliest stages of an as yet undetected pregnancy: many radiologists avoid all procedures requiring pelvic irradiation after the tenth day following the beginning of menstruation. This is not necessary in non-sexually active women or in those using effective contraceptive measures. Early therapeutic abortion can be discussed in women submitted to radiological examinations before pregnancy was known; the law and attitudes regarding this vary according to country and individuals (Moreau et al. 1988).

Normal intravenous urography

On the plain film, the skeleton is normal. There is no abnormal gas filling, abnormal hydric mass, or abnormal metallic opacity. After bolus injection of concentrated contrast medium, there are no vascular abnormalities. Total body opacification shows no opaque mass. The opacification of both pelvicalyceal systems occurs 2.5 min after the injection is started. Both kidneys are located normally in the lumbar fossae, with their longitudinal axes parallel to the psoas lines, and at distances proportional to the amount of retroperitoneal fat. The bipolar distances are between three and four lumbar vertebrae; the left kidney is slightly longer (about 10 mm) and situated at a higher level than the right kidney, but both are of similar width (about half of the length). The renal margins are well defined all along their perimeters, and corticomedullary differentiation is also well defined and without defects or hyperdensities. Tubular nephrography is homogeneous. The ratio between the bipolar distances and the upper–lower calyx distances is near 0.5. The radiolucency of intrasinusal fat is homogeneous.

Each calyceal cup is concave, with normal papilla, and all calyces located in the same plane feature Hodson's lines parallel to the renal outlines; minor calyces drain into normally located

1.6.2 Conventional uroradiology and contrast media

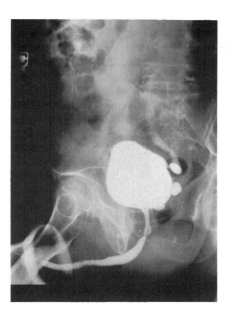

Fig. 5 Stenosis of the distal penile urethra after renal transplantation of the left iliac fossa, demonstrated by suprapubic cystography.

major calyces. The renal pelvis is triangular on both sides; the lower margins are smoothly concave above the lower lip of the renal sinus. The volumes of the pelvicalyceal systems are normal and the densities of both pyelograms are compatible with a normal glomerular filtration rate and with an average osmotic diuresis rate.

The lumbar ureters cross the transversal apophyses of the three lower lumbar vertebrae and are opacified early; their function is shown by transient asymmetrical peristaltic waves and rapid bladder filling. The iliac ureters run inward toward from the sacroiliac joints; the pelvic ureter curves are convex outward and nearly parallel to the margin of pelvis fossa.

The posterior face of the bladder is also opacified early. The inferior part of the retroureteric fossa is symmetrically convex above the intramural tip of the ureters. The V-shape of the bladder neck, projected on the coccyx, is located symmetrically along the longitudinal axis of the body, above the inferior part of the bladder contour. The latter is slightly above and parallel to the upper margins of the pubic bones. The bladder horns are symmetrical.

When the bladder is filled, opacity extends into the abdomen and projects on the sacral bone. Micturating urethrocystography shows normal opening of the bladder neck and active urine flow, which dilates the urethra uniformly from the neck to the meatus. In a normal male, the supramontanal, montanal, and inframontanal thirds of the posterior urethra are of equal length. The oval radiolucency of the veru montanum regularly appears on the posterior face of the montanal urethra. There is no evidence of vesicoureteric reflux or postvoiding residual urine.

Radiological signs and syndromes

Renal mass

The urographic signs result from the size and the shape of the mass, which is usually spherical or oval (never crescentic), and from its localization. The major signs are deformity of the renal contour; displacement and/or compression and/or elongation of the pelvicalyceal system. The bulge of the external margin extends contrary to the pelvicalyceal displacement. When a compressed cavity is parallel to the radiographic film, it looks enlarged with a lower radio-opacity of the urine; conversely, when it is perpendicular to the film, the cavity is narrow and densely opaque.

Ultrasonography and computed tomography have decreased the use of intravenous urography in the investigation of the renal masses. The urographic diagnosis of benign masses should be discouraged: malignancy is suspected when a mass is seen as an invasion of the cavities (calyceal amputation or filling defect). Whatever their features, calcifications are initially suggestive of carcinoma, and of hydatic cyst in endemic areas.

Obstructive syndrome

Ultrasonography is usually diagnostic in acute renal colic. The usefulness of intravenous urography during the painful episode is controversial, although it has the best accuracy: the occlusion is assessed on the homolateral delayed opacification of the distended urinary tract. Low-osmolality contrast medium injection decreases the risk of urine leakage through ruptured fornices; this complication is, however, benign.

There are several patterns of chronic obstruction. Minor obstructions can induce the so-called persisting 'standing ureter' or an intermittent hydronephrosis; there is no delayed opacification; depending on the degree of urinary flow, dilatation of the excretory duct above the obstacle is total or segmental. Investigations can benefit from a frusemide wash-out test. In more severe obstructions, the prognosis is closely related to the rate of atrophy of the papillae, which makes the calyceal base flat or convex; the distended excretory duct is enlarged and elongated. Delayed opacification of the calyces and the 'crescent sign' suggest a distension; these abnormalities vanish when the treatment is successful, whereas the dilatation often remains more or less unchanged (hypotonic dilatation). End-stage obstruction destroys the kidney; the urographic pattern of the obstructed 'silent' kidney shows a thin peripheral slice of renal parenchyma surrounding an enormous radiolucent dilatation of pelvicalyceal system.

Large and small kidneys

The causes of large and small kidneys are summarized in Tables 1 and 2.

Filling defects in the collecting system

According to the location of the cause, filling defects are seen as radiolucencies within the column of opacified urine or as localized impressions on the outline. Only intraluminal gas is visible on the plain film and on the urogram. The other forms are detected on urograms only; they are better studied with low osmolality contrast medium (Fig. 6). Three mechanisms can be suggested; (i) intraluminal radiolucent foreign bodies, all surrounded by radiopaque urine such as radiolucent calculi, blood clots, or sloughed papillae; (ii) localized thickening of the wall of the urinary tract, for example primary or secondary tumours, pseudotumours, intramural oedema, or parasitic granulomas; (iii) extrinsic compression, usually by normal or abnormal blood vessels compressing on the upper tract, or by prostatic lesions affecting the lower tract. In practice, the major difficulties arise from the inability to differentiate transitional cell carcinoma

Table 1 Enlarged kidneys

1. Normal long and narrow kidney(s), hypertrophied column(s) of Bertin (duplex kidney)
2. Thickened parenchyma
   2.1 Unilateral harmonious
       Compensatory hypertrophy
       Renal vein thrombosis
       Unilateral parenchymal infiltration (oedema, infiltrating tumours, lymphoma, Hodgkin's disease)
   2.2 Unilateral inharmonious
       Pseudotumoral compensatory hypertrophy (for example in reflex nephropathy)
       Renal masses (tumours, pseudotumours, cysts)
   2.3 Bilateral harmonious
       Constitutional large kidneys
       Diffuse renal infiltration (oedema, fat, intratubular obstruction, for example by uric acid crystals, malignant process such as lymphoma)
       Polycystic kidney disease
   2.4 Bilateral inharmonious (multiple masses)
       Polycystic kidney disease
       Multiple cysts
       Multiple renal metastases
       Malignant haemopathies
       Multiple angiomyolipomas
3. Distended pelvicalyceal system
4. Enlargement of the sinus
   Intrasinusal masses (such as parapelvic cyst)
   Pseudotumuoral fibrolipomatosis of the sinus

Table 2 Small kidneys

1. False-positive small kidney (rotation on the transverse axis)
2. Unilateral
   2.1 Harmonious proportional
       Congenital renal hypoplasia
       Stenosis of the renal artery
       End-stage thrombosis of the renal vein
       Radiation nephritis
   2.2 Harmonious non-proportional
       Obstructive pyelonephritis
       Chronic interstitial nephritis with papillary necrosis
   2.3 Inharmonious
       Congenital renal aplasia
       Reflux nephropathy, Ask-Upmark kidney, segmental congenital hypoplasia
       Ectopic kidney
       Aseptic papillary necrosis
       Non-obstructive pyelonephritis (infection, urolithiasis, staghorn calculus)
       Focal obstructive pyelonephritis
       Segmental renal infarction
       Post-traumatic or postsurgical renal atrophy
       Renal tuberculosis
3. Bilateral
   Symmetrical: chronic nephritides
   Asymmetric: bilateral involvement by lesions described in 2

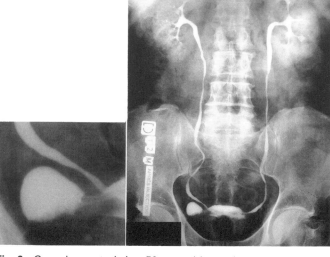

Fig. 6 Gross haematuria in a 50-year-old man. Intravenous urogram with 100 ml Hexabrix 320. Normal upper urinary tract. Transitional cell carcinoma of the lower margin of the right diverticulum of the bladder (see insert). Note that the intradiverticular lesion could be missed with cystoscopy.

from a uric acid stone—ultrasonography and computed tomography may be helpful. If these investigations are negative, it is wise to repeat high-contrast intravenous urography after a week's course of alkalinization of the urine, which dissolves pure uric acid stones, before considering an antegrade or retrograde ureteropyelography.

Extraluminal collections of iodinated urine.

Abnormal collections of iodinated urine result from opacification of congenital or acquired cavities communicating with the collecting system. When they develop on the kidney, they project laterally to Hodson's interpapillary line. There are two major mechanisms: when the lesion extends from the duct to the outer space the cause is diverticulum or fistula. When the lesion extends in the opposite direction the cause is an aseptic or infectious necrotic focus (Fig. 7(a,b)).

1.6.2 Conventional uroradiology and contrast media

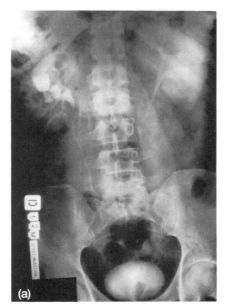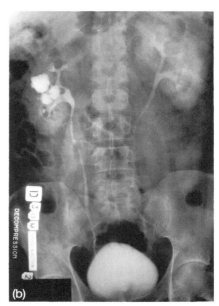

Fig. 7 Tubercular sequelae of the right kidney. (a) Low contrast urogram with 60 ml of Telebrix 380 (hyperosmolar compound). (b) Extemporaneous enhancement of the contrast by addition of 60 ml of Hexabrix 320 (low osmolarity compound).

References

Abrams, H.L. (1971). Introduction and historical notes. In *Angiography*, Vol. 1, (ed. H.L. Abrams), pp. 3–13. Little, Brown, Boston.

Belleville, J., Baguet, J., Paul, J., Clendinnen, G., and Eloy, R. (1985). In vitro study of the inhibition of coagulation induced by different radio-contrast molecules. *Thrombosis Research*, **38**, 149–62.

Berns, A.S. (1989). Nephrotoxicity of contrast media. *Kidney International*, **36**, 730–40.

Brezis, M. and Epstein, F.H. (1989). A closer look at radiocontrast-induced nephropathy. *New England Journal of Medicine*, **320**, 179–81.

Cochran, S.T., Wong, W.S., and Roe, D.J. (1983). Predicting angiography induced acute renal function impairment: clinical risk model. *American Journal of Roentgenology*, **14**, 1027–33.

Collins, R.D. Jr. (1989). Catastrophic arterial thrombosis associated with an unsuspected anticardiolipin antibody, following pyelography. *Arthritis Rheumatism*, **32**, 1490–1.

Dawson, P., Hewitt, P., Mackie, I.J., Machin, S.J., Amin, A., and Bradshaw, A. (1986). Contrast, coagulation, and fibrinolysis. *Investigative Radiology*, **21**, 248–52.

Deray, G., et al. (1990). Néphrotoxicité comparée des produits de contraste iodés de haute et basse osmolalité: aspects expérimentaux et cliniques. In *Séminaires d'uro-néphrologie*, No. 16, (eds. C. Chatelain and C. Jacobs), pp. 193–206. Masson, Paris.

Derobert, L., Wolfromm, R., Dehouve, A., and Lange, R. (1964). Les accidents graves par injection intraveineuse de substances iodées pour urographie. *Annales de Médecine Légale*, **44**, 330–45.

Errfmeyer, J.E., Siegle, R.L., and Lieberman, P. (1986). Anaphylactoid reactions to radiocontrast material. *Journal of Allergy and Clinical Immunology*, **75**, 401–10.

Harkonen, S. and Kjellstrand, C. (1981). Contrast nephropathy. *American Journal of Nephrology*, **7**, 69–77.

Hartman, G.W., Hattery, R.R., Witten, D.M., and Williamson, B. Jr. (1982). Mortality during excretory urography: Mayo Clinic experience. *American Journal of Roentgenology*, **139**, 919–22.

Harvey, E.B., Boice, J.T. Jr, Honeyman, M., and Flannery, J.T. (1985). Prenatal X-ray exposure and childhood cancer in twins. *New England Journal of Medicine*, **312**, 541–45.

Heyman, S.N., Weinstein, J.M., Rosen, S., and Brezis, M. (1990). Current concepts on the pathophysiology of radiocontrast nephrotoxicity. In *Séminaires d'uro-néphrologie*, No. 16, (eds. C.Chatelain and C. Jacobs), pp. 188–92. Masson, Paris.

Hopton, P.A., et al. (1985). X-rays in pregnancy and the risk of childhood cancer. *Lancet*, **ii**, 773.

Jacobson, P.D. and Rosenquist C.J. (1988). The introduction of low-osmolar contrast agents in radiology: medical, economic, legal and public policy issues. *Journal of the American Medical Association*, **260**, 186–92.

Katayama, H., et al. (1988). Clinical survey on adverse reactions of iodinated contrast media. The Japanese Committee on Safety of Contrast Media. Scientific Poster. *Radiological Society of North America Annual Meeting*, Chicago, Illinois, U.S.A.

Lalli, A.F. (1980). Contrast media reactions: data analysis and hypothesis. *Radiology*, **134**, 1–12.

Lasser, E.C., et al. (1987). Pretreatment with corticosteroids to alleviate reactions to intravenous contrast material. *New England Journal of Medicine*, **317**, 845–9.

McClennan, B.L.(1987). Low-osmolality contrast media: premises and promises. *Radiology*, **162**, 1–8.

Michel, J.R. (1989). *Radiologie de l'urètre*. Masson, Paris.

Moreau, J.F. and Affre, J. (1979). *L'urographie intraveineuse*. Flammarion Médecine, Paris.

Moreau, J.F. and Mazzara, L. (1983). *Intravenous Urography*. Wiley, New York.

Moreau, J.F., Droz, D., Noel, L.H., Leibowitch, J., Jungers, P., and Michel, J.R. (1980). Tubular nephrotoxicity of water-soluble iodinated contrast media. *Investigative Radiology*, **15**, S54–60.

Moreau, J.F., Lesavre, P., de Luca H., Hennessen, U., Fischer, A.M., and Giwerc, M. (1988a). General toxicity of water-soluble iodinated contrast media. Pathogenic concepts. *Investigative Radiology*, **23** (Suppl 1), S75–78.

Moreau, J.F., Droz, D, Giwerc, M., and Chabriais, J. (1988b). Nephrotoxicity of the iodinated contrast media. In *Contrast Media* (ed. D. Carr), pp. 66-77, Churchill Livingstone, Edinburgh.

Moreau, J.F., Pasco, A., Morel P., and Richard, O. (1988c). Radiologie diagnostique: risques chez la femme enceinte. In *Pathologies médicales et grossesse*. (eds Weschler, B., Janse-Marec, J., and Péchère, J.C.), Medsi-McGraw Hill, Paris.

Neoh, S.H., Sage, M.R., Willis, R.B., Robert-Thompson, P., and Bradley, J. (1981). The in vitro activation of complement by radiologic contrast media and its inhibition with ε aminocaproic acid. *Investigative Radiology*, **12**, 152–62.

Parfrey, P.S., et al. (1989). Contrast material-induced renal failure in patients with diabetes mellitus, renal insufficiency, or both: a prospective controlled study. *New England Journal of Medicine*. **320**, 143–9.

Pinet, A., Lyonnet, D, Maillet, P., and Groleau, J.M. (1982). Adverse reactions to intravenous contrast media in urography. Result of a national survey. In *Contrast media in radiology* (ed M. Amiel), pp. 14-15. Springer, Berlin.

Rice, M.C., Lieberman, P., Siegle, R.L., and Mason, J. (1983). In vitro histamine release induced by radiocontrast media and various analogs in reactor and control subjects. *Journal of Allergy and Clinical Immunology*, **72**, 180–6.

Robertson, H.J.F. (1989). Thromboembolic complications in coronary angiography associated with the use of nonionic contrast medium. *Catheterization and Cardiovascular Diagnosis*, **16**, 145–6.

Robertson, H.J.F. (1987). Blood clot formation in angiographic syringes containing nonionic contrast media. *Radiology*, **163**, 621–2.

Schwab, S.J., *et al*. (1989). Contrast nephrotoxicity: a randomized controlled trial of a nonionic and an ionic radiographic contrast agent. *New England Journal of Medicine*, **320**, 149–53.

Shehadi, W.H. and Toniolo, G. (1980). Adverse reactions to contrast media: a report from the Committee on Safety of Contrast Media of the International Society of Radiology. *Radiology*, **137**, 299–302.

Shieh, S.D., *et al*. (1982). Low risk of contrast media-induced acute renal failure in nonazotemic type-2 diabetes mellitus. *Kidney International*, **21**, 739–43.

Sovak, M. (ed.) (1984). *Radiocontrast Agents*. Springer, Berlin.

Strain, W.H. (1971). Historical development of radiocontrast agents. In *International encyclopedia of Pharmacology and Therapeutics, Section 76, Vol. 1. Radiocontrast Agents*, pp. 1–22, Pergamon, Oxford.

Vari, C.R., Natarajan, L.A., Whitescarver, S.A., Jackson, B.A., and Ott, C.E. (1988). Induction, prevention and mechanisms of contrast media-induced acute renal failure. *Kidney International*, **33**, 699, 702.

Vix, V.A. (1966). Intravenous pyelography in multiple myeloma. Review of 52 studies in 40 patients. *Radiology*, **87**, 896.

Westaby, S., Dawson, P., Turner, M.W., and Pridie, R.B. (1985). Angiography and complement activation. Evidence for generation of C3a anaphylotoxin by intravascular contrast agents. *Cardiovascular Research*, **19**, 85–8.

Yamagushi, K., Katayama, H., Kozuka, T., Takashima, T., and Matsuura, K. (1989). An evaluation of presenting in the problem of severe adverse reactions to ionic and non-ionic contrast media in Japan. Communication at the *Contrast Media Research 1989 Symposium*, Sydney, Australia, No. AR5.

1.6.3 Renal ultrasonography (including Doppler echography)

PAUL A. DUBBINS

The use of high-frequency sound waves for the diagnosis of renal disease was first described for the differentiation of solid from cystic renal lesions. There have been many advances in technology and thereby in the applications of ultrasound since this time. While it is beyond the scope of this chapter to describe the physics of ultrasound, it is important to introduce some basic principles, although for the purposes of brevity this section will resemble perhaps an incomplete glossary of terms.

Ultrasonography: basic principles and aspect of the normal kidney

Basic principles of ultrasound

In contrast to most imaging modalities which rely upon the transmission of electromagnetic wave forms, ultrasound is a pulse-echo technique, reflections of sound at acoustically different tissues being returned to the transmitting transducer. Sound is attenuated in its passage through tissue, and therefore, in order to maintain uniformity of signal throughout an organ, there is differential amplification of sound which increases with depth. This explains one of the diagnostically important artefacts that occur with ultrasound imaging. When sound passes through a structure which attenuates sound only poorly, such as a cyst, the intensity of sound being reflected from structures distal to the cyst is over-compensated for by the amplifier producing through transmission, or 'bright up', beyond the cyst. Certain substances such as bone, gas, and calcium will attenuate the sound beam so effectively that no sound passes through and an acoustic shadow is produced obscuring all diagnostic information distal to the structure; this acoustic shadow is also of diagnostic importance (see below).

The image produced by an ultrasound machine is a tomographic slice, whose thickness is determined by the thickness of the ultrasound beam, whose length is determined by the traverse of the ultrasound probe or transducer, and whose depth of penetration is determined by the acoustic impedance of the tissues and the frequency of sound transmitted. Higher frequencies of sound produce better resolution of structures along the sound beam but poorer tissue penetration. The frequencies used in adult renal sonography are between 2.5 and 5 MHz and up to 7.5 or even 10 MHz in the sonography of neonate kidneys.

The Doppler effect is the change in frequency of sound that occurs when a sound source is moving. This effect is regularly encountered by the changing pitch of an ambulance siren both towards and away from you. The change in frequency is proportional to the velocity of motion of the sound source. The Doppler effect can be used with ultrasound particularly to assess blood flow velocity because of the reflection of sound from the moving red blood cells. The frequency shift that results is in the audible range but although a simple auditory assessment may distinguish between flow and absence of flow, computer processing of the signal will allow, for example, either a graphical representation of frequency change with time (the Doppler sonogram) or may allow the demonstration of a colour-coded flow image superimposed upon the real time two-dimensional ultrasound image (colour flow mapping). It is possible to calculate peak and mean velocity of blood flow from these data, and

1.6.3 Renal ultrasonography (including Doppler echography)

thereby volume flow if the diameter of the vessel is known. However, errors of measurement of angle of incidence and of diameter may cause errors of calculation of volume flow in excess of 40 per cent. Therefore, for the purposes of clinical evaluation, velocity and volume calculations are not currently performed.

Technique

A knowledge of surface anatomy is important for the proper performance of an ultrasound examination. The kidneys are situated in the retroperitoneum to either side of midline with axes slightly deviated from the longitudinal. The lower poles are more anteriorly and laterally placed than the upper poles. Scanning technique takes account of these positions. The right kidney is demonstrated in long axis by a scan plane parallel to the expected long axis of the kidney with the probe placed on the right hypochondrium. The kidney is imaged posterior to the right lobe of the liver which acts as an acoustic window (a sound passage free of gas and bone). Rotating the probe at right angles to this produces an axial plane of scan. On the left there is no such acoustic window unless there is splenomegaly and there is therefore usually no counterpart of this view for the left kidney. Both kidneys however, may be demonstrated by alignment of the probe to the expected long axis of the kidney placing the probe paraspinally with the patient prone; 90° rotation produces an axial plane of scan. Coronal planes of scan may be achieved with the probe placed on each flank in turn with the patient lying on his side. The ureters are rarely visualized distal to the pelviureteric junction and proximal to the ureterovesical junction except when significantly dilated: an anterior approach following the surface markings of the course of the ureters with abdominal compression to displace gas will then often demonstrate the dilated ureter and associated pathology. Visualization of the distal ureter is via the pelvis using the full urinary bladder as an acoustic window.

Sonographic anatomy of the kidney (Cook et al. 1977)

The appearances of the kidney will vary depending upon the plane of scan. In the sagittal plane the kidney appears as an oval structure with several gross anatomical regions identifiable. It measures between 10 and 12 cm. in long axis (Fig. 1). The renal capsule is a thin brightly reflective structure whose margins are often undiscernable from the surrounding perirenal fat. The renal cortex returns midlevel grey echoes while the medullary pyramids are less echoic or darker. The renal sinus which contains the collecting system, vessels, nerves, and connective tissue as well as a variable amount of fat is usually highly reflective and may attenuate the sound beam sufficiently to produce slight distal acoustic shadowing. High resolution ultrasound may show the branching vascular pattern of interlobar and arcuate vessels. Individual calyces are not usually identified separately within the central sinus but the renal pelvis can usually be seen as an echo-free structure passing medially and anteriorly from the kidney. The proximal ureter is not usually seen. Short segments of the distal ureter in the pelvis may occasionally be demonstrated through a full urinary bladder as tubular structures on longitudinal oblique sections in the line of the ureter or as small ring-like structures just proximal to the ureterovesical junction. The ureteric orifices can often be identified as two discrete elevations at the base of the bladder. (Dubbins *et al.* 1984).

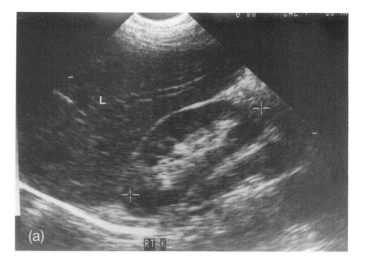

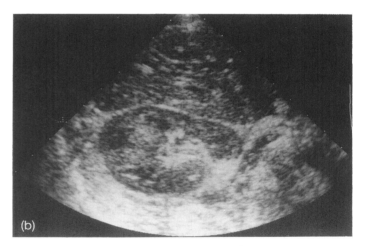

Fig. 1(a) Long-axis scan through the right upper quadrant demonstrating longitudinal section of the right lobe of the liver (L) and the right kidney (between the crosses). The right renal capsule, the low-level grey echoes of the renal cortex, the darker medullary pyramids and the bright right renal sinus are demonstrated. (b) Transverse axis scan through the right upper quadrant demonstrating right lobe of liver and right kidney.

Anatomy of the neonate kidney (Hayden et al. 1984)

The kidney of the neonate (Fig. 2) shows certain important anatomical differences from that of the adult. The relative size of the kidney is greater, (long axis measurement approximately 4 cm): there are charts available relating kidney size to age (Fitzsimons 1983). The size of the medullary pyramids is also relatively increased. There is increased differentiation between cortex and medulla; the renal cortex is more reflective, the medulla less so. The size and reflectivity of the renal sinus are decreased due to paucity of sinus fat.

Characteristics of the normal Doppler sonogram

The kidney is an extremely vascular organ receiving 29 per cent of the cardiac output. The complex network of vessels supplying the glomerular tufts supply little resistance to blood flow and this

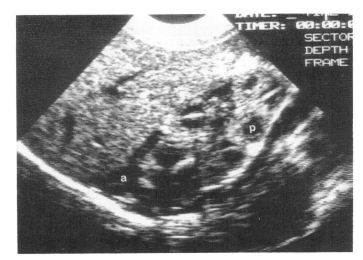

Fig. 2 Long-axis scan through the right upper quadrant of a neonate showing the right lobe of the liver and the right kidney. The margination between the right kidney and the liver is less well defined because of the similar reflectivity of the cortex and lack of prominence of the renal capsule. The relatively large and prominent echo-poor medullary pyramids are noted (p). The renal sinus which contains little fat is not well seen. a = adrenal gland.

is reflected in the characteristics of the Doppler frequency spectrum. High velocities of blood flow are recorded in systole but there is only a gradual decline of low velocities in early diastole and flow continues throughout diastole. This pattern has been linked to a 'ski slope' (Fig. 3). There are many parameters used to quantify the Doppler frequency spectrum. Of these the pulsatility index of Gosling (1975) is perhaps the best known. This is given by the difference between maximum and minimum frequency divided by the mean. For the main renal artery the pulsatility index is usually between 1.2 and 1.6. Similar values and similar Doppler sonograms although of smaller amplitude are seen within the branch arteries of the kidney (Dubbins 1989).

Ultrasound in the evaluation of renal pathology

Renal mass lesions

In spite of the advent of newer imaging techniques such as computerized tomography and magnetic resonance imaging ultrasound remains the cornerstone of diagnosis of a mass lesion either detected clinically or at plain abdominal radiography or intravenous urography.

Cyst versus carcinoma

A renal cyst has several well-defined characteristics. It is smooth in outline with well-demarcated borders. Its fluid content is devoid of echoes except occasionally for those adjacent to its anterior margin due to reverberation of sound. There is increased brightness of echoes beyond the cyst known as through transmission due to inappropriate amplification (Fig. 4) (Pollack *et al*. 1982).

By contrast, carcinoma of the kidney usually demonstrates an irregular outline, lobulated margins, and a variable content of high and low-level echoes which are at least partly dependent on

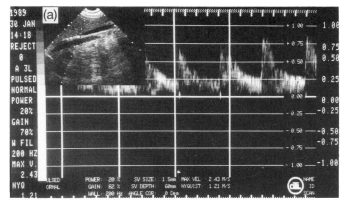

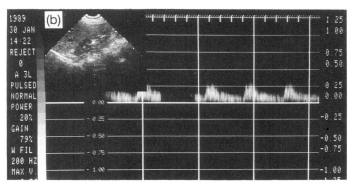

Fig. 3(a) Doppler sonogram scan from the main renal artery. In the left upper corner there is an image of the sampling site showing the parallel echo-free tubular structures of inferior vena cava and aorta. The Doppler signal is being sampled from the right renal artery which courses from the aorta underneath the inferior vena cava. The resulting signal is a classical pattern of a normal renal artery similar to a 'ski-slope'. (b) Doppler sonogram sampled more peripherally within the kidney (interlobar arteries). It shows similar features but lower amplitude.

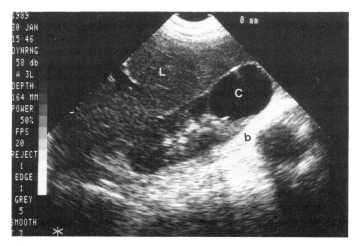

Fig. 4 Long-axis scan through right upper quadrant. The liver (L) is demonstrated. At the lower pole of the right kidney is a renal cyst (C) which expands the outline of the kidney but which is echo free apart from a few reverberation echoes within the anterior portion of the cyst, and has clearly defined walls. Beyond the cyst structures appear unusually bright due to increased through transmission of sound by the cyst (acoustic enhancement) (b).

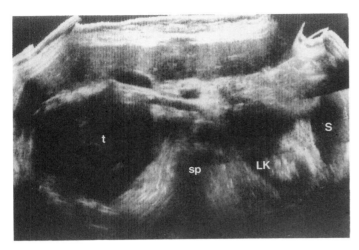

Fig. 5 Transverse scan through the abdomen. There is a large right renal tumour (t) which contains a mixture of echoes but which transmits sound only poorly. The spleen (S), left kidney (LK) and spine are identified (sp).

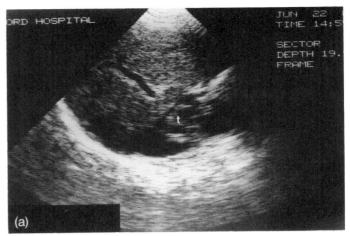

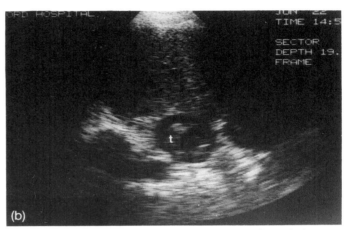

Fig. 7(a) and (b) Long-axis and transverse scans through the inferior vena cava, demonstrating solid material within the inferior vena cava representing tumour invasion (t). The liver is seen immediately adjacent.

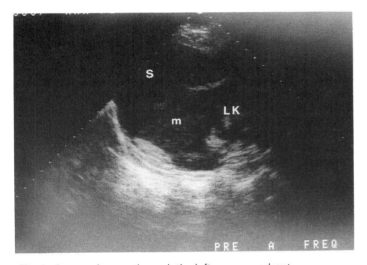

Fig. 6 Long-axis scan through the left upper quadrant demonstrating the upper pole of the left kidney (LK) and the spleen (S), but with a mass at the upper pole (m) whose margins blend with those of the spleen. When the reflectivity of a structure is similar to that of an adjacent structure the margins are difficult to define and invasion difficult to assess. This is termed the ultrasound-masking sign and is similar to the silhouette sign in chest radiology. This was a renal adenoma, completely enclosed by renal capsule.

the degree of tumour necrosis (Fig. 5) (Coleman et al. 1980). There is little or no through transmission of sound. Although ultrasound allows the differentiation of cyst from solid this is the limit of tissue characterization except by the recognition of ancillary features. Primary and secondary tumours may show the same echographic characteristics and indeed benign and malignant tumours of the kidney are indistinguishable unless ancillary features are observed (Fig. 6) (Kurtz et al. 1981). These include breach of the renal capsule, invasion of peri- and pararenal structures and most particularly evidence of tumour within the renal vein and inferior vena cava (Fig. 7) (Plainfosse et al. 1983; Schwerk et al. 1985). This is recognized by the presence of low-level echoes within the renal vein or inferior vena cava usually causing distension of the vessel lumen. Characteristic Doppler signals deriving both from the tumour itself and from the supplying renal artery may provide more information (Fig. 8). Vascular renal tumours with arteriovenous shunting cause increased velocity of flow both in the main renal artery and particularly in the vessels supplying and surrounding the tumour. (Taylor et al. 1988; Dubbins and Wells 1986). These velocity changes are most noticeable in diastole. In hypovascular tumours unless there is complete replacement of normal renal substance there is no effect upon the normal renal Doppler sonogram scan. When there is renal vein occlusion by tumour this causes increased peripheral resistance to arterial blood flow with resultant loss of detectable blood flow in diastole (Fig. 9).

Because of the poor visualization of individual calyces ultrasound is an insensitive investigation for the detection of transitional cell tumours of the kidney except when large (Fig. 10). However, ultrasound may be of value in distinguishing the non-opaque calculus demonstrated by intravenous urography from a small transitional cell carcinoma. Even non-opaque calculi will produce distal acoustic shadowing whereas no such shadowing

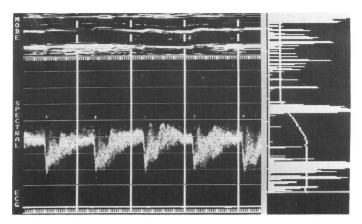

Fig. 8 Doppler sonogram from a renal artery supplying a renal tumour. Blood flow is away from the transducer and therefore the sonogram is negative. There is increased blood-flow velocity throughout diastole with much irregularity of blood flow indicating turbulence.

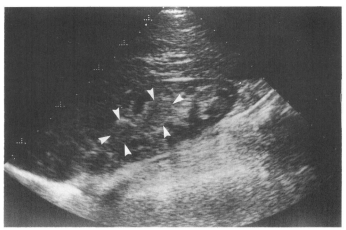

Fig. 10 Long-axis scan through the right upper quadrant showing a large transitional carcinoma (arrowheads); the renal sinus echoes are separated by echogenic material of slightly lesser reflectivity than the renal sinus itself. Transitional cell tumours are difficult to identify but can be easily distinguished from renal calculi by the absence of acoustic shadowing (see Fig. 21); blood clot however, produces similar appearances.

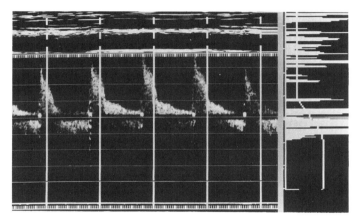

Fig. 9 Doppler sonogram to renal tumour with renal vein invasion but not complete renal vein occlusion: there is reduction in the velocity of blood flow in diastole indicating increased peripheral resistance to blood flow.

derives from either transitional cell carcinoma or clot (Charboneau *et al.* 1983; Joseph *et al.* 1986).

Lymphoma of the kidney (Heiken *et al.* 1983)

Although the kidney involved by lymphoma may be echographically normal the most common feature is the demonstration of ill-defined, often multiple, echo-poor masses. There may be distortion of the renal outline, loss of normal corticomedullary differentiation, and irregularity of the renal sinus echoes.

Pseudotumours (Mahoney *et al.* 1983)

Pseudotumours such as the splenic hump or column of Bertin do not usually provide diagnostic problems since their echographic texture is exactly the same as the rest of the kidney. However, occasionally other investigative techniques such as isotope imaging are required to show the functioning nature of the pseudomass. Colour flow mapping will demonstrate normal blood-vessel appearances in these anatomical variants (unpublished observations).

Cyst aspiration

Because of the increasing specificity of ultrasound in the demonstration of renal cysts, cyst aspiration is now reserved for those patients in whom there is either disagreement between ultrasound and the intravenous urogram with respect to the nature of the renal lesion, or where ultrasound demonstrates an irregularity of cyst margin, or the contents of the cyst are not uniformly anechoic. Where cyst puncture is deemed necessary ultrasound can be used to guide the aspirating needle into the centre of the cyst because the needle tip is visible as a brightly reflecting echo on the ultrasound image. As a general rule the injection of contrast into the cyst cavity is no longer used (Pollack *et al.* 1982).

Cystic disease of the kidney (Grossman *et al.* 1983; Joseph *et al.* 1986)

It is arguable as to whether this topic should be considered under renal mass lesions or in the section on renal failure. However, not all cystic diseases of the kidney progress to renal failure, nor will all cystic diseases present with uraemia.

Autosomal dominant polycystic disease

Although polycystic renal disease may present with renal failure, with hypertension, with haematuria, or as a renal mass lesion or lesions, patients are increasingly referred for ultrasound because of a strong family history of adult-type polycystic kidney disease. The appearances of established polycystic renal disease are characteristic with multiple cysts of varying size replacing normal renal parenchyma (Fig. 11). Good images of the affected kidneys are difficult because the multiple cysts produce complex features of multiple areas of increased through-transmission. If patients are referred before established disease in childhood or in their teens, the signs are often very subtle. Early features of polycystic kidney disease are several small, sometimes even isolated, cysts within the renal parenchyma. It must be remembered that very

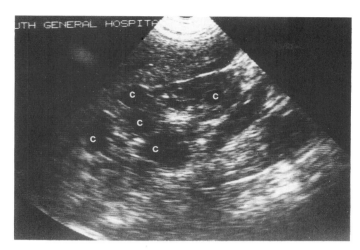

Fig. 11 Long-axis scan through the right upper quadrant demonstrating kidney largely replaced by cysts of varying size (c). The appearances are those of polycystic renal disease.

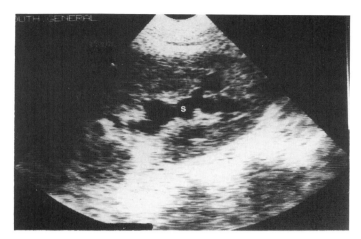

Fig. 13 Long-axis scan through the left upper quadrant, demonstrating mild hydronephrosis. There is slight separation of the echoes from the renal sinus by echo-free fluid (s).

small cysts will be beyond the resolution of ultrasound and thus a normal ultrasound examination particularly before the age of 20 will not exclude subsequent development of polycystic renal disease (see Chapter 18.2.2).

While cyst complications such as haemorrhage, infection, or even tumour may be demonstrated on ultrasound it is frequently very difficult to distinguish between these because all will simply show increase in the echoes of the cyst contents.

Multicystic dysplastic disease

The diagnosis of a multicystic or dysplastic kidney in the neonate is becoming less common as this diagnosis is made antenatally with increasing frequency. The disease if bilateral is invariably fatal but when unilateral and when unassociated with other congenital anomalies is of little prognostic moment. Ultrasound appearances are those of multiple cysts, usually fewer in number than with adult-type polycystic renal disease, completely replacing the renal substance (Fig. 12). Follow up of infants with multi-

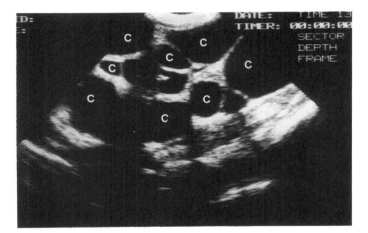

Fig. 12 Long-axis scan through the left upper quadrant, demonstrating multiple cysts of varying size, totally replacing renal tissue. These are characteristic appearances of a multicystic dyplastic kidney in a neonate.

cystic dysplastic kidneys usually demonstrates progressive reduction in size of the cysts. Eventually no cyst or residual renal tissue can be demonstrated on the affected side and the appearances then mimic unilateral renal agenesis.

Infantile polycystic renal disease (Stapleton *et al.* 1983)

This diagnosis is being increasingly made with antenatal ultrasound, although not usually until the third trimester. Termination of pregnancy is therefore unusual. Ultrasound features of infantile polycystic kidneys are increased size with a diffuse increase in reflectivity as a result of the multiple reflections from the walls of the diffusely scattered tiny cysts. Individual cysts are not resolved by ultrasound (Mahoney *et al.* 1986).

Obstructive renal disease (Green *et al.* 1986)
Diagnosis of obstruction

Unlike other imaging modalities such as intravenous urography and isotope renography ultrasound relies entirely upon anatomical features for the diagnosis of obstruction rather than a functional element. In situations therefore, when there is acute obstruction such as ureteric calculus disease, where there may be little or no calyceal dilatation, ultrasound is an insensitive indicator of obstruction (Morin *et al.* 1979). When the obstruction is more long standing, for whatever cause, ultrasound features become characteristic. The central renal sinus echoes become separated by distension of the calyces and pelvis with echo-free urine (Fig. 13). As the hydronephrosis becomes more marked the degree of separation of the central sinus echoes is also increased. In the coronal and transverse scan plane the typical pattern of dilated calyces communicating with a dilated pelvis can be seen (Fig. 14). Gross hydronephrosis may simply be represented by a large irregular echo-free sac.

Pyonephrosis (Yoder *et al.* 1983)

In an infected obstructed kidney the appearances may not differ from those seen in hydronephrosis. Frequently however, the fluid contained within the central sinus may contain low-level echoes as a result of the presence of pus. Occasionally if the patient is immobile there may be layering of the cellular content with a resultant 'fluid/fluid' level (Fig. 15).

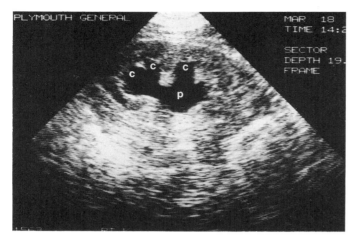

Fig. 14 Moderate hydronephrosis: coronal scan through the left flank, demonstrating dilated calyces (c) communicating with a central echo-poor dilated pelvis (p).

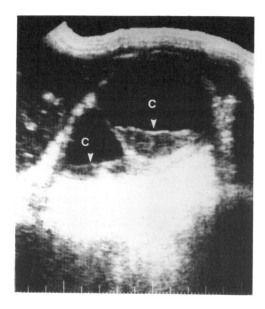

Fig. 15 Right pyonephrosis: the right kidney is represented by two large cystic (c) spaces with a fluid-fluid level (arrowheads) representing layered pus.

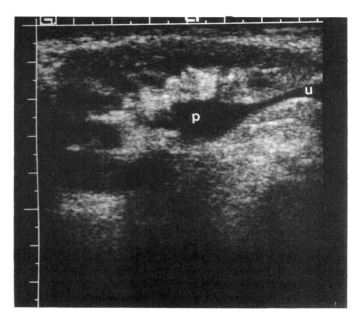

Fig. 16 Coronal scan through the left kidney demonstrating a dilated renal pelvis (p) and proximal ureter (u) in a patient with vesicoureteric reflux.

Level of obstruction

Although the ureters are poorly seen on ultrasound both the level and cause of obstruction can frequently be demonstrated. In pelviureteric junction obstruction, there is a rounded contour to the distal part of the pelvis and no demonstrable proximal ureter. Obstruction of the midureter is more difficult to assess unless there is significant ureteric dilatation (Fig. 16). However, it may be possible to demonstrate causative pathology not visible at the time of intravenous urography: e.g. an echo-poor mantle enveloping the anterior aorta suggesting retroperitoneal fibrosis, and lobulated masses around the aorta and inferior vena cava indicating para-aortic lymphadenopathy.

Although ultrasound is commonly used for evaluation of the pelvic organs it relies upon the presence of a full urinary bladder for adequate demonstration of pelvic structures. If the bladder can be filled either from above if the obstruction is not complete and bilateral, or from below via catheter, ultrasound will demonstrate uterine and ovarian tumours as well as transitional cell bladder carcinomas, or ureteroceles. The transitional cell tumour is characterized by a lobulated mass of midrange echoes usually projecting into the bladder lumen. Transabdominal ultrasound will frequently show the extent, and help in the diagnosis of stage 3 and 4 bladder tumours (Fig 17). More recently the passage of a fine real-time transducer *per urethram* has allowed the more accurate differentiation of stages 1 and 2. A ureterocele has the appearance on ultrasound of a 'cyst within a cyst'.

Benign prostatic hypertrophy

Ultrasound is being increasingly used in the preoperative evaluation of the patient with prostatism. The initial enthusiasm for extensive investigation with intravenous urography and urodynamic studies has been replaced by a more critical assessment of the value of these techniques. There appears to be incomplete correlation between symptoms and surgical outcome with results of urodynamic studies and scant justification to screen just those elderly males with symptoms of prostatism for renal neoplastic disease. Ultrasound will demonstrate features of upper tract obstruction and the residual volume of urine in the bladder, as well as demonstrating most of the renal mass lesions without the need for further evaluation by other imaging techniques. Therefore, if any other preoperative imaging investigation is required in patients with benign prostatic hypertrophy ultrasound fulfils this role best (Dubbins 1989).

Percutaneous nephrostomy

In renal obstruction there are a number of indications for acute decompression of a dilated system. Using ultrasound to guide initial needle placement before guidewire exchange, track dilatation and catheter placement reduces significantly the number

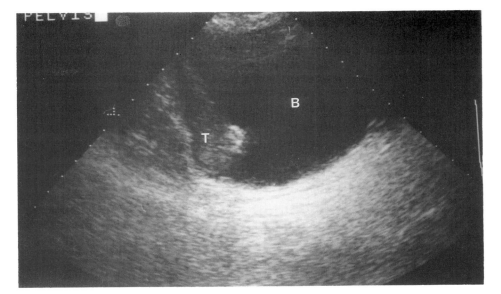

Fig. 17 Long-axis scan through the bladder (B). There is irregular thickening of the fundus of the bladder with a bright, possibly calcified focus, but the appearances are those of a transitional cell carcinoma of the bladder (T).

of needle passes required for satisfactory placement of a nephrostomy drainage catheter, thereby reducing the morbidity of the technique. The procedure is performed under direct ultrasound visualization when the highly reflective needle tip can be seen entering the dilated collecting system (Zegel et al. 1981).

Renal infection (Kuligowska 1986; Morehouse et al. 1984)

Acute pyelonephritis usually produces no significant abnormality of the ultrasound image although occasionally there may be generalized enlargement. Occasionally focal pyelonephritis may produce a focal enlargement of the kidney with reduction in echoes. Gas-producing organisms may result in areas of acoustic shadowing. Chronic pyelonephritis, usually as a result of reflux, may produce cortical scarring with associated calyceal dilatation (Fig. 18).

Renal abscesses have similar appearances to those of a renal cyst when fully liquefied although there may be echogenic material within the fluid and this may layer. In early evolution the renal abscess may just appear as a focal area of reduced reflectivity within the kidney, and extension of the infection will give rise to a collection of fluid in a subcapsular, perirenal, or pararenal location.

Xanthogranulomatous pyelonephritis may result in focal or diffuse enlargement of the kidney. There may be mixed reflectivity, often associated with focal or widespread calyceal dilation. Occasionally the obstructing calculus is imaged as a brightly reflective focus with distal shadowing.

Diffuse parenchymal renal disease (Green et al. 1986)

Ultrasonographic findings

The patient with diffuse renal disease may present for ultrasound for a variety of causes. There may be a history of an acute nephritis or of a condition with known renal sequelae such as systemic lupus erythematosus or the patient may present with renal failure. There are no specific features on ultrasound that allow a particular diagnosis and many of the acute nephritides produce similar ultrasound findings. There is a progressive increase in the reflectivity or brightness of the renal cortex with increasing pro-

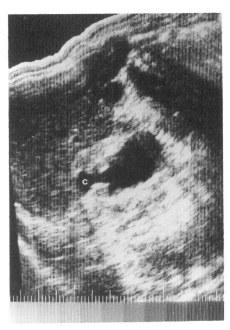

Fig. 18 Chronic pyelonephritis: transverse scan through the right upper quadrant. The kidney is small, and the cortex and medulla are of equivalent brightness. The renal pelvis is dilated and there is a dilated clubbed calyx adjacent to marked cortical thinning (c).

minence of the medullary pyramids by contrast with the cortex but with maintenance of the normal gross anatomical appearance of the kidney. This has been termed Type 1 renal disease (Fig. 19). It may be seen in the glomerulonephritides, diabetic nephropathy, and hypertensive nephrosclerosis, as well as in the many connective tissue disorders. Acute tubular necrosis by contrast is usually characterized by kidneys of normal appearance. If the renal disease is progressive then the morphological changes on ultrasound are also progressive. The renal cortex becomes progressively more reflective but eventually the increased reflectivity extends to the medulla so that there is eventual loss of corticomedullary differentiation. As this progression occurs so there is reduction in renal size. The kidney of end-stage renal

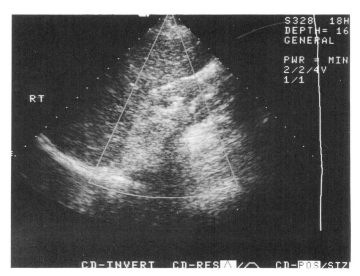

Fig. 19 Longitudinal scan in the right upper quadrant. The right kidney is of increased reflectivity although the medullary pyramids are still just visible. The appearances are those of parenchymal renal disease.

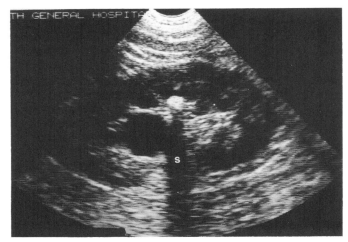

Fig. 21 Long-axis scan through left upper quadrant. There is a bright highly reflective focus within the middle of the kidney which demonstrates an acoustic shadow (s). This is characteristic of a calculus.

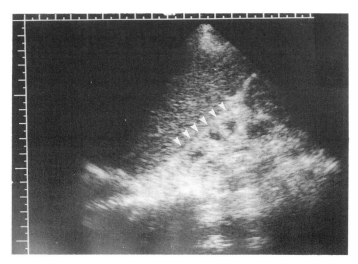

Fig. 20 End-stage renal disease: the right kidney is small and highly reflective and almost indistinguishable from perirenal fat (arrowheads).

disease is very small and uniformly highly reflective, blending almost completely with the surrounding perirenal fat. It therefore becomes almost invisible on ultrasound (Fig. 20).

Doppler findings

The changes in the Doppler sonogram scan in diffuse parenchymal renal disease are interesting. In the acute phase there may be either a normal scan or some increase in the diastolic component of flow with a consequent reduction in the pulsatility index. This may be due to intrarenal shunting. As the condition progresses and becomes chronic and as azotaemia appears so there is a reduction in the diastolic component of flow and a consequent increase in the pulsatility index (unpublished data). There is then close correlation with a rising blood urea and the pulsatility index. There is currently inadequate information to determine whether a measure of the pulsatility index might predict the reversibility or otherwise of the renal disease.

Uraemic cystic disease of the kidney (Green et al. 1986)

The development of cysts and occasionally tumours is a well-known complication of renal dialysis and renal transplantation. Ultrasound can be used to demonstrate the development of renal cysts within end-stage kidneys. Although renal tumours can be demonstrated, the detection of a small solid tumour within small highly reflective kidneys is difficult (see Chapter 10.3). It is possible that assessment of changing Doppler flow patterns within the renal arteries might provide this information in the future, especially with colour flow mapping.

Renal calcification

Renal calculous disease (or nephrolithiasis)

Ultrasound is not the investigation of first choice in the evaluation of renal calculous disease. Because of its essentially tomographic nature small stones might be missed unless an exhaustive study is performed. Rather it should be reserved for differentiation of calculus from tumour suggested by a filling defect on intravenous urography (Mulholland et al. 1979). Occasionally it may be used for confirmation of renal calculous disease when intravenous urography is difficult, inconclusive, or contraindicated. The appearances on ultrasound of renal calculi are those of a brightly echogenic focus either within the renal substance or within the renal sinus, which produces a distal acoustic shadow (Fig. 21). When the calculus occurs in the renal sinus it is occasionally difficult to confirm an acoustic shadow because of the normal shadowing effect of the renal sinus itself. In contrast to renal calculi (even those not obviously calcified on the plain film) transitional cell tumours and blood clots within the collecting system produce no acoustic shadowing (Joseph et al. 1986).

Multiple and staghorn calculi produce such extensive acoustic shadowing that they may obscure much of the renal substance: while the diagnosis of a staghorn calculus may be suggested on

1.6.3 Renal ultrasonography (including Doppler echography)

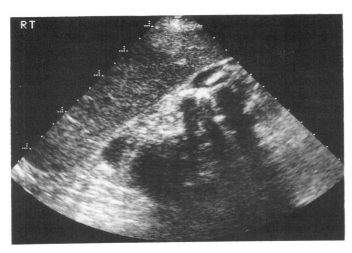

Fig. 22 Longitudinal scan in the right upper quadrant: the kidney is obscured by multiple foci of calcification, all of which produce an acoustic shadow. These appearances are those of a staghorn calculus.

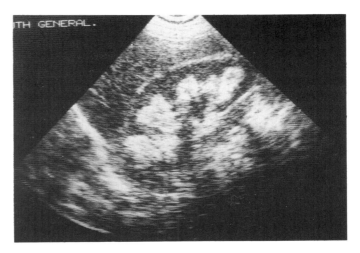

Fig. 24 Long-axis scan through the left upper quadrant in a patient with nephrocalcinosis. The medullary calcification produces little if any shadowing but marked increase in reflectivity.

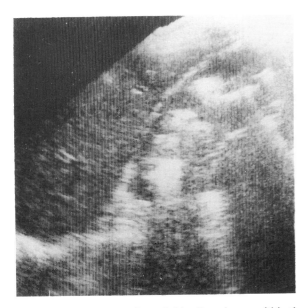

Fig. 23 Emphysematous pyelonephritis: there is gas within the renal substance which produces areas of increased reflectivity with distal acoustic shadowing which is 'dirty'.

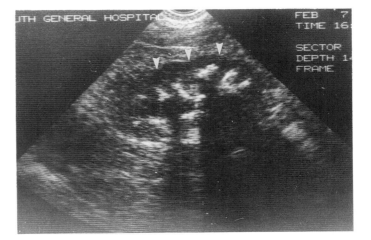

Fig. 25 Renal papillary necrosis: small, highly reflective areas throughout the renal substance in the region of the renal papillae in association with irregularity of the margin of the kidney as a result of scarring (arrowheads).

ultrasound little information about the size of the staghorn can be achieved and care must be taken when using ultrasound to evaluate the amount of residual cortex (Fig. 22).

The extensive acoustic shadowing derived from staghorn calculi may be difficult to differentiate from gas within the collecting system. This may occur as a result of emphysematous pyelonephritis or as a result of tumour invasion either from gut into kidney or vice versa (Fig. 23).

Nephrocalcinosis

Diffuse calcification of the kidney parenchyma may be recognized with ultrasound. Most commonly this is medullary in location. There is increase in reflectivity confined to the medullary pyramids (Fig. 24). At first there is only subtle acoustic shadowing but this is progressive with increasing density of calcification. It is usually possible to differentiate medullary calcification from papillary necrosis where the foci of calcification more closely mimic renal calculi. The small papillae are associated with variable distal acoustic shadowing and focal irregularity of the renal sinus (Fig. 25).

Parathyroid adenomata (Krudy et al. 1984)

In patients with chronic renal disease, and particularly those with nephrocalcinosis or other signs of hyperparathyroidism, it is possible with ultrasound to examine the thyroid gland and to demonstrate parathyroid adenomata. These appear as enlarged, echo-poor oval masses usually situated related to the posterior aspect of the thyroid. The limitations of use of ultrasound for the

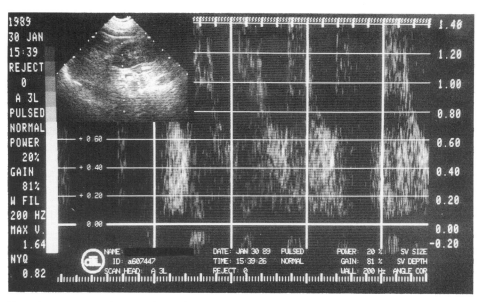

Fig. 26 Doppler sonogram from the renal artery in a patient with renal artery stenosis. There is massive increase in the peak frequency with significant spectral broadening and increased velocity of flow in diastole.

diagnosis of parathyroid adenomata usually relate to ectopic parathyroid glands. Where glands are not in close relationship to the thyroid ultrasound is a poor diagnostic tool compared with isotope imaging. Furthermore in the presence of a multinodular thyroid goitre the differentiation of a parathyroid tumour becomes difficult.

Renovascular disease (Dubbins 1986; Robertson et al. 1988)

Renal artery disease

The mainstay for the investigation of hypertension has long been intravenous urography. However, the sensitivity and specificity of this technique has been brought into question. It appears that ultrasound imaging combined with Doppler echography may allow the more accurate detection of renal artery disease. Ultrasound imaging may demonstrate a small kidney or indeed a kidney progressively decreasing in size on sequential examinations together with evidence of atheroma at the ostium of the renal artery. These findings are, however, inconstant and are therefore insensitive for the diagnosis of renal artery stenosis if taken in isolation. However, there are characteristic Doppler flow patterns which occur in significant renal artery stenosis. Sampling blood flow with Doppler echography at the origin of the renal artery shows that in the presence of a stenosis there is a significant increase in peak Doppler frequency in systole. In diastole there is a broad spread of frequencies of high intensity termed spectral broadening (Fig. 26). These appearances of Doppler sonography reflect the increased velocity and flow disturbance that occur at the site of a stenosis. In cases of vascular occlusion no flow is detected. The combination of Doppler spectral findings and ultrasound morphologic features appears to represent a sensitive method for the detection of renovascular disease and is suggested as a method of screening in arterial hypertension.

Renal vein thrombosis

Venous thrombosis may be diagnosed by ultrasound purely on morphological grounds. There is increase in the size of the lumen of the vessel because of the distension by blood clot. Moreover there is usually increased reflectivity of the contents of the vessel occasioned by the presence of clot. The affected kidney becomes enlarged, with loss of normal echo pattern. There may be evidence of perirenal fluid (Ezzat 1986). Doppler studies may confirm this original impression, first by demonstrating the absence of blood flow within the affected renal vein but also demonstrating abnormal blood flow characteristics within the supplying renal artery. Renal vein thrombosis will increase the peripheral resistance to blood flow, thereby increasing the pulsatility index within the main and branch renal arteries (unpublished observations).

Renal trauma (Monstrey et al. 1988)

Ultrasound is often used in the assessment of solid viscus injury following blunt abdominal trauma. Indeed abnormalities of the kidney may be identified during the course of a general abdominal survey. However, management of renal trauma depends upon control of haemorrhage and preservation of renal function. The decision to intervene is therefore based either on clinical grounds or upon the results of a procedure which allows functional assessment of the kidney, usually intravenous urography.

There remain significant roles for ultrasound in renal trauma. Morphological abnormalities of a kidney shown by intravenous urography to be non-functioning can be characterized. Intrarenal haematomata produce either focal or generalized enlargement of the kidney with alteration of the normal echo pattern. Perirenal or pararenal fluid collections may be either echo free or may contain echoes depending upon their content of blood clot (Fig. 27). The size and evolution of fluid collections may be followed with ultrasound. Occasionally unrelated renal abnormalities (e.g. congenital hydronephrosis, multicystic disease) may be demonstrated.

It is possible that Doppler echography may allow the demonstration of vascular compromise as a result of damage to the renal pedicle, but this hypothesis remains to be tested.

The renal transplant

Ultrasound occupies a central role in the assessment of complications of renal transplantation (Fig. 28) (Hricak and Hoddick 1986). While many of the difficulties of surgical technique have been resolved complications of the renal transplant are still broadly divisible into surgical and functional groups.

1.6.3 Renal ultrasonography (including Doppler echography)

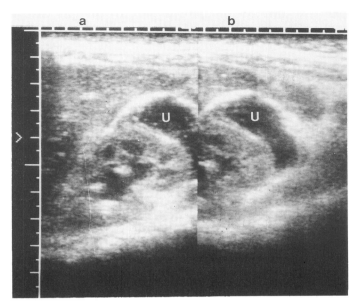

Fig. 27 Long-axis scans through the right upper quadrant showing in (a) neonate kidney with an echo-poor collection at the lower pole (u). In (b) the collection is slightly better shown at the lower pole of the kidney (u). This was a urinoma following trauma.

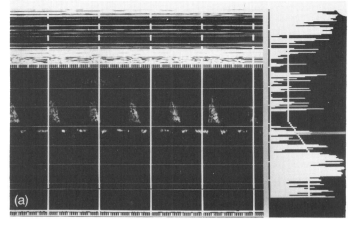

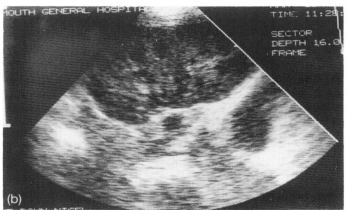

Fig. 29(a) Doppler sonogram in acute rejection demonstrating forward blood flow in systole but with reverse flow in diastole. (b) Ultrasound image of the same patient showing enlarged renal transplant with prominent medullary pyramids and loss of renal sinus echoes characteristic of acute rejection.

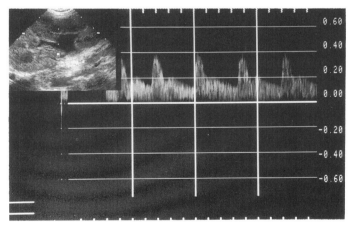

Fig. 28 Doppler sonogram of the hilar artery of a renal transplant showing normal flow characteristics.

Surgical complications

Fluid collections related to the renal transplant are well demonstrated by ultrasound usually as echo-free or echo-poor areas. While there are no specific features that allow the differentiation of different types of fluid certain appearances will suggest a particular type of collection. For example, lymphoceles have a multilocular appearance. However, ultrasound can also be used to guide needle placement for aspiration and/or drainage of fluid collections, thus achieving accurate diagnosis and therapy with a single needle puncture.

Fluid collections are but one cause of obstruction to the renal transplant and the demonstration of separation of the renal sinus echoes as evidence of calyceal dilatation will identify the obstructed transplant kidney. Moreover ultrasound-guided antegrade pyelography and subsequent percutaneous nephrostomy will allow the diagnosis of site, and cause of obstruction and its temporary relief.

Vascular complications

The ability to demonstrate blood flow characteristics in the artery of supply and the vein of drainage of the transplant kidney allows the early diagnosis of vascular complications. The appearances of renal artery stenosis are similar to those described for the native kidney although kinking of the renal artery and disproportionate size between donor and recipient vessels may also produce abnormal flow characteristics. Occasionally venous stenoses may occur and these are reflected by increased flow velocities focally within the draining renal vein (Murphy et al. 1987; Taylor et al. 1988). Arteriovenous fistulae as a result of renal biopsy are best demonstrated with colour flow mapping as focal areas of prominent, high-intensity colour. These fistulae are normally short lived, apparently closing spontaneously and usually of no haemodynamic or functional significance.

Transplant failure

Poor function in the renal transplant may be due to a number of causes, such as, in the early stages, acute tubular necrosis and hyperacute rejection, subsequently acute cellular and vascular rejection and in the longer term cyclosporin A toxicity. Eventually chronic rejection may ensue or there may be a recurrence in the transplanted kidney of the original disease. There was much initial enthusiasm for the use of ultrasound and of Doppler echography for the diagnosis of complications producing transplant failure. Initially it was thought that there were characteristics specific for acute vascular rejection. However, it is clear that there is significant overlap between the different conditions. Classically the patient with acute vascular rejection will show a transplant kidney of increased size with prominent medullary pyramids, patchy cortical reflectivity with focal areas of reduced reflectivity and a Doppler sonographic scan of high pulsatility index (Fig. 29). However, the morphologic features of the kidney in acute rejection frequently lag behind changes in the Doppler signal. Moreover the patient with acute tubular necrosis will frequently display an abnormal Doppler scan. The role of duplex ultrasound in cyclosporin toxicity has yet to be established. The combination of grossly abnormal renal morphology and reverse flow in diastole indicate irreversible rejection allowing prompt surgical removal when indicated.

In the right clinical setting duplex ultrasound can provide absolute confirmatory evidence of transplant rejection, but it does not represent the specific discriminant that was at first hoped (Murphy *et al.* 1987; unpublished observations).

Interventional ultrasound (Brooke Jeffrey and Kuligowska 1986)

Because of a particular ability to image the tip of a needle with ultrasound it has become increasingly used to guide interventional procedures, such as renal cyst and abscess aspiration and drainage, using direct guidance of the needle. There are several probes available in which the path of the needle is predetermined using a guide of the probe itself. However, freehand guidance is often less cumbersome and still allows the continuous imaging of the organ, the abnormality, and the needle tip.

References

Brooke Jeffrey, R. Jr. and Kuligowska, E. (1986). Interventional ultrasound. In *Genitourinary Ultrasound*. Clinics in Diagnostic Ultrasound No. 18. (ed. H. Hricak), pp. 113–134, Churchill Livingstone, Edinburgh.

Charboneau, J.W., Hattery, R.R., Ernst, E.C. 3d, James, E.M., Williamson, B. Jr., and Hartman, G.W. (1983). Spectrum of sonographic findings in 125 renal masses other than benign simple cyst. *American Journal of Roentgenology*, **140**, 87–94.

Coleman, D.G., Arger, P.H., and Mulhern, C.B. (1980). Grey scale sonographic spectrum of hypernephromas. *Radiology*. **137**, 757.

Cook, J.H., Rosenfield, A.T., and Taylor, K.J.W., (1977). Ultrasonic demonstration of intrarenal anatomy. *American Journal of Roentgenology*, **129**, 831–5.

Dubbins, P.A. (1986). Renal artery stenosis—Doppler evaluation. *British Journal of Radiology*, **59**, 225–9.

Dubbins, P.A. (1989a). Intravenous urography or ultrasound in prostatism. (abstract). In *Proceedings of the British Institute of Radiology Conference*, Eastbourne, p. 123.

Dubbins, P.A. (1989b). Doppler ultrasound in renal disease. *Current Imaging*, **1**, 32–3.

Dubbins, P.A. and Wells, I. (1986). Renal carcinoma: duplex Doppler evaluation. *British Journal of Radiology*, **59**, 231–6.

Dubbins, P.A., Kurtz, A.B., Derby, J., and Golberg, B.B. (1984). Ureteric jet effect – the echographic appearance of urine entering the bladder. *Radiology*. **140**, 513–5.

Ezzat, M. (1986). Acute renal vein thrombosis in adults. *International Urology and Nephrology*, **18**, 243–53.

Fitzsimons, S. (1983). Kidney length in the newborn measured by ultrasound. *Acta Paediatrica Scandinavica*, **72**, 885–7.

Gosling, R.G. and King, D.H. (1975). In *Arteries and Veins* (eds. A.W. Harcus and L. Adamson) pp. 61–3. Churchill Livingstone, Edinburgh.

Green, D. and Carroll, B. (1986). Ultrasound of renal failure. In *Genitourinary Ultrasound*. Clinics in Diagnostic Ultrasound No. 18. (ed. H. Hricak), pp. 55–88, Churchill Livingstone, Edinburgh.

Grossman, H., Rosenberg, E.R., Bowie, J.D. Ram, P., and Merten, D.F. (1983). Sonographic diagnosis of renal cystic diseases. *American Journal of Roentgenology*. **140**, 81–5.

Hayden, C.K. Jr., Santa-Cruz, F.R., Amparo, E.G. Brouchard, B., Swischuk, L.E., and Ahrendt, D.K. (1984). Ultrasonic evaluation of the renal parenchyma in infancy and childhood. *Radiology*, **152**, 413–7.

Heiken, J.P., Gold, R.P., and Schnur, M.J. (1983). Computed tomography of renal lymphoma with ultrasound correlation. *Journal of Computer Assisted Tomography*, **7**, 245–50.

Hricak, H. and Hoddick, W.K. (1986). Ultrasound in renal transplantation. In *Genitourinary Ultrasound*. Clinics in Diagnostic Ultrasound No. 18. (ed. H. Hricak), pp. 161–80, Churchill Livingstone, Edinburgh.

Joseph, N., Neiman, H.L., and Vogelzang, R.L. (1986). Renal Masses. In *Genitourinary Ultrasound*. Clinics in Diagnostic Ultrasound No. 18. (ed. H. Hricak), pp. 135–60, Churchill Livingstone, Edinburgh.

Krudy, A.G., *et al.* (1984). Ultrasonic parathyroid localisation in previously operated patients. *Clinical Radiology*, **35**, 113–8.

Kuligowska, E. (1986). Renal infections. In *Genitourinary Ultrasound*. Clinics in Diagnostic Ultrasound. No. 18. (ed. H. Hricak), pp. 89–112, Churchill Livingstone, Edinburgh.

Kuriano, G., Williams, B., and Dana, K. (1984). Kidney length correlated with age: normal values in children. *Radiology*, **150**, 703–4.

Kurtz, A.B., *et al.* (1981). Echogenicity: analysis, significance and masking. *American Journal of Roentgenology*, **137**, 471–6.

Mahoney, B.S. and Filly R.A. (1986). The genitourinary system in utero. In *Genitourinary Ultrasound*. No. 18. (ed. H. Hricak), pp. 1–22, Churchill Livingstone, Edinburgh.

Mahoney, B.S., Jeffrey, R.B., and Laing, F.C. (1983). Septa of Bertin, a sonographic pseudotumour. *Journal of Clinical Ultrasound*, **11**, 317–9.

Monstrey, S.J.M., Werken, C.H.R., and de Bruyne, F.M.J., (1988). Rational guidelines in renal trauma assessment. *Urology*, **31**, 469–73.

Morehouse, H.T., Weiner, S.N., and Hoffman, J.C. (1984). Imaging in inflammatory disease of the kidney. *American Journal of Roentgenology*. **143**, 135–41.

Morin, M.E. and Baker, D.A. (1979). The influence of hydration and bladder distention on the sonographic diagnosis of hydronephrosis. *Journal of Clinical Ultrasound*, **7**, 192–4.

Mulholland, S.G., Arger, P.G., Goldberg, B.B., and Pollack, H.M. (1979). Ultrasonic differentiation of renal pelvic filling defects. *Journal of Urology*, **122**, 14–16.

Murphy, A.A., Robertson, R., and Dubbins, P.A. (1987). Duplex ultrasound in the assessment of renal transplant complications. *Clinical Radiology*, **38**, 229–34.

Plainfosse, M.C., Delecoeullerie, G., Vital, J.L., Paty, E., and Morran, S. (1983). 165 Renal carcinomas: Accuracy of imaging for diagnosis and spread-cost efficiency. *European Journal of Radiology*, **2**, 95–172.

Pollack, H.M., Banner, M.P., Arger, P.H., Peters, J., Mulhern, C.B.Jr., and Coleman, B.G., (1982). The accuracy of grey scale renal ultrasonography in differentiating cystic neoplasms from benign cysts. *Radiology*, **143**, 741–5.

Robertson, R., Murphy, A., and Dubbins, P.A. (1988). Renal artery stenosis—the use of duplex ultrasound as a screening technique. *British Journal of Radiology*, **61**, 196–201.

1.6.4 Computed tomography and magnetic resonance imaging of the kidney and upper urinary tract

A.N. DARDENNE AND G.C. DOOMS

Introduction

Computed tomography of the kidney might not seem to be of much importance compared with conventional radiology, which can detect small areas of calcification and provide extremely detailed images of the opacified excretory pathway as well as clinically useful information on the state of the renal parenchyma. Intravenous pyelography (IVP+tomography of the parenchyma) remains the most complete morphological examination of the urinary system, and also provides information about function. Nevertheless, the relative weakness of the conventional procedures lies in their poor ability to discriminate between the soft tissues, which are of fairly similar densities.

Computed tomography overcomes this failing by allowing more detailed study of the parenchyma and non-opacifying (or poorly opacified) excretory pathway and, most importantly, by showing abnormal features in the perirenal space and retroperitoneum—abnormalities that are hard to explore with conventional techniques. CT scanning thus makes it possible, on the one hand, to answer, without contrast enhancement, certain questions that would formerly have required the injection of a contrast medium, and, on the other hand, to solve equivocal diagnostic problems, with injection of opacifying material if necessary without trauma.

The contrast media used in CT are those used in intravenous pyelography with the same concentration, dose and procedure. It thus seems logical to consider CT as part of a diagnostic procedure involving administration of intravascular contrast media and, depending on the case, including other means of image collection, e.g., conventional, digital, and digital subtraction films.

CT scanning is particularly indicated in conditions that alter the morphology or physiological opacification of the parenchyma, in certain cases of urinary obstruction and vascular disease, and in urinary tract injuries.

Parenchymatous masses

Simple renal cysts

Simple renal cysts are commonly discovered incidentally during ultrasound or CT scanning; There are often several cysts within

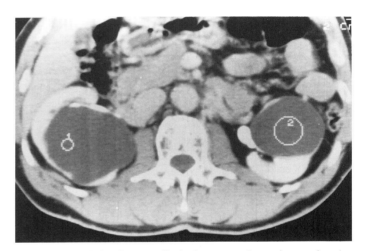

Fig. 1 Bilateral simple cysts displacing parenchyma (arrows) and calyces (arrowheads).

one or both parenchymata. Fortunately, these techniques afford great diagnostic reliability.

The sites and, above all, sizes of these cysts are highly variable, and they can cause bulging, displacement, and compression of the renal architecture (Fig. 1).

The CT features of an uncomplicated cyst are, without contrast medium: no calcification and relatively low (close to 0 Hounsfield unit) homogeneous density; with contrast medium: lack of change in the spontaneous density, very thin (often indiscernible) wall, and sharp contour at all levels. Adhering strictly to these criteria will guarantee practically error-free diagnosis (McClennan et al. 1979a; Bosniak 1986) (Fig. 2).

Some cysts have an atypical CT appearance, which may make their diagnosis harder (Balfe et al. 1982). Cysts with thick walls or parietal calcifications may be seen, particularly following haemorrhagic incidents, and these can have a very similar appearance to necrotic malignant tumors (Fig. 3).

Occasionally, a benign cyst may have a high density on an

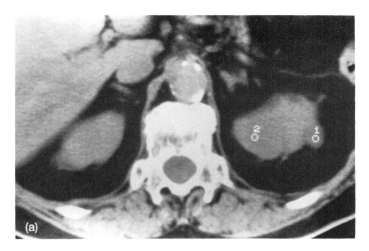

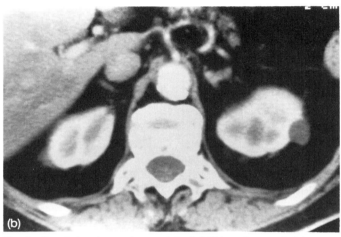

Fig. 2 Small simple renal cyst of the left kidney (arrows): (a) without contrast; (b) following intravenous contrast injection: low density, lack of enhancement, indiscernible wall, sharp contour.

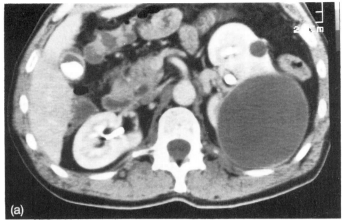

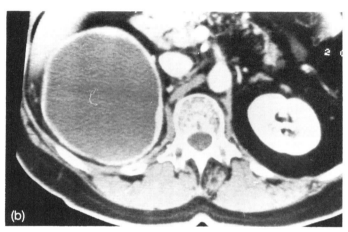

Fig. 3 (a) Atypical simple cyst: thick though regular wall; (b) necrotic malignant tumour: even thicker wall with irregular thickness (arrows).

unenhanced scan (Fishman *et al.* 1983; Pealstein 1983; Sussman *et al.* 1984); this is frequently due to recent intracystic haemorrhage. The other CT signs of an uncomplicated simple cyst, including the absence of density enhancement after injection of the contrast medium, will be present. Rarely, renal cysts, (mainly haemorrhagic cysts) rupture and disappear (Papanicolaou *et al.* 1986; Davis and McLaughlin 1987; Friedland 1987).

Chronic haemodialysis patients with renal failure due to parenchymatous nephropathy, especially glomerulonephritis, usually develop multiple parenchymatous cysts and, sometimes, renal clear-cell carcinoma (Jabour *et al.* 1987; Soffer *et al.* 1987) (Fig. 4).

Adult dominant polycystic kidney disease (see Chapter 18.2.1)

Adult polycystic renal disease has a characteristic CT appearance at an advanced stage of its development, when both kidneys are enlarged and deformed by countless cysts of variable size. The invasion of the parenchyma is diffuse, generalized, and without normal patches of renal tissue, unlike the pattern of scattered cysts and areas of normal renal parenchyma characteristic of multiple simple cysts (Levine and Grantham 1981; Segal and Spataro 1982) (Fig. 5).

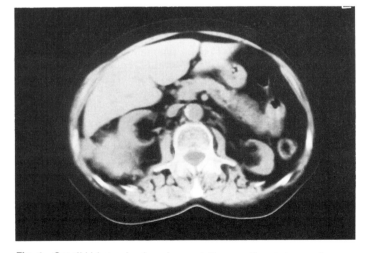

Fig. 4 Small kidneys in chronic renal disease. Renal clear cell carcinoma on the right of the patient.

Very rarely it is impossible to differentiate between bilateral multiple simple cysts and polycystic disease. The discovery of associated hepatic cysts will swing the balance strongly in favour of the second diagnosis (Rosenfield *et al.* 1977) (Fig. 6).

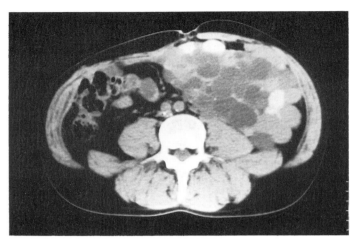

Fig. 5 Advanced adult dominant polycystic kidney disease. Scan through the left lower pole. Many cysts with no normal parenchyma; several hyperdense cysts.

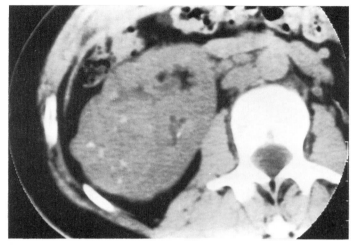

Fig. 7 Right renal clear cell carcinoma (arrows) with small central calcifications. Scan without intravenous opacification.

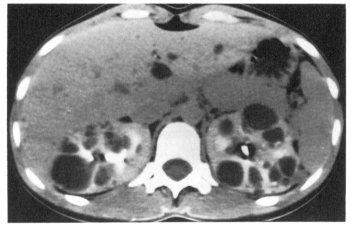

Fig. 6 Early stage adult dominant polycystic kidney disease. Several hepatic cysts.

CT is the most sensitive method for detecting early stages of polycystic disease, since it can detect cysts that are less than 0.5 cm in diameter (after intravenous opacification). The local complications of polycystic disease in the kidneys are revealed best by CT. Haemorrhagic or infected cysts are characterized by thickened walls, an increase in the intracystic density (Fig. 5), thickening of the perirenal Gerota's fascia, and sometimes, in the case of infection, by the presence of intracystic gas bubbles. CT is also the best technique for detecting complications such as calculi, obstructions, ruptured haemorrhagic cysts, and perinephric haemorrhage (Segal and Spataro 1982; Meziane *et al.* 1986; Levine and Grantham 1981).

Tumours of the renal parenchyma (see Section 19)
Renal cell carcinoma

Renal cell carcinoma is the most frequent tumour of the renal parenchyma in adults. CT is the method of choice for detecting renal tumours (Hricak 1987) since it allows one make a diagnosis of a solid tumour and also to determine its local and regional extension (Lemaître and Delambre 1983). Many more small renal tumours are being detected with the use of CT and ultrasound. The cure rate of renal cell carcinoma undoubtedly increases if these tumours are detected when they are small and asymptomatic. Partial nephrectomy will increasingly be used in the management of such small tumours (Amendola *et al*: 1988; Smith *et al.* 1989).

CT findings The degree to which the tumour changes the morphology of the kidney and its environment depends on tumour size, site, and its localized or invasive nature.

The anatomical changes produced by a well-defined tumour may be identical to those observed in the case of a simple cyst, i.e., bulging of a parenchymal margin, displacement of the kidney, and backward thrusting and compression of the pyelocalyceal cavities. A locally more aggressive tumour may produce signs of involvement of the excretory pathway, renal vein, perirenal space, and neighbouring organs. In non-contrast-enhanced scans the attenuation values of the tumour tissue are generally comparable to those of the normal parenchyma. Hypodense centres, corresponding to necrotic tissue, are sometimes seen. The detection of central calcification is strongly suggestive of malignancy. In contrast, the fine calcifications found at the edges of abnormal areas are of no diagnostic value, since they are observed in 'atypical' cysts and malignant tumours alike. In one series, approximately 40 per cent of calcified renal masses were found to be malignant. These must be surgically explored or followed up closely with computed tomography (Patterson *et al.* 1987) (Fig. 7).

Since renal cell carcinomas are very frequently hypervascularized, the density of a typical carcinoma will rise dramatically after rapid intravenous injection of contrast medium. This intense opacification is heterogeneous and anarchical, but transient—the plasma contrast medium concentration falls rapidly as the tubule nephrogram of the parenchyma develops, so that the mass exhibits delayed relative hypodensity (Weyman *et al.* 1980; Wadsworth *et al.* 1982; Lemaître and Delambre 1983) (Fig. 8).

Carcinomas may also be hypovascular, and these are either large, thick-walled, cyst-like, sometimes septate, highly necrotic tumours (Fig. 3(b)) or very poorly vascularized, usually small, solid tumours that resemble small cysts, so slight is the increase

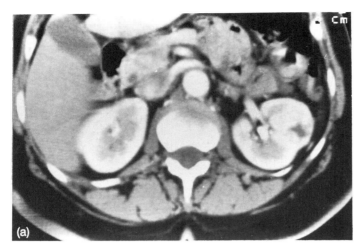

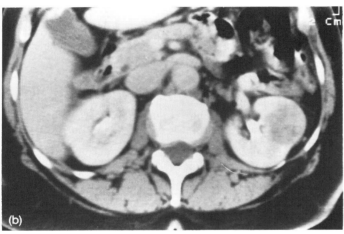

Fig. 8 Small typical renal clear cell carcinoma of the left kidney (arrows). (a) After fast intravenous injection of contrast medium (20 s): intense heterogeneous opacification of the tumour. (b) After opacification (10 min): hypodensity of the tumour in contrast with the high density of the renal parenchyma.

Table 1 Staging of renal cell carcinoma

Stage I:	Tumour confined to renal capsule
Stage II:	Tumour extends through renal capsule and perinephric fat but is enclosed by perirenal fascia. Adrenal involvement is possible
Stage IIIA:	Tumour involves the renal vein and/or the inferior vena cava
Stage IIIB:	Tumour involves adjacent lymphatic structures
Stage IIIC:	Tumour involves both renal venous structures and adjacent lymphatics
Stage IVA:	Tumour extends directly to adjacent organs outside perirenal fascia
Stage IVB:	Tumour has metastasized to distant sites

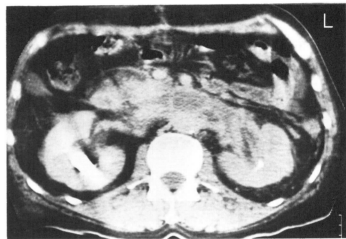

Fig. 9 Malignant renal vein thrombosis of the left kidney (tumour not shown at this level): enlargement of the vessel and central hypodensity.

in their density. In such cases, ultrasound must be used to determine whether these small masses contain fluid.

About 15 per cent of renal cell carcinomas are cyst-like on radiological and pathological examination. There are four basic growth patterns: intrinsic multiloculated growth, intrinsic unilocular growth (cystadenocarcinoma), cystic necrosis, and origin in the wall of a simple cyst. These tumours resemble a vascularized nodule bulging into the cyst lumen (Hartman *et al.* 1986; 1987).

Local and regional extension Assessing the extension of renal cell carcinomas allows a prognosis to be assessed and possible surgical approach to be established (Karp *et al.* 1981; Probst *et al.* 1981; Cronan *et al.* 1982; Hata *et al.* 1983; Benson *et al.* 1989).

Renal cell carcinomas have been staged by Robson *et al.* (1969) (Table 1). The spread of tumour tissue into the renal vein is the first thing to look for in checking extension of the mass (Vasile *et al.* 1981). The right renal vein is more difficult to explore than the left, because of its shorter, more oblique path. Nevertheless, CT usually proves to be the best technique for confirming or ruling out the presence of a tumour thrombus in the renal vein.

Renal vein involvement may appear as an increase in vessel diameter, an endoluminal lacuna (Fig. 9), retroperitoneal collateral veins, and, sometimes, hypervascularization of the tumour thrombus. Tumour thrombi spreading to the inferior vena cava are usually easy to detect, since the blood vessel lumen is enlarged and its centre is hypodense after opacification.

Perirenal extension of the tumour tissue is characterized by thickening of the tissue streaking the perirenal fat. This is not a specific sign, however, and may simply reflect an inflammatory reaction. Pararenal extension is easy to detect when it occurs in the direction of the planes of the section, especially in the pararenal fat and nearby muscles. It is more difficult to ascertain in the other directions (Lemaître *et al.* 1986).

Lymph node involvement can only be detected when the nodes are more than 2 cm in diameter—CT scans cannot reveal microscopic involvement of the nodes. In one recent series, CT correctly staged lesions in 90 per cent of patients (Zeman *et al.* 1988). In von Hippel-Lindau's disease one may come across associations of renal cell carcinomas, multiple simple cysts, and very rare types of solid tumours in the renal parenchymata. (Levine *et al.* 1979; Das *et al.* 1981).

Less common tumours of the renal parenchyma
Secondary tumours: *metastases, lymphomas, and leukaemias*
Several masses are often present; they are often poorly

1.6.4 Computed tomography and magnetic resonance imaging

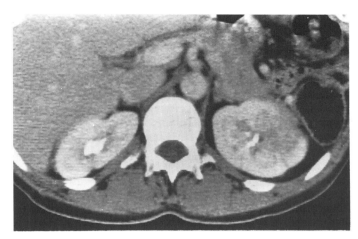

Fig. 10 Metastasis of a bronchial carcinoma (left kidney): hypodensity and poor margins (after opacification).

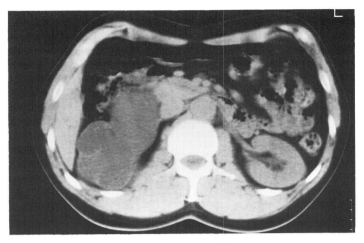

Fig. 12 Long-term severe obstruction of the right kidney. Non-enhanced scan. Extremely atrophic parenchyma and hypodense urine contents.

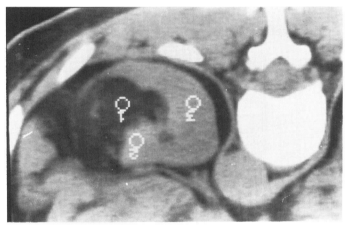

Fig. 11 Angiomyolipoma (arrows). Scan without opacification. Large dark area of low density representing fat within the tumour.

vascularized, poorly defined, and infiltrating (Fig. 10). Contrast-enhanced CT is the best radiographic technique currently available for detecting renal metastases, lymphomas and leukaemias (Hartman et al. 1982; Bruneton et al. 1984; Lemaître et al. 1986; Choyke et al. 1987).

Benign tumours The tumour most frequently seen in clinical practice is the angiomyolipoma, which originates in the mesenchyma. On a CT scan it may be recognized by the extremely hypodense areas corresponding to its lipid components (Bosniak 1981; Hartman et al. 1981; Kolmannskog et al. 1981; Bosniak et al. 1988) (Fig. 11). In Bourneville's disease (tuberous sclerosis complex), both kidneys may be the seats of angiomyolipomas and multiple cysts.

The following points should be remembered with regard to the other benign tumours of epithelial origin: a small adenoma cannot be differentiated from a carcinoma in the early stages of development (Lemaître et al. 1986); a particular form of adenoma, the oncocytoma, may exhibit a highly recognizable hypodense stellate central scar if it is fairly large (more than 3 cm) (Lautin et al. 1981; Hara et al. 1982; Cohan et al. 1984; Velas-

quez et al. 1984); the characteristics of multilocular cystic nephroma, another benign tumour of epithelial origin, are midway between those of a simple, septate cyst and a necrotic malignant tumour (Parienty et al. 1981; Slasky and Wolfe 1982; Hartman et al. 1987).

Rare malignant tumours These include sarcomas of mesenchymal origin. Given their often large size at the time of discovery, it is often difficult to tell whether they arose in or outside the kidney. CT is useful in suggesting a diagnosis: a well-encapsulated solid neoplasm originating in the renal capsule or renal sinus is suggestive of renal sarcoma. Negative attenuation of the tumour suggests the diagnosis of liposarcoma. When a sarcoma arises within the renal parenchyma, it is difficult to differentiate the mass from a renal cell carcinoma. However, sarcomas usually do not involve the renal vein or the inferior vena cava (Shirkhoda and Lewis 1987). Wilms' tumour, or nephroblastoma, is an embryonal tumour that affects children almost exclusively. Its CT features are not very specific.

Tumour-like masses produced by inflammation

These are acute parenchymal abscesses whose CT features sometimes resemble those of renal cell carcinomas. Xanthogranulomatous pyelonephritis may also mimic renal cell carcinoma (Subramanyam et al. 1982; Goldman et al. 1984).

Urinary tract obstruction

CT scanning allows the aetiology of ureteral obstructions already detected by other techniques to be determined (O'Reilly 1982). It can also be used for the primary diagnosis of an obstruction. In unenhanced scans, the dilated excretory pathways are seen as hypodense structures with attenuation coefficients below that of the renal parenchyma (Fig. 12). The dilated ureter, for example, appears as a round structure with a liquid density, and may or not be edged with a denser fringe representing the ureteral wall, which is more or less visible, depending on its thickness (Fig. 13). Opacification of the parenchyma after intravenous injection of contrast material is delayed, progressive, and, in the case of an acute obstacle, is eventually of higher-than-normal intensity. Unless the kidney is almost non-functioning, progressive opaque sedimentation develops in the sloping areas of the

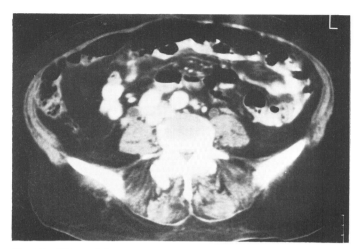

Fig. 13 Bilateral dilated ureters.

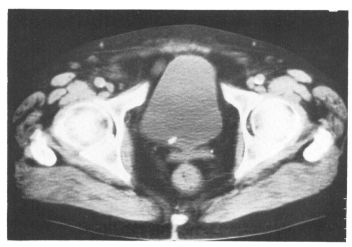

Fig. 15 Small ureteral stone at the ureterovesical junction on the right; non-opacified bladder.

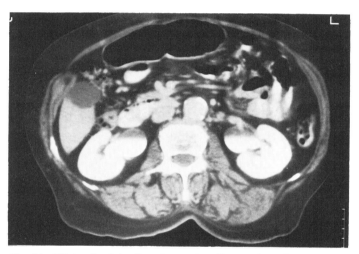

Fig. 14 Bilateral pelvic dilatation and sedimentation of contrast medium.

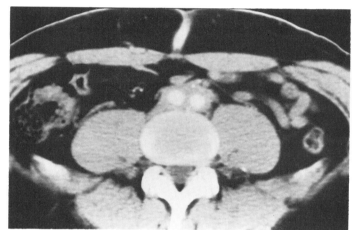

Fig. 16 Retroperitoneal fibrosis. Fibrous plaque in front of the proximal iliac vessels.

dilated excretory pathway lumen on contrast-enhanced scans after a variable delay (Fig. 14).

Severe obstructions in an almost non-functioning kidney may benefit most from CT scanning, which reliably evaluates residual parenchymal thickness and often reveals the origin of the obstruction (Lemaître *et al*. 1986).

Like other imaging techniques, CT scanning is of limited value in obstructive syndromes with minimal dilatation of the collecting system. In these cases, visualizing the functional abnormalities is not always enough to make a diagnosis with certainty, unless an obvious abnormality, such as a ureteral calculus, is present (Curry *et al*. 1982; Maillet *et al*. 1986) (Fig. 15).

CT scanning is primarily useful in determining the aetiologies of ureteral obstructions, e.g., retroperitoneal fibrosis, lymphadenopathy, retroperitoneal tumour, differential diagnosis of radiotransparent stones and tumours of the ureter (Fig. 16). A complete diagnosis of obstruction of the upper pole moiety in renal duplication can be achieved by CT of the kidney and of the pelvis (Cronan *et al*. 1986). All urinary calculi, including those that are radiotransparent on conventional X-rays, are radio-paque on unenhanced CT scans, with densities ranging from 250 to 1600 Hounsfield units depending on their chemical composition (Hillman *et al*. 1984; Newhouse *et al*. 1984) (Fig. 15). In some rare cases, including very small, infiltrating tumours of the ureter, the cause of the obstruction may not be detected.

Kidney infections (see Section 13)

Chronic pyelonephritis, renal tuberculosis, and papillary necrosis (of infectious or non-infectious origin) do not often benefit from the use of CT, although acute, especially severe infections, are good indications for CT.

Acute parenchymal infections

It is now well established that CT is more sensitive than intravenous pyelography or ultrasonography for the detection of acute inflammatory disease in the kidney. CT defines the extent of the disease, guiding early and accurate therapy (Hoddick *et al*. 1983; Soulen *et al*. 1989).

The diffuse infections are classified, somewhat arbitrarily, as acute pyelonephritis and diffuse acute bacterial nephritis, based

1.6.4 Computed tomography and magnetic resonance imaging

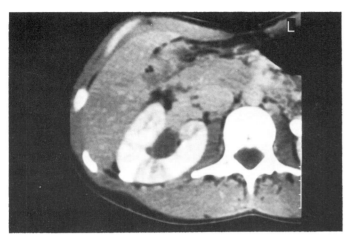

Fig. 17 Acute pyelonephritis. Scan after opacification. Radiating hypo- and hyperdense streaks.

on the severity of renal involvement. The former is clinically unimportant, and responds quickly and favourably to medical treatment of the infection. CT scans in such cases show, in addition to frequent kidney enlargement, radiating streaks of alternating dense and hypodense lines perpendicular to the parenchymal surface (Fig. 17). Gold *et al.* (1983) describe these, after injection of contrast media, as hypodense wedge-shaped zones with distinct borders stretching from the sinus to the capsule, scattered among the areas of normal parenchyma and may show internal striation. They show delayed enhancement on CT performed some hours after injection (Ishikawa *et al.* 1985). Diffuse acute bacterial nephritis is used when these signs are particularly marked and clinically resolution is not achieved quickly. This may occur in diabetics, in whom the risk that the condition will evolve to emphysematous pyelonephritis is ever present (Hoffman *et al.* 1980).

Focal acute bacterial nephritis is the term used for acute focal infectious involvement of the parenchyma in a non-suppurative stage. In unenhanced CT scans this appears as an isodense, focal, parenchymal area. Very rarely this area is of increased or decreased density resulting from the presence of haemorrhage at different stages, associated with focal infection (Rigsby *et al.* 1986). After contrast enhancement it shows up as a poorly-defined area for which the density enhancement is lower than that of the parenchyma. It is either triangular, lobar with a peripheral base, or rounded in shape (Lee *et al.* 1980). Late enhancement of this area is possible (Ishikawa *et al.* 1985).

A renal abscess appears as a hypodense, rounded area which is not enhanced by the injection of a contrast material. According to the literature, it is always round (Rauschkolb *et al.* 1982). Thickening of the perirenal (Gerota's) fascia, as well as opaque septa in the perirenal fat, is frequently observed. A renal abscess may eventually break through the capsule to become a perirenal abscess (Hoddick *et al.* 1983). Focal acute bacterial nephritis is cured by appropriate antibiotic therapy, whereas renal abscesses that have reached a certain size must be drained.

Pyonephrosis is the septic complication of a urinary obstruction. In addition to the signs of urinary obstruction one often sees thickening of the excretory pathway and signs of parenchymal and perirenal infection.

Emphysematous pyelonephritis is a variety of acute bacterial nephritis with a particularly poor prognosis and is seen primarily in diabetics. It is characterized by the presence of gas bubbles in the parenchyma, excretory pathway, or perirenal space (Hoffman *et al.* 1980)

Xanthogranulomatous pyelonephritis

Xanthogranulomatous pyelonephritis is an extremely mutilating form of a chronic infection of the renal parenchyma whose differential diagnosis is difficult. It may be diffuse or focal and often combines staghorn calculi, hydrocalyces, focal loss of parenchymal substance, and severely altered renal function. It must be differentiated from extensive kidney abscesses, advanced pyonephrosis, renal tuberculosis, and even necrotic renal cell carcinoma (Subramanyam *et al.* 1982; Goldman *et al.* 1984).

Parenchymal consequences of vascular disease

The major arterial disorders which have a chronic effect on renal parenchymal function and morphology are atheromatous and, to a lesser extent, dysplastic stenosis, while embolisms are the most frequent cause of acute ischaemia. Extrinsic compression of the vascular pedicle, aneurysmal lesions, and arterial trauma are much less common. Renal vein thrombosis is, in practice, the major venous disorder that can alter the renal parenchyma anatomy or physiology.

The main CT feature of acute arterial ischaemia is isolated opacification (after injection of a contrast material) of a narrow peripheral parenchymal cortical rim (cortex corticis) (the 'rim sign') which is thought to be due to the conserved permeability of the capsular arteries (Harris and Dorros 1981; Glazer and London 1981; Demos *et al.* 1982; Ishikawa *et al.* 1982; Glazer *et al.* 1983). When this sign is absent, one must rely on the absence of parenchymatous opacification and, if ischaemia is total, the lack of opacification of the excretory pathway. The rim sign may also occur in cases of acute renal vein thrombosis and acute tubular necrosis. Absence of parenchymal opacification associated with retrograde opacification of the renal vein may be a sign of renal artery avulsion (Cates *et al.* 1986). If the ischaemia is segmental, the parenchymal abnormalities will be limited and the opacification of the excretory pathway is normal (Parker 1981).

Chronic ischaemia is expressed primarily by parenchymal atrophy that is frequently non-specific in CT scans, as the underlying intactness of the calyces and any functional changes (delayed opacification of the calyces) that may have occurred show up much more poorly on CT scans than on intravenous pyelography.

The CT image of renal vein thrombosis varies according to the speed of development and the extent of the venous obstruction, as well as the development of a collateral venous network. In the acute stage, the kidney is enlarged. After opacification, the nephrogram intensity is either diminished or comparable to that observed in acute obstruction of the excretory pathway or acute arterial ischaemia. The opacity of the pelvis-calyces complex is either normal, low, or nil. Venous thrombi often show up as clear central areas in enlarged veins (Fig. 9). In the chronic stage, the kidney seems normal or small and is frequently opacified normally (Coleman *et al.* 1980; Adler *et al.* 1981; Winfield *et al.* 1981; Greene *et al.* 1982; Gatewood *et al.* 1986).

Renal failure

In the investigation of renal failure, ultrasound, radiographs without contrast materials, and intravenous pyelography can determine whether the renal failure is acute or chronic and whether it is obstructive or not (Fig. 4). Ultrasound also makes it easy to diagnose polycystic renal disease that has reached the renal failure stage. Computed tomography is reserved for patients in whom previous examinations have failed to answer these two questions (McClennan 1979b; Jeffrey and Federle 1983). In the case of a polycystic kidney complicated by a urinary obstruction, for example, ultrasound will differentiate poorly between the dilated calyces and surrounding cysts (Thomsen et al 1981; Segal and Spataro 1982). In bilateral vascular, infectious, or obstructive diseases one can obviously find the indices described for each disease on both sides.

The CT pattern is often of little use in determining the aetiology of parenchymatous disease other than polycystic disease, with a few rare exceptions. However, after intravenous contrast injection denser opacification of a band of subcapsular cortex similar to the rim sign seen in acute renal ischaemia has been described in some cases of tubular necrosis. Similarly, in cortical necrosis, the sections imaged after opacification can show a hypodense peripheral parenchymal band located just within a fine border (1–2 mm) of peripheral cortex that is vascularized by the capsular arteries (Goergen et al. 1981).

The detection of residual opacification of the glomerular cortex greater than 140 Hounsfield units 24 h after a radiological examination that includes intravenous injection of a contrast material seems to be an early indicator of contrast nephropathy (Older et al. 1980; Slasky and Lenkey 1981; Love et al. 1989). Computed tomography is the most sensitive technique for revealing faintly discernible nephrocalcinosis, such as occurs in some cases of oxalosis (Luers and Siegler 1980).

Miscellaneous diseases

The most frequent diseases of the renal sinus are lipomatosis, parapelvic cysts, and pyelocalyceal urothelial tumours. Lipomatosis (fibrolipomatosis) of the renal sinus is an abnormal proliferation of fatty (or fatty fibrous) tissue in the renal sinus which is usually easy to recognize due to its negative tissue attenuation value on unenhanced scans (Subramanyam et al. 1983; Thierman et al. 1983; Kullendorf et al. 1987). Parapelvic cysts, which are often multiple, show the same intrinsic pattern of signs as the simple renal cyst, but without parenchymatous attachments (Amis et al. 1982; Downey et al. 1982) (Fig. 1). Pyelocalyceal urothelial tumours are found most often in their vegetative form, which appear after opacification, as poorly vascularized, endoluminal defects of tissue density (Fig. 18). The parenchyma-infiltrating form appears as hypodense patches on the nephrogram that are difficult to differentiate from various other types of malignant renal tumours (poorly vascularized adenocarcinomas, lymphomas, and metastases) or even from certain focal infections. In some very rare cases this tumour may contain calcifications (Pollack et al. 1981; Baron et al. 1982; Dinsmore et al. 1988). CT scanning is by far the best technique for obtaining an accurate picture of parenchymal involvement—both morphological and functional—and for demonstrating blood or urinary perirenal effusions in renal trauma. It will also reveal lesions in other organs, such as the liver and spleen (Federle et al. 1981, 1987; Sandler and Toombs 1981; Federle 1983; Pollack and Wein 1989). CT also has advantages over the other imaging techniques in the diagnosis of renal agenesis and revealing non-functioning very small kidneys. It is also excellent for visualizing specific vascular abnormalities such as retro-aortic left renal vein and left

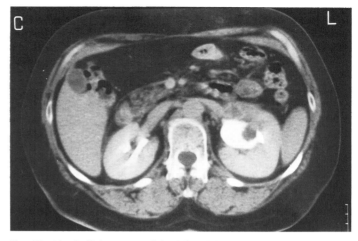

Fig. 18 Urothelial tumour of the left pelvis. Bulging of a soft tissue density mass into the opacified pelvis.

inferior vena cava (Royal and Callen 1979; Buschi et al. 1980; Parikh et al. 1981; Reed et al. 1982).

Retroperitoneal diseases

The retroperitoneal abnormalities that may affect the urinary tract include diseases of the large blood vessels and psoas muscle, primary tumors, lymphadenopathy, retroperitoneal fibrosis, and collections. The last two affect the urinary system more than any of the others: urinary obstruction is the main complication of fibrous plaque, while the urinary system is the most frequent source of retroperitoneal collections behind the anterior pararenal space.

Retroperitoneal fibrosis

Idiopathic retroperitoneal fibrosis

In idiopathic retroperitoneal fibrosis there is gradual transformation of the retroperitoneal fatty tissues into a retractile fibrosclerous plaque, which is centred around the aorta. Both very discreet and voluminous, tumour-like forms are seen. The fibrous plaque usually starts in front of the vessels at the level of L4–L5, below the aortic bifurcation and extends upward, enveloping the aorta, vena cava, and ureters. Its downward extension follows the path of the iliac vessels (Fig. 16). On unenhanced sections this plaque appears as a prespinal opacity blurring the contours of the aorta and inferior vena cava. The aorta lies against the vertebral body. After injection of the contrast material, the fibrous plaque density increases from 30 to 80 Hounsfield units. The ureters are almost always positioned anterolaterally or laterally with respect to the plaque, to which they often adhere. They are sometimes enveloped by the plaque over a short distance.

The major clinical consequence of this process is ureteral obstruction, which is often bilateral by the time it is discovered (Brun et al. 1981; Dalla-Palma et al. 1981).

Secondary retroperitoneal fibrosis

Retroperitoneal fibrosis can occur following medication (methysergide, ergotamine), retroperitoneal collections of blood, urine, or contrast material, surgical history (surgery of the aorta) or radiotherapy, diseases of the pancreas, intestines, kidneys and adjacent muscles and bones, metastases and aneurysms. The fibrous plaque appearance and location are similar to those of

1.6.4 Computed tomography and magnetic resonance imaging

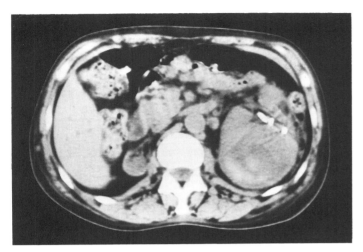

Fig. 19 Subcapsular haematoma of the left kidney. Scan without contrast. Areas of high density. Thickened capsule.

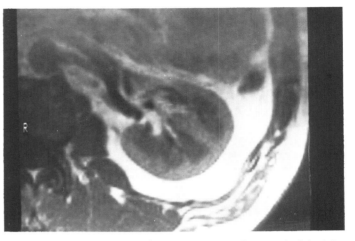

Fig. 20 T_1 weighted image (TR=415 ms and TE=20 ms) of the left kidney (transverse plane). Distinction between cortex (intermediate signal intensity) and medulla (low signal intensity) is readily achieved without contrast medium.

idiopathic retroperitoneal fibrosis (Tracy *et al.* 1979; Allibone and Saxton 1980; Lemaître *et al.* 1984).

Differential diagnosis

Unlike retroperitoneal fibrosis, lymphadenopathies readily displace the aorta and inferior vena cava forward and often have polycyclic contours. Retroperitoneal sarcomas usually give a stronger impression of mass, with anterior development. Secondary lymph node tumour involvement, sarcomas, and lymphomas may lead to dislocation of the psoas or bone involvement, neither of which has ever been reported in retroperitoneal fibrosis (Jing *et al.* 1982).

Retroperitoneal fluid collections

Renal subcapsular collections are the result of an accumulation of blood or pus. Trauma, either external or internal (e.g., renal needle biopsy), is the most frequent cause of subcapsular haematomas, and these collections exhibit the highest densities (80 Hounsfield units) during the acute haemorrhagic phase (Fig. 19). Those of infectious origin result from the peripheral extension of the parenchymal abscesses.

The subcapsular collections tend to be crescent shaped, and flatten or even depress the renal parenchyma. The non-parenchymal surface of the fluid is surrounded by the renal capsule, which is often thickened and hyperdense, either on unenhanced scans or after intravenous opacification (Fig. 19).

Perirenal collections, whether of inflammatory origin (extension of a kidney infection), haematic (following rupture of the capsule), or urinary (through a ruptured excretory duct) do not change the parenchymal surface but, given their volume, can shift the positions of the urinary tract components (Alexander *et al.* 1981; Hilton *et al.* 1981; Morettin and Kummar 1981; Pummil 1981; Renouard *et al.* 1984).

MAGNETIC RESONANCE IMAGING OF THE KIDNEY

Introduction

The multiplanar imaging capability and excellent soft tissue contrast of magnetic resonance imaging (MRI) initially suggested

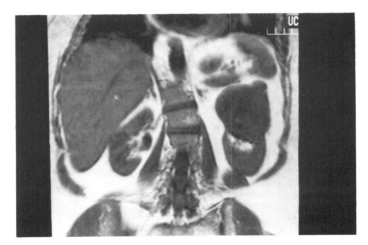

Fig. 21 T_1 weighted image (TR=415 ms and TE=20 ms) of the kidneys (coronal plane). Huge simple cyst at the upper pole of the left kidney is imaged with homogeneous low signal intensity.

that it would replace CT for imaging the urinary tract. For example, cortex and medulla can be distinguished by means of T_1 weighted images, without need for contrast medium (Fig. 20). Comparison between MRI and CT findings have been used mainly in the diagnosis of renal masses and staging of malignant tumours.

Detection of renal masses

On spin echo images, renal cell carcinomas vary from hypointense to hyperintense compared with surrounding renal parenchyma. This depends to some extent on the pulse sequence used (Hricak *et al.* 1985b). Since, in some instances, the signal intensity of renal cell carcinoma may be similar to that of normal surrounding renal parenchyma on both T_1 and T_2 weighted images (Fig. 21) MRI cannot be used to screen for renal tumours (Hricak *et al.* 1988; Quint *et al.* 1988). Tumours with an exophytic component are easy to detect because of the deformation of the

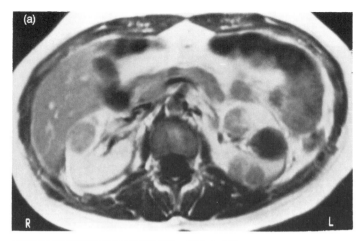

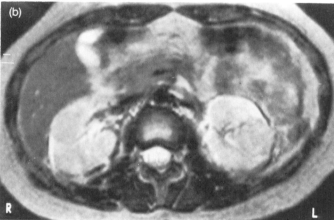

Fig. 22 (a)T_1 weighted image (TR=415 ms and TE= 20 ms) of the kidneys (transverse plane) with Gd-DTPA intravenous injection of contrast medium. A left-sided simple renal cyst is imaged with low signal intensity (no contrast uptake) while multiple solid bilateral renal tumours are imaged with intermediate signal intensity compared to the high signal intensity of the normal renal parenchyma. Biopsy-proven multiple renal metastases from thymus carcinoma. (b) T_2 weighted image (TR=1650 ms and TE=100 ms) at the same level. Simple cyst, multiple solid tumours, and normal renal parenchyma are imaged with almost the same signal intensity.

renal contours (even if the tumour is imaged with similar signal intensity); however, a small tumour which does not distort the renal contour may be very difficult to detect.

Tissue characterization of renal masses

The excellent soft tissue contrast of MRI makes distinction between simple renal cysts, complex renal cysts, and solid renal masses relatively easy (Marotti et al. 1987). Simple renal cysts appear hypointense on T_1 weighted images and hyperintense on T_2 weighted images because of the very long T_1 and T_2 relaxation times of simple fluid (Fig. 22). Haemorrhagic renal cysts are usually imaged with a very high signal intensity on both T_1 and T_2 weighted images, because of the very short T_1 relaxation time of blood (in the subacute phase). The signal intensity of solid renal masses may vary considerably on both T_1 and T_2 weighted images (Hricak et al. 1985): the lesion often has irregular margins and may show a non-homogeneous appearance due to partial haemorrhage and/or necrosis.

Use of intravenous contrast media (for example Gd-DTPA) will probably improve the differential diagnosis between these various entities (Fig. 22(a)).

Staging of renal neoplasms

MRI is accurate and particularly useful in detecting venous tumour extension (Hricak et al. 1985a). Rapidly flowing blood in the renal and inferior vena cava does not normally produce a detectable MR signal. Venous tumour thrombus causes an intraluminal signal which is easily demonstrated against the zero signal intensity of the residual free lumen (Hricak et al. 1985a)). Direct sagittal MRI is particularly helpful in evaluating the superior extent of caval tumour thrombus relative to the diaphragm, hepatic veins, and right atrium.

MR demonstration of lymph nodes is very easy, again because of the excellent soft tissue contrast between lymph nodes (intermediate signal intensity), surrounding fat (high signal intensity), and vascular structures (zero signal because of rapidly flowing blood) on T_1 weighted images (Dooms et al 1984). However, characterization of lymphadenopathy using MR relaxation times is not better than with CT (Dooms et al. 1985) and detection of lymphadenopathy either by CT or MRI is primarily based on the lymph-node size.

Conclusion

At present, MRI cannot be used as a screening modality for renal diseases, particularly tumours. However, it is an excellent staging modality that should be used when the CT findings are equivocal (mainly for venous tumour extension and lymphadenopathy).

References

Alder, J., Greweldinger, J., Hallac R., and Frier S. (1981). Computed tomographic findings in a case of renal vein thrombosis with nephrotic syndrome. *Urologic Radiology*, 3, 181–3.

Alexander, E.S., Colley, D.P., and Clark R.A. (1981). Computed tomography of retroperitoneal fluid collections. *Seminars in Roentgenology*, 16, 268–276.

Allibone, G.W. and Saxton, H.M. (1980). The association of aorto-iliac aneurysms with ureteral obstruction. *Urological Radiology* 1, 205–10.

Amendola, M.A., et al. (1988). Small renal cell carcinomas: resolving a diagnostic dilemma. *Radiology*, 166, 637–41.

Amis, E.S., Cronan, J.J., and Pfister, R.C. (1982). Pseudohydronephrosis on noncontrast tomography. *Journal of Computed Assisted Tomography*, 6, 511–13.

Balfe, D.M., McClennan, B.L., Stanley, R.J., Weyman, P.J., and Sagel, S.S. (1982). Evaluation of renal masses considered indeterminate on computed tomography. *Radiology*, 142, 421–8.

Baron, R.L., McClennan, B.L., Lee, J.K.T., and Lawson, J.L. (1982). Computed tomography of transitional cell carcinoma of the renal pelvis and ureter. *Radiology*, 144, 125–30.

Benson, A., Haaga, J.R., and Resnick, M.I. (1989). Staging renal carcinoma: what is sufficient? *Archives of Surgery*, 123, 71–3.

Bosniak, M.A. (1981). Angiomyolipoma (hamartoma) of the kidney; a preoperative diagnosis is possible in virtually every case. *Urologic Radiology*, 3, 135–42.

Bosniak, M.A. (1986). The current radiological approach to renal cysts. *Radiology*, 158, 1–10.

Bosniak, M.A., Megibow, A.J., Hulnick, D.H., Horii, S., and Raghavendra B.N. (1988). CT diagnosis of renal angiomyolipoma: the

1.6.4 Computed tomography and magnetic resonance imaging

importance of detecting small amounts of fat. *American Journal of Radiology*, **151**, 497–501.

Brun, B., Laursen, K., Sorenson, I.N., Lorentzen, J.E., and Kristensen, J.K. (1981). CT in retroperitoneal fibrosis. *American Journal of Roentgenology*, **137**, 535–8.

Bruneton, J.M., Drouillard, J., Caramella, E., and Manzino, J.J. (1984). Lymphomes du rein, Intérêt de l'échographie et de la scanographie. *Journal de Radiologie*, **65**, 755–60.

Buschi, A.J., et al. (1980). Distended left renal vein: CT/and sonographic normal variant. *American Journal of Roentgenology*, **135**, 339–42.

Cates, J.D., Foley, W.D., and Lawson, T.L. (1986). Retrograde opacification of the renal vein: a CT sign of renal artery avulsion. *Urological Radiology*, **8**, 92–4.

Choyke, P.L., White, E.M., Zeman, R.K., Jaffe, M.H., and Clark, L.R. (1987). Renal metastases: clinicopathologic and radiologic correlation. *Radiology*, **162**, 359–63.

Cohan, R.H., Dunnick, N.R., Degesys, G.E., and Korobkin, M. (1984). Computed tomography of renal oncocytoma. *Journal of Computed Assisted Tomography*, **8**, 284–87.

Coleman, C.C., Saxena, K., and Johnson, K.W. (1980). Renal vein thrombosis in a child with the nephrotic syndrome: CT diagnosis. *American Journal of Roentgenology*, **135**, 1285–6.

Cronan, J.J., Zeman, R.K., and Rosenfield, A.T. (1982). Comparison of computerized tomography, ultrasound and angiography in staging renal cell carcinoma. *Journal of Urology*, **127**, 712–14.

Cronan, J.J., Amis, E.S., Zeman, R.K., and Dorfman, G.S. (1986). Obstruction of the upper-pole moiety in renal duplication in adults: CT evaluation. *Radiology*, **161**, 17–21.

Curry, N.S., Gobien, R.P., and Schabel, S.I. (1982). Minimal dilatation obstructive nephropathy. *Radiology*, **143**, 531–4.

Dalla-Palma, L., Rocca-Rossetti, S., Pozzi-Mucelli, R.S., and Rozzato, G. (1981). Computed tomography in the diagnosis of retroperitoneal fibrosis. *Urologic Radiology*, **3**, 77–83.

Das, S., Egan, R.M., and Amar, A.D. (1981). Von Hippel-Lindau syndrome with bilateral synchronous renal cell carcinoma. *Urology*, **18**, 599–600.

Davis, J.M. and McLaughlin, A.P. (1987). Spontaneous renal hemorrhage due to cyst rupture: CT findings. *American Journal of Roentgenology*, **148**, 763–4.

Demos, T.C., Gadwood, K., Love, L., and Engel, G. (1982). The rim nephrogram in renovascular compromise. *Urologic Radiology*, **4**, 227–30.

Dinsmore, B.J., Pollack, H.M., and Banner M.P. (1988). Calcified transitional cell carcinoma of the renal pelvis. *Radiology*, **167**: 401–4.

Dooms G.C., Hricak H., and Crooks L.E. (1984). Magnetic resonance imaging of the lymph nodes: comparison with CT. *Radiology*, **153**, 719–728.

Dooms G.C., Hricak H., and Moseley M.E. (1985). Characterization of lymphadenopathy by magnetic resonance relaxation times: preliminary results. *Radiology*, **155**, 691–7.

Downey, E.F., Friedman, A., Hatman, D.S., Pyatt, R.S., Thane, T.T., and Warnock, G.R. (1982). Pseudocystic CT pattern of renal sinus lipomatosis. *Radiology*, **144**, 840.

Federle, M.P., Kaiser, J.A., and McAninch, J.W. (1981). The role of computed tomography in renal trauma. *Radiology*, **41**, 455–60.

Federle, M.P. (1983). Computed tomography of blunt abdominal trauma. *Radiologic Clinics of North America*, **21**, 461–75.

Federle, M.P., Brown, T.R., and McAninch, J.W. (1987). Penetrating renal trauma: CT evaluation. *Journal of Computed Assisted Tomography*, **11**, 1026–30.

Fishman, M.C., Pollack, H.M., Arger, P.H., and Banner, M.P. (1983). High protein content: another cause of CT hyperdense benign renal cyst. *Journal of Computed Assisted Tomography*, **7**, 1103–6.

Friedland, G.W. (1987). Shrinking and disappearing renal cysts. *Urologic Radiology*, **9**, 21–25.

Gatewood, O.M., Fishman, E.K., Burrow, C.R., Walker, W.G., Goldman, S.M., and Siegelman, S.S. (1986). Renal vein thrombosis in patients with nephrotic syndrome: CT diagnosis. *Radiology*, **159**, 117–22.

Glazer, G.M. and London, S.S. (1981). CT appearance of global renal infarction. *Journal of Computed Assisted Tomography*, **5**, 847–50.

Glazer, G.M., Francis, I.R., Brady, T.M., and Teng, S.S. (1983). Computed tomography of renal infarction: clinical and experimental observation. *American Journal of Roentgenology*, **140**, 721–7.

Goergen, T.G., Lindstrom, R.R., Tan, H., and Lilley, J.J. (1981). CT appearance of acute renal cortical necrosis. *American Journal of Roentgenology*, **137**, 176–7.

Gold, R.P., McClennan, B.L., and Rottenberg, R.R. (1983). CT appearance of acute inflammatory disease of the renal interstitium. *American Journal of Roentgenology*, **141**, 343–9.

Goldman, S.M., Hartman, D.S., Fishman, E.K., Finizio, J.P., Gatewood, O.M.B., and Siegelman, S.S. (1984). CT of xanthogranulomatous pyelonephritis, **141**, 963–9.

Greene, A., Cromie, W., and Goldman, M. (1982). Computerized body tomography in neonatal renal vein thrombosis. *Urology*, **19**, 213–15.

Hara, M., Yoshida, K., Tomita, M., Akimoto, M., Kawai, H., and Fukuda, Y. (1982). A case of bilateral renal oncocytoma. *Journal of Urology*, **127**, 576–8.

Harris, R.D., and Dorros, S. (1981). Computed tomographic diagnosis of renal infarction. *Urology*, **17**, 287–9.

Hartman. D.S., Goldman, S.M., Friedman, A.C., Davis, Ch.J., Madewell, J.E., and Sherman, J.L. (1981). Angiomyolipoma: ultrasonic-pathologic correction. *Radiology*, **139**, 451–8.

Hartman, D.S., Davis, C.J., Jr. Goldman, S.M., Friedman, A.C., and Fritsche, P. (1982). Renal lymphoma: radiologic-pathologic correlation of 21 cases. *Radiology*, **144**, 759–66.

Hartman, D.S., Davis, H.J., Johns, T., and Goldman S.M. (1986). Cystic renal cell carcinoma. *Urology*, **28**; 145–53.

Hartman, D.S., Davis, C.J., Sanders, R.C., Johns, T.T., Smirniotopoulos, J., and Goldman, S.M. (1987). The multiloculated renal mass: considerations and differential features. *Radiographics*, **7**, 29–52.

Hata, Y., Tada, S., Kato, Y., Onishi, T., Masuda, F., and Machida, T. (1983). Staging of renal cell carcinoma by computed tomography. *Journal of Computed Assisted Tomography*, **7**, 828–32.

Hillman B.J., Drach, G.W., Tracey, P., and Gaines, J.A. (1984). Computed tomographic analysis of renal calculi. *American Journal of Roentgenology*, **142**, 549–52.

Hilton, S., Bosniak, M.A., Megibow, A.J., and Ambos, M.A. (1981). Computed tomographic demonstration of a spontaneous subcapsular hematoma due to a small renal cell carcinoma. *Radiology*, **141**, 743–44.

Hoddick, W., Jeffrey, R.B., Goldberg, H.I., Federle, M.P., and Laing F.C. (1983). CT and sonography of severe renal and perirenal infections. *American Journal of Roentgenology*, **140**, 517–20.

Hoffman, E.P., Mindelzun, R.E., and Anderson, R.U. (1980). Computed tomography in acute pyelonephritis associated with diabetes. *Radiology*, **135**, 691–5.

Hricak, H. (1987). Urologic cancer. Methods of early detection and future developments. *Cancer*, **60**, 677–85.

Hricak H., Amparo E., and Fisher, M.R. (1985a). Abdominal venous system: assessment using MR. *Radiology*, **156**, 415–22.

Hricak, H., Demas, B.E., and Williams, R.D. (1985b). Magnetic resonance imaging in the diagnosis and staging of renal and perirenal neoplasms. *Radiology*, **154**, 709–15.

Hricak, H., Thoemi R., and Caroll P. (1988). Detection and staging of renal neoplasms: a reassessment of MR imaging. *Radiology*, **166**, 643–9.

Ishikawa, I., Matsuura, H., Onouchi, Z., and Susuki, M. (1982). CT appearance of the kidney in traumatic renal artery occlusion. *Journal of Computed Assisted tomography*, **6**, 1021–4.

Ishikawa, I., et al. (1985). Delayed contrast enhancement in acute focal bacterial nephritis: CT features. *Journal of Computed Assisted tomography*, **9**, 894–7.

Jabour, B.A., et al. (1987). Acquired cystic disease of the kidneys. *Investigative Radiology*, **22**, 728–32.

Jeffrey, R.B. and Federle, M.P. (1983). CT and ultrasonogrpahy of acute renal abnormalities. *Radiologic Clinics of North American*, 21, 515–25.

Jing, B.S., Wallace, S., and Zornoza, J. (1982). Metastases to retroperitoneal and pelvic lymph nodes. Computed tomography and lymphangiography. *Radiologic Clinics of North America*, 20, 511–30.

Karp, W., Ekelund, L., Olafsson, G., and Olsson, A. (1981). Computed tomography, angiography and ultrasound in staging of renal carcinoma. *Acta Radiologica*, 22, 625–33.

Kolmannskog, F., Kolbenstvedt, A., Nakstad, P.H., and Aakhus, T. (1981). Computer tomography and angiography in renal angiomyolipoma. *Acta Radiologica*, 22, 635–9.

Kullendorff, B., Nyman, U., and Aspelin, P. (1987). Computed tomography in renal replacement lipomatosis. *Acta Radiogica*, 28, 447–50.

Lautin, E.M., *et al.* (1981). Radionuclide imaging and computed tomography in renal oncocytoma. *Radiology*, 142, 185–90.

Lee, J.K.T., Mc Clennan, B.L., Melson, G.L., and Stanley, R.J. (1980). Acute focal bacterial nephritis: emphasis on gray scale sonography and computed tomography. *American Journal of Roentgenology*, 135, 87–92.

Lemaître, G. and Delambre, Y. (1983). Place de la tomodensitométrie dans le diagnostic et le bilan d'extension des tumeurs malignes du rein. *Journal de Radiologie*, 64, 91–8.

Lemaître, G., Renound, O., and Kasbarian, M. (1984). Usefulness of computed tomography in the diagnosis and follow-up of retroperitoneal fibrosis. *Journal Belge de Radiologie*, 67, 13–19.

Lemaître, G., Dardenne, A.N., and Caron-Poitreau, C. (1986). Tomodensitométrie du rein et du rétropéritoine. Collection de monographies du traité de radiodiagnostic. Masson, Paris, 119 pp.

Levine, E., Kyo Rek Lee, P.D., Weigel, J.W., and Farber, B. (1979). Computed tomography in the diagnosis of renal carcinoma complicating Hippel-Lindau syndrome. *Radiology*, 130, 703–6.

Levine, E. and Grantham, J.J. (1981). The role of computed tomography in the evaluation of adult polycystic kidney disease. *American Journal of Kidney Disease*, 1, 99.

Love, L., Lind, J.A., and Olson, M.C. (1989). Persistent CT nephrogram: significance in the diagnosis of contrast nephropathy. *Radiology*, 172, 125–9.

Luers, P.R. and Siegler, R.L. (1980). CT demonstration of cortical nephrocalcinosis in congenital oxalosis. *Pediatric Radiology*, 10, 116–18.

Maillet, P.J., Pelle-Francoz, D., Laville, M., Gay, F., and Pinet, A. (1986). Nondilated obstructive acute renal failure: diagnostic procedures and therapeutic management. *Radiology*, 160, 659–62.

Marotti M., Hricak H., Fritzsche P. (1987). Complex and simple renal cysts: comparative evaluation with MR imagiang. *Radiology*, 162, 679–84.

McClennan, B.L. (1979b). Current approaches to the azotemic patient. *Radiologic Clinics of North America*, 17, 197–211.

McClennan, B.L., Stanley, R.J., Melson, G.L., Levitt, R.G., and Sagel, S.S. (1979a). CT of the renal cyst: is cyst aspiration necessary? *American Journal of Roentgenology*, 133, 671–5.

Meziane, M.A., Fishman, E.K., Goldman, S.M., Friedman, A.C., and Siegelmann, S.S. (1986). Computed tomography of high density renal cysts in adult polycystic kidney disease. *Journal of Computed Assisted Tomography*, 10, 767–70.

Morettin, L.B. and Kumar, R. (1981). Small renal carcinoma with large retroperitoneal hemorrhage; diagnosis considerations. *Urologic Radiology*, 3, 143–8.

Newhouse, J.H., Prien, E.L., Arnis, E.S., Dretler, S.P., Pfister, R.C. (1984). Computed tomographic analysis of urinary calculi. *American Journal of Roentgenology*, 142, 545–8.

Older, R.A., Korobkin, M., Cleeve, D.M., Schaaf, R. and Thompson, W. (1980). Contrast induced acute renal failure; persistent nephrogram as clue to early detection. *American Journal of Roentgenology*, 134, 339–42.

O'Reilly, P.H. (1982). Role of modern radiological investigations in obstructive uropathy. *British Medical Journal*, 284, 1847–51.

Papanicolaou, N., Pfister, R.C., and Yoder, I.C. (1986). Spontaneous and traumatic rupture of renal cysts: diagnosis and outcome. *Radiology*, 160, 99–103.

Parienty, R.A., Pradel, J., Imbert, M.C., Picard, J.D., and Savart, P. (1981). Computed tomography of multilocular cystic nephroma. *Radiology*, 140, 135–9.

Parikh, S.J., Peeters, J.C., and Kihm, R.H. (1981). The anormalous left renal vein. Computed tomographic appearance and clinical implication. *Computed Tomography*, 5, 529–33.

Parker, M.D. (1981). Acute segmental renal infarction: difficulty in diagnosis despite multimodality approach. *Urology*, 18, 523–6.

Patterson, J., Briscoe G., Lohr, D., and Flanigan R.C. (1987). Calcified renal masses. *Urology*, 29, 353–6.

Pearlstein, A.E. (1983). Hyperdense renal cysts. *Journal of Computed Assisted Tomography*, 7, 1029–31.

Pollack, H.M. and Wein, A.J. (1989). Imaging of renal trauma. *Radiology*, 172 297–308.

Pollack, H.M., Arger, P.H., Banner, M.P., Mulhern, C.B. Jr., and Coleman, B.G. (1981). Computed tomography of renal pelvic filling defects. *Radiology*, 138, 645–51.

Probst, P., Hoogewoud, H.M., Haertel, M., Zingg, E., and Fuchs, W.A. (1981). Computerized tomography versus angiography in the staging of malignant renal neoplasm. *British Journal of Radiology*, 54, 744–53.

Pummill, CH.L. (1981). Lesions involving the perirenal space. *Seminars in Roentgenology*, 16, 237–8.

Quint, L.E., Glazer G.M., Chenevert, Th.L. (1988). In vivo and in vitro MR imaging of renal tumors: histopatholgic correlation and pulse optimization. *Radiology*, 169, 359–62.

Rauschkolb, E.N., Sandler, C.M., Patel, S., and Childs, T.L. (1982). Computed tomography of renal inflammatory disease. *Journal of Computed Assisted Tomography*, 6, 502–6.

Reed, M.D., Friedman, A.C., and Nealey, P. (1982). Anomalies of the left renal vein. Analysis of 433 CT scans. *Journal of Computed Assisted Tomography*, 6, 1124–6.

Renouard, O., Mazeman, E., Locquet, Ph., and Lemaître, G. (1984). Apports de la tomodensitométrie et de l'échogaphie dans le diagnostic des hématomes sous-capsulaires du rein. *Journal de Radiologie*, 65, 645–51.

Rigsby, C.M., Rosenfield, A.T., Glickman, M.G., and Hodson, J. (1986). Hemorrhagic focal bacterial nephritis: findings on gray-scale sonography and CT. *American Journal of Roentgenology*, 146, 1173–7.

Robson, Ch.J., Churchill, B.M., and Anderson, W. (1969). The results of radical nephrectomy for renal carcinoma. *Journal of Urology*, 101, 297–301.

Rosenfield, A.T., *et al.* (1987). Gray scale ultrasonography, computerized tomography and nephrotomography in evaluation of polycystic kidney and liver diseases. *Urology*, 9, 436–8.

Royal, S.A. and Callen, P.W. (1979). Computed tomographic evaluation of anomalies of the inferior vena cava and left renal vein. *American Journal of Roentgenology*, 132, 759–63.

Sandler, C.M. and Toombs, B.D. (1981). Computed tomographic evaluation of blunt renal injuries. *Radiology*, 141, 461–6.

Segal, A.J. and Spataro, R.F. (1982). Computed tomography of adult polycystic disease. *Journal of Computer Assisted Tomography*, 6, 777–80.

Shirkhoda, A. and Lewis, E. (1987). Renal sarcoma and sarcomatoid renal cell carcinoma: CT and angiographic features. *Radiology*, 162, 353–7.

Slasky, B.S. and Lenkey, J.L. (1981). Acute renal failure, contrast media, and computer tomography. *Urology*, 18, 309–13.

Slasky, B.S., and Wolfe, P.W. (1982). Cross-sectional imaging of multilocular cystic nephroma. *Journal of Urology*, 128, 128–31.

Smith, S.J., Bosniak, M.A., Megibow, A.J., Hulnick, D.H., Horii,

S.C., and Raghavendra, B.N. (1989). Renal cell carcinoma: earlier discovery and increased detection. *Radiology*, **170**, 699–703.
Soffer, O., Miller, L.R., and Lichtman, J.B. (1987). CT findings in complications of acquired renal cystic disease. *Journal of Computer Assisted Tomography*, **11**, 905–8.
Soulen, M.C., Fishman G.K., Goldman S.M., and Gatewood O.M. (1989). Bacterial renal infection: role of CT. *Radiology*, **171**, 703–7.
Subramanyam, B.R., Megibow, A.J., Raghavendra, B.N., and Bosniak M.A. (1982). Diffuse xanthogranulomatous pyelonephritis: analysis by computed tomography and sonography. *Urologic Radiology*, **4**, 5–9.
Subramanyam, B.R., Bosniak, M.A., Horii, S.C., Megibow, A.J., and Balthazar, E.J. (1983). Replacement lipomatosis of the kidney: diagnosis by computed tomography and sonography. *Radiology*, **148**, 791–92.
Sussman, S., *et al.* (1984). Hyperdense renal masses: a CT manifestation of hemorrhagic renal cysts. *Radiology*, **150**, 207–211.
Thierman, D., Haaga, J.R., Anton, P., and Lipuma, J. (1983). Renal replacement lipomatosis. *Journal of Computer Assisted Tomography*, **7**, 241–3.
Thomsen, H.S., Madsen, J.K., Thaysen, J.H., and Damgaard-Petersen, K. (1981). Volume of polycystic kidneys during reduction of renal function. *Urological Radiology*, **3**, 85–9.
Tracy, D.A., Eisenberg, R.L., and Hedgcock, M.W. (1979). Ureteral obstruction resulting from vascular prosthetic graft surgery. *American Journal of Roentgenology*, **132**, 415–18.
Vasile, N., Lacrosniere, L., Abbou, C., and Larde, D. (1981). Extension veineuse des cancers du rein, rôle de la TDM. *Journal de Radiologie*, **62**, 615–20.
Velasquez, G., Glass, T.A., D'Souza, V.J., and Formanek, A.G. (1984). Multiple oncocytomas and renal carcinoma. *American Journal of Roentgenology*, **142**, 123–4.
Wadsworth, D.E., McClennan, B.L., and Stanley, R.J. (1982). CT of renal mass. *Urological Radiology*, **4**, 85–94.
Weyman, P.J., McClennan, B.L., Stanley, R.J., Levitt, R.G., and Sagel, S.S. (1980). Comparison of computed tomography and angiography in the evaluation of renal cell carcinoma. *Radiology*, **137**, 417–24.
Winfield, A.C., Gerlock, A.J., and Shaff, M.I. (1981). Perirenal cobwebs: a CT sign of renal vein thrombosis. *Journal of Computer Assisted Tomography*, **5**, 705–8.
Zeman, R.K., Cronan J.Y., Rosenfield A.T., Lynch, J.H., Jaffe, M.H., and Clark Z.R. (1988). Renal cell carcinoma: dynamic thin-section CT assessment of vascular invasion and tumor vascularity. *Radiology*, **167**, 393–6.

1.6.5 Renal angiography

F. JOFFRE

Renal arteriography

In the 1960s and 1970s, renal arteriography was considered one of the most valuable techniques for assessing renal masses, renovascular disease, renal trauma and renal bleeding. Since the 1980s, the use of arteriography has been challenged by new techniques, such as renal ultrasonography and computed tomography (CT). Concomitantly, the techniques of arterial catheterization and contrast media have improved, and digital techniques have greatly simplified collection of the image. Thus arteriography is being used less, whereas endovascular interventional radiology has developed rapidly in the last decade.

Technical considerations

Opacification of the renal arteries can be obtained either by intra-arterial injection or by peripheral intravenous injection of the radiological contrast medium.

Intra-arterial injection

Intra-arterial injection (total arteriography in front position) is usually performed by retrograde femoral catheterization. It is essential to be able to visualize the renal artery ostia: this may require additional injections in an oblique position in renovascular disease, particularly when there are severe aortic atheromatous lesions. Selective catheterization of one or both renal arteries is often needed to evaluate the intrarenal vasculature, particularly in renal tumours.

This technique is currently performed under local anaesthesia, after simple premedication. The narrow catheters now used carry a decreased risk of femoral artery puncture, allowing the investigation to be performed in 1 day. The safety of the procedure has been widely improved: local complications after puncture (haematoma, arterial thrombosis) are now rare (less than 0.5 per cent). In addition, the use of new, low osmolality contrast media may contribute to reducing the renal toxicity observed with classical contrast media, although this has not so far been demonstrated. However, these new contrast media have the advantage of making the aortic injection almost painless.

Peripheral intravenous injection

Intravenous injection, combined with digital data collection, allows satisfactory opacification of the aorta and the renal artery trunks in 95 per cent of cases (Fig. 1). The nephrogram thus obtained allows the size and contours of the kidneys to be determined. Complementary urography is needed to evaluate the morphology of the urinary tract. These investigations can be combined with selective renal vein sampling and can be accomplished on an outpatient basis. The only contraindications are those normally associated with the use of contrast materials (Chapter 1.6.2). This technique has some disadvantages: the quality of the opacification obtained is not sufficient to allow precise evaluation of the renal artery branches and the intrarenal vasculature. Morphological evaluation of trunk lesions may not be sufficiently precise for choice of therapy: in approximately 5 per cent of cases, the patient's anatomic and haemodynamic con-

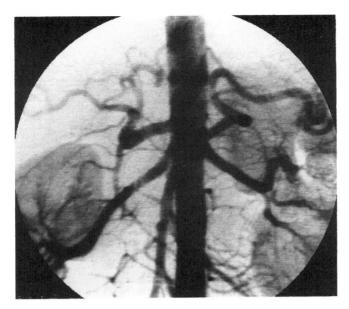

Fig. 1 Intravenous digital subtraction angiogram: showing renal arteries and their branches.

ditions do not allow correct evaluation of the renal artery trunks (Hillman et al. 1982). Anatomic conditions may also necessitate repeated injections, which require the use of high doses of contrast media (60 ml per injection) to be efficient. Intravenous digital angiography is a good screening test but may need to be complemented by direct arteriography.

Image collecting

Rapid seriography is the standard method for producing an angiographic image. Its main advantage is its high definition, which reveals the renal arteries as far as the interlobar and arcuate arteries. In the last few years, digital techniques, which assign a numerical value to each point of the collected image on an intensifier, have progressively replaced standard radiology. This digital processing, which replaces the standard photographic and analogue process, allows almost instantaneous collection of information from multiple operations, such as subtraction (removal of some superimpositions) and contrast enhancement to be performed (Harrington et al. 1982; Hillman 1985). Digital angiography has the advantages of allowing intravenous investigations to be performed, using intra-arterial doses of contrast medium, which are 50 per cent lower than conventional doses (this does not apply to peripheral intravenous administration). In addition, the injection flow rate can be reduced, along with the diameter of the catheter. Patient and operator exposure to irradiation is reduced due to the shorter examination time and small area photographed. There are, however, disadvantages regarding the quality of the image, which is inferior to that obtained by conventional radiologic films. Obesity, the presence of intestinal gas, and incomplete apnoea may also reduce the quality of the image and create misleading interference images or artefacts (Fig. 2).

Normal arteriogram

Whatever the injection method, arteriographic investigation of the renal vasculature takes place in three successive phases: arterial, parenchymatous, and venous. The arterial phase is very short, about 2 s, and allows evaluation of the renal pedicle and the intrarenal branches. Thirty per cent of patients will have abnormalities of the renal artery, which arise from the lateral aorta. Before they enter the hilum of the kidney, these arteries give rise to extrarenal branches, which can be seen by selective injection of contrast medium. These arteries which supply the perirenal arterial circle and the most superficial zone of the renal cortex (cortex corticis) through the perforans arteries, undergo hypertrophy when the parenchyma is ischaemic.

Although the intrarenal vasculature is extremely variable from one side to the other and from one patient to another, the most typical scheme is the subdivision into one prepyelic branch to the anterior half of the kidney and the inferior pole, and one retropyelic branch which reaches the posterior half and the superior pole of the kidney. Each of these arteries gives rise to lobar and interlobar branches which pass through the Bertin's columns and form the arcuate arteries by curving at the basis of the Malpighian pyramids. The arcuate artery branches (interlobular arteries to the glomeruli and the vasa recta of the medulla) are not visualized because they are masked by the parenchymatous opacification, which occurs within 2 s of the contrast medium being injected. This opacification (nephrography) is mainly cortical and induced by the opacification of the glomerular and peritubular capillaries. The cortex and its internal radiated extensions, which correspond to Bertin's columns, are clearly demonstrated, whereas the Malpighian pyramids are not opacified. The returning venous phase appears after 10 s with opacification of the renal vein which is usually faint and depends on the injected dose. Usually, only the renal vein trunk is visualized; this allows the patency of this vein to be confirmed.

Indications

Apart from endoluminal techniques of interventional radiology which are considered below, current indications for renal arteriography are limited (Michel 1980); it is chiefly used in the

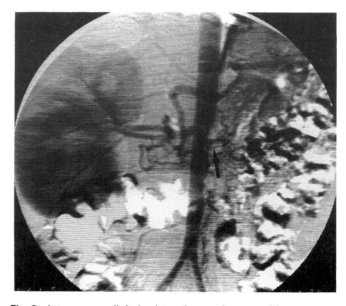

Fig. 2 Intravenous digital subtraction angiogram with poor visualization of the right renal artery. The left renal artery (arrow) and the left kidney cannot be evaluated because of artefacts created by superimpositions of gastrointestinal gas.

1.6.5 Renal angiography

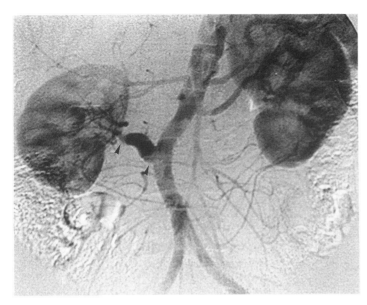

Fig. 3 Intravenous digital angiography: control of the patency of right aortorenal bypass. Excellent visualization of the bypass (arrow) and of the nephrography.

screening of renovascular disease (Hillman *et al.* 1983). Intravenous digital angiography is a good screening method, the limits of which have been discussed above (Smith *et al.* 1982), and which has replaced the inadequate method of intravenous urography (Thornbury 1982). The choice of investigations in the screening and detection of renal artery stenosis in hypertensives (as well as the cost/effectiveness ratio) is discussed in Chapter 17.7. When intravenous angiography yields insufficient evidence and additional angiographic information is necessary, conventional renal arteriography should be performed (Wilms *et al.* 1986). This is often the first phase of transluminal angioplasty. Follow-up after surgical or radiologic repair of renal artery stenosis represents an excellent indication for intravenous digital angiography (Novick *et al.* 1982) (Fig. 3).

Rapid arteriography is indicated in conditions where there is acute ischaemia of the renal parenchyma, including renal artery embolism, acute thrombosis, aortic dissection extending to the renal arteries, or traumatic obstruction of the renal artery. The results allow the institution of adequate management, including surgical repair if needed. Renal arteriography may also be used to screen for intrarenal arterial aneurysms, frequently associated with stenosis or obstruction of the intrarenal branches, when classical macroscopic polyarteritis nodosa is suspected (Helenon 1977). In these cases, it must be performed before renal biopsy.

Renal arteriography is of limited use in the investigation of renal masses (Mauro *et al.* 1982; Prager *et al.* 1984). It may, however, be helpful when carcinoma develops in a functional or anatomically solitary kidney and partial surgical removal is considered (Engelmann *et al.* 1984). In unexplained recurrent or persistent gross haematuria, renal arteriography may reveal vascular malformations such as cirsoid aneurysm, angioma, or congenital arteriovenous fistula. When an intrarenal vascular lesion is suspected, for example following renal needle biopsy, renal arteriography can establish the nature of the lesion (false aneurysm or arteriovenous fistula) and evaluate the possibility of endovascular therapy. Vascular lesions of transplanted kidneys including acute arterial obstruction in the immediate post-transplant period, cortical necrosis due to hyperacute rejection and hypertension induced by renal artery stenosis can also be detected. In these cases, renal arteriography is preferred to intravenous digital angiography (Irving and Khoury 1983; Tessier *et al.* 1986). The indication for arteriography in renal trauma patients is restricted to those in whom conservative surgery is being considered.

Endovascular interventional radiology

Selective catheterization of the renal artery allows two major categories of endovascular therapy to be performed: vascular occlusion, of which embolization is the most important, and vascular dilatation, mainly as percutaneous transluminal angioplasty. Although procedures represent a definite therapeutic advance, their use must result from close collaboration between radiologists and the medical and surgical team and the advantages and risks of each method must be considered (Mitty and Gribetz 1982).

Renal embolization

Renal embolization involves the introduction of materials which tend to occlude the vascular lumen of the renal arterial tree, via an appropriate catheter. The site, size, and permanence of vascular obstruction can be varied according to the clinical indication. Materials which can be used in different clinical situations include, spongel fragments, lyophized dura mater, polymer paste, absolute alcohol, balloon release, and metallic spirals. Injection of drugs can be combined with the embolization aims at procedure (chemoembolization).

Preoperative embolization

This aims to decrease the blood supply to a tumour, facilitating surgery by reducing blood loss and making dissection easier (Almgard *et al.* 1973). There is controversy over the usefulness of this technique, which should be restricted to large, highly vascular tumours.

Symptomatic management by embolization

Palliative embolization aims at relieving haematuria or pain, usually caused by inoperable tumours (Nurmi *et al.* 1987). The effects of such therapeutic intervention are often temporary; however, embolization represents a particularly efficient and attractive treatment for post-traumatic haematuria, since the catheterization allows the maximum amount of intact parenchyma to be preserved. It is particularly well suited in haematuria caused by needle biopsy.

'Medical nephrectomy'

It is possible to destroy the renal parenchyma almost totally, particularly if there is an existing major impairment, by sclerosing the whole renal vasculature. This procedure may be an efficient alternative to bilateral nephrectomy (Pinet and Lyonnet 1981), particularly in patients with malignant refractory hypertension (McCarron *et al.* 1976) or severe nephrotic syndrome with major protein loss.

The risks of embolization are moderate. Adverse effects related to renal infarction are constant and transitory, and consist of lumbar pain, and sometimes fever and hyperleucocytosis. The migration of embolic fragments into the aortic lumen must be avoided by careful radioscopic control during the injection.

An abscess may develop within tumoural necrosis, requiring antibiotic therapy (Lammer *et al* 1985).

Percutaneous transluminal angioplasty

Percutaneous transluminal angioplasty was first performed on renal arteries in 1978 by Gruntzig, since when it has been increasingly considered as the initial approach to renal artery stenosis whenever possible (Sos *et al.* 1983). It is indicated in patients with isolated hypertension due to renal artery stenosis (renovascular hypertension) and in those with artery stenosis jeopardizing renal function (renovascular disease due to bilateral renal artery stenosis or stenosis in a solitary kidney). In the latter case, the main aim is to preserve the renal vasculature.

Technique

An appropriate metallic guide wire is introduced through the stenosis after previous arteriography. A catheter with a balloon at its end is placed on this guide wire. The size and length of the balloon are selected to be appropriate to the lesion and, when expanded, the arterial wall is stretched and the atheromatous plaque is crushed, resulting in dilation of the stenosis. This procedure is done under fluoroscopic and manometric controls.

Results

Angioplasty provides good long-term anatomical results in approximately 85 per cent of patients (Fig. 4). In rare cases dilatation cannot be achieved, e.g. when the stenosis is hard to pass through, impossible to dilate or elastic, recurring immediately after dilatation (Martin *et al.* 1986). Failures are particularly encountered in ostial stenoses secondary to an atheromatous plaque of the aortic wall (Cicuto *et al.*1981). Complications are rare and usually minor (haematoma at the puncture site, distal intrarenal thrombosis), although more severe complications can occur, including renal insufficiency due to overdose of contrast medium, renal artery thrombosis, or retroperitoneal haematoma. Immediate surgery is required in less than 1 per cent of patients. Mid-term recurrences are observed in 10 to 15 per cent of cases, and are closely related to the quality of the immediate result. It has been shown that the presence of more than 30 per cent residual stenosis and/or longitudinal dissection within the stenotic area are the main factors associated with recurrence. Clinical results depend mainly on the indication for treatment (Geyskes 1988). In renovascular hypertension, the aim is to cure or to make arterial hypertension more responsive to medical therapy. Clinical results are not always related to anatomic results: angiosclerotic changes in the contralateral kidney can maintain arterial hypertension in spite of successful dilatation. Clinical benefit is obtained in approximately 50 per cent of all atheromatous stenoses and in 85 per cent of stenoses due to fibromuscular dysplasia. When angioplasty is performed in an attempt to preserve renal function, results are less good. Patients in this group tend to be older hypertensives with atheromatous ostial lesions, often with renal insufficiency. A satisfactory clinical result (improvement or stabilization of renal failure) is obtained in approximately 50 per cent of cases (Pickering *et al.* 1986).

Indications

Indications for percutaneous trausnluminal angioplasty depend on the clinical condition of the patient and on the type of lesion (Bernadet *et al.* 1989). Angioplasty is the treatment of choice for

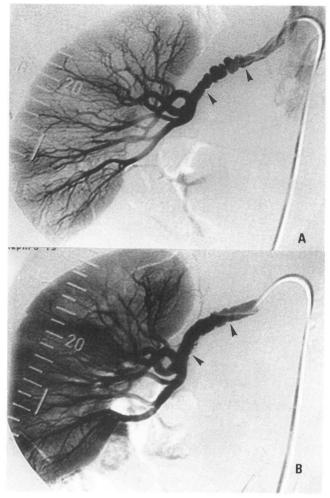

Fig. 4 A. Digital angiography of right renal artery shows multiple stenoses of the distal part of the renal artery. This suggests fibromuscular dysplasia by fibroplasia of the media (arrows).
B. After right renal artery angioplasty there is good patency, with persistence of some wall irregularity (arrows).

most renal artery stenoses. Its unquestionable advantages over surgery include simplicity, lower morbidity, almost no mortality, nearly similar results, and markedly lower cost (Miller *et al.* 1985). It can be repeated and does not prevent subsequent surgery. Contraindications usually include complete renal artery occlusions, complex and distal stenoses inaccessible to dilatation, renal artery dissections, and stenoses with poor downstream circulation. Patients with ostial stenoses should be considered individually, in view of frequent failures. Surgery, when it is possible, is restricted to those in whom angioplasty is contraindicated, or those who have had complications or failed angioplasty. Immediate recurrences may respond to the placement of an expandable endoprosthesis at the dilated area (Sigvart *et al.* 1987). This keeps the lumen open and restores the contours of the vessel wall (Fig. 5). Such endoprostheses are made of a cylinder of wire netting which becomes incorporated into the arterial wall within a few weeks. Preliminary results are encouraging—some patients have been successfully treated by the procedure with more than 1 year of follow-up (Joffre *et al.* 1989).

1.6.5 Renal angiography

versial, since newer imaging techniques—particularly MRI—allow safer and easier detection of renal vein thrombosis, whether of primary or of neoplastic origin. Only opacification of the inferior vena cava through an iliocavogram retains some indications, in cases where caval extensions of a renal vein thrombus (primary or neoplastic) cannot be diagnosed with certainty using other imaging methods. Selective catheterization of the renal veins for venous sampling is still an indication in the pretherapeutic evaluation of renovascular hypertension (Hillman 1989).

References

Abrams, H.L. (1961). Renal venography. In *Angiography* (ed. H.L. Abrams), pp. 915–25, Little Brown & Co., Boston.

Almgard, L.E., Fernstrom, J., and Haverlin, M. (1973). Treatment of renal adenocarcinoma by embolic occlusion of renal circulation. *British Journal of Urology*, **45**, 474–9.

Bernadet, P., Durand, D., Joffre, F., Cerene, A. and Suc, J.M. (1989). Sténose de l'artère rénale: nouvelles stratégies de diagnostic et de traitement. *La Revue du Praticien*, **10**, 855–7.

Cicuto, K.P., McLean, G.K., Oleaga, J.A., Ring, E.J., Freiman, D.B., and Grossman, R.A. (1981). Renal artery stenosis: anatomic classification for percutaneous transluminal angioplasty. *American Journal of Roentgenology*, **137**, 599–601.

Englemann, U., Schaub, T., and Schweden, F. (1984). Digital subtraction angiography in staging renal cell carcinoma: comparison with computerized tomography and histopathology. *Journal of Urology*, **137**, 248–51.

Geyskes, G.G. (1988). Treatment of renovascular hypertension with percutaneous transluminal angioplasty. *American Journal of Kidney Diseases*, **4**, 253–65.

Gruntzig, A., Kuhlmann, K., Verter, W., Lutolf, K., Meier B., and Siegenthaler, W. (1978). Treatment of renovascular hypertension with percutaneous transluminal dilatation of renal artery stenosis. *Lancet*, **i**, 801–2.

Harrington, D.P., Boxt, L.M., and Murray, P.D. (1982). Digital subtraction angiography: overview of technical principles. *American Journal of Roentgenology*, **139**, 781–6.

Helenon, C., Bigot, J.M., Jacri, P., Sraer, J.D., Mignon, F., and Richet, C. (1977). Aspects de l'artériographie au cours de l'évolutivité des lésions de la péri-artérite noueuse. *Journal de Radiologie et d'Electrologie*, **58**, 45–50.

Hillman, B.J. (1985). Renal digital subtraction angiography. *Urologic Clinics of North America*, **12**, 699–713.

Hillman, B.J., Ovitt, T.W., and Capp, M.P. (1982). The potential impact of digital video subtraction angiography on screening for renovascular hypertension. *Radiology*, **142**, 577–9.

Hillman, B.J., Sceley, G.W., Roeehnig, II., Tracey, P., and Raynaud, A. (1983). Digital nephropyelography subtraction angiography. *Radiology*, **149**, 405–10.

Irving, J.D. and Khoury, G.A. (1983). Digital subtraction angiography in renal transplant recipients. *Cardiovascular and Interventional Radiology*, **6**, 224–30

Joffre, F. (1980). Renal venography. In *Venography of the Inferior Vena Cava and its Branches* (eds. J. Chermet, J.M. Bigot), pp. 139–56. Springer Verlag, Berlin.

Joffre, F., et al. (1989). The usefulness of an endovascular prosthesis for treatment of renal artery stenosis. Preliminary experience. *Diagnostic and Interventional Radiology*, **1**, 15–21.

Lammer, J., Justich, E., Soreyer, H., and Pettek, R. (1985). Complications of renal tumor embolisation. *Cardiovascular and Interventional Radiology*, **8**, 31–3.

Martin, L.G., Casarella, W.J., Alspauh, J.P., and Chuang, U.P. (1986). Renal artery angioplasty: increased technical success and decreased complication in the second 100 patients. *Radiology*, **159**, 631–4.

Mauro, M.A., Wandsworth, D.E., and Stanley, R.J. (1982). Renal cell carcinoma: angiography in the CT era. *American Journal of Roentgenology*, **139**, 1135–9.

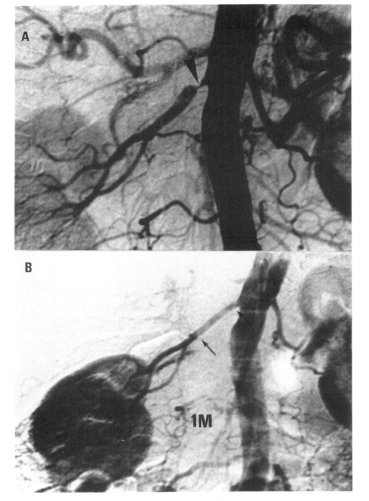

Fig. 5 Renal artery endoprosthesis. A. Digital abdominal aortography showing a tight stenosis near the ostium (arrows) in a patient who had already undergone angioplasty 2 months earlier. B. Aortographic control 1 month after placing an endoprosthesis (arrows). Excellent artery patency.

Renal venography

Selective catheterization of the renal veins for venous blood sampling is one of the few remaining indications for renal venography. Since most patients who require renal venography have thrombotic disease, an iliocavogram should be performed first, through percutaneous catheterization of both femoral veins (Abrams 1961), in order to ascertain that the inferior vena cava is patent.

A catheter is inserted at the site of convergence of the intrarenal veins (usually next to the renal hilum), through which contrast medium is injected against the bloodstream and under high pressure (Joffre 1980). This usually opacifies the proximal tributaries of the renal vein, although false-positive intraluminal filling defects may be caused by the flow of unopacified blood. The presence of more than one renal vein and the possibility of anomalies of the left renal vein should be taken into consideration when performing and interpreting the venogram; intrarenal anastomoses between interlobar veins are frequently present.

At present, many indications for renal venography are contro-

McCanon, D.A., Rubin, F.J., and Baines, D.A. (1976). Therapeutic bilateral renal infarction in end-stage renal disease. *New England Journal of Medicine*, **294**, 652–8.
Michel, J.R. (1980). Etat actuel des explorations vasculaires en pathologie urinaire. *Journal d'Urologie*, **86**, 545–9.
Miller, C.A., *et al.* (1985). Percutaneous transluminal angioplasty vs surgery for renovascular hypertension. *American Journal of Roentgenology*, **144**, 447–50.
Mitty, H.A. and Gribetz, M.E. (1982). The status of interventional uroradiology. *Journal of Urology*, **127**, 2–9.
Novick, A.C., Buonocore, E., and Meaney, T.E. (1982). Digital subtraction angiography for postoperative evaluation of renal arterial reconstruction. *Journal of Urology*, **127**, 14–7.
Nurmi, M., Satokari, K., and Puntala, P. (1987). Renal artery embolisation in the palliative treatment of renal adenocarcinoma. *Scandinavian Journal of Urology and Nephrology*, **21**, 93–6.
Pickering, T.G., Sos, T.A., and Saddekni, S. (1986). Angioplasty in patients with azotemia and renovascular hypertension. *Journal of Hypertension*, **5**, 667–9.
Pinet, A. and Lyonnet, D. (1981). Embolisation des affections bénignes non tumoraies du rein. *Annales de Radiologie*, **24**, 420–5.
Prager, P., Hoevels, J., and Georgi, M. (1984). Digital subtraction angiography in the preoperative evaluation of patients with renal tumor. *Acta Radiologica (Diagnosis) (Stockholm)*, **25**, 101–5.
Sigvart, U., Puel, J., Mirkovitch, U., Joffre, F., and Kappenberger, L. (1987). Intravascular stents to prevent occlusion and restenosis after transluminal angioplasty. *New England Journal of Medicine*, **316**, 701–6.
Smith, C.W., Windfield, A.C., and Price, R.R. (1982). Evaluation of digital venous angiography for the diagnosis of renovascular hypertension. *Radiology*, **144**, 462–5.
Sos, T., *et al.* (1983). Percutaneous transluminal renal angioplasty in renovasular hypertension due to atheroma or fibromuscular dysplasia. *New England Journal of Medicine*, **309**, 274–9.
Tessier, J.P., Teyssou, H., and Marchal, M. (1986). Angiographie numérisée par voie intraveineuse dans l'étude de l'artère du rein transplanté. *Journal de Radiologie*, **67**, 373–7.
Thornbury, J.R., Stanley, J.C., and Fryback, D.G. (1982). Hypertensive urogram: a non-discriminatory test for renovascular hypertension. *American Journal of Roentgenology*, **138**, 43–9.
Wilms, G.E., Baert, A.L., and Staessen, J.A. (1986). Renal artery stenosis: evaluation with intravenous digital subtraction angiography. *Radiology*, **160**, 713–5.

1.6.6 Nuclear imaging in nephrology

MICHAEL MAISEY AND K.E. BRITTON

Renal radiopharmaceuticals

Radioactively labelled compounds can provide a very wide range of information about renal function and renal pathophysiology. This information may be related to total renal function but, more importantly, it provides images and measurements of the distribution of renal function both between the two kidneys and within a single kidney; in some cases this information is critical in making clinical decisions. To obtain the best use of the wide choice of techniques available it is essential to have a complete understanding of the physiological and functional basis of the investigations, to perform them carefully, and to understand the limitations and possible sources of error. In almost all instances they are complementary to the predominantly structural information provided by other imaging methods and should be interpreted with the full knowledge of the morphological abnormalities as well as the clinical context.

Renal handling of radiolabelled tracers

Inulin, ^{51}Cr-labelled ethylenediaminetetraacetate (EDTA), or pure ^{99}Tc^m-labelled diethylenetriaminepentaacetate (DTPA) are only excreted from the kidney through glomerular filtration and, being neither reabsorbed nor metabolized by the kidney, may be used to measure glomerular filtration rate (GFR). *p*-Aminohippurate (PAH), radioiodine-labelled *o*-iodohippurate (OIH) and ^{99}Tc^m-labelled mercaptoacetyltriglycine (MAG3) are primarily secreted by the tubules and may be used to provide an index of renal plasma flow. However, an overall look at blood—kidney exchange of compounds shows that the amount per unit time taken up by the kidneys of all the compounds mentioned above depends on their supply rate to the kidneys, and the kidneys' extraction efficiency for each, (E). The amount in the plasma supplied per unit time, (t, the supply rate) is given by the renal plasma flow (RPF) multiplied by the plasma concentration, (P_t), at unit time. Thus the amount taken up by the kidneys, Q, per unit time is given by

$$Q_t = \text{RPF} \times P_t \times E$$

For OIH and PAH, $\text{RPF} \times E$, is usually referred to as the effective renal plasma flow, to take account of the fact that these compounds are not 100 per cent extracted—about 80 per cent extraction is the generally accepted figure. For agents filtered in the glomerulus, the fraction filtered (the filtration fraction, FF) is given by the fraction of renal plasma flow that is glomerular filtration rate (GFR). Thus E for these compounds equals GFR/RPF. Thus

$$Q_t = \text{RPF} \times P_t \times \text{GFR}/\text{RPF}$$
$$= \text{GFR} \times P_t$$

for glomerular filtered compounds.

The very short time taken for ^{99}Tc^m-DTPA to pass the glomerular filter and the slight difference in timing between this and the secretion of ^{99}Tc^m-MAG3 or ^{123}I-OIH into the proximal tubular lumen are quite lost in consideration of the nephron transit times when the data sampling interval is 10 s (Fig 1). The mean parenchymal transit time is made up of a minimum transit time, which is common to all nephrons, and parenchymal transit time index (PTTI). The mean parenchymal transit time (MPTT) is the time taken for a compound to pass through the whole of the paren-

1.6.6 Nuclear imaging in nephrology

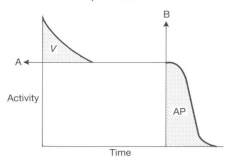

Fig. 1 Parenchymal impulse retention function. A, extrapolation of the plateau of the impulse retention to exclude vascular component, V. B, time of the change from plateau to a falling activity which delineates the minimum transit time. The parenchymal transit time index is given by the area divided by the height of the impulse retention function for transit times longer than the minimum and is measured in seconds.

chyma and when added to the pelvic transit time (PVTT) equals the whole kidney transit time (WKTT). The whole kidney transit time corrected for the minimum transit time is called the whole kidney transit time index (WKTTI).

Normal ranges for these parameters are: PTTI, 10–156 s; MPTT, 100–240 s; WKTTI, 20–170 s.

The normal pelvis has effectively laminar flow with a mean residence time of 5 to 10 s. However, if the pelvis is dilated, whether floppy walled and unobstructed, or with its outlet obstructed, then eddying and mixing of tracer results and the mean residence time may be over 3 min. Any method which increases the flow rate of urine will reduce the pelvic transit time in the absence of obstruction to outflow and this forms the basis for the use of diuretics such as frusemide or mannitol in making this distinction.

Radiopharmaceuticals for renal imaging

$^{99}Tc^m$-DTPA

$^{99}Tc^m$-DTPA is a chelating agent able to bind reduced technetium firmly, and it has been the routine renal radiopharmaceutical for gamma-camera studies for 20 years. Pure $^{99}Tc^m$-DTPA is a glomerular filtered agent; however, certain preparations may have variable protein binding which may invalidate their use for the measurement of the true glomerular filtration rate.

The disadvantage of $^{99}Tc^m$-DTPA is its low extraction efficiency. It has a fourfold lower extraction efficiency than ^{123}I-OIH, and the tissue and blood background is, therefore, always higher for $^{99}Tc^m$-DTPA than for ^{123}I-OIH. When renal function is poor, inferior quality images are obtained and data analysis is limited by the background levels and the poor statistical quality of the data. The advantages of $^{99}Tc^m$-DTPA are the ready availability of the $^{99}Tc^m$-generator and a robust kit that is simple to prepare.

^{131}I-o-Iodohippurate

^{131}I-OIH has been an ideal tubule-secreted compound for probe renography for 30 years. It is bought ready prepared and may be diluted from stock with sterile 0.9 per cent physiological saline. For probe renography and for measuring effective renal plasma flow 1.5 MBq (55 μCi), are usually sufficient, but for gamma-camera studies up to 20 MBq (500 μCi) may be used.

^{131}I-OIH has the advantage of high extraction efficiency and cheapness, but ^{131}I is a poor radiolabel for the modern gamma-camera. The beta emission and long half-life limit its activity, so that image quality and the statistical quality of the data obtained is poor, and ^{131}I should no longer be used for routine renal gamma-camera studies.

^{123}I-o-Iodohippurate

^{123}I-OIH is a better renal imaging agent than ^{131}I-OIH, and has a gamma-ray energy, 159 keV, suitable for the gamma-camera and a 13-h half-life. The main problems of ^{123}I-OIH are related to its availability and cost. Excellent quality images may be obtained even when renal function is poor, provided that the administered dose is related to that usually used for $^{99}Tc^m$-DTPA: 2.5 mCi (100 MBq) ^{123}I OIH will give similar count rate images to 10 mCi (400 MBq) $^{99}Tc^m$-DTPA, but because of its more rapid excretion, the background count of the ^{123}I-OIH image is lower, giving a better kidney signal to noise ratio.

Compounds secreted from the tubules are associated with physiological problems. Many drugs can interfere with their excretion including the penicillins, probenecid and most diuretics, and red cell uptake of OIH is about 45 per cent at equilibrium. Weak protein binding, of around 70 per cent in blood, reduces the glomerular filtration of OIH from 20 to 6 per cent and the tubular extraction is between 80 and 87 per cent. Hippurates are secreted from postglomerular capillary blood into the proximal tubules of both cortical and juxtamedullary nephrons.

$^{99}Tc^m$-mercaptoacetyltriglycine

The disadvantages of the current renal imaging agents have led many workers to search for a better alternative. An ideal substitute for I-OIH should be labelled with $^{99}Tc^m$- and have an extraction efficiency three times greater than that of $^{99}Tc^m$-DTPA (20 per cent) and preferably better than ^{123}I-OIH. It should be a stable compound easily prepared from a robust kit, cheap to purchase and easy to use, with weak or no protein binding to improve glomerular filtration and tubular secretion. Compounds based on a triamide monomercaptide (N_3S ligand) were investigated by Fritzberg et al. (1981,1986), and lead to the development of MAG3, which was shown by probenecid inhibition studies to be tubule-secreted and equivalent to ^{123}I-OIH.

The three nitrogens and the sulphur of MAG3 are arranged in a ring structure which is able to bind reduced technetium in a stable way. It also contains a combination of polar and non-polar groups which makes it suitable for proximal tubular uptake and secretion, and a potential technetium-labelled replacement for OIH. The MAG3 kit preparation contains a benzoyl group to protect the ring structure; this must be displaced by boiling to enable the binding of technetium in the ring.

The volume of distribution of MAG3 is 65 per cent smaller than that of ^{131}I-hippuran, and the blood clearance half-lives are similar (slow component ratio $^{99}Tc^m$-MAG3: ^{123}I-hippuran, 1.09:1). The clearance relationship is:

ERPF hippuran = 1.5 × ERPF $^{99}Tc^m$-MAG3 + 40 ml/min.

Although there is a linear relationship between effective renal plasma flow and clearance, $^{99}Tc^m$-MAG3 is not yet proposed as a substitute for the absolute measurement of glomerular filtration rate or effective renal plasma flow.

Clinical studies have shown that MAG3 is a successful radiopharmaceutical for routine renal work, combining the physiological advantages of OIH with the benefits of ^{99}Tc labelling. It has been used in the whole range of clinical studies without side effects (Jafri et al. 1988). A dose of 100 MBq (2.5 mCi) gives better results than 400 MBq (10 mCi) ^{99}Tc^m-DTPA and equivalent to 80 MBq (2 mCi) of ^{123}I-OIH for routine renal imaging, for relative function measurements, for renal transit time analysis and for frusemide diuresis studies. It is suitable for the measurement of intrarenal plasma flow distribution.

^{99}Tc^m-dimercaptosuccinate (DMSA)

Succinate is an important substrate for proximal tubules, which possess an uptake system to accumulate this substrate. DMSA labelled with ^{99}Tc^m is therefore taken up by the kidneys and retained in the proximal tubules, with less than 5 per cent being excreted. In order to avoid urinary excretion and liver uptake, careful preparation is required: poor preparations show high urinary loss which may interfere with the measurement of renal function. A dose of 100 MBq (2.7 mCi) is administered intravenously and static images are taken at 1 h if renal function is good, or 3 to 6 if renal function is poor. The compound is filtered by the glomeruli and taken up by the kidney tubules. If renal function is poor, uptake by the liver may be seen. Acidosis also affects renal uptake.

Many other radiopharmaceuticals have been used to evaluate renal function, but those described above will meet all routine clinical requirements.

Uptake and output components of radiopharmaceuticals

The activity—time curve recorded over a organ using an externally placed detector or gamma-camera represents the variation with time in the quantity of radiation arriving at the detector from radioactive material within its field of view. When the organ of interest is the kidney and a probe detector is used then the activity—time curve has been called the 'renogram' and the technique 'renography'. When the gamma camera is used, the term renal radionuclide study is preferred for the technique and 'renogram' is used loosely for the activity—time curve recorded from a region of interest.

This 'renal' activity—time curve is a complex and composite curve, $R(t)$. One component represents the variation with time of the quantity of activity in non-renal tissue in the chosen region of interest, mainly in front of, and behind, the kidney; another represents activity in the renal vasculature not taken up by the kidney. The component of interest is the variation of radiotracer activity taken up by the kidney over time; this is the kidney activity—time curve, $K(t)$, and may be considered to have been derived or obtained theoretically as in an idealized situation free of blood and tissue background activity. In practice the kidney curve is obtained as the resultant activity time curve after appropriate assumptions or corrections for tissue and blood background activity, depth, attenuation and size of region of interest have been applied to the composite activity—time curve $R(t)$ obtained from the gamma-camera. The supply curve, that is the blood clearance curve $B(t)$, which is usually obtained from a region of interest posteriorly over the left ventricle in the field of view of the camera, decreases with time as DTPA is taken up by the kidneys.

Taking ^{99}Tc^m-DTPA as an example, it is evident that the amount of DTPA in the kidney before any leaves the parenchyma to enter the pelvis in urine depends on the amount which has been supplied to it in the blood and on the renal extraction efficiency (for a filtered agent the extraction efficiency is equal to the filtration fraction, GFR/RPF). The early part of the kidney activity—time curve during the minimum time of transit of DTPA from its glomerular uptake site along the nephron, before any urinary loss of DTPA activity occurs, represents the accumulation of DTPA in the kidney over time. The kidney is in effect integrating the uptake of the supplied and extracted DTPA. The DTPA supplied to the kidney in renal arterial blood can take one of two paths: either it remains in the intrarenal blood circulation, returning via the renal vein, or it is filtered at the glomerulus. The renal artery to renal vein transit time is of the order of 4 s, with a range from about 3 to 20 s. As the rate of loss of DTPA from the intrarenal blood by glomerular filtration is clearly equivalent to the glomerular filtrate rate, then the fraction of blood activity taken up per second (designated U; the uptake constant) will be proportional to the blood activity and the total uptake (during the period of the minimum parenchymal transit time, t) will be proportional to the integral of the blood curve. This is the uptake component $Q(t)$ of $K(t)$.

Since during the minimum transit time period $K(t)$ is proportional to $\int B(t) dt$, consider a situation where no activity appears in the urine for a long time as with complete outflow obstruction. The kidney continues to integrate the supply for this period and $K(t)$ becomes a 'zero output' curve. The falling phase (third phase) of a kidney curve is not an excretory phase but an 'amount of tracer left behind in the kidney' phase.

Applications to clinical nephrology

The measurement of individual renal function

The relative contribution of each kidney to total function can be reliably determined using an excreted radiopharmaceutical.

From the regions of interest over the two kidneys, activity—time curves, corrected for the background, are generated. From these, a normal peak occurs at about 150 to 180 s at which time the tracer starts to leave the kidney. No analysis of percentage function can be taken after this time. During the first 90 s or so there are unstable conditions with mixing effects: it is best to limit the consideration of relative renal function to the period when uptake is unaffected by these factors, usually between 90 and 150 s in kidneys without outflow disorder. This period may occur later, e.g. 3 to 4 min in kidneys with an outflow disorder. The potential errors in measuring $Q(t)$ are not negligible. The most important assumption is that the kidneys are at equal depth so that the attenuation of ^{99}Tc^m or radioiodine may be considered to be equal. This requires that the kidneys are prevented from falling forward during the study: the patient should be reclining back or even supine, although placing the camera is then more difficult. In such a position 75 per cent of kidneys have less than 1 cm difference in depth and the count rate loss is about 10 per cent/cm for ^{99}Tc^m or ^{123}I. This leads to an error of about +3.5 per cent for the measured percentage relative function. Corrections for depth may be undertaken using true lateral views of the kidney, ultrasound or from a height/weight formula.

Errors are also due to (1) variations of distribution of the radiopharmaceutical in the kidney which are probably small, (2) counting errors due to Poisson statistics, and (3) non-renal background activity. This gives reliable results in moderately well functioning kidneys but may lead to errors when the contribution of one kidney is less than 15 per cent of total function.

Fig. 2 $^{99}Tc^m$-DMSA scan showing the effect of reflux into a lower moiety of a duplex kidney on the left and a lower pole scar in the right kidney.

The normal range for relative uptake function determined in practice and thus incorporating both biological variation and physical sources of error is 42.5 to 57.5 per cent for the contribution of one kidney. The accuracy of a particular figure depends on the overall renal function and the number of nephrons in a kidney but ranges from ±4.5 per cent at 40 per cent of total uptake with a creatinine clearance of 90 ml/min to ±7.5 per cent at 20 per cent of total uptake at a creatinine clearance of 20 ml/min. This accuracy is sufficient for most clinical applications since the usual decision is whether to perform nephrectomy or a restorative operation.

The same considerations apply to the relative function determined by $^{99}Tc^m$-MAG3 or by the use of $^{99}Tc^m$-DMSA. The preferred method for determining relative function with $^{99}Tc^m$-DMSA is to use the geometric mean of background corrected anterior and posterior, left and right, renal counts taken between 2 and 3 h after injection.

Some of the clinical situations when knowledge of distribution of renal function is helpful in clinical management will be considered.

Reflux nephropathy

The measurement of divided renal function is now an essential part of the investigation and follow-up of the patient with ureteric reflux, as decisions about surgical management may depend on serial measurements of renal function. Urinary reflux may differentially affect parts of a single kidney; for example in a duplex system with reflux up one ureter only, damage may occur only to that part (usually the lower moiety) of the duplex system. In this situation it is essential to know the proportion of function in each of the two moieties in order to make rational decisions about the need for partial nephrectomy, total nephrectomy, or other surgery. A $^{99}Tc^m$-DMSA scan performed 3 h after injection of the radiolabel will clearly display the regional variations in functional damage brought about by reflux and associated infection (Fig. 2).

Nephrolithiasis

The contribution of each kidney to total function should always be measured in the assessment and follow-up of patients with intrarenal stone formation. Knowledge of the distribution of function within the kidney will also enable the surgical approach to be planned so as to minimize surgical damage to the residual renal tissue and also contribute to the decision between total and partial nephrectomy or simple stone removal. Postoperative repeat assessment will be used to determine how successful the operation has been in preserving renal function and to monitor subsequent progress. Large masses of calcium lying between the kidney and the gamma-camera may result in misinterpretation of the distribution of function so it is advisable to make anterior as well as routine posterior images.

Hypertension due to local ischaemia

Renal scanning with DMSA, DTPA, OIH or MAG3 may be helpful in defining an abnormal renal segment with poor function which is responsible for hypertension and which can be treated by partial nephrectomy or segmented angioplasty. The delayed DMSA scan will show a focal area of decreased function (Fig. 3) and the DTPA scan may show an early uptake defect at 2 min, with an increase later due to local delayed transit times secondary to local ischaemia and water reabsorption. This assessment will usually be performed together with regional venography for the measurement of segmental renin secretion and arteriography. The theoretical basis for these abnormalities is discussed under renovascular hypertension.

Renal localization

Although renal localization prior to biopsy is routinely performed under X-ray fluoroscopy control, localization using $^{99}Tc^m$-DMSA may occasionally be preferable, for example when renal function is impaired so that localization of contrast media is poor or when there is a particular region of the kidney which is functionally abnormal. It may also be valuable in localizing the renal outline on the skin surface when planning radiotherapy of the abdomen in order to avoid unnecessary renal radiation.

Pyelonephritic scarring

Chronic pyelonephritis is one of the most common causes of end-stage renal failure and in children is often associated with reflux. To follow up and treat children and young adults appropriately with long-term antibiotics or ureteric reimplantation it is necessary to identify the presence or absence and of renal parenchymal scarring, and its progression. The identification of scars using urography is good when there is impeccable technique and good preparation, especially with nephrotomography. However, bowel gas frequently complicates the picture, too low a dose of

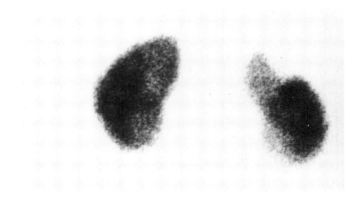

Fig. 3 Loss of function at the upper pole of the right kidney ($^{99}Tc^m$-DMSA) due to a segmental stenosis confirmed at arteriography.

whereas a small pelvic kidney, especially if it is poorly functioning, cannot always be seen against the background of the pelvic bone. Renal abnormalities and the assessment of divided function associated with neurological abnormalities such as meningomyelocele can easily be documented.

Absolute measurement of each kidney's individual function

The difficulties in measuring absolute individual kidney function include the rates of change of the renal input, the blood and tissue background and renal uptake, the effects of different kidney depths on the attenuation of the renal count rate, the problem of relating the detected activity to the injected radiolabel, the sensitivity of the gamma-camera and the dead time losses in count rate.

Many attempts have been made to overcome these problems so that an absolute measurement of the uptake of an intravenously injected radiopharmaceutical by each kidney may be made. Most published methods correct for some, but not all, of the above problems (Fleming et al. 1987; Russell et al. 1985). Some methods require one or more blood samples to calibrate the activity time curves, but measurements with or without blood sampling tend to be made when the rate of change of activity in the blood is high. While these apparently direct methods give a pleasing computer printout in terms of the GFR or effective renal plasma flow of each kidney in ml/min, it is more accurate and reliable in principle (and probably in practice) to measure the blood clearance properly in ml/min and apply the relative uptake function percentage measurements to it to give the individual renal clearance.

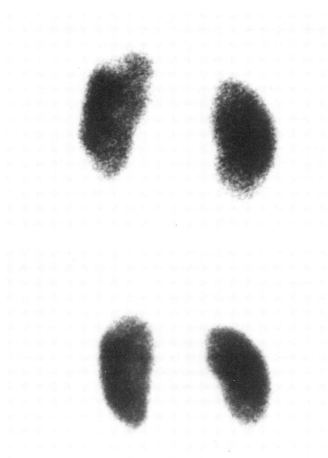

Fig. 4 Focal loss of function at the upper pole of the left kidney on a DMSA scan during an infection and resolving a few months later with no scar formation.

contrast is used, and nephrotomography is not employed. Merrick et al. (1980) have shown that excretion urography has a sensitivity of 86 per cent and a specificity of 92 per cent whereas radionuclide imaging with $^{99}Tc^m$-DMSA has a sensitivity of 96 per cent and a specificity of 98 per cent. Renal scanning is, therefore, an important adjunct in the identification of scars as well as in the measurement of divided function in children and young adults with reflux or urinary infections. Care must be taken in the presence of current infection as focal 'scars' may be due to focal nephritis and resolve with time, whereas true scars cannot resolve. (Fig. 4). Recent advances in single photon emission computerized tomography (SPECT) are increasing the accuracy of the detection of focal abnormalities.

Congenital abnormalities

Renal imaging with $^{99}Tc^m$-DMSA is valuable for the proper assessment of many congenital abnormalities of the renal tract (Fig. 5). Reflux and duplex kidneys with 'seesaw' reflux have already been mentioned; other examples include the assessment of horseshoe kidneys where the function of the 'bridge' can often be assessed very much more easily than with urography. Ectopic kidneys, for example pelvic kidney, can usually be easily identified and investigated since once the radiopharmaceutical has been given a whole-body search can be undertaken if necessary,

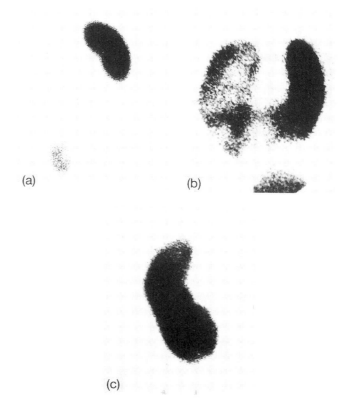

Fig. 5 $^{99}Tc^m$-DMSA scans showing (a) posterior view of a pelvic kidney, (b) anterior view of a horseshoe kidney with obstruction, and (c) posterior view of crossed renal ectopia.

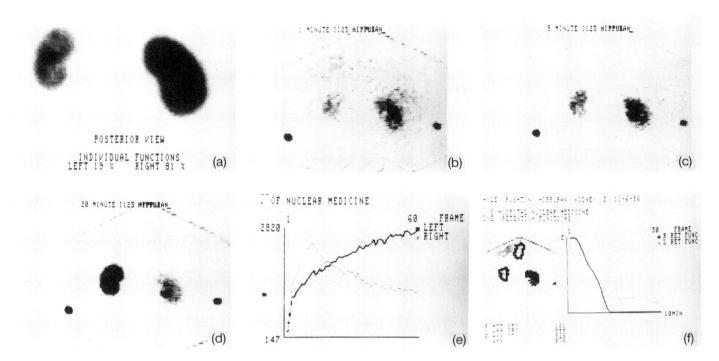

Fig. 6 An example of left renal artery stenosis. The ^{99}Tc^m-DMSA scan shows decreased uptake of the left kidney. (a) The ^{123}I-hippuran study (b,c,d) shows decreased uptake on the left with prolonged parenchymal transit. Time–activity curves from the hippuran study (e) show a delayed renogram peak on the left and the deconvolution study (f) a prolonged transit time (333 s versus 161 s).

The use of ^{99}Tc^m-DMSA is an alternative and more reliable approach to a direct measurement of absolute individual kidney function since the conditions are much closer to a steady state at 2 or 3 h than they are during the first 30 min. The geometrical mean method overcomes some of the problems of count attenuation due to different kidney depths. In order to determine the absolute uptake the amount injected must be known either by preparing a proper standard, or by counting the active syringe before and after injection. Kidney depth may be measured from true lateral images of the kidneys or by ultrasound. Kidney phantom studies using different thicknesses of tissue equivalent material give the relationship between depth, count rate, and activity in the kidney, or a linear attenuation coefficient taking scatter into account may be incorporated into the calculation (Groshar et al. 1989).

The normal value for absolute uptake at 2 h is 27.0 ± 6.0 per cent, at 3 h 25.4 ± 8.9 per cent, and at 6 h 30.0 ± 9.2 per cent (mean ± SD) injected activity in each kidney (Zananin et al. 1987).

Renovascular hypertension

Uptake function and transit times

The effect of renal artery stenosis on renal function has two features relevant to radionuclide studies. Because there is an increased proximal tubular water and salt reabsorption related to the reduction in peritubular capillary pressure caused by the occlusive arterial lesion, a non-reabsorbable solute such as ^{131}I-hippuran will become relatively concentrated within a pool of fluid that travels more slowly along the nephron. The time to peak of the renogram, which represents a crude measure of the mean transmit time of hippuran entering and leaving the kidney, will be prolonged compared with a normal kidney. In the absence of outflow system disorder and in a normally hydrated person a difference of over a minute between peak times in the two renograms signifies prolonged hippuran transit. The blood supply, and therefore delivery, of the radiopharmaceutical to a kidney with renovascular disorder will be reduced below normal due to the occlusive arterial lesions. An uptake function less than 42 per cent of the total is abnormal, assuming that the function of the concentrated kidney is normal.

Functionally significant branch artery stenosis may be identified as delayed parenchymal uptake at one pole of a kidney. Such appearances are due to local prolongation of the tracer's parenchymal transit time.

The demonstration of an abnormally prolonged parenchymal transit time reflects the increased salt and water reabsorption due to a drop in peritubular capillary pressure relative to the intratubular luminal pressure in the proximal tubules and to increased generation of intrarenal angiotensin II. The flow of fluid in the nephrons is thus reduced, and the mean parenchymal transit time is prolonged in renovascular disorder because of both the prolonged minimum transit time and a prolonged parenchymal transit time index. The separation of the mean parenchymal transit time from the whole kidney transit time is important since it obviates the effect of changes in pelvic transit time by which the specificity of the whole kidney transit time or activity time curve is reduced.

In a normally hydrated hypertensive patient, a mean parenchymal transit time over 240 s or, in a small kidney, more than 60 s longer than that of a normal contralateral kidney, together with an uptake function of less than 42 per cent, is strongly suggestive of functionally significant renovascular disorder. An example of renal artery stenosis is shown in Fig. 6.

Such a finding does not distinguish between large-vessel and small-vessel disease. Bilateral functionally significant renal artery stenosis is rare and causes bilateral prolongation of mean parenchymal transmit time (MPTT). Renal activity time curves and MPTT findings often demonstrate that one stenosis is significant at the time of the study when arteriography shows bilateral stenoses.

Congenitally small kidney and the small kidney due to unilateral pyelonephritis will have normal parenchymal transit times if they occur incidentally in a hypertensive patient.

Captopril test

The 'captopril test' has been introduced as an alternative approach to improving the specificity of the changes in renal activity—time curves associated with renovascular disorder (Wenting *et al.* 1984). Captopril reduces angiotensin II formation, and thus relaxes the efferent arterioles, causing a fall in glomerular capillary pressure and filtration. For this test a baseline dynamic scan is performed using either ^{123}I-hippuran, ^{99}Tc^m-DTPA or ^{99}Tc^m-MAG3. After 1 h, 25 mg of captopril is given orally, followed 30 min later by a second injection of radiopharmaceutical and a second dynamic scan. Normally there should be no effect on the relative GFR or transit times; in renal artery stenosis the captopril will cause a drop in GFR and prolongation of transit times in the affected kidney (Fig.7). Dondi *et al.* (1989) used this test to investigate 105 hypertensive patients; 34 of 37 patients were diagnosed as having unilateral or bilateral stenosis of over 50 per cent, and in the 55 with no stenosis or less than 50 per cent stenosis there were only two false positives. A single dose of captopril can cause potentially dangerous hypotension when renin levels are high and in salt-depleted patients. The latter is typically related to diuretic therapy, which must be stopped for at least 2 days before the use of captopril. Good hydration is important and blood pressure must be monitored during the test.

The success of the mean parenchymal transit time measurement in predicting the outcome of angioplasty of the renal artery stenosis has been recently demonstrated by Gruenewald *et al.* (1988), who evaluates 32 patients with hypertension and radiologically important renal artery stenosis. Nine patients with a normal MPTT all failed to show improvement in blood pressure, whereas 20 of 23 with a prolonged MPTT showed improvement in, or achieved normal blood pressure. The importance of using MPTT in the follow-up to detect restenosis was also demonstrated. In a study of 60 patients Geyskes *et al.* (1986) found that the uptake of hippuran was a good predictor of response to angioplasty. These radionuclide studies are complementary to angiography in the detection of renovascular hypertension, provide invaluable baseline data for follow-up, and allow prediction of response and early detection of restenosis.

Vesicoureteral reflux

The micturating cystogram remains the preferred method and is the 'gold standard' for the diagnosis of vesicoureteral reflux. This investigation is usually necessary to establish the diagnosis and to achieve the anatomical information necessary for patient management. The micturating cystogram does, however, have the disadvantages of being an unpleasant investigation to perform, increasing the risk of infection due to bladder catheterization and requiring a relatively high radiation dose. Radionuclide methods are now used therefore, particularly in the follow-up of children with established reflux in whom the anatomical information has been obtained from a micturating cystogram.

There are two methods in routine use. The indirect method, (Fig.8) uses a standard dynamic ^{99}Tc^m-DTPA renal scan, with the generation of appropriate images, renogram curves, and measurements of total and individual renal function. At the end of the study, instead of emptying the bladder, the child is encouraged to drink until the bladder is full and there is a desire to micturate. At this point the child stands (boys) or sits (girls) with the back against the gamma camera with the field of view including both kidneys and the bladder. After a baseline period of data acquisition the data are obtained while the bladder is emptied and for several minutes afterwards. Regions of interest are placed over each kidney and over the bladder. The time—activity curves from the renal regions of interest will show peaks in the renal area if radiolabelled urine refluxes up as far as the kidney. Lesser degrees of reflux in the ureters can be detected but this is generally more prone to error. The bladder curve will show the rapid emptying phase if it is a good study and a partial refilling phase if the refluxed volume is large enough. The time—activity curves should always be correlated with the images obtained during reflux to avoid false positive reports. This method has been shown to be sensitive and accurate and can be

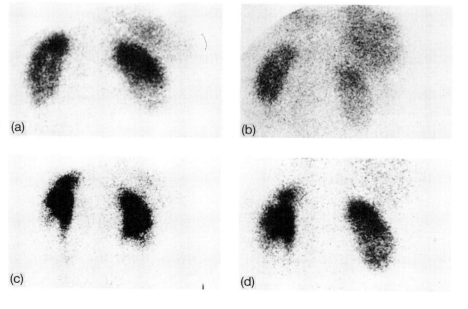

Fig. 7 An example of a positive captopril test in a hypertensive patient. The ^{99}Tc^m-DTPA baseline study (a,b), is normal with a right individual GFR of 51 per cent. After 25 mg of captopril the right GFR drops to 42 per cent with a delayed transit time (c,d).

1.6.6 Nuclear imaging in nephrology

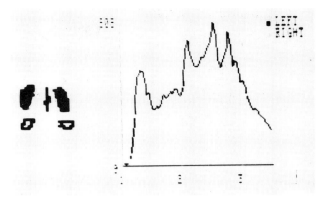

Fig. 9 Time–activity curve from the third phase of the renogram curve showing intermittent reflux into the left renal area during bladder filling.

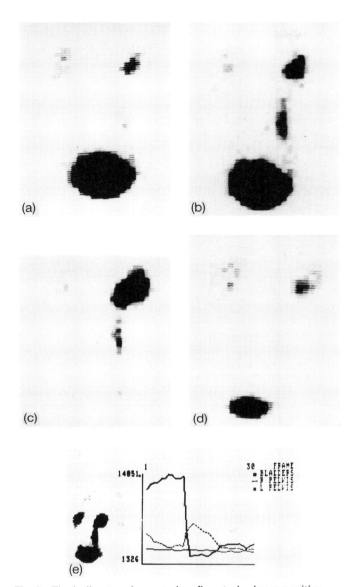

Fig. 8 The indirect vesicoureteric reflex study shows positive reflux on the right side using $^{99}Tc^m$-DTPA. (a) Premicturation; (b) right-sided reflux at the initiation of micturition; (c) further reflux as the bladder empties and (d) rapid drainage partially to refill the bladder. (e) Time–activity curves from the two pelves and the bladder.

used as a follow-up procedure. In addition to being less traumatic than the micturating cystogram, avoiding catheterization and having a lower radiation dose, it has the added advantage of providing functional renal information from the early part of the DTPA scan. The disadvantage, apart from lack of anatomical detail, is that a true filling phase reflux is not seen unless it occurs spontaneously during the renogram phase, when it shows up as multiple spikes in the third phase (Fig.9). The correlation of reflux detection using these methods is good. Hedman (1978) showed 90/102 renal units in agreement and Kogan (1986) suggested that about 75 per cent of micturating cystograms could be dispensed with. The degree of reflux can also be graded satisfactorily using this method. Zhang et al. (1987) showed an 80 per cent correlation when all five grades were compared and 100 per cent correlation when grades IV and V were combined into one high-grade reflux. Measurement of the actual volume of refluxed urine is also possible and can be very accurate when performed with careful attenuation to detail and quantitation (Godley et al. (in press)).

This method can also be used in the screening of siblings with reflux and the screening of patients presenting with urinary tract infection.

The retrograde (direct) method is used for children less than 3 or 4 years of age, who are usually not able to co-operate with the indirect method. $^{99}Tc^m$-DTPA diluted in saline is run into the catheterized bladder during imaging with the gamma camera. Although similar to a micturating cystogram, this investigation does not yield anatomical detail; on the other hand there is a much lower radiation dose and it can be made more quantitative. A refinement of this method is to include $^{99}Tc^m$-colloid in the radiopharmaceutical and image the kidney after 20 h, when residual radioactivity in the kidney indicates the presence of intrarenal reflux (Rizzoni et al. 1986).

Obstruction to outflow
Effect on the renogram
An obstructing process may affect structure, as demonstrated by intravenous urography, renal function, as demonstrated by radionuclide studies; and pressure/flow relationships, as demonstrated by antegrade perfusion pressure measurements in the upper urinary tract and by urodynamic studies of the lower urinary tract.

In the context of renal radionuclide studies, obstruction affects the movement of radio tracer from the renal input to the renal output in the following ways. DTPA, MAG3, and I-OIH are non-reabsorbable solutes; their times of transfer from renal blood to the early proximal tubules are similar and very short, compared to a typical data sampling interval of 10 s. The time taken for these non-reabsorbable solutes to travel along the nephron from the early proximal tubule to the papillary ducts is of the order of 180 s under normal conditions of hydration. When there is an increased resistance to flow, as in obstruction, the intratubular pressure gradient changes, and there is greater pressure difference between the lumen of the nephron and the peritubular capillary. As a consequence the passive component of salt and water reabsorption is increased, and the non-

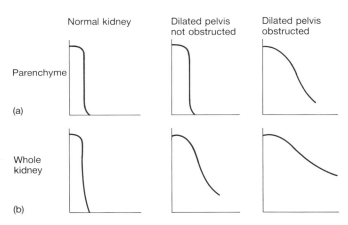

Fig. 10 Impulse retention functions from which 16 min transit times have been subtracted for the parenchyma and whole kidney, a dilated non-obstructed pelvis and a dilated obstructed pelvis.

reabsorbable tracer is concentrated in the tubular luminal fluid, reducing its flow rate and prolonging its transit time along the nephron. Thus DTPA, MAG3 and I-OIH are concentrated in a smaller volume of fluid in the lumen of the nephron, the intranephron fluid flow rate is reduced, and the parenchymal transit time of these tracers is increased.

Diuretic renography

Clinical tests based on the use of the diuretic frusemide (diuretic renography) may help to determine the strength of the resistance to outflow. Frusemide acts directly on the thick ascending limb of the loop of Henle, inhibiting the active Na–K–Cl transport system. In order for frusemide to work, there must be a sufficient number of nephrons to produce a diuresis and the patient should not be chloride or sodium depleted. Thus frusemide diuresis is an unreliable test of outflow resistance when renal function is poor and when the patient has major electrolyte disturbances. Conversely, if renal function is good there may be such a diuretic response that a nephrologically important, but relatively slight outflow resistance may be overcome, leading to a false diagnosis of lack of obstruction. The response may also be lost in a grossly dilated pelvis and an inappropriately good response obtained if the pelvis is rigid, as has been shown by modelling the system (Zechmann 1988). Variable correlations have been shown between the frusemide diuresis technique and antegrade perfusion pressure measurements, from poor (Hay et al. 1984) to good (O'Reilly et al. 1987).

The frusemide test is usually performed using $^{99}Tc^m$-DTPA but ^{123}I-OIH or $^{99}Tc^m$-MAG3 may be used. If there is significant delay of tracer in the collecting system, intravenous frusemide is given 30 min after the tracer, at a dose of 0.5mg/kg body weight. Data are collected for a further 15 min and time–activity curves are obtained from regions of interest over the renal areas and may be divided into three categories: (i) normal, when there is steepening of the rate of fall with a concave washout curve; (ii) obstructed, when there is no washout, and (iii) intermediate, which is more difficult to interpret (Fig. 10). Quantitatively, results are expressed as the percentage washout in 15 min and the washout rate (measured as half-life). Results are considered normal if washout is greater than 50 per cent, with a half-life faster than 10 min. In obstruction less than 25 per cent washes out over 15 min with a half-life slower than 20 min; results between the two are considered intermediate.

Since the measured change in activity with time is the basis of determining the response to frusemide, this response is also crucially dependent on the amount of radiolabel taken up by the kidney and on the rate of uptake. Thus the rate of fall of the activity–time curve of the kidney in response to frusemide is dependent on its previous rate of rise and one has to judge whether the rate of fall is appropriate for a given rate of rise. A moderately poorly functioning kidney would have a moderately impaired rate of rise and a moderately impaired, but appropriate, rate of fall in response to frusemide in the absence of obstructing uropathy. An inappropriately slow rate of fall in response to frusemide would support a diagnosis of obstructing uropathy.

The frusemide response must, therefore, be evaluated not only in terms of 'good' or 'poor' but as appropriate or inappropriate. A poor response is appropriate to a poorly functioning kidney and a nil response is appropriate to a non-functioning kidney but in neither case does the frusemide response indicate obstructing nephropathy. Poorly functioning kidneys often provide a real problem to the surgeon evaluating possible outflow obstruction and it is difficult to rely on the frusemide diuresis either visually, graphically, or on numerical indices applied to the third phase unless they show unequivocal absence of obstruction.

A fall in renal function is a typical response to chronic outflow obstruction. This is due partly to a normal control loop, which causes a reduction in glomerular filtration rate and renal plasma flow, and partly to a reduction in nephron population which typically occurs in the presence of infection. The extent to which this response is reversible depends upon the degree on which it is a physiological consequence of the nephron control loop. Recovery of renal function is often much greater than expected, particularly when the appearance of the kidney or an estimate of cortical thickness was determined from intravenous urography. Prolongation of the parenchymal transit time occurs before reduction in uptake function and shortening of the parenchymal transit time towards normal precedes an increase in uptake function after relief of the obstruction.

Renal failure

Renal failure is present when the urinary output of metabolites and waste products is insufficient to maintain normal body composition without alteration in body fluids. This may be chronic, resulting in progressive renal insufficiency, or may be an acute clinical presentation. The possible role of radionuclide studies in these conditions will be reviewed.

Acute renal failure

The mechanism of acute tubular necrosis is controversial, as is the terminology, since it is now well established that necrosis is by no means always present or necessary for the clinical picture. The initial insult is probably decreased renal perfusion which results in a decrease in renal blood flow and renal ischaemia. As GFR is acutely dependent on renal blood flow, the GFR falls, resulting in a decrease in urine flow (oliguria). The renal ischaemia then becomes self-perpetuating, involving mechanisms such as arteriolar vasoconstriction due to angiotensin, microcirculatory obstruction due to swelling of the vascular endothelium or, possibly, a decrease in secretion of intrarenal vasodilators such as prostaglandins. Certainly if the ischaemia is severe or nephro-

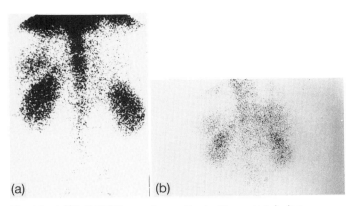

Fig. 11 A $^{99}Tc^m$-DTPA scan in a patient with acute tubular necrosis. (a) Good perfusion to both kidneys but no concentration of DTPA; (b) the image fades as the tracer distributes into the larger extracellular space.

toxins are present actual tubular cell necrosis will occur, with consequent back leakage of filtrate and tubule obstruction by debris.

The first problem is to decide upon initial investigations and whether any additional studies are necessary. Ultrasound examination combined with a radionuclide functional study will usually provide most of the information necessary for clinical management, with no risk of the toxicity and further deterioration of renal function which might be associated with the use of contrast agents. A dynamic $^{99}Tc^m$-DTPA or $^{99}Tc^m$-MAG3 scan is the method of choice, occasionally with the addition of a $^{99}Tc^m$-complex radiopharmaceutical such as DMSA.

The radionuclide study provides information about first-pass renal perfusion, the handling of the glomerular filtered agent and whether there is sufficient urine flow through the collecting systems. The frequency with which the kidneys can be visualized by radionuclide study when intravenous urography fails is variable, but we have found this to be quite high, especially if $^{99}Tc^m$-DMSA is also used. Radionuclide imaging also provides prognostic information. The single most useful role of the dynamic radionuclide scan is to make a firm diagnosis of acute tubular necrosis which, if treated with adequate dialysis, has a good prognosis, compared to other causes of acute renal failure.

It may also be possible to differentiate the early onset phase of acute tubular necrosis, which may be reversed by fluid and electrolye correction, from the established phase. The typical images seen in acute tubular necrosis are shown in Fig. 11. A practically normal perfusion phase during the first transit is followed by a moderately good visualization of the kidneys at 90 to 180 s which represents a blood pool image of the kidneys; as the tracer diffuses into the larger extracellular space from the vascular space the renal image diminishes without the appearance of tracer into the collecting system. Early signs of recovery are increasing retention of tracer in the kidney, as the GFR returns to normal and is superimposed on the blood pool image; progressive concentration as glomerular filtration continues to improve but intrarenal transit remains grossly prolonged; and excretion as at the onset of the diuretic phase of acute tubular necrosis.

Acute obstruction is characterized by a decreased perfusion image during first transit and poor early uptake image with progressive parenchymal accumulation, due to the grossly delayed intrarenal transit. Dilated calyces are frequently seen on the early images as negative photon-deficient areas which progressively accumulate tracer over several hours. These findings, although characteristic, should be confirmed with ultrasound prior to surgical treatment and the absence of evidence of obstruction on a $^{99}Tc^m$-DTPA scan should never be used to exclude the diagnosis without ultrasound confirmation.

Prerenal failure associated with acute oliguria due to, for example, dehydration but before the onset of established acute tubular necrosis is identical to phase 2 or 4 of the acute tubular necrosis recovery pattern. Radionuclide imaging shows well-perfused kidneys with a significant secretory peak representing glomerular filtration, markedly delayed parenchymal transit, but with some excretion of tracer which may only appear in the calyces at the end of the normal 20 to 30 min period of imaging.

Acute nephritis and most other renal parenchymal diseases usually show significantly worse perfusion than acute tubular necrosis, with markedly decreased uptake at 2 min and either no accumulation or slow progressive accumulation in the renal parenchyma, but without significant excretion.

These patterns represent the majority of cases of acute renal failure, but other causes, such as acute loss of perfusion unilaterally associated with renal artery embolus, may also be diagnosed. In aortic obstruction (Fig. 12) failure to visualize the aorta accompanies grossly diminished renal perfusion; aneurysm, if associated with a large lumen, will be identified as a large abdominal blood pool; venous thrombosis decreases perfusion but to a lesser extent than arterial occlusion.

When the visualization is very poor or absent using $^{99}Tc^m$-DTPA or MAG3 a repeat image following administration $^{99}Tc^m$-DMSA may give better results, and be a better guide to eventual prognosis. However, no systematic studies of this are available at the present time.

Chronic renal failure

Ultrasound examination remains the initial investigation in chronic renal failure and may reveal the aetiology. This may include

Fig. 12 An example of severe loss of perfusion which is complete on the left and almost complete on the right due to vascular complications during aortic surgery for aneurysm.

irregularly scarred kidneys due to pyelonephritis or reflux; the enlarged kidneys of polycystic disease; the small, uniformly contracted kidneys of chronic glomerulonephritis, or dilated ureters, collecting system, and enlarged kidneys arising from chronic obstruction. Although radionuclide investigation certainly shows chronically damaged kidneys in this situation and, to a very limited extent, predicts the degree of recoverable function, it is only occasionally of any clinical value and then is best performed with ^{123}I-hippuran, ^{99}Tc^m-DTPA or MAG3, or ^{99}Tc^m-DMSA.

Renal mass lesions

Renal parenchyma that has been replaced by a space-occupying lesion, whether tumour, cyst, infarct, or scar, can be identified by a loss of functioning tissue compared to the surrounding normal parenchyma, with the presence or absence of changes in local perfusion during the angiographic phase. Early studies with ^{99}Tc^m-DTPA and ^{99}Tc^m-DMSA concentrated on the detection of tumours and cysts of the kidney but with the development of improved structural imaging techniques (high dose nephrotomography, ultrasonography, and CT scanning) and more widespread use of diagnostic cyst puncture with cytological examination, the role of nuclear medicine techniques have changed markedly.

The vast majority of patients receive the appropriate investigative sequence and treatment. Radionuclide scans, which for the purpose of this particular problem are always ^{99}Tc^m-DMSA scans, usually with first-pass radionuclide angiography, are used in three clinical situations. The first is when the ultrasound examination does not confirm a mass lesion or is equivocal. This is not often a problem with quality ultrasound examination, but when it does arise the renal scan is the next best way of confirming or excluding a tumour. The scan may also help to determine whether there is a single or multiple lesion. Lateral and oblique views of the kidney should always be performed, and the results should be reviewed in conjunction with the intravenous urogram, in order to correlate abnormalities seen on the two examinations and to identify the site and size of the calyces on the renal scan, because, with current high-resolution images of the renal parenchyma, normal calyces may be confused with a space-occupying lesion.

Renal radionuclide scanning may also be helpful in the further evaluation of a lesion seen on intravenous urography which is solid on ultrasound, but could possibly be a pseudotumour, such as the lump from splenic pressure, fetal lobulation, compensatory hypertrophy, and prominent columns of Bertin. It is an advantage to be able to demonstrate that the 'tumour' is functionally normal renal tissue without resorting to arteriography.

When arteriography and surgery are contraindicated the renal scan, with particular emphasis on the first-pass arteriographic phase, may help to establish a diagnosis. Increased blood flow in the lesion indicates a high probability of tumour; no flow in a large lesion increases the likelihood of a cyst, but there are a significant number of cases in which the tumour has a relatively poor supply and differentiation between benign and malignant is not possible. It must be emphasized that this is a small part of the investigation of such patients; radionuclide scans must not be used to document the lesion if it is not going to influence the management of the patient.

Renal transplantation

Radionuclide investigations can contribute significantly to the management of renal transplant patients. Complications are dis-

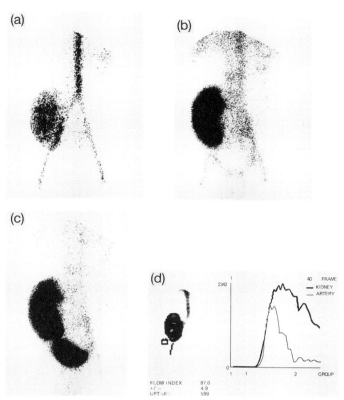

Fig. 13 A normal ^{99}Tc^m-DTPA transplant study showing: (a) good perfusion during the first-pass study; (b) good function with DTPA filtered at 2 min with (c) rapid drainage into the bladder. The time–activity curves (d) show the iliac and renal flow curves superimposed, calculated normal flow index and uptake index.

cussed in more detail in Chapter 11.2. Although divided into immediate, intermediate, and delayed, they are not confined to those times. Most patients who receive a cadaveric graft will have some degree of postoperative acute tubular necrosis, which may cause prolonged primary non-function. Acute rejection remains important and may be superimposed on primary non-function; the use of cyclosporin complicates the picture since it may have functionally similar effects to those of acute rejection.

Most of the important and common complications result in changes in renal function without morphological abnormalities; consequently techniques which essentially provide images of renal function, are well suited to the monitoring of the graft. They are most effective when performed in a very reproducible way and with accurate quantification, early after surgery (within 3 days) and repeated regularly until function is normal.

The dynamic ^{99}Tc^m-DTPA scan is most widely used for monitoring graft function (Fig. 13). The patient lies supine with a LFOV gamma-camera over the anterior pelvis and a bolus of 400 MBq of ^{99}Tc^m-DTPA is given intravenously. A series of images of the first pass of tracer through the kidney and vessels (vascular phase) is obtained, followed by 1 min images for a total of 20 min as the tracer concentrates in the parenchyma and is excreted into the collecting system. Delayed images may be necessary if there is delayed excretion, suspected obstruction, or a urinary leak. Data are stored on the computer and time–activity curves are generated from the regions of interest over the kidney, the iliac artery and background. Quantitative measurement may be made according to a number of published protocols (Baillet *et al.*

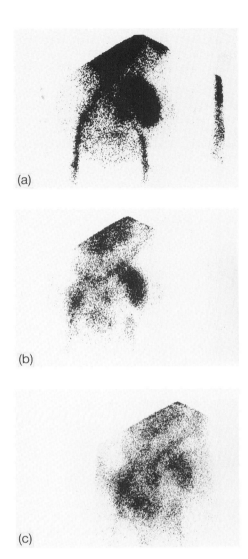

Fig. 14 Twenty-four hours after transplantation the cause of anuria is acute tubular necrosis with good perfusion (a), a blood pool and no GFR (b), and fading of the image (c).

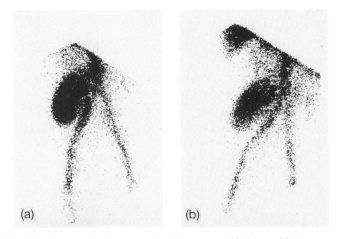

Fig. 15 Acute rejection occurring in an anuric kidney with acute tubular necrosis. (a) Before rejection, (b) decreased flow due to rejection superimposed on acute tubular necrosis.

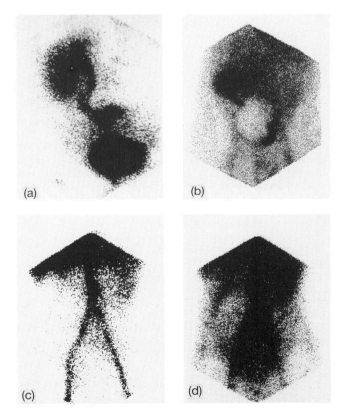

Fig. 16 Complications of renal transplantation may be shown such as urinary leak (a), lymphocele (b), or acute loss of perfusion (c,d).

1986), one of which generates a flow index, which is a measure of perfusion, and an uptake index, which is a measure of GFR (Hilson et al. 1987). Absolute uptake measurements, which enable GFR to be calculated, can also be obtained. This method will diagnose acute tubular necrosis (Fig. 14), and demonstrate decreased perfusion and function of various aetiologies, e.g. acute rejection (Fig. 15), cyclosporin A toxicity and renal arterial disorders. In most cases, unless the results are very unequivocal, temporal changes in the scan over a period of days and the clinical background are important. Cyclosporin A toxicity can be particularly difficult to distinguish from acute rejection (Gedroyc et al. 1986). Many other complications (lymphocele, leaks, obstruction etc.) can also be detected using the scan, often before they are clinically apparent (Fig. 16).

Although $^{99}Tc^m$ is the most common radiopharmaceutical used for transplant evaluation, others have been introduced and are used in some centres. These include ^{131}I-OIH or ^{123}I-OIH which are used with imaging protocols similar to DTPA and a variety of different quantitative procedures. They may also be combined with serum sampling to measure effective renal plasma flow. Recently, $^{99}Tc^m$-MAG3 has been used as a DTPA replacement with encouraging results, the higher clearance giving better quality images, especially in patients with impaired function and in children. Some radiopharmaceuticals have been directed at the diagnosis of rejection, these include ^{67}Ga-citrate,

$^{99}Tc^m$-sulphur colloid, and ^{111}In-labelled platelets and white cells. However, none of these has been widely accepted as a routine clinical tool.

Bibliography

Bischof-Delaloye, A. and Blaufox, M.D. (1987). Radionuclides in Nephrology. *Contributions to Nephrology*, **86**, 77–81.

Blaufox, M.D. (1987). The current status of renal radiopharmaceuticals. *Contributions to Nephrology*, **56**, 31–7.

Dubovsky, E.V. and Russell, C.D. (1988). Radionuclide evaluation of renal transplants. *Seminars in Nuclear Medicine*, **18**, 181-98.

Fine, E.J. and Sarkar, S. (1989). Differential diagnosis and management of renovascular hypertension through nuclear medicine techniques. *Seminars in Nuclear Medicine*, **19**, 101–15.

Fogelman, I. and Maisey, M.N. (1988). *An Atlas of Clinical Nuclear Medicine*, Martin Dunitz, London.

Gordon, I. (1986). Use of TC-99m DMSA and Tc-99m DTPA in reflux. *Seminars in Urology*, **4**, 99–108.

Heyman, S. (1989). An update of radionuclide renal studies in paediatrics. In *Nuclear Medicine Annual* (eds. L.M. Freeman and H.S. Weissman), p. 179. Raven Press, New York.

O'Reilly, P.H., Shields, R.A., and Testa, H.J. (1987). In *Nuclear Medicine in Urology and Nephrology* (eds. H.J. Testa, R.A. Shields, and P.H. O'Reilly). Butterworths: London.

Sfakianakis, G.N. and Sfakianaki, E.D. (1988). Nuclear-medicine in paediatric urology and nephrology. *Journal of Nuclear Medicine*, **29**, 1287–300.

References

Baillet, G., et al. (1986). Evaluation of allograft perfusion by radionuclide first-pass study in renal failure following renal transplantation. *European Journal of Nuclear Medicine*, **11**, 463–9.

Dondi, M., et al. (1989). Evaluation of hypertensive patients by means of captopril enhanced renal scintigraphy with technetium ^{99m}-DTPA. *Journal of Nuclear Medicine*, **30**, 615–21.

Fleming, J.S., Keast, C.M., Waller, D.G., and Ackery, D. (1987). Measurement of glomerular filtration rate with ^{99m}Tc-DTPA: a comparison of gamma camera methods. *Journal of Nuclear Medicine*, **13**, 250–3.

Fritzberg, A.R., Kasina, S., Eshima, D., and Johnson, D.L. (1981). Chemical and biological studies of Tc-^{99m} N, N-bis-(mercaptoacetamide)-ethylenediamine: a potential replacement for I-131 iodohippurate. *Journal of Nuclear Medicine*, **22**, 258–63.

Fritzberg, A.R., et al. (1986). Synthesis and biological evaluation of Tc-99m MAG3 as a hippuran replacement. *Journal of Nuclear Medicine*, **27**, 111–16.

Gedroyc, W., Taupe, D., Fogelman, I., Neild, G., Cameron, J.S., and Maisey, M.N. (1986). Tc-99m DTPA scans in renal allograft rejection and cyclosporine nephrotoxicity. *Transplantation*, **42**, 494–7.

Geyskes, G.G., Oei, H.Y., and Faber, J.A. (1986). Renography: prediction of blood pressure after dilatation of renal artery stenosis. *Nephron*, **44**, 54–9.

Godley, M.L., Ransley, P.G., Parkhouse, H.F., Gordon, I., Evans, K., and Peters, A.M. (1990). Quantitation of vesicoureteric reflux by radionuclide cystography and urodynamics. *Paediatric Nephrology*, in press.

Groshar, D., et al. (1989). Quantitation of renal uptake of technetium-^{99m}-DMSA using SPECT. *Journal of Nuclear Medicine*, **30**, 246–50.

Gruenewald, S.M., et al. (1988). Quantitative renography in patient follow-up following treatment of renal artery stenosis. *Proceedings of the Royal Australian College of Physicians Golden Jubilee Meeting*, A 156.

Hay, A.M., Norman, W.J., Rice, M.L., and Steventon, R.D. (1984). A comparison between diuresis renography and the Whitaker test in 64 kidneys. *British Journal of Urology*, **56**, 561–64.

Hedman, P.J.K., Kempi, V., and Voss, H. (1978). Measurement of vesicoureteral reflux with intravenous $^{99}Tc^m$-DTPA compared to radiographic cystography. *Radiology*, **126**, 205–8.

Hilson, A.J., Maisey, M.N., Brown, C.B., Ogg, C.S., and Bewick, M.S. (1978). Dynamic renal transplant imaging with $^{99}Tc^m$-DTPA (Sn) supplemented by a transplant perfusion index in the management of renal transplants. *Journal of Nuclear Medicine*, **19**, 994–1000.

Jafri, R.A., et al. (1988). $^{99}Tc^m$-MAG3: a comparison with I-123 and I-131 orthoiodohippurate in patients with renal disorder. *Journal of Nuclear Medicine*, **29**, 147–58.

Kogan, S.J., Sigler, L., Levitt, S.B., Reda, E.F., Weiss, R., and Greifer, I. (1986). Elusive vesicoureteral reflux in children with normal contrast cystograms. *Journal of Urology*, **136**, 325–8.

Merrick, M.V., Uttley, W.S., and Wild, S.R. (1980). Detection of pyelonephritic scarring in children by radioisotope imaging. *British Journal of Radiology*, **53**, 544–6.

O'Reilly, P.H., Shields, R.A., and Testa, H.J. (1987). Nuclear Medicine in Urology and Nephrology. 2nd ed. pp. 91–108. Butterworths, London.

Rizzoni, G., et al. (1986). Radionuclide voiding cystography in intrarenal reflux detection. *Annals of Radiology*, **29**, 415–20.

Russell, C.D., et al. (1985). Measurement of glomerular filtration rate using $^{99}Tc^m$-DTPA and the gamma camera: a comparison of methods. *European Journal of Nuclear Medicine*, **10**, 519–21.

Wenting, G.J., Tan-Tjiong, H.L., Derkx, F.H.M., de Bruyn, J.H.B., Man in t'Veld, A.J., and Schalekamp, M.A.D.H. (1984). Split renal function after captopril in unilateral renal artery stenosis. *British Medical Journal*, **288**, 886–90.

Zananiri, M.C., Jarritt, P.H., Sarfarazi, M., and Ell, P.J. (1987). Relative and absolute $^{99}Tc^m$-DMSA uptake measurements in normal and obstructed kidneys. *Nuclear Medicine Communications*, **8**, 869–80.

Zechmann, W. (1988). The experimental approach to explain some misinterpretations of diuresis renography. *Nuclear Medicine Communications*, **9**, 283–94.

Zhang, G., Day, D.L., Loken, M., and Gonzalez, R. (1987). Grading of reflux by radionuclide cystography. *Clinical Nuclear Medicine*, **12**, 106–9.

1.7.1 The renal glomerulus—the structural basis of ultrafiltration

MARLIES ELGER AND WILHELM KRIZ

The correct name for the structure to be described is 'renal corpuscle'; 'glomerulus' strictly refers only to the tuft of glomerular capillaries (glomerular tuft). However, the use of the term glomerulus for the entire corpuscle is widely accepted.

A renal corpuscle is made up of a tuft of specialized capillaries supplied by an afferent arteriole, drained by an efferent arteriole, and enclosed in Bowman's capsule (Figs. 1 and 2). The entire tuft of capillaries is covered by epithelial cells (podocytes), representing the visceral layer of Bowman's capsule. At the vascular pole, i.e. at the place where the afferent arteriole enters and the efferent arteriole leaves the capillary tuft, the visceral layer of Bowman's capsule becomes the parietal layer, which is a simple squamous epithelium. At the urinary pole the parietal epithelium of Bowman's capsule abruptly transforms to the high epithelium of the proximal tubule. The space between both layers of Bowman's capsule is called the urinary space; at the urinary pole this passes into the tubule lumen.

The glomerular basement membrane lies at the interface between the glomerular capillaries and the epithelial (visceral) layer of Bowman's capsule. During developement this membrane originates from both the epithelial podocytes and the capillary endothelial cells (Sariola *et al.* 1984; Abrahamson 1987). At the vascular pole the glomerular basement membrane becomes the basement membrane of Bowman's capsule. Through the opening of Bowman's capsule, bordered by the glomerular basement membrane/Bowman's capsule basement membrane the afferent arteriole enters and the efferent arteriole leaves the glomerular tuft, respectively.

Renal corpuscles are the starting points of the nephrons. Thus, the number of nephrons exactly correlates with the number of renal corpuscles—in man, this is about one million in each kidney, in the rat about 30 000 per kidney. Renal corpuscles are roughly spherical in shape, with a diameter of approximately 120 μm in the rat, and about 200 μm in man. Juxtamedullary renal corpuscles are generally somewhat larger (by about 20 per cent) than midcortical and superficial corpuscles (Kriz and Kaissling 1985; Tisher and Brenner 1989).

The glomerular tuft—a 'wonder net'

The glomerular tuft derives from an afferent arteriole, which divides into several (three to five) primary capillary branches at the entrance level (Figs. 1 and 3) (Yang and Morrison 1980). Each of these branches gives rise to a superficially located, anastomosing capillary network which runs toward the urinary pole, thereby establishing a subdivision of the glomerular tuft, which is often called a glomerular lobule. These lobules are generally not strictly separated from each other; within deeper layers of the tuft anastomoses occur between the lobules. The efferent arteriole develops deep within the centre of the glomerular tuft from the confluence of tributaries from all lobules. Thus, in contrast to the afferent arteriole, the efferent arteriole has an intraglomerular segment. The efferent arteriole leaves the tuft at the vascular pole; two efferent arterioles are sometimes observed.

Topography of a glomerular lobule—glomerular capillaries are unique

Glomerular capillaries are a specific type of blood vessel (Fig. 4) whose wall is made up of an endothelial tube. A small strip of the outer circumference of this tube is in touch with the mesangium, which constitutes the axis of a glomerular lobule. The glomerular basement membrane and the visceral epithelium of Bowman's capsule (podocytes) do not form complete envelopes around individual glomerular capillaries; together they constitute a common surface cover wrapping an entire glomerular lobule (i.e. the mesangium with its attached capillaries). Consequently, the wall

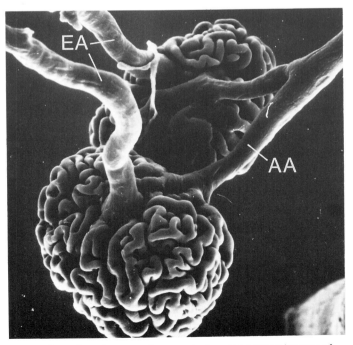

Fig. 1 Scanning electron micrograph showing a vascular cast of two juxtamedullary glomeruli (rat). Each capillary tuft is supplied by an afferent arteriole (AA) which, on the surface of the tuft, immediately divides into several branches. Efferent arterioles (EA) emerge out of the centre of the tuft (× 400).

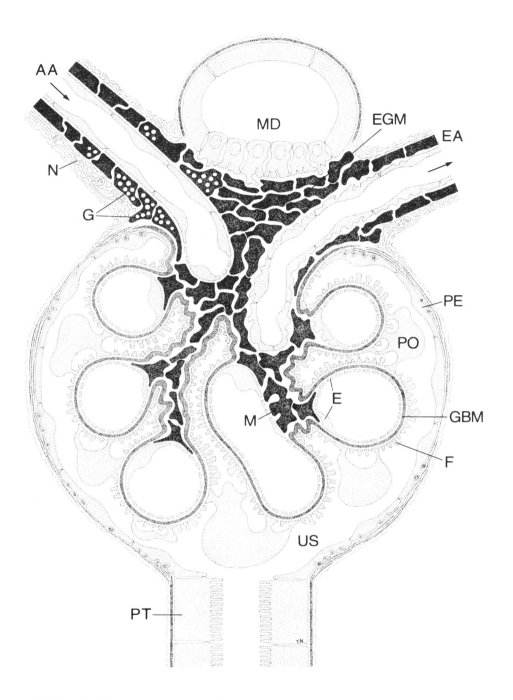

Fig. 2 Diagram of a longitudinal section through a renal corpuscle and the juxtaglomerular apparatus (JGA). The capillary tuft consists of a network of specialized capillaries, which are outlined by a fenestrated endothelium (E). At the vascular pole an afferent arteriole (AA) enters and an efferent arteriole (EA) leaves the tuft. The capillary network is surrounded by Bowman's capsule comprising two different epithelia: the visceral epithelium consisting of highly branched podocytes (PO) directly follows—together with the glomerular basement membrane (GBM)—the surface of the capillaries. At the vascular pole, the visceral epithelium and the GBM are reflected into the parietal epithelium (PE) of Bowman's capsule, which passes over into the epithelium of the proximal tubule (PT) at the urinary pole. Mesangial cells (M) are situated in the axes of glomerular lobules. At the vascular pole mesangial cells are continuous with extraglomerular mesangial cells (EGM); together with the granular cells (G) of the afferent arteriole and the macula densa (MD) they establish the JGA. All cells which are suggested to be derived from smooth muscle cells are shown in black. F, foot processes; N, sympathetic nerve terminals; US, urinary space.

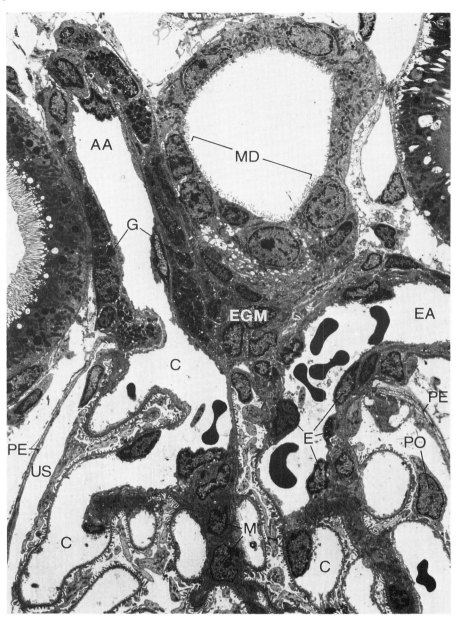

Fig. 3 Longitudinal section through the glomerular vascular pole showing both arterioles and the juxtaglomerular apparatus (JGA) (rat). At the entrance into the glomerulus, the afferent arteriole (AA) immediately branches into capillaries (C). The efferent arteriole (EA) usually arises deeper in the tuft and can be distinguished by the high number of endothelial cells (E) at the exit from the tuft. The macula densa (MD) of the thick ascending limb is in contact with the extraglomerular mesangial cells (EGM) and the glomerular arterioles. The media of the AA contains granular cells (G). M, mesangial cells; PE, parietal epithelium; PO, podocytes; US, urinary space. Electron micrograph ($\times$ 1700).

of glomerular capillaries has two portions: a urinary portion, which bulges into Bowman's space and is covered by the glomerular basement membrane and by the visceral epithelium of Bowman's capsule (podocytes), and a much smaller juxtamesangial portion, which only consists of the endothelium (without an underlying glomerular basement membrane) directly abutting the mesangium.

At the points where the capillaries come into contact with the mesangium, the glomerular basement membrane and the podocyte layer deviate from a pericapillary course and cover the mesangium; these points have been called mesangial angles. Therefore two parts of the glomerular basement membrane and the visceral epithelium can be distinguished: a pericapillary part and a perimesangial part. The pericapillary part of the glomerular basement membrane is smooth and follows the outline of the capillary, whereas its perimesangial part is irregular in thickness and is frequently wrinkled. The epithelial layer of Bowman' capsule (podocyte layer) forms the interdigitating processes which are generally more perfectly developed in the pericapillary part than in the perimesangial part. In the mesangium, two parts must also be distinguished: a juxtacapillary region which directly borders the juxtamesangial portion of the glomerular capillaries and the axial region which is bounded by the perimesangial part of the glomerular basement membrane (Sakai and Kriz 1987).

Capillary endothelium—a perforated highly charged structure

The capillary endothelium is made up of flat endothelial cells whose cell bodies usually occupy a juxtamesangial position, while the thin and fenestrated peripheral parts of these cells line the urinary portions of the capillaries (Fig. 4). The cell bodies contain the usual organelles such as mitochondria, lysosomes, Golgi bodies, and some rough and smooth endoplasmic reticulum. The endothelial skeleton comprises intermediate filaments

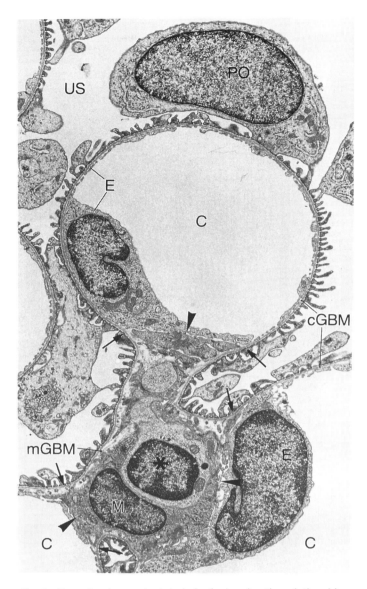

Fig. 4 Part of a glomerular lobule (rat), showing the relationship of glomerular capillaries to the lobule centre. The capillary (C) is outlined by a flat fenestrated endothelium (E). The podocyte layer (PO) and the glomerular basement membrane (GBM) do not encircle the capillary completely; they form a common surface cover around the entire lobule. At the urinary portion of the capillary the filtration barrier is formed (see also Fig. 5). Two subdomains of the GBM are delineated from each other by mesangial angles (arrows): the pericapillary GBM (cGBM) faced by the podocyte and endothelial layer, and the perimesangial GBM (mGBM) bordered by the podocyte layer and the mesangium. Within the mesangium two types of mesangial cells are shown: contractile mesangial cells proper (M) and a cell (*) which is probably a macrophage which has taken residence in the mesangium. Note the intimate relationships between the endothelium and the mesangium (arrowheads). US, urinary space (× 6100).

and microtubules; individual fenestrations are lined by clusters of microfilaments (Vasmant et al. 1984).

The fenestrated regions are extremely attenuated and characterized by round to oval pores 50 to 100 nm in diameter (Larsson and Maunsbach 1980) (Fig. 5). Compared to fenestrated endothelia at other sites, most pores are fully open; thus they lack a diaphragm. In the rat, fenestrations are occasionally covered by diaphragms in man (Larsson and Maunsbach 1980); these have also been described (Jorgensen 1966). In the rat, about 60 per cent of the capillary surface is covered by the fenestrated regions; the total area of fenestrae occupies about 13 per cent of the capillary surface (Bulger et al. 1983). The total area of all pores in a glomerulus amounts to about 22 mm^2 in rat (Larsson and Maunsbach 1980). In contrast to other capillaries of the body, micropinocytotic vesicles are very rare in glomerular endothelial cells.

The luminal membrane of endothelial cells bears a negatively charged cell coat due to the presence of several polyanionic glycoproteins (Horvat et al. 1986), including a sialoprotein called podocalyxin, which is considered to be the major surface polyanion of glomerular endothelial as well as epithelial cells (Sawada et al. 1986).

Renal endothelial cells share an antigen system with cells of the monocyte/macrophage lineage; they express surface antigens of the class II histocompatibility antigens. Like platelets, glomerular endothelial cells contain components of the coagulation pathway, and are capable of binding factors IXa and Xa, and synthesize, release and bind von Willebrand factor (factor VIII) (Wiggins et al. 1989).

Visceral epithelium—filtration slits are formed by interdigitating foot processes

The visceral epithelial layer of Bowman's capsule consists of highly differentiated cells, the podocytes (see Kriz and Kaissling 1985; Bulger and Hebert 1988; Tisher and Brenner 1989). Mitotic divisions never occur in the adult. They have a voluminous cell body, which bulges into the urinary space (Figs. 6 and 7). Long primary processes emerge from the cell body and extend towards the capillaries, to which they are affixed by numerous secondary processes (foot processes or pedicles) in a very specific pattern. The foot processes of neighbouring podocytes regularly interdigitate with each other, leaving between them meandering slits, which are called filtration slits. These are narrowest at their floor, ranging between 25 and 65 nm (Fig. 5).

The filtration slits are bridged by an extracellular structure, called slit membrane or slit diaphragm, which has a regular substructure. Like a zipper, the slit membrane consists of two rows of subunits that extend from both sides of the slit to its centre where they are linked by a central bar. Between the subunits on both sides rectangular spaces occur, 4 times 14 nm in area, which is approximately the size of an albumin molecule. The total area of the slit membrane represents approximately 2 to 3 per cent of the total surface area of glomerular capillaries (Rodewald and Karnovsky 1974).

The presence of the filtration slits means that podocytes are not connected to each other. Each podocyte is individually fixed to the glomerular basement membrane by its own foot processes; thus the major part of the cell floats within the filtrate of the urinary space. Two cell surface domains (separated from each other at the level of the slit membrane) must be distinguished: the luminal cell membrane facing Bowman's space and the abluminal membrane at the bases of the foot processes which anchor the cells to the glomerular basement membrane.

The luminal podocyte membrane is covered by a thick surface

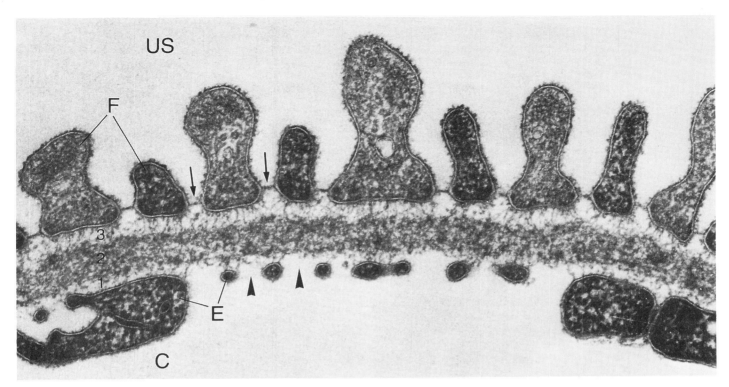

Fig. 5 Glomerular capillary wall. The filtration barrier comprises the fenestrated endothelial layer (E), the glomerular basement membrane, and the podocyte layer with foot processes (F) and filtration slits bridged by diaphragms (long arrows). Arrowheads point to the endothelial pores. The glomerular basement membrane consists of three layers, the lamina rara interna (1), the lamina densa (2), and the lamina rara externa (3). In this picture tannic acid staining allows differentiation between the alternating foot processes of two neighbouring podocytes: the more densely stained processes belong to one cell, and the others to the neighbouring cell. C, capillary lumen (× 87 000).

coat (glycocalyx) which is rich in sialic acid, demonstrated by a variety of techniques including colloidal iron, alcian blue, ruthenium red, and polycationized ferritin (see Andrews 1988). Podocalyxin has been identified as the major sialoprotein of this glycocalyx (Sawada et al. 1986). The high concentration of polyanionic sialic acid components is responsible for the high negative surface charge of podocytes.

The foot processes are anchored within the glomerular basement membrane to a depth of about 60 nm. The abluminal membrane at the bases of the foot processes contains quite different glycocalix components, including, for example, N-acetyl-D-galactosamine residues of glycoconjugates (Roth et al. 1983). The occurrence of a fibronectin receptor within the abluminal membrane underlines the apparent role of fibrillar surface glycoproteins such as fibronectin, laminin, or entactin, in the attachment of the podocytes to the glomerular basement membrane (Kerjaschki et al. 1989).

Other membrane proteins, of which the C3b-receptor (Kazatchkine et al. 1982) and Heymann's antigen (Kerjaschki and Farquhar 1983) are extremely important, are contained in the entire surface of podocytes, i.e. within the luminal and abluminal cell membrane. Heymann's antigen, a glycoprotein of 330 kDa (gp 330) is generally associated with coated pits at the cell surface; within the cytoplasm it is found in the endoplasmic reticulum and the Golgi apparatus (Kerjaschki and Farquhar 1983).

The cell body contains a prominent nucleus situated peripherally, close to the urinary surface. On the opposite site, i.e. between the nucleus and the surface which points towards the capillary tuft, a large cytoplasmic compartment contains a conspicuously well-developed Golgi apparatus, Cisternal profiles of rough and smooth endoplasmic reticulum, many free ribosomes, prominent lysosomes, and a large number of mitochondria (Fig. 7(a)). In contrast to the cell body, the cell processes contain only a few organelles.

The organelles contained within the cytoplasm of the cell body indicate a high level of anabolic as well as catabolic activity. In addition to the work necessary to sustain the structural integrity of these complicated cells, it is widely agreed that most (if not all) components of the glomerular basement membrane are synthesized by the podocytes (Striker and Striker 1985; Abrahamson 1987).

The maintenance of the complicated shape of podocytes requires a well developed cytoskeleton: microfilaments, intermediate filaments and microtubules are abundant. In the cell body and the primary processes, microtubules and intermediate filaments dominate, as demonstrated by routine electron microscopy (Fig. 7(b)) and by immunocytochemical localization of their main protein subunits, tubulin and vimentin (Bachmann et al. 1983; Andrews 1988; Vasmant et al. 1984; Drenckhahn and Franke 1988). Within the foot processes microfilaments, generally arranged in bundles extending in the long axis of the foot processes, are most abundant. Immunocytochemical studies show that actin, myosin and α-actinin are associated with these microfilament bundles (Trenchev et al. 1976; Andrews 1988; Drenckhahn and Franke 1988).

The cytoskeletal equipment of the foot processes suggest a

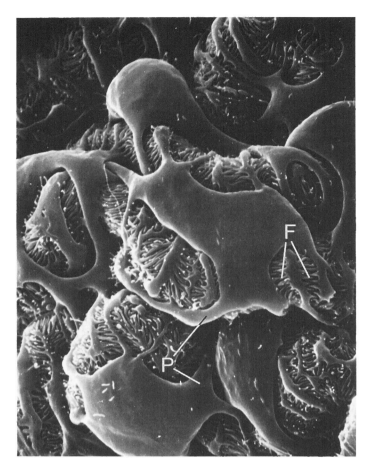

Fig. 6 A scanning electron micrograph of a rat glomerular capillary. The urinary side of the capillary is covered by the highly branched cells of the podocyte layer. The cell bodies give rise to primary processes (P) embracing adjacent capillaries. The foot processes (F) of neighbouring cells interdigitate regularly, sparing the filtration slits in between (× 3300).

contractile ability, whose relevance, however, is still unclear. It has often been suggested, but so far never proven, that the width of the filtration slits may be changed by the contractile activity of the podocyte foot processes (Andrews 1988). In this context it has to be mentioned that the surface charge contributes decisively to sustain the interdigitating pattern of the foot processes and the width of their filtration slits. In response to neutralization of the surface charge by cationic substances (e.g. protamin sulphate, poly-*l*-lysin), the glomerular epithelium undergoes a series of changes including flattening and retraction of foot processes, narrowing of filtration slits, and formation of tight junctions between adjacent foot processes (Seiler *et al.* 1977; Andrews 1988).

Glomerular basement membrane—the backbone of the glomerular tuft

The glomerular basement membrane represents the skeletal backbone of the glomerular tuft. It is divided into two different domains, the pericapillary or peripheral glomerular basement membrane and the perimesangial or axial region. The border

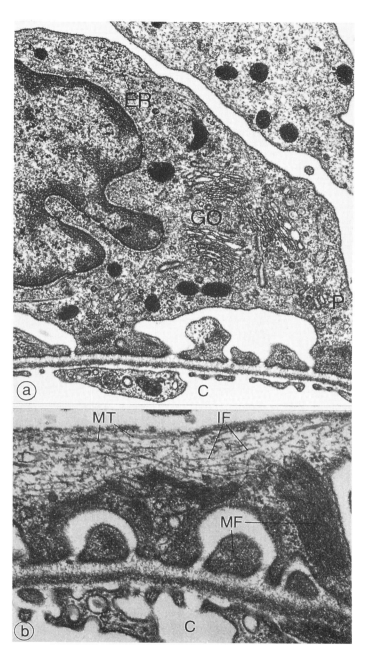

Fig. 7 (a) Electron micrograph showing part of a podocyte cell body anchored via primary processes (P) and foot processes to the glomerular basement membrane. Note the prominent Golgi apparatus (GO); also rough endoplasmic reticulum (ER) is fairly abundant. C, capillary lumen (× 25 000). (b) Primary and secondary processes of podocytes showing cytoskeletal elements. Intermediate filaments (IF) and microtubules (MT) are abundant in the primary processes, whereas thick bundles of microfilaments (MF) are located in the foot processes. C, capillary lumen (× 46 000).

between both domains is represented by a change from a convex pericapillary into a concave perimesangial course; the turning points are called mesangial angles (Fig. 4) (Sakai and Kriz 1987). Routine electron micrographs show that the glomerular basement membrane is composed of three layers of different electron density (Fig. 5). The most dense, middle, layer is called lamina

1.7.1 The renal glomerulus—the structural basis of ultrafiltration

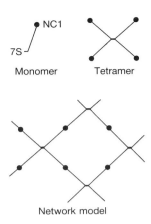

Fig. 8 Schematic diagram for the formation of macromolecular structures of type IV collagen. After Timpl and Dziadek 1986, modified after Glanville (1987).

densa and is bounded by two, more electron lucent layers, the lamina rara interna and lamina rara externa. The lamina rara interna attaches the lamina densa to the capillary endothelium; at the perimesangial domain it blends with the mesangial matrix. The lamina rara externa attaches the lamina densa to the bases of the foot processes of podocytes and fills the gap between two foot processes up to the slit membrane.

The width of the glomerular basement membrane differs between man and experimental animals. In man the thickness ranges between 300 and 370 nm (Steffes *et al.* 1983), in rat and other experimental animals its thickness ranges from 110 to 190 nm (Bulger and Herbert 1988; Rasch 1979). The lamina densa is not only the most dense layer but also the thickest layer. Generally both laminae rarae have a thickness of about one-half to two-thirds of the lamina densa.

The biochemical composition of the glomerular basement membrane is similar to that of basement membranes at other sites (Mohan and Spiro 1986; Timpl and Dziadek 1986). The major components include type IV collagens, heparan sulphate proteoglycans, and laminin. Type V collagen, fibronectin, and entactin has also been demonstrated. The arrangement and location of these substances in the glomerular basement membrane is still controversial (Kanwar 1984; Abrahamson 1987).

Current models picture the lamina densa as a mat of collagen type IV. Monomers of type IV collagen consist of a triple helix of length 400 nm which, at its carboxy-terminal end, has a large non-collagenous globular domain, called NC 1. At the amino-terminus the helix possesses a triple helical rod of length 60 nm, the 7S domain. Interactions between the 7S domain and the NC 1 domain allow collagen type IV monomers to form dimers and tetramers which, by lateral association of triple helical strands, assemble into a three-dimensional network (Fig. 8) (Timpl and Dziadek 1986; Yurchenko and Ruben 1987). This network provides mechanical stability to the basement membrane and serves as a basic structure upon which other matrix components attach.

The laminae rarae obviously serve to connect the adjacent cells to the lamina densa. Whether there is a basic substructure of collagen type IV in the laminae rarae is still a matter of debate (Laurie *et al.* 1984; Kerjaschki *et al.* 1986). In routine electron micrographs thin fibrils are generally seen to extend from the meshwork of the lamina densa into both laminae rarae. The affixation to the adjacent cell membranes is obviously effected by glycoproteins such as fibronectin, laminin, and entactin, of which laminin is the most abundant. It is a non-collagenous glycoprotein consisting of three polypeptide chains, two of which are glycosylated and cross-linked by disulphide bridges (Timpl and Dziadek 1986). Laminin is thought to bind to specific sites on the polymerized meshwork of type IV collagen as well as to the basal endothelial and epithelial surfaces, thereby connecting these cells to the glomerular basement membrane.

In common with the endothelial and epithelial cell layers, the glomerular basement membrane has an electronegative charge, mainly due to polyanionic proteoglycans. The major proteoglycan of the glomerular basement membrane is heparan sulphate which is composed of a core protein and four heparan sulphate chains (glycosaminoglycans) (Kanwar 1984). Aggregation of proteoglycan molecules results in the formation of a meshwork that is kept highly hydrated by water molecules trapped in the interstices. In some studies, the localization of heparan sulphate proteoglycans was confined to the laminae rarae (Kerjaschki *et al.* 1986), whereas other authors concluded that proteoglycans are present throughout the glomerular basement membrane (Laurie *et al.* 1984). Heparan sulphate proteoglycans within the glomerular basement membrane may act as an anticlogging agent to prevent hydrogen bonding and adsorption of anionic plasma proteins and maintain an efficient flow of water through the membrane (Kanwar 1984).

Filtration barrier—filtration occurs along an extracellular pathway

Filtration in the glomerulus occurs across the urinary wall portion of glomerular capillaries. The filtration barrier is composed of (1) the endothelium with large open pores, (2) the pericapillary portion of the glomerular basement membrane, and (3) the slits between the podocyte foot processes which are bridged by the slit membrane. Thus, filtration is obviously effected along an extracellular route through the glomerular capillary wall.

Compared with the barrier established in capillaries elsewhere in the body, there are at least two outstanding characteristics of the filtration barrier in the glomerulus: the permeability for water, small solutes and ions is extremely high, while the permeability for plasma proteins and other macromolecules is very low.

The selective permeability is obviously rooted in the fact that filtration occurs along extracellular routes. All components of this route, the endothelium, the highly hydrated glomerular basement membrane, and the slit membrane can be expected to be quite permeable for water and small solutes (including small charged solutes). Whether the major hydraulic barrier is established by the glomerular basement membrane or by the filtration slits is in debate (see Olivetti *et al.* 1985). However, it can readily be assumed that in the end the entire filtrate has to go through the filtration slits.

The barrier function for macromolecules is selective for size and charge. The charge selectivity arises from the dense accumulation of negatively charged molecules throughout the entire depth of the filtration barrier, including the surface coats of endothelial and epithelial cells and the glomerular basement membrane. Polyanionic macromolecules in the plasma, such as plasma proteins, are repelled by these assemblies of negative charges establishing an electronegative shield. This mechanism obviously is of great importance in minimizing the clogging of the filter. Removal or blocking of the negative charge in experimental models results in proteinuria (Brenner et al. 1978).

The size selectivity of the filtration barrier is established by the dense network of the glomerular basement membrane. Uncharged macromolecules up to an effective radius of about 2.0 nm freely pass through the filter. Larger compounds are more and more restricted (indicated by their fractional clearances which progressively decrease) and are totally restricted at effective radii of more than 4.0 nm. The term effective radius is an empirical value, measured in artificial membranes, which takes into account the shape of macromolecules and attributes a radius to non-spherical molecules. Plasma albumin has an effective radius of 3.6 nm; without the repulsion due to the negative charge, plasma albumin would pass through the filter in considerable amounts (Deen et al. 1979).

Mesangium—maintenance of the structural integrity of the tuft

The mesangium occupies the axial region of a glomerular lobule and consists of mesangial cells and the surrounding mesangial matrix (Figs. 4 and 9), first described by Zimmermann (1929). Since the ultrastructural characterization of the mesangium in the early sixties (Latta et al. 1960; Farquhar and Palade 1962), mesangial cells have been the focus of glomerular research. They are generally believed to form a supporting frame work which maintains the structural integrity of the glomerular tuft. Recent evidence, mainly from experiments with cultured mesangial cells has suggested that mesangial cells also play an important role in the regulation of glomerular haemodynamics and filtration (see Kreisberg et al. 1985; Schlondorff 1987).

Mesangial cells are considered to be contractile cells. They are irregular in shape with numerous cytoplasmic processes extending from the cell body, and contain assemblies of microfilaments, particularly within the processes (Fig. 9). This, along with their many gap junctions (Pricam et al. 1974) suggests that they have a common origin with smooth muscle cells, and resemble pericytes.

Immunocytochemistry and light microscopy has shown that mesangial cells contain actin, myosin, and α-actinin (Pease 1968; Kreisberg et al. 1985; Drenckhahn and Franke 1988); electron microscopy reveals that these contractile proteins are contained in the microfilament bundles (Drenckhahn et al., unpublished work). Thus, there is good evidence that mesangial cells can contract not only in culture but also within the glomerular tuft in the living animal. Mesangial cells possess receptors for angiotensin II, vasopressin, atrial natriuretic factor, and prostaglandins (among others) (Dworkin et al. 1983; Sraer et al. 1974). They also share a surface antigen, Thy-1, with T lymphocytes (Paul et al. 1984).

In addition to their contractile ability mesangial cells are phagocytic, serving to unblock the glomerular basement membrane. They contain some lysosomal elements and have been shown to take up particulate tracers (Farquhar and Palade 1962) as well as immune complexes which may accumulate within the mesangial region (see Schlondorff 1987). It appears, however, that the majority of mesangial cells (all of which have the contractile properties and may therefore be considered as mesangial cells proper) are not bona fide phagocytic cells. A small subpopulation (3 to 7 per cent) of mesangial cells is bone marrow

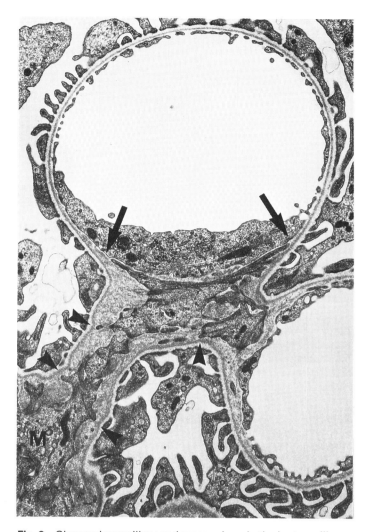

Fig. 9 Glomerular capillary and mesangium. In the juxtacapillary region, long mesangial cell processes extend between opposite mesangial angles, where they are fixed to the glomerular basement membrane (arrows). In the axial region, finger-like processes of different length connect the mesangial cells to the perimesangial glomerular basement membrane (arrowheads). Note bundles of microfilaments in the cell processes. M, mesangial cell ($\times$ 13 000).

derived; they can be regarded as macrophages, which have taken up residence in the mesangium (Kreisberg et al. 1985).

The main function of mesangial cells therefore arises from their contractile abilities. The glomerular basement membrane has recently been shown (Sakai and Kriz 1987) to be the primary effector site of mesangial cell contraction. Mesangial cells are extensively connected with this membrane either by direct apposition of mesangial cell processes or by microfibrils (Fig. 9). These connections, which appear to be mechanically strong, are found throughout the mesangial region.

The mesangial matrix basically consists of densely interwoven microfibrils which connect the entire perimesangial part of the glomerular basement membrane to mesangial cells. Microfibrils are a major component of the mesangial matrix, as has been shown by transmission electron microscopy after tannic acid staining (Mundel *et al.* 1988) and by immunocytochemistry using antibodies against microfibrillar proteins derived from elastic tissue (Gibson *et al.* 1989). The high levels of fibronectin within the mesangium (Madri *et al.* 1980) are related to need for firm connections between the different structures. The wrinkles frequently found in the perimesangial glomerular basement membrane appear to be the result of mesangial cell contraction by which the membrane is drawn towards the mesangial axis of a glomerular lobule.

The mesangial cell-glomerular basement membrane connections are most prominent at mesangial angles and are of a specific character. The juxtamesangial portion of each capillary is consistently associated with tongue-like mesangial cell processes (emerging from a trunk process) which run towards the mesangial angles (Figs. 9 and 10). They frequently extend for some distance into the narrowing space between the endothelium and the glomerular basement membrane being connected to the membrane either directly or indirectly via microfibrils (Sakai and Kriz 1987).

These juxtacapillary mesangial cell processes are densely filled with assemblies of contractile filaments which run in a tangential direction to the circular circumference of the capillaries. When a process is connected to the glomerular basement membrane by long microfibrils, the direction of the intracellular microfilaments is continued extracellularily into that of microfibrils which finally insert into the glomerular basement membrane in an oblique direction. Intracellular microfilaments and extracellular microfibrils are consistently arranged in such a way that contraction of mesangial cells brings the membrane from opposing mesangial angles closer together, maintaining the typical narrow necks of glomerular capillaries (Fig. 11) (Sakai and Kriz 1987).

It must be stressed that the relevance of this arrangement is not fully known. It is thought that the major role of the contractile apparatus of mesangial cells is static in nature, operating by isometric or minute isotonic contractions. The large pressure gradient across their walls is obviously a crucial challenge to glomerular capillaries. To withstand this pressure gradient a correspondingly high tension has to be developed, and the contractile apparatus of mesangial cells and the glomerular basement membrane apparently represent a biomechanical unit capable of developing such tension in capillaries, and adapting this tension to changes in distending forces.

Parietal epithelium

The parietal layer of Bowman's capsule consists of squamous epithelial cells resting on a basement membrane. The flat cells are polygonal, with a central cilium and few microvilli. Parietal cells are filled with bundles of actin filaments running in all directions (Pease 1968; Newstead 1971), especially prominent at the vascular pole, where they are located within cytoplasmic ridges that run in a circular fashion around the glomerular entrance. They appear to be functionally connected via fibronexus to fibrils within the underlying basement membrane. This basement membrane has a very specific structure which is established at the transition from the glomerular basement membrane at the vascular pole. At this point the thick membrane splits into several layers which together establish the basement membrane of Bowman's capsule. Thus, this basement membrane is a multilayered structure composed of several dense layers which are

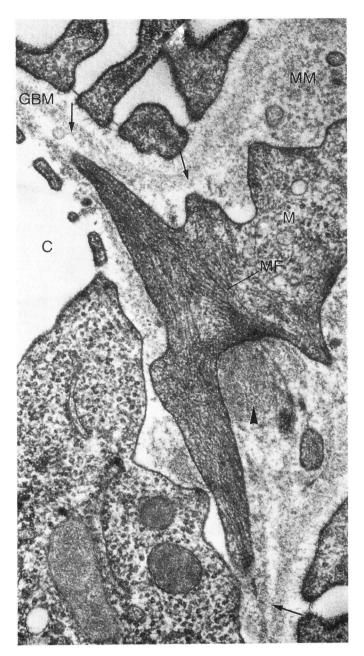

Fig. 10 Juxtacapillary portion of mesangium showing tongue-like mesangial processes fixed to the glomerular basement membrane (GBM) at the mesangial angles (arrows). Note the rich equipment of mesangial processes with bundles of microfilaments (MF) which are attached to the cell membrane. The mesangial matrix (MM) contains abundant microfibrils (arrowhead). C, capillary lumen ($\times$ 61 200).

separated by several translucent layers (Mbassa *et al.* 1988). It is thickest near the vascular pole; in more peripheral parts it may be found to be much thinner. As already mentioned, the transition from the glomerular basement membrane into the basement membrane of Bowman's capsule borders the glomerular entrance. The transitional region is mechanically connected to the smooth muscle cells of the afferent and efferent arterioles as well as to extraglomerular mesangial cells (see below).

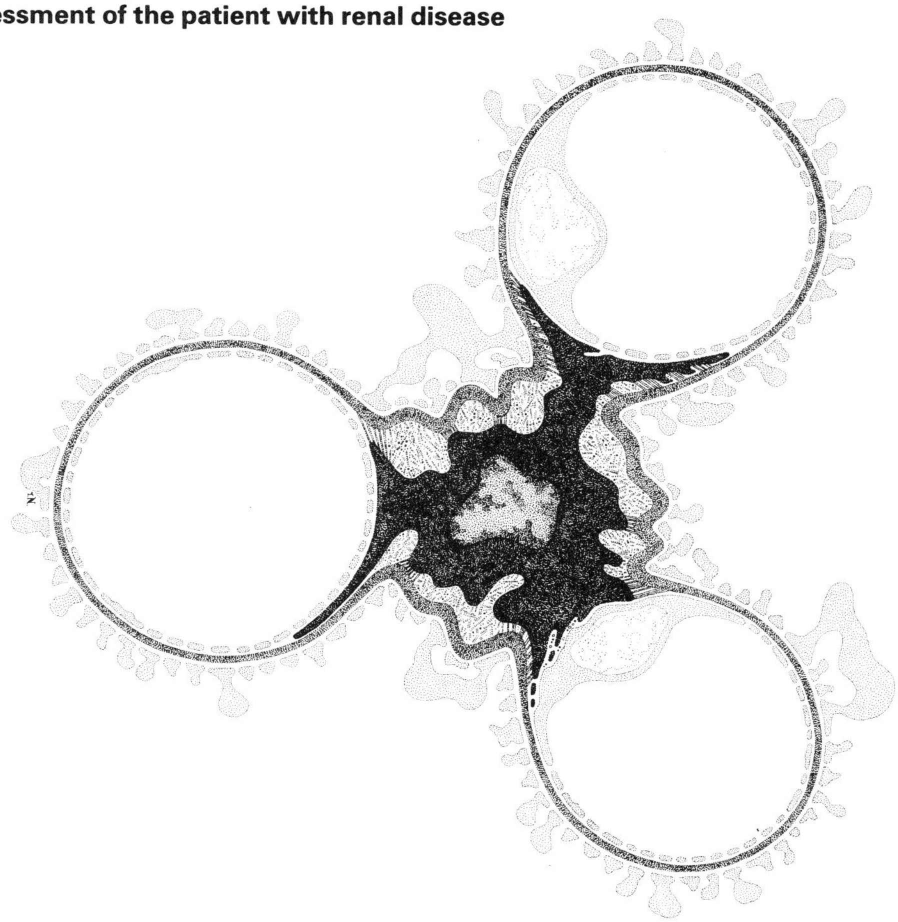

Fig. 11 Schematic showing the centrolobular position of a mesangial cell and its relationships to the glomerular capillaries and to the glomerular basement membrane. Connections between mesangial cell processes and the membrane are prominent at mesangial angles. In addition, numerous connections with the perimesangial glomerular basement membrane are found. Many of these connections are mediated by microfibrils. Thus, a mechanical firm linkage of the perimesangial glomerular basement membrane to the contractile apparatus of the mesangial cells is established. (Modified after Sakai and Kriz 1989.)

Extraglomerular mesangium—a closure device of the glomerular entrance

At the glomerular entrance the glomerular mesangial cells pass out of the glomerular stalk to become extraglomerular mesangial cells (Figs. 2 and 3). Together with their surrounding matrix they form the extraglomerular mesangium (Barajas et al. 1989), which is often also called the polar cushion or Goormaghtigh cell field. The extraglomerular mesangium represents a solid cell complex that is not penetrated by blood vessels or lymphatic capillaries. Nerves pass on both sides from the afferent to the efferent arteriole but do not enter this cell complex.

The extraglomerular mesangium is located within the triangular space bordered by the two glomerular arterioles and the macula densa of the thick ascending limb. Reconstruction of extraglomerular mesangial cells has shown them to be flat and elongated, separating into bunches of long cell processes at their poles (Spanidis and Wunsch 1979). They are arranged in several layers, parallel to the base of the macula densa. The cells nearest to the glomerular stalk and filling the deepest portion of the triangle loose this grouping in parallel to the macula densa base: they extend—either with part of their cell body or at least with some processes—into the stalk of the glomerular tuft, mixing with mesangial cells. The conspicuous extraglomerular mesangial matrix is clearly different from the mesangial matrix. Even if microfibrils can be found in the extraglomerular matrix, a dense microfibrillar network has never been seen.

Although direct evidence is lacking, extraglomerular mesangial cells can be expected to be contractile for several reasons. First, they contain a moderate amount of contractile microfilaments, mainly in their processes. Second, intimate structural similarities are found among arteriolar smooth muscle cells, granular cells, and intra- and extraglomerular mesangial cells suggesting that they have the same origin. Third, they are extensively coupled by gap junctions (Pricam et al. 1974; Taugner et al. 1978).

The contractile processes of extraglomerular mesangial cells are connected to the basement membrane of Bowman's capsule and to the walls of both glomerular arterioles. As a whole, the extraglomerular mesangium represents a spider-like contractile clamp sitting above and interconnecting all structures of the glomerular entrance. The extraglomerular mesangium can be regarded as a closure device of the glomerular entrance, maintaining its structural integrity against the distending forces exerted to the entrance by the high intraglomerular pressure.

Juxtaglomerular apparatus—intersection of local and systemic regulations

The juxtaglomerular apparatus comprises (1) the macula densa of the thick ascending limb, (2) the extraglomerular mesangium (already described above), and (3) the renin producing granular cells of the afferent arteriole. Because of the intimate connection with the juxtaglomerular apparatus, and probably due to their functional relationships, the mesangial cells and the ordinary smooth muscle cells of the afferent and efferent arterioles are sometimes considered as part of this apparatus.

The macula densa is a plaque of specialized cells within the thick ascending limb at the site where it is affixed to the extraglomerular mesangium of the corresponding glomerulus. The most obvious structural feature, from which the cell plaque gets the name 'macula densa', is the large narrowly packed cell nuclei (Zimmermann 1933).

In contrast to the surrounding cells of the thick ascending limb the macula densa cells do not interdigitate with each other but have a fairly polygonal outline. Basically, the cells display numerous infoldings of the plasma membrane. Narrow basal plasma membrane folds are anchored to the underlying basement membrane, blending with the matrix of the extraglomerular mesangium (Kriz and Kaissling 1985).

The lateral membranes of macula densa cells bear folds and finger-like villi which are frequently connected to those of neighbouring cells by desmosomes. The luminal cell membrane is densely studded by stubby microvilli and bears one cilium. The cells are joined by tight junctions consisting of several junctional strands, similar to those of the thick ascending limb. The cells contain the usual cytoplasmic organelles, comprising some small mitochondria, Golgi apparatus, and smooth endoplasmic reticulum; free ribosomes are abundant but rough endoplasmic reticulum is infrequent. Few small lysosomes may be encountered in the apical cell portion.

The lateral intercellular spaces are a prominent feature of the macula densa. They are different from those in most nephron segments, but resemble those between collecting duct cells. Electron microscopic studies, and studies on isolated macula densa segments *in vitro*, have shown that the width of the lateral interspaces varies under different functional conditions (Kaissling and Kriz 1982; Kirk *et al.* 1985). In agreement with the suggestion that water flow through the macula densa epithelium is secondary to active sodium reabsorption, compounds such as frusemide, that block sodium transport by macula densa cells, as well as high osmolalities of impermeable solutes such as mannitol are associated with narrowing of the intercellular spaces (Kaissling and Kriz 1982; Alcorn *et al.* 1986). The spaces are apparently dilated under most physiological conditions, usually regarded as normal control conditions.

The granular cells are assembled in clusters within the terminal portion of the afferent arteriole, replacing ordinary smooth muscle cells (Fig. 3). Their name refers to their specific cytoplasmic granules which are dark, membrane-bound, and irregular in size and shape. Immunocytochemical studies have clearly shown that they contain renin which is synthesized by the cells and stored in granular form. Renin release occurs by exocytosis into the surrounding interstitium (see Taugner and Hackenthal 1989).

Granular cells are modified smooth muscle cells. In situations that require enhanced renin synthesis (e.g. volume depletion and stenosis of the renal artery) additional smooth muscle cells located upstream in the wall of the afferent arteriole transform into granular cells (see Taugner and Hackenthal 1989). It has already been mentioned that granular cells are connected to the extraglomerular mesangial cells and probably also to adjacent smooth muscle cells by gap junctions, and that granular cells are densely innervated by sympathetic nerve terminals (Taugner and Hackenthal 1989).

The structural organization of the juxtaglomerular apparatus strongly suggests a regulatory function. Goormaghtigh (1937) was the first to propose that some component of the distal urine is sensed by the macula densa and this information is used to adjust the tonus of the glomerular arterioles, thereby producing a change in glomerular blood flow and filtration rate. Even if many details of this mechanism are still subject to debate, the essence of this hypothesis has been verified by many studies and is known as the tubular glomerular feedback mechanism (Briggs and Schnermann 1987). In addition, however, the juxtaglomerular apparatus seems to be the main site of regulation of renin secretion; this is not only of local but obviously also of systemic relevance. It is widely believed that the extraglomerular mesangium receives signals from the macula densa, modulating and transferring them to the granular cells and the ordinary smooth muscle cells of the glomerular arterioles. It makes sense (as described above) for the basal tonic contractility of the extraglomerular mesangium to reflect the degree of distension brought about by the intraglomerular pressure.

References

Abrahamson, D.R. (1987). Structure and development of the glomerular capillary wall and basement membrane. *American Journal of Physiology*, **253**, F783–F794.

Alcorn, D., Anderson, W.P., and Ryan, G.B. (1986). Morphological changes in the renal macula densa during natriuresis and diuresis. *Renal Physiology*, **9**, 335–47.

Andrews, P. (1988). Morphological alterations of the glomerular (visceral) epithelium in response to pathological and experimental situations. *Journal of Electron Microscopy Technique*, **9**, 115–144.

Bachmann, S., Kriz, W., Kuhn, C., and Franke, W.W. (1983). Differentiation of cell types of the mammalian kidney by immunofluorescence microscopy using antibodies to intermediate filament proteins and desmoplakins. *Histochemistry*, **77**, 365–94.

Barajas, L., Salido, E.C., Smolens, P., Hart, D., and Stein, J.H. (1989). Pathology of the juxtaglomerular apparatus including Bartter's syndrome. In *Renal Pathology*, (eds. C. C. Tisher and B. M. Brenner), pp. 877–912, Lipincott, Philadelphia.

Brenner, B.M., Hostetter, T.H., and Humes, H.D. (1978). Glomerular permselectivity: barrier function based on discrimination of molecular size and charge. *American Journal of Physiology*, **234**, F455–F460.

Briggs, J.P. and Schnermann, J. (1987). The tubuloglomerular feedback mechanism: functional and biochemical aspects. *Annual Review of Physiology*, **49**, 251–73.

Bulger, R.E., Eknoyan, G., Purcell, D.J., and Dobyan, D.C. (1983). Endothelial characteristics of glomerular capillaries in normal, mercuric chloride-induced, and gentamicin-induced acute renal failure in the rat. *The Journal of Clinical Investigation*, **72**, 128–41.

Bulger, R.E. and Hebert, S.C. (1988). Structural-functional relationships in the kidney. In *Diseases of the Kidney*, (eds. R.W. Schrier and C.W. Gottschalk), pp. 3–63, Little, Brown and Company, Boston/Toronto.

Deen, W.M., Bohrer, M.P., and Brenner, B.M. (1979). Macromolecule transport across glomerular capillaries: Application of pore theory. *Kidney International*, **16**, 353–65.

Drenckhahn, D. and Franke, R.P. (1988). Ultrastructural organization of contractile and cytoskeletal proteins in glomerular podocytes of chicken, rat, and man. *Laboratory Investigation*, **59**, 673–682.

Dworkin, L.D., Ichikawa, I., and Brenner, B.M. (1983). Hormonal modulation of glomerular function. *American Journal of Physiology*, **244**, F95–F104.

Farquhar, M.G. and Palade, G.E. (1962). Functional evidence for the existence for a third cell type in the renal glomerulus. *Journal of Cell Biology*, **13**, 55–87.

Gibson, M.A., Kumaratilake, J.S., and Cleary, E.G. (1989). The protein components of the 12-nanometer microfibrils of elastic and nonelastic tissues. *Journal of Biological Chemistry*, **264**, 4590–4598.

Glanville, R.W. (1987). Type IV collagen. In *Structure and function of collagen types*, (ed. R. Mayne), pp. 43–80. Academic Press, London.

Goormaghtigh, N. (1937). L'appareil neuro-myo-artériel juxta-glomérulaire du rein: ses réactions enpathologie et ses rapports avec le tube urinifère. *Comptes Rendus des Sciences de la Société de Biologie et de ses Filiales (Paris)*, **124**, 293–6.

Horvat, R., Hovorka, A., Dekan, G., Poczewski, H., and Kerjaschki, D. (1986). Endothelial cell membranes contain podocalyxin—the major sialoprotein of visceral glomerular epithelial cells. *Journal of Cell Biology*, **102**, 484–91.

Jorgensen, F. (1966). *The ultrastructure of the normal human glomerulus*, Ejnar Munksgaard, Copenhagen.

Kaissling, B. and Kriz, W. (1982). Variability of intercellular spaces between macula densa cells: A transmission electron microscopic study in rabbits and rats. *Kidney International*, **12**, S9–S17.

Kanwar, Y.S. (1984). Biology of disease. Biophysiology of glomerular filtration and proteinuria. *Laboratory Investigation*, **51**, 7–21.

Kazatchkine, M.D., Fearon, D.T., Appay, M.D., Mandet, C., and Bariety, J. (1982). Immunohistochemical study of the human glomerular C3b receptor in normal kidney and in seventy-five cases of renal diseases. *Journal of Clinical Investigation*, **69**, 900–12.

Kerjaschki, D. and Farquhar, M.G. (1983). Immunohistochemical localization of the Heymann antigen (gp 330) in glomerular epithelial cells of normal Lewis rats. *Journal of Experimental Medicine*, **157**, 667–86.

Kerjaschki, D., Sawada, H., and Farquhar, M.G. (1986). Immunoelectron microscopy in kidney research: some contributions and limitations. *Kidney International*, **30**, 229–45.

Kerjaschki, D., *et al.* (1989). A β1-integrin receptor for fibronectin in human kidney glomeruli. *American Journal of Pathology*, **134**, 481–9.

Kirk, K.L., Bell, P.D., Barfuss, D.W., and Ribadeneira, M. (1985). Direct visualization of the isolated and perfused macula densa. *American Journal of Physiology*, **248**, F890–F894.

Koushanpour, E. and Kriz, W. (1986). *Renal Physiology. Principles, Structure and Function*, pp. 162–175, Springer, New York.

Kreisberg, J.I., Venkatachalam, M.A., and Troyer, D. (1985). Contractile properties of cultured glomerular mesangial cells. *American Journal of Physiology*, **249**, F457–F463.

Kriz, W. and Kaissling, B. (1985). Structural organization of the mammalian kidney. In *The Kidney: Physiology and Pathophysiology*, (eds. D.W. Seldin and G. Giebisch), pp. 265–306. Raven Press, New York.

Larsson, L. and Maunsbach, A.B. (1980). The ultrastructural development of the glomerular filtration barrier in the rat kidney: a morphometric analysis. *Journal of Ultrastructural Research*, **72**, 392–406.

Latta, H., Maunsbach, A.B. and Madden, S.C. (1960). The centrolobular region of the glomerulus studied by electron microscopy. *Journal of Ultrastructural Research*, **4**, 455–72.

Laurie, G.W., Leblond, C.P., Inoue, S., Martin, G.R., and Chung, A. (1984). Fine structure of the glomerular basement membrane and immunolocalization of five basement membrane components to the lamina densa (basal lamina) and its extensions in both glomeruli and tubules of the rat kidney. *American Journal of Anatomy*, **169**, 463–81.

Madri, J.A., Roll, F.J., Furtmayr, H., and Foidart, J.M. (1980). Ultrastructural localization of fibronectin and laminin in the basement membranes of the murine kidney. *Journal of Cell Biology*, **77**, 837–52.

Mbassa, G., Elger, M., and Kriz, W. (1988). The ultrastructural organization of the basement membrane of Bowman's capsule in the rat renal corpuscle. *Cell and Tissue Research*, **253**, 151–63.

Mohan, P.S. and Spiro, R.G. (1986). Macromolecular organization of basement membranes. *Journal of Biological Chemistry*, **261**, 4328–36.

Mundel, P., Elger, M., Sakai, T., and Kriz, W. (1988). Microfibrils are a major component of the mesangial matrix in the glomerulus of the rat kidney. *Cell and Tissue Research*, **254**, 183–7.

Newstead, J.D. (1971). Filaments in the renal parenchyma and interstitial cells. *Journal of Ultrastructural Research*, **343**, 16–28.

Olivetti, G., Giacomelli, F., and Weiner, J. (1985). Morphometry of superficial glomeruli in acute hypertension in the rat. *Kidney International*, **27**, 31–8.

Paul, L.C., Rennke, H.G., Milford, E.L., and Carpenter, C.B. (1984). Thy-1.1 in glomeruli of rat kidneys. *Kidney International*, **25**, 771–7.

Pease, D.C. (1968). Myoid features of renal corpuscles and tubules. *Journal of Ultrastructural Research*, **23**, 304–20.

Pricam, C., Humbert, F., Perrelet, A., and Orci, L. (1974). Gap junctions in mesangial and lacis cells. *Journal of Cell Biology*, **63**, 349–54.

Rasch, R. (1979). Prevention of diabetic glomerulopathy in streptozotocin diabetic rats by insulin treatment. Glomerular basement membrane thickness. *Diabetologia*, **16**, 319–24.

Rodewald, R. and Karnovsky, M.J. (1974). Porous substructure of the glomerular slit diaphragm in the rat and mouse. *Journal of Cell Biology*, **60**, 423–33.

Roth, J., Brown, D., and Orci, L. (1983). Regional distribution of N-acetyl-D-galactosamine residues in the glycocalyx of glomerular podocytes. *The Journal of Cell Biology*, **96**, 1189–96.

Sakai, T. and Kriz, W. (1987). The structural relationship between mesangial cells and basement membrane of the renal glomerulus. *Anatomy and Embryology*, **176**, 373–86.

Sariola, H., *et al.* (1984). Dual origin of glomerular basement membrane. *Developmental Biology*, **101**, 86–96.

Sawada, H., Stukenbrok, H., Kerjaschki, D., and Farquhar, M.G. (1986). Epithelial polyanion (podocalyxin) is found on the sides but not the soles of the foot processes of the glomerular epithelium. *American Journal of Pathology*, **125**, 309–18.

Schlondorff, D. (1987). The glomerular mesangial cell: an expanding role for a specialized pericyte. *Federation of American Societies for Experimental Biology Journal*, **1**, 272–81.

Seiler, M.R., Rennke, H.G., Venkatachalam, M.G., and Cotran, R.S. (1977). Pathogenesis of polycation-induced alteration (fusion) of glomerular epithelium. *Laboratory Investigation*, **36**, 48–61.

Spanidis, A. and Wunsch, H. (1979). Rekonstruktion einer Goormaghtigh'schen und einer Epitheloiden Zelle der Kaninchenniere. *Dissertation*, University of Heidelberg.

Sraer, J.D., Sraer, J., Ardaillou, R., and Mimoune, O. (1974). Evidence for renal glomerular receptors for angiotensin II. *Kidney International*, **6**, 241–6.

Steffes, M.W., Barbosa, J., Basgen, J.M., Sutherland, D.E.R., Najarian, J.S., and Mauer, S.M. (1983). Quantitative glomerular morphology of the normal human kidney. *Laboratory Investigation*, **49**, 82–6.

Striker, G.E. and Striker, L.J. (1985). Biology of disease. Glomerular cell culture. *Laboratory Investigation*, **53**, 122–31.

Taugner, R., Schiller, A., Kaissling, B., and Kriz, W. (1978). Gap junctional coupling between the JGA and the glomerular tuft. *Cell and Tissue Research*, **186**, 279–85.

Taugner, R. and Hackenthal, E. (1989). *The Juxtaglomerular Apparatus*, Springer-Verlag, Berlin.

Timpl, R. and Dziadek, M. (1986). Structure, development, and molecular pathology of basement membranes. *International Review of Experimental Pathology*, **29**, 1–112.

Tisher, C.C. and Brenner, B.M. (1989). Structure and function of the glomerulus. In *Renal Physiology*, (eds. C.C. Tisher and B.M. Brenner), pp. 92–110. Lippincott, Philadelphia.

Trenchev, P., Dorling, J., Webb, J., and Holborrow, E.J. (1976). Localization of smooth muscle-like contractile protein in kidney by immunoelectron microscopy. *Journal of Anatomy*, **121**, 85–95.

Vasmant, D., Maurice, M., and Feldmann, G. (1984). Cytoskeleton ultrastructure of podocytes and glomerular endothelial cells in man and in the rat. *Anatomical Record*, **210**, 17–24.

Wiggins, R.C., Fantone, J., and Phan, S.H. (1989). Mechanisms of vascular injury. In *Renal Pathology*, (ed. C.C. Tisher and B.M. Brenner), pp. 965–93. Lippincott, Philadelphia.

Yang, G.C.H. and Morrison, A.B. (1980). Three large dissectable rat glomerular models reconstructed from wide-field electron micrographs. *Anatomical Record*, **196**, 431–40.

Yurchenko, P.D. and Ruben, G.C. (1987). Basement membrane structure in situ: evidence for lateral associations in the Type IV collagen network. *Journal of Cell Biology*, **105**, 2559–68.

Zimmermann, K.W. (1929). Über den Bau des Glomerulus der menschlichen Niere. *Zeitschrift fuer Mikroskopische und Anatomische Forschung*, **18**, 520–52.

Zimmermann, K.W. (1933). Über den Bau des Glomerulus der Säugerniere. *Zeitschrift fuer Mikroskopische und Anatomische Forschung*, **32**, 176–278.

1.7.2 Renal biopsy: performance and interpretation

CLAUDIO PONTICELLI, M.J. MIHATSCH, AND ENRICO IMBASCIATI

The introduction of renal biopsy in clinical practice (Iversen and Brun 1951) has represented one of the most important advances in the field of clinical nephrology. Renal biopsy has greatly contributed to a rational classification of intrinsic renal diseases and to a better knowledge of the pathogenetic mechanisms involved in these diseases. Even today, in spite of the flood of new and less invasive tests renal biopsy is still considered by most nephrologists as an irreplaceable tool in assessing diagnosis and prognosis and guiding treatment of many renal diseases.

Procedure

Percutaneous renal biopsy

This is done with the patient in prone position upon a firmly rolled sheet compressing the upper abdomen and lower ribs and fixing the kidney. Local anaesthesia and sedation are sufficient in most patients but children under the age of 7 or 8 may require general anaesthesia. The lower pole of the kidney is chosen for biopsy. This area presents several advantages: it is furthest away from the renal pelvis and the large renal vessels; it contains the smallest number of large vessels; and it falls in a natural anatomic triangle where the kidney is almost subcutaneous.

Localization of the kidney may be carried out by means of either plain film of the abdomen or intravenous pyelography. The operator localizes the midspinal line and the lowest rib and then marks the proposed biopsy site in the left lower pole. Direct fluoroscopic visualization, with the use of an image intensifier or television monitoring can localize the kidney rather better (Kark 1968). Ultrasonography is now widely used to locate the kidney and to guide the needle. After ultrasonic delineation of the kidney, the calyceal system is noted and avoided. A location on the lower pole is chosen so that variations in respiration will not result in passage of the needle into the calyceal system. The angle of entry of the needle can be determined by observing the transducer angle with the skin. With the patient prone it is generally more comfortable for a righthanded operator to biopsy the left kidney. The risk of hitting the great vessels is negligible. Once the biopsy site is mapped, local anaesthesia is administered. The patient is asked to hold his breath and an exploring needle is gently guided through the tissues to the area to be biopsied. As the operator feels the tip entering the kidney he stops. If the needle is in the kidney it moves in unison with respiration. After measuring the depth of the exploring needle the operator repeats the manoeuvre with the biopsy needle, usually a Trucut®. When the needle is in the kidney the patient is asked to take a deep breath and the biopsy is done. Transillumination of biopsy specimens is useful to confirm the presence of adequate glomeruli for evaluation by light microscopy, immunofluorescence, and/or electron microscopy (Hefter and Brennan 1981).

Open renal biopsy

The kidney can be exposed by surgical opening of the triangle formed by the lateral border of the sacrospinalis muscle, the inferior border of the last rib, and the upper border of the quadratus lumborum muscle. Incision of the lumbodursal fascia and separations of the latissimus dorsi expose the aponeurosis of the transversus abdominus. The aponeurosis is incised and the inferolateral surface of the kidney is exposed by incising Gerota's fascia and separating the perirenal fat. A wedge of cortex can be removed with a lancet or scalpel. Some clinicians prefer to use a biopsy needle (Kark *et al.* 1968) or the Kawamura-modified forceps (Nomoto *et al.* 1987) to obtain medullary tissue also. Moreover, squeezing artefacts are seen less often with a biopsy needle than with any other procedure. Accurate haemostasis can be achieved by stitching or compression. The transverse aponeurosis and the lumbodursal fascia are then sutured and the skin is closed with nylon sutures. A small vertical opening (2–5 cm) is more than sufficient and results in a short, uncomplicated course. Large exposures are unnecessary, painful, and favour wound complications. In co-operative patients open renal biopsy can be done under local anaesthesia.

In a study which compared open surgical to percutaneous biopsy, tissue was adequate for diagnosis in 92 per cent of patients with biopsy localized at a site by fluoroscopy and in 100 per cent of surgically-operated patients (Bolton and Vaughn 1977). One patient required exploration after closed biopsy and none following surgical biopsy. In choosing between the two procedures the slightly higher yield of tissue by open biopsy should be weighed against its higher cost and longer stay in hospital. It is also difficult to perform a second biopsy in the same area as an open first biopsy.

Table 1 Complications of renal biopsy

Complication	Incidence (%)
Gross haematuria	5–7
Perirenal haemorrhage	
Severe	0.2–1.4
Mild	85
Arteriovenous fistula	15
Aneurysms	Rare
Renal dysfunction	Exceptional
Puncture of other organs	Exceptional

Preoperative and postoperative care

Before performing biopsy it is necessary to know the size and the localization of the kidneys as well as the possible presence of morphologic abnormalities, cysts, or neoplasms. Ultrasonographic or intravenous pyelographic examinations are therefore required. In hypertensive patients blood pressure should be lowered to normal values by adequate therapy, before biopsy. Coagulation disorders should be screened by determining the blood platelet count, the partial thromboplastin time, the prothrombin time, and the Howell time. However the most reliable haemostatic test in renal patients is probably the bleeding time, which reflects the platelet-plug formation at the transected ends of blood vessels and is an indirect indicator of the platelet function. Template bleeding time should be measured by the special devices available. Normal values range between 3 and 7 min. Bleeding time is generally prolonged in uraemia. A transient shortening of bleeding time can be obtained in uraemic patients by giving cryoprecipitate, which may increase factor VIII: von Willebrand factor (Janson et al. 1980). More recently, 1-deamino-8-D-arginine vasopressin, a synthetic derivate of the antidiuretic hormone without smooth-muscle activity, proved to shorten the bleeding time significantly while increasing factor VIII-coagulant activity in uraemic patients (Mannucci et al. 1983). Infusion over 30 min of 0.3 μg/kg body weight diluted in 50 ml of physiologic saline may normalize the bleeding time for 4 to 8 h in most uraemic patients, so allowing renal biopsy.

Following the biopsy, the patient must remain in the prone position upon a rolled sheet or pillow, for at least 1 h to improve compressive haemostasis. Strict bedrest is recommended for at least 24 h in uncomplicated patients. Vital signs, haematocrit values, and urine samples should be checked frequently. Activity should be restricted for several days after discharge and intensive physical exercise should be avoided for at least 4 weeks after renal biopsy.

Complications of renal biopsy

Several complications may occur after biopsy (Table 1). They are more frequent with the percutaneous technique than with open renal biopsy but severe and even lethal complications have been (exceptionally) encountered even with surgical biopsy (Zawada et al. 1987). Patients with haemorrhagic disorders, renal insufficiency, arterial hypertension, amyloidosis, and/or infection are particularly prone to complications. The skill of the operator is also of great importance. Although the mortality rate is less than 0.1 per cent (Wickre and Golper 1982) renal biopsy remains a serious procedure and should be performed only by experienced nephrologists. A complete preoperative assessment of the patient and careful postoperative assistance are necessary to minimize the complications of renal biopsy.

Haematuria

Gross haematuria occurs in 5 to 7 per cent of patients (Wickre and Golper 1982). Bleeding frequently stops within a few hours and in 50 per cent of patients stops within a week. Macroscopic haematuria may be associated with clots, colicky pain, urinary obstruction, and in the most severe cases with haemorrhagic shock. Puncture of the renal calyceal system, large vessel damage, aneurysms, and arteriovenous fistulae are the main causes. Strict bedrest is recommended in patients with gross haematuria. High urine flow rates may help to prevent urinary obstruction. Some patients may benefit from antifibrinolytic agents, such as aminocaproic acid. In uraemic patients a course with conjugated oestrogens can be useful (Liu et al. 1984). An intravenous infusion over 40 min of Emopremarin (Ayerst, New York), at a dose of 0.6 mg/kg body weight per day for 5 days, significantly shortens the bleeding time in uraemic patients. The maximum effect is reached between days 5 and 7 and lasts for 14 days (Livio et al. 1986).

Perirenal haemorrhage

A prevalence of 85 per cent of perirenal haematomas has been observed in a series of patients investigated by computerized tomography after renal biopsy (Rosenbaum et al. 1978). Fortunately, in most cases these haematomas were asymptomatic and completely resolved within 1 to 3 months. The incidence of clinically significant haematomas is far less, ranging between 0.2 and 1.4 per cent (Brun and Raaschou 1958; Diaz-Buxo and Donadio 1975). Perirenal bleeding usually occurs immediately after biopsy but can be delayed for some days or even weeks. A consistent decrease in haematocrit and loin tenderness occur in patients with large perirenal haematomas. Tachycardia with collapse may accompany heavy bleedings. Computerized tomography and/or ultrasonography can be helpful for supporting the diagnosis, localizing the haematoma, and following its outcome. Severe anaemia, impairment of renal function due to compressive ischaemia, and perinephric abscesses are the most severe complications of haematomas. In most patients bedrest is suf-

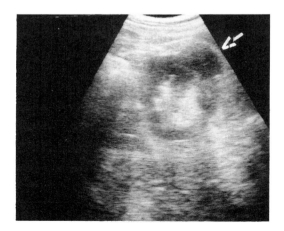

Fig. 1 Postbiopsy haematoma. At ultrasonography it appears to involve partially the renal parenchyma (arrow).

1.7.2 Renal biopsy: performance and interpretation

ficient but in more severe cases surgical evacuation of the haematoma and local compressive haemostasis or reparation of capsular lacerations are needed.

Arteriovenous fistulae

Angiographic studies have documented a 15 per cent incidence of arteriovenous fistulae after biopsy. Hypertensive patients are more prone to this complication (Ekelund and Lindholm 1971). In most patients arteriovenous fistula is asymptomatic and heals spontaneously within some months. Other patients present persistent haematuria, hypertension or high output cardiac failure. Fistula occurs more often in patients with massive renal haemorrhage after biopsy. Arteriography is the diagnostic procedure of choice (Wickre and Golper 1982). In severe cases, occlusion of the fistula by arterial embolization or partial nephrectomy is required.

Aneurysms

The formation of arteriolar aneurysms is a rare complication of renal biopsy. The involvement of arcuate arteries may be responsible for arterial hypertension and/or gross haematuria (Wickre and Golper 1962).

Renal dysfunction

Rare cases of temporary loss of renal function have been reported in patients with perinephric haemorrhage after percutaneous renal biopsy. Oligoanuria may occur as a result of posthaemorrhagic hypotension. Another possible mechanism of renal dysfunction is restriction of renal perfusion brought about by the compression of the perirenal haematoma (Wijeyesinghe et al. 1987).

Puncture of other organs

Inadvertent puncture of the spleen, pancreas, liver, intestine, gall-bladder, and adrenal glands as well as laceration of renal, mesenteric, and subcostal arteries have been reported as exceptional complications (Zollinger and Mihatsch 1978). Today with the use of ultrasonography it is almost impossible for such complications to occur for an expert operator.

Contraindications

Renal mass

The presence of polycystic kidney disease is a formal contraindication to renal biopsy. Large cysts and renal neoplasms are not an absolute contraindication if they can be well localized. Renal biopsy under the guidance of ultrasonography or an open surgical biopsy should be done in these cases.

Solitary kidney

This condition is generally considered as an almost absolute contraindication to percutaneous renal biopsy. Open biopsy should be considered in patients with solitary kidney. An exception is the transplanted kidney which is commonly biopsied both because it is easy to puncture, being almost subcutaneous, and because compressive haemostasis can be carried out.

Chronic renal failure

Patients with decreased renal function have a high rate of complications. Moreover useful information cannot be obtained from the biopsy of small, contracted kidneys. For these reasons, renal biopsy should be considered only in patients with normal-sized kidneys after careful control of hypertension and correction of coagulation disorders.

Urinary infection

The presence of an active, untreated urinary tract infection is considered a contraindication to renal biopsy, in view of the potential communication between the collecting system and a possible perirenal haematoma, with consequent catastrophic infection of the haematoma (Rosenbaum et al. 1978).

Hypertension

The risk of complications after biopsy is directly related to the degree of elevation of blood pressure (Diaz-Buxo and Donadio 1975). Although normalization of blood pressure with antihypertensive agents may reduce the risks of renal biopsy, nevertheless the transection of sclerotic vessels is more likely to produce severe haemorrhage. Uncontrolled hypertension should be considered as a high-risk factor for complications.

Coagulation disorders

Haemorrhagic diathesis is a formal contraindication to renal biopsy. Many patients with renal dysfunction present a prolongation of the skin bleeding time associated with normal results on coagulation tests. The risk of haemorrhagic complications after biopsy is high in patients with very prolonged bleeding time. However, as reported above, in many cases preoperative infusion of 1-deamino-8-D-arginine vasopressin can achieve normal values for the bleeding time for some hours and allow renal biopsy.

Miscellaneous

Renal artery aneurysm, marked calcific arteriosclerosis, perinephric abscess, and horseshoe kidney are generally considered as contraindications to percutaneous renal biopsy. In particular cases, however, an open biopsy may be considered if the clinical situation warrants the risk. Patients with amyloidosis may have a factor X deficiency due to binding of this factor to the amyloid tissue deposit (Furie et al. 1977). Moreover, vascular amyloid deposits may impair vascular occlusion and vasoconstriction after transection of vascular structures by biopsy. In view of the increased risk of haemorrhage, patients with amyloidosis should be carefully assessed to rule out possible haemostatic defects.

Indications for renal biopsy

Idiopathic nephrotic syndrome

Until recently it was uniformly agreed that renal biopsy was an essential procedure in patients with idiopathic nephrotic syndrome, not only in terms of diagnosis but also in selecting treatment (Kark 1968). Today, however, some clinicians restrict their use of renal biopsy in this condition. Paediatricians, for example, think that most nephrotic children up to age 8 should first receive an 8-week course of high-dose prednisone before considering renal biopsy (Habib et al. 1979). In fact, in childhood lipoid nephrosis is by far the most common cause of the nephrotic syndrome and usually responds with a complete remission of proteinuria within a few days or some weeks of corticosteroid therapy. These clinicians would reserve renal biopsy for steroid-resistant patients or for children with clinical and laboratory features (such as hypertension, renal insufficiency, multiorgan involvement, haematuria, unselective proteinuria) which suggest an underlying renal disease other than minimal change nephropathy.

A similar approach has also been proposed by some investigators for adults with idiopathic nephrotic syndrome. Using decision analysis Hlatky (1982) concluded that an initial trial with alternate-day prednisone given for 2 months in nephrotic adults may yield as many remissions, fewer complications, and lower mortality as routine renal biopsy. The analysis assumed that corticosteroid therapy would be administered after renal biopsy only for lesions in which previous randomized clinical trials had proved the efficacy of alternate-day prednisone, namely minimal change and membranous nephropathy. Levey et al. (1987) also used decision analysis to compare the conventional strategy of biopsy-tailored therapy with alternative strategies which do not use renal biopsy. They evaluated data on alternate-day steroids given for 8 weeks plus long-term platelet-inhibitor therapy for resistant nephrotic syndrome, and concluded that renal biopsy is not necessary in caring for nephrotic adults; empiric short-term alternate-day prednisone was equally effective as biopsy-guided therapy in minimal change and membranous nephropathy. Sequential therapy with steroids followed by long-term platelet inhibitors was also equally effective for patients with membranoproliferative glomerulonephritis.

These studies are very elegant but some recent clinical trials indicate that the responses to treatment in adults with idiopathic nephrotic syndrome may differ from those accepted a few years ago. This might invalidate the assumptions of the decision analysis. There is, for example, evidence that only 60 per cent of adults with minimal change nephrotic syndrome are in complete remission after an 8 week course of prednisone but more than 80 per cent are without proteinuria after 16 weeks of treatment (Nolasco et al. 1986). Even more striking, the experience of Korbet et al. (1988) shows that only 32 per cent of patients older than 40 years achieved a complete remission within 8 weeks of corticosteroids but 77 per cent had responded by 16 weeks of treatment and 94 per cent became free of proteinuria with more prolonged therapy. Using the short-term approach a large number of late-responding patients would be considered as steroid-resistant and would not be treated further. Some recent papers (Pei et al. 1987; Griswold et al. 1987) have reported that 40 to 60 per cent of patients with focal glomerulosclerosis and nephrotic syndrome can respond with a complete remission of proteinuria, when treated with aggressive and prolonged corticosteroid and/or cytotoxic therapy. Without biopsy these patients would not receive potentially beneficial treatment.

An English multicentre trial has shown that an 8-week course of alternate-day prednisone does not interfere with the natural course of membranous nephropathy (Cameron et al. 1990) which is contrary to results reported by the American Collaborative Study Group (1979). On the other hand, another controlled trial in idiopathic membranous nephropathy showed that a 6-month course with methylprednisolone and chlorambucil favours remission of the nephrotic syndrome and protects renal function in the long term (Ponticelli et al. 1984, 1989). Without renal biopsy, patients with membranous nephropathy would receive an useless course of prednisone while being excluded from a potentially helpful treatment. The value of renal biopsy in the idiopathic nephrotic syndrome is further stressed, without the use of decision analysis, by prospective studies showing that about 50 per cent of clinically predicted diagnosis and prognosis changed after renal biopsy (Paone and Meyer 1981; Turner et al. 1986; Cohen et al. 1989). Moreover, in some cases biopsy can show that what appears clinically to be an idiopathic nephrotic syndrome is actually secondary to a systemic disease such as amyloidosis, myeloma, lupus.

Table 2 Indications for renal biopsy in patients with acute renal failure

Gradual onset of acute renal failure
No obvious cause of acute renal failure
Heavy proteinuria
Significant haematuria
Clinical evidence or history of systemic disease
Significant hypertension
Prolonged oliguria

We conclude that since there is no way to diagnose the underlying disease correctly in patients with idiopathic nephrotic syndrome and since there are data emerging to suggest a role for treatment in several instances, most adults should undergo renal biopsy in order better to assess their diagnosis and prognosis and to decide on possible treatment, which today may be tailored using the results of biopsy.

Acute renal failure

The most frequent cause of acute renal failure is tubular necrosis. In about 80 per cent of cases the diagnosis of acute tubular necrosis is made correctly on clinical grounds (Mustonen et al. 1984). In these cases biopsy adds little to prognosis and nothing to choice of therapy. Thus, in anuric patients, renal biopsy should be reserved for doubtful cases in order to detect causes other than acute tubular necrosis which may benefit from specific treatment. Biopsy may also be useful in those patients who fail to regain renal function after 4 or more weeks (Table 2).

Rapidly progressive renal failure

In patients with acute glomerulonephritis, interstitial nephritis, vasculitis, or systemic disease renal biopsy is generally needed. In these instances the clinical diagnosis may be difficult and incomplete. Idiopathic or lupus glomerulonephritis may cause acute renal failure which is sometimes associated with diffuse crescents, sometimes with exudative lesions, sometimes with a pure acute tubular necrosis. The prognosis and therapeutical approach are clearly different in each of these settings. In rare instances, isolated renal failure, due to granulomatous nephritis, is the only manifestation of an underlying sarcoidosis (King et al. 1976; Allegri et al. 1980). In these cases kidney biopsy can allow correct diagnosis and treatment which usually results in complete recovery. Interstitial nephritis is a histological diagnosis. Fever, rash, and eosinophilia are present only in one-third of cases. New stains have improved the recognition of eosinophiluria but false results are frequent (Nolan et al. 1986). Some vasculitides may be difficult to recognize, although the recent technique of antibodies to components of neutrophil cytoplasm has allowed an earlier diagnosis (Van der Woude et al. 1985). Multiple myeloma may present with an acute renal failure with no symptoms of the underlying disease (Border and Cohen 1980) and the same may occur for primary amyloidosis (Zawada et al. 1987). Thus, in most cases of acute renal failure other than tubular necrosis renal biopsy can provide useful and irreplaceable information.

Chronic renal insufficiency

In patients with small, contracted kidneys renal biopsy is of no help and is often followed by severe complications. However, in the few patients with normal or mildly-reduced sized kidneys and especially when there is a change from slowly progressive to rapidly deteriorating renal insufficiency, biopsy is indicated to identify the cause of functional impairment (such as drug-related interstitial nephritis, superimposed extracapillary glomerulonephritis, atheroembolic disease), which may sometimes be reversed with appropriate treatment.

Asymptomatic urinary findings

There is no rule for the indications to renal biopsy in patients with isolated haematuria or asymptomatic proteinuria. Some nephrologists prefer not to submit these patients to renal biopsy. In most cases patients with asymptomatic urinary changes do not show renal dysfunction, at least for several years, and therefore are not candidates for any type of specific therapy. These patients should be monitored over the years and if deterioration of renal function, increase in proteinuria, or hypertension develops they should be reconsidered for renal biopsy. Some nephrologists choose to perform renal biopsy even in patients with asymptomatic urinary changes, to rule out the presence of progressive and/or potentially treatable renal disease which can be clinically silent at presentation. A renal biopsy may also be indicated for eugenic reasons in women in childbearing age with mild proteinuria and microhaematuria to detect the presence of Alport's syndrome or other hereditary diseases, or biopsy may be required by those patients who want a firm diagnosis and information about the long-term prognosis.

Systemic diseases

In patients with diabetes mellitus, long-standing proteinuria, and typical course, a renal biopsy is seldom indicated. However biopsy can be helpful in patients whose course is complicated by rapid development of nephrotic syndrome and/or renal insufficiency. A multitude of glomerulopathies may be associated with diabetic nephropathy, including membranous nephropathy (Silva et al. 1983) and more rarely minimal change nephropathy (Urizar et al. 1969) or dense-deposit disease (Couser et al. 1977).

Renal biopsy has been extensively used in lupus nephritis. A WHO workshop identified five main patterns of renal involvement in patients with systemic lupus, i.e., normal kidneys, mesangial changes, focal and segmental proliferative glomerulonephritis, diffuse proliferative glomerulonephritis, and membranous glomerulonephritis (Appel et al. 1978). This classification has been widely accepted, although several subclasses have been added to the original divisions by different investigators. There is a trend for distinctive clinical features and long-term outcome in the different histological classes (Leaker et al. 1987; Appel et al. 1987), but in view of the frequent discrepancy between clinicobiological features and renal histology, some authorities think that renal biopsy is of little help in lupus nephritis (Fries et al. 1978). In particular, in some patients with lupus and clinically silent kidney disease renal biopsy may show glomerular changes. Usually these changes are mild, mesangial, or occasionally membranous patterns. However, in a minority a diffuse proliferative glomerulonephritis can be seen at renal biopsy (Eiser et al. 1979). Since in these patients the outcome is usually benign the value of biopsy might be challenged but even in these instances some investigators advocate prednisone treatment (Leehey et al. 1982).

In clinically active lupus nephritis renal biopsy can help in predicting the prognosis and in taking therapeutic decisions. For example, the presence of severe interstitial fibrosis, tubular atrophy, and diffuse glomerular sclerosis indicates irreversible lesions, treatment for which is useless. On the other hand, lesions such as fibrinoid deposits, florid crescents, increased cellularity, or segmental necrosis, although with a possibility of becoming sclerotic, are potentially reversible and warrant aggressive therapy. The Bethesda group has particularly stressed the prognostic value of renal biopsy, suggesting the use of a chronicity and an activity index, which are obtained by summing the semiquantitative scores of certain histological features (Table 3). Both indices would have predictive value (Austin et al. 1984) although some investigators pointed out that the chronicity index applies only to diffuse proliferative glomerulonephritis (Lewis et al. 1987).

In patients with vasculitis, renal biopsy may be useful both to assess the diagnosis and to decide whether to reduce treatment in patients who are over the acute phase.

In some patients with multiple myeloma, amyloidosis or light chain deposition disease the clinical diagnosis may be particularly difficult and renal biopsy is invaluable. The biopsy findings may be startling and may suggest a much more long-standing lesion than has been apparent clinically (Mallick and Williams 1988).

Renal transplantation

Biopsy can give direct information on the events occurring in the graft. There are several clinical settings in which it is indicated (Table 4). These include post-transplant oligoanuria, graft dysfunction, failure to respond to antirejection therapy, and development of proteinuria. The introduction of the new immunosuppressant cyclosporin has further extended the indications for renal biopsy. Because of its nephrotoxicity this agent may be responsible for renal dysfunction which poses difficult problems of differential diagnosis with rejection, both in early and in long-term follow-up.

A new invasive technique for assessing intragraft events is the fine-needle aspiration biopsy. This is carried out using a spinal needle of 22 G $3\frac{1}{2}$. The aspirate is prepared with the May-Grünwald-Giemsa stain after cytocentrifugation and several

Table 3 Renal pathology scoring system in lupus nephritis*

Activity index	Chronicity index
Glomerular abnormalities	
1. Cellular proliferation	1. Glomerular sclerosis
2. Fibrinoid necrosis, karyorrhexis	2. Fibrous crescents
3. Cellular crescents	
4. Hyaline thrombi, wire loops	
5. Leucocyte infiltration	
Tubulointerstitial abnormalities	
1. Mononuclear-cell infiltration	1. Interstitial fibrosis
	2. Tubular atrophy

Each factor is scored from 0 to 3.
* Fibrinoid necrosis and cellular crescents are weighted by a factor of 2. Maximum score of activity index is 24, and of chronicity index is 12. (Taken from Austin, H.A. et al. (1983). Prognostic factors in lupus nephritis. American Journal of Medicine, 75:382–91 with permission.)

Table 4 Indications for renal biopsy in kidney transplantation

Clinical event	Differential diagnosis
Post-transplant anuria	Acute tubular necrosis Hyperacute rejection Accelerated rejection Superimposed acute rejection Renal infarction
Acute renal dysfunction	Acute rejection Cyclosporin toxicity Drug toxicity Acute interstitial nephritis Pyelonephritis Obstruction of the lower urinary tract
Chronic renal dysfunction	Late rejection Cyclosporin toxicity Transplant glomerulopathy
Proteinuria	Recurrence of glomerulonephritis de novo Glomerulonephritis Late rejection Transplant glomerulopathy
Failure to respond to antirejection therapy	Cellular rejection Vascular rejection Reversibility of the lesions Glomerulonephritis Obstruction of the lower urinary tract

washings. A peripheral blood sample is obtained at the same time to correct the cell count by subtracting the peripheral white cell count (Häyry et al. 1981). The technique is simple, safe, can give information within 1 h, and can be repeated many times in the same patient. It is particularly useful for evaluating early events and in diagnosing rejection, tubular necrosis, and cyclosporin toxicity; its role in late events is more limited (Egidi et al. 1985).

Biopsy specimen processing

To allow good evaluation of a renal biopsy, sufficient cortical tissue should be available for light microscopy and immunofluorescence. Ideally small fragments of cortex should also be fixed for electron microscopy, although in most cases a diagnosis can be achieved with the combination of light microscopy and immunofluorescence (Morel-Maroger 1982). In practice it is preferable to have two adequate biopsy specimens, one for light microscopy and the other for immunohistology. For electron microscopy, two or three small pieces of cortical tissue are cut from one of the two specimens and put in the fixative solution as quickly as possible. The light microscopy specimen is laid on a strip of heavy filter paper and plunged into the fixative. The tissue cylinder chosen for immunohistology is placed on a blockholder previously covered by a supporting substance and is immersed in isopentane precooled by liquid nitrogen. If only one specimen is available the fragment for immunohistology can be obtained by cutting the cylinder of tissue longitudinally. Alternatively, with insufficient material, immunohistology could be performed in formalin-fixed and paraffin-embedded tissue (Bolton and Mesnard 1982; Fogazzi et al. 1989), but the technique on snap-frozen specimens is more sensitive.

The choice and the optimization of the methods used for tissue processing in the three techniques are of paramount importance in obtaining maximum information. The reader can find detailed descriptions elsewhere of these techniques (Zollinger and Mihatsch 1978). We summarize here the guidelines for optimal tissue processing.

Light microscopy

Fixation in neutral phosphate-buffered 4 per cent formalin is recommended for routine examination. Good results can also be obtained with fixation in Dubosq-Brazil and postfixation in neutral formalin. However, specimens processed with this fixative cannot be used for immunohistology. Alcoholic dehydration and embedding in paraffin or paraplast is commonly used to process renal biopsies. Numbered slides are then obtained from the paraffin block by cutting serial 2 μm sections. A regular thickness is important for semiquantitative or morphometric examination. To obtain thinner (0.5–1 μm) and more regular sections, methyl methacrylate as embedding medium has been proposed (Striker et al. 1978). Sections should be stained routinely with haematoxylin–eosin, PAS, silver methenamine, Masson trichrome stain, or AFOG stain (Zollinger and Mihatsch 1978). If amyloidosis is suspected sections should be stained by Congo red and examined under polarized light.

Immunohistology

Immunohistological techniques are usually performed in renal biopsies in order to recognize and characterize deposits of immunoglobulin and complement factors. Immunofluorescence is more commonly used, but immunoperoxidase is also applied since it offers advantages with stability of the stain and with use of light microscopy for examination. A wide series of immunesera is commercially available for both techniques: anti-IgG, -IgA, -IgM, -C3, -C4, -C1q, -fibrin(ogen) sera. Antilight chain sera should be used routinely in aged patients. Additional antisera, such as those specific for other complement components, for coagulation factors, or for viral antigens, may be useful in selected cases.

The intensity of staining of the individual sera is semiquantitatively scored. Standardization of the techniques is necessary to obtain reproducible data (Nairn 1976). Artefacts should be avoided carefully. Poorly preserved specimens during the snap-freezing procedure, thick sections, unspecific fixation of the sera or of the free-labelling material are the most common causes of misinterpretation. Immunohistology is generally performed on cryostat sections from frozen specimens. Formalin-fixed and paraffin-embedded specimens can be used by applying conjugated antisera on sections previously treated by proteolytic digestion with trypsin or pronase. This method is more complex and time consuming than that on frozen sections. It gives comparable results for immunoglobulins, fibrin, and light chain detection but is less sensitive for complement factors (Bolton and Mesnard 1982; Fogazzi et al. 1989).

Electron microscopy

Ultrastructural studies are expensive and time consuming. In some centres, therefore, small tissue specimens are routinely embedded in resin, but are examined by electron microscopy only when diagnosis has not been obtained by light and immunofluorescence microscopy.

Standard processing for electron microscopy includes fixation in glutaraldehyde, postfixation in osmium, embedding in epon, and preparation of semithin sections and ultrathin sections which

1.7.2 Renal biopsy: performance and interpretation

Table 5 Stepwise evaluation of renal biopsy (from Zollinger and Mihatsch (1978), modified)

Evaluation of clinical and laboratory data
Study at low power magnification (adequacy and preliminary examination)
Analytical study of light microscopy features (glomeruli, tubules, interstitium, vessels)
Presumptive diagnosis and correlation with clinical data
   Evaluation of immunohistological and electron microscopy findings
Confirmation of diagnosis or:
   Revision of light microscopy and clinical data
Final diagnosis, staging of lesions, and evaluation of prognostic indices

---

are stained by uranyl acetate and lead citrate. Silver impregnation of ultrathin sections is an additional staining procedure seldomly used for evaluation of basement membrane abnormalities. Scanning electron microscopy, immunoelectron microscopy, and histochemistry on electron microscopy are further possible applications of this technique, which however requires special preparation of the specimens.

General guidelines for renal biopsy evaluation

Evaluation of the biopsy specimen by the different techniques is included in a stepwise diagnostic process where clinical and laboratory data are integrated and correlated with morphologic findings (Table 5).

The first step in a renal biopsy analysis is low power evaluation of the light microscopy slides. This allows assessment of the proportion of cortical and medullary tissue and detection of rough abnormalities such as infarcted areas or large scars. The adequacy of the specimen is evaluated from the number of glomeruli observed, although this criterion is not absolute but depends on the underlying disease and on the purpose of the study. When the lesion is confined to the glomeruli and is diffuse, even a single glomerulus may be sufficient for diagnosis. Clearly, however, other associated interstitial and vascular changes or the degree of glomerular obsolescence cannot be assessed. When the lesion is focally distributed the representativity of the specimen is directly proportional to the number of glomeruli. The accuracy of biopsy evaluation in these conditions can be quantitatively assessed considering the actual number of abnormal glomeruli and the size of the specimen (Corwin *et al.* 1988). In general, for the evaluation of glomerular lesions a biopsy should contain at least five glomeruli. For tubulointerstitial lesions a biopsy size of six to 10 glomeruli is necessary (Oberholzer *et al.* 1983). In order to improve the accuracy of histological evaluation, it can be useful to cut a large number of sections and to study semithin sections also.

The second step consists of an analytical examination of features of glomeruli, tubules, interstitium, and vessels by light microscopy. For each of these structures several histological parameters must be considered (Table 6). Careful examination at high-power magnification, especially of sections stained by silver-methenamine and trichrome techniques, is needed for identifying the lesions.

The third step consists of the interpretation of immunohistology and electron microscopy findings. Immunohistological findings should be evaluated by considering for each antiserum the site of fixation on different structures, the extension and the distribution of positively stained material, the intensity of the staining and the morphological characteristics of deposits. These findings are very important in diagnosis especially in glomerular diseases (Morel-Maroger *et al.* 1973). For each histological type of glomerular disease a well-defined immunohistological pattern can be identified on the basis of the antigen composition, conformation, and allocation of deposits on glomerular structures (Fig. 2). In addition some diseases such as IgA mesangial deposits nephropathy, antibasement membrane glomerulonephritis, type III crescentic glomerulonephritis cannot be correctly classified without immunohistology.

There are many peculiar abnormalities which may be recognized only by electron microscopy (Table 7). Ultrastructural studies are essential for the diagnosis in thin-basement-membrane nephropathy (Tiebosch *et al.* 1989); Alport's syndrome and other hereditary diseases; in the initial phase of diabetic nephropathy; in dense-deposit disease; in cryoglobulinaemic nephropathy; in some cases of amyloidosis, and in the so-called immunotactoid nephropathy (Korbet *et al.* 1985). In glomerular disease an ultrastructural study provides detailed information about glomerular basement membrane changes, the characteristics of the deposits, and of some cellular abnormalities. These findings may allow recognition of morphological subtypes of membranous nephropathy (Ehrenreich and Churg 1968), membranoproliferative glomerulonephritis (Habib *et al.* 1973; Jackson *et al.* 1987), and lupus nephropathy (Appel *et al.* 1978). Finally some findings related to dissolution of the deposits or accumulation of basement membrane-like material may give pointers to the duration of the pathological phenomena.

The final step is a comprehensive evaluation of histological, clinical, and laboratory findings. Discrepancies between clinical and pathological data should stimulate both a critical revision of biopsy data or further laboratory investigations. Renal biopsy is not only irreplaceable in the recognition of a well-defined pathological entity to place the patient in a nosological framework but may also provide information on the activity or chronicity of the lesions, as well as on staging or subclassification of pathological patterns. It is seldom that the inadequacy of the biopsy specimen, the presence of extensive sclerosing lesions, the presence of an unclassifiable pattern, or the superimposition of two different pathologic processes (Bertani *et al.* 1986) preclude a firm classification. However even when the final morphologic diagnosis remains uncertain, renal biopsy may suggest a diagnostic hypothesis.

Analytic approach to biopsy findings

Identification of some basic morphological changes plays a key role in renal biopsy interpretation. This type of analysis may be schematic and is not always applicable. However it represents an effort to establish a correlation between morphologic changes and pathogenetic mechanisms underlying renal lesions. These changes may be isolated or may coexist with other lesions. They will be described briefly considering together light and electron microscopy.

Glomerular lesions

Glomerular changes can be divided into three categories (Fig. 3): (1) changes of cellular components; (2) changes resulting from deposition of immune material; and (3) changes of extracellular components.

Changes of cellular components include intracapillary (endothelial or mesangial) proliferation, infiltration of capillary loops (by polymorphonuclear and/or mononuculear cells), extracapillary or crescentic proliferation. The use of monoclonal antibodies may be helpful in characterizing the cells causing hypercellularity, e.g. monocytes, lymphocytes (Hancock and Atkins 1984; Hooke et al. 1987; Nolasco et al. 1987, Castiglione et al. 1988).

Glomerular deposits can be identified in mesangial, subendothelial, intramembranous, and subepithelial positions. Two types of the last deposits can be distinguished: (1) those typical of poststreptococcal glomerulonephritis, termed humps, which are scattered, globular, or elongated and lean on the lamina rara externa; and (2) those typical of membranous nephropathy, which are in contact with the lamina densa.

A great variety of changes can be identified in extracellular structures. These changes may be the consequence of immune injury, of haemodynamic abnormalities, of biochemical abnormalities in diabetes, of direct endothelial injury in haemolytic uraemic syndrome or of hereditary abnormalities in Alport's syndrome. A reciprocal effect between cellular and extracellular changes of the glomerular structure exists and it has been demonstrated that platelets, mononuclear cells, and mesangial cells may produce factors that promote mesangial proliferation, basement membrane and matrix formation, and eventually sclerosis (Klahr et al. 1988). A peculiar type of change related to extracellular structures is due to an accumulation of foreign material such as amyloid fibrils, light chain deposits (Ganeval et al. 1984) and lipoproteins (Imbasciati et al. 1986).

Tubular lesions

Since renal biopsy is seldom performed in patients with acute tubular necrosis, tubular changes reflecting acute injury are rarely encountered in biopsies as isolated findings. Conversely tubular cell swelling, vacuolation, and necrosis are frequently found in association with acute and severe glomerulonephritis, interstitial nephritis, vasculitis, and graft rejection. In these conditions tubular lesions are probably the consequence of intrarenal haemodynamic changes. Severe acute tubular lesions consisting of protein and/or necrotic cells casts may also be observed in patients with minimal change nephrotic syndrome complicated by acute renal failure (Imbasciati et al. 1981). A peculiar type of casts composed of dense and fractured material surrounded by multinucleated giant cells may be seen in patients with multiple myeloma and renal dysfunction. Erythrocyte casts

Table 6 Items for analytical evaluation of light microscopy

Glomeruli
Number of glomeruli
Number of obsolescent glomeruli
Size and shape of glomerular tuft
Bowman capsule
Bowman space
Capsular adhesion
Crescents (number, size and type)
Leucocytes in capillary loops
Mesangial cell proliferation
Mesangial matrix increase
Capillary lumen patency
Deposits
Basement membrane features
Segmental sclerosis
Glomerular capillary necrosis
Glomerular capillary thrombosis
Juxtaglomerular apparatus

Vessels
Arteries/arterioles
　Intimal fibrosis
　Intimal proliferation
　Mucoid transformation (onion-skin)
　Elastic reduplication
　Hyaline or fibrinoid deposits
　Medial hypertrophy
　Necrosis
　Thrombosis
　Leucocytes infiltration
Peritubular capillaries and venules
　Congestion
　Leucocytes infiltration
Mononuclear cells in vasa recta

Tubules
Epithelia cell (proximal)
　Swelling or flattening
　Vacuolization (size of vacuoles)
　Brush border loss
　Inclusion (hyaline, crystals)
　Foamy transformation
　Nuclear changes
　Mitoses
　Cell necrosis
Tubular lumen
　Size and patency
　Casts (type and size)
　Cells (erythrocytes, leucocytes, epithelial) and cell debris
　Crystals
Basement membrane

Interstitium
Oedema
Leucocyte infiltration
Granuloma formation
Red blood cell extravasation
Fibrosis
Foam cells
Lymph casts
Crystals or calcified aggregates

1.7.2 Renal biopsy: performance and interpretation

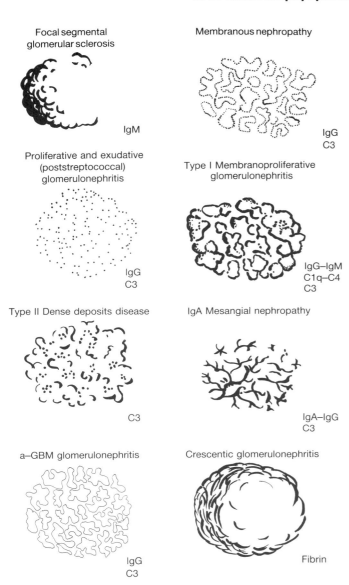

Fig. 2 Characteristic immunohistological patterns of the different types of primary glomerulonephritis.

may be seen in acute glomerulonephritis and in some rare cases of acute renal failure in patients with IgA nephropathy (Praga *et al.* 1985).

Tubular lesions may assume particular relevance in renal transplantation. Acute tubular necrosis is relatively frequent in cadaveric renal transplantation. In acute rejection tubular lesions are frequent and may be severe especially in the presence of vascular lesions. When acute rejection occurs early in the posttransplant period, tubular lesions may reflect both immunological and ischaemic insults. The use of cyclosporin as an immunosuppressive agent in transplantation has introduced another cause of tubular damage. Some tubular lesions, i.e. isometric vacuolization, giant mitochondria, and calcification of individual necrotic cells, although non-specific have been considered highly suggestive of cyclosporin toxicity (Mihatsch *et al.* 1983).

A tubular change observed in patients with chronic and long-lasting renal diseases is tubular atrophy. A significant correlation has been found between tubular atrophy and impairment of glomerular filtration rate (Mackensen-Haen *et al.* 1981). Two patterns of this lesion may be observed. The first is characterized by a reduction of the outer diameter of the lumen, which becomes virtual. Epithelial cells are decreased in size and show a clear cytoplasm and loss of brush border. In the second pattern the outer diameter may be unchanged or even increased, although the epithelium is flattened and dedifferentiated. The lumen is widened and may contain hyaline casts. Tubular basement membranes may be thickened in both types of changes. Tubular atrophy develops in a variety of renal disorders associated with glomerular obsolescence and interstitial fibrosis. Inflammatory destruction of parts of the nephron, narrowing of major blood vessels, or obliteration of peritubular capillaries with consequences of reduced blood supply to tubular cells are probably the cause of tubular atrophy. Tubular atrophy alone or in association with interstitial fibrosis has been found to be a reliable index of long-term outcome in membranous nephropathy (Ponticelli *et al.* 1989; Wehrmann *et al.* 1989), lupus nephritis (Austin *et al.* 1983), and in IgA nephropathy (D'Amico *et al.* 1986).

Interstitial tissue

Diffuse interstitial infiltration of leucocytes is characteristic of drug-induced interstitial nephritis. Inflammatory cells, mainly mononuclear and scattered eosinophils and polymorphs or even granulomas fill interstitial spaces, which appear expanded and oedematous. Cells tend to accumulate in peritubular capillaries and around vascular structures and even infiltrate into the tubules. Other causes of acute interstitial nephritis are bacterial and viral infections, sarcoidosis, systemic lupus erythematosus, and other immune-mediated renal diseases. In many cases the cause of interstitial nephritis remains unknown (Cameron 1988).

Acute interstitial rejection is another condition characterized by diffuse interstitial infiltration of mononuclear cells. These cells may fill peritubular capillaries and penetrate into the tubules. Studies with monoclonal antibodies have characterized the infiltrating cells and have suggested a delayed hypersensitivity reaction as mechanism of cellular infiltration (Hall *et al.* 1984).

Mononuclear cells infiltrating interstitial spaces may be seen, independently of nephron loss, in primary glomerulonephritis, in lupus nephritis, and in essential mixed cryoglobulinaemia (D'Amico 1988). Mild to severe focal interstitial inflammation is frequently found. Small foci of interstitial infiltration are generally localized near obsolescent glomeruli and atrophic tubules. In lupus nephritis interstitial inflammation has been found to correlate with the activity of glomerular lesions and to have prognostic significance (Park *et al.* 1986). Deposits of immune material in peritubular capillaries or along tubular basement membrane are rarely found in case of interstitial inflammation (Park *et al.* 1986). An interstitial nephritis may be an isolated finding in rare cases of systemic lupus (Cunningham *et al.* 1978).

According to these findings it has been suggested that interstitial infiltration may be an active reaction which parallels glomerular injury. It may be mediated by mechanisms in some aspects independent of those producing glomerular lesions and might play a role in the progression of glomerular diseases (D'Amico 1988).

Interstitial fibrosis may be the consequence of inflammation, tubular obstruction, glomerular obsolescence and ischaemia. It is produced by stimulation of fibroblasts through mechanisms which remain largely unexplored (Cameron 1988). Interstitial fibrosis, usually in association with tubular atrophy is an obvious parameter of chronicity in any type of renal disease, and may be predictive of the outcome (D'Amico 1988).

Vascular lesions

Microvascular changes are the main characteristic of several diseases such as malignant hypertension, progressive systemic sclerosis, thrombotic thrombocytopenic purpura, haemolytic uraemic syndrome, micropolyarteritis, and acute and chronic graft rejection. Vascular lesions may also appear in chronic renal diseases as a consequence of arterial hypertension, immune deposits, and/or adaptative changes of intrarenal haemodynamics. A peculiar form of vascular damage can be seen in patients treated with cyclosporin. The lesion which is similar to that observed in haemolytic uraemic syndrome, has been described in kidneys of patients treated with cyclosporin for renal transplantation (Mihatsch et al. 1983), autoimmune disease (Mihatsch et al. 1988), and heart or bone marrow transplantation (Nizze et al. 1988).

Recognition and interpretation of renal vascular lesions may be difficult. Although each pathological condition is associated with some morphologic characteristics, there is a considerable overlap in the morphologic patterns of injury. Moreover, the pattern may depend on the severity and on the stage of the lesion. In the same biopsy different stages may be observed. The more advanced the lesion the more difficult is the recognition of the original morphologic characteristics. Lesions related to age or with concomitant metabolic disorders should also be considered. Finally vascular lesions are usually focally distributed and the sample may be inadequate for reliable evaluation of the vascular tree. In cases of vascular lesions, the whole biopsy cylinder should be cut to provide maximum information.

Arterial and arteriolar lesions are commonly grouped into a few main categories (Fig. 4), although the criteria for differential diagnosis among pathological groups are not clearly established. It is useful to recognize some basic changes which alone or in combination may contribute to the lesion. They include endothelial swelling or necrosis, proliferation of myofibroblasts, new basement membrane formation, accumulation of immune granular or filamentous material (mucoid transformation) (Hsu and Churg 1980), reduplication of the elastic membranes, fibrinoid necrosis of vascular wall, and leucocyte infiltration and thrombotic obliteration of the lumen. Unfortunately the pathogenesis of the diseases which involve renal microvasculature is largely unknown. Endothelial injury produced by different aetiological factors may initiate a variety of pathological events within the vessel wall and the lumen: within the vessel wall accumulation of fine fibrillar material on the subendothelial side, overproduction of basement membrane, insudation of proteinaceous material, necrosis and/or proliferation of myofibroblasts; in the vascular lumen endothelial damage promotes platelet aggregation, fibrin deposition, and thrombosis. Considering that endothelial injury is the common event of different disorders, it seems reasonable to group within the same histopathological pattern haemolytic uraemic syndrome, thrombotic thrombocytopenic purpura, malignant nephrosclerosis, and perhaps progressive systemic sclerosis (Neild 1987). The demonstration that these disorders share some basic lesions (Sinclair et al. 1976) confirms this assumption.

Semiquantitative evaluation and morphometry

The degree of severity and the distribution (in focal lesions) of the different morphologic parameters can be arbitrarily scored from 0 to 3 or 4+. Relatively objective criteria can be established for the assessment of some lesions such as tubular atrophy or interstitial fibrosis, where the percentage of tissue area covered by this change may be evaluated approximately. Obsolescent glomeruli, glomeruli focally involved by proliferative or sclerosing lesions, crescents, and glomerulocapsular adhesions can be numerically evaluated. Semiquantitative evaluation of renal biopsy has been demonstrated to be a reproducible method

Table 7 Electron microscopy findings relevant for diagnosis

Electron microscopy findings	Disease
Glomerular cell	
Foot process effacement of podocytes	Minimal change nephropathy
Microtubular inclusions (virus-like structures)	Lupus nephritis
Cytoplasmic inclusions of storage material	Lecithin-cholesterol acyltransferase deficit disease—Fabry's disease
Mesangial cell interposition	Membranoproliferative glomerulonephritis
Leucocytes (polymorphonuclear, lymphocytes and monocytes) *and platelets in capillary loops*	Acute glomerulonephritis, lupus nephritis Cryoglobulinaemia
Extracellular glomerular structures	
Mesangial matrix increase and interposition	Glomerulonephritis
Lamina densa abnormalities (thickening, thinning, splitting, reticulation)	Alport syndrome
Lamina rara interna widening	Haemolytic uraemic syndrome, scleroderma, transplant glomerulopathy
Basement membrane new formation	Glomerulonephritis
Collagen deposition	Nail patella syndrome
Fibrin in loops	IV coagulation, haemolytic uraemic syndrome
Dense deposits allocation (subepithelial, subendothelial, mesangial, intramembranous) and characterization (size and shape, osmiophilia, structure)	Glomerulonephritis
Microfibrils (amyloid and non-amyloid)	Amyloidosis—fibrillosis

1.7.2 Renal biopsy: performance and interpretation

Cell changes Deposits Extracellular structures

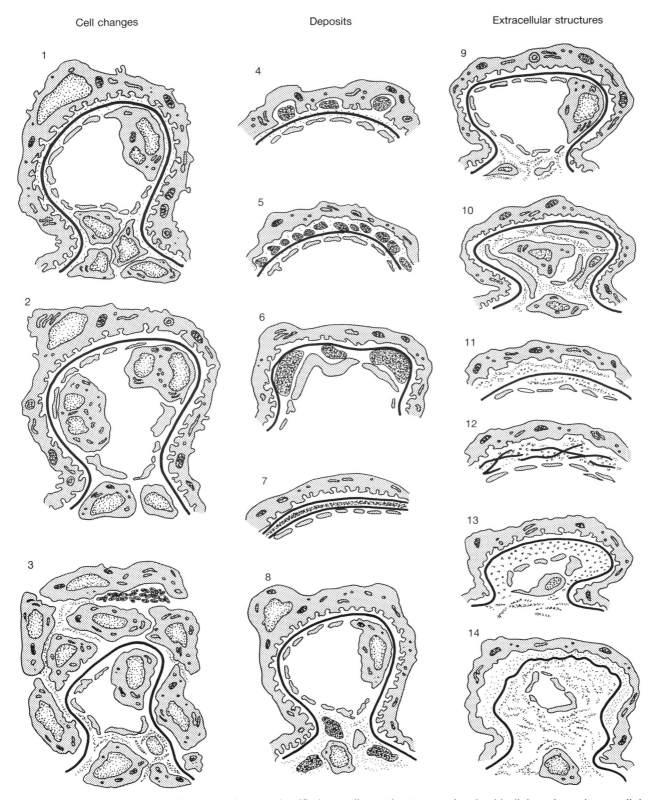

Fig. 3 Schematic representation of glomerular changes classified according to the structure involved (cell deposits and extracellular structures). (1) Mesangial cell proliferation; (2) leucocyte infiltration; (3) extracapillary proliferation with crescent formation; (4) subepithelial deposits (hump-like); (5) subepithelial deposits (membranous nephropathy); (6) subendothelial deposits (membranoproliferative glomerulonephritis type I); (7) intramembranous dense deposits (membranoproliferative glomerulonephritis type II); (8) mesangial deposits; (9) mesangial matrix increase; (10) mesangial interposition; (11) new formation of basement membrane; (12) lamina densa reticulation (Alport's syndrome); (13) lamina rara internal widening with fine granular material (haemolytic uraemic syndrome, systemic sclerosis, malignant hypertension); (14) glomerular sclerosis with capillary lumen obliteration.

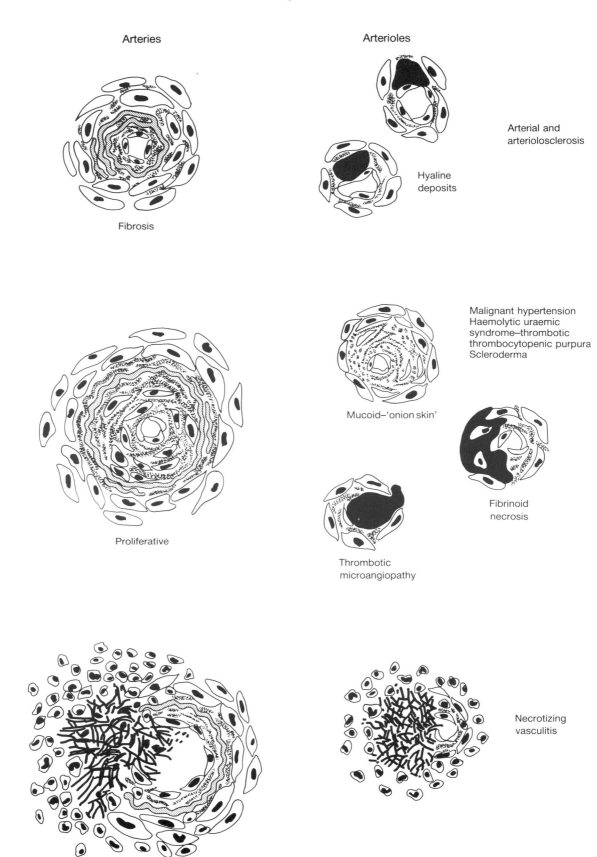

Fig. 4 Schematic representation of the main pathological patterns of vascular diseases involving the kidney.

(Pirani *et al.* 1964). This method increases the accuracy of evaluation of the biopsy and provides useful information in predicting the outcome and in making therapeutic decisions. If a limited number of cases is studied, the ranking technique is recommended. The least severe lesion is ranked 1, the most severe lesion is ranked highest, and this may correspond to the number of cases studied. This method allows a better correlation with laboratory values than the scoring system from 0 to +++ (Solez and Whelton 1983).

Morphometry

Different morphometric techniques have been described for light and electron microscopy. For basic information on morphometry the reader is referred to standard monographs (Torhost 1974; Bohman *et al.* 1979; Jensen 1979; Weibel 1979, 1980; Gundersen and Osterby 1982).

The prerequisites for morphometric analysis by light microscopy are: (1) absolute standardized preparation of material from fixation to embedding; (2) sufficient sample size (at least six to 10 glomeruli) (Oberholzer *et al.* 1983); (3) uniform section thickness; and (4) evaluation of identically stained sections.

The point counting technique, which is very time consuming, is still widely used (Torhost 1974; Weibel 1979, 1980). Alternatively, different semiautomatic systems are available.

In electron microscopy morphometric techniques are mainly used for basement membrane thickness measurements (Jensen *et al.* 1979; Gundersen and Osterby 1982).

References

Allegri, L., Olivetti, G., David, S., Concari, G.M., Dascola, G., and Savazzi, G. (1980). Sarcoid granulomatous nephritis with isolated and reversible renal failure. *Nephron*, 25, 207–8.

Appel, G.B., Silva, F.G., Pirani, C.L., Meltzer, J.I., and Estes, D. (1978). Renal involvement in systemic lupus erythematosus (SLE): A study of 56 patients emphasizing histologic classification. *Medicine* 57, 371–410.

Appel, G.B., Cohen, D.J., Pirani, C.L., Meltzer, J., and Estes, D. (1987). Long-term follow-up of patients with lupus nephritis. *American Journal of Medicine*, 83, 877–85.

Austin, H.A. III, *et al.* (1983). Prognostic factors in lupus nephritis. Contribution of renal histologic data. *American Journal of Medicine*, 75, 382–91.

Austin, H.A. III, Muenz, L.R., Joyce, K.M., Antonovich, T.T., and Balow, J.E. (1984). Diffuse proliferative lupus nephritis: identification of specific pathologic features affecting renal outcome. *Kidney International*, 25, 689–95.

Bertani, T., Mecca, G., Sacchi, G., and Remuzzi, G. (1986). Superimposed nephritis: a separate entity among glomerular diseases. *American Journal of Kidney Diseases*, 3, 205–12.

Bohman, S.O., Deguchi, N., Gundersen, H.J., Hesbech, J., Maunsbach, A.B., and Olsen, S. (1979). Evaluation of a procedure for systematic semiquantitative analysis of glomerular ultrastructure in human renal biopsies. *Laboratory Investigation*, 40, 433–44.

Bolton, W.K. and Mesnard, R.M. (1982). New technique of kidney tissue processing for immunofluorescence microscopy. Formol sucrose/gum sucrose/paraffin. *Laboratory Investigation*, 47, 206–13.

Bolton, W.K. and Vaughn, E.D., Jr. (1977). A comparative study of open surgical and percutaneous renal biopsies. *Journal of Urology*, 117, 696–98.

Border, W.A. and Cohen, A.H. (1980). Renal biopsy diagnosis in clinically silent multiple myeloma. *Annals of Internal Medicine*, 93, 43–6.

Brun, C. and Raaschou, F. (1958). Kidney biopsies. *American Journal of Medicine*, 24, 676–91.

Cameron, J.S. (1988). Allergic interstitial nephritis: clinical features and pathogenesis. *Quarterly Journal of Medicine*, 66, 97–116.

Cameron, J.S., Healey, M.J.R., and Adu, D. (1990). The Medical Research Council Trial of short-term high-dose alternate day presnisone in idiopathic membranous nephropathy with a nephrotic syndrome in adults. *Quarterly Journal of Medicine*, 74, 133–56.

Castiglione, A., Bucci, A., D'Amico, G., and Atkins, R.C. (1988). The relationship of infiltrating renal leukocytes to disease activity in lupus and cryoglobulinaemic glomerulonephritis. *American Journal of Nephrology*, 50, 14–23.

Cohen, A.H., Nast, C.C., Adler, S.G., and Kopple, J.D. (1989). Clinical utility of kidney biopsies in the diagnosis and management of renal disease. *American Journal of Nephrology*, 9, 309–15.

Collaborative Study of the Adult Idiopathic Nephrotic Syndrome (1979). A controlled study of short-term prednisone treatment in adults with membranous nephropathy. *New England Journal of Medicine*, 301., 1301–6.

Corwin, H.L., Schwartz, M.N., and Lewis, E.J. (1988). The importance of sample size in the interpretation of the renal biopsy. *American Journal of Nephrology*, 8, 85–9.

Couser, W.G, Stilmant, M.M., Idelson, B.A., and Ooi, Y.M. (1977). Ultrastructural dense deposit disease in diabetes mellitus. *Archives of Pathology*, 101, 221 (letter).

Cunningham, E., Provost, T., Brentjens, J., Reichlin, M., and Venuto, R.C. (1978). Acute renal failure secondary to interstitial lupus nephritis. *Archives of Internal Medicine*, 138, 1560–1.

D'Amico, G. (1988). Role of interstitial infiltration in glomerular disease. *Nephrology Dialysis and Transplantation*, 3, 596–600.

D'Amico, G., *et al.* (1986). Prognostic indicators in idiopathic IgA mesangial nephropathy. *Quarterly Journal of Medicine*, 59, 363–78.

Diaz-Buxo, J.A. and Donadio, J.V. (1975). Complications of percutaneous renal biopsy: an analysis of 1,000 consecutive biopsies. *Clinical Nephrology*, 4, 223–7.

Egidi, F., De Vecchi, A., Tarantino, A., Imbasciati, E., and Ponticelli, C. (1985). Comparison, of renal biopsy and fine needle aspiration biopsy in renal transplantation. *Transplantation Proceedings*, 17, 61–3.

Ehrenreich, T. and Churg, J. (1968). Pathology in membranous nephropathy. In *Pathology Annual* (ed. S.C. Sommers), pp. 145–86, Appleton Century-Crofts, New York.

Eiser, A.R., Katz, S.M., and Swartz, C. (1979). Clinically occult diffuse proliferative lupus nephritis. An age-related phenomenon. *Archives of Internal Medicine*, 139, 1022–5.

Ekelund, L. and Lindholm, T. (1971). Arteriovenous fistulae following percutaneous renal biopsy. *Acta Radiologica*, 11, 38–48.

Fogazzi, G.B., Bajetta, M.T., Banfie G., and Mihatsch M. (1989). Comparison of immunofluorescent findings in kidney after snap-freezing and formalin fixation. *Pathology Research and Practice*, 185, 225–30.

Fries, J.F., Porta, J., and Loang, M.H. (1978). Marginal benefit of renal biopsy in systemic lupus erythematosus. *Archives of Internal Medicine*, 138, 1386–9.

Furie, B., Greene, E., and Furie, B.C. (1977). Syndrome of acquired factor X deficiency and systemic amyloidosis. *New England Journal of Medicine* 297, 81–5.

Ganeval, D., Noël, L.H., Preud'Homme, J.L., Droz, D., and Grünfeld, J.P. (1984). Light chain deposition disease. Its relation with AL type amyloidoses. *Kidney International* 26, 1–9.

Griswold, W.R., *et al.* (1987). Treatment of childhood prednisone-resistant nephrotic syndrome and focal segmental glomerulosclerosis with intravenous methylprednisolone and oral alkylating agents. *Nephron*, 46, 73–7.

Gundersen, H.J.G. and Osterby, R. (1982). Optimizing sampling efficiency of stereological studies in biology: or 'Do more less well!' *Journal of Microscopy*, 121, 65–73.

Habib, R., Kleinchnecht, C., Gubler, M.C., and Levy, M. (1973). Idiopathic membranoproliferative glomerulonephritis in children. Report of 105 cases. *Clinical Nephrology*, 1, 194–214.

Habib, R., Levy, M., and Gubler, M.C. (1979). Clinicopathologic correlations in the nephrotic syndrome. *Pediatrician*, **8**, 325–48.

Hall, B.M., *et al.* (1984). Identification of the cellular subpopulations infiltrating rejecting cadaver renal allografts. *Transplantation*, **37**, 564–70.

Hancock, W.W. and Atkins, R.C. (1984). Cellular composition of crescents in human rapidly progressive glomerulonephritis identified using monoclonal antibodies. *American Journal of Nephrology*, **4**, 177–81.

Häyry, P., von Willebrand, E., and Ahonen, J. (1981). Monitoring of organ allograft rejection by transplant aspiration cytology. *Annals of Clinical Research*, **13**, 264–87.

Hefter, L.G. and Brennan, G.G. (1981). Transillumination of renal biopsy specimens for rapid identification of glomeruli. *Kidney International*, **20**, 411–5.

Hlatky, M.A. (1982). Is renal biopsy necessary in adults with nephrotic syndrome? *Lancet*, **2**, 1264–8.

Hooke, D.H., Gee, D.C., and Atkins, R.C. (1987). Leukocyte analysis using monoclonal antibodies in human glomerulonephritis. *Kidney International*, **31**, 964–72.

Hsu, H.C. and Churg, J. (1980). The ultrastructure of mucoid 'onionskin' intimal lesions in malignant nephrosclerosis. *American Journal of Pathology*, **99**, 67–80.

Imbasciati, E., *et al.* (1981). Acute renal failure in idiopathic nephrotic syndrome. *Nephron*, **28**, 186–91.

Imbasciati, E., Pasties, C., Scarpioni, L., and Mihatsch, M.J. (1986). Renal lesions in familial lecithin-cholesterol acyltransferase deficiency. Ultrastructural heterogeneity of glomerular changes. *American Journal of Nephrology*, **6**, 66–70.

Iversen, P. and Brun, C. (1951). Aspiration biopsy of the kidney. *American Journal of Medicine*, **11**, 324–30.

Jackson, E.C.M.C., Adams, A.J., Strife, C.F., Forrestal, J., Welch, T.R., and West, C.D. (1987). Differences between membranoproliferative glomerulonephritis type I and III in clinical presentation, glomerular morphology, and complement perturbation. *American Journal of Kidney Diseases*, **9**, 115–20.

Janson, P.A., Jubelirer, S.J., Weinstein, M.S., and Deykin, D. (1980). Treatment of bleeding tendency in uremia with cryoprecipitate. *New England Journal of Medicine*, **303**, 1318–22.

Jensen, E.B., Gundersen, H.J.G., and Osterby, R. (1979). Determination of membrane thickness distribution from orthogonal intercepts. *Journal of Microscopy*, **115**, 190–33.

Kark, R.M. (1968). Renal biopsy. *Journal of American Medical Association*, 205:80–6.

King, B.P., Esparza, A.R., Kahn, S.J., and Garella, S. (1976). Sarcoid granulomatous nephritis occurring as isolated renal failure. *Archives of Internal Medicine*, **136**, 241–5.

Klahr, S., Schreiner, G., and Ichikawa, I. (1988). The progression of renal disease. *New England Journal of Medicine*, **318**, 1657–66.

Korbet, S.M., Schwartz, M.M., Rosemberg, B.F., Sibley, R.K., and Lewis, E.J. (1985). Immunotactoid glomerulopathy. *Medicine*, 64, 228–43.

Korbet, S.M., Schwartz, M.M., and Lewis, E.J. (1988). Minimal-change glomerulopathy of adulthood. *American Journal of Nephrology* **8**, 291–7.

Leaker, B., Fairley, K.F., Dowling, J., and Kincaid-Smith, P. (1987). Lupus nephritis: clinical and pathological correlations. *Quarterly Journal of Medicine*, **238**, 163–79.

Leehey, D.J., Katz, A.I., Azaran, A.H., Aronson, A.J., and Spargo, B.H. (1982). Silent diffuse lupus nephritis: long-term follow-up. *American Journal of Kidney Diseases*, **2**(S), 188–196.

Levey, A.S., Lau, J., Pauker, S.G., and Kassirer, J.P. (1987). Idiopathic nephrotic syndrome. Puncturing the biopsy myths. *Annals of Internal Medicine*, **107**, 697–713.

Lewis, E.J., Kawala, K., and Schwartz, M.M. (1987). Histologic features that correlate with the prognosis of patients with lupus nephritis. *American Journal of Kidney Diseases*, **10**, 192–7.

Liu, Y.K., Kosfeld, R.E., and Marcum, S.G. (1984). Treatment of uraemic bleeding with conjugated oestrogens. *Lancet*, **2**, 887–90.

Livio, M., *et al.* (1986). Conjugated estrogens for the management of bleeding associated with renal failure. *New England Journal of Medicine*, **315**, 731–5.

Mackensen-Haen, S., Bader, R., Grund, K.E., and Bohle, A. (1981). Correlation between renal cortical interstial fibrosis, atrophy of the proximal tubules and impairment of the glomerular filtration rate. *Clinical Nephrology*, **15**, 167–71.

Mallick, N.P. and Williams, G. (1988). Glomerular and associated tubular injury by light chains. The spectrum of damage and effect of treatment. In *The Kidney in Plasma Cell Dyscrasias* (eds. L. Minetti, G. D'Amico and C. Ponticelli), pp. 77–84, Kluwer Academic, Dordrecht.

Mannucci, P.M., *et al.* (1983). Deamino-8-D-arginine vasopressin shortens the bleeding time in uremia. *New England Journal of Medicine*, **308**, 8–12.

Mihatsch, M.J., *et al.* (1983). Morphological findings in kidney transplants after treatment with cyclosporine. *Transplantation Proceedings*, **15**, 2821–36.

Mihatsch, M.J., *et al.* (1988). Cyclosporin associated nephropathy in patients with autoimmune diseases. *Klinische Wochenschrift*, **66**, 43–7.

Morel-Maroger, L. (1982). The value of renal biopsy. *American Journal of Kidney Diseases*, **1**, 244–8.

Morel-Maroger, L., Mery J.P., and Richet, G. (1973). Lupus nephritis immunofluorescence study of 54 cases. In *Glomerulonephritis: Morphology, Natural History, and Treatment*, Part II. (eds. P. Kincaid-Smith and T.H. Mathew and E.L. Becker, pp. 1183–6, Wiley, New York.

Mustonen, J., Pasternack, A., Helin, H., Pystynen, S., and Tuominen, T. (1984). Renal biopsy in acute renal failure. *American Journal of Nephrology*, **4**, 27–31.

Nairn, R.C. (1976). *Fluorescent protein tracing*. Churchill Livingstone, Edinburgh.

Neild, G. (1987). The haemolytic uraemic syndrome: a review. *Quarterly Journal of Medicine*, **63**, 367–76.

Nizze, H., *et al.* (1988). Cyclosporin associated nephropathy in patients with heart and bone marrow transplants. *Clinical Nephrology* **30**, 248–60.

Nolan, C.R. III, Anger, M.S., and Kelleher, S.P. (1986). Eosinophiluria – a new method of detection and definition of the clinical spectrum. *New England Journal of Medicine*, **315**, 1516–9.

Nolasco, F., Cameron, J.S., Heywood, E.F., Hicks, J., Ogg, C., and Williams, D.G. (1986). Adult-onset minimal change nephrotic syndrom: A long-term follow-up. *Kidney International*, **29**, 1215–23.

Nolasco, F.E.B., Cameron, J.S., Hartley, B., Coelho, A., Hildreth, G., and Reuben, R. (1987). Intraglomerular T cells and monocytes in nephritis: study with monoclonal antibodies. *Kidney International*, **31**, 1160–6.

Nomoto, Y., *et al.* (1987). Modified open renal biopsy: results in 934 patients. *Nephron*, **45**, 224–8.

Oberholzer, M., Torhorst, J., and Mihatsch, M.J. (1983). Minimum sample size of kidney biopsies for semiquantitative and quantitative evaluation. *Nephron*, **34**, 192–5.

Paone, D.B. and Meyer, L.R.E. (1981). The effect of biopsy on therapy in renal disease. *Archives of Internal Medicine*, **141**, 1039–41.

Park, M.H., D'Agati, V., Appel, G.B., and Pirani, C.L. (1986). Tubulointerstitial disease in lupus nephritis: relationship to immune deposits, interstitial inflammation, glomerular changes, renal function and prognosis. *Nephron*, **44**, 309–19.

Pei, Y., Cattran, D., Delmore, T., Katz, A., Lang, A., and Rance, P. (1987). Evidence suggesting undertreatment in adults with idiopathic focal segmental glomerulosclerosis. Regional Glomerulonephritis Registry Study. *American Journal of Medicine*, **82**, 938–44.

Pirani, C.L., Pollack, V.E., and Schwartz, F.D. (1964). The reproducibility of semiquantitative analyses of renal histology. *Nephron*, **1**, 230–46.

Ponticelli, C., *et al.* (1984). Controlled trial of methylprednisolone and chlorambucil in idiopathic membranous nephropathy. *New England Journal of Medicine*, **310**, 946–50.

Ponticelli, C., *et al.* (1989). A randomized trial of methylprednisolone and chlorambucil in idiopathic membranous nephropathy. *New England Journal of Medicine*, **320**, 8–13.

Praga, M., *et al.* (1985). Acute worsening of renal function during episodes of macroscopic haematuria in IgA nephropathy. *Kidney International* **28**, 69–74.

Rosenbaum, R., Hoffstein, P.E., Stanley, R.J., and Klahr, S. (1978). Use of computerized tomography to diagnose complications of percutaneous renal biopsy. *Kidney International*, **14**, 87–92.

Silva, F., Pace, E.H., Burns, D.K., and Krous, H. (1983). The spectrum of diabetic nephropathy and membranous glomerulopathy: report of two patients and review of the literature. *Diabetic Nephropathy*, **2**, 28–32.

Sinclair, R.A., Antonovych, T.T., and Mostofi, F.K. (1976). Renal proliferative arteriopathies and associated glomerular changes: A light and electron microscopic study. *Human Pathotholoy*, **7**, 565–88.

Solez, K. and Whelton, A. (1983). *Acute Renal Failure: Correlations between Morphology and Function*, Dekker, New York.

Striker, G.E., Quadracci, L.J., and Cutler, R.E. (1978). *Use and Interpretation of Renal Biopsy*. Saunders, Philadelphia.

Tiebosch, A.T.M.G., *et al.* (1989). Thin-basement-membrane nephropathy in adults with persistent haematuria. *New England Journal of Medicine*, **320**, 14–18.

Torhorst, J. (1974). Studies on the pathogenesis and morphogenesis of glomerulonephritis. Application of a newly developed morphometric method. *Current Topics in Pathology*, **59**, 1–68.

Turner, M.W., Hutchinson, T.A., Barré, P.E., Prichard, S., and Jothy, S. (1986). A prospective study on the impact of the renal biopsy in clinical management. *Clinical Nephrology*, **26**, 217–21.

Urizar, R.E., Schwartz, A., Top, F. Jr., and Vernier, R.L. (1969). The nephrotic syndrome in children with diabetes mellitus of recent onset: report of five cases. *New England Journal of Medicine*, **281**, 173–81.

Van der Woude, F.J., *et al.* (1985). Autoantibodies against neutrophils and monocytes: tool for diagnosis and marker of disease activity in Wegener's granulomatosis. *Lancet*, **1**, 425–9.

Wehrman, M., *et al.* (1989). Long-term prognosis of chronic idiopathic membranous glomerulonephritis. *Clinical Nephrology*, **31**, 67–76.

Weibel, E.R., (1979). *Stereological methods*. Vol. I. *Practical Methods for Biological Morphometry*. Academic Press, London.

Weibel, E.R. (1980). *Stereological methods*. Vol. II. *Theoretical foundations*. Academic Press, London.

Wickre, C.G. and Golper, T.A. (1982). Complications of percutaneous needle biopsy of the kidney. *American Journal of Nephrology*, **2**, 173–8.

Wijeyesinghe, E.C.R., Richardson, R.M.A., and Uldall, P.R. (1987). Temporary loss of renal function: an unusual complication of perinephric hemorrhage after percutaneous renal biopsy. *American Journal of Kidney Diseases*, **10**, 314–17.

Zawada, E.T., Jensen, R., Hicks, D., Putnam, W., and Ramirez, D. (1987). Nephrotic syndrome and renal failure in an elderly man. *American Journal of Nephrology*, **7**, 482–9.

Zollinger, H.U. and Mihatsch, J.M. (1978). *Renal Pathology in Biopsy* Springer-Verlag, Berlin.

SECTION 2
Pharmacology and drug use in kidney patients

2.1   Drug-induced nephropathies
2.2   Handling of drugs in kidney disease
2.3   Action and clinical use of diuretics

2.1 Drug-induced nephropathies

J.P. FILLASTRE AND M. GODIN

Introduction

Drug-induced nephropathies are common, numerous and often underdiagnosed. Little known only 30 years ago, they now have an important place in nephrology. In fact, that drugs may be responsible should now be considered in every type of renal failure, whether acute or chronic, glomerular, tubular, interstitial, or vascular. The kidney is a particularly vulnerable organ. It is highly vascularized, receiving 25 per cent of cardiac output, and contact surfaces between the nephron structures and a xenobiotic are extensive, both for the glomerular endothelium and the surface of the tubular epithelium. The kidney has many functions, one of which is the excretion of waste. It also plays a determinant role in homeostasis of the *milieu interieur*, and it is capable of diluting or concentrating urine in relation to the plasma. The concentrating capacity of the kidney may have deleterious consequences when it leads to the accumulation of toxic substances in the renal interstitium. Being the site of numerous metabolic transformations, the kidney requires normal perfusion, sufficient energy-producing substances, and constant oxygen supply. The kidney is also sensitive to insult from the exterior.

A large number of drugs may induce renal lesions, and an exhaustive review would be impossible. Rather than analysing the toxic effect of individual drugs, we will consider first the major syndromes. To begin with, we will give some examples of functional involvement, then consider the nephropathies according to their effect on various parts of the nephron, i.e., glomerular, tubular, interstitial, and vascular. We will only briefly comment on nephropathies that are discussed in other chapters of this work, such as those induced by non-steroidal anti-inflammatory drugs (NSAIDs) (Chapter 6.2), lithium (Chapter 6.6), antibiotics (Chapter 8.10), and contrast products used in radiology (Chapter 1.6.2). We will not discuss renal lesions secondary to analgesic abuse (Chapter 6.1), or to cyclosporin, which are discussed elsewhere (see Chapter 1.5).

Functional renal failure

Cases of functional renal failure are frequently seen in clinical practice. We will consider only those that modify the regulation of glomerular filtration by the kidney. This regulation is independent of renal blood flow and is mainly determined by variations in pre- and postglomerular resistance in the arterioles. Functional renal failure is now mainly found during treatment with inhibitors of angiotensin conversion enzyme and with NSAIDs.

Angiotensin conversion enzyme inhibitors

Inhibitors of angiotensin converting enzyme are usually well tolerated and are highly effective drugs. Angiotensin II induces vasoconstriction of the efferent arteriole of the glomerulus and may thus prevent a drop in GFR. Consequently, in some circumstances, inhibitors of the converting enzyme can lead to an abrupt drop in glomerular filtration, if such filtration is maintained by vasoconstriction of the postglomerular efferent arteriole. The first such observations were made in patients with bilateral stenosis of the renal artery, or stenosis of a single functioning kidney. Renal failure often occurs during administration of this type of drug in cases where, due to dehydration or hypovolaemia, the renin–angiotensin system is vital to the maintenance of glomerular filtration.

Such renal failure can also occur in nephroangiosclerosis (Davin and Mahieu 1985) and in cirrhotics (Wood *et al.* 1985) receiving conversion enzyme inhibitors.

Non-steroidal anti-inflammatory drugs (see also Chapter 6.2)

Whatever their chemical nature, NSAIDs inhibit the formation of prostaglandins through inhibition of cyclo-oxygenase. Prostaglandins have a net vasodilatory effect on the kidney. They do not seem to be greatly involved in autoregulation of the kidney in normal circumstances. On the other hand, in some circumstances, such as hypovolaemia, they can modulate the vasoconstrictor action of angiotensin II. The experimental induction of vasoconstriction immediately leads to increased synthesis and liberation of vasodilator prostaglandins. The administration of NSAIDs in patients with hypovolaemia (from cardiac insufficiency, haemorrhage, nephrotic syndrome, or ascitic cirrhosis) can lead to kidney failure. Such temporary renal failure is reflected only by increased urea and blood creatinine, without abnormalities of proteinuria or urinary sediment, and is rapidly reversible after stopping the treatment (see Chapter 6.2).

We will not describe the anomalies of tubular function that may be induced by drugs. They include urinary concentrating and diluting disorders (see Chapter 5.5), and involvement of the proximal portion of the nephron, leading to a Fanconi syndrome (see Chapter 5.1).

Organic renal failure

All parts of the nephron may be injured by drugs. Drugs can induce various renal lesions, of which tubulointerstitial are the most common and glomerular less frequent.

Drug-induced glomerulonephritis

These usually present as proteinuria or a full nephrotic syndrome in which renal function is normal. Membranous nephropathy is the most frequent type of glomerular lesion encountered; nevertheless, it is important to remember that a given drug can be associated with various different glomerular lesions. The reader

Gold salts

The incidence of proteinuria in patients with rheumatoid arthritis treated with gold salts was once considered to be low (Hartfall *et al.* 1937; Empire Rheumatism Council 1960). From recent studies, however, proteinuria and nephrotic syndrome are known to be much more frequent, occurring in from 6 to 17 per cent and 2.6 to 5.3 per cent of cases, respectively (Silverberg *et al.* 1970; Kean *et al.* 1983).

We have traced 27 accounts, comprising 122 patients with rheumatoid arthritis who underwent renal biopsies after the development of proteinuria (Fillastre *et al.* 1988). Fifty of them (41 per cent) presented with nephrotic syndrome. In eight cases, the renal biopsy was difficult to interpret because neither electron microscopic nor immunofluorescence studies were done. Among the 114 remaining cases, 102 (89.5 per cent) had a membranous nephropathy, whereas only 11 (9.6 per cent) had minimal glomerular changes. The clinical picture was similar in both situations. A full nephrotic syndrome was present in 40 of 69 patients (57.9 per cent) with membranous glomerulonephritis and in 3 of the 11 patients with minimal glomerular changes. High blood pressure, microhaematuria, and renal insufficiency were rare. No death related to renal disease had been reported.

Glomeruli from patients with membranous glomerulonephritis were most often considered normal by light microscopy. Subepithelial electron-dense deposits, typical of stage I, membranous glomerulonephritis, were usually scarce and focal, which we consider to be suggestive of drug-induced membranous glomerulonephritis. Granular IgG deposits were always seen along glomerular capillary walls and were usually also scarce and focal; C3 and/or IgM deposits were occasionally associated. Gold-induced membranous glomerulonephritis is usually diagnosed at stage I, probably because proteinuria is repeatedly looked for and renal biopsies are therefore taken very early. In some patients from whom sequential renal biopsies were taken, it was noted that electron-dense deposits had a rather similar course to that described in patients with idiopathic membranous glomerulonephritis.

Minimal glomerular changes have been found in 11 patients. This seems to be very rare but as glomeruli appeared normal by light microscopy, and as no subepithelial electron-dense deposits were seen and immunofluorescence studies were negative, it is clear that gold salts can be responsible for nephrotic syndrome with this type of lesion.

Mesangial glomerulonephritis was also noted in gold-treated patients who had renal biopsy for proteinuria without haematuria. The incidence of this type of nephritis has been reviewed by Hall (1988). It appears in 6 per cent of patients with the nephrotic syndrome (Helin *et al.* 1986), in 10 per cent with proteinuria of any severity (Hall 1988), and in 25 per cent with mild, asymptomatic proteinuria (Helin *et al.* 1986). It is significant that, when gold-treated patients underwent renal biopsy for microscopic haematuria with or without proteinuria, mesangial glomerulonephritis was found in 60 per cent (Helin *et al.* 1986). The histological appearances were an increase in mesangial cells and matrix, granular mesangial deposits of immunoglobulins and complement by immunofluorescence, and mesangial electron-dense deposits by electron microscopy. The immunoglobulin found most frequently was IgM, followed by IgG and IgA. Hall (1988) suggests that in gold-treated patients with rheumatoid arthritis, when urinary infection, urothelial carcinoma, and analgesic nephropathy have been excluded, the presence of microscopic haematuria is suggestive of mesangial rather than membranous glomerulonephritis.

Information from the follow-up of gold-induced, membranous glomerulonephritis after the withdrawal of the drug was available in 49 patients (Fillastre *et al.* 1988). Twenty-four were not treated; the remaining 25 received low-dose steroids (5–15 mg/day), high-dose steroids (1 mg/kg.day), or immunosuppressive agents. Proteinuria disappeared in 23 (95.8 per cent) of the untreated patients and in 21 (84 per cent) of the treated. Treatment did not hasten a favourable outcome, and proteinuria disappeared both in treated and untreated patients within 4 to 18 months. A similar, favourable outcome occurred in patients with minimal glomerular changes and nephrotic syndrome. There is little information about the clinical course and prognosis of those with gold-induced, mesangial glomerulonephritis. The haematuria usually resolves when treatment is stopped (Horden *et al.* 1984; White *et al.* 1984) but persists whilst treatment is continuing (White *et al.* 1984). Thus the prognosis of this glomerulopathy is spontaneously favourable after withdrawal of the drug. Treatment with corticosteroids is therefore not indicated in gold-associated nephrotic syndromes. Because gold has always been stopped when proteinuria was detected, it is not known whether the proteinuria could disappear spontaneously if treatment is continued, as is the case with some other drugs (see below).

Two studies (Wooley *et al.* 1980, Bardin *et al.* 1982) have shown that patients with rheumatoid arthritis who have HLA B8 or DRW3 antigens have an increased incidence of proteinuria.

Ayes *et al.* (1987) reported that many patients who developed proteinuria during gold treatment had a low capacity for sulphoxidation of carbocysteine, but this point is still unclear.

The development of gold nephropathy is not correlated with the cumulative dosage of gold (which varied from 200 to 1000 mg) nor with the duration of treatment (which varied from 5 to 12 months). Burger *et al.* (1979) have suggested that gold nephropathy is more frequent in patients receiving sodium aurothiomalate. Though aurothiomalate was the gold salt used in about 75 per cent of the cases we reviewed, all the gold compounds that have been given parenterally (sodium aurothiopropanol sulphonate, sodium aurothiosulphate, and aurothioglucose) have been responsible for some cases of membranous glomerulonephritis. Preliminary findings with auranofin, a recently introduced oral gold-containing medication, show that it has an efficacy similar to that of gold sodium thiomalate and appears less nephrotoxic. The prevalence of proteinuria has been retrospectively analysed in 1283 patients treated with auranofin (Katz *et al.* 1984). A mild proteinuria (< 1 g/day) occurred in 1.1 per cent of them, and a moderate or heavy proteinuria (> 1 g/day) in 2 per cent. A renal biopsy, taken from four patients, showed membranous glomerulonephritis. Proteinuria disappeared in most patients after withdrawal of the drug and reappeared in only one of the eight patients who were rechallenged. In a controlled trial, Ward *et al.* (1983) found that proteinuria appeared in 2 per cent of the 75 patients treated with parenteral gold and in none of the 72 treated with auranofin. The low incidence of proteinuria allows us to conclude that patients treated with auranofin are less liable to renal complications than those receiving gold parenterally.

The pathogenesis of gold-induced, membranous glomerulonephritis is still unclear. It is tempting to speculate that gold induces a disease of immune-complex type in which circulating immune complexes are either deposited in glomeruli or formed *in situ*. The fact that gold has not been found in the subepithelial deposits suggests that it is not deposited as a hapten bound to proteins. It has been suggested that tubular lesions induced by gold could be responsible for the release of a tubular antigen, which could then provoke a Heymann-type nephritis. However, such an antigen has not been demonstrated in glomeruli.

Mercury compounds

Few mercury-containing drugs are still in use. Between 1947 and 1968, about 20 cases of nephrotic syndrome were described after treatment with organomercurial diuretics. Antiseptics, laxatives, and vaginal contraceptives containing mercury were used until recently or are still in use in some countries (Lipsky and Ziff 1977; Husby *et al.* 1981). Nephrotic syndrome has been observed in patients using topical ammoniated mercury for treatment of psoriasis (Fillastre *et al.* 1988). The clinical presentation was heavy proteinuria, normal renal function, and normal blood pressure. The outcome was favourable after withdrawal of the drug, whether or not the patient had received steroids; membranous glomerulonephritis was generally found.

A skin-lightening cream containing mercury seems to have been responsible for several cases of nephrotic syndrome in east Africa. Indeed, Barr *et al.* (1972) found that 32 of 60 adult Africans presenting with that syndrome used such creams. Renal biopsies were taken in 16 patients: minimal glomerular changes were found in 14 and membranous glomerulonephritis in the other two. The prognosis was good after the use of the cream was discontinued.

Nephrotic syndrome occurred after occupational or environmental exposure to mercury in 11 patients (Fillastre *et al.* 1988). A renal biopsy showed membranous glomerulonephritis in seven, with IgG and C3 deposits. In the four other patients, only minimal glomerular changes were found. The outcome was favourable in all of these patients. The pathogenesis of mercury-induced glomerulonephritis is discussed in Chapter 3.1.

Silver

Silver can be deposited as an organic silver compound in most organs of the body; the concentration of silver is highest in the kidney. In rats, silver given in the drinking water has been found in the basement membrane of several tissues. Whether such deposits in the kidney induce proteinuria is unknown. Similar dense deposits, demonstrated at the light microscopic and ultrastructural levels without any counterstaining, have been reported by Zech *et al.* (1973) and Leroux-Robert *et al.* (1982) in two patients exposed to silver salts for a long time. These deposits were located exclusively in the lamina densa. Immunofluorescence analysis was done in one case and was negative; this confirmed that the glomerular disease was not a membranous glomerulonephritis.

D-Penicillamine

Proteinuria has been detected in 7 per cent, 10 per cent (Stein and Smythe 1968), and up to 20 per cent (Day *et al.* 1974) of patients with rheumatoid arthritis treated with D-penicillamine. We have reviewed 21 accounts comprising 190 cases (Fillastre *et al.* 1988). Proteinuria varied from 0.3 g/day to more than 20 g/day. Sixty-seven of the 135 (49.6 per cent) cases described in detail had nephrotic syndrome. Proteinuria rarely occurred within the first 3 months of treatment and usually appeared after 7 to 18 months. However, it occurred after 4 years in a patient with Wilson's disease (Walshe 1962), and after 5 and 12 years in two of our patients. It is important that in four patients (Neild *et al.* 1979), proteinuria first appeared one to four months after D-penicillamine was stopped. The incidence and the degree of proteinuria seem similar, whether patients were taking a low dose (500–750 mg/day) or a high dose (1000–1800 mg/day). Hypertension, haematuria, and renal insufficiency are rare.

Renal biopsies were taken in 147 patients (Fillastre *et al.* 1988). In 12 cases, the histological findings were difficult to interpret. Among the 135 remaining cases, 115 had membranous glomerulonephritis (85.2 per cent); 14 were described as having a mesangioproliferative glomerulonephritis, three had minimal glomerular changes and the three remaining had focal, necrotizing glomerulonephritis. The clinical presentation was similar in these different cases. More details of renal histology are given by Fillastre *et al.* (1988).

Follow-up information was available in 121 patients (Fillastre *et al.* 1988). Proteinuria usually decreased progressively after withdrawal of D-penicillamine and disappeared within 6 to 8 months. It is significant that proteinuria can decrease or even disappear despite continued treatment. Kirby *et al.* (1979) found that proteinuria resolved in an average of 14 months in ten patients who continued D-penicillamine, compared to 15 months in seven who stopped it. In cases studied by Jaffe *et al.* (1968), and by Stein and Smythe (1968), those who received a second course of D-penicillamine had a recurrence of proteinuria after a similar delay and from a similar, cumulative dosage. In contrast, it did not recur in the five patients described by Hill *et al.* (1978), who were retreated with an initial low dosage of D-penicillamine, which was then slowly increased; the follow-up ranged from 13 to 24 months. It is thus difficult to know whether D-penicillamine should be withdrawn where proteinuria occurs during treatment. This, together with the benign course of D-penicillamine-induced glomerulonephritis, which never results in chronic renal failure, led Huskinsson and Dudley Hart (1972) to recommend maintenance of the drug. In contrast, Bacon *et al.* (1976) preferred to withdraw D-penicillamine as soon as proteinuria was found. Hill *et al.* (1978) adopted an intermediate attitude and recommended discontinuation of treatment only when proteinuria exceeds 3 g per day. We agree with Bacon and recommend withdrawal of D-penicillamine as soon as proeinuria is found.

Two special circumstances should be mentioned: the appearance of a lupus-like syndrome, and of one resembling Goodpasture's syndrome. In some patients treated for various diseases with D-penicillamine, a lupus-like syndrome has been described (Harkcom *et al.* 1978; Kirby *et al.* 1979; Chalmers *et al.* 1982). Renal involvement was present in six, who all presented with proteinuria (Fillastre *et al.* 1988). A membranous glomerulonephritis or a focal, necrotic glomerulonephritis was found. Proteinuria and systemic symptoms rapidly subsided after D-penicillamine was discontinued.

To our knowledge, in nine patients, a most severe and rapidly progressive glomerulonephritis has been described under the name of Goodpasture's syndrome (Sternlieb *et al.* 1975; Banfi 1983; Fillastre *et al.* 1988). The patients presented suddenly with dyspnoea, haemoptysis, pleural effusions, gross haematuria, and oliguria. Diffuse extracapillary proliferations with necrosis and diffuse crescents were found. It is of note that a linear pattern of C3 deposition was present in only one patient, whereas the pattern was granular in the others and no circulating antibodies to glomerular basement membrane were detected in three cases in whom they were sought. Five patients from some time ago, who received neither steroids nor immunosuppressive agents, died rapidly. In four more recent cases, intensive treatment with high doses of prednisone, cyclophosphamide, or azathioprine and heparin together with plasmapheresis, was beneficial (Swainson *et al.* 1982).

A genetic predisposition to nephrotoxicity by D-penicillamine is suggested by several studies (Wooley *et al.* 1980), which have shown that those who developed proteinuria when treated with gold are at higher risk when subsequently treated with D-penicillamine. The frequency of D-penicillamine-induced proteinuria is higher in DRW3-positive patients.

The mechanism by which D-penicillamine induces glomerulonephritis is still unknown. There have been a few experimental studies. Batsford *et al.* (1976) and Seelig *et al.* (1977) detected proteinuria in rats fed very high doses of D-penicillamine (1.5–2.0 g/day). A mesangial proliferation without subepithelial deposits was found by Batsford, and a focal glomerulonephritis with granular IgG and C3 deposits by Seelig. These deposits disappeared totally or partially within a few weeks of D-penicillamine being discontinued. Antibodies to glomerular basement membrane, occasionally associated with intravascular coagulation, were induced in Brown Norway rats fed with low dosages of D-penicillamine (20 mg/day) but not in Lewis or Sprague-Dawley rats (Donker *et al.* 1984). *In-vitro* studies have shown that D-penicillamine can interfere with T-cell function (Lipsky and Ziff 1978, 1980). An immunological mechanism is very likely in this type of glomerulonephritis.

Thiopronine, pyrithioxine, and 5-thiopyridoxine

These sulphydryl-containing drugs are also used mostly in the treatment of rheumatoid arthritis.

About 30 patients developed nephrotic syndrome after receiving thiopronine (Amor *et al.* 1982; Pasero *et al.* 1982; Fillastre *et al.* 1988); minimal glomerular changes were detected in five and membranous glomerulonephritis in one. In some patients, reintroduction of thiopronine at a lower dosage did not induce a relapse of the syndrome, which suggests that the reaction of the kidney may be dose-related.

The nephrotic syndrome was found in three patients with rheumatoid arthritis treated with pyrithioxine (Amor *et al.* 1979; Segond *et al.* 1979; Prouvost-Danon *et al.* 1981); there was a membranous glomerulonephritis in all of them. Proteinuria appeared in 3 of 12 patients treated with 5-thiopyridoxine (Huskisson and Dudley Hart 1972).

Captopril

Captopril, like D-penicillamine, thiopronine, pyrithioxine and 5-thiopyridoxine, contains a sulphydryl group. A nephrotic syndrome has also been observed in patients treated with high dosages of captopril (450–600 mg/24 h). In 1982, the overall incidence of proteinuria in such cases was 1 per cent, but this is certainly less now since lower doses of captopril are used. In the *Safety report on captopril in the treatment of hypertension* (Pigott 1980) there are 65 patients with proteinuria, and of these, 13 had a nephrotic syndrome. A renal biopsy was taken in 19 patients: 18 had membranous glomerulonephritis. The rate of disappearance of proteinuria was the same whether captopril was discontinued or not; indeed it disappeared in 18 of 33 patients in whom captopril was maintained without any reduction in dosage, compared to 22 of 32 patients in whom it was withdrawn (Pigott 1980). Thus, captopril-induced proteinuria is not a serious problem.

Non-steroidal anti-inflammatory drugs

The use of NSAIDs may lead to a number of renal and electrolyte disorders. They may cause sodium and fluid retention, diminished renal blood flow, a decreased GFR, and hyperkalaemia (see Chapter 6.2). Many of these changes have been attributed to inhibition of prostaglandin synthesis in the kidney and are reversible on stopping the drug. In addition, NSAIDs have been incriminated in chronic interstitial nephritis, renal papillary necrosis and, more frequently, in acute renal failure (see Chapter 8.9) due to acute interstitial nephritis. Such reactions are, however, very rare when one considers the wide use of NSAIDs.

We emphasize that in some patients treated with NSAIDs, renal failure is not only due to acute interstitial nephritis but also to nephrotic syndrome. The first two cases of acute renal failure with nephrotic syndrome were reported in 1979 (Brezin *et al.* 1979) and more than 100 cases have been described since.

Acute renal failure and nephrotic syndrome generally appeared after several months of exposure to the drug. The severity of renal failure varied greatly: it was moderate in some patients or severe enough in others to require peritoneal or haemodialysis. Eosinophilia was present in two of the 20 cases in which counts of eosinophils were made (Fillastre *et al.* 1988). There were no extrarenal manifestations of hypersensitivity. When performed, immunological tests, including complement components, LE clot test, antinuclear and antiDNA antibodies, cryoglobulins, rheumatoid factor, and circulating immune complexes were always normal or negative.

The histological findings combined acute interstitial nephritis and minimal-change glomerulopathy. By light microscopy, glomeruli had no significant changes whereas interstitial lesions were prominent. The interstitium was oedematous and infiltrated by lymphocytes, plasma cells, and eosinophils. There were eosinophils in the interstitial infiltration in at least 40 per cent of cases. Tubular damage of varying degrees was usually an associated finding. Fluorescent staining of glomeruli, tubules, and interstitium for immunoglobulins and complement was almost always negative. There was extensive fusion of foot processes in all the cases studies by electron microscopy, the glomeruli being otherwise normal. The lymphocyte subpopulations within the inflammatory interstitial infiltrates were studied in some cases. T lymphocytes were predominant, with cytotoxic/suppressor T cells predominating over helper/inducer T cells (Solez *et al.* 1983; Stachura *et al.* 1983).

Females and elderly patients were particularly at risk. Renal function returned to normal and proteinuria disappeared in the majority of patients; however, a few retained some degree of renal insufficiency and/or proteinuria for several months.

Antitumour agents

These drugs induce interstitial or tubular toxicity more often than glomerular changes. Nevertheless, mitomycin in man and Adriamycin in animals may induce glomerulopathies.

Mitomycin

The incidence of human nephrotoxicity caused by mitomycin varies in different reports from 4 to 10 per cent. We have reviewed this condition and found 106 reported cases (Fillastre *et al.* 1988). Most reports mention a haemolytic–uraemic syndrome as the essential complication of mitomycin. The majority of patients were receiving mitomycin for adenocarcinoma of the gastrointestinal tract or of the breast.

Renal failure was preceded by haemolytic anaemia and subsequently a full haemolytic–uraemic syndrome occurred. Nephrotoxicity was rarely noted after the first course and generally appeared after the fourth or sixth course. The dosage of mitomycin was the usual 10 to 15 mg/m^2. Clinical manifestations included abrupt severe haemolysis, thrombocytopenia, elevated lactic dehydrogenase, decreased or absent haptoglobin, reticulocytosis, and nucleated erythrocytes and schizocytosis, together with a negative Coombs' test. Proteinuria was detected, lower than 2 g/24 h; nephrotic syndrome was never reported. Hypertension occurred in about half of the cases. Renal failure was of variable severity; the serum creatinine rose to a final concentration of 10 mg/100 ml over a period of approximately 3 weeks; in 12 cases, the severity of renal failure required immediate haemodialysis. Progression to death was frequent, and renal insufficiency was reported as the cause in many cases. Associated pulmonary or neurological manifestations were of poor prognosis. Blood transfusions could precipitate death.

By light microscopy, the kidneys had fibrin thrombi in afferent arterioles and glomerular loops, thickened glomerular basement membranes, and fibrin exudation into Bowman's capsule. By electron microscopy, the arterioles and glomerular capillaries contained aggregates of platelets and fragments of red blood cells. Immunofluorescence staining revealed IgM and fibrin deposition in damaged arteriole walls, with fibrin exudation into Bowman's capsule. Tubules had more or less extensive secondary ischaemic damage with loss of proximal tubules and associated interstitial damage. In some cases, mesangiolysis was seen.

Mitomycin was stopped when renal insufficiency was detected. Spontaneous regression was very rare: usually haemolytic episodes, progressive renal failure, and pulmonary or neurological complications provoked terminal changes and eventual death. Many therapeutic manoeuvres were proposed: corticosteroids, platelet antiaggregants, anticoagulants, and plasmapheresis. Some treatments benefited the haematological abnormalities but had little effect on the renal disease. The best results were obtained with combination of repeated plasmapheresis and high dosage of steroids (Lyman *et al*. 1983).

Doxorubicin (Adriamycin)
Doxorubicin hydrochloride is an anthracycline antibiotic widely used as an antitumour agent; it induces a nephrotic syndrome in the rat and rabbit (Fajardo *et al*. 1980; Bertani *et al*. 1982) Giroux *et al*. 1984). No apparent changes in glomeruli or tubules were seen by light microscopy except vacuolization of visceral epithelial cells; no immune deposits were found. Electron microscopy revealed loss of the normal architecture of foot processes. In humans, to date, there is no evidence of nephrotoxicity.

Other drug-induced glomerulopathies
Hydralazine-induced lupus is considered in Chapter 4.6.2. The association between lithium and renal disease has been discussed in Chapter 6.6. The development of nephrotic syndrome in patients taking lithium is extremely rare. In our review, we collected only 10 patients (for more details see Fillastre *et al*. 1988). Minimal glomerular changes constituted the most frequent histological picture.

Many other drugs have been occasionally associated with proteinuria and/or nephrotic syndrome (Fillastre and Morel Maroger 1969). In some instances these associations are poorly documented and may have been coincidental. The drugs include procainamide (Whittingham *et al*. 1970), antiepileptics, i.e., oxazolidinedione derivatives (Bar Khayim *et al*. 1973), hydantoin derivatives (Barnett *et al*. 1948), probenecid (Ferris *et al*. 1961), phenindione (Tait 1960), carbutamide (Csapo and Hobi 1966), tolbutamide (Schnall and Wiener 1958), chlorpropamide (Appel *et al*. 1983), quinidine (Yudis and Meehan 1976), dapsone (Belmont 1967), psoralen (Eyanson *et al*. 1979), potassium perchlorate (Lee *et al*. 1961), methimazole (Bahner *et al*. 1969; Schaum *et al*. 1973, Reynolds and Bhathena 1979), and levamisole (Hansen *et al*. 1978). All these drugs may induce glomerular lesions.

The association of nephrotic syndrome and acute renal failure, as described for NSAIDs earlier in this chapter, has been reported in association with three other drugs: ampicillin (Rennke *et al*. 1980), rifampicin (Neugarten *et al*. 1983), and interferon (Averbuch 1984), each of which was responsible for one case. In the case attributed to interferon the patient had been treated with recombinant leucocyte α-interferon for mycosis fungoides.

Drug-induced acute toxic tubular nephropathies
Antibiotics are the most frequent drugs to induce acute tubular nephropathies and of these aminoglycosides are the group most frequently cited. Tubular lesions are due to a direct action on the tubule and it is the proximal portion that is principally altered.

Aminoglycosides
Nephrotoxicity and ototoxicity constitute the major secondary effects of aminoglycosides. The incidence of renal involvement has been reduced by better laboratory surveillance during administration and by shorter periods of treatment. The incidence, however, is still high (4–10 per cent; Hou *et al*. 1983). All aminoglycosides in current use are nephrotoxic to different degrees. Renal involvement is characterized by a non-oliguric acute renal failure; in fact, the first sign is usually polyuria due to a defect in the urinary concentrating capacity (Bennett *et al*. 1978). Other signs are not clinically evident and comprise abnormal tubular function resulting in proteinuria, leucocyturia, and elevated urinary excretion of certain tubular enzymes (Mondorf *et al*. 1978). Toward the seventh to the tenth day of treatment, reduced glomerular filtration and in consequence, elevated blood urea and creatinine, occur. The episode is reversible and leaves no sequelae. Biopsy shows varying severity of lesions according to the time it was taken. The first lesions are rarification of the brush border, followed by its disappearance. Soon thereafter, intracellular modification appears, with an increase in size of lysosomes, myeloid bodies in the lysosomes, and mitochondrial swelling (Fillastre *et al*. 1980). Later, cellular necrosis is seen, alternating with regeneration (Luft *et al*. 1977).

Aminoglycosides accumulate in the kidney such that the concentration in the cortex is higher than that in the medulla and reaches 10 times that in blood (Fabre 1980). They adhere to phospholipids on the brush border of the proximal tubular epithelium and enter the cell mainly by endocytosis. They interfere with many intracellular activities, leading to inhibition of phospholipases, accumulation of lipids, modification of the activities of enzymes (alanineaminopeptidase, *N*-acetyl-β-D-glucosaminidase, alkaline phosphatase, cathepsin, sphingomyelinase) (Kaloyanides and Pastoriza-Munoz 1980), reduction of the permeability of the lysosomal membrane (Morin *et al*. 1980), and reduction of mitochondrial respiration (Bendirdjian *et al*. 1982) and the energy potential of the cell; they also modify the activity of sodium–potassium ATPase (Williams et al. 1981).

Many factors contribute to the potential nephrotoxicity of aminoglycosides (Lietman and Smith 1983). Two are very important: the daily quantity injected and the length of treatment. It is imperative to respect the recommended dosages. Aminoglycosides are typical of drugs with a narrow therapeutic margin; however, the length of treatment can now be reduced because numerous other antibiotics are available to replace them. Treatment with aminoglycosides should not exceed 10 to 12 days. Hydroelectrolyte changes are also important: sodium depletion, any hypovolaemic state, and potassium depletion all favour aminoglycoside nephrotoxicity. Conversely sodium and potassium loading have the reverse effect, and increased dietary calcium supplements are protective. Drug interactions frequently play a part in the appearance of aminoglycoside nephrotoxicity. The deleterious effects of a diuretic–aminoglycoside association, and that of cephalothin or cephaloridine with an aminoglycoside, are well known. It is also possible that nephrotoxic effects are potentiated when an aminoglycoside is associated with cyclosporin, vancomycin, amphotericin B, or angiotensin converting enzyme inhibitors (Fillastre et al. 1989). In contrast, calcium antagonists may provide protection from nephrotoxic effects of gentamicin.

Lastly, the risk varies with the clinical state. On the one hand, nephrotoxic risk is increased in septicaemia with endotoxaemia and acute pyelonephritis (Bergeron et al. 1982); on the other, it has been shown experimentally that rats rendered diabetic resist the toxic effects of gentamicin. It is not known whether diabetic patients also benefit from such protection.

Thus, the nephrotoxicity of aminoglycosides persists as a clinical problem. However, if the rules for their use are respected (see Chapter 2.2), if the risk factors are known, and if susceptible subjects are followed more closely, it should be possible to reduce the number of renal incidents.

Cephalosporins

Many antibiotics belong to this class of compounds. It is unfortunate that because the first two introduced into clinical practice, cephaloridine and cephalothin, were nephrotoxic, all of the others have also been presumed to be so. In fact, renal involvement has been found only rarely after treatment with cephazolin, cephalexin, cephradine, cefoxitin, or cephaloglycin (Moellering and Swartz 1976; Barza 1978; Giamarellou et al. 1982). Most cephalosporins, those of the second and third generations, are not nephrotoxic. Cephaloridine accumulates in the proximal tubular cells and induces necrosis (Tune et al. 1974). Cephalosporins inhibit mitochondrial respiration, induce the formation of reactive metabolites, interfere with cytochrome P-450 and favour lipid peroxidation. We recommend monitoring of renal function during the use of cephalosporins.

Other antibiotics

Polymyxins can be nephrotoxic but they are now rarely used (Fillastre and Godin 1989). Vancomycin was thought to be nephrotoxic but, since impurities have now been eliminated during its manufacture, it no longer appears to be the origin of renal lesions. However, it may potentiate aminoglycoside nephrotoxicity. Some cases of ciprofloxacin-induced acute renal failure have now been reported. Gerritsen et al. (1987) described one such case, treated with ciprofloxacin for several weeks: in the renal biopsy, light microscopic examination was normal, but electron microscopy showed epithelial vacuolization similar to that found in aminoglycoside nephrotoxicity. Hootkins et al. (1989) described three cases of acute renal failure after ciprofloxacin given from several days to several weeks. All patients were non-oliguric and the renal failure was completely reversed after discontinuing the drug. No histological studies were made. More recently, ciprofloxacin has been implicated in the development of an allergic interstitial nephritis. Rippelmeyer and Symhavsky (1988) reported a case with interstitial oedema, lymphocytic infiltration, and occasional eosinophils.

Non-steroidal anti-inflammatory drugs

As we noted above, NSAIDs can be responsible for several types of renal involvement.

Acute reversible renal failure occurring relatively quickly (8–15 days) after the beginning of treatment suggests a haemodynamic disorder, which is related to inhibition of prostaglandin synthesis. Such incidents occur most often in the elderly and in patients who are hypovolaemic (see above). We have mentioned acute interstitial nephropathies, with or without nephrotic syndrome (see also Chapter 8.9).

Finally, it is also of significance that these drugs can induce acute tubular nephropathies. In contrast to the haemodynamic renal failure mediated by NSAIDs, there appear to be no clear-cut risk factors for this acute renal failure.

Contrast media

Radiopaque products are responsible for a large number of renal lesions (see Chapter 6.2). The frequency is difficult to assess, as it depends on the subjects exposed: it may be less than 1 per cent in apparently healthy subjects but can affect 80 to 90 per cent of elderly subjects and those with diabetes, arteriosclerosis, and chronic renal failure (and possibly those with hypovolaemia). The manifestations of renal involvement are highly variable, ranging from a modest increase in blood creatinine to oligoanuria.

Oligoanuria appears soon after injection of a contrast medium, disappears rapidly, and in most cases does not leave sequelae. All water-soluble tri-iodide contrast media have similar renal nephrotoxicity. In most reports, no relationship could be established between the dose of iodine injected and the occurrence of renal failure; the threshold dose that causes nephrotoxicity cannot be defined. The susceptibility of the patient is the determining factor (see above).

There have been many suggestions concerning the mechanism of action, including direct toxicity to renal tubular cells and intracellular modification with aggregation of blood cells. Emphasis has been placed on disturbances of intrarenal haemodynamics: contrast media increase secretion of renin and angiotensin, and thus induce intense vasoconstriction and reduction or even interruption of glomerular filtration.

Such incidents can be prevented now that the risks have been identified, and physicians must find substitutes for investigations requiring the injection of contrast media for subjects at high risk. Alternatives have been made available by progress in ultrasonic and isotopic techniques. When the use of a contrast medium cannot be avoided, sufficient hydration must be maintained, the injections must not be repeated and, if several examinations are necessary, they should be carried out at intervals of several days (see Chapter 1.6.2).

2.1 Drug-induced nephropathies

Anticancer drugs
Cisplatin

Cisplatin (*cis*-diammine dichloroplatinum or CDDP) is a potent anticancer drug, used intensively in man, being effective against testicular carcinoma, ovarian and bladder carcinoma, osteosarcoma, neuroblastoma, and head and neck carcinoma. The clinical use of cisplatin is often complicated by nephrotoxicity, ototoxicity, gastrointestinal disturbances, and myelosuppression.

Early clinical trials of cisplatin in cancer patients showed a striking incidence of persistent azotaemia and acute renal failure (Rossof *et al* 1972; Lippman *et al* 1973; Dentino *et al.* 1978). It was concluded in the early years of therapeutic use of cisplatin that for high doses, greater than 100 mg/m^2, there was a high incidence of nephrotoxicity, including irreversible toxicity. The incidence of renal side-effects has been largely reduced in recent years because of preventive measures, as will be discussed below.

Some studies have now been made of the primary alterations of renal function after cisplatin. Offerman *et al.* (1984) measured GFR and renal plasma flow in 10 patients receiving cisplatin infusion at a dose of 20 mg/m^2 over 4 h. These measurements were made before, during and after the administration of cisplatin on the first day of treatment. The patients were hydrated with intravenous saline during the whole study period. In all of them there was a decrease in renal plasma flow during, or within 3 h of cisplatin infusion with no important changes in GFR. Thus, an increase in the corresponding filtration fraction was found in all patients. Subsequently, a decreased GFR was also detected. The fall in renal plasma flow that occurred before any change in GFR was suggestive of an increase in renal vascular resistance. Vascular effects of cisplatin are perhaps more widespread: Raynaud's phenomenon was seen in many patients (Vogelzang *et al.* 1985); severe coronary artery diseases were also encountered (Edwards *et al.* 1979); and systemic hypertension was detected during intra-arterial infusion of cisplatin (Kletzel and Jaffe 1981).

Tubular dysfunctions have also been demonstrated very early after cisplatin administration. Alanine aminopeptidase, a brush-border membrane enzyme of proximal tubules, and *N*-acetyl-β-D-glucosaminidase, a lysosomal enzyme, were elevated in many patients treated with cisplatin, as were leucine-aminopeptidase and β-glucuronidase (Kuhn *et al.* 1978; Jones *et al.* 1980). These increases in urinary enzymes did not correlate with a fall in creatinine clearance. Excretion of β_2-microglobulin was also shown to rise after cisplatin (Jones *et al.* 1980): this excretion increased in most patients within 5 days of treatment and then fell; the rise in excretion did not correlate with a fall in creatinine clearance.

Electrolyte disturbances appeared very frequently in cisplatin-treated patients. Hypomagnesaemia, hypocalcaemia, and hypokalaemia were those most often detected. Many patients developed significant hypomagnesaemia with serum levels below 1.4 mmol/l. Hypomagnesaemia persisted for as long as 20 months after cisplatin had been discontinued in about half of the patients (Schilsky and Anderson 1979). This high incidence of hypomagnesaemia has been confirmed: it was found in 87 per cent (Vogelzang *et al.* 1985) and 76 per cent of patients (Buckley *et al.* 1984). Urinary excretion of magnesium was determined in only a few patients and it was found to be inappropriately high. Hypokalaemia and hypocalcaemia associated with hypomagnesaemia were considered to be an effect of magnesium deficiency.

After prolonged administration of cisplatin, chronic renal failure appeared, without any particular characteristics. The histological appearances of the kidney in the early phase of the disease are focal acute tubular necrosis affecting primarily the distal and collecting tubules, with dilatation of convoluted tubules and the formation of casts (Gonzalex-Vitale *et al.* 1977; Dentino *et al.* 1978). No glomerular or vascular lesions were reported. It is important that prolonged renal damage, lasting more than 12 months, has been reported (Dentino *et al.* 1978), which suggests that cisplatin could cause irreversible injury.

Cisplatin undergoes rapid distribution to nearly all organs after intravenous administration; the highest levels appear in kidney, liver, ovary, uterus, skin, and bone (Jacobs *et al.* 1980). The elevated tissue/plasma ratios are maintained for a long time. The principal route of cisplatin excretion is the kidney. By 2 to 4 weeks after treatment, significant levels of platinum were still found in the kidney (Litterst *et al.* 1979); the largest portion of this platinum was recovered from the cytosol fraction.

The mechanisms of cisplatin nephrotoxicity remain unclear and many hypotheses have been suggested. The effect is certainly not related to the platinum atom. The delay in the appearance of nephrotoxicity suggests that a metabolite of the cisplatin complex produces the effect. The stereospecificity of cisplatin determines its toxic potential. *Cis*- and *trans*-dichlorodiammine platinum result in similar renal concentrations of platinum, and the *trans* isomer does not produce renal toxicity, indicating that the geometry of these complexes, rather than the presence of the platinum atom, plays a crucial role (Leonard *et al.* 1971).

Biotransformation of cisplatin may also play a part in the toxic renal effects. *In vitro*, the chloride ligands of the complex become labile in aqueous media: by extrapolation, it has been suggested that cisplatin exists *in vivo* as a neutral complex in extracellular fluid, because the chloride concentration is sufficiently high to stabilize the complex and prevent hydrolysis. In fact, the markedly lower intracellular concentration of chloride facilitates the displacement of chloride by water molecules yielding a positively charged, hydrated and hydroxylated complex. Hydration of cisplatin induces formation of monochloromonoaquodiammineplatinum or diaquodiammineplatinum. These agents alkylate the purine and pyrimidine bases of nuclear material. There is evidence that the hydration products of cisplatin induce the nephrotoxic effect of the drug. Litterst *et al.* (1979) showed that cisplatin, 9 mg/kg bodyweight, prepared and administered in distilled water, resulted in acute renal failure and death in all rats so treated. When the same dose was prepared and administered in 0.9 per cent saline, the deaths fell to 66 per cent. The antitumour potency of cisplatin in tumour-bearing rats was not affected by the solvent vehicle.

Other proposed explanations of the nephrotoxicity of cisplatin include the possibility that it could generate reactive metabolites that bind covalently to tissue macromolecules. The nephrotoxic effects might also be due to sulphydryl binding of heavy metals. A fall in sulphydryl groups in the rat renal cortex has been demonstrated; this occurred before any significant change in renal function could be detected, suggesting that this biochemical change may be a primary event. Cell fractionations have shown that the greatest decline of sulphydryl groups occurs in the mitochondrial and cytosol fractions; these also had the highest concentrations of platinum.

Because of the substantial antitumour effects of cisplatin, much effort has been directed toward preventing its nephrotoxicity (see Fillastre *et al.* 1987 for more details). The rate of cisplatin infusion has been manipulated in an attempt to minimize the renal toxicity; nephrotoxicity occurred significantly less often when the infusion rate did not exceed 1 mg/kg body weight. Fractionation of the total dose over 5 days, or weekly administration of 40 mg/m^2, did not reduce the incidence of nephrotoxicity. Many investigators have now confirmed the important role of hydration in reducing the nephrotoxicity of cisplatin. A large hydration (150–200 ml/h) during and for 6 h after cisplatin is obligatory. One proposal made recently was to give verapamil, a calcium entry blocker, or captopril to cisplatin-treated patients in order to reduce nephrotoxicity. Cisplatin can increase the generation of prostaglandin E$_2$; it can also stimulate angiotensin II through stimulation of prostaglandin synthesis or through proximal tubular damage with a decreased reabsorption of sodium (Nittscke *et al.* 1981). Captopril did reduce the initial decrease in renal plasma flow, but the subsequent decrease in GFR could not be prevented. Calcium entry blockers are known to induce vasodilation of the renal vascular bed by inhibiting local activity of angiotensin II. Verapamil did prevent the initial changes in renal function during the first cisplatin administration but failed to reduce the subsequent fall in GFR (Offerman *et al.* 1984).

Nitrosoureas
Streptozotocin is the most nephrotoxic drug in the group of nitrosoureas, and has a marked effect. It is excreted rapidly in the urine: within 4 h, 72 per cent of the drug is found therein. The liver, intestine, and kidney have consistently greater concentrations of streptozotocin than that in plasma (Bhuyan *et al.* 1974). Renal toxicity is the limiting factor for the use of streptozotocin: in many cases, it is mild and reversible, with proteinuria appearing in 2 to 3 weeks, sometimes associated with glycosuria, hyperphosphaturia, or renal tubular acidosis. In 20 to 30 per cent of treated patients, a decrease of GFR has appeared (Sadoff 1970; Stolinsky *et al.* 1972; Schein *et al.* 1974); the histological findings were tubular atrophy and interstitial fibrosis.

BCNU, CCNU, and methyl CCNU are the nitrosoureas most commonly used in chemotherapy. These are rapidly metabolized; antitumour and toxic effects seem to be due to metabolites, not all of which have been identified.

In none of the first clinical trials were signs of nephrotoxicity noted; the first observations of this were reported in 1978. There are two reviews of the nephrotoxicity of these drugs (Weiss *et al.* 1983; Cluzel *et al.* 1985). We found 11 other case reports, and thus 51 observations are now known (Fillastre *et al.* 1988). The most frequent and usually only manifestation of nephrotoxicity was a slow rise in the serum creatinine. Most patients had only sporadic urinalysis and it is not possible to say whether there was proteinuria or haematuria. Small kidneys were found clinically in many patients and in all cases at autopsy. The first indications of renal abnormalities appeared at a median of 24 months after the start of treatment. In more than half of the cases, the renal dysfunction appeared from 2 to 49 months after discontinuing the treatment. The clinical course was variable and reversal of the renal dysfunction was possible; nevertheless for many patients, renal insufficiency was progressive even if treatment with nitrosoureas was discontinued. Twenty-four patients out of 51 required haemodialysis. The histological appearances of the kidney were of tubular and interstitial lesions; interstitial fibrosis, glomerulosclerosis and endarteritis were also seen, especially when the examination was made late in the course of the condition. It is of significance that much cellular atypia was found. Thus, for these nitrosoureas, nephrotoxicity is chronic and dose-dependent (Harmon *et al.* 1979). The incidence of nephrotoxicity was very low or even absent in patients receiving cumulative doses lower than 1.200 mg/m^2. A specific mechanism of nephrotoxicity has not yet been found for any of the nitrosoureas, nor is it known if the different drugs are likely to cause renal injury by the same process. It is conceivable that alkylation and carbamoylation could have antitumour actions and also lead to the toxicity.

Methotrexate
This drug is widely used in cancer treatment. It is excreted into the urine both unchanged and as a metabolite, 7-hydroxy-methotrexate (Jacobs *et al.* 1976). Methotrexate may be nephrotoxic when doses higher than 1000 mg/m^2 are used; when given in low, conventional doses, nephrotoxicity is only an occasional problem. Pitman and Frei (1977) found that nearly two-thirds of a group of 33 patients receiving high doses of methotrexate sustained rises in their serum creatinine of more than half, and that most of these patients had this rise after just one dose. Similar observations were made by Jaffe and Traggis (1975).

Intraluminal precipitation of methotrexate and of 7-hydroxymethotrexate has been demonstrated. It is necessary, in controlling the nephrotoxicity of the drug, to correct high tubular concentrations of methotrexate and the possibly acidic urine by full hydration and alkalinization; these methods have been successful in reducing nephrotoxicity to very low levels. Thyss *et al.* (1986) demonstrated the interaction between methotrexate and ketoprofen; acute renal failure and medullary aplasia were the cause of death in two patients. Ketoprofen modifies the pharmacokinetics of methotrexate; the peak level and area under the curve were significantly increased when methotrexate and ketoprofen were combined.

Other antitumour drugs
Many other antitumour agents may cause nephrotoxicity; these include mithamycin, 5-azacytidine, cytosine arabinoside, ifosfamide, and cyclosphosphamide (for more details see the review by Fillastre *et al.* 1988).

Drug-induced interstitial nephritis

Drug-induced acute interstitial nephritis is recognized with increasing frequency, although its true frequency is unknown and probably underestimated (see Chapter 8.10) In a review of acute renal failure, drug-induced interstitial nephritis represented 4 per cent of all cases studied histologically (Richet *et al.* 1978). Linton *et al.* (1980) estimated the frequency at 8 per cent. More than 70 drugs have been implicated in interstitial nephritis, of which the most common are antibiotics and NSAIDs (see Chapter 8.10).

In brief, it is known that methicillin is the agent most likely to cause such reactions (Kleinknecht *et al.* 1982). At the beginning, macroscopic haematuria, fever, and a skin eruption appear. Eosinophilia is present in blood and urine. Renal failure occurs a few days later. Diuresis is often maintained, and only an increased blood urea or blood creatinine reflect the renal failure. The kidneys are often enlarged and, on pathological examination, interstitial oedema with an abundant infiltration of lymphocytes, plasmocytes, and eosinophils is seen. Epithelioid granulomas are sometimes present. The prognosis is usually favourable, but renal failure may persist in approximately 10 per cent of cases. A penicillin derivative and linear IgG and C3 deposits are sometimes observed along the tubular basement membrane.

Penicillin G, which is used widely, is less often responsible for such incidents. When they do occur, the clinical picture is identical to that described for methicillin. Similar observations have also been reported after treatment with ampicillin, carbenicillin, oxacillin, nafcillin, and amoxicillin (Kleinknecht *et al.* 1982).

Many cases of acute interstitial nephritis have also been reported after treatment with rifampicin (Kleinknecht and Adhemar 1977). The clinical picture was similar to that found with methicillin.

NSAIDs have been discussed above in the context of glomerulopathies; many other drugs that may be responsible for tubular interstitial nephritis are considered in Chapter 8.10.

Abuse of laxatives

Long-continued use of purgatives and laxatives, particularly in women, causes excessive loss of potassium. Until potassium deficiency is advanced, signs and symptoms are vague. The characteristic features of advanced potassium deficiency are thirst, polyuria, muscular weakness, irritability, and constipation. In these cases, the urine contains only minimal amounts of potassium. The plasma bicarbonate is usually raised in the absence of severe renal impairment. The ability to concentrate the urine is lost at an early stage, although the ability to dilute persists for a long time. Polyuria and nocturia are distinctive features of the renal failure associated with potassium deficiency. Then, progressively, the renal failure increases (Schwartz and Relman 1953).

The most characteristic histological change found in potassium deficiency in man is extensive vacuolation of the cells of the proximal tubule (Schwartz and Relman 1953). This change can be rapidly reversed if laxative abuse is stopped and potassium given. Nevertheless, if abuse is prolonged, interstitial nephritis appears and is irreversible. It has been suggested that potassium depletion may predispose to urinary infections or to chronic pyelonephritis. Many continue to doubt whether chronic renal disease is directly related to potassium depletion.

Drug-induced renal vasculitis

Drug-induced vasculitis is infrequently observed and is suggested only in the presence of extrarenal manifestations. Angiitis involving the small vessels and capable of inducing acute renal failure has mainly been described during treatment with penicillin and sulphonamides (see Chapter 4.5.1). Some reports have now also incriminated amphetamine abuse although it is not known if the damage to the kidney is caused directly by the drug or by some other aspect of parenteral drug abuse (Bennett *et al.* 1977). Renal involvement is characterized by proteinuria, microscopic haematuria, hypertension, and a more or less elevated blood creatinine. Extrarenal manifestations can be cutaneous, such as eruptions or focal necrosis, or abdominal pain, pancreatic involvement, and joint or pulmonary changes. Histological examination shows fibrinoid necrosis of the intima and media of the small and medium arteries with periarterial cellular infiltration. The vascular lumen is partly obstructed by fibrin deposits. Hypertension is very frequently found, and may be severe and associated with rapidly progressive renal failure. Administration of high doses of corticosteroids may be beneficial.

Drug-induced lithiasis

The incidence of calculi in the urinary tract is high and is greater in men than in women. It is estimated at 7 cases per 1000 men aged from 40 to 60 years; its incidence is three times less in women. Réveillaud and Daudon (1986), who analysed 4000 urinary stones, found 58 cases of drug-induced calculus; for further details, see their general review.

Sulphonamides

Sulphonamides are known to provoke urinary crystallization and calculus formation because of their poor solubility. Those in current use that may be responsible for renal stones are sulphaperine, sulphaguanidine, and sulphadiazine. The stones result from urinary oversaturation due to the poor solubility of the *N*-acetylated metabolites of these sulphonamides.

Allopurinol

Through its metabolic action, allopurinol induces xanthinuria, and a few cases of xanthine stones have been reported. Generally, in these cases, the medical recommendations for urinary alkalinization had not been followed, and the subjects were either children with the Lesch–Nyhan syndrome or patients suffering from haematological diseases during chemotherapy. Allopurinol is metabolized to oxypurinol, and excretion and accumulation of this poorly soluble metabolite may have led to oversaturation, crystallization, and stone formation. Cases of stones composed of xanthine and oxypurinol have also been reported; allopurinol increases urinary xanthine and hypoxanthine to 30 times its normal value. These two derivatives are poorly soluble, and such an increase can induce crystallization by urinary oversaturation.

Glafenine

Urinary stones have developed during treatment with glafenine; Réveillaud and Daudon (1986) identified 14 cases in their study of 4000 urinary stones. They are formed either of free glafenic acid or of hydroxylated derivatives; they are often of mixed composition (calcium oxalate and glafenine), and therefore usually radiopaque. These calculi occur after long-term intake of glafenine, for example two to three tablets per day over several years.

Carbonic anhydrase inhibitors

When used for long-term treatment, acetazolamide can lead to urinary lithiasis. Its development may be explained by the increased urinary pH that favours oversaturation of phosphates on the one hand, and, on the other a clearly decreased urinary citrate and magnesium, which are physiological inhibitors of urinary crystallization.

Triamterene

Triamterene can induce renal lithiasis when given for a long time, in general, for several years, at doses of 150 to 300 mg/day. When pure, such stones are radiotransparent. Triamterene and its metabolites can favour crystallization of monohydrated calcium oxalate. Spontaneous nucleation of triamterene and its metabolites can also occur in saturated urine. It is also possible that the level of sodium chloride in the urine may be a predisposing factor; the solubility of triamterene is decreased in the presence of high concentrations of urinary sodium chloride.

Quinolones

Cases of urinary stone have been reported with flumequine and oxolinic acid. Given the wide prescription of these agents, the incidence of such stones is very low, and no case has been reported with the new quinolones.

Piridoxilate

Piridoxilate is used to treat peripheral arterial and coronary insufficiency. It is a salt of glyoxylic acid and pyridoxine. Calcium oxalate stone can occur in these patients; in fact, oxaluria can be increased up to three times normal during treatment with piridoxilate. Réveillaud and Daudon (1986) identified 14 such stones in their study of 4000.

Frusemide

In general, frusemide does not induce urinary stone. However, in premature infants, it would appear that high doses of this diuretic may induce micronephrocalcinosis. This could be due to the hypercalciuria induced by the diuretic and perhaps to the immature renal function of the premature infant.

Silicates and aluminium or magnesium hydroxides

Silicates can induce radiopaque stones; the same is true for long-term treatment with aluminium or magnesium hydroxides, or with carbonates.

Phenazopyridine

Phenazopyridine is used as a urinary antibiotic and analgesic. Phenazopyridine-induced urinary stone appears to be rare (four reported cases).

Comments

Incidence

The incidence of acute renal failure due to drugs has increased during the last 20 years. The number of drugs that may be responsible has reached at least 200, and a quarter of all acute renal failures are from drug toxicity (Baldwin *et al*. 1968). In France, a prospective study made over 1 year in 58 nephrology units found 398 cases of acute renal failure from drugs (Kleinknecht *et al*. 1986). The major cause of these renal failures was antibiotics, with 136 observations (34.2 per cent), including 107 cases of aminoglycoside toxicity (78.7 per cent of the cases). Glafenine and its derivatives, antrafenine and floctafenine, were the second most frequent cause, with 79 cases (19.8 per cent). This drug is not available in the United Kingdom or the United States. NSAIDs were third, with 62 cases (15.6 per cent), and fourth were radiological contrast media with 50 observations (12.6 per cent).

It has often been stated that renal failure due to drugs is abrupt and that renal function returns to normal after the treatment is stopped. This notion should be reconsidered, with some tempering of this optimism. In Kleinknecht's study, 259 out of the 398 patients regained normal renal function, but 93 subsequently suffered from chronic renal failure and 50 died, which is a death rate of 13 per cent. Risk factors were advanced age, the initial severity of acute renal failure, and the previous presence of cardiac, liver, or kidney failure.

The reader is advised that, for the sake of clarity in this chapter, renal lesions due to drugs have been artificially organized into glomerular, tubular, interstitial, and vascular; it should be remembered, that some drugs can induce multiple renal lesions—NSAIDs are a good example of these. By inhibiting prostaglandin synthesis they favour renal haemodynamic disorders, have an antinatriuretic effect, favour dilution hyponatraemia, reduce renal excretion of potassium, and as well induce tubular necrosis, interstitial nephritis, nephrotic syndrome, vasculitis, and even papillary necrosis. The same can be said of cyclosporin, which can induce functional renal failure, oliguria, and chronic renal failure with tubular, interstitial, glomerular, and vascular lesions.

Mechanisms involved in renal lesions

Knowledge of the mechanisms involved in drug-induced renal failure should lead to treatments adapted to preventing such complications. Drug-induced renal lesions are caused by two major types of mechanisms, a direct toxic effect and an action that is essentially immunological.

Nephropathies due to toxic mechanisms

Drugs attack renal cells via many mechanisms and at multiple sites. The membrane is frequently involved, with the drug acting on the phospholipids. These phospholipids play an important role in ensuring membrane permeability and permitting normal function of membrane enzymes such as Na, K-ATPase. Toxins may directly affect membrane permeability, as has been demonstrated for amphotericin and for other polyene antibiotics (Clejan and Bittman 1985). Drugs can act by increasing the activity of membrane phospholipase and by inhibiting normal reconstruction of the membrane. The addition of mercuric chloride to cultures of fibroblasts leads to increased hydrolysis of phospholipids (Shier and Dubourdieu 1983). Cultures of LLCPK1 cells exposed to mercuric chloride become depleted of phosphatidylethanolamine and phosphatidylcholine, whereas their content of free fatty acids and lysophospholipids is increased. Phospholipid degradation products, lysophospholipids, and free fatty acids, have membrane detergent properties. Aminoglycosides displace membrane-bound Ca^{2+} and may influence the permeability and enzymic properties of the membrane (Laughry and McKeown 1984). Even in absence of major changes in membrane permeability, the failure of plasma membrane pumps would be enough to allow major, potentially injurious changes in the cation homeostasis of the cell (Weinberg 1988). The most important and well defined of these pumps are Na,K-ATPase and Ca-ATPase. The activity of each may be affected by limitations of ATP, compromise of functions of the enzyme protein, or changes in the phospholipid microenvironment surrounding the enzyme and required for optimal function and regulation (Weinberg 1988). Toxins may also, through direct or indirect effects, lead to remodelling of the surface of the renal tubular cell, thus changing the area available for transportation.

Numerous experiments have shown that during cellular insult, whether toxic or ischaemic, an early and common change is the accumulation of calcium in the cell (Farber 1981). An increase in intracellular calcium is found at the plasma membrane, in the mitochondria and endoplasmic reticulum, as well as in the cytoplasm. Changes in mitochondrial function are deleterious to the survival of the cell. An increase in intracellular calcium can modify the permeability of the internal membrane of the mitochondria and thus change the electrochemical gradient across it. This perturbation will decrease the oxidative/phosphorylative capacity of the mitochondria. Disordered permeability of this internal membrane leads to loss of enzymes and nucleotides.

Careful studies have correlated calcium concentrations in the kidney and mitochondria with functional and structural modifications of the tubular epithelial cell and the mitochondria (Weinberg and Humes 1983). Three hours after injection of 5 mg/kg of $HgCl_2$ in the rat, the proximal tubular cells were intact and there was no change in the content of calcium, sodium, potassium, magnesium, or water in the renal cortex. The potassium and magnesium content of the mitochondria was slightly decreased. On the other hand, between the sixth and the twelfth hour after the injection, cell necrosis appeared, the concentrations of sodium and calcium were greatly increased in the renal cortex, and the mitochondria no longer functioned normally and had a very high calcium concentration. In rats given gentamicin, the appearance of cellular necrosis and renal failure is well correlated with an increase in calcium in the renal cortex and mitochondria. The degradation products of membrane phospholipids lyse the internal membrane of the mitochondria and play a role in the accumulation of calcium in those organelles.

The intracellular metabolism of drugs, particularly under the effect of the mixed-function oxidase system of the endoplasmic reticulum, leads to the formation of reactive metabolites. These metabolites are toxic for the cell, as are free radicals. The superoxide ion is normally formed during oxidation reactions, but superoxide dismutase catalyses its conversion into hydrogen peroxide and oxygen, then to water and oxygen by the catalase effect. The superoxide ion, especially if it is formed within the membranes, forms hydroxyl radicals by the Haber–Weiss reaction (Humes and Weinberg 1983); the result is lipid peroxidation. Because the membranes of the mitochondria and the endoplasmic reticulum are rich in non-saturated fatty acids, lipid peroxidation leads to oxidative deterioration of polyunsaturated lipids, and the result is a dramatic modification of the membrane structure and function (McCord and Fridovich 1978). These changes are greater than might be expected because the toxic agent reduces the concentration of glutathione in the renal cortex, which is a protective system that can detoxify active metabolites. The deleterious action of reactive metabolites has been clearly demonstrated in several circumstances. The mixed-function oxidase system can activate acetaminophenone and lead to the production of an electrophile metabolite, which will then bind covalently to cellular macromolecules (McMurtry et al. 1978). Diacetylation of acetaminophenone to p-aminophenol then leads to the production of free radicals, which can in turn bind covalently to tissue macromolecules (Newton et al. 1982). These two reactions occur in the cortical part of the kidney, whereas activation of acetaminophenone to an aryl metabolite, through the effect of the synthetase–endoperoxide–prostaglandin systems, can occur in the renal medulla. Glutathione depletion greatly favours the toxic effect of acetaminophenone in the kidney (McMurtry et al. 1978). Lipid peroxidation could be one of the mechanisms of cephaloridine toxicity (Kuo et al. 1983). Reactive metabolites are also at the origin of the nephrotoxicity induced by dibromochloropropane, paraquat, and bromobenzine.

The toxic substance can also have its own effect on mitochondrial respiration without the intervention of reactive oxygen metabolites or free radicals: for example, gentamicin, cephaloridine, cephaloglycine, and cephalexin inhibit state 3 of mitochondrial respiration, and platinum, cadmium, copper, silver, gold, and arsenic can have the same effects (Weinberg 1988). The changes in the lysosomal membrane also participate in cellular toxicity. Membrane phospholipids can be modified, resulting in an increase in lysosomal permeability, as has been shown for aminoglycosides. The increase in permeability causes leakage of lysosomal enzymes, which can be highly toxic to the other cell components. Accumulation of drugs or toxic agents in the lysosomes through endocytosis can modify the activities of cathepsins, phospholipases, and sphingomyelinases, favouring thesaurismoses (phospholipidoses). This has been well shown for gentamicin.

Thus, multiple mechanisms of action are possible for a single nephrotoxic drug. These mechanisms are still only partly elucidated. The production of reactive metabolites or of free radicals, for example, is known to be deleterious to cell function, but the exact mechanism of this action is not known. Nonetheless, it is our increased knowledge of the mechanisms of direct toxicity that has already made it possible to propose some measures to prevent drug nephrotoxicity.

Nephropathies due to an immunological mechanism

Renal lesions induced by drugs can be also due to immunological mechanisms. Experiments have greatly contributed to our knowledge of the side-effects of drugs, and have shown that glomerular, interstitial, tubular, and vascular lesions can arise immunologically. Several experimental models of immunological, glomerular nephropathies have been devised; of these, glomerulonephritis induced by mercury salts is by far the most studied (Druet et al. 1986). Repetitive subcutaneous injections of mercuric chloride ($HgCl_2$) in the Brown Norway rat induce a glomerulopathy that evolves in two phases. From the 8th day after starting the injections, antibasement membrane antibodies appear along the various basement membranes, mainly glomerular, as shown by the linear fixation of rat anti-IgG fluorescent antiserum (Sapin et al. 1977). Two months, later the aspect has changed. From linear, the deposit has become granular (Druet et al. 1978). Granular deposits of IgG are also found in many organs and can be detected in the spleen as from the 16th day.

The pathogenesis of this disorder has now been determined. Antibasement membrane antibodies are present in the serum of the animals from the 11th day (Bellon et al. 1982); they have antilaminin and antiprocollagen IV activity. These circulating antibodies subsequently disappear, despite continuing injections. Circulating immune complexes are concomitantly detected, and the decrease in complement (activated by the classical pathway as shown by the decrease in C2 and C4) is no doubt related to their formation (Bellon et al. 1982).

The antigen that forms part of these immune complexes is still unknown. The disease is manifested by the appearance of proteinuria and a transient nephrotic syndrome (Druet et al. 1978). Intravascular coagulation can also appear, and is responsible for the death of some animals (Michaud et al. 1980). The importance of genetic control in this experimental model has been demonstrated. Lewis rats submitted to the same experimental conditions as the Brown Norway do not develop glomerular disease. However, studies of segregants bred from Brown Norway and Lewis rats (first and second generation hybrids and cross-bred animals) have shown that the sensitivity to development of the disease depends on one of three genes, one of which is linked to the major histocompatibility complex (Druet et al. 1977). Investigation of other rat lines has shown that four breeds having the haplotype 1 of the Lewis rat in the major histocompatibility complex, as well as those having haplotype u, were resistant. Other breeds with various other haplotypes develop glomerular nephropathy with granular deposits of IgG, without detectable antibasement membrane antibodies, but with antibodies directed against denatured DNA (Druet *et al.* 1982). These findings clearly show the importance of genetic control and of certain genes linked to the major histocompatibility complex in sensitivity to mercury-induced glomerulonephritis. They probably explain why glomerulonephritis is observed inconstantly when non-consanguineous animals are used.

Weening *et al.* (1978) showed that $HgCl_2$ induces glomerulonephritis with appearance of antinuclear antibodies in PVC/c rats. Roman Franco *et al.* (1978) produced in the rabbit a disease that is very similar to that induced in the Brown Norway rat. Makker and Aikawa (1979) induced a glomerulonephritis that had mesangial IgG deposits in the Wistar rat. These various experimental models have led to a better understanding of the mechanism of action of $HgCl_2$. The absence of mercury in the immune deposits (Druet *et al.* 1978), as well as the absence of anti-brush border antibodies (Druet *et al.* 1978; Druet *et al.* 1986), are arguments against the hypothesis that pictures mercury as a hapten binding to serum or tissue proteins. On the other hand, several lines of evidence strongly suggest that it could act directly on immunocompetent cells, provoking dysregulation of the immune system. In the Brown Norway rat, $HgCl_2$ acts as a polyclonal activator (Hirsch *et al.* 1982); *in vivo* it provokes hyperplasia of splenic lymph nodes, the appearance of diverse antibodies (antiglomerular basement membrane, anti-DNA, anti-sheep erythrocytes, and anti-trinitrophenyl), as well as a major polyclonal increase in non-specific IgE (Prouvost-Danon *et al.* 1981). When added *in vitro* to cultures of lymphocytes from the Brown Norway rat, $HgCl_2$ can provoke synthesis of various antibodies (Hirsch *et al.* 1982). A final important factor is that the anomalies induced by $HgCl_2$ are transient despite its continuing administration. This finding strongly suggests that the events of suppression occur after the phase of polyclonal activation. This should be compared with the observation that certain drug-induced glomerulonephritides in man, as we have seen, can disappear despite continuing treatment.

Other drugs also induce dysregulation of the immune system. Gleichmann (1981) has demonstrated fixation of diphenylhydantoin to certain murine lymphocytes, so modifying the recognition systems between syngeneic lymphocyte subpopulations, which then became capable of stimulating each other. Such a mechanism perhaps could explain the autoimmune manifestations observed after exposure to various toxic agents.

Tubulointerstitial nephropathies can also be of immunological origin and in man can complicate various autoimmune diseases (Mery and Kenouch, 1988; and Chapter 4.6.2). Several experimental protocols reproduce the renal tubulointerstitial lesions observed in man (Druet 1988; Neilson 1989). The antigen responsible may be a structural antigen of the kidney, a component of basement membrane, or exogenous (for example, a drug). With methicillin, immunofluorescence studies have shown the presence of the dimethoxyphenylpenicilloyl hapten and linear deposits of IgG along the tubular basement membranes in some cases (Baldwin *et al.* 1968). Antibodies to tubular basement membrane have been found in the blood of patients with interstitial nephropathy due to methicillin (Border *et al.* 1974); these react with tubular basement membranes of human and monkey kidney. Elution studies suggest that the methicillin hapten is linked to proteins of the kidney structure rather than to antibody–antigen complexes. It is possible that the hapten, dimethoxyphenylpenicilloyl, when eliminated by tubular secretion, combines with proteins making up the tubular basement membrane, giving rise to an immunogenic complex. This complex could lead to the formation of IgG or IgE antibodies, which, in the presence of complement, could be responsible for tubulointerstitial lesions.

Acute tubulointerstitial renal disease can be induced experimentally by immunizing guinea pigs with rabbit tubular basement membrane and complete Freund's adjuvant (the classic nephropathy described by Steblay). The pathogenic role of antibasement membrane antibodies has been confirmed by transfer experiments: the serum of a diseased animal can in fact induce the nephropathy when injected into a healthy animal. The response to tubular basement membrane is controlled by a gene localized in the major histocompatibility complex, which explains why certain lines of the guinea-pig are sensitive and others are resistant. The same phenomenon was mentioned above in the case of experimental glomerulonephritis induced by mercury injections. Van Zwieten *et al.* (1977) showed that guinea-pigs given gold salts produced antibodies to tubular basement membrane and suggested that gold salts may modify that membrane and render it antigenic.

In the rat and mouse, induction and especially transfer of anti-tubular basement membrane disease is much more difficult. This suggests that nephrotoxicity is not induced by antibody production, but depends upon a cell-mediated immune process. For example, in the mouse, the capacity to develop tubulointerstitial nephritis after immunization with rabbit tubular basement membranes is not linked to the capacity to produced antibodies, nor to the degree of IgG deposition along the tubular membranes, but is rather due to the aggression of cytotoxic T lymphocytes. There is no reliable experimental model that reproduces the findings in acute drug-induced tubulointerstitial nephropathy in humans. Often there is extensive cellular infiltration of the renal interstitium, sometimes with the development of granulomas. These infiltrates are composed of eosinophils, plasmocytes, and mononuclear cells. The phenotype of these mononuclear cells has been identified in several studies of interstitial nephropathy due to NSAIDs or to cimetidine. They are mostly formed of T cells, usually a majority of T4 and a minority of T8, although the reverse may be true in acute cases. It is possible that the T cells could be activated under the influence of the drug or one of its metabolites. Severe interstitial nephropathy can be induced in the Lewis rat by immunization with a homogenate of synergic kidney and adjuvant; the cells of the infiltrate are mononuclear (Sugisaki *et al.* 1980). In the guinea-pigs or rat immunized with an antigen, focal interstitial nephritis can be induced by intrarenal injection of the same antigen into the sensitized animals. Subsequently, interstitial infiltration of mononuclear cells and tubular destruction without immunoglobulin deposition occur. Similar events have also been found in numerous drug-induced, interstitial nephropathies. This model may be close to imitation of lesions occurring in response to the presence of drug fixation on renal structures.

Can drug-induced nephrotoxicity be prevented?

The occurrence of lesions of an immunological nature is unpredictable. A first manifestation requires definitive stopping of the drug; a second course usually leads to renewed renal involvement that can be more severe than the first. For renal lesions due to direct toxicity, predisposing factors have been demonstrated and preventive action is possible. The state of hydration is a major element influencing the occurrence of functional deterioration of the kidney due to nephrotoxic insult. The protective effect of sodium chloride has been shown in many experimental models of acute toxic renal failure. For cisplatin, glycerol, mercuric chloride, and uranyl nitrate, a high-salt diet is more protective than a normal diet, and a low-salt diet increases the renal functional defect. The mechanism of this preventive effect is not entirely clear. Increased urinary flow favours the excretion of cellular debris accumulated in the tubular lumen and may decrease the importance of the tubular obstruction factor. Sodium chloride may decrease the activity of the renin–angiotensin system, which sometimes plays an important role in the appearance of acute renal failure. It may also preserve renal blood flow. The role of such hydration is essential in preventing the appearance of cisplatin nephrotoxicity.

Diuretics have also been prescribed to prevent the appearance of acute renal failure. Mannitol was the first to be used. Cvitkovic et al. (1977) showed that the administration of mannitol prevents cisplatin nephrotoxicity in the dog. In fact, it prevents an increase in urea and blood creatinine levels, but does not modify the intensity of the microscopic lesions. The hypertonicity of mannitol leads to increased renal blood flow and decreased oedema of the tubular cell. It is not known whether the protective effect of mannitol is superior to that of the salt solution because few prospective studies have been made. Frusemide has also been widely used, with divergent results. It was not shown to be more beneficial than hydration alone in preventing toxic acute renal failure, although it increases urinary volume, increases renal blood flow, decreases tubular reabsorption of sodium, and thus diminishes energy consumption and could protect the tubular cell from the effects of lack of oxygen. As for other drugs with preventive action, frusemide may be capable of preventing renal functional defects without improving structural alterations. More recently, a protective effect of calcium channel inhibitors has been shown in both experimental and clinical situations. Verapamil can reduce the nephrotoxicity of gentamicin and cyclosporin, diltiazem that of glycerol, and nifedipine that of cisplatin (Deray et al. 1988). Inhibition of the renin–angiotensin system by enzyme conversion inhibitors or by antagonists of angiotensin II receptors also can prevent acute toxic renal failure.

References

Amor, B., Cherot, A., and Delbarre, F. (1979). Syndrome néphrotique lié à l'emploi de la pyrithioxine dans le traitement de la polyarthrite rhumathoïde. *Nouvelle Presse Médicale*, 8, 2023–4.

Amor, B., Mery, C., and De Gery, A. (1982). Tiopronin (*N*-[2-mercapto-propionyl]glycin) in rheumatoid arthritis. *Arthritis and Rheumatism*, 25, 698–703.

Appel, G.B., D'Agati, V., Bergman, M., and Pirani, C.I. (1983). Nephrotic syndrome and immune complex glomerulonephritis associated with chlorpropamide therapy. *American Journal of Medicine*, 74, 337–42.

Averbuch, A. (1984). Acute interstitial nephritis with the nephrotic syndrome following recombinant leucocyte interferon therapy for mycosis fungoidis. *New England Journal of Medicine*, 310, 32–54.

Ayes, R., Mitchell, C.S., and Waring, R.H. (1987). Sodium aurothiomalate toxicity and sulphoxidation capacity in rheumatoid arthritic patients. *British Journal of Rheumatology*, 26, 197.

Bacon, P.A., Tribe, C.R., Mackenzie, J.C., Verrier Jones, J., Cumming, R.H., and Amer, B. (1976). Penicillamine nephropathy in rheumatoid arthritis. *Quarterly Journal of Medicine*, 45, 661–84.

Bahner, V.B., Fritzer, W., and Obiditsch-Mayer L. (1969). Nephrotisches syndrom nach methyl mercaptoimidazol therapie. *Wien Z Intern Medische*, 50, 280–8.

Baldwin, D.S., Levine, B.B., McCluskey, R.T., and Gallo, G.R. (1968). Renal failure and interstitial nephritis due to penicillin and methicillin. *New England Journal of Medicine*, 279, 1245–9.

Banfi, G. (1983). Extracapillary glomerulonephritis with necrotizing vasculitis. *Nephron*, 33, 56–60.

Bardin, L., Dryll, A., and Debeyre, N. (1982). HLA system and side effects of gold and D-penicillamine treatment of gold of rheumatoid arthritis. *Annals of Rheumatic Diseases*, 41, 599–601.

Bar Khayim, Y., Teplitz, C., Garella, S., and Chazan, J.A. (1973). Trimethadione (Tridione®)-induced nephrotic syndrome. A report of a case with unique ultrastructural renal pathology. *American Journal of Medicine*, 54, 272–80.

Barnett, H.L., Simons, D.J., and Wells, R.C. (1948). Nephrotic syndrome occurring during Tridione therapy. *American Journal of Medicine*, 4, 760–4.

Barr, R.D., Rees, P.H., Cordy, P.E., Kungu, A., Woodger, B.A., and Cameron, H.M. (1972). Nephrotic syndrome in adult Africans in Nairobi. *British Medical Journal*, 2, 131–4.

Barza, M. (1978). The nephrotoxicity of cephalosporins. An overview. *Journal of Infectious Disease*, 137, 500–73.

Batsford, S.R., Rohrbach, R., Riede, U.N., Sandritter, W., and Kluthe, R. (1976). Effects of D-penicillamine administration to rats, induction of renal changes; preliminary communication. *Clinical Nephrology*, 6, 394–7.

Bellon, B., et al. (1982). Mercuric chloride-induced autoimmune disease in Brown Norway rats. Sequential search for anti-basement membrane antibodies and circulating immune complexes. *European Journal of Clinical Investigation*, 12, 127–33.

Belmont, A. (1967). Dapsone-induced nephrotic syndrome. *Journal of American Medical Association*, 200, 262–3.

Bendirdjian, J.P., Fillastre, J.P., and Foucher, B. (1982). Mitochondria modifications with the aminoglycosides. In *The aminoglycosides, microbiology, clinical use and toxicology*, (ed. A. Whelton and H. Neu). Dekker, New York.

Bennett, W.M., Plamp, C., and Porter, G.A. (1977). Drug-related syndromes in clinical nephrology. *Annals of Internal Medicine*, 87, 582–90.

Bennett, W.M., Plamp, C., Reger, K., McClung, M., and Porter, G.A. (1978). The concentrating defect in experimental gentamicin nephrotoxicity. *Clinical Research*, 26, 540.

Bergeron, M.G., Trottier, S., Lessard, C., Beauchamp, D., and Gagnon, P.M. (1982). Disturbed intrarenal distribution of gentamicin in experimental pyelonephritis due to *Escherichia coli*. *Journal of Infectious Disease*, 146, 436–9.

Bertani, T., et al. (1982). Adriamycin-induced nephrotic syndrome in rats. Sequence of pathologic events. *Laboratory Investigation*, 46, 16–23.

Bhuyan, B.K., Kuentzel, S.L., Gray, L.G., Fraser, T.J., Wallach, D., and Neil, G.L. (1974). Tissue distribution of streptozotocin (NSC 25998). *Cancer Chemotherapy Reports*, 58, 157–65.

Border, W.A., Lehman, D.H., Egan, J.D., Saas, H.J., Glode, J.E., and Wilson, C.B. (1974). Antitubular basement–membrane antibodies in methicillin-associated interstitial nephritis. *New England Journal of Medicine*, 291, 381–4.

Brezin, J.H., Katz, S.M., Schwartz, A.B., and Chinitz, J.L. (1979). Reversible renal failure and nephrotic syndrome associated with nonsteroidal anti-inflammatory drugs. *New England Journal of Medicine*, 301, 1271–3.

Buckley, J.E., Clark, V.L., Meyer, T.J., and Pearlman, N.W. (1984). Hypomagnesemia after cisplatin combination chemotherapy. *Archives of Internal Medicine*, 144, 2347–8.

Burger, H.R., Briner, J., and Spycher, M.A. (1979). Auffalige Haüfung membranoser glomerulonephritiden nach goldtherapie beichronischer polyarthritis. *Schweizerische Medizinische Wochenschrift*, 105, 31.

Chalmers, A., Thompson, D., Stein, H.E., Reid, G., and Patterson, A.C. (1982). Systemic lupus erythematosus during penicillamine therapy for rheumatoid arthritis. *Annals of Internal Medicine*, **97**, 659–63.

Clejan, S. and Bittman, R. (1985). Rates of amphotericin B and fillipin association with sterols. A study of changes in sterol structure and phospholipid composition of vesicles. *Journal of Biology and Chemistry*, **260**, 2884–91.

Cluzel, P., Philippe, P., Alphonse, J.C., and Marcheix, J.C. (1985). Néphrotoxicité des nitrosourées. Une nouvelle observation. *Semaine des Hôpitaux de Paris*, **61**, 1847–52.

Csapo, G. and Hodi, M. (1966). Carbutamide. *The Lancet*, **i**, 324.

Cvitkovic, E., Spaulding, J., and Bethune, V. (1977), Improvement of cis–dichlorodiammine platinum (NSC 119875) therapeutic index in an animal model. *Cancer*, **39**, 1357–61.

Davin, J.C. and Mahieu, P.R. (1985). Captopril–associated renal failure with endarteritis but not renal artery stenosis in transplant recipient. *Lancet*, **i**, 820.

Day, A.T., Golding, J.R., Lee, P.N., and Butterworth, D. (1974). Penicillamin in rheumatoid disease. A long term study. *British Medical Journal*, **1**, 180–3.

Dentino, M., Luft, F.C., Moo Nahm Yum, Williams, S.D., and Einhorn, L.H. (1978). Long term effect of cis-diamminedichloride platinum (CDDP) on renal function and structure in man. *Cancer*, **41**, 1247–81.

Deray, G., Baumelou, A., Cacoub, P., Beaufils, H., Anouar, M., and Jacobs, C. (1988). Néphropathies interstitielles d'origine médicamenteuse avec syndrome néphrotique. In *Seminaire d'Uronéphrologie*, (ed. C. Chatelain and C. Jacobs), pp. 90–8. Masson, Paris.

Donker, A.J., Brentjens, J.R., Pawlowski, I.B., Venuto, R.C., and Andres, G.A. (1984). Development of auto-antibodies and diffuse intravascular coagulation in D-penicillamin treated Brown Norway rats. *Clinical Immunology and Immunopathology*, **30**, 152–5.

Druet, Ph. (1988). Physiopathologie des néphropathies interstitielles expérimentales immunes. In *Seminaire uronéphrologie*, (ed. C. Chatelain and Cl. Jacobs), pp. 41–8. Masson, Paris.

Druet, E., Sapin, C., Gunther, E., Feingold, N., and Druet, Ph. (1977). Mercuric chloride-induced antiglomerular basement membrane antibodies in the rat. Genetic control. *European Journal of Immunology*, **7**, 348–51.

Druet, Ph., Druet, E., Podevin, F., and Sapin, C. (1978). Immune type glomerulonephritis induced by $HgCl_2$ in the Brown Norway rat. *Annales d'Immunologie*, (Paris), **129C**, 777–92.

Druet, E., Sapin, C., Fournie, G., Mandet, C., Gunther, E., and Druet, Ph. (1982). Genetic control of susceptibility to mercury-induced immune nephritis in various strains of rats. *Clinical Immunology and Immunopathology*, **25**, 203–12.

Druet, Ph., Baran, D., Pelletier, L., Hirsch, H., Druet, E. and Sapin, C. (1986). Experimental models of drug induced immunologically mediated glomerulonephritis. In *Drugs and kidney*, (ed. T. Bertani, G. Remuzzi and S. Garattini), Vol. 33, pp. 61–74. Raven Press, New York.

Edwards, G.S., Lane, M., and Smith, F.E. (1979). Long-term treatment with cisdichlorodiammineplatinum (II)–vinblastine–bleomycin: possible association with severe coronary artery disease. *Cancer Treatment Report*, **63**, 551.

Empire Rheumatism Council. (1961). Gold therapy in rheumatoid arthritis. *Annals of the Rheumatic diseases*, **20**, 315–34.

Eyanson, S., Greist, M.C., Brandt, K.D., and Skinner, B. (1979). Systemic lupus erythematosus. Association with psoralen–ultraviolet A treatment of psoriasis. *Archives of Dermatology*, **115**, 54–6.

Fajardo, L.F., Eltringham, J.R., Stewart, J.R., and Klauber, M.R. (1980). Adriamycin nephrotoxicity. *Laboratory Investigation*, **43**, 242–53.

Farber, J.L. (1981). The role of calcium in cell death. *Life Sciences*, **29**, 1289–93.

Fabre, J. (1980) La néphrotoxicité des aminoglycosides. I. Clinique, morphologie, pharmacocinétique tissulaire. *Néphrologie*, **1**, 37–43.

Ferris, T.F., Morgan, W.S., and Levitin, H. (1961). Nephrotic syndrome caused by probenecid. *New England Journal of Medicine*, **265**, 381–3.

Fillastre, J.P. and Godin, M. (1989). Complications rénales des médicaments. *Encyclopédie Médico-Chirurgicale* (Paris) (Rein, organes génito-urinaire). 18066, **5**, 22.

Fillastre, J.P., Morel Maroger, L., Mignon, F., and Mery, J.P. (1969). Les néphropathies médicamenteuses. In *Actualités néphrologiques de l'Hôpital Necker*, (ed. J. Hamburger, J. Crosnier and J. L. Funck-Brentano), pp. 155–193. Flammarion, Paris.

Fillastre, J.P., Morin, J.P., Bendirdjian, J.P., Viotte, G., and Godin, M. (1980), La néphrotoxicité des aminoglycosides. III. Modification des éléments subcellulaires. *Néphrologie*, **1**, 145–52.

Fillastre, J.P., Viotte, G., Morin, J.P., and Moulin, B. (1987). Néphrotoxicité des antitumoraux in *Actualités néphrologiques de l'Hôpital Necker*, (ed. J. Crosnier, J. L. Funck Brentano, J. F. Bach, and J. P. Grünfeld), pp. 161–200. Flammarion, Paris.

Fillastre, J.P., Druet, P., and Mery, J.P. (1988). Proteinuric nephropathies associated with drugs and substances of abuse. In *The nephrotic syndrome*, (ed. J.S. Cameron and R.J. Glassock), pp. 697–744. Dekker, New York.

Fillastre, J.P., Moulin, B., and Josse, S. (1989). Aetiology of nephrotoxic damage to the renal interstitium and tubuli. *Toxicology Letters*, **46**, 45–54.

Gerritsen, W.R., Peters, A., Henny, F.C., and Brouwers, J.R.B. (1987). Ciprofloxacin-induced nephrotoxicity. *Nephrology, Dialysis and Transplantation*, **2**, 382.

Giamarellou, H., Metzikoff, C.H., Papachristophorou, S., Petrikkos, G., Doudoulaki, P., and Daikos, G.K. (1982). Do cephalosporins enhance gentamicin nephrotoxicity. In *Nephrotoxicity-ototoxicity of drugs*, pp. 321–32. Université de Rouen, France.

Giroux, L., Smeeters, C., Boury, P., Faure, M.P., and Jean, G. (1984). Adriamycin and adriamycin DNA-nephrotoxicity in rats. *Laboratory*, **50**, 190–6.

Gleichmann, H. (1981). Studies on the mechanism of drug sensitization: T-cell dependent popliteal lymph node reaction to diphenylhydantoin. *Clinical Immunology and Immunopathology*, **18**, 203–11.

Gonzalez-Vitale, J.C., Hayes, D.M., Cvitkovic, E., and Sternberg, S.S. (1977). The renal pathology in clinical trials of cis-platinum (II) diammine-dichloride. *Cancer*, **39**, 1362–71.

Hall, C.L. (1988). Gold nephropathy, *Nephron*, **50**, 265–72.

Hansen, T.M. *et al.* (1978). Levamisole induced nephropathy. *The Lancet*, **ii**, 737.

Harkcom, T.M., Conn, D.L., and Holley, K.E. (1978). D-penicillamine and lupus erythematosus-like syndrome. *Annals of Internal Medicine*, **89**, 1012.

Harmon, W.E., Cohen, H.J., and Schneeberger, E.E. (1979). Chronic renal failure in children treated with methyl CCNU. *New England Journal of Medicine*, **300**, 1200–3.

Hartfall, S.J., Garland, H.G. and Goldie, W. (1937). Gold treatment of arthritis, *Lancet*, **2**, 838–42.

Helin, H., Korpela, M., Mustonen, J., and Pasternack, A. (1986). Mild mesangial glomerulopathy, a frequent finding in rheumatoid arthritis patients with haematuria or proteinuria. *Nephron*, **42**, 224–31.

Hill, H., Hill, A., and Davison, A.M. (1978). Can penicillamine be used again in rheumatoïd arthritis after it has induced proteinuria, *Annals of the Rheumatic Diseases*, **37**, 566–7.

Hirsch, F., Couderc, J., Sapin, C., Fournie, G., and Druet, P. (1982). Polyclonal effect of $HgCl_2$ in the rat, its possible role in an experimental immune disease. *European Journal of Immunology*, **12**, 620–5.

Hootkins, R., Fenves, A.Z., and Stephens, M.K. (1989). Acute renal failure secondary to oral ciprofloxacin therapy: a presentation of three cases and a review of the literature. *Clinical Nephrology*, **32**, 75–8.

Horden, L.D., Sellars, L., Morley, A.R., Wilkinson, R., Thompson, M., and Griffiths, I.D. (1984). Haematuria in rheumatoid arthritis: an association with mesangial glomerulonephritis. *Annals of the Rheumatic Diseases*, **43**, 440–7.

Hou, S.H., Bashinsky, D.A., Wish, J.B., Cohen, J.J., and Harrington, J.T. (1983). Hospital-acquired renal insufficiency. A prospective study. *American Journal of Medicine*, **74**, 243–51.

Humes, H.D. and Weinberg, J.M. (1983). Alterations in renal tubular cell metabolism in acute renal failure. *Mineral Electrolyte and Metabolism*, **9**, 290–305.

Husby, G., Tung, K.S.K., and Williams, R.L. (1981). Characterization of renal tissue lymphocytes in patients with interstitial nephritis. *American Journal of Medicine*, **70**, 31–8.

Huskisson, E.C. and Dudley Hart, F. (1972). Penicillamine in the treatment of rheumatoid arthritis. *Annals of the Rheumatic Diseases*, **31**, 402–4.

Huskisson, E.C., Jaffe, I.A., Scott, J., and Dieppe, P.A. (1980). 5-thiopyridoxine in rheumatoid arthritis: clinical and experimental studies. *Arthritis and Rheumatism*, **23**, 106–10.

Jacobs, S.A., Stoller, R.G., Chabner, B.A., and Johns, D.G. (1976). 7-hydroxymethotrexate as a urinary metabolite in human subjects and rhesus monkeys receiving high dose methotrexate. *Journal of Clinical Investigation*, **57**, 534–8.

Jacobs, Ch., Kalman, S.M., Tretton, M., and Weiner, M.W. (1980). Renal handling of *cis*-diamminedichloroplatinum (II). *Cancer Treatment Reports*, **64**, 1223–6.

Jaffe, N. and Traggis, D. (1975). Toxicity of high dose-methotrexate (NSC-140) and citrovorum factor (NSC-3590) in osteogenic sarcoma. *Cancer Chemotherapy Reports*, **6**, 61–6.

Jaffe, I.A., Treser, G., Suzuki, Y., and Ehrenreich, T. (1968). Nephropathy-induced by D-penicillamine. *Annals of Internal Medicine*, **69**, 549–56.

Jones, B.R. *et al.* (1980). Comparison of methods of evaluating nephrotoxicity of *cis*-platinum. *Clinical Pharmacology and Therapeutics*, **27**, 557–62.

Kaloyanides, G.J. and Pastoriza-Munoz, E. (1980). Aminoglycoside nephrotoxicity. *Kidney International*, **18**, 571–82.

Katz, W.A., Blodgett, R.C., and Pietrusko, R.G. (1984). Proteinuria in gold-treated rheumatoid arthritis *Annals of Internal Medicine*, **101**, 176–9.

Kean, W.F., Bellamy, N., and Brooks, P.M. (1983). Gold therapy in the elderly rheumatoid arthritis patient. *Arthritis and Rheumatism*, **26**, 705–11.

Kincaid-Smith, P. and Whitworth, J.A. (1983). Hydralazine-associated glomerulonephritis. *The Lancet*, **ii**, 348.

Kirby, J.D., Dieppe, P.A., Huskisson, E.C., and Smith, B. (1979). D-penicillamine and immune complex deposition. *Annals of the Rheumatic Diseases*, **38**, 344–6.

Kleinknecht, D. Adhemar, J.P. (1977). Les insuffisances rénales aiguës dues à la rifampicine. *Médecine et Maladies Infectieuses*, **7**, 117–21.

Kleinknecht, D. *et al.* (1982). Les néphrites interstitielles aiguës immuno-allergiques d'origine médicamenteuses. Aspects actuels. In *Actualités néphrologiques de l'Hôpital Necker*, (ed. J. Hamburger, J.Crosnier and J.L. Funck Brentano), pp. 111–45. Flammarion, Paris.

Kleinknecht, D., Landais, P., and Goldfarb, B. (1986). Les insuffisances rénales aiguës associées à des médicaments ou à des produits de contraste iodés. Résultats d'une enquête coopérative multicentrique de la société de néphrologie. *Néphrologie*, **7**, 41–6.

Kletzel, M. and Jaffe, N. (1981). Systemic hypertension: a complication of intra-arterial cis-diammine-dichloroplatinum (II) infusion. *Cancer*, **47**, 245–7.

Kuhn, J.A., Argy, W.R., Hakowski, T.A., Schreiner, G.E., and Schein, P.S. (1978). Nephrotoxicity of *cis*-diammine dichloroplatinum as measured by urinary beta glucuronidase. *Clinical Research*, **26**, 776.

Kuo, C.H., Maita, K., Sleight, S.D., and Hook, J.B. (1983). Lipid peroxidation: A possible mechanism of cephaloridine-induced nephrotoxicity. *Toxicology and Applied Pharmacology*, **67**, 78–92.

Laughry, E.G. and McKeown, J.W. (1984). Effects of luminal addition of gentamicin on fluid and phosphate transport in rabbit proximal convoluted tubules. *Kidney International*, **25**, 147.

Lee, R.E., Vernier, R.L., and Ulstrom, R.A. (1961). The nephrotic syndrome as a complication of perchlorate treatment of thyrotoxicosis. *New England Journal of Medicine*, **264**, 12–21.

Leonard, B.J., Eccleston, E., Jones, D., Todd, P., and Walpoles, A. (1971). Antileukemic and nephrotoxic properties of platinum compounds. *Nature*, **234**, 43–5.

Leroux-Robert, C., Benevent, J., and Benevent, D. (1982). Argyrie rénale découverte lors d'un syndrome néphrotique. *Néphrologie*, **3**, 101.

Lietman, P.S. and Smith, C.R. (1983). Aminoglycoside nephrotoxicity in humans. *Review of Infectious Diseases*, **5**, 284–93.

Linton, A.L., Clark, W.F., Driedger, A.A., Turnbull, D.I., and Lindsay, R.M. (1980). Acute interstitial nephritis due to drugs. *Annals of Internal Medicine*, **93**, 735–41.

Lippman, A.J., Helson, C., Helson, L., and Krakoff, I.H. (1973). Clinical trials of cisdiammine dichloroplatinum (NSC 119875). *Cancer Chemotherapy Reports*, **57**, 191–200.

Lipsky, P.E. and Ziff, M. (1977). Inhibition of antigen- and nitrogen-induced human lymphocyte proliferation by gold compounds. *Journal of Clinical Investigation*, **59**, 455–66.

Lipsky, P.E. and Ziff, M. (1978). The effect of D-penicillamine on mitogen-induced human lymphocyte proliferation: synergistic inhibition by D-penicillamine and copper salts. *Journal of Immunology*, **120**, 1006–13.

Lipsky, P.E. and Ziff, M. (1980). Inhibition of human helper T cell function *in vitro* by D-penicillamine and Cu SO$_4$. *Journal of Clinical Investigation*, **65**, 1069–76.

Litterst, C.I., Leroy, A.F., and Guarino, A.M. (1979). Disposition and distribution of platinum following parenteral administration of *cis*-dichlorodiammine platinum (II) to animals. *Cancer Treatment Reports*, **63**, 1485–92.

Luft, F.C., Yum, M.N., Walker, P.D., and Kleit, S.A. (1977). Gentamicin gradient patterns and morphologic changes in human kidneys. *Nephron*, **18**, 167–72.

Lyman, N.W., Michaelson, R., Viscuso, R.L., Winn, R., Mulgoaonkar, S., and Jacobs, M.G. (1983). Mitomycin-induced hemolytic–uremic syndrome. Successful treatment with corticosteroids and intense plasma exchange. *Archives of Internal Medicine*, **143**, 1617–18.

Makker, S.P. and Aikawa, M. (1979). Mesangial glomerulopathy with deposition of IgG, IgM and C$_3$ induced by mercuric chloride. A new model. *Laboratory Investigation*, **41**, 45–50.

McCord, J.M. and Fridovich, I. (1978). The biology and pathology of oxygen radicals. *Annals of Internal Medicine*, **82**, 122–7.

McMurtry, R.J., Snodgrass, W.R. and Mitchell, J.R. (1978). Renal necrosis, glutathione depletion and covalent binding after acetaminophen. *Toxicology and Applied Pharmacology*, **46**, 87–98.

Mery, J.P. and Kenouch, S. (1988). Les atteintes de l'interstitium rénal au cours des maladies systémiques. In *Séminaires d'Uro-néphrologie*, (ed. C. Chatelain and Cl. Jacobs), pp. 57–89. Masson, Paris.

Michaud, A., Sapin, C, Aiach, M., and Druet, Ph. (1980). Clotting abnormalities during the course of immune type glomerulonephritis induced by HgCl$_2$ in the Brown Norway rat. In *Abstract 14th Meeting European Society for Clinical Investigation*, p.25. Salzburg, Austria.

Moellering, R.C. and Swartz, M.N. (1976). The newer cephalosporins. *New England Journal of Medicine*, **294**, 24–8.

Mondorf, A.W. *et al.* (1978). Effect of aminoglycoside on proximal tubular membranes of the human kidney. *European Journal of Clinical Pharmacology*, **13**, 133–47.

Morin, J.P., Viotte, G., Vandewalle, A., Van Hoof, F., Tulkens, P., and Fillastre, J.P. (1980). Gentamicin induced nephrotoxicity: a cell biology approach. *Kidney International*, **18**, 583–90.

Neilson, E.G. (1989). Pathogenesis and therapy of interstitial nephritis. *Kidney International*, **35**, 1257–70.

Neugarten, J., Gallo, G.R., and Baldwin, R.S. (1983). Rifampicin-induced nephrotic syndrome and acute interstitial nephritis. *American Journal of Nephrology*, **3**, 38–42.

Newton, J.F., Kuo, C.H., Gemborys, M.W., Mudge, G.H., and Hook, J.B. (1982). Nephrotoxicity of p–aminophenol, a metabolite of acetaminophen, in the Fisher 344 rat. *Toxicology and Applied Pharmacology*, **65**, 336–44.

Nield, G.H., Gartner, H.V., and Bohle, A. (1979). Penicillamine-induced membranous glomerulonephritis. *Scandinavian Journal of Rheumatology*, **28** (suppl.), 79–90.

Nitschke, R., Starling, K., and Land, V. (1976). *Cis*-platinum in childhood malignancies. *Proceedings of the American Association for Cancer Research*, **17**, 310.

Offerman, J.J.G. *et al.* (1984). Acute effects of *cis*-diammine dichloroplatinum (CDDP) on renal function. *Cancer Chemotherapy and Pharmacology*, **12**, 36–8.

Pasero, G. *et al.* (1982) Controlled multicenter trial of tiopronin and D-penicillamine for rheumatoid arthritis. *Arthritis and Rheumatism*, **25**, 923–9.

Pigott, P.V. (June 1980). *Side effects induced by captopril multicenter results.* Communication at the captopril satellite symposium of the 8th European Congress of Cardiology, pp. 46, Paris.

Pitman, S.W. and Frei, E. (1977), Weekly methotrexate calcium leucovorin rescue: effect of alkalinization on nephrotoxicity: pharmacokinetics in the CNS; and use in CNS non-Hodgkin's lymphoma. *Cancer Treatment Reports*. **61**, 695–701.

Prouvost-Danon, A., Abadie, A., Sapin, C., Bazin, H., and Druet, P. (1981). Induction of IgE synthesis and potentiation of anti-albumin IgE antibody response by HgCl$_2$ in the rat. *Journal of Immunology*, **126**, 699–702.

Rennke, H.G., Roos, P.C., and Wall, S.G. (1980). Drug-induced interstitial nephritis with heavy glomerular proteinuria. *New England Journal of Medicine*, **302**, 691–2.

Réveillaud, R.J. and Daudon, M. (1986). Les lithiases urinaires. In *Séminaires d'uronéphrologie*, (ed. C. Jacobs and C. Chatelain), Vol 12, pp. 14–39. Masson, Paris.

Reynolds, L.R. and Bhathena, D. (1979). Nephrotic syndrome associated with methimazole therapy. *Archives of internal medicine*, **139**, 236–7.

Richet, G., Sraer, J., and Kourilsky, O. (1978). La ponction biopsie rénale dans l'insuffisance rénale aiguë. *Annales de Médecine Interne*, (Paris), **129**, 445–7.

Rippelmeyer, D.J. and Synhavsky, A. (1988). Ciprofloxacin and allergic interstitial nephritis. *Annals of Internal Medicine*, **109**, 170.

Roman-Franco, A.A., Turiello, M., Albini, B., Ossi, E., Milgrom, F., and Andres, G.A. (1978). antibasement membrane antibodies and antigen–antibody complexes in rabbits injected with mercuric chloride. *Clinical Immunology and Immunopathology*, **9**, 464–81.

Rossof, A.H., Slayton, R.E., and Perlia, C.P. (1972). Preliminary clinical experience with *cis*-diamminedichloroplatinum (II) NSC 119875 CACP. *Cancer*, **30**, 1451–62.

Sadoff, L. (1970). Nephrotoxicity of streptozotocin (NSC 85998). *Cancer Chemotherapy Reports*, **54**, 457–9.

Sapin, C. Druet, E., and Druet, P. (1977). Induction of anti-glomerular basement membrane antibodies in the Brown-Norway rat by mercuric chloride. *Clinical Experimental Immunology*, **28**, 173–9.

Schaum, E. Kriegel, W., and Eickenbush, N. (1973). Nephrotisches syndrom nach faviatan behanlung einer hyperthyreose. *Munchen Medizinische Woschenschrift*, **11**, 2243–4.

Schein, P. *et al.* (1974). Clinical antitumor activity and toxicity of streptozotocin (NSC 85988). *Cancer*, **34**, 993–1000.

Schilsky, R.L. and Anderson, T. (1979). Hypomagnesemia and renal magnesium wasting in patients receiving cisplatin. *Annals of Internal Medicine*, **90**, 929–31.

Schnall, C. and Wiener, J.S. (1958). Nephrosis occurring during tolbutamide administration *Journal of the American Medical Association*, **167**, 214–5.

Schwartz, W.B. and Relman, A.S. (1953). Metabolic and renal studies in chronic postassium depletion resulting from overuse of laxatives. *Journal of Clinical Investigation*, **32**, 258–67.

Seelig, H.P., Seelig, R., Fischer, A., Maurer, R., Safer, A., and Schnitzlein, W. (1977). Immun komplex-Glomerulonephritis nach peroraler Penicillamin-applikation bei Ratten. *Review of Experimental Medicine*, **170**, 35–55.

Segond, P., Dellas, J.A., Massias, P., and Delfraissy, J.F. (1979). Syndrome néphrotique au cours d'une polyarthrite rhumatoïde traitée par la pyrithioxine. *Revue du Rhumatisme et des Maladies Ostéo-articulaires*, **46**, 509–10.

Shier, W.T. and Dubourdieu, D.J. (1983). Stimulation of phospholipid hydrolysis and cell death by mercuric chloride: evidence for mercuric ion acting as a calcium-mimetic agent. *Biochemistry Biophysic Research Communications*, **110**, 758–65.

Silverberg, D.S., Kidd, E.G., Shnitka, K., and Ulan, R.A. (1970). Gold nephropathy: a clinical and pathologic study. *Arthritis and Rheumatism*, **13**, 812–25.

Solez, K., Beschorner, W.L., Hall-Craggs, M., and Whelton, A. (1983). The 'unique' lesion of fenoprofen nephropathy, an over simplification. *Kidney International*, **23**, 198.

Stachura, I., Jayakumar, S., and Bourke, E. (1983). T and B lymphocyte subsets in fenoprofen nephropathy. *American Journal of Medicine*, **75**, 9–16.

Stein, J., and Smythe, H.A. (1968). Nephrotic syndrome induced by penicillamine. *Canadian Medical Association Journal*, **98**, 505–7.

Sternlieb, I., Bennett, B., and Scheinberg, H.I. (1975). D-penicillamine induced goodpasture's syndrome in Wilson's disease. *Annals of Internal Medicine*, **82**, 673–6.

Stolinsky, D.C., Sadoff, L., Braun-Wald, J., and Bateman, J.R. (1972) Streptozotocin in the treatment of cancer: phase II study. *Cancer*, **30**, 61–7.

Sugisaki, T., Yoshida, T., McCluskey, R.T., Andres, G.M., and Klassen, J. (1980). Autoimmune cell mediated tubulointerstitial nephritis induced in Lewis rats by renal antigens. *Clinical Immunology and Immunopathology*, **15**, 33–42.

Swainson, C.P., Thompson, D., Short, A.K., and Winney, R.J. (1982). Plasma exchange in the successful treatment of drug-induced renal disease. *Nephron*, **30**, 244–9.

Tait, G.B. (1960). Nephropathy during phenindione therapy. *Lancet*, **ii**, 1198–9.

Thyss, A., Milano, G., Kubar, J., Namer, M., and Schneider, M. (1986). Clinical and pharmacokinetic evidence of a life-threatening interaction between methotrexate and ketoprofen. *The Lancet*, **i**, 256–8.

Tune, B.M., Fernholt, M., and Schwartz, A. (1974). Mechanism of cephaloridine transport in the kidney. *Journal of Pharmacology and Experimental Therapeutics*, **191**, 311–17.

Van Zwieten, M.J., Leber, P.D., Bhan, A.K., and McCluskey, R.T. (1977). Experimental cell mediated interstitial nephritis induced with exogenous antigens. *Journal of Immunology*, **118**, 589–96.

Vogelzang, N.J., Torkelson, J.L., and Kennedy, B.J. (1985). Hypomagnesemia, renal dysfunction, and Raynaud's phenomenon in patients treated with cisplatin, vinblastine, and bleomycin. *Cancer*, **56**, 2765–70.

Walshe, J.M. (1962). Penicillamine and nephrotoxicity. (Letter). *British Medical Journal*, **1**, 1009.

Walshe, J.M. (1968). Toxic reaction to penicillamine in patients with Wilson's disease. *Postgraduate Medical Journal*, **44**, 6–8.

Ward, J.R., Williams, H.J., Egger, M.J., Reading, J.C., and Boyce, E. (1983). Comparison of auranofin, gold sodium thiomalate, and placebo in the treatment of rheumatoid arthritis. A controlled clinical trial. *Arthritis and Rheumatism*, **26**, 1303–15.

Weening, J.J., Fleuren, G.J., and Hoedemaeker, P.J. (1978). Demonstration of antinuclear antibodies in mercuric chloride induced glomerulopathy in the rat. *Laboratory Investigation*, **39**, 405–11.

Weinberg, J.M. (1988). The cellular bases of nephrotoxicity. In *Diseases of the kidney*, (ed. R.W. Schrier, and C.W. Gottschalk), Vol 2, pp. 1137–96. Little Brown, Boston.

Weinberg, J.M. and Humes, H.D. (1983). In vivo origin of mitochondrial calcium overload during nephrotoxic renal failure. *Kidney International*, **23**, 209-17.

Weinberg, J.M., Hardin, P.G., and Humes, H.D. (1983). Alterations in renal cortex cation homeostasis during mercuric chloride and gentamicin nephrotoxicity. *Experimental Molecular Pathology*, **39**, 43.

Weiss, R.B., Posada, J.G. Jr., Kramer, R.A., and Boyd, M.R. (1983). Nephrotoxicity of semustine. *Cancer Treatment Reports*, **67**, 1105–12.

White, E.G., Smith, D.H., and Zaphiropoulos, G.C. (1984). Haematuria occurring during antirheumatoid therapy. *British Journal of Rheumatology*, **1**, 57–62.

Whittingham, S. and Mackay, I.R. (1970). Systemic lupus erythematosus by procaine amide. *Australasian Annals of Medicine*, **4**, 358–61.

Williams, P.D. and Bridge, H.G.T. (1958). Nephrotic syndrome after the application of mercury ointment. *The Lancet*, **ii**, 602.

Williams, P.D., Holohan, P.D., and Ross, C.R. (1981). Gentamicin nephrotoxicity. I. Acute biochemical correlates in rat. *Toxicology and Applied Pharmacolqogy*, **61**, 234–42.

Wilson, C.B. (1989). Study of the immunopathogenesis of tubulointerstitial nephritis using model systems. *Kidney International*, **35**, 938–53.

Wood, L.J., Goergen, S., Stockigt, J.R., Powell, L.W., and Dudley, F.J. (1985). Adverse effects of captopril in treatment of resistant ascites, a state of functional bilateral renal artery stenosis. *The Lancet*, **ii**, 1008–9.

Wooley, P.H., Griffin, J., Panayi, G.S., Batchelor, J.R., Welsch, K.I., and Gibson, T.J. (1980). HLA-DR antigens and toxic reaction to sodium aurothiomalate and D-penicillamine in patients with rheumatoid arthritis. *New England Journal of Medicine*, **303**, 300–2.

Yudis, M. and Meehan, J.J. (1976). Quinidine-induced lupus nephritis *Journal of the American Medical Association*, **235**, 2000.

Zech, P., Colon, S., Labeeuw, R., Blanc-Brunat, M., Richard, P., and Perol, M. (1973). Syndrome néphrotique avec dépôt d'argent dans les membranes basales glomérulaires au cours d'une argyrie. *Nouvelle Presse Médicale*, **2** 161–4.

2.2 Handling of drugs in kidney disease

D.J.S. CARMICHAEL

Introduction

Many drugs and their metabolites are excreted via the kidney by glomerular filtration, tubular secretion or in some cases both. Renal impairment thus has a significant effect on the clearance of these drugs, with important clinical consequences. These are most obvious in patients with overt renal failure; more subtle forms of renal dysfunction may also be important and are extremely common, most notably as an accompaniment of ageing (see Chapter 1.5). Patients on haemodialysis are frequently prescribed multiple drugs, up to eight per day in the study of Anderson et al. (1982), and patients both with acute and chronic renal failure are often on complex medical therapy. Although in theory changes in dosage and dose interval of all drugs that are affected by renal impairment need to be considered, in practice dose adjustment is important for relatively few specific drugs with a narrow therapeutic index or adverse effects related to drug or metabolite accumulation. The corollary is that drugs are often prescribed in higher doses than those necessary to produce therapeutic plasma concentrations. This is less likely to occur, even in the presence of abnormal renal function, when the starting dose of a drug is low and the dose increased to achieve the desired effect (e.g. antihypertensive agents).

Although renal impairment has its most important effect upon excretion, other aspects of pharmacokinetics—absorption, distribution (including protein binding), metabolism, and renal haemodynamics may be affected, as may pharmacodynamics.

The major determinant of alteration in dosage is the change in drug clearance. This can be estimated by measurement of glomerular filtration rate (GFR) (Dettli 1983) although this does not always change in parallel with tubular secretion of drugs (Reidenberg 1985) and alterations in pharmacokinetics due to extrarenal factors (Gibaldi 1977). There are many handbooks (Brater 1985; Bennett et al. 1987; British National Formulary 1990) which provide guidelines for the adjustment of dosage in renal impairment. Many of these data are derived from measurement or estimation of changes in clearance, half-life ($t_{\frac{1}{2}}$) and volume of distribution (V_d). The determination of these pharmacokinetic parameters is very model-dependent and their application has limitations (Chennevasin and Brater 1981; Michael et al. 1985); consequently the guidelines should only be regarded as useful approximations. They cannot be expected to overcome the major errors made when prescribing in renal disease:

1. ignorance of renal impairment before a drug is prescribed
2. ignorance of how a drug is cleared from the body
3. failure to monitor therapeutic and adverse effects

This chapter considers the basic pharmacokinetic changes which occur in the presence of impaired renal function and how they affect drug usage. Pharmacokinetic models tend to be cumbrous, and the more elaborate ones can be criticized in that their multiple variables are arbitrary and permit a degree of precision of curve fitting that is unrealistic. We have used a one-compartment model, which despite its evident oversimplifications provides a useful framework for clinical decisions. The effects of haemodialysis, haemofiltration, and peritoneal dialysis are considered. Common therapeutic problems in patients with impaired renal function are illustrated, with additional reference to the elderly and patients with the nephrotic syndrome.

Pharmacokinetics

Gastrointestinal absorption

Ammonia production in the stomach occurs in chronic renal failure concomitant with urea accumulation and hydrolysis. Ammonia buffers hydrochloric acid, causing a rise in gastric pH. Consequently there may be reduced absorption of drugs such as ferrous sulphate, chlorpropamide, folic acid, pindolol, and cloxacillin (Bennett 1980) whose absorption is greater at acid pH. Aluminium hydroxide, which is used as a phosphate binding agent, may bind several drugs including iron, aspirin, and ciprofloxacin and therefore should not be administered simultaneously with these agents. Most drugs are absorbed in the proximal small intestine; D-xylose absorption is reduced by renal failure (Craig et al. 1983), but malabsorption of drugs does not appear to be a major problem in uraemia.

Diuretic resistance is reported to occur in the nephrotic syndrome, and this has been attributed to poor absorption of loop diuretics from the oedematous intestine (Huang et al. 1974; Odlind and Beermann 1980). Intravenous administration gives rise to a high peak concentration and can overcome this 'resistance' (Editorial 1979). Factors other than malabsorption that may reduce the efficacy of diuretics in nephrotic states include protein binding within the tubular lumen (Bowman 1975), coexisting renal impairment (Keller et al. 1982) and an increase in volume of distribution (Tilstone and Fine 1978). The efficacy of a combination of an oral loop diuretic with a thiazide in this circumstance argues against an explanation for diuretic resistance based on intestinal malabsorption.

Peritoneal absorption

The peritoneum is used as an absorptive surface during peritoneal dialysis in patients with peritonitis or for the administration of insulin to uraemic diabetics. The transfer of antibiotics is bidirectional for some antibiotics, although it is primarily from peritoneum to the circulation (Albin et al. 1985; Bonati et al. 1987); gentamicin absorption however, is unidirectional from peritoneum to plasma (Somani et al. 1982). Balducci et al. (1981) observed that insulin was better absorbed from an empty peritoneum, and concluded that dialysate reduced absorption by lowering insulin concentration in the peritoneal fluid. During peritonitis, insulin requirements may fall as absorption increases with mesothelial damage (Henderson et al. 1985).

Absorption from subcutaneous and intramuscular depositions

In patients with acute renal failure who are critically ill with shock or hypotension drug absorption from subcutaneous and intramuscular injections may be retarded.

Distribution

Protein binding is affected by renal impairment, which is accompanied by increased concentrations in plasma of a number of acidic compounds that compete for binding sites on albumin and other plasma proteins (Tillement et al. 1983; Reidenberg 1983). Some of these have been identified: indoxyl sulphate and 2-hydroxyhippuric acid (Bowmer and Lindup 1982); 3-carboxy-4-methyl-5-propyl-2-furanpropranoic acid (which inhibits phenytoin binding (Mahuchi and Nakahashi 1988)); 2-hydroxybenzoylglycine (Lichtenwalner and Suh 1983) and free fatty acids (Bowmer and Lindup 1982). α_1-Acid glycoprotein binds basic drugs such as propranolol, oxprenolol and disopyramide (Belpaire et al. 1988; Pedersen et al. 1987). The binding of a variety of drugs has been evaluated in renal impairment in vivo and in vitro (Table 1) (Biaski 1980; Gulyassy and Depner 1983; Reidenberg 1983; Tillement et al. 1983; Webb et al. 1986; Kapstein et al. 1987; Vanholder et al 1988).

Serum albumin concentration is low in patients with the nephrotic syndrome and may also fall in cachectic patients and the elderly, reducing the number of drug binding sites (see Chapter 3.2). As a consequence the proportion of free to bound drug is increased, and there are greater fluctuations in the free drug concentration following administration of each dose. This could be responsible for an increased susceptibility to adverse drug reactions (Lewis et al. 1971; Gugler and Azarnoff 1983). In the elderly, effects related to plasma albumin concentration have been noted during treatment with warfarin, phenytoin, sulphonylureas, and salicylates (Cook 1979; Bliss 1981), but these appear to be of minor therapeutic importance. Indeed, few of the changes in pharmacokinetics that are caused by altered protein binding are of clinical significance (Reidenberg 1983). Phenytoin therapy is an exception and illustrates one potential pitfall of drug monitoring by measurement of plasma concentration. Routine measurement of plasma drug concentration includes both bound and unbound (free) drug. In renal impairment binding of phenytoin to plasma protein is reduced in direct proportion to the fall in GFR (Brater 1985). Since protein binding is reduced the proportion of free (active) drug increases for a fixed total plasma concentration. The therapeutic range estimated by total plasma concentration, therefore has to be adjusted to lower concentrations: if phenytoin dose is adjusted to give concentrations within the usual therapeutic range, toxicity will occur.

Digoxin tissue binding (and consequently its volume of distribution) falls in renal failure and a smaller loading dose is needed. The effect of reduced digoxin clearance as GFR falls is even more important, and a lower maintenance dose is required than in patients with normal renal function. Monitoring by plasma concentration measurement is often useful, especially if digoxin is being used for its inotropic action (which is relatively difficult to gauge clinically), in contrast to its effect on ventricular response in patients with atrial fibrillation. In the anephric patient elimination of digoxin by non-renal mechanisms usually permits a dose of approximately 0.125 mg on alternate days.

Volume of distribution may alter in renal failure because of fluid retention and expansion of the circulating blood volume, alteration in protein and tissue binding and alterations in the

Table 1 Alteration in drug binding in renal impairment

Reduced	Unaltered	Increased
Theophylline	Indomethacin	Imipramine
Phenytoin	Metoclopramide	
Methotrexate	Trimethoprim	
Diazepam	D-Tubocurarine	
Prazosin	Quinidine	
Frusemide	Dapsone	
Dicloxacillin		
Warfarin		
Barbiturates		
Clofibrate		
Salicylates		
Morphine		

2.2 Handling of drugs in kidney disease

Table 2 Parent drugs, metabolites, and possible adverse effects

Drug	Metabolite	Effect of metabolite
Allopurinol	Oxypurinol	? Cause of rashes,
Clofibrate	Chlorophenoxyisobutyric acid	Muscle damage, Neuropathy
Nitroprusside	Thiocyanate	Toxic symptoms
Primodone	Phenobarbitone	Active drug
Procainamide	N-Acetyl procainamide	Antiarrhythmic
Sulphonamides	Acetylsulphonamides	Rashes
Pethidine	Norpethidine	Causes seizures
Morphine	Morphine-6-glucuronide	Prolongs analgesia and respiratory depression
Codeine	Morphine	
Propoxyphene	Norpropoxyphene	Cardiotoxic
Acebutalol	n-Acetyl analogue	Confers selectivity
Nitrofurantoin	Metabolite	Peripheral neuropathy

proportion of fat and muscle in the body. In the elderly body fat increases as a proportion of total body weight by approximately 25 per cent in men and 40 per cent in women and this leads to an increase in volume of distribution for water-soluble drugs.

Metabolism

The majority of drugs are excreted by the kidney either as the original compound or after metabolism in the liver to more polar (water-soluble) substances. Uraemia may affect drug metabolism (Gibson 1986) and reduces non-renal clearance of drugs such as acyclovir, aztreonam, moxolactam, cefotaxime, captopril, cimetidine, and metoclopramide. This alteration in clearance is minor in comparison to the retention in renal impairment of metabolites which have therapeutic or adverse effects (Table 2) (Palmer and Lasseter 1975; Verbeeck et al. 1981; Drayer 1983). Measurements of plasma drug concentrations, for therapeutic monitoring, must be interpreted cautiously. Not only is there the possibility of the free fraction of the drug (as discussed above in the case of phenytoin), but assays of imperfect specificity may detect accumulating metabolites (active or inactive) confounding interpretation.

Renal failure may reduce drug metabolism: for instance the conversion of sulindac to its active sulphide metabolite is reduced in uraemia (Gibson et al. 1987). This has been invoked as a partial explanation for the lower incidence of adverse effects as compared to indomethacin (Berg and Talseth 1985).

The kidney synthesizes $1,25\text{-}(OH)_2$-vitamin D_3 from its precursor $25\text{-}(OH)D_3$. In chronic renal impairment this metabolism is reduced (Coburn and Slatopolsky 1986). Metabolism of $24,25\text{-}(OH)_2D_3$ is also affected. The synthetic analogue $1\alpha\text{-}(OH)D_3$ is converted to $1,25\text{-}(OH)_2D_3$ in vivo and is used therapeutically.

Presystemic ('first pass') metabolism by the liver of some drugs such as propranolol and cimetidine may be reduced in renal impairment causing increases in plasma concentrations. Terao and Shen (1985) studied an in-vitro model of acute renal failure in rats and confirmed that an unidentified property of uraemic blood reduced the liver extraction of 1-propranolol. However, this is probably of little importance for the majority of drugs.

Renal excretion

Renal excretion of drugs depends upon

1. filtration
2. active tubular secretion and reabsorption
3. passive diffusion

Renal clearance of drugs can be expressed as a function of GFR, the fraction of the drug that is unbound or free in plasma, secretion, and reabsorption (Gibaldi 1984).

$$Cl_R = (f_u \times GFR) + \text{secretion} - \text{reabsorption}$$

where Cl_R is renal clearance and f_u is fraction of unbound drug (available for filtration).

Therefore if Cl_R is lower than $GFR \times f_u$, reabsorption (usually passive) must be taking place and if it is greater than $GFR \times f_u$ then secretion must be taking place.

Compounds with a molecular weight below 60 000 Da are filtered to a variable extent depending on molecular size through the glomerulus unless they are protein bound when only the unbound portion is filtered. Non-polar (lipid-soluble) drugs diffuse readily across tubular cells whereas polar (water-soluble) compounds do not. Since less than 1 per cent of the volume filtered is usually excreted as urine, drugs in tubular fluid become concentrated relative to plasma as water is reabsorbed. Polar drugs generally remain in the tubular fluid and are excreted in the urine, while non-polar drugs are reabsorbed by passive diffusion down their concentration gradient into plasma (Reidenberg 1985). Some polar drugs are eliminated in the urine as a result of active or facilitated transport mechanisms that transport organic acids or bases. Many drugs are metabolized, primarily in the liver, to produce more polar compounds which cannot be passively reabsorbed and so are eliminated in the urine.

Examples of drugs that are actively secreted into the tubule are given in Table 3. In addition, some drugs interact to inhibit tubular secretion of others (e.g. probenicid with penicillin, with cephalosporins, and with frusemide). Elimination of organic acids (AH) or bases (B) is affected by the H^+ ion concentration

Table 3 Examples of drugs with active tubular secretion

Organic acids	Organic bases
Penicillins	Amiloride
Cephalosporins	Procainamide
Sulphonamides	Quinidine
Frusemide	
Thiazides	
Salicylates	
Probenecid	

of the tubular fluid with any change of urinary pH that favours ionization leading to more drug excretion:

$$H^+ + B \overset{pK_B}{=} BH^+$$
$$AH = H^+ + A^-$$
$$pK_A$$

The amount of ionized drug at any particular pH is determined by its pK. The pK is the pH at which 50 per cent of the drug is ionized. If an organic acid has a pK_A below 7.5, making the urine alkaline (i.e. increasing its pH) increases the amount of ionized drug (A^-) and therefore its excretion. The converse is true for organic bases with a pK_B above 7.5, which are eliminated as the charged (BH^+) form favoured by acid pH. The excretions of salicylates (weak acids) and amphetamines (weak bases) exemplify these principles. Co-trimoxazole contains trimethoprim, which has increased elimination in acid urine, and sulphamethoxazole, which has increased elimination in alkaline urine (Craig and Kunin 1973). Marked variations in urinary pH could therefore, in principle, disturb the usual ratio of trimethoprim to sulphamethoxazole after administration of this drug combination. Synergy between these components as regards their antimicrobial effect depends on their presence in appropriate proportions, and may therefore be lost if urinary pH is altered. This is unlikely to be important in situations where trimethoprim alone is effective (urinary tract infection) but could be important in the treatment of *Pneumocystis carinii* infection. This remains to be investigated. Drugs present in tubular fluid may affect the excretion of other compounds; aspirin and paracetamol reduce methotrexate excretion (Reidenberg 1985). A low protein diet reduces the acidity of the urine and may thereby lead to increased reabsorption of oxpurinol, a metabolite of allopurinol which is thought to be responsible for some of the adverse side-effects of allopurinol (Berlinger *et al.* 1985; Kitt *et al.* 1989).

Although it is a simplification to disregard the tubular handling of drugs in renal impairment both filtration and secretion of drugs appear to fall in parallel and in proportion to GFR (Dettli 1983; Gibaldi 1984). It is paramount to understand that, irrespective of the assumptions made about the handling of drugs by the kidney, the most important aspect of prescribing in renal disease is awareness of the existence of renal impairment and of changes in renal function. These may occur either as a consequence of progression of the underlying renal pathology, general clinical condition, or effects of drug therapy.

Measurement of glomerular filtration

Some measure of GFR is needed (see Chapter 1.3). Isotope or inulin clearance measurements are for the most part impractical. In practice measurement of endogenous creatinine clearance or estimation of creatinine clearance from a single plasma or serum creatinine measurement, with adjustment for body weight, age and sex, is usually sufficient, provided their limitations are noted. Both the former (Gabriel 1986) and the latter method (Dodge *et al.* 1987; Cockcroft and Gault 1976) have their advocates. Timed urine collections are not always accurate even in hospital and very often patients have already been started on drug therapy before the result is known. Estimated creatinine clearance may be inaccurate in very obese patients (Salazar and Corcoran 1988) or in the very ill, in whom lean body weight may be lower either because of oedema or reduced muscle mass. Brater (1985) provides formulae to correct for both lean body mass and body surface area calculated from body weight and height. The main limitation of plasma creatinine as an estimate of GFR is in the presence of changing renal function (e.g. acute renal failure) since there is a considerable lag between the change in GFR and the consequent rise in plasma creatinine concentration. An increased protein intake may lead to overestimation of endogenous creatinine production and GFR (Laville *et al.* 1989). Similarly tubular secretion of creatinine, which increases as GFR falls, may lead to an overestimation of GFR (Shemesh *et al.* 1985). In the elderly, in particular, some measurement or estimate of GFR is important, as creatinine production is reduced and therefore the serum creatinine may seriously underestimate the degree of renal impairment (see Chapter 1.7 for a detailed discussion). Both GFR and effective renal plasma flow decline with age (Watkin *et al.* 1950; Rowe *et al.* 1976; O'Malley and Meagher 1985). Despite the potential inaccuracy of the measurement of GFR, it is essential that some estimate is made before drugs are prescribed to patients with renal impairment.

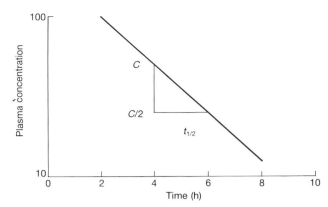

Fig. 1 Exponential decline in drug concentration.

Drug kinetics

Most drugs that are eliminated by the kidney display first-order kinetics (Dettli 1983). This means that the rate of removal is proportional to the concentration of the drug. The elimination rate constant k_e is the proportion of total amount of drug removed per unit time. This can be expressed by the equation

$$\frac{dC}{dt} = -k_e C$$

Integration for $t = 0$ to $t = t$ yields

$$C_t = C_0 e^{-k_e t}$$

where k_e is the elimination constant, C_0 is the drug concentration at t_0, and C_t is the drug concentration at t_t.

Since this is a simple exponential decline in concentration it gives rise to a straight line when the concentration is plotted on a semilogarithmic scale against time (Fig. 1).

The half-life ($t_{1/2}$) of a drug is the time for its plasma concentration to fall by half after absorption and distribution are complete. It is useful in determining dosage interval, drug accumulation (both extent of accumulation and time taken to reach steady state), and persistence of drug after dosing is stopped. It is inversely related to k_e:

$$t_{1/2} = 0.693/k_e \quad (0.693 = \ln 2)$$

The clearance of a drug depends upon $t_{1/2}$ (k_e) and volume of distribution (V_d). Volume of distribution does not usually correspond to a real volume, although for a drug confined exclusively within the plasma it would approximate the plasma volume. In practice, drugs distribute through a larger space than this and volume of distribution is a constant with units of volume that relate the amount of drug in the body (units of mass) to the measured plasma concentration (units mass/ volume):

$$V_d = \frac{m}{C_p}$$

where m is the mass of the drug in the body and C_p is plasma concentration.

Volume of distribution thus represents an apparent volume which the drug would have distributed in to produce the measured plasma concentration.

Clearance of a drug is related to k_e and V_d:

$$\text{Clearance} = k_e \times V_d = \frac{0.693 V_d}{t_{1/2}}$$

Clearance can be used to estimate the steady state plasma concentration (C_{ss}) that can be anticipated in response to any particular dosage regimen. At steady state the rate that a drug enters the plasma is equal to the rate at which it leaves.

$$\text{Rate in} = \frac{F \times \text{dose}}{\text{dose interval}}$$

where F is fraction reaching the circulation (assumed to be 1.0 (100 per cent) for IV administration)

$$\text{Rate out} = C_{ss} \times Cl$$

where C_{ss} is plasma concentration at steady state. By substituting the above equation we obtain

$$C_{ss} = \frac{F \times \text{dose} \times t_{1/2}}{\text{dose interval} \times 0.693 \times V_d}$$

From these relationships it can be seen that C_{ss} will increase with a longer $t_{\frac{1}{2}}$ and a smaller volume of distribution. C_{ss} can be reduced either by lowering the dose or by increasing the dose interval.

The relationship between the elimination of a drug in the presence of renal impairment and normal renal function can be estimated. This estimate can be used to gauge the appropriate reduction in dose or increase in dose interval compared to a standard regimen.

$$Q_r = \frac{k_e}{k_n}$$

where Q_r is the elimination rate fraction, k_e is actual elimination rate constant, and k_n is elimination rate constant if the GFR was 100 ml/min.

The difference between k_e and k_n is only clinically important in renal impairment when non-renal elimination accounts for less than 50 per cent of total elimination, provided that the non-renal clearance is not reduced markedly by other means.

The expected reduction in dose of a drug necessitated by renal impairment can be calculated in terms of Q_r:

$$D_r = D_n \times Q_r$$

where D_r is dose in renal failure and D_n is dose assuming normal renal function. Alternatively the dosage interval can be prolonged by the same proportion:

$$I_r = \frac{I_n}{Q_r}$$

where I_r is dose interval in renal failure and I_n is dose interval assuming normal renal function.

Either way, the dose/unit time is the same and is less than that in patients with normal renal function. In practice the decision whether to reduce dose or prolong dose interval is not, however, always equivalent. An example is the use of aminoglycoside antibiotics which must achieve a threshold peak concentration in order to kill bacteria effectively. In consequence small but frequent doses may fail to achieve efficacy whereas the same total dose delivered less frequently may achieve the desired therapeutic effect without leading to accumulation and toxicity. With respect to nephrotoxicity, this principle is illustrated by Verpooten et al. (1989) who demonstrated that a continuous infusion of gentamicin reaching the same plateau concentration as an intravenous bolus produced more cortical binding in nephrectomy specimens and by inference more tubular toxicity. The Q_r value has been estimated for many drugs and nomograms have been constructed as guides to correct dosages. These assume that other factors such as non-renal elimination and volume of distribution remain constant in the presence of reduced or changing renal function, limiting their precision. However they can be extremely useful at the start of therapy which can be modified subsequently on the basis of actual, i.e. individual, plasma concentrations.

Renal haemodynamics

Renal blood flow has little effect on the excretion of most drugs independent of that on the GFR. However, once GFR is reduced, renal clearance will be impaired. Dehydration or saline depletion induced by diuretic therapy, or as a consequence of tetracycline therapy in pre-existing renal failure, is important. Tetracycline is antianabolic and will produce a rise in blood urea which initiates a cycle of osmotic diuresis and a further rise in blood urea. All the tetracyclines can cause this, but it is less likely to occur with doxycycline and minocycline, which are not excreted by the kidney (Heaney and Eknoyan 1978) and therefore do not accumulate in renal impairment. In addition demeclocycline which is used as a physiological antagonist of arginine vasopressin may alter intrarenal haemodynamics by lowering prostaglandin biosynthesis (Oster et al. 1976, Carvillo 1977; Perez-Ayuso et al. 1984). Non-steroidal anti-inflammatory agents may also alter GFR by inhibiting prostaglandin synthesis

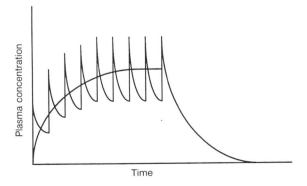

Fig. 2 Drug accumulation during repeated dosing and drug elimination are both related to half-life.

Table 4 Factors influencing clearance of drugs

Properties of the drug	Molecular weight
	Protein binding
	Volume of distribution
Delivery of drug to the filter	Blood flow to filter
	Blood flow within filter
	Volume of distribution
Properties of the filter	Pore size
	Surface area

(Schrier and Conger 1986). This may cause deterioration in renal function and in the presence of infection, particularly in acute pyelonephritis, acute renal failure may develop. Antibiotic treatment in these circumstances, often before renal function has been assessed, may produce additional difficulties. Angiotensin converting enzyme (ACE) inhibitors may lower GFR by lowering glomerular efferent arteriolar tone under circumstances, such as dehydration or renovascular hypertension, when GFR is dependent upon the balance between afferent/efferent arteriolar tone. In acute renal failure, in patients with hypovolaemia, changes in GFR may cause a rapid fall in drug clearance.

Dialysis and haemofiltration

The clearance of drugs by haemodialysis and haemofiltration follows first-order kinetics. The rate of removal can be calculated from the law of conservation of matter using the equation below:

$$Cl_d = \frac{Q \times (A - V)}{V}$$

where Cl_d is dialysis or haemofilter clearance, Q is blood flow, A is blood concentration entering the dialyser, and V is blood concentration leaving the dialyser.

Modification of this equation is needed for drug concentrations measured in plasma as blood flow represents whole blood. Since it is not always possible to obtain measurements of drug plasma concentrations in individual cases, an estimate of drug clearance can be obtained by use of the sieving coefficient (Bickley 1988). This coefficient is the proportion of the drug (or solute) which will cross the membrane and should be constant for a particular drug and membrane.

The measurement of the sieving coefficient has been simplified.

$$S = \frac{C_{uf}}{C_{arterial}}$$

where S is sieving coefficient, C_{uf} is concentration in filtrate, and $C_{arterial}$ is concentration in blood entering the filter.

This value can be used to calculate the clearance and is more applicable to haemofiltration as ultrafiltrate volumes are known.

$$Cl = S \times UFR_{avg}$$

where UFR_{avg} is the average ultrafiltration rate per unit time.

This value can be substituted into the formula for calculating drug dosages in renal impairment discussed above including, where necessary, non-renal elimination.

The factors given in Table 4 should be considered. The clearance of a drug depends on its molecular weight (and size) and protein binding (Gibson and Nelson 1983). Haemofilters have a pore size of 0.01 µm; artificial kidneys for haemodialysis 0.001 µm. In haemofiltration drugs with a molecular weight below that of inulin (5200 Da) will pass through, in haemodialysis most drugs with a molecular weight below 500 Da (which includes most antibiotics) will be cleared but drugs such as vancomycin (1800 Da), amphotericin (960 Da) and erythromycin (734 Da) behave differentially with respect to haemofiltration/haemodialysis. Heavily protein-bound drugs, even with a lower molecular weight (propranolol 259 Da) will not be filtered but drugs which are displaced from binding sites in the presence of renal impairment will become available for filtration. Water-soluble drugs pass through filters more readily than those which are fat soluble. A list of some commonly used drugs and their clearance by haemodialysis or peritoneal dialysis is shown in Table 5.

Digoxin and antidepressants are examples of drugs with a large volume of distribution which will have low plasma concentrations and therefore little of the drug available for filtration.

There are differences, other than pore size, between haemodialysis and haemofiltration. In haemodialysis clearance is by diffusion with a concentration gradient between the dialysate and plasma. Therefore drugs of a low molecular weight will be cleared more than in haemofiltration where there is no concentration gradient and clearance is by convection. Drug clearance will be optimal (whichever system is used) the greater the blood

Table 5 Dialysis of drugs

Dialysed	Slightly dialysed	Not dialysed
Gentamicin	Ciprofloxacin	Acebutalol
Netilmicin	Co-trimoxazole	Amphoteracin B
Amikacin	Erythromycin	Bretylium
Tobramycin	Enalapril	Chloroquine
Streptomycin	Tetracyclines	Clonidine
Lithium	Procainamide	Cyclosporin
Metronidazole	Nadolol	Diazoxide
Cefotaxime	Lorazepam	Digoxin
Ceftazidime	Methylprednisolone	Disopyramide
Cefuroxime	Ethosuximide	Flecainide
Cephalexin	Ethambutol	Glutethamide
Cephradine	Amantidine	Heparin
Aztreonam		Methicillin
Moxolactam		Methotrexate
Acyclovir		Metoprolol
Allopurinol		Miconazole
Ampicillin		Oxazepam
Amoxycillin		Pethidine
Atenolol		Phenytoin
Flucytosine		Prednisone
Isoniazid		Procainamide
Mecillinam		Propoxyphen
Penicillin		Propranolol
Ranitidine		Valproic acid
Theophylline		Vancomycin
Ticarcillin		Warfarin
Carbenicillin		
Trimethoprim		

Drugs cleared by peritoneal dialysis

Gallamine	Aspirin	Phenobarbitone
Ranitidine	Lithium	Quinidine
Ethambutol	Aminoglycosides	Isoniazid
Ceftazidime	Cephalexin	Theophylline
Aztreonam	Flucytosine	

2.2 Handling of drugs in kidney disease

Table 6 Penicillin dose adjustment

Drug	Dose reduction	Comments	Reference
Flucloxacillin	Nil		Nauta and Mattie 1976
Ampicillin		Dose seldom adjusted	Fabre et al. 1983
Amoxycillin + clavulanic acid	<20 ml/min	One-half normal dose after haemodialysis	Davies et al. 1988 Franke et al. 1979
Benzylpenicillin	<20 ml/min	Daily dose should not exceed 20 mU	Fabre et al. 1983
Piperacillin	20–50 ml/min One-half normal dose <20 ml/min	One-half normal dose after haemodialysis One-third to one-half normal dose	Bennett et al. 1987 Heim-Duthoy 1986

flow to the artificial filter, provided the filter does not become clotted or the membrane pores obstructed by protein which may occur in prolonged continuous haemofiltration. Haemodialysis is performed for short periods and it is usually sufficient to administer drugs, which are cleared by dialysis, at the end of treatment in a dose adjusted to take into consideration absent or diminished renal function. These principles apply if automated, high blood flow (300–400 ml/min), large filtrate volume (up to 80 ml/min), and short duration haemofiltration (with more efficient clearance of higher molecular weight drugs than haemodialysis) is employed. In the more usual form of haemofiltration (either arteriovenous or venous to venous) blood flow rates and therefore filtration rates are much lower and drug clearance is very slow. However because it is continuous and may be used for several days, net clearance may be high and supplemental doses may be needed.

Peritoneal dialysis clears drugs very much less efficiently than haemodialysis/filtration. However some drugs (particularly antibiotics) need readjustment in treatment of CAPD peritonitis.

Drugs

Drug prescribing

Once the clinician has identified that renal impairment is present, and that a drug which he plans to administer has clinically important renal excretion, he can adjust the dose in two main ways. Either the size of each dose, or the frequency of administration can be reduced. Plasma drug concentrations can be used to confirm that the initial adjustment of dosage is correct in that particular individual. Steady-state concentrations of anticonvulsants, digoxin, and theophylline can be measured after the equivalent of five half-lives of the drug. For antibiotics such as gentamicin the peak and trough level taken after the first day's administration is essential and continued monitoring is required if renal function is impaired or changing.

The combination of reduction in dosage and less frequent administration is suitable for most drugs. Dosage reduction alone is more likely to lead to subtherapeutic plasma concentrations. Unfamiliar dosages and administration of drugs at odd times may result in errors. Dose adjustments must be kept simple and clear.

Antimicrobials

Many antimicrobial agents are excreted by the kidney. With the exception of aminoglycosides and vancomycin, most have a wide therapeutic index and little or no dose adjustment is usually made until the GFR is below 20 ml/min. Although larger doses than strictly necessary are frequently used, this is not unreasonable in seriously ill patients in whom the priority must be to provide adequate plasma concentrations. Special consideration should be given to patients with acute changes in renal function, as adverse effects may occur before it is realized that renal function is impaired. Antimicrobials that are removed by dialysis should be administered after dialysis, or a supplemental dose given at that time.

Penicillins

Adjustments of dose for the commonly used pencillins are given in Table 6. Fractions of usual doses are shown rather than the absolute dose. Piperacillin is cited as an example of a broad-spectrum injectable penicillin with antipseudomonas activity; azlocillin (Aletta et al. 1980), ticarcillin (Parry and Neu 1976), and carbenicillin (Latos et al. 1975) require similar dose adjustment. Carbenicillin and ticarcillin solutions contain approximately 5 mmol Na^+/g, and caution is needed in the presence of salt and water retention. Mezlocillin (Bergan et al. 1979) requires similar dose reduction in renal impairment, but unlike the other penicillins it is not dialysed. Liver disease reduces mezlocillin clearance; in combined renal and hepatic dysfunction clearance is markedly reduced and should be generally avoided in this circumstance.

In patients with normal renal function penicillin elimination can be deliberately reduced by simultaneous administration of probenecid. This still has some therapeutic applications, e.g. in the treatment of gonorrhoea.

Cephalosporins

There are few absolute indications for this multifarious class of drugs, and only a few representative examples will be discussed. Cephalosporins are particularly useful as broad-spectrum antibiotics in patients with impaired or changing renal function as less toxic alternatives to aminoglycosides. Cephalexin and cephradine, which are active when given by mouth, can be used in normal dose until the GFR is below 10 ml/min when the daily dose is halved. The combination of loop diuretic and the first-generation cephalosporin, cephaloridine caused nephrotoxicity; the later generation drugs are safer but caution is still needed with cefuroxime when used with loop diuretics. In this situation ceftazidime appears to be a safer alternative (Walstad et al. 1988). Dose adjustments are shown in Table 7. Dosage

Table 7 Cephalosporin dose adjustment

Drug	Dose reduction	Comments	References
Cefotaxime	GFR 20–50 ml/min	One-half to one-fifth normal dose	Albin et al. 1985
Ceftazidine	<20 ml/min	One-tenth to one-fifth normal dose	Leroy et al. 1984
		After haemodialysis one-half normal dose	Walstad 1988
Cefuroxime			Local et al. 1981
			Walstad 1983

requirements for patients with renal failure are theoretically very small (< 100 mg/day) (Walstad et al. 1983; Bennett et al. 1987) but in practice only modest reductions in dose are appropriate (Cantu and Fantozzi 1988; Aronoff and Bennett 1988). Cephalosporins are one of several acceptable alternatives for the treatment of CAPD peritonitis and are readily absorbed via the peritoneum, but have been shown to be a cause of sometimes fatal *Clostridium difficile* gut superinfection in such cases. (See Chapter 10.6 for treatment of peritonitis in CAPD patients.)

Aminoglycosides

Aminoglycosides need dose adjustment in mild renal impairment, which may not have been discovered. Furthermore they are inherently nephrotoxic, and their use may worsen renal impairment and cause ototoxicity.

Several factors make nephrotoxicity more likely: prior or prolonged treatment, hypovolaemia, dehydration, concomitant administration of diuretics, hypokalaemia, and hypomagnesaemia (Humes 1988). Obstructive jaundice also increases the risk (Desai and Tsang 1988). The simplest way to prevent aminoglycoside toxicity is to avoid their use altogether in patients with any suspicion of renal impairment. Whelton (1988) suggests aztreonam and moxalactam as newer alternatives. Of the available aminoglycosides, tobramycin is marginally less nephrotoxic than gentamicin (Smith et al. 1980) and netilmicin appears to be less ototoxic than tobramycin (Lerner et al. 1983).

There are many nomograms and other guidelines for dose adjustment of aminoglycoside for patients with renal impairment (Kunin 1967; Sarubbi and Hull 1978; Brater 1985; Bennett et al. 1977), but each method has drawbacks. Whether reduction in dose, or reduction in frequency of administration, is used a loading dose is required. Since the volume of distribution of aminoglycosides is not materially affected by renal impairment, an adequate loading dose of 1 to 1.5 mg/kg of gentamicin is required. Dose adjustment may take two forms. Reduction of dose (without alteration in frequency of administration) may lead to an increased likelihood of subtherapeutic peak plasma levels (see above). If the frequency of administration, alone, is reduced then subtherapeutic plasma concentrations are more likely to occur over longer periods. A combination of both methods with frequent peak measurements (taken 1 h after intravenous dosing) and trough measurements (immediately before the next dose) is optimal. Such measurements should be made daily when alterations in renal function are anticipated and two to three times a week under other circumstances. The dose regimen is adjusted to produce peak plasma concentrations of gentamicin between 6 and 10 mg/l and trough concentrations not exceeding 2 mg/l. Doses and therapeutic concentrations are shown in Table 8.

Aminoglycosides are often used in the treatment of CAPD peritonitis (Chapter 10.6), and are readily absorbed from the peritoneum. Absorption however is unidirectional (Somani et al. 1982).

Vancomycin

Vancomycin is used extensively in patients with staphylococcal infections. It is indicated in the treatment of epidemic methicillin resistant strains and is used in the treatment of CAPD peritonitis which is often caused by *Staphylococcus epidermidis*. It is used for intravenous line sepsis, particularly in patients in the intensive care unit. The target steady-state plasma concentration is approximately 15 mg/l (Brown and Mauro 1988). Vancomycin is excreted by the kidney and is not dialysed so in patients with end-stage renal failure on dialysis, therapeutic concentrations can be maintained for 5 days or more after a single intravenous dose. It is usually given in a loading dose of 15 mg/kg over a period of 1 h using an intravenous filter. Further doses are given on the basis of plasma concentration. The incidence of the 'red man syndrome' which is seen with too-rapid infusion does not appear to be increased in renal failure. Since it is not removed by haemodialysis or haemofiltration (except at extremely high flow rates) it can be given at a rate of 500 mg/h during the last few hours of dialysis (Edell et al. 1988) or during haemofiltration.

In the treatment of CAPD peritonitis a single dose of 500 mg intravenously can be given every fifth day to produce satisfactory therapeutic levels (Whitby et al. 1987). Alternatively the drug can be administered intraperitoneally (Beaman et al. 1989) to produce adequate therapeutic concentrations and satisfactory treatment of peritonitis.

Teicoplanin

This antibiotic is a glycopeptide related to vancomycin. It is used intravenously and can also be given intraperitoneally and reportedly enters the bloodstream but does not cross back into the peritoneal cavity. In renal failure the half-life is approximately

Table 8 Doses and therapeutic plasma concentrations of aminoglycosides

Drug	Usual daily dose	Therapeutic concentration	
		Peak (mg/l)	Trough (mg/l)
Gentamicin	2–5 mg/kg	5–10	<2.5
Tobramycin	2–5 mg/kg	5–10	<2.5
Netilmicin	2–5 mg/kg	5–10	<2.5
Amikacin	10–30 µg/kg	20–30	<10
Kanamycin	10–30 µg/kg	20–30	<10

* Peak concentration measured 1 h after injection.
Trough concentration immediately before next dose.
Usual daily dose is administered 8-hourly.
Netilmicin dose can be increased to 7.5 mg/kg.day in severe infections but only with careful therapeutic monitoring.

three times that in normal subjects (Bonati et al. 1987). It can be given in a loading dose of 400 mg with virtually no clearance by peritoneal or haemodialysis.

Aztreonam

Aztreonam is a monocyclic β-lactam. It has a wide spectrum of activity against Gram-negative organisms. It is usually given in a dose of 1 g 8-hourly. However the dose should be reduced in renal impairment (Gerig et al. 1984; Fillastre et al. 1985). If the GFR is below 10 ml/min a loading dose of 1 g is given followed by a maintenance dose of 250 mg 8-hourly. Aztreonam is haemodialysed and half the usual dose is given as a supplement after dialysis. It can be used in CAPD peritonitis (Gerig et al. 1984) (1 g intravenously followed by 500 mg intraperitoneally 6-hourly).

Imipenem/cilastin

Imipenem is a carbapenen and is a new broad-spectrum antibiotic but without activity against Pseudomonas species. It is combined with cilastatin (an inhibitor of dipeptidase) to confer protection against renal toxicity. The usual dose is 0.5 to 1 g 6-hourly and this should be reduced to 0.5 g 12-hourly. It is cleared by haemodialysis and a supplementary dose is required (Schroder et al. 1989).

Erythromycin

Erythromycin is often used in upper respiratory tract infections (including mycoplasma, psittacosis, and legionnaires' disease) and soft tissue infections. It is particularly useful in patients who are allergic to penicillin. It is an important drug in the treatment of patients with *Legionella pneumophilia* infection, which can produce acute renal failure. Erythromycin should be given in full dosage to patients with renal failure. It is not haemodialysed to any significant extent (Kanfer et al. 1987; Periti et al. 1989). It inhibits the metabolism of cyclosporin and may therefore cause cyclosporin toxicity in transplant recipients. Its use can often be avoided in these patients.

Tetracyclines

All the tetracyclines, with the exception of minocycline and doxycycline, are excreted renally. Plasma half-lives are markedly prolonged (up to 100 h) in renal impairment. Tetracyclines are antianabolic and cause a concentration-related increase in blood urea, setting up a vicious cycle leading to deterioration in renal function (see above). This emphasizes the importance of a measure of GFR independent of blood urea (e.g. plasma creatinine) in patients with potential renal impairment before prescribing these drugs. Doxycycline or minocycline can be used cautiously in patients with renal impairment, but the other tetracyclines are contraindicated. Doxycycline does have some renal clearance but hepatic clearance increases with renal impairment, partly because there is a reduction in binding to plasma proteins and red blood cells (Houin et al. 1983). It is not dialysed and no dose adjustments are needed in patients on dialysis. Demeclocycline (used in the treatment of the syndrome of inappropriate ADH secretion) has particular effects.

Metronidazole

Metronidazole is used against anaerobic bacteria and against protozoa including *Trichomonas* spp. and *Entamoeba histolytica*. It is given in usual doses to patients with renal impairment. It is dialysed (Somogyi et al. 1983) and a supplemental dose (half the usual dose) is required after dialysis.

Sulphonamides and co-trimoxazole

The use of sulphonamides as single agents has largely been superseded. They are eliminated by acetylation followed by renal excretion, and acetylated metabolites (which have no antibacterial activity) cause crystalluria and tubular damage. Most sulphonamide usage is currently accounted for by co-trimoxazole (sulphamethoxazole 400 mg, trimethoprim 80 mg). However it is now appreciated that trimethoprim alone is effective for most urinary and respiratory infections which were previously treated with co-trimoxazole. The use of trimethoprim as a single agent avoids the toxicity of the sulphonamide. Sulphamethoxazole and trimethoprim display similar renal excretion, except at extremes of urinary pH (see above). In the presence of renal impairment (GFR <20 ml/min) one-half the usual dose may be used. A supplement of one-half the normal dose is given after dialysis.

Much higher doses of co-trimoxazole are needed in the treatment of *Pneumocystis carinii* infection, the risk of adverse effects being balanced against the seriousness of the condition. Such patients often have impaired renal function. The dose is trimethoprim 20 mg/sulphamethoxazole 100 mg/kg body weight. day divided into two or more doses. The plasma concentration should be maintained at approximately 5 to 8 μ/l measured after five doses. Full dosage should be given initially and then reduced if necessary.

Trimethoprim

It is now appreciated that for many indications, other than for *Pneumocystis carinii* infection, trimethoprim alone is as effective as co-trimoxazole and lacks the toxicity of the sulphonamide component of the combination drug. Considerations regarding its use in renal impairment are as shown above for co-trimoxazole.

Pentamidine

Pentamidine is used in the treatment and prophylaxis of *Pneumocystis carinii* infection and in the treatment of resistant cases of leishmaniasis. The usual dose is 4 mg/kg given by slow intravenous infusion over 90 min; this will reduce the incidence of hypotension and hypoglycaemia. There is considerable tissue binding and the drug is excreted in the urine over long periods. The dose should be reduced in patients with renal impairment and since the drug is nephrotoxic the dose should be reduced by 30 to 50 per cent if the serum creatinine increases by 88 μmol/l (1 mg/dl) (Sattler et al. 1988). Inhaled nebulized pentamidine is used prophylactically to prevent *Pneumocystis carinii* infection without producing renal impairment (Golden et al. 1989), although Miller et al. (1989) describe acute renal failure in one case.

Ciprofloxacin

This is a quinoline, and although related to nalidixic acid differs from this in that it can be prescribed for urinary infections in patients with renal impairment. It is particularly useful in the treatment of antibiotic resistant Gram-negative organisms, including *Pseudomonas aerogenosa*. Unlike other antibiotics effective against pseudomonas, it is available as both oral and intravenous preparations. It is also acquiring a role in the treatment of CAPD peritonitis (Fleming et al. 1987) for which it can be given orally (500 mg 6-hourly or with each bag exchange is an effective and relatively non-toxic regimen). Renal excretion exceeds GFR, and in patients with normal renal function approximately 60 per cent is cleared by the kidneys (Roberts and

Table 9 Drug dosage in tuberculosis

Drug	Chronic renal failure	Renal transplants
Rifampicin	Normal dosage	Avoid in cyclosporin therapy
Isoniazid	Normal dosage	Normal dosage
Ethambutol	GFR <30 ml/min	10–15 mg/kg.day
	GFR <10 ml/min	5 mg/kg.day
Pyrazinamide	GFR <30 ml/min	10–15 mg/kg.day
Capreomycin	GFR <10 ml/min	100 mg/day
Streptomycin	500 mg single dose: monitor levels	Avoid in cyclosporin therapy

Williams 1989). It is recommended that the dose should be reduced in renal impairment (500–750 mg/day maximum); however the proportion that is eliminated by the kidney is reduced in renal failure (Singlas et al. 1987) as a result of an increase in hepatic clearance and of secretion through the bowel wall (Roberts and Williams 1989). It is not significantly removed by haemodialysis and a supplemental dose is not required.

Antimicrobials in urinary tract infections

It is essential that agents reach therapeutic concentrations in the urine or renal parenchyma. Nitrofurantoin and nalidixic acid are used in the treatment of cystitis but are of limited use in pyelonephritis. If there is renal impairment nalidixic acid will not reach sufficient urinary concentrations. In addition nitrofurantoin causes peripheral neuropathy in patients with impaired renal function. There is little value in the use of urinary antiseptics such as hexamine.

Mycobacterial infections

Tuberculosis

Tuberculosis can be a difficult therapeutic problem in patients with renal failure (see Table 9). Rifampicin (Fabre et al. 1983) and isoniazid (Gold et al. 1976) can be given in the usual dosage. Neither is cleared significantly by dialysis. Pyridoxine should be given with isoniazid to prevent peripheral neuropathy (Cuss et al. 1986). The plasma half-life of ethambutol is, however, prolonged in renal impairment (Andrew 1980; Lee et al. 1980). If GFR is below 30 ml/min, the dose should be 10 to 15 mg/kg.day with a further reduction to 5 mg/kg daily if GFR is below 10 ml/min. Although 10 mg/kg daily has been used at these levels of GFR, cases of optic atrophy have been reported (Andrew 1980). Ethambutol is not dialysed to any significant extent. Pyrazinamide should be given at reduced dose (10–15 mg/kg) (Cuss et al. 1986). Capreomycin (Lehmann et al. 1988), a second-line drug, can be used in isoniazid or streptomycin resistance but is itself nephrotoxic. The dose should be reduced to 100 mg/day (from 1 g/day) if GFR is below 10 ml/min; it is dialysed. Streptomycin is also nephrotoxic and ototoxic but can be given in a single dose of 500 mg with repeated monitoring of plasma concentration until a further dose is required. In transplant patients on cyclosporin, rifampicin cannot be used as it reduces cyclosporin concentration very substantially as a result of hepatic enzyme induction (Allen et al. 1985).

The necessity to use more than three agents will be dictated by the nature and severity of the infection. Treatment may need to be prolonged to between 9 and 12 months in uraemic or immunosuppressed patients in contrast to the shorter courses that are preferred in patients with normal renal function.

Leprosy

Rifampicin (see above) is used in combination with dapsone and clofazimine. No alteration in dosage in patients with impaired renal function is required (Zuiderma et al. 1986).

Antiviral agents

Acyclovir and ganciclovir

Acyclovir and ganciclovir are both eliminated by the kidney. The main use of acyclovir in renal transplant patients is for the treatment of herpes and cytomegalovirus infections. Ganciclovir is also indicated in treatment of severe cytomegalovirus infection. High doses of acyclovir are needed in cytomegalovirus infections, and although acyclovir in these doses has been reported to cause renal impairment, dose reduction can prevent adverse effects (Blum et al. 1982; Laskin et al. 1982; Fletcher et al. 1988; Lake et al. 1988). These include cerebral irritation, ataxia, and myoclonus. The dose should be reduced stepwise from 800 mg five times daily at GFR above 25 ml/min to 800 mg daily at GFR below 10 ml/min. The drug is dialysed and between 60 and 100 per cent of the dose should be given postdialysis (Laskin et al. 1982).

Zidovudine

Zidovudine is used in patients with human immunodeficiency virus. It is a basic substance and undergoes glucuronidation and is eliminated by the kidney by tubular secretion. Although experience is limited, the dose should be reduced in patients with renal impairment. Probenecid will block the glucuronidation of the drug and will partially block the tubular excretion of this metabolite (Kornhauser et al. 1989) but will not block the tubular excretion of zidovudine itself. This property of probenecid may be used to allow reduction in zidovudine dosage with ensuing reduction in cost. However, the toxicity of this combination has yet to be fully evaluated.

Antifungal agents

Amphotericin

Amphotericin is nephrotoxic, and should only be used with great caution in patients who already have renal impairment. Toxicity may be ameliorated by sodium supplementation either as intravenous sodium chloride or by other means and it is worth noting that the high sodium content of carbenicillin can be an advantage in this context (Branch 1988). Protein binding, to lipoproteins, is reduced in renal impairment and low plasma concentrations need interpreting accordingly. No dose adjustments are needed and it is not removed by dialysis (Block and Bennett 1974).

Flucytosine

Flucytosine clearance follows GFR (Daneshmend and Warnock 1983) and the dose should be reduced progressively from roughly one-half the usual dose with GFR of 50 ml/min to one-quarter with GFR below 20 ml/min, with individual plasma concentration monitoring. It is removed by dialysis and a supplemental dose is required after dialysis (Block and Bennett 1974). It attains high urine concentrations and is therefore useful in fungal pyelonephritis or urine infections. It should be given in combination with another agent because of resistance.

Imidazoles

Ketoconazole, miconazole

The oral absorption of ketoconazole is reduced in patients with renal impairment (Daneshmend and Warnock 1983). Both miconazole and ketoconazole are extensively metabolized and can be used in usual dosage despite renal impairment (Lewi 1976). Ketoconazole, given by mouth, achieves therapeutic concentration in CAPD peritonitis (Johnson et al. 1985). Neither ketoconazole nor miconazole should be used with cyclosporin as they induce metabolizing enzymes causing cyclosporin plasma concentrations to fall substantially.

Fluconazole
Fluconazole is used in candidiasis, and is particularly valuable in oesophageal disease in immunocompromised patients. Unlike the other imidazoles, its elimination is dependent upon GFR; thus it should be reduced to one-half the usual dose after the first 2 days of therapy in patients with a GFR below 50 ml/min. It achieves high concentrations in the urine.

Griseofulvin
Griseofulvin can be given in usual dosage to patients with renal impairment. It does not interfere with cyclosporin.

Antiprotozoal agents
Malaria
Up-to-date information as to the likely pattern of resistance in different parts of the world may be sought from local reference laboratories.

Treatment This is summarized below.
Severe and complicated malaria Quinine may be given by slow intravenous infusion to patients severely ill with *Plasmodium falciparum* infection. It should be given in usual dosage (White 1985) to patients with chronic renal impairment. If acute renal failure develops as a complication of the disease then the dose may need to be reduced after 2 to 3 days. Quinidine is also active and available in parenteral form. It is given as an alternative to quinine in similar dose by slow intravenous infusion with monitoring of ECG and blood pressure.

Uncomplicated malaria *Plasmodium falciparium* infection is treated with quinine in usual dosage (White 1985). Tetracycline is often given to patients with normal renal function to enhance the curative effect of quinine, but only doxycycline is appropriate in this regard in patients with renal impairment.

Infection with *Plasmodium vivax*, *Plasmodium ovale*, and *Plasmodium malariae* can be treated with chloroquine. The dose is reduced by one-half if GFR is below 50 ml/min and to one-quarter if GFR is below 10 ml/min. Primaquine can be given in usual doses to achieve a radical cure with elimination of hepatic forms.

Prophylaxis Chloroquine (usual dose 300 mg/week) can be given to patients with renal impairment. Proguanil (usual dose 200 mg daily) should be given at one-half the usual dose if GFR is below 10 ml/min.

Antihelminthics
Mebendazole and thiobendazole
The most important helminth infestation in patients with chronic renal disease (particularly recipients) is *Strongyloides stercoralis*. This is endemic in several areas including parts of the West Indies (Guayana and Jamaica), East Africa and south-east Asia. Patients receiving immunosuppression may develop serious infestation of the gut, lung, or brain. Mebendazole is the drug of choice for treatment (and also as prophylaxis in high-risk patients); thiobendazole is an alternative but can cause vomiting.

Mebendazole is highly protein bound, not haemodialysed, and is metabolized in the liver (Allgayer et al. 1984). It is given in usual dosage to patients with renal impairment (100 mg twice daily for each 3-day course, with 2 weeks between courses). Mebendazole can be used in other infestations including those of *Ascaris lumbricoides*, *Trichosis trichuria*, and threadworms, and in hydatid disease.

Bilharzia
Praziquantel is the drug of choice for *Schistosoma haematobium*, *S. mansonii*, or *S. japonicum* infestation and can be given in usual doses to patients with impaired renal function. Oxamniquine is an alternative for *Schistosoma mansoni* with no dose adjustment.

Leishmaniasis
Sodium stibogluconate requires no alteration in dosage in patients with renal failure. Pentamidine and amphotericin are alternatives which have been discussed above.

Filariasis
Diethylcarbamazine is used for patients with *Wuchereria bancrofti* infestation. Its renal clearance is reduced in alkaline urine. This may be significant in vegetarian patients (Edwards et al. 1981). The overall excretion is reduced in the presence of renal impairment and a stepwise reduction in dosage (2 mg/kg three times daily for 2–4 weeks) is required (Adjepon-Yamoah et al. 1982).

Trypanosomiasis
Suramin or pentamadine, which is less effective, are used in haemolymphatic state and sleeping sickness. A high proportion of suramin is protein bound and a small amount is excreted renally. Collins et al. (1986) found that there was no accumulation in a study which included patients with impaired renal function. However, caution is advised as suramin can produce proteinuria and haematuria.

In *Trypanosoma cruzi* infections (Chagas disease) benznidazole, nifurtimox, or ketoconazole are given in normal doses to patients with renal impairment.

Drugs acting on the central nervous system

Drugs acting on the central nervous system may have a prolonged effect not only because of changes in pharmacokinetics but also because of increased sensitivity as a consequence of uraemia.

Drugs used in anaesthesia
Anaesthetic agents

Inhalation anaesthetics and injectable anaesthetics such as propofol are used in the usual dosage in patients with renal impairment. Fentanyl and alfentanyl also require no dose adjustment (Chauvin et al. 1987a; Meuldermans et al. 1988) and are used in combination with neuroleptics. The potency of thiopentone (Dundee and Richards 1954) and other barbiturates is increased in uraemia (Danhof et al. 1984; Lynam et al. 1988; Dingemanse et al. 1988). This appears to be due to a direct effect of uraemia upon the central nervous system; ethanol behaves similarly (Hisaoka et al. 1985). Although there are minor pharmacokinetic changes, primarily in volume of distribution, thiopentone is cleared by non-renal means. Despite the pharmacodynamic effect precise dose adjustments are unnecessary.

Neuromuscular blocking agents

Depolarizing
Succinylcholine is rapidly hydrolysed by plasma cholinesterase, and dose adjustment is not required in patients with renal impairment.

Non-depolarizing
Most of these nicotinic receptor antagonists are quaternary ammonium compounds. They are highly polar and consequently are eliminated by the kidney. Some are best avoided altogether: tubocurarine, gallamine, alcuronium, pancuronium, pipercuronium (Caldwell et al. 1988), and vecuronium (Lynam et al. 1988). Atracurium is the drug of choice in anaesthesia and prolonged mechanical ventilation (Granistad 1987; Ward et al. 1987; Parker et al. 1987). It is degraded by non-enzymatic Hofmann elimination which is independent of kidney and liver function. A metabolite, laudanosine, accumulates in renal impairment but the significance of this is unclear (Parker et al. 1987; Ward et al. 1988). It is removed by dialysis and haemofiltration.

Aminoglycosides, which accumulate in renal impairment without monitoring and appropriate dose adjustment (see above) are themselves weak non-depolarizing neuromuscular blocking agents (particularly if there is hypokalaemia) and may interact to produce prolonged paralysis.

Narcotic analgesics
Analgesic agents may be needed by patients with renal impairment for pain relief (including postoperative pain) or as analgesia and sedation in ventilated patients. Opiates are affected by renal failure and retention of metabolites can produce adverse effects. Reduced intermittent doses, epidural administration, or low dose continuous infusions reduce the incidence of adverse effects. Diamorphine is metabolized into morphine and then to morphine-3-glucuronide and morphine-6-glucuronide (Guay et al. 1988) both of which accumulate in renal failure (Chauvin et al. 1987b; Sawe et al. 1987; Sear 1988). These metabolites prolong both analgesia (Wolff et al. 1988) and respiratory depression (Chauvin et al. 1987b).

Pethidine (meperidine) is converted to norpethidine (normeperidine) which accumulates and can cause seizures (Szeto et al. 1977). The elderly appear to be more sensitive to pethidine but this is not an effect of altered kinetics (Herman et al. 1985) and the dose can be adjusted according to response in the usual manner. Papaveretum is a mixture of alkaloids of opium including morphine, codeine, noscapine, and papaverene: its use is not recommended in renal impairment.

Codeine and dihydrocodeine, although weaker analgesics, still have the potential to cause severe respiratory depression in some patients, and should be used with caution in renal impairment.

Dextropropoxyphene (combined with paracetamol as coproxamol) can produce drowsiness and respiratory depression and its metabolite norproxyphene, which accumulates in renal failure, sometimes causes cardiac toxicity (Drayer 1983).

Buprenorphine is metabolized in the liver and does not appear to have any important toxic metabolites.

Non-narcotic analgesics
Paracetamol (which in overdosage is an important cause of acute renal failure (see Chapters 8.10 and 8.11)) is excreted in small amounts by glomerular filtration with some passive tubular reabsorption (Prescott et al. 1989). The majority of the drug is metabolized and the glucuronide and sulphide metabolites, which are subject to active tubular secretion, accumulate in renal impairment. There is some regeneration of the parent compound. Despite this, paracetamol is used in usual doses for pain relief in patients with renal impairment.

Aspirin which is an alternative mild analgesic has the disadvantage of causing gastric damage and increasing the bleeding diathesis of patients with renal failure (see Chapter 9.10). Renal elimination of its metabolite salicylate is enhanced in alkaline urine (see above). Other non-steroidal anti-inflammatory drugs, which like aspirin inhibit cyclo-oxygenase, are discussed below (and in Chapter 6.2).

Antidepressants
Clinically significant effects of renal impairment on antidepressants have not been described. Tricyclics can be prescribed in usual dosage (Lieberman et al. 1985); there is a reduction in protein binding of imipramine (Vanholder et al. 1988) but this is not clinically significant.

Lithium
Lithium is used primarily in affective disorders. It is filtered and then reabsorbed, mainly in the proximal tubule. The dose should be reduced in renal impairment with careful monitoring of plasma concentration. In sodium depletion (e.g. with chronic thiazide diuretic use (Petersen et al. 1974)) tubular reabsorption of lithium is increased, leading to higher plasma concentrations and toxicity (Kristensen 1983).

Major tranquillizers
No dose change is required when phenothiazines or butyrophenones are used in patients with renal impairment (Forsman and Ohman 1976).

Minor tranquillizers
Benzodiazepines can be prescribed in usual dosage. Diazepam and chlordiazepoxide have active metabolites which may accumulate (Brater 1985), and drugs without active metabolites such as nitrazepam and temazepam (Morrison et al. 1984; Kroboth et al. 1985) may avoid hangover the morning after use as night sedation. Chronic benzodiazepine use should be avoided.

Midazolam is used extensively as an intravenous sedative for minor operations or procedures. Patients with renal impairment are more sensitive to midazolam and the dose should be reduced to between one-quarter and one-third when GFR is below 10 ml/min (Vinik et al. 1983).

Heminevrin can be used in patients with impaired renal function who are suffering alcohol withdrawal, with careful monitoring of conscious state and respiration.

Anticonvulsants
Phenytoin and valproic acid are both highly protein bound. Phenytoin protein binding falls in proportion to GFR (Brater 1985; Mauchi and Nakahashi 1988). The unbound (free or active components) will be proportionately higher for a given plasma concentration as the GFR falls and this will have the effect of lowering the therapeutic range (see above). Changes in protein binding of valproic acid are not clinically important (Vanholder et al. 1988). Both drugs are prescribed in usual dosage to patients with renal impairment and neither is dialysed (Adler et al. 1975). Carbamazepine is prescribed in usual doses (Gulyassy and Depner 1983). Primodone is converted to phenobarbitone; this process is not affected by renal impairment. However both pri-

modone and phenobarbitone accumulate (Lee et al. 1982) in patients with renal impairment because an appreciable fraction of their elimination depends upon renal excretion. Additional doses are not required after dialysis. Therapy with phenobarbitone or primodone can be monitored by measuring plasma concentrations of phenobarbitone together with clinical observations of efficacy and toxicity, especially nystagmus and ataxia.

All anticonvulsants other than sodium valproate and bromide (which is now seldom used) are potent inducers of cytochrome P-450 and because of this immunosuppression using glucocorticoids and/or cyclosporin is difficult to achieve in epileptic patients. Indeed, even very large doses of cyclosporin may fail to achieve therapeutic concentrations because of this enzyme induction (see Chapter 11.5).

Antihistamines

H_1 antagonists are frequently prescribed to patients with renal failure for pruritus although their efficacy is uncertain. Both terfenadine (Carter et al. 1985) and prochlorperazine (Paton and Webster 1985) are used in usual dosage.

Cardiac drugs

Antiarrhythmics

Patients with renal disease may develop a variety of cardiac arrhythmias. In patients with abnormal renal function it is advisable to keep treatment simple, and most antiarrhythmic drugs are used without dose modification, e.g. lignocaine (Thomson et al. 1973; Collingsworth et al. 1975), amiodarone (Latini et al. 1984), flecainide (Forland et al. 1988), mexilitene (Wang et al. 1985) and verapamil (Mooy et al. 1985). Digoxin is a notable exception (Koup et al. 1975).

Acute arrhythmias

Hyperkalaemia causes many of the severe arrhythmias that occur in the setting of renal impairment. Emergency treatment should include intravenous calcium gluconate, glucose and insulin, salbutamol, and dialysis (see Chapter 8.4). Ion exchange resins may be used in this situation as a temporizing measure when dialysis facilities are unavailable. Sodium bicarbonate will correct acidosis and lower potassium concentrations. However, patients with chronic renal failure and hypocalcaemia may develop tetany; some patients will not tolerate the sodium load.

Acute supraventricular tachycardia

Supraventricular tachycardia in patients with renal failure can be treated by DC cardioversion, vagal manoeuvres, or drugs, including intravenous β-blockers or calcium-channel blockers. Verapamil can be used in usual doses in patients with renal impairment (Mooy et al. 1985), and is the drug of first choice. β-Blockers or oral verapamil are used in prophylaxis of recurrent supraventricular tachycardia. Propranolol can be given in usual doses (Fabre et al. 1983).

Arrhythmias in acute myocardial infarction

These may require treatment by DC cardioversion or cardiac pacing (bradycardias may also be treated with atropine or isoprenaline in similar doses to those used in patients with normal renal function). Serious ventricular arrhythmias may respond to intravenous lignocaine. This is used in usual doses in patients with renal impairment (Collingsworth et al. 1975; Thomson et al. 1973), since it is eliminated by hepatic metabolism. As with patients with normal renal function, accumulation and toxicity may occur as a result of reduced hepatic blood flow secondary to acute myocardial dysfunction. In the longer term β-blockers may reduce cardiac mortality, although this has not been specifically addressed in patients with renal failure.

Atrial fibrillation

Acute atrial fibrillation may require DC cardioversion. In chronic atrial fibrillation digoxin is the drug of first choice for controlling the ventricular rate. Tissue protein binding and volume of distribution are somewhat reduced in renal failure, and a modest reduction in loading dose (e.g. 1 mg rather than 1.5 mg) is appropriate. Since renal excretion is the main mode of elimination of digoxin (85 per cent in individuals with normal renal function) the maintenance dose also needs to be reduced in patients with renal failure. Dose of 0.125 mg/day or less are appropriate depending on plasma concentration (therapeutic levels 0.7–2 ng/ml) and ventricular response (Koup et al. 1975). Since $t_{\frac{1}{2}}$ is increased (to approximately 4.5 days) in anephric patients the possibility of accumulation and toxicity needs to be considered. Digoxin toxicity is more likely to occur in the presence of hypokalaemia. Digoxin is not dialysed. Amiodarone may be valuable in resistant cases of atrial fibrillation and is also useful in ventricular arrhythmias; in both circumstances the dose need not be modified (Latini et al. 1984).

Ventricular arrhythmias

As in patients with normal renal function the treatment of ventricular arrhythmias is problematical. On the one hand severe arrhythmias may give rise to recurrent hypotension or sudden death; on the other ventricular arrhythmias may be asymptomatic markers of underlying cardiac disease. Antiarrhythmic drugs may reduce the number and nature of ectopics without improving prognosis (or even worsening outcome).

Amiodarone is a highly effective antiarrhythmic drug but has significant toxicity; the dose is unchanged in renal failure (see above). Other orally active antiarrhythmics (e.g. quinidine, procainamide, mexilitene, encainide, flecainide, disopyramide) are of unproven efficacy in terms of reduction in death rate. Disopyramide is renally excreted and in maintenance needs to be reduced to one-half the usual dose when GFR is below 20 ml/min (Burk and Peters 1983). Disopyramide is not dialysed. Antimuscarinic side-effects are even more of a problem in patients with renal impairment because metabolites with potent atropine-like effects accumulate.

Quinidine can be used in usual dosage although there are minor reductions in volume of distribution and minor increases in protein binding (Ueda et al. 1975).

Procainamide is metabolized to n-acetylprocainamide (Gibson et al. 1977) which is retained in renal impairment. This metabolite has antiarrhythmic activity in its own right. However, this is unpredictable; many assays are non-specific so plasma concentration measurement is unreliable. Cimetidine reduces renal procainamide clearance (Christian et al. 1984).

Angina

Nitrates are used in normal dosages. β-Blockers should be started in low doses and the dose increased, monitoring symptoms and heart rate. Atenolol is eliminated by the kidney unlike propranolol, metoprolol, oxprenolol, and pindolol (Fabre 1983). Atenolol is dialysed and one-half the regular normal dose may be given as a supplement after dialysis.

Calcium-channel blockers are extensively protein bound and not renally excreted. Verapamil (Mooy et al. 1985), nifedipine (Martre et al. 1985), and diltiazem (Grech-Belanger et al. 1988) can all be used in usual dosage and are not dialysed.

Hypertension

Diuretics

Thiazide diuretics lose their potency when the GFR is below 25 ml/min. In consequence loop diuretics such as frusemide are used in hypertensive patients with impaired renal function. Larger doses of loop diuretics may be needed as the GFR falls (Huang et al. 1974).

Spironolactone, triamterene, and amiloride, all potassium sparing diuretics, should be avoided in renal impairment because of the danger of hyperkalaemia. The same applies for combination diuretics such as 'Moduretic' (amiloride and hydrochlorothiazide) or 'Dyazide' (triamterene and hydrochlorothiazide).

Severe sodium and water depletion can occur in diuretic therapy. This may affect renal haemodynamics with secondary effects upon renal function and concomitant drug therapy.

β-Blockers

These have been discussed in the section on angina. They may be used to treat hypertension in patients with renal impairment, although they may reduce renal blood flow necessitating close monitoring of renal function.

Angiotensin converting enzyme inhibitors

Starting doses of captopril, enalapril, or lisinopril should be low, and the dose increased slowly with careful monitoring of serum creatinine and potassium. Particular caution is necessary if these drugs are used in combination with diuretics or in other high renin states (e.g. volume depletion), when marked hypotension ('first dose effect') may be anticipated. Caution is also necessary when there is (or may be) a possibility of renal artery stenosis, both because of the risk of hypotension and also because of reduced GFR in the affected kidney(s). They should generally not be used with potassium sparing diuretics because of the added risk of hyperkalaemia: this is particularly important in patients with impaired renal function. All are eliminated by the kidney which accounts for the reduced dose usually required in the elderly (Laher et al. 1988). Captopril has a shorter half-life than enalapril and lisinopril and dose adjustments can therefore be made more readily (Duchin et al. 1984). Both enalapril and the active metabolite enalaprilat have reduced clearance and volume of distribution in renal impairment (Hockings et al. 1986). Lisinopril (which is the lysine analogue of enalaprilat) is not a prodrug and therefore does not require metabolism to its active form. Its clearance is affected, in addition, by cardiac failure (Thomson et al. 1989). Converting enzyme inhibitors are haemodialysed (Kelly et al. 1988) and in patients on haemodialysis drug accumulation occurs between dialyses.

Vasodilators

Hydralazine, α_1-blockers (e.g. prazosin, terazosin, and doxazisin) and calcium-channel blockers may be used in patients with renal impairment. Minoxidil, a potassium channel activator, is particularly valuable in the most severely affected and most refractory patients. Usual doses are used. Diuretics are almost invariably needed, usually in high doses, as fluid retention occurs. Minoxidil is removed by both haemodialysis and peritoneal dialysis.

Centrally acting agents

Alpha-methyldopa (Bennett et al. 1987) and clonidine are occasionally useful in patients with renal impairment who have contraindications to other drugs. They are given in usual doses.

Cardiac failure

Diuretics

The use of diuretics in hypertension has been described above. Loop diuretics in higher doses than usual are required in patients with renal impairment. Metolazone, a quinolone which is often considered with the thiazides, is effective in patients with resistant oedema when used in combination with loop diuretics, although severe sodium and water depletion with consequent effects on the kidney may occur, necessitating close monitoring of renal function. Diuretic resistance and poor absorption in oedema and nephrotic states have been discussed in the sections on absorption and distribution.

Other drugs

Other drugs affecting cardiac preload include organic nitrates which are unaffected by renal impairment. Converting enzyme inhibitors which reduce both pre- and afterload have been discussed in the section on hypertension: the same precautions apply even more strongly in patients with heart failure.

Digoxin is most effective in patients with atrial fibrillation, and is discussed in the section on antiarrhythmics. It is also effective as a positive inotrope.

Endocrine

Diabetes mellitus

Patients with chronic renal failure due to diabetes mellitus require frequent monitoring of blood glucose and glycosylated haemoglobin both before dialysis, during CAPD, or haemodialysis, and subsequent transplantation. The dose of insulin and oral hypoglycaemic agents can be adjusted accordingly.

Insulin

Insulin requirements fall with declining renal function, probably as a consequence of reduced metabolism of insulin by the kidney in both acute renal failure (Naschitz et al. 1983) and chronic renal failure (Rabkin et al. 1970). In patients on haemodialysis it is often necessary to give supplemental insulin during treatment and this can be best achieved by use of an infusion pump and hourly titration against the blood sugar. The same situation applies in patients on haemofiltration in acute renal failure, particularly if they are being fed parenterally, and in continuous arteriovenous haemodialysis when the dialysate is a glucose-based solution. Non-diabetic patients may require insulin temporarily under similar circumstances.

Patients on CAPD may need a change in insulin preparation and adjustment in the frequency and route of administration. Insulin can be given subcutaneously on a normal twice or three times a day regimen with combinations of long- and short-acting insulins. Alternatively (and usually preferably) insulin can be given intraperitoneally with each bag change. The intraperitoneal requirement is approximately 50 per cent of intravenous requirements. Soluble insulin can be (and usually is) injected into the bag, but absorption from the peritoneum is better as a bolus down the catheter before the dialysate is run in (Henderson et al. 1985).

Oral hypoglycaemic agents

Chlorpropamide has a greatly increased half-life (>36 h) and should be avoided if there is renal impairment as the risk of severe and prolonged hypoglycaemia is considerable. In the elderly renal impairment may be covert (see Chapter 1.8) and hypoglycaemia may go unrecognized. Tolbutamide and glibenclamide are both extensively metabolized with no active metabolites. The former is highly protein bound, and displacement may occur in advanced uraemia. Tolbutamide is used in usual dosage until the GFR is below 10 ml/min when the dose may need to be reduced. Glibenclamide should be started at a low dose (2.5–5 mg daily) and increased slowly. Glicazide is being used more frequently than glibenclamide and tolbutamide. It is cleared by the kidneys but if started at a low dose (40 mg daily) hypoglycaemia can be avoided and the dose increased to the usual maintenance level (160 mg daily as a single dose; up to 320 mg in divided doses).

Biguanides

Metformin is excreted unchanged in the urine and there is an increased risk of lactic acidosis in renal impairment. It should be avoided when GFR is below 20 ml/min. It is eliminated by dialysis and this can be used in the treatment of metformin-induced lactic acidosis (Lalau *et al.* 1989).

Thyroid disease

Thiouracils are used in usual doses. Thyroxine is highly protein bound and there may be alterations in protein binding in hypoalbuminaemic states (Kapstein *et al.* 1987). These changes do not appear to be clinically important.

Erythropoietin

For a discussion of erythropoietin see Chapter 9.6.8.

Asthma

β-Agonists administered by inhalation, oral, or parenteral routes need no adjustment in patients with renal impairment. Aminophylline and theophylline can be given in usual doses (Kraan *et al.* 1988), but in rats the seizure threshold is reduced (Ramzan and Levy 1987). Plasma concentration should be monitored but concentrations may be overestimated by the recently described dip-stick test (Nanji and Greenway 1988).

Gastrointestinal drugs

H$_2$-antagonists

There is an increased risk of confusional states with cimetidine in patients with impaired renal function. Cimetidine is cleared by the liver but metabolites accumulate if GFR is below 20 ml/min (Bjaeldager *et al.* 1980). Ranitidine is preferable in this situation, but it should be noted that it interferes with creatinine secretion and raises plasma creatinine (see Chapter 1.3). It is partly cleared by the kidneys (Roberts 1984) and the dose should be halved when GFR is below 10 ml/min. It is dialysed, and a supplemental dose is needed after dialysis. Approximately one-third of a 50-mg dose is cleared by a single CAPD exchange (Sica *et al.* 1987) and 150 mg twice daily can be used safely. It is commonly given to patients requiring artificial ventilation, to reduce the risk of stress ulceration. In these circumstances 25 mg twice or thrice daily intravenously is sufficient. No supplemental dose is needed during haemofiltration even though 20 per cent may be removed during 20 l of exchange (Gladziwa *et al.* 1988).

Antacids

Alginates, magnesium trisilicate mixture (but not magnesium trisilicate powder), and sodium bicarbonate all have high sodium content. The use of aluminium-containing compounds, such as aluminium hydroxide or sulcralfate in patients with severe renal impairment or those on dialysis is controversial because of the potential risks of aluminum retention with deleterious effects on bone, bone marrow, and the central nervous system. Calcium carbonate should not be used as an antacid in patients with severe renal impairment or end-stage renal failure, although it is used as a phosphate binder with careful monitoring of serum calcium and phosphate.

Newer antiulcer drugs

Experience with proton pump inhibitors (e.g. omeprazole, which has been used without dose reduction in patients with renal failure) and prostaglandin analogues (e.g. misoprostol) is limited to date.

Laxatives

Constipation is frequent in dialysis patients, and can be a major problem in those on CAPD. Drugs based on ispaghula husk have a high sodium content. Lactulose or senna are indicated for constipation in patients with renal failure. Sodium picosulphate is also useful in severe constipation, but its use in patients with renal impairment prior to radiological procedures should be avoided as dehydration and the increased risk of contrast induced renal failure may occur.

Antidiarrhoeal agents

Compounds containing opioids or their derivatives should be used with the same caution as in their use as analgesics. Loperamide can be given in usual doses.

Antiemetics

Metoclopramide is not significantly affected by renal impairment despite a small reduction in clearance (Wright *et al.* 1988) with no effect on volume of distribution or protein binding (Webb *et al.* 1986). It is not appreciably dialysed.

Hyperuricaemia
Allopurinol

Allopurinol is metabolized to oxypurinol which is retained in renal impairment and may be responsible for some of the adverse effects including rashes, bone marrow depression, and gastrointestinal upset. The dose should be reduced to 100 mg/day when GFR is below 20 ml/min. The dose should be given after haemodialysis. Allopurinol interferes with the metabolism of 6-mercaptopurine (an active metabolite of azathioprine) causing accumulation and toxicity (e.g. leucopenia). In transplant patients with severe gout it may be necessary to change from azathioprine to cyclosporin therapy for this reason, or to give probenecid, because cyclosporin raises uric acid (see Chapters 11.5 and 11.6).

Uricosuric agents

Uric acid is excreted most effectively in an alkaline urine. This may be achieved with potassium citrate, with care to avoid hyperkalaemia, but in advanced renal failure the urine is relentlessly acid in pH.

Probenecid

Probenecid inhibits secretion of acids in the proximal tubule and prevents reabsorption of urate from the tubular lumen. It prolongs the effect of penicillins, cephalosporins, naproxen, indomethacin, methotrexate, and sulphonylureas (all of which are weak acids), causing accumulation and the potential for toxicity. It also inhibits tubular secretion (and hence activity) of frusemide and bumetanide. It also inhibits liver uptake and hence glucuronidation of several drugs including zidovudine (see section on antivirals).

Colchicine

Colchicine has been partly replaced by NSAIDs for the treatment of acute gout. However it remains valuable in patients in whom NSAIDs are undesirable (e.g. peptic ulcer disease, cardiac failure, renal impairment). It is also sometimes used in low dose in prophylaxis. It is mainly metabolized by the liver and no adjustment is needed in renal impairment if it is used to treat acute gout.

Anti-inflammatory agents (see Chapter 6.2)

NSAIDs, including aspirin, inhibit prostaglandin synthesis by inhibition of cyclo-oxygenase. The principal renal prostaglandins in man are prostaglandin E_2 and prostaglandin I_2 each of which is vasodilator and natriuretic (Dunn 1983). In addition to effects on renal blood flow prostaglandins also influence tubular ion transport directly. In healthy individuals inhibition of cyclo-oxygenase has no detectable effect on renal function, but in patients with cardiac failure, nephrotic syndrome, liver disease, glomerulonephritis, and other renal disease cyclo-oxygenase inhibitors predictably cause a reversible fall in GFR which can be severe. They also cause fluid retention. They may also cause hyperkalaemia. There is evidence that sulindac causes less inhibition of renal cyclo-oxygenase than a dose of ibuprofen that is equi-effective on extrarenal tissues; sulindac may cause less renal impairment than other NSAIDs. Aspirin may also spare cyclo-oxygenase in the kidney to some extent. The clinical relevance of these observations remains uncertain and caution is needed in severe renal impairment for all of this group of drugs.

Indomethacin, azapropazone, and diflunisal have important renal excretion, whereas most other NSAIDs are eliminated by metabolism. Diflunisal has lowered protein binding in renal failure although the clinical importance of this is slight (Verbeeck and de Schepper 1980; Eriksson *et al.* 1989). Sulindac has an active sulphide metabolite (Gibson *et al.* 1987) but in renal impairment this metabolism is reduced (Diskin *et al.* 1988; Nesher *et al.* 1988). It has been reported to have fewer side-effects than indomethacin (Berg and Talseth 1985; Nesher *et al.* 1988). The NSAIDs are highly protein bound and are not removed by dialysis.

Drug therapy in patients with renal impairment in intensive care

Neuromuscular and analgesic agents have been dealt with above. Cardiac inotropes are used in normal dosage although renal vasoconstriction may be deleterious so the minimum effective dose of adrenaline, dobutamine, or dopamine should be used, and dopamine (as a vasodilator in low doses (up to 2 μg/kg.min) is preferable. Intravenous nitrates are given in normal dosage. Sodium nitroprusside may be used in the management of hypertension or left ventricular failure. It is metabolized to sodium thiocyanate (Palmer and Lasseter 1975) which is eliminated by the kidney, and so may accumulate in renal failure, causing toxicity. It is removed by haemodialysis or peritoneal dialysis. In liver failure the detoxification of cyanide to thiocyanate may be impaired so the use of sodium nitroprusside should be avoided in this circumstance. Plasma concentrations above 10 mg/dl produce nausea, anorexia, and fatigue. Levels above 20 mg/dl may be fatal.

Anticoagulants

Warfarin is used in normal dosage (Van Peer *et al.* 1978) and its effect monitored by measuring prothrombin time in the usual way. It is highly protein bound and there may be slight displacement and consequent reduction in volume of distribution. In nephrotic patients (see Chapter 3.2) hypoalbuminaemia leads to increased sensitivity to warfarin. It is not dialysed. Heparin is used in normal dosage.

Corticosteroids and immunosuppressive agents

Prednisone and prednisolone are not eliminated by the kidney. However, in theory the dose should be reduced when dialysis is indicated in patients with end-gage renal failure (Bergrem 1983) as uraemia may reduce the clearance of prednisolone by the liver (Bianchetti *et al.* 1976). Bergrem confirms that normal doses of prednisolone can be used in patients with the nephrotic syndrome. Hypoalbuminaemia reduces the number of binding sites on plasma protein, but the increase in steroid side-effects reported by Lewis *et al.* (1971) failed to account for other factors such as underlying disease and renal dysfunction. Methylprednisolone is cleared by haemodialysis, and should be given after dialysis (Sherlock and Letteri 1977). Azathioprine accumulates in renal impairment and the dose should be reduced from a maximum of 3 mg/kg per day to 1 mg/kg.day if the GFR falls below 10 ml/min (Schusziarra *et al.* 1976). Allopurinol prevents the metabolism of 6-mercaptopurine, the active metabolite of azathioprine, and the combination should be avoided as described above.

Cyclosporin is discussed in detail in Chapters 11.5 and 11.9 but its important drug interactions will be discussed here. It is a highly lipid-soluble drug which is extensively bound to plasma proteins and has a large volume of distribution. It is metabolized by the liver via the cytochrome P-450 system by mono- and dihydroxylation as well as N-demethylation. Only minor amounts are excreted as parent drug or metabolites in the urine. Renal impairment does not affect the metabolism. However since many other drugs may be prescribed to patients on cyclosporin therapy several important interactions may occur. These will both increase plasma concentration and therefore increase the risk of nephrotoxicity or to reduce plasma concentrations to increase the risk of transplant organ rejection. Aminoglycosides may have an additive effect upon the nephrotoxicity itself. The common interactions are listed in Table 10.

Cancer chemotherapy

Cytotoxic drugs and antimetabolites are frequently administered in single doses or short courses intended to destroy cancer cells at a particular stage of the cell cycle, with periods to allow the bone marrow to recover. High peak concentrations rather than lower steady state concentrations are required. Most anticancer drugs are cleared by metabolism (see Table 11) but the fate of

Table 10 Drugs affecting the cytochrome P-450 system

Inhibitors	Inducers
Ketoconazole	Rifampicin
Erythromycin	Phenytoin
Oral contraceptives	Phenobarbitone
Methylprednisolone	Carbamazepine
Diltiazem	Sodium valproate
Nicardipine	
Verapamil	
Cimetidine	

the metabolites is largely unknown. Nephrotoxicity can occur as a direct effect on the kidney or an indirect effect via increased urate production as tumour cells are destroyed: hydration and pretreatment with allopurinol (with suitable dose adjustment in patients with renal impairment (see above) is indicated when rapid breakdown of cells is anticipated.

Cisplatin (testicular, ovarian, bladder cancer) and mitomycin (lymphomas) are directly nephrotoxic. Cisplatin displays cumulative nephrotoxicity. The drug is given three to four times weekly and repeat doses should only be given if the serum creatinine is below 130 μmol/l. The risk of nephrotoxicity is reduced if the drug is given with a fluid load and if necessary mannitol to maintain a high urine flow. Mitomycin may give rise to a haemolytic uraemic syndrome (see Chapter 8.7).

Cyclophosphamide is metabolized and the metabolites which are excreted in the urine are responsible for the chemical haemorrhagic cystitis. A high urine flow rate is essential when high bolus doses of the drug are used. Methotrexate is acidic and is eliminated by proximal tubule secretion. This is blocked by salicylate and reduced by NSAIDs; alkalinization of the urine increases elimination. In theory this could be applied therapeutically in methotrexate overdosage but the volume of distribution of the drug is so large that this is ineffectual. The drug is not haemodialysed significantly and toxicity from larger doses (whether accidental or deliberate) can be limited by the use of folinic acid rescue.

Table 11 Clearance of cancer chemotherapy agents

Metabolized	Renal elimination
Alkylating agents	
Busulphan	
Chlorambucil	
Cyclosphosphamide	
Thiotepa	
Cytotoxic antibiotics	
Doxorubicin hydrochloride	Actinomycin D
Mitomycin	Bleomycin
Antimetabolites	
Cytarabine	Methotrexate
Fluorouracil	Cisplatin
Mercaptopurine	Hydroxyurea
(metabolism inhibited by allopurinol)	
Procarbazine	
Vinblastine	
Vincristine	
Mitozantrone	Mitozantrone

Melphalan, which is used in myeloma where there is frequently impairment of renal function, is largely metabolized. However, increased bone marrow toxicity seems to occur in patients with renal impairment and the usual dose may need to be halved if GRF is below 25 ml/min (Alberts *et al.* 1979).

Bleomycin is renally eliminated, but since it is given in a single dose no dose adjustment is needed in patients with renal impairment.

Miscellaneous drugs

Acetazolamide

Acetazolamide is used in the treatment of glaucoma. It can produce serious acidosis and other electrolyte disturbances in patients with renal impairment, even though its diuretic potency is reduced. In the elderly in particular this is a common problem (Chapron *et al.* 1989) and renal function and electrolytes should be monitored carefully. As a sulphonamide of limited solubility, it may crystallize out in the tubules and cause acute renal failure. If possible, acetazolamide should be avoided in patients with renal impairment.

Radiocontrast agents (see Chapter 8.2)

Brezis and Epstein (1989) reviewed the problem of nephrotoxicity of contrast agents and noted that sodium and water depletion, prostaglandin inhibition (particularly by drugs), and a single kidney were all risk factors. The damage occurs in the medullary thick ascending limb and is either caused by vasoconstriction or by a direct toxic action. Pre-existing renal impairment and diabetes mellitus, alone or in combination, produce an increased risk whichever contrast media was used (Parfrey *et al.* 1989). However a non-ionic compound, iopanidol, rather than the ionic diatrizoate, was less nephrotoxic in cardiac catheter cases (Schwab *et al.* 1989). Intravenous saline loading is beneficial in patients with pre-existing renal impairment. Dehydration before intravenous urography in patients with pre-existing renal impairment or multiple myeloma is contra-indicated; an infusion IVU will suffice.

Lipid lowering agents

Hyperlipidaemia is found in patients with the nephrotic syndrome, in most patients with chronic renal failure, and in many patients with successful renal transplants. Diet should be used to control these abnormalities but drug treatment may be needed. The anion-exchange resins, cholestyramine and colestipol are given in usual doses. The clofibrate group of drugs (clofibrate, bezafibrate, and gemfibrizol) should be avoided as dose reductions are needed and the myopathy seen with clofibrate may occur. The HMG-CoA reductase simvastatin is metabolized in the liver and dose adjustment is not required in patients with impaired renal function. However close monitoring (as in other patients) of liver function and muscle enzymes is necessary. This is particularly important in patients on cyclosporin as cases of reversible rhabdomyolysis have been reported.

Practical notes for use of drugs in renal failure

1. Use few drugs and get to know them well.
2. In prescribing for any patient with renal disease, ask yourself how much you know about their renal function. A normal plasma creatinine may NOT be enough.

3. Beware of prescribing in elderly patients.
4. What do you know about the elimination of the drug?
5. If you do not know find out.
6. If you cannot find out do not prescribe!
7. Do not prescribe and go away; MONITOR clinically and when necessary (and if available) by blood concentration.

Bibliography

Anderson, R.J., Gambertoglio, J.G., and Schrier, R.W. (1976). *Clinical Use of Drugs in Renal Failure*. Charles C. Thomas, Springfield, Illinois.

Aronson, J.K., Dengler, H.J., Dettli, L., and Follath, F. (1988). Standardization of symbols in clinical pharmacology. *European Journal of Clinical Pharmacology*, **35**, 1–7.

Balant, L.P., Dayer, P., and Fabre, J. (1983). Consequences of renal insufficiency on the hepatic clearance of some drugs. *International Journal of Clinical Pharmacology Research*, **3**, 459.

Drayer, D.E. (1977). Active drug metabolites and renal failure. *American Journal of Medicine*, **62**, 486–9.

Duchin, K.L. and Schrier, R.W. (1983). Inter-relationship between renal haemodynamics, drug kinetics and drug action. In *Handbook of Clinical Pharmacokinetics* (eds. M. Gibaldi and L. Prescott), Section I, pp. 183–97. ADIS Health Science Press, Balgowlah, Australia.

Gibson, T.P. and Nelson, H.A. (1983). Drug kinetics and artificial kidneys. In *Handbook of Clinical Pharmacokinetics* (eds. M. Gibaldi and L. Prescott), Section III, pp. 301–24. ADIS Health Science Press, Balgowlah, Australia.

Grahame-Smith, D.G. and Aronson, J.K. (1984). The pharmacokinetic process. In *The Oxford Textbook of Clinical Pharmacology and Drug Therapy*, pp. 25–29. Oxford University Press, Oxford.

Park, G.D. (1987). Pharmacokinetics. In *The Scientific Basis of Clinical Pharmacology* (ed. R. Spector), pp. 67–102. Little, Brown and Co., Boston.

Reynolds, J.E.F., ed. (1989). *Martindale. The Extra Pharmacopoeia*. The Pharmaceutical Press, London.

Rowland, M. and Tozer, T.N. (1980). *Clinical Pharmacokinetics*. Philadelphia, Lea and Febiger.

References

Adjepon-Yamoah, K.K., Edwards, G., Breckenridge, A.M., Orme, M.L'E., and Ward, S.A. (1982). The effects of renal disease on the pharmacokinetics of diethylcarbamazine in man. *British Journal of Clinical Pharmacology*, **13**, 829–34.

Adler, D.S., Martin, E., Gambertoglio, J.G., Tozer, T.N., and Spire, J-P. (1975). Haemodialysis of phenytoin in a uremic patient. *Clinical Pharmacology and Therapeutics*, **18**, 65–9.

Alberts, D.S., *et al*. (1979). Kinetics of intravenous melphalan. *Clinical Pharmacology and Therapeutics*, **26**, 73–80.

Albin, H.C., Demotes-Mainard, F.M., Bouchet, J.L., Vincon, G.A., and Martin-Dupont, C. (1985). Pharmacokinetics of intravenous and intraperitoneal cefotaxime in chronic ambulatory peritoneal dialysis. *Clinical Pharmacology and Therapeutics*, **38**, 285–9.

Aletta, J.M., Francke, E.F., and Neu, H.C. (1980). Intravenous azlocillin kinetics in patients on long-term haemodialysis. *Clinical Pharmacology and Therapeutics*, **27**, 563–6.

Allen, R.D.M., Hunnisett, A.G., and Morris, P.J. (1985). Cyclosporin and rifampicin in renal transplantation. *Lancet*, **i**, 980.

Allgayer, H., Zahringer, J., Bach, P., and Bircher, J. (1984). Lack of effect of haemodialysis on mebendazole kinetics: studies in a patient with echinococciosis and renal failure. *European Journal of Clinical Pharmacology*, **27**, 243–5.

Anderson, R.J., Gambertoglio, J.G., and Schrier, R.W. (1982). Prescriber medication in long-term dialysis units. *Archives of Internal Medicine*, **142**, 1305–8.

Andrew, O.T. (1980). Tuberculosis in patients with end-stage renal failure. *American Journal of Medicine*, **68**, 59–65.

Aronoff, G.R. and Bennett, W.J. (1988). Cefuroxime dosage in renal failure. *Annals of Internal Medicine*, **109**, 990.

Balducci A., Slama, D.G., Rottembourg, J., Baumelou, A., and Delage, A. (1981). Intraperitoneal insulin in uraemic diabetics undergoing continuous ambulatory peritoneal dialysis. *British Medical Journal*, **283**, 1021–3.

Belpair, F.M., Van de Velde, E.J., Fraeyman, N.H., Bogaert, M.G., and Lameire, N. (1988). Influence of continuous ambulatory peritoneal dialysis on serum alpha$_1$-acid glycoprotein concentration and drug binding. *European Journal of Clinical Pharmacology*, **35**, 339–43.

Bennett, W.M. (1980). Drug therapy in renal failure: dosing guidelines for adults. Parts I and II. *Annals of Internal Medicine*, **93**, 62–89.

Bennett, W.M., Aronoff, G.R., Golper, T.A., Morrison, G., Singer, I., and Brater, D.C. (1987). *Drug Prescribing in Renal Failure: Dosing Guidelines for Adults*. Philadelphia, American College of Physicians.

Beaman, M., Soalro, L., McGonigle, R.J.S., Michael, A., and Adu, D. (1989). Vancomycin and ceftazidime in the treatment of CAPD peritonitis. *Nephron*, **51**, 51–5.

Berg, J.K. and Talseth, T. (1985). Acute renal effects of sulindac and indomethacin in chronic renal failure. *Clinical Pharmacology and Therapuetics*, **37**, 325–9.

Bergan, T., Broadwall, E.K., and Wilk-Larsen, E. (1979) Mezlocillin pharmacokinetics in patients with normal and impaired renal functions. *Antimicrobial Agents and Chemotherapy*, **16**, 651–4.

Bergrem, H. (1983). Pharmacokinetics and protein binding of prednisolone in patients with nephrotic syndrome and patients undergoing haemodialysis. *Kidney International*, **23**, 876–81.

Berlinger, W.G., Park, G.D., and Spector, R. (1985). The effect of dietary protein and the clearance of allopurinol and oxypurinol. *New England Journal of Medicine*, **313**, 771–6.

Bianchetti, G., *et al.* (1976). Pharmacokinetics and effects of propranolol in terminal uraemic patients and in patients undergoing regular dialysis treatment. *Clinical Pharmacokinetics*, **1**, 373–84.

Biaski, K.M. (1980). Disease induced changes in the plasma binding of basic drug. *Clinical Pharmacokinetics*, **5**, 246–62.

Bickley, S.K. (1988). Drug dosing during continuous arteriovenous hemofiltration. *Clinical Pharmacy*, **7**, 198–206.

Bjaeldager, P.A.L., Jensen, J.B., and Larsen, N-E. (1980). Elimination of oral cimetidine in chronic renal failure and during haemodialysis. *British Journal of Clinical Pharmacology*, **9**, 585–92.

Bliss, M.R. (1981). Prescribing for the elderly. *British Medical Journal*, **283**, 203–6.

Block, E.R. and Bennett, J.E. (1974). Flucytosine and amphotericin B: haemodialysis effects on plasma concentration and clearance. *Annals of Internal Medicine*, **80**, 613–7.

Blum, M.R., Liao, S.H.T., and de Miranda, P. (1982). Overview of acyclovir pharmacokinetic disposition in adults and children. *American Journal of Medicine*, **73**, 186–92.

Bonati, M., *et al.* (1987). Teicoplanin pharmacokinetics in patients with chronic renal failure. *Clinical Pharmacokinetics*, **12**, 292–301.

Bowman, R.H. (1975). Renal secretion of [S^{35}] furosemide and its depression by albumen binding. *American Journal of Physiology*, **229**, 93–8.

Bowmer, C.J. and Lindup, W.E. (1982). Decreased drug binding in uraemia: effect of indoxyl sulphate and other endogenous substances on the binding of drugs and dyes to human albumin. *Biochemical Pharmacology*, **31**, 319–23.

Branch, R.A. (1988). Prevention of amphotericin B-induced renal impairment. A review on the use of sodium supplementation. *Archives of Internal Medicine*, **148**, 2389–94.

Brater, D.C. (1985). *Handbook of Drug Use in Patients with Renal Disease*, 2nd edn. Improved Therapeutics, Lancaster, USA.

Brezis, M. and Epstein, F.H. (1989). A closer look at radiocontrast induced nephropathy. *New England Journal of Medicine*, 320, 179–81.
British Medical Association and Royal Pharmaceutical Society of Great Britain (1990). Prescribing in renal impairment. *British National Formulary*, 17th edn. pp. 19–27. The Bath Press, United Kingdom.
Brown, D.L. and Mauro, L.S. (1988). Vancomycin dosing chart for use in patients with renal impairment. *American Journal of Kidney Diseases* 11, 15–19.
Burk, M. and Peters, U. (1983). Disopyramide kinetics in renal impairment: determinants of inter individual variability. *Clinical Pharmacology and Therapeutics*, 34, 331–40.
Caldwell, J.E., *et al.* (1988). Pipercuronium and pancuronium: comparison of pharmacokinetics and duration of action. *British Journal of Anaesthesia*, 61, 693–7.
Cantu, T.G. and Fantozzi, D. (1988). Cefuroxime dosage in renal failure. *Annals of Internal Medicine*, 109, 989–90.
Carter, C.A., Wojcieckowski, N.J., Hayes, J.M., Skoutakis, V.A., and Rickman, L.A. (1985). Terfenadine, a non-sedating antihistamine. *Drug Intelligence and Clinical Pharmacy*, 19, 812–7.
Carvillo, F., Bosch, J., Arroyo, V., Mas, A., Viver, J., and Rodes, J. (1977). Renal failure associated with demeclocycline in cirrhosis. *Annals of Internal Medicine*, 87, 195–7.
Chapron, D.J., Gomolin, I.H., and Sweeney, K.R. (1989). Acetazolamide blood concentrations are excessive in the elderly: Propensity for acidosis and relationship to renal function. *Journal of Clinical Pharmacology*, 29, 348–53.
Chauvin, M., Lebrault C., Levron J.C., and Duvaldestin P. (1987a). Pharmacokinetics of alfentanil in chronic renal failure. *Anaesthetics and Analgesia*, 66, 53–6.
Chauvin, M., Sandouk, P., Scherrmann, J.M., Farinotti, R., Strumza, P., and Duvaldestin, P. (1987b). Morphine pharmacokinetics in renal failure. *Anaesthiology*, 66, 327–31.
Chennavasin, P. and Brater, D.C. (1981). Nomograms for drug use in renal disease. *Clinical Pharmacokinetics*, 6, 193–214.
Christian, C.D., Meredith, C.G., and Speeg, K.V. (1984). Cimetidine inhibits renal procainamide clearance. *Clinical Pharmacology and Therapeutics*, 36, 221–7.
Coburn, J.W. and Slatopolsky, E. (1986). Vitamin D, parathyroid hormone, and renal osteodystrophy. In *The Kidney* (eds. B.M. Brenner and F.C. Rector), 3rd edn. pp. 1657–729. W.B. Saunders, Eastbourne.
Cockcroft, D.W. and Gault, B.H. (1976). Prediction of creatinine clearance from serum creatinine. *Nephron*, 16, 31–41.
Collingsworth, K.A., Strong, J.M., Atkinson, A.J., Winkle, R.A., Perlroth, F., and Harrison, D.C. (1975). Pharmacokinetics and metabolism of lidocaine in patients with renal failure. *Clinical Pharmacology and Therapeutics*, 18, 59–64.
Collins, J.M., *et al.* (1986). Clinical pharmacokinetics of suramin in patients with HTLV-III/LAV inefection. *Journal of Clinical Pharmacology*, 26, 22–6.
Cook, P. (1979). How drug action is altered in the elderly. *Geriatric Medicine*, 34, 45–6.
Craig, W.A. and Kunin, C.M. (1973). Trimethoprim-sulfamethoxazole: pharmacodynamic effects of urinary pH and impaired renal function. *Annals of Internal Medicine*, 78, 491–7.
Craig, R., Murphy, T., and Gibson, T.P. (1983). Kinetic analysis of D-xylose absorption in normal subjects and in patients with chronic renal insufficiency. *Journal of Laboratory and Clinical Medicine*, 101, 496–506.
Cuss, F.M.C., Carmichael, D.J.S., Allington, A., and Hulme, B. (1986). Tuberculosis in renal failure: a high incidence in patients born in the third world. *Clinical Nephrology*, 25, 129–33.
Daneshmend, T.K. and Warnock, D.W. (1983). Clinical pharmacokinetics of antifungal drugs. *Clinical Pharmacokinetics*, 8, 17–42.
Danhof, M., Hisaoka, M., and Levy, G. (1984). Kinetics of drug action in disease states II. Effect of experimental renal dysfunction on phenobarbital concentration in rats at onset of loss of righting reflex. *Journal of Pharmacology and Experimental Therapeutics*, 230, 627–31.
Davies, B.E., Boon, R., Horton, R., Reubi, F.C., and Descoeudres, C.E. (1988). Pharmacokinetics of amoxicillin and clavulanic acid in haemodialysis patients following intravenous administration of augmentin. *British Journal of Clinical Pharmacology*, 26, 385–90.
Deray, G., *et al.* (1988). Pharmacokinetics of zidovudine in a patient on maintenance haemodialysis. *New England Journal of Medicine*, 319, 1606–7.
Desai, T.K. and Tsang, T-K. (1988). Aminoglycoside nephrotoxicity in obstructive jaundice. *American Journal of Medicine*, 85, 47–50.
Dettli, L. (1983). Drug dosage in renal disease. In *Handbook of Clinical Pharmacokinetics* (eds. M. Gibaldi and L. Prescott), Section III, pp. 261–76. ADIS Health Science Press, Balgowlah, Australia.
Dingemanse, J., Polhuijs, M., and Danhof, M. (1988). Altered pharmacokinetic-pharmacodynamic relationship of heptabarbital in ecperimental renal failure in rats. *Journal of Pharmacology and Experimental Therapeutics*, 246, 371.
Diskin, C.J., Ravis, W., Campagna, K.D., and Clark, C.R. (1988). Pharmacokinetics of sulindac in ESRD. *Nephron*, 50, 397.
Dodge, W.F., Travis, L.B., and Daeschner, C.W. (1967). Comparison of endogenous creatinine clearance with inulin clearance. *American Diseases of Childhood*, 113, 683–92.
Drayer, D.E. (1983). Pharmacologically active drug metabolites: therapeutic and toxic activities, plasma and urine data in man, accumulation in renal failure. In *Handbook of Clinical Pharmacokinetics* (eds. M. Gibaldi and L. Prescott), pp. 114–32. New York, Raven Press.
Drummer, O.H., Workman, B.S., Miach, P.J., Jarrett, B., and Lovis, W.J. (1987). The pharmacokinetics of captopril and captopril disulfide conjugates in uraemic patients on maintenance dialysis: comparison with patients with normal renal function. *European Journal of Clinical Pharmacology*, 3, 267–71.
Duchin, K.L., Pierides, A.M., Heald, A., Singhvi, S.M., and Rommel, A.J. (1984). Elimination kinetics of captopril in patients with renal failure. *Kidney International*, 25, 942–8.
Dundee, J.W. and Richards, R.K. (1954). Effect of azotemia upon the action of intravenous barbiturate anesthesia. *Anesthesiology*, 15, 333–46.
Dunn, M.J. (1983). Renal prostaglandins. In *Renal Endocrinology* (ed. M.J. Dunn), pp. 1–74. Williams and Wilkins, Baltimore/London.
Edell, L.S., Westby, G.R., and Gould, S.R. (1988). An improved method of vancomycin administration to dialysis patients. *Clinical Nephrology*, 29, 86–7.
Editorial (1979). Diuretic resistance? *Lancet*, i, 253–4.
Edwards, G., Breckenridge, A.M., Adjepon-Yamoah, K.K., Orme, M.L'e., and Ward, S.A. (1981). The effects of variations in urinary pH on the pharmacokinetics of diethylcarbamazine. *British Journal of Clinical Pharmacology*, 12, 807–12.
Eriksson, R.K., Wahlin-Boll, E., Odar-Cederlof, I., Lindholm, L., and Melander, A. (1989). Influence of renal failure, rheumatoid arthritis and old age on the pharmacokinetics of diflunisal. *European Journal of Clinical Pharmacology*, 36, 165–74.
Fabre, J., Fox, H.M., Dayer, P., and Balant, L. (1983). Differences in kinetic properties of drugs: implications as to the selection of a particular drug for use in patients with renal failure, with special emphasis on antibiotics and beta adrenoceptor blocking agents. In *Handbook of Clinical Pharmacokinetics* (eds. M. Gibaldi and L. Prescott), Section III, pp. 233–60. ADIS Health Science Press, Balgowlah, Australia.
Fillastre, J.P., *et al.* (1985). Pharmacokinetics of aztreonam in patients with chronic renal failure. *Clinical Pharmacokinetics*, 10, S91–100.
Fleming, L.W., Morland, T.A., Scott, A.C., Stewart, W.K., and White, L.O. (1987). Ciprofloxacin in plasma and peritoneal dialysate after oral therapy in patients on continuous ambulatory peritoneal dialysis. *Journal of Antimicrobial Chemotherapy*, 19, 493–503.
Fletcher, C.V., Chinnock, B.J., Chace, B., and Balfour, H.H. (1988). Pharmacokinetics and safety of high-dose oral acyclovir for

suppression of cytomegalovirus disease after renal transplantation. *Clinical Pharmacology and Therapeutics*, **44**, 158–63.

Forland, S.C., et al. (1988). Oral flecainide pharmacokinetics in patients with impaired renal function. *Journal of Clinical Pharmacology*, **28**, 259–67.

Forsman, A. and Ohman, R. (1976). Pharmacokinetic studies on haloperidol in man. *Current Therapeutic Research*, **20**, 319–36.

Francke, E.L., Appel, G.B., and Neu, H.C. (1979). Kinetics of intravenous amoxycillin in patients on long-term dialysis. *Clinical Pharmacology and Therapeutics*, **26**, 31–5.

Gabriel, R. (1986). Time to scrap creatinine clearance? *British Medical Journal*, **293**, 1119–20.

Gerig, J.S., Bolton, N.D., Swabb, E.A., Scheld, M., and Bolton, W.K. (1984). Effect of haemodialysis and peritoneal dialysis on aztreonam pharmacokinetics. *Kidney International*, **26**, 308–18.

Gibaldi, M. (1977). Drug distribution in renal failure. *American Journal of Medicine*, **62**, 471–4.

Gibaldi, M. (1984). *Biopharmaceutics and Clinical Pharmacokinetics*. Lea and Febiger, Philadelphia.

Gibson, T.P. (1986). Renal disease and drug metabolism: an overview. *American Journal of Kidney Diseases*, **8**, 7–17.

Gibson, T.P., et al. (1977). Kinetics of procaine and N-acetyl procainamide in renal failure. *Kidney International*, **12**, 422–9.

Gibson, T.P., Dobrinska, M.R., Entwhistle, L.A., and Davies, R.O. (1987). Biotransformation of sulindac in end-stage renal disease. *Clinical Pharmacology and Therapeutics*, **42**, 82–8.

Gladziwa, U., et al. (1988). Pharmacokinetics of ranitidine in patients undergoing haemofiltration. *European Journal of Clinical Pharmacology*, **35**, 427–30.

Gold, C.H., Buchanan, N., Tringham, V., Viljoen, M., Strickwold, B., and Moodley, G.P. (1976). Isoniazid pharmacokinetics in patients in chronic renal failure. *Clinical Nephrology*, **6**, 365–9.

Golden, J.A., Chernoff, D., Hollander, H., Fiegal, D., and Conte, J.E. (1989). Prevention of *Pneumocystis carinii* pneumonia by inhaled pentamidine. *Lancet*, **i**, 654–7.

Granistad, L. (1987). Atracurium, vecuronium and pancuronium in end-stage renal failure. *British Journal of Anaesthesia*, **59**, 995–1003.

Grech-Belanger, O., Langlois, S., and LeBoeuf, E. (1988). Pharmacokinetics of diltiazem in patients undergoing continuous ambulatory peritoneal dialysis. *Journal of Clinical Pharmacology*, **284**, 73–80.

Guay, D.R., et al. (1988). Pharmacokinetics and pharmacodynamics of codeine in end-stage renal disease. *Clinical Pharmacology and Therapeutics*, **43**, 63–71.

Gugler, R. and Azarnoff, D.L. (1983). Drug protein binding and the nephrotic syndrome. In *Handbook of Clinical Pharmacokinetics* (eds. M. Gibaldi and L. Prescott), Section III, pp. 96–108. ADIS Health Science Press, Balgowlah, Australia. 96–108.

Gulyassy, P.F. and Depner, T.A. (1983). Impaired binding of drugs and ligands in renal diseases. *American Journal of Kidney Diseases*, **2**, 578–601.

Heaney, D. and Eknoyan, G. (1978). Minocycline and doxycycline kinetics in chronic renal failure. *Clinical Pharmacology and Therapeutics*, **24**, 233–9.

Heim-Duthoy, K.L. (1986). The effect of haemodialysis on piperacillin pharmacokinetics. *International Journal of Clinical Pharmacology, Therapeutics, and Toxicology*, **24**, 680–4.

Henderson, I.S., Patterson, K.R., and Leung, A.C.T. (1985). Decreased intraperitoneal insulin requirements during peritonitis on continuous ambulatory peritoneal dialysis. *British Medical Journal*, **290**, 1474.

Herman, R.J., McAllister, C.B., Branch, R.A., and Wilkinson, G.R. (1985). Effects of age on meperidine disposition. *Clinical Pharmacology and Therapeutics*, **37**, 19–24.

Hisaoka, M., Danhof, M., and Levy, G. (1985). Kinetics of drug action in disease states VII. Effect of experimental renal dysfunction on the pharmacodynamics of ethanol in rats. *Journal of Pharmacology and Experimental Therapeutics*, **232**, 717–21.

Hockings, N., Ajayi, A., and Reid, J.L. (1986). Age and pharmacokinetics of angiotensin converting enzyme inhibitors enalapril and enalaprilat. *British Journal of Clinical Pharmacology*, **21**, 341–8.

Houin, G., Burnner, F., Nebout, T., Cherfasni, M., Lagrue, G., and Tillement, J.P. (1983). The effects of chronic renal insufficiency on the pharmacokinetics of doxycycline in man. *British Journal of Clinical Pharmacology*, **16**, 245–52.

Huang, C.M., Atkinson, A.J., Levin, M., Levin, N.W., and Quintanilla, A. (1974). Pharmacokinetics of furosemide in advanced renal failure. *Clinical Pharmacology and Therapeutics*, **16**, 659–66.

Humes, H.D. (1988). Aminoglycoside nephrotoxicity. *Kidney International*, **33**, 900–11.

Johnson, R.J., Blair, A.D., and Ahmed, S. (1985). Ketoconazole kinetics in chronic peritoneal dialysis. *Clinical Pharmacology and Therapeutics*, **37**, 325–9.

Kanfer, A., Stamatakis, G., Torlotin, J.C., Fredj, G., Kenouch, S., and Mery, J.P. (1987). Changes in erythromycin pharmacokinetics induced by renal failure. *Clinical Nephrology*, **27**, 147–50.

Kapstein, E.M., Change, E.I., Egodage, P.M., Nicoloff, J.T., and Massry, S.G. (1987). Thyroxine transfer and distribution in critical nonthyroidal illnesses, chronic renal failure and chronic ethanol abuse. *Journal of Clinical Endocrinology and Metabolism*, **65**, 606–16.

Keller, E., Hoppe-Seyler, G., and Schoolmeyer, P. (1982). Disposition and diuretic effect of furosemide in the nephrotic syndrome. *Clinical Pharmacology and Therapeutics*, **32**, 442–9.

Kelly, J.G., Doyle, G.D., Carmody, M., Gover, D.R., and Cooper, W.D. (1988). Pharmacokinetics of lisinopril, enalapril and enalaprilat in renal failure: Effects of haemodialysis. *British Journal of Clinical Pharmacology*, **26(6)**, 781–6.

Kitt, T.M., Park, G.D., Spector, R., and Tsalikian E. (1989). Reduced renal clearance of oxypurinol during a 400 calorie protein-free diet. *Journal of Clinical Pharmacology*, **29**, 65–71.

Kornhauser, D.M., et al. (1989). Probenecid and zidovudine metabolism. *Lancet*, **ii**, 473–5.

Koup, J.R., Jusko, W.J., Elwood, C.M., and Kohli, R.K. (1975). Digoxin pharmacokinetics: role of renal failure in dosage regimen design. *Clinical Pharmacology and Therapeutics*, **18**, 9–21.

Kraan, J., et al. (1988). The pharmacokinetics of theophylline and enprofylline in patients with liver cirrhosis and in patients with chronic renal disease. *European Journal of Clinical Pharmacology*, **35**, 357–62.

Kristensen, M.B. (1983). Drug interactions and clinical pharmacokinetics. In *Handbook of Clinical Pharmacokinetics* (eds. M. Gibaldi and L. Prescott), Section I, pp. 242–63. ADIS Health Science Press, Balgowlah, Australia.

Kroboth, P.D., et al. (1985). Effects of end-stage renal disease and aluminium hydroxide on temazepam kinetics. *Clinical Pharmacology and Therapeutics*, **37**, 453–9.

Kunin, C.M. (1967). A guide to the use of antibiotics in renal disease. *Annals of Internal Medicine*, **67**, 151–8.

Laher, M.S., Donohoe, J.F., Kelly, J.G., and Doyle, G.D. (1988). Antihypertensive and renal effects of lisinopril in older patients with hypertension. *American Journal of Medicine*, **85**, 38–43.

Lake, K.D., Fletcher, C.V., Love, K.R., Brown, D.C., Joyce, L.D., and Pritzker, M.R. (1988). Ganciclovir pharmacokinetics during renal impairment. *Antimicrobial Agents and Chemotherapy*, **32**, 1899–900.

Lalau, J.D., et al. (1989). Hemodialysis in the treatment of lactic acidosis in diabetics treated by metformin: a study of metformin elimination. *Clinical Pharmacology, Therapy and Toxicology*, **27**, 285–8.

Laskin, O.I., et al. (1982). Acyclovir kinetics in end-stage renal disease. *Clinical Pharmacology and Therapeutics*, **31**, 594–601.

Latini, R., Tognoni, G., and Kates, R.E. (1984). Clinical pharmacokinetics of amiodarone. *Clinical Pharmacokinetics*, **9**, 136–56.

Latos, D.S., Bryan, C.S., and Stone, W.J. (1975). Carbenicillin therapy in patients with normal and impaired renal function. *Clinical Pharmacology and Therapeutics*, **17**, 692–700.

Laville, M., Hadj-Aissa, A., Pozet, N., Le Bras, J-H., Labeeuw, M., and Zech, P. (1989). Restrictions on use of creatinine clearance for measurement of renal functional reserve. *Nephron*, **51**, 233–6.

Lee, C.S., Marbury, T.C., and Benet, L.Z. (1980). Clearance calculations in haemodialysis: application to blood, plasma and dialysate measurements for ethambutol. *Journal of Pharmacokinetics and Biopharmaceuticals*, **8**, 69–81.

Lee, C.S., Marbury, T.C., Perchalskin, R.T., and Wilder, B.J. (1982). Pharmakokinetics of primodone elimination by uremic patients. *Journal of Clinical Pharmacology*, **22**, 301–8.

Lehmann, C.R., *et al.* (1988). Capreomycin kinetics in renal impairment and clearance by hemodialysis. *American Review of Respiratory Disease*, **138**, 1312–13.

Lerner, A.M., *et al.* (1983). Randomized, controlled trial of the comparative efficacy, auditory toxicity and nephrotoxicity of tobramycin and netilmycin. *Lancet*, **ii**, 1123–6.

Leroy, A., Leguy, F., Borsa, F., Spencer, G.R., Fillastre, J.P., and Humbert, G. (1984). Pharmacokinetics of ceftazidime in normal and uremic subjects. *Antimicrobial Agents and Chemotherapy*, **25**, 638–42.

Lewi, P.J., *et al* (1976). Pharmacokinetic profile of intravenous miconazole in man. Comparison of normal subjects and patients with renal insufficiency. *European Journal of Clinical Pharmacology*, **10**, 49–54.

Lewis, G.P., Jusko, W.J., and Burke, C.W. (1971). Prednisolone side-effects and serum protein levels. *Lancet*, **ii**, 778–80.

Lichtenwalner, D.M. and Suh, B. (1983). Isolation and chemical characterization of 2-hydroxybenzoylglycine as a drug binding inhibitor in uraemia. *Journal of Chemical Investigation*, **71**, 1289–96.

Lieberman, J.A., *et al.* (1985). Tricyclic antidepressant and metabolite levels in chronic renal failure. *Clinical Pharmacology and Therapeutics*, **37**, 301–7.

Local, F.K., *et al.* (1981). Pharmacokinetics of intravenous and intraperitoneal cefuroxime in patients undergoing peritoneal dialysis. *Clinical Nephrology*, **16**, 1640–3.

Lynam, D.P., *et al.* (1988). The pharmacokinetics of vecuronium in patients anesthetized with isoflurane with normal renal function or with renal failure. *Anesthesiology*, **69**, 227–31.

Mahuchi, H. and Nakahashi, H. (1988). A major inhibitor of phenytoin binding to serum protein in uraemia. *Nephron*, **48**, 310–4.

Martre, H., Sari, R., Taburet, A.M., Jacobs, C., and Singlas, E. (1985). Haemodialysis does not affect the pharmacokinetics of nifedipine. *British Journal of Clinical Pharmacology*, **20**, 155–8.

Meuldermans, W., *et al.* (1988). Alfentanil pharmacokinetics and metabolism in humans. *Anesthesiology*, **69**, 527–34.

Michael, K.A., *et al.* (1985). Failure of creatinine clearance to predict gentamycin half-life in a renal transplant with diabetes mellitus. *Clinical Pharmacology*, **4**, 572–5.

Miller, R.F., Delany, S., and Semple, S.J.G. (1989). Acute renal failure after nebulized pentamidine. *Lancet*, **i**, 1271–2.

Mooy, J., Schols, M., v Baak, M., v Hoof, M., Muytjents, A., and Rahn, K.H. (1985). Pharmacokinetics of verapamil in patients with renal failure. *European Journal of Clinical Pharmacology*, **28**, 405–10.

Morrison, G., Chiang, S.T., Koepke, H.H., and Walker, B.R. (1984). Effect of renal impairment and haemodialysis on lorazepam kinetics. *Clinical Pharmacology and Therapuetics*, **35**, 646–52.

Nanji, A.A. and Greenway, D.C. (1988). Falsely raised plasma theophylline concentrations in renal failure. *European Journal of Clinical Pharmacology*, **34**, 309–10.

Naschitz, J.E., Barak, C., and Yeshurun, D. (1983). Reversible diminished insulin requirement during acute renal failure. *Postgraduate Medical Journal*, **59**, 269–71.

Nauta, E.H. and Mattie, H. (1976). Dicloxacillin and cloxacillin; pharmacokinetics in healthy and haemodialysis subjects. *Clinical Pharmacology and Therapeutics*, **20**, 98–108.

Nesher, G., Zimran, A., and Hershko, C. (1988). Reduced incidence of hyperkalaemia and azotaemia in patients receiving sulindac compared with indomethacin. *Nephron*, **48**, 291–5.

O'Malley, K. and Meagher, F. (1985). Pharmacological aspects of therapeutics. In *Practical Geriatric Medicine* (eds. A.N. Exton-Smith and M.E. Weksler), Churchill Livingstone, Edinburgh.

Odlind, B.G. and Beerman, B. (1980). Diuretic resistance reduced bioavailability and effect of oral frusemide. *British Medical Journal*, **280**, 1577.

Oster, J.R., Epstein, M., and Ulano, H.B. (1976). Deterioration of renal function with demeclocycline administration. *Current Therapeutics Research*, **20**, 794–801.

Palmer, R.F. and Lasseter, K.C. (1975). Sodium nitroprusside. *New England Journal of Medicine*, **292**, 294–7.

Parfrey, P.S., *et al.* (1989). Contrast material-induced renal failure in patients with diabetes mellitus, renal insufficiency, or both. *New England Journal of Medicine*, **320**, 143–9.

Parker, C.J.R., Jones, J.E., and Hunter, J.M. (1988). Disposition of infusions of atracurium and its metabolite, laudanosine, in patients in renal and respiratory failure in an ITU. *British Journal of Anaesthetics*, **61**, 531–40.

Parry, M.F. and Neu, H.C. (1976). Pharmacokinetics of ticarcillin in patients with abnormal renal function. *Journal of Infectious Diseases*, **133**, 46–9.

Paton, D.M. and Webster, D.R. (1985). Clinical pharmacokinetics of H1-receptor antagonists (the antihistamines). *Clinical Pharmacokinetics*, **10**, 477–97.

Pedersen, L.E., Booude, J., Graudal, N.A., Backer, A.V.V., Hausen, J.E.S., and Kampmann, J.P. (1987). Quantitative and qualitative binding characteristics of disopyramide in serum from patients with decreased renal and hepatic function. *British Journal of Pharmacology*, **23**, 41–6.

Perez-Ayuso, R.M., *et al.* (1984). Effect of demeclocycline on renal function and urinary prostaglandin E_2 and kallikrein in hyponatraemic cirrhosis. *Nephron*, **36**, 30–7.

Periti, P., Mazzei, T., Mini, E., and Norelli, A. (1989). Clinical pharmacokinetic properties of the macrolide antibiotics. Effects of age and various pathophysiological states. Part 1. *Clinical Pharmacokinetics*, **16**, 193–214.

Petersen, V., Hvidt, S., Thomsen, S., and Schon, M. (1974). Effect of prolonged thiazide treatment on renal lithium clearance. *British Medical Journal*, **3**, 143–5.

Prescott, L.F., Speirs, G.C., Critchley, J.A.J.H., Temple, R., and Winney, R.J. (1989). Paracetamol disposition and metabolite kinetics in patients with chronic renal failure. *European Journal of Clinical Pharmacology*, **36**, 291–7.

Rabkin, R., Simon, N.M., Steiner, S., and Collwell, J.A. (1970). Effect of renal disease on renal uptake and excretion of insulin in man. *New England Journal of Medicine*, **282**, 182–7.

Ramzan, I.M. and Levy, G. (1987). Kinetics of drug action in disease states XVIII. Effect of experimental renal failure on the pharmacodynamics of theophylline-induced seizures in rats. *Journal of Experimental Pharmacology and Therapeutics*, **240**, 584–8.

Reidenberg, M.M. (1983). The binding of drugs to plasma proteins from patients with poor renal function. In *Handbook of Clinical Pharmacokinetics* (eds. M. Gibaldi and L. Prescott), Section III, pp. 89–95. ADIS Health Science press, Balgowlah, Australia.

Reidenberg, M.M. (1985). Kidney function and drug action. *New England Journal of Medicine*, **313**, 816–8.

Roberts, C.J.C. (1984). Clinical pharmacokinetics of ranitidine. *Clinical Pharmacokinetics*, **9**, 211–21.

Roberts, D.E. and Williams, J.D. (1989). Ciprofloxacin in renal failure. *Antimicrobial Agents and Chemotherapy*, **23**, 820–3.

Rowe, J.W., Andres, R.A., Tobin, J.D., Norris, A.H., and Shock, N.W. (1976). Age-adjusted normal standards for creatinine clearance in man. *Annals of Internal Medicine*, **84**, 567–9.

Salazar, D.E. and Corcoran, G.B. (1988). Predicting creatinine clearance and renal drug clearance in obese patients from estimated fat-free body mass. *American Journal of Medicine*, **84**, 1053–60.

Sarubbi, F.A. and Hull, J.H. (1978). Amikacin serum concentrations: prediction of levels and dosage guidelines. *Annals of Internal Medicine*, **89**, 612–8.

Sattler, F.R., et al. (1988). Trimethoprim-sulphamethoxazole compared with pentamidine for treatment of *Pneumocystis carinii* pneumonia in the acquired immunodeficiency syndrome. *Annals of Internal Medicine*, **109**, 280–7.

Sawe, J., et al. (1987). Kinetics of morphine in patients in renal failure. *European Journal of Pharmacology*, **32**, 377–82.

Schrier, R.W. and Conger, J.D. (1986). Acute renal failure: pathogenesis, diagnosis and management. In *Renal and Electrolyte Disorders* (ed. R.W. Schrier), 3rd edn. pp. 423–60. Little, Brown, and Company, New York.

Schusziarra, V., et al. (1986). Pharmacokinetics of azathioprine under haemodialysis. *International Journal of Clinical Pharmacology and Therapeutics*, **14**, 298–302.

Schroeder, S.A., Krupp, M.A., Tierney, L.M., and McPhee, S.J., eds. (1989). *Current Medical Diagnosis and Treatment*. Appleton and Lange, Connecticut.

Schwab, S.J., et al. (1989). Contrast nephrotoxicity: a randomized controlled trial of a nonionic and an ionic radiographic contrast agent. *New England Journal of Medicine*, **320**, 149–53.

Sear, J.W. (1988). Drug metabolites in anaesthetic practice—are they important? *British Journal of Anaesthesia*, **61**, 525–6.

Shemesh, O., Goldbetz, H., Kriss, J.P., and Myers, B.D. (1985). Limitations of creatinine as a filtration marker in glomerulopathic patients. *Kidney International*, **28**, 830–8.

Sherlock, J.E. and Letteri, J.M. (1977). Effect of haemodialysis on methylprednisolone plasma levels. *Nephron*, **18**, 208–11.

Sica, D.A., Conistock, T., Harford, A., and Eshelman, F. (1987). Ranitidine pharmacokinetics in continuous ambulatory peritoneal dialysis. *European Journal of Clinical Pharmacology*, **32**, 587–91.

Singlas, E., Taburet, A.M., Landru, I., Albin, H., and Ryckelinck, J.P. (1987). Pharmacokinetics of ciprofloxacin tablets in renal failure: influence of haemodialysis. *European Journal of Clinical Pharmacology*, **31**, 589–93.

Smith, C.R., et al. (1980). Comparison of the nephrotoxicity and auditory toxicity of gentamicin and tobramycin. *New England Journal of Medicine*, **302**, 1106–9.

Somani, P., Shapiro, R.S., Stockard, H., and Higgins, J.T. (1982). Unidirectional absorption of gentamicin from the peritoneum during continuous ambulatory peritoneal dialysis. *Clinical Pharmacology and Therapeutics*, **32**, 113–21.

Somogyi, A., Kong, C., Sabto, J., Gurr, F.W., Spicer, W.J., and McLean, A.J. (1983). Disposition and removal of metronidazole in patients undergoing haemodialysis. *European Journal of Clinical Pharmacology*, **25**, 683–7.

Szeto, H.H., Inturissi, C.E., Honde, R., Saal, S., Cheigh, J., and Reidenberg, M.M. (1977). Accumulation of normeperidine, and active metabolite of meperidine in patients with renal failure. *Annals of Internal Medicine*, **86**, 738–41.

Terao, N. and Shen, D.D. (1985). Reduced extraction of 1-propranolol by perfused rat liver in the presence of uraemic blood. *Journal of Pharmacology and Experimental Therapeutics*, **233**, 277–84.

Thomson, A.H., Kelly, J.G., and Whiting, B. (1989). Lisinopril population pharmacokinetics in elderly and renal disease patients with hypertension. *British Journal of Clinical Pharmacology*, **27**, 57–65.

Thomson, P.D., et al. (1973). Lidocaine pharmacokinetics in advanced heart failure, liver disease and renal failure in humans. *Annals of Internal Medicine*, **778**, 499–508.

Tillement, J.P., Lhoste, F., and Guidicelli, J.F. (1983). Diseases and drug protein binding. In *Handbook of Clinical Pharmacokinetics* (eds. M. Gibaldi and L. Prescott), Section III, pp. 57–69. ADIS Health Science Press, Balgowlah, Australia.

Tilstone, W.J. and Fine, A. (1978). Furosemide kinetics in renal failure. *Clinical Pharmacology and Therapeutics*, **23**, 644–50.

Ueda, C.T., et al. (1975). Disposition kinetics of quinidine. *Clinical Pharmacology and Therapeutics*, **19**, 30–6.

Vanholder, R., Van Landschoot, N., De Swet, R., Schoots, A., and Ringoir, S. (1988). Drug protein binding in chronic renal failure: evaluation of nine drugs. *Kidney International*, **33**, 996–1004.

Van Peer, A., Belpaire, F., and Bogaert, M. (1978). Warfarin elimination and responsiveness in patients with renal dysfunction. *Journal of Clinical Pharmacology*, **18**, 84–8.

Verbeeck, R.K. and De Schepper, P.J. (1980). Influence of chronic renal failure and hemodialysis on diflunisal plasma protein binding. *Clinical Pharmacology and Therapeutics*, **27**, 628–35.

Verbeeck, R.K., Branch, R.A., and Wilkinson, G.R. (1981). Drug metabolites in renal failure: pharmacokinetic and clinical implications. *Clinical Pharmacokinetics*, **6**, 329–45.

Verpooten, G.A., Giuliano, R.A., Verbist, L., Eestermans, G., and De Broe, M.E. (1989). Once-daily dosing decreases renal accumulation of gentamicin and netilmicin. *Clinical Pharmacology and Therapeutics*, **45(1)**, 22–7.

Vinik, H.R., Reeves, J.G., Greenblatt, D.J., Abernethy, D.R., and Smith, L.R. (1983). The pharmacokinetics of midazolam in chronic renal failure patients. *Anesthesiology*, **59**, 390–4.

Walstad, R.A., Nilson, O.G., and Berg, K.J. (1983). Pharmacokinetics and clinical effects of cefuroxime in patients with severe renal insufficiency. *European Journal of Clinical Pharmacology*, **24**, 391–8.

Walstad, R.A., Dahl, K., Hellum, K.B., and Thurmann-Nielsen, E. (1988). The pharmacokinetics of ceftazidime in patients with impaired renal function and concurrent frusemide therapy. *European Journal of Clinical Pharmacology*, **35**, 273–9.

Wang, T., Wuellner, D., Woolsley, R.L., and Stone, W.J. (1985). Pharmacokinetics and nondialysability of mexiletene in renal failure. *Clinical Pharmacology and Therapeutics*, **37**, 649–53.

Ward, S., Boheimer, N., Weatherley, B.C., Simmonds, R.J., and Dopson, T.A. (1987). Pharmacokinetics of atracurium and its metabolites in patients with normal renal function, and patients in renal failure. *British Journal of Anaesthesia*, **59**, 697–706.

Watkin, D.M. and Shock, W.W. (1950). Age changes in glomerular filtration rate, effective renal plasma flow and tubular excretory capacity in adult males. *Journal of Clinical Investigation*, **29**, 496.

Webb, D., Buss, D.C., Fifield, R., Bateman, N., and Routledge, P.A. (1986). The plasma protein binding of metoclopramide in health and renal disease. *British Journal of Clinical Pharmacology*, **21**, 334–6.

Whelton, A. (1988). Treatment of Gram-negative infection in patients with renal impairment: new alternatives to aminoglycosides. *Journal of Clinical Pharmacology*, **28**, 866–78.

Whitby, M., Edwards R., Aston E., and Finch, R.G. (1987). Pharmacokinetics of single dose of intravenous vancomycin in CAPD peritonitis. *Journal of Antimicrobial Chemotherapy*, **19**, 351–7.

White, N.J. (1985). Clinical pharmacokinetics of antimalarial drugs. *Clinical Pharmacokinetics*, **10**, 187–215.

Wolff, J., Bigler, D., Christensen, C.B., Rasmussen, S.N., Andersen, H.B., and Tonnesen, K.H. (1988). Influence of renal function on the elimination of morphine and morphine glucuronides. *European Journal of Clinical Pharmacology*, **34**, 353–7.

Wright, M.R., et al. (1988). Effect of haemodialysis on metoclopramide kinetics in patients with severe renal failure. *British Journal of Clinical Pharmacology*, **26**, 474–7.

Ziuderma, J., Hilbers-Modderman, E.S.M., and Merkus, F.W.H.M. (1986). Clinical pharmacokinetics of dapsone. *Clinical Pharmacokinetics*, **11**, 299–315.

2.3 Action and clinical use of diuretics

R. GREGER AND A. HEIDLAND

Cellular mechanisms of action of diuretics

Studies of the mechanisms by which diuretics exert their effects have been performed over the past decades, and they have characterized these drugs as specific blockers of ion transporters. Different types of diuretics interact with the different ion transporters present in the various nephron segments, and diuretics can, therefore, be conveniently classified either according to their main target site in the nephron, or according to the ion transporter with which they interact. The former classification is more traditional, since the main target sites of the various groups of diuretics have been known for many years. The latter classification is probably more appropriate, however, since it takes into account the pharmacological interaction.

The aim of this section is to outline the mechanisms of water, Na^+, and Cl^- reabsorption in the tubule, and to relate these to the mode of action of the osmotic diuretics, which act throughout the water-permeable sites of the nephron; the carbonic anhydrase inhibitors, which act mainly in the proximal nephron; the loop diuretics; the thiazides acting in the early distal tubule and the K^+-sparing diuretics which act in the distal tubule and collecting duct.

Osmotic diuretics such as mannitol

The osmotic diuretics are freely filtered at the glomerulus and are not reabsorbed by the tubule; hence their tubular concentration increases as water is reabsorbed. The presence of these drugs in the tubule lumen inhibits the reabsorption of NaCl and water throughout the nephron, both in the proximal tubule and distal tubule.

Carbonic anhydrase inhibitors—proximal diuretics

The proximal nephron reclaims some 60 to 70 per cent of filtered water, Na^+, K^+, Cl^-, even more HCO_3^-, and almost 100 per cent of organic solutes such as glucose and amino acids. During recent years a variety of increasingly sophisticated techniques has been developed and used to characterize the processes involved in the absorption of water and many solutes (Ullrich and Greger 1985), although several aspects remain unresolved. These controversial aspects will not be addressed here. Instead, we base our further consideration of the topic on the generally accepted, but simplified, concept shown in Fig. 1(a) (Burckhardt and Greger 1991).

It is apparent that transport work is coupled to the action of the Na^+, K^+ pump which, as in most mammalian cells, expels three Na^+ ions from the cell in exchange for two K^+ ions, the process being fuelled by the hydrolysis of ATP. The proximal tubule cell, like all the different cell types in the nephron, is polarized, the pump being present only in the basolateral plasma membrane. The extrusion of Na^+ to the capillary side (also referred to as the blood, contraluminal, or basolateral side) provides the driving force for Na^+ uptake across the luminal membrane. This occurs chiefly via Na^+/H^+ exchange, mediated by a transport (carrier) protein. The rate of exchange is determined by the driving forces, and is regulated by a pH sensing site which increases the turnover rate whenever the cytosolic pH becomes acidic (Kinsella and Aronson 1980).

A smaller proportion of Na^+ uptake occurs through cotransport with solutes such as D-glucose, amino acids, or phosphate. The brush border (luminal) membrane is equipped with many such cotransporters, which all possess a Na^+ binding site and another site which is highly substrate specific. A small amount of Na^+ is taken into the lumen via selective ion channels (Gögelein and Greger 1986).

A very important feature of the proximal nephron is the extremely high permeability of the paracellular shunt pathway. The tight junctions and lateral spaces are highly permeable to water and small ions; the epithelium cannot therefore sustain any substantial ionic or osmotic gradient. Similarly, the transepithelial electrical potential is only about 1 mV (Frömter and Gessner 1974). These properties mean that very small osmotic gradients across the epithelium will lead to substantial fluxes of water. As Fig. 1(a) shows, the combination of large transcellular reabsorptive fluxes and a high paracellular permeability is specifically suited to 'bulk' transport, but it will not enable the epithelium to generate large ionic gradients.

We shall now briefly expand on this point by following the routes of water, Na^+, Cl^-, and HCO_3^- movement through the proximal tubule (Fig. 1(a)). HCO_3^- is titrated by the H^+ ions in the luminal fluid, which are continuously replaced by Na^+/H^+ exchange. The carbonic acid so formed is then dehydrated by carbonic anhydrase present in the luminal membrane (Lang et al. 1978), with the production of CO_2 which can easily penetrate the plasma membrane. Rehydration within the cell regenerates carbonic acid which dissociates to produce H^+ and HCO_3^- ions. The latter leave the cell at the basolateral side via a specific transport system (Boron 1983). The Na^+ taken up in exchange for H^+ is pumped out by the Na^+,K^+-ATPase. The net effect is that $Na^+HCO_3^-$ has been reabsorbed.

As a result of the ion movements described above, the Cl^- concentration in the tubule fluid increases to the same extent as the loss of HCO_3^-. On the other side of the epithelium, and more specifically in the lateral spaces, HCO_3^- accumulates. This separation of anions, due to the selective permeability properties of the paracellular pathway, with a high permeability for Cl^- and a low permeability for the larger HCO_3^-, generates a small lumen-positive transepithelial voltage. The HCO_3^- produces an effective hyperosmolality on the blood side (Schafer et al. 1974), and thus induces a reabsorptive water flux. Although much of the Na^+ and Cl^- reabsorption occurs together with the water flux

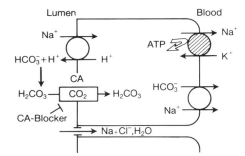
(a) Proximal tubule

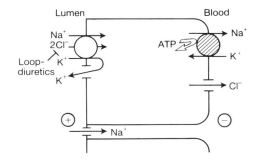
(b) Thick ascending limb

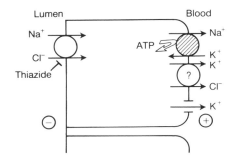
(c) Distal tubule

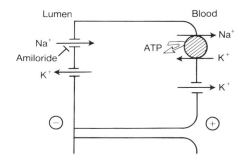
(d) Collecting tubule

Fig. 1 Simplified schemes for the mechanisms of Na^+, Cl^- and H_2O transport in the nephron segments. ◯ = carrier, ◍ = primary active pump, ⇌ = ion channel. (a) Proximal tubule. Note that transcellular transport involves mostly Na^+ and HCO_3^-. Much Na^+, and most of the Cl^- and H_2O is reabsorbed via the paracellular shunt pathway. CA-blocker = blocker of the carbonic anhydrase. (b) Thick ascending limb of the loop of Henle. The entire Cl^- flux is across the cell. Half of the Na^+ is transported across the cell, the other half is reabsorbed across the paracellular shunt pathway. The driving force for this latter component comes from the lumen positive transepithelial voltage. Loop diuretics such as frusemide bind to one of the Cl^--binding sites of the Na^+2Cl^-, K^+-cotransporter. (c) Distal tubule. The entire NaCl-reabsorption occurs across the cell. Thiazides block the NaCl-cotransporter. (d) Principal cell of the collecting tubule. The entire Na^+-reabsorption occurs across the cell. The extent of K^+ secretion depends on the Na^+ reabsorption inasmuch as Na^+ depolarizes the luminal cell membrane and increases the driving force for K^+ exit into the lumen. Amiloride blocks the luminal Na^+ channel. The reabsorption of Na^+ exceeds the secretion of K^+. Electroneutrality is maintained by Cl^- reabsorption. The route of Cl^--reabsorption is not clear. It does not, however, proceed across the principal cell.

(solvent drag), some diffuses across the paracellular pathway due to the lumen-positive electrical potential difference, in the case of Na^+, and the concentration gradient which exists with respect to Cl^-. Thus, a large proportion of the reabsorptive fluxes of water, Na^+, and Cl^- take the paracellular rather than the transcellular route. Energetically, this is advantageous, since ATP is only consumed during transcellular transport. In a very simplified analysis (Greger 1987) one can deduce that with such a scheme and with a Na^+,K^+-ATPase stoichiometry of three Na^+/ATP, up to 9 mol of Na^+ can be reabsorbed with the breakdown of only 1 mol of ATP.

The above concept suggests that inhibition of the transcellular reabsorption of $NaHCO_3$ would lead to a very strong diuretic and saluretic effect. All one needs to do is to block the Na^+/H^+ exchanger, the Na^+,K^+-ATPase or the carrier mediated exit of HCO_3^-. However, inhibition of the Na^+,K^+-ATPase is hardly feasible: it would probably have profound effects in many other cells, and would lead to K^+ depletion, and to uptake of Na^+, Cl^-, and water by the tubule cells. No high affinity blocker of the Na^+/H^+ exchanger has yet been described. Drugs such as amiloride, and its derivative ethylisopropylamiloride, block this exchanger *in vitro*, but the required concentrations ($>10^{-4}$ and $>10^{-5}$ mol/l) are so high that this effect cannot be exploited *in vivo*. Carbonic anhydrase inhibitors (Fig. 2(a)) such as acetazolamide inhibit the cytosolic as well as the membrane bound enzyme, and thus reduce the reabsorption of HCO_3^-. In the proximal nephron they inhibit the generation of CO_2 at the luminal membrane, the production of carbonic acid in the cell, and the carrier mediated exit of HCO_3^- across the basolateral membrane, reducing HCO_3^- and Na^+ reabsorption. The inhibition of HCO_3^- reabsorption is, however, never complete, and in addition, the proximal natriuretic and diuretic response is blunted by augmented reabsorption in the more distal nephron segments. Hence, the diuretic and natriuretic effect of carbonic anhydrase inhibitors is very modest. The increased urinary losses of HCO_3^- produce a metabolic acidosis. The increase in tubule fluid pH caused by these substances may reduce the reabsorption of weak organic acids, and thus increase their renal elimination.

Proximal saluresis and diuresis can also be induced by osmotic diuretics such as mannitol or any other non-reabsorbable osmolyte (such as sugar or high plasma bicarbonate). As stated above, manoeuvres directed at inducing a proximal diuresis are of limited effect, since much of the diuretic response is blunted further downstream.

Inhibitors of the $Na^+2Cl^-K^+$ cotransporter—loop diuretics

The thick ascending limb of the loop of Henle reabsorbs some 30 per cent of the filtered load of Na^+ and Cl^-. Some K^+ is also reabsorbed at this site, along with substantial amounts of Ca^{2+} and Mg^{2+}. Due to the low water permeability of this nephron segment, however, very little water is reabsorbed (Greger 1985a). This means that the tubule fluid at the end of the thick ascending limb is hypotonic while the reabsorbed fluid is hypertonic. This hypertonic reabsorptive process feeds the countercurrent system, which enables the kidney to excrete concentrated urine. During water diuresis, this same mechanism generates a tubular fluid with low Na^+ and Cl^- concentrations, making it possible to excrete a dilute urine.

The basic mechanism responsible for Na^+ and Cl^- reabsorption in the thick ascending limb of Henle's loop (Greger 1985a) is depicted in Fig. 1(b). The reabsorptive process is fuelled by the Na^+,K^+-ATPase localized in the basolateral membrane. The uptake of Na^+ across the luminal membrane occurs via a cotransport system which couples the uptake of Na^+ to that of Cl^- and K^+. Hence, after determination of its stoichiometry, this cotransporter was labelled the $Na^+2Cl^-K^+$ carrier (Geck and Heinz 1986, Greger and Schlatter 1981). The K^+ taken up across the luminal membrane almost completely recycles across the same membrane through K^+ channels (Bleich et al. 1990), but may be reduced in volume contraction and is regulated by cytosolic pH, being increased in cytosolic alkalosis and reduced in acidosis (Greger 1988; Bleich et al. 1990). Cl^- leaves the cell through selective ion channels (Greger et al. 1990) and via a KCl cotransport system, and a lumen-positive transepithelial potential difference of some 6 to 15 mV is generated. The paracellular shunt pathway in this nephron segment is far less permeable than that of the proximal nephron. The tight junctions are more complex, with more strands, are only permeable to small ions, and are also cation-selective (Greger 1985a). The lumen positive potential leads to a substantial reabsorptive flux in Na^+ via the paracellular route. In total, six Cl^- ions are moved across the cell, while three Na^+ ions are extruded by the Na^+, K^+-ATPase and another three Na^+ ions are reabsorbed through the paracellular pathway. Given the poor water permeability of this nephron segment, little water will follow the reabsorptive flux of NaCl, thus generating a dilute luminal fluid.

By binding at (one?) Cl^- binding site of the $Na^+2Cl^-K^+$ carrier, frusemide-type loop diuretics (Fig. 2(b)) stop the luminal uptake of these ions, and, therefore, the reabsorption of NaCl (Greger and Schlatter 1983; Greger 1985b). The structure–activity relationship of many of these compounds has been explored and, as a result, the minimum requirements of a molecule which will bind to and inhibit the $Na^+2Cl^-K^+$ carrier have been defined (Schlatter et al. 1983; Wittner et al. 1987). All diuretics belonging to this group (e.g. frusemide, bumetanide, piretanide, torasemide, azosemide) act from the luminal side of the membrane and have the same mechanism of action. Inspection of Fig. 2(b) indicates that the thick ascending limb cells have a grossly reduced Na^+, K^+-ATPase activity, and thus reduced ATP consumption, in the presence of loop diuretics. This effect can be used in vitro to preserve this tissue and it may be useful clinically in the prophylaxis of acute renal failure (Brezis et al. 1984; Greger 1985a, 1987; Greger and Wangemann 1987) (see Chapter 8.4).

The diuresis and natriuresis produced by loop diuretics is profound: up to 20 or 30 per cent of the filtered load may be excreted. If the K^+ recycling in the thick ascending limb is incomplete in the normal state, loop diuretics will increase the load of luminal K^+ delivered to the distal tubule. In addition to this effect the increased distal load of NaCl will lead to an increased secretion of K^+ and protons in the distal nephron (see below): hypokalaemia is, therefore, a predictable effect of all loop diuretics. Since at least part of the Ca^{2+} and Mg^{2+} reabsorption in the thick ascending limb of the loop of Henle is paracellular and voltage driven (Greger 1985a), loop diuretics will, by abolishing the transepithelial potential difference, reduce the reabsorption of these divalent cations and induce their excretion.

The onset of the natriuretic and diuretic effect of these drugs is very rapid and the losses of salt and water may be substantial. If these volume losses are not matched by adequate repletion, hyper-reninaemia and a reduction in glomerular filtration rate are predictable consequences. When volume contraction is induced by loop diuretics, urate reabsorption is increased in the loop of Henle (Lang et al. 1977), thus leading to hyperuricaemia.

A feedback response normally reduces single nephron filtration rate in response to increased NaCl delivery to the macula densa by the tubular fluid: this is paralysed by loop diuretics which block the $Na^+2Cl^-K^+$ carrier in the apical membrane of the macula densa cells (Schlatter et al. 1989a). Loop diuretics have little effect in nephron segments other than the thick ascending limb of the loop of Henle, although some proximal effect has been reported, probably due to the weak inhibition of carbonic anhydrase by this class of substances.

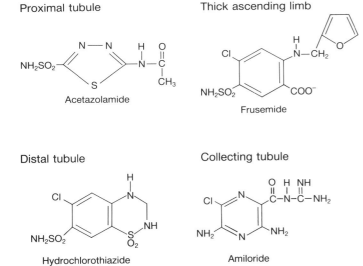

Fig. 2 Structural formula of several diuretics. The carbonic anhydrase inhibitor acetazolamide acts mainly in the proximal tubule. Its target is the carbonic anhydrase in the luminal membrane and in the cell, as well as the HCO_3^- exit system across the basolateral membrane (cf. Fig. 1). Frusemide is one example of an inhibitor of the $Na^+2Cl^-K^+$-cotransporter in the thick ascending limb of the loop of Henle. It binds to one of the Cl^- binding sites of this carrier. Hydrochlorothiazide acts by binding to the putative NaCl-carrier in the luminal membrane of the distal tubule. Amiloride blocks the Na^+-channels in the principal cell of the collecting duct.

Inhibitors of early distal NaCl reabsorption—thiazide diuretics

The early distal tubule, i.e. the nephron segment which follows the thick ascending limb, reabsorbs some 10 per cent of the filtered load of the NaCl by mechanisms which are only partially understood. A tentative scheme of how the reabsorption in this nephron segment might proceed (Greger 1988) is depicted in Fig. 1(c). There is no doubt that this reabsorption is also fuelled by the basolateral Na^+, K^+-ATPase, and the uptake of Na^+ probably proceeds via a NaCl cotransport system. The transepithelial electrical potential difference at this nephron site is slightly lumen negative.

Thiazide diuretics (Fig. 2(c)) apparently interact with the NaCl cotransporter, which is clearly distinct from the $Na^+2Cl^-K^+$ cotransporter, and hence inhibit luminal uptake of both ions (Velázquez and Wright 1986; Fanestil et al. 1990). Loop diuretics and thiazides do not inhibit the NaCl cotransporter (Schlatter et al. 1983). The natriuretic and diuretic effect of thiazide diuretics is smaller than that of loop diuretics, but they still produce a substantial increase in distal tubule NaCl load and hence cause kaliuresis. In addition to their effect in the early distal tubule these substances all have some inhibitory effect in the proximal nephron, probably due to the inhibition of carbonic anhydrase; it should be kept in mind that this class of substances was originally discovered in the search for more potent carbonic anhydrase inhibitors. Thiazides reduce the glomerular filtration rate even if in the presence of adequate hydration, although the mechanism of this is not known. It is well known that these substances, unlike loop diuretics, reduce the excretion of Ca^{2+} (see below), but again the cellular mechanism of this effect is unknown (Costanzo et al. 1978).

Sodium channel blockers in the collecting duct—K^+-sparing diuretics

Only a very small percentage of the filtered Na^+ is reclaimed in the collecting duct system, and the cellular mechanism of this process is shown in Fig. 1(d). The active extrusion of Na^+ again occurs via the Na^+,K^+-ATPase, while the uptake of Na^+ across the luminal membrane proceeds via Na^+ selective channels (Palmer and Frindt 1987; Garty and Benos 1988). The same cell membrane possesses selective ion channels (Hunter et al. 1986; Frindt and Palmer 1987; Schlatter et al. 1989) through which K^+ leaks into the tubule fluid. The extent to which Na^+ is reabsorbed will regulate the K^+ secretory flux, simply because the uptake of Na^+ depolarizes the luminal membrane, and depolarization increases the driving force for K^+ secretion (Schafer et al. 1990). Hence, K^+ losses will be augmented in hyperaldosteronism or whenever the distal delivery of NaCl is increased. There is also some recent evidence to suggest that the K^+ channels, like those in the luminal membrane of the thick ascending limb, are gated by cytosolic alkalinity. The paracellular pathway in this nephron segment is rather tight, and enables this nephron segment to establish large ionic and electrical gradients. Unlike the transport mechanism in the proximal nephron or the thick ascending limb of the loop of Henle, one molecule of ATP is consumed for the reabsorption of only three Na^+ ions, making this transport mechanism energetically inferior to that in the other nephron segments (Burckhardt and Greger 1991). The mechanism of Cl^- reabsorption in this nephron segment is still a matter of debate. It is clear, however, that it does not proceed through the principal cell (Schlatter and Schafer 1988). By exclusion, it should occur between the cells, or through the intercalated cells. The regulation of Na^+ reabsorption in the principal cell is under the control of mineralocorticoids and, at least in some species, also under the more acute control of arginine vasopressin. Both hormones increase the reabsorption of Na^+ (Schlatter 1989).

The Na^+ reabsorption in the collecting duct can be blocked by amiloride and triamterene. There is direct evidence that amiloride blocks the Na^+ channel (Li and Lindemann 1983; Palmer and Frindt 1987); the only data available for triamterene have been obtained using the frog-skin Na^+ channel. By blocking the Na^+ channel, these substances indirectly reduce K^+ secretion through the luminal K^+ channels; they are, therefore, natriuretic and antikaliuretic. The natriuresis is only moderate, and the specific importance of these substances is their antikaliuretic effect. In hyperaldosteronism, a comparable effect can also be achieved with aldosterone antagonists. Clinical doses of amiloride or triamterene act specifically on the collecting duct. Very high concentrations are required to inhibit other transporters such as the Na^+/H^+ exchanger (see above).

Organ specificity and drug interactions of diuretics

The above section has dealt with the cellular mechanisms of Na^+ and water reabsorption and the mechanisms by which the different groups of diuretics interfere with this reabsorption. In the following discussion we shall briefly address the questions of (i) why diuretics are kidney specific, (ii) how other drugs may modify the diuretic action, and (iii) how the different groups of diuretics may be combined.

Why are diuretics organ specific?

The organ specificity of diuretics is surprising, since they act on transport proteins, many of which are present in organs other than the kidney. For example, the $Na^+2Cl^-K^+$ cotransporter has been found in intestine, excretory glands, inner ear epithelium, red blood cells, neuronal cells, and smooth muscle cells, Na^+ channels occur in the intestine, respiratory epithelium, and glandular ducts, and carbonic anhydrase is present in many polar and apolar cells. Why should inhibitors of these transport proteins have effects on the kidney but little if any systemic action? The reason for this organ specificity is the accumulation of high concentrations of the diuretic in the luminal fluid due to water reabsorption and to secretion by the proximal nephron. The basolateral cell pole of the proximal tubule possesses transporters which take up organic anions such as PAH and drugs such as diuretics, causing them to accumulate in the tubule cell (Ullrich et al. 1989). Other transporters then transfer the drugs into the luminal fluid. The combination of water reabsorption and this secretory mechanism lead to a drug concentration in the tubule fluid which may be 10 times (or more) higher than that of the plasma. Clinically used doses, will, therefore, affect the transporters in the luminal membrane of the tubule before those of any other organ. In some disease states, where secretion in the proximal tubule may be defective, much higher doses of, for example, a loop diuretic are required to achieve a diuretic response, and serious side effects such as ototoxicity may occur, since endolymph secretion in the inner ear is inhibited (Marcus et al. 1987).

Drugs interfering with the secretion of diuretics

Many drugs can interfere with the secretion of diuretics. Amongst those, probenecid has been studied in some detail. Probenecid markedly reduces the diuretic response to a previously effective dose of a loop diuretic. All loop diuretics are affected (Braitsch et al. 1990) and the secretion of these different diuretics in the proximal tubule is equally inhibited by a given dose of probenecid. One would predict that all drugs which are weak organic acids would potentially reduce the effectiveness of loop diuretics and produce a reduced renal excretion rate. On the other hand, loop diuretics might reduce the renal clearance of these weak organic acids. Unfortunately very few experimental data are available regarding this issue.

Combination of diuretics

Since different diuretics act at different nephron sites, one might predict that the combinations of two diuretics from different groups would be additive. The response is usually more than additive; a fraction of the diuretic response exerted at the level of the loop of Henle will be blunted by increased reabsorption in more distal tubule segments. If the loop diuretic is now combined with a thiazide, this compensatory reabsorption is eliminated by the diuretic response to the second drug. The same phenomena can be looked at from a different perspective: if a small load is delivered to a certain tubule segment the diuretic response in this segment can only be minimal. If for example the flow rate at the end of the proximal tubule is reduced by volume contraction, the effect of a loop diuretic will be much smaller than it is in volume expansion, when the end-proximal flow rate is increased.

Choice and combination of diuretics will be discussed more explicitly in the next sections; it is sufficient to state here that the adverse effects of threatening hypokalaemia, which may be caused by loop diuretics and thiazides, can be compensated for by concomitant use of potassium-sparing diuretics such as amiloride or triamterene.

Clinical use of diuretics

Arterial hypertension

Thiazide diuretics have long been considered to be the 'cornerstone' in the treatment of arterial hypertension. Accordingly, in the recent recommendations of the Joint National Committee on Detection, Evaluation and Treatment of High Blood Pressure (JNC 1988) diuretics still represent one option for the first step of therapy, along with β-blockers, calcium antagonists, and angiotensin converting enzyme inhibitors. As monotherapy they are particularly effective in elderly hypertensives (Hulley et al. 1985), in black patients (Veterans Administrative Cooperative Study Group of Antihypertensive Agents 1982) and perhaps in individuals with a low renin activity (Bühler et al. 1985a, 1982b). If not chosen as the first drug, diuretics are an ideal second antihypertensive agent to be added to an inadequately effective drug.

Mechanisms of antihypertensive action

Thiazides initially reduce blood pressure by reducing intra- and extravascular volume and cardiac output. With continued treatment, however, cardiac output and intravascular volume return to pretreatment levels, and there is a marked decrease in total peripheral vascular resistance (Tarazi et al. 1970). This reaction is probably caused by a 'reversed autoregulatory response' in the presence of a sustained decrease in extra- and intravascular fluid volume (Tobian 1974; Shah et al. 1978). The following antihypertensive mechanisms have also been discussed: (i) decreased sodium content, particularly of the vascular tree (Kusumoto et al. 1974). The results of such studies are contradictory (Tobian et al. 1962; Daniel 1962). (ii) Increased formation of vasodilating prostaglandins (Scherer and Weber 1979); (iii) decreased vascular responsiveness to vasoconstrictors such as angiotensin II and noradrenaline (Weidmann et al. 1983); and (iv) reduction of increased free cytosolic calcium activity (Erne et al. 1984).

Dosage considerations—combined therapy

Diuretics have often been used in larger doses than necessary; the dose–response curve for the antihypertensive effect of thiazides is flatter than that for their diuretic effect. To minimize unwanted metabolic complications the lowest effective dose should be used (Beerman and Groschinsky-Grind 1978; Campbell et al. 1985); e.g. bendroflumethiazide and indapamide 2.5 mg daily. In mild to moderate hypertension, normalization of mean arterial blood pressure can be achieved in 29 per cent of patients with only 25 mg of hydrochlorothiazide daily. A dose of 50 mg daily is adequate in 40 per cent of patients (Hollifield 1986). Low dose loop diuretics are also an effective antihypertensive monotherapy.

The antihypertensive action of diuretics per se probably is limited by two factors: (i) potassium depletion can raise blood pressure (Krishna et al. 1989), probably via constriction of resistance vessels (Haddy 1975), sodium retention (Welt et al. 1960), and/or elevation of plasma renin (Luke et al. 1982; Tannen 1983). Correction of diuretic-induced hypokalaemia can lower the blood pressure (Kaplan et al. 1985). However, the combined administration of potassium-losing and potassium-sparing diuretics does not always produce additive effects (Myers 1987). (ii) Any extracellular volume contraction stimulates renin secretion, resulting in an increased formation of angiotensin II. This potent vasoconstrictor counteracts the antihypertensive effect of diuretics as has been shown by the enhanced hypotensive effect of additional angiotensin converting enzyme inhibitors (Brunner et al. 1988). A similar, although smaller, effect can also be exerted by the coadministration of β-blockers which reduce renin secretion. Finally, diuretics augment the antihypertensive efficacy of many drugs, especially vasodilators and sympathicolytic agents, at least in part by preventing or correcting their sodium-retaining actions (pseudotolerance).

Effects of diuretics on morbidity and mortality during long-term therapy of hypertension

The effectiveness of thiazide diuretics has been proven in numerous long-term trials involving more than 40 000 hypertensive patients. All studies revealed a striking reduction in stroke rate in severe, moderate, and in mild hypertension.

However, different results emerged concerning coronary artery disease. In the Multiple Risk Factor Intervention Trial (MRFIT 1985), a post-hoc analysis of subgroups showed that hypertensive patients with resting ECG abnormalities has a higher mortality rate after diuretic treatment, whereas hypertensive subjects with a normal ECG seemed to benefit from diuretic therapy. These results are in contrast to the findings of the

European Working Party on Hypertension in the Elderly (Amery *et al.* 1985) and in the Hypertension Detection Follow-Up Program (1982) in which a decline in cardiac deaths was achieved. In comparison to β-blockers, diuretic treatment showed similar results so far as stroke and cardiac events are concerned, both in the International Primary Prospective Preventive Study in Hypertension (IPPPSH 1985) and in the Heart Attack Primary Prevention in Hypertension (HAPPHY 1987) Trial. Taken together, the epidemiological studies suggest that thiazide treatment does not increase the incidence of coronary events when compared to placebo or β-blocker therapy, although the long-term results of cardiac mortality are disappointing in some, but not all studies. This lack of 'cardioprotection' despite effective blood pressure reduction is possibly due to adverse effects of diuretics on electrolyte, carbohydrate, and lipid metabolism, as well as to the activation of the renin-angiotensin system.

Consequences for therapy

Since the negative metabolic effects of diuretics can be minimized by the use of lower doses without reducing their antihypertensive effects, they are still the antihypertensive agents of choice. Low doses of diuretics are particularly effective in hypertensive patients already being treated with an angiotensin converting enzyme inhibitor. They will also continue to be important in those conditions in which volume expansion contributes to the rise in blood pressure (i.e. subgroups of essential hypertension, renoparenchymal disease, and steroid-dependent hypertension) as well as in some specific groups of patients (e.g. black patients and the elderly). Finally, diuretics are indispensible drugs in overcoming the sodium-retaining effects of other antihypertensive agents such as the classical vasodilators (minoxidil, dihydralazine).

Congestive heart failure and pulmonary oedema

Chronic diuretic therapy is of proven efficacy in the therapy of congestive heart failure (Coodley *et al.* 1979; Schäfer and Sievert 1987; Reyes 1988). Haemodynamic measurements show that left ventricular end-diastolic pressure is reduced due to a fall in preload. If patients are on the steep portion of the Frank-Starling curve, cardiac output may be reduced (Parmley 1985), but cardiac output is unaltered in the majority of patients, and may even be slightly enhanced by diuretics, since they may reduce the afterload by diminishing total peripheral resistance (Nishijama *et al.* 1984). The clinical improvement of patients with congestive heart failure is mostly caused by a fall in respiratory work, due to a decrease in pulmonary water and improved pulmonary compliance (Biddle and Yu 1979; Mathur *et al.* 1984). Successful treatment of congestive heart failure by diuretics is associated with a progressive decline of elevated plasma concentrations of atrial natriuretic peptide to normal values (Anderson *et al.* 1988).

Direct effects of diuretics

Intravenous administration of loop diuretics immediately improves ventricular filling pressure and pulmonary congestion in left ventricular failure following myocardial infarction (Dikshit *et al.* 1973). This early effect precedes the diuretic action and is associated with a 50 per cent increase in venous capacitance, which is believed to be mediated by the kidney, perhaps through the release of prostaglandins (Bourland *et al.* 1977). This 'internal phlebotomy' is further augmented by the ensuing diuresis. An opposite effect may occur in patients with severe chronic heart failure (NYHA class III–IV), in whom intravenous administration of frusemide promotes a reversible acute vasoconstriction and aggravated haemodynamic deterioration. The underlying mechanism seems to be an activation of the 'neurohormonal axis' with an increase in plasma renin, adrenaline, and arginine vasopressin activities (Francis *et al.* 1985).

Recommendations for clinical use

The majority of patients with mild congestive heart failure can be managed by dietary sodium restriction and the administration of thiazide diuretics, either on a daily or an alternate day basis. However, in moderate congestive heart failure the administration of diuretics in combination with vasodilators or angiotensin converting enzyme inhibitors is indicated. Since thiazides are rarely effective in this stage, the more potent loop diuretics should be used. If the clinical response is suboptimal, the dose of the loop diuretics should be increased or a blockade of sodium reabsorption in different nephron segments should be attempted. Coadministration of potassium sparing drugs (triamterene, amiloride, or spironolactone) may prove useful, since these inhibit sodium reabsorption and potassium as well as hydrogen ion secretion in the late distal tubule and collecting duct. The additional administration of a thiazide diuretic completes the distal nephron blockade (Puschett and Greenberg 1984).

In severe congestive heart failure loop diuretics should be combined with a proximally acting diuretic, since the delivery of sodium from the proximal tubule to the loop of Henle may be markedly reduced in such patients. A very useful regimen is the combination of metolazone and frusemide in increasing dosages.

Repeated haemofiltration may be necessary in the small group of patients with refractory cardiac failure. Alternatively, regular haemodialysis or peritoneal dialysis may extend the lives of these patients from several months up to 2 or 3 years.

Liver cirrhosis and ascites

Several local and systemic factors are involved in the pathogenesis of cirrhotic ascites, including portal hypertension, arteriovenous fistula, increased lymph production in the liver and in the intestine (Witte 1980), and decreased plasma oncotic pressure due to hypoalbuminaemia (Cherrick *et al.* 1960). During formation of ascites, urinary sodium excretion is markedly reduced, due to an enhanced tubular sodium reabsorption and a reduced glomerular filtration rate.

Two theories have been put forward to explain the increased fractional reabsorption of sodium. The 'overflow' theory (Liebermann *et al.* 1970) suggests a primary renal sodium retention with the overfilling of the circulation and the consequent formation of ascites. This interpretation fits well with the increased plasma volume and high cardiac output observed in cirrhotics. On the other hand, the theory of 'underfilling' (i.e. reduced effective arterial blood volume) claims that stimulated volume receptors induce sodium retention (Epstein 1982). This concept is supported by the activation of the 'neurohormonal axis' with elevated concentrations of noradrenaline, vasopressin, and angiotensin II (Bichet *et al.* 1982*a*, 1982*b*; Epstein *et al.* 1977*b*), a normal concentration of plasma atrial natriuretic factor (Gerbes *et al.* 1985), and the diuretic effect of both head out of water immersion (Epstein 1978) and the implantation of a peritoneal-venous shunt (Epstein 1982). The theory of underfilling was recently confirmed by direct measurements which showed that, on average, central blood volume was reduced by 25 per cent with lowest volumes in patients with gross ascites and a reduced systemic vascular resistance (Hennksen *et al.* 1989). Decreased effective central blood volume may be due to arteriolar vasodilation, portosystemic collateral flow, and/or sequestration of fluid in the peritoneal cavity. Hyperaldosteronism is not an invariable finding in cirrhosis: elevated, normal, and even reduced concentrations of aldosterone have been reported (Epstein *et al.* 1977*b*).

Therapeutic approach

Diuretic treatment is indicated in patients with tense ascites or in those suffering from ascites-induced dyspnoea, abdominal pain, anorexia, gastro-oesophageal reflux, and abdominal hernia (Maharaj 1988). However, before diuretics are introduced a rigid reduction of dietary sodium (<0.5 g/day), with fluid restriction in those with hyponatraemia (1–1.5 l/day), combined with bed rest for at least 3 days is recommended. About 5 to 15 per cent of patients benefit from this regimen. Ascitic fluid is poorly mobilized during diuretic therapy and therefore the daily weight loss should not exceed 300 to 500 g. If there is additional peripheral oedema a weight loss of up to 1 kg per day may be acceptable (Pockros *et al.* 1986).

Aldosterone antagonists such as spironolactone have provided the most favourable results. Increasing the doses from 50 to 300 mg/day within 10 days will induce a sustained diuresis in about 75 per cent of patients (Perez-Ayuso *et al.* 1983). Amiloride (15–30 mg) has been suggested as an alternative to spironolactone (Yamada and Reynolds 1970), and triamterene may be considered, although its rapid metabolism in the normal liver has to be taken into account: in liver cirrhosis this metabolism may be impaired (Knauf *et al.* 1978). If potassium-sparing diuretics alone fail, loop diuretics may have a synergistic effect. Generally, the natriuretic effects of diuretics are significantly greater in the supine, as compared to the upright, position (Ring-Larson *et al.* 1986). Resistance to loop diuretics is pharmacodynamic in nature, largely due to increased proximal and/or distal tubule sodium reabsorption (Brater 1985; Brater *et al.* 1986). This effect occurs particularly when renal failure is present (Shear *et al.* 1965).

A peritoneovenous shunt has been proposed for patients resistant to diuretics. This improves systemic and renal haemodynamics (Stassen and McCullough 1985), but is unfortunately, associated with a high rate of severe complications such as thrombosis, infections, and disseminated intravascular coagulation. Recent studies have shown that large volume paracentesis combined with the administration of salt-poor albumin is an effective alternative therapy devoid of such complications (Quintero *et al.* 1985).

Specific complications of diuretic therapy

Cirrhotic patients with ascites may develop a multitude of complications. In the classical study of Sherlock *et al.* (1966) 75 per cent of the patients developed serum electrolyte disturbances (hyponatraemia, hypokalaemia, hypochloraemic alkalosis), particularly after the administration of loop diuretics. Hypotension, renal failure and hepatic encephalopathy are also frequent complications (Naranjo *et al.* 1979). With regard to these side effects, Sherlock (1981) stated pithily that the patient is better 'wet and wise rather than dry and demented'.

In summary, the therapeutic goal in hepatic ascites is a slow, gradual diuresis, since only a limited amount of ascitic fluid can be mobilized each day. Any contraction of the plasma volume is harmful, and enhances the risk of hepatorenal syndrome and hepatic encephalopathy. Great care should be taken in the monitoring of plasma potassium: any additional potassium loss caused by therapy may induce profound hypokalaemia (Epstein 1984). As shown by Gabazda and Hall (1966), hypokalaemia enhances renal ammonia production, thus contributing to hepatic encephalopathy.

Nephrotic syndrome—response to diuretics

Various factors are involved in the pathogenesis of sodium and fluid retention in patients with nephrotic syndrome (see Chapter 3.3). Micropuncture studies in rats with experimentally induced glomerulonephritis showed a striking heterogeneity in the function of cortical nephrons (Rocha *et al.* 1973): nephrons with reduced single nephron filtration rate showed an increased reabsorption of fluid. In nephrotic patients, as well as decreased GFR, a contraction of effective arterial blood volume and increased blood concentrations of noradrenaline (Kelsch *et al.* 1972) and vasopressin (Bichet and Manzini 1984) contribute to sodium and fluid retention. On the other hand, reduced peritubular capillary protein concentration decreases sodium reabsorption in the proximal tubule, suggesting a major role of the distal nephron in the sodium retention of the nephrotic syndrome. There are some data which indicate that fractional sodium excretion in response to loop diuretics is enhanced in nephrotic patients with very low serum albumin concentrations (<2 g/l) as compared to healthy subjects (Grausz *et al.* 1972). In certain patients with minimal change glomerulopathy and complicating renal insufficiency frusemide-induced diuresis was shown to improve renal function, possibly due to the reduction in interstitial oedema (Lowenstein *et al.* 1981).

Resistance to frusemide in nephrotic syndrome is not caused by altered pharmacokinetics since the total amount of diuretic delivered to the urine and the time course of urinary diuretic excretion are both identical to those in healthy subjects (Rane *et al.* 1978; Keller *et al.* 1982). Experiments in the rat suggest that the most important cause for impaired response is the binding of the diuretic to filtered proteins (Green and Mirkin 1980). In microperfusion experiments, addition of albumin to the perfusate blunted the inhibitory effect of frusemide (Kirchner *et al.* 1989). As a consequence, in patients with resistant oedema, the dose of a loop diuretic has to be increased to two or three times the normal in order to provide sufficient unbound diuretic at its site of action. Another cause of an impaired pharmacodynamic response may be an increased proximal and/or distal tubule fluid reabsorption: this may respond to spironolactone and/or the proximally acting thiazide metolazone. In patients with severe hypoalbuminaemia (serum albumin less than 20 g/l) infusion of hypertonic albumin has been shown to increase the diuretic response (Puschett and Greenberg 1984) (see Chapter 3.3)

To achieve a negative sodium balance in patients with nephrotic oedema, dietary sodium intake should be reduced to 2 g (5 g NaCl) per day, which essentially corresponds to a diet with no added salt. Generally it is clinical practice to employ diuretics such as frusemide, and the effective dose of up to 500 mg orally may be repeated during the day if necessary. If these doses are inadequate, a combination with the proximally-acting drug metonazole should be attempted: intravenous administration, starting with low doses, may increase the diuretic effect. Only in resistant cases should intravenous hypertonic albumin be tried, always in combination with intravenous loop diuretics. The effect of this therapy is limited, since the albumin is lost rapidly in the urine. In order to prevent too rapid a volume reduction, body weight has to be measured daily: orthostatic hypotension, prerenal azotaemia, and, occasionally, even acute renal failure may occur as a consequence of aggressive diuretic therapy.

Chronic renal failure

Most patients suffering from chronic renal failure have increased total body sodium and water content, even if this is not apparent clinically. Fluid overload is the major cause of hypertension in these patients and predisposes to congestive heart failure and pulmonary oedema. Since most thiazide diuretics, except metolazone, lose their effectiveness if GFR is below 30 ml/min (Reubi and Cottier 1961), loop diuretics are the drugs of choice in these patients. Interestingly, in patients with a mean GFR of 20 ml/min, intravenous administration of frusemide produces an exaggerated response in the remaining nephrons, as measured by fractional fluid and sodium excretion (Heidland et al. 1972; Brater et al. 1986). This response may be caused by an increased fluid delivery to the thick ascending limb of the loop of Henle either by glomerular hyperfiltration and/or decreased reabsorption in the proximal tubule.

In advanced renal failure (GFR <10 ml/min) resistance to loop diuretics develops with a need for increased doses. This is due to an extreme reduction of filtered NaCl and a reduction of proximal secretion of diuretics. A number of endogenous acids accumulate in advanced renal failure (Spustova et al. 1988), and these compete with diuretics for their secretion by the proximal tubule (Ullrich et al. 1989; Rose et al. 1976; Tilstone et al. 1977), prolonging their half-lives (Cutler et al. 1974; Beermann et al. 1977, Kampf and Baethke 1980). Only 10 per cent of intravenously administered frusemide reaches the urine, compared to 50 per cent in subjects with normal renal function (Brater et al. 1986). In addition, both the non-renal clearance of frusemide and its glucuronidation are diminished to about 50 per cent of the normal value, probably due to a blockade of hepatic frusemide uptake. In contrast, the non-renal elimination of bumetanide appears to be normal in patients with renal failure (Voelker et al. 1987).

High doses of frusemide

Patients with advanced renal failure require much higher doses of frusemide than normal in order to achieve adequate levels at its site of action. A sufficient response to large doses of diuretic may be achieved even at a GFR as low as 3 ml/min (Heidland et al. 1969): for example, intravenous infusion of 1000 mg frusemide over 45 min increases fractional fluid and sodium excretion from 43 to 68 per cent and 26 to 61 per cent, respectively, and increases inulin clearance by 50 per cent. A significant rise in GFR was also reported by Fairley and Laver (1971), but not by Verenstraeten et al. (1975) or by Keeton and Morrison (1981). When different doses of frusemide were used in patients with terminal renal failure a maximal response was achieved with (100, 200, 500, and 1000 mg intravenously) a single dose of 500 mg (Heidland et al. 1972). Brater et al. (1986) postulate that even lower doses of frusemide (up to 160 mg intravenously) are sufficient to reach the upper plateau of the dose–response curve, although the highest doses were not administered in these studies. Animal experiments suggest that the maximally effective dose of frusemide is much higher (5–25 mg/kg; Muschawek and Hadju 1964). The risk of ototoxic side effects limits the dose of frusemide which can be given to no more than 4 mg/min intravenously, with a maximum daily dose of 1 g. This corresponds to an oral dose of 2 g/day, since the bioavailability of frusemide is about 50 per cent.

Chronic treatment

Oral frusemide, in doses of up to 2 g has been effective in the treatment of oedema, volume overload and hypertension (Muth 1967). Although the GFR rises after acute administration of high doses of frusemide, this effect does not persist during chronic administration, even if the urinary losses of salt and water are replaced, and prolonged administration of loop diuretics in chronic renal failure is, therefore, without any benefit on renal function. A negative sodium balance is followed by a rise in blood urea nitrogen: this may be due to both an enhanced reabsorption of urea in the distal part of the nephron (Dal Canton et al. 1985) and, occasionally, an enhanced urea production (Kamm et al. 1983).

Frusemide treatment may play a role in the control of fluid balance in selected dialysis patients with residual renal function (Verenstraeten et al. 1975). A significant increase in urinary output can be achieved, reducing the fluid gain between dialyses.

High doses of other loop diuretics, such as bumetanide, torasemide, and piretanide (12–60 mg/day) are promising alternatives to frusemide in chronic renal failure (Kampf and Baethke 1980; Marone et al. 1984). Unfortunately, long-term therapy with high doses of muzolimine in advanced renal failure may be complicated by severe neurological disturbances such as encephalomyelopathy (Daul et al. 1987).

In summary, volume overload and oedema in chronic renal failure have to be controlled by salt restriction and, if necessary, with progressive doses of frusemide (up to 500 mg twice daily orally), or other loop diuretics in corresponding doses. In unresponsive patients additional administration of metolazone may be helpful.

Diuretics in acute renal failure
Mannitol

Mannitol is a volume expanding osmotic diuretic whose basic mechanism of action has been briefly discussed above.

Experimental studies and mechanism of action

Mannitol has been demonstrated to have an unequivocally beneficial effect in numerous experimental models of acute renal failure (for review see Levinsky and Berhard 1988). Prophylactic and in some instances, therapeutic administration of mannitol has improved kidney function after partial and total renal artery obstruction, as well as in noradrenaline-induced acute renal failure (Kashgarian et al. 1976; Patak et al. 1979; Burk et al. 1983; Zager et al. 1985). Mannitol was also effective in the treatment of experimental acute renal failure induced by glycerol (Wilson et al. 1967), methaemoglobin (Parry et al. 1963), myoglobin (Mason et al. 1961), mercuric chloride (Vanholder et al. 1984), cisplatin (Pera et al. 1979) and amphotericin B (Said et al. 1980): renal function improved, but tubular necrosis was not prevented.

The beneficial effect of mannitol seems to arise from vascular as well as tubular effects. Mannitol may minimize the osmotic swelling of endothelial and tubule cells (Flores et al. 1972) and may protect mitochondrial function (Schrier et al. 1984). It may also act as a scavenger for free oxygen radicals (Freeman and Crapo 1982), and hence reduce post-ischaemic injury of tubule cells. The increased tubule flow rate induced by mannitol may relieve tubular obstruction by debris and protein precipitates (Burke *et al.* 1980). Mannitol also causes vasodilation of small vessels, by increasing prostaglandin synthesis (Johnston *et al.* 1981) and inhibiting renin release (Vander and Miller 1964). Recently mannitol has been shown to increase the release of atrial natriuretic petide (Gianello *et al.* 1989) which is known to improve acute renal failure in experimental animals.

Clinical studies (see also Chapter 8.4)

The prophylactic effect of mannitol has been examined in a number of uncontrolled and controlled studies. High-risk patients undergoing open-heart or vascular surgery showed no significant benefit when mannitol was compared to standard volume repletion (Berman *et al.* 1964). In patients with obstructive jaundice the results were different: postoperative GFR was markedly higher in the mannitol-treated group (Dawson 1965; Untura 1979), although in a randomized trial conducted by Gurbern *et al.* (1988) the results were disappointing. In uncontrolled studies a beneficial effect of mannitol was observed in patients with toxic nephropathy due to cisplatin and amphotericin B (Hayes *et al.* 1977). The incidence of acute renal failure induced by radiocontrast media was also reduced by mannitol in uncontrolled studies (Old and Lehrner 1980; Anto *et al.* 1981).

Acute renal failure is a frequent complication in patients with severe rhabdomyolysis and consequent myoglobinuria, occurring in at least 50 per cent of cases (Gabow *et al.* 1982). These patients usually respond to mannitol therapy combined with urinary alkalinization (Eneas *et al.* 1979). In seven cases with extensive crush injuries and severe rhabdomyolysis, immediate therapy, inducing a diuresis of about 8 l/day, prevented acute renal failure in all patients (Ron *et al.* 1984).

The incidence of acute renal failure in recipients of cadaveric kidney transplants was reduced from 55 per cent to 14 per cent by mannitol in a controlled study (Weimar *et al.* 1983), and in another study this form of treatment was successful if combined with hydration (Tiggeler *et al.* 1985). On the other hand, 75 per cent of recipients in another study developed acute renal failure when given mannitol with or without frusemide (Lachance and Barry 1985).

Early acute loss of renal function is often reversible: in several uncontrolled studies infusion of mannitol reversed oliguria and improved renal function in two-thirds of the patients (Barry and Malloy 1962; Eliahou 1964; Luke *et al.* 1965; Scheer 1965). Immediate intervention seems to be crucial.

During mannitol therapy urine flow rate, plasma osmolality, and central venous pressure must be carefully monitored. The initial dose should not exceed 12 to 25 g: if renal function does not improve, therapy should be stopped to prevent hypervolaemia and acute left ventricular overload. Acute renal failure after administration of mannitol has, paradoxically, been reported in neurosurgical patients (Dorman *et al.* 1988).

Loop diuretics

The mechanism of the action of loop diuretics has been discussed explicitly above. Their possible beneficial effect in acute renal failure is summarized below.

Animal experiments

The rationale underlying the use of this group of diuretics in the prevention of deterioration of thick ascending limb function has already been presented. Several studies in intact animals support its use under the respective conditions (Kramer *et al.* 1980*a, b*). Frusemide, like mannitol, improved renal function in dogs with acute renal failure induced by noradrenaline (Patak *et al.* 1979), but the data obtained in models of ischaemic renal failure are controversial. In partial occlusion models, no clear beneficial effect was seen, but protection against acute renal failure was observed in animals with total occulsion (Kramer *et al.* 1980*a*). Results are also variable in models of nephrotoxicity: mercuric chloride nephrotoxicity was attenuated (Thiel *et al.* 1976; Brunner *et al.* 1985), although histological findings were not improved. Frusemide was only protective against uranyl nitrate nephrotoxicity if combined with dopamine (Bailey *et al.* 1973). Frusemide aggravated glycerol-induced acute renal failure (Greven and Klein 1976). Its beneficial effect was only short-lived in rats with methaemoglobin-induced acute renal failure (Montoreano *et al.* 1971), and no effect was demonstrable in dogs injected with haemolysed blood (Mohring and Madsen 1973). Only a moderate beneficial effect was seen in rats with cisplatin-induced nephrotoxicity (Pera *et al.* 1979), but frusemide reduced necrosis of the thick ascending limb of the loop of Henle in acute renal failure induced by radio-contrast media administered in the presence of indomethacin (Heyman *et al.* 1989). Hence, frusemide and related diuretics have a prophylactic effect in some animal models, but the same substances are ineffective in other models, for unknown reasons.

Pathophysiology of acute renal failure and the rational basis for the use of loop diuretics

As has been stated above the thick ascending limb of the loop of Henle seems to be damaged early in the course of acute renal failure induced by toxins and ischaemia. This damage may be caused by an imbalance between oxygen supply and demand on the one hand, and substrate demand and supply on the other (Heyman *et al.* 1988): the close link between substrate supply and transport work has been shown directly in experiments with isolated perfused thick ascending limbs (Wittner *et al.* 1984). Loop diuretics appear to have a direct protective effect: as transport work is reduced, the oxygen and substrate demand are lowered (Brezis *et al.* 1984, 1985; Greger and Wangemann 1987). Other effects may also be beneficial: frusemide-induced prostaglandin release may cause renal vasodilatation (Ludens *et al.* 1968; Stein *et al.* 1972; Abe *et al.* 1977), and loop diuretics may inhibit mitochondrial respiration (Eknoyan *et al.* 1975; Brezis *et al.* 1986). High tubule flow rates may also prevent or release tubular obstruction (Steinhausen and Parekh 1984).

Clinical studies with loop diuretics in acute renal failure

In uncontrolled studies prophylactic treatment of open heart and vascular surgery patients has been ineffective in some studies (Yeboah *et al.* 1972; Ray *et al.* 1974; Abel *et al.* 1976) but beneficial in others (Mahoney 1976). Some beneficial effect on postoperative glomerular filtration rate was also reported in controlled studies (Nuutinen *et al.* 1978), and in an uncontrolled study of radiocontrast media-induced acute renal failure.

In the early stages of acute renal failure loop diuretics can reverse oliguria in a large fraction of patients, with a consequent higher rate of survival. However, the better prognosis of this group may simply reflect a less severe course of acute renal failure.

Few controlled trials on the usefulness of loop diuretic therapy have been carried out in established acute renal failure. Cantarovich *et al.* (1971) reported a reduction in the duration of oliguria in the group treated with large doses of frusemide, but the mortality rate was not different in the treatment group. Other studies along these lines (Cantarovich et al. 1973; Karayannopoulos 1974) indicated that less frequent dialysis was required in the treatment group, but again this had no effect on the rate of mortality. In other studies no beneficial effect was seen, even on urinary flow rate (Keinknecht et al. 1976). Brown (1981) compared the course of acute renal failure in patients treated with frusemide at a dose of 1 or 3 g/day: oliguria improved in the high dose group but there was no advantage with respect to the mortality rate. Several uncontrolled studies have been inconclusive: some report beneficial effects in patients treated with high doses of loop diuretics (Anderson et al. 1977) whilst others deny such an effect (Fries et al. 1971; Minuth et al. 1976). Even infusions of frusemide into the renal artery had no beneficial effect on one study (Epstein et al. 1975).

Recently, the addition of dopamine has been suggested for patients not responding to high doses of diuretics. This regimen is supported by experimental (Linder et al. 1979) and clinical studies (Graziani et al. 1980, 1984; Henderson et al. 1980; Hörl et al. 1989). In clinical use, dopamine should be administered during the first day of oliguria. If there is no immediate response, prolonged trials are of no benefit and may even be harmful. This may be due to the formation of 6–OH-dopamine, a neurotoxin capable of destroying peripheral sympathetic nerve terminals (Rollema et al. 1988).

In summary, numerous experimental and some clinical studies have shown that diuretics have a 'limited value to prevent, reverse or speed recovery from acute renal failure' (Levinsky and Bernhard 1988).

Volume expansion should be performed first in the prophylaxis of postischaemic or toxic acute renal failure. Just prior to the potential renal insult intravenous mannitol (12.5–25 g) or frusemide (20–40 mg) should be administered, with adequate replacement of the urinary losses of fluid and electrolytes. In patients with severe traumatic rhabdomyolysis, Better and Stein (1990) recommend aggressive volume replacement followed by forced solute alkaline diuresis. Isotonic saline solution (1.5 l/h) should be infused immediately after the accident. Once the systemic circulation has been stabilized and as soon as urine flow is achieved, forced mannitol–alkaline diuresis should be initiated using hypotonic sodium chloride and sodium bicarbonate (NaCl 40 mmol/l, $NaHCO_3$ 110 mmol/l) in 50 g/l glucose with 10 g/l mannitol added. This solution should be infused initially at a rate of approximately 12 l/day, forcing a diuresis of approximately 8 l/day and maintaining an urine pH of above 6.5, until myoglobinuria disappears. The positive fluid balance is explained by the sequestration of fluid in the injured muscles. Up to now, however, the usefulness of this aggressive therapy has not been confirmed by other centres.

In patients with early renal failure due to septicaemia, surgery or administration of radiocontrast dye intravenous frusemide (up to 250 mg) other loop diuretics or single doses of mannitol in increasing doses should be tried within the first 24 to 48 h after onset of oliguria. In patients who do not respond to this regimen, a combination with a low dose of dopamine (1–3 μg/kg.min) for 24 h is reasonable.

It is uncertain whether induction of diuresis in oliguric patients with established acute renal failure shortens the course of the acute renal failure. It is, however, easier to manage patients with non-oliguric acute renal failure, particularly since fluid and sodium overload and hyperkalaemia are less severe.

Although the risks of diuretic therapy are small in experienced centres, it should be stressed that mannitol therapy may lead to volume overload and should be avoided in elderly subjects and patients with congestive heart failure. Frusemide may potentiate the nephrotoxicity of various drugs such as aminoglycosides, and may cause ototoxicity. If the urinary losses of fluid and electrolytes are not replaced quantitatively serious volume and electrolyte deficits may occur, resulting in worsening renal function.

Other indications for diuretic therapy
Therapeutic use of carbonic anhydrase inhibitors
Glaucoma
This is the most common indication for the use of carbonic anhydrase inhibitors, which reduce intraocular hypertension and re-establish normal pressure by inhibiting fluid secretion into the anterior chamber (Wistrand 1984). A dose of 0.25 to 1 g of acetazolamide daily is required. The treatment may be complicated by metabolic acidosis and hypokalaemia, and occasionally crystalluria (for review see Berger and Warnock 1986).

Urinary alkalinization
Acetazolamide-induced urinary bicarbonate excretion may be beneficial in acute urate and pigment nephropathy. Furthermore, alkalinization of the tubule fluid helps to increase the renal clearance of weak acids such as aspirin and barbiturates due to the fact that non-ionic backdiffusion decreases with alkaline tubule pH. The effect of acetazolamide may be increased by additional bicarbonate infusion (Conger et al. 1977). Carbonic anhydrase inhibitors are dangerous in salicylate overdose since they can increase salicylate toxicity: during the metabolic acidosis resulting from renal bicarbonate loss, the protonated and lipophilic form of salicylic acid prevails and is accumulated in the central nervous system (Gabow et al. 1978).

Treatment of metabolic alkalosis
Acetazolamide may be used to treat metabolic alkalosis caused by diuretics and glucocorticoids. Another indication may be metabolic alkalosis following hypercapnia in patients with left ventricular dysfunction or cor pulmonale, in whom saline administration is contraindicated (for review see Berger and Warnock 1986). Acetazolamide may also be used in patients suffering from chronic obstructive lung disorders with complicating metabolic alkalosis due to steroid and/or diuretic treatment (Miller and Berns 1977) (see Chapter 7.3)

Acute mountain sickness
Untrained lowlanders may develop post-hypercapnic alkalosis, associated with fluid retention and pulmonary and cerebral oedema, when they climb rapidly to altitudes of 3000 m and higher. The symptoms can be treated (Larson et al. 1982) or prevented (Greene et al. 1981) by acetazolamide.

Hyperphosphataemia
Hyperphosphataemia due to acute phosphate load (rhabdomyolysis, phosphate laxative abuse, chemotherapy, and hypoparathyroidism) can be treated with acetazolamide (Agus et al. 1981) since this drug is phosphaturic (Beck and Goldberg, 1973).

Seizures
Acetazolamide therapy has been claimed to be beneficial in patients suffering from periodic paralysis due to hyperkalaemia as well as hypokalaemia and from petit mal type seizure disorders (Griggs et al. 1970; Streeten et al. 1971).

2.3 Action and clinical use of diuretics

Therapeutic use of thiazides
Calcium nephrolithiasis (see also Chapter 14.5)
Thiazide diuretics and indapamide have been shown to reduce stone formation in patients with idiopathic hypercalciuria (Coe 1981; Mortensen et al. 1986, Lemieux 1986) and in normocalciuric individuals (Yendt and Cohanim 1978), with a decrease in urinary calcium excretion of about 25 to 50 per cent. The response is increased by low, and reduced by high sodium intake (Yendt and Cohanim 1978). In contrast to renal calciuria, some patients with absorptive hypercalciuria become resistant to long-term thiazide treatment (Preminger and Pak 1987). Intestinal calcium reabsorption is also reduced by thiazides, although this effect is less marked than the effect on the kidney. Thiazides therefore cause calcium retention, although this has no impact on serum calcitonin and parathyroid hormone concentrations (Coe et al. 1988).

Thiazides have little effect on oxalate excretion (Ahlstrand et al. 1984) except in patients with absorptive hyperoxaluria (Yendt and Cohanim 1986). Thiazides may be effective in primary calcium nephrolithiasis (Baggio et al. 1986). The effect of thiazides on calcium reabsorption occurs in the early distal tubule (Costanzo and Windhager 1978). Parathyroid hormone is probably not required for the hypocalciuric effect of thiazides in man, since thiazides are anticalciuric in hypoparathyroid patients (Porter et al. 1978). The incidence of hypercalcaemia in patients during chronic thiazide therapy is three times that observed in the normal population (Christensson et al. 1977).

Thiazides and bone mineralization
Long-term thiazide treatment has been associated with enhanced bone mineralization in elderly male hypertensives (Wasnich et al. 1983), and a reduced incidence of femoral neck fractures has also been noted (Hale et al. 1984; Rashiq and Logan 1986). Some of the studies on the ability of thiazides to prevent osteoporosis suffer from the fact that the number of observations is small (Condon et al. 1978). Further studies are required to settle this issue.

Diuretics and diabetes insipidus
In patients with diabetes insipidus (pituitary and renal type) polyuria and polydipsia can be reduced by thiazides and salt restriction (Crawford et al. 1960; Lant and Wilson 1971). The beneficial effect appears to be due to reduction in glomerular filtration rate, reduced extracellular volume, and inhibition of the diluting effect of the distal tubule. In nephrogenic diabetes insipidus caused by lithium therapy, thiazides will increase serum lithium concentrations (Petersen et al. 1974; Singer 1981), and the administration of amiloride has also recently been advocated. Amiloride is claimed to inhibit entry of lithium into principal cells of the distal nephron, hence reducing the resistance to antidiuretic hormone caused by lithium (Batlle et al. 1982). This therapy carries the risk of renal tubular acidosis (Arruda et al. 1980).

Therapeutic use of loop diuretics
Treatment of hypercalcaemia
Loop diuretics inhibit NaCl as well as Mg^{2+} and Ca^{2+} reabsorption in the thick ascending limb of the loop of Henle (Duarte et al. 1971), and reduce plasma Ca^{2+} concentrations in patients with marked hypercalcaemia (Suki et al. 1970). It is important to couple this treatment with adequate volume and salt repletion: volume contraction will otherwise occur, and this will reduce the calciuric effect, due to increase in proximal tubule Ca^{2+} reabsorption.

Loop diuretics and thiazides in renal tubular acidosis
Distal renal tubular acidosis (type I) is a congenital or acquired acidification defect in which urine pH cannot be lowered because no pH gradient can be generated in the distal tubule (see Chapter 5.3). Loop diuretics have been used successfully to increase proton excretion in this disease (Györy and Edwards 1971; Heidbreder et al. 1982; Rastogi et al. 1985; Hropot et al. 1985). This beneficial effect of loop diuretics is explained by the increased NaCl load to the distal nephron, leading to increased NaCl reabsorption at this distal site and generation of a negative voltage in the lumen. The increased lumen-negative voltage is thought to increase distal proton secretion.

Thiazide diuretics may also be useful in type II renal tubular acidosis, which is due to proximal tubule bicarbonate wasting. They probably exert their effect by reducing glomerular filtration rate and increasing proximal tubule bicarbonate reabsorption. Correction of bicarbonaturia, of metabolic acidosis and hypophosphataemia with rickets can be achieved by thiazide treatment, even in patients with the full expression of type II renal tubular acidosis (de Toni-Debre-Fanconi-Syndrome) (Rampini et al. 1968; Santos-Atherton and Frank 1986).

Patients with hyperkalaemic renal tubular acidosis (type IV) show a generalized distal tubule dysfunction, which may be caused by aldosterone deficiency (hyporeninaemic hypoaldosteronism), frequently secondary to diabetes mellitus (DeFronzo 1980), or by resistance to the action of aldosterone (pseudohypoaldosteronism type I and II; Schambelan et al. 1981), or by a hydrogen secretion defect. Loop diuretics are beneficial in this disorder through their induction of aldosterone secretion due to volume contraction, and due to the increased distal fluid load (Rastogi et al. 1984). In some patients hydrogen secretion may not respond to frusemide: this is called 'irreversible voltage-dependent defect' (Batlle et al. 1981). The effectiveness of frusemide in patients with type IV renal tubular acidosis can easily be assessed by monitoring the change in urine pH after drug administration (Tarka et al. 1983).

Diuretics and the syndrome of inappropriate antidiuretic hormone (ADH) secretion
The basic treatment for chronic inappropriate ADH secretion is fluid restriction. In some cases ADH antagonists such as demeclocycline or lithium carbonate may be effective. In the long-term management of this condition oral frusemide and generous dietary sodium intake has been recommended (Decaux et al. 1981). Frusemide prevents ADH-induced concentration of the urine, by blocking NaCl reabsorption in the thick ascending limb of the loop of Henle. When acute hyponatraemia develops due to inappropriate ADH secretion the combination of frusemide and hypertonic saline may be used (Hantman et al. 1973)—but only after careful assessment of the volume status.

Diuretics in asthma
The rationale behind the use of loop diuretics comes from the knowledge that frusemide inhibits salt and fluid secretion in airway epithelia (Widdicombe et al. 1983; Welsh 1983) and from the experience that diuretic inhalation may prevent certain forms of bronchoconstriction (Bianco et al. 1988; Robuschi et al. 1988). A beneficial effect of inhaled frusemide was seen in a recent

double-blind, placebo controlled study in patients with asthma induced by specific allergen challenge (Bianco et al. 1989). Further studies on this issue are urgently needed.

Diuretics in intoxications

Forced diuresis induced by loop diuretics with adequate volume and electrolyte repletion is of value in many cases of intoxication. Since bromide and iodide may, to some extent, be reabsorbed by the thick ascending limb of the loop of Henle (Walser and Rahill 1966) halide intoxication can be treated by loop diuretics. Together with administration of Prussian blue and haemodialysis, loop diuretics are also recommended in thallium intoxication, because they increase renal excretion of thallium (Koch et al. 1972; Lameijer and van Zwieten 1978; Pedersen et al. 1978).

Interactions

Non-steroidal anti-inflammatory drugs and diuretics

Non-steroidal anti-inflammatory drugs interfere with both the pharmacokinetics and pharmacodynamics of loop diuretics. As organic anions, these compounds compete for the proximal tubular secretion of diuretics, and shift the dose–response curve of diuretics to the right (Chennavasin et al. 1980; Petrinelli et al. 1980). Apart, perhaps, from the selective extrarenal cyclo-oxygenase inhibitor sulindac (Ciabattoni et al. 1980), the non-steroidal anti-inflammatory agents impair the natriuresis induced by loop diuretics (Patak et al. 1975), which has a tubular as well as a haemodynamic component. According to Higashikara et al. (1979) prostaglandins inhibit sodium chloride reabsorption in the thick ascending limb of the loop of Henle, and they also are involved in the increase in renal blood flow caused by loop diuretics (Attallah 1979; Mackay et al. 1984). Conflicting data have been published concerning the effects of indomethacin on thiazide-induced diuresis and no clear effect (Fanelli et al. 1980; Favre et al. 1982, 1983) or reduction of the saliuretic response (Kramer et al. 1980a) has been reported.

Simultaneous use of non-steroidal anti-inflammatory agents and potassium-sparing diuretics may cause renal failure in healthy subjects (Anonymous 1986) and elderly patients (Favre et al. 1983; Lynn et al. 1985). The risk of hyperkalaemia is particularly high in patients with reduced GFR.

Control of hypertension by diuretics is impaired by non-steroidal anti-inflammatory drugs. Pretreatment with indomethacin attenuates the blood pressure lowering action of thiazides (Watkins et al. 1980), and additional use of naproxen and piroxicam in hydrochlorothiazide-treated patients may even cause an elevation of blood pressure (Wong et al. 1986).

Angiotensin converting enzyme inhibitors and diuretics

Acute renal failure complicated by severe hyponatraemia has been reported after combined administration of diuretics and angiotensin converting enzyme inhibitors (Watson et al. 1983; Hogg and Hillis 1986). The acute deterioration of renal function may be explained by both impaired renal autoregulation with a reduced tone of the glomerular efferent arteriole (Blythe 1983), and by a decreased production of glomerular prostaglandins in the presence of sodium depletion (Podjarny et al. 1988). As would be expected, simultaneous administration of angiotensin converting enzyme inhibitors and potassium-sparing diuretics can cause severe abrupt hyperkalaemia, leading in some cases to complete heart block (Lo and Cryer 1986; Lakhani 1986).

Allopurinol and diuretics

Adverse reactions to allopurinol (toxic epidermal necrolysis and hypersensitivity syndrome) seem to occur more often in patients taking thiazides (Aubock and Fritsch 1969; Anonymous 1985). The net excretion of the active metabolite, oxipurinol (although not of the parent allopurinol) is very sensitive to circulating volume, and thus retention may occur in diuretic-treated patients.

Lithium and diuretics

Diuretics have a direct influence on the renal clearance of lithium. Loop diuretics acutely increase the renal clearance of lithium by reducing lithium reabsorption in the thick ascending limb of the loop of Henle. Similarly, potassium-sparing diuretics have been shown to increase the renal clearance of lithium. On the other hand, diuretics can reduce the renal clearance of lithium if they induce volume contraction (Petersen et al. 1974; Greger 1990).

Cardiac glycosides and diuretics

Potassium-losing diuretics enhance the risk of cardiac arrhythmias in patients on glycoside therapy. This can probably be explained by competition between K^+ and glycosides for the same binding site on the Na^+, K^+–ATPase. Hence, hypokalaemia shifts the dose-response to glycosides to the left. Norgaard et al. (1981) have also shown that hypokalaemia decreases the number of ouabain binding sites in skeletal muscle.

Antibiotics and diuretics

The ototoxic effects of aminoglycosides (gentamicin, kanamycin) and the nephrotoxic potential of the first and second generation of cephalosporin may be enhanced by loop diuretics (Tilstone et al. 1977; Brummet et al. 1981).

Adverse reactions to diuretics

Metabolic disturbances

Hyponatraemia

Hyponatraemia is common after administration of thiazide diuretics, but it is rarely of clinical significance. Severe hyponatraemia once it occurs may, however, be life threatening (Ashraf et al. 1981). Diuretic-induced hyponatraemia is frequently observed in patients with congestive heart failure, cirrhosis of the liver, in the elderly, and after volume contraction (Gross et al. 1988). The mechanisms of diuretic-induced hyponatraemia are not well understood (see Chapter 7.1), but the following pathogenetic factors may be involved.

1. Excessive urinary sodium loss in the presence of a low salt intake (Ashraf et al. 1981).

2. A shift of sodium from extracellular to intracellular compartments (Fuizs et al. 1962; Fichman et al. 1971).

3. Impaired free water excretion, e.g. due to loss of the urinary dilution capacity caused by the diuretic (Kennedy and Earley 1970) as well as a reduced renal blood flow.

4. Inappropriate (enhanced) secretion of ADH or AVP (Horowitz et al. 1972; Luboshitzky et al. 1978). Increased AVP concentrations have been described in congestive heart failure, liver cirrhosis and in the elderly (Gross et al. 1988). Additional non-osmotic AVP release may occur in response

to diuretic-induced contraction of effective arterial blood volume.

5. Unrestricted intake of hypotonic fluid. According to Schrier and Berl (1980) and Gross et al. (1988) the fluid intake in patients with hyponatraemia is surprisingly high, suggesting that enhanced fluid intake is an important contributor to hyponatraemia. A dipsogenic effect of thiazides has been shown in certain elderly (Fuizs 1963; Kone et al. 1986) or psychotic patients (Hariprasad et al. 1980; Emsley and Gledhill 1984). According to Friedman et al. (1989) high-risk patients can be identified using a single challenge dose of the thiazide diuretic. Within 6 h a decrease in serum sodium concentration is observed, combined with a weight gain.

6. Concomitant drug therapy. Water intake may be increased by drugs which stimulate thirst (tricyclic antidepressants) or which have AVP-like effects (chlorpropamide, non-steroidal anti-inflammatory agents).

In patients with volume contraction, circulatory underfilling may cause postural hypotension (especially in the elderly), a decline in GFR, and elevated concentrations of catecholamines, renin, angiotensin II, and aldosterone. The latter may result in potassium wastage and in metabolic alkalosis.

The management of diuretic-induced hyponatraemia involves withdrawal of the diuretic, restriction of fluid intake (<1 l per day) and, if necessary, sodium and potassium repletion (Ayus 1986).

Hypokalaemia

Prevalance of hypokalaemia (see also Chapter 7.2)

Long-term administration of thiazide and loop diuretics is associated with a reduction in serum and total body potassium levels (for review see Tannen 1985): the degree of hypokalaemia is generally mild and asymptomatic. Approximately 50 per cent of patients treated with thiazides have potassium concentrations below 3.5 mmol/l, and 7 per cent have levels below 3.0 mmol/l (Morgan and Davidson 1980). Long-acting diuretics such as chlorthalidone cause a more severe hypokalaemia (Tweeddale et al. 1977), while potassium depletion is less pronounced with short-acting loop diuretics such as frusemide, ethacrynic acid, and bumetanide (Venkata et al. 1981). The fall in serum potassium is generally proportional to the dose (Hollifield 1986), and potassium concentration reaches its nadir 1 to 4 weeks after initiation of diuretic therapy. Females seem to be more susceptible to hypokalaemia than males (Krakauer and Lauritzen 1978), and the severity of potassium depletion is also related to sodium intake. Very low and high sodium intakes aggravate hypokalaemia, whereas renal potassium loss is minimal if sodium intake is moderate (ca. 100 mmol/day) (Ram et al. 1975; Landmann-Suter and Struyvenberg 1978).

There is considerable variability in the degree of total body potassium depletion during long-term diuretic therapy. Various techniques have estimated the potassium deficits to be between 0 and 400 mmol, with an average loss of 200 mmol, corresponding to 5 per cent of total body potassium. It is assumed that this potassium loss corresponds to a fall in serum potassium concentration by 0.6 mmol/l (Tannen 1985).

Pathophysiology

As already mentioned, any diuretic acting in the thick ascending limb of the loop of Henle (loop diuretics) or in the distal tubule (thiazides) causes kaliuresis, and hence potassium losses through several mechanisms. Firstly, enhanced potassium secretion is caused by an increased delivery of water and sodium to the collecting duct. The rate of sodium reabsorption and consequent potassium secretion at this nephron site is controlled by aldosterone, and secondary hyperaldosteronism, which follows diuretic-induced volume depletion, therefore augments potassium losses. Secondly, diuretic-induced losses of protons (metabolic alkalosis) aggravate and maintain hypokalaemia (Evans et al. 1954). Diuretic-induced losses of chloride, arising via an increased lumen negative voltage, enhance potassium secretion (Velázquez et al. 1982), and potassium losses are also induced by a reduction in bicarbonate reabsorption, which is caused by carbonic anhydrase inhibitors and all sulphamoyl diuretics.

Clinical consequences of diuretic-induced hypokalaemia

There is an ongoing discussion of whether hypokalaemia has direct arrhythmogenic effect (Hollifield 1981; Holland 1984; for review see Tannen 1985). On the one hand no such effect was seen in four prospective studies encompassing 64 patients (Lief et al. 1983; Madias et al. 1984; Papademetrio 1984; Whelton 1984). On the other hand, three other prospective studies with a total of 50 patients, and also the chronic phase of the MRC study with 74 patients (Whelton 1984) clearly demonstrated the occurrence of ventricular premature beats and complex ventricular premature beats during diuretic therapy. After correction of hypokalaemia the incidence of arrhythmias returned to baseline values, even though the diuretic therapy was continued. According to Caralis et al. (1984) only a subgroup of hypertensives who had pre-existing cardiac disease showed enhanced arrhythmias during diuretic therapy.

The coincidence of diuretic-induced hypokalaemia and arrythmias has been demonstrated especially in patients with acute myocardial infarction (Hulting 1981; Nordehaug and Vonder Lippe 1983; Solomon 1984). In this regard it is of interest that catecholamines, by enhancing cellular uptake of potassium, can further aggravate hypokalaemia (Struthers et al. 1983). It is well known that hypokalaemia-induced arrhythmias are more likely in patients treated with digitalis glycosides. In hypokalaemia more K^+-binding sites of the Na^+, K^+–ATPase seem to be occupied by glycosides (Norgaard et al. 1981). Other diuretic-induced electrolyte disturbances, such as magnesium depletion, may contribute to the occurence of arrhythmias.

In summary, all diuretics (except the potassium-sparing agents) can decrease plasma potassium concentration: this is related to drug dose and duration of action. Diuretic-induced hypokalaemia may potentiate arrhythmias due to ischaemic heart disease and/or myocardial infarction and may have negative metabolic effects: potassium depletion should therefore be prevented, or corrected once it develops. Changes in serum potassium can be minimized or avoided by administering a reduced dose of diuretics (if the GFR is higher than 50 ml/min) or by administration of angiotensin converting enzyme inhibitors. Correction of hypokalaemia may protect against ventricular arrhythmias and minimize carbohydrate intolerance. There are also some recent provocative data supporting a 'vasculoprotective' effect of potassium in man (Khaw and Barrett-Conner 1987) and in rats (Tobian 1988).

Treatment of hypokalaemia

The decision to treat mild hypokalaemia (3.0–3.5 mmol/l) must be weighed against the dangers of potential hyperkalaemia.

Patients with impaired renal function (GFR <50ml/min), elderly subjects, and diabetics are at especially high risk of developing hyperkalaemia. However, even mild hypokalaemia should be prevented or treated in patients receiving cardiac glycosides or in those predisposed to arrhythmias and congestive heart failure. Potassium chloride supplements are not often effective (Ramsay *et al.* 1980) and they carry the risk of mucosal ulceration if passage through the intestine is slow. Amiloride, triamterene, and spironolactone have both a potassium and magnesium sparing effect (Heidland *et al.* 1970; Dykner and Wester 1984). Another attractive method for correcting diuretic-induced hypokalaemia is the concomitant administration of angiotensin converting enzyme inhibitors (Griffing *et al.* 1983). Prevention of hypokalaemia by the use of low doses of diuretics should, however, be the first principle.

Hyperkalaemia

Concomitant use of diuretics and a potassium-sparing agent (or potassium supplements) may occasionally induce hyperkalaemia (Mor *et al.* 1983; Akbarpour *et al.* 1985), particularly in patients suffering from impairment of renal function (George 1980). Potassium supplements should therefore not be used in patients with a GFR below 50ml/min. Diabetic patients, especially those on insulin therapy, appear to develop hyperkalaemia more often than other subjects, in part at least because of hypoaldosteronism.

Magnesium depletion

The pathophysiology of diuretic-induced magnesium depletion is discussed above. Long-term therapy with thiazides may induce magnesium deficiency; this is directly related to the dosage (Lim and Jacob 1972; Heidland *et al.* 1973; Hollifield 1986). Decreased magnesium concentrations have also been observed in skeletal muscle after long-term thiazide therapy (Dykner and Wester 1984). Interestingly, potassium deficiency may fail to respond to supplementation if magnesium depletion is not corrected (Dykner and Wester 1979), and an inter-relationship between magnesium and hyponatraemia has been described (Dykner and Wester 1981). Potassium-sparing agents such as amiloride, triamterene, and spironolactone reduce magnesium losses, and this, at least initially, is not the result of a decline in GFR (Heidland *et al.* 1970; Ryan *et al.* 1984).

Effects of diuretics on calcium balance

The pathophysiology has been described above. Calcium reabsorption is increased by thiazides and decreased by loop diuretics. The effect of thiazides on calcium is probably a direct effect on the distal tubule, and this has been exploited in the prevention of recurrent formation of calcium containing urinary calculi (Yendt and Cohanim 1978) (see also Chapter 14.5). Hypercalcaemia caused by thiazide therapy may occasionally unmask autonomous (primary) hyperparathyroidism. In other cases, thiazides may affect the interaction of PTH and bone. There are some data indicating a beneficial effect of thiazide-induced calcium retention in osteoporosis (Wasnich *et al.* 1983).

Loop diuretics, on the other hand, reduce calcium reabsorption in the thick ascending limb of the loop of Henle, and hence increase urinary calcium excretion. This effect is partially compensated for by enhanced intestinal calcium reabsorption, and diminishes with long-term use of loop diuretics (Lant 1985*a*, 1985*b*).

Metabolic alkalosis

As described above, loop diuretics and thiazides cause increased renal loss of protons, and hence induce metabolic alkalosis. On the other hand, potassium-sparing diuretics such as amiloride or triamterene conserve protons: if administered together with loop diuretics or thiazides they tend to counteract metabolic alkalosis (Hyams 1981).

Carbohydrate intolerance

Predisposing factors

The appearance of impaired glucose tolerance and the worsening of diabetes is common during long-term diuretic therapy. Although usually clinically silent, hyperglycaemia can be severe and may induce a non-ketotic hyperosmolar state (Dollery 1973). Worsening of carbohydrate tolerance is particularly frequent after the use of long-acting thiazide diuretics, but may also occur after moderate or high doses of loop diuretics. Combinations of thaizides and β-blockers exert a more deleterious effect than that of either drug alone (Helgeland *et al.* 1984; Dornhorst *et al.* 1985). Impaired glucose tolerance is found more frequently in type II diabetics, in the elderly (Amery *et al.* 1978; Goldner *et al.* 1960) and in vulnerable patients with essential hypertension (Bengtsson *et al.* 1988). In a prospective study, lasting 9 years on average, 12 of 73 hypertensives developed overt diabetes mellitus, compared to only two of 65 non-treated normotensive controls (Skarfors *et al.* 1989). Whether thiazides and loop diuretics can also induce glucose intolerance in previously glucose tolerant subjects is a matter of discussion. Insulin resistance is, however, a frequent finding in essential hypertension *per se* (Ferrannini *et al.* 1987; Ferrari and Weidmann 1990). Fortunately, the diabetogenic effect of diuretics appears to be reversible (Murphy *et al.* 1982). In the MRC Trial (1981) about 60 per cent of patients with glucose intolerance during bendroflumethiazide treatment returned to normal tolerance within a year after stopping the drug.

Pathogenesis of diuretic-induced glucose intolerance

The present evidence points to a multifactorial pathogenesis, involving increased hepatic glucose production, inhibition of glucose uptake and utilization in skeletal muscle, and inhibition of insulin release by a direct and/or indirect effect (hypokalaemia).

In rats thiazides increase hepatic glucose production by enhancing glycogenolysis (Senft *et al.* 1966). This seems to be mediated by an inhibition of hepatic phosphodiesterase and a consequent increase in cyclic AMP, which then increases glycogenolysis (Hoskins and Jackson 1978).

Recently, direct evidence for other mechanisms of insulin resistance has been accumulated. Glucose utilization by rat diaphragms is drastically reduced during incubation with various diuretics (Dzurik *et al.* 1990). Frusemide has also been shown to inhibit glucose transport in isolated rat adipocytes *via* direct inactivation of a carrier protein (Jacobs *et al.* 1984). Pancreatic insulin production is impaired *in vitro* and *in vivo* by thiazides and frusemide (Frumann 1981): in the latter case the mechanism of inhibition was localized in chloride channels (Sandström and Sehlin 1988).

In man, the main mechanism underlying thiazide-induced glucose intolerance is reduced insulin sensitivity of the tissue (Beardwood *et al.* 1965); basal glucose and insulin levels are both elevated during long-term diuretic therapy. The mechanism of diuretic-induced insulin resistance was characterized by Pollare *et al.* (1989) using the euglycaemic hyperinsulinaemic clamp

2.3 Action and clinical use of diuretics

Table 1 Changes in blood lipids after short-term hydrochlorothiazide therapy

	Total triglycerides	Total cholesterol	VLDL–cholesterol	LDL–cholesterol	HDL–cholesterol
Range of increase (%)	4–37	4–30	7–65	7–29	
Mean increase (%)	23	6.5	28	15	0

Data of Ames (1986).

technique. Insulin secretion by the β-cell of the pancreas may, at least in part, be disturbed by local prostaglandin activation (Giugliano et al. 1979).

Although it has been suggested that diuretics per se can interfere with glucose metabolism (Berglund and Andersen 1981), many authors favour the view that impaired carbohydrate tolerance is related to diuretic-induced hypokalaemia (Amery et al. 1978; Sundberg et al. 1982). The central role of hypokalaemia is further emphasized by the clinical observation that glucose intolerance can be prevented by potassium supplementation (McFarland and Carr 1977; Helderman et al. 1983) and avoidance of hypokalaemia (Berglund and Anderson 1981; Sundberg et al. 1982). However, it should be stressed that during hydrochlorothiazide therapy small reductions of plasma potassium (-0.28 mmol/l), within the normal range, may be associated with decreased insulin sensitivity (Pollare et al. 1989). Potassium depletion not only decreases the insulin sensitivity of tissues, but may also cause an inhibition of insulin secretion (Rapoport and Heard 1965; Grünfeld and Chappell 1983). Increased levels of catecholamines associated with diuretic treatment (Heidland et al. 1969; Luke et al. 1979) may also contribute to reduced insulin sensitivity.

Therapeutic consequences

Prospective population studies indicate that disturbances in carbohydrate metabolism cause both a higher incidence of coronary artery disease (Fuller et al. 1980) and a higher mortality rate (Jarrett et al. 1982). Basal hyperinsulinaemia is also considered to be a risk factor for cardiovascular disease (Pyörälä 1979), and the benefits of lowering blood pressure by chronic diuretic therapy may be offset by the diabetogenic effect. To prevent these complications the lowest effective diuretic dose should be used.

Hyperlipidaemia

Incidence of hyperlipidaemia during diuretic therapy

Potassium-losing diuretics may cause unfavourable alterations in blood lipids. In short-term studies of hypertensive patients thiazide diuretics increased concentrations of total triglycerides, total cholesterol, very low density lipoprotein cholesterol, and low density lipoprotein cholesterol, whereas the concentration of high density lipoprotein cholesterol was not altered (Ames and Hill 1976; Pollare et al. 1989) or even fell (Lasser et al. 1984; Greenberg 1984). Table 1 summarizes data obtained by Ames (1986).

The changes in lipid metabolism caused by various diuretics are not uniform. Indapamide, for instance, has no effect on total triglycerides or low density lipoprotein cholesterol (Weidmann et al. 1981) but may increase high density lipoprotein cholesterol (Meyer-Saballek et al. 1985). Whether the effects of diuretics on lipid metabolism disappear with chronic use is controversial. After long-term therapy cholesterol concentrations have been reported to be close to baseline values (Helgeland 1978; Hypertension Detection and Follow-up Program 1979; Veteran Administration Cooperative Study 1982; Amery et al. 1982; Greenberg 1984; MRFIT study 1985; HAPPHY 1987; Moser 1988). Other authors, however, report increases in serum lipid concentrations after 1 year of diuretic treatment (Weinberger et al. 1985; Middecke et al. 1987). Similarly Skarfors et al. (1989) found that total triglyceride levels, measured over 9 years, were higher in treated hypertensives than in untreated normotensives. The MRFIT study also provided some evidence to suggest that diuretics may blunt the beneficial effects of a low fat diet on lipid concentrations (Shekelle et al. 1984).

Pathophysiology of diuretic-induced hyperlipidaemia

Possible mechanisms include potassium depletion, insulin resistance and hyperinsulinaemia, and other hormonal factors.

A causal role for hypokalaemia seems likely: spironolactone and indapamide have no effect on total cholesterol, and the combination of thiazides and triamterene is also without effect on lipid concentrations (Amery et al. 1982). Insulin resistance and hyperinsulinaemia, which themselves are aggravated by hypokalaemia, may also be causally involved. Hence, it is conceivable that potassium sparing therapy may prevent the lipid changes which may be caused by long-term diuretic therapy. Insulin not only modulates glucose uptake, but also inhibits lipolysis. Hyperinsulinaemia can result in lipolysis and hence in enhanced concentrations of free fatty acids and in a rise in synthesis of very low density lipoproteins. Insulin resistance may also induce non-enzymatic glycosylation and slow the clearance of glycosylated lipids and lipoproteins from the circulation (Steinbrechen and Witztum 1984). Hormonal disturbances may also be involved: premenopausal, but not postmenopausal women, for instance, show no lipid changes during diuretic therapy (Boehringer et al. 1982). In addition, catecholamines may play a role, since α_1-receptors are involved in the regulation of lipoprotein-lipase, synthesis of very low density lipoproteins, and in low density lipoprotein-receptor mediated catabolism.

Therapeutic consequences

The results of the Framingham Study suggested that altered lipid metabolism, and increases in low density lipoprotein cholesterol concentrations in particular, are important risk factors for coronary heart disease (Kannel et al. 1979). Diuretic-induced hyperlipidaemia may therefore offset the cardiovascular benefits to be derived from lowering blood pressure. Nevertheless, diuretics are still recommended as first line therapy in arterial hypertension, but the lowest effective doses should be used (JNC 1984). Hyperlipidaemia should also be controlled by a low fat, low cholesterol, and low calorie diet.

In summary, long-term antihypertensive therapy with high doses of diuretics may cause changes in serum lipoproteins. These alterations are generally mild and may return to pretreatment values after 2 to 3 years, although they may persist for 10 years or longer.

Hyperuricaemia and gouty arthritis

Urate retention is a common and dose-dependent complication of long-term diuretic therapy, and 50 to 75 per cent of treated hypertensives have been reported to develop hyperuricaemia (Shutske *et al.* 1984), while the incidence of clinical gouty arthritis is no more than 2 per cent (Beevers *et al.* 1971). In the MRC Trial, gouty arthritis occurred during bendrofluazide therapy with an incidence of 12.2 cases in men and 1.1 cases in women per 1000 treatment years. Currently there is no evidence suggesting that hyperuricaemia may be an additional and independent cardiovascular risk factor (Bloxham and Beevers 1979; Bulpitt *et al.* 1979).

Diuretic-induced hyperuricaemia occurs through increased renal urate reabsorption, which itself is caused by volume contraction (Steele 1971; Lang *et al.* 1977, 1980). Reduced urate secretion may be another causal factor: diuretics and urate compete for proximal secretion (Lang *et al.* 1977).

Diuretic-induced changes in renal function

Short-term administration of thiazides and carbonic anhydrase inhibitors results in a decline of glomerular filtration rate, while renal plasma flow remains constant (Corcoran *et al.* 1955; Crosley *et al.* 1960; Heidland *et al.* 1964). These alterations have been related to extracellular fluid volume depletion and an activation of the tubuloglomerular feedback system (Tucker *et al.* 1978). After prolonged antihypertensive therapy with thiazides the fall in plasma volume, GFR, and renal plasma flow is only transient (van Brummelen *et al.* 1979; Loon *et al.* 1989). Continued therapy with diuretics leads to a reduced natriuretic response (Loon *et al.* 1989). Studies in rats have shown that prolonged frusemide infusion causes hypertrophy of the distal convoluted tubule, connecting tubule, and principal cells of the collecting tubule (Kaissling and Stanton 1987; Ellison *et al.* 1989).

In patients with severe hypertension treated by long-term administration of diuretics (Veterans Administration Study 1972) the decline in GFR was slowed. Similarly, treatment with conventional antihypertensive drugs combined with frusemide slowed the decline of GFR in patients with diabetic nephropathy (Parving *et al.* 1985). In contrast to these results, Walker *et al.* (1987) showed that diuretics worsen the natural course of diabetic nephropathy in hypertensive insulin-dependent and non-insulin-dependent subjects. The EWPHE-Trial (Amery *et al.* 1985) found an increase in renal impairment (2 cases per 1000 treatment years) in patients receiving the combination of hydrochlorothiazide and triamterene. Lynn *et al.* (1985) described 19 elderly patients with renal failure due to treatment with potassium-sparing diuretics and hydrochlorothiazide.

There is no doubt that vigorous diuretic treatment, with consequent hypovolaemia, may enhance the risk of acute renal failure, particularly in the elderly. In rats diuretic-induced sodium depletion results in azotaemia which seems to arise from both reduced urea clearance and enhanced urea nitrogen production (Kamm *et al.* 1987). An increase in urea nitrogen has also been observed in uraemic patients during diuretic therapy (Dal Canton *et al.* 1985). The underlying mechanism for this increase in urea production may be enhanced proteolysis of skeletal muscle due to volume contraction. Recently, the combination of thiazides and angiotensin converting enzyme-inhibitors was also found to have this effect (Hogg and Hillis 1986). Furthermore, interstitial nephropathy may develop in patients with pre-existing glomerular disease after administration of frusemide and thiazides (Abt and Gordon 1985) (see Chapters 2.1 and 8.10). Long-term frusemide treatment of idiopathic oedema causes a decline in GFR and tubulointerstitial changes which are clearly dose-dependent (Schichiri *et al.* 1984).

Diuretic-induced ototoxicity

Clinical observations

Several authors have reported transient hearing losses in patients with renal failure treated with ethacrynic acid (Maher and Schreiner 1965; Schneider and Becker 1966); and permanent deafness has also been seen after ethacrynic acid therapy of uraemic patients (Pillay *et al* 1969). Acute hearing losses of up to 30 dB and more were observed following intravenous administration of high doses of frusemide (1 g/40 min): these hearing losses were reversible within some hours in all cases (Heidland and Wigand 1970; Wigand and Heidland 1971). A lower dose of 0.6 g over 40 min exerted smaller and transient effects. In the recent past three cases of permanent deafness due to frusemide therapy have been reported: one of these followed oral administration (Quick and Hoppe 1975; Gallagher and Jones 1979). In a review of the cases reported to the FDA six patients were reported to have developed deafness with daily doses of 200 mg or less of frusemide (Gallagher and Jones 1979). Loop diuretics are particularly harmful if they are given together with aminoglycosides.

Pathophysiology of diuretic-induced ototoxicity

Currently, most experimental evidence favours the view that loop diuretics act in the stria vascularis by inhibiting the $Na^+2Cl^-K^+$ cotransporter, and hence reduce KCl secretion into the endolymph (Bosher *et al.* 1973; Syka and Melichar 1981; Marcus *et al.* 1987). In addition, and similar to the effects of aminoglycosides, loop diuretics cause a depletion of glutathione, resulting in a limited capacity to detoxify free radicals (Hoffman *et al.* 1988).

Prevention of ototoxicity

The ototoxic potential of equipotent doses of bumetanide and piretanide has been shown to be less than that of frusemide and ethacrynic acid in dogs (Brown 1981), guinea-pigs (Jung and Schön 1979; Brummet *et al.* 1981), cats (Goettl *et al.* 1985), and in clinical studies (Tazel 1981).

Urinary voiding

In elderly patients, whose urinary bladder has reduced capacity, brisk diuresis should be avoided, because both urinary retention and urinary incontinence may occur. Due to the mechanisms of action loop diuretics are more likely to produce this problem than thiazides.

Diuretic induced impotence

A high incidence of male impotence was reported in the MRC Trial (1981), although it should be stressed that the diuretic dose used in this study (bendrofluazide 10 mg daily) was excessive by current standards. The prevalence of impotence after 3 months was 16 per cent in the bendrofluazide group, compared to 9 per cent in the placebo group. After 2 years the numbers were 23 per cent versus 10 per cent. A much lower incidence was found in the MRFIT Study (Grimm 1985): after 6 years the incidence of decreased sexual activity was 25 per cent in the treated group compared to 16 per cent in the control group. Interestingly, 80 per cent of the decrease in sexual activity could not be attributed to antihypertensive therapy. Similar results were obtained in a double-blind short-term (4 months) study by Helgeland *et al.* (1986) comparing hydrochlorothiazide (25–50 mg daily), atenolol (50–100 mg daily), and enalapril (20–40 mg daily). The prevalence of impotence was 1.7, 2.3, and 1 per cent respectively. Hence, the relevance of this side-effect was probably overestimated in the MRC Trial (1981).

Diuretic-induced skin reactions

Robinson *et al.* (1985) described four cases of chronic photosensitivity attributed to hydrochlorothiazide, which persisted after withdrawal of the drug. Reed *et al.* (1985) reported a new photosensitive skin eruption resembling subacute lupus erythematosus related to hydrochlorothiazide therapy.

Specific toxicities

Acetazolamide

Metabolic acidosis is a frequent complication of therapy with carbonic anhydrase inhibitors. Examination of blood gases in 27 elderly patients taking acetazolamide (250–1000 mg daily) revealed severe acidosis (pH 7.15) in one patient, moderate acidosis (pH 7.21–7.29) in 35 per cent, and mild acidosis (pH 7.3–7.31) in 15 per cent of the patients (Heller *et al.* 1985). Carbonic anhydrase inhibitors may cause hepatic encephalopathy in patients with hepatic cirrhosis, due to increased ammonia levels in the blood. Prolonged use of these agents is occasionally associated with the development of nephrocalcinosis and apatite stones (Gordon and Sheldon 1957). This complication has been related to the combined effects of phosphaturia and hypercalciuria, caused by systemic acidosis in the presence of an alkaline urine, in which calcium phosphate salts are only poorly soluble.

Spironolactone

Painful gynaecomastia is a frequent complication of long-term therapy with spironolactone. Its incidence seems to be lower after chronic administration of potassium canrenoate (Dupont 1985; Bellati and Ideo 1986; Overdiek and Merkus 1986). Spironolactone is also implicated in menstrual disturbances (Hutchinson 1985) and male impotence. This may be caused both by a decreased biosynthesis of testosterone and by an antiandrogenic effect (Overdiek and Merkus 1986).

In the rat, long-term administration of potassium canrenoate, but not spironolactone, caused a dose-related increase in myelocytic leukaemia, breast cancer, and testicular and thyroid adenomas (Lumb *et al.* 1978). Recent investigations suggest that one or more metabolites of canrenoate are mutagenic and, presumably, responsible for the myelotic leukaemia. These mutagenic metabolites are not formed from spironolactone (Oppermann *et al.* 1988). Specific high performance liquid chromatography showed that, in contrast to earlier investigations with non-specific fluorimetry, canrenone accounts for only 20 to 25 per cent of the fluorimetric metabolites of spironolactone in man (Abshagen *et al.* 1990).

Due to the various side-effects and the potential carcinogenic risk the use of spironolactone in the United Kingdom is generally restricted to the treatment of Conn's syndrome and of oedema in hepatic cirrhosis.

Triamterene

Triamterene is poorly water soluble and may precipitate in the urine. Urinary abnormalities (red-brown deposits, red-brown crystals, and casts) are frequently found during triamterene treatment, and are assumed to be due to precipitation of this diuretic. Occasionally, triamterene-induced nephrolithiasis has been reported (Ettinger *et al.* 1979). In most cases, however, triamterene and its metabolites seem to be simply incorporated into stones as innocent bystanders (Anonymous 1985). A controlled study failed to show any increased incidence of renal calculi in association with triamterene therapy (Jick *et al.* 1982).

References

Abe, K., Yasiuma, M., and Cheiba, L. (1977). Effect of furosemide on urinary excretion of prostaglandin E in normal volunteers and patients with essential hypertension. *Prostaglandins*, **14**, 513–21.

Abel, R.M., Buckley, M.J., and Austen, W.G. (1976). Etiology, incidence, and prognosis of renal failure following cardiac operations. Results of a prospective analysis of 500 consecutive patients. *Journal of Thoracic and Cardiovascular Surgery*, **71**, 323–37.

Abshagen, U., Besenfelder, E., Endele, R., Koch, K., and Neubert, B. (1990). Kinetics of canrenone after single and multiple doses of spironolactone. *European Journal of Clinical Pharmacology*, **16**, 255–62.

Abt, A.B. and Gordon J.A. (1985). Drug-induced interstitial nephritis: coexistence with glomerular disease. *Archives of Internal Medicine*, **145**, 1063–7.

Agus, Z.S., Goldfarbs, S., and Wasserstein, A. (1981). Disorders of calcium and phosphate metabolism. In *The Kidney*, 2nd edn. (eds. B. M. Brenner and F. C. Rector, Jr.), p. 985. W. B. Saunders Company, Philadelphia.

Ahlstrand, C., Tiselius, H.G., Larsson, L., and Hellgren, E. (1984). Clinical experience with long-term bendroflumethiazide treatment in calcium oxalte stone formers. *British Journal of Urology*, **56**, 255–62.

Akbarpour, F., Afrasiabi, A., and Vaziri, N.D. (1985). Severe hyperkalemia caused by indomethacin and potassium supplementation. *Southern Medical Journal*, **78**, 756–7.

Amery, A., *et al.* (1978). Glucose intolerance during diuretic therapy. *Lancet*, **i**, 681–3.

Amery, A., *et al.* (1982). Influence of antihypertensive therapy on serum cholesterol in elderly hypertensive patients. *Acta Cardiologica*, **37**, 235–44.

Amery, A., *et al.* (1985). Mortality and morbidity results from the European Working Party on High Blood Pressure in the Elderly trial. *Lancet*, **i**, 1349–54.

Ames, R.P. (1986). The effect of antihypertensive drugs on serum lipids and lipoproteins. I. Diuretics. *Drugs*, **32**, 260–78.

Ames, R.P. and Hill, P. (1976). Elevation of serum lipid levels during diuretic therapy of hypertension. *American Journal of Medicine*, **61**, 748–57.

Anderson, J.V., Woodruff, P.W., and Bloom, S.R. (1988). The effect of treatment of congestive heart failure on plasma atrial natriuretic peptice concentration: a longitudinal study. *British Heart Journal*, **59**, 207–11.

Anderson, R.J., *et al.* (1977). Nonoliguric acute renal failure. *New England Journal of Medicine*, **296**, 1134–38.

Anonymous (1985). Treatment of gout. *Drug and Therapeutics Bulletin*, **23**, 47–8.

Anonymous (1986). Triamterene and the kidney. *Lancet*, **1**, 424.

Anto, H.R., Chou, S.Y., Porush, J.G., and Shapiro, W.G. (1981). Infusion intravenous pyelography and renal function-effects of hypertonic mannitol in patients with chronic renal insufficiency. *Archives of Internal Medicine*, **141**, 1652–6.

Arruda, J.A.L., Subbarayudu, K., Dytko, G., and Kurtzman, N.A. (1980). Voltage dependent distal acidification defect induced by amiloride. *Journal of Laboratory and Clinical Medicine*, **95**, 407–16.

Ashraf, N., Locksley, R., and Arieff, A.I. (1981). Thiazide-induced hyponatremia associated with death or neurologic damage in outpatients. *American Journal of Medicine*, **70**, 1163–8.

Attallah, A.A. (1979). Interaction of prostaglandins with diuretics. *Prostaglandins*, **18**, 369–75.

Aubock, J. and Fritsch, P. (1985). Asymptomatic hyperuricaemia and allopurinol induced toxic epidermal necrolysis. *British Medical Journal*, **290**, 1969.

Ayus, J.C. (1986). Diuretic-induced hyponatremia. *Archives of Internal Medicine*, **146**, 1295–6.

Baggio, B., et al. (1986). An inheritable anomaly of red-cell oxalate transport in 'primary' calcium nephrolithiasis correctable with diuretics. *New England Journal of Medicine*, **314**, 599–604.

Bailey, R.R., Natale, R., Turnbull, D.L., and Linton, A.L. (1973). Protective effect of furosemide in acute tubular necrosis and acute renal failure. *Clinical Science and Molecular Medicine*, **45**, 1–17.

Barry, K.G. and Malloy, J.P. Oliguric renal failure. Evaluation and therapy by the intravenous infusion of mannitol. (1962). *Journal of the American Medical Association*, **179**, 510–13.

Batlle, C., Arruda, J.A.L., and Kurtzman, N.A. (1981). Hyperkalemic distal renal tubular acidosis associated with obstructive uropathy. *New England Journal of Medicine*, **304**, 373–80.

Batlle, D., Gavira, M., Grupp, M., Arradu, J.A.L., Wynn, J., and Kurtzman, N.A. (1982). Distal nephron function in patients receiving chronic lithium therapy. *Kidney International*, **21**, 477–85.

Beardwood, D.M., Alden, J.S., Graham, C.A., Beardwood, J.T., and Marble, A. (1965). Evidence for a peripheral action of chlorothiazide in normal men. *Metabolism*, **14**, 561–7.

Beck, L.M. and Goldberg, M. (1973). Effects of acetazolamide and parathyroidectomy on renal transport of sodium, calcium and phosphate. *American Journal of Physiology*, **233**, 1136–46.

Beermann, B., Dalen, E., and Lindstrom, B. (1977). Elimination of furosemide in healthy subjects and in those with renal failure. *Clinical Pharmacology and Therapeutics*, **22**, 70–8.

Beermann, B. and Groschinsky-Grind, M. (1978). Antihypertensive effect of various doses of hydrochlorothiazide and its relation to the plasma level of the drug. *European Journal of Clinical Pharmacology*, **13**, 195–201.

Beevers, D.G., Hamilton, M., and Harpur, J.E. (1971). The long-term, treatment of hypertension with thiazide diuretics. *Postgraduate Medical Journal*, **47**, 639–43.

Bellati, G. and Ideo, G. (1986). Gynaecomastia after spironolactone and potassium canrenoate. *Lancet*, **1**, 626.

Bengtsson, C., Blohme, C., Lapidus. L., and Lundgren, H. (1988). Diabetes in hypertensive women: an effect of antihypertensive drugs or of the hypertensive state *per se*? *Diabete et Métabolisme*, **5**, 261–4.

Berger, B.E. and Warnock, D.G. (1986). Mechanism of action and clinical use of diuretics. In: *The Kidney*, 3rd edn. (eds. Brenner, B.M., Rector, F.C.) pp. 433–455, Saunders Company, Philadelphia.

Berglund, G. and Andersson, O. (1981). Beta-blockers or diuretics in hypertension? A six year follow-up of blood pressure and metabolic side effects. *Lancet*, **1**, 744–7.

Berman, L.B., Smith, L.L., Chisholm, G.D., and Weston, R.E. (1964). Mannitol and renal function in cardiovascular surgery. *Archives of Surgery*, **88**, 239–42.

Better, O.S. and Stein, J.H. (1990). Early management of shock and prophylaxis of acute renal failure in traumatic rhabdomyolysis. *New England Journal of Medicine*, **322**, 825–8.

Bianco, S., Vaghi, A., Robuschi, M., and Pasargiklian, M. (1988). Prevention of exercise-induced bronchoconstriction by inhaled furosemide. *Lancet*, **ii**, 252–5.

Bianco, S., Pieroni, M.G., Refini, R.M., Rottoli, L., and Sostini, P. (1989). Protective effect of inhaled furosemide on allergen-induced early and late asthmatic reactions. *New England Journal of Medicine*, **321**, 1069–73.

Bichet, D. and Manzini, C. (1984) Role of vasopressin (AVP) in the abnormal water excretion in nephrotic patients (NP). *Kidney International*, **25**, 160.

Bichet, D., Szatalowicz, V., Chainovitz, C., and Schrier, R.W. (1982a). Role of vasopressin in abnormal water excretion in cirrhotic patients. *Annals of Internal Medicine*, **96**, 413–7.

Bichet, D.G., Van Putten, V.J., and Schrier, R.W. (1982b). Potential role of increased sympathetic activity in impaired sodium and water excretion in cirrhotic patients. *New England Journal of Medicine*, **307**, 1552–7.

Biddle, T.L. and Yu, P.N. (1979). Effect of furosemide on hemodynamics and lung water in actue pulmonary edema secondary to myocardial infarction. *American Journal of Cardiology*, **43**, 86–90.

Bleich, M., Schlatter, E., and Greger, R. (1990). The luminal K^+ channel of the thick ascending limb of Henle's loop. *Pflügers Archiv*, **415**, 449–60.

Bloxham, C.A. and Beevers, D.G. (1979). The effects of thiazide diuretics on coronary risk factors. *Postgraduate Medical Journal*, **55**, 9–13.

Blythe, W.B. (1983). Captopril and renal autoregulations. *New England Journal of Medicine*, **308**, 390–1.

Boehringer, K., Weidmann, P., Mordasini, R., Schiffl, H., Bachmann, C., and Risen, U. (1982). Menopause-dependant plasma lipoprotein alterations in diuretic-treated women. *Annals of Internal Medicine*, **97**, 206–9.

Boron, W.F. (1983). Transport of H^+ and ionic weak acids and bases. *Journal of Membrane Biology*, **72**, 1–16.

Bosher, S.K., Smith, C., and Warren, R.L. (1973). The effects of ethacrynic acid upon the cochlear endolymph and stria vascularis. *Acta Oto-Laryngica*, **75**, 184–91.

Bourland, W.A., Day, D.K., and Williamson, H.E. (1977). The role of the kidney in the early nondiuretic action of furosemide to reduce elevated left atrial pressure in the hypervolemic dog. *Journal of Pharmacology and Experimental Therapeutics*, **201**, 221–9.

Braitsch, R., Lohrmann, E., and Greger, R. (1990). Effect of probenecid on loop diuretic induced saluresis and diuresis. In *Diuretics III, Chemistry, Pharmacology, and Clinical applications*. (ed. J. Puschett), pp. 137–9. Elsevier, New York.

Brater, D.C. (1985). Resistance to loop diuretics: Why it happens and what to do about it. *Drugs*, **30**, 427–43.

Brater, D.C., Anderson, S.A., and Brown-Cartwright, D. (1986). Response to furosemide in chronic renal insufficiency. Rationale for limited doses. *Clinical Pharmacology and Therapeutics*, **40**, 134–9.

Brezis, M., Rosen, S., Silva, P., and Epstein, F.H. (1984a). Selective vulnerability of the thick ascending limb to anoxia in the isolated perfused kidney. *Journal of Clinical Investigation*, **73**, 182–9.

Brezis, M., Rosen, S., Silva, P., and Epstein, F.H. (1984b). Renal ischaemia: A new perspective. *Kidney International*, **26**, 375–83.

Brezis, M., et al. (1985). Disparate mechanisms for hypoxic cell injury in different nephron segments. Studies in the isolated perfused rat kidney. *Journal of Clinical Investigation*, **76**, 1796–1806.

Brezis, M., Rosen, S., and Silva, P. (1986). Mitochondrial activity: A possible determinant of anoxic injury in renal medulla. *Experientia*, **42**, 570–2.

Brown, R.D. (1981). Comparisons of the acute effects of i.v. furosemide and bumetanide on the cochlear action potential (N1) and on the AC cochlear poten 6 kHz in cats, dogs and the guninea pig. *Scandinavian Audiology*, **14 (Suppl.)**, 71–84.

Brummett, R.E., Bendrick, T., and Himes, D. (1981). Comparative ototoxicity of bumetanide and furosemide when used in combination with kanamycin. *Journal of Clinical Pharmacology*, **21**, 628–36.

Brunner, F.P., deRoughemont, D., Robbiani, M., Seiler, H., and Thiel, G. (1985). Renal mercury content in $HgCl_2$-induced acute renal failure in furosemide/saline-protected and nonprotected rats. *Nephron*, **41**, 94–9.

Brunner, H.R., Waeber, B., and Nussberger, J. (1988). Renin secretion responsiveness: understanding the efficacy of renin-angiotensin inhibition. *Kidney International*, **34 (Suppl. 26)**, S80–S85.

Bühler, F.R., Bolli, P., Kiowski, W., and Müller, F.B. (1985). Relevance of the renin-angiotensin-aldosterone system for effectiveness and adversity of antihypertensive treatment. *Progress in Pharmacology*, **6**, 33–44.

Bulpitt, C.J., *et al.* (1979). Risk factors for death in treated hypertensive patients. *Lancet*, **ii**, 134–7.

Burckhardt, G. and Greger, R. (1991). Principles of electrolyte transport across plasma membranes of renal tubular cells. In *Handbook of Physiology Section on Renal Physiology*, (ed. E. E. Windhager), American Physiological Society, Rockville.

Burke, T.J., Arnold, P.E., and Schrier, R.W. (1983). Prevention of ischemic acute renal failure with impermeant solutes. *American Journal of Physiology*, **244**, F646–F649.

Burke, T.J., Cronin, R.E., Duchin, K.L. Peterson, R.L., and Schrier, R.W. (1980). Ischaemia and tubule obstruction during acute renal failure in dogs–mannitol protection. *American Journal of Physiology*, **238**, F305–F314.

Campbell, D.B., Lowe, S., Taylor, D., Turner, P., and Walsh, N. (1985). A fresh approach to the evaluation of antihypertensive agents. *Hypertension*, **7 (Supple. II)**, 143–51.

Cantarovich, F., Locatelli, A., and Fernandez, J.C. (1971). Furosemide in high doses in the treatment of acute renal failure. *Postgraduate Medical Journal*, **47 (Supple.)**, 13–16.

Cantarovich, F., Galli, C., and Benedetti, L. (1973). High dose furosemide in established acute renal failure. *British Medical Journal*, **4**, 449–50.

Caralis, P.C., Materson, B.J., and Perez-Stable, E. (1984). Potassium and diuretic-induced ventricular arrhythmias in ambulatory hypertensive patients. *Mineral Electrolyte Metabolism*, **10**, 148–54.

Chennavasin, P., Seiwell, R., and Brater, D.C. (1980). Pharmacokinetic-dynamic analysis of the indomethacin-furosemide interaction in man. *Journal of Pharmacology and Experimental Therapeutics*, **215**, 77–81.

Cherrick, G., Kerr, D.N., Read, A.E., and Sherlock, S. (1960). Colloid osmotic pressure and hydrostatic pressure relationships in the formation of ascites in hepatic cirrhosis. *Clinical Science*, **19**, 361–75.

Christensson, T., Hellstrom, K., and Wengle, B. (1977). Hypercalcemia and primary hyperparathyroidism. Prevalence in patients receiving thiazides as detected in a health screen. *Archives of Internal Medicine*, **137**, 1138–42.

Ciabattoni, G., Pugliese, F., Cinotti, G.A., and Patrono, C. (1980). Renal effects of anti-inflammatory drugs. *European Journal of Rheumatology and Inflammation*, **3**, 210–21.

Coe, F.L. (1981). Prevention of kidney stones. *American Journal of Medicine*, **71**, 514–16.

Coe, F.L., Parks, J.H., Bushinsk, D.A., Langman, C.B., and Favus, M.J. (1988). Chlorthalidone promotes mineral retention in patients with idiopathic hypercalciuria. *Kidney International*, **33**, 1140–6.

Condon, J.P., Nassim, J.R., Dent, C.F., Hilb, A., and Stainthorpe, E.M. (1978). Possible prevention of treatment of steroid-induced osteoporosis. *Postgraduate Medical Journal*, **54**, 249–52.

Conger, J.D. (1977). Intrarenal dynamics in the pathogenesis and prevention of acute urate nephropathy. *Journal of Clinical Investigation*, **59**, 786.

Coodley, E.L., Nandi, P.S., and Chiotellis, P. (1979). Evaluation of a new diuretic, diapamide, in congestive heart failure. *Journal of Clinical Pharmacology*, **19**, 127–36.

Corcoran, A.C., Macleod, C., Dustan, H.P., and Page, I.H. (1959). Effects of chlorothiazide on specific renal functions in hypertension. *Circulation*, **19**, 355–9.

Costanzo, L.S. and Windhager, E.E. (1978). Calcium and sodium transport by the distal convoluted tubule of the rat. *American Journal of Physiology*, **235**, F492–F506.

Crawford, J.D., Kennedy, G.C., and Hill, L.E. (1960). Clinical results of treatment of diabetes insipidus with drugs of the chlorothiazide series. *New England Journal of Medicine*, **262**, 737–43.

Crosley, A.P., Jr., Cullen, R.C., White, D., Fresman, J.F., Castillo, C.A., and Rowe, G.G. (1960). Studies on the mechanism of action of chlorothiazide in cardiac and renal diseases. I. Acute effects on renal and systemic hemodynamics and metabolism. *Journal of Laboratory and Clinical Medicine*, **55**, 182–90.

Cutler, R.E., Forrey, A.W., Christopher, T.G., and Kimpel, B.M. (1974). Pharmacokinetics of furosemide in normal subjects and functionally anephric patients. *Clinical Pharmacology and Therapeutics*, **15**, 588–96.

Dal Canton, A., *et al.* (1985). Mechanism of increased plasma urea after diuretic therapy in uremic patients. *Clinical Science*, **68**, 255–61.

Daniel, E.E. (1962). On the mechanism of the antihypertensive action of hydrochlorothiazide in rats. *Circulation Research*, **11**, 941–54.

Daul, A., Graben, N., and Bock, K.D. (1987). Neuromyeloenzephalopathie nach hochdosierter Muzolimintherapie bei Dialysepatienten. *Münchener Medizinische Wochenschrift*, **129**, 542–3.

Dawson, J.L. (1965). Post-operative renal function in obstructive jaundice. Effect of a mannitol diuresis. *British Medical Journal*, **1**, 82–6.

Decaux, G., Waterlot, Y., Genette, F., and Mockel, J. (1981). Treatment of the syndrome of inappropriate secretion of antidiuretic hormone with furosemide. *New England Journal of Medicine*, **304**, 329–30.

DeFronzo, R.A. (1980). Hyperkalemia and hyporeninemic hypoaldosteronism. *Kidney International*, **17**, 118–34.

Dikshit, K., Vyden, J.K., Forrester, J.S., Chatterjee, K., Prakash, R., and Swan, H.J.C. (1973). Renal and extra-renal hemodynamic effects of furosemide in congestive heart failure after myocardial infarction. *New England Journal of Medicine*, **288**, 1087–90.

Dollery, C.T. (1973). Diabetogenic effect of diuretics: In *Modern Diuretic Therapy in the Treatment of Cardiovascular and Renal Disease*. (eds. A. F. Lant and G. M. Wilson) pp. 320–7, Excerpta Medica, Amsterdam.

Dorman, H., Sondheimer, J., and Cadnapaphornchai, P. (1988). Characteristics of mannitol-induced (MI) acute renal failure (Abstract). *Kidney International*, **33**, 188.

Dornhorst, A., Powell, S.H., and Penskv, J. (1985). Aggravation by propranolol of hyperglycemic effect of hydrochlorothiazide in type II diabetics without alteration of insulin secretion. *Lancet*, **i**, 687–90.

Duarte, C.G., Winnacker, J.L., Becker, K.L., and Pace, A. (1971). Thiazide-induced hypercalcemia. *New England Journal of Medicine*, **284**, 828–30.

Dupont, A. (1985). Disappearance of spironolactone-induced gynaecomastia during treatment with potassium canrenoate. *Lancet*, **ii**, 731.

Dykner, T. and Wester, P.O. (1979), Ventricular extrasystoles and intracellular electrolytes before and after potassium and magnesium infusion in hypokalemic patients on diuretic treatment. *American Heart Journal*, **97**, 12–18.

Dykner, T. and Wester, P.O. (1981). Effects of magnesium infusions in diuretic-induced hyponatremia. *Lancet*, **ii**, 585–6.

Dyckner, T. and Wester, P.O. (1984). Intracellular magnesium loss after diuretic administration. *Drugs*, **28 (Suppl. 1)**, 161–6.

Dzau, V.J. (1987). Renal effects of angiotensin-converting enzyme inhibition in cardiac failure. *American Journal of Kidney Diseases*, **10, (Suppl. 1)**, 74–80.

Dzurik, R. (1990). In, *International Symposium on Hypertension* (eds. W. H. Birkenhäger, F. Halberg, and P. Prikryl) *Proceedings BRNP* 98–106.

Dzurik, R., Krirosikova, Z., Spastova, V., and Dzurikova, V. (1990) Recent development in diuretics for the treatment of hypertension, *in press*

Eknoyan, G., Sawa, H., and Hyde, S. III. (1975). Effect of diuretics on oxidative phosphorylation of dog kidney mitochondria. *Journal of Pharmacology and Experimental Therapeutics*, **194**, 614–23.

Eliahou, H.E. (1964). Mannitol therapy in oliguria of acute onset. *British Medical Journal*, **1**, 807–9.

Ellison, D.H., Velazquez, H., and Wright, F.S. (1989). Adaptation of the distal convoluted tubule of the rat: Structural and functional effects of dietary salt intake and chronic diuretic infusion. *Journal of Clinical Investigation*, **83**, 113–26.

Emsley, R.A. and Gledhill, R.F. (1984). Thiazides, compulsive water drinking and hyponatraemic encephalopathy. *Journal of Neurology Neurosurgery and Psychiatry*, **47**, 886–7.

Eneas, J.F., Schoenfeld, P.Y., and Humphreys, M.H. (1979). The effect of infusion of mannitol-solium bicarbonate on the clinical course of myoglobinuria. *Archives of Internal Medicine*, **139**, 801–5.

Epstein, M. (1978). Renal effects of head-out water immersion in man: Implications for an understanding of volume homeostasis. *Physiological Review*, **58**, 529–81.

Epstein, F.H. (1982a). Underfilling versus overflow in hepatic ascites. *New England Journal of Medicine*, **307**, 1577–9.

Epstein, M. (1982). Peritonvenous shunt in the management of ascites and the hepatorenal syndrome. *Gastroenterology*, **82**, 790–9.

Epstein, M. (1984). Therapy in renal disorders in liver disease. In *Therapy of Renal Diseases and Related Disorders* (eds. W. N. Suki, and S. G. Massry) pp. 335–46. Martinus Nijhoff Publishers, Boston.

Epstein, M., Schneider, N.S., and Befeler, B. (1975). Effect of intrarenal furosemide on renal function and intrarenal hemodynamics in acute renal failure. *American Journal of Medicine*, **58**, 510–16.

Epstein, M., Lepp, B.A., Hoffman, D.S., and Levinson, R. (1977a). Potentiation of furosemide by metolazone in refractory edema. *Current Therapeutic Research*, **21**, 656–67.

Epstein, M., Levinson, R., Sancho, J., Haber, E., and Re, R. (1977b). Characterization of the renin-aldosterone system in decompensated cirrhosis. *Circulation Research*, **41**, 818–29.

Erne, P., Bolli, P., Bürgisser, E., and Bühler, F.R. (1984). Correlation of platelet calcium with blood pressure. Effect of antihypertensive therapy. *New England Journal of Medicine*, **310**, 1084–8.

Ettinger, B. and Weil, E. (1979). Triamterene-induced nephrolithiasis. *Annals of Internal Medicine*, **91**, 745–6.

Ettinger, B., Weil, E., Mandel, N.S., and Darling, S. (1979) Triamterene-induced nephrolithiasis. *Annals of Internal Medicine*, **91**, 745–6.

Evans, B.M., Hughes-Jones, N.C., Milne, M.D., and Steiner, S. (1954). Electrolyte excretion during experimental potassium depletion in man. *Clinical Science*, **13**, 305–16.

Fairley, K.F. and Laver, M. (1971). High dose furosemide in chronic renal failure. *Postgraduate Medical Journal*, **47** (**Suppl.**), 29–35.

Fanelli, G.M. Jr., Bohn, D.L., Camp, A.E., and Shum, W.K. (1980). Inability of indomethacin to modify hydrochlorothiazide diuresis and natriuresis in the chimpanzee kidney. *Journal of Pharmacology and Experimental Therapeutics*, **213**, 596–9.

Fanestil, D.D., Tran, J.M., Vaughn, D.A. Maciejewski, A.R., and Beaumont, K. (1990). Investigations of the metazolone receptor. In *Diuretics III, Chemistry, Pharmacology, and Clinical Applications*, (ed. J. Puschett) pp. 195–204, New York: Elsevier.

Favre, L., Glasson, P., and Vallotton, M.B. (1982). Reversible acute renal failure with combined triamterene and indomethacin. A study in healthy subjects. *Annals of Internal Medicine*, **96**, 317–20.

Favre, L., Glasson P.H., Riondel, A., and Vallotton, M.B. (1983). Interaction of diuretics and non-steroidal anti-inflammatory drugs in man. *Clinical Science*, **64**, 407–11.

Ferranini, E., *et al*. (1987). Insulin resistance in essential hypertension. *New England Journal of Medicine*, **317**, 350–7.

Ferrari, P. and Weidmann, P. (1990). Insulin sensitivity in humans: alterations during drug administration and in essential hypertension. *Mineral Electrolyte Metabolism*, **16**, 16–24

Fichman, M.P., Vorherr, H., Kleeman, C.R., and Telfer, N. (1971). Diuretic-induced hyponatremia. *Annals of Internal Medicine*, **75**, 853–63.

Flores, J., DiBona, D.R., Beck, C.H., and Leaf, A. (1972). The role of cell swelling in ischemic renal damage and the protective effect of hypertonic solute. *Journal of Clinical Investigation*, **51**, 118–26.

Francis, G.S., Siegel, R.M., Goldsmith, S.R., Olivari, M.T., Levine, B., and Cohn, J.N. (1985). Acute vasoconstrictor response to intravenous furosemide in patients with chronic congestive heart failure. *Annals of Internal Medicine*, **103**, 1–6.

Freeman, B.A. and Crapo, J.D. (1982). Free radicals and tissue injury. *Laboratory Investigation*, **47**, 412–26.

Friedman, E., Shadel, M., Halkin, H., and Farfel, Z. (1989). Thiazide-induced hyponatraemia. Reproducibility by single dose rechallenge and an analysis of pathogenesis. *Internal Medicine*, **110**, 24–30.

Fries, D., Pozet, N., Dubois, N., and Traeger, J. (1971). The use of large doses of furosemide in acute renal failure. *Postgraduate Medical Journal*, **47** (**Suppl. 18–20**).

Frindt, G. and Palmer, L.G. (1987). Ca-activated K channels in apical membrane of mammalian CCT, and their role in K secretion. *American Journal of Physiology*, **252**, F458–F467.

Frumann, B.L. (1981). Impairment of glucose tolerance produced by diuretics and other drugs. *Pharmacology and Therapeutics*, **12**, 613–49.

Fuisz, R.E., Lauler, D.P., and Cohen, P. (1962). Diuretic-induced hyponatremia and sustained antidiuresis. *American Journal of Medicine*, **33**, 783–91.

Fuisz, R.E. (1963). Hyponatremia. *Medicine*, **42**, 149–70.

Fuller. J.H., Shipley, M.J., Rose, G., Jarrett, R.J., and Keen, H. (1980). Coronary heart disease risk and impaired glucose tolerance. The Whitehall Study. *Lancet*, **i**, 1373–6.

Gabazda, G.J. and Hall, P.W. (1966). Relation of potassium depletion to renal ammonium metabolism and hepatic coma. *Medicine*, **45**, 481–90.

Gabow, P.A., Anderson, R.J., Potts, D.E., and Schrier, R.W. (1978). Acid-base disturbances in the salicylate-intoxicated adult. *Archives of Internal Medicine*, **138**, 1481–93.

Gabow, P.A., Kaehny, W.D., and Kelleher, S.P. (1982). The spectrum of rhabdomyolysis. *Medicine*, **61**, 141–52.

Gallagher, K.L. and Jones, J.K. (1979). Furosemide induced ototoxicity. *Annals of Internal Medicine*, **91**, 744–5.

Garty, H. and Benos, D.J. (1988). Characteristics and regulatory mechanisms of the amiloride-blockable Na^+ channel. *Physiological Reviews*, **68**, 309–73.

Geck, P. and Heinz, E. (1986). The Na-K-2Cl cotransport system. *Journal of Membrane Biology*, **91**, 97–105.

George, C.F. (1980). Amiloride handling in renal failure. *British Journal of Clinical Pharmacology*, **9**, 94–5.

Gerbes, A.L., Arendt, R.M., Ritter, D., Jungst, D., Zahringer, J., and Paumgartner, G. (1985). Plasma atrial natriuretic factor in patients with cirrhosis. *New England Journal of Medicine*, **313**, 1609–10.

Gianello, P., *et al.* (1989). Evidence that atrial natriuretic factor is the humoral factor by which volume loading or mannitol infusion produces an improved renal function after acute ischemia. *Transplantation*, **48**, 9–14.

Gifford, R.W. (1989). Review of the long-term controlled trials of usefulness of therapy for systemic hypertension. *American Journal of Cardiology*, **63B**, 8–16.

Giugliano, D., Torella, R., and Sgambato, S. (1979). Acetyl-salicylic acid restores acute insulin response reduced by furosemide in man. *Diabetes*, **28**, 841–5.

Goettl, K.H., Roesch, A., and Klinke, R. (1985). Quantitative evaluation of ototoxic side-effects of furosemide, piretanide, bumetanide, azosemide, and ozolinone in the cat. A new approach to the problem of ototoxicity. *Naunyn-Schmiedebergs Archiv of Pharmacology*, **331**, 275–82.

Gögelein, H. and Greger, R. (1986). Na^+ selective channels in the apical membrane of rabbit late proximal tubules (pars recta). *Pflüger's Archiv*, **406**, 198–203.

Goldner, M.G., Zarowitz, H., and Akgun, S. (1960). Hyperglycemia and glucosuria due to thiazide derivates administered in diabetes mellitus. *New England Journal of Medicine*, **262**, 403–5.

Gordon, E.E. and Sheldon, S.G. (1957). Effect of acetazolamide on citrate excretion and formation of renal calculi. Report of a case and study of five normal subjects. *New England Journal of Medicine*, **256**, 12–15.

Grausz, H., Lieberman, R., and Earley, L.E. (1972). Effect of plasma albumin on sodium reabsorption in patients with nephrotic syndrome. *Kidney International*, **1**, 47.

Graziani, G., Cairo, G., Tarantino, F., and Ponticelli, C. (1980). Dopamine and furosemide in acute renal failure. *Lancet*, **ii**, 1301–3.

Graziani, G., *et al.* (1984). Dopamine and furosemide in olilguric acute renal failure. *Nephron*, **37**, 39–42.

Green, T.P. and Mirkin, B.L. (1980). Resistance of proteinuric rats to furosemide: Urinary drug protein binding as a determinant of drug effect. *Life Science*, **26**, 623–30.

Greenberg, G., Brennan, P.J., and Miall, W.E. (1984). Effects of diuretic and beta-blocker therapy in the Medical Research Council Trial (MRC). *American Journal of Medicine*, **76**, 45–51.

Greene, M.K., Kerr, A.M., McIntosh, I.B., and Prescott, R.J. (1981). Acetazolamide in prevention of acute mountain sickness: a double-blind controlled cross-over study. *British Medical Journal*, **283**, 811–18.

Greger, R. (1985a). Ion transport mechanisms in thick ascending limb of Henle's loop of mammalian nephron. *Physiological Reviews*, **65**, 760–97.

Greger, R. (1985b). Wirkung von Schleifendiuretika auf zellulärer Ebene. *Nieren Hochdruckkrankheiten*, **14**, 217–20.

Greger, R. (1987). Pathophysiologie der renalen Ischämie. *Zeitschrift für Kardiologie*, **76**, 81–6.

Greger, R. (1988). Chloride transport in thick ascending limb, distal convolution, and collecting duct. *Annual Review of Physiology*, **50**, 111–22.

Greger, R. (1990). Possible sites of lithium transport in the nephron. *Kidney International*, **37**, S26–30.

Greger, R. and Schlatter, E. (1981). Presence of luminal K^+, a prerequisite for active NaCl transport in the thick ascending limb of Henle's loop of rabbit kidney. *Pflügers Archiv*, **392**, 92–4.

Greger, R. and Schlatter, E. (1983). Cellular mechanism of the action of loop diuretics on the thick ascending limb of Henle's loop. *Klinische Wochenschrift*, **61**, 1019–27.

Greger, R. and Wangemann, Ph. (1987). Loop diuretics. *Renal Physiology*, **10**, 174–83.

Greger, R., Bleich, M., and Schlatter, E. (1990). Ion channels in the thick ascending limb of Henle's loop. *Renal Physiology and Biochemistry*, **13**, 37–50.

Greven, J. and Klein, H. (1976). Renal effects of furosemide in glycerol-induced acute renal failure of the rat. *Pflügers Archiv*, **365**, 81–7.

Griffing, G.T., Sindler, B.H., Aurecchia, S.A., and Melby, J.C. (1983). Reversal of diuretic-induced secondary hyperaldosteronism and hypokalemia by enalapril (MK 421): A new angiotensin-converting enzyme inhibitor. *Metabolism*, **32**, 711–16.

Griggs, R.C., Engel, W.K., and Resnick, J.S. (1970). Acetazolamide treatment of hypokalemic periodic paralysis. Prevention of attacks and improvement of persistent weakness. *Annals of Internal Medicine*, **73**, 39–48.

Grimm, R.H., Cohen, J.D., McFate Smith, W., Falvo-Gerard, L., and Neaton, J.D. (1985). Hypertension management in the multiple risk factor intervention trial (MRFIT) (1985). *Archives of Internal Medicine*, **145**, 1191.

Gross, P., *et al.* (1988). Role of diuretics, hormonal derangements, and clinical setting of hyponatremia in medical patients. *Klinische Wochenschrift*, **66**, 662–9.

Grünfeld, C. and Chappell, D.A. (1983). Hypokalemia and diabetes mellitus. *American Journal of Medicine*, **75**, 553–4.

Gubern, J.M., Sancho, J.J., Simo, J., and Sitges-Serra, A. (1988). A randomized trial of the effect of mannitol on postoperative renal function in patients with obstructive jaundice. *Surgery*, **103**, 39–44.

Györy, A.Z. and Edwards, K.D.C. (1971). Effect of mersalyl, ethacrynic acid and sodium sulfate infusion on urinary acidification in hereditary renal tubular acidosis. *Medical Journal of Australia*, **2**, 940–5.

Haddy, F.J. (1975). Potassium and blood vessels. *Life Sciences*, **16**, 1489–98.

Hale, W.E., Stewart, R.B., and Marks, R.G. (1984). Thiazide and fractures of bone. *New England Journal of Medicine*, **310**, 926–7.

Hantman, D., Rossier, B., Zohlman, R., and Schrier, R. (1973). Rapid correction of hyponatremia in the syndrome of inappropriate secretion of antidiuretic hormone. *Annals of Internal Medicine*, **78**, 870–5.

HAPPHY (1987). Beta-blockers versus diuretics in hypertensive men: main results from the HAPPHY trial. *et al. Journal of Hypertension*, **5**, 561–72.

Hariprasad, M.K., Eisinger, R.P., Nadler, I.M., Padmanabhan, C.S., and Nidus, B.D. (1980). Hyponatremia in psychogenic polydipsia. *Archives of Internal Medicine*, **140**, 1639–42.

Hayes, D.M., Cvitkovic, E., Colbey, R.G. (1977). High dose cis-platinum diammine dichloride. Amelioration of renal toxicity by mannitol diuresis. *Cancer*, **39**, 1372–4.

Heidbreder, E., Hennemann, H., Krempien, B., and Heidland A. (1982). Salidiuretische Behandlung der renal-tubulären Acidose und ihrer Komplikationen. *Deutsche Medizinische Wochenschrift*, **97**, 1504–7.

Heidland, A. and Wigand, M.E. (1970) Einfluß hoher Furosemiddosen auf die Gehörfunktion bei Urämie. *Klinische Wochenschrift*, **48**, 1052–6.

Heidland, A., Klütsch, K., Schneider, K.W., and Suzuki, F. (1964). Thiaziddiuretika und Nierenfunktion bei Hypertonie und kardialer Dekompensation. *Klinische Wochenschrift*, **42**, 831–3.

Heidland, A., Klütsch, K., Moormann, A., and Hennemann, H. (1969). Möglichkeiten und Grenzen hochdosierter Diuretikatherapie bei hydropischer Niereninsuffizienz. *Deutsche Medizinische Wochenschrift*, **94**, 1568–74.

Heidland, A., Rockel, A., Maidhof, R., Klutsch, K., and Hennemann, H. (1970). Divergenz der renalen Natrium- und Magnesiumexkretion nach Trometanol (Trispuffer), Amilorid-HCl, Acetazolamid und Parathormon—ein Effekt der Harnalkalisierung? *Klinische Wochenschrift*, **48**, 371–4.

Heidland, A., Hennemann, H., and Wigand, M.E. (1972). Dosiswirkung und optimale Infusionsgeschwindigkeit hoher Furosemidgaben bei terminaler Niereninsuffizienz. In: *Medikamentöse Therapie bei Nierenerkrankungen*. (ed. R. Kluthe) pp. 221–228, Georg Thieme Verlag, Stuttgart.

Heidland, A., Hennemann, H.M., and Rockel, A. (1973). The role of magnesium and substances promoting the transport of electrolytes. *Acta Cardiologica*, **Suppl. 17**, 52–75.

Helderman, J.H., *et al.* (1983). Prevention of the glucose intolerance of thiazide diuretics by maintenance of body potassium. *Diabetes*, **32**, 106–11.

Helgeland, A., Hjermann, I., Leren, P., Enger, S., and Holms, I. (1978) High density lipoprotein cholesterol and antihypertensive drugs: the Oslo Study. *British Medical Journal*, **2**, 403–8.

Helgeland, A., Leren, P., Per Foss, O., Hjermann, J., Holme, J., and Lund-Larsen, P.G. (1984). The Oslo Study. Serum glucose levels during long-term observation of treated and untreated men with mild hypertension. *American Journal of Medicine*, **76**, 802–5.

Helgeland, A., Strommen, R., Hagelund, C.H., and Tretli, S. (1986). Enalapril, atenolol and hydrochlorothiazide in mild to moderate hypertension. *Lancet*, **1**, 872–5.

Heller, I., Halvey, J., Cohen, S., and Theodor, E. (1985). Significant metabolic acidosis induced by acetazolamide. *Archieves of Internal Medicine*, **145**, 1815–17.

Henderson, I.S., Beattie, T.J., and Kennedy, A.C. (1980). Dopamine hydrochloride in oliguric states. *Lancet*, **ii**, 827–8.

Henriksen, J.H., Bendtsen, F., Sorasen, T.A., Stadenger, C., and Rhing-Labon, H. (1989). Reduced central blood volume in cirrhosis. *Gastroenterology*, **97**, 1506–13.

Heyman, S.N., Brezis, M., and Reubinoff, C.A. (1988). Acute kidney failure with selective renal medullary injury. *Journal of Clinical Investigation*, **82**, 401–12.

Heyman, S.N., Brezis, M., Greenfeld, Z., and Rosen, S. (1989). Protective role of furosemide and saline in radiocontrast-induced acute renal failure in the rat. *American Journal of Kidney Disease*, **4–5**, 377–85.

Higashihara, E., Stokes, J.B., Kokko, J.P., Campbell, W.B., and Dubose, T.D. (1979). Cortical and papillary micropuncture examination of chloride transport in segments of the rat kidney during inhibition of prostaglandin production—possible role of postaglandins in the chloruresis of acute volume expansion. *Journal of Clinical Investigation*, **64**, 1277–87.

Hoffman, D.W., Jones-King, K.L., Whitworth, C.A., and Rybak, L.P. (1988) Potentiation of ototoxicity by gluthathione depletion. *Annals of Otology Rhinology Laryngology*, **57**, 36–41.

Hogg, K.J., and Hillis, W.S. (1986). Captopril/metolazone induced renal failure. *Lancet*, **i**, 501–2.

Holland, O.B. (1984). Diuretic induced hypokalemia and ventricular arrhythmias. *Drugs*, **28 (Suppl. 1)**, 86–92.

Hollifield, J.W. (1986). Thiazide treatment of hypertension: effects of thiazide diuretics on serum potassium, magnesium and ventricular ectopy. *American Journal of Medicine*, **80 (Suppl. 4A)**, 8–12.

Hollifield, J.W. and Slaton, P.E. (1981). Thiazide diuretics, hypokalemia and cardiac arrhythmias. *Acta Medica Scandinavica*, **647 (Suppl.)**, 67–73.

Hörl, W.H., Müller, V., and Heidbreder, E. (1989). Verlauf und Prognose bei postoperativem und posttraumatischem akuten Nierenversagen. *Medizinische Welt*, **40**, 616–19.

Horowitz, J., Keynan, A., and Ben-Ishay, D. (1972). A syndrome of inappropriate ADH secretion induced by cyclothiazide. *Journal of Clinical Pharmacology, New Drugs*, **12**, 337–41.

Hoskins, B. and Jackson, C.M. (1978). The mechanism of chlorothiazide-induced carbohydrate intolerance. *Journal of Pharmacology and Experimental Therapeutics*, **206**, 423–30.

Hropot, M., Fowler, N., Karlmark, B., and Giebisch, G. (1985). Tubular action of diuretics: Distal effects on electrolyte transport and acidification. *Kidney International*, **28**, 477–89.

Hulley, S.B., *et al*. (1985). Systolic hypertension in the elderly program (SHEP): antihypertensive efficacy of chlorthalidone. *American Journal of Cardiology*, **56**, 913–20.

Hulting, J. (1981). In-hospital ventricular fibrillation and its relation to serum potassium. *Acta Medica Scandinavica*, **647 (Suppl.)**, 109–16.

Hunter, M., Lopes, A.G., Boulpaep, E., and Giebisch, G.H. (1986). Regulation of single potassium ion channels from apical membrane of rabbit collecting tubule. *American Journal of Physiology*, **251**, F725–F733.

Hutchinson, W.D. (1985). Amenorrhoea after treatment with spironolactone and hydroflumethiazide. *British Medical Journal*, **291**, 1094.

Hyams, D.E. (1981). Amiloride: a review. In *Arrhythmias and Myocardial Infarction: the Role of Potassium*. (eds. Wood and Somerville), Royal Society of Medicine International Congress and Symposium Series, No. 44, pp. 65–73. Academic Press, London, Grune and Stratton, New York.

Hypertension Detection and Follow-up Program Cooperative Group (1979). Five year findings of the Hypertension Detection and Follow-up Program: II. Mortality by race, sex and age. *Journal of the American Medical Association*, **42**, 2572–7.

Hypertension Detection and Follow-up Program cooperative Group (1982). The effect of treatment on mortality in "mild" hypertension. *New England Journal of Medicine*, **307**, 976–80.

IPPPSH Collaborative Group (1985). Cardiovascular risk and risk factors in a randomized trial of treatment based on the beta-blocker oxprenolol: the international prospective primary prevention study in hypertension (IPPPSH). *Journal of Hypertension*, **3**, 379–92.

Jacobs, D.B., Mookerja, B.K., and Jang, C.Y. (1984). Furosemide inhibits glucose transport in isolated rat adipocytes via direct inactivation of carrier proteins. *Journal of Clinical Investigation*, **74**, 1679–85.

Jarrett, R.J., McCartney, P., and Keen, H. (1982). The Bedford survey: Ten year mortality roles in newly diagnosed diabetics and normoglycaemic controls and risk of indices for coronary heart disease in borderline diabetics. *Diabetologia*, **22**, 79–84.

Jick, H., Dinan, B.J., and Hunter, J.R. (1982). Triamterene and renal stones. *Journal of Urology*, **27**, 224–5.

JNC (1984). The 1984 Report of the Joint National Committee on the Detection, Evaluation, and Treatment of High Blood Pressure 1984. *Archives of Internal Medicine*, **44**, 1045–57.

JNC (1988). The 1988 Report of the Joint National Committee on the Detection, Evaluation and Treatment of High Blood Pressure. *Archives of Internal Medicine*, **148**, 1023–38.

Johnston, P.A., Bernard, D.B., Perrin, N.S., and Levinsky, N.G. (1981). Prostaglandins mediate the vasodilatory effect of mannitol in the hypoperfused rat kidney. *Journal of Clinical Investigation*, **68**, 127–33.

Jung, W. and Schön, F. (1979). Effects of modern loop diuretics on the inner ear. A quantitative evaluation using computer techniques. *Archives of Otorhinolaryngology*, **224**, 143–7.

Kaissling, B. and Stanton, B.A. (1987). Adaptation of distal tubuli and collecting duct to increased sodium delivery. I. Ultrastructure. *American Journal of Physiology*, **225**, F910–F915.

Kamm, E.E., Genin, M., Kuschmy, P., and Hollander, J. (1983). Diuretic induced acotemia: unmasking the role of increased peripheral catabolism through eviseration hepatectomy. *Kidney International*, **24**, 58–60.

Kamm, D.E., Wu, L., and Kuchmy, B.L. (1987). Contribution of the urea appearance rate to diuretic-induced azotemia in the rat. *Kidney International*, **32**, 47–56.

Kampf, V.D. and Baethke, R. (1980). The diuretic acitivity of bumetanide in a controlled comparison with furosemide in patients with various degrees of impaired renal function. *Arzneimittelforschung*, **30**, 1015–8.

Kannel, W.B., Castelli, W.P., and Gordon, T. (1979). Cholesterol in the prediction of artherosclerotic diseases. *Annals of Internal Medicine*, **90**, 85–91.

Kaplan, N.M., Carnegie, A., Raskin, P., Heller, J.A., and Simmons, M. (1985). Potassium supplementation in hypertensive patients with diuretic-induced hypokalemia. *New England Journal of Medicine*, **312**, 746–9.

Karayannopoulos, S. (1974). High-dose furosemide in renal failure. *British Medical Journal*, **2**, 278.

Kashgarian, M., Siegel, N.J., Ries, A.L., Di Meola, H.J., and Hayslett, J.P. (1976). Hemodynamic aspects in development and recovery phases of experimental postischemic acute renal failure. *Kidney International*, **10**, 160–68.

Keeton, G.F. and Morrison, S. (1981). Effects of furosemide in chronic renal failure. *Nephron*, **28**, 169–73.

Keller, E., Hoppe-Seyler, G., and Schollmeyer, P. (1982). Disposition and diuretic effect of furosemide in the nephrotic syndrome. *Clinical Pharmacology and Therapeutics*, **32**, 442–9.

Kelsch, R.C., Light, G.S., and Oliver, W.J. (1972). The effect of albumin infusion upon plasma norepinephrine concentration in nephrotic children. *Journal of Laboratory and Clinical Medicine*, **79**, 516–25.

Kennedy, R.M. and Earley, L.E. (1970). Profound hyponatremia resulting from a thiazide-induced decrease in urinary diluting capacity in a patient with primary polydipsia. *New England Journal of Medicine*, **282**, 1185–6.

Khaw, K.T. and Barrett-Connor, E. (1987). Dietary potassium and stroke-associated mortality: A 12 year prospective population study. *New England Journal of Medicine*, **316**, 235–40.

Kinsella, J.L. and Aronson, P.S. (1980). Properties of the Na^+-H^+ exchanger in renal microvillus membrane vesicles. *American Journal of Physiology*, **238**, F461–F469.

Kirchner, K.A., Voelker, J.R., and Brater, D.C. (1989). Tubular albumin blunts the response to furosemide. A mechanism for diuretic resistance, (Abstract). *Kidney International* **35**, 432.

Kleinknecht, D., Ganeval, D., Gonzalez-Duque, L.A., and Fermanian, J. (1976). Furosemide in acute oliguric renal failure: A controlled trial. *Nephron*, **17**, 51–8.

Knauf, H., Mutschler, E., Vogler, K.D., and Wais, U. (1978). Pharmacological effects of phase I and phase II metabolites of triamterene. *Arzneimittel-Forschüng*, 28, 1417–20.

Koch R., Winter, R., Tillmann P., and Wiessmann, B. (1972). Forcierte Diurese bei Thalliumvergiftung. *Medizinische Welt*, 23, 649–52.

Kone, B., Gimenez, L., and Watson, A.J. (1986). Thiazide-induced hyponatremia. *Southern Medical Journal*, 79, 1456–7.

Krakauer, R. and Lauritzen, M. (1978). Diuretic therapy and hypokalemia in geriatric outpatients. *Danish Medical Bulletin*, 25, 126–32.

Kramer, H.G., *et al.* (1980a) Interactions of conventional and antikaliuretic diuretics with the renal prostaglandin system. *Clinical Science*, 59, 67–70.

Kramer, H.J., Schumann, J., Wasserman, C., and Dusing, R. (1980b). Protaglandin-independent protection by furosemide from oliguric ischemic renal failure in conscious rats. *Kidney International*, 17, 455–64.

Krishna, G.G., Müller, E., and Kapoor, S. (1989). Increased blood pressure during potassium depletion in normotensive men. *New England Journal of Medicine*, 320, 1177–82.

Kusumoto, M., *et al.* (1974). Effects of hydrochlorothiazide on water cation and norepinephrine contents of cardiovascular tissues in renovascular hypertensive drugs. *Proceedings of the Society of Experimental Biology and Medicine*, 147, 769–74.

Lachance, S.L. and Barry, J.M. (1985). Effect of furosemide on dialysis requirement following cardaveric kidney transplantation. *Journal of Urology*, 133, 950–3.

Lakhani, M. (1986). Complete heart block induced by hyperkalaemia associated with treatment with a combination of captopril and spironolactone. *British Medical Journal*, 293, 271.

Lameijer, W. and van Zwieten, P.A. (1978). Accelerated elimination of thallium in rat due to subchronic treatment with furosemide. *Archives of Toxicology*, 40, 7–16.

Landmann-Suter, R. and Struyvenberg, A. (1978). Initial potassium loss and hypokalemia during chlorthalidone administration in patients with essential hypertension: the influence of dietary sodium restriction. *European Journal of Clinical Investigation*, 8, 155–64.

Lang, F., Greger, R., and Deetjen, P. (1977). Effect of diuretics on uric acid metabolism and excretion. In *Diuretics in Research and Clinics*, (eds. W. Siegenthaler, R. Beckerhoff, and W. Vetter) pp. 213–24. Thieme, Stuttgart.

Lang, F., Quehenberger, P., Greger, R., and Oberleithner, H. (1978). Effect of benzolamide on luminal pH in proximal convoluted tubules of the rat kidney. *Pflügers Archiv*, 375, 39–43.

Lang, F., Greger, R., Oberleithner, H., Griss, E., Lang, K., and Pastner, D. (1980). Renal handling of urate in healthy man in hyperuricaemia and renal insufficiency: circadian fluctuation, effect of water diuresis and of uricosuric agents. *European Journal of Clinical Investigation*, 10, 285–92.

Lant, A. (1985a) Diuretics: clinical pharmacology and therapeutic use (Part I). *Drugs*, 29, 58–87.

Lant, A. (1985b) Diuretics: clincial pharmacology and therapeutic use (Part II). *Drugs*, 29, 162–8.

Lant, A.F. and Wilson, G.M. (1971). Long-term therapy of diabetes insipidus with oral benzothiadizine and phthalmidine diuretics. *Clinical Science*, 40, 497–511.

Larson, E.B., Roach, R.C., Schoene, R.B., and Hornbein, T.F. (1982). Acute mountain sickness and acetazolamide: clincial efficacy and effect on ventilation. *Journal of the American Medical Association*, 248, 328–32.

Lasser, N.L., *et al.* (1984). Effects of antihypertensive therapy on plasma lipids and lipoproteins in the multiple risk factor intervention trial. *American Journal of Medicine*, 76, 52–66.

Lemieux, G. (1986). Treatment of idiopathic hypercalciuria with indapamide. *Canadian Medical Association Journal*, 135, 119–21.

Levinsky, N.G. and Bernard, D.B. (1988). Mannitol and loop diuretics in acute renal failure. In *Acute Renal Failure*, (eds. B. Brenner, and J. M. Lazarus), pp. 841–856, Churchill Livingstone, Edinburgh.

Li, H.-Y. and Lindemann, B. (1983). Competitive blocking of epithelial sodium channels by organic cations: the relationship between macroscopic and microscopic inhibition constants. *Journal of Membrane Biology*, 76, 235–51.

Lieberman, F.L., Denison, E.K., and Reynolds, T.B. (1970). The relationship of plasma volume, portal hypertension, ascites and renal soldium retention in cirrhosis: the overflow theory of ascites formation. *Annals of the New York Academy of Sciences*, 170, 202–12.

Lief, P.D., Belizon, I., Matos, J., and Bank, N. (1983). Diuretic induced hypokalemia does not cause ventricular ectopy in uncomplicated essential hypertension (abstract). *Proceedings of the American Society of Nephrology*, 16, A66.

Lim, P. and Jacob, E. (1972). Magnesium deficiency in patients on long-term diuretic therapy for heart failure. *British Medical Journal*, 3, 620–2.

Lindner, A., Cutler, R.E., and Goodman, W.G. (1979). Synergism of dopamine plus furosemide in preventing actue renal failure in the dog. *Kidney International*, 16, 158–66.

Lo, T.C.N. and Cryer, R.J. (1986). Complete heart block induced by hyperkalaemia associated with treatment with a combination of captopril and spironolactone. *British Medical Journal*, 292, 1672.

Loon, N.R., Wilcox C.S., and Unwin, R.J. (1989) Mechanism of impaired natriuretic response to furosemide during prolonged therapy. *Kidney International*, 36, 682–9.

Lowenstein, J., Schacht, R.G., and Baldwin, D.S. (1981). Renal failure in minimal change nephrotic syndrome. *American Journal of Medicine*, 70, 227–33.

Luboshitzky, R., Tal-Or, Z., and Barzilai, D. (1978). Chlorthalidone-induced syndrome of inappropriate secretion of antidiuretic hormone. *Journal of Clinical Pharmacology*, 18, 336–9.

Ludens, J.H., Hook, J.B., Brody, M.J., and Williams, H.E. (1968). Enhancement of renal blood flow by furosemide. *Journal of Pharmacological and Experimental Therapeutics*, 163, 456–68.

Luke, C.R., Ziegler, M.G., Coleman, M.D., and Kopin, I.J. (1979). Hydrochlorothiazide induced sympathetic hyperactivity in hypertensive patients. *Clinical Pharmacology and Therapeutics*, 26, 428–32.

Luke, R.G., Linton, A.L., Briggs, J.D., and Kennedy, A.C. (1965). Mannitol therapy in acute renal failure. *Lancet*, i, 980–2.

Luke, R.G., Lyerly, R.H., Anderson, J., Galla, J.H., and Kotchen, T.A. (1982). Effect of potassium depletion on renin release. *Kidney International*, 21, 14–19.

Lumb, G., Newborne, P., Rust, J.H., and Wagner, B. (1978). Effects in animals of chronic administration of spironolactone—a review. *Journal of Environmental Pathology and Toxicology*, 1, 641–60.

Lynn, K.L., Bailey, R.R., Swainson, C.P., Saintsbury, R., and Low, W.J. (1985). Renal failure with potassium-sparing diuretics. *New Zealand Medical Journal*, 98, 629–31.

Mackay, G., Miur, A.L., and Watson, M.L. (1984). Contribution of prostaglandins to the systemic and renal vascular responses to furosemide in normal man. *British Journal Clinical Pharmacology*, 17, 513–9.

Madias, J.E., Madias, N.E., and Gavras, H.P. (1984). Non-arrhythmogenicity of diuretic-induced hypokalemia. *Archives of Internal Medicine*, 144, 2171–6.

Maharaj, B. (1988). Diuretics in cirrhotic ascites. *Progress in Pharmacology*, 63, 245–66.

Maher, J.F. and Schreiner, G.E. (1965). Studies on ethacrynic acid in patients with refractory edema. *Annals of Internal Medicine*, 62, 15–29.

Mahony, J.F. (1976). Treatment of acute renal failure. *Drugs*, 12, 381–7.

Marcus D.C., Marcus, N.Y., and Greger, R. (1987). Sidedness of action of loop diuretics and ouabain on nonsensory cells of utricle: a micro-Ussing chamber for inner ear tissues. *Hearing Research*, 30, 55–64.

Marone, C., Reubi, F.C., and Lahn, W. (1984). Comparison of the short-term effects of loop diuretics prietanide and furosemide in patients with renal insufficiency. *European Journal of Clinical Pharmacology*, 26, 413–18.

Mason, A.D., Bowler, E., and Brown, W. (1961). Experimental acute renal failure. *Clinical Research*, **9**, 205–7.

Mathur, P.N., Pugsley, S.C., Powles, P., McEwan, M.P., and Campbell, J.M. (1984). Effect of diuretics on cardiopulmonary performance in severe chronic airway obstruction. *Archives of Internal Medicine*, **144**, 2154–7.

McFarland, K.F. and Carr, A.A. (1977). Changes in fasting blood glucose sugar after hydrochlorothiazide and potassium supplementation. *Journal of Clinical Pharmacology*, **17**, 13–17.

MRC (1981). Medical Research Council Working Party on Mild to Moderate Hypertension 1981. Adverse reactions to bendrofluazide and propranolol for the treatment of mild hypertension. *Lancet*, **ii**, 539–43.

Meyer-Sabellek, W., Gotzen, R., Heitz, J., Arntz, H.R., and Schulte, K.L. (1985). Serum lipoprotein levels during long-term treatment of hypertension with indapamide. *Hypertension*, **7**, 170–4.

Middecke, M., Weisweiler, P., Schwandt, P., and Holzgreve, H. (1987). Serum Lipoproteins during antihypertensive therapy with beta-blockers and diuretics: a controlled long-term comparative trial. *Clinical Cardiology*, **10**, 94–8.

Miller, P.D. and Berns, A.S. (1977). Acute metabolic alkalosis perpetuating hypercarbia: a role for acetazolamide in chronic obstructive pulmonary disease. *Journal of the American Medical Association*, **238**, 2400–1.

Minuth, A.N., Terrell, J.B., and Suki, W.N. (1976). Acute renal failure: A study of the course and prognosis of 104 patients and of the role of furosemide. *American Journal of Medical Sciences*, **271**, 317–24.

Mohring, K. and Madsen, P.O. (1973). The effect of dextrans, furosemide, and heparin on renal function changes caused by hemolysis in the dog. *Investigative Urology*, **10**, 404–7.

Montoreano, R., Cunarro, J., Mouzet, M.T., and Ruiz-Guinazu, A. (1971). Prevention of the initial oliguria of acute renal failure by administration of furosemide. *Postgraduate Medical Journal*, **47 (Suppl.)**, 7–10.

Mor, R., Pitlik, S., and Rosenfeld, J.B. (1983). Indomethacin- and moduretic-induced hyperkalemia. *Israeli Journal of Medical Sciences*, **19**, 535–7.

Morgan, D.B. and Davidson, C. (1980). Hypokalemia and diuretics: an analysis of publications. *British Medical Journal*, **280**, 905–8.

Mortensen, J.T., Schultz, A., and Ostergaard, A.H. (1986). Thiazides in the prophylactic treatment of recurrent idiopathic kidney stones. *International Urology and Nephrology*, **18**, 265–9.

Moser, M. (1988). Some highlights of the 4th joint national committee report on the detection, evaluation, and treatment of high blood pressure. *Hypertension*, **11**, 560–2.

Moser, M. (1989). Relative efficacy of and some adverse reactions to different antihypertensive regiments. *American Journal of Cardiology*, **63**, 2B–7B.

MRFIT (1985). Multiple Risk Factor Intervention Trial Research Group. Baseline rest electrocardiographic abnormalities, antihypertensive treatment, and mortality in the Multiple Risk Factor Intervention trial. *American Journal of Cardiology*, **55**, 1–15.

Murphy, M.B., Lewis, P.J., Kohner, E., Schumer, B., and Dollery, C.T. (1982) glucose intolerance in hypertensive patients treated with diuretics; a fourteen-year-follow up. *Lancet*, **ii**, 1293–5.

Muschaweck, R. and Hajdu, P. (1964). Die salidiuretische Wirksamkeit der 4-Chlor-N-(2-furylmethyl-5-sulfamoyl-anthranilsaure. *Arzneimittelforschung*, **14**, 44–7.

Muth, R.G. (1967). Diuretic properties of furosemide in renal disease. *Annals of Internal Medicine*, **69**, 249–61.

Myers, M.G. (1987). Hydrochlorothiazide with or without amiloride for hypertension in the elderly. A dose titration study. *Archives of Internal Medicine*, **147**, 1026–30.

Naranjo, C.A., Pontigo, E., Valdenegro, C., Gonzalez, G., Ruiz, I., and Busto, U. (1979). Furosemide-induced adverse reactions in cirrhosis of the liver. *Clinical Pharmacology and Therapeutics*, **25**, 154–60.

Navarro, J.M., Bosch, R., Casado, S., Lopez Novoa, J.M., Hernando, L., and Caramelo, C. (1986). Mechanism of enhancement of furosemide (F)-induced water excretion by converting-enzyme inhibition (CEI) in congestive heart failure (Abstract). *Fundation Jimenez Diaz, Madrid/Spain*, **54 A**.

Nishijma, J., *et al.* (1984). Acute and chronic hemodynamic effects of the basic therapeutic regimen for congestive heart failure. Diuretics, low salt diet and bed rest. *Japanese Heart Journal*, **25**, 571–85.

Nordehaug, J.E. and von der Lippe, G. (1983). Hypokalemia and ventricular fibrillation in acute myocardial infarction. *British Heart Journal*, **50**, 525–9.

Norgaard, A., Kjeldsen K., and Clausen, T. (1981). Potassium depletion decreases the number of ^{3}H-ouabain binding sites and the active Na-K transport in skeletal muscle. *Nature*, **293**, 739–41.

Nuutinen, L.S., Kairaluoma, M., Tuononen, S., and Larmi, T.K.I. (1978). The effect of furosemide on renal function in open heart surgery. *Journal of Cardiovascular Surgery*, **19**, 471–9.

Old, C.W. and Lehrner, L.M. (1980). Prevention of radiocontrast induced acute renal failure with mannitol. *Lancet*, **i**, 885.

Old, C.W., Duarte, C.M., Lehrner, L.M., Henry, A.R., and Simmott, R.C. (1981). A prospective evaluation of mannitol in the prevention of radiocontrast acute renal failure. *Clinical Research*, **29**, A472.

Oppermann, J.A., Piper, Ch., and Gardiner, P. (1988). Spironolactone and potassium canrenoate despite chemical similarities, differing metabolism accounts for different toxicological findings in animals. In *Therapie mit Aldosteronantagonisten*. (ed. E. Mutschler) pp. 3–9, Urban und Schwarzenberg, München.

Overdiek, J.W.P.M. and Merkus, F.W.H.M. (1986). Spironolactone metabolism and gynaecomastia. *Lancet*, **i**, 1103.

Palmer L.G. and Frindt, G. (1987). Effects of cell Ca and pH on Na channels from rat cortical collecting tubule. *American Journal of Physiology*, **253**, F333–F339.

Papademetriou, V., Price, M.B., Notargiacomo, A., Fletcher, R.D., and Freis, E.D. (1984). Effect of thiazides on ventricular arrhythmias in patients with uncomplicated systemic hypertension. *Clinical Research*, **32**, 337A.

Parmley, W.W. (1985). Pathophysiology of congestive heart failure. *American Journal of Cardiology*, **56**, 7A–11A.

Parry, W.L., Schaefer, J.A., and Mueller, C.B. (1963). Experimental studies of actue renal failure. 1. The protective effect of mannitol. *Journal of Urology*, **89**, 1–6.

Parving, H-H., Andersson, A.R., Hommel, E., and Smidt, U. (1985). Effects of long-term antihypertensive treatment on kidney function in diabetic nephropathy. *Hypertension*, **7 (Suppl. 2)**, 114–7.

Patak, R.V., Mookerjee, B.K., Bentzel, C.J., Hysert, P.E., Babej, M., and Lee, J.B. (1975). Antagonism of the effects of furosemide by indomethacin in normal and hypertensive man. *Prostaglandins*, **10**, 649–58.

Patak, R.V., Fadem, S.Z., Lifschitz, M.D., and Stein, J.H. (1979). Study of factors which modify the development of norepinephrine-induced acute renal failure in the dog. *Kidney International*, **15**, 227–37.

Pedersen, R.S., Olesen, A.S., Freund, L.G., Solgaard, P., and Larsen, E. (1978). Thallium intoxication with long-term hemodialysis, forced diuresis and Prussian blue. *Acta Medica Scandinavica*, **204**, 429–33.

Pera, M.F., Zook, B.G., and Harder, H.C. (1979). Effects of mannitol or furosemide diuresis on the nephrotoxicity and physiological disposition of *cis*-dichlorodiammineplatinum-(II) in rats. *Cancer Research*, **39**, 1269–78.

Perez-Ayuso, R.M., *et al.* (1983). Randomized comparative study of efficacy of furosemide versus spironolactone in nonazotemic cirrhosis with ascites. *Gastroenterology*, **84**, 961–8.

Petersen, V., Hvidt, S., Thompson, K., and Schou, M. (1974). Effect of prolonged thiazide treatment on renal lithium clearance. *British Medical Journal*, **3**. 143–5.

Petrinelli, R., Magagna, A., and Arizilli, F. (1980). Influence of indomethacin on the natriuretic and renin-stimulating effect of bumeta-

nide in essential hypertension. *Clinical Pharmacology and Therapeutics*, **28**, 722–31.
Pillay, V.K.G., Schwartz, F.D., Aimi, K., and Kark, R.M. (1969). Transient and permanent deafness following treatment with ethacrynic acid in renal failure. *Lancet*, **i**, 77–9.
Pockros, P.J. and Reynolds, T.B. (1986). Rapid diuresis in patients with ascites from chronic liver disease: The importance of peripheral edema. *Gastroenterology*, **90**, 1827–33.
Podjarny, E., et al. (1988). Prostanoids in renal failure induced by converting enzyme inhibition in sodium-depleted rats. *American Journals of Physiology*, **254**, F358–F363.
Pollare, T., Lithell, M., and Berne, C. (1989). A comparison of the effects of hydrochlorothiazide and captopril on glucose and lipid metabolism in patients with hypertension. *New England Journal of Medicine*, **321**, 868–73.
Porter, R.H., Cox, B.G., Heaney, D., Hostetter, T.H., Stinebaugh, B.J., and Suki, W.N. (1978). Treatment of hypoparathyroid patients with chlorthalidone. *New England Journal of Medicine*, **298**, 577–81.
Preminger, G.M. and Pak, C.Y. (1987). Eventual attenuation of hypocalciuric response to hydrochlorothiazide in absorptive hypercalciuria. *Journal of Urology*, **137**, 1104–9.
Puschett, J.B. and Greenberg, A. (1984). Treatment of edematous states. In: Therapy of renal diseases and related disorders. (eds. W. N. Suki, and S. G Massry) pp. 3–45, Martinus Nijhoff Publishing, Boston.
Pyörälä, K. (1979) Relationship of glucose tolerance and plasma insulin to the incidence of coronary heart disease: results from two population studies in Finland. *Diabetes Care*, **2**, 131–41.
Quick, C.A. and Hoppe, W. (1975) Permanent deafness associated with furosemide administration. *Annals of Otology Rhionology & Laryngology*, **84**, 94–101.
Quintero, E., et al. (1985). Paracentesis versus diuretics in the treatment of cirrhotics with tense ascites. *Lancet*, **i**, 611–2.
Ram, C.V.S., Garrett, B.N., and Kaplan, N.M. (1975). Moderate sodium restriction and various diuretics in the treatment of hypertension. *Archives of Internal Medicine*, **141**, 1015–19.
Rampini, S., Fanconi, A., Illig, R., and Prader, A. (1968). Effect of hydrochlorothiazide on proximal renal tubular acidosis in a patient with idiopathic 'de Toni-Debre-Fanconi syndrome'. *Helvetica Paediatrica Acta*, **23**, 13–15.
Ramsay, L.E., Hettiarachshi, J., and Morton, J.J. (1980). Amiloride, spironolactone and potassium chloride in thiazide-treated hypertensive patients. *Clinical Pharmacology and Therapeutics*, **27**, 543–53.
Rane, A., Villeneuve, J.P., Stone, W.J., Nies, A.S., Wilkinson, G.R., and Brauch, R.A. (1978). Plasma binding and disposition of furosemide in the nephrotic syndrome and in uremia. *Clinical Pharmacology and Therapeutics*, **24**, 119–207.
Rapoport, M.J. and Heard, H.F. (1965). Thiazide induced glucose intolerance treated with potassium. *Archives of Internal Medicine*, **113**, 405–11.
Rashiq, S. and Logan, R.F.A. (1986). Role of drugs in fractures of the femoral neck. *British Medical Journal*, **292**, 861–3.
Rastogi, S., Crawford, C., Wheeler, R., Flanigan, W., and Arruda, J.A.L. (1984). Effect of furosemide on urinary acidification in patients with distal renal tubular acidosis. *Journal of Laboratory and Clinical Medicine*, **104**, 271–82.
Rastogi, S., Bayliss, J.M., Nascimento, L., and Arruda, J.A.L. (1985). Hyperkalemic renal tubular acidosis: Effect of furosemide in humans and in rats. *Kidney International*, **28**, 801–7.
Ray, J.F., Winemiller, R.H., and Parker, J.P. (1974). Postoperative renal failure in the 1970's. A continuing challenge. *Archives of Surgery*, **108**, 576–83.
Reed, B.R., Huff, J.C., and Jones, S.K. (1985). Subacute cutaneous lupus erythematosus associated with hydrochlorothiazide therapy. *Annals of Internal Medicine*, **103**, 49.

Reubi, F. and Cottier, P. (1961). Effects of reduced glomerular filtration rate on responsiveness to chlorothiazide and mercurial diuretics. *Circulation*, **23**, 200–10.
Reyes, A.J. (1988). Therapy with diuretics in congestive heart failure. *Progess in Pharmacology*, **6**, 167–82.
Ring-Larsen, H., Henriksen, J.H., Wilken, C., Clausen, J., Pals, H., and Christensen, N.J. (1986). Diuretic treatment in decompensated cirrhosis and congestive heart failure: effect of posture. *British Medical Journal*, **292**, 1351–3.
Robinson, H.N., Morison, W.L., and Hood, A.F. (1985) Thiazide diuretic therapy and chronic photosensitivity. *Archives of Dermatology*, **121**, 522–3.
Robuschi, M., Vaghi, A., Gambaro, G., Spagnotto, S., and Bianco, S. (1988). Inhaled furosemide (F) is highly effective in preventing ultrasonically nebulized water (UNH_2O) bronchoconstriction. *American Review of Respiratory Diseases*, **137**, 412 (abstract).
Rocha, A., Marcondes, M., and Malnic, G. (1973). Micropuncture study in rats with experimental glomerulonephritis. *Kidney International*, **3**, 14–18.
Rollema, H., Booth, R.G., and Castagno, N. (1988). In vivo dopaminergic neurotoxicity of the 2-beta-methyl-carbolinum ion. A potential endogenous MPP+ analog. *European Journal of Pharmacology*, **153**, 131–4.
Ron, D., Taitelman, U., Michaelson, M., Sar-Joseph, G., Bursztein, S., and Better, O.S. (1984). Prevention of acute renal failure in traumatic rhabdomyolysis. *Archives of Internal Medicine*, **144**, 277–80.
Rose, H.J., O'Malley, K., and Pruitt, A.W. (1976). Depression of renal clearance of furosemide in man by azotemia. *Clinical Pharmacology and Therapeutics*, **21**, 141–6.
Ryan, M.P., Devane, J., Ryan, M.F., and Counihan, T.B. (1984). Effects of diuretics on the renal handling of magnesium. *Drugs*, **28 (Suppl. 1)**, 167–81.
Said, R., Marin, P., Anicama, H., Quintanilla, A., and Lewin, M.L. (1980). Effect of mannitol on acute amphotericin B nephrotoxicity. *Research in Experimental Medicine (Berlin)*, **177**, 85.
Sandström, P.E. and Sehlin, J. (1988). Furosemide reduces insulin release by inhibition of Cl^- and Ca^{2+} fluxes in beta cells. *American Journal of Physiology*, **255**, E591–E596.
Santos-Atherton, D. and Frank, S. (1986). The functional response to furosemide in a case of de Toni–Debre–Fanconi disease. *Acta Endocrinologica Supplementum (Copenhagen)*, **279**, 452–7.
Schafer, J.A., Troutman, S.L., and Andreoli, T.E. (1974). Volume reabsorption, transepithelial potential differences, and ionic permeability properties in mammalian superficial proximal straight tubules. *Journal of General Physiology*, **64**, 582–607.
Schafer, J.A., Troutman, S.L., and Schlatter, E. (1990). Vasopressin and mineralocorticoid increase apical membrane driving force for K^+ secretion in rat CCD. *American Journal of Physiology*, **258**, F199–F210.
Schäfer, G.E. and Sievert, H. (1987). Treatment of chronic congestive heart failure: effect of hydrochlorothiazide and triamterene on cardiac performance. In *Diuretics: Basic Pharmacological and Clinical Aspects*. (eds. V. E. Andreucci, and A. Dal Canton), pp. 350–2 Martinus Nijhoff Publishing, Boston.
Schambelan, M., Sebastian, A., and Rector, F.C., Jr. (1981). Mineralocorticoid-resistant renal hyperkalemia without salt wasting (type II pseudohypoaldosteronism): role of increased chloride reabsorption. *Kidney International*, **19**, 716–27.
Scheer, R.L. (1965). The effect of hypertonic mannitol on oliguric patients. *American Journal of Medical Science*, **250**, 483–6.
Scherer, B. and Weber, P.C. (1979). Time dependent changes in prostaglandin excretion in response to furosemide in man. *Clinical Science*, **56**, 77–81.
Schlatter, E. (1989). Antidiuretic hormone regulation of electrolyte transport in the distal nephron. *Renal Physiology and Biochemistry*, **12**, 65–84.

Schlatter, E. and Schafer, J.A. (1988). Intracellular chloride activity in principal cells of rat collecting ducts (CCT) (Abstract). *Pflügers Archiv*, **410**, R86.

Schlatter, E., Greger, R., and Weidtke, C. (1983). Effect of 'high ceiling' diuretics on active salt transport in the cortical thick ascending limb of Henle's loop of rabbit kidney. Correlation of chemical structure and inhibitory potency. *Pflügers Archiv*, **396**, 210–17.

Schlatter, E., Salomonsson, M., Persson, A.E.G., and Greger, R. (1989). Macula densa cells sense luminal NaCl concentration via the furosemide sensitive Na-2Cl-K cotransporter. *Pflügers Archiv*, **414**, 286–90.

Schlatter, E., Bleich, M., and Greger, R. (1990). Properties of the luminal K^+-channel of isolated perfused cortical collecting ducts (CCT) of the rat (Abstract). Proceedings of the American Society of Nephrology. *Kidney International*, **37**.

Schneider, W.J. and Becker, E.L. (1966). Acute transient hearing loss after ethacrynic acid therapy. *Archives of Internal Medicine*, **117**, 715–17.

Schrier, R.W. and Berl, T. (1980). Disorders of water metabolism. In *Renal and Electrolyte Disorders*. (ed. R. W. Schrier) pp. 1–64, Little Brown, Boston.

Schrier, R.W., Arnold, P.E., Gordon, J.A., and Burke, T.J. (1984). Protection of mitochondrial function by mannitol in ischemic acute renal failure. *American Journal of Physiology*, **247**, F365.

Senft, G., Losert, W., and Schultz, G. (1966). Ursachen der Störungen im Kohlenhydratstoffwechsel unter dem Einfluβ sulfonamidierter Diuretika. *Naunyn-Schmiedebergs Archiv für Pharmakologie und Experimentelle Pathologie*, **255**, 369–82.

Shah, S., Khatri, J., and Freis, E.D. (1978). Mechanism of antihypertensive effect of thiazide diuretics. *American Heart Journal*, **95**, 611–18.

Shear, L., Kleinerman, J., and Gabuzda, G.J. (1965). Renal failure in paitents with cirrhosis of the liver. 1. Clinical and pathologic characteristics. *American Journal of Medicine*, **39**, 184–98.

Shekelle, R.B., Caggiula, A.W., and Grimm, R.H. (1984). Diuretic treatment of hypertension and changes in plasma lipids over 6 years in the Multiple Risk Intervention Trial. *Atherosclerosis*, **12**, 113–27.

Sherlock, S. (1981). Ascites. In *Diseases of the Liver and Biliary System*. 6th edn. (ed. S. Sherlock), pp. 116–33. Blackwell Scientific Publications, Oxford.

Sherlock, S., Senewiratne, B., Scott, T.A., and Walker, J.G. (1966). Complications of diuretic therapy in hepatic cirrhosis. *Lancet*, **i**, 1049–53.

Shichiri, M., Shiigai, T., and Takeuchi, J. (1984). Long-term furosemide treatment in idiopathic edema. *Archives of Internal Medicine*, **144**, 2161–4.

Shutske, G.M. and Allen, R.C. (1984). Diuretics in hypertension: whence and whither. In *Hypertension: Physiological Basis and Treatment*. (eds. Ong and Lewis) pp. 123–92. Academic Press, Inc., New York.

Singer, I. (1981). Lithium and the kidney. *Kidney International*, **19**, 374–87.

Skarfors, E.T., Lithell, H.O., Selinus, I., and Aberg, H. (1989). Do antihypertensive drugs precipitate diabetes in predisposed men? *British Medical Journal*, **298**, 1147–52.

Solomon, R.J. (1984). Ventricular arrhythmias in patients with myocardial infarction and ischemia. Relationships to serum potassium and magnesium. *Drugs*, **28 (Suppl. 1)**, 66–76.

Spustova, V., Gerykova, M., and Dzurik, R. (1988). Serum hippurate accumulation and urinary excretion in renal insufficiency. *Biochemia Clinica Bohemoslovaka*, **17**, 205–12.

Stassen, W.N. and McCullough, A.J. (1985). Management of ascites. *Seminars in Liver Disease*, **5**, 291–307.

Steele, T.H. (1971). Control of uric acid secretion. *New England Journal of Medicine*, **284**, 1193–6.

Stein, J.H., Mauk, R.C., and Bonjarern, S. (1972). Difference in the effect of furosemide and chlorothiazide on the distribution of renal cortical blood flow in the dog. *Journal of Laboratory and Clinical Medicine*, **79**, 995–1003.

Steinbrechen, U.P. and Witztum, J.C. (1984). Glycosylation of low-density lipoproteins to an extent comparable to that seen in diabetes slows their catabolism. *Diabetes*, **33**, 130–4.

Steinhausen, M. and Parekh, N. (1984). Principles of acute renal failure. In *Nephrology*. (eds. R. R. Robinson, V. W. Dennis, T. F. Ferris, R. J. Glassock, J. P. Kokko, and C. C. Tisher) pp. 702–10, Springer, New York.

Streeten, D.H.P., Dalakos, T.G., and Fellerman, H. (1971). Studies on hyperkalemic periodic paralysis. Evidence of changes in plasma Na and Cl and induction of paralysis by adrenal glucocorticoids. *Journal of Clinical Investigation*, **50**, 142–55.

Struthers, A.D., Whitesmith, R., and Reid, J.L. (1983). Prior thiazide diuretic treatment increases adrenaline-induced hypokalemia. *Lancet*, **i**, 1358–60.

Suki, W.N., Yuin, J.J., and VonMinden, M. (1970). Acute treatment of hypercalcemia with furosemide. *New England Journal of Medicine*, **283**, 836–40.

Sundberg, S., Salo, H., and Gordin, A. (1982) Effect of low dose diuretics on plasma and blood cell electrolytes, plasma uric acid and blood glucose. *Acta Medica Scandinavica*, **668**, 95–101.

Syka, J. and Melichar, I. (1981). Comparison of the effects of furosemide and ethacrynic acid upon the cochlear function in guinea pig. *Scandian Audiology*, **Suppl. 14**, 63–9.

Tannen, R. (1985) Diuretic-induced hypokalemia. *Kidney International*, **28**, 988–1000.

Tannen, R.L. (1983). Effects of potassium on blood pressure control. *Animals Internal Medicine*, **98**, 773–80.

Tarazi, R.C., Dustan, H.P., and Fröhlich, E.D. (1970). Long-term thiazide therapy in essential hypertension. *Circulation*, **41**, 709–17.

Tarka, J., Kurtzman, N.A., and Batlle, D.C. (1983). Clinical assessment of urinary acidification using a short test with oral furosemide (abstract). *Kidney International*, **23**, 737.

Tazel, J.H. (1981) Comparison of adverse reactions to bumetanide and furosemide. *Journal of Clinical Pharmacology*, **21**, 615–19.

Thiel, G., et al. (1976). Protection of rat kidney against HgCl2-induced acute renal failure by induction of high urine flow without renin suppression. *Kidney International*, **10**, S191–S200.

Tiggeler, R.G., Berden, J.H., Hoitsma, A.J., and Koene, R.A. (1985). Prevention of acute tubular necrosis in cadaveric kidney transplantation by the combined use of mannitol and moderate hydration. *Annals of Surgery*, **201**, 246–9.

Tilstone, W.J., Semple, P.F., Lawson, D.H., and Boyle, J.A. (1977). Effects of furosemide on glomerular filtration rate and clearance of practolol, digoxin, cephaloridine, and gentamicin. *Clinical Pharmacology and Therapeutics*, **22**, 389–96.

Tobian, L. (1974). Hypertension and the kidney. *Archives of Internal Medicine*, **133**, 959–67.

Tobian, L. (1988). The Volhard lecture: Potassium and sodium in hypertension. *Journal of Hypertension*, **6 (Suppl. 4)**, 12–24.

Tobian, L., Janecek, J., Foker, J., and Ferreira, D. (1962). Effect of chlorothiazide on renal juxtaglomerular cells and tissue electrolytes. *American Journal of Physiology*, **202**, 905–8.

Tucker, B.J., Steiner, R.W., Gushwa, L.C., and Blantz, R.C. (1978). Studies on the tubuloglomerular feedback system in the rat. The mechanism of reduction in filtration rate with benzolamide. *Journal of Clinical Investigation*, **62**, 993–1004.

Tweeddale, M.G., Ogilvie, R.I., and Ruedy, J. (1977). Antihypertensive and biochemical effects of chlorthalidone. *Clinical Pharmacology and Therapeutics*, **22**, 519–27.

Ullrich, K.J. and Greger, R. (1985). Approaches to the study of tubule transport functions. In *The Kidney, Physiology and Pathophysiology*. (eds. D. W. Seldin and G. Giebisch). pp.427–69, New York: Raven Press.

Ullrich, K.J., Rumrich, G., and Klöss, S. (1989). Contraluminal organic anion and cation transport in the proximal renal tubule: V. Interaction with sulfamoyl- and phenoxy diuretics, and with β-lactam antibiotics. *Kidney International*, **36**, 78–88.

Untura, A. (1979). Incidence and prophylaxis of acute postoperative renal failure in obstructive jaundice. *Reviews of Medical Chir. Soc. Med. Nat. IASI*, **83**, 247–50.

Van Brummelen, P., Woerlee, M., and Schalekamp, M.A.D.H. (1979). Long-term versus short-term effects of hydrochlorothiazide on renal haemodynamics in essential hypertension. *Clinical Science*, **56**, 463–9.

Vander, A.J. and Miller, R. (1964). Control of renin secretion in the anesthetized dog. *American Journal of Physiology*, **207**, 537–46.

Vanholder, R., Leusen, I., and Lameire, N. (1984). Comparison between mannitol and saline infusion in HgCl$_2$-induced acute renal failure. *Nephron*, **38**, 193–201.

Velázquez, H. and Wright, F.S. (1986). Effects of diuretic drugs on Na, Cl and K transport by rat renal distal tubule. *American Journal of Physiology*, **250**, F1013–23.

Velázquez, H., Wright, F.S., and Good, D.W. (1982). Luminal influences on potassium secretion: chloride replacement with sulfate. *American Journal of Physiology*, **242**, F44–F55.

Venkata, C., Ram, S., Garrett, N., and Kaplan, N.M. (1981). Moderate sodium restriction and various diuretics in the treatment of hypertension: Effects of potassium wastage and blood pressure control. *Archives of Internal Medicine*, **141**, 1015–19.

Verenstraeten, P.J.C., Dupis, F., and Toussaint, C. (1975). Effects of large doses of furosemide in endstage chronic renal failure. *Nephron*, **14**, 333–8.

Veterans Administration Cooperative Study Group on Antihypertensive Agents (1972). Effects of treatment on morbidity in hypertension: III. Influence of age, diastolic pressure and prior cardiovascular disease: Further analysis of side effects. *Circulation*, **45**, 991–1004.

Veterans Administration Cooperative Study Group on Antihypertensive Agents (1982a). Comparison of propranolol and hydrochlorothiazide for the initial treatment of hypertension II. Results of long-term therapy. *Journal of the American Medical Association*, **248**, 2004–11.

Veterans Administration Cooperative Study Group on Antihypertensive Agents (1982b) Comparison of propranolol and hydrochlorothiazide for the initial treatment of hypertension. I. Results of short-term titration with emphasis on racial differences in response. *Journal of the American Medical Association*, **248**, 1996–2003.

Voelker, J.R. *et al.* (1987). Comparison of loop diuretics in patients with chronic renal insufficiency. *Kidney International*, **32**, 572–8.

Walker, W.G., Hermann, J., Yin, D.P., Murphy, R.P., and Patz, A. (1987). Diuretics accelerate diabetic nephropathy in hypertensive insulin-dependent and non-insulin-dependent subjects. *Clinical Research*, **35**, 663A.

Walser, M. and Rahill, W.J. (1966). Renal tubular reabsorption of bromides compared with chlorides. *Clinical Science*, **30**, 191–208.

Wasnich, R.D., Benfante, R.J., Yanok, K., Heilburn, L., and Vogel, J.M. (1983). Thiazide effect on the mineral content of bone. *New England Journal of Medicine*, **309**, 344–7.

Watkins, J., Carl-Abbot, E., Hensby, C.N., Webster, J., and Dollery, C.T. (1980) Attenuation of hypotensive effect of propranolol and thiazide diuretics by indomethacin. *British Medical Journal*, **281**, 702–5.

Watson, M.L., Bell, G.M., Muir, A.L., Buist, T.A.S., Kellett, R.J., and Padfield, P.L. (1983). Captopril/diuretic combinations in severe renovascular disease. A cautionary note. *Lancet*, **ii**, 404–5.

Weidmann, P., Berettap, C., Link, L., Blanchet, M.G., and Boehring, K. (1983). Cardiovascular counterregulation during sympathetic inhibition in normal subjects and patients with mild hypertension. *Hypertension*, **5**, 873–80.

Weidmann, P., Meier, A., Mordasani, R., Riesen, W., Bachmann, C., and Peheim, E. (1981). Diuretic treatment and serum lipoproteins: effect of tienilic acid and indapamide. *Klinische Wochenschrift*, **59**, 343–6.

Weimar, W., *et al.* (1983). A controlled study on the effect of mannitol on immediate renal function after cadaver donor kidney transplantation. *Transplantation*, **35**, 99–101.

Weinberger, M.H. (1985) Antihypertensive therapy and lipids: evidence, mechanisms, and implications. *Archives of Internal Medicine*, **145**, 1102–5.

Welsh, M.J. (1983). Inhibition of chloride secretion by furosemide in canine tracheal epithelium. *Journal of Membrane Biology*, **72**, 219–26.

Welt, L.G., Hollander, W., Jr., and Blythe, W.B. (1960). The consequences of potassium depletion. *Journal of Chronic Diseases*, **11**, 213–54.

Whelton, P.K. (1984). Diuretics and arrhythmias in the Medical Research Council Trial. *Drugs*, **28 (Suppl. 1)**, 54–65.

Widdicombe, J.H., Nathanson, I.T., and Higland, E. (1983). Effects of 'loop' diuretics on ion transport by dog tracheal epithelium. *American Journal of Physiology*, **245**, C388–C396.

Wigand, M.E. and Heidland, A. (1971). Ototoxic side-effects of high doses of furosemide in patients with uraemia. *Postgraduate Medical Journal*, **(47 Suppl.)** 54–6.

Wilson, D.R., Thiel, G., Arce, M.L., and Oken, D.E. (1967). Glycerol-induced hemoglobinuric acute renal failure in the rat. III. Micropuncture study of the effects of mannitol and isotonic saline on individual nephron function. *Nephron*, **4**, 337–55.

Wistrand, P.J. (1984). The use of carbonic anhydrase inhibitors in ophtalminology and clinical medicine. *Proceedings of the New York Academy of Science*, **429**, 609–19.

Witte, C.L., Witte, M.H., and Dumont, A.E. (1980). Lymph imbalance in the genesis and perpetuation of the ascites syndrome in hepatic cirrhosis. *Gastroenterology*, **78**, 1059–68.

Wittner, M., Weidtke, C., Schlatter, E., DiStefano, A., and Greger, R. (1984). Substrate utilization in the isolated perfused cortical thick ascending limb of rabbit nephron. *Pflügers Archiv*, **402**, 52–62.

Wittner, M., Di Stefano, A., Wangemann, P., Delarge, J., Liegeois, J.F., and Greger, R. (1987). Analogues of torasemide—structure function relationships. Experiments in the thick ascending limb of the loop of Henle of rabbit nephron. *Pflügers Archiv*, **408**, 54–62.

Wong, D.G., Spence, J.D., Lamki, L., Freemen, D., and McDonald, J.W. (1986). Effect of non-steroidal anti-inflammatory drugs on control of hypertension by beta-blockers and diuretics. *Lancet*, **i**, 997–1001.

Yamada, S. and Reynolds, T.B. (1970). Amiloride (MK–870), a new antikaliuretic diuretic. *Gastroenterology*, **59**, 833–41.

Yeboah, E.D., Petrie, A., and Pead, J.L. (1972). Acute renal failure and open heart surgery. *British Medical Journal*, **1**, 415–18.

Yendt, E.R. and Cohanim, M. (1978). Prevention of calcium stones with thiazides. *Kidney International*, **13**, 397–409.

Yendt, E.R. and Cohanim, M. (1986). Absorptive hyperoxaluria: a new clinical entity-successful treatment with hydrochlorothiazide. *Clinical and Investigative Medicine*, **9**, 44–50.

Zager, R.A., Mahan, J., and Merola, A.J. (1985). Effects of mannitol on the postischemic kidney. *Laboratory Investigation*, **53**, 433–42.

SECTION 3
The patient with glomerular disease

3.1 The clinical approach to haematuria and proteinuria
3.2 Immune mechanisms of glomerular damage
3.3 Fluid retention in renal disease: the genesis of renal oedema
3.4 Clinical consequences of the nephrotic syndrome
3.5 Minimal changes and focal segmental glomerular sclerosis
3.6 IgA nephropathies
3.7 Membranous nephropathy
3.8 Mesangiocapillary glomerulonephritis
3.9 Acute endocapillary glomerulonephritis
3.10 Crescentic glomerulonephritis
3.11 Antiglomerular basement membrane disease
3.12 Infection-associated glomerulonephritis
3.13 Malignancy-associated glomerular disease
3.14 Glomerular disease in the tropics

1

3.1 The clinical approach to haematuria and proteinuria

N.P. MALLICK AND COLIN D. SHORT

Introduction

The presence of either protein or blood in the urine (see Chapter 1.2) is regarded as a significant marker of disease in the kidneys or urinary tract. Although this is usually the case, proteinuria sometimes reflects systemic overproduction of a particular protein, with its consequent loss by filtration through normal kidneys. Protein is also secreted by the distal nephron (Tamm–Horsfall glycoprotein), by the urothelial lining of the urinary tract (secretory IgA) and by the prostate (acid phosphatase). Thus, proteinuria may originate from postglomerular sources, just as haematuria may arise from anywhere within the urinary tract, due to inflammatory, infective, neoplastic, or vascular lesions.

This chapter concentrates on the glomerular causes of proteinuria and haematuria, and offers guidelines for their identification, investigation, and management in the context of our present knowledge.

Proteinuria

The source of proteinuria

The efficiency of the glomerulus as a filter means that most of the protein content of the plasma escapes filtration; up to 900 g of albumin might pass through the glomerular barrier at a glomerular filtration rate of 120 ml/min if this protein was filtered freely.

Smaller proteins such as insulin, β_2-microglobulin and free immunoglobulin light chains are filtered freely, and the proximal renal tubule is an important catabolic site for these smaller proteins, as it is for albumin; only 1 per cent of the filtered protein escapes into the urine (Sumpio and Hayslett 1985).

Some proteins are secreted by the renal tubular system. In addition, the tubular system may release lysosomal and other cellular enzymes into the tubular lumen, and their presence in the urine has been taken to indicate tubular damage (see Section 8).

Thus, normal urine contains small amounts of various proteins from different sources. The physicochemical characteristics of each of these proteins can be determined, and its presence accounted for, but this is a complex exercise rarely undertaken in clinical practice.

The detection and analysis of proteinuria (see also Chapter 1.2)

The easiest and most reliable screening method is the test strip which sensitively records the presence of most proteins, especially albumin. It may give a false-positive colour change in very alkaline urine and may not detect monoclonal free immunoglobulin light chains, the overflow of which into urine may be the only overt feature of a monoclonal gammopathy (see Chapter 1.2). These are highly unusual circumstances and the test strip is otherwise a sensitive monitor of proteinuria. Older screening methods such as boiling the urine or detecting urine turbidity by eye after adding salicylsulphonic or trichloracetic acid, are much less reliable than the test strip. Turbidometric methods which utilize nephelometry (the detection of light scattering in a cuvette) are often used in laboratories and are reasonably accurate except at low, or very high, levels of proteinuria, when the turbidity is either very faint or so heavy that analysis is difficult even in a diluted specimen. The Kjeldahl method for nitrogen remains the standard by which other tests are considered. Many laboratories use automated Biuret analysis to detect protein. This is specific for peptide bonds and accurate, if not highly sensitive, for protein.

An outline of the available screening methods is given in Table 1. The subject of testing for proteinuria is dealt with in detail in Pesce and First (1979).

In studying glomerular disease it is sometimes worthwhile quantitating the relative loss of proteins of different size (urine protein 'selectivity') (White 1981). Such qualitative analysis can be valuable in assessing whether or not there is overflow proteinuria, or in determining if proteinuria is likely to have arisen by glomerular leakage or by failure of tubular reabsorption.

Variations in urine protein loss in health

A 24-h urine collection normally contains only small amounts of protein. Berggård (1970) gives figures of 80 ± 24 mg (mean ± SD), with 128 mg/24 h as the upper limit of normal. Albumin and other plasma proteins account for about 90 mg of this, which represents an albumin excretion rate of less than 30 μg/min. The remainder is principally Tamm–Horsfall protein (Hoyer and Seiler 1979) and secretory IgA (Bienenstock and Tomasi 1968) with a small amount of small molecular weight proteins. Albumin appears to be excreted predominantly during periods of upright posture and in some people this is exaggerated, with up to 160 μg/min being excreted after a period of quiet walking and up to 2 mg/min, transiently, after severe exercise, such as a marathon run. Since the dipstick test for protein becomes positive at about 100 μg/ml, many healthy individuals thus show 'trace' positive readings occasionally, and frank positives may be seen after exercise.

Orthostatic proteinuria

Orthostatic (postural) proteinuria (proteinuria present above normal limits only when the patient is upright and active) is not always purely physiological. It is observed during the recovery

Table 1 Available tests for detection of urine protein

Method	Description	Detection limit (mg/l)	Comment
Kjeldahl	Measures nitrogen following digestion of protein, after removal of non-protein nitrogen sources	10–20	Technically demanding and time-consuming. Not suitable for routine use
Test strip	Reagent strip impregnated with indicator which changes colour in the presence of protein	100	More sensitive to albumin than globulins. False-positive results may occur in alkaline urine. Useful for screening purposes
Turbidimetric methods (a) Trichloracetic acid (b) Sulphosalicyclic acid (c) Benzethomicin	Alteration of colloid properties of proteins causing steadily increasing turbidity (precipitation) providing concentration not too high. Turbidity is measured by light absorbance	50–100	All methods show differing reactivity for albumin and globulin; (c) shows most consistent bias with different globulin fractions and is more sensitive and precise than (a) or (b). False-positive results have been reported
Dye-binding methods (a) Ponceau S (b) Coomassie brilliant blue	Indicator changes colour in the presence of protein	50–100	Variable response to different proteins and in some cases poor linearity
Biuret and Folin/Lowry methods	Essentially use copper protein binding. There is a colour change which can be measured quantitatively. Folin is more sensitive	50	Usually requires precipitation of protein from urine to remove interfering substances and improve sensitivity

from acute glomerular disease as initially heavy proteinuria declines, and may also occur in the presence of mild glomerular changes.

Many healthy individuals undergo urine tests for protein for life insurance or employment purposes, or as a result of health screening, and a proportion of adolescents and young adults, most frequently males, exhibit proteinuria in excess of the limits outlined above. This should be considered in the context of the clinical history, and the occupation of the subject and their coincident physical activity are also relevant. Urinalysis must be carried out after recumbency and then after standardized activity, for example, by collecting urine after overnight rest and a further specimen after 2 h of quiet walking and standing. The first morning urine may be free of protein. A qualitative analysis of the urine protein pattern is helpful.

Pathological proteinuria

By definition this is any protein loss which exceeds the normal limits, when factors such as posture and exercise have been taken into account, or the presence of any protein not found in normal urine, however small the amount. A quantitative test may not detect very small amounts of a given protein, or even larger amounts of unusual constituents, if the method does not identify such proteins (see Table 1).

'Overflow' proteinuria

This rather loose term encompasses conditions in which proteinuria occurs as a result of the increased plasma level of a protein whose physicochemical characteristics result in its filtration through a normal glomerular barrier at a rate which exceeds the tubular reabsorptive capacity for the protein.

The increased plasma level of a protein may be due to its overproduction or abnormal production (e.g. free light chains secreted by a B cell clone, or lysosome from leukaemic cells) or its release from a normal intracellular site after tissue damage (e.g. haemoglobin, myoglobin, or amylase). A summary of the findings in such cases is given in Table 2.

Proteinuria due to tubular dysfunction

Proteinuria of up to 1 or 1.5 g/24 h is readily detected, but it should not be assumed that this always indicates glomerular damage, since protein losses of this magnitude may be the result of impaired reabsorption of protein by the proximal renal tubule. The hallmark of 'tubular' proteinuria is the presence of a number of proteins of a lower molecular weight than albumin, which together equal or exceed it in quantity. This contrasts with glomerular proteinuria, in which albumin is the predominant species.

The consensus of clinical and experimental studies is that the proximal renal tubule is a major catabolic site for several plasma proteins which, by virtue of their small size and favourable isoelectric point, are filtered freely by the normal glomerulus. When there is proximal tubular dysfunction due to congenital or acquired disorders (see Chapter 5.1) the reabsorption of these proteins is impaired, and they appear in increased amounts in the urine. Albumin excretion is also increased in these circumstances, because the small amount which escapes the normal glomerular filter is reabsorbed in bulk by a low affinity, high capacity system, which results in escape of some albumin into the urine even at low tubular fluid concentrations (Park and Maack 1984). The reabsorption of mixtures of filtered proteins is non-selective, at least in the rat (Bernard et al. 1988).

Proteins such as β_2-microglobulin, which has a molecular weight of 11.9 kDa and a glomerular sieving coefficient approaching 1 compared to water, are efficiently reabsorbed by the proximal tubule, and can be used as a marker of tubular dysfunction (Wibell and Evrin 1979). Normally β_2-microglobulin is present in the urine at a concentration of <0.4 mg/l. It is unstable in acid urine and other small molecular weight proteins, such as retinol-binding protein or lysosyme (molecular weight 17/kDa) may also be used as markers. For both of these, as for

Table 2 Representative 'overflow' proteins in urine

Protein	Source	Size (kDa)	Isoelectric point/sieving coefficient	Fate of filtered protein	Detection in urine	Effects on kidney	Significance of raised urinary levels
Monoclonal light chain	B-cell monoclone	22 (monomer) 44 (dimer)	4 to 7 or more Varies, may approach 1	Principal catabolic site is proximal tubule	Test strip is unreliable. Electrophoresis and immunofixation of concentrated urine are helpful	Determined by physicochemical characteristics of the light chain. May affect tubular system at different sites or cause casts. Many have no apparent nephrotoxic effects	Indicates the presence of an unstable B-cell monoclone and the potential for renal damage
Lysozyme	Ubiquitous lysosomal enzyme. High plasma levels found in some leukaemias	17	11 Approaches 1	75% of normal daily output is reabsorbed from glomerular filtrate in the proximal tubule	Chemical or immunochemical methods	None reported	Indicates increased release from tissue especially white blood cells
Haemoglobin	Erythrocytes	68	6.99 0.33	Limited tubular reabsorption	Red to dark brown urine (the latter due to formation of haemoglobin in acid urine) with characteristic absorption spectrum	Accumulates in tubular cells. Forms casts. Acute renal failure may occur concurrently. Direct link not established	Indicates intravascular haemolysis and postglomerular red cell lysis
Myoglobin	Muscle	17	6.8 0.7	Limited tubular reabsorption	Light to dark-brown urine (the latter due to formation of metmyoglobin in acid urine) with characteristic absorption spectrum	Similar to those associated with haemoglobinuria	Indicates muscle damage
Amylase	Pancreas and salivary gland	50	7 (pancreatic) 6 (salivary) 0.03 (pancreatic) 0.01 (salivary)	Reabsorbed	Enzymatic or immunological methods	None reported	Indicates damage to relevant tissue

Table 3 Examples of conditions in which there is evidence of proximal tubular dysfunction (aminoaciduria, phosphaturia, bicarbonate wasting, metabolic bone disorders) and 'tubular' proteinuria (see also Chapter 5.1)

Congenital disorders	Cystinosis; hepatolenticular degeneration (Wilson); oculo-cerebral-renal syndrome (Lowe)
Acquired disorders	Heavy metal poisoning (e.g. cadmium, lead) Free monoclonal light chains Autoimmune disease (Sjögren's syndrome)

other low molecular weight proteins such as insulin, the catabolic function of the proximal renal tubule has been established (Sumpio and Hayslett 1985). Plasma levels of all these small molecular weight proteins rise if glomerular filtration rate is impaired. If there is selective proximal renal tubular impairment, it is the urinary, not the plasma levels, which rise.

Low molecular weight proteinuria is best detected by electrophoresis of concentrated urine, usually on cellulose acetate, and individual proteins can be identified and quantitated by specific immunoassays.

A representative list of conditions in which isolated proximal renal tubular proteinuria arises is given in Table 3 (see also Chapter 5.1).

Pathological proteinuria due to a damaged glomerular filter

Glomerular damage is the most common cause of pathological proteinuria. The mechanisms by which the glomerular filter prevents protein from entering the urinary space has been studied extensively and only a summary is given here.

The filter operates as both a size-selective and a charge-selective barrier to the passage of larger molecules (Brenner et al. 1978). Physical impedance to the passage of larger molecules through the filter arises through the ordered arrangement of type IV collagen and its glycoprotein matrix in the outer and inner laminae rarae and by the more tightly structured, principally type IV, collagen of the central lamina densa of the glomerular basement membrane. It is less certain to what extent the slit pore structure of the interdigitating, lateral extensions of the epithelial cell podocytes on the outer, or urinary space, aspect of the glomerular basement membrane contribute to the physical barrier (see Chapter 1.7.1). The net effect is a rate dependent exclusion of proteins of mean cross-section from 1.5 nm to 4.4 nm: for a given size, globular proteins appear to be excluded more readily than those of elongated or tubular shape. These findings have been interpreted using the Pappenheimer model, which suggests the presence of water-filled 'pores' of 4.7 nm diameter occupying about 10 per cent of the filtering surface (Deen *et al.* 1985), but it must be emphasized that the physical existence of such 'pores' is doubtful (Simpson 1986).

To this structural barrier is added that due to the net negative charge on the sialoglycoproteins, particularly heparan sulphate and carboxyl residues, which constitute the matrix of the glomerular basement membrane (Bertolatus and Hunsicker 1987). Those plasma proteins with isoelectric point (pI) equal to or higher than that of the glomerular basement membrane pass through the barrier more readily than those of lower pI since at the pH of the plasma they carry a net positive charge.

The physical and electrical constituents of the glomerular barrier may have separate effects, but are interdependent. If the structure of the glomerular basement membrane is altered, as in diabetes mellitus for example, the ordered arrangement of the negatively charged sialoglycoproteins will be disrupted. Since the physical structure adopted by collagen in the membrane is itself dependent on the ordered, negatively charged sialoglycoproteins of its matrix, it follows that, if these are destroyed, the physical barrier itself is affected.

How is the glomerular barrier damaged in different diseases?
We have no satisfactory answers to this central question. Damage may occur with or without histologically detectable changes in glomerular structure, and in a given case it may heal (that is proteinuria ceases) without any immediate change in the histological appearance. One can infer that it is predominantly the physical (size determined), or the electrical (charge determined), element of the barrier which is affected but we do not know how such damage has been brought about.

It is tempting to suggest that in some proteinuric nephropathies proteinuria arises from loss or absence of anionic glycoproteins (Cotran and Rennke 1983). The most likely candidates are, firstly, the Finnish form of congenital nephrotic syndrome (see Section 20), in which it seems established that there is a reduction in glomerular capillary polyanion (Vernier *et al.* 1983), and secondly, the nephrotic syndrome with few or no histological changes and an absence of immune deposits and immune reactants in the glomeruli. In this 'minimal change' disease in humans (Carrie *et al.* 1981) (see also Chapter 3.5) and in experimental puromycin aminonucleoside nephropathy in rats, it has been alleged that there is a decreased amount of polyanionic glycoprotein stainable with cationic probes such as colloidal iron or polyethyleneimine, and also that injection of plasma from patients with minimal change nephrotic syndrome would induce 'loss' of polyanion (Wilkinson *et al.* 1989).

These results must be viewed with caution however, since apparent decreases at an optical microscopic level can result from the simplification ('fusion') of the podocyte foot processes, with consequent reduction in the area of charged surface of the cell (Alcorn and Ryan 1981). Also, even though dextran sieving data can be interpreted to indicate alterations in charge structure (Winetz *et al.* 1981), excretion of variously charged albumins remains selective in minimal change disease, suggesting normal electrostatic function of the barrier (Ghiggeri *et al.* 1987). Nevertheless, the idea of a decrease in anionic charge of the glomerulus in response to a substance, possibly released from activated lymphocytes and thus curable by steroids, cytotoxic agents, and cyclosporin is an attractive one. In some experimental nephropathies, including some models of membranous nephropathy, proteinuria cannot be induced in animals congenitally deficient in C5 or C6, and an absolute requirement for the C5b-9 membrane attack complex alone in inducing proteinuria is established. Polymorphonuclear leucocytes or monocytes are required in other models of mainly proliferative nephritis (see Chapter 3.2).

The metabolic consequences of glomerular proteinuria

These are considered in detail in Chapters 3.2 and 3.4, and are summarized here only in relation to their influence on clinical or laboratory assessment of the patient. If urine albumin loss exceeds 10 g/24 h for over 1 week, stores of preformed albumin in the plasma and interstitial space are depleted, and hepatic albumin synthesis increases several fold. Hepatic lipoprotein and fibrinogen synthesis seems to be linked inextricably to the raised albumin production rate, and so there is hyperfibrinogenaemia and hyperlipoproteinaemia, which itself may be clinically significant (see Chapter 3.4). Peripheral protein anabolism is compromised by the utilization of available nitrogen for increased hepatic protein production, and there is tissue wasting, especially in muscle and skin.

Concurrently, there is urinary loss of other proteins whose size or charge permit their escape through the now less efficient glomerular barrier. These include vitamin D-binding and thyroid-binding proteins which are important in homeostasis, immunoglobulins and antithrombin III. Many of the clinical features of the nephrotic syndrome are attributable to these losses (see Chapter 3.4).

Haematuria

It takes only 0.5 ml of blood visibly to redden 100 ml of urine, or about 10 ml of blood to ensure that there is macroscopic haematuria throughout the day: gross bleeding from the kidneys, or, directly into the urinary tract, is therefore detected rapidly. Macroscopic haematuria usually arises from a site of tumour or trauma, only rarely from the glomerulus. Other degrees of blood loss occur from all sites, including the nephron, but is easy to overlook the significance of the chance finding on testing of blood in otherwise clear urine of a symptomless person.

Detection of haematuria

Normally fewer than 30×10^6 red blood cells are excreted each day, at least some of which originate from the lower urinary tract. Red cell excretion can be significantly increased and remain invisible (microscopic haematuria), detected only by biochemical analysis or by one of the microscopic techniques. The methods used are listed in Table 4. The dipstick test is a useful screening tool, but the results should be interpreted with caution (see Chapter 1.2, in which differentiation between haematuria and haemoglobinuria is discussed). While false-positive results are unusual, it is a very sensitive test and may detect the haemoglobin from red cell excretion within the limit of the normal range or transiently increased excretion occurring after vigorous exertion ('jogger's nephritis') (Kincaid-Smith 1982). It will also be positive if there is any contamination of urine by menstrual loss or from urethral or perineal lesions. It is always wise to examine directly a fresh, and preferably a timed, urine specimen for red blood cells.

Reliable results are obtained by calculating the red cell excretion rate in a fresh 2-h urine collection. Equivocal red cell excretion rates are unusual; whereas normal urine will contain less

3.1 The clinical approach to haematuria and proteinuria

Table 4 The detection of haematuria

Method	Sensitivity	Specificity	Normal values	Comment
Test strip (impregnated with o-tolidine)	1000 cells/ml unspun urine	Positive reaction with free haemoglobin, myoglobin and oxidating agents	< 1000 cells/ml unspun urine	Convenient screening test
Direct microscopy of fresh urine	1 cell/high-power field of spun urine	Good in experienced hands, but dysmorphic cells may be difficult to identify with confidence	< 3 cells/high-power field of spun urine	Not useful if urine hypo-osmolar (red cell lysis occurs). Read promptly
Direct microscopy of fresh (2-h) timed urine in ambulant subject			< 20 000 cells/min	Read promptly. Glomerular haematuria is quantitatively usually well above normal range
Phase contrast microscopy		Experienced observer can identify dysmorphic cells		Red cells lysed if urine hypo-osmolar. Dysmorphic cells may occur normally and increase after vigorous exercise
Automated red cell analysis		Channel width adjusted for red cell size		Detects a population of dysmorphic cells, if present. Less helpful if dysmorphic cell numbers are few or if total red cell loss mild

than 20 000 red blood cells/min, this is greatly exceeded in nearly all cases of glomerular haematuria.

It has been claimed that red blood cells are distorted by passage through the glomerulus and that the detection of such distorted cells in a freshly collected urine specimen is pathognomonic of a glomerular lesion. Two techniques have been used to assess distortion of red cells (see Chapter 1.2); the first employs phase contrast microscopy to assess morphology directly (Birch and Fairley 1979). In the second, suitably prepared urine is passed through a red cell analyser which produces both a frequency histogram for erythrocyte volume, and also the modal urinary red cell volume (Shichiri et al. 1988). The sensitivity and specificity of each method is still debated. Clear-cut distinctions can be made in cases of macroscopic haematuria attributable to a glomerular or non-glomerular source, but it is more difficult confidently to localize the source of microscopic haematuria by either method. Phase contrast microscopy is time and labour intensive and requires a skilled observer. Automated red cell analysis does not have such drawbacks, but it depends entirely on the detection of a population of red cells sufficiently distorted to register a distinctively abnormal pattern in the red blood cell profile. Subtle, and perhaps diagnostic, morphological changes to a minority of red blood cells cannot be defined in this way (see Chapter 1.2 for details). Finally, the basis for the distortion of the cells is under debate: it seems likely that this results from osmotic stresses in traversing the hypertonic medulla in tubular fluid (Schramek et al. 1989).

Even with such sophisticated approaches, it remains mandatory to examine a spun urine deposit for tubular casts: their presence points firmly to the renal parenchyma, rather than the lower urinary tract, as the source of the haematuria.

The detection and quantitation of cast excretion can be carried out conveniently while determining the red cell excretion rate on a fresh, timed specimen of urine. These investigations are far better performed on a fresh specimen in the ward side room or outpatients' department than on a stale specimen examined in a distant laboratory. Casts may have lysed completely in many dilute or alkaline urines during transit and storage (see Chapter 1.2).

In glomerular disease with established proteinuria, the presence of macroscopic or microscopic haematuria has diagnostic but little prognostic significance, since it may result from either serious glomerular inflammation or from resolving acute nephritis of one form or another. However, the development of proteinuria in a patient whose glomerular lesion was first signalled only by isolated haematuria may be important, and often leads to the performance of a renal biopsy. This serves to underscore the need to establish the prognosis of a glomerular lesion by sequential study, rather than from single 'snapshot' findings, however detailed these may be.

Origin of haematuria

Usually it has been assumed that in glomerular diseases, red cells make their passage into the urine through the capillary walls of the inflamed glomeruli. It is easy to imagine this as the mechanism in severe glomerulonephritis, in which actual rupture of capillary walls leads to fibrin exudation and crescent formation. However, there are remarkably few observations to support this thesis in less active glomerular diseases (Makino et al. 1986), and since many diseases associated with glomerular inflammation also show interstitial inflammation, the peritubular capillaries deserve consideration as sources of haematuria. It is also difficult to see how (for example) purely mesangial deposits of IgA can sometimes lead to the prominent and often macroscopic haematuria common in this condition (see Chapter 3.3), even though red cells and red cell casts can be seen in the lumina of proximal tubules. Certainly we know even less about the entry of red cells into the urine than we do about the entry of protein.

An algorithm for screening patients with haematuria is presented in Fig. 1. It need hardly be mentioned that glomerular causes form only a small part of the spectrum of haematuria,

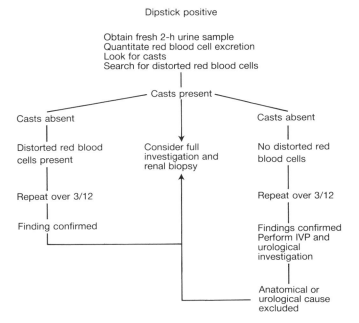

Fig. 1 An algorithm for screening patients with haematuria.

which may arise from anticoagulant treatment or sickle-cell disease, through polycystic kidneys to tumours, stones, and even urethral caruncles. In family practice the most common cause of haematuria, both micro- and macroscopic, is almost certainly urinary tract infections. Finally the significance of persistent microscopic haematuria differs in patients at various ages, from infancy to the elderly (see Chapter 1.1).

The symptom patterns of glomerular disease

Postural proteinuria

As indicated above the exact mechanism of postural proteinuria eludes us. What is clear, however, is that the very long-term (20–40 year) prognosis of these 'patients', who are of course in excellent health by definition, is also excellent (Levitt 1967; Rytand and Spreiter 1981; Springberg 1982). Renal biopsy is not necessary in such patients, but the proteinuria may persist for decades, or even become persistent; it is always, however, without haematuria.

Symptomless isolated proteinuria

This is by definition, a chance finding, usually as part of a routine medical assessment for employment, insurance, or at an antenatal visit (see Section 21). The urine is positive for protein on repeated dipstick testing and is negative for blood. A 'trace' reading is of little use clinically, since it detects about 10 mg/100 ml protein, and so may be positive in normal persons on occasion.

It is essential to exclude the influence of posture or of vigorous exercise by direct questioning and appropriate retesting. If orthostatic proteinuria is present (see above) it should be quantitated and reviewed in the light of the clinical and laboratory findings (see below).

Fixed (as opposed to orthostatic) proteinuria is not eliminated by recumbency, is present on repeated testing, does not exceed 1.5 g/day or thereabouts in the presence of a normal glomerular filtration rate, and the kidneys are normal radiographically.

Even if such proteinuria continues in isolation over months or years, long-term studies have shown little histological evidence of renal damage either in adults (Antoine et al. 1969), or in children (Dodge et al. 1976), and a good outcome. Proteinuria of this nature appears to arise as the sum of minor, posture-related variations in glomerular loss, variations in tubule reabsorption of proteins and tubular or urothelial secretion, rather than representing active renal disease. Normally, in persisting but isolated proteinuria of $\leq 1.5–2.0$ g/24 h a renal biopsy does not give useful information, either for prognosis (which is excellent) or treatment.

Such a finding is said to be more common in men than women but it should be remembered that disciplined long-term studies have been conducted principally in such 'captive' groups as young male Army recruits! The authors have seen fixed and isolated proteinuria in young women referred from obstetric practice because proteinuria has been noted from early pregnancy and has persisted postnatally.

A simple algorithm, presented in Fig. 2, provides for the screening of subjects with documented proteinuria in order to define those for whom inpatient study, perhaps including a renal biopsy, is indicated.

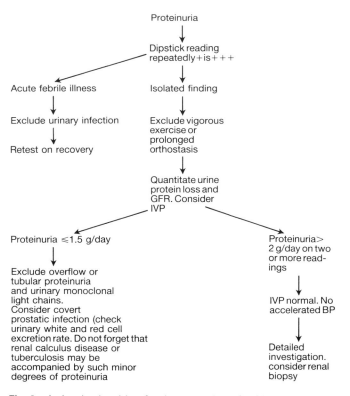

Fig. 2 A simple algorithm for the screening of subjects with proteinuria.

3.1 The clinical approach to haematuria and proteinuria

Table 5 Settings for the development of proteinuria, oedema, and visible haematuria (acute nephritic syndrome)

In an idiopathic glomerulopathy	e.g. IgA-associated mesangial proliferation;* mesangiocapillary glomerulonephritis
As a postinfective lesion	Bacterial, e.g. nephritogenic group A or G streptococci, some pneumococci, or staphylococci Viral, e.g. some enteroviruses (Coxsackie), varicella, cytomegalovirus Parasitic, e.g. Schistosoma
In autoimmune and multisystem disease	e.g. Antiglomerular basement membrane disease* Systemic vasculitides* Henoch-Schönlein disease*†

* May develop aggressive, crescentic glomerular damage.
† Has laboratory and histopathological features in common with those of IgA nephropathy.

There is some debate surrounding the choice of radiographic technique. Modern ultrasound methods permit the partial definition of renal anatomy and detect renal calculi, but do not define precisely the pelvicalyceal system, ureters or bladder, and provide no information on the blood supply to, or urine flow from, each kidney separately. These features can be determined, if indirectly, by intravenous pyelourography which also shows anatomical detail; the technique is invasive, however, and potentially harmful in some settings (see Chapter 1.6.2). It also takes longer and is more costly. For these reasons renal ultrasound appears to be the better screening tool; if the kidneys are shown to be of normal and equal size without cysts, calculi, or other distortion and the proteinuria is an isolated finding, further radiology may be withheld, but the limitations of ultrasonography must be remembered.

Proteinuria with oedema

When a non-renal cause for oedema has been excluded and hypoalbuminaemia accompanies proteinuria the triad of symptoms defining the nephrotic syndrome has been completed, and detailed inpatient investigation is required. In patients over the age or 10 years or so, this will include a renal biopsy (see Section 1 and Chapter 3.5). One diagnostic trap is the patient with congestive heart failure (Albright et al. 1983) and proteinuria, but normal serum albumin and raised neck veins. This proteinuria usually reverses within 2 weeks of successful treatment of the cardiac problem.

Proteinuria, oedema, and visible haematuria

The abrupt onset of these features, commonly in association with oliguria, hypertension and sometimes mild uraemia, is widely recognized to result from an acute inflammatory lesion of the glomerulus, referred to as 'acute glomerulonephritis' or 'acute nephritic syndrome'. The aetiology is varied (see Table 5): this syndrome may arise in the setting of a pre-existing but less obvious glomerulopathy, or as a new, but not original, presentation of one. Clinical and laboratory evidence will not always be sufficient to decide whether this is a new illness, and a renal biopsy may show evidence of long-standing disease. It is important to determine as precisely as possible both the aetiology and the acuteness of the lesion, since these dominate the prognosis.

In the worst cases there is severe uraemia and proteinuria, and the urine is full of red cell casts and tubular debris as well as red cells. This type of acute nephritic syndrome has been called 'rapidly progressive glomerulonephritis' and is usually (but not always) associated with crescent formation in the renal biopsy. The development of this syndrome in an otherwise healthy individual usually carries an excellent prognosis if it is a consequence of an acute bacterial or viral disease. The outlook must be treated more cautiously in patients with a pre-existing autoimmune or idiopathic glomerulopathy, or if there is a multisystem disease. Aggressive glomerular damage with extensive crescent formation carries an urgent need for diagnosis and for decisive intervention (see Chapter 3.6).

Proteinuria with visible haematuria

This finding poses problems: it is the macroscopic haematuria which attracts attention, and the proteinuria may only be recognized later. Presentation may masquerade as a glomerular lesion: the authors have seen proteinuria of up to 5 g/day due to bleeding from a vascular anomaly of the renal pelvis (the equivalent of the loss of about 600 ml of blood on the relevant day), and tuberculosis of the urinary tract should also be considered in the diagnosis. Lesser degrees of haematuria from a neoplasm, from local damage produced by a renal calculus or, even from an infected site may cause sufficient proteinuria to be detected by dipstick. There may be benefit in determining red cell morphology, and the proteinuria should be analysed both qualitatively and quantitatively, to establish whether the loss is commensurate with the degree of haematuria and its pattern is, or is not, glomerular.

This presentation may be due to a variety of glomerular lesions. Full inpatient investigation is required, usually including renal biopsy which rarely shows normal glomeruli.

Visible haematuria alone (isolated macroscopic haematuria)

Pure or isolated macroscopic haematuria is unusual simply because of the ability of the dipstick to detect very small amounts of protein. Protein losses of 70 to 300 mg/day may arise from the loss of 10 to 40 ml of blood per day and may not represent a change in the glomerular barrier to protein.

The analysis of red cell morphology and the detection of urinary red blood cell casts may confirm an active glomerular lesion, often an IgA-associated nephropathy, if biopsy is performed.

Microscopic haematuria and proteinuria

The ability to detect changes in red cell morphology has increased awareness of the spectrum of glomerular lesions in which the combination of microscopic 'glomerular' haematuria and proteinuria occurs; in addition, IgA-dominant mesangial proliferation is now recognized as an important lesion worldwide. This type of presentation is common in this disease and may also be an early marker of insidiously progressive disease (see Chapter 3.6). Many such patients may be seen first by a general physician, or a surgeon, rather than a nephrologist, and the true nature of the lesion may not be determined quickly. Furthermore, since the patients are symptomless, the majority never receive medical attention. Some patients may be moderately hypertensive with minor degrees of haematuria and proteinuria: it may not be clear whether the latter results from the hypertension, or whether the hypertension is the result of glomerular disease. Treating the blood pressure may sometimes restore the urine findings to normal.

Isolated microscopic haematuria

This is a common and difficult finding to interpret. Since only a few red cells are needed to produce macroscopic haematuria, the microscopic detection of even smaller amounts of blood in the urine suggests either a small localized source, opening onto the urothelium, or loss from sources as far apart as the renal parenchyma or prostate. Investigation of such cases can be time consuming and often leads to no firm conclusion. Until the past 5 to 10 years, a renal parenchymal origin for isolated microscopic haematuria was thought to be rare; there was little evidence that it marked progressive glomerular disease, and few renal biopsies were performed for such an indication. By contrast, many patients underwent extensive urological investigation, which frequently proved unrewarding.

Recent attention has altered the clinical perception of the significance which might attach to the finding of microscopic haematuria. In population studies, it has been demonstrated clearly that persistent isolated microscopic haematuria may be the only detectable abnormality in well established and even progressive glomerulopathies, especially those with IgA-dominant mesangial proliferation. With this awareness more biopsy studies are being undertaken, and hitherto undetected cases are coming to light (Turi *et al.* 1989).

At present it seems prudent to proceed cautiously in individual patients. The red cell excretion rate should be determined, and a search made for distorted red cells, although current techniques may not be diagnostic in the presence of so little blood in the urine.

Full urological investigation is almost always needed, especially in the middle-aged or elderly; this should include a search for renal neoplasms, calculi, urothelial tumours or vascular anomalies, and prostatic disease (Froom *et al.* 1984; Britton *et al.* 1989). If these investigations are negative and the haematuria persists for over 6 months, full medical investigation, including renal biopsy, may be indicated (see Fig. 1).

Secondary nephropathies

An isolated 'renal' presentation suggests that one is dealing with an idiopathic disease, but this can be a diagnostic pitfall, since most glomerulopathies arise in the course of other diseases, which are not obvious or even detectable at first renal presentation.

Secondary nephropathies with a proximate but not obvious cause

Drug-induced glomerulopathies

This is a subject in itself, and is dealt with in Chapter 2.1. A wide variety of drugs is involved and the mechanism by which each drug induces differing, if sometime characteristic, histopathology remains largely unclarified. With some agents, such as gold or pencillamine, the associated glomerulopathy is well known and the relationship is usually detected promptly. This is not so with other drugs especially if, as with the non-steroidal anti-inflammatory drugs, they have been obtained for pain relief from a local chemist or even from prescriptions for other family members rather than on medical prescription. Glomerulopathy may only appear after many months ingestion, and the cause may only come to light after careful and direct questioning. An increasing problem in some communities is the use of intravenous psychotropic drugs which can induce glomerulopathies (see Chapter 2.1).

Infection-associated glomerulopathies (see also Chapter 3.12)

The diffuse proliferative glomerulopathy associated with a restricted number of group A, or occasionally group G, β-haemolytic streptococci, still occurs in epidemic form, especially in countries with poor public health facilities (see Chapter 3.9).

Skin-related streptococcal lesions are recognized as a predisposing source for acute poststreptococcal nephritis, particularly when streptococcal infection is superimposed on epidemic scabies. Acute nephritic syndrome with its typical renal presentation is also caused by bacterial infections other than nephritogenic streptococci and by viruses such as coxsackie B. It may also occur as a manifestation of autoimmune or drug associated glomerulopathy.

The uncommon, and less acute, sequelae of persisting infection with hepatitis B, HIV, or Schistosoma are more difficult to differentiate from apparently idiopathic glomerular disease.

Paraprotein-associated glomerulopathy (see Chapter 4.3)

Overt myeloma presents frequently as renal disease, and the underlying primary illness should be diagnosed quickly. Greater care is required to establish that an apparently idiopathic glomerular lesion has been caused by monoclonal immunoglobulin or light-chain deposition. Occasionally, light chain-associated amyloid may also be sparsely deposited in glomeruli, despite heavy glomerular proteinuria—the converse also applies.

Secondary nephropathies with an occult primary lesion

Autoimmune disease

The paradigm is systemic lupus erythematosus, in which glomerular involvement is frequent. However, only in recent years has it become clear that an isolated nephropathy may be the presenting feature of systemic lupus erythematosus in the absence of laboratory evidence of autoimmune disease. There is mounting evidence that treating the nephropathy of lupus nephritis is beneficial (see Chapter 4.6.2), although the best regimens are not yet defined. It is therefore important to establish the diagnosis: this can only be achieved by sequential studies over time, and it may be 1 or 2 years before other clinical or immunological evidence of systemic lupus erythematosus appears.

Neoplastic diseases (see also Chapter 3.13)

A carcinoma or a lymphoma may underlie the development of a glomerulopathy; the putative pathogenetic mechanisms differ. It seems probable that the lesion associated with carcinoma, which may take several histopathological forms, is caused by the glomerular localization of a circulating neoantigen on which immune complexes develop and complement or polymorphonuclear-induced reactions occur. Alternatively, circulating immune complexes with inflammatory potential may be deposited in the kidney. A few tumour neoantigens have been identified in glomeruli, and there are sporadic reports of the cure of a glomerular lesion when the incriminated carcinoma is excised (Alpers and Cotran 1986).

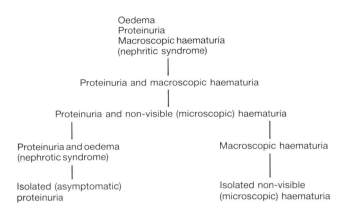

Fig. 3 The 'presentation pyramid' for glomerular disease. The most dramatic presentation—the nephritic syndrome—is at the apex. It is found far less often than the symptomless findings of isolated proteinuria or isolated microscopic haematuria.

The glomerulopathies which occur in lymphomata do not appear to be caused only by neoantigens released from tumour tissue; some probably arise because distortion of normal immunological surveillance by the lymphoma facilitates glomerular damage by other agencies. A variety of glomerular pathology is seen; the renal lesions are cured by effective treatment of the lymphoma.

The glomerulopathy associated with lymphoma or carcinoma may precede detection of the underlying disease by up to 2 years, making the long-term follow up of all apparently idiopathic glomerulopathies essential for full diagnosis effective management.

The clinical syndromes in relation to the history of glomerular diseases

It is necessary to distinguish clearly between the onset of a glomerular lesion and its observed clinical course. The former is documented only rarely, whereas the latter commences at a random point in the natural history of the lesion. Symptomless proteinuria is, by definition, a random finding and may be documented at any time during its course. However, the level of urinary protein loss varies with time, and in our experience up to 50 per cent of subjects in whom proteinuria was initially symptomless and less than 5 g per day developed heavier, potentially symptomatic (that is oedema-producing) proteinuria later. This usually remitted eventually, as does proteinuria which initially presents with much heavier loss.

It is very uncommon for end-stage renal failure to develop if proteinuria remains below 3 g (and probably even below 5 g) per day. Above this level, perhaps 20 per cent of patients develop end-stage renal failure within 3 to 5 years of the apparent onset of such a phase of their disease—and importantly, about 15 per cent die of other causes over the same period.

It is necessary to record the findings in detail at the first assessment, but this should not lull the clinician into believing that no more needs to be done. Vigilant follow-up is mandatory whatever the level of protein loss recorded initially. The only exception may be in those subjects in whom protein loss remains consistently less than 1 to 1.5 g/24 h (see above).

The clinical syndromes associated with glomerular disease are outlined in Fig. 3. It should be noted that the more dramatic the findings, the less frequently the syndrome occurs. Also, the prognosis is not closely related to the severity of the initial or current illness. Despite efforts to refine the information that can be obtained from a renal biopsy, at present the histopathology does not tend to predict the clinical findings (see Chapter 1.7.2), nor does it offer more than a broad prediction of outcome.

Apart from defining the nature and course of the proteinuria, there is the ever present question of prognosis. This requires longitudinal assessment over months or years since a glomerular lesion may already be recovering or may develop further, with consequent heavier protein loss. In the light of present knowledge we consider the approach outlined in the Appendix a reasonable one with which to address this problem.

Initial assessment and subsequent follow-up of patients with proteinuria and/or haematuria

It must always be remembered that the presentation of glomerular disease does not necessarily coincide with its onset. This is self-evident for symptomless proteinuria or microscopic haematuria, since these are clearly chance findings. There is good evidence that in most histopathologically defined groups of patients the nephrotic syndrome may occur after months—or even years—of asymptomatic proteinuria. This is apparently not true in minimal change nephropathy, in which very heavy proteinuria develops abruptly, rapidly followed by the nephrotic syndrome (see Chapter 3.5). Equally important is the observation that macroscopic haematuria may occur in a patient with already documented, symptomless microscopic haematuria.

Similarly, slight proteinuria or haematuria may be detected for many months, or even years, after an overt glomerular illness (see Chapter 3.9).

Taken together, these observations point to the need for caution in determining the natural history of a given lesion. It takes time and repeated analysis to develop a sense of the progress of an individual glomerulopathy and so to separate out the minority in whom healing is not occurring and in whom active intervention may be worthwhile. Furthermore, relapse has been recorded in all histopathological groups, often abruptly, and even after some years of freedom from the proteinuria.

For the individual patient with suspected glomerular disease (see Figs. 1 and 3) the following pattern is suggested:

1. Full inpatient assessment including renal biopsy (see Chapter 1.7.2);
2. Regular follow-up for 3 years, or until proteinuria has disappeared. This should be at 4-monthly intervals; more frequent assessment may be necessary at times;
3. Triennial detailed inpatient investigation for those with persisting proteinuria. Consideration should be given to repeating the renal biopsy at this time.

Some notes of value on the individual investigations suggested, together with a format which has proved useful in tabulating results, is given in the Appendix.

Conclusion

The study of glomerular disease is fascinating, but requires considerable effort over months, or even years, to be fully productive. A careful assessment of the natural history of each patient, together with detailed evaluation of all of the histopathological and laboratory tests, is needed to assess whether or not the patient is going to do well, which is the rule, or badly. It is, of course, crucially important to determine the latter as early as possible. We hope that the approach we have outlined will help the reader construct his own studies and define groups of patients which now, and in the future, will require active treatment if their disease is not to progress to renal failure.

References

Albright, P., Brensilver, J., and Cortell, S. (1983). Proteinuria in congestive heart failure. *American Journal of Nephrology*, **3**, 272–5.

Alcorn, D. and Ryan, G.B. (1981). Distribution of anionic groups in the glomerular capillary wall in rat nephrotoxic nephritis and aminonucleoside nephrosis. *Pathology*, **13**, 37–50.

Alpers, C.E. and Cotran, R.S. (1986). Neoplasia and glomerular injury. *Kidney International*, **30**, 465–473.

Antoine, B., Symvoulides, A., and Dardenne, M. (1969). La stabilité évolutive des états de protéinurie permanente insolée. *Nephron*, **6**, 526–536.

Berggård, I. (1970). Plasma proteins in normal urine. In *Proteins in normal and pathological urine* (eds. H. Manuel, H. Betuel, J.P. Revillard), p. 719. Karger, Basel.

Bernard, A., Amor, A.O., Viau, C., and Lauwerys, R. (1988). The renal uptake of proteins: a non-selective process in conscious rat. *Kidney International*, **34**, 175–185.

Bertolatus, J.A., Abuyousef, M., and Hunsicker, L.G. (1987). Glomerular sieving of high molecular weight proteins in proteinuric rats. *Kidney International*, **31**, 1257–1266.

Birch, D.F. and Fairley, K.F. (1979). Haematuria: glomerular or non-glomerular. *Lancet*, **ii**, 845–846.

Brenner, B.M., Hostetter, T.H., and Humes, H.D. (1978). Molecular basis of proteinuria of glomerular origin. *New England Journal of Medicine*, **298**, 826–833.

Britton, J.P., Dowell, A.C., and Whelan, P.O. (1989). Dipstick haematuria and bladder cancer in men over 60: results of a community study. *British Medical Journal*, **299**, 1010–1012.

Carrie, B.J., Salyer, W.R., and Myers, B.D. (1981). Minimal change nephropathy: an electrochemical disorder of the glomerular membrane. *American Journal of Medicine*, **70**, 262–268.

Cotran, R.S. and Rennke, H.G. (1983). Anionic sites and the mechanisms of proteinuria. *New England Journal of Medicine*, **309**, 1050–1052.

Deen, W.M., Bridges, C.R., Brenner, B.M., and Myers, B.D. (1985). Heteroporous model of glomerular size selectivity: application to normal and nephrotic humans. *American Journal of Physiology*, **249**, F374–F389.

Dodge, W.G., West, E.F., Smith, E.H., and Bunce, H.B. III (1976). Proteinuria and haematuria in schoolchildren: epidemiology and early natural history. *Journal of Paediatrics*, **88**, 327–347.

Froom, P., Ribak, H., and Benbassat, J. (1984). Significance of microhaematuria in young adults. *British Medical Journal*, **288**, 20–22.

Ghiggeri, G.M., *et al.* (1987). Renal selectivity properties towards endogenous albumin in minimal change nephropathy. *Kidney International*, **32**, 69–77.

Hoyer, J. and Seiler, M.W. (1979). Pathophysiology of Tamm–Horsfall protein. *Kidney International*, **16**, 279–289.

Kaysen G.A., Gambertoglio, J., Jimenes, I., Jones, H., and Hutchinson, F.N. (1986). Effect of dietary protein intake. *Kidney International*, **29**, 572–577.

Kincaid-Smith, P. (1982). Haematuria and exercise-related haematuria. *British Medical Journal*, **285**, 1595–1597.

Levitt, J.I. (1967). The prognostic significance of proteinuria in young college students. *Annals of Internal Medicine*, **66**, 685–696.

Makino, H., Nishimura, S., Takaoka, M., and Ota, Z. (1988). Mechanism of haematuria. II. A scanning electron microscopic demonstration of the passage of blood cells through a glomerular capillary wall in rabbit Masugi nephritis. *Nephron*, **50**, 142–150.

Park, C.H. and Maack, T. (1984). Albumin absorption and catabolism by isolated perfused convoluted tubules of the rabbit. *Journal of Clinical Investigation*, **73**, 767–777.

Pesce, A.J. and First, M.R. (1979). *Proteinuria*, Marcel Dekker, New York.

Rytand, D.A. and Spreiter, S. (1981). Prognosis in postural (orthstatic) proteinuria. Forty to fifty year follow-up of six patients after diagnosis by Thomas Addis. *New England Journal of Medicine*, **305**, 618–621.

Schramek, P., Moritsch, A., Haschkowitz, H., Binder, B.R., and Maier, M. (1989). *In vitro* generation of dysmorphic erythrocytes. *Kidney International*, **36**, 72–77.

Schichiri, M., *et al.* (1988). Red-cell-volume distribution curves in diagnosis of glomerular and non glomerular haematuria. *Lancet*, **i**, 90.

Simpson, F.O. (1986). Is current research into basement membrane chemistry and ultrastructure providing any new insights into the way the glomerular basement membrane functions? *Nephron*, **43**, 1–4.

Springberg, P.D., *et al.* (1982). Fixed and reproducible orthostastic proteinuria: results of a 20-year follow-up study. *Annals of Internal Medicine*, **97**, 516–519.

Sumpio, B.E. and Hayslett, J.P.H. (1985). Renal handling of proteins in normal and diseased states. *Quarterly Journal of Medicine*, **57**, 611–635.

Turi, S., *et al.* (1989). Long term follow-up of patients with persistent/recurrent, isolated haematuria: a Hungarian multi-centre study. *Paediatric Nephrology*, **3**, 235–239.

Vernier, R.W., Klein, D.J., Sisson, S.P., Mahan, J.D., Ogema T.R., and Brown, D.B. (1983). Heparin sulfate-rich anionic sites in the human glomerular basement membrane. Decreased concentration in congenital nephrotic syndrome. *New England Journal of Medicine*, **309**, 1001–1009.

White, R.H.R. (1981). The clinical applications of selectivity of proteinuria. *Contributions to Nephrology*, **24**, 63–71.

Winetz, J.A., *et al.* (1981). The nature of the glomerular injury in minimal change and focal sclerosing glomerulopathies. *American Journal of Kidney Diseases*, **1**, 91–98.

Wilkinson, A.H., Gillespie, C., Hartley, B., and Williams, D.G. (1989). Increase in proteinuria and reduction in number of anionic sites on the glomerular basement membrane in rabbits by infusion of human nephrotic plasma *in vivo*. *Clinical Science*, **77**, 43–48.

Appendix

A proforma suitable for recording both initial and follow-up data is given below.

Basic data (measurements should be repeated every 3 years)

Date of investigation
Name
Record number
Sex
Date of birth

Height	cm
	To be recorded initially, but only on follow-up if patient is less than 18 years
Weight	kg
Blood pressure	Supine, diastolic phase V
Plasma creatinine	μmol/l
Plasma albumin	g/l
Plasma calcium	mmol/l
Plasma phosphate	mmol/l

Serum urate	mmol/l
Blood sugar	mmol/l
24-h Urinary protein	g
24-h Urinary creatinine	mmol
Timed urine	
1. Time	min
2. Volume	ml
3. Red blood cells	+ 1000 per ml (if technique is available, record red blood cell morphology)
Serum immunoglobulins	g/l
Fibrinogen	g/l
SCAT	Negative or positive
ANF	Negative or positive
Anti-DNA	titre
C3	mg/dl
C4	mg/dl
Cryoglobulins	Negative or positive
Cholesterol	mmol/l
Triglycerides	mmol/l
HDL	mmol/l
HBsAg	Negative or positive
Selectivity	Ratio
Serum (immuno)electrophoresis	Normal, abnormal, or paraprotein
Urine (immuno)electrophoresis	Normal, abnormal, or paraprotein (consider immunoprecipitation in selected cases (see text))
HLA	Tissue type

Follow-up (4-monthly (or more frequently if required clinically))

Height (if child or adolescent)
Weight
Blood pressure
Oedema
Plasma creatinine
Plasma albumin
Urinary protein
Urinary creatinine
Diuretics
Antihypertensives
Prednisolone
Other immunoactive drug
Diet
Event since last entry (denotes potentially important events such as infections, trauma, or operations which may trigger relapse or deterioration, or transiently affect otherwise stable biochemical/pathological variables)

Protocol for studying a patient with glomerular disease

Full clinical details will be recorded. Particular attention should be paid in the history to previous illnesses especially if there has been any evidence of renal disease. It is always worth enquiring about previous employment-related or insurance examinations and, in females, about any previous gynaecological or obstetric assessment, since proteinuria may have been recorded at such times.

Direct questioning regarding drug ingestion, both prescribed and casual, is mandatory.

Since some glomerular diseases run in families (see Section 18), a detailed family history is always important. In practice it is very difficult to establish satisfactorily whether or not a member of the family has had renal disease since this may be, or may have been, present but symptomless. Only a minority of apparently idiopathic disease appears to have a familial context but such cases are always instructive and should be followed through. There are, of course, a few conditions such as Alport's syndrome (see Chapter 18.4.1) in which a familial association is already well documented, but it is in the other glomerulopathies, such as membranous nephropathy, that pathogenetic clues may lie hidden.

Age, height and weight should be documented. Height is usually recorded in children but it is a useful parameter in determining ideal as opposed to actual weight in adults, and is necessary in determining body surface area, a valuable baseline when comparing data from different patients.

Blood pressure

This important variable is poorly annotated. It is difficult to compare with confidence either intra- or interindividual data unless baseline conditions are defined and it is clearly stated as to whether the fourth or fifth Korotkoff point was selected consistently. Since a single reading in hospital conditions is notoriously unreliable as the index blood pressure, only broad principles can be deduced from less precise data, for example, on the influence of raised blood pressure (itself needing careful definition) on a given renal disease (see Chapter 17.1).

The presence of oedema

This may be noted suitably according to Fig. 4.

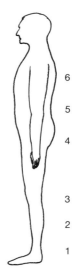

A useful scoring system is:

1. Ankle oedema (+)
2. Ankle and leg oedema, but not to knee ++
3. Ankle and leg oedema, to knee +++
4. Oedema to thighs
5. Oedema to thighs with ascites and/or sacral fluid
6. Oedema to thighs with ascites and with pleural effusion

Fig. 4 A scoring system for oedema.

Body mass

The degree of cachexia in a patient with nephrotic syndrome is masked by oedema, but can be revealed by such measurements as skin-fold thickness and mid-arm circumference. The nutritional status of patients with prolonged heavy proteinuria needs more careful attention than it often receives. The active co-operation of an informed dietitian improves the health and mobility of such patients but requires time and skill. The diet must be palatable and balanced for protein and calorie content to ensure that nitrogen anabolism is maintained. The authors suggest that a diet of 1 g protein/kg ideal weight and 1 g protein/g daily urine protein loss, balanced with 32 kcal/g dietary protein restores nitrogen anabolism and is beneficial. A slightly lower dietary regimen (0.8 g/kg body weight) with a similar carbohydrate intake has been recommended (Kaysen *et al.* 1986). However, the benefits claimed may well have been attributable to the increased calorie content of the diet over the 2-week study period; our own extended studies have shown that the diet we suggest is effective in the longer term.

Data can be stored in a form suitable for future analysis as shown.

Explanatory notes
Biochemistry
Plasma creatinine

Calculation of the predicted glomerular filtration rate using the formula. This method offers a useful check if there is severe cachexia and urine collections have been erratic (see Chapter 1.4 for more details).

$$\frac{(140 - \text{age (years)}) \times \text{weight (kg)}}{\text{Serum creatinine } (\mu\text{mol/l})} \times (1.25 \text{ (males) or } 1.03 \text{ (females)})$$

Blood urea

This should be treated with caution as it is the result of several processes which are not easily separable. Protein intake, both alone and in relation to calorie intake, may be insufficient to ensure maximal protein anabolism even if there is normal liver function. The balance of tissue protein anabolism and catabolism is itself influenced by heavy albuminuria, since the consequent increase in liver albumin synthesis makes increased demands on all available nitrogen sources, including body tissues. The 24-h urea output (in mmol) is more useful than the blood urea level since it reflects the overall degree of protein catabolism. This figure, multiplied by 0.216, gives an approximate measure of the equivalent protein catabolized in g/24 h.

Serum albumin

There are various methods for determining this, the most accurate being (relatively expensive) immunochemical analysis. Other routine methods invariably use a dye-binding technique which may overestimate low results by about 4 to 5 g/l. At equilibrium, plasma albumin reflects the total albumin pool in the extracellular space, which is 3 to 4 g/kg ideal body weight. Of this 200 to 300 g of albumin, less than 50 per cent is in the plasma. In the presence of albuminuria the total pool is depleted and plasma albumin falls only as a reflection of this depletion. Similarly, if protein anabolism is restored in the hypoalbuminaemic subject, the total albumin pool increases gradually and this is reflected by the plasma albumin concentration. It follows that changes in plasma albumin concentration are not a reliable marker of the prevailing state of albumin synthesis or catabolism, but rather of the preceding balance of these metabolic events. If the extracellular albumin pool is halved (with a fall of plasma albumin to say about 20 g/l) it may be several weeks after albumin loss has ceased before the pool is restored to normal and for the plasma albumin to return to its premorbid level.

Other plasma proteins

Analysis of other plasma proteins rarely yields valuable information, although some general conclusions may be drawn. Providing dietary intake of protein/calories is adequate, albuminuria of less than 5 g/24 h is usually compensated for by increased liver albumin synthesis so that the plasma albumin level is maintained. However, increased hepatic protein synthesis appears not to be confined to albumin but necessarily involves a parallel increase in the synthesis of fibrinogen, β-microglobulin, and the lipoproteins. On the other hand, the increased glomerular permeability to protein sometimes results in the loss of vitamin D, thyroxine-carrying proteins and the immunoglobulins, especially IgG (see Chapter 3.4).

The resultant plasma protein pattern reflects these changes; the more marked the proteinuria the greater are these changes.

Lipoproteins

Lipoprotein abnormalities are a frequent, but not universal, accompaniment to the nephrotic syndrome. Triglycerides and total cholesterol are usually raised, the latter predominantly due to an increase in the cholesterol content of the low-density lipoprotein and very low-density lipoprotein fractions (see Chapter 3.4 for detailed discussion). The high-density lipoprotein fraction in nephrotic sera has been variously reported as raised, normal or lowered, but few studies have addressed the question of which high-density lipoprotein subfractions are particularly affected. Furthermore, it is possible that different profiles may be produced with different histological types. The genesis of lipoprotein abnormalities appears to stem from increased lipoprotein production by the liver. It remains uncertain whether the stimulus is a fall in plasma oncotic pressure or increased plasma albumin synthesis with a non-specific rise in production of many hepatically produced proteins. This phenomenon is compounded by a fall in lecithin-cholesterol acetyltransferase and lipoprotein lipase activity and loss of certain lipoprotein moieties in the urine.

The relevance of such perturbations in lipoprotein metabolism is two-fold. First, it is likely that such potentially atherogenic lipoprotein profiles at least contribute to the increased incidence of coronary artery disease often observed in patients with heavy, prolonged proteinuria. Secondly, and perhaps more importantly, it is now proposed that deposition of lipid within the glomeruli may contribute to the progression of renal disease associated with persistent proteinuria, by inducing glomerulosclerosis (see Chapter 9.3). Both these hypotheses, however, await proof.

Abnormal plasma protein constituents in the urine

Both myoglobin and haemoglobin may appear in the urine, causing characteristic discoloration, which for haemoglobin may vary from a pink tinge to frankly black urine (the latter is only a question of density; with a very strong light the red colour is obvious). The presence of either may be associated with renal tubular damage. Identification in each case is by spectrometry, electrophoresis, or by direct immunochemical analysis.

Monomeric free immunoglobulin light chains have a molecular weight of 22 000 kDa, but they may be present as dimers or even tetramers. The isoelectric point of light chains varies widely; some are as ionic as albumin, with a p*I* of 4.6, while others have a p*I* exceeding 7. The proximal renal tubule is a predominant site for light chain catabolism and in proximal tubule disorders a mixture of light chains is detected in urine. This can be related to the daily catabolic rate, indicating how important tubular catabolism is for these small proteins (see Chapter 4.3 and Sumpio and Hayslett 1985).

The detection of a monoclonal light chain has a different connotation, since it implies the existence of a B-cell monoclone; such a monoclone is usually malignant, producing myeloma. However, more benign monoclones may produce free light chains and a wide variety of glomerular and tubular lesions (see Chapter 4.3). Unless a careful search is made for such free light chains the cause of a glomerulopathy may be missed entirely.

The best screening test for free light chains in the urine is electrophoresis of concentrated urine. Immunofixation will confirm the results and determine the light chain type, if clinical suspicion that there might be a covert B-cell monoclone exists. It may be appropriate to use immunofixation when electrophoresis shows no abnormality (see Chapter 4.3).

Immunological studies

The complement cascade

Some complement components are known to be under genetic control and their complete absence (say of C3) may provide a diathesis on which glomerular disease arises more frequently than in the general population (see Chapter 3.2 and 4.6.1).

C4 and Factor B each have several alleles coded within the major histocompatibility complex on the short arm of chromosome 6. In some communities, particular alleles may be found more frequently, or may be present only rarely in a given glomerulopathy.

These primary, genetically determined complement patterns are of interest only as part of the understanding of the pathogenetic mechanisms underlying glomerular disease, and are worth recording only as part of a structured study addressing such questions. Random recordings are of little benefit. To determine that such defects are genuinely persistent, repeat analysis is clearly required and there should be no evidence of complement consumption, e.g. complement breakdown products should not be detectable.

Complement consumption does occur in glomerulopathies, but if this is only a mild feature it is difficult to detect by analysing the complement cascade components, since production may balance utilization. Measurement of stable products of complement activation, such as C3d, is more valuable; recently there has been interest in the detection of this product and in the presence in the urine of the membrane attack complex formed by C5b-9 as markers of local intraglomerular complement activation.

For routine purposes it is sufficient to measure serum C3 and C4 levels, together with C3d. The total haemolytic complement (CH_{50}) is not usually contributory except to screen for congenital absence of other components, e.g. C1 or C2. Factor B levels need only be measured if there is reason to suppose that the alternate complement pathway is activated. Measurement of urinary C3d or the membrane attack complex should be considered if intraglomerular complement activation is suspected, and if these tests are available locally; serial measurements are more useful than 'spot' samples.

It is important to remember that complement utilization is an active process; it may be observed only intermittently and it may occur for a prolonged period with persisting low levels of complement cascade components. However, these may eventually return to normal even after many months. Regular assays are always required to determine such patterns, and random findings are unlikely to assist in diagnosis or management.

C3 nephritic factor

The association of one of this family of globulins with mesangiocapillary glomerular disease is a recognized feature, especially when this glomerulopathy occurs in patients with partial lipodystrophy (see Chapter 3.8). The most common C3 nephritic factor acts by stabilizing C3 convertase and thus generates consumption of C3, leading to low serum levels. It should be sought by crossed electrophoresis of mixtures of normal and nephritic sera if there is a persistingly low C3 but maintained C4 concentrations in the serum.

Antiglomerular basement membrane antibodies (see Chapter 3.11)

Traditionally, the presence of antiglomerular basement membrane antibodies has been demonstrated on renal biopsies by continuous linear deposits of IgG along the basement membrane. In recent years the nature of the antigen has been sought, and it is clear that the epitope resides in the NC1 domain of type IV collagen. The pathogenicity of these autoantibodies has been demonstrated by the induction of nephritis in experimental animals by the passive transfer of antibody eluted from diseased kidneys. Assays which will detect serum levels of circulating antibodies against the glomerular basement membrane are now detectable. These enzyme-linked immunosorbent assays, and radioimmunassays, are more sensitive than an indirect immunofluorescence method, are quantitative, have a very high degree of specificity for such autoimmune disease and are rapid, results being available the same day. Immunoassay for antiglomerular basement membrane antibody is therefore of great diagnostic value, especially when renal biopsy is inappropriate for clinical reasons, and also has an important role in monitoring treatment. The assay will plot the reduction in circulating antibody which occurs as a result of plasmaphoresis and immunosuppression; when the antibody is undetectable, treatment may be terminated.

Antineutrophil cytoplasm antibodies

Antibodies specific for neutrophil cytoplasmic antigen(s) have been described in patients with systemic vasculitis and necrotizing glomerulonephritis (see Chapters 4.5.1 and 4.5.3). Early interest centred on the relationship of these antibodies to Wegener's granulomatosis in both diagnosis and management of the disease. However, subsequent clinical and laboratory research has shown this field to be very complex; several cytoplasmic antigens are clearly involved, and can be defined by variable clinical criteria. Following an International Workshop in Copenhagen, the standard method of detecting antineutrophil cytoplasmic antibodies is the indirect immunofluorescence assay using ethanol-fixed human leucocytes.

Two types of staining pattern have been described. The classical pattern is a coarse, granular staining of the cytoplasm, often accentuated in the centre of the cell but showing no nuclear binding. Lymphocytes do not show this staining. Current evidence suggests that about 70 per cent of sera showing this staining pattern react with an α-granule serine proteinase, of molecular weight 29 kDa. This staining occurs in 75 to 90 per cent of patients with biopsy-proven, active Wegener's granulomatosis and in up to 50 per cent of patients with microscopic polyarteritis nodosa. The perinuclear staining pattern, sometimes described as a granulocyte-specific antinuclear factor, appears as a linear nuclear membrane staining of the multilobed nucleus of neutrophils. There is general agreement that the antigen detected here is, in fact, of cytoplasmic origin and that it migrates during fixation to produce the perinuclear pattern. Current research implicates myeloperoxidase as the predominant perinuclear antigen.

This staining pattern is reported to be present in the majority of patients with idiopathic crescentic, necrotizing glomerulonephritis, and in a variable number of cases of microscopic polyarteritis nodosa. Antinuclear cytoplasmic antibodies have not yet been shown to be pathogenic, and may simply be secondary markers of the disease process. Reports differ on their usefulness for monitoring treatment. This rapidly developing field of research offers the clinician the hope of a diagnostic test, which may be helpful in the management of these very variable vasculitic glomerulonephritides.

3.2 Immune mechanisms of glomerular damage that affect the kidney

DENIS GLOTZ AND PHILIPPE DRUET

Introduction

Understanding immunological disorders that affect the kidney, whether glomerular or tubulointerstitial, requires some understanding of the normal functioning of the immune system. As well as helping to elucidate the mechanisms of immunologically mediated renal disease, such understanding is also relevant to treatments which are being introduced.

After a brief overview of the immune system, we will consider how the induction of autoimmune responses leads to glomerulopathies, whether glomerular-specific or in the context of polyclonal activation. We will then look at the glomerular deposition of immune reactants, antibodies as well as cells, and finally focus on the mediators of glomerular damage such as antibodies, cells, cytokines, reactive oxygen species, and complement.

Overview of the immune system

This overview will briefly describe the main phases of the immune response, that is, antigen processing and presentation to T cells, T-cell activation and differentiation into subsets, the biology of immunoglobulins, and the theoretical concepts on self-recognition, tolerance, and autoimmunity. All of these processes, whose normal function is the elimination of antigen, may sometimes be deleterious to the host, as exemplified by the process of healing through scarring.

Antigen presentation and processing

The central event in the immune response is the presentation of processed antigen to T cells (reviewed in Kourilsky and Clavery 1989) resulting in the generation of cytotoxic T cells or helper T cells. Both types of T cells use the same antigen receptor: this receptor is only capable of recognizing 'processed antigen' bound to a major histocompatibility complex molecule, is in obvious contrast to antibodies, which recognize native antigens.

Resolution of the three-dimensional structure of two HLA-class I molecules by X-ray crystallography (Björkman et al. 1987) and the extrapolation of these results to class II molecules has greatly increased understanding of these processes. Both MHC class I and class II molecules appear to have an antigen (peptide) binding groove bordered by α helices and with a β-pleated floor. The amino acids which determine the polymorphism of these molecules face this antigen-binding groove and influence its shape and, therefore, the peptides that bind to it. These interactions between antigen, MHC molecule and T-cell receptor determine the specificity of the immune response and also tolerance to self antigens.

Unlike antibodies, MHC molecules bind to antigens only after their enzymatic fragmentation: this has been elegantly illustrated using fixed cells bearing MHC molecules pulsed with synthetic peptides based on the primary amino acid sequence of antigens. The principles underlying processing by intact cells, whether *in vitro* or *in vivo*, are clear though the details have yet to be worked out.

There appear to be two main pathways for processing antigen. Exogenous antigens are internalized by antigen-presenting cells: these may be non-specific, such as macrophages and dendritic cells, or specific, such as immunoglobulin-bearing B cells. The internalized antigen is digested and the resultant peptides bind to MHC class II molecules in endosomes before being expressed as a complex on the cell surface. Antigenic peptides bound to class II molecules are recognized by T-cell receptors on T cells expressing the CD4 molecule; these usually function as T-helper cells. Endogenous molecules, such as those derived from viruses, are processed by a different route before combining with MHC class I molecules expressed on the cell surface. Peptides bound to MHC class I molecules are recognized by CD8-positive T cells, which are usually cytotoxic.

Antigen-presenting cells continuously take up proteins native to the host, and it is perhaps inevitable that the peptide-binding grooves of most MHC molecules expressed on the cell surface contain processed self antigens. Individuals do not usually respond to these, since potentially autoreactive T cells are either eliminated or inactivated in the thymus during ontogeny. The

3.2 Immune mechanisms of glomerular damage

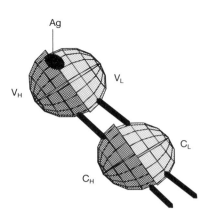

Fig. 1 Schematic representation of the tertiary structure of the antigen receptor. The globular domains of the heavy and light chain and their association are shown. This holds true for the B-cell antigen receptor (immunoglobulin) as well as for the T-cell antigen receptor.

processes responsible will be described in more detail later. It is important to realize that the ability to respond to exogenous antigens and yet remain tolerant to self antigens is determined by the specificities of T-cell receptors.

T cells

T-cell receptor

There are several structural similarities between immunoglobulin molecules and the T-cell receptor (Fig. 1), and they both belong to the immunoglobulin supergene family. Two different T-cell receptors have been described (Mak 1988). They are present at the cell surface as heterodimers made of two polypeptide chains (α and β, molecular weight 43–49 and 38–44 kDa, respectively, or γ and δ, molecular weight 40–55 and 38–42 kDa, respectively). The $\gamma\delta$ receptor is present on a minority of T cells, whose function is unknown. In mice, these cells are mainly found in the skin, uterus, and gut. The $\alpha\beta^+$ T cells recognize antigens bound to MHC molecules.

Each T-cell receptor chain consists of two extracellular domains, one of which is variable, with a V and J sequence (α,γ) or V, D, and J sequence (β,δ), and one constant (C sequence), a transmembrane portion, and a short intracytoplasmic tail. The separate chains are linked together by an interchain disulphide bond (Table 1). Most peripheral blood T cells, whether $CD4^+$ or $CD8^+$, have an $\alpha\beta$ receptor, while most gut-associated T cells bear $\gamma\delta$ receptors. The diversity of T-cell receptors is brought about by germ-line diversity coupled to gene rearrangements; however, in contrast to immunoglobulin genes in B cells somatic mutation has not been detected in T-cell receptor V genes.

Each of the above sequences is encoded by a different exon, and rearrangements which occur during intrathymic cell differentiation arise from recombination between them. This is probably identical to the immunoglobulin gene rearrangements which are described later. The T-cell repertoire is smaller than the B-cell repertoire because of intrathymic selection (see below) and the absence of somatic mutation and thus is 'incomplete'. The main characteristics of the T-cell receptors are given in Table 2.

The CD3 complex associated with the T-cell receptor is made up of three peptide chains ($\gamma,\delta,\varepsilon$). Two additional chains have been described in mice. This complex is responsible for transducing the signal initiated by antigen interaction with the receptor: tyrosine kinase and phospholipase C are activated, resulting in an increase in intracellular Ca^{2+} and the opening of ion channels. There may also be alternatives means by which T cells are activated which are less dependent on antigen.

Ontogeny and thymic selection

T-cell subsets may be characterized by the expression of various T-cell markers, such as CD25 (interleukin-2 receptor), CD2, CD4 and CD8, which can be recognized by various monoclonal antibodies. Three main stages of T-cell development in the thymus have been described using such antibodies. Pre-T cells leaving the fetal liver are CD2, CD3, CD4, CD8, and CD25 negative. After 8 weeks (stage I), T cells penetrate the outer cortex of the thymus, and become $CD25^+$, $CD2^+$, and $CD45$ (T 200)$^+$. During the tenth week of gestation (stage II) cells become double positive for both $CD4^+$ and $CD8^+$, and genes coding for the T-cell receptor are rearranged. Rearrangement of the γ and δ genes occurs first; a secondary rearrangement involving the α and β genes occurs in most cells, leading to the expression of only an $\alpha\beta$ receptor. By the 15th week of gestation (stage III), fetal T cells, like mature T cells, are $CD3^+$, $CD4^+$ or $CD3^+$, $CD8^+$ (Table 3): these two populations of T cells become inducer/helper ($CD4^+$) T cells and suppressor/cytotoxic ($CD8^+$) T cells (reviewed in Ferrick 1989).

About 90 per cent of T cells die in the thymus, indicating a powerful selection process. During the last 2 to 3 years much progress has been made in understanding the processes involved in this selection. (Kappler *et al.* 1987; Kisielow *et al.* 1988; reviewed in Owen and Lamb 1988). The basis for selection is thought to be related to reactions to self MHC, and the presence of T cells specific for such molecules (DR, DQ, DP, etc.) in different parts of the thymus has been studied. Another approach is to use mice transfected with a given T-cell receptor gene, and such experiments demonstrated that both positive and negative selection for T cells occurred within the thymus (Fig. 2).

During the first step (positive selection) T cells specific for self MHC are selected following recognition by the receptor of class I or II antigens on thymic epithelial cells. During the second step (negative selection) autoreactive T cells with a high affinity for self antigens are eliminated. Such processed self antigens would be presented by (bone marrow-derived) dendritic cells. It must be stressed that clonal deletion, which seems to be valid for self antigens present in the thymus, probably does not hold true for all antigens since autoreactive T-cell clones specific for various antigens (gp330, GBM, myelin etc.) may be induced at the periphery in normal animals.

Tolerance at the B-cell level seems to proceed via different

Table 1 Genes of the human T-cell receptors

Genes	Chain of the TCR			
	α (14)	β (7)	γ (7)	δ (14)
V	≈ 100	≈ 50	6–8	7–15
D	0	2	0	2
J	≈ 100	12–13	3–5	2–3
C	1	2	2	1

Numbers in parentheses refer to chromosome number.

Table 2 Main characteristics of the T-cell receptors

	TCR $\alpha\beta$	TCR $\gamma\delta$
Distribution	60–70% of peripheral blood lymphocytes	1–10% of peripheral blood lymphocytes
Phenotype*	CD_2, CD_3, CD_4 (60%)	CD_2, CD_3, CD_4 (< 1%)
	CD_2, CD_3, CD_8 (35%)	CD_2, CD_3, CD_8 (20–50%)
	CD_2, CD_3 (< 1%)	CD_2, CD_3 (50–80%)
Ontogeny	Thymic	Thymic and extrathymic
Function	Recognizes foreign peptides plus MHC class I or II	Recognizes heat shock proteins and MHC or MHC-like structures

* Only positive markers are indicated (modified from Ferrick et al. 1989).

Table 3 Intrathymic differentiation

Bone marrow	Stem cell
	Pre-T-cell
Thymus	
Outer cortex	CD25, CD2, CD45*†
Inner cortex	CD4, CD8
	$\alpha+$ $\beta+$
Junction	CD4, CD8
	T_3–Ti
Medulla	CD4 CD8
	T_3–Ti T_3–Ti
Periphery	CD4 CD8
	T_3–Ti T_3–Ti
	(T helper–T DTH) (T cytotoxic–T suppressor)

* Only positive markers are indicated.
† These markers are present on all the subsequent subsets and will not be mentioned.

mechanisms, since B-cell clones specific for self are not deleted but rendered anergic.

T-cell subsets and functions

Most T cells at the periphery (peripheral lymphoid organs or blood) are $CD3^+$ and either $CD4^+$, $CD8^-$ or $CD4^-$, $CD8^+$. The $CD4^+$, $CD8^-$ cells recognize either a peptide generated from exogenous antigens by antigen-presenting cells (monocyte/macrophage or B cells) and associated with class II molecules, or allogeneic class II molecules themselves. The $CD4^-$, $CD8^+$ cells recognize either a viral determinant associated with class I molecules or allogeneic class I molecules alone.

$CD4^+$ T cells may be either regulatory or effector cells. Regulatory $CD4^+$ T cells collaborate with B cells to initiate antibody production. Two different regulatory $CD4^+$ T cells have recently been described in mice (Mosman and Coffman 1989); these produce different cytokines and induce B cells to produce different isotypes. TH1 cells produce mainly interleukin-2 and γ-interferon and induce the production of IgM and IgG2a antibodies; whereas TH2 cells produce interleukin-4 and interleukin-5 and induce the production of IgG1, IgE, and IgA. There is good, though indirect, evidence that the equivalents of TH1 and TH2 cells exist in humans. The effector $CD4^+$ T cells also produce IL-2 and γ-interferon and are responsible for delayed type hypersensitivity reactions.

The $CD8^+$ T cells are also usually divided into regulatory (suppressor) and effector (cytotoxic) T cells. Such suppressor T cells have rarely been cloned but, as will be described, there is good evidence that such cells are present in Heymann's nephritis and in $HgCl_2$-injected Lewis rats. The effector $CD8^+$ T cells are classical cytotoxic T cells that recognize either viral determinants associated with class I MHC molecules, or allogeneic class I determinants. These cells do not produce IL-2 and rely on IL-2 produced by $CD4^+$ T cells to enable them to proliferate.

Accessory molecules for T cells

Several of the proteins expressed on the membrane of T cells and antigen presenting cells are involved in lymphocyte adhesion

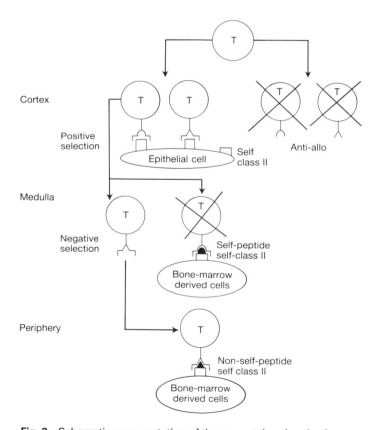

Fig. 2 Schematic representation of the current thymic selection theory. The first step involves positive selection of those autoreactive T cells which recognize self class II molecules on thymic epithelial cells. Alloreactive T cells are deleted. The second step occurs in the medulla and is characterized by the elimination of T cells binding with high affinity to self-peptides associated to self-class II molecules presented on bone marrow-derived cells. Thus, only T cells reactive with non-self-peptides associated with self-class II molecules are found in the periphery.

Table 4 Accessory molecules

Antigen presenting cell	T lymphocyte
LFA–3	CD2
Class I MHC	CD8
Class II MHC	CD4
ICAM–1	LFA–1

(Table 4). There is good evidence that these molecules are responsible for the cellular contact which is essential for regulatory and effector functions of T cells (reviewed in Owen and Lamb 1988). CD2 (also called lymphocyte function antigen-2 (LFA2)), and intercell adhesion molecule-1 (ICAM-1) molecules interact with LFA-3 and LFA-1, respectively. CD4 and CD8 are also adhesive molecules that recognize MHC class II and class I antigens. Several experiments clearly show the importance of such determinants: anti-LFA-1 antibodies, for example, abrogate T-cell mediated lysis. The role of such molecules in immunologically mediated nephritis has not yet been considered.

Immunoglobulins

Structure

Antibodies are unique bifunctional molecules which can bind antigen in one region of the molecule and initiate effector pathways through another region of the polypeptide chain.

Immunoglobulins are heterogeneous glycoproteins, 96 per cent polypeptide and 4 per cent carbohydrate, whose basic unit comprises four chains, two heavy (H) and two light (L) (Fig. 3). The two heavy and the two light chains of an immunoglobulin are identical, and all immunoglobulins can be represented as a $(H_2 L_2)_n$ polymer. The chains are linked through covalent (disulphide bridges) and non-covalent bonds. Enzymatic digestion by papain or pepsin produces Fab and F(ab)'$_2$ fragments respectively (Fig. 3).

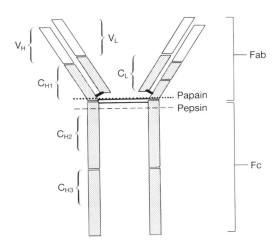

Fig. 3 The antibody molecule. Schematic representation of an IgG molecule, with its variable regions (V_H for the heavy chain and V_L for the light chain) and constant regions (C_H for the heavy chain and C_L for the light chain). The Fab and Fc fragments are shown, together with the cleavage sites of papain and pepsin.

Primary structure

Amino-acid sequencing of immunoglobulins shows that each chain is composed of a constant region which is identical in many different immunoglobulins, and a variable (V) region which is very heterogeneous and contains hypervariable regions. These account for the fine antigenic specificity of the molecule. This observation, made in 1965, was only fully explained in 1976 by the demonstration of different genes coding for a single polypeptide chain (Hozumi and Tonegawa 1976).

Secondary structure

Each heavy and light chain folds into globular regions, called domains, separated by short linear stretches of amino acids. All V regions have a similar appearance, irrespective of differences in their heavy and light chains and this secondary structure is very similar to that of MHC antigens. This lead to the theory of a superfamily of genes being involved in every aspect of immune regulation (Hood et al. 1985).

Tertiary structure

The three-dimensional structure of immunoglobulins has been studied by crystallography, which shows a close physical relationship between the corresponding V and C domains of the heavy and light chains (Fig. 1). Furthermore, the conformation of the V domains allows for a 'hole', bordered by the hypervariable regions, where the antigen makes contact with the antibody molecule. Residues from both the heavy and the light chain are involved in this interaction.

Isotypes

Antigenic differences in the constant regions of heavy chains define five classes of immunoglobulins: IgG, containing the γ heavy chain; IgM, containing the μ heavy chain; IgA, containing the α heavy chain; IgD, containing the δ heavy chain; and IgE, containing the ε heavy chain. Similarly, light chains are divided into two types, $\varkappa$ and λ. Every normal individual possesses all the isotypes of the species. Table 5 summarizes some of the properties of the different classes of immunoglobulins.

Allotypes

Unlike the isotypes, some antigenic determinants are expressed in some but not all individuals of a given species. These determinants, or allotypes, have a Mendelian pattern of inheritance, and most are borne on the constant regions of the heavy chains: they are thus restricted to a class of immunoglobulin. Three sets of allotypes are known in man: Gm (1)-(25) is localized on the γ chain, Am (1), (2) is localized on the α chain, and Km (1)-(3) is expressed on the $\varkappa$ chain.

Biological functions

Binding of antigen to the variable region of an immunoglobulin initiates a number of biological phenomena which vary depending on the type of the antibody. We will thus review separately the different classes of immunoglobulins and their biological properties.

IgG

IgG antibodies are composed of a monomer of the basic $H_2 L_2$ unit, and comprise about 75 per cent of the serum immunoglobulins. IgG can be divided into four subclasses, each of which has slightly different properties (Table 6). Maternal IgG antibodies cross the placenta and are readily found in the newborn, where they represent the major part of the immunoglobulins for

Table 5 Properties of different classes of immunoglobulins

Properties	IgG	IgM	IgA	IgE	IgD
H chain subclasses	G1, 2, 3, 4	M1, 2	A1, 2	–	–
Sedimentation coefficient	6 to 7	19	7 to 11	8	6 to 7
Molecular weight	150	900	160 to 400	190	180
Serum concentration (mg/ml)	12	1	2.5	0.003	0.03
Serum half-life (days)	23	5	6	2.5	3
Synthesis rate (mg/kg.day)	33	7	25	0.02	0.5
Complement fixation	Yes	Maximal	No	No	No
Antibacterial activity	Yes	Maximal	Yes	?	?
Antiviral activity	Yes	Yes	Maximal	?	?
Placental transfer	Yes	No	No	No	No

the first few weeks of life. IgG can activate the complement system, and macrophages and monocytes possess an IgG Fc receptor.

IgA
IgA represents 15 per cent of the total immunoglobulin in serum and is also found in the body secretions. Serum IgA is mainly monomeric and of the IgA_1 subclass. Secretory IgA is produced at a rate of 40 mg/kg.day, and is mainly dimeric, with a joining (J) piece linking the heavy chains: it usually belongs to the IgA_2 subclass. They are present in secretions and are the primary defence against mucosal infections.

IgM
Ten per cent of the total serum immunoglobulins is pentameric IgM, composed of five units of two heavy and two light chains, with an approximate molecular weight of 900 000 kDa. IgM is mainly seen in the early (primary) phase of the immune response, and most 'natural' autoantibodies belong to this class. Most membrane-bound immunoglobulins at the surface of B cells are also of the IgM class. They bind complement very efficiently.

IgD
Minute quantities of IgD are found in the serum, most on the surface of B cells. The biological activity is unknown, but IgD_2 may be involved in differentiation of B cells and development of clonal anergy.

IgE
These immunoglobulins have a unique role in mediating immediate hypersensitivity ('allergy'). The binding of their Fc region to mast cells leads to the release of pharmacologically active mediators, such as histamine and serotonin.

Table 6 Properties of IgG subclasses

Properties	IgG_1	IgG_2	IgG_3	IgG_4
Percentage of total IgG	60–70	15–20	4–8	2–6
Serum concentration (mg/ml)	5–10	2–6	0.5–1.5	0.1–0.8
Serum half-life (days)	23	23	11	21
Complement fixation	+++	++	++++	+/–
Monocyte receptor	+	–	+	–
Staph.–A binding	+	+	–	+

Molecular biology and genetics
The bifunctional nature of the immunoglobulin molecule, as well as the variability of the N-terminal end of the molecule and the conservation of the C-terminal end implies that more than one gene could code for a single polypeptide chain. This was demonstrated by Hosumi and Tonegawa in 1976.

Recombination events and the generation of diversity
The assembly of an immunoglobulin heavy or light chain necessitates genetic recombination which starts at the pre-B cell stage and always begins with the heavy chain genes. Once an effective recombination has occurred, recombination of the light chain genes takes place, and the mature plasma cell synthesizes and expresses at its membrane a complete IgM molecule. Cytokines induce another rearrangement, or switch of the heavy chain constant genes, resulting in the production of an immunoglobulin of the same antigenic specificity, but of a different isotype (reviewed in Alt et al. 1986).

Heavy chains
Four genes are involved in the construction of a heavy chain: V_H codes for variable region, D_H for the diversity segment, J_H for the joining region, and C_H for the constant part. In man, 300 V_H genes, four D_H genes, and six J_H genes have been identified. Sequence homology allows the V_H genes to be ordered in a number of families: these are well defined in the mouse but less so in man. All of the heavy chain genes are located on chromosome 14, but are separated by long stretches of DNA (Fig. 4).

The first recombination event leads to the formation of a $D_H J_H$ segment with deletion of the intervening portion of DNA. A second rearrangement leads to the formation of a $V_H D_H J_H$ segment coding for the entire variable region of the polypeptide chain. The genes coding for the constant region of the heavy chain lie further downstream, and it is during their transcription that an RNA coding for the $V_H D_H J_H C_H$ sequence is created. At this stage, the C_H used is always the $C\mu$, and thus the corresponding molecule is an IgM.

Light chains
Following recombination of the heavy chain genes, similar recombination events generate the light chain. Only three genes code for the light chain, namely V_L for the variable region, J_L for the joining region, and C_L for the constant region (Fig. 5). It is important to note that there are two sets of each genes, located on different chromosomes, one for the $\varkappa$ allotype (chromosome 2), the other for the λ allotype (chromosome 22). Thus, there is no allotype switching for the light chain.

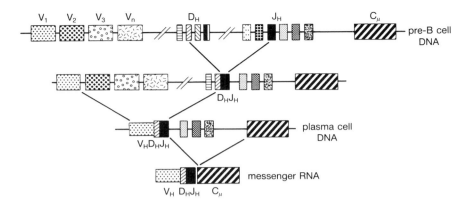

Fig. 4 The immunoglobulin heavy chain genes rearrangement (see text for details).

The mechanisms responsible for the stoichiometric production of heavy and light chains as well as for their assembly into a complete Ig molecule are still poorly understood.

Switching
A secondary rearrangement at the DNA level is necessary to produce IgG, IgA and IgE. As shown in Fig. 6, the corresponding constant region genes have been located downstream from the V, D, and J genes. The $C\mu$ gene lies closest to the rearranged $V_H D_H J_H$ complex; this explains how an IgM molecule can be transcribed without further rearrangement. The switch phenomenon corresponds to another rearrangement involving deletion of the intervening C region genes; secretion of an IgA molecule requires the deletion of all constant region genes located between the $V_H D_H J_H$ complex and the $C\alpha$ gene.

The generation of antibody diversity
The incredible diversity of immunoglobulin molecules results from two different processes: recombination between immunoglobulin genes and somatic mutations. If one accepts that any given D gene can recombine with any J and V gene, about 10^5 different heavy chains can be generated by recombination (300 V × 4 D × 6 J). One can probably multiply this number by 10 since during DJ joining some base pairs are added or subtracted through poorly understood mechanisms. For the light chains, about $1-2 \times 10^3$ different $\varkappa$ chains are possible. Therefore, if any light chain could combine with any heavy chain the number of different immunoglobulin molecules directly arising from the germ line is in excess of 10^8. However, there are only 10^6 lymphocytes present in any individual at a given time (Gearhart 1982). In addition, somatic mutations modify the antigenic specificity of the immunoglobulins. These mechanisms explain how an antibody response can be mounted against nearly all antigens presented to the immune system. The B-cell repertoire is therefore 'complete'.

Idiotypes
Definition
An idiotype is a unique antigenic determinant borne on the variable portion of an immunoglobulin of a given specificity in a single individual (Oudin and Michel 1963; Kunkel et al. 1963). The presence of certain idiotypes has been correlated with a particular antigenic specificity of the immunoglobulin on the one hand, and with transcription of a specific V_H gene or group of genes on the other (Monestier et al. 1986). Idiotypes can therefore be considered as serum markers of the clonality and genetic origin of immunoglobulins.

Idiotype network and the regulation of the immune response
Immunization of an animal with a given immunoglobulin (called Ab_1) recognizing an antigen leads to the synthesis of an anti-idiotypic antibody (called Ab_2) which binds specifically to the idiotype of Ab_1. Ab_2 can, in turn, induce the synthesis of a specific antibody (Ab_3) and so on (Fig. 7). Moreover, two different types of anti-idiotypes are found; $Ab_2\alpha$, which binds the idiotype of the Ab_1, and $Ab_2\beta$, whose idiotype is bound by Ab_1 (Jerne et al. 1982).

The idiotype network theory (Jerne 1974) postulates that anti-idiotypes modulate the synthesis of idiotypes and are themselves controlled by anti-anti-idiotypes. In many experimental models, administration of such anti-idiotypes produces profound changes in the immune response directed towards the antigen (reviewed in Zanetti et al. 1986a). Thus immunization with idiotype prior to the administration of thyroglobulin generates anti-idiotypic antibodies which abrogate the response to thyroglobulin (Zanetti et al. 1986b). The physiological and therapeutic relevance of such a network has yet to be fully explored. Anti-idiotypes have been shown to appear spontaneously in animals (Strosberg 1983) as well as in human beings (Abdou et al. 1981; Geha 1983), but the ability of such anti-idiotypes to down-regulate the production of the idiotype has yet to be proven.

Self recognition, tolerance, and autoimmunity (see also Chapter 4.6.1)
Self-recognition is a physiological process which can lead to a pathological situation termed 'autoimmunity'. At the beginning of the century, Ehrlich introduced the concept that the body normally does not react against its own constituents. He coined the term 'horror autotoxicus' meaning that self-recognition was

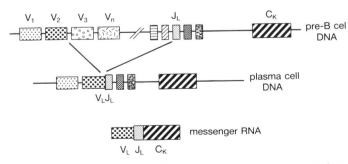

Fig. 5. The immunoglobulin light chain genes rearrangement (see text for details).

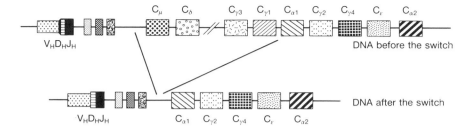

Fig. 6 The immunoglobulin switch mechanism (see text for details).

synonymous with autoimmune disease (Ehrlich and Morgenroth 1959). Since then, an ever-growing body of evidence has emerged to show that recognition of self moieties such as MHC molecules, classical self antigens (DNA, thyroglobulin) or V regions of antibodies is an essential phenomenon necessary for the economy of the immune system.

Normal human serum contains antibodies that react with self-antigens such as collagen, actin, and myosin (Guilbert *et al.* 1982). In neonatal mice, a number of B cells spontaneously produce antibodies reactive with DNA, actin, myosin, and thyroglobulin, in the absence of autoimmune disease (Dighiero *et al.* 1983; Glotz *et al.* 1988). Similarly, self-reactive T cell clones have been characterized in normal human beings and mice (Nagarkatti *et al.* 1985). Spontaneous anti-idiotype antibodies may be seen in normal human beings following immunization with tetanus toxoid (Geha 1983).

If self-reactive B and T cells exist, then active mechanisms must down-regulate the immune response toward self, preventing the occurrence of autoimmune disease. Such mechanisms seem to be very different for B and T cells, and these can now be explored using recombinant DNA technology. For example, mice have been transfected with a T-cell receptor that reacts with the minor histocompatibility complex antigen HY (Kisielow *et al.* 1988), which is expressed in male, but not female mice. Transfected females have normal thymic T-cell populations and their peripheral T cells proliferate in response to HY antigen. In contrast, transfected males have lower levels of mature T cells in the peripheral blood and are unable to mount a response against HY antigens. This shows that thymic deletion of autoreactive clones therefore plays an important role in the generation of T cell tolerance by minimizing the number of self-reactive mature T cells in the peripheral circulation.

Other mechanisms could contribute to tolerance. Although the majority of T cells that emerge from the thymus are either $CD4^+$ or $CD8^+$, a small percentage are double negative cells which possess the $\alpha\beta$ T-cell receptor and which could become autoreactive under certain circumstances. Clonal anergy of T cells may also occur; this has been shown for B cells. Autoreactive cells could also be controlled by regulatory mechanisms involving suppressor cells or anti-idiotypes. Lastly, accessory molecules and costimulatory factors from other cells are required for T cells to proliferate and tolerance may be due to a defect in such signals.

Clonal deletion does not appear to be an important factor in the generation of B-cell tolerance. This was demonstrated by an experiment in which double-transgenic mice were transfected with the hen egg lysozyme (*HEL*) gene and the rearranged anti-*HEL* immunoglobulin gene. High levels of anti-*HEL* B cells that were unable to secrete immunoglobulins were present in the circulation (Goodnow 1988), suggesting that clonal deletion of B cells does not occur and that down-regulation must be achieved by other means. Clonal anergy, lack of a costimulatory signal or suppression could be at play, possibly mediated by suppressor T cells, as suggested by studies of $HgCl_2$-induced autoimmunity in rats. In this model, injection of $HgCl_2$ to Brown Norway rats induces a T-cell dependent B-cell activation which results in an autoimmune disease characterized by anti-self antibodies such as those directed against the glomerular basement membrane or DNA (Hirsch *et al.* 1982), as well as the appearance of self-MHC reactive T cells (Pelletier 1988*b*). A defect of the suppressor T cells has been demonstrated in this model (Weening *et al.* 1981), and this could play a major role in the induction of autoimmunity.

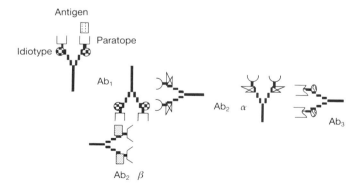

Fig. 7 Idiotypic network. The antibody reacting with the antigen is called Ab_1. Two different anti-idiotypic antibodies can be found, one reacting with the idiotype of Ab, called $Ab_2\ \alpha$, and another one whose idiotype is recognized by Ab_1, called $Ab_2\ \beta$. Each Ab_2 can in turn induce the formation of anti-anti-idiotypes, termed Ab_3, and so on . . .

Induction of autoimmune responses leading to glomerulopathies

Autoimmune glomerulonephritis is attributable to autoantibodies and/or effector autoreactive T cells that recognize glomerular autoantigens.

Autoimmune diseases can be classified in two ways: firstly, whether they occur spontaneously or are intentionally induced, or secondly by whether they are organ specific or are systemic diseases. These classifications also apply to autoimmune glomerulonephritides (Table 7).

Glomerulus-specific autoimmunity
Antiglomerular basement membrane-mediated glomerulonephritis

Glomerulonephritis caused by autoantibodies directed against the glomerular basement membrane was one of the first autoimmune diseases to be studied (reviewed in Wilson and Dixon 1986). The first model was described by Steblay (1962) who immunized sheep with heterologous or homologous glomerular basement membrane. Severe autoimmune crescentic glomerulonephritis developed in some animals, causing renal failure and death.

More recently, several models have been developed in the rat, and these have allowed more precise studies of the mechanisms involved and of the determinants of susceptibility to this disease, which varies greatly with different strains. The genetic control of autoantibody production has been shown to depend upon several genes, one of which is MHC-linked (Stuffers-Heimann et al. 1979; Sado et al. 1984). Some strains, such as Brown-Norway rats produce autoantibodies directed against the glomerular basement membrane following immunization with bovine collagenase-digested glomerular basement membrane, without the development of either proteinuria or crescentic glomerulonephritis. In contrast, WKY/NCry rats immunized with homologous or isologous glomerular basement membrane and complete Freund's complete adjuvant produce autoantibodies which are deposited along the glomerular basement membrane, and exhibit significant proteinuria and a focal proliferative glomerulonephritis. In a study of the antibody response of various strains of rats to homologous or isologous collagenase-digested glomerular basement membrane, Pusey et al. (1991) found that Brown-Norway rats produce circulating antiglomerular basement membrane autoantibodies, even when adjuvant was not administered. The magnitude of the response was higher in rats immunized with homologous antigen and CFA, and these animals developed albuminuria. A proportion also had focal proliferative glomerulonephritis with segmental lesions, and pulmonary lesions were also apparent in some animals. The other strains studied (PVG/c, DA, LEW, WAG) also produced autoantibodies following immunization with homologous glomerular basement membrane and complete Freund's adjuvant; PVG/c and DA rats produced higher amounts of antibody and kidney-bound antibodies were only found in those two strains. The autoantibody response was transient in all but Brown-Norway rats: in this strain it was sustained for 12 months. Neither the autoantigen nor the mechanism of the induction of autoimmunity in this model are known.

Brown-Norway rats are extremely prone to develop autoantibodies directed against glomerular basement membrane, when compared to other strains of rats. Such antibodies are found not only after immunization with the specific antigen but also (see below) when T cell-dependent polyclonal activation of B cells occurs following exposure to drugs or toxins, or in chronic graft-versus-host disease. This suggests a low level of tolerance to native glomerular basement membrane antigens in these animals. It is tempting to explain the enhanced susceptibility of DR2 positive patients to antiglomerular basement membrane nephritis by a similar lack of tolerance (Rees et al. 1978). Whether the autoantigen has to be modified (by collagenase in the experimental situation or by environmental agents in humans) for the immune response to occur remains to be shown, although this is unlikely in the chronic graft-versus-host disease model.

In all of these models the autoimmune disorder is thought to be mediated by antibody, and although a number of arguments support this view, a cell-mediated effect cannot be excluded especially as effector T cells are important in the generation of heterologous nephrotoxic nephritis (Tipping et al. 1985). Chickens immunized with heterologous glomerular basement membrane develop a classical autoimmune glomerulonephritis with antibodies deposited along the glomerular basement membrane (Bolton et al. 1984). Bursectomized chickens, which cannot mount antibody responses, still develop autoimmune glomerulonephritis, despite the absence of antibodies against the glomerular basement membrane. Mononuclear cells infiltrate the glomeruli of these birds and the disease can be transferred to naive recipients by mononuclear cells from immunized animals (Bolton et al. 1988), indicating that T cells directed at the glomerular basement membrane are responsible for this form of autoimmune glomerulonephritis.

Heymann's nephritis

This form of autoimmune membranous glomerulonephritis occurs in susceptible strains of rats (LEW or PVG/c) following immunization with a rat renal tubular antigen and complete Freund's adjuvant (Heymann et al. 1959) and provides a convincing model for membranous nephropathy. The crude antigenic fraction originally used, called FX1A, has been shown to contain the nephritogenic antigen, a 330 kDa glycoprotein (gp 330) which is expressed on both the brush border of convoluted proximal tubules and the epithelial cells of the glomerular capillary wall (Kerjaschki and Farquhar 1982; Chatelet et al. 1986). The gp 330 antigen is concentrated into the clathrin coated pits that are responsible for endocytosis, where it is irregularly distributed at the cell surface. Fixation of free circulating autoantibodies on this discontinuously distributed autoantigen leads to the formation of *in-situ* immune complexes which give rise to the typical granular IgG deposits (Couser et al. 1978; Van Damme et

Table 7 Autoimmune glomerulonephritis

	Organ-specific	Systemic
Spontaneous	Membranous glomerulonephritis (rabbit)	Lupus nephritis
Induced	Anti-GBM Heymann's nephritis (and related)	Lipopolysaccharide (Parasites) Allogeneic reactions (Graft-versus-host disease—transplantation tolerance) Toxic-induced

al. 1978) along the capillary wall observed by immunofluorescence. Other antigen–antibody complexes may also participate in the pathogenesis of the disease, including gp 90 (dipeptidyl peptidase IV) or as yet unknown basement membrane components and their corresponding antibodies (Verroust et al. 1987; Hogendoorn et al. 1988).

The mechanisms responsible for the induction and regulation of Heymann's nephritis have not been studied as extensively as the target antigen. Antibody response to FX1A plus adjuvant is T-cell dependent (Cheng et al. 1988) and requires the presence of antigen-specific $CD4^+$ T cells (of the helper/inducer subset). Neonatal thymectomy, or adult thymectomy associated with irradiation and reconstitution with bone marrow B cells, abrogates the antibody response and prevents the occurrence of the disease (Cheng et al. 1988); Heymann's nephritis can also be induced in T-cell depleted adult animals following reconstitution with normal syngeneic $CD4^+$ T cells (Cheng et al. 1988). The immune response, as assessed by the detection of autoreactive B cells, is limited to the lymph node draining the site of immunization (De Heer et al. 1985), where $CD4^+$ T cells are likely to be found. Such experiments show that, at least in some strains, autoreactive T and B cells are present and can be triggered following appropriate immunization. They also show that tolerance is broken only locally.

The cellular mechanisms responsible for down-regulation of the autoimmune response have also been studied in two different situations. The amount of autoantibody produced by B cells in the draining lymph node decreases after the 10th week in rats with active Heymann's nephritis, but the level of circulating autoantibodies does not decrease until later. De Heer et al. (1986) have shown that $CD8^+$ T cells appear in the spleen of rats with active Heymann's nephritis from week 12, and that transfer of these cells into syngeneic recipients markedly depresses the antibody response to subsequent challenge with FX1A and adjuvant. This is consistent with the increased renal IgG deposits seen following splenectomy. Interestingly, transfer of $CD4^+$ T cells from the spleen of nephritic animals to normal rats increases the immune response to subsequent immunization (De Heer et al. 1986). These experiments clearly demonstrate the presence of autoreactive B cells and $CD4^+$ helper T cells in the draining lymph nodes of rats after immunization, and antigen-specific $CD4^+$ helper T cells and $CD8^+$ suppressor/cytotoxic T cells in the spleen. These may enhance or down-regulate autoimmunity, respectively. The fact that adult thymectomy enhances autoantibody production has been considered to be due to depletion of short-lived suppressor T cells.

Immunization with high doses of antigen plus incomplete Freund's adjuvant induces tolerance in susceptible LEW and PVG/c rats (IFA) (Litwin et al. 1979; De Heer et al. 1985; Cheng et al. 1988). This tolerance can be transferred into syngeneic recipients by lymph node cells. Splenic $CD8^+$ T cells from tolerant animals also transfer resistance into syngeneic recipients subsequently challenged while $CD8^-$ T cells enhanced autoimmunity (De Heer et al. 1985; Cheng et al. 1988). The role of $CD8^+$ T cells in the resistance of Brown-Norway or DA rats to immunization with antigen has not been studied: in Brown-Norway rats, resistance may be overcome by immunization with the appropriate adjuvant (Stenglein et al. 1978). The effect of in vivo $CD8^+$ T cell depletion remains an open question.

T-lymphocyte subsets thus play an important role in the regulation of Heymann's nephritis. $CD4^+$ helper T cells are preferentially triggered following immunization with complete Freund's adjuvant and $CD8^+$ suppressor/cytotoxic T cells are produced following immunization with incomplete adjuvant. $CD8^+$ T cells are also involved in the down-regulation of the active disease. Although the role of $CD8^+$ T cells in autoimmunity has recently been questioned, they clearly play a role in Heymann's nephritis. Generation of gp 330-specific T-cell clones may help the understanding of the cellular basis of this important autoimmune disorder.

Autoimmune glomerulonephritis in the context of polyclonal activation

In the situations described above, antigen-specific T- and B-cell clones are triggered selectively. In other situations numerous T- and/or B-cell clones are activated non-selectively, leading to a polyclonal activation (reviewed in Goldman et al. 1988b) which will produce an autoimmune glomerulonephritis if some of the stimulated clones have kidney-specific reactivity.

Autoimmune glomerulonephritis induced by bacterial products

Bacterial lipopolysaccharide is composed of two polysaccharidic structures and of the lipid A region. The latter probably binds to a receptor on B lymphocytes and is responsible for polyclonal activation of B cells. Bacterial lipopolysaccharide triggers the differentiation of between 10 and 50 per cent of all resting normal B cells into antibody-producing cells with a wide variety of specificities. Antibodies to DNA and rheumatoid factor, as well as antibodies to exogenous antigens and haptens, are produced and the total serum immunoglobulin level (mainly IgM) increases. Some other non-bacterial agents have the same effect (Goldman et al. 1988b).

Injection of these B-cell mitogens alone induces glomerular deposition of immunoglobulins (Izui et al. 1977). In mice, exposure to lipopolysaccharide induces prominent glomerular IgM and IgG deposits in the mesangial areas and in the capillary loops, which are associated with slight glomerular lesions at the light microscope level. About 40 per cent of the immunoglobulins eluted from such kidneys are directed against DNA (Izui et al. 1977). DNA is released following lipopolysaccharide injections (Fournie et al. 1974), this DNA may combine with anti-DNA antibodies to form circulating immune complexes that can be trapped in the kidney. Alternatively, DNA may bind to collagen in the glomerular basement membrane, and circulating antibodies then become fixed to this 'planted' antigen (Izui et al. 1976).

Since only 40 per cent of the kidney-bound immunoglobulins are DNA-specific, other antibodies must be involved. Rheumatoid factors may combine with the deposited anti-DNA antibodies, but many other systems could also be involved: antiphosphoryl choline antibodies bearing a particular idiotype (T15) are also produced, along with antibodies to the T15 idiotype. Such antibodies have been detected in the kidney (Goldman et al. 1982).

Experimental exposure of mice to lipopolysaccharide has demonstrated that strains lacking the lipopolysaccharide receptor do not respond, while the autoimmune disease in lupus-prone mice is enhanced following lipopolysaccharide injections (Hang et al. 1983).

Autoimmune glomerulonephritis induced by parasites

The association between parasitic infections and glomerulonephritis is well known in man (Goldman et al. 1988b), although the mechanisms involved are poorly understood. Although the induction of a specific immune response is important for parasite control, hyperimmunoglobulinaemia is a common feature of parasitic infections and suggests that polyclonal activation occurs. Evidence from studies in mice supports this, and a B-cell mitogen has been recovered from *Plasmodium* spp. (Goldman *et al.* 1988*b*). Animals infected with schistosomes (Goldman *et al.* 1988*b*) or trypanosomes (Rose *et al.* 1982; Bruijn *et al.* 1987) may also develop glomerulonephritis, and autoantibodies directed against DNA, rheumatoid factor, and laminin have been described in them. This again strongly suggests that glomerulonephritis associated with parasitic infection is at least in part, due to polyclonal activation of B cells.

Autoimmune glomerulonephritis induced by allogeneic reactions

Graft-versus-host disease

A graft-versus-host reaction occurs when T cells from a mouse or rat of strain X are transferred into a semi-allogeneic (X × Y) F_1 hybrid (Gleichmann *et al.* 1984). The reaction is acute and lethal if the parental strains X and Y differ at both class I and class II MHC antigens, whereas a chronic reaction is induced if they only differ at class II antigens.

Acute graft-versus-host disease is caused by the development of alloreactive T cells from the parental (X) strain which recognize Y class II or class I antigens in the F_1 hybrid. These anti-class II T cells induce the proliferation of parental cytotoxic anti-class I T cells that will ultimately destroy all of the F_1 hybrid cells leading to bone marrow aplasia, atrophy of the lymphoid organs, severe skin and gastrointestinal lesions, and finally death.

Chronic graft-versus-host disease is seen in three different situations (Gleichmann *et al.* 1984; Via and Shearer 1988). First, when class I MHC antigens are compatible and cytotoxic T cells are not involved, T cells are directed against the allogeneic class II antigens and polyclonally activate the class II alloantigen-bearing B cells. Autoantigens are thought to be very important in B-cell stimulation and only antigens bearing repetitive determinants appear to be able to stimulate B cells (Gleichmann *et al.* 1984). The various autoantibodies produced following polyclonal stimulation are responsible for a lupus-like syndrome which includes autoimmune glomerulonephritis with granular IgG deposits along the glomerular capillary walls (Goldman *et al.* 1983). Anti-DNA antibodies have been eluted from the glomeruli of animals with this type of graft-versus-host disease. Secondly, chronic graft-versus-host disease may also be observed between some strain combinations with incompatibility at class I and class II antigens in spite of depletion of parental donor $CD8^+$ T cells by monoclonal antibodies. This leaves only $CD4^+$ T cells from which the anti-class II alloantigen T cell clones will be generated. Lastly, a similar situation is seen when T cells from the parent strains are constitutively deficient in $CD8^+$ T cells.

Transplantation tolerance model

Mice from a given strain (Y) can become tolerant to alloantigens of another strain (X) if they are injected at birth with spleen cells from (X × Y) F_1 hybrids (Goldman *et al.* 1983; Luzuy *et al.* 1986). Such mice do not reject skin grafts from X parents and do not generate cytotoxic T cells against the X alloantigens. The transferred F_1 T and B cells persist in the recipient for a long time, and these tolerant mice become chimaeric (Luzuy *et al.* 1986). However, T cells directed against class II alloantigens do not become tolerant: these are able to stimulate B cells from the F_1 hybrid (Abramowicz *et al.* 1987) to produce autoantibodies, some of which are deposited in the glomeruli of the host and cause immune complex glomerulonephritis with extramembranous deposits. The specificity of the kidney-bound antibodies has been only partly characterized and contains rheumatoid factor activity. It is interesting that the autoantibodies produced are of the IgG_1 isotype and that total serum IgE level increases (Goldman *et al.* 1988*a*). This suggests a role for TH2 cells ($CD4^+$ T cells that produce IL-4 and IL-5) because IL-4 is the cytokine responsible for IgE and IgG_1 production (Mosmann and Coffman 1989). There is also additional evidence for a role of TH2 cells and IL-4 in this phenomenon. An increase in MHC class II antigen expression by F_1 B cells and parental T cells is able to increase expression of class II antigen on normal F_1 B cells (Abramowicz *et al.* 1990). This effect is known to depend upon IL-4, and is inhibited by incubation with monoclonal against this interleukin. Treatment of mice with such monoclonal antibody completely abrogates the increase in IgE and IgG_1 production. Interestingly, other isotypes (IgG_{2a}, IgG_3) are still produced, suggesting that TH1 cells were now involved and implicating a reciprocal control of TH1 cells by TH2 cells (M. Goldman *et al.* personal communication).

Drug-induced autoimmune glomerulonephritis

Although many drugs are known to induce autoimmune glomerulonephritides in humans (reviewed in Fillastre *et al.* 1988), the mechanisms responsible have yet to be defined. Two of the most plausible hypotheses are either that drugs act as haptens fixed on glomerular antigens or, alternatively, they release or modify autoantigens. Drugs such as gold salts and D-penicillamine also induce non-renal autoimmune disorders (Fillastre *et al.* 1988), indicating that at least some drugs are able to perturb the immune system.

Extensive experimental studies have been performed in the Brown-Norway rat, principally using $HgCl_2$ as the inducing agent (Pelletier *et al.* 1987*a*), but there are some studies using D-penicillamine and gold salts (Donker *et al.* 1984; Tournade *et al.* 1989). The mechanisms are probably quite similar to those responsible for chronic graft-versus-host disease (Tournade *et al.* 1990*a*).

$HgCl_2$-induced autoimmunity in Brown-Norway rats

Non-toxic amounts (100 μg/100 g body weight) of $HgCl_2$ induce autoimmune disease in Brown-Norway rats, rabbits, and in H-2 susceptible B10 S and A.SW strains of mice (H-2) (the H-2 complex in mice is the equivalent of the human HLA system). Manifestations of disease appear in Brown-Norway rats in the second week of treatment and peak during the third and fourth weeks; they include lymphoproliferation, mainly due to an increase in the number of $CD4^+$ T cells and B cells in the spleen and lymph nodes (Pelletier *et al.* 1988*a*), and an increase in the total serum immunoglobulin concentration, mainly due to an increase in the IgE, IgG_1 and $IgG2_b$ isotypes (Pelletier *et al.* 1988*a*). A number of autoantibodies that recognize glomerular basement membrane components, including laminin, collagen IV, and fibronectin are produced (Bellon *et al* 1982; Fukatsu *et al.* 1987) as well as antibodies against nuclear antigens, IgG, collagen II, and thyroglobulin (Hirsch *et al.* 1982; Pusey *et al.* 1990). Antibodies against non-self autoantigens (trinitrophenyl, sheep red blood cells) are also produced (Hirsch *et al.* 1982) and the level of circulating natural polyreactive antibodies is increased.

There is a biphasic autoimmune glomerulonephritis; in the

first phase, IgG is deposited along the glomerular basement membrane in a linear pattern from day 8 (Sapin et al. 1977). From the third week, a typical membranous glomerulonephritis is also seen (Druet et al. 1978). The mechanisms of formation of subepithelial deposits is still unknown, but immunoglobulins eluted from kidneys with predominant granular deposits bind in a linear pattern when incubated on normal kidney cryostat sections (Druet et al. 1978). Hypotheses proposed to explain the formation of granular deposits include the deposition of circulating immune complexes containing glomerular basement membrane antigen and antibody, or the rearrangement of linear IgG deposits due to the deposition of anti-idiotypic antibodies. It is also possible, as suggested by Aten et al. (1988b), that antilaminin antibodies react against glomerular epithelial cells. Other autoimmune abnormalities (Sjögren's syndrome, dermatitis) have been reported (Aten et al. 1988b).

These data suggest that $HgCl_2$ induces a polyclonal activation of B cells in Brown-Norway rats, and this has been supported by the observation that T cells from $HgCl_2$-injected rats are able to collaborate with normal B cells to induce antibody production in vitro (Hirsch et al. 1982). In-vivo experiments have shown that T cells are required for this polyclonal activation to occur (Pelletier et al. 1987b). Other experiments have shown that T cells from mercury-injected Brown-Norway rats are able to recognize B cells from both mercury-injected rats and from normal syngeneic rats, and that this interaction is blocked by an anticlass II monoclonal antibody. These self-reactive T cells are common; about 1/5000 at the height of the disease (Rossert et al. 1988). Finally T cells from rats exposed to $HgCl_2$ transfer autoimmunity to normal syngeneic rats that have been previously depleted of $CD8^+$ T cells (Pelletier et al. 1988b).

These experiments suggest that $HgCl_2$ promotes the emergence of self reactive, anticlass II, $CD4^+$ T cells. Recent data have demonstrated unequivocally that the vast majority of self-reactive T cells are clonally deleted in the thymus (Kappler et al. 1987; Kisielow et al. 1988), but there are other experiments which show that some T-cell receptor $\alpha\beta^+$, double negative ($CD4^-$, $CD8^-$) T cells may escape thymic deletion and, under certain circumstances may become $CD4^+$ and self reactive (deTalance et al. 1986). The exact mechanisms of the action of mercury remain to be elucidated. Transfer experiments suggest that autoreactive T cells from mercury-injected rats transferred into normal recipients provoke the appearance of $CD8^+$ T cells, which may prevent expansion of these autoreactive T cells (Pelletier et al. 1988b). The nature of these cells (cytotoxic ?, anti-idiotypic ?) is not yet known, but such $CD8^+$T cells cannot be detected in mercury-injected rats, supporting the idea that mercury acts by inhibiting such cells. It has been shown previously that $HgCl_2$ induces a defect in T suppressor cells in PVG/c rats (Weening et al. 1981).

Similarities between chronic graft-versus-host disease and mercury-induced autoimmunity are striking (Fig. 8); anticlass II T cells are produced in both situations, and a defect at the $CD8^+$ T-cell level seems to be required. To confirm these similarities normal Brown-Norway T cells were transferred into (LEW × Brown-Norway) F_1 hybrids. The F_1 hybrids produced antilaminin and antinuclear autoantibodies, total serum IgE level increased, and the rats developed autoimmune glomerulonephritis characterized by linear IgG deposits along the glomerular capillary walls (Tournade et al. 1990a). More interestingly, experiments using monoclonal antilaminin antibodies and rabbit anti-idiotypic antibodies obtained in the mercury model have

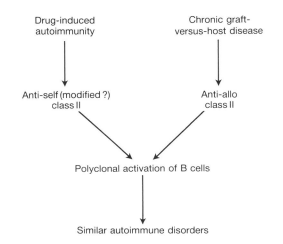

Fig. 8 Mechanisms of drug-induced autoimmunity and chronic graft-versus-host reaction.

demonstrated that antilaminin antibodies share a cross-reactive idiotype which is also present on circulating IgG and on kidney-bound antibodies in chronic graft-versus-host disease (Guéry et al. 1990a). These data strongly support the idea that both of these autoimmune diseases are due to anticlass II T cells which activate the same B cell clones.

The genetic control of susceptibility has also been studied in this model (Druet et al. 1977; Sapin et al. 1984). Several strains of rats with the RTI-l or −u haplotype at the major histocompatibility complex are resistant (Brown-Norway rats bear the RT1−n haplotype). Classical breeding experiments between resistant (LEW) rats and Brown-Norway rats, as well as between congenic rats demonstrated that both MHC and non-MHC linked genes are involved; about three genes have been implicated in the control of susceptibility (Druet et al. 1982).

About 50 per cent of the rats die during the third and fourth weeks of the disease. The remaining rats recover and all of the autoimmune abnormalities progressively disappear. After 2 months, IgE serum levels return to normal and most autoantibodies can no longer be detected. At least two possible mechanisms for this have been proposed. Several arguments favour a role for T-suppressor (Ts) cells: firstly, rats injected with low doses of $HgCl_2$ do not develop autoimmunity and become resistant to further challenge with high doses of $HgCl_2$ (Pusey et al. 1983); secondly, transfer of spleen cells from convalescent Brown-Norway rats to naive syngeneic recipients attenuates the severity of $HgCl_2$-induced disease (Bowman et al. 1984); thirdly, the number of $CD8^+$ T cells in the blood increases significantly during the regulation phase, even though they were reduced during the induction phase, (Bowman et al. 1987). However treatment with an anti-CD8 monoclonal antibody does not significantly affect the induction or regulation phases (Pelletier et al. 1990). These observations are compatible with an initial defect in suppressor T cells, but they also show that $CD8^+$ T cells alone are not responsible for the regulation phase.

Anti-idiotypic antibodies have been proposed as another mechanism to explain the regulation phase. Chalopin et al. (1984) obtained indirect evidence for anti-idiotypic autoantibodies from the capacity of serum from $HgCl_2$-injected animals to enhance a plaque forming cell assay for antiglomerular basement membrane antibodies; however other factors (such as immune complexes containing glomerular basement membrane antigens and antibodies) could also explain this effect. A monoclonal anti-idiotypic autoantibody which has recently been obtained from a Brown-Norway rat injected with $HgCl_2$ has recently been shown to recognize monoclonal antilaminin antibodies produced in the mercury model (Guery et al. 1990c). Although this confirms that such antibodies are synthesized during the disease, their regulatory and/or pathogenic effect is unknown.

Cytokines may also be produced during the two phases of the disease: IL-4 during the former phase and γ-interferon during the latter. It has been shown that these cytokines may down regulate TH1 and TH2 cells respectively (Mosmann and Coffman 1989).

$HgCl_2$-induced autoimmunity in other species
New Zealand rabbits injected with $HgCl_2$ develop an autoimmune glomerulonephritis similar to that seen in Brown-Norway rats (Roman-Franco et al. 1978). Various strains of mice have been tested, and those with the $H-2$ haplotype (B10 S, A.SW) seem to be particularly susceptible to the induction of autoimmune glomerulonephritis, which appears with several other autoimmune manifestations (Robinson et al. 1986; Pietsch et al. 1989). Anti-S3 autoantibodies can be recovered from kidney eluates of these mice, and are similar to those found in patients with scleroderma. As in Brown-Norway rats, total serum IgE level increases, and recent results have demonstrated that IL-4 plays a major role in this. Treatment of B10 S mice with an anti-IL-4 monoclonal antibody abrogates the increase in total serum IgE level and the production of autoantibodies bearing the IgG_1 isotype (E. Gleichmann, personal communication).

Other drug-induced autoimmune glomerulonephritis
Several models of autoimmune glomerulonephritis have been described following administration of D-penicillamine, gold salts and other drugs, but the mechanisms of induction have not been studied. Several recent experiments suggest analogies between autoimmune glomerulonephritis associated with D-penicillamine (Donker et al. 1984; Tournade et al. 1990b) or gold (Guéry et al. 1990b) exposure in Brown-Norway rats. Most of the abnormalities encountered in Hg-induced autoimmunity were also observed in these models, but the disease was less severe than that induced by mercury. The cross-reactive idiotype described in the mercury model (Guéry et al. 1990a) and in chronic graft-versus-host disease has also been found in D-penicillamine and gold-induced disease (Guéry et al. 1990b); anti-class II T cells have also been observed (Fig. 8). Lewis rats do not develop autoimmune glomerulonephritis after D-penicillamine (Donker et al. 1984) or gold treatment (Tournade in preparation). These data strongly support the idea of a common mechanism of drug-induced autoimmunity in all three models and show that the Brown-Norway rat is the most appropriate strain in which to test the potential of drugs to induce systemic autoimmunity.

Murine lupus nephritis
There are good recent reviews of the spontaneous autoimmune disease that develops in female (NZB × NZW) F_1 hybrids, in MRL 1pr/1pr, or in male B × SB mice (Theofilopoulos and Dixon 1985; Theofilopoulos et al. 1989). These models, and the mechanisms leading to the in-situ formation and/or to the deposition of immune complexes responsible for these diseases will not be described.

There is a general agreement that the major abnormality is a generalized polyclonal B-cell activation (Theofilopoulos and Dixon 1985) but clearly this does not rule out a role for autoantigens in causing a switch from IgM to IgG or IgA production (Dziarski 1988). It is also recognized that T cells are required for the polyclonal activation to occur and that both genetic and non-genetic factors play a major role in inducing or accelerating autoimmunity.

More recent studies have investigated the molecular genetics of the B- and T-cell antigen receptors in these models (Dziarski 1988). The immunoglobulin germ line genes in normal and lupus-prone mice do not appear to differ, and no peculiar V_H gene or gene family seems to be responsible for autoantibody production (Theofilopoulos et al. 1986). In addition, no differences in the degree of somatic mutation of immunoglobulin genes have been defined. There does not seem to be any special T cell receptor gene or family of genes associated with autoimmunity, in spite of the unusual nature of the NZW T cell receptor β chain (Kotzin et al. 1985; Kotzin and Palmer 1987), and the T-cell receptor repertoire is similar to that of normal mice.

Self-reactive T cells appear to be deleted normally in the thymuses of (NZB × NZW) F_1 mice and of MRL 1pr/1pr mice (Theofilopoulos et al. 1986). A role for T-cell receptor $\alpha\beta^+$ double negative T cells in these models of autoimmunity has recently been proposed (Theofilopoulos et al. 1989). These cells form a very low percentage (5 per cent) of T cells in normal BALB/c mice and may become $CD4^+$ and anti-self class II reactive following stimulation with concanavalin A (Morisset et al. 1988). The lymphoproliferation observed in MRL 1pr/1pr mice mainly involves double negative T cells, which contribute to the helper activity for anti-DNA production in (NZB × NZW) F_1 mice. It is possible, therefore, that TcR $\alpha\beta^+$, $CD4^-$, $CD8^-$ T cells represent potentially autoreactive T cells that have escaped clonal elimination in the thymus and may become autoreactive under different situations. It is interesting that $CD4^+$ anticlass II T cells have been obtained from MRL 1pr/1pr (Rosenberg et al. 1983) and from (NZB × NZW) F_1 mice (Ando et al. 1987). The role of double negative T cells, their relationship to anticlass II T cells, and their role in autoimmunity require further studies. The immunopathogenesis of human lupus is discussed in Chapter 4.6.1.

Glomerular localization of immune reactants

Whatever the mechanism(s) responsible for the induction and regulation of the humoral and the cellular immune responses, the immune reactants that appear as a consequence may have a deleterious effect on the glomerulus. We will now consider how these reagents may become localized within the glomerulus (Table 8).

Antibody-mediated glomerulonephritis
Antibodies become deposited in the glomerulus either because they circulate in the form of immune complexes and are passively entrapped within the glomerulus, or because they bind to targets fixed in the glomerulus. Traditionally, the first mechanism is considered to be of major importance in the development of serum sickness models.

Circulating immune complexes
Serum sickness
Large amounts of heterologous proteins such as bovine serum albumin injected intravenously into rabbits can cause serum

Table 8 Glomerular localization of immune reactants

Antibody-mediated
  Circulating immune complexes:
    Serum sickness (?), IgA nephropathy
  Free circulating antibodies specific for:
    Glomerular antigens (GBM, epithelial cell: gp 330, endothelial cell, mesangial cell)
    Planted glomerular antigens
      Immune reactants
      Cationic antigens (native, modified)
      Lectins
      DNA (?)
Cell-mediated
  GBM specific (chicken)
  TNP specific

sickness (Wilson and Dixon 1986). Acute or chronic forms of the disease can be induced depending upon the number of injections, and circulating immune complexes can be detected in both models. These complexes circulate for a long period of time when antigen is present in excess, and granular deposits consisting of rabbit IgG, bovine serum albumin and complement components can be observed by immunofluorescence along the glomerular capillary walls and also in mesangial areas. Electron-dense deposits are located on the subepithelial aspect of the glomerular basement membrane and in the mesangial areas. Different forms of glomerulonephritis (proliferative, crescentic, membranous, or mesangioproliferative) may develop, depending upon the experimental conditions. Besides the antigen/antibody ratio, factors such as size of the immune complexes, antibody avidity, blood flow, blood pressure, complement and ability of the mononuclear-phagocyte system to clear immune complexes have all been shown to influence immune complex deposition. However, in spite of numerous attempts, it has been extremely difficult to obtain subepithelial deposits following injection of preformed immune complexes (Couser 1985).

Formation of immune complexes in the subepithelial space mainly depends upon the electric charge of the antigen or antibody. This was demonstrated following the crucial observation that several alternate perfusions of bovine serum albumin and antibovine serum albumin could lead to subepithelial deposit formation (Fleuren *et al.* 1980). Numerous studies have since established that extramembranous deposits are readily observed when cationic bovine serum albumin, or cationic immune complexes are used; it is now widely accepted that these appear as a consequence of fixation of cationic bovine serum albumin on the anionic sites of the glomerular basement membrane, where it acts as a planted antigen (Border *et al.* 1982). It is of interest that cationic antibody to bovine serum albumin may bind first to these anionic sites; circulating bovine serum albumin then binds to the planted antibody. Anionic antigens usually induce the formation of cationic antibodies and vice versa (Couser 1985).

Although it is now clear that subepithelial deposits rarely occur as a consequence of circulating immune complex deposition, deposits at mesangial sites, and probably also in a subendothelial position, can develop from trapping of circulating immune complexes. This has been shown following passive injection of preformed immune complexes or of heat-aggregated IgG. However this usually does not result in glomerular inflammation (Couser 1985).

IgA nephropathy

Berger's disease represents the most common form of glomerulonephritis in humans. Although its pathogenesis is not yet clear, there is good evidence that it results from polyclonal activation of IgA-secreting B cells leading to the deposition of immune complexes containing IgA within the mesangial area.

Several experimental models have been described (Rifai 1987); in most of them, IgA deposition is not associated with proteinuria or haematuria. Injection of immune complexes containing dinitrophenyl-specific mouse monoclonal polymeric IgA complexed to dinitrophenyl-bovine serum albumin results in glomerular IgA deposits within the mesangium and transient haematuria. Such deposits are also seen when BABL/c mice bearing MOPC–315 myeloma, which produces monoclonal anti-dinitrophenyl IgA are injected with dinitrophenyl-bovine serum albumin. Passive injection of a dextran-specific monoclonal IgA antibody complexed with dextrans with different sizes also results in glomerular mesangial IgA deposits (Rifai 1987).

Interestingly, glomerular IgA deposits have also been observed after oral immunization (Emancipator *et al.* 1983), indicating a role for the mucosal IgA immune system, as in the human situation. These mesangial IgA deposits contain the J chain, and are, at least in part, made of polymeric IgA. Monomeric IgA can also bind provided that immune complexes containing polymeric IgA have previously been deposited in the mesangium (Chen *et al.* 1988).

Oral immunization of mice with lactalbumin, combined with blockade of the reticuloendothelial system leads to the co-deposition of IgA, IgG, and C3 in the mesangium, occasionally with proteinuria (Sato *et al.* 1986).

Several models have been described in mice or rats in which bile duct ligation or cirrhosis of the liver leads to an increase in total serum IgA level and to glomerular IgA deposits (Rifai 1987). These models again suggest that circulating immune complexes containing polymeric IgA may become deposited within the mesangium, but they are of little relevance to the understanding of IgA nephropathy in humans with cirrhosis of the liver since the biliary transport of IgA differs in rats and man.

Finally, ddY mice spontaneously develop mesangial deposits of IgA and IgG which are probably related to the presence of gp70–anti-gp70 immune complexes (Imai *et al.* 1985). The presence of the gp70 antigen in these mice is related to infection by the murine leukaemia virus. Interestingly these mice develop a proliferative glomerulonephritis with occasional crescents which mimics human IgA nephropathy.

3.2 Immune mechanisms of glomerular damage

Free circulating antibodies

There is now good evidence to suggest that most experimental glomerulonephritides are due to circulating antibodies that become fixed on glomerular structural or planted antigens.

Structural glomerular basement membrane antigens

Since the classical descriptions by Lindemann (1900), and later by Masugi (1934), numerous models of so-called nephrotoxic nephritides have been reported.

In the heterologous form, antibodies against, for example, rat glomerular basement membrane are produced in rabbits and, when injected intravenously into rats, become fixed, together with complement, in a linear pattern along the glomerular basement membrane. The host then produces antibodies against the heterologous IgG which are also deposited along the glomerular basement membrane. The degree of proteinuria that results depends on the amount of antibody injected and the species studied. It may or may not be associated with various degrees of proliferative glomerulonephritis including, occasionally, the crescentic form. Induction of inflammation depends initially on complement activation and on the influx of neutrophils, and later on monocyte infiltration (reviewed by Wilson and Dixon 1986). When the recipient has been preimmunized against IgG from the species that in which the heterologous antibodies were produced, an accelerated form of the disease is observed (Unanue et al. 1965; Schreiner et al. 1978). Several models of autoimmune glomerulonephritis in sheep have also been described (Steblay 1962). Autoimmune glomerulonephritis can also be produced in rats after immunization with heterologous or homologous glomerular basement membrane (Stuffers-Heimann et al. 1979; Pusey et al. 1990b); the antigen(s) involved has not been characterized. Glomerulonephritis mediated by antibodies directed against the glomerular basement membrane is also observed in the context of polyclonal activation in the chronic graft-versus-host model (Tournade et al. 1990a), in drug-induced autoimmunity (Tournade et al. 1989, 1990b), and in rats with trypanosomiasis (Bruijn et al. 1987).

Structural glomerular cellular antigens

Following the description of serum sickness models and of the antiglomerular basement membrane mediated glomerulonephritis, glomerular granular immunoglobulin deposits observed by immunofluorescence were considered to be the consequence of deposition of circulating immune complexes, while a linear pattern of fixation was considered to be due to the fixation of antibodies to the glomerular basement membrane.

Experiments performed in the Heymann's nephritis model showed that this dogma had to be reconsidered. The granular IgG deposits seen in this model were first thought to result from the deposition of circulating immune complexes. Later, it became clear that these deposits were due to the fixation of free anti-gp 330 antibodies to gp330 antigen irregularly distributed on visceral epithelial cells (Kerjaschki and Farquhar 1982; Chatelet et al. 1986; Verroust et al. 1987). These antibodies bound in a granular pattern to the glomerular capillary wall when injected intravenously (Van Damme et al. 1978), or when infused into the renal artery (Couser et al. 1978). The antigen responsible (gp 330) is associated with clathrin in coated pits of the podocytes. Immune complexes thus formed *in situ* are then released and accumulate on the subepithelial side of the glomerular basement membrane, leading to the formation of electron-dense deposits. In this situation, granular deposits are therefore due to fixation of antibody to an irregularly distributed glomerular antigen. It is, however, difficult to rule out the additional participation of circulating immune complex deposition.

Other antigens such as dipeptidyl peptidase IV (gp 70–90), which are mainly found on glomerular endothelial cells, may also participate in the formation of granular IgG glomerular deposits in the rat or in mice (Verroust et al. 1987; Brentjens and Andres 1989). Such antibodies, as well as antibodies to glomerular basement membrane components (Hoedemaeker and Weening 1989) might facilitate the access of gp 330-specific antibodies to the epithelial cells in the active model of Heymann nephritis.

Further convincing evidence of the role of free antibodies in the pathogenesis of experimental membranous glomerulopathy comes from the model of the kidney transplantation developed by Thoenes et al. (1979), who grafted a LEW.1N rat kidney into a Brown-Norway histocompatible recipient. This did not lead to rejection but, after a few weeks, to development of typical membranous glomerulonephritis with granular deposits in the grafted kidney but not in the recipient's own kidney. Antibodies were found that recognized a brush border antigen on the LEW.1N kidney, but not on the native kidney. It is likely that these antibodies, rather than circulating immune complexes, were responsible for the disease. This antigen has not been further characterized but is likely to represent an alloantigen expressed both on the brush border of proximal convoluted tubules and on the glomeruli.

Membranous glomerulopathy can also be caused by other mechanisms. Antiangiotensin converting enzyme antibodies injected intravenously first bind to the endothelial cells which express this enzyme. Three days later, granular deposits are found in a subepithelial position. Although these could result from circulating immune complex deposition, they could also appear following shedding of antibodies from the endothelium and their relocalization at the subepithelial site (Brentjens and Andres 1989). In the mercury model, granular IgG deposits are found after an initial phase which is characterized by linear deposits (Druet et al. 1978). These granular deposits could be due to the deposition of circulating immune complexes. They could also result from either the redistribution of antiglomerular basement membrane antibodies, for example following the production of anti-idiotypic antibodies, or from the fixation of anti-laminin antibodies to cell-bound laminin or to laminin produced in excess by the epithelial cells (Aten et al. 1988b). This mechanism would explain the appearance of granular subepithelial deposits in classical antiglomerular basement membrane-mediated glomerulonephritis. If this holds true, the boundary between antiglomerular basement membrane-mediated nephritis and membranous glomerulopathy would become very tenuous. The different potential mechanisms of subepithelial deposit formation are shown in Fig. 9.

Monoclonal antibodies that recognize glomerular epithelial cells have been produced in the rat (Orikasa et al. 1988; Brentjens and Andres 1989). These fix in a granular pattern along the glomerular capillary wall when injected intravenously, but do not produce electron-dense deposits. Some of them provoke a transient proteinuria. Finally, antibodies that recognize epitopes shared between mesangial cells and thymocytes induce a complement dependent mesangiolysis when injected intravenously (Bagchus et al. 1984; Yamamoto and Wilson, 1987).

There are, therefore, numerous experimental examples of antibodies which interact with glomerular components and may consequently induce glomerular injury. It is quite likely that

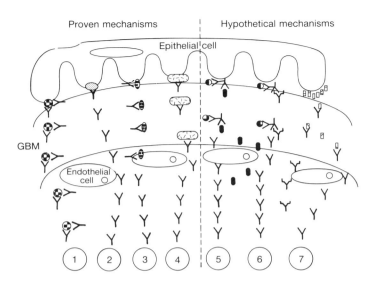

Fig. 9 Mechanisms of formation of subepithelial deposits. 1. Deposition of circulating immune complexes. 2. Fixation of circulating antibodies on irregularly distributed epithelial cell antigen such as gp 330. 3. Shedding of immune complexes formed *in situ* at the membrane of endothelial cells. This is the case for angiotensin converting enzyme and its antibodies. 4. Fixation of antibodies to cationic 'planted' antigens. 5. Polyreactive antibodies could first bind a GBM antigen and then bind a circulating autoantigen. These antibodies could be secondarily redistributed. 6. Same mechanism as in (5) but redistribution is induced by anti-idiotypic (or anti-isotypic) antibodies. 7. Antilaminin antibodies are fixed to the GBM but could also fix laminin produced in excess by epithelial cells.

similar mechanisms occur in man but, except for the Goodpasture antigen, no target antigen has been identified.

Planted glomerular antigens
A major advance in the understanding of pathogenesis of glomerulonephritis occurred with the discovery that molecules can bind to glomerular structures for non-immune reasons and then serve as targets for circulating antibodies. These non-renal molecules have been called 'planted' antigens (Fig. 9).

The first demonstration for a role of planted antigen came from Mauer *et al.* (1973), who demonstrated that heat-aggregated human IgG administered intravenously to rats localizes within the mesangium without inducing any renal injury. However, when a kidney from such a rat is transplanted into syngeneic recipients preimmunized against human IgG, circulating antihuman IgG antibodies rapidly bind to the planted IgG, resulting in a focal proliferative glomerulonephritis. It must be stressed that immune reactants previously fixed to glomerular structures for immune reasons may also function as planted antigens. The heterologous antibodies used in antiglomerular basement membrane-mediated nephritis (Wilson and Dixon 1986) or in passive Heymann's nephritis (Van Damme *et al.* 1978) induce autologous antibodies which bind to these fixed heterologous antibodies. More interestingly anti-idiotypic antibodies produced during the normal immune response may also bind autologous IgG in any model of immune-mediated glomerulonephritis and perpetuate glomerular injury. This has already been shown in the serum sickness model (Zanetti and Wilson 1983).

Numerous other antigens may behave as planted antigens. Cationic molecules have been carefully studied as they bind to the anionic sites in the glomerular basement membrane. Binding of such antigens is normally transient, but can be long lasting if antibodies to the planted antigens are given (Oite *et al.* 1982; Border *et al.* 1982). These cationic molecules (pI usually >9) may be either artificially cationic molecules, such as ferritin or bovine serum albumin, or natural cationic proteins. As mentioned earlier, the serum sickness model probably results from such a mechanism. Electron-dense deposits are first seen on the internal and then on the external aspect of the glomerular basement membrane. This change in the site of deposition probably results from redistribution or rearrangement of *in-situ* immune complexes. Endogenous cationic molecules, such as lyzozyme, platelet factors or cationic proteins from neutrophils, may also serve as planted antigens. Interestingly, cationic IgG may bind to the glomerular basement membrane for non-immune reasons (Oite *et al.* 1983), the corresponding antigen being fixed in a second step.

There is some evidence that cationic molecules also play a role in human glomerular disease: these have been isolated from streptococci and are probably involved in the pathogenesis of poststreptococcal glomerulonephritis (Vogt *et al.* 1983), and in the formation of extramembranous deposits observed in this condition.

Planted antigens may also be composed of lectins such as concanavalin A (Golbus and Wilson 1979), which has an affinity for glucose and mannose present within the glomerular basement membrane. Again, injection of antibodies specific for the bound lectin will initiate a glomerulonephritis with immune deposits present mainly in a subendothelial position (Johnson *et al.* 1989).

Anti-DNA antibodies have been eluted from lupus kidneys, suggesting a role for immune complex deposition (Koffler *et al.* 1974) (see Chapter 4.6.1). However DNA may bind *in vitro* to components of the glomerular basement membrane, and could possibly act as a planted antigen (Izui *et al.* 1976). In one study, a monoclonal anti-DNA antibody injected intravenously bound to the glomerular basement membrane *in vivo* (Madaio *et al.* 1985). Other studies have suggested that the binding of anti-DNA antibodies to glomerular basement membrane antigens is due to cross-reactivity between DNA and heparan sulphate. More recently Jacob *et al.* (1987) reported that a monoclonal antibody specific for DNA recognized a protein expressed at the membrane of several cell types (Jacob *et al.* 1987), including glomerular epithelial cells. Finally Schmiedeke *et al.* (1989) have demonstrated that cationic histones bind to the glomerular basement membrane and that DNA bound to fixed histones can be recognized by anti-DNA or antihistones antibodies. This is a new and attractive hypothesis to explain how glomerular deposits are formed in lupus nephritis.

From a theoretical point of view, the process of immune complex deposition could be much more complicated. It is now widely accepted that numerous autoantibodies are polyreactive, which means that they recognize not only the expected antigen but other haptens or autoantigens as well (Schoenfeld *et al.* 1983; Matsiota *et al.* 1987). It is therefore conceivable that antibodies fixed in the glomerulus because of one specificity could, for one reason or another, bind circulating antigens of another specificity (Fig. 9). If this hypothesis holds true it could be extremely difficult to define the composition of immune deposits in human membranous glomerulopathy and to understand the underlying mechanisms.

Cell-mediated glomerulonephritis

Glomerular immunoglobulins deposits are found in the vast majority of experimental models of glomerulonephritis and many investigators have considered that cells within glomeruli have effector functions but no role in the induction process. It is now clear that cells of the immune system may, themselves, be responsible for experimental glomerulonephritis and may be important in rapidly progressive glomerulonephritis without immunoglobulin deposits.

Bahn et al. (1979) showed that proteinuria develops when both antiglomerular basement membrane antibodies and cells from rats sensitized to the glomerular basement membrane are passively transferred into naïve recipients. Proteinuria does not occur when either antibodies or cells are omitted. Tipping et al. (1985) observed CD4$^+$ T cells within the glomeruli of rats with anti-glomerular basement membrane glomerulonephritis before the macrophage influx.

A definitive demonstration of the role of cell-mediated immunity in experimental glomerulonephritis has recently been provided. As discussed above, mononuclear cells (presumably T cells) are involved in proliferative glomerulonephritis which occurs in bursectomized chickens immunized with glomerular basement membrane (Bolton et al. 1988).

Other models have been described using planted antigens. Kagami et al. (1988) initially described classical antibody-mediated, trinitrophenyl-specific, glomerulonephritis in rats. Immunization with trinitrophenyl coupled to bovine serum albumin as a carrier and Freund's complete adjuvant produced, as expected, antibodies to trinitrophenyl. Trinitrophenyl, coupled to cationic human IgG, was then injected into the left renal artery of these animals. Circulating antibodies bonded to planted trinitrophenyl and produced glomerulonephritis. The same group (Oite et al. 1989) then took advantage of the fact that only a CD4$^+$ T cell-mediated response, the so-called delayed hypersensitivity reaction, occurs following application of the hapten alone, without carrier, on the skin. Following immunization with trinitrophenyl, trinitrophenyl coupled to cationic BSA was injected in a renal artery. Rats developed proliferative glomerulonephritis and adhesion of glomerular capillary loops to Bowman's capsule with transient proteinuria. They produced very small amounts of circulating antibodies to trinitrophenyl, and no glomerular deposits of IgG were seen even though trinitrophenyl could be demonstrated in the glomeruli. The phenotype of cells infiltrating the glomeruli has not yet been described. More recently, Rennke et al. (1990) reported that rats immunized with azabenzenarsonate coupled to keyhole limpet haemocyanin develop a crescentic glomerulonephritis associated with periglomerular granulomas and an interstitial nephritis when challenged with azabenzenarsonate alone via the renal artery. Antibody deposition was not observed and passive transfer of T cells from sensitized animals also produced glomerulonephritis.

These experimental models are certainly extremely valuable, and show clearly that specific T cells can localize in the glomeruli and react with fixed antigens.

Effector mechanisms and mediators of glomerular damage

Direct antibody effect

Heterologous antibodies may cause glomerular injury by a number of mechanisms, depending on the species and isotype of

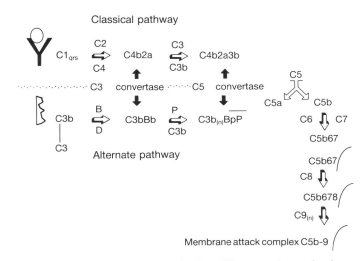

Fig. 10 The complement cascade. Two different pathways lead to the formation of a C3 convertase, then to the formation of a C5 convertase. The classical pathway is initiated through the binding of the early complement components C1q and C1r to an antigen–antibody complex. The alternate pathway is initiated through the binding of C3 on an 'activator' surface. A single cascade then induces the formation of the membrane attack complex.

the antibody. A direct effect of the antibody itself, independent of complement or recruitment of cells, has been demonstrated in a few instances, for example, nephrotoxic nephritis induced in guinea-pigs by heterologous anti-GBM F(ab')$_2$ fragment (reviewed in Couser 1985).

Complement
Activation

Two different pathways can lead to the formation of C5a and the membrane attack complex, which are mainly responsible for the complement-mediated lesions of glomerulonephritis (Fig. 10).

The classical pathway is initiated through binding of the C1q component to the Fc portion of an IgG or IgM molecule. After activation of C1r and C1s, the C1qrs molecule cleaves C4 and C2, leading to the formation of the classical pathway C3 convertase, C4b2a. C3 is then cleaved into C3b and C3a, and C3b binds to this C3 convertase to form the classical pathway C5 convertase, C4b2a3b.

The alternative pathway of complement is initiated through binding of the soluble C3b to 'activator' surfaces, where the B factor joins C3b and is cleaved by factor D, which leads to the formation of the alternative pathway C3 convertase C3bB, which is stabilized by properdin (P). More C3b is then generated, and the alternative pathway C5 convertase C3b(n)BbP is formed.

Both pathways lead to the cleavage of C5 into C5a, a potent neutrophil chemoattractant, and C5b, which initiates the formation of the C5b–9 complex, or membrane attack complex.

Cell-mediated action

Nephrotoxic serum nephritis follows intravenous administration of serum containing antibodies to glomerular basement membrane. Neutrophils infiltrate the glomeruli 15 to 30 min after deposition of these antibodies, immediately before the onset of massive proteinuria. The neutrophils degranulate, and neutrophil-derived proteolytic enzymes as well as fragments of

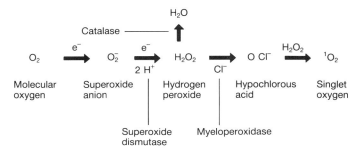

Fig. 11 The reactive oxygen species (see text for details).

glomerular basement membrane appear in the urine (Gang et al. 1970). Neutrophil depletion prevents the appearance of proteinuria and structural alterations of the glomerular basement membrane (Cochrane et al. 1965). Decomplementation, through administration of cobra venom factor, also greatly reduces the proteinuria, and abolishes the glomerular neutrophil infiltrate (Cochrane et al. 1970). In this model complement is primarily activated through the classical pathway, and, although direct evidence is lacking, the chemotactic properties of C5a may be partly responsible for the generation of the neutrophil infiltrate. Neutrophils could alter immune adherence mechanisms involving C3b-receptors or interact with the Fc receptors of deposited immunoglobulins.

Direct effect

Passive Heymann's nephritis occurs following administration of heterologous anti-FX1A antibody (see above) to naive rats. Complement deposits, along with antibodies, are formed and proteinuria occurs in the absence of any cellular infiltrate. Neutrophil depletion has no effect, but decomplementation by cobra venom factor totally abolishes proteinuria without modifying the glomerular deposition of immunoglobulins. This demonstrates a direct involvement of complement in glomerular injury (Salant et al. 1980). Decomplementation using antibodies against C6 or utilization of C6-deficient animals produces the same results, suggesting a role of the membrane attack complex in the development of the glomerular lesions responsible for the proteinuria (Groggel et al. 1983). Immunofluorescence demonstrates antigens of the membrane attack complex in the capillary wall deposits of rats with serum sickness (Koffler et al. 1983). Similar results have been obtained in other models, such as chronic serum sickness, antiglomerular basement membrane-mediated nephritis and active Heymann's nephritis.

Membrane attack complex antigens, together with IgG and C3, have been demonstrated in the capillary walls of humans with many different types of glomerulonephritis, including lupus nephritis, membranous nephropathy, poststreptococcal nephritis, IgA nephritis, membranoproliferative nephritis, and antiglomerular basement membrane-mediated nephritis (Falk et al. 1983).

Reactive oxygen species

Metabolism

Reduction of molecular oxygen within the cell, if complete, leads to the formation of water and, if incomplete, to reactive oxygen species, including superoxide anion, hydrogen peroxide, hypochlorous acid, and singlet oxygen (Fig. 11). Under normal conditions, kidney cells produce minute quantities of reactive oxygen species particularly superoxide anion. This anion is detoxified by superoxide dismutase, an enzyme which transforms it into hydrogen peroxide, which is in turn converted to water by a catalase. There are two types of superoxide dismutase in the kidney, one located in the cytoplasm and the other in the mitochondria. Catalase is found in the subcellular organelles, such as the peroxisomes of the kidney.

Toxic effects of reactive oxygen species

Reactive oxygen species can react with polyunsaturated fatty acids to form peroxide derivatives. Such reactions alter the lipid bilayer of the cell membrane and interfere with ion transport, resulting, for example, in an increase of the K^+ permeability of cells. Other toxic effects include a decrease in ATP levels and inactivation of intracellular proteins.

Reactive oxygen species and glomerulonephritis

Injection of phorbol myristate acetate, a potent inducer of reactive oxygen species formation, into the renal artery results in proteinuria and glomerular damage (Rehan et al. 1985) which is abolished by prior depletion of polymorphonuclear leucocytes. Early administration of catalase in nephrotoxic serum nephritis greatly reduces proteinuria and endothelial cell damage, whereas superoxide dismutase has no effect, indicating that hydrogen peroxide is more toxic in this model (Rehan et al. 1984). Both infiltrating polymorphonuclear leucocytes and the resident glomerular cells are capable of producing reactive oxygen species and polymorph depletion lowers but does not abolish proteinuria. A role for reactive oxygen species in immune complex nephritis is suggested by the fact that injection of hydrogen peroxide and myeloperoxidase into the renal artery induces proteinuria and glomerular alterations similar to those observed in nephritis induced by concanavalin A antigen-antibody complexes (Johnson et al. 1987). In this model myeloperoxidase and superoxide anion are likely to be produced by neutrophils following the complement-dependent glomerular localization of platelets (Johnson et al. 1989).

Monocytes and macrophages

Cells of the monocyte/macrophage lineage are responsible for the glomerular hypercellularity observed in models of nephrotoxic nephritis, such as these induced by mercuric chloride or serum sickness, and their presence in these models has been shown by electron microscopy or immunohistochemistry. The pathogenetic role of these cells has been directly demonstrated in the rabbit nephrotoxic nephritis model (Cochrane et al. 1965) as well as in serum sickness models, in which treatment with an antimacrophage antiserum prevents proteinuria.

Other cells may also be involved; antibody-dependent cell-mediated cytotoxicity has recently been found to play a role in models of antiglomerular basement membrane-mediated nephritis and in a model of passive Heymann's nephritis.

Arachidonic acid derivatives

Arachidonic acid is synthesized from cell membrane phospholipids and gives rise, through two different metabolic pathways, to the families of prostaglandins (catalysed by cyclooxygenase) and leukotrienes (catalysed by lipoxygenase) (Fig. 12). Polymorphonuclear leucocytes synthesize leukotrienes as well as prostaglandins, whilst vascular endothelium synthesizes mainly prostaglandins, especially prostacyclin. Glomeruli are able to

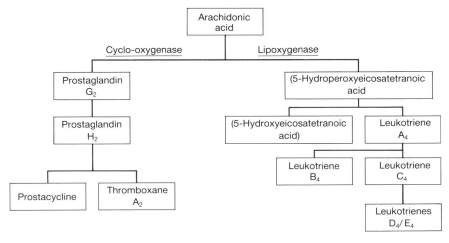

Fig. 12 The arachidonic acid metabolism. Two main pathways are shown, depending upon the enzyme used, cyclo-oxygenase for the prostaglandin pathway, lipoxygenase for the leukotriene pathway.

synthesize thromboxane A_2 as well as leukotrienes. Although the role of leukotrienes in ischaemic renal disease, is well established, little is known of their role in immune nephritis. Some reports described an increased level of leukotriene B4 in nephrotoxic serum nephritis, Heymann's nephritis, and in experimental membranous nephropathy. Reactive oxygen species are potent activators of leukotriene synthesis, and both compounds could act synergistically to induce glomerular damage (reviewed in Ardaillou et al. 1987).

Cytokines

Interleukin-1 and tumour necrosis factor have major effects on the vascular endothelium and play an important role in promoting vascular injury. Both are secreted by activated macrophages and promote coagulation through the increased secretion of tissue factor and plasminogen activator inhibitor, and by inhibition of thrombomodulin and plasminogen activator secretion. They also exacerbate inflammatory reactions by increasing leucocyte adhesion to the endothelial cells through the induction of expression of endothelial leucocyte adhesion molecule ELAM 1, by the stimulation of intercellular adhesion molecule 1 at the cell surface (reviewed in Cotran and Pober 1989) and by promoting transendothelial passage of neutrophils (Moser et al. 1989). They also stimulate the expression of HLA class I molecules, thus augmenting the targets of an ongoing immune response. Tomosugi et al. (1989) have shown that IL-1 and tumour necrosis factor participate in glomerular injury. These molecules, as well as lipopolysaccharide, which stimulates their production, are indeed able to potentiate proteinuria in a model of anti-glomerular basement membrane γ-antibody-mediated glomerulonephritis.

Interferon is mainly secreted by activated T cells and has a unique effect on vascular endothelial cells by promoting the expression of HLA class II molecules on the endothelial cell surface, thus allowing those cells to act as antigen-presenting cells and initiate an immune response.

Growth factors

Growth factors may contribute to the pathogenesis of glomerulonephritis though their action on glomerular cell proliferation and differentiation. T-cell growth factor β stimulates the production of proteoglycans by mesangial cell (Border et al. 1990a), and administration of antibodies to this growth factor prevents mesangial proliferation in a model of anti-thy-1.1 mediated glomerulonephritis (Yamamoto and Wilson 1987, Border et al. 1990b).

Coagulation factors

Glomerular fibrin deposition is a common finding in a number of glomerulopathies and can be the consequence of either systemic or local activation of the coagulation system. Many glomerular structures can interfere with the coagulation fibrinolysis: the glomerular basement membrane comprises both collagen, a potent inducer of platelet aggregation, and heparan sulphate, a natural antithrombin agent; glomerular endothelial cells secrete both von Willebrand factor VIII, which activates the coagulation system, and thrombomodulin, which binds to thrombin and enhances its ability to activate the anticoagulant protein C. Moreover, both tissue plasminogen activator and the urokinase-type plasminogen activator are readily found in glomeruli, together with inhibitors or plasminogen activation (reviewed in Kanfer 1989).

Fibrin deposition is seen in many animal models, including nephrotoxic, antiglomerular basement membrane-mediated, and mercuric chloride-induced nephritis. In the latter model, enhanced glomerular procoagulant activity has been demonstrated and correlates with proteinuria and the occurrence of fibrin deposits (Kanfer et al. 1987). In nephrotoxic nephritis, defibrination with ancrod results in the improvement of renal function and disappearance of fibrin deposits. (Thomson et al. 1976). In man, systemic hypercoagulability, marked by elevated platelets counts, hyperfibrinogenaemia, and decreased antithrombin III, has been reported in many patients with severe nephritis.

Conclusion

Although they are not always the exact counterpart of human autoimmune glomerulonephritis, experimental models have helped advance our understanding of glomerular autoimmunity. However much more information is needed in many areas. Better definition of autoantigens is crucial: the only substantial work in this area concerns those antigens involved in anti-glomerular basement membrane-mediated nephritis (Wieslander et al. 1987) and Heymann's nephritis (reviewed in Verroust et al. 1987). T-cell mediated immunity is difficult to study and its role in the pathogenesis of autoimmune glomerulonephritis has undoubtedly been underestimated. Very little is known about the generation of autoreactive T cells, their fine antigenic specificity, and their preferential recognition of some MHC alleles. Autoreactive T cells are induced in several experimental models of autoimmune glomerulonephritis: T cell lines or clones are needed to enable the T cell receptor to be characterized and the regulatory circuits that prevent activation of autoreactive T cells to be elucidated (see Chapter 4.6.1.). In this respect, evidence that suppressor T cells, idiotypic network, and cytokines may be of major importance in models of autoimmune glomerulonephritis may point the way to new therapeutic strategies. Lastly, an important potential mechanism lies in the fact that resident glomerular cells might act as antigen-presenting cells, providing a major issue for future research (Mendrick *et al.* 1990).

Understanding of human autoimmune glomerulonephritis has greatly improved during the last few years due to a better understanding of experimental models and to major advances in basic immunology. Future contributions in these areas will certainly bring further insights into the mechanisms of glomerular autoimmunity in humans.

References

Abdou, N.I., Wall, H., Lindsley, H.B., Halsey J.F., and Suzuki, T. (1981). In vitro suppression of serum anti-DNA antibody binding to DNA by anti-idiotypic antibody in systemic lupus erythematosus. *Journal of Clinical Investigation*, **67**, 1297–304.

Abramowicz, D., Goldman, M., Bruyns, C., Lambert, P.H., Thoua, Y., and Toussaint, C. (1987). Autoimmune disease after neonatal injection of semi-allogeneic spleen cells in mice; involvement of donor B and T cells and characterization of glomerular deposits. *Clinical and Experimental Immunology*, **70**, 61–7.

Abramowicz, D., Doutrelepont, J.M., Lambert, P., Van Der Vorst, P., Bruyns, C., and Goldman, M. (1990). Increased expression of Ia antigens on B cells after neonatal induction of lymphoid chimerism in mice: role of interleukin-4. *European Journal of Immunology*, **20**, 469–76.

Alt, F.W., Blackwell, T.K., Depinho, R.A., Reth, M.G., and Yancopoulos, G.D. (1986). Regulation of genome rearrangement events during lymphocyte differentiation. *Immunological Review*, **89**, 5–29.

Ando, D.G., Sercarz, E.E., and Hahn, B.V. (1987). Mechanisms of T and B cell collaboration in the *in vitro* production of anti-DNA antibodies in the NZB/NZW F1 murine SLE model. *Journal of Immunology*, **87**, 3185–90.

Ardaillou, R., Baud, L., and Sraer, J. (1987). Role of arachidonic acid metabolites and reactive oxygen species in glomerular immune-inflammatory process. *Springer Seminars in Immunopathology*, **9**, 371–85.

Aten, J., Bruijn, J.A., Veringa, A., De Heer, E., and Weening, J.J. (1988b). Anti-laminin autoantibodies and mercury-induced membranous glomerulopathy. *Kidney International*, **33**, 309.

Bagchus, W.M., Donga, J., Hoedemaeker, P.J., and Bakker, W.W. (1984). The specificity of nephritogenic antibodies. II Immune complex glomerulopathy in rats induced by heterologous antithymocyte serum. *Transplantation*, **38**, 165–9.

Bagchus, W.M., Hoedemaeker, P.J., Rozing, J., and Bakker, W.W. (1987). Glomerulonephritis induced by monoclonal anti-Thy 1.1 antibodies. A sequential histological and ultrastructural study in the rat. *Laboratory Investigation*, **55**, 680–7.

Bahn, A.K., Collins, A., Schneeberger, E., and Mc Cluskey, R. (1979). A cell-mediated reaction against glomerular-bound immune complexes. *Journal of Experimental Medicine*, **150**, 1410–17.

Bellon, B., *et al.* (1982). Mercuric chloride induced auto-immune disease in Brown-Norway rats: sequential search for antibasement membrane antibodies and circulating immune complexes. *European Journal of Clinical Investigation*, **12**, 127–33.

Bjorkman, P.J., Saper, M.A., Samroui, B., Bennet, W.S., Strominger, J.L., and Willey, D.C. (1987). Structure of the human class I histocompatibility antigen, HLA-A2. *Nature*, **329**, 506–12.

Bolton, W.K., Tucker, F.L., and Sturgill, B.C. (1984). A new avian model of experimental glomerulonephritis consistent with mediation by cellular immunity. *Journal of Clinical Investigation*, **73**, 1263–76.

Bolton, W.K., Chandra, M., Tyson, T.M., Kirkpatrick, P.R., Sadovnic, M.J., and Sturgill, B.C. (1988). Transfer of experimental glomerulonephritis in chickens by mononuclear cells. *Kidney International*, **34**, 598–610.

Border, W.A., Ward, H.J., Kamil, E.S., and Cohen, A.H. (1982). Induction of membranous nephropathy in rabbits by administration of an exogenous cationic antigen: demonstration of a pathogenic role for electrical charge. *Journal of Clinical Investigation*, **69**, 451–61.

Border, W.A., Okuda, S., Languino, L.R., and Ruoslahti, E. (1990a). Transforming growth factor–β regulates production of proteoglycans by mesangial cells. *Kidney International*, **37**, 689–95.

Border, W.A., Okuda, S., Languino, L.R., Sporm, M.B. and Ruoslahti, E. (1990b). Suppression of experimental glomerulonephritis by antiserum against transforming growth factor β_1. *Nature*, **346**, 371–4.

Bowman, C., Mason, D.W., Pusey, C.D. and Lockwood, C.M. (1984). Autoregulation of antibody synthesis in mercuric chloride nephritis in the Brown-Norway rat. I. A role for T suppressor cells. *European Journal of Immunology*, **14**, 464–70.

Bowman, C., Green, C., Borysiewicz, L., and Lockwood, C. (1987). Circulating T-cell population during mercuric chloride-induced nephritis in the Brown-Norway rat. *Immunology*, **61**, 515–20.

Brentjens, J.R. and Andres, G. (1989). Interaction of antibodies with renal cell surface antigens. *Kidney International*, **35**, 954–68.

Bruijn, J.A., Oemar, B.S., Ehrich, J.H.H., Foidart, J.M., and Fleuren, G.J. (1987). Anti-basement membrane glomerulopathy in experimental trypanosomiasis. *Journal of Immunology*, **139**, 2482–8.

Chalopin, J.M. and Lockwood, C.M. (1984). Autoregulation of autoantibody synthesis in mercuric chloride nephritis in the Brown-Norway rat. II. Presence of antigen-augmentable plaque-forming cells in the spleen is associated with humoral factors behaving as auto-anti-idiotypic antibodies. *European Journal of Immunology*, **14**, 470–5.

Chatelet, F., Brianti, E., Ronco, P., Roland, J., and Verroust, P. (1986). Ultrastructural localization by monoclonal antibodies of brush border antigens expressed by glomeruli. I. Renal distribution. *American Journal of Pathology*, **122**, 500–11.

Chen, A., Wrong, S.S., and Rifai, A. (1988). Glomerular immune deposits in IgA nephropathy. A continuum of circulating and *in situ* formed immune complexes. *American Journal of Pathology*, **130**, 216–22.

Cheng, I.K.P., Dorsch, S.E., and Hall, B.M. (1988). The regulation of autoantibody production in Heymann's nephritis by T lymphocyte subsets. *Laboratory Investigation*, **59**, 780–8.

Cochrane, C.G., Unanue, E.R., and Dixon, F.J. (1965). A role of polymorphonuclear leukocytes and complement in nephrotoxic nephritis. *Journal of Experimental Medicine*, **122**, 99–119.

Cochrane, C.J., Muller-Eberhard, H.J., and Aikin, B.S. (1970). Depletion of plasma complement *in vivo* by a protein of cobra venom: its effect on various immunologic reactions. *Journal of Immunology*, **105**, 55–69.

Cotran, R.S. and Pober, J.S. (1989). Effects of cytokines on vascular endothelium: their role in vascular and immune injury. *Kidney International*, **35**, 969–75.

Couser, W.G., Steinmuller, D.R., Stilmant, M.M., Salant, D.J., and Lowenstein, L.M. (1978). Experimental glomerulonephritis in the isolated perfused rat kidney. *Journal of Clinical Investigation*, **62**, 1275–87.

Couser, W.G. (1985). Mechanisms of glomerular injury in immune-complex disease. *Kidney International*, **28**, 569–83.

De Heer, E., Daha, M.R., and Van Es, L.A. (1985). The autoimmune response in active Heymann's nephritis in Lewis rats is regulated by T lymphocyte subsets. *Cellular Immunology*, **92**, 254–64.

De Heer, E., Daha, M.R., Burger, J., and Van Es, L.A. (1986). Re-establishment of self tolerance by suppressor T-cells after active Heymann's nephritis. *Cellular Immunology*, **98**, 28–33.

3.2 Immune mechanisms of glomerular damage

deTalance, A., Regnier, D., Spinella, S., Morisset, J., and Seman, M. (1986). Origin of autoreactive T helper cells. I. Characterization of Thy-1$^+$, Lyt$^-$/L3T4$^-$ precursors in the spleen of normal mice. *Journal of Immunology*, **137**, 1101–8.

Dighiero, G., et al. (1983). Murine hybridomas secreting natural monoclonal antibodies reacting with self-antigens. *Journal of Immunology*. **131**, 2267–72.

Donker, Ab.J., Venuto, R.C., Vladutio, A.O., Brentjens, J.R., and Andres, G.A. (1984). Effects of prolonged administration of D-penicillamine or captopril in various strains of rats. *Clinical Immunology and Immunopathology*, **30**, 142–55.

Druet, E., Sapin, C., Gunther, E., Feingold, N., and Druet, P. (1977). Mercuric chloride induced antiglomerular basement membrane antibodies in the rat. Genetic control. *European Journal of Immunology*, **7**, 348–51.

Druet, P., Druet, E., Potdevin, F., and Sapin, C. (1978). Immune type glomerulonephritis induced by HgCl$_2$ in the Brown-Norway rat. *Annales d'Immunologie*, **129C**, 777–92.

Druet, E., Sapin, C., Fournie, G., Mandet, C., Gunther, E., and Druet, P. (1982). Genetic control of susceptibility to mercury-induced immune nephritis in various strains of rat. *Clinical Immunology and Immunopathology*, **25**, 203–12.

Dziarski, R. (1988). Autoimmunity: Polyclonal activation or antigen induction? *Immunology Today*, **9**, 340–2.

Emancipator, S.N., Gallo, G.R., and Lamm, M.E. (1983). Experimental IgA nephropathy induced by oral immunization. *Journal of Experimental Medicine*, **157**, 572–82.

Ehrlich, P. and Morgenroth, J. (1959). *The Collected Papers of Paul Ehrlich – 1900*. (ed. F. Himmelweit), pp. 205–5. Pergamon Press, Oxford.

Falk, R.J., et al. (1983). Neoantigen of the polymerised ninth component of complement. Characterization of a monoclonal antibody and immunohistochemical localization in renal disease. *Journal of Clinical Investigation*, **72**, 560–73.

Ferrick, D.A., Ohashi, P.S., Wallace, V., Schilham, M., and Mak, T.W. (1989). Thymic onteny and selection of $\alpha\beta$ and δ T cells. *Immunology Today*, **10**, 403–7.

Fillastre, J.P., Druet, P., and Mery, J.PH. (1988). Drug-induced glomerulonephritis. In *The Nephrotic Syndrome*, (eds. J. S. Cameron and R. J. Glassock), pp. 697–744. Marcel Dekker Inc., New York.

Fleuren, G., Grond, J., and Hoedemaeker, P.J. (1980). In situ formation of subepithelial glomerular immune complexes in passive serum sickness. *Kidney International*, **17**, 631–7.

Fournie, G.J., Lambert, P.H., and Miescher, P.A. (1974). Release of DNA in circulating blood and induction of anti-DNA antibodies after injection of bacterial lipopolysaccharides. *Journal of Experimental Medicine*, **140**, 1189–206.

Fukatsu, A., Brentjens, J.R., Killen, P.D., Kleinman, K.D., Martin, G.R., and Andres, G.A. (1987). Studies on the formation of glomerular immune deposits in Brown-Norway rats injected with mercuric chloride. *Clinical Immunology and Immunopathology*, **45**, 35–47.

Gang, N.F., Trachtenberg, E., Allerhand, J., Kalant, N., and Mautner, W. (1970). Nephrotoxic serum nephritis III: correlation of proteinuria, excretion of glomerular basement-like protein and changes in the ultrastructure of the glomerular basement membrane as visualized with lanthanum. *Laboratory Investigation*, **23**, 436–41.

Geha, R.S. (1983). Presence of circulating anti-idiotype bearing cells after booster immunization with tetanus toxoid (TT) and inhibition of anti-TT antibody synthesis by auto-anti-idiotypic antibodies. *Journal of Immunology*, **130**, 1634–9.

Gearhart, P.J. (1982). Generation of immunoglobulin variable gene diversity. *Immunology Today*, **3**, 107–12.

Gleichmann, E., Pals, S.T., Rolink, A.G., Radaszkiewicz, T., and Gleichmann, H. (1984). Graft-versus-host reactions: clues to the etiopathology of a spectrum of immunological diseases. *Immunology Today*, **5**, 324–32.

Glotz, D., Sollazo, M., Riley, S., and Zanetti, M. (1988). Isotype, V$_H$ genes, and antigen binding analysis of hybridomas from newborn normal Balb/c mice. *Journal of Immunology*, **141**, 383–90.

Goodnow, C.C., et al. (1988). Altered immunoglobulin expression and functional silencing of self-reactive B lymphocytes in transgenic mice. *Nature*, **334**, 676–82.

Golbus, S.M. and Wilson, C.B. (1979). Experimental glomerulonephritis induced by in situ formation of immune complexes in the glomerular capillary wall. *Kidney International*, **16**, 148–57.

Goldman, M., Rose, L.M., Hochmann, A., and Lambert, P.H. (1982). Deposition of idiotype-anti-idiotype immune complexes in renal glomeruli after polyclonal B cell activation. *Journal of Experimental Medicine*, **155**, 1385–99.

Goldman, M., Feng, H.M., Engers, H., Hochmann, A., Louis, J., and Lambert, P.H. (1983). Autoimmunity and immune complex disease after neonatal induction of transplantation tolerance in mice. *Journal of Immunology*, **131**, 251–8.

Goldman, M., Abramowicz, D., Lambert, P., Vandervorst, P., Bruyns, C., and Toussaint, C. (1988a). Hyperactivity of donor B cells after neonatal induction of lymphoid chimerism in mice. *Clinical and Experimental Immunology*, **72**, 79–83.

Goldman, M., Baran, D., and Druet, P. (1988b). Polyclonal activation and experimental nephropathies. *Kidney International*, **34**, 141–50.

Groggel, G.C., Adler, S., Rennke, H.G., Couser, W.G., and Salant, D.J. (1983). Role of the terminal complement pathway in experimental membranous nephropathy in the rabbit. *Journal of Clinical Investigation*, **72**, 1948–57.

Guéry, J.C., et al. (1990a). Specificity and cross-reactive idiotypes of anti-glomerular basement membrane autoantibodies in HgCl$_2$-induced autoimmune glomerulonephritis. *European Journal of Immunology*, **20**, 93–100.

Guéry, J.C., Tournade, H., Pelletier, L., Druet, E., and Druet, P. (1990b). Rat anti-glomerular basement membrane antibodies in toxin-induced autoimmunity and in chronic graft-versus-host reaction share recurrent idiotypes. *European Journal of Immunology*, **20**, 101–5.

Guéry, J.C. and Druet, P. (1990c). A spontaneous hybridoma producing autoanti-idiotypic antibodies that recognize a Vx-associated idiotype in mercury-induced autoimmunity. *European Journal of Immunology*, **20**, 1027–31.

Guilbert, B., Dighiero, G., and Avrameas, S. (1982). Naturally occuring antibodies against nine common antigens in human sera. I. Detection, isolation and characterization. *Journal of Immunology*, **128**, 2779–87.

Hang, L.M., Slack, J.H., Amundson, C., Izui, S., Theofilopoulos, A.N., and Dixon, F.J. (1983). Induction of murine autoimmune disease by chronic polyclonal B cell activation. *Journal of Experimental Medicine*, **157**, 874–83.

Heymann, W., Hackel, D.B., Harwood, S., Wilson, S.G.F., and Hunter, J.L.P. (1959). Production of nephrotic syndrome in rats by Freund's adjuvants and rat kidney suspensions. *Proceedings of The Society of Experimental Biology and Medicine*, **100**, 660–4.

Hirsch, F., Couderc, J., Sapin, C., Fournie, G., and Druet, P. (1982). Polyclonal effect of HgCl$_2$ in the rat, its possible role in an experimental auto-immune disease. *European Journal of Immunology*, **12**, 620–5.

Hoedemaeker, Ph.J. and Weening, J.J. (1989). Relevance of experimental models for human nephropathology. *Kidney International*, **35**, 1015–25.

Hogendoorn, P.C.W., et al. (1988). Antibodies to purified renal tubular epithelial antigens contain activity against laminin, fibronectin and type IV collagen. *Laboratory Investigation*, **58**, 278–86.

Hood, L., Kronenberg, M., and Hunkapiller, T. (1985). T cell antigen receptor and the immunoglobulin supergene family. *Cell*, **40**, 225–9.

Hozumi, N. and Tonegawa, S. (1976). Evidence for somatic rearrangement of immunoglobulin genes coding for variable and constant regions. *Proceedings of the National Academy of Science USA*, **73**, 3628–32.

Imai, H., Nakamoto, Y., Asakura, K., Mini, K., Yasuda, T., and Miura, A.B. (1985). Spontaneous glomerular IgA deposition in ddY mice: an animal model of IgA nephritis. *Kidney International*, **27**, 756–61.

Izui, S., Lambert, P.H. and Miescher, P.A. (1976). *In vitro* demonstration of a particular affinity of glomerular basement membrane and collagen for DNA. A possible basis for a local formation of DNA-anti-DNA complexes in systemic lupus erythematosus. *Journal of Experimental Medicine*, **144**, 428–43.

Izui, S., Lambert, P.H., Fournie, G.J., Turler, H., and Miescher, P.A. (1977). Features of systemic lupus erythematosus in mice injected with bacterial lipopolysaccharide. Identification of circulating DNA and renal localization of DNA-anti-DNA complexes. *Journal of Experimental Medicine*, **145**, 1115–30.

Jacob, L., *et al.* (1987). Presence of antibodies against a cell-surface protein, cross-reactive with DNA in systemic lupus erythematosus: a marker of the disease. *Proceedings of the National Academy of Science USA*, **84**, 2956–9.

Jerne, N. (1974). Towards a network theory of the immune system. *Annales d'Immunologie, Paris*, **125c**, 373–89.

Jerne, N.K., Roland, J. and Cazenave, P.A. (1982). Recurrent idiotypes and internal images. *EMBO Journal*, **1**, 243–7.

Johnson, R.J., *et al.* (1987). Participation of the myeloperoxidase–H_2O_2–halide system in immune-complex nephritis. *Kidney International*, **32**, 342–9.

Johnson, R.J., *et al.* (1989). Mechanisms and kinetics for platelet and neutrophil localisation in immune complex nephritis. *Kidney International*, **36**, 780–9.

Kagami, S, Miyao, M., Shimizu, F., and Oite, T. (1988). Active *in situ* immune complex glomerulonephritis using the hapten-carrier system: role of epitope density in cationic antigens. *Clinical Experimental Immunology*, **74**, 121–5.

Kanfer, A., *et al.* (1987). Enhanced glomerular procoagulant activity and fibrin deposition in rats with mercuric chloride induced autoimmune glomerulonephritis. *Laboratory Investigation*, **57**, 138–43.

Kanfer, A. (1989). Role of coagulation in glomerular injury. *Toxicology Letters*, **46**, 83–92.

Kappler, J.W., Roehm, N. and Marrack, P. (1987). T cell tolerance by clonal elimination in the thymus. *Cell*, **49**, 273–80.

Kerjaschki, D. and Farquhar, M.G. (1982). The pathogenic antigen of Heymann nephritis is a membrane glycoprotein of the renal proximal tubular brush border. *Proceedings of the National Academy of Science USA*, **79**, 5557–61.

Kisielow, P., Bluthmann, H., Staerz, U.D., Steinmetz, M., and Von Boehmer, H. (1988). Tolerance in T-cell-receptor transgenic mice involves deletion of nonmature $CD4^+8^+$ thymocytes. *Nature*, **333**, 742–6.

Klouda, P.T., *et al.* (1979). Strong association between idiopathic membranous nephropathy and HLA-DRw3. *Lancet*, **ii**, 770–1.

Koffler, D., Agnello, V., and Kunkel, H.G. (1974). Polynucleotide immune complexes in serum and glomeruli of patients with systemic lupus erythematosus. *American Journal of Pathology*, **74**, 109.

Koffler, D., Biesecker, G., Noble, B., Andres, G.A., and Martinez-Hernandez, A. (1983). Localization of the membrane attack complex (MAC) in experimental immune complex nephritis. *Journal of Experimental Medicine*, **157**, 1885–905.

Kotzin, B.L., Barr, V.L., and Palmer, E. (1985). A large deletion within the T-cell receptor beta-chain gene complex in New Zealand white mice. *Science*, **229**, 167–71.

Kotzin, B.L. and Palmer, E. (1987). The contribution of NZW genes to lupus-like disease in (NZB × NZW) F_1 mice. *Journal of Experimental Medicine*, **165**, 1237–51.

Kourilsky, P. and Claverie, J.M. (1989). MHC-antigen interaction: what does the T cell receptor see? *Advances in Immunology*, **45**, 107–93.

Kunkel, H.G., Mannik, M., and Williams, R.C. (1963). Specificity of isolated antibodies. *Science*, **140**, 1218–20.

Lindemann, W. (1900). Sur le mode d'action de certains poisons renaux. *Annales de l'Institut Pasteur (Paris)*, **14**, 49–58.

Litwin, A., Bash, J.A., Adams, L.E., Donovan, R.J., and Hess, E.V. (1979). Immunoregulation of Heymann's nephritis. I. Induction of suppressor cells. *Journal of Immunology*, **122**, 1029–34.

Luzuy, S., Merino, J., Engers, H., Izui, S., and Lambert, P.H. (1986). Autoimmunity after induction of neonatal tolerance to alloantigens: role of B cell chimerism and F1 donor B cell activation. *Journal of Immunology*, **136**, 4420–6.

Madaio, M.P., Carlson, J.A., and Hodder, S. (1985). Monoclonal anti-DNA antibodies bind directly to intrinsic glomerular antigens and form immune deposits. *Kidney International*, **27**, 216a.

Mak, T.W. (1988). *The T-cell Receptors*. Plenum Press, New York.

Masugi, M. (1934). Uber die experimentelle glomerulonephritis auch des specifische antinierenserum. Ein beitrag zur pathogene der diffusen glomerulonephritis. *Beiträge zur Pathologie*, **92**, 429–42.

Matsiota, P., Druet, P., Dosquet, P. Guilbert, B., and Avrameas, S. (1987). Natural antibodies in systemic lupus erythematosus. *Clinical and Experimental Immunology*, **69**, 79–88.

Mauer, S.M., *et al.* (1973). The glomerular mesangium: III. Acute immune mesangial injury: a new model of glomerulonephritis. *Journal of Experimental Medicine*, **137**, 553–70.

Mendrick, D.L., Kelly, D., and Rennke, H. (1990). Antigen presentation by rat glomerular epithelial cells. *Kidney International*, **37**, 428a.

Monestier, M., *et al.* (1986). Shared idiotypes and restricted immunoglobulin variable region heavy chain genes characterize murine autoantibodies of various specificities. *Journal of Clinical Investigation*, **78**, 753–9.

Morisset, J., *et al.* (1988). Genetics and strain distribution of concanavalin-A reactive $Ly-2^-$, $L3T4^-$ peripheral precursors of autoreactive T cells. *European Journal of Immunology*, **18**, 387–94.

Moser, R., Schleiffenbaum, B., Groscurth, P., and Fehr, J. (1989). Interleukin 1 and tumor necrosis factor stimulate human vascular endothelial cells to promote transendothelial neutrophil passage. *Journal of Clinical Investigation*, **83**, 444–55.

Mosmann, T.R. and Coffman, R.L. (1989). Heterogeneity of cytokine secretion patterns and functions of helper T cells. *Advances in Immunology*, **46**, 111–47.

Nagarkatti, P.S., Nagarkatti, M., and Kaplan, A.M. (1985). Normal $Lyt-1^+2^-$ T cells have the unique capacity to respond to syngeneic autoreactive T cells. *Journal of Experimental Medicine*, **162**, 375–80.

Oite, T., Batsford, S.R., Mihatsch, M.J., Takamiya, H., and Vogt, A. (1982). Quantitative studies of *in situ* immune complex glomerulonephritis in the rat induced by planted cationized antigen. *Journal of Experimental Medicine*, **155**, 460–74.

Oite, T., Shimizu, F., Kihara, I., Batsford, S.R., and Vogt, A. (1983). An active model of immune complex glomerulonephritis in the rat employing cationized antigen. *American Journal of Pathology*, **112**, 185–94.

Oite, T., Shimizu, F., Kagami, S., and Morioka, T. (1989). Hapten-specific cellular immune response producing glomerular injury. *Clinical and Experimental Immunology*, **76**, 463–8.

Orikasa, M., Matsui, K., Oite, T., and Shimizu, F. (1988). Massive proteinuria induced in rats by a single intravenous injection of a monoclonal antibody. *Journal of Immunology*, **141**, 807–14.

Oudin, J. and Michel, M. (1963). Une nouvelle forme d'allotypie des globulines du sérum de lapin, apparemment liée à la fonction et à la spécificité anticorps. *Compte-rendu Hebdomadaire de l'Académie des Sciences*, **257**, 805–8.

Owen, M.J. and Lamb, J.R. (1988). The T-cell antigen receptor. In *Immune recognition*. (ed. D. Male), pp. 37–52. IRL Press, Oxford.

Pelletier, L., Hirsch, F., Rossert, J., Druet, E., and Druet, P. (1987a). Experimental mercury-induced glomerulonephritis. *Springer Seminars of Immunopathology*, **9**, 359–69.

Pelletier, L., *et al.* (1987b). Mercury-induced autoimmune glomerulonephritis: requirement for T-cells. *Nephrology, Dialysis, Transplantation*, **1**, 211–18.

Pelletier, L., et al. (1988a). HgCl₂ induces T and B cell to proliferate and differentiate in BN rats. *Clinical and Experimental Immunology*, **71**, 336–42.

Pelletier, L., Pasquier, R., Rossert, J., Vial, M.C., Mandet, C., and Druet, P. (1988b). Autoreactive T cells in mercury-induced autoimmunity. Ability to induce the autoimmune disease. *Journal of Immunology*, **140**, 750–4.

Pelletier, L., Rossert, J., Pasquier, R., Vial, M.C., and Druet, P. (1990). Role of CD8⁺ T cells in mercury-induced autoimmunity or immunosuppression in the rat. *Scandinavian Journal of Immunology*, **31**, 65–74.

Pietsch, P., Vohr, H.W., Degitz, K., and Gleichmann, E. (1989). Immunological alterations inducible by mercury compounds. II. HgCl₂ and gold sodium thiomalate enhance serum IgE and IgG concentrations in susceptible mouse strains. *International Archives of Allergy and Applied Immunology*, **90**, 47–53.

Pusey, C.D., Bowman, C., Peters, D.K., and Lockwood, C.M. (1983). Effects of cyclophosphamide on autoantibody synthesis in the Brown-Norway rat. *Clinical and Experimental Immunology*, **54**, 697–704.

Pusey, C.D., Bowman, C., Morgan, A., Weetman, A.P., Hartley, B., and Lockwood, C.M. (1991). Kinetics and pathogenicity of autoantibodies induced by mercuric chloride in the Brown-Norway rat. *Clinical and Experimental Immunology*, **81**, 76–82.

Pusey, C.D., et al. (1990b). Autoimmunity to glomerular basement membrane induced by homologous and isologous in Brown-Norway rats. *Nephrology, Dialysis, Transplantation*, in press.

Rehan, A., Johnson, K.J., Wiggins, R.C., Kunkel, R.G., and Ward, P.A. (1984). Evidence for the role of oxygen radicals in acute nephrotoxic nephritis. *Laboratory Investigation*, **51**, 398–403.

Rehan, A., Johnson, K.J., Kunkel, R.G., and Wiggins, R.C. (1985). Role of oxygen radicals in phorbol myristate-induced glomerular injury. *Kidney International*, **27**, 501–3.

Rennke, H.G., Klein, P.S., and Mendrick, D.L. (1990). Cell-mediated immunity in hapten-induced interstitial nephritis and glomerular crescent formation in the rat. *Kidney International*, **37**, 428a.

Rees, A.J., Peters, D.K., Compston, D.A.S., and Batchelor, J.R. (1978). Strong association between HLA DRw2 and antibody mediated Goodpasture's syndrome. *Lancet*, **i**, 966–8.

Rifai, A. (1987). Experimental models for IgA-associated nephritis. *Kidney International*, **31**, 1–7.

Robinson, C.J.G., Balazs, T., and Egorov, I.K. (1986). Mercuric chloride−, gold sodium thiomalate−, and D-penicillamine-induced antinuclear antibodies in mice. *Toxicology and Applied Pharmacology*, **86**, 159–69.

Roman-Franco, A.A., Turiello, M., Albini, B., Ossi, E., Milgrom, F., and Andres, G.A. (1978). Anti-basement membrane antibodies and antigen-antibody complexes in rabbits injected with mercuric chloride. *Clinical Immunology and Immunopathology*, **9**, 464–81.

Rose, L.M., Goldman, M., and Lambert, P.H. (1982). Simultaneous induction of an idiotype, of corresponding anti-idiotypic antibodies and of immune complexes during African trypanosomiasis in mice. *Journal of Immunology*, **128**, 79–85.

Rosenberg, Y.J., Steinberg, A.D., and Santoro, T.J. (1983). The basis of autoimmunity in mrl-1pr/1pr mice: a role for self Ia-reactive T cells. *Immunology Today*, **5**, 64–7.

Rossert, J., Pelletier, L., Pasquier, R., and Druet, P. (1988). Autoreactive T cells in mercury-induced autoimmunity. Demonstration by limiting dilution analysis. *European Journal of Immunology*, **18**, 1761–6.

Sado, Y., Okigaki, T., Takamiya, H., and Seno, S. (1984). Experimental autoimmune glomerulonephritis with pulmonary hemorrhage in rats. The dose-effect relationship of the nephritogenic antigen from bovine glomerular basement membrane. *Journal of Clinical and Laboratory Immunology*, **15**, 199–205.

Salant, D.J., Belok, S., Madaio, M.P., and Couser, W.G. (1980). A new role for complement in experimental membranous nephropathy in rats. *Journal of Clinical Investigation*, **66**, 1339–50.

Sapin, C., Druet, E., and Druet, P. (1977). Induction of anti-glomerular basement membrane antibodies in the Brown-Norway rat by mercuric chloride. *Clinical and Experimental Immunology*, **28**, 173–9.

Sapin, C., Hirsch, F., Delaporte, J.P., Bazin, H., and Druet, P. (1984). Polyclonal IgE increase after HgCl₂ injections in BN and LEW rats. A genetic analysis. *Immunogenetics*, **20**, 227–36.

Sato, M., Ideurat, T., and Koshikawa, S. (1986). Experimental IgA nephropathy in mice. *Laboratory Investigation*, **54**, 377–84.

Schmiedeke, T.M.J., Stockl, F.W., Weber, R., Sugisaki, Y., Batsford, S.R., and Vogt, A. (1989). Histones have affinity for the glomerular basement membrane. Relevance for immune complex formation in lupus nephritis. *Journal of Experimental Medicine*, **169**, 1879–94.

Schoenfeld, Y., et al. (1983). Polyspecificity of monoclonal lupus autoantibodies produced by human-human hybridomas. *New England Journal of Medicine*, **308**, 414–20.

Schreiner, G.F., Cotran, R.S., Pardo, V., and Unanue, E.R. (1978). A mononuclear cell component in experimental immunological glomerulonephritis. *Journal of Experimental Medicine*, **147**, 369–84.

Steblay, R.W. (1962). Glomerulonephritis induced in sheep by injections of heterologous glomerular basement membrane and Freund's complete adjuvant. *Journal of Experimental Medicine*, **116**, 253–72.

Steinglein, B., Thoenes, G.H., and Günther, E. (1978). Genetic control of susceptibility to autologous immune complex glomerulonephritis in inbred rat strains. *Clinical and Experimental Immunology*, **33**, 88–94.

Strosberg, A.D. (1983). Anti-idiotype and anti-hormone receptor antibodies. *Springer Seminars of Immunopathology*, **6**, 67–78.

Stuffers-Heimann, M., Günther, E., and Van Es, L.A. (1979). Induction of autoimmunity to antigens of the glomerular basement membrane in inbred Brown-Norway rats. *Immunology*, **36**, 759–67.

Theofilopoulos, A.N. and Dixon, F.J. (1985). Murine models of systemic lupus erythematosus. *Advances in Immunology*, **37**, 269–390.

Theofilopoulos, A.N., Kofler, R., Noonan, D., Singer, P., and Dixon, F.J. (1986). Molecular aspects of murine systemic lupus erythematosus. *Springer Seminars of Immunopathology*, **9**, 121–42.

Theofilopoulos, A.N., Kofler, R., Singer, P.A., and Dixon, F.J. (1989). Molecular genetics of murine lupus models. *Advances in Immunology*, **46**, 61–109.

Thoenes, G.H., Pielsticker, H., and Schubert, G. (1979). Transplantation-induced immune complex kidney disease in rats with unilateral manifestation in the allografted kidney. *Laboratory Investigation*, **41**, 321–33.

Thomson, N.P., Moran, J., Simpson, I.J., and Perters, D.K. (1976). Defibrination with ancrod in nephrotoxic serum nephritis in rabbits. *Kidney International*, **10**, 343–7.

Tipping, P.G., Neale, T.J., and Holdsworth, S.R. (1985). T lymphocyte participation in antibody-induced experimental glomerulonephritis. *Kidney International*, **27**, 530–7.

Tournade, H., Pelletier, L., Glotz, D., and Druet, P. (1989). Toxiques et auto-immunité. *Médecine/Sciences*, **5**, 303–10.

Tournade, H., Pelletier, L., Pasquier, R., Vial, M.C., Mandet, C., and Druet, P. (1990a). Graft-versus-host reactions in the rat mimic toxin-induced autoimmunity. *Clinical and Experimental Immunology*, **81**, 334–8.

Tournade, H., Pelletier, L., Pasquier, R., Vial, M.C., Mandet, C., and Druet P., (1990b). D-penicillamine-induced autoimmunity in Brown-Norway rats: similarities with HgCl₂−induced autoimmunity. *Journal of Immunology*, **144**, 2985–91.

Unanue, E.R. and Dixon, F.J. (1965). Experimental glomerulonephritis. VI. The autologous of nephrotoxic serum nephritis. *Journal of Experimental Medicine*, **121**, 715–25.

Tomosugi, N.I., et al. (1989). Modulation of antibody-mediated glomerular injury in vivo by bacterial lipopolysaccharide, tumor necrosis factor and I₂−1. *Journal of Immunology*, **142**, 3083–90.

Van Damme, B.J.C., Fleuren, G.J., Bakker, W.W., Vernier, R.L., and Hoedemaeker, Ph.J. (1978). Experimental glomerulonephritis in the rat induced by antibodies against tubular antigens. V. Fixed

glomerular antigens in the pathogenesis of heterologous immune complex glomerulonephritis. *Laboratory Investigation*, **38**, 502–10.
Verroust, P., Ronco, P., and Chatelet, F. (1987). Antigenic targets in membranous glomerulonephritis. *Springer Seminars of Immunopathology*, **9**, 341–58.
Via, C.S. and Shearer, G.M. (1988). T-cell interactions in autoimmunity: insights from a murine model of graft-versus-host disease. *Immunology Today*, **9**, 207–13.
Vogt, A., Batsford, S.R., Rodriguez-Iturbe, B., and Garcia, R. (1983). Cationic antigens in poststreptococcal glomerulonephritis. *Clinical Nephrology*, **20**, 271–9.
Weening, J.J., Fleuren, G.J., and Hoedemaeker, P.J. (1981). Immunoregulation and antinuclear antibodies in mercury-induced glomerulopathy in the rat. *Clinical and Experimental Immunology*, **45**, 64–71.
Wieslander, J., Kataja, M., and Hudson, B.G. (1987). Characterisation of the human Goodpasture antigen. *Clinical and Experimental Immunology*, **69**, 332–40.

Wilson, C.B. and Dixon, F.J. (1986). The renal response to immunological injury. In *The Kidney*, (eds. Brenner, B.M. and Rector, F.C.) p. 800. W.B. Saunders, Philadelphia.
Yamamoto, T. and Wilson, C.B. (1987). Quantitative and qualitative studies of antibody-induced mesangial cell damage in the rat. *Kidney International*, **32**, 514–25.
Zanetti, M. and Wilson, C.B. (1983). Participation of auto-anti-idiotypes in immune complex glomerulonephritis in rabbits. *Journal of Immunology*, **131**, 2781–83.
Zanetti, M., Glotz, D., and Rogers, J. (1986a). A regulatory idiotype on autoantibodies. In *Idiotypes*, (eds. J. D. Capra and M. Reichlin), pp. 591–6. Academic Press, New York.
Zanetti, M., Glotz, D., and Rogers, J. (1986b). Perturbation of the autoimmune network II: immunization with isologous idiotype induces auto-anti-idiotypic antibodies and suppresses the autoantibody response elicited by antigen: a serologic and cellular analysis. *Journal of Immunology*, **10**, 3140–6.

3.3 Fluid retention in renal disease: the genesis of renal oedema

E.J. DORHOUT MEES

Introduction

The most frequent disturbance of renal function is insufficient excretion of salt and water, which results in oedema. The word (which means swelling) is a clinical term and as such ill-defined: it is used when the eyes of a patient look 'puffy' or when persistent 'pitting' can be demonstrated, usually in the lower parts of the body. Important amounts of fluid retention (up to 5 l) may not be detected in this way. This chapter will deal with 'fluid excess' rather than oedema in the strict sense. In a way, every oedema is 'renal', because the excess of fluid must have been retained by the kidney, even when the primary disturbance is elsewhere (e.g. lymphatic obstruction). As we will show later, distinction between primary and secondary renal fluid retention proves to be very difficult. This arises because the main task of the kidney is to keep the composition and volume of the body fluids within normal limits. The latter, volume regulation, is eventually aimed at maintenance of the blood circulation, and closely related to regulation of the blood pressure. It is of such importance that its control involves multiple mechanisms, outside as well as inside the kidney itself. Nature apparently has built many safeguards against disturbance, and as a result made it extremely difficult for investigators to unravel the processes of normal regulations.

Regulation of salt and water balance

Volume- and osmoregulation

The extracellular fluid (of which the blood is an integral part) is mainly a solution of sodium chloride and bicarbonate. It is essential to realize that osmolality and volume are regulated separately by the kidney. Osmoregulation is accomplished by varying intake and excretion of water while volume is primarily dependent on the sodium content of the body (Peters 1948). Consequently the afferent (extrarenal) part of volume regulation is triggered by changes in volume, but the efferent (renal) part is accomplished by modulating Na (Cl) excretion.

Normally the two mechanisms are performed within the kidney simultaneously, but they can be dissociated in pathological conditions: one of these is the so-called 'inappropriate ADH secretion syndrome'. When this is experimentally produced (Fig. 1) by giving water to a normal subject while preventing the kidney excreting it by giving antidiuretic hormone, a 'paradoxical' natriuresis occurs, which aggravates the hyponatraemia (Leaf *et al.* 1953). When (as often happens) a physician tries to correct the hyponatraemia in such a patient by infusing NaCl solution, this provokes a natriuresis, which may strengthen the false impression that the patient is 'salt losing'. In fact, he is slightly overhydrated, and his kidney reacts in a normal way as far as one part is concerned. This example shows that the kidney actually regulates 'volume' by changing NaCl excretion. Restriction of water in the hope of decreasing oedema is an erroneous 'treatment' because it primarily increases osmolality and thirst. To give water without salt for the purpose of correcting loss of extracellular fluid (dehydration) is equally illogical. Because cell membranes are freely permeable to water, 1 l of water will expand the extracellular space with ± 0.4 l, while 0.6 l will go into the cells. Conversely, giving salt to an overhydrated patient because of hyponatraemia is inadvisable. It can be calculated that for a man of 70 kg who has oedema of about 10 l, 500 mmol (± 30 g) NaCl would be required to increase plasma sodium con-

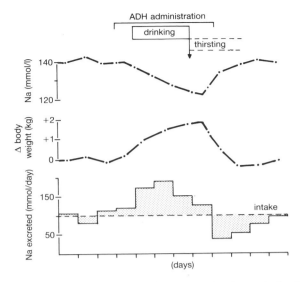

Fig. 1 Schematic presentation of sequence leading to induction and correction of experimental 'inappropriate antidiuretic hormone (ADH)' syndrome (see text).

centration by 10 mmol, while interstitial fluid would increase 3.6 l by a shift from intra- to extracellular water.

Volume receptors

It is logical to consider intravascular volume as the variable which is primarily regulated, interstitial volume being changed indirectly via the 'Starling forces' (see below). It has proved difficult to localize areas in the vascular tree where changes are sensed, or to identify the signals that tell the kidney to change its sodium output. The problem starts by defining 'normal' volume.

Because blood may be 'pooled' in some areas or 'shunted' without achieving adequate tissue perfusion, a more dynamic concept of 'effective circulating volume' has been proposed. 'Volume' receptors which actually sense pressure or stretch are present in the arterial as well as in the venous side. Baroreceptors in the carotid sinus and aortic arch react on decreased pressure and protect against deficient circulation. They induce increased sympathetic activity, which at the same time helps to maintain blood pressure and to retain sodium and water (Kon et al. 1985). If the circulation is severely compromised, non-osmotic stimulation of thirst mechanism and antidiuretic hormone (ADH) by angiotensin occurs. This induces hyponatraemia and may also increase blood pressure by vasoconstriction (Schrier 1988). It is unlikely that high ADH levels increase sodium retention, but they are certainly a marker of pathological volume deregulation.

A second site where arterial pressure is sensed lies in the kidney itself: the juxtaglomerular apparatus. Deficient volume directly (as well as via sympathetic impulses) stimulates release of renin. This generates the powerful vasoconstrictor angiotensin II which raises the blood pressure. In addition it causes sodium retention by two mechanisms: it increases Na reabsorption in the proximal nephron, and induces aldosterone secretion by the adrenal cortex, which in turn stimulates Na reabsorption in the distal nephron (see below).

On the venous side of the heart, stretch receptors in the right atrium have been found which release atrial natriuretic factor (ANF) in case of overhydration. ANF (otherwise known as ANP) is a peptide hormone which increases sodium excretion by the kidney. It is probable that it has an important physiological function in acute changes, but its role in long-term regulation is less certain. Nevertheless the plasma concentration of ANF is a good marker for the state of hydration, and varies more or less inversely with renin activity. It is clear that if the heart fails, this regulation will be completely deranged and both volume excretion (ANF) and retention (low arterial pressure, renin–angiotensin) simultaneously stimulated. Apparently the latter will prevail in heart failure. It seems probable at present that not all factors involved are yet delineated.

Normal renal salt and water handling (see also Chapters 1.4 and 7.1)

General remarks

The mammalian kidney functions in such a way that hardly any NaCl is excreted by normal animals, and current human consumption is a habit, with features of addiction. Sodium reabsorption occurs all along the tubular system in a stepwise fashion and is primarily accomplished through active transport. Passive transport of NaCl also contributes, notably in the proximal tubules. Water follows sodium passively, but in a very selective way in different segments. Despite a tremendous amount of research, there still remains much uncertainty as to the quantitative contribution in the different tubular segments. It is important to consider two general features of this system.

Reabsorption in every tubular segment is dependent on delivery. Thus, with increasing loads, absolute reabsorption rises, while 'fractional' reabsorption (as part of the delivered amount) may decrease. When glomerular filtration rate (GFR) is increased, proximal reabsorption increases. When delivery to the thick ascending limb (TAL) of Henle's loop increases, more NaCl is reabsorbed there. The same applies to the distal and collecting tubular segments. In this way multiple possibilities exist for the kidney to 'compensate' for a given disturbance at a particular level. In addition a special regulatory system exists at the macula densa in the first part of the distal convoluted tubule: the tubuloglomerular feedback (TGF).

The (passive) water transport is dependent on the hydraulic permeability of the tubules and varies in the different segments: It is very high in the proximal tubules and thin descending limb, low in the thick ascending loop. In the distal convoluted tubules and the collecting duct it depends on the presence of antidiuretic hormone (ADH). Despite this separation in salt and water handling, it is safe to conclude that no single segment is completely impermeable to water in any circumstances. In addition the ability to generate 'free water' is dependent on the delivery of filtrate to the 'diluting sites': when there is avid sodium retention, water excretion is impaired.

Renal haemodynamics

Sympathetic activity and angiotensin II increase resistance mainly in the efferent glomerular capillaries, decreasing renal blood flow while filtration is maintained. This leads to changes in physical forces in the peritubular capillaries (increased oncotic and decreased hydrostatic pressure) which promote salt and water reabsorption in the proximal tubules. Concomitantly, local prostaglandin synthesis is activated opposing afferent constriction. Thus, inhibition of the latter may precipitate a fall in GFR and further fluid retention. Conversely, inhibition of angiotensin generation by converting enzyme blockade may promote sodium excretion, also by a direct effect on the proximal tubules.

Much emphasis has been laid in the past on the fact that changes in GFR are not important determinants of NaCl excretion. However, it cannot be denied that a normal filtration per nephron (SNGFR) is a prerequisite for normal sodium and water handling and that important decreases in SNGFR are bound to impair excretion. Moreover, nearly all manoeuvres that change sodium excretion lead to parallel changes in GFR.

Blood pressure is a factor which at first sight has little to do with sodium excretion because both may vary independently. Yet, when blood pressure can be isolated from simultaneous changes in other factors it proves to be an important determinant of volume regulation. Conversely, the sodium excretion threshold of the kidney is the major factor determining the set point for chronic blood pressure regulation. Whereabouts in the kidney blood pressure affects sodium reabsorption is not known.

Tubular reabsorption

Apart from the changes induced indirectly by glomerular haemodynamics, both sympathetic activity (Di Bona 1977) and angiotensin II directly stimulate proximal reabsorption.

Reabsorption in the loop of Henle and distal convoluted tubules is very 'load dependent'. There is no indication that they contribute independently to regulation of normal balance (Barton et al. 1972). Prostaglandins decrease both sodium and water reabsorption in this region and their inhibition may cause retention in some individuals.

Despite the relatively small fraction of filtered load that is reabsorbed in the late distal and collecting duct, the critical location of these tubular segments makes them of paramount importance in the maintenance of sodium balance. It is here that aldosterone, the well-known antinatriuretic hormone, stimulates sodium reabsorption, while facilitating potassium excretion. Without aldosterone, it is impossible to conserve sodium and severe dehydration may result. Yet this loss is a small fraction of filtered load and can be compensated by liberal supplementation. Atrial natriuretic factor (ANF) probably has its main action also in these parts of the nephron. This makes it understandable why its effect is much decreased in conditions of sodium retention where delivery of fluid of these segments is reduced.

Limits of renal performance

While the normal kidney is able to produce practically sodium-free urine, its capacity to excrete sodium is limited. A few studies have investigated the normal upper limit. It appears that the increased excretion after acute infusions of saline is, for unknown reasons, quite variable among individuals. Less acute studies (days rather than hours or minutes) have shown that most normal subjects are able to excrete up to 1500 mmol NaCl per day (only 6 per cent of filtered load), but sometimes at the cost of considerable signs of overhydration.

It must be remembered that with longer periods of time better adaptation occurs. However, it is remarkable that a single salt-retaining stimulus does not usually lead to progressive retention. The best known example is the 'escape' from prolonged mineralocorticoid action. If this syndrome is reproduced experimentally by giving daily aldosterone or an analogue (DOCA) to a normal person, sodium excretion decreases only temporarily but comes back to control levels after 150 to 300 mmol NaCl is retained, because natriuretic mechanisms are activated and compensate the sodium-retaining stimulus. Other escape phenomena can be observed with inappropriate ADH secretion and after sodium retention by non-steroidal anti-inflammatory drugs. They illustrate the extreme complexity of volume regulation.

Pathophysiology of oedema formation

Distribution of body fluids ('Starling forces')

The extracellular fluid consists of a vascular (plasma) and an extravascular (interstitial) compartment. These two fluid spaces are in dynamic equilibrium, which is maintained by the 'Starling forces' (Guyton et al. 1975). The vascular compartment also contains a large erythrocyte mass, which is relatively constant and determined by factors unrelated to volume changes. Thus, for practical purposes, blood volume is regulated by changes in plasma volume. Because the capillaries are permeable to water and electrolytes but not to large (colloidal) particles, ultrafiltration at this level can be represented by the formula

$$J_v = k(P_c - P_i) - (\pi_c - \pi_i)$$

where P_c and P_i are capillary and interstitial hydrostatic pressure, and π_c and π_i are capillary and interstitial colloid osmotic pressure respectively, while k is the ultrafiltration coefficient. In the steady state, the resulting flow (J_v) must return to the blood and thus be equal to the lymph flow, which is about 2 to 4 l per 24 h in normal conditions. In fact, much more fluid leaves the early capillaries but is taken up again more distally, as hydrostatic pressure in the capillary falls along its length. It is estimated that this 'tissue perfusion fluid' amounts to 40 to 60 l per day. Although permeability for proteins is negligible in relation to these large fluxes, it is sufficient for an amount equal to the total intravascular albumin mass to be filtered within 24 h. It has only recently been established that the interstitial fluid contains large amounts of albumin (40 to 60 per cent of the plasma concentration in different tissues), which is responsible for a value of π of about 14 mmHg. In addition, it is important that interstitial hydrostatic pressure (p) is normally negative, (estimated at -2 to -5 mmHg) (Fig. 2(a)).

Consequences of disturbances in 'Starling forces'

Let us consider what happens to these forces when the blood is expanded with a saline solution (Schad and Brechtelsbauer 1978). This will cause a drop in π, leading to increased filtration (J_v). If nothing else changes, this would go on until all of the administered fluid has left the plasma. However, the increased J_V value will cause some expansion of the interstitial space. Because the compliance of this space is low in normal condition, this small increase in volume will raise hydrostatic pressure (P_i) which in turn increases lymph flow. This lymph, containing rather too much albumin, is returned to the blood. At the tissue level it is replaced by ultrafiltrate of very low albumin content. Thus intravascular albumin increases while tissue π will drop. The gradient ($\pi_p - \pi_i$) will return to its original level, and equilibrium is restored while the administered fluid is proportionally distributed between plasma and tissues. The same mechanisms were shown to operate in the lung circulation (Erdmann et al. 1975).

Similar changes will occur when there is a primary drop in plasma (hypoalbuminaemia) or an increase in intracapillary hydrostatic pressure (venous congestion). At the expense of only

3.3 Fluid retention in renal disease: the genesis of renal oedema

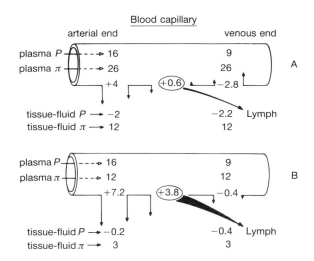

Fig. 2 Estimate of 'Starling forces' in peripheral capillaries. Numbers represent pressure in mmHg. $P=$ hydrostatic pressure, π = osmotic pressure. A = normal condition, B = nephrotic syndrome. At the lower border of each schematic capillary, net ultrafiltration pressure at the beginning and end are given. Encircled: mean net pressure difference, leading to a more or less proportional fluid movement, which is carried away with lymph.

a small amount of plasma fluid, which can easily be replenished by retaining some sodium, a small increase in interstitial volume (undetectable by clinical means) will cause relatively large changes in π_i which soon restore the balance (Fig. 2(b)). The three factors interstitial hydrostatic pressure (P_i), osmotic pressure π_i, and lymph flow have been called 'oedema preventing factors'. It would be equally justifiable to denote them as 'blood volume preserving factors'. When P_i reaches zero, interstitial compliance increases rather abruptly. Oedema appears, and further increase in fluid will change P_i very little. Lymph flow has also reached its maximum of $\pm$ 30 l per 24 h) (Fig. 3) and tissue albumin will have been almost washed out, resulting in a very low π_i. Thus the 'preventing factors' are exhausted, and any retained or administered fluid will be shifted completely to the tissues. There is no more protection against decrease in blood volume. If a further decrease in π_c occurs in that situation, the only way to prevent plasma fluid from escaping to the tissues would be a decrease in capillary (and arterial) hydrostatic pressure. These considerations make it clear why plasma volume can be maintained within normal limits even with severe hypoproteinaemia, but on the other hand is relatively independent of changes in extracellular fluid in that condition. When plasma proteins are normal however, large changes can be expected in plasma and blood volume when extracellular volume is changed. These facts are illustrated in Fig. 4.

Capillary leak

Little is known about possible changes in the other factors, k and P_c, which may conceivably influence fluid partition between blood and tissues. It has been suggested that an increase in the permeability coefficient k plays a role in oedema formation. In view of the fact that the actual amount of filtrate greatly exceeds lymph flow, but is reabsorbed again by the same capillaries, it is theoretically unlikely that changes in k would be important in changing net ultrafiltration. Another factor is, of course, the capillary reflection coefficient for colloids which normally is very

high. Decrease of this coefficient would indeed greatly endanger maintenance of the intravascular volume, soon leading to a situation incompatible with life. Anaphylactic shock and extensive burns may represent examples of such a condition.

Localized oedema

When venous pressure is raised and outflow impaired in a limb of an otherwise normal subject, all oedema-preventing factors will be mobilized: lymph flow will increase, π_i will drop, and interstitial pressure will increase. No appreciable oedema will occur, as is indeed the case in artificial arteriovenous fistulae ('Cimino shunts'). However, when lymph flow is impaired, gross oedema will result, because wash-out of interstitial albumin is stopped and π_i will rise, attracting more fluid. Balance may only be reached when interstitial pressure rises to very high positive values, i.e. 'tense' oedema.

Primary and secondary oedema

It follows from the foregoing discussion that it will be not as easy as was originally believed to define exactly the sequence of events leading to, and the factors responsible for, the maintenance of, generalized oedema. Traditionally a sharp distinction is made between oedematous states which result from pathologic inability of the kidney to excrete the normally ingested amounts of NaCl, and those conditions in which this retention is secondary, due to a physiological response of a normal kidney to the (apparent) need of the body to restore an 'effective' blood volume. In the latter situation, the retained fluid fails for various reasons to reach this goal, perpetuating the retention and causing oedema. However in some forms of presumed secondary

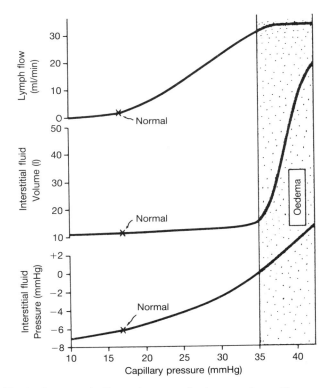

Fig. 3 Computed effects of progressive increase in capillary pressure on interstitial fluid pressure, interstitial fluid volume, and lymph flow. (Taken from Guyton (1975) with permission.)

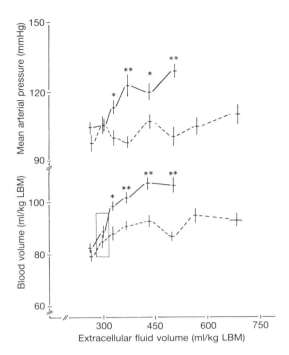

Fig. 4 Changes in blood volume and blood pressure in relation to induced changes in extracellular volume in patients with nephrotic syndrome (stippled line) and with renal failure but normal plasma albumin levels (continuous line). (Taken from Koomans *et al.* (1987) with permission.)

retention it is difficult to prove the existence of a 'deficient circulation', while there are also arguments that the kidney is not 'normal' as regards salt and water handling. In particular this 'hypovolaemia concept' has been challenged in patients with the nephrotic syndrome, to which we will devote most of our attention. Because of the related problems existing for the role of the kidney in any kind of overhydration, we will also briefly review other forms of oedema.

Renal fluid retention

Acute nephritic syndrome

This syndrome, most often caused by poststreptococcal glomerulonephritis, is the prototype of primary renal oedema. The sudden decrease in urine and sodium output in a patient who did not feel ill and continued his normal (salt-rich) diet leads to fluid retention and generalized oedema. It has been suggested that periorbital oedema reliably distinguishes 'nephritic' oedema from other forms, because of increased capillary permeability. As discussed in the preceding section, there are no grounds for believing that such an abnormality operates in any form of renal oedema. The cause of periorbital oedema is simply recumbency at night; it may be seen in other oedematous conditions as well. Further clinical features in these patients are hypertension, increased cardiac output (Binak *et al.* 1975), elevated venous pressure (to be seen in the neck veins), and cardiac enlargement. As a result of overhydration, plasma renin activity is decreased (Powell *et al.* 1974) and the elevated blood pressure must be considered as a 'volume hypertension'. In severe cases pulmonary congestion and oedema occur. Because it is unusual for 'heart failure' to occur in a young individual with a normal heart, some kind of cardiac damage has been suggested to accompany poststreptococcal nephritis. However, this has not been confirmed, and it is quite conceivable that pulmonary oedema will occur despite a normal heart during a large fluid overload, while a normal protein content of blood and interstitium lead to a proportional increase in blood volume. Pulmonary oedema is particularly apt to occur after acute increases in volume.

Renal mechanism for the fluid retention should be looked for primarily in the glomeruli, where microscopical analysis shows intense inflammation and proliferation. However, it is not exactly clear how this leads to sodium retention. In animal models comparable with human acute nephritis, micropuncture studies confirmed the presence of a decrease in ultrafiltration coefficient and hypofiltration, largely compensated by an increase in filtration pressure, due to decrease in afferent resistance. In most studies, fractional proximal reabsorption was normal or low, because the decrease in filtered load was balanced by a decrease in peritubular π and increase in peritubular hydrostatic pressure (Bennet *et al.* 1979). Thus glomerulotubular balance was preserved. Although the reported amounts of filtrate delivered from the proximal tubules varied, the main cause of sodium retention was attributed to relatively elevated reabsorption in the distal nephron (Wen and Wagnild 1976). The latter was not related to plasma aldosterone, nor was it responsive to volume expansion. Less acute ('rapidly progressive') forms of glomerulonephritis may give rise to fluid retention with basically similar pathophysiology.

Other forms of acute renal failure

These are mainly various forms of tubular necrosis and acute interstitial (drug-induced) nephritis. This is usually seen in very ill patients and oedema is not the first sign to be noticed. The amount of overhydration is determined by the balance between infusions and mechanical ultrafiltration with or without dialysis, which has usually to be applied.

The nephrotic syndrome
History and definition

The term 'nephrosis' was introduced in 1905 in an attempt to distinguish between the several disease entities that were hitherto collectively called 'Morbus Bright'. It was applied to those proteinuric patients whose kidneys showed no signs of glomerular inflammation ('nephritis') but in whom only tubular 'degenerative' abnormalities were seen. Because the concept of a glomerular disease without visible abnormalities did not appeal, it was considered to be a tubular disease. Later it was shown that the high fat content of the tubular cells (which had led to the term 'lipoid nephrosis') was not degenerative but an expression of excessive protein reabsorption secondary to glomerular leakage. Because a similar clinical picture could be seen with various glomerular abnormalities, the term 'nephrotic syndrome' gained general acceptance (Bradley and Tyson 1948). This constituted a conceptual change from a histological to a pathophysiological entity. Although the nephrotic syndrome is usually defined as the combination of proteinuria (exceeding an arbitrary amount), hypoalbuminaemia and oedema, it is more or less tacitly assumed that the oedema is 'the predictable consequence' of proteinuria which, by causing hypoalbuminaemia, sets into motion pathophysiological processes leading to oedema (Glassock *et al.* 1981). As will be shown in the following pages, this pathophysiology is much less evident and uniformly present than formerly believed.

The hypovolaemia concept

This concept, which traditionally forms the basis of the diagnosis of 'nephrotic' (as opposed to 'nephritic') oedema, can be summarized as follows. As a result of the massive protein loss (mainly the albumin fraction) with the urine, plasma albumin concentration drops, because synthetic capacity of the liver is not sufficient to compensate. This means that plasma colloid osmotic pressure π (which is mainly determined by albumin concentration) decreases in parallel. Because π is the main driving force in keeping plasma fluid from escaping to the tissue, the drop in plasma π causes a fluid shift from the blood to the interstitium. The resulting decrease in plasma and blood volume (hypovolaemia) will stimulate the kidneys to retain salt and water by the same mechanisms as in other hypovolaemic states. These mechanisms include increased sympathetic tone and elevated renin–angiotensin–aldosterone levels. As long as the disequilibrium of the capillary fluid exchange remains, the retained fluid will continue to accumulate in the interstitial spaces, leading to the well-known state of massive oedema.

It is clear that this concept is founded on two postulates: the first is that the peripheral circulation shows the characteristics of an insufficient circulating volume and the second is that the kidney function is normal as far as salt and water handling is concerned. In the following paragraphs we will review the evidence for and against these two assumptions.

Clinical (physiological) findings

Blood volume It has been assumed for a long time that blood volume is low in the nephrotic syndrome, and evidence to the contrary was neglected or ascribed to technical errors. More recently it was shown that blood volume is normal or even slightly elevated in the majority of the patients (Geers *et al.* 1984) although low volumes have been reported in from 7 to 38 per cent of the cases in different studies (Dorhout Mees *et al.* 1984). Some of this variability may be due to differences between the groups studied, children presenting more often with extreme hypoproteinaemia and a low blood volume. On the other hand, normal values are subject to biological variation depending on body build. Oedema will increase the value of body weight used to calculate normal reference values. Failure to take this into account alone was responsible for some reports of decreased plasma volumes. Use of before disease weight (or ideal weight from height) may still overestimate normal values because adult patients may lose up to 10 kg 'real weight' during the nephrotic episode. Investigators using measurements obtained soon after recovery from a nephrotic episode found an elevated blood volume in the majority of patients (Kelsch *et al.* 1972; Dorhout Mees *et al.* 1984).

Volume determinations have generally been performed during recumbency. On standing, patients with the nephrotic syndrome show a larger drop in blood volume (approximately −9 per cent) than normal subjects (approximately −5 per cent). Thus the normal stimulus for volume retention in that position is probably exaggerated. Acute reduction of plasma volume results in haemoconcentration. An elevated haematocrit is, however, not the rule in the nephrotic syndrome, values often being on the low normal side. Occasionally, however, an elevated haematocrit may be found, particularly at the onset or in acute exacerbations, especially in children.

It has been argued that the degree of hypovolaemia needed to cause renal sodium retention may be too delicate to be detected with conventional methods, and that arterial filling is indeed low. This problem may be obviated by looking for other signs expected in reduced effective circulating volume.

Blood pressure Absence of hypertension was considered to be a criterion for 'uncomplicated' nephrotic syndrome. Yet blood pressure is often mildly elevated even in 'minimal change' disease (Cameron *et al.* 1974), and subnormal values are uncommon. Occasionally, the blood pressure may even be severely elevated. Patients with oedema due to cirrhosis of the liver (to which the nephrotic oedema is often compared), usually present with subnormal blood pressures. This implies that the pathogeneses of oedema in cirrhosis of the liver and nephrotic syndrome are essentially different.

Plasma renin activity (PRA) and aldosterone These usually reflect 'effective circulating volume'. Elevated values would be expected if blood volume were decreased, even in the presence of extravascular fluid excess. Yet a wide range of low to high values of PRA and aldosterone have been found in patients with definite histological lesions as well as in patients with 'minimal change' actively retaining sodium. While there is no doubt that in the nephrotic syndrome sodium retention can exist in the absence of a stimulated renin–aldosterone system (Danielsen 1985), an inverse relationship between plasma aldosterone and sodium excretion is reported in several studies. Thus, according to the renin–aldosterone profile, volume retention in some patients seems to be due to a primary renal impairment in excreting sodium, whereas in others a hypovolaemic factor plays a role.

Several observations suggest that the role of the renin–angiotensin system in the pathogenesis of nephrotic oedema is more complex. For instance, no relationship was found between blood volume and PRA in individual patients, (Geers *et al.* 1984). Often PRA was high, despite an elevated blood pressure and a normal or slightly increased blood volume. A retarded fall in PRA was found after albumin infusion, values still falling when blood volume had long passed its maximum (Koomans *et al.* 1984) and PRA often does not decrease after volume expansion (Chonko *et al.* 1977). It cannot be excluded therefore that, for unknown reasons in some patients with the nephrotic syndrome at least renin release is inappropriate, and influenced by intrarenal changes.

Other neurohumoral factors Some studies have reported increased antidiuretic hormone (ADH) concentrations in the nephrotic syndrome. This was associated with increased as well as normal renin, aldosterone, and catecholamines, but often also with an elevated blood pressure. Increased urinary noradrenaline excretion and exaggerated rise in plasma noradrenaline upon tilt have been found in children (Kelsch *et al.* 1972) and in a minority of adults with the nephrotic syndrome. Atrial natriuretic factor plasma levels have been reported to be either normal or increased (Tulassay *et al.* 1987).

Albumin distribution and Starling forces

Total albumin mass in the nephrotic syndrome is not only reduced, but the partition between intra- and extravascular fraction is also changed. While normally 60 per cent of albumin is distributed in the interstitial fluid this proportion is decreased to 50 per cent or less in nephrotics, despite an increased interstitial volume. It has recently been confirmed that interstitial albumin

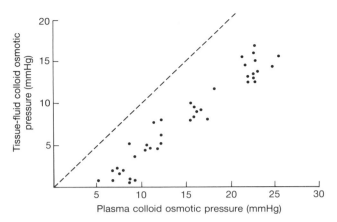

Fig. 5 Relationship between plasma and tissue fluid colloid osmotic pressure in 12 patients with nephrotic syndrome and during follow up with partial or complete recovery. (Taken from Koomans et al. (1985) with permission.)

concentration and colloid osmotic pressure in the tissue fluid (π_i), is decreased in the nephrotic syndrome, and proportionally more than that in the plasma (π_c) (Koomans et al. 1985). Thus the gradient ($\pi_c - \pi_i$) remains close to the normal value (Fig. 5), and there is no driving force for a drop in blood volume. That lymph flow is greatly increased has been shown long ago by Hollander (1961). When oedema is present, tissue compliance becomes very high and interstitial pressure will hardly change when tissue fluid volume changes (Guyton et al. 1975). Thus oedema-preventing factors are largely spent, and when the kidney for any reason retains fluid, this will accumulate preferentially in tissue, without a change in blood volume.

In contrast, in normal conditions the volume retention would elevate both blood volume and interstitial volume (Koomans et al. 1986). This explains the different course of blood volume and blood pressure during removal of oedema in normoproteinaemic patients with renal failure, and in hypoproteinaemic patients with nephrotic syndrome (Fig. 4). Note that in the nephrotic group the blood volume changed little over a large range of fluid excess, and only dropped below normal in dehydration. The mean plasma π in these patients was 11 mmHg. Joles et al. (1988) showed that dogs with experimental hypoproteinaemia were able to excrete sodium normally, and that only when plasma π dropped to as low as 7 mmHg did retention occur. Because tissue π is at a minimal level, it can be expected that when plasma π drops below this value, blood volume would no longer be maintained. In that case, however, the only way to prevent progressive decrease would be a decrease in capillary hydrostatic pressure by a drop in arterial pressure. Thus at this critical level there is only little margin from a shock-like state. Recently, Kaysen et al. (1985) showed that very low plasma π of 9 mmHg in rats with a model of membranous nephritis did not cause oedema, unless kidney mass was reduced. In the latter animals, plasma volume was higher than in those without oedema. It should be pointed out that the time needed to redistribute interstitial albumin to the blood is probably several hours, and a drop in blood volume may occur transiently when plasma π drops very fast.

In summary, adaptations in tissue-fluid Starling forces explain why blood volume in patients with hypoproteinaemia due to the nephrotic syndrome is not necessarily low in the oedema-free state, and also why during oedema blood volume is not elevated. Thus, although the hypoproteinaemia cannot explain the fluid retention in the nephrotic syndrome, it does explain its distribution. Only when plasma π drops very fast, or when it drops to a very low level, would a truly low blood volume be present.

Renal function

Because a variety of histological changes give rise to a nephrotic syndrome it is conceivable that primary renal volume retention may be caused by obvious glomerular lesions, but not in case of 'minimal change'. We will therefore concentrate our discussion on the latter, because in this condition renal function with regard to water and salt handling is most controversial.

Glomerular permselectivity

Investigators have long been puzzled why with 'minimal lesions' plasma proteins can permeate a capillary wall that looks completely normal in light microscopy, and even shows obliteration of the normal splits between the epithelial cells when viewed under the electron microscope. Studies of fractional clearances of dextran molecules in animals have resolved this problem (Brenner et al. 1977). Normally, permeability for albumin is only about 1 per cent of similarly-sized dextran. In contrast, the clearance of dextran sulphate which, like albumin, is negatively charged, approaches that of albumin. Thus, the permeability of the normal glomerulus for proteins is not only determined by their size but also by their charge (Vehaskari et al. 1982), because normal glomerular basement membrane carries negative charges. In rats with experimental nephritis and proteinuria, including models resembling minimal change disease (Bridges et al. 1982), the permeability for dextran sulphate molecules was found to be increased, apparently due to loss of charge of the basement membrane, but that of neutral dextran was decreased, suggesting that structural changes were also present (Fig. 6).

In accordance with these data, decreased rather than increased permeability for synthetic macromolecules has been found in children with minimal lesions (Robson et al. 1974). In addition, depending on the pathological lesions, the number of large 'non-selective pores' may be increased causing leakage of large molecules (Deen et al. 1985). It thus appears likely that loss of negative charges of the glomerular basement membrane is the common denominator causing proteinuria in the nephrotic syndrome, while structural changes manifested by a decreased number of pores account for the decrease in ultrafiltration in certain models.

Renal haemodynamics

It has been assumed that GFR is hardly decreased, and even supranormal values have been reported occasionally in children. Nevertheless, moderately decreased GFR is the rule in most patients with minimal change disease, and GFR is generally lower in the same patient during a nephrotic episode than in remission. Severe drops in GFR have been attributed to extreme hypovolaemia, but there are few arguments for this assumption, as discussed in Chapter 8.6. Even a normal GFR, however, is remarkable because the low plasma π would be expected to increase GFR and filtration fraction (FF). In patients with the rare syndrome of congenital hypoalbuminaemia filtration fraction is markedly elevated as predicted (Bennhold et al. 1960). In contrast, patients with the nephrotic syndrome usually have a low to very low filtration fraction, while the renal plasma flow

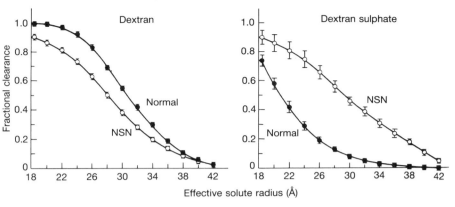

Fig. 6 Fractional clearances of neutral dextran and dextran sulphate (negatively charged) in normal rats and in rats with severe proteinuria due to nephrotic serum nephritis (NSN). Similar results were obtained with models closely resembling human minimal change disease. (Taken from Brenner *et al.* (1977) with permission.)

(RPF) is normal or elevated (Berg and Bohlin 1982) and decreases after recovery (Dorhout Mees *et al.* 1979).

These findings argue strongly against the presence of hypovolaemia in the nephrotic syndrome, since in cases of insufficient circulation, such as salt depletion or congestive heart failure, moderate to marked reduction of renal plasma flow with increased filtration fraction are seen (Kon *et al.* 1985). The low filtration fraction in nephrotic patients can only be explained by primary impairment of glomerular filtration, as also suggested by the macromolecule clearance studies. Micropuncture studies in various models of the nephrotic syndrome in the rat have shown a reduced filtration coefficient (K_g), which is partly compensated by an increase in glomerular perfusion pressure. The latter is due to afferent arteriolar dilatation, probably mediated by prostaglandins. Indeed, increased urinary prostaglandin excretion is reported in the nephrotic syndrome. Experiments with unilateral aminonucleoside nephrosis (Ichikawa *et al.* 1983) are particularly relevant. Sodium retention was present only in the diseased kidney, showing directly that in this model, which in function resembles minimal change nephropathy rather closely, it is not due to hypovolaemia but to intrarenal changes.

Excretion of sodium and water

Because in the nephrotic syndrome there appears to exist a decrease in ultrafiltration coefficient, the salt and water retention may conceivably be due to the same (not adequately understood) mechanisms which are operative in acute nephritis. Indeed, micropuncture studies in different experimental models of glomerular damage, including those resembling 'minimal change' yielded similar results, although increased proximal reabsorption was found in one study (Lewy 1976).

Investigations in humans with the nephrotic syndrome are necessarily confined to indirect clearance methods. In one study 'total distal blockade' with diuretics was applied (Grausz *et al.* 1972). They were interpreted as showing decreased proximal reabsorption, while a similar study in patients with heart failure suggested increased distal reabsorption. Other studies reported a lowered glucose threshold, which also would indicate reduced proximal reabsorption. On the other hand, all investigators using maximal free water clearance found reduced values suggesting increased proximal reabsorption (Gur *et al.* 1976). This method is also based on some unproven assumptions, notably that ADH is completely suppressed. Finally fractional lithium clearance, a marker of distal delivery, was found normal or moderately decreased (Koomans *et al.* 1987). Taken together it seems that the results of animal and human studies are inconclusive, and cannot differentiate between primary and secondary sodium retention.

Effects of induced or spontaneous changes

If hypovolaemia were the cause of sodium retention, one would expect that manoeuvres which correct it or its consequences would cause a massive natriuresis. The best known is infusion of hyperosmotic albumin solutions, which may indeed provoke a natriuresis, but only rarely of great quantity. In about half of nephrotic patients, little or no effect is produced, despite expansion of the blood volume to well above normal levels. In one study free-water clearance increased without natriuresis, which is in agreement with increased distal reabsorption (Koomans *et al.* 1984). Accordingly, natriuresis after loop diuretics may be greatly enhanced by albumin infusion. Another way to expand effective circulating volume is 'head out of water immersion'. This manoeuvre induces natriuresis together with suppression of renin and aldosterone in normal subjects. In nephrotic subjects natriuresis was also reported, but much more variable in amount (Krishna and Danovitch 1982). This can be interpreted as supporting the hypovolaemia concept, but also could be interpreted to indicate that some intrinsic renal effect was overcome by the expansion. The fact that natriuresis did not correlate with PRA, but was proportional to the blood volume, suggests that more than one mechanism is involved in this type of oedema formation, a conclusion probably also applicable to the albumin infusion experiments.

It should be borne in mind that a comparable control condition for these expansion studies does not exist. Because hypovolaemia acts partly on the kidney by stimulating the renin–aldosterone system, it is interesting that blocking the angiotensin production with a converting enzyme inhibitor did not increase sodium excretion, while blood pressure decreased less than in normal subjects (Brown *et al.* 1982).

Observations during initiation and recovery

In experimental aminoglucoside nephrosis, sodium retention precedes marked proteinuria, leading to an increase in blood volume (Kalant *et al.* 1962). Such observations have not been described in humans. Sequential changes in recovery from a nephrotic episode can be summarized as follows (Oliver 1963; Koomans *et al.* 1987). First, proteinuria starts to fall, followed after one day by increased sodium excretion. At that stage, plasma albumin and plasma volume have hardly changed (Fig. 7). There is however, a marked fall in PRA and aldosterone, which rises again after the peak of the natriuresis. The maximal daily sodium excretion often exceeds 500 mmol. These changes apparently are associated with the repair of the glomerular lesion, as the filtration fraction also increases towards normal during the natriuresis. These observations suggest that the fall in PRA during natriuresis results from intrarenal changes.

Comparison of patients with minimal change and patients with histological lesions

It is not unlikely that the presence of a histological lesion may be an important determinant in the pathogenesis of oedema formation. In particular, it has been suggested by Meltzer *et al.* (1979) that in patients with histological lesions, these are the primary cause of a lowered GFR and renal sodium excretion impairment. In contrast, a 'hypovolaemic–vasoconstrictive' pattern with low blood volume and blood pressure was postulated to be characteristic only for patients with minimal lesions.

However, systematic comparison of blood volume measurements and other parameters in both groups of patients failed to find any systematic differences (Geers *et al.* 1984). A reduced filtration fraction is as characteristic for minimal change nephrotics as it is for patients with histological lesions (Fig. 8). Thus there is no basically different pathophysiology of oedema in these populations, but some subsets may be distinguished. High renin and aldosterone levels are probably found mainly in patients with minimal lesions in whom plasma albumin and sodium excretion are very low, even when blood volume and blood pressure are not below normal. It is likely that they represent one end of the spectrum. Such a pattern has been mainly reported in children, in whom minimal lesions often occur, frequently with very low plasma protein concentrations. While in both populations a primary renal disturbance is operative, very severe hypoproteinaemia may add a hypovolaemic component. This explains why clear-cut correlations between haemodynamic parameters, the renin–aldosterone system and sodium excretion are often absent.

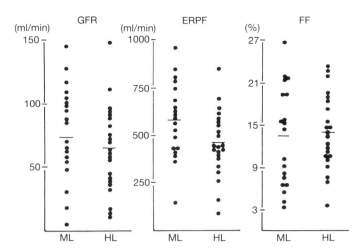

Fig. 8 Glomerular filtration rate (GFR) inulin clearance, effective renal plasma flow (ERPF), PAH clearance, and calculated filtration fraction (FF) in 19 patients with minimal change disease (ML) and in 22 nephrotic patients with various histological abnormalities. (Taken from Geers *et al.* (1984) with permission.)

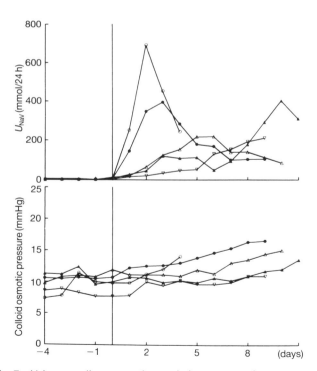

Fig. 7 Urinary sodium excretion and plasma osmotic pressure during prednisone-induced remission from minimal change disease. Symbols correspond to five different patients. Vertical line marks start of increase in sodium excretion. (A.J. Rabelink 1989, unpublished observations.)

Treatment of oedema in the nephrotic syndrome

When cure or improvement of the renal lesion itself is not possible (see other chapters in this section) symptomatic treatment of the oedema is of great importance for the patient, even though the oedematous condition is not life threatening. However, complete removal of all the fluid excess is not necessary, and may even be dangerous by increasing plasma and whole blood viscosity, thus promoting thrombosis (see Chapter 3.4). Basically sodium restriction in the diet should be applied, which in less severe retentive states will restore sodium balance. A sodium content of less than 50 mmol/day is seldom practicable, and the amount allowed also depends on the motivation of the patient. Bed rest may have a diuretic effect and enhance other diuretic measures. However, it should be realized that it may increase the danger of thromboembolism to which nephrotic syndrome predisposes (see Chapter 3.4).

When sodium restriction alone fails, diuretics (see Chapter 2.3) should be added. In some patients thiazide may suffice to keep them oedema free. However, more often the more potent loop diuretics like frusemide (furosemide) (40–500 mg/24 h), and bumetanide (1–5 mg/24 h) are needed. Because of their short time of action (3–6 h) multiple doses are preferable. In contrast to thiazides, they have a steep dose–activity relation, and doses up to five times those normally applied may be necessary. Intravenous administration has no advantages, and should be avoided because of increased danger of ototoxicity, particularly with ethacrynic acid, of which the dose should never exceed 150 mg orally at a time. If Na^+ retention is extreme, with practically sodium-free urine, a combination of diuretics acting on different tubular levels is logically the next step, because their combined effect proves larger than the sum of each drug separately. Hydrochlorothiazide 50 mg twice daily, chlorthalidone 50 to 100 mg in one dose, or metolazone 5 to 20 mg may be added. As hypokalaemia is likely to result from such a strong combination triamterene (100–300 mg) amiloride (5–10 mg) each twice daily or potassium supplementation as well should be considered. Addition of acetazolamide (250–500 mg) twice daily will further potentiate this regimen, but will encourage more loss of potassium, and perhaps should generally be avoided. Such rather extreme combinations should only be carried out under close supervision in the hospital.

Finally in some extremely oedematous patients mechanical ultrafiltration may then be a more profitable treatment. It is noteworthy that spironolactone is more effective than frusemide in cirrhosis of the liver (Pérez-Arguso et al. 1983) but not in the nephrotic syndrome. This also suggests a different pathophysiology between these two diseases.

The presence of hypoproteinaemia may change the pharmacokinetics of some diuretics markedly. Frusemide is highly protein-bound, and the unbound fraction is rapidly cleared from the blood by extrarenal routes. Thus plasma levels are lower and duration of action shorter in nephrotic patients (Keller et al. 1982). Because the natriuretic effect is more dependent on the time that an effective drug concentration is present, rather than on the height of the peak level, intravenous administration is less efficient than oral use (Brater 1988). The danger of ototoxicity by the elevated unbound fraction also argues against intravenous use, as already mentioned.

Because of the large leak of proteins by the glomeruli, infused concentrated albumin is rapidly lost. It may however temporarily increase natriuretic action of diuretics and permit rapid removal of large volumes of oedema with safety. Only rarely do albumin infusions provoke considerable diuresis when given alone. Sustained albumin infusions may increase blood volume, cause hypertension and rarely even pulmonary oedema, which can be fatal on occasion.

Nephrotic patients usually are catabolic and striking loss of lean body mass becomes apparent once they are oedema free. High protein diet may improve nitrogen balance (Blainey 1955). However, prescribing high-protein diets has recently been shown to give no benefit for protein concentration. As discussed in a previous section, a given amount of infused albumin will expand blood volume temporarily more in hypoproteinaemic than in normal subjects. After 1 day, this effect will be abolished, not only because albumin is lost in the urine, but also because of a shift into the tissue fluid (Koomans et al. 1987). Albumin synthesis by the liver cannot be stimulated beyond a rather limited rate. Feeding more protein only results in elevations of GFR and increased proteinuria, finally causing no increase or even a reduction in plasma albumin (Kaysen 1988). Finally, several drugs can decrease protein loss without affecting the disease process itself. Non-steroidal anti-inflammatory drugs such as indomethacin may induce substantial decrease in proteinuria (Vriezendorp et al. 1986). This is due to prostaglandin inhibition, and is accompanied by some decrease in GFR but a much larger decrease in permeability to proteins by the 'shunt pathway' (Shemesh et al. 1986). This effect is most apparent when strong sodium restriction is applied. Although some elevation of plasma albumin level may be achieved, this treatment cannot be considered 'diuretic'. Moreover, acute renal failure is occasionally induced, which may be either due to excessive dependancy of the glomerular circulation on prostaglandins or to a toxic effect of the drug (interstitial nephritis). On the other hand in a retrospective study of patients with various histological diagnoses prolonged treatment was shown to retard development of renal failure significantly (Vriezendorp et al. 1986).

Recently, it has been shown that converting enzyme inhibition also can reduce proteinuria apparently by a non-specific alteration of glomerular dynamics (Kaysen 1988).

Even more dramatic symptomatic improvement can be achieved with cyclosporin treatment which may stop proteinuria, although often it only induces a reduction in proteinuria and in GFR. As with NSAIDS, proteinuria soon returns to its previous levels when the drug is stopped, or relapse is seen in fully responsive patients. This treatment, probably symptomatic, is still in its experimental stage (see Chapter 3.5).

Chronic renal failure
General remarks

When renal function slowly deteriorates, as it often does in renal diseases whatever their nature, increasing demands are laid upon the remaining nephrons. As far as sodium and volume balance are concerned, this implies excretion of a much larger fraction of the filtered load. A patient with 10 per cent of functioning renal tissue on a diet containing the amount of salt in an average western diet (approximately 150 mmol) has a Na^+ excretion of 6 per cent of the filtered load, i.e. around the upper limit of the amount which can be excreted by a normal person (see above). That most patients are capable of excreting this, and even larger amounts, is remarkable, and it is not surprising when an occasional patient fails to do so. The mechanisms involved in adaptation of the kidney are probably the same in health and disease, but it is evident that the impulses to enable the remaining nephrons capable of achieving such relatively larger performance, have to be proportionally stronger. This view has been popularized by Bricker as 'magnification phenomena' (Bricker et al. 1978). It was also suggested that such extreme activation of physiological mechanisms might have some unwanted side-effects ('trade off'). It is not difficult to imagine that in order to raise a maximal natriuretic stimulus, an expansion of extra- and intravascular volume up to, or beyond, their upper limits will be present, with hypertension as accompanying feature (Blumberg et al. 1967).

Overfilling and hypertension

As shown in Fig. 4, fluid retention in patients with chronic renal insufficiency is accompanied by a proportional increase in blood volume and blood pressure. Salt balance is often preserved at the cost of a chronically expanded blood volume and hypertension. Indeed the 'salt sensitivity' of blood pressure in patients with chronic renal insuffiency is inversely related to the degree of renal failure (Koomans 1982). Without visible oedema, extracellular volume may be expanded by 30 per cent, while the expanded blood volume and hypertension inevitably lead to cardiac dilatation, no matter how healthy the heart may be. Finally such conditions result in generalized and/or pulmonary oedema, depending on subtle changes which usually escape clinical recognition.

When patients are being treated by regular dialysis, this regulation, which is normally performed by their own kidneys, must be taken over by adjustment of intake and subtraction by ultrafiltration. Some 'non-compliant' patients may gain 5 to 6 kg in a few days without apparent harm, thanks to the compliance of their tissues. Nevertheless, such changes in the long run are deleterious for the heart and arterial system. Now that our technology is able to prevent patients dying from uraemia, their life expectancy is largely determined by the damage to heart and arteries which (unlike kidneys) cannot be replaced. While the hyperlipidaemia which is often seen with chronic renal failure may aggravate atherosclerosis, the main cause is hypertension, prevention of which is dependent on adequate and timely prevention of chronic overhydration.

Although in normal conditions glomerular filtration is largely independent of blood pressure, it is probable that filtration per nephron in most chronic renal patients varies more or less with arterial pressure, because autoregulation is limited. High arterial pressure induces high intraglomerular pressure and supranormal filtration, which accelerates glomerular damage. Conversely, lowering of elevated blood pressure often induces (sometimes temporary) decrease in creatinine clearance. This feature may deter the doctor from vigorously treating hypertension and its main cause, chronic expansion of intravascular volume.

Thus it is logical to restrict sodium intake in all renal patients with hypertension, in proportion to the decrease in GFR. If this is not feasible, or sodium-excreting capacity is particularly low, diuretics should be added. When volume is normal and hypertension still persists, additional hypotensive drugs should be given. Because small degrees of volume expansion up to 5 l often are not readily apparent to clinical examination, it is difficult to decide when this is needed. One way to investigate this problem is to test the effect of a converting enzyme blocker (e.g. captopril 12.5 mg). As long as this does not induce a decrease in blood pressure, the volume probably is not too low.

Patients with renal failure and chronic fluid retention often develop pulmonary congestion which may unexpectedly present as acute pulmonary oedema. The lungs often show impressive, unusual pictures on radiography, suggesting infiltrates ('uraemic lung'). This led earlier investigators to assume abnormal permeability of the lung capillaries, due to the uraemia. However, with modern techniques such an abnormality was not found (Rocker et al. 1988). Another feature of fluid retention may be the presence of ascites, particularly in patients treated by long-term dialysis. The term 'dialysis ascites' has been applied, but in my opinion there is no proof that it is caused by any other abnormality than general fluid retention. While it has been reported frequently in some dialysis units, others have never observed a single case among hundreds of patients.

Treatment

It can be expected that the effect of diuretics (see Chapter 2.3), which block tubular reabsorption, will be reduced in proportion to the decreased number of functioning nephrons. Thus, reduced natriuretic effect need not be due to 'resistance' of the kidney. However, weaker diuretics will have little effect and for clinically useful results higher doses of strong (i.e. 'loop') diuretics are required. Nevertheless it has been shown that the amount of frusemide reaching its site of action in the loop of Henle is diminished in azotaemic subjects. This is due to competition at its transport site by organic acids which are retained (Rose 1976). Therefore, the maximal effective dose varies and should be 'titrated' in the individual patient by doubling the individual dose until the effect has been achieved (Brater 1988). This can be done at intervals of a few hours, because of the short duration of action. The same maximal natriuresis of approximately 20 per cent of filtered load as in normal renal function can be reached. There is no rationale for increasing the single dose of frusemide to 200 mg IV or 400 mg orally. Because in uraemia extrarenal clearance of frusemide is decreased but not of bumetanide, the potency ratio on weight basis of these two drugs is lower than in normal subjects. The net effect over 24 h can be increased by increasing frequency up to three or four times daily. If, due to severely decreased overall renal function, the desired result cannot be reached in this way, adding diuretics with a different site of action may be tried as outlined in the previous section. However potassium-sparing diuretics are contraindicated in these patients, because sudden hyperkalaemia may appear, and may lead to cardiac arrest. This arises from the need to increase the secretion of K^+ 20-fold per nephron in advanced renal failure to maintain a normal plasma K^+; gut secretion of K^+ also rises. Blockage of this K^+ secretion is therefore absolutely undesirable.

Presumed forms of secondary fluid retention

Congestive heart failure

Oedema in the course of heart disease is 'by definition' caused by secondary renal salt and water retention. Although blood volume is never low, 'insufficient filling of the arterial system' somehow induces the kidney to retain fluid, which increases blood volume as described earlier. If the resulting increase in central venous pressure is not sufficient to improve cardiac performance, a vicious circle is initiated which may in turn even adversely affect the heart function. Further fluid retention by the kidney causes oedema, while further increase in blood volume may be prevented by increased venous pressure and P_c, shifting the fluid to the interstitium. The efferent signals inducing the kidney to retain NaCl are probably temporary decreases in blood pressure and increased sympathetic drive (Zambraski 1989), both resulting in increased renin–angiotensin and aldosterone. Renal blood flow is severely impaired while GFR is maintained by an increased filtration fraction, and micropuncture studies show increased proximal reabsorption of sodium (Bennett et al. 1973). In this stage, converting enzyme blockade may improve cardiac function and favour natriuresis. It does not follow from these findings that converting enzyme blockade is also beneficial in heart failure (congestion) due to inability to excrete sodium in renal disease. To my knowledge this point has not yet been investigated. Generally, to combat fluid overload by diuretics is helpful, in particular to prevent pulmonary oedema.

Cirrhosis of the liver (see Chapter 8.11)

It is usually considered that, as in heart failure, renal fluid retention in this condition is secondary to a decrease in 'effective circulating volume' (Schrier 1988).

Signs that the circulation is indeed inadequate are the low blood pressure and the often extremely high values of plasma renin and aldosterone. Micropuncture studies showed increased proximal reabsorption (Lopez-Novoa et al. 1977). Nevertheless, several lines of evidence suggest that in the early stages of cirrhosis these signs are not present, yet patients may still retain sodium and water (overfilling theory). This concept is favoured by the finding of increased ANF levels. It is not clear how their kidneys would be induced to retain sodium. Possibly this is accomplished by hepatorenal nervous reflexes. In that case the retention would still in a way be 'secondary'.

Pregnancy (see Chapter 15.1)

Pregnancy is a normal condition associated with considerable fluid retention and sometimes oedema. The blood volume is increased, evidently to accommodate the enlarged increased vascular bed, because blood pressure is lower than normal. There is a generalized vasodilatation as a result of smooth muscle relaxing substances such as progesterone and prostaglandins. These same substances block the actions of renin–angiotensin and aldosterone, which are several times higher than normal, without causing hypertension or hypokalaemia. Thus peripheral resistance and response of peripheral vessels to vasoconstriction stimuli are both low. The kidney shares this vasodilatation, resulting in a 20 to 30 per cent increase of renal blood flow and glomerular filtration rate. Unfortunately there have been few investigations in this area, and many details of the mechanisms of these 'extremes of normal physiology' still await elucidation.

Toxaemia of pregnancy (see Chapter 15.4)

Toxaemia of pregnancy is an oedematous condition in which sodium retention is most likely due to a renal abnormality. The glomeruli of these patients show unique changes consisting of swelling of endothelial cells and fibrin deposition, which probably account for the often extreme sodium retention, with practically sodium-free urine. The mechanism of oedema and sodium retention may be glomerulotubular imbalance, similar to that of acute proliferative glomerulonephritis. Other expressions of the glomerular lesion are profuse proteinuria and at least a relative decrease in GFR. Toxaemia is accompanied by striking changes in haemodynamics suggesting imbalance between vasoconstrictor and relaxing factors, as hypertension occurs despite a decrease in blood volume and plasma renin activity. Many symptoms are probably due to vasospasm, which may be accompanied by intravascular coagulation. The high arterial pressure contrasts with a low central venous pressure (Wallenburg 1988) suggesting disturbance of the Starling forces with a volume shift from intra- to extravascular compartment. Analysis of these problems with modern non-invasive techniques would be of great help not only in establishing a rational treatment but also to elucidate the role of Starling forces in general pathophysiology.

Idiopathic oedema

This condition has excited the interest of many nephrologists but until now has not been clearly understood. It occurs only in women, often obese and usually of childbearing age, and is characterized by periodic (usually not 'cyclic') increase in weight up to 5 kg and more with swelling of legs, hands, and face. It appears that in some at least the normal weight gain during the day related to upright posture is exaggerated, and excretion during the night insufficient (Kuchel et al. 1970). Therefore the syndrome is elusive because it may disappear when the patient is admitted to a hospital for investigation because she will be recumbent most of the day. There is no doubt that renal fluid retention is secondary to some circulatory disturbance. Abnormalities in capillary permeability, excessive pooling of blood in the legs with orthostatic decrease in blood pressure, elevated renin–aldosterone levels and the female hormones have been variously claimed to be responsible (Streeten 1978).

An interesting observation was made by de Wardener (1981) who found that most women with this syndrome used (or abused) diuretics, upon whose interruption delayed decreases of renin and aldosterone with fluid retention occurred. Though it goes too far to attribute all cases to diuretic abuse, there is no doubt that a similar syndrome can be provoked in this way, and that many patients aggravate and perpetuate their salt-retaining tendency with diuretics. There is also no doubt that many such patients are overconcerned about their weight, their body image, and the fat which they often insist is fluid. The syndrome may represent one extreme of physiological variations in fluid balance, which affects many women who do not seek medical attention. Although it is limited to women, there is no relation to the menstrual cycle (Edwards and Bayliss 1975). Treatment should be non-pharmacologic and include psychological support, weight loss, regular exercise (in particular swimming), and no or minimal (certainly not short-acting) diuretics. Other drugs like amphetamines and converting-enzyme blockers have been advocated but not proven to be successful.

Drugs

Many drugs are known to induce some sodium retention, although this does not always lead to manifest oedema because of 'escape' reaction. Individual susceptibility varies markedly.

Non-steroidal anti-inflammatory drugs (see Chapter 6.2) These probably induce sodium retention by decreasing synthesis of prostaglandin within the kidney, which stimulates reabsorption in the loop of Henle. For phenylbutazone an aldosterone-like effect has been claimed.

Antihypertensive drugs Drugs with strong vasodilating action such as dixazoide and minoxidil may cause sodium retention and oedema. This is usually attributed to increased vascular capacity and relative underfilling of the circulation, (thus the hypovolaemia concept), but may perhaps be better described as a drop in renal perfusion pressure below the sodium excretion threshold. In order to preserve their hypotensive effect, these drugs have to be combined with diuretics.

Oestrogens These have a weak sodium retaining effect on the kidney, and may cause oedema in susceptible individuals. The mechanism is unknown.

Liquorice Liquorice alone or as the base of some sweets that are popular in a few countries, particularly in Holland, may be a puzzling source of oedema. Their salt-retaining and hypertensive action is due to glycyrrhizic acid, which by inhibiting 11-β-OH-dehydrogenase sensitizes renal tubular cells to mineralocorticoid action.

Potassium depletion It has long been known that potassium salts are natriuretic, and that they may be helpful if oedema is associated with hypokalaemia. On the other hand K^+ depletion and chronic hypokalaemia induce sodium retention, not only substituting intracellular K^+ loss, but actual volume expansion. The mechanism, which is certainly renal, remains to be elucidated. Although patients with this condition (usually from laxative abuse and psychiatric abnormalities) are rare, the phenomenon may be of more general pathophysiological importance.

Overtreatment

Since the dangers of sodium and volume depletion, in particular in association with large operations and traumatic conditions have become common knowledge, it is to be expected that many such patients will be overtreated. In fact the most extreme examples of extracellular volume expansion can be seen in intensive care wards. As a rule, these patients had to be treated for hypotension because of blood and fluid loss as well as extreme enlargement of the vascular bed resulting from septicaemia. However it is not probable that, once oedema appears, further expansion with non-colloidal solutions will have any influence on intravascular volume, because the 'oedema-preventing factors' are exhausted. Many of these patients develop renal failure. Complete immobilization probably also impairs excretion of the excess fluid. Positive-pressure artificial respiration itself induces hypotension and fluid retention, but one of the reasons for continuing this treatment is the threat of pulmonary oedema, which is in turn increased by the overhydration. Although these critically-ill patients form a very difficult 'material' to study, it is clear that further research in this field is needed urgently.

In the past, much emphasis has been laid on renal 'salt wasting'. In particular patients recovering from tubular necrosis (see Chapter 8.2) and subjects in whom a urinary obstruction has been relieved (see Chapter 16.2) may produce large amounts of urine, excrete much sodium, and lose weight. This tempts physicians to substitute these losses in order to stop the negative balance. On close scrutiny, it is not unusual to find some oedema and hypertension in these patients, proving that they were not usually losing, but simply excreting accumulated fluid, which were obscured by the concomitant muscle and fat wasting. It appears that renal sodium wasting is extremely rare except in some tubulointerstitial diseases, either primary such as nephronophthisis (see Chapter 18.3), or secondary as in lithium poisoning (see Chapter 6.6).

Bibliography

Bradley, S.E. and Tyson, C.J. (1948). The 'nephrotic syndrome'. *New England Journal of Medicine*, **238**, 223–2 and 260–6.

Brater, D.C. (1988). Use of diuretics in renal insufficiency and nephrotic syndrome. *Seminars in Nephrology*, **8**, 333–41.

Brenner, B.M. (1977). Determinants of glomerular permselectivity: Insights derived from observations *in vivo*. *Kidney International*, **12**, 229–37.

Dorhout Mees, E.J., Geers, A.B., and Koomans, H.A. (1984). Blood volume and sodium retention in the nephrotic syndrome: a controversial pathophysiological concept. *Nephron*, **36**, 201–11.

Glassock, R.J., Cohen, A.H., Bennett, C.M., and Martinez-Maldonado, M. (1981). Primary glomerular diseases. In *The Kidney*, (eds. B.M. Brenner, and F.C. Rector), 2nd edn. W.B. Saunders, Philadelphia.

Kaysen, G.A. (1988). Albumin metabolism in the nephrotic syndrome: The effect of dietary protein intake. *American Journal of Kidney Diseases*, **12**, 461–80.

Peters, J.P. (1948). The role of sodium in the production of oedema. *New England Journal of Medicine*, **239**, 353–62.

Schrier, R.W. (1988). Pathogenesis of sodium and water retention in high-output and low-output cardiac failure, nephrotic syndrome, cirrhosis, and pregnancy. *New England Journal of Medicine*, **319**, 1065–72 and 1127–34.

Zambraski, E. (1989). Renal nerves in renal sodium-retaining states: cirrhotic ascites, congestive heart failure, nephrotic syndrome. *Mineral Electrolyte Metabolism*, **15**, 88–96.

References

Barton, L.J., Lackner, L.H., Rector, F.C., Jr., and Seldin, D.W. (1972). The effect of volume expansion on sodium reabsorption in the diluting segment of the dog kidney. *Kidney International*, **1**, 19–26.

Bennett, C.M., Thompson, G.R., and Glassock, R.J. (1979). Sodium homeostasis in acute glomerulonephritis. *Mineral Electrolyte Metabolism*, **2**, 63–9.

Bennett, W.M., Bagby, G.C., Jr., Antonovic, J.N., and Porter, G.A. (1973). Influence of volume expansion on proximal tubular sodium reabsorption in congestive heart failure. *American Heart Journal*, **85**, 55–64.

Bennhold, H., Klaus, D., and Scheurlen, P.G. (1960). Volume regulation and renal function in analbuminaemia. *Lancet*, 1169–70.

Berg, U. and Bohlin, A.B. (1982). Renal hemodynamics in minimal change nephrotic syndrome in childhood. *International Journal of Pediatric Nephrology*, **3**, 187–92.

Blainey, J.D. (1955). High protein diets in the treatment of the nephrotic syndrome. *Clinical Science*, **13**, 567–81.

Binak, K., Sirmaci, N., Uqak, D., and Harmanci, N. (1975). Circulatory changes in acute glomerulonephritis at rest and during exercise. *British Heart Journal*, **37**, 833–9.

Blumberg, A., Nelp, W.B., Hegstrom, R.M., and Scribner, B.H. (1967). Extracellular volume in patients with chronic renal disease treated for hypertension by sodium restriction. *Lancet*, 69–72.

Bricker, N.S., Fine, L.G., Kaplan, A.M., Epstein, M., Bourgoignie, J.J., and Licht, A. (1978). 'Magnification phenomenon' in chronic renal disease. *New England Journal of Medicine*, **299**, 1287–93.

Bridges, Ch.R., Myers, B.D., Brenner, B.M., and Deen, W.M. (1982). Glomerular charge alterations in human minimal change nephropathy. *Kidney International*, **22**, 677–84.

Brown, E.A., Markandu, N.D., Sagnella, G.A., Jones, B.E., Squires, M., and MacGregor, G.A. (1982). Evidence that some mechanism other than the renin system causes sodium retention in nephrotic syndrome. *Lancet*, 1237–9.

Cameron, J.S., Turner, D.R., Ogg, C.S., Sharpstone, P., and Brown, C.B. (1974). The nephrotic syndrome in adults with minimal change glomerular lesions. *Quarterly Journal of Medicine*, **171**, 461–87.

Chonko, A.M., Bay, W.H., Stein, J.H., and Ferris, T.H. (1977). The role of renin and aldosterone in the salt retention of oedema. *American Journal of Medicine*, **63**, 881–9.

Danielsen, H., Pedersen, E.B., Madsen, M., and Jensen, T. (1985). Renal sodium excretion in the nephrotic syndrome after furosemide. *Acta Medica Scandinavica*, **217**, 513–18.

Deen, W.M., Bridges, Ch.R., Brenner, B.M., and Myers, B.D. (1985). Heteroporous model of glomerular size selectivity application to normal and nephrotic humans. *American Journal of Physiology*, **249**, F374–89.

Di Bona, G.F. (1977). Neurogenic regulation of renal tubular sodium reabsorption. *American Journal of Physiology*, **233**, F73–81.

Dorhout Mees, E.J., Roos, J.C., Boer, P., Oei, Y.H., and Simatupang, T.A. (1979). Observations on oedema formation in the nephrotic syndrome in adults with minimal lesions. *American Journal of Medicine*, **67**, 378–84.

Edwards, O.M. and Bayliss, R.I.S. (1975). Postural fluid retention in patients with idiopathic oedema: Lack of relationship to the phase of the menstrual cycle. *Clinical Science*, **48**, 331–3.

Erdmann, A.J., Vaughan, T.R., Brigham, K.L., Jr., Woolverton, W.C., and Staub, N.C. (1975). Effect of increased vascular pressure on lung fluid balance in unanesthetized sheep. *Circulation Research*, **37**, 1.

Geers, A.B., Koomans, H.A., Roos, J.C., Boer, P., and Dorhout Mees, E.J. (1984). Functional relationship in the nephrotic syndrome. *Kidney International*, **26**, 324–30.

Grausz, H., Lieberman, R., and Early, L.E. (1972). Effect of plasma albumin on sodium reabsorption in patients with nephrotic syndrome. *Kidney International*, **1**, 47–54.

Gur, A., Adefuin, P., Siegel, M., and Hayslet, J. (1976). A study of renal handling of water in lipid nephrosis. *Pediatric Research*, **10**, 197–201.

Guyton, A.C., Taylor, A.E., and Granger, H.J. (1975). *Dynamics and Control of the Body Fluids*, W.B. Saunders, Philadelphia.

Hollander, W., Reilly, P., and Burrows, B.A. (1960). Lymphatic flow in human subjects as indicated by the disappearance of labeled albumine from the subcutaneous tissue. *Journal of Clinical Investigation*, **40**, 222–34.

Ichikawa, J., Rennke, J.R., Hoyer, J.R., Badr, K.E., Schor, N., and Troy, J.L., (1983). Role of intrarenal mechanism in the impaired salt excretion of experimental nephrotic syndrome. *Journal of Clinical Investigation*, **71**, 91–103.

Joles, J.A., Koomans, H.A., Kortlandt, W., Boer, P., and Dorhout Mees, E.J. (1988). Hypoproteinemia and recovery from oedema in dogs. *American Journal of Physiology*, **254**, F887–94.

Kalant, N., Gupta, D.D., Despointes, R., and Giroud, C.J.P. (1962). Mechanisms of oedema in experimental nephrosis. *American Journal of Physiology*, **202**, 91–6.

Kaysen, G.A., Paukert, Th.T., Menke, D.J., Gouser, W.G., and Humphreys, H. (1985). Plasma volume expansion is necessary for oedema formation in the rat with Heymann nephritis. *American Journal of Physiology*, **248**, F247–53.

Keller, E., Hoppe-Seyler, G., and Schollmeyer, P. (1982). Disposition and diuretic effect of furosemide in the nephrotic syndrome. *Clinical Pharmacology and Therapeutics*, **32**, 442–9.

Kelsch, R.C., Light, G.S., Oliver, W.J., and Mich, A.A. (1972). The effect of albumin infusion upon plasma norepinephrine concentration in nephrotic children. *Clinical Medicine*, **79**, 516–25.

Kon, V., Yared, A., and Ichikawa, I. (1985). Role of renal sympathetic nerves in mediating hypoperfusion of renal cortical microcirculation in experimental congestive heart failure and acute extracellular fluid volume depletion. *Journal of Clinical Investigation*, **76**, 1913–20.

Koomans, H.A., Roos, J.C., Boer, P., Geyskes, G.G., and Dorhout Mees, E.J. (1982). Salt sensitivity of blood pressure in chronic renal failure. *Hypertension*, **4**, 190–7.

Koomans, H.A., Geers, A.B., vd Meiracker, H., Roos, J.C., Boer, P., and Dorhout Mees, E.J. (1984). Effects of plasma volume expansion on renal salt handling in patients with the nephrotic syndrome. *American Journal of Nephrology*, **4**, 2–234.

Koomans, H.A., Kortlandt, W., Geers, A.B., and Dorhout Mees, E.J. (1985). Lowered protein content of tissue fluid in patients with the nephrotic syndrome: observations during disease and recovery. *Nephron*, **40**, 391–5.

Koomans, H.A., Braam, B., Geers, A.B., Roos, J.C., and Dorhout Mees, E.J. (1986). The importance of plasma protein for blood volume and blood pressure homeostasis. *Kidney International*, **30**, 730–5.

Koomans, H.A., Boer, W.H., and Dorhout Mees, E.J. (1987). Renal function during recovery from minimal lesions nephrotic syndrome. *Nephron*, **47**, 173–8.

Krishna, G.G. and Danovitch, G.M. (1982). Effects of water immersion on renal function in the nephrotic syndrome. *Kidney International*, **21**, 395–401.

Kuchel, O., *et al.* (1970). Inappropriate response to upright posture: a precipitating factor in the pathogenesis of idiopathic edema. *Annals of Internal Medicine*, **73**, 245–52.

Leaf, A., Bartter, F.C., Sautor, R.F., and Wrong, O. (1953). Evidence in man that urinary electrolyte loss induced by pitressin is a function of water retention. *Journal of Clinical Investigation*, **32**, 868–78.

Lewy, J.E. (1976). Micropuncture study of fluid transfer in aminonucleoside nephrosis in the rat. *Paediatric Research*, **10**, 30–4.

Lopez-Novoa, J.M., Rengel, M.A., Rodicio, J.L., and Hernando, L. (1977). A micropuncture study of salt and water retention in chronic experimental cirrhosis. *American Journal of Physiology*, **232**, F315–18.

Meltzer, J.I., Keim, H.J., Laragh, J.H., Sealey, J.E., Kung-Ming, J.D., and Chien, S. (1979). Nephrotic syndrome: vasoconstriction and hypervolemic types indicated by renin—sodium profiling. *Annals of Internal Medicine*, **91**, 688–96.

Oliver, W.J. and Mich, A.A. (1963). Physiologic responses associated with steroid-induced diuresis in the nephrotic syndrome. *Journal of Laboratory and Clinical Medicine*, **62**, 419–63.

Pérez-Ayuso, R.M., *et al.* (1983). Randomized comparative study of efficacy of furosemide versus spironolactone in nonazotemic cirrhosis with ascites. *Gastroenterology*, **84**, 961–8.

Powell, H.R., Rotenberg, E., Williams, A.L., and McCredie, D.A. (1974). Plasma renin activity in acute poststreptococcal glomerulonephritis and the hemolytic-uraemic syndrome. *Archives of Disease in Childhood*, **49**, 802–7.

Robson, A.M., Giangiacomo, J., Keinstra, R.A., Naqvi, S.T., and Ingelfinger, J.R. (1974). Normal glomerular permeability and its modification by minimal change nephrotic syndrome. *Journal of Clinical Investigation*, **54**, 1190–9.

Rocker, G.M., Morgan, A.G., and Shale, D.J. (1988). Pulmonary oedema and renal failure. *Nephrology Dialysis Transplantation*, **3**, 244–6.

Rose, H.J., O'Malley, K., and Pruitt, A.W. (1976). Depression of renal clearance by azotemia. *Clinical Pharmacology and Therapeutics*, **21**, 141–6.

Schad, H. and Brechtelsbauer, H. (1978). The effect of saline loading and subsequent anaesthesia on thoracic duct lymph, transcapillary protein escape and plasma protein of conscious dogs. *European Journal of Physiology*, **378**, 1–133.

Shemesh, O., Ross, J.C., Deen, W.M., Grant, G.W., and Myers, B.D. (1986). Nature of the glomerular capillary injury in human membranous glomerulopathy. *Journal of Clinical Investigation*, **77**, 868–77.

Streeten, D.H.P. (1978). Idiopathic edema: pathogenesis, clinical features, and treatment. *Metabolism*, **27**, 353–82.

Tulassay, T., Rascher, W., Lang, R.E., Seyberth, H.W., and Schärer, K. (1987). Atrial natriuretic peptide and other vasoactive hormones in nephrotic syndrome. *Kidney International*, **31**, 1391–5.

Vehaskari, V.M., Root, E.R., Germuth, F.G., Jr., and Robson, A.M. (1982). Glomerular charge and urinary protein excretion: effects of systemic and intrarenal polycation infusion in the rat. *Kidney International*, **22**, 1–135.

Vriezendorp, R., Donker, A.J.M., de Zeeuw, D., de Jong, P.E., and van der Hem, G.K. (1986). Effects of nonsteroidal anti-inflammatory drugs on proteinuria. *American Journal of Medicine*, **81**, 84–94.

Wallenburg, H.C.S. (1988). Hemodynamics in hypertensive pregnancy. In *Hypertension in pregnancy*, (ed. P.C. Rubin), pp. 66–101, Amsterdam, Elsevier Science Publishers.

de Wardener, H.E. (1981). Idiopathic oedema: role of diuretic abuse. *Kidney International*, **19**, 881–92.

Wen, S.F. and Wagnild, J.P. (1976). Acute effect of nephrotoxic serum on renal sodium transport in the dog. *Kidney International*, **9**, 243–51.

3.4 Clinical consequences of the nephrotic syndrome

J. STEWART CAMERON

Introduction

The nephrotic syndrome, mainly through alterations in concentration of plasma proteins, affects every cell and every tissue in the body, just as the uraemic syndrome does (see Chapters 9.6–9.12). In the nephrotic syndrome, there is a general overproduction of all hepatically synthesized proteins, with selective loss of low molecular weight protein in the urine (Fig. 1). Thus low molecular weight proteins tend to be depleted in the plasma, with accumulation of the higher molecular weight protein species; the size of protein for which the plasma concentration tends to be approximately normal is about 180–200 kDa; for proteins below this size, levels tend to be lowered, and above this limit, raised. Another factor is the rate at which the protein is normally replaced, proteins with fast turnovers in health being more resistant to depletion than those with slower normal synthetic rates.

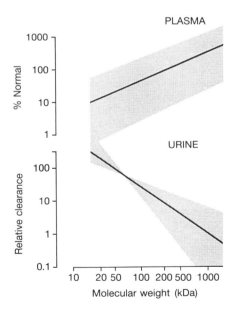

Fig. 1 The relationship between (lower panel) the loss of proteins into the urine in relation to molecular weight in nephrotic patients, and (upper panel) the concentrations of proteins in the plasma in relationship to molecular weight. It can be seen that there is a reciprocal relationship between the two sets of measurements. One hypothesis which would explain this relationship is that all proteins are synthesized in excess in the nephrotic syndrome, and those of low molecular weight lost predominantly in the urine. Thus, those of low molecular weight are depleted, and those of high molecular weight accumulate in the plasma.

Most of the abnormalities found in the nephrotic syndrome result from the effects of these profound alterations in the protein environment, either directly as a result of the altered concentrations (for example in the coagulation system) or as a secondary result of alterations in cellular function induced by the change in protein environment (for example binding of increased quantities of LDL to cells).

Infection in nephrotics

Incidence and clinical features

Early reports emphasized the high incidence and considerable mortality of bacterial infections in patients with a nephrotic syndrome, especially children (Leiter 1931; Schwartz and Cohn 1935). In Arneil's data (1961) 28 of 35 deaths within 1 year of onset during 1928–57 were from sepsis; overall, even after 10 years of follow up, deaths from infections exceeded those from renal failure (35 vs. 28). Others such as Lawson *et al.* (1960) reported similar figures for the preantibiotic era, but noted that even from 1945 to 1957, only five of nephrotic children who died did so from sepsis. However, sepsis remains an important cause of the much smaller number of deaths in nephrotic children: recently, the International Study of Kidney Disease in Children (1984) noted that of 389 children with minimal change nephrotic syndrome, only six died from sepsis, compared with four from other causes, and Gorensek *et al.* (1988) show that peritonitis remains a threat even today. In the Third World, sepsis is still a major problem in nephrotics (Choudhry and Gai 1977; Futrakul 1978; Elidrissy 1982)

Primary peritonitis

Primary peritonitis is particularly characteristic of nephrotic children; in our own series, the oldest case was aged 20. The onset may be insidious but is usually sudden, and should be suspected in any nephrotic child who develops abdominal pain. Unfortunately this is a common symptom associated with hypovolaemia, and the diagnosis must be confirmed by direct microscopic examination of a Gram stain, or an immunochemical search for bacterial antigens, on ascitic fluid removed by needle. Signs of systemic infection as well make the diagnosis easier. Blood cultures are usually positive also, but of course take much longer to perform. Hypotension, shock, and even acute renal failure may follow rapidly, sometimes with disseminated intravascular coagulation (see below).

In the past the organism was almost always *Strep. pneumoniae* (pneumococcus), but more recent reports suggest that other organisms are becoming relatively more common (Speck *et al.* 1974; Gorensek *et al.* 1988); these include β-haemolytic streptococci (Bannatyne *et al.* 1979), Haemophilus and Gram-negative bacteria (Wilfert and Katz 1968; Speck *et al.* 1974; Rubin *et al.* 1975; O'Regan *et al.* 1980). Even so, the pneumococcus remains the most important organism: Krensky *et al.* (1982) reported 12 of 24 episodes in 1970–80 from pneumococcus, with *E.coli* accounting for six; in three of these six however, there were additional causes: appendicitis, renal biopsy, and intussusception. Gorensek *et al.* (1988) similarly noted 11 cases of *Strep. pneumoniae* amongst 37 patients seen during 1979–86. There is some suggestion that *Strep. pneumoniae* peritonitis may be more common in black than white children (Rubin *et al.* 1975; O'Regan *et al.* 1980). Although now treatable, primary pneumococcal peritonitis is still a major illness, and remains an important cause of death in nephrotics in the Third World (Choudhry and Gai 1977; Futrakul 1978; Elidrissy 1982).

Cellulitis

This is equally characteristic, especially in severely oedematous nephrotics. Many such infections arise from skin punctures, either spontaneous or as a result of venepuncture, or (in the past) attempts at mechanical drainage by puncture or scarification. Organisms responsible include β-haemolytic streptococci and a variety of Gram-negative bacteria. Usually the clinical diagnosis is clear, with obvious demarcation of the infected area, which may spread rapidly within hours to include most of the limb; the patient may be very toxic, febrile, or even become hypotensive. Other patients run a more indolent course with a well-demarcated area of infection which remains localized. It is difficult to stain or culture organisms from fluid aspirated from the area, but as in primary peritonitis blood cultures are usually positive.

Miscellaneous infections

Urinary tract infections

It has been suggested that nephrotic children are more prone to urinary tract infections than others. Sonnenschein and Joos (1970) also suggested that these infections might inhibit response to corticosteroids. However, McVicar and Nicastri (1973) found seven episodes of infection amongst five of 30 nephrotic children over a 4-year period. These episodes occurred both in remission and in relapse. There was no evidence that the infections either precipitated relapse, or prevented response to treatment.

Viral infections

The relationship between the nephrotic syndrome and viral infection is an interesting one. First, relapses of the minimal change nephrotic syndrome often follow viral upper respiratory infections. There is, in contrast, some evidence that viral infections may benefit the minimal change nephrotic syndrome. Measles may lead to remission (Blumberg and Cassidy 1947), and the disease was even used as a treatment for the condition before the advent of corticosteroid (Janeway *et al.* 1948). Viral infections do not, in the absence of chemical immunosuppression, seem to pose a major threat to the nephrotic patient. Both varicella and measles do, however, present a major danger to children receiving either corticosteroid and cytotoxic agents (see Chapter 3.5).

Other infections

Particularly in the Third World, infections (including peritonitis) may arise from a variety of other organisms, such as Alkaligenes (Choudhry and Gai 1977), Bacteroides and Aerobacter (Tanphaichitr and Chesdavanijkul 1978), and *Strep. viridans* (Choudhry and Gai 1977).

Causes of infection in nephrotics

A number of aspects of immune response to, and defence against, invading micro-organisms are defective in nephrotic patients.

Clinical clues

Several clinical features of these infections give us clues to possible pathogenesis. First, they are most common in children, reports of severe infections in adults being rare (Rusthoven and Kabins 1978). Second, the predominance of pneumococcal infections is striking. Although minimal change is the most frequently associated condition, this may merely reflect the fact that it is the most common form of underlying disease in children; we have seen pneumococcal peritonitis in children with mesangiocapillary glomerulonephritis and in Henoch-Schönlein purpura. The organisms are nearly always encapsulated; some others, such as Staphylococci are notable for their almost complete absence.

Physical factors

There are features of the nephrotic state which predispose to the spread of infection. The first of these are the large fluid collections in abdomen and pleural cavity, which as in liver disease are tempting sites of growth for bacteria. The second is the fragility of nephrotic skin, which may even rupture spontaneously under the pressure of oedema. Local humoral defences may be diluted by oedema, which as in peritoneal dialysis may be an important factor. However, it seems likely that immunologic factors play the major role in the susceptibility of nephrotic patients to infection.

Low IgG concentrations

Immunoglobulin G provides the main defence in tissues against infection, and low serum IgG concentrations are characteristic of nephrotic patients (Giangiacomo *et al.* 1975). The reasons for these low IgG concentrations are still not clear. Obviously urinary losses and tubular catabolism are important in many patients, but these cannot alone account for the very depressed levels noted in nephrotics. Giangiacomo *et al.* (1978) suggested that patients, especially those with minimal change disease, had a defect in IgG synthesis, perhaps as a result of a failure to switch adequately from IgM synthesis (IgM concentrations are raised in nephrotic patients) to IgG synthesis, the result in turn of a failure of T-cell help. Ooi and colleagues (1980) and Heslan *et al.* (1982) confirmed that *in-vitro* IgG production by B cells from nephrotics with membranous nephropathy, minimal change disease, or mesangiocapillary glomerulonephritis (MCGN) was indeed reduced. In minimal change patients, in remission, the latter study showed that IgG production was normal.

However, in patients with inherited common hypogammaglobulinaemia, levels of IgG below 2 g/l are required before serious infections are seen. These levels are very rarely reached in nephrotic patients. Also, again in patients with congenital deficiencies of IgG, infections tend to be chronic and pulmonary, and primary peritonitis exceptional. Thus the low IgG concentration is probably not critical in determining the high incidence of infection in nephrotics, although it may have an accessory role.

Complement abnormalities

Pneumococcal infections are a particular problem in patients with some congenital deficiencies of the complement system (Agnello 1978). McLean *et al.* (1977) pointed out that factor B, crucial in the alternative pathway of the complement system, which has a molecular weight of only 55 000, is lost in the urine and that plasma concentrations of this protein are very low in nephrotics. Thus alternative pathway activity is likely to be low also in nephrotics, proportionally to their urinary protein losses. The alternative pathway is of particular significance on the opsonization of encapsulated organisms, such as *Strep. pneumoniae*.

McLean *et al.* (1977) also found defective opsonization of *E.coli* by nephrotic sera in 11 of 20 sera from 14 patients. The addition of IgG had no effect, whilst addition of purified factor B resulted in improvement in three to five samples tested. These data have been confirmed by Anderson *et al.* (1979) and so it seems that factor B deficiency is the main defect responsible for pneumococcal infections.

By 20 years of age, most adults have acquired antibodies against a variety of pneumococcal capsular antigens. Thus it is only during childhood that there is this peculiar vulnerability to *Strep. pneumoniae*. In accordance with this idea, susceptibility to encapsulated organisms is noted also in other conditions with defective alternative pathway complement activity, such as sickle haemoglobinopathy.

Lowered transferrin and zinc concentrations

Transferrin is essential for normal lymphocyte function, and acts as a carrier for a number of metals, including zinc. One study (Warshaw *et al.* 1984) suggested that, *in vitro*, added transferrin could restore defective lymphocyte function in nephrotics to normal, and Bensman and colleagues (1984) suggest that loss of zinc into the urine in nephrotics results in insufficient production of the zinc-dependent thymic hormone, thymulin. Urinary losses of zinc are discussed further below.

Impaired white cell function

There is abundant—but confusing—evidence that lymphocyte function is depressed in nephrotic patients, especially those with minimal change disease (see Chapter 3.5). This depression of T-cell function appears to arise from serum factor(s) which may or may not persist in remission (Moorthy *et al.* 1976; Taube *et al.* 1981) and which may or may not be specific for minimal change disease, independent of treatment (Chapman *et al.* 1982; Taube *et al.* 1984; Taube and Williams, 1988).

There are fewer data on polymorph function, but Yetgkin and colleagues (1980) demonstrated impaired phagocytosis of *Staph. aureus* and *E.coli* by polymorphonuclear leucocytes from nephrotics, even when the assays were performed in normal serum. Similar results were obtained by Sillix *et al.* (1983).

Treatment and prophylaxis
Prevention and antibiotic prophylaxis

Obviously prevention of the nephrotic state itself, or prompt induction of remission of oedema or proteinuria, are the most important goals. Probably one of the main reasons for the fall in death rate from infection in nephrotic children is that so few patients now labour for prolonged periods in a severely nephrotic state. Care must be taken over asepsis in (for example) taking blood from nephrotics, and indwelling central or peripheral cannulae should be avoided unless essential. Breaks in the skin, spontaneous or induced, should be carefully dressed and kept dry—not always easy if the patient is severely oedematous.

In children there is a good case for using prophylactic penicillin to avoid pneumococcal infections, at least whilst the child is oedematous. This is especially true for children who have already experienced pneumococcal infection, since recurrences are well known (Anderson *et al.* 1979; Moore *et al.* 1980). The case for prophylactic penicillin treatment in those known to suffer relapses of their nephrotic syndrome whilst they are in remission is weaker, but still persuasive.

Antipneumococcal vaccines

For some time vaccines to promote active immunization against pneumococcal capsular antigens have been available, and their use in nephrotic children seems logical (Klein and Mortimer 1978; Fikrig *et al.* 1978; Tejani *et al.* 1985). The response when given in remission is adequate (except perhaps against antigen 19F) (Wilkes *et al.* 1982; Hogg 1983), but not always protective when given during steroid therapy, and frankly defective when given shortly after cytotoxic therapy (Fikrig *et al.* 1978). However, recurrent pneumococcal sepsis has been reported in children previously immunized after a first attack (Primack *et al.* 1979; Moore *et al.* 1980). Persistence of the antibody is low in relapsing cases (Spika *et al.* 1986), and it seems that, as in pneumonia in the elderly and other at-risk groups (Forrester *et al.* 1987; Noah 1988), pneumococcal vaccine may be helpful, but its effectiveness should not be trusted too far.

Treatment of established sepsis

Much the most important feature of treatment for sepsis in a nephrotic patient is that it should begin quickly. This in turn rests with anticipation, suspicion, and rapid diagnosis. The ESR is useless in nephrotic children, and the white cell count may be misleading, especially in those taking corticosteroids. Eskola and colleagues (1986) suggest that the level of C-reactive protein may be useful in the diagnosis of serious sepsis, but this is not always available, or is available too late.

Parenteral antibiotics should always be used even when the infection appears to be localized because septicaemia is often already present. They should be begun as soon as cultures have been taken and not await results of sensitivity tests. In children benzylpenicillin should always be a component of the antibiotic therapy; conversely when pneumococcus is observed, backup broad-spectrum therapy is needed, because it may not be the only organism present. A broad-spectrum cephalosporin, perhaps with an aminoglycoside, may be used as initial 'blind' treatment in adults. Blood pressure, peripheral temperature, and venous pressure should be watched carefully, colloid given as necessary and anticoagulate heparin for anticoagulation because of the danger of secondary thrombosis (see Futrakul 1978, and next section). Finally, do not forget the need for supplementary corticosteroids in those taking these drugs, or those who have just stopped them.

Thromboembolic complications

Introduction

Thrombosis in both the arterial and the venous circulation is a relatively frequent and serious complication of the nephrotic syndrome (Cameron 1984; Cameron *et al.* 1988). After the intro-

3.4 Clinical consequences of the nephrotic syndrome

Table 1 Prevalence of deep vein thrombosis in nephrotic adults

Diagnostic technique	Authors	n	Total	Percentage
Doppler	Andrassy et al. (1980)	21	84	25
Clinical	Combined series*	12	191	6
	Llach (1982)	3	118	2.5
	Pohl et al. (1984)	3	59	5.5
	Cameron et al. (1988)†	11	90	12
Total		29	458	6

* Kanfer et al. (1970); Kendall et al. (1971); Thompson et al. (1974); Bernard et al. (1977); Kauffman et al. (1978). All these papers reported concentrations of coagulation factors in nephrotics.
† All minimal changes.

duction of antibiotics, corticosteroids, and powerful diuretics, reports of this complication became more frequent, but at the same time survival dramatically improved and the apparent increase may simply have been a result of increased survival. However, both corticosteroids and diuretics may contribute to thrombosis, the one from effects on coagulation proteins, the other through effects of blood viscosity (see below). Many clinicians including our own group, have the impression that thrombotic complications have become less common in recent years, but there are no secure data on this point. However they remain a serious danger in many patients, and occasionally may be fatal.

Peripheral venous thrombosis and pulmonary embolism

Deep vein thrombosis

Thrombosis of the deep calf veins is common in nephrotic patients, and Table 1 summarizes data on its clinical diagnosis, suggesting that overt deep venous thrombosis is evident in about 6 per cent of nephrotic adults. Thrombi may be detectable in as many as 25 per cent if Doppler utrasonography is used (Andrassy et al. 1980). In our series of adult-onset patients with minimal changes, 11 of 90 (12 per cent) showed clinical deep vein thrombosis, but other histological groups are affected by the problem. It is much less common in childhood; less than 1 per cent of the 3377 children studied in Egli's survey (1974) showed deep vein thrombi, and recently Mehls et al. (1987) reported only two cases amongst 204 nephrotic children.

Pulmonary emboli

Not surprisingly, in view of the frequent finding of deep vein thrombosis in adult nephrotics, pulmonary emboli are also common. At a clinical level, Table 2 summarizes data which indicate an incidence of clinically diagnosed pulmonary emboli of 6 per cent, and in our series of those with minimal changes, seven of 90 (8 per cent) had emboli. However, if ventilation—perfusion scanning is done routinely, 12 per cent of adult nephrotics show evidence of pulmonary emboli, often symptomless. However, only a single patient in our own series up to 1982, followed for 2100 patient-years, actually died of pulmonary embolism. As with deep-vein thrombosis, children seem to be much less affected: Egli's data (1974) discovered only a single child out of 3377 with an embolus, although two of 204 cases of the series of Mehls' (1987) had pulmonary emboli. However, no systematic studies of scans have been done in nephrotic children, and the incidence of both emboli and pelvic or deep venous thrombi may be underestimated.

Other venous thrombi

Other venous thrombi are much less common, with the exception of renal vein thrombosis, discussed below. However subclavian or axillary, jugular, iliac, portal, splenic, hepatic, and mesenteric vein thrombosis have all been described (see Cameron (1984) and Cameron et al. (1988) for detailed references). Sagittal sinus thrombosis has been described both in children and adults, and we have seen a fatal case (unpublished) in which the thrombus extended to include superficial cortical veins. We have also seen a nephrotic child with priapism.

Arterial thrombosis

In all series reported in adults arterial thrombosis is much less common than venous thrombosis (Cameron 1984). In contrast, children show about an equal incidence of arterial and venous thrombosis, although at a much lower total level than in adults (lower because of the lower incidence of deep venous thrombosis (Egli et al. 1973; Egli 1974)). Even so, the incidence of arterial thrombosis in adults is still twice that in children. Thrombosis of various arteries has been recorded, and is summarized in detail in Cameron (1984) and Cameron et al. (1988). Almost every artery has been involved: aorta, coronary, pulmonary, mesenteric, cerebral, renal, ophthalmic, and carotid arteries, together with major intracardiac thrombi.

Thromboses at only a few of these sites have been recorded with any frequency. One is thrombosis of the femoral artery (Kanfer et al. 1970; Cameron et al. 1971). This is usually found in

Table 2 Prevalence of pulmonary emboli in adult nephrotics

Method	Authors	n	Total	Percentage
V/Q scans	Ogg et al. (1982)	7	37	19
	Llach (1982)	7	70	10
	Kuhlmann et al. (1981)	4	26	26
	Andrassy et al. (1980)	7	80	9
Total		25	213	12
Clinical	Ogg et al. (1982)	2	37	5
	Cameron et al. (1988)†	7	90	8
	Llach (1982)	0	70	0
	Kuhlmann et al. (1981)	4	26	15
	Combined series*	15	191	8
Total		28	414	7

* Collected data from five papers on coagulation in nephrotics, as in Table 1.
† All minimal changes.

Table 3 Frequency of renal venous thrombosis in nephrotic adults with membranous nephropathy

Method	Authors	n	Total	Percentage	Remarks
Clinical	Ehrenreich and Churg (1968)	5	60	8	Autopsy
	Trew et al. (1978)	6	90	7	20 had venograms
	Noel et al. (1979)	5	116	4	16 venograms
	Cameron et al. (1988)	5	86	6	24 venograms
Total		21	352	6	
Venography	Wagoner et al. (1983)	13	27	48	Only 2 clinical
	Llach (1982)	20	69	29	Only 2 clinical
	Andrassy et al. (1980)	4	20	20	
	Ogg et al. (1982)	2	15	13	
	Pohl et al. (1984)	4	47	9	

Table 4 Prevalence of renal venous thrombosis in adult nephrotics with histological changes other than those of membranous nephropathy

Method	Authors	n	Total	Percentage	Remarks
Venography*	Llach (1982)	13	82	16	Only 4 clinical
	Bennett et al. (1975)	1	11	9	
	Pohl et al. (1977)	0	34	0	
	Ogg et al. (1982)	5	40	12.5	
Clinical	Cameron et al. (1988)	5	322	1.5	
	Andrassy et al. (1980)	5	64	3	
	Llach (1982)	4	82	5	

* All consecutive cases investigated by venography.

children (10 cases in Egli's survey of 1973) often in association with attempts at femoral vein puncture in hypovolaemic subjects, but has been noted in adults also (Kanfer et al. 1970), including one of our own patients with minimal change; this was the only arterial thrombosis in the 90 adult patients studied. Thrombosis of the pulmonary artery has been commonly recorded (Symchych and Perrin, 1965; Habib et al. 1968). In the survey by Egli et al. (1973) it was the single most common thrombosis, 34 cases having been recorded in children as opposed to 25 cases of deep venous thrombosis. Again this has almost always been confined to children (although it has been recorded in an adult (Kanfer et al. 1970)), and now seems almost to have disappeared, the great majority of cases having been recorded in the 1960s. An association here with haemoconcentration seems likely; the only other condition in which pulmonary artery thrombosis is seen is congenital cyanotic heart disease.

Renal vein thrombosis in nephrotic subjects
Introduction
Few topics have generated such controversy in nephrology, and the subject has been extensively reviewed (Llach et al. 1980; Harrington and Kassirer 1982; Llach 1985). The early assumptions that the thrombosis was a cause of the nephrotic syndrome gave way during the 1960s to the idea that it was a complication of the hypercoagulable state of the nephrotic syndrome (Cameron et al. 1988). However, controversy exists about the incidence of the condition, how energetically it should be sought, and how it should best be managed, especially when symptomless. All agree, however, that it was commonly found in association with membranous nephropathy, usually idiopathic but also in lupus membranous nephropathy (Appel et al. 1976), even after gold therapy (Nelson and Birchmore 1979), and after transplantation (Arruda et al. 1973; Liaño et al. 1988), as well as in the Heymann model of membranous nephropathy in rats. The reason for this predilection for membranous nephropathy has so far eluded explanation.

Prevalence
The apparent prevalence of renal vein thrombosis varies according to whether a clinical or a venographic diagnosis is considered (Table 3), the latter showing a much higher prevalence. Also, if renal vein thrombosis and membranous nephropathy are associated, those series with a high proportion of patients with membranous nephropathy will show a higher incidence of renal vein thrombosis. In several American series, more than 50 per cent of adult nephrotics have shown membranous nephropathy in renal biopsies (Llach 1982) whereas in the United Kingdom the proportion in our own series is only 22 per cent (Cameron et al. 1988) and in the United Kingdom Glomerulonephritis Registry, only 13 per cent (A.M. Davison, personal communication). Nevertheless, there appear to be inexplicable geographical variations in the incidence of renal venous thrombosis, even when these factors are taken into account.

The prevalence of clinical renal venous thrombosis in adult nephrotics with membranous nephropathy appears to be between 5 and 10 per cent (Table 3) and there is little disagreement over this figure. In contrast, in other forms of nephrotic syndrome (Table 4), the incidence is much lower; an analysis of 322 nephrotics in our own series showed only five with renal venous thrombosis (1.5 per cent).

The prevalence of venographic thrombosis without symptoms is however a subject of controversy. Some authors, as Tables 3 and 4 illustrate have found considerable incidences of symptomless thrombosis, in up to half their cases of membranous nephropathy; others (Pohl *et al.* 1984), including ourselves (Ogg *et al.* 1982), have found much lower figures (approximately 10 per cent).

In nephrotic children, with the exception of the Finnish form of congenital nephrotic syndrome in infants, renal venous thrombosis is rare (Trygstad *et al.* 1970). In Egli's survey of 3371 nephrotic children, 20 had renal vein thrombosis; in 10 of these the vena cava was also affected.

Clinical diagnosis and evaluation

Clinically evident renal venous thrombosis may be present acutely, with loin pain, haematuria, renal enlargement, and deterioration in renal function, or as a slower fall off in renal function without dramatic signs or symptoms. Otherwise, the thrombosis may be quite silent clinically and be found incidentally by venography or other evaluation (Table 3). Leg oedema increases if the vena cava is involved, although caval thrombosis can be surprisingly silent. Thrombosis of renal veins is frequently bilateral, but may be unilateral. Men are more commonly affected than women. About 35 per cent of patients with renal venous thrombosis will have pulmonary emboli which are clinically evident or show up on scanning (see Table 5, and Cameron *et al.* (1988) for detailed discussion).

Despite the figures of Llach and colleagues (1982) and Wagoner *et al.* (1983), few if any nephrologists have adopted a policy of doing renal venous venography in all nephrotic patients, or even in all those with membranous nephropathy (Harrington and Kassirer 1982). Venography, or venous phase arteriography are invasive tests which are expensive and have some morbidity. In general those worth investigating are those with clinical signs suggestive of renal venous thrombosis (flank pain, otherwise unexplained deterioration in renal function, sudden onset of haematuria) and those with pulmonary emboli. It has yet to be established that seeking symptomless renal venous thrombi is useful, since their prognosis appears to be benign (see below), and how frequently or at what intervals rescreening must be undertaken is not established.

The position would be much easier if a reliable non-invasive diagnostic test were available. Ultrasound (Rosenfeld 1980; Avasthi *et al.* 1983) is sometimes useful, but the veins cannot always be visualized and the technique does not always detect venographically evident thrombi. CAT scanning (Zerhouni *et al.* 1980; Gatewood *et al.* 1986) has not been systematically evaluated, nor has MRI scanning (Tobias *et al.* 1986) although either could in theory be the answer. The last authors found that CAT scanning missed some thrombi detectable by both MRI and venography.

Prognosis

Early reports on the prognosis of renal venous thrombosis were biased by the inclusion of postmortem cases and these data are of little value today. The prognosis of untreated renal venous thrombosis in the absence of anticoagulation, whether clinically evident or silent, is no longer clear since virtually all patients now receive heparin and/or warfarin. In particular, the prognosis of radiographically diagnosed symptomless renal venous thrombosis in the absence of anticoagulation is not known. Functional renal impairment is an adverse prognostic sign (Laville *et al.* 1988) but in many instances recovery of renal function is possible and recanalization of the veins usually occurs.

Treatment

It has been usual to begin anticoagulation in all patients with renal venous thrombosis (Ross and Lubowitz, 1978), irrespective of how the diagnosis was reached, or whether clinical signs are present; in the face of an incidence of overt or occult pulmonary emboli of 35 per cent it is difficult to argue against this course. There appears to be no advantage either to thrombectomy (Laville *et al.* 1988) or local fibrinolytic therapy (Balabanian *et al.* 1973; Harrington 1984; Laville *et al.* 1988). In most cases in which warfarin has been used and subsequent venography performed, the vein had recanalized (Kassirer 1973; Llach 1982), but in some it did not (Laville *et al.* 1988). Five of 27 patients in Laville's series had bleeding complications from anticoagulation (see below on use of warfarin in nephrotics)

One problem which arises is when—and on what grounds—anticoagulation can be stopped. Because of the large body of data relating to deep vein thrombosis in other situations (Petiti *et al.* 1986), 6 months of treatment has been usual. However, stopping the warfarin in the continuing presence of a nephrotic

Table 5 Concentrations of proteins important in coagulation in the nephrotic syndrome in relation to molecular weight

Protein	MW (Da)	Concentration in nephrotic plasma
Zymogens and cofactors		
von Willebrand factor	840 000	Raised
Factor V	350 000	Raised
Factor I (fibrinogen)	330 000	Raised
Factor VII	200 000	Raised–normal
Factor XI	160 000	Normal–reduced
Factor XII	79 000	Normal–reduced
Factor II (prothrombin)	72 000	Normal–reduced
Factor X	56 000	Normal–reduced
Factor IX	55 400	Normal–reduced
Regulator proteins		
α_1-Macroglobulin	840 000	Raised
Plasminogen	81 000	Normal–reduced
Protein S	75 000	Normal–reduced
α_2-Antiplasmin	70 000	Normal–reduced
Antithrombin III	68 000	Normal–reduced
Protein C	65 000	Raised–normal
α_1-Antitrypsin	54 000	Reduced

syndrome may lead to late rethrombosis (Briefel et al. 1978), and a reasonable policy is to continue warfarin until the nephrotic syndrome remits, or at least until the serum albumin is above 25 g/l (see below)

Abnormalities of coagulation in nephrotics

It is not clear exactly what underlies this increased tendency to thromboembolism in nephrotic patients. No single protein factor appears overwhelmingly important, since many proteins involved in the initiation or regulation of clotting show altered concentrations, most of which might be expected to lead to enhanced coagulability. These are summarized in Table 5, and reviewed in greater detail in Vaziri (1983), Paniucci (1983), Cameron (1984), Cameron et al. (1988), and Livio (1988). Therefore only very recent references will be given to details of the actual concentrations of clotting and regulator proteins in the discussion below.

Physical factors

Nephrotic patients are relatively immobile and may have haemoconcentrated blood from hypovolaemia, especially in childhood. Whole blood viscosity is increased by both the increase in haematocrit and as a result of increased plasma viscosity, in turn related to high fibrinogen concentrations (McGinley et al. 1983; Ozanne et al. 1983). Thus, even without alterations in coagulation proteins or in platelet function, nephrotic patients are likely to suffer thromboses.

Alterations in zymogens and cofactors

Several studies of adults (Kanfer et al. 1971; Kendall et al. 1971; Thompson et al. 1974; Bernard et al. 1977; Robert et al. 1987) and children (Honig and Lindley 1971; Loirat et al. 1979; Alkjaersig et al. 1987) have surveyed concentrations and/or activities of zymogens and their cofactors (as well as regulator proteins, see next section) in nephrotic patients. In general, factors of the intrinsic pathway are reduced, and occasional patients with a bleeding tendency from this have been reported (e.g. Handley and Lawrence 1967) but all other zymogens or cofactors are either present in normal amounts, or (more usual) raised, particularly von Willebrand factor, fibrinogen, and factors V, X, and VII. Concentrations of prothrombin (factor II) are in general normal. Whether in fact increases in zymogen concentrations, over and above the great excess in which all are present in normal circumstances will induce a 'prethrombotic' state is not clear; none are rate-limiting at the concentrations found in normal plasma. However, it is worth noting that in non-nephrotic individuals fibrinogen and factor VIIc concentrations are independent variables predicting arterial thrombosis.

Alterations in fibrinolytic and regulator proteins

Whilst zymogens and their cofactors are all promoters of thrombosis, the situation is more complicated with regard to regulator proteins, since some of them are concerned both with inhibiting fibrinolysis (which limits thrombosis) and also thrombin generation (which promotes thrombosis). It is therefore more difficult to predict the effects of abnormal concentrations of individual components. In addition, there are several modulators of plasmin and thrombin which act in concert, concentrations of some of which go down, and some up in the nephrotic patient. Thus even though levels of antithrombin III are very low (Kauffman et al. 1978) total antithrombin activity is normal in the nephrotic syndrome, as described by Kanfer and colleagues (1970), because of the rise in α_1-macroglobulin concentration (Boneu et al. 1981). Rydzewski et al. (1986) noted low concentrations of α_1-antitrypsin, but felt that the increases in α_1-macroglobulin did not compensate for losses of antithrombin III. Consistently, plasminogen concentrations have been found to be low in nephrotics, but the role of plasmin inhibitors (Du et al. 1985) is not yet clear.

There has been much interest in a possible role in nephrotic hypercoagulability of the recently described vitamin K-dependent natural anticoagulants protein C and protein S (Clouse and Comp 1985). Protein C, which is activated by the thrombin—thrombomodulin complex, antagonizes the action of activated factors V and VIII and stimulates release of plasminogen activator from endothelial cells. Protein S is a cofactor of activated protein C, and is responsible for its binding to platelets and endothelial cells where anticoagulant activity is expressed. Patients with hereditary deficiencies of either protein have recurrent thromboses. Protein S exists in the plasma in a free active form, and also in an inactive form, bound to the C4-binding protein. Despite a rather low molecular weight (62 000) even though urinary losses occur (Cosio et al. 1985; Sala et al. 1985; Mannucci et al. 1986) there is agreement that levels of antigenic protein C remain normal (Soff et al. 1985) or raised (Cosio et al. 1985; Sorensen et al. 1985; Mannucci et al. 1986; Rostoker et al. 1987) in nephrotics; the same is true of protein C functional activity (Pabinger-Fasching et al. 1985; Vaziri et al. 1988). The position is more complicated with regard to protein S. Total levels of protein S are misleading, since only the free protein is active. Nevertheless, raised (Vaziri et al. 1988), approximately normal (Gouault-Heilman et al. 1988), and reduced (Vigano-D'Angelo et al. 1987) protein S levels have been reported in nephrotics, although the reduction is modest compared with those of antithrombin III, whose concentration may fall to only 10 to 20 per cent of normal.

Summary of alterations in coagulation proteins

Thus the immensely complex balance of coagulant, fibrinolytic, and regulator proteins is much disturbed in the dysproteinaemia of the nephrotic state. However, it seems unlikely that the increases in zymogens and other promoters of coagulation will enhance coagulation in view of the excess normally present in plasma. If the alterations in proteins do promote hypercoagulability, it is probably through urinary losses of regulator proteins; even here there is no convincing evidence that total antithrombin activity is reduced in nephrotics, when all the changes are summed up. Fibrinolysis has been less studied in nephrotics (Thompson et al. 1974), and it may be here that the effects of dysproteinaemia, if any, are to be found. In fact conventional tests of coagulation such as the prothrombin time are usually normal in nephrotics (Andrassy et al. 1980). The partial thromboplastin time may be prolonged in the few patients with low intrinsic pathway activities (factors XII, XI, IX). However it is safe to biopsy these patients without fresh frozen plasma, as it is in those with a lupus 'anticoagulant' (antiphospholipid antibody) (Kant et al. 1982).

Alterations in platelet function

In contrast, it is possible that platelet hyperaggregability plays a significant role in the thrombotic state of nephrotic patients (Livio 1988; Cameron 1989) even though there is as yet no firm evidence that this operates in vivo. There are three possible mechanisms by which platelet hyperaggregability may arise: effects of lowered serum albumin on platelet prostaglandin metabolism, effects of hyperlipidaemia on platelet membrane lipids, and effects of raised von Willebrand factor concentrations.

The platelet count in nephrotic patients has variously been reported as either normal or raised (see Cameron (1984) and Livio (1988) for review) but the elevation, if present, is usually mild.

Many workers have demonstrated abnormalities of aggregation and thromboxane production in platelets from nephrotic patients studied *ex vivo*. Hyperaggregability to ADP stimulation when in their own plasma, with occasional spontaneous aggregation, is seen. Bang *et al.* (1973) showed that this hyperaggregability could be reversed by adding concentrated urine protein to the *in-vitro* system. Thus it seemed that some factor or factors present in plasma modulate platelet function, and are lost in the urine in nephrotics.

Albumin binds exogenous, added arachidonic acid, which can induce aggregation *in vitro*, and addition of pure albumin either *in vitro* (Yoshida and Aoki 1978; Remuzzi *et al.* 1979; Stuart *et al.* 1980), or *in vivo* by infusion (Remuzzi *et al.* 1979) to nephrotic platelets restores this hyperaggregability to normal. Those patients with a plasma albumin below 20 g/l showed the greatest effect. The production of thromboxane A_2 (or malonaldehyde, as a general indicator of prostaglandin production) by platelets *in vitro* is greater at lower ambient albumin concentrations (Yoshida and Aoki 1978; Stuart *et al.* 1980; Jackson *et al.* 1982; Schiepatti *et al.* 1984). It seems possible, therefore, that binding sites for arachidonate on the albumin compete with the cyclo-oxygenase which leads to production of thromboxane A_2, for arachidonate.

However, endogenous arachidonate released from membrane lipids by phospholipase A is available only intracellularly, and it is difficult to see how this competition for arachidonate could operate *in vivo*. Hypoalbuminaemia itself reduces the production of endogenous arachidonate from the membrane lipid (Stuart *et al.* 1980), and collagen (which leads to endogenous arachidonate release) leads to only minor hyperaggregability *in vitro* in nephrotic platelets when compared to the effect of exogenously added arachidonate (Remuzzi *et al.* 1979; Jackson *et al.* 1982). Thus a major role for the lowered albumin binding of arachidonate in nephrotic platelet hyperaggregability remains unproven.

Hyperlipidaemia is also a prominent feature of the nephrotic syndrome, either hypercholesterolaemia alone or together with hypertriglyceridaemia. Lipoproteins bind to platelets, LDL enhancing and HDL inhibiting aggregation; these effects may be mediated through effects on adenylate cyclase. Hyperlipidaemia is associated, in nephrotics, with spontaneous aggregation (Jackson *et al.* 1982). Dietary lipids alter the composition of platelet membrane lipids—for example, if n_3-unsaturated fatty acids are fed, these replace arachidonic acid in the platelet membrane, with a decrease in platelet aggregability and an increase in bleeding time. There are also opposing effects of LDL and HDL on endothelial cell production of prostacyclin, which is a powerful modulator of platelet activity. The effects of nephrotic hyperlipidaemia on *in-vitro* hyperaggregability and especially on *in-vivo* thrombosis deserve more exploration.

It has been suggested also that the charge on platelets may be important. Like all cells, platelets have a net negative charge on their surface, as can be demonstrated directly by electrophoresis. Using an indirect assessment of charge using the binding of the cationic dye Alcian blue, Levin and colleagues (1985) suggested that in children with the steroid-responsive nephrotic syndrome there was a reduced net negative charge on the platelets and red cells. This might lead to more ready platelet aggregation, and spontaneous red cell aggregation (Ozanne *et al.* 1983) as well as spontaneous platelet aggregation (Jackson *et al.* 1982; Levin *et al.* 1985) has been described in nephrotics. Levin and colleagues suggested that this phenomenon might relate specifically to the pathogenesis of minimal change nephrotic syndrome, but there are doubts both about the specificity for this disease and the technique. Boulton-Jones *et al.* (1986) found the same effect in all nephrotics of whatever histology, and Feehally *et al.* (1985) could not confirm the data at all even using Alcian blue. Recently, direct measurements of platelet charge in nephrotics using electrophoresis has shown no difference in charge from normal subjects (Cohen *et al.* 1988), so the significance of the data based on studies using Alcian blue binding is doubtful.

However we (Bennett *et al.* 1987) have found hyperaggregability to ristocetin (which is a cationic compound) in nephrotic adults irrespective of underlying histology. This hyperaggregability appeared to be independent of the depletion of albumin, or the availability of arachidonate, which of course is not needed for aggregation using ristocetin. Hyperaggregability to ristocetin probably relates most to the elevated levels of von Willebrand factor which are found in almost all nephrotics (Cameron *et al.* 1984; Bennett *et al.* 1987), and to which ristocetin binds in a triple complex with the platelet. However, in addition, the amount of IgG present on (and in) the platelets was reduced in proportion to the reduction in serum IgG found in nephrotics. This reduction in surface protein would reduce the negative charge, and this could also affect aggregability and adhesion. Obviously this is an area which deserves further study.

Role of drugs
Corticosteroids
The description of thromboses in nephrotic patients coincided with the introduction of corticosteroids in their treatment (Cosgriff 1951; Calcagno and Rubin 1961); in Egli's paediatric survey, 26 of 59 patients who developed thrombosis were taking steroids at the time. Corticosteroids raise the concentration of some zymogens, particularly factor VIII (Ozsolylu *et al.* 1962), and prothrombin times and APTT may be shortened under steroid treatment (Ueda *et al.* 1987). Mukherjee *et al.* (1956) noted an increase in thromboplastin generation when three nephrotic patients were started on corticosteroids; the action of warfarin is also antagonized by corticosteroids (Menczel and Dreyfuss 1960). On the positive side, steroids tend to raise concentrations of antithrombin III (Thaler and Lechner 1978) and inhibit platelet aggregation (Glass *et al.* 1981), at least in large doses.

Diuretics
Most patients who develop thrombosis are receiving diuretics, since almost all are oedematous; some develop thrombosis during diuresis. The reduced plasma volume and consequent increased haematocrit induced by diuretics can only promote thrombosis; whole blood viscosity rises steeply with haematocrit, and nephrotic patients already have a high plasma viscosity which probably arises in the main from the very increased

concentrations of fibrinogen (Ozanne *et al.* 1983; McGinley *et al.* 1983). It may be that the more judicious use of diuretics during the 1970s and 1980s has led to a decrease in the frequent thromboses, for example of the pulmonary artery in children, which we no longer seem to observe. Another factor may the common practice of giving albumin infusions at the same time as powerful diuretics, especially when the latter are given intravenously.

Treatment and prophylaxis of thrombosis in nephrotics

Some 10 per cent of adult nephrotics and 1.8 per cent of children will suffer some clinically-evident thrombotic episode during their course. The thromboses in the children, being more frequently arterial will on average be more serious clinically. These data ignore the subclinical deep venous thrombosis, pulmonary embolism, and renal vein thrombosis discussed above. First, there are general factors which can be avoided: patients should be mobilized, sepsis avoided or treated promptly, dehydration from incidental causes (e.g. diarrhoea) treated, diuretics used with care, and haemoconcentration minimized.

Anticoagulation treatment would seem to be the answer, but carries the usual risks and presents additional difficulties in nephrotic patients. Heparin acts mainly (although by no means exclusively) through the activation of antithrombin III, whose concentration may be grossly diminished in nephrotics (Kauffman *et al.* 1978). Higher doses of heparin are required in nephrotics to achieve anticoagulation, although there are probably additional explanations for this finding (Vermylen *et al.* 1987). Sie *et al.* (1988), however, describe normal plasma concentrations of heparin cofactor II (MW 65 600) in nephrotics, and heparin binds also to α_1-macroglobulin and to endothelial cell surfaces, as well as promoting platelet aggregation. The role of heparin therefore is not clear in nephrotic patients.

Warfarin similarly presents problems (Ganeval *et al.* 1986), mainly because it is albumin bound and the levels of albumin may change therapeutically or spontaneously. However, warfarin also raises antithrombin III levels (Andrassy *et al.* 1980). The problem still remains, however, of where the balance of benefit lies in treating some or all nephrotics with anticoagulants.

Clearly those with symptomatic thromboses should receive anticoagulation treatment at least for the normal time of about 6 months (Petiti *et al.* 1986). At this point their serum albumin should be assessed, and if this is under 25 or especially 20 g/l, then anticoagulation should probably be continued until it exceeds this level, because of the risks of rethrombosis. Whether nephrotic patients with symptomless deep venous thrombosis should receive anticoagulation treatment has never been studied; probably if they also have symptomless pulmonary emboli most physicians would anticoagulate. If the physician does wish to anticoagulate all nephrotics with symptomless deep vein thromboses, then one in four will require anticoagulation treatment. Despite the fact that most physicians will anticoagulate patients with symptomless renal venous thrombosis, there is in fact no basis for doing this. Finally, very few nephrologists use prophylactic warfarin in all nephrotic patients, and only a few in those with very low albumin concentrations (less than 20 g/l). There is a case for giving this at-risk subgroup low-dose aspirin (75 mg daily, which is safe with steroids) and dipyridamole (100–200 mg three times a day), as suggested by Andrassy's data (1980).

Alterations in lipid and carbohydrate metabolism

Possible effects of lipid alterations on ischaemic vascular disease

It has been known for 160 years that fasting nephrotic plasma may be lipaemic, and much of the recent interest in the alterations in lipid metabolism in the nephrotic syndrome has centred on possible effects on the induction of atheromatous vascular disease, and what might be learnt both from this secondary hyperlipidaemia about risks of primary hyperlipidaemia, and vice versa.

It has become standard teaching (Mallick and Short 1981) that nephrotics have a higher incidence and prevalence of vascular disease, in particular myocardial ischaemia and infarction. This idea requires more detailed examination.

Early descriptions noted atheroma even in nephrotic children (Leiter 1931; Schwartz and Cohn, 1935) and several small series of nephrotic patients with myocardial infarction were published (Berlyne and Mallick, 1969; Alexander *et al.* 1974) with the suggestion that the incidence of this was much higher than in controls. Other studies (Curry and Roberts 1977) and anecdotes (Kallen *et al.* 1977) supported this concept. However, examination of these reports shows that apart from the small numbers (15 and 17 patients respectively in the studies of Berlyne *et al.* and Alexander *et al.*), and the problem of ascertainment artefact, the 'controls' were often inappropriate, given that the incidence of myocardial disease differs sharply with race, sex, age, blood pressure, and geographical area. Also, the disparity of numbers (more than 300 deaths/million population from ischaemic heart disease, compared with an incidence of only 10 new adult nephrotic patients/million population each year), the inclusion of patients with diabetes and lupus (conditions known to show excess ischaemic disease) in the paper of Curry and Roberts (1977), and the ignoring of other major risk factors such as smoking and hypertension render these comparisons almost valueless.

On the other hand, several observers (Hopper *et al.* 1970; Gilboa 1976), as well as our own group (Wass *et al.* 1979; Wass and Cameron 1981) noted an incidence of ischaemic heart disease indistinguishable from a control population, and most of the cases noted were in middle-aged males as was to be expected. Our careful epidemiological survey (Wass *et al.* 1979) failed to show in a group of nephrotics of mixed aetiology (but excluding diabetes and lupus) that the death rate from, or incidence of, ischaemic heart disease was higher than appropriate sex- and age-matched local controls.

However, the numbers involved even in this regional survey are small, and this is not to deny that a subgroup of nephrotic patients with a particularly severe or prolonged course may be at extra risk (Mallick and Short 1981; Kallen *et al.* 1977). An important feature of our study was how briefly the majority of nephrotic adults remained in that state, most going into renal failure or remitting within 3 to 4 years. However, the numbers of patients with (for example) membranous nephropathy and focal segmental sclerosis resistant to treatment with severe prolonged nephrotic syndromes are very small in any series, and it is likely that it will be impossible to conduct an appropriate epidemiological survey.

Obviously these data and their interpretation have considerable importance when considering possible treatments, and this topic is dealt with below.

Alterations in lipids in nephrotic patients

It has also become received dogma that the possibly increased incidence of ischaemic vascular disease in nephrotics is in turn the result, in the main, of alterations in the levels of circulating lipids. Again, this proposition requires careful examination. Space precludes a summary of lipid and lipoprotein metabolism; the recent excellent description of Deckelbaum (1987) may be consulted. Nor, of course, are circulating lipids the only possible risk factor in the genesis of vascular disease in the population as a whole.

Cholesterol and triglycerides

Raised concentrations of total cholesterol (both free cholesterol and cholesterol esters) are present in almost all nephrotic patients (Baxter 1962; Chopra *et al.* 1971; Newmark *et al.* 1975), which is greater in the more severe nephrotic syndromes, especially when judged by the concentration of serum albumin, with which there is a strong negative statistical correlation. Phospholipids increase also (Baxter 1962), but to a lesser extent than the cholesterol, so that the cholesterol:phospholipid ratio falls as the nephrotic syndrome becomes worse. The phospholipid also shows a reduced amount of lecithin (Nye and Waterhouse, 1961) with increased lysolecithin and sphingomyelin.

Increases in fasting triglyceride levels are less common than cholesterol, and are usually found in the more severe nephrotic syndromes with a serum albumin of 20 g/l or less; below 10 g/l, hypertriglyceridaemia is almost invariable (Baxter 1962; Chopra *et al.* 1971; Newmark *et al.* 1975) and may reach four times normal levels. There is a negative correlation between triglycerides and serum albumin (Newmark *et al.* 1975) but this is less strong than that observed for cholesterol, probably because of additional effects of diet. Free fatty acid levels are reduced, probably because of their binding to albumin (Shafrir 1958) although the concomitant increase in lipids results in normal total fatty acid concentrations.

Lipid and cholesterol concentrations, although depending upon the severity of the nephrotic syndrome, do not seem to vary with different histological types (Chopra *et al.* 1971; Newmark *et al.* 1975), and in particular diabetics and those with lupus were (perhaps surprisingly) no different from those with glomerulonephritis of various types. Diminished renal function does of course affect triglyceride levels profoundly, but this is only important in a minority of nephrotics.

Also, there is some evidence (Zilleruelo *et al.* 1984) that at least in nephrotic children, the lipid abnormalities may persist into remission, or between relapses.

Lipoproteins

Since lipids such as cholesterol, cholesterol esters, phospholipids, and triglycerides are insoluble in water, they circulate as mixtures of carrier protein (apoprotein) and lipid, which together form lipoproteins. These are usually classified by their ultracentrifugal properties into very low density lipoproteins (VLDL), intermediate density lipoprotein (IDL), low density lipoproteins (LDL) and high density lipoproteins (HDL). Figure 2 summarizes the relationships between these three species of lipoprotein, and the alterations of the nephrotic state. The pro-

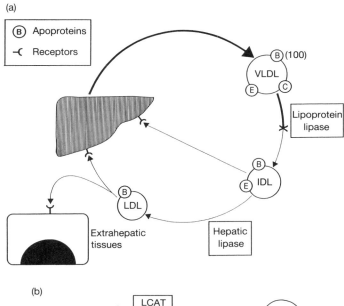

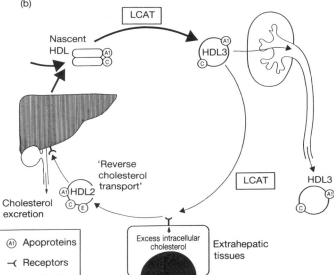

Fig. 2 (a) (left) Abnormalities of very low density lipoprotein (VLDL), IDL and LDL in the nephrotic syndrome. Hepatic overproduction of VLDL leads to an increase in plasma concentration of this lipoprotein and its products, intermediate (IDL) and low (LDL) density lipoprotein. In addition the activity of endothelial lipoprotein lipase is reduced, so that catabolism of VLDL is retarded. Excess VLDL may deposit in vascular structures (see the text for more details). (b) (right) Abnormalities of high density lipoprotein (HDL) in the nephrotic syndrome. Secretion of nascent HDL by the liver is increased. HLD3 tends to be lost in the urine since its molecular weight is higher than that of HDL2. LCAT activity is reduced, which also inhibits conversion of HDL3 to HDL2. Removal of cholesterol from tissues to the liver is inhibited by this defective maturation of HDL (see text for additional details). Taken from Wheeler D.C., Varghese, Z., and Moorhead, J.F. (1989). Hyperlipidemia in the nephrotic syndrome. *American Journal of Nephrology*, **9**, (suppl 1), 78–84, with permission of the authors and publishers.

tein moiety of lipoproteins, the apolipoproteins, are synthesized in the liver, and are generally divided into apoproteins A-I, A-II, B, C-I, C-II, and E.

Low-density and very low-density lipoproteins (LDL and VLDL)
VLDL are increased in moderate and severe nephrotic syndromes but may be normal in mild cases (Baxter 1962; Chopra *et al.* 1971; Gherardi *et al.* 1977). The lipoprotein fraction contains much of the excess triglyceride, and the cholesterol:phospholipid ratio is increased. Kashyap *et al.* (1980) showed reduced apolipoprotein C II in VLDL from nephrotics, which may be important since Apo C II is an activator of lipoprotein lipase (see below). Like cholesterol and triglyceride concentrations, VLDL concentrations vary inversely with the serum albumin concentration (Gherardi *et al.* 1977).

LDL concentrations are also consistently raised in nephrotics (Baxter 1962; Chopra *et al.* 1971; Gherardi *et al.* 1977) in parallel with the VLDL concentrations (Gherardi *et al.* 1977).

High-density lipoproteins (HDL)
In contrast to the unanimity of data on VLDL and LDL, data on HDL concentrations in nephrotics are conflicting: low concentrations (Baxter 1962; Lopes-Virella *et al.* 1979; Appel *et al.* 1985), normal levels (Baxter 1962; Lewis and Page 1953; Widhalm *et al.* 1980; Lopes Virella *et al.* 1979; Chan *et al.* 1981; Short *et al.* 1986), and high concentrations (Gherardi *et al.* 1977; Oetliker *et al.* 1980) of HDL have all been reported. In most of these studies, the reduced HDL concentrations have been noted only in the most severe nephrotic syndromes, as in our own data (V.J. Wass, unpublished observations). However, these differences probably arise from the mixture of nephrotic patients of various ages, renal function, treatments (Sokolovskaya and Nikiforova 1984), duration, and severity included in the studies. In addition true control populations studied in parallel are lacking in many of these papers.

Joven *et al.* (1987) examined the composition of the HDL in both children and adults with nephrotic syndrome, none of whom were uraemic. Apolipoprotein A-I concentrations were raised in both children and adults compared with controls, HDL cholesterol was normal in both, but HDL phospholipids were raised in children and low in adults. Others have observed normal or low Apo AI concentrations in adult nephrotics (Short *et al.* 1986).

Gherardi *et al.* (1977) examined HLD2 and HDL3 separately, finding that only HLD2 was reduced, the levels of HDL3 being normal or increased, except in the most severe cases. The data of Short *et al.* (1986) are similar. The HDL2 concentrations correlated with the serum albumin concentrations in Gherardi's data, and as HDL2 decreased the VLDL concentrations rose. Since HDL3 molecules are smaller than HDL2 (molecular weight 175 000 *vs.* 360 000), one would expect HDL3 to be lost more readily into the urine (Short *et al.* 1986), and a good explanation is lacking for these apparently paradoxical observations. Unfortunately many of the patients in the study of Gherardi *et al.* were also uraemic, so the contributions of the nephrotic and uraemic states to these changes is not clear.

There is some evidence (Oetliker *et al.* 1980; Widhalm *et al.* 1980; V.J. Wass, unpublished results) that the raised HDL concentrations noted in minimal change nephrotic children in relapse persist during remission. The significance of this provocative observation is not yet clear.

Lipiduria
Lipiduria is well recognized in the nephrotic syndrome, and fatty urinary casts are a characteristic feature (see Chapter 1.2). Cholesterol, phospholipid, free fatty acids, and triglyceride are all present in nephrotic urine. HDL is also present (Felts and Mayerle 1974; Lopes-Virella *et al.* 1979; Kashyap *et al.* 1979; Short *et al.* 1986) and Gitlin *et al.* (1958) showed that labelled LDL did not appear in the urine when injected into nephrotics, whereas HDL did so. This is not surprising when one considers that HDL molecules are little larger than albumin, whilst LDL molecules are much larger.

It seems that little catabolism of lipoproteins occurs in the kidney in the nephrotic syndrome (Kashyap *et al.* 1979). Shore *et al.* (1982) showed that ApoE and ApoC-I may be dissociated from lipoprotein molecules in the kidney, but ApoB and ApoA-I remain relatively unaffected. Thus it seems that only easily degradable or surface proteins are catabolized in passage through the kidney.

Cardiovascular risk profiles in nephrotic patients

It seems that plasma concentrations of total cholesterol are raised in almost all nephrotics, and triglycerides also in the more severe cases. VLDL and LDL lipoproteins rise in even a mild nephrotic syndrome, whilst HDL and ApoA-I remain normal except in all but the most severe cases, in whom urinary losses of this lipoprotein become significant. Thus the total:HDL and LDL:HDL lipoprotein ratios are increased in all nephrotics to some extent. All these factors have been shown, in general non-nephrotic populations, to be amongst the several risk factors for the appearance of ischaemic heart disease. Many would argue, therefore, that the secondary hyperlipidaemia of the nephrotic patient must be a major risk factor for vascular disease. As in the general population, however, other factors which are even stronger generators of risk are at work, particularly hypertension. Only a few nephrotic patients remain nephrotic long enough for the hyperlipidaemia to have much effect, and despite the strong probability that the hypothesis which suggests hyperlipidaemia induces vascular disease is true for some nephrotics, there is no justification in these data for treating all nephrotics with hypolipidaemic agents (see below).

Causes of lipid alterations in the nephrotic syndrome

The causes of the raised plasma lipids in the nephrotic syndrome are still not understood but both a major increase in production and a relatively minor decrease in removal of lipoproteins seem to be present (Wheeler *et al.* 1989), and abnormalities on almost every step in production and removal have been described. The production of lipoproteins and the relation to the various apoproteins is summarized in Fig. 2(a) (see Deckelbaum (1987) and Wheeler *et al.* (1989) for further details). Important in understanding the physiology and pathology of lipoproteins is that LDL is removed largely through high-affinity receptor-mediated pathways, principally into hepatocytes, but also via low-affinity receptor-independent pathways. Monocytes can assimilate LDL by both mechanisms, and develop into the foam cells infiltrating fatty streaks in early atheromatous lesions. LDL may become oxidized (Steinberg *et al.* 1989) and the atherogenic potential of these molecules is believed to be much greater than for normal LDL.

Effects of hypoalbuminaemia
Hypoalbuminaemia was thought for some time to be the non-specific stimulus inducing increased liver apoprotein (and thus

lipoprotein) synthesis. Not only are there strong negative correlations between serum albumin and lipid levels as outlined above, but infusion of albumin into nephrotic patients reduces the concentrations of both cholesterol and triglyceride (Soothill and Kark 1956; Baxter 1962) and Kekki and Nikkilä (1971) showed that increased triglyceride synthesis returned to normal. However, there is evidence that this is not so in the long term, since alterations in albumin synthesis by changing diet were not accompanied by alterations in lipids (Kaysen et al. 1987). Also, Allen et al. (1961) showed that dextran infusions would lower cholesterol levels in nephrotic rats, although the same experiment seems not to have been done in humans. Appel et al. (1985) found that plasma osmotic pressure correlated with lipid concentrations, which agrees with observations that perfusate osmolarity altered albumin synthesis in isolated perfused livers (Davis et al. 1980). Yedgar et al. (1984) suggested that plasma viscosity was the signal for cultured hepatocytes, but Appel et al. (1985) found no correlation in intact human nephrotics. At the moment, therefore, the precise signal is unknown.

Hepatic lipoprotein synthesis

Hepatic synthesis of both the lipid and protein moieties of lipoproteins is markedly increased in the human nephrotic syndrome (Wheeler et al. 1989). Synthesis of cholesterol (McKenzie and Nestel 1968), total triglyceride (Kekki and Nikkilä 1971), VLDL triglyceride (McKenzie and Nestel 1968), and LDL (Scott et al. 1970) have all been shown to be increased, and in the case of VLDL triglyceride (McKenzie and Nestel 1968; Kekki and Nikkilä 1971), to return to normal on remission. This increased synthesis, however triggered, (see previous paragraph) probably underlies most of the major increase in lipids in the disease.

Removal of LDL: receptor mediated removal and lipoprotein lipase (LPL)

Decreased catabolism of lipoproteins has also been described; however. Warwick et al. (1989) used metabolism of native and chemically-altered LDL to study receptor and non-receptor mediated removal of LDL in adult nephrotics, and showed that the major defect was in receptor-mediated catabolism. They found that the fractional catabolic rate of LDL was markedly reduced, in agreement with data of Gitlin et al. (1958) in nephrotic children; Vega and Grundy (1988), however, found a normal catabolic rate.

Lipoprotein lipase (LPL), mainly localized in muscle and adipose tissue, performs a crucial step in the transformation of lipids, which hydrolyses VLDL and LDL and (eventually) HDL. This enzyme requires ApoC-II for activation, which is transported in HDL and transferred to VLDL during activation of the enzyme, then back to HDL after the reaction is complete. The studies previously cited (McKenzie and Nestel 1968; Kekki and Nikkilä 1971) showed also a marked decrease in the rate of degradation of triglyceride in nephrotics, and most studies of LPL activity in nephrotics have shown reduced levels (Hyman et al. 1969; Kashyap et al. 1980).

The cause of this reduction in LPL activity in nephrotics has not been clarified (Chan et al. 1981; Wheeler et al. 1989); in some series of nephrotics, patients with reduced renal function were included, and in uraemia reduced LPL activity is well known. However, free fatty acids, normally bound to albumin, accumulate in nephrotic plasma and can inhibit LPL (Nikkilä and Pykäpistö 1968). Also, the apoC-II already mentioned may play a role. Loss of HDL3 in the urine (see above and Fig. 2(b)) would result in loss of associated ApoC-II, bound to it and essential for LPL activation (Kashyap et al. 1980). However, plasma concentrations of ApoC-II were normal in this study. Straprans et al. (1981) also identified glycoprotein cofactors of LPL present in nephrotic urine, and Vermylen et al. (1987a) showed inhibition of LPL by sera from nephrotic children. The inhibition of LPL in nephrotics may therefore be multifactorial.

Lecithin cholesterol acyl transferase (LCAT)

HDL are secreted by the liver and intestines (Fig. 2(a)), and require the action of lecithin cholesterol acyl transferase (**LCAT**) for their maturation. HDL form a means whereby lipid is returned to the liver. Reduced activity of LCAT has been noted in nephrotics (Cohen et al. 1980) which could reduce HDL production. Again, the reasons for LCAT deficiency in nephrotics, which would reduce ability to mobilize lipid from endothelium or other peripheral tissue, are obscure. The enzyme (molecular weight 65 000 Da) may be lost in the urine, but loss of albumin may play a part, since low albumin concentrations inhibit LCAT activity *in vitro*.

Treatment of lipid alterations in nephrotics

Because of the possibility that nephrotic hyperlipidaemia is a risk factor for accelerated vascular disease, as outlined above, attempts have been made to modify the lipid environment of the nephrotic patient, in the hope that, as in several studies of lipid modification in the general population, there might be a reduction in vascular disease now or in the future.

It seems reasonable to give dietary advice to nephrotic as to any other hyperlipidaemic patients, but surprisingly no studies have been done on compliance or effects on circulating lipids in nephrotics, despite the availability of data from similar studies in uraemic patients (see Chapter 9.6.1).

As mentioned above, there seems to be no justification in treating all nephrotic patients with hypolipidaemic agents. Most nephrotics remit spontaneously, are treated, or go on to develop chronic renal failure within 3 to 5 years of onset. The decision to treat hyperlipidaemia in a patient with a nephrotic syndrome can therefore be postponed until a year or so has passed, or in the light of the probable outcome, and the response to treatment or lack of it, which seems likely in the light of the renal biopsy appearances.

Several possible modes of treatment are available (Table 6). As noted above, no reports exist of studies on attempted dietary modification of plasma lipids, or the administration of n_3 unsaturated fish oils, which might affect platelet function favourably, also as noted above. The actions of hypolipidaemic drugs are different; there is no indication which may be the most useful in treating nephrotics, and no long-term studies of this situation. Now that more effective treatments for lowering hyperlipidaemia are available, interest is reviving in the treatment of this aspect of the nephrotic syndrome, but it must be emphasized that as yet there is no way of knowing whether, in any individual, reduction of the lipids towards normal levels has achieved benefit and that therefore the risks and side-effects of any drugs must be considered very carefully.

Resins (cholestyramine and colestipol)

These bile-acid binding resins act by inducing increased expression of LDL receptors on hepatic cells, and lower LDL

Table 6 Treatment of nephrotic hyperlipidaemia

Type of agent	Effect on total cholesterol	Effect on triglycerides	Disadvantages
Bile-acid binding resins	Reduction	Increase	Increase thyroglobulin gastrointestinal side-effects
Fibric acid derivatives	Reduction	Strong reduction	Myonecrosis, gallstones
Probucol	Reduction	No effect	Reduces HDL
Nicotinic acid	Reduction	Reduction	Flushing
HMG coA reductase inhibitors	Strong reduction	Reduction	?Myonecrosis, no long-term studies

Fibric acid derivatives = clofibrate, benzafibrate, gemfibrozil.
Bile-acid binding resins = cholestyramine, colestipol.
Nicotinic acid derivatives = nicotinamide, Bradilan®, Acipimax®.
HMG coA reductase inhibitors = lovastatin, simvastatin, pravastatin.

cholesterol modestly (approximately 13 per cent) in non-nephrotic subjects. However, not only do they not reduce triglycerides but may actually increase them (Crouse 1987). Resins have been shown to reduce cardiovascular mortality in major trials (LRC-CPPT 1984). They must be taken in quite large quantities (16–24 g/24 h of cholestyramine and 15–30 g/24 h of colestipol), are unpleasant to take and frequently produce abdominal symptoms such as bloating and constipation. Not all patients will tolerate them over a long period. They are more palatable when made up in fruit juice and refrigerated overnight before consumption to disperse the drug. In addition the absorption of fat-soluble vitamins may be impaired, and they should not be given at the same time as other drugs because of possible effects on absorption.

Edwards et al. (1973) reported a decrease in cholesterol concentrations of 46 per cent in seven nephrotics treated with cholestyramine. More recently Valeri et al. (1986) treated seven patients with colestipol and lowered total cholesterol by 20 per cent (although not to normal levels) and LDL cholesterol by 32 per cent. There was no effect on HDL levels. Rabelink et al. (1988) compared cholestyramine treatment with simvastatin (see below) in 10 nephrotic adults. There was only a modest (20 per cent) reduction in total and LDL cholesterol concentrations in the cholestyramine group.

Fibric acid derivatives (clofibrate, benzafibrate, and gemfibrozil)

Cholesterol levels are lowered even less by fibrates than by resins, but they have a powerful effect on triglycerides. Clofibrate has now been withdrawn in the United Kingdom, because, in a WHO-sponsored trial, the incidence of cancer deaths was increased in the clofibrate patients, but it is still available in the United States and elsewhere. Clofibrate, like gemfibrozil in the Helsinki heart study (Frick et al. 1987), has been shown to reduce cardiovascular mortality in general populations with hyperlipidaemia.

Clofibrate effectively reduced triglyceride levels in Edwards' study in nephrotics (1973), and gemfibrozil has been used in Finland (Eisalo et al. 1976). Five patients showed a mean reduction in triglycerides from 5.6 to 1.4 mmol; cholesterol fell from 11.1 to 7.6 mmol/l on gemfibrozil 800 mg daily. Similarly Groggel et al. (1989) reported a 51 per cent fall in triglycerides but only a 15 per cent fall in cholesterol concentrations. Since in nephrotics (unlike in renal failure—see Chapter 9.6.1) the main need is to reduce cholesterol, it does not seem that fibrates are the best drugs to use. Also, clofibrate produces muscle necrosis in the early phase after starting the drug in some nephrotics, because of the increased amounts of free drug not bound to albumin (Bridgman et al. 1972). The risk of gallstones is increased by both clofibrate and benzafibrate.

Probucol

This drug does not affect triglyceride concentrations at all, but is effective in lowering cholesterol levels. Unfortunately it reduces HDL cholesterol concentrations as well as LDL cholesterol in many subjects. Its exact mode of action remains uncertain, and no large population studies in cardiovascular disease are available. However, it is also an antioxidant, and since oxidized LDL may be a factor in intimal injury (Steinberg et al. 1989), this property might be valuable in preventing atheroma in nephrotics, as indeed it does in hyperlipidaemic rabbits.

Probucol was used in five nephrotics in the trial of Valeri et al. (1986) in a dose of 1 g/24 h. Cholesterol was reduced by 23 per cent, LDL by 24 per cent; HDL cholesterol fell by 12 per cent, but with a fall in LDL:HDL ratio. Iida et al. (1987) studied 12 patients: total cholesterol fell from mean 10.2 to 7.2 mmol/l, and triglyceride from 3.3 to 2.5 mmol/l. No fall in HDL levels was reported in this study.

HMG coenzyme A reductase inhibitors (lovastatin, simvastatin, pravastatin)

These drugs (Grundy 1988), which interfere with a crucial step in cholesterol synthesis (3-hydroxy-3-methylglutaryl coenzyme A reductase), are now in general use. However, they seem to act to reduce lipidaemia principally by increasing LDL receptors on hepatocytes, secondary to the inhibition of cholesterol synthesis. These drugs will reduce both cholesterol and triglyceride levels, although the latter to a smaller extent. There are no long-term studies in cardiovascular disease available. Like clofibrate, lovastatin has been reported to cause rhabdomyolysis (Corpier et al. 1988) with renal failure, but perhaps as a result of interaction with cyclosporin, and commonly symptomless increases in muscle enzymes may be noted (Israeli et al. 1989); these should be monitored regularly, and the drug stopped if creatinine kinase rises over 500 i.u./l.

Lovastatin was given to eight nephrotics by Golper et al. (1987) in a dose of 40 to 80 mg/24 h. LDL and total cholesterol were reduced by a mean of 35 per cent; HDL cholesterol remained unchanged. One patient was unresponsive. Vega and Grundy (1988) studied four nephrotic patients: cholesterol, VLDL-ApoB and triglycerides were reduced in all, and HDL cholesterol levels rose. Simvastatin (20 mg twice daily) produced a similar decrease in total cholesterol of 30 per cent and a decrease in LDL cholesterol of 39 per cent in 10 nephrotics studied by Rabelink et al. (1988).

3.4 Clinical consequences of the nephrotic syndrome

Table 7 Loss of binding proteins in urine in the nephrotic syndrome

Protein	MW (Da)	Ligand
Cortisol binding globulin	52 000	Hydrocortisone
Vitamin D binding globulin	59 000	25-Hydroxyvitamin D, 1,25-dihydroxyvitamin D
Albumin	65 000	Drugs (prednisolone), Zinc, arachidonic acid
Thyroid binding globulin	85 000	T_4, T_3
Transferrin	87 000	Iron, Zinc
Caeruloplasmin	151 000	Copper

Nicotinic acid

Nicotinic acid (1–6 g/24 h) has been used as a hypolipidaemic agent for two decades, but there have been no studies in nephrotics since that of Edwards in 1973. It has been shown to reduce cardiovascular mortality in controlled trials (Canner: Coronary drug project 1986), and reduces both cholesterol and triglycerides. Few patients will tolerate it in full doses because of its vasodilator actions, and it may be best used in combination with other drugs such as colestipol; this combination has been used in general studies. Nicofuranose is a nicotinic acid derivative marketed as *Bradilan* ®; it also is effective alone and in combination but has not been used in nephrotics. *Acipimax* ® (Carlo Erba) has also been used. Inhibition of side-effects of nicotinic acid by NSAIDs has been reported, but their use in nephrotics has obvious potential dangers (see Chapter 6.2). It has also been suggested that nicotinic acid increases the frequency of myopathy seen with HMG CoA reductase inhibitors, but this is not established.

Summary of treatment of nephrotic hyperlipidaemia

Dietary advice seems prudent but is not known to affect nephrotic hyperlipidaemia. At the moment there is no clear message as to whether, and with which drug, nephrotic hyperlipidaemia should or can be treated. Both probucol and resins have disadvantages and are less effective than other drugs. Nevertheless, resins have been the only widely available and relatively safe drugs to use for some time. Perhaps the HMG CoA reductase inhibitors will be most effective, but data on detailed long-term studies in larger numbers of nephrotics are needed before they can be generally recommended. Whether this will affect morbidity and mortality either during the nephrotic phase, or during subsequent renal failure in those who progress, is quite unknown. (Possible effects of hyperlipidaemia on rates of progression into renal failure are discussed below and in Chapter 9.3.)

Alterations in carbohydrate metabolism

Unlike the more comprehensively studied alterations in lipid metabolism, there appear to be no direct clinical consequences of the changes in carbohydrate metabolism in the nephrotic syndrome, and thus they have received rather little attention (Spitz *et al.* 1970; Bridgman *et al.* 1975). The higher rate of protein synthesis in the liver probably accounts for the accelerated glycogenolysis found (Drabkin *et al.* 1955; Malmendier *et al.* 1970). Data after glucose loading have been controversial. Kleinknecht *et al.* (1972) found an exaggerated insulin response to intravenous glucose, tolbutamide, and glucagon in nephrotic children, and flattened glucose tolerance curves in response to oral glucose; however, surprisingly the insulin response was in this case blunted also. Spitz *et al.* (1970) noted elevated resting insulin and glucose levels, whilst Bridgman *et al.* (1975) found a poor insulin response to intravenous glucose. Finally Loschiavo *et al.* (1983) found an impaired response to oral glucose loading, with a diabetic-like plasma glucose curve.

These differences can be explained in part by the fact that some of these studies (e.g. that of Bridgman *et al.*) included patients with reduced renal function, in which insulin resistance is well documented, but Loschiavo's study was confined to those with normal renal function. All studies (Spitz *et al.* 1970; Bridgman *et al.* 1975; Loschiavo *et al.* 1983) showed raised basal levels of plasma growth hormone, with a paradoxical rise in concentrations following insulin rather than the usual suppression. These findings resemble those found in the low albumin state of chronic malnutrition (Becker *et al.* 1971).

Losses of binding transport proteins in the urine

A number of plasma proteins important in the transport of metals, hormones, and drugs are of relatively small molecular weight and thus are lost easily into the urine of nephrotic patients (Table 7).

Metal-binding proteins

Transferrin—iron

Cartwright *et al.* first reported low concentrations of iron and copper in nephrotic plasma in 1954, and Rifkind *et al.* (1961) and Dagg *et al.* (1966) losses of transferrin (molecular weight 87 000) and iron into the urine. Jensen *et al.* (1968) studied transferrin turnovers in nephrotics, and found increased synthesis and degradation (presumably in renal tubules after reabsorption). Iron deficiency is however rare in nephrotics, although adults with anaemia alleged to be a result have been described (Hancock *et al.* 1976; Brown *et al.* 1984). However, since losses of iron are at most 0.5 to 1.0 mg/24 h even with the heaviest proteinuria other factors must operate to produce iron deficiency. The patient of Ellis (1977) had a normochromic anaemia in association with mesangiocapillary glomerulonephritis, a well-described conjunction.

In addition the transferrin-iron complex binds to many proliferating cells, and the effects of transferrin depletion on immunity may not depend entirely upon zinc deficiency (Brock and Mainou-Fowler 1983; Warshaw *et al.* 1984).

Caeruloplasmin—copper

The copper-binding protein, caeruloplasmin, has a molecular weight of 151 000 Da, and although Cartwright originally described low red-cell and plasma copper concentrations, most other authors have reported normal levels of the metal, even though the binding protein is lost in the urine, and its plasma concentration is low (Jensen 1967; Brown et al. 1984). No clinical consequences of copper losses have been reported.

Albumin/transferrin—zinc

Zinc circulates bound mainly to albumin and also to transferrin, and thus the reported reduction in plasma, hair, and white cell zinc concentrations in nephrotics is not surprising (Reimold 1980; Temes Montes et al. 1980; Cameron et al. 1988). The clinical significance of this finding is unclear. One group reported hypogeusia in nephrotics (Mahajan et al. 1982), and the possible effects in reducing cell-mediated immunity have been discussed above (Dardenne et al. 1982; Bensman et al. 1984; Warshaw et al. 1984).

Vitamin D binding protein and calcium metabolism

A number of abnormalities of calcium and vitamin D metabolism have been described in children and adults with nephrotic syndromes, not all of which can be the result of the undoubted losses of vitamin D binding protein (molecular weight 59 000 Da) and its associated vitamin in the urine (Schmidt Gayk et al. 1977; Barrary et al. 1977). Clinically, both osteomalacia and hyperparathyroidism have been described in nephrotic children more often than adults (Malluche et al. 1979; Alon et al. 1980; Tessitore et al. 1984; Freundlich et al. 1986), although they are rarely of major clinical significance and bone biopsies are often normal (Lim et al. 1977; Korkor et al. 1983). An important finding in the study of Tessitore et al. (1984) was that nephrotics with reduced renal function readily develop bone disease, and early treatment with vitamin D may have a place.

The biochemical abnormalities described include hypocalcaemia, both total (protein bound) and ionized (Malluche et al. 1979), hypocalciuria (Lim et al. 1977), reduced intestinal absorption of calcium and negative calcium balance in both adults (Lim et al. 1977) and children (Emerson and Beckman 1945), reduced plasma 25-hydroxy- and 24,25-dihydroxy-cholecalciferol (Goldstein et al. 1977) and surprisingly also 1,25-dihydroxy-cholecalciferol (Malluche et al. 1979; Goldstein et al. 1981; Auwerx et al. 1986), and a blunted response to administration of PTH (Goldstein et al. 1981).

There have been many inconsistencies in these findings and a clear picture has yet to emerge. Not all workers have found reduced plasma 1,25-dihydroxycholecalciferol (Freundlich et al. 1986), and this finding is surprising since plasma levels are normal in nutritional vitamin D deficiencies. Low 25-hydroxy-cholecalciferol concentrations would be expected because this metabolite is bound to vitamin D binding globulin, and 25-hydroxycholecalciferol is lost in nephrotic (but not normal) urine (Saito et al. 1982). Thus the significance of low total vitamin D concentrations is obscured by the low level of the binding protein, as in all the protein-bound substances discussed in this section. Levels of free vitamin D are obviously more important, but data are lacking. There is general agreement, however, that plasma PTH levels are raised in nephrotic subjects with normal renal function (Goldstein et al. 1977; Schmidt-Gayk et al. 1977; Malluche et al. 1979; Saito et al. 1982), and that the effects of exogenous PTH are blunted, as noted above (Goldstein et al. 1981).

Thyroid-binding globulin

The puffy face and low metabolic rate when uncorrected for oedema led many early observers to suppose that all nephrotic patients were hypothyroid. In fact despite losses of thyroid-binding globulin (molecular weight 85 000 Da) in the urine (Recant and Riggs 1952; Rassmussen and Rapp 1956; Robbins et al. 1957) proportional to total proteinuria and accompanied by bound T_3 and T_4 (Afrasiabi et al. 1975; Gavin et al. 1978), plasma concentrations of T_3, T_4, and TSH are usually normal in nephrotic patients, although the T_3 may be rather low and the T_4 somewhat elevated, with increased reverse T_3 (Gavin et al. 1978), perhaps because of the loss of T_3 in the urine. These findings, although confusing, are not of clinical significance except that they must be remembered when pursuing a diagnosis of incidental hypothyroidism in a nephrotic patient. Obviously the TSH level is the important investigation.

Cortisol-binding protein

Cortisol-binding protein (molecular weight 52 000 Da) is lost in nephrotic urine and in one study (Musa et al. 1967) plasma concentration was reduced, but not in another (Loschiavo et al. 1983). More important is binding of prednisolone to albumin (see next section).

Drug binding in nephrotics

Despite the large number of drugs normally bound in part to albumin, the gross reductions in serum albumin in nephrotics give rise to remarkably few problems (Gugler et al. 1975; Gugler and Arzanoff 1976). Those concerning warfarin (Ganeval et al. 1986) and clofibrate (Bridgman et al. 1972) have been discussed above. Prednisolone is frequently given to nephrotics, especially those with minimal glomerular changes, and is normally bound to albumin. Despite this fact, even severe hypoalbuminaemia does not render dose modification necessary (Wagner et al. 1981; Bergrem 1983; Frey and Frey 1984) since the levels of free drug equilibrate rapidly to nearly normal concentrations, although in the Boston drug surveillance programme there was a direct relation between side-effects of steroid treatment and serum albumin concentrations.

Renal effects of the nephrotic syndrome

Hypovolaemia and acute renal failure

This is discussed in detail in Chapters 3.3 and 8.9. However, it is worth noting here that although acute renal failure is exceptionally rare in children with nephrotic syndrome (Steele et al. 1982), hypovolaemia seems to be relatively common compared with adult nephrotics, even in the absence of diuretic therapy. Thus a nephrotic child in the initial attack or relapse may have a cold periphery, collapsed peripheral veins and a low central venous pressure, a high haematocrit, a variable but sometimes very low blood pressure (perhaps because of variable renin secretion), and oliguria with urine almost devoid of sodium; all of these reverse on infusion of albumin. The frequency of hypovolaemia in children and the rarity of acute renal failure suggests that the acute renal failure described in Chapter 8.8 does not result from hypovolaemia; in addition, in nephrotic children acute renal failure is almost always described with either sepsis or thrombosis: two cases of each in our own series (H. Georgaki et al., unpublished data).

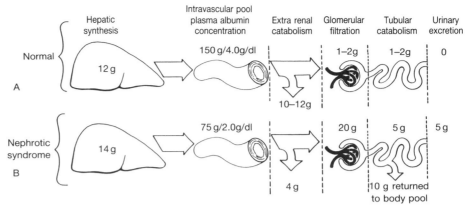

Fig. 3 A summary of albumin metabolism in the nephrotic syndrome. From Bernard (1982) with permission of the author and publishers.

Progression of renal failure

Two possible consequences of profuse proteinuria may contribute to progression of renal failure, and explain in part the correlation between relentless proteinuria and decline of renal function into renal failure (discussed further in Chapter 9.3). These relate to hypercoagulability and hyperlipidaemia.

Hypercoagulability has been implicated in the progression of remnant kidney renal failure in rats (Klahr et al. 1988). Hypercoagulability in the nephrotic state results principally from platelet hyperaggregability, as discussed above. If a persistent nephrotic syndrome is present, this could contribute through intraglomerular thrombosis and sclerosis induced by release of platelet-derived factors. Thus treatment with antiplatelet agents, on this hypothesis, would be expected to slow progression of renal failure in persistent nephrotic syndromes.

Hyperlipidaemia has a complex relation to glomerulosclerosis (Moorhead et al. 1982; Keane et al. 1988). The analogy has been drawn between the macrophage infiltration and foam-cell formation of atheroma, and events in the glomerulus in progressive glomerulosclerosis (Keane et al. 1988). Certainly, mesangial cells bear LDL receptors, and there is evidence from *in-vitro* studies that binding of excess LDL can damage these cells (Moorhead et al. 1989). The renal disease seen in inherited LCAT deficiency is well recognized (see Chapter 18.5). Thus control of hyperlipidaemia in persistent nephrotic syndromes could have the aim not only of preventing generalized atheroma, but also of preventing entry into renal failure, or at least slowing its progression. These points are discussed further in Chapter 9.3.

Renal tubular dysfunction in nephrotics

Renal tubular dysfunction has been described in a rather small number of nephrotic patients since the first descriptions by Teglaers and Tiddens (1955) and Woolf and Giles (1956). These usually take the form of multiple proximal tubular deficiencies in reabsorption to give a complete or partial Fanconi syndrome (see Chapter 5.1; Stanbury and Macaulay 1957). Whilst undoubtedly some of these cases have had reversible tubular defects dependent upon their nephrotic syndrome (Hooft and Vermassen 1960; Shioji et al. 1974) (presumably as a result of tubular damage from excess protein reabsorption), in many others the tubular defect arises from tubular damage as part of the underlying disease, often focal segmental glomerulosclerosis. Indeed in children the presence of glucosuria or other tubular abnormalities (McVicar et al. 1980; Boissou et al. 1980) or β_2-microglobulinuria (Portman et al. 1986) can help differentiate children with focal segmental glomerulosclerosis from those with minimal change lesions only.

Protein malnutrition in the nephrotic syndrome

Although nitrogen balances have rarely been done in nephrotic patients, the wasting of muscle mass is often obvious as oedema recedes. Turnover of albumin is greatly increased in nephrotics, up to 50 per cent of the albumin pool/24 h (Fig. 3), probably because of increased tubular catabolism of filtered albumin (Kaysen et al. 1987, 1988; Bernard 1988) rather than absolute urinary losses, which are, in comparison to turnover, rather small. There may also be gastrointestinal losses of albumin. Equally, feeding extra protein leads to an increase in albumin synthesis (Kaysen et al. 1986) but also to an increase in proteinuria, and vice versa.

Thus the optimum protein intake in nephrotic patients remains a source of controversy. Probably a relatively normal protein intake is adequate, about 75 to 100 g (1–1.5 g/kg)/24 h of mainly first-class (animal) protein, but a high-protein diet (2–3.5 g/kg.24 h) as usually recommended (Blainey 1954; Bernard 1988) may have advantages, and may even be essential in some individuals to maintain nitrogen balance (Blainey 1956) A low-protein diet, even if it reduces proteinuria in the short term, runs the risk of negative nitrogen balance in the longer term. A high-calorie intake seems appropriate to permit optimum usage of the protein, as in uraemia, although there are no data on this point in nephrotics.

Bibliography

Bernard, D.B. (1988). Extrarenal complications of the nephrotic syndrome. (Nephrology Forum). *Kidney International*, **33**, 1184–1202.

Brenner, B.M. and Stein, J.H. (eds.) (1982). *The nephrotic syndrome*. Churchill Livingstone, New York.

Cameron, J.S., Ogg, C.S., and Wass, V.J. (1988). Complications of the nephrotic syndrome. In *The nephrotic syndrome* (eds. J.S. Cameron, and R.J. Glassock,) pp. 849–920. Marcel Dekker, New York.

Cameron, J.S. (1987). The nephrotic syndrome. *American Journal of Kidney Diseases*, **10**, 157–71.

References

Afrasiabi, M.A., et al. (1975). Thyroid function in the nephrotic syndrome. *Annals of Internal Medicine*, **90**, 335–8.

Agnello, V. (1978). Complement deficiency states. *Medicine*, **57**, 1–23.

Alexander, J.H., Schapel, G.J., and Edwards, K.D.G. (1974). Increased incidence of coronary heart disease associated with combined elevation of triglyceride and cholesterol concentrations in the nephrotic syndrome in man. *Medical Journal of Australia*, **2**, 119–22.

Alkjaersig, N., Fletcher, A.P., Narayanan, M., and Robson, A.M. (1987). Course and resolution of the coagulopathy in nephrotic children. *Kidney International*, **31**, 772–80.

Allen, J.C. Baxter, J.H., and Goodman, H.C. (1961). Effects of dextran, polyvinylpyrrolidone and gamma globulin on the hyperlipidemia of experimental nephrosis. *Journal of Clinical Investigation*, **40**, 499–508.

Alon, U. and Chan, J.C.M. (1980). Calcium and vitamin D metabolism in nephrotic syndrome. *International Journal of Pediatric Nephrology*, **4**, 115–8.

Anderson, D.C., York, T.L., Rose, G., and Smith, C.W. (1979). Assessment of factor B, serum opsonins, granulocyte chemotaxis and infection in nephrotic syndrome of children. *Journal of Infectious Diseases*, **140**, 1–11.

Andrassy, K., Ritz., E., and Bommer, J. (1980). Hypercoagulability in the nephrotic syndrome. *Klinische Wochenschrift*, **58**, 1029–36.

Appel, G.B., Blum, C.B., Chien, S., Kunis, C., and Appel, A. (1985). The hyperlipidemia of the nephrotic syndrome. Relation to plasma albumin concentration, oncotic pressure and viscosity. *New England Journal of Medicine*, **312**, 1544–8.

Appel, G.B., Williams, G.S., Melzer, J.I., and Pirani, C.L. (1976). Renal vein thrombosis, nephrotic syndrome and systemic lupus erythematosus. An association in four cases *Annals of Internal Medicine*, **85**, 310–7.

Arneil, G.C. (1961). 164 children with nephrosis. *Lancet*, **ii**, 1103–10.

Arruda, J.A.L., Jonasson, O., Gutierrez, L.F., Pillay V.K.G., and Kurtzman, N.A. (1973). Renal-vein thrombosis in kidney allografts. *Lancet*, **ii**: 585–6.

Auwerx, J., de Keyser, L., Bouillon, R., and de Moor, P. (1986). Decreased free 1,25-dihydroxycholecalciferol index in patients with the nephrotic syndrome. *Nephron*, **42**, 231–5.

Avasthi, P.S., Greene, E.R., Scholler, C., and Fowler, C.R. (1983). Noninvasive diagnosis of renal vein thrombosis by ultrasonic echo-Doppler flowmetry. *Kidney International*, **23**, 882–7.

Balabanian, M.B., Schenzler, D.E., and Kaloyanides, G.J. (1973). Nephrotic syndrome, renal vein thrombosis and renal failure. Report of a case with recovery of renal function, loss of proteinuria on dissolution of thrombus after anticoagulant therapy. *American Journal of Medicine*, **54**, 768–76.

Bang, N.U., Trygstad, C.W., Schoeder, J.E., Heidenreich, R.O., and Csiscko, B.M. (1973). Enhanced platelet function in glomerular renal disease. *Journal of Laboratory and Clinical Medicine*, **81**, 651–5.

Bannatyne, R.M., Stringel, G., and Simpson J.S. (1979). Spontaneous peritonitis due to group B streptococci. *Canadian Medical Association Journal*, **121**, 442–3.

Barragry, J.M., *et al*. (1977). Vitamin D metabolism in nephrotic syndrome. *Lancet*, **ii**, 629–31.

Baxter, J.H. (1962). Hyperlipoproteinemia in nephrosis. *Archives of Internal Medicine* **109**, 742–57.

Becker, D.J., Pimstone, B.L., Hansen, J.D.L., and Hendricks, S. (1971). Serum albumin and growth hormone relationships in kwashiorkor and the nephrotic syndrome. *Journal of Laboratory and Clinical Medicine*, **78**, 865–71.

Bennett, W.M. (1975). Renal vein thrombosis and nephrotic syndrome; comments and corrections. *Annals of Internal Medicine*, **183**, 577–8.

Bennett, A. and Cameron, J.S. (1987). Hyperaggregability of platelets to ristocetin in the nephrotic syndrome not dependent upon thromboxane production or serum albumin concentration. *Clinical Nephrology*, **27**, 182–8.

Bensman, A., Dardenne, M., Murnaghan, K., Vasmant, K., and Bach, J-F. (1984). Decreased biological activity of serum thymic hormone (thymulin) in children with nephrotic syndrome. *International Journal of Pediatric Nephrology*, **5**, 201–4.

Bergrem, H. (1983). Pharmacokinetics and protein binding of prednisolone in patients with nephrotic syndrome and patients undergoing hemodialysis. *Kidney International*, **23**, 876–81.

Berlyne, G.M. and Mallick, N.P. (1969). Ishaemic heart disease as a complication of nephrotic syndrome. *Lancet*, **ii**, 399–400.

Bernard, D., Govault-Heilmann, M., Ansquer, J-C., Levenet, M., Ladieu, D., and Lagrue G. (1977). Étude des facteurs de l'hémostase au cours des syndromes néphrotiques. *Annales de Médicine Interne*, **128**, 325–33.

Blainey, J.D. (1954). High protein diets in the treatment of the nephrotic syndrome. *Clinical Science*, **13**, 567–81.

Blumberg, R.W. and Cassidy, H.A. (1947). Effects of measles on the nephrotic syndrome. *American Journal of Diseases of Children*, **73**, 151–66.

Boissou, F., Barthe, F.L., and Pierragi, M. Th. (1980). Severe idiopathic nephrotic syndrome with tubular dysfunction (report of nine pediatric cases). *Clinical Nephrology*, **14**, 135–41.

Boneu, B., Boissou, F., Abbal, M., Sie, P., Caranobe, C., and Barthe, P. (1981). Comparison of progressive antithrombin activity and concentration of three thrombin inhibitors in nephrotic syndrome. *Thrombosis and Haemostasis*, **46**, 623–25.

Boulton-Jones, J.M., McWilliams, G., and Chardrachud, L. (1986). Variation in care on red cells of patients with different glomerulopathies. *Lancet* **ii**, 186–189.

Bridgman, J.F., Rosen, S.M., and Thorp, J.M. (1972). Complications during clofibrate treatment of nephrotic syndrome hyperlipidaemia. *Lancet*, **ii**, 506–9.

Bridgman, J.F., Summerskill, J.F., Buckler, J.M.H., Hillman, B., and Rosen, S.M. (1975). Insulin and growth hormone secretion in the nephrotic syndrome. *Quarterly Journal of Medicine*, **44**, 115–23.

Briefel, G.R., Marris, T., Gordon, D.H., Nicastri, A.D., and Friedman, E.A. (1978). Recurrent renal vein thrombosis consequent to membranous glomerulonephritis. *Clinical Nephrology*, **10**, 32–7.

Brock, J.H. and Mainou-Fowler, T. (1983). The role of iron and transferrin in lymphocyte transformation. *Immunology Today*, **4**, 347–51.

Brown, E.A., Sampson, B., Muller, B.R., and Curtis J.R. (1984). Urinary iron loss in the nephrotic syndrome—an unusual case of iron deficiency with a note on the excretion of copper. *Postgraduate Medical Journal*, **160**, 125–8.

Calcagno, P.L. and Rubin, M.I. (1961). Physiologic considerations concerning corticosteroid therapy and complications in the nephrotic syndrome. *Journal of Pediatrics*, **58**, 686–706.

Cameron, J.S. (1984). Thromboembolic complications of the nephrotic syndrome. *Advances in Nephrology*, **13**, 75–114.

Cameron, J.S. (1990). Platelets in renal disease. In *Platelets in health and disease* (ed. C.P. Page), in press. Blackwell, Oxford.

Cameron, J.S., Ogg, C.S., Ellis, F.G., and Salmon, M.A. (1971). Femoral artery thrombosis in nephrotic syndrome. *Archives of Disease in Childhood*, **42**, 836–7.

Canner, P.L., *et al*. (1986). Fifteen year mortality in coronary drug project patients: long term benefit with niacin. *Journal of the American College of Cardiologists*, **8**, 1245–55.

Cartwright, G.E., Gubler, C.J., and Wintrobe, M.M. (1954). Studies on copper metabolism XI: copper and iron metabolism in the nephrotic syndrome. *Journal of Clinical Investigation*, **33**, 685–98.

Chan, M.K., Persaud, J.W., Ramdial, L., Varghese, Z., Sweny, P., and Moorhead, J.M. (1981). Hyperlipidemia in untreated nephrotic syndrome, increased production or decreased removal? *Clinica Chimica Acta*, **117**, 317–23.

Chapman, S.M., Taube, D., Brown, Z., and Williams, D.G. (1982). Impaired lymphocyte transformation in minimal change nephropathy in remission. *Clinical Nephrology*, **18**, 34–8.

Chopra, J.S., Mallick, N.P., and Stone, M.C. (1971). Hyperlipoproteinaemias in the nephrotic syndrome. *Lancet*, **i**, 317–21.

Choudhry, V.P. and Ghai, O.P. (1977). Peritonitis in nephrotic syndrome. *Indian Paediatrician*, **14**, 405–8.

Clouse, L.H. and Comp, P.C. (1986). The regulation of hemostasis: the protein C system. *New England Journal of Medicine*, **314**, 1298–1304.

Cohen, H.T., Singh, A.K., Kasinath, B.S., and Lewis, E.J. (1988). Red-cell surface charge in patients with nephrotic syndrome. *Lancet* **i**, 1459 (letter).

Cohen, S.L., Caramps, D.G., Lewis, A.D., and Tickner, T.R. (1980). The mechanism of hyperlipidaemia in nephrotic syndrome—role of low albumin and the LCAT reaction. *Clinica Chimica Acta*, **104**, 393–400.

Corpier, C.L., *et al.* (1988). Rhabdomyolysis and renal injury with lovastatin use. Report of two cases in cardiac allograft recipients. *Journal of the American Medical Association* **260**: 239–41.

Cosgriff, S.W. (1951). Thromboembolic complications associated with ACTH and cortisone therapy. *Journal of the American Medical Association*, **147**, 924–6.

Cosio, F.G., Harker, C., Batard, M.A., Brandt, J.T., and Griffin, J.H. (1985). Plasma concentrations of the natural anticoagulants protein C and protein S in patients with proteinuria. *Journal of Laboratory and Clinical Medicine*, **106**, 218–22.

Crouse, J.R. III (1987). Hypertriglyceridemia: a contraindication to the use of bile acid binding resins. *American Journal of Medicine*, **83**, 243–8.

Curry, R.C. and Roberts, W.C. (1977). Status of the coronary arteries in the nephrotic syndrome (analysis of 20 necropsy patients aged 15 to 35 years to determine if coronary atherosclerosis is accelerated). *American Journal of Medicine*, **63**, 183–92.

Dagg, J.H., Smith, J.A., and Goldberg A. (1966). Urinary excretion of iron. *Clinical Science*, **30**, 495–503.

Dardenne, M., *et al.* (1982). Contribution of zinc and other metals to the biological activity of the serum thymic factor. *Proceedings of the National Academy of Science of the USA*, **79**, 5370–3.

Davis, R.A., Engelhorn, S.C., Wienstein, D.B., and Steinberg D. (1980). Very low density lipoprotein secretion by cultured rat hepatocytes: inhibition by albumin and other macromolecules. *Journal of Biological Chemistry*, **255**, 2039–45.

Deckelbaum, R.J. (1987). Structure and composition of human plasma lipoproteins. In *Atherosclerosis, Biology and Clinical Science* (ed. A.G. Olsson) pp. 251. Churchill Livingstone, New York.

Drabkin, D.L. and Marsh J.B. (1955). Metabolic channeling in experimental nephrosis, protein and carbohydrate metabolism. *Journal of Biological Chemistry*, **212**, 623–31.

Du, X.H., Glas-Greenwalt, P., Kant, K.S., Allen, C.M., Hayes, S., and Pollak, V.E. (1985). Nephrotic syndrome with renal vein thrombosis: pathogenic importance of plasmin inhibitor (alpha-2-antiplasmin). *Clinical Nephrology*, **24**, 186–191.

Edwards, K.D.G., Schapel, G.J., and Neale, F.C. (1973). Efficacy of cholestyramine, L-tryptophan and clofibrate in the treatment of nephrotic hyperlipidemia, studied by factorial analysis. *Australian and New Zealand Journal of Medicine*, **3**, 441–8.

Egli, F., Eimiger, P., and Stalder, G. (1973). Thromboembolism in the nephrotic syndrome. *Pediatric Research*, **8**, 803 (abstract).

Egli, F. (1974). Thromboembolism beim nephrotischer Syndrom in Kindesalter. Thesis, University of Basel.

Ehrenreich, T. and Churg, J. (1968). Pathology of membranous nephropathy. In *Pathology Annual, vol. 3* (Sommers, S.C. ed.) pp. 145–86. Appleton-Century-Crofts, New York.

Eisalo, A., Manninen, V., Malkonen, M., and Kuhlback, B. (1976). Hypolipidaemic action of gemfibrozil in adult nephrotics. *Proceedings of the Royal Society of Medicine*, **69 (suppl 2)**, 47–8.

Elidrissy, A.T.H. (1982). Primary peritonitis and meningitis in nephrotic syndrome in Riyadh. *International Journal of Pediatric Nephrology*, **3**, 9–12.

Ellis, D. (1977). Anemia in the course of the nephrotic syndrome secondary to transferrin depletion. *Journal of Pediatrics*, **90**, 953–5.

Emerson, K. and Beckman, W.W. (1945). Calcium metabolism in children with nephrosis I. A description of an abnormality in calcium metabolism in children with nephrosis. *Journal of Clinical Investigation*, **24**, 564–572.

Eskola, J., Holmberg, C., and Koskimies, O. (1986). C-reactive protein—an indicator of generalized bacterial infection in children with nephrotic syndrome. *Acta Paediatrica Scandinavica*, **75**, 846–8.

Feehally, J., Samanta, S., Kinghorn, H., Burden, A.C., and Walls, J. (1986). Red cell surface charge in glomerular disease. *Lancet* **ii**, 635 (letter).

Felts, J.M. and Mayerle, J.A. (1974). Urinary loss of plasma high density lipoprotein—a possible cause of hyperlipidemia of the nephrotic syndrome. *Circulation*, **50, suppl III**, 263 (abstract).

Fikrig, S.M., Schiffman, G., Phillipp, J.C., and Moel, D.I. (1978). Antibody response to capsular polysaccharide vaccine of *Streptococcus pneumoniae* in patients with nephrotic syndrome. *Journal of Infectious Diseases*, **137**, 818–21.

Forrester, H.L., Jahnigen, D.W., and LaForce, F.M. (1987). Inefficacy of pneumococcal vaccine in a high risk population. *American Journal of Medicine*, **83**, 425–30.

Freundlich, M., Bourgognie, J.J., Zilleruelo, G., Jacobs, A.I., Canterbury, J.M., and Strauss, J. (1986). Bone modulating factors in nephrotic children with normal glomerular filtration rate. *Pediatrics*, **76**, 280–5.

Frey, F.J. and Frey, B.M. (1984). Altered plasma protein-binding of prednisolone in patients with the nephrotic syndrome. *American Journal of Kidney Diseases*, **3**, 339–48.

Frick, M.H., *et al.* (1987). Helsinki heart study: primary prevention trial with gemfibrozil in middle aged men with dyslipidemia: safety of treatment, changes in risk factors, and incidence of coronary heart disease. *New England Journal of Medicine*, **317**, 1237–45.

Futrakul, P. (1978). Primary peritonitis syndrome: a state of hypercoagulability in the nephrotic syndrome. *Journal of the Medical Association of Thailand*, **61**, 268–72.

Ganeval, D., *et al.* (1986). Pharmacokinetics of warfarin in the nephrotic syndrome and effect on vitamin K dependent clotting factors. *Clinical Nephrology*, **25**, 75–80.

Gatewood, O.M.B., Fishman, E.K., Burrow, C.R., Walker, W.G., Goldman, S.M., and Siegelman, S.S. (1986). Renal vein thrombosis in patients with nephrotic syndrome: CT diagnosis. *Radiology*, **159**, 117–22.

Garin, L.A., McMahon, F.A., Castle, J.N., and Caralreri, R.R. (1978). Alterations in serum thyroid hormones and thyroxine binding globulin in patients with nephrosis. *Journal of Clinical Endocrinology and Metabolism*, **46**, 125–30.

Gherardi, E., Rota, E., Calandra, S., Genova, R., and Tamborino, A. (1977). Relationship among the concentrations of serum lipoproteins and changes in their chemical composition in patients with untreated nephrotic syndrome. *European Journal of Clinical Investigation*, **7**, 563–70.

Giangiacomo, J., Cleary, T.G., Cole, B.R., Hoffsten, P., and Robson, A.M. (1975). Serum immunoglobulins in the nephrotic syndrome. *New England Journal of Medicine*, **293**, 8–12.

Gilboa, N. (1976). Incidence of coronary heart disease associated with nephrotic syndrome. *Medical Journal of Australia*, **1**, 207–8 (letter).

Gitlin, D., Cornwell, D.G., Makasato, D., Oncley J.L., Hughes, W.L., and Janeway, C.A. (1958). Studies on the metabolism of plasma proteins in the nephrotic syndrome. II The lipoproteins. *Journal of Clinical Investigation*, **37**, 172–84.

Glass, F., Lippton, H., and Kadowitz, P.J. (1981). Effects of methylprednisolone and hydrocortisone on aggregation of rabbit platelets induced by arachidonic acid and other aggregating substances. *Thrombosis and Haemostasis*, **46**, 676–79.

Goldstein, D.A., Oda, W.Y., Kurokawa, K., and Massry, S.G. (1977). Blood levels of 25-hydroxy vitamin D in nephrotic syndrome. studies in 26 patients. *Annals of Internal Medicine*, **87**, 664–7.

Goldstein, D.A., Haldiman, B., Sherman, D., Norman, A.W., and Massry, S.G. (1981). Vitamin D metabolites and calcium metabolism in patients with nephrotic syndrome and normal renal function. *Journal of Clinical Endocrinology and Metabolism*, **52**, 116–21.

Golper, T.A., Illingworth, D.R., Morris, C.D., and Bennett, W.M. (1989). Lovastatin in the treatment of multifactorial hyperlipidemia associated with proteinuria. *American Journal of Kidney Diseases*, **13**, 312–28.

Gorensek, M.J., Lebel, M.H., and Nelson, J.D. (1988). Peritonitis in children with nephrotic syndrome. *Pediatrics* **81**, 849–56.

Gouault-Heilmann, M., Gadelha-Parenete, T., Levent, M., Intrator, L., Rostocker, G., and Lagrue, G. (1988). Total and free protein S in nephrotic syndrome. *Thrombosis Research*, **49**, 37–42.

Groggel, G.C., Cheung, A.K., Ellis-Benigni, K., and Wilson, D.E. (1989). Treatment of nephrotic hyperlipoproteinemia with gemfibrozil. *Kidney International*, **36**, 266–71.

Grundy, S.M. (1988). HMG-CoA reductase inhibitors for treatment of hypercholesterolemia. *New England Journal of Medicine*, **319**, 24–33.

Gugler, R., Showman, D.W., Huffman, D.H., Cholmia, J.B., and Azarnoff, D.L. (1975). Pharmacokinetics of drugs in patients with the nephrotic syndrome. *Journal of Clinical Investigation*, **55**, 1182–9.

Gugler, R. and Azarnoff, D.L. (1976). Drug protein binding and the nephrotic syndrome. *Clinical Pharmacokinetics*, **1**, 25–35.

Habib, R., Courtecuisse, V., and Bodaghi, E. (1968). Thrombose des arteres pulmonaires dans les syndromes nephrotiques de l'enfant. *Journal d'Urologie et Nephrologie*, **74**, 349–52.

Hancock, D.E., Onstad, J.W., and Wolf, P.L. (1976). Transferrin loss in the urine with hypochromic microcyctic anaemia. *American Journal of Clinical Pathology*, **6**, 73–8.

Handley, D.A. and Lawrence, J.R. (1967). Factor IX deficiency in the nephrotic syndrome. *Lancet*, **i**, 1079–81.

Harrington, J.T. (1984). Thrombolytic therapy in renal vein thrombosis. *Archives of Internal Medicine*, **144**, 33–4.

Harrington, J.T. and Kassirer, J.P. (1982). Renal venous thrombosis. *Annual Review of Medicine*, **33**, 255–62.

Heslan, J.M., Lautie, J.P., Intrator, L., Blanc, C., Lagrue, G., and Sobel, A.T. (1982). Impaired IgG synthesis in patients with the nephrotic syndrome. *Clinical Nephrology*, **18**, 144–7.

Hogg, R.J. (1983). Protective antibody concentrations in children with nephrotic syndrome four years after pneumococcal vaccination. *European Journal of Pediatrics*, **140**, 189–93.

Honig, G.A. and Lindley, A. (1971). Deficiency of Hageman factor (factor XII) in patients with the nephrotic syndrome. *Journal of Paediatrics*, **78**, 633–37.

Hooft, C. and Vermassen, A. (1960). De Toni-Debre-Fanconi syndrome in nephrotic children. *Annales de Pédiatrie*, **194**, 193–216.

Hopper, J., Ryan, P., Lee, J.C., and Rosenau, W. (1970). Lipoid nephrosis in 31 adult patients: renal biopsy study by light, electron and fluorescence microscopy with experience in treatment. *Medicine (Baltimore)*, **49**, 321–41.

Hyman, L.R., Wong, P.K.W., and Grossman, A. (1969). Plasma lipoprotein lipase in children with idiopathic nephrotic syndrome. *Pediatrics*, **44**, 1021–4.

Iida, H., Asaka, M., Fujita, M., Nishino, A., and Sasayama, S. (1987). Effect of probucol on hyperlipidemia in patients with nephrotic syndrome. *Nephron*, **47**, 280–3.

International Study of Kidney Disease in Children (1984). Minimal change nephrotic syndrome in children: deaths occurring during the first five to fifteen years. *Pediatrics*, **73**, 497–501.

Israeli, A., Raveh, D., Arnon, R., Eisenberg, S., and Stein, Y. (1989). Lovastatin and elevated creatine kinase: results of rechallenge. *Lancet*, **i**, 725 (letter).

Jackson, C.A., Greaves, M., Patterson, A.D., Brown, C.B., and Preston, F.E. (1982). Relationship between platelet aggregation, thromboxane synthesis and albumin concentration in nephrotic syndrome. *British Journal of Haematology*, **52**, 69–77.

Janeway, C.A., Moll, G.H., Armstrong, S.H., Wallace, W.M., Hallman, N., and Barness, L.A. (1948). Diuresis in children with nephrosis. Comparison of response to injection of normal human serum albumin and to infection, particularly measles. *Transactions of the Association of American Physicians*, **61**, 108–11.

Jensen, H. (1967). Plasma protein and lipid patterns in the nephrotic syndrome. *Acta Medica Scandinavica*, **182**, 465–73.

Jensen, H., Bro-Jorgensen, K., Jarnum, S., Oelsen, H., and Yssing, M. (1968). Transferrin metabolism in the nephrotic syndrome and in protein-losing nephropathy. *Scandinavian Journal of Clinical and Laboratory Medicine*, **21**, 293–304.

Joven, J., Rubies-Prat, J., Espinel, E., Ras, M.R., and Piera, L. (1987). High density lipoproteins in untreated nephrotic syndrome without renal failure. *Nephrology Dialysis Transplantation*, **2**, 149–53.

Kallen, R.J., Bryner, R.K., Aronson, A.J., Lichtig, C., and Spargo, B.H. (1977). Premature coronary atherosclerosis in a 5-year old with corticosteroid- refractory nephrotic syndrome. *American Journal of Diseases in Children*, **131**, 976–80.

Kanfer, A., Kleinknecht, D., Broyer, M., and Josso, F. (1970). Coagulation studies in 45 cases of nephrotic syndrome without uremia. *Thrombosis Diathesis Haemorrhagica*, **24**, 562–71.

Kant, S.K., Pollak, V.E., Weiss, M.A., Glueck, H.I., Miller, M.A., and Hess, E.V. (1981). Glomerular thombosis in systemic lupus erythematosus: prevalence and significance. *Medicine (Baltimore)*, **60**, 71–86.

Kashyap, H.L., Ooi, B.S., Hynd, B.A., Glueck, C.J., Pollack, V.E., and Robinson, K. (1979). Sequestration and excretion of high density and low density lipoproteins by the kidney in human nephrotic syndrome. *Artery*, **6**, 108–21.

Kashyap, H.L., et al. (1980). Apolipoprotein CII and lipoprotein lipase in human nephrotic syndrome. *Atherosclerosis*, **35**, 29–40.

Kassirer, J.P. (1979). Thrombosis and embolism of the renal vessels. In *Strauss and Welt's diseases of the kidney*, (Earley, L.E. and Gottschalk, C.W., eds.) pp 1385–1402. Little Brown, Boston.

Kauffmann, R.H., Veltkamp, J.J., van Tilburg, N.H., and van Es, L.A. (1978). Acquired antithrombin III deficiency and thrombosis in the nephrotic syndrome. *American Journal of Medicine*, **65**, 607–613.

Kaysen, G.A. (1988). Albumin metabolism in the nephrotic syndrome: the effect of dietary protein intake. *American Journal of Kidney Diseases*, **12**, 461–80.

Kaysen, G.A., Gambertoglio, J., Felts, J., and Hutchison, F.N. (1987). Albumin synthesis, albuminuria and hyperlipidemia in nephrotic patients. *Kidney International*, **31**, 1368–76.

Keane, W.F., Kasiske, B.L., and O'Donnell, M.P. (1988). Lipids and progressive glomerulosclerosis. A model analogous to atherosclerosis. *American Journal of Nephrology*, **8**, 261–71.

Kekki, M. and Nikkila, E.A. (1971). Plasma triglyceride metabolism in the adult nephrotic syndrome. *European Journal of Clinical Investigation*, **1**, 345–51.

Kendal, A.F., Lohmann, R.C., and Dossetor, J.B. (1971). Nephrotic syndrome: a hypercoagulable state. *Archives of Internal Medicine*, **127**, 1021–7.

Klahr, S., Schreiner, G., and Ichikawa, I. (1988). The progression of renal disease. *New England Journal of Medicine*, **318**, 1657–66.

Klein, J.D. and Mortimer, E.A. (1978). Use of pneumococcal vaccine in children. *Pediatrics*, **61**, 321–2.

Kleinknecht, D., Laudat, M.H., Strauch, G., Jungers, P., and Laudat, P. (1972). Lipoprotein and carbohydrate disorders in nephrotic syndrome and uraemia. *Revue Européene d'Études Cliniques et Biologiques*, **17**, 27–37.

Korkor, A., et al. (1983). Absence of metabolic bone disease in adult patients with the nephrotic syndrome and normal renal function. *Journal of Clinical Endocrinology and Metabolism*, **56**, 496–500.

Krensky, A.M., Ingelfinger, J.M., and Grupe, W.E. (1982). Peritonitis in childhood nephrotic syndrome. *American Journal of Diseases of Children*, **136**, 732–6.

Kuhlmann, U., Steurer, J., Bollinger, A., Pouliadis, G., Briner, J., and Sieganthaler, W. (1981). Inzidenz und klinische Bedeutung von Thrombosen und thromboembolischem Komplikationen bei Pazienten mit nephrotischen Syndrom. *Schweitzer Medizinische Wochenschrift*, **111**, 1034–40.

Laville, M., Aguilera, D., Maillet, P.J., Labeeuw, M., Madonna, O., and Zech, P. (1988). The prognosis of renal vein thrombosis: a re-evaluation of 27 cases. *Nephrology Dialysis Transplantation*, **3**, 247–56.

Lawson, D., Moncreiff, A.W., and Payne, W.W. (1960). Forty years of

nephrosis in childhood. *Archives of Disease in Childhood*, **35**, 115–26.
Leiter, L. (1931). Nephrosis. *Medicine (Baltimore)*, **10**, 135–242.
Lewis, L.A. and Page, I.H. (1953). Electrophoretic and ultracentrifugal analysis of serum lipoproteins of normal, nephrotic and hypertensive persons. *Circulation*, **7**, 707.
Levin M., Smith, C., Walters, M.D.S., Gascoine, P., and Barratt, T.M. (1985). Steroid-responsive nephrotic syndrome: a generalised disorder of membrane negative charge. *Lancet* ii, 239–242.
Liaño, F., et al. (1988). Allograft membranous glomerulonephritis and renal-vein thrombosis in a patient with a lupus anticoagulant factor. *Nephrology Dialysis Transplantation*, **3**, 684–9.
Lim, P., Jacob, E., Tock, E.P.C., and Pwee, H.S. (1977). Calcium and phosphorus metabolism in nephrotic syndrome. *Quarterly Journal of Medicine*, **46**, 327–38.
Lipid research clinics program: the lipid research clinics coronary primary prevention trial results I. Reduction in incidence of coronary heart disease; II The relation of reduction in incidence of coronary heart disease to cholesterol lowering. *Journal of the American Medical Association*, **251**, 351–64, 365–74.
Livio, M. (1988). 'Hypercoagulability' in the nephrotic syndrome. In *Haemostasis and the kidney* (eds. G. Remuzzi and E.C. Rossi) pp. 145–52, Butterworth, London.
Llach, F. (1982). Nephrotic syndrome: hypercoagulability, renal vein thrombosis and other thromboembolic complications. In *The nephrotic syndrome* (eds. B.M. Brenner, and J.H. Stein,) pp. 121–44. Churchill-Livingstone, New York.
Llach, F. (1985). Hypercoagulability, renal vein thrombosis, and other thrombotic complications of nephrotic syndrome (Nephrology Forum). *Kidney International* **28**, 429–39.
Llach, F., Papper, S., and Massry, S.G. (1980). The clinical spectrum of renal vein thrombosis. *American Journal of Medicine*, **69**, 819–27.
Loirat, C., Pillion, G., Schleyel, N., Dovenias, R., and Mahieu, H. (1979). Étude des protéines de la coagulation et de la fibrinolyse au cours des syndromes néphrotiques de l'enfant. *Archives Françaises de Péiatrie*, 36 (suppl 1), 56–63.
Lopes-Virella, M., Virella, G., Debeukelaer, M., Owens, C.J., and Colwell, J.A. (1979). Urinary high density lipoprotein in minimal change glomerular disease and chronic glomerulonephritis. *Clinica Chimica Acta*, **94**, 73–81.
Loschiavo, C., et al. (1983). Carbohydrate metabolism in patients with nephrotic syndrome and normal renal function. *Nephron*, **33**, 257–61.
McKenzie, I.F.C. and Nestel, P.J. (1968). Studies on the turnover of triglyceride and esterified cholesterol in subjects with the nephrotic syndrome *Journal of Clinical Investigation*, **47**, 1689–95.
McGinley, E., Lowe, G.D.O., Boulton-Jones, M., Forbes, C.D., and Prentice, C.R.M. (1983). Blood viscosity and haemostasis in the nephrotic syndrome. *Thrombosis and Haemostasis*, **49**, 155–7.
McLean, R.H., Forsgren, A.B., Jorksten, B., Kim, Y., Quie, P.G., and Michael, A.F. (1977). Decreased serum factor B concentration associated with decreased opsonisation of *Escherichia coli* in the nephrotic syndrome. *Pediatric Research*, **11**, 910–16.
McVicar, M., Exeni, R., and Susin, M. (1980). Nephrotic syndrome and multiple tubular defects in children: an early sign of focal glomerulosclerosis. *Journal of Pediatrics*, **87**, 918–22.
McVicar, M. and Nicastri, A.D. (1973). Incidence of bacteriuria in the nephrotic syndrome in children. *Pediatrics*, **82**, 166–7.
Mahajan, S., et al. (1982). Zinc metabolism in nephrotic syndrome. *Kidney International*, **23**, 129. (abstract).
Mallick, N.P. and Short, C.D. (1981). The nephrotic syndrome and ischaemic heart disease. *Nephron*, **27**, 54–7.
Malluche, H.H., Goldstein, D.A., and Massry, S.G. (1979). Osteomalacia and hyperparathyroid bone disease in patients with nephrotic syndrome. *Journal of Clinical Investigation*, **63**, 494–500.
Malmendier, C.L., Delacroix, C., and van den Bagen, C.J. (1970). In vivo demonstration of differences in substrate oxidation in normal and nephrotic rats. *American Journal of Physiology*, **219**, 911–8.

Mannucci, P.M., Valsecchi, C., Bottasso, B., D'Angelo, A., Casati, S., and Ponticelli, C. (1985). High plasma levels of protein C activity and antigen in the nephrotic syndrome. *Thrombosis and Haemostasis*, **55**, 31–3.
Mehls, O., Andrassy, K., Koderisch, J., Herzog, U., and Ritz, E. (1987). Hemostasis and thromboembolism in children with nephrotic syndrome: differences from adults. *Journal of Pediatrics*, **110**, 862–7.
Menczel, J. and Dreyfuss, F. (1960). Effect of prednisone on blood coagulation time in patients on dicoumarol therapy. *Journal of Laboratory and Clinical Medicine*, **56**, 14–20.
Moore, D.H., Shackelford, P.G., Robson, A.M., and Rose, G.M. (1980). Recurrent pneumococcal sepsis and defective opsonisation after pneumococcal polysaccharide vaccine in a child with nephrotic syndrome. *Journal of Pediatrics*, **96**, 882–85.
Moorhead, J.F., El Nahas, M., Chan, M.K., and Varghese, Z. (1982). Lipid nephrotoxicity in chronic progressive glomerular and tubulointerstitial disease. *Lancet*, ii, 1309–10.
Moorhead, J.F., Wheeler, D.C., Fernando, R., Sweny, P., and Varghese, Z. (1989). Injury to rat mesangial cells in culture by low density lipoproteins. *Kidney International*, **35**, 433 (abstract).
Moorthy, A.V., Zimmerman, S.W., and Burkholder, P.M. (1976). Inhibition of lymphocyte blastogenesis by plasma of patients with minimal change nephrotic syndrome. *Lancet*, i, 1160–2.
Mukherjee, A.P., Tog, B.H., Chan, G.L., Lau, K.S., and White, J.C. (1956). Vascular complications in nephrotic syndrome: relationship to steroid therapy and accelerated thromboplastin generation. *British Medical Journal*, **4**, 273–6.
Musa, B.J., Seal, U.S., and Doe, R.P. (1967). Excretion of corticosteroidbinding globulin, thyroxine binding globulin and total protein in adult males with nephrosis. Effect of sex hormone. *Journal of Clinical Endocrinology* **27**, 768–74.
Nelson, D.C. and Birchmore, D.A. (1979). Renal vein thrombosis associated with nephrotic syndrome and gold therapy in rheumatoid arthritis. *Southern Medical Journal*, **72**, 1616–8.
Newmark, S.R., Anderson, C.F., Donadio, J.V., and Ellefson, R.D. (1975). Lipoprotein profiles in adult nephrotics. *Mayo Clinic Proceedings*, **50**, 359–64.
Nikkilä, E.A. and Pykäpistö, O. (1968). Regulation of adipose tissue lipoprotein lipase synthesis by intracellular free fatty acid. *Life Sciences*, **7**, 1303–9.
Noah, N.D. (1988). Vaccination against pneumcoccal infection. Only for selected cases. *British Medical Journal*, **297**, 1351–2.
Noel, L.H., Zanetti, M., Droz, D., and Barbanel, C. (1979). Long term prognosis of idiopathic membranous glomerulonephritis: study of 116 untreated patients. *American Journal of Medicine*, **66**, 82–90.
Nye, W.H.R. and Waterhouse, C. (1961). The phosphatides of human plasma; II Abnormalities encountered in the nephrotic syndrome. *Journal of Clinical Investigation*, **40**, 1202–7.
Oetliker, O.H., Mordasisi, R., Lutschig, J., and Riesen, W. (1980). Lipoprotein metabolism in nephrotic syndrome of childhood. *Pediatric Research*, **14**, 64–6.
Ogg, C.S., et al. (1982). Renal vein thrombosis in the nephrotic syndrome. In *Controversies in Nephrology* (eds. G.E. Schreiner, and J.F Winchester, pp. 160–8. Masson, New York.
Ooi, B.S., Ooi, Y.M., Hsu, A., and Hurtebise, P.E. (1980). Diminished synthesis of immunoglobulin by peripheral lymphocytes of patients with idiopathic membranous glomerulonephritis. *Journal of Clinical Investigation*, **65**, 789–97.
O'Regan, S., Mongeau, J., and Robitaille, P. (1980). Primary peritonitis in the nephrotic syndrome. *International Journal of Pediatric Nephrology*, **1**, 216–7.
Ozanne P., Francis, R.B., and Meiselman, H.J. (1983). Red blood cell aggregation in nephrotic syndrome. *Kidney International*, **23**, 519–25.
Ozsosylu, S., Strauss, H.S., and Diamond, L.K. (1962). Effects of corticosteroids on coagulation of the blood. *Nature*, **195**, 1214–5.
Pabinger-Fasching, I., Lechner, K., Niessner H., Schmidt, P., Balzar,

E., and Mannhalter, C. (1985). High levels of protein C in nephrotic syndrome. *Thrombosis and Haemostasis*, **53**, 3–5.

Paniucci, F., *et al.* (1983). Comprehensive study of haemostasis in nephrotic syndrome. *Nephron*, **33**, 9–13.

Petiti, D.B., Strom, B.L., and Melmon, K.L. (1986). Duration of warfarin anticoagulant therapy and the probabilities of recurrent thromboembolism and haemorrhage. *American Journal of Medicine*, **81**, 255–9.

Pohl, M.A., Maclaurin, J.P., and Alfredi, R.J. (1977). Renal vein thrombosis and the nephrotic syndrome. *Kidney International*, **12**, 472 (abstract).

Pohl, M.A., DeSio, F., MacLaurin, J., Alfredi, R., and Zelch, M. (1984). Renal vein thrombosis in membranous and membranoproliferative glomerulonephritis. Abstracts, IXth Congress of the International Society of Nephrology, Los Angeles. p. 119A.

Portman, R.J., Kissane, J.M., and Robson, A.M. (1986). Use of β-microglobulin to diagnose tubulo-interstitial lesions in children. *Kidney International*, **30**, 91–8.

Primack, W.A., Rosel, M., Thirumorrthi, M.C., Fleischmann, L.E., and Schiffman, G. (1979). Failure of pneumococcal vaccine to prevent *Streptococcus pneumoniae* sepsis in nephrotic children. *Lancet*, **ii**, 1192.

Rabelink, A.J., Hene, R.J., Erekelns, D.W., Joles, J.A., and Koomans, H.A. (1988). Effects of simvastatin and cholestyramine on lipoprotein profile in hyperlipidaemia of nephrotic syndrome. *Lancet*, **ii**, 1335–8.

Rasmussen, H. and Rapp, B. (1956). Thyroxine metabolism in the nephrotic syndrome. *Journal of Clinical Investigation*, **35**, 792–9.

Recant, L. and Riggs, D.S. (1952). Thyroid function in nephrosis. *Journal of Clinical Investigation*, **31**, 789–97.

Reimold, E.W. (1980). Changes in zinc metabolism during the course of the nephrotic syndrome. *American Journal of Diseases of Children*, **134**, 46–50.

Remuzzi, G., *et al.* (1979). Platelet hyperaggregability and the nephrotic syndrome. *Thrombosis Research*, **16**, 345–54.

Rifkind, D., Kravetz, H.K., Knight, V., and Schade, A.L. (1961). Urinary excretion of iron-binding protein in the nephrotic syndrome. *New England Journal of Medicine*, **265**, 115–8.

Robbins, J., Rall, J.E., and Peterman, H.L. (1957). Thyroxine binding by serum and urinary proteins in nephrosis. Qualitative aspects. *Journal of Clinical Investigation*, **36**, 1333–42.

Robert, A., Olmer, M., Sampol, J., Gugliotta, J-E., and Casanova, P. (1987). Clinical correlation between hypercoagulability and thromboembolic phenomena. *Kidney International*, **31**, 830–5.

Rosenfeld, A.T., Zeman, R.K., Cronana, J., and Taylor, K.J.W. (1980). Ultrasound in experimental and clinical renal vein thrombosis. *Radiology*, **137**, 735–41.

Ross, D.L. and Lubowitz, H. (1978). Anticoagulation in renal vein thrombosis. *Archives of Internal Medicine*, **178**, 1349–51.

Rostoker, G., Goualt-Heilmann, M., Levent, M., Robeva, R., Lang, P., and Lagrue, G. (1987). High level of protein C and S in nephrotic syndrome. *Nephron*, **46**, 20–1.

Rubin, H.M., Blau, E.B., and Michaels, R.H. (1975). Hemophilus and pneumococcal peritonitis in children with nephrotic syndrome. *Pediatrics*, **56**, 598–601.

Rusthoven, J. and Kabins, S.A. (1978). *Hemophilus influenzae* cellulitis with bacteremia, peritonitis and pleuritis in an adult with the nephrotic syndrome. *Southern Medical Journal*, **71**, 1433–5.

Rydzewski, A., Mysilwiec, M., and Soszka, J. (1986). Concentration of three thrombin inhibitors in the nephrotic syndrome in adults. *Nephron*, **42**, 200–203.

Saito, K.W., Gray, R.W., and Lemann, J. (1982). Urinary excretion of 25-hydroxy vitamin D in health and the nephrotic syndrome. *Journal of Laboratory and Clinical Medicine*, **99**, 325–30.

Sala, N., Olivier, N., Estivill, X., Moreno, R., Felez, J., and Rutllant, M. (1985). Plasmatic and urinary protein C levels in nephrotic syndrome. *Thrombosis and Haemostasis*, **54**, 900.

Schiepatti, A., *et al.* (1982). The metabolism of arachidonic acid by platelets in the nephrotic syndrome. *Kidney International*, **25**, 671–6.

Schmidt-Gayk, H., *et al.* (1977). 25-hydroxy-vitamin D in nephrotic syndrome. *Lancet*, **ii** 105–8.

Schwartz, H. and Cohn, J.R. (1935). Lipoid nephrosis: clinical and pathologic study based on 15 years' observation with special reference to prognosis. *American Journal of Diseases in Children*, **44**, 579–93.

Scott, P.J., White, B.M., Winterbourne, C.C., and Hurley, P.J. (1970). Low density lipoprotein peptide metabolism in nephrotic syndrome. *Australasian Annals of Medicine*, **12**, 23–7.

Shafrir, E. (1958). Partition of unesterified fatty acids in normal and nephrotic serum and its effect on serum electrophoretic pattern. *Journal of Clinical Investigation*, **37**, 1775–82.

Shore, V.G., Forte, T., Licht, H., and Lewis, S.B. (1982). Serum and urinary lipoproteins in the human nephrotic syndrome: evidence for renal catabolism of lipoproteins. *Metabolism*, **31**, 258–68.

Short, C.D., Durrington, P.J., Mallick, N.P., Hunt, L.P., Tetlow, N.P. and Ishola, M. (1986). Serum and urinary high density lipoproteins in glomerular disease with proteinuria. *Kidney International*, **29**, 1224–8.

Shioji, R., Sasaki, Y., Siato, H., and Furuyama, T. (1974). Reversible tubular dysfunction associated with chronic renal failure in an adult patient with the nephrotic syndrome. *Clinical Nephrology*, **2**, 76–80.

Sié, P., Meguira, B., Bouissou, F., Boneu, B., and Barthe, Ph. (1988). Plasma levels of plasma co factor II in nephrotic syndrome of children. *Nephron*, **48**, 175–6.

Sillix, D., Francis, J., Mahajan, S., and Briggs, W. (1983). Impaired granulocyte function in the nephrotic syndrome. *Kidney International*, **23**, 135 (abstract).

Soff, G.A., Sica, D.A., Marlar, R.A., Evans, H.J., and Qureshi, G.D. (1985). Protein C levels in nephrotic syndrome: use of a new enzyme-linked immunoadsorbent assay for protein C. *American Journal of Hematology*, **22**, 43–9.

Sokolovskaya, I. and Nikiforova, N. (1984). High-density lipoprotein cholesterol in patients with untreated and treated nephrotic syndrome. *Nephron*, **37**, 49–53.

Sonnenschein, H. and Joos, H.A. (1979). Observations on infections of the urinary tract and childhood nephrosis. *Clinical Pediatrics*, **9**, 419–21.

Soothill, J.F. and Kark, R.M. (1956). The effects of infusion of salt-poor serum albumin on serum cholesterol, cholinesterase and albumin levels in healthy subjects and in patients ill with the nephrotic syndrome. *Clinical Research Proceedings*, **4**, 140–1.

Sorensen, P.J., Knudsen, F., Nielsen, A.H., and Dyerberg, J. (1985). Protein C activity in renal disease. *Thrombosis Research*, **38**, 243–9.

Speck, W.T., Dresdale, S.S., and McMillan, R.W. (1974). Primary peritonitis and the nephrotic syndrome. *American Journal of Surgery*, **127**, 267–9.

Spika, J.S., *et al.* (1986). Decline of vaccine-induced antipneumococcal antibody in children with nephrotic syndrome. *American Journal of Kidney Diseases*, **7**, 466–70.

Spitz, I.M., Rubinstein, A.H., Bersohn, I., Abrahams, C., and Lowy, C. (1970). Carbohydrate metabolism in renal disease. *Quarterly Journal of Medicine*, **39**, 201–26.

Stanbury, S.W. and Macauley, D. (1957). Defects of renal tubular function in the nephrotic syndrome. *Quarterly Journal of Medicine*, **26**, 7–30.

Steele, B.T., Bacheyie, G.S., Baumal, R., and Rance, C. Ph. (1982). Acute renal failure of short duration in minimal changes lesion of childhood. *International Journal of Pediatric Nephrology*, **3**, 59–62.

Steinberg, D., Parsatharthy, S., Carew, T.E., Khoo, J.C., and Witzum, J.L. (1989). Beyond cholesterol: modifications of low-density lipoprotein that increase its atherogenicity *New England Journal of Medicine*, **320**, 915–23.

Straprans, I., Garon, S.J., Hopper, J., and Felts, J.M. (1981). Characterisation of glycoasminoglycans in urine from patients with nephrotic syndrome and control subjects, and their effects on lipoprotein lipase. *Biochimica Biophysica Acta*, **678**, 414–22.

Stuart, M.J., Gerrard, J.M., and White, J.G. (1980). The influence of albumin and calcium on human platelet arachidonic acid metabolism. *Blood*, **55**, 418–23.

Symchych, P.J. and Perrin, E.V. (1965). Thrombosis of the main renal artery in nephrosis. *American Journal of Diseases of Children*, **110**, 636–42.

Tanphaichitr, P. and Chesdavanijkul, W. (1978). An approach to the management of primary peritonitis in children with nephrotic syndrome. *Journal of the Medical Association of Thailand*, **61**, 369–73.

Taube, D., Chapman, S., Brown, Z., and Williams, D.G. (1981). Depression of normal lymphocyte transformation by sera from patients with minimal change nephropathy and other forms of nephrotic syndrome. *Clinical Nephrology*, **15**, 286–90.

Taube, D., Brown, Z., and Williams, D.G. (1984). Impaired lymphocyte and suppressor cell function in minimal change nephropathy, membranous nephropathy and focal segmental glomerulosclerosis. *Clinical Nephrology*, **22**, 176–82.

Taube, D. and Williams, D.G. (1988). Pathogenesis of minimal change nephropathy. In *The nephrotic syndrome* (eds. J.S. Cameron, and R.J. Glassock, pp. 193–218. Marcel Dekker, New York.

Teglaers, W.H.H. and Tiddens, H.W. (1955). Nephrotic-glucosuricaminoaciduric dwarfism and electrolyte metabolism. *Helvetica Paediatrica Acta*, **10**, 269–78.

Tejani, A., Fikrig, S., Schiffman, G., and Gurumurthy, K. (1985). Persistence of protective pneumococcal antibody following vaccination in patients with a nephrotic syndrome. *American Journal of Nephrology* **4**, 32–7.

Temes Montes, X.L., *et al*. (1980). Valores sericos y urinarios de cinc en nefropatias exteriorizadas por sindrome nefrotico o como proteinuria en rango nefrotico. *Revista Clinica Espanola*, **159**, 159–61.

Tessitore, N., *et al*. (1984). Bone histology and calcium metabolism in patients with nephrotic syndrome and normal or reduced renal function. *Nephron*, **37**, 153–9.

Thaler, E. and Lechner, K. (1978). Thrombophilie bei erworbenem Antithrombin III- Mangel von Patienten mit nephrotischem Syndrom. In *Niere und Hämostase* (eds. R. Marx, and H. Thiess, **5**, 123–9. Schattauer, Stuttgart.

Thompson, C., Forbes, C.D., Prentice, C.R.M., and Kennedy, A.C. (1974). Changes in blood coagulation and fibrinolysis in the nephrotic syndrome. *Quarterly Journal of Medicine*, **43**, 399–407.

Tobias, J.A., Cain, J.R., Soila, K.P., and Sheldon, J.P. (1986). Usefulness of magnetic resonance imaging (MRI) in the diagnosis of renal vein thrombosis. *Kidney International*, **31**, 206 (abstract).

Trew, P.A., Biava, C.C., Jacobs, R.P., and Hopper, J.Jr. (1978). Renal vein thrombosis in membranous glomerulonephritis: incidence and association. *Medicine (Baltimore)*, **57**, 69–82.

Trygstad, C.W., McCabe, E., Fancyk, W.P., and Crummy, W.P. (1970). Renal vein thrombosis and the nephrotic syndrome. A case report with protein selectivity studies. *Journal of Pediatrics*, **76**, 861–6.

Ueda, N., *et al*. (1987). Effect of corticosteroids on coagulation factors in children with nephrotic syndrome. *Pediatric Nephrology*, **1**, 286–9.

Valeri, A., Gelfand, J., Blum, C., and Appel, G.B. (1986). Treatment of the hyperlipidemia of the nephrotic syndrome: a controlled trial. *American Journal of Kidney Diseases*, **8**, 388–96.

Vaziri, N. (1983). Nephrotic syndrome and coagulation and fibrinolytic abnormalities. *American Journal of Nephrology*, **3**, 1–6.

Vaziri, N.D., Alikhani, S., Patel, B., Nguyen, Q., Barton, C.H., and Gonzales, E.V. (1988). Increased levels of protein C activity, protein C concentration, total and free protein S in nephrotic syndrome. *Nephron*, **49**, 20–3.

Vega, G.L. and Grundy, S.M. (1988). Lovastatin therapy in nephrotic hyperlipidemia: effects on lipoprotein metabolism. *Kidney International*, **33**, 1160–68.

Vermylen, C., Levin, M., Barratt, T.M., and Muller, D.P.R. (1987*a*). Inhibition of lipoprotein lipase by plasma from children with the steroid responsive nephrotic syndrome. *Pediatric Research*, **22**, 197–200.

Vermylen, C.G., Levin, M., Lanham, J.G., Hardisty, R.M., and Barratt, T.M. (1987*b*). Decreased sensitivity to heparin in vitro in steroid-responsive nephrotic syndrome. *Kidney International*, **31**, 1396–1401.

Vigano-D'Angelo, S., D'Angelo, A., Kaufman, C.E., Scholer, C., Esmon, C.T., and Comp, P.C. (1987). Protein S deficiency occurs in the nephrotic syndrome. *Annals of Internal Medicine*, **107**, 42–7.

Wagner, J.G., Agebeyoglu, T., Bergstrom, R.F., Sakmar, E., and Kary D.R. (1981). Plasma protein binding parameters of prednisolone in immune disease patients receiving long term prednisone therapy. *Journal of Laboratory and Clinical Medicine*, **97**, 487–501.

Wagoner, R.D., Stanson, A.W., Holley, K.E., and Winter, C.S. (1983). Renal vein thrombosis in idiopathic membranous glomerulopathy and nephrotic syndrome: incidence and significance. *Kidney International*, **23**, 368–74.

Warshaw, B.L., Check, I.J., Hymes, L.C., and DiRusso, S.C. (1984). Decreased serum transferrin concentrations in children with nephrotic syndrome: Effect on lymphocyte proliferation and correlation with serum transferrin levels. *Clinical Immunology and Immunopathology*, **33**, 210–19.

Warwick, G., Boulton-Jones, J.M., Caslake, M., Dagen, M., Packard, C., and Shepherd, J. (1989). Changes in low density lipoprotein (LDL) apoprotein metabolism in the nephrotic syndrome. *Nephrology Dialysis Transplantation*, **4**, 831 (abstract).

Wass, V.J., Jarrett, R.J., Chilvers, C., and Cameron, J.S. (1979). Does the nephrotic syndrome increase the risk of cardiovascular disease? *Lancet*, **ii**, 664–7.

Wass, V.J. and Cameron, J.S. (1981). Cardiovascular disease and the nephrotic syndrome. The other side of the coin. *Nephron*, **27**, 58–61.

Wheeler, D.C., Varghese, Z., and Moorhead, J.F. (1989). Hyperlipidemia in nephrotic syndrome. *American Journal of Nephrology*, **9 (suppl 1)**, 78–84.

Widhalm, K., Singar, P., and Balzar, E. (1980). Serum lipoproteins in children with nephrotic syndrome and chronic renal failure. *Artery*, **2**, 191–8.

Wilfert, C.M. and Katz, S.L. (1968). Etiology of bacterial sepsis in nephrotic children 1963–67. *Pediatrics*, **42**, 840–42.

Wilkes, J.C., Nelson, J.D., and Worthen, H.G. (1982). Response to pneumococcal vaccine in children with nephrotic syndrome. *American Journal of Kidney Diseases*, **11**, 43–6.

Woolfe, L.E. and Giles, H. McC. (1956). Urinary excretion of amino acids and sugar in the nephrotic syndrome. *Acta Paediatrica (Uppsala)*, **45**, 489–500.

Yedgar, S., Weinstein, D.B., Patsch, W., Schonfeld, G., Casanada, F.E., and Steinberg, D. (1984). Viscosity of culture medium as a regulator of synthesis and secretion of very low density lipoproteins by cultured hepatocytes. *Journal of Biological Chemistry*, **257**, 2188–92.

Yetgkin, S., Guy, Y., and Saatci, U. (1980). Non-specific immunity in nephrotic syndrome. *Acta Paediatrica Scandinavica*, **69**, 21–4.

Yoshida, N. and Aoki, N. (1978). Release of arachidonic acid from human platelets. A key role for the potentiation of platelet hyperaggregability in normal subjects as well as those with a nephrotic syndrome. *Blood*, **52**, 969–77.

Zerhouni, E.A., Barth, K.H., and Sigelman, S. (1980). Demonstration of venous thrombosis by computed tomography. *American Journal of Radiology*, **134**, 753–8.

Zilleruelo, G., Hisua, S.L., Freundlich, M., Gorman, H.M., and Strauss, J. (1984). Persistence of serum lipid abnormalities in children with idiopathic nephrotic syndrome. *Journal of Pediatrics*, **104**, 61–4.

3.5 Minimal changes and focal segmental glomerular sclerosis

MICHEL BROYER, ALAIN MEYRIER, PATRICK NIAUDET AND RENÉE HABIB

History and definition

Historically, the term 'nephrosis' was first used by Müller in 1905, and later by Volhard and Fahr (1914), to denote a renal condition in which an inflammatory component was absent from the kidney. The term 'lipoid nephrosis' was coined by Munk (1913) to describe a group of patients with oedema, heavy proteinuria, hypoproteinaemia, and hyperlipidaemia, in whom microscopic examination of the kidney showed normal glomeruli but lipid droplets in the cells of the proximal tubule. The same clinical symptoms were shown to occur in patients with various systemic diseases, including lupus, diabetes, and amyloidosis, and appeared as a consequence of heavy proteinuria whatever the cause (Hamburger et al. 1968): the term 'nephrotic syndrome' was therefore considered to be preferable. Routine use of renal biopsy and immunofluorescence staining was helpful in recognizing the underlying pathology (Habib et al. 1961). After excluding diffuse glomerular lesions a large group of patients—mainly children—exhibited minimal changes on light microscopy in all or part of their glomeruli. This group of patients has been referred to as having 'nil' disease, minimal change nephrotic syndrome, or idiopathic primary nephrotic syndrome.

By definition, idiopathic nephrotic syndrome is a clinicopathological entity characterized on the one hand by massive proteinuria, hypoalbuminaemia, hyperlipidaemia, and oedema, and on the other by non-specific histological abnormalities of the kidney including minimal changes, focal and segmental glomerular sclerosis, and diffuse mesangial proliferation. All of these features are associated with fusion of the foot processes of epithelial cells on electron microscopy and insignificant deposition of immunoglobulins or complement.

This unitary view appears to give a better account of the facts and also seems more appropriate for the clinician who has to take decisions on behalf of patients. It is also compatible with the possibility of varied aetiologies and/or physiopathological mechanisms. At the same time we are well aware of the fact that a number of nephrologists consider minimal change nephrosis and focal and segmental glomerular sclerosis as separate entities, and that they are presented as such in the majority of nephrology textbooks.

There is no doubt that patients who exhibit focal and segmental glomerular sclerosis on renal biopsy generally have more severe disease, are often resistant to corticosteroid treatment and have a significant propensity for progression to renal failure. In the early stages, however, focal and segmental glomerular sclerosis and minimal change nephrosis are indistinguishable (Kashgarian et al. 1974). Several other facts strongly favour the single disease concept: a significant number of patients with focal and segmental glomerular sclerosis respond to steroids (Habib 1973), while some steroid resistant patients have no sclerotic changes visible on adequate biopsies (Siegel et al. 1972; Habib 1973). In addition, a benign steroid-sensitive minimal change nephrotic syndrome has been noted in transplanted kidneys in recipients whose primary renal disease was severe corticosteroid-resistant focal and segmental glomerular sclerosis (personal observation). Experimental models support the hypothesis of a single disease. Rats given a single dose of the aminonucleoside of puromycin develop a minimal change nephrotic syndrome, while repeated exposure results in progressive glomerular sclerosis (Wilson et al. 1958). Another argument is the non-specificity of the lesion in focal and segmental glomerular sclerosis: this may be observed in a number of conditions associated with proteinuria, and may be the consequence of mesangial overload arising from proteins escaping through the glomerular basement membrane.

For all these reasons, and in the absence of a known aetiology and pathogenesis, it seems to us justified to consider the different histological features of minimal change nephritis and focal and segmental glomerular sclerosis as different stages or variants in severity of a single disease: the idiopathic nephrotic syndrome.

Epidemiology

Idiopathic nephrotic syndrome is most common in children, but also occurs in adults of all ages: the features of the disease are similar in all age groups. The incidence of idiopathic nephrotic syndrome varies with age, race and geographical location: the annual incidence in children in three areas of the United States has been estimated to be 2 to 2.7 per 100 000 (Rothenberg et al. 1957; Schlesinger et al. 1986; McEnery and Strife 1982), with a cumulative prevalence of 16 per 100 000. An English survey found an incidence of three per million adults. Whereas idiopathic nephrotic syndrome accounts for only 25 per cent of patients with nephrotic syndrome in adults (Hamburger et al. 1968; Cameron et al. 1974b; Glassock et al. 1976) it is by far the most common cause of nephrotic syndrome in children: figures of 63 per cent (Habib and Kleinknecht 1971), 87 per cent (Churg et al. 1970), and 93 per cent (White et al. 1970) of nephrotic children have been quoted. Almost all nephrotic children under 6 years of age in Western countries have idiopathic nephrotic syndrome.

Geographical and/or ethnic differences are well known. In the United Kingdom the incidence of idiopathic nephrotic syndrome is six-fold higher in Asian than in European children (Sharples et

3.5 Minimal changes and focal segmental glomerular sclerosis

Table 1 Factors reported to be related to onset of idiopathic nephrotic syndrome

---

Drugs
  Non-steroidal anti-inflammatory drugs
    Tolmetin, indomethacin
    Sulindac
    5-Aminosalicylic acid, salsalate
    Zomepirac
    Pirprofen, naproxen, fenoprofen
    Ibuprofen
    Diclofenac
  D-Penicillamine
  Lithium
  Mercury, gold
  Trimethadione (see Chapter 2.1)
Allergy
  Pollens
  Food allergy (milk, pork)
  House dust
  Bee stings
  Medusa stings
  Contact dermatitis, poison ivy, and oak
Malignancies (adult patients)
  Carcinoma, sarcoma, lymphocytic leukaemia, Hodgkin's disease, mycosis fungoides, Kimura's disease
Others
  Schistosoma haematobium infection
  EBV primary infection
  AIDS (?) (see Chapter 3.13)
Immunization

---

al. 1985): this is also true for Indians (Srivastava et al. 1975) and for Japanese, south-west Asians and Hungarian gipsies (Horwarth and Sulyok 1986). Idiopathic nephrotic syndrome is rare in Africa and most of the cases of nephrotic syndrome seem to be related to structural glomerular lesions unresponsive to steroids (Coovadia et al. 1979). Only 13.5 per cent of a series of nephrotic children examined in the Ivory Coast had idiopathic nephrotic syndrome (de Paillerets et al. 1972).

A high male prevalence is found in children, with a male: female ratio of 2:1 (Hayslett et al. 1973; ISKDC 1978) but both sexes are similarly affected in adolescents and adults (Cameron et al. 1974b; Hopper et al. 1970).

The nephrotic syndrome tends to be more severe in black populations (Schlesinger et al. 1968), but it would be interesting to know whether this observation is due to race or to environment. Indians and white South African children with minimal change disease seem to have similar long-term outcome to that reported from Europe, but black children are generally resistant to corticosteroid treatment (Coovadia and Adhikari 1989).

Aetiology and inheritance

Idiopathic nephrotic syndrome is, by definition, a primary disease. Nevertheless, in a number of cases an upper respiratory tract infection, an allergic reaction, or another factor may immediately precede the development or a relapse of the disease.

The factors (many extremely rare) which are regularly cited in textbooks as possible causes of idiopathic nephrotic syndrome includes infectious diseases, drugs, allergies, immunizations, and some malignancies (Table 1). The question remains whether these factors are a real cause, a simple coincidence, or a precipitating agent. Among these factors, allergy and malignancies are of particular interest, and are discussed in more detail below.

Clinical manifestations of allergy have been associated with up to 30 per cent of cases in some series (Lagrue and Laurent 1984), and several reports of improvement after desensitization or exclusion of an allergen provide evidence for their involvement in the disease (see below).

The association with malignancies mainly concerns lymphomatous disorders (see Chapter 3.13): 26 of 33 patients with a cancer-related minimal change nephrotic syndrome had Hodgkin's disease and a further two had non-Hodgkin's lymphomas (Eagen and Lewis 1977). In another study (Cale et al. 1982) 36 of 44 patients who had Hodgkin's disease and the nephrotic syndrome had minimal change disease, while only two had membranous nephropathy. The nephrotic syndrome may be the presenting feature of the disease, and disappear after its successful treatment, but there are patients who show dissociation between the evolution of the malignancy and the nephrotic syndrome. Interestingly, minimal change nephropathy has also been reported in a case of mycosis fungoides, a cutaneous T-cell malignant lymphoma (Allon et al. 1988).

Other types of neoplasia may also be associated with nephrotic syndrome (see also Chapter 3.13). Lee et al. (1966) reported a 54-year-old woman with minimal change disease in whom nephrotic syndrome disappeared after surgical ablation of a colon carcinoma. It is not absolutely certain that this patient did not suffer from membranous glomerulopathy. However, almost two decades later, Moorthy (1983) published two case reports of minimal change glomerular disease in patients with bronchogenic, small cell carcinoma. In both, chemotherapy-induced reduction of tumour mass reduction was accompanied by a significant diminution of proteinuria. Since this report, several other cases of minimal change nephropathy associated with diverse solid tumours, including mesothelioma, pancreatic and prostatic carcinomas, etc. have been described (Whelan and Hirszel 1988). One of us has observed three (unpublished) cases in adults, one of whom had small cell bronchogenic carcinoma and nephrotic syndrome which remitted when chemotherapy produced regression of the tumour. Relapse of the nephrotic syndrome was seen when the tumour recurred. Interestingly, treatment had not included corticosteroid therapy. In another case, massive nephrotic syndrome with normal glomeruli was the presenting feature of cancer of the breast with bone metastases. The last case was a female in whom steroid-resistant nephrotic syndrome preceded the discovery of a cancer of the endocervix. The pathophysiology of such paraneoplastic minimal change nephrosis is obscure. However, it should be considered, in the light of the frequency of proteinuria (of non-nephrotic range) in large series of patients with cancer (Gabriel and Wignen 1986).

Paraneoplastic minimal change nephrotic syndrome is rare—almost anecdotal. However, this now well-documented association leads to two considerations. First, minimal change nephrotic syndrome occurring in a patient over 50 years of age should suggest an underlying tumour as a possible aetiology. Second, occurrence of a solid tumour during the course of minimal change nephrosis treated with cytostatics should not systematically be ascribed to the oncogenic hazards of treatment, as it may occasionally be the actual cause of nephrosis.

Eosinophilic lymphoid granuloma in Orientals (Kimura's disease) has also been reported as a possible association with corticosteroid-responsive minimal change nephrotic syndrome (Matsumoto et al. 1988).

Several cases of minimal change disease have been reported in association with onset of insulin dependent diabetes mellitus (Dornan et al. 1988; Peces et al. 1987) (see also Chapter 4.2). The disease is usually responsive to corticosteroids and follows a relapsing course. There may be a pathogenetic link between these two diseases, which are known to have an immunological basis, and to be associated with genes coded within the major histocompatibility complex.

Inheritance

The familial occurrence of idiopathic nephrotic syndrome, first described by Fanconi in 1951, has since been noted by several authors (Bader et al. 1974; Vernier et al. 1957). White 1973 found that 3.3 per cent of 1877 patients with idiopathic nephrotic syndrome (excluding congenital nephrotic syndrome) had affected family members, mainly siblings. This incidence was higher than that among the general population and was probably not due to coincidence. Idiopathic nephrotic syndrome has also been reported in identical twins (Roy and Pitcock 1957). In one series of familial cases (Moncrief et al. 1973) the disease tended to develop in affected siblings at the same age, with the same renal morphology and the same outcome. Another study from the Hôpital des Enfants Malades reported 34 patients in 15 families; the response to therapy was identical within members of individual families (Gonzalez et al. 1977). The age and sex distribution, the relapsing course and the absence of renal failure in 15 steroid-responsive familial cases was similar to that in sporadic steroid responders. By contrast, the non-responder group was characterized by an early onset, often during the first year of life, a high percentage of affected males with constant concordance of sex within individual families, and a high incidence of focal scler-

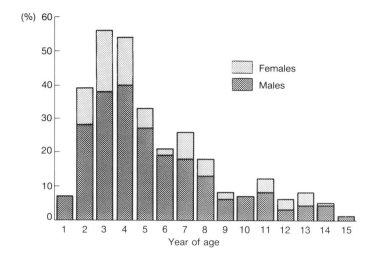

Fig. 1 Histogram of age at onset of idiopathic nephrotic syndrome (personal observation).

osis and/or mesangial proliferation and of renal failure. These cases represented 15 per cent of the steroid-resistant patients with nephrosis in this institution. Steroid-resistant nephrotic syndrome with focal and segmental glomerular sclerosis occurred within a 4-month period in three of five siblings aged 18 to 20.5 years, but these patients also had sickle-cell anaemia or thalassaemia (Chandra et al. 1981a). There are also examples of steroid-sensitive familial groups in which one of the parents and one or several children develop the disease (Schwartz and Cornfield 1979).

In conclusion, there is a possible genetic predisposition to the development of idiopathic nephrotic syndrome, and this is reinforced by the reported association between the disease and certain histocompatibility complex antigens, mainly HLA B12 and DRW7 (see below).

Finally, some familial cases of minimal change nephrotic syndrome are associated with other abnormalities, such as microcephaly and hiatus hernia as in the syndrome of Galloway and Mowat (1968). Other syndromes, including microcephaly (Roos et al. 1987) or disturbances of neuronal migration (Palm et al. 1986), have also been reported in association with early onset nephrotic syndrome.

Clinical features

The disease may appear in the first year of life, and as early as the neonatal period, but it is more common after this age, with a peak incidence (74 per cent) between the ages of 2 and 7 years. At this age the male to female ratio is 2:1 (ISKDC 1978) (Fig. 1), while in adolescents and adults the sex ratio is almost 1:1 (Cameron et al. 1974b). In 30 to 60 per cent of patients, the first attack of nephrosis and further relapses follows benign infections of the upper respiratory tract (Habib and Kleinknecht 1971; Rubin 1975): inapparent viral infections may be even more commonly involved. Oedema is the major presenting symptom, and an almost constant sign. It increases gradually, only becoming clinically detectable when retention of fluid exceeds 3 to 5 per cent of body weight. Initially, facial oedema may be taken for an allergic symptom. The oedema is gravity dependent, becoming localized preferentially to the lower extremities in the upright position, and to the dorsal part of the body in reclining position. This oedema is classically said to be white, soft, and pitting, keeping the marks of cloth creases or finger pressure. When more marked, serious effusions develop: although initially clinically latent they may be revealed by radiography or on ultrasonography. Later still, frank anasarca develops with patent ascites, and pleural and pericardial effusions. Although there may also be abdominal distension dyspnoea or respiratory embarrassment are rare. At this stage, an impressive oedema of the scrotum and penis, or labiae, may be seen and the eyelids are swollen and shut in a puffy face. In children, the rapid formation of ascites is often associated with abdominal pain and malaise: these symptoms may also be related to the concomitant hypovolaemia. Abdominal pain is occasionally due to a complication such as peritonitis, thrombosis or, exceptionally, pancreatitis. Cardiovascular shock is not unusual, even in the absence of any infectious complication, presumably following the sudden loss of plasma albumin in the urine. Blood pressure however is usually normal and even sometimes elevated. The question of blood volume in adult and childhood nephrosis is discussed further in Chapters 3.1 and 18.2.

The illness may be discovered on routine urinalysis, and in some series this accounts for up to 10 per cent of cases. In some rare cases macroscopic haematuria may be observed. The disease may also be revealed by a complication, particularly infections due to *Streptoccocus pneumoniae*. 'Medical' peritonitis is a classical mode of onset (Krensky *et al.* 1982), but other sites of infection are seen, including meningitis, cellulitis, and pneumonia. Deep vein or artery thromboses are classical complications of nephrotic syndrome (see Chapter 3.4), and are mainly observed in persistent non-responsive nephrosis, but they may also be seen during the first attack of the disease or during a relapse. Pulmonary embolism must be suspected if there is any thoracic or cardiovascular symptom; the diagnosis of this complication is ascertained by angiography or angioscintigraphy.

Changes may be noted in the subungual areas, with inversion of colours of the nailbed, the lunula becoming pink and the rest white, with transverse white bands in the nail beds.

Acute renal failure

Renal insufficiency with marked oliguria has been reported as a complication of idiopathic nephrotic syndrome, either with minimal changes or focal and segmental glomerular sclerosis (see Chapter 8.2). This complication occurs mainly in adults, particularly in middle-aged or older patients (Cameron *et al.* 1974b; Esparza *et al.* 1981), but may also occur in children (Springate *et al.* 1977). Oliguric renal failure may be the presenting symptom of idiopathic nephrotic syndrome (Hulter and Bonner 1980) but has usually been reported in association with a relapse.

Several causes may underly such an event: a role for a hypovolaemia with poor renal perfusion, the first hypothesis to be proposed, is generally not supported since hypertension is often present and plasma volume not significantly decreased (see also Chapter 3.1). The possibility of bilateral renal vein thrombosis is recognized easily by sonography. Interstitial nephritis is a possible cause having been reported in several cases, especially in association with frusemide administration (Lyons *et al.* 1973); other symptoms such as skin rash and peripheral blood eosinophilia are suggestive of this diagnosis (see Chapters 2.1 and 8.2.).

Acute renal failure in idiopathic nephrotic syndrome is usually reversible, often by producing diuresis with a high dose of frusemide, with or without cautious intravenous albumin infusion. In some cases, where there is no renal vein thrombosis, and where glomerular structure is normal on initial histological examination, renal failure may be irreversible (Raij *et al.* 1976).

Histopathology

In addition to renal vein thrombosis, and interstitial nephritis, diffuse and severe tubular damage with dilatation and vacuolization of tubular cells, and casts of proteinaceous material in the lumina have been reported in several series of oligoanuric idiopathic nephrotic syndrome (Imbasciati *et al.* 1981; Esparza *et al.* 1981). It is possible that massive precipitation of casts in the collecting ducts obstructs urine flow, leading to a back leak of filtrate and/or a reflux vasoconstriction of the affluent arterioles. Tubular injury due to obstruction and/or ischaemia seems to be the most frequent association with acute renal failure in this disease.

Pathophysiology

Apart from renal vein thrombosis and interstitial nephritis, other mechanisms have been suggested. Lowenstein *et al.* (1981) reported a remarkably low filtration fraction in a series of patients with minimal change nephrotic syndrome and acute renal failure. This reduction was not due to hypovolaemia, which is usually accompanied by an increase in the filtration fraction, and has also been reported (but to a lesser extent) in minimal change nephrotic syndrome without acute renal failure (Dorhout Mees *et al.* 1979; Metcoff and Janeway 1961) (see Chapter 3.1). Several experimental studies (Bohrer *et al.* 1977; Baylis *et al.* 1977) in rats with aminonucleoside nephrosis have related the low GFR to a reduction of the glomerular capillary ultrafiltration coefficient (K_f), which is also reduced by hypoproteinaemia, but the rapid improvement of GFR which may follow a diuresis induced with diuretics does not support a decreased K_f as the main explanation for acute renal failure in man.

Lowenstein *et al.* (1981) also suggested that increased hydrostatic pressure in the proximal tubule and Bowman's space, as a consequence of renal interstitial oedema, finally produces the reduced GFR and low filtration fraction. However, no observations of interstitial or tubular pressures have yet been made in this condition.

Usual laboratory abnormalities

Urinalysis

In massive proteinuria the urine becomes foamy on voiding: patients sometimes report this spontaneously or on direct questioning. Proteinuria is also readily identified by dipstick testing 3 or 4+ (see Chapter 1.4). Quantitative evaluation of proteinuria gives figures ranging from less than 1g to several tens of grams per day. Proteinuria is, not surprisingly, heavier in adults than in children, and it is better expressed with reference to body weight or body surface area. The nephrotic range of proteinuria is defined as >50 mg/kg.day or 40 mg/h.m^2 but the mean value during the first days of an attack may be four or five times higher. This arises because proteinuria is also dependent on the plasma protein concentration, and may decrease as the serum albumin concentrations fall to below 10 g/l.

It has to be remembered that the amount of protein excreted in the urine does not equal the quantity of protein crossing the glomerular basement membrane since a significant (but largely unknown) amount is reabsorbed by the proximal tubule. As a consequence of massive proteinuria the specific gravity of the urine is typically high, usually reaching 1025 to 1030, but urine osmolality is hardly changed since the molecules are very large, and (comparatively) few in number.

The nature of the protein passed in the urine is important since it may depend on the type and/or severity of the renal abnormality responsible for the disease. Typically, in minimal change corticosteroid-sensitive nephrotic syndrome, proteinuria consists mainly of albumin and other low molecular weight proteins, while in severe nephrotic syndrome with glomerular lesions and resistance to steroid treatment, the urine contains not only albumin but also higher molecular weight proteins such as gamma-globulins. This can be easily seen on a polyacrylamide gel electrophoresis and can also be quantified by the means of the selectivity index, although the latter is much more expensive to perform. The selectivity index which is currently used is the ratio of IgG to albumin or transferrin clearance. A 'good' selectivity index would be below 0.10, or better below 0.05; a poor index is above 0.15 or 0.20. There is a considerable overlap in results from benign and severe idiopathic nephrotic syndrome and the test has a limited value, especially in adult patients.

The urine sediment of patients with idiopathic nephrotic syndrome often contains fat bodies, which give the typical 'Maltese' cross appearance when viewed with a polarized light microscope. Hyaline casts are also usually found in patients with massive proteinuria, but granular casts are not present unless there is associated acute renal failure and acute tubular necrosis.

Macroscopic haematuria is rare in idiopathic nephrotic syndrome, occurring in 1 per cent of steroid responders, but in 3 per cent of non-responders (Kleinknecht and Gubler 1983). Microscopic haematuria is more common, and may be observed up to 30 per cent of patients. It is with usually inconstant in minimal change disease, but more frequent, more constant, and more important in patients with steroid-resistant idiopathic nephrotic syndrome. Granular casts are also often associated with the haematuria, again suggesting structural renal damage.

Urinary sodium is usually low (1–2 mmol/24 h), associated with sodium retention and oedema. In steroid-responsive patients an onset of natriuresis is observed 2 or 3 days after the fall of proteinuria, before any significant change of plasma protein concentration is seen (Koomans et al. 1987). Kaliuresis is usually higher than natriuresis, but the urinary output of potassium may be reduced in oliguric patients.

Blood chemistry

By definition serum proteins are markedly reduced and serum lipid usually increased in nephrotic syndrome. Proteinaemia is below 50 g/l in 80 per cent of patients, and below 40 g/l in 40 per cent (Habib and Kleinknecht 1971). Albumin concentration usually falls below 20 g/l and may be less than 10 g/l. Electrophoresis of the plasma proteins shows a typical pattern since, besides the low albumin levels, α_1-Globulins are not modified but α_2-globulins and, to a lesser extent, β-globulins are markedly increased, whilst γ-globulins are decreased. Assays of the different classes of immunoglobulins also show typical patterns: IgG is considerably decreased, IgA slightly reduced, IgM is increased, and IgE is normal or increased. Hyperlipidaemia is almost always seen within a few days of the onset of massive proteinuria, and is usually considered to be part of the definition of nephrotic syndrome. Serum total cholesterol is increased, averaging 5 g/l but sometimes in excess of 10 g/l. Levels of serum lipids exceeding 15 g/l are observed in 30 per cent of patients (Habib and Kleinknecht 1971): a detailed analysis of the hyperlipidaemia of the nephrotic syndrome and its related complications is given in Chapter 3.4.

Serum electrolytes are usually within the normal range. A low sodium level may be related to dilution from inappropriate renal retention of water, but there is usually a proportionate retention of sodium and water and the mild reduction of plasma sodium concentration may be an artefact related to hyperlipidaemia, true sodium concentration in the plasma water remaining normal. Serum potassium may be high in oliguric patients. Serum calcium is consistently low as a result of the hypoproteinaemia, since 1g of albumin binds 0.8 mg of calcium. Ionized calcium is usually normal but may be decreased in persistent nephrotic syndrome due to urinary loss of 25-hydroxyvitamin D_3 Sato et al. 1982) and normal but inappropriate levels of calcitriol (Freundlich et al. 1986) (see Chapter 3.4). BUN and creatinine concentrations are usually within the normal range, or slightly increased in relation to a modest reduction in the GFR (see below)

Haematology

Haemoglobin levels and haematocrit may be increased in patients with reduced circulating plasma volume, which is sometimes associated with the first attack or a relapse with a sudden and heavy loss of albumin in the urine. Anaemia with microcytosis may be observed in chronic steroid-resistant nephrotic syndromes, probably related to a urinary loss of siderophilin. Erythrocytosis has been reported in idiopathic nephrotic syndrome, but is exceptional (Myers et al. 1979). Thrombocytosis is common and may reach 5.10^5 to $10^6/mm^3$: this is related to lipid abnormalities (Chapter 3.4).

Renal function

Renal function is usually within normal limits, at least during the first attack of idiopathic nephrotic syndrome (Barnett et al. 1978). Some studies, however, have reported reduction of the glomerular filtration rate, attributed to hypovolaemia (White et al. 1970). A variable reduction of GFR has also been reported in some cases of minimal change disease with a complete return to normal after remission (Bohlin 1984). A reduced GFR may also be found in the proteinuric phase of the idiopathic nephrotic syndrome, despite normal effective plasma flow (Dorhout Mees et al. 1979; Berg and Bohlin 1982; Bohlin 1984). Bohman et al. (1984) showed a close relationship between the degree of foot process fusion and both GFR and filtration fraction, suggesting that fusion of foot processes could lead to a reduction of glomerular filtering area and/or of permeability to water and small solutes. This reduction is transitory, with a rapid return to normal of GFR and filtration fraction after disappearance of proteinuria (Koomans 1987). The renal functional reserve after a protein load is conserved (Mizuiri et al. 1988).

Tubular function is sometimes modified in idiopathic nephrotic syndrome with some symptoms of the Fanconi syndrome: glycosuria, aminoaciduria, loss of bicarbonate, and hypokalaemia. A defect in urinary acidification has also been reported, corrected by administration of frusemide (Rodriguez-Soriano et al. 1982). These symptoms were reported in children with focal and segmental glomerular sclerosis and a poor subsequent outcome (McVicar et al. 1980): they may be related to tubular lesions, but may also disappear after disappearance of proteinuria.

Histology

Indications for renal biopsy
Since the definition of idiopathic nephrotic syndrome is both clinical and histological, renal biopsy is theoretically required to make the diagnosis. In fact, the indications for a renal biopsy depend on age and clinical features.

Children
A renal biopsy is not indicated at onset of the disease in a child 1 to 8 years old with typical symptoms, and in these cases initial treatment with corticosteroids is usually prescribed. A clear-cut, complete remission obtained following treatment is a major support for the diagnosis and confirms that a renal biopsy was not necessary. Renal biopsy is, however, indicated at onset in circumstances which suggest another type of glomerular disease (Habib and Kleinknecht 1971), including a moderate nephrotic syndrome or a long antecedent course of minor proteinuria and the presence of macroscopic haematuria and/or marked hypertension and/or renal insufficiency. A decreased plasma C3 fraction is also an indication for a biopsy. An age of under 12 months and, to a lesser extent, above 8 to 10 years is another indication for biopsy, even in patients with typical symptoms. However the main indication is failure to respond to a 4-week course of prednisone given in adequate dosage, thus demonstrating initial steroid non-responsiveness.

Renal biopsy is sometimes also proposed for steroid responders who relapse frequently. However the risk/benefit ratio is not favourable in these circumstances, since the therapeutic approach does not depend on histology (see below). A biopsy might be indicated before starting treatment with a cytotoxic drug, but this point of view is controversial (Schulman et al. 1988). Biopsy is necessary, however, before starting cyclosporin-treatment to allow possible nephrotoxicity to be assessed with a later biopsy. Finally, a renal biopsy is indicated in secondary acquired steroid unresponsiveness and in patients with a sudden deterioration in renal function not obviously explained by haemodynamic changes.

Adults
Since idiopathic nephrotic syndrome is much less common in adults, and since there is a higher risk of the presence of vascular disease a renal biopsy is usually performed before any treatment, although this remains controversial (Gault and Muehrcke 1983; Levey et al. 1987).

Light and electron microscopy
Light microscopy shows three morphological patterns in biopsies from patients with idiopathic nephrotic syndrome: minimal change disease, focal and segmental glomerular sclerosis, and diffuse mesangial proliferation. The relative incidence of these three patterns is difficult to determine, since it varies according to the bias in patient selection, the indications for renal biopsy, age of patients etc. Minimal changes are found in the majority of unselected children, and focal sclerosis occurs in only 5 to 7 per cent (Southwest Paediatric Nephrology study group 1985b). Mesangial proliferation, not always recognized in the past, is now regularly reported in a small number (3–5 per cent) of patients. These proportions are quite different in adult patients, half of whom have minimal change disease, and half of whom have focal and segmental glomerular sclerosis.

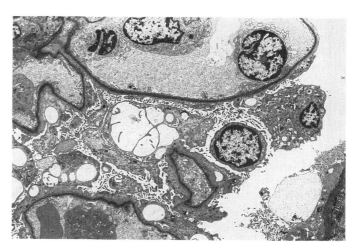

Fig. 2 Minimal change disease. The glomerular basement membrane is normal. The cytoplasm of the podocytes is vacuolated with effacement of foot processes and microvilli formation (electron microscopy. Silver methamine ($\times$ 2800)).

Minimal change disease
This is the most frequent pattern of idiopathic nephrotic syndrome in children. Glomeruli may be completely normal with normal capillary walls and normal cellularity. Mild glomerular changes, including swelling and vacuolation of epithelial cells, together with a slight increase in mesangial matrix, are usually present: however, a mild and focal mesangial hypercellularity may be noted (Churg et al. 1970; Cameron et al. 1974; Habib and Kleinknecht 1971) and morphometric methods consistently show proliferation of intercapillary cells (Bohle et al. 1974). Lipid vacuoles and degenerative changes of proximal tubules are rare. Scattered foci of tubular lesions and interstitial fibrosis may be observed: obstruction by large hyaline casts, dilatation with thinning of epithelial cells, atrophy with thickening of tubular basement membrane, interstitial foam cells, and calcium deposits. Vascular changes are absent in children (White et al. 1970; Habib and Kleinknecht 1971), whilst in adults they are age-related (Cameron 1973; Jao et al. 1973).

Electronic microscopic studies
In contrast to the paucity of glomerular changes seen on light microscopy, ultrastructural changes are always present, mainly involving podocytes and mesangial stalks. Podocyte foot process fusion, first described by Farquhar et al. in 1957, has been emphasized (Fig. 2); its extent is closely related to the degree of proteinuria (Jao et al. 1973; Powell 1976). In fact, as demonstrated by scanning electron microscopic study of rats with aminonucleoside nephrosis (Arakawa 1970; Carroll et al. 1973) there is no actual fusion, but swelling and retraction, leading to effacement of foot processes in such a way that an almost continuous cytoplasmic layer covers the glomerular basement membrane. Other epithelial changes consist of microvillus formation, the presence of numerous protein reabsorption droplets and (less frequently) increased organelles and lysosomes. The glomerular basement membranes are mostly normal: the lamina densa is of normal width and structure, the lamina rara interna is irregularly thickened by an electronlucent, fluffy material (Grishman and Churg 1973; Jao et al. 1973). No parietal deposits are present. The endothelial cells are often swollen (Kawano et

al. 1971). Intracapillary fibrin formation and platelet aggregates may occasionally be seen (Jao *et al.* 1973).

Mesangial alterations include mesangial cell hyperactivity (Kawano *et al.* 1971) increased mesangial matrix, and occasionally finely granular, osmiophilic deposits located along the internal side of the basement membrane. These ultrastructural alterations are non-specific and are probably related to massive proteinuria. Repeat biopsies demonstrate their progressive disappearance after remission (Hopper *et al.* 1970).

Focal glomerular sclerosis

This is characterized by the presence of glomerular lesions of focal distribution affecting only some of the glomeruli. It has been found as early as one or a few months after clinical onset of nephrosis (Churg *et al.* 1970; Habib and Gubler 1971, 1975; Habib 1973).

Two main patterns may be seen in the affected glomeruli: in the first, the focal changes are limited to only part of the involved tuft, the other capillary loops showing no modification. The term 'focal and segmental glomerular hyalinosis and/or sclerosis' has been proposed for this type (Habib and Kleinknecht 1971; Habib and Gubler 1971). In the other, the whole of the involved tuft is completely sclerosed: the terms focal and global sclerosis or 'focal fibrosis' seem appropriate. In both these forms diffuse mesangial proliferation may be associated with the focal lesion (Hayslett *et al.* 1969; Habib and Gubler 1971; 1975; Habib 1973).

Focal and segmental glomerular sclerosis and/or hyalinosis

Different names have been proposed for this histological type, including focal glomerular sclerosis and focal sclerosing glomerulonephritis (or glomerulopathy) with segmental hyalinosis (McGovern 1964; Churg *et al.* 1970; Hyman and Burkolder 1974).

The segmental lesion of focal and segmental glomerular sclerosis affects a few capillary loops which stick together either at the hilum or at the periphery of the tuft, often at both. Recent reports (Howie and Brewer 1984; Ito *et al.* 1984) have stressed that the location of these sclerotic lesions has prognostic significance: the clinical course is considered more benign when they are peripheral (the so-called tip-lesion), although this opinion is not shared by others (Poucell *et al.* 1985). Hyaline material is often present within the sclerosed areas, appearing either as a peripheral rim or as rounded deposits obstructing lumens (Fig. 3). In addition, foamy endothelial cells and lipid inclusions may be found in some instances. At the periphery of these sclerotic segments there is, in most cases, a clear 'halo' zone (Fig. 4). The segmental lesion has a different aspect depending on whether it affects a group of capillary loops free in Bowman's space or is adherent to Bowman's capsule. The 'free' sclerotic segments are always surrounded by a 'crown' of severely altered podocytes, which are either flat or hypertrophied with a blueish cytoplasm often containing vacuoles. The podocytes are closely packed together forming a continuous layer overlying the damaged areas of the tuft and in close apposition to the clear 'halo'. When the sclerosing lesion is adherent to Bowman's capsule, podocytes are no longer indentifiable and there is a direct synechia between the collapsed capillary loops covered by the amorphous 'halo' and Bowman's basement membrane.

In some instances, all types of lesions and all grades of involvement, from discrete segmental hyalinosis to complete glomerular obliteration, are observed. The lesions always predominate in

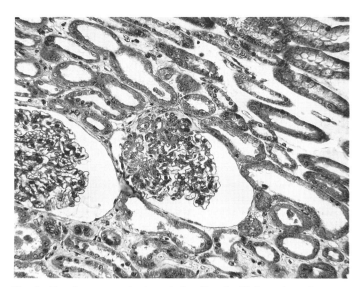

Fig. 3 Focal segmental sclerosis/hyalinosis. Obliteration of capillary lumens at the vascular pole of the glomerulus by a combination of sclerosis and hyalinosis (light microscopy. Trichrome (× 180)).

Fig. 4 Focal segmental sclerosis/hyalinosis. Segmental lesion of the tuft characterized by the deposition of hyaline material in the inner side of glomerular basement membrane, and a ring of detached podocytes separated from the glomerular basement membrane by a clear 'halo' (light microscopy. Trichrome (× 370)).

the deeper cortex and juxtamedullary glomeruli are mainly affected (Rich 1957). The rest of the tuft and the other glomeruli show only 'minimal changes'. However, as already mentioned, moderate diffuse mesangial hypercellularity may be present, and tubular atrophy and interstitial fibrosis are often present: this seems to be proportional to the glomerular damage (Habib and Gubler 1971; Hyman and Burkholder 1973; Janis *et al.* 1974; Newman *et al.* 1976). Focal glomerular lesions should therefore be suspected when focal tubular and interstitial changes are found associated with minimal glomerular changes. Foam cells may be seen in the interstitium, and occasionally in the glomeruli. In rare instances tubular lesions predominate (Hayslett *et al.* 1969; Habib and Gubler 1971). Vascular changes are rare in children, although subendothelial hyalinosis of afferent arterioles may be observed (Habib and Gubler 1971; 1975; Hyman and Burkholder 1974). In adults, they increase in severity with the age of patients (Jao *et al.* 1973; Cameron *et al.* 1973; Saint Hillier *et al.* 1975).

Electron microscopy studies

Electron microscopy demonstrates that the capillary obliteration seen in the affected segments is mainly due to the presence of paramesangial and subendothelial, finely granular, osmiophilic deposits (Nagi *et al.* 1971; Velosa *et al.* 1975; Hyman and Burkholder 1974; Janis *et al.* 1974; Newman *et al.* 1976) with either disappearance or swelling of endothelial cells, and an increase in mesangial matrix material. Fatty vacuoles may be seen, either in the middle of the abnormal deposit or in the cytoplasm of endothelial and mesengial cells. The peripheral synechia, located between podocytes and basement membrane, is formed by the apposition of a cloudy acellular material in which thin and irregular layers of newly formed basement membranes are visible (Fig. 5). Sclerosing lesions result from a marked increase in basement membrane-like material with wrinkling of capillary walls, leading to capillary collapse (Rumpelt and Thoenes 1974). Collagen fibres are identifiable within some segmental areas. Diffuse effacement of foot processes is confirmed by all investigators, and diffuse mesangial alterations are always present. Modifications of the podocytes consist of focal cytoplasmic degeneration, breakdown of cell membranes, and detachment of epithelial cells from basement membranes, with filling of resulting space by cell debris and new membranes (Grishman and Churg 1975). Such podocyte damage has been reported in heroin-associated nephropathy, but is also observed with a milder degree in some patients with nephrosis and focal and segmental glomerular sclerosis.

Another strong argument for the single-disease concept of idiopathic nephrotic syndrome is the appearance of the non-sclerotic glomeruli on electron microscopy. Yoshikawa *et al.* (1982) found no difference in the non-sclerotic glomeruli of patients with idiopathic nephrotic syndrome, whatever the light microscopic pattern—whether minimal change disease, focal and segmental glomerular sclerosis or focal global fibrosis, with the exception of clusters of electron-dense round microparticles and microfilaments, which were found in all three types, but were more common in patients with focal segmental sclerosis. These observations, in agreement with previous studies (Rumpelt and Thoenes 1974; Velosa *et al.* 1975; Newman *et al.* 1976), are consistent with common pathogenetic factors operating at different intensity in minimal change disease and focal and segmental glomerular sclerosis.

Focal and segmental glomerular sclerosis is an irreversible scarring process in the glomeruli, as demonstrated by morphological analysis of repeat biopsies (McGovern 1964; Grishman and Churg 1973; Hayslett *et al.* 1973; Hyman and Burkholder 1973; Jao *et al.* 1973; Janis *et al.* 1974; Velosa *et al.* 1975; Nash *et al.* 1976). It is usually regarded as a progressive lesion leading to chronic end-stage glomerulonephritis. But, except for Saint Hillier *et al.* (1975) who observed histological aggravation in a few apparently cured patients, extension of segmental lesions has always been documented in association with a persistent nephrotic syndrome.

Focal and segmental glomerular sclerosis is not a specific histopathological lesion: similar alterations may be seen in persistent idiopathic proteinuria, heroin-associated nephropathy (Rao *et al.* 1974; Grishman *et al.* 1976), Alport's syndrome, hypertension, pyelonephritis, etc. It has also been reported in renal hypoplasia with oligo-meganephronia (Habib and Gubler 1975) and in ageing rats (Couser and Stilmant 1975; Elema and Arenda 1975) or after partial nephrectomy (Lalich *et al.* 1975; Shimamura and Morrison 1975).

Studies in experimental animals (Couser and Stilmant 1975) as well as in nephrotic patients have shown that proteinuria precedes development of focal sclerotic lesions. Therefore, focal and segmental glomerular sclerosis appears not to be the cause of proteinuria, but rather a complication.

Focal glomerular fibrosis

This pattern is less common than focal and segmental glomerular sclerosis and is characterized by the coexistence of optically normal glomeruli and sclerosed glomeruli. The affected glomeruli are shrunken into small wafer-like formations of low cellularity, positive for collagen stains, and irregularly disseminated over the renal parenchyma. Only those patients in whom at least 15 to 20 per cent of glomeruli are sclerosed, in association with conspicuous interstitial and tubular damage, should be considered as significant (Habib and Gubler 1975).

In these cases it is possible to differentiate such atrophic glomeruli from 'congenital glomerulosclerosis' which is a developmental anomaly frequently found in the kidneys of infants and young children (Friedli 1966; Kohaut *et al.* 1976) and which is not associated with tubulointerstitial changes.

In normal adults, the number of sclerosed glomeruli increases with age (Kaplan *et al.* 1975) and thus their presence in biopsies from nephrotic patients older than 40 years is difficult to interpret (Cameron *et al.* 1974*b*).

Diffuse mesangial proliferation

Although it is not always easy to know where to draw the line between 'mild' and 'marked' mesangial hypercellularity, it appears from a number of published series (Churg *et al.* 1970; White *et al.* 1970; Habib and Kleinknecht 1971; Waldherr *et al.*

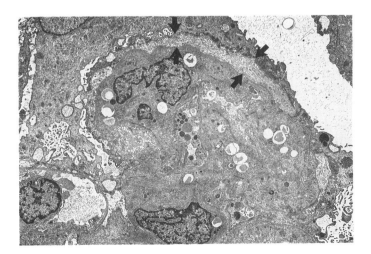

Fig. 5 Focal segmental sclerosis/hyalinosis. Note the presence around the collapsed capillary-loops of multilayered basement membrane material in a subepithelial location (arrows) (electron microscopy. Uranyl acetate-lead citrate ($\times$ 2380)).

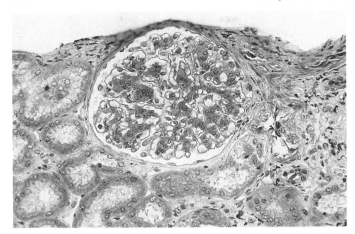

Fig. 6 Diffuse mesangial proliferation. The glomerular basement membrane is normal but there is an increase in the number of mesangial cells and in the amount of mesangial matrix (light microscopy. Trichrome (× 212)).

1978) that there is a group of patients presenting with all the features of idiopathic nephrotic syndrome who show a marked increase in mesangial matrix associated with hypercellularity (Fig. 6), a pattern usually associated with subsiding acute glomerulonephritis. However peripheral capillary walls are normal, and neither electron microscopy nor immunofluorescence shows the presence of humps or other immune aggregates.

Immunofluorescence studies

Classically, immunofluorescent microscopic examination of renal biopsy specimens from patients with idiopathic nephrosis reveals no immune material (Hopper et al. 1970; Hayslett et al. 1973; Jao et al. 1973; Michael et al. 1973; Lim et al. 1974), except in the focal lesions of segmental sclerosis and/or hyalinosis, which may strongly fix anti-IgM serum and weakly anti-C3 and anti-IgC (Rumpelt and Thoenes 1973, 1974; Janis et al. 1974; Habib and Gubler 1975; Velosa et al. 1975; Newman et al. 1976). The focal lesion often (but not always) binds anti-β-lipoprotein serum. In some reports minimal to moderate granular deposits of IgM, IgG, C3, and, more rarely, IgA may be observed in patients with minimal change disease without any relationship with response to therapy, suggesting that these deposits are non-specific and have no pathogenetic role (Prasad et al. 1977).

Fibrin/fibrinogen was present in 10 of the 31 patients studied by Hyman et al. (1973), but to a much more limited extent than immunoglobulins or C3. Matalon et al. (1974) found a continuous linear staining of all glomerular basement membranes for IgG in his 10 patients, but no deposits were seen on electron microscopy, and no circulating antiglomerular basement membrane antibodies were present. During the proteinuric phase, droplets of albumin are present within the cytoplasm of podocytes and proximal tubular cells. In some cases IgM and/or C3 was found in small amounts in the mesangial stalk (Fig. 7).

IgM associated nephropathy

Cohen et al. (1978) followed by Bhasin et al. (1978), Lawler et al. (1980), Helin et al. (1982), and Kopolovic et al. (1987) proposed that patients with IgM in the mesangium should be considered as a separate entity—IgM-associated nephropathy. Based on the small numbers of patients reported, they concluded that mesangial deposits of IgM were a marker for either poor response to corticosteroids, or progression to renal failure. A large number of other studies however, have not confirmed this concept (Murphy et al. 1979; Cavallo et al. 1981; Mampaso et al. 1981; Papadopoulos et al. 1982; Vilches et al. 1982; Hirszel et al. 1984; Hsu et al. 1984; Ji-Yun et al. 1984; Pardo et al. 1984; Gonzalo et al. 1985; Bhuyan and Srivastava 1987; Andal et al. 1989). Recently, immunofluorescence studies in a large series of children with idiopathic nephrotic syndrome confirmed that even though IgM was the immunoglobulin most frequently found in the glomeruli (54 of 222), there was no correlation between these deposits and initial response to therapy or final outcome (Habib et al. 1988). IgM deposits have also been described in association with diffuse mesangial proliferation (Bhasin et al. 1978), but the study of Habib et al. (1988) showed that this immunofluorescence pattern was not correlated with a specific histopathological category, and was not associated with electron-dense deposits on electron microscopy. These results do not support the concept that these deposits represent immune complexes. In fact, IgM is a large molecule that could be trapped in an injured glomerulus, and it is likely that mesangial IgM deposits, which are a frequent finding in a variety of situations (Ji-Yun et al. 1984) represent an epiphenomenon which is likely to be the consequence of altered mesangial function (Michael et al. 1980).

IgG 'deposits'

The second most frequently found immunoglobulin in the mesangium is IgG, with or without C1q. Again, Habib et al. (1988) found no correlation between IgG immunofluorescence and response to steroids or final outcome. In these particular cases, electron-dense deposits were observed and it is conceivable that these might be related to the IgG immune complexes described by Levinsky et al. (1978) in idiopathic nephrotic syndrome.

The scattered granules of isolated C3 reported in 15 of 222 patients by Habib et al. (1988) were associated with a non-significant poorer outcome, especially in patients with focal lesions.

The lack of specificity of these various immunofluorescence patterns in idiopathic nephrotic syndrome is further emphasized by the results of sequential biopsies. Ten of 22 repeat biopsies reported by Habib et al. (1988) exhibited a transformation of the initial pattern.

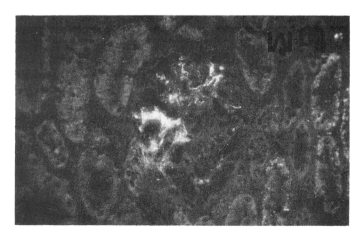

Fig. 7 Minimal change disease. Mesangial deposits of IgM are present in most mesangial areas of the glomerulus (immunofluorescence microscopy (× 165)).

IgA

A number of patients with nephrotic syndrome and IgA in the mesangium have been reported (see Chapter 3.6). They were classified by some authors as Berger's disease (Abreo and Wen 1983; Mustonen *et al.* 1983; Furuse *et al.* 1985; Lai *et al.* 1989). Other authors believed that nephrotic syndrome and IgA nephropathy were associated, especially when macroscopic haematuria developed (Southwest Pediatric Nephrology Study Group 1985*a*; Rambausek *et al.* 1987).

Habib *et al.* (1988), Sinnassamy and O'Regan (1985), Barbiano di Belgiojoso *et al.* (1986) and Hervé *et al.* (1984) consider the presence of mesangial IgA in this setting to be an incidental finding.

In conclusion immunofluorescence studies in patients with idiopathic nephrotic syndrome are usually negative, and if a particular mesangial fixation pattern is found there is no evidence to suggest that it represents a distinct clinicopathological entity, and no reason to modify the standard treatment protocol.

Relationship between the different histological patterns

All important series of idiopathic nephrotic syndrome include cases with morphological transition between the three main histological patterns in repeat biopsies. A number of patients with minimal changes on a first biopsy containing adequate glomeruli and a view of the corticomedullary area, had focal and segmental glomerular sclerosis on a second biopsy (Habib and Gubler 1975). These patients usually have more severe disease, either steroid-resistant or highly steroid dependent. A retrospective morphometric study of the first renal biopsy of 10 patients with initial apparent minimal change disease and who subsequently developed focal and segmental glomerular sclerosis found that these patients had significant hypertrophy of their glomeruli in comparison with age-matched patients and controls (Fogo *et al.* 1990). Transition from minimal change to focal and segmental glomerular sclerosis has also been observed in one case of Hodgkin's lymphoma with nephrotic syndrome, while the association between minimal change disease and Hodgkin's disease is well established (Watson *et al.* 1983).

Focal and segmental glomerular sclerosis developed in 60 per cent of 48 patients, in association with aggravation of symptoms (Tejani 1985). The progression of minimal change to sclerosis may occur directly or through stages of mesangial proliferation (Waldherr *et al.* 1978; Hirszel *et al.* 1984).

Conversely, some patients who have a diffuse mesangial proliferation pattern, whether or not associated with focal sclerosis on the first biopsy, may lose their hypercellularity with time, and show minimal change or focal and segmental glomerular sclerosis on the second biopsy (Southwest Pediatric Nephrology Study Group 1985).

In conclusion it may be considered that, at least in children, minimal change disease, focal and segmental glomerular sclerosis, and diffuse mesangial proliferation represent histological variations of idiopathic nephrotic syndrome which may be found alone or in any combination on sequential biopsies in the same patient (Fig. 8).

Differential diagnosis

From a clinical point of view, a frequent mistake is to relate eyelid or facial oedema to an allergic reaction, thus postponing a search for proteinuria. The diagnosis may be more difficult when allergy and nephrosis develop simultaneously.

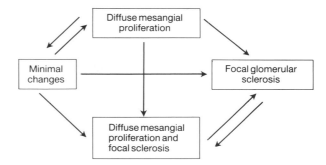

Fig. 8 The different histological patterns of idiopathic nephrotic syndrome and their possible sequential evolution.

Idiopathic nephrotic syndrome may occur in the first year of life, but is rare, and it may be difficult to distinguish this (at least at onset) from other glomerular diseases occurring at this age, particular the congenital nephrotic syndrome of Finnish type and diffuse mesangial sclerosis.

Congenital nephrotic syndrome (see Chapter 18.4.4) is commonly observed in Finland but also in other countries and is an inherited autosomal recessive disease which is present before birth. Prenatal diagnosis is possible, based on the non-specific increase of alpha-fetoprotein in amniotic fluid and in maternal plasma, and/or an abnormally large placenta. Profuse proteinuria and a severe nephrotic syndrome are present from the first days of life. There is no specific histological pattern, and although microcystic dilatation of proximal tubules usually develops after several weeks or months, this seems to be a secondary phenomenon. There are no definite histopathological criteria which allow certain differentiation of the idiopathic nephrotic syndrome from the congenital nephrotic syndrome of Finnish type (Sibley *et al.* 1985), which seems to be related to a defect in the structural proteins of glomerular basal membranes associated with a decrease of heparan sulphate (Vernier *et al.* 1983) and a loss of negative charges. This has also been suspected in idiopathic nephrotic syndrome (Vermylen *et al.* 1989) (see below).

Diffuse mesangial sclerosis is another glomerular disease which, in young children, is associated with a severe nephrotic syndrome. This clinical entity can be confused with congenital nephrotic syndrome of Finnish type because of its early onset, but differs in its rapid progression to end-stage renal failure and by the characteristic histopathology. However, it may be also be confused with idiopathic nephrotic syndrome when it develops after several months of life, and if renal biopsy is performed at the very beginning of the disease. In the early stages, light microscopy shows glomerular lesions which are characterized by a fibrillar increase. The mesangial matrix expands with no increase in number of mesangial cells and no apparent abnormality of the capillary walls, but with a widespread hypertrophy of the podocytes. The fully developed lesion is characterized by a combination of thickened basement membranes and a spongy appearance of expanded mesangial zones. The mesangial cells are embedded in a delicate network which stains positive with periodic acid-Schiff and silver stains. The capillary lumens are more or less obliterated by the accumulation of mesangial matrix. In advanced stages the tuft appears as a solidified mass in a urinary space which is often dilated, and a layer of hypertrophied and vacuolized podocytes surrounds the tuft. These various stages may coexist in the same specimen with a corticomedullary gradient, the deepest glomeruli being least altered, in contrast to focal and segmental glomerular sclerosis. Impressive tubular lesions are always present.

Electron microscopy shows mild hypertrophy of endothelial cells, with markedly hypertrophic mesangial cells surrounded by an abundant mesangial matrix which contains collagen fibrils. There is a segmental splitting of the glomerular basement membrane with a wavy appearance of its external contour. Immunofluorescence reveals mesangial deposits of IgM and C1q in the least affected glomeruli, and deposits of IgM and C3 outline the periphery of the sclerosed glomeruli.

The nephropathy is usually discovered in the first year of life, but some patients are diagnosed between 1 and 3 years. In more than one-third of patients reported by Habib et al. (1985), a diagnosis of Drash's syndrome was made due to the association of male pseudohermaphrotidism and or a Wilms' tumour with the nephropathy (see Chapter 3.13), and this suggested that children with isolated diffuse mesangial sclerosis still have Drash's syndrome with either absent or undiagnosed gonadal dysgenesis and with a potential risk of developing a Wilms' tumour (Habib et al. 1989) (see Chapters 3.13 and 19.2).

Several infectious or parasitic diseases may cause nephrotic syndrome in the first month of life: congenital syphilis must be systematically eliminated by the specific serology, but is usually recognized on other clinical symptoms. Secondary syphilis is associated with a membranous nephropathy, not always obvious by light microscopy, but recognized as electron-dense deposits on the epithelial side of the glomerular capillary wall.

Nephrotic syndrome occurring in African children may be related to a peculiar nephropathy which has been associated with quartan malaria in Uganda and Nigeria (Hendrickse and Adenyi 1979), but not in Ivory Coast and Senegal (de Paillerets et al. 1972; Morel-Maroger et al. 1975) (see Chapters 3.6 and 3.12). Histological examination shows a characteristic thickening of the glomerular capillary wall, segmental in the early stage and affecting more and more glomeruli with time. The thickening involves the subendothelial area, consisting of a double contour or plexiform arrangement of argyrophilic fibrils and immunofluorescence reveals either coarse granular deposits of IgG, IgM, and C3 along the capillary wall, or very small deposits of IgA homogeneously distributed along the capillary wall. This nephropathy does not usually respond to steroids and has a poor prognosis. The aetiology and pathogenesis of this nephropathy are not clear at the present time.

At any age, if there is no response to corticosteroids a renal biopsy must be performed which easily permits distinction between idiopathic nephrotic syndrome and other glomerular diseases which may be associated with heavy proteinuria and nephrotic syndrome, such as membranous nephropathy, idiopathic membranoproliferative glomerulonephritis, or other type of glomerulopathy, such as amyloidosis, diabetic nephropathy, or systemic lupus.

Focal segmental glomerular sclerosis may be observed in a number of clinical situations, including hypoplastic kidneys and any other states where there is a reduction in nephron numbers, including reflux nephropathy, obstructive uropathy or advanced glomerulonephritis. This lesion has also been reported in heroin addicts and, independently, in association with AIDS. Focal segmental glomerular sclerosis has also been reported in patients with a chronic proteinuria below the nephrotic range and a possible evolution to end-stage renal failure (Saint Hillier et al. 1975; Yoshikawa et al. 1986). This nephropathy seems to be more frequent in black people (Bakir et al. 1989). We do not know whether any of these situations share common features with the focal and segmental glomerular sclerosis which is observed in idiopathic nephrotic syndrome.

In adults, the nephrotic syndrome has been described in association with renal vein thrombosis, and it was formerly believed that the latter produced the nephrotic syndrome. A number of careful and fully documented observations have definitely shown that thrombosis is the consequence and not the cause of nephrotic syndrome (Heptinstall 1983) (see Chapters 3.4 and 3.7).

Massive proteinuria, not usually associated with nephrotic syndrome, has been reported in adults with extreme obesity and in association with renal artery stenosis. Proteinuria disappears after weight reduction or relief of the stenosis.

Pathophysiology

Mechanisms of proteinuria

As described elsewhere (Chapter 3.1), the impermeability of the glomerular capillary wall to serum albumin rests in part upon the presence of anionic (negative) charges on the endothelium, the glomerular basement membrane and the foot processes. These negative charges repulse the negatively charged serum albumin molecules, whose isoelectric point is of the order of 4.6. Theoretically, albuminuria in the absence of visible lesions of the glomerular capillary walls might be due to loss of the glomerular polyanion, less anionic serum albumin, or both. In any of these situations the electrochemical derangement could be due to a circulating factor capable of increasing the positive charge of the structures discussed above. Each of these hypotheses has been studied.

Loss of glomerular polyanion in minimal change disease

In minimal change disease the contrast between virtual absence of histological lesions and greatly increased glomerular permeability to serum albumin has long suggested that the disease is mostly functional, and due to an electrochemical disorder of the glomerular membrane. This hypothesis has been substantiated by the demonstration that the glomerular K_f is diminished despite an increased permeability to serum albumin. Robson et al. (1974) used dispersed polyvinylpyrrolidone as a test macromolecule in nephrotic children and found a reduction in fractional clearances, indicating that in this condition the notional 'pore' size of the barrier was smaller than normal. Carrie et al. (1981) investigated adult patients with minimal change disease by infusing uncharged dextrans with Einstein-Stokes radii between 2.0 and 4.8 nm. They confirmed that the fractional clearance of macromolecules similar in size to albumin (that is, with a molecular radius of 3.6 nm) was reduced. The discrepancy between this apparent reduction in 'pore' size and massive albuminuria again suggested loss of glomerular negative charges. This was confirmed indirectly by renal histopathology: Carrie et al. studied renal biopsy sections stained by the colloidal iron reaction and showed that glomerular uptake of colloidal iron was severely reduced in five cases and less reduced in three. They considered that this was evidence for reduced sialic acid content in the glomerular basement membrane, as sialic acid residues are the major chemical group responsible for glomerular negative charges (Blau and Haaz 1973).

More recently, Vermylen *et al.* (1989) found evidence tending to indicate that diminished glomerular basement membrane content of heparan sulphate, more than sialic acid alterations, could be responsible for the abnormal glomerular permeability in congenital nephrotic syndrome. (In fact, heparan sulphate constitutes 59 per cent of the glycosaminoglycans in the normal infant.) No such studies are available in minimal change disease focal and segmental glomerular sclerosis, which precludes extrapolation of Vermylen's observations to acquired idiopathic nephrotic syndrome, but whatever the intimate biochemical basis of the phenomenon, the weight of the evidence is in favour of diminished glomerular polyanion as the basis of massive albuminuria in minimal change disease (see also Chapter 3.1).

Cationic serum albumin fractions in minimal change disease

Two reports (Ghiggeri *et al.* 1987a, b) showed that at the onset of disease in proteinuric children urinary albumin is detectable as one band with an isoelectric point (pI) of 4.7 and by several others with a pI between 4.8 and 5.5, accounting for about 50 per cent of the total amount of urinary albumin. The investigators examined an index of conformation of the albumin molecules by determining the fluorescence quantum yield of tyrosine and tryptophan. In the proteinuric phase the fluorescence quantum yield of tryptophan of urinary albumin was markedly quenched, but returned to near normal values after steroid-induced remission of proteinuria. The authors concluded that charge and conformation of albumin were modified in idiopathic nephrosis, explaining both modifications. They amplified these observations by studying the fatty acid content of the cationic fractions of albumin in nephrotic children. Their data indicated that the pI of urinary albumin was a function of its fatty acid content, and that only partially defatted isoalbumins were present in the urine. They suggested that the inability of fatty acids to react with cationic isoalbumins was brought about by competitive mechanisms with some substance(s), or an 'inaccessibility' of the fatty acid-containing pocket of the protein.

In 1989, Levin *et al.*, using a method to detect charge neutralizing substances based on inhibition of binding of the cationic dye Alcian blue 8GX to heparin, plasma, and urine from children with steroid-responsive nephrotic syndrome, found a charge-neutralizing factor. This substance was purified by ion exchange chromatography, gel filtration and FPLC and found to be protease sensitive, as pronase reduced its heparin neutralizing activity. Levin *et al.* postulated that the basic defect in steroid-responsive nephrotic syndrome might be an excessive or uncontrolled release of cationic proteins from some cellular component of the immune or acute phase response system.

These fascinating observations suggested that serum albumin could be the villain, as much as the victim of glomerular basement membrane hyperpermeability—a return to the idea of 'dyscrasic proteinuria' itself, popular earlier in this century. Nonetheless, loss of glomerular polyanion and increased positive charge of some fractions of serum albumin are not mutually exclusive: both could be due to a humoral factor inducing a generalized electrochemical disorder responsible for the nephrotic syndrome. The hunt for such a factor was started 15 years ago, but although tracks of the game have been found, no hunter has yet cornered the fox.

The search for a 'vascular permeability factor' in minimal change disease

The quest for a humoral factor responsible for loss of anionic charges on glomerular structures (and perhaps on serum albumin molecules) was undertaken with particular conviction after the suggestion by Shalhoub (1974) that a lymphokine might be this factor. Lagrue *et al.* (1975) injected a supernatant of lymphocytes cultured from patients with nephrotic syndrome into the renal artery of rats. The rats became proteinuric and the authors concluded that the lymphocytes had elaborated a cytokine capable of increasing glomerular permeability. In another experiment, Wilkinson *et al.* (1989) injected the plasma from patients with minimal change disease into the aorta of rabbits, and observed a decrease of anionic sites in the glomerular basement membrane and the simultaneous occurrence of proteinuria. The Ovary test was also used, in which the supernatant of lymphocyte cultures is injected intradermally into the shaved abdominal skin of guinea pigs. The animals are then injected with Evans' blue. The diameter of the blue area surrounding each injection site is measured and its surface area is considered proportional to vascular permeability and hence to the presence of vascular permeability factor. Using this test, Heslan *et al.* (1986) and Tomizawa *et al.* (1985) showed increased vascular permeability in patients with untreated idiopathic nephrotic syndrome. Tomizawa *et al.* also showed that plasma from patients with active minimal change nephrotic syndrome markedly inhibited the test, compared with plasma taken from 'inactive' cases. Their conclusion was that lymphocytes from these patients indeed secreted a substance responsible for increased glomerular permeability. Unfortunately, apart from its rather elementary nature, the test was also positive in other sorts of glomerulopathies, including focal and segmental glomerular sclerosis, but also membranous glomerulonephritis, membranoproliferative glomerulonephritis, and IgA nephropathy (Bakker *et al.* 1982; Heslan *et al.* 1986).

Two series of experiments using other models were more convincing, and showed directly that cytokines secreted by lymphocytes of patients with idiopathic nephrosis, or even the white cells themselves, were able to suppress negative charges of glomerular structures.

Boulton-Jones *et al.* (1983) infused supernatants of lymphocyte cultures into the renal circulation of rats and immediately processed the kidneys for light and electron microscopy. The quality of the glomerular polyanion was estimated by colloidal iron staining. Infusion of lymphokine from six of seven patients reduced colloidal iron staining, and electron microscopy confirmed loss of negative charges, with fusion of epithelial cell foot processes.

Bakker *et al.* (1986) studied 17 nephrotic patients with minimal change disease aged from 5 to 57 years, in parallel with 14 patients with other nephrotic syndromes and with control subjects without nephropathy. Monocytes, stimulated with concanavalin A or unstimulated, were cultured with slices of rat kidneys. Glomerular sialoproteins were then stained by colloidal iron. The kidney slices cultured with concanavalin A-stimulated monocytes from minimal change disease patients exhibited diminished fixation, reflecting reduced glomerular negative charges. This technique seemed clearly more specific than the Ovary test, as positive results were observed in 15 of 17 (88 per cent) patients with minimal change disease, but only four of 14 (28 per cent) patients with other types of glomerulonephritis and three of 18 (16 per cent) controls. The fact that direct contact of monocytes with renal tissue yielded such convincing modification of the glomerular polyanion suggested that the 'circulating factor' which modifies selective glomerular permeability in minimal change disease is more likely to be derived from activated lymphocyte subpopulations than a humoral substance present in the serum.

Minimal change disease: a generalized disorder of membrane negative charges?

Since the beginning of this century, every clinician faced with a patient with nephrosis and generalized oedema, from puffy face to swollen feet has wondered whether this condition is different from other oedematous states, in that fluid seems to ooze from every capillary in the body. Attempts to demonstrate increased protein concentration in fluids other than urine, such as cerebrospinal fluid or saliva, have yielded inconclusive results. Such a concept of a generalized disorder was revived when Levin et al. (1985) and Boulton-Jones et al. (1986) presented data which indicated that the loss of negative charges was not restricted to the glomeruli but could also be found on other cell membranes. Levin et al. (1985) used a simple chemical test based on the binding of the cationic dye Alcian blue to measure negative charge on blood cell membranes. They studied 17 children with relapsing nephrotic syndrome responsive to steroids, during 26 relapses, and six children with steroid-resistant nephrotic syndrome due to focal and segmental glomerular sclerosis. Normal children, normal adults, and children with other renal diseases were used as controls. Alcian blue binding to red blood cells and platelets was significantly less in the nephrotic children than in controls. However, the sialic acid content of the red blood cells measured after neuraminidase treatment was similar in both nephrotic and normal children. These data suggested that a generalized loss of membrane negative charges occurred in nephrosis and that this was due to neutralization, rather than absence, of anionic groups. Boulton-Jones et al. (1986) carried out a similar study on red cells of patients with different glomerulopathies. They found that the charge was significantly less in patients with minimal change nephropathy and membranous nephropathy than in patients with IgA nephropathy. An interesting point was that the charge was still less when blood was taken during clinical remission. Hence, the authors inferred that the differences probably represented a 'premorbid' state and postulated that the red cell charge reflected the glomerular polyanion.

Two groups of investigators failed to reproduce Levin's experiments and cast doubt on the above data, claiming that some methodological problems existed in the assay used for determining the red cell charge, namely partial insolubility of Alcian blue (Feehally et al. 1986; Sewell and Brenchley 1986). Cohen et al. (1988) could show no differences in charge on nephrotic platelets by direct electrophoretic measurement. In order to clarify the issue, Ginevri et al. 1989 studied the structural composition of erythrocyte ghosts sampled from children with either steroid-responsive or steroid-resistant nephrotic syndrome. They found that the composition of inner layer phospholipids was altered in steroid-responsive nephrosis. Signs of peroxidative damage were present in ghosts from both steroid-responsive and steroid-resistant children. Ginevri et al. concluded that, taken together, these results supported the concept that erythrocytes are target cells for peroxidative damage in nephrosis. One cannot but accept the final statement of this astute group that 'whether similar alterations also involve renal structures and induce proteinuria is a fascinating topic for future studies'.

The immune system in idiopathic nephrotic syndrome

There are several arguments which incriminate immunological abnormalities in the pathogenesis of the disease. It must be kept in mind that it is often not known whether a given immunological abnormality is specific for the disease or rather a result from the patient's nephrotic state, and may also be present in other forms of nephrotic syndrome—for example that arising from membranous nephropathy.

Role of cellular immunity

As noted above, Shalhoub (1974) postulated that idiopathic nephrotic syndrome might be secondary to a disorder of T-lymphocyte function. This proposal was consistent with the dramatic response of the disease to corticosteroids and alkylating agents, the remission occurring in association with measles, which depresses cell mediated immunity (Lin and Hsu 1986), and the occurrence of minimal change nephrotic syndrome in patients with Hodgkin's disease. Several authors have looked for abnormalities of cellular immunity in order to address Shalhoub's hypothesis. Peripheral blood T-lymphocyte subpopulations, defined using monoclonal antibodies against differentiation antigens CD3, CD4 and CD8, do not generally show any abnormalities, and the total number of T and B lymphocytes is not different from controls (Cagnoli et al. 1982; Herrod et al. 1983). However, an increased percentage of suppressor T cells has been reported in patients with minimal change nephrotic syndrome (Sasdelli et al. 1981; Dall'Aglio et al. 1982).

More important are studies of the activity of these cells. Lymphocytes from patients with idiopathic nephrotic syndrome have impaired responses to polyclonal activators such as concanavalin A or phytohaemagglutinin (Schulte-Wisserman et al. 1977; Iitaka and West 1979; Minchin et al. 1980; Sasdelli et al. 1981; Taube et al. 1981a; Taube et al. 1984). This decreased response is partly reversible when normal human serum is added to the culture medium instead of autologous serum, suggesting both an intrinsic defect of the cells and the presence of inhibitory factors in the serum. Macrophages may be responsible for the impaired response of mononuclear cells to mitogens. Taube et al. (1982) showed that when monocytes were removed by physical means, the response of lymphocytes to concanavalin A was identical to that of monocyte-depleted mononuclear cells from normal donors. The addition of indomethacin also partially restores mitogen responsiveness, suggesting a role for prostaglandins secreted by macrophages. The proliferative response of mononuclear cells returns to normal when the patients are in remission.

Abnormalities of T-suppressor cell function have been described. Two studies found an increase in T-suppressor cell activity in patients during relapses using an assay with concanavalin A-activated lymphocytes (Wu and Moorthy 1982; Matsumoto et al. 1982). This abnormality was reversible after steroid treatment (Osakabe and Matsumoto 1981). However, Taube et al. (1981b) found a prolonged impairment of suppressor cell activity in patients treated with cyclophosphamide. Interestingly, suppressor cell activity increased in patients who relapsed following cyclophosphamide therapy, suggesting that this abnormality may have a pathogenic role. However, Feehally *et al.* (1983) did not find the same impairment of suppressor cell activity following cyclophosphamide treatment.

Other abnormalities of cell-mediated immunity have been described in minimal change nephrotic syndrome. Mallick *et al.* (1972) reported that peripheral blood lymphocytes from patients with idiopathic nephrotic syndrome inhibit leucocyte migration when they are incubated with kidney antigens. Eyres *et al.* (1976) found that lymphocytes from patients with minimal change nephrotic syndrome display a cytotoxic activity when they are incubated with kidney tubular cells. Decreased skin reactivity to common antigens has also been found in patients during relapses, with a return to normal during remission (Fodor *et al.* 1982; Matsumoto *et al.* 1981). Graft-versus-host reactions induced in immunosuppressed rats by intradermal injection of lymphocytes from patients in relapse are decreased (Matsumoto 1982).

Role of humoral immunity

Patients with minimal change nephrotic syndrome have depressed levels of IgG. This abnormality is more pronounced during relapses, but persists during remission which suggests that the low IgG levels are not only due to urinary loss (Giangiacomo *et al.* 1975; Brouhard *et al.* 1981; Heslan *et al.* 1982). Conversely, serum IgM levels are elevated (Giangiacomo *et al.* 1975; Ganguly *et al.* 1979). Altered serum levels of IgG and IgM may be secondary to abnormal T-cell regulation of immunoglobin synthesis (Yokoyama *et al.* 1985 1987). It has been postulated that increased concentrations of serum IgM in patients during relapse or in remission could be due to a fault in switching from IgM to IgG during the immune response (Giangiacomo *et al.* 1975). IgM synthesis by peripheral blood lymphocytes stimulated *in vitro* with pokeweed mitogen is normal but spontaneous IgM synthesis is low. Similar findings were observed for IgA and IgG (Brouhard *et al.* 1981; Beale *et al.* 1983). These data could indicate a state of spontaneous activation of patient lymphocytes. The increased susceptibility to infections, particularly to peritonitis, might be related to the low IgG concentrations (Krensky *et al.* 1982) (see Chapter 3.4). Varying titres of specific antibody against bacterial or viral antigens have been observed (Fikrig *et al.* 1978; Wilkes *et al.* 1979; Fodor *et al.* 1982; Spika *et al.* 1982, 1986). Interestingly, specific antibodies to pneumococcal or streptococcal antigens are reduced in patients up to 20 years after remission (Lange *et al.* 1981). This suggests that patients with minimal change nephrotic syndrome have a defect of the immune response which is not directly related to the nephrotic state. However it does not imply that this abnormality is pathogenic (see also Chapter 3.4 for susceptibility to infection).

Role of immune complexes

Circulating immune complexes have been found in the serum of patients with minimal change disease by a variety of techniques (Poston *et al.* 1978; Cairns *et al.* 1982; Solling 1982). In a review of the literature, Border (1979) reported that 45 per cent of patients with minimal changes had circulating immune complexes, and Abrass *et al.* (1980) found immune complexes in 79 per cent of adults with idiopathic nephrotic syndrome. Levinsky *et al.* (1978) found IgG-containing immune complexes that did not bind to C1q in 94 per cent of patients. The pathogenic role of such immune complexes is questionable, since immunofluorescence microscopy shows no deposits in renal biopsies from most patients. Furthermore, there has been no analysis of the composition of the complexes, which may be secondary to the nephrotic syndrome since they are found during relapses and disappear after remission.

Role of complement

There is no clear evidence to suggest that the complement system plays a role in the pathogenesis of the disease. Complement levels, including C4 and C3, are usually normal. Elevated levels of immunoconglutinin may suggest *in-vivo* complement activation (Ngu *et al.* 1970). Low concentrations of C1q were found in 47 of 132 patients with minimal change disease by Lewis *et al.* (1971). Factor B levels are decreased during relapses but are normal in remission (McLean *et al.* 1977). Similarly, levels of Factor D may be decreased (Ballow *et al.* 1982) and these abnormalities in the alternate pathway of complement could be partly responsible for the increased susceptibility to infections (see Chapter 3.4). It is likely that most of the abnormalities of the complement system are secondary to the nephrotic state, particularly to the urinary loss of proteins of lower molecular weight.

Circulating factors

A number of immunological abnormalities could be secondary to circulating factors present in patients' serum. Moorthy *et al.* (1976) showed that the plasma from patients with minimal change nephrotic syndrome in relapse was able to inhibit the proliferative response of normal human lymphocytes to phytohaemagglutinin. However, this inhibition also occurred with the plasma from patients with other forms of nephrotic syndrome (Beale *et al.* 1980; Taube *et al.* 1981*a*). The mixed lymphocyte reaction and the proliferative response of lymphocytes to concanavalin A are also inhibited by sera from patients with minimal change nephrotic syndrome and other forms of nephrotic syndrome (Moorthy *et al.* 1976; Iitaka and West 1979; Beale *et al.* 1980). Different factors may be responsible for this effect of serum such as α_2-macroglobulin, lipoproteins, or immune complexes.

One of the most convincing arguments for a role of circulating factors in the pathogenesis of the disease is the possible recurrence of the nephrotic syndrome after renal transplantation. Since Shalhoub (1974) postulated that lymphokines may be responsible for the increased permeability of glomerular basement membrane, several authors have looked for lymphokines in the supernatant of cultured lymphocytes from patients. The data of Lagrue *et al.* (1975) and others were discussed in the previous section. Interestingly, a vascular permeability factor, which induced proteinuria when injected into the renal artery of rats, was found in the serum from one transplanted patient who had recurrence of the nephrotic syndrome (Zimmerman 1984), but many similar attempts to isolate such a factor have failed. Tomizawa *et al.* (1985) reported that the plasma from patients in relapse was able to inhibit the production of vascular permeability factor *in vitro* by stimulated lymphocytes, whereas the plasma from patients in remission had no such effect.

Schnaper and Aune (1985, 1987) have reported the presence of another lymphokine, which they have called soluble immune response suppressor, in the urine and serum from patients with steroid-responsive nephrotic syndrome, including minimal change nephrotic syndrome and other histopathological forms of nephrotic syndrome (membranoproliferative glomerulonephritis, membranous glomerulonephritis, or immune complex glomerulonephritis). This lymphokine is produced by suppressor T lymphocytes stimulated by concanavalin A, and inhibits both delayed hypersensitivity reactions and antibody responses. Soluble immune response suppressor was found to disappear from the urine of patients in remission after steroid therapy, although not always (Cheng et al. 1989). This suppressor factor, although not directly responsible for proteinuria, may contribute to the decreased immune responsiveness of patients with minimal change nephrotic syndrome. Recently, Schnaper (1990) found that the sera from patients with steroid-responsive nephrotic syndrome stimulated normal CD8+ cells to produce soluble immune response suppressor. Supernatants of cultured CD4+ lymphocytes from patients also activated normal CD8+ cells to produce this factor. The molecular weight of these factors was estimated to be 13 000 to 18 000 Da.

Other lymphokines may play a pathogenic role (Bakker and Van Luijk 1989). For example, increased interleukin 2 levels have been found in lymphocyte culture supernatants from patients with idiopathic nephrosis (Lagrue et al. 1988, and interleukin-2 can induce proteinuria when injected into the rat kidney (Kawaguchi et al. 1987; Heslan et al. 1988).

Histocompatibility antigens and idiopathic nephrotic syndrome

The human major histocompatibility complex on chromosome 6 consists of a series of polymorphic class I (HLA, A, B, C) and class II (HLA DR, -DQ, -DL) genes that have a major role in the immune response. Many of these antigens are linked together in haplotypes, which are inherited with varying degrees of linkage. Histocompatibility antigens are involved in the presentation of antigens to T lymphocytes.

The relationship between minimal change nephrotic syndrome and the major histocompatibility complex has been studied by different authors. An increased incidence of HLA-DR7 has been reported in France (de Mouzon-Cambon et al. 1981; Ansquer et al. 1983), in Spain (Nunez-Roldan et al. 1982), and in Australia (Alfiler et al. 1980). These authors found that HLA-DR7 is three to four times more frequent in children with the disease compared to controls and that the relative risk of having the disease was between 4.5 and 6.8. HLA DR3 has been associated with steroid-resistant nephrosis in children with a relative risk of three. Conversely, in adult patients, an association between DR4 and idiopathic nephrotic syndrome with focal sclerosis and end-stage renal failure was found in a group of 57 patients (Glicklich et al. 1988). An association with DR8 and DQW3 has been found in adult Japanese patients (Kobayashi et al. 1985).

An association with HLA-B8 was reported in England (Thomson et al. 1976), and the same study found that children with atopy and HLA-B12 had a 13-fold increased risk of developing nephrosis. An association with HLA-B8 was found in Ireland (O'Reagan et al. 1980) and in Germany (Noss et al. 1981). Other workers did not find such associations (Meadow et al. 1981).

These differences may be explained if the disease segregates with different alleles in different populations (Schnaper 1989). Apart from single allele associations, possible associations with certain haplotypes have been studied. The incidence of the phenotype DR3-DR7 is increased in steroid-resistant patients (30 per cent versus 4 per cent) with a high relative risk of 9.3 (Yokoyama et al. 1985). This phenotype is associated with early onset and lesions of focal sclerosis (Cambon Thomsen et al. 1986). Clark et al. (1990) investigated the frequencies of the major histocompatibility complex class II alleles using analysis of restriction fragment length polymorphisms DNA. They found a strong association between HLA DR7 and the DQB1 gene of HLA DQW2 and steroid-sensitive nephrotic syndrome. These authors suggest that the beta chains of DR7 and DQW2 contribute to disease susceptibility. Lagueruela et al. (1990) also found a strong association with DQW2 and steroid-responsive nephrotic syndrome. They demonstrated that two extended haplotypes (HLA-A1, B8, DR3, DRW52, SCO1) and (HLA-B44, DR7, DRW53, FC31) occur with a significantly increased incidence in these patients. It should be remembered that class I and class II glycoproteins play a key role in the presentation of the antigen to the T-cell receptor. Some antigens reacting with particular class II glycoproteins and not with others may play a role in the disease. From these studies, it can be concluded that there are probably major genetic determinants residing in the major histocompatibility region on chromosome 6 which confer a susceptibility to idiopathic nephrotic syndrome.

Allergy, IgE

The role of allergic episodes in triggering relapses of idiopathic nephrotic syndrome has long been known, and it is not surprising that some nephrologists have suggested that at least some cases of nephrosis are indeed due to allergy (Fontana 1956). Several observations indicated that respiratory allergens such as pollen may induce nephrosis. Hardwicke et al. (1959) reported the case of a patient whose nephrotic syndrome relapsed each year, corresponding to pollen-induced hayfever. In this patient renal biopsy showed normal glomeruli by light microscopy but also an interstitial infiltrate containing polymorphonuclear eosinophils. Dr James Hopper Jr from San Francisco (personal communication) mentioned a similar patient in whom the disease relapsed after each contact with poison ivy. Among a list of anecdotal cases, the respiratory allergens reported include fungi, ragweed pollen, house dust, medusa stings, and cat fur (Wittig and Goloman 1970; Reeves et al. 1975; De Montis et al. 1976; Lagrue 1982; Cochat and Larbre 1988). Relapses elicited by bee stings are also classical, although the total of reported cases hardly exceeds 20 (Cuoghi 1988).

In these cases the allergen was easily identified, but it is probably more interesting to focus on observations where a food allergen was suspected and was apparently responsible for relapsing episodes of steroid-sensitive nephrosis. Cow's milk (Sandberg 1977) and egg (Richards 1977) were shown to be responsible for skin and respiratory allergic phenomena together with massive proteinuria. The introduction of an 'elemental diet' induced remission and in some cases challenge with the allergen provoked a relapse. Lagrue et al. (1982) and Laurent et al. (1987) studied 40 cases of nephrosis and stressed the frequency of an 'allergic background'. In three-quarters of the cases there was a family and personal history of allergy. Half of these patients had high eosinophil counts, elevated IgE serum levels, and/or positive basophil degranulation tests.

More recently Laurent et al. (1987) evaluated the effect of an oligoantigenic diet given for 10 to 15 days to 13 patients with idiopathic nephrotic syndrome with unsatisfactory response to corticosteroids. Eight of them had a history of allergy, including six with increased serum IgE levels. Skin tests to food antigens showed an immediate reaction in five cases (three times to cow's milk, twice to beef, and once to wheat flour, rice, chicken, and pork). Four of five patients reacted to two foods. The oligoantigenic diet coincided with definite improvement of proteinuria in nine, including complete remission in five. Most patients relapsed when a normal diet was reintroduced.

Increased IgE serum levels were also considered to be an argument for the role of allergy, especially when associated with positive basophil degranulation tests (Laurent *et al.* 1987). Nonetheless, increased serum IgE levels have been found in various primary glomerular diseases (Laurent *et al.* 1986; Shu 1988), including not only minimal change disease and focal and segmental glomerular sclerosis, but also membranous glomerulonephritis and Berger disease. Shu (1988) indicated that higher serum IgE levels were associated with more frequent relapse or steroid resistance in minimal change disease, IgA nephropathy, and focal and segmental glomerular sclerosis.

A possible association between atopy, nephrosis, and some HLA haplotypes has been already stressed.

The importance of such observations in understanding the pathophysiology of idiopathic nephrosis is debatable. Their relationship with the overall immunological background of nephrosis can be the subject of innumerable hypotheses, including the release of vascular permeability factors by mast cells, platelets, basophils, and so on. Conversely, Lagrue and Laurent (1982) wisely emphasized the discrepancy between the frequency of allergy in a broad sense, which is in the order of 15 per cent in the general population, and the rarity of nephrosis, which is in the order of 7/100 000. Nonetheless, the notion that, at least in patients with a genetic predisposition, a food allergen may induce relapses of steroid-sensitive nephrotic syndrome and that a diet based on elimination of one specific component might suppress the need for steroids should certainly lead to further systematic research.

Pathogenesis of glomerular sclerosis

Is focal and segmental glomerular sclerosis passively induced by increased glomerular basement membrane permeability and excessive mesangial traffic of macromolecules?

In some glomerular capillary loops mesangial cells are separated from the blood only by the highly permeable endothelial cells, and in others they are in direct contact with the circulation. There is a continuous flow of plasma through the lacis region. Circulation of macromolecules can be demonstrated experimentally by uptake and transport of intravenously injected colloidal carbon (Elema *et al.* 1976) or the iron compound imposil (Leiper *et al.* 1977), and immune complexes may also be taken up into the mesangium after transiting through endothelial fenestrae (Keane and Raij 1981). The frequent presence of immunoglobulins, especially IgM, within lesions of glomerular sclerosis has been cited as supporting a similar causal phenomenon. In fact, such a pathophysiology for focal fibrosis and hyalinosis would only be acceptable when this lesion complicates non-immunological glomerular basement membrane hyperpermeability (such as, for instance, in Alport's syndrome), or primary glomerular hyperfiltration (as in oligomeganephronia). More generally, the first phenomenon which calls for explanation in nephrosis is the increased hyperpermeability of the glomerular basement membrane, which precedes fibrosis.

Does focal and segmental glomerular sclerosis begin in the mesangial cells? Is it an equivalent of atheroma?

The mesangial cell shares numerous properties with smooth muscle cells. This observation has led to the interesting hypothesis that a lipid disorder induces mesangial cell disturbances analogous to the vascular smooth muscle lesions of atherosclerosis. In fact, mesangial cells of patients with focal and segmental glomerular sclerosis resemble the foam cells found in atheromatous plaques (Brown and Goldstein 1983; Roos 1986). Grond *et al.* (1982; 1986) demonstrated lipid accumulation in sclerosed glomeruli of aminonucleoside-treated rats and noted the relationship of glomerular lesions and serum cholesterol concentrations. Kasiske *et al.* (1987) showed that a cholesterol synthesis inhibitor reduced glomerular injury in a rat model of focal and segmental glomerular sclerosis. In man Verani and Hawkins (1986) studied the early lesion of recurrent focal and segmental glomerular sclerosis approximately 2 months after renal transplantation by electron microscopy. One of the earliest lesions observed was the presence of lipid vacuoles and foam cells in the mesangium. Nevertheless, the mesangial cell lesion was concomitant with epithelial cell lesions which might play a dominant part in the initial events leading to focal sclerosis.

There is, therefore, an increasing body of information which suggests that lipids may cause or aggravate the focal fibrotic mesangial lesions of focal and segmental glomerular sclerosis. Unfortunately, this does not explain the generalized loss of sieving properties which affects all glomerular basement membranes, and induces nephrotic syndrome.

Recent observations of rapidly progressive focal and segmental glomerular sclerosis leading to end-stage renal failure in the acquired immunodeficiency syndrome (see Chapter 3.13) have identified viral particles in mesangial cells (Chander *et al.* 1987; Alpers *et al.* 1988).

Is focal and segmental glomerular sclerosis primarily a disease of glomerular epithelial cells?

The work of Verani and Hawkins (1986) and that of Korbet *et al.* (1988) point to the epithelial cell as the first victim of the process which leads to focal scarring. Both groups took advantage of a natural experimental model—serial histology after renal transplantation in man—and confirmed previous histological observations in patients with focal and segmental glomerular sclerosis (Grishman and Churg 1975; Schwartz and Lewis 1985). Their findings implied that glomerular abnormalties are not limited to segmental scars but stem from injury and proliferation of podocytes. The cause of this primary cellular lesion is unknown, but its early appearance after post-transplantation recurrence of proteinuria suggests the same pathophysiology as the generalized disorder of glomerular basement membrane permeability. It is tempting to speculate that the same mechanism also explains the early tubular and even the renal hyaline arteriolosclerosis which characterize focal and segmental glomerular sclerosis with a poor prognosis (Lee and Spargo 1986). This recalls the electron microscopy observations of Grishman *et al.* (1976) which showed that a toxic substance, heroin, induces nephrotic syndrome with focal and segmental glomerular sclerosis with the initial lesions of glomerular podocytes.

Treatment and outcome in children

Overview

Antibiotics represented the first major advance in the treatment of idiopathic nephrotic syndrome, reducing the mortality, which was mainly related to infection, by more than 50 per cent (Lawson *et al.* 1960; Cornfeld and Schwartz 1966). In spite of their obvious effectiveness, corticosteroids did not change patient survival so much, only reducing mortality by 3 to 7 per cent (White et al. 1970; Habib and Kleinknecht 1971) probably because of their adverse side-effects. With time and experience of steroid administration, these drugs became the basic treatment of idiopathic nephrotic syndrome.

Excluding spontaneous remissions, it could be claimed that prognosis of idiopathic nephrotic syndrome is a function of the response to high dose corticosteroid therapy (Habib and Kleinknecht 1971; Lim et al. 1974; Makker and Heymann 1974; Schwartz et al. 1974). Patients who respond to steroid therapy may relapse, but the majority continue to respond throughout the subsequent course of the disease (Pollak et al. 1968; Habib and Kleinknecht 1971). Only 1 to 3 per cent of initially steroid-sensitive patients become steroid resistant later (Habib and Kleinknecht 1971; Siegel et al. 1974; Trainin et al. 1975). Conversely, patients who do not respond to initial steroid therapy given in an adequate dose will usually remain non-responders. Response to corticosteroids is therefore the best prognostic index, not only because non-responders are more exposed to the complications of persistent nephrotic syndrome, but also (and mainly) because they may develop end-stage renal failure after several years of disease. The response to corticosteroids allows us also to foresee, more surely than histology, the response to other drugs used in this disease. Corticosteroid responders and non-responders will, therefore, be considered separately in this chapter, this distinction being more helpful for the clinician and for therapeutic decisions than the usual contrast made between 'minimal change disease' and 'focal segmental sclerosis'. A subset of partial responders also has to be considered; this group seems to be intermediate between the two main groups of patients and correspondingly has an intermediate prognosis.

Initial treatment of idiopathic nephrotic syndrome

There are several reasons for not starting specific treatment too quickly in a child with the initial symptoms of idiopathic nephrotic syndrome. Firstly, there is the possibility of spontaneous remission within the first 8 to 15 days. Before the corticosteroid era this was reported to occur in around 25 per cent of patients (Cornfeld and Schwartz 1966). More recently, spontaneous remissions were reported in 4 to 6 per cent of cases (Habib and Kleinknecht 1971; Cameron et al. 1974b). Some of these early spontaneous remissions are definitive, others are not. Secondly, the clinician should be sure that idiopathic nephrotic syndrome is not related to allergy or malignancy, both of which need specific treatment before starting steroids. Finally, any infection has to be diagnosed and treated effectively before starting steroid treatment, not only to prevent the risk of an overwhelming infective process during treatment, but also because occult infection may be responsible for non-responsiveness to steroids (McEnery and Strife 1982). Infection may be obvious, but occult infections such as those in sinuses or teeth may have to be sought.

Corticosteroids

Even if one regrets the fact that no controlled study was performed in children when corticosteroids were first used for the treatment of idiopathic nephrotic syndrome, the effectiveness of these drugs is so obvious that a controlled trial may not now be appropriate.

When the diagnosis of idiopathic nephrotic syndrome is made in a child, with or without a kidney biopsy, and whatever the histological pattern, specific treatment with corticosteroids should be started after the short delay mentioned above. There are several possible drugs, variable time schedules and modes of administration, and none has been shown definitively to be the most appropriate. However, it is possible to propose a standard protocol from the day to day experience and available data in the published literature.

First of all, it must be said that prednisone remains the reference drug to date. Prednisolone has the advantage of being soluble in water, making its administration easier in young children, but may be unable to induce remission in some patients who respond quickly to the same dosage of prednisone (personal observation). There is no special advantage in using other steroids. Oxazacort ® has been proposed as being associated with fewer side effects (Balsan et al. 1987), but controlled studies are not yet available.

The standard dosage regimen is 60 mg/m^2.day and it is better not to give more than 80 mg/day. This dose is best prescribed once daily in the morning (Warshaw and Hymes 1989). A response occurs in most of the cases within 10 to 15 days (median 11 days). Several studies have shown relapses to be less frequent if the initial treatment is not stopped too rapidly: it is usually continued daily for 4 weeks, and then switched to every other day. According to the ISKDC (1981), approximately 90 per cent of responders are in remission 4 weeks after starting steroids, less than 10 per cent going into remission after 2 to 4 additional weeks using a daily regimen. A few patients are also reported to go into remission after 8 to 12 weeks of daily steroids (Schlesinger et al. 1968; ISKDC 1978), but prolongation of daily steroid therapy beyond 4 or 5 weeks increases the risk of side-effects dramatically. An alternative for patients who are not in remission after the first 4 weeks of steroid therapy is to give three to four boluses of methyl prednisolone (1 g/1.73 m^2): this additional regimen seems to be associated with fewer side-effects than the prolongation of daily high dose steroids and probably produces remission more rapidly in the few patients who would have entered the second month of daily therapy (Murnaghan et al. 1984).

The definition of 'steroid-resistant' idiopathic nephrotic syndrome is somewhat arbitrary, but there is a good agreement for considering as resistant those children who have received at least 4 weeks of adequate high dose corticosteroid, and yet remained nephrotic with heavy proteinuria. The marginal benefit obtained with additional steroids has to be evaluated according to the possible additional side-effects in each individual patient.

Boluses of intravenous methylprednisolone followed by low dose daily corticosteroid therapy have also been proposed as the initial specific treatment: this regimen seems to be marginally (Imbasciati et al. 1985) or clearly (Yeung et al. 1983) less effective than the standard protocol.

After the fourth week, or in the days following remission when this occurs later, one can recommend switching corticosteroid administration to an alternate day schedule. The Arbeitgemenshaft für Pädiatrische Nephrologie (1979; 1981) showed unequivocally in a prospective controlled trial that at this stage of the treatment alternate day was better than intermittent steroids (three consecutive days out of every seven), with a significantly lower number of relapses during and after the treatment period.

The length of time during which the alternate-day corticosteroid regimen has to be continued for the initial treatment of idiopathic nephrotic syndrome depends on the preference of individual centres. However, the shorter schedule of 4 to 6 weeks seems to be associated with a higher number of relapses after stopping steroids (ISKDC 1982; Arbeitgemenschaft für Pädiatrische Nephrologie 1988). Our practice is to continue with a dose of 60 mg/m^2 every other day for 6 to 8 weeks, then to taper the dosage and stop corticosteroids completely within the next 6 weeks. This regimen was associated with significantly fewer relapses than shorter schedules in at least two studies (Kleinknecht et al. 1982; Ueda et al. 1988).

It does not seem useful to continue with an alternate day regimen beyond 4 to 5 months, as was demonstrated in another study from the French Club of Pediatric Nephrology which compared a 1-year versus a 4-month alternate-day regimen (Kleinknecht et al. 1982).

Corticosteroid responsive nephrosis in children

The vast majority of patients with idiopathic nephrotic syndrome are steroid responsive: 89 per cent of children studied by White et al. (1970) and 98 per cent of the patients with minimal changes on histology. In a study of 665 cases Kleinknecht and Gubler (1983) reported 87 per cent of patients to be steroid responsive. The incidence of steroid responders—78 to 94 per cent—in patients with minimal change disease is similar in adults (Hopper et al. 1970; Jao et al. 1973; Cameron et al. 1974; Lim et al. 1974).

After the first remission the clinical course is predictable: the proportion of patients who had only one attack varies greatly from one series to another, from 6 per cent to 30 per cent. This variation seems mainly to be due to bias in patient selection (White et al. 1970; Habib and Kleinknecht 1971; McCrory et al. 1973; Glassock and Bennett 1976). In a series of 77 unselected patients Kleinknecht and Gubler (1983) reported 40 per cent of children who were definitively cured after the first attack. After 18 to 24 months in persistent remission after stopping adequate steroid treatment, it is likely that the cure is definitive, and the risk of later relapses is low, although possible. Such cases almost always remain steroid-sensitive (Habib and Kleinknecht 1971; Schärer and Minges 1973; Makker and Haymann 1974). In 10 to 20 per cent of cases relapses occur several months after stopping treatment, and a cure may take place after two or three episodes which respond quickly to a standard course of steroids. The remaining 40 to 50 per cent of patients who responded initially to steroids experience frequent relapses (Barnett et al. 1978). An early relapse following the first attack is predictive of a frequent relapsing course (Habib and Kleincknecht 1971; Siegel et al. 1972; ISKDC 1982). It is these patients who raise difficult therapeutic problems.

Relapses

Relapse occurs in 60 to 70 per cent of unselected patients with steroid-sensitive idiopathic nephrotic syndrome (Koskimies 1982; Kleinknecht and Gubler 1983) and the majority of these—85 per cent—had multiple relapses. According to an ISKDC study (1982), no correlation was found between the frequency of relapse and the detailed histological pattern, the clinical or laboratory findings at onset, the time needed for initial response, or the time elapsed between the initial response and the first relapse. However, the number of relapses during the first 6 months in patients receiving the ISKDC corticosteroid protocol was highly predictive of the subsequent disease course.

The role of upper respiratory tract infection in exacerbation of nephrotic syndrome has been underlined in all series: 71 per cent of relapses were preceded by such an event in a prospective study, but only 45 per cent of respiratory infections were followed by an exacerbation of proteinuria (McDonald et al. 1986). Varicella was consistently followed by relapse in this study, as well as in the Hôpital des Enfants Malades series.

A critical threshold for an efficient steroid dose can sometimes be approximately defined. Of 112 patients with at least three relapses, 18 had infrequent relapses occurring after more than 6 months without treatment, 45 relapsed within 3 months after cessation of treatment or when alternate-day therapy was below 0.5 mg/kg every other day, 23 relapsed during treatment with prednisone 1.5 to 2 mg/kg every other day, and 26 relapsed under too variable a dose to be classified. Even in patients showing a clear critical dose, changes in the effective dose may be observed. Thus a threshold in treatment dosage cannot be determined definitely from one or two relapses (Kleinknecht and Gubler 1983).

In many cases, exacerbations of proteinuria are only transient, and spontaneous remissions are quite frequent. This was observed in 23 per cent of frequently relapsing patients and in 10 per cent of steroid-dependent patients, with disappearance of proteinuria in 4 to 14 days in 79 per cent of cases (Wingen et al. 1985).

Leisti et al. (1977a, b) suggested a role for post-therapeutic hypocortisolism in triggering relapses, and some reports have suggested possible prevention of early relapse by low dose maintenance hydrocortisone administration (Leisti et al. 1978; Schoeneman 1983); this endocrine abnormality seems to have been present in these patients before steroid treatment. In fact, the prescription of low-dose hydrocortisone after stopping corticosteroids did not prevent the risk of becoming a frequent relapser (Arant and Bernstein 1982).

The risk of relapse depends on age at onset, being higher in patients whose nephrotic syndrome started at a young age, and on sex, being higher in males, and on time elapsed from onset, with a plateau after 4 years from the last relapse (Lewis et al. 1989). It is impossible, however, to set a definite end-point. Late relapses from 14 to 29 years after the last exacerbation have been reported, usually with a benign evolution (Pru et al. 1984).

Pregnancy may be a precipitating factor for relapse in female patients who had steroid-sensitive idiopathic nephrotic syndrome in childhood followed by prolonged remission (Makker and Heymann 1972; Wynn et al. 1988).

Basic corticosteroid treatment in relapsing nephrosis

Two main approaches may be proposed when the first or subsequent relapse occurs while decreasing steroid dosage, or within 2 months after stopping steroid therapy.

The first approach, followed by the ISKDC group (1974), recommends a short course of prednisone therapy of 60 mg/m^2.day for 4 weeks followed by alternate day therapy for 4 weeks, after which the treatment is stopped. In this protocol, by definition frequent relapsers have had at least two relapses during a 6-month period, and steroid-dependent patients have either two consecutive relapses during a period of steroid taper, or two consecutive relapses occurring within 2 weeks of ending a course of cortisteroid therapy.

The second approach, proposed by our group at the Hôpital des Enfants Malades (Broyer and Kleinknecht 1978; Broyer 1988), recommends treatment of the relapse with prednisone every day at a dose of 40 to 60 mg/m^2 until proteinuria has been absent for a week. Thereafter, the regimen is switched to alternate days for 6 weeks. The dosage is then progressively tapered

Table 2 Results of cyclophosphamide treatment of relapsing idiopathic nephrotic syndrome in children

Reference	Type of idiopathic nephrotic syndrome	Patients (n)	Dose (mg/kg.day)	Duration (week)	Cumulative rate of sustained remission
Barratt and Soothill (1970)	NS	10	3	8	80% at 1 year
Barratt et al. (1973)	NS	16	3	8	93% at 1 year
McDonald et al. (1974)	NS	34	2–5–5	4–15	64% at 5 years
Cameron et al. (1974a)	NS	58	0–5–5	3–30	36% at 5 years
Chiu and Drummond (1974)	NS	36	2–5	16	50% at 5 years
ISKDC (1974)	NS	27	5 then 1–3	6	52% at 2 years
Barratt et al. (1975)	NS	15	3	8	67% at 1 year
					44% at 4 years
Rance et al. (1976)	NS	NS	3	12	66% at 5 years
Pennisi et al. (1976)	NS	29	3–5	6–8	58% at 1 year
		24*	2	+4	92% at 1 year
Garin et al. (1978)	Steroid dependent	13	2	10	20% at 3 years
	Frequent relapser	10	2	12	70% at 3 years
Siegel et al. (1981)	FSGS	11	2	12	27% at 6 years
	Minimal change	18	2	12	78% at 6 years
	DMP	9	2	12	45% at 6 years
APN (1982)	Steroid dependent	8	2–5	8	28% at 2 years
	Frequent relapser	18	2–5	8	62% at 2 years
APN (1987)	Steroid dependent	18	2	12	67% at 2 years
	Frequent relapser	18	2	8	22% at 2 years
Shohet et al. (1988)	Frequent relapser	11	2–5	8	64% at 1 year
	Steroid dependent	8	2–5	8	12% at 1 year

NS, not specified; DMP, diffuse mesangial proliferation; FSGS, focal segmental glomerular sclerosis. * These patients received the same regimen as the 29 mentioned above plus an additional 4-week course.
APN: Arbeitsgemeinschaft für Pädiatrische Nephrologie.

within the next 6 weeks to 15 mg/mg.m² every other day, the dosage being intentionally continued for at least 12 to 18 months after the last relapse. In this protocol the maintenance dose is later modified according to the apparent threshold necessary to obtain a sustained remission. Even in patients receiving 15 to 30 mg/m² every other day, there was no or little change in appearance, except for an increase in weight for height. In contrast with daily therapy, growth rate is usually normal, but no catch up is obtained in previously stunted children.

The ISKDC approach allows a better definition in terms of number of relapses by year. However it is associated in many cases with many more relapses. In the 'Enfants Malades' long-term/low-dose/alternate-day schedule, patients are less well defined but the number of relapses is lower and the cumulative dose of corticosteroids is also lower. A similar protocol was used by Elzouki and Jaiswal (1988), who reported a dramatic decrease of relapse rate and frequent spontaneous remissions of exacerbations of nephrotic syndrome associated with episodes of infection.

Immunosuppressive drugs

Whatever the basic corticosteroid treatment used in patients with frequent relapses treatment with effective dose cannot be continued indefinitely in all the cases. If symptoms of steroid toxicity appear, including a decrease in growth velocity, it is time to try and withdraw corticosteroids. In these cases, immunosuppressive drugs need to be used: these are not without risk and should be kept in reserve. The decision to prescribe immunosuppressive drugs in idiopathic nephrotic syndrome does not depend on the number of relapses, but rather on the side-effects of steroids and sometimes also on sociopsychological reasons, including the fatigue of the family and the child.

Alkylating agents

Alkylating agents have been used to achieve long lasting remission in idiopathic nephrotic syndrome for almost 40 years (Kelley and Panos 1952). The cytotoxic effect of these drugs arises from the ability of their alkyl chains link to purine bases of DNA, preventing the normal transcription process and possibly leading to cell death. Three drugs have been used in the treatment of idiopathic nephrotic syndrome: cyclophosphamide, chlorambucil, and mechlorethamine.

Cyclophosphamide The efficacy of cyclophosphamide for preventing further relapse of idiopathic nephrotic syndrome was reported more than 20 years ago (Coldbeck 1963), and was proven by a prospective study by Barratt and Soothill (1970) who compared an 8-week course of cyclophosphamide to prednisone alone. Identical results were obtained in an international study of kidney disease in children trial (ISKDC 1974). However, it soon became apparent that remission induced by this treatment varied in duration from patient to patient and was in some cases very short. Table 2 shows data from a number of reports describing a persisting remission rate of 67 to 93 per cent after 1 year, and of 36 to 66 per cent after 5 years. However, even in patients who relapsed after cyclophosphamide treatment, the time between stopping steroids and relapse was shorter than in the period before cyclophosphamide with a reduced relapse rate (Bergstrand et al. 1973, Chiu and Drummond 1974).

Several trials have addressed the relationship between dose and duration of treatment and therapeutic efficacy. Barratt et al. (1973) obtained significantly better results with a dose of 3 mg/kg.day for 8 weeks than with the same dose for 2 weeks. Similarly, Pennisi et al. (1976) recorded a relapse rate of 42 per cent in the first year after a treatment of 6 to 8 weeks at a dose of 3 to 5 mg/kg.day, compared to a relapse rate of only 8 per cent if this treatment was followed by 2 mg/kg.day for an additional 4 weeks. The same findings were reported by Rance *et al.* (1976): two-thirds of their patients were in remission 5 years after a 3-month course of cyclophosphamide. Treatment for 12 weeks at a dose of 2 mg/kg.day was recently shown to be much more effective than an 8-week course, with 67 per cent remaining in remission after 2 years versus 22 per cent (Fig. 9) (Arbeitsgemeinschafte für Pediatrische Nephrologie 1987). Conversely Cameron *et al.* found that treatment for 12 or 30 weeks gave identical results (1974*a*).

Another factor which may influence the result of cyclophosphamide treatment is the pattern of response to prednisone (Garin *et al.* 1978*b*). These authors found only 20 per cent of steroid-dependent patients in remission after 3 years following an 8-week course of cyclophosphamide, whereas the remission rate in patients with frequent relapses was 70 per cent. Tejani *et al.* (1983) and Shohet *et al.* (1988) reached a similar conclusion, which was also borne out in a multicentre prospective trial (Arbeitgemenshaft für Pädiatrische Nephrologie 1982). In fact, the results of cyclophosphamide treatment in steroid-dependent patients also depends on the duration of treatment, being much better with a 12-week course than with an 8-week course (Arbeitgemenshaft für Pädiatrische Nephrologie 1987). A subset of steroid-dependent patients with atopy and HLA B12 histocompatibility antigen had a poor response to cyclophosphamide (Trompeter *et al.* 1980).

Kidney histopathology is associated with different success rates. The incidence of relapse after cyclophosphamide was significantly greater in patients with focal and segmental sclerosis (73 per cent) and mesangial proliferation, than in children with minimal change disease (22 per cent) in a retrospective study by Siegel *et al.* (1981).

Finally, response to cytotoxic therapy for childhood idiopathic nephrotic syndrome depends more on the type of response to corticosteroid therapy than on histopathological pattern. Schulman *et al.* (1988) showed that the actuarial remission rate after cyclophosphamide was similar in patients with minimal changes, IgM deposits, or global focal sclerosis and only 10 to 15 per cent lower in patients with focal and segmental glomerular sclerosis who responded initially to steroids. These data suggest that biopsy is not really useful, at least in those who respond to corticosteroids, in making therapeutic decisions.

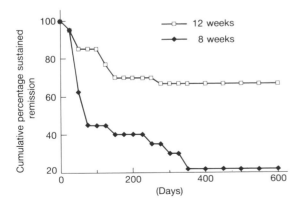

Fig. 9 Actuarial remission rate after treatment with cyclophosphamide in steroid-dependent nephrotic syndrome (according to Arbeitsgemeinschaft für Pädiatrische Nephrologie 1987 and with permission of the publisher).

Cyclophosphamide alone has been compared with steroids as initial treatment in a randomized trial (Tsao and Yeung 1971) with a clear-cut decrease in relapse rate. This approach has also been reported to be effective in adult patients (Al-Khader *et al.* 1979) and may be indicated in elderly patients who are more prone to develop complications with corticosteroid treatment. As a rule it seems wise to reserve this treatment for patients who cannot tolerate steroid treatment. It has been suggested that cyclophosphamide administered after steroid-induced remission (Cameron *et al.* 1974*a*) may reduce toxicity by maintaining a high diuresis through the bladder. The cumulative dose has to be assessed in order to remain below the threshold of gonadal toxicity, usually for a duration of 12 weeks. Steroids should be stopped at the same time in order to avoid corticosteroid toxicity in the sometimes short period of remission produced by cyclophosphamide.

Chlorambucil Beneficial results have also been achieved with chlorambucil in steroid-responsive nephrosis (Table 3). Leveque *et al.* (1969) reported an increased incidence of prolonged remissions with chlorambucil, but the duration of treatment in this study (6–18 months) must be emphasized. Grupe (1973) demonstrated the efficacy of shorter courses of treatment, lasting 2.5 to 12 weeks, with a relapse rate of only 13 per cent. Grupe *et al.* (1976) later published the results of a controlled trial of chlorambucil at a dose of 0.2 to 0.4 mg/kg.day for 6 to 12 weeks: this induced prolonged remissions. During a 1 to 3-year follow up in 11 patients, none had suffered a further relapse by the end of the trial. Comparing these results with those published for cyclophosphamide, the authors concluded that chlorambucil was more effective. Baluarte *et al.* (1978) obtained identical results in relapsing steroid-responsive nephrosis with a dose of 0.2 mg/kg.day chlorambucil for 2 months and with a higher dose of 0.6 mg/kg.day for the same duration. Williams *et al.* (1980) have shown that low daily doses are preferred: 91 per cent of patients on a dose of 0.3 mg/kg and 80 per cent of those on 3 mg/kg were still in remission 4 years later.

Contrary to the situation with cyclophosphamide, the response to chlorambucil does not seem to be related so much to the clinical pattern of relapse: Grupe (1982) found similar results in frequent relapsers and in steroid-dependent patients. However, this point of view was not shared by the cooperative German group (Arbeitgemeinschaft für Pädiatrische Nephrologie 1982), who clearly showed the same relationship between corticosteroid dependence/multiple relapse and subsequent stability of remission following chlorambucil treatment.

Mechlorethamine Mechlorethamine was one of the first drugs used to treat nephrosis but was gradually abandoned after the introduction of steroids and the other alkylating agents that could be administered orally. However, several groups have returned to mechlorethamine therapy, because of the side-effects of cyclophosphamide and chlorambucil.

Schoeneman (1983) recently used mechlorethamine in 12 children with frequently relapsing steroid-responsive nephrosis. Actuarial analysis showed a remission rate of 46 per cent after 27 months. Six children relapsed, all during the first 7 months following treatment. The authors concluded that mechlorethamine was only slightly less effective than cyclophosphamide or chlorambucil. Maisin *et al.* (1985) reported long lasting remissions in 16 of 27 patients with steroid-dependent or partially steroid-responsive nephrosis. Mechlorethamine induced rapid remission, within an average of 7 days. Eleven children relapsed,

Table 3 Results of chlorambucil treatment of relapsing idiopathic nephrotic syndrome in children

Reference	Type of idiopathic nephrotic syndrome	Patients (n)	Dose (mg/kg.day)	Duration (week)	Cumulative rate of sustained remission
Grupe (1973)	NS	23	0.3	2.5–12	87% at 1 year
Kleinknecht (1973)	NS	36	0.2	12–24	72% at 5 years
		32	0.2	24–48	91% at 5 years
Baluarte et al. (1978)	NS	10	0.2	8	80% at 3 years
		11	0.2–0.6	6–11	80% at 3 years
Williams et al. (1980)	NS	25	< 0.3	5–15	91% at 4 years
		31	> 3	5–15	80% at 2 years
APN (1982)	Frequent relapser	8	0.15	8	87% at 2 years
	Steroid dependent	16	0.15	8	31% at 4 years
Grupe (1982)	Frequent relapser	13	0.26	10	74% at 4 years
	Steroid dependent	36	0.27	10	82% at 4 years

NS, not specified.

most during the first 7 months, but the disease profile improved in five of these children.

Our own experience with mechlorethamine (Broyer et al. 1985) consisted of treating children who each received a total dose of 0.8 mg/kg in two courses of four injections 1 month apart. Thirty children had relapsing steroid-responsive nephrosis and major corticosteroid side-effects. At the time of treatment, 20 were in remission and 10 were in relapse. Following mechlorethamine treatment, most children remained in remission without corticosteroid therapy or with lower corticosteroid doses than previously. Nevertheless, actuarial analysis showed a sharp decrease in the cumulative remission rate: 43 per cent after 1 year, 30 per cent after 2 years, and only 15 per cent after 3 years. Although these results do not seem encouraging, mechlorethamine therapy allowed the signs of corticosteroid toxicity to resolve in most children and the course of the illness was altered by reducing the level of corticosteroid dependence.

The effect of mechlorethamine on proteinuria has been studied in the 10 children who received the treatment while in relapse. Proteinuria resolved during the following month in all children, and it did so rapidly—in less than 5 days—in half the cases. The time for mechlorethamine to take effect in these cases was unrelated to corticosteroid dose or to the duration of higher dose corticosteroid therapy.

Cyclosporin

Hoyer et al. (1986) first reported cyclosporin to be effective in steroid-responsive nephrotic syndrome. These authors reported five patients showing frequent relapses, despite treatment with cytotoxic agents, who had fewer relapses and needed much less prednisone during cyclosporin treatment at a daily dose of 150 to 200 mg/m^2.day. Capodicasa et al. (1986) reported similar results in six frequent relapsers with long lasting remission after stopping cyclosporin in five cases. Niaudet et al. (1987) reported 20 steroid-dependent children with corticosteroid-induced side effects who were given cyclosporin at a dose of 6 mg/kg.day. Seventeen of the 20 either remitted or did not relapse despite prednisone withdrawal, but 10 of the 12 children in whom cyclosporin had been tapered off, and who had initially responded to this drug, relapsed. Tejani et al. (1988) reported 13 patients with relapsing nephrotic syndrome, of whom 11 achieved remission with cyclosporin, but seven of them also relapsed after stopping the drug.

The Enfants Malades group reported recently an update of the results obtained in steroid-dependent nephrotic syndrome (Niaudet et al. 1989). Thirty-six of 45 patients remained in remission without steroids whilst taking cyclosporin at a mean dose of 150 mg/m^2.day, whilst seven relapsed after stopping prednisone and two relapsed when switching to alternate day prednisone. All of the patients in remission with a follow-up longer than 1 month after stopping cyclosporin relapsed: most of them were again in remission after resuming cyclosporin alone or cyclosporin plus steroids, but a few patients failed to respond immediately, recovering cyclosporin sensitivity after several months' treatment with corticosteroids alone or on symptomatic treatment. In this study some patients were receiving cyclosporin for periods as long as 3 years with a yearly kidney biopsy.

The preliminary results of a prospective randomized trial in relapsing patients comparing cyclosporin and chlorambucil given at a dose which avoids gonadal toxicity showed a clear-cut advantage for chlorambucil: almost all patients who received cyclosporin relapsed within 3 months of stopping the drug, while more than half of those who received chlorambucil remained in remission (Niaudet et al. 1989).

In conclusion, these data indicate that although almost 90 per cent of patients with corticosteroid-responsive idiopathic nephrotic syndrome respond to cyclosporin, most patients relapse after tapering or stopping the drug, and the treatment should be continued as needed in order to prevent new relapses. Thus cyclosporin may be useful in the treatment of idiopathic nephrotic syndrome, especially when there is a contraindication to continuing corticosteroid treatment and after failure of a non-toxic cumulative dose of alkylating agent. The precise indications and duration of this treatment will need further study, especially with regard to the risk of side-effects. The main risk appears to be nephrotoxicity, which seems much less marked in steroid-responsive than in steroid-resistant nephrosis (see below).

Azathioprine

After a controlled multicentric trial in children, Abramowicz et al. (1970) concluded that 6-months' treatment with azathioprine had no effect in preventing relapse in idiopathic nephrotic syndrome. In this study five of 18 treated patients relapsed within 6 months after treatment, versus eight out of 18 control patients. A recent report of successful treatment with azathioprine in adults with steroid-resistant idiopathic nephrotic syndrome (Cade et al. 1986) could lead to other trials of longer treatment with this drug in frequent relapsers.

Levamisole

The use of *l*-tetramisole (levamisole) in relapsing idiopathic nephrotic syndrome was first described by Tanphaichitr *et al.* (1980), who reported remission using this drug alone (1.5 to 3.9 mg/kg twice a week) in association with an increase of T cells, assessed by E rosette formation. Another uncontrolled study was less optimistic but reported a subgroup of patients relapsing disease who responded to this drug at a dose of 2.5 mg/kg twice a week. These patients had a reduced relapse rate and received fewer corticosteroids than in the period before levamisole treatment, and their blood leucocyte counts were also lower (Niaudet *et al.* 1984). Similar results were obtained with levamisole administered at a dose of 2.5 mg/kg every other day (Mongeau *et al.* 1988). In two other studies of children with relapsing idiopathic nephrotic syndrome levamisole was effective in patients with a dysfunction of cell-mediated immunity (Mehta *et al.* 1986; Drachman *et al.* 1988).

Non-steroidal anti-inflammatory drugs

The study of Garin *et al.* (1978*a*) seems to have removed any indication for the use of non-steroidal anti-inflammatory drugs in steroid-sensitive idiopathic nephrotic syndrome. In this study, oral indomethacin for 5 weeks produced only a limited reduction of proteinuria, and none of the patients was really improved.

Steroid-resistant idiopathic nephrotic syndrome in children

Overview

As has been stated several times, the response to corticosteroids determines prognosis in idiopathic nephrotic syndrome. This information is more useful for the clinician than the histopathological pattern. Unfortunately there are few reported series of 'non-responsive idiopathic nephrotic syndrome': most of the literature on this subject refers to 'focal sclerosis'. Another difficulty is the quite arbitrary and variable definitions of 'non-responsiveness' to steroids. In this chapter, we will consider as a non-responder a patient who had an unchanged nephrotic syndrome after at least 4 weeks of high-dose prednisone treatment with additional corticosteroid therapy either as boluses of methylprednisolone, or 2 to 4 more weeks of oral steroids. Histopathological criteria are also mandatory for accepting the diagnosis of idiopathic nephrotic syndrome. According to this definition, less than 10 per cent of children with the disease are non-responders (ISKDC 1978). This proportion is even lower in non-selected series (Koskimies 1982; Kleinknecht and Gubler 1983).

The natural history of steroid-resistant idiopathic nephrotic syndrome is difficult to describe, since a variable proportion of the most difficult cases are referred to specialized centres. There is no way of knowing in advance whether a child presenting with nephrotic syndrome will be responsive to corticosteroids. Nevertheless, in the study of Kleinknecht and Gubler (1983), although the proportion of patients with macroscopic haematuria, severe nephrotic syndrome, hypertension, and azotaemia at onset was similar in both groups, a significant difference was noted for several other parameters, including sex, age at onset, microscopic haematuria, selectivity of proteinuria, and histology (Table 4).

In a child with a nephrotic syndrome who fails to respond to adequate high-dose steroid treatment, after the kidney biopsy needed for the diagnosis of idiopathic nephrotic syndrome, steroids have to be withdrawn progressively and another treatment proposed, whatever the histological pattern, unless the patient has irreversible renal insufficiency with diffuse lesions of the kidney.

The results obtained with different drugs are often contradictory and difficult to analyse for several reasons, including the lack of a uniform definition of steroid non-responsiveness, the mixing of complete and partial steroid resistance, and the possibility of spontaneous remission. An interesting point is the good prognosis generally reported in cases of complete remission, who rarely relapsed.

Immunosuppressive and other drugs

Alkylating agents

Cyclophosphamide Cyclophosphamide is the alkylating agent used most frequently in this group of patients. The rate of full or partial remission achieved with the drug differs from trial to trial and appears higher in patients with partial steroid resistance, those with late steroid resistance, or those in whom there is biopsy evidence of minimal change disease, when compared with those showing focal and segmental glomerular sclerosis and/or complete resistance to steroids. The International Study of Kidney Disease in Children (1983) reported the largest trial series, consisting of 37 children with steroid-resistant nephrosis and focal segmental glomerular sclerosis on renal biopsy. The full or partial remission rate on cyclophosphamide was 32 per cent, which was practically identical to that of patients receiving steroid therapy alone. Geary *et al.* (1984) reported full or partial response to cyclophosphamide in 12 of 29 steroid-resistant patients with focal disease on biopsy.

Renal failure developed less frequently in partial responders (one of nine) than in those who did not respond at all (seven of

Table 4 Symptoms and signs of steroid non-responders versus responders according to Kleinknecht and Gubler (1983)

	Non-responder (%)	Responder (%)	*p*
Macroscopic haematuria	3	1	NS
Initially severe nephrotic syndrome	+ + +	+ + +	NS
Hypertension	11	6	NS
Azotaemia at onset	9	10	NS
Female/male	45/55	24/76	< 0.001
Onset before 1 year of age	11	2	< 0.01
Microscopic haematuria	47	20	< 0.01
Non-selectivity of proteinuria	83	10	< 0.001
Focal segmental glomerular sclerosis at first biopsy	50	10	< 0.01

eight). Siegel et al. (1975) achieved full remission in six steroid-resistant patients with minimal change disease, three of whom relapsed after cyclophosphamide but then became steroid responsive. Similarly, Bergstrand et al. (1973) reported that patients with steroid-resistant nephrosis treated with cyclophosphamide became steroid-responsive but also steroid-dependent at a high dosage level.

Other investigators have had much less clear results. White and Glasgow (1971) observed no improvement after cyclophosphamide treatment in 15 steroid-resistant children with focal sclerosis. Cameron et al. (1974) reported only one responder out of 13 children with steroid-resistant nephrosis and focal and segmental glomerular sclerosis. Similarly, Tejani et al. (1983) reported no remission with cyclophosphamide in 10 steroid-resistant children.

Some interesting results were reported by Griswold et al. (1987) using both cyclophosphamide and a series of methylprednisolone boluses at a dose of 30 mg/kg, given every other day for 1 week, then weekly for 2 months and biweekly for the next 2 months. With this protocol two of seven patients had long-term, nearly complete remissions.

Chlorambucil   Chlorambucil has been used to treat steroid-resistant nephrosis less often than cyclophosphamide. Baluarte et al. (1980) treated 17 cases of steroid-resistant nephrosis with chlorambucil, achieving remission in 10 cases, three of whom relapsed. Williams et al. (1980) treated six children who all went into remission with a follow up of 1.3 to 9.4 years.

Our own experience with chlorambucil in steroid-resistant nephrosis is greater than that with cyclophosphamide. Patients received chlorambucil, 0.2 mg/kg, for between 2 and 6 months. In our initial trial in 1980, comprising 20 children, the response rate was 20 per cent (four full and four partial remissions) and our present results scarcely differ. Thus, of 74 steroid-resistant children who received a course of chlorambucil, 14 went into full or partial remission, during or shortly after treatment.

Mechlorethamine   Fine et al. (1976) used mechlorethamine in seven patients with steroid-resistant nephrosis at a cumulative dose of 1 mg/kg per course of treatment. Histological examination of renal biopsy specimens had shown minimal change disease in four cases, and focal and segmental glomerular sclerosis in two. The seven patients received a total of nine courses of treatment. Complete remission was achieved eight times and partial remission once. Response to treatment occurred between days 3 and 21, and the mean duration of remission was 14.9 months. Three patients were still in remission 20 to 47 months after receiving mechlorethamine. Our own experience with the drug has been less encouraging (Broyer et al. 1985): only three of 13 steroid-resistant children went into remission, and this was transient in all cases, lasting less than 1 year.

Combined immunosuppression
Trompeter (1987) proposed an aggressive immunosuppression for a 'malignant' subgroup of patients with focal and segmental glomerular sclerosis, refractory nephrotic syndrome, and/or rapid progression to renal failure. The regimen included vincristine 1.5 mg/m^2.week in eight weekly intravenous doses, cyclophosphamide 3 mg/m^2.day for 8 weeks and prednisolone 2 mg/kg.day initially, and progressively tapering after 2 weeks. Lasting remission was observed in seven of 21 children, starting 6 months to 3 years from the commencement of combined immunosuppression. This delay may question the relationship between treatment and response. In addition, side-effects of this treatment may be significant.

Azathioprine
Azathioprine was considered to be ineffective in steroid-resistant patients after the report of Abramowicz et al. (1970) who found no difference between a 3-month course of azathioprine and a placebo. Recently, the study of Cade et al. (1986) which reported complete remission in 13 adult patients with steroid-resistant idiopathic nephrotic syndrome drew attention to the possible efficacy of this drug.

Cyclosporin
Cyclosporin has been given to steroid resistant patients in a few small, uncontrolled trials. Capodicasa et al. (1986) reported that two of four patients who received a combination of cyclosporin and steroids over a 6-month period went into remission. The four patients treated by Brandis (1987) with cyclosporin alone went into remission within 8 weeks. The results were not so impressive in other series: Waldo and Kohaut (1987) reported one partial and transient remission in six patients; Niaudet et al. (1987) had one full and two partial remissions in 10 patients treated for 2 to 7 months; Brodehl et al. (1987) obtained two partial remissions in 7 patients treated 6 to 29 months; and Tejani et al. (1985) had 3 remissions in seven patients. In summary, only 26 per cent of 38 steroid-resistant patients went into complete remission, and in most cases the remission has been long lasting. A partial response was observed in 13 per cent of cases, but this response was usually transient.

Response to cyclosporin correlated with the initial steroid responsiveness better than with the histological category: 89 per cent of steroid responders versus 29 per cent of non-responders went into remission with cyclosporin treatment, while 72 per cent of patients with minimal changes and 35 per cent of those with focal and segmental glomerular sclerosis did so (Niaudet and Broyer 1989).

Preliminary results of a prospective study in corticosteroid non-responders treated with cyclosporin 150 mg/m^2 and prednisone 30 mg/m^2.day during the first month and then every other day, are more encouraging. Eight of 18 patients in this protocol went into remission, most within the first month (Niaudet et al. 1989).

Non-steroidal anti-inflammatory drugs
These drugs are known to be able to decrease proteinuria (Conte et al. 1967; Michielsen and Lambert 1967). Several authors have used indomethacin in the treatment of idiopathic nephrotic syndrome, as well as other glomerulopathies, with variable results. Donker et al. (1978) found an effective reduction of proteinuria in patients with focal sclerosis as a result of simultaneous reduction of GFR. This effect was enhanced by a sodium restricted diet, but was immediately reversible when the drug was discontinued. A clear reduction in proteinuria, with an increase in plasma albumin, was also reported in patients treated with meclofenamate in an uncontrolled study (Velosa et al. 1985). The antiproteinuric effect of these drugs was associated with reduction of urinary excretion of prostaglandin E$_2$ (Vriesendorp et al. 1986). Apart from these encouraging reports, a high incidence of irreversible renal failure was observed in a prospective randomized study (Kleinknecht et al. 1980).

In summary, even if non-steroidal anti-inflammatory drugs may help to reduce proteinuria in some patients with steroid-resistant idiopathic nephrotic syndrome, these drugs must be administered with caution, taking into account the possible side-effects. Effectiveness and the reversibility of decreased GFR must be assessed regularly by stopping the drug for some days.

Other approaches

Antiplatelet therapy with dipyridamole was used together with anticoagulation in a group of children with focal and segmental glomerular sclerosis who were also receiving cyclophosphamide and steroids: seven of nine went into remission (Futrakul 1980). This high rate of remission has not been confirmed by other groups. Plasma exchange has been used in some patients with steroid-resistant idiopathic nephrotic syndrome, with a lack of consistent results.

The angiotensin converting enzyme inhibitor captopril was reported to decrease dramatically nephrotic range proteinuria due to renovascular hypertension (Martinez-Vea et al. 1987). A decrease in proteinuria, and even complete remission, have also been observed in patients with chronic glomerulopathies with or without hypertension (Ferder et al. 1990). Further studies are needed to confirm these encouraging, but anecdotal observations. The role of converting enzyme inhibitors in preventing deterioration of renal function which has been reported in experimental animals with reduced renal mass (see Section 9) is another reason to consider this prescription in patients with refractory idiopathic nephrotic syndrome.

Practical recommendations for the treatment of children

Steroid-responsive idiopathic nephrotic syndrome

There are still no definitive recommendations for the treatment of relapsing idiopathic nephrotic syndrome, and the multiplicity of drugs and strategies reviewed above does not make the prescription easier. The following recommendations are not universal. They apply to patients who suffer relapses.

When relapse occurs during steroid treatment or within the first 2 to 3 months after stopping steroids, we recommend resuming corticosteroids at an effective dose to obtain remission and to continue at a low dose on alternate days for at least 12 months before attempting to withdraw the drug. A new relapse indicates resumption of steroids for another period of at least 1 year, and so on, as long as this regimen is tolerated with a normal growth velocity. Levamisole may help reduce the steroid dose in some patients, but is without effect in those who are highly steroid-dependent.

When symptoms of steroid toxicity develop, or in patients with psychological intolerance, we would prescribe an alkylating agent at a dose not exceeding the threshold of gonadotoxicity. Chlorambucil (cumulative dose <8 mg/kg) or cyclophosphamide (cumulative dose <150 mg/kg) are valuable and equal alternatives, but intravenous mechlorethamine is probably the most appropriate if remission is needed rapidly.

The main problem comes later, with patients who remain highly steroid-dependent, and who develop symptoms of drug toxicity. In those patients, cyclosporin administered under careful surveillance, allows maintenance of remission, lowering or stopping of corticosteroid and disappearance of these symptoms. For those who continue to relapse and remain steroid-dependent but with side-effects, cyclosporin should be stopped and a second course of alkylating agent considered, being aware of the risk of irreversible gonadotoxicity. At this time it may be better to use a different alkylating agent.

One of the most important points in such a chronic disease, which could last more than 10 years, is to try and maintain a normal life style, interrupting the child's schooling and sports as little as possible, avoiding hospital admission, useless blood sampling, and limiting the follow up to strictly needed outpatient consultations. Such follow up is made easier if the parents keep a diary and record each day's proteinuria and treatment.

Steroid non-responders

Again there is no universally accepted scheme of treatment for this group of patients, after having clearly proved steroid-resistance, and diagnosed idiopathic nephrotic syndrome from a kidney biopsy. Even if some clinicians are tempted not to give any further specific treatment, due to the poor results already mentioned, our current approach is to prescribe a course of alkylating agent at a non-gonadotoxic cumulative dose for 8 to 12 weeks in association with alternate-day corticosteroids. In some cases proteinuria disappears progressively, and corticosteroids may be continued every other day at a lower dose for 3 to 6 months; when there is complete unresponsiveness to this treatment we usually prescribe a 'wash-out' period of 2 months before the next step, which remains under study and at the present time is represented by cyclosporin in association with corticosteroids for at least 6 months, with careful monitoring of renal function. If there is no response after this attempt, we think that no further specific treatment is currently justified, and symptomatic treatment, which should be started in any case from the onset, is needed. An angiotensin converting enzyme inhibitor may be useful for controlling arterial hypertension and may also have beneficial effects on the level of proteinuria and also on the associated process of progressive renal damage.

Symptomatic treatments

The short duration of nephrotic syndrome in steroid-responsive patients does not often render it necessary to prescribe symptomatic treatment. This is the reverse in the case of non-responders with a persisting nephrotic syndrome.

The treatment of oedema with diuretics and albumin infusion is discussed in Chapters 2.2 and 3.1. Refractory oedema with serious effusions may require cautious removal of ascites and/or pleural effusions by repeat tapping. Immersion of the body up to the neck in a bath may be helpful in these cases (Rascher et al. 1986). In extreme cases resistant oedema can be treated by mechanical ultrafiltration (Fauchald et al. 1985).

Any arterial hypertension has to be carefully controlled, using for preference an angiotensin converting enzyme inhibitor.

Calcium metabolism may be altered by the urinary loss of 25-hydroxycholecalciferol and its carrier protein (see Chapter 3.4). Preventive treatment with vitamin D supplements is sometimes useful.

Dietary therapy should include a 'normal' or moderately increased protein intake, of around 130 to 150 per cent of the recommended daily allowance, avoiding protein overload which may increase protein losses and consequently worsen nephrotic syndrome. Carbohydrates should be given preferentially as starch or dextrin-maltose, avoiding sucrose which increases lipid abnormalities.

Thyroxine substitution may be indicated, but only in patients with documented hypothyroidism due to urinary loss of iodinated proteins (see Chapter 3.4).

Lipid abnormalities are often present and may merit treatment (see Chapter 3.4).

Anticoagulants have been proposed as preventive treatment for thrombosis in some cases but are usually only given in patients with clinical symptoms or suspicion of thrombosis (see Chapter 3.4).

Complications related to treatment

Corticosteroids

There is a clear relationship between the cumulative dose, and the side-effects of these drugs, with a variable individual component, some subjects being more prone than others to develop these effects with similar cumulative dosage.

Corticosteroid side-effects must be looked for carefully in patients undergoing long-term treatment, in order to take the appropriate therapeutic decision in time and to avoid further development of these effects. Growth retardation in children, a major cause of concern, is discussed elsewhere.

The most apparent symptom of steroid toxicity is the Cushingoid syndrome with facial and truncal obesity and skin alterations. The cutaneous alterations include reversible symptoms such as hirsutism, acne and a tendency to atrophy, but also to striae, which are irreversible. Poor wound healing is usual. Facial changes and obesity may have serious psychological consequences in children and adolescents for whom body image is so important. Changes in mood and activity, including hyperactivity, increased aggressiveness, inattention, and sleep abnormalities are frequently seen, but more serious psychiatric disturbances are much less common.

Hypertension is a common side-effect and blood pressure has to be checked regularly after starting corticosteroid treatment. Gastrointestinal complications which may develop are mainly peptic ulceration, and exceptionally pancreatitis, the risk of which is increased by other factors such as hyperlipidaemia. A decreased resistance to infections is also associated with corticosteroid toxicity.

Osteoporosis is a long-term complication which can be assessed by annual measurement of bone mineral content. Aseptic osteonecrosis may be also observed. Enamel hypoplasia has been reported in the long term. Posterior lenticular cataracts are probably more frequent than realized when systematically looked for, but are usually asymptomatic (Limaye *et al.* 1988).

Multiple metabolic disturbances are related to steroid treatment, including increased protein catabolism with a tendency to muscle atrophy, abnormalities of carbohydrate and lipid metabolism, and hypercalciuria with the risk of urinary lithiasis. Blunting of the hypothalamoadrenal axis with absence of appropriate response to stress and risk of cardiovascular collapse must be remembered when tapering and stopping long-term steroid treatment. 'Benign' intracranial hypertension with papilloedema and convulsions are also possible complications, especially when steroids are tapered too quickly in patients undergoing this treatment for several years.

Alkylating agents
Cyclophosphamide

The acute toxic effects of cyclophosphamide includes bone marrow depression, haemorrhagic cystitis, gastrointestinal tract symptoms, alopecia, and infection. Leukopenia is frequently observed, but weekly haematological monitoring should limit its severity and concomitant steroids help blunt the marrow depression. Haemorrhagic cystitis can cause concern, but rarely occurs with the doses used in these patients, particularly when treatment is begun only after remission is achieved with corticosteroid therapy and is combined with forced diuresis (Levine and Richie 1989). Alopecia, which varies in degree, remits fully a few weeks after treatment has been stopped: it also could be avoided with a cooled helmet. Viral infections can be overwhelming if cyclophosphamide therapy is not stopped in time, and are most common in patients taking concomitant corticosteroids.

Long-term toxic effects include the risk of developing cancer, pulmonary fibrosis, ovarian fibrosis, and sterility. Gonadal toxicity is now well established and the risk of sterility is greater in boys than in girls. The cumulative threshold dose above which oligo/azoospermia may be expected lies between 150 and 250 mg/kg (Penso *et al.* 1974; Trompeter *et al.* 1981; Hsu *et al.* 1979). Such azoospermia is reversible in some patients (Buchanan *et al.* 1975). In female patients the minimal cumulative dose associated with risk of sterility is much higher, probably more than several times but not well defined. Pregnancies have been reported after treatments exceeding 18 months (Watson *et al.* 1986).

Cyclophosphamide is undoubtedly oncogenic in animals, but the risk of cancer is difficult to define with the short courses used in idiopathic nephrotic syndrome. One case of malignant mixed Mullerian tumour of the cervix was reported in a patient who received a 22-month course of cyclophosphamide (Bashour *et al.* 1973); Cameron *et al.* (1974b) reported a case of leiomyosarcoma in a 62-year old patient with relapsing minimal change disease treated with cyclophosphamide for over 1 year. A decrease of T-helper cells persisting for 6 to 12 months has been reported after stopping cyclophosphamide (Feehally *et al.* 1984); the extent to which this could be responsible for an increased susceptibility to infection is not known.

Chlorambucil

Acute toxic effects are less frequent with chlorambucil than with cyclophosphamide. Leukopenia and thrombocytopenia may occur, and are reversible within 1 to 3 weeks, and for this reason haematological monitoring should be performed weekly. Severe microbial and viral infections have been reported, including malignant hepatitis and measles encephalitis, in patients who contracted the infection during treatment (Lenoir *et al.* 1977). Latent EEG abnormalities were described in one-fifth of children given chlorambucil (Ichida *et al.* 1985): this may help explain the occurrence of clinical neurologic disorders, such as focal seizures (Williams *et al.* 1978).

Long-term toxic effects include the risk of developing cancer or leukaemia, which has only been reported in patients who had prolonged courses of treatment (Lenoir *et al.* 1977; Muller and Brandis 1981). Gonadal toxicity, as with cyclophosphamide, essentially affects boys. Azoospermia is total and probably irreversible at cumulative doses above 10 to 20 mg/kg. No case of azoospermia was reported in patients given less than 8 mg/kg (Guesry *et al.* 1978; Callis *et al.* 1980).

Mechlorethamine

Several types of side-effect occur with mechlorethamine treatment. In our experience (Broyer *et al.* 1985) one-third of children suffered from vomiting, despite parenteral antiemetics. Bacterial infections were seen in 11 per cent of children who had a persistent nephrotic syndrome. Haematological changes were noted in one-third of patients: lymphocytopenia (30 per cent), neutropenia (5 per cent), pancytopenia (1 case), most proving reversible in 4 to 5 weeks. Local perivenous reactions were prevented in all but one case by careful intravenous injection, followed by adequate flushing of the cannula.

Levamisole

Allergic reactions are possible, but the main complication is marrow suppression leading to leukopenia, with consequent risks of bacterial infection, avoided by regular control of blood counts.

Indomethacin and other non-steroidal anti-inflammatory drugs

Risks associated with non-steroidal anti-inflammatory agents in corticosteroid-resistant nephrotic syndrome have been reviewed by Velosa and Torres (1986). Their detrimental effect on renal function is well known, and patients with renal disease seem more vulnerable to this side-effect (Bennet 1983; Clive and Stoff 1984) (see also Chapter 6.2). Positive sodium balance, increased oedema and risk of arterial hypertension are recognized complications. In addition, non-steroidal anti-inflammatory agents antagonize the effects of frusemide (Tiggeler et al. 1977). The decrease of GFR observed with non-steroidal anti-inflammatory drugs is usually reversible, and marked only in salt-sodium depleted patients. However irreversible renal failure has been reported in children with non-responsive nephrotic syndrome and focal sclerosis (Kleinknecht et al. 1980). Renal damage was also observed in experimental nephrotic syndrome (Kleinknecht et al. 1983).

Several reports have described a syndrome including nephrotic syndrome with minimal change in glomeruli and more or less severe interstitial nephritis (Feinfeld et al. 1984) occurring after the prescription of non-steroidal anti-inflammatory drugs, e.g. fenoprofen, phenylbutazone, but there are no reports indicating that these drugs increase proteinuria in patients with established nephrotic syndrome. Other side-effects include allergic reactions with dermatitis and pruritus, gastrointestinal symptoms such as diarrhoea, and complications such as gastrointestinal perforation.

Cyclosporin

The main unwanted effect of cyclosporin is nephrotoxicity which is generally shown by deteriorating renal function (Myers 1986). However, renal insufficiency may occur either in relapsing steroid-sensitive or in steroid-resistant idiopathic nephrotic syndrome, and it is not always possible to determine whether deterioration of renal function is due to drug-induced nephrotoxicity, or to the renal disease itself, especially as the two events can, at least in theory, be superimposed on one another.

Impairment of renal function was reported in 26 of 83 patients treated with cyclosporin (Niaudet and Broyer 1989). In 19 of these, the rise of serum creatinine was transient, with a rapid return to pretreatment values after cessation of cyclosporin therapy. Some patients experienced a transient increase of serum creatinine during a relapse of a steroid-sensitive nephrotic syndrome. Impairment of renal function was irreversible or only partially reversible in seven patients, all with steroid-resistant nephrotic syndrome: in these patients the deterioration of renal function could be related to the natural course of the disease. Most patients who experienced nephrotoxicity had persistent nephrotic syndrome or were in relapse when the glomerular filtration rate was noted to be decreased, which suggests that factors other than cyclosporin nephrotoxicity may have been operative.

Several authors have found changes in renal histology in cyclosporin-treated patients with idiopathic nephrotic syndrome. Tejani et al. (1987) reported on three patients with repeat renal biopsy after 8 weeks of cyclosporin therapy and noted no changes in two and tubular atrophy in one. Niaudet et al. (1989) performed repeat renal biopsies in 30 patients, 27 with steroid-dependent and three with steroid-resistant nephrotic syndrome. There were no significant changes in 14, and small areas of interstitial fibrosis with clusters of atrophic tubules in 13 cases, but these lesions were difficult to interpret in six cases due to the presence of focal and segmental glomerular sclerosis in the initial biopsy. After 2 to 7 months of treatment one of the three steroid-resistant patients exhibited more severe tubulointerstitial nephritis with calcified tubular necrosis.

A third biopsy was performed in 10 children with steroid-dependent nephrosis, 20 to 37 months after starting cyclosporin. Of these, three showed normal kidneys, four had mild tubulointerstitial lesions and three had more severe lesions, clearly related to cyclosporin, and already present to a lesser extent in the previous biopsy. Renal function was not affected in two of these patients.

Other side-effects give less cause for concern. Moderate elevation of blood pressure was reported in 13 patients, hyperkalaemia in three patients, hypertrichosis in 27 patients, gum hypertrophy in 11 patients, and hypomagnesaemia in 32 patients.

Long-term outcome in children with idiopathic nephrotic syndrome

Long-term outcome of those initially steroid responsive

General evolution

Early reports on the long-term outcome of idiopathic nephrotic syndrome before the antibiotic and corticosteroid era (Lawson et al. 1960; Cornfeld and Schwartz 1966) showed that approximately 50 per cent of children recovered completely, while the mortality rate was high. Several other long term follow-up studies, including patients treated with corticosteroids and also immunosuppressive drugs, provide us with a more precise idea of the outcome. Schwartz et al. (1974) found that 20 per cent of patients had only one attack, and they underlined the major difference in outcome between corticosteroid responders and non-responders. In the steroid-sensitive group, 78 per cent achieved a persisting remission for 10 years and 85 per cent were in remission for 5 years. In the non-responders the patient survival was only 45 per cent after 5 years, 20 per cent after 10 years, and 16 per cent after 15 years. Similar results were found by Schärer and Minges (1973) with 22 per cent of patients with only one attack, 35 per cent of the relapsing patients continuing to relapse after 10 years, and a high mortality rate of 85 per cent in those not responsive to corticosteroids. In another study (ISKDC 1984) dealing with death during the first 5 to 15 years in 389 children with minimal change nephrotic syndrome, the mortality rate was 0.4 per cent in the 283 patients who were not early relapsers, but 6.3 per cent in the 63 who had relapse during the initial 8 weeks' treatment, death being mainly due to infection.

More recently, Trompeter et al. (1985) reported on the late outcome of 152 cases of steroid-responsive minimal change disease after a follow up of 14 to 19 years: 127 (83 per cent) were in remission, four had hypertension, 10 were still having nephrotic relapses requiring corticosteroids and 11 had died, seven from avoidable complications of the disorder. The duration of the disorder was inversely related to age at onset of symptoms, those starting before age 6 being predisposed to a lengthy relapsing course. In this study, there was no relationship between cure of the disease and pubertal development.

Wynn et al. (1988), found that 15 per cent of 132 patients with a mean follow up of 27.5 years had a persistent relapsing course; in this series three of 56 children with minimal change disease died from uraemia. Lewis et al. (1989) reported on 63 patients with corticosteroid-sensitive minimal change disease followed for between 10 and 21 years: two died accidentally but all the survivors had normal renal function and blood pressure. The proportion of patients relapsing during a given year after onset fell rapidly over the first 4 years and then reached a plateau up to 10 years, at around 40 per cent for those less than 6 at start, and around 10 per cent for the others; this proportion was 10 to 20 per cent higher in boys than in girls. This study found an inverse linear relationship between the length of remission and the proportion of patients relapsing in the subsequent 5 years, with a risk of 5 per cent after 9 years, and without the possibility of defining an end-point to the disease.

Late outcome must be analysed according to the mode of treatment. In the Enfants Malades series (Broyer 1988) 60 per cent of patients starting treatment before 1975 received large cumulative doses of chlorambucil over a long period of time (6 to 18 months). After 1975, only 20 per cent received cytotoxic drugs, and these were administered for only a short period of time, in order to avoid gonadal toxicity. In the first group the proportion of patients in persistent remission was 55 per cent, 75 per cent and 84 per cent at 5, 10, 15 years after onset, respectively; in the second group only 27 per cent were in remission at 5 years, and 50 per cent at 10 years.

It must be stressed that all of the above studies were performed in specialized centres, with a bias towards the more difficult cases. The study of Koskimies et al. (1982) reporting all patients with nephrotic syndrome occurring in Finland from 1967 to 1976, is less open to criticism from this point of view: 94 of 114 cases responded to corticosteroids: 24 per cent of these had no relapse, 22 per cent infrequent relapse and 54 per cent frequent relapses, more than two-thirds being in long remission at time of report. None of these patients developed renal insufficiency.

It is noteworthy that this generally good long-term prognosis was also observed in all steroid responders, even those with patterns other than minimal change on renal biopsy: the 19 steroid-sensitive patients with focal sclerosis reported by Arbus et al. (1982) remained responders, and none had renal insufficiency after a mean follow up of 10 years. Similar observations have been made in the Enfants Malades group. In other series, however almost all patients with focal and segmental glomerular sclerosis had a poor outcome, even if they were responders initially (Tejani et al. 1983; Southwest Pediatric Nephrology Study Group 1985b). These series include mainly black children, who may have different genetic or environmental backgrounds to European children.

Late non-responders

Some patients who initially respond to steroids become unresponsive after a variable period of time, from some months to more than 10 years. This course was noted in 3 to 5 per cent of patients with steroid-sensitive nephrotic syndrome (Trainin et al. 1975; Srivastava et al. 1985) but in no more than 2 per cent of the Enfants Malades series (Kleinknecht and Gubler 1983). The long term outcome was reported as good by Trainin et al. (1975), with a return to the responder state, either after another course of steroid treatment or, more often, after cyclophosphamide. Late non-responsiveness may, however also lead to end-stage renal failure. This outcome is usually anecdotal, after a long period of relapses (Nash et al. 1982) but was observed in four of 12 late non-responders in spite of repeat treatment with cyclophosphamide (Srivastava et al. 1985). End-stage renal failure due to late non-responsiveness occurred in eight patients of the large Enfants Malades series (Kleinknecht and Gubler 1983), all of whom were steroid sensitive to begin with. It is noteworthy that five received treatment for a short period of time and were then lost to follow up, with persistent proteinuria for 10 years or more; three others became corticosteroid-unresponsive after 1, 14, and 42 months, despite adequate treatment, and developed end-stage renal failure in less than 3 years. The series reported by Tejani (1985) was notably more pessimistic, with 12 of 48 initially steroid-responsive patients developing end-stage renal failure after transition from minimal change to focal and segmental glomerular sclerosis on a repeat biopsy. This series however is drawn from a different population where black subjects are preponderant, as already noted.

Agressive treatment has to be applied in these patients, whatever the histological pattern, since complete remission has been reported in a significant number of cases, using either cyclophosphamide (Trainin et al. 1975; Srivastava et al. 1985) or cyclosporin (Niaudet et al. 1989).

In summary, the long-term outcome of steroid-sensitive nephrotic syndrome may be considered as excellent for the majority of patients, in spite of the fact that approximately 50 per cent experience multiple relapses, the duration of the disease being longer with younger age at onset. After 10 to 15 years, 10 to 20 per cent of patients continue to relapse. Very few patients become late non-responders, and fewer—probably less than 1 per cent—go into end-stage renal failure, with the exception of some groups which include a majority of black American children.

Growth and adult height

Statural growth is *a priori* a major cause for concern, since it is known to be impaired by corticosteroid drugs, which form the basic treatment for idiopathic nephrotic syndrome and are often necessary for long periods of time.

Early reports showed that corticosteroid treatment for periods of 6 to 18 months was frequently associated with a reduction of stature (Lam and Arneil 1968). Use of alternate-day steroids was shown to minimize growth retardation in children with relapsing idiopathic nephrotic syndrome (Broyer and Kleinknecht 1978), and this regimen may even be associated with a gain of height in relation to the mean normal for age in prepubertal children (Polito et al. 1986). A recent study showed, nevertheless, that low-dose alternate-day steroids were able to inhibit growth, especially in boys more than 10 years of age, with a loss of 1 SD from the mean and a delay in pubertal development. This effect was not related to the relapse rate or the use of cyclophosphamide and was associated with a blunting of the expected overnight pulsatile release of growth hormone and gonadotrophins (Rees et al. 1988) in addition to the peripheral effects on chondrocyte metabolism. It must be emphasized that even if there is

3.5 Minimal changes and focal segmental glomerular sclerosis

Table 5 Outcome in 84 cases of primary corticosteroid non-responsive idiopathic nephrotic syndrome

	End-stage renal disease	Chronic renal failure	Nephrotic syndrome	Proteinuria + hypertension	Remission	
					Partial	Complete
n	37	5	6	2	3	31
Mean duration of evolution (years)	$5\frac{9}{10}$	$10\frac{9}{10}$	$10\frac{7}{10}$	13	$8\frac{4}{10}$	$8\frac{6}{10}$

some subsequent catch up, this delay may be associated with more or less severe emotional disturbance caused by loss of height and physical immaturity relative to peers.

Growth curves must be carefully determined in children with idiopathic nephrotic syndrome, with height being measured by the same person every 3 months in order to check tolerance of steroid treatment. A loss of 0.5 to 1 SD from the mean in a period of 1 to 2 years has to be taken into account when considering a change in the therapeutic approach with the aim of stopping steroids. Alkylating agents allow a variable period of remission without corticosteroids. Cyclosporin, which is able to maintain a remission state in 80 per cent of highly steroid-dependent patients without corticosteroids, is certainly one of the best ways to cope with the problem of growth retardation in pubertal boys who have already received alkylating agents and who continue to relapse and would otherwise need the continuation of corticosteroids (Niaudet and Broyer 1989).

In spite of the real stunting effect of long-term steroid therapy seen in a number of children with idiopathic nephrotic syndrome, adult height seems to be quite satisfactory. Foote et al. (1985) reported the height attainment in patients measured 5 to 24 years after diagnosis of steroid-responsive nephrotic syndrome. The mean height standard deviation score was −0.22, equivalent to a height on the 40th centile; the cumulative dose of corticosteroids prescribed correlated only weakly with the height SD. Trompeter et al. (1985) reported a mean height of 167 ± 1.2 cm in six male patients and of 159 ± 2 cm in four female patients who had had idiopathic nephrotic syndrome since childhood and who continued to relapse into adult life. Similar data have been observed in the Enfants Malades unit.

Long-term outcome of primary non-responders

The main difference between responders and non-responders is the tendency of the latter to develop end-stage renal failure, which is seen in less than 3 per cent of responders, even in the highly-selected series. Non-responders have also a higher risk of extrarenal complications of nephrotic syndrome.

In recent study the Enfants Malades group (Table 5) analysed 84 patients defined by the criteria already given, and who were followed up for at least 5 years. Fourteen of these patients were aged less than 1 year at onset, and nine had a sib affected by the same disease. The first renal biopsy showed minimal change in 33 cases (39 per cent), focal sclerosis in 34 cases (40 per cent) and diffuse mesangial proliferation in 17 cases (21 per cent); five of these were associated with focal sclerosis. At last examination 5 to 25 years after onset, 50 per cent of these patients had renal insufficiency and 40 per cent were in complete or partial remission, six remaining nephrotic and two having proteinuria and hypertension. Seventeen of the 34 remissions occurred during immunosuppressive treatment, whilst other patients entered progressively into remission. Six of the 12 patients who did not receive another treatment after the initial steroid treatment spontaneously went into complete remission afterwards, although the true proportion of such patients is unknown. Actuarial survival of kidney function in this series was 76 per cent at 5 years and 60 per cent at 10 years, with 50 per cent of patients on renal replacement therapy 11.5 years after onset (Fig. 10). There was some relationship between the pattern of the initial kidney biopsy and the evolution (Table 6). Chronic renal failure occurred in 38 per cent of those with minimal changes, 48 per cent of those with focal and segmental glomerular sclerosis, and 66 per cent of those with diffuse mesangial proliferation. Two subsets of patients had a more severe evolution: those with an onset below 1 year of age, and the familial cases, in whom there was no case of complete remission, and an incidence of 50 per cent of end-stage renal failure and 50 per cent of persisting nephrotic syndrome.

The data reported in several other series on non-responders with focal and segmental glomerular sclerosis are difficult to compare. Nevertheless, most of the clinical features described in the non-responders from the Enfants Malades series are similar to those reported by the Southwest Pediatric Nephrology Study group (1985). In this latter study, no statistically significant relationship was found between the histological pattern and the outcome, but glomerular hyalinosis tended to be more frequent in the initial biopsy specimen of patients who subsequently developed renal failure. Children who had minimal changes in the initial biopsy and who develop focal and segmental glomerular sclerosis had a longer period between onset of the disease and end-stage renal failure.

Growth may be more or less severely stunted in children with persistent nephrotic syndrome. Depletion of hormones due to urinary losses represents a possible cause of stunting. Hypothy-

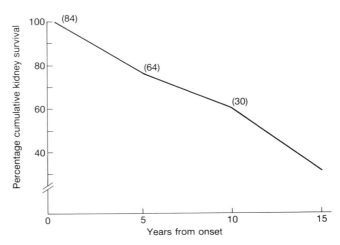

Fig. 10 Actuarial survival of kidney function in non-responsive idiopathic nephrotic syndrome (Enfants Malades Hôpital series).

Table 6 Relationship between histology in the first year and outcome more than 5 years later in primary steroid-resistant idiopathic nephrotic syndrome

	Minimal changes	Focal sclerosis	Diffuse mesangial proliferation + focal sclerosis
Complete or partial remission	17	14	4
Nephrotic syndrome	3	1	1
Chronic renal failure and end-stage renal disease	12(38%)	14(48%)	10(66%)

roidism related to urinary loss of iodinated proteins has been observed and may be corrected. A low plasma IgF1 and IgF2 level associated with a urinary loss of the carrier proteins has also been reported recently (Garin et al. 1989). It is not known whether recombinant growth hormone treatment could overcome the negative effects of these losses.

Recurrence of idiopathic nephrotic syndrome in transplanted kidneys

The development of nephrotic syndrome following transplantation (see Chapter 11.6) may be related to multiple causes, including de-novo membranous nephropathy, allograft nephropathy, and also chronic rejection (Cheigh et al. 1980). Focal and segmental glomerular sclerosis has been reported in kidneys transplanted into patients in whom this was not the original disease (Ettenger et al. 1977; Cheigh et al. 1981). Thus strict criteria have to be present before the diagnosis of recurrence is made in a transplanted patient with heavy proteinuria. We suggest the following criteria: past history of idiopathic nephrotic syndrome in the recipient associated with any of the morphological variants of this entity; the demonstration in the graft of one of three patterns observed in patients with idiopathic nephrotic syndrome, and especially focal and segmental glomerular sclerosis on late kidney graft biopsy. The rapid development of nephrotic syndrome after grafting is an additional criterion, but not mandatory. Recurrence of nephrotic syndrome with focal and segmental glomerular sclerosis was first reported by Hoyer et al. (1972). Several single centre (Malekzadeh 1979; Maizel et al. 1981; Cheigh et al. 1981; Morzycka et al. 1982; Habib et al. 1982; Axelsen et al. 1984; Korbet et al. 1988b; Senggutuvan et al. 1989) or multicentre (Zimmerman 1978; Leuman et al. 1981; Pinto et al. 1981) studies confirmed this risk in more than 200 patients with this primary renal disease. Efforts have been made to determine risk factors in order to predict the recurrence in individual patients.

The overall risk of recurrence was estimated to be around 25 per cent, but it is certainly different in children and in adult patients, as was shown in a recent single-centre study reporting eight recurrences in 16 children versus three in 27 adults (Senggutuvan et al. 1990). In children, recurrence appeared more frequent when the disease started after 6 years of age than before (Habib et al. 1982; Gagnadoux et al. 1987).

Rapid development of end-stage renal failure seems to be a major factor associated with recurrence. In most of the series a duration of disease shorter than 3 years was associated with a 50 per cent risk of recurrence, (Leumann et al. 1980; Pinto et al. 1981; Habib et al. 1982). In none of the reported series was the relationship between good HLA matching and recurrence reported by Zimmerman (1978) confirmed, especially after living-related donor transplantation.

The histopathological pattern observed on the first biopsy performed during the initial course of the disease is probably the most important predictive factor (Leumann et al. 1980; Maizel et al. 1981; Habib et al. 1982; Striegel et al. 1986; Gagnadoux et al. 1987; Senggutuvan et al. 1990). Nephrosis recurred in 50 to 80 per cent of patients in whom mesangial proliferation was present on initial biopsy, while patients with minimal changes on the first biopsy had a recurrence rate of 20 to 25 per cent. When renal histology was considered simultaneously with other risk factors: diffuse mesangial proliferation on the first biopsy and development of end-stage renal failure within 3 years: this was associated with recurrence risk of 80 per cent or more (Donckerwolke et al. 1982; Gagnadoux et al. 1987).

Recurrence is usually noted as soon as urine is produced by the graft with massive proteinuria (Maizel et al. 1981). Proteinuria and nephrotic syndrome may occur after several months or years, with a histological pattern suggesting recurrence (Gagnadoux et al. 1987).

The high frequency of prolonged oligoanuria after grafting in immediate recurrence was underlined by Striegel et al. (1986) and found in 80 per cent of cases by Gagnadoux et al. (1987), a proportion much higher than that observed in the non-recurring patients.

Evolution of the disease in the graft is not uniform. Sixty per cent of those with recurrence lost their graft, versus 23 per cent of those without recurrence (Gagnadoux et al. 1987). Similar failure rates were reported in other series (Pinto et al. 1981; Striegel et al. 1986). All of the patients with immediate recurrence lost their graft in the series of Senggutuvan et al. (1990). Some patients entered into remission either after additional treatment or spontaneously (Gagnadoux et al. 1987). Other patients may show good renal function for many years despite proteinuria and nephrotic syndrome (Leumann et al. 1980; Pinto et al. 1981; Gagnadoux et al. 1987).

Pulse methylprednisolone was effective in one of six cases (Gagnadoux et al. 1987). A partial remission was observed in one patient given meclofenamate (Torres et al. 1984) but heavy proteinuria was seen when the drug was stopped transiently. Plasma exchange was used in several patients (Solomon et al. 1981; Zimmerman 1985), with partial or transient results. More recently, cyclosporin has given variable results. Burke and Rigby (1986) reported complete remission with this drug, but complete failure was also reported (Voets et al. 1986; Pirson et al. 1986). In the Enfants Malades group, six patients with recurrence were given cyclosporin; two had a complete remission within the first months, two a partial remission, and two did not respond at all.

Early graft biopsy showed typical electron microscopic changes of idiopathic nephrotic syndrome, with fusion of epithelial cell foot processes as early as 12 days postgrafting (Maizel et al. 1981). Focal sclerosis was regularly observed in patients who had immediate or early recurrence, at least on repeat biopsy.

This lesion was preceded in two cases by focal segmental epithelial proliferation (Korbet *et al.* 1988*a*). It was remarkable that massive proteinuria preceded these lesions in one patient given sequential biopsy (Morales *et al.* 1988), which is another argument for considering focal and segmental glomerular sclerosis more as a consequence of heavy proteinuria than its cause.

Subsequent management of patients who lose their graft due to recurrence is difficult. An incidence of 75 per cent multiple recurrence in successive allografts has been reported (Chandra *et al.* 1981*b*; Hosenpud *et al.* 1985; Senggutuvan *et al.* 1990) and the tendency to recur may persist for a long period, as shown in the series of Senggutuvan *et al.* (1990) with a patient whose disease recurred in the fifth graft 9 years after he first started on dialysis.

Recurrence of idiopathic nephrotic syndrome after grafting has implications for the pathogenesis of the condition: it clearly suggests that humoral factor(s) are involved in the induction of glomerular proteinuria. The presence of such factor(s) was indirectly shown in a patient who had a recurrence in his graft (Zimmerman 1984).

In some patients recurrence of idiopathic nephrotic syndrome supports the unitary concept of the disease. Steroid-sensitive minimal change disease may also recur in a kidney graft (Mauer *et al.* 1979), again suggesting the role of humoral factor(s). Transplantation allows us to observe the reverse transition from focal and segmental glomerular sclerosis to minimal change disease: one patient with steroid-resistant focal and segmental glomerular sclerosis and a rapid development of end-stage renal failure had an early recurrence in a second graft which responded to boluses of methylprednisolone. Relapse occurred several times until he received cyclophosphamide followed by persistent remission: a graft biopsy showed only minimal changes (personal observation partly reported by Gagnadoux *et al.* 1987).

Treatment and outcome in adults

There are very few hints (apart from the data given above on differing HLA typing) that adult nephrosis with minimal changes or focal and segmental glomerular sclerosis is essentially different from idiopathic nephrotic syndrome in children. It may, therefore, seem unnecessary to discuss the issue of evolution and treatment on adults separately. However, treatment schedules in adults are less standardized than those used in children, making definitions of steroid sensitivity and resistance less precise. In addition, relapse rates vary according to age in minimal change disease, progressively decreasing in older patients. The third reason is that response to steroids and cytotoxic drugs in equivalent doses is slower in adults than in children, and courses lasting much longer than 4 weeks are needed to establish the presence of resistance (Nolasco *et al.* 1986; Korbet *et al.* 1988*b*).

Response to corticosteroid therapy

Considerable differences in treatment protocols between adults and children (Meyrier and Simon 1988) are reported in the literature. Nephrologists treating adults usually prescribe lower doses than paediatricians, from 0.33 to 2 mg/kg.day. In papers published over the last 25 years, the length of treatment varied from 3 weeks to 3 years, and dosage varied, with some authors starting at a maximum, and others increasing stepwise by 20 mg/day. Drugs have been given daily or on alternate days, and after obtaining remission, cessation of therapy may be either progressive or abrupt, the latter perhaps occasionally explaining short-term relapses due to a rebound effect. These inconsistencies in treatment modes used in adult nephrotics obviously make the definitions of steroid sensitivity, dependence, resistance, and multiple relapses most imprecise.

Table 7 Response of adult patients with idiopathic nephrotic syndrome to corticosteroids (1961–1986)

	Patients (*n*)	Complete remissions (%)	Partial remissions (%)	Failures (%)
Minimal changes	301	226 (75.8)	21 (7)	55 (18.2)
Focal sclerosis	153	24 (15.6)	31 (20.2)	98 (64.2)
Biopsy not done	48	31 (64.6)	0 (0)	17 (35.4)
Total	503	281 (55.8)	52 (10.3)	170 (33.9)

Despite these shortcomings, an analysis of the literature (Meyrier and Simon 1988) shows that response to steroid treatment does not seem to differ greatly from that reported in children (Table 7). Differences between adults and children stem rather from the relapse rate after a first course of treatment. In some studies, however, the response rate appears to be lower in adults than in children (Nolasco *et al.* 1986; Korbet *et al.* 1988*b*). Nephrosis due to minimal change disease often follows a frequently relapsing course.

The trend toward a diminished number and frequency of relapses with greater age is apparent in adults (provided nephrosis did not start during childhood). Stable remissions were reported in 20 to 70 per cent of patients (Grupe 1982*b*). Nolasco *et al.* (1986) followed 89 patients with adult onset minimal change, 58 of whom responded to corticosteroid treatment: 24 per cent never relapsed, 56 per cent relapsed on a single occasion or infrequently, and only 21 per cent were frequent relapsers. Korbet *et al.* (1988*b*) followed 40 adults with minimal change disease; 34 were treated with prednisone, and 31 (91 per cent) achieved remission, of whom only three suffered multiple relapses. Relapses were infrequent in 99 adult nephrotics with minimal change disease followed by Wang *et al.* (1982) for 3 to 102 months. In 85 patients whose urine was protein-free for at least 6 months, four relapsed; of 46 who were protein-free for 24 months, three relapsed; of 37 followed for 36 to 96 months, only three relapsed.

Thus, the experience of nephrologists treating adult nephrosis is comparable in Europe, America, and the Far East, with multiple relapse rate in the order of 10 per cent, as opposed to a probably higher percentage in children.

Apart from age, insufficient treatment might also explain the frequency of relapses after a first treatment course. In four of six reports published between 1971 and 1988, patients were treated with a long course of steroids; the other two used a short regimen. The remission rate was comparable with the two treatment modes, but duration of remission was superior when patients received more than 8 months of prednisone. To test the hypothesis that relapses might be due to a rebound effect after inadequate initial treatment, two groups of adult nephrotics of the same age (Simon and Meyrier 1989) were compared. One group, comprising 28 patients, was treated for 2 to 4 months with 1 mg/kg.day prednisone. Another group of 26 was treated with 2 mg/kg every other day for 3 months, after which the dosage was slowly tapered over the ensuing 9 months. Initial remission rate was similar but subsequent relapse rate differed. In the first

Table 8 Response of adult patients with idiopathic nephrotic syndrome to conventional immunosuppressive drugs (1966–1986)

	Patients (n)	Complete remissions (%)	Partial remissions (%)	Failures (%)
Minimal change disease	152	123 (80.9)	13 (8.5)	16 (10.5)
Focal sclerosis	63	16 (25.4)	8 (12.7)	36 (61.9)
Total	215	139 (64.6)	21 (9.8)	55 (25.6)

group, there was one relapse in nine patients, two in seven patients, three in two patients, and multiple relapses in 10. In the second group, there was one relapse in 19 patients, two in two patients, three in none, and multiple relapses in five. In retrospect, the cumulative dosage of corticosteroids had been less in the second group. The study confirmed that long initial alternate-day corticosteroid therapy followed by slow tapering is effective in obtaining sustained remission in adult minimal change disease, as it is in children.

Response to conventional immunosuppressive therapy

The immunosuppressive regimens used in the treatment of adult nephrosis are similar to those used by paediatricians. Experience with various immunosuppressive regimens using cyclophosphamide, chlorambucil, and mechlorethamine published between 1966 and 1986 has recently been reviewed (Meyrier and Simon 1988) and is summarized in Table 8. The indications were the same as in children, namely steroid-resistant, steroid-dependent, or relapsing idiopathic nephrotic syndrome. There were no significant differences in response to such drugs between adults and children.

Azathioprine deserves special mention, since it has occasionally been administered to adult nephrotics, with a few anecdotal successful results (Meyrier and Simon 1988). Cade et al. (1986) studied the effect of azathioprine in 13 adult patients, including 11 in whom nephrosis appeared after the age of 14. The diagnosis was focal and segmental glomerular sclerosis in four and minimal change disease in nine. Six were steroid-resistant from the outset; the others had multirelapsing nephrotic syndrome and four eventually developed resistance. The results of prolonged treatment with azathioprine were favourable: the 12 patients who complied to follow up had progressively attained complete remission. Such observations should be an incentive to undertake large-scale prospective controlled studies on the effect of azathioprine in adult nephrosis. Such a protocol should be applied for at least 18 months before conclusions are reached with regard to efficacy.

Response to treatment with cyclosporin

Treatment of idiopathic nephrotic syndrome with cyclosporin was undertaken in 1986 and at the time of writing the experience exceeds 300 cases (Meyrier et al. 1989). The rationale, the indications, and the results of cyclosporin treatment in childhood nephrosis have been analysed in the corresponding section of this chapter. We shall focus on some minor differences observed between children and adults.

In patients who are responders to corticosteroids but who suffer multiple relapses, cyclosporin induced complete remission or maintained remission after dose reduction or withdrawal of steroids slightly less frequently in adults (77 per cent) than in children (89 per cent). The success rate was lower in steroid-resistant patients, whether adults or children (Figs. 11 and 12). The main difference stemmed from the interpretation of 'partial remission', the definition of which is vague. Apparently more adults than children achieved such partial remission.

Cyclosporin alone or cyclosporin + corticosteroids?

In adults the trial of the Société de Néphrologie (Meyrier et al. 1989), consisted of treatment with cyclosporin alone for the first 3 months in 51 patients. Among 23 patients with minimal change disease, cyclosporin induced complete remission in 13, partial remission in two, and was a failure in eight. Cyclosporin treatment failed in 20 of 33 patients with focal and segmental glomerular sclerosis. These figures differed from the somewhat better results obtained in other trials where a small dosage of corticosteroids was maintained during cyclosporin treatment (Lagrue et al. 1986). The impression that cyclosporin plus low-dose prednisone produced better results than cyclosporin alone was con-

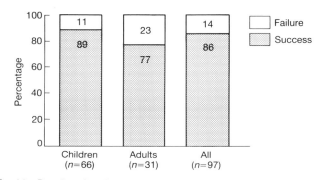

Fig. 11 Results of cyclosporin treatment in corticosensitive idiopathic nephrotic syndrome (minimal-change disease and focal segmental glomerulosclerosis). 'Success' regarded as complete remission or patterns of remission after dose reduction or withdrawal of steroids (from Meyrier 1989, with permission of the publisher).

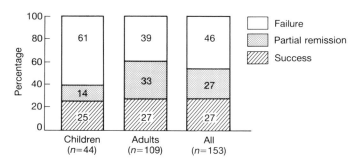

Fig. 12 Results of cyclosporin treatment in corticoresistant idiopathic nephrotic syndrome (minimal-change disease and focal segmental glomerulosclerosis). 'Success' regarded as complete remission or maintenance of remission after dose reduction or withdrawal of steroids (from Meyrier 1989, with permission of the publisher).

firmed by the subsequent development of the trial: half of the patients in sustained remission seemed to depend on the drug combination. Finally, the experience of nephrologists treating adults was similar to that of paediatricians: cyclosporin had the major advantage of reducing the threshold of corticosensitivity and allowing a reduction in steroids, or their withdrawal.

References

Abramowicz, M., et al. (1970). Controlled trial of azathioprine in children with nephrotic syndrome. *Lancet*, **i**, 959–61.

Abrass, C.K., Hall, C.L., Border, W.A., Brown, C.A., Glassock, R.J., and Coggins, C.H. (1980). Circulating immune complexes in adults with idiopathic nephrotic syndrome. *Kidney International*, **17**, 545–53.

Abreo, K. and Wen, S.-F. (1983). A case of IgA nephropathy with an unusual response to corticosteroid and immunosuppressive therapy. *American Journal of Kidney Diseases*, **3**, 54–7.

Alfiler, C.A., Roy, L.P., Doran, T., Sheldon, A., and Bashir, H. (1980). HLA-DRW7 and steroid responsive nephrotic syndrome of childhood. *Clinical Nephrology*, **14**, 71–4.

Al-Khader, A.A., Lien, J.W., and Aber, G.M. (1979). Cyclophosphamide alone in the treatment of adult patients with minimal change glomerulonephritis. *Clinical Nephrology*, **11**, 26–30.

Allon, M., Campbell, W.G., Nasr, S.A., Bourke, E., Stoute, J., and Guntupalli, J. (1988). Minimal change glomerulonephropathy and interstitial infiltration with mycosis fungoides. *American Journal of Medicine*, **84**, 756–9.

Alpers, C.E., Harawi, S., and Rennke, M.S. (1988). Focal glomerulosclerosis with tubuloreticular inclusions: possible predictive value for AIDS. *American Journal of Kidney Diseases*, **12**, 240–2.

Andal, A., Saxena, S., Chellani, H.K., and Sharma, S. (1989). Pure mesangial proliferative glomerulonephritis. A clinico-morphologic analysis and its possible role in morphologic transition of minimal change lesion. *Nephron*, **51**, 314–9.

Ansquer, J.C., Laurent, J., Lagrue, G., Cambon de Mouzon, A., and Bracq, G. (1983). HLA DR7 in adult lipoid nephrosis patients. *Nephron*, **34**, 270.

Arakawa, M. (1976). A scanning electron microscopy of the glomerulus of normal and nephrotic rats. *Laboratory Investigation*, **23**, 489–94.

Arant, B.S., Singer, S.A., and Bernstein, J. (1982). Steroid dependent nephrotic syndrome. *Journal of Pediatrics*, **100**, 328–33.

Arbeitsgemeinschaft für Pädiatrische Nephrologie. (1979). Alternate-day vs. intermittent prednisone in frequently relapsing nephrotic syndrome. *Lancet*, **i**, 401–3.

Arbeitsgemeinschaft für Pädiatrische Nephrologie. (1981). Alternate-day prednisone is more effective than intermittent prednisone in frequently relapsing nephrotic syndrome. *European Journal of Pediatrics*, **135**, 229–37.

Arbeitsgemeinschaft für Pädiatrische Nephrologie. (1982). Effect of cytotoxic drugs in frequently relapsing nephrotic syndrome with or without steroid dependance. *New England Journal of Medicine*, **306**, 451–4.

Arbeitsgemeinschaft für Pädiatrische Nephrologie. (1987) Cyclophosphamide treatment of steroid dependant nephrotic syndrome: comparison of eight week with 12 week course. *Archives of Diseases in Childhood*, **62**, 1102–1106.

Arbeitsgemeinschaft für Pädiatrische Nephrologie. (1988). Short versus standard prednisone therapy for initial treatment of idiopathic nephrotic syndrome in children. *Lancet*, **i**, 380–3.

Arbus, G.S., Poucell, S., Bacheyie, G.S., and Baumal, R. (1982). Focal segmental glomerulosclerosis with idiopathic nephrotic syndrome: Three types of clinical response. *Journal of Pediatrics*, **101**, 40–5.

Axelsen, R.A., Seymour, A.E., Mathew, T.H., Fisher, G., Canny, A., and Pascoe, V. (1984). Recurrent focal glomerulosclerosis in renal transplants. *Clinical Nephrology*, **21**, 110–4.

Bader, P.I., Grove, J., Trygstad, C.W., and Nance, W.E., (1974). Familial nephrotic syndrome. *American Journal of Medicine*, **56**, 34–43.

Bakir, A.A., Bazilinski, N.G., Rhee, H.L., Ainis, H., and Dunea, G. (1989). Focal glomerular sclerosis, a common entity in nephrotic black adults. *Archives of Internal Medicine*, **149**, 1802–4.

Bakker, W.W. and Van Luijk, W.H.J. (1989). Do circulating factors play a role in the pathogenesis of minimal change nephrotic syndrome? *Pediatric Nephrology*, **3**, 341–9.

Bakker, W.W., Beukhof, J.R., Van Luijk, W.H.J., and Van der Hem, G.K. (1982). Vascular permeability increasing factor (VPF) in IgA nephropathy. *Nephron*, **18**, 165–7.

Bakker, W.W., Van Luijk, W.H.J., Hené, R.J., Desmit, E.M., Van der Hem, G.K., and Vos, J.T. (1986). Loss of glomerular polyanion *in vitro* induced by mononuclear blood cells from patients with minimal-change nephrotic syndrome. *American Journal of Nephrology*, **6**, 101–6.

Ballow, M., Kennedy, T.L., Gaudio, K.M., Siegel, N.J., and McLean, R.H. (1982). Serum hemolytic factor D values in children with steroid-responsive idiopathic nephrotic syndrome. *Journal of Pediatrics*, **100**, 192–6.

Balsan, S., Lenoir, G., Steru, D., Bourdeau, A., and Grimberg, R. (1987). Effects of long term deflazacort maintenance therapy on mineral metabolism and statural growth in children. *Calcified Tissue International*, **40**, 303–9.

Baluarte, H.J., Hiner, L., and Gruskin, A.B. (1978). Chlorambucil dosage in frequently relapsing nephrotic syndrome: a controlled clinical trial. *Journal of Pediatrics*, **92**, 295–8.

Baluarte, H.J., Gruskin, A.B., Polinski, M.S., Prebis, J.W., and Rosenblum, H. (1980). Chlorambucil therapy in the nephrotic syndrome. In *Pediatric Nephrology. Proceedings of the 5th International Pediatric Nephrology Symposium*. (eds. A. B. Gruskin and H. E. Norman), pp. 423–429. Martinus Nijhoff, The Hague.

Barbiano di Belgiojoso, G., Mazzucco, G., Casanova, S., Radaelli, L., Monga, G., and Minetti, L. (1986). Steroid-sensitive nephrotic syndrome with mesangial IgA deposits: a separate entity? *American Journal of Nephrology*, **6**, 141–5.

Barnett, H.L., Schoeneman, M., Bernstein, J., and Edelmann, C.M. Jr. (1978). Minimal change nephrotic syndrome. In *Pediatric Kidney Disease*. (ed. C. M. Edelmann Jr.), pp. 695–711. Little, Brown and Co. Boston.

Barratt, T.M. and Soothill, J.F. (1970). Controlled trial of cyclophosphamide in steroid sensitive relapsing nephrotic syndrome of childhood. *Lancet*, **ii**, 479–82.

Barratt, T.M., Cameron, J.S., Chantler, C., Ogg, C.S., and Soothill, J.F. (1973). Comparative trial of 2 weeks and 8 weeks cyclophosphamide in steroid sensitive relapsing nephrotic syndrome in childhood. *Archives of Diseases in Childhood*, **48**, 286–90.

Barratt, T.M., Osofsky, F.G., Bercowsky, A., Soothill, J.F., and Kay, R. (1975). Cyclophosphamide treatment in steroid sensitive nephrotic syndrome of childhood. *Lancet*, **i**, 55–8.

Bashour, B.N., Mancer, K., and Rance, C.P. (1973). Malignant mixed Müllerian tumour of the cervix following cyclophosphamide therapy for nephrotic syndrome. *Journal of Pediatrics*, **82**, 292–3.

Baylis, C., Ichikawa, I., Willis, W.T., Wilson, C.B., and Brenner, B.M. (1977). Dynamics of glomerular ultrafiltration IX effects of plasma protein concentration. *American Journal of Physiology*, **232**, F58–F71.

Beale, M.G., Hoffsten, P.E., Robson, A.M., and MacDermott, R.P. (1980). Inhibitory factors of lymphocyte transformation in sera from patients with minimal change nephrotic syndrom. *Clinical Nephrology*, **13**, 271–6.

Beale, M.G., Nash, G.S., Bertovich, M.J., and MacDermott, R.P. (1983). Immunoglobulin synthesis by peripheral blood mononuclear cells in minimal change nephrotic syndrome. *Kidney International*, **23**, 380–6.

Bennett, W.M. (1983). The adverse renal effects of nonsteroidal anti-inflammatory drugs: increasing problems or overrated risk. *American Journal of Kidney Diseases*, **2**, 477.

Berg, U. and Bohlin, A.B. (1982). Renal hemodynamic in minimal change nephrotic syndrome in childhood. *International Journal of Pediatric Nephrology*, **3**, 187–92.

Bergstrand, A., Bollgren, I., and Samuelson, A. (1973). Idiopathic nephrotic syndrome of childhood. Cyclophosphamide induced conversion from steroid refractory to highly steroid sensitive disease. *Clinical Nephrology*, **1**, 302–6.

Bhasin, H.K., Abuelo, J.G., Nayak, R., and Esparza, A.R. (1978). Mesangial proliferative glomerulonephritis. *Laboratory Investigation*, **39**, 21–9.

Bhuyan, U.N. and Srivastava, R.N. (1987). Incidence and significance of IgM mesangial deposits in relapsing idiopathic nephrotic syndrome of childhood. *Indian Journal of Medical Research*, **86**, 53–60.

Blau, E.B. and Haaz, J.E. (1973). Glomerular sialic acid and proteinuria in human renal disease. *Laboratory Investigation*, **28**, 477–81.

Bohle, A., Fischbach, H., Werner, H., Woerz, U., Edel, H.H., Kluthe, R., and Scheller, F. (1974). Minimal change lesion with nephrotic syndrome and focal glomerular sclerosis. *Clinical Nephrology*, **2**, 52–8.

Bohlin, A.B. (1984). Clinical course and renal function in minimal change nephrotic syndrome. *Acta Paediatrica Scandinavica*, **73**, 631–6.

Bohman, S.O., Jaremko, G., Bohlin, A.B., and Berg, U. (1984). Foot process fusion and glomerular filtration rate in minimal change nephrotic syndrome. *Kidney International*, **25**, 696–700.

Bohrer, M.P., Baylis, C., Robertson, C.R., and Brenner, B.M. (1977). Mechanism of the puromycin-induced defects in the transglomerular passage of water and macromolecules. *Journal of Clinical Investigation*, **60**, 152–61.

Border, W.A. (1979). Immune complex detection in glomerular disease. *Nephron*, **24**, 105–13.

Boulton-Jones, J.M., Tulloch, I., Dore, B., and McLay, A. (1983). Changes in the glomerular capillary wall induced by lymphocyte products and serum of nephrotic patients. *Clinical Nephrology*, **20**, 72–7.

Boulton-Jones, J.M., McWilliams, G., and Chandrachud, L. (1986). Variation in charge on red cells of patients with different glomerulopathies. *Lancet*, **ii**, 186–9.

Brandis, M., Burchard, R., Leititis, J., Zimmerhackl, B., Hildebrandt, F., Helmcher, V. (1987). Cyclosporine A for treatment in nephrotic syndrome (abstract). *Pediatric Nephrology*, **1**, C42.

Brodehl, J., Ehrich, J.J.H., Hoyer, P.F., Lee, I.J., Oemar, B.S., and Wonigeit, K. (1987). Cyclosporine A treatment of minimal change nephrotic syndrome and focal segmental glomerulosclerosis in children. *Korean Journal of Nephrology*, **6**, 26–33.

Brouhard, B.H., Goldblum, R.M., Brunce, H. III, and Cunningham, R.J. (1981). Immunoglobulin synthesis and urinary IgG excretion in the idiopathic nephrotic syndrome of children. *International Journal of Pediatric Nephrology*, **2**, 163–9.

Brown, M. and Goldstein, J. (1983). Lipoprotein metabolism in the macrophage: Implications for cholesterol deposition in atherosclerosis. *Annual Review of Biochemistry*, **52**, 223–61.

Broyer, M. (1988). La néphrose idiopathique de l'enfant et son traitement. *Archives Francaises de Pédiatrie*, **45**, 1–4.

Broyer, M. and Kleinknecht, C. (1978). Traitement des néphroses corticosensibles par la corticothérapie discontinue prolongée. In *Journées Parisiennes de Pédiatrie* (eds. H.E. Brissaud, *et al.*), pp. 158–170. Flammarion Médecine-Science, Paris.

Broyer, M., Meziane, A., Kleinknecht, C., and Niaudet, P. (1985). Nitrogen mustard therapy in idiopathic nephrotic syndrome of childhood. *International Journal of Pediatric Nephrology*, **6**, 29–34.

Buchanan, J.D., Fairley, K.F., and Barris, J.U. (1975). Return of spermatogenesis after stopping cyclophosphamide therapy. *Lancet*, **ii**, 156–7.

Burke, J.R. and Rigby, R.J. (1986). Ciclosporine and prednisolone: do they prevent recurrence of focal segmental glomerulosclerosis? *Nephron*, **43**, 236–7.

Cade, R., *et al.* (1986). Effect of long-term azathioprine administration in adults with minimal change glomerulonephritis and nephrotic syndrome resistant to corticosteroids. *Archives of Internal Medicine*, **146**, 737–41.

Cairns, S.A., London, R.A., and Mallick, N.P. (1982). Circulating immune complexes in idiopathic glomerular diseases. *Kidney International*, **21**, 507–12.

Cagnoli, L., Tabacchi, P., Pasquali, S., Cenci, M., Sasdelli, M., and Zucchelli, P. (1982). T cell subset alterations in idiopathic glomerulonephritis. *Clinical and Experimental Immunology*, **50**, 70–6.

Cale, W.F., Ulrich, I.H., and Jenkins, J.J. (1982). Nodular sclerosing Hodgkin's disease presenting as nephrotic syndrome. *Southern Medical Journal*, **75**, 604–6.

Callis, L., Nieto, J., Vila, A., and Rende, J. (1980). Chlorambucil treatment in minimal lesion nephrotic syndrome: a reappraisal of its gonadal toxicity. *Journal of Pediatrics*, **97**, 653–6.

Cambon-Thomsen, A., *et al.* (1986). HLA et Bf dans le syndrome néphrotique de l'enfant: différences entre les formes corticosensibles et corticorésistantes. *Pathologie et Biologie*, **34**, 725–30.

Cameron, J.S., Ogg, C.S., Turner, D.R., and Weller, R.O. (1973). Focal glomerulosclerosis. In *Glomerulonephritis* (eds. P. Kincaid-Smith, T. H. Mathew, and E. L. Becker), pp. 249–261, Part I John Wiley and Sons, New York.

Cameron, J.S., Chantler, C., Ogg, C.S., and White, R.H.R. (1974*a*). A long term stability of remission in nephrotic syndrome after treatment with cyclophosphamide. *British Medical Journal*, **4**, 7–11.

Cameron, J.S., Turner, D.R., Ogg, C.S., Sharpstone, R., and Brown C.B. (1974*b*). The nephrotic syndrome in adults with minimal change glomerular lesions. *Quarterly Journal of Medicine*, **43**, 461–88.

Capodicasa, G., De Santo, N.G., Nuzzi, F., and Giordano, C. (1986). Cyclosporine A in nephrotic syndrome of childhood a 14 months experience. *International Journal of Pediatric Nephrology*, **7**, 69–72.

Carrie, B.J., Salyer, W.R., and Myers, B.D. (1981) Minimal change nephropathy: An electrochemical disorder of the glomerular membrane. *American Journal of Medicine*, **70**, 262–8.

Carroll, N., Crock, G.W., Funder, C.C., Green, C.R., Ham, K.N., and Tange, J.D. (1973). Scanning electron microscopy of aminonucleoside nephrosis. *Journal of Pathology*, **111**, 37–42.

Cavallo, T. and Johnson, M.P. (1981). Immunopathologic study of minimal change glomerular disease with mesangial IgM deposits. *Nephron*, **27**, 281–4.

Chandler, P., Soni, A., Suri, A., Bhagwat, R., Yoo, J., and Treser, G. (1987). Renal ultrastructural markers in AIDS associated nephropathy. *American Journal of Pathology*, **126**, 513–25.

Chandra, M., Mouradian, J., Hoyer, J.R., and Lewy, J.E. (1981*a*). Familial nephrotic syndrome and focal segmental glomerulosclerosis. *Journal of Pediatrics*, **98**, 556–60.

Chandra, M., Lewy, J.E., Mouradian, J., Susin, M., and Hoyer, J.R. (1981*b*). Reccurent nephrotic syndrome with three successive renal allografts. *American Journal of Nephrology*, **1**, 110–4.

Cheigh, J.S., *et al.* (1980). Kidney transplant nephrotic syndrome: relationship between allograft histopathology and natural course. *Kidney International*, **18**, 358–65.

Cheigh, J.S., *et al.* (1981). Focal segmental glomerulosclerosis in kidney transplants. *Transplantation Proceedings*, **13**, 125–7.

Cheng, I.K.P., Jones, B.M., Chan, P.C.K., and Chan, M.K. (1989). The role of soluble immune response suppressor lymphokine in prediction of steroid responsiveness in idiopathic nephrotic syndrome. *Clinical Nephrology*, **32**, 168–72.

Chiu, J. and Drummond, K.N. (1974). Long-term follow-up of cyclophosphamide therapy in frequent relapsing minimal lesion nephrotic syndrome. *Journal of Pediatrics*, **84**, 825–30.

Churg, J., Habib, R., and White, R.H.R. (1970). Pathology of the nephrotic syndrome in children. *Lancet*, **i**, 1299–1302.

Clark, A.G.B., Vaughan, R.W., Stephens, H.A.F., Chantler, C., Williams, D.G., and Welsh, K.I. (1990). Genes encoding the β-chains of HLA-DR7 and HLA-DQW2 define major susceptibility determinants for idiopathic nephrotic syndrome. *Clinical Science*, **78**, 391–7.

Clive, D.M. and Stoff, J.S. (1984). Renal syndrome associated with nonsteroidal antiinflammatory drugs. *New England Journal of Medicine*, **310**, 563–72.

Cochat, P. and Larbre, F. (1988). Syndrome nephrotique après piqûre de méduse. *Pediatrie*, **43**, 419–20.

Cohen, A.H., Border, W.A., and Glassock, R.J. (1978). Nephrotic syndrome with glomerular mesangial IgM deposits. *Laboratory Investigations*, **38**, 610–9.

Cohen, H.T., Singh, A.K., Kasinath, B.S., and Lewis, E.J. (1988). Red cell surface charge in patients with nephrotic syndrome. *Lancet*, **i**, 1459.

Coldbeck, J.H. (1963). Experience with alkylating agents in the treatment of children with the nephrotic syndrome. *Medical Journal of Australia*, **2**, 987–9.

Conte, J., Suc, J.M., and Mignon-Conte, M. (1967). Effet antiprotéinurique de l'indométacine dans les glomerulopathies. *Journal d'Urologie et de Néphrologie*, **73**, 850–6.

Coovadia, H.M., Adhikari, M., and Morel-Maroger, L. (1979). Clinicopathological features of the nephrotic syndrome in South African children. *Quarterly Journal of Medicine*, **48**, 77–91.

Coovadia, H.M. and Adhikari, M. (1989). Outcome of childhood minimal change disease. *Lancet*, **i**, 1199–1200.

Cornfeld, D. and Schwartz, M.W. (1966). Nephrosis: a long-term study of children treated with corticosteroids. *Journal of Pediatrics*, **68**, 507–15.

Couser, W.G. and Stilmant, M.M. (1975). Mesangial lesion and focal glomerular sclerosis in the aging rat. *Laboratory Investigation*, **33**, 491–501.

Cuoghi, D., Venturi, P., and Cheli, E. (1988). Bee sting and relapse of nephrotic syndrome. *Childhood Nephrology and Urology*, **9**, 82–3.

Dall'Aglio, P., Chizzolini, C., and Brigati, C. (1982). Minimal change glomerulonephritis and focal glomerulosclerosis markers and "*in vitro*" activity of peripheral blood monoclear cells. *Proceedings of the European Dialysis and Transplant Association*, **19**, 673–8.

De Montis, G., Chabrolle, J., Berman, D., and Rossier, A. (1976). Syndrome néphrotique et hypersensibilité aux pollens. *Annales de Médecine Interne*, **127**, 387–92.

De Mouzon-Cambon, A., Bouissou, F., Dutau, G., Barthe, P., Parra, M.T., Sevin, A., and Ohayon, E. (1981). HLA-DR7 in children with idiopathic nephrotic syndrome. Correlation with atopy. *Tissue Antigens*, **17**, 518–24.

De Paillerets, F., Habib, R., Loubières, B., Clerc, M., Chapuis, Y., and Assi-Adou, J. (1972). The nephrotic syndrome in children in the Ivory Coast. *Guigoz Scientific Review*, **89**, 2–8.

Donckerwolcke, R.A., et al. (1982). Combined report on regular dialysis and transplantation of children in Europe, XI, 1981. *Proceedings of the European Dialysis and Transplant Association*, **19**, 61–91.

Donker, A.J.M., Brentjens, J.R.H., Van der Hem, G.K. (1978). Treatment of the nephrotic syndrome with indomethacin. *Nephron*, **22**, 374–81.

Dorhout Mees. E.J., Roos, J.C., and Boer, P. (1979). Observations on edema formation in the nephrotic syndromes in adults with minimal lesion. *American Journal of Medicine*, **67**, 378–84.

Dornan, T.L., Jenkins, S., Cotton, R.T., Tattersall, R.B., and Burden, R.P. (1988). The nephrotic syndrome at presentation of insulin dependent diabetes mellitus: cause or coincidence? *Diabetic Medicine*, **5**, 387–90.

Drachman, R., et al. (1988). Immunoregulation with levamisole in children with frequently relapsing steroid responsive nephrotic syndrome. *Acta Paediatrica Scandinavica*, **77**, 721–6.

Eagen, J.W. and Lewis, E.J. (1977). Glomerulopathies of neoplasia. *Kidney International*, **11**, 297–303.

Elema, J.D. and Arends, A. (1975). Focal and segmental glomerular hyalinosis and sclerosis in the rat. *Laboratory Investigation*, **33**, 554–61.

Elema, J.D., Hoyer, J.R., and Vernier, R.L. (1976). The glomerular mesangium: Uptake and transport of intravenously injected colloidal carbon in rats. *Kidney International*, **9**, 395–406.

Elzouki, A.Y. and Jaiswal, O.P. (1988). Long-term small dose prednisone therapy in frequently relapsing nephrotic syndrome of childhood. *Clinical Pediatrics*, **27**, 387–92.

Esparza, A.R., Kahn, S.I., Carella, S., and Abuelo, J.G. (1981). Spectrum of acute renal failure in nephrotic syndrome with minimal (or minor) glomerular lesions role of hemodynamic factors. *Laboratory Investigation*, **45**, 510–21.

Ettenger, R.B., Heuser, E.T., Malekzadeh, M.H., Pennisi, A.J., Uittenbogaart, C.H., and Fine, R.N. (1977). Focal glomerulosclerosis in renal allografts. *American Journal of Diseases of Children*, **131**, 1347–52.

Eyres, K.E., Mallick, N.P., and Taylor, G. (1976). Evidence for cell mediated immunity to renal antigens in minimal change nephrotic syndrome. *Lancet*, **i**, 1158–60.

Farquhar, M.G., Vernier, R.L., and Good, R.A. (1957). An electron microscopic study of the glomerulus in nephrosis, glomerulonephritis and lupus erythematosus. *Journal of Experimental Medicine*, **106**, 649–60.

Fauchald, P., Noddeland, H., and Norseth, J. (1985). An evaluation of ultrafiltration as treatment of diuretic resistent oedema in nephrotic syndrome. *Acta Medica Scandinavica*, **217**, 127–31.

Feehally, J., et al. (1983). T lymphocyte subpopulations and lymphocyte function in minimal change nephropathy during long term remission. *European Journal of Pediatrics*, **140**, 158.

Feehally, J., et al. (1984). Modulation of cellular immune function by cyclophosphamide in children with minimal change nephropathy. *New England Journal of Medicine*, **310**, 415–20.

Feehally, J., Samanta, A., Kinghom, H., Burden, A.C., and Walls, J. (1986). Red-cell surface charge in glomerular disease. *Lancet*, **ii**, 635.

Feinfeld, A.D., Olesnicky, L., and Pirani, C.L. (1984). Nephrotic syndrome associated with use of the nonsteroidal antiinflammatory drugs. Case report and review of the literature. *Nephron*, **37**, 174–9.

Ferder, L.F., Inserra, F., Daccordi, H., and Smith, R.D. (1990). Enalapril improved renal function and proteinuria in chronic glomerulopathies. *Nephron*, **55**, (Suppl. 1), 90–5.

Fikrig, S.M., Schiffman, G., Phillipp, J.C., and Moel, D.L. (1978). Antibody response to capsular polysaccharide vaccine of streptococcus pneumonia in patients with nephrotic syndrome. *Journal of Infectious Diseases*, **173**, 818–21.

Fine, B.P., Munoz, R., Uy, C.P., and Ty, A. (1976). Nitrogen mustard therapy in children with nephrotic syndrome unresponsive to corticosteroid therapy. *Journal of Pediatrics*, **89**, 1014–16.

Fodor, P., Saitua, M.T., Rodriguez, E., Gonzales, B., and Schesinger, L. (1982). T-cell dysfunction in minimal change nephrotic syndrome of childhood. *American Journal of Diseases in Children*, **136**, 713–7.

Fogo, A., et al. (1990). Glomerular hypertrophy in minimal change disease predicts subsequent progression to focal glomerular sclerosis. *Kidney International*, **38**, 115–23.

Fontana, V.J., Spain, W.C., and Desanctis, A.G. (1956). The role of allergy in nephrosis. *New York State Journal of Medicine*, **56**, 3927–10.

Foote, K.D., Brocklebank, J.T., and Meadow, S.R. (1985). Height attainment in children with steroid responsive nephrotic syndrome. *Lancet*, **ii**, 917–9.

Freundlich, M., Bourgoignie, J.J., Zilleruelo, G., Abitbol, C., Canterbury, J., and Strauss, J. (1986). Calcium and vitamin D metabolism in children with nephrotic syndrome. *Journal of Pediatrics*, **108**, 383–7.

Friedli, B. (1966). Le glomérule hyalin du nouveau né. Étude morphologique. *Biologia Neonatorum*, **10**, 359–67.

Furuse, A., Hiramatsu, M., Adachi, N., Karuschima, S., Hattori, S., and Matsuda, I. (1985). Dramatic response to corticosteroid therapy of nephrotic syndrome associated with IgA nephropathy. *International Journal of Pediatric Nephrology*, **6**, 205–8.

Futrakul, P. (1980). A new therapeutic approach of nephrotic syndrome associated with focal segmental glomerulosclerosis. *International Journal of Pediatric Nephrology*, **1**, 18–21.

Gabriel, R. and Wignen, M. (1986). Proteinuria and gut tumors. *Nephron*, **42**, 348.

Gagnadoux, M.F., Broyer, M., and Habib, R. (1987). Transplantation in children with idiopathic nephrosis. In *Recent advances in pediatric nephrology*. (eds. K. Murakami, T. Kitagawa, K. Yabuta, and T. Sakai), pp. 351–356. Excerpta Medica, International Congress Series 733, Amsterdam, New York, Oxford.

Galloway, W.H. and Mowat, A.P. (1968). Congenital microcephaly and nephrotic syndrome in two sibs. *Journal of Medical Genetics*, **5**, 319–21.

Ganguly, N.K., *et al.* (1979). Serum immunoglobulins in glomerulonephritis with a special reference to minimal lesion of glomerulonephritis. *Journal of the Association of Physicians of India*, **27**, 1003–8.

Garin, E.H., Williams, R.L., and Fennell, R.S. (1978*a*). Indomethacin in the treatment of idiopathic minimal lesion nephrotic syndrome. *Journal of Pediatrics*, **93**, 138–40.

Garin, E.H., Pryor, N.D., Fennell, R.S., and Richard, G.A. (1978*b*). Pattern of response to prednisone in idiopathic minimal lesion nephrotic syndrome as a criterion in selecting patients for cyclophosphamide therapy. *Journal of Pediatrics*, **92**, 304–8.

Garin, E.H., Grant, M.B., and Silverstein, J.H. (1989). Insulin-like growth factors in patients with active nephrotic syndrome. *American Journal of Diseases of Children*, **43**, 865–7.

Gault, M.H. and Muehrcke, R.C. (1983). Renal biopsy: current views and controversies. *Nephron*, **34**, 1–34.

Geary, D.F., Farine, M., Thorner, P., and Baumal, R. (1984). Response to cyclophosphamide in steroid resistant focal segmental glomerulosclerosis. A reappraisal. *Clinical Nephrology*, **22**, 109–13.

Ghiggeri, G.M., *et al.* (1987*a*). Renal selectivity properties towards endogenous albumin in minimal change nephropathy. *Kidney International*, **32**, 69–77.

Ghiggeri, G.M., *et al.* (1987*b*). Characterization of cationic albumin in minimal change nephropathy. *Kidney International*, **32**, 547–53.

Giangiacomo, J., Cleary, T.G., Cole, B.R., Hoffsten, P., and Robson, A.M. (1975). Serum immunoglobulins in the nephrotic syndrome. A possible cause of minimal change nephrotic syndrome. *New England Journal of Medicine*, **293**, 8–12.

Ginevri, F., *et al.* (1989). Peroxidative damage of the erythrocyte membrane in children with nephrotic syndrome. *Pediatric Nephrology*, **3**, 25–32.

Glassock, R.Y. and Bennett, C.M. (1978). Lipoid nephrosis. In *The Kidney* (eds. B.M. Brenner and F.C. Rector), p. 994. Saunders and Co, Philadelphia.

Glassock, R.J., Cohen, A.H., Bennett, C.M., and Martinez-Maldonado, M. (1976). Primary glomerular diseases. In *The Kidney*. (ed. B. M. Brenner and F. C. Rector), pp. 1351–1492. Saunders, Philadelphia.

Glicklich, D., Haskell, L., Senitzer, D., and Weiss, R.A. (1988). Possible genetic predisposition to idiopathic focal segmental glomerulosclerosis. *American Journal of Kidney Diseases*, **12**, 26–30.

Gonzalez, G., Kleinknecht, C., Gubler, M.C., and Lenoir, G. (1977). Nephrose familiale. *La Revue de Pédiatrie*, **13**, 427–33.

Gonzalo, A., Mampaso, P., Gallego, N., Quereda, C., Fierro, C., and Ortuno, J. (1985). Clinical significance of IgM mesangial deposits in the nephrotic syndrome. *Nephron*, **41**, 246–9.

Grishman, E. and Churg, J. (1973). Pathology of nephrotic syndrome with minimal or minor glomerular changes. In *Glomerulonephritis part I*, (ed. P. Kincaid-Smith, T. H. Mathew, and E. L. Becker), pp. 165–181, John Wiley and Sons, New York.

Grishman, E. and Churg, J. (1975). Focal glomerular sclerosis in nephrotic patients: an electron microscopy study of glomerular podocytes. *Kidney International*, **7**, 111–22.

Grishman, E., Churg, J., and Porush, J.G. (1976). Glomerular morphology in nephrotic heroin addicts. *Laboratory Investigation*, **35**, 415–24.

Griswold, W.R., *et al.* (1987). Treatment of childhood prednisone-resistant nephrotic syndrome and focal segmental glomerulosclerosis with intravenous methylprednisolone and oral alkylating agents. *Nephron*, **46**, 73–7.

Grond, J., Schilthuis, M., Koudstaal, J., and Elema, J. (1982). Mesangial function and glomerular sclerosis in rats after unilateral nephrectomy. *Kidney International*, **22**, 338–43.

Grond, J., Van Goor, H., Erkelens, D., and Elema, J. (1986). Glomerular sclerotic lesions in the rat. Histological analysis of their macromolecular and cellular composition. *Virchow's Archives, Abteilung B Zell-Pathologie*, **51**, 521–34.

Grupe, W.E. (1973). Chlorambucil in steroid dependent nephrotic syndrome. *Journal of Pediatrics*, **82**, 598–604.

Grupe, W.E. (1982*a*). Cytotoxic drugs for nephrotic syndrome. *New England Journal of Medicine*, **307**, 313.

Grupe, W.E. (1982*b*). Minimal change disease. *Seminars in Nephrology*, **2**, 241–50.

Grupe, W.E., Makker, S.P., and Ingelfinger, J.R. (1976). Chlorambucil treatment of frequently relapsing nephrotic syndrome. *New England Journal of Medicine*, **295**, 746–9.

Guesry, P., Lenoir, G., and Broyer, M. (1978). Gonadal effects of chlorambucil given to prepubertal and pubertal boys for nephrotic syndrome. *Journal of Pediatrics*, **92**, 299–303.

Habib, R. (1973). Focal glomerulosclerosis. *Kidney International*, **4**, 355–61.

Habib, R. and Gubler, M.C. (1971). Les lésions glomerulaires focales des syndromes néphrotiques idiopathiques de l'enfant. A propos de 49 observations. *Nephron*, **8**, 382–401.

Habib, R. and Gubler, M.C. (1975). Focal glomerular sclerosis associated with idiopathic nephrotic syndrome in children. In *Pediatric Nephrology* (ed. M. Rubin, and T. M. Barratt), pp. 499–514. Williams and Wilkins, Baltimore.

Habib, R. and Kleinknecht, C. (1971). The primary nephrotic syndrome in childhood. Classification and clinicopathologic study of 406 cases. In *Pathology Annual*. (ed. S. C. Sommers), pp. 417–74. Appleton Century Crofts, New York.

Habib, R., Michielsen, P., de Montera, H., Hinglais, N., Galle, P., and Hamburger, J. (1961). Clinical microscopic and electron microscopic data in the nephrotic syndrome of unknown origin. In *Ciba Foundation Symposium on Renal Biopsy*. (ed. G. Wolstenholme and S. Cameron), pp. 70–92. Churchill Ltd., London.

Habib, R., Hebert, D., Gagnadoux, M.F., and Broyer, M. (1982). Transplantation in idiopathic nephrosis. *Transplantation Proceedings*, **14**, 489–95.

Habib, R., *et al.* (1985). The nephropathy associated with male pseudohermaphroditism and Wilm's tumor (Drash syndrome): a distinctive glomerular lesion. Report of 10 cases. *Clinical Nephrology*, **24**, 269–78.

Habib, R., Girardin, E., Gagnadoux, M.F., Hinglais, N., Levy, M., and Broyer, M. (1988). Immunopathological findings in idiopathic nephrosis: clinical significance of glomerular "immune deposits". *Pediatric Nephrology*, **2**, 402–8.

Habib, R., Gubler, M.C., Niaudet, P., and Gagnadoux, M.F. (1989). Congenital/infantile nephrotic syndrome with diffuse mesangial sclerosis: relationship with Drash syndrome. In *Genetics of Kidney disorders* (ed. C. S. Bartsocas pp. 193–197. R. Alan Liss, New York.

Hamburger, J., *et al.* (1968), *Nephrology*, vol. 1 pp. 215–274. Saunders Co, Philadelphia.

Hardwicke, J., Soothill, J.F., Squire, J.R., and Holti, G. (1959). Nephrotic syndrome with pollen hypersensitivity. *Lancet*, 500–2.

Hayslett, J.P., Krassner, L.S., Klausse, G., Bensch, M.D., Kashgarian, M., and Epstein, F.H. (1969). Progression of lipoid nephrosis to renal insufficency. *New England Journal of Medicine*, **281**, 181–7.

Hayslett, J.P., Kashgarian, M., Bensch, K.G., Spargo, B.H., Freedman, L.R., and Epstein, F.H. (1973). Clinicopathological correlation in the nephrotic syndrome due to primary renal disease. *Medicine*, **52**, 93–120.

Helin, H., Mustonen, J., Pasternack, A., and Antonen, J. (1982). IgM-associated glomerulonephritis. *Nephron*, **31**. 11–16.

Hendrickse, R.G. and Adeniyi, A. (1979). Quartan malarial nephrotic syndrome in children. *Kidney International*, **16**, 64–74.

Heptinstall, R.H. (1983). The nephrotic syndrome. In *Pathology of the Kidney*, (ed. R. H. Heptinstall) pp. 637–740. Little Brown and Co., Boston.

Herrod, H.G., Stapleton, F.B., Trouy, R.L., and Roys, S. (1983). Evaluation of T lymphocyte subpopulations in children with nephrotic syndrome. *Clinical and Experimental Immunology*, **52**, 581–5.

Hervé, J.P., Cledes, J., Leroy, J.P., Simon, P., Ramée, M.P., and Legall, E. (1984). Syndromes néphrotiques corticosensibles avec dépots mésangiaux d'IgA: 4 observations. *Néphrologie*, **5**, 46.

Heslan, J.M., Lautie, J.P., Intrator, L., Blanc, C., Lagrue, G., and Sobel, A.T. (1982). Impaired IgA synthesis in patients with the nephrotic syndrome. *Clinical Nephrology*, **18**, 144–7.

Heslan, J.M., Branellec, A.I., Laurent, J., and Lagrue, G. (1986). The vascular permeability factor is a T lymphocyte produce. *Nephron*, **42**, 187–8.

Heslan, J.M., Pirotsky, E., Branellec, A.I., and Lagrue, G. (1988). Recombinant interleukin-2 induced proteinuria into isolated perfused rat kidney. In *Progress in basement membrane reasearch. Renal and related aspects in health and disease*. (ed. M. C. Gubler, and M. Sternberg), pp. 277–280. Libbey, London.

Hirszel, P., *et al.* (1984). Mesangial proliferative glomerulonephritis with IgM deposits. Clinicopathologic analysis and evidence for morphologic transitions. *Nephron*, **38**, 100–8.

Hopper, J., Ryan, P., Lee, J.C., and Rosenau, W. (1970). Lipoid nephrosis in 31 adult patients renal biopsy study by light, electron and fluorescence microscopy with experience in treatment. *Medicine*, **49**, 321–41.

Horwarth, M. and Sulyok, E. (1986). Steroid responsive nephrotic syndrome in Asians (letter). *Archives of Diseases in Childhood*, **61**, 528.

Hosenpud, J., Piering, W.F., Garancis, J.C., and Kaufman, H.M. (1985). Successful second kidney transplantation in a patient with focal glomerulosclerosis, a case report. *American Journal of Nephrology*, **5**, 299–304.

Howie, A.J. and Brewer, D.B. (1984). The glomerular tip lesion: a previously undescribed type of segmental glomerular abnormality. *Journal of Pathology*, **142**, 205–20.

Hoyer, J.R., Raij, L., Vernier, R.L., Simmons, R.L., Najarian, J.S., and Michael, A.F. (1972). Recurrence of idiopathic nephrotic syndrome after renal transplantation. *Lancet*, **ii**, 343–8.

Hoyer, P.F., Krull, F., and Brodelh, J. (1986). Cyclosporin in frequently replasing minimal change nephrotic syndrome. *Lancet*, **ii**, 335.

Hsu, A.C., Folami, A.O., Bain, J., and Rance, C.P. (1979). Gonadal functions in males treated with cyclophosphamide for nephrotic syndrome. *Fertility and Sterility*, **31**, 173–7.

Hsu, H.C., *et al.* (1984). Clinical and immunopathologic study of mesangial IgM nephropathy: report of 41 cases. *Histopatholgoy*, **8**, 435–46.

Hulter, H.N. and Bonner, E.L. (1980). Lipoid nephrosis appearing as acute oliguric renal failure. *Archives of Internal Medicine*, **140**, 403–5.

Hyam, J.S. and Carrey, D.E. (1988). Corticosteroids and growth. *Journal of Pediatrics*, **113**, 249–54.

Hyman, L.R. and Burkholder, P.M. (1973). Focal sclerosing glomerulonephropathy with segmental hyalinosis. A clinicopathologic analysis. *Laboratory Investigation*, **28**, 533–44.

Hyman, L.R. and Burkholder, P.M. (1974). Focal sclerosing glomerulonephropathy with hyalinosis. *Journal of Pediatrics*, **84**, 217–25.

Ichida, F., *et al.* (1985) Chlorambucil central nervous toxicity: a significant side effect of chlorambucil therapy in childhood nephrotic syndrome. *European Journal of Pediatrics*, **144**, 283–6.

Iitaka, K. and West, C.D. (1979). A serum inhibitor of blastogenesis in idiopathic nephrotic syndrome transferred by lymphocytes. *Clinical Immunology and Immunopathology*, **12**, 121–6.

Imbasciati, E., *et al.* (1981). Acute renal failure in idiopathic nephrotic syndrome. *Nephron*, **28**, 186–91.

Imbasciati, E., *et al.* (1985). Controlled trial of methylprednisone pulses and low dose oral prednisone for the minimal change nephrotic syndrome. *British Medical Journal*, **291**, 1305–8.

International Study of Kidney Disease in Children. (1974). Prospective controlled trial of cyclophosphamide therapy in children with the nephrotic syndrome. *Lancet*, **ii**, 423–7.

International Study of Kidney Disease in Children. (1978). Nephrotic syndrome: prediction of histopathology from clinical and laboratory characteristics at time of diagnosis. *Kidney International*, **13**, 159–65.

International Study of Kidney Disease in Children. (1981). The primary nephrotic syndrome in children. Identification of patients with minimal change nephrotic syndrome from initial response to prednisone. *Journal of Pediatrics*, **98**, 561–4.

International Study of Kidney Disease in Children. (1982). Early identification of frequent relapsers among children with minimal change nephrotic syndrome. *Journal of Pediatrics*, **101**, 514–8.

International Study of Kidney Disease in Children. (1983). Cyclophosphamide therapy in focal segmental glomerular sclerosis. A controlled clinical trial. *European Journal of Pediatrics*, **140**, 149.

International Study of Kidney Disease in Children. (1984). Minimal change nephrotic syndrome in children: deaths during the first 5 to 15 years observation. *Pediatrics*, **73**, 497–501.

Ito, I., *et al.* Twenty seven children with focal segmental glomerulosclerosis: correlation between the segmental location of the glomerular lesions and prognosis. *Clinical Nephrology*, **22**, 9–14.

Jao, W., Pollak, V.E., Norris, S.H., Lewy, P., and Pirani, C.L. (1973). Lipoid nephrosis. *Medicine*, **52**, 445–68.

Jenis, E.H., *et al.* (1974). Focal segmental glomerulosclerosis. *American Journal of Medicine*, **57**, 695–701.

Ji-Yun, Y., Melvin, T., Sibley, R., and Michael, A.F. (1984). No evidence for a specific role of IgM in mesangial proliferation of idiopathic nephrotic syndrome. *Kidney International*, **25**, 100–6.

Kaplan, C., Pasternak, B., Shah, H., and Gallo, G. (1975). Age related incidence of sclerotic glomeruli in human kidneys. *American Journal of Pathology*, **80**, 227–34.

Kashgarian, M., Hayslett, J.P., and Seigel, N.J. (1974). Lipoid nephrosis and focal sclerosis: distinct entities or spectrum of disease. *Nephron*, **13**, 105–8.

Kasiske, B.L., O'Donnell, M.P., and Keane, W.F. (1987). Cholesterol synthesis inhibition reduces glomerular injury in obese Zucker rats. *Kidney International*, **31**, 387.

Kawaguchi, H., Yamaguchi, Y., Nagata, M., and Itoh, K. (1987). The effects of human recombinant interleuckin-2 on the permeability of glomerular basement membranes in rats. *Japanese Journal of Nephrology*, **29**, 1–11.

Kawano, K., Wenzl, J., McCoy, J., Porch, J. and Kimmelstiel, P. (1971). Lipoid nephrosis. A multifold blind study including quantitation. *Laboratory Investigation*, **24**, 499–503.

Keane, W.F. and Raij, L. (1981). Determinants of glomerular mesangial localization of immune complexes. Role of endothelial fenestrae. *Laboratory Investigation*, **45**, 366–71.

Kelley, V.C. and Panos, T.C. (1952). Nephrotic syndrome in children: clinical response to nitrogen mustard therapy. *Journal of Pediatrics*, **41**, 505–17.

Kleinknecht, C. and Gubler, M.C. (1983). Nephrose. In *Néphrologie Pédiatrique* 3rd. Ed. (ed. P. Royer, H. Mathieu, R. Habib, and M. Broyer). pp. 274–293. Flammarion Médecine Science, Paris.

Kleinknecht, C., Broyer, M., and Gubler, M.C. (1980*a*). Irreversible renal failure after indomethacin in steroid-resistant nephrosis. *New England Journal of Medicine*, **302**, 691.

Kleinknecht, C., Broyer, M., Loirat, C., Nivet, H., Palcoux, J.B., and Parchoux, B. (1982). Comparison of short and long term treatment at onset of steroid sensitive nephrosis. Preliminary results of a multicentric controlled trial from the French Society of Pediatric Nephrology. *International Journal of Pediatric Nephrology*, **3**, 45.

Kleinknecht, C., Laouari, D., and Gubler, M.C. (1983). Adverse effects of indomethacin in experimental chronic nephrosis. *International Journal of Pediatric Nephrology*, **4**, 83–8.

Kobayashi, Y., Chen, X.M., Hiki, Y., Fujii, K., and Kashiwagi, N. (1985). Association of HLA-DRW8 and DQW3 in minimal change nephrotic syndrome in Japanese adults. *Kidney International*, **28**, 193–7.

Kohaut, E.C., Singer, B., and Leighton, H. (1976). The significance of focal glomerular sclerosis in children who have nephrotic syndrome. *American Journal of Clinical Pathology*, **66**, 545–50.

Koomans, H.A., Boer, W.H., and Dorhout Mees, E.J. (1987). Renal function during recovery from minimal lesion nephrotic syndrome. *Nephron*, **47**, 173–8.

Kopolovic, J., Shvil, Y., Pomeranz, A., Ron, N., Rubinger, D. and Oren, R. (1987). IgM nephropathy: morphological study related to clinical findings. *American Journal of Nephrology*, **7**, 275–80.

Korbet, S.M., Schwartz, M.M., and Lewis, E.J. (1988*a*). Recurrent nephrotic syndrome in renal allografts. *American Journal of Kidney Diseases*, **11**, 270–6.

Korbet, S.M., Schwartz, M.M., and Lewis, E.J. (1988*b*). Minimal-change glomerulopathy of adulthood. *American Journal of Nephrologyy*, **8**, 291–7.

Koskimies, O., Vilska, J., Rapola, J., and Hallman, N. (1982). Long-term outcome of primary nephrotic syndrome. *Archives of Diseases in Childhood*, **57**, 544–8.

Krensky, A.M., Ingelfinger, J.R., and Grupe, W.E. (1982). Peritonitis in childhood nephrotic syndrome. *American Journal of Diseases of Children*, **136**, 732–6.

Kunin, C.M., et al. (1959). Antibody response to influenza virus vaccine in children with nephrosis: effect of cortisone. *Pediatrics*, **23**, 54–62.

Lagrue, G., et al. (1975). A vascular permeability factor in lymphocyte culture supernatants from patients with nephrotic syndrome. *Biomédecine*, **23**, 73–5.

Lagrue, G. and Laurent, J. (1982) Is lipoïd nephrosis an allergic disease? *Transplantation Proceedgins*, **4**, 485–8.

Lagrue, G. and Laurent, J. (1984). Role de l'allergie dans la néphrose lipoïdique. *Nouvelle Presse Médicale*, **11**, 1465–6.

Lagrue, G. and Laurent, J., and Belghiti, D. (1982). Immunopathologie de la néphrose lipoïdique. *Néphrologie*, **3**, 40–5.

Lagrue, G., Laurent, J., Belghiti, D., and Robeva, R. (1986). Cyclosporin and idiopathic nephrotic syndrome. *Lancet*, **ii**, 692–3.

Lagrue, G., et al. (1988). Increased interleukin-2 levels in lymphocyte culture supernatants from patients with idiopathic nephrotic syndrome. In *Progress in Basement Membrane Research. Renal and Related Aspects in Health and Disease* (eds. M.C. Gubler and M. Sternberg), pp. 281–4. Libbey, London.

Lagueruela, C.C., Buettner, T.L., Cole, B.R., Kissane, J.M., and Robson, A.M. (1990). HLA extended haplotypes in steroid-sensitive nephrotic syndrome of childhood. *Kidney International*, **38**, 145–50.

Lai, K.N., Lai, F.M., Chan, K.W., Ho, C.P., Leung, A.C., and Vallance-Owen, J. (1989). An overlapping syndrome of IgA nephropathy and lipoid nephrosis. *American Journal of Clinical Pathology*, **86**, 716–23.

Lalich, J.J., Burkholder, P.M., and Paik, W.C.W. (1975). Protein overload nephropathy in rats with unilateral nephrectomy. *Archives of Pathology*, **99**, 72–9.

Lam, C.N. and Arneil, G.C. (1968). Long term dwarfing effects of corticosteroid treatment for childhood nephrosis. *Archives of Diseases in Childhood*, **43**, 589–94.

Lange, K., Ahmed, U., Seligson, G., and Grover, A. (1981). Depression of endostreptosin, streptolysin O and streptizyme antibodies in patients with idiopathic nephrosis with and without a nephrotic syndrome. *Clinical Nephrology*, **15**, 279–85.

Laurent, J., Lagrue, G., Belghiti, D., and Kersterbaum, S. (1986). Elévation des IgE sériques totales dans la glomérulonéphrite à dépôts extramembraneux. *Néphrologie*, **4**, 171–4.

Laurent, J., Rostoker, G., Robeva, R., Bruneau, C., and Martin-Govantes, J. (1987). Is adult idiopathic nephrotic syndrome food allergy? *Nephron*, **47**, 7–11.

Lawler, W., Williams, G., Tarpey, P., and Mallick, N.P. (1980). IgM associated primary diffuse mesangial proliferative glomerulonephritis. *Journal of Clinical Pathology*, **33**, 1029–38.

Lawson, D., Moncrieff, A., and Payne, W.W. (1960). Fourty years of nephrosis in childhood. *Archives of Diseases in Childhood*, **35**, 115–26.

Lee, H.S. and Spargo, B.H. (1986). Significance of renal hyaline arteriolosclerosis in focal segmental glomerulosclerosis. *Nephron*, **41**, 86–93.

Lee, J.C., Yamauchi, H., and Hopper, J., Jr. (1966). The association of cancer and the nephrotic syndrome. *Annals of Internal Medicine*, **64**, 41–51.

Leiper, J.M., Thomson, D., and McDonald, M.K. (1977). Uptake and transport of imposil by the glomerular mesangium in the mouse. *Laboratory Investigation*, **37**, 526–33.

Leisti, S., Koskimies, O., Rapola, J., Hallman, H., Perheentupa, J., and Vilska, J. (1977*a*). Association of post-medication hypocortisolism with early first relapse of idiopathic nephrotic syndrome. *Lancet*, **ii**, 795–6.

Leisti, S., Vilska, J., and Hallman, N. (1977*b*). Adrenocortical insufficiency and relapsing in the idiopathic nephrotic syndrome of childhood. *Pediatrics*, **60**, 334–42.

Leisti, S., Koskimies, O., Perheentupa, J., Vilska, J., and Hallman, N. (1978). Idiopathic nephrotic syndrome: prevention of early relapse. *British Medical Journal*, **1**, 892.

Lenoir, G., Guesry, P., Kleinknecht, C., Gagnadoux, M.F., and Broyer, M. (1977). Complications extragonadiques du chlorambucil chez l'enfant. *Archives Françaises de Pédiatrie*, **34**, 798–807.

Leumann, E.P., Briner, J., Donckerwolcke, R.A., Kuijten, R., and Largiader, F. (1980). Recurrence of focal segmental glomerulosclerosis in the transplant kidney. *Nephron*, **25**, 65–71.

Leveque, B., Debauchez, C., Deflandre, L., and Marie, J. (1969). Le chlorambucil dans le traitement du syndrome néphrotique idiopathique sans lesions glomérulaires chez l'enfant: à propos de 30 observations. *Annales de Pédiatrie*, **16**, 13–23.

Levey, A.S., Lau, J., Pauker, S.G., and Kassirer, J.P. (1987). Idiopathic nephrotic syndrome: puncturing the biopsy myth. *Annals of Internal Medicine*, **107**, 697–713.

Levin, M., Walters, M.D.S., Smith, C., Gascoine, P., and Barratt, T.M. (1985). Steroid-responsive nephrotic syndrome: a generalized disorder of membrane negative charge. *Lancet*, **ii**, 239–42.

Levin, M., Gascoine, P., Turner, M.W., and Barratt, T.M. (1989). A highly cationic protein in plasma and urine of children with steroid responsive nephrotic syndrome. *Kidney International*, **36**, 867–77.

Levine, L.A. and Richie, J.P. (1989). Urological complications of cyclophosphamide. *Journal of Urology*, **141**, 1063–9.

Levinsky, R.J., Malleson, P.N., Barratt, T.M., and Soothill, J.F. (1978). Circulating immune complexes in steroid responsive nephrotic syndrome. *New England Journal of Medicine*, **298**, 126–9.

Lewis, E.J., Carpenter, C.B., and Schur, P.H. (1971). Serum complement levels in human glomerulonephritis. *Annals of Internal Medicine*, **75**, 555–60.

Lewis, M.A., Davis, N., Baildom, E., Houston, I.B., and Postlethwaite, R.J. (1989). Nephrotic syndrome from toddlers to twenties. *Lancet*, **i**, 255–9.

Lim, W.S., Sibley, R., and Spargo, B. (1974). Adult lipoid nephrosis: clinicopathological correlations. *Annals of Internal Medicine*, **81**, 314–20.

Limaye, S.R., Pillai, S., and Tina, L.U. (1988). Relationship of steroid dose to degree of posterior subcapsular cataracts in nephrotic syndrome. *Annals of Opththalmology*, **20**, 225–7.

Lin, C.Y. and Hsu, H.C. (1986). Histopathological and immunological studies in spontaneous remission of nephrotic syndrome after intercurrent measles infection. *Nephron*, **42**, 110–5.

Lowenstein, J., Schacht, R.G., and Baldwin, D.S. (1981). Renal failure in minimal change nephrotic syndrome. *American Journal of Medicine*, **70**, 227–33.

Lyon, H., Pinn, V.N., Cortell, S., Cohen, J.J., and Harrington, J.J. (1973). Allergic interstitial nephritis causing reversible renal failure in four patients with idiopathic nephrotic syndrome. *New England Journal of Medicine*, **288**, 124–8.

Maisin, A., Loirat, C., Pillon, G., Macher, M.A., and Mathieu, M. (1985). Intérêt de la chlormétine chez les enfants atteints de néphrose cortico-dépendante ou partiellement cortico-sensible en intoxication stéroïdienne. *Archives Francaises de Pédiatrie*, **42**, 635–8.

Maizel, S.E., Sibley, R.K., Horstman, J.P., Kjellstrand, C.M., and Simmons, R.L. (1981). Incidence and significance of recurrent focal segmental glomerulosclerosis in renal allograft recipients. *Transplantation*, **32**, 512–6.

Makker, S.P. and Heymann, W. (1972). Pregnancy in patients who have had the idiopathic nephrotic syndrome in childhood. *Journal of Pediatrics*, **81**, 1140–4.

Makker, S.P. and Heymann, W. (1974). The idiopathic nephrotic syndrome of childhood. *American Journal of Diseases of Children*, **127**, 830–7.

Malekzadeh, M.H., et al. (1979). Focal sclerosis and renal transplantation. *Journal of Pediatrics*, **95**, 249–54.

Mallick, N.P., Williams, R.J., McFarlane, H., Orr, W.M., Taylor, G., and Williams, G. (1972). Cell mediated immunity in nephrotic syndrome. *Lancet*, **i**, 507–9.

Mampaso, F., et al. (1981). Mesangial deposits of IgM in patients with the nephrotic syndrome. *Clinical Nephrology*, **16**, 230–4.

Martinez-Vea, A., Garciaruiz, C., Carrera, M., Oliver, J.A., and Richart, C. (1987). Effect of captopril in nephrotic range proteinuria due to renovascular hypertension. *Nephron*, **45**, 162–3.

Maruyama, K., Tomizawa, S., Shimabukoro, N., Fukuda, T., and Kuroume, T. (1989). Effect of supernatants derived from T Lymphocyte culture in minimal change nephrotic syndrome on rat kidney capillaries. *Nephron*, **51**, 73–6.

Matalon, R., Kartz, L., Gallo, G., Waldo, E., Cabaluna, C., and Eisinger, R.P. (1974). Glomerular sclerosis in adults with nephrotic syndrome. *Annals of Internal Medicine*, **80**, 488–95.

Matsumoto, K. (1982). Impaired local graft versus host reaction in lipoid nephrosis. *Nephron*, **31**, 281–2.

Matsumoto, K., Osakabe, K., Harada, M., and Hatano, M. (1981). Impaired cell-mediated immunity in lipoid nephrosis. *Nephron*, **29**, 190–4.

Matsumoto, K., Osakabe, K., Katayama, H., and Hatano, M. (1982). In vitro lymphocyte dysfunction in lipoid nephrosis mediated by suppressor cells. *Nephron*, **32**, 270–2.

Matsumoto, K., Katayama, H., and Hatano, M. (1988). Minimal change nephrotic syndrome associated with subcutaneous eosinophilic lymphoid granuloma (Kimura's disease). *Nephron*, **49**, 251–4.

Mauer, S.M., Hellerstein, S., Cohn, R.A., Sibley, R.K., and Vernier, R.L. (1979). Recurrence of steroid responsive nephrotic syndrome after renal transplantation. *Journal of Pediatrics*, **95**, 261–4.

McCrory, W.W., Shibuya, M., Lu, W.H., and Lewy, J.E. (1973). Therapeutic and toxic effects observed with different dosage programs of cyclophosphamide in treatment of steroid-responsive but frequently relapsing nephrotic syndrome. *Journal of Pediatrics*, **82**, 614–8.

McDonald, N.E., Wolfish, N., McLaine, P., Phipps, P., and Rossier, E. (1986). Role of respiratory viruses in exacerbation of primary nephrotic syndrome. *Journal of Pediatrics*, **108**, 378–82.

McDonald, J., Murphy, A.V., and Arneil, G.C. (1974). Long term assessment of cyclophosphamide therapy for nephrosis. *Lancet*, **ii**, 980–2.

McEnery, P.T. and Strife, C.F. (1982). Nephrotic syndrome in childhood. Management and treatment in patients with minimal change disease, mesangial proliferation, or focal glomerulosclerosis. *Pediatric Clinics of North America*, **29**, 875–94.

McGovern, V.J. (1964). Persistant nephrotic syndrome: a renal biopsy study. *Australian Annals of Medicine*, **13**, 306–12.

McLean, R.H., Forsgren, A., Björksten, B., Kim, Y., Quie, P.G., and Michael, A.F. (1977). Decreased serum factor B concentration associated with decrease opsonization of *Escherichia coli* in idiopathic nephrotic syndrome. *Pediatric Research*, **11**, 910–16.

McVicar, M., Exeni, R., and Susin, M. (1980). Nephrotic syndrome and multiple tubular defects in children: an early sign of focal segmental glomerulosclerosis. *Journal of Pediatrics*, **97**, 918–22.

Meadow, S.R., Sarsfield, J.R., Scott, D.G., and Rajah, S.M. (1981). Steroid responsive nephrotic syndrome and allergy: immunological studies. *Archives of Diseases in Childhood*, **56**, 517–24.

Mehta, K.P., Ali, U., Kutty, M., and Kolhatkar, U. (1986). Immunoregulatory treatment for minimal change nephrotic syndrome. *Archives of Diseases in Childhood*, **61**, 153–8.

Metcoff, J. and Janeway, C.A. (1961). Studies on the pathogenesis of nephrotic edema. *Journal of Pediatrics*, **58**, 640–85.

Meyrier, A. and Simon, P. (1988). Treatment of corticoresistant idiopathic nephrotic syndrome in the adult: minimal change disease and focal segmental glomerulosclerosis. In *Advances in Nephrology*, (ed J. P. Grünfeld, et al.) pp. 127–150. Year Book, Chicago.

Meyrier, A. and Collaborative Group of the Société de Néphrologie (1989). Ciclosporin in the treatment of nephrosis. *American Journal of Nephrology*, **9 (Suppl, 1)**, 65–71.

Michael, A.F., et al. (1973). Immunologic aspects of the nephrotic syndrome. *Kidney International*, **3**, 105–15.

Michael, A.F., Keane, W.F., Raij, L., Vernier, R.L., and Mauer, S.M. (1980). The glomerular mesangium. *Kidney International* **17**, 141–54.

Michielsen, P. and Lambert, P.P. (1967). Effets du traitement par les corticostéroides et l'indomethacine sur la proteinurie. *Bulletin et Mémoirs de la Société Médicales des Hôpitaux de Paris*, **118**, 217–21.

Minchin, M.A., Turner, K.J., and Bower, G.D. (1980). Lymphocyte blastogenesis in nephrotic syndrome. *Clinical and Experimental Immunology*, **42**, 241–46.

Miziuri, S., Hayashi, I., Ozawa, T., Hirata, K., Takano, M., and Sasaki, Y. (1988). Effects of an oral protein load on glomerular filtration rate in healthy controls and nephrotic patients. *Nephron*, **48**, 101–6.

Moncrieff, M.W., White, R.H.R., Glasgow, E.F., Winterborn, M.H., Cameron, J.S. and Ogg, C.S. (1973). The familial nephrotic syndrome II. A clinicopathological study. *Clinical Nephrology*, **1**, 220–9.

Mongeau, J.G., Robitaille, P.O., and Roy, F. (1988). Clinical efficacy of levamisole in the treatment of primary nephrosis in children. *Pediatric Nephrology*, **2**, 398–401.

Moorthy, A.V., Zimmerman, S.W., and Burkholder, P.M. (1976). Inhibition of lymphocyte blastogenesis by plasma of patients with minimal change nephrotic syndrome. *Lancet*, **1**, 1160–3.

Moorthy, A.V. (1983). Minimal change glomerular disease: a paraneoplastic syndrome in two patients with bronchogenic carcinoma. *American Journal of Kidney Diseases*, **3**, 58–62.

Morales, J.M., et al. (1988). Clinical and histological sequence of recurrent focal segmental glomerulosclerosis. *Nephron*, **48**, 241–2.

Morel-Maroger, L., et al. (1975) "Tropical nephropathy" and "tropical extra-membranous glomerulonephritis" of unknown aetiology in Senegal. *British Medical Journal*, **1**, 541–6.

Morzycka, M., Crocker, B.P., Jr, Seigler, H.F., and Tisher, C.C. (1982). Evaluation of recurrent glomerulonephritis in kidney allografts. *American Journal of Medicine*, **72**, 588–98.

Müller, F. (1905). Morbus Brighti. *Verhandl-ung der Deutsche Pathologie Gesellschaft*, **9**, 64–99.

Müller, W. and Brandis, M. (1981). Acute leukemia after cytotoxic treatment for non maligant disease in childhood. *European Journal of Pediatrics*, **136**, 105–8.

Munk, F. (1913). Klinische Diagnostick der degenerativen Nierenkrankungen. *Klinische Medizin*, **78**, 1–52.

Murnaghan, W.M., Vasmant, D., and Bensman, A. (1984). Pulse methylprednisolone therapy in severe idiopathic childhood nephrotic syndrome. *Acta Paediatrica Scandinavica*, **73**, 733–9.

Murphy, W.M., Jukkola, A.F., and Roy, S. (1979). Nephrotic syndrome with mesangial-cell proliferation in children—a distinct entity? *American Journal of Clinical Pathology*, **72**, 42–7.

Mustonen, J., Pasternack, A., and Rantala, I. (1983). The nephrotic syndrome in IgA glomerulonephritis: response to corticosteroid therapy. *Clinical Nephrology*, **20**, 172–6.

Myers, B.D. (1986). Cyclosporine nephrotoxicity. *Kidney International*, **30**, 964–76.

Myers, D.I., Ciuffo, A.A., and Cooke, C.R. (1979). Focal glomerulosclerosis and erythrocytosis. *John Hopkins Medical Journal*, **145**, 192–5.

Nagi, A.H., Alexander, F., and Lanigan, R. (1971). Light and electron miscroscopical studies of focal glomerular sclerosis. *Journal of Clinical Pathology*, **24**, 846–50.

Nash, M.A., Greifer, I., Olbing, H., Bernstein, J., Bennett, B., and Spitzer, A. (1976). The significance of focal sclerotic lesions of glomeruli in children. *Journal of Pediatrics*, **88**, 806–13.

Nash, M.A., Bakare, M.A., D'Agati, V., and Pirani, C.L. (1982). Late development of chronic renal failure in steroid-responsive nephrotic syndrome. *Journal of Pediatrics*, **101**, 411–4.

Newman, W.J., *et al.* (1976). Focal glomerular sclerosis: contrasting clinical patterns in children and adults. *Medicine*, **55**, 67–87.

Ngu, J., Barratt, T.M., and Soothill, J.F. (1970). Immunoconglutinin and complement changes in steroid sensitive relapsing nephrotic syndrome of children. *Clinical and Experimental Immunology*, **6**, 109–16.

Niaudet, P. (1986). Traitement d'attaque de la néphrose: resultats d'une enquête coopérative du Club Français de Néphrologie Pédiatrique. *Néphrologie*, **7**, 86–7.

Niaudet, P. and Broyer, M. (1989). Cyclosporin in the therapy of idiopathic nephrotic syndrome in children. In *International Yearbook of Nephrology*, (ed. V. E. Andreucci), pp. 155–168. Kluwer Academic Publication

Niaudet, P., Drachman, R., Gagnadoux, M.F., and Broyer, M. (1984). Treatment of idiopathic nephrotic syndrome with levamisole. *Acta Paediatrica Scandinavica*, **73**, 637–41.

Niaudet, P., Habib, R., Tête, M.J., Hinglais, N., and Broyer, M. (1987). Cyclosporin in the treatment of idiopathic nephrotic syndrome in children. *Pediatric Nephrology*, **1**, 566–73.

Niaudet, P., Habib, R., and Broyer, M. (1989). Cyclosporine et néphrose idiopathique chez l'enfant. In *Journée Parisienne de Pédiatrie*, (eds. D. Alagille, *et al.*), pp. 143–8. Flammarion Medecine Science. Paris.

Nolasco, F., Cameron, J.S., Heywood, E.F., Hicks, J., Ogg C. and Williams, D.G. (1986). Adult–onset minimal change nephrotic syndrome: a long-term follow-up. *Kidney International*, **29**, 1215–23.

Noss, G., Bachmann, H.J., and Olbing, H. (1981). Association of minimal change nephrotic syndrome (MCNS) with HLA-B8 and B13. *Clinical Nephrology*, **15**, 172–4.

Novis, B.H., Korzets, Z., Chen, P., and Bernheim J. (1988). Nephrotic syndrome after treatment with 5 aminosalicylic acid. *British Medical Journal*, **296**, 1442.

Nunez-Roldan, A., Villechenous, E., Fernandez-Andrado, C., and Marin-Govantes, J. (1982). Increased HLA-DR7 and decreased DR2 in steroid-responsive nephrotic syndrome. *New England Journal of Medicine*, **306**, 366–7.

O'Reagan, E., O'Callaghan, V., and Dundon, S. (1980). HLA antigens and steroid-responsive nephrotic syndrome of childhood. *Tissue Antigens*, **16**, 147–51.

Osakabe, K. and Matsumoto, K. (1981). Concanavalin A-induced suppressor cell activity in lipoid nephrosis. *Scandinavian Journal of Immunology*, **14**, 161–6.

Palm, L., Hagerstrand, I., Kristofferson, U., Blennow, G., Brun, A., and Jorgensen, C. (1986). Nephrosis and disturbances of neuronal migration in male siblings: a new hereditary disorder? *Archives of Diseases in Childhood*, **61**, 545–8.

Papadopoulou, Z.L., Jenis, E.H., Tina, A.C., Jose, P.A., and Calcagno, P.L. (1982). Chronic relapsing minimal change nephrotic syndrome with or without mesangial deposits: long term follow-up. *International Journal of Pediatric Nephrology*, **3**, 179–86.

Pardo, V., Riesgo, I., Zilleruelo, G., and Strauss, J. (1984). The clinical significance of mesangial IgM deposits and mesangial hypercellularity in minimal change nephrotic syndrome. *American Journal of Kidney Diseases*, **3**, 264–9.

Peces, R., Riera, J.R., Lopez-Larrea, C., and Alvarez, J. (1987). Steroid responsive relapsing nephrotic syndrome associated with early diabetic nephropathy in a child. *Nephron*, **46**, 78–82.

Pennisi, A.J., Grushkin, C.M., and Lieberman, E. (1976). Cyclophosphamide in the treatment of idiopathic nephrotic syndrome. *Pediatrics*, **57**, 948–51.

Penso, J., Lippe, B., Ehrlich, R., and Smith, F.C. (1974). Testicular function in prepubertal and pubertal male patients treated with cyclophosphamide for nephrotic syndrome. *Journal of Pediatrics*, **84**, 831–6.

Pinto, J., Lacerda, G., Cameron, J.S., Turner, D.R., Bewick M., and Ogg, C.S. (1981). Recurrence of focal segmental glomerulosclerosis in renal allografts. *Transplantation*, **32**, 83–9.

Pirson, Y., Squifflet, J.P., Marbaix, E., Alexandre, G.P., and Van Ypersele de Strihou, C. (1986). Recurrence of focal glomerular sclerosis despite cyclosporin treatment after renal transplantation. *British Medical Journal*, **292**, 1336.

Polito, C., Oporto, M.R., Totino, S.F., Lamanna, A., and Di Toro, R. (1986). Normal growth of nephrotic children during long term alternate day prednisone therapy. *Acta Paediatrica Scandinavica*, **75**, 245–50.

Pollak, V.E., Rosen, S., Pirani, C.L., Muehrcke, R.C., and Kark, R.M. (1968). Natural history of lipoid nephrosis and of membranous glomerulonephritis. *Annals of Internal Medicine*, **69**, 1171–96.

Poston, R.N., Cerio, R., and Cameron, J.S. (1978). Circulating immune complexes in minimal change nephritis. *New England Journal of Medicine*, **298**, 1089.

Poucell, S., Baumal, R., Farine, M., and Arbus, G.A. (1985). Location of glomerular lesions in focal segmental glomerulosclerosis. *Archives of Pathology and Laboratory Medicine*, **109**, 482–3.

Powell, H.R. (1976). Relationship between proteinuria and epithelial cell changes in minimal lesion glomerulopathy. *Nephron*, **16**, 310–7.

Prasad, D.R., Zimmerman, S.W., and Burkholder, P.M. (1977). Immunohistologic features of minimal change nephrotic syndrome. *Archives of Pathology and Laboratory Medicine*, **101**, 345–8.

Pru, C., Kjellstrand, C.M., Cohn, R.A., and Vernier, R.L. (1984). Late recurrence of minimal lesion nephrotic syndrome. *Annals of Internal Medicine*, **100**, 69–72.

Raij, L., Keane, W.F., Leonard, A., and Shapiro, F.L. (1976). Irreversible acute renal failure in idiopathic nephrotic syndrome. *American Journal of Medicine*, **61**, 207–14.

Rao, T.K.S., Nicastri, A.D., and Friedman, E.D. (1974). Natural history of heroin associated nephropathy. *New England Journal of Medicine*, **290**, 19–23.

Rambausek, M., Waldherr, R., Rauterberg, W., Andrassy, K., and Ritz, E. (1987). Mesangial IgA nephropathy and idiopathic nephrotic syndrome. *Nephron*, **47**, 190–3.

Rance, C.P., Arbus, G.S., and Balfe, J.W. (1976). Management of the nephrotic syndrome in children. *Pediatric Clinics of North America*, **23**, 735–50.

Rascher, W., Tulassay, T., Seyberth, H.W., Himbert, U., Lang, U., and Schärer, K. (1986). Diuretic and hormonal responses to head out water immersion in nephrotic syndrome. *Journal of Pediatrics*, **109**, 609–14.

Rees, L., *et al.* (1988). Growth and endocrine function in steroid sensitive nephrotic syndrome. *Archives of Diseases in Childhood*, **63**, 484–90.

Reeves, W.G., Cameron, J.S., and Johansson, S.G. (1975). Seasonal nephrotic syndrome, description and immunological findings. *Clinical Allergy*, **5**, 121–7.

Rich, A.R. (1957). A hitherto undescribed vulnerability of the juxta medullary glomeruli in the lipoid nephrosis. *Bulletin of the John Hopkins Hospital*, **100**, 173–9.

Richards, W., Olson, D., and Church, J.A. (1977). Improvement of idiopathic nephrotic syndrome following allergy therapy. *Annals of Allergy*, **39**, 332–3.

Robson, A.M., Giangiacomo, J., Kienstra, R.A., Naqui, S.T., and Ingelfinger, J.R. (1974). Normal glomerular permeability and its modification by minimal change nephrotic syndrome. *Journal of Clinical Investigation*, **54**, 1190–9.

Rodriguez-Soriano, J., Vallo, A., and Castillo, G. (1982). Defect in urinary acidification in nephrotic syndrome and its correction by furosemide. *Nephron*, **32**, 308–13.

Roos, R. (1986). The pathogenesis of atherosclerosis—an update. *New England Journal of Medicine*, **314**, 488–500.

Roos, R.A.C., Maaswinkel-Mooy, P.D., Loo, E.M., and Kanhai, H.H.H. (1987). Congenital microcephaly, infantile spasms, psychomotor retardation and nephrotic syndrome in two sibs. *European Journal of Pediatrics*, **146**, 532–6.

Rothenberg, M.B. and Heyman, W. (1957). The incidence of the nephrotic syndrome in children. *Pediatrics*, **19**, 446–52.

Roy, S. and Pitcock, J.A. (1971). Idiopathic nephrosis in identical twins. *American Journal of Diseases of Children*, **121**, 428–39.

Rubin, M.I. (1975). Nephrotic syndrome. In *Pediatric Nephology*. (ed. M. I. Rubin, and J. M. Barratt), pp. 454–499. Williams and Wilkins, Baltimore.

Rumpelt, H.J. and Thoenes, W. (1973). Intraglomerular immune deposits in focal and segmental sclerosing glomerulopathy. *Clinical Nephrology*, **1**, 367–71.

Rumpelt, H.J. and Thoenes, W. (1974). Focal and segmental sclerosing glomerulopathy: a pathomorphological study. *Virchow's Archives A: Pathology Anatomy and Histology*, **362**, 265–82.

Saint-Hillier, Y., Morel-Maroger, L., Woodrow, D., and Richet, G. (1975). Focal and segmental hyalinosis. In *Advances in Nephrology* (ed. J. Hamburger, J. Crosnier, and M. H. Maxwell). pp. 67–88. Yearbook, Chicago.

Sandberg, D.H., MacIntosh, R.M., Bernstein, C.W., Carr, R.K., and Strauss, J. (1977). Seven steroid responsive nephrosis associated with hypersensitivity. *Lancet*, **i**, 388–90.

Sasdelli, M., Cagnoli, L., Candi, P., Mandreoli, M., Bettrandi, E., and Zuchelli, P. (1981). Cell mediated immunity in idiopathic glomerulonephritis. *Clinical and Experimental Immunology*, **46**, 27–34.

Sato, K.A., Gray, R.W., and Lemann, J. (1982). Urinary excretion of 25-hydroxy-vitamin D in health and the nephrotic syndrome. *Journal of Laboratory and Clinical Medicine*, **99**, 325–30.

Schärer, K. and Minges, U. (1973). Long term prognosis of the nephrotic syndrome in childhood. *Clinical Nephrology*, **1**, 182–7.

Schlesinger, E.R., Sultz, H.A., Mosher, W.E., and Feldman, J.G. (1968). The nephrotic syndrome. Its incidence and implications for the community. *American Journal of Diseases of Children*, **116**, 623–32.

Schnaper, H.W. and Aune, T.M. (1985). Identification of the lymphokine soluble immune response suppressor in urine of nephrotic children. *Journal of Clinical Investigation*, **76**, 341–9.

Schnaper, J.W. and Aune, T.M. (1987). Steroid-sensitive mechanism of soluble immune response suppressor production in steroid-responsive nephrotic syndrome. *Journal of Clinical Investigation*, **79**, 257–64.

Schnaper, H.W. (1989). The soluble immune response suppressor pathway in nephrotic syndrome. *Seminars in Nephrology*, **9**, 107–11.

Schnaper, H.W. (1990). A regulatory system for soluble immune response suppressor production in steroid-responsive nephrotic syndrome. *Kidney International*, **38**, 151–9.

Schoeneman, M.J. (1983). Minimal change nephrotic syndrome: treatment with low doses of hydrocortisone. *Journal of Pediatrics*, **102**, 791–3.

Schoeneman, M.J., Spitzer, A., and Greifer I. (1983). Nitrogen mustard therapy in children with frequent relapsing nephrotic syndrome and steroid toxicity. *American Journal of Kidney Diseases*, **2**, 526–9.

Schulman, S.L., Kaiser, B.A., Polinsky, M.S., Srinivasan, R., and Baluarte, H.J. (1988). Predicting the response to cytotoxic therapy for childhood nephrotic syndrome: superiority of response to corticosteroid therapy over histopathologic patterns. *Journal of Pediatrics*, **113**, 996–1001.

Schulte-Wisserman, H., Lemmel, E.M., Reitz, M., Beck, J. and Straub, E. (1977). Nephrotic syndrome of childhood and disorders of T-cell function. *European Journal of Pediatrics*, **124**, 121–6.

Schwartz, M.W., Schwartz, G.J., and Cornfeld, D.A. (1974). A 16 years follow-up study of 163 children with nephrotic syndrome. *Pediatrics*, **54**, 547–52.

Schwartz, M.W. and Cornfeld, D.A. (1979). An unusual variation of familial nephrosis. *American Journal of Diseases of Children*, **133**, 216–7.

Schwartz, M.W. and Lewis, E.J. (1985). Focal segmental glomerular sclerosis: the cellular lesion. *Kidney International*, **28**, 968–74.

Senggutuvan, P., et al. (1990). Recurrence of focal segmental glomerulosclerosis in transplanted kidneys: analysis of incidence and risk factors in 59 allografts. *Pediatric Nephrology*, **4**, 21–8.

Sewell, R.F. and Brenchley, P.E. (1986). Red-cell surface charge in glomerular disease. *Lancet*, **ii**, 635–6.

Shalhoub, R.J. (1974). Pathogenesis of lipoid nephrosis: a disorder of T cell function. *Lancet*, **ii**, 556–9.

Sharples, P.M., Poulton, J., and White, R.H.R. (1985). Steroid responsive nephrotic syndrome is more common in Asians. *Archives of Diseases in Childhood*, **60**, 1014–7.

Shimamura, T. and Morrison, A.B. (1975). A progressive glomerulosclerosis occuring in partial five-sixths nephrectomized rats. *American Journal of Pathology*, **79**, 95–106.

Shohet, I., Meyerovitch, J., Aladjem, M., and Boichis, H. (1988). Cyclophosphamide in treatment of minimal change nephrotic syndrome. *European Journal of Pediatrics*, **147**, 239–41.

Shu, K., Lian, J., Yang, Y., Lu, Y., and Wang, J. (1988). Serum IgE in primary glomerular diseases and its clinical significance. *Nephron*, **49**, 24–8.

Sibley, R.K., Mahan, J., Mauer, S.M., and Vernier, R.L. (1985). A clinicopathologic study of forty eight infants with nephrotic syndrome. *Kidney International*, **27**, 544–52.

Siegel, N.J., Goldberg, B., Krassner, L.S., and Hayslett, J.P. (1972). Long term follow-up of children with steroid responsive nephrotic syndrome. *Journal of Pediatrics*, **81**, 251–8.

Siegel, N.J., Kashgarian, M., Spargo, B.H., and Hayslett, J.P. (1974). Minimal change and focal sclerotic lesions in lipoid nephrosis. *Nephron*, **13**, 125–37.

Siegel, N.J., Gur, A., and Krassner, L.S. (1975). Minimal lesion nephrotic syndrome with early resistance to steroid therapy. *Journal of Pediatrics*, 377–80.

Siegel, N.J., Gaudio, K.M., Krassner, L.S., MacDonald, B., Anderson, F.P., and Kashgarian, M. (1981). Steroid dependent nephrotic syndrome in children: histopathology and relapses after cyclophosphamide treatment. *Kidney International*, **19**, 454–9.

Simon, P. and Meyrier, A. (1989). One-year, alternate day dosage, corticosteroid Rx reduces rate of further relapses in adult minimal change nephrosis. *Kidney International*, **35**, 201 (abstract).

Sinnassamy, P. and O'Regan, S. (1985). Mesangial IgA deposits with steroid responsive nephrotic syndrome: probable minimal lesion nephrosis. *American Journal of Kidney Diseases*, **5**, 267–9.

Sobel, A.T., Branellec, A.L., Blanc, C.J., and Lagrue, G.A. (1977). Physiochemical characterization of a vascular permeability factor, produced by Con A stimulated human lymphocytes. *Journal of Immunology*, **119**, 1230–4.

Solling, J. (1982). Molecular weight of immune complexes in patients with glomerulonephritis. *Nephron*, **30**, 137–42.

Solomon, L.R., Cairns, J.A., Lawler, W., Johnson, R.W., and Mallick, N.P. (1981). Reduction of post transplant proteinuria due to recurrent mesangial proliferative (IgM) glomerulonephritis following plasma exchange. *Clinical Nephrology*, **16**, 44–50.

Southwest Pediatric Nephrology Study Group (1985a). Association of IgA nephropathy with steroid-responsive nephrotic syndrome. *American Journal of Kidney Diseases*, **5**, 157–64.

Southwest Pediatric Nephrology Study Group (1985b). Focal segmental glomerulosclerosis in children with idiopathic nephrotic syndrome. *Kidney International*, **27**, 442–9.

Spika, J.S., *et al.* (1982). Serum antibody response to pneumococcal vaccine in children with nephrotic syndrome. *Pediatrics*, **69**, 219–23.

Spika, J.S., *et al.* (1986). Decline of vaccine-induced antipneumococcal antibody in children with nephrotic syndrome. *American Journal of Kidney Diseases*, **7**, 466–70.

Springate, J.E., Coyne, J.F., Karp, M.P., and Feld, L.G. (1987). Acute renal failure in minimal change nephrotic syndrome. *Pediatrics*, **80**, 946–8.

Srivastava, R.N., Mayekar, G., Anand, R., Choudhry, V.P., Gnaî, O.P., and Tandon, H.D. (1975). Nephrotic syndrome in Indian children. *Archives of Diseases in Childhood*, **50**, 626–80.

Srivastava, R.N., Agarwal, R.K., Moudgil, A., and Bhuyan, U.N. (1985). Late resistance to corticosteroids in nephrotic syndrome. *Journal of Pediatrics*, **107**, 66–70.

Striegel, J.E., Sibley, R.K., Fryd, D.S., and Mauer, S.M. (1986). Recurrence of focal segmental sclerosis in children following renal transplantation. *Kidney International*, **30**, S44–S50.

Tanphaichitr, P., Tanphaichitr, D., Sureetanan, J., and Chatasigh, S. (1980). Treatment of nephrotic syndrome with levamisole. *Journal of Pediatrics*, **96**, 490–3.

Taube, D., Chapman, S., Brown, Z., and Williams, D.G. (1981a). Depression of normal lymphocyte transformation by sera of patients with minimal change nephropathy and other forms of nephrotic syndrome. *Clinical Nephrology*, **15**, 286–90.

Taube, D., Brown, Z., and Williams, D.G. (1981b). Long term impairment of suppressor cell function by cyclophosphamide in minimal change nephropathy and its association with therapeutic response. *Lancet*, **i**, 235–8.

Taube, D., Brown, Z., and Williams, D.G. (1982). Increased plastic-adherent mononuclear suppressor cell activity in patients with minimal change nephropathy. *Kidney International*, **22**, 581.

Taube, D., Brown, Z., and Williams, D.G. (1984). Impaired lymphocyte and suppressor cell function in minimal change nephropathy membranous nephropathy and focal glomerulosclerosis. *Clinical Nephrology*, **22**, 176–82.

Tejani, A. (1985). Morphological transitions in minimal change nephrotic syndrome. *Nephron*, **39**, 157–9.

Tejani, A., *et al.* (1983). Long term evaluation of children with nephrotic syndrome and focal segmental glomerular sclerosis. *Nephron*, **35**, 225–32.

Tejani, A., *et al.* (1985). Cyclosporine induced remission of relapsing nephrotic syndrome in children. *Kidney International*, **29**, 206.

Tejani, A., Butt, K., Trachtman, H., Suthanthiran, M., Rosenthal, C.J., and Khawar, M.R. (1987). Cyclosporine induced remission of relapsing nephrotic syndrome in children. *Journal of Pediatrics*, **111**, 1056–62.

Tejani, A., Butt, K., Trachtman, H., Suthanthiran, M., Rosenthal, C.J., and Khawar, M.R. (1988). Cyclosporine A induced remission of relapsing nephrotic syndrome in children. *Kidney International*, **33**, 729–34.

Thomson, P.D., Barratt, T.M., Stokes, C.R., Turner, M.W., and Soothill, J.F. (1976). HLA antigens and atopic features in steroid-responsive nephrotic syndrome of childhood. *Lancet*, **ii**, 765–8.

Tiggeler, R.G., Koene, R.A., and Wijdeveld, P.G. (1977). Inhibition of furosemide-induced natriuresis by indomethacin in patients with the nephrotic syndrome. *Clinical Science and Molecular Medicine*, **52**, 149–51.

Tomizawa, S., Maruyama, K., Nagasawa, N., Suzuki, S., and Kuroume, T. (1985). Studies of vascular permeability factor derived from T lymphocytes and inhibitory effect of plasma on its production in minimal change nephrotic syndrome. *Nephron*, **41**, 157–60.

Torres, V.W., Velosa, J.A., Holley, K.E., Frohnert, P.P., Zincke, H., and Sterioff, S. (1984). Meclofenamate treatment of recurrent idiopathic nephrotic syndrome with focal segmental glomerulosclerosis after renal transplantation. *Mayo Clinic Proceedings*, **59**, 146–52.

Trainin, E.B., Boichis, H., Spitzer, A., Edelmann, C.M., and Greifer, I. (1975). Late non responsiveness to steroids in children with the nephrotic syndrome. *Journal of Pediatrics*, **87**, 519–23.

Trompeter, R.S. (1987). Steroid resistant nephrotic syndrome: a review of the treatment of focal segmental glomerulosclerosis in children. In *Recent advances in pediatric nephrology*. (ed. K. Murakami, T. Kitagawa, K. Yabuta, and T. Sakai), pp. 363–371. Excerpta medica, Amsterdam.

Trompeter, R.S., Barratt, T.M., Kay, R., Turner, M.W., and Soothill, J.F. (1980). HLA. atopy and cyclophosphamide in steroid-responsive childhood nephrotic syndrome. *Kidney International*, **17**, 113–7.

Trompeter, R.S., Evans, P.R., and Barratt, T.M. (1981). Gonadal function in boys with steroid responsive nephrotic syndrome treated with cyclophosphamide for short periods. *Lancet*, **i**, 1177–80.

Trompeter, R.S., Hicks, J., Lloyd, B.W., White, R.H.R., and Cameron, J.S. (1985). Long term outcome for children with minimal change nephrotic syndrome. *Lancet*, **i**, 368–70.

Tsao, Y.C. and Yeung, C.H. (1971). Paired trial of cyclophosphamide and prednisone in children with nephrosis. *Archives of Diseases in Childhood*, **46**, 327–31.

Turner, I., Ibels, L.S., Alexander, J.H., Harrisson, A., and Moir, D. (1987). Minimal change glomerulonephritis associated with schistosomia hematobium infection: resolution with praziquantel treatment. *Australian and New Zeland Journal of Medicine*, **17**, 596–8.

Ueda, N., *et al.* (1988). Intermittent versus long term tapering prednisolone for initial therapy in children with idiopathic nephrotic syndrome. *Journal of Pediatrics*, **112**, 122–6.

Velosa, J.A. and Torres, V.E. (1986). Benefits and risks of nonsteroidal antiinflammatory drugs in steroid resistant nephrotic syndrome. *American Journal of Kidney Diseases*, **8**, 345–50.

Velosa, J.A., Donadio, J.V., and Holley, K.E. (1975). Focal sclerosing glomerulopathy. *Mayo Clinic Proceedings*, **50**, 121–32.

Velosa, J.A., *et al.* (1985). Treatment of severe nephrotic syndrome with meclofenamate: A uncontrolled pilot study. *Mayo Clinic Proceedings*, **60**, 586–92.

Verani, R.R. and Hawkins, E.P. (1986). Recurrent focal segmental glomerulosclerosis: a pathological study of the early lesion. *American Journal of Nephrology*, **6**, 263–70.

Vermylen, C., Levin, M., Mossman, J., and Barratt, T.M. (1989). Glomerular and urinary heparan sulphate in congenital nephrotic syndrome. *Pediatric Nephrology*, **3**, 122–9.

Vernier, R.L., Brunson, J., and Good, R.A. (1957). Studies on familial nephrosis. *American Journal of Diseases of Children*, **93**, 469–85.

Vernier, R.L., Klein, D.J., Sisson, S.P., Mahon, J.D., Oegema, T.R., and Brown, D.M. (1983). Heparan sulfate-rich anionic sites on the glomerular basement membrane. Decreased concentration in congenital nephrotic syndrome. *New England Journal of Medicine*, **309**, 1001–9.

Vilches, A.R., Turner, D.R., Cameron, J.S., Ogg, C.S., Chantler, C., and Williams, D.G. (1982). Significance of mesangial IgM deposition in "minimal change" nephrotic syndrome. *Laboratory Investigation*, **46**, 10–15.

Voets, A.J., Hoitsma, A.J., and Koene, R.A.P. (1986). Recurrence of nephrotic syndrome during cyclosporin treatment after renal transplantation. *Lancet*, **i**, 266–7.

Volhard, F. and Fahr, T. (1914). *Die brightsche nierenkrankeit*. Springer, Berlin.

Vriesendorp, R., De Zeeuw, D., De Jong, P.E., Donker, A.J.M., Pratt, J.T., and Van der Hem, G.K. (1986). Reduction of urinary protein and prostaglandin E2 excretion in the nephrotic syndrome by non steroidal antiinflammatory drugs. *Clinical Nephrology*, **25**, 105–10.

Waldherr, R., Gubler, M., Levy, M., Broyer, M., and Habib, R. (1978). The significance of pure mesangial proliferation in idiopathic nephrotic syndrome. *Clinical Nephrology*, **10**, 171–9.

Waldo, F.B. and Kohaut, E.C. (1987). Therapy of focal segmental glomerulosclerosis with cyclosporine A. *Pediatric Nephrology*, **1**, 180–2.

Wang, F., Loo, L.M., and Chua, C.T. (1982). Minimal change glomerular disease in Malaysian adults and use of alternate day steroid therapy. *Quarterly Journal of Medicine*, **51**, 312–28.

Warshaw, B.L. and Hymes, L.C. (1989). Daily single dose and daily reduced dose prednisone therapy for children with the nephrotic syndrome. *Pediatrics*, **83**, 694–9.

Watson, A., Stachura, I., Fragola, J., and Bourke, E. (1983). Focal segmental glomerulosclerosis in Hodgkin's disease. *American Journal of Nephrology*, **3**, 228–32.

Watson, A.R., Taylor, J., Rance, C.P., and Bain, J. (1986). Gonadal function in women treated with cyclophosphamide for childhood nephrotic syndrome: a long term follow-up study. *Fertility and Sterility*, **46**, 331–3.

Whelan, T.W. and Hirszel, P. (1988). Minimal change nephropathy associated with pancreatic carcinoma. *Archives of Internal Medicine*, **148**, 975–6.

White, R.H.R. (1973). The familial nephrotic syndrome. A European survey. *Clinical Nephrology*, **1**, 215–9.

White, R.H.R. and Glasgow, E.F. (1971). Focal glomerulosclerosis. A progressive lesion associated with steroid resistant nephrotic syndrome. *Archives of Diseases in Childhood*, **46**, 877–86.

White, R.H.R., Glasgow, E.F., and Mills, R.J. (1970). Clinicopathological studies of nephrotic syndrome in childhood. *Lancet*, **i**, 1353–9.

Wilkes, J.C., Nelson, J.D., Worthen, H.G., and Hogg, R.J. (1979). Pneumococcal vaccination in nephrotic syndrome. *Kidney International*, **16**, 914.

Wilkinson, A.H., Gillespie, C., Hartley, B., and Williams, D.G. (1989). Increase in proteinuria and reduction in number of anionic sites on the glomerular basement membrane in rabbits by infusion of human nephrotic plasma *in vivo*. *Clinical Science*, **77**, 43–8.

Williams, S.A., Makker, S.P., and Grupe, W.E. (1978). A significant side effect of chlorambucil therapy in children. *Journal of Pediatrics*, **93**, 516–8.

Williams, S.A., Makker, S.P., Ingelfinger, J.R., and Grupe, W.E. (1980). Long term evaluation of chlorambucil plus prednisone in idiopathic nephrotic syndrome of childhood. *New England Journal of Medicine*, **302**, 929–33.

Wilson, S.G.F., Hackel, D.B., Horwood, S., Nash, G., and Heymann, W. (1958). Aminonucleoside nephrosis in rats. *Pediatrics*, **21**, 963–73.

Wingen, A.M., Muller-Wiefel, D.E., and Scharer, K. (1985). Spontaneous remissions in frequently relapsing and steroid dependent idiopathic nephrotic syndrome. *Clinical Nephrology*, **23**, 35–40.

Wittig, J. and Goldman, A. (1970). Nephrotic syndrome associated with inhaled allergens. *Lancet*, **1**, 542–3.

Wu, M.J. and Moorthy, A.V. (1982). Suppressor cell fucntion in patients with primary glomerular disease. *Clinical Immunology and Immunopathology*, **22**, 442–7.

Wynn, S.R., Stickler, G.B., and Burke, E.C. (1988). Long term prognosis for children with nephrotic syndrome. *Clinical Pediatrics*, **27**, 63–8.

Yeung, C.K., Wong, K.L., and Ng, W.L. (1983). Intravenous methylprednisolone pulse therapy in minimal change nephrotic syndrome. *Australian and New Zedland Journal Medicine*, **13**, 349–51.

Yokoyama, H.M., *et al.* (1985). Immunodynamics of minimal change nephrotic syndrome in adults. T and B Lymphocyte subsets and serum immunoglobulin levels. *Clinical and Experimental Immunology*, **61**, 601–7.

Yokoyama, H., Kida, H., Abe, T., Koshino, Y., Yoshimura, M., and Hattori, N. (1987). Impaired immunoglobulin G production in minimal change nephrotic syndrome in adults. *Clinical and Experimental Immunology*, **70**, 110–5.

Yoshikawa, N., Cameron, A.H., and White, R.H.R. (1982). Ultrastructure of the non sclerotic glomeruli in childhood nephrotic syndrome. *Journal of Pathology*, **136**, 133–47.

Yoshikawa, N., *et al.* (1986). Focal segmental glomerulosclerosis with and without nephrotic syndrome in children. *Journal of Pediatrics*, **109**, 65–70.

Zimmerman, C.E. (1978). Renal transplantation for focal segmental glomerulosclerosis. *Transplantation*, **29**, 172.

Zimmerman, S.W. (1984). Increased urinary protein excretion in the rat produced by serum from patient with recurrent focal glomerular sclerosis after renal transplantation. *Clinical Nephrology*, **22**, 32–8.

Zimmerman, S.W. (1985). Plasmapheresis and dipyridamole for recurrent focal glomerular sclerosis. *Nephron*, **40**, 241–5.

3.6 IgA nephropathies

F. PAOLO SCHENA

Definition of IgA nephropathies

IgA nephropathies are characterized pathologically by the presence of diffuse mesangial deposits of IgA in the glomeruli. IgA is frequently detected in selected pathological entities such as Berger's disease, Schönlein-Henoch purpura and glomerulonephritis associated with hepatic cirrhosis. Some clinical and laboratory findings, such as the clinical progression of Berger's disease to Schönlein-Henoch purpura in some patients, the increased serum levels of polymeric IgA, in part as immune complexes, and the deposition of IgA in the glomeruli and in the walls of blood vessels in the skin support the hypothesis that these diseases may represent different manifestations of a single syndrome.

Primary IgA nephropathy (pIgAN) or Berger's disease is a disease characterized by recurrent gross haematuria, microhaematuria and/or proteinuria in the absence of any recognizable systemic disease (lupus erythematosus, Schönlein-Henoch purpura, cryoglobulinaemia), liver disease, or lower urinary tract diseases. Berger *et al.* (1967) first reported this disease at the International Nephrology Congress in Washington in 1966 and described the immunological and clinical findings in more detail in two later reports (Berger and Hinglais 1968; Berger 1969). Their patients had macro- or microhaematuria and/or proteinuria with normal renal function. The macrohaematuria appeared mainly in association with recurrent episodes of upper respiratory tract infections. Since most of these patients had normal renal function, this form was named benign recurrent haematuria. In the next few years long-term studies demonstrated impaired renal function in 20 to 50 per cent of adult patients with this type of nephritis, so the definition of pIgAN was changed to include indolent or slowly progressive disease.

Mesangial IgA deposits are observed in a variety of other diseases, most commonly but not exclusively in Schönlein-Henoch purpura, systemic lupus erythematosus, and liver diseases. Schönlein-Henoch purpura is described in Chapter 4.5.2, and Systemic lupus erythematosus in Chapters 4.6.1 and 4.6.2.

Glomerulonephritis associated with liver disease is usually clinically silent and characterized by mild microhaematuria, proteinuria and glomerular lesions. IgA is almost always the immunoglobulin deposited in the glomeruli.

Several diseases in which an abnormal response of the IgA system is present may be associated with IgA nephritis. The nephropathy may occur before and during the clinical course of the disease and a wide variety of glomerular lesions may be present.

Primary IgA nephropathy (Berger's disease)

Clinical features

At the onset

The occurrence of glomerular haematuria and heavy mesangial IgA deposits in the renal biopsy specimens is very high in subjects with asymptomatic microhaematuria and/or slight proteinuria (Simon et al. 1984; Kurokawa et al. 1985; Propper et al. 1987).

Age at the time of the first clinical manifestations of pIgAN ranges between 15 and 30 years and this is often 7 to 10 years earlier than the age when diagnosis is made by biopsy, i.e. the onset of pIgAN is usually in the teenage years or earlier. Affected children do not present symptoms and/or urinary signs before the age of 3 years; thereafter they are evenly spread through childhood (Levy et al. 1985).

The clinical signs and symptoms at the onset of the disease are those of microscopic haematuria/proteinuria, acute glomerulonephritis, and the nephrotic syndrome. In the majority of cases the first signs are chance findings of microhaematuria/proteinuria. Microscopic haematuria is persistent, and sometimes intermittent. In many cases microhaematuria is detected accidentally as a result of pre-employment or blood donor screening, or of investigations for other causes.

Episodes of macroscopic haematuria can be associated with any of the events reported in Table 1, but the majority occur after infections, usually of the upper respiratory tract and less often of other sites (Fig. 1); rarely they occur after vaccination or heavy physical exercise. The interval between the precipitating event and the appearance of macrohaematuria is very short (24–72 h), compared with 1 or 2 weeks in postinfectious acute glomerulonephritis. The macrohaematuria persists less than 3 days and is sometimes accompanied by flank and loin pain and, occasionally, fever. The colour of the urine is red or brown (coke-coloured); rarely, it contains blood clots. The number of recurrent episodes of gross haematuria is variable in adults, whereas they are a characteristic feature of the clinical picture of the disease in 80 to 95 per cent of children, declining in fre-

Table 1 Precipitating factors for macrohaematuria

Upper respiratory tract infections
  Tonsillitis
  Faringitis
  Bronchitis
Acute gastroenteritis
Hepatitis A/B
Periostitis
Staphylococcal osteomyelitis
Septic arthritis
Peritonitis
Lobar pneumonia
Erysipelas
Erythema polymorphus
Staphylococcal sepsis
Typhoid fever
Brucellosis
Infectious mononucleosis
Influenza-like syndromes
Rubella
Mumps
Herpes zoster
Tonsillectomy
Tooth extraction
Appendicectomy
Heavy physical exercise
Vaccine
BCG overdose

quency with increasing age in both sexes (Southwest Pediatric Nephrology Study Group 1982; Levy et al. 1985). In the past patients have had their tonsils removed because of recurrent infections and macrohaematuria, but this does not reduce the number of haematuric episodes unless the tonsils are diseased.

Occasionally the only manifestation of pIgAN is asymptomatic hypertension or proteinuria. Hypertension may be present with variable frequency (4–7 per cent). Its incidence is higher in patients with microhaematuria, with or without proteinuria.

In a few cases there is an acute nephritic syndrome, similar to poststreptococcal glomerulonephritis, at the onset of the disease. In these patients macrohaematuria is associated with increased serum creatinine and BUN, and also hypertension. Furthermore, in a few patients acute oliguric renal disease, usually spontaneously reversible, accompanies the episodes of macrohaematuria (Lupo et al. 1987). Finally, pIgAN may recur 1 to 4 years after transplantation with the appearance of microhaematuria and mild proteinuria (Berger et al. 1975; Cameron et al. 1977).

At the time of biopsy

Haematuria

The episodes of macrohaematuria are more frequent during the first few years of the disease and then tend to disappear. The relative incidence of this symptom is higher in children and young adults than in older patients. This difference varies greatly in various geographical areas because of differences in attitude to submitting patients with this symptom to renal biopsy (Emancipator and Lamm 1989).

In some countries of the Asian Pacific area (Japan, Hong Kong, Singapore) microscopic haematuria together with asymptomatic proteinuria is the main clinical finding at the time of biopsy in children and in young adults because most of the cases

3.6 IgA nephropathies

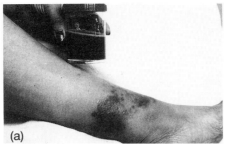
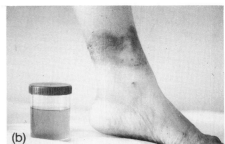

Fig. 1 Coke-coloured urine in patient with previous diagnosed pIgAN during an episode of erysipelas (a). Urine shows variation in colour a few days after the infectious episode (b).

are detected in school screening programmes or during the routine or regular medical examinations of male army recruits. By contrast, macroscopic haematuria is an unusual presenting symptom. The prevalence of microscopic and macroscopic haematuria in three different areas of the world is shown in Fig. 2. Gross haematuria occurs more frequently in North America and in European patients than in those living in the Asian-Pacific countries (Asia and Australia): the reverse is true for microhaematuria.

In a small number of patients presentation of the disease is characterized clinically by macroscopic haematuria, proteinuria, and rapid deterioration of renal function. In these cases the histological examination of renal biopsy samples reveals prominent crescent formations that usually involve more than 30 to 50 per cent of the glomeruli. This condition has been called 'rapidly progressive glomerulonephritis' (Abuelo *et al.* 1984) or 'malignant IgA nephropathy' (Nicholls *et al.* 1985).

The incidence of IgA nephropathy is variable among subjects with asymptomatic microhaematuria. Two reports from the United Kingdom (Propper *et al.* 1987; Michael *et al.* 1976) record an incidence of 37 and 54 per cent, respectively. In contrast, Pardo *et al.* (1979) and Kupor *et al.* (1975) report a lower incidence of IgA nephropathy (22.7 and 20 per cent, respectively) in the United States.

French investigators (Berthoux *et al.* 1986) have shown that microscopic haematuria with more than 5000 RBC/min is an indicator of the disease.

Proteinuria

Proteinuria may be mild (less than 1 g/day), moderate (more than 1 g/day and less than 3.0 g/day) or heavy (more than 3.0 g/day). In asymptomatic patients detected at a routine medical examination the associated proteinuria is usually less than 1 g/24 h and exceeds 3 g/day in only a few cases. The level of proteinuria tends to remain fairly stable. A transient increase in proteinuria occurs in subjects with a history of fever, sore throat, pain in the lumbar or loin region, dysuria, and transient ankle oedema.

Nephrotic syndrome

The nephrotic syndrome rarely appears at the onset of the disease but the prevalence is higher in Asian-Pacific countries than in Europe (Fig. 3). The nephrotic syndrome is more frequent later in the course of the disease, and the incidence differs throughout the world (Fig. 3); the renal lesions may be minimal or severe.

Spontaneous remission of the nephrotic syndrome has been reported (Wu *et al.* 1985) in patients with pIgAN, in whom the renal biopsy revealed a minimal change disease, with mesangial deposits of IgA. Similar cases have been observed in children (Southwest Pediatric Nephrology Study Group 1985). The

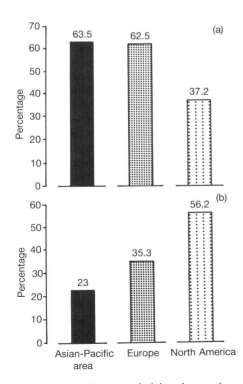

Fig. 2 Incidence of microhaematuria (a) and macrohaematuria (b) observed during the clinical presentation at the time of renal biopsy in patients with pIgAN from three different areas of the world.

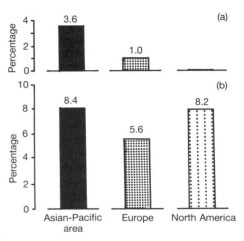

Fig. 3 Incidence of nephrotic syndrome at onset of the disease (a) and at renal biopsy (b) in patients with pIgAN from different areas of the world.

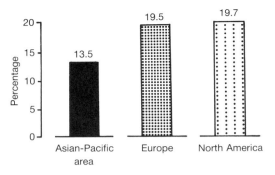

Fig. 4 Percentage of IgA nephropathy patients with decreased renal function (serum creatinine 1.5 mg/dl; clearance creatinine 80 ml/min) at the time of renal biopsy.

occurrence of mild glomerular lesions, normal renal function, good response to corticosteroids and a tendency to relapse is similar to the behaviour of minimal change disease, and it is possible that, in some cases, the development of the pIgAN was superimposed on preceding but undiagnosed minimal change disease. This suggestion is supported by the finding of nephrotic children whose first renal biopsies were negative for IgA but in whom a subsequent biopsy was positive (Southwest Pediatric Nephrology Study Group 1985). Alternative explanations for this particular aspect of pIgAN may be the possible intermittent appearance of mesangial IgA deposits and failure to identify IgA at the initial biopsy, or the presence of IgAN with IgA deposits in a focal distribution affecting only some glomeruli.

Renal insufficiency

The frequency of renal insufficiency in patients with pIgAN at the time of renal biopsy is higher in European and North American patients than in patients from the Asian-Pacific area (Fig. 4). The difference may be accounted for by the absence of screening programmes in schools in Europe and North America and by the different approaches adopted to renal biopsy in eastern and western areas of the world. There is general agreement that the severity of the histological lesions correlate well with the progression of renal failure. The relative risk of progression towards renal insufficiency is 10 times lower in patients with minor abnormalities (Noel *et al.* 1987).

Hypertension

The frequency of hypertension in IgAN patients without renal insufficiency is higher than that found in the healthy population, matched for sex and age, and living in the same geographical area (Zucchelli *et al.* 1984; D'Amico and Vendemmia 1987). Nevertheless, it is less frequent at presentation than in other types of primary glomerulonephritis (D'Amico 1983). Hypertension often develops during the course of the disease, sometimes before serum creatinine is appreciably raised. It is found in 20 per cent of adult patients at the time of biopsy but less commonly in children (5 per cent). It occurs more frequently in patients who present the disease after the age of 35 and in those with decreased renal function (Noel *et al.* 1987). In some subjects hypertension accelerates the impairment of renal function. Clarkson *et al.* (1977), Rambausek *et al.* (1982), and Subias *et al.* (1987) have all reported that malignant hypertension occurs in 7 to 15 per cent of patients and is associated with a rapid decline in renal function. In these cases the renal biopsy reveals epithelial crescents in the majority of glomeruli.

Several factors, such as increased plasma renin activity, hypersympathetic activity, renal prostaglandins, natriuretic hormones and histological lesions may be responsible for the development of hypertension.

Hypertension may aggravate the glomerular damage via an increased arteriolar sclerosis effect. Alternatively hypertension may cause glomerular damage directly through the elevated hydrostatic pressure in the glomerular capillaries. Clarkson *et al.* (1977) and Feiner *et al.* (1982) found a correlation between arteriolar sclerosis and glomerular sclerosis in pIgAN whilst D'Amico *et al.* (1986) found a relationship between arterial hypertension and glomerular sclerosis, interstitial fibrosis, arteriolar hyalinosis, and diffuse mesangial proliferation. It is possible that the development of mesangial injury and relative sclerosis favour the appearance of hypertension through modifications of the normal glomerular microcirculation. The presence of vascular damage, independent of its severity, may not be accompanied by high plasma renin activity in normotensive or in hypertensive patients. Therefore, the high plasma renin activity found in a third of patients may be due to other mechanisms, such as a generalized increase in sympathetic tone, intrarenal sympathetic overactivity, or a β-receptor hyper-responsiveness (Zucchelli *et al.* 1984).

In both normotensive and hypertensive patients Valvo *et al.* (1987) found an increased plasma and blood volume associated with fluid retention which implies a decreased capacity to excrete sodium and water despite normal renal function. They also observed a high plasma renin activity which could be due to the inflammatory process in the mesangium. Finally the vasoconstriction could be due to the release of other vasoactive substances that can be present at high levels, such as noradrenaline and adrenaline (Ishii *et al.* 1983).

Recently Manno *et al.* (1988) found high values of human atrial natriuretic factor in patients with pIgAN. This hormone was able to control the blood pressure by natriuresis and by reducing peripheral resistance in normotensive patients, whereas its compensatory mechanism was not sufficient in hypertensive patients, in whom the prolonged clinical course could be responsible for increased peripheral vascular resistance and the consequent hypertension.

Extrarenal manifestations

Extrarenal manifestations involving mainly the respiratory and gastrointestinal tracts and the skin have been observed in a high percentage of patients with pIgAN in some series (Mustonen 1984; Makdassy *et al.* 1984; Berthoux and Alamartine 1986), but there is no agreement over the role of IgA in these cases and whether they have true IgAN. Alternative explanations are that extrarenal manifestation is a pure coincidence or that IgAN is the result of a systemic disorder that involves the kidney. Recently Churg and Sobin (1982) suggested that Berger's disease was classified as a glomerulonephritis in the group of systemic diseases but until there is a better understanding of the significance of IgAN in these cases, it seems more appropriate to deal with them in the chapter on IgA glomerulonephritis and associated diseases.

Clinical outcome

There is a marked variation in the clinical outcome of pIgAN in all geographical areas, with a spectrum from total disappearance of blood and protein in the urine to the development of chronic renal failure requiring dialysis. The disappearance of microhae-

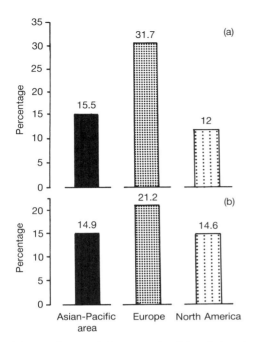

Fig. 5 Incidence of systemic hypertension (a) and chronic renal insufficiency (b) after a short mean follow-up (2–4 years) in patients with pIgAN from different areas of the world.

maturia reported by many investigators (De Werra *et al.* 1973; Schmekel *et al.* 1981; Wyatt *et al.* 1984b; Rodicio 1984; Nicholls *et al.* 1984; Noel *et al.* 1987) occurs in 3 to 25 per cent of the patients, but the regression of renal lesions and disappearance of mesangial IgA deposits have never been observed. In a small number of patients in whom the presentation is characterized by the presence of macroscopic haematuria, with loin pain and a normal serum creatinine, the progression of the renal damage to end-stage renal disease proceeds rapidly over 2 to 3 years. In these cases the renal lesions are characterized by persisting crescents and progressive loss of glomeruli. Furthermore, patients with crescentic IgAN have a higher prevalence of hypertension, renal insufficiency and nephrotic-range proteinuria.

Differences in the prevalence of hypertension (12–31 per cent) and impairment of renal function (15–21 per cent) among patients from different geographical areas present problems in determining the likelihood of progression of the glomerular lesion (Fig. 5). These differences probably reflect dissimilar glomerular diseases that have the finding of mesangial IgA deposits in common.

The overall clinical outcome shows that 69 to 80 per cent of patients continue to have normal renal function after a weighted mean follow-up of 2 to 5 years (Schena 1989a). Renal biopsies from these patients usually reveal either no change or only minimal glomerular lesions. Stable renal functions is generally present in subjects with a mild form of mesangial proliferative glomerulonephritis, while deterioration of renal function occurs in patients with advanced histological lesions. The prevalence of renal impairment is higher in European patients than in subjects from the Asian-Pacific area and from North America (Fig 5). Renal insufficiency is more common in males than in females, but the rate of deterioration varied considerably from patient to patient. Some reports (Van der Peet *et al.* 1977; Woo *et al.* 1986; Kobayashi *et al.* 1983) have identified two groups of patients by their clinical course: one with slowly progressive disease over several years and another with a more rapid course progressing to renal insufficiency with a few years. Patients in the first group represent the natural history of the nephritis, whereas many patients in the second group have unfavourable prognostic factors such as severe uncontrolled hypertension.

Patients with normal or stable renal function have mild proteinuria, normal blood pressure and mild histological lesions. In contrast, patients with progressive renal impairment show heavy proteinuria, hypertension, proliferative lesions, and crescents. Patients with a proteinuria of more than 3.5 g/day and deteriorated renal function develop chronic renal insufficiency within a short time of the appearance of clinical signs. Furthermore, a few cases have a rapid course to terminal renal failure and they are referred to as having 'malignant IgA nephropathy'.

Pregnancy does not influence the clinical course of the disease in patients with pIgAN in whom the renal function is normal or mildly impaired and when hypertension is minimal or absent at conception. More crucial is the adverse effect of pregnancy on the disease in patients with moderate renal insufficiency or hypertension. Patients with a serum creatinine more than 2.3 mg/dl (Jungers *et al.* 1987), should be advised of the risk that pregnancy will provide a rapid deterioration of the renal function and the fetus may be lost. A renal biopsy study (Kincaid-Smith *et al.* 1987) in these patients has shown a greater degree of glomerular proliferation and crescent formation as well as focal and segmental hyalinosis and sclerosis.

Renal survival

It has been suggested that pIgAN should be regarded as an indolent, chronic nephritis that may be benign in childhood and may progress slowly to chronic renal failure in adults. Nevertheless, the actuarial curves of renal survival from numerous series of patients that have been followed up have shown 5 to 15 per cent of patients with end-stage renal disease after 5 years from the apparent onset of the disease, 10 to 20 per cent after 10 years, 15 to 30 per cent after 15 years, and 20 to 50 per cent after 20 years (Fig. 6). The percentage of patients with normal renal function during the course of the disease is low, as chronic renal failure develops frequently, but renal insufficiency usually progresses slowly. The predictive power of the first renal biopsy must be emphasized, since the histological and immunofluorescence grading are reliable prognostic indicators. The renal survival rate in patients with diffuse proliferative glomerulonephritis

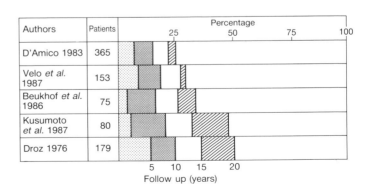

Fig. 6 Percentages of IgA nephropathy patients with end-stage kidney disease after 5, 10, 15, and 20 years of mean follow-up.

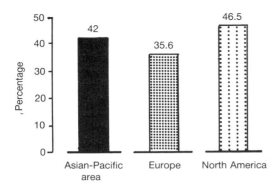

Fig. 7 Prevalence of IgA nephropathy patients with high serum levels of IgA.

and focal crescents is less than that of patients with minimal glomerular lesions, focal proliferative glomerulonephritis and mild to moderate proliferative glomerulonephritis. The patients who develop chronic renal failure have marked capsular adhesions, fibrocellular crescents, glomerular hyalinization, and sclerosis. The extracapillary lesion is especially important since long-term follow-up studies have shown that the renal survival rate decreases as the percentage of crescents present increases (Nicholls et al. 1984a; Abe et al. 1986). Renal survival is 40 per cent 5 years after the initial biopsy in patients with more than 50 per cent of crescents and 25 per cent after 10 years, whereas the 5-year survival rate is 91 per cent in patients with less than 50 per cent of crescents and 81 per cent after 10 years. These findings suggest that diffuse proliferative lesions and crescents indicate a poor prognosis. Children have been reported to have a better prognosis in various European countries, probably because severe mesangial proliferation, sclerotic glomeruli, severe interstitial changes and renal arteriosclerosis are less frequent.

There is a significant difference in renal survival rates between patients with moderate proteinuria and heavy proteinuria. The renal survival rates (specified by a serum creatinine of 2 mg/dl or less), have been reported to be 100 per cent 5 years after diagnosis in patients with mild proteinuria, 87 per cent in those with moderate proteinuria and 69 per cent in patients with heavy proteinuria (Neelakantappa et al. 1988). The relative risk of renal insufficiency was six times greater in patients with heavy proteinuria, and 2.5 times greater in those with moderate proteinuria than in those with mild proteinuria.

The renal survival rate in hypertensive patients, both children and adults, is lower than that of non-hypertensive patients (Payton et al. 1988). Furthermore, the renal survival rate is lower in patients with persistent microscopic haematuria than in those with macroscopic episodes of haematuria (Beukhof et al. 1986; D'Amico et al. 1986). Finally, the renal survival rate is lower in males than in females (Wyatt et al. 1984b; Nicholls et al. 1984) and in adults than in children (Kusumoto et al. 1987). No difference in renal survival rate between patients with elevated and normal levels of serum IgA has been observed.

Laboratory findings

Serum IgA

Elevated serum IgA concentrations are observed in a proportion (range 9–70 per cent) of patients with a higher incidence being reported in North America (46.5 per cent) than in the Asian-Pacific area (42 per cent) and Europe (35.6 per cent) (Fig. 7).

This serum marker may suggest the possible presence of the disease. IgA levels do not change during the clinical course of the disease, and initial high values persist in some patients, and fall to normal levels in others. IgA is unique amongst immunoglobulins in its ability to form multimers (a mixture of monomers and polymers). Several clinical studies (Lopez-Trascasa et al. 1980; Valentijn et al. 1984; Sinico et al. 1986) have reported the presence of high serum concentrations of polymeric IgA which may have an anti IgG (rheumatoid factor) activity. The occurrence of polymeric IgA rheumatoid factor in these patients is an example of an autoimmune phenomenon, and implies a general disturbance of IgA metabolism.

Serum IgA antibodies

Elevated IgA antibody titres against antigens from viruses (Herpes simplex, cytomegalovirus, Epstein-Barr virus), bacteria (Pneumococcus, *Streptococcus mutans*, *E.coli*), and food proteins (bovine serum albumin, ovoalbumin, casein, wheat gliadin) have been found by many investigators (Woodroffe et al. 1980; Sancho et al. 1983; Nagy et al. 1984b; Russel et al. 1986; Drew et al. 1987; Davin et al. 1987) in patients with pIgAN. These may reflect the persistence of viral/bacterial and food antigens, leading to the formation of immune complexes with excess antibody similar to those thought to be deposited in the mesangium in some experimental models of nephritis.

Immune complexes

High titres of IgA immune complexes and a low frequency of IgG immune complexes have been found consistently in the blood of patients with pIgAN using a variety of techniques. The former are present both in the acute phase of the disease (fever, upper respiratory tract infection and macrohaematuria) and in remission; the latter appear only in relapses. Some investigators (Schena et al. 1986, 1989c) have suggested that IgG immune complexes are responsible for activation of the complement system and progression of the renal lesions. The immune complexes may contain antibodies against bacterial or viral and food antigens. Complexes containing food antigen have been found in the blood of patients after food challenge and some investigators (Coppo et al. 1986) have suggested a relationship between diet and the high levels of IgA immune complexes in the blood, although this has not been confirmed by others (Yagama et al. 1988).

Proteinuria

The presence of proteinuria is closely correlated with impairment of renal function. When proteinuria exceeds 2 g/day, there is a higher incidence of renal failure as well as an increased incidence of crescents in renal biopsy. Hence, a more severe proteinuria (above 3g/day) seems to be related to more severe histological lesions. As heavy proteinuria precedes the appearance of hypertension, its identification is a prognostic marker. A protein selectivity study in pIgAN patients (Woo et al. 1986a) has shown that patients with poorly selective proteinuria are more likely to have poor prognostic features, such as renal impairment, hypertension, and severe glomerulosclerosis. In contrast, selective proteinuria is associated with mild histological changes. In addition, selective proteinuria is steroid responsive whereas a non-selective proteinuria does not respond well.

In addition to the glomerular lesions, morphological and functional tubulointerstitial damage occurs in patients with pIgAN. There is a highly significant positive correlation between low molecular weight urinary proteins and the severity of the tubulointerstitial changes found at renal biopsy (Nagy et al. 1987a).

Other laboratory features

Antistreptolysin O titres are elevated in a small number of patients. The presence of cryoglobulins and cryofibrinogen in serum and plasma samples has been reported by a few investigators (Agrofiotis et al. 1981; Nagy et al. 1987). Most of these patients had a history of long exposure to cold in the course of their work. The cryoprecipitate was composed of one or more proteins, but IgA was present in only a few cases. There was a good correlation between the presence of cryofibrinogen in the plasma samples and the tubulointerstitial fibrin/fibrinogen deposits. These findings suggest that renal deposition of cryofibrinogen might be responsible for the tubulointerstitial fibrocellular changes, which are of prognostic importance.

Routine tests have indicated that the serum complement levels of C3, C4, and factor B are normal. Nevertheless, sensitive techniques able to detect small amounts of some neoantigens of C3, such as iC3b and C3b, provide evidence of complement activation in patients with pIgAN, both in adults and in children (Wyatt et al. 1987b). No association has been found between the raised levels of the C3 fragments and a history of macroscopic haematuria, chronic renal failure, or the degree of proteinuria. In addition, a partial deficiency of a complement component, such as factor H which controls the C3b amplification loop has been observed in some patients with pIgAN and also in their relatives (Wyatt et al. 1982).

High serum values of β_2-microglobulin have been observed in patients with pIgAN, mainly in those with glomerular sclerosis (Woo et al. 1984; Lai et al. 1986a) and so it may be a marker of a poor prognosis, although there was no correlation between elevated levels of β_2-microglobulin and the degree of glomerular sclerosis.

Renal biopsy

Berger's disease is included in the large histological group of primary mesangial proliferative glomerulonephritides; mesangial proliferation may occur in other glomerulopathies such as minimal change disease, and glomerulonephritis with C3 and/or IgM deposits (Cohen and Border 1982). In all these cases the immunohistochemical findings and the presence of electron-dense deposits may differentiate between the various forms.

Light microscopy

The first renal biopsy study on pIgAN was performed by Galle and Berger in 1962. They recognized a specific form of glomerulonephritis characterized by intercapillary fibrinoid deposits, which were studied by electron microscopy and described in detail by Galle in 1964. These deposits were also reported by others, but no relationship between their size and appearance, and the duration of the disease was demonstrated.

The light microscopic picture of glomeruli is characterized by apparently normal glomeruli interspersed with others with areas of increased cellularity. The proportion of normal and abnormal glomeruli varies and a wide spectrum of proliferative changes may be observed.

The histological changes in the glomeruli consist of an increased mesangial matrix and hypercellularity in the mesangium. After the first descriptions (Berger et al. 1967, 1968; Druet et al. 1970), the spectrum of glomerular abnormalities has been enlarged; virtually all morphological manifestations of glomerular damage, such as adhesions, segmental sclerosis, and crescents, may be observed. For this reason different morphological classifications have been used by renal pathologists.

The grading of histological damage is based principally on the severity of the cellular proliferation of glomerulosclerosis and the number of crescents as well as the presence or absence of tubular atrophy and interstitial cellular infiltration or fibrosis. On the basis of similar renal pathological and immunohistological findings in Berger's disease and in Henoch-Schönlein nephritis, some investigators (Lee et al. 1982; Sinniah 1985) used the classification of Meadow et al. (1972) for the morphological distinction of renal lesions in Berger's disease. Finally, Churg and Sobin (1982) devised a classification of the renal lesions which has been officially accepted by the World Health Organization.

This morphological classification includes five classes of renal lesions (Table 2). These are:

Grade I. The majority of glomeruli are normal by light microscopy. Small areas of slight mesangial thickening with or without hypercellularity are present. These lesions are defined as minimal changes. Tubular and interstitial lesions are absent.

Grade II. Less than 50 per cent of glomeruli show mesangial proliferation, and sclerosis, adhesions, and small crescents are rare. These lesions are defined as minor changes. There are no tubular and interstitial lesions.

Grade III. Diffuse mesangial proliferation and thickening with focal and segmental variations. Adhesions and small crescents are occasionally present. This histological picture is defined as focal and segmental glomerulonephritis. Focal interstitial oedema and mild infiltrate are occasionally present.

Grade IV. All the glomeruli show marked diffuse mesangial proliferation and sclerosis with varying degrees of hypercellularity and irregular distribution. Obsolescent glomeruli are frequently found, in variable numbers. Up to 50 per cent of glomeruli contain adhesions and crescents. These lesions characterize diffuse mesangial proliferative glomerulonephritis. Tubular atrophy and interstitial inflammation are evident.

Grade V. This grade is similar to grade IV but more severe. Segmental and/or global sclerosis, hyalinosis, and capsular adhesions are present. More than 50 per cent of glomeruli show crescents. These lesions characterize diffuse sclerosing glomerulonephritis. In addition, tubular and interstitial changes are more severe than those observed in grade IV.

In biopsy specimens taken within the first year of the onset of disease, the changes are minor. In contrast, focal lesions, diffuse mesangial cell proliferation, and diffuse sclerosing glomerulonephritis are more frequent in patients examined by renal biopsy 3 years or more after the clinical onset of disease. In general the clinical course of the disease corresponds with the histological course; therefore renal biopsy is a good prognostic indicator, as the changes in morphology and the distribution of renal lesions can be followed. Patients with asymptomatic microscopic haematuria and proteinuria detected at a routine medical examination most frequently have minimal or minor changes on the initial renal biopsy. Renal biopsies from patients with episodes of gross haematuria reveal histological changes of grades I to III, whilst patients with the nephrotic syndrome mainly show grade IV and V lesions, similar to those found in patients with renal insufficiency. Hypertension is more frequent in patients with grade IV and V histological lesions. Biopsies performed immediately after the episodes of macrohaematuria reveal foci of active epithelial proliferation that disappear thereafter. Patients

Table 2 Morphological renal lesions in patients with primary IgA nephropathy

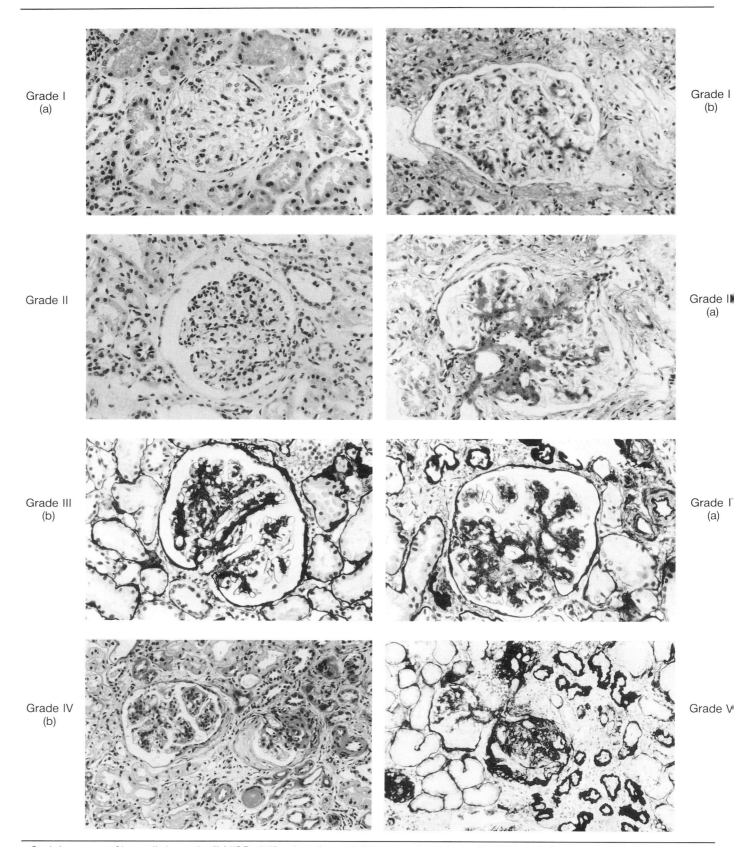

Grade I: presence of 'normal' glomerulus ((a) H&E ×250) and small area of slight mesangial thickening ((b) H&E ×250). Grade II: local area of mesangial hypercellularity in the upper part of the glomerulus (Masson ×250). Grade III: segmental sclerosis with glomerular adhesion ((a) and (b) Masson ×250). Grade IV: marked diffuse mesangial thickening ((a) PASM ×250); lobulated glomerulus with sclerotic mesangium ((b) PASM ×250). Grade V: segmental and global mesangial sclerosis, glomerular hyalinosis and capsular thickening ((a) and (b) PASM ×150).

admitted with rapidly progressive glomerulonephritis have true crescents in many glomeruli and a diffuse sclerosing glomerulonephritis. Despite the presence of occasional small crescents in pIgAN, there are a small number of patients in whom the histological lesions consist mainly of crescents; this form has been called 'crescentic IgA nephropathy' (Abuelo et al. 1984). In these cases the rapid formation of crescents is thought to be due to the passage of plasma and cells into Bowman's space through a ruptured glomerular capillary wall. The presence of fibrin and other biologically active substances stimulates the accumulation of monocytes and triggers the local proliferation of visceral and parietal epithelium.

Tubulointerstitial changes are present in a varying percentage of cases (Nagy et al. 1984a; Hotta et al. 1987). Studies designed to characterize interstitial cell infiltrates have demonstrated the presence of an increased number of monocyte/macrophage and T lymphocytes, mainly T-helper/inducer cells, while the amount of T-suppressor/cytotoxic cells was within the normal range (Alexopoulos et al. 1989).

There are a few reports (De Werra et al. 1973; Droz 1976; Lee et al. 1982; Nicholls et al. 1984; D'Amico et al. 1985) which describe the renal lesions observed at repeat biopsies. The proliferative lesions change little and in some cases mesangial cell proliferation is replaced by an increase in mesangial matrix. Glomerular sclerosis, interstitial fibrosis, and vessel changes increase with tir   although there is a large variation in the rate of increase among patients.

A strong correlation has been found between serum creatinine at the time of biopsy and the degree of global sclerosis, arteriosclerosis, and interstitial fibrosis. The parameters of serum creatinine, global sclerosis, interstitial fibrosis, and arteriosclerosis correlate with age at biopsy and with age at the onset of the disease.

The most constant abnormality in children with IgA nephropathy is a widening of the glomerular mesangium. The mesangial widening is produced by a variable combination of mesangial hypercellularity and increased mesangial matrix. Histological examination of Japanese children with pIgAN (Yoshikawa et al. 1987) revealed that predominant mesangial hypercellularity was almost exclusively seen in the initial biopsy whereas predominant mesangial matrix was usually seen at the follow-up biopsy. Thus mesangial hypercellularity is characteristic of the early lesion of pIgAN in children, but progression of the disease leads to a gradual decrease in mesangial cellularity and an increase in matrix with sclerosis. Focal tubular atrophy, together with interstitial cellular infiltration and fibrosis are frequently observed. A predominance of mesangial hypercellularity is rare in adult patients. This difference between adults and children may be due to the short interval between onset of the disease and the renal biopsy. It is usual to find a very long asymptomatic period preceding the apparent onset of the disease in adults.

Children with pIgAN characterized by a predominance of matrix are resistant to treatment; they have persistent proteinuria at follow up and do not show a decrease in matrix at the second biopsy. This matrix increase is an irreversible glomerular change and produces glomerular sclerosis.

Immunofluorescence

As immunofluorescent deposits are present in all mesangial regions of both normal and affected glomeruli, the descriptive histological term of focal and segmental glomerulonephritis is incorrect, since all glomeruli are affected by the immunological process. In fact, since the 1970s more accurate light and electron microscopy studies have indicated that there is a wide spectrum of glomerular lesions. In the first immunofluorescence studies Berger and Hinglais (1968) reported mesangial IgA and IgG deposits, but it became evident from subsequent reports that in the immunofluorescent mesangial deposits the predominant immunoglobulin is IgA. An early report (André et al. 1980) suggested that IgA2 deposits were predominant but numerous subsequent studies (Conley et al. 1980; Tomino et al. 1981; Murakami et al. 1983) using monoclonal antisera, have shown clearly that the renal deposits are composed almost exclusively of the IgA1 subclass. Furthermore, the composition of IgA consists mainly of IgA1 with lambda light chain (Lai et al. 1988).

By definition, IgA is the sole or dominant immunoglobulin present in all glomeruli, C3 has the same distribution as IgA. IgG is present in 50 to 70 per cent of the renal biopsies (Schena 1989a), but it often has the same intensity of staining as IgA, a feature that explains why this disease was initially called IgA–IgG nephropathy. IgM deposits are also found but less commonly (31–66 per cent). The early complement components, such as C1q and C4, are rarely present, but when they do occur are invariably in association with IgG and/or IgM. Fibrin/fibrinogen deposits have been observed in 30 to 40 per cent of cases; their deposits are more frequent in the crescents.

The presence of renal vascular C3 was reported to be associated with renal hyaline arteriolosclerosis by Clarkson et al. (1977). This finding has been confirmed by others (Mustonen et al. 1985), who noted that the presence of both renal arteriolosclerosis and C3 seemed to precede the appearance of hypertension. Similar findings were reported by Cameron et al. (1982b) in other forms of glomerulonephritis.

IgA deposits may be confined to the mesangium (mesangial type) or be present in the glomerular capillary walls as well as in the mesangium (capillary type) (Table 3). Changes in the pattern of deposition may occur, so that at repeat biopsy a change from the capillary to the mesangial type or vice versa may be observed. In the latter case, renal function often shows a progressive deterioration. Acid-urea treatment of renal tissue sections has been used to demonstrate that the IgA deposits contain the J chain which is a subunit of dimeric and polymeric immunoglobulins (Monteiro et al. 1985) and this has been used to argue that the IgA deposited in the kidney of patients with pIgAN is mainly polymeric IgA. The antigen specificity of IgA in mesangial deposits is unknown. Gregory et al. (1988) examined the possible role of infective agents and reported that the deposits contained antibodies against cytomegolovirus; this finding has not been confirmed (Waldo et al. 1989). Glomerular deposition of food antigens has also been detected by immunofluorescent techniques (Russel et al. 1986; Sato et al. 1988). IgA mesangial deposits recur very frequently in pIgAN patients with a transplanted kidney (Berger et al. 1975; Cameron et al. 1977). In contrast they disappear from the IgA containing renal allografts that have occasionally been transplanted into patients with renal failure from other causes (Sanfilippo et al. 1982; Silva et al. 1982).

The immunopathological pattern of pIgAN in children is characterized by diffuse deposits of IgA in the mesangial areas, often extending into adjacent capillary walls. IgA deposits may be associated with IgG or with IgM and C3 which have a similar distribution.

Table 3 Immunofluorescent deposits in mesangial proliferative glomerulonephritis

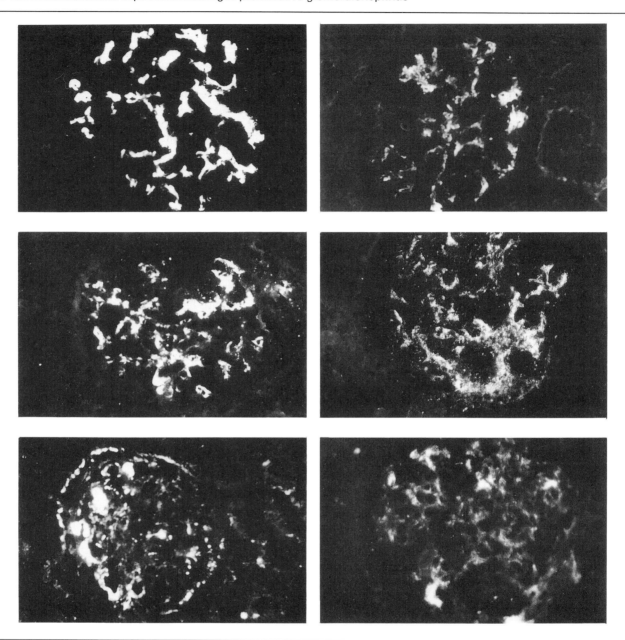

Mesangial deposits of IgA, IgG, IgM, C3, and fibrin. Parietal deposition of IgG and C3. Deposits of C3, along the Bowman's capsule (×250).

Electron microscopy

The most characteristic change on electron microscopy is the presence of dense deposits in mesangial and paramesangial areas. The quantity, size, shape, and density of these deposits vary from glomerulus to glomerulus and from one mesangial zone to another.

In proliferative forms of pIgAN the mesangium contains an increased number of cells which have notched nuclei, abundant cytoplasm, numerous free ribosomes, well developed Golgi apparatus, and smooth-surfaced endoplasmic reticulum. Phagocytic mesangial cells with numerous spinous cytoplasmic processes, lipid inclusions, and numerous dense bodies are commonly present in the sclerosing form of pIgAN. Mesangial cells with irregular nuclei, a small amount of cytoplasm with reduced processes, and poorly developed Golgi apparatus and smooth-surfaced endoplasmic reticulum probably represent resting cells occurring in minimal or focal sclerosing glomerular lesions (Ng 1981).

Another constant finding is the increase in the mesangial matrix of basement membrane-like material. The mesangial matrix is more dense in patients with diffuse mesangial proliferative glomerulonephritis, and minor changes with superimposed focal lesions, than in cases with minimal lesions alone. In biopsy samples taken after the onset of clinical disease, and showing sclerotic lesions on light microscopy, the mesangial matrix is very dense and there is atrophy of the mesangial cells. The widening of mesangial channels with the presence of deposits suggests functional obstruction of the mesangial drainage system by the deposited material.

The mesangial deposits are often hemispherical and, when present in the paramesangial area, are accompanied by narrowing of the capillary lumen. They are usually homogeneous in appearance, mixed with vesicular debris but devoid of any organized structure. The extent and size of the mesangial deposits is variable. The amount of mesangial electron-dense deposits is not related to the severity of the mesangial proliferation or sclerosis.

The glomerular basement membrane shows local abnormalities, such as thinning, splitting, or duplication of the lamina densa in one-third of cases (Shigematsu et al. 1982, 1983). Local splitting of the lamina densa causes swelling of the basement membrane, and the endothelial cytoplasm is irregularly pushed up into the side of the lamina. Tiny intramembranous electron-dense deposits can be seen in a few cases. Abnormal areas of the glomerular basement membrane are covered by flattened epithelial foot processes which show cross-striated fibrils. The cytoplasm of epithelial cell covers the abnormal basement membrane from both sides, indicating an interruption of the membrane, while the cytoplasm of endothelial cell protrudes irregularly into the abnormal basement membrane. These findings suggest that glomerular loop abnormalities are repaired by the epithelial and endothelial cell response. The occurrence of membrane disruption causes the deposition of electron-dense material, accumulation of fibrin, the escape of intralaminal material into Bowman's space with a monocyte reaction around the efflux material with consequent formation of multiple small crescents and adhesions.

Peripheral glomerular capillary wall deposits may also be found in the subendothelial and subepithelial region (Davies et al. 1973; Zimmerman et al. 1975; Clarkson et al. 1977). The subendothelial deposits are usually small and scanty. They occur most frequently in the capillary wall adjacent to the mesangium, and have also been reported in the peripheral part of the loops. They are usually restricted to a small number of capillary loops of the glomeruli (Hara et al. 1980). Subendothelial deposits are similar to mesangial deposits in granularity and in electron density, and have no structure suggestive of fibrin. The basement membrane, where the deposits are located, is either normal or also shows only mild thickening of the lamina rara interna. Patients with subendothelial dense deposits have moderate proteinuria and decreased renal function, and the prognosis differs from that of patients without subendothelial deposits. Minimal subepithelial deposits may also be present and are often surrounded by areas of thinning, lamination, and replication of the lamina densa, and expansion of the lamina rara externa (Yoshikawa et al. 1987).

Electron-dense deposits may occur in extraglomerular vascular structures, such as the intrarenal arterioles (Hulette et al. 1985). The significance of these arteriolar deposits is not clear but it has been postulated that arterioles are areas of relative turbulence which facilitate the deposition of circulating immune complexes. The focal nature of the arteriolar deposits may support this hypothesis.

Dysart et al. (1983) and Laasonen et al. (1984) have demonstrated by immune electron microscopy that deposits of IgA/IgG and C3 may exist in the mesangial, paramesangial, and capillary deposits, and in the mesangial channels. They are usually homogeneous and without any organized structure. However, Mannik et al. (1983) used immunofluorescence to demonstrate glomerular deposits of immunoglobulins, not apparent by electron microscopic examination. Deposits could be detected in the glomeruli by immunofluorescence at least 1 week before they were visible by electron microscopy. Rearrangement of immune complexes in early deposits appears to be necessary before reaching the threshold of electron density resolvable by electron microscopy.

Clinical and histological prognostic markers

PIgAN has a variable course. The overall prognosis of patients however, varies between series (D'Amico 1983). This variation is probably dependent on the criteria applied to biopsy. Initial short-term studies of pIgAN indicated a favourable prognosis, but long follow-up reports have shown progression of renal damage to end-stage kidney disease in 20 to 50 per cent of the patients 20 years after the apparent onset of the disease. Apparent clinical resolution has been reported by some workers (De Werra et al. 1973; Rodicio 1984; Nicholls et al. 1984a; Mustonen et al. 1985) in a few cases, most of which had mild histological lesions (grade I and II). Nevertheless, some of these showed persisting histological lesions and mesangial IgA deposits when submitted to second biopsy. Recent reports have indicated that some clinical and histological findings are of value for predicting the clinical course (Hood et al. 1981; Kobayashi et al. 1983; Bennet and Kincaid-Smith 1983; Croker et al. 1983; Droz et al. 1984; Nicholls et al. 1984a; Abe et al. 1986; D'Amico et al. 1986). These findings are shown in Table 4.

Haematuria

Van der Peet et al. (1977), Burkholder et al. (1978) and Gärtner et al. (1979) noted that the degree of haematuria, whether gross or microscopic had no apparent bearing on the severity or likelihood of progression of IgAN. Others (Sissons et al. 1975; Hood et al. 1981; Kobayashi et al. 1983; Droz et al. 1984; D'Amico et

Table 4 Clinical and histological prognostic markers in patients with primary IgA nephropathy at the time of renal biopsy

Severe microscopic haematuria
Heavy proteinuria
Blood hypertension
Renal insufficiency
Severity of renal lesions (grade IV and V)
IgA deposits in the peripheral capillary walls
Male sex
Older age at the onset of the disease

al. 1986) have reported that isolated or recurrent gross haematuria is more common in patients with a benign course. In contrast, Linné et al. (1982), the Southwest Pediatric Nephrology Study Group (1982), and Bennet and Kincaid-Smith (1983) have reported that gross haematuria is a poor prognostic sign in these cases, since crescents and impaired renal function are often present. Nicholls et al. (1984) have demonstrated a strong correlation between haematuria (RBC>10^5/ml) and the presence of crescents.

Proteinuria

The absence of proteinuria at diagnosis is generally a good prognostic sign, although it does not confer immunity against subsequent progression of the disease.

The presence of moderate or heavy proteinuria at the time of biopsy is considered to be an indication of the eventual development of renal insufficiency (Hood et al. 1981; Kobayashi et al. 1983; Katz et al. 1983; Croker et al. 1983; Chida et al. 1985; D'Amico et al. 1986). In fact, its presence correlates with global sclerosis, interstitial fibrosis, and serum creatinine. Therefore, segmental and global proliferation, glomerular sclerosis, tubulointerstitial damage, and vessel sclerosis increase with increasing levels of proteinuria. Persisting or increasing proteinuria during the course of the disease is associated with an impairment of renal function and sometimes with end-stage renal failure.

The nephrotic syndrome is uncommon, but several groups (Croker et al. 1983; Hattori et al. 1985; Lai et al. 1985) consider that this clinical aspect of the disease carries a poor prognosis. Lai et al. (1986b), in a large number of patients with pIgAN associated with proteinuria, demonstrated that presence of the nephrotic syndrome was associated with a varying prognosis because patients with mild renal lesions responded to steroids, whereas those with advanced changes did not.

Lai et al. (1986a) demonstrated that an increase in the excretion of β_2-microglobulin in the urine was correlated with the degree of tubulointerstitial changes, suggesting that this marker may be a good indicator of the long-term prognosis.

Hypertension

Many studies (Hood et al. 1981; Kobayashi et al. 1983; Nicholls et al. 1984; D'Amico et al. 1986; Lee et al. 1987) have demonstrated a progressive decrease in renal function in patients with hypertension. Persistent hypertension develops most frequently in patients who have had heavy proteinuria for a long period. Uncontrolled, severe hypertension is a possible cause of rapid deterioration to end-stage kidney disease. Hypertension and renal insufficiency at the time of biopsy are unfavourable prognostic signs.

The development of hypertension is more common in patients with vascular sclerosis (Rambausek et al. 1987b), the appearance of which may be due to intrarenal haemodynamics, afferent vasodilatation, and exposure of afferent vessels to higher wall stress.

The presence of renal hyaline arteriosclerosis and renal vascular C3 seems to precede the appearance of hypertension. The propensity to develop hypertension may be related to the high renin levels observed in normotensive IgAN patients. The degree of hypertension increases according to the length of time that symptoms have been apparent, and the age of patients at the time of diagnosis. Patients with macrohaematuria have a lower mean blood pressure than those presenting with proteinuria. Malignant hypertension develops in about 5 per cent of patients.

Kusumoto et al. (1987) have demonstrated that the effect of hypertension in children, who developed signs of the disease at a mean age of 12 years, appeared 21 years or more after the onset of the disease (mean age 30 years). In adults hypertension influenced kidney survival rate during the first 12 years after the onset of hypertension (mean age 40 years). Therefore, hypertension affects renal function at a mean age of 30 to 40 years, but the effect of arteriosclerosis and ageing increases after age 40 years.

Renal insufficiency

End-stage kidney disease occurs in 20 to 50 per cent of patients after 20 years after the apparent onset of the disease, while chronic renal insufficiency (serum creatinine above 1.5 mg/dl) is present in 50 to 70 per cent of patients. In addition, differences in age and in the severity of the clinical findings play an important role.

Hood et al. (1981) and Kobayashi et al. (1983) noted that, as the disease progressed, renal function decreased further in patients with initially impaired renal function (serum creatinine 2–4 mg/dl), whereas it remained unchanged in patients who initially had normal renal function. These findings show clearly that in some patients with pIgAN the renal damage progresses slowly. Van der Peet et al. (1977) and Kobayashi et al. (1983) stated that patients with a decrease in the glomerular filtration rate were, on average, 10 years older than patients with a preserved glomerular filtration rate and that the known duration of the disease in the former was longer. Kobayashi et al. (1983) observed that patients with decreased renal function were more frequently hypertensive.

However, the rate of progression to renal insufficiency is quite different from case to case, since some patients show a rapid decline of renal function over a few years, while in others there is good preservation of the renal function for over 30 years.

Histological lesions

The best predictive index in pIgAN is the histological type of the glomerular lesions present in the renal biopsy samples. Therefore, severe mesangial proliferation, frequent sclerotic glomeruli, crescents, a high proportion of glomerular adhesions, vascular sclerosis, and marked interstitial fibrosis are considered histological markers of a poor prognosis. Clarkson and Woodroffe (1977) and Feiner et al. (1982) suggested that glomerular sclerosis and progressive renal damage are mediated by vascular sclerosis caused by ischaemia. Chida et al. (1985) found a significant decrease in renal survival rate in patients with proteinuria of more than 1 g/day, hypertension, severe diffuse proliferative glomerulonephritis, focal crescents, and glomerular deposition of IgM and/or fibrinogen.

Abe et al. (1986), in a follow-up study, demonstrated that patients with more extracapillary lesions had more severe proteinuria, lower renal function, and higher blood pressure, Moreover, a progressive reduction of renal function occurred in patients with recurrent extracapillary lesions. Such lesions are a very important factor for determining the progression of pIgAN and are an index of poor prognosis.

Serial renal biopsies performed in patients followed for 2 years after the second biopsy showed that changes in mesangial areas, evaluated by quantitative analysis, were a valuable prognostic indicator. Patients with decreased mesangial sclerosis at the second biopsy had stable renal function and a favourable outcome, whereas those in whom mesangial sclerosis increased had a deterioration in renal function during the follow-up period of approximately 5 years. In the latter group, there were glomeruli which showed varying degrees of sclerosis including global sclerosis (Tateno et al. 1987).

There is a good correlation between the activity of the lesions, their severity, and prognosis. Patients with mild lesions have a benign course, whereas patients with more severe lesions of grade IV or V develop end-stage kidney disease.

Sclerosis reflects renal function more closely than proliferative changes, and mesangial sclerosis is the most important histological prognostic factor.

Mesangial matrix increases with the duration of the disease in adult patients (Clarkson et al. 1977; Hara et al. 1980). A predominance of matrix is, therefore, characteristic of the late lesion.

Interstitial sclerosis is highly suggestive of an unfavourable prognosis. Tubulointerstitial infiltrates occur mainly in patients with high serum creatinine levels. The observation of a significant correlation between these tubulointerstitial changes and serum creatinine levels suggests a local cell-mediated immune response, since T-helper cells are mainly present in patients with impaired renal function. Furthermore, the liberation of cytokines by T-helper cells may be responsible for tubular damage. The extent of T-cell infiltration (T-helper and T-suppressor cells) is more marked in patients with impaired renal function at the time of renal biopsy or in those who have a deterioration in renal function during the follow-up period after biopsy (Sabadini et al. 1988; Alexopoulous et al. 1989). A correlation between the severity of interstitial lymphocyte infiltration and the extent of proteinuria has been found. A good correlation between the tubulointerstitial damage and the glomerular filtration rate has been found in all types of glomerulonephritis, except for minimal change disease (Cameron 1987). Recently D'Amico et al. (1986) demonstrated that interstitial sclerosis is a significant prognostic factor, whether or not there is concomitant glomerular sclerosis. This would explain why progression of renal damage differs in patients with the same degree of glomerular involvement but with different degrees of tubulointerstitial inflammatory infiltrate or sclerosis.

Immunofluorescent deposits

IgA, IgG, and C3 tend to be deposited in the capillary walls in patients with crescents or adhesions of the capillary loop to Bowman's capsule. The descriptions of many investigators (Kobayashi et al. 1983; Abuelo et al. 1984; Andreoli et al. 1986; Yoshimura et al. 1987) suggest that capillary IgA deposits, especially IgA–IgG subepithelial deposits, may induce the rupture or increased permeability of the glomerular capillary walls, in association with thinning and splitting of the lamina densa, and cause the formation of crescents of various sizes, resulting in segmental lesions. These lesions contribute to the formation of small crescents and their development plays an important role in the progression of pIgAN. The capillary IgA deposits can be useful as a marker of insidious progression.

Kobayashi et al. (1983) stated that the presence of large amounts of IgA and C3 in association with IgG and fibrinogen in the glomeruli has some bearing on the subsequent progression of the disease.

Deposition of IgG and/or IgM appears to coincide with a more severe form of the disease (McCoy et al. 1974; Sinniah et al. 1976; Clarkson et al. 1977). The spontaneous disappearance of IgA deposits in repeat biopsy samples has been described only in renal transplants.

Serum IgA

Clarkson et al. (1977), Van der Peet et al. (1977), Gärtner et al. (1979), and D'Amico et al. (1986) have shown that high serum levels of IgA had no influence on the prognosis of the disease whereas other reports (Droz 1976; Hood et al. 1981) have emphasized the poor prognosis in patients with high serum IgA values.

Age

The better prognosis in children is attributed to the less frequent occurrence of hypertension and to the lesser degree of glomerular injury found at the initial renal biopsy. Older age at the time of detection of the disease is a marker of poor prognosis. In fact, D'Amico et al. (1986) demonstrated that the risk of developing renal failure is higher for older patients.

In general, renal function decreases according to the age of the patient at presentation. However, creatinine clearance tends to fall during follow up in those patients with proteinuria and/or microhaematuria.

Sex

Men have a worse prognosis than women (Hood et al. 1981; Wyatt et al. 1984b; D'Amico et al. 1985; Lee et al. 1987), since it is more frequently associated with progression of renal damage. No differences between the sexes have been found when haematuria, proteinuria, fibrosis, and arteriosclerosis in the kidney were considered. However, creatinine levels were higher in males.

Epidemiology and immunogenetics

Geographical distribution, race and sex ratio

PIgAN is the most common type of glomerulonephritis worldwide with differences in prevalence in the various geographical areas (Schena 1989a). Immunofluorescent studies of renal biopsy samples have shown that pIgAN is more frequent in the Asian-Pacific area (Asia and Australia) (24.7 per cent) than in Europe (10.5 per cent) and North America (4.8 per cent) (Fig. 8). Furthermore, it is common in patients with primary glomerulonephritis examined by renal biopsy; the prevalence is higher in the Asian-Pacific area (34.1 per cent) than in Europe (19.4 per cent) and it is particularly common in Singapore, Japan, and China, as well as in Australia and in southern Europe (France, Spain and Italy) (D'Amico 1985). Furthermore, a high incidence of the disease (37 per cent) has been observed (Propper et al. 1988) among patients with asymptomatic microhaematuria or recurrent proteinuria in Scotland, apparently in contrast to the incidence elsewhere in the United Kingdom. This may be due to substantial underestimation of the incidence of pIgAN in England and Wales (Julian et al. 1988). Although it is accepted that this

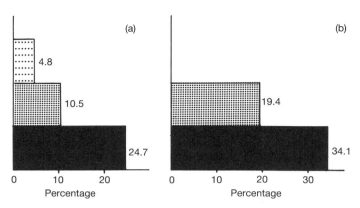

Fig. 8 Distribution of pIgAN in the world. (a) Incidence of pIgAN in all patients who were examined by renal biopsy. (b) Incidence of pIgAN among all patients with primary glomerulonephritis who underwent biopsy. ⬚ North America; ▦ Europe, ■ Asian-Pacific area.

disease affects individual patients randomly, recent papers have reported a regional and familial clustering of pIgAN in Europe and the United States (Egido *et al.* 1985; Julian *et al.* 1985). These reports suggest that genetic factors may influence the distribution of the disease. Nevertheless, the different incidence of pIgAN in the world could have other causes.

1. Ethnic factors and geographical situations.
2. Different social and national attitudes to the management of renal diseases in different countries; for example, the high incidence of pIgAN in Singapore is due to a high selection of patients with asymptomatic microhaematuria or recurrent proteinuria, discovered during the regular medical examination of army recruits. In Japan school children receive a yearly urinalysis, because of the School Health Law of 1973 and the persistence of microhaematuria or macrohaematuria associated with proteinuria leads to a renal biopsy and may be the reason why pIgAN is the most common primary glomerulonephritis in Japanese children attending renal units (Kitagawa 1987).
3. There are also differences in the interest and judgement of clinicians on the diagnostic value of renal biopsy.
4. Differences in the indications for and background to renal biopsy.

The disease occurs in both Caucasians and Orientals, but only rarely in the black population (Julian *et al.* 1983; Jennette *et al.* 1985; Seedat *et al.* 1988). The incidence of pIgAN in the United States is approximately six times greater in the white than in the black population. Some racial groups (the Japanese and some American Indians) appear to have a higher incidence of pIgAN than do the white population. The prevalence of pIgAN among American Indians, such as the Navajos and Zunis in New Mexico, may be the result of a combination of both genetic and environmental factors (Smith *et al.* 1985, 1989; Hughson *et al.* 1989).

Males have a much higher incidence of pIgAN in different geographical areas, except for Hong Kong, and the male-to-female ratio has been reported to be 2:1 to 6:1. However, when the disease occurs in the black population, females are much more often affected. In some Indian tribes of New Mexico, such as the Zunis, pIgAN occurs with the same frequency in the two sexes.

The disease is most common in patients aged between 16 and 40 years, and is less frequent in subjects over 40 years of age.

Genetic studies

The occurrence of the disease in siblings (Sabatier *et al.* 1978; Tolkoff-Rubin *et al.* 1978) and in renal allografts (Berger *et al.* 1975; Cameron *et al.* 1982*a*) of patients originally affected with pIgAN has suggested that genetic factors, determining a host immune response, may be related to the onset or the progression of the disease. Research on healthy relatives of patients with pIgAN (Egido *et al.* 1985) has encouraged this idea, but the invasive nature of renal biopsies has limited genetic studies in family members with haematuria and/or mild proteinuria. Familial pIgAN may, therefore, be inadvertently overlooked.

Genetic studies on chromosome 6

Interest in the role of HLA antigens in pIgAN began in 1976, when MacDonald *et al.* described a strong association between pIgAN and the HLA-Bw35 antigen. Numerous other studies followed from many parts of the world (Table 5) with inconsistent results. The frequency of the HLA antigen association ranged between 6 and 66 per cent and there are no consistently positive associations (Nagy *et al.* 1979; Savi *et al.* 1979; Chan *et al.* 1981; Arnaiz-Villena *et al.* 1981).

Some investigators have described a worse prognosis in a positive subgroup, Bw35, of IgAN patients (Noel *et al.* 1978; Berthoux *et al.* 1979; Bignon *et al.* 1980; Hiki *et al.* 1982). In contrast, Kashiwabara *et al.* (1980) found a high frequency of Bw35 in cases with a favourable course. Wyatt *et al.* (1987*a*) also did not confirm the high frequency of Bw35 in a series from the United States with chronic renal insufficiency. Hiki *et al.* (1982) described a strong association of HLA Bw5 and HLA DR4 with patients with a stable course. In contrast, Kashiwabara *et al.* (1982) stated that the frequency of DR4 increased more prominently as renal function deteriorated. These discrepancies may be due to many factors including geographical situation, biased selection of patients, and heterogeneity of the disease.

A recent study on the restriction fragment length polymorphisms of the HLA-D region genes has demonstrated that disease susceptibility genes are localized within or near the DQ and DP subregions (Moore *et al.* 1989).

The frequencies of the phenotypes of the polymorphic C4 and Bf genes, which are located with the HLA complex, have been studied in patients with pIgAN. An association between pIgAN and a homozygous C4 null phenotype (McLean *et al.* 1984; Wyatt *et al.* 1987*b*), and an excess of the BfFF phenotype (Rambausek *et al.* 1987*a*) have been reported. In addition, a higher risk of an adverse outcome was found in patients with C4A deficiency and in BfFF homozygotes.

Genetic studies on chromosome 19

Gene and phenotype frequencies for C3 have been investigated by Wyatt *et al.* (1984*a*, 1987*b*) in pIgAN patients from Kentucky and Alabama (Table 6). They found a high gene frequency for C3*F and its related phenotype. A recent study in German and Dutch patients demonstrated a significant excess of the homozygous phenotype C3FF; in addition this phenotype was associated with an adverse clinical course of the disease (Rambausek *et al.* 1987*a*).

Genetic studies on chromosome 14

Genetic and amino-acid sequence studies (Gally *et al.* 1973; Williamson 1976) have established that three gene families control

Table 5 Genetic studies on the sixth chromosome in patients with pIgAN

Authors	Year	Country	No. of patients	No. of controls	Antigen association	Incidence in Patients (%)	Controls (%)	p
HLA A, B, C								
MacDonald et al.	1976	Australia	13		Bw35	46		
Noel et al.	1978	France	29		Bw35	48	18	0.005
Berthoux et al.	1978	France	43	105	Bw35	39	13	0.02
Brettle et al.	1978	UK	17	210	Cw1	29.4	4	0.05
Richman et al.	1979	USA	17	100	B12	59	20	0.05
Kashiwabara et al.	1980	Japan	24	139	Bw44	33	9	0.01
HLA D								
Komori et al.	1979	Japan	40	115	DEn	46	18	0.02
Hors et al.	1980	France	37					
Fauchet et al.	1980	France	45	113	DR4	49	19	0.001
Feehally et al.	1984	UK	41		DR1			
Kashiwabara et al.	1980	Japan	24	64	DR4	66	29	0.001
Kashahara et al.	1982	Japan	75		DR4			
Hiki et al.	1982	Japan	80	884	DR4	66	41	0.001
Simon et al.	1983	France	58		DR4	41	19	0.01
HLA complement								
McLean et al.	1984	USA	67	102	C4	18	4	0.003
Wyatt et al.	1987b	USA	141	117	C4A	6	2	
Rambausek et al.	1987a	W. Germany	67	50	BfFF	10	0	0.05

Table 6 Genetic studies on the 19th chromosome in patients with pIgAN

Authors	Year	Country	No. of patients	No. of controls	Gene	Phenotype	Incidence Patients (%)	Controls (%)	p
Wyatt et al.	1987b	USA	160	485	C3*F	C3F	7	2	0.01
Rambausek et al.	1987a	West Germany	67	204		C3FF	10	3	0.05

the organization and expression of the immunoglobulins. There is a gene family for the λ light chains, another for the ϰ light chains and a third for the various types of heavy chains (μ, δ, $\gamma 1$, $\gamma 2b$, $\gamma 2a$, $\gamma 3$, ε and α). The three gene families are localized on chromosomes 2, 14, and 22 in man. In each family the variant gene region is separated by a constant region. The consequences of polymorphism in the Ig heavy chain switch region for Ig function have not been defined in man. Since the polymorphism of the switch region may be associated with the differences in the variable heavy chain region gene, it is possible that differences in the $C\mu$ and $C\alpha$ gene regions might influence the Ig heavy chain switch region. The first report on the Ig ϰ light chain in pIgAN showed a significant increase of Km 1 in this disease (Le Petit et al. 1982). The Km 1 gene frequency modification might indicate that the genetic control of the disease is related to immunoregulation involving idiotype and anti-idiotype interaction. Beukhof et al. (1984) found a specific γ heavy chain allotype associated with macrohaematuria in young patients with normal renal function. A subsequent study (Demaine et al. 1988) on the switch region sequences showed differences for the $S\mu$ and $S\alpha 1$ genotypes, and phenotypes that control the antigen recognition repertoire. A significant increase in the $S\alpha 1$ genotype number was observed in English and German patients with pIgAN. These results suggest that genes within the Ig heavy chain loci may play an important role in the pathogenesis of pIgAN. Nevertheless, a recent study (Hitman et al. 1989) on the restriction fragment length polymorphisms of the immunoglobulin heavy chain region has not confirmed significant differences in the genotypic frequencies of $S\mu$ and $S\alpha 1$.

Family studies

Many investigators have reported the presence of pIgAN in one or more family members of patients with IgAN (Julian et al. 1985a, 1985b; Egido et al. 1987). Some members may suffer from mild urinary abnormalities, such as microhaematuria or proteinuria, whereas others show recurrent haematuria or chronic renal failure. In the latter group it is easy to perform a renal biopsy, while in the former it is not possible for a variety of reasons to confirm the presence of pIgAN.

The more distant family relations of 92 patients with pIgAN were investigated in the United States by Julian et al. (1985b) and Wyatt et al. (1987a). They demonstrated that 60 per cent of patients from Lexington, Kentucky were related to at least one other patient. Epidemiological investigation did not reveal a common environmental factor since occupations, types of residence, and foods were different. However, this study on family relations suggests that a genetic mechanism may be important in the pathogenesis of pIgAN in some patients.

Renal biopsies were not thought to be indicated in family members presenting only minimal urinary findings. Futhermore, the resolution of microhaematuria and/or proteinuria in a certain percentage of patients (15–30 per cent) and the lack of opportunity to perform renal biopsy on patients with past episodes of micro- or macrohaematuria clearly reduce the possibility of carrying out genetic studies on family members. Nevertheless, the occurrence of pIgAN in first-degree relatives has been reported by many investigators (Egido *et al.* 1987). Although these reports seem to indicate an autosomal dominant, single-locus mode of inheritance in these particular families it is possible that separate pathophysiological processes account for the clinical and immunological heterogeneity of pIgAN. This hypothesis is supported by the environmental and clinical heterogeneity which Julian *et al.* (1985*b*) found in affected families in eastern Kentucky. Furthermore, there have been conflicting results in studies on the segregation of HLA haplotypes in familial pIgAN.

Immunological abnormalities have been observed in relatives of patients with pIgAN; these include high serum levels of IgA (Sakai 1984), increased number of IgA-bearing B lymphocytes (Nomoto *et al.* 1979), and the production of large amounts of IgA (mainly polymeric) by peripheral blood mononuclear cells after pokeweed-mitogen stimulation (Egido *et al.* 1983*a*; Waldo *et al.* 1986). Thus, it is probable that at least some determinants of pIgAN must be under genetic control. The fact that pIgAN patients from eastern Kentucky were descended from a limited number of ancestors (the earliest settlers of the Kanawha section of the Allegheny Plateau) suggests the occurrence of a common gene pool that confers a particular susceptibility to the development of the disease. All these subjects carry this trait for susceptibility to pIgAN but the disease appears in only a certain percentage, probably as a result of environmental factors.

Pathogenesis

Since Berger's original report the immunological mechanisms implicated in the pathogenesis of pIgAN have been extensively described (Woodroffe *et al.* 1982; Egido *et al.* 1982*a*; Clarkson *et al.* 1984; Feehally 1988; Emancipator and Lamm 1989; Schena and Emancipator 1989*e*): nevertheless many points of the pathogenesis remain controversial. It is evident that improved comprehension of the implicated immune mechanisms will derive from basic studies. However, a link between the experimental data and the clinical studies can be extrapolated from comparative analysis of the factors which cause the renal damage.

Immunological cellular abnormalities

The total number of T cells is normal in peripheral blood of patients with pIgAN but the incidence of lymphocytes with the Fcα receptor on the surface has been reported to be high (Nomoto *et al.* 1979; Garcia-Hoyo *et al.* 1986). The distribution of T-cell subsets is altered because of the presence of increased numbers of Tα cells which have an IgA specific helper activity on human B cells (Sakai *et al.* 1982). Furthermore, the T cells have a high density of Fcα-receptor in contrast to the depressed synthesis of this receptor found in patients with selective IgA deficiency (Adachi *et al.* 1983).

The recent report of a spontaneous overproduction of interleukin-2 (IL-2) *in vitro* by T lymphocytes from IgAN patients, together with increased expression of IL-2 receptors on their surfaces (Schena *et al.* 1989*b*), suggests the presence of a continuous stimulation of T cells, especially as they also express other activation antigens such as HLA DR. Furthermore, the overactivity of helper T cells could be explained by the low number of true suppressor T cells that have been reported (Adachi *et al.* 1983; Egido *et al.* 1983*b*; Rothschild *et al.* 1984; Antonaci *et al.* 1989). The activity of IgA-specific suppressor T cells has been found to be very low when it was studied *in vitro* by measuring the quantity of immunoglobulins produced by pokeweed mitogen-stimulated B cells in the presence of T-cell supernatant obtained from concanavalin A-stimulated T cells (Sakai *et al.* 1979). These findings indicate that the higher spontaneous IL-2 production in IgAN determines a putative, rapid acquisition of IL-2 receptors by helper T cells, which are responsible for the continued production or increased production of polymeric IgA by B lymphocytes (Lopez-Trascasa *et al.* 1980; Egido *et al.* 1982*b*) (Fig. 9).

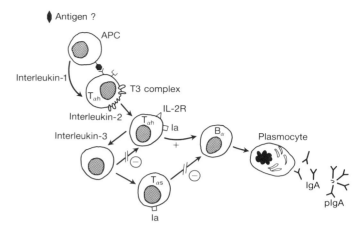

Fig. 9 Possible immune defect in patients with primary IgA nephropathy. The higher spontaneous production of interleukin-2 by Tα-helper cells and the decreased function of Tα-suppressor cells may be responsible for the increased production of monomeric and polymeric IgA.

A local cell-mediated immunological mechanism may be implicated in the renal damage; D'Amico (1988) found a high number of T-helper cells in the tubulointerstitial infiltrates that occur frequently in patients with impairment of renal function during the postbiopsy follow-up period. It has been postulated that T-helper cells may intervene by two mechanisms, either through the cytolytic activity of the lymphocytes, which may release or unmask endogenous antigens capable of eliciting cell-mediated hypersensitivity in the interstitium, or by cytokine liberation, which enhances other immunoinflammatory cells. The renal damage is characterized by interstitial fibrosis, which is responsible for a reduction of glomerular blood flow through a narrowing of intertubular capillaries; the resultant atrophy of the proximal tubules may disturb their function and modify the tubuloglomerular feedback mechanism.

A heterogeneous immunological response has been observed in patients with pIgAN, when spontaneous immunoglobulin synthesis by peripheral mononuclear cells or mitogen-stimulated T and B cells were studied *in vitro*. All reports (Cosio *et al.* 1982; Cagnoli *et al.* 1985; Linné *et al.* 1985; Casanueva *et al.* 1986; Hale *et al.* 1986; Waldo *et al.* 1986; Schena *et al.* 1986; Feehally *et al.* 1986; Williams *et al.* 1987; Lai *et al.* 1987*a*; Wyatt *et al.* 1988) describe a large, though not always statistically significant, production of IgA by peripheral mononuclear cells that occurs both spontaneously and after mitogen stimulation, while an increase in IgG and/or IgM synthesis has only been observed occasionally (Linné and Wasserman 1985; Williams *et al.* 1987). These contrasting findings could be due to variations in techniques or to different phases of the disease. Recently, some investigators (Schena *et al.* 1986) demonstrated that patients with active pIgAN (infection of the upper respiratory tract, fever, and gross haematuria) have increased numbers of cells actively secreting both IgA and IgG in their peripheral blood, whereas, in remission, only IgA secreting cells were increased. Moreover, they also found a significant increase in the serum levels of IgG and IgM and circulating IgG immune complexes. Therefore, the overproduction of IgG and IgG immune complexes may be more important in relapse than persistent overproduction of IgA, and the circulating IgG immune complexes may be responsible for the gross haematuria. Emancipator et al. (1987b, 1987c), demonstrated the presence of haematuria only in mice with IgG and/or IgM and C3 deposited in glomeruli.

Immunological serum abnormalities

Increased serum levels of IgA, mainly in the polymeric form, have been detected in patients with pIgAN. Study of these IgA forms has revealed the occurrence of IgA antibodies against bacteria, viruses and food antigens. Furthermore, a high production of polymeric IgA by stimulated peripheral blood lymphocytes and the inhibition of neutrophil migration by polymeric IgA have been observed in vitro in these patients (Egido et al. 1982c). These findings could explain the persistence of polymeric IgA in patients with pIgAN. A study of the charge on the serum IgA in these patients has demonstrated high levels of IgA binding to cationized bovine serum albumin (BSA) irrespective of whether the IgA was of a mono- or polymeric nature (Monterio et al. 1988). Moreover, the presence of negatively charged IgA is not correlated with the serum levels of IgA, this alteration being, therefore, a qualitative rather than quantitative feature of the serum IgA. Increased levels of negatively charged serum IgA have also been found in transplanted patients with previous pIgAN, often in association with high values of IgA rheumatoid factor. These findings suggest that both negatively charged IgA and IgA-rheumatoid factor may play a role in the recurrence of mesangial IgA deposits.

Various techniques have demonstrated circulating immune complexes in patients with IgAN. The first report (Woodroffe et al. 1980) showed the presence of circulating immune complexes in 43.6 per cent of cases; this proportion increased during the episodes of macrohaematuria; the results of serial determinations in individual patients showed that immune complex levels correlated with clinical evidence of disease activity. In that study solid phase C1q assay was employed. This study suggested the important role of these immune complexes but it should be noted that this technique detects IgG and/or IgM immune complexes preferentially.

In the following years many investigators oriented their studies to the detection of IgA immune complexes by different methods (Stachura et al. 1981; Mustonen et al. 1981; Coppo et al. 1982; Nagy et al. 1982; Lesavre et al. 1982; Valentijn et al. 1983; Hall et al. 1983; Yagame et al. 1987; Imai et al. 1987; Cederholm et al. 1988). Since IgA immune complexes are present in a high percentage of patients, many investigators suggested that they play an important role in the pathogenesis of the disease. In fact, some years ago Egido et al. (1984c) detected the presence of high serum levels of IgA immune complexes by two different techniques, but did not explain the presence of the high serum levels of IgG immune complexes found in 64 per cent of their cases. Recently, Czerkinsky et al. (1986) have reported the presence of increased levels of circulating immune complexes containing IgG and IgA1. More recently, other investigators (Schena et al. 1989c) using a solid-phase jacalin ELISA were able to detect IgA1 immune complexes containing IgG and/or IgM in patients with pIgAN; in addition a significant increase in IgA1–IgG immune complexes was observed in relapses. These findings support the hypothesis that IgG immune complexes may play an important role. On the other hand, 10 years ago Levinsky and Barratt (1979) suggested that only IgG class immune complexes are nephritogenic in children with Henoch-Schönlein purpura. It is possible, therefore, that mesangial IgA deposits derived from polymeric or aggregated IgA and IgA immune complexes or IgA reacting with 'planted antigens' cause small lesions, like that shown by Rifai et al. (1979) experimentally, while IgG or IgM immune complexes due to environmental factors or nuclear antigens (Nomoto et al. 1986) localize in the renal tissue and activate the complement system and are responsible for renal damage. The pronounced loss of C3 receptors in the mesangial area, particularly in those patients who develop severe crescentic and sclerosing pIgAN, implicates complement in the evolution of the disease. Furthermore, the high serum levels of polymeric IgA may alter the clearance of IgG immune complexes which could deposit secondarily in the mesangium (Schena et al. 1988b).

Autoimmune phenomena in which the IgG system is implicated have been mentioned recently by some investigators. Jackson et al. (1987) observed high serum levels of IgG antibody specific for the Fab region of the autologous IgA in IgAN patients. Ballardie et al. (1988) reported the occurrence of IgG autoantibodies with specificity for mesangial antigenic determinants in IgAN patients with more frequent IgG glomerular deposits. Furthermore, the levels of circulating IgG autoantibodies rose during the episodes of macroscopic haematuria and in IgAN patients with high urinary protein excretion. These findings provide evidence for serum Fab α fragments and glomerular antigens which are autoantigenic in pIgAN.

Other immunological abnormalities

The persistence of immune complexes in the peripheral blood may be caused by several factors, including reduced phagocytic activity of the polymorphonuclear leucocytes (Sato et al. 1983; Tomino et al. 1984b) and a depressed activity of Fc and C3b receptor function in the reticuloendothelial system (Lawrence et al. 1983; Nicholls and Kincaid-Smith 1984b; Roccatello et al. 1985). This last finding has been demonstrated by using radiolabelled autologous red blood cells coated with either IgG antibody to the rhesus D antigen or C3b; these results show clearly a preferentially defective clearance of IgG immune complexes. In addition, recent studies have demonstrated that the clearance kinetics and the fate of polymeric IgA and IgA oligomers, which resemble IgA immune complexes, occur regularly in patients with pIgAN (Rifai et al. 1989).

The persistence of immune complexes in the blood and renal tissue (Tomino et al. 1983b) may be due to the low capacity of the serum to solubilize immune complexes. Recently Schena et al. (1987) demonstrated this reduced property of serum using BSA-antiBSA (IgG) immune complexes preformed at equivalence in vitro. Later they found that the impaired capacity was due to the presence of macromolecules in the blood (Schena et al. 1988a, 1988b), since high levels of polymeric IgA and/or IgA rheumatoid factor are present in these patients (Valentijn et al. 1983; Endoh et al. 1985; Czerkinsky et al. 1986; Sinico et al. 1986).

Furthermore, the slower clearance of IgA immune complexes in relation to those of IgG class (Egido et al. 1982a) might favour the persistence of IgA immune complexes in both the remission and relapse stages of the disease, thus providing ideal conditions for producing anti-idiotypic antibodies.

The role of the mucosa and the bone marrow

The tonsils are important organs of contact with bacteria and viruses and, hence, are sites of clonal expansion and cellular differentiation. An immunoregulatory dysfunction of the secretory immune system has been demonstrated at this level in patients with pIgAN. It consists of an increased CD4/CD8 cell ratio, which is responsible for a significant increase in IgA-bearing lymphocytes and cells producing polymeric IgA (Egido et al. 1984b); in fact, a high production of IgA by B cells from these patients is found when they are cocultured with allogeneic T cells from controls (Lozano et al. 1984). The immunohistomorphometrical analysis of tonsillar plasma cells has confirmed the increase in the number of dimeric IgA-secreting cells associated with decreased production of IgG (Bené et al. 1983). An inversion of the IgG/IgA-secreting plasma cell ratio is present only in patients with pIgAN. Nevertheless, it is important to note that an increased number of IgG-secreting plasma cells has also been found in patients with recurrent episodes of gross haematuria without renal deposits of IgA (Bené et al. 1986).

The altered distribution of IgG and IgA-secreting cells in patients with pIgAN is responsible for a high concentration of proteins and IgA in saliva and pharyngeal and nasal secretions (Finlayson et al. 1975; Whitworth et al. 1976; Tomino et al. 1986a). Recent studies using extracts obtained by freezing and thawing pharyngeal cells from these patients have shown a potent cytopathic effect on fibroblasts. The effect is also present in the supernatant from fibroblasts previously cocultured with these extracts (Tomino et al. 1986b). These findings suggest that an antigen is located in the infected cells of the upper respiratory tract of patients with pIgAN which may originate from viral or bacterial infections or may be the expression of the presence of structurally altered proteins. IgA antibodies eluted from the renal tissue of these patients bind specifically to the nuclear region of tonsillar cells and fibroblasts previously cocultured with extracts obtained by freezing and thawing pharyngeal cells (Tomino et al. 1981b, 1983a). It seems possible, therefore, that a common antigenic site in both the mesangial area of the glomeruli and the nuclear region of tonsillar and pharyngeal cells may be present in these patients. Serum IgA antibodies circulating in the blood of these patients may also bind to the nuclear region of cells from the mucosa of the upper respiratory tract and to cocultured fibroblasts; some antigenic substances may thus be transferred from the pharyngeal cells to fibroblasts *in vitro* (Tomino et al. 1985).

The participation of the mucosa is confirmed by the demonstration of high values of IgA in saliva (Whitworth et al. 1976) and nasal secretions (Finlayson et al. 1975) from patients with pIgAN. Furthermore, tonsillectomy in patients with recurrent episodes of macroscopic haematuria following upper respiratory tract infections has resulted in a decrease in the synthesis of polymeric IgA by peripheral blood mononuclear cells (Lozano et al. 1985); unfortunately this was not associated with an improvement in the renal lesions.

Controversial findings concerning the role of the intestinal mucosa have been obtained in man. Studies performed on the mucosa of the jejunum and the small intestine of patients with pIgAN have shown a normal number of plasma cells producing IgA and its subclass A1 (Westberg et al. 1983; Hené et al. 1988). In contrast, high values of IgA antibodies against gliadin (Coppo et al. 1986; Rostoker et al. 1988; Rodriguez-Soriano et al. 1988) and other food proteins (Woodroffe et al. 1977; Russel et al. 1986) have been observed in the blood of IgAN patients, findings confirmed by others in a smaller number of patients (Fornasieri et al. 1987; Yap et al. 1987). The participation of food antigens could be possible in a small number of patients who should be perhaps considered to be a subgroup of pIgAN.

From these observations, it can be argued that polymeric IgA molecules are produced essentially by mucosal plasma cells; it is thus probable that in some subjects, bearing particular genetic characteristics, viral or bacterial stimuli could enhance the production of dimeric and polymeric IgA by the lymphoid cells present in the mucosa. Normally the secretory component binds to IgA as it moves through the epithelium to reach the lumen. An excessive production of IgA or impaired secretory component function in these subjects could result in the rapid and abnormal passage of dimeric and/or polymeric IgA into the blood and impaired clearance contribute to their deposition in the kidney (Lamm and Emancipator 1987). Therefore, in patients with IgAN it is more likely that mucosal IgA passes directly into the circulation, without entering the epithelial cells, and consequently without binding the secretory component. In fact, the absence of the secretory component in the kidney indicates that the deposited IgA is not of the secretory type, and the IgA is not bound to the mesangium via the secretory component-mediated mechanism. Although IgA antibodies eluted by the glomeruli are specific to mesangial areas, they show some heterogeneity among patients with this disease (Tomino et al. 1981b).

Nevertheless, the occurrence of IgA1 deposited in the mesangium of glomeruli, the possible binding of a part of the deposited IgA to the secretory component, the recurrence of the disease in renal transplants, and the presence of increased serum levels of polymeric IgA1 suggest that the glomerular IgA may be derived from the circulation. Recent data obtained from Van Es and coworkers (van der Wall Bake et al. 1988) have demonstrated an increased percentage of plasma cells containing IgA1 in the bone marrow of patients with pIgAN. Furthermore, increased synthesis of IgA, preferentially IgA1, from bone marrow cells of these patients has been observed. These findings suggest the participation of the bone marrow in the pathogenesis of the disease. Further studies are needed to elucidate the relation between the bone marrow and mucosa in pIgAN. The immunological reactions to oral polio vaccine (Leinikki et al. 1987), food challenge (Sancho et al. 1983; Sato et al. 1987), and the subcutaneous infection of inactivated mumps virus vaccine (Pasternack et al. 1986) have always shown an abnormal response of IgA.

In conclusion, the widely divergent clinical and pathological manifestations of pIgAN may be due to differing contributions by several possible pathogenic components (Emancipator et al. 1985). Therefore a classification by the predominant pathogenic mechanism may result in the recognition of several syndromes which may have a wider spectrum of manifestations and natural history.

Experimental IgA nephropathy models

About 10 years after Berger's publication, Rifai et al. (1979) developed the first animal model of IgAN using IgA immune complexes obtained with an IgA myeloma protein (MOPC 315) with specific reactivity for the hapten dinitrophenol (DNP) enabling bovine serum albumin (BSA) conjugated with DNP (DNP-BSA) as the antigen. The clinical and pathological findings showed haematuria associated with glomerular IgA and C3 deposition, and increased mesangial matrix. The striking feature of this study was the nephritogenic property showed by the polymeric form of IgA immune complexes. In fact, monomeric IgA immune complexes, prepared *in vitro* or formed *in vivo*, did not localize in the kidney; in contrast polymeric IgA immune complexes were present. In addition, Rifai *et al.* (1979) showed that the antigen–antibody ratio did not play a critical role but that the concentration of the immune reactants was responsible for inducing the renal lesions. The same group confirmed the nephritogenic role of polymeric IgA by another *in-vivo* study using the same monomeric or polymeric IgA (MOPC 315) and DNP-Ficoll. They underlined the renal deposition of preformed circulating polymeric IgA immune complexes and speculated about the possibility of *in-situ* monomeric IgA mediated complex formation analysis of kinetic studies (Chen *et al.* 1988). These studies demonstrated that (a) IgAN is an immune complex glomerulonephritis; (b) antibody predominance is the IgA class; and (c) renal deposition of circulating polymeric IgA immune complexes and monomeric IgA immune complexes formed *in situ* are possible immunopathogenetic mechanisms.

Comparison with clinical findings suggests the possibility of nephritogenic large-size immune complex formation during a respiratory or gastrointestinal infection, frequently present in the history of patients with pIgAN, their mesangial deposition and the consequent haematuria.

Some years after Rifai's original report, Isaacs and Miller (1981, 1982, 1983) described passive and active models of IgAN using different forms of dextran. These workers emphasized the importance of the size and charge of the antigen. On the other hand, while Rifai's group had developed a passive model of IgAN using immune complexes obtained in antibody excess, Isaacs and Miller used IgA immune complexes formed in 40-fold antigen excess.

In the active model, IgA nephropathy was induced by an intraperitoneal injection of dextran followed by an intravenous injection of different types of dextran, varying in size and charge. The workers showed that the different forms of dextran can induce an IgA and IgM response and outlined the importance of the role of the antigen. Bright, granular, and diffuse IgA deposition occurred in 100 per cent of the glomeruli of mice treated with the cationic and anionic forms of antigen, but there was no relationship to the molecular size. All different sizes of neutral dextran elicited a more focal segmental IgA pattern of deposition. About 80 per cent of the immunized animals showed C3 deposits in the same distribution of IgA. A similar distribution pattern was observed for IgM, but in smaller amounts. The light microscopic findings emphasized the relationship between the antigen characteristics and the renal lesions: mild proliferative lesions were observed in mice treated with neutral dextran while a striking substantial increase in mesangial matrix and considerable mesangial proliferation were present in mice treated with diethylaminoethyldextran and in those who received dextran sulphate. In conclusion, high molecular weight dextran had a more aggressive activity. Finally, the model used in these studies may have very practical aspects since carbohydrate antigens, originating from bacterial cell walls may play an important role in the pathogenesis of IgA nephropathy.

Recently, using the same immunization protocol, Gesualdo *et al.* (1990) showed a significant increase in renal expression of platelet-derived growth factor mRNA. Since this factor promotes mesangial cell proliferation and matrix synthesis (Ross *et al.* 1986) it has the potential role to mediate glomerulonephritis.

In a recent study, Rifai *et al.* (1987a) demonstrated the important role played by the antigen in complement activation. They used two different types of antigens, DNP-Ficoll and DNP-phosphorylcholine (PC). Administration of cross-linked IgA anti-DNP alone showed mesangial IgA without C3 staining by immunofluorescence. Mice that received immune complexes prepared with IgA anti-PC and PC/DNP-BSA or with IgA-anti DNP and DNP-Ficoll showed IgA and C3 deposits in the glomeruli, but mice treated with immune complexes formed with IgA anti-PC and PC-BSA failed to show glomerular C3 deposits. This study underlined the fact that the composition of the antigen, in this case DNP-conjugated, could have an important and central role in the activation and co-deposition of C3 by IgA immune deposits.

Many investigators, although not all, believe IgA nephropathy originates from a mucosal immune response possibly provided by an infectious agent or dietary antigen. Since IgA is derived predominantly from the mucosal associated lymphoid tissue, Emancipator *et al.* (1983a) tested this hypothesis experimentally on mice by oral immunization with different proteins. This novel experimental model showed that mice continuously immunized orally with bovine gammaglobulins for 14 weeks had mesangial deposits of IgA. No haematuria or histological lesions were found. Later, the same group (Emancipator *et al.* 1987c) demonstrated that a protracted oral immunization not only promoted IgA antibody synthesis but also activated a mechanism called 'oral tolerance' which induced active suppression of IgG and IgM antibody response. These findings underlined the potential importance of immunoregulation in the expression of the disease (Emancipator 1988b). In fact, these workers observed microhaematuria associated with IgA, IgG, IgM, and C3 deposition in mice continuously immunized orally for 6 weeks and challenged parenterally with the same antigen. The same mice showed glomerular C3 deposits and high serum titres of IgA, IgG, and IgM anti-immunogen. Between 6 and 14 weeks of continuous oral immunization, they observed suppression of serum IgG and IgM antibody against the oral immunogen and confirmed that a pure IgA antibody response does not efficiently activate complement (Emancipator 1988a). Recently Gesualdo *et al.* (1989) demonstrated that when mucosal tolerance was abrogated by cyclophosphamide and oestradiol, mice had persistently high serum levels of IgG and perhaps IgM antibody against the oral immunogen, mesangial deposition of IgG/IgM and C3, and microhaematuria after 14 weeks of oral immunization. These data suggest that the mucosal immune response can markedly influence nephritogenesis by altering oral tolerance.

The role of the mucosal immune system is suggested by other experimental models of IgAN. Genin *et al.* (1986) demonstrated that selected C3H/HeJ mice develop a high non-specific serum IgA response after oral immunization with ferritin for 30 days, which is associated with mesangial deposits of IgA. However, the occurrence of low levels of IgA antibody against ferritin and the absence of ferritin and its antibody in the glomeruli suggest that a strong oral immunization with an antigen leads to high serum levels of non-specific IgA and their deposition in the mesangium. These data were not confirmed by Biewenga *et al.* (1987) who demonstrated that long-term oral administration of TNP ovalbumin was not sufficient to induce an IgA nephropathy in mice. In contrast, intraperitoneal injection of anti-TNP producing tumour cells and followed by intravenous administration of antigen resulted in IgA immune complexes in the kidney.

Table 7 Therapeutic approaches in primary IgA nephropathy

Drugs	References
Corticosteroids	Lai *et al.* 1985, 1986*b*; Kobayashi *et al.* 1986, 1987, 1988; Mustonen *et al.* 1983, 1987
Immunosuppressives (cyclophosphamide, azathioprine, chlorambucil)	Lagrue *et al.* 1975; Woo *et al.* 1987
Cyclosporin A	Lai *et al.* 1987*c*
Antiplatelet drugs	Woo *et al.* 1987; Chan *et al.* 1987
NSAIDs	Kobayashi *et al.* 1986, 1987, 1988
Phenytoin	Clarkson *et al.* 1980; Houssin *et al.* 1983; Egido *et al.* 1984*a*; Coppo *et al.* 1984
Danazol	Tominao *et al.* 1984*a*
Eicosapentanoic acid	Hamazaki *et al.* 1984; Bennett *et al.* 1989
Diaminodiphenylsulphone-dapsone	Deteix *et al.* 1984
Urokinase	Sakai *et al.* 1984
Mediterranean diet	Coppo *et al.* 1986
Plasma exchange	Genin *et al.* 1984; Coppo *et al.* 1985*b*
Tonsillectomy	Lagrue *et al.* 1980

Recent experimental data from Sato *et al.* (1986) have demonstrated an important role for the reticuloendothelial system; IgA nephropathy was induced after administration of lactalbumin for a long period only in mice whose reticuloendothelial system had been blocked.

During the last 10 years animal models of IgA nephritis caused by viral infections have also been described. Portis and Coe (1979) reported the appearance of IgA deposits in the kidney of mink affected with the Aleutian disease; Imai *et al.* (1985) observed spontaneous and prominent accumulation of IgA in the glomeruli of ddY mice at some points in their lives. The natural history of the renal disease in these mice revealed mesangial cell proliferation associated with IgA, IgG, IgM, and C3 deposits after 16 weeks and progressive proteinuria after 28 weeks. Chino *et al.* (1981) had already reported that retroviral infections may be related to the high incidence of malignant tumours in these mice. Recently Takeuchi *et al.* (1989) demonstrated that circulating gp70m, a retroviral glycoprotein, and circulating gp70-anti-gp70 immune complexes deposit in the mesangium of glomeruli of ddY mice over the age of 24 weeks. Jessen *et al.* (1986) have developed an animal model of IgA nephropathy in which an inactivated infectious pathogen, Sendai virus, was introduced intranasally under light ether anaesthesia followed by intravenous injection of virus.

In conclusion, the various experimental models of IgA nephropathy, developed in the last 10 years have been extensively reviewed (Emancipator *et al.* 1987; Rifai 1987), but there are differences between the models and between the models and the human disease. Therefore, extrapolation of animal models to disease in humans must always be accompanied by caution.

Therapy

PIgAN is classed as a slowly progressive renal disease since 20 to 50 per cent of patients in long-term follow-up studies develop end-stage kidney failure 20 years or more after the apparent onset of the disease. The first reaction of many nephrologists, after the initial histological assessment, is to monitor the clinical signs, such as the frequency of macrohaematuria episodes, the degree of proteinuria, hypertension, and renal function, without starting drug treatment (Clarkson and Woodroffe 1987). Tonsillectomy and antibiotics have been reported to reduce the number of episodes of macrohaematuria, but no renal histological data were given and very few patients were studied (Lagrue *et al.* 1980). The occurrence of high values of dietary antibodies, their participation in circulating IgA immune complexes, and their return to the normal range after placing patients on a gluten-free diet (Coppo *et al.* 1986) suggest a possible therapy for pIgAN. Nevertheless, dietary therapy would only be helpful in selected cases, because antibodies to food antigens and IgA immune complexes have been detected in only a small proportion of patients (Fornasieri *et al.* 1988).

Even though the disease is progressive in a minority of patients various investigators have attempted treatment with various drugs (Table 7). Long-term treatment with phenytoin, antiplatelet drugs, or eicosapentenoic acid has no effect on the course of the disease and does not arrest the evolution of the renal lesions. Short-term therapy with cyclosporin A has had a transient effect on proteinuria, but it increases again soon after the drug is stopped (Lai *et al.* 1987); there was transient impairment of renal function in this trial during administration of the drug. Retrospective studies (Lai *et al.* 1985; Mustonen *et al.* 1987; Kobayashi *et al.* 1988) have shown a beneficial effect of corticosteroids in patients with proteinuria of 1 to 2 g/day; when the same patients were evaluated after 10 or more years, renal function was preserved in subjects being treated with corticosteroids at the time of renal biopsy. In contrast, patients with a previous deterioration in renal function had a poor clinical course, whether or not steroids had been given. Thus, corticosteroids seem to have no effect on the course of pIgAN with the possible exception of patients with heavy proteinuria or the nephrotic syndrome. Similarly, review of randomized controlled trials (Schena *et al.* 1989*d*) has shown that corticosteroids and cytotoxic drugs have beneficial effects in the group of patients with heavy proteinuria or the nephrotic syndrome but they do not influence the course of the disease in patients with moderate proteinuria.

The presence of mild glomerular lesions and normal renal function in many cases with heavy proteinuria indicate that there will probably be a good response to drugs. However, the immediate response to corticosteroids and the subsequent tendency to relapse in some cases may suggest that pIgAN is superimposed on existing asymptomatic minimal change disease in some patients. This is supported by the fact that in nephrotic children, initial renal biopsies were negative for IgA deposits, but repeat biopsies were positive. In addition, some patients may have a spontaneous remission of their nephrotic syndrome.

In conclusion, steroid therapy can be beneficial in patients with heavy proteinuria, a creatinine clearance greater than 70 ml/min, and mild histological lesions. Tubular atrophy, interstitial fibrosis, and vascular changes are poor prognostic signs which contraindicate the administration of corticosteroids.

Liver glomerulonephritis

This condition is usually a clinically silent disease characterized by mild microhaematuria, proteinuria, and glomerular lesions associated with mesangial IgA deposits and smaller amounts of other immunoglobulins and complement components. It is found most commonly in patients with alcoholic cirrhosis, but sometimes occurs in other liver diseases not related to alcohol, such as viral and cryptogenic chronic hepatitis, and chronic hepatosplenic schistosomiasis (Nakamoto et al. 1981). Postmortem studies (Bloodworth and Somers 1959; Fisher and Hellstrom 1959; Jones et al. 1961) and renal biopsies (Newell 1987) have demonstrated that glomerular abnormalities in patients with liver cirrhosis occur in 50 to 100 per cent of cases. Many studies (Masugi et al. 1979; Nakamoto et al. 1981; Fukuda 1982) have confirmed that liver glomerulonephritis is a distinct entity but the mesangial deposits of IgA observed in hepatic glomerulonephritis are similar to those of Berger's disease and Henoch-Schönlein purpura.

Clinical features

Liver glomerulonephritis is usually a clinically silent disease discovered unexpectedly by urinalysis. Macrohaematuria occurs but with much less frequency than in patients with Berger's disease and renal insufficiency is rare and does not correlate with the glomerular lesions. Few cases have heavy proteinuria which may or may not be associated with the nephrotic syndrome. Arterial hypertension, which is present in some patients, may indicate a liver glomerulonephritis since increases of systemic blood pressure are rare in cirrhotics with portal hypertension.

Laboratory findings

Urinalysis reveals microhaematuria and proteinuria; this is in the nephrotic range in a small percentage of cases, particularly in those with membranoproliferative glomerulonephritis. In general there are mild urine abnormalities in patients who have minimal glomerular lesions without proliferation and heavy proteinuria, cylindruria, and microhaematuria with increasing severity and proliferation.

Increased serum levels of IgA, mainly in the polymeric form, are present in 77 to 91 per cent of the patients, although high values are also found in 54 per cent of subjects with liver cirrhosis without nephritis. The high serum IgA level is due to increased synthesis of antibodies to bacterial and dietary antigens and the highest values of IgA are found in cirrhotic patients with portacaval shunts. In addition, reduced catabolism of polymeric IgA, unrelated to hepatobiliary IgA transport, has been demonstrated in patients with alcoholic cirrhosis. These mechanisms are partly responsible for the hypergammaglobulinaemia that occurs in these patients.

Antibodies against dietary and bacterial antigens have often been found (Triger et al. 1972; André et al. 1978; Woodroffe et al. 1980). They are always of the IgA class, but are sometimes associated with those of the IgG and IgM classes. Normally, dietary and bacterial antigens reach the liver, where they are cleared by the reticuloendothelial system. Consequently general immunization (IgG and IgM responses) is rare in subjects with normal gut liver axis; furthermore the intestinal IgA response is blocked by the liver through the hepatocytes and the bile. Hepatic clearance of intestinal antigens is less effective in patients with liver cirrhosis which leads to immunization and consequently to increased concentrations of IgG and IgM antibodies. The IgA antibody response must be high enough to overcome liver trapping of intestinal IgA and the overspill of IgA antibodies will be enhanced both by liver bypass through the portosystemic collaterals and by insufficiency of the reticuloendothelial system. As a consequence of those abnormalities, high titres of IgA antibodies against dietary antigens are frequent in cirrhotic patients.

Low levels of C3 (and rarely of C4) may be observed in these patients. It is not known whether hypocomplementaemia is due to reduced hepatic synthesis or to increased catabolism from the activation of the complement system but high levels of C3d are present in the blood. However, no correlation between hypocomplementaemia and the degree of liver abnormalities has been observed and low C3 levels occur in both mesangial and proliferative glomerular lesions.

Circulating immune complexes have been detected using various techniques in the blood of cirrhotic patients (Woodroffe et al. 1980; Sancho et al. 1981; Coppo et al. 1985d). They are mainly IgA but of both subclasses and do not correlate with renal abnormalities; this does not support the suggestion that they have an important pathogenic role in the renal disease.

Mixed cryoglobulins containing IgA have also been found in such patients (Nochy et al. 1976).

Renal biopsy

The first description of glomerular lesions in patients with liver cirrhosis appeared between 1946 and 1947. Patek et al. (1947) described the coexistence of cirrhosis and glomerulonephritis in 14 cases. Since then mild proliferative glomerular lesions have been reported in 25 per cent of unselected patients with liver cirrhosis; similarly moderate or severe glomerulosclerotic lesions have been reported in 53 per cent of cirrhotic patients at autopsy (Sakaguchi 1968; Berger et al. 1977; Nakamoto et al. 1981; Fukuda 1982; Sinniah 1984; Nochy et al. 1984). The lesions were thought to be specific by some investigators (Patek et al. 1947; Sakaguchi 1968) but non-specific by others (Fisher and Perez-Stable 1968). After the introduction of immunofluorescence techniques, immunoglobulin deposits were described in the mesangium of cirrhotic patients with nephritis (Sakaguchi et al. 1965; Salomon et al. 1965) and later Manigand et al. (1970) reported that IgA and C3 deposits were present in 75 per cent of biopsy specimens from patients with liver cirrhosis.

The characteristic glomerular lesion in the kidneys of patients with liver disease is glomerulosclerosis with widening of the mesangial matrix, thickening of the capillary wall, a small increase in mesangial cellularity, and electron-dense deposits in the mesangium. Thickening of the glomerular basement membrane with areas of rarefaction is occasionally present. Proliferative lesions such as mesangiocapillary and rapidly progressive glomerulonephritis in some patients indicate that there are two forms of the disease. However, it is not clear whether

Table 8 Factors involved in the pathogenesis of liver glomerulonephritis

Factor	Authors
Arteriolar sclerosis	Baxter et al. 1946
Choline deficiency	Allich et al. 1949
Alcohol	Chaikoff et al. 1948
Hyperoestrogenaemia	Trevan et al. 1956
Hyperaldosteronism	Bloodworth and Somers 1959
Toxic substances	Jones et al. 1961
Lecithin cholesterol Acyltransferase deficiency	Hovig et al. 1978
Circulating immune complexes	Woodroffe et al. 1980; Sancho et al. 1981; Coppo et al. 1985

non-proliferative lesions evolve into proliferative disease or if the two forms represent different pathological responses to the same insult.

Immunofluorescence reveals IgA deposits associated with IgG and/or IgM and/or C3. They are present both in proliferative lesions and in non-proliferative glomerulonephritis. IgA is the main immunoglobulin present. It contains the J chain and is able to bind the secretory component (Sancho et al. 1981; Donini et al. 1982). IgA2 is the major component of the mesangial deposits (André et al. 1980). This finding is in agreement with the fact that IgA2 is the main immunoglobulin formed by the intestinal mucosa. However, the classical immunohistological picture of liver glomerulonephritis is characterized by the presence of mesangial IgA deposits, occasionally with increased cellularity and mesangial interposition.

Pathogenesis

Various mechanisms, listed in Table 8, have been suggested for the pathogenesis of liver glomerulonephritis, but there is no evidence for any of them. The interaction between IgA produced by intestinal plasma cells and food antigens which cross the intestinal mucosa damaged by alcohol causes the formation of immune complexes. The impaired liver catabolism of the polymeric IgA and immune complexes could be responsible for the persistence of immune complexes in the blood of patients with liver cirrhosis (Woodroffe et al. 1980; Sancho et al. 1981; Coppo et al. 1985a). Nevertheless, abnormal levels of serum IgA immune complexes occur only after alcohol-induced hepatic damage, since chronic alcoholics without biochemical evidence of liver damage have normal levels of IgA immune complexes. High values of monomeric and polymeric IgA are present in patients with alcoholic cirrhosis (Sancho et al. 1981; Kutteh et al. 1982). In normal subjects the selective transfer of polymeric IgA from serum to bile has been reported. It has been suggested that interruption of this pathway might explain the elevated levels of IgA observed in patients with liver disease. Nevertheless, the increase of monomeric IgA cannot explain the selective loss of clearance of polymeric IgA. The presence of spontaneous, enhanced synthesis of IgA by peripheral lymphocytes (Kalsi et al. 1983), rather than, or in addition to, decreased catabolism, may explain the high levels of IgA found in alcoholic cirrhosis.

Other investigators (André and André) 1976; Triger et al. 1973) have suggested that the high value of circulating immune complexes may be secondary to portacaval shunting of bacterial or food antigens which bypass the hepatic Kupffer cells. This hypothesis is supported by the presence of high serum levels of antibodies against bacteria and food and by an abnormal antibody production by peripheral blood lymphocytes after stimulation with these antigens.

Glomerular lesions associated with liver cirrhosis have been reproduced in Lewis rats rendered cirrhotic by the administration of carbon tetrachloride (Gormly et al. 1981). Immunoglobulin deposits, especially IgA, and complement were present in the mesangium of the glomeruli. The rats also had high serum levels of circulating immune complexes and IgA and it was suggested that defective hepatic clearance of dietary antigens in the cirrhotic rats allowed them to pass into the systemic circulation and cause the serological abnormalities and mesangial deposits. As most of the deposited immunoglobulin is IgA2 this may support this mechanism in patients with cirrhotic glomerulonephritis. Other experimental situations in which IgA is deposited in the mesangium include experimental hepatosplenic schistosomiasis (Van Mark et al. 1977) and after bile duct ligation in rats (Melvin et al. 1983) and mice (Emancipator et al. 1983b).

IgA nephropathy and associated diseases

Since first described by Berger (1969), pIgAN has been found typically in association with liver diseases and Henoch-Schönlein purpura, but in the last few years other associations have been reported, mainly with systemic disorders, such as lupus erythematosus, cryoglobulinaemia, rheumatic diseases, and other illnesses involving the upper respiratory tract, gastrointestinal tract, skin, and other organs. Table 9 lists the disease associated with IgA nephropathy. Patients may develop these diseases before or during the clinical course of the nephritis. In reviewing papers describing these cases it is possible to draw some general conclusions; these are discussed below. The incidence is very high since more than half of the patients with IgAN have another pathological condition (Mustonen 1984; Makdassy et al. 1984). Some associations have been reported in isolated cases and may be purely coincidental whilst others appear to be the result of concomitant mucosal infection, as in bronchial and pulmonary disease, disseminated tuberculosis, *Mycoplasma pneumoniae* pneumonia, sarcoidosis, silicosis, enteric fever, or coeliac disease. Finally, the appearance of IgA nephropathy after the placement of nickel-alloy base dental crowns suggests that the prosthesis implantation might stimulate the oral mucosa to produce IgA with resultant immune complex formation and renal deposition. The description of remission after the removal of the prosthesis supports this hypothesis.

Proliferative and/or sclerosing glomerulonephritis is considered very unusual in patients with neoplastic diseases. IgA nephropathy has been reported in patients with neoplasia and the cases are listed in Table 8. It occurs in older age groups, often as neoplasms in the mucosae, and the presence of high serum levels of IgA suggest that an abnormal IgA response to chronic mucosal irritation or to specific tumour antigen might be responsible for the onset of IgA nephropathy.

In many patients the IgA system is involved, since the presence of high serum levels of IgA in the polymeric form, or of circulating IgA immune complexes, is characterized by the secondary deposition in the kidney. In these cases the IgA nephropathy may be considered a complication rather than a disease, as extrarenal vessels are also involved, as shown by the deposits of IgA in the wall of these vessels. These conditions occur mainly in immunological and systemic disorders, such as ankylosing spondylitis, psoriatic arthritis, juvenile rheumatoid arthritis, and arthritis after Yersinia infection. Furthermore, the association of various rheumatic disorders with scleritis may explain the appearance of IgA nephropathy in patients with scleritis. An immunofluorescence study (Bené et al. 1984) demonstrated numerous infiltrating plasma cells secreting dimeric IgA. Raised serum levels of IgA may occur in the presence of other immunological and proliferative disorders, such as mixed cryoglobulinaemia, the Sjögren syndrome, and immunothrombocytopenia.

Concluding remarks

IgA nephropathies are characterized by the presence of mesangial deposits of IgA in the glomeruli. IgA is detected in selected pathological entities such as Berger's disease, Henoch-Schönlein purpura, and glomerulonephritis associated with hepatic cirrhosis.

Berger's disease or primary IgA nephropathy (pIgAN) is characterized by recurrent gross haematuria, microhaematuria, and/or proteinuria in the absence of systemic disease (lupus erythematosus, Henoch-Schönlein purpura, cryoglobulinaemia), liver disease, and lower urinary tract diseases. Onset of the disease is usually in the teenage years. The occurrence of end-stage kidney failure in 20 to 50 per cent of adult patients 20 years after the apparent onset of the disease has caused this disease to be named indolent or slowly progressive glomerulonephritis. Since pIgAN is the most common type of glomerulonephritis throughout the world, it is the reason for periodic haemodialytic treatment in many of these patients; this aspect of the disease poses an important social problem because the disease begins at a very early age without obvious symptoms. The frequency of renal insufficiency at the time of renal biopsy is higher in European and North American patients than in patients from the Asian-Pacific area. Several reasons may account for this difference such as the absence of screening programmes in schools in Europe and North America and the different approaches adopted to renal biopsy in eastern and western areas of the world. Some clinical and histological findings at the time of biopsy are of value for predicting the clinical course, such as the presence of moderate or heavy proteinuria, hypertension, serum creatinine more than 1.5 mg/dl, severe mesangial proliferation, frequent sclerotic glomeruli, crescents, a high proportion of glomerular adhesions, vascular sclerosis, marked interstitial fibrosis, IgA and IgG/C3 deposits in the capillary walls, older age at the time of detection of the disease, and male sex.

The occurrence of pIgAN in siblings and in renal transplants of patients originally affected with pIgAN has suggested that genetic factors, determining a host immune response, may be related to the onset or the progression of the disease. Study of healthy relatives of patients with pIgAN has encouraged this idea even though preliminary studies on gene and phenotype frequencies have given contrasting findings in different countries. However, the invasive nature of renal biopsy in family members of patients with IgA nephropathy when they have haematuria and/or mild proteinuria has limited genetic studies.

Since circulating immune complexes occur in patients with pIgAN, immunological mechanisms have been studied extensively. Increased number of Tα-helper cells, reduced number of true suppressor T cells, spontaneous overproduction of interleukin-2 associated with increased expression of interleukin-2 receptors on the surface of T cells may be responsible for the continued production or increased production of polymeric IgA by B lymphocytes. This abnormal immunological response causes high synthesis of other immunoglobulins and antibodies which participate in the formation of immune complexes. A local cell-mediated immunological mechanism may be implicated in the renal damage since high number of T-helper cells occurs in the tubulointerstitial infiltrates in patients with impaired renal function during follow up after biopsy.

Even though the disease is progressive in some patients many tentative studies have been done with drugs; however they do not influence the course of the disease. Some studies have recently shown that steroid therapy is beneficial only in patients with heavy proteinuria, a creatinine clearance greater than 70 ml/min, and mild histological lesions. Tubular atrophy, interstitial fibrosis, and vascular changes are poor prognostic signs which contraindicate the administration of corticosteroids.

Mesangial IgA deposits are observed in liver diseases and other disorders. In these conditions the glomerulonephritis is usually a clinically silent disease discovered unexpectedly by urinanalysis which reveals mild microhaematuria and/or proteinuria. Macrohaematuria occurs, but much more infrequently than in patients with Berger's disease; renal insufficiency is rare and does not correlate with the glomerular lesions.

In many patients with mucosal diseases, neoplasia, and systemic disorders the IgA system is involved, since the presence of high serum levels of IgA in the polymeric form or of circulating IgA immune complexes is characterized by secondary deposition in the kidney. In these cases IgA nephropathy may be

Table 9 Diseases associated with IgA nephropathy

Mucosal diseases
    Bronchial and pulmonary diseases
    Enteric fever
    Coeliac disease
    Scleritis
    Oral mucosa sensitization
    Cystic fibrosis
Neoplasia
    Bronchial carcinoma
    Mycosis fungoides
    Multiple myeloma
Immunological and systemic disorders
    Ankylosing spondylitis
    Psoriatic arthritis
    Juvenile rheumatoid arthritis
    Reiter's disease
    Rheumatoid arthritis
    Sjögren's syndrome
    Immunothrombocytopenia
    Mixed cryoglobulinaemia

considered a complication rather than a disease, since extrarenal vessels are also involved as shown by deposits of IgA in the walls of these vessels.

Bibliography

Berger, J. (1969). IgA glomerular deposits in renal disease. *Transplantation Proceedings*, **1**, 939–44.

Berthoux, F.C. and Alamartine, E. (1986). Les glomérulonéphrites à dépots mésangiaux d'immunoglobuline A.*Encyclopédie médicochirurgicale*, rein-organes génito-urinaires, 18052 H25, **7**, pp. 1–8, Paris.

Clarkson, A.R., Woodroffe, A.J., Bannister, K.M., Lomax-Smith, J.D., and Aarons, I. (1984). The syndrome of IgA nephropathy. *Clinical Nephrology*, **21**, 7–14.

Clarkson, A.R. and Woodroffe, A.J. (1987). Treatment tentatives in IgA nephropathy. *Seminars in Nephrology*, **7**, 393–8.

Cohen, A.H. and Border, W.A. (1982). Mesangial proliferative glomerulonephritis. *Seminars in Nephrology*, **2**, 228–40.

D'Amico, G. (1985a). Idiopathic IgA mesangial nephropathy. *Nephron*, **41**, 1–13.

D'Amico, G. (1987a). The commonest glomerulonephritis in the World: IgA nephropathy. *Quarterly Journal of Medicine*, **64**, 709–27.

Egido, J., Sancho, J., Blasco, R., Rivera, F., and Hernando, L. (1982a). Aspects immunopathogéniques de la glomerulonéphrite à dépôts mésangiaux d'IgA. pp. 147–73. *Actualités Néphrologiques de l'Hôpital Necker*, Flammarion, Paris.

Egido, J., Julian, B.A., and Wyatt, R.J. (1987). Genetic factors in primary IgA nephropathy. *Nephrology, Dialysis, Transplantation*, **2**, 134–42.

Emancipator, S.N., Gallo, G.R., and Lamm, M.E. (1985). IgA nephropathy: perspectives on pathogenesis and classification. *Clinical Nephrology*, **24**, 161–79.

Emancipator, S.N., Gallo, G.R., and Lamm, M.E. (1987a). Animal models of IgA nephropathy. In *Topics in renal medicine: IgA nephropathy* (eds. V. Andreucci and A.R. Clarkson), pp. 188–203. Martinus Nijhoff, Boston.

Emancipator, S.N. and Lamm, M.E. (1989). IgA nephropathy: pathogenesis of the most common form of glomerulonephritis. *Laboratory Investigation*, **60**, 168–83.

Feehally, J. (1988). Immune mechanisms in glomerular IgA deposition. *Nephrology, Dialysis, Transplantation*, **3**, 361–78.

Julian, B.A., Wyatt, R.J., and Quiggins, P.A. (1985a). The immunogenetics of IgA nephropathy. *Plasma Therapy Transfusion Technology*, **6**, 687–704.

Julian, B.A., Waldo, F.B., Rifai, A., and Mestecky, J. (1988). IgA nephropathy, the most common glomerulonephritis worldwide. A neglected disease in the United States? *American Journal of Medicine*, **84**, 129–32.

Kincaid-Smith, P. and Nicholls, K. (1983). Mesangial IgA nephropathy. *American Journal of Kidney Diseases*, **2**, 90–102.

Kurokawa, K., et al. (1985). IgA nephropathy in Japan. *American Journal of Nephrology*, **5**, 127–37.

Lamm, M.E. and Emancipator, S.N. (1987). The mucosal immune system and IgA nephropathy. *Seminars in Nephrology*, **7**, 280–2.

Lévy, M., et al. (1985). Berger's disease in children. Natural history and outcome. *Medicine*, **64**, 157–80.

Makdassy, R., Beaufils, M., Meyrier, A., Mignon, F., Moulonquet-Doleris L., and Richet, G. (1984). Pathologic conditions associated with IgA mesangial nephropathy: Preliminary results. *Contributions to Nephrology*, **40**, 292–5.

Mustonen, J. (1984). IgA glomerulonephritis and associated diseases. *Annals of Clinical Research*, **16**, 161–6.

Newell, G.C. (1987). Cirrhotic glomerulonephritis: incidence, morphology, clinical features, and pathogenesis. *American Journal of Kidney Diseases*, **9**, 183–90.

Rifai, A. (1987). Experimental models for IgA-associated nephritis. *Kidney International*, **31**, 1–7.

Rodicio, J.L. (1984). Idiopathic IgA nephropathy. *Kidney International*, **25**, 717–29.

Schena, F.P. and Emancipator, S.N. (1989e). Primary IgA-associated nephropathy. A clue to pathogenesis. *Journal of Nephrology*, **2**, 135–43.

Sinniah, R. (1985). IgA mesangial nephropathy: Berger's disease. *American Journal of Nephrology*, **5**, 73–83.

Woodroffe, A.L., Clarkson, A.R., Seymour, A.E., and Lomex-Smith, J.D. (1982). Mesangial IgA nephritis. *Springer Seminars in Immunopathology*, **5**, 321–32.

References

Abe, T., et al. (1986). Participation of extracapillary lesions (ECL) in progression of IgA nephropathy. *Clinical Nephrology*, **25**, 37–41.

Abuelo, J.G., Esparza, A.R., Matarese, R.A., Endreny, R.G., Carvalho, J.S., and Allegra, S.R. (1984). Crescentic IgA nephropathy. *Medicine*, **63**, 396–406.

Adachi, M., Yodoi, J., Masuda, T., Takatsuki, K., and Uchino, H. (1983). Altered expression of lymphocyte Fc$_\alpha$-receptor in selective IgA deficiency and IgA nephropathy. *Journal of Immunology*, **131**, 1246–51.

Agrofiotis, A., Gluckman, C., Baumelou, A., Beaufils, H., Jacob, N., and Legrain, M. (1981). Is cryoglobulin detection of clinical significance in chronic glomerulonephritis not related to systemic disease? *Clinical Nephrology*, **16**, 146–50.

Alexopoulos, E., Seron, D., Hartley, R.B., Nolasco, F., and Cameron, J.S. (1989). The role of interstitial infiltrates in IgA nephrology: A study with monoclonal antibodies. *Nephrology, Dialysis, Transplantation*, **4**, 187–95.

Allich, J.J., Kline, B.E., and Rush, H.P. (1949). Degenerative renal lesions induced by prolonged choline deficiency. *Archives of Pathology*, **48**, 583.

André, C., Berthoux, F.C., André F., Gillon, J., Genin, C., and Sabatier, J.C. (1980). Prevalence of IgA2 deposits in IgA nephropathies. A clue to their pathogenesis. *New England Journal of Medicine*, **203**, 1343–6.

André, F. and André, C. (1976). Cirrhotic glomerulonephritis and secretory immunoglobulin A. *Lancet*, **i**, 197.

André, F., Druguet, M., and André, C. (1978). Effect of food in circulating antigen–antibody complexes in patients with alcoholic liver cirrhosis. *Digestion*, **17**, 554.

Andreoli, S.P., Yum, M.N., and Bergstein, J.M. (1986). Significance of glomerular basement membrane deposition of IgA. *American Journal of Nephrology*, **6**, 28–33.

Antonaci, S., Serlenga, E., Garofalo, A.R., Colizzi, M., and Schena, F.P. (1989). Imbalance of T cell immunoregulatory subsets in primary IgA nephropathy. *Cytobios*, **59**, 95–100.

Arnaiz-Villena, A., Gonzalo, A., Mampaso, F., Teruel, J.L., and Ortunno, J. (1981). HLA and IgA nephropathy in Spanish population. *Tissue Antigens*, **17**, 549–50.

Ballardie, F.W., Brenchley, P.E.C., Williams, S., and O'Donoghue, D.J. (1988). Autoimmunity in IgA nephropathy. *Lancet*, **ii**, 588–92.

Baxter, J.H. and Ashworth, C.T. (1946). Renal lesions in portal cirrhosis. *Archives of Pathology*, **41**, 476–7.

Bené, M.C., Faure, G., Hurault de Ligny, B., Kessler, M., and Duheille, J. (1983). Quantitative immunohistomorphometry of the tonsillar plasma cells evidences an inversion of the immunoglobulin A versus immunoglobulin G secreting cell balance. *Journal of Clinical Investigation*, **71**, 1342–7.

Bené, M.C., Hurault de Ligny, B., Sirbat, D., Faure, G., Kessler, M., and Duheille, J. (1984). IgA nephropathy: dimeric IgA-secreting cells are present in episcleral infiltrate. *American Journal of Clinical Pathology*, **82**, 608–11.

Bené, M.C., Hurault de Ligny, B., Faure, G., Kessler, M., and Duheille, J. (1986). Histoimmunological discrepancies in primary IgA nephropathy and anaphylactoid purpura sustain relationship between mucosa and kidney. *Nephron*, **43**, 214–6.

Bennet, W.M. and Kincaid-Smith, P. (1983). Macroscopic haematuria in mesangial IgA nephropathy: correlation with glomerular crescents and renal dysfunction. *Kidney International*, **23**, 393–400.

Bennet, W.M., Walker, R.G., and Kincaid-Smith, P. (1989). Treatment of IgA nephropathy with eicosapentanoic acid (EPA): a two-year prospective trial. *Clinical Nephrology*, **31**, 128–31.

Berger, J. and Hinglais, N. (1968). Les dépôts intercapillaires d'IgA-IgG. *Journal d'Urologie et Nephrologique*, **74**, 694–5.

Berger, J., De Montera, H., and Hinglais, N. (1967). Classification des gloméruo-néphrites en pratique biopsique. *Proceedings of the Third International Congress on Nephrology*, Washington 1966, **2**, 198–221.

Berger, J., Yaneva, H., Nabarra, B., and Barbanel C. (1975). Recurrence of mesangial deposition of IgA after renal transplantation. *Kidney International*, **7**, 232–41.

Berger, J., Yaneva, H., and Nabarra, B. (1977). Glomerular changes in patients with cirrhosis of the liver. *Advances in Nephrology*, **7**, 3–14.

Berthoux, F.C., et al. (1978). HLA-Bw35 and mesangial IgA glomerulonephritis. *New England Journal of Medicine*, **298**, 1034.

Berthoux, F.C., Genin, C., Gagne, A., Le Petit, J.C., and Sabatier, J.C. (1979). HLA-Bw35 antigen and mesangial IgA glomerulonephritis: a poor prognosis marker? *Proceedings of the European Dialysis and Transplant Association*, **16**, 551–5.

Beukhof, J.R., et al. (1984). Subentities within adult primary IgA-nephropathy. *Clinical Nephrology*, **22**, 195–9.

Beukhof, J.R., et al. (1986). Toward individual prognosis of IgA nephropathy. *Kidney International*, **29**, 549–56.

Biewenga, J., Noordhoek, G.T., Donker, A.J.M., and Beukhof, J.R. (1987). IgA nephropathy is not induced in mice by oral administration of TNP-conjugated ovalbumin. *Nephron*, **47**, 295–8.

Bignon, J.D., Houssin, A., Soulillou, J.P., Denis J., Guimbretiere, J., and Guenel, J. (1980). HLA antigens and Berger's disease. *Tissue Antigens*, **16**, 108–11.

Bloodworth, J.M.B. Jr. and Somers, S.C. (1959). Cirrhotic glomerulosclerosis. A renal lesion associated with hepatic cirrhosis. *Laboratory Investigation*, **8**, 962–78.

Brettle, R., Peters, D.K., and Batchelor, J.R. (1978). Mesangial IgA glomerulonephritis and HLA antigens. *New England Journal of Medicine*, **299**, 200.

Burkholder, P.M., Zimmerman, S.W., and Moorthy, A.V. (1978). A clinicopathologic study of the natural history of mesangial IgA nephropathy. In *Glomerulonephritis*. (eds. Y. Yoshitoshi, and Y. Ueda), pp. 143–66. Proceedings of the International Symposium on Glomerulonephritis. Progression and Regression (6–8 December 1977). Baltimore, University Press.

Cagnoli, L., et al. (1985). B and T cell abnormalities in patients with primary IgA nephropathy. *Kidney International*, **28**, 646–51.

Cameron, J.S. (1982a). Glomerulonephritis in renal transplants. *Transplantation*, **34**, 237–245.

Cameron, J.S. (1982b). Glomerulonephritis: current problems and understanding. *Journal of Laboratory and Clinical Medicine*, **99**, 755.

Cameron, J.S. (1987). Interstitial changes in glomerulonephritis. In *Prevention in Nephrology* (ed. G. Buccianti), pp. 8–18. Masson, Milano.

Cameron, J.S. and Turner, D.R. (1977). Recurrent glomerulonephritis in allografted kidneys. *Clinical Nephrology*, **7**, 47.

Casanueva, B., Rodriguez-Valverde, V., Arias, M., Vallo, A., Garcia-Fuentes, M., and Rodriguez-Soriano, J. (1986). Immunoglobulin-producing cells in IgA nephropathy. *Nephron*, **43**, 33–7.

Cederholm, B., Wieslander, J., Bygren, P., and Heinegard, D. (1988). Circulating complexes containing IgA and fibronectin in patients with primary IgA nephropathy. *Proceedings of the National Academy of Sciences of the USA*, **85**, 4865–8.

Chaikoff, H., et al. (1948). Pathologic reactions in the livers and kidneys of dogs fed alcohol while maintained on a high protein diet. *Archives of Pathology*, **45**, 435.

Chan, S.H., Ku, G., and Sinniah, R. (1981). HLA and Chinese IgA mesangial glomerulonephritis. *Tissue Antigens*, **17**, 351–2.

Chan, M.K., Kwan, S.Y.L., Chan, K.W., and Yeung, C.K. (1987). Controlled trial of antiplatelet agents in mesangial IgA glomerulonephritis. *American Journal of Kidney Diseases*, **9**, 417–21.

Chen, A., Wong, S.S., and Rifai, A. (1988). Glomerular immune deposits in experimental IgA nephropathy. *American Journal of Pathology*, **130**, 216–22.

Chida, Y., Tomura, S., and Takeuchi, J. (1985). Renal survival rate of IgA nephropathy. *Nephron*, **40**, 189–90.

Chino F., Sato, F., and Sasaki, S. (1981). Retrovirus particles in spontaneously occurring and radiation-induced tumours in ddY mice. *Acta Pathologica Japonica*, **31**, 233–47.

Churg, J. and Sobin, L.H. (1982). In *Renal disease. Classification and atlas of glomerular disease*. Igaku-Shoin, Tokyo, New York.

Clarkson, A.R., Seymour, A.E., Thompson, A.J., Haynes, W.D.G., Chan, Y-L., and Jackson, B. (1977). IgA nephropathy: a syndrome of uniform morphology, diverse clinical features and uncertain prognosis. *Clinical Nephrology*, **8**, 459–71.

Clarkson, A.R., Seymour, A.E., Woodroffe, A.J., McKenzie, P.E., Chan, Y.L., and Wootton, A.M. (1980). Controlled trial of phenytoin therapy in IgA nephropathy. *Clinical Nephrology*, **13**, 215–8.

Conley, M.E., Cooper, M.D., and Michael, A.F. (1980). Selective deposition of immunoglobulin A1 in immunoglobulin A nephropathy, anaphylactoid purpura nephritis, and systemic lupus erythematosus. *Journal of Clinical Investigation*, **66**, 1432–6.

Coppo, R., et al. (1982). Circulating immune complexes containing IgA, IgG and IgM in patients with primary IgA nephropathy and with Henoch-Schönlein nephritis. Correlation with clinical and histologic signs of activity. *Clinical Nephrology*, **18**, 230–9.

Coppo, R., Basolo, B., Bulzomi, M.R., and Piccoli, G. (1984). Ineffectiveness of phenytoin treatment on IgA-containing circulating immune complexes in IgA nephropathy. *Nephron*, **36**, 275–6.

Coppo, R., et al. (1985a). Presence and origin of IgA1 and IgA2 containing circulating immune complexes in chronic alcoholic liver diseases with and without glomerulonephritis. *Clinical Immunology and Immunopathology*, **35**, 1–8.

Coppo, R., et al. (1985b). Plasmapheresis in a patient with rapidly progressive idiopathic IgA nephropathy: Removal of IgA-containing circulating immune complexes and clinical recovery. *Nephron*, **40**, 488–90.

Coppo, R., et al. (1986). Mediterranean diet and primary IgA nephropathy. *Clinical Nephrology*, **26**, 72–82.

Cosio, F.G., Lam, S., Folami, A.O., Conley, M.E., and Michael, A.F. (1982). Immune regulation of immunoglobulin production in IgA nephropathy. *Clinical Immunology and Immunopathology*, **23**, 430–6.

Croker, B.P., Dawson, D.V., and Sanfilippo, F. (1983). IgA nephropathy. Correlation of clinical and histologic features. *Laboratory Investigations*, **48**, 19–24.

Czerkinsky, C., et al. (1986). Circulating immune complexes and immunoglobulin A rheumatoid factor in patients with mesangial immunoglobulin A nephropathies. *Journal of Clinical Investigation*, **77**, 1931–8.

D'Amico, G. (1983). Idiopathic mesangial IgA nephropathy. In *Glomerular Injury 300 Years After Morgagni*, (eds. T. Bertani and G. Remuzzi), pp. 205–28. Wichtig, Milano.

D'Amico, G., et al. (1985b). Idiopathic IgA mesangial nephropathy. Clinical and histological study of 374 patients. *Medicine*, **64**, 49–60.

D'Amico, G., et al. (1986). Prognostic indicators in idiopathic IgA mesangial nephropathy. *Quarterly Journal of Medicine*, **59**, 363–78.

D'Amico, G. and Vendemia, F. (1987b). Hypertension in IgA nephropathy. *Contribution in Nephrology*, **54**, 113–8.

D'Amico, G. (1988). Role of interstitial infiltration of leukocytes in glomerular diseases. *Nephrology, Dialysis, Transplantation*, **3**, 596–600.

Davies, D.R., Tighe, J.R., Jones, N.F., and Brown, G.W. (1973).

Recurrent haematuria and mesangial IgA deposition. *Journal of Clinical Pathology*, **26**, 672–7.

Davin, J.C., Malaise, M., Foidart, J., and Mahieu, P. (1987). Anti-α-galactosyl antibodies and immune complexes in children with Henoch-Schönlein purpura or IgA nephropathy. *Kidney International*, **31**, 1132–39.

Demaine, A.G., Rambausek, M., Knight, J.F., Williams, D.G., Welsh, K.I., and Ritz, E. (1988). Relation of mesangial IgA glomerulonephritis to polymorphism of immunoglobulin heavy chain switch region. *Journal of Clinical Investigation*, **81**, 611–4.

Deteix, P., *et al.* (1984). Prospective controlled therapeutic trial with diaminodiphenylsulfone-dapsone in primitive IgA nephropathy. Abstract. 9th International Congress of Nephrology, Los Angeles, 11–19 June, p. 83A.

De Werra, P., Morel-Maroger, L., Leroux-Robert, C., and Richet, G. (1973). Glomérulites à dépôts d'IgA diffus dans le mésangium. Etude de 96 cas chez l'adulte. *Schweizerishe Medizinische Wochenschrift*, **103**, 761–8.

Donini, U., Casanova, S., Zini, N., and Zucchelli, P. (1982). The presence of J chain in mesangial immune deposits of IgA nephropathy. *Proceedings of the European Dialysis and Transplant Association*, **19**, 655–61.

Drew, P.A., Nieuwhof, W.N., Clarkson, A.R., and Woodroffe, A.J. (1987). Increased concentration of serum IgA antibody to pneumococcal polysaccharides in patients with IgA nephropathy. *Clinical and Experimental Immunology*, **67**, 124–9.

Droz, D. (1976). Natural history of primary glomerulonephritis with mesangial deposits of IgA. *Contributions in Nephrology*, **2**, 150–7.

Droz, D., Kramar, A., Nawar, T., and Noel, L.H. (1984). Primary IgA nephropathy: prognostic factors. *Contributions in Nephrology*, **40**, 202–7.

Druet, P., Bariety, J., Bernard, D., and Lagrue, G. (1970). Les glomérulopathies primitives a dépôts mésangiaux d'IgA et d'IgG. Etude clinique et morphologique de 52 cas. *La Presse Medicale*, **78**, 583–7.

Dysart, N.K. Jr., Sisson, S., and Vernier, R.L. (1983). Immunoelectron microscopy of IgA nephropathy. *Clinical Immunology and Immunopathology*, **29**, 254–70.

Egido, J., Blasco, R., Sancho, J., Lozano, L., Sanchez-Crespo, M., and Hernando, L. (1982b). Increased rate of polymeric IgA synthesis by circulating lymphoid cells in IgA mesangial glomerulonephritis. *Clinical and Experimental Immunology*, **47**, 309–16.

Egido, J., Sancho, J., Lorente, F., and Fontan, G. (1982c). Inhibition of neutrophil migration by serum IgA from patients with IgA nephropathy. *Clinical and Experimental Immunology*, **49**, 709–16.

Egido, J., Blasco, R., Sancho J., Lozano, L., and Gutierrez-Millet, V. (1983a). Immunological studies in a familial IgA nephropathy. *Clinical and Experimental Immunology*, **54**, 532–38.

Egido, J., Blasco, R., Sancho, J., and Lozano, L. (1983b). T-cell dysfunctions in IgA nephropathy: specific abnormalities in the regulation of IgA synthesis. *Clinical Immunology and Immunopathology*, **26**, 201–12.

Egido, J., Rivera, F., Sancho, J., Barat, A., and Hernando, L. (1984a). Phenytoin in IgA nephropathy: a long-term controlled trial. *Nephron*, **38**, 30–9.

Egido, J., Blasco, R., Lozano, L., Sancho, J., and Garcia-Hoyo, R. (1984b). Immunological abnormalities in the tonsils of patients with IgA nephropathy: inversion in the ratio of IgA: IgG bearing lymphocytes and increased polymeric IgA synthesis. *Clinical and Experimental Immunology*, **57**, 101–6.

Egido, J., Sancho, J., Rivera, F., and Hernando, L. (1984c). The role of IgA and IgG immune complexes in IgA nephropathy. *Nephron*, **36**, 52–9.

Egido, J., Blasco, R., Sancho, J., and Hernando, L. (1985). Immunological abnormalities in healthy relatives of patients with IgA nephropathy. *American Journal of Nephrology*, **5**, 14–20.

Emancipator, S.N. (1988a). Experimental models of IgA nephropathy. *American Journal of Kidney Diseases*, **12**, 415–9.

Emancipator, S.N. (1988b). Oral tolerance as a mechanism of protection for immune mediated injury. *Monographs in Allergy*, **24**, 244–50.

Emancipator, S.N. and Lamm, M.E. (1987b). The role of IgG, IgM and C3 in experimental murine IgA nephropathy. *Seminars in Nephrology*, **7**, 286–8.

Emancipator, S.N., Gallo, G.R., and Lamm, M.E. (1983a). Experimental IgA nephropathy induced by oral immunization. *Journal of Experimental Medicine*, **157**, 572–82.

Emancipator, S.N., Gallo, G.R., Razaboni, R., and Lamm, M.E. (1983b). Experimental cholestasis promotes the deposition of glomerular IgA immune complexes. *American Journal of Pathology*, **113**, 19–26.

Emancipator, S.N., Ovary, Z., and Lamm, M.E. (1987c). Mesangial complement deposition causes haematuria in murine IgA nephropathy. *Laboratory Investigation*, **57**, 269–76.

Endoh, M., Suga, T., and Sakai, H. (1985). IgG, IgA and IgM rheumatoid factors in patients with glomerulonephritis. *Nephron*, **39**, 330–5.

Fauchet, R., *et al.* (1980). HLA-DR4 antigen and IgA nephropathy. *Tissue Antigens*, **16**, 405–10.

Feehally, J., Dyer, P.A., Davidson, J.A., Harris, R., and Mallick, N.P. (1984). Immunogenetics of IgA nephropathy: experience in a U.K. centre. *Disease Markers*, **2**, 493–500.

Feehally, J., Beattie, T.J., Brenchley, P.E.C., Coupes, B.M., Mallick, N.P., and Postlethwaite, R.J. (1986). Sequential study of the IgA system in relapsing IgA nephropathy. *Kidney International*, **30**, 924–31.

Feiner, H.D., Cabili, S., Baldwin, D.S., Schacht, R.G., and Gallo, G.R. (1982). Intrarenal vascular sclerosis in IgA nephropathy. *Clinical Nephrology*, **18**, 183–92.

Finlayson, G., *et al.* (1975). Immunoglobulin a glomerulonephritis. A clinicalopathologic study. *Laboratory Investigation*, **32**, 140–8.

Fisher, E.R. and Hellstrom, H.R. (1959). The membranous and proliferative glomerulonephritis of hepatic cirrhosis. *American Journal of Pathology*, **32**, 48–55.

Fisher, E.R. and Perez-Stable, E. (1968). Cirrhotic (hepatic) lobular glomerulonephritis. Correlation of ultrastructural and clinical features. *American Journal of Pathology*, **52**, 869–90.

Fornasieri, A., Sinico, R.A., Maldifassi, P., Bernasconi, P., Vegni, M., and D'Amico, G. (1987). IgA-antigliadin antibodies in IgA mesangial nephropathy (Berger's disease). *British Medical Journal*, **295**, 78–80.

Fornasieri, A., *et al.* (1988). Food antigens, IgA-immune complexes and IgA mesangial nephropathy. *Nephrology, Dialysis, Transplantation*, **3**, 738–43.

Fukuda, Y. (1982). Renal glomerular changes associated with liver cirrhosis. *Acta Pathologica Japonica*, **32**, 561–74.

Galle, P. (1964). Sur une variété inédite de dépôts dans le glomérule. *Revue Francoise d'Etude Cliniques Biologiques*, **9**, 49–60.

Galle, P. and Berger, J. (1962). Dépôts fibrinoides intercapillaires. *Journal d'Urologie et Nephrologique*, **68**, 123–7.

Gally, J.A. (1973) in *The Antigens* (ed., M. Seal), vol. 1, pp. 162–298. Academic Press, New York.

Garcia-Hoyo, R., Lozano, L., Blasco, R., Sancho, J., Egido, J., and Hernando, L. (1986). T and B cells alterations in IgA nephropathy. *Kidney International*, **30**, 134.

Gärtner, H.-V., Hönlein, F., Traub, U., and Bohle, A. (1979). IgA-nephropathy (IgA-IgG-nephropathy/IgA-nephritis)—a disease entity? *Virchows Archives. A Pathological Anatomy and Histology*, **385**, 1–27.

Genin, C., Sabatier, J.C., Laurent, P., Gonthier, R., Laurent, B., and Berthoux, F.C. (1984) Efficacy of high dose steroids and plasma exchanges in severe mesangial IgA glomerulonephritis. p. 89A. Abstract. 9th International Congress of Nephrology, Los Angeles, 11–19 June.

Genin, C., Laurent, B., Sabatier, J.C., Colon, S., and Berthoux, F.C. (1986). IgA mesangial deposits in C3H/HeJ mice after oral immuni-

zation with ferritin or bovine serum albumin. *Clinical and Experimental Immunology*, **63**, 385–94.
Gesualdo, L., Emancipator, S.N., Lamm, M.E., and Schena, F.P. (1989a). Estradiol and cyclophosphamide blunt oral tolerance and promote nephritogenesis in experimental IgA nephropathy. Abstract. *Kidney International*, **35**, 756.
Gesualdo, L., et al. (1990). Platelet-derived growth factor (PDGF) expression is increased in murine IgA nephropathy. Abstract. *Kidney International*, **37**, 414.
Gormly, A.A., Smith, P.S., Seymour, A.E., Clarkson, A.R., and Woodroffe, A.J. (1981). IgA glomerular deposits in experimental cirrhosis. *American Journal of Pathology*, **104**, 50–4.
Gregory, M.C., Hammond, M.E., and Brewer, E.D. (1988). Renal deposition of cytomegalovirus antigen in immunoglobulin-A nephropathy. *Lancet*, **i**, 11–4.
Hale, G.M., McIntosh, S.L., Hiki, Y., Clarkson, A.R., and Woodroffe, A.J. (1986). Evidence for IgA-specific B cell hyperactivity in patients with IgA nephropathy. *Kidney International*, **29**, 718–24.
Hall, R.P., Stachura, I., Cason, J., Whiteside, T.L., and Lawley, T.J. (1983). IgA-containing circulating immune complexes in patients with IgA nephropathy. *American Journal of Medicine*, **74**, 56–63.
Hamazaki, T., Tateno, S., and Shishido, H. (1984). Eicosapentaenoic acid and IgA nephropathy. *Lancet*, **i**, 1017–8.
Hara, M., Endo, Y., Nihei, H., Hara, S., Fukushima, O., and Mimura, N. (1980). IgA nephropathy with subendothelial deposits. *Virchows Archives. A. Pathological Anatomy and Histology*, **386**, 249–63.
Hattori, S., Karashima, S., Furuse, A., Terashima, M., Murakami M., and Matsuda, I. (1985). Clinicopathological correlation of IgA nephropathy in children. *American Journal of Nephrology*, **5**, 182.
Hené, R.J., Schuurman, H.J., and Kater, L. (1988). Immunoglobulin A subclass-containing plasma cells in the jejunum in primary IgA nephropathy and in Henoch-Schönlein purpura. *Nephron*, **48**, 4–7.
Hiki, Y., Kobayashi, Y., Tateno, S., Sada, M., and Kashiwagi, N. (1982). Strong association of HLA-DR4 with benign IgA nephropathy. *Nephron*, **32**, 222–6.
Hitman, G., et al. (1989). Immunoglobulin heavy chain switch region gene polymorphisms in idiopathic membranous nephropathy and IgA nephropathy. *Kidney International*, **35**, 349.
Hood, S.A., Velosa, J.A., Holley, K.E., and Donadio, J.V. Jr. (1981). IgA–IgG nephropathy: predictive indices of progressive disease. *Clinical Nephrology*, **16**, 55–62.
Hors, J., et al. (1980). *Strong association between HLA DRw4 and mesangial IgA nephropathy (Berger's disease)*. Abstract 8.5.26, in 4th International congress of Immunology (eds. J.L. Preud'homme and V.A.L. Harven), Paris.
Hotta, O., et al. (1987). Significance of renal hyaline arteriolosclerosis and tubulointerstitial change in IgA glomerulonephropathy and focal glomerular sclerosis. *Nephron*, **47**, 262–5.
Houssin, A., Riberi, P., and Hurez, D. (1983). Incidence de la phénytoïne sur les immunoglobulines au cours de la maladie de Berger. *La Presse Médicale*, **12**, 2001.
Hovig, T., et al. (1978). Plasma lipoprotein alterations and morphologic changes with lipid deposition in the kidney of patients with hepatorenal syndrome. *Laboratory Investigation*, **3**, 540–9.
Hughson, M.D., Megill, D.M., Smith, S.M., Tung, K.S.K., Miller, G., and Hoy, W.E. (1989). Mesangiopathic glomerulonephritis in Zuni (New Mexico) Indians. *Archives of Pathololy and Laboratory Medicine*, **113**, 148–57.
Hulette, C. and Carstens, P.H.B. (1985). Electron-dense deposits in extraglomerular vascular structures in IgA nephropathy. A retrospective study. *Nephron*, **39**, 179–83.
Imai, H., Nakamoto, Y., Asakura, K., Miki, K., Yasuda, T., and Miura, A.B. (1985). Spontaneous glomerular IgA deposition in ddY mice: an animal model of IgA nephritis. *Kidney International*, **27**, 756–61.
Imai, H., Chen, A., Wyatt, R.J., and Rifai, A. (1987). Composition of IgA immune complexes precipitated with polyethylene glycol. *Journal of Immunological Methods*, **103**, 239–45.

Isaacs, K.L. Miller, F., and Lane, B. (1981). Experimental model for IgA nephropathy. *Clinical Immunology and Immunopathology*, **20**, 419–26.
Isaacs, K.L. and Miller, F. (1982). Role of antigen size and charge in immune complex glomerulonephritis. *Laboratory Investigation*, **47**, 198–205.
Isaacs, M.L. and Miller, F. (1983). Antigen size and charge in immune complex glomerulonephritis. *American Journal of Pathology*, **111**, 298–306.
Ishii, M., et al. (1983). Elevated plasma catecholamines in hypertensives with primary glomerular diseases. *Hypertension*, **5**, 545–51.
Jackson, S., Montgomery, R.I., Julian, B.A., Galla, J.H., and Czerkinsky, C. (1987). Aberrant synthesis of antibodies directed at the Fab fragment of IgA in patients with IgA nephropathies. *Clinical Immunology and Immunopathology*, **45**, 208–13.
Jennette, J.C., Wall, S.D., and Wilkman, A.S. (1985). Low incidence of IgA nephropathy in blacks. *Kidney International*, **28**, 944–50.
Jessen, R.H., Nedrud, J.G., and Emancipator, S.N. (1987). A murine model of IgA nephropathy induced by a viral respiratory pathogen. *Advances in Experimental Medicine and Biology*, **216**, 1609–18.
Jones, W.T., Rao, D.R.G., and Braunstein, H. (1961). The renal glomerulus in cirrhosis of the liver. *American Journal of Pathology*, **39**, 393–404.
Julian, B.A., Wyatt, R.J., McMorrow, R.G., and Galla, J.H. (1983). Serum complement proteins in IgA nephropathy. *Clinical Nephrology*, **20**, 251–58.
Julian, B.A., Quiggins, P.A., Thompson, J.S., Woodford, S.Y., Gleason, K., and Wyatt, R.J. (1985b). Familial IgA nephropathy. Evidence of an inherited mechanism of disease. *New England Journal of Medicine*, **312**, 202–8.
Jungers, P., Forget, D., Hovillier, P., Henry-Amar, M., and Grünfeld, J.P. (1987). Pregnancy in IgA nephropathy, reflux nephropathy and focal glomerular sclerosis. *American Journal of Kidney Diseases*, **9**, 334–8.
Kalsi, J., Delacroix, D.L., and Hodgson, H.J. (1983). IgA in alcoholic cirrhosis. *Clinical and Experimental Immunology*, **52**, 499–504.
Kashahara, M., et al. (1982). Role of HLA in IgA nephropathy. *Clinical Immunology and Immunopathology*, **25**, 189–95.
Kashiwabara, H., Shishido, H., Yokoyama, T., and Miyajima, T. (1980). HLA in IgA nephropathy. *Tissue Antigens*, **16**, 411–2.
Kashiwabara, H., Shishido, H., Tomura, S., Tuchida, H., and Miyajima, T. (1982). Strong association between IgA nephropathy and HLA-DR4 antigen. *Kidney International*, **22**, 377–82.
Katz, A., Walker, J.F., and Landy, P.J. (1983). IgA nephritis with nephrotic range proteinuria. *Clinical Nephrology*, **20**, 67–71.
Kincaid-Smith, P. and Fairley, K.F. (1987). Renal disease in pregnancy. Three controversial areas: mesangial IgA nephropathy, focal glomerular sclerosis (focal and segmental hyalinosis and sclerosis), and reflux nephropathy. *American Journal of Kidney Diseases*, **9**, 328–33.
Kitagawa, T. (1987). A multicenter clinical study on the epidemiology of chronic glomerulonephritis in children in Japan. *Recent Advances in Pediatric Nephrology*, 13–20.
Kobayashi, Y., Tateno, S., Hiki, Y., and Shigematsu, H. (1983). IgA nephropathy: Prognostic significance of proteinuria and histological alterations. *Nephron*, **34**, 146–53.
Kobayashi, Y., Fujii, K., Hiki, Y., and Tateno, S. (1986). Steroid therapy in IgA nephropathy: A prospective pilot study in moderate proteinuric cases. *Quarterly Journal of Medicine*, **61**, 935–43.
Kobayashi, Y., Hiki, Y., Fujii, K., Kurokawa, A., Kamiyama, M., and Tateno, S. (1987). IgA nephropathy: heterogeneous clinical pictures and steroid therapy in progressive cases. *Seminars in Nephrology*, **7**, 382–5.
Kobayashi, Y., Fujii, K., Hiki, Y., Tateno, S., Kurokawa, A., and Kamiyama, M. (1988). Steroid therapy in IgA nephropathy: a retrospective study in heavy proteinuric cases. *Nephron*, **48**, 12–7.

Komori, K., Nose, Y., Inouye, H., Tsuji, K., Nomoto, Y., and Sakai, H. (1979). Study on HLA system in IgA nephropathy. *Tissue Antigens*, **14**, 32–6.

Kupor, L.R., Mullins, J.D., and McPhaul, J.J. (1975). Immunopathologic findings in idiopathic renal haematuria. *Archives of Internal Medicine*. **135**, 1204–11.

Kusumoto, Y., Takebayashi, S., Taguchi, T., Harada, T., and Naito, S. (1987). Long-term prognosis and prognostic indices of IgA nephropathy in juvenile and in adult Japanese. *Clinical Nephrology*, **28**, 118–24.

Kutteh, W.H., et al. (1982). Properties of immunoglobin A in serum of individuals with liver disease and in hepatic bile. *Gastroenterology*, **82**, 1847.

Laasonen, A., Rantala, I., Mustonen, J., and Pasternack, A. (1984). Immunoelectron microscopic localization of immune deposits in IgA glomerulonephritis. *Acta Pathologica, Microbiologica et Immunologica Scandinavica. Sect. A*, **92**, 249–56.

Lagrue, G., Bernard, D., Bariety, J., Druet, P., and Guenel, J. (1975). Traitement par le chlorambucil et l'azathioprine dans les glomerulonéphrites primitives. Résultats d'une étude 'contrôlée'. *Journal d'Urologie et de Néphrologie*, **9**, 655–72.

Lagrue, G., Sadreux, T., Laurent, J., and Hirbec, G. (1980). Is there a treatment of mesangial IgA glomerulonephritis? *Clinical Nephrology*, **14**, 161.

Lai, K.N., Ho, C.P., Chan, K.W., Yan, K.W., Lai, F.M., and Vallance-Owen, J. (1985). Nephrotic range proteinuria—a good predictive index of disease in IgA nephropathy? *Quarterly Journal of Medicine*, **57**, 677–88.

Lai, K.N., Mac-Maune Lai F., and Vallance-Owen, J. (1986a). The clinical use of serum beta-2-microglobulin and fractional beta-2-microglobulin excretion in IgA nephropathy. *Clinical Nephrology*, **25**, 260–5.

Lai, K.N., Lai, F.M., Ho, C.P., and Chan, K.W. (1986b). Corticosteroid therapy in IgA nephropathy with nephrotic syndrome: a long-term controlled trial. *Clinical Nephrology*, **26**, 174–80.

Lai, K.N., et al. (1987a). Studies of lymphocyte subpopulations and immunoglobulin production in IgA nephropathy. *Clinical Nephrology*, **28**, 281–7.

Lai, K.N., Lai, F.M., Leung, A.C.T., Ho, C.P., and Vallance-Owen, J. (1987b). Plasma exchange in patients with rapidly progressive idiopathic IgA nephropathy: a report of two cases and review of literature. *American Journal of Kidney Diseases*, 1987; **10**, 66–70.

Lai, K.N., Lai, F.M.M., Li, P.K., and Vallance-Owen, J. (1987c). Cyclosporin treatment of IgA nephropathy: a short term controlled trial. *British Medical Journal*, **295**, 1165–8.

Lai, K.N., Lai, F.M.M., Lo, S.T.H., and Lam, C.W.K. (1988). Light chain composition of IgA in IgA nephropathy. *American Journal of Kidney Diseases*, **11**, 425–9.

Lawrence, S., Pussell, B.A., and Charlesworth, J.A. (1983). Mesangial IgA nephropathy: detection of defective reticulo-phagocytic function *in vivo*. *Clinical Nephrology*, **16**, 280–3.

Lee, S-M., K., et al. (1982). IgA nephropathy: morphologic predictors of progressive renal disease. *Human Pathology*, **13**, 314–22.

Lee, H.S., Koh, H.I., Lee, H.B., and Park, H.C. (1987). IgA nephropathy in Korea: a morphological and clinical study. *Clinical Nephrology*, **27**, 131–40.

Leinikki P.O., Mustonen, J., and Pasternack, A. (1987). Immune response to oral polio vaccine in patients with IgA glomerulonephritis. *Clinical and Experimental Immunology*, **68**, 33–8.

Le Petit, J.C., et al. (1982). Genetic investigation in mesangial IgA nephropathy. *Tissue Antigens*, **19**, 108–14.

Lesavre, P., Digeon, M., and Bach, J.F. (1982). Analysis of circulating IgA and detection of immune complexes in primary IgA nephropathy. *Clinical and Experimental Immunology*, **48**, 61–9.

Levinsky, R.J. and Barratt, T.M. (1979). IgA immune complexes in Henoch-Schönlein purpura. *Lancet*, **ii**, 1100–3.

Linné, T., Aperia, A., Broberger, O., Bergstrand, A., Bohman S.O., and Rekola, S. (1982). Course of renal function in IgA glomerulonephritis in children and adolescents. *Acta Paediatrica Scandinavica* **71**, 735–43.

Linné, T. and Wasserman, J. (1985). Lymphocyte subpopulations and immunoglobulin production in IgA nephropathy. *Clinical Nephrology*, **23**, 109–11.

Lopez-Trascasa, M., Egido, J., Sancho, J., and Hernando, L. (1980). IgA glomerulonephritis (Berger's disease): evidence of high serum levels of polymeric IgA. *Clinical and Experimental Immunology*, **42**, 247–54.

Lozano, L., Blasco, R., Sancho, J., Garcia-Hoyo, R., and Egido, J. (1984). Abnormalities of the immune regulation of IgA by tonsil lymphocytes of patients with IgA nephropathy. *Kidney International*, **26**, 223.

Lozano, L., Garcia-Hoyo, R., Egido, J., Blasco, R., Sancho, J., and Hernando, L. (1985). Tonsillectomy decreases the synthesis of polymeric IgA by blood lymphocytes and clinical activity in patients with IgA nephropathy. *Proceedings of the European Dialysis and Transplant Association–European Renal Association*, **22**, 800–4.

Lupo, A., Rugiu, C., Cagnoli, L., Zucchelli, P., and Maschio, G. (1987). Acute changes in renal function in IgA nephropathy. *Seminars in Nephrology*, **7**, 359–62.

MacDonald, I.M., Dumble, L.J., and Kincaid-Smith, P. (1976). HLA and glomerulonephritis. In *HLA and Disease*, p. 203 INSERM Paris.

McCoy, R.C., Abramowsky, C.R., and Tisher, C.C. (1974). IgA nephropathy. *American Journal of Pathology*, **76**, 123–44.

McLean, R.H., Wyatt, R.J., and Julian, B.A. (1984). Complement phenotypes in glomerulonephritis: Increased frequency of homozygous null C4 phenotypes in IgA nephropathy and Henoch-Schönlein purpura. *Kidney International*, **26**, 855–60.

Manigand, G., et al. (1970). Lesiones glomerulaires de la cirrhose du foie. Note preliminaire sur de lesiones histologiques du rein au cours des cirrhose hepatiques, d'après 20 prelevements biopsieques. *Revue Européen d'Etude Biologiques*, **15**, 898–996.

Mannik, M., Agodoa, L.Y.C., and David, K.A. (1983). Rearrangement of immune complexes in glomeruli leads to persistence and development of electron-dense deposits. *Journal of Experimental Medicine*, **157**, 1516–27.

Manno, C., et al. (1988). Increased plasma levels of immunoreactive human atrial natriuretic factor in primary IgA nephropathy. *Clinical Nephrology*, **6**, 308–14.

Masugi, Y., et al. (1979). Immunopathologic studies of renal glomerular change in liver cirrhosis with special reference to its pathogenesis. *Acta Pathologica Japonica*, **29**, 571–83.

Meadow, S.R., Glasgow, E.F., White, R.H.R., Moncrieff, M.W., Cameron, J.S., and Ogg, C.S. (1972). Schönlein-Henoch nephritis. *Quarterly Journal of Medicine*, **41**, 241–58.

Melvin, T., Burke, B., Michael, A.F., and Kim, Y. (1983). Experimental IgA nephropathy in bile duct ligated rats. *Clinical Immunology and Immunopathology*, **27**, 369–77.

Michael, J., Jones, N.F., Davies, D.R., and Tighe, J.R. (1976). Recurrent haematuria: role of renal biopsy and investigative morbidity. *British Medical Journal*, **1**, 686–8.

Monteiro, R.C., Halbwachs-Mecarelli, L., Roque-Barreira, M.C., Nöel, L.H., Berger, J., and Lesavre, P. (1985). Charge and size of mesangial IgA in IgA nephropathy. *Kidney International*, **28**, 666–71.

Monteiro, R.C., Chevailler, A., Noel, L.H., and Lesavre, P. (1988). Serum IgA preferentially binds to cationic polypeptides in IgA nephropathy. *Clinical and Experimental Immunology*, **73**, 300–6.

Moore, R., et al. (1989), HLA-D region gene polymorphisms associated with IgA nephropathy. *Kidney International*, **35**, 356.

Murakami, T., Furuse, A., Hattori, S., Kobayashi, K., and Matsuda, I. (1983). Glomerular IgA1 and IgA2 deposits in IgA nephropathies. *Nephron*, **35**, 120–3.

Mustonen, J., et al. (1981). Circulating immune complexes, the concentration of serum IgA and the distribution of HLA antigens in IgA nephropathy. *Nephron*, **29**, 170–5.

Mustonen, J., Pasternack, A., and Rantala, I. (1983). The nephrotic syndrome in IgA glomerulonephritis: response to corticosteroid therapy. *Clinical Nephrology*, **20**, 172–6.

Mustonen, J., Pasternack, A., Helin, H., and Nikkilä, M. (1985). Clinicopathologic correlations in a series of 143 patients with IgA glomerulonephritis. *American Journal of Nephrology*, **5**, 150–7.

Mustonen, J., Pasternack, A., and Helin, H. (1987). The course of disease in patients with nephrotic syndrome and IgA nephropathy. *Seminars in Nephrology*, **7**, 374–6.

Nagy, J., Hamori, A., Ambrus, M., and Hernadi, E. (1979). More on IgA glomerulonephritis and HLA antigens. *New England Journal of Medicine*, **300**, 92.

Nagy, J., Füst, G., Ambrus, M., Trinn, C., Paàl, M., and Burger, T. (1982). Circulating immune complexes in patients with IgA glomerulonephritis. *Acta Medica Academiae Scientiarum Hungaricae*, **39**, 211–8.

Nagy, J., Trinn, C., Deàk, G., Schmelzer, M., and Burger, T. (1984a). The role of the tubulointerstitial changes in the prognosis of IgA glomerulonephritis. *Klinische Wochenschrift*, **62**, 1094–6.

Nagy, J., Uj, M., Szücs, G., Trinn, C., and Bürger, T. (1984b). Herpes virus antigens and antibodies in kidney biopsies and sera of IgA glomerulonephritic patients. *Clinical Nephrology*, **21**, 259–62.

Nagy, J., Miltenyi, M., Dobos, M., and Burger, T. (1987a). Tubular proteinuria in IgA glomerulonephritis. *Clinical Nephrology*, **27**, 76–8.

Nagy, J., Ambrus, M., Paal, M., Trinn, C.S., and Burger, T. (1987b). Cryoglobulinaemia and cryofibrinogenaemia in IgA nephropathy: a follow-up study. *Nephron*, **46**, 337–42.

Nakamoto, Y., *et al.* (1981). Hepatic glomerulonephritis. Characteristics of hepatic IgA glomerulonephritis as the major part. *Virchows Archives. A. Pathological Anatomy*, **392**, 45–54.

Neelakantappa, K., Gallo, G.R., and Baldwin, D.S. (1988). Proteinuria in IgA nephropathy. *Kidney International*, **33**, 716–21.

Ng, W.L. (1981). The ultrastructural morphology of the mesangial cell in IgA nephropathy. *Journal of Pathology*, **134**, 209–17.

Nicholls, K.M., Fairley, K.F., Dowling, J.P., and Kincaid-Smith, P. (1984a). The clinical course of mesangial IgA associated nephropathy in adults. *Quarterly Journal of Medicine*, **53**, 227–50.

Nicholls, K. and Kincaid-Smith, P. (1984b). Defective *in vivo* Fc and C3b-receptor function in IgA nephropathy. *American Journal of Kidney Diseases*, **4**, 128–34.

Nicholls, K., Walker, R.G., Dowling, J.P., and Kincaid-Smith, P. (1985). 'Malignant' IgA nephropathy. *American Journal of Kidney Diseases*, **5**, 42–6.

Nochy, D., *et al.* (1976). Association of overt glomerulonephritis and liver diseases. A study of 34 patients. *Clinical Nephrology*, **6**, 422–27.

Nochy, D., Druet, P., and Bariety, J. (1984). IgA nephropathy in chronic liver disease. *Contributions in Nephrology*, **40**, 268–75.

Noel, L.H., *et al.* (1978). HLA antigens in three types of glomerulonephritis. *Clinical Immunology and Immunopathology*, **10**, 19–23.

Noel, L.H., Droz, D, Gascon, M., and Berger, J. (1987). Primary IgA nephropathy: from the first-described cases to the present. *Seminars in Nephrology*, **7**, 351–4.

Nomoto, Y., Sakai, H., and Arimori, S. (1979). Increase of IgA-bearing lymphocytes in peripheral blood from patients with IgA nephropathy. *American Journal of Clinical Pathology*, **71**, 158–60.

Nomoto, Y. and Sakai, H. (1979). Cold reacting antinuclear factor in sera from patients with IgA nephropathy. *Journal of Laboratory and Clinical Medicine*, **94**, 76–82.

Nomoto, Y., Suga, T., Miura, N., Nomoto, H., Tomino, Y., and Sakai, H. (1986). Characterization of an acid nuclear protein recognized by auto antibodies in sera from patients with IgA nephropathy. *Clinical and Experimental Immunology*, **65**, 513–9.

Pardo, V., *et al.* (1979). Benign primary haematuria: Clinicopathologic study of 65 patients. *American Journal of Medicine*, **67**, 817–22.

Pasternack, A., Mustonen, J., and Leinikki, P. (1986). Humoral immune response in patients with IgA and IgM glomerulonephritis. *Clinical and Experimental Immunology*, **63**, 228–33.

Patek, A.J., Segal, D., and Bevans, M. (1947). The coexistence of cirrhosis of the liver and glomerulonephritis. Report of 14 cases. *American Journal of Medicine*, **221**, 77–85.

Payton, C.D., McLay, A., and Boulton Jones, J.M. (1988). Progressive IgA nephropathy: The role of hypertension. *Nephrology, Dialysis, Transplantation*, **2**, 138–42.

Portis, J.L. and Coe, J.E. (1979). Deposition of IgA in renal glomeruli of mink affected with aleutian disease. *American Journal of Pathology*, **96**, 227–36.

Propper, D.J., Power, D.A., Simpson, J.G., Edward, N., and Catto, G.R.D. (1987). The incidence, mode of presentation, and prognosis of IgA nephropathy in Northeast Scotland. *Seminars in Nephrology*, **7**, 363–6.

Propper, D.J., Power, D.A., Simpson, J.G., Edward, N., and Catto G.R.D. (1988). Incidence of IgA nephropathy in U.K. *Lancet*, **i** 190.

Rambausek, M., Seelig, H.P., Andrassy, K., Waldherr, R., Lenhard, V., and Ritz, E. (1982). Clinical and serological features of mesangial IgA glomerulonephritis. *Proceedings of the European Dialysis and Transplant Association*, **19**, 663–6.

Rambausek, M., *et al.* (1987a). Genetic polymorphism of C3 and Bf in IgA nephropathy. *Nephrology, Dialysis, Transplantation*, **2**, 208–11.

Rambausek, M., Rauterberg, E.W., Waldherr, R., Demaine, A., Krupp, G., and Ritz, E. (1987b). Evolution of IgA glomerulonephritis: Relation to morphology, immunogenetics, and BP. *Seminars in Nephrology*, **7**, 370–3.

Richman, A.V., Mahoney, J.J., and Fuller, T.J. (1979). Higher prevalence of HLA-B12 in patients with IgA nephropathy. *Annals of Internal Medicine*, **90**, 201.

Rifai, A. (1987b). Experimental models for IgA-associated nephritis. *Kidney International*, **31**, 1–7.

Rifai, A., Small, P.A. Jr., Teague, P.O., and Ayoub, E.M. (1979). Experimental IgA nephropathy. *Journal of Experimental Medicine*, **150**, 1161–73.

Rifai, A., Chen, A., and Imai, H. (1987a). Complement activation in experimental IgA nephropathy: an antigen-mediated process. *Kidney International*, **32**, 838–44.

Rifai, A., *et al.* (1989). Clearance kinetics and fate of macromolecular IgA in patients with IgA nephropathy. *Laboratory Investigation*, **61**, 381–8.

Roccatello, D., *et al.* (1985). Circulating Fc-receptor blocking factors in IgA nephropathies. *Clinical Nephrology*, **23**, 159–68.

Rodriguez-Soriano, J., Arrieta, A., Vallo, A., Sebastian, M.J., Vitoria, J.C., and Masdevall, M.D. (1988). IgA antigliadin antibodies in children with IgA mesangial glomerulonephritis. *Lancet*, **i**, 1109–10.

Ross, R., Raines, E.W., and Bowen-Pope, D.F. (1986). The biology of platelet-derived growth factor. *Cell*, **46**, 155–69.

Rostoker, G., Laurent J., André, C., Cholin, S., and Lagrue, G. (1988). High levels of IgA antigliadin antibodies in patients who have IgA mesangial glomerulonephritis but not coeliac disease. *Lancet*, **i**, 356–9.

Rothschild, E. and Chatenoud, L. (1984). T cell subset modulation of immunoglobulin production in IgA nephropathy and membranous glomerulonephritis. *Kidney International*, **25**, 557–64.

Russel, M.W., Mestecky, J., Julian, B.A., and Galla, J.H. (1986). IgA-associated renal diseases: antibodies to environmental antigens in sera and deposition of immunoglobulins and antigens in glomeruli. *Journal of Clinical Immunology*, **6**, 74–86.

Sabadini, E., *et al.* (1988). Characterization of interstitial infiltrating cells in Berger's disease. *American Journal of Kidney Diseases*, **12**, 307–15.

Sabatier, J.C., Genin, C., Assenat, H., Colon, S., Ducret, F., and Berthoux, F.C. (1978). Mesangial IgA glomerulonephritis in HLA-identical brothers. *Clinical Nephrology*, **11**, 35–38.

Sakaguchi, H. (1968). Hepatic glomerulosclerosis – Light microscopic study of autopsy cases. *Acta Pathologica Japonica*, **18**, 407–15.

Sakaguchi, H., et al. (1965). Hepatic glomerulosclerosis. An electron microscopic study of renal biopsies in liver disease. *Laboratory Investigation*, **14**, 533–45.

Sakai, H. (1984a). T-cell function. *Contribution in Nephrology*, **40**, 124–29.

Sakai, H., Nomoto, Y., and Arimori, S. (1979). Decrease of IgA-specific suppressor T cell activity in patients with IgA nephropathy. *Clinical and Experimental Immunology*, **38**, 243–8.

Sakai, H., Endoh, M., Tomino, Y., and Nomoto, Y. (1982). Increase of IgA specific helper Tα cells in patients with IgA nephropathy. *Clinical and Experimental Immunology*, **50**, 77–82.

Sakai, H., Nomoto, Y., Tomino, Y., Endoh, E., Suga, T., and Miura, M. (1984b). *Effects of urokinase and danazol on proteinura in patients with IgA nephropathy.* p. 72A. Abstract. 9th International Congress of Nephrology, Los Angeles, 11–19 June.

Salomon, M., et al. (1965). Renal lesions in hepatic disease: a study based on kidney biopsies. *Archives of Internal Medicine*, **115**, 704–9.

Sancho, J., Egido, J., Sanchez-Crespo, M., and Blasco, R. (1981). Detection of monomeric and polymeric IgA containing immune complexes in serum and kidney from patients with alcoholic liver disease. *Clinical and Experimental Immunology*, **47**, 327–35.

Sancho, J., Egido, J., Rivera, F., and Hernando, L. (1983). Immune complexes in IgA nephropathy: presence of antibodies against diet antigens and delayed clearance of specific polymeric IgA immune complexes. *Clinical and Experimental Immunology*, **54**, 194–202.

Sanfilippo F., Crocker, B.P., and Bollinger, R.R. (1982). Fate of four cadaveric donor renal allografts with mesangial IgA deposits. *Transplantation*, **33**, 370–6.

Sato, M., Kinugasa, E., Ideura, T., and Koshikawa, S. (1983). Phagocytic activity of polymorphonuclear leukocytes in patients with IgA nephropathy. *Clinical Nephrology*, **19**, 166–71.

Sato, M., Ideura, T., and Koshikawa, S. (1986). Experimental IgA nephropathy in mice. *Laboratory Investigation*, **54**, 377–83.

Sato, M., Takayama, K., Wakasa, M., and Koshikawa, S. (1987). Estimation of circulating immune complexes following oral challenge with cow's milk in patients with IgA nephropathy. *Nephron*, **47**, 43–8.

Sato, M., Kojima, H., Takayama, K., and Koshikawa S. (1988). Glomerular deposition of food antigens in IgA nephropathy. *Clinical and Experimental Immunology*, **73**, 295–9.

Savi, M., Neri, T.M., Silvestri, M.G., Allegri, L., and Migone, L. (1979). HLA antigens and IgA mesangial glomerulonephritis. *Clinical Nephrology*, **11**, 45–56.

Schena, F.P. (1989a). The natural history of primary IgA nephropathy in the world by prospective analysis. *American Journal of Medicine*, in press.

Schena, F.P., Mastrolitti, G., Fracasso, A.R., Pastore, A., and Ladisa, N. (1986). Increased immunoglobulin-secreting cells in the blood of patients with active idiopathic IgA nephropathy. *Clinical Nephrology*, **26**, 163–8.

Schena, F.P., Pastore, A., Sinico, R.A., Ladisa, N., Montinaro, V., and Fornasieri, A. (1987). Studies on the mechanism producing solubilization of immune precipitates in the serum of patients with primary IgA nephropathy. *Seminars in Nephrology*, **7**, 336–40.

Schena, F.P., Pastore, A., Sinico, R.A., Montinaro, V., and Fornasieri, A. (1988a). Polymeric IgA and IgA rheumatoid factor decrease the capacity of serum to solubilize circulating immune complexes in patients with primary IgA nephropathy. *Journal of Immunology*, **141**, 125–30.

Schena, F.P., Pastore, A., and Montinaro, V. (1988b). The role of polymeric IgA in complement-medicated solubilization of IgG and IgA immune complexes. *American Journal of Kidney Diseases*, **12**, 433–6.

Schena, F.P., et al. (1989b). Increased production of interleukin 2 and IL-2 receptor in primary IgA nephropathy. *Kidney International*, **35**, 875–9.

Schena, F.P., Pastore, A., Ludovico, N., Sinico, R.A., Benuzzi, S., and Montinaro, V. (1989c). Increased serum levels of IgA1-IgG immune complexes and anti-F(ab') 2 antibodies in patients with primary IgA nephropathy. *Clinical and Experimental Immunology*, **77**, 15–20.

Schena, F.P., Montenegro, M., and Scivittaro, V. (1989d). Metaanalysis of randomized controlled trials with patients with primary IgA nephropathy or Berger's disease. *Nephrology, Dialysis, Transplantation*, in press.

Schmekel, B., Svalander, C., Bucht, H., and Westberg, N.G. (1981). Mesangial IgA glomerulonephritis in adults – clinical and histopathological observations. *Acta Medica Scandinavica*, **210**, 363–72.

Seedat, Y.K., Nathoo, B.C., Parag, K.B., Naiker, I.P., and Ramsaroop, R. (1988). IgA nephropathy in blacks and Indians of Natal. *Nephron*, **50**, 137–41.

Shigematsu, H., Kobayashi, Y., Tateno, S., Hiki, Y., and Kuwao, S. (1982). Ultrastructural glomerular loop abnormalities in IgA nephritis. *Nephron*, **30**, 1–7.

Shigematsu, H., Kobayashi, Y., Tateno, S., Hiki, Y., and Kuwao, S. (1983). Glomerular tissue injury in IgA nephritis. *Acta Pathologica Japonica*, **33**, 367–80.

Silva, F.G., Chanter, P., Pirani, C.L., and Hardy, M.A. (1982). Disappearance of mesangial IgA deposits after renal allograft. *Transplantation*, **33**, 214–6.

Simon, P., Ramée, M.P., Ang, K.S., Genetet, B., and Fauchet, R. (1983). Minimum prevalence and evolution of IgA nephropathy in a French region with 250,000 inhabitants: seven year follow-up. *Kidney International*, **24**, 137.

Simon, P., Ang. K.S., Bavay, P., Cloup, C., Mignard, J.P., and Ramée, M.P. (1984). Glomérulonéphrite à immunoglobulines. A. Epidémiologie dans une population de 250.000 habitants. *La Presse Médicale*, **13**, 257–60.

Sinico, R.A., Fornasieri, A., Oreni, N., Benuzzi, S., and D'Amico, G. (1986). Polymeric IgA rheumatoid factor in idiopathic IgA mesangial nephropathy (Berger's disease). *Journal of Immunology*, **137**, 536–41.

Sinniah, R. (1984). Heterogenous IgA glomerulonephropathy in liver cirrhosis. *Histopathology*, **8**, 947–62.

Sinniah, R., Pwee, H.S., and Lim, C.H. (1976). Glomerular lesions in asymptomatic microscopic haematuria discovered on routine medical examination. *Clinical Nephrology*, **5**, 216–28.

Sissons, J.G.P., et al. (1975). Isolated glomerulonephritis with mesangial IgA deposits. *British Medical Journal*, **3**, 611–4.

Smith, S.M. and Tung, K.S.K. (1985). Incidence of IgA-related nephritides in American Indians in New Mexico. *Human Pathology*, **16**, 181–4.

Smith, S.M., Hoy, W.E., Pathak, D., Megill, D.M., Tung, K.S.K., and Hughson, M.D. (1989). Pathologic findings in mesangiopathic glomerulonephritis in Navajo Indians. *Archives of Pathology Laboratory Medicine*, **113**, 158–63.

Southwest Pediatric Nephrology Study Group (1982). A multicenter study of IgA nephropathy in children. *Kidney International*, **22**, 643–52.

Southwest Pediatric Nephrology Study Group (1985). Association of IgA nephropathy with steroid-responsive nephrotic syndrome. *American Journal of Kidney Diseases*, **3**, 157–64.

Stachura, I., Singh, G., and Whiteside, T.L. (1981). Immune abnormalities in IgA nephropathy (Berger's disease). *Clinical Immunology and Immunopathology*, **20**, 373–88.

Subìas, R., Botey, A., Darnell, A., Montoliu, J., and Revert, L. (1987). Malignant or accelerated hypertension in IgA nephropathy. *Clinical Nephrology*, **27**, 1–7.

Takeuchi, E., Doi, T., Shimada, T., Muso, E., Maruyama, N., and Yoshida, H. (1989). Retroviral gp70 antigen in spontaneous mesangial glomerulonephritis of ddY mice. *Kidney International*, **35**, 638–46.

Tateno, S. and Kobayashi, Y. (1987). Quantitative analysis of mesangial areas in serial biopsied patients with IgA nephropathy. *Nephron*, **46**, 28–33.

Tolkoff-Rubin, N.E., Cosimi, A.B., Fuller, T., Rubin, R.H., and Colvin, R.B. (1978). IgA nephropathy in HLA-identical siblings. *Transplantation*, **26**, 430–3.

Tomino, Y., Endoh, M., Nomoto, Y., and Sakai, H. (1981*a*). Immunoglobulin A1 in IgA nephropathy. *New England Journal of Medicine*, **305**, 1159–60.

Tomino, Y., Endoh, M., Nomoto, Y., and Sakai, H. (1981*b*). Specificity of IgA antibody in IgA nephropathy. *Nephron*, **29**, 103–04.

Tomino, Y., *et al*. (1983*a*). Cross-reactivity of IgA antibodies between renal mesangial areas and nuclei of tonsillar cells in patients with IgA nephropathy. *Clinical and Experimental Immunology*, **51**, 605–10.

Tomino, Y., *et al.* (1983*b*). Impaired solubilization of glomerular immune deposits by sera from patients with IgA nephropathy. *American Journal of Kidney Diseases*, **3**, 48–53.

Tomino, Y., Sakai, H., Miura, M., Suga, T., Endoh, M., and Nomoto, Y. (1984*a*). Effect of danazol on solubilization of immune deposits in patients with IgA nephropathy. *American Journal of Kidney Diseases*, **4**, 135–40.

Tomino, Y., Miura, M., Suga, T., Endoh, M., Nomoto, Y., and Sakai, H. (1984*b*). Detection of IgA1-dominant immune complexes in peripheral blood polymorphonuclear leukocytes by double immunofluorescence in patients with IgA nephropathy. *Nephron*, **37**, 137–9.

Tomino, Y., *et al.* (1985). Specific binding of circulating IgA antibodies in patients with IgA nephropathy. *American Journal of Kidney Diseases*, **6**, 149–53.

Tomino, Y., Endoh, M., Kaneshige, H., Nomoto, Y., and Sakai, H. (1986*a*). Increase of IgA in pharyngeal washings from patients with IgA nephropathy. *American Journal of Medical Science*, **286**, 15–21.

Tomino, Y., *et al.* (1986*b*). Cytopathic effects of antigens in patients with IgA nephropathy. *Nephron*, **42**, 161–6.

Trevan, D. (1956). Glomerular changes induced by stilbesterol: preliminary communication. *Lancet*, **ii**, 22–3.

Triger, D.R., Alp, M.H., and Wright, R. (1972). Bacterial and dietary antibodies in liver disease. *Lancet*, **i**, 60–3.

Triger, Dr. R., Alp, M.H., and Wright, R. (1973). Hyperglobulinemia in liver diseases. *Lancet*, **i**, 1494–6.

Valentijn, R.M., Kauffmann, R.H., Brutel de la Riviere, G., Daha, M.R., and Van Es., L.A. (1983). Presence of circulating macromolecular IgA in patients with haematuria due to primary IgA nephropathy. *American Journal of Medicine*, **74**, 375–81.

Valentijn, R.M., *et al.* (1984). Circulating and mesangial secretory component-binding IgA1 in primary IgA nephropathy. *Kidney International*, **26**, 760–66.

Valvo, E., *et al.* (1987). Hypertension in primary immunoglobulin A nephropathy (Berger's disease): hemodynamic alterations and mechanisms. *Nephron*, **45**, 219–23.

van der Peet, J., Arisz, L., Brentjens, J.R.H., Marrink, J., and Hoedemaeker, P.J. (1977). The clinical course of IgA nephropathy in adults. *Clinical Nephrology*, **8**, 335–40.

van der Wall Bakke, A.W.L., *et al.* (1988). The bone marrow as production site of the IgA deposited in the kidneys of patients with IgA nephropathy. *Clinical and Experimental Immunology*, **72**, 321–5.

Van Mark, E.A.E., Deelder, A.M., and Gigase, P.L.J. (1977). Effect of partial portal vein ligation on immune glomerular deposits in *Schistosoma mansoni* infected mice. *British Journal of Experimental Pathology*, **58**, 412–7.

Velo, M., Lozano, L., Egido, J., Gutierrez-Millet, V., and Hernando, L. (1987). Natural history of IgA nephropathy in patients followed up for more than ten years in Spain. *Seminars in Nephrology*, **7**, 346–50.

Waldo, F.B., Beischel, L., and West, C.D. (1986). IgA synthesis by lymphocytes from patients with IgA nephropathy and their relatives. *Kidney International*, **29**, 1229–33.

Waldo, F.B., Britt, W.J., Tomana, M., Julian, B.A., and Mestecky, J. (1989). Non-specific mesangial staining with antibodies against cytomegalovirus in immunoglobulin. A nephropathy. *Lancet*, **i**, 129–31.

Westberg, N.G., Baklien, K., Schmekel, B., Gillberg, R., and Brandtzaeg, P. (1983). Quantitation of immunoglobulin-producing cells in small intestinal mucosa of patients with IgA nephropathy. *Clinical Immunology and Immunopathology*, **26**, 442–5.

Whitworth, J.A., Leibowitz, S., Kennedy, M.C., Cameron, J.S., and Chantler, C. (1976). IgA and glomerular disease. *Clinical Nephrology*, **5**, 33–6.

Williams, D.G., Perl, S.J., Knight, J.F., and Harada, T. (1987). Immunoglobulin production *in vitro* in IgA nephropathy and Henoch-Schönlein purpura. *Seminars in Nephrology*, **7**, 322–4.

Williamson, A. (1976). The biological origin of antibody diversity. *Annual Review Biochemistry*, **45**, 467–500.

Woo, K.T., Tan, Y.O., Yap, H.K., Lau, Y.K., Tay, J.S.H., and Lim, C.H. (1984). Beta-2-microglobulin in mesangial IgA nephropathy. *Nephron*, **37**, 78–81.

Woo, K.T., Edmondson, R.P.S., Wu, A.Y.T., Chiang, G.S.C., Pwee, H.S., and Lim, C.H. (1986*a*). The natural history of IgA nephritis in Singapore. *Clinical Nephrology*, **25**, 15–21.

Woo, K.T., *et al.* (1986*b*). Protein selectivity in IgA nephropathy. *Nephron*, **42**, 236–9.

Woo, K.T., *et al.* (1987). Effects of triple therapy on the progression of mesangial proliferative glomerulonephritis. *Clinical Nephrology*, **27**, 56–64.

Woodroffe, A.J., *et al.* (1977). Detection of circulating immune complexes in patients with glomerulonephritis. *Kidney International*, **12**, 268–78.

Woodroffe, A.J., *et al.* (1980). Immunologic studies in IgA nephropathy. *Kidney International*, **18**, 366–74.

Wu, G., Katz, A., Cardella, C., and Oreopoulos, D.G. (1985). Spontaneous remission of nephrotic syndrome in IgA glomerular disease. *American Journal of Kidney Diseases*, **6**, 96–9.

Wyatt, R.J., Julian, B.A., Weinstein, A., Rothfield, N.F., and McLean, R.H. (1982). Partial H (β1H) deficiency and glomerulonephritis in two families. *Journal of Clinical Immunology*, **2**, 110–17.

Wyatt, R.J., Julian, B.A., Galla, J.H., and McLean, R.H. (1984*a*). Increased frequency of C3 fast alleles in IgA nephropathy. *Disease Markers*, **2**, 419–28.

Wyatt, R.J., Julian, B.A., Bhathena, D.B., Mitchell, B.L., Holland, N.H., and Malluche, H.H. (1984*b*). IgA nephropathy: presentation, clinical course and prognosis in children and adults. *American Journal of Kidney Diseases*, **4**, 192–200.

Wyatt, R.J., *et al.* (1987*a*). Regionalization in hereditary IgA nephropathy. *American Journal of Human Genetics*, **41**, 36–50.

Wyatt, R.J., *et al.* (1987*b*). Complement activation in IgA nephropathy. *Kidney International*, **31**, 1019–23.

Wyatt, R.J., *et al.* (1988). Immunoregulatory studies in patients with IgA nephropathy. *Journal of Clinical Laboratory Immunology*, **25**, 109–14.

Yagama, M., *et al.* (1987). Detection of IgA-class circulating immune complexes in sera from patients with IgA nephropathy using a solid-phase anti-C3Facb enzyme immunoassay. *Clinical and Experimental Immunology*, **67**, 270–6.

Yagama, M., *et al.* (1988). Levels of circulating IgA immune complexes after gluten-rich diet in patients with IgA nephropathy. *Nephron*, **49**, 104–6.

Yap, H.K., Sakai, R.S., Woo, K.T., Lim, C.H., and Jordan, S.C. (1987). Detection of bovine serum albumin in the circulating IgA immune complexes of patients with IgA nephropathy. *Clinical Immunology and Immunopathology*, **43**, 395–402.

Yoshikawa, N., *et al.* (1987). Mesangial changes in IgA nephropathy in children. *Kidney International*, **32**, 585–9.

Yoshimura, M., Kida, H., Abe, T., Takeda, S., Katagiri, M., and Hattori, N. (1987). Significance of IgA deposits on the glomerular capillary walls in IgA nephropathy. *American Journal of Kidney Diseases*, **9**, 404–9.

Zimmerman, S.W. and Burkholder, P.M. (1975). Immunoglobulin A nephropathy. *Archives of Internal Medicine*, **135**, 1217–23.

Zucchelli, P., Zuccalà, A., Santoro, A., Degli Esposti, E., Sturani, A., and Chiarini, C. (1984). Characteristics of hypertension in primary IgA glomerulonephritis. *Contributions in Nephrology*, **40**, 174–81.

3.7 Membranous nephropathy

PIETRO ZUCCHELLI AND SONIA PASQUALI

Definition

Membranous nephropathy (MN) is a glomerular disease well-defined by its histopathological picture. Its typical morphological appearance is characterized by a uniform thickening of the glomerular capillary wall due to a diffuse subepithelial localization of immune aggregates in the absence of inflammatory or proliferative changes. In most patients the clinicopathological entity of membranous nephropathy does not appear to be associated with any known aetiological factors (idiopathic membranous nephropathy IMN), while in a small minority of patients the disease appears related to antigens or environmental agents derived from various sources (secondary membranous nephropathies). The synonymous terms membranous glomerulonephritis, epimembranous glomerulonephritis, extramembranous nephropathy are also used.

Epidemiology

The incidence of membranous nephropathy in the community at large is unquantified owing to the high renal centre variability (Mallick *et al.* 1983). Its overall frequency in all renal biopsies varies from 3.3 to 14.9 per cent (Zollinger and Mihatsch 1978). Its frequency in the British Medical Research Council (MRC) registry was 10 per cent of all renal biopsies over 4 years, corresponding to an annual incidence of 0.39 per million adult population. In our series of 3380 renal biopsies performed over 20 years until the end of 1988, membranous nephropathy amounted to 244 (7.2 per cent).

Membranous nephropathy is the most frequent cause of nephrotic syndrome in adults, accounting for 20 to 40 per cent of adult nephrotic patients (Glassock 1984; Lewis 1988), while it is rare in infants and children with a frequency of less than 1 per cent of nephrotic children under the age of 16 (Glassock 1984).

Idiopathic membranous nephropathy accounts for 62 to 86 per cent of all membranous nephropathies (Ehrenreich *et al.* 1976; Zollinger and Mihatsch 1978; Noel *et al.* 1979; Abe *et al.* 1986; Honkanen 1986). In our series, 39 cases were secondary forms corresponding to a frequency of 16 per cent. Table 1 lists the associated conditions that have been linked to membranous nephropathy (secondary membranous nephropathies).

Clinical features

In many idiopathic membranous nephropathy series, there is a consistent male preponderance: 61 per cent of 1046 patients discussed by Mallick *et al.* (1983) were males. Males accounted for 70 per cent of the series of Honkanen (1986) and Davison (1984), 55 per cent of that of Murphy *et al.* (1988) and Gluck *et al.* (1973) but only 51 per cent of the series of Ramzy *et al.* (1981) and Abe *et al.* (1986). Membranous nephropathy mainly occurs in adults: in a review of eight studies up to 1971, Rosen (1971) recorded a mean age range of 28 to 47 years. Subsequent studies have confirmed a peak age in the 36 to 40 age group (Noel *et al.* 1979; Davison *et al.* 1984; Honkanen *et al.* 1986; Kida *et al.* 1986; Zucchelli *et al.* 1987; Murphy *et al.* 1988). In children, it may be seen in the neonatal period or in the first year of life

Table 1 Secondary membranous nephropathies

Association with	Cross-reference
Infections	
Hepatitis B (active or without overt liver disease)	3.12
Syphilis (congenital or secondary)	3.12
Quartan malaria	3.12, 3.14
Streptococcal infection	3.9, 3.12
Leprosy	3.14
Filariasis	3.14
Schistosomiasis	3.14
Neoplasia	
Solid tumours (lung, colon, rectum, kidney, breast, stomach)	3.13
Hodgkin's disease	3.13
Non-Hodgkin's lymphoma	3.13
Other rare tumours (phaeochromocytoma, carotid body tumour)	
Multisystem diseases	
Systemic lupus erythematosus	4.6.2
Autoimmune thyroiditis	
Dermatitis herpetiformis	
Sarcoidosis	4.4
Sjögren's syndrome	4.10
Mixed connective tissue disease	4.10
Drugs or toxic agents	
Captopril	2.1
Organic gold compounds	2.1, 4.8
Mercury	2.1
D-Penicillamine	2.1, 4.8
Thiola	2.1
Probenicid	2.1
Tridione, paramethadione	2.1
Miscellaneous diseases and events	
Renal transplantation	11.6
Sickle-cell disease	4.11
Diabetes mellitus	4.1
Gardner-Diamond syndrome	
Fanconi syndrome	
Guillain-Barré syndrome	
Bullous pemphigoid	
Renal vein thrombosis*	3.4

* Associated abnormalities.

Table 2 Profile of symptoms and clinical findings in 205 cases of idiopathic membranous nephropathy at time of renal biopsy

Age (years)		
Mean±SD	50.1±15	
Range	5–79	
Number of males (%)	139	(67)
Delay between apparent onset and renal biopsy (months)		
Mean±SD	9.5±7.1	
Range	1–37	
Symptoms		
Episodic macrohaematuria	3	(1.4)
Microhaematuria	110	(53.6)
Asymptomatic proteinuria	40	(19.5)
Hypertension (blood pressure ≥160/95 mmHg)	47	(22.9)
Oedema	151	(73.6)
Oligoanuria	0	–
Serum creatinine ≥ 2 mg/dl	18	(8.8)
AST ↑	0	–
C3 ↓	0	–
Circulating immune complexes	14*	(14.8)

Percentages in parentheses.
* Out of 95 patients.

(Kleinknecht and Habib 1987). The clinical manifestations of the disease are often so insidious than in many patients membranous nephropathy is discovered accidentally. In many other cases, progressive oedema is the most important finding which points to the disease. A very rare early symptom of the disease may be macrohaematuria, mainly seen in children: only one case in Honkanen's series (1986), and three cases in the series of Ramzy et al. (1981) were reported, but it was seen in several children of the series of Habib et al. (1973). At renal biopsy the clinical picture is dominated by a nephrotic syndrome in most cases. Eighty-four per cent of 515 patients reported by Mallick et al. (1983) had nephrotic syndrome at the time of renal biopsy. Nephrotic syndrome was present in 75.8 per cent of the series of Noel et al. (1979), 74 per cent of that of Ramzy et al. (1981), 85 per cent of that of Davison et al. (1984), in virtually all of the patients of Gluck et al. (1973), but only 54 per cent of the series of Murphy et al. (1988).

Asymptomatic proteinuria is usually present in 10 to 25 per cent of patients with idiopathic membranous nephropathy. Proteinuria, which demonstrates wide fluctuations from day to day, is usually of a non-selective type (Mallick et al. 1983). Microhaematuria varies from 20 to 55 per cent (Noel et al. 1979; Murphy et al. 1988). Data relating to hypertension also exhibit considerable variations, from 20 to 40 per cent of patients being affected (Noel et al. 1979; Murphy et al. 1988). Hypertension was as frequently encountered in the early morphologic as in the later stages (Zollinger and Mihatsch 1978). Hypertension is rare in children (Kleinknecht and Habib 1987). Impaired renal function at the time of renal biopsy was often found: 7.5 per cent (Honkanen et al 1986), 4 per cent (Murphy et al 1988), and 8 per cent (Gluck et al. 1973) of patients. The urinary sediment is typically benign, commonly with oval fat bodies and hyaline casts. Radiographically, the kidneys are of a normal or slightly increased size.

The clinical data of our idiopathic membranous nephropathy patients at the time of renal biopsy are reported in Table 2. It is interesting to note that the mean age is higher than that reported previously. We feel this demonstrates a tendency towards a progressive increase in membranous nephropathy in the elderly, but could equally result from more active investigation of older patients.

Patients with secondary membranous nephropathy and underlying covert solid tumours, SLE, or hepatitis B infection may initially appear to be indistinguishable from those with idiopathic disease (Glassock 1984). In a recent paper, membranous nephropathy was reported to differ clinically from idiopathic membranous nephropathy only in that the mean age was higher in one group as opposed to the other (Cahen et al. 1989).

Serum levels of the C3 and C4 components of complement are nearly always normal in idiopathic membranous nephropathy while they may be reduced in the presence of underlying SLE or hepatitis B infection. Circulating immune complexes are found in 10 to 25 per cent of patients with idiopathic membranous nephropathy, depending on the method employed; levels are generally only moderately above the upper normal limit. They have no apparent relationship with disease activity (Solling 1983; Mallick et al. 1983).

Infrequently, antibody to glomerular basement membrane, tubular basement membrane, or renal tubular epithelial antigens may be found in the circulation.

Patients with membranous nephropathy are subject to several complications which may lead to a sudden deterioration in renal function. These complications include an acute hypersensitivity interstitial nephritis secondary to diuretics (see Chapters 2.1) acute volume depletion (see Chapters 3.4 and 8.2), superimposed crescentic glomerulonephritis (type I or II) (see Chapter 3.10) and acute or chronic renal vein thrombosis (see Chapter 3.4).

The association of renal-vein thrombosis with idiopathic membranous nephropathy has a reported frequency varying from 5 to 62 per cent of cases (Llach et al. 1975; Murphy et al. 1988). No single risk factor discriminates adequately between patients with and without thrombosis. A significant discrimination seems to be offered by low antithrombin III, a low plasminogen level associated with high haematocrit, high fibrinogen and elevated 2PI (α_2-antiplasmin) Llach 1985). These complications are dealt with in Chapter 3.4. Interestingly, no case of renal vein thrombosis is noted in the 200 cases reported in children in the literature (Kleinknecht and Habib 1987).

Immunogenetic abnormalities

Recent data indicate a strong link between MHC antigens and membranous nephropathy. In 1979, Klouda et al. found that the frequency of HLA-DR3 was significantly higher in idiopathic membranous nephropathy patients than in a control population (75 versus 20 per cent, with a relative risk of 12). HLA A1-B8-B18 were also significantly more frequent in patients than in controls.

Subsequently, Berthoux et al. (1984), Müller et al. (1981) and many other authors (Garavoy 1980; Papiha et al. 1987) confirmed the strong association between HLA B8 and DR3 and membranous nephropathy in other European countries, and we have recorded similar results in an Italian population (see Table 4).

Contrary to findings in the European population, a significant increase in DR2 antigen, and relatively low values of DR3 (and DR4) antigens have been reported in the Japanese population

(Tomura et al. 1984; Hiki et al. 1984); DR3 is in any case very rare in healthy Japanese. Finally, patients with rheumatoid arthritis and DR2 and/or DR3 antigens are likely to develop either penicillamine or gold-related membranous renal disease (Panayi et al. 1978).

There is also a very strong association in idiopathic membranous nephropathy with the haplotype B18-BfF1-DR3, which is very rare in the Caucasoid population. This association seems to be stronger than with B8-DR3 alone, thus suggesting that the locus which determines vulnerability to the disease is not DR itself but is closely related to HLA-DR (Mallick et al. 1983; Berthoux et al. 1984). The strong association of the major histocompatibility complex antigens with membranous nephropathy suggests that chromosome 6 in humans plays an important role in the pathogenetic mechanisms of this disease.

The widely reported close association between membranous nephropathy and the B8-DR3 haplotype may be associated with a number of immunological aberrations. These include

1. A primary defect in reticuloendothelial function. A selective defect in splenic Fc-receptor function in idiopathic membranous nephropathy has been recorded by some workers (McGinley et al. 1984; Berthoux et al. 1984) but not by others (Solomon et al. 1981)

2. An impaired lymphocyte proliferation response to the T-cell mitogen PHA (Greenberg and Yunis 1978)

3. An impaired IgG3 response to tetanus toxoid (Feehally et al. 1986).

In addition, other immunological abnormalities have been reported to occur frequently in idiopathic membranous nephropathy

1. An impairment in delayed skin reactivity to tuberculin (PPD) (Matsumoto et al. 1978)

2. A defective antibody response of peripheral blood lymphocytes to pokeweed mitogen, a polyclonal B–cell activator (Ooi et al. 1980)

3. A significant increase in the helper/suppressor T-cell ratio due to a reduction in CD8 T-cell subset (Zucchelli et al. 1988).

Pathology

Membranous nephropathy is a histological entity showing a combination of subepithelial electron-dense deposits on the outer surface or within the glomerular basement membrane (GBM), and thin, argyrophilic projections (usually called 'spikes') rising from the lamina densa of the GBM. The deposits are discrete, continuous, usually homogeneous and separated by the spikes (Figs. 1 and 2). The podocytes over the deposits show oedema, loss of foot processes, and microvilli. The tubules may contain protein droplets, lipid vacuoles in the cytoplasm, and numerous proteinaceous casts in the lumen. These changes are not static, but seem to evolve in a regular manner over time, according to experimental observations and studies based on serial biopsies from patients with gold-induced membranous nephropathy. These data suggest that projections and irregular thickening of the GBM develop secondary to the formation of immune deposits on the external surface. Spike formation is due to the increased synthesis of GBM-like material by the visceral epithelial cells. Moreover, the deposits are gradually displaced towards the inner side of the GBM, while undergoing considerable mor-

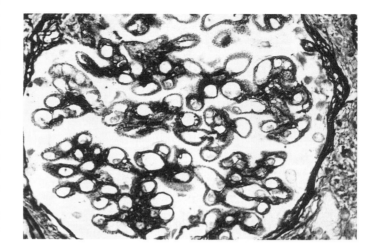

Fig. 1 Membranous nephropathy (stage II). The optical microscopic picture shows obvious thickening of the glomerular basement membrane with easily recognizable 'spikes' on the outer surface of the glomerular basement membrane (× 800, PASM).

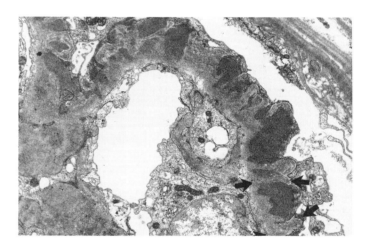

Fig. 2 Membranous nephropathy (stage II) on electron microscopy. The glomerular basement membrane is markedly thickened. There are prominent 'spikes' (arrow) between the subepithelial deposits (× 9600).

phological changes, thus suggesting a slow, continuous synthesis and turnover of the GBM (Törnroth and Skrifvars 1975).

On the basis of these observations many authors have used a 'staging' of the glomerular morphology to classify the histopathological evolution of membranous nephropathy. It is unfortunate that the stages have differed somewhat amongst groups: Ehrenreich and Churg (1968) used a four-point scale, Zollinger and Mihatsch (1978) five categories, while Bariéty et al. (1970) used only three.

However, the five-stage development scale could be used by modifying the schemes proposed by Ehrenreich and Churg (1968), and Gärtner et al. (1977). These stages seem to correlate with the data offered by electron scanning microscopy (Weidner and Lorentz 1986), experimental data (Törnroth and Skrifvars 1975; Törnroth et al. 1987), and clinical observations (Zucchelli

3.7 Membranous nephropathy

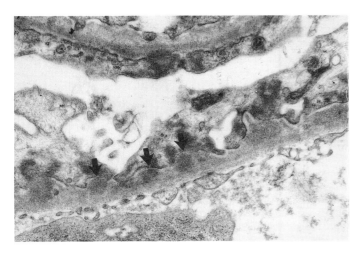

Fig. 3 Membranous nephropathy (stage I). On electron miscroscopy subepithelial deposits are clearly recognizable (arrows) (× 27 200).

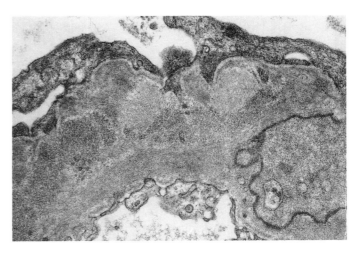

Fig. 4 Electron micrograph. Membranous nephropathy (stage III). Slightly osmiophilic deposits are covered by a thin rim of a newly-formed layer of glomerular basement membrane (× 32 500).

et al. 1986). It should be stressed that a specific stage represents the sum of the predominant lesions seen in the glomeruli. The grading system, based on a combination of light (semithin section) and electron microscopic changes, may be outlined as follows.

Stage I (subepithelial deposits) Optical microscopy shows normal glomeruli. Numerous small, usually flat and superficially placed, subepithelial osmiophilic deposits are present at electron microscopy or can be seen with Masson's trichrome stains. There is no evidence of spike formation or only very scattered, small projections from the GBM. Interstitium and vessels are virtually normal (Fig. 3).

Stage II (spike formation) The osmiophilic deposits are increased both in number and size and are flanked by very prominent (PASM-positive) spikes in virtually every capillary loop. The GBM is focally thickened (Figs. 1, 2).

Stage III (incorporation stage) The capillary wall is highly irregular and thickened and seems to be split into two thin layers. It has a 'moth-eaten' appearance. Large dense deposits are within the GBM because they are extensively incorporated by the adjacent spikes (endomembranous position). A few isolated synechiae and capillary loop obliterations can be seen (Fig. 4).

Stage IV (disappearing deposit stage) The GBM is quite irregular and thickened with the deposits either completely absent or only a few in number and isolated with a granular and vacuolated appearance. Numerous electron-lucent vacuolated areas are present within the thickened membrane (Fig. 5).

Stage V (reparation stage, according to Gärtner et al. 1977) Extensive normalization of the GBM occurs which appears very delicate and only partially thickened or transformed into a chain-like appearance. Some lucent areas within the apparently normal GBM may be seen (Fig. 6).

An acute relapse, characterized by fresh, small and partially short-linear deposits may be superimposed upon the lesions of various stages. Finally, an endstage with progressive glomerular obsolescence, adhesions, crescents, and sclerosis may be reached

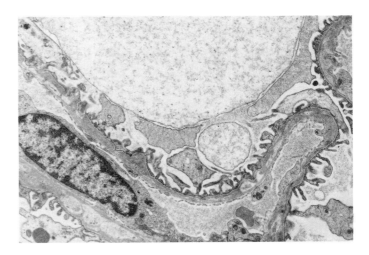

Fig. 5 Electron micrograph showing a case of membranous nephropathy in stage IV. The glomerular basement membrane is still irregular; areas of former deposits appear as small, clear lacunae along the glomerular basement membrane (× 8750).

(Stage V according to Zollinger and Mihatsch (1978) or Donadio *et al*. 1988, Stage IV according to Gluck *et al*. 1973).

Immunofluorescence studies reveal similar findings in Stage I, II, and III, with diffuse, uniform IgG granular deposits in all cases and usually deposits of C3 along the capillary loops as well (Fig. 7). In Stage I the deposits may have a 'pseudolinear' appearance because they are small and numerous (Grishman *et al*. 1988). IgM, and especially IgA and fibrin(ogen), are less frequently demonstrable with a frequency varying from 6 to 50 per cent (Kusonoki *et al*. 1989). The immunofluorescent deposits correspond to the osmiophilic deposits seen in electron microscopy. Reports of the frequency of C1q and C4 differ: seven of 51 cases for C1q, and 27 of 68 cases for C4 in the series of Noel *et al*. (1979), 18 of 78 and 38 of 108 cases in that of Jennette *et al*. (1983)). Although C1q and C4 deposits have rarely been found, C4d and C4bp antigens have been identified in 92 per cent of

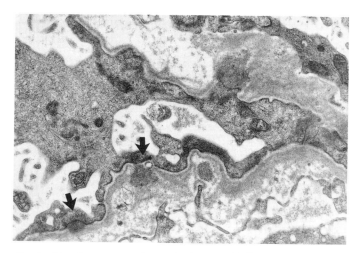

Fig. 6 Membranous nephropathy in stage V. Old deposits are replaced by empty-looking spaces in the basement membrane. Small, newly-formed osmiophilic deposits are closely overlaid on the epithelial covering (arrows) (× 28 000).

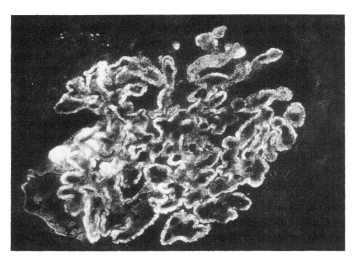

Fig. 7 Immunofluorescence staining with antibodies anti-IgG. The granular deposits along the glomerular capillary walls are intensively stained (× 360).

cases (Kusonoki et al. 1989), thus suggesting that complement activation usually occurs in idiopathic membranous nephropathy, via the classic pathway. In Stage V, following a complete and persistent remission of proteinuria, the IgG deposits may become scanty or entirely negative (Zucchelli et al. 1986).

Even though these stages may evolve gradually and often overlap, morphometric techniques have demonstrated significant correlations between GBM thickening and the duration of disease, proteinuria, and renal function (Aparicio et al. 1986).

Lesions superimposed upon membranous nephropathy have been reported, and are usually associated with a deterioration in renal function. Such lesions comprise superimposed crescentic glomerulonephritis with anti-GBM antibody (see Chapter 3.10), membranous nephropathy and acute or chronic interstitial nephritis, membranous nephropathy and focal sclerosis, membranous nephropathy with antitubular basement membrane (anti-TBM) antibodies, membranous nephropathy and IgA glomerulonephritis.

The presence of interstitial infiltrates in patients with membranous nephropathy is well documented although the role of such lesions in the pathogenesis and clinical outcome of the disease are not yet clear (Cameron 1979). T cells and monocytes/macrophages constitute the predominant cell types in the interstitium. In a recent study (Alexopoulos et al. 1989), total leucocytes, T cells, monocytes/macrophages, and B cells all showed positive correlations with plasma creatinine and inverse correlations with GFR at the time of biopsy. A significantly higher number of interstitial T cells and monocytes/macrophages was present in the patients who had a more rapid decline in renal function during the subsequent 4-year follow-up period, while no relationship was demonstrated between histologically-assessed tubulointerstitial damage and the progression of the disease. Analysing their results, Alexopoulos et al. (1989) suggested that both cellular and humoral immune mechanisms may be important in the initiation of the disease, whereas progression towards renal failure seems to be determined mainly by cell-mediated immunity.

Several cases have been described in which an association is recognized between membranous nephropathy and focal sclerosis (Cameron 1979; Grishman et al. 1988). In a retrospective study of 107 patients with idiopathic membranous nephropathy, Ehrenreich and Churg (1977) reported 32 cases of focal and segmental glomerular sclerosis. Their outcome was considerably worse when compared to the remaining 75 cases without focal sclerosis.

Levy et al. (1978) described a rare case of a 3-year-old boy who had severe tubular damage as a consequence of the development of anti-TBM antibodies, shown by the linear fluorescence pattern along the tubular basement membranes. The case was also remarkable as antialveolar antibodies were demonstrated in the lung.

The simultaneous presence of membranous changes together with IgA glomerulonephritis has occasionally been reported (Jennette et al. 1987).

Secondary forms of membranous nephropathy sometimes have a similar histomorphologic appearance to idiopathic membranous nephropathy. However, a number of histopathological features are significantly more frequent in SLE-related membranous nephropathy than in idiopathic membranous nephropathy. Mesangial deposits, revealed by toluidine-blue staining or electron microscopy, are frequent in SLE, but occur in only 3 to 11 per cent of patients with idiopathic membranous nephropathy (Gaffney and Panner 1981; Jennette et al. 1983). A few small subendothelial deposits can be identified in 61 per cent of biopsy specimens in SLE, but they are absent in idiopathic membranous nephropathy. In addition, endothelial tubuloreticular inclusions, TBM deposits and intense C1q deposition can be found in many SLE-related cases of membranous nephropathy (Jennette et al. 1983; Weidner and Lorentz 1986). The presence of mesangial deposits of immunoglobulin and/or complement should lead to other secondary forms. Although the presence of renal tubular antigen in glomerular lesions has been reported in a few cases (Naruse et al. 1973) suggesting the possibility of a Heymann nephritis-like mechanism in human membranous nephropathy, subsequent attempts in larger series to identify renal tubular antigens in immunofluorescent deposits have been largely unsuccessful (Whitworth *et al.* 1976). Finally, DNA, tumour-associated antigen and hepatitis B-associated antigens have been identified in the glomeruli (Glassock 1984) in occasional cases.

Aetiology

As shown in Table 1 many pathological conditions have been linked to membranous nephropathy. The frequency of so-called secondary membranous nephropathies in adults varies from 15 to 42 per cent (mean 22 per cent) (Row *et al.* 1975; Kingswood *et al.* 1984). In a recent study analysing 82 consecutive Caucasian adults with membranous nephropathy, the incidence of secondary forms was 21 per cent (Cahen *et al.* 1989). In our series, secondary membranous nephropathy accounted for 16 per cent of 244 patients. In children, about one-third of cases have associated conditions (Kleinknecht and Habib 1987). The principal causes of secondary membranous nephropathy appear to be drug therapy, malignancy, SLE, and hepatitis B. Moreover membranous nephropathy seems to be a part of the spectrum of autoimmune disease.

In many series drugs appear to be the cause of secondary membranous nephropathy, accounting for 6 to 9 per cent of all cases (Cahen *et al.* 1989) (see Chapter 2.1). In our series, five patients of 39 with secondary membranous nephropathy (13 per cent) had gold-induced disease. D-penicillamine, captopril, non-steroidal anti-inflammatory drugs may be related to membranous nephropathy (Schillinger *et al.* 1987).

Another common condition associated with membranous nephropathy in adulthood is malignancy (see Chapter 3.13). In the mid-1960s, Lee *et al.* (1966) reported 11 patients with underlying cancer associated with 101 patients with nephrotic syndrome; eight of the 11 had membranous nephropathy. In the intervening period, until 1986, more than 100 patients with nephrotic syndrome associated with carcinoma have been described, and about 60 per cent of these had membranous nephropathy (Ponticelli 1986). The prevalence of malignancy in patients with membranous nephropathy was reported to vary from 5 per cent (Cahen *et al.* 1989) to 11 per cent (Row *et al.* 1975). Most patients with neoplasia-associated membranous nephropathy were over 40 years of age. Seven patients in our series (3 per cent of all the 244 patients, 18 per cent of those with secondary forms) had malignancy-associated membranous nephropathy. The most common tumours associated with membranous nephropathy are bronchogenic lung cancers followed, in order of frequency, by tumours of the colon, rectum, kidney, breast and stomach (see Chapter 3.13).

The association of membranous nephropathy with infection has been clearly documented (Takekoshi *et al.* 1978) (see Chapter 3.12). Of these hepatitis B virus has been most studied. Hepatitis B virus-related antigens are involved in the development of glomerulonephritis, particularly membranous nephropathy in children. Many observations support the theory that the disease is caused by the formation or deposition of HBe antigen-anti HBe immune complexes (Takekoshi *et al.* 1979; Hattori *et al.* 1988; Milner *et al.* 1988). Circulating HB viral DNA and the delta antigen do not appear to be involved in immune complex formation. Other infections leading to membranous nephropathy (Table 1) include syphilis, both secondary and congenital (see Chapter 3.12), quartan malaria (see Chapters 3.12, 3.14) and infections from the tropics (leprosy, filariasis and schistosomiasis) (see Chapter 3.14). Syphilis-associated nephrotic syndrome is well documented but very uncommon, at least in temperate zones (Cahen *et al.* 1989). Identification of antitreponemal antibody and treponemal antigen in glomerular deposits has provided direct evidence for the immunopathogenesis of the glomerular lesion (Cahen *et al.* 1989). *Plasmodium malariae* has been suggested as a major cause of the nephrotic syndrome in children in Nigeria and Uganda (Seggie and Adu 1988). Several histological lesions, including membranous nephropathy, have been described in these patients (Seggie and Adu 1988). However, it is very difficult to assess the precise incidence of infection-associated membranous nephropathy in tropical areas since most reports concern selected groups of patients and, in many cases, present several infections concomitantly.

Membranous nephropathy accounts for 16 to 27 per cent of SLE (Baldwin *et al.* 1977; Austin *et al.* 1983), as discussed in Chapter 4.6.2. Patients with autoimmune thyroiditis may develop membranous nephropathy with the nephrotic syndrome (Weetman *et al.* 1981). Thyroglobulin and microsomal antigens have been identified in epimembranous deposits and in circulating immune complexes (Cahen *et al.* 1989). Membranous glomerulonephritis has also been reported in several patients with sarcoidosis (see Chapter 4.4), but a fortuitous association cannot be ruled out (Torres and Donadio 1988). The association of membranous nephropathy with diabetes mellitus is variably reported depending on the indications for renal biopsy (see Chapter 4.1). In a study of 150 cases with membranous nephropathy (Ehrenreich *et al.* 1974) coexistent diabetes mellitus was noted in 30 (20 per cent), while Cameron (1979) described only one diabetic patient in 100 cases of membranous nephropathy. Membranous glomerulonephritis can occur in a variety of dermatological diseases such as bullous pemphigoid and dermatitis herpetiformis, but there are few detailed descriptions (Torres and Donadio 1988). A case of membranous nephropathy with glomerular deposition of erythrocyte antigen-antibody complexes has been reported in a patient affected by Gardner-Diamond syndrome, a rare disorder characterized by recurrent crops of painful pruritic lesions (Torres and Donadio 1988).

Sometimes membranous nephropathy is the first manifestation of an underlying disease, preceding any other evidence of the systemic process by months or even years. A delayed diagnosis occurs most commonly in SLE and neoplasia (Lee *et al.* 1966; Row *et al.* 1975; Adu *et al.* 1983; Ponticelli 1986).

de novo Membranous nephropathy in allografted kidneys

Finally, *de novo* membranous nephropathy (see Chapter 11.6) represents the most commonly reported *de novo* glomerulonephritis in transplanted kidneys with an incidence of 1–2 per cent (Matthew 1988). The time to clinical evidence of *de novo* membranous nephropathy ranges from 3 months to 6 years, which is longer than the time to clinical recurrence (1 week to 3.8 years) (Cameron 1982). The outstanding clinical finding is proteinuria, usually of nephrotic dimensions, which is persistent in the majority of patients (Matthew 1988). The outcome of *de novo* membranous nephropathy is not well defined because it is difficult to be sure whether rejection or *de novo* glomerulonephritis is the cause of the failure. Some authors (Berger *et al.* 1983) have described a long and stable course while others (Dische *et al.* 1981) have reported a decline to renal failure 2/26 months after the detection of proteinuria secondary to glomerulonephritis. There are no pathological features to distinguish *de novo* membranous nephropathy in native kidneys or its post-transplantation recurrence. The association of membranous change with evidence of rejection is frequent (Matthew 1988).

Pathogenesis

Although in most patients with membranous nephropathy the aetiology remains unknown and they are thus classified as having 'idiopathic' disease, membranous nephropathy appears to be a disease arising from a combination of environmental stimuli and genetic factors. A strong association between idiopathic membranous nephropathy and the major histocompatibility complex antigens has been described above. Susceptibility to the autoimmune form of membranous nephropathy in rats (Heymann nephritis) is also linked to the major histocompatibility complex antigens.

Equally, the pathogenesis of membranous nephropathy in man remains unknown, although the ability to produce virtually identical glomerular lesions in animal models has provided considerable insight into the mechanisms which presumably underlie this disease. In experimental animals, as discussed in Chapter 3.1, two general immunologic mechanisms responsible for the formation of diffuse subepithelial deposits have been described.

The first mechanism involves the electrostatic interaction of a cationic protein with the anionic network of the glomerular capillary wall (Border 1985). Any component of the immune complex, such as antigen, antibody, or immune complex that possesses a positive charge, is capable of direct glomerular binding. Subepithelial immune complex formation can in fact occur in serum sickness with cationic bovine serum albumin (Border 1985, 1988) or with the administration of antibody-directed against GBM heparan sulphate-proteoglycan into a presensitized rat (Makino *et al.* 1988).

The second mechanism occurs in Heymann nephritis, in which a circulating antibody reacts with the structural antigen of the epithelial cell wall. Heymann nephritis in rats accurately simulates the clinical features and pathological characteristics of the human disorder. Active Heymann nephritis is induced by immunizing rodents with an antigen derived from the luminal brush border of tubular epithelial cells (Fx 1A) (Couser 1988). The *in situ* formation of the subepithelial immune deposits is suggested by the absence of detectable circulating immune complexes (Cattran and Chodirker 1982) and by the rapid production of an identical lesion after the administration of heterologous antibody to Fx 1a (passive Heymann nephritis). In the passive model, recent studies have established that glomerular immune deposit formation in a subepithelial distribution results from the binding of an antibody to an antigen system expressed in coated pits on the glomerular epithelial cell membrane as well as on the proximal tubular brush border (gp 330, gp 108 etc.) (Couser 1988). The reaction of antibodies with the antigen system is followed by complement activation and then by patching and capping of immune aggregates on the cell membrane, with subsequent shedding from the cell membrane into the GBM to produce a discontinuous granular subepithelial immune complex (Andres *et al.* 1986).

The immunohistochemical pattern of membranous nephropathy and the presence of several serologic abnormalities together suggest that humoral immunologic mechanisms, similar to these described in experimental models, are involved in the pathogenesis of the human disease. On the basis of our present knowledge, there are indications that human disease may arise from all three of these main pathogenetic pathways (Glassock 1984, Glassock 1988).

Glomerular subepithelial deposition of a circulating immune complex composed of non-glomerular, exogenous antigen, and endogenous low-affinity antibody

This mechanism can be implicated in secondary membranous nephropathy, such as the one related to hepatitis B, quartan malaria, neoplastic antigens, and in patients who develop this lesion as a form of lupus nephritis (Friend and Michael 1978). However, there is little evidence of a direct role for circulating immune complexes in idiopathic membranous nephropathy. Their presence in 10 to 25 per cent of patients with the idiopathic disease is probably a marker of an impairment in immune control (Mallick *et al.* 1983), as reported in association with the B8-DR3 haplotype.

Glomerular subepithelial deposition of circulating autoantibody reacting with an intrinsic, native glomerular antigen (similar to Heymann nephritis)

Although an autologous, renal-derived antigen has been found in only a few cases (Niles *et al.* 1987), this mechanism may be suggested during polyclonal B cell activation (Goldman *et al.* 1988) as happens during captopril or penicillamine administration, or associated with autoimmune diseases (Mallick *et al.* 1983).

Many autoimmune diseases, including Sjögren's syndrome, type I diabetes mellitus, Graves's disease, and gluten enteropathy, are frequently associated with the DR3-B8 haplotype, as happens in idiopathic membranous nephropathy, suggesting that unknown genes, closely related to HLA DR, may be associated to alterations in the B-cell and T-cell network response. In all the conditions listed above, membranous nephropathy has also been reported (Mallick *et al.* 1983, Shoenfeld and Schwartz 1984).

Glomerular subepithelial deposition of circulating antibody to extrinsic antigen 'planted' in the subepithelial space by virtue of a biochemical or electrostatic affinity

Numerous endogenous proteins are present (e.g. cationic proteins of human neutrophils) which are capable of direct glomerular binding as well as being immunogenic (Mallick *et al.* 1983; Gauthier *et al.* 1984). Inherited properties of the glomerulus may determine an individual's susceptibility to this nephropathy (Chandrachud and Boulton-Jones 1988).

In experimental models, whatever pathways are involved, the subepithelial immune deposit formation is associated both with a reduction in the glomerular ultrafiltration coefficient and to the onset of abnormal proteinuria (Gabbai *et al.* 1987). Because of the lack of cell infiltrates, it is unlikely that a cellular mechanism is involved in the production of the permeability defect (Glassock 1988). The terminal complement complex appears to be directly involved in mediating tissue damage and causing proteinuria (Salant *et al.* 1980). C5b-9 appears to be essential for the subepithelial deposits to induce proteinuria in experimental animals, because it can be abrogated completely by manoeuvres which prevent the formation of the C5b-9 membrane attack complex of the complement system by the immune deposits (Salant *et al.* 1980; Gabbai *et al.* 1988). Although the precise mechanism by which the C5b-9 membrane attack complex induces proteinuria remains unknown, it is possible to postulate an altered syn-

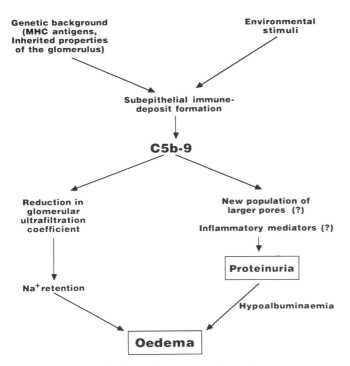

Fig. 8 Pathogenesis of membranous nephropathy.

thesis of the normal component of the filtration barrier causing the development of a population of larger pores.

Another mechanism may involve a modification of glomerular epithelial cell metabolism induced by the production of inflammatory mediators by the epithelial cells (reactive oxygen species, interleukin 1, prostaglandins) (Adler *et al.* 1986; Couser 1988). Activation of the terminal complement pathway and the formation of membrane attack complexes have also been documented in human membranous nephropathy (Falk *et al.* 1983). Moreover, the clearance of inulin and the fractional clearance of uncharged dextrans of a radius of 2.8 to 4.8 nm were significantly below control values in 20 patients with idiopathic membranous nephropathy studied by Shemesh *et al.* (1986), documenting a significant reduction in the glomerular ultrafiltration coefficient during the disease. In contrast, the fractional clearance of dextrans of radius 5 nm was markedly elevated in idiopathic membranous nephropathy patients compared to controls, suggesting an increment in the fraction of filtrate volume permeating the non-discriminatory shunt pathway (new population of large pores).

The reduction of glomerular hydraulic permeability which coincides with the onset of proteinuria, probably reflecting an altered epithelial cell surface area due to foot process effacement (Gabbai *et al.* 1987), may explain the sodium retention and the development of generalized oedema. Figure 8 summarizes all the mechanisms involved in the pathogenesis of membranous nephropathy including the recent theories of oedema formation (Dorhout Mees *et al.* 1979; Brown *et al.* 1982, 1984; Schrier 1988) (see Chapter 3.2).

Natural history

There are remarkably few reports (Noel *et al.* 1979; Davison *et al.* 1984, MacTier *et al.* 1986; Donadio *et al.* 1988) on the natural history of untreated idiopathic membranous nephropathy, and there is still controversy concerning its long-term prognosis, while the outcome of this nephropathy is often deduced from untreated control populations of retrospective or prospective studies on treatment. In general, the natural history of idiopathic membranous nephropathy is considered as rather indolent, chronic, and slowly progressive (Cameron 1979; Ramzy *et al.* 1981). It can also be seen as dichotomous, in that some patients maintain normal renal function and even achieve a spontaneous remission of proteinuria, while others progress to terminal renal failure, or die following complications related to the persisting nephrotic syndrome.

In the uncontrolled studies (Table 3), the proportion of untreated patients with a poor prognosis has varied. Noel *et al.* (1979) followed 116 untreated patients, 88 of whom had nephrotic syndrome, for an average period of 54 months. At the end of the study 19 per cent of the patients had renal insufficiency with end-stage renal failure in 9.5 per cent. Honkanen *et al.* (1986) reported that four of the 26 (15 per cent) untreated patients progressed to renal insufficiency after a mean period of 6.7 years. A less favourable outcome has been observed in the other series reported in Table 3, with a rate of renal failure ranging from 23 to 71 per cent. An Italian multicentre retrospective study analysed the natural history of 38 untreated patients with idiopathic membranous nephropathy (Gruppo di Immunologia Renale 1983). After an average observation period of 7 years, 34 per cent had progressive renal failure. In the Ehrenreich series (Ehrenreich *et al.* 1976), 21 of 44 (48 per cent) untreated patients presented a poor renal outcome. More recently, Davison *et al.* (1984) reported that 43 per cent of 64 patients, who had never received corticosteroids or cytotoxic agents, reached a serum creatinine of at least 400 μmol/l (4.5 mg/dl) within 32 months and a further 8 per cent showed a deterioration in renal function during the same period. Furthermore, Franklin *et al.* (1973), found that 10 of their 32 patients (31 per cent) developed uraemia in an average of 2.7 years, while Gluck *et al.* (1973) reported that 24 of 38 patients (63 per cent) progressed to renal failure. In our own experience (Zucchelli *et al.* 1987), out of 49 untreated patients with nephrotic syndrome and followed for at least 10 years, 22 died or underwent regular dialysis, while 20 presented persistent renal disease.

The survival curves of untreated patients reported in the literature also differ considerably in different series, probably reflecting the heterogeneity of the sample population (Zucchelli *et al.* 1987). The 10–year survivals reported have ranged from as high as 90 per cent in the series of Kida (1986) in a Japanese population, in which nephrotic patients formed only 59 per cent, down to 52 per cent in our own series (Zucchelli *et al.* 1987) in which all the patients studied had a nephrotic syndrome. In two recent studies by Honkanen *et al.* (1986) and Donadio *et al.* (1988), the 10-year survival was 83 and 75 per cent respectively, with no differences between treated and untreated patients.

The series studied by Noel *et al.* (1979) and Murphy *et al.* (1988), for a similar period (4.6 and 4.3 years, respectively) both presented similar characteristics (an unusually high number of women (48 and 45 per cent, respectively) and of patients with asymptomatic proteinuria (25 and 46 per cent, respectively)); both of these are associated with a good prognosis (see below). Nevertheless, the survival rates diverged greatly at 15 years (Fig. 9), even though they had been virtually the same at 10 years (approximately 76 per cent).

It is well known that in the course of idiopathic membranous

Table 3 Uncontrolled studies: outcome of untreated patients with idiopathic membranous nephropathy

Authors	No. of patients	Mean follow-up (years)	Complete remission (%)	Renal failure (%)
Erwin et al. (1973)	20	4.7	1 (5)	9 (45)
Ehrenreich et al. (1976)	44	6.5	0	21 (48)
Pierides et al. (1977)	17	4.4	3 (18)	4 (23)
Noel et al. (1979)	116	4.6	27 (23)	22 (19)
Ramzy et al. (1981)	6	14	3 (50)	2 (33)
Hopper et al. (1981)	21	7.8	3 (14)	15 (71)
Gruppo di Immunologia Renale (1983)	38	7.0	15 (39)	13 (34)
Davison et al. (1984)	64	8.5	–	32 (50)
Honkanen et al. (1986)	26	6.7	8 (31)	4 (15)
MacTier et al. (1986)	54	5.3	8 (15)	15 (28)
Zucchelli et al. (1987)	49	10.6	7 (14)	22 (45)
Totals	455	*	75 (19)	159 (35)

* Average period ranging from 4.4 to 14 years.

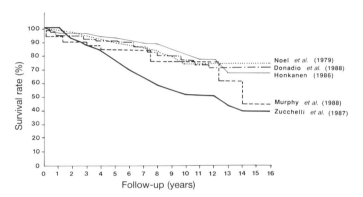

Fig. 9 Survival rates of untreated patients with idiopathic membranous nephropathy.

nephropathy there is a possibility of improvement in the nephrotic syndrome, and there may even be a spontaneous complete disappearance of proteinuria. Reviewing several uncontrolled studies (Table 3), we found that at the end of periods of varying lengths, 75 of 391 patients (19 per cent, excluding the Davison series which did not report the incidence of complete remission) who had never been previously treated with corticosteroids and/or cytotoxic agents spontaneously presented a complete disappearance of proteinuria. In greater detail, Ehrenreich et al. (1976) observed no such cases in 44 untreated patients, while in the series of Erwin (1973) proteinuria disappeared in only one of 20 cases (5 per cent). At the other extreme, such studies as the Italian Multicentre Retrospective Study (Gruppo di Immunologia Renale 1983) or the series of Noel (1979) reported an incidence of complete remission in 39 and 23 per cent of their patients respectively. In one of our recent studies (Zucchelli et al. 1987), 14 per cent of the patients reached complete remission of proteinuria; the data are similar to those described by Pierides et al. (1977), Hopper et al. (1981) and MacTier et al. (1986). Analysis of the control populations of the prospective, randomized trials (see Table 7) compared with untreated patients of the uncontrolled studies, suggests that renal failure occurs less frequently in the former groups (20 vs. 35 per cent) while the incidence of complete remission is virtually the same (18 vs. 19 per cent).

It seems clear that about 20 per cent of idiopathic membranous nephropathy patients have a chance of achieving a complete remission of proteinuria during a period of several months or a few years, while progression to chronic renal failure is highly variable. The percentage of renal insufficiency increases over the study period which is longer in most of the uncontrolled compared to the controlled studies. It must be stressed, however, that the effects of therapy on the progression to renal failure are generally harder to assess than the different rates of complete remission of proteinuria. Given this background of extraordinary clinical variability, it is not surprising that most of the literature has been devoted to analysing the factors which appear to affect the natural course of idiopathic membranous nephropathy.

Factors which appear to affect prognosis of idiopathic membranous nephropathy

Age

The outcome of idiopathic membranous nephropathy is generally better in children than in adults. Olbing et al. (1973) and Trainin et al. (1976) found no cases of renal insufficiency in children with idiopathic membranous nephropathy under the age of 10. Habib and coworkers (1973) reported an actuarial renal survival rate of 90 per cent after 10 years of observation. A poorer prognosis in adults than children with idiopathic membranous nephropathy was also found by Pierides et al. (1977) and by Row et al. (1975). Davison et al. (1984) reported that approximately 50 per cent of the untreated patients between 30 and 60 years of age developed renal failure, whereas it was less frequent in subjects below 30 years of age using univariate analysis. By contrast, Tu et al. (1984) in a multivariate analysis did not find that age was an independent predictor of terminal renal failure in adults with idiopathic membranous nephropathy. In our own retrospective analysis (Zucchelli et al. 1987), untreated patients with spontaneous remission were on the whole younger than non-treated patients who developed chronic renal failure and renal death (36.7 vs. 48.1 years), although the difference did not reach significance at a 1:20 level.

Sex

Hopper et al. (1981) reported that a deterioration in renal function was significantly more common in males than females. After a mean period of 99 months, renal function had deteriorated in just 17 per cent of women compared with 55 per cent of men. The women in this study fared better than the men whether or not they had been given prednisone. These data, confirmed by some reports (Davison et al. 1984, Donadio et al. 1988), have not been noted in all series (Cameron 1979). In our own experience of 82 idiopathic membranous nephropathy patients followed for at least 10 years, (Zucchelli et al. 1987) sex did not appear to influence the outcome. Over the whole series complete remission was seen in seven of the 24 women, and in 13 of the 58 men.

Nephrotic syndrome

There is widespread agreement that patients with idiopathic membranous nephropathy and nephrotic proteinuria are more likely to progress to uraemia than patients with less severe proteinuria (Erwin et al. 1973; Mallick et al. 1983; Davison et al. 1984). Indeed, patients with non-nephrotic proteinuria appear to have a significantly more benign course and rarely progress to renal failure (Row et al. 1975; Pierides et al. 1977). Davison et al. (1984) reported that in 50 per cent of untreated patients with nephrotic syndrome at presentation, renal function deteriorated within 5 years. Reviewing some larger series, Mallick and co-workers found that after a mean period of 92 months, only 7.7 per cent of patients with idiopathic membranous nephropathy and non-nephrotic proteinuria developed end-stage renal failure, and 30.8 per cent reached complete remission. Donadio et al. (1988) found a good association between progression to end-stage renal disease and nephrotic syndrome, but only with an average urine protein excretion of 10 g or more per 24 h. In contrast to all the previous studies, Murphy et al. (1988) have more recently reported from Australia that the survival analysis of initially nephrotic patients was not significantly different from the survival of the total patient population.

Stages of glomerular lesion

It is widely accepted that stages I to stage IV do not represent a progressively worse manifestation of the disease, although stages I to III may represent a histological evolution (Cameron 1979). Some investigators have found that patients whose renal biopsy disclosed stages I or II are less likely to progress to uraemia than those with stages III to IV (Ehrenreich et al. 1976, Noel et al. 1979). Others have however found no correlation between staging and eventual outcome (Mallick et al. 1983, Donadio et al. 1989). Our own observations (Zucchelli et al. 1986) have shown that a relationship exists between histological lesions and clinical course. We studied 25 patients with idiopathic membranous nephropathy and nephrotic syndrome who had participated in the Italian Randomized Trial on a 6 month steroid and chlorambucil schedule (Ponticelli et al. 1984), who underwent repeated renal biopsy after a mean interval of 41 months. Extensive normalization of the basement membrane (stage V) was observed in six (five in the treated group and one in the untreated group) of the nine patients with complete and sustained remission. Moreover, only patients in stages I or II at admission seemed to reach reparation stage V, while a persistence of nephrotic syndrome or partial remission was usually associated with the progression of the capillary wall lesion to stages III or IV.

Genetic and immunological factors

Some HLA antigens seem to influence the long-term course of the disease. A better prognosis has recently been recorded in a Japanese population even though they presented with more severe disease as shown histologically than Caucasian patients (Abe et al. 1986; Kida et al. 1986). this has been at least partially attributed to the strong association of the disease with HLA DR2. We have recently shown in Italian patients who participated at our randomized multicentre trial (Zucchelli et al. 1988) that those with HLA DR3+-B8− were more likely to have a complete remission than those with HLA DR3+B8+ (Table 4). Similar results have recently been reported by Papiha et al. (1987). Among 55 patients with idiopathic membranous nephropathy they found a subset of six individuals carrying DR3, B8 and BfSS antigens who had a significantly worse clinical course compared to the remainder of the patients. On the other hand,

Table 4 (a) Phenotype frequency of Class I and II antigens in patients with idiopathic membranous nephropathy

	No.	Class I		Class II	
		A1	B8	DR2	DR3
Patients	55	16 (29.1%)	18 (32.7%)	6 (10.9%)	30 (54.5%)
Controls	425	92 (21.6%)	43 (10.1%)	66 (16%)	100 (23.6%)
p		NS	0.002	NS	<0.01
Relative risk		1.48	4.32	0.60	3.7

Distribution of DR3 and B8 antigens in 28 treated and 24 untreated patients with idiopathic membranous nephropathy

	Treated group		Untreated group		Total
	Remission (complete or partial)	Unchanged or worsened	Remission (complete or partial)	Unchanged or worsened	
DR3+B8+	5	2	1	8	16
DR3+B8−	7	3	3	1	14
DR3−B8+	1	−	−	1	2
DR3−B8−	8	2	4	6	20
Totals	21	7	8	16	52

association with HLA antigens and clinical outcome has not been observed in the patients of the British MRC Trial (Cameron et al. 1990).

Finally, a higher helper–inducer/suppressor–cytotoxic T-cell ratio before therapy may be a good prognostic index. In fact, this ratio seems to identify a subpopulation of patients who more frequently respond to treatment (Zucchelli et al. 1988).

Two other predictors of the outcome have been studied: renal function and interstitial lesions. In general, there is fair agreement that impaired renal function at the time of renal biopsy is a predictor of terminal renal failure during the period (Davison et al. 1984; Tu et al. 1984). Moreover, the severity of the interstitial lesions at renal biopsy (Ramzy et al. 1981; Ponticelli et al. 1989) seems to be correlated with the subsequent development of renal insufficiency.

As a general conclusion, it should be stressed that the adequate assessment of the evolution of this slowly progressive disease requires a long follow-up period, certainly more than the 5 years or so which is typical of most of the studies in the literature.

Treatment

Before considering the treatment of idiopathic membranous nephropathy we shall briefly sum up the therapy of secondary forms. Since several underlying conditions may be associated with membranous nephropathy, as previously described, in the presence of a nephrotic syndrome due to membranous nephropathy, it is important to search for the most common aetiological factors such as medication, HBV infection, and malignancy. In our opinion taking a careful history focused on a long exposure to the drugs listed in Table 1 can achieve a correct diagnosis. The study of hepatitis B markers must be included in the routine evaluation of every patient affected by membranous nephropathy. The search for an underlying cancer presents, on the other hand, some difficulties since tumours may appear later in the follow-up period. However, by recalling the types of neoplasia most frequently associated with membranous nephropathy such as lung, kidney and colorectal carcinomas, the clinician can make a correct diagnosis by a careful history and physical examination and by radiological or endoscopic studies. In fact, lung cancer is usually evident on routine chest radiography while the presence of a renal tumour is generally identified by intravenous pyelography and/or kidney ultrasonography, part of the routine evaluation of a nephrotic patient. A patient with membranous nephropathy who is over the age of 50, and who complains of changes in bowel habits, or changes in upper gastrointestinal function should be evaluated with radiography and/or endoscopic study of the stomach and of the bowel together with repeated tests of the stool for occult bleeding. Unless other clinical findings suggest a specific malignancy, we do not think that the mere presence of membranous nephropathy warrants a more extensive evaluation.

Once a causative agent has been identified (Table 1), therapeutic intervention is directed to treating the primary cause. In membranous nephropathy secondary to infection, the nephropathy may be resolved with antimicrobial therapy. For instance, renal disease often improves with penicillin in patients affected by syphilis, although spontaneous remission has even been noted before the institution of such treatment (Levy 1988). Antimalarial treatment is often associated with a remission of the nephrotic syndrome, yet some patients will develop progressive renal deterioration despite the apparent removal of the antigen (Seggie and Adu 1988). Although the prognosis of HBV-related membranous nephropathy is considered good and spontaneous remissions have been described, many patients have been treated with steroids and/or immunosuppressive drugs (Levy 1988). Ito et al. (1981) suggested that the condition of the patients may remit as they seroconvert from HBeAg to anti-HBe, and that steroids may suppress the tumoural antibody response retarding the seroconversion to anti-HBe. However, in the study reported by Wiggenlinkhuizen et al. (1983), remission of the renal disease did not correlate with seroconversion to HBe.

Generally the treatment of neoplasia associated with membranous nephropathy by the resection of solid tumour or chemotherapy in haematological malignancies, may allow a complete or partial remission to be achieved (Cahen et al. 1989).

Drug-induced membranous nephropathy is usually resolved when treatment is stopped. There is no doubt that a specific association between gold treatment and membranous nephropathy exists, given that the lesion develops during the therapy and resolves when the treatment is withdrawn (Hall 1988). Large doses of corticosteroid have been used to treat gold nephropathy either alone or in combination with immunosuppressive drugs (Skrifvars et al. 1977). However, there is no reliable evidence to show that this treatment either shortens the duration of proteinuria or leads to a more rapid or more complete resolution of the renal lesion (Hall 1988) (see Chapter 4).

We now consider the management of idiopathic membranous nephropathy. Ideally, the treatment should be based on an understanding of its pathogenesis and a knowledge of drug kinetics. In fact however, although such aspects are becoming clearer, they remain largely undefined. As a consequence, the therapeutic schedules employed in the treatment mainly comprise corticosteroids and/or cytotoxic agents in an attempt to modify the abnormal immune regulation which is currently believed to be a likely cause of the glomerulopathy. These drugs do indeed seem to interfere with immune responses mediated by B and T cells and appear to be preferentially effective in suppressing hyperactivity of the immune system (Balow et al. 1987). Attempts to reduce proteinuria in patients with idiopathic membranous nephropathy by non-steroidal anti-inflammatory agents have to date had only limited success (Arisz et al. 1976).

In general it is very difficult to assess the role of a particular regimen in a disease which is characterized by such a variable natural course, especially in the long run. Moreover, in most reports the distribution of the main prognostic factors mentioned above is not even considered: in particular patients with nephrotic syndrome and asymptomatic proteinuria have often been grouped together. When these difficulties are combined with the heterogeneity of the treatment schedules used in terms of dose, duration, stop point and response definitions, it becomes apparent why the role of therapy in idiopathic membranous nephropathy is still a matter of controversy. We shall now briefly analyse the results of some major retrospective uncontrolled studies, and the few prospective controlled trials.

Uncontrolled studies

A number of uncontrolled trials showed no benefit from corticosteroids either in inducing remission or in preserving renal function, whereas other retrospective analyses reported an

3.7 Membranous nephropathy

Table 5 Uncontrolled studies: outcome of 406 treated patients with idiopathic membranous nephropathy

Study	No. of patients	Mean follow-up (years)	Complete remission (%)	Renal failure (%)	Therapy
Erwin et al. (1973)	21	4.7	1 (5)	6 (28)	Steroids
Ehrenreich et al. (1976)	59	6.5	17 (29)	9 (15)	Steroids, cytotoxic drugs
Pierides et al. (1977)	20	4.4	5 (25)	5 (25)	Steroids, cytotoxic drugs
Bolton et al. (1977)	24	3.4	8 (33)	2 (8)	Steroids
Hopper et al. (1981)	51	7.8	15 (29)	17 (33)	Steroids
Ramzy et al. (1981)	29	14.0	10 (34)	11 (38)	Steroids, cytotoxic drugs
Suki et al. (1981)	19	3.0	16 (84)	0	Steroids, cytotoxic drugs
Gruppo di Immunologie Renale (1983)	61	7.0	31 (51)	17 (28)	Steroids, cytotoxic drugs
Blainey et al. (1986)	57	–	6 (10)	28 (49)	Steroids
Honkanen et al. (1986)	32	6.7	12 (37)	8 (25)	Steroids, cytotoxic drugs
Zucchelli et al. (1987)	33	10.6	13 (39)	8 (24)	Steroids, cytotoxic drugs
Totals	406	*	134 (33)	111 (27)	

* Average period ranging from 3 to 14 years.
Percentages in parentheses.

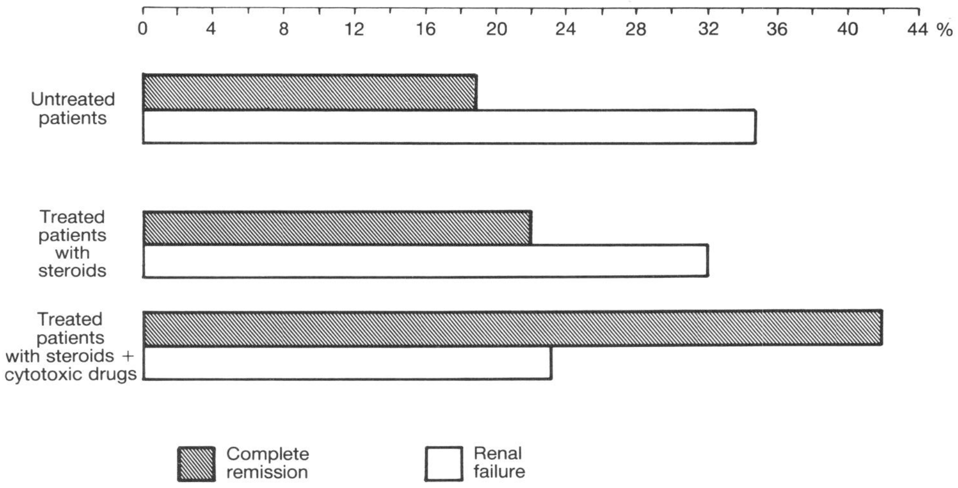

Fig. 10 The incidence of complete remission and renal failure of the untreated patients (Table 3) is compared with the percentage of therapeutic results in the series reported in Table 5.

impressively high rate of remission of nephrotic syndrome in patients treated with corticotherapy alone or associated with cytotoxic agents (Table 5). Erwin et al. (1973) compared the results from 21 patients treated by steroids with 20 patients who were historical controls. They concluded that the outcome in patients who had received therapy was essentially identical to that in untreated patients. A few years later, Pierides et al. (1977) reached the same conclusions in a series of 37 patients. More recently, Honkanen et al. (1986) have shown no significant differences between 32 treated patients and 26 control patients as regards renal function, the outcome of proteinuria, and survival rate.

In contrast, the benefits of therapy have been emphasized by Ehrenreich et al. (1976) who treated 59 patients with prednisone plus immunosuppressive agents. At the end of a mean period of 6.5 years, 17 patients (29 per cent) experienced complete remission and only nine (15 per cent) reached renal failure. Bolton and coworkers (1977) achieved eight cases of complete remission in 24 patients (33 per cent) treated with alternate-day prednisone from a mean period of 3 years. Hopper et al. (1981) observed complete remission in 15 (29 per cent) of 51 treated patients while 17 (33 per cent) developed renal failure. Suki et al. (1981) reported 19 patients treated with steroids plus cytotoxic agents. In 16 cases proteinuria decreased to less than 1 g/24 h, and normal renal function was maintained in all patients. In the Italian Multicentre Retrospective Study (Gruppo di Immunologia Renale 1983) 51 per cent of the treated patients achieved complete remission and 28 per cent went to renal insufficiency. In our own experience (Zucchelli et al. 1987), treatment increased the likelihood of obtaining complete remission (14 per cent of non-treated vs. 39 per cent of treated patients) and reduced the chances of reaching renal failure (45 per cent of non-treated vs. 24 per cent of treated patients).

Summarizing the available reports on the effects of therapy in idiopathic membranous nephropathy (Table 5), 134 of 406 (33 per cent) patients were in complete remission at the end of the follow-up period. This proportion is higher than that reported for untreated patients (19 per cent). If the influence of therapy on renal insufficiency is considered, it appears that treatment may prevent progression to renal failure in patients with idiopathic membranous nephropathy and normal or slightly impaired renal function (27 per cent of renal failure in treated vs. 35 per cent in untreated). These data become more evident if the patients treated with corticosteroids alone, and those treated with corticosteroids plus immunosuppressive agents are analysed separately, as shown in Fig. 10.

Table 6 Controlled prospective studies

Corticosteroids	No. of patients	Therapy	Conclusions
Medical Research Council (1970)	19	Prednisone 32 mg/24 h for 3 weeks, followed by 20 mg/24 h for at least 6 months	No difference between treated and untreated patients
Collaborative Study of the Adult Idiopathic Nephrotic Syndrome (1979)	72	Prednisone 100–150 mg on alternate days for 8 weeks, then gradually tapered off	More rapid deterioration of renal function in untreated patients
Multicentre Canadian Trial (Cattran et al. 1989)	158	Prednisone 45 mg/m^2 on alternate days for 6 months	No difference between treated and untreated patients
Medical Research Council Trial (Cameron et al. 1990)	103	Prednisone 100–150 mg on alternate days for 8 weeks, then gradually tapered off	No difference between treated and untreated patients
Cytotoxic agents			
Donadio et al. (1974)	22	Cyclophosphamide 1.5–2.5 mg/kg/24 h for 12 months	No differences between treated and placebo group
Lagrue et al. (1975)	38	Chlorambucil 0.2 mg/kg/24 h for 6 months, then 0.1 mg/kg/24 h for 6 months or azathioprine 3 mg/kg/24 h for 6 months	More remission of proteinuria in chlorambucil than in azathioprine or in placebo groups
Australian Multicentre Trial (Tiller et al. 1981)	54	Cyclophosphamide 1.5 mg/kg.24 h and warfarin and dipyridamole 400 mg for 36 months	Significantly less proteinuria in the treated group
Cytotoxic agents plus corticosteroids			
Medical Research Council Working Party (1971)	14	Prednisone 20 mg/24 h + azathioprine 2.5 mg/kg.24 h for 8 weeks	No difference between treated and untreated patients
Italian Collaborative Study (1984, 1989)	81	Six-month course of methylprednisolone (1 g intravenously) for 3 days followed by 0.4 mg/kg.24 h for 27 days alternated with chlorambucil (0.2 mg/kg.24 h) every other month	More remission and preservation of renal functions in treated patients

Controlled studies

Over the last few years several randomized prospective trials have been carried out in patients affected by idiopathic membranous nephropathy, as summarized in Table 6.

Prednisolone

Four of these studies have evaluated the role of corticosteroid treatment. The first controlled trial of prednisone was performed in the United Kingdom (Black et al. 1970), and a small but statistically insignificant benefit from treatment was observed. However, there were too few patients to support negative findings, and the doses of steroids were much lower than those used in later trials. The subsequent United States Collaborative Study (1979) compared the clinical course of 34 adult patients with nephrotic syndrome, given 100 to 150 mg of prednisone every other day for at least 8 weeks, with the course of 38 similar patients given a placebo. Patients were followed for a mean period of 23 months. By the end of this period, there was an equal number of patients in complete remission in both groups, but there was a significant effect on the development of renal insufficiency. Indeed, at the latest assessment 10 of 38 patients given placebo were in renal failure compared with only one of the treated patients. This difference persisted for a further 5 years of observation (Glassock, R.J. personal communication).

In complete contrast, the recent Canadian Study (Cattran et al. 1989) reported no significant therapeutic effects of a 6-month course of prednisone (45 mg/m^2 every other day) in 81 patients, compared with 77 untreated patients. After a mean observation time of 48 months, there were no significant differences between the two groups in the rate of decline of renal function, or in the numbers of patients with complete or partial remission of proteinuria. It is difficult to reconcile the results of the last two studies, since the two untreated control groups differed widely in terms of disease severity. In fact, the annual median decline in renal function differed dramatically upon comparing the Canadian Study (-2.4 per cent) and the United States Collaborative Study (-10 per cent). It should also be underlined that in the Canadian Study the prednisone group on average had a higher serum creatinine and urinary protein excretion, but lower serum albumin levels than the control group. Moreover, the number of subjects lost to the study were significantly higher in the control group than in the prednisone group (17 of 77 control vs. 10 of 81 prednisone patients). In the American study, the median known duration of disease before randomization was fractionally less then 6 months, while in the Toronto Study it was between 15 and 17 months.

The situation is complicated by the results of the British MRC Trial (Cameron et al. 1990) in which 52 patients with idiopathic membranous nephropathy and nephrotic syndrome were randomized to the treatment group and 51 to the control group.

Although the treatment regimen was identical to the United States Collaborative Study and the analysis was similar, the British study showed no differences between treated and untreated patients after a mean follow-up of 4.1 years. Cameron et al. (1990) tried to provide an explanation for the conflicting results of the two studies by analysing the different characteristics of the patients in the United States and in the United Kingdom. In the United States, the proportion of adult nephrotics with idiopathic membranous nephropathy is much higher than in the United Kingdom (64 versus 12 per cent) and in the United Kingdom, as in the rest of Europe, there is a very strong association with the HLA haplotype B8-DR3 which has not been found in the United States.

Cytotoxic drugs

Three further controlled studies have analysed the effects of cytotoxic drugs alone in idiopathic membranous nephropathy. In 1974 Donadio et al. randomly allocated 22 adults to receive either no drugs or cyclophosphamide at a mean dose of 1.8 mg/kg.24 h for 1 year. After treatment, a downward trend in proteinuria was noted while renal function remained stable. There were no significant differences between the two groups for either parameter. Of 38 patients studied by Lagrue et al. (1975), after a mean study period of at least 2 years, nine subjects taking chlorambucil and two taking placebo were in complete remission while none of the azathioprine-treated patients responded. In the Australian Multicentre Trial (Tiller et al. 1981) 54 patients with idiopathic membranous nephropathy were randomly allocated to receive supportive treatment, or a combination of cyclophosphamide with warfarin and dipyridamole. After 36 months, the treated patients had less proteinuria and higher levels of serum albumin than the controls, whereas the levels of serum creatinine in the two groups did not differ significantly.

Lastly, there are two randomized prospective trials in which cytotoxic agents were used in combination with steroids. In the 1971 study (Medical Research Council Working Party, 1971), only 14 patients with idiopathic membranous nephropathy entered a trial comparing prednisone plus azathioprine with an untreated control group. There were no significant differences between the two groups, but owing to the small samples used the results appear to be inconclusive.

The Italian Collaborative Study (Ponticelli et al. 1984, 1989) was started in 1976 to assess the effects of a therapy consisting of methylprednisolone and chlorambucil, each given every other month for 6 months. Eighty-one patients with proteinuria more than 3.5 g/day and biopsy-proven membranous nephropathy were randomly assigned to receive either supportive therapy alone or a specific treatment. The following schedule was chosen for patients allocated to receive specific therapy: 1 g of methylprednisolone given intravenously over 20 to 30 min on 3 consecutive days, followed by oral prednisone (0.5 mg/kg.24 h) or methylprednisolone (0.4 mg/kg.24 h) given in a single morning dose for 27 days. At the end of the first month, steroid treatment was stopped and the patients were given chlorambucil at a dosage of 0.2 mg/kg.24 h for 1 month. Steroid administration was then repeated with the same schedule followed by chlorambucil, then methylprednisolone again and chlorambucil, each one for a month at a time so that the period of treatment lasted 6 months in total.

At randomization, the treatment and control groups were similar with regard to age, sex, duration of disease before admission, plasma creatinine levels, urinary protein excretion, hypertension, histological stages of glomerular lesions, and the presence of vascular or tubulointerstitial lesions. The median duration of observation in both groups was 5 years. During the entire follow-up period, seven of the 39 control patients and 24 of the 42 treated patients had a complete remission of proteinuria (urinary protein excretion less than 0.2 g/24 h). In particular, the efficacy of the treatment for proteinuria was assessed after 2 and 5 years, only taking into account the patients who had actually been observed for those periods. At 2 years 17 of 42 treated patients were in complete remission, compared with none of the 39 control patients. Likewise at 5 years the number of patients in complete remission was higher in the treated group (14 out of 30) than in the control group (four out of 25). With reference to renal function, 19 of the 39 controls had an increase in plasma creatinine levels up to at least 50 per cent over the basal values (13 doubled their plasma creatinine levels, and four had to undergo regular dialysis) while among the 42 treated patients, only four had deteriorated at the last examination (two increased plasma creatinine levels up to at least 50 per cent over the basal values, one had a doubling of plasma creatinine while the last has been undergoing regular dialysis).

To summarize the results from the controlled studies (Table 7), we believe that treatment may increase the chances of achieving complete remission (17 per cent in untreated patients vs. 26 per cent in treated patients) and it may prevent progression to renal failure (22 per cent in untreated patients vs. 8 per cent in treated patients). It should be emphasized that complete and sustained remission of proteinuria and stable renal function may occur with no histological improvement (Mallick et al. 1983), but in some patients remission is associated with the histological appearance of reparative stage V (Zucchelli et al. 1986).

Delayed treatment in patients with deteriorating renal function

The 'controlled' study by West et al. (1987) deserves to be mentioned first, although it is based on a retrospective comparison between matched historical controls and a treated group. These workers analysed 26 patients affected by idiopathic membranous nephropathy with serum creatinine concentrations exceeding 135 mmol (1.4 mg/dl) and 24-h urine protein excretion of at least 3.5 g/24 h. Nine patients received cyclophosphamide (mean 1.5 mg/kg.24 h) for 23 months, associated with prednisone in six cases. These nine patients were compared with 17 concurrent controls. At most recent observation, serum creatinine values exceeded 400 μmol/l (4.5 mg/dl) in eight controls but in none of the treated patients. The mean serum creatinine level was significantly lower in the treated group than in the controls. Moreover, there were four complete remissions among the treated patients, while none of the controls presented a disappearance of proteinuria.

Other uncontrolled reports have emphasized the favourable effects of therapy in patients affected by idiopathic membranous nephropathy whose renal function had been deteriorating. Short et al. (1987) have reported that high-dose prednisolone may be beneficial in some of these patients, but the effect was variable and in most cases remissions were not sustained. Williams and Bone (1989) reported that pulse methylprednisolone therapy (1 g × 3 days) followed by oral prednisolone (30 mg/daily) and azathioprine (50 mg/daily) reversed the deterioration in renal function. In fact, this treatment in 10 patients with severe nephrotic syndrome, who had developed progressive renal failure,

Table 7 Controlled studies: outcome of patients with idiopathic membranous nephropathy

	Mean follow-up (years)	Treated				Untreated		
		Therapy	No. of patients	Complete remission (%)	Renal failure (%)	No. of patients	Complete remission (%)	Renal failure (%)
Donadio et al. (1974)	1.0	Cytotoxic agents	11	0	1 (9)	11	0	2 (18)
Lagrue et al. (1975)	2.0	Cytotoxic agents	24	9 (37)	0	14	2 (14)	0
United States Collaborative Study (1979)	1.9	Steroids	34	4 (12)	1 (3)	38	4 (11)	10 (26)
Multicentre Canadian Trial (1989) (Cattran et al. 1989)	4.0	Steroids	81	16 (20)	3 (4)	77	19 (25)	4 (5)
Italian Collaborative Study (1989) (Ponticelli et al. 1989)	5.0	Steroids, cytotoxic agents	42	24 (57)	4 (9)	39	7 (18)	19 (49)
MRC Trial (1989) (Cameron et al. 1990)	4.1	Steroids	52	10 (19)	11 (21)	51	7 (14)	15 (29)
Totals	*	–	244	63 (26)	20 (8)	230	39 (17)	50 (22)

* Average period ranging from 1 to 5 years.
Percentages in parentheses.

obtained a significant improvement in renal function and renal stabilization in a further three patients.

Finally, Mathieson et al. (1988) treated eight patients with nephrotic syndrome due to idiopathic membranous nephropathy and declining renal function with six alternating monthly cycles of prednisolone and chlorambucil, as proposed by the Italian Collaborative Study (Ponticelli et al. 1984). Proteinuria was reduced in all eight from a mean of 15.3 g/24 h at the start of treatment to 2 g/24 h at follow-up (mean 8.4 months). Creatinine clearance increased in six and the rate of decline was reduced in another two patients (group mean 51.6 ml/min. at the start of treatment and 81.4 ml/min. at follow-up).

Side-effects of treatment

The possible deleterious consequences of the use of steroids and cytotoxic drugs are dealt with in Section 2. However, in the literature on treatment of idiopathic membranous nephropathy, symptoms or signs of steroid toxicity have been reported in only a few patients. In particular, the investigators of the United States Collaborative Study (1979) were impressed by the near-total absence of side-effects with steroid treatment. In very few cases the development of peptic ulcer, which responded to H_2 inhibitors, was observed during the treatment (Collaborative Study 1979; Ponticelli et al. 1984, 1989). Patients sometimes suffered from tremor, and anxiety and had a Cushingoid appearance. In all cases these side-effects were resolved within a few weeks after the steroids had been tapered off or stopped.

The myelosuppressive action, cancerogenic effect and infertility induced by cytotoxic agents have been thoroughly described (Guesry et al. 1978; Chaplin 1982). West et al. (1987), utilizing cyclophosphamide at a mean dosage of 1.5 mg/kg.day for a mean of 23±4 months, did not report any serious toxicity. In our opinion, however, caution is required because of the possible risk of sterilizing young male patients. In the Italian Collaborative Study (Ponticelli et al. 1988) chlorambucil was given on alternate months with steroids at a cumulative dose ranging between 0.9 and 1.4 g. The oncogenic risk of this drug has been reported with a cumulative dose ranging between 7 and 8 g (Chaplin 1982). More recently, however, Mathieson et al. (1988) utilizing the Italian Collaborative schedule (Ponticelli et al. 1984) in eight patients with deteriorating renal function, observed far more side-effects and reported that only one patient was able to tolerate the full dose of chlorambucil throughout.

Conclusions

Idiopathic membranous nephropathy is the most frequent cause of nephrotic syndrome in adults. Its pathogenesis is still unclear, even if it appears to be related to immunological abnormalities induced by a genetic predisposition, and by environmental stimuli.

The natural history of idiopathic membranous nephropathy is likely to be influenced by numerous clinical variables. Upon reviewing the copious literature devoted to the condition, it seems clear that 20 per cent of patients have a chance of achieving a complete remission of proteinuria. Although the evolution to end-stage renal failure varies greatly from series to series, the disease seems to be responsible for approximately 8 per cent of cases of end-stage renal disease due to primary glomerulopathy (Jacobs et al. 1977). Unfortunately, there are as yet no clearly-defined criteria to select cases with an unfavourable outcome (Garattini et al. 1987), even though males and those with severe nephrotic syndromes forms the majority of this subgroup.

Until these new criteria are available, what is our *modus operandi* in the therapy of membranous nephropathy? After searching for a primary cause, we believe that corticosteroid and/or cytotoxic treatment is not justified if the patient has an idiopathic membranous disease with a subnephrotic proteinuria. In the presence of a nephrotic syndrome, we observe the patients for a least 6 months with careful clinical and laboratory monitoring. During this period the patient is maintained on a low-sodium and moderately low-protein diet (approximately 0.9–1.0 g/kg.body weight) only without excessive saturated fats, and associated with a restriction on water intake. After this 'waiting policy' period, during which a strict control of the 'dry' body weight is usually obtained, if the nephrotic syndrome persists, we recommend specific treatment, excluding only patients over 70, with diabetes mellitus, or with systemic vascular disorders. Although it is still uncertain which is the most effective and the least toxic among the various forms of therapy, we believe that the favourable effects of the methylprednisolone/chlorambucil regimen, as outlined above, are the best. In our patients, this therapy is able significantly to improve the chance of a complete remission, and to reduce the incidence of renal failure, without important and long-lasting deleterious side-effects. We believe that to achieve maximum benefit the treatment must be given early in the course of the disease; even so some recent studies have reported good therapeutic results in patients with membranous nephropathy and deteriorating renal function.

References

Abe, S., Amagasaki, Y., Konishi, K., Kato, E., Iyori, S., and Sakaguchi, H. (1986). Idiopathic membranous glomerulonephritis: aspects of geographical differences. *Journal of Clinical Pathology*, **39**, 1193–8.

Adler, S., *et al.* (1986). Complement membrane attack complex stimulates production of reactive oxygen metabolites by cultured rat mesangial cells. *Journal of Clinical Investigation*, **77**, 762–7.

Adu, D., *et al.* (1983). Late onset systemic lupus erythematosus and lupus-like disease in patients with apparent idiopathic glomerulonephritis. *Quarterly Journal of Medicine*, **208**, 471–87.

Alexopoulos, E., Seron, D., Hartley, B., Nolasco, F., and Cameron, J.S. (1989). Immune mechanisms in idiopathic membranous nephropathy: the role of the interstitial infiltrates. *American Journal of Kidney Diseases*, **13**, 404–12.

Andres, G., Brentjens, J.R., Caldwell, P.R.B., Camussi, G., and Matsuo, S. (1986). Formation of immune deposits and disease. *Laboratory Investigation*, **55**, 510–20.

Aparicio, S.R., Woolgar, A.E., Aparicio, S.A.J.R., Walkins, A., and Davison, A.M. (1986). An ultrastructural morphometric study of membranous glomerulonephritis. *Nephrology, Dialysis and Transplantation*, **1**, 22–30.

Arisz, L., Donker, A.J.M., Brentjens, J.R.H., and Van Der Hem, G.K. (1976). The effect of indomethacin on proteinuria and kidney function in the nephrotic syndrome. *Acta Medica Scandinavica*, **199**, 121–5.

Austin, H.A., *et al.* (1983). Prognostic factors in lupus nephritis. Contribution of renal histological data. *American Journal of Medicine*, **75**, 382–91.

Baldwin, D.S., Gluck, M.C., Lowenstein, J., and Gallo, G.R. (1977). Lupus nephritis clinical course as related to morphologic forms and their transitions. *American Journal of Medicine*, **62**, 12–30.

Balow, J.E., Austin, H.A., Tsokos, G.C., Antonovych, T.T., Steinberg, A.D., and Klipper, J.H. (1987). Lupus Nephritis. *Annals of Internal Medicine*, **106**, 79–94.

Bariéty, J., Druet, P., Lagrue, G., Samarcq, P., and Milliez, P. (1970). Les glomérulopathies "extra-membraneuses" (GEM). Etude morphologique en microscopie optique, électronique et en immunofluorescence. *Pathologie et Biologie*, **18**, 5–32.

Berger, B.E., Vincenti, F., Biava, C., Amend, W.J., Feduska, N., and Salvatierra, O. (1983). De novo and recurrent membranous glomerulopathy following kidney transplantation. *Transplantation*, **35**, 315–9.

Berthoux, F.C., *et al.* (1984). Immunogenetics and immunopathology of human primary membranous glomerulonephritis: HLA-A, B, DR antigens; functional activity of splenic macrophage Fc receptors and peripheral blood T-lymphocyte subpopulations. *Clinical Nephrology*, **22**, 15–20.

Black, D.A.K., Rose, G., and Brewer, D.B. (1970). Controlled trial of prednisone in adults patients with the nephrotic syndrome. *British Medical Journal*, **3**, 421–6.

Blainey, J.D., Brewer, D.B., and Hardwicke, J. (1986). Proteinuric glomerular disease in adults: cumulative life tables over twenty years. *Quarterly Journal of Medicine*, **59**, 557–67.

Bolton, W.K., Atuk, N.O., Sturgill, B.C., and Westervelt, F.B. (1977). Therapy of the idiopathic nephrotic syndrome with alternate day steroids. *American Journal of Medicine*, **62**, 60–70.

Border, W.A. (1985). Role of antigen and antibody charge in immune complex disease. In *Nephrology Proceedings of IXth International Congress of Nephrology*, Vol. 1, (ed. R.R. Robinson), pp. 550–559. Springer-Verlag, New York.

Border, W.A. (1988). Experimental membranous nephropathy: the pathogenetic role of cationic proteins. In *Nephrology Proceedings of the Xth International Congress of Nephrology*, Vol. 2, (ed. A.M. Davison), pp. 689–694. Bailliére Tindall, London.

Brown, E.A., Markandu, N.D., Roulston, J.E., Jones, B.E., Squires, M., and MacGregor, G.A. (1982). Is the renin-angiotensin-aldosterone system involved in the sodium retention in the nephrotic syndrome? *Nephron*, **32**, 102–7.

Brown, E.A., Markandu, N.D., Saguella, G.A., Jones, B.E., and MacGregor, G.A. (1984). Lack of effect of captopril on the sodium retention of nephrotic syndrome. *Nephron*, **37**, 43–8.

Cahen, R., Francois, B., Trolliet, P., Gilly, J., and Parchoux, B. (1989). Aetiology of membranous glomerulonephritis: a prospective study of 82 adult patients. *Nephrology, Dialysis and Transplantation*, **4**, 172–80.

Cameron, J.S. (1979). Pathogenesis and treatment of membranous nephropathy. *Kidney International*, **15**, 88–103.

Cameron, J.S. (1982). Glomerulonephritis in renal transplants. *Transplantation*, **34**, 237–45.

Cameron, J.S., Healy, M.J.R., and Adu, D. (1990). The Medical Research Council Trial of short-term high dose alternate day prednisolone in idiopathic membranous nephropathy with a nephrotic syndrome in adults. *Quarterly Journal of Medicine*, **74**, in press.

Cattran, D.C. and Chodirker, W.B. (1982). Experimental membranous glomerulonephritis. The relationship between circulating free antibody and immunecomplexes to subsequent pathology. *Nephron*, **31**, 260–5.

Cattran, D.C., *et al* (1989). A randomized controlled trial of prednisone in patients with idiopathic membranous nephropathy. *New England Journal of Medicine*, **320**, 210–5.

Chandrachud, L. and Boulton-Jones, J.M. (1988). Inherited glomerular properties and their role in the expression of various forms of experimental glomerular injury. *Clinical Science*, **74**, 249, 254.

Chaplin, H. (1982). Lymphoma in primary cold hemagglutinin disease treated with chlorambucil. *Archives of Internal Medicine*, **142**, 2119–23.

Collaborative Study of the Adult Idiopathic Nephrotic Syndrome: a controlled study of short term prednisone treatment in adults with membranous nephropathy (1979). *New England Journal of Medicine*, **301**, 1301–6.

Couser, W.G. (1988). Pathogenesis and theoretical basis for treating membranous nephropathy. In *Nephrology. Proceedings of the Xth International Congress of Nephrology*, Vol. 2, (ed. A.M. Davison), pp. 701–713. Bailliere Tindall, London.

Davison, A.M., Cameron, J.S., Kerr, D.N.S., Ogg, C.S., and Wilkinson, R.W. (1984). The natural history of renal function in untreated idiopathic membranous glomerulonephritis in adult. *Clinical Nephrology*, **22**, 61–7.

Dische, F.E., Herbertson, B.M., Melcher, D.H., and Morley, A.R. (1981). Membranous glomerulonephritis in transplant kidneys: recurrent or de novo disease in four patients. *Clinical Nephrology*, **15**, 154–63.

Donadio, J.V., Holley, K.E., Anderson, C.F., and Taylor, W.F. (1974). Controlled trial of cyclophosphamide in idiopathic membranous nephropathy. *Kidney International*, **6**, 431–439.

Donadio, J.V., *et al.* (1988). Idiopathic membranous nephropathy: the natural history of untreated patients. *Kidney International*, **33**, 708–15.

Dorhout Mees, E.J., Roos, J.C., Boer, P., Yoe, O.H., and Simatupang, T.A. (1979). Observations on edema formation in nephrotic syndrome in adults with minimal lesions. *American Journal of Medicine*, **67**, 378–84.

Ehrenreich, T. and Churg, J. (1968). Pathology of membranous nephropathy. In *Pathology Annual*, (ed. S.C. Sommers), pp. 145–86. Appleton-Century-Crofts, New York.

Ehrenreich, T., Grishman, E., and Churg, J. (1974). Glucose dysmetabolism and membranous nephropathy. *American Journal of Pathology*, **74**, 330.

Ehrenreich, T. et al. (1976). Treatment of idiopathic membranous nephropathy. *New England Journal of Medicine*, **295**, 741–6.

Ehrenreich, T. and Churg, J. (1977). Focal sclerosis in membranous nephropathy. *American Journal of Pathology*, **86**, 37a.

Erwin, D.T., Donadio, J.V., and Holley, K.E. (1973). The clinical course of idiopathic membranous nephropathy. *Mayo Clinic Proceedings*, **48**, 697–712.

Falk, R.J., et al. (1983). Neoantigen of the polymerized ninth component of complement. Characterization of a monoclonal antibody and immunohistochemical localization in renal disease. *Journal of Clinical Investigation*, **72**, 560–73.

Feehally, J., Brenchley, P.E.C., Coupes, B.M., Mallick, N.P., Morris, D.A., and Short, C.D. (1986). Impaired IgG response to tetanus toxoid in human membranous nephropathy: association to HLA-DR3. *Clinical and Experimental Immunology*, **63**, 376–84.

Franklin, W.A., Jennings, R.B., and Earle, D.P. (1973). Membranous glomerulonephritis: long-term serial observations on clinical course and morphology. *Kidney International*, **4**, 36–56.

Friend, P.S. and Michael, A.F. (1978). Hypothesis: immunologic rationale for the therapy of membranous lupus nephropathy. *Clinical Immunology and Immunopathology*, **10**, 35–40.

Gabbai, F.B., Gushwa, L.C., Wilson, C.B., and Blantz, R.C. (1987). An evaluation of the development of experimental membranous nephropathy. *Kidney International*, **31**, 1267–78.

Gabbai, R.B., Mundy, C.A., Wilson, C.B., and Blantz, R.C. (1988). An evaluation of the development of experimental membranous nephropathy in the rat. *Laboratory Investigation*, **58**, 539–48.

Gaffney, E.F. and Panner, B.J. (1981). Membranous glomerulonephritis: clinical significance of glomerular hypercellularity and parietal epithelial abnormalities. *Nephron*, **29**, 209–15.

Garattini, S., Bertani, T., and Remuzzi, G. (1987). What is the basis for the use of steroids in the treatment of idiopathic membranous nephropathy? *Nephron*, **45**, 1–6.

Garavoy, M.R. (1980). Idiopathic membranous glomerulonephritis: an HLA-associated disease. In *Histocompatibility Testing*, pp.673–680. UCLA Tissue Typing Lab., Los Angeles.

Gärtner, H.W., et al. (1977). Correlations between morphologic and clinical features in idiopathic perimembranous glomerulonephritis. *Current Topics in Pathology*, **65**, 1–29.

Gauthier, V.J., Striker, G.E., and Mannik, M. (1984). Glomerular localization of preformed immune complexes prepared with anionic antibodies or with cationic antigens. *Laboratory Investigation*, **50**, 636–44.

Glassock, R.J. (1984). Membranous glomerulopathy. In *Textbook of Nephrology*, **Vol. 1**, (ed. S.G. Massry and R.J. Glassock), pp. 6.42–6.46. Williams & Wilkins, Baltimore.

Glassock, R.J. (1988). Pathogenesis of the nephrotic syndrome in humans. In *The Nephrotic Syndrome*, (ed. J.S. Cameron and R.J. Glassock), pp. 163–92. Marcel Dekker, New York.

Gluck, M.C., Gallo, G., Lowenstein, J., and Baldwin, D.S. (1973). Membranous glomerulonephritis. Evolution of clinical and pathologic features. *Annals of Internal Medicine*, **78**, 1–12.

Goldman, M., Baran, D., and Druet, P. (1988) Polyclonal activation and experimental nephropathies. *Kidney International*, **34**, 141–50.

Greenberg, L.J. and Yunis, E.J. (1978). Histocompatibility determinants, immune responsiveness and ageing in man. *Federation Proceedings*, **37**, 1258–62.

Grishman, E., Ehrenreich, T., and Churg, J. (1988). The morphologic spectrum of primary and secondary nephrotic syndrome in man. In *The Nephrotic Syndrome*, (ed. J.S. Cameron and R.J. Glassock), pp. 285–372. Marcel Dekker, New York.

Gruppo di Immunologia Renale. Società Italiana di Nefrologia. (1983). Evoluzione della nefropatia membranosa. Studio multicentrico in 186 casi. In *Nefrologia, Dialisi, Trapianto*, (ed A.Albertazzi, D. Brancaccio, P. Cappelli, G. Del Rosso, B. Di Paolo, P.F. Palmieri, and C. Spiani) pp. 209–13, Wichtig, Milano.

Guesry, B., Lenoir, H.G., and Broyer, M. (1978). Gonadal effects of chlorambucil given to prepubertal and pubertal boys for nephrotic syndrome. *Journal of Pediatrics*, **92**, 299–303.

Habib, R., Kleinknecht, C., and Gubler, M.C. (1973). Extramembranous glomerulonephritis in children. Report of 50 cases. *Journal of Pediatrics*, **82**, 754–66.

Hall, C.L. (1988). Gold nephropathy. *Nephron*, **50**, 265–72.

Hattori, S., Furuse, A., and Matsuda, I. (1988). Presence of HBe antibody in glomerular deposits and in membranous glomerulonephritis is associated with hepatitis B virus infection. *American Journal of Nephrology*, **8**, 384–7.

Hiki, Y., Kobayashi, Y., Itoh, I., and Kashiwagi, N. (1984). Strong association of HLA-DR2 and MT1 with idiopathic membranous nephropathy in Japan. *Kidney International*, **25**, 953–67.

Honkanen, E. (1986). Survival in idiopathic membranous glomerulonephritis. *Clinical Nephrology*, **25**, 122–8.

Hopper, J., Trew, P.A., and Biava, C.G. (1981). Membranous nephropathy: the relative benignity in women. *Nephron*, **29**, 18–24.

Ito, H., et al (1981). Hepatitis Be antigen-mediated membranous glomerulonephritis. Correlation of ultrastructural changes with HBeAg in the serum and glomeruli. *Laboratory Investigation*, **44**, 214–20.

Jacobs, C., et al. (1977). Combined report on regular dialysis and transplantation in Europe. In *Dialysis, transplantation and nephrology*, (ed. B. Robinson, J. Hawkins, and P. Verstraeten), pp. 3–69. Pitman, London.

Jennette, J.C., Iskandar, S.S., and Dalldorf, F.G. (1983). Pathologic differentiation between lupus and non lupus membranous glomerulonephritis. *Kidney International*, **24**, 377–85.

Jennette, J.C., Newman, W.J., and Diaz-Buxo, J.A. (1987). Overlapping IgA membranous nephropathy. *American Journal of Clinical Pathology*, **88**, 74–78.

Kida, H., Asamoto, T., Yokojama, H., Tomosugi, N., and Hattori, N. (1986). Long-term prognosis of membranous nephropathy. *Clinical Nephrology*, **25**, 64–9.

Kingswood, J.C., Banks, R.A., Tribe, R., Owen-Jones, J., and MacKenzie, J.C. (1984). Renal biopsy in the elderly: clinicopathological correlations in 143 patients. *Clinical Nephrology*, **22**, 183–187.

Kleinknecht, C. and Habib, R. (1987) Membranous glomerulonephritis. In *Pediatric Nephrology* (2nd edn) (ed. M.A Holliday, T.M. Barratt, and R.L. Vernier) pp. 462–70. Williams and Wilkins, Baltimore.

Klouda, P.T., et al. (1979). Strong association between idiopathic membranous nephropathy and HLA-DRw3. *Lancet*, **ii**, 770–1.

Kusonoki, Y., Itami, N., Tochimarn, H., Takekoshi, Y., Nagasawa, S., and Yoshiki, T. (1989). Glomerular deposition of C4-cleavage fragment (C4 d) and C4 binding protein in idiopathic membranous glomerulonephritis. *Nephron*, **51**, 17–19.

Lagrue, G., Bernard, J., Bariéty, P., Druet, P., and Guenel, J. (1975). Traitement par le chlorambucil et l'azathioprine dans les glomérulonéprites primitives. Résultats d'une étude contrôllée. *Journal d'Urologie et de Néphrologie*, **9**, 655–72.

Lee, J.C., Yamauchi, H., and Hopper, J. (1966). The association of cancer and nephrotic syndrome. *Annals of Internal Medicine*, **64**, 41–51.

Levy, M., Gagnadoux, M.F., Beziau, A., and Habib, R. (1978). Membranous glomerulonephritis associated with anti-tubular and anti-alveolar basement membrane antibodies. *Clinical Nephrology*, **10**, 158–62.

Levy, M. (1988). Infection-related proteinuric syndromes. In *The nephrotic syndrome*, (ed. J.S. Cameron and R.J. Glassock), pp. 745–804. Marcel Dekker, New York.

Lewis, E.J. (1988). Management of the nephrotic syndrome in adults. In *The nephrotic syndrome*, (ed. J.S. Cameron and R.J. Glassock), pp. 461–521. Marcel Dekker, New York.

Llach, F., Arieff, A.T., and Massry, S.G. (1975). Renal vein thrombosis and nephrotic syndrome: a prospective study of 36 adult patients. *Annals of Internal Medicine*, **83**, 9–14.

Llach, F. (1985). Hypercoagulability, renal vein thrombosis and other thrombotic complications of nephrotic syndrome. *Kidney International*, **28**, 429–39.

McGinley, E., Martin, W., Henderson, N., and Boulton-Jones, J.M. (1984). Defective splenic reticuloendothelial function in idiopathic membranous nephropathy. *Clinical and Experimental Immunology*, **56**, 295–301.

MacTier, R., Boulton Jones, J.M., Payton, C.D., and McLay, A. (1986). The natural history of membranous nephropathy in the West of Scotland. *Quarterly Journal of Medicine*, **60**, 793–802.

Makino, H., Lelongt, B., and Kanwar, S. (1988). Nephritongenicity of proteoglycans. III Mechanisms of immune deposit formation. *Kidney International*, **34**, 209–19.

Mallick, N.P., Short, C.D., and Manos, J. (1983). Clinical membranous nephropathy. *Nephron*, **34**, 209–19.

Mathew, T.H. (1988). Nephrotic syndrome in patients with transplanted kidneys. In *The Nephrotic syndrome*, (ed. J.S. Cameron and R.J. Glassock), pp. 805–847. Marcel Dekker, New York.

Mathieson, P.W., Turner, A.N., Maidment, C.G.H., Evans, D.J., and Rees, A.J. (1988). Prednisolone and chlorambucil treatment in idiopathic membranous nephropathy with deteriorating renal function. *Lancet*, **ii**, 869–72.

Matsumoto, K., Yoshizawa, N., and Hatano, M. (1978). Studies of cell mediated immunity in human glomerulonephritis by macrophage migration inhibition test. *Nephron*, **21**, 192–200.

Medical Research Council Working Party (1971). Controlled trial of azathioprine and prednisone in chronic renal disease. *British Medical Journal*, **2**, 239–41.

Milner, L.S., *et al.* (1988). Biochemical and serological characteristics of children with membranous nephropathy due to hepatitis B virus infection: correlation with hepatitis Be antigen, hepatitis B DNA and hepatitis D. *Nephron*, **49**, 184–9.

Müller, G.A., Müller, C., Liebau, G., Kompf, J., Ising, H., and Wernet, P. (1981). Strong association of idiopathic membranous nephropathy (IMN) with HLA-DR3 and MT-2 without involvement of HLA-B18 and no association of BfF1. *Tissue Antigens*, **17**, 332–7.

Murphy, B.F., Fairley, K.F., and Kincaid-Smith, P.J. (1988). Idiopathic membranous glomerulonephritis: long-term follow-up in 139 cases. *Clinical Nephrology*, **30**, 175–81.

Naruse, T., Kitamura, D., Miyakawa, Y., and Shibata, S. (1973). Deposition of renal tubular epithelial antigen along the glomerular capillary walls of patients with membranous glomerulonephritis. *Journal of Immunology*, **110**, 1163–6.

Niles, J., *et al.* (1987). Antibodies reactive with a renal glycoprotein and with deposits in membranous nephritis. (Abstract) *Kidney International*, **31**, 338.

Noel, L.H., Zanetti M., Droz, D., and Barbanel, C. (1979). Long-term prognosis of idiopathic membranous glomerulonephritis. *American Journal of Medicine*, **66**, 82–90.

Olbing, H., Greifer, I., Bennett, B.P. Bernstein, J., and Spitzer, A. (1973). Idiopathic membranous nephropathy in children. *Kidney International*, **3**, 381–90.

Ooi, B.S., Ooi, J.M., Hsu, A., and Hurtubise, P.E. (1980). Diminished synthesis of immunoglobulin by peripheral lymphocytes of patients with idiopathic membranous glomerulonephritis. *Journal of Clinical Investigation*, **65**, 789–97.

Panayi, G.S., Wooley, P., and Batchelor, I.R. (1978). Genetic basis of rheumatoid disease, HLA antigens, disease manifestations, and toxic reactions to drugs. *British Medical Journal*, **2**, 1326–8.

Papiha, S.S., *et al.* (1987). HLA-A, B, Dr and Bf allotypes in patients with idiopathic membranous nephropathy (IMN). *Kidney International*, **31**, 130–4.

Pierides, A.M., Malasit, P., Morley, A.R., Wilkinson, R., Uldall, P.R., and Kerr, D.N.S. (1977). Idiopathic membranous nephropathy. *Quarterly Journal of Medicine*, **46**, 163–78.

Ponticelli, C., *et al.* (1984). Controlled trial of methylprednisolone and chlorambucil in idiopathic membranous nephropathy. *New England Journal of Medicine*, **310**, 946–50.

Ponticelli, C. (1986). Prognosis and treatment of membranous nephropathy. *Kidney International*, **29**, 927–40.

Ponticelli, C., *et al.* (1989). A randomized trial of methylprednisolone and chlorambucil in idiopathic membranous nephropathy. *New England Journal of Medicine*, **320**, 8–13.

Ramzy, M.H., Cameron, J.S., Turner, D.R., Neild, G.H., Ogg, C.S., and Hicks, J. (1981). The long-term outcome of idiopathic membranous nephropathy. *Clinical Nephrology*, **16**, 13–19.

Rosen, S. (1971). Membranous glomerulonephritis: current status. *Human Pathology*, **2**, 209–31.

Row, P.G., *et al.* (1975). Membranous nephropathy. *Quarterly Journal of Medicine*, **44**, 207–39.

Salant, D.J., Darby, C., and Couser, W.G. (1980). Experimental membranous glomerulonephritis in rats. Quantitative studies of glomerular immune deposit formation in isolated glomeruli and whole animals. *Journal of Clinical Investigation*, **66**, 71–81.

Schillinger, F., Montagnac, R., and Milcent, T. (1987). Glomérulonéphrite extra-membraneuse réversible secondaire à un traitement par diclofénac. *Semaine des Hôpiteaux*, **63**, 1831–2.

Schrier, R.M. (1988). Pathogenesis of sodium and water retention in high-output and low-output cardiac failure, nephrotic syndrome, cirrhosis, and pregnancy. *New England Journal of Medicine*, **319**, 1065–72.

Seggie, J.L. and Adu, D. (1988). Nephrotic syndrome in the Tropics. In *The nephrotic syndrome*, (ed. J.S. Cameron and R.J. Glassock), pp. 653–695. Marcel Dekker, New York.

Shemesh, O., Ross, J.C., Deen, W.M., Grant, G.W., and Meyers, B.D. (1986). Nature of the glomerular capillary injury in human membranous glomerulopathy. *Journal of Clinical Investigation*, **77**, 868–77.

Shoenfeld, J. and Schwartz, R.S. (1984). Immunological and genetic factors in autoimmune diseases. *New England Journal of Medicine*, **311**, 1019–29.

Short, C.D., Solomon, L.R., Gokal, R., and Mallick, N.P. (1987). Methylprednisolone in patients with membranous nephropathy and declining renal function. *Quarterly Journal of Medicine*, **247**, 929–40.

Skrifvars, N., Törnroth, T., and Tallqvist, G. (1977). Gold induced immune complexes nephritis in seronegative rheumatoid arthritis. *Annals of Rheumatic Disease*, **36**, 549–56.

Solling, J. (1983). Circulating immune complexes in glomerulonephritis: a longitudinal study. *Clinical Nephrology*, **20**, 177–89.

Solomon, L.R., Rawlinson, V., Howarth, S., and Mallick, N.P. (1981). A study of reticulo-endothelial clearance in glomerulonephritis and SLE. *Abstracts. 8th International Congress of Nephrology*, **IM 86**, p. 146.

Suki, W.N. and Chavez, A. (1981). Membranous nephropathy: response to steroids and immunosuppression. *American Journal of Nephrology*, **1**, 11–16.

Takekoshi, Y., Tanaka, M., Shida, N., Satake, Y., Saheki, Y., and Matsumoto, S. (1978). Strong association between membranous nephropathy and hepatitis B surface antigenemia in Japanese children. *Lancet*, **ii**, 1065–8.

Takekoshi, Y., Tanaka, M., Miyakawa, J., Yoshizawa, H., Takahashi, K., and Mayumi, M.(1979). Free 'small' and IgG associated 'large' hepatitis Be antigen in the serum and glomerular capillary walls of two patients with membranous glomerulonephritis. *New England Journal of Medicine*, **300**, 814–19.

Tiller, D.J., *et al* (1981). A prospective randomized trial in the use of cyclophosphamide, dipyridamole and warfarin in membranous and mensangiocapillary glomerulonephritis. In *Nephrology. Proceedings of 8th International Congress of Nephrology*, (ed. W. Zurukzoglu, M. Papadimitriou, M. Pyrpasopoulos, M. Sion, and C. Zamboulis). pp. 345–354. University Studio, Thessaloniki and Karger, Basel.

Tomura, S., et al. (1984). Strong association of idiopathic membranous nephropathy with HLA-DR2 and MT1 in Japanese. *Nephron*, **36**, 242–5.

Törnroth, T. and Skrifvars, B. (1975). The development and resolution of glomerular basement membrane changes associated with subepithelial immune deposits. *American Journal of Pathology*, **79**, 219–36.

Törnroth, T., Honkanen, E., and Pettersson, E. (1987). The evolution of membranous glomerulonephritis reconsidered: new insights from a study on relapsing disease. *Clinical Nephrology*, **28**, 107–17.

Torres, V.E. and Donadio, G.V. (1988). Clinical aspects of the nephrotic syndrome in systemic diseases. In *The nephrotic syndrome*, (ed. J.S. Cameron and R.J. Glassock), pp. 555–561, Marcel Dekker, New York.

Trainin, E.B., Boichis, H., Spitzer, A., and Greifer, I. (1976). Idiopathic membranous nephropathy. *New York State Journal of Medicine,* **76**, 357–60.

Tu, W.H., Petitti, D.B., Biava, C.G., Tulunay, O., and Hopper, J. (1984). Membranous nephropathy: predictors of terminal renal failure. *Nephron*, **36**, 118–24.

Weetman, A.P., Pinchin, A.J., Pussel, B.A., Evans, D.J., Seny, P., and Rees, A.J. (1981). Membranous glomerulonephritis and autoimmune thyroid disease. *Clinical Nephrology*, **15**, 50–51.

Weidner, N. and Lorentz, W.B. (1986). Scanning electron microscopy of the acellular glomerular basement membranes in idiopathic membranous glomerulopathy. *Laboratory Investigation*, **54**, 84–92.

West, M.L., Jindal, K.K., Bear, R.A., and Goldstein, M.B. (1987). A controlled trial of cyclophosphamide in patients with membranous glomerulonephritis. *Kidney International*, **32**, 579–84.

Whitworth, J.A., et al. (1976). Absence of glomerular renal tubular epithelial antigen in membranous glomerulonephritis. *Clinical Nephrology*, **5**, 159–62.

Wiggelinkhuizen, J., Sinclair-Smith, C., Stannard, L.M., and Smuts, H. (1983). Hepatitis B virus associated membranous glomerulonephritis. *Archives of Disease in Childhood*, **58**, 488–96.

Williams, P.S. and Bone, J.M. (1989). Immunosuppression can arrest progressive renal failure due to idiopathic membranous glomerulonephritis. *Nephrology, Dialysis and Transplantation*, **4**, 181–6.

Zollinger, H.U. and Mihatsch, M.J. (1978). Epimembranous glomerulonephritis. In *Renal pathology in biopsy*, pp. 261–78. Springer-Verlag, Berlin.

Zucchelli, P., Cagnoli, L., Pasquali, S., Casanova, S., and Donini, U. (1986). Clinical and morphologic evolution of idiopathic membranous nephropathy. *Clinical Nephrology*, **25**, 282–8.

Zucchelli, P., Ponticelli, C., Cagnoli, L., and Passerini, P. (1987). Long-term outcome of idiopathic membranous nephropathy with the nephrotic syndrome. *Nephrology, Dialysis and Transplantation*, **2**, 73–8.

Zucchelli, P., Ponticelli, C., Cagnoli, L., Aroldi, A., and Beltrandi, E. (1988). Prognostic value of T lymphocyte subset ratio in idiopathic membranous nephropathy. *American Journal of Nephrology*, **8**, 15–20.

3.8 Mesangiocapillary glomerulonephritis

D. GWYN WILLIAMS

Introduction

The term mesangiocapillary glomerulonephritis (MCGN) defines a histological type of nephritis characterized by mesangial proliferation and thickening of the capillary wall, partly due to extension of the mesangium into the wall. It has a range of clinical presentations which do not distinguish it from other histological types of nephritis and has no unique serological markers. Most cases are idiopathic, although a minority are clearly associated with infections (see Chapter 3.12). Histology is therefore mandatory to make the diagnosis of mesangiocapillary glomerulonephritis. Recognition of the current histological descriptions of mesangiocapillary glomerulonephritis dates from the early 1960s, although credit must be given to Volhard and Fahr, Ellis, and Allen for their light-microscopic descriptions of 'lobular' glomerulonephritis, which equate with mesangiocapillary glomerulonephritis (Jones 1977).

Despite its lack of defining features, apart from its microscopic appearance, mesangiocapillary glomerulonephritis has several important features for nephrologists. Foremost among these are its association in many cases with prolonged hypocomplementaemia, which is characterized by evidence of activation of the alternative pathway potentiated by the activity of the nephritic factors, the most prevalent of which is now known to be an autoantibody directed against assembled components of the complement system. Another intriguing feature is the unique appearance of the osmophilic electron-dense material which defines the Type II (dense deposit) variety of mesangiocapillary glomerulonephritis. The nature and genesis of this material remain unknown. There has been much interest in the apparent decrease in incidence of mesangiocapillary glomerulonephritis in north-west Europe and North America during the last decade, although it remains a common type of nephritis in poorer and non-industrialized countries (see Chapter 3.14). Finally, on a practical note, the histologically clearly defined entity of mesangiocapillary glomerulonephritis allows a clinical prognosis to be given with some certainty in the idiopathic cases.

Nomenclature

Mesangiocapillary glomerulonephritis has been known by several other names. Those most commonly used are membranoproliferative glomerulonephritis, lobular glomerulonephritis, chronic mesangioproliferative glomerulonephritis, all of which are histological descriptions, and chronic or persistent hypocomplementaemic glomerulonephritis, which highlights the strongly associated, but not unique complement abnormality. The term mesangiocapillary glomerulonephritis will be used in this chapter, being that most frequently used and the most accurately descriptive.

Mesangiocapillary glomerulonephritis has been subdivided on histological grounds into two main types, Type I, which is characterized by subendothelial capillary wall deposits, and Type II, with its electron-dense material within the glomerular basement membrane. Although these features totally distinguish between the two types, raising the question of whether they should be considered the same disease, they have many features in common, making it practicable to discuss them together from a clinical point of view. Where the two types differ other than histologically this will be stated in the text. Other types, e.g. Type III, have also been described, and these will be mentioned as necessary when they diverge from the general picture.

Pathology

Histology

The all-embracing features of the various types of mesangiocapillary glomerulonephritis are an increase in mesangial cells and matrix and thickening of the capillary walls. The mesangial increase, when generalized throughout the glomeruli, causes an exaggeration of their lobular form, giving rise to one of the alternative names of 'lobular glomerulonephritis'. Sclerotic nodules are usually found in the centre of the lobules. The non-lobular type has been termed 'simple' or 'pure' mesangiocapillary glomerulonephritis. Types I and II are not differently distributed between the simple and lobular forms (Bohle *et al.* 1974).

Type I (subendothelial mesangiocapillary glomerulonephritis; classic mesangiocapillary glomerulonephritis) (Plate 1)

The typical mesangial cellular proliferation and increase in mesangial matrix extend into the capillary walls to contribute to their thickening. The cellular increase is due to mesangial cells and infiltrating leucocytes, mainly neutrophils. Eosinophilic and, less frequently, basophilic deposits occur in the mesangium.

The thickening of the capillary walls is not necessarily continuous—some may be of normal thickness—and is due to interposition of mesangial matrix and cells and polymorphonuclear leucocytes into the capillary wall. A 'double contour' in the capillary walls is a typical finding, but not unique. This is most easily seen with a silver stain (Fig. 1), and is due to staining of the true basement membrane of the capillary and a false basement membrane arising from the mesangial interposition; it is not due to splitting of the true basement membrane, as it is sometimes erroneously described. The capillary lumina are reduced, due not only to thickening of the wall but also to the increase in endothelial cells. Although these changes are usually diffuse, focal and segmental abnormalities have been described.

As with any chronic glomerulonephritis, there may be increase in interstitial cells and eventual fibrosis.

Electron microscopy confirms the above glomerular changes, with the additional finding of electron-dense deposits in the subendothelial sites (Fig. 2), and also, less frequently, in the glomerular basement membrane itself and in the mesangium. The deposits are featureless or finely granular.

Immunohistology reveals deposition of immunoglobulins and complement components. The typical picture is a granular deposition of C3 in the capillary walls, outlining the lobular structure of the glomerulus (Fig. 3). IgG and, less frequently, IgM and IgA and the complement components Clq, C4, factor B, and properdin are found in the same distribution; serial biopsies have shown IgG on some occasions and not others. Occasionally C3 may be found alone; absence of IgG is correlated with absence of the early complement components Clq and C4. Fibronectin has been found in capillary loops. All these components may also be found in the mesangium, but less prominently than in the capillary walls.

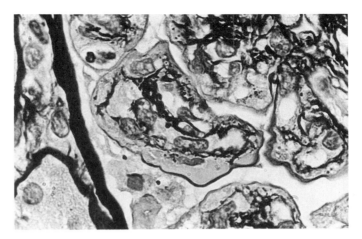

Fig. 1 Mesangiocapillary glomerulonephritis Type I. A high power view of the appearances in the capillary wall, showing the thin silver positive true basement membrane on the outer surface of the capillary lobule, with a more irregular inner deposit of silver staining basement membrane-like material secreted by interposed mesangial cells, within the lobule, which creates the 'double contour' appearance. Between the two are mesangial cell cytoplasm and nuclei, and silver-negative immune aggregates (silver methenamine counterstained with haematoxylin/eosin; original magnification × 500. (By courtesy of Dr Barrie Hartley.)

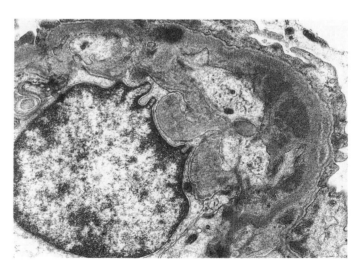

Fig. 2 Mesangiocapillary glomerulonephritis Type I. This electron micrograph of the capillary wall shows the complex thickening to consist of the normal-looking basement membrane on the outside, with subendothelial electron-dense material representing immune aggregates in a subendothelial position. Within this are electron-lucent areas of interposed mesangial cell cytoplasm, and new basement membrane-like material abutting the capillary lumen, filled by a large mononuclear cell (original magnification × 9747). (By courtesy of Dr Barrie Hartley.)

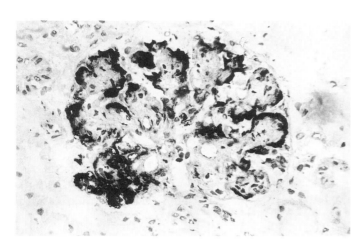

Fig. 3 Mesangiocapillary glomerulonephritis Type I. Frozen section stained with immunoperoxidase conjugated anti-IgM serum to demonstrate the lumpy aggregates of immune material present around the capillary walls (original magnification × 200). (By courtesy of Dr Barrie Hartley.)

Type II (dense deposit disease) (Fig. 4)

Light microscopy shows that, as in Type I, there is proliferation of mesangial cells and increase of mesangial matrix, with similar effects on the appearance of the glomerulus. The degree of proliferation can vary widely. The cardinal feature of Type II is the structural change in the basement membrane which is thickened by osmophilic electron-dense material (hence the descriptive term) but which in many, but not all, cases is visible on light microscopy. The basement membranes appear as a darker or abnormally prominent ribbon when stained with particular histochemical agents such as periodic acid-Schiff, thioflavine T and eosin, and notably do not react strongly with silver, which may produce a double line either side of the altered structure. These changes may be seen in the basement membranes of the glomeruli, Bowman's capsule, the tubules, and peritubular capillaries. From a practical point of view it is important to remember that the diagnosis of Type II mesangiocapillary glomerulonephritis may be missed on light microscopy as the diagnostic basement membrane changes may be detectable only on electron microscopy. This point is also relevant when assessing reports of series describing clinical features, response to treatment, etc. in which electron microscopy (or an alternative technique of toluidine blue staining) has not been performed on all cases. Another type of deposit may be seen in approximately one-third of patients in relation to the capillary loops—subepithelial 'humps' similar to those seen in poststreptococcal glomerulonephritis. The capillary lumina are reduced by a combination of the basement membrane structural changes, the mesangial interposition into the capillary wall and leucocytes within the lumen.

Apart from the increase in matrix and cellular proliferation, the mesangium is abnormal in that it contains electron-dense deposits similar to those in the basement membrane, which are, however, less easily seen on light microscopy. Polymorphonuclear leucocytes may infiltrate the glomerular tuft. The interstitium may show changes similar to those in Type I.

Electron microscopy (Fig. 5) is of especial interest in Type II mesangiocapillary glomerulonephritis as it reveals the unique alteration in the glomerular basement membrane, which is diagnostic and focuses attention on the as yet unanswered question of the nature of the altered structure. The electron-dense abnormalities are not continuous, and may change abruptly so that in some glomeruli the capillary walls may appear normal, particularly if mesangial interposition has not occurred. Electron microscopy confirms similar alterations in the basement membranes of the tubules, Bowman's capsule, and, infrequently, in the peritubular capillaries. The deposits in the glomerular basement membrane may be extremely wide, extending across the entire width of the lamina densa and widening the membrane to 1500 or 2000 nm. Even with high magnification the deposits appear amorphous and contrast therefore with the deposits seen in or around the glomerular basement membrane in other diseases, especially those thought to be typical of or associated with deposition of immune complexes. If the cells are removed from the glomeruli and then examined by transmission electron microscopy, the dense deposits remain *in situ*, whereas deposits of mesangiocapillary glomerulonephritis Type I and other forms of glomerulonephritis are removed (Nishimura *et al.* 1989).

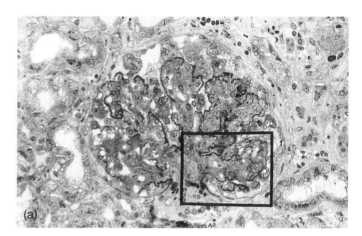

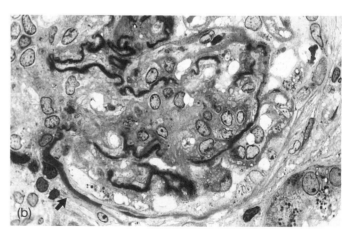

Fig. 4 Mesangiocapillary Type II (dense deposit disease). (a) Plastic embedded section cut at 1μm thickness and stained with toluidine blue. The linear material following the capillary wall basement membrane is also evident in Bowman's capsule (bottom) (toluidine blue; original magnification × 400). (By courtesy of Dr Barrie Hartley.) (b) A higher magnification of a portion of the field in (a). The reflection of the dense material into Bowman's capsule is seen on the left (arrow). 'Breaks' in the continuity of the material within the capillary wall basement membrane can be seen also (toluidine blue; original magnification × 500).

3.8 Mesangiocapillary glomerulonephritis

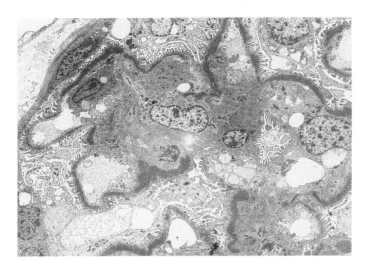

Fig. 5 Mesangiocapillary Type II (dense deposit disease). The capillary basement membranes are replaced with a rather irregular, electron-dense, osmiophilic material which occupies the lamina densa continuously with only a few breaks (electron microphotograph; original magnification × 1965). (By courtesy of Dr Barrie Hartley.)

Although the chemical and structural reasons for this difference is not known, the contrast between the deposits in Types I and II is further emphasized.

The nature of this electron-dense material is not known. The use of the word 'deposit' to describe the abnormality gives the impression that some extraneous material, usually in this context an antibody or circulating antigen–antibody complexes, has been laid down in the glomerular basement membrane. There is no evidence that this is so, either from the fine detail of the abnormal membrane or from the immunofluorescence findings, nor do serological studies provide any good circumstantial evidence. It is just as likely that the basement membrane has undergone a chemical change, and there is some suggestion that alteration in galactosyl residues has occurred. The reaction with thioflavine T which stains amyloid has an obvious implication, and staining of the dense deposit with antibody to serum amyloid A protein has been reported.

Immunohistology shows deposition of C3 in the capillaries, less often accompanied by C1q, C4, factor B, and properdin than in Type I. Immunoglobulins are also seen less often, with IgG being most frequently detected. The C3 may be deposited along the borders of the electron-dense material, not within it, giving a double or parallel appearance, or as granular deposits (Plate 2). In the mesangium, C3 is found as discrete deposits or outlining the deposits seen on electron microscopy, giving the appearance of rings.

Although typically associated with mesangiocapillary glomerulonephritis, dense deposits of a similar nature have also been described in a small number of patients with the light microscopic appearances of focal segmental necrotizing glomerulonephritis and focal sclerosing crescentic proliferative glomerulonephritis (Davis et al. 1977). A distinguishing feature is that the dense deposits are predominantly in the paramesangial area of the glomerular basement membrane and in the mesangium itself. Mazzucco et al. (1980) have also described a small number of cases of dense-deposit disease without the mesangial proliferative element, leading them to postulate strongly that Type II is a separate entity from Type I. One case of typical dense deposit disease associated with systemic lupus erythematosus has been described (Friedman et al. 1983).

Type III mesangiocapillary glomerulonephritis (mixed membranous and proliferative glomerulonephritis) (Fig. 6)

This is a variant of Type I in which there are changes in the capillary wall similar to those of membranous nephropathy, i.e. epimembranous deposits with interspersed projections of basement membrane. Electron microscopy confirms the deposits and immunohistology shows granular deposition of C3, IgG, and IgM predominantly in the capillary walls (Strife et al. 1982).

Type IV mesangiocapillary glomerulonephritis

This is typified by a disruption of the basement membranes and a layering appearance of the lamina densa. Subepithelial and subendothelial deposits occur.

Other variants

Attention has been drawn to a variant of Type I with focal proliferative changes; its importance lies in its relatively good clinical prognosis (Davis et al. 1977).

Crescentic mesangiocapillary glomerulonephritis

Crescents, which may be diffuse, are found in 10 to 15 per cent of cases (see Chapter 3.10).

Clinical interpretation of histology

The histological appearances described above can be confused with changes seen in entirely different clinical disorders, and it is therefore necessary for the nephrologist to be aware that a

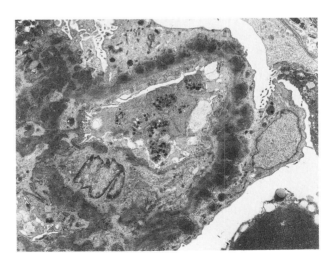

Fig. 6 Mesangiocapillary glomerulonephritis Type III. This electron micrograph shows the irregular appearance of electron-dense, presumed immune aggregates within, outside and under the glomerular capillary basement membrane. Many observers regard this as a late variant of Type I mesangiocapillary glomerulonephritis, but others as a distinct histopathological entity. This appearance may be easily confused with a late membranous nephropathy (original magnification × 3822). (By courtesy of Dr Barrie Hartley.)

Table 1 Clinical associations of mesangiocapillary glomerulonephritis (MCGN)

Disease	MCGN Type	Cross reference
Infections		
Streptococcal (serological)	I and II	Chapter 3.12
Infective endocarditis (mainly streptococcal)	I	Chapter 3.12
Shunt nephritis (staphylococcal)	I	Chapter 3.12
Abscesses	I	Chapter 3.12
Buckley's syndrome, (raised serum IgE, dermatitis, sinusitis)	I	
Cystic fibrosis	I	
Tuberculosis	I	Chapter 3.14
Leprosy	I	Chapter 3.14
Mycoplasma	I	
Hepatitis B	I	Chapter 3.12/3.14
Filariasis	I	Chapter 3.12/3.14
Malaria	I	Chapter 3.12/3.14
Schistosomiasis	I	Chapter 3.12/3.14/13.5
Candidiasis	II	
Autoimmunity		
Systemic lupus erythematosus	I (II)	Chapter 4.6
Systemic sclerosis	I	Chapter 4.7
Mixed cryoglobulinaemia	I	Chapter 4.9
Neoplasia/dysproteinaemia		
Carcinoma	I	Chapter 3.13
Leukaemia	I	Chapter 3.13
Lymphoma (Hodgkin's and non-Hodgkin's)	I	Chapter 3.13
Myeloma	I	Chapter 4.3
Light-chain nephropathy	I	Chapter 4.3
Waldenstrom's macroglobulinaemia	I	Chapter 4.3
Metabolic		
α_1-Antitrypsin deficiency	I	
Complement deficiency	I and II	Chapter 3.1
Miscellaneous		
Sickle-cell disease	I	Chapter 4.12
Cyanotic heart disease	I	
Sarcoidosis	I	Chapter 4.4
Partial lipodystrophy (with complement deficiency)	I and II	

histological diagnosis of mesangiocapillary glomerulonephritis can not only indicate the presence of the diseases shown in Table 1, but if misinterpreted may not lead to the diagnosis of other following conditions.

Diabetes mellitus
Confusion can arise because the small nodular lesions and lobular enlargement seen in diabetes mellitus are similar to the lobular form of mesangiocapillary glomerulonephritis which may contain nodular lesions. Distinguishing features are the presence of nodular lesions in most, if not all, glomeruli in mesangiocapillary glomerulonephritis and their relative infrequency in diabetes, and the immunohistology (see Chapter 4.1).

Amyloidosis
Nodular deposits of amyloid can cause an appearance which resembles mesangiocapillary glomerulonephritis on staining with haematoxylin and eosin. However, amyloid is not argyrophilic, and immunohistological techniques, specific amyloid stains, and electron microscopy distinguish totally between amyloid and mesangiocapillary glomerulonephritis (see Chapter 4.2).

Light chain nephropathy
Light microscopy of light chain nephropathy and mesangiocapillary glomerulonephritis can appear remarkably similar; the main differences are the non-staining with silver of the nodules in the mesangium and the immunohistology (see Chapter 4.3).

Systemic lupus erythematosus
Systemic lupus erythematosus can produce most histological types of nephritis, mesangiocapillary glomerulonephritis Type I being no exception. Indeed, it is more correct to include lupus as a 'cause' of mesangiocapillary glomerulonephritis than to consider it, like diabetes and light chain nephropathy, as a mistaken diagnosis, as it can be truly indistinguishable from mesangiocapillary glomerulonephritis. A strong indication for a diagnosis of systemic lupus erythematosus is the presence of each of IgG, IgM, IgA, C1q, C4, and C3, a combination which is rare in mesangiocapillary glomerulonephritis but common in SLE.

Henoch-Schönlein purpura
A variant of mesangiocapillary glomerulonephritis (pseudomembranoproliferative glomerulonephritis) occurs in Henoch-Schönlein purpura. This is distinguished from the idiopathic variety by the mesangial deposition of IgA.

Postinfectious nephritis
It can sometimes be difficult to distinguish between Type I mesangiocapillary glomerulonephritis and postinfectious nephritis. Several features are common, such as proliferation, poly-

morphonuclear cells, and subepithelial humps. This diagnostic blurring may account for a proportion of patients with mesangiocapillary glomerulonephritis Type I who undergo spontaneous improvement. On occasion, classic endocapillary glomerulonephritis may evolve into mesangiocapillary glomerulonephritis (see Chapter 3.9).

Evolution of histological changes

Although sequential changes have been described in mesangiocapillary glomerulonephritis it must be borne in mind that the data from these studies may be selected and skewed because repeat biopsies are usually performed to assess therapy, to establish an unclear diagnosis, or to explain a deterioration in renal function. An overall picture is therefore not readily apparent, but certain points may be made. Since the clinical outcome of mesangiocapillary glomerulonephritis is poor, with approximately 50 per cent renal survival at 10 years from diagnosis, the majority of cases must progress, increasing glomerular sclerosis leading to end-stage renal failure. Taguchi and Bohle (1989) have reported their experience of serial biopsies in 33 patients, and reviewed the work of others. It appears that most patients of Types I and II who have more than 50 per cent of glomeruli affected remain in this category, although a decrease can occur. Most patients with less than 50 per cent of glomeruli involved show an increase, but again reversal may occur. In mesangiocapillary glomerulonephritis Type I focal changes may become diffuse and vice versa; both may remit. The majority of cases have diffuse changes on presentation, and these remain. Proliferation and lobularity tend to diminish with time. The number and distribution of deposits in Type I varies with time, but Type II deposits tend to persist.

Spontaneous improvements resulting in a return of glomeruli to normal have been noted, particularly in those cases associated with infections which regress, but also in apparently idiopathic cases of Type I. Clinical observation shows that most patients with Type II mesangiocapillary glomerulonephritis have continuing and progressive disease; recorded histological regression of the dense deposits is rare (Habib *et al.* 1973).

The complement system (see also Chapter 3.1)

Mesangiocapillary glomerulonephritis is accompanied by chronic hypocomplementaemia in a large proportion of cases (Gotoff *et al.* 1965; West *et al.* 1965): this observation makes this particular form of nephritis even more distinctive, as the only other forms of glomerular disease regularly associated with hypocomplementaemia are acute poststreptococcal nephritis and the nephritis of systemic lupus erythematosus. Not only do the clinical features and production of antibodies differ between these three diseases, but there are distinctive features pertaining to the hypocomplementaemia of mesangiocapillary glomerulonephritis. Although there has been much interest in the abnormalities of the complement system in mesangiocapillary glomerulonephritis, the part played by the hypocomplementaemia in the pathogenesis of the disease remains unknown.

General description

Point prevalence studies show that hypocomplementaemia occurs in 50 to 60 per cent of patients, and is mainly due to a low serum concentration of C3 and, consequently, a low serum level of haemolytic complement (CH_{50}) (Herdman *et al.* 1970; Habib *et al.* 1973, 1975; Williams *et al.* 1974; Ooi *et al.* 1976; Schena *et al.* 1982; Cameron *et al.* 1983). Levels of C3 can be profoundly depressed, with median values of 30 per cent of normal with a lower limit of the range of 2 to 5 per cent. C3, as well as being the bulk component of the complement system, is the focal point at which the alternative and classical pathways meet, C3 itself forming the alternative pathway convertase with factor B (C3bBb) (see Chapter 3.2). Levels of the early classical pathway components C1q and C4 are clearly low in some, but not all, patients, whereas the alternative pathway components properdin and factor B are low in others; in some patients levels of all of these components are below normal. C3 breakdown products are detectable in the circulation, indicating activation of C3. Activation of the complement cascade beyond C3 is suggested by low serum concentrations of C5 and a correlation of low C3 with the lower values of the range of C7 in some patients.

The finding of circulating C3 breakdown products, plus the circumstantial evidence that mesangiocapillary glomerulonephritis may be an immune complex disease, makes activation of complement and therefore hypercatabolism an obvious explanation for the hypocomplementaemia. This has indeed been confirmed by metabolic studies of patients with mesangiocapillary glomerulonephritis using radioactive human C3 (Charlesworth *et al*, 1974). These studies showed the expected increased turnover of C3 due to increased catabolic rates, but somewhat surprisingly also demonstrated a decreased rate of synthesis of C3. The cause for this seemingly paradoxical behaviour of a protein, i.e. reduced synthesis in the face of increased consumption, has not been explained, but it has been suggested that it may be due to a negative feedback effect of the C3 breakdown products on C3 synthesis, with the biological advantage of reducing production of a protein causing tissue damage. There is increased catabolism of factor B in patients with a low serum C3 concentration, adding further evidence to involvement of the alternative pathway in mesangiocapillary glomerulonephritis (Charlesworth *et al.* 1974).

Activators of complement in mesangiocapillary glomerulonephritis

The observation that the addition of hypocomplementaemic serum from patients with mesangiocapillary glomerulonephritis to normal human serum *in vitro* caused conversion of the C3 in the latter excited much interest (Spitzer *et al.* 1969), and although the study of this phenomenon has not actually led to increased understanding of the pathogenesis of mesangiocapillary glomerulonephritis, it has given considerable insight into the working of the complement system.

This property of hypocomplementaemic serum of patients with mesangiocapillary glomerulonephritis was given the name of nephritic factor, or C3 nephritic factor. It is now known to be an IgG autoantibody directed against determinants of the C3bBb complex, the C3 converting enzyme of the alternative pathway (Scott *et al.* 1978) (see Chapter 3.2). C3 nephritic factor, although found in both Type I and Type II mesangiocapillary glomerulonephritis, is more frequent in the latter. C3 nephritic factor combines with the convertase, preventing its degradation by its normal inactivators, factors H, I and Cr1, and consequently enhancing the rate and extent of C3 breakdown. The reason for the genesis of this autoantibody is not known; it is not seen in all cases of hypocomplementaemic mesangiocapillary glomerulonephritis, and can also occur in other diseases, e.g. systemic lupus erythematosus. Variant forms have been described in functional terms, including C3 nephritic factor which does not cause C3 breakdown in the parent nephritic serum, and C3 nephritic factor which does not cause detectable C3 breakdown *in vitro* but nonetheless causes activation of the complement cascade so that C5 is activated and is present in low concentrations in the serum (Mollnes *et al.* 1986; Ng *et al.* 1986).

C3 convertase enzymes are not the only activators in the complement system to give rise to autoantibodies: an IgG anti-C4b has also been described in mesangiocapillary glomerulonephritis and systemic lupus erythematosus. C3 activators which are not immunoglobulins have also been described in various hypocomplementaemic disorders including mesangiocapillary glomerulonephritis.

Correlation of complement abnormalities with disease

The presence of hypocomplementaemia, measured as CH_{50} or any individual component, its degree, and the presence of C3 nephritic factor are, according to most reports, not related to the severity of the nephritis, its clinical course or outcome. Complement abnormalities are more commonly found in Type II mesangiocapillary glomerulonephritis than Type I. Bilateral nephrectomy does not alter the hypocomplementaemia or result in the disappearance of C3 nephritic factor.

Evolution of complement abnormalities with time

Sequential observations have shown that many patients with hypocomplementaemia lose their abnormalities of the complement system (Cameron *et al.* 1983) Levels of C3, C4, C1q, factor B, and properdin may rise towards or even attain normal values, with consequent normal CH_{50} levels, and C3 nephritic factor may disappear. These changes tend to occur over several years, and are rarely seen in the reverse order—normocomplementaemic patients rarely become hypocomplementaemic. As discussed below, neither the absolute complement values nor their fluctuations are related to clinical evolution of the nephritis, and absence of abnormalities cannot be used as a negative diagnostic point—some patients are normocomplementaemic throughout their observed course.

Variations in complement abnormalities between types of mesangiocapillary glomerulonephritis

The general picture given above has distinct variations according to the histological types of mesangiocapillary glomerulonephritis (Habib *et al.* 1973, 1975; Williams *et al.* 1974; Cameron *et al.* 1983). In Type I mesangiocapillary glomerulonephritis the incidence of hypocomplementaemia at diagnosis is 33 to 50 per cent; 25 to 30 per cent of patients have low C1q and C4, and a similar proportion have low C5. Levels of the alternative pathway factors properdin and factor B are reduced in 15 to 20 per cent of patients. By contrast, low C3 occurs in 70 per cent of patients with Type II mesangiocapillary glomerulonephritis and C1q, C4 and C5 are usually normal, whereas properdin and factor B are reduced in a higher proportion of patients. These differences indicate activation of the classical pathway in a proportion of patients with Type I, with consumption of the early classical components C1q and C4, and of the alternative pathway in Type II, bypassing these early components. The low concentrations of properdin and factor B point to activation of the C3b feedback mechanism. C3 nephritic factor occurs in a higher proportion of Type II patients—60 per cent compared with 20 per cent in Type I, and when quantitated is more active than in Type I, contributing to the greater depression of serum C3 seen in Type II disease. Nephritic factor occurs in secondary cases of Type I mesangiocapillary glomerulonephritis (e.g. shunt nephritis) as well as in idiopathic cases. The return towards normal of the individual complement components and disappearance of C3 nephritic factor is much less common in Type II. Type III is characterized by a low incidence of complement abnormalities and C3 nephritic factor; the abnormalities are similar to those of Type I (Jackson *et al.* 1987).

The histological distribution of complement components and their variations according to the type of mesangiocapillary glomerulonephritis were described earlier.

Complement deficiency and mesangiocapillary glomerulonephritis

A small number of cases of mesangiocapillary glomerulonephritis has occurred in individuals with an inherited deficiency of the complement system, usually C3 or C2 (Coleman *et al.* 1983). Some of these patients also have an increased incidence of infection, but no other evidence of systemic disease. The self same complement deficiencies are associated with other forms of nephritis or disease, implying that factors other than the complement deficiency govern the nature of the association. The pathogenic implications of the association of mesangiocapillary glomerulonephritis with complement deficiency are discussed below.

Partial lipodystrophy, hypocomplementaemia and mesangiocapillary glomerulonephritis.

An association between renal disease and partial lipodystrophy was recognized 30 years ago. It was subsequently found that the most common association of partial lipodystrophy was with mesangiocapillary glomerulonephritis, with the great majority of cases being Type II (Eisinger *et al.* 1972; Peters *et al.* 1973; Sissons *et al.* 1976). The observation that partial lipodystrophy could occur with hypocomplementaemia due to alternative pathway activation and in the presence of C3 nephritic factor without mesangiocapillary glomerulonephritis, as well as with this form of nephritis, provoked much interest and speculation. The numerous clinical observations may be summarized as follows:

1. The majority of patients with partial lipodystrophy have hypocomplementaemia and C3 nephritic factor indistinguishable from mesangiocapillary glomerulonephritis.
2. Partial lipodystrophy occurs with and without mesangiocapillary glomerulonephritis; the mesangiocapillary glomerulonephritis is Type II in 90 per cent of cases.
3. Partial lipodystrophy almost always precedes the development of nephritis, implying that hypocomplementaemia, accompanying the partial lipodystrophy, has similarly predated the nephritis.
4. Partial lipodystrophy is associated with recurrent infections without nephritis, i.e. analogous to the isolated complement component deficiencies described above.

These observations have contributed to the hypothesis, discussed in detail below, that hypocomplementaemia precedes and predisposes to the development of mesangiocapillary glomerulonephritis. Patients with partial lipodystrophy and hypocomplementaemia but without clinical evidence of nephritis have had renal biopsy performed with the demonstration of normal kidneys, (Sissons *et al.* 1976) although a patient with partial lipodystrophy, hypocomplementaemia and clinically occult mesangiocapillary glomerulonephritis was found to have a clinically 'silent' nephritis on biopsy (Bennett et al. 1977). The pathogenesis of partial lipodystrophy and its relationship, if any, to the accompanying hypocomplementaemia are unknown, so no links can be made between this and the pathogenesis of mesangiocapillary glomerulonephritis. It is worth noting however that the onset of partial lipodystrophy has been related to infections, particularly measles, allowing a tenuous link between a common and perhaps persistent antigen in partial lipodystrophy and mesangiocapillary glomerulonephritis to be made, although measles itself is not a recognized antecedent of mesangiocapillary glomerulonephritis.

Immune complexes

A variety of techniques has demonstrated an increase in the level of circulating immune complexes in mesangiocapillary glomerulonephritis Types I, II, and III (Stuhlinger et al. 1976; Pussell et al. 1978; Davis et al. 1981). The increase is small, being much less when compared to a disease such as systemic lupus erythematosus, is not significantly different between the various types, and is due to a positive test in 30 to 50 per cent of patients, i.e. a considerable proportion of cases do not have detectable immune complexes. No analysis of these complexes has been performed in idiopathic mesangiocapillary glomerulonephritis. The presence of complexes is not related to disease activity, serum concentrations of C3, C4, and C3 nephritic factor, or to serial biopsies showing disappearing deposits. An interesting observation that circulating complexes are more frequently detectable in early mild mesangiocapillary glomerulonephritis was interpreted as evidence that the complexes are not nephritogenic (Davis et al. 1981).

These data suggest that circulating immune complexes are not relevant to the pathogenesis of mesangiocapillary glomerulonephritis, or that immune complex deposition in the kidneys, if it occurs, may be by initial antigen deposition followed by antibody fixation in situ in the kidney.

Platelets

Evidence for platelet activation in mesangiocapillary glomerulonephritis has been provided by the detection of reduced concentrations of serotonin within platelets, accompanied by increased serum serotonin concentrations (Parbtani et al. 1980) and decreased in-vivo survival of platelets (George et al. 1974). Immunohistology has shown platelet antigens in glomeruli in mesangiocapillary glomerulonephritis (Miller et al. 1980).

Family studies and genetics

Mesangiocapillary glomerulonephritis has been reported in two unrelated pairs of siblings (Berry et al. 1981) and in successive generations of a family, showing either X-linked recessive (Stutchfield et al. 1986) or autosomal dominant modes of inheritance (Sherwood et al. 1987). Hypocomplementaemia and C3 nephritic factor were not detected.

C3 nephritic factor not associated with overt mesangiocapillary glomerulonephritis has been demonstrated in other siblings (Lopez–Trascasa et al. 1986; Teisberg et al. 1973). An interesting family in which a mother and two of her four children had C3 nephritic factor and partial lipodystrophy has been described. The mother, one child with C3 nephritic factor and another without had mesangiocapillary glomerulonephritis Type I (Power et al. 1990).

These observations, in addition to those relating to inherited complement deficiency, strongly suggest that there is some linkage of a genetic susceptibility to mesangiocapillary glomerulonephritis, C3 nephritic factor, and partial lipodystrophy, or an underlying pathogenetic mechanism common to each of them. On the other hand, reports of identical twins, one of whom had mesangiocapillary glomerulonephritis, C3 nephritic factor, and partial lipodystrophy whereas the other was normal, indicate that any genetic predisposition requires an environmental factor to make these abnormalities manifest (Reichel et al. 1976).

Tissue typing studies in a white family showed that susceptibility to mesangiocapillary glomerulonephritis Type I was associated with HLA B8, DR3 SC01, G402 (Welch et al. 1986). This is of interest in view of the general association of HLA B8/DR3 with autoimmune diseases.

Aetiology and pathogenesis

Despite the clearly defined histological appearances of the two main types of mesangiocapillary glomerulonephritis and the unique hypocomplementaemia of Type II, the strongly associated abnormalities of the complement system, and the related clinical studies, the pathogenesis of this disease remains unknown. The aetiology may be related to antecedent infection in a minority of cases but otherwise it also is largely unknown.

Animal models

Unfortunately the traditional method of injecting known antigen does not produce reliable experimental models of mesangiocapillary glomerulonephritis, although a proportion of rabbits who developed chronic glomerulonephritis induced by bovine albumin did show a histological picture similar to that of mesangiocapillary glomerulonephritis Type I (Kuriyama 1973): this was related to the high precipitating nature of the antibody. In-vivo activation of the complement system by administration of substances such as cobra venom factor, a potent activator of the alternative pathway of complement, has not caused mesangiocapillary glomerulonephritis.

Spontaneous Type I mesangiocapillary glomerulonephritis has been reported in cats, dogs and Finnish-Landrace lambs. Only the last have been studied in depth, and are of particular interest in that they exhibit not only a picture identical to the human disease on light and electron microscopy and on immunohistology, but similar complement abnormalities are also found. This condition also occurs congenitally (Angus et al. 1980).

Two negative points of interest are that Type II mesangiocapillary glomerulonephritis has not been reported in any animal, and mesangiocapillary glomerulonephritis has not yet been found spontaneously in non-human primates, although histological changes similar to mesangiocapillary glomerulonephritis Type I have been reported in an experimental model of nephritis produced by trypanosomiasis in monkeys (Nagle et al. 1974).

Other diseases associated with mesangiocapillary glomerulonephritis

Mesangiocapillary glomerulonephritis of both types has been associated with a wide range of conditions, summarized in Table 1. It is probably no accident that the largest group of conditions, and indeed most cases, are associated with infectious diseases.

These range through viruses, bacteria, and parasites, and an obvious explanation is that these are providing the antigens in an immune–complex mediated disease to provide the Type I picture. Although the histological descriptions of these associated Type I lesions differ in some instances from the typical picture, this may represent no more than variations introduced by differences between the organisms themselves. Type II mesangiocapillary glomerulonephritis is associated with infections, particularly those caused by streptococci. Two interesting points arise from the association of mesangiocapillary glomerulonephritis with streptococci infection. First, the well known cross-reactivity between streptococcal and kidney antigens may result in antibody-mediated renal damage without deposition of exogenous antigens in the kidney. Second, the reduction in streptococcal infections with serious sequelae, i.e. poststreptococcal nephritis and rheumatic fever, in industrialized countries may account for the decreasing incidence of mesangiocapillary glomerulonephritis which is not seen in poorer countries, where streptococcal infection remains common.

Other infectious diseases associated with mesangiocapillary glomerulonephritis, such as malaria and schistosomiasis, produce autoantibodies such as anti-DNA, which can cross-react with glomerular constituents such as heparan sulphate.

The autoimmune diseases noted in Table 1 may cause mesangiocapillary glomerulonephritis by a similar mechanism as they are associated with production of anti-DNA antibodies, or, as with the malignant disorders and pentazoline, may be responsible for an antigen–antibody complex mediated nephritis.

Deficiency of α-antitrypsin could cause nephritis by a failure to inhibit neutrophil elastase activity, with consequent damage to the glomeruli, analogous to the lung damage seen in emphysema: neutrophils are seen quite commonly in the glomeruli in mesangiocapillary glomerulonephritis. Alternatively, since α-antitrypsin was demonstrated by immunofluorescence in subendothelial deposits and along the capillary wall in one case, it has been suggested that the enzyme, which accumulates in cells in the ZZ homozygote because of failure in the final stages of its production, may act as antigen in an immune complex disease (Strife *et al.* 1983).

The hypocomplementaemia

The association of mesangiocapillary glomerulonephritis with congenital deficiencies of complement along with evidence deduced from the association of partial lipodystrophy hypocomplementaemia, and mesangiocapillary glomerulonephritis that hypocomplementaemia precedes mesangiocapillary glomerulonephritis, has led to the general hypothesis that it is in some way due to complement deficiency.

The hypocomplementaemia could operate by allowing persistent subclinical infection due to inefficient opsonization and continuous immune complex formation, and/or by a relative failure of the complement-mediated solubilization of immune complexes, either in the circulation or those deposited in the kidney. This notion fits well with the associations of overt repeated infections and immune complex disease with congenital complement deficiency. However, the immunohistological features of Type II mesangiocapillary glomerulonephritis, which has a stronger association with hypocomplementaemia than Type I, are quite atypical of immune complex disease in that there is a relative paucity of immunoglobulin deposition: in most cases C3 deposition is the sole abnormality. Although it has been suggested that complement activation itself is primarily pathogenetic and causes the glomerular injury, experimental attempts to mimic this mechanism in animals have failed to cause glomerular damage similar to mesangiocapillary glomerulonephritis of any type. However, it is not possible to produce prolonged complement activation as an isolated factor in animals for a period analogous to that occurring in human mesangiocapillary glomerulonephritis, so the correct experiment may not yet have been performed. A third possibility, applicable to Type II, is that the primary abnormality is a structural and chemical change in the basement membranes, i.e. the dense deposits, which activate complement. Complement activation by particulate substances such as yeast is well recognized, and such activation usually favours the alternative pathway. This hypothesis would explain the distribution of the C3 deposits apparently on the 'surface' of the dense deposits, giving the tram-line effect, the chronic and prolonged complement activation in the circulation (remembering the large proportion of circulating blood volume which passes through the kidneys), and would suggest that C3 nephritic factor then arises secondarily to the continuously present alternative pathway convertase C3bBb. Against this notion is the failure of bilateral nephrectomy to alter complement activation and hypocomplementaemia in mesangiocapillary glomerulonephritis (Vallota *et al.* 1972).

Clearly, the above mechanisms do not operate in normocomplementaemic patients with mesangiocapillary glomerulonephritis, unless their normocomplementaemia was preceded by hypocomplementaemia before they presented with their disease, which has now become self-perpetuating. Certainly the sequence of hypocomplementaemia and normocomplementaemia is well documented, while the converse is not (Cameron *et al.* 1983).

Histology

The nature of the 'dense deposit' in Type II mesangiocapillary glomerulonephritis, unique to this disease and to man, could provide important clues as to pathogenesis; however, little is known of its structure, either physically or chemically. The sialic acid content of the glomerular basement membrane is increased and that of cystine is reduced, and by implication this could be due to the presence of the electron-dense material (Galle and Mahieu 1975). There is no detectable staining of the electron-dense material with antisera to complement, immunoglobulins, laminin, fibronectin, entactin, type IV collagen, and glycoproteins derived from human renal tumours. On the other hand it does bind to a lectin known as wheat-germ agglutinin, but not other lectins. Since wheat-germ agglutinin binds to N-acetylglucosamine, it is suggested that this glycoprotein is present in the dense deposits. The peculiar pattern of C3 deposition, shown by immunohistology to outline the dense deposits but not to be within them, suggests that the material comprising the deposits, whether they are due to altered amount and/or quality of carbohydrate and/or amino-acid molecules, is capable of activating complement at a surface. This is analogous to complement activation by such naturally occurring substances as pneumococci or yeast particles.

Recurrence after transplantation

Finally, a clue to the aetiology of mesangiocapillary glomerulonephritis must lie in the high rate of recurrence, particularly of Type II, in patients receiving a renal transplant (see also Chapter 11.6). Whether this points to a role for circulating antigen–

antibody complexes, antibody alone, or a biochemical defect which alters glomerular constituents remains to be seen.

Having gleaned any possible clues from the presumed aetiological associations discussed above, and adding the knowledge from histological and immunological studies, what can we state about the pathogenesis of mesangiocapillary glomerulonephritis? No single mechanism emerges, and indeed the findings in the two main types are so different that this would not be expected. Three main possibilities may be considered.

Firstly, the association with infections and the site of the deposits in Type I, with the associated immune reactants, suggest an antigen–antibody type of disease, either with circulating complexes or with deposition of antigen alone in the kidney and subsequent deposition of antibody. However, circulating antigen–antibody complexes are detected in only a minority of cases, a significant proportion do not have detectable deposition of immunoglobulin (particularly Type II), and no exogenous antigens derived from potential infectious causes have been demonstrated in the glomeruli.

Secondly, the profound and chronic hyposynthesis and activation of complement may cause mesangiocapillary glomerulonephritis. This may be through activation of the complement cascade and its attendant inflammatory sequelae occurring in the kidney. Against this notion is the general inability to induce nephritis in experimental animals by complement activation, with the exception of a mild glomerulitis with C3 deposition in one report using levan in rabbits, and the occurrence of mesangiocapillary glomerulonephritis, particularly in transplanted kidneys, without evidence of complement activation. Alternatively, hypocomplementaemia may predispose to infection, the failure of efficient elimination of exogenous antigen and the persistence of antigen–antibody complexes, manifest as nephritis. This latter explanation obviously relates to the first explanation.

Thirdly, the abnormalities in the glomerular tubular and capillary basement membranes in Type II may be structural, arising from a biochemical defect, due either to some primary defect in the tract, or possibly induced by a foreign antigen deposited in the kidney. The altered basement membrane, in common with other surface activators, may be able to activate complement, predominantly via the alternative pathway. In Type II particularly it is conceivable that the mesangial proliferation of cells and matrix and their extension into the glomerular loops is secondary to complement activation and its inflammatory sequelae.

None of these suggested pathogenetic mechanisms is mutually exclusive, nor is any one particularly favoured by the genetic and hereditary data. These could be associated with a predisposition for any of the mechanisms, either directly or by linkage to other as yet undefined genes.

Clinical features

Incidence

It is difficult to be sure of the true incidence of any form of glomerulonephritis because of variations in patterns of referral to a renal unit, and differences in biopsy policy between units. When expressed as a percentage of biopsies, mesangiocapillary glomerulonephritis accounted for 10 to 20 per cent of primary glomerulonephritis in two European reports during the 1970s (Belgiojoso et al. 1985; Simon et al. 1987); however, these authors report a decreasing incidence to 6 and 2 per cent, respectively, by the mid-1980s. This decrease is due to a reduction in Type I alone; the proportion of Type II rose progressively from 7 to 21 per cent, and there was a parallel fall in poststreptococcal nephritis. The reason for the falling incidence of mesangiocapillary glomerulonephritis over such a short time is best explained by a change in environmental factors, particularly infectious organisms, and circumstantial evidence points again to nephritogenic streptococci. The incidence of Type II mesangiocapillary glomerulonephritis is much higher in children, being 43 to 48 per cent, compared to 22 per cent in adult populations studied at the same time (Habib et al. 1973; Cameron et al. 1983). As a cause of nephrotic syndrome, mesangiocapillary glomerulonephritis is found in 5 per cent of childhood cases, and in 10 per cent of adults.

In non-industrialized countries, mesangiocapillary glomerulonephritis accounts for a higher proportion of cases of nephritis, for example, 40 per cent in Mexico (Mota-Hernandez et al. 1985). Because of the association of infection with Type I, its incidence is considerably higher than that of Type II. Thus Date et al. (1983) reporting from southern India, found mesangiocapillary glomerulonephritis in 7 per cent of biopsied cases of nephritis; of these, 6 per cent were Type II.

Clinical presentation

The general picture of the clinical features of mesangiocapillary glomerulonephritis will be discussed first, followed by the particular features associated with the histological types, and ages and geographical location of the patients. Several groups in Europe and North America have published large series of patients of Type I and II together, and provide a reasonably concordant picture of the clinical characteristics of the idiopathic disease in children and adults (Herdman et al. 1970; Mandalenakis et al. 1971; Habib et al. 1973; West et al. 1976; Belgiojoso et al. 1977; Davis et al. 1978; Donadio et al. 1979; Cameron et al. 1983; Swainson et al. 1983).

Mesangiocapillary glomerulonephritis can present with any of the whole gamut of clinical syndromes of glomerular disease, ranging from symptomless proteinuria and haematuria through nephrotic syndrome, chronic renal failure, and acute renal failure. Asymptomatic proteinuria and microscopic haematuria and the nephrotic syndrome account for the majority of presentations, reported as 23 to 30 per cent and 42 to 67 per cent, respectively. Patients presenting with chronic or even end-stage renal failure and an acute nephritic syndrome are not uncommon. The latter is often associated with an antecedent infection and/or with crescentic glomerulonephritis and accounts for 16 to 30 per cent of patients. Macroscopic haematuria, which is intermittent and in some patients associated with sore throats and chest infections, is reported in 10 to 20 per cent of patients.

Anaemia may accompany mesangiocapillary glomerulonephritis; affected patients present with complaints of tiredness, breathlessness, and pallor which are not thought to be explicable by the degree of renal failure. This anaemia has been related to the presence of activated complement on the surface of the red cells. Hypertension is frequent, present in 80 to 90 per cent of patients at presentation, as is, according to some reports, a preceding infection, particularly in non-industrialized countries. In Western countries 40 to 60 per cent of patients have raised antistreptolysin antibody titres.

Reference to Table 1 will emphasize that at presentation patients with mesangiocapillary glomerulonephritis may have clinical features of a wide range of associated disorders, and these may clinically overshadow the nephritis itself. This is particularly the case when mesangiocapillary glomerulonephritis is associated with infections. The clinical features change as the disease evolves and progresses. Since the majority of patients develop end-stage renal failure symptoms of uraemia will inevitably supervene; many patients may become nephrotic after presentation, as many as 80 per cent in some series. The development of crescentic nephritis or the rapid stepwise deterioration which characterizes the course of many patients can produce an acute nephritic syndrome or acute renal failure.

Histological types and clinical features

Type I
The mean age of these patients is 24 ± 16 years, and this type has been more predominantly associated with the symptoms of anaemia and fatigue. Type I is also more frequently associated with infections and complement deficiency than Types II and III. There are differing accounts of the sex incidence, none of which have markedly favoured one sex, indicating that the male/female ratio does not deviate much from unity.

Type II
In addition to the studies describing Types I and II together, noted above, several groups have reported on their experience of Type II mesangiocapillary glomerulonephritis alone (Antoine and Faye 1972; Habib *et al.* 1975; Vargas *et al.* 1976; Droz *et al.* 1977; Lamb *et al.* 1977; Bennett *et al.* 1989). These patients are younger, mean age 15 ± 11 years, again with differing accounts of the predominant sex. There is a higher incidence of acute renal failure and crescentic nephritis, and the association between mesangiocapillary glomerulonephritis and partial lipodystrophy is mainly with this type.

Type III (Strife *et al.* 1984; Jackson *et al.* 1987)
This variety of mesangiocapillary glomerulonephritis is characterized by a higher incidence of symptomless presentation with microscopic haematuria and proteinuria. One series reported that 66 per cent of patients presented in this way. There is also a predilection for the tiredness–pallor–anaemia cluster of symptoms.

Age groups and clinical features
Although mesangiocapillary glomerulonephritis can affect any age group, it predominantly affects children and young adults. Reports focusing on Types I and II in children have been published by Habib *et al.* (1973) and Davis *et al.* (1978), while Watson *et al.* (1984) have described Type I and The Southwest Paediatric Nephrology Study Group (USA) (1985) described Type II. Particularly important are two reports from single centres comparing Type I mesangiocapillary glomerulonephritis alone in adults and in children (Magil *et al.* 1979) and comparing Types I and II in both age groups (Cameron *et al.* 1983). The bulk of patients are between 5 and 30 years of age, but rare patients as young as 2 years old have been reported. Since Type II mesangiocapillary glomerulonephritis is more common in younger patients, accounting for 45 per cent of patients, their presenting features, associated conditions, and clinical course are influenced by the factors noted above. Studies of paediatric populations have shown a high incidence of nephrotic syndrome in mesangiocapillary glomerulonephritis: e.g. 67 per cent at presentation rising to 86 per cent during the whole of the disease (Habib *et al.* 1973). Mesangiocapillary glomerulonephritis accounts for 12 to 15 per cent of cases of childhood nephrotic syndrome.

Children with mesangiocapillary glomerulonephritis presenting with nephrotic syndrome differ from those with minimal change nephropathy in that there is no male preponderance, the median age of onset is greater, and there is more risk of developing hypertensive encephalopathy on daily high-dose steroids.

In the elderly (patients aged more than 60 years) mesangiocapillary glomerulonephritis is a rare cause of nephritis, with an incidence of 3 to 4 per cent (Moorthy and Zimmerman 1977; Cameron *et al.* 1983), which is half its incidence in the under 60s. Consequently it accounts for only a small proportion (6 per cent) of nephrotic syndromes in the elderly, in whom membranous nephropathy, amyloidosis, and minimal change nephropathy account for the majority of cases (Bolton 1984).

The majority of cases reported in the elderly have been Type I, which may reflect the increased susceptibility of this group to infections.

Geography and clinical features
Since the bulk of published information comes from the Western industrialized nations, there has been inadequate recognition of the different picture of mesangiocapillary glomerulonephritis seen in non-industrialized countries. Reports from India, Mexico, and Africa exemplify the less frequent occurrence of Type II and the higher incidence of associated infections, predominantly with Type I (Date *et al.* 1983).

Diagnosis
The diagnosis of mesangiocapillary glomerulonephritis is of course histological, since the clinical features due directly to the renal disease are common to virtually all other glomerular disorders. A clinical diagnosis is therefore not very practicable, but certain features are suggestive. For example, suspicions should be aroused by a patient who is a child or young adult, with episodes of macroscopic haematuria, possibly with an anaemia out of proportion to the impairment of glomerular filtration rate. If the patient has one of the associated diseases, or for some reason is known to have chronic hypocomplementaemia in the absence of symptoms and signs suggestive of systemic lupus, then the likelihood of mesangiocapillary glomerulonephritis is increased. The urinary findings of non-selective proteinuria, haematuria, and casts typical of glomerulonephritis are not diagnostic for mesangiocapillary glomerulonephritis.

Outcome and prognostic features
Several centres, virtually all of them from Western industrialized countries, have reported on the clinical course of mesangiocapillary glomerulonephritis. Some authors have included a minority of patients who have been treated with anti-inflammatory and immunosuppressive drugs and have analysed them with the untreated group; it is unlikely that this is a confounding feature as the proportion of treated patients is small, and it is debatable whether these drugs have an effect on outcome. The following is a precis of these reports, which deal with idiopathic mesangiocapillary glomerulonephritis: Types I and II: Herdman *et al.* 1970; Mandezanakis *et al.* 1971; Habib *et al.* 1973; West *et al.* 1976; Belgiojoso *et al.* 1977; Davis *et al.* 1978; Donadio *et al.* 1979; Cameron *et al.* 1983; Swainson *et al.* 1983. Type II: Antoine and Faye, 1972; Habib *et al.* 1975; Vargas *et*

al. 1976; Droz et al. 1977; Lamb et al. 1977; Bennett et al. 1989. Type I Magil et al. 1979.

General picture

Mesangiocapillary glomerulonephritis carries a poor prognosis. The 50 per cent survival to death or end-stage renal failure is closely agreed at 10 to 11 years, with 10 per cent survival at 20 years. The period from presentation to death or end-stage renal failure varies from 3 to 20 years. Of those surviving, 60 to 70 per cent, are in chronic renal failure. These figures put idiopathic mesangiocapillary glomerulonephritis amongst the forms of idiopathic glomerulonephritis with the worst prognosis. Spontaneous remission has been reported in a range of 2 to 20 per cent of patients, with a median of 5 to 6 per cent.

The progression through uraemia to end-stage renal failure is characterized by an increasing incidence of hypertension and nephrotic syndrome. A particular feature of the progression of mesangiocapillary glomerulonephritis is the stepwise and relatively sudden deterioration in renal function seen in some patients. Although repeat biopsy may reveal the development of crescentic nephritis in a proportion of patients, others decline seemingly without explanation.

Variations according to types of mesangiocapillary glomerulonephritis

In comparisons of Types I and II from single centres it should be emphasized that few have included more than 10 individual cases of Type II, thus making the comparisons somewhat uncertain.

The general conclusion, supported by one of the largest studies (Cameron et al. 1983) is that there is no difference in outcome between Types I and II, although there is a much lower rate of remission in Type II. The exception is provided by a study in which children with Type I were compared with children with Type II (Habib et al. 1975); the former achieved 50 per cent survival at 18 years and the latter at 10 years. The authors related the difference to the higher incidence of crescentic nephritis in their patients with Type II mesangiocapillary glomerulonephritis, but if the cases of crescentic nephritis were excluded, the outcomes of Types I and II were the same.

Variation according to age at presentation

There is a paucity of studies from single centres comparing mesangiocapillary glomerulonephritis in adults and in children. Magil et al. (1979) presented data on 46 patients with Type I only, and Cameron et al. (1983) reported on 104 patients with Types I and II disease (Type III being analysed with Type I). The main difference between adults and children is that the younger patients have a slower rate of progression to end-stage renal failure. Thus, when compared to adults over a 10-year period, children had a lower incidence of renal failure. This difference was seen for Type I and Type II considered both together and separately, although statistical significance was not achieved for the latter alone. However, after 10 years children catch up with adults, and the final results are not different. Since there is a higher incidence of Type II in children than in adults, the remission rate is lower in the younger group.

Variation according to sex

There is a tendency for adult males with Type I to fare worse than females, but there are no data for Type II (Magil et al. 1979; Cameron et al. 1983).

Geographical variations

As mentioned above, few long-term studies of mesangiocapillary glomerulonephritis have come from outside Europe and North America. A recent study from Hong Kong shows a remarkable similarity in outcome for Type I in 46 adult Chinese patients studied (Chan et al. 1989). Twenty per cent of these patients were positive for HBsAg, but the outcome for this group was no different to that of the negative group.

Prognosis

Analyses of the clinical course and outcome of mesangiocapillary glomerulonephritis have tried to define features which can indicate prognosis. In practical terms, because of the overall poor outcome, this means features which define those patients who will die or develop end-stage renal failure more rapidly than average.

Histology

The differences between Types I and II have been discussed above: there is little effect on outcome, although Type II carries a lower incidence of remission. A comparison of the simple and lobular types in one study showed no difference in outcome (Bohle et al. 1974), although another group found that the lobular form was associated with a worse prognosis (Belgiojoso et al. 1977).

It is generally agreed that the presence of crescents or sclerosis worsens the prognosis, and, surprisingly, one study found that this effect was not related to an increasing proportion of crescents (Cameron et al. 1983). The main effect of these changes is in the first decade of follow up; thereafter the prognosis becomes similar, implying that the crescents govern the rate of loss of renal function rather than eventual development of end-stage renal failure. The amount of interstitial fibrosis correlates well with the reduction in GFR, in common with other forms of nephritis (Schmitt et al. 1987).

Clinical features

Most authors agree that presentation with a nephrotic syndrome worsens the prognosis by an approximate factor of two, although one study found that the outcome was not related to this feature (Davis et al. 1970). Remission despite the presence of a nephrotic syndrome has been noted. Not surprisingly, uraemia at presentation has been linked with a worse prognosis, although in one study this was true of Type I only (Cameron et al. 1983), and so too has an acute nephritic syndrome. Hypertension has been variously reported as being a good or poor guide to outcome.

Most reports state that macroscopic haematuria has no bearing on prognosis, but a study of children alone found that its presence was related to a worse prognosis (Habib et al. 1973).

There are no features predicting remission, although there is an increased likelihood of remission in patients with a treatable cause of infection.

Immunological features

The presence or absence of hypocomplementaemia, its severity when present, and the presence or absence of C3 nephritic factor are, according to most reports, irrelevant to the prognosis. The exceptions are the studies by Swainson et al. (1983), which related decreased serum C3 concentrations to diminished survival, and by Klein et al. (1980), whose patients with a benign course were found to have normal serum concentrations of C3

Table 2 Controlled trials of various drugs in patients with mesangiocapillary glomerulonephritis

Reference	Drugs	Patients n; Types I/II	Renal function			Follow-up (years)	Controls n; Types I/II	Renal function		
			Increased	Stable	Decreased			Increase	Stable	Decrease
Kincaid-Smith (1972)	Cyclophosphamide + Dipyridamole + Warfarin	16; NK	10	3	3	3	13; NK	0	0	13
Lagrue (1975)	Chlorambucil Azathioprine	25; NK	1	4	20	2	9; NK	1	2	6
Donadio (1984)	Dipyridamole + Acetylsalicylic acid	21; 21/2	0	18	3	7	19; 19/0	0	10	9
Zimmerman* (1983)	Dipyridamole + Warfarin	13; NK	0	13	0	1	13; NK	0	9	4
Cattran (1985)	Cyclophosphamide + Dipyridamole + Coumadin	22; 17/5	4	12	6	1.5	25; 18/5	6	13	6
Mota-Hernandez (1985)	Prednisolone	10; 10/0	1	9	0	6.5	8; 7/1	1	3	4

NK, not known.
* Crossover trial.

and factor B, and no C3 nephritic factor, in contrast to the progressive group; the latter study was of Type II patients only. Schena et al. (1982) related a more rapid rate of deterioration in renal function to the presence of C3 nephritic factor, but found that its serum levels were not helpful for monitoring the clinical course.

Pregnancy

Pregnancy and its outcome in patients with mesangiocapillary glomerulonephritis is dealt with in Chapter 15.3.

Treatment

The problems attending the choice and evaluation of treatment of patients with idiopathic mesangiocapillary glomerulonephritis epitomize the difficulties which have beset the treatment of chronic nephritis. The disease lasts many years, it may remit spontaneously, and even when progressive it may fluctuate quite markedly and quickly. There is no identified pathogenetic mechanism or aetiological agent which may be monitored, in contrast to, for example, antiglomerular basement membrane disease. Unless the cases are defined clearly, the varying mix of Types I, II, and III may influence the outcome of therapeutic trials because of possible individually different prognoses. Most trials have studied small numbers of patients and have been uncontrolled. Comparative trials using historical control groups are flawed because treated groups have survived from clinical onset to treatment, a condition not pertaining to the untreated group. Despite these difficulties (or by dint of ignoring them) several studies have been reported and are summarized in Tables 2 and 3. Consideration of crescentic mesangiocapillary glomerulonephritis is excluded from this chapter (see Chapter 3.10).

Rationale of treatments used

In those cases of mesangiocapillary glomerulonephritis associated with a known infectious organism it is obvious that attempts should be made to eradicate it. The approach to treatment of idiopathic mesangiocapillary glomerulonephritis has been coloured by the view that it is an immune complex disease with reactive proliferation in the glomeruli. Therapy, limited anyway by the types of drugs available, has been rationalized by its description as being immunosuppressive, anti-inflammatory, and antiproliferative. The agents used have been steroids, mainly for their anti-inflammatory effect, azathioprine, cyclophosphamide, and chlorambucil for their antimitotic, antiproliferative, and immunosuppressive effects, anticoagulants and antiplatelet drugs to prevent fibrinogen deposition (although it appears that fibrinogen is not laid down via the normal coagulation cascade) and platelet activation, and non-steroidal anti-inflammatory drugs. More recently there have been reports of the use of plasma exchange, to remove circulating antigens, antibodies, antigen–antibody complexes, and mediators of inflammation, and inevitably cyclosporin A, to interrupt the immune response by its inhibition of interleukin 2 mRNA production. There is no clear evidence that there is any overall successful treatment, and the few controlled trials have not confirmed the positively interpreted uncontrolled results of steroid therapy. Some trials of therapy have been confused by the introduction in some cases of other treatments in addition to the basic regimen.

Steroids

Over the years much attention has been paid to the continuing series of patients, who were children or teenagers, treated with 60 mg/kg prednisolone on alternate days for prolonged periods by the group at Cincinnati (McAdams et al. 1976; McEnery et al. 1980, 1986; West 1986). Several patients were also pretreated with a continuous course of steroids for several weeks, or received courses of methchlorethamine or dipyridamole and aspirin during the alternate day steroid treatment. By comparison with series of cases reported from France and the United Kingdom, it was concluded that patient/renal survival was improved, that the overall incidence of nephrotic syndrome was reduced, and that repeat renal biopsies showed evidence of reduced proliferation, although there had been inevitable progression of some glomeruli to sclerosis. The authors strengthened their case by dividing their treated patients into two groups according to the length of known disease at the time of beginning treatment. Significant differences for renal and patient survival and incidence of nephrotic syndrome in favour of those with less evolved disease were found. In a separate report of the treatment of Type III mesangiocapillary glomerulonephritis, uncontrolled results showed that 60 per cent of patients without nephrotic syndrome improved, and 65 per cent of those without nephrotic syndrome did not develop renal failure. Repeat renal biopsy showed evidence of improvement in 50 per cent.

The only controlled trial of steroids reported selected children with Type I disease, and analysed the data in two ways (International Study of Kidney Diseases in Children, 1980, 1982). Fifty per cent reduction of GFR at 5 years was taken as the endpoint. It was avoided in 95 per cent of treated patients, and in 57 per cent of controls. Costing the treatment by assessing major toxic effects, such as fits or hypertension, and by no reduction in GFR, showed no difference in the groups—55 per cent treated and 53 per cent placebo. Six patients had to be withdrawn because of toxicity of steroids. It would seem therefore that steroids used in this way can have a beneficial effect on the rate of progression of mesangiocapillary glomerulonephritis Type I, but at a cost which negates the benefit.

Anticoagulants/antiplatelet drugs

Controlled trials have examined mixtures of these drugs; none has been performed on one agent alone. A combination of dipyridamole and aspirin in 40 patients with Type I mesangiocapillary glomerulonephritis produced a decreased progression of renal failure at 1 year in the treated group, associated with prolongation of the reduced platelet survival time. The benefit persisted over the succeeding 7 years, during which the treated group continued taking their drugs and suffered a lower rate of end-stage renal failure (Donadio et al. 1984), but analysis at 10 years showed no significant difference between the two groups (Donadio and Offord, 1989).

A combination of dipyridamole and warfarin in 17 Type I and 1 Type II patients was succeeded at 1 year by a cross-over trial alternating treatment with placebo. The results of the first stage showed better preservation of function in the treated group during the first phase, but the advantage was not evident in the cross-over trial. Again, therapeutic cost was high, one-third of patients suffering significantly complications due to haemorrhage, with one death (Zimmerman et al. 1983).

It seems therefore that a combination of antiplatelet and anticoagulant drugs may slow deterioration in mesangiocapillary glomerulonephritis, but the renal benefits are not overwhelming, and may be bought dearly.

Using a rigorous selection of published data, Schena and Cameron (1988) analysed the data from Kincaid-Smith (1972), Lagrue et al. (1975), Vanrenterghem et al. (1975), Abreo et al. (1982), Donadio et al. (1982), Zimmerman et al. (1983), Cameron et al. (1983), Cattran et al. (1985), and Blainey et al. (1986). They found that the percentage of unchanged and improved function was not statistically different for treated and untreated patients, either in all the studies combined, or when the prospective controlled trials were analysed separately.

Antimitotic/immunosuppressive drugs

Published reports have rarely used these drugs alone, which makes their assessment difficult. They have mostly been used with steroids, not in controlled trials, and the data available show little evidence of an effect. An early report in 1972 found that cyclophosphamide with dipyridamole and warfarin in Types I and II produced improvement in GFR in 60 per cent of patients, and histological improvement in 25 per cent (Kincaid-Smith 1972). This study was not controlled; retrospective analysis of the treated group with an untreated group showed improved survival in the former. Two controlled trials, one in the United States and one in Australia, examined this combination of drugs and showed no beneficial effects; again there was a high cost manifest by bleeding complications and marrow toxicity (Tiller et al. 1981; Cattran et al. 1985).

Used in combination with the non-steroidal anti-inflammatory drug indomethacin, cyclophosphamide reduced proteinuria and decreased rate of deterioration in renal function; only nine patients were observed and the study was uncontrolled (Vanrenterghem et al. 1975). Another uncontrolled study of patients using a mixture of non-steroidal anti-inflammatories showed improvement in renal function.

Table 3 Uncontrolled trials of various drugs in patients with mesangiocapillary glomerulonephritis

Reference	Drugs	Patients n; Types I/II	Renal function			Follow-up (years)
			Increased	Stable	Decreased	
Habib (1973)	Prednisone	28; NK	3	13	12	
	Chlorambucil	66; NK	20	33	13	
Vanrenterghem (1975)	Non-steroidal anti-inflammatory drug + Cyclophosphamide	9; NK	4	3	2	1.7
McEnery (1980)	Prednisolone	27; 15/5 + 7–III	7	16	4	6
Abreo (1982)	Prednisone Dipyridamole Warfarin Cyclophosphamide	9; NK	3	4	2	7.8
Cameron (1983)	Prednisone Azathioprine Cyclophosphamide	38;	7	19	12	—
Strife (1984)	Prednisone	16; all III	1	9	6	5
Warrady (1985)	Prednisone	6; 6/0	4	2	0	5
Blainey (1986)	Prednisone Cyclophosphamide	69; NK	7	20	42	6.8

Plasma exchange

Sporadic reports of plasma exchange in conjunction with other treatments have shown some improvement, although this was not long-lasting in some cases. In one small study of six patients, three with Type I and three with Type II, plasma exchange was used alone. In four patients with Type I mesangiocapillary glomerulonephritis and one with normocomplementaemic Type II, there was a reduction in plasma creatinine with treatment which reversed when the treatment ceased. The other two patients, with hypocomplementaemic Type II, did not respond (McGinley *et al.* 1985).

Summary

The therapeutic picture at present may be summarized by saying that the evidence from controlled trials indicate that only a combination of anticoagulants and antiplatelet drugs may have produced some benefit. Steroids alone seem to be without effect, despite the strong case made in their favour. However the patients treated, although classified as idiopathic, could represent an aetiological mixture, some of which may be steroid responsive. They were also of differing times of onset, with a probable reduction in likely response with a later start of treatment in the course of the disease. There may, therefore, be a subset of steroid-responsive patients in the whole group of reported cases of treated mesangiocapillary glomerulonephritis. The roles of non-steroidal anti-inflammatory drugs, plasma exchange, and cyclosporin A must await further evaluation as the numbers reported are very small, and no controlled studies have been performed.

With the overall disappointing results of treatment and the relative and increasing rarity of idiopathic mesangiocapillary glomerulonephritis, it is not easy to advise a strategy for the individual patient. Excluding the case with rapidly progressive/crescentic nephritis, a reasonable plan may be as follows. If the patient is nephrotic a course of steroids may be tried to reduce the proteinuria. Since the available evidence indicates that anticoagulants plus antiplatelet drugs may improve the prognosis, these can be employed with monitoring of renal function; in the absence of response they may be discontinued. Since the main benefit of therapy would seem to be to delay end-stage renal failure, it should be remembered that the treatment of the latter in the young—the main sufferers from mesangiocapillary glomerulonephritis—has good results. Where treatment of end-stage renal failure is easily available there is, therefore, a case to be made for not treating mesangiocapillary glomerulonephritis with the advantages of avoiding high doses of steroids during growth and the risks of anticoagulation in an age-group prone to trauma.

Mesangiocapillary glomerulonephritis—how many diseases?

Having reviewed the pathological, immunological and clinical features of mesangiocapillary glomerulonephritis, we should now address the question of whether we are studying a single disease or not. Since 'mesangiocapillary glomerulonephritis' is a histological diagnosis and Types I and II are designated by major histological differences, it would seem self-evident that it represents more than one disease, especially in view of the profound and totally definitive electron microscopic difference between the two types. Enhancing this difference are the contrasts in immunohistology, the higher incidence of and more marked hypocomplementaemia with the increased frequency of C3 nephritic factor in Type II, and its much stronger association with partial lipodystrophy. On the other hand, there are many similarities in the light microscopic appearances and the clinical features which are in favour of regarding the two types as the same disease.

As with any disease in which the pathogenesis is unknown, it is not possible to answer the question satisfactorily. Perhaps the best approach at present is to compromise by regarding mesangiocapillary glomerulonephritis as an all-embracing term for clinical practice, but to separate Types I and II for the purposes of investigative pathology until there is convincing evidence that they are identical.

References

Abreo, K. and Moorthy, V. (1982). Type 3 mebranoproliferative glomerulonephritis. *Archives of Pathology and Laboratory Medicine*, **106**, 413–7.

Angus, K.W., Gardiner, A.C., Mitchell, B., and Thomson, D. (1980). Mesangiocapillary glomerulonephritis in lambs: the ultra-structure and immunopathology of diffuse glomerulonephritis in newly born Finnish Landrace lambs. *Journal of Pathology*, **131**, 65–74.

Antoine, B. and Faye, C. (1972). The clinical course associated with dense deposits in the kidney basement membranes. *Kidney International*, **1**, 420–7.

Belgiojoso, B.G., Tanantino, A., Colasanti, G., Bazzi, C., Guerra, L., and Durante, A. (1977). The prognostic value of some clinical and histological parameters in membranoproliferative glomerulonephritis. *Nephron*, **19**, 250–8.

Belgiojoso, G.B., *et al* (1985). Is membranoproliferative glomerulonephritis really decreasing. *Nephron*, **40**, 380–1.

Bennett, W.M., *et al.* (1977). Partial lipodystrophy, C3 nephritic factor and clinically inapparent mesangiocapillary glomerulonephritis. *American Journal of Medicine*, **62**, 757–60.

Bennett, W.M., Fassett, R.G., Walker, R.G., Fairley, K.F., D'Apice, A.J.F., and Kincaid-Smith, P. (1989). Mesangiocapillary glomerulonephritis Type II (dense deposit disease): clinical features of progessive disease. *American Journal of Kidney Diseases*, **13**, 469–76.

Berry, P.L., McEnery, P.T., McAdams, A.J., and West, C.D. (1981). Membranoproliferative glomerulonephritis in two shibships. *Clinical Nephrology*, **16**, 101–6.

Blainey, J.D., Brewer, D.B., and Hardwick, J. (1986). Proteinuric glomerular disease in adults: cumulative life tables over twenty years. *Quarterly Journal of Medicine*, **59**, 555–67.

Bohle, A., *et al.* (1974). The morphological and clinical features of membranoproliferative glomerulonephritis in adults. *Virchow's Archiv (A): (Pathological Anatomy)*, **363**, 213–24.

Bolton, W.K. (1984). Nephrotic syndrome in the aged. *Proceedings of Ninth International Congress of Nephrology*, **72A**.

Cameron, J.S., *et al.* (1983). Idiopathic mesangiocapillary glomerulonephritis. Comparison of Types I and II in children and adults and long-term prognosis. *American Journal of Medicine*, **74**, 175–92.

Cattran, D.C., *et al.* (1985). Results of a controlled drug trial in membranoproliferative glomerulonephritis. *Kidney International*, **27**, 436–41.

Chan, M.K., Chan, K.W., Chan, P.C.K., Fang, G.X., and Cheng, I.K.P., (1989). Adult-onset mesangiocapillary glomerulonephritis: a disease with a poor prognosis. *Quarterly Journal of Medicine*, **72**, 599–608.

Charlesworth, J.A., Williams, D.G., Sherington, E., Lachmann, P.J., and Peters, D.K. (1974). Metabolic studies of the third component of complement and the glycine-rich B glycoprotein in patients with hypocomplementaemia. *Journal of Clinical Investigation*, **53**, 1578–87.

Coleman, T.H., Forristal, J., Kosaka, T., West, C.D. (1983). Inherited complement component deficiencies in membranoproliferative glomerulonephritis. *Kidney International*, **24**, 681–90.

Date, A., Neela, P., Shastry, J.C.M. (1983). Membranoproliferative glomerulonephritis in a tropical environment. *Annals of Tropical Medicine and Parasitology*, **3**, 279–85.

Davis, C.A., Marder, H., and West, C.D. (1981). Circulating immune complexes in membranoproliferative glomerulonephritis. *Kidney International*, **20**, 728–32.

Davis, A.E., Schneeberger, E., McCluskey, R., and Grupe, W.E. (1970). Mesangial proliferative glomerulonephritis with irregular intramembranous deposits. Another variant of hypocomplementaemic nephritis. *American Journal of Medicine*, **63**, 481–7.

Davis, A.E., Schneeberger, E.E., Grupe, W.E., and McCluskey, R.T. (1978). Membranoproliferative glomerulonephritis (MPGN Type I) and dense deposit disease (DDD) in children. *Clinical Nephrology*, **9**, 184–93.

Donadio, J.V. and Offord, K.P. (1989). Reassessment of treatment results in membranoproliferative glomerulonephritis, with emphasis on life-table analysis. *American Journal of Kidney Disease*, **14**, 445–51.

Donadio, J.V., Slack, T.K., Holley, K.E., and Ilstrup, D.M. (1979). Idiopathic membranoproliferative (mesangiocapillary) glomerulonephritis. A clincopathologic study. *Mayo Clinic Proceedings*, **54**, 141–50.

Donadio, J.V., Anderson, F., Mitchell, J.C., Fuster, V., Holley, K.E., and Ilstrup, D. (1984). Membranoproliferative glomerulonephritis. A prospective trial of platelet-inhibitor therapy. *New England Journal of Medicine*, **310**, 1421–6.

Droz, D., Zanetti, M., and Noel, L.H. (1977). Dense deposits disease. *Nephron*, **19**, 1–11.

Eisinger, A.J., Shortland, J.R., and Moorhead, P.J. (1972). Renal disease in partial lipodystrophy. *Quarterly Journal of Medicine*, **56**, 343–54.

Friedman, A.L., Chesney, R.W., and Oberley, T.D. (1983). Clinical systemic lupus erythematosus with dense deposit disease. *International Journal of Pediatric Nephrology*, **4**, 171–5.

Galle, P. and Mahieu, P. (1975). Electron dense alteration of kidney basement membranes. *American Journal of Medicine*, **58**, 749–64.

George, C.R.P., Slichter, S.J., and Quadracci, L.J. (1974). A kinetic evaluation of hemostasis in renal disease. *New England Journal of Medicine*, **291**, 1111–7.

Gotoff, S.P., Fellers, F.X., Vawter, G.F., Janeway, C.A., and Rosen, F.S. (1965). The B1C globulin in childhood nephrotic syndrome. *New England Journal of Medicine*, **237**, 524–9.

Habib, R., Kleinknecht, C., Gubler, M.C., Levy, M. (1973). Idiopathic membranoproliferative glomerulonephritis in children. Report of 105 cases. *Clinical Nephrology*, **1**, 194–213.

Habib, R., Gubler, M.C., Loirat, C., Ben Maiz, H., and Levy, M. (1975). Dense deposit disease: a variant of membranoproliferative glomerulonephritis. *Kidney Internationl*, 204–15.

Herdman, R.C., *et al.* (1970). Chronic glomerulonephritis associated with low serum complement activity (chronic hypocomplementaemic glomerulonephritis). *Medicine (Baltimore)*, **49**, 207–26.

ISKDC (1980). Membranoproliferative glomerulonephritis (MPGN). A double trial of alternate day prednisone (ADP). *Pediatric Research*, **17**, 100.

ISKDC (1982). Alternate day steroid treatment in membranoproliferative glomerulonephritis: a ramdomized controlled clinical trial. *Kidney International*, **21**, 150.

Jackson, E.C., McAdams, A.J., Strife, F., Forristal, J., Welch, T.R., and West, C.D. (1987). Differences between membranoproliferative glomerulonephritis Types I and III in clinical presentation, glomerular morphology and complement perturbation. *American Journal of Kidney Diseases*, **9**, 115–20.

Jones, D.B. (1977). Membranoproliferative glomerulonephritis. One or many diseases? *Archives of Pathology and Laboratory Medicine*, **101**, 457–61.

Klein, M., *et al.* (1980). Characteristics of a benign sub-type of dense deposit disease: companion with the progressive form of this disease. *Clinical Nephrology*, **20**, 163–71.

Kincaid-Smith, P. (1972). The treatment of chronic mesangiocapillary (membranoproliferative) glomerulonephritis with impaired renal function. *Medical Journal of Australia*, **2**, 589–92.

Kuriyama, T. (1973). Chronic glomerulonephritis induced by prolonged immunization in the rabbit. *Laboratory Investigation*, **28**, 224–35.

Lagrue, G., Bernard, D., Bariety, J., Druet, P., and Guevel, J. (1975). Trâitement par le chlorambucil et l'azathiorpine dans les glomerulonephritis primitives. Resultats d'une étude controleé. *Journal d'Urologie et Nephrologie*, **9**, 655–72.

Lamb, V., Tisher, C.G., McCoy, R.C., and Robinson, R.R. (1977). Membranoproliferative glomerulonephritis with dense intramembranous alterations. *Laboratory Investigation*, **36**, 607–17.

Lopez-Trascasa, M., Vicario, J.L., Martin, J.M., and Martin, M.A. (1986). Familial incidence of C3 nephritic factor. *Complement*, **3**, 204–5.

Magil, A.B., Price, J.D.E., Bower, G., Rance, C.P., Huber, J., and Chase, W.I. (1979). Membranoproliferative glomerulonephritis Type I in children: comparison of natural history in children and adults. *Clinical Nephrology*, **11**, 234–44.

Mandalenakis, N., Mendoza, N., Pirani, C.L., and Pollak, V.E. (1971). Lobular glomerulonephritis and membranoproliferative glomerulonephritis. *Medicine (Baltimore)*, **50**, 319–55.

Mazzucco, G., Belgiojoso, G.B., Confalonieri, R., Coppo, R., and Monga, G. (1980). Glomerulonephritis with dense deposits: a variant of membranoproliferative glomerulonephritis or a separate morphological entity. *Virchows Arch (A): Pathological Anatomy and Histology*, **387**, 17–29.

McAdams, A.J., McEnery, P.T., and West, D. (1976). Mesangiocapillary glomerulonephritis: changes in glomerular morphology with long-term alternate-day prednisone therapy. *Journal of Pediatrics*, **86**, 23–31.

McEnery, P.T., McAdams, A.I., and West, C.D. (1980). Membranoproliferative glomerulonephritis: improved survival with alternate day prednisone therapy. *Clinical Neprhology*, **13**, 117–24.

McEnery, P.T., McAdams, A.I., and West, C.D. (1986). The effect of prednisone in a high-dose alternate-day regimen on the natural history of idiopathic membranoproliferative glomerulonephritis. *Medicine*, **64**, 401–23.

McGinley, E., Watkins, R., McLay, A., and Jones, J.M.B. (1985). Plasma exchange in the treatment of mesangiocapillary glomerulonephritis. *Nephron*, **40**, 385–90.

Miller, K., Dresner, I.G., and Michael, A.F. (1980). Localization of platelet antigens in human kidney disease. *Kidney International*, **18**, 472–7.

Mollnes, T.E., Ng, Y.C., Peters, D.K., Lea, T., Tschopp, J., and Harboe, M. (1986). Effect of nephritic factor on C3 and the terminal pathway of complement *in vivo* and *in vitro*. *Clinical and Experimental Immunology*, **65**, 73–9.

Moorthy, A.V. and Zimmerman, S.W. (1977). Renal disease in the elderly: clinicpathologic analysis of renal disease in 115 elderly patients. *Clinical Nephrology*, **14**, 223–9.

Mota-Hernandez, F., Gordillo-Paniagua, G., Munoz-Arizpe, R., Lopez-Arriaga, J.A., and Barboza-Madueno, L. (1985). Prednisone versus placebo in membranoproliferative glomerulonephritis: long term clinicopathological correlations. *International Journal of Pediatric Nephrology*, **6**, 25–8.

Nagle, R.B., *et al.* (1974). Experimental infections with African trypanosomes VI. Glomerulonephritis involving the alternate pathway of complement activation. *American Journal of Tropical and Medical Hygiene*, **23**, 15–26.

Ng, Y.C. and Peters, D.K. (1986). C3 nephritic factor (C3NeF): dissociation of cell-bound and fluid phase stablization of alternative pathway C3 convertase. *Clinical and Experimental Immunology*, **65**, 450–7.

Nolasco, F.E.B., Cameron, J.S., Hartley, B., Coelho, A., Hildreth, G., and Reuben, R. (1987). Intraglomerular T cells and monocytes in nephritis: study with monoclonal antibodies. *Kidney International*, **31**, 1160–6.

Ooi, Y.M., Vallota, E.H., and West, C.D. (1976). Classical complement pathway activation in membranoproliferative glomerulonephritis. *Kidney International*, **9**, 46–53.

Parbtani, A., Frampton, G., and Cameron, J.S. (1980). Platelet and plasma serotonin concentrations in glomerulonephritis. II. *Clinical Nephrology*, **14**, 112–7.

Peters, D.K., et al. (1973). Mesangiocapillary nephritis, partial lipodystrophy and hypocomplementaemia. *Lancet*, **2**, 535–8.

Power, A.D., Ng, Y.C., and Simpson, J.G. (1990). Familial incidence of C3 nephritic factor, partial lipodystrophy and membranoproliferative glomerulonephritis. *Quarterly Journal of Medicine*, **75**, 387–98.

Pussell, B.A., Lockwood, C.M., and Scott, D.M. (1978). Value of immune complex assays in diagnosis and management. *Lancet*, **2**, 359–62.

Reichel, W., Kobberling, J., Fischbach, H., and Scheler, F. (1976). Membranoproliferative glomerulonephritis with partial lipodystrophy: discordant occurence in identical twins. *Klinische Wochenschrift*, **54**, 75–81.

Schena, F.P., Pertosa, G., and Stanziale, P. (1982). Biological significance of the C3 nephritic factor in membranoproliferative glomerulonephritis. *Clinical Nephrology*, **18**, 240–7.

Schmitt, H., Cavalcanti de Oliveira, V., and Bohle, A. (1987). Tubulointerstitial alterations in Type I membranoproliferative glomerulonephritis. An investigation of 259 cases. *Pathology Research and Practice*, **182**, 6–10.

Scott, D.M., Amos, N., Sissons, J.G.P., Lachmann, P.J., and Peters, D.K. (1978). The immunoglobulin nature of nephritic factor (NeF). *Clinical and Experimental Immunology*, **32**, 12–17.

Sherwood, M.C., Pincott, J.R., Goodwin, F.J., and Dillon, M.J. (1987). Dominantly inherited glomerulonephritis and an unusual skin disease. *Archives of Disease in Childhood*, **62**, 1278–80.

Sibley, R.K. and Kim, Y. (1984). Dense intramembranous deposit disease. New pathologic features. *Kidney International*, **25**, 660–70.

Simon, P., Ramee, M.P., Ang, K.S., and Cam, G. (1987). Variations of primary glomerulonephritis incidence in a rural area of 40,000 inhabitants in the last decade. *Nephron*, **45**, 171.

Sissons, J.G.P., et al. (1976). The complement abnormalities of lipodystrophy. *New England Journal of Medicine*, **294**, 461–5.

Spitzer, R.E., et al. (1969). Serum C3 lytic system in patients with glomerulonephritis. *Science*, **164**, 436–7.

Strife, C.F., McAdams, A.J., and West, C.D. (1982). Membranoproliferative glomerulonephritis characterized by focal, segmental proliferative lesions. *Clinical Nephrology*, **18**, 9–16.

Strife, C.F., Hug, G., Chuck, G., McAdams, A.J., Daus, C.A., and Kline, J.K. (1983). Membranoproliferative glomerulonephritis and alpha-1 antitrypsin deficiency in children. *Prediatrics*, **71**, 88–92.

Strife, C.F., Jackson, E.C., and McAdams, A.J. (1984). Type III membranoproliferative glomerulonephritis: long-term clinical and morphological evaluation. *Clinical Nephrology*, **21**, 323

Stuhlinger, W.D., Verroust, P.J., and Morel-Maroger, L. (1976). Detection of circulating soluble immune complexes in patients with various renal disease. *Immunology*, **30**, 43–7.

Stutchfield, P.R., White, R.H.R., Cameron, A.H., Thompson, R.A., McKintosh, P., and Wells, L. (1986). X-linked mesangiocapillary glomerulonephritis. *Clinical Nephrology*, **26**, 150–6.

Swainson, C.P., Robson, J.S., Thomson, D., and MacDonald, M.K. (1983). Mesangiocapillary glomerulonephritis. A long-term study of 40 cases. *Journal of Pathology*, **141**, 449–68.

Taguchi, T. and Bohle, A. (1989). Evaluation of change with time of glomerular morphology in membranoproliferative glomerulonephritis: a serial biopsy study of 33 cases. *Clinical Nephrology*, **31**, 297–306.

Teisberg, P., Grottum, K.A., Myhre, E., and Flatmark, A. (1973). In vivo activation of complement in hereditary nephropathy. *Lancet*, **2**, 356–8.

The Southwest Paediatric Nephrology Group. (1985). Dense deposit disease in children: prognostic value of clinical and pathologic indicators. *American Journal of Kidney Diseases*, **6**, 161–9.

Tiller, D.J., Clarkson, A.R., and Mathew, T. (1981). A prospective randomized trial in the use of cyclophosphamide, dipyridamole and warfarin in membranous and mesangiocapillary glomerulonephritis. Proceedings of the Eighth International Congress of Nephrology, Basel. Karger. p. 345–351.

Valotta, E.H., Forristal, J., and Davis, N.C. (1972). The C3 nephritic factor and membranoproliferative nephritis: correlation of serum levels of the nephritic factor with C3 levels, with therapy, and with progression of the disease. *Journal of Pediatrics*, **80**, 947–87.

Vanrenterghem, Y., Roels, L., Ver Berchmoes, R., and Michielsen, P. (1975). Treatment of chronic glomerulonephritis with a combination of indomethacin and cyclophosphamide. *Clinical Nephrology*, **4**, 218–22.

Vargas, A.R., et al. (1976). Mesangiocapillary glomerulonephritis with dense deposits in the basement membranes of the kidney. *Clinical Nephrology*, **5**, 73–82.

Warady, B.A., Guggenheim, S., Sedman, A., and Lum, G.M. (1985). Prednisone therapy of membranoproliferative glomerulonephritis in children. *Journal of Pediatrics*, **107**, 702–7.

Watson, A.R., Poucell, S., Thorner, P., Arbus, G.S., Rance, C.P., and Baumal, R. (1984). Membranoproliferative glomerulonephritis Type I in children: correlation of clinical features with pathologic subtypes. *American Journal of Kidney Diseases*, **4**, 141–6.

Welch, T.R., Beischel, L., Balakrishman, K., Quinlan, M., and West, C.D. (1986). Major histocompatibility—complex extended haplotypes in membranoproliferative glomerulonephritis. *New England Journal of Medicine*, **314**, 1476–81.

West, C.D. (1986). Childhood membranoproliferative glomerulonephritis: an approach to management. *Kidney International*, **29**, 1077–93.

West, C.D., McAdams, A.J., McConville, J.M., Davis, N.C., and Holland, N.H. (1965). Hypocomplementaemic and normocomplementaemic persistent (chronic) glomerulonephritis; clinical and pathologic charcteristics. *Journal of Pediatrics*, **67**, 1089–112.

Williams, D.G., Peters, D.K., Fallows, J., Morel-Maroger, L., and Cameron, J.S. (1974). Studies of serum complement in the hypocomplementaemic nephritides. *Clinical and Experimental Immunology*, **18**, 391–405.

Zimmerman, S.W., Moorthy, A.V., Dremer, W.H., Friedman, A., and Varanasi, U. (1983). Prospective trial of warfarin and dipyridamole in patients with membranoproliferative glomerulonephritis. *American Journal of Medicine*, **76**, 920–7.

3.9 Acute endocapillary glomerulonephritis

BERNARDO RODRIGUEZ-ITURBE

Definition

Acute endocapillary glomerulonephritis is characterized by an increased number of cells within the glomerular tuft. Endocapillary proliferation may occur without apparent cause (idiopathic), but is often a feature of primary renal diseases, such as IgA nephropathy, and of the glomerulonephritis observed in systemic diseases such as lupus erythematosus and cryoglobulinaemia. In addition, endocapillary glomerulonephritis is associated with a large number of bacterial, viral, and fungal infections, as well as with parasitic infestations (Table 1). Poststreptococcal glomerulonephritis is the prototype disease for this disease entity.

Glomerular endocapillary proliferation may be present alone or in association with pathological changes in the basement membrane, mesangial sclerosis or with extracapillary crescent formation (discussed in detail in Chapter 3.10). It should be emphasized that diseases which are manifested only by endocapillary proliferation in their typical presentation may, at times, show a mixed pathological appearance.

Epidemiology

In Western countries the incidence of endocapillary glomerulonephritis represents only 10 to 15 per cent of all glomerular disease, and is now very rare in children as reported in international studies. In contrast, it is the most common histological presentation of primary renal disease in Africa, India, Pakistan, Uganda, Malaysia, and Papua New Guinea. It accounts for more than 40 per cent of the cases of nephrotic syndrome in tropical Africa (reviewed by Rastegar et al. 1988). Aetiological associations with poststreptococcal infections and malaria are strongly suspected in some countries but not in others.

The incidence of poststreptococcal glomerulonephritis is decreasing in the United States, the United Kingdom, and Central Europe. The reasons are not entirely clear but are assumed to be related to improved living conditions and increased natural resistance. Nevertheless, poststreptococcal glomerulonephritis continues to have a worldwide distribution, as indicated by reports from Alaska, Africa, the Middle East, the Caribbean islands, and South America (reviewed in Rodriguez-Iturbe et al. 1985a).

Sporadic and epidemic cases follow upper respiratory or skin infections, but postimpetigo cases have become more common in recent years. The disease is most common between the ages of 2 and 12 years, but can occur at any age; in large series, 5 per cent of the patients are younger than 2 years and 5 to 10 per cent are older than 40 years. Acute poststreptococcal glomerulonephritis is twice as common in males as in females, but the male predominance disappears when asymptomatic cases are taken into account (Rodriguez-Iturbe 1984).

Epidemic outbreaks occur in closed communities or in densely populated areas with low socioeconomic status and poor hygienic conditions. These epidemics tend to be cyclic in certain areas, such as in the Red Lake Indian Reservation in Minnesota (Anthony et al. 1969), in Trinidad (Poon-King et al. 1967), and in Maracaibo (Rodriguez-Iturbe et al. 1985a). Isolated outbreaks also occur in rugby team members with infected skin abrasions, when the disease has the colourful name of 'scrum kidney' (Ludlam and Cookson 1986).

Table 1 Infectious agents associated with endocapillary glomerulonephritis

1. Associated with infectious syndromes
   Skin and throat (Streptococcus group A)
   Bacterial endocarditis (*Staphylococcus aureus, Streptococcus viridans*)
   Pneumonia (*Diplococcus pneumoniae, Mycoplasma*)
   Meningitis (*Meningococcus pneumoniae,* Staphylococcus)
   Visceral abscesses and osteomyelitis (*Staphylococcus aureus, Escherichia coli, Pseudomona aeruginosa, Proteus mirabilis, Klebsiella, Clostridium perfringens*)
   'Shunt' nephritis (*Staphylococcus aureus, Staph. albus, Streptococcus viridans*)
   Infected vascular prosthesis (*Staphylococcus aureus*)
   Guillain-Barré syndrome (G-B infectious agent?)
2. Associated with specific bacterial diseases
   Typhoid fever (*Salmonella typhi*)
   Leprosy
   Yersiniosis
   Brucellosis
   Leptospirosis
3. Associated with viral infections
   Hepatitis B
   Epstein-Barr virus
   Cytomegalovirus
   Measles
   Mumps
   Varicella
   Coxsackie virus
4. Associated with parasitic infestation
   Malaria (*Plasmodium falciparum, P. malariae*)
   Schistosomiasis (*Schistosoma haematobium, S. mansoni*)
   Toxoplasmosis
   Filariasis
5. Associated with other infectious organisms
   Rickettsiae (*Coxiella*)
   Fungi (*Candida albicans, Coccidioides immitis*)

Clinical aspects

The infection preceding nephritis

As emphasized in Table 1, certain infections appear to be more commonly associated with glomerulonephritis. The clinical characteristics of infection with nephritogenic group A haemolytic streptococcus are well known. The usual site of infection is the skin or throat, but other locations are possible. Streptococcal pharyngitis may cause trivial symptoms but is usually accompanied by tonsillar exudate, fever, and cervical lymphadenopathy. The likelihood of a streptococcal aetiology is less than 3 per cent when none of these is present (Komaroff et al. 1986). Scarlatinal rash, if present, is due to an erythrogenic toxin which is thought to be produced by the bacteria; vomiting may be a prominent symptom in scarlet fever.

Streptococcal impetigo is characterized by groups of small vesicles which appear in exposed skin areas. They break rapidly and leave lesions covered with a thick crust; this distinguishes them from staphylococcal impetigo which usually starts with larger vesicles that have a thin friable crust when they break. Regional lymphadenopathy is present in 90 per cent of the patients (Dillon 1968). Streptococcal impetigo does not affect deeper tissues and heals without scarring, unless it originates as a superinfection of lesions produced by the scratching associated with scabies. A history of intense itching, particularly if it is also present in other family members, is a diagnostic clue and careful inspection of the skin in the interdigital spaces and flexor surface of the wrists may reveal scabiosis.

Nephritis follows throat infections after a latent period of 1 or 2 weeks; the latent period after skin infections typically 3 to 6 weeks. Microscopic haematuria may be found during the latent period, and this may predict the development of clinically overt nephritis (Stetson et al. 1955).

Clinical features of acute nephritis

Acute endocapillary glomerulonephritis may be asymptomatic or may present with clinical haematuria and other features of the acute nephritic syndrome. Occasionally, patients present with massive proteinuria with or without the complete nephrotic syndrome. Even more rarely, rapidly increasing uraemia may be present, a clinical course that is characteristic of so-called 'rapidly progressive glomerulonephritis', usually associated with widespread crescentic extracapillary proliferation.

Subclinical or asymptomatic glomerulonephritis occurs during epidemics of poststreptococcal glomerulonephritis (Anthony et al. 1969), in children presenting to outpatient clinics with streptococcal sore throats (Sagel et al. 1973), and in the families of index cases (Dodge et al. 1967; Rodriguez-Iturbe et al. 1981b). Most studies have shown that subclinical cases are far more frequent than those with symptoms. However, Sharrett et al. (1971) found that clinical cases were 2 to 33 times more frequent than subclinical nephritis in household members of index cases; however, serial testing of individuals at risk was not performed and so their data probably underestimate the true incidence of subclinical cases. Repeated follow up of children with streptococcal upper respiratory infections in an outpatient clinic detected 19 subclinical cases for each case with symptoms (Sagel et al. 1973), and Anthony et al. (1969) have suggested that patients with subclinical nephritis outnumber the patients with clinical disease in epidemics by a factor of 1.5. Two prospective studies in families of index cases revealed a remarkably similar ratio of subclinical to clinical disease of 5.3 (Dodge et al. 1967) and 4.0 (Rodriguez-Iturbe et al. 1981b).

The acute nephritic syndrome was first described as a feature of the convalescent period of scarlet fever in the epidemics of the 18th century (von Plenciz 1792). It remains the most characteristic clinical presentation of acute poststreptococcal glomerulonephritis and is generally referred to as acute glomerulonephritis. A typical patient is a boy 2 to 14 years of age who suddenly develops puffiness of the eyelids and facial oedema; his urine becomes dark and scanty and his blood pressure is found to be elevated.

Forty per cent of children have the full clinical picture of the acute nephritic syndrome: oedema, haematuria, hypertension and oliguria, and about 96 per cent have at least two of these symptoms (Rodriguez-Iturbe 1984). In a typical case of poststreptococcal nephritis, the urine volume increases 4 to 7 days after admission, followed rapidly by resolution of oedema and return of the blood pressure to normal levels.

Haematuria is a universal finding and may be the only indication of glomerulonephritis. Microscopic haematuria is found in two-thirds of patients and dark red urine is observed in the remainder. Red blood cell casts are usually present in the urine sediment. Gross haematuria usually disappears when urine output increases; microscopic haematuria usually persists for many months after the acute attack. Microscopic examination of the urine should be performed as soon as possible after voiding, since red blood cells are destroyed rapidly, particularly in alkaline urine. Erythrocytes make their way through the glomeruli and become distorted, fragmented and smaller during their transit through the nephrons; phase-contrast microscopy is particularly useful in evaluating the shape of erythrocytes (Fairley and Birch 1982). Dysmorphism of 80 per cent or more of the red blood cells in the urine is characteristic of haematuria of glomerular origin. In laboratories where urine flow cytometry is used, a shift towards a smaller size should raise the suspicion of glomerular haematuria (Shichiri et al. 1986). However, the diagnostic usefulness of these findings has been questioned by Raman et al. (1986), who could not confirm a close correlation between erythrocyte dysmorphism and renal parenchymal disease. They also found that two observers differed in their interpretation of red cell morphology on 38 per cent of occasions.

Oedema is the chief complaint in most patients; in fact, the earliest accounts of acute glomerulonephritis deal with the 'dropsy which succeeds scarlet fever' (Wells 1812). Oedema is usually confined to the face and legs in adolescents, while generalized oedema is more common in younger children. Fluid retention in acute glomerulonephritis has long been associated with a decreased glomerular filtration rate, which is the functional expression of pathological damage localized primarily in the glomerulus. The inflammatory reaction in the glomerulus decreases the glomerular filtration rate by reducing the useful filtration area and shunting the circulation between glomerular capillaries. Renal blood flow is normal or even increased and, consequently the filtration fraction is severely depressed. Traditionally the reduced glomerular filtration rate is thought to be responsible for loss in the glomerulotubular 'balance', urine output decreases and fluid retention follows as tubular reabsorption is not correspondingly down-regulated. This is clearly an oversimplification because a reduced glomerular filtration rate is seen in a variety of other conditions where fluid retention does not occur. Profound sodium retention can occur in patients with a mildly reduced glomerular filtration rate and spontaneous diuresis can occur before glomerular filtration rate has improved (Glassock 1980).

Tubular function is also affected and studies have shown proximal tubular reabsorption is to be decreased in proportion to the fall in glomerular filtration rate, which has been interpreted as a mechanism to compensate for the reduction in the volume of filtrate. Nevertheless, there is a diminished distal delivery of filtrate, which, in association with a constant or even relatively increased distal reabsorption, results in extracellular fluid expansion. This is aggravated if sodium intake is maintained. The reasons for increased distal sodium reabsorption are not clear, but there is evidence against a major role for aldosterone. Consistent with this sequence of events is the observation that fractional sodium reabsorption is less than 1 per cent and the urine plasma creatinine ratio is more than 40.

Hypertension is present in more than 80 per cent of patients, but only 50 per cent require antihypertensive treatment; hypertensive encephalopathy may occasionally complicate the clinical course but this is unusual. When there is evidence of central nervous system involvement, such as somnolence or convulsions, the possibility of systemic lupus erythematosus or haemolytic uraemic syndrome should be considered.

Several authors have reported haemodynamic measurements made in patients with acute glomerulonephritis (reviewed by Rodriguez-Iturbe and Parra 1987). Their findings are consistent and show that all the patients have increased plasma volume, 85 per cent have an increased cardiac output, and that 64 per cent have increased peripheral vascular resistance. Thus it appears that both the increased cardiac output and elevated vascular resistance contribute to the hypertension in these patients; this is consistent with the volume-dependent nature of arterial hypertension in the nephritic syndrome. The severity of hypertension is directly related to fluid retention, as judged by the correlation between the increment in blood pressure in the acute phase and the magnitude of weight lost during resolution (Birkenhager et al. 1970; Rodriguez-Iturbe et al. 1981a).

Congestive heart failure may complicate the clinical course of the acute nephritic syndrome and the severity of hypertension and fluid retention are determining factors. Nevertheless, there is a long-standing observation that patients can develop heart failure with little increase in blood pressure and a minor degree of oedema (Murphy and Murphy 1954), and close monitoring of even seemingly mild disease is advisable, particularly in adult patients.

Proteinuria of 3 g/day or more, with or without other features of the nephrotic syndrome is seen in 4 per cent of symptomatic cases of acute poststreptococcal glomerulonephritis, which contrasts with the prevalence of nephritic syndrome in other diseases that present with endocapillary glomerulonephritis, such as systemic lupus erythematosus, 'shunt' nephritis and nephritis associated with visceral abscess. Proteinuria is a hallmark of glomerular disease and it is the result of damage to the size and charge-dependent filtration barrier. Inflammatory damage to the glomerular basement membrane diminishes the number of pores per unit area, but the new pores are larger and have reduced anionic charge (Myers 1985). The charge-selective filtration barrier may be altered without concomitant changes in pore density or size, as has been shown in minimal change disease, and the modification in charge can result in albuminuria. In plasma IgG molecules are large and uncharged; therefore, increased urinary gammaglobulin excretion may be used as an indication of the severity of the damage to the size filtration barrier (Myers 1985).

Table 2 Clinical manifestations of acute poststreptococcal glomerulonephritis in children and elderly adults

	Children (%)	Elderly patients (%)
Haematuria	100	100
Proteinuria	80	92
Oedema	90	75
Hypertension	60–80	83
Oliguria	10–50	58
Dyspnoea/heart failure	<5	43
Nephrotic proteinuria	4	20
Azotaemia	25–40	83
Early mortality	<1	25

Data in elderly patients taken from the review of Melby et al. (1987). Data in children taken from several studies reviewed in the text.

Non-specific symptoms such as general malaise, weakness, and nausea, with or without occasional vomiting, are often present in addition to the clinical features of nephritis. About 5 to 10 per cent of the patients complain of a dull lumbar pain which is assumed to be related to distension of the renal capsule by parenchymal oedema.

The course of acute glomerulonephritis in elderly patients is different from that in children. Table 2 compares the clinical characteristics of acute poststreptococcal glomerulonephritis in children and in patients of 60 years of age or more. Dyspnoea and pulmonary congestion, oliguria, massive proteinuria, azotaemia, and early mortality are substantially more common in the adult population (reviewed by Melby 1987). Prognosis is poorer and may be related to the coexistence of diabetes, cardiovascular or liver disease rather than to the severity of the renal disease itself and it is important to be aware of the frequency of a complicated clinical course in these patients.

Hormonal systems

Hormonal systems involved in the regulation of extracellular volume are appropriately modified in response to the extracellular expansion which occurs in poststreptococcal glomerulonephritis. Plasma renin activity and plasma aldosterone levels are depressed in the acute oedematous phase and the degree of fluid retention correlates with the suppression of the plasma renin activity (Rodriguez-Iturbe et al. 1981a). Atrial natriuretic factor increases three- to five-fold during the acute oedematous phase (unpublished observations). The hormonal changes suggest a secondary response to primary fluid retention; it is also possible that suppression of the renin–angiotensin system could be inappropriately incomplete. This would explain why converting enzyme inhibitors induce a transient increase in the glomerular filtration rate in patients with acute glomerulonephritis a finding that suggests that intrarenal angiotensin II activity is responsible for some degree of preglomerular vasoconstriction (Parra et al. 1988).

Urinary prostaglandins E_2 and F_2 and kallikrein excretion are reduced in the acute nephritic syndrome, presumably because of decreased renal synthesis; however, the reasons for this are not clear at the present time (Colina-Chourio et al. 1983; Cumming and Robson 1985). The histological characteristics of renal disease are not the only contributing factor since other forms of glomerulonephritis with endocapillary proliferation, such as lupus nephritis are frequently associated with increased urinary prostaglandin excretion. Reduced renal prostaglandin synthesis

Table 3 Serum complement and proteinuria in diseases that may present with features of acute glomerulonephritis

	Serum complement	
	Normal (>90%)	Low (>90%)
Nephrotic proteinuria		
Likely (70–90%)	Visceral abcesses	Membranoproliferative glomerulonephritis I and II, Lupus nephritis, Shunt nephritis
Unlikely (<10%)	IgA nephritis, Henoch-Schönlein purpura, Microscopic polyarteritis nodosa	Acute poststreptococcal glomerulonephritis Subacute endocarditis Cryoglobulinaemia
No help (40–60%)	Antiglomerular basement membrane	

Numbers refer to percentage of cases with the corresponding finding.

might be involved in the sodium retention associated with acute glomerulonephritis, but its relative importance has not been defined.

In contrast to the findings in acute nephritic syndrome, patients with the nephrotic syndrome show increased urinary excretion of prostaglandin E_2 and thromboxane B_2 which correlates with the severity of proteinuria and microhaematuria, respectively (Niwa et al. 1987) and with increased urinary kallikrein excretion (Cumming and Robson 1985). The role of these hormones in compensatory responses to impaired renal haemdynamics or as aggravating factors in these renal conditions has been postulated but not proven (Niwa et al. 1987).

Diagnosis

Subclinical disease is characterized by a transient drop in complement levels, microscopic haematuria and, at times, arterial hypertension. Detection of subclinical disease is only possible if children at risk are placed under prospective surveillance; this may be justified in the siblings of index cases or in 'closed' populations in epidemic conditions. Renal biopsy is not indicated in subclinical cases since practically all patients at risk who develop haematuria and a fall in serum complement have endocapillary proliferation (Sagel et al. 1973) and because subclinical cases have an excellent immediate and long-term prognosis.

The bedside differential diagnosis between nephrotic syndrome and acute nephritic syndrome in an oedematous child can be helped by examination of the ear cartilage and nailbeds of the patients. The consistency of the auricular cartilage is affected by long-standing severe hypoalbuminaemia and its attendant hypocalcaemia; as a consequence, ear cartilage acquires a very soft, almost paper-like, consistency in young children (Heyman et al. 1963). A pair of white oedematous bands may also be present in nail beds in patients with the nephrotic syndrome (Muerckhe 1956). Softening of ear cartilages and subungual oedema are absent in patients with acute glomerulonephritis and the rarity of ascites nephritis was noted almost two centuries ago (Wells 1812).

The clinical presentation of a variety of renal diseases may include features of the acute nephritis syndrome. It is useful, therefore, to consider from the outset whether the patient may have a primary renal disease or the initial manifestation of an otherwise silent systemic disease. Systemic lupus erythematosus, Goodpasture's syndrome, essential cryoglobulinaemia, subacute and acute bacterial endocarditis, 'shunt' nephritis, visceral abscess, 'microscopic' polyarteritis glomerulonephritis, and Wegener's granulomatosis may occasionally present as an isolated nephritis. This is particularly important in nephritis associated with chronic infections: if they are unrecognized, nephritis usually persists and may become chronic, but if the infection is eradicated the acute inflammatory phase is likely to resolve, with complete recovery.

Measurement of serum complement has been suggested as a first-line test in evaluation of acute glomerulonephritis (Madaio and Harrington 1983), because the low serum complement (C3) is a feature of 90 per cent or more of the cases of acute poststreptococcal glomerulonephritis, type II membranoproliferative glomerulonephritis, diffuse proliferative lupus nephritis, subacute bacterial endocarditis, 'shunt' nephritis, and essential cryoglobulinaemia (although in the last of these C3 usually remains normal and only C4 and C2 are depressed). The finding of a normal serum complement would make any of these diagnoses unlikely, and a low serum complement level excludes IgA nephropathy and antiglomerular basement disease.

Serial determinations of serum complement are of clinical value in glomerulonephritis associated with bacterial infections. Persistence of low complement levels may indicate a failure to control the infection (Neugarten and Baldwin 1984). Serum complement returns to normal in less than 2 months in 96 per cent of patients with uncomplicated poststreptococcal glomerulonephritis (Cameron et al. 1973), and persistence of low levels should raise the possibility of membranoproliferative glomerulonephritis or lupus nephritis.

It is useful to consider the presence or absence of nephrotic proteinuria (>3.5 g/day), in conjunction with the serum complement levels, in the differential diagnosis of patients presenting with features of acute glomerulonephritis. As shown in Table 3, massive proteinuria is characteristic of mesangioproliferative glomerulonephritis types I and II, lupus nephritis, shunt nephritis, and nephritis associated with visceral abscess, but is unusual in other renal diseases.

Indications for renal biospy

From a clinical view point, it is pertinent to ask when a renal biopsy contributes to the management of poststreptococcal glomerulonephritis. The acute disease has a self-limited course with an excellent prognosis and the risks of a biopsy, however small, are only justified on clinical grounds if unusual features are present that cast substantial doubt on the presumptive diagnosis. For example, renal biopsy is justified when massive proteinuria is present in the acute stage, or if the serum complement is normal; in both these circumstances the possibility of a diagnosis other than acute poststreptococcal glomerulonephritis should be entertained. A biopsy should also be considered when the serum creatinine rises progressively, because a biopsy showing crescentic glomerulonephritis might lead to other modalities of treatment, such as immunosuppressive or anticoagulant drugs. Persistence of a low serum complement after a month of the acute nephritic syndrome is another circumstance when a biopsy is justified, because this course is more characteristic of lupus nephritis and membranoproliferative glomerulonephritis.

Aetiopathogenesis of poststreptococcal glomerulonephritis

The vast majority of acute endocapillary nephritides are thought to be immunologically mediated. The nature of the antigens involved, the characteristics of the antibody response, and the site of the immune reaction as well as the role of cellular immune mechanisms are of obvious relevance and are considered generally elsewhere; and so only those aspects directly relevant to poststreptococcal glomerulonephritis will be discussed here.

It is usual to start with the work of Schick (1907) who noted the latent period between scarlet fever and nephritis and suggested pathogenetic similarities between postscarlatinal nephritis and experimental serum sickness. When the streptococcal aetiology of scarlet fever was established (Dochez and Sherman 1924), the nephritis that followed was ascribed to an 'allergic' reaction to the bacterium. Since acute rheumatic fever and glomerulonephritis are both complications of streptococcal infection, but have epidemiological and biological differences but only rarely occur in the same patient, Seegal and Earle (1941) postulated the existence of rheumatogenic and nephritogenic strains of the bacterium. It is well recognized that streptococcal M-types 1, 2, 4, 12, 18, 25, 49, 55, 57, and 60 have been cultured from the throat or skin of patients with nephritis, and it is likely that they share a common nephritogenic antigen. Numerous attempts to induce nephritis with streptococcal products and to localize putative antigens in the glomeruli of patients with acute poststreptococcal glomerulonephritis have yielded inconsistent results, possibly due to differences in the timing of the renal biopsy. Treser et al. (1970) have suggested that all free antigenic sites of the bacterium will be covered by antibody after the first few days of the disease, making it inaccessible to detection by standard immunofluorescence methods.

Streptococcal antigens

At the present time, three antigenic fractions are actively investigated; these have been isolated from nephritogenic streptococci and are found in the glomeruli of patients with the disease. In addition, elevated serum antibody titres to each of these fractions has been detected in most patients with poststreptococcal glomerulonephritis. The New York Medical College group (Yoshizawa et al. 1973; Lange et al. 1976) have isolated an antigen with a molecular weight between 40 000 and 50 000 Da and a pI of 4.7 from nephritogenic streptococci. They have named it endostreptosin or water-soluble preabsorbing antigen. This antigen is also found in lesser amounts in some group C and G streptococci (Seligson 1985). Villareal et al. (1979) have described a protein fraction present only in nephritogenic strains (nephritis strain associated protein, which has a very similar molecular weight and pI to that isolated by the New York Medical College group. This protein is antigenically related to streptokinase, in particular to a variant produced at 32°C and is found in some group C and G streptococcal strains (Johnston and Zabriskie 1986). Elevated antibody titres to the New York Medical College antigen (Seligson et al. 1985) and to the nephritis-associated antigen (Ohkuni et al. 1983) have been detected in 70 per cent and 96 per cent of patients with acute poststreptococcal glomerulonephritis respectively.

Finally, Vogt et al. (1983) used supernatants of nephritogenic streptococcal cultures to isolate a cationic proteinase that has epitopes in common with the glomerular basement membrane. (Bohus et al. 1987). Cationic antigens are attractive nephritogens because they easily penetrate the negatively charged filtration barrier, becoming localized in the subepithelial regions (Vogt 1984), not unlike the situation in poststreptococcal glomerulonephritis in which the most characteristic lesions are the electron-dense subepithelial deposits known as 'humps'. An antibody response to cationic streptococcal proteinase (most frequently directed against its precursor zymogen) is found in 82 per cent of patients with poststreptococcal glomerulonephritis and is a better marker for the disease than anti-DNAse B, antihyaluronidase or antistreptokinase (Zaum et al. 1987).

Autoimmune reactivity

Endogenous antigen–antibody complexes may play a role in poststreptococcal nephritis. McIntosh et al. (1972) showed experimentally that streptococcal neuraminidase (sialidase) could react with IgG, and that autologous sialic acid-depleted IgG could induce the formation of antiglobulins and subsequent nephritis in rabbits. These experimental findings were followed by observations that serum rheumatoid factor activity was increased in 32 to 43 per cent of patients with poststreptococcal nephritis (Rodriguez-Iturbe 1984; Sesso et al. 1986). The possibility of an autoimmune response to altered Ig has received additional support from the finding of anti-IgG reactivity in the antibody eluted from the kidney of a fatal case of poststreptococcal glomerulonephritis and in the immune deposits detected by immunofluorescence in renal biopsies from such patients (reviewed by Rodriguez-Iturbe 1988). Neuraminidase activity and free sialic acid have been detected in the serum of some patients with acute poststreptococcal glomerulonephritis, findings that were initially questioned but have been subsequently confirmed independently by Asami et al. (1985).

Streptococci isolated from patients with glomerulonephritis produce neuraminidase which appears to be particularly active on human IgM (Mosquera et al. 1985). Further evidence for a role for neuraminidase activity comes from the finding that peanut agglutinin binds to the glomeruli of biopsies taken in the first few weeks of clinical disease. This reagent identifies free acetylgalactosamine radicals, presumably exposed by the loss of sialic acid from the immunoglobulin deposited within the glomeruli or, conceivably, from glomerular structures (Mosquera and Rodriguez-Iturbe 1986).

Another possible mechanism for the production of antiglobulins is suggested by the work of Kronvall (1973) and Myhre and Kronvall (1977) who indentified and isolated a receptor in the bacterial wall of group A streptococci with affinity for the Fc fragment of IgG. Antiglobulin reactivity can be induced in rabbits by minimal amounts of IgG bound to this receptor, and immunization with strepococci cultured in a media containing horse or autologous serum induces anti-IgG reactivity (Schalén et al. 1985); a similar mechanism may play a role in human poststreptococcal nephritis.

Additional evidence of autologous immune complex disease in streptococcal infections has been supplied by the demonstration of DNA–anti–DNA complexes in patients with nephritis (Vilches and Williams 1984).

Site of immune reaction

The site of immune reactivity is also a subject of controversy. The role of circulating immune complexes has been suggested on the basis of the similarities between acute poststreptococcal nephritis and experimental models of serum sickness (Dixon et al. 1961; Germuth and Rodriguez 1973; Wilson and Dixon 1971). These and other investigators have postulated that circulating immune complexes deposit in the glomeruli, in ways that depend on a variety of factors, including their intrinsic properties, and that the result is acute glomerulonephritis following activation of the complement and coagulation systems.

In fact, human acute poststreptococcal glomerulonephritis and acute serum sickness share the characteristics of rising antibody response, transient fall of serum complement, and immune deposits in the kidney and nephritis; in both conditions there is complete resolution in a short period of time. Almost 90 per cent of the patients with acute poststreptococcal nephritis, have increased levels of IgG and IgM, normal levels of IgA and depressed serum complement (Rodriguez-Iturbe 1984). This contrasts with another complication of streptococcal infection, acute rheumatic fever, in which serum complement is not decreased and serum IgA levels are high (Potter et al. 1982b). Elevated levels of circulating antigen–antibody complexes, measured by a variety of methods, have been found in as many as two-thirds of patients in the first week of nephritis (reviewed by Mezzano et al. 1986).

However, in glomerulonephritis, as in serum sickness, electron-dense deposits may be found at the subendothelial, intramembranous, and subepithelial regions of the glomerular basement membrane, as well as within the glomerular mesangium. As discussed by Vogt (1984), theories postulating that deposits originate from circulating immune complexes do not explain why large subepithelial deposits are not excluded by the size filtration barrier. Since the original studies by Vogt and his associates (reviewed by Vogt 1984), numerous investigators have shown that the charge of the antigen, antibody, or immune complex at physiological pH has an important influence on its penetration through the glomerular basement membrane. Neutral or negatively charged molecules are excluded if their size exceeds that of serum albumin (effective radius of 35.5 Å), whereas cationic molecules penetrate the glomerular basement membrane freely, attracted by its negative electrostatic charge. It has been shown that in-situ reactivity between cationic (positive) antigen and antibody takes place initially in the subendothelial site. The immune complex, separated or in smaller units, penetrates through the lamina densa and recombines at subepithelial sites, where it persists for some time.

In contrast to the difficulties encountered in producing significant nephritis by administration of preformed immune complexes using anionic antigens which are deposited in the subendothelial and mesangial regions, small quantities of cationic antigen consistently produce massive proteinuria in association with subepithelial deposits. These findings have particular relevance in acute poststreptococcal glomerulonephritis in view of the typical subepithelial location of immune deposits in this disease.

The charge of the antibody is also important. Immunoglobulins have a variable charge and subpopulations are cationic with a pI higher than 9; these could become fixed to anionic sites in the glomerular basement membrane and act as a target for circulating antigen. McIntosh et al. (1972) showed that streptococcal neuraminidase removed sialic acid from immunoglobulin molecules, making them more immunogenic and increasing their pI. This would favour penetration of the glomerular basement membrane. The potential relevance of change in immunoglobulin charge in the pathogenesis of acute poststreptococcal glomerulonephritis is obvious, and recently, Mosquera and Rodriguez-Iturbe (1986) found glomerular deposits of sialic acid-depleted material, as evidenced by their lectin binding specificity, in early biopsies of these patients.

Nephritogenicity is not restricted to cationic molecules, since the electrostatic repulsion may be also blocked or minimized by mechanisms that reduce the negative charge of the capillary wall. Removal of sialic acid from the glomerular basement membrane by neuraminidase causes a charge as well as a size permeability defect (Kanwar 1984), and this could be a factor in poststreptococcal glomerulonephritis: bacterial neuraminidase may remove sialic acid from epithelial and endothelial cell surfaces in the glomerular basement membrane. Furthermore, as noted by Couser (1985), non-immune cationic proteins which result from inflammatory reactions involving neutrophils, complement, platelets, and lymphocytes may interact with the glomerular basement membrane and reduce its anionic sites. These mechanisms may be important in the penetration of streptococcal antigens of negative (anionic) charge, as discussed above, and in fact, Lange et al. (1983) suggested that in poststreptococcal nephritis the antigenic fraction studied by their group (endostreptosin) could be acting as a fixed target antigen.

The role of intraglomerular infiltrating cells

Inflammatory mononuclear infiltration is increasingly recognized in human proliferative glomerulonephritis. The work of Ferrario et al. (1985) and Hooke et al. (1987) showed particularly intense infiltration of the glomeruli by monocytes and T lymphocytes in postinfectious nephritis: this correlated with the severity of proteinuria. In these studies, all types of proliferative glomerulonephritis had significant glomerular and interstitial infiltration whereas non-proliferative forms of glomerulonephritis showed minimal cellular accumulation. Increased numbers of intraglomerular monocytes and lymphocytes were also found in proliferative glomerulonephritis (Nolasco et al. 1987), and there was a strong correlation between the number of T cells and the number of monocytes in biopsies from patients with this disease.

The involvement of lymphocytes in immune complex nephritis was first suggested by the experiments of Bhan et al. (1979), who showed that deposition of antigen–antibody complexes in the rat kidney could only promote glomerular hypercellularity if the animals were injected with sensitized lymphocytes. In acute serum sickness, the infiltrating cells inside the glomeruli were shown to be monocytes, and infiltration by these cells correlated with the appearance of proteinuria (Hunsicker et al. 1979). The relationship between monocyte infiltration and nephritis was convincingly demonstrated by Holdsworth et al. (1981), who were able to prevent proteinuria in the acute serum sickness model by treatment with antimacrophage serum.

Following this lead, Parra *et al.* (1984) have shown a three-fold increase in the number of glomerular monocytes in early biopsies of acute poststreptococcal glomerulonephritis. Monocyte accumulation could be due to the attachment of these cells to the Fc fragment of the immunoglobulins deposited in the glomeruli, or it could be the result of lymphokine activity. There is evidence in favour of both mechanisms: the administration of the Fab fragment of antiglomerular basement membrane antibody is not associated with monocyte infiltration or proteinuria in animal models of acute serum sickness (Holdsworth 1983), whereas lymphokine-mediated attraction is suggested by the work of Neild *et al.* (1984), who showed that treatment with cyclosporin A prevents acute serum sickness nephritis.

In early biopsies of patients with poststreptococcal glomerulonephritis, monocyte accumulation is associated with infiltration of $CD4^+$ lymphocytes (i.e. putative helper/inducer cells). In contrast, glomerular monocytes and $CD4^+$ lymphocytes are very rare or absent in biopsies taken more than 1 month after the acute attack. It should be kept in mind that the CD4 or CD8 receptors of lymphocytes do not define their function as helper or suppressor but, rather, their recognition of class II or class I MHC antigens; therefore, pathogenetic speculations cannot be derived from these indentifications. Nevertheless, the demonstration of such cells, even in a small number, emphasizes the possibility of cell-mediated immune responses. $CD4^+$ cells (T4 lymphocytes) may have a variety of functions of potential relevance, such as initiation of the delayed hypersensitivity reaction and induction of antibody production or suppressor cells.

Cellular immunity may also modulate the long-term progression of renal damage: Reid *et al.* (1984) showed a depressed cellular response to streptococcal antigen in patients who developed streptococcal nephritis after the age of 10, and suggested that this finding could be related to the worse prognosis of the disease in adults.

Direct nephritogenicity of terminal complement components

Complement activation has long been accepted as playing a role in immune complex-induced nephritis, since glomerular immune deposits are capable of activating the complement system and components of the complement system are localized in the renal tissue. It is well recognized that complement deposits may, occasionally, be present without demonstrable immunoglobulin deposits, raising the question of direct complement activation by unbound fixed antigen, but this possibility remains unproven.

The traditional view of complement mediated damage postulated the need for recruitment of inflammatory cells. Indeed, the strongly chemotactic properties of components of the complement cascade, such as C5a, make it likely that a major part of complement induced injury is mediated by neutrophils. However, it is increasingly apparent that cell-independent mechanisms may be responsible for some pathogenic properties of the complement system. For instance, the production of C3a, C5a, and anaphylatoxins cause histamine release and increased capillary permeability, and the terminal components of complement, the C5b–C9 complex (membrane attack complex) may have a direct effect on the glomerular capillary membrane. Non-lytic effects of the C5b–C9 components in glomerulonephritis may be mediated by their potential to stimulate platelets to secrete serotonin and thromboxane B_2, macrophages to secrete phospholipids and arachidonic acid, and mesangial cells to secrete substances such as prostaglandins, proteases, phospholipases and oxygen reactive radicals.

A direct action of the membrane attack complex is an attractive explanation for glomerular damage in conditions characterized by little or no cellular infiltration within the glomeruli, as is the case in human membranous nephropathy and in some cases of acute poststreptococcal glomerulonephritis in which dense deposits of complement and IgG outline the glomerular capillary in a 'garland pattern' (Sorger *et al.* 1982); these patients characteristically have massive proteinuria.

This mechanism is of considerable potential importance and several workers have identified the membrane attack complex inside the glomeruli in a variety of primary renal and systemic diseases. Membrane attack complex is deposited prominently along the capillary wall, in proliferative endocapillary glomerulonephritis of poststreptococcal aetiology (Parra *et al.* 1984), as well as in IgA and lupus nephritis and Henoch-Schönlein purpura, membranoproliferative glomerulonephritis, antiglomerular basement membrane nephritis, and membranous nephropathy (Falk *et al.* 1983).

In acute endocapillary glomerulonephritis, with preferential activation of the alternate pathway, Endre *et al.* (1984) have recently found that C3 depletion correlated with C5 levels, a finding interpreted to indicate that complement activation occurs via a convertase bound to the glomerular capillary, capable of cleaving both C3 and C5.

Genetic aspects

Associations with major histocompatibility complex antigens have been reported in a variety of types of proliferative glomerulonephritis. Strong associations have been found in autoimmune conditions, such as the link between HLA–DR4 and the risk for hydralazine induced lupus, but there is much less evidence in other types of endocapillary glomerulonephritis.

The increased familial incidence of acute poststreptococcal glomerulonephritis has been noted since 1806, when Wells read to the Society for the Improvement of Medical and Surgical Knowledge his observations on postscarlatinal nephritis and wrote, 'when one child of a family has been attacked with this disease, the other children of the same family, who have lately passed through the scarlet fever are more liable to become dropsical, than the children of another family, who have also lately laboured under that fever, but among whom no instance of dropsy has yet occurred . . . (due) in part, to similarity of constitution derived from common parents' (Wells 1812). Dodge *et al.* (1967) also noted an increased familial incidence of poststreptococcal glomerulonephritis. In prospective family studies, we found that poststreptococcal glomerulonephritis developed in almost 38 per cent of the siblings of index cases with sporadic disease (Rodriguez-Iturbe *et al.* 1981*b*). This incidence is higher than the attack rate of children at risk in epidemics, estimated to range between 28.3 per cent for throat infections and 4.5 per cent for skin infections (Anthony *et al.* 1969).

Nevertheless, definite associations between acute poststreptococcal glomerulonephritis and HLA antigens have been elusive, and only DR4 was found to be more common in unrelated patients with the disease (Layrisse *et al.* 1983). A recent investigation from Japan has reported associations of DR1 and poststreptococcal nephritis, and the finding that patients with HLA

Bw48 and DRw8 are less susceptible to primary glomerulonephritis (Naito et al. 1987).

Pathology

Acute endocapillary glomerulonephritis is typical of poststreptococcal nephritis. The glomerular tuft shows an increased number of cells and appears to fill up the Bowman's space (Fig. 1(a)). The basement membrane is normal. Both mesangial and endothelial cells proliferate and, in addition, there is infiltration of cells from the circulation. Polymorphonuclear leucocyte infiltration may be intense in early biopsies. Monocytes and T lymphocytes also infiltrate glomeruli (Fig. 1(b and c)) and the interstitium in acute proliferative nephritis (Parra et al. 1984; Hooke et al. 1987; Ferrario et al. 1985; Nolasco et al. 1987). The interstitial cellular infiltrate may be of considerable importance since tubulointerstitial damage shows a recognized correlation with depression of the glomerular filtration rate. Recently, Hooke et al. (1987) reported a strong association between interstitial leucocyte accumulation and reduction of creatinine clearance.

Immune deposits in the glomerular basement membrane and mesangium can be seen by immunofluorescence microscopy. Granular deposits of C3 (Fig. 1(d)) and other complement components, including the C5b–C9 and terminal membrane attack complex (Fig. 1(e)) are prominent, as are deposits of immunoglobulins. Findings of immune deposits and infiltrating cells in acute poststreptococcal glomerulonephritis are shown in Table 4.

Ultrastructural studies demonstrate subepithelial, subendothelial and intramembranous deposits as well as proliferating and infiltrating cells. Electron-dense subepithelial deposits (humps) are typical of acute poststreptococcal glomerulonephritis (Fig. 1(f)), but also occur in glomerulonephritis of other aetiologies, such as systemic lupus erythematosus, bacterial endocarditis, and cryoglobulinaemia. Immunoperoxidase staining has shown the presence of IgG inside the humps in acute poststreptococcal glomerulonephritis (Yoshizawa et al. 1973).

Clinicopathological correlations have emerged from the careful studies of Sorger et al. (1982, 1983) who noted three distinct immunofluorescence patterns in postinfectious glomerulonephritis. The finely granular 'starry sky' pattern of immune deposition in the capillary walls, and to a lesser degree in the mesangium, may be seen in the first week of the disease. These desposits consist primarily of C3 and IgG and sometimes of IgM and C4. The mesangial pattern is characterized by lumpy deposits of C3; immunoglobulins are usually absent, and this pattern is frequently found in biopsies taken 4 to 6 weeks later. The garland pattern is formed by heavy confluent deposits of IgG and C3 which tend to outline the glomerular capillaries. The garland pattern correlates with a large number of humps seen on electron microscopy and is frequently associated with massive proteinuria. From a prognostic viewpoint, patients who present the garland pattern have a higher incidence of persistent renal damage. Heavy proteinuria has also been correlated with the severity of the histological damage in the initial biopsy (Clark et al. 1988).

Transformation of histological characteristics may occur, and is the rule in cases which do not resolve completely. In acute poststreptococcal glomerulonephritis, endocapillary proliferation in an early biopsy may change to membranoproliferative morphology after some weeks (unpublished observations), or may rapidly progress to a crescentic form of nephritis (reviewed in Fairley et al. 1987). Similar observations have been made by Beaufils et al. (1981) in glomerulonephritis associated with severe infections. These authors noted that the proliferative morphology is typical of cases occurring within 2 months of infection, while cases showing mesangioproliferative changes and crescent formation appear after a longer delay.

In follow-up studies performed years after the initial episode of acute poststreptococcal glomerulonephritis, immune deposits and a variable degree of mesangial sclerosis and obliteration is frequently observed, even in the absence of clinical manifestations of renal disease (Baldwin et al. 1980; Gallo et al. 1980).

Treatment

The treatment of endocapillary proliferative glomerulonephritis will depend on the aetiology and the clinical presentation. We shall discuss only the treatment of the acute nephritic syndrome and streptococcal infection that may result in nephritis.

The acute nephritic syndrome should be treated with restriction of sodium and fluid intake. Our ususal practice is to withhold oral intake in the first 24 h, which may produce an early significant loss of weight. As discussed by De Wardener (1979), it is necessary to determine the urine output, information that is usually imprecise in the history obtained on admission, before deciding on fluid allowance. Loop diuretics may be administered from the outset in patients who require admission to hospital, since these drugs increase the urine flow several fold in many of the patients and decrease central venous pressure with beneficial effects on volume dependent hypertension and cardiovascular congestion; fortunately, the diuretic induced rise in plasma renin activity is not observed in these patients (Powell et al. 1980).

Hypertensive encephalopathy is rare, but if present, may need treatment with intravenous diazoxide, hydralazine or nitroprussate. The occasional patient who presents with repeated convulsions may need deep sedation and intubation. If hypertension is less severe, captopril in doses of 75 to 100 mg should decrease the blood pressure in less than 1 h in most cases. The results obtained with short-term captopril treatment have been attributed to blocking of inappropriate angiotensin II activity, even though levels of plasma renin activity are depressed; it also decreases the activity of the prostaglandin and kallikrein systems (Parra et al. 1988). Converting enzyme inhibition carries the risk of hyperkalaemia and these patients usually have a low urine output; serum potassium should be followed closely.

Pulmonary oedema may complicate the clinical course and should be treated with ancillary methods of morphine, oxygen, loop diuretics, and rotating tourniquets and, if necessary venesection. Digitalis is ineffective and carries an increased risk of intoxication. Dialysis should be used early if there is rising azotaemia or hyperkaleamia. Extra care should be taken with peritoneal dialysis in dyspnoeic patients; it should either be avoided or started with a reduced quantity of liquid and shorter equilibration periods until a significant negative balance is achieved and the patient feels better.

Enforced bed rest is of doubtful value. If the patient is not feeling well he will keep to bed of his own accord.

Removal of the antigen is of obvious importance in nephritis of immune aetiology. In the case of poststreptococcal glomerulonephritis, treatment and, in certain circumstances, prevention of streptococcal infection is often possible. Oral phenoxymethyl or phenoxyethyl penicillin G, 125 mg, every 6 h for 7 to 10 days should eradicate the bacterium from the throat and skin. Oral

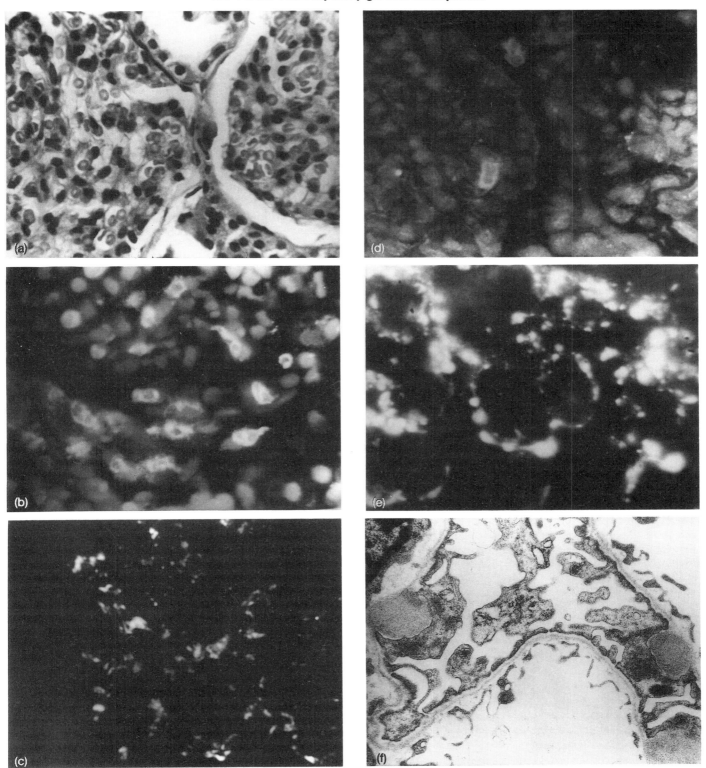

Fig. 1 Acute endocapillary glomerulonephritis of poststreptococcal aetiology. (a) Intense endocapillary proliferation in a biopsy taken 2 weeks after the beginning of symptoms. Basement membrane appears normal. Numerous red blood cells inside and outside the glomerular capillaries. (b) Intraglomerular infiltration of monocytes indentified with the peripheral fluorescence surrounding a central darkness (staining with fluorescein-labelled monoclonal antibody OKMI (Ortho Pharmaceutical Corporation) and counterstaining with ethydium bromide). Non-specific global fluorescence is observed in polymorphonuclear leucocytes. (c) Intraglomerular $CD4^+$ T lymphocyte demonstrated with fluorescein-labelled monoclonal OKT4 antibody and similar counterstaining as in (b).
(d) Immunofluorescent C3 deposits in the mesangium and glomerular capillaries. (e) Heavy deposits of the membrane attack complex of the complement in the mesangium and in the glomerular capillary walls. (f) Electron-dense subepithelial deposits (humps) shown by electron-microscopy. (Figs. 1(b), 1(c), 1(e), and 1(f) reproduced by courtesy of Dr G. Parra.)

Table 4 Immune deposits and infiltrating cells in acute poststreptococcal glomerulonephritis

Immune deposits	Positive/total biopsies
Complement C3	53/53
C5b–C9*	6/7
IgG	32/52
IgM	38/50
IgA	1/40
IgE	0/34
Antihuman IgG	16/55
Infiltrating cells*	**Positive cells per 100 glomerular cells (normal)**
Polymorphonuclear neutrophils	4.94 (0.1)
Monocytes	3.35 (0)
$CD4^+$ lymphocytes	0.25 (0)
$CD8^+$ lymphocytes	0.07 (0.06)
CD4/CD8 ratio	3.57 (—)

Data obtained from biopsies taken less than 1 month after the beginning of symptoms. * Data from Parra et al. (1984).

benzylpenicillin (200 000 units 4 times daily for 7 to 10 days) is equally effective. Better compliance and similar results may be obtained with a single injection of 1.2 million units of benzathine benzylpenicillin in adults and half this dose in small children. In persons allergic to penicillin, erythromycin in doses of 250 mg every 6 h in adults, and 40 mg per kg in children, given for 7 to 10 days is an appropriate alternative.

All patients with acute poststreptococcal glomerulonephritis should be treated as if they had active infection. The challenge for the clinician resides in the diagnosis of streptococcal pharyngitis in a patient without nephritis who consults with a sore throat. Since only 10 to 15 per cent of throat infections in the adult are streptococcal in origin, certain guidelines may be useful when deciding whether or not to use antibiotics. The throat cultures give 10 per cent false-negative and 30 to 50 per cent false-positive results; the latter arise in streptococcal carriers.

Certain clinical clues are helpful. Fever, tonsillar exudate, and cervical adenitis occur more frequently in streptococcal infection and the absence of all three makes this diagnosis unlikely. In contrast, 40 to 50 per cent of the cases who present with all three symptoms have a streptococcal aetiology. Because of the rate of false-positive and false-negative results from throat cultures, patients with a probability higher than 40 per cent should be treated, and those with a probability lower than 3 per cent should not be treated, irrespective of culture results. Outside this range of probability the culture is not helpful. In the vast majority of patients the culture should be done and treatment given to those patients with a positive result (Komaroff et al. 1986).

It is important to remember that streptococcal impetigo frequently complicates scabiosis. If this is present both the patient and the family group will need to be treated, in addition to treatment of the streptococcal infection.

Preventive antibiotic treatment is justified in populations at risk during epidemics, and in siblings of index cases. The latter show evidence of streptococcal infection within 2 to 3 weeks after the presentation of the index case in a vast majority of cases, and over one-third will develop nephritis (Rodríguez-Iturbe et al. 1981b).

Prognosis

The early mortality of acute endocapillary glomerulonephritis is very low. Cardiovascular complications, such as hypertensive encephalopathy and acute pulmonary oedema cause death in less than 1 per cent of patients who present with the acute nephritic syndrome. Irreversible renal failure may follow acute glomerulonephritis if widespread crescentic extracapillary proliferation with permanent obliteration of most glomeruli is superimposed on the initial histological damage.

The natural history of poststreptococcal glomerulonephritis has been extensively studied, but remains controversial. Several reviews have discussed the long follow up of this disease (Roy et al. 1976; Baldwin et al. 1980; García et al. 1981; Vogl et al. 1986). Some of the discrepancies may be explained by differing criteria for defining chronicity; evaluation of the glomerular filtration rate and histological studies were not performed in all studies. Not surprisingly, the incidence of abnormal findings during follow up varies greatly ranging from 3.5 per cent (Potter et al. 1982a) to 60 per cent (Baldwin et al. 1980). Discrepancies may also result when the prognosis of children and adults are considered together. Studies by Hinglais et al. (1974), García et al. (1981) and Vogl et al. (1986), which have analysed the course in children and adults separately, suggest that 30 to 55 per cent of adults have an abnormal urine sediment, proteinuria or a low creatinine clearance up to 15 years after the acute episode. These values were significantly higher than those for children at similar intervals. A poorer long-term prognosis has also been associated with the clinical presentation of nephrotic syndrome which is more common in adults.

If all published studies are taken into account, it may be concluded that one to two decades after acute poststreptococcal glomerulonephritis, about 20 per cent of the patients have an abnormal urine analysis or creatinine clearance, but less than 2 per cent of the patients have azotaemia. The incidence of glomerular sclerosis and fibrosis is considerably higher (Gallo et al. 1980), but their clinical relevance is uncertain. The incidence of chronic renal failure may be as high as 77 per cent in a subset of adult patients who were nephrotic at the initial presentation (Vogl et al. 1986).

Recently, we have found a decreased renal functional reserve in patients 10 to 15 years after an acute attack of poststreptococcal glomerulonephritis (Rodríguez-Iturbe et al. 1985b). In these studies, we stimulated the glomerular filtration rate with an oral protein load and assumed, after Bosch et al. (1983), that the difference between stimulated and unstimulated glomerular filtration rate represented the functional reserve of the kidney. We have extended our studies and the results are shown in Table 5. Apparently healthy individuals who appear to have recovered from an attack of acute glomerulonephritis showed a poor functional response to the protein meal, with a 60 per cent reduction of renal functional reserve compared to normal subjects (Molina et al. 1988). In contrast, four patients who still had subclinical nephritis responded in a manner similar to control groups. If these findings are confirmed in a larger number of subclinical patients, they would suggest that renal functional reserve is better preserved in patients who had a milder initial attack. It remains to be shown whether a diminished renal reserve capacity has any clinical significance. Similar reductions of the glomerular filtration rate response have also been demonstrated years after single nephrectomy (Rodríguez-Iturbe et al. 1985b; ter Wee et al. 1987), but has not been associated with progressive renal failure

Table 5 Renal functional reserve in postacute glomerulonephritis patients

	n	Unstimulated C_{cr} (ml/min)	Stimulated C_{cr} (ml/min)	Renal functional reserve (ml/min)
Normal school children	260	84.7 ± 2.08	143.1 ± 2.24	60.2 ± 2.51
Normal adults	55	92.6 ± 4.27	150.7 ± 6.25	57.9 ± 6.19
10–15 years postacute				
Epidemic acute poststreptococcal glomerulonephritis	34	76.9 ± 3.96	89.6 ± 5.45*	13.1 ± 5.22*
Endemic acute poststreptococcal glomerulonephritis	6	78.8 ± 9.93	93.5 ± 13.5	14.7 ± 8.79*
Subclinical poststreptococcal glomerulonephritis	4	81.9 ± 13.3	122.1 ± 21.8	40.1 ± 34.5

Values represent mean ± SEM. *$p < 0.001$ vs. normal. n = number of patients. Data corrected for 1.73 m² surface area. Data in school children from Molina et al. (1988). Renal functional reserve (stimulated C_{cr} − unstimulated C_{cr}).

even though these patients show a higher incidence of mild proteinuria and hypertension; longer follow up of these patients is obviously needed.

References

Anthony, B.F., Kaplan, E.L. Wannamaker, L.W., Briese, F.W., and Chapman, S.S. (1969). Attack rates of acute nephritis after type 49 streptococcal infection of the skin and of the respiratory tract. *Journal of Clinical Investigation*, **48**, 1697–1704.

Asami, T., Tanaka, T., Gunji, T., and Sakai, K. (1985). Elevated serum and urine sialic acid levels in renal diseases of childhood. *Clinical Nephrology*, **23**, 112–9.

Baldwin, D.S., Schacht, R.G., Gallo, G., Gluck, M.C., and Feiner, H.D. (1980). Natural history of poststreptococcal glomerulonephritis. In *Streptococcal diseases and the immune response* (eds. J.B. Zabriskie and S.E. Read). pp. 563–579, Academic Press, New York.

Beaufils, M. (1981). Glomerular disease complicating abdominal sepsis (Nephrology Forum). *Kidney International*, **19**, 609–18.

Bhan, A.K., Collins, A.B., Schneeberger, E., and McClusky, R. (1979). A cell-mediated reaction against glomerular bound immune complexes. *Journal of Experimental Medicine*, **150**, 1410–20.

Birkenhager, W.H., Schalekamp, M.A., Schalekamp-Kuyen, M.A.P., Kolsters, G., and Kraus, H.K. (1970). Interrelations between arterial pressure, fluid volumes and plasma renin concentration in the course of acute glomerulonephritis. *Lancet*, **i**, 1086–7.

Bohus, M., Batsford, S., and Vogt, A. (1987). Cationic streptococcal proteinase and human renal basement membrane have common epitopes. Xth Lancefield International Symposium on Streptococcal and Streptococcal Diseases. Cologne, Germany, p. 79.

Bosch, J.P., Saccaggi, A., Lauer, A., Ronco, C., Belledonne, M., and Glabman, S. (1983). Renal functional reserve in humans: effect of protein intake in glomerular filtration rate. *American Journal of Medicine*, **75**, 943–50.

Cameron, J.S., Vick, R.M., Ogg, C.S., Seymour, W.W., Chantler, C., and Turner, D.R. (1973). Plasma C3 and C4 concentrations in the management of glomerular nephritis. *British Medical Journal*, **3**, 668–72.

Clark, G., White, R.H.R., Glasgow, E.F., Chantler, C., and Cameron, J.S. (1988). poststreptococcal glomerulonephritis in children: clinicopathological correlations and long-term prognosis. *Pediatric Nephrology*, **2**, 381–8.

Colina-Chourio, J.A., Rodriguez-Iturbe, B., Baggio, B., García, R., and Borsatti, A. (1983). Urinary excretion of prostaglandins (PGE2 and PGF2a) and kallikrein in acute glomerulonephritis. *Clinical Nephrology*, **20**, 217–24.

Couser, W.G. (1985). Mechanisms of glomerular injury in immune-complex disease (Nephrology Forum). *Kidney International*, **28**, 569–83.

Cumming, A.D. and Robson, J.S. (1985). Urinary kallikrein excretion in glomerulonephritis and nephrotic syndrome. *Nephron*, **39**, 206–10.

De Wardener, H.E. (1979). Treatment of acute glomerular nephritis. *American Heart Journal*, **98**, 523.

Dillon, H.C. (1968). Impetigo contagiosa: suppurative and non-suppurative complications. I. Clinical, bacteriologic and epidemiologic characteristics of impetigo. *American Journal of Diseases of the Children*, **115**, 530–41.

Dixon, F.J., Feldman, J.D., and Vasquez, J.J. (1961). Experimental glomerulonephritis: the pathogenesis of a laboratory model resembling the spectrum of human glomerulonephritis. *Journal of Experimental Medicine*, **113**, 889–920.

Dochez, A.R. and Sherman, L. (1924). The significance of Streptococcus hemolyticus in scarler fever and the preparation of a specific antiscarlatinal serum by immunization of the horse to Streptococcus hemolyticus scarlatinae. *Journal of American Medical Association*, **2**, 542–4.

Dodge, W.F., Spargo, B.F., and Travis, L.B. (1967). Occurence of acute glomerulonephritis in sibling contacts of children with sporadic acute glomerulonephritis. *Pediatrics*, **40**, 1028–30.

Endre, Z.H., Pusell, B.A., Charlesworth, J.A., Coovadia, H.M., and Seedat, Y.K. (1984). C3 metabolism in acute glomerulonephritis: implications for sites of complement activation. *Kidney International*, **25**, 937–41.

Fairley, C., Mathew, D.C., Mathews, D.C., and Becker, G.J. (1987). Rapid development of diffuse crescents in poststreptococcal glomerulonephritis. *Clinical Nephrology*, **28**, 256–60.

Fairley, K. and Birch, D.F. (1982). Hematuria: a simple method for identifying glomerular bleeding. *Kidney International*, **21**, 105–8.

Falk, R.J., et al. (1983). Neoantigen of the polymerized ninth component of complement characterization of a monoclonal antibody and immuno-histochemical localization in renal disease. *Journal of Clinical Investigation*, **72**, 560–73.

Ferrario, F., et al. (1985). The detection of monocytes in human glomerulonephritis. *Kidney International*, **28**, 513–9.

Gallo, G.R., Feiner, H.D., Steel, J.M., Jr., Schacht, R.G., Gluck, M.C., and Baldwin, D.S. (1980). Role of intrarenal vascular sclerosis in progression of poststreptococcal glomerulonephritis. *Clinical Nephrology*, **13**, 449–57.

García, R., Rubio, L., and Rodriguez-Iturbe, B. (1981). Long-term prognosis of epidemic poststreptococcal glomerulonephritis in Maracaibo: Follow-up studies 11–12 years after the acute episode. *Clinical Nephrology*, **15**, 291–8.

Germuth, F.J. and Rodríguez, E. (1973). *Immunopathology of the human glomerulus*. Boston: Little, Brown.

Glassock, R.J. (1980). Sodium homeostasis in acute glomerulonephritis and nephrotic syndrome. *Contributions in Nephrology*, **23**, 181–93.

Heyman, W., Harwood, S., Isaacs, E.V., and Cuppage, F. (1963). Decreased elasticity of auricular cartilage in the nephrotic syndrome of children. *Journal of Pediatrics*, **62**, 74–6.

Hinglais, N., García-Tores, R., and Kleinknecht, D. (1974). Long-term prognosis in acute glomerulonephritis. The predictive value of early clinical and pathologic features observed in 65 patients. *American Journal of Medicine*, **56**, 52–60.

Holdsworth, S.R. (1983). Fc dependence of macrophage accumulation and subsequent injury in experimental glomerulonephritis. *Journal of Immunology*, **130**, 735–9.

Holdsworth, S.R., Neale, T.J., and Curtis, C.B. (1981). Abrogation of macrophage dependent injury in experimental glomerulonephritis in the rabbit. Use of antimacrophage serum. *Journal of Clinical Investigation*, **68**, 686–96.

Hooke, D.H., Gee, D.C., and Atkins, R.C. (1987). Leucocyte analysis using monoclonal antibodies in human glomerulonephritis. *Kidney International*, **31**, 964–72.

Hunsicker, L.G., Shearer, T.P., Plattner, S.B., and Weisenburger, D. (1979). The role of monocytes in serum sickness nephritis. *Journal of Experimental Medicine*, **150**, 413–25.

Johnston, K.H. and Zabriskie, J.B. (1986). Purification and partial characterization of the nephritis strain-associated protein from streptococcus pyogenes. *Journal of Experimental Medicine*, **163**, 697–712.

Kanwar, Y. (1984). Biophysiology of glomerular filtration and proteinuria. *Laboratory Investigation*, **58**, 7–21.

Komaroff, A.L., Pass, T.M., and Aronson, M.D. (1986). The prediction of streptococcal pharyngitis in adults. *Journal of General Internal Medicine*, **1**, 1–7.

Kronvall, G. (1973). A surface component of A, C, and G. streptococci with non-immune reactivity for immunoglobulin G. *Journal of Immunology*, **111**, 1401–06.

Lange, K., Ahmed, U., Kleinberger, H., and Treser, G. (1976). A hitherto unknown streptococcal antigen and its probable relation to acute poststreptococcal glomerulonephritis. *Clinical Nephrology*, **5**, 207.

Lange, K., Seligson, G., and Cronin, W. (1983). Evidence for the *in-situ* origin of poststreptococcal glomerulonephritis: glomerular localization of endostreptosin and clinical significance of the subsequent antibody response. *Clinical Nephrology*, **19**, 3–10.

Layrisse, Z., Rodriguez-Iturbe, B., García, R., Rodríguez, A., and Tiwari, J. (1983). Family studies of the HLA system in acute poststreptococcal glomerulonephritis. *Human Immunology*, **7**, 177–85.

Ludlam, H. and Cookson, B. (1986). Scrum kidney: epidemic pyoderma caused by a nephritogenic streptococcus pyogenes in a rugby team. *Lancet*, **ii**, 331–3.

Madaio, M.P. and Harrington, J.T. (1983). The diagnosis of acute glomerulonephritis. *New England Journal of Medicine*, **309**, 1299–1302.

McIntosh, R.M., Kaufman, D.B., McIntosh, J.R., and Griswold, W. (1972). Glomerular lesions produced in rabbits by autologous serum and autologous IgG modified by treatment with a culture of hemolytic streptococcus. *Journal of Medical Microbiology*, **5**, 1–5.

Melby, P.C., Musick, W.D., Luger, A.M., and Khanna, R. (1987). Poststreptococcal glomerulonephritis in the elderly. *American Journal of Nephrology*, **7**, 235–40.

Mezzano, S., Olavarria, F., Ardiles, L., and Lopez, M.I. (1986). Incidence of circulating immune complexes in patients with acute glomerulonephritis and in patients with streptococcal impetigo. *Clinical Nephrology*, **26**, 61–5.

Molina, E., Herrera, J., and Rodriguez-Iturbe, B. (1988). The renal functional reserve in health and renal disease in school age children. *Kidney International*, **34**, 809–16.

Mosquera, J.A., Katiyar, V.N., Coello, J., and Rodriguez-Iturbe, B. (1985). Neuraminidase production by streptococci isolated from patients with glomerulonephritis. *Journal of Infectious Diseases*, **151**, 259–63.

Mosquera, J.A. and Rodriguez-Iturbe, B. (1986). Glomerular binding sites for peanut agglutinin in acute poststreptococcal glomerulonephritis. *Clinical Nephrology*, **26**, 227–34.

Meuhrcke, R.C. (1956). The fingernails in chronic hypoalbuminemia. A new physical sign. *British Medical Journal*, **1**, 1327–8.

Murphy, T.R. and Murphy, F.D. (1954). The heart in acute glomerulonephritis. *Annals of Internal Medicine*, **41**, 510–32.

Myers, B. (1985). *In-vivo* evaluation of glomerular permselectivity in normal and nephrotic man. In *Proteinuria* (ed. M. Avram). pp. 17–35. Plenum Press, New York.

Myhre, E.B. and Kronvall, G. (1977). Heterogenicity of nonimmune immunoglobulin Fc reactivity among Grampositive cocci; description of three major types of receptors of human immunoglobulin G. *Infection and Immunity*, **17**, 475–82.

Naito, S., Hohara, M., and Arakawa, K. (1987). Associations of class II antigens of HLA with primary glomerulopathies. *Nephron*, **45**, 111–4.

Neild, G.H., Ivory, K., and Williams, D.G. (1984). Cyclosporin A inhibits acute serum sickness in rabbits. *Clinical and Experimental Immunology*, **52**, 586–94.

Neugarten, J. and Baldwin, D.S. (1984). Glomerulonephritis in bacterial endocarditis. *American Journal of Medicine*, **77**, 297–304.

Niwa, T., Maeda, K., and Shibata, M. (1987). Urinary prostaglandins and thromboxane in patients in chronic glomerulonephritis. *Nephron*, **46**, 281–7.

Nolasco, F.E.B., Cameron, J.S., Hartley, B., Coelho, A., Hildreth, G., and Reuber, R. (1987). Intraglomerular T cells and monocytes in nephritis: study with monoclonal antibodies. *Kidney International*, **31**, 1160–6.

Ohkuni, H., Freidman, J., van de Rijn, I., Fischetti, V.A., Poon-King, T., Zabriskie, J.B. (183). Immunological studies of post-streptococcal sequelae: serological studies with an extracellular protein associated with nephritogenic streptococci. *Clinical and Experimental Immunology*, **54**, 185–93.

Parra, G., Platt, J.L., Falk, R.J., Rodriguez-Iturbe, B., and Michael, A.F. (1984). Cell populations and membrane attack complex in glomeruli and patients with poststreptococcal glomerulonephritis: indentification using monoclonal antibodies by indirect immunofluorescence. *Clinical Immunology and Immunopathology*, **33**, 324–32.

Parra, G., Rodriguez-Iturbe, B., Colina-Chourio, J., and García, R. (1988). Short-term treatment with captopril in hypertension due to acute glomerulonephritis. *Clinical Nephrology*, **29**, 58–62.

Poon-King, T., *et al.* (1967). Recurrent epidemic nephritis in South Trinidad. *New England Journal of Medicine*, **277**, 728–33.

Potter, E.V., Lipschultz, S.A., Abidh, S., Poon-King, T., and Earle, D.P. (1982*a*). Twelve to seventeen-year follow-up of patients with poststreptococcal acute glomerulonephritis in Trinidad. *New England Journal of Medicine*, **307**, 725–9.

Potter, E.V., Shaugnessy, M.A., Poon-King, T., and Earle, D.P. (1982*b*). Serum immunoglobulin A and antibody to M associated protein in patients with acute glomerulonephritis or rheumatic fever. *Infection and Immunity*, **37**, 227–34.

Powell, H.T., McCredie, D., and Rotenber, E. (1980). Response to furosemide in acute renal failure: dissociation of renin and diuretic responses. *Clinical Nephrology*, **15**, 55–9.

Raman, G.V., Pead, L., Lee, H.A., and Maskell, R. (1986). A blind controlled trial of phase contrast microscopy by two observers for evaluating the source of haematuria. *Nephron*, **44**, 304–8.

Rastegar, A., Sitprija, V., and Rocha, H. (1988). Tropical Nephrology. In *Diseases of the kidney*, 4th edn. (ed. R.W. Schrier and C.W. Gottschalk). pp. 2583–613. Boston, Little, Brown and Co.

Reid, H.F.M., Read, S.E., Zabriskie, J.B., Ramkisson, R., and Poon-King, T. (1984). Suppression of cellular reactivity in group A streptococcal antigens in patients with acute poststreptococcal glomerulonephritis. *Journal of Infectious Diseases*, **149**, 841–50.

Rodriguez-Iturbe, B. (1984). Epidemic poststreptococcal glomerulonephritis. (Nephrology Forum). *Kidney International*, **25**, 129–36.

Rodriguez-Iturbe, B. (1988). Poststreptococcal glomerulonephritis. In *Diseases of the kidney* (ed. R.W. Schrier and C. W. Gottschalk), pp. 1929–47. Boston, Little and Brown Co.

Rodriguez-Iturbe, B., and Parra, G. (1987). Loop diuretics and angiotensin enzyme inhibitors in the acute nephritic syndrome. In *Diuretics II. Chemistry, pharmocology and clinical applications* (ed. J. B. Puschett and A. Greenberg), pp. 536–541. Amsterdam, Elsevier Publishers.

Rodriguez-Iturbe, B., et al. (1981a). Studies on the renin-aldosterone system in the acute nephritic syndrome. *Kidney International*, 19, 47–55.

Rodriguez-Iturbe, B., Rubio, L., and Garcia, R. (1981b). Attack rate of poststreptococcal glomerulonephritis in families. A prospective study. *Lancet*, i, 401–3.

Rodriguez-Iturbe, B., García, R., Rubio, L., and Cuenca, L. (1985a). Características clínicas y epidemiológicas de la glomerulonefritis postestreptocóccica en la región Zuliana. *Investigación Clínica*, 26, 191–211.

Rodriguez-Iturbe, B., Herrera, J., and Garcia, R. (1985b). Response to acute protein load in kidney donors and in apparently normal post-acute glomerulonephritis patients. Evidence for glomerular hyperfiltration. *Lancet*, ii, 461–4.

Roy, S., Pitcock, J.A., and Ettledorf, J.N. (1976). Prognosis of acute poststreptococcal glomerulonephritis in childhood: prospective study and review of the literature. *Advances in Pediatrics*, 23, 35–69.

Sagel, I., et al. (1973). Occurrence and nature of glomerular lesions after group A streptococcal infection in children. *Annals of Internal Medicine*, 79, 492–9.

Schalén, C., Burova, L.A., Christensen, P., Grubb, R., Svensson, L., and Totolian, A.A. (1985). Induction and specificity of anti-IgG following immunization with streptococci. In *Recent Advances in Streptococci and Streptococcal Diseases* (ed. Y. Kimura, S. Kotami, and Y. Shiokawa). pp. 94–5. Reedbooks Ltd. Publishers, Windsor.

Schick, B. (1907). Die Nachkrankheiten des Scharlach. *Jahrbuch der Kinderheilkunde*. 65(Suppl.), 132–73.

Seegal, D. and Earle, D.P. (1941). A consideration of certain biological differences between glomerulonephritis and rheumatic fever. *American Journal of Medical Sciences*, 201, 528–39.

Sesso, R.C.C., Ramos, O.L., and Periera, A.B. (1986). Detection of IgG-rheumatoid factor in sera of patients with acute poststreptococcal glomerulonephritis and its relation with circulating immunocomplexes. *Clinical Nephrology*, 26, 55–60.

Seligson, G., Lange, K., Majeed, H.A., Deol, Hl, Cronin, W., and Bovie, R. (1985). Significance of endostreptosin antibody titers in poststreptococcal glomerulonephritis. *Clinical Nephrology*, 24, 69–75.

Sharrett, A.R., Poon-King. T., Potter, E.V., Finklea, J.F., and Earle, D.P. (1971). Subclinical nephritis in South Trinidad. *American Journal of Epidemiology*, 94, 231–45.

Shichiri, M., Dowada, A., Nishio, Y., Tomita, K. (1986). Use of autoanalyzer to examine urinary red-cell morphology in the diagnosis of glomerular haematuria. *Lancet*, ii, 781–2.

Sorger, K., et al. (1982). Subtypes of acute post-infectious glomerulonephritis. Synopsis of clinical and pathological features. *Clinical Nephrology*, 17, 114–28.

Sorger, K., et al. (1983). The garland type of acute postinfectious glomerulonephritis: morphological characteristics and follow-up studies. *Clinical Nephrology*, 20, 17–26.

Stetson, C.A., Rammelkamp, C.H., Krause, R.M., Cohen, R.J., and Perry, W.A. (1955). Epidemic acute nephritis: Studies on etiology, natural history and prevention. *Medicine*, 34, 431–50.

ter Wee, P.M., Tegless, A.M., and Donker, A.J.M. (1987). Effect of low-dose dopamine on renal function in uninephrectomized patients. Special emphasis on kidney donors before and after nephrectomy. *Clinical Nephrology*, 28, 211–6.

Treser, G., Semar, M., Ty, A., Sagel, I., Franklin, M.A., and Lange, K. (1970). Partial characterization of antigenic streptococcal plasma membrane components in acute glomerulonephritis. *Journal of Clinical Investigation*, 49, 762–8.

Vilches, A.R., and Williams, D.G. (1984). Persistent anti-DNA antibodies and DNA-anti complexes in poststreptococcal glomerulonephritis. *Clinical Nephrology*, 22, 97–101.

Villareal, H., Jr., Fischetti, V.A., van de Rijn, I., and Zabriskie, J.B. (1979). The occurence of a protein in the extracellular products of streptococci isolated from patients with acute glomerulonephritis. *Journal of Clinical Investigation*, 149, 459–72.

Vogl, W., Renke, M., Mayer-Eichberger, D., Schmitt, H., and Bohle, A. (1986). Long-term prognosis for endocapillary glomerulonephritis of poststreptococcal type in children and adults. *Nephron*, 44, 58–65.

Vogt, A., Bastford, S., Rodriguez-Iturbe, B., and García, R. (1983). Cationic antigens in poststreptococcal glomerulonephritis. *Clinical Nephrology*, 20, 271–9.

Vogt, A. (1984). New aspects of the pathogenesis of immune complex glomerulonephritis: formation of subepithelial deposits. *Clinical Nephrology*, 21, 15–20.

Von Plenciz, M.A. (1792). Tractatus III de Scarlatina. Vienna: J.A. Trattner. Cited by Becker, C.G., and Murphy, G.E. (1968). The experimental induction of glomerulonephritis like that in man by infection with group A streptococcus. *Journal of Experimental Medicine*, 127, 1–23.

Wells, W.C. (1812). Observation on the dropsy which succeeds scarlet fever. *Transactions of the Society for the Improvement in Medical and Chirurgical Knowledge*, 3, 167–86.

Wilson, C.B., and Dixon, F.J. (1971). Quantitation of acute and chronic serum sickness in the rabbit. *Journal of Experimental Medicine*, 134, 7s–18s.

Yoshizawa, N., Treser, G., Sagel, I., Ty, A., Ahmed, U., and Lange, K. (1973). Demonstration of antigenic sites in glomeruli of patients with acute poststreptococcal glomerulonephritis by immunofluorescein and immunoferritin technique. *American Journal of Pathology*, 70, 131–50.

Zaum, R., Vogt, A., Rodriguez-Iturbe, B. (1987). Analysis of the immune response to streptococcal proteinase in poststreptococcal disease. *Xth Lancefield International Symposium on Streptococci and Streptococcal Diseases*. Cologne, Germany, p. 88.

3.10 Crescentic glomerulonephritis

A.J. REES AND J. STEWART CAMERON

Introduction and definition

Volhard and Fahr, in 1914, provided the classic description of what has come to be known as 'crescentic glomerulonephritis' (Fig. 1). They described autopsy specimens with severe glomerular destruction, in which the tuft was surrounded by a mass of cells filling Bowman's space. This was referred to as extracapillary proliferation, but its appearance in cross-section has determined that this type of nephritis is now usually referred to by the much more evocative name of crescentic nephritis. It was appreciated that patients with crescentic nephritis usually died of renal failure within weeks of the apparent onset of the disease. Later, Ellis (1942) approached the same topic from a clinical standpoint and introduced the term 'rapidly progressive glomerulonephritis type I', later reduced to rapidly progressive glomerulonephritis, for patients with nephritis who developed renal failure within weeks or months. Most patients with rapidly progressive glomerulonephritis were shown to have crescentic nephritis and were thought to have severe poststreptococcal disease.

In 1948, Davson, Ball and Platt described a group of patients with systemic vasculitis involving small arteries in whom 'epithelial crescent formation was widespread and often associated with fibrinoid necrosis' of the glomerular tuft. They called the condition the microscopic form of periarteritis nodosa, nowadays more usually microscopic polyarteritis. Davson et al. distinguished microscopic polyarteritis from poststreptococcal disease and discussed the question as to whether some of Ellis' patients had it rather than poststreptococcal nephritis; later they, and others, made detailed comparisons of the two conditions (Davson and Platt 1949; Harrison et al. 1964). Harrison et al. also discussed the difficulty of diagnosing microscopic polyarteritis from renal biopsies, and described patients with crescentic nephritis on needle biopsy who were subsequently shown to have microscopic polyarteritis at autopsy. This topic has, of course, been much discussed recently (Serra et al. 1984; Velosa 1987; Couser 1988).

Although much less accurate than autopsies for making the diagnosis of microscopic polyarteritis, the widespread use of renal biopsies enabled clinicopathological correlations to be made during life. Throughout the 1960s, numerous studies of crescentic nephritis were published and have been summarized in several reviews (Pollak and Mendoza 1971; Couser 1982; Heaf et al. 1983; Glassock 1985). During this time, Bacani et al. (1968) clearly distinguished between streptococcal and non-streptococcal forms of the disease, and Scheer and Grossman (1964) demonstrated the association between antibodies to glomerular basement membrane and crescentic nephritis. However, the observation that had the greatest practical implications was that less than 10 per cent of untreated patients survived and were independent of dialysis within months or a few years (Couser 1982; Glassock 1985), the only apparent exception being those with poststreptococcal nephritis. It now became clear that 'crescents' were in fact a superadded feature found, more or less commonly, in many forms of nephritis and that 'crescentic' nephritis was a large and heterogeneous group.

The lack of effective treatment meant that the principal purpose of many of the larger series of patients treated during the 1960s and 1970s was to define prognostic features more accurately. However, they revealed the variety of clinical contexts in which crescentic nephritis occurred (Whitworth et al. 1976; Beirne et al. 1977; McLeish et al. 1978; Morrin et al. 1978; Davis et al. 1979; Stilmant et al. 1979; Neild et al. 1983) and enabled crescentic nephritis to be categorized into three groups depending on the immunofluorescent findings:

1. patients with linear staining implicit of antiglomerular basement membrane antibodies;
2. those in whom glomerular immunoglobulins were scanty or absent;
3. those with prominent granular deposits.

At the time these differences had little effect on prognosis,

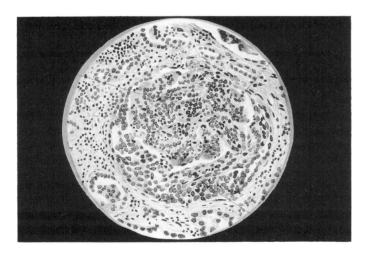

Fig. 1 Crescentic nephritis as illustrated by Volhard and Fahr (1914). The glomerular tuft shows obvious proliferation, and material, probably at a subendothelial site, can be seen peripherally in the lower half of the glomerulus in the original coloured drawings. A cellular crescent occupies the lower segment of Bowman's capsule. Thus the illustration shows proliferative glomerulonephritis complicated by crescents, probably mesangiocapillary glomerulonephritis type I in contemporary terms.

3.10 Crescentic glomerulonephritis

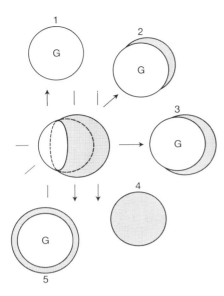

Fig. 2 The effect of the direction of section on the appearance of a single 'crescent'. The centre figure represents a glomerulus partially surrounded by a crescent. The arrows indicate various sections, and point to the appearance seen in each section. G = glomerular tuft; the shaded area represents the crescent. The crescent can appear to be absent, segmental or circumferential depending on which plane the section takes.

and so attention was concentrated on better ways to assess the severity. The severity of renal failure at presentation was immediately recognized as indicating a bad prognosis, whether assessed by the presence of oliguria or the need for dialysis (Whitworth *et al.* 1976; Beirne *et al.* 1977; Morrin *et al.* 1978). Patients with preceding infection (Whitworth *et al.* 1976; Beaufils *et al.* 1976), or with systemic evidence of vasculitis, were said to fare better. Proliferation within the glomerular tuft was also said to be associated with a better prognosis; but the most important factor appeared to be the proportion of glomeruli surrounded by crescents (Elfenbein *et al.* 1975; Whitworth *et al.* 1976; Morrin *et al.* 1978). Patients with less than half of glomeruli affected by crescents fared much better than those with 50 to 79 per cent affected, who in turn had a better prognosis than those with more than 80 per cent crescents.

The obvious importance of crescent scores for prognosis led to detailed discussions of how crescents should be defined, and how crescent scores should to be calculated. This focused on three main issues: (a) whether the denominator used to calculate a crescent score should be the total number of glomeruli in the biopsy or only those that were not sclerosed; (b) the minimum number of glomeruli in a biopsy needed to calculate a crescent score; and (c) the extent of extracapillary proliferation that was needed before it was appropriate to use the term 'a crescent'. With hindsight, these discussions seem irrelevant, especially as needle biopsies take such small samples, and because the plane through which glomeruli were cut can have a tremendous influence on the apparent extent of extracapillary proliferation (Fig. 2).

It was also at this stage that terminology became thoroughly confused, especially with regard to so-called 'idiopathic' crescentic glomerulonephritis. This term has been used to describe at least three subtly different sets of patients: (a) those with crescentic glomerulitis without extrarenal disease (Glassock 1985); (b) those with crescentic glomerulonephritis, irrespective of the pathogenesis, in whom another specific diagnosis such as vasculitis, mesangial IgA disease, or poststreptococcal nephritis cannot be made (Beirne *et al.* 1977); and (c) crescentic nephritis with either granular or scanty immune deposits (Couser 1982). Although confusing, this had no practical consequences until the development of treatments whose effectiveness was shown to be influenced by the pathogenesis and the associated diagnosis. Thus severe antiglomerular basement membrane disease was reported to have a worse prognosis than other types of crescentic nephritis of comparable severity (Beirne *et al.* 1977; McLeish *et al.* 1978; Hind *et al.* 1983), a fact still not recognized in the protocols of some trials of treatment in crescentic nephritis (Glockner *et al.* 1988; Keller *et al.* 1989).

The confusion over terminology was easy to resolve in patients with antiglomerular basement membrane disease who could be diagnosed unequivocally by serology or by immunofluorescence. Using this approach, Goodpasture's syndrome (nephritis and pulmonary haemorrhage) and isolated antiglomerular basement membrane antibody-induced nephritis were shown to be different clinical expressions of the same disease (see Chapter 3.11). It is becoming increasingly likely that an analogous situation exists in the case of crescentic nephritis with scanty immune deposits, also known as 'no immune deposits' (Couser 1982) or as 'pauci-immune' (Jennette *et al.* 1989). Comparisons of clinical (Serra 1984; Couser 1988), histological (Velosa 1987), and serological data (Falk and Jennette 1988; Jennette *et al.* 1989; Cohen Tervaert *et al.* 1990) suggest that microscopic polyarteritis and idiopathic, 'pauci-immune', crescentic nephritis are the same or very similar diseases. Until this is known for certain, we would favour the descriptive name of renal microscopic vasculitis for these conditions, rather than one which implied a pathogenesis. This term could then be qualified by a more specific diagnosis, such as Wegener's granulomatosis or microscopic polyarteritis, where appropriate.

The group of patients with obvious glomerular immune deposits is clearly heterogeneous. Most have pre-existing nephritis, which appears to have been complicated by the development of crescents. Thus it is more important to consider patients in pathogenetically defined groups when considering the response to trials of therapy, rather than to rely on similarities of crescent scores.

There are other reasons why crescent scores are an inadequate guide to the need for treatment. Clinical (Rees *et al.* 1977; Gill *et al.* 1977; Juncos *et al.* 1979; Fairley *et al.* 1987) and experimental (Wilson and Dixon 1986) studies have shown that glomerular inflammation can evolve very rapidly, and that an individual with 20 per cent crescents one day can have 80 per cent as little as 3 days later. Obviously, effective treatment should be introduced at the earliest possible stage for those patients who need it, even though the crescent score might justify a diagnosis of crescentic glomerulonephritis by traditional criteria. For this reason, some (Parfrey *et al.* 1985; Weiss and Crissman 1985; Furlong *et al.* 1987) have used the name 'focal necrotizing glomerulonephritis', whilst others have rediscovered the term renal microscopic polyarteritis (Serra *et al.* 1984; Savage *et al.* 1985; Coward *et al.* 1986; Croker *et al.* 1987).

Despite these developments, there are enough similarities between different types of crescentic nephritis to justify a common approach to management, and especially with regard to the urgent need for treatment. The purpose of this chapter is to

Table 1 The main forms of crescentic glomerulonephritis

	Serology	Immunohistology†
From anti-GBM* antibodies	Anti-GBM positive	Linear immunofluorescence
Arising from vasculitis**	ANCA positive	Negative immunofluorescence
Complicating proliferative glomerulonephritis		Granular immunofluorescence
Common:		
Henoch-Schönlein purpura		Mesangial IgA
Post-streptococcal glomerulonephritis		Granular IgG, C3
Mesangiocapillary glomerulonephritis type I		Granular IgG, C3
Mesangiocapillary glomerulonephritis type II		C3 only, IgM
Less common:		
Cryoglobulinaemia		IgM, IgG, C3
Systemic lupus erythematosus		All immunoglobulins and complement
IgA nephropathy		Mesangial IgA
Membranous nephropathy		Granular IgG, C3
Endocarditis/shunts/abscesses		IgM, IgG, C3
Behçet's syndrome		
Monoclonal gammopathy		
Drug ingestion (e.g. penicillamine, hydralazine)		
Neoplasia		

* GBM, glomerular basement membrane.
** Including microscopic polyarteritis, Wegener's granulomatosis, Churg–Strauss syndrome and ANCA positive, apparently idiopathic, glomerulonephritis.
† Always accompanied by fibrin(ogen) in crescents.

consider this approach, but not to provide detailed descriptions of individual diseases, which can be found in the appropriate chapters.

Varieties of crescentic nephritis

Crescents occur when breaks in glomerular capillaries allow leakage of cells and plasma proteins into Bowman's space, and so it is not surprising that occasional crescents have been described in most chronic renal diseases, including those that do not primarily affect the glomerulus. Widespread crescent formation, however, requires active and specific attack on glomerular capillaries; predictably this can occur in a variety of settings (Table 1). At present, crescentic nephritis is categorized using a variety of clinical, morphological, immunohistological, and serological criteria. None of the systems is entirely satisfactory, largely because crescent formation is the result of a disease rather than a disease of itself.

Three broad groups of crescentic nephritis can be distinguished:

1. antiglomerular basement membrane disease, characterized by circulating antiglomerular basement membrane antibodies and linear deposition of antibody along the membrane;
2. renal microscopic vasculitis, characterized by scanty glomerular deposits of immunoglobulin and circulating antineutrophil cytoplasmic antibodies (ANCA);
3. a more heterogeneous group, usually associated with obvious granular deposits of immunoglobulin, in which crescent formation complicates an identifiable form of nephritis, usually proliferative in type.

In adults, including the elderly (Potvliege *et al.* 1975; Montoliu *et al.* 1984; Kingswood *et al.* 1984), antiglomerular basement membrane disease accounts for 10 to 20 per cent of patients with severe crescentic nephritis, whilst the remainder is divided equally between the other two groups (Whitworth *et al.* 1976; Stilmant *et al.* 1979; Neild *et al.* 1983; Velosa 1987). Fewer children with crescentic nephritis have antiglomerular basement disease and a greater proportion has had granular immune deposits, as will be described later.

Antiglomerular basement membrane disease

Antiglomerular basement membrane disease presents no problems in diagnosis, despite differences in clinical presentation (see Chapter 3.11). This is because the pathogenesis is relatively well understood, and can be assessed serologically.

Renal microscopic vasculitis

Renal microscopic vasculitis is much more difficult (see Chapters 4.5.1 and 4.5.3). The clinical presentation is diverse and some patients present with systemic disease, whilst others appear to have nephritis alone. Most patients with small-vessel vasculitis and crescentic nephritis can be classified clinically as having Wegener's granulomatosis or microscopic polyarteritis. Exceptionally there is definite evidence of another vasculitic syndrome such as relapsing polychondritis (Neild *et al.* 1978; Dalal *et al.* 1978; reviewed by Chang-Miller *et al.* 1987); Churg–Strauss syndrome (Clutterbuck *et al.* 1990); Takayasu's disease (Hellmann *et al.* 1987); giant-cell arteritis (Droz *et al.* 1979); rheumatoid vasculitis (Breedveld *et al.* 1985; Naschitz *et al.* 1989); and vasculitis associated with narcotics abuse (Citron *et al.* 1970). There have been isolated case reports of crescentic glomerulonephritis in juvenile polymyositis (Kamata *et al.* 1982), acute rheumatic fever (Mustonen *et al.* 1983), and, rather unconvincingly, in Behçet's disease (Landwehr *et al.* 1980; Olsson *et al.* 1980; Donnelly *et al.* 1989).

It is apparent that many patients who are diagnosed in Europe as having systemic vasculitis on clinical grounds are considered to have 'idiopathic' crescentic glomerulonephritis in the United States (Couser 1988). In addition, there is another group who have identical glomerular appearances to those of systemic vasculitis but without evidence of extrarenal disease. This group was emphasized by Serra *et al.* (1984), who described 53 patients with focal necrotizing glomerulonephritis and either histological evidence of extraglomerular vasculitis, or clinical evidence that included constitutional symptoms and evidence of multisystem disease. Seventeen of the patients (32 per cent) had clinical evidence of systemic vasculitis without histological proof, and eight (15 per cent) had histological evidence of extraglomerular vasculitis in a renal biopsy without signs or symptoms of extrarenal disease. Broadly similar findings have been reported by others (Heilman *et al.* 1987).

Jennette *et al.* (1989) and Cohen Tervaert *et al.* (1990) have also correlated the clinical and morphological findings with the presence of ANCA. These autoantibodies are found in patients with systemic vasculitis and are discussed extensively in Chapters 4.5.1 and 4.5.3. The important point to mention here is that over 80 per cent of patients with crescentic nephritis with scanty glomerular deposits have detectable ANCA, regardless of whether there is evidence of extrarenal disease (Jennette *et al.* 1989; Nassberger *et al.* 1989; Pusey and Lockwood 1989; Cohen Tervaert *et al.* 1990). There is a suggestion that inheritance of HLA DR2 increases the susceptibility to these diseases (Muller *et al.* 1984); but this needs to be confirmed. Thus it seems reasonable to consider these patients as an homogeneous group. Certainly it can be impossible to distinguish clinically between the patients with and without systemic vasculitis. This point is emphasized by descriptions of patients who have developed typical Wegener's granulomatosis shortly or even many years after presenting with 'idiopathic' crescentic nephritis (Woodworth *et al.* 1987).

Crescentic nephritis with granular immune deposits

The third group of patients with crescentic nephritis is much more heterogeneous. They have obvious granular deposits of immunoglobulins in the mesangium, the capillary wall, or both. Most have identifiable forms of proliferative nephritis, sometimes against a background of infection or other systemic disease. Most of the patients fall into one of three broad groups: (a) those with infection; (b) those with systemic immune complex disease; and (c) those with pre-existing nephritis. But there is a small number of patients who appear to have a variety of other associated conditions.

Systemic infections

Poststreptococcal nephritis

The prognosis is excellent for the vast majority of patients with poststreptococcal nephritis (see Chapter 3.9), but occasional patients have a rapidly progressive clinical course. Typically these patients present with severe crescentic nephritis from the outset (Lohlein 1907; Jennings and Earle 1961; McCluskey and Baldwin 1965; Chugh *et al.* 1981; Lewy *et al.* 1971; Roy *et al.* 1981; Southwest Pediatric Nephrology Study Group 1985; Niaudet and Levy 1987). Occasionally patients with poststreptococcal disease develop vasculitis (Ingelfinger *et al.* 1977; Chugh *et al.* 1981). However, at least six cases have been described in which diffuse endocapillary proliferative nephritis evolved into severe crescentic disease over periods from 10 days to 5 weeks (Morel-Maroger *et al.* 1974; Gill *et al.* 1977; Ferraris *et al.* 1983; Old *et al.* 1984; Modai *et al.* 1985; Fairley *et al.* 1987). One of these patients also developed antitubular basement membrane antibodies (Morel-Moroger *et al.* 1974).

Infective endocarditis

Typically, infective endocarditis causes a focal or a diffuse proliferative nephritis, depending on the organism (see Chapter 3.12). It is rarely associated with severe renal failure and even in this situation there are usually few if any crescents (Boulton-Jones *et al.* 1974; Beaufils *et al.* 1976). However, exceptional patients develop severe crescentic nephritis (Beaufils *et al.* 1976; Neugarten *et al.* 1984; Rovzar *et al.* 1986).

Infected atrioventricular shunts

Renal biopsies from patients with shunt nephritis usually show membranoproliferative nephritis (see Chapter 3.12), but occasionally this is complicated by severe crescentic disease (Southwest Pediatric Nephrology Study Group 1985; Niaudet and Levy 1983; A.J. Rees, unpublished).

Visceral abscesses

Crescentic nephritis has been reported in patients with deep-seated infections (Beaufils *et al.* 1976; Whitworth *et al.* 1976; Connelly and Gallacher 1987), but it is difficult to know whether the infections were responsible for nephritis, or whether they merely exacerbated it, as described by Rees *et al.* (1977). Four of the eight patients described by Beaufils *et al.* (1976) had pure extracapillary proliferation, three had membranoproliferative glomerulonephritis, and one focal proliferative nephritis. In retrospect it seems likely that three of the patients may have had Wegener's granulomatosis (three had pulmonary abscesses and another sinusitis), which can also be exacerbated by infection (Pinching *et al.* 1980).

Other infections

Crescentic nephritis has been reported during the course of a number of other viral and bacterial infections. These include, Ross river virus (Davies *et al.* 1982), legionella (Wegmuller *et al.* 1985), syphilis (Walker *et al.* 1984), and leprosy (Singhal *et al.* 1977; Nigam *et al.* 1986). Crescentic nephritis has also been attributed to rifampicin during treatment for tuberculosis (Hirsch *et al.* 1983; Murray *et al.* 1987).

Systemic immune complex disease

Systemic lupus erythematosus

Small numbers of crescents are found in biopsies from many patients with severe focal or diffuse proliferative nephritis but only rarely is a substantial proportion of glomeruli affected (see Chapter 4.5.4). This type of lupus nephritis has been described in detail by Yeung *et al.* (1984).

Henoch-Schönlein purpura

Moderate numbers of crescents are commonly found in renal biopsies from patients with Henoch-Schönlein purpura, and the implications they have for prognosis are discussed in Chapter 4.5.2. There are small numbers of patients with this disease in most large series of children with crescentic nephritis (Neild *et al.* 1983; Niaudet and Levy 1985; Southwest Pediatric Nephrology Study Group 1985).

Monoclonal gammopathies

Monoclonal synthesis of immunoglobulins can cause renal disease in a variety of ways. Mixed essential cryoglobulinaemia is usually associated with membranoproliferative glomerulonephritis but is occasionally associated with severe crescentic disease (see Chapter 4.8; Weber *et al.* 1985). More recently, it has been appreciated that monoclonal immunoglobulins without these special properties can also be associated with crescentic nephritis. Some of these patients have had myeloma (Kaplan and Kaplan 1970; Lapenas et al. 1983; Meyrier et al. 1984; Kebler et al. 1985; Cadnapaphornchai and Sillix 1989), whilst others have had Waldenström's macroglobulinaemia (Meyrier et al. 1984; Ogami et al. 1989). Possibly related are the patients with familial Mediterranean fever (Said et al. 1989) and amyloidosis who have developed crescentic nephritis (Panner 1980; Harada et al. 1984; Vernier et al. 1987).

Chronic glomerulonephritis
Mesangial IgA disease
Biopsies taken from patients with mesangial IgA disease frequently contain small numbers of crescents (or glomeruli with extracapillary proliferation), especially when macroscopic haematuria is present (see Chapter 3.6). Severe nephritis has been described regularly, both in series selected because they had IgA disease (Abuelo et al. 1984; Nicholls et al. 1984; Boyce et al. 1986; Lai et al. 1987; Welch et al. 1988) or because of crescentic nephritis (Whitworth et al. 1976; Morrin et al. 1978; Neild et al. 1983; Niaudet and Levy 1983).

Membranoproliferative glomerulonephritis
Between 10 and 15 per cent of patients with membranoproliferative nephritis have crescents; these affect large numbers of glomeruli in a small proportion of patients (McCoy et al. 1975; Davis et al. 1978; Chapman et al. 1980; Kleinknecht et al. 1980; Neild et al. 1983; Niaudet and Levy 1983; Southwest Pediatric Nephrology Study Group 1985; Korzets and Bernheim 1987).

Membranous nephropathy
At least 13 patients have been described with membranous nephropathy complicated by crescentic nephritis late in the course of the disease (Klassen et al. 1974; Nicholson et al. 1975; Moorthy et al. 1976; Hill et al. 1978; Taylor et al. 1979; Tateno et al. 1981; Mitas et al. 1983; Kurki et al. 1984; Koethe et al. 1986; Abreo et al. 1986; Nguyen et al. 1988). Interestingly, three of the patients also had antiglomerular basement membrane antibodies. One patient, without such antibodies, developed recurrent crescentic nephritis in a renal transplant (Hill et al. 1978).

Hypersensitivity to drugs
Penicillamine
Penicillamine is a common cause of nephrotoxicity (see Chapter 2.1) and is the only drug to have been shown unequivocally to cause crescentic nephritis. Its use has been associated with crescentic nephritis in patients with rheumatoid arthritis (Gibson et al. 1976; Gavaghan et al. 1981; Swainson et al. 1982; Banfi et al. 1983; Sadjadi et al. 1985; Devogelaer et al. 1987; Peces et al. 1987), Wilson's disease (Sternleib et al. 1975), primary biliary cirrhosis (Matloff and Kaplan 1980), and systemic sclerosis (Ntoso et al. 1986). Many of the patients have had the syndrome of pulmonary haemorrhage and crescentic nephritis without antiglomerular basement membrane antibodies and are indistinguishable clinically from microscopic polyarteritis. Penicillamine is known to activate B cells polyclonally, and provoke synthesis of pathogenic autoantibodies. Recently, we studied a patient with penicillamine-induced nephritis in whom antimyeloperoxidase antibodies with perinuclear ANCA were detectable (G. Gaskin and C.D. Pusey, unpublished observations). Thus, it is possible that crescentic nephritis is another reflection of the propensity of penicillamine to provoke autoantibody synthesis.

Other drugs
There are two reports suggesting that hydralazine may provoke crescentic nephritis (Bjork et al. 1985; Mason and Lockwood 1986), but the findings are too preliminary for definite conclusions. There have been single case reports apparently implicating phenylbutazone (Leung et al. 1985), enalapril (Bailey and Lynn 1986), and rifampicin (Hirsch et al. 1983; Murray et al. 1987). Crescentic nephritis has also been attributed to the use of streptokinase (Murray et al. 1986) but the development of acute renal failure in this context is not necessarily associated with crescentic change (Davies et al. 1990).

Miscellaneous causes
Malignancy
A small number of patients in most large series with crescentic nephritis has had, or has later developed, malignancies (Whitworth et al. 1976; Beirne et al. 1977; Neild et al. 1983). However, it is difficult to be sure there is a causal link because many of the patients are usually elderly (but see Chapter 3.13). Even so, there are grounds for suspicion of a genuine, causal link in the case of carcinomas (Biava et al. 1984) and especially for lymphoid malignancies (Petzel et al. 1979; Pollak et al. 1988).

Other possible associations
It has been suggested that silicosis may predispose to crescentic nephritis (Arnalich et al. 1989; Sherson and Jorgensen 1989), possibly as a form of IgA nephropathy (Bonnin et al. 1987). α_1-Antitrypsin deficiency has also been described in two patients with crescentic nephritis (Lewis et al. 1985; Levy et al. 1986).

Two patients with Alport's syndrome have presented with crescentic nephritis (Harris et al. 1978).

Crescentic nephritis in children
Rapidly progressive glomerulonephritis is rare in childhood and relatively few large series have been collected (Heilman et al. 1970; Anand et al. 1975; Cunningham et al. 1980; Niaudet and Levy 1983; Miller et al. 1984; Southwest Pediatric Research Group 1985; Walters et al. 1988). It affects children of all ages but is exceptional under the age of 4 years. The range of diseases responsible is broadly similar to that for crescentic nephritis in adults but the proportions are different. Three diseases are clearly more common: acute poststreptococcal nephritis, which accounts for between 10 (Walters et al. 1988) and 55 per cent of cases (Cunningham et al. 1980); mesangiocapillary nephritis of both types; and Henoch-Schoenlein purpura. Antiglomerular basement membrane disease is rare but does occur (Cunningham et al. 1980; Walters et al. 1988). Systemic vasculitis (Niaudet and Levy 1983) and ANCA-positive microscopic renal vasculitis (Walters et al. 1988) have both been reported, as have crescentic phases of mesangial IgA disease, shunt nephritis, and subacute infective endocarditis (see Chapter 3.12).

The approach to diagnosis in children should be the same as in adults but there are small differences in clinical presentation and obviously practical differences in the management. The one point that should be made here is that, irrespective of the cause, children with crescentic nephritis appear to have a greater propensity to hypertension.

Epidemiology

Crescentic glomerulonephritis can affect patients of all races and of all ages, excepting perhaps infants. The diversity of conditions that cause crescentic nephritis means that descriptions of demography, epidemiology, and clinical presentation should be interpreted cautiously. Nevertheless the proportion of patients with crescentic nephritis in unselected series of renal biopsies is remarkably consistent throughout the world. An incidence of 2 to 5 per cent has been reported from France (Whitworth *et al.* 1976), the United Kingdom (Neild *et al.* 1983), the United States (Heilman *et al.* 1987), Africa (Dilma *et al.* 1981; Parag *et al.* 1988), India (Bhuyan *et al.* 1982; Date *et al.* 1987) and in the Chinese (Woo *et al.* 1986). There is an overall male predominance of approximately two to one.

Crescentic glomerulonephritis has been described in all age groups from children as young as 2 years to adults as old as 87 years. The diseases responsible vary with the age of onset; poststreptococcal nephritis and Henoch-Schönlein purpura are more common in childhood, and systemic vasculitis is in the elderly. There is relatively little information on the influence of race on susceptibility, but there is a strong clinical impression that the black races are less susceptible, at least to some types of crescentic nephritis. In a series of 110 patients with Wegener's and microscopic polyarteritis studied at Guy's Hospital (London), only one patient was black, compared with 6 to 11 per cent of black patients in other categories of glomerular disease (J.S. Cameron, unpublished). Support for this idea comes from Parag *et al.* (1988) in a study of 24 patients in Natal. They provided population statistics for blacks, caucasoids, and Indians, which showed that poststreptococcal nephritis occurs in exactly the expected frequencies for the three populations, whereas antiglomerular basement membrane disease and 'idiopathic' crescentic nephritis were more common than expected in caucasoids.

Clinical aspects

The clinical features of crescentic nephritis are determined in part by the underlying disease. Thus, for example, patients with poststreptococcal nephritis have oedema, hypertension, and give a history of a sore throat. Antiglomerular basement membrane disease, systemic vasculitis, and lupus have equally characteristic signs and these are discussed in other chapters. Nevertheless, many features are common to patients with crescentic nephritis as a group.

The rate of progression to renal failure is variable and may be as short as hours (Rees *et al.* 1978) or many months (Baldwin *et al.* 1987). Roughly half the patients have constitutional symptoms including fever, weight loss, and malaise. These occur most prominently in patients with renal microscopic polyarteritis and other types of vasculitis, but they are also found in antiglomerular basement membrane disease, and in nephritis associated with infections. Patients with severe injury may have loin pain with tenderness. As first pointed out by Berlyne and Baker (1964), hypertension is relatively uncommon, except in patients with diffuse endocapillary proliferative nephritis, chronic glomerulonephritis, or when fluid overload is severe. Serial measurements of serum creatinine may provide evidence of recent deterioration of renal function, but many patients present acutely with a raised serum creatinine. The urine usually contains blood and protein on stick testing, but nephrotic-range proteinuria is rare except when there is pre-existing glomerular disease. Urine microscopy reveals numerous dysmorphic erythrocytes, and counts of greater than 1 million/ml have been reported to correlate well with acute crescent formation (Bennett and Kincaid-Smith 1983; Nicholls *et al.* 1984), at least in crescentic IgA disease. The urine also contains increased numbers of leucocytes, which have been studied quantitatively by Segasothy *et al.* (1989). He showed that patients with crescentic nephritis had more than 30 000/ml, 30 per cent of which were polymorphs, 10 per cent monocytes, and roughly 5 per cent each of T-helper and T-suppressor cells. Red cell and granular casts are common.

The kidneys are usually swollen and there is often loss of the corticomedullary junction on ultrasonography. Intravenous urography gives poor pictures but the vessels are usually of normal calibre and outline on renal arteriography, even in the patients with systemic vasculitis (Travers *et al.* 1979; Serra *et al.* 1984; Savage *et al.* 1985) because the size of vessel involved in patients with crescentic nephritis is usually beyond the resolution of the technique. Isotope renograms show reduced renal perfusion and filtration, and DMSA scans may show areas of non-functioning cortex.

Anaemia is common and so are leucocytosis and thrombocytosis. They, together with a raised concentration of C-reactive protein, provide evidence of a vigorous, acute-phase response. Serum albumin concentrations are usually low, but immunoglobulins are normal or increased (Haworth 1983; Almroth *et al.* 1989). Antiglomerular basement membrane antibodies are a specific serological marker for antiglomerular basement membrane disease and the correlation between ANCA and renal microscopic vasculitis is increasingly close (Savage *et al.* 1987; Andrassy *et al.* 1989; Jennette *et al.* 1989; Nassberger *et al.* 1989; Cohen Tervaert *et al.* 1990), but their involvement in pathogenesis is speculative; this is discussed in more detail in Chapters 4.5.1 and 4.5.3. Immune complexes have been described in 40 to 100 per cent of patients with crescentic nephritis (Lockwood *et al.* 1977; Cohen *et al.* 1981; Sjoholm *et al.* 1983; Solling 1983) but their presence does not correlate with disease activity and their significance is uncertain.

Pathology of crescentic nephritis

Histopathology

In the acute stage, kidneys from patients with crescentic nephritis are usually swollen, but may be of normal size. Characteristically, the surface is studded with petechial haemorrhages and these may also be seen along the core of a renal biopsy (Fig. 3). The most obvious feature on light microscopy is the presence of extracapillary proliferation or infiltration. In some glomeruli the extracapillary accumulation of cells is discrete and localized to an individual lobule of the glomerulus. This is often referred to as a 'segmental' crescent, and is particularly characteristic of focal segmental proliferative glomerulonephritides, such as IgA nephropathy and Henoch-Schönlein purpura (see Chapter 4.5.2). In other patients, the accumulation of cells is severe and fills Bowman's space (Fig. 4), but the effect of geometry on the appearance of crescents mentioned previously and illustrated in Fig. 2 must be borne in mind. This effect makes definitions of crescents contentious, and, as they evolve into fibrous scars (see below), whether fibrous crescents should be taken into account, even if not very cellular, also arises. Even so, there must be some distinction between adhesions with a mild capsular reaction and

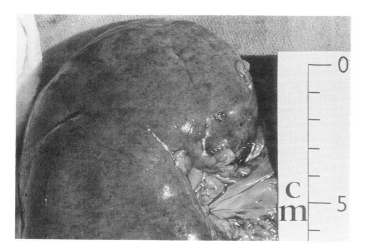

Fig. 3 The postmortem appearance of a kidney affected by severe crescentic nephritis. The kidney is congested, and the surface is covered in petechiae.

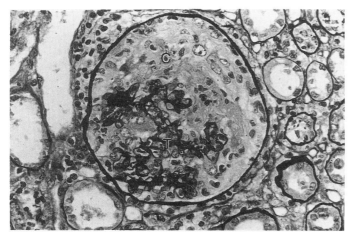

Fig. 4 A large crescent (C) is seen occupying Bowman's space and surrounding a peripheral section of the glomerular tuft (T) (methenamine silver/haematoxylin and eosin; × 150).

until the crescent is entirely composed of collagen. Nevertheless, there is convincing evidence that crescents can resolve completely without scarring (Heptinstall 1983). This phenomenon is probably more common in children, and in patients in whom the glomerular tuft is relatively intact and Bowman's capsule not ruptured. In some biopsies, all the crescents are at the same stage of their evolution, which suggests a single, devastating, immunopathological event, and this is often the case in antiglomerular basement membrane disease. In other biopsies, crescents are at many different stages, which suggests a more repetitive process; this is more typical of renal microscopic vasculitis and those with granular immune deposits.

The glomerular tuft itself may be relatively normal or show varying degrees of necrosis, proliferation, and scarring. Focal and segmental glomerular necrosis is usual in patients with antiglomerular basement membrane disease and renal microvascular vasculitis (Serra et al. 1984; Savage et al. 1985; Jennette et al. 1989). It is also common in crescentic lupus nephritis and Henoch–Schönlein purpura. Necrosis is less prominent in crescentic poststreptococcal nephritis, and in those with an underlying chronic nephritis. Diffuse, endocapillary proliferation is seen especially in patients with poststreptococcal nephritis, and is much less prominent in patients with antiglomerular basement membrane disease and microscopic polyarteritis; there is considerable overlap, however (Bacani et al. 1968). Completely sclerosed glomeruli and segmental scars are common in patients with microscopic vasculitis. Crescents do not obscure the appropriate features of membranous nephropathy and mesangiocapillary nephritis when present.

There is usually an extensive interstitial infiltrate (Fig. 6), often containing neutrophils in the acute stage as well as more chronic inflammatory cells. Immunohistological studies have shown that the interstitial infiltrates contain large numbers of macrophages as well as lymphocytes of both helper and suppressor phenotypes (Hooke et al. 1987; Nolasco et al. 1987a; Boucher et al. 1987; Muller et al. 1988). Occasionally, large numbers of eosinophils are seen, and this suggests a diagnosis of Churg–Strauss syndrome (Clutterbuck and Pusey 1990). Patients with more chronic disease, including those with microscopic polyarteritis, often have interstitial fibrosis and tubular atrophy. The

a true crescent, and it is not always clear what definition has been used when using the word 'crescent'. Most would require at least two or three layers of cells to be present to distinguish hyperplasia of the capsular cells (Neild et al. 1983); the apparent extent of the crescent depends so much upon the angle of the cut that it is probably not useful to attempt to define any minimum circumference of the tuft that must be occupied, below which it is to be considered 'segmental'.

Severe crescents compress the glomerular tuft except in patients with underlying chronic nephritis, when the contour of the tuft is maintained by the expanded mesangial matrix. In extreme cases, the glomerular tuft appears to dissolve into the mass of inflammatory cells and the basement membrane of Bowman's capsule may also rupture (Fig. 5). Initially, crescents consist exclusively of densely packed polygonal cells, whose origin cannot be ascertained by light microscopy. Within 2 weeks, some of the cells develop a histiocytic appearance and reticulin can be detected with appropriate stains. This process continues

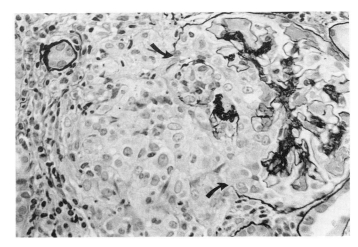

Fig. 5 Breaches in Bowman's capsule (arrows) and fusion of crescentic cells with periglomerular infiltrate (Methenamine silver haematoxylin and eosin).

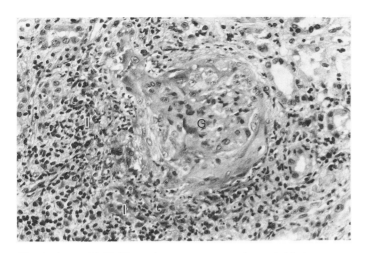

Fig. 6 Interstitial infiltrate in severe crescentic nephritis from a patient with Wegener's granuloma. The glomerulus (G) is surrounded by a dense infiltrate (I) of mononuclear cells, shown by phenotyping using monoclonal antibodies to consist mainly of CD4- and CD8-positive T cells and monocytes.

tubules in some patients may also show evidence of acute injury and regeneration, but attempts to correlate these changes with renal function have not usually been successful (Hind *et al.* 1983). The tubular cells also express increased amounts of HLA Class II antigens (Muller *et al.* 1988).

Immunohistology

Immunohistology has demonstrated the heterogeneity of crescentic nephritis and provided more accurate evidence of the pathogenesis. Linear staining of the basement membrane of glomerular capillaries, Bowman's capsule, and distal tubules is characteristic of antiglomerular basement membrane disease (see Chapter 3.11). Most commonly the antibodies deposited are IgG, but IgA and IgM are sometimes found. However, caution is needed when interpreting negative results in patients with severely disrupted glomeruli, as multiple levels of sectioning may be needed to find the few short segments of intact basement membrane that remain. Immunofluorescence findings, except for fibrin (Fig. 7), are unimpressive in most patients with renal microscopic vasculitis and this is the usual situation in patients with other types of systemic vasculitis, or when ANCA are detectable (Jennette *et al.* 1989).

The remaining patients show granular deposits of immunoglobulin and complement. This occurs in crescentic nephritis complicating poststreptococcal nephritis, in which there are prominent deposits of IgG and C3 in capillary loops. Obvious deposits of IgA in the mesangium suggests either Berger's disease or Henoch-Schönlein purpura, depending on the clinical context. There are strong deposits of C3 with little or no immunoglobulin in patients with type II mesangiocapillary nephritis. Substantial deposits of all three immunoglobulin classes together with complement suggest systemic lupus erythematosus or perhaps bacterial endocarditis, in which IgM may be particularly prominent.

Fibrin is invariably found in cellular crescents (Fig. 7) and sometimes in capillary loops and small vessels. Small deposits of C3 and IgM are sometimes found in areas of necrosis.

Electron microscopy

Electron microscopy has been used extensively in crescentic nephritis to look for deposits, and to identify the nature of the cells within crescents, as well as for more general descriptions of glomerular morphology. Electron-dense areas consistent with immune deposits are unusual in antiglomerular basement disease (Duncan *et al.* 1965; Beirne *et al.* 1977; Poskitt 1970) and in patients with systemic vasculitis and renal microscopic polyarteritis (Stilmant *et al.* 1979; Serra *et al.* 1984; Jennette *et al.* 1989). They are found in a subepithelial position in patients with crescentic postinfective nephritis, in some as typical humps whilst in others as smaller and more widely distributed deposits (Richardson *et al.* 1970; Beirne *et al.* 1977; McLeish *et al.* 1978; Morrin *et al.* 1978); some patients also have subendothelial deposits (Beirne *et al.* 1977; Neild *et al.* 1983). The location of deposits in other types of crescentic nephritis varies considerably depending on the underlying nephritis; they may be found in mesangium or in peripheral capillary loops (Bohman *et al.* 1974).

Other common electron microscopic findings include collapse of capillary walls with wrinkling of the glomerular basement membrane and accompanying areas of mesangiolysis (Bonsib 1988). Numerous breaks in the capillary basement membrane can be seen (Burkholder 1969; Stejskal *et al.* 1973). Bonsib has shown, using scanning electron microscopy of acellular preparations, that the breaks really are holes in the glomerular basement membrane (Bonsib 1985, 1988; Fig. 8). Fibrin can be found both within capillary loops and in Bowman's space in patients with cellular crescents, but this remains detectable only for a short time.

Pathogenesis

The crucial event in crescent formation appears to be the development of breaches in the glomerular basement membrane, as first suggested by Burkholder *et al.* (1969). Bonsib (1985; 1988;

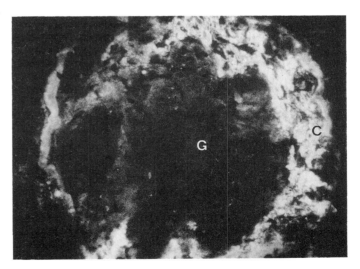

Fig. 7 Glomerulus from a patient with crescentic glomerulonephritis stained with fluorescein-conjugated antifibrin (-ogen) antibody. The glomerulus (G) contains almost no fibrin (-ogen), but the crescent (C) stains intensely. Cryostat section of snap-frozen material viewed using ultraviolet light; × 240.

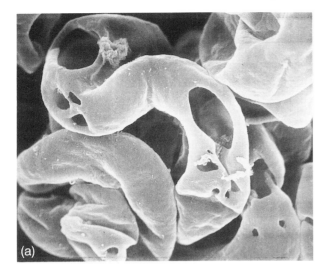

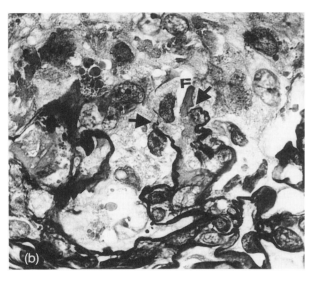

Fig. 8(a) Holes in the basement membrane in crescentic glomerulonephritis. The scanning electron microscopic preparation is a glomerulus from a renal biopsy which has been treated to remove all cells (from Bonsib 1985, with permission). (b) Shows in conventional sectioning the efflux of fibrin (F) through a gap in the silver-stained basement membrane (arrows), with cells around the exuded fibrin. Silver methenamine/haematoxylin and eosin, × 167.

Fig. 8(a)) have shown that these breaks, or more accurately holes, measure about 5×10 μm in diameter. Intravascular contents can be seen leaking into Bowman's space through them, and they are presumed to be the cause of crescent formation (Fig. 8(b)). Experimental studies by Vassali and McCluskey (1964) suggested that coagulation was important for this process by demonstrating the deposition of fibrin in Bowman's space before the development of crescents; they also showed protection from crescent formation by high-dose heparin in rabbits with experimental crescentic nephritis; lower doses of heparin, in subsequent experiments by others, proved to be ineffective (Thomson et al. 1975). Later, Naish et al. (1975) and Thomson et al. (1976) demonstrated the role of fibrin formation definitively in experiments in which prior defibrination with ancrod prevented fibrin deposition and crescent formation; it also appeared to be effective in reversing crescent formation. The fibrin in crescents is probably generated by the extrinsic pathway, because tissue factor can be demonstrated in crescentic glomeruli (Tipping et al. 1988), and because factor VIII of the intrinsic pathway cannot (Hoyer et al. 1974). It is not clear why fibrin should elicit the response it does, but, interestingly, cleavage of fibrinogen releases fibrinopeptides that are chemotactic for monocytes; however, these are not the only cells in crescents. This work has been reviewed by Salant (1987).

The traditional view, first advanced by Langhans (1885), was that crescents were composed of proliferating epithelial cells. This was supported by early studies using electron microscopy, which were interpreted as showing increased numbers of visceral and parietal epithelial cells (Morita et al. 1973; Bohman et al. 1974; Min et al. 1974; Gabbert and Thoenes 1977). This belief was strengthened by reports that cells within crescents were able to synthesize basement membrane (Gabbert and Thoenes 1977; Fig. 9). Later studies with monoclonal antibodies have shown that only parietal epithelial cells participate in crescent formation (Jennette et al. 1985; Boucher et al. 1987; Muller et al. 1988), and it is interesting that basement membrane produced by some of these cells contains the Goodpasture antigen (S.J. Cashman, A.J. Rees, and D.J. Evans, unpublished observations).

The idea that crescents were composed exclusively of proliferating epithelial cells was not supported by the experiments of Kondo et al. (1972), who pointed out that many of the cells were macrophages. Occasional cells, thought to be monocytes, were also described in early clinical studies using electron microscopy (Morita et al. 1973). Schiffer and Michael (1978) emphasized the role of macrophages in crescents in renal biopsies from two male patients with renal transplants from female donors. These showed that very few cells in the crescents contained Barr bodies, which suggested that they were of host rather than donor origin. Later Atkins and his colleagues used cell culture to show that macrophages were abundant in cultured glomeruli from patients with crescentic nephritis (Atkins et al. 1976; 1980). With these studies, the consensus shifted to favour the idea that most cells within the crescent were macrophages. Immunohistochemical studies and monoclonal antibodies capable of recognizing specific cell types facilitated quantitative studies of the cellular composition of crescents, which confirmed that monocytes were

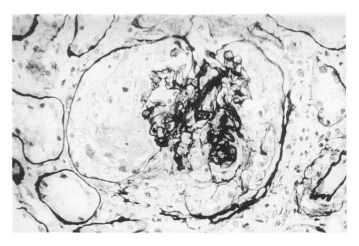

Fig. 9 Late crescentic nephritis with formation of silver-positive strands of basement membrane (arrows) within the crescent (silver methenamine/haematoxylin and eosin, × 135).

often abundant (Monga et al. 1981; Laopahand et al. 1983; Hooke et al. 1984; Stachura et al. 1984; Ferrario et al. 1985; Magil 1985; Bolton et al. 1987; Boucher et al. 1987; Nolasco et al. 1987b), and that small numbers of T cells also were present (Stachura et al. 1984; Bolton et al. 1987; Boucher et al. 1987; Nolasco et al. 1987a; Muller et al. 1989). Some of these studies have also demonstrated considerable numbers of epithelial cells in cresents (Jennette et al. 1985; Magil 1985; Boucher et al. 1987), and in agreement with this, we have noted that most crescents express the transferrin receptor, whilst not expressing major histocompatibility complex (MHC) antigens, despite the statement of the contrary by Muller et al. (1988). Glomerular expression of the CR-1 receptor for complement is also depressed on podocytes (Nolasco et al. 1987b).

Boucher et al. (1987), in one of the most detailed studies to date, used a panel of monoclonal antibodies capable of distinguishing between parietal and visceral epithelial cells, monocytes, and lymphocytes. They reported great heterogeneity in the cellular composition of crescents, even between those found in the same biopsy. The results depended on whether the basement membrane of Bowman's capsule remained intact. Crescents in which Bowman's capsule was intact largely comprised parietal epithelial cells, whereas macrophages predominated in crescents in which Bowman's capsule had ruptured. The conclusion drawn by Boucher was that the state of Bowman's capsule was critical to the composition of crescents, but this is unlikely to be the whole explanation. Both intuitively and from direct observation of the published pictures, it seems much more likely that epithelial cells predominate in less severe crescents, when neither the glomerular basement membrane nor Bowman's capsule has been extensively breached, and that extensive damage to either structure is associated with a greater proportion of macrophages.

There is direct experimental evidence that parietal epithelial cells proliferate during crescent formation (Cattell and Jamieson 1978; Sterzl and Pabst 1982), and there are many potential stimuli to epithelial cell proliferation in this situation. Inflammatory macrophages release large amounts of polypeptide growth factors capable of stimulating epithelial cell division (Nathan 1987; Sporn and Roberts 1988) including interleukin 1, and platelet-derived growth factor. The latter is released *in vitro* from renal microvascular (possibly glomerular) endothelium exposed to thrombin (Daniel et al. 1986), and this could happen, *in vivo*, in thrombosed glomerular capillaries. Interestingly, neither platelets themselves nor platelet-related antigens can be demonstrated within crescents (P. Duffus and F.E.B. Nolasco unpublished). Glomeruli from patients with crescentic glomerulonephritis release interleukin 1 when cultured for 24 h *in vitro* (Matsumoto et al. 1988). But it is too soon to say whether any of these factors is important in crescent formation.

Cells within crescents rapidly start to lay down collagen, and expression of collagen genes occurs within two days of the onset of experimental nephrotoxic nephritis in rabbits (Downer et al. 1988). The integrity of Bowman's capsule may influence the rate crescents evolve to the fibrotic stage. It has been suggested that rupture of the capsule allows inward migration of fibroblasts from the interstitium (Striker et al. 1973; Olsen 1974; Southwest Pediatric Nephrology Study Group 1985), but this remains controversial. Nevertheless, the Southwest Group reported a correlation between gaps in Bowman's capsule and fibrous transformation of crescents (Fig. 10). The question of whether crescents can resolve without scarring is even more contro-

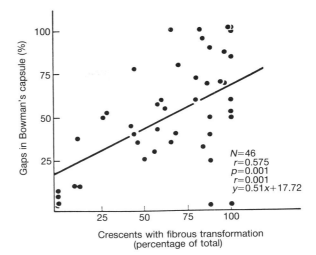

Fig. 10 Relationship between frequency of gaps in Bowman's capsule and the percentage of total crescents showing fibrous or fibrocellular changes (from Southwest Pediatric Nephrology Study Group (1985). *Kidney International*, 27, 456, with permission).

versial. It is difficult to envisage this happening in glomeruli with extensive injury to the capillary walls and multiple breaks in capillary membranes. Nevertheless, some glomeruli with circumferential crescents have relatively minor damage to capillary walls, and there are very convincing anecdotes concerning serial biopsies to suggest that some crescents resolve (McCluskey and Baldwin 1963; Faarup et al. 1978; Hepinstall 1983; Southwest Pediatric Nephrology Study Group 1985), at least in children with poststreptococcal nephritis. The genesis and fate of crescents can thus be summarized as shown in Fig. 11.

Treatment

General considerations

A common approach to the management of crescentic nephritis was entirely reasonably in the 1960s, before the heterogeneity of the condition was appreciated and before any treatments were effective. This is illustrated by the fact that all the patients reported by Berlyne and Baker (1964), Bacani et al. (1968), and Lewis et al. (1971) died. Later, survival figures improved, probably because of increased availability of dialysis, and also because of the more widespread use of steroids. Even so, less than 25 per cent of patients escaped the need for dialysis (Leonard et al. 1970; Striker et al. 1973; Beirne et al. 1977; Whitworth et al. 1976). Couser (1988) reviewed the early literature and calculated that 73 per cent of the 339 patients with idiopathic rapidly progressive glomerulonephritis reported at the time either died or went on to chronic dialysis. The only groups with a more encouraging prognosis were those with poststreptococcal nephritis (Leonard et al. 1970), and possibly with crescentic nephritis associated with other types of infection (Beaufils et al. 1976; Whitworth et al. 1976) There are two possibilities why this might be: first, because treatment of infection eradicated the stimulus to crescent formation (i.e. the immunopathology); and second, because the patients tended to be younger. Nevertheless, the observation did encourage the hope that renal function might recover in other types of crescentic nephritis, provided

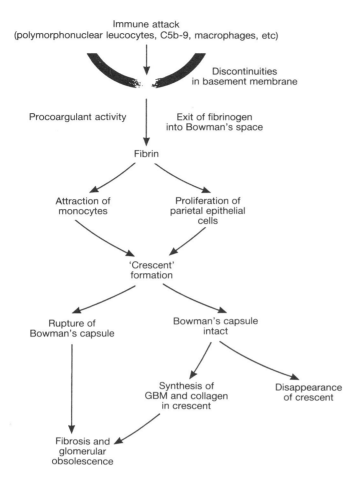

Fig. 11 Summary of the proposed pathogenesis of crescent formation.

that treatments to abort the immunopathological process could be found.

Treatment in the acute stage has three objectives: (a) to control the immunopathology; (b) to suppress acute inflammation; and (c) to limit scarring. Such treatment would need to be introduced early to be successful, as even the most effective control of the immunopathology would not improve renal function if the kidney had already been destroyed. This principle is illustrated by the effects of immunosuppression and plasma exchange on patients with antiglomerular basement membrane disease (see Chapter 3.11). Concentrations of circulating antiglomerular basement membrane antibody can be reduced rapidly both in dialysis-dependent patients and in those who still have reasonable renal function, but renal function improves only in the latter group (Pusey and Lockwood 1984). Similar principles are likely to apply to other forms of crescentic nephritis, but are more difficult to demonstrate because of the uncertainties about pathogenesis, and because they seem to have a higher threshold above which treatment becomes ineffective (Hind et al. 1983). This emphasizes the need for early diagnosis, which has been helped considerably by the more widespread use of serological assays. This may explain in part the improved prognosis now being reported for both the kidney and the patient (Coward et al. 1986; Pusey and Lockwood 1984; Furlong et al. 1987; Bruns et al. 1989; Bolton and Sturgill 1989).

The indifferent results from using immunosuppressive drugs in the 1960s have been reviewed by Cameron (1971). Better immunosuppressive and anti-inflammatory regimens are almost certainly an important contributing factor to the better results being reported, and it should now be an exception for renal function to continue to deteriorate once treatment has been started. Various treatments have been used in the past including anticoagulants and antiplatelet drugs (Cameron 1973; Brown et al. 1974; Suc et al. 1976). Occasional patients with severe renal failure survived (Fischer and Blumberg 1978) but this was unusual. Most of the modern immunosuppressive regimens used for crescentic nephritis are 'cocktails' consisting of oral steroids and a cytotoxic drug, usually azathioprine or cyclophosphamide. These are often supplemented with pulses of methylprednisolone (Cole et al. 1976; Bolton and Couser 1979; O'Neill et al. 1979; Oredugba et al. 1980; Adu et al. 1987; Bolton and Sturgill 1989) or plasma exchange (Lockwood et al. 1976; Becker et al. 1977; Thysell et al. 1983; Pusey and Lockwood 1984; Walker et al. 1986; Burran et al. 1986). It is not easy to ascertain which components of these regimens are critical to success, because most studies contain relatively small numbers of patients with disease of varying aetiologies and treated with a variety of regimens. A few investigators have used a consistent approach, and have shown that the response to an individual therapy is influenced more by the immunopathogenesis than by crescent scores or other morphological markers of prognosis (Hind et al. 1983; Bolton and Sturgill 1989).

There are also suggestions of different responses to treatment within the group of patients with renal microscopic polyarteritis. Patients with isolated nephritis have been reported to have a much worse prognosis than those with either systemic disease or proven systemic vasculitis (Weiss and Crissman 1985; Heilman et al. 1987); but these results have been contradicted by others (Coward et al. 1986), and so the claims need to be examined critically. Heilman et al. reported that survival at 3 years was 80 per cent in the patients with evident vasculitis compared to 15 per cent for those without. Because of these results, Velosa (1987) had advised making rigorous efforts to distinguish between these two groups. However, there was at least one other obvious difference between the groups in Heilman's study; the patients with vasculitis had less severe renal injury at presentation when judged by serum creatinine or the proportion of patients on dialysis (9 of 15 without vasculitis, compared to none of 13 with vasculitis), even though the crescent scores were similar. This alone could account for the difference in prognosis reported by Heilman. There were differences in treatment in the patients reported by Weiss and Crissman (1985): more patients with vasculitis were treated with cyclophosphamide, and the investigators' own conclusion was that this was likely to be responsible for the better prognosis in patients with vasculitis rather than intrinsic differences in the response to treatment. This conclusion is supported by recent reports of patients treated with cyclophosphamide, whose responses in renal microscopic vasculitis are excellent regardless of the presence of systemic signs (Coward et al. 1986; Adu et al. 1987; Furlong et al. 1987; Fuiano et al. 1988; Bruns et al. 1989).

The studies already cited allude to the difficulties of assessing the prognosis of crescentic nephritis by morphology. The traditional approach has been to use the crescent score, which used to be a good guide (Whitworth et al. 1976; Neild et al. 1983) but is now ineffective (Hind et al. 1983; Heilman et al. 1987), possibly because current treatments have changed the outlook. Evans et al. (1986), in a detailed analysis of a large number of patients, found that outcome could still be predicted from the renal biopsy in antiglomerular basement membrane disease but not in other forms of crescentic nephritis. Reports by Neild et al. (1983), Parfrey et al. (1984), Heilman et al. (1987), and Hind et al. (1983) suggest that overall measures of renal function are better guides to immediate prognosis, i.e., to the response to treatment; and they have the additional advantage that they can be assessed much more easily. Parfrey et al. (1984) and Heilman et al. (1987) all reported that serum creatinine, at presentation, was a good prognostic indicator. Others have used the need for dialysis as the yardstick (Hind et al. 1983; Stevens et al. 1983; Serra et al. 1983; Bolton and Sturgill 1989). It now seems sensible to categorize patients by renal function as well as disease when considering trials of therapy in crescentic nephritis.

Most recent attention has concentrated on finding out whether methylprednisolone and plasma exchange confer benefits above those obtained by steroids and cytotoxic drugs alone, but the choice of cytotoxic drugs should not be underestimated. Cyclophosphamide has been reported to be more effective than azathioprine for treating Wegener's granulomatosis (Fauci et al. 1983) and microscopic polyarteritis (Fauci et al. 1979), and it is used much more often to treat crescentic nephritis than before. Many of the groups now reporting the highest remission rates in renal microscopic vasculitis routinely use cyclophosphamide or chlorambucil (Johnson et al. 1973; Hind et al. 1983; Coward et al. 1986; Furlong et al. 1987; Bruns et al. 1989), but the importance of this change can only be guessed at. The use of intravenous pulses of cyclophosphamide has been advocated (Fort et al. 1988) but, as yet, there is no evidence to suggest it is better than oral cyclophosphamide for treatment of the acute stage; the opposite may even be true (Hoffman et al. 1990; Falk et al. 1990).

The doses of methylprednisolone that have been used to treat nephritis have been fairly consistent (Bolton and Couser 1979; Cole et al. 1976; O'Neill et al. 1979; Oredugba et al. 1980; Stevens et al. 1983; Adu et al. 1987; Bruns et al. 1989; Bolton and Sturgill 1989), but those for plasma exchange have not. Some groups have used daily regimens of whole plasma volume (Lockwood et al. 1976, 1977; Becker et al. 1977; Stevens et al. 1983; Muller et al. 1986), whereas others have preferred much less intensive treatment (Johnson et al. 1978; Glockner et al. 1988); clearly these differences should be considered when assessing the results, and when comparing the results reported by different groups.

Patients with renal failure being treated with immunosuppressive drugs need to be monitored very closely if infective complications are to be avoided. Neither methylprednisolone (Adu et al. 1987) nor plasma exchange (Rondeau et al. 1989) are without risk. This aspect has been reviewed in detail by Cohen et al. (1982). There is another important aspect to the treatment of crescentic nephritis, which has only became relevant when the acute stage could be controlled effectively. Increasingly, patients are being reported who have made early responses to immunosuppression but who go on to develop progressive renal injury, sometimes with, and sometimes without, overt evidence of relapse (Serra et al. 1984; Savage et al. 1985; Bolton and Sturgill 1989; Bruns et al. 1989; Keller et al. 1989). Clearly many of these patients could have progressive renal scarring as a direct result of the severity of their original injury; alternatively they may have continuing, though partially suppressed, activity of the crescentic nephritis. Distinguishing between these possibilities is important though not easy. In the case of antiglomerular basement membrane disease, circulating autoantibody concentrations can be used as a guide to the need for immunosuppression. Recent evidence suggests that ANCA may provide similar help in patients with microscopic vasculitis, but only in a negative sense as relapses in patients with undetectable ANCA are very rare (Nassberger et al. 1989; see Chapter 4.5.3).

Nevertheless, complete long-term follow-up studies of patients with crescentic nephritis have shown that those with both renal microscopic polyarteritis and Wegener's can relapse (Belghiti et al. 1987; Bruns et al. 1989; see Chapter 4.5.2) and many need long-term treatment with immunosuppressive drugs. This necessity carries the dangers of long-term immunosuppression, especially for the bladder and bone marrow in the case of alkylating agents (Boitard and Bach 1989), and of cataracts and avascular necrosis of the hips for patients continued on high-dose steroids (Bolton and Sturgil 1989). This places a premium on the continued careful assessment of individual patients and comparisons of the costs and benefits of individual regimens. This can only be done by considering different diseases separately.

Treatment of the acute stage
Antiglomerular basement membrane disease

The treatment of antiglomerular basement membrane disease is discussed fully in Chapter 3.11 and only the comparative aspects need be considered here. Conventional regimens of immunosuppressive drugs are ineffective. The disease is rare, and it is unlikely that convincing controlled trials will ever be mounted. Regimens that combine daily plasma exchanges with prednisolone and cyclophosphamide rapidly reduce circulating titres of antiglomerular basement membrane antibody (Johnson et al. 1978; Peters et al. 1983) and shorten the duration of their synthesis (Pusey and Lockwood 1986). Renal function also improves coincident with starting treatment in most patients with serum creatinine less than 600 μmol/l. Plasma exchange is much less effective when the creatinine is higher than this at presentation, but a few patients have recovered even when already on dialysis (see Chapter 3.11 for details). There is little tendency for the disease to recur once antibodies have become undetectable, and no need for long-term immunosuppressive therapy. However, long-term follow-up, including serial measurements of the antibody concentrations, is needed for optimal mangement. Serial measurements of antiglomerular basement membrane antibody are essential when considering a patient for renal transplant.

Renal microscopic vasculitis

Interpretation of the results of treating renal microscopic vasculitis is much more difficult. Small numbers of patients, even with severe disease, recover without treatment (Maxwell et al. 1979; Coward et al. 1986), and rather more respond when treated with steroids alone (Barlow 1985). The prognosis for patients in series reported since 1980 is considerably better than before. Data extracted from 13 recent large series of patients diagnosed as having crescentic, or focal necrotizing, glomerulonephritis or renal microscopic vasculitis show that the overall renal survival of 414 patients was 64 per cent. A wide variety of treatments was used in these series, and many patients did well without methylprednisolone or plasma exchange. The capacity for improvement of crescentic nephritis is shown even more strikingly in patients who were already on dialysis before treatment was started (Table 2). Over two-thirds of patients came off dialysis in the

Table 2 Response to therapy of dialysis-dependent patients with crescentic nephritis not caused by antiglomerular basement membrane antibodies

	Number of patients	Improved (%)	Dialysis-dependent	Death	Therapy*
Hind et al. (1983)	27	17(63)	0	10	P, Cy, PE mainly
Neild et al. (1983)	19	5(26)	3	11	P, Az mainly
Serra et al. (1984)	16	2(12)	1	13	P, Az mainly
Coward et al. (1986)	18	9(50)	7	2	P, Cy, some PE
Heilman et al. (1987)	9	1(11)	7	1	P mainly
Fuiano et al. (1988)	6	3(50)			
Glockner et al. (1988)	12	8(66)	3	1	P, Cy, Az, PE
Bolton and Sturgill (1989)	9	0(0)	?	?	P, Cy
Bolton and Sturgill (1989)	23	16(70)	6	1	MP, P, Cy
Hammersmith Hospital (unpublished)	42	32(76)	6	4	P, Cy, PE

* Az = azathioprine; Cy = cyclophosphamide; P = prednisolone; MP = pulse methylprednisolone; PE = plasma exchange.

five recent large series in which either methylprednisolone or plasma exchange were used.

The principal question is whether methylprednisolone and plasma exchange are really beneficial, or whether the improved survival is due to other, less well-defined changes in management. The Hammersmith group has attempted to answer this question for plasma exchange by a controlled trial (Pusey and Lockwood 1986; C.D. Pusey et al., unpublished). The results showed that all but one of the 29 patients who were not dialysis dependent improved, irrespective of whether they were treated with oral steroids and cytotoxic drugs alone or in combination with plasma exchange. Thus there appears to be little or no need for additional measures in this group of patients. In contrast, there was a significant benefit of plasma exchange in dialysis-dependent patients; 10 of 11 treated with plasma exchange improved, compared to three of eight given drugs alone ($p<0.04$) (see Chapter 4.5.3). These findings are similar to the preliminary results of a controlled trial reported by Rifle et al. (1981), but not to that by Glockner et al. (1988), which failed to show benefit from plasma exchange. However, several difficulties complicate the interpretation of Glockner's study, including the small numbers, the lack of stratification for renal function, or for disease (the study included patients with systemic lupus erythematosus and scleroderma), and, most particularly, because three patients in the control group were treated by plasma exchange and responded, having failed to improve during the first month with drug therapy alone.

There are no controlled trials of methylprednisolone for treating crescentic nephritis, but a comparative, though not strictly controlled, study by Stevens et al. (1983) suggests that both plasma exchange and methylprednisolone are valuable in dialysis-dependent patients with microscopic vasculitis. Methylprednisolone gave broadly similar results in another series of patients (Bolton and Sturgill 1989). In both these series, 70 per cent of dialysis-dependent cases came off dialysis, which provides a strong argument for the aggressive management of this group of patients.

Poststreptococcal nephritis

The natural history of crescentic poststreptococcal nephritis is much less clearly documented than that of either antiglomerular basement membrane disease or renal microscopic polyarteritis. There were strong suggestions from early series that complete recovery and resolution of crescents occurred in some patients given supportive treatment only; possibly as many as half of these with severe crescentic disease (Nakamoto et al. 1965; Leonard et al. 1970; Anand et al. 1975; Whitworth et al. 1976). Even so, the mortality was high, but it is unclear whether this would have been reduced with modern supportive treatment. Later studies have confirmed that severely affected patients can recover renal function (Roy et al. 1981; Niaudet and Levy 1983; Southwest Paediatric Nephrology Study Group 1985). Many, but not all, of these patients have received corticosteroids, or other forms of immunosuppressive therapy.

Substantial recovery has occurred spontaneously in some children with severe renal failure and over 80 per cent crescents. In fact, Roy et al. (1981) could demonstrate no difference in outcome when patients given supportive measures were compared with those treated with a 'quintuple' immunosuppressive regimen. Similar spontaneous recoveries have been reported by the Southwest Paediatric Nephrology Study Group (1985). Follow-up biopsies from these patients showed various combinations of normal, partially, or completely sclerosed glomeruli and some fibrosed crescents. Nevertheless, the proportion of intact and largely intact glomeruli suggested that many of the crescents had resolved completely (Roy et al. 1981). This accords with previous observations on children with poststreptococcal nephritis (McCluskey and Baldwin 1965; Faarup et al. 1978). Small numbers of patients develop progressive deterioration of renal function many months after the initial recovery (Roy et al. 1981; Niaudet and Levy 1983). Repeat biopsies in these patients show glomerular and interstitial scarring, presumptively related to the severity of the original injury.

It is not certain whether the relatively good prognosis in children can be extrapolated to adults with poststreptococcal nephritis, or the patients in whom acute endocapillary nephritis evolves to crescentic glomerulonephritis over a period of weeks (Morel-Maroger et al. 1974; Gill et al. 1977; Ferraris et al. 1983; Old et al. 1984; Modai et al. 1985; Fairley et al. 1987). Certainly, these patients have been reported to improve immediately after the introduction of immunosuppressive therapy, but too few have been studied to know what the natural history would have been if they had been left untreated. It would be interesting to know whether the endocapillary injury and the crescentic phase have the same pathogenesis, or, alternatively, whether the original glomerular injury provokes a second, presumably autoimmune, phase as appeared to be the case in a patient (described by

Morel-Maroger *et al.* 1974) who developed antitubular basement membrane antibodies.

Given the present uncertainties it seems reasonable to control the streptococcal infection first, and then give a brief course of high-dose steroids or methylprednisolone to patients presenting with severe crescentic poststreptococcal nephritis, and to reserve full immunosuppressive regimens for those patients whose crescentic disease evolves more slowly.

Infective endocarditis

The situation with crescentic nephritis in patients with infective endocarditis is very similar to that found in poststreptococcal disease. Control of infection is paramount and may necessitate early valve replacement. There is insufficient evidence to be dogmatic about the role of immunosuppressive drugs. The three patients with crescentic disease were included in the review by Neugarten *et al.* (1984) and all developed renal failure. More recently, Rovzar *et al.* (1986) treated a patient with immunosuppressive drugs and plasma exchange with apparent benefit. This seems a reasonable approach to the most severely affected patients after control of sepsis.

Crescentic lupus nephritis

This is usually managed in identical fashion to other patients with severe lupus nephritis (see Chapter 4.6.1).

Henoch-Schönlein purpura

The implications of crescents for the management of this purpura are discussed in Chapter 5.2.3.

Crescentic nephritis complicating chronic nephritis

Crescentic nephritis associated with membranous nephropathy is rare and none of the 13 cases described in the literature has recovere renal function, irrespective of the presence of antiglomerular basement membrane antibodies, and despite the use of modern immunosuppressive regimens (Abneo *et al.* 1986). Crescentic mesangiocapillary glomerulonephritis is also regarded as having a bad prognosis. None of nine patients reported by Niaudet and Levy (1983) recovered, despite treatment with various combinations of steroids, alkylating agents, anticoagulants, and plasma exchange. However, four of the six patients reported by Neild *et al.* (1983) had improved renal function after 'quadruple therapy' with steroids, azathioprine, anticoagulants, and dipyramidole. However, later all but one went into renal failure (J.S. Cameron, unpublished observations). A few patients have been subjected to plasma exchange but the results are not clear. Patients with severe, crescentic mesangiocapillary glomerulonephritis show a striking tendency to recurrence in allografts (Eddy *et al.* 1984; see Chapter 11.6).

Long-term outcome

The long-term outcome for patients with severe crescentic nephriris has been transformed by more effective early treatment, but many need continuous treatment and follow up if the improvement is to be maintained. This is dealt with more fully in Chapters 3.11, 4.5.3 and 4.5.2 for antiglomerular basement membrane disease, renal microscopic vasculitis associated with Wegener's granulomatosis and microscopic polyarteritis, and Henoch-Schönlein purpura, respectively. This presents the biggest problem for patients with ANCA-associated, renal microscopic vasculitis. Many of the patients who responded initially relapse if treatment is discontinued, irrespective of whether the original diagnosis was Wegener's granulomatosis (Fauci *et al.* 1983; Pinching *et al.* 1980; Fuiano *et al.* 1988; Weiss and Crissman 1985; Gaskin *et al.* 1989), microscopic polyarteritis (Savage *et al.* 1985; Coward *et al.* 1986; Fuiano *et al.* 1988; Heilman *et al.* 1987), or idiopathic rapidly progressive glomerulonephritis (Bruns *et al.* 1989; Keller *et al.* 1989; Stevens and Bone 1985; Belghiti *et al.* 1987; Bolton and Sturgill 1989). These findings strongly suggest that immunosuppression should be maintained in these patients well after the acute disease has been controlled.

Long-term immunosuppression carries inevitable risks (reviewed by Boitard and Bach 1989) and it is desirable to limit its use as much as possible. Unfortunately, there are no certain ways to do this but serial assays for ANCA may help, at least in the negative sense. Many treated patients continue to have detectable ANCA for many years without evidence of reactivation of nephritis or systemic vasculitis (see Chapter 4.5.2), but relapses are very rare in patients in whom ANCA are no longer detectable. Thus, it is reasonable to reduce maintenance therapy in patients with negative ANCA but to be cautious about reducing the dose in these in whom ANCA remains positive. Obsessional follow up to look for relapses is essential and should include objective measurements, such as C-reactive protein, platelet counts, and urinary microscopy, as well as purely clinical assessment (Hind *et al.* 1984).

It is even more difficult to determine the most effective combination of maintenance treatment. Relatively high doses of steroids carry the risk of cataracts and avascular necrosis, which were prevalent in the series of Bolton and Sturgill (1989), whereas the continuous use of cyclophosphamide, even in low doses, carries a considerable risk of haemorrhagic cystitis and bone marrow dysplasia (Fauci *et al.* 1983). At present, we favour the approach of the use of cyclophosphamide for induction treatment and then a switch to azathioprine for routine maintenance after 2 to 3 months. This regimen is described fully in Chapter 4.5.2, but a proportion of patients relapse on this regimen and can be controlled only with long-term cyclophosphamide.

Data for long-term survival are now available for patients treated in this general fashion. They show 5-year survival with independent renal function of 60 to 80 per cent in recent series (Savage *et al.* 1985; Coward *et al.* 1986; Furlong *et al.* 1987; Fuiano *et al.* 1988; Bruns *et al.* 1989; Gaskin *et al.* 1989). This is the standard that needs to be improved upon.

Conclusions

Over the past 20 years, developments in treatments have transformed the outlook for patients with crescentic nephritis in ways only equalled for those with systemic lupus. In doing so, many of the guiding principles have changed; for example, crescent scores now have little value in determining the prognosis of most patients with crescentic nephritis, although they may have some use in determining the intensity of initial treatment.

It is no longer possible to assert that glomeruli surrounded by crescents are inevitably destroyed; many will be, but the example of the degree of recovery possible in poststreptococcal nephritis is proof against this. This suggests that effective control of the underlying immunopathology may allow resolution without scarring, and this perhaps underlies the effectiveness of current regimens for renal microscopic polyarteritis. Unfortunately, it is impossible to tell as there is no certain way to assess the immunopathology.

The second cause for encouragement comes from studies in antiglomerular basement membrane disease, which have shown the value of being able to relate injury directly to the presence of antiglomerular basement membrane antibodies, so that treatment can be given and monitored in a rational way. One hopes that the recent realization that ANCA are closely associated with injury in most, possibly all, patients with renal microscopic polyarteritis will provide the necessary insight into the pathogenesis of these diseases to enable them to be approached as rationally as antiglomerular basement membrane disease.

There remains a clear need for more specific treatments. These should follow naturally on improved understanding of the pathogenesis. It is of interest that this applies to the inflammatory response as well as the underlying immunopathology. Recent advances in understanding of the control of inflammation by cytokines and the demonstration of abnormal cytokine concentrations in vasculitis (Grau *et al.* 1989) may indicate new areas for study.

References

Abreo, K., Abreo, F., Mitchell, B., and Schloemer, G. (1986). Idiopathic crescentic membranous glomerulonephritis. *American Journal of Kidney Diseases*, **8**, 257–61.

Abuelo, J.G., Esparaza, A.R., Matarese, R.A., Endreny, R.G., Carvalho, J.S., and Allagra, S.R. (1984). Crescentic IgA nephropathy. *Medicine*, **63**, 396–406.

Adu, D., Howie, A.J., Scott, D.G.I., Bacon, P.A., McGonigle, R.J.S., and Michael, J. (1987). Polyarteritis and the kidney. *Quarterly Journal of Medicine*, **239**, 221–37.

Almroth, G., Sjostrom, P., Svalander, C., and Danielsson, D. (1989). Serum immunoglobulins and IgG subclasses in patients with glomerulonephritis. *Journal of Interal Medicine*, **225**, 3–7.

Anand, S.K., Trygstad, C.W., Sharma, H.M., and Northway, J.D. (1975). Extracapillary proliferative glomerulonephritis in children. *Pediatrics*, **56**, 434–42.

Andrassy, K., Koderisch, J., Rufer, M., Erb, A., Waldherr, R., and Ritz, E. (1986). Detection and clinical implication of anti-neutrophil cytoplasmic antibodies in Wegener's granulomatosis and rapidly progressive glomerulonephritis. *Clinical Nephrology*, **32**, 159–67.

Arnalich, F., *et al.* (1989). Polyarteritis nodosa and necrotizing glomerulonephritis associated with long-standing silicosis. *Nephron*, **51**, 544–7.

Atkins, R.C., Holdsworth, S.R., Glasgow, E.F., and Mathews, F.E. (1976). The macrophage in human rapidly progressive glomerulonephritis. *Lancet*, **i**, 830–2.

Atkins, R.C., Glasgow, E.F., Holdsworth, S.R., Thomson, N.M., and Hancock, W.W. (1980). Tissue culture of isolated glomeruli from patients with glomerulonephritis. *Kidney International*, **17**, 515–27.

Bacani, R.A., Velasquez, F., Kanter, A., Pirani, C.L., and Pollak, V.E. (1968). Rapidly progressive (non-streptococcal) glomerulonephritis. *Annals of Internal Medicine*, **69**, 463–85.

Bailey, R.R. and Lynn, K.L. (1986). Crescentic glomerulonephritis developing in a patient taking enalapril (letter). *New Zealand Medical Journal*, **99**, 958–9.

Baldwin, D.S., Neugarten, J., Feiner, H.D., Gluck, M., and Spinowitz, B. (1987). The existence of a protracted course in crescentic glomerulonephritis. *Kidney International*, **31**, 790–4.

Balow, J.E. (1985). Renal Vasculitis (clinical conference). *Kidney International*, **27**, 954–64.

Banfi, G., Imasciati, E., Guerra, L., Mihatsch, M.J., and Ponticelli, C. (1983). Extracapillary glomerulonephritis with necrotizing vasculitis in D-penicillamine-treated rheumatoid arthritis. *Nephron*, **33**, 56–60.

Beaufils, M., Morel-Maroger, L., Sraer, J-D., Kanfer, A., Kourilsky, O., and Richet, G. (1976). Acute renal failure of glomerular origin during visceral abcesses. *New England Journal of Medicine*, **295**, 185–9.

Becker, G., Kincaid-Smith, P., D'Apice, A., and Walker, R.G. (1977). Plasmaperesis in the treatment of glomerulonephritis. *Medical Journal of Australia*, **2**, 693–6.

Beirne, G.J., Wagnild, J.P., Zimmerman, S.W., Mackem, P.D., and Burkholder, P.M. (1977). Idiopathic crescentic glomerulonephritis. *Medicine*, **56**, 349–81.

Belghiti, D., *et al.* (1987). Relapses of idiopathic diffuse crescentic glomerulonephritis without immune deposits: report of 6 cases. *American Journal of Nephrology*, **7**, 22–7.

Bennett, W.M. and Kincaid-Smith, P. (1983). Macroscopic haematuria in mesangial IgA nephropathy: clinical pathologic correlations. *Kidney International*, **23**, 393–400.

Berlyne, G. and Baker, J.S. (1964). Acute anuric glomerulonephritis. *Quarterly Journal of Medicine*, **33**, 105–15.

Bhuyan, U.N., Dash, S.C., Srivastava, R.N., Sharma, R.K., and Malhotra, K.K. (1982). Immunopathology, extent and course of glomerulonephritis with crescent formation. *Clinical Nephrology*, **18**, 280–5.

Biava, C.G., Gonwa, T.A., Naugton, J.L., and Hopper, J.Jr. (1984). Crescentic glomerulonephritis associated with nonrenal malignancies. *American Journal of Nephrology*, **4**, 208–14.

Bjorck, S., Svalander, C., and Westberg, G. (1985). Hydralazine-associated glomerulonephritis. *Acta Medica Scandinavica*, **218**, 261–9.

Bohman, S.O., Olsen, S., and Petersen, V.P. (1974). Glomerular ultrastructure in extracapillary glomerulonephritis. *Acta Pathologica et Microbiologica Scandinavica (A)*, **249**, 29–54.

Boitard, C. and Bach, J-F. (1989). Long term complications of conventional immunosupressive treatment. *Advances in Nephrology*, **18**, 335–54.

Bolton, W.K. and Sturgill, B.C. (1989). Methylprednisolone therapy for acute crescentic rapidly progressive glomerulonephritis. *American Journal of Nephrology*, **9**, 368–75.

Bolton, W.K. and Couser, W.G. (1979). Intravenous pulse methylprednisolone therapy of acute crescentic rapidly progressive glomerulonephritis. *American Journal of Medicine*, **66**, 495–502.

Bolton, W.K., Innes, D.J.Jr., Sturgill, B.C., and Kaiser, D.L. (1987). T-cells and macrophages in rapidly progressive glomerulonephritis: clinicopathologic correlations. *Kidney International*, **32**, 869–76.

Bonnin, A., Mousson, C., Justrabo, E., Tanter, Y., Chalopin, J.M., and Rifle, G. (1987). Silicosis associated with crescentic IgA mesangial nephropathy. *Nephron*, **47**, 229–30.

Bonsib, S.M. (1985). Glomerular basement membrane discontinuities. Scanning electron microscopic study of acellular glomeruli. *American Journal of Pathology*, **119**, 357–60.

Bonsib, S.M. (1988). Glomerular basement membrane necrosis and crescent organization. *Kidney International*, **33**, 966–74.

Boucher, A., Droz, D., Adafer, E., and Noel, L.H. (1987). Relationship between the intergrity of Bowman's capsule and the composition of cellular crescents in human crescentic glomerulonephritis. *Laboratory Investigation*, **56**, 526–33.

Boulton-Jones, J.M., Sissions, J.G.P., Evans, D.J., and Peters, D.K. (1974). Renal lesions of subacute infective endocarditis. *British Medical Journal*, **2**, 11–14.

Boyce, N.W., Holdsworth, S.R., Thomson, N.M., and Atkins, R.C. (1986). Clinicopathological associations in mesangial IgA nephropathy. *American Journal of Nephrology*, **6**, 246–52.

Breedveld, F.C., Valentin, R.M., Westdt, M.L., and Weening, J.J. (1985). Rapidly progressive glomerulonephritis with glomerular crescent formation in rheumatoid arthritis. Rapidly progressive renal failure in a child (clinical conference). *Scottish Medical Journal*, **30**, 184–9.

Brown, C.B., *et al.* (1974). Combined immunosuppression and anticoagulation in rapidly progressive glomerulonephritis. *Lancet*, **ii**, 1166–72.

Bruns, F.J., Adler, S., Fraley, D.S., and Segel, D.P. (1989). Long-term follow-up of aggressively treated idiopathic rapidly progressive glomerulonephritis. *American Journal of Medicine*, **86**, 400–6.

Burkholder, P.M. (1969). Ultrastructure demonstrations of injury and perforation of the glomerular capillary basement membrane in acute proliferative glomerulonephritis. *American Journal of Pathology*, **56**, 251–65.

Burran, W.P., Avasthi, P., Smith, K.J., and Simon, T.L. (1986). Efficacy of plasma exchange in severe idiopathic rapidly progressive glomerulonephritis: A report of ten cases. *Transfusion*, **26**, 382–7.

Cadnapaphornchai, P. and Sillix, D. (1989). Recurrence of monoclonal gammopathy-related glomerulonephritis in renal allograft. *Clinical Nephrology*, **31**, 156–9.

Cameron, J.S. (1971). Immunosuppressant agents in the treatment of glomerulonephritis. 2. Cytotoxic drugs. *Journal of the Royal College of Physicians*, **5**, 301–22.

Cameron, J.S. (1973). Are anticoagulants beneficial in the treatment of rapidly progressive glomerulonephritis. *Proceedings of the European Dialysis and Transplant Association*, **10**, 57–90.

Cattell, V. and Jamieson, S.W. (1978). The origin of glomerular crescents in experimental nephrotoxic serum nephritis in the rabbit. *Laboratory Investigation*, **391**, 589–90.

Chang-Miller, A., *et al.* (1987). Renal involvement in relapsing polychondiritis. *Medicine*, **66**, 202–17.

Chapman, S.J., Cameron, J.S., Chantler, C., and Turner, D. (1980). Treatment of mesangiocapillary glomerulonephritis in children with combined immunosuppression and anticoagulation. *Archives of Diseases of Childhood*, **55**, 446–51.

Chugh, K.S., Gupta, V.K., Singhal, P.C., and Sehgal, S. (1981). Case report: Poststreptococcal crescentic glomerulonephritis and pulmonary hemorrhage simulating Goodpasture's syndrome. *Annals of Allergy*, **47**, 104–6.

Citron, B.P., *et al.* (1970). Necrotising angiitis associated with drug abuse. *New England Journal of Medicine*, **283**, 1003–11.

Clutterbuck, E.J., Evans, D.J., and Pusey, C.D. (1990). Renal involvement in Churg–Strauss syndrome. *Nephrology Dialysis and Transplantation* (in press).

Cohen, A.H., Border, W.A., Shankel, E., and Glassock, R.J. (1981). Crescentic glomerulonephritis: Immune versus non-immune mechanisms. *American Journal of Nephrology*, **1**, 78–83.

Cohen, J., Pinching, A.J., Rees, A.J., and Peters, D.K. (1982). Infection and immunosuppression: A study of the infective complications in 75 patients with immunologically medicated disease. *Quarterly Journal of Medicine*, **201**, 1–15.

Cohen Tervaert, J.W., *et al.* (1990). Autoantibodies against myeloid lysosomal enzymes in crescentic glomerulonephritis. *Kidney International*, **37**, 799–806.

Cole, B.R., Brocklebank, T.J., Keinstra, R.A., Kissane, J.M., and Robson, A.M. (1976). 'Pulse' methylprednisolone therapy in the treatment of severe glomerulonephritis. *Journal of Pediatrics*, **88**, 302–14.

Connelly, C.E. and Gallacher, B. (1987). Acute crescentic glomerulonephritis as a complication of a *Staphylococcus aureus* abscess of hip joint prosthesis (letter). *Journal of Clinical Pathology*, **40**, 1486.

Couser, W.G. (1982). Idiopathic rapidly progressive glomerulonephritis. *American Journal of Nephrology*, **2**, 57–69.

Couser, W.G. (1988). Rapidly progressive glomerulonephritis: Classification, pathogenetic mechanisms, and therapy. *American Journal of Kidney Disease*, **11**, 449–64.

Coward, R.A., Hamdy, N.A.T., Shortland, J.S., and Brown, C.B. (1986). Renal micropolyarteritis: A treatable condition. *Nephrology Dialysis and Transplantation*, **1**, 31–7.

Croker, B.P., Lee, T., and Gunnells, J.C. (1987). Clinical and pathologic features of polyarteritis nodosa and its renal-limited variant: Primary crescentic and necrotizing glomerulonephritis. *Human Pathology*, **18**, 38–44.

Cunningham, R.J., *et al.* (1980). Rapidly progressive glomerulonephritis in children: A report of thirteen cases and a review of the literature. *Pediatric Research*, **14**, 128–32.

Dalal, B.I., Wallace, A.C., and Slinger, R.P. (1988). IgA nephropathy in relapsing polychondritis. *Pathology*, **20**, 85–9.

Daniel, T.O., Gibbs, V.C., Milfay, D.F., Garovoy, M.R., and Williams, L.T. (1986). Thrombin stimulates c-*sis* gene expression in microvascular endothelial cells. *Journal of Biological Chemistry*, **261**, 9579–82.

Date, A., Raghavan, R., John, J., Richard, J., Kirubakaran, M.G., and Shastry, J.G. (1987). Renal disease in adult Indians: A clinicopathological study of 2827 patients. *Quarterly Journal of Medicine*, **64**, 729–37.

Davies, D.J., Moran, J.E., Niall, J.F., and Ryan, G.B. (1982). Segmental necrotising glomerulonephritis with antineutrophil antibody: possible arbovirus aetiology? *British Medical Journal*, **285**, 606.

Davies, K.A., Marthieson, P., Winearls, C.G., Rees, A.J., and Walport, M.J. (1990). Serum sickness and acute renal failure after Streptokinase therapy for myocardial infarction. *Clinical and Experimental Immunology* (in press).

Davis, C.A., McEnery, P.T., Maby, S., McAdams, A.J., and West, C.D. (1978). Observations on the evolution of idiopathic rapidly progressive glomerulonephritis. *Clinical Nephrology*, **9**, 91–101.

Davis, C.A., McAdams, A.J., Wyatt, R.J., Forristal, J., and McEnery, P.T. (1979). Idiopathic rapidly progressive glomerulonephritis with C3 nephritic factor and hypocomplementemia. *Journal of Pediatrics*, **94**, 559–63.

Davson, J. and Platt, R. (1949). A clinical and pathological study of renal disease: I glomerulonephritis. *Quarterly Journal of Medicine*, **18**, 149–71.

Davson, J., Ball, J., and Platt, R. (1948). The kidney in periarteritis nodosa. *Quarterly Journal of Medicine*, **67**, 175–202.

Devogelaer, J.P., Pirson, Y., Vandenbroucke, J.M., Cosyns, J.P., Brichard, S., and Nagant de Deuxchaisnes, C. (1987). D-penicillamine induced crescentic glomerulonephritis: Report and review of the literature. *Journal of Rheumatology*, **14**, 1036–41.

Dilma, M.G., Adhikari, M., and Coovadia, H.M. (1981). Rapidly progressive glomerulonephritis in black children: A report of 4 cases. *South African Medical Journal*, **60**, 829–32.

Donnelly, S., Jothy, S., and Barre, P. (1989). Crescentic glomerulonephritis in Behçet's syndrome—results of therapy and review of the literature. *Clinical Nephrology*, **31**, 213–18.

Downer, G., Phan, S.H., and Wiggins, R.C. (1988). Analysis of renal fibrosis on a rabbit model of crescentic nephritis. *Journal of Clinical Investigation*, **82**, 998–1006.

Droz, D., Noel, L.H., Leibowitch, M., and Barbanel, C. (1979). Glomerulonephritis and necrotizing angiitis. *Advances in Nephrology*, **8**, 343–63.

Duncan, D.A., Drummon, K.N., Michael, A.F., and Vernier, R.L. (1965). Pulmonary haemorrhage and glomerulonephritis: Report of six cases and study of the renal lesion by the fluorescent antibody technique and electron microscopy. *Annals of Internal Medicine*, **62**, 920–38.

Eddy, A., Sibley, R., Mauer, S.M., and Kim Y. (1984). Renal allograft failure due to recurrent dense intramembranous deposit disease. *Clinical Nephrology*, **21**, 305–13.

Elfenbein, I.B., Baluarte, H.J., Cubillos-Rohas, M., Gruskin, A.B., Cote, M., and Cornfeld, D. (1975). Quantitative morphometry of glomerulonephritis with crescents diagnostic and predictive value. *Laboratory Investigation*, **32**, 56–64.

Ellis, A. (1942). Natural history of Brights' disease. Clinical, histological and experimental observations. *Lancet*, **i**, 34–36.

Evans, D.J., Savage, C.O.S., Winearls, C.G., Rees, A.J., Pusey, C.D., and Peters, D.K. (1986). Renal biopsy in prognosis of treated 'glomerulonephritis with crescents'. *Abstract of the X International Congress of Nephrology*, p. 60.

Faarup, P., Norgaard, T., Elling, F., and Jensen, H. (1978). Structural changes in the kidney of patients with oliguric extracapillary glomerulonephritis during immunosupressive therapy. *Acta Pathologica et Microbiologica Scandinavica (A)*, **86**, 409–14.

Fairley, C., Mathewson, D.C., and Becker, G.J. (1987). Rapid development of diffuse crescents in post-streptococcal glomerulonephritis. *Clinical Nephrology*, **28**, 256–60.

Falk, R.J. and Jennette, J.C. (1988). Anti-neutrophil cytoplasmic autoantibodies with specificity for myeloperoxidase in patients with systemic vasculitis and idiopathic necrotizing and crescentic glomerulonephritis. *New England Journal of Medicine*, **318**, 1651–7.

Falk, R.J., Hogan, S., Carey, T.S., and Jennette, C. (1990). Clinical Course of anti-neutrophil cytoplasmic autoantibody-associated glomerulonephritis and vasculitis. *Annals of Internal Medicine*, **113**, 656–63.

Fauci, A.S., Katz, P., Haynes, B.F., and Wolff, S.M. (1979). Cyclophosphamide therapy of severe systemic necrotizing vasculitis. *New England Journal of Medicine*, 235–8.

Fauci, A.S., Haynes, B.F., Katz, P., and Wolff, S.M. (1983). Wegener's granulomatosis: Prospective clinical and therapeutic experience with 85 patients for 21 years. *Annuals of Internal Medicine*, **98**, 76–85.

Ferrario, F., *et al.* (1985). The detection of monocytes in human glomerulonephritis. *Kidney International*, **28**, 513–19.

Ferraris, J.R., Gallo, G.E., Ramirez, J., Iotti, R., and Gianantonio, C. (1983). 'Pulse' methylprednisolone therapy in the treatment of acute crescentic glomerulonephritis. *Nephron*, **34**, 207–8.

Fischer, E. and Blumberg, A. (1978). Prolonged anuria in Wegener's granulomatosis: Recovery of renal function. *Journal of the American Medical Association*, **240**, 1174–5.

Fort, J.G., and Abnuzzo, J.L. (1988). Reversal of progressive necrotizing vasculitis with intravenous pulse cyclophosphamide and methylprednisolone. *Arthritis and Rheumatism*, **31**, 1194–8.

Fuiano, G., Cameron, J.S., Raftery, M., Hartley, B.H., Williams, D.G., and Ogg, C.S. (1988). Improved prognosis of renal microscopic polyarteritis in recent years. *Nephrology, Dialysis and Transplantation*, **3**, 383–91.

Furlong, T.J., Ibels, L.S., and Eckstein, R.P. (1987). The clinical spectrum of necrotizing glomerulonephritis. *Medicine*, **66**, 192–201.

Gabbert, H. and Thoenes, W. (1977). Formation of basement membrane in extracapillary proliferates in rapidly progressive glomerulonephritis. *Virchows Archives (Cell Pathology)*, **25**, 265.

Gaskin, G., Bateson, K., Evans, D.J., Rees, A.J., and Pusey, C.D. (1989). Response to treatment and long-term follow-up in 60 patients with Wegener's granulomatosis (abstract). *Nephrology, Dialysis and Transplantation*, **4**, 832–3.

Gavaghan, T.E. *et al.* (1981). Penicillamine-induced "Goodpasture Syndrome"; successful treatment of a fulminant case. *Australian and New Zealand Journal of Medicine*, **11**, 261–5.

Gibson, T., Burry, H.C., and Ogg, C.S. (1976). Goodpasture's syndrome. *Annals of Internal Medicine*, **84**, 100–1.

Gill, D.G., *et al.* (1977). Progression of acute proliferative poststreptococcal glomerulonephritis to severe epithelial crescent formation. *Clinical Nephrology*, **8**, 449–52.

Glassock, R.J. (1985). Natural history and treatment of primary proliferative glomerulonephritis: A review. *Kidney International suppl.* **17**, 136–42.

Glockner, W.M., *et al.* (1988). Plasma exchange and immunosuppression in rapidly progressive glomerulonephritis: A controlled, multicenter study. *Clinical Nephrology*, **29**, 1–8.

Grau, G.E., *et al.* (1989). Serum cytokine changes in systemic vasculitis. *Immunology*, **68**, 196–8.

Harada, A., *et al.* (1984). Renal amyloidosis associated with crescentic glomerulonephritis. *American Journal of Nephrology*, **4**, 52–5.

Harris, J.P., Rakowski, T.A., Argy, W.P.Jr., and Schreiner, G.E. (1978). Alport's syndrome representing as crescentic glomerulonephritis: A report of two siblings. *Clinical Nephrology*, **10**, 245–9.

Harrison, C.V., Loughridge, L.W., and Milne, M.D. (1964). Acute oliguric renal failure in acute glomerulonephritis and polyarteritis. *Quarterly Journal of Medicine*, **129**, 39–55.

Haworth, S.J. (1983). Renal involvement in Wegener's granulomatosis. The Hammersmith experience. *Nephrology*, 33–43.

Heaf, J.G., Jorgensen, F., and Neilsen, L.P. (1983). Treatment and prognosis of extracapillary glomerulonephritis. *Nephron*, **35**, 217–24.

Heilman, R.L., Offord, K.P., Holley, K.E., and Velosa, J.A. (1987). Analysis of risk factors for patient and renal survival in crescentic glomerulonephritis. *American Journal of Kidney Disease*, **9**, 98–107.

Hellmann, D.B., Hardy, K., Lindenfeld, S., and Ring, E. (1987). Takayasu's arteritis associated with crescentic glomerulonephritis. *Arthritis and Rheumatism*, **30**, 451–4.

Heptinstall, R.H. (1983). *Pathology of the Kidney*, Little Brown and Company, Boston.

Hill, G.S. *et al.* (1978). An unusual variant of membranous nephropathy with abundant crescent formation and recurrence in the transplanted kidney. *Clinical Nephrology*, **10**, 114–20.

Hind, C.R., Paraskevakou, H., Lockwood, C.M., Evans, D.J., Peters, D.K., and Rees, A.J. (1983). Prognosis after immunosuppression of patients with crescentic nephritis requiring dialysis. *Lancet*, **i**, 263–5.

Hind, C.R., Savage, C.O.S., Winearls, C.G., and Pepys, M.B. (1984). Objective monitoring of disease activity in polyarteritis by measurement of serum C reactive protein concentration. *British Medical Journal*, **288**, 1027–30.

Hirsch, D.J., Bia, F.J., Kashgarian, M., and Bia, M.J. (1983). Rapidly progressive glomerulonephritis during antituberculous therapy. *American Journal of Nephrology*, **3**, 7–10.

Hoffman, G.S., Leavitt, R.Y., Fleisher, T.A., Minor, J.R., and Fauci, A.S. (1990). Treatment of Wegener's granulomatosis with intermittent high-dose intravenous cyclophosphamide. *American Journal of Medicine*, **89**, 403–10.

Hooke, D.H., Hancock, W.W., Gee, D.C., Kraft, N., and Atkins, R.C. (1984). Monoclonal antibody analysis of glomerular hypercellularity in human glomerulonephritis. *Clinical Nephrology*, **22**, 163–8.

Hooke, D.H., Gee, D.C., and Atkins, R.C. (1987). Leukocyte analysis using monoclonal antibodies in human glomerulonephritis. *Kidney International*, **31**, 964–72.

Hoyer, J.R., Michael, A., and Hoyer, L. (1974). Immunofluorescent localisation of anti-hemophilic factor antigen and fibrinogen in human renal disease. *Journal of Clinical Investigation*, **53**, 1375–84.

Ingelfinger, J.R., McClusky, R.T., Schneeberger, E.E., and Grupe, W.E. (1977). Necrotizing arteritis in acute poststreptococcal glomerulonephritis: report of a recovered case. *Journal of Pediatrics*, **91**, 228–32.

Jennette, J.C. and Hipp, C.G. (1985). The epithieal cell antigen phenotype of glomerular crescent cells. *American Journal of Clinical Pathology*, **86**, 274–80.

Jennette, J.C., Wilkman, A.S., and Falk, R.J. (1989). Anti-neutrophil cytoplasmic autoantibody—associated glomerulonephritis and vasculitis. *American Journal of Pathology*, **135**, 921–30.

Jennings, R.B. and Earle, D.P. (1961). Post-streptococcal glomerulonephritis: Histopatholgical and clinical studies of the acute, subsiding acute and early chronic latent phase. *Journal of Clinical Investigation*, **40**, 1525–95.

Johnson, J.P., Whitman, W., Briggs, W.A., and Wilson, C.B. (1978). Plasmapheresis and immunosuppressive agents in antibasement membrane antibody-induced Goodpasture's syndrome. *American Journal of Medicine*, **64**, 354–9.

Johnson, J.R., McGovern, V.J., May, J., and Lauer, C. (1973). Focal necrotizing glomerulonephritis: Treatment with prednisone and chlorambucil. *England Journal of Medicine*, **289**, 628–34.

Juncos, L.I., Alexander, R.W., and Marbury, T.C. (1979). Intravascular clotting preceding crescent formation in a patient with Wegner's granulomatosis and rapidly progressive glomerulonephritis. *Nephron*, **24**, 17–20.

Kamata, K., Kobayashi, Y., Shigematsu, H., and Saito, T. (1982). Childhood type polymyositis and rapidly progressive glomerulonephritis. *Acta Pathologica Japanica*, **32**, 801–6.

Kaplan, N.G. and Kaplan, M.C. (1970). Monoclonal gamopathy, glomerulonephritis, and the nephrotic syndrome. *Archives of Internal Medicine*, **125**, 696–700.

Kebler, R., Kithier, K., McDonald, F.D., and Cadnapaphornchai, P. (1985). Rapidly progressive glomerulonephritis and monoclonal gammopathy. *American Journal of Medicine*, **78**, 133–8.

Keller, F., Oehlenberg, B., Kunzendorf, U., Schwarz, A., and Offerman, G. (1989). Long-term treatment and prognosis of rapidly progressive glomerulonephritis. *Clinical Nephrology*, **31**, 190–7.

Kingswood, J.C., Banks, R.A., Tribe, C.R., Owen Jones, J., and MacKenzie, J.C. (1984). Renal biopsy in the elderly: clinicopathological correlations in 143 patients. *Clinical Nephrology*, **22**, 183–7.

Klassen, J. et al. (1974). Evolution of membranous nephropathy into anti-glomerular-basement-membrane glomerulonephritis. *New England Journal of Medicine*, **290**, 319–25.

Klienknecht, D. et al. (1980). Dense deposit disease with rapidly progressive renal failure in a narcotic addict. *Clinical Nephrology*, **14**, 309–12.

Koethe, J.D., Gerig, J.S., Glickman, J.L., Sturgill, B.C., and Bolton, W.K. (1986). Progression of membranous nephropathy to acute crescentic rapidly progressive glomerulonephritis and response to pulse methylprednisolone. *American Journal of Nephrology*, **6**, 224–8.

Kondo, Y., Shigemastu, H., and Kobayashi, (1972). Cellular aspects of rabbit Masugi nephritis. II. Progressive glomerular injuries with crescent formation. *Laboratory Investigation*, **27**, 620–31.

Korzets, Z. and Bernheim, J. (1987). Rapidly progressive glomerulonephritis (crescentic glomerulonephritis) in the course of type I idiopathic membranoproliferative glomerulonephritis. *American Journal of Kidney Disease*, **10**, 56–61.

Kurki, P. et al. (1984). Transformation of membranous glomerulonephritis into crescentic glomerulonephritis with glomerular basement membrane antibodies: serial determinations of anti-GBM before the transformation. *Nephron*, **38**, 134–7.

Lai, K.N., Lai, F.M., Leung, A.C., Ho, C.P., and Vallance Owen, J. (1987). Plasma exchange in patients with rapidly progressive idiopathic IgA nephropathy: a report of two cases and review of literature. *American Journal of Kidney Diseases*, **10**, 66–70.

Landwehr, D.M., Cooke, C.L., and Rodriguez, G.E. (1980). Rapidly progressive glomerulonephritis in Beçhet's syndrome. *Journal of the American Medical Association*, **244**, 1709–11.

Langhans, T. (1885). Ueber die entzundlichen Veranderungen der Glomeruli und die acute Nephritis. *Virchows Archives (Pathology and Anatomy)*, **99**, 193–204.

Lapenas, D.J., Drewny, S.J., Luke, R.L. and Leeber, D.A. (1983). Crescentic light chain glomerulopathy. *Archives of Pathology*, **107**, 319–23.

Leonard, C.D., Nagle, R.B., Striker, G.E., Cutler, R.E., and Scribner, B.H. (1970). Acute glomerulonephritis with prolonged oliguria: An analysis of 29 cases. *Annals of Internal Medicine*, **73**, 703–11.

Laopahand, R.T., Cattell, V., and Gabriel, J.R.T. (1983). Monocyte infiltration in glomerulonephritis: alpha-1-antitrypsin as a marker for mononuclear phagocytes in renal biopsies. *Clinical Nephrology*, **19**, 309–16.

Leung, A.C., McLay, A., Dobbie, J.W., and Boulton-Jones, J.M. (1985) Phenylbutazone-induced systemic vasculitis with crescentic glomerulonephritis associated with deficiency of alpha 1-antitrypsin. *Archives of Internal Medicine*, **145**, 685–7.

Levy, M. (1986). Severe alpha-1 antitrypsin deficiency presenting with cutaneous vasculitis, rapidly progressive glomerulonephritis, and colitis (letter). *American Journal of Medicine*, **81**, 363–4.

Lewy, J.E., Salinas-Madrigal, L., Hendron, P.B., Pirani, C.L., and Metcaff, J. (1977). Clinico-pathologic correlations in acute poststreptococcal nephritis. *Medicine*, **50**, 453–501.

Lewis, E. et al. (1971). An immunopathological study of rapidly progressive glomerulonephritis in the adult. *Human Pathology*, **2**, 185–208.

Lewis, M. et al. (1985). Severe deficiency of alpha 1-antitrypsin associated with cutaneous vasculitis, rapidly progressive glomerulonephritis, and colitis. *American Journal of Medicine*, **79**, 489–94.

Lockwood, C.M., Rees, A.J., Pearson, T., Evans, D.J. and Peters, D.K. (1976). Immunosupression and plasma exchange in the treatment of Goodpasture's syndrome. *Lancet*, **i**, 711–15.

Lockwood, C.M. et al. (1977). Plasma-exchange and immunosuppression in the treatment of fulminating immune-complex crescentic nephritis. *Lancet*, **i**, 63–7.

Magil, A.B. (1985). Histogenesis of glomerular crescents. Immunohistochemical demonstration of cytokeratin in crescent cells. *American Journal of Pathology*, **120**, 222–9.

Mason, P.D. and Lockwood, C.M. (1986). Rapidly progressive nephritis in patients taking hydralazine. *Journal of Clinical Laboratory Immunology*, **20**, 151–3.

Matloff, D.S. and Kaplan, M.M. (1980). D-penicillamine-induced bilary cirrhosis: successful treatment with plasmapheresis and immunosupressives. *Gastroenterology*, **78**, 1046–9.

Matsumoto, K., Dowling, J., and Atkins, R.C. (1988). Production of interleukin 1 in glomerular cell cultures from patients with rapidly progressive crescentic glomerulonephritis. *American Journal of Nephrology*, **8**, 463–70.

Maxwell, D.R., Ozawa, T., Nielsen, R.L., and Luft, F.C. (1979). Spontaneous recovery from rapidly progressive glomerulonephritis. *British Medical Journal*, **2**, 1447.

McCluskey, R.I. and Baldwin, D.S. (1963). The natural history of acute glomerulonephritis. *American Journal of Medicine*, **35**, 213–30.

McCoy, R., Clapp, J., and Seigler, H.F. (1975). Membranoproliferative glomerulonephritis. Progression from the pure form to the crescentic form with recurrence after transplantation. *American Journal of Medicine*, **59**, 288–92.

McLeish, K.R., Yum, M.N., and Luft, F.C. (1978). Rapidly progressive glomerulonephritis in adults: Clinical and histologic correlations. *Clinical Nephrology*, **10**, 43–50.

Meyrier, A., Simon, P., Mignon, F., Striker, L., and Ramee, M.P. (1984). Rapidly progressive ('crescentic') glomerulonephritis and monoclonal gammapathies. *Nephron*, **38**, 156–62.

Miller, M.N., Baumal, R., Poucell, S., and Steele, B.T. (1984). Incidence and prognostic importance of glomerular crescents in renal diseases of childhood. *American Journal of Nephrology*, **4**, 244–7.

Min, K.W., Gyorkey, F., Gyorkey, P., Yium, J.J., and Eknoyan, G. (1974). The morphogenesis of glomerular crescents in rapidly progressive glomerulonephritis. *Kidney International*, **5**, 47–56.

Mitas, J.A., Frank, L.R., Swerdlin, A.R., Johnson, D.L., and Rabetoy, G.M. (1983). Crescentic glomerulonephritis complicating idiopathic membranous glomerulonephropathy. *Southern Medical Journal*, **76**, 664–7.

Modai, D. et al. (1985). Biopsy proven evolution of post streptococcal glomerulonephritis to rapidly progressive glomerulonephritis of a post infectious type. *Clinical Nephrology*, **23**, 198–202.

Monga, G., Mazzucco, G., Belgiojoso, B., and Bushach, G. (1981). Monocyte infiltration and glomerular hypercellularity in human acute and persistant glomerulonephritis: light and electron microscopic, immunofluorescent and histochemical investigation on twenty eight patients. *Laboratory Investigation*, **44**, 381–7.

Montoliu, J., Darnell, A., Torras, A., and Revert, L. (1981). Acute and rapidly progressive forms of glomerulonephritis in the elderly. *Journal of the American Geriatric Society*, **29**, 108–16.

Moorthy, A.V., Zimmerman, S.W., Burkholder, P.M., and Harrington, A.R. (1976). Association of crescentic glomerulonephritis with membranous glomerulonephropathy: a report of these cases. *Clinical Nephrology*, **6**, 319–25.

Morel-Maroger, L., Kourilsky, O., Midnon, F., and Richet, G. (1974). Antitubular basement membrane antibodies in rapidly progressive poststreptococcal glomerulonephritis: Report of a case. *Clinical Immunology and Immunopathology*, **2**, 185–94.

Morita, T., Suzuki, Y., and Churg, J. (1973). Structure and development of the glomerular crescent. *American Journal of Pathology*, **72**, 349–68.

Morrin, P.A., Hinglais, N., Nabarra, B., and Kreis, H. (1978). Rapidly progressive glomerulonephritis. A clinical and pathologic study. *American Journal of Medicine*, **65**, 446–60.

Muller, G.A., Gebhardt, M., Kompf, J., Baldwin, W.M., Ziegenhagen, D., and Bohle, A. (1984). Association between rapidly progressive glomerulonephritis and the properdin factor BfF and different HLA-D region products. *Kidney International*, **25**, 115–18.

Muller, G.A., Seipel, L., and Risler, T. (1986). Treatment of non anti-GBM-antibody mediated, rapidly progressive glomerulonephritis by plasmapheresis and immunosuppression. *Klinische Wochenschrift*, **64**, 231–8.

Muller, G.A., Muller, C.A., Markovic-Lipkovski, J., Kilper, R.B., and Risler, T. (1988). Renal, major histocompatibility complex antigens and cellular components in rapidly progressive glomerulonephritis identified by monoclonal antibodies. *Nephron*, **49**, 132–9.

Murray, A.N., Cassidy, M.J., and Templecamp, C. (1987). Rapidly progressive glomerulonephritis associated with rifampicin therapy for pulmonary tuberculosis. *Nephron*, **46**, 373–6.

Murray, N., Lyons, J., and Chappell, M. (1986). Crescentic glomerulonephritis: A possible complication of streptokinase treatment for myocardial infarction. *British Heart Journal*, **56**, 483–5.

Mustonen, J., Helin, H., Pasternack, A., and Vanttinen, T. (1983). Acute rheumatic fever with extracapillary glomerulonephritis and the nephrotic syndrome. *Annals of Clinical Research*, **15**, 92–4.

Naish, P.F., Evans, D.J., and Peters D.K. (1975). The effects of defibrination with ancrod in experimental allergic glomerular injury. *Clinical and Experimental Immunology*, **20**, 303–9.

Nakamoto, S., Dunea, G., Kolff, W., and McCormack, L. (1965). Treatment of oliguric glomerulonephritis with dialysis and steroids. *Annals of Internal Medicine*, **63**, 359–68.

Naschitz, J.E., Yushurun, D., Scharf, Y., Sajrawi, I., Lazarov, N.B., and Boss, J.H. (1989). Recurrent massive alveolar hemorrhage, crescentic glomerulonephritis, and necrotizing vasculitis in a patient with rheumatoid arthritis. *Archives of Internal Medicine*, **149**, 406–8.

Nassberger, L., Sjoholm, A.G., Bygren, P., Thysell, H., Hojer-Madsen, M., and Rasmussen, N. (1989). Circulating anti-neutrophil cytoplasm antibodies in patients with rapidly progressive glomerulonephritis and extracapillary proliferation. *Journal of Internal Medicine*, **255**, 191–6.

Nathan, C.F. (1987). Secretory products of macrophages. *Journal of Clinical Investigation*, **79**, 319–26.

Neild, G.H., Cameron, J.S., Lessof, M.H., Ogg, C.S., and Turner, D.R. (1978). Relapsing polychondritis with crescentic glomerulonephritis. *British Medical Journal*, **1**, 743–5.

Neild, G.H., et al. (1983). Rapidly progressive glomerulonephritis with extensive glomerular crescent formation. *Quarterly Journal of Medicine*, **52**, 395–416.

Neugarten, J., Gallo, G.R., and Baldwin, D.S. (1984). Glomerulonephritis in bacterial endocarditis. *American Journal of Kidney Disease*, **3**, 371–9.

Nguyen, B.P., Reisen, E., and Rodriguez, F.H.J.R. (1988). Idiopathic membranous glomerulopathy complicated by crescentic glomerulonephritis and renal vein thrombosis. *American Journal of Kidney Disease*, **12**, 326–8.

Niaudet, P. and Levy, M. (1983). Glomerulonephritis à croissants diffuse. In *Nephrologie Pediatrique* (ed. P. Royer, R. Habib, H. Mahieu, and M. Broyer pp. 381–94. Flammarion, Paris.

Nicholls, K.M., Fairley, K.F., Dowling, J.P., and Kincaid-Smith, P. (1984). The clinical course of mesangial IgA nephropathy. *Quarterly Journal of Medicine*, **53**, 227–250.

Nicholson, G.D., Amin, U.F., and Alleyne, G.A. (1975). Membranous glomerulonephropathy with crescents. *Clinical Nephrology*, **4**, 198–201.

Nigam, P., et al. (1986). Rapidly progressive (crescenteric) glomerulonephritis in erythema nodosum leprosum: Case report. *Hansenology International*, **11**, 1–6.

Nolasco, F.E.B., Cameron, J.S., Hartley, B., Coelho, A., Hildreth, G., and Reuben, R. (1987a). Intraglomerular T cells and monocytes in nephritis: study with monoclonal antibodies. *Kidney International*, **31**, 1160–6.

Nolasco, F.E., Cameron, J.S., Hartley, B., Coelho, R.A., and Hildredth, G. (1987b). Abnormal podocyte CR-1 expression in glomerular diseases: Association with glomerular cell proliferation and monocyte infiltration. *Nephrology Dialysis and Transplantation*, **2**, 304–12.

Ntoso, K.A., Tomaszewski, J.E., Jimenez, S.A., and Neilson, E.G. (1986). Penicillamine-induced rapidly progressive glomerulonephritis in patients with progressive systemic sclerosis: Successful treatment of two patients and a review of the literature. *American Journal of Kidney Disease*, **8**, 159–63.

Ogami, Y., et al. (1989). Waldenstrom's macroglobulinemia associated with amyloidosis and crescentic glomerulonephritis. *Nephron*, **51**, 95–8.

Old, C.W., Herrera, G.A., Reimann, B.E., and Latham, R.D. (1984). Acute post-streptococcal glomerulonephritis progressing to rapidly progressive glomerulonephritis. *Southern Medical Journal*, **77**, 1470–2.

Olsen, S. (1974). Extracapillary glomerulonephritis. A semiquantitative lightmicroscopical study of 59 patients. *Acta Pathologica et microbiologica Scandinavica (A)*, **249**, 7–19.

Olsson, P.J., Gaffney, E., Alexander, R.W., Mars, D.R., and Fuller, T.J. (1980). Proliferative glomerulonephritis with crescent formation in Beçhet's syndrome. *Archives of Internal Medicine*, **140**, 713–14.

O'Neill, W.M.Jr., Etheridge, W.B., and Bloomer, H.A. (1979). High-dose corticosteroids: their use in treating idiopathic rapidly progressive glomerulonephritis. *Archives of Internal Medicine*, **139**, 514–18.

Oredugba, O., Mazumdar, D.C., Meyer, J.S., and Lubowitz, H. (1980). Pulse methylprednisolone therapy in idiopathic, rapidly progressive glomerulonephritis. *Annals of Internal Medicine*, **92**, 504–6.

Panner, B.J. (1980). Rapidly progressive glomerulonephritis and possible amyloidosis. *Archives of Pathology and Laboratory Medicine*, **104**, 603–9.

Parag, K.B., Naran, A.D., Seedat, Y.K., Nathoo, B.C., Naicker, I.P., and Naicker, S. (1988). Profile of crescentic glomerulonephritis in Natal—a clinicopathological assessment. *Quarterly Journal of Medicine*, **68**, 629–36.

Parfrey, P.S., Hutchinson, T.A., Jothy, S., Cramer, B.C., Martin, J., and Seely, J.F. (1985). The spectrum of diseases associated with necrotizing glomerulonephritis and its prognosis. *American Journal of Kidney Disease*, **6**, 387–96.

Peces, R., Riera, J.R., Arboleya, L.R., Lopez-Larrea, C., and Alvarez, J. (1987). Goodpasture's syndrome in a patient receiving penicillamine and carbimazole. *Nephron*, **45**, 316–20.

Peters, D.K., Rees, A.J., Lockwood, C.M., and Pusey, C.D. (1983). Treatment and prognosis in anti-basement membrane antibody-mediated nephritis. *Transplant Proceedings*, **14**, 513–21.

Petzel, R.A., Brown, D.C., Staley, N.A., McMillen, J.J., Sibley, R.K., and Kjellstrand, C.M. (1979). Crescentic glomerulonephritis and renal failure associated with malignant lymphoma. *American Journal of Clinical Pathology*, **71**, 728–32.

Pinching, A.J., Rees, A.J., Pussell, B.A., Lockwood, C.M., Mitcheson, R.S., and Peters D.K. (1980). Relapse in Wegener's granulomatosis: the role of infection. *British Medical Journal*, **281**, 836–8.

Pollak, V.E. and Mendoza, N. (1971). Rapidly progressive glomerulonephritis. *Medical Clinics of North America*, **55**, 1397–415.

Pollak, C.A., Ibels, L.S., Levi, J.A., Eckstein, R.P., and Wakeford, P. (1988). Acute renal failure due to focal necrotizing glomerulonephritis in a patient with non-Hodgkin's lymphoma. Resolution with treatment of lymphoma. *Nephron*, **48**, 197–200.

Poskitt, T.R. (1970). Immunogic and electron microscopic findings in Goodpasture's syndrome. *American Journal of Medicine*, **49**, 250–7.

Potvliege, P.R., De Roy, G., and Dupuis, F. (1975). Necropsy study on glomerulonephritis in the elderly. *Journal of Clinical Pathology*, **28**, 891–8.

Pusey, C.D. and Lockwood, C.M. (1984). Plasma exchange for glomerular disease. In *Nephrology* (ed. R.R. Robinson), pp. 1474–85. New York, Springer-Verlag.

Pusey, C.D. and Lockwood, C.M. (1989). Autoimmunity in rapidly progressive glomerulonephritis. *Kidney International*, 35, 929–37.

Rees, A.J., Lockwood, C.M., and Peters, D.K. (1977). Enhanced allergic tissue damage in Goodpasture's syndrome by intercurrent bacterial infection. *British Medical Journal*, 2, 723–6.

Rees, A.J., Lockwood, C.M., and Peters, D.K. (1978). Nephritis caused by antibodies to GBM. In *Progress in Glomerulonephritis* (ed. P. Kincaid-Smith, A.J.F. D'Apice, R.C. Atkins) pp. 347–66. New York, Wiley.

Richardson, J.A., Rosenau, W., Lee, J.C., and Hopper, J. (1970). Kidney transplantation for rapidly progressive glomerulonephritis. *Lancet*, ii, 180–2.

Rifle, G., *et al.* (1981). Treatment of idiopathic acute crescentic glomerulonephritis by immunodepression and plasma-exchanges: A prospective randomised study. *Proceedings of the European Dialysis and Transplantation Association*, 18, 493–502.

Rondeau, E., Levy, M., Dosquet, P., Ruedin, P., Mougenot, B., and Kanfer, A. (1989). Plasma exchange and immunosuppression for rapidly progressive glomerulonephritis: prognosis and complications. *Nephrology Dialysis and Transplantation*, 4, 196–200.

Rovzar, M.A., Logan, J.L., Ogden, D.A., and Graham, A.R. (1986). Immunosuppressive therapy and plasmapheresis in rapidly progressive glomerulonephritis associated with bacterial endocarditis. *American Journal of Kidney Disease*, 7, 428–33.

Roy, S., Murphy, W.M., and Arant, B.S. (1981). Post-streptococcal glomerulonephritis in children: Comparison of qunitruple therapy versus supportive care. *Pediatrics*, 98, 403–10.

Rubinger, D., *et al.* (1986). Combined cyclophosphamide and corticosteroid-induced remission in severe glomerulopathy associated with systemic vasculitis. *American Journal of Nephrology*, 6, 346–52.

Sadjadi, S.A., Seelig, M.S., Berger, A.R., and Milstoc, M. (1985). Rapidly progressive glomerulonephritis in a patient with rheumatoid arthritis during treatment with high-dosage D-penicillamine. *American Journal of Nephrology*, 5, 212–16.

Said, R., Hamzeh, Y., Tarawneh, M., El Khateeb, M., Abdeen, M., and Shaheen, A. (1989). Rapid progressive glomerulonephritis in patients with familial Mediterranean fever. *American Journal of Kidney Disease*, 14, 412–16.

Salant, D.J. (1987). Immunopathogenesis of crescentic glomerulonephritis and lung purpura. *Kidney International*, 32, 408–25.

Savage, C.O.S., Winearls, C.G., Evans, D.J., Rees, A.J., and Lockwood, C.M. (1985). Microscopic polyarteritis: presentation, pathology and prognosis. *Quarterly Journal of Medicine*, 56, 467–83.

Savage, C.O.S., Winearls, C.G., Jones, S., Marshall, P., and Lockwood, C.M. (1987). Prospective study of radioimmunoassy for antibodies against neutrophil cytoplasm in diagnosis of systemic vasculitis. *Lancet*, i, 1389–93.

Scheer, R.L. and Grossman, M.A. (1964). Immune aspects of glomerulonephritis associated with pulmonary haemorrhage. *Annals of Internal Medicine*, 60, 1009–21.

Schiffer, M.C. and Michael, A. (1978). Renal cell turnover studied by the Y chromosome (Y body) staining of the transplanted human kidney. *Journal of Laboratory and Clinical Medicine*, 92, 841–8.

Segasothy, M., Fairley, K.F., Birch, D.F., and Kincaid Smith, P. (1989). Immunoperoxidase identification of nucleated cells in urine in glomerular and aute tubular disorders. *Clinical Nephrology*, 31, 281–91.

Serra, A., *et al.* (1984). Vasculitis affecting the kidney: Presentation, histopathology and long-term outcome. *Quarterly Journal of Medicine*, 210, 181–207.

Sherson, D. and Jorgensen, F. (1989). Rapidly progressive crescenteric glomerulonephritis in a sandblaster with silicosis. *British Journal of Industrial Medicine*, 46, 675–6.

Singhal, P.C., Chugh, K.S., Kaur, S., and Malik, A.K. (1977). Acute renal failure in leprosy. *International Journal of Leprosy and Other Mycobacterial Diseases*, 45, 171–4.

Sjoholm, A.G., Brun, C., Larsen, S., and Thysell, H. (1983). Circulating immune complexes and C1 activation in patients with rapidly progressive glomerulonephritis, before and after treatment with immunosuppression and plasma exchange. *International Archives of Allergy and Applied Immunology*, 72, 9–15.

Solling, J. (1983). Circulating immune complexes in glomerulonephritis: A longitudinal study. *Clinical Nephrology*, 20, 177–89.

Southwest Pediatric Nephrology Study Group (1985). A clinicopatholgical study of crescentic glomerulonephritis in 50 children. *Kidney International*, 27, 450–8.

Sporn, M.B. and Roberts, A.B. (1988). Peptide growth factors and multifunctional. *Nature*, 332, 217–9.

Stachura, I., Si, L., and Whiteside, T.L. (1984). Mononuclear-cell subsets in human idiopathic crescentic glomerulonephritis (ICGN): analysis in tissue sections with monoclonal antibodies. *Journal of Clinical Immunology*, 4, 202–8.

Stejskal, J. *et al.* (1973). Discontinuites (gaps) of the glomerular capillary wall basement membrane in renal disease. *Laboratory Investigation*, 29, 149–65.

Sternlieb, I., Bennett, B., and Scheinberg, I.H. (1975). D-penicillamine induced Goodpasture's syndrome in Wilson's disease. *Annals of Internal Medicine*, 82, 673–6.

Sterzel, R.B. and Pabst, R. (1982). The temporal relationship between glomerular cell proliferation and monocyte infiltration in experimental glomerulonephritis. *Virchows Archivs (Cell Pathology)*, 38, 337–50.

Stevens, M.E. and Bone, J.M. (1985). Follow-up prednisolone dosage in rapidly progressive crescentic glomerulonephritis successfully treated with pulse methylprednisolone or plasma exchange. *Proceedings of the European Dialysis and Transplantation Association*, 21, 594–9.

Stevens, M.E., McConnell, M., and Bone, J.M. (1983). Aggressive treatment with pulse methylprednisolone or plasma exchange is justified in rapidly progressive glomerulonephritis. *Proceedings of the European Dialysis and Transplantation Association*, 19, 724–31.

Stilmant, M.M., Bolton, W.K., Sturgill, B.C., Schmitt, G.W., and Couser, W.G. (1979). Crescentic glomerulonephritis without immune deposits: clinicopathologic features. *Kidney International*, 15, 184–95.

Striker, G., Cutler, R.E., Haung, T., and Benditt, E. (1973). Renal failure, epithelial cell hyperplasia. In *Glomerulonephritis* (ed. P. Kincaid-Smith, R. Mathew, and E. Becker, pp. 657–75. Wiley, New York.

Suc, J.M., *et al.* (1976). The use of heparin in the treatment of idiopathic rapidly progressive glomerulonephritis. *Clinical Nephrology*, 5, 9–13.

Swainson, C.P., *et al.* (1982). Plasma exchange in the successful treatment of drug induced renal disease. *Nephron*, 30, 244–9.

Tateno, S., Sakai, T., Kobayashi, Y., and Shigematsu, H. (1981). Idiopathic membranous glomerulonephritis with crescents. *Acta Pathologica Japanica*, 31, 211–19.

Taylor, T.K., *et al.* (1979). Membranous nephropathy with epilthelial crescents in a patient with pulmonary sarcoidosis. *Archives of Internal Medicine*, 139, 1183–5.

Thomson, N.M., Simpson, I.J., and Peters, D.K. (1975). A quantitiative evaluation of anticoagulants in experimental nephrotoxic nephritis. *Clinical and Experimental Immunology*, 19, 301–8.

Thomson, N.M., Moran, J., Simpson, I.J. and Peters, D.K. (1976). Defibrination with Ancrod in nephrotoxic nephritis in rabbits. *Kidney International*, 10, 343–347.

Thysell, H., *et al.* (1983). Improved outcome in rapidly progressive glomerulonephritis by plasma exchange treatment. *International Journal of Arificial Organs*, 6, 11–14.

Tipping, P.G., Dowling, J.P., and Holdsworth, S.R. (1988). Glomerular procoagulant activity in human proliferative glomerulonephritis. *Journal of Clinical Investigation*, 81, 119–25.

Travers, R.L., Allison, D.J., Brettle, R.P., and Hughes, G.R.V. (1979). Polyarteritis nodosa: A clinical and angiographic analysis of 17 cases. *Seminars in Arthritis and Rheumatism*, **8**, 184–99.

Vassalli, P. and McCluskey, R. (1964). The pathogenetic role of the coagulation process in rabbit Masugi nephritis. *American Journal of Pathology*, **45**, 653–77.

Velosa, J.A. (1987). Idiopathic crescentic glomerulonephritis or systemic vasculitis. *Mayo Clinic Proceedings*, **62**, 145–7.

Vernier, I., Pourrat, J.P., Mignon Conte, M.A., Hemery, M., Dueymes, J.M., and Conte, J.J. (1987). Rapidly progressive glomerulonephritis associated with amyloidosi: Efficacy of plasma exchange. *Journal of Clinical Apheresis*, **3**, 226–9.

Volhard, F. and Fahr, T. (1914). *Die Brightsche Nierenkrankheit*. Springer, Berlin.

Walker, P.D., Deeves, E.C., Sahaba, G., Wallin, J.D., and O'Neill, W.M.Jr. (1984). Rapidly progressive glomerulonephritis in a patient with syphilis Identification of antitreponemal antibody and treponemal antigen in renal tissue. *American Journal of Medicine*, **76**, 1106–12.

Walker, R.G., Becker, G.J., D'Apice, A.J., and Kincaid-Smith, P. (1986). Plasma exchange in the treatment of glomerulonephritis and other renal diseasees. *Australian and New Zealand Journal of Medicine*, **16**, 828–38.

Walters, M.D., Savage, C.O.S., Dillon, M.J., Lockwood, C.M., and Barratt, T.M. (1988). Antineutrophil cytoplasm antibody in crescentic glomerulonephritis. *Archives of Diseases of Childhood*, **63**, 814–17.

Weber, M., Kohler, H., Fries, J., Theones, W., Meyer Zum Buschenfelde, K.H. (1985). Rapidly progressive glomerulonephritis in IgA/IgG cryoglobulinemia. *Nephron*, **41**, 258–61.

Wegmuller, E., Weidmann, P., Hess, T., and Reubi, F.C. (1985). Rapidly progressive glomerulonephritis accompanying Legionnaires' disease. *Archives of Internal Medicine*, **145**, 1711–13.

Weiss, M.A. and Crissman, J.D. (1985). Segmental necrotizing glomerulonephritis: diagnostic, prognostic, and therapeutic significance. *American Journal of Kidney Disease*, **6**, 199–211.

Welch, T.R., McAdams, A.J., and Berry, A. (1988). Rapidly progressive IgA nephropathy. *American Journal of Diseases of Childhood*, **142**, 789–93.

Whitworth, J.A., Morel-Maroger, L., Mignon, F., and Richet, G. (1976). The significance of extracapillary proliferation. Clinicopathological review of 60 patients. *Nephron*, **16**, 1–19.

Wilson, C.B. and Dixon, F.J. (1986). The renal response to immunological injury. In (ed. B.M. Brenner and F.C. Rector) *The Kidney* (pp. 800–900. Saunders, Philadelphia.

Woo, K.T., Chiang, G.S., Edmondson, R.P., Wu, A.Y., Lee E.J., and Pwee, H.S. (1986). Glomerulonephritis in Singapore: An overview. *Annals of Academy of Medicine, Singapore*, **15**, 20–31.

Woodworth, T.G., Abuelo, J.G., Austin, H.A., and Esparza, A. (1987). Severe glomerulonephritis with late emergence of classic Wegener's granulomatosis. Report of 4 cases and review of the literature. *Medicine*, **66**, 181–91.

Yeung, C.K., Wong, K.L., Wong, W.S., Ng, M.T., Chan, K.W., and Ng, W.L. (1984). Crescentic lupus glomerulonephritis. *Clinical Nephrology*, **21**, 251–8.

3.11 Antiglomerular basement membrane disease

NEIL TURNER AND A.J. REES

Definition

Human antiglomerular basement membrane (anti-GBM) disease is an autoimmune disorder characterized by the presence of autoantibodies directed against a target restricted to the glomerular and a few other specialized basement membranes. Its classical manifestations are rapidly progressive nephritis and lung haemorrhage, a syndrome commonly known as 'Goodpasture's syndrome'. This terminology is confusing though, as the eponym has been used in two different ways. In this chapter: Goodpasture's syndrome is a clinical syndrome of rapidly progressive glomerulonephritis (RPGN) and pulmonary haemorrhage. Ernest Goodpasture originally described an 18-year-old man with these findings at autopsy during an epidemic of influenza (Goodpasture, 1919). Nearly 40 years later, Stanton and Tange (1958) described a group of nine patients with similar findings, and applied Goodpasture's name to it before the immunological heterogeneity of the syndrome was appreciated. In man, Goodpasture's disease is used synonymously with human anti-GBM disease, which is one of the causes of Goodpasture's syndrome (Table 1). The classical linear staining of the GBM on direct immunofluorescence was first recognized by Scheer and Grossman (1964). This is now known to be caused by the formation of antibodies directed against a specific basement membrane antigen, referred to as the Goodpasture antigen. By indirect immunofluorescence, the antigen is found not only in GBM, but also in the basement membranes of a limited number of extrarenal tissues (Table 2).

Epidemiology

Incidence

Goodpasture's disease is rare, with an estimated incidence of about 0.5 cases per million per annum in Great Britain, based on the identification of anti-GBM antibodies by radioimmunoassay and renal biopsy. It has been estimated to cause up to 5 per cent of glomerulonephritis (Wilson and Dixon, 1973; New Zealand glomerulonephritis study group, 1989), and data from Australia and New Zealand shows it as causing 2 per cent of end-stage renal failure (Disney, 1986). The disease is more common in Caucasoids, and it seems to be particularly rare in Asians and Afro-Caribbeans. A New Zealand study (Teague et al. 1978) reported a relatively high proportion of Maoris (five of 29 cases,

3.11 Antiglomerular basement membrane disease

Table 1 (a) Causes of rapidly progressive glomerulonephritis with alveolar haemorrhage (Goodpasture's syndrome)

Goodpasture's (anti-GBM) disease	
Wegener's granulomatosis Microscopic polyarteritis Systemic lupus erythematosus	Types of systemic vasculitis commonly causing Goodpasture's syndrome
Churg-Strauss syndrome Henoch-Schönlein purpura Behçet's disease Essential mixed cryoglobulinaemia Rheumatoid vasculitis	Types of systemic vasculitis occasionally causing Goodpasture's syndrome
Penicillamine	Usually in association with a systemic vasculitis

(b) Other important causes of acute renal and respiratory failure—the 'pulmonary-renal syndrome'

Pulmonary oedema secondary to hypervolaemia in acute renal failure of any aetiology
Opportunistic or other infection in patients with rapidly progressive glomerulonephritis treated with immunosuppressive agents
Severe cardiac failure with pulmonary oedema
Severe pneumonia (including legionella pneumonia)
Paraquat poisoning
Thrombosis of renal vein/inferior vena cava with pulmonary emboli

Leatherman 1987; Leatherman et al. 1984; Holdsworth et al. 1985; Clutterbuck and Pusey 1987.

Table 2 Distribution of the Goodpasture antigen in human organs by indirect immunofluorescence using serum or antibodies eluted from kidneys of affected patients, Steblay nephritis eluate, or monoclonal antibodies (Cashman et al. 1988; Kleppel et al. 1989)

Kidney	GBM, distal TBM, Bowman's capsule
Lung	Alveolar basement membrane
Eye	Bruch's and Descemet's membranes, basement membranes of retinal capillaries, lens capsule, and cornea
Ear	Cochlear basement membrane
Brain	Basement membrane of choroidal epithelium
Other	Basement membranes of adrenal, breast, pituitary, thyroid

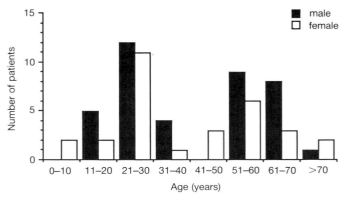

Fig. 1 Age and sex of 68 patients with Goodpasture's disease treated at Hammersmith Hospital, 1972–88.

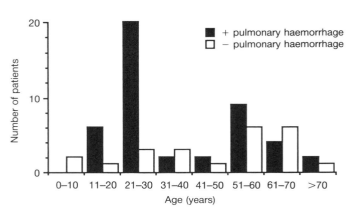

Fig. 2 Pulmonary haemorrhage in 68 patients with Goodpasture's disease treated at Hammersmith Hospital, 1972–88.

17 per cent, while they only made up 7 per cent of the population).

The disease may occur from childhood to old age. Early series showed a marked preponderance of young men. The wider use of radioimmunoassays and direct immunofluorescence, and greater awareness of the disease, have led to later series showing greater proportions of women and of older patients, although it is still more common in men (see Table 6 and Fig. 1). Our data shows a second peak in incidence in the sixth and seventh decades. It is interesting that this age and sex distribution is in contrast to those of other organ-specific autoimmune disorders. Lung haemorrhage is more common in younger patients and in men (Fig. 2). Glomerulonephritis alone is more common in older patients and in women. The sex ratio (males/females) in 71 cases from Great Britain in 1980–84 (Savage et al. 1986b) was 1.4, but for those with pulmonary haemorrhage as well as glomerulonephritis it was 3.0, and for those with nephritis alone it was 0.9.

Clustering of cases has been described in three reports: 15 cases in 3 years from Auckland, New Zealand (Simpson et al. 1982), four cases within a 25-mile radius during one winter in Connecticut, U.S.A. (Perez et al. 1974), and 13 cases within 8 months in north-west England (Williams et al. 1988). Such reports are notoriously difficult to analyse, but do tend to favour the hypothesis that exposure to an infective or other exogenous agent is involved in the pathogenesis. Influenza virus has been specifically mentioned in a number of reports, including Goodpasture's original description, but infection with the virus is common. Wilson studied patients with influenza but did not find evidence of anti-GBM antibodies, and many have looked for influenza antibodies in patients with Goodpasture's disease, rarely with any success (Wilson and Smith 1972; Wilson and Dixon 1973; Clark et al. 1978). A wider survey looking for antibodies to 13 viruses in 22 patients found no evidence of a disease association with any particular infection (Wilson et al. 1973). A nearly insuperable problem in analysing anecdotal reports is that an infection may non-specifically exacerbate tissue injury caused by the disease, leading to its clinical presentation in association with an infection that is not causally related to it.

A seasonal variation in incidence was noted in Great Britain in 1980–1982 (Savage et al. 1986b), but there was no discernible seasonal variation in 140 positive samples between 1985 and 1988 (unpublished data).

Disease associations

Although Goodpasture's disease is not commonly associated with the other organ-specific autoimmune diseases, it has been associated with a limited number of other disorders (Table 3). The best documented example is idiopathic membranous nephropathy which has evolved into a rapidly progressive nephritis. Circulating and fixed anti-GBM antibodies have been demonstrated in six of 10 published cases. The repeated demonstration of this phenomenon suggests that the relationship is likely to be of pathogenetic significance, but it could be either through an effect on the autoantigenicity of the GBM, or through a common defect in the immune system. An additional fascinating observation is that none of the six cases of membranous-associated anti-GBM disease in the literature has been described as having pulmonary haemorrhage. Closer investigation of these cases is needed.

Goodpasture's disease has also been associated with lymphoma, which may also hold clues to aetiology. The two patients on whom details are available developed Goodpasture's disease at a time when their lymphoma was in remission or cured, suggesting that it may have been an effect of the treatment rather than the disease. One had received both chemotherapy and full mantle radiotherapy, whereas the other had received upper mantle radiotherapy alone.

There is a single case report (Judge et al. 1977) of a multisystem disease associated with linear fixation of IgG to basement membranes of skeletal muscle fibres and GBM. The patient presented with a progressive myopathy of some years' duration, but developed RPGN and then a fulminant and fatal vasculitis affecting lungs, skin, and spleen. Elution studies were not performed and circulating anti-GBM antibodies were not sought, so it is not clear what the target antigen was in this disorder. It may well be an example of 'anti-GBM disease' in which the target antigen is different from that of Goodpasture's disease.

Clinical features

General manifestations

There is often a history of malaise and sometimes arthralgia and weight loss, but these are not generally marked, in contrast to many cases of systemic vasculitis, where in other aspects the clinical picture at presentation may be identical. Symptoms due to anaemia are common, even when there has been little or no haemoptysis. This is probably due to subclinical pulmonary hae-

Table 3 Reported associations of Goodpasture's disease with other disorders

Disorder	Number of cases	References
Other renal diseases		
Membranous nephropathy*	6	Beirne et al. 1977; Klassen et al. 1974; Moorthy et al. 1976; Richman et al. 1981; Kurki et al. 1984
Cortical necrosis	1	Hume et al. 1970
Other autoimmune disturbances		
Microscopic polyarteritis	1	Relman et al. 1971
Wegener's granulomatosis	1	Wahls et al. 1987
Coeliac disease	1	Savage et al. 1986
Hashimoto's thyroiditis	1	Kalderon et al. 1973
Myasthenia gravis	1	A.J. Rees et al., unpublished observation
Penicillamine therapy	1	R. Lechler, personal communication
Partial lipodystrophy	1	Blake et al. 1980
Malignant disorders		
Hodgkin's disease/lymphoma	6	Kleinknecht et al. 1980; Ma et al. 1978; Wilson and Dixon 1981
Congenital/inherited disorders		
Nail-patella syndrome	1	Curtis et al. 1976
Alport's syndrome	7†	
After renal transplantation		
Recurrence of original disease	Many (see text)	
Alport's syndrome	7†	McCoy et al. 1982; Milliner et al. 1982; Shah et al. 1988; Fleming et al. 1988; Querin et al. 1986
de novo	(Theoretical only)	

* Cases where there was only histological evidence for pre-existing membranous nephropathy have been excluded.
† Patients with Alport's syndrome may develop antibodies to the Goodpasture antigen present in the transplanted kidney (see text).

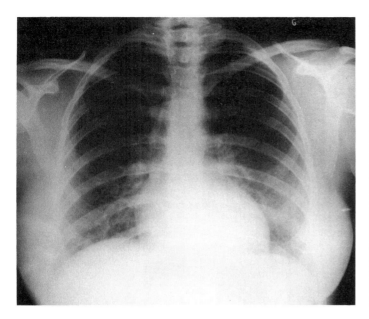 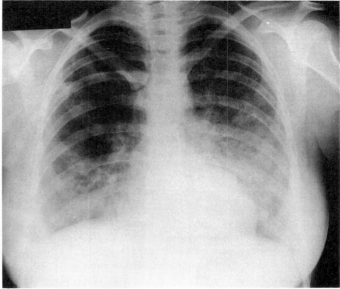

Figs. 3 and 4 Chest radiographs of a 23-year-old woman who presented with pulmonary haemorrhage. Figure 3, 1 week before, and Fig. 4, on admission with 2-week history of breathlessness and three minor episodes of haemoptysis over 3 months, the last 4 days previously. She had a haemoglobin of 5.4 g/dl with indications of iron deficiency. Her renal biopsy is shown in Fig. 7; she had microscopic haematuria but her serum creatinine was normal at 69 μmol/l. Circulating anti-GBM antibodies were detected by radioimmunoassay and direct immunofluorescence of the renal biopsy showed linear fixation of IgG and C3 to the GBM. The radiological changes resolved on treatment with cyclophosphamide, prednisolone, and plasma exchange, but 2 weeks later she had a relapse of pulmonary haemorrhage 1 day after resuming smoking.

morrhage, which may cause marked iron deficiency. In some of our patients, the degree of anaemia has led to an initial diagnosis of gastrointestinal disease or chronic renal failure, with consequent delays in appropriate treatment. The anaemia is usually microcytic, but in common with other types of RPGN there may be some, though usually minor, microangiopathic features. Anaemia is common in systemic vasculitis even without pulmonary haemorrhage, but it is usually normochromic and normocytic.

Pulmonary manifestations

Pulmonary haemorrhage occurs in about two-thirds of patients in most recent series. There is commonly a history of episodic or continuous breathlessness for weeks or months. Haemoptysis may be trivial, absent, or torrential, but bears little relation to the actual quantity of pulmonary bleeding. Occasional patients give a history of similar episodes, resolving spontaneously, over many years (sometimes over more than a decade). Even when the history is short, iron deficiency anaemia often suggests that subclinical haemorrhage may have been occurring for longer. Pulmonary haemorrhage shows a marked tendency to remit and relapse, a fact which has complicated the interpretation of various therapeutic interventions. A number of factors may bear on this, notably whether or not the patient is a current cigarette smoker. Pulmonary haemorrhage is extremely rare in non-smokers, although exposure to other inhaled toxins, fluid overload, and local or remote infections may also influence its occurrence. The reasons for this are discussed later.

Examination in mild to moderate cases is usually normal apart from tachypnoea. Severely affected patients are tachypnoeic and cyanosed, and often expectorating some fresh blood. Rather dry-sounding inspiratory crackles are commonly audible over the lower lung fields, and there may be areas of bronchial breathing.

Remitting and relapsing lung disease alone may be confused with idiopathic pulmonary haemosiderosis, and other causes of isolated pulmonary haemorrhage (Morgan and Turner-Warwick, 1981; Leatherman et al. 1984). The transition to rapidly progressive disease may occur even in patients with a long history, and it can develop over hours rather than days (Bergrem et al. 1981; Briggs et al. 1979; Rees et al. 1979); the factors responsible remain poorly understood. It is the rapidity of this deterioration, and the rapidly changing prognosis for renal recovery, that makes prompt diagnosis and treatment of Goodpasture's disease imperative (see Fig. 5).

Even severe lung lesions seem to have the capacity to resolve almost completely. Significant pulmonary impairment is not seen in patients who have survived even life-threatening pulmonary haemorrhage; nor is there radiological evidence of fibrosis.

Radiological changes

Most episodes of pulmonary haemorrhage are associated with changes in the chest radiograph (Bowley et al. 1979a; Bowley et al. 1979b). Usually the shadows involve the central lung fields, with peripheral and upper lobe sparing (Figs. 3 and 4). The abnormalities are generally symmetrical but can be markedly asymmetrical. Changes range from 1 to 4 mm ill-defined nodules to confluent consolidation with an air bronchogram, though usually some nodular shadows are still visible at the edge of confluent areas. Shadowing limited or entirely confined by a fissure, or at the apex of a lung, is strongly suggestive of infection, either alone or as a complication. Shadows caused by bleeding are usually clearing within 48 h of an episode of haemorrhage, longer if it has been confluent. Residual minor changes are

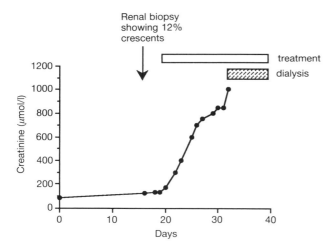

Fig. 5 Very rapid progression of renal injury in one patient in association with infection, despite treatment with cyclophosphamide, prednisolone, and plasma exchange.

usually gone within 2 weeks. Interpretation may be considerably complicated by the coexistence of infection or fluid overload—which may themselves precipitate haemorrhage.

Diagnosis of pulmonary haemorrhage

Pulmonary haemorrhage frequently occurs without haemoptysis, and subclinical bleeding may be observed quite frequently by close observation and the use of sensitive techniques to detect it. Indicators of pulmonary haemorrhage include fresh haemoptysis; a sudden drop in haemoglobin without other explanation; new shadows on the chest radiograph; or a rise in K_{CO}, the rate of carbon monoxide uptake by the lung. Haemoglobin binds carbon monoxide (CO) avidly, and the measurement of CO uptake by the lungs is used clinically as a measure of the efficiency of gas exchange. The total lung CO uptake, TL_{CO}, divided by the alveolar volume, is known as the K_{CO}. It must be corrected for the patient's haemoglobin. This measurement is usually about 30 per cent lower than predicted in patients with renal failure (Lee and Stretton 1975), even in most cases of pulmonary oedema, but pulmonary haemorrhage leads to additional free haemoglobin being available to bind to CO within the alveoli, and an increase in K_{CO}. The rise is usually abrupt and often precedes changes in the chest radiograph. It may demonstrate haemorrhage undisclosed by other techniques (Ewan et al. 1976; Bowley et al. 1979b). The test is usually performed by the single-breath method, and is easily repeated daily if necessary. It can be done by a bedside rebreathing method in patients who require artificial ventilation or are too ill to be moved to the pulmonary function laboratory (Clark et al. 1978).

Renal manifestations

Approximately one-third of patients with Goodpasture's disease have no evidence of pulmonary haemorrhage at presentation. Some of these develop clinical or subclinical pulmonary haemorrhage later, but most never do. The renal lesion, like the lung lesion, can improve spontaneously when mild, but usually deteriorates inexorably once significant renal damage has occurred. As already described, this deterioration may be extremely rapid (Fig. 5). The renal disease appears to have much less tendency to remit and relapse, although this difference may be largely illusory, and to do with the stage of the disease. Patients rarely die of pulmonary haemorrhage without evidence of advanced renal disease, although this may have developed from a mild renal lesion in days or less (Bergrem et al. 1980; Simpson et al. 1982; Rees et al. 1979). It may simply be that minor tissue injury is more evident in the lungs than the kidney—for example, bleeding may occlude unaffected alveoli, and haemoptysis is a dramatic symptom, whereas microscopic haematuria is not; or it may be that with the right cofactors the lung can suffer significant immunologically-mediated damage at an earlier stage in the disease than the kidneys. Very occasional patients have a history and findings suggestive of subacutely or chronically progressive renal disease (McPhaul and Mullins 1976).

Abnormalities of the urine sediment, usually microscopic haematuria, are the earliest sign of renal damage. At this stage, before the serum urea or creatinine are abnormal, the changes may remit and relapse without specific treatment. Later the urine contains numerous crenated red blood cells and usually red cell casts. Proteinuria is generally modest (<5 g/24 h), but it may occasionally be heavy.

The clinical manifestations of established renal disease are common to other causes of RPGN. Macroscopic haematuria is common, and is frequently accompanied by loin pain. Hypertension is uncommon at presentation in the absence of hypervolaemia. Oliguria is a late feature and is a bad prognostic sign. However the chance of superimposed acute tubular necrosis in hypoxic and severely ill patients is always high.

Kidney size is usually normal. There are no specific morphological or other abnormalities on any type of renal imaging.

Other manifestations possibly related to immunopathology

Anti-GBM antibodies fixed to other basement membranes have been shown in individual cases of Goodpasture's disease, but rarely appear to cause much injury (McPhaul and Dixon, 1970; McIntosh et al. 1975; Cashman et al. 1988). An exception is a report (Jampol et al. 1975) describing two cases of retinal detachment in patients with Goodpasture's disease. Antibody fixation to Bruch's membrane and the basement membranes of choroidal vessels were demonstrated in one of their cases.

Ocular abnormalities may occasionally be seen in other patients (Fig. 6). Similarly, fixation of antibody to the choroid plexus has been reported, and it has been suggested that this could be related to the convulsions that occur in some patients. However convulsions may occur in acute renal failure of any aetiology for a variety of possible reasons, and it is not clear that there is an aetiological relationship.

The presentation of Goodpasture's disease after renal transplantation is discussed in the final section of this chapter.

Diagnosis

Immunological features

The specific diagnosis of Goodpasture's disease rests on the demonstration of anti-GBM antibodies in the circulation or fixed to the kidney. Circulating antibodies were originally detected by indirect immunofluorescence using sections of normal kidney. However this method is relatively insensitive, difficult to quantify, and can be difficult to interpret. Radioimmunoassay (RIA) or ELISA methods are now more commonly employed. These

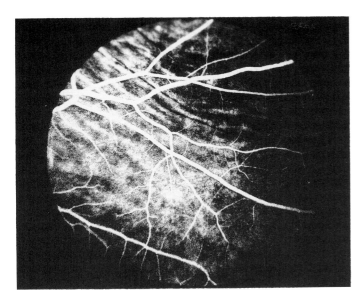

Fig. 6 Fluorescein angiogram at presentation of the retinal vessels of a 50-year-old woman with Goodpasture's disease. Ocular fundoscopy showed choroidal folds, confirmed on fluorescein angiography, which also shows retinal capillary leakage in the same region.

Table 4 Causes of linear staining on direct immunofluorescence of renal tissue. In most cases the appearances are of weaker binding than in Goodpasture's disease, but confirmation of specificity can only be achieved by elution studies (Klahr et al. 1988; Wilson and Dixon 1974; Querin et al. 1986)

Goodpasture's disease
Systemic lupus erythematosus
Diabetes mellitus
Normal autopsy kidneys
Cadaver kidneys after perfusion
Transplant biopsies (1.4 per cent in the study of Querin et al.)

are both highly specific and sensitive when an adequate substrate is used (Wilson and Dixon, 1981; Bowman and Lockwood 1985; Wheeler et al. 1988; Saxena et al. 1989). There have been no reports of circulating antibodies in the absence of disease attributable to them, and false-negative results are extremely rare. The antibodies have not been found in relatives of patients, in those with other autoimmune disorders, or in normal individuals, though it is fair to say that large surveys on this subject have not been published. Direct immunofluorescence on the renal biopsy is the most sensitive technique of all, if adequate renal tissue is obtained and glomerular destruction is not severe; it is the main method of diagnosis in many centres. There may be occasional false-positives, as linear fluorescence not attributable to Goodpasture's disease has been noted in a number of circumstances (Table 4). These observations are also pertinent to the interpretation of indirect immunofluorescence on 'normal' kidney sections; control sera must always be compared. Direct immunofluorescence is not useful for following disease activity as linear antibody fixation may be demonstrable for a year or more after diagnosis, after circulating antibodies have become undetectable and in the absence of clinical disease (Teague et al. 1978).

Circulating anti-GBM antibodies are predominantly IgG1 and IgG4 (Bowman et al. 1987; Weber et al. 1988; Noel et al. 1988), although there is one reported case in which only IgA antibodies could be detected (Border et al. 1979). It seems reasonable to rely on techniques that detect only circulating IgG antibodies for most cases, but to look routinely for IgG, IgA, and IgM, as well as C3 on renal biopsies of all patients with RPGN.

There are no consistent changes in the concentration of complement factors or serum immunoglobulins, but plasma exchange will deplete both. Rheumatoid factor, antinuclear antibodies, and cryoglobulins are not usually found. Circulating immune complexes have been found only exceptionally, usually in patients with intercurrent infections.

It is interesting that a proportion of patients may also have antineutrophil cytoplasm antibodies (ANCA) (Pusey and Lockwood 1989). There are single clear case reports of the near-simultaneous occurrence of Goodpasture's disease and microscopic polyarteritis (Relman et al. 1971), an eosinophilic systemic vasculitis (Komadina et al. 1988), and Wegner's granulomatosis (Wahls et al. 1987), but it is not yet clear whether the majority of 'double-positive' patients have any distinctive clinical features. Some data suggest that most such patients have low titres of anti-GBM antibodies and clinical features more suggestive of vasculitic disease (Jayne et al. 1989, and Chapter 4.5.3). The anti-GBM antibodies may disappear promptly after initiation of treatment. From a diagnostic point of view it is perhaps equally interesting to know whether there are many 'double-negative' patients among the group of patients with idiopathic RPGN, and if so, whether they have unusual clinical or pathological features. Experience suggests that such patients are rare, and that most RPGN can be attributed to Goodpasture's disease, to the ANCA-positive, systemic vasculitis group, or to another specific underlying disease (see Chapter 3.10 on crescentic nephritis).

Renal features

An adequate renal biopsy should be considered an essential part of the assessment of patients with this disease, because of its prognostic as well as diagnostic importance. The glomeruli are generally abnormal even in patients with no clinical evidence of renal involvement (Heptinstall 1983). The earliest and mildest changes consist of segmental mesangial matrix expansion and hypercellularity, progressing to a more generalized but still focal and segmental proliferative glomerulonephritis with increased numbers of neutrophils in the glomeruli. Later, glomeruli show a diffuse nephritis with segmental or total necrosis and extensive crescent formation. (Figs. 7 and 8). Glomerular capillary thrombi may be seen; this, and minor degrees of schistocytosis, may be seen in RPGN of any aetiology, and they probably do not justify a separate categorization or imply a close relationship to haemolytic uraemic syndrome or thrombotic thrombocytopenic purpura (Chapter 8.7), as has been suggested (Stave and Croker 1984). Glomerular giant cells have been reported in occasional patients (Sabnis et al. 1988). Interstitial inflammation is usually present and may be severe; it has been correlated with the extent of antibody fixation to the basement membrane of the distal convoluted tubules (Andres et al. 1978).

Vasculitis of small intrarenal vessels has been described in the renal biopsies of occasional patients with otherwise typical

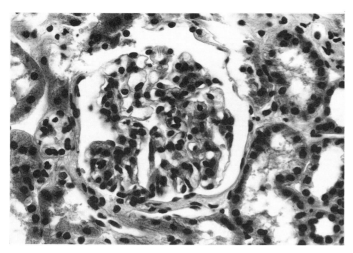

Fig. 7 A typical glomerulus from the renal biopsy of a patient with very mild renal disease (microscopic haematuria but normal serum creatinine). Abnormalities are minimal, although direct immunofluorescence showed linear fixation of IgG and C3 as in Fig. 9.

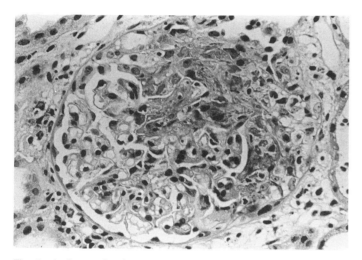

Fig. 8 A glomerulus from a patient with severe renal disease (requiring dialysis at presentation). There is segmental necrosis with extracapillary proliferation. Seventy per cent of glomeruli showed crescent formation. However, there are still patent capillary loops in this and many other glomeruli, and the crescents were generally cellular; renal function improved with intensive treatment, to a serum creatinine of 255 μmol/l at 8 weeks and 168 μmol/l at 1 year.

Goodpasture's disease (Wu *et al.* 1980a). ANCA assays were not available at the time of these reports, but one similar case in our series was negative for ANCA by immunofluorescence.

The picture of linear binding of antibody to GBM visualized by direct immunofluorescence is something of a pathological classic (Fig. 9). Such deposits are found in all patients, however mild the renal disease (Mathew *et al.* 1975; Saraf *et al.* 1978). Antibody can also usually be seen on the basement membrane of the distal convoluted tubule, but not proximal tubule, collecting duct, or other site in renal tissue. IgG is almost universally detected, accompanied by other immunoglobulins (IgA, IgM, or a combination of these) in about a third of cases. Linear C3 is demonstrable in 60 to 70 per cent of cases, and bright in about 30 per cent. Occasional reports mention IgM alone, IgA alone, or C3 alone (sometimes associated with typical circulating antibodies (Savage *et al.* 1986b)). There is one report of a patient with linear IgA alone, circulating antibodies of IgA alone, and typical disease with lung haemorrhage and RPGN (Border *et al.* 1979).

Additional granular deposits have been noted in some patients, associated with subepithelial or sometimes intramembranous or subendothelial deposits ultrastructurally (Jennette *et al.* 1982; Kurki *et al.* 1984; Pettersson *et al.* 1984; Rajaraman *et al.* 1984). In one case the granular deposits appeared during resolution (Agodoa *et al.* 1976). Usually the clinical picture is entirely typical, and there has been little reason to suspect pre-existing membranous nephropathy. However as described earlier, there are clear-cut examples of patients with membranous glomerulonephritis devloping Goodpasture's disease. It is not clear whether these two groups are related.

Ultrastructurally the GBM usually shows a widespread irregular broadening, often with mottled thickening of the lamina rara interna. Breaks in the GBM are common. Endothelial and epithelial cells are swollen, and epithelial foot processes may be fused.

Pulmonary features

At autopsy of patients with pulmonary haemorrhage the lungs are characteristically heavy, showing patchy congestion or haemorrhage. Histologically, intra-alveolar haemorrhage is accompanied by haemosiderin-containing macrophages, deposits of fibrin, and alveolar cell hyperplasia. Electron microscopy shows thickening of the alveolar basement membrane, often with defects (Donald *et al.* 1975). Thickened alveolar walls may show oedema, fibrosis, and modest inflammatory cell infiltration, mainly with polymorphs and lymphocytes.

Immunofluorescence investigations are more difficult in the lung, but linear fixation of immunoglobulin can be seen by direct immunofluorescence in most patients with lung haemorrhage. Binding is often patchy, and although a high success rate in obtaining diagnostic material by transbronchial biopsy has been

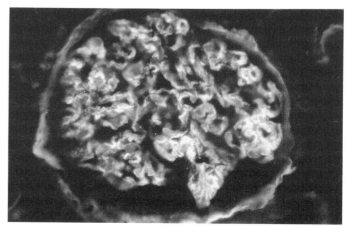

Fig. 9 Direct immunofluorescence for IgG in a glomerulus from a patient with Goodpasture's disease, showing linear fixation of antibody to the GBM.

Table 5 Incidence of various HLA-DR haplotypes in 45 patients with Goodpasture's disease versus normal Caucasoids (Burns et al., in preparation)

HLA-DR allele	Percentage incidence in patients	Percentage incidence in controls	χ^2	Relative risk	Aetiological fraction
2	69	30	26.4	4.9	0.55
4	60	37	10.3	2.55	0.36
2, 4	38	11	32.4	4.8	0.29
Neither 2 nor 4	8	56			

reported, the technique is too unreliable for diagnostic use (Johnson et al. 1985).

Aetiology

There is unlikely to be a 'simple' cause for any human autoimmune disorder. Most are now believed to be the product of an environmental influence on a genetically susceptible background. Cases of Goodpasture's disease occurring in first-degree relatives are mentioned repeatedly in the literature, but in recent years attention has been more closely focused on the specific role of major histocompatibility antigens, especially class II molecules, in susceptibility to all autoimmune disorders. In contrast, the role of environmental factors remains an area for speculation.

Immunogenetics of Goodpasture's disease

Goodpasture's disease has been reported in four sibling pairs (Stanton and Tange, 1958; Gossain et al. 1972), and in two sets of identical twins (d'Apice et al. 1978; Simonsen et al. 1982). The two reported sets of identical twins are particularly interesting as the onset of the disease in both twins was 5 months and 6 weeks apart respectively. The twins in the first report had not been in contact for 18 months, but had inhaled petrol and turpentine fumes shortly before presentation. Such reports provide strong circumstantial evidence that environmental factors might provoke the disease. The importance of non-genetic factors is emphasized by the observation that two pairs of identical twins in our series, and others in the literature (Almkuist et al. 1981) were discordant for the disease.

Like other human autoimmune disorders, Goodpasture's disease has been associated with specific class II HLA antigens. An association with HLA-DR2 (Rees et al. 1978), carried by 88 per cent of patients in the original study, has been confirmed by series from other countries (Perl et al. 1981; Garovoy 1982) (87 and 50 per cent versus 25 and 23 per cent in controls). More recent data obtained by analysing restriction fragment length polymorphisms of genomic DNA have suggested additional associations with HLA-DR4 (Table 5) and the HLA-DQ β chain genes DQwIb and DQw3. Because DQwIb is in tight linkage disequilibrium with HLA-DR2, and DQw3 likewise with HLA-DR4, it is not possible to say whether the DR or the DQ alleles are more likely to be primarily responsible for the association.

A further association has been shown between a class I antigen and severity of the nephritis in patients carrying HLA-DR2 (Rees et al. 1984a). Because HLA-B7 and HLA-DR2 are in linkage disequilibrium, the incidence of HLA-B7 is increased in Goodpasture's disease. However patients carrying both antigens had significantly higher creatinines and crescent scores at presentation than patients with HLA-DR2 but not HLA-B7. Formal genetical analysis of the data confirmed that DR2 was related to susceptibility to the disease and B7 to its severity.

A third genetic influence on Goodpasture's disease has been described (Rees et al. 1984b). This concerns the Gm allotypes of the IgG heavy chain constant region, which are coded for at three loci which are in strong linkage disequilibrium, so they are usually inherited as fixed haplotypes. Different combinations of Gm allotypes have been associated with both susceptibility to and severity of various autoimmune disorders, and the magnitude of antibody responses in health. In a group of British patients, the Gm haplotype 1,2,21 (axg) was carried by 54 per cent of patients and 17 per cent of controls, giving a relative risk of 5.75.

Environmental influences

The reports of clustering of cases, and the evidence from identical twins, as well as various anecdotal reports of associations with inhaled toxins and infections, provide evidence for the importance of environmental factors in causing or precipitating Goodpasture's disease. It is impossible to identify any such agents definitively at present, although a number of candidates are repeatedly mentioned. One of the difficulties in analysing reports on this subject is the problem of distinguishing agents that initiate the disease from those that reveal it. Some substances may have a non-specific irritant or toxic effect on the lungs and precipitate pulmonary haemorrhage in patients who already have circulating antibodies, but who have not yet developed overt disease. Such a mechanism is suggested by reports of remission and relapse of pulmonary haemorrhage in association with exposure to a putative toxic agent (Bernis et al. 1985). There is strong evidence that cigarette smoking may act in this way, as discussed in the section on pathogenesis. Alternatively, some influences may aggravate tissue injury in a non-specific way, again 'revealing' underlying disease without actually causing it. Remote or systemic infections may do this. Finally, some agents may alter either the immune system or the autoantigen in some way that facilitates the breakdown of normal tolerance, so that an autoimmune response is generated; they may have a truly causal role.

Exposure to organic solvents and hydrocarbons has been linked persistently with glomerulonephritis in case reports and reviews. Most of the reports refer to Goodpasture's disease, and the number is impressive. However the level of exposure mentioned is extremely varied, as are the agents implicated in different reports, and the level of reporting has probably been increased by a general awareness of the possible association. Nevertheless case-control studies (Daniell et al. 1988) have tended to support a link between solvent exposure and nephritis. Some case reports describe nephritis without pulmonary haemorrhage, suggesting that the mechanism might be more than just 'revealing' the disease by precipitating pulmonary haemorrhage (Beirne and Brennan 1972; Kleinknecht et al. 1980; Daniell et al. 1988). The rarity of glomerulonephritis makes a prospective study of exposed populations unlikely to succeed, but there is scope for better case-control studies focusing on Goodpasture's disease and comparing it with other acute nephritides.

Pathogenesis

As Goodpasture's disease appears to be so clearly defined by clinical, pathological, and immunological features, it has been the subject of a great deal of investigation in an attempt to reveal more about the mechanisms underlying a human nephritis.

The Goodpasture antigen
Glomerular basement membrane (GBM)

Basement membranes are specialized extracellular structures that are usually found at the boundary between cells and connective tissue stroma. The bulk of the components of their matrix is composed of molecules that are common to all basement membranes, and several have been characterized (Martinez-Hernandez and Amenta 1983; Timpl et al. 1987). Other components have a more restricted distribution, presumably related to specific functional roles. These include the antigens of at least two human autoimmune diseases, bullous pemphigoid and Goodpasture's disease.

The GBM is a highly specialized basement membrane and has unique morphological and biochemical features. Structurally, it lacks the lamina reticularis which anchors many basement membranes to the interstitium, as it separates the endothelial cells of the glomerular capillaries from the visceral glomerular epithelial cells, serving as basement membrane to both. Uniquely for a basement membrane it is exposed to circulating blood at 'fenestrae' in its endothelial covering. When injected intravenously, antibodies to basement membrane components bind to GBM but not to most other basement membranes, including alveolar basement membrane (Lerner et al. 1967; McPhaul and Dixon 1970). The major biochemical constituents of the GBM are the same as those of other basement membranes, but their individual analysis has been hampered by more extensive cross-linking than is usually found in other membranes, which makes extraction of intact components difficult.

Of the major basement membrane components, polyanionic heparan sulphate proteoglycans carry the negative charge which gives the GBM some of its characteristic filtration properties. Laminin is believed to be the major cell-binding protein in basement membranes, but also shows interations with various molecules found in the basement membrane matrix. Type IV collagen is only found in basement membranes, and is believed to form their structural skeleton by associating into covalently cross-linked networks. It makes up 25 to 90 per cent of their dry weight. The majority of the molecule is typical collagenous triple helix, previously thought to consist of two $\alpha1$ and one $\alpha2$ chains, both of which have now been cloned and sequenced. Several groups now believe that there may be one or more additional species of type IV collagen chain, with restricted tissue distribution and probably specialized function, one of which may carry the major antigenic determinant of human anti-GBM disease, the Goodpasture antigen (Kleppel et al. 1986; Wieslander et al. 1987; Saus et al. 1988; Hudson et al. 1989).

In animal models of renal disease injection of antibodies to GBM components such as these can produce linear staining on direct immunofluorescence and induce pathological effects (Yaar et al. 1982; Feintzeig et al. 1986; Wick et al. 1986). Similarly, antibodies to various GBM components have been described in some human nephritides (Fillit et al. 1985; Kefalides et al. 1986; Bygren et al. 1989). However there is only good evidence that injury mediated by such antibodies is of pathogenic importance in one human disease—Goodpasture's, or human anti-GBM disease.

Distribution of the Goodpasture antigen

Studies using antibodies to localize basement membrane components to specific parts of the GBM have produced conflicting results (Timpl et al. 1987), probably because of differences in fixation techniques and reagents. By indirect immunofluorescence and immunoelectron microscopy, the Goodpasture antigen has been localized to the endothelial aspect of the lamina densa (Sisson et al. 1982). As shown in Table 2, indirect immunofluorescence studies using the serum of patients, eluted antibody from affected kidneys, or monoclonal antibodies, show the antigen to have a limited extrarenal distribution too. It is interesting to note that many of these sites are ones in which the basement membrane is formed by fusion of epithelial and endothelial cell basement membranes.

It has been stated that the antigen can be detected in the basement membrane of other tissues after treatment of sections with acid and urea (Yoshioka et al. 1985). We have not been able to repeat these observations, but there is other evidence that the antigen may be present in small quantities in tissues that are usually negative by indirect immunofluorescence (Weber et al. 1987; Kleppel et al. 1986; Langeveld et al. 1988).

Anti-GBM antibodies

Pathogenicity of circulating anti-GBM antibodies in Goodpasture's disease was suggested by their very close association with the disease, and also by the correlation between antibody titre and the severity of renal damage (Simpson et al. 1982; Savage et al. 1986b). Classic experiments (Lerner et al. 1967) unequivocally established the pathogenicity of kidney-fixed antibodies in Goodpasture's disease. They showed that antibody eluted from kidneys of patients with Goodpasture's disease could fix to the GBM of squirrel monkeys in vivo and cause pathological glomerular changes. This is perhaps not surprising, as human anti-GBM antibodies are known to be mainly IgG1 and complement fixing, and similar antibodies in animal models exert pathological effects.

An additional feature of the antibody response in the disease, and one which originally led to hope for the possible success of a short but intensive course of treatment, is its transient nature. As discussed below, treatment reduces the time to disappearance of the antibody, but even without treatment, most patients no longer have detectable circulating antibody after 18 to 24 months (Wilson and Dixon 1981).

Cell-mediated immunity

For many years Goodpasture's disease was considered to be an archetypal antibody-mediated disease. However it would be naive to think now that B lymphocytes could be responsible for an autoimmune disease without a role for T lymphocytes. Few antigens can elicit an antibody response without the assistance of helper T cells, and the association of Goodpasture's disease with HLA class II antigens is strong evidence for the importance of class II-restricted lymphocytes in pathogenesis. In a model of anti-GBM disease in chickens (the target antigen is not known, but it is probably different from the Goodpasture antigen), it can be shown that complete removal of B cells does not prevent the disease, though antibody is no longer involved, and that the nephritis can be transferred to other animals by T lymphocytes (Bolton *et al.* 1988). Transfer of antibody alone leads to linear fixation of immunoglobulin to the GBM, but no disease. This provides evidence that as well as providing help for B cells, T cells can be involved directly in at least one model of anti-GBM disease. It is becoming clear that T lymphocytes also play a pivotal role in those models of anti-GBM disease where autoantibody synthesis is associated with polyclonal activation (Hoedemaeker and Weening, 1989) (and Chapter 3.2).

A number of studies have attempted to show lymphocyte responses to glomerular antigens in human nephritides (Fillit and Zabriskie 1982). However the study of cell-mediated mechanisms in human autoimmune disease is in its infancy, largely because of difficulty in identifying the antigens recognized by the lymphocytes involved (Oliveira and Peters 1989). The presence of T lymphocytes in the inflammatory infiltrate in the renal cortex (Nolasco *et al.* 1987) is circumstantial evidence that cell-mediated mechanisms might be important in Goodpasture's disease. In support of the idea that cell-mediated mechanisms are important in the disease is the one case in which an apparently unaffected baby was delivered shortly before the mother died of severe Goodpasture's disease (Wilson and Dixon 1981), and it is well recognized that there can be GBM-fixed antibody and even significant levels of circulating antibody with minimal evidence of tissue damage (Wilson and Smith 1972; Dahlberg *et al.* 1978; Wilson and Dixon 1981; Bailey *et al.* 1981; Hind *et al.* 1984; Bernis *et al.* 1985), although there could be explanations other than those that invoked cell-mediated mechanisms. At presentation, however, there is a correlation between antibody titre and the severity of renal damage (Simpson *et al.* 1982; Savage *et al.* 1986b), and as already mentioned, the association between antibodies and disease is exceptionally close in Goodpasture's disease, and in contrast to the situation in most other human autoimmune disorders.

It seems likely that any significant advances in the explanation of the pathogenesis of Goodpasture's disease are certain to involve a better understanding of the cell-mediated mechanisms involved. This also seems likely to be the major hope for the introduction of more specific immunotherapy than can be attempted from our present understanding.

Other influences on tissue injury

A number of investigators have noted the phenomenon (Rees *et al.* 1977) of a worsening of clinical condition in association with infection, at a time when the disease seems to be controlled or coming under control (Fig. 10; Fig. 5 also shows a rapid deterioration in renal function in association with injection). The ability to monitor renal function constantly, to detect lung damage equally readily, and to correlate these with circulating antibody titres, makes Goodpasture's disease an exceptionally useful one for the study of this phenomenon in man. Presumably similar effects might occur in other immunologically-mediated diseases, but they are usually more difficult to detect.

In animal models of anti-GBM disease, infection has been shown to amplify the tissue damage caused by a given dose of heterologous anti-GBM antibody (Van Zyl Smit *et al.* 1983). This has been further refined to show that endotoxin alone, with-

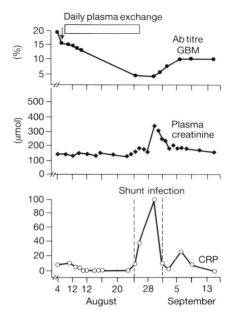

Fig. 10 The effect of intercurrent infection on tissue injury in Goodpasture's disease. Anti-GBM antibody titres (upper panel), creatinine (middle panel), and C-reactive protein (lower panel) in a patient with disease controlled by immunosuppressive agents and plasma exchange, who then developed an infection at the site of an arteriovenous shunt. (From Donaghy and Rees 1985, with permission.)

out an actual infection, and some cytokines known to be produced in response to endotoxin, such as tumour necrosis factor and interleukin 1, can achieve the same effect (Tomosugi *et al.* 1989).

The pathogenesis of lung haemorrhage

Lung haemorrhage is a variable feature of Goodpasture's disease and, in contrast to renal disease, lung disease shows a very poor correlation with antibody titre (Wilson and Dixon 1981; Simpson *et al.* 1982). Nevertheless immunofluoresence studies using monoclonal antibodies as well as serum of patients or antibody eluted from affected kidneys (Cashman *et al.* 1988; Kleppel *et al.* 1989) have suggested that the autoantigen in alveolar and glomerular basement membranes is the same. Immunizing sheep with alveolar basement membrane produces a nephritis indistinguishable from that caused by immunization with GBM, Steblay nephritis, in which it has been shown that antibodies to the Goodpasture antigen are formed (Steblay and Rudofsky 1983b; Steblay and Rudofsky 1983a; Jeraj *et al.* 1982; Bygren *et al.* 1987). Interestingly, the sheep do not get lung disease after immunization with antigen from either source. Some workers (McPhaul and Dixon 1970) have suspected differences in the properties of antibody eluted from lung and kidney, and differences have been noted in the immunoblotting pattern of alveolar basement membrane (Weber *et al.* 1987; Wieslander and Heinegard 1985; Yoshioka *et al.* 1988). However antibody eluted from lungs binds to kidneys in the same way as glomerular eluates (McPhaul and Dixon 1970), and the results of immunoblotting using a monoclonal antibody also suggest that the same epitopes are present (unpublished results). Rather than being a different antigen, it is possible that the same antigen is differently cross-linked or displayed in the alveolar basement membrane.

There is an obvious structural difference to account for the dissociation of lung and kidney disease. In normal circumstances there is a lack of direct contact between circulating antibodies and the matrix of the alveolar basement membrane. The endothelium of the alveolar capillaries is not fenestrated, so the antigen is usually concealed from both circulating antibodies and lymphocytes by the endothelial layer. It has been suggested that pulmonary haemorrhage may occur when this barrier is disturbed by any of a variety of non-specific insults to the lung, including smoking, inhalation of toxic agents (particularly organic solvents), and infection. The patchy distribution of antibody binding to alveolar basement membranes by direct immunofluorescence in affected patients is in keeping with fixation being dependent on some such exogenous influence, and there is some additional support from animal models. Passively injected anti-GBM antibodies bind to glomerular but not to alveolar basement membranes in monkeys (Lerner *et al*. 1967; McPhaul and Dixon 1970), sheep (Steblay and Rudofsky 1983*a*), rabbits, and rats.

However it has been shown in the latter two species that toxic pulmonary insults which increase pulmonary capillary permeability, such as exposure to 100 per cent oxygen for three days, the installation of petrol, or exposure to cigarette smoke for 2 weeks (A.J. Rees, unpublished observations) leads to fixation of the antibody to the alveolar basement membrane (Yamamoto and Wilson 1987; Jennings *et al*. 1981; Downie *et al*. 1982). These experiments did not examine the effects of antibodies to the specific antigen of Goodpasture's disease, but the conclusions may reasonably be expected to extend to them too. It is puzzling that some animal models of anti-GBM disease (probably involving different antigens) do include lung damage, from both passive and active immunization protocols (for example Willoughby and Dixon 1970; Wick *et al*. 1986). In some circumstances this appears to be a highly variable phenomenon (C.D. Pusey *et al*., manuscript submitted). Possible explanations for this might be that the animals have subclinical infections or exposure to toxic inhaled substances, or that the administered antigen or antibody contains agents that may modify the endothelial covering in the alveoli. Antiendothelial antibodies may be present in some sera, as may endotoxin or cytokines. One of these, tumour necrosis factor (TNF), has been shown to cause endothelial cells to contract in culture (Brett *et al*. 1989).

The strongest evidence for any exogenous agent having an effect on the occurrence of lung haemorrhage comes from the epidemiological data on smoking (Donaghy and Rees 1983). This study of 51 patients with Goodpasture's disease found that all of the 37 current smokers had pulmonary haemorrhage, compared with only two of the 10 non-smokers. Our experience since then has confirmed this relationship, although another survey has failed to do so. This may have been because smoking history was ascertained retrospectively in that series (Leaker *et al*. 1984), but there is a need for prospective data from other centres. On an anecdotal, but persuasive level, a prompt recurrence of pulmonary haemorrhage has been noted immediately following resumption of smoking in patients in whom the disease seemed to be coming under control (Heale *et al*. 1969; Donaghy and Rees 1983). Fortunately most patients with pulmonary haemorrhage stop smoking without too much prompting, but we have seen a further example of the phenomenon. Many reports suggest that exposure to various other inhaled agents might precipitate pulmonary haemorrhage by disrupting the endothelial barrier in a similar way, 'revealing' the underlying disease in some circumstances. It is interesting that there seems to be no relation between smoking and pulmonary haemorrhage in patients with Goodpasture's syndrome caused by systemic vasculitis of the Wegener's or microscopic polyarteritis type (Haworth *et al*. 1985), presumably reflecting differences in the way in which tissue damage is caused in the two types of disorder. More specifically, it might suggest that the target antigen in these disorders is not normally shielded from antibodies or lymphocytes by the pulmonary endothelium.

Fluid overload (Briggs *et al*. 1979; Simpson *et al*. 1982; Rees *et al*. 1979) and exposure to various other inhaled agents, particularly organic solvents or hydrocarbons, may act to disrupt the alveolar endothelium in the same way.

Lung infections may precipitate pulmonary haemorrhage (Bailey *et al*. 1981; Briggs *et al*. 1979; Simpson *et al*. 1982; Rees *et al*. 1979), perhaps in a similar way. However so may remote or systemic infection (Rees *et al*. 1977; Johnson *et al*. 1979) (and Fig. 10), probably by amplifying slight lung injury by mechanisms of the type discussed in the previous section. It is known that antibody can be bound to alveolar basement membrane without pulmonary haemorrhage occurring (Wilson and Dixon 1981).

Although difficult to prove in man, the animal evidence seems to make it prudent to try to avoid any damage from oxygen toxicity. This applies particularly to patients who already have severe lung haemorrhage: it is worth being more than usually careful to keep inspired oxygen concentrations as low as is consistent with good tissue oxygenation.

Treatment and outcome

Goodpasture's disease has an appalling prognosis if it remains untreated once renal impairment has developed (early series in Table 6); most patients die shortly after diagnosis of pulmonary haemorrhage or renal failure. Historically various authors have ascribed improvement of pulmonary haemorrhage to bilateral nephrectomy or to corticosteroid therapy, while others reported failures of the same treatment. As pulmonary haemorrhage has a tendency to spontaneous remission and relapse, such reports are hard to interpret. Renal function was not improved (Benoit *et al*. 1964). In a later series, retrospective analysis of survivors with preserved renal function suggested that they were more likely to have received corticosteroids than those who had developed end-stage renal failure (Wilson and Dixon 1973). Anecdotal reports also suggested that the addition of immunosuppressive agents might confer additional benefit (Briggs *et al*. 1979; Cohen *et al*. 1976; Wu *et al*. 1980*b*), although numerous failures were also noted (Wilson and Dixon 1973; Wilson and Dixon 1981). However as more effective treatments became available, it became increasingly clear that the severity of disease at presentation was a very powerful influence on outcome, particularly on whether there would be any recovery of renal function. At one extreme, patients with normal serum creatinine levels quite frequently showed spontaneous resolution of microscopic haematuria. At the other, those who were oliguric or dialysis-dependent at presentation never regained renal function, and indeed neither did most patients with renal impairment of any severity.

Current treatment regimes

The evidence for the pathogenicity of circulating antibodies in Goodpasture's disease has already been discussed, and it seems logical that an effective treatment regime might be one that removed circulating antibodies as rapidly as possible, and kept

Table 6 Results of treatment for Goodpasture's disease in larger series since 1964. The first two series included patients with Goodpasture's syndrome of other aetiologies, as the distinction was not then possible

Series	Number of patients	Male/female ratio	Pulmonary haemorrhage (%)	At one year (%)		Remarks
				Mortality	Preserved renal function	
Benoit et al. 1964	52	9.4	100	96	4	43 cases from previous series. Mortality figures for 3 years
Proskey et al. 1970	56	3.6	100	77	≤23*	51 cases from series 1964–9. *Maximum; figures not given
Wilson and Dixon 1973	53	3.6	60	25	23	Only 11.5% (six patients) were still off dialysis by the time of analysis
Beirne et al. 1977	26	2.2	54	54	15	Three anti-GBM antibody-negative patients excluded from their data
Teague et al. 1978	29	2.2	100	38	31	This series excluded patients without lung haemorrhage
Briggs et al. 1979	18	8.0	61	17	11	
Peters et al. 1982	41	1.2	56	24	39	
Johnson et al. 1985	17	7.5	94	6	45	
Walker et al. 1985	22	2.1	62	41	45	
Savage et al. 1986	108	1.4	52	21	22	Data from a number of British centres over 10 years

their titre depressed as effectively as possible. This was the rationale for the introduction of the combination of plasma exchange with more powerful immunosuppressive therapy in the mid 1970s (Lockwood et al. 1976) (and Table 7). Anecdotally this regime produced much better results, an experience that was confirmed by others (Johnson et al. 1978, and series in Table 8). Most centres now use very similar protocols to that introduced over a decade ago, and find that pulmonary haemorrhage is usually arrested in 24 to 48 h, and that even quite severely impaired renal function can be recovered in most patients. The regime of Lockwood et al. included azathioprine as well as cyclophosphamide, but other groups have shown that the use of two immunosuppressive agents is not essential, and cyclophosphamide is generally used alone. Plasma exchange regimes have differed more substantially, but if removal of antibody is indeed the mode of action, one would expect the more intensive regimes to be most effective. Our current regime is shown in Table 7; similar regimes are widely used. The improvement in mortality in series over 25 years is illustrated in Table 6, although improvements in and greater availability of dialysis, artificial ventilation, and other means of support have undoubtedly made significant contributions to this. A more detailed comparison of recent series using 'modern' treatment is made in Table 8.

Unfortunately advanced renal impairment is still not generally salvaged by any current treatment (Fig. 11), and it is interesting to contrast these results with those in rapidly progressive glomerulonephritis caused by systemic vasculitis (microscopic polyarteritis or Wegener's disease, for practical purposes), in which oliguric renal failure with a high proportion of crescents is arrested by similar treatment to that used in Goodpasture's disease (Hind et al. 1983; Bolton and Sturgill 1989). It is not clear whether this reflects some subtle and undetermined difference in the damage inflicted by the disease, or whether the results in Goodpasture's disease could be brought up to this level by future advances in treatment.

The use of bolus doses of methylprednisolone (10 mg/kg intravenously once daily for 1 to 3 days) has been advocated when there is severe pulmonary haemorrhage or very rapidly declining renal function (Johnson et al 1985). While this is simpler and cheaper than plasma exchange, it has not been compared adequately with the rational and proven treatment already described, and it may carry an increased risk of later infection. In the face of considerable immunosuppression, secondary infections are a major concern in these patients, and may exacerbate the renal and pulmonary injury caused by the disease, as discussed above.

The specific role of plasma exchange

Plasma exchange hastens the disappearance of circulating antibody when used in combination with immunosuppressive agents

Table 7 Protocol for treatment of acute Goodpasture's disease (adapted from Lockwood et al. 1976)

1. Prednisolone 1 mg/kg.24 h orally
2. Cyclophosphamide 3 mg/kg.24 h orally, rounded down to the nearest 50 mg
3. Daily exchange of 4 l of plasma for 5 per cent human albumin for 14 days or until the circulating antibody is suppressed. In the presence of pulmonary haemorrhage, or within 48 h of an invasive procedure, 300–400 ml of fresh frozen plasma are given at the end of each treatment

Table 8 Results of treatment in recent series using immunosuppression and plasma exchange. Untreated patients have been excluded. Treated patients are divided into two groups according to their creatinine at the time treatment commenced, or at presentation if this is not available (number in each group in parentheses). The percentage of patients who were alive and not requiring dialysis at 1 year is shown

Series	Percentage with independent renal function at 1 year according to initial serum creatinine level		Notes on regime used
	≤ 600 µmol/l	> 600 µmol/l	
Briggs et al. 1979 (n = 15)	36 (11)	0 (4)	Only 4/15 received plasma exchange
Simpson et al. 1982 (n = 12)	70 (10)	0 (2)	8/12 received plasma exchange
Johnson et al. 1985 (n = 17)	69 (13)	0 (4)	Less cyclophosphamide than given in Table 7. Half received plasma exchange, but only every third day, and using frozen plasma
Walker et al. 1985 (n = 22)	82 (11)	18 (11)	Slightly less cyclophosphamide and plasma exchange than given in Table 7
Hammersmith 1976–88 (n = 56)	90 (21)	11 (35)	As Table 7, except some also received azathioprine 1 mg/kg.day

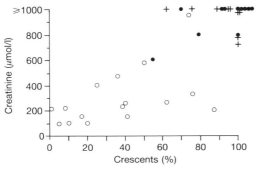

○ Independent renal function at 1 year
● On dialysis at 1 year
+ Dead at 1 year

Fig. 11 Graph showing creatinine at presentation and proportion of glomeruli with crescents in 38 patients treated at Hammersmith Hospital between 1976 and 1988. Those who did not receive the combination of plasma exchange, cyclophosphamide, and prednisolone, or who had less than 10 glomeruli in their renal biopsies, have been excluded. Illustrating: (1) the correlation between the creatinine at presentation and histological evidence of glomerular damage, except in one patient with acute tubular necrosis; (2) the close relationship between severity of renal damage at presentation and outcome; and (3) that death from pulmonary haemorrhage occurs only in those with severe renal disease.

(Johnson et al. 1985; Savage et al. 1986b). There is a strong clinical impression of its efficacy, but proving that it is an essential component in the treatment of severe Goodpasture's disease has been more difficult. The rarity of the disease and the strong influence of severity of renal damage at presentation is bound to make obtaining and matching of adequately sized treatment groups difficult.

Evidence from the use of plasma exchange in other disorders suggests that its maximum effect should be in fulminant acute disease, perhaps before immunosuppression or anti-inflammatory therapy have had time to exert their full effects. For example, muscle strength improves rapidly after plasma exchange in myasthenia gravis, another disease thought to be caused by circulating autoantibodies (Newsom-Davis et al. 1978). However, when used without other agents in Goodpasture's disease it has only a transient effect on antibody levels (Proskey et al. 1970), and there is no case for using plasma exchange alone. The role of plasma exchange was examined in the only controlled trial of treatments in Goodpasture's disease (Johnson et al. 1985). The group receiving plasma exchange achieved a better outcome than those receiving immunosuppressive agents and prednisolone alone, although a less intensive plasma exchange regime and lower doses of cyclophosphamide were used than many workers would employ. However, the authors were cautious about accepting that plasma exchange had been responsible for the improved outcome, as the group receiving it appeared to have a lesser degree of renal damage when crescent scores were compared. They did show, however, that patients with mild renal disease (<30 per cent crescents and plasma creatinine <300 µmol/l) did well on cyclophosphamide and prednisolone whether or not they received plasma exchange, and confirmed that those with severe renal disease (>70 per cent crescents and plasma creatinine >600 µmol/l) generally fared badly, as in Fig. 11.

The use of staphylococcal protein A to adsorb circulating IgG antibodies has been reported in Goodpasture's disease (Bygren et al. 1985). Although it has a number of theoretical advantages (complement and clotting factors are not depleted, expensive and potentially dangerous replacement colloid solutions are not required), it is a more time-consuming and complicated technique, and it is not likely to be generally available in the immediate future.

The best approach currently available is to use intensive plasma exchange in all patients with significant renal impairment that is thought to be retrievable, and in all patients with continuing or recurrent pulmonary haemorrhage. It should be continued until the antibody titre has reached background levels, and reintroduced if there is a later rise in creatinine or recurrence of pulmonary haemorrhage in association with an elevated antibody titre.

Deciding not to treat

The renal prognosis of patients with severe nephritis on biopsy, or severely impaired renal function, or oliguria at presentation, is extremely poor. In the absence of pulmonary haemorrhage it has been suggested that they should not be given potentially dangerous treatment (Flores et al. 1986). However there are occasional reports of patients who recovered despite acute renal failure requiring dialysis (Cohen et al. 1976; Johnson et al. 1978; Walker et al. 1985), and we have treated similar cases. Their hallmark is that they have particularly short histories, rapid renal deterioration, and either surprisingly mild, or very recent changes on renal biopsy, with cellular crescents without evidence of fibrosis, and patent capillary loops (D.J. Evans, unpublished observations). Some have relatively mild glomerular changes with superimposed acute tubular necrosis. Such patients are rare, but emphasize the value of an urgent renal biopsy even in patients where the diagnosis has been established serologically.

Duration and monitoring of treatment

Fortunately there are relatively simple ways of monitoring the consequences of the disease on both of its major target organs. Serum creatinine is the most useful guide to renal function, as urea will rise considerably with high-dose steroid therapy. It may continue to rise in the first few days of treatment, but usually stabilizes or begins to fall within a week if useful renal function is to be recovered. Haemoptysis, chest radiograph changes, changes in K_{CO}, and falls in haemoglobin are guides to the occurrence of pulmonary haemorrhage. Of these, the K_{CO} measurement is the most sensitive and earliest guide. A fall in the level of anti-GBM antibodies is taken to the an indication of effective therapy, but it does not preclude continuing or worsening disease.

Repeated measurement of the blood count is important both for the detection of haemorrhage (gastrointestinal as well as lung) and to ensure that the cytotoxic agents used are not causing leucopenia. Dose adjustments may be required to allow recovery. The commonest time for leucopenia to occur is 2 to 3 weeks with daily oral cyclophosphamide, and it is rarely seen before 10 days.

If, by these criteria, the disease is coming under control, we generally reduce prednisolone weekly, aiming to stop all treatment at about 3 months in the absence of relapses or resurgences in antibody production.

Relapse and recurrence

Relapse can be defined as a worsening of signs or symptoms of the disease before antibody titre has been totally suppressed. Early in the hospital course, severity of renal and pulmonary injury may increase either because of inadequate therapy with a rising antibody titre, or in the face of a falling titre and another reason for increased tissue damage, such as infection or fluid overload. The rational way to deal with this phenomenon is to attack both arms of the damaging mechanisms: to continue or increase therapy aimed at reducing the antibody titre, and to treat the complicating infection aggressively. However, it is better to prevent such episodes, by keeping the risk of infection to a minimum. Intravascular catheters and shunts are the most common source of significant infection, and attention to them needs to be rigorous.

Detectable antibody production may recur months to years

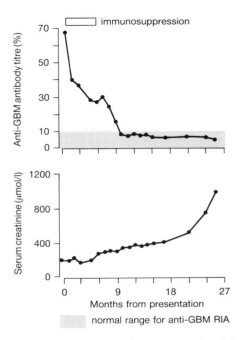

Fig. 12 Late deterioration in renal function despite clinically inactive disease in a 27-year old woman with Goodpasture's disease.

after diagnosis, with or without clinical disease, and in the presence or absence of continuing immunosuppression. Recurrences of pulmonary haemorrhage over many years are well described in the literature, the episodes usually accompanied by minimal renal disease and resolving spontaneously (Dahlberg et al. 1978, Mehler et al. 1987). Recurrences of circulating antibodies may occur with or without signs of tissue damage (Hind et al. 1984). Renal transplantation is the best characterized precipitant of recurrences of antibody production with or without clinical disease (Wilson and Dixon 1973), but many individual case histories are suggestive of infection or exposure to toxic agents at the onset of the illness or of an episode in it.

Treatment of recurrences is identical to that of initial disease, but probably stands a better chance of success as the patients and their physicians make the correct diagnosis more quickly.

Late deterioration in renal function

Patients who are left with significant renal impairment after recovery from the acute disease are at risk of the inexorable decline in renal function that has been described in renal disease of widely diverse types (Klahr et al. 1988) (and Chapter 9.3) (Fig. 12). This process is probably unrelated to the immunological disorder that caused the original disease. In such cases, but particularly if the deterioration is rapid or deviating from a linear reciprocal creatinine plot, it is important to exclude active disease. This is easily done by examination of urine sediment, assay for circulating anti-GBM antibodies, and, if necessary, renal biopsy. If there is no evidence for continuing disease activity, continuation of immunosuppressive therapy is more dangerous than helpful, and management should focus on rigorous control of hypertension, and planning for renal replacement therapy if appropriate.

Goodpasture's disease after renal transplantation

In patients with previous Goodpasture's disease

Patients with previous Goodpasture's disease have been shown to be suitable for transplantation, although cases of recurrence of the disease with graft loss have been noted, especially where circulating antibodies are present at the time of transplantation (Lerner et al. 1967; Wilson and Dixon, 1973; Beleil et al. 1973; Briggs et al. 1979; Simpson et al. 1982). Pulmonary haemorrhage in these circumstances seems to be rare. Recommendations to avoid recurrence of disease have ranged from simply monitoring the disappearance of circulating antibodies to bilateral nephrectomy. Neither of these precautions is adequate protection, and there is certainly now no case for performing a nephrectomy for immunological reasons. A patient is described (Almkuist et al. 1981) who had no detectable circulating antibody for 18 months, and a bilateral nephrectomy, before receiving the kidney of her identical twin 2 years after her original illness. No immunosuppression was given after the transplant. She developed microscopic haematuria at 2 weeks, and at 6 months circulating antibodies were detectable again, and she had crescentic nephritis. Renal function was not seriously impaired however, and the abnormalities resolved after treatment with azathioprine and corticosteroids.

Renal transplantation should be postponed until at least 6 months of the disappearance of circulating antibodies as shown by RIA or ELISA. In patients with persisting antibodies in whom transplantation is urgent, similar treatment to that used at presentation (Table 7) can be given in preparation for transplantation. With modern immunosuppression and after taking these precautions clinical relapse should be rare, although linear fixation of IgG to the GBM without overt disease can occur (Wilson and Dixon 1973; Couser et al. 1973). In any case, the renal function and urine sediment should be closely monitored in all such patients, and anti-GBM titres should be obtained at intervals.

In patients with hereditary nephritis

When studied by indirect immunofluoresence, the Goodpasture antigen is absent or greatly diminished in most patients with strictly-defined hereditary nephritis of the Alport type (Chapter 18.4.1). The transplantation of a normal kidney may allow the development of anti-GBM antibodies. Linear IgG fixation to the GBM probably occurs quite commonly, circulating anti-GBM antibodies may develop, and a small proportion of patients have developed crescentic nephritis. Lung haemorrhage has not been reported. The graft has been lost in all cases where significant nephritis has occurred (references in Table 3, epidemiology section).

The exact nature of the abnormality in Alport's syndrome is not yet clear. Western blotting studies with Alport kidneys (Kleppel et al. 1987; Savage et al. 1989), though difficult as they are usually chronically diseased and shrunken, show that the antigen is not totally absent, but its appearance is simplified. Some of the bands that are present in Western blots of normal kidneys are absent in patients with Alport's syndrome. This may indicate that the abnormality is not in the molecule of the Goodpasture antigen itself, but perhaps in a supporting or modifying structure or enzyme.

Some cases of *de-novo* anti-GBM nephritis after transplantation may occur in patients who have this type of hereditary nephritis but who lack a clear family history of renal disease, or deafness or eye changes. If earlier biopsies of native kidneys are available, it is worth re-examining them by immunohistochemical techniques for evidence of the Goodpasture antigen (Savage et al. 1986a).

References

Agodoa, L.C., Striker, G.E., George, C.R., Glassock, R., and Quadracci, L.J. (1976). The appearance of nonlinear deposits of immunoglobulins in Goodpasture's syndrome. *American Journal of Medicine*, **61**, 407–13.

Almkuist, R.D., Buckalew, V.M. Jr., Hirszel, P., Maher, J.F., James, P.M., and Wilson, C.B. (1981). Recurrence of anti-glomerular basement membrane antibody mediated glomerulonephritis in an isograft. *Clinical Immunology and Immunopathology*, **18**, 54–60.

Andres, G., et al. (1978). Histology of human tubulo-interstitial nephritis associated with antibodies to renal basement membranes. *Kidney International*, **13**, 480–91.

Bailey, R.R., Simpson, I.J., Lynn, K.L., Neale, T.J., Doak, P.B., and McGiven, A.R. (1981). Goodpasture's syndrome with normal renal function. *Clinical Nephrology*, **15**, 211–15.

Beirne, G.J. and Brennan, J.T. (1972). Glomerulonephritis associated with hydrocarbon solvents: mediated by antiglomerular basement membrane antibody. *Archives of Environmental Health*, **25**, 365–9.

Beirne, G.J., Wagnild, J.P., Zimmerman, S.W., Macken, P.D., and Burkholder, P.M. (1977). Idiopathic crescentic glomerulonephritis. *Medicine (Baltimore)*, **56**, 349–81.

Beleil, O.M., Coburn, J.W., Shinaberger, J.H., and Glassock, R.J. (1973). Recurrent glomerulonephritis due to anti-glomerular basement membrane antibodies in two successive allografts. *Clinical Nephrology*, **1**, 377–80.

Benoit, F.L., Rulon, D.B., Theil, G.B., Dooland, P.D., and Watten, R.H. (1964). Goodpasture's syndrome. A clinicopathological entity. *American Journal of Medicine*, **37**, 424–44.

Bergrem, H., Jervell, J., Brodwall, E.K., Flatmark, A., and Mellbye, O. (1980). Goodpasture's syndrome. A report of seven patients including long-term follow-up of three who received a kidney transplant. *American Journal of Medicine*, **68**, 54–8.

Bernis, P., Hamels, J., Quoidbach, A., Mahieu, P., and Bouvy, P. (1985). Remission of Goodpasture's syndrome after withdrawal of an unusual toxic. *Clinical Nephrology*, **23**, 312–17.

Blake, D.R., Rashid, H., McHugh, M., and Morley, A.R. (1980). A possible association of partial lipodystrophy with anti-GBM nephritis (Goodpasture's syndrome). *Postgraduate Medical Journal*, **56**, 137–9.

Bolton, W.K., Chandra, M., Tyson, T.M., Kirkpatrick, P.R., Sadovnic, M.J., and Sturgill, B.C. (1988). Transfer of experimental glomerulonephritis in chickens by mononuclear cells. *Kidney International*, **34**, 598–610.

Bolton, W.K. and Sturgill, B.C. (1989). Methylprednisolone therapy for acute crescentic rapidly progressive glomerulonephritis. *American Journal of Nephrology*, **9**, 368–75.

Border, W.A., Baehler, R.W., Bhathena, D., and Glassock, R.J. (1979). IgA antibasement membrane nephritis with pulmonary haemorrhage. *Annals of Internal Medicine*, **91**, 21–5.

Bowley, N.B., Steiner, R.E., and Chin, W.S. (1979a). The chest X-ray in antiglomerular basement membrane antibody disease (Goodpasture's syndrome). *Clinical Radiology*, **30**, 419–29.

Bowley, N.B., Hughes, J.M., and Steiner, R.E. (1979b). The chest X-ray in pulmonary capillary haemorrhage: correlation with carbon monoxide uptake. *Clinical Radiology*, **30**, 413–17.

Bowman, C., Ambrus, K., and Lockwood, C.M. (1987). Restriction of human IgG subclass expression in the population of auto-antibodies to glomerular basement membrane. *Clinical and Experimental Immunology*, **69**, 341–9.

Bowman, C. and Lockwood, C.M. (1985). Clinical application of a radio-immunoassay for auto-antibodies to glomerular basement membrane. *Journal of Clinical and Laboratory Immunology*, **17**, 197–202.

Brett, J., Gerlach, H., Nawroth, P., Steinberg, S., Godman, G., and Stern, D. (1989). Tumour necrosis factor/cachectin increases permeability of endothelial cell monolayers by a mechanism involving regulatory G proteins. *Journal of Experimental Medicine*, **169**, 1977–91.

Briggs, W.A., Johnson, J.P., Teichman, S., Yeager, H.C., and Wilson, C.B. (1979). Antiglomerular basement membrane antibody-mediated glomerulonephritis and Goodpasture's syndrome. *Medicine (Baltimore)*, **58**, 348–61.

Bygren, P., Freiburghaus, C., Lindholm, T., Simonsen, O., Thysell, H., and Wieslander, J. (1985). Goodpasture's syndrome treated with staphylococcal protein A immunoadsorption (letter). *Lancet*, **ii**, 1295–6.

Bygren, P., Wieslander, J., and Heinegard, D. (1987). Glomerulonephritis induced in sheep by immunization with human glomerular basement membrane. *Kidney International*, **31**, 25–31.

Bygren, P., Cederholm, B., Heinegard, D., and Wieslander, J. (1989). Non-Goodpasture anti-GBM antibodies in patients with glomerulonephritis. *Nephrology Dialysis and Transplantation*, **4**, 254–61.

Cashman, S.J., Pusey, C.D., and Evans, D.J. (1988). Extraglomerular distribution of immunoreactive Goodpasture antigen. *Journal of Pathology*, **155**, 61–70.

Clark, E.H., Jones, H.A., and Hughes, J.M.B. (1978). Bedside rebreathing technique for measuring carbon-monoxide uptake by the lung. *Lancet*, **i**, 791–3.

Clutterbuck, E.J. and Pusey, C.D. (1987). Severe alveolar haemorrhage in Churg-Strauss syndrome. *European Journal of Respiratory Diseases*, **71**, 158–63.

Cohen, L.H., Wilson, C.B., and Freeman, R.M. (1976). Goodpasture syndrome: recovery after severe renal insufficiency. *Archives of Internal Medicine*, **136**, 835–7.

Couser, W.G., Wallace, A., Monaco, A.P., and Lewis, E.J. (1973). Successful renal transplantation in patients with circulating antibody to glomerular basement membrane: Report of two cases. *Clinical Nephrology*, **1**, 381–8.

Curtis, J.J., Bhathena, D., Leach, R.P., Galla, J.H., Lucas, B.A., and Luke, R.G. (1976). Goodpasture's syndrome in a patient with the Nail-Patella syndrome. *American Journal of Medicine*, **61**, 401–6.

d'Apice, A.J., Kincaid Smith, P., Becker, G.H., Loughhead, M.G., Freeman, J.W., and Sands, J.M. (1978). Goodpasture's syndrome in identical twins. *Annals of Internal Medicine*, **88**, 61–2.

Dahlberg, P.J., *et al.* (1978). Recurrent Goodpasture's syndrome. *Mayo Clinic Proceedings*, **53**, 533–7.

Daniell, W.E., Couser, W.G., and Rosenstock, L. (1988). Occupational solvent exposure and glomerulonephritis. A case report and review of the literature. *Journal of the American Medical Association*, **259**, 2280–3.

Disney, A.P.S. (1986). *Tenth report of the Australian and New Zealand combined dialysis and transplant registry (ANZ-DATA)*. Queen Elizabeth Hospital, Adelaide, South Australia.

Donaghy, M. and Rees, A.J. (1983). Cigarette smoking and lung haemorrhage in glomerulonephritis caused by autoantibodies to glomerular basement membrane. *Lancet*, **ii**, 1390–3.

Donald, K.J., Edwards, R.L., and McEvoy, J.D. (1975). Alveolar capillary basement membrane lesions in Goodpasture's syndrome and idiopathic pulmonary haemosiderosis. *American Journal of Medicine*, **59**, 642–9.

Downie, G.H., Roholt, O.A., Jennings, L., Blau, M., Brentjens, J.R., and Andres, G.A. (1982). Experimental anti-alveolar basement membrane antibody-mediated pneumonitis. II. Role of endothelial damage and repair, induction of autologous phase, and kinetics of antibody deposition in Lewis rats. *Journal of Immunology*, **129**, 2647–52.

Ewan, P.W., Jones, H.A., Rhodes, C.G., and Hughes, J.M. (1976). Detection of intrapulmonary haemorrhage with carbon monoxide uptake. Application in Goodpasture's syndrome. *New England Journal of Medicine* **295**, 1391–6.

Feintzeig, I.D., Abrahamson, D.R., Cybulsky, A.V., Dittmer, J.E., and Salant, D.J. (1986). Nephritogenic potential of sheep antibodies against glomerular basement membrane laminin in the rat. *Laboratory Investigation*, **54**, 531–42.

Fillit, H., Damle, S.P., Gregory, J.D., Volin, C., Poon-King, T., and Zabriskie, J. (1985). Sera from patients with poststreptococcal glomerulonephritis contain antibodies to glomerular heparan sulfate proteoglycan. *Journal of Experimental Medicine*, **161**, 277–89.

Fillit, H.M. and Zabriskie, J.B. (1982). Cellular immunity in glomerulonephritis. *American Journal of Pathology*, **109**, 227–43.

Fleming, S.J., *et al.* (1988). Antiglomerular basement membrane antibody-mediated nephritis complicating transplantation in a patient with Alport's syndrome. *Transplantation*, **46**, 857–9.

Flores, J.C., Taube, D., Savage, C.O.S., Cameron, J.S., Lockwood, C.M., and Ogg, C.S. (1986). Clinical and immunological evolution of oligoanuric anti-GBM nephritis treated by haemodialysis. *Lancet*, **i**, 5–8.

Garovoy, M.R. (1982). Immunogenetic associations in nephrotic states. *Contemporary Issues in Nephrology*, **9**, 259–82.

Goodpasture, E.W. (1919). The significance of certain pulmonary lesions in relation to the etiology of influenza. *American Journal of Medical Science*, **158**, 863–70.

Gossain, V.V., Gerstein, A.R., and Janes, A.W. (1972). Goodpasture's syndrome: a familial occurrence. *American Review of Respiratory Diseases*, **105**, 621–4.

Haworth, S.J., Savage, C.O.S., Carr, D., Hughes, J.M.B., and Rees, A.J. (1985). Pulmonary haemorrhage complicating Wegener's granulomatosis and microscopic polyarteritis. *British Medical Journal*, **290**, 1775–8.

Heale, W.F., Matthiesson, A.M., and Niall, J.F. (1969). Lung haemorrhage and nephritis (Goodpasture's syndrome). *Medical Journal of Australia*, **2**, 355–7.

Heptinstall, R.H. (1983). Schönlein-Henoch syndrome; lung haemorrhage and glomerulonephritis. In *Pathology of the kidney*, (ed. R. H. Heptinstall), pp. 761–791. Little, Brown & Co, Boston.

Hind, C.R., Paraskevakou, H., Lockwood, C.M., Evans, D.J., and Peters, D.K. (1983). Prognosis after immunosuppression of patients with crescentic nephritis requiring dialysis. *Lancet*, **i**, 263–265.

Hind, C.R., Bowman, C., Winearls, C.G., and Lockwood, C.M. (1984). Recurrence of circulating anti-glomerular basement membrane antibody three years after immunosuppressive treatment and plasma exchange. *Clinical Nephrology* **21**, 244–6.

Hoedemaeker, P.J. and Weening, J. (1989). Relevance of experimental models for human nephropathology. *Kidney International* **35**, 1004–14.

Holdsworth, S., Boyce, N., Thomson, N.M., and Atkins, R.C. (1985). The clinical spectrum of acute glomerulonephritis and lung haemorrhage (Goodpasture's syndrome). *Quarterly Journal of Medicine*, **55**, 75–86.

Hudson, B.G., Wieslander, J., Wisdom, B.J., and Noelken, M.E. (1989). Goodpasture syndrome: molecular architecture and function of basement membrane antigen. *Laboratory Investigation*, **61**, 256–69.

Hume, D.M., Sterling, W.A., Weymouth, R.J., Siebel, H.R., Madge, G.E., and Lee, H.M. (1970). Glomerulonephritis in human renal homotransplants. *Transplantation Proceedings*, **2**, 361–412.

Jampol, L.M., Lahov, M., Albert, D.M., and Craft, J. (1975). Ocular clinical findings and basement membrane changes in Goodpasture's syndrome. *American Journal of Ophthalmology*, **79**, 452–63.

Jayne, D.R.W., Marshall, P.D., Jones, S.J., and Lockwood, C.M. (1989). Autoantibodies to glomerular basement membrane and neutrophil cytoplasm in rapidly progressive glomerulonephritis. *Kidney International*, in press.

Jennette, J.C., Lamanna, R.W., Burnette, J.P., Wilkman, A.S., and Iskander, S.S. (1982). Concurrent antiglomerular basement membrane antibody and immune complex mediated glomerulonephritis. *American Journal of Clinical Pathology*, **78**, 381–6.

Jennings, L., Roholt, O.A., Pressman, D., Blau, M., Andres, G.A., and Brentjens, J.R. (1981). Experimental anti-alveolar basement membrane antibody-mediated pneumonitis. I. The role of increased permeability of the alveolar capillary wall induced by oxygen. *Journal of Immunology*, **127**, 129–34.

Jeraj, K., Michael, A.F., and Fish, A.J. (1982). Immunologic similarities between Goodpasture's and Steblay's antibodies. *Clinical Immunology and Immunopathology*, **23**, 408–13.

Johnson, J.P., Whitman, W., Briggs, W.A., and Wilson, C.B. (1978). Plasmapheresis and immunosuppressive agents in antibasement membrane antibody-induced Goodpasture's syndrome. *American Journal of Medicine*, **64**, 354–9.

Johnson, J.P., Moore, J.,Jr., Austin, H.A., Balow, J.E., Antonovych, T.T., and Wilson, C.B. (1985). Therapy of antiglomerular basement membrane antibody disease: analysis of prognostic significance of clinical, pathologic and treatment factors. *Medicine (Baltimore)* **64**, 219–27.

Judge, D.M., McGlynn, T.J., Abt, A.B., Luderer, J.R., and Ward, S.P. (1977). Immunologic myopathy. Linear IgG deposition and fulminant terminal episode. *Archives of Pathology and Laboratory Medicine*, **101**, 362–5.

Kalderon, A.E., Bogaars, H.A., and Diamond, I. (1973). Ultrastructural alterations of the follicular basement membrane in Hashimoto's thyroiditis. *American Journal of Medicine*, **55**, 485–91.

Kefalides, N.A., Pegg, M.T., Ohno, N., Poon King, T., Zabriskie, J., and Fillit, H. (1986). Antibodies to basement membrane collagen and to laminin are present in sera from patients with poststreptococcal glomerulonephritis. *Journal of Experimental Medicine*, **163**, 588–602.

Klahr, S., Schreiner, G., and Ichikawa, I, (1988). The progression of renal disease. *New England Journal of Medicine*, **318**, 1657–66.

Klassen, J., et al. (1974). Evolution of membranous nephropathy into anti-glomerular-basement-membrane glomerulonephritis. *New England Journal of Medicine*, **290**, 1340–4.

Kleinknecht, D., Morel Maroger, L., Callard, P., Adhemar, J.P., and Mahieu, P. (1980). Antiglomerular basement membrane nephritis after solvent exposure. *Archives of Internal Medicine*, **140**, 230–2.

Kleppel, M.M., Michael, A.F., and Fish, A.J. (1986). Antibody specificity of human glomerular basement membrane type IV collagen NC1 subunits. Species variation in subunit composition. *Journal of Biological Chemistry* **261**, 16547–52.

Kleppel, M.M., Kashtan, C.E., Butkowski, R.J., Fish, A.J., and Michael, A.F. (1987). Alport familial nephritis – absence of 28 kilodalton non-collagenous monomers of type IV collagen in glomerular basement membrane. *Journal of Clinical Investigation*, **80**, 263–6.

Kleppel, M.M., Santi, P.A., Cameron, J.D., Wieslander, J., and Michael, A.F. (1989). Human tissue distribution of novel basement membrane collagen. *American Journal of Pathology*, **134**, 813–25.

Komadina, K.H., Houk, R.W., Vicks, S.L., Desrosier, K.F., Ridley, D.J., and Boswell, R.N. (1988). Goodpasture's syndrome associated with pulmonary eosinophilic vasculitis. *Journal of Rheumatology*, **15**, 1298–1301.

Kurki, P., et al. (1984). Transformation of membranous glomerulonephritis into crescentic glomerulonephritis with glomerular basement membrane antibodies. Serial determinations of anti-GBM before the transformation. *Nephron*, **38**, 134–7.

Langeveld, J.P.M., et al. Structural heterogeneity of the noncollagenous domain of basement membrane collagen. *Journal of Biological Chemistry* **263**, 10481–8.

Leaker, B., Walker, R.G., Becker, G.J., and Kincaid Smith, P. (1984). Cigarette smoking and lung haemorrhage in anti-glomerular-basement-membrane nephritis (letter). *Lancet*, **ii**, 1039.

Leatherman, J.W., Davies, S.F., and Hoidal, J.R. (1984). Alveolar haemorrhage syndromes: diffuse microvascular lung haemorrhage in immune and idiopathic disorders. *Medicine (Baltimore)*, **63**, 343–61.

Leatherman, J.W. (1987). Immune alveolar haemorrhage. *Chest*, **91**, 891–7.

Lee, H.Y. and Stretton, T.B. (1975). The lungs in renal failure. *Thorax*, **30**, 46–53.

Lerner, R.A., Glassock, R.J., and Dixon, F.J. (1967). The role of antiglomerular basement membrane antibody in the pathogenesis of human glomerulonephritis. *Journal of Experimental Medicine*, **126**, 989–1004.

Lockwood, C.M., Rees, A.J., Pearson, T.A., Evans, D.J., and Peters, D.K. (1976). Immunosuppression and plasma-exchange in the treatment of Goodpasture's syndrome. *Lancet*, **i**, 711–15.

Ma, K.W., Golbus, S.M., Kaufman, R., Staley, N., Londer, H., and Brown, D.C. (1978). Glomerulonephritis with Hodgkin's disease and herpes zoster. *Archives of Pathology and Laboratory Medicine*, **102**, 527–9.

Martinez-Hernandez, A. and Amenta, P.S. (1983). The basement membrane in pathology. *Laboratory Investigation*, **48**, 656–77.

Mathew, T.H., Hobbs, J.B., Kalowski, S., Sutherland, P.W., and Kincaid Smith, P. (1975). Goodpasture's syndrome: normal renal diagnostic findings. *Annals of Internal Medicine*, **82**, 215–18.

McCoy, R.C., Johnson, H.K., Stone, W.J., and Wilson, C.B. (1982). Absence of nephritogenic GBM antigen(s) in some patients with hereditary nephritis. *Kidney International*, **21**, 642–52.

McIntosh, R.M., Copack, P., Chernack, W.B., Griswold, W.R., and Weil, R., and Koss, M.N. (1975). The human choroid plexus and autoimmune nephritis. *Archives of Pathology*, **99**, 48–50.

McPhaul, J.J., Jr., and Dixon, F.J. (1970). Characterization of human anti-glomerular basement membrane antibodies eluted from glomerulonephritic kidneys. *Journal of Clinical Investigation*, **49**, 308–17.

McPhaul, J.J. Jr., and Mullins, J.D. (1976). Glomerulonephritis mediated by antibody to glomerular basement membrane. Immunological, clinical, and histopathological characteristics. *Journal of Clinical Investigation*, **57**, 351–61.

Mehler, P.S., Brunvand, M.W., Hutt, M.P., and Anderson, R.J. (1987). Chronic recurrent Goodpasture's syndrome. *American Journal of Medicine*, **82**, 833–5.

Milliner, D.S., Pierides, A.M., and Holley, K.E. (1982). Renal transplantation in Alport's syndrome: anti-glomerular basement membrane glomerulonephritis in the allograft. *Mayo Clinic Proceedings*, **57**, 35–43.

Moorthy, A.V., Zimmerman, S.W., Burkholder, P.M., and Harrington, A.R. (1976). Association of crescentic glomerulonephritis with membranous glomerulonephropathy: a report of three cases. *Clinical Nephrology*, **6**, 319–25.

Morgan, P.G. and Turner-Warwick, M. (1981). Pulmonary haemosiderosis and pulmonary haemorrhage. *British Journal of Diseases of the Chest*, **75**, 225–42.

New Zealand Glomerulonephritis Study Group (1989). The New Zealand glomerulonephritis study: introductory report. *Clinical Nephrology*, **31**, 239–46.

Newsom-Davis, J., Pinching, A.J., Vincent, A., and Wilson, S.G. (1978). Function of circulating antibody to acetylcholine receptor in myasthenia gravis: Investigation by plasma exchange. *Neurology*, **28**, 266–72.

Noel, L.H., Aucouturier, P., Monteiro, R.C., Preud Homme, J.L., and Lesavre, P. (1988). Glomerular and serum immunoglobulin G subclasses in membranous nephropathy and anti-glomerular basement membrane nephritis. *Clinical Immunology and Immunopathology*, **46**, 186–94.

Nolasco, F.E.B., Cameron, J.S., Hartley, B., Coelho, A., Hildreth, G., and Reuben, R. (1987). Intraglomerular T cells and monocytes in nephritis: Study with monoclonal antibodies. *Kidney International*, **31**, 1160–6.

Oliveira, D.B.G. and Peters, D.K. (1989). Autoimmunity and the kidney. *Kidney International*, **35**, 923–8.

Perez, G.O., Bjornsson, S., Ross, A.H., Aamato, J., and Rothfield, N. (1974). A mini-epidemic of Goodpasture's syndrome clinical and immunological studies. *Nephron*, **13**, 161–73.

Perl, S.I., Pussell, B.A., Charlesworth, J.A., Macdonald, G.J., and Wolnizer, M. (1981). Goodpasture's (anti-GBM) disease and HLA-DRw2 (letter). *New England Journal of Medicine*, **305**, 463–4.

Peters, D.K., Rees, A.J., Lockwood, C.M., and Pusey, C.D. (1982). Treatment and prognosis in antibasement membrane antibody-mediated nephritis. *Transplantation Proceedings*, **14**, 513–21.

Petterson, E., Tornroth, T., and Miettinen, A. (1984). Simultaneous anti-glomerular basement membrane and membranous nephritis: case report and literature review. *Clinical Immunology and Immunopathology*, **31**, 171–80.

Proskey, A.J., Weatherbee, L., Easterling, R.E., Greene, J.A., Jr., and Weller, J.M. (1970). Goodpasture's syndrome. A report of five cases and review of the literature. *American Journal of Medicine*, **48**, 162–73.

Pusey, C.D., and Lockwood, C.M. (1989). Autoimmunity in rapidly progressive glomerulonephritis. *Kidney International*, **35**, 929–37.

Querin, S., Noel, L.H., Grunfeld, J.P., Droz, D., Mahieu, P., and Berger, J. (1986). Linear glomerular IgG fixation in renal allografts: incidence and significance in Alport's syndrome. *Clinical Nephrology*, **25**, 134–40.

Rajaraman, S., Pinto, J.A., and Cavallo, T. (1984). Glomerulonephritis with coexistent immune deposits and antibasement membrane activity. *Journal of Clinical Pathology*, **37**, 176–81.

Rees, A.J., Lockwood, C.M., and Peters, D.K. (1977). Enhanced allergic tissue injury in Goodpasture's syndrome by intercurrent bacterial infection. *British Medical Journal*, **2**, 723–6.

Rees, A.J., Peters, D.K., Compston, D.A., and Batchelor, J.R. (1978). Strong association between HLA-DRW2 and antibody-mediated Goodpasture's syndrome. *Lancet*, **i**, 966–8.

Rees, A.J., Lockwood, C.M., and Peters, D.K. (1979). Nephritis due to antibodies to GBM. In *Progress in Glomerulonephritis* (eds. P. Kincaid-Smith, A. J. F. d'Apice, and R. C. Atkins), pp. 347–370. Wiley, New York.

Rees, A.J., Peters, D.K., Amos, N., Welsh, K.I., and Batchelor, J.R. (1984a). The influence of HLA-linked genes on the severity of anti-GBM antibody-mediated nephritis. *Kidney International*, **26**, 445–50.

Rees, A.J., Demaine, A.G., and Welsh, K.I. (1984b). Association of immunoglobulin Gm allotypes with antiglomerular basement membrane antibodies and their titer. *Human Immunology*, **10**, 213–20.

Rehan, A., Johnson, K.J., Wiggins, R.C., Kunkel, R.G., and Ward, P.A. (1984). Evidence for the role of oxygen radicals in acute nephrotoxic nephritis. *Laboratory Investigation*, **51**, 396–403.

Relman, A.S., Dvorak, H.F., and Colvin, R.B. (1971). Case records of the Massachusetts General Hospital. Weekly clinicopathological exercises. Case 46–1971. *New England Journal of Medicine*, **285**, 1187–96.

Richman, A.V., Rifkin, S.I., and McAllister, C.J. (1981). Rapidly progressive glomerulonephritis. Combined antiglomerular basement membrane antibody and immune complex pathogenesis. *Human Pathology*, **12**, 597–604.

Sabnis, S.G., Nandedkar, M.A., and Antonovych, T.T. (1988). Antiglomerular basement membrane antibody-induced glomerulonephritis with glomerular multinucleated giant cell reaction: a case study. *American Journal of Kidney Diseases*, **12**, 544–7.

Saraf, P., Berger, H.W., and Thung, S.N. (1978). Goodpasture's syndrome with no overt renal disease. *Mount Sinai Journal of Medicine (New York)*, **45**, 451–4.

Saus, J., Wieslander, J., Langeveld, J.P.M., Quinones, S., and Hudson, B.G. (1988). Identification of the Goodpasture antigen as the alpha 3(IV) chain of collagen IV. *Journal of Biological Chemistry*, **263**, 13374–80.

Savage, C.O.S., *et al.* (1986a). The Goodpasture antigen in Alport's syndrome: studies with a monoclonal antibody. *Kidney International*, **30**, 107–12.

Savage, C.O.S., Pusey, C.D., Bowman, C., Rees, A.J., and Lockwood, C.M. (1986b). Antiglomerular basement membrane antibody mediated disease in the British Isles 1980–4. *British Medical Journal*, **292**, 301–4.

Savage, C.O.S., Noel, L.H., Crutcher, E., Price, S.R., and Grunfeld, J.P. (1989). Hereditary nephritis: immunoblotting studies of the glomerular basement membrane. *Laboratory Investigation*, **60**, 613–18.

Saxena, R., Isaksson, B., Bygren, P., and Wieslander, J. (1989). A rapid assay for circulating anti-glomerular basement membrane antibodies in Goodpasture syndrome. *Journal of Immunological Methods*, **118**, 73–8.

Scheer, R.L. and Grossman, M.A. (1964). Immune aspects of the glomerulonephritis associated with pulmonary haemorrhage. *Annals of Internal Medicine*, **60**, 1009–21.

Shah, B., First, M.R., Mendoza, N.C., Clyne, D.H., Alexander, J.W., and Weiss, M.A. (1988). Alport's syndrome: risk of glomerulonephritis induced by anti-glomerular-basement-membrane antibody after renal transplantation. *Nephron*, **50**, 34–8.

Simonsen, H., Brun, C., Thomsen, O.F., Larsen, S., and Ladefoged, J. (1982). Goodpasture's syndrome in twins. *Acta Medica Scandinavica*, **212**, 425–8.

Simpson, I.J., *et al.* (1982). Plasma exchange in Goodpasture's syndrome. *American Journal of Nephrology*, **2**, 301–11.

Sisson, S., Dysart, N.K.,Jr., Fish, A.J., and Vernier, R.L. (1982). Localization of the Goodpasture antigen by immunoelectron microscopy. *Clinical Immunology and Immunopathology*, **23**, 414–29.

Stanton, M.C. and Tange, J.D. (1958). Goodpasture's syndrome (pulmonary haemorrhage associated with glomerulonephritis). *Australia and New Zealand Journal of Medicine*, **7**, 132–44.

Stave, G.M. and Croker, B.P. (1984). Thrombotic microangiopathy in anti-glomerular basement membrane glomerulonephritis. *Archives of Pathology and Laboratory Medicine*, **108**, 747–51.

Steblay, R.W. and Rudofsky, U.H. (1983a). Experimental autoimmune antiglomerular basement membrane antibody-induced glomerulonephritis. I. The effects of injecting sheep with human, homologous or autologous lung basement membranes and complete Freund's adjuvant. *Clinical Immunology and Immunopathology*, **27**, 65–80.

Steblay, R.W. and Rudofsky, U.H. (1983b). Experimental autoimmune glomerulonephritis induced by anti-glomerular basement membrane antibody. II. Effects of injecting heterologous, homologous, or autologous glomerular basement membranes and complete Freund's adjuvant into sheep. *American Journal of Pathology*, **113**, 125–33.

Teague, C.A., Doak, P.B., Simpson, I.J., Rainer, S.P., and Herdson, P.B. (1978). Goodpasture's syndrome: an analysis of 29 cases. *Kidney International*, **13**, 492–504.

Timpl, R., Paulsson, M., Dziadek, M., and Fujiwara, S. (1987). Basement membranes. *Methods in Enzymology*, **145**, 363–91.

Tomosugi, N.I., *et al.* (1989). Modulation of antibody-mediated glomerular injury in vivo by bacterial lipopolysaccharide, tumor necrosis factor, and IL-1. *Journal of Immunology* **142**, 3083–90.

Van Zyl Smit, R., Rees, A.J., and Peters, D.K. (1983). Factors affecting severity of injury during nephrotoxic nephritis in rabbits. *Clinical and Experimental Immunology* **54**, 366–72.

Wahls, T.L., Bonsib, S.M., and Schuster, V.L. (1987). Coexistent Wegener's granulomatosis and anti-glomerular basement membrane disease. *Human Pathology*, **18**, 202–5.

Walker, R.G., Scheinkestel, C., Becker, G.J., Owen, J.E., Dowling, J.P., and Kincaid Smith, P. (1985). Clinical and morphological aspects of the management of crescentic anti-glomerular basement membrane antibody (anti-GBM) nephritis/Goodpasture's syndrome. *Quarterly Journal of Medicine*, **54**, 75–89.

Weber, M., Kohler, H., Manns, M., Baum, H.P., and Meyer zum Buschenfelde, K.H. (1987). Identification of Goodpasture target antigens in basement membranes of human glomeruli, lung, and placenta. *Clinical and Experimental Immunology*, **67**, 262–9.

Weber, M., Lohse, A.W., Manns, M., Meyer zum Buschenfelde, K.H., and Kohler, H. (1988). IgG subclass distribution of autoantibodies to glomerular basement membrane in Goodpasture's syndrome compared to other autoantibodies. *Nephron*, **49**, 54–7.

Wheeler, J., Simpson, J., and Morley, A.R. (1988). Routine and rapid enzyme linked immunosorbent assays for circulating anti-glomerular basement membrane antibodies. *Journal of Clinical Pathology*, **41**, 163–70.

Wick, G., von der Mark, H., Dietrich, H., and Timpl, R. (1986). Globular domain of basement membrane collagen induces autoimmune pulmonary lesions in mice resembling human Goodpasture disease. *Laboratory Investigation*, **55**, 308–17.

Wieslander, J. and Heinegard, D. (1985). The involvement of type IV collagen in Goodpasture's syndrome. *Annals of the New York Academy of Sciences*, **460**, 363–74.

Wieslander, J., Kataja, M., and Hudson, B.G. (1987). Characterization of the human Goodpasture antigen. *Clinical and Experimental Immunology*, **69**, 332–40.

Willoughby, W.F. and Dixon, F.J. (1970). Experimental haemorrhagic pneumonitis produced by heterologous anti-lung antibody. *Journal of Immunology*, **104**, 28–37.

Wilson, C.B. and Dixon, F.J. (1973). Anti-glomerular basement membrane antibody-induced glomerulonephritis. *Kidney International*, **3**, 74–89.

Wilson, C.B. and Smith, R.C. (1972). Goodpasture's syndrome associated with influenza A2 virus infection. *Annals of Internal Medicine*, **76**, 91–4.

Wilson, C.B., Dixon, F.J., Evans, A.S., and Glassock, R.J. (1973). Antiviral antibody responses in patients with renal diseases. *Clinical Immunology and Immunopathology* **2**, 121–32.

Wilson, C.B. and Dixon, F.J. (1974). Diagnosis of immunopathologic renal disease. *Kidney International*, **5**, 389–401.

Wilson, C.B. and Dixon, F.J. (1981). The renal response to immunological injury. In *The Kidney* (eds. B. M. Brenner and F. C. Rector), pp. 1237–1350. W. B. Saunders, Philadelphia.

Wu, M.J., Rajaram, R., Shelp, W.D., Beirne, G.J., and Burkholder, P.M. (1980a). Vasculitis in Goodpasture's syndrome. *Archives of Pathology and Laboratory Medicine*, **104**, 300–2.

Wu, M.J., Moorthy, A.V., and Beirne, G.J. (1980b). Relapse in antiglomerular basement membrane antibody mediated crescentic glomerulonephritis. *Clinical Nephrology*, **13**, 97–102.

Yaar, M., Foidart, J.M., Brown, K.S., Rennard, S.I., Martin, G.R., and Liotta, L. (1982). The Goodpasture-like syndrome in mice induced by intravenous injections of anti-type IV collagen and anti-laminin antibody. *American Journal of Pathology*, **107**, 79–91.

Yamamoto, T. and Wilson, C.B. (1987). Binding of anti-basement membrane antibody to alveolar basement membrane after intratracheal gasoline instillation in rabbits. *American Journal of Pathology*, **126**, 497–505.

Yoshioka, K., Michael, A.F., Velosa, J., and Fish, A.J. (1985). Detection of hidden nephritogenic antigen determinants in human renal and nonrenal basement membranes. *American Journal of Pathology*, **121**, 156–65.

Yoshioka, K., Iseki, T., Okada, M., Morimoto, Y., Eryu, N., and Maki, S. (1988). Identification of Goodpasture antigens in human alveolar basement membrane. *Clinical and Experimental Immunology*, **74**, 419–24.

3.12 Infection-associated glomerulonephritis

A.M. DAVISON

General introduction

The effects on the kidney of infections are variable and depend on a number of factors relating both to the host and the infecting organism. For example the well-recognized association between a group 'A' β-haemolytic streptococcal infection and the subsequent development of an acute nephritic illness has been used for many years to explain certain important immunological principles. However, in current nephrological practice, particularly in the developed countries, poststreptoccal glomerulonephritis is relatively uncommon, although it has become apparent that other infections may produce a similar clinical picture and similar glomerular lesions. In this situation it has been postulated that the infecting organism releases antigens to which there is a subsequent antibody response. Local or systemic immune complex formation, complement activation, and a resulting inflammatory response may occur in the glomerulus, leading to glomerulonephritis.

It is becoming increasingly recognized that infections may be associated with many other lesions in the kidney. Septicaemia is frequently associated with significant toxin production which can result in an acute tubular necrosis (Mactier and Dobbie 1984) and the development of acute renal failure. Severe *Escherichia coli* septicaemia may be associated with verocytotoxin production and the development of haemolytic uraemic syndrome, particularly in children (see Chapter 8.7). Other infections have a major influence on the interstitium: acute tubulointerstitial nephritis is well-recognized in patients with Hantavirus infection, and a wide variety of lesions have been reported in AIDS patients, including glomerular changes, tubular necrosis, tubulointerstitial nephritis, granulomatous lesions due to opportunistic infections, nephrocalcinosis, and direct involvement of the kidney by Kaposi's sarcoma and lymphoma. Other viral infections may precipitate or cause a relapse of nephritis caused by antiglomerular basement membrane antibodies (Goodpasture's syndrome) or may result in an haemolytic uraemic syndrome with renal involvement. Vasculitis has been described in association with hepatitis B virus infections. Chronic bacterial infections such as subacute bacterial endocarditis, infected ventriculoatrial shunts, and visceral abscesses have been associated with the development of glomerulonephritis, particularly of the mesangiocapillary form. Other chronic bacterial infections, particularly bronchiectasis, oesteomyelitis, and tuberculosis, are associated with a chronic stimulation of the immune system which results in amyloid deposition. More recently, the complications of antibiotic therapy have been shown to result in the development of an allergic type eosinophilic tubulointerstitial nephritis (Pusey *et al.* 1983). Thus all compartments of the kid-

3.12 Infection-associated glomerulonephritis

Table 1 Recognized renal consequences of infections

Blood vessels	Vasculitis
	Haemolytic–uraemic syndrome
Glomeruli	Glomerulonephritis
	Intravascular coagulation
	Amyloidosis
Tubules	Acute tubular necrosis
Interstitium	Tubulointerstitial nephritis
	Infection-associated
	Antibiotic-induced nephrocalcinosis
	Granuloma formation
	Kaposi's sarcoma
	Lymphoma

ney may be involved as an indirect effect of infections (Table 1). It is particularly important to remember that the same organism may produce different morphological appearances in different patients. For instance, membranous glomerulonephritis, mesangiocapillary glomerulonephritis, and IgA nephropathy have all been described in patients with persistent hepatitis B antigenaemia. In addition, different infections may produce the same glomerular morphological appearance in different patients (Boulton-Jones and Davison 1986).

Furthermore, similar infections from different geographical regions may present completely different clinical features (see Chapters 3.14 and 8.13). It would appear that the glomerular response to infection depends on both patient factors and characteristics of the infecting organism. The immune response to infection may determine the resulting glomerular lesion through the type, quantity, and duration of antibody response, and localization of the antigen–antibody immune complexes in a subendothelial, subepithelial, and/or mesangial position. The immune response may be influenced by genetic factors, and this may explain the geographical differences which have been reported for certain infections. Some forms of glomerular disease, such as IgA nephropathy, are common and this will increase the possibility of a coincidental association with an infecting organism. Furthermore, organisms differ in their virulence, and this may be further affected by passage, resulting in either increased virulence or attenuation. Thus the consequences of an infection in any given population is dependent on a number of variables, and this may well explain the wide variety of responses that are seen even in epidemic infections.

The clinical situation is further complicated by socioeconomic factors. With increasing affluence, the general health of a community improves, and this is associated with an increased ability to deal with infection. Poststreptococcal glomerulonephritis is now relatively uncommon in the United Kingdom, although it is still epidemic in certain areas of the world. Other factors such as a reduction in virulence of the streptococcus may also in part explain the phenomenon. However, it is interesting to note that in a recent study of infection as a cause of renal disease leading to renal biopsy carried out in the United Kingdom, 21 per cent of patients with infection-associated renal disease were immigrants (Boulton-Jones and Davison 1986).

Many other factors can be implicated in the changing epidemiology of infection associated glomerular disease. The eradication of *Plasmodium malariae* from British Guiana has been associated with a reduction in the incidence of nephropathy (see Chapter 3.14). However the AIDS epidemic has been accompanied by an increasing number of reports describing an associated nephropathy and renal disease. Although many diseases cannot be eradicated, the development of effective antibiotic therapy has been associated with a reduction in infection-related complications. It is now relatively uncommon to encounter patients with renal amyloidosis secondary to tuberculosis or chronic osteomyelitis in the United Kingdom. Thus socioeconomic status and environmental health are significant factors in explaining the changing clinical pattern of glomerular disease associated with infections.

A causal relationship between infection and nephropathy was initially established by epidemiological studies in which epidemic or endemic infections were associated with specific glomerular abnormalities. The link has been strengthened further by the clinical observation that in many instances the renal lesion resolves following effective treatment of the underlying infection; malaria and schistosomiasis are exceptions. In addition, there have been many experimental animal studies in which glomerular lesions have been induced by bacterial, viral, or parasitic infections. A direct causal relationship has been suggested by the detection of antigen arising from the infecting organism within the glomerulus, and from elution studies demonstrating specific antibody directed against bacterial or viral antigens. A role for the immune system has been suggested by demonstration of immunoglobulins and complement components within glomerular deposits. Although many early reports are of doubtful significance, due to the poor specificity of the antisera used, recent reports (particularly those using monoclonal antibodies) have demonstrated specific immunoglobulins and complement within the glomerulus, thus confirming the immunological nature of the nephropathy.

Reports from the preantibiotic era frequently associated nephropathy with chronic infections such as tuberculosis. It is interesting that a recent re-examination of three of the kidneys which formed the original report by Richard Bright in the early nineteenth century revealed that one has amyloidosis, possibly associated with an underlying chronic tuberculosis (Weller and Nester 1972). At about the same time the association between an infectious illness, scarlatina, and renal abnormalities was reported (Miller 1849). Detailed studies, of course, could not be undertaken until the nature of the infecting organism was elucidated. Many early investigators relied on postmortem material, and it was not until the development of percutaneous renal biopsy 35 years ago that it became possible to understand renal pathology, particularly of acute illnesses. Since then numerous studies have shown an association between bacterial, parasitic, and viral infections and nephropathy.

Infections do not always produce an adverse effect on the glomerulus. It has been noted that measles may produce a remission in proteinuria in children with minimal changed nephrotic syndrome, and primary measles vaccination has been used therapeutically (Yuceoglu *et al.* 1969) (see Chapter 3.5). It is possible that measles virus infection has an inhibitory effect on T cells, suppressing lymphokine secretion and resulting in glomerular capillary wall changes which reduce proteinuria. Infections with varicella-zoster virus, *Salmonella typhi*, Plasmodium, and pneumococcal infections have also occasionally been reported to have a similar effect. There has also been an increased interest in the role of cellular immunity in infection-related glomerular disease following the reports of AIDS-associated nephropathy.

Many associations between infections and nephropathy are

well accepted and well documented (see Levy 1988 for a comprehensive review). However, in many instances there are few or sporadic adequately documented reports. It is particularly difficult to establish a causal relationship if specific antigen has not been detected within the glomerulus and/or elution studies have not been undertaken. There is frequently more than one potential cause for the nephropathy. Patients with parasitic infections are frequently malnourished and have other bacterial infections. Many of those with AIDS-associated nephropathy are intravenous drug abusers, and have opportunistic infections; there is also the possibility of antibiotic-induced nephrotoxicity. Caution is needed before a particular infection is associated with a glomerular lesion unless there has been (1) a specific antigen detected within the glomerulus, (2) elution studies have identified an immunoglobulin which reacts specifically with antigens of the affecting organism, and (3) epidemiological studies have excluded other compounding factors.

In general, management is directed towards the identification of the infection and its subsequent treatment. In the majority of patients, treatment of the renal disease is symptomatic, aiming to control oedema, hypertension, and impaired kidney function. It must always be remembered that drugs, particularly antibiotic and diuretic agents, may be associated with a tubulointerstitial nephritis.

Bacterial infections

General remarks

The association of bacterial infections with glomerulopathy has been accepted for many years. It is well established that the antibody response to bacterial antigens can result in the presence of immune complexes in the circulation and/or tissues. Whereas it was previously assumed that these immune complexes formed in the circulation and then became trapped in the glomerulus as a secondary event, it is now generally accepted that in the majority of cases the complexes are formed locally within the glomerular capillary wall or mesangium. Here they activate the complement and coagulation systems and induce an inflammatory response (see Chapter 3.1).

The pattern of glomerular response is not uniform: even in epidemics not everyone who is infected develops glomerular disease, and the clinical picture and histological appearances of those who do are seldom uniform. This suggests that many other factors, possibly genetic and nutritional, have an important influence on the development of the glomerular lesion. The widespread use of antibiotic therapy, which has resulted in the prompt treatment of most infectious diseases, also has an effect. Chronic infections are much less of a problem than in the past and, as a result, complications such as amyloidosis are decreasing in frequency.

Streptococcus (see also Chapter 3.9)

The relationship between acute glomerulonephritis and epidemics of scarlatina has been recognized for over 100 years (Miller 1849). There is no doubt, however, that the epidemiology of poststreptococcol glomerulonephritis has changed significantly in the past century, in that there has been a significant decline in the incidence in developed countries, although sporadic cases continue to be reported. Epidemic poststreptococcal glomerulonephritis appears to occur mainly in developing countries such as Africa, the West Indies, and the Middle East.

Reasons for this changing epidemiology relate to the nutritional status of the community, the more liberal use of antibiotics, and possibly, also to changes in the nephritogenic potential of streptococcus.

Streptococcal infection may be followed by an acute nephritic illness or by acute rheumatic fever. It is uncommon, although not unknown, for these two conditions to coexist. There are interesting epidemiological differences between rheumatic fever and poststreptococcal glomerulonephritis in that acute rheumatic fever only follows group A streptococcal infection of the upper respiratory tract, whereas glomerulonephritis may follow either pharyngeal or cutaneous infections. In addition, rheumatic fever may recur, whereas glomerulonephritis normally does not, although there are a few reports of more than one attack. Some streptococcal serotypes have been considered to be particularly nephritogenic, while others may have the potential to induce both rheumatic fever and glomerulonephritis. This is supported by the studies of Russell (1962), who found that 38 per cent of 246 patients with rheumatic heart disease had evidence of 'glomerulitis'. Grishman *et al.* (1967) performed renal biopsies in 22 patients with rheumatic fever and found a focal glomerulonephritis in two and diffuse proliferative glomerulonephritis in four patients. More recently, Kujala *et al.* (1989) have reported the simultaneous appearance of acute rheumatic fever and acute glomerulonephritis occurring in an 8-year-old girl shortly after an episode of scarlet fever.

Poststreptococcal glomerulonephritis usually develops 10 days after an upper respiratory tract infection or some 3 weeks after a skin infection such as pyoderma (impetigo). In tropical countries, streptococcal infection may complicate insect bites. Glomerulonephritis has also been reported following middle-ear infections and in association with endocarditis due to both α-haemolytic and β-haemolytic streptococci, as well as *Strep. viridans*, *Strep. mitis* and *Strep. mutans*.

Poststreptococcal glomerulonephritis is associated with infection with group A β-haemolytic streptococcus of types 1, 2, 4, 12, 18, 25, 49, 55, 57, and 60. The incidence of nephritis varies with the prevalence of the nephritogenic streptococcus in the community. In Chicago 76 per cent of children with acute poststreptococcal glomerulonephritis had a preceding upper respiratory tract infection (Lewy *et al.* 1971), whereas in the southern parts of the United States 66 per cent of children had evidence of a preceding skin infection (Sanjad *et al.* 1977). Not all patients infected with the same serotype develop glomerular abnormalities. In a prospective study of children, only 22 per cent of infected patients develop urinary abnormalities and/or a low C3 (Sagel *et al.* 1973).

Patients who develop poststreptococcal glomerulonephritis have a very variable clinical picture. Most commonly, it occurs in children 3 to 8 years of age, and is extremely rare before the age of 2 years. The male:female ratio is approximately 2:1, and signs of renal disease become apparent in 50 per cent of patients within 1 to 2 weeks of an upper respiratory tract infection. In patients with streptococcal skin infection the latent interval is difficult to determine. The classical picture consists of macroscopic haematuria, which may last for up to 2 weeks and may remain microscopic for many months thereafter, oedema, which is variable but may be severe enough to produce pleural effusions and ascites, and hypertension, which is usually mild, but may be severe. Some patients have cerebral symptoms, such as headaches, encephalopathy, and convulsions, whilst others have cardiovascular complications in the form of congestive cardiac failure; approximately one-third will have gastrointestinal symptoms with nausea, vomiting, and abdominal pain. Purpura, arthralgia, and arthritis are rare, but may cause diagnostic confusion with lupus erythematosus, polyarteritis, Henoch-Schönlein purpura, or endocarditis.

At clinical presentation, 75 per cent of patients will have some evidence of impaired renal function, and this will be mild in 50 per cent. Haematuria is universal and the urine is described as 'smoky' in appearance. Red cell casts are evident on urine microscopy. Proteinuria is usually moderate, and does not exceed 2 g/day in 85 per cent of patients. Chest radiographs may reveal pulmonary oedema. Isolation of streptococcus from the pharynx is possible in about 20 per cent of patients and culture from the skin is very variable. A raised antistreptolysin titre is present in 90 per cent of patients, but care must be taken with interpretation as not all types of streptococcus are nephritogenic. Concentrations of complement (C3) are frequently depressed but return to normal in approximately 8 weeks.

The nephritis is immune-complex mediated. The bacterium produces numerous products, including toxin and enzymes such as streptolysin, hyaluronidase, and streptokinase. The immune complex pathogenesis is supported by the finding of a reduction in C3 values, the presence of circulating immune complexes and cryoglobulins, and the demonstration of IgG and complement within glomerular deposits. The precise nature of the nephritogenic antigen is not known; it is, however, possible that the initiating antigen localizes in the mesangium and subendothelial space of the glomerulus, subsequently reacting with antibody to form in-situ complexes (Lange et al. 1983). Streptococcal antigens are principally found in deposits in the mesangium of the glomerulus in the early stages of the disease, but can rarely be detected later than 2 weeks after the clinical presentation.

An alternative explanation, is that streptococcal neuraminidase exposes antigenic sites on host IgG molecules, inducing an autoimmune reaction leading to immune complex formation and deposition. McIntosh et al. (1978) detected glomerular deposition of anti-IgG antibody (rheumatoid factor) in 19 of 22 patients with acute poststreptococcal glomerulonephritis. In addition, cationic antigen, which can enter the negatively charged basement membrane, has been isolated from nephritogenic streptococci; this may act as a 'planted' antigen, with the subsequent formation of immune complexes in situ (Vogt et al. 1983).

Proliferative glomerulonephritis involving both mesangial and endothelial cells is the most common finding on renal biopsy. Glomerular infiltration with polymorphs is common in the early stages of the illness: immunofluorescence microscopy reveals deposition of IgG, and less commonly IgM, in association with C3 and C1q. Dense deposits within the mesangium are detected on electron microscopy, and there may also be large subepithelial deposits usually described as 'humps'. Other histological appearances which have been reported in a few patients include type 1 mesangiocapillary glomerulonephritis. (McCluskey and Baldwin 1965), crescentic glomerulonephritis (Roy et al. 1981; Old et al. 1984; Parag et al. 1988), and IgA nephropathy (Rekola et al. 1985).

The exudative changes last for approximately 2 weeks, but a return of normal glomerular appearances may take years. Glomerulosclerosis may develop in patients with crescentic lesions.

Patient management depends on the severity of the illness, but is predominantly symptomatic. Renal failure is usually mild and self-limiting, although, dialysis may occasionally be necessary. Peripheral oedema should be treated with appropriate diuretic therapy and hypertension controlled to within acceptable limits. Antibiotic therapy is only required if signs of the initiating infection are still present. In the majority of patients macroscopic haematuria, oedema, oliguria, hypertension, and CNS symptoms usually subside within 1 week of clinical presentation. A diuresis is usually complete by 2 weeks, with consequent resolution of the oedema. Renal function improves over 2 to 4 weeks and hypertension rapidly subisdes. C3 usually returns to normal within 8 weeks. Microscopic haematuria may persist for many years and minor proteinuria may last even longer.

There is considerable debate about long-term prognosis. There is general agreement that the short-term prognosis is excellent, particularly in children, although the early mortality has been reported as between 0.5 and 0.8 per cent (Rodriguez-Iturbe 1984). Doubt has been cast on the long-term prognosis by Baldwin (1977) who followed 168 patients for up to 18 years, of whom 95 per cent were followed for more than 10 years. In these patients, proteinuria, hypertension, and impaired renal function appeared to resolve within 2 years of the initial illness, but during follow up there appeared to be an increasing incidence of these three clinical features. In contrast, a study undertaken 10 years after the Red Lake epidemic found no difference between patients and controls with respect to haematuria and biopsy changes (Perlman et al. 1965). A further long-term study also concluded that the majority of patients had a good prognosis (Lien et al. 1979). Clark et al. (1988) concluded that although 20 per cent of children with biopsy proven acute poststreptococcal glomerulonephritis remained with some urinary abnormality, the long-term outcome was excellent. The debate regarding long-term prognosis may be confounded because the various reported series have not been comparable. However, the prognosis appears to be better in children than in adults, and epidemic cases may have a more mild disease than sporadic cases. Although, histological changes, in particular glomerular sclerosis, may be present it is difficult to determine whether these lesions are residual or represent progressive disease. However, it is likely that the chance of progressing to end-stage renal failure is probably less than 1 per cent. In the small group of patients with progressive disease, the mechanism of progression is unknown, but it may be related to the development of cell-mediated immunity acting against cross-reacting antigens of the streptococcus and the glomerular basement membrane, or as a direct consequence of the induced hypertension.

Staphylococcus

Glomerular lesions associated with staphylococcal infections arise in many different clinical situations, but most commonly in patients with infective endocarditis. Many organisms have been implicated in infective endocarditis, but the most common is *Staphylococcus aureus*. The first reports (Lohlein 1910) suggested that the glomerular lesion was due to bacterial embolization from the infected valve. This view is no longer supportable and the lesion is considered to be immunologically mediated. Infective endocarditis is associated with high titres of circulating immune complexes, hypocomplementaemia, and glomerular immunoglobulin deposition, and staphylococcal antigens are present within involved glomeruli. It is also possible that staphylococcal cell wall antigens may directly activate complement by the alternate pathway, resulting in glomerular damage, and this may explain the occurrence of glomerulonephritis after a short duration of endocarditis, in the absence of immune complexes and with no glomerular immunoglobulin deposition. On light microscopy patients with acute fulminating endocarditis tend to have a diffuse proliferative glomerulonephritis, whereas those with a subacute form tend to have a focal segmental proliferative glomerulonephritis, which, in the late stages, may show sclerosis.

Staphylococcal septicaemia is associated with proliferative glomerulonephritis; an autopsy study demonstrated glomerular changes in 35 per cent of patients who died from staphylococcal septicaemia (Powell 1961). Septicaemia may also occur in association with intravenous drug abuse and such patients have a high incidence of associated proliferative glomerulonephritis (Treser et al. 1974).

Staphylococcal colonization of ventriculoatrial and ventriculoperitoneal shunts, used in the treatment of hydrocephalus, is a relatively common problem. These patients present with symptoms suggestive of recurrent bacteraemia such as fever, arthralgia, hepatosplenomegaly, and oedema. Infection most commonly occurs within 2 months of catheter insertion. Renal involvement is manifest by haematuria, which is occasionally macroscopic, and proteinuria, which results in a nephrotic syndrome in 25 per cent of patients. Cryoglobulinaemia, hypocomplementaemia, and high titres of antistaphylococcal antibodies are well described. Renal biopsy most commonly reveals Type 1 mesangiocapillary glomerulonephritis which may be associated with minor crescent formation. A few patients have diffuse exudative proliferative glomerulonephritis, minor mesangial proliferation or segmental lesions.

Occasionally, other staphylococcal infections, such as osteomyelitis (Boonshaft et al. 1970) and impetigo (Takaue and Tokumaru 1982) have been associated with glomerulonephritis, and very occasionally, IgA nephropathy has been associated with staphylococcal infection (Levy et al. 1985).

Streptococcus pneumonia (pneumococcus)

Historically, renal disease was a common sequel to pneumococcal infections, but since the introduction of antibiotics only sporadic case reports have appeared. Clinical presentation is with haematuria, minimal proteinuria, and only mild impairment of renal function. Renal biopsy shows acute glomerulonephritis, with subepithelial 'humps' and intramembranous deposits. Pneumococcal antigen has been detected within the mesangium (Hyman et al. 1975; Kaehny et al. 1978). In addition, cryoglobulins were detected in one patient and pneumococcal capsular antigen was demonstrated within the cryoprecipitate (Kaehny et al. 1978).

Disseminated intravascular coagulation may occur following pneumococcal infection, and one case has been described with focal membranous glomerulonephritis and granular deposition of pneumococcal capsular antigen along glomerular capillary walls (Rytel et al. 1974). Disseminated intravascular coagulation is thought to be precipitated by neuraminidase released by the bacterium.

Salmonella

Typhoid fever is relatively common but is rarely associated with glomerulonephritis, even in endemic areas, presumably due to the predominance of the gastrointestinal symptoms. Renal involvement in typhoid fever may be manifest as cystitis or pyelonephritis. Acute renal failure may result from dehydration or as a result of intravascular coagulation and haemolysis, as in haemolytic uraemic syndrome; this occurs more commonly in patients with glucose 6-phosphate dehydrogenase deficiency. Glomerular involvement occurs in approximately 2 to 4 per cent of patients in endemic areas.

Typhoid fever rarely presents with renal manifestation although renal involvement may be relatively common. Microscopic, and occasionally macroscopic haematuria, associated with moderate proteinuria and normal or slightly diminished renal function are found on presentation. In a series of children reported from South Africa oedema had been present for more than 1 month in 50 per cent of patients, and pyrexia and splenomegaly were common (Buka and Coovadia 1980). *Salmonella typhi* may be isolated from blood, stool, or urine culture, and there may be serological evidence of infection with the detection of O and H antibodies. C3 concentrations are frequently reduced as is IgG and IgM, although IgA values are significantly elevated, presumably due to antigenic stimulation of plasma cells within the lamina propria of the gastrointestinal tract, the site of entry of the bacteria.

Renal biopsy has shown both mesangial proliferative glomerulonephritis (Musa et al. 1981) and IgA nephropathy (Amerio et al. 1972; Indrapasit et al. 1985); acute diffuse proliferative glomerulonephritis has also been described (Buka and Coovadia 1980). Most commonly, there is a proliferation of mesangial cells and matrix and minor crescents may be present in a few glomeruli. Mesangial deposition of IgA, IgG, and C3 are common, and there are two reports of *Salmonella* V_i antigen being detected within glomeruli (Sitprija et al. 1974; Indrapasit et al. 1985). There is frequently a focal interstitial fibrosis. In patients in whom haemolytic uraemic syndrome develops as a consequence of typhoid fever, histology reveals normal glomeruli, but there is widespread tubular degeneration and interstitial mononuclear cell infiltration similar to that found in acute tubular necrosis.

Several reports from Egypt and Sudan have implicated chronic salmonella bacteraemia in the pathogenesis of acute nephrotic syndrome in patients with hepatosplenic schistosomiasis (Barsoum et al. 1977). Biopsy in these patients reveals acute diffuse proliferative glomerulonephritis. Resolution occurs on effective treatment of the typhoid fever, suggesting that salmonella, rather than the schistosomes, is responsible for the nephropathy in these patients.

The glomerular lesion is probably immune complex mediated. The significantly elevated serum IgA may result from antigenic stimulation within the lamina propria of the gastrointestinal tract at the site of entry of the bacteria. The association of mesangial IgA deposition and the detection of the V_i antigen within the mesangium is further support for the conception of immune mediation.

The glomerular disease is frequently mild and transient. Prognosis clearly depends on many factors: many patients are frequently malnourished. In the study of Buka and Coovadia the overall mortality of children with typhoid fever was between 3 and 8 per cent, but was 20 per cent in those complicated by glomerulonephritis. In adults, the mortality in typhoid glomerulonephritis may be as much as 30 per cent.

Mycobacteria (see also Chapter 3.14)
Tuberculosis

Glomerular involvement in tuberculosis is usually due to chronic infection which results in amyloidosis (see Chapter 4.2). The decreasing incidence of tuberculosis associated with effective antituberculous chemotherapy has resulted in a significant decline in the incidence of such amyloidosis.

Chronic tuberculosis may also be associated with other types of glomerular disease. One patient with a dense deposit type mesangiocapillary glomerulonephritis was described by Hariprasad *et al.* (1979), and a further case report described a patient with focal proliferative glomerulonephritis (Shribman *et al.* 1983).

Leprosy (see also Chapter 3.13)

A wide variety of renal lesions has been associated with leprosy. Glomerulonephritis (Johny *et al.* 1975), amyloidosis (Ozaki and Furuta 1975), interstitial nephritis, pyelonephritis and acute tubular necrosis have all been described. Renal disease is most commonly found in lepromatous leprosy, although it has occasionally been described in the tuberculoid form. Episodes of erythema nodosum leprosum are associated with acute glomerulonephritis, while recurrent episodes are associated with amyloidosis. In a review of the literature by Ng *et al.* (1981), 58 of 187 patients reported had evidence of glomerulonephritis, and although these are selected patients it can be assumed that glomerular involvement is uncommon but not rare.

Immune complexes can be detected in many patients with lepromatous leprosy, but to date antigens to *Mycobacterium leprae* have not been identified in glomerular deposits.

Clinically, patients present with minor proteinuria which, in some instances, may be sufficient to cause nephrotic syndrome. Microscopic haematuria is present but renal functional impairment is uncommon, although there are occasional reports of rapidly progressive disease due to crescentic nephritis.

The most common finding on renal biopsy is diffuse proliferative change in which there is mesangial cell proliferation, with little increase in mesangial matrix. There may be a significant inflammatory cell infiltrate in the glomeruli, giving an exudative appearance in some patients; focal proliferative glomerulonephritis and, rarely, crescentic nephritis, have also been described. Immunofluorescence microscopy may reveal IgM, IgG and C3 but IgA is detected only occasionally. Electron microscopy shows electron-dense deposits in a subendothelial and mesangial position; a number of patients have typical subepithelial 'humps'.

The nephropathy is mild and self limiting in the majority of patients and progression to renal failure is uncommon.

Renal amyloidosis is also well recognized in long-standing lepromatous leprosy, although there is a marked geographical variation in the incidence; it commonly occurs in America, but is uncommon in India and Japan. The amyloid is typically of the AA type with the fibrils being composed of serum acute phase reactant amyloid-associated protein (see Chapter 3.2). Clinically there is proteinuria, frequently sufficient to produce nephrotic syndrome. Progression to renal failure is common and death is frequently due to uraemia.

Treponema

The association between syphilis and dropsy was reported early in the 19th century (Blackall 1818), although it is recognized as uncommon. Nephropathy may develop either in congenital syphilis or in the secondary phase of an acquired infection.

Nephropathy is uncommon in congenital syphilis, but may become evident in the first few months of life usually with other clinical signs of syphilis. Very occasionally, clinical evidence of congenital syphilis may be absent and it is wise to undertake serological tests for syphilis in all nephrotic neonates. The most common mode of presentation is with nephrotic syndrome, which may be particularly severe. A few cases present as a nephritic illness. The typical findings on renal biopsy are membranous glomerulonephritis (Losito *et al.* 1979) with subepithelial deposition of IgG and complement (C1q and C3). Mesangial proliferation is variable and in some cases small crescents may be present. The glomerular lesion is immunologically mediated: treponema antigens have been identified in the glomerular capillary wall deposits (O'Regan *et al.* 1976) and antitreponemal antibodies have been detected by elution studies (Losito *et al.* 1979). Associated tubulointerstitial nephritis is common and may be the dominant feature. Treatment with penicillin is accompanied by resolution of the proteinuria, usually within 2 to 6 weeks, and by recovery of the glomerular lesion. Further courses of penicillin may be required for patients in whom significant proteinuria persists. Follow-up biopsies frequently demonstrate a significant percentage of hyalinized glomeruli.

Nephropathy is an uncommon complication of acquired syphilis but may become clinically apparent during the secondary phase of the infection. The majority of patients present with a nephrotic syndrome (O'Reagan *et al.* 1976) which becomes clinically manifest about the same time as the typical skin lesions develop. Microscopic haematuria, hypertension, and/or impairment of renal function only rarely occur. On light microscopy there is a membranous glomerulonephritis, frequently accompanied by a slight mesangial proliferation. Immunofluorescence demonstrates subepithelial electron-dense deposits containing IgG and complement (C3 and C1q). Following the institution of penicillin therapy there is a rapid resolution of the nephropathy. Spontaneous improvement in nephrotic syndrome prior to penicillin therapy is also well recognized (Sterzel *et al.* 1974). This is presumably because the secondary phase of syphilis is a self-limiting immune complex-mediated disease.

Leptospirosis (see also Chapter 8.13)

Leptospirosis (Weil's disease) is most commonly associated with acute oliguric renal failure with evidence of hepatic involvement. The renal lesion is usually acute tubulointerstitial nephritis, although some patients develop a mesangial proliferation associated with mesangial subepithelial and intramembranous deposits which contain immunoglobulins and complement (Lai *et al.* 1982).

Other bacterial infections

There are many reports of glomerulonephritis in patients with other bacterial infections. However, the frequency of the infections and the relative paucity of the reports must cast doubt on a causal relationship. In the majority, detection of an appropriate bacterial antigen has not been possible, and it is likely that glomerular disease is coincidental in a number of cases.

Brucellosis may be complicated by proteinuria and haematuria, particularly during acute infection. There is a case report of a patient with acute *Brucella melitensis* infection who had haematuria and proteinuria: a renal biopsy revealed IgA nephropathy (Nunan *et al.* 1984).

Infection with *Escherichia coli* is common and a significant number of patients have septicaemia but glomerular disease has been reported only twice. In both patients, renal involvement became manifest after *E.coli* septicaemia; in one as a severe nephrotic syndrome (Jacquot and Bariéty 1980) and in the other acute renal failure due to acute proliferative glomerulonephritis (Zappacosta and Ashby 1977). Following antibiotic therapy renal disease completely resolved in both patients.

Klebsiella pneumoniae-associated glomerulonephritis has been described in two patients who developed a microscopic haematuria and minor proteinuria in association with severe glomerulonephritis. *Klebsiella pneumoniae* antigens were detected along glomerular basement membranes on immunofluorescence microscopy. In addition, immunoglobulins containing capsular polysaccharide antigens were eluted from the glomeruli.

Meningococcal meningitis has been associated with a proliferative glomerulonephritis associated with glomerular 'humps'. Unfortunately meningococcal antigen was not identified in these deposits.

There is also an association between infection and haemolytic uraemic syndrome (see Chapter 8.7).

Yersinia entercolitica has been associated with glomerulonephritis, particularly in reports from Scandinavia (Denneberg *et al.* 1981; Friedberg *et al.* 1981). Renal involvement is usually characterized by a variable degree of proteinuria, with occasional microscopic haematuria and less frequently, impairment of renal function. Renal biopsy shows either proliferative or membranous glomerulonephritis. Yersinia antigen has been demonstrated in both mesangium and in the glomerular capillary wall, particularly in those patients with a membranous glomerulonephritis. One patient has been described whose renal biopsy revealed an IgA nephropathy three weeks after an acute *Yersinia entercolitica* infection (Cusack *et al.* 1983).

There are a number of sporadic case reports of glomerular lesions in association with *Mycoplasma pneumoniae* infection. Renal biopsy has shown proliferative glomerulonephritis, both type 1 and type 2 mesangiocapillary glomerulonephritis, and IgA nephropathy. Mycoplasma antigen has been demonstrated in the mesangium and in the capillary wall (Cochat *et al.* 1985).

The most usual renal disease associated with legionnaires' disease is tubulointerstitial nephritis (see Chapter 8.9), but one patient has been reported in whom acute renal failure was associated with mesangial proliferative glomerulonephritis (Hariprasad *et al.* 1985).

Rickettsial infections

Rickettsial infections have a predilection for endothelium. Initial swelling of infected endothelial cells is followed by perivascular and intramural infiltration by lymphocytes and macrophages resulting in a vasculitis.

The mortality from Rocky Mountain spotted fever remains at about 7 per cent despite effective early antibiotic therapy. Death is associated with multiorgan vasculitis involving the central nervous system, myocardium, and kidney. Acute renal failure is common in fatal cases, and is a consequence of systemic rickettsial vasculitis (Walker and Mattern 1979.) In these patients the kidneys show a multifocal perivascular interstitial nephritis of the intertubular capillaries and vessels of the other medulla and corticomedullary junction. Immunofluoresence studies reveal *R. rickettsii* antigens in intertubular capillaries and in the endothelium of arteries and veins. There are reports of an acute diffuse glomerulonephritis (Allen and Spitz 1945) while others describe a slight glomerular endothelial cell swelling associated with mesangial prominence (DeBrito *et al.* 1968). However, none of the 10 cases reported by Walker and Mattern (1979) had evidence of glomerulonephritis.

Haemolytic uraemic syndrome has been described in a rickettsial infection in which the organism has been provisionally classified as a microtatobiote (Mettler 1969).

Viral infections

General remarks

Viral infection may produce glomerular effects by three different mechanisms: a direct cytopathic effect of the virus on glomerular cells, by stimulation of an antibody response resulting in immune complex formation, and by a direct effect on T cells altering the helper/suppressor ratio and affecting humoral immunity.

Viral infections resulting in immune complex-mediated glomerular diseases occur naturally and in experimental animal diseases, including Aleutian disease in mink, spontaneous Gross leukaemia virus infection, and lymphocytic choriomeningitis virus infection. Hepatitis B virus, cytomegalovirus, and human immunodeficiency virus have all been reported to cause glomerular lesions.

Viral infections are particularly common in man and it is surprising that so few patients with acute infections develop evidence of renal disease. There are many sporadic case reports of glomerulonephritis following viral diseases: varicella (Yuceoglu *et al.* 1968), measles virus (Lin and Hsu 1983), mumps (Hughes *et al.* 1966), subacute sclerosing encephalitis (O'Regan *et al.* 1979), Echo virus infection (Huang and Weigenstein 1977), and Epstein-Barr virus infection (Lee and Kjellstrand 1978). Viral infections may also precipitate haemolytic uraemic syndrome, particularly in children (see Chapter 8.7).

Chronic infections, particularly those involving hepatitis B virus, cytomegalovirus, and human immunodeficiency virus, have been associated with glomerular changes. Chronic hepatitis B infection has been associated with both glomerulonephritis (Kohler *et al.* (1974) and polyarteritis nodosa (Michalak 1978). Cytomegalovirus infection has been most commonly reported in patients following transplantation, but there is considerable doubt over whether this lesion is specific or represents a form of rejection. The human immunodeficiency virus is associated with a focal segmental glomerulosclerosis, but the case reports contain large numbers of intravenous drug abusers and patients with opportunistic infections, and a direct causal relationship is difficult to establish.

Hepatitis B (see also Chapter 3.14)

Renal involvement in hepatitis B has been widely reported: polyarteritis nodosa, membranous glomerulonephritis, type 1 mesangiocapillary glomerulonephritis, IgA nephropathy, and essential mixed cryoglobulinaemia have all been described. The first reported association was with polyarteritis nodosa (Gocke *et al.* 1970), and in the following year the associated membranous lesion was reported by Combes *et al.* (1971). Since then there have been numerous reports confirming these associations, and whilst glomerulonephritis is well recognized arteritis is considered to be uncommon. There is widespread geographical variation in the incidence of hepatitis B infection, which may account for the varying reports in the literature.

Clinically there appears to be considerable difference in the nephropathy seen in adults and children. In children, the disease most frequently presents as a nephrotic syndrome accompanied by microscopic haematuria; only rarely is there evidence of renal functional impairment. In such patients there is only minor evidence of hepatic disease and proteinuria frequently remits within 1 year. In adults, however, the major clinical feature is hepatic disease, which may have been present for many years prior to the development of the nephropathy. It is difficult to explain these differences, but most reported cases in children arise from areas where hepatitis B is endemic and the effects of infection acquired in childhood from either infected mothers or siblings may be considerably different from that acquired in later life. Only very rarely is an unsuspected hepatitis B antigenaemia detected in adult patients undergoing renal biopsy for suspect glomerular disease, although there is evidence that the incidence of antigenaemia is greater in patients undergoing renal biopsy than in the general community (Maggiore et al. 1981).

All of the available evidence suggests that hepatitis B nephropathy is immune complex mediated. Many patients do not have evidence of circulating immune complexes, thus supporting the concept of 'in-situ' complex formation within the glomerulus. Nearly all patients have circulating HBsAg and HBsAb while some 60 per cent will also have HBeAg. Detailed studies have shown evidence of HbsAg HbcAg, and HbeAg in glomeruli, although not all studies have found all three antigens. Surface and core antigens are predominantly detected in the mesangium, whereas HBe antigen is usually present in subepithelial deposits. This distribution is probably accounted for by differences in the molecular weight of these antigens. Surface and core antigens are large, whereas the e antigen, which is part of nucleoprotein, exists in two forms with molecular weights of 19 000 and 300 000. This may further explain why some patients are described as having membranous glomerulonephritis whereas others have been reported with mesangiocapillary glomerulonephritis or IgA nephropathy. Although there is general agreement that HbsAg and HbeAg are present, the role of HbcAg is still unclear. Caution is required in the interpretation of these studies as Lai et al. (1989) detected deposition of HbsAg and HbcAg using polyclonal antisera, whereas only HbsAg and not HbcAg could be detected when monoclonal antibodies were used on the same tissue. In contrast, a study from Taiwan (Hsu et al. 1989), using monoclonal antibodies, detected only HBeAg in glomerular deposits in 41 of 43 children with hepatitis B-associated membranous nephropathy; no HBsAg or HBcAg was detected in any of these cases. The immune complex nature of the associated nephropathy is further supported by the detection of HbsAb in the kidney using elution studies.

Renal biopsy most commonly reveals membranous glomerulonephritis (Combes et al. 1971; Kohler et al. 1974; Slusarczyk et al. 1980), particularly in children, in whom the idiopathic form is rare (Takekoshi et al. 1978, Yoshikawa et al. 1985). Interestingly, 14 of 33 children with a biopsy diagnosis of membranous glomerulonephritis were carriers of HBsAg, compared to three of 170 children with other glomerular diseases (Kleinknecht et al. 1979); similar findings were reported by La Manna et al. (1985). Immunofluorescence studies of renal biopsies have detected IgG, IgA, IgM, and C3. Viral antigens, particularly HbsAg and HbeAg, have also been described. Electron microscopy reveals numerous subepithelial and intramembranous deposits, but, unlike the idiopathic form of membranous nephropathy, there may also be some subendothelial and/or mesangial deposits.

In a number of patients, particularly adults, renal biopsy shows type 1 mesangiocapillary glomerulonephritis (Takeda et al. 1988) and this seems to progress and may result in terminal renal failure. Associated IgA nephropathy has been reported from Hong Kong, where hepatitis B is endemic (Lai et al. 1987). There are high titres of HBcAg and HBsAg in such patients indicating that they are persistent carriers. These antigens were present in the mesangium in eight of 10 patients suggesting that they may play a part in the pathogenesis of IgA nephropathy. Lai et al. (1988) have also reported surface antigen in the serum of 17.2 per cent of patients with IgA nephropathy in Hong Kong, an incidence significantly greater than the prevalence of HbsAg carriers in the general population.

The prognosis of membranous nephropathy associated with hepatitis B is variable, with up to 50 per cent of children having a spontaneous remission within the first 6 months of the illness. In HBeAg-positive patients, proteinuria appears to remit, with regression of the membranous lesion, when they produce HBe antibody (Ito et al. 1981; Wiggelinkhuizen et al. 1983). However, in another study (Hsu et al. 1989) remission was more closely related to HBsAg seroconversion; although HBeAg seroconversion was accompanied by remission, many patients had already entered remission prior to seroconversion. Patients who remit following HBeAg seroconversion do not subsequently relapse. Thus although the exact role of the three antigens has not been precisely defined it appears that the e antigen plays an important part in the pathogenesis of membranous nephropathy and that, although circulating HBeAg is necessary for the maintenance of the glomerular lesion, its presence does not exclude complete remission.

Progressive disease is more usual in adults, with the development of chronic renal failure in addition to the chronic active hepatitis. Garcia et al. (1985) described the use of leucocyte interferon in two children with nephrotic syndrome and hepatitis B-associated membranous nephropathy. Interferon therapy was associated with a loss of HbsAg and HbeAg and the associated development of anti-Hbe in one patient, accompanied by significant diminution in proteinuria. There was a significant reduction in proteinuria in the second patient, although HbsAg remained. Another patient with hepatitis B-associated membranous nephropathy with deposition of HBcAg and HBeAg was treated with interferon and acyclovir (de Man et al. 1989). One year later the child was asymptomatic, and although a mild membranous appearance was still visible on biopsy HBcAg and HBeAg were no longer detectable. The mechanism of action of interferon is unknown, but is probably related to its immunostimulatory properties.

Hepatitis B-associated polyarteritis nodosa

Although the first reported association of hepatitis B and polyarteritis nodosa was made in 1970 (Gocke et al. 1970) it is more than likely that earlier reports of this disease developing after vaccination (Paull 1947) were also due to hepatitis B antigen deposition. The incidence of hepatitis-associated polyarteritis nodosa has been variously reported from zero to 54 per cent; most reports originate from Europe and America and involve adults, in whom infection is usually acquired parenterally. The patients present within a few months of a clinically mild hepatitis with fever, arthralgia, and urticaria which progresses to vasculitis after several weeks. Renal disease is manifest by haematuria, variable proteinuria, hypertension or impaired renal function, in a manner similar to 'classical' polyarteritis (see Chapter 4.5.3).

Studies of liver function reveal a mild increase in enzymes; HbsAg and HBcAg are usually present in blood. There is focal inflammation of medium sized arteries, characterized by fibrinoid necrosis, cellular infiltration, fibrin deposition, and, at a later stage, aneurysm formation. Glomeruli usually show ischaemic changes; the appearances of 'microscopic polyarteritis' (focal segmental necrotizing glomerulonephritis) are rare. Circulating immune complexes containing HBsAg and HBsAb can be demonstrated in the majority of patients; vascular lesions may be due to the deposition of these antigens in vessel walls, with subsequent activation of the complement and coagulation systems resulting in inflammation. HBsAg, immunoglobulins, and complement have been detected in vessel walls by immunofluorescence, and while this may only represent 'trapping' of such reactants in already damaged vasculature it is more likely that they are playing an active role.

Prognosis is difficult to determine. As yet there are no reported studies of treatment for hepatitis B-associated polyarteritis; only uncontrolled observations in small numbers of patients are available (Sergent et al. 1976; Fauci et al. 1979; Oriente et al. 1986). Most would suggest the use of corticosteroids, with or without cytotoxic drugs such as cyclophosphamide. This therapy, however, is likely to result in prolongation of the carrier state with the attendant risk of developing chronic liver disease. Remission has been reported following the clearance of HBsAg in a patient developing HBsAg (Trepo et al. 1974).

Cytomegalovirus

Cytomegalovirus infection has been associated with glomerular changes in two groups of patients; children with severe congenital or neonatal infection, and immunosuppressed patients, particularly following transplantation. Circulating immune complexes are common in children with severe disease, but evidence of nephropathy is rare. In such patients, although granular deposits of IgG and C3 have been demonstrated in glomerular capillary walls, no cytomegalovirus antigen has been detected within glomeruli, casting doubt on the pathogenesis and specificity of the lesion.

In adults the situation remains unclear. Although most individuals will eventually become infected with cytomegalovirus, symptomatic infection is rare. Ozawa and Stewart (1979) reported a patient who developed proteinuria and haematuria during the course of fatal cytomegalovirus pneumonitis. The kidney showed mesangial proliferative glomerulonephritis with mesangial deposition of IgG, IgA, C3, and C4. Cytomegalovirus antigen was present in a distribution similar to that of the immunoglobulin, and antibody was detected by elution studies. This seemed to confirm the causal relationship between cytomegalovirus infection and glomerulonephritis.

Cytomegalovirus infection is particularly common following transplantation and can arise through reactivation of a latent infection, reinfection, or primary infection in a seronegative recipient receiving tissue from a seropositive donor. Approximately 66 per cent of renal transplant patients have serological evidence of cytomegalovirus infection and half of these patients will chronically excrete the virus in their urine. Although reactivation may be virtually asymptomatic, primary infection in transplant recipients is frequently associated with a variable clinical picture including pyrexia, pneumonia, hepatitis, retinitis, leucopenia, and thromocytopenia. Reinfection due to transmission of a different strain from that of the primary infection is usually symptomatic (Grundy et al. 1988). Richardson and colleagues (1981) studied 14 renal transplant recipients and concluded that there was evidence that cytomegalovirus infection could result in an acute allograft dysfunction associated with distinctive glomerular changes which were unlike those of rejection. The glomerular capillary loops contained large eosinophilic cells which were considered to be damaged endothelial cells. There was also slight mesangial proliferation and minor focal interstitial changes. Electron microscopy revealed subendothelial accumulation of amorphous material, but no subepithelial deposits and only occasional mesangial deposits. Immunofluorescence demonstrated IgM and C3 and cytomegaloviral antigens were detected in cells in glomerular capillaries, thought possibly to be leucocytes in the one case examined. It was postulated that the infection precipitated rejection: such patients frequently require a reduction in immunosuppression due to the infection-associated leucopenia and thrombocytopenia and thus it has been difficult to confirm a causal role for the virus. However, in a study which compared autopsy material from children who received bone-marrow transplants, children with severe disseminated cytomegalovirus infection, and renal transplant recipients (Herrera et al. 1986) there was no direct evidence for a specific cytomegalovirus nephropathy and it was concluded that the changes were those of an unusual form of rejection involving endothelial cells. Furthermore in a study of 298 renal allograft recipients Boyce et al. (1988) reported that 33 had an acute transplant glomerulopathy, but there was no correlation with cytomegalovirus infection.

In transplant recipients, therefore, it is difficult to determine whether the lesions described are due to cytomegalovirus infection or represent glomerular endothelial cell changes associated with rejection. The balance of opinion at present would seem to favour this latter explanation.

AIDS-associated nephropathy

The acquired immune deficiency syndrome is a disorder of immune regulation in which there is a marked abnormality in cellular immunity, resulting in the development of Kaposi's sarcoma and opportunistic infections. Renal involvement in patients infected with the human immunodeficiency virus (HIV) include a focal segmental glomerulosclerosis, tubulointerstitial disease, acute tubular necrosis, drug-related interstitial nephritis, pyelonephritis, nephrocalcinosis, electrolyte disorders, neoplasia (Kaposi's sarcoma and lymphoma), and opportunistic infections. There has been considerable debate regarding the specificity of glomerular lesions, since the majority of patients have multiple infections and/or intravenous drug abuse.

There are reports of acute glomerulonephritis in AIDS patients which may be secondary to bacterial infections and membranous glomerulonephritis has also been associated with coincidental hepatitis B virus infection.

There are considerable geographical, racial, and other socio-economic influences on HIV nephropathy. Reports from the east coast of the United States of America report a high incidence of renal involvement in a population which is predominantly black in origin. D'Agati et al. (1989), working in New York, concluded that the prevalence of nephropathy is 3.3 per cent in autopsies of AIDS patients, and that black patients and intravenous drug addicts were particularly at risk of developing this complication. On the other hand reports from the west coast of America describe a low incidence of nephropathy in a population which is mainly white. Reports from Europe, another predominantly white population, also indicate that renal involvement is rare (Brunkhorst *et al.* 1989; De Meyer *et al.* 1989).

The first reported association of nephropathy in an AIDS patient was made by Collins *et al.* (1983) who reported hepatitis B-associated nephropathy in a homosexual man with AIDS. The following year Rao *et al.* (1984) reported from Brooklyn on a study of 92 AIDS patients, of whom nine had a nephrotic syndrome and two had chronic renal failure. Renal histology was obtained by biopsy in eight patients, and at autopsy in a further three. On light microscopy, 10 patients exhibited a focal segmental glomerulosclerosis and immunofluorescence microscopy showed IgM and C3. One patient had mesangial proliferative glomerulonephritis with deposition of IgG and C3. In the patients with focal segmental glomerulosclerosis electron microscopy showed focal loss of foot processes, shrinkage of glomerular tufts and occasional dense mesangial deposits. Five of the patients were known intravenous drug abusers and all had opportunistic infections including herpes, oesophageal candidiasis, *Mycobacterium avium intracellulare* infection, Pneumocystis infection, and cytomegalovirus infection. This study concluded, however, that although the lesion was similar to that found in intravenous drug abusers, it probably represented a discrete entity since drug abuse did not appear to be a factor in six of the 11 patients. The incidence of nephropathy in AIDS patients is much greater than could be expected from the known prevalence of nephropathy in intravenous drug abusers. In addition, the AIDS-related nephropathy was not similar to that associated with cytomegalovirus or hepatitis B virus infections.

The pathogenesis was considered to be altered cell-mediated immunity, with reversal of the normal helper/suppressor ratio and the polyclonal hypergammaglobulinaemia associated with circulating immune complexes. The patients developing renal functional impairment progressed rapidly to end-stage renal failure and, although the cause for this was not obvious, it was considered to be due to the presence of multiple infections and the adverse effects of drug treatment.

Pardo *et al.* (1984) reported a prospective study of 75 patients from Miami, 32 of whom had proteinuria in excess of 0.5 g/day. Seven of these had proteinuria in excess of 3 g daily. In addition they reported an autopsy study of 36 patients, 18 black and 18 white, of whom 18 were homosexual and 10 were intravenous drug abusers. Proteinuria was present in nine of these 36 patients. At autopsy, 17 of the 36 patients had glomerular abnormalities; five had focal segmental glomerulosclerosis, of whom four also had tubulointerstitial disease, five had focal segmental proliferative glomerulonephritis, four had mesangial proliferative glomerulonephritis, and three had diffuse proliferation. The glomeruli appeared normal in 19 patients. In addition to these findings cytomegalovirus was present in the glomeruli and there was evidence within the kidney of Kaposi's sarcoma and a cryptococcal granuloma. There was a high incidence of acute renal failure and changes of acute tubular necrosis were seen. On electron microscopy, electron-dense deposits were found in the mesangium of patients with glomerular abnormalities, and also in a number of patients in whom the glomeruli appeared normal by light microscopy. Immunofluorescence showed IgM and C3 deposition. Although the focal segmental glomerulosclerosis seen in these patients was no different from that seen in intravenous drug abusers, only 10 of the patients examined were known drug addicts. Thirty-five of the 36 patients in this group had evidence of more than one opportunistic infection, pneumocystis and cytomegalovirus being most common.

In an attempt to determine whether AIDS nephropathy was a distinct clinical entity Chander *et al.* (1987) from New York studied three groups of patients; AIDS patients with proteinuria in excess of 2 g daily, AIDS patients with no proteinuria and a matched group of intravenous heroin addicts with proteinuria; all but one of the AIDS patients were also heroin abusers. There was a high incidence of opportunistic infections in the AIDS patients and all reported one or more such infections, particularly with pneumocystis. This study concluded that there were many similarities between the groups, particularly with respect to focal segmental glomerulosclerosis and global sclerosis. However four of seven AIDS patients with significant proteinuria had a hypocellular mesangium, whereas the other two groups had either a normal or hypercellular mesangium. In addition, there was a paucity of interstitial infiltrates and evidence of tubular ectasia in the AIDS patients. Electron microscopy revealed an increased frequency, size, complexity and heterogeneity of nuclear bodies. These are usually associated with drugs or viral or tumour stimulation.

Rousseau *et al.* (1988) from Montreal studied the renal complications of AIDS in children. A low incidence of renal disease was noted, the most common being acute renal failure due to dehydration or drug nephrotoxicity. Four patients, however, had focal glomerulosclerosis, and one of these patients had overt nephrotic syndrome. As these patients had no associated heroin addiction, this supports the concept of an AIDS-related nephropathy; however, as with adult patients, there were frequent associated and multiple opportunistic infections. These findings were confirmed by Connor *et al.* (1988), from the New Jersey, who found clinically apparent renal disease in five of 175 children with symptomatic HIV infection. Two had a focal glomerulosclerosis and three had mesangial proliferation; all had evidence of opportunistic infections.

Reports from Europe have not confirmed the North American findings. Brunkhorst *et al.* (1989) investigated 203 patients with HIV infection; 11 had proteinuria in excess of 0.5 g/l and none developed nephrotic syndrome or a sustained increase in serum creatinine. Proteinuria and transient increases in serum creatinine were always associated with severe opportunistic infections or high doses of potentially nephrotoxic antibiotics. A further European study (De Meyer *et al.* 1989) reported on 38 patients, of whom 19 were black. In only two was there a history of intravenous drug addiction. Twenty-six patients had no clinical evidence of renal involvement; the remainder showed acute renal failure (four patients), chronic renal failure (one patient), proteinuria (five patients), and microscopic haematuria (two patients). Although light microscopy showed acute tubulointerstitial disease in 14 patients and chronic interstitial nephritis in six, mesangial hypercellularity was observed in only two patients. It was concluded that glomerular lesions are relatively uncommon and that, by contrast, interstitial lesions due to intercurrent infection or nephrotoxic drugs are frequent.

The question of a specific HIV-associated nephropathy distinct from AIDS-associated nephropathy remains unresolved. It would appear that many factors play a part, including intravenous drug addiction, opportunistic infections and drug reactions. Recently proviral HIV DNA has been demonstrated in glomerular epithelial cells by *in-vitro* DNA hybridization studies using biopsy and autopsy material from patients with HIV-associated nephropathy (Cohen *et al.* 1979). However, in another study of 370 HIV infected patients Genderini *et al.* (1989) found 12 had clinical and histological signs of HIV nephropathy. Seven biopsies were tested for specific HIV antigens, core P18 and P24, envelope GP41 and GP110 by means of monoclonal antibodies. No specific binding was detected despite ultrastructural evidence of viral involvement in the renal lesions: three patients had microtubular virus-like particles, five patients showed nucleogranular transformation of tubular cells and two patients had nuclear bodies. Thus the causal relationship of HIV infection to nephropathy must still remain open; however, immune complexes do not appear to be involved in the lesion.

Electrolyte disorders

Hyponatraemia is common (Glassock et al. 1990), occurring in 40 per cent of patients with AIDS or AIDS-related complex, and may be due to a variety of causes, including gastrointestinal fluid loss, adrenal insufficiency, inappropriate ADH secretion associated with pulmonary and central nervous system infections and acute renal failure. Hyponatraemia is mild in the majority of patients, but in a few it is severe and symptomatic. Hyperkalaemia is less common but may arise due to adrenal insufficiency or acute renal failure, whereas hypokalaemia may result from gastrointestinal loss due to diarrhoea or metabolic alkalosis due to prolonged vomiting.

Epstein-Barr virus

The Epstein-Barr virus is associated with infectious mononucleosis and Burkitt's lymphoma. Renal complications are common in infectious mononucleosis, with proteinuria in 14 per cent and haematuria in 11 per cent of patients (Lee and Kjellstrand 1978). Some patients have macroscopic haematuria and there may be mild impairment of renal function (Woodroffe et al. 1974); some patients have been reported with severe impairment of renal function (Lee and Kjellstrand 1978). Renal biopsy shows a variety of glomerular changes, including IgA nephropathy and proliferative glomerulonephritis. Renal abnormalities appear to resolve with resolution of the infections. Glomerulonephritis is uncommon in Burkitt's lymphoma, although a mesangiocapillary glomerulonephritis has been described.

Influenza

Although influenza is common and epidemics frequently occur associated renal disease has only been reported occasionally. Goodpasture (1919) described a young man who died 6 weeks after influenza and who had pulmonary haemorrhage and glomerulonephritis. Wilson and Smith (1972) described a patient who developed proteinuria and haematuria associated with haemoptysis following influenza. Renal biopsy revealed proliferative glomerulonephritis and a linear deposition of immunoglobulins on the glomerular capillary basement membrane. There was no impairment of renal function and the pulmonary and renal abnormalities resolved completely. Perez et al. (1974), however, reported a number of patients with rapidly progressive renal failure and a typical Goodpasture's syndrome during an outbreak of influenza.

Haemolytic uraemic syndrome has also been reported in association with influenza A virus infection (Davison et al. 1973). Clinical presentation was with acute renal failure requiring haemodialysis. Renal biopsy revealed endothelial cell changes and glomerular capillary fibrin deposition typical of disseminated intravascular coagulation. One patient recovered completely, whilst the other died from massive intrapulmonary haemorrhage.

Measles

Measles is common and is frequently severe in malnourished children. Urinary epithelial cells from infected patients can be shown to contain measles antigen. Glomerular involvement has, however, only rarely been reported.

Lin and Hsu (1983) described a child who developed acute nephritis 2 weeks after exposure to measles and 4 days after the appearance of clinical measles. Renal biopsy showed mesangial proliferative glomerulonephritis and electron microscopy revealed discrete granular electron-dense deposits within the basement membrane and also in a subepithelial position. Immunofluorescence revealed C3 and measles virus antigen within the mesangium. There was rapid resolution of the acute nephritis and at follow up 20 months later renal function, urine analysis and serum complement were all within normal limits.

Fungal infections

Candida albicans

Candida infection is relatively common, but there has only been one report convincingly demonstrating an association with glomerular disease. Chesney et al. (1976) described a patient with mucocutaneous candidiasis associated with a type 2 mesangiocapillary glomerulonephritis. C. Albicans antigen was detected within both the mesangium and tubular basement membranes. A causal link between the candidiasis and the renal involvement was inferred from the improved renal function and diminution in proteinuria which followed improvement in the mucocutaneous candidiasis.

Histoplasma capsulatum

Renal involvement may occur in approximately one-third of patients with disseminated histoplasmosis and is thought to be immune-complex mediated. Bullock et al. (1979) described a patient with disseminated histoplasmosis in whom the renal biopsy revealed mild diffuse mesangial proliferation with IgA, IgM, and C3 deposition within the mesangium. Unfortunately, histoplasma antigen could not be detected by indirect immunofluorescence.

Parasitic infections (see also Chapter 3.14)

General remarks

Parasitic infections are widespread, particularly in tropical areas. The many published reports linking such infections with specific glomerular lesions must be interpreted with caution since these patients frequently have multiple infections—parasitic, bacterial, and viral—and there is commonly associated malnutrition. However, malarial and schistosomal antigens have been detected within glomerular lesions, strongly supporting a causal link for these two parasitic infections.

Malaria

Although the association between nephrotic syndrome and malaria was reported more than 70 years ago (McFie and Ingram 1917), and may even have been recognized in the ancient world, the reason for this association remains to be elucidated. It is not clear why some patients with *Plasmodium malariae* infection develop glomerular lesions, and although this infection is endemic throughout the tropics, the associated nephrotic syndrome has only been reported from certain geographical areas. Early reports from West Africa and British Guiana implicated *Plasmodium malariae*, although there was a greater incidence of *Plasmodium vivax* and *Plasmodium falciparum* infections in these communities. There has been a marked reduction in the incidence of nephritis since the eradication of malaria from British Guiana. (Giglioli 1962), supporting the concept of a causal relationship between malaria and glomerular disease, although it must be remembered that there have been many other changes in that community with respect to both nutrition and health.

Plasmodium malariae

Reports from West African countries have recently produced conflicting information. In Nigeria, 88 per cent of nephrotic children are known to have *Plasmodium malariae* parasitaemia (Gilles and Hendrickse 1963); however, Adu *et al*. (1981) were not able to confirm this finding in a study of children in Ghana, where *P. malariae* is endemic. This suggests that not all strains of *P. malariae* are nephritogenic, or there may be other factors, such as nutrition and genetic predisposition, which influence the immune response and/or the development of glomerular disease.

The association between *P. malariae* infection and nephrotic syndrome is well accepted, although the pathogenesis remains unclear. The glomerular changes may not be specific for *P. malariae* infection, although the term malarial nephropathy has been widely accepted. Clinically, patients who present with nephrotic syndrome are between the ages of 5 and 8 years, and there appears to be an equal sex incidence. Hypoalbuminaemia is frequently marked and ascites common. Microscopic haematuria occurs in the majority of patients, but hypertension is rare.

On renal biopsy, the most common glomerular lesion detected is focal and segmental glomerular capillary wall thickening involving the subendothelial aspect and producing, in some areas, a double contour appearance. This progresses in time to segmental, and ultimately global, sclerosis. Cellular proliferation is minimal and only occasionally are small crescents found. Immunofluorescence shows deposition of IgG, IgM, C3, and antigens of *P. malariae*, but not *P. falciparum*, have been described. Electron microscopy reveals irregular glomerular capillary wall thickening due to an increase in basement membrane material, subendothelial electron-dense aggregates, and small intramembranous lacunae. Although these appearances were originally described as 'quartan malarial nephropathy' similar appearances have been described by Morel-Maroger *et al*. (1975) in nephrotic children from Senegal, and *P. malariae* could not be implicated. Furthermore, immunofluorescence studies demonstrating *P. malariae* antigens may be of doubtful significance as there is cross-reactivity of antibodies to different plasmodial strains and patients with chronic malaria produce anti-IgG rheumatoid factors and other antibodies, including autoantibodies directed against single-stranded DNA.

Other glomerular lesions have been described, including minimal change nephropathy, minor proliferative or focal lesions, and mesangiocapillary glomerulonephritis. A direct link with malarial infection should, however, be interpreted with caution in view of the endemic nature of malaria in tropical areas and the frequency of other infectious illnesses.

The management of patients is mainly symptomatic. Antimalarial treatment does not produce a remission of the nephrotic syndrome and steroids—with or without immunosuppressive drugs—are unlikely to be beneficial and may be associated with unacceptable complications due to malnutrition and increased risk of infection in affected children.

Plasmodium falciparum

P. falciparum infection may be associated with immune complex-mediated glomerulonephritis or with acute renal failure from massive haemolysis and haemoglobinuria (blackwater fever). Glomerulonephritis is most commonly mild and transient in falciparum malaria; although proteinuria is common during episodes of fever, subsequent remission is usual as the fever subsides. Acute glomerulonephritis with renal failure has been described (Hartenbower *et al*. 1975), as has nephrotic syndrome associated with proliferative glomerulonephritis (Berger *et al*. 1967). Light microscopy usually reveals mesangial proliferative glomerulonephritis with some thickening of the basement membrane. Immunofluorescence studies reveal IgM and complement deposition, while on electron microscopy there may be mesangial and subendothelial deposits. Resolution of the glomerular lesion usually occurs within 6 weeks of effective antimalarial treatment (Boonpucknavig and Sitprija 1979).

Schistosomiasis (see also Chapters 3.14 and 13.5)

Schistosomal infection is widespread throughout many areas in the tropics. A number of species have been described, but *S. haematobium* is most common in Africa, the Middle East and India, *S. mansoni* in the Nile Valley and Near East, while *S. japonicum* is found in the Far East. Although four species of schistosomes are known to affect man nephropathy has been associated with only three; *S. haematobium*, (Ezzat *et al*. 1974), *S. mansoni* (Andrade and Rochal 1979), and *S. japonicum* (Chandra Shekkar and Pathmanathan 1987).

Nephropathy usually complicates hepatosplenic schistosomiasis: hepatic fibrosis develops about 10 years after infection and nephropathy appears some 5 years later. The disease usually affects males in their second or third decade and is characterized by proteinuria, hypertension, and progressive renal failure. There are marked geographical differences in the prevalence of nephropathy, and schistosomes of the same species seem to differ with respect to infectivity and pathogenicity. The two most common lesions on renal biopsy are mesangiocapillary glomerulonephritis and focal segmental glomerulosclerosis. Immunofluorescence microscopy reveals IgG, IgM, and C3 deposition, while electron microscopy shows mesangial and subendothelial electron-dense deposits. Schistosomal gut antigens have been demonstrated in glomerular deposits.

Long-standing schistosomiasis may result in secondary amyloidosis with glomerular involvement.

Treatment of the underlying schistosomiasis does not appear to have a marked effect on the nephropathy and there is no evidence that corticosteroids or immunosuppression affect the progressive nature of the renal disease.

Toxoplasmosis

Toxoplasma gondii is an intracellular parasite which has a predilection for cells of the reticuloendothelial and nervous systems. Infection may be congenital or acquired, and in the latter case patients present with a clinical picture similar to that of infectious mononucleosis.

One child with congenital toxoplasmosis and glomerular disease presented with the nephrotic syndrome and a renal biopsy revealed mesangial proliferative glomerulonephritis in which there was also a segmental area of sclerosis and a capsular adhesion. Immunofluorescence revealed IgM and toxoplasma antigen. Following remission a further biopsy was performed, and this revealed a significant number of hyalinized glomeruli; although IgM was still present, toxoplasma antigen could not be identified.

Renal involvement is extremely rare in acquired toxoplasmosis. Ginsburg et al. (1974) examined 150 biopsy specimens and detected only one patient in whom there was a simultaneous increase in circulating antitoxoplasma antibodies and glomerular deposition of toxoplasma antigen.

Treatment with corticosteroids produced a gradual improvement in symptoms and subsequent resolution. A repeat renal biopsy revealed segmental sclerosis, but the toxoplasma antigen was no longer detectable.

Leishmaniasis (kala-azar)

Leishmaniasis is a tropical disease which occurs in two clinical forms, visceral and cutaneous. A number of patients with visceral leishmaniasis, caused by *Leishmania donovani*, have been reported to have associated nephropathy. De Brito et al. (1975) described increased mesangial matrix and cellularity and paramesangial electron-dense deposits in biopsy tissue from such patients, and mesangial IgG and C3 deposition have also been reported. The exact relationship between these findings and leishmaniasis is difficult to determine as the majority of patients have considerable malnutrition and intercurrent infections are common. Clinically, the nephropathy is manifest by microscopic haematuria and minor proteinuria.

Filariasis (see also Chapter 3.14)

A number of nematodes of the family Filariidae may infect man and there is a recognized association between loiasis (*loa loa* infection) and onchocerciasis (*Onchocerca volvulus* infection) and glomerulonephritis. Membranous glomerulonephritis and mesangiocapillary glomerulonephritis have been described in patients with loiasis (Ngu et al. 1985). Glomerular deposition of onchocercal antigen has also been described in some patients. In spite of this it is difficult to establish a causal relationship in view of the frequency of coexisting infections such as malaria.

Infections at particular sites

Infective endocarditis

Renal involvement in bacterial endocarditis is the result of immune complex-mediated damage to glomeruli and small blood vessels. Tubulointerstitial changes are also common and may arise as a direct consequence of the infective endocarditis or through a reaction to antibiotic therapy. The association between endocarditis and renal disease was reported early this century (Lohlein 1910) and originally thought to be embolic in nature; the descriptive term of 'flea-bitten' kidney was applied. The embolic nature of the glomerular lesion however, was questioned by Bell (1932), who suggested an immunological mechanism. This was further supported by the work of Barker (1949) in a study of endocarditis confined to the right side of the heart. However, the immune nature of the condition was not confirmed until the finding of hypocomplementaemia (Williams and Kunkel 1962), the demonstration of glomerular immune complex deposition (Gutman et al. 1972), the identification of circulating

Table 2 Organisms reported to be involved in endocarditis and associated glomerulonephritis

β-Haemolytic streptococci	Gutman et al. 1972
α-Haemolytic streptococci	Perez et al. 1976
Strep. viridans	Boulton-Jones et al. 1974
Strep. mitis	Keslin et al. 1973
Strep. mutans	Allen et al. 1982
Staph. aureus	Tu et al. 1969; Gutman et al. 1972*
Staph. epidermidis	Boulton-Jones et al. 1974
Enterococcus	Levy and Hong 1973
Gonococcus	Ebright and Komorowski 1980
Pseudomonas aeruginosa	Beaufils et al. 1978
Coxiella burnetti	Hall et al. 1975
Chlamydia psittaci	Boulton-Jones et al. 1974
Fungi	Roberts and Rabson 1962

* Plus many more subsequent reports.

immune complexes in 97 per cent of patients (Bayer et al. 1976), and the elution from the kidney of antibody which specifically reacted with bacteria isolated from the blood (Levy and Hong 1973).

Clinically, endocarditis results from infection of valves which are congenitally abnormal or have been previously damaged by rheumatic heart disease or hypertension. Normal valves may also be infected, particularly following septicaemia with virulent bacteria such as *Staphylococcus aureus*, group A β-haemolytic streptococci, or *Streptococcus pneumoniae*. Such infections may arise as a complication of central intravenous catheters or through the use of non-sterile needles in intravenous drug addicts.

Clinically, there is a fulminant illness leading to rapid destruction of the valves, associated with myocarditis and evidence of infection elsewhere, such as phlebitis, meningitis, and pneumonia. Less virulent organisms such as *Streptococcus viridans* produce a subacute form of endocarditis which has an insidious clinical onset and is more common in patients with previously damaged valves. Many different organisms have been isolated in patients with endocarditis and associated glomerulonephritis (Table 2).

The frequency with which renal disease occurs in endocarditis is not clear. O'Connor et al. (1978) showed abnormal urine analysis in 13 of 24 patients with infective endocarditis; a more recent postmortem study (Neugarten et al. 1984) detected glomerulonephritis in 22 per cent of patients, most of whom had right-sided endocarditis complicating intravenous drug abuse.

The clinical presentation depends on the virulence of the infecting organism and the degree of cardiac involvement. Commonly, patients present with fever, malaise, arthralgia, and intermittent arthritis. The diagnosis can be made on clinical grounds by the finding of a changing cardiac murmur, splenomegaly, 'splinter' haemorrhages in the nail beds, and microscopic haematuria. Retinal and subconjunctival haemorrhages may occur, but are uncommon. There is also a normochromic, normocytic anaemia associated with significant leucocytosis. Non-specific evidence of inflammation is provided by the finding of an increased ESR and increased C-reactive protein. On urine microscopy, haematuria is universal and red blood cell casts are seen, with varying amounts of proteinuria. Blood culture is positive in approximately 70 per cent of patients.

Rheumatoid factor, cryoglobulinaemia, and circulating immune complexes are variably present. During the acute phase of the illness, C3 and C4 are decreased, which is an indicator of (but not diagnostic of) renal involvement. The blood urea and serum creatinine values depend on the extent of the renal failure, which is frequently mild, but occasionally there is acute renal failure with a rapid deterioration in renal function.

The introduction of effective antibiotic therapy has produced a significant change in the pattern of the disease. *Staphylococcus aureus* appears to be increasing in incidence, particularly in drug abusers, whereas *Streptococcus viridans* is becoming less common. The pathogenesis involves immune complex formation with complement activation, apparently by the classical pathway although there are reports of alternate pathway activation by *Staph. aureus* (O'Connor et al. 1978). The immune complexes arise as a result of prolonged antigenaemia and are frequently present in high titres in the circulation. The role of immune complexes has been confirmed by the detection of antibodies to infecting organisms in glomeruli (Perez et al. 1976). Further confirmation comes from the demonstration of staphylococcal antigen (Pertschuk et al. 1976; Yum et al. 1978) and streptococcal antigen (Perez et al. 1976) in glomeruli.

Renal biopsy reveals two distinct patterns of glomerular damage. The findings in acute fulminating infections are similar to those in acute poststreptococcal glomerulonephritis, with an acute diffuse exudative proliferation with or without crescents (Neugarten et al. 1984). In such patients immunofluorescence demonstrates IgG, IgM, and C3, deposition, and dense subendothelial and subepithelial deposits may be present on ultrastructural examination. In the subacute form, focal segmental proliferative glomerulonephritis is more common; those with a prolonged history may have evidence of sclerosis. Although the changes seen on light microscopy are focal in nature, immunofluorescence shows the deposits to be more diffuse and predominantly mesangial. As with many other forms of infection-associated glomerulonephritis, a wide variety of glomerular changes have been noted, and there are reports of type 1 mesangiocapillary glomerulonephritis and crescentic glomerulonephritis. Tubulointerstitial changes are common, consisting of a mononuclear cell infiltrate, tubular atrophy and fibrosis (Morel-Maroger et al. 1972). The interstitial changes may be aggravated by antibiotic therapy prescribed for the cardiac lesion.

The course and prognosis are very variable. No specific treatment is required for the renal lesion, which commonly improves following effective treatment of the endocarditis. In patients with significantly impaired renal function, recovery depends more on the outcome of the endocarditis than on the type of renal disease. Antibiotic therapy is essential and must be given in a sufficient dose and for a sufficient length of time to eradicate the endocarditis completely. Although it has been advocated that antibiotic therapy should be given intravenously for 6 weeks, this is now disputed, and currently there is no consensus regarding the length of intravenous therapy or of overall treatment.

Shunt nephritis

Surgical treatment for hydrocephalus consists of the insertion of a catheter into the distended ventricle and drainage to the atrium or the peritoneal cavity, either via the jugular vein or superior vena cava. A common complication of this procedure is colonization of the catheter by bacteria of low virulence. About 30 per cent of juguloatrial catheters become infected and 70 per cent of these infections occur within 2 months of insertion. Although

Table 3 Micro-organisms responsible for shunt nephritis

Staphylococcus epidermidis (*albus*) (Lam et al. 1969)
Staphylococcus aureus (Stickler et al. 1968)
Corynebacterium bovis (Bolton et al. 1975)
Bacillus subtilis (Schoenbaum et al. 1975)
Bacillus cereus (Campbell et al. 1981)
Listeria monocytogenes (Strife et al. 1976)
Proprionibacterium acnes (Beeler et al. 1976)
Pseudomonus aeruginosa (Zunin et al. 1977)
Serratia (Levy et al. 1981)
Peptococcus (Caron et al. 1979)
Diphtheroids (O'Regan and Makker 1979)

infection seems to be relatively common the subsequent development of glomerulonephritis may occur in less than 5 per cent of infected patients. A variety of shunts is available, but it appears that superadded infection can complicate all types. The first report of shunt nephritis (Black et al. 1965) described two patients who developed haematuria and a nephrotic syndrome after prolonged bacteraemia following the insertion of Spitz-Holter valves. Since then there have been many further reports confirming the association between ventriculoatrial shunts and the development of glomerulonephritis. A few reports indicate that ventriculoperitoneal shunts can also be associated with this complication, but this occurs much less frequently. The causal organism is *Staphylococcus epidermidis* (albus) in approximately 70 per cent of patients and *Staphylococcus aureus* in another 20 per cent. A wide variety of organisms has been reported from the remaining 10 per cent (see Arze et al. 1983 for a review), in some instances as single case reports (Table 3).

Colonization of the shunt probably occurs at insertion or shortly thereafter. It appears that some subtypes of *Staph. epidermidis* produce an excess of a mucoid substance ('slime') which promotes adherence of the bacteria to smooth surfaces and may provide a protective layer against the action of lysozymes (Schena 1985). The glomerular lesion arises as a consequence of prolonged low grade bacteraemia leading to antibody production and the formation of immune complexes. Such complexes within the glomerulus activate complement by the classical pathway and lead to glomerular inflammation. A role for the bacterial antigen has been confirmed by the detection of specific bacterial antigens, including *Staph. epidermidis*, (Kaufman and McIntosh 1971; Dobrin et al. 1975) *P. acnes* (Groenveld et al. 1982), diphtheroids (O'Regan and Makker 1979), *C. bovis*, (Bolton et al. 1975) within glomerular deposits; and also within circulating cryoprecipitates in some cases.

Evidence of infection always precedes the renal manifestations. Fever is universal and may be low grade or spiking. It is associated with symptoms such as arthralgia, purpura, rash, anorexia, malaise, and occasionally signs of increasing intracranial pressure. Very occasionally a patient may be asymptomatic, but have a profound anaemia. Anaemia and hepatosplenomegaly are common; haematuria is usually microscopic, but may occasionally be macroscopic. Proteinuria is variable, but is sufficient to result in a nephrotic syndrome in approximately 25 per cent of patients. There is evidence of infection with a leucocytosis associated with increased ESR, plasma viscosity, and C-reactive protein. Complement studies may indicate a reduced C3, C4, and C1q. Some patients produce rheumatoid factor, and mixed cryoglobulins may be detected. Blood cultures are positive and whilst in normal clinical practice, *Staph. epidermidis* would be considered as a contaminant, it must always be taken as clinically significant in a patient with a ventriculoatrial shunt.

The most common appearance on renal biopsy is that of type 1 mesangiocapillary glomerulonephritis, in some cases with occasional crescent formation. Subendothelial electron-dense deposits are common and immunofluorescence microscopy usually reveals IgM and C3. Some patients have a diffuse exudative proliferative glomerulonephritis, while others may have a minor mesangial proliferation or segmental lesions.

Antibiotic therapy alone is usually inadequate and patients require shunt removal. Repeat renal biopsies have been performed in a number of successfully treated patients, and show a reduction in glomerular cellularity with disappearance of the electron-dense deposits (Zunin *et al.* 1977; Levy *et al.* 1981). The prognosis is variable and whilst adequate antibiotic therapy and shunt removal will allow urinary abnormalities to resolve, these will persist in a few patients (Dobrin *et al.* 1975; Levy *et al.* 1981). There was progression to end-stage renal failure in one reported case. Repeat renal biopsies have only been reported in few instances (see Wakabayashi *et al.* 1985 for a review); these usually show residual mild mesangial proliferative glomerulonephritis with resolution of deposits previously identified by immunofluorescence and electron microscopy.

There are two reports of nephritis complicating infected ventriculoperitoneal shunts. Noe and Roy (1981) described a child who presented with microscopic haematuria, nephrotic syndrome, and deteriorating renal function; these resolved following removal of the shunt. Unfortunately, no specific organisms were isolated in this case. Partiarca and Lauer (1980) described a child with fever in whom *Haemophilus influenzae* was isolated; this organism was considered to be responsible for the subsequent development of proteinuria.

Visceral abcesses

Glomerulonephritis has been associated with visceral abcesses in a variety of clinical situations. Multiple cutaneous abcesses (Pertshuk *et al.* 1976), severe pulmonary infection (Salyer and Salyer 1974; Danovitch *et al.* 1979), chronic granulomatous disease (Van Rhenen *et al.* 1979; Cottin *et al.* 1982), infected arterial prostheses (Beaufils 1981; Coleman *et al.* 1983), and a variety of deep seated abscesses (Beaufils *et al.* 1976) have all been described. A common feature is that of a severe infection being present for many months, unlike the situation in amyloidosis, where the infection has been present for many years.

Clinical presentation is extremely variable and can range from mild urinary abnormalities (Salyer and Salyer 1974) to severe fulminating renal failure (Nydahl and Hall 1965; Beaufils *et al.* 1976). Impaired renal function is common (Pertshuk *et al.* 1976; Spector *et al.* 1980), but interestingly, the nephrotic syndrome has been rarely described (Danovitch *et al.* 1979).

In visceral abscesses the most common infecting organism is *Staphylococcus aureus* (Pertshuk *et al.* 1976; Beaufils *et al.* 1976), although there are instances of infections due to *Pseudomonas aeruginosa, Escherichia coli, Proteus mirabilis*, and *Candida albicans*.

As with many other forms of infection-related glomerular disease, a wide variety of histological appearances has been described. Proliferative glomerulonephritis (Nydahl and Hall 1965; Danovitch *et al.* 1979) or mesangial proliferative glomerulonephritis (Beaufils *et al.* 1976; Spector *et al.* 1980) are most common. Others have reported mesangiocapillary glomerulonephritis (Beaufils 1981), membranous glomerulonephritis (Cottin *et al.* 1982), a focal proliferative glomerulonephritis (Becaufils *et al.* 1976), and glomerulosclerosis (Van Rhenen *et al.* 1979). Thus no single histological entity is associated with visceral abscesses, although in the majority there are marked proliferative changes.

The management of such patients is directed predominantly at the underlying infection, and culture of the infecting organism is particularly important. The majority of patients have some impairment of renal function and antibiotic therapy may need to be adjusted appropriately. Management of the renal failure is along conventional lines, although it is frequently complicated by the fact that many patients are malnourished because of their long-standing sepsis. The overall outcome depends on the severity of the underlying infection and whether it can be eradicated or controlled. Recovery with effective treatment occurs in a number of patients (Beaufils *et al.* 1976). Patients presenting with acute renal failure clearly have a poor prognosis, although recovery has been described. A small number of patients may progress to a chronic impairment of renal function.

General conclusions

Infection-associated glomerulonephritis is well established and has been recognized for almost two centuries. In that time there has been many significant advances including the identification of micro-organisms, the development of effective specific therapy in the form of antibiotics, antiviral drugs, and antifungal agents, the complete eradication of certain infections such as smallpox, and eradication of other infections from certain geographical regions, e.g. malaria from Guyana. There have also been very significant changes in the socioeconomic status of many communities, which have had an influence on the development of infection-associated nephropathy. There is no doubt that changes will continue to take place and it is likely that bacterial and parasitic infections will be less prominent, while viral infections are likely to be of more importance.

In the general management of a patient with glomerulonephritis careful consideration should be given to the possibility of an associated underlying infection. In many, but not all, instances this will be obvious at clinical presentation. A search into potential infective causes will depend to a large extent on local factors, particularly where certain infections are endemic. A causal link can only be established by identification of antigen and elution of specific antibody and so, in the majority of instances, the link will be inferred from the satisfactory resolution of the renal abnormalities after successful eradication of the infection. Although this is the usual course, malaria and schistosomiasis are striking examples of progression of the renal disease in spite of effective treatment of the infection.

References

Adu, D., *et al.* (1981). The nephrotic syndrome in Ghana: clinical and pathological aspects. *Quarterly Journal of Medicine*, **50**, 297–306

Allen, A.C. and Spitz, S. (1945). A comparative study of the pathology of scrub typhus (tsutsugamushi disease) and other rickettsial diseases. *American Journal of Pathology*, **21**, 603–82

Allen, D.B., Friedman, A.L., Croker, B., and Osofsky, S.G. (1982). Glomerulonephritis associated with infective endocarditis in a pediatric patient. *North Carolina Medical Journal*, **43**, 113–7.

Amerio, A., Campese, V., Coratelli, P., and Schena F.P. (1972). Glomerulonephritis in typhoid fever. 5th International Congress of Nephrology. Abstract book p. 62.

Andrade, Z.A. and Rocha H. (1979). Schistosomal glomerulopathy. *Kidney International*, **16**, 23–9.

Arze, R.S., Rashid, H., Morley, R., Ward, M.R., and Kerr D.N.S. (1983). Shunt-nephritis. Report of two cases and review of the literature. *Clinical Nephrology*, **19**, 48–53.

Baldwin, D.S. (1977). Post-streptococcal glomerulonephritis, a progressive disease? *American Journal of Medicine*, **62**, 1–11.

Barker, P.S. (1949). A clinical study of subacute bacterial infection confined to the right side of the heart. *American Heart Journal*, **37**, 1054–60.

Barsoum, R.S., et al. (1977). Renal disease in hepatosplenic schistosomiasis a clinicopathological study. *Transactions of the Royal Society for Tropical Medicine and Hygiene*, **71**, 287–91.

Bayer, A.S., Theofilopoulos A.N., Eisenberg R., Dixon F.J., and Guze L.B. (1976). Circulating immune complexes in infective endocarditis. *New England Journal of Medicine*, **295**, 1505–11.

Beaufils, M., Morel-Maroger, L., Sraer, J.D., Kanfer, A., Kourilsky, O., and Richet, G. (1976). Acute renal failure of glomerular origin during visceral abscesses. *New England Journal of Medicine*, **295**, 185–9.

Beaufils, M., et al. (1978). Glomerulonephritis in severe bacterial infections with and without endocarditis. *Advances in Nephrology*, **7**, 217–34.

Beaufils, M. (1981). Glomerular disease complicating abdominal sepsis. *Kidney International*, **19**, 609–18.

Beeler, B.A., Crowder, J.G., Smith, J.W., and White, A. (1976). Propionbacterium acnes: Pathogen in central nervous system shunt infection. Report of 3 cases including immune complex glomerulonephritis. *American Journal of Medicine*, **61**, 935–8.

Bell, E.T. (1932). Glomerular lesions associated with endocarditis. *American Journal of Pathology*, **8**, 639–64.

Berger, M., Birch L.M., and Conte, N.F. (1967). The nephrotic syndrome secondary to acute glomerulonephritis during falciparum malaria. *Annals of International Medicine*, **67**, 1163–71.

Black, J.A., Challacombe, D.N., and Ockenden, B.G. (1965). Nephrotic syndrome associated with bacteraemia after shunt operations for hydrocephalus. *Lancet*, **ii**, 921–4.

Blackhall, J. (1818). *Observations on the nature and cure of dropsies, and particularly on the presence of the coagulable part of the blood in dropsied urine*. London, Longman Green.

Bolton, W.K., et al. (1975). Ventriculojugular shunt nephritis with *Corynebacterium bovis*. *American Journal of Medicine*, **59**, 417–23.

Boonpucknavig, V. and Sitprija, V. (1979). Renal disease in acute *Plasmodium falciparum* infection in man. *Kidney International*, **16**, 44–52.

Boonshaft, B., Maher, J.F., and Schreiner, G.E. (1970). Nephrotic syndrome associated with osteomyelitis without secondary amyloidosis. *Archives of Internal Medicine*, **125**, 322–7.

Boulton-Jones, J.M., Sissons, J.G.P., Evans, D.J., and Peters, D.K. (1974). Renal lesions of subsequent infective endocarditis. *British Medical Journal*, **2**, 11–14.

Boulton-Jones, J.M. and Davison, A.M. (1986). Persistent systemic infection as a cause of renal disease in patients submitted to renal biopsy: a report from the Glomerulonephritis Registry of the United Kingdom MRC. *Quarterly Journal of Medicine*, **58**, 123–32.

Boyce, N.W., Hayes, K., Gee, D., Holdsworth, S.R., Thomson, N.M., Scott, D., and Atkins, R.C. (1988). Cytomegalovirus infection complicating renal transplantation and its relationships to acute transplant glomerulopathy. *Transplantation*, **45**, 706–9.

Brunkhorst, R., Brunkhorst, U., Eisenbach, G., Schedel, I., and Koch, K. (1989). Lack of clinical evidence for a glomerulopathy in 203 patients with HIV infection. *Nephrology Dialysis Transplantation*, **4**, 433.

Buka, I. and Coovadia, H.M. (1980). Typhoid glomerulonephritis. *Archives Diseases of Childhood*, **55**, 305–7.

Bullock, W.E., Artz, R.P., Bhathena, D., and Tung, K.S.K. (1979). Histoplasmosis. Association with circulating immune complexes, eosinophilia and mesangiopathic glomerulonephritis. *Archives of Internal Medicine*, **139**, 700–2.

Campbell, A., Futrakul, P., Musgrave, J.E., Surapathana, L.O., and Eisenberg, C.S. (1981). *Bacillus cereus* prosthesis nephritis. Second International Symposium of Pediatric Nephrology, Paris.

Caron, C. Luneau, C., Gervais, M.H., Plante, G.E., Sanchez G., and Blain, G. (1979). La glomerulonéphrite de shunt: manifestations cliniques et histopathologiques. *Canadian Medical Association Journal*, **120**, 557–61.

Chander, P., Soni A., Suri A., Bhagwat R., Yoo J., and Treser, G. (1987). Renal ultrastructural markers in AIDS-associated nephropathy. *American Journal of Pathology*, **126**, 513–26.

Chandra Shekkar, K. and Pathmanathan, R. (1987). Schistosomiasis in Malaysia. *Reviews of Infectious Diseases*, **9**, 1026–37.

Chesney, R.W., O'Regan, S., Guyda, H.J., and Drummond, K.N. (1976). *Candida* endocrinopathy syndrome with membranoproliferative glomerulonephritis: demonstration of glomerular *Candida* antigen. *Clinical Nephrology*, **5**, 232–8.

Clark, G., et al. (1988). Post-streptococcal glomerulonephritis in children: clinicopathological correlations and long-term prognosis. *Pediatric Nephrology*, **2**, 381–8.

Cochat, P., Colon S., Bosshard S., Zech P., and Traeger J. (1985). Glomérulonéphrite membranoproliferative ét infection à *Mycoplasma pneumoniae*. *Archives Françaises de Pédiatrie*, **42**, 29–31.

Cohen, A.M., Sun, N.C., Shapshak, P., and Imagwa, D.T. (1989). Demonstration of human immunodeficiency virus in renal epithelium in HIV-associated nephropathy. *Modern Pathology*, **2**, 125–8.

Coleman, M., Burnett, J., Barrat, L.J., and Dupont, P. (1983). Glomerulonephritis associated with chronic bacterial infection of a Dacron arterial prosthesis. *Clinical Nephrology*, **20**, 315–20.

Collins B., et al. (1983). Hepatitis B immune complex glomerulonephritis: simultaneous glomerular deposition of hepatitis B surface and e antigens. *Clinical Immunology and Immunopathology*, **26**, 137–53.

Combes B., et al. (1971). Glomerulonephritis with deposition of Australia antigen–antibody complexes in glomerular basement membrane. *Lancet*, **ii**, 234–7.

Connor, E., et al. (1988). Acquired immunodeficiency syndrome associated renal disease in children. *Journal of Pediatrics*, **113**, 39–44.

Cottin, X., Chopard, P., Cotton, J.B., and Larbre, F. (1982). Glomerulonéphrite extra-membraneuse, au cours d'une granulomatose septique chronique. *Pédiatrie*, **37**, 299–304.

Cusack, D., Martin, P., Schinittger, T., McCafferty M., Keane C., and Keogh B. (1983). IgA nephropathy in association with *Yersinia enterocolitica*. *Irish Journal of Medical Science*, **152**, 311–2.

D'Agati, V., Suh, J.-I., Carbone, L., Cheng, J.-T., and Appel, G. (1989). Pathology of HIV-associated nephropathy: a detailed morphologic and comparative study. *Kidney International*, **35**, 1358–70.

Danovitch, G.M., Nord, E.P., Barki, Y., and Krugliak, L. (1979). Staphylococcal lung abscess and acute glomerulonephritis. *Israel Journal of Medical Science*, **10**, 840–3.

Davison, A.M., Thomson, D., and Robson J.S. (1973). Intravascular coagulation complicating influenza A virus infection. *British Medical Journal*, **1**, 654–5.

De Brito, T., Tiriba, A., and Godoy, C.V.F. (1968). Glomerular response in human and experimental rickettsial disease (Rocky Mountain Spotted Fever Group). A light and electron microscopy study. *Pathologia et Microbiologia*, **31**, 365–77.

De Brito, T., Hoshino-Schimaza, S., Amato Neto, V., Duarte I.S., and Penna, D.O. (1975). Glomerular involvement in human Kala-azar. *American Journal of Tropical Medicine and Hygiene*, **24**, 9–18.

De Man, R.A., Schalm, S.W., van der Heijden A.J., ten Kate, F.W.J., Wolff, E.D., and Heijtink, R.A. (1989). Improvement of hepatitis B—associated glomerulonephritis after antiviral combination therapy. *Journal of Hepatology*, **8**, 367–72.

De Meyer, M., Cosyns J.P., and Van Ypersele De Strihou, C. (1989). Does HIV-related nephropathy exist in AIDS patients? *Nephrology Dialysis and Transplantation*, **4**, 434.

Denneberg, T., Friedberg, M., Samuelsson, T., and Winblad, S. (1981). Glomerulonephritis in infections with *Yersinia enterocolitica* O-serotype 3. *Acta Medica Scandanavia*, **209**, 97–101.

Dobrin, R.S., et al. (1975). The role of complement, immunoglobulin and bacterial antigen in coagulase—negative staphylococcal shunt nephritis. *American Journal of Medicine*, **59**, 660–73.

Ebright, J.R. and Komorowski, R. (1980). Gonococcal endocarditis associated with immune complex glomerulonephritis. *American Journal of Medicine*, **68**, 793–6.

Ezzat, E., Osman, R., Ahmed, K.Y., and Soothill, J.F. (1974). The association between Schistosoma haematobium infection and heavy proteinuria. *Transactions of the Royal Society of Tropical Medicine and Hygiene*, **68**, 315–7.

Fauci, A.S., Katz, P., Haynes, B.F., and Wolff, S.M. (1979). Cyclophosphamide therapy of severe systemic necrotising vasculitis. *New England Journal of Medicine*, **301**, 235–9.

Friedberg, M., Denneberg, T., Brun, C., Hannover Larsen, J., and Larsen, S. (1981). Glomerulonephritis in infection with *Yersinia enterocolitica* O-serotype 3. *Acta Medica Scandanavia*, **209**, 103–10.

Gamble, C. and Reardan, J.S. (1975). Immunopathogenesis of syphilitic glomerulonephritis. *New England Journal of Medicine*, **292**, 449–54.

Garcia, G., et al. (1985). Preliminary observation of hepatitis B associated membranous nephropathy treated by leucocyte interferon. *Hepatology*, **5**, 317–20.

Genderini A., et al. (1989). Absence of viral antigens in renal tissue in patients with HIV nephropathy. *Nephrology Dialysis and Transplantation*, **4**, 426.

Giglioli, G. (1962). Malaria and renal disease with special reference to British Guiana. *Annals of Tropical Medicine and Parasitology*, **56**, 225–41.

Gilles, H.M. and Hendrickse, R.G. (1963). Nephrosis in Nigerian children. Role of *Plasmodium malariae* and effect of antimalarial treatment. *British Medical Journal*, **2**, 27–31.

Ginsburg, B.E., Wasserman, J., Huldt, G., and Bergstrand, A., (1974). Case of glomerulonephritis associated with acute toxoplasmosis. *British Medical Journal*, **3**, 664–5.

Glassock, R.J., Cohen, A.H., Danovitch, G. and Parsa, K.P. and (1990). Human immunodeficiency virus (HIV) infection and the kidney. *Annals of Internal Medicine*, **112**, 35–49.

Gocke, D.J., Hsu, K., Morgan, C., Bombardieri, S., Lockshin, M., and Christian, C.L. (1970). Association between polyarteritis and Australia antigen *Lancet*, **ii**, 1149–53.

Goodpasture, E.W. (1919). The significance of certain pulmonary lesions in relation to the etiology of influenza. *American Journal of Medical Science*, **158**, 863–70.

Gorevic, P.D., et al. (1980). Mixed cryoglobinemia: clinical aspects and long term follow up to 40 patients. *American Journal of Medicine*, **69**, 287–308.

Grishman, E., Cohen, S., Salomon, M.D., and Churg, H. (1967). Renal lesions in acute rheumatic fever. *American Journal of Pathology*, **51**, 1045–61.

Groeneveld, A.B.J., Nommensen, F.E., Mullink, M., Ooms, E.C.M., and Bode, W.A. (1982). Shunt nephritis associated with *Propionibacterium acnes* with demonstration of the antigen in the glomeruli. *Nephron*, **32**, 365–9.

Grundy, J.E., et al. (1988). Symptomatic cytomegalovirus infections in seropositive kidney recipients: re-infection with donor virus rather than re-activation of recipient virus. *Lancet*, **ii**, 132–5.

Gutman, R.A., Striker, G.E., Gilliland, B.C., and Cutler, R.E. (1972). The immune complex glomerulonephritis of bacterial endocarditis. *Medicine*, **57**, 1–25.

Hall, G.H., Hart, R.J.C., Davies, S.W., George, M., and Head A.C. (1975). Glomerulonephritis associated with *Coxiella burnetti* endocarditis. *British Medical Journal*, **ii**, 275.

Hariparsad, D., Ramsaroop, R. Seedat, Y.K., and Patel, P.L. (1985). Mesangial proliferative glomerulonephritis with Legionnaire's disease. A case report. *South African Medical Journal*, **67**, 649–650.

Hariprasad, M.K., Dodelson, R., Eisinger, R.P., and Gary, N.E. (1979). Dense deposit disease in tuberculosis. *New York State Journal of Medicine*, **79**, 2084–5.

Hartenbower, D.L., Kantor, G.L., and Rosen, V.J. (1972). Renal failure due to acute glomerulonephritis during falciparum malaria: case report. *Military Medicine*, **137**, 74–6.

Hassan, A.A. (1982). Schistosomal nephropathy and HLA association. Thesis, Cairo University.

Herrera, G.A., et al. (1986). Cytomegalovirus glomerulopathy: a controversial lesion. *Kidney International*, **29**, 725–33.

Hsu, H.-C., et al. (1989). Membranous nephropathy in 52 hepatitis B surface antigen (HBsAg) carrier children in Taiwan. *Kidney International*, **36**, 1103–7.

Huang, T.W. and Wiegenstein, L.M. (1977). Mesangiolytic glomerulonephritis in an infant with immune deficiency and echovirus infection. *Archives of Pathology*, **101**, 125–8.

Hughes, W.T., Steigman, A.J., and Delong, H.F. (1966). Some implications of fatal nephritis associated with mumps, *American Journal of Disease in Children*, **111**, 297–301.

Hyman, L.R., Jenis, E.H., Hill, G.S., Zimmerman, S.W., and Burkholder, P.M. (1975). Alternate C3 pathway activation in pneumococcal glomerulonephritis. *American Journal of Medicine*, **58**, 810–4.

Indrapasit, S., Boonpucknavig, V., and Boonpucknavig, S. (1985). IgA nephropathy associated with enteric fever. *Nephron*, **40**, 219–22.

Ito, H., et al. (1981). Hepatitis Be antigen-mediated membranous glomerulonephritis. *Laboratory Investigation*, **44**, 214–20.

Jacquot, C. and Bariéty, J. (1980). Syndromes glomérulaires aïgues révélateurs d'une infection bactérienne grave. *Nouvelle Press Médicale*, **9**, 1629–31.

Johny, J.V., Karat, A.B.A., Rao, P.S.S., and Date, A. (1975). Glomerulonephritis in leprosy—a percutaneous biopsy study. *Leprosy Review*, **46**, 29–37.

Kaehny, W.D., Ozawa, T., Schwartz, M.I., Standford, E.E., Kohler, P.F., and McIntosh, R.M. (1978). Acute nephritis and pulmonary alveolitis following pneumococcal pneumonia. *Archives of Internal Medicine*, **138**, 806–8.

Kark, R.M. and Muehrcke, R.C., (1954). Biopsy of the kidney in the prone position. *Lancet*, **i**, 1047–49.

Kaufman, M.B. and McIntosh, R. (1971). The pathogenesis of the renal lesion in a patient with steptococcal disease, infected ventriculoatrial shunt, cryoglobulinaemia and nephritis. *American Journal of Medicine*, **50**, 262–8.

Keslin, M.H., Messner, R.P., and Williams, R.C. (1973). Glomerulonephritis with subacute bacterial endocarditis. *Archives of Internal Medicine*, **132**, 578–81.

Kleinknecht, C., Levy, M., Peix, A., Broyer, M., and Courtecuisse, V. (1979). Membranous glomerulonephritis and hepatitis surface antigen in children. *Journal of Pediatrics*, **95**, 946–52.

Kohler, P.F., Cronin, R.E., Hammond, W.S., Olin, D., and Carr, R.I. (1974). Chronic membranous glomerulonephritis caused by hepatitis B antigen-antibody immune complexes. *Annals of Internal Medicine*, **81**, 448–51.

Kujala, G.A., Doshi, H., and Brick, J.E. (1989). Rheumatic fever and poststeptococcal glomerulonephritis: a case report. *Arthritis and Rheumatism*, **32**, 236–9.

Lai, K.N., Arrons, I., Woodroffe, A.J., and Clarkson, A.R. (1982). Renal lesions in leptospirosis. *Australian and New Zealand Journal of Medicine*, **12**, 276–9.

Lai, K.N., Lai, F.M-M., Lo, S., Ho, C.P., and Chan, K.W. (1987). IgA nephropathy associated with hepatitis B virus antigenaemia. *Nephron*, **47**, 141–3.

Lai, K.N., Lai, F.M-M., Tam, J.S., and Vallance-Owen, J. (1988). Strong association between IgA nephropathy and hepatitis B surface antigenemia in endemic area. *Clinical Nephrology*, **29**, 229–34.

Lai, K.N., Lai, F.M-M., and Tam, J.S. (1989). IgA nephropathy associated with chronic hepatitis B infection in adults: the pathogenic role of HBsAg. *Journal of Pathology*, **157**, 321–7.

Lam, C.N., McNeish, A.S., and Gibson, A.M.M. (1969). Nephrotic syndrome associated with complement deficiency and staphylococcal albus bacteraemia. *Scottish Medical Journal*, **14**, 86–8.

La Manna, A., Polito, C., Delgado, R., Olivieri, P.N., and Di Toro, R. (1985). Hepatitis B surface antigenaemia and glomerulonephritis in children. *Acta Paediatrica Scandanavia*, **74**, 122–5.

Lange, K., Seligson, G., and Cronin, W. (1983). Evidence for the *in-situ* origin of post-streptococcal glomerulonephritis. Glomerular localisation of endostreptosin and clinical significance of subsequent antibody responses. *Clinical Nephrology*, **19**, 3–10.

Lee, S. and Kjellstrand, C.M. (1978). Renal disease in infectious mononucleosis. *Clinical Nephrology*, **9**, 236–40.

Levy M. (1988). Infection related proteinuria syndromes. In: Cameron J.S., and Glassock, R.J. (eds), The Nephrotic Syndrome. New York: *Marcel Dekker*, pp. 745–804.

Levy, M., Gubler, M.C., and Habib, R. (1981). Pathology and immunopathology of shunt nephritis in children: report of 10 cases. Proceedings 8th International Congress of Nephrology, Athens. Zurukyoglu, W., Papadimitriou, M., Pyrpasopoules, M., Sion, M., and Zamboulis, C. (eds) Basel, Karger, pp. 290–96.

Levy, M., *et al.* (1985). Berger's disease in children. *Medicine*, **64**, 157–80.

Levy, R.L. and Hong, R. (1973). The immune nature of subacute bacterial endocarditis. *American Journal of Medicine*, **54**, 645–52.

Lewy, J.E., Salinas-Madrigal, L., Herdson, P.B., Pirani, C.L., and Metcoff, J. (1971). Clinico-pathologic correlations in acute poststreptococcal glomerulonephritis. *Medicine*, **50**, 453–501.

Lien, J.W.K., Mathew, T.H., and Meadows, R. (1979). Acute poststreptococcal glomerulonephritis in adults: a long-term study. *Quarterly Journal of Medicine*, **48**, 99–111.

Lin, C.-Y. and Hsu, H.-C. (1983). Measles and acute glomerulonephritis. *Pediatrics*, **71**, 398–401.

Löhlein, M. (1910). Ueber hammorrhagische Nierenaffektionen bei chronischen ulzerözer Endocarditis. *Medinizische Klinik*, **10**, 375–9.

Losito, A., Bucciarelli, E., Massi-Benedetti, F., and Lato, M. (1979). Membranous glomerulonephritis in congenital syphilis. *Clinical Nephrology*, **12**, 32–7.

Mactier, R.A. and Dobbie, J.W. (1984). Acute infectious disease presenting as acute renal failure. *Scottish Medical Journal*, **28**, 96–100.

Maggiore, Q., Bartolomeo, F., L'Abbate, A., and Misefari, V. (1981). HBsAg glomerular deposits in glomerulonephritis: fact or artifact. *Kidney International*, **19**, 579–86.

McCluskey, R.T. and Baldwin D.S. (1965). The natural history of acute glomerulonephritis. *American Journal of Medicine*, **35**, 213–30.

McFie, J.W.S. and Ingram, A. (1917). Observation on malaria in the Gold Coast Colony, West Africa. *Annals of Tropical Medicine and Parasitology*, **11**, 1–26.

McIntosh, R.M., Garcia, R., Rubio, L., Rabideau, D., and Rodriguez-Iturbe, B. (1978). Evidence of an autologous immune complex pathogenic mechanism in acute post streptococcal glomerulonephritis. *Kidney International*, **14**, 501–10.

Mettler, N.E. (1969). Isolation of a microtatobiote from patients with hemolytic-uremic syndrome and thrombotic thrombocytoponeic purpura and from mites in the United States. *New England Journal of Medicine*, **281**, 1023–7.

Michalak, T. (1978). Immune complexes of hepatitis B surface antigen in the pathogenesis of polyarteritis nodosa. A study of seven necropsy cases. *American Journal of Pathology*, **90**, 619–32.

Miller, J. (1849). Observations on scarlatinal albuminuria and a short notice on the sequela proper to that infection. *Lancet*, **i**, 127–8.

Morel-Maroger, L., Sraer, J.D., Herrman, G., and Godeau, P. (1972). Kidney in subacute endocarditis. *Archives of Pathology*, **94**, 205–13.

Morel-Maroger, L., *et al.* (1975). 'Tropical nephropathy' and 'Tropical extramembranous glomerulonephritis' of unknown aetiology in Senegal. *British Medical Journal*, **1**, 541–6.

Musa, A.M., Saleh, S.T., and Abu Asha, H. (1981). Transient nephritis during typhoid fever in five Sudanese patients. *Annals of Tropical Medicine and Parasitology*, **75**, 181–4.

Neugarten, J., Gallo, G.R., and Baldwin, D.S. (1984). Glomerulonephritis in bacterial endocarditis. *American Journal of Kidney Diseases*, **5**, 371–9.

Ng, W.L., Scollard, D.M., and Hua, A. (1981). Glomerulonephritis in leprosy. *American Journal of Clinical Pathology*, **76**, 321–9.

Ngu, J.L., Chatelanat, F., Leke, R., Ndumbe, P., and Youmbissi, J. (1985). Nephropathy in Cameroon: evidence of filarial derived immune-complex pathogeneses in some cases. *Clinical Nephrology*, **24**, 128–34.

Noe, N.H. and Roy, S. (1981). Shunt nephritis. *Journal of Urology*, **125**, 731–3.

Nunan, T.O., Eykyn, S.J., and Jones, N.F. (1984). Brucellosis with mesangial IgA nephropathy: successful treatment with doxycyline and rifampicin. *British Medical Journal*, **288**, 1802.

Nydahl, B.C. and Hall, W.H. (1965). The treatment of staphylococcal infection with nafcillin with a discussion of staphylococcal nephritis. *Annals of Internal Medicine*, **63**, 27–43.

Old, C.W., Herrera, G.A., Reimann, B.E.F., Latham, R.D. (1984). Acute post streptococcal glomerulonephritis progressing to rapidly progressive glomerulonephritis. *Southern Medical Journal*, **77**, 1470–2.

O'Connor, D.T., Weisman, M.H., and Fierer, J. (1978). Activation of the alternate complement pathway in *Staph. aureus* infective endocarditis and its relationship to thrombocytopaenia, coagulation abnormalities and acute glomerulonephritis. *Clinical and Experimental Immunology*, **34**, 179–87.

O'Regan, S. and Makker, S.P. (1979). Shunt nephritis: demonstration of diptheroid antigen in glomeruli. *American Journal of Medical Science*, **278**, 161–5.

O'Regan, S., Fong, J.S.C., de Chadarevian, J.P., Rishikof, J.R., and Drummond K.N. (1976). Treponemal antigens in congenital and acquired syphilis. *Annals of Internal Medicine*, **85**, 325–7.

O'Regan, S., Turgeon-Knaack, E., Mongeau, J.G., Lapointe, N., Nolin, L., and Robitaille, P.O. (1979). Subacute sclerosing panencephalitis associated glomerulopathy. *Nephron*, **23**, 304.

Oriente, P., *et al.* (1986). Cyclophosphamide treatment in polyarteritis nodosa. *Clinical Rheumatology*, **5**, 193–200.

Ozaki, M. and Furuta, M. (1975). Amyloidosis in leprosy. *International Journal of Leprosy*, **43**, 116–124.

Ozawa, T. and Stewart, J.A. (1979). Immune-complex glomerulonephritis associated with cytomegalovirus infection. *American Journal of Clinical Pathology*, **72**, 103–7.

Parag, K.B., *et al.* (1988). Profile of crescentic glomerulonephritis in Natal—a clinicopathological assessment. *Quarterly Journal of Medicine*, **68**, 629–36.

Pardo, V., *et al.* (1984). Glomerular lesions in the acquired immunodeficiency syndrome. *Annals of Internal Medicine*, **101**, 429–34.

Patriarca, P.A. and Lauer, B.A. (1980). Ventriculoperitoneal shunt-associated infection due to *Haemophilus influenzae*. *Pediatrics*, **65**, 1007–9.

Paull, R. (1947). Periarteritis nodosa (panarteritis nodosa) with report of four proven cases. *Californian Medicine*, **67**, 309–14.

Perez, G.O., Bjornsson, S., Ross, A.H., Amato, J., and Rothfield, N. (1974). A mini epidemic of Goodpastures' syndrome. *Nephron*, **13**, 161–73.

Perez, G.O., Rothfield, N., and Williams, R.C. (1976). Immune complex nephritis in bacterial endocarditis. *Archives of Internal Medicine*, **136**, 334–6.

Perlman, L., Herdman, R.C., Kleiman, H., and Vernier, R.L. (1965). Post streptococcal glomerulonephritis: a ten year follow-up of an epidemic. *Journal of the American Medicine Association*, **194**, 175–82.

Pertschuk, L.P., Vuletin, J.C., Sutton, A.L., and Vesasquez, L.A. (1976). Demonstration of antigen and immune complex in glomerulonephritis due to *Staphylococcus aureus*. *American Journal of Clinical Pathology*, **66**, 1027.

Powell, D.E.B. (1961). Non-suppurative lesions in staphylococcal septicaemia. *Journal of Pathology and Bacteriology*, **82**, 141–9.

Pusey, C.D., Saltissi, D., Bloodworth, L., Rainford, D.J., and Christie, J.L. (1983). Drug associated acute interstitial nephritis: clinical and pathological features and response to high dose steroids. *Quarterly Journal of Medicine*, **52**, 194–211.

Rao, T.K.S., Nicastri, A.D., and Friedman, E.A. (1974). Natural history of heroin associated nephropathy. *New England Journal of Medicine*, **290**, 19–23.

Rao, T.K.S., et al. (1984). Associated focal and segmental glomerulosclerosis in the acquired immunodeficiency syndrome. *New England Journal of Medicine*, **310**, 669–73.

Rekola, S., Bergstrand, A., Bucht, H., and Lindberg, A. (1988). Are beta-haemolytic streptococci involved in the pathogenesis of mesangial IgA nephropathy. *Proceedings European Dialysis and Transplant Association*, **21**, 698–702.

Richardson, W.P., et al. (1981). Glomerulopathy associated with cytomegalovirus viremia in renal allografts. *New England Journal of Medicine*, **305**, 57–63.

Roberts, W.C. and Rabson, A.S. (1975). Focal glomerular lesions in fungal endocarditis. *Annals of Internal Medicine*, **71**, 963–70.

Rodriguez-Iturbe, B. (1984). Epidemic post-streptococcol glomerulonephritis. *Kidney International*, **25**, 129–36.

Rosseau, E., Russo, P., Lapointe, N., and O'Regan, S. (1988). Renal complications of acquired immunodeficiency syndrome in children. *Americal Journal of Kidney Disease*, **11**, 48–50.

Roy, S., Murphy, W.M., and Arant, B.S. (1981). Post streptococcal crescentic glomerulonephritis in children: comparison of quintuple therapy verses supportive care. *Journal of Pediatrics*, **98**, 403–10.

Russell, D.S. (1962). The kidney in rheumatic heart disease. *Journal of Clinical Pathology*, **15**, 414–20.

Rytel, M.W., Dee, T.H., Ferstenfeld, J.E., and Hensley, G.T. (1974). Possible pathogenetic role of capsular antigens in fulminant pneumococcal disease with disseminated intravascular coagulation. *American Journal of Medicine*, **57**, 889–96.

Sadigursky, M., Andrade, Z.A., Danner, R., Cheever, A.W., Kamel, I.A., and Elwi, A.M. (1976). Absence of schistomsomal glomerulopathy in *Schistosoma haematobium* infection in man. *Transactions of the Royal Society of Tropical Medicine and Hygiene*, **70**, 322–3.

Sagel, I., et al. (1973). Occurrence and nature of glomerular lesions after Group A streptococcal infections in children. *Annals of Internal Medicine*, **79**, 492–9.

Salyer, W.R. and Salyer, D.C. (1974). Unilateral glomerulonephritis. *Journal of Pathology*, **113**, 247–57.

Sanjad, S., Tolaymat, A., Whitworth, J., and Levin, S. (1977). Acute glomerulonephritis in children: a review of 153 cases. *Southern Medical Journal*, **70**, 1202–6.

Schena, F.P. (1985). Renal Manifestations in bacterial infections. *Contributions to Nephrology*, **48**, 125–34.

Schoenbaum, S.C., Gardner, P., and Shillito, J. (1975). Infections of cerebrospinal fluid shunts: epidemiology, clinical manifestations and therapy. *Journal of Infectious Diseases*, **131**, 543–51.

Schoeneman, M., Bennett, B., and Greifer, I. (1982). Shunt nephritis progressing to chronic renal failure. *American Journal of Kidney Disease*, **11**, 375–7.

Scragg, J. (1976). Typhoid fever and its management. *South African Journal of Hospital Medicine*, **2**, 556–60.

Sergent, J.S., Lockshin, M.D., Christian, C.L., and Gocke, D.J. (1976). Vasculitis with hepatitis B antigenaemia: long term observations in some patients. *Medicine*, **55**, 1–17.

Shahin, B., Papadopoulou, Z.L., and Jenis, E.H. (1974). Congenital nephrotic syndrome associated with congenital toxoplasmosis. *Journal of Pediatrics*, **85**, 366–70.

Shribman, J.H., Eatwood, J.B., and Uff, J. (1983). Immune complex nephritis complicating miliary tuberculosis. *British Medical Journal*, **287**, 1593–4.

Sitprija, V., Pipatanagul, V., Boonpucknavig, V., and Boonpucknavig, S. (1974). Glomerulonephritis in typhoid fever. *Annals of Internal Medicine*, **81**, 210–3.

Slusarczyk, J., Michalak, T., Mezer, T.M., Krawczynski, K., and Nowoslawski, A. (1980). Membranous glomerulopathy associated with hepatitis B core antigen immune complexes in children. *American Journal of Pathology*, **98**, 29–43.

Sobh, M.A., Moustafa, F. El-Arbagy, A., and Shehab El-din, M. (1988). Nephropathy in asymptomatic patients with active *Schistosoma mansoni* infection. Proceedings 1st African Kidney and Electrolyte Conference, Cairo, pp. 314–323.

Spector, D.A., Millan, J., Zauber, N., and Buton, J. (1980). Glomerulonephritis and *Staphylococcus aureus* infections. *Clinical Nephrology*, **14**, 256–61.

Sterzel, R.B., Krause, P.H., Zobl, H., and Kuhn, K. (1974). Acute syphilitic nephrosis: a transient glomerular immunopathy. *Clinical Nephrology*, **2**, 164–8.

Stickler, G.B., Shin, M.H., Burke, E.C., Holley, K.E., Miller, R.H., and Segar, W.E. (1968). Diffuse glomerulonephritis associated with infected ventriculoatrial shunt. *New England Journal of Medicine*, **279**, 1077–82.

Strife, C.F., McDonald, B.H., Ruley, E.J., McAdams, A.J., and West, C.D. (1976). Shunt nephritis: the nature of the serum cryoglobulins and their relation to the complement profile. *Journal of Pediatrics*, **88**, 403–13.

Takaue, Y. and Tokumaru, M. (1982). Staphylococcal skin lesions and acute glomerulonephritis. *New England Journal of Medicine*, **307**, 1213–4.

Takeda, S., Kida, H., Katagiri, M., Yokoyama, H., Abe, T., and Hattori, N. (1988). Characteristics of glomerular lesions in hepatitis B virus infections. *American Journal of Kidney Disease*, **11**, 57–62.

Takekoshi, Y., Tanaka, M., Shida, N., Satake, Y., Shaeki, Y., and Matsumoto, S. (1978). Strong association between membranous nephropathy and hepatitis B surface antigenaemia in Japanese children. *Lancet*, **ii**, 1065–8.

Trepo, C.G., Zuckerman, A.J., Bird, R.C., and Prince, A.M. (1974). The role of circulating hepatitis B antigen/antibody immune complexes in the pathogenesis of vascular and hepatic manifestations in polyarteritis nodosa. *Journal of Clinical Pathology*, **27**, 863–8.

Treser, G., et al. (1974). Renal lesions in narcotic addicts. *American Journal of Medicine*, **57**, 687–94.

Tu, W.H., Shearn, M.A., and Lee, J.C. (1969). Acute diffuse glomerulonephritis in acute staphylococcal endocarditis. *Annals Internal Medicine*, **71**, 335–41.

van Rhenen, D.J., Koolen, M.I., Feltkamp Vroom, Th.M., and Weening, R.S. (1976). Immune complex glomerulonephritis and chronic gramulomatous disease. *Acta Medica Scandinavica*, **206**, 33–7.

Vogt, A., Batsford, S., Rodriguez-Itrube, B., and Garcia, R. (1983). Cationic antigens in post streptococcal glomerulonephritis. *Clinical Nephrology*, **20**, 271–9.

Wakabayashi, Y., Kobayashi, Y., and Shigematsu, M. (1985). Shunt Nephritis: histological dynamics following removal of the shunt. *Nephron*, **40**, 111–17.

Walker, D.H. and Mattern, W.D. (1979). Acute renal failure in Rocky Mountain spotted fever. *Archives of Internal Medicine*, **139**, 443–8.

Weller, R.O. and Nester, B. (1972). Histological re-assessment of three kidneys originally described by Richard Bright in 1827–1836. *British Medical Journal*, **2**, 761–3.

Wiggelinkhuizen, J., Sinclair-Smith, C., Stannard, L.M., and Smuts, H. (1983). Hepatitis B virus associated membranous glomerulonephritis. *Archives of Diseases in Childhood*, **58**, 488–96.

Williams, R.L. and Kunkel, H.C. (1962). Rheumatoid factor, complement and conglutinin aberrations in patients with subacture bacterial endocarditis. *Journal of Clinical Investigation*, **41**, 666–75.

Wilson, C.B. and Smith, R.C. (1972). Goodpastures' syndrome associated with influenza A2 virus infection. *Annals of Internal Medicine*, **76**, 91–4.

Woodroffe, A.J., Row, P.G., Meadows, R., and Lawrence, J.R. (1974). Nephritis in infectious mononucleosis. *Quarterly Journal of Medicine*, **43**, 451–60.

Yoshikawa, N., *et al.* (1985). Membranous glomerulonephritis associated with hepatitis B antigen in children: a comparison with idiopathic membranous glomerulonephritis. *Clinical Nephrology*, **23**, 28–34.

Yuceoglu, A.M., Berkovich, S., Chiu, J. (1969). Effect of live measles vaccine on childhood nephrosis. *Journal of Pediatrics*, **74**, 291–4.

Yuceoglu, A.M., Berkovich, S., and Minkowitz, S. (1967). Acute glomerulonephritis as a complication of varicella. *Journal of the American Medical Association*, **202**, 879–81.

Yum, M.N., Wheat, L.J., Maxwell, D., and Edwards, J.L. (1978). Immunofluorescent localisation of *Staphylococcus aureus* antigen in acute bacterial endocarditis nephritis. *American Journal of Clinical Pathology*, **70**, 832–5.

Zappacosta, A.R. and Ashby, B.L. (1977). Gram-negative sepsis with acute renal failure. *Journal of the American Medical Association*, **238**, 1389–90.

Zunin, C., Castellani, A., Olivetti, G., Marini, G., and Gabriele, P.W. (1977). Membranoproliferative glomerulonephritis associated with infected ventriculoatrial shunt; report of two cases recovered after removal of shunt. *Pathologica*, **69**, 297–305.

3.13 Malignancy-associated glomerular disease

A.M. DAVISON AND D. THOMSON

General introduction

Both benign and malignant tumours, including solid tumours, most commonly adenocarcinoma of the lung and gastrointestinal tract, and lymphoproliferative and myeloproliferative disorders, have been associated with nephropathy. This chapter is devoted to the glomerular consequences of solid tumours, lymphoproliferative, and myeloproliferative disorders: those associated with monoclonal gammopathies are detailed in Chapter 4.3.

The first report of an association between malignancy and glomerular disease was made by Galloway in 1922, who described a patient with Hodgkin's disease who subsequently developed a nephrotic syndrome, and in 1931 Volhard described a patient with nephrotic syndrome associated with gastrointestinal cancer. Apart from these two anecdotal reports, the association was either unrecognized or ignored until the classic paper by Lee *et al.* (1966) which reported carcinoma in 11 of 101 patients with nephrotic syndrome. This first series suggested a high prevalence of carcinoma in adult patients with nephrotic syndrome, and indicated that the most common glomerular histological appearance associated with neoplasia was that of membranous nephropathy. Since then, many further series and anecdotal reports have been published which confirm the initial suggestions. It is now generally accepted that glomerular disease associated with malignancy usually presents as a nephrotic syndrome, that the majority of such patients have membranous nephropathy and that most patients are adults; only very rarely are children involved.

Abnormal urinary findings are common in patients with neoplasia and there are many possible reasons for this. Proteinuria is more common in patients with malignancy than in controls (Sawyer *et al.* 1988) and, interestingly, this indicates a poor prognosis with respect to the neoplasms. In addition, some studies suggest that mild glomerular pathology without extensive proteinuria is relatively common in patients with malignancy (Pascal *et al.* 1976; Helin *et al.* 1980).

Although the most common histological finding is a membranous nephropathy, minimal change disease, mesangiocapillary glomerulonephritis, IgA nephropathy, crescentic nephritis, and amyloidosis have all been reported. Similarly, although the most widely recognized association with Hodgkin's disease is a minimal change nephropathy, proliferative, mesangiocapillary, and crescentic glomerulonephritis have also been described. Particular glomerular appearances are not specific for a particular tumour type, and there is no association between tumour load, as assessed by tumour size or the presence of metastases, and glomerulopathy.

Evidence for a causal relationship between malignancy and nephropathy

A number of conditions should be satisfied to establish a causal relationship between malignancy and nephropathy (Table 1).

In a literature review of 93 patients, Keur *et al.* (1989) found that the glomerulopathy was clinically apparent in the 6 months before or after recognition of the tumour in 63 per cent of the reported cases, and that the nephrotic syndrome and the tumour were recognized simultaneously in nearly 40 per cent. However, 15 patients developed the nephrotic syndrome more than 12 months before the tumour was diagnosed, and the nephrotic syndrome developed more than 12 months after tumour diagnosis in seven patients. Thus there is a reasonable temporal association between diagnosis of tumour and nephropathy in most patients.

Table 1 Evidence of a causal relationship

1. Close temporal relationship between the clinical appearance of the renal lesion and the tumour
2. Remission or complete removal of tumour should be associated with a remission of the nephropathy
3. Recurrence of the tumour should be accompanied by recurrence of the nephropathy
4. Demonstration of tumour antigen and appropriate antibody in glomeruli

Improvement or remission of glomerulonephritis has been described after removal or treatment of a wide variety of tumours (Cantrell 1969; Barton *et al.* 1980; Yamauchi *et al.* 1985), and most reports indicate a very prompt response (Young *et al.* 1985; Beauvais *et al.* 1989). However, few studies of follow-up renal biopsies to determine whether the glomerular appearances have returned to normal have been reported. Such reports need to be interpreted with caution as many patients have had therapy for the tumour, such as corticosteroids and immunosuppressive or cytotoxic drugs, which may have had a direct effect on the glomerular pathology. Some patients fail to improve after remission or removal of the tumour (Couser *et al.* 1974; Row *et al.* 1975).

Recurrence of the tumour is usually associated with recurrence of the nephropathy, but this is not invariable. In a case reported by Cairns *et al.* (1978) there was a rapid resolution of proteinuria and haematuria with an increase in creatinine clearance following removal of a squamous cell carcinoma of the bronchus, but a local recurrence of the tumour was not associated with evidence of any recurrence of renal disease.

Glomerular deposition of tumour antigen and antitumour antibodies have been described in a number of patients. Olson *et al.* (1979) and Weksler *et al.* (1974) both described melanoma antigen and antibody within the glomerulus. Haskell *et al.* (1990) described positive immunoperoxidase staining for prostate-specific acid phosphatase and prostate-specific antigen, and detected immune deposits on electron microscopy in samples from a patient with metastatic prostate carcinoma. In addition, eluates from kidney have been shown to react with tumour cells in patients with bronchogenic carcinoma (Lewis *et al.* 1971, DaCosta *et al.* 1974) and in three patients with gastric carcinoma (Wakashin *et al.* 1980). These findings do not, however, excluded the possibility of passive localization of tumour-associated proteins on pre-existing glomerular deposits, as has been described in Heymann nephritis (Bellon *et al.* 1982).

The criteria for a causal relationship set out in Table 1 seem, therefore, to have been satisfied, and it can be concluded that in a number of patients with malignancy there is a causally related nephropathy.

Renal involvement in carcinoma

Malignant disease may involve the kidney in a variety of different ways (Table 2), either having a direct effect on the kidney or by indirect systemic effect of the tumour. Renal disease may also be a result of therapy prescribed for either the tumour or its consequences (see Chapter 21.5).

Direct effects

Metastatic spread of a solid tumour to the kidney is relatively uncommon, despite the magnitude of renal blood flow. However when present, metastases are usually multiple and often bilateral. They are usually manifest by either loin pain or haematuria; renal functional impairment is very rare. Tumours at the hilum of the kidney, either metastatic or from lymphoma involving the para-aortic nodes, may compress the renal artery, resulting in ischaemia with the subsequent development of hypertension. Urinary obstruction may result from lesions arising within the walls of the urinary tract or from external compression, usually of the ureters, by retroperitoneal or pelvic tumours. Although retroperitoneal tumours rarely cause such obstruction, pelvic tumours, particularly of cervix or the recto-sigmoid colon, may produce ureteric compression and eventually occlusion.

Obstructive uropathy is most frequently produced by prostatic hypertrophy, which is benign in the majority of patients—less than 10 per cent of cases are due to carcinoma. Bladder cancer, particularly involving the trigone, can also lead to ureteric obstruction with subsequent obstructive uropathy.

Indirect effects

Malignancy may have many indirect effects on the kidney, particularly in patients with advanced disease. Hypokalaemia may develop either from potassium loss, for example in patients with prolonged diarrhoea or in those with a villous adenoma of the rectum (Roy and Ellis 1959), or from the profound metabolic alkalosis that accompanies prolonged vomiting. Hyponatraemia may occur in association with diarrhoea and vomiting, or from inappropriate ADH secretion in patients with oat cell carcinoma of the lung. Hypercalcaemia arises more commonly from marrow infiltration with myeloma than from metastatic disease, and may also be caused by release of PTH-like substances from certain tumours, such as oat-cell tumours of the lung, and by secretion of prostaglandins from tumours (Seyberth *et al.* 1975). Release of skeletal calcium following immobilization and, rarely, as a result of hormonal therapy for breast carcinoma can also cause hypercalcaemia (Swaroop and Krant 1973). In some patients the hypercalcaemia causes nephrocalcinosis when calcium becomes deposited in tubular cell mitochondria, resulting in cell death. Deposition in tubular basement membranes leads to tubulointerstitial inflammation and fibrosis. Hypercalcaemia can also impair urinary concentration, by reducing the medullary concentration gradient through the inhibition of sodium transport in the ascending limb of the loop of Henle, and through inhibition of the effect of antidiuretic hormone in collecting ducts. The net result is polyuria and polydipsia, with the risk of dehydration if the patient is unable to maintain an adequate fluid intake.

Acute renal failure may arise as an indirect effect of malignancy. This may be prerenal, as in patients with hypovolaemia from prolonged diarrhoea and vomiting associated with hepatic failure due to metastatic tumours, or may be due to acute tubular necrosis caused by septicaemia as a consequence of the general-

Table 2 Renal involvement in carcinoma

Direct effects
   Metastases
   Infiltration
   Ischaemia
   Obstruction
Indirect effects
   Electrolyte disorders
      Hypokalaemia, hyponatraemia, hypercalcaemia
   Acute renal failure
   Disseminated intravascular coagulation
   Renal vein thrombosis
   Amyloidosis
   Glomerulopathy
   Nephrocalcinosis
Treatment associated
   Drug nephrotoxicity
   Tumour lysis syndrome
   Radiation

ized immunosuppression of advanced malignancy. Impaired renal function may also occur in patients who develop disseminated intravascular coagulation. This has been described in carcinoma of the pancreas, lung, and in trophoblastic tumours. Renal vein thrombosis may also arise (De Swit and Wells 1957) due to the generalized hypercoagulable state of the malignant patient.

A variety of glomerular changes (Table 3) have been reported in association with malignancy (see below).

Table 3 Glomerular appearances most commonly associated with carcinoma

Membranous nephropathy
Mesangiocapillary glomerulonephritis
IgA nephropathy
Mesangial proliferative glomerulonephritis
Crescentic nephritis
Amyloidosis

Effects of therapy

Renal disease can arise as a direct consequence of anticancer therapy. Drug nephrotoxicity may occur due to drugs such as cisplatinum prescribed for treatment of the malignancy (see Chapter 21.5) or from drugs such as antibiotics and analgesics prescribed to treat consequences of the malignancy. Dosik et al. (1978) described three patients with malignant melanoma treated with C. parvum who developed an acute renal failure and in whom renal biopsy showed a mesangial proliferative glomerulonephritis. Although subendothelial deposits containing immunoglobulin and complement were found, no C. parvum antigen could be detected. A causal link is clearly tenuous, but the renal failure resolved on stopping C. parvum therapy. Although radiotherapy is widely used in the treatment of malignancies, radiation nephropathy (see Chapter 6.6) is now uncommon: recognition of the susceptibility of renal tubular cells to irradiation injury has led to the development of effective shielding during therapy. In some circumstances it is impossible to shield the whole kidney and acute and chronic radiation nephritis may result. Acute effects usually become manifest 6 to 12 months after irradiation, whereas chronic effects develop years later and may not have been preceded by an acute syndrome. Proteinuria, hypertension, and impaired renal function are recognized clinical effects.

Pathogenesis of glomerulopathy

Glomerular injury in malignancy may be immunologically mediated, or may result from disseminated intravascular coagulation or amyloidosis (Table 4). Eagen and Lewis (1977) observed that the immunopathological features of the glomerular lesions associated with carcinoma were similar to those found in immune complex disease and suggested that they might result from (1) tumour associated antigens; (2) re-expressed fetal antigens; (3) viral antigens; or (4) the development of autoimmunity to autologous non tumour antigens.

Although malignancy can be associated with impaired cell mediated immunity, tumours are able to stimulate both cellular and humoral responses.

Table 4 Pathogenesis of malignancy-associated glomerulonephritis

1. Tumour associated antigen
2. Re-expressed fetal antigen
3. Viral antigen
4. Autologous non-tumour antigen
5. Disseminated intravascular coagulation
6. Amyloidosis

Circulating immune complexes have been described in up to 80 per cent of patients with malignancy (Rossen et al. 1976). The prevailing view of the 1970s, that deposition of entire complexes in glomeruli leads to the development of glomerulonephritis, is now regarded with considerable scepticism, except as a nidus for in-situ complex formation. Three factors should be borne in mind. Firstly, although circulating immune complexes are common in patients with malignancy, only a small proportion develop nephropathy; this could, however, be due to variation in the size and/or charge of the complexes. Secondly, the most common histological appearance in malignancy associated nephropathy is a membranous lesion, and patients with idiopathic membranous glomerulonephritis rarely have circulating immune complexes. Finally, electron microscopy commonly shows sub-epithelial deposits, whereas in patients with glomerular disease associated with circulating immune complexes, and also in experimental animals given intravenous injections of preformed immune complexes, deposits are usually found within the subendothelial area of the glomerular capillary wall or in the mesangium. All observations suggest that the immune deposits are more likely to have been formed in situ from antigen and antibody deposited separately.

Melanoma antigen and antibody have been demonstrated both directly and indirectly within glomerular deposits (Olson et al. 1979; Weksler et al. 1974), and eluates from glomeruli have been shown to cross-react with antigens on bronchogenic carcinoma (Lewis et al. 1971), colonic carcinoma (Couser et al. 1974), and gastric carcinoma (Wakashin et al. 1980) cells. One recent report describes a patient with metastatic prostatic carcinoma who developed acute renal failure and was found to have a crescentic glomerulonephritis at autopsy. This was associated with glomerular deposition of prostate-specific acid phosphatase and prostate-specific antigen in a focal and segmental distribution (Haskell et al. 1990). Unfortunately, there have been no large studies, and there is a surprising lack of recent information. Nonetheless, carefully conducted studies involving small numbers of patients provide convincing evidence that the malignancy associated nephropathy is immune complex mediated in a number of patients.

Re-expressed fetal antigens have been demonstrated, both on tumours and within glomerular deposits. Carcinoembryonic antigen has been demonstrated in glomerular deposits (Costanza et al. 1973; Couser et al. 1974) and Pascal and Slovin (1980) described a patient with gastric carcinoma in whom carcinoembryonic antigen and antibodies could be detected within the glomerular capillary wall in a subendothelial and paramesangial position. Thus, in a number of patients the immune response is accompanied by the production of a re-expressed fetal antigen and appropriate antibody.

Viruses are being implicated as causal agents in an increasing number of malignancies, and particularly in lymphoproliferative diseases: no such association has yet been demonstrated with human solid tumours. Pascal *et al.* (1980) described virus like particles adjacent to the plasma membranes of numerous mesangial and epithelial cells in a patient with prostatic cancer and membranoproliferative glomerulonephritis. The significance of these findings and, in particular, their relationship to the pathogenesis of the glomerular findings is unknown. Glomerular deposition of viral antigen and antibody has been demonstrated in mice with virus-induced mammary tumours (Pascal *et al.* 1975). Thus, although viruses may play a part in the glomerular lesions associated with lymphoma and leukaemia, there is currently little evidence that they play any part in the nephropathy associated with carcinoma.

Patients with malignancy seem more prone than controls to the development of autoimmunity, with the formation of non-specific autoantibodies, and there are reports of glomerular deposition of autologous non-tumour antigens. Higgins *et al.* (1974) described a patient with oat cell carcinoma of the bronchus in whom large quantities of extracellular DNA in necrotic tumour tissue were associated with serum antinuclear antibody and the presence of Feulgen-positive material within glomerular capillary walls. The authors suggested that, in this patient, there was antibody formation against nuclear antigen released by tumour necrosis. Ozawa *et al.* (1975) described three patients in whom glomerular and tumour eluates reacted with normal proximal tubular brush border antigens as well as antigens on the tumour membrane. This finding supports the concept that the renal tubular epithelial antigen is shed from renal cell carcinoma and that subsequent antibody leads to immune complex formation.

Local intravascular coagulation may be responsible for glomerular lesions in a small number of cases. Young *et al.* (1985) described two patients with distinctive glomerular abnormalities consisting of occlusive eosinophilic deposits in glomerular capillary lumina in which immunofluorescence revealed fibrin-related antigens and IgM. It is possible that this lesion was initiated by the release of coagulant proteins from tumour cells.

Amyloidosis is a well recognized complication of malignancy and is discussed fully in Chapter 4.2; it has been described particularly in association with renal cell carcinoma (Kimball 1961). The mechanism by which renal cell carcinoma produces amyloidosis is not clear, but there is speculation that the tumour secretes a precursor of amyloid proteins, which is later cleaved and precipitates as amyloid. Alternatively the tumour may secrete an enzyme which cleaves a precursor of amyloid already present in the circulation (Couser 1980).

Clinical features

The most common presentation of tumour-associated nephrosis in adults is the nephrotic syndrome, but this is influenced by whether the series is being reported by a nephrologist or an oncologist. Nephrotic syndrome is manifest before the tumour in about 40 per cent of patients (Keur *et al.* 1989), and these are therefore likely to present to a nephrologist (Fig. 1). There is, however, no indication of the true incidence of nephropathy in malignancy since many patients may have only minor degrees of proteinuria. In a prospective study of 150 patients with various malignancies there was a 12 per cent incidence of proteinuria and/or haematuria (Puolijoki *et al.* 1989).

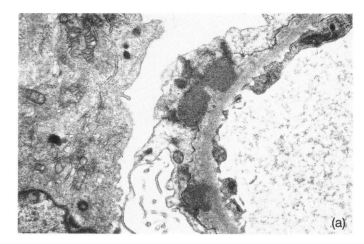
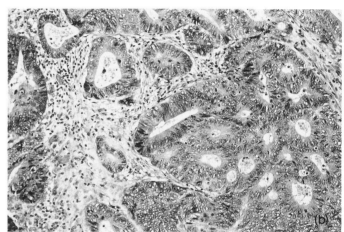

Fig. 1 A 60-year-old man who presented with nephrotic syndrome. Renal biopsy revealed a membranous type lesion. (a) Electron micrograph illustrating three finely granular subepithelial electron-dense deposits in the glomerular capillary wall ($\times$ 1325 reduced for publication). (b) Ten months later a well differentiated adenocarcinoma of the sigmoid colon was excised (haematoxylin and eosin, $\times$ 212 reduced for publication) with subsequent resolution of the nephrotic syndrome. Illustrations by kind permission of Dr S.R. Aparicio, Department of Pathology, St James' University Hospital.

Microscopic haematuria or asymptomatic proteinuria are common, but a number of patients have proteinuria severe enough to produce a nephrotic syndrome. There is seldom significant renal functional impairment: Puolijoki *et al.* (1989) reported that three of 70 patients had a serum creatinine above 150 μmol/1, but in none was this above 190 μmol/1. Nevertheless there are a number of reports of patients whose reduced creatinine clearance at presentation increased after removal of the tumour, with subsequent improvement in the nephropathy (Jermanovich *et al.* 1982; Cairns *et al.* 1978). No reports have documented the rate of deterioration of renal function, but it is assumed that progression of the tumour is faster than progression of the renal impairment.

Other less common forms of presentation have included steroid-resistant nephrotic syndrome in a child (Beauvais *et al.* 1989) and Henoch-Schönlein purpura (Cairns *et al.* 1978).

Incidence of glomerulopathy in neoplasia

The prevalence of tumour-associated nephropathy is unknown, but in view of the frequency of malignancy in the general population the development of clinically apparent nephropathy must be rare. Subclinical glomerular involvement is probably much more common. The majority of patients with advanced or incurable malignancy are not subjected to invasive renal investigations and in many the rate of progression of the malignancy may well exceed that of the nephropathy. In an autopsy study of 303 patients, Sutherland *et al.* (1974) found that two of 124 patients with solid tumours had deposition of IgG and complement in histologically normal glomeruli. The antigens were not identified, clinical details were not given and some of the patients had been treated with irradiation and chemotherapy. In another autopsy study, glomerular immune deposits were detected in 22 of 129 patients (17 per cent) with solid tumours, mainly of the gastrointestinal tract (Beaufils *et al.* 1985). The deposits were predominantly mesangial and/or subendothelial: none were subepithelial. Twenty of the patients had a mild increase in mesangial matrix, with occasional hypercellularity on light microscopy, and two had no glomerular changes. A greater incidence of immune deposits was reported by Pascal *et al.* (1976) who found mesangial and subendothelial deposits in 30 per cent of patients at autopsy. In a further autopsy study, Pascal (1980) detected overt glomerulonephritis in five of 314 patients with malignancy, giving an incident of glomerular disease of about 1.5 per cent among patients with solid tumours. Thus it appears that although immune deposits may be common, occurring in 17 to 40 per cent of patients with malignancy coming to autopsy, histologically obvious glomerular changes are rare.

Two studies have attempted to determine the incidence of urinary abnormalities in patients with carcinoma. Sawyer *et al.* (1988) found that 58 per cent of 504 patients with a variety of malignancies had urinary abnormalities consisting of either proteinuria and/or haematuria. In addition, actuarial analysis showed that patients with proteinuria had significantly shorter median survival than those without. A further retrospective study from Finland (Puolijoki *et al.* 1989) found proteinuria and haematuria in 15 per cent of 450 patients with lung cancer. This group also reported a prospective study of 150 patients, 23 of whom had proteinuria and/or haematuria. Exclusion of patients with renal metastases, anticoagulant therapy, nephrolithiasis, and hypertension gave an incidence of 12 per cent.

It is difficult to reconcile the differences between autopsy and clinical studies. It is possible that the clinical studies detected a significant number of patients with mild proteinuria associated with little in the way of glomerular pathology. This is supported by Pascal *et al.* (1976) and Helin *et al.* (1980), who concluded that mild glomerulopathy without extensive proteinuria was relatively common in patients with solid tumours. It is likely that the autopsy series underestimated the frequency of deposits, since both electron microscopy and immunological tests are notoriously difficult to perform in postmortem material.

Incidence of carcinoma in patients with glomerulopathy

Adults

The initial report of Lee *et al.* (1966) recorded that 11 of 101 patients with nephrotic syndrome had cancer. This high incidence has not been reported in other series or reviews. Kaplan *et al.* (1976) reviewed 14 series, comprising a total of 1643 patients with nephrotic syndrome and found only six who were reported to have malignancy. Details of the 14 series are not recorded, but they could not have included the 11 patients reported by Lee *et al.* (1966). In a study of 76 patients with nephrotic syndrome and over the age of 60, Zech *et al.* (1982) showed that 7.9 per cent had an associated malignancy. Keur *et al.* (1989) reviewed nine reported series and identified 52 of 763 patients with an associated malignancy. This gives an overall incidence of 7 per cent, while the incidence in the individual series varied between 3 and 13 per cent.

It is likely that the incidence increases with increasing age, but this may be due to the fact that secondary causes of nephrotic syndrome become more common and malignancy increases in frequency with age.

Children

Glomerular disease has only very rarely been reported in children with malignancy. There were no cases of malignancy in 121 children with membranous glomerulonephritis (Kleinknecht *et al.* 1979; Latham *et al.* 1982; Ramirez *et al.* 1982). Cameron and Ogg (1975), however, recorded that three of 350 children with nephrotic syndrome had evidence of neoplasia, and one with Wilms' tumour and nephrotic syndrome was reported by Row *et al.* (1975). A 1-year-old child with an abdominal neuroblastoma and membranous nephropathy was also described: the nephropathy apparently resolved after tumour removal and chemotherapy (Zheng *et al.* 1979). Similarly, steroid-resistant nephrotic syndrome in a 14-year-old girl with a benign ovarian tumour resolved after tumour removal (Beauvais *et al.* 1989).

Diffuse mesangial sclerosis has been reported in association with the Drash syndrome, (Wilms' tumour, pseudohermaphroditism, and nephropathy) (Drash *et al.* 1970; Habib *et al.* 1985). Since the original two patients (Drash *et al.* 1970) there have been a number of reports confirming the association. Clinically, the patients present at a young age (2 months to 2 years old) with proteinuria which may be sufficient to cause the nephrotic syndrome, hypertension, and progressive renal failure. Mesangial sclerosis has been reported in a few children without either male pseudohermaphroditism or Wilms' tumour, and the clinical detection of two features of the triad should therefore lead to an active search for the third. The prognosis is poor with death being due to renal failure or as a consequence of the tumour.

Thus the incidence of malignancy in patients with nephrotic syndrome is likely to be less than 7 per cent, but it may increase with increasing age. Nonetheless, it is advisable to seek an underlying malignancy in adults presenting with nephrotic syndrome due to a membranous glomerulonephritis, especially in those aged more than 50 years, as tumour removal may be curative. Simple screening tests such as chest radiography, faecal occult blood, and carcinoembryonic antigen should be performed since the nephrotic syndrome is the initial presenting complaint in at least 40 per cent of patients. Gastrointestinal contrast studies however, are not justified as a routine investigation.

Types of malignancy involved

A very wide variety of malignant and benign tumours has been implicated. The most common are adenocarcinomata of the lung and gastrointestinal tract (stomach, colon, rectum), but many

Table 5 Types of malignancy associated with nephropathy

Site	Type	Reference
Lung	Adenocarcinoma	Lewis et al. 1971
	Squamous cell carcinoma	Cairns et al. 1978
		Heaton et al. 1975
	Oat cell carcinoma	Mustonen et al. 1981
Oesophagus	Squamous cell carcinoma	Heckerling 1985
Stomach	Adenocarcinoma	Cantrell 1969
		Wakashin et al. 1980
Colon	Adenocarcinoma	Costanza et al. 1973
Rectum	Adenocarcinoma	Kitano 1984
Bile duct	Adenocarcinoma	Brueggemeyer et al. 1987
Skin	Basal cell	Lee et al. 1966
	Squamous cell	Lee et al. 1966
	Melanoma	Weksler et al. 1974
Breast	Adenocarcinoma	Lewis et al. 1971
		Barton et al. 1980
Thyroid	Adenocarcinoma	Castleman et al. 1963
Adrenal	Phaeochromocytoma	Endoh 1984
Ovary	Teratoma	Beauvais et al. 1989
	Adenocarcinoma	Lee et al. 1966
Cervix	Adenocarcinoma	Lee et al. 1966
Uterus	Trophoblastic	Young et al. 1985
Kidney	Renal cell carcinoma	Ozawa et al. 1975
		Kimball 1961
	Wilms' tumour	Drash et al. 1970
Prostate	Adenocarcinoma	Stuart et al. 1986
Bladder	Transitional cell carcinoma	Brueggemeyer et al. 1987
Palate	Adenocarcinoma	Borochovitz et al. 1982

other tumours have been associated with nephropathy, usually on the basis of single case reports (Table 5). Interestingly, the nephropathy is relatively rarely associated with carcinoma of the breast, despite the relatively high incidence of this tumour (Barton et al. 1980).

Benign tumours are only occasionally implicated, but there are two particularly striking reports. Membranous glomerulonephritis was associated with a benign teratoma in a child and the associated nephrotic syndrome resolved on removal of the tumour (Beauvais et al. 1989). In addition a carotid body tumour was reported in association with a membranous glomerulonephritis in a 16-year-old girl (Lumeng and Moran 1966).

Glomerular appearances (Table 5)

Membranous nephropathy

Membranous glomerulonephritis is by far the most common histological type observed, occurring in approximately 70 per cent of patients. It has been associated with both benign and malignant tumours, but most commonly with bronchogenic carcinoma, although there are also reports of colorectal, renal (Kerpen et al. 1978), and gastric adenocarcinomas. Other tumours have only been described in small numbers, the majority being single case reports. The histological appearances of membranous nephropathy associated with malignancy on light microscopy, immunofluorescence, and electron microscopy are no different from those of idiopathic membranous nephropathy.

The frequency of neoplasia in patients with membranous nephropathy is variable. Hopper (1974) found an associated carcinoma in 6 per cent of a series of patients with membranous glomerulonephritis, whereas in the series reported by Row et al. (1975) there was evidence of malignancy in 10.9 per cent. Not all series or reviews have recorded such incidences; in a review of 14 reported series of patients with nephrotic syndrome (Kaplan et al. 1976) only six patients with an associated malignancy were detected from a total of 1643 patients.

Mesangiocapillary glomerulonephritis

A number of reports describe an association between malignancy and mesangiocapillary glomerulonephritis. Lee et al. (1966) describe one patient as having a 'lobular glomerulonephritis'. A subendothelial type mesangiocapillary glomerulonephritis was reported in a patient with carcinoma of the breast (Lewis et al. 1971) and in a patient with bronchogenic carcinoma (Heaton et al. 1975). Similar findings were reported by Robinson et al. (1984) in a patient with adenocarcinoma of the stomach. In addition, Olson et al. (1979) reported three patients with malignant melanoma in whom glomerular deposits contained antigen which was also identified on the tumour: the appearance on light microscopy was that of mesangiocapillary glomerulonephritis. Walker et al. (1981) described a patient with squamous cell carcinoma of the oesophagus and membranoproliferative glomerulonephritis.

IgA nephropathy

There are two reports of typical IgA nephropathy associated with malignancy. Cairns et al (1978) reported two patients with bronchial squamous cell carcinoma who presented with clinical features of Henoch-Schönlein purpura and mesangial IgA deposition. Mustonen et al. (1981) also described two patients, one with recurrent haematuria during febrile illnesses who was found to have a bronchogenic carcinoma associated with IgA nephropathy, the second with purpura in a distribution similar to that of Henoch-Schönlein purpura associated with microscopic haematuria and mesangial IgA deposition. Although the association may have been coincidental, Mustonen et al. (1981) thought this unlikely taking into consideration the expected incidence of this association in his local population and the lack of any other associated possible causes such as alcoholic liver disease.

Although IgA nephropathy is rarely associated with malignancy, Endo and Hara (1986) found that 17 per cent of patients with lung carcinoma had glomerular IgA deposits.

Mesangial proliferative glomerulonephritis

There is a single case report of a mesangial proliferative glomerulonephritis associated with an anaplastic small cell carcinoma of the lung (Jermanovich et al. 1982).

Crescentic nephritis

A number of reports have suggested an association between malignancy and crescentic glomerulonephritis (Whitworth et al. 1975; Hopper et al. 1976; Eagan and Lewis 1977; Haskell et al. 1990). Biava et al. (1984) reviewed 80 patients with crescentic glomerulonephritis and found seven (9 per cent) to have a co-existent non-renal malignancy—six with carcinoma and one with lymphoma. This is a similar incidence to that reported by Whitworth et al. (1975) who found malignancy in four of 60 patients (7 per cent) with crescentic glomerulonephritis. The clinical course appears to be similar to that of idiopathic crescentic glomerulonephritis, with the exception that tumour removal may

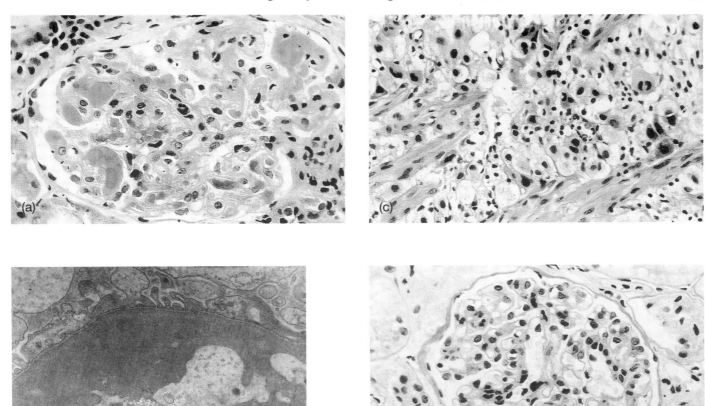

Fig. 2 A 26-year-old girl who presented with acute renal failure 2 weeks after delivery. The cause of the acute renal failure was not obvious and a renal biopsy was undertaken; this showed on light microscopy large intracapillary amorphous deposits which in some, occluded the capillary lumina ((a) (haematoxylin and eosin × 400). Electron microscopy revealed large subendothelial deposits which had a granular character ((b) × 8160). A uterine curettage suggested the presence of a choriocarcinoma and consequently hysterectomy was undertaken. On light microscopy strands of placental trophoblastic tumour were seen to be invading the myometrium ((c) haematoxylin and eosin × 256). Renal function promptly returned to normal and 6 months later a follow-up renal biopsy was undertaken. This showed resolution of the deposits although a minor increase in mesangial cells remained ((d) haematoxylin and eosin × 400).

be associated with improvement in renal function (Biava et al. 1984).

Minimal change nephropathy

Minimal change nephropathy has only rarely been associated with solid tumours, in contrast to its frequent association with Hodgkin's disease (see below). One of the 11 patients reported by Lee et al. (1966) had minimal change nephropathy and an adenocarcinoma of the colon. Other single cases have been reported, and the associated malignancies include carcinoma of the colon (Caruana and Griffin 1980), small cell carcinoma of the lung (Moorthy 1983), renal cell carcinoma (Dupond et al. 1980), and a mesothelioma (Schroeter et al. 1986).

Amyloidosis

Vanatta et al. (1983) reviewed the literature and found 40 renal cell carcinoma patients with systemic amyloidosis, presumed to be of the AA type (see Chapter 4.2). The incidence of renal amyloidosis in carcinoma is not known, but in a study of 4033 autopsies, 16 patients were identified with amyloidosis, seven of whom also had carcinoma (Kimball 1961). A review by Glenner (1980) concluded that 24 to 33 per cent of all tumour-associated amyloidosis was associated with renal cell carcinoma. It is possible that circulating precursors are produced by epithelial cells in the renal cell carcinoma and lead to subsequent amyloid deposition (see Chapter 4.2).

Other glomerular pathology

Placental trophoblastic tumour has been associated with glomerular appearances identical to those of disseminated intravascular coagulation (Young et al. 1985).

Prognosis

In general, the prognosis of malignancy-associated nephropathy is determined more by the malignancy than by the glomerular findings. There are as noted above, several reports of improvement in the nephropathy or a remission of glomerulonephritis after removal or adequate treatment of the tumour (Cantrell 1969; Walker et al. 1981; Moorthy 1983; Yamauchi et al. 1985) (Fig. 2). There is one interesting case report of a patient with renal cell carcinoma in whom systemic amyloidosis, as assessed by serial liver biopsies, showed a progressive reduction and was undetectable 22 months after nephrectomy (Paraf *et al.* 1970).

Few reports contain information obtained at follow-up biopsy, but Walker *et al.* (1981) described a patient with a marked membranoproliferative glomerulonephritis who showed a marked regression of mesangial hypercellularity, a reduction in immunofluorescence and a disappearance of electron microscopic evidence of mesangial interposition in the glomerular capillary walls 5 months after oesophagectomy and 3 months after initial biopsy. Although some studies report no improvement after remission/removal of the tumour (Couser *et al.* 1974; Row *et al.* 1975), it is generally believed that complete resection or remission of the malignancy is associated with long-term remission of the nephropathy. If complete removal of the tumour is not possible there may be improvement in the nephropathy with reduction in tumour mass. Robinson *et al.* (1984) reported a patient in whom the nephrotic syndrome remitted on two occasions following irradiation of metastatic disease; the patient died from gastric cancer nearly 4 years after presentation with nephrotic syndrome. In other patients, however, in whom the tumour is unresponsive to treatment 50 per cent are dead within 3 months, usually from disseminated carcinomatosis.

Renal involvement in lymphoma and leukaemia

Introduction

Renal disease can arise in patients with lymphoma and leukaemia in a number of ways (Table 6). Obstructive uropathy may result from enlargement of hilar nodes or para-aortic nodes. Hydronephrosis due to ureteral obstruction was reported in 10 per cent of patients with lymphoma (Richmond *et al.* 1962), although it is likely that this is much less in present-day practice, because of earlier detection of lymphadenopathy by CT scanning and more effective treatment règimens which limit lymph node enlargement.

Renal infiltration occurs in approximately one-third of patients with lymphoma and in 50 per cent of patients with leukaemia. The infiltration is usually nodular in patients with Hodgkin's lymphoma, but more diffuse in non-Hodgkin's lymphoma. Although all types of leukaemia may infiltrate the kidney, this most commonly occurs with lymphoblastic leukaemia, when it is usually bilateral and diffuse throughout the cortex (Norris and Wiener 1961). Renal infiltration only rarely leads to significant impairment of renal function.

Amyloidosis is now rarely associated with lymphoma. Earlier reports described amyloidosis developing insidiously after many years of active disease. This late and previously frequently fatal complication is now uncommon due to the more recent development of successful therapeutic règimens.

As in patients with carcinoma, renal disease may arise as a complication of therapy, either directly, as in the tumour lysis syndrome (see below), or indirectly due to the development of intercurrent infection.

Hyperuricaemia is common in diseases associated with a rapid turnover of cells. Urinary uric acid excretion is increased by approximately 80 per cent in patients with leukaemia: this is sparingly soluble at a low pH and urinary precipitation may occur in the distal tubules where concentration and acidification (see Chapter 6.3) are maximal. Effective chemotherapy increases the breakdown of cells and therefore increases nucleic

Table 6 Renal involvement in lymphoma and leukaemia

1. Obstructive uropathy
2. Infiltration of renal parenchyma
3. Amyloidosis
4. Therapy associated
5. Urate nephropathy
6. Glomerulopathy
7. Disseminated intravenous coagulation

acid catabolism; severe uric acid nephropathy may follow chemotherapy in up to 10 per cent of patients with acute lymphocytic leukaemia. The so-called, 'tumour lysis syndrome' is, in part, an acute urate nephropathy which occasionally occurs in patients with lymphoproliferative disorders treated with cytotoxic drugs (see Chapter 6.4). However, many other substances which could affect renal function are released from disintegrating tumour cells.

Glomerulopathy is a rare complication of lymphoma or leukaemia. In a survey of the literature, Keur *et al.* (1989) identified only 77 patients with Hodgkin's lymphoma and an associated nephropathy, most commonly minimal change nephropathy; an additional 20 patients had chronic lymphocytic leukaemia, most commonly associated with a mesangiocapillary glomerulonephritis. There is probably a direct association between Hodgkin's lymphoma and nephropathy, but the rarity of nephropathy and the diversity of lesions reported in association with leukaemias must cast doubt on a causal relationship.

Pathogenesis

The pathogenesis of the minimal change nephropathy in Hodgkin's lymphoma is thought to be due to a T-cell disorder in which there is an abnormal T4/T8 ratio with a reduction in T4 (helper cells) and an increase in T8 (suppressor cells). It has been suggested that increased release of proteases or lymphokines by T cells increase glomerular permeability, resulting in proteinuria (see Chapter 3.2).

An alternative hypothesis implicates a tumour related virus or an unrelated intercurrent viral infection as the cause of the glomerular lesion. Experimentally, many New Zealand mice chronically infected with murine leukaemia virus develop lymphomas and glomerulonephritis, with immune deposits containing viral antigens and antibodies. Similarly, AKR mice develop leukaemia due to the Gross murine leukaemia virus and have a high incidence of glomerular deposition of immune complexes containing viral antigens.

Clinically, glomerular deposits of antibodies to Epstein-Barr virus have been demonstrated in Burkitt's lymphoma (Oldstone *et al.* 1974), and an oncornavirus has been implicated in two patients with an acute myelomonocytic leukaemia (Sutherland and Mardinay 1973). The involvement of an intercurrent viral infection is supported by the report of Hyman *et al.* (1973) who detected Herpes zoster viral antigen in a young child with Hodgkin's disease, Herpes zoster infection, and nephrotic syndrome.

Hodgkin's lymphoma

The association between nephropathy and Hodgkin's lymphoma, although well recognized, is rare: Kramer *et al.* (1981) and Plager and Stutzman (1971) report glomerular disease in only seven of more than 1700 patients with Hodgkin's lymphoma. In a review of the literature, Keur et al. (1989) identifies 77 patients with Hodgkin's lymphoma and nephropathy. Although a direct link has not been established there is considerable circumstantial evidence to link the two conditions: they can both present at the same time, they may relapse together, and effective treatment of the lymphoma is associated with remission of the nephrotic syndrome, even when this is confined to local radiotherapy to upper body glandular involvement only.

Hodgkin's lymphoma presents before or at the same time as the nephrotic syndrome in 75 per cent of patients. It may occur at any age and has a male: female ratio of 2:1. Severe proteinuria with hypoproteinaemia and hypercholesterolaemia is common, but haematuria, hypertension, and impaired renal function are rare.

Prior to effective therapy for Hodgkin's lymphoma amyloidosis was the most commonly associated glomerular lesion. All recent reports, however, indicate that minimal change nephropathy is now the most common histological finding, present in 41 of the 77 patients reviewed by Keur et al. (1989). Other lesions which have been reported include focal glomerulonephritis (Watson et al. 1983), membranous nephropathy (Egan and Lewis 1977) and mesangiocapillary glomerulonephritis (Morel-Maroger Striker et al. 1984). A number of patients have been described who progressed from a minimal lesion nephropathy to a focal glomerulosclerosis (Hyman et al. 1973; Watson et al. 1983). Immunofluorescence microscopy may show IgG, IgM, and complement C3. Deposits have been described as being subendothelial (Hyman et al. 1973), intramembranous (Lokich et al. 1973) and/or subepithelial (Froom et al. 1972).

Effective treatment of Hodgkin's lymphoma is almost always associated with remission of the nephrotic syndrome (Froom et al. 1972) and recurrence of proteinuria may signal relapse of the Hodgkin's lymphoma.

Non-Hodgkin's lymphoma

Nephropathy has been described in patients with non-Hodgkin's lymphoma, but the association is extremely rare. In a review of the literature, Gonzalez et al. (1986) were able to identify only 22 patients with non-Hodgkin's lymphoma and a variety of glomerular diseases. The average age of the patients was 47 years (range 7–74) and there was a male preponderance. The nephropathy and lymphoma became apparent simultaneously in seven cases, and the majority had a nephrotic syndrome. The glomerular appearances have included minimal change nephropathy (Ghosh and Muehrcke 1970; Gagliano et al. 1976; Muggia and Ultman 1971), mesangiocapillary glomerulonephritis, membranous nephropathy (Rabkin et al. 1973; Gluck et al. 1973), focal glomerulosclerosis (Belghiti et al. 1981), and crescentic glomerulonephritis (Biava et al. 1984). In addition to the wide variety of glomerular histological changes, the associated lymphomas have been diverse, including lymphosarcoma (Ghosh and Muehrcke 1970; Gluck et al. 1973), and lymphocytic lymphoma (Gonzalez et al. 1986). It is thus very much open to question as to whether these represent coincidental occurrences or whether there is any direct causal link.

Immunoblastic lymphadenopathy is a prelymphomatous lymphoproliferative disorder that has been associated with acute renal failure in two patients (Wood and Harkins 1979).

Mycosis fungoides, a variant of malignant lymphoma involving the skin, has been associated with IgA nephropathy in two patients (Ramirez et al. 1982).

Leukaemia

Chronic lymphatic leukaemia has been associated with the nephrotic syndrome in 30 patients (reviewed by Touchard et al. 1989). Most commonly, the nephropathy, usually manifest as nephrotic syndrome, and leukaemia have been detected simultaneously. A variety of histological appearances have been described, including amyloidosis (Scott 1957; Leonard 1957), atypical membranous nephropathy (Brodovsky et al. 1968; Dathan et al. 1974), mesangiocapillary glomerulonephritis (Mandalenakis et al. 1971; Feehally et al. 1981; Touchard et al. 1989) and minimal change nephropathy (Kerkhoven et al.) 1973; Gagliano et al. 1976). Improvement of the nephropathy after treatment for the leukaemia is well described (Brodovsky et al. 1968; Touchard et al. 1989). Unfortunately, however, the therapy prescribed for the leukaemia (e.g. prednisolone or chlorambucil) may often have had a direct effect on the renal lesion.

An AA type amyloidosis has been described in association with hairy cell leukaemia (Linder et al. 1982).

Disseminated intravascular coagulation has been associated with a variety of neoplasms, particularly acute progranulocytic leukaemia.

Conclusions

An association between nephropathy and neoplasia is now generally accepted and has been reported for benign and malignant tumours, including carcinoma, lymphoma, leukaemia, and myeloma. In carcinoma, the pathogenesis in the majority of patients is thought to involve the formation, most commonly *in situ*, of immune complexes in a subepithelial position in the glomerular capillary wall. On light microscopy the appearances are those of a membranous nephropathy. The association has been demonstrated most commonly in adults and in patients with adenocarcinoma, particularly of the lung and gastrointestinal tract. Many other malignancies have been implicated, but usually on the basis of anecdotes. In adult patients with apparently idiopathic membranous glomerulonephritis, routine cancer screening tests are advisable to detect underlying malignancy, especially in those aged over 50. The prognosis depends more on the underlying malignancy than the type of renal disease. If tumour removal or therapeutic remission is achieved the nephropathy sometimes resolves, but when the tumour is uncontrollable the prognosis is particularly poor.

Glomerular disease also occurs in patients with lymphoma or leukaemia. The majority of patients with Hodgkin's lymphoma have an associated minimal change nephropathy, possibly mediated by concurrent T cell abnormalities; the nephropathy resolves on successful treatment of the lymphoma. The situation is far from clear. In non-Hodgkin's lymphoma there are only a few reported cases, involving a wide variety of glomerular appearances and of lymphoma types, suggesting that the association may be fortuitous. In leukaemia, particularly chronic lymphatic leukaemia, a number of patients have been reported, particularly with a mesangiocapillary or membranous nephropathy. While this association is probably not fortuitous, it is apparently rare.

References

Barton, C.H., Vaziri, N.D., and Spear, G.S. (1980). Nephrotic syndrome associated with adenocarcinoma of the breast. *American Journal of Medicine*, **68**, 308–12.

Beaufils, H., Jouanneau, C., and Chomette, G. (1985). Kidney and cancer: results of immunofluorescence microscopy. *Nephron*, **40**, 303–8.

Beauvais, P., Vaudour, G., Boccon Gibod, L., and Levy, M. (1989). Membranous nephropathy associated with ovarian tumour in a young girl: recovery after removal. *European Journal of Paediatrics*, **148**, 624–5.

Belghiti, D., Vernant, J.P., Hirbec, G., Gubler, M.C., Andre, C., and Sobel A., (1981). Nephrotic syndrome associated with T-cell lymphoma. *Cancer*, **47**, 1878–81.

Bellon, B., Belair, M.F., Kuhn, J., Druet, P., and Bariety, J. et al. (1982). Trapping of circulating proteins in immune deposits of Heymann nephritis. *Laboratory Investigation*, **46**, 306.

Biava, C.G., Gonwa, T.A., Naughton, J.L., and Hopper, J., Jr. (1984). Crescentic glomerulonephritis associated with non-renal malignancies. *American Journal of Nephrology*, **4**, 208–14.

Borochovitz, D., Kam, W.K., Nolte, M., Graner, S., and Kiss, J. (1982). Adenocarcinoma of the palate associated with nephrotic syndrome and epimembranous carcinoembryonic antigen deposition. *Cancer*, **49**, 2097–102.

Brodovsky, M.H., Samuels, M.L., Migliore, P.J., and Howe, C.D. (1968). Chronic lymphocytic leukaemia, Hodgkin's disease and the nephrotic syndrome. *Archives of Internal Medicine*, **121**, 71–5.

Brueggemeyer, C.D. and Ramirez, G. (1987). Membranous nephropathy: a concern for malignancy. *American Journal of Kidney Diseases*, **9**, 23–6.

Cairns, S.A., Mallick, N.P., Lawler, W., and Williams, G. (1978). Squamous cell carcinoma of bronchus presenting with Henoch-Schönlein purpura. *British Medical Journal*, **2**, 174–5.

Cameron, J.S. and Ogg, C.S. (1975). Neoplastic disease and the nephrotic syndrome. *Quarterly Journal of Medicine*, **44**, 630–1.

Cantrell, E.G. (1969). Nephrotic syndrome cured by removal of gastric carcinoma. *British Medical Journal*, **2**, 739–40.

Caruana, R.J. and Griffin, J.W. (1980). Nephrotic syndrome in a patient with ulcerative colitis and colonic carcinoma. *American Journal of Gastroenterology*, **74**, 525–8.

Castleman, B., Nichols, G., and Roth, S.I. (1963). Case records of the Massachusetts General Hospital. *New England Journal of Medicine*, **268**, 943–53.

Costanza, M.E., Pinn, V., Schwartz, R.S., and Nathanson, L. (1973). Carcinoembryonic antigen-antibody complexes in a patient with colonic carcinoma and nephrotic syndrome. *New England Journal of Medicine*, **289**, 520–2.

Couser, W.G. (1980). Renal cell cancer with amyloid disease. Case records of the Massachussets General Hospital. *New England Journal of Medicine*, **303**, 985–95.

Couser, W.G., Wagonfeld, J.B., Spargo, B.H., and Lewis, E.J. (1974). Glomerular deposition of tumour antigen in membranous nephropathy associated with colonic carcinoma. *American Journal of Medicine*, **57**, 962–70.

Da Costa, C.R., Dupont, E., Hamers, R., Hooghe, R., Dupuis, F., and Potvliege, R. (1974). Nephrotic syndrome in bronchogenic carcinoma: report of two cases with immunochemical studies. *Clinical Nephrology*, **2**, 245–50.

Dathan, J.R.E., Heyworth, M.F., and MacIver, A.G. (1974). Nephrotic syndrome in chronic lymphocytic leukaemia. *British Medical Journal*, **3**, 655–75.

De Swiet, J. and Wells, A.L. (1957). Nephrotic syndrome associated with renal venous thrombosis and bronchial carcinoma. *British Medical Journal*, **1**, 1341–3.

Dosik, G.M., Guterman, J.V., Hersh, E.M., Akhtar, M., Sonoda, T., and Horn, R.G. (1978). Nephrotoxicity from cancer immunotherapy. *Annals of Internal Medicine*, **89**, 41–6.

Drash, A., Sherman, F., Hartman, W., and Blizzard, R.M. (1970). A syndrome of pseudohermaphroditism, Wilm's tumour hypertension and degenerative renal diseases. *Journal of Pediatrics*, **76**, 585–93.

Dupond, J.L., et al. (1980). Renal cancer disclosed by a nephrotic syndrome with minimal glomerular lesion. *Presse Medicale*, **9**, 884–5.

Eagen, J.W. and Lewis, E.J. (1977). Glomerulopathies of neoplasia. *Kidney International*, **11**, 297–306.

Endo, Y. and Hara, M. (1986). Glomerular IgA deposition in pulmonary diseases. *Kidney International*, **29**, 557–62.

Endoh, M. (1984). Focal segmental glomerulonephritis associated with pheochromocytoma. *Tokai Journal of Experimental and Clinical Medicine*, **9**, 191–7.

Feehally, J., Hutchinson, R.M., MacKay, E.H., and Walls, J. (1981). Recurrent proteinuria and chronic lymphatic leukaemia. *Clinical Nephrology*, **16**, 51–4.

Friou, G.J. (1974). Current knowledge and concepts of the relationships of malignancy autoimmunity and immunologic disease. *Annals of the New York Academy of Science*, **230**, 23.

Froom, D.W., Franklin, W.A., Hano, J.E., and Potter, E.V. (1972). Immune deposits in Hodgkin's disease with nephrotic syndrome. *Archives of Pathology*, **94**, 547–53.

Gagliano, R.G., Costanzi, J.J., Beathard, G.A., Sarles, H.E., and Bells, J.D. (1976). The nephrotic syndrome associated with neoplasia: an unusual paraneoplastic syndrome. *American Journal of Medicine*, **60**, 1026–31.

Galloway, J. (1922). Remarks on Hodgkin's disease. *British Medical Journal*, **2**, 1201.

Ghosh, L. and Muehrcke, R.C. (1970). The nephrotic syndrome: a prodrome to lymphoma. *Annals of Internal Medicine*, **72**, 379–82.

Glenner, G.G. (1980). Amyloid deposits and amyloidosis. *New England Journal of Medicine*, **302**, 1283–92.

Gluck, M.C., Gallo, G., Lowenstein, J., and Baldwin, D.S. (1973). Membranous glomerulonephritis: evolution of clinical and pathologic features. *Annals of Internal Medicine*, **78**, 1–12.

Gonzalez, J.A.G., Bango, M.Y., Morales, F.M., Caro, J.L.M., Marin, F.J.D., and Martinez, A.D. (1986). The association of non-Hodgkin's lymphoma with glomerulonephritis. *Postgraduate Medical Journal*, **62**, 1141–5.

Habib, R., et al. (1985). The nephropathy associated with male pseudohermaphroditism and Wilm's tumour (Drash syndrome): a distinctive glomerular lesion—report of 10 cases. *Clinical Nephrology*, **24**, 269–78.

Haskell, L.P., Fusco, M.J., Wadler, S., Sablay, L.B., and Mennemeyer, R.P. (1990). Crescentic glomerulonephritis associated with prostatic carcinoma: evidence of immune-mediated glomerular injury. *American Journal of Medicine*, **88**, 189–92.

Heaton, J.M., Menzin, M.A., and Carney, D.N. (1975). Extra-renal malignancy and the nephrotic syndrome. *Journal of Clinical Pathology*, **28**, 944–6.

Heckerling, P.S. (1985). Oesophageal carcinoma with membranous nephropathy. *Annals of Internal Medicine*, **103**, 474.

Helin, M., Pasternack, A., Hakala, J., Penttinen, K., and Wager, O. (1980). Glomerular electron-dense deposits and circulating immune complexes in patients with malignant tumours. *Clinical Nephrology*, **14**, 23–30.

Higgins, M.R., Randall, R.E., and Still, W.J.S. (1974). Nephrotic syndrome with Oat-cell carcinoma. *British Medical Journal*, **3**, 450–1.

Hopper, J. (1974). Tumour-related renal lesions. *Annals of Internal Medicine*, **81**, 550–1.

Hopper, J., Biava, C.B., and Naughton, J.L. (1976). Glomerular extracapillary proliferation, E.P. (crescentic glomerulonephritis) associated with non-renal malignancies. *Kidney International*, **10**, 544.

Hyman, L.R., Burkholder, P.M., Joo, P.A., and Segar, W.E. (1973). Malignant lymphoma and nephrotic syndrome. *Journal of Pediatrics*, **82**, 207–17.

Jermanovich, N.B., Giammarco, R., Ginsberg, S.J., Tinsley, R.W., and Jones, D.B. (1982). Small cell anaplastic carcinoma of the lung with mesangial proliferative glomerulonephritis. *Archives of Internal Medicine*, **142**, 397–9.

Kaplan, B.S., Klassen, J., and Gault, M.H. (1976). Glomerular injury in patients with neoplasia. *Annual Review of Medicine*, **27**, 117–25.

Kerkhoven, P., Briner, J., and Blumberg, A. (1973). Nephrotisches Syndrome als Erstmanifestation maligner Lymphoma. *Schweizer Medizinische Wochenschrift*, **103**, 1706–9.

Kerpen, H.O., Ganesh Bhat, J., Felner, H.D., and Baldwin, D.S. (1978). Membranous nephropathy associated with renal cell carcinoma. *American Journal of Medicine*, **64**, 863–7.

Keur, I., Krediet, R.T., and Arisz, L. (1989). Glomerulopathy as a paraneoplastic phenomenon. *Netherlands Journal of Medicine*, **34**, 270–84.

Kimball, K.G. (1961). Amyloidosis in association with neoplastic disease: report of an unusual case and clinicopathological experience at the Memorial Center for Cancer and Allied Diseases during eleven years (1948–1958). *Annals of Internal Medicine*, **55**, 58–74.

Kitano, S. (1984). Poorly differentiated adenocarcinoma of rectum in a nephrotic patient with focal segmental glomerulosclerosis. *Japanese Journal of Surgery*, **14**, 155–8.

Kleinknecht, C., Levy, M., Gagnadoux, M.F., and Habib, R., (1979). Membranous glomerulonephritis with extra-renal disorders in children. *Medicine*, **58**, 219–28.

Kramer, P., Sizoo, W., and Twiss, E.E. (1981). Nephrotic syndrome in Hodgkin's disease. *Netherlands Journal of Medicine*, **24**, 114–19.

Latham, P., Poucell, D., Koresaar, A., Arbus, G., and Baumal, R. (1982). Idiopathic membranous glomerulopathy in Canadian children: clinicopathologic study. *Journal of Pediatrics*, **101**, 682–5.

Lee, J.C., Yamauchi, H., and Hopper, J. (1966). The association of cancer and the nephrotic syndrome. *Annals of Internal Medicine*, **64**, 41–51.

Leonard, B.J. (1957). Chronic lymphatic leukaemia and the nephrotic syndrome. *Lancet*, **i**, 1356–7.

Lewis, M.G., Loughbridge, L., and Phillips, T.M. (1971). Immunological studies in nephrotic syndrome associated with extra-renal malignant disease. *Lancet*, **ii**, 134–5.

Linder, J., Silberman, H.R., and Croker, B.P. (1982). Amyloidosis complicating hairy cell leukaemia. *American Journal of Clinical Pathology*, **78**, 864–7.

Lokich, J.J., Galvanek, E.G., and Moloney, W.C. (1973). Nephrosis of Hodgkin's disease. *Archives of Internal Medicine*, **132**, 597–600.

Lumeng, J., and Moran, J.F. (1966). Carotid body tumour associated with mild membranous glomerulonephritis. *Annals of Internal Medicine*, **65**, 1266–70.

Mandalenakis, N., Mendoza, N., Pirani, C.L., and Pollak, V.E. (1971). Lobular glomerulonephritis and membranoproliferative glomerulonephritis: a clinical and pathologic study based on renal biopsies. *Medicine*, **50**, 319–55.

Moorthy, A.V. (1983). Minimal change glomerular disease: a paraneoplastic syndrome in two patients with bronchogenic carcinoma. *American Journal of Kidney Diseases*, **3**, 58–62.

Morel-Maroger Striker, L., Mignon, F., Dabbs, D., and Striker, G.E. (1984). Glomerular lesions in lymphomas and leukaemias In *Nephrology* Robinson R.R. (ed., et al.) New York. Springer-Verlag, p.905–15.

Muggia, F.M., and Ultmann, J.E. (1971). Glomerulonephritis or nephrotic syndrome in malignant lymphoma, reticulum-cell type. *Lancet*, **i**, 805.

Mustonen, J., Henn, H., and Pasternack, A. (1981). IgA nephropathy associated with bronchial small cell carcinoma. *American Journal of Clinical Pathology*, **76**, 652–6.

Norris, H.J., and Weiner, J. (1961). The renal lesions in leukaemia. *American Journal of Medical Science*, **241**, 512–7.

Oldstone, M.B.A., Theofilopoulos, A.N., Gunven, P., and Klein, G. (1974). Immune complexes associated with neoplasia: Presence of Epstein-Barr virus antigen-antibody complexes in Burkitt's lymphoma. *Intervirology*, **4**, 292–302.

Olson, J.L., Philips, T.M., Lewis, M.G., and Solez, K. (1979). Malignant melanoma with renal dense deposits containing tumour antigens. *Clinical Nephrology*, **12**, 74–82.

Ozawa, T., et al. (1975). Endogenous immune complex nephrology associated with malignancy, 1 Studies on the nature and immunopathogenic significance of glomerular bound antigen and antibody. Isolation and characterisations of tumour specific and antibody circulating immune complexes. *Quarterly Journal of Medicine*, **44**, 523–41.

Pascal, R.R. (1980). Renal manifestations of extra-renal neoplasms. *Human Pathology*, **11**, 7–17.

Pascal, R.R., et al. (1975). Glomerular immune complex deposits associated with mouse mammary tumour. *Cancer Research*, **35**, 302–4.

Pascal, R.R., Finney, R.P., Rifkin, S.I., and Kahana, L. (1980). Glomerulonephritis and virus like particles associated with prostatic cancer. *Human Pathology*, **11**, 391–5.

Pascal, R.R., Innaccone, P.M., and Rollwagen, F.M. (1976). Electron microscopy and immunofluorescence of glomerular immune complex deposits in cancer patients. *Cancer Research*, **36**, 43–7.

Pascal, R.R., and Slovin, S.G. (1980). Tumour directed antibody and carcinoembryonic antigen in the glomeruli of a patient with gastric carcinoma. *Human Pathology*, **11**, 679–82.

Paraf, A., Coste, T., Rautureau, J., and Texier, J. (1970). La régression de l'amylose: disparition d'une amylose hépatique massive après nephrectomie pour cancer. *Presse Medicale*, **78**, 547–8.

Plager, J., and Stutzman, J. (1971). Acute nephrotic syndrome as a manifestation of active Hodgkin's disease. *American Journal of Medicine*, **50**, 56–66.

Puolijoki, H., Mustonen, J., Pettersson, E., Pasternack, A., and Lamdensuo, A. (1989). Proteinuria and haematuria and frequently present in patients with lung cancer. *Nephrology Dialysis Transplantation*, **4**, 947–50.

Rabkin, R., Thatcher, G.N., Diamond, L.H., and Eales, L. (1973). The nephrotic syndrome, malignancy and immunosuppression *South African Medical Journal*, **14**, 605–6.

Ramirez, F., Brouhard, B.H., Travis, L.B., and Ellis, E.N. (1982). Idiopathic membranous nephropathy in children. *Journal of Pediatrics*, **101**, 677–81.

Richmond, J., et al. (1962). Renal lesions associated with malignant lymphomas. *American Journal of Medicine*, **32**, 184–207.

Robinson, W.L., Mitas, J.A., Haerr, R.W., and Cohen, I.M. (1984). Remission and exacerbation of tumour-related nephrotic syndrome with treatment of the neoplasm. *Cancer*, **54**, 1082–4.

Rossen, R.D., Reisberg, R.A., Hersh, E.M., and Gutterman, J.U. (1976). Measurement of soluble immune complexes: A guide to prognosis in cancer patients. *Clinical Research*, **24**, 462A.

Row, P.G., et al. (1975). Membranous nephropathy, long term follow up and association with neoplasia. *Quarterly Journal of Medicine*, **44**, 207–39.

Roy, A.D. and Ellis, M. (1959). Potassium secreting tumour of the large intestine. *Lancet*, **i**, 759–60.

Sagel, J., Muller, H., and Logan, E. (1971). Lymphoma and the nephrotic syndrome. *South African Medical Journal*, **45**, 79–80.

Sawyer, N., Wadsworth, J., Winnen, M., and Gabriel, R. (1988). Prevalence, concentration and prognostic importance of proteinuria in patients with malignancies. *British Medical Journal*, **296**, 295–8.

Schroeter, N.J., Rushing, D.A., Parker, J.P., and Beltaos, E. (1986). Minimal change nephrotic syndrome associated with malignant mesothelioma. *Archives of Internal Medicine*, **146**, 1834–6.

Scott, R.B. (1957). Chronic lymphatic leukaemia. *Lancet*, **i**, 1162–7.

Seyberth, K.W., Segre, G.V., Morgan, J.L., Sweetman, B.J., Potts, J.T., and Oaes, J.A. (1975). Prostaglandins as mediators of hypercalcaemia associated with certain types of cancer. *New England Journal of Medicine*, **293**, 1278–83.

Stuart, K., Fallon, B.G., and Cardi, M.A. (1986). Development of the nephrotic syndrome in a patient with prostatic carcinoma. *American Journal of Medicine*, **80**, 295–8.

Sutherland, J.C. and Mardiney, M.R. (1973). Immune complex disease in the kidneys of lymphoma-leukaemia patients: the presence of an oncornavirus-related antigen. *Journal of the National Cancer Institute*, **50**, 633–44.

Sutherland, J.C., Markham, R.V., and Mardiney, M.R. (1974). Subclinical immune complexes in the glomeruli of kidneys postmortem. *American Journal of Medicine*, **57**, 536–41.

Swaroop, S. and Krant, M.J. (1973). Rapid estrogen-induced hypercalcaemia. *Journal of the American Medical Association*, **223**, 913–4.

Touchard, G., et al. (1989). Nephrotic syndrome associated with chronic lymphocytic leukaemia: an immunological study. *Clinical Nephrology*, **31**, 107–16.

Vanatta, P.R., Silva, F.G., Taylor, W.E., and Costa, J.C. (1983). Renal cell carcinoma and systemic amyloidosis. *Human Pathology*, **14**, 195–201.

Volhard, F. (1931). (Cited by Revol L., et al. (1964)). Protéinurie associé à des manifestations paranéoplastiques au course d'un cancer bronchogenique. *Lyon Médecine*, **212**, 907–16.

Wakashin, M., Wakashin, Y., and Iesato, K. (1980). Association of gastric cancer and nephrotic syndrome. An immunologic study in three patients. *Gastroenterology*, **78**, 749–56.

Walker, J.F., O'Neill, S., and Campbell, E. (1981). Carcinoma of the oesophagus associated with membranoproliferative glomerulonephritis. *Postgraduate Medical Journal*, **57**, 592–6.

Watson, A., Stachura, I., Fragola, J., and Bourke, E. (1983). Focal and segmental glomerulosclerosis in Hodgkin's disease. *American Journal of Nephrology*, **3**, 228–32.

Weksler, T.C., Day, N., Susin, M., Sherman, R., and Becker, C. (1974). Nephrotic syndrome in malignant melanoma: demonstration of melanoma antigen-antibody complexes in kidney. *Kidney International*, **6**, 112A.

Whitworth, J.A., Unge, A., and Cameron, J.S. (1975). Carcinoembryonic antigen in tumour-associated membranous nephropathy. *Lancet*, **ii**, 611.

Wood, W.G. and Harkins, M.M. (1979). Nephropathy in angioimmunoblastic lymphadenopathy. *American Journal of Clinical Pathology*, **71**, 58–63.

Yamauchi, M., Linsey, M.S., Biava, C.G., and Hopper, J. (1985). Cure of membranous nephropathy after resection of carcinoma. *Archives of Internal Medicine*, **145**, 2061–3.

Young, R.H., Scully, R.E., and McCluskey, R.T. (1985). A distinctive glomerular lesion complicating placental site trophoblastic tumour: report to two cases. *Human Pathology*, **16**, 35–42.

Zech, P., Colon, S., Pointet, P.L., Deteiz, P., Labeevu, M., and Leitienne, P.L. (1982). The nephrotic syndrome in adults aged over 60: etiology, evolution and treatment of 76 cases. *Clinical Nephrology*, **17**, 232–6.

Zheng, H.L., Maruyama, T., Matsuda, S. and Satomura, K. (1979). Neuroblastoma presenting with the nephrotic syndrome. *Journal of Paediatric Surgery*, **14**, 414–19.

3.14 Glomerular disease in the tropics

K.S. CHUGH AND VINAY SAKHUJA

Glomerular disease in the tropics differs considerably from that seen in temperate zones in its incidence, aetiology, and natural history. In some parts of Africa, the hospital admission rate for nephrotic syndrome is 100 times higher than that in the United Kingdom, Europe, and United States (Table 1). It has also been estimated that in the tropics, nearly 100 young adults per million population reach end-stage renal failure due to glomerulonephritis each year (Editorial 1980). In rural areas, the syndrome may be underdiagnosed because anasarca may be attributed to hypoproteinaemia resulting from malnutrition. Moderate proteinuria on the other hand is more likely to manifest itself by the development of oedema in malnourished subjects.

The prevalence of primary and secondary glomerular disease in adults in different countries is shown in Table 2. In most areas, primary glomerular disease is more common and accounts for more than two-thirds of all patients with glomerulonephritis. However, in Jamaica, more than 50 per cent of patients with a nephrotic syndrome have secondary glomerular disease (Morgan et al. 1984). In Zimbabwe, while 60 per cent of adult nephrotics have primary glomerular disease (Seggie et al. 1984), 80 per cent of children have secondary glomerular disease associated with hepatitis B or streptococcal infection. The prevalence of various histological types of primary glomerular disease in adults with a nephrotic syndrome is shown in Table 3, and in children in Table 4.

Table 1 Incidence of nephrotic syndrome as a percentage of hospital admissions

Country	Authors	Incidence (%)
United Kingdom	Kibukamusoke et al. (1967)*	0.04
United States	Kibukamusoke et al. (1967)	0.03–0.15
China	Kibukamusoke et al. (1967)	0.02
South Africa	Seedat (1978)	0.02
Zimbabwe	Seggie et al. (1984)	0.05
Uganda	Kibukamusoke et al. (1967)	2.0
Nigeria	Kibukamusoke et al. (1967)	2.4
Papua New Guinea	Powell et al. (1977)	4.0
Senegal	Verroust et al. (1979)	0.85
Yemen	Kibukamusoke (1984)	3.0
Guyana	Giglioli (1962)	0.05

* Data of J.S. Cameron et al., unpublished.

Primary glomerular disease

The clinical and histological features of various primary glomerulopathies will not be discussed here. Only the prevalence and features specific to these conditions in the tropical countries are highlighted.

Minimal change nephropathy

A striking feature is the rarity with which minimal change is encountered on the African continent. In Nigeria, Tunisia and

3.14 Glomerular disease in the tropics

Table 2 Prevalence of primary and secondary glomerular disease in adults in different countries

Country	Authors	Primary (%)	Secondary (%)
Jamaica	Morgan et al. (1984)	46	54
Ghana	Adu et al. (1981)	86	14
Sudan	Musa et al. (1980)	78	22
South Africa	Seedat (1978)	93	7
Zimbabwe	Seggie et al. (1984)	60	40
Pakistan	Sadiq et al. (1978)	78	22
Singapore	Sinniah and Khoo (1979)	87	13
India	Chugh and Sakhuja (1984)	69	31
Developed countries	Pesce and First (1979)	70	30

Table 3 Prevalence of histological types of primary glomerular disease in adults with the nephrotic syndrome in different countries

	Indonesia (Sidabutar et al. 1986)	Jamaica (Morgan et al. 1984)	Ghana (Adu et al. 1981)	Sudan (Musa et al. 1980)	South Africa (Seedat 1978)	Pakistan (Sadiq et al. 1978)	Papua New Guinea (Powell et al. 1977)	Singapore (Sinniah and Khoo 1979)	North India*	Developed countries (Pesce and First 1979)
Minimal change	16	26	10	17	4	15	4	17	24	27
Membranous	4	4	16	33	54	26	10	6	13	19
Mesangial proliferation	34	39	–	–	2	–	16	69	4	31
Diffuse proliferation	11	18	16	38	29	41	25	–	10	–
Mesangiocapillary	14	9	13	–	11	–	8	–	18	8
Focal glomerulosclerosis	7	4	36	12	–	–	7	–	16	7
Others	14	–	9	–	–	18	30	8	15	8

Numbers are percentages.
* Data of K.S. Chugh and V. Sakhuja (unpublished).

Table 4 Prevalence of various forms of glomerulonephritis in children with a nephrotic syndrome

	South India (n = 120) (Johny 1978)	Papua New Guinea (Powell et al. 1977)	North India (n = 195) (Srivastava et al. 1975)	Zimbabwe (n = 17) (Seggie et al. 1984)	South African black patients (n = 74) (Coovadia et al. 1979)	South African Indian patients (n = 56) (Coovadia et al. 1979)	Nigeria (n = 40) (Abdurrahman and Edington 1982)
Minimal change	38	15	77	6	13.5	75	17.5
Membranous	1	15	1.5	41*	28.4	3.6	
Mesangial proliferative	35		2	12*			
Diffuse proliferative	7	62	5	17†	19.2	8.8	7.5
Mesangiocapillary	3		4		8.1	3.6	35
Focal glomerulosclerosis	9		5.5	12	5.4	1.8	
Others	7	8	5	12	26	7.2	40‡

Numbers are percentages.
* All associated with hepatitis B surface antigenaemia.
† All poststreptococcal in aetiology.
‡ Mostly patients with quartan malarial nephropathy.

amongst the South African black population, minimal change is observed in less than 4 per cent of adults while it is not seen at all in Senegal and Malawi (Morel-Maroger et al. 1975; Brown et al. 1977). In a study of 130 South African children of whom 74 were black and 56 were Indians, contrasting clinicopathological patterns were noted. While 86 per cent of black children had structural glomerular lesions (proliferative or membranous nephropathy), and 13.5 per cent had minimal change disease, 75 per cent of Indians had minimal change nephropathy (Coovadia et al. 1979). Similarly, in Nigeria, Abdurrahman and Edington (1982) found minimal lesions in only 17.5 per cent of children. The spectrum of glomerular disease in children with a nephrotic syndrome in India is not different from that seen in Europe and North America (Srivastava et al. 1975). As a result the prognosis of idiopathic nephrotic syndrome in African patients is likely to be poorer than in those elsewhere.

Focal segmental glomerulosclerosis

An unusually high incidence of focal segmental glomerulosclerosis has been reported amongst adult nephrotics in Ghana (Adu et al. 1981) and in children in Senegal (Morel-Maroger et al. 1975). The latter authors observed that the condition seen in Senegal children has distinctive histological features comprising focal segmental sclerosis and fibrillary splitting of the capillary walls with intrusion of basement membrane-like material into the capillary lumen. The lesion appeared identical to that seen in Nigerian children with quartan malarial nephropathy; however,

Membranous nephropathy

A wide variation in the incidence of membranous nephropathy in various regions is apparent from the figures given in Table 3. It is a rare cause of the nephrotic syndrome in adults in Jamaica, where it accounts for only 4 per cent of those with primary glomerular disease (Morgan et al. 1984). Membranous nephropathy in Africa and south-east Asia is frequently associated with hepatitis B infection. From Senegal a tropical membranous nephropathy of unknown aetiology associated with hypocomplementaemia has been reported (Morel-Maroger et al. 1975).

Poststreptococcal nephritis

While poststreptococcal nephritis has become a rare disease in developed countries (Meadow 1975), it remains common in several tropical and subtropical areas. In Zimbabwe, it is the commonest cause of nephrotic syndrome in young adults (Seggie et al. 1984) and accounts for 28 per cent of all patients with a nephrotic syndrome. In South Africa, it is the commonest renal problem encountered in children, with 2 per cent of all paediatric admissions being affected. In Trinidad, four epidemics of poststreptococcal nephritis have occurred between 1952 and 1973 (Poon-King et al. 1973). In India too, it continues to be a common problem (Johny 1978; Chugh et al. 1987). In the tropics, streptococcal infection is predominantly present in the skin, often in association with lesions of scabies (Whittle et al. 1973; Poon-King et al. 1973), although pharyngeal infection remains the major site of infection in some regions (Chugh et al. 1987; Date et al. 1987).

IgA nephropathy

Data on the incidence of IgA nephropathy in the tropical countries have recently been reviewed by Levy and Berger (1988). IgA nephropathy is the most common primary glomerular disease in Hong Kong, Singapore, and Taiwan where it accounts for 25 to 39 per cent of all patients (Sinniah and Khoo 1979; Chen et al. 1986). In Malaysia and Thailand, 18 and 8 per cent of patients with primary glomerular disease respectively have IgA nephropathy. Data on the prevalence of this problem on the African continent are scanty. In Tunisia, only 1 per cent of all biopsies for glomerular disease reveal IgA nephropathy; in South Africa only 0.8 per cent of black and as many as 13.3 per cent of Indian patients with primary glomerular disease have IgA nephropathy. A low incidence in black patients has also been observed in North America. In India, 4 to 7 per cent of adults and 10 per cent of children with renal disease who undergo renal biopsy have IgA nephropathy. IgA nephropathy appears to be uncommon in Mexico where only 1.4 per cent of all biopsies show this lesion. In Brazil 10 per cent of all biopsies and in Chile 10 per cent of biopsies in those with primary glomerular disease reveal IgA nephropathy.

Secondary glomerular diseases

Systemic lupus erythematosus (see Chapter 4.6.2)

Differences in racial susceptibility may account for the uneven distribution of systemic lupus erythematosus. It is common amongst Chinese patients in Malaysia (Frank 1980; Prathap and Looi 1982) and Singapore (Feng 1986). The admission rates for SLE in six hospitals in China have been observed to range from 1.6 to 3.3 per cent. In a hospital-based study of different ethnic groups in Hawaii, the prevalence rates for SLE per 100 000 patient population were white patients 5.8, Chinese 24.1, Filipino 19.9, Japanese 18.2, and part-Hawaiian 20.4 (Serdula and Rhoads 1979). In Jamaica, it is the commonest cause of nephrotic syndrome resulting from secondary glomerular disease and accounts for 38 per cent of all nephrotics (Morgan et al. 1984). However, the disease does not appear to be common in West Africa, the origin of the forebears of most Jamaicans (Adu et al. 1981). Black patients in America have a three-fold greater prevalence of the disease than white patients (Fessel 1974). In other African countries also, SLE appears to be uncommon. Kanyerezi et al. (1980) saw only 21 patients over a period of 11 years at a referral centre in Uganda while Seedat and Pudifin (1977) encountered only 13 black patients in Natal in South Africa over a 6-year period. Seboxa et al. (1984) reported on 16 patients in Ethiopia over a 4-year period. In India, systemic lupus is not uncommon, and accounts for 11 per cent of all patients with secondary glomerular disease (Chugh and Sakhuja 1984). In Zimbabwe, 2 per cent of all patients with a nephrotic syndrome have systemic lupus erythematosus (Seggie et al. 1984).

Amyloidosis

Amyloidosis (see Chapter 4.2) is one of the commonest causes of secondary glomerular disease in India. The incidence of amyloidosis at autopsy has ranged from 0.5 to 5.1 per cent at various centres (Mathur and Jhala 1964; Chitkara et al. 1965; Date and Job 1974). In a series of 1980 renal biopsies carried out in our unit for glomerular disease, amyloidosis was encountered in 8.4 per cent (Chugh et al. 1981). Most Indian patients with renal amyloidosis have secondary amyloidosis: among 233 patients, 87 per cent had secondary amyloid (Chugh et al. 1981). Tuberculosis was the commonest cause of secondary amyloid in the latter study, whereas rheumatoid arthritis is now the most frequent cause in the West (Kyle and Bayrd 1975) (see Chapter 4.2).

In Papua New Guinea, leprosy has been observed to be the predisposing disease in 60 per cent of cases of secondary amyloidosis (Cooke and Champness 1970; McAdam et al. 1975). It is surprising to note that amyloidosis has seldom been encountered in various African surveys of the nephrotic syndrome, despite the high incidence of tuberculosis and suppurative disease. The prevalence in Nigeria and Uganda was only 0.3 and 0.57 per cent of biopsy series respectively (Edington and Mainwaring 1964; James and Owor 1975). Amyloidosis was found in only 3 per cent of adult black patients with a nephrotic syndrome in South Africa (Seedat 1978) and 2 per cent of adult nephrotics in Jamaica (Morgan et al. 1984).

Schistosomal glomerulopathy (see also Chapter 13.5)

Schistosomiasis (bilharziasis) is one of the most widespread parasitic diseases in the tropics. More than 200 million people in Africa, the Near East, Asia, and South America are affected. *Schistosoma haematobium*, *S. mansoni*, and *S. japonicum* are the three main species infecting man from contact with water containing cercariae liberated by infected snails and usually contracted in childhood or adolescence. The pathological lesions are characterized by granuloma formation as a result of egg deposition by the adult worm, or ova travelling by the bloodstream to various organs.

S. haematobium involves the urinary bladder, ureters, and the genital tract. The bladder capacity is reduced and the ureters become stenosed causing hydroureter, hydronephrosis, and insidious renal failure (Chugh *et al.* 1986). Glomerular disease has been reported only in isolated case reports (Greenham and Cameron 1980).

S. mansoni and *S. japonicum* affect mainly the gut and liver. Granulomata with fibrosis develop in the liver around branches of the portal vein (pipestem fibrosis). Portal hypertension, hepatomegaly, splenomegaly, ascites, dilated abdominal veins, and oesophageal varices may result.

Though both *S. mansoni* and *S. japonicum* have been shown to cause glomerular lesions in experimental studies, clinical glomerular disease has been reported only in association with *S. mansoni* infection. Most reports of glomerulopathy associated with *S. mansoni* have emanated from Brazil (Andrade *et al.* 1971; Falcao and Gould 1975; Rocha *et al.* 1976) and Egypt (Sobh *et al.* 1987). Prospective studies in patients with *S. japonicum* infection in the Philippines have failed to show increased incidence of renal disease in these patients compared with that in a control population without schistosomiasis (Watt *et al.* 1987). The reasons for the difference in prevalence of nephropathy in *S. mansoni* and *S. japonicum* infection are not clear.

Incidence

Lopes (1964) reported abnormal proteinuria and leucocyturia in 26.7 per cent of patients with hepatosplenic schistosomiasis and 3.8 per cent of patients with the milder intestinal form of the disease. In another study on 100 consecutive patients with hepatosplenic disease, 15 were found to have overt renal involvement (Rocha *et al.* 1976). At the same centre, an incidence of glomerulonephritis of 12 per cent was recorded amongst patients at autopsy (Andrade *et al.* 1971). However, these figures may not reflect the true prevalence of the condition since the surveys were conducted in a hospital with a special interest in schistosomal nephropathy. A significant number of patients may also have subclinical renal involvement. Of 15 patients with hepatosplenic disease who had no clinical evidence of glomerulonephritis, six showed glomerular lesions when subjected to open renal biopsy at the time of splenectomy (Rocha *et al.* 1976). Although earlier investigators had observed renal involvement primarily in patients with hepatosplenic disease, a recent study has shown that renal lesions could occur even in the hepatointestinal phase (Sobh *et al.* 1988).

Clinical manifestations

The majority of the infected individuals acquire only a light worm load and may be asymptomatic. However, about 4 to 6 per cent carry a heavy worm load and develop hepatosplenic disease over several years. These patients often present with hepatosplenomegaly, ascites, and haematemesis. Among those with severe hepatosplenic schistosomiasis, glomerulopathy has been documented in 12 to 15 per cent (Andrade and Rocha 1979).

Sixty per cent of patients with renal involvement have a nephrotic syndrome (Rocha *et al.* 1976) while the remainder have non-nephrotic proteinuria. Hypertension is seen in less than 50 per cent of these patients. One-third of patients have hyper-

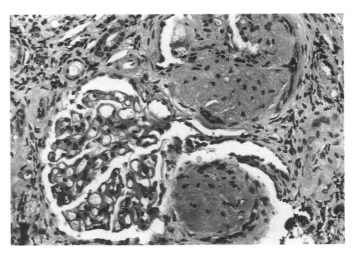

Fig. 1 Microphotograph showing membranous nephropathy in a patient with end-stage renal failure associated with *Schistosoma mansoni* infection (PAS× 345). (Reproduced with permission of *Kidney International* (data of Sobh *et al.* 1987).)

globulinaemia and a normal serum cholesterol level. Proteinuria is either non-selective or poorly selective (Queiroz *et al.* 1973) resulting in excretion of IgM, α_2-macroglobulins and lipoproteins. Serum complement (C3) levels are low and circulating immune complexes are present.

The presence of viable ova in stool specimens is diagnostic of the parasitic infection. Stool examination is rarely negative and in such cases, rectal and/or liver biopsies may be required to confirm the diagnosis. Histological examination of the tissues may show granulomas containing the eggs.

Renal histology
Light microscopy

The majority of patients show a chronic membranoproliferative glomerulonephritis, frequently with lobular accentuation. The mesangium is expanded and this is most obvious at the glomerular vascular pole. The mesangial expansion is due to the presence of amorphous and a fibrillar, PAS-positive material. There is mild to moderate increase in cellularity (Andrade and Rocha 1979). In early schistosomal nephropathy, Sobh *et al.* (1988a) observed mesangial proliferation as the most frequent lesion.

Focal glomerular sclerosis is almost as common. Other glomerular lesions encountered in these patients are those of membranous nephropathy (Fig. 1), acute diffuse proliferative glomerulonephritis, crescentic glomerulonephritis, end-stage kidney disease, and occasionally amyloidosis (Omer and Wahab 1976; Andrade and Rocha 1979). Progression from a predominantly membranous lesion to a lobular mesangiocapillary glomerulonephritis has also been seen in a few instances (Andrade and Rocha 1979).

Immunofluorescence

Glomeruli show heavy deposits of IgM and to a lesser extent IgG, IgA, IgE, and complement components C3 and C1q. The deposits are granular and localized to mesangium and the capillary walls.

Indirect immunofluorescence shows deposits of circulating anodic (Fig. 2) and cathodic antigen in the mesangium and along the capillary wall (Sobh *et al.* 1987; Sobh *et al.* 1988a).

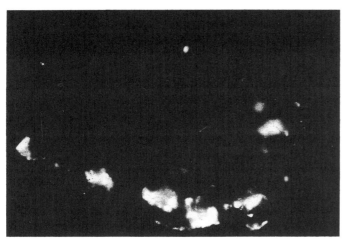

Fig. 2 Granular deposits of circulating anodic antigen [CAA] along the capillary basement membrane. (Same patient as Fig. 1, reproduced with permission.)

Electron microscopy

Electron-dense deposits and laminar bodies have been demonstrated in the mesangial area. Changes in the basement membrane are less prominent and consist of subepithelial, subendothelial, and intramembranous deposits. In some cases schistosomal pigment can be found in the mesangial matrix.

Pathogenesis

There is adequate evidence that schistosomal glomerulopathy is an immune complex-mediated glomerulonephritis since both antigen and antibody have been demonstrated in the glomerular lesions (Hoshino-Shimizu et al. 1976; Morierty and Brito 1977). However, the exact antigenic component responsible has not been characterized. About 60 immunogenic substances have been extracted from adult worms, cercariae and schistosomula (Waksman and Cook 1975). However, only the few that are present on exposed sites may be important in inciting antibody formation.

Berggren and Weller (1967) demonstrated the presence of a 'circulating schistosomal antigen' in the sera and urine of hamsters and mice infected with S. mansoni. This antigen was later characterized by Nash et al. (1974) as a polysaccharide with a molecular weight of 100 000 Da, and could be demonstrated in the primordial oesophagus of cercariae, caecum of the schistosomula, and vomitus of the adult worm. Another polysaccharide antigen of molecular weight of 30 000 Da, which presumably originates from the gut of the adult worm has been demonstrated by Deelder et al. (1976) in the serum and urine of infected hamsters.

Apart from adult worm antigen, a soluble egg antigen has been described by Houba et al. (1976), which appears to cause the periovular granulomatous reaction. However, its role in the pathogenesis of the glomerular lesion has not been defined.

The load of antigen seems to be important in the pathogenesis of schistosomal glomerulopathy. In less severe infestation the antigen excreted by the worm is cleared from the portal circulation by the reticuloendothelial cells in the liver and does not reach the systemic circulation. In heavily infested subjects with hepatosplenic disease the antigen reaches the systemic circulation because of the excess load as well as a portasystemic bypass in the liver as a result of presinusoidal portal vein obstruction. The importance of this bypass is shown by the fact that glomerulonephritis is rarely seen in humans infected with S. haematobium where liver damage is seldom severe. In baboons infected with S. mansoni, liver fibrosis does not develop and in this species, even heavy infection does not result in glomerular lesions. In experimental animals infected with S. mansoni, if the portal vein is ligated, the immune complexes bypass the liver, reach the systemic circulation, and induce changes of immune complex glomerulonephritis (Van Marck et al. 1977). The recent demonstration of antibodies against Schistosoma mansoni adult worm antigen in the eluates from the kidney-biopsy tissue of patients of schistosomal nephropathy and their absence in control specimens by the ELISA technique as well as the presence of granular deposits of circulating anodic antigen in the ELISA positive cases, is clear proof that nephropathy in these cases is schistosomal specific (Sobh et al. 1987).

S. japonicum and S. haematobium

Whereas experimentally infected monkeys have been shown to develop glomerular lesions associated with the presence of schistosomal antigen and deposition of immunoglobulins and complement (Tada et al. 1975), no human case of S. japonicum nephropathy has been reported.

In S. haematobium infection, although a nephrotic syndrome was observed in some patients by Ezzat et al. (1974), it was felt that this was a consequence of concomitant chronic salmonella infection and was not due to schistosomiasis itself (Farid et al. 1972). However, Greenham and Cameron (1980) have reported two instances of nephrotic syndrome due to mesangiocapillary glomerulonephritis in the absence of salmonellosis. Although the association may have been purely coincidental, remission of the nephrotic state following a course of niridazole suggested that the glomerular lesion was causally related to S. haematobium infection.

Management

Schistosomal glomerulopathy represents a type of glomerulonephritis associated with a potentially treatable parasitic infestation. Therefore it would be reasonable to expect that elimination of the worms and thereby the antigen load would lead to resolution of the renal lesions. However, the results of treatment with both immunosuppressive drugs and antiparasitic agents have been disappointing.

Of the 14 nephrotics treated by Queiroz et al. (1973) with cyclophosphamide with or without prednisolone, only two had a complete remission, four had a partial remission, and eight showed no response.

Dutra et al. (1979) treated 27 patients with antischistosomal drugs and immunosuppressives. The former had no effect on the degree of proteinuria or renal dysfunction but with immunosuppressive therapy, almost one-third of patients had a complete remission. Martinelli et al. (1987) on the other hand observed no benefit in 16 patients treated with a combination of antiparasitic drugs (oxamniquine or hycanthone) and prednisolone with or without cyclophosphamide. Similarly, a combination of praziquantel and oxamniquine proved ineffective in inducing remission in 21 patients with schistosomal nephropathy (Sobh et al. 1988b). Sooner or later, most patients with schistosomal nephropathy progress to end-stage renal failure.

Glomerular disease in leprosy

Leprosy is a widespread disease in Africa, India, Brazil, China, and many other countries of south-east Asia and South America. According to recent estimates, there are about 12 million cases of leprosy throughout the world, of which 3 million are in India.

Mitsuda and Ogawa (1937) were the first to recognize that significant renal lesions can be associated with leprosy, and also observed that renal failure was a common cause of death amongst these patients. A variety of glomerular lesions have since been identified. The lesions encountered include secondary amyloidosis and several types of glomerulonephritis. Interstitial nephritis, leproma formation, and isolated functional renal tubular defects have also been documented.

Clinical manifestations

Most patients with renal involvement present with asymptomatic urinary abnormalities. Less frequently, patients may develop nephrotic syndrome or a nephritic illness. The clinical course in some cases is consistent with rapidly progressive glomerulonephritis manifesting as oliguric renal failure (Bedi et al. 1977; Singhal et al. 1977). Hypertension is observed only rarely. Progression to chronic renal failure has been well documented in patients with renal amyloidosis (Shuttleworth and Ross 1956; Desikan and Job 1968) and in crescentic glomerulonephritis; however, the natural history of other forms of glomerulonephritis associated with leprosy is not clear.

Renal amyloidosis

Incidence

The incidence of amyloidosis (see Chapter 4.2) in leprosy has been observed to vary widely in different geographical areas. In earlier autopsy studies from the United States and Argentina, renal amyloidosis complicating leprosy was reported to vary from 30 to 55 per cent (Powell and Swan 1955; Shuttleworth and Ross 1956; Bernard and Vazquez 1973). In another study in American patients in which gingival biopsies were used for the diagnosis of amyloid disease, 31 per cent of leprosy patients were confirmed to have amyloidosis (Williams et al. 1965). However, the incidence has been considerably lower in other areas. In India, renal amyloidosis has been detected in only 2 to 8 per cent of patients (Junnarkar 1957; Krishnamurthy and Job 1966; Job 1968; Johny and Karat 1971; Mittal et al. 1972; Sainani and Rao 1974; Gupta 1977, 1981; Chugh et al. 1983).

Similarly, the incidence of systemic amyloidosis in leprosy has been observed to be 3.3 per cent in Mexico (Williams et al. 1965), 2.4 per cent in Malawi (Editorial, Lancet, 1975), 8.4 per cent in Papua New Guinea (McAdam et al. 1975), and 15 per cent in Japanese patients (Ozaki and Furuta 1975). The lower frequency of amyloidosis can only partly be accounted for by the fact that patients with non-lepromatous leprosy were also included in some of these studies. Of 203 consecutive patients of secondary amyloidosis seen in a north Indian centre, only 2.5 per cent were due to lepromatous leprosy (Chugh et al. 1981). However, in Papua New Guinea, 60 per cent of all cases of amyloidosis are secondary to leprosy (Cooke and Champness 1967).

Pathogenesis

Although secondary amyloidosis has occasionally occurred in patients with tuberculoid leprosy and suppurating chronic trophic ulcers (McAdam et al. 1975; Chugh et al. 1983), it is more

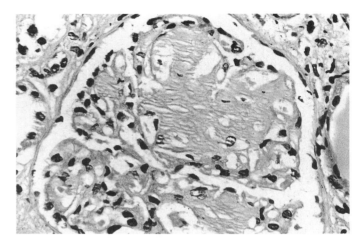

Fig. 3 Glomerulus shows marked involvement by amyloid deposits expanding the mesangium to form nodules, and thickening of the capillary loops (haematoxylin and eosin × 600).

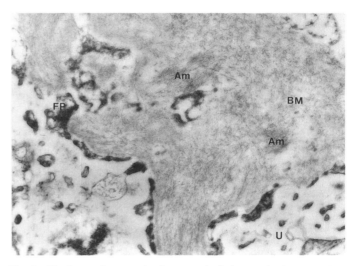

Fig. 4 Electron micrograph of segment of glomerular basement membrane shows typical fibrillar structure of amyloid (Am) deposits. BM = basement membrane; FP = foot process; U = urinary space (× 47 800).

frequent in those with lepromatous (Figs. 3 and 4), and borderline leprosy. There seems to be no correlation between the duration of disease and the development of amyloidosis which may supervene as early as 2 to 3 years after the diagnosis of leprosy. A definite correlation, however, has been documented between erythema nodosum leprosum and amyloidosis. McAdam et al. (1975) noted that all patients who had developed amyloidosis had suffered more than 20 episodes of erythema nodosum leprosum.

The amyloid fibril protein in leprosy has been confirmed to be of the AA type (see Chapter 4.2), as in secondary amyloidosis due to other causes. During each episode of erythema nodosum leprosum there is a marked elevation in the blood levels of serum amyloid A-related protein (SAA), the precursor of the amyloid fibril, and the elevation persists for several months after a severe reaction (McAdam et al. 1975). Erythema nodosum leprosarium

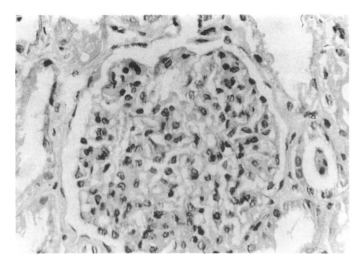

Fig. 5 Leprosy: glomerulus showing diffuse mesangial cell proliferation (haematoxylin and eosin × 345).

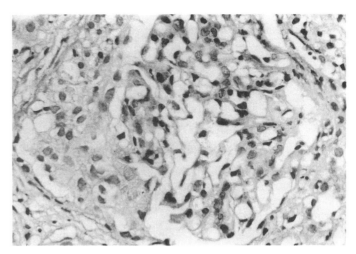

Fig. 6 Glomerulus showing mesangial proliferation and a cellular crescent in a patient with lepromatous leprosy (haematoxylin and eosin × 345).

is not a feature of non-lepromatous leprosy and this appears to be the reason why amyloidosis is uncommon in this type of disease.

Glomerulonephritis

Incidence

Recent studies have shown that glomerulonephritis is commoner than amyloidosis in patients with leprosy (Mittal et al. 1972; Johny et al. 1975; Cologlu 1979; Chugh et al. 1983). The frequency with which glomerulonephritis has been encountered varies widely, ranging from 6 to 50 per cent in studies utilizing renal biopsies (Shwe 1972; Sainani and Rao 1974; Johny et al. 1975; Cologlu 1979; Chugh et al. 1983). However, Drutz and Gutman (1973) found urinary abnormalities suggestive of glomerulonephritis in only 11 of 636 leprosarium inhabitants in Taiwan.

Pathology

Light microscopy

All morphological types of glomerulonephritis except focal glomerulosclerosis have been reported in leprosy. The two most common types are a diffuse endocapillary proliferative and a mesangial proliferative glomerulonephritis (Fig. 5) (Kean and Childress 1942; Shwe 1972; Mittal et al. 1972; Sainani and Rao 1974; Johny et al. 1975; Peter et al. 1981; Gupta et al. 1981; Ng et al. 1981; Chugh et al. 1983). Other glomerular lesions reported include focal proliferative, chronic sclerosing and in rare instances, membranous, mesangiocapillary, and crescentic glomerulonephritis (Fig. 6) (Bullock et al. 1974; Date et al. 1974; Date et al. 1977; Singhal et al. 1977; Peter et al. 1981).

Immunofluorescence

Granular deposits of IgG and C3, and less commonly IgM, IgA and fibrin are seen in the mesangium and along capillary walls on immunofluorescent microscopy (Shwe 1972; Iveson et al. 1975; Cologlu 1979; Ng et al. 1981).

Electron microscopy

Dense deposits have been identified in the mesangial-subendothelial region (Bullock et al. 1974; Date et al. 1977; Cologlu 1979; Date 1982) or subepithelially (Date and Johny 1975; Date et al. 1977; Ng et al. 1981). Other ultrastructural abnormalities include neutrophil leucocyte infiltration, focal foot process fusion, basement membrane thickening, and mesangial proliferation, expansion and interposition.

Pathogenesis

Glomerulonephritis has been documented in all forms of leprosy. Although some studies have reported a higher frequency in lepromatous leprosy (Johny et al. 1975; Chugh et al. 1983), an analysis of 186 patients who underwent biopsy (pooled data from several studies) has shown a similar incidence in lepromatous and non-lepromatous cases (Ng et al. 1981). Erythema nodosum leprosum is an immune complex disease involving mycobacterial antigens released by breakdown of *Mycobacterium leprae* following chemotherapy. During an episode of erythema nodosum leprosum, glomerulonephritis (Drutz and Gutman 1973; Iveson et al. 1975; Date et al. 1977; Cologlu 1979), and abnormalities of renal function (Thomas et al. 1970; Drutz and Gutman 1973; Bajaj et al. 1981) have often been noted. Whereas 76 per cent of patients have revealed circulating immune complexes during an erythema nodosum leprosum reaction, only 33 per cent of those not in reaction showed positive results (Moran et al. 1972). Hypocomplementaemia has been described during the reaction (Shwe 1972b; Drutz and Gutman 1973). These observations coupled with the demonstration of electron-dense deposits and presence of immunoglobulins and complement in diseased glomeruli suggest that glomerulonephritis is immune-complex mediated. However, the exact nature of the antigen–antibody complexes has not as yet been determined. The antigens involved could be mycobacterial and specific to leprosy or non-mycobacterial and exogenous such as streptococcal, staphylococcal, and hepatitis B virus antigens which are commonly present in leprosy patients and are known to cause glomerulonephritis in patients without leprosy. Even dapsone: antidapsone antibodies in the circulating immune complexes have been suspected (Das et al. 1980). Autoantibodies, of which the best documented are IgG–IgM cryoglobulins are also associated with leprosy and may be involved in the pathogenesis of glomerular lesions (Iveson et al. 1975). Glomerulonephritis in leprosy may therefore have a multifactorial origin and this is probably reflected in the variety of its morphological expressions.

Management
Whether drugs given for leprosy influence the course of glomerular lesions is unknown. Since erythema nodosum leprosum reactions predispose to the development of renal complications and are less frequent when dapsone and clofazimine are used in combination (than when dapsone is used alone), this combined drug regimen should preferably be used in lepromatous leprosy (Yawalkar and Vischer 1979). Although dapsone is largely metabolized in the liver, its metabolites are excreted in the urine and its dose therefore needs modification in those with chronic renal failure (Israili et al. 1973).

With the availability of dialysis and transplantation in endemic areas some leprosy patients are likely to be considered for these therapies. Lepromatous leprosy is associated with deficient cell-mediated immunity and is known to prolong the survival of allogeneic skin grafts (Bullock 1968). Experience with renal transplantation in leprosy is scanty. In one patient with tuberculoid leprosy, the disease showed continued healing despite immunosuppressive therapy (Date et al. 1982). However, in another patient with lepromatous leprosy, recrudescence of disease occurred 2 years after transplantation along with repeated lepra reactions (Adu et al. 1973). If transplantation is done for amyloidosis, recurrence of the latter could occur in some patients particularly if leprosy has not been adequately treated. A careful evaluation of cardiac status before transplantation is also mandatory because cardiac amyloidosis would be a contraindication (Kennedy and Castro 1977).

Glomerular disease in malaria
About 300 million people in the world develop malaria each year and about a million die in tropical Africa alone (Wyler 1983). Although glomerular disease had been suspected in malaria as early as the last century (Atkinson 1884), it was only in 1930 that Giglioli, working in British Guiana, obtained sufficient evidence to postulate a cause and effect relationship between *Plasmodium malariae* infection and a nephrotic syndrome. The relationship, however, remained unproven until Gilles and Hendrickse (1963) showed a highly significant increase in *P. malariae* parasitaemia in nephrotic children (88 per cent) compared to healthy controls. In *Plasmodium falciparum* infection, though the most frequent renal manifestation is acute renal failure associated with acute tubular necrosis seen in less than 1 per cent of patients (Chugh and Sakhuja 1986; Sitprija 1988), glomerular lesions have also been documented.

Incidence
The exact incidence of glomerular disease associated with malaria is not known. However, the incidence of nephrotic syndrome amongst hospital admissions in regions where malaria is endemic is 20 to 60 times greater than that observed in non-endemic areas. In Nigeria, Uganda, and Yemen, the nephrotic syndrome accounts for 2 to 5 per cent of all hospital admissions compared to only 0.02 to 0.03 per cent in the People's Republic of China and the United States (Kibukamusoke 1984). In Northern Nigeria, quartan malarial nephropathy is found in one-third of children with nephrotic syndrome (Abdurrahman et al. 1983). The admission rate for the nephrotic syndrome fell dramatically in Guyana after the elimination of endemic *P. malariae* (Giglioli 1962). Because of differences in the type of renal lesions, their progression, and response to therapy, nephropathy associated with *P. falciparum* and *P. malariae* is described separately.

Falciparum malaria
A transient glomerulonephritis often occurs in association with *P. falciparum* infection but mostly remains undetected because of the absence of hypertension, oedema, or any decline in renal function (Bhamarapravati et al. 1973). In a few instances, however, a nephrotic syndrome and renal insufficiency have been described (Berger et al. 1967; Hartenbower et al. 1972). Proteinuria is non-selective and less than 1 g per 24 h. Microscopic haematuria and cylindruria are seen in 20 to 50 per cent of patients (Boonpucknavig and Sitprija 1979). There is no correlation between the height of fever, the degree of proteinuria, or the type of urinary sediment. During the acute phase, serum complement levels (C3 and C4) are decreased and *P. falciparum* soluble antigen, antibodies to *P. falciparum*, and circulating immune complexes can be demonstrated (McGregor et al. 1968; Bhamarapravati et al. 1973).

Pathologic findings
Light microscopy
Only a few reports have documented the renal histological lesions associated with falciparum malaria. In a study of 50 consecutive autopsies, Spitz (1946) found glomerular lesions in 18 per cent of patients who died of a falciparum infection. There is widening of the mesangial area, mesangial and endothelial cell proliferation, and irregular thickening of the glomerular basement membrane. Eosinophilic granular material and pigment-laden macrophages are present in the capillary lumina (Bhamarapravati et al. 1973; Futrakul et al. 1974). Parasitized erythrocytes are frequently seen in peritubular capillaries and interlobular veins but are rarely detected in the glomerular capillaries. In patients who develop disseminated intravascular coagulation, giant nuclear masses may be seen in some capillaries (Churg et al 1988).

Immunofluorescence
Immunofluorescent studies show deposits of IgM and C3, mainly in the mesangium and along some capillary walls. IgG and IgA are occasionally seen. Fibrin is not detectable in the glomeruli. Falciparum antigen can be demonstrated in the mesangium and capillary walls in some instances.

Electron microscopy
Electron microscopy reveals electron-dense deposits in the mesangial and subendothelial areas and widening of the foot processes. Many deformed erythrocytes may be seen entrapped in spaces formed by endothelial cytoplasm.

The morphological findings are similar to those seen in mice infected with *Plasmodium berghei* (Boonpucknavig et al. 1972; Parbtani and Cameron 1979) and *Aotus* monkeys infected with *P. falciparum* (Hutt et al. 1975) and provide evidence in support of an immune complex-mediated glomerulonephritis.

Glomerulonephritis associated with falciparum malaria is transient. Lesions resolve within 4 to 6 weeks after antimalarial treatment.

Quartan malarial nephropathy

Nephropathy in quartan malaria has been reported from British Guiana, Uganda, Nigeria, Kenya, the Ivory Coast, Yemen, Sumatra, and New Guinea (Giglioli 1930; Surbek 1931; Carothers 1934; James 1939; Gilles and Hendrickse 1963; Kibukamusoke and Hutt 1967; Kibukamusoke et al. 1967; Thuriaux 1971; Paillerets et al. 1972).

Quartan malarial nephropathy differs from the acute transient glomerulonephritis associated with falciparum infection in its potential for progression to chronic end-stage disease and unresponsiveness to antimalarial therapy in the majority of cases.

Clinical features

This condition affects predominantly children and young adults with a peak incidence at the age of 5 years. *P. malariae* infection has been shown to be more frequent among malnourished children (Hendrickse and Adeniyi 1979). Fever is present only in the early stages of the illness and may be of the quartan type with spikes every 72 h. Over the next few weeks, a nephrotic state with generalized oedema and ascites develops. In adolescents and adults a combined picture of renal failure and the nephrotic syndrome is the rule. Blood pressure is usually normal but may become elevated in those with advanced renal disease.

Urinalysis reveals non-selective proteinuria in the majority. However, highly selective proteinuria occurs in about 20 per cent of patients (Adeniyi et al. 1970; Hendrickse et al. 1972; Adeniyi et al. 1976). Microscopic haematuria may also be present. Serum complement (C3) is usually within the normal range (Soothill and Hendrickse 1967). Serum cholesterol concentrations tend to be lower than those observed in European children with the nephrotic syndrome, because of the low dietary lipid intake, but are significantly elevated by local standards.

Renal histology

Light microscopy

Children in general show mild changes. Advanced mesangial sclerosis may be seen in late adolescence and early adulthood. In early cases, there is focal and segmental thickening of the capillary walls which at times may show double contour. The thickening is due to the presence of PAS-positive argyrophilic fibrils arranged in plexiform fashion in the subendothelial region. As the disease progresses, more capillaries become affected and eventually complete obliteration of capillary lumina occurs. The mesangium shows sclerosis which is initially segmental but later becomes diffuse. Cellular proliferation is usually absent but occasionally small fibroepithelial crescents may be seen. Tubular atrophy and interstitial infiltration by lymphomononuclear cells is also observed.

The severity of histological lesions may be graded as follows (Hendrickse et al. 1972): Grade I, up to 30 per cent of glomeruli show definite evidence of involvement with localized capillary-wall thickening and segmental sclerosis; Grade II, 30 to 75 per cent of glomeruli are affected, often with diffuse capillary wall thickening and sclerosis producing a 'honeycomb' appearance; Grade III, more than 75 per cent of glomeruli are affected with prominent tubular atrophy and interstitial inflammation.

Immunofluorescence microscopy

Three patterns of deposits have been observed.

1. Coarse, medium-sized granular deposits along the capillary walls—this is the most common appearance. A coarse granular pattern is positive for subclass IgG_3, either alone or in combination with other subclasses of IgG, IgM, and complement.

2. Diffuse and very fine deposits which are homogeneously distributed along the capillary walls. A fine granular pattern is more commonly associated with presence of IgG alone, predominantly of IgG_2 subclass, and absence of complement.

3. A mixture of granular and diffuse deposits in different proportions. Immunoglobulins of the IgG and IgM type, either alone or in combination are found in 96 per cent, complement in 66 per cent, and *P. malariae* antigen in about 25 per cent of patients (Ward and Conran 1966; Ward and Kibukamusoke 1969; Houba et al. 1971). The presence of IgA has been noted occasionally but always in combination with IgG or IgM.

Immunoglobulins G and M are also demonstrated in the proximal tubules. In one Nigerian study, 11 of 86 nephrotic children showed the presence of *P. malariae* antigen in the proximal tubules.

Electron microscopy

Electron microscopy reveals thickening of the capillary basement membrane with deposition of basement membrane-like material in the subendothelial zone. Small lacunae containing islands of electron-dense material similar in density to the basement membrane are often observed (Hendrickse et al. 1972.)

Course and prognosis

Spontaneous remission of established quartan malarial nephrotic syndrome is very rare. The natural history of this disease is one of slowly progressive renal damage leading to renal failure in 3 to 5 years. Death is usually due to hypertension, renal failure or intercurrent infection.

Antimalarial therapy has not been found valuable. In a controlled trial, during which 44 patients were given chloroquine followed by weekly pyrimethamine, 45 patients were given no antimalarials, and 24 patients received chloroquine and primaquine, there was no difference in outcome 6 months later amongst the three groups (Gilles and Hendrickse 1963).

Prednisolone is ineffective in inducing a remission in most patients, and its use is associated with a high rate of serious complications, including severe infections and hypertension (Adeniyi et al. 1970; Adeniyi et al. 1976). Of the small minority who have highly selective proteinuria, less than 50 per cent show a good response to prednisolone and a trial of therapy is, therefore, justified only in this category of patients (Adeniyi et al. 1970). Azathioprine has also been found to be ineffective and is not recommended. Following administration of azathioprine, a significant reduction in 2-year survival rate was observed in one study (Adeniyi et al. 1979). Cyclophosphamide has occasionally been associated with complete remission or reduction in proteinuria, but has not led to improved survival (Adeniyi et al. 1979). It may therefore be tried in those who fail to respond to prednisolone.

Histological grading has also been found valuable in predicting response to therapy (Hendrickse et al. 1972). All patients who have responded to prednisolone or cytotoxic drugs have Grade I lesions, while those with Grade II or III lesions have been unresponsive to any form of treatment. However, the majority of patients in the Grade I group show no differences in clinical behaviour or response to therapy from those graded II and III.

Of the three patterns of immunofluorescence, the coarse granular pattern is most often associated with a good response to therapy.

Normal immunofluorescent patterns have been shown in second biopsies after 12 months in the minority of children who showed good response to steroids, azathioprine, or cyclophosphamide.

Pathogenesis

In mice infected with *Plasmodium berghei*, malarial antigen first appears as the free-floating or corpuscular form in erythrocytes within blood vessels 3 days after infection. From the seventh day, the antigen can be detected by immunofluorescent microscopy in granular deposits along the walls of glomerular capillaries and in mesangial areas, together with immunoglobulins and complement (Boonpucknavig *et al.* 1973). These findings suggest that the parasitic antigen released from erythrocytes is processed and changed into another form which is deposited in glomeruli and causes local immune complex formation. Alternatively, it enters the circulation where it binds with antibodies and these antigen–antibody complexes localize in the glomeruli. A similar sequence of events has been observed in rhesus monkeys infected with *P. cynomolgi* (Ward and Conran 1966) and in *Aotus* monkeys with *P. malariae* infection (Voller *et al.* 1971).

Studies in quartan malaria in man have shown the binding of IgG specific antibody with sera of the host *in vitro*, suggesting the presence of either free antigens or soluble immune complexes with antigen excess in the circulation (Houba *et al.* 1976). Malarial antigen can be demonstrated in the kidneys of about one-third of patients. These data strongly support the belief that quartan malarial nephropathy is immune-complex mediated.

However, several issues still remain unresolved. Firstly, it is not clear why only a small proportion of patients with *P. malariae* infection develop renal lesions, and secondly why the lesion of quartan malarial nephropathy is persistent and progressive. Since *P. malariae* infection can persist for long periods in the liver, it has been suggested that this constant supply of antigen may contribute to perpetuation of renal lesions. However, this seems unlikely, since the progress of the disease is unaffected by antimalarial therapy and the detection of malarial antigen decreases with the duration of the disease (Houba 1975, 1979). Formation of autoantibodies stimulated by release of autologous antigens as a result of plasmodium-inflicted damage has also been postulated. Tubular antigens could be involved in such a process (Houba *et al.* 1971) but their role remains unproven.

Rheumatoid factor-like antiglobulins have been found in the sera of some patients but evidence that they aggravate renal lesions is lacking. Cross-reactivity of plasmodial antigens with autologous antigen has also been suspected. The increased incidence of antinuclear factor in sera of individuals living in malarious areas may represent a cross-reacting antibody induced by malarial nuclear material (Voller 1974) but the significance of these is not clear.

Filarial glomerulopathy

An association between filariasis and glomerular disease has been reported from India, Cameroon, and some other areas (Bariety *et al.* 1967; Pillay *et al.* 1973; Chugh *et al.* 1978; Date *et al.* 1979; Ngu *et al.* 1985). Renal involvement has been reported with *Onchocerca volvulus*, *Wuchereria bancrofti*, and loa-loa infections. Loiasis is prevalent in West and Central Africa and characteristically manifests with localized areas of allergic inflammation called Calabar swellings. Subconjunctival presence of adult worms may result in intense lacrimation and pain.

Onchocerciasis (river blindness) is characterized by subcutaneous nodules, a pruritic skin rash, sclerosing lymphadenitis, and ocular lesions and is seen in South America, tropical Africa, Saudi Arabia, and Yemen.

Bancroftian filariasis causes febrile episodes associated with acute lymphangitis and lymphadenitis. In advanced cases, hydrocele and elephantiasis occur. This form of filariasis is endemic in Africa and south-east Asia.

Renal involvement may occasionally cause an acute nephritic syndrome but is more frequently associated with a nephrotic syndrome and varying degrees of renal insufficiency. In *W. bancrofti* infection, chyluria may be associated with glomerular disease (Date *et al.* 1979; Ormerod *et al.* 1983).

Renal histology
Light microscopy

In patients with Bancroftian filariasis, mesangial proliferative, diffuse proliferative, and acute eosinophilic glomerulonephritis have been reported (Chugh *et al.* 1978; Date *et al.* 1979; Ormerod *et al.* 1983). Microfilariae may sometimes be seen in the glomerular capillaries.

In onchocercal infection, minimal change, mesangial proliferative, mesangiocapillary, and chronic sclerosing glomerulonephritis have been observed (Ngu *et al.* 1985). In those with loa-loa infection, membranous, mesangiocapillary, and chronic sclerosing glomerulonephritis have been documented (Pillay *et al.* 1973; Ngu *et al.* 1985).

Immunofluorescence microscopy

Deposits of IgM, IgG, and C3 are seen in the glomerular mesangium and capillary walls (Waugh *et al.* 1980; Ngu *et al.* 1985). Onchocercal antigen was identified in glomerular deposits in nine of 18 patients with *O. volvulus* infection in one study (Ngu *et al.* 1985). Specific antigens have not been looked for in patients with loa-loa and *Wuchereria bancrofti* infections.

Electron microscopy

Focal foot process fusion is present and electron-dense deposits are seen in the mesangium (Ormerod *et al.* 1983).

Pathogenesis

An epidemiologic survey of 1101 subjects from a hyperendemic area in Cameroon and 890 controls showed a significantly higher prevalence of proteinuria amongst subjects from the onchocercal zone (Ngu *et al.* 1985). The presence of immunoglobulins, complement, and filarial (onchocercal) antigen in glomeruli is suggestive of an immune-complex glomerulonephritis. In experimental animals also, an immune-complex glomerulonephritis has been seen following infection with *Dirofilaria immitis* or *Dipetalonema viteae* (Klei *et al.* 1974). The administration of diethylcarbamazine is at times associated with the occurrence of proteinuria in patients with filariasis (Greene *et al.* 1980; Ngu *et al.* 1980). This too could be the result of an immune response to release of antigens into the circulation following death of the parasite.

Management

In patients with a nephritic presentation, treatment with diethylcarbamazine may result in resolution of renal manifestations (Chugh et al. 1978; Date et al. 1979). However in those with a nephrotic syndrome, therapy does not usually influence the course of renal disease.

Glomerulonephritis associated with hepatitis B virus (HBV) infection (see also Chapter 3.12)

An association between persistent hepatitis B surface antigenaemia and a nephrotic syndrome was first reported by Combes et al. in 1971. Since then a variety of glomerulonephritides have been documented in patients with HBV infection. The most common histologic types reported with this infection are membranous nephropathy in children (Takekoshi et al. 1978; Levy and Kleinknecht 1980 ; Hsu et al. 1983; Seggie et al. 1984) and membranoproliferative glomerulonephritis in adults (Brozosko et al. 1974; Amemiya et al. 1983; Lee et al. 1988). The markers which have been used to identify HBV-related glomerulopathy include the presence of virus-associated antigens HBs, HBe, or HBc in the glomeruli and in the sera of these patients. The presence of immunoglobulins, complement components, and viral antigens in glomerular deposits indicates an immune-complex pathogenesis.

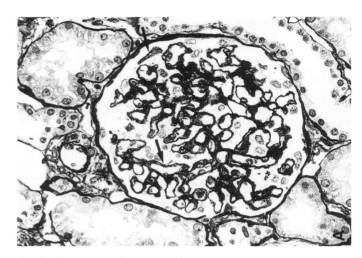

Fig. 7 Glomerulus showing uniform thickening of basement membrane with argyrophilic spike and chain-like pattern (arrows) in a patient with HBV-associated membranous nephropathy (periodic acid–methenamine silver stain × 300). (Reproduced with permission from *Virchows Archives of Pathological Anatomy* (data from Lai *et al.* 1989).)

Prevalence

Although HBV infection is a global problem, its prevalence varies considerably. Low endemic areas with a carrier rate of less than 0.1 per cent in the general population include North America, Northern Europe, and Australia. In Eastern Europe, South America, West Asia, Japan, and India the prevalence rate is not greater than 5 per cent. High endemic areas with a prevalence rate of more than 5 per cent include China, Taiwan, south-east Asia, the Pacific region, and 'tropical Africa. Some pockets of high endemicity have also been detected in Southern India (Thyagarajan et al. 1989).

The association between HBV and glomerular disease is most apparent in areas of high endemicity. Lai et al. (1987) found that 22 per cent of their patients with primary glomerular disease had hepatitis B surface (HBs) antigenaemia and the prevalence of a carrier state in these patients was significantly higher than that in the general population of Hong Kong. Vos et al. (1973) reported a prevalence of HBs antigenaemia of 20 per cent in South African Bantus with chronic glomerulonephritis. In Italy, only 2.7 per cent of a population in hospital had HBs antigenaemia compared with 8 per cent of patients with a variety of glomerular lesions (Maggiore et al. 1981). In South Korea, however, frequency of antigenaemia in those with glomerulonephritis is not higher than that in the general population (Lee et al. 1988). In Nigerian children with a nephrotic syndrome, glomerular deposits of HBs antigen were found in 12 of 50 patients, even though there was no difference in frequency of HBs antigenaemia between patients and a control group (Abdurrahman et al. 1983).

Though HBV-related antigens have been reported to be involved in the development of several types of glomerulonephritis, the most striking association is with membranous and membranoproliferative glomerulonephritis. In Zimbabwe, Japan, and Taiwan, HBs antigenaemia has been recorded in 80 to 100 per cent of children with membranous lesions (Takekoshi et al. 1978; Hsu et al. 1983; Seggie et al. 1984; Adhikari et al. 1985; Lai et al. 1987). In South Korea, Lee et al. (1988) observed that 87.5 per cent of adults with membranoproliferative glomerulonephritis were HBsAg carriers. The reason for a different predilection of glomerular lesions in children and adults is not clear.

The association of IgA nephropathy and HBs antigenaemia reported by some authors is intriguing. Nagy and coworkers (1979) reported a HBs carrier rate of 16 per cent amongst Hungarian patients with IgA nephropathy although the prevalence rate in the population is only 1 per cent. These workers observed glomerular deposits of HBsAg in 31 per cent of their patients. Recently Lai et al. (1987) have demonstrated glomerular deposits of surface or core antigen in 61 per cent of patients with IgA nephropathy and persistent HBs antigenaemia. In rare instances, combined features of membranous and IgA nephropathy have been recorded (Doi et al. 1983; Magil et al. 1986). The development of polyarteritis nodosa in patients with HBV infection has been well documented (see Chapter 3.5).

Renal histology

The morphological features of HBV associated membranous nephropathy (Fig. 7) are indistinguishable from idiopathic membranous nephritis except for the frequent presence of small mesangial deposits in the former. In some patients subendothelial deposits may also be seen (Yoshikawa et al. 1985; Southwest Pediatric Nephrology Study Group 1985; Lee et al. 1988). The features of membranoproliferative glomerulonephritis associated with HBV are similar to those of idiopathic type I disease (Swainson et al. 1983). Besides these two types, mesangial proliferative glomerulonephritis has also been documented (Stratta et al. 1975; Takeda et al. 1988). On electron microscopy, subendothelial electron-dense deposits and mesangial interposition are frequently observed especially in HBeAg positive patients, thereby suggesting the involvement of HBeAg in the causation of capillary wall lesions (Takeda et al. 1988). Microtubular virus-like structures are observed in glomerular endothelial cells in

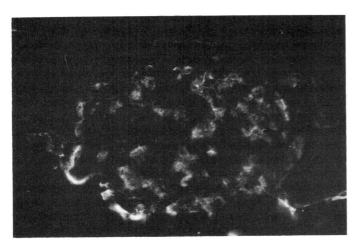

Fig. 8 Glomerular mesangial staining for HBsAg in a patient with IgA nephropathy (indirect immunofluorescence, rabbit anti-human HBsAg antiserum × 420). (By courtesy of K.N. Lai.)

some patients with membranous and membranoproliferative lesions (Lee et al. 1988).

Immunofluorescence

Of the three hepatitis B-associated antigens (HBs, HBc, and HBe), that most consistently seen in the glomeruli is the last (Hirose et al. 1984). The presence of HBe antigen, IgG, and C3 in a granular fashion in the glomerular capillary basement membrane of patients with membranous nephritis suggests that the glomerulopathy associated with HBV antigenaemia is an HBe antigen–anti-HBe immune-complex disease (Takekoshi et al. 1978; Hattori et al. 1988). In patients with IgA nephropathy associated with chronic hepatitis B virus infection, Lai et al. (1989a) have shown mesangial deposits of HBsAg (Fig. 8), suggesting that the latter may be playing a pathogenetic role in these patients. These authors have further suggested that since polyclonal anti-HBcAg antiserum reacts both with HBcAg and HBeAg, only monoclonal antibodies should be used for immunofluorescence studies for the purpose of identifying viral antigens in the glomeruli (Lai et al. 1989b).

Therapy

In HBV-associated membranous nephropathy, spontaneous regression of nephrotic syndrome has been reported in 30 to 60 per cent of patients (Kleinknecht et al. 1979; Ito et al. 1981; Hsu et al. 1983). Most patients, however, remain symptomatic for 12 months or longer and there are reports of progression to renal insufficiency (Kohler et al. 1974; Hsu et al. 1983). Steroid therapy has therefore been tried in a few patients. Although six of 11 patients in one study developed remission of proteinuria following administration of steroids, treatment was associated with increased serum concentrations of HBeAg, HBV DNA, and serum alanine aminotransferase, and the appearance of virus-like particles in the glomeruli suggesting active viral replication despite the absence of overt clinical hepatitis (Lai et al. 1989).Thus, the available evidence indicates that whereas steroids reduce the duration of nephrotic syndrome, they may potentiate the risk of viral replication.

The use of interferon in patients with persistent hepatitis B virus infection has been accompanied by eradication of markers of hepatitis B virus (HBeAg, HBsAg, hepatitis B virus DNA polymerase) in some cases. This treatment has recently been extended to patients with extrahepatic manifestations. Preliminary observations on the use of interferon in patients with hepatitis B-associated membranous glomerulonephritis have shown a resolution of proteinuria in isolated cases (Garcia et al. 1985; Mizushima et al. 1987). Whether the decrease in proteinuria in patients with membranous lesions was due to a spontaneous remission as a part of natural history or was the result of treatment with interferon is not certain.

Further prospective trials to evaluate the potential benefits and risks of steroid and interferon therapy in HBV-associated nephropathy are warranted.

Concluding remarks

The prevalence and pattern of glomerular disease vary widely in different geographical regions and are greatly influenced by environmental, nutritional and socioeconomic conditions. The tropical countries with vast populations and poor economic resources are not only exposed to diseases common to temperate and tropical climates alike but are also heavily burdened with diseases caused by aetiologic agents specific to those regions. Among the agents which have been identified as contributing to a higher prevalence of glomerular disease in the tropics are P. malariae, S. mansoni, M. leprae, and hepatitis B virus. Thus, glomerular disease appears to be far more common in the tropics than in countries with temperate climates. In some hospitals in Africa, patients with chronic renal failure resulting from chronic glomerulonephritis constitute more than 2 to 3 per cent of the total admissions. The patterns are only now being accurately described. Of the primary glomerular diseases, for reasons which are still not clear, IgA nephropathy and minimal change disease are rare among the African black population.

During the last two decades, the industrially advanced countries of the world have witnessed a significant change in disease patterns because of marked improvement in socioeconomic conditions and rapidly expanding medical facilities. Most tropical countries with poor resources and vast populations living in rural areas continue to be faced with malnutrition, poverty, illiteracy, lack of consciousness of personal hygiene, poor health care facilities, and delays in hospitalization—conditions which epitomize underdevelopment. Early signs are often missed; for example, micturition in a drain or a field provides no opportunity to the patient for recognition of haematuria. Nephrotic syndrome is not infrequently disguised by concomitant malnutrition or anaemia. Patients often seek medical advice for the first time when the disease is far too advanced. Altered immunological responses brought about by malnutrition are now being recognized as important determinants of the pathogenesis and morbidity associated with glomerular disease in the tropics.

It is also obvious that the distribution of glomerular disease associated with tropical infections varies considerably from country to country and a decline in the prevalence can be expected only with eradication of these infections. A striking example is provided by the decline in deaths from chronic renal failure which followed eradication of malaria in British Guyana. Schistosomiasis and malaria which are often associated with serious glomerular disease continue to be endemic in Africa and many other parts of the tropics. The available evidence also suggests that hepatitis B virus plays a significant role in the development of membranous and membranoproliferative glomerulonephritis and the incidence of these lesions is high in areas with high endemicity. Homozygous sickle-cell disease is another important cause of glomerulopathy in Africa and the West Indies (see Chapter 4.12). A significant decline in the prevalence of membranoproliferative glomerulonephritis has been noted recently in some European countries and it is suspected that this is related to a decrease in the incidence of bacterial infections. Acute diffuse proliferative glomerulonephritis following throat infections, which has become a rarity in temperate climates, continues unabated, with impetigo complicating scabies as an additional source of streptococcal infection in tropical countries. Secondary amyloidosis associated with tuberculosis, leprosy, and other chronic infections contributes significantly to the incidence of nephrotic syndrome and a decline can be expected only with eradication of these primary conditions still rampant in these countries. The prevention and eradication of glomerular diseases associated with tropical infections is undoubtedly a gigantic task, and accurate knowledge of patterns and prevalence of such diseases is essential for launching effective measures to achieve these goals.

References

Abdurrahman, M.B. and Edington, G.M. (1982). Correlation between proteinuria selectivity index and kidney histology of nephrotic children in Northern Nigeria. *Journal of Tropical Paediatrics*, **28**, 124–6.

Abdurrahman, M.B., Fakunle, Y.M., and Whittle, H.C. (1983). The role of hepatitis B surface antigen in Nigerian children with nephrotic syndrome. *Annals of Tropical Paediatrics*, **3**, 13–16.

Abdurrahman, M.B., et al. (1983). The role of malaria in childhood nephrotic syndrome in Northern Nigeria. *East African Medical Journal*, **59**, 467–71.

Adeniyi, A., Hendrickse, R.G., and Houba, V. (1970). Selectivity of proteinuria and response to prednisolone or immunosuppressive drugs in children with malarial nephrosis. *Lancet*, **i**, 644–8.

Adeniyi, A., Hendrickse, R.G., and Soothill, J.F. (1976). Differential protein clearances and response to treatment in Nigerian nephrotic children. *Archives of Disease in Childhood*, **51**, 691–6.

Adeniyi, A., Hendrickse, R.G., and Soothill, J.F. (1979). A controlled trial of cyclo-phosphamide and azathioprine in Nigerian children with the nephrotic syndrome and poorly selective proteinuria. *Archives of Disease in Childhood*, **54**, 204–7.

Adhikari, M., Coovadia, H.M., and Chrystal, V. (1985). Extramembranous nephropathy in black South African children. *Annals of Tropical Paediatrics*, **5**, 19–22.

Adu, D., Evans, D.B., Millard, P.R., Calne, R.Y., Shwe, T., and Jopling, W.H. (1973). Renal transplantation in leprosy. *British Medical Journal*, **2**, 280–1.

Adu, D., et al. (1981). The nephrotic syndrome in Ghana: Clinical and pathological aspects. *Quarterly Journal of Medicine*, **50**, 297–306.

Amemiya, S., Ito, H., Kato, K., Sakaguchi, H., Hasegawa, O., and Hajikano, H. (1983). A case of membrano-proliferative glomerulonephritis type III (Burkholder) with the deposition of both HBeAg and HBsAg. *International Journal of Pediatric Nephrology*, **4**, 267–73.

Andrade, Z.A. and Rocha, H. (1979). Schistosomal glomerulopathy. *Kidney International*, **16**, 23–9.

Andrade, Z.A., Andrade, S.G., and Sadigursky, M. (1971). Renal changes in patients with hepatosplenic schistosomiasis. *American Journal of Tropical Medicine and Hygiene*, **20**, 77–83.

Atkinson, I.E. (1984). Bright's disease of malarial origin. *American Journal of Medical Sciences*, **88**, 149–66.

Bajaj, A.K., Gupta, S.C., Sinha, S.N., Govil, D.C., Gaur, U.C., and Kumar, R. (1981). Renal functional status in lepromatous leprosy. *International Journal of Leprosy*, **49**, 37–41.

Bariéty, J., et al. (1967). Proteinurie et loase. Étude histologique, optique et électronique d'un cas. *Bulletin et Mémoires de la Société Médicale des Hôpitaux de Paris*, **118**, 1015–25.

Bedi, T.R., Kaur, S., Singhal, P.C., Kumar, B., and Banerjee, C.K. (1977). Fatal proliferative glomerulonephritis in lepromatous leprosy. *Leprosy India*, **49**, 500–3.

Berger, M., Birch, L.M., and Conte, N.F. (1967). The nephrotic syndrome secondary to acute glomerulonephritis during falciparum malaria. *Annals of Internal Medicine*, **67**, 1163–71.

Berggren, W.L. and Weller, T.H. (1967). Immunoelectrophoretic demonstration of specific circulating antigen in animals infected with *S. mansoni*. *American Journal of Tropical Medicine and Hygiene*, **16**, 606–12.

Bernard, J.C. and Vazquez, C.A.J. (1973). Visceral lesions in lepromatous leprosy. *International Journal of Leprosy*, **41**, 94–101.

Bhamarapravati, N., Boonpucknavig, S., Boonpucknavig, V., and Yaemboonruang, C. (1973). Glomerular changes in acute plasmodium falciparum infection: an immunologic study. *Archives of Pathology*, **96**, 289–93.

Boonpucknavig, V. and Sitprija, V. (1979). Renal disease in acute *Plasmodium falciparum* infection in man. *Kidney International*, **16**, 44–52.

Boonpucknavig, S., Boonpucknavig, V., and Bhamarapravati, N. (1972). Immunopathological studies of *Plasmodium berghei*-infected mice. *Archives of Pathology*, **94**, 322–30.

Boonpucknavig, V., Boonpucknavig, S., and Bhamarapravati, N. (1973). *Plasmodium berghei* infection in mice. An ultrastructural study of immune complex nephritis. *American Journal of Pathology*, **70**, 89–99.

Brown, K.G.E., Abrahams, C., and Meyers, A.M. (1977). The nephrotic syndrome in Malawian blacks. *South African Medical Journal*, **52**, 275–8.

Brozosko, W.J., Krawczynski, K., Nazarewicz, T., Morzycka, M., and Nowoslawski, A. (1974). Glomerulonephritis associated with hepatitis B surface antigen immune complexes in children. *Lancet*, **ii**, 477–82.

Bullock, W.E. (1968). Studies of immunologic mechanisms in leprosy. I. Depression of delayed allergic response to skin antigens. *New England Journal of Medicine*, **278**, 298–304.

Bullock, W.E., Callerame, M.L., and Panner, B.J. (1974). Immunohistologic alterations of skin and ultrastructural changes of glomerular basement membranes in leprosy. *American Journal of Tropical Medicine and Hygiene*, **23**, 81–6.

Carothers, J.C. (1934) An investigation of the etiology of subacute nephritis as seen among the children of North Kavirondo. *East African Medical Journal*, **10**, 335–6.

Chen, W., et al. (1986). Clinical study on primary IgA glomerulonephritis in adults. *Journal of the Formosan Medical Association*, **85**, 209–26.

Chitkara, N.L., Chugh, T.D., Chhuttani, P.N., and Chugh, K.S. (1965). Secondary amyloidosis. *Indian Journal of Pathology and Bacteriology*, **8**, 285–8.

Chugh, K.S. and Sakhuja, V. (1984). Renal disease in northern India. In *Tropical Nephrology* (ed. J.W. Kibukamusoke), pp. 428–40, Citforge, Canberra.

Chugh, K.S. and Sakhuja, V. (1986). Renal involvement in malaria. *International Journal of Artificial Organs*, **9**, 391–2.

Chugh, K.S., Singhal, P.C., and Tewari, S.C. (1978). Acute glomerulonephritis associated with filariasis. *American Journal of Tropical Medicine and Hygiene*, **27**, 630–1.

Chugh, K.S., Singhal, P.C., Sakhuja, V., Datta, B.N., Jain, S.K., and Dash, S.C. (1981). Pattern of renal amyloidosis in Indian patients. *Postgraduate Medical Journal*, **57**, 31–5.

Chugh, K.S., et al. (1983). Renal lesions in leprosy amongst North Indian patients. *Postgraduate Medical Journal*, **59**, 707–11.

Chugh, K.S., et al. (1986). Urinary schistosomiasis in Maiduguri, North East Nigeria. *Annals of Tropical Medicine and Parasitology*, **80**, 593–9.

3.14 Glomerular disease in the tropics

Chugh, K.S., *et al.* (1987). Progression to end-stage renal disease in poststreptococcal glomerulonephritis. *International Journal of Artificial Organs*, **10**, 189–94.

Churg, J., *et al.* (1988). Protozoal infections. In *Renal Disease—Classification and Atlas of Infectious and Tropical Renal Diseases* (ed. WHO) pp. 13–40, ASCP Press, Chicago.

Cologlu, A.S. (1979). Immune complex glomerulonephritis in leprosy. *Leprosy Review*, **50**, 213–22.

Combes, B., *et al.* (1971). Glomerulonephritis with deposition of Australia antigen–antibody complexes in the glomerular basement membrane. *Lancet*, **ii**, 234–7.

Cooke, R.A. and Champness, L.T. (1967). Amyloidosis in Papua and New Guinea. *Papua and New Guinea Medical Journal*, 10, 43–8.

Coovadia, H.M., Adhikari, M.A., and Morel Maroger, L. (1979). Clinicopathological features of the nephrotic syndrome in South African children. *Quarterly Journal of Medicine*, **48**, 77–91.

Das, P.K., *et al.* (1980). Dapsone and anti-dapsone antibody in circulating immune complexes in leprosy patients. *Lancet*, **i**, 1309–10 (letter).

Date, A. (1982). The immunological basis of glomerular disease in leprosy—a brief review. *International Journal of Leprosy*, **50**, 351–4.

Date, A. and Job, C.K. (1974). The prevalence of amyloidosis in autopsy material. *Journal of the Indian Medical Association*, **62**, 287–9.

Date, A. and Johny, K.V. (1975). Glomerular subepithelial deposits in lepromatous leprosy. *American Journal of Tropical Medicine and Hygiene*, **24**, 853–6.

Date, A., Thomas, A., Mathai, R., and Johny, K.V. (1977). Glomerular pathology in leprosy. An electron microscopic study. *American Journal of Tropical Medicine and Hygiene*, **26**, 266–72.

Date, A., Gunasekaran, V., Kirubakaran, M.G., and Shastry, J.C.M. (1979). Acute eosinophilic glomerulonephritis with Bancroftian filariasis. *Postgraduate Medical Journal*, **55**, 905–7.

Date, A., Shastry, J.C.M., and Johny, K.V. (1979). Ultrastructural glomerular changes in filarial chyluria. *Journal of Tropical Medicine and Hygiene*, **82**, 150–5.

Date, A., Mathai, R., Pandey, A.P., and Shastry, J.C.M.(1982). Renal transplantation in leprosy. *International Journal of Leprosy*, **50**, 56–7.

Date, A., Raghavan, R., John, T.J., Richard, J., Kirubakaran, M.G., and Shastry, J.C.M. (1987). Renal disease in adult Indians: a clinicopathological study of 2827 patients. *Quarterly Journal of Medicine*, **64**, 729–37.

Deelder, A.M., Klappe, H.T.M., Aardweg, G.J.M.J., and Meerbeke, E.H.E.M. (1976). *Schistosoma mansoni*: demonstration of two circulating antigens in infected hamsters. *Experimental Parasitology*, **40**, 189–97.

Desikan, K.V. and Job, C.K. (1968). A review of postmortem findings in 37 cases of leprosy. *International Journal of Leprosy*, **36**, 32–44.

Doi, T., Kanatsu, K., Nagai, H., Kohrogi, N., and Hamashima, Y. (1983). An overlapping syndrome of IgA nephropathy and membranous nephropathy. *Nephron*, **35**, 24–30.

Drutz, D.J. and Gutman, R.A. (1973). Renal manifestations of leprosy: glomerulonephritis, a complication of erythema nodosum leprosum. *American Journal of Tropical Medicine and Hygiene*, **22**, 496–502.

Dutra, M., de Carvalho Gilho, E.M., and Gusmâo, E.A. (1979). Tratamento da glomerulopatia da Esquistossomose mansonica: E feito de corticosteroides, ciclofosfamida e esquistossomicidas. Cited by Andrade, Z.A., and Rocha, H. (1979) Schistosomal glomerulopathy. *Kidney International*, **16**, 23–9.

Edington, G.M., and Mainwaring, A.R. (1964). Amyloidosis in Western Nigeria. *Pathologia et Microbiologia*, **27**, 841–5.

Editorial. (1975). Amyloidosis and leprosy. *Lancet*, **ii**, 589–90.

Editorial. (1980). Nephrotic syndrome in tropics. *Lancet*, **ii**, 461–2.

Ezzat, E., Osman, R.A., Ahmet, K.Y., and Soothill, J.F. (1974). The association between *Schistosoma haematobium* infection and heavy proteinuria. *Transactions of the Royal Society of Tropical Medicine and Hygiene*, **68**, 315–18.

Falcao, H.A. and Gould, D.B. (1975). Immune complex nephropathy in schistosomiasis. *Annals of Internal Medicine*, **83**, 148–54.

Farid, Z., Higashi, G.I., Bassily, S., Young, S.W., and Sparks, H.A. (1972). Chronic salmonellosis, urinary schistosomiasis and massive proteinuria. *American Journal of Tropical Medicine and Hygiene*, **21**, 578–81.

Feng, P.H. (1986). Lupus nephritis—the Asian connection. *Philippines Journal of Nephrology*, **1**, 30–1.

Fessel, J.W. (1974). Systemic lupus erythematosus in the community—incidence, prevalence, outcome and first symptoms. The high incidence in black women. *Archives of Internal Medicine*, **134**, 1027–34.

Frank, A.O. (1980). Apparent predisposition to systemic lupus erythematosus in Chinese patients in West Malaysia. *Annals of Rheumatic Diseases*, **39**, 266–9.

Futrakul, P., Boonpucknavig, V., Boonpucknavig, S., Mitrakul, C., and Bhamarapravati, N. (1974). Acute glomerulonephritis complicating *Plasmodium falciparum* infection. *Clinical Paediatrics*, **12**, 281–3.

Garcia, G., *et al.* (1985). Preliminary observations of hepatitis B associated membranous glomerulonephritis treated with leukocyte interferon. *Hepatology*, **5**, 317–20.

Giglioli, G. (1930). Malarial nephritis: epidemiological and clinical notes on malaria. *Blackwater fever, Albuminuria and Nephritis in the Interior of British Guiana, based on Seven Years' Continual Observation*. Churchill, London.

Giglioli, G. (1962). Malaria and renal disease with special reference to British Guiana II. The effect of malaria eradication on the incidence of renal disease in British Guiana. *Annals of Tropical Medicine and Parasitology*, **56**, 225–41.

Gilles, H.M. and Hendrickse, R.G. (1963). Nephrosis in Nigerian children: role of *Plasmodium malariae*, and effect of anti-malarial treatment. *British Medical Journal*, **1**, 27–31.

Greene, B.M., Taylor, H.R., and Humphrey, R.L. (1980). Proteinuria associated with diethylcarbamazine treatment of onchocerciasis. *Lancet*, **i**, 254–5.

Greenham, R. and Cameron, A.H. (1980) *Schistosoma haematobium* and nephrotic syndrome. *Transactions of the Royal Society of Tropical Medicine and Hygiene*, **74**, 609–13.

Gupta, J.C., Divakar, R., Singh, S., Gupta, D.K., and Panda, P.K. (1977). A histopathological study of renal biopsies in 50 cases of leprosy. *International Journal of Leprosy*, **45**, 167–70.

Gupta, J.C., Bajaj, A.K., Govil, D.C., Sinha, S.N., and Kumar, R. (1981). A study of percutaneous renal biopsy in lepromatous leprosy. *Leprosy India*, **53**, 179–84.

Hartenbower, D.L., Kantor, G.L., and Rosen, V.J. (1972). Renal failure due to acute glomerulonephritis during falciparum malaria. *Military Medicine*, **137**, 74–6.

Hattori, S., Furuse A., and Matsuda, I. (1988). Presence of HBe antibody in glomerular deposits in membranous glomerulonephritis is associated with hepatitis B virus infection. *American Journal of Nephrology*, **8**, 384–7.

Hendrickse, R.G. and Adeniyi, A. (1979). Quartan malarial nephrotic syndrome in children. *Kidney International*, **16**, 64–74.

Hendrickse, R.G., Adeniyi, A., Edington, G.M., Glasgow, E.F., White, R.H.R., and Houba, V. (1972). Quartan malarial nephrotic syndrome: collaborative clinicopathological study in Nigerian children. *Lancet*, **i**, 1143–9.

Hirose, H., *et al.* (1984). Deposition of hepatitis Be antigen in membranous glomerulonephritis: identification by F[ab]2 fragments of monoclonal antibody. *Kidney International*, **26**, 338–41.

Hoshino-Shimizu, S., *et al.* (1976). Human schistosomiasis: *Schistosoma mansoni* antigen detection in renal glomeruli. *Transactions of the Royal Society of Tropical Medicine and Hygiene*, **70**, 492–6.

Houba, V. (1975). Immunopathology of nephropathies associated with malaria. *Bulletin of the World Health Organization*, **52**, 199–207.

Houba, V. (1979). Immunologic aspects of renal lesions associated with malaria. *Kidney International*, **16**, 3–8.

Houba, V., Allison, A.C., Adeniyi, A., and Houba, J.E. (1971). Immunoglobulin classes and complement in biopsies of Nigerian children with the nephrotic syndrome. *Clinical and Experimenal Immunology*, **8**, 761–74.

Houba, V., Koech, D.K., Sturrock, R.F., Butterworth, A.E., Kusel, J.R., and Mahmoud, A.A.F. (1976). Soluble antigens and antibodies in sera from baboons infected with *Schistosoma mansoni*. *Journal of Immunology*, **177**, 705–7.

Hsu, H.C., Lin, G.H., Chang, M.H., and Chen, C.H. (1983). Association of hepatitis B surface antigenemia and membranous nephropathy in children in Taiwan. *Clinical Nephrology*, **20**, 121–9.

Hutt, M.S.R., Davies, D.R. and Voller, A. (1975). Malarial infections in aotus trivirgatus with special reference to renal pathology. II. *P. falciparum* and mixed malaria infections. *British Journal of Experimental Pathology*, **56**, 429–38.

Israili, Z.H., Cucinell, S.A., Vaught, J., Davis, E., Lesser, J.M., and Dayton, P.G. (1973). Studies of the metabolism of dapsone in man and experimental animals: formation of *N*-hydroxy metabolites. *Journal of Pharmacology and Experimental Therapeutics*, **187**, 138–42.

Ito, H., *et al.* (1981). Hepatitis Be antigen-mediated membranous glomerulonephritis. Correlation of ultrastructural changes with HBeAg in the serum and glomeruli. *Laboratory Investigation*, **44**, 214–20.

Iveson, J.M.I., McDougall, A.C., Leathem, A.J., and Harris, H.J. (1975). Lepromatous leprosy presenting with polyarthritis, myositis and immune complex glomerulonephritis. *British Medical Journal*, **3**, 619–21.

James, C.S. (1939). Malarial nephritis (nephrosis) in the Solomon Islands and Mandated Territory of New Guinea. *Medical Journal of Australia*, **1**, 759–61.

James, P.D. and Owor, R.D. (1975). Systemic amyloidosis in Uganda—an autopsy study. *Transactions of the Royal Society of Tropical Medicine and Hygiene*, **69**, 480–3.

Johny, K.V. (1978). Nephrotic syndrome. In *Progress in Clinical Medicine in India*, (Ed. M.M.S. Ahuja), pp. 372–403, Arnold Heinemann, New Delhi.

Johny, K.V. and Karat, A.B.A. (1971). Renal biopsy studies in leprosy (Abstr.) *Journal of Association of Physicians of India*, **19**, 117–18.

Johny, K.V., Karat, A.B.A., Rao, P.S.S., and Date, A. (1975). Glomerulonephritis in leprosy. A percutaneous renal biopsy study. *Leprosy Review*, **46**, 29–37.

Junnarkar, R.V. (1957). Late lesions in leprosy. *Leprosy India*, **29**, 148–54.

Kanyerezi, B.R., Lutalo, S.K., and Kigonya, E. (1980). Systemic lupus erythematosus—clinical presentation among Ugandan Africans. *East African Medical Journal*, **57**, 274–8.

Kean, B.H. and Childress, M.E. (1942). A summary of 103 autopsies on leprosy patients on the Isthmus of Panama. *International Journal of Leprosy*, **10**, 51–9.

Kennedy, C.L. and Castro, J.E. (1977). Transplantation for renal amyloidosis. *Transplantation*, **24**, 382–6.

Kibukamusoke, J.W. (1984). Quartan malarial nephropathy. In *Tropical Nephrology*, (ed. J.W. Kibukamusoke), pp. 58–75, Citforge, Canberra.

Kibukamusoke, J.W. and Hutt, M.S.R. (1967). Histological features of the nephrotic syndrome associated with quartan malaria. *Journal of Clinical Pathology*, **20**, 117–23.

Kibukamusoke, J.W., Hutt, M.S.R., and Wilks, N.E. (1967). The nephrotic syndrome in Uganda and its association with quartan malaria. *Quarterly Journal of Medicine*, **36**, 393–407.

Klei, T.R., Cornell, W.A., and Thompson, P.E. (1974). Ultrastructural glomerular changes associated with filariasis. *American Journal of Tropical Medicine and Hygiene*, **23**, 608–18.

Kleinknecht, C., Levy, M., Peix, A., Broyer, M., and Courtecuisse, V. (1979). Membranous glomerulonephritis and hepatitis B surface antigen in children. *Journal of Pediatrics*, **95**, 946–52.

Kohler, P.F., Chronin, R.E., Hammond, W.S., Olin, D., and Carr, R.I. (1974). Chronic membranous glomerulonephritis caused by hepatitis antigen–antibody immune-complexes. *Annals of Internal Medicine*, **81**, 448–51.

Krishnamurthy, S. and Job, C.K. (1966). Secondary amyloidosis in leprosy. *International Journal of Leprosy*, **34**, 155–8.

Kyle, R.A. and Bayrd, E.D. (1975). Amyloidosis: review of 236 cases. *Medicine (Baltimore)*, **54**, 271–85.

Lai, F.M., Tam, J.S., Li, P.K.T., and Lai, K.N. (1989). Replication of hepatitis B virus with corticosteroid therapy in hepatitis B virus related membranous nephropathy. *Virchows Archives of Pathological Anatomy*, **414**, 279–84.

Lai, K.N., Lai, F.M., Chan, K.W., Chow, C.B., Tong, K.L., and Vallance-Owen, J. (1987). The clinicopathologic features of hepatitis B virus-associated glomerulonephritis. *Quarterly Journal of Medicine*, **63**, 323–33.

Lai, K.N., Lai, F.M., and Tam, J.S. (1989a). IgA nephropathy associated with chronic hepatitis B virus infection in adults: the pathogenetic role of HBsAg. *Journal of Pathology*, **157**, 321–7.

Lai, K.N., Lai, F.M., and Tam, J.S. (1989b). Comparison of polyclonal and monoclonal antibodies in determination of glomerular deposits of hepatitis B virus antigens in hepatitis B virus associated glomerulonephritides. *American Journal of Clinical Pathology*, in press.

Lee, H.S., Choi, Y., Yu, S.H., Koh, H.I., Kim, M.J., and Ko, K.W. (1988). A renal biopsy study of hepatitis B virus associated nephropathy in Korea. *Kidney International*, **34**, 537–43.

Levy, M. and Berger, J. (1988). Worldwide perspective of IgA nephropathy. *American Journal of Kidney Diseases*, **12**, 340–7.

Levy, M. and Kleinknecht, C. (1980). Membranous glomerulonephritis and hepatitis B virus infection. *Nephron*, **26**, 259–65.

Lopes, M. (1964). Aspectos renais da sindrome hepatoesplenica da esquistossomose mansonica. Thesis, University of Minas Gerais School of Medicine, Belo Horizonte, Brazil.

Maggiore, Q., Bartolomeo, F., L'Abbate, A., and Misefari, V. (1981). HBsAg glomerular deposits in glomerulonephritis: fact or artifact? *Kidney International*, **19**, 579–86.

Magil, A., Weber, D., and Chan, V. (1986). Glomerulonephritis associated with hepatitis B surface antigenemia. *Nephron*, **42**, 335–99.

Martinelli, R., Pereira, L.J., and Rocha, H. (1987). Influence of antiparasitic therapy on the course of the glomerulopathy associated with schistosomiasis mansoni. *Clinical Nephrology*, **27**, 229–32.

Mathur, B.L. and Jhala, C.I. (1964). Amyloidosis: an emphasis on increasing incidence in India. *Indian Journal of Pathology and Bacteriology*, **7**, 133–5.

McAdam, K.P.W.J., Anders, R.F., Smith, S.R., Russell, D.A., and Price, M.A. (1975). Association of amyloidosis with erythema nodosum leprosum reactions and recurrent neutrophil leucocytosis in leprosy. *Lancet*, **ii**, 572–5.

McGregor, I.A., Turner, M.W., Williams, K., and Hall, P. (1968). Soluble antigens in the blood of African patients with severe *Plasmodium falciparum* malaria. *Lancet*, **i**, 881–4.

Meadow, S.R. (1975). Poststreptococcal nephritis—a rare disease. *Archives of Disease in Childhood*, **50**, 379–81.

Mitsuda, K. and Ogawa, M. (1937). A study of one hundred and fifty autopsies on cases of leprosy. *International Journal of Leprosy*, **5**, 53–60.

Mittal, M.M., Aggarwal, S.C., Maheshwari, H.B., and Kumar, S. (1972). Renal lesions in leprosy. *Archives of Pathology*, **93**, 8–12.

Mizushima, N., Kanai, K., and Matsuda, H. (1987). Improvement in proteinuria in a case of hepatitis B-associated glomerulonephritis after treatment with interferon. *Gastroenterology*, **92**, 524–6.

Moran, C.J., Turk, J.L., Ryder, G., and Waters, M.F.R. (1972). Evidence for circulating immune complexes in lepromatous leprosy. *Lancet*, **ii**, 572–3.

Morel-Maroger, L., *et al.* (1975). Tropical nephropathy and tropical extramembranous glomerulonephritis of unknown aetiology in Senegal. *British Medical Journal*, **1**, 541–4.

Morgan, A.G., Shah, D.J., Williams, W., and Forrester, T.E. (1984). Proteinuria and glomerular disease in Jamaica. *Clinical Nephrology*, **21**, 205–9.

Morierty, P.L. and Brito, E. (1977). Elution of renal anti-schistosome antibodies in human schistosomiasis mansoni. *American Journal of Tropical Medicine and Hygiene*, 26, 717–22.

Musa, A.R.M., Veress, B., Kordofani, A.M., Asha, H.A., Satir, A., and Hassan, A.M.E. (1980). Pattern of the nephrotic syndrome in the Sudan. *Annals of Tropical Medicine and Parasitology*, 74, 37–42.

Nagy, J., Bajtai, G., Brasch, H., Sule, T., Ambrus, M., Deak, G., and Hamori, A. (1979). The role of hepatitis B surface antigen in the pathogenesis of glomerulonephritis. *Clinical Nephrology*, 12, 109–16.

Nash, T.E., Prescott, B., and Neva, F.A. (1974). The characteristics of a circulating antigen in schistosomiasis. *Journal of Immunology*, 112, 1500–7.

Ng, W.L., Scollard, D.M., and Hua, A. (1981). Glomerulonephritis in leprosy. *American Journal of Clinical Pathology*, 76, 321–9.

Ngu, J.L., Adam, M., Leke, R., and Titanji, V. (1980). Proteinuria associated with diethylcarbamazine treatment of onchocerciasis. *Lancet*, i, 254–5.

Ngu, J.L., Chatelanat, F., Leke, R., Ndumbe, P., and Youmbissi, J. (1985). Nephropathy in Cameroon: evidence for filarial derived immune complex pathogenesis in some cases. *Clinical Nephrology*, 24, 128–34.

Omer, H.O. and Wahab, S.M.A. (1976). Secondary amyloidosis due to *Schistosoma mansoni* infection. *British Medical Journal*, 1, 375–7.

Ormerod, A.D., Petersen, J., Hussey, J.K., Weir, J., and Edward, N. (1983). Immune complex glomerulonephritis and chronic anerobic urinary tract infection: complication of filariasis. *Postgraduate Medical Journal*, 59, 730–3.

Ozaki, M. and Furuta, M. (1975). Amyloidosis in leprosy. *International Journal of Leprosy*, 43, 116–24.

Paillerets, F. de., Habib, R., Leubieres, R., Clerc, M., Chapus, Y., and Assi-Adeu, J. (1972). The nephrotic syndrome in children in the Ivory Coast. *Guigoz Science Review*, 89, 2.

Parbtani, A. and Cameron, J.S. (1979). Experimental nephritis associated with plasmodium infection in mice. *Kidney International*, 16, 53–63.

Pesce, A.J. and First, M.R. (1979). *Proteinuria: An Integrated Review*, p. 131, Marcel Dekker, New York.

Peter, K.S., Vijayakumar, T., Vasudevan, D.M., Leena Devi, K.R., Mathew, M.T., and Gopinath, T. (1981). Renal involvement in leprosy. *Leprosy India*, 53, 163–78.

Pillay, V.K.G., Kirch, E., and Kurtzman, N.A. (1973). Glomerulopathy associated with filarial loiasis. *Journal of the American Medical Association*, 255, 179 (letter).

Poon-King, T., Svartman, M., Mohammed, I., Potter, E.V., Achong, J., and Cox, R. (1973). Epidemic acute nephritis with reappearance of M-type 55 streptococci in Trinidad. *Lancet*, i, 475–79.

Powell, C.S. and Swan, L.L. (1955). Leprosy: pathologic changes observed in fifty consecutive necropsies. *American Journal of Pathology*, 31, 1131–47.

Powell, K.C., Meadows, R., Anders, R., Draper, C.C., and Lauer, C. (1977). The nephrotic syndrome in Papua New Guinea: aetiological, pathological and immunological findings. *Australian and New Zealand Journal of Medicine*, 7, 243–8.

Prathap, K. and Looi, L.M. (1982). Morphological patterns of glomerular disease in renal biopsies from 1000 Malaysian patients. *Annals of Academy of Medicine (Singapore)*, 21, 52–6.

Queiroz, F.P., Brito, E., Martinelli, R., and Rocha, H. (1973). Nephrotic syndrome in patients with *Schistosoma mansoni* infections. *American Journal of Tropical Medicine and Hygiene*, 22, 622–8.

Rocha, H., Cruz, T., Brito, E., and Susin, M. (1976). Renal involvement in patients with hepatosplenic schistosomiasis mansoni. *American Journal of Tropical Medicine and Hygiene*, 25, 108–15.

Sadiq, S., Jafrey, N.A., and Naqvi, S.A.J. (1978). An analysis of percutaneous renal biopsies in fifty cases of nephrotic syndrome. *Journal of Pakistan Medical Association*, 28, 121–4.

Sainani, G.S. and Rao, K.V.N. (1974). Renal changes in leprosy. *Journal of Association of Physicians of India*, 22, 659–64.

Seboxa, T., Teklu, B., and Van der Meulen, J. (1984). A four year review of systemic lupus erythematosus. *Ethiopian Medical Journal*, 22, 13–16.

Seedat, Y.K. (1978). Nephrotic syndrome in the Africans and Indians of South Africa. A ten year study. *Transactions of the Royal Society of Tropical Medicine and Hygiene*, 72, 506–12.

Seedat, Y.K. and Pudifin, D. (1977). Systemic lupus erythematosus in black and Indian patients in Natal. *South African Medical Journal*, 51, 335–7.

Seggie, J., Davies, P.G., Ninin, D., and Henry, J. (1984). Pattern of glomerulonephritis in Zimbabwe. Survey of disease characterised by nephrotic proteinuria. *Quarterly Journal of Medicine*, 53, 109–18.

Seggie, J., Nathoo, K., and Davies, P.G. (1984). Association of hepatitis B antigenemia and membranous glomerulonephritis in Zimbabwean children. *Nephron*, 38, 115–19.

Serdula, M.K. and Rhoads, G.G. (1979). Frequency of systemic lupus erythematosus in different ethnic groups in Hawaii. *Arthritis and Rheumatism*, 22, 328–33.

Shuttleworth, J.S. and Ross, S.H. (1956). Secondary amyloidosis in leprosy. *Annals of Internal Medicine*, 45, 23–38.

Shwe, T. (1972a) Immune complexes in glomeruli of patients with leprosy. *Leprosy Review*, 42, 282–9.

Shwe, T. (1972b) Serum complement [C_3] in leprosy. *Leprosy Review*, 42, 268–72.

Sidabutar, R.P., et al. (1986). Glomerulonephritis in Indonesia. *Proceedings of the 3rd Asian Pacific Congress of Nephrology*, pp. 282–291, Singapore.

Singhal, P.C., Chugh, K.S., Kaur, S., and Malik, A.K. (1977). Acute renal failure in leprosy. *International Journal of Leprosy*, 45, 171–4.

Sinniah, R. and Khoo, O.T. (1979). The pathology and immunopathology of glomerulonephritis in Singapore. *Proceedings of the 1st Asian Pacific Congress of Nephrology*, p. 114, Tokyo.

Sitprija, V. (1988). Nephropathy in falciparum malaria. *Kidney International*, 34, 867–77.

Sobh, M.A., Moustafa, F.E., El-Housseini, F., Basta, M.T., Deelder, A.M., and Ghoniem, M.A. (1987). Schistosomal specific nephropathy leading to end stage renal failure. *Kidney International*, 31, 1006–11.

Sobh, M.A., Moustafa, F.E., Sally, S.M., Deelder, A.M., and Ghoniem, M.A. (1988a). Characterisation of kidney lesions in early schistosomal-specific nephropathy. *Nephrology Dialysis Transplantation*, 3, 392–8.

Sobh, M.A., Moustafa, F.E., Sally, S.M., Deelder, A.M., and Ghoniem, M.A. (1988b). Effect of anti-schistosomal treatment of schistosomal specific nephropathy. *Nephrology Dialysis Transplantation*, 3, 744–51.

Soothill, J.F. and Hendrickse, R.G. (1967). Some immunological studies of nephrotic syndrome of Nigerian children. *Lancet*, ii, 629–32.

Southwest Pediatric Nephrology Study Group. (1985). Hepatitis B surface antigenemia in North American children with membranous glomerulo-nephropathy. *Journal of Pediatrics*, 106, 571–8.

Spitz, S. (1946). The pathology of acute falciparum malaria. *Military Surgeon*, 99, 555–72.

Srivastava, R.N., Mayekar, G., Anand, R., Choudhry, V., Ghai, O.P., and Tandon, H. (1975). Nephrotic syndrome in Indian children. *Archives of Disease in Childhood*, 50, 626–30.

Stratta, P., Camussi, G., Ragni, R., and Vercellone, A. (1975). Hepatitis B antigenemia associated with active chronic hepatitis and mesangioproliferative glomerulonephritis. *Lancet*, ii, 179 (letter).

Surbek, K.E. (1931). A striking case of quartan-nephrosis. *Transactions of the Royal Society of Tropical Medicine and Hygiene*, 25, 201–4.

Swainson, C.P., Robson, J.S., Thomson, D., and MacDonald, M.K. (1983). Mesangiocapillary glomerulonephritis: A long term study of 40 cases. *Journal of Pathology*, 141, 449–68.

Tada, T., Kondo, Y., Okumura, K., Sano, M., and Yokogawa, M. (1975). *S. japonicum*: immunopathology of nephritis in Macaca fascicularis. *Experimental Parasitology*, 38, 291–302.

Takeda, S., Kida, H., Katagiri, M., Yokoyama, H., Abe, T., and Hattori, N. (1988). Characteristics of glomerular lesions in hepatitis B virus infection. *American Journal of Kidney Diseases*, **11**, 57–62.

Takekoshi, Y., Tanaka, M., Shida, N., Satake, Y., Saheki, Y., and Matsumoto, S. (1978). Strong association between membranous nephropathy and hepatitis B surface antigenemia in Japanese children. *Lancet*, **ii**, 1065–8.

Thomas, G., Karat, A.B.A., Rao, P.S.S., and Prathapkumar, C. (1970). Changes in renal function during reactive phases of lepromatous leprosy. *International Journal of Leprosy*, **38**, 170–6.

Thuriaux, M.C. (1971). The nephrotic syndrome and *Plasmodium malariae* in the Yemen Arab Republic. *Journal of Tropical Medicine and Hygiene*, **74**, 36–8.

Thyagarajan, S.P., *et al.* (1989). Serum and tissue positivity for hepatitis B virus markers in histopathologically proven glomerulonephropathies. *Journal of Medical Microbiology*, **29**, 243–50.

Van Marck, E., Deelder, A.M., and Gigase, P.L.J. (1977). Effect of partial portal vein ligation on immune glomerular deposits in *Schistosoma mansoni* infected mice. *British Journal of Experimental Pathology*, **58**, 412–17.

Verroust, P., *et al.* (1979). A clinical and immunopathological study of 304 cases of glomerulonephritis in Tunisia. *European Journal of Clinical Investigation*, **9**, 75–9.

Voller, A. (1974). Immunopathology of malaria. *Bulletin of the World Health Organization*, **50**, 177–86.

Voller, A., Draper, C.C., Shwe, T., and Hutt, M.S.R. (1971). Nephrotic syndrome in monkey infected with human quartan malaria. *British Medical Journal*, **4**, 208–10.

Vos, G.H., Grobbelaar, G., and Milnet, L.V. (1973). A possible relationship between persistent hepatitis B antigenemia and renal disease in South African Bantus. *South African Medical Journal*, **47**, 911–12.

Waksman, B.H. and Cook, J.A. (1975). A report of a conference on newer immunologic approaches to schistosomiasis. *American Journal of Tropical Medicine and Hygiene*, **24**, 1037–9.

Ward, P.A. and Conran, P.B. (1966). Immunopathology of renal complications in simian malaria and human quartan malaria. *Military Medicine*, **134**, 1228–36.

Ward, P.A. and Kibukamusoke, J.W. (1969). Evidence for soluble immune complexes in the pathogenesis of the glomerulonephritis in quartan malaria, *Lancet*, **i**, 281–5.

Watt, G., Long, G.W., Calubaquib, C., and Ranoa, C.P. (1987). Prevalence of renal involvement in *Schistosoma japonicum* infection. *Transactions of the Royal Society of Tropical Medicine and Hygiene*, **81**, 339–42.

Waugh, D.A., Alexander, J.H., and Ibels, L.H. (1980). Filarial chyluria associated glomerulonephritis and therapeutic consideration in the chyluric patient. *Australian and New Zealand Journal of Medicine*, **10**, 559–62.

Whittle, H.C., Abdullahi, M.T., Fakunle, F., Parry, E.H.O., and Rajkovic, A. (1973). Scabies pyoderma and nephritis in Zaria, Nigeria. *Transactions of the Royal Society of Tropical Medicine and Hygiene*, **67**, 349–63.

Williams, R.C., Cathcart, E.C., Calkins, E., Fite, G.L., Rubio, J.B., and Cohen, A.S. (1965). Secondary amyloidosis in lepromatous leprosy. Possible relationships of diet and environment. *Annals of Internal Medicine*, **62**, 1000–7.

Wyler, D.J. (1983). Malaria: resurgence, resistance and research. *New England Journal of Medicine*, **308**, 875–8.

Yawalker, S.J. and Vischer, W. (1979). Lamprene (Clofazimine) in leprosy. *Lancet Review*, **50**, 135–44.

Yoshikawa, N., *et al.* (1985). Membranous glomerulonephritis associated with hepatitis B antigen in children. A comparision with idiopathic membranous glomerulonephritis. *Clinical Nephrology*, **23**, 28–34.

SECTION 4
The patient with systemic disease affecting the kidney

4.1 The patient with diabetes mellitus

4.2 Amyloidosis

4.3 Kidney involvement in plasma-cell dyscrasias

4.4 Sarcoidosis

4.5 Vasculitis

    4.5.1 Pathogenesis of angiitis

    4.5.2 The nephritis of Henoch–Schönlein purpura

    4.5.3 Systemic vasculitis

4.6 Systemic lupus erythematosus

    4.6.1 Recent advances in the pathogenesis of systemic lupus erythematosus

    4.6.2 Systemic lupus erythematosus (clinical)

4.7 Scleroderma—systemic sclerosis

4.8 Rheumatoid arthritis, mixed connective tissue disease, and polymyositis

4.9 Essential mixed cryoglobulinaemia

4.10 Sjögren's syndrome and overlap syndromes

4.11 Sickle-cell disease and the kidney

4.1 The patient with diabetes mellitus

JOSÉ REIMÃO PINTO AND GIANCARLO VIBERTI

Introduction

Renal disease is a relatively common microvascular complication of both insulin-dependent and non-insulin-dependent diabetes mellitus. It is defined clinically as the presence of persistent proteinuria (>0.5 g/24 h) in a diabetic patient with concomitant retinopathy and elevated blood pressure, but without urinary tract infection, other renal disease or heart failure.

Signs of renal disease in diabetics were recognized as early as the 18th century by Cotunnius (1764) and Rollo (1798), who in his textbook on diabetes reported the presence of protein in the urine of some diabetic patients. In 1836 Richard Bright gave the first full account of albumin in the urine as a sign of serious renal disease: some of his patients had diabetes, presumably of the non-insulin-dependent type. It was Rayer who, in 1840, postulated that diabetes might cause a form of 'Bright's disease'. One hundred years later, in 1936, Kimmelstiel and Wilson described nodular glomerular intercapillary lesions developing in the diabetic kidney, and related this to the clinical syndrome of profuse proteinuria and renal failure accompanied by arterial hypertension.

The size of the problem of diabetic kidney disease became clear in the 1950s, as insulin-dependent diabetic patients began to survive for longer periods, following the discovery of insulin in 1921. Recent studies indicate that approximately 600 cases of end-stage diabetic renal failure occur every year in the United Kingdom (just under 10 cases per million population) (Joint Working Party on Diabetic Renal Failure 1988); in 1985 about one-third of all patients beginning renal replacement therapy in the United States were diabetic. The cost for caring for these diabetics in renal failure approached $1 billion in 1985 (Eggers 1988).

Epidemiology

The epidemiology of diabetic nephropathy has been predominantly, though not exclusively, studied in insulin-dependent patients. Until recently, non-insulin-dependent patients had been investigated less intensively, but it is now becoming clear that within this type of diabetes major differences exist between different ethnic groups. While waiting for more information about the latter group of patients, it is prudent to treat the epidemiology of nephropathy separately in the two types of diabetes.

Insulin-dependent diabetes

Two major cohort studies have described the prevalence and incidence of diabetic nephropathy, as defined by persistent clinical proteinuria, in insulin-dependent patients who developed diabetes before the age of 31 years (Andersen *et al.* 1983; Krolewski *et al.* 1985; Kofoed-Enevoldsen *et al.* 1987). The prevalence of nephropathy increases with duration of diabetes to a peak of 21 per cent after 20 to 25 years, and then declines to about 10 per cent in patients who have had diabetes for 40 years or more. Only a small proportion of patients (4 per cent) develops nephropathy within 10 years of diabetes. After a lag time of approximately 5 years, the annual incidence of nephropathy rises rapidly over the next 10 years to a peak after 15 to 17 years of about 3 per cent per year, and then declines to around 1 per cent per year in patients with 40 years or more of diabetes. Nephropathy develops in only 4 per cent of those with diabetes duration of more than 35 years, who therefore have a low risk of this complication.

This pattern of risk indicates that accumulation (i.e. intensity times duration) of exposure to diabetes is not sufficient to explain the development of clinically manifest kidney disease, and suggests that only a subset of patients are susceptible to renal complications. The paucity of new cases of nephropathy among long-standing diabetic patients would support the view that this complication occurs in most of the susceptible individuals earlier in the course of diabetes. The cumulative incidence clearly indicates that only a proportion of juvenile diabetic patients will ever develop nephropathy.

This proportion has changed over the years (Krolewski *et al.* 1985; Kofoed-Enevoldsen *et al.* 1987). In cohorts of patients diagnosed before 1942, the cumulative risk after 25 to 30 years of diabetes was approximately 41 per cent but this has declined to around 25 per cent in patients diagnosed after 1949 (Fig. 1). The reason(s) for the lower frequency of diabetic nephropathy in recent decades is not entirely clear. Cohort differences may be related to changes in diabetes care and control, to more intensive and early treatment of concomitant conditions such as hypertension, or, less likely, to dietary changes. An alterna-

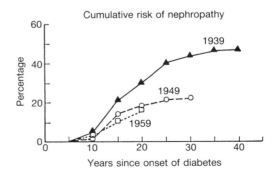

Fig. 1 Cumulative incidence of clinical proteinuria in insulin-dependent diabetic patients according to duration of diabetes and calendar year of diagnosis. Reproduced with permission from Krolewski *et al.* 1987b.

tion is that before 1942 persistent proteinuria in diabetic patients may have been, in a proportion of cases, the manifestation of some other form of renal disease, particularly glomerulonephritis, the incidence of which has declined during this century (Cameron 1979). This interpretation would explain why a sharp decrease in incidence occurred in patients diagnosed in the late 1940s, with no further decline demonstrable thereafter.

There is a clear male preponderance in the development of proteinuria, with a male to female ratio of around 1.7:1. After 40 years of diabetes, the cumulative incidence is 46 per cent in male but only 32 per cent in female diabetics. This sex difference in the incidence of renal disease is also found in non-diabetic subjects (Finn and Harmer 1979; Pasternack et al. 1985). Interestingly, a similar male preponderance has also been reported in the development of diabetic proliferative retinopathy (Danielsen et al. 1983; Klein et al. 1986). It has been suggested that sex steroids may be of importance, since castrated diabetic rats seem less prone to late diabetic complications than non-castrated control animals (Williamson et al. 1986).

Age at diagnosis significantly influences incidence of nephropathy. The time to development of microvascular complications is not influenced by prepubertal duration of the disease (Kostraba et al. 1989), and nephropathy develops more slowly in individuals who develop diabetes before the age of 10 than in those diagnosed after puberty (Krolewski et al. 1985). The highest incidence, 44 per cent, is seen in subjects who develop diabetes between the age of 11 and 20 years (Kofoed-Enevoldsen et al. 1987). Patients who develop diabetes after the age of 20 have a lower cumulative incidence of nephropathy, at around 35 per cent. Current age has been found by some authors (Krolewski et al. 1985) but not by others (Kofoed-Enevoldsen et al. 1987) to influence the incidence of proteinuria, with a maximal risk in the age interval 18 to 35 years and a rapid decline in incidence after age 35 regardless of duration of diabetes (Derby et al. 1988). This discrepancy in findings may be partly related to the different age groups of the cohorts studied.

In one study the level of hyperglycaemia during the first 15 years of diabetes was found to be positively related to the risk of persistent proteinuria (Krolewski et al. 1985). However, the incidence of nephropathy rapidly declined after 15 years of diabetes, even though there was no improvement in the control of glycaemia, suggesting that other non-metabolic factors (probably genetic) are involved in the genesis of this condition. The complex metabolic disturbance of diabetes appears therefore necessary, but not sufficient, for the clinical manifestation of nephropathy.

The development of end-stage renal failure and the mortality associated with it closely conform with the occurrence of persistent proteinuria (Krolewski et al. 1985). The range of survival after the onset of persistent proteinuria is wide, varying between 1 and 24 years. Median survival has been reported to be around 7 (Andersen et al. 1983) or 10 years (Krolewski et al. 1985), with 25 per cent of the patients developing end-stage renal failure within 6 years and 75 per cent within 15 years of onset of proteinuria. Progression to end-stage renal failure seems to take longer in those diabetic patients diagnosed before puberty. Median time between onset of persistent proteinuria and development of end-stage renal failure was 14 years in the group of subjects with onset of diabetes before age 12, and 8 years in those with onset of diabetes between age 12 and 20 years (Krolewski et al. 1985).

The ominous significance of renal involvement in insulin-dependent diabetes mellitus is clearly shown by the comparison of long-term outcome in patients with and without nephropathy.

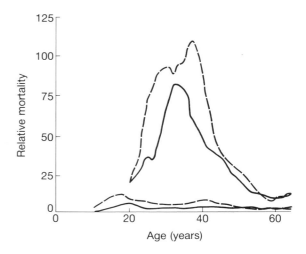

Fig. 2 Relative mortality of insulin-dependent diabetics with (upper curves) and without (lower curves) clinical proteinuria as a function of age. (Dashed lines women, solid lines, men.) Reproduced with permission from Borch-Johnsen et al. 1985a.

Only 10 per cent of patients with proteinuria survive after 40 years of diabetes, in contrast to more than 70 per cent of those without proteinuria (Andersen et al. 1983). Almost all of these have normal renal function, and up to 70 per cent are clinically well in all respects, only about 10 per cent being seriously disabled (i.e., blind, amputated, hemiparetic, infarcted) (Deckert et al. 1978).

Before renal replacement therapy was widely available, the main cause of death (approximately 60 per cent) in insulin-dependent patients with diabetic nephropathy was uraemia, but a substantial proportion died on the way to terminal renal failure from ischaemic heart disease (19 per cent) or stroke (5 per cent). The proportion of cardiovascular deaths seems to increase significantly, to around 41 per cent, in those patients who develop proteinuria after more than 20 years of diabetes (Andersen et al. 1983). The advent of renal replacement therapy has postponed uraemic death but has increased the pool of cardiovascular deaths, which now occur after institution of treatment for end-stage renal disease. There is general agreement that the development of persistent proteinuria in insulin-dependent diabetic patients increases early mortality from cardiovascular disease approximately 9-fold (Borch-Johnsen and Kreiner 1987; Jensen et al. 1987a). Similarly, it has been estimated that the risk of developing coronary artery disease is 15 times higher in those with compared to those without proteinuria (Krolewski et al. 1987a), and its cumulative incidence 5 years after onset of proteinuria is eight times higher than in matched controls without renal disease (Jensen et al. 1987a). Whatever the cause, mortality by age 45 in insulin-dependent diabetic patients with proteinuria has been reported to be 20 to 40 times higher than that of patients without proteinuria, who experience a relative mortality only double that of the non-diabetic population (Borch-Johnsen et al. 1985a) (Fig. 2). Recent studies suggest that more effective antihypertensive treatment during the past decade has significantly improved the prognosis of diabetic nephropathy. Ten-year survival after onset of persistent proteinuria has risen from 30 to 50 per cent (Andersen et al. 1983; Krolewski et al. 1985) to the present 80 per cent (Parving and Hommel 1989; Mathiesen et al. 1989) (Fig. 3).

Non-insulin-dependent diabetes

Data on the prevalence and incidence of nephropathy, as indicated by persistent proteinuria, were relatively scarce in this type of diabetes until recent years.

In European diabetic patients diagnosed after age 40, the overall prevalence of proteinuria in excess of 500 mg/24 h was reported to be approximately 16 per cent (Fabre et al. 1982). Other studies in slightly differently selected populations have reported lower prevalences, at around 12 per cent (Klein et al. 1988) and 14 per cent (Parving et al. 1990), with males being affected more frequently (19 per cent) than females (4 per cent). Unlike in insulin-dependent diabetics, the prevalence of proteinuria increases steadily with duration of diabetes, from 7 to 10 per cent in patients with less than 5 years of diabetes to 20 to 35 per cent in patients who have had diabetes for more than 20 to 25 years (Fabre et al. 1982; Klein et al. 1988). The lowest prevalence of proteinuria, 2.4 per cent, was reported in diabetic patients in Hong Kong (WHO Multinational Study 1985), but prevalences higher than those seen in European diabetic patients have been found in American Indians (Rate et al. 1983; WHO Multinational Study 1985; Nelson et al. 1989), Mexican Americans (Haffner et al. 1989), black Americans (Cowie et al. 1989), Japanese (WHO Multinational Study 1985), Nauruan from Central Pacific islands (Collins et al. 1989), and Asian Indians in the United Kingdom (Samanta et al. 1986), in whom an increased prevalence of microalbuminuria has also been described (Allawi et al. 1988). In these non-European ethnic groups there is also a tendency for the prevalence to increase with duration of dia-

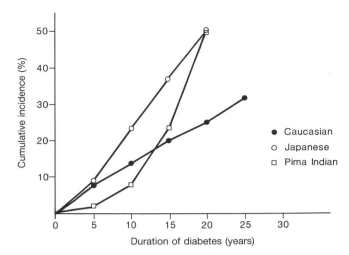

Fig. 4 Cumulative incidence of clinical proteinuria by duration of diabetes in non-insulin-dependent subjects of Caucasian, Japanese, and Pima Indian origin. Data taken from Ballard et al. 1988, Sasaki et al. 1986 and Kunzelman et al. 1989.

betes, though this does not seem to apply to all populations. A male preponderance in the prevalence of proteinuria has been reported by some authors (Ballard et al. 1988). Significant relationships have been found between prevalence of proteinuria and both blood glucose control and level of arterial blood pressure. These findings seem to apply across ethnic differences and suggest that cumulative exposure to diabetes and perhaps raised arterial blood pressure are important determinants of renal involvement in non-insulin-dependent diabetes. Of interest is the observation that as many as 62 per cent of European diabetic patients have been found to maintain normal protein excretion even after 16 years of diabetes. Those who developed proteinuria had a higher prevalence of hypertension before onset of persistent proteinuria than those who did not (Hasslacher et al. 1988).

Incidence data in Caucasian non-insulin-dependent diabetic patients of European origin in a population based study in Rochester showed a cumulative risk of persistent proteinuria of 25 per cent after 20 years of diabetes (Ballard et al. 1988). Recent data from Europe shows a 57 per cent cumulative frequency of proteinuria after 25 years of diabetes (Hasslacher et al. 1989). Reports in non-insulin-dependent diabetic patients of Japanese (Sasaki et al. 1986) and of Pima Indian (Kunzelman et al. 1989) origin indicate that the incidence rate of nephropathy, as measured by persistent proteinuria, rises with duration of diabetes, with a cumulative risk of proteinuria of 50 per cent after 20 years of diabetes (Fig. 4). Of particular importance are the recent observations in the Pima that elevated blood pressure before the onset of diabetes predicts abnormal albuminuria (Knowler et al. 1988), and that the risk of development of diabetic nephropathy in the offspring of a diabetic parent is increased when the parent also has diabetic renal disease (Pettitt and Saad 1988). Thus in both types of diabetes susceptibility factors, either genetic or shared environmental appear to be critical in the pathogenesis of nephropathy.

Progression to end-stage renal failure in proteinuric non-insulin-dependent diabetic patients is variable, and has been reported to be infrequent in individuals of European origin

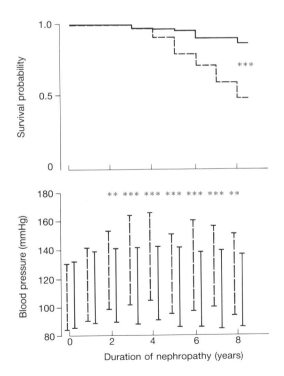

Fig. 3 Survival probability and blood pressure in two cohorts of insulin-dependent diabetic patients with clinical nephropathy who either received (cohort II) or did not receive (cohort I) antihypertensive treatment. Treatment of hypertension resulted in improved survival (48 vs. 87 per cent at 8 years) (**$p < 0.01$; ***$p < 0.001$). Reprinted with permission from Mathiesen et al. 1989.

(Fabre *et al.* 1982). In a period prevalence survey of diabetic renal failure in the United Kingdom in 1985, only 3.5 cases per million population were found, and some of these patients were of non-European origin. This figure compares with 6.5 cases per million population of end-stage renal failure in insulin-dependent diabetic patients (Joint Working Party 1988). Given the approximately 10-fold higher prevalence of non-insulin-dependent diabetes, this observation suggests that end-stage renal failure is about 20 times less frequent than in insulin-dependent diabetes in patients of European origin.

This generally held view has recently been challenged by the finding of a cumulative incidence of chronic renal failure in non-insulin-dependent diabetic patients who had persistent proteinuria at the time of diagnosis of 12 per cent by 15 years. In those patients developing proteinuria after the diagnosis of diabetes, the cumulative risk was 17 per cent 15 years after the onset of persistent proteinuria (Humphrey *et al.* 1989). In another longitudinal study in Europeans followed for a median time of 20 years, the cumulative frequency of renal failure (diagnosed by a serum creatinine greater than 124 μmol/l) 3 and 5 years after onset of persistent proteinuria was even higher, at 41 and 63 per cent, respectively, not different from the cumulative risk in patients with insulin-dependent diabetes (Hasslacher *et al.* 1989).

The discrepancy between frequency of proteinuria and frequency of end-stage renal disease in some studies of European subjects may arise for different reasons. Many non-insulin-dependent patients are taking insulin by the time they reach terminal renal failure, which may lead to misclassification of type of diabetes. Many of these patients are old and have associated cardiovascular disease, and they may simply not be considered for renal replacement therapy (Joint Working Party 1989) or they may die before reaching end-stage renal failure. Proteinuria of non-diabetic origin, which occurs in close to 30 per cent of proteinuric non-insulin-dependent diabetic patients (Grenfell *et al.* 1988; Parving *et al.* 1990) may also contribute to this disparity.

In a prospective study of a cohort of 503 non-insulin-dependent diabetic subjects, 10-year survival was significantly poorer in patients with elevated urinary albumin excretion—around 30 per cent, compared to 55 per cent in patients with normal albumin excretion rates. The majority of deaths (58 per cent) were due to cardiovascular causes, only 3 per cent being ascribed to uraemia (Schmitz and Vaeth 1988). These data in persistently proteinuric patients agree with previous findings of prospective studies of European non-insulin-dependent diabetics, in whom lesser elevation of urinary albumin excretion markedly increased the risk of cardiovascular death (Jarrett *et al.* 1984; Mogensen 1984).

In other ethnic groups the incidence of end-stage renal failure is strongly associated with the presence of proteinuria. In non-insulin-dependent diabetic Pima Indians, the incidence of end-stage renal failure increases with duration of diabetes to about 41 cases per 1000 person-years at risk after 20 years of diabetes, the cumulative risk of this complication reaching 15 per cent after this time (Nelson *et al.* 1988*a*). In comparison to non-insulin-dependent patients of European origin, an excess incidence of end-stage renal disease has also been reported in Mexican Americans (Mexican American/White ratio about 6) (Pugh *et al.* 1988) and in black diabetic subjects (Rostand *et al.* 1982; Cowie *et al.* 1989). In the United Kingdom, an excess of Afro-Caribbean and Asian patients has been found among non-insulin-dependent patients treated for end-stage renal disease (Grenfell *et al.* 1988), and the incidence of this complication has also been reported to be higher in African countries (Abdullah 1978; Adetuyibi 1976).

In diabetic Pima Indians, proteinuria confers a 3.5-fold higher risk of premature mortality, and the concomitant presence of arterial hypertension increases this relative risk to 7.1 (Nelson *et al.* 1988*b*). Interestingly, the mortality rate of the diabetic Pima Indians without proteinuria is similar to that of the non-diabetic subjects. Of the excess mortality associated with non-insulin-dependent diabetes in this population, 97 per cent is found in patients with proteinuria. Sixteen per cent of deaths were ascribed to uraemia, while 22 per cent were due to cardiovascular disease. These mortality data are slightly higher than the two-fold excess early mortality observed in the European non-insulin-dependent patients with proteinuria (Schmitz and Vaeth 1988).

Clinical course and natural history

The natural history of diabetic nephropathy has been better defined in insulin-dependent than in non-insulin-dependent diabetes, partly because the usually acute onset of diabetes in the former allows a more precise timing of observed clinical and physiological events.

Early phase

After diagnosis of diabetes a clinically silent phase of variable duration occurs. During this period, however, important abnormalities of renal function and structure take place.

Glomerular filtration rate is found to be an average of 20 to 40 per cent above that of age-matched normal subjects, both in adults and children with insulin-dependent diabetes (Cambier 1934; Fiaschi *et al.* 1952; Stalder and Schmid 1959; Ditzel and Schwartz 1967; Mogensen 1971*a*). Data are controversial in non-insulin-dependent patients, with some groups reporting normal and others supranormal GFR values (Schmitz *et al.* 1989*a*; Palmisano and Lebowitz 1989; Vora *et al.* 1990; Loon *et al.* 1990). Approximately 25 per cent of patients with insulin-dependent diabetes have a GFR exceeding the upper limit of the normal range (Wiseman *et al.* 1984*a*). The frequency distribution of GFR seems to be shifted to the right in these patients (Fig. 5), suggesting that the GFR of all insulin-dependent subjects (even those in the normal range) are probably higher than their putative non-diabetic GFR. Hyperfiltration is related to the degree of blood glucose control, at least within the range of moderate hyperglycaemia (i.e. up to 14 mmol/l), whereas higher blood glucose values tend to be associated with normal or low GFR (Wiseman *et al.* 1984*a*). Intensified insulin-treatment and good metabolic control reduce the GFR toward normal levels after a period of weeks to months in both insulin-dependent (Christiansen *et al.* 1982*a*; Wiseman *et al.* 1985*a*) and non-insulin-dependent diabetes (Schmitz *et al.* 1989*b*). Renal plasma flow has been reported as elevated, normal or reduced in insulin-dependent diabetes (Mogensen 1971*a*; Christiansen *et al.* 1981*a*; Ditzel and Junker 1972), although more recent work shows an elevation of renal plasma flow ranging between 9 and 14 per cent (Christiansen 1984).

The increased GFR and renal plasma flow are accompanied by an approximately 20 per cent increase in kidney size, and a good correlation between GFR and kidney volume has been described in insulin-dependent diabetic patients (Mogensen and Andersen 1973; Puig *et al.* 1981; Christiansen *et al.* 1981*a*; Wiseman and Viberti 1983). Approximately 40 per cent of insulin-dependent

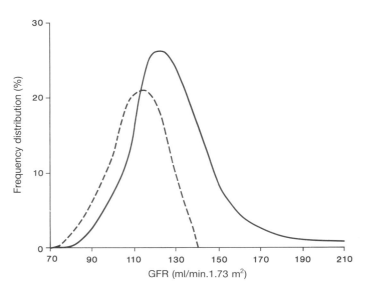

Fig. 5 GFR distribution in 127 insulin-dependent diabetic patients (solid line) and in 40 age-matched (32 (16–58) vs. 35 (18–56) years) controls (dashed line).

patients have kidneys larger than normal; a large kidney is a prerequisite for the occurrence of a GFR above the upper limit of the normal range, but a normal GFR can be found in patients with large kidneys (Wiseman and Viberti 1983). In one autopsy study of non-insulin-dependent subjects, increased kidney volumes were found only on those patients with normal serum creatinine (Dumler et al. 1987).

The relationship of kidney size to glycaemic control is less clear cut than that of GFR. Three months of insulin treatment has been claimed to reduce kidney size in newly diagnosed insulin-dependent (Mogensen and Andersen 1975) and non-insulin-dependent (Schmitz et al. 1989b) patients. This observation has not, however, been confirmed in longer term insulin-dependent subjects treated by intensified insulin treatment for periods of up to 1 year. In these patients, in spite of a reduction in GFR, kidney volume remained unchanged (Christiansen et al. 1982a; Wiseman et al. 1985a; Christensen et al. 1986). Nephromegaly in the animal model can also be reversed only if insulin treatment is started soon after induction of diabetes (Rasch 1979a): correction of hyperglycaemia after 4 weeks fails to return kidney volume to normal (Seyer-Hansen 1976). Indeed, contrary to the situation in most forms of renal disease, large kidneys can persist in diabetes, even in the presence of advanced renal failure (Kahn et al. 1974; Ellis et al. 1985). It is not known why renal enlargement becomes irreversible, but it has been speculated that this irreversibility may have prognostic implications for the development of clinically significant renal damage (Kleinman and Fine 1988). If hydronephrosis and duplex kidneys are excluded, diabetes is the most common cause of nephromegaly associated with renal disease (Segel et al. 1984).

The prognostic significance of glomerular hyperfiltration in humans also remains uncertain. A positive correlation between initial hyperfiltration and subsequent increase in albuminuria and development of clinical nephropathy was described in two retrospective studies (Mogensen and Christensen 1984; Mogensen 1986), but these findings were not confirmed by a more recent study spanning an observation period of 18 years (Lervang et al. 1988). In all of these reports, a number of potentially important confounding variables, such as levels of albumin excretion rate, blood pressure, or protein intake, which can all affect outcome, were not controlled for. The 5-year results of an ongoing prospective case-control study of insulin-dependent diabetic patients with and without hyperfiltration (Jones et al. 1987) seem to suggest that diabetics with hyperfiltration, though showing a faster rate of GFR decline, do not show an increased frequency of rising albuminuria or blood pressure.

Although patients in this silent phase do not, by definition, demonstrate clinically detectable proteinuria, a sensitive radioimmunoassay for urinary albumin has shown that in several circumstances a proportion of patients with relatively early diabetes have elevated supranormal rates of albumin excretion. A significant increase in albumin excretion rate to three or four times above the normal level was first demonstrated in non-insulin-dependent diabetics newly detected in a population survey (Keen et al. 1969). Young newly diagnosed insulin-dependent subjects, or short-term diabetics in poor control, often demonstrate an elevated albumin excretion rate (Mogensen 1971c, 1976a; Parving et al. 1976). Subclinical elevation of both albumin and IgG excretion is also found in longer-term insulin-dependent diabetic patients (Viberti et al. 1982a). These abnormal, though subclinical, increases in albumin excretion rate have been termed microalbuminuria. Albumin excretion in healthy individuals ranges between 1.5 and 20 μg/min, with a geometric mean around 6.5 μg/min: these levels have been termed normoalbuminuria. Diabetic patients with dipstick-positive proteinuria generally have albumin excretion rates in excess of 200 μg/min, and these levels are defined as persistent or clinical albuminuria. 'Microalbuminuria' thus defines the wide subclinical range of albumin hyperexcretion ranging from 20 to 200 μg/min.

The albumin excretion rate in both normal individuals and diabetic patients tends to be about 25 per cent higher during the day than during the night, and has an average day-to-day variation of about 40 per cent (Mogensen 1971c; Feldt-Rasmussen and Mathiesen 1984; Mathiesen et al. 1984; Rowe et al. 1985; D.L. Cohen et al. 1987; Chachati et al. 1987). Similar coefficients of variation are found for albumin/creatinine ratios, suggesting that this variability is a true biological phenomenon and not due to inadequate urine collection (Mogensen 1971c; Mathiesen et al. 1984; Rowe et al. 1985; D.L. Cohen et al. 1987). An important practical implication of this variability is that the diagnosis of microalbuminuria should depend on finding increased rates of excretion in at least three collections over a 6-month period. The prevalence of microalbuminuria in insulin-dependent diabetes has been reported to vary from 5 to 37 per cent in different population- and diabetic clinic-based studies (Mathiesen et al. 1986; Gardete et al. 1986; Close et al. 1987; Gatling et al. 1988; Parving et al. 1988a; Marshall and Alberti 1989). These rather large differences in prevalence are probably ascribable to patient selection. No correlation is found between albumin excretion rate and age, but there is a male preponderance among insulin-dependent microalbuminuric patients, who also tend to have an earlier onset and longer duration of diabetes (Parving et al. 1988a; Marshall and Alberti 1989). Persistent elevation of albumin excretion rate is exceptional in the first 5 years of diabetes (Close 1987; Marshall and Alberti 1989), and microalbuminuria has not been detected in children less than 15 years of age (Mathiesen et al. 1986). The phase of persistent microalbuminuria has been termed by some authors 'incipient nephropathy' (Mogensen et

Table 1 Predictive value of albumin excretion rate for persistent clinical proteinuria in insulin-dependent diabetics

Investigator (no. of subjects)	Baseline albumin excretion rate (μ/min)	Follow-up (years)	Type of collection	Sensitivity (%)	Specificity (%)	Predictive value of a positive test	
Viberti et al. (n = 63)	> 30	14	Overnight	77	98	7/8	(82%)
Parving et al. (n = 71)	> 70	6	24 hr	70	100	7/10	(70%)
Mogensen et al. (n = 43)	> 15	10	Short term	100	94	12/14	(86%)
Jerums et al. (n = 53)	> 30	8	24 hr	50	94	1/4	(25%)

Modified from Bennett 1989.

al. 1983). Since albumin excretion rate is influenced by glycaemic control, correction of hyperglycaemia can reduce microalbuminuria, not only in short-term but also in insulin-dependent diabetes of long duration (Parving et al. 1976; Mogensen 1976a; Viberti et al. 1979, 1982a; Mathiesen et al. 1984). Moderately strenuous exercise may also provoke an exaggerated rise in albumin excretion rate in diabetic patients with normal resting values (Mogensen and Vittinghus 1975; Viberti et al. 1978). The severity of this exercise-induced albuminuria seems related to duration of diabetes, and is modulated by the level of blood glucose control (Koivisto et al. 1981; Viberti et al. 1981a; Vittinghus and Mogensen 1982).

Prevalences of microalbuminuria between 8 and 46 per cent have been reported in Europeans with non-insulin-dependent diabetes (Fabre et al. 1982; Damsgaard and Mogensen 1986; Gatling et al. 1988; Schmitz and Vaeth 1988; Marshall and Alberti 1989), with a prevalence of 47 per cent in the Pima Indians (Nelson et al. 1989). Its correlation with glycaemic levels is apparent not only in diabetic patients, but also in patients with impaired glucose tolerance (Keen et al. 1969; Damsgaard and Mogensen 1986; Nelson et al. 1989). No correlation is found with sex, but there are significant associations with diastolic blood pressure levels and resting heart rate (Allawi and Jarrett 1990). Microalbuminuria has been reported to be more frequent in non-insulin-dependent patients treated with insulin (Damsgaard and Mogensen 1986; Nelson et al. 1989).

The prognostic significance of microalbuminuria for development of persistent proteinuria and overt nephropathy has been demonstrated by four longitudinal studies of cohorts of insulin-dependent patients (Viberti et al. 1982b; Parving et al. 1982; Mathiesen et al. 1984; Mogensen and Christensen 1984), all of which suggested the existence of a threshold of albumin excretion rate above which the risk of progression to clinical nephropathy increases by about 20-fold. The overall findings of these studies are remarkably similar, the differences in methods of urine collection and length of follow-up probably being responsible for the different risk levels associated with different albumin excretion rates (Table 1). Microalbuminuria is unlikely to be a marker of susceptibility to the development of clinical nephropathy as it is undetectable in the first 5 years of diabetes, but is more likely a sign of early disease. This interpretation has recently been corroborated by the finding that patients with persistent microalbuminuria have more severe renal histological lesions (Chavers et al. 1989). There is no evidence, at present, that detection of exercise-induced microalbuminuria improves the predictive power of resting microalbuminuria. A good correlation has been found between albumin excretion rate and albumin/creatinine ratios, particularly in first morning urine samples, and albumin excretion rates greater than 30 μg/min correspond to albumin/creatinine ratios greater than 2.5 (D.L. Cohen et al. 1987).

Three retrospective studies examined the prognostic value of microalbuminuria in cohorts of non-insulin-dependent patients (Jarrett et al. 1984; Mogensen 1984; Schmitz and Vaeth 1988). Over an interval of 10 to 14 years they have all shown an increased risk of cardiovascular death in these patients. A recent 3-year prospective study has confirmed the greater incidence of premature mortality due to cardiovascular events in non-insulin-dependent diabetics with microalbuminuria (Mattock et al. 1990); microalbuminuria has also been found to be predictive of risk of cardiovascular events in the non-diabetic population (Yudkin et al. 1988).

Late phase: clinical nephropathy

The onset of the clinical phase of diabetic nephropathy is signalled, by convention, by the appearance of persistent proteinuria (total protein excretion of 0.5 g/day or more), which corresponds to an albumin excretion rate greater than 200 μg/min (about 300 mg/day). A phase of intermittent dipstick-positive urine for protein precedes persistent proteinuria (Ireland et al. 1982), but this probably corresponds to the late phases of high microalbuminuria, with values intermittently breaking through the clinical threshold (Bending et al. 1986; Jerums et al. 1987): with modern quantitative techniques for the measurement of urinary albumin, the term and concept of intermittent proteinuria should probably be abandoned. The development of persistent proteinuria is followed by progressive decline of the GFR to end-stage renal failure (Mogensen 1976b; Jones et al. 1979; Parving et al. 1981; Viberti et al. 1983a). Whether the fall in GFR starts at the time or after the appearance of the microalbuminuric phase, is a matter for debate (Feldt-Rasmussen et al. 1986; Bending et al. 1986). The fall in GFR appears to be linear with time in all patients, but the rate of decline varies over an approximately five-fold range in individual patients (Fig. 6).

The reasons for these different rates of progression are not entirely clear and do not seem to be related to age, sex, duration of diabetes, or degree of proteinuria at onset. Adequacy of blood glucose control has a limited impact on progression (Viberti et al. 1983b; Hasslacher et al. 1985), even though some authors have reported a correlation between glycosylated haemoglobin levels and rate of fall of GFR (Nyberg et al. 1987). These data have, however, not been confirmed (Viberti et al. 1987a). The level of diastolic blood pressure has been found to be correlated with the rate of progression of established diabetic nephropathy, and serum creatinine concentrations have been reported to rise sooner in those proteinuric patients with the higher blood pressure levels (Mogensen 1976b; Hasslacher et al. 1985; Laffel et al. 1987). The failure of other authors (Jones et al.

1979; Parving et al. 1981; Viberti et al. 1983a) to confirm a relationship with blood pressure may result from the current more widespread treatment of mild hypertension in these patients. Caution must be applied when serum creatinine or creatinine clearance are used as indices of glomerular function (Walser et al. 1988) (see Chapter 1.5): the asymptotic relationship between serum creatinine and GFR means that a rise in serum creatinine may not occur until more than 50 per cent of the GFR has been lost. A linear decline in inverse creatinine is not seen for serum creatinine concentrations below 200 μmol/l (Jones et al. 1979). The use of creatinine clearance also tends to overestimate GFR, because of the tubular secretion of creatinine which takes place in advanced renal failure (Shemesh et al. 1985).

The average time between onset of persistent proteinuria and end-stage renal failure has been calculated as 7 years in the past; this is likely to have been considerably extended in recent years (probably more than doubled) with the early use of more intensive treatment for hypertension (Parving and Hommel 1989; Mathiesen et al. 1989) and the early prescription of low protein diets (Walker et al. 1989; Zeller et al. 1990).

Elevations of arterial blood pressure almost invariably occur in diabetic patients with established nephropathy. A small proportion (about 25 per cent) of patients may still have blood pressures within the so-called normal range at the onset of persistent proteinuria, although it may have risen within the normal range from previous lower preproteinuric levels (Wiseman et al. 1984b; Mathiesen et al. 1984; Jensen et al. 1987b), but virtually all patients become hypertensive with progression to renal insufficiency (Ireland et al. 1982; Hasslacher et al. 1985; Parving and Hommel 1989). Several studies suggest that the excess of arterial hypertension described in insulin-dependent patients is almost entirely accounted for by patients with persistent proteinuria (Keen et al. 1975; Drury 1983; Hasslacher et al. 1985): long-term insulin-dependent patients without proteinuria have lower blood pressures than age-matched controls (Borch-Johnsen et al. 1985b). The degree of proteinuria in diabetic patients is often in the subnephrotic range, but heavy protein excretion and nephrotic syndrome may occur, and this has been related to a poorer renal outcome (Watkins et al. 1972). Although the level of proteinuria is thought to relate roughly to the severity of the glomerular lesions, severe glomerular damage has been reported in patients without proteinuria (Thomsen 1965; Watkins et al. 1972). The clinical significance of this latter finding remains unclear, as in our personal experience of more than 90 insulin-dependent patients with diabetic nephropathy, we have not met a single case of progressive renal failure in the absence of some degree of proteinuria.

Non-renal complications

Diabetic retinopathy is present in virtually all insulin-dependent diabetic patients with nephropathy (Thomsen 1965; Malins 1968; Deckert and Poulsen 1968; Parving et al. 1988a). Indeed, the absence of retinopathy should lead to careful consideration of other non-diabetic causes for proteinuria and renal disease (see below). In advanced renal disease retinopathy is usually severe with new vessel formation. While all patients with nephropathy have retinopathy, the reverse is not true: retinopathy, even of the proliferative kind, may occur in the absence of proteinuria and renal disease (Bilous et al. 1985). Up to one-third of patients with proliferative retinopathy may be free of proteinuria (Root et al. 1959; Feldman et al. 1982). The exact reasons for this discrepancy in manifestations of microvascular disease are not clear, but epidemiological evidence suggests that retinopathy and nephropathy are associated with different environmental determinants, the former being more closely related to a history of poor blood glucose control, and the latter showing a stronger association with blood pressure. Retinopathy is present in 47 to 63 per cent of non-insulin-dependent diabetics with persistent proteinuria (West et al. 1980; Schmitz and Vaeth 1988; Parving et al. 1990), this discordance being probably related to the observation that about 30 per cent of proteinuric cases are of non-diabetic origin.

Urinalysis in proteinuric diabetic patients shows in a substantial number of cases the presence of microhaematuria: those with microhaematuria have a higher prevalence of concomitant non-diabetic renal disease (Hommel et al. 1987a). However, microhaematuria can occur in proteinuric diabetic patients as part of the diabetic nephropathy syndrome and in the absence of other renal conditions. It has been reported in 66 per cent of all cases in a large series of 136 consecutive renal biopsies from insulin-dependent and non-insulin-dependent patients, and it has been claimed to be of little use in distinguishing non-diabetic renal disease (Taft et al. 1990). Red cell casts are unusual in diabetic nephropathy and call for further evaluation. Though still controversial, the presence of red cells in the urinary sediment must alert the physician to the possibility of another renal disease.

End-stage renal failure

The development of uraemia in diabetic patients is compounded by a number of other complications. Fluid retention and oedema occur relatively early in the development of renal failure, and their cause may not be readily apparent. They are often noted in the absence of hypoalbuminaemia (Hatch and Parrish 1961), and the variable contributions, particularly in older patients, of cardiac insufficiency and of vasomotor defects secondary to neuropathy and peripheral vascular disease were recognized long ago (Rifkin et al. 1948; Bell 1953; Gellman et al. 1959). Depressed renal function further compromises disposal of water and solutes and impairs osmotic diuresis: these limitations become more

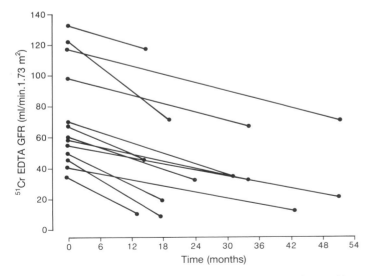

Fig. 6 ^{51}Cr EDTA GFR in insulin-dependent diabetic patients with clinical nephropathy. Individual linear rates of fall varied from 0.63 ml/min.month to 2.43 ml/min.month. Adapted from Viberti et al. 1983a.

significant in the face of rapid compartmental shifts secondary to brisk variations in glycaemia (Axelroth 1975). Pulmonary oedema may follow and the prognosis at this stage is poor. Hyperkalaemia may develop, partly due to the hyporeninaemic hypoaldosteronism common in patients with diabetic nephropathy (DeFronzo 1980), which may also aggravate the metabolic acidosis of chronic renal failure.

Peripheral neuropathy affects the majority of diabetics with renal failure. Uraemia, itself a cause of neuropathy, is likely to contribute to the severity of symptoms in a number of cases. Foot sepsis leading to amputation occasionally occurs, probably due to a combination of neural and arterial disease. Autonomic neuropathy, notably postural hypotension, can be a problem as it makes treatment of arterial hypertension particularly difficult. Good control of blood pressure in the supine position often results in standing blood pressures which prove incompatible with the maintenance of the upright position by the patient. Diabetic diarrhoea (Malins 1968) and gastroparesis (Campbell et al. 1977) causing nausea and vomiting are sometimes hard to distinguish from the gastrointestinal symptoms of uraemia. Impotence and profuse sweating are also common manifestations of autonomic neuropathy in uraemic patients (Watkins 1973). Neurogenic bladder, the most serious problem of all, will be treated separately (see below). Arterial disease and medial calcification of larger arteries (Monckeberg's sclerosis) is present in almost all diabetics with advanced renal disease (White and Graham 1981). Disturbances of lipid metabolism of diabetes and uraemia, combined with arterial hypertension, may all contribute to the development of severe sclerotic damage. Coronary artery disease is the major cause of death in these patients and peripheral artery disease may contribute to gangrene and amputation both before and after renal replacement therapy.

The treatment of end-stage renal failure by dialysis and transplantation is discussed in Chapter 12.3.

Pathophysiology

Early functional changes
Glomerular hyperfiltration

Several studies have used accurate techniques to confirm that the GFR in insulin-dependent diabetics is elevated by 20 to 40 per cent above the normal range. The intrarenal haemodynamic basis for hyperfiltration has been investigated in humans, but particularly in animal models of insulin-dependent diabetes.

Renal plasma flow, one of the major determinants of GFR, has been reported as elevated, normal or reduced in human diabetes (Mogensen 1971a, b; Ditzel and Junker 1972). However, most recent work shows an elevation of renal plasma flow in insulin-dependent diabetic patients ranging between 9 and 14 per cent (Christiansen 1984). Several studies have shown that there is a good correlation between the increases in renal plasma flow and GFR in diabetic patients (Mogensen and Andersen 1973; Christiansen et al. 1981a; Puig et al. 1981), suggesting that at least part of the increased GFR is accounted for by an elevation of the renal plasma flow.

Increased renal plasma flow, however, does not explain the whole rise in GFR. In human studies where GFR and renal plasma flow were measured simultaneously, the latter accounted for about 50 to 60 per cent of the GFR increase (Mogensen and Andersen 1973; Christiansen et al. 1981a). Findings of an increased filtration fraction, sometimes associated with an elevation of urinary albumin excretion, in short-term insulin-dependent diabetics have been taken to suggest that intraglomerular pressure may be elevated under these circumstances (Mogensen 1971b, 1976a).

Direct measurements of hydraulic filtration pressure are not obtainable in the human kidney, but micropuncture studies in moderately hyperglycaemic rats have shown a significant increase in the transglomerular pressure gradient (Hostetter et al. 1981; Zatz et al. 1985; Jensen et al. 1986), though this has not always been confirmed (Jensen et al. 1981; Michels et al. 1981). Severely hyperglycaemic rats show no elevation in intraglomerular pressure (Hostetter et al. 1981; Michels et al. 1981; O'Donnell et al. 1986). In the rat model, analysis of the arteriolar vascular resistance has shown that an increase in both flow and pressure is achieved by a reduction of total arteriolar vascular resistance, more marked at the afferent than at the efferent end of the arteriole (Hostetter et al. 1981). These findings have a counterpart in diabetic humans with glomerular hyperfiltration, in whom reduced calculated renal vascular resistance has been found (Wiseman et al. 1987a). However, the renal lesion in the rat model is that of focal segmental glomerulosclerosis, not of diabetic glomerulopathy, and not all strains of rats develop hyperfiltration and increased glomerular pressure (see under pathogenesis).

The elevation in glomerular transcapillary hydraulic pressure difference, the second determinant of GFR, appears therefore to account for a further proportion of the rise in GFR, which can be estimated to be approximately 25 per cent. The third determinant of GFR is systemic oncotic pressure, which has been reported to be normal both in human (Christiansen et al. 1981b) and animal diabetes (Hostetter et al. 1981).

The fourth determinant of GFR is the glomerular ultrafiltration coefficient, the product of the capillary hydraulic conductivity and the capillary surface area available for filtration. Although in animal studies the calculated glomerular ultrafiltration coefficient did not account for the increased filtration rate (Michels et al. 1981; Jensen et al. 1981; Hostetter et al. 1981), filtration surface area has been found to be increased in insulin-dependent diabetic patients (Kroustroup et al. 1977) as well as in diabetic rats (Gøtzsche et al. 1981), and a significant correlation has been described between GFR and filtration surface area in young insulin-dependent diabetic patients (Hirose et al. 1980). Thus, an increase in the surface area available for filtration (a component of the ultrafiltration coefficient) may be a determinant of the elevation of the GFR. However, significant changes in GFR can occur upon rapid improvement of metabolic control (Parving et al. 1976, 1979; Christiansen et al. 1982a), over a period that does not appear sufficient to reduce the filtration surface area (Kroustroup et al. 1977). A close correlation has been found between GFR and kidney size. Kidney size, which may be increased in diabetes, has been found to be related to changes in at least two of the four determinants of GFR, namely transglomerular pressure gradient and filtration surface area.

Metabolic and hormonal mediators of GFR elevation

The elevation of GFR appears to be multifactorial. Hyperfiltration is a phenomenon which occurs under conditions of moderate hyperglycaemia: hyperglycaemia exceeding 14 to 16 mmol/l is associated not with an elevated GFR, but with normal or reduced glomerular filtration (Mogensen 1971d; Wiseman et al. 1984a). In unselected groups of diabetics with baseline GFR ranging from normal to high, increasing blood glucose concentration to an average of 16 mmol/l by intravenous glucose infusion leads to an average GFR rise of approximately 5 per cent (Christiansen et al. 1981b). Other studies show that only those diabetics with baseline glomerular hyperfiltration increase their GFR significantly (on average by 12 per cent) in response to intravenous glucose (Wiseman et al. 1987a). In normal individuals, blood glucose elevations are associated with either an increase (Fox et al. 1964; Brøchner-Mortensen 1973; Christiansen et al. 1981b) or no change in the GFR (Wiseman et al. 1987a). High blood glucose levels may induce vasodilation: this has been shown to occur in the retinal circulation (Atherton et al. 1980) and a similar mechanism might apply to the glomerular arterioles. In rats, hyperglycaemia and glycosuria, both separately and additively, affect the tubuloglomerular feedback mechanism by blunting the GFR reduction in response to increased proximal tubular flow and reduced sodium delivery to the distal tubule and macula densa, thus producing relative hyperfiltration (Blantz et al. 1982). An increase in glucose-coupled reabsorption of sodium and water in the proximal tubule may also increase GFR (Leyssac 1976). Recently, infusion of ketone bodies in supraphysiological doses in diabetics has been shown to cause an increase in GFR by about 33 per cent. This is the single most powerful effect so far described, and is accompanied by a concomitant rise of about 16 per cent in renal plasma flow and a 14 per cent increase in the filtration fraction (Trevisan et al. 1987).

In addition to the altered level of glucose and other metabolites found in diabetes, a number of changes in circulating levels of, and vascular responsiveness to, metabolic fuel-related and vasoactive hormones may be seen. Acute reduction of blood glucose in diabetic patients by insulin infusion reduces the GFR within 30 to 60 min, but the effect is small (about 6 per cent) (Mogensen et al. 1978). This does not occur if blood glucose is maintained at euglycaemic levels by concomitant glucose infusion (Christiansen et al. 1981c). Euglycaemic insulin infusion in normal man is not associated with changes in either GFR or renal plasma flow (DeFronzo et al. 1975; Skøtt et al. 1989). These findings have been interpreted as indicating that insulin *per se* has no effect on GFR and renal plasma flow. Daily administration of growth hormone for some days increases GFR and renal plasma flow in normal subjects (Corvilain and Abramow 1962; Christiansen et al. 1981d) and in insulin-dependent diabetic patients (Christiansen et al. 1982b) by 7 and 6 per cent, respectively, in the face of unchanged glycaemic control. However, a correlation between plasma growth hormone concentration and GFR has not been found in unselected diabetic patients (Lundbaeck et al. 1970), and hyperfiltering diabetics have diurnal profiles of growth hormone not dissimilar from those of diabetics with a normal GFR (Wiseman et al. 1985b). Intravenous infusion of glucagon to plasma levels found in poorly controlled diabetics raises GFR both in normal (Parving et al. 1977) and diabetic (Parving et al. 1980) subjects. In diabetics, this increase is accompanied by an equivalent rise in renal plasma flow, with a significant relationship between this and GFR. Peripheral plasma levels of glucagon have not been found to be higher throughout the day in hyperfiltering diabetics than in those patients with a normal GFR (Wiseman et al. 1985b). It is controversial whether glucagon infused into the renal artery causes consistent modification of GFR, but intraportal infusion of the hormone induces a significant rise (Ueda et al. 1977; Uranga et al. 1979). Recent evidence suggests that the effect of glucagon on renal haemodynamics may be mediated through renal prostaglandin production (Fioretto et al. 1990). Small elevations of plasma renin activity, coupled with increased urinary excretion of 6-keto prostaglandin $F_{1\alpha}$, have been described in insulin-dependent diabetic patients with glomerular hyperfiltration (Viberti et al. 1989). Other authors, however, have found significantly lower levels of plasma renin activity and normal levels of urinary prostaglandin excretion in similar patients (Esmatjes et al. 1985).

In streptozotocin diabetic rats, both the circulating levels of renin and the binding of angiotensin II to glomeruli were found to be abnormal (Ballerman et al. 1984). The finding of reduced numbers of angiotensin II receptors in this model suggests that the expected upregulation of receptors with suppression of renin did not occur. Other workers have provided evidence in the animal model that renal prostaglandin production may be important in the genesis of glomerular hyperfiltration.

In the isolated perfused kidney, the increased GFR and the vasodilation in response to hyperglycaemia have been found to be blocked by indomethacin infusion, implicating renal prostaglandins as possible mediators of this effect (Kasiske et al. 1985). Administration of indomethacin to diabetic animals has been found to cause a significant reduction in GFR and renal plasma flow (Jensen et al. 1986), and an increased synthesis of prostaglandins in isolated glomeruli taken from diabetic rats has been described (Schambelan et al. 1985). Administration of aspirin to streptozotocin diabetic rats prevented early hyperfiltration and also the decrease in GFR and accompanying increase in glomerular basement membrane thickness seen after 16 weeks in the control animals (Moel et al. 1987). These findings, taken together, can be interpreted as reflecting an imbalance between the vasoconstrictive and vasodilatory systems contributing to the regulation of glomerular vascular resistance, in favour of the vasodilatory arm.

The predominance of the prostaglandin system is further supported by the finding that inhibition of prostaglandin production by acetylsalicylate reduces GFR in hyperfiltering insulin-dependent diabetic subjects (Esmatjes et al. 1985). Whereas short-term administration of indomethacin to recently diagnosed patients does not seem to have the same effect (Christiansen et al. 1985), a reduction in GFR was seen in persistently proteinuric subjects (Hommel et al. 1987b). Some authors have postulated that a hepatic product could be responsible for the increased GFR. This substance, which has been named glomerulopressin, is thought to be a glucuronic acid conjugate (Del Castillo et al. 1977), and raised levels have been found in dogs with diabetes (Uranga et al. 1979). In diabetic rats, high levels of atrial natriuretic peptide have been found to be associated with glomerular hyperfiltration and intraglomerular hypertension; reduction of these levels by antibody blockade led to a parallel fall in GFR. It has been postulated that diabetes-induced volume expansion (as seen on average in these animals) may mediate the rise in atrial natriuretic peptide, which in turn would elevate the GFR (Ortola et al. 1987). Atrial natriuretic peptide levels have been reported to be higher in hyperfiltering diabetics (Rave et al. 1987; Sawicki et al. 1988; Solerte et al. 1987), but these findings have not been confirmed by other authors (Jones et al. 1988), and conflicting results in patients with microalbuminuria and persistent proteinuria have also been produced (Solerte et al. 1987; Sawicki et al. 1988; Hommel et al. 1989). Similarly, extracellular fluid volume has been found to be normal or expanded by different authors (Brøchner-Mortensen and Ditzel 1982;

Feldt-Rasmussen *et al.* 1987; Jones *et al.* 1988; Hommel *et al.* 1989).

In conclusion, metabolic and hormonal mechanisms, acting both systemically and locally, are likely to be involved in the altered GFR seen in diabetes. The effects of a number of mediators, themselves small, may combine to induce the observed GFR elevations. These disturbances are not only related to the diabetic state, but seem to affect specifically the function of susceptible kidneys in diabetes.

Renal hypertrophy

Animals made diabetic develop kidney enlargement within days of the induction of diabetes. Whole kidney weight increases by an average of 15 per cent within the first 3 days of diabetes, and parallel increases in protein and RNA content of the kidney are observed. DNA synthesis remains stable for the first 3 days but increases thereafter (Seyer-Hansen 1976; Cortes *et al.* 1980). Insulin treatment started soon after the induction of diabetes is capable of preventing the increase in kidney weight (Rasch 1979*a*), but if such treatment is started after 28 days kidney enlargement is irreversible (Seyer-Hansen 1976). When near normoglycaemia is achieved by islet transplantation after 4 weeks of diabetes, there is a partial reduction in kidney size with a return of the RNA/DNA ratio to normal (Gøtzsche *et al.* 1981). This situation resembles the findings in human insulin-dependent diabetics, who generally have larger kidneys than controls. Good metabolic control at onset of diabetes appears capable of reducing kidney size (Mogensen and Andersen 1973), whereas strict metabolic control for 1 year in longer-term diabetics with established kidney enlargement does not seem to reduce kidney volume (Christiansen *et al.* 1982*a*; Wiseman *et al.* 1985*a*; Christensen *et al.* 1986).

In the animal model, glomerular growth is prominent during the first 4 days following induction of diabetes. Tubular growth then catches up, and eventually exceeds the glomerular growth over the first 6 weeks after onset of diabetes (Seyer-Hansen *et al.* 1980). Glomerular capillary length appears to be the earliest change, followed in the succeeding weeks by increases in the radial cross-sectional area of the capillary loop, which eventually becomes the dominant change for the increased glomerular size (Østerby and Gundersen 1980). Both proximal and distal tubule length increases by about 20 per cent, but whereas proximal tubular cells retain their normal appearance, the distal cells in the cortex and outer medullary stripe appear laden with glycogen-like granules, and show a marked reduction in the number of organelles and basal infoldings (Rasch 1984). Some authors believe that these morphological changes precede the early haemodynamic alterations and may, in fact, be contributory to them (Seyer-Hansen 1983).

Microalbuminuria

In a percentage of insulin-dependent diabetics urinary albumin excretion rate is increased above the upper limit of normal but still undetectable by routine clinical tests. Exaggerated excretion of albumin can also be induced by periods of poor metabolic control (Parving *et al.* 1976) or moderately strenuous exercise (Mogensen and Vittinghus 1975; Viberti *et al.* 1978). This microalbuminuria is not, however, accompanied by changes in β_2-microglobulin excretion rates. β_2-Microglobulin, a small molecular weight protein with an Einstein-Stokes radius of approximately 1.6 nm, is essentially freely filtered across the glomerular capillary barrier and taken up by proximal tubular cells. Its excretion reflects the degree of tubular uptake of the filtered protein (Peterson *et al.* 1969; Wibell 1974). The consistency of β_2-microglobulin excretion rate in the face of augmented albumin excretion rate suggests that the excess albuminuria is not the result of a change in tubular reabsorption of protein, but is more likely to derive from increased glomerular leakage. Thus microalbuminuria is likely to be glomerular in origin.

The glomerular capillary blood–urine barrier can be regarded functionally as a membrane perforated by pores of an average size of 5.5 nm and uniformly coated by a negative electrical charge (Pappenheimer *et al.* 1951; Brenner *et al.* 1978; Venkatachalam and Renke 1978; Deen and Satvat 1981; Myers *et al.* 1982). Both the size and charge of the circulating molecule, as well as the set of haemodynamic forces operating across the capillary wall, will therefore determine the passage of protein across the glomerular membrane. In microproteinuric diabetic patients, the clearances of albumin, a polyanion with molecular radius of about 3.6 nm, and of IgG, a larger but electrically neutral molecule with radius of around 5.5. nm, are both increased (Viberti *et al.* 1983*c*; Viberti and Keen 1984). These early increases are likely to be the consequence of alterations in glomerular haemodynamics. Increases in filtration fraction have been reported in poorly controlled diabetics (Mogensen 1971*b*; Mogensen and Andersen 1975; Parving *et al.* 1976). An increased filtration fraction would result in a greater concentration of protein within the glomerular capillary wall, and would provide a driving force for protein diffusion into the Bowman's space. In humans, the absolute urinary clearance of neutral dextran is raised over a wide range of molecular weights in parallel with elevations in the glomerular filtration rate (Parving *et al.* 1979). Micropuncture studies in moderately hyperglycaemic diabetic rats confirm an elevation of transglomerular pressure gradient (Hostetter *et al.* 1981; Zatz *et al.* 1985; Jensen *et al.* 1986).

As microalbuminuria becomes persistent and increases in degree, the selectivity index, i.e. the clearance of IgG over the clearance of albumin, starts to fall, reaching its lowest values when albumin excretion is approximately 90 μg/min or more. This is due to a disproportionate increase in the filtration of albumin compared with that of IgG, and it marks a new stage of selective glomerular loss of polyanionic albumin (Viberti *et al.* 1983*c*; Viberti and Keen 1984). Experiments with clearance of neutral dextran have shown that medium size pores are unchanged at this stage (Mogensen 1971*b*; Myers *et al.* 1982). A likely reason for the increased glomerular filtration of albumin is a loss of the fixed negative electrical charge on the membrane (Westberg and Michael 1973; Winetz *et al.* 1982; Parthasarathy and Spiro 1982; Schober *et al.* 1982). This would permit increased permeation of anionic albumin, but would have little influence on IgG, a neutral molecule, the filtration of which is regulated by pore radius or number, and by glomerular pressures and flows. The mechanism of this transition from low to high levels of microalbuminuria is unknown, but it may result from a combination of haemodynamic abnormalities and the cumulative metabolic derangement of synthesis of the electronegative membrane glycosialoprotein proteoglycans (Hostetter *et al.* 1982; Kanwar *et al.* 1983; Viberti and Keen 1984; Brenner 1985). Recent studies suggest, however, that preferential filtration on the basis of charge discrimination due to loss of glomerular polyanion does not explain the facilitated clearance of anionic proteins. Permeation of both large and medium sized molecules through a non-size-discriminatory shunt pathway could entirely account for the observed renal clearances of albumin, IgG and dextran probe molecules (Nakamura and Myers 1988). The facilitated clearance of anionic proteins, including IgG_4 (Deckert et al. 1988), may reflect an increased expression of cationic sites on the glomerular filter, rather than a loss of glomerular polyanion (Bertolatus et al. 1987). Indeed, linear deposition of both albumin and IgG in the glomerular basement membrane is a consistent finding in diabetic kidneys, and it has been shown that only anionic species bind to the diabetic glomerular capillary wall (Melvin et al. 1984).

Exposure of structural and circulating proteins to high glucose concentrations will increase the rate of their non-enzymatic glycosylation (Guthrow et al. 1979; Cohen et al. 1981). Recent studies indicate that microvessels isolated from rat epididymal fat pads preferentially take up glycosylated rat albumin by endocytosis; glycosylation of endothelial membrane components seems to enhance this process of pinocytosis even further (Williams et al. 1981). Two recent reports seem to indicate that glycosylated proteins (including albumin) may undergo preferential transport across the glomerular barrier (Ghiggeri et al. 1985; Williams and Siegel 1985). The reason for this facilitated flux of glycosylated macromolecules through the glomerular membrane barrier remains unknown, but it is possible that conformational changes induced by glycosylation are important (Shaklai et al. 1984). The transition to high selectivity proteinuria signals the advent of heavier losses; this may indicate the critical importance of the loss of the charge barrier in the unfolding sequence of pathogenic events.

The concomitants of microalbuminuria

The positive association between urine flow and albumin excretion (Pillay et al. 1972; Jarrett et al. 1976) has led to the suggestion that glucose-induced diuresis could impair proximal tubular reabsorption of albumin, as it does for several other solutes. However, rat albumin excretion by the kidney is unaffected by osmotic diuresis, whereas it is increased after volume expansion, probably through a change in the GFR (First et al. 1978). In the diabetic human, no correlation has been found between glycosuria and urinary albumin excretion (Mogensen 1971c; Hegedüs et al. 1980). The transient microalbuminuria seen in normal subjects after water loading is likely to be mediated, as in the volume expanded rat, by transient changes in glomerular filtration rate (Viberti et al. 1982c). Whether acute worsening of glycaemia by glucose ingestion or infusion increases albumin excretion rate remains somewhat controversial. Glucose ingestion has been reported to increase albuminuria in normal subjects but not in diabetic patients (Hegedüs et al. 1980). A number of other studies have failed to show any acute effect of glucose, oral or intravenous, on urinary albumin excretion (Viberti et al. 1981b; Christiansen et al. 1981b). Metabolic acidosis has no effect on urinary albumin excretion but increases the excretion of β_2-microglobulin (Trevisan et al. 1987).

A bolus intravenous injection of insulin was reported to increase urinary albumin excretion (Mogensen et al. 1978), but these findings have not been reproduced (Viberti et al. 1980; Christiansen et al. 1981c). Neither glucagon nor growth hormone affect urinary albumin excretion in normal or in diabetic subjects (Parving et al. 1977, 1980; Christiansen et al. 1981d, 1982b).

There is an association between microalbuminuria and higher levels of arterial pressure (Table 2). A positive, linear, and independent correlation between arterial pressure and albumin excretion rate has been confirmed by several investigators (Wiseman et al. 1984b; Mathiesen et al. 1984). This association is much closer than that between albumin excretion rate and blood glucose, and is independent of a number of other variables including age, duration of diabetes, body mass index, sex, and blood glucose itself. The observation of raised arterial pressures in microalbuminuric patients who still have perfectly maintained levels of glomerular function speaks against the assumption that blood pressure is a consequence of renal dysfunction, and argues in favour of a more complex relationship. This makes it possible to postulate that the rise in blood pressure could be contributory to the renal disease, or alternatively, that microalbuminuria and high blood pressure may recognize a common determinant. It is of interest that microalbuminuric patients with elevation of arterial pressure show significantly more marked histological lesions of mesangial expansion compared with patients with a similar duration of diabetes but lower levels of albumin excretion rate and arterial pressure (Chavers et al. 1989).

Table 2 Blood pressure in age, sex and duration matched insulin-dependent diabetic patients with albumin excretion rates above and below 30 μg/min

	Albumin excretion rates	
	> 30 μg/min	< 30 μg/min
No. of patients	12	28
Blood pressure (mmHg)		
Systolic	136 ± 5	118 ± 2**
Diastolic	87 ± 3	76 ± 2*

Mean ± SE. * $p < 0.005$; ** $p < 0.002$.

Tubular function

Changes in tubular function take place early in insulin-dependent diabetes and seem to be largely related to the degree of metabolic control. Maximum rates of glucose reabsorption are elevated in these patients and absolute rates of sodium reabsorption are also increased (Mogensen 1971d). The latter is probably due, at least in part, to the proximal tubular cotransport of sodium with glucose (Kokko 1973). It has been suggested that enhanced proximal tubular reabsorption could diminish distal sodium delivery, and thereby trigger tubuloglomerular feedback mechanisms which would lead to augmentation of glomerular filtration rate (Ditzel et al. 1982). Insulin has been shown to directly increase distal sodium reabsorption in both normal (DeFronzo et al. 1975) and diabetic patients (Skøtt et al. 1989). A number of tubular proteins such as N-acetyl-β-D-glucosaminidase have also been found to be increased in insulin-dependent diabetic patients, the increase being related to the degree of glycaemic control (Watanabe et al. 1987; Gibb et al. 1989). On the other hand, phosphate absorption is diminished in insulin-dependent diabetics and this also appears to be related to the blood glucose concentration. It is believed that there is a net competition in the tubular reabsorption of these two solutes. This competitive interaction may represent competition for a common driving force for reabsorption (Ditzel and Brochner-Mortensen 1983). Insulin has been shown to reduce the renal clearance of phosphate by stimulating its proximal reabsorption (DeFronzo et al. 1975; Skøtt et al. 1989). The prospective significance for overall renal function of all these alterations in sodium, glucose, phosphate, and enzyme transport and excretion, is largely unknown. It is of note that these abnormalities are rapidly corrected by improvement of blood glucose control.

Late functional changes
Glomerular filtration rate

In established diabetic nephropathy GFR declines relentlessly towards end-stage renal failure, accompanied by a reduction in renal plasma flow, while filtration fraction remains relatively constant (Winetz *et al.* 1982). Progressive mesangial expansion and capillary occlusion can be seen as renal disease advances. This process leads to a reduction in the filtering surface area and would imply changes in the ultrafiltration coefficient as a possible cause for the decline in GFR (Mauer *et al.* 1984). Although direct measurements of these variables are not possible in humans, indirect calculations, using neutral dextran sieving curves, support the view of a reduction in the ultrafiltration coefficient due to a diminution in the glomerular capillary surface area available for filtration (Winetz *et al.* 1982; Tomlanovitch *et al.* 1987). It is believed that, in the setting of established renal disease, the surviving glomeruli filter at their maximal capacity. This assumption seems to be supported by studies in which administration of an oral protein load was not capable of producing any further expansion of GFR over baseline values in diabetic patients with impaired renal function (Pinto *et al.* 1988). This loss of 'renal functional reserve' has been suggested as one of the possible deleterious mechanisms which would lead to further renal damage.

This notion has, however, been recently questioned, as stimuli other than protein loading have been shown to have profound effects on the GFR in patients with established diabetic nephropathy and reduced GFR. Hyperglycaemia, for instance, has been shown to induce a rise of about 30 per cent in the GFR in these renal failure patients (Remuzzi *et al.* 1990). This glucose-induced effect seems to be mediated by changes in renal prostaglandin production, as it can be significantly blunted by cyclo-oxygenase inhibition (De Cosmo *et al.* 1989). Whether in the setting of renal failure, hyperglycaemia therefore becomes a mechanism to maintain GFR, or whether hyperglycaemia-induced higher GFR would in the end be deleterious, remains an open question. It is worth noting, however, that correction of hyperglycaemia at this advanced stage of renal disease has no significant impact on the progression of renal failure (Viberti *et al.* 1983b; Bending *et al.* 1986).

Clinical proteinuria

As the degree of proteinuria progresses, a change from high to low selectivity proteinuria takes place. For GFR below 10 to 20 ml/min, more IgG is filtered relative to albumin, the selectivity index rising from 0.12 to approximately 0.6 (Viberti *et al.* 1983c). Studies with neutral dextran sieving curves have demonstrated that the proteinuria of the late stages of overt nephropathy is probably the result of a defect in size selectivity properties of the glomerular membrane. The fractional clearance of neutral molecules with radii greater than 4.6 nm is elevated, and a mathematical analysis of the phenomenon indicates that this increase is consistent with the appearance of a 'shunt pathway' within the glomerular capillary wall (Fig. 7). The development of a small population of unselective pores would allow the unrestricted movement of very large plasma proteins into the urine. Whether the size selectivity defect totally explains the proteinuria of advanced nephropathy remains to be clearly established (Myers *et al.* 1982; Tomlanovitch *et al.* 1987; Nakamura and Myers 1988). It is likely that charge selectivity defects, as well as abnormal renal haemodynamics, persist at this stage of advanced

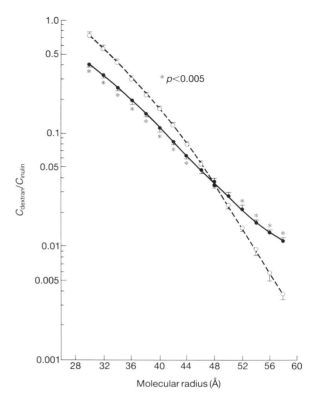

Fig. 7 Mean (±SE) fractional clearances of graded size neutral dextrans in patients with diabetic nephropathy (●) and in normal controls (O). A reduction in K_f and therefore possibly in filtering surface area leads to a reduced fractional clearance of small-size dextrans. The increased fractional clearance of larger-size dextrans is best explained by an increase in the 'shunt pathway'. Reproduced with permission from Tomlanovitch *et al.* 1987.

nephropathy. It is of interest that the sialic acid component of the glomerular barrier has been found to be reduced in patients with long-standing diabetes (Westberg and Michael 1973; Kefalides 1974; Wahl *et al.* 1982). A reduction in the synthesis of heparan sulphate within the glomerular basement membrane has been described in diabetic animals (Parthasarathy and Spiro 1982; Rohrbach *et al.* 1983; Kanwar *et al.* 1983; Cohen and Surma 1984; Wu *et al.* 1987), and a loss of heparan sulphate from the glomerular basement membrane has been demonstrated in insulin-dependent diabetic patients with nephropathy (Shimomura and Spiro 1987). This glycosaminoglycan is a major constituent of the glomerular capillary wall fixed negative charge.

As renal failure advances the proteinuria becomes of mixed tubular and glomerular origin. At a GFR of 40 ml/min or below there is a significant increase in the urinary excretion of β_2-microglobulin (Viberti and Keen 1984). It is unclear at present whether this represents tubular damage or is the consequence of saturation of proximal tubular reabsorption capacity for β_2-microglobulin, with overspill into the urine. With advancing renal failure, plasma concentrations of β_2-microglobulin increase and there is a greater filtered load of β_2-microglobulin per functioning nephron.

Hyperlipidaemia

Clear disturbances of plasma lipoproteins develop with renal disease. Increases in cholesterol, LDL cholesterol, total triglycerides, VLDL triglyceride, and apolipoprotein B1 have been described, and HDL_2 cholesterol has been found to be decreased (Vannini *et al.* 1984; Winocour *et al.* 1987; Jensen *et al.* 1988). The pathogenic role of these lipid disturbances in the progression of renal failure (see Chapter 9.3) remains uncertain in human diabetes. However, studies in the diabetic animal model seem to indicate that hyperlipidaemia may contribute to late glomerular sclerotic changes (Kasiske *et al.* 1988). The recent finding that these lipid changes may not entirely be secondary to heavy proteinuria and advancing renal disease is of great interest. Both insulin-dependent (Jones *et al.* 1989*a*; Dullaart *et al.* 1989) and non-insulin-dependent diabetics (Mattock *et al.* 1988) with normal renal function have been found to have a similar pattern of lipid abnormalities, though of a lesser degree. It is clearly important now to establish whether correction of lipid disturbances at this early stage of the natural history of the disease may prevent deterioration of renal function and renal histological damage.

Pathogenesis

Metabolic pathways

Hyperglycaemia and non-enzymatic glycosylation

There is no doubt that there is a positive relationship between the abnormal glycaemic milieu of diabetes and microvascular complications. The view that small vessel disease, and in particular abnormalities in capillary basement membrane thickness, is primarily an inherited phenomenon (Siperstein *et al.* 1968) has been negated by an overwhelming body of evidence linking hyperglycaemia to diabetic complications (Williamson and Kilo 1977; Steffes *et al.* 1985). Small vessel complications can be found in secondary diabetes in humans (Becker and Miller 1960; Ireland *et al.* 1967*a*) and in a variety of animal models of diabetes, both chemically induced and genetically determined (Brown *et al.* 1982). Established histological lesions such as mesangial expansion may be reversed by the transplantation of a diabetic kidney into a normal animal (Lee *et al.* 1974) or by curing diabetes with islet cell transplantation (Mauer *et al.* 1975; Steffes *et al.* 1980; Gøtzsche *et al.* 1981). In the rat model renal histological lesions (Rasch 1979*a, b*) and albuminuria (Rasch 1980) can be prevented by maintenance of near-normal blood glucose with intensified insulin treatment from induction of diabetes.

There are, however, two important aspects to consider. The animal model most frequently used for the study of diabetic nephropathy is the rat, which does not develop advanced renal failure or the severe histological lesions seen in the human diabetic. There are also considerable species differences in the glomerular haemodynamic responses to different glycaemic levels, making extrapolation of animal data to man difficult (Brown *et al.* 1982; O'Donnell *et al.* 1988). In addition, highly inbred strains of rat are used as models of chemically induced or genetic diabetes, a condition which is unrepresentative of the genetic heterogeneity of human diabetes.

In man, the evidence for a straightforward causal relationship between hyperglycaemia and renal disease becomes less compelling. The development of clinically overt renal disease is not linearly related to the duration of diabetes, and affects only between 35 and 50 per cent of patients, depending on the type of diabetes. The majority of diabetic patients do not develop renal failure, and although some histological damage occurs in their kidneys, renal function remains essentially normal until death. Kidneys from non-diabetic human donors develop typical lesions of diabetic glomerulopathy when they are transplanted into a diabetic recipient (Mauer *et al.* 1983), but the rate of development of the lesions varies widely in different kidneys, independently of the history of blood glucose control (Mauer *et al.* 1989). It would therefore appear that in man hyperglycaemia is necessary but not sufficient to cause the renal damage which leads to kidney failure, and that other, possibly non-environmental, factors are needed for the manifestation of the clinical syndrome.

Non-enzymatic reactions between glucose and the lysine amino-terminus of circulating and structural proteins give rise to two major classes of glycation products (Brownlee *et al.* 1984). Relatively short-lived proteins form a Schiff base which undergoes an Amadori rearrangement, with the formation of a stable, but still chemically reversible, sugar–protein adduct. Structural proteins that turn over at a much slower rate such as collagen, myelin, crystallin, and elastin, accumulate different products derived from slow reactions of dehydration, degradation and rearrangement of the Amadori adducts to form chemically irreversible advanced glycation end-products (Brownlee *et al.* 1984, 1988). Non-enzymatic glycation is thus likely to affect the glomerular basement membrane and other matrix components in the glomerulus.

The pathophysiological consequences of this process are unsettled, but several possibilities have been suggested. Excess glomerular basement membrane glycation, as seen in diabetes, may lead to an increase in the degree of disulphide bridge cross-linking between collagen components via an increased oxidation of sulphydryl groups. This process may induce molecular rearrangement and has been implicated in cataract formation in the lens (Stevens *et al.* 1978). Similar cross-linking by disulphide bonds might affect the assembly and architecture of the glomerular basement membrane and mesangial matrix. Advanced glycation end-products are also capable of extensive cross-linking throughout the collagen molecule. Rotary shadowing electron microscopy of glycated basement membrane components has revealed increased collagen IV cross-linking and altered molecular morphology (Brownlee *et al.* 1988). It is noteworthy that in Lewis alloxan-diabetic rats the compound aminoguanidine, which blocks the formation of advanced glycation end-products, prevents increased aortic collagen cross-linking as well as increased cross-linking of collagen to lipoproteins, besides preventing thickening of glomerular basement membrane and glomerular trapping of IgG molecules (Brownlee *et al.* 1988).

The pathogenetic consequences of enhanced cross-linking in the kidney remain however largely obscure. The glycation reaction may theoretically enhance the binding of circulating plasma proteins to structural components in the glomerular basement membrane and mesangial matrix through the presence of the reactive carbonyl group on the glucose attached to these structures. It has been suggested that this increased binding may account for the linear deposition of albumin and IgG observed along glomerular and tubular basement membranes in diabetes (Brownlee *et al.* 1983*a*). Glycation of structural proteins or of circulating proteins trapped in the glomerular structures may interfere with their degradation: degradation of fibrin by plasmin has been found to be reduced by glycation (Brownlee *et al.* 1983*b*). Reduced degradation of glomerular components may result in their accumulation in mesangial matrix and glomerular basement membrane. Other glycoproteins, such as fibronectin, which are found in the mesangial matrix and are involved in the control of cell growth, replication and adhesion, are also susceptible to non-enzymatic glycation. *In vitro* this process interferes with the binding characteristics of fibronectin, inhibiting its adhesion to matrix components (Cohen and Ku 1984) and enhancing its binding to glomerular basement membrane (M.P. Cohen *et al.* 1987). *In vivo* this phenomenon may alter the integrity and adhesive properties of the matrix and promote the development of selectivity defects in the capillary barrier. It has been claimed that an alteration of the charge distribution of glycated albumin could provide a mechanism for the abnormal flux of modified albumin across the glomerular membrane (Ghiggeri *et al.* 1985). These results were not, however, confirmed by Nakamura and Myers (1988), who were unable to show differences in the isoelectric point distribution of glycated and non-glycated albumin. A membrane-associated macrophage receptor that specifically recognizes advanced glycation end-products in proteins has recently been identified (Vlassara *et al.* 1986). This receptor enables the selective removal of senescent cross-linked denatured proteins. The binding of this receptor to protein with advanced glycation end products induces synthesis of macrophage monokines (interleukin-1 and tumour necrosis factor), which in turn stimulate nearby mesenchymal cells to synthesize extracellular proteases. These monokines also initiate a cascade of stimuli and interactions whose end result is an increased protein synthesis and cell proliferation, with an increased vascular permeability in endothelium (Brownlee *et al.* 1988).

In conclusion, excessive formation of glycation products in the glomerulus may lead to enhanced deposition of basement membrane-like material and circulating proteins in the mesangium, interfere with mesangial clearance mechanisms and alter the macrophage removal system, and contribute to mesangial expansion and glomerular occlusion.

The polyol pathway
Aldose reductase is present in the papilla, glomerular epithelial cells, distal tubular cells, and most probably mesangial cells of the normal kidney (Ludvigson and Sorenson 1980; Kikkawa *et al.* 1987). The physiological significance of this enzyme is difficult to define in many tissues, but, in the renal medullary cells of the kidney, its primary role seems to be in the generation of sorbitol, an organic osmolyte, from glucose in response to the high salinity in the medullary interstitium. Sorbitol would aid in preventing osmotic stress (Burg 1988). It has been argued that, in diabetes more glucose becomes available for reduction by aldose reductase in tissues where glucose uptake is insulin-independent. An increased sorbitol and/or a reduced intracellular myoinositol concentration would therefore contribute to diabetic complications, probably by upsetting osmoregulation in these cells (Cogan 1984; Kador *et al.* 1985).

A series of trials has been carried out with various aldose reductase inhibitors, with the aim of blocking the intracellular conversion of glucose to sorbitol and thus preventing some complications of diabetes. In glomeruli obtained from diabetic rats, the increased flux through the polyol pathway, with accumulation of sorbitol, depletion of myoinositol, and reduced Na^+, K^+-ATPase activity, has been shown to be preventable by the administration of the aldose reductase inhibitor sorbinil. The increased GFR and proteinuria seen in rats with streptozotocin-induced diabetes have also been reported to be reduced by inhibitors of aldose reductase or by supplementation of myoinositol (Beyer-Mears *et al.* 1984, 1986; Goldfarb *et al.* 1986). These studies, however, were not confirmed by a more recent report, and have been criticized on methodological grounds because large volumes of saline were infused during clearance studies, kidney haemodynamics were factored for body weight, diabetes was of short duration, and total urinary protein rather than albumin was measured (Daniels and Hostetter 1989). There is some evidence that low molecular weight proteins may be affected to a greater degree by sorbinil than albumin or larger proteins, which reflect glomerular barrier permeability function more directly (Beyer-Mears *et al.* 1986). It has also been suggested that sorbinil may act as a vasoconstrictor (Coco *et al.* 1988; Mower *et al.* 1989) and lower renal prostaglandin synthesis (Frey *et al.* 1989), thereby modifying renal haemodynamics independently of the aldose reductase pathway. It is of interest that sulindac, a potent inhibitor of prostanoid synthesis, also blocks aldose reductase, suggesting that certain aldose reductase inhibitors may have similar mode of action to non-steroidal anti-inflammatory drugs (Jacobson *et al.*1983). Although sorbinil has been shown to prevent renal hypertrophy in galactose-fed rats (Beyer-Mears *et al.* 1983), all authors concur that aldose reductase inhibitors have no effect on the increased kidney weight seen in streptozotocin-induced diabetes. Moreover, the histological lesions of glomerular disease after 6 months of diabetes in the rat were unaffected by the administration of statil, another aldose reductase inhibitor (Rasch and Østerby 1989).

Despite the relative lack of effect of statil on renal function and structure, Tilton *et al.* (1989) induced a reduction in the clearance of ^{51}Cr-EDTA, in albuminuria and in the permeation of ^{125}I-BSA into the vascular wall by administering any one of three aldose reductase inhibitors to Sprague-Dawley rats with streptozotocin-induced diabetes. These authors concluded that 'virtually all of the early functional and structural renal and vascular changes associated with diabetes in animals are aldose reductase-linked phenomena'. The same group has also shown that myoinositol supplemented diets that raise plasma myoinositol levels by five-fold reduce GFR, renal blood flow, urinary protein excretion, and serum albumin permeation of blood vessels, and may even return them to normal (Pugliese *et al.* 1990).

This diversity of effects of aldose-reductase inhibitors can also be seen in man although few studies have been performed. Reduced GFR (Pedersen *et al.* 1989) and albumin excretion rates (Blohmé and Smith 1989) have been claimed in insulin-dependent diabetics, but no effects have been reported in a controlled study of microalbuminuric non-insulin dependent diabetic patients (Cohen *et al.* 1989).

Biochemical abnormalities of extracellular matrix
Diabetic glomerulopathy is characterized by an excessive accumulation of glomerular basement membrane and mesangial matrix. Studies in diabetic animals suggest that the rates of matrix and glomerular basement membrane synthesis are significantly accelerated. Collagen represents a major component of extracellular membranes, and its biosynthesis, measured by the incorporation of radiolabelled amino acids, has been shown to be increased in diabetic rats (Brownlee and Spiro 1979). The activity of lysyl-hydroxylase, an enzyme involved in the hydroxylation of peptide-bound lysine during collagen biosynthesis (Khalifa and Cohen 1975), has been found to be increased in the glomeruli of diabetic rats. These abnormalities can be prevented if insulin therapy is started at the time of induction of diabetes.

The non-collagenous moieties of the glomerular basement membrane and glomerular basement membrane-like material can be measured by the rate at which glucosamine, galactose, and sulphate are incorporated into sialoglycoproteins and glycosaminoglycans, the two major carbohydrate constituents of the glomerular basement membrane. Glycosaminoglycans account for approximately 90 per cent of the total carbohydrate component of glomerular basement membrane; the principal glycosaminoglycan is heparan sulphate which, together with sialic acid, contributes to the negative charge of the glomerular capillary wall, and, therefore, the charge selectivity properties of the filtration barrier (Brenner et al. 1978; Farquhar 1981). In diabetes there is reduced *de novo* synthesis of glomerular heparan sulphate, and the total glycosaminoglycan content of the glomerulus and glomerular basement membrane is reduced (Parthasarathy and Spiro 1982; Kanwar et al. 1983; Rohrbach et al. 1983; Cohen and Surma 1984; Wu et al. 1987). The heparan sulphate content of the glomerular basement membrane has been found to be decreased in patients with insulin-dependent diabetic nephropathy (Shimomura and Spiro 1987), and studies of both diabetic humans and experimental diabetic animals have consistently reported a reduction in sialic acid components (Westberg and Michael 1973; Kefalides 1974; Wahl et al. 1982). Sialoglycoproteins are highly negatively charged and coat glomerular epithelial cells, their foot processes and the epithelial slit diaphragm. A loss of negative charge in the glomerular membrane may be responsible for foot process fusion, with consequent obliteration of the slit diaphragm, and could partly explain the albuminuria of diabetic nephropathy.

The existence of abnormalities of the carbohydrate components of the glomerular basement membrane in diabetes remains more controversial. An increase in the hydroxylysine level, as well as an elevation of the glucose and galactose disaccharide units attached to hydroxylysine residues, have been described. Increased activity of the enzyme glycosyltransferase, responsible for the attachment of glucose to the glycoprotein, has also been reported. None of these findings have, however, been confirmed by a number of other studies (Spiro and Spiro 1971; Beisswenger and Spiro 1972; Westberg and Michael 1973; Beisswenger 1976; Kefalides 1974; Wahl et al. 1982).

Glucotoxicity

Direct pathogenetic effects of glucose itself have only recently been described. Lorenzi et al. (1985, 1987a) have demonstrated convincingly that prolonged, though not acute, exposure to high ambient glucose concentrations produces consistent alterations in replication and maturation of cultured human endothelial cells, which cannot be ascribed to abnormalities of the polyol pathway (Lorenzi et al. 1987b). These abnormalities are associated with evidence of damage to DNA (Lorenzi et al. 1986), which has also been demonstrated in peripheral blood lymphocytes from poorly controlled diabetic patients, but not from those with better control (Lorenzi et al. 1987c) (Fig. 8). High glucose levels also enhance the expression in cultured endothelial cells of those glycoproteins characteristically increased in the diabetic basement membrane (Cagliero et al. 1988). These cells also show abnormal expression of tissue factor mRNA in response to thrombin and interleukin-1 after prolonged exposure to high glucose concentrations (Boeri et al. 1989).

It is not yet known whether glucose exerts a direct toxic effect on human endothelial cells *in vivo*, but abnormalities of endothelial cell function have been implicated in the increased frequency of cardiovascular disease which is a feature of diabetic nephropathy. Such abnormalities, evidenced by raised plasma von Willebrand factor (Jensen 1989) and decreased release of tissue plasminogen activator in response to exercise (Jensen et al. 1989), are present even before overt nephropathy develops.

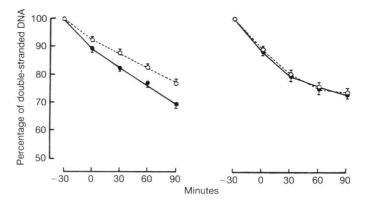

Fig. 8 Kinetics of unwinding of DNA from lymphocytes of control (O) and diabetic subjects (●) with high (12.9 ± 2.4 per cent; left panel) or normal (7.4 ± 1.5 per cent; right panel) glycated haemoglobin. Unwinding is expressed as a percentage of double-stranded DNA recovered after incubation with alkali and is increased in proportion to single-stranded DNA breaks. The difference between lymphocytes of high glycated haemoglobin subjects and normal controls was highly significant ($p < 0.01$). Reproduced with permission from Lorenzi et al. 1987c.

Haemodynamic and hypertrophic pathways

Haemodynamic disturbances within the glomerulus, with raised flows and pressures, occur early in the course of diabetes and have been suggested to be directly responsible for the development of glomerulosclerosis and its attendant proteinuria (Hostetter et al. 1982). The notion that increases in GFR, renal plasma flow, and glomerular capillary hydraulic pressure produce diabetic glomerular injury is based on several observations. Mesangial expansion and mesangial accumulation of circulating plasma proteins were found to be greater in uninephrectomized diabetic rats than in control animals without unilateral nephrectomy (Steffes et al. 1978). Although intrarenal haemodynamics were not measured in this study, it was assumed that altered microcirculatory dynamics may affect the rate of development of glomerular lesions. The induction of systemic hypertension, using the two-kidney, one-clip Goldblatt hypertension method, resulted in the development of much more severe glomerular lesions in the unclipped kidney of diabetic rats than in the kidneys of normotensive diabetic control animals (Mauer et al. 1978). The clipped kidney showed less severe glomerular damage than the kidney of diabetic control rats. Autopsy findings in two diabetic men with hypertension and unilateral renal artery stenosis showed nodular glomerulosclerotic lesions to be confined to the kidney with the patent renal artery, while the contralateral kidney was spared (Berkman and Rifkin 1973; Béroniade et al. 1987). Finally, in the diabetic animal model, manoeuvres that lessen disturbed renal haemodynamics, such as low protein diet or converting enzyme inhibition, have been shown to prevent the increase in urinary albumin excretion and the glomerular histological lesions which occur in the untreated diabetic control animal (Zatz et al. 1985, 1986; Anderson et al. 1986).

How alterations in glomerular haemodynamics lead to mesangial expansion, glomerular basement membrane thickening, and eventually sclerosis, is uncertain, but several suggestions have been made. Elevated intraglomerular pressure may lead to an increase in mesangial cells and matrix production, and to basement membrane thickening via an increase in arteriolar or capillary wall tension, as is seen in smooth muscle cells exposed to supranormal pressures (Ausiello et al. 1980; Webb and Bohr 1981; Hostetter et al. 1982). Physical stress and shear forces may damage endothelial and epithelial surfaces and disrupt the normal glomerular barrier; this is analogous to the situation which is believed to occur at sites of turbulence in larger arteries in systemic hypertension (Leung et al. 1976; Ross and Glomset 1976; Olson et al. 1982). Proteinuria associated with disruption of the normal glomerular barrier would lead to accumulation and deposition of plasma proteins and lipoproteins in the mesangial area. Reduced clearance of these proteins in diabetes leads to their persistence (Mauer et al. 1979), and they become a local stimulus for more mesangial matrix production and accumulation (Velosa et al. 1977; Mauer et al. 1981; Grond et al. 1982).

The evidence that alterations in local physical forces within the glomerulus lead to later sclerotic changes in diabetes is, however, debatable and has been recently questioned (Fogo and Ichikawa 1989). A dissociation between the haemodynamic changes and the subsequent sclerosis has been reported by a number of authors. Severely hyperglycaemic rats develop renal sclerotic changes even though there is no evidence of raised pressures and flows early in the course of diabetes (Steffes et al. 1980; Orloff et al. 1986). Bank et al. (1987) found no relationship between levels of glomerular hyperfiltration and pressure and subsequent degree of glomerular sclerosis in two strains of diabetic rat. Lowering high lipid levels in the Zucker rat model protects against renal sclerosis without affecting the haemodynamic pattern (Kasiske et al. 1988). Moreover, manoeuvres which affect renal haemodynamics also have marked effects on other cell functions. Unilateral nephrectomy, for instance, is a potent stimulus for glomerular hypertrophy and hyperplasia, and Goldblatt hypertension leads to (compensatory) hypertrophy in the unclipped kidney (Neugarten et al. 1982): hypertrophic changes in the glomerulus are invariably related to subsequent glomerular sclerosis (Fogo and Ichikawa 1989). It is of interest that marked renal hypertrophy is an early event in diabetes. Treatment which modifies renal haemodynamics and the degree of glomerular sclerosis also affects the accompanying glomerular hypertrophy. Low protein diets and angiotensin converting enzyme inhibitors significantly attenuate the renal hypertrophy associated with nephrectomy or streptozotocin-induced diabetes (Halliburton and Thomson 1965; Hostetter et al. 1981; Anderson et al. 1985). In certain animal models of renal disease, hypertrophic and sclerotic changes have been reduced independently of modification of renal haemodynamics, and agents which affect glomerular and mesangial cell proliferation, but are not known to have haemodynamic effects, have been shown to ameliorate glomerular sclerosis (Purkerson et al. 1976, 1982, 1985, 1988; Olson 1984; Fogo et al. 1988; Ichikawa et al. 1988; Castellot et al. 1985).

It has been argued that hyperplastic and hypertrophic changes in the diabetic kidney precede the haemodynamic abnormalities (Cortes et al. 1987). Hypertrophic factors may activate mesangial cell proliferation and augment mesangial matrix formation or suppress matrix degradation, giving rise to the histological alterations which are pathognomonic of diabetic glomerulopathy. Perhaps one of the clearest demonstrations linking glomerular hypertrophy to subsequent sclerosis has been obtained using transgenic mice with chronic overexpression of growth hormone and growth hormone releasing factor. These animals develop early enlargement of the glomerulus which is followed by glomerulosclerosis (Doi et al. 1988).

Familial/genetic pathways

A central question which remains to be answered in human diabetes is why do only a proportion of diabetic patients develop renal failure? If a diabetes-induced abnormality in systemic or local growth promoters, or for that matter haemodynamic forces, were sufficient to cause renal sclerosis leading to renal insufficiency, one would expect that, given time, all patients would develop overt renal disease. This is not the case.

Diabetes induces important metabolic, hormonal, and growth factor changes. These changes, which are partly related to the degree of glycaemic control, occur in virtually all patients, but, to date, it has been impossible to isolate a subset of individuals in whom their severity is convincingly linked to the development of renal complications. On the contrary, there is a growing body of evidence to suggest that the degree of diabetic control is a necessary component in, but is not linearly related to, the development of renal failure. Early renal hypertrophic and haemodynamic changes consistently occur in only a subgroup of subjects, and to explain their susceptibility to renal failure it is therefore necessary to formulate an alternative hypothesis which takes into account the host response to diabetes-induced environmental disturbances.

Familial clustering of diabetic kidney disease has been reported. In insulin-dependent diabetes, diabetic siblings of probands with diabetic nephropathy have evidence of nephropathy significantly more frequently than the diabetic siblings of probands without nephropathy (83 versus 17 per cent) (Seaquist et al. 1989). A genetic influence on the development of nephropathy has similarly been described in Pima Indians with non-insulin-dependent diabetes (Pettitt and Saad 1988). The findings of these studies are consistent with the postulate that inherited factors play an important role in determining susceptibility to diabetic nephropathy, but they provide no insight into the nature of these factors. Two independent studies suggested that a familial predisposition to raised arterial pressure may be a possible contributing factor to susceptibility to nephropathy in diabetes. Parents of insulin-dependent diabetics with proteinuria were found to have significantly higher arterial pressure than matched parents of non-proteinuric diabetic patients (Viberti et al. 1987b; Krolewski et al. 1988).

Further insights into the predisposition to and mechanisms of diabetic renal disease, and possibly the attendant cardiovascular disease, has come from studies of cell membrane cation transport systems. Rates of red cell sodium–lithium countertransport, which are largely genetically determined and associated with the risk of essential hypertension (Boerwinkle et al. 1986; Dadone et al. 1984), have been found to be higher in proteinuric diabetic patients than in matched long-term normoalbuminuric controls (Mangili et al. 1988; Krolewski et al. 1988). The risk of nephropathy seems to be increased by the combination of a previous history of poor glycaemic control and a high sodium–lithium countertransport activity (Krolewski et al. 1988) (Fig. 9). Microalbuminuric diabetic patients, a group at increased risk of overt nephropathy, have also been found to have higher rates of sodium–lithium countertransport (Jones et al. 1990). Moreover, an association between plasma lipoprotein levels and Na^+/Li^+

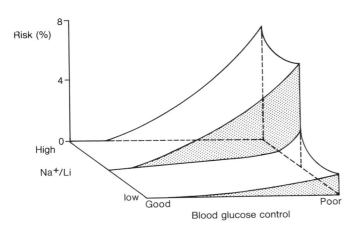

Fig. 9 Tridimensional diagram plotting levels of part blood glucose control and rates of Na^+/Li^+ countertransport against risk of development of diabetic nephropathy in insulin-dependent diabetics. The combination of a history of poor glycaemic control and high Na^+/Li^+ gives the highest risk of nephropathy. Adapted from data of Krolewski et al. 1988.

countertransport activity in insulin-dependent patients without persistent clinical proteinuria has been described. Higher rates of Na^+/Li^+ countertransport were associated with elevated LDL cholesterol, total and VLDL triglycerides and with reduced HDL_2-cholesterol levels (Jones et al. 1990b).

The mechanisms underlying the association between sodium–lithium countertransport activity, hypertension, and lipid abnormalities in the context of susceptibility to diabetic renal and vascular disease are unclear, but could be related to factors involved in the control of insulin sensitivity (Bunker and Mallinger 1985). In a study of short-term diabetic patients without clinical proteinuria but with arterial hypertension (blood pressure >140/90 mmHg), those with higher rates of sodium–lithium countertransport were more insulin resistant, and had higher albumin excretion rates, increased total body exchangeable sodium, enlarged kidneys, and left ventricular hypertrophy (Nosadini et al. 1989). These associations were independent of the actual level of blood pressure or the duration of arterial hypertension. It seems to be the diabetic hypertensive patient with high sodium–lithium countertransport who displays the albuminuria, left ventricular and renal hypertrophy, and insulin resistance that have been related to renal and vascular injury (Chavers et al. 1989; Foster 1989; Silberberg et al. 1989; Sampson et al. 1990).

Sodium–lithium countertransport is believed to be a mode of operation of the physiological sodium–hydrogen antiport, a crucial system in the control of intracellular pH, cell growth and the renal reabsorption of sodium, and thus in the regulation of blood pressure (Mahnensmith and Aronson 1985). Recently, leucocytes and cultured skin fibroblasts from insulin-dependent diabetic patients with albuminuria have been reported to have elevated Na^+–H^+ antiport activity (Li et al. 1990; Ng et al. 1990). An increase in ^3H-thymidine incorporation into DNA of skin fibroblasts of diabetic patients with nephropathy has also been documented (Li et al. 1990). These findings are consistent with the view that cells of diabetic patients who develop nephropathy have an intrinsic enhanced capacity to proliferate, and that this phenomenon is associated with high rates of Na^+/H^+ exchange. Phases of growth of the whole body, such as puberty in humans,

are associated with insulin resistance (Amiel et al. 1986); thus the activity of the sodium–hydrogen antiport seems to act as an indicator of some mechanism, possibly genetically determined, controlling cell growth and hypertrophy on the one hand, and intracellular sodium homeostasis on the other. The environmental changes brought about by diabetes could lead to dysregulation of these mechanisms in susceptible individuals, and induce cell hypertrophy and hyperplasia, contributing to glomerular hypertrophy, mesangial expansion, and tubular hypertrophy and hyperplasia. An increased renal sodium reabsorption would augment systemic and renal perfusion pressure to maintain sodium balance: this increased perfusion pressure would be readily transmitted to the glomerular capillaries because of the general vasodilation present in diabetes (Parving et al. 1983). The resulting increased intraglomerular pressure is responsible, at least in part, for an increase in GFR, and may be responsible for the disruption of glomerular membrane permeability properties which result in proteinuria. Progressive mesangial expansion would lead to glomerulosclerosis and further disruption of the selective permeability of the glomerular basement membrane. The insulin resistance associated with excessive growth and the consequent hyperinsulinaemia may cause lipid abnormalities that, in the setting of the vascular hyperpermeability characteristic of diabetic microvascular disease (Feldt-Rasmussen 1986), would further aggravate the renal damage (Moorhead et al. 1982) and contribute to the accelerated atherosclerosis of diabetic renal failure. An association between insulin resistance and an increased incidence of cardiovascular events has been reported in the general population (Zavaroni et al. 1989). The sequence described above would trigger a vicious cycle of events producing reduction in renal function, more hypertension, more proteinuria, more severe glomerulosclerosis, more hyperlipidaemia, and eventually renal failure and cardiovascular death (Fig. 10).

Pathology

The renal morphological changes associated with diabetes mellitus were first described in 1936 by Kimmelstiel and Wilson (1936) and soon confirmed and extended by other workers (Fahr 1942; Spühler and Zollinger 1943). The early autopsy and biopsy series and later studies (Kamenetzky et al. 1974; Taft et al. 1990) have shown that the histological changes found in kidneys of insulin-dependent and non-insulin-dependent diabetic patients are similar.

Light microscopy

The nodular lesion described by Kimmelstiel and Wilson in 1936 has long been considered virtually specific for diabetes (Bell 1953; Gellman et al. 1959). Reports that it could be found in the absence of diabetes have not stood critical review (Tchobroutsky 1979), and some patients may have had the nodular form of light chain nephropathy (see Chapter 4.3).

The nodules are well-demarcated, eosinophilic, and periodic acid-Schiff positive hard masses, located in the central regions of peripheral glomerular lobules (Fig. 11). When not acellular they contain pyknotic nuclei, and foam cells frequently surround them. They are relatively homogeneous on haematoxylin staining, and have a laminated structure when viewed with periodic acid-Schiff or reticulin stains. The irregularity of their size and distribution, within and between glomeruli, and their location away from the hilus is characteristic. A rim of mesangial cells can sometimes be seen between them and the adjoining capillary,

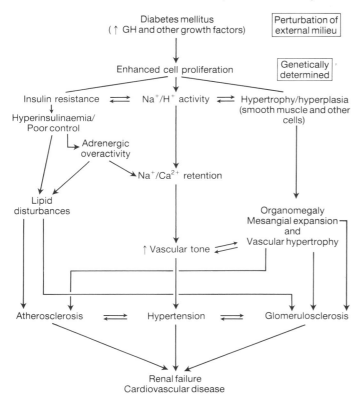

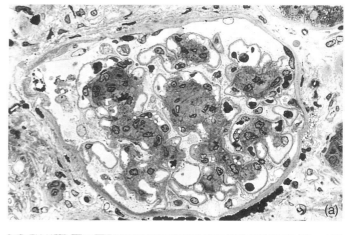

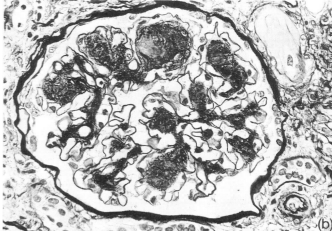

Fig. 10 Hypothetical diagrammatic representation of a sequence of events leading a susceptible subset of insulin-dependent diabetic patients to renal and cardiovascular disease.

which is often distended. The location and the morphogenesis of the nodules has been the subject of long discussion, as reflected by their description as 'intracapillary' and 'intercapillary'. Recent evidence (Saito *et al.* 1988) seems to establish the mesangium as their site of origin, and extends the original suggestion that mesangial disruption and lysis of the lobule centre is related to prior microaneurysmal dilatation of the respective capillary, followed by a laminar reorganization of the mesangial debris (Bloodworth 1978).

Although nearly pathognomonic for diabetes, this lesion is not always present. Its incidence varies considerably from 12 to 46 per cent in different series, which included both insulin-dependent and non-insulin-dependent patients, probably due to differences in selection (Heptinstall 1983). Nodules were found in 55 per cent of an autopsy series of non-insulin-dependent Pima patients (Kamenetzky *et al.* 1974). It is generally accepted that nodules are not seen in the absence of the diffuse lesion, and this reflects its appearance only after a long period of disease (14 years in the series of Gellman *et al.* 1959).

The diffuse glomerular lesion arises from an increase of the mesangial area and capillary wall thickening, with the mesangial matrix extending to involve the capillary loops (Fig. 12). The staining properties of the accumulated material are similar to those of the nodules. In its early stages it may be difficult to distinguish minor mesangial expansion from changes which occur with ageing or other glomerular pathology (Watkins *et al.* 1972). In more severe cases, the capillary wall thickening and the mesangial expansion lead to capillary narrowing and eventual complete hyalinization. In this advanced state, periglomerular

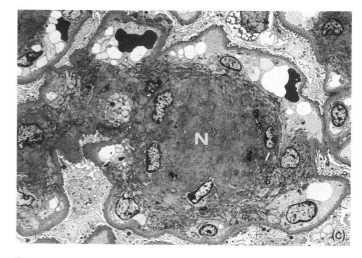

Fig. 11 (a) A nodular lesion in a 54-year-old insulin-dependent diabetic. The mesangium shows enormous expansion with relatively accelular masses of material, with preservation of the patency of surrounding capillary loops and the glomerular capillary walls (toluidine blue × 200). (b) Staining with silver methanamine reveals the silver-positive, laminar nature of the nodules (silver methanamine/haematoxylin and eosin × 200). (c) Electron microscopy shows again the complex texture of the mesangial nodule and its relative accelularity; only three nuclei are clearly seen, as usual round the periphery of the mass. A developing foam cell can be seen on the left (impregnated with osmium tetroxide and uranyl acetate (× 4833)). (By courtesy of Dr Barrie Hartley, Department of Pathology, Guy's Campus, UMDS, London.)

fibrosis is often present. The distribution of the diffuse lesion is non-uniform, both among lobules of the same glomerulus and between different glomeruli, leading to appearances suggestive of transition to nodule formation. The thickening of the capillary walls also tends to be non-uniform, particularly when the histological changes are not very severe. This lesion is more frequent than the nodules, but again its incidence varies in different series (Heptinstall 1983). In a large autopsy study, however, changes compatible with diabetic glomerulosclerosis were found in up to 90 per cent of insulin-dependent diabetic patients with disease duration of more than 10 years (Thomsen 1965). Renal morphological changes have been reported in 25 to 51 per cent of non-insulin-dependent patients (Bell 1953; Heptinstall 1983; Dumler *et al.* 1987).

The exudative lesions are highly eosinophilic, rounded homogeneous structures seen in the capsular space (Fig. 13), overlying a capillary loop (fibrin cap) or lying on the inside of Bowman's capsule (capsular drop). They are non-specific, containing various proteins and sometimes lipid material, and similar lesions are seen in a variety of other renal conditions.

An increase in glomerular volume and in luminal volume occurs 1 to 6 years after diagnosis of insulin-dependent diabetes (Østerby and Gundersen 1975). This initial glomerular enlargement is not accompanied by hyperplasia, but the number of nuclei is increased later in the course of the disease, and the volume of the patent glomeruli increases further. This late hypertrophy has been considered to be distinct from the early one, and may be an expression of compensatory growth in the face of progressive glomerular loss (Gundersen and Østerby 1977; Østerby *et al.* 1987). Although atrophic ischaemic glomeruli are present, some of the non-functioning obsolete glomeruli seem filled up with solid periodic acid-Schiff positive material, and preserve their increased size. This hypertrophy has not been found in subjects with non-insulin-dependent diabetes (Schmitz *et al.* 1988), although hyalinized non-atrophic glomeruli can be seen in these patients (Thomsen 1965).

Arteriolar lesions are prominent in diabetes, with hyaline

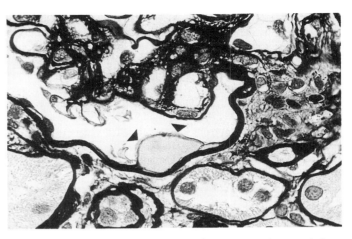

Fig. 13 A 'capsular drop' (arrowheads) on Bowman's capsule in a diabetic patient. The lesion is eosinophilic and homogenous (silver methenamine/haemotoxylin and eosin × 500). (By courtesy of Dr Barrie Hartley, Department of Pathology, Guy's Campus, UMDS, London).

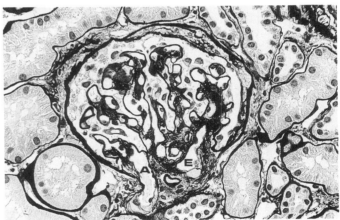

Fig. 14 Cross-section of a glomerulus showing the hilar region from a diabetic patient. The afferent arteriole (A) and the efferent arteriole (E) show hyaline masses within their walls, shown as clear non-argyrophilic spaces. This dual lesion affecting also the efferent arteriole is believed to be relatively specific for diabetes, and may be present in the absence of proteinuria and/or obvious glomerular changes (silver methenamine/haemotoxylin and eosin × 250). (By courtesy of Dr Barrie Hartley, Department of Pathology, Guy's Campus, UMDS, London.)

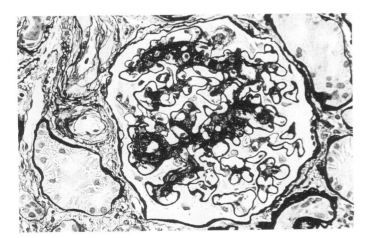

Fig. 12 The increase in mesangium seen in patients with diffuse capillary wall thickening can be observed in the upper part of this glomerulus, whilst in the lower right there is relatively little change. The capillary walls have taken the silver stain with avidity and are clearly thickened (silver methenamine/haemotoxylin and eosin × 200). (By courtesy of Dr Barrie Hartley, Department of Pathology, Guy's Campus, UMDS, London.)

material progressively replacing the entire wall structure. Bell (1953) first underlined the fact that both afferent and efferent arterioles could be affected (Fig. 14). He also pointed out that these lesions were often present in the absence of hypertension, and that involvement of the efferent vessel was highly specific for diabetes. These arteriolar changes may be the first change detectable by light microscopy in the diabetic kidney, as judged by their recurrence at 2 years in non-diabetic kidneys transplanted into diabetic patients (Mauer *et al.* 1976).

The tubules and interstitium may show a variety of non-specific changes which are similar to those seen in other forms of progressive renal disease. The Armanni-Ebstein lesion is the result of accumulation of glycogen in tubular cells of the corticomedullary region in patients with profound glycosuria, and was common in insulin-dependent diabetics before the advent of insulin. More subtle tubular changes, consisting of vacuolization, a decrease in the intercellular spaces normally present between the macula densa cells, and a significant increase in the contact area between these cells and the extraglomerular mesangial cells of Goormatigh, have been described at the ultrastructural level in the streptozotocin model of rat diabetes. It was suggested that these changes may represent a morphological counterpart to the disturbed tubuloglomerular feed-back and the hyperfiltration of diabetes (Rasch and Holck 1988).

The histological changes described, although individually non-specific, are highly suggestive when present simultaneously. The degree of arteriolar hyalinization seen in diabetes is uncommon outside this condition, especially when hypertension is mild or absent, and its presence in the efferent arteriole may be unique to diabetes (Bell 1953). Nodular masses similar to the diabetic nodule may be present in other conditions: the nodules in mesangiocapillary glomerulonephritis (see Chapter 3.8) tend to be much more evenly distributed, and there are characteristic loop changes and hypercellularity. Similar nodular masses can be seen in light chain nephropathy (see Chapter 4.3), and the immunofluorescence findings should establish the diagnosis. In amyloid nephropathy (see Chapter 4.2) the nodules may be unevenly distributed, and the use of specific stains may be necessary to make the diagnosis. When nodules are not present, glomerular changes may have to be distinguished from those present in a number of other renal diseases. The specific stains for amyloid may again be of help. In membranous nephropathy (see Chapter 3.7) the thickening of the capillary loops and the degree of glomerular involvement are even; this is rarely the case in diabetes. The differential diagnosis from non-specific arteriolosclerosis and with the changes associated with ageing may be difficult, but in both of these conditions, the obsolete glomeruli tend to be reduced in size, whereas in diabetes large hyalinized glomeruli are often seen.

Immunopathology

Westberg and Michael (1972) confirmed previous observations of thin linear staining of the glomerular basement membrane for IgG, IgM, albumin, and fibrinogen in kidneys of insulin-dependent diabetic patients, and concluded that it represented a non-specific consequence of increased permeability rather than an expression of specific binding. Their findings were later extended, and the presence of thin linear positivity for IgG and albumin, along not only the glomerular basement membrane but also the Bowman's capsule and especially the outer aspect of the tubular basement membrane, was considered specific for diabetes (Miller and Michael 1976). This assertion has been supported by the demonstration of similar staining in non-diabetic kidneys transplanted into insulin-dependent diabetic patients (Mauer et al. 1976).

Immunofluorescence studies have shown increased mesangial amounts of type IV collagen, as well as of type V collagen, laminin and fibronectin, and the presence of antigens normally only expressed in fetal glomeruli (Falk et al. 1983). Immunochemical analysis confirmed the increase in type IV collagen, but showed reduced laminin levels and markedly decreased heparan sulphate proteoglycan, whereas fibronectin levels were not different from those in normal controls (Shimomura and Spiro 1987).

Collagens type I (Glick et al. 1990) and type VI, but not type III (Ikeda et al. 1990) have also been described in the diabetic mesangium. Levels of type VI collagen, which is probably synthesized only by mesangial cells and is, therefore, absent from the basement membrane, have been found to be increased by about three-fold in kidneys from patients with long-standing diabetes when compared to diabetic controls (Mohan et al. 1990). The possibility of anomalies in the contractile properties of the mesangium has been considered following the demonstration of increased amounts of actomyosin (Scheinman et al. 1974). The role of fibrinogen present along the endothelial aspect of the glomerular basement membrane and in the mesangium has been discussed (Farquhar et al. 1972; Westberg and Michael 1972), and fibrin products as well as IgG and complement components have also been described in the exudative lesions of the glomerulus and of the arterioles (Heptinstall 1983).

Electron microscopy

Glomerular basement membrane thickening is prominent in the kidneys of patients with diabetes. It has been studied particularly in insulin-dependent diabetics, in whom it is absent at diagnosis (Østerby-Hansen 1965) becoming detectable after 2 to 5 years of disease (Østerby 1975). The increased membrane synthesis which takes place during this period is reflected in the 80 per cent increase of peripheral capillary wall area found in young insulin-dependent subjects within months of diagnosis (Kroustroup et al. 1977). In advanced stages of the disease, the basement membrane can become very thick (Fig. 15), particularly when associated with nodular lesions (Kimmelstiel et al. 1966). In early disease the increase in thickness seems even, but there is marked irregularity in membrane width in later disease (Østerby et al. 1983). Localized areas of electron-lucent material can account for up to 10 per cent of total basement membrane volume, and between 1 and 6 per cent of the capillary length is lined by abnormally thin membrane (Østerby et al. 1987). Throughout all the stages up to sclerosis, aspects suggestive of immune deposits are absent, although subendothelial fibrillar electron dense material interpreted as fibrin–fibrinogen products has been described (Farquhar et al. 1972) (Fig. 13).

The foot processes of the epithelial cells in diabetes remain discrete after proteinuria has become persistent (Ireland et al. 1967b) and renal function has declined to 20 per cent of normal (Østerby et al. 1987) (Fig. 15). At this stage, epithelial cell cytoplasm preserves a healthy appearance, with prominent organelles (rough endoplasmic reticulum, mitochondria, Golgi vesicles) that may be the expression of continued basement membrane synthesis. As the disease progresses further the foot processes appear wider in cross-section, and the length of the filtration slits tends to decrease, especially after the onset of persistent proteinuria (Ellis et al. 1987). Aspects suggestive of degenerative changes have also been noted, accompanied by detachment from the underlying basement membrane, which is left exposed (Cohen et al. 1977). Eventually, podocyte effacement and effusion are seen (Kimmelstiel et al. 1966).

The fractional volume of the mesangium, although normal at diagnosis of insulin-dependent diabetes (Ireland 1970), increases, initially due to an expansion of the matrix component following the thickening of the glomerular basement membrane (Østerby 1973). Its delicate strands anastomose and widen, and collagen fibrils become detectable with progression of the disease (Fig. 13). The core of the typical nodules seems to corre-

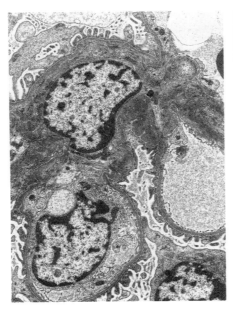 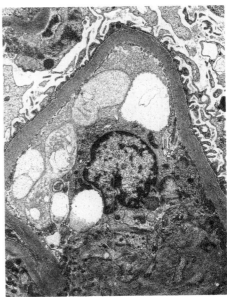

Fig. 15 The diffuse capillary membrane lesion of diabetic nephropathy. On the left a normal capillary wall thickness (250–350 nm) is shown. On the right, a capillary loop from a diabetic patient with severe capillary basement membrane thickening: the basement membrane is at least twice as thick as normal, with a homogenous texture. Despite proteinuria, the architecture of the podocytes outside the capillary wall is relatively well preserved. (impregnated with osmium tetroxide and uranyl acetate × 4475). (By courtesy of Dr Barrie Hartley, Department of Pathology, Guy's Campus, UMDS, London.)

spond to local accumulation of these components (Kimmelstiel *et al.* 1962; Dachs *et al.* 1964), while their laminated periphery may derive from expanded and folded basement membrane (Ikeda *et al.* 1990). In more advanced stages of the disease, the number of mesangial nuclei is sometimes increased, and the fractional volume of the mesangium can become more than double that found in normal controls (Mauer *et al.* 1984).

The exudative lesions appear as finely granular electron-dense material on electron microscopy. In the arterioles, this material spreads from an initial subintimal location towards the media, sometimes producing near replacement. The 'capsular drop' corresponds to accumulation of this material between the epithelial cells and Bowman's capsule basement membrane, whereas the 'fibrin cap' represents its presence along the endothelial side of the glomerular basement membrane (Heptinstall 1983).

Endothelial cell changes are not prominent in diabetes.

Structure–function relationships

Understanding the relationship between abnormal morphological appearances and altered function may help clarify the pathogenesis of diabetic kidney disease. Such information is almost totally restricted to patients with insulin-dependent diabetes.

Increased glomerular volume is a salient feature in the kidneys of insulin-dependent diabetic patients, and is part of a generalized hypertrophic process which also involves the tubules and results in increased kidney volume. Larger glomeruli are probably a necessary condition for the early hyperfiltration of insulin-dependent diabetes, and large kidneys can persist with advancing renal failure in diabetes mellitus, unlike in most other renal diseases.

Gellman *et al.* (1959) were the first to show that the nodular lesions are of little functional significance, and that it is the degree of diffuse glomerulosclerosis, encompassing changes in the glomerular capillary and the mesangial region, that correlates with the clinical manifestations of worsening renal function. In the early phases of insulin-dependent diabetes, the increase in luminal volume and filtering surface area shown by light and electron microscopy morphometric studies offer a structural counterpart to the higher GFR (Kroustroup *et al.* 1977). With advancing renal disease, no correlation is seen between the width of the thickened basement membrane and any of the parameters of renal function, but a close association has been reported between functional changes and mesangial expansion. A fractional mesangial volume in excess of 37 per cent was shown to discriminate between patients with and without clinical proteinuria (Mauer *et al.* 1984). Mesangial expansion also correlates inversely with the capillary filtering area, a variable closely associated with glomerular filtration rate, from levels of hyperfiltration to markedly reduced renal function (Ellis *et al.* 1986; Østerby *et al.* 1988). It has therefore been suggested that it is the expansion of the mesangium, with the attendant reduction in glomerular filtration surface area, that is responsible for the progressive loss of renal function in insulin-dependent diabetes. It should be remembered, however, that there is considerable overlap of fractional mesangial volume between patients with and without proteinuria (Mauer *et al.* 1984; Thomsen *et al.* 1984; Steffes *et al.* 1989), making the prognostic significance of this variable open to question.

One study has attempted to correlate morphological changes with microalbuminuria in insulin-dependent diabetic patients (Chavers *et al.* 1989), and showed no changes in morphometric measurements in patients with albumin excretion rates below the at risk level of 40 mg/day (about 30 μg/min). In patients with higher albumin excretion rates, however, the fractional volume of mesangium was on average significantly increased, and a minor reduction in creatinine clearance and elevation in blood pressure was observed. Similar findings have recently been reported in non-insulin-dependent diabetic patients (Inomata *et al.* 1989). These data confirm the reliability of persistent microalbuminuria and the attendant blood pressure elevation as clinical markers of significant renal damage (Wiseman *et al.* 1984*b*).

Indications for renal biopsy in diabetics

It can be argued a patient with diabetes and urinary abnormalities has no need of a renal biopsy, which will almost certainly show changes consistent with diabetic nephropathy. While it is true that proteinuria and/or haematuria are the result of diabetic nephropathy in the great majority of patients with clinical diabetes, this is not always the case.

The first step in the investigation of a proteinuric diabetic is to obtain information on the number, size and position of the kidneys by ultrasonography. Should there be a suspicion of papillary necrosis or tuberculosis of the renal tract, conditions that have an increased incidence in diabetic patients and may be associated with modest proteinuria, a full intravenous urogram should be performed.

There may be several clues to the existence of disease other than diabetic nephropathy. First, absence of retinopathy, which is almost invariably present in patients with proteinuria of diabetic origin. A concordance rate varying between 63 per cent in non-insulin-dependent diabetics (Parving *et al.* 1990) and up to 85 to 99 per cent in insulin-dependent diabetics (Thomsen 1965; Malins 1968; Deckert and Poulsen 1968; Parving *et al.* 1988*a*) makes the absence of retinopathy a strong argument for performing a biopsy.

Secondly, the presence of haematuria and red cell casts. The frequency of haematuria is variable depending on the type of diabetics studied and on the definition of haematuria. In a large series of 136 consecutive biopsies in insulin-dependent and non-insulin-dependent patients, Taft *et al.* (1990) found haematuria in excess of 10 000 cells/ml in 66 per cent of patients, and this was of little help in distinguishing diabetic and non-diabetic renal disease. In a smaller biopsy series, 69 per cent of insulin-dependent diabetic patients with proteinuria and haematuria had a concomitant non-diabetic renal disease. Red cell casts have been noted in 4 per cent of cases of biopsy-proven diabetic disease (Kincaid-Smith and Whitworth 1988). Although still controversial, the presence of frank haematuria, especially when accompanied by red cell casts, should suggest the possibility of other renal disease.

Thirdly, the duration of diabetes at onset of proteinuria. Proteinuria develops in only 4 per cent of insulin-dependent patients within 10 years of onset of diabetes; early onset of proteinuria is much more likely to be the result of disease other than diabetic nephropathy (Amoah *et al.* 1988). Conversely, the likelihood that a young or adolescent diabetic with onset in childhood who develops proteinuria 20 years later has diabetic nephropathy approaches 100 per cent. Proteinuria can be present at diagnosis in as many as 8 per cent of non-insulin-dependent patients (Marshall and Alberti 1989), and the number presenting with proteinuria increases with increasing age at presentation (Malins 1968). In these older patients the duration of diabetes is uncertain, and the presence of other renal disease more likely (Amoah *et al.* 1988). Thus early proteinuria is of less value in differentiating between renal injury of diabetic and non-diabetic origin.

Almost every form of glomerular disease has been shown to occur in diabetic patients: this is not surprising considering that about 3 per cent of the general population is diabetic. The diagnosis of a coincident glomerular disease by renal biopsy may be helpful from both the therapeutic and prognostic standpoints. Besides detecting a potentially treatable non-diabetic renal disease, renal biopsy may be useful in the future for monitoring the effect of treatment of the early potentially reversible phases of diabetic nephropathy.

Other glomerular diseases in diabetics

IgA nephropathy and other forms of mesangial glomerulonephritis, acute glomerulonephritis, lupus, amyloidosis, mesangiocapillary glomerulonephritis types I and II, crescentic glomerulonephritis and steroid responsive minimal change disease in both children and adults (Kasinath *et al.* 1983; Silva *et al.* 1983; Chihara *et al.* 1986; Taft *et al.* 1990) have all been reported in diabetics. In a few patients, two additional glomerulopathies have been noted as well as diabetic nephropathy (Bertani *et al.* 1983); in our own unit we have seen a 71-year-old patient who had diabetic nephropathy, membranous nephropathy and amyloidosis, probably arising from coincident rheumatoid arthritis.

The overall incidence of non-diabetic glomerular disease was 8 per cent in 50 biopsied diabetics (Rao and Crosson 1980), 9 per cent of 122 patients (Kasinath *et al.* 1983), 22 per cent of 164 cases (Chihara *et al.* 1986), and 10 per cent of 136 diabetic patients (Taft *et al.* 1990). When only non-insulin-dependent patients were examined, 27 per cent of 33 were found to have non-diabetic renal disease (Parving *et al.* 1990). In a retrospective study of renal biopsies, 12 per cent of 49 insulin-dependent and 28 per cent of 60 non-insulin-dependent patients were found to have another disease (Amoah *et al.* 1988). These figures will undoubtedly be an overestimate of the true rate, since routine renal biopsy of proteinuric diabetics is not practised in most units, and those submitted to biopsy will contain an augmented number of patients who do have non-diabetic glomerulopathies.

Membranous nephropathy is the glomerular disease which has most often been reported in association with diabetes. Since the first description by Churg's group (Dachs *et al.* 1964) at least 60 cases have been reported, and many other similar patients seen more recently have not been reported since the association is now well recognized (see Rao and Crosson 1980; Kobayashi *et al.* 1981; Kasinath *et al.* 1983; Silva *et al.* 1983 for reviews of published cases). The mode age at presentation was 40 to 60 years, similar to that of both membranous nephropathy (Chapter 3.7), and of maturity onset diabetes mellitus. However, patients as young as 13 years have been reported, together with several in their 20s. Interestingly, most patients had had diabetes for 10 years or more; two-thirds of the patients were using insulin. Only 25 per cent of the patients had retinopathy, although not all were studied by fluorescent angiography. Most of the patients were nephrotic, with proteinuria as high as 35 g/24 h. Histological appearances ranged from typical membranous nephropathy to an almost typical diabetic glomerulosclerosis complicated by mild or occasional membranous changes.

It is difficult to be sure whether this association is simply the coincidence of one of the more common glomerulopathies of middle age with diabetes of either type, or whether it represents a specific pathogenetic association. Certainly the excess of membranous cases over other glomerular pathologies noted in nephrotic patients aged 30 to 60 years is striking. It is conceivable that diabetic alterations of the glomerular structure might predispose to one of the several mechanisms thought to lead to subepithelial deposits, and thus to the pattern of membranous nephropathy (see Chapter 3.2).

Although Chihara *et al.* (1986) reported that coincidence of membranous nephropathy and diabetes did not affect the prognosis for kidney function in man, the superimposition of Heymann membranous nephropathy on streptozotocin-induced diabetes in rats has been found to lead to a more severe nephrotic syndrome and a higher mortality than either disease alone (Okuda *et al.* 1984).

Treatment

Blood pressure control

Up to 10 years ago, hypertension was viewed as a late phenomenon of diabetic nephropathy in insulin-dependent diabetes, and was thought to be a consequence of it (Ireland et al. 1982); there was no consensus regarding its frequency or pathogenic importance in non-insulin-dependent diabetes (Jarrett et al. 1982). It has, however, become apparent that rises in blood pressure occur in proteinuric insulin-dependent diabetes even in the face of normal, although declining, GFR (Parving et al. 1981). Microalbuminuric insulin-dependent patients have also been shown to have higher arterial pressures than matched normoalbuminuric patients (Wiseman et al. 1984b; Mathiesen et al. 1984). These observations, coupled with evidence for a central role of blood pressure level in the progression of diabetic renal disease (Hasslacher et al. 1985), have led to a different perception of the importance of arterial hypertension.

Mogensen (1976b) was the first to show that reduction in blood pressure slowed the rate of decline of GFR and checked the increasing albuminuria in insulin-dependent diabetics. A prospective self-controlled study of 6 years' duration (Parving et al. 1987) demonstrated that effective blood pressure treatment reduced the rate of fall of GFR, from 0.94 ml/min.month before therapy to 0.29 ml/min.month during the first 3 years, and to 0.1 ml/min.month in the second 3 years. This change was accompanied by a 50 per cent reduction in albuminuria, from 1038 µg/min to 504 µg/min (Fig. 16). Despite the small number of patients and the lack of a randomized controlled design, these results were taken as evidence of the significant impact of antihypertensive treatment on the progression of diabetic kidney disease. The authors speculated that, should the slowing of progression be maintained, the renal survival of these patients would be extended from the current 7 to 10 years to more than 20 years. Recent longitudinal cohort studies seem to support this prediction: a remarkable reduction in cumulative mortality, from over 50 per cent to 18 per cent at 10 years, has been noticed in patients receiving antihypertensive treatment (Parving and Hommel 1989; Mathiesen et al. 1989) (Fig. 3).

Multiple drug therapy for hypertension was used in the earlier studies, including a β-blocker, a diuretic, and a vasodilator. During the last few years, great interest has surrounded the use of angiotensin converting enzyme inhibitors, not only in established diabetic nephropathy but also in non-hypertensive patients with microalbuminuria. Studies in experimentally diabetic animals (Zatz et al. 1986; Anderson et al. 1986) have led to the suggestion that these drugs may be specifically protective of renal function because of their ability to reduce efferent arteriolar vasoconstriction, and thus reduce intraglomerular hypertension. This 'specific' effect has been claimed to be separate from any effect on systemic blood pressure. A reduction in proteinuria and in the rate of fall of GFR was shown to occur independently of significant blood pressure changes in two uncontrolled studies (Taguma et al. 1985; Bjørck et al. 1986). In more recent controlled trials, however, the reduction of proteinuria and the decreasing rate of GFR decline were found to be associated with a reduction of systemic blood pressure, even when non-hypertensive proteinuric patients were investigated (Hommel et al. 1986; Parving et al. 1988b, 1989). The decrease in the rate of decline of GFR obtained during treatment with angiotensin converting enzyme inhibitors does not appear to be of greater magnitude than that achieved with conventional antihypertensive therapy. A physiological approach to the study of the antiproteinuric effect of converting enzyme inhibitors has employed the fractional clearance of graded size neutral dextrans. Three studies using different designs (Pinto et al. 1990; Ruggenenti et al. 1990; Morelli et al. 1990) have demonstrated that treatment of proteinuric diabetic patients with these drugs significantly reduces the augmented clearance of large-sized dextrans (molecular radii 5.6 to 7.4 nm) and of large, essentially neutral plasma proteins such as IgG. This improvement in glomerular membrane size-selective properties seems to be unique to angiotensin converting enzyme inhibitors and independent of systemic blood pressure changes, since other antihypertensive drugs (such as diuretics or clonidine) do not produce the same effect in spite of similar reductions in systemic arterial pressure. The prognostic significance of these changes remains unknown.

The effect of inhibition of angiotensin converting enzyme has also been studied prospectively in non-hypertensive insulin-dependent patients with microalbuminuria. In a randomized study, Marre et al. (1988) showed a reduction in albumin excretion rate, with a return to normal in 50 per cent of cases after 1 year of enalapril treatment (Fig. 17). Mean blood pressure was reduced by about 10 mmHg throughout the study and GFR remained unchanged. By contrast, in the placebo group mean blood pressure rose and GFR decreased significantly, with 30 per cent of patients becoming persistently proteinuric by the end of the study. A decrease in the fractional clearance of albumin, in the absence of changes in GFR or renal plasma flow, was also found in a short-term cross-over study of the effects of enalapril

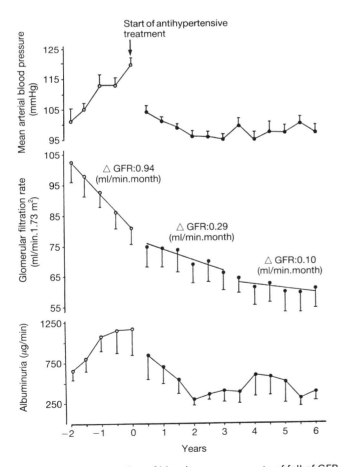

Fig. 16 Effect of reduction of blood pressure on rate of fall of GFR and on albumin excretion rate in nine insulin-dependent diabetic patients with clinical nephropathy. Reprinted with permission from Parving et al. 1987.

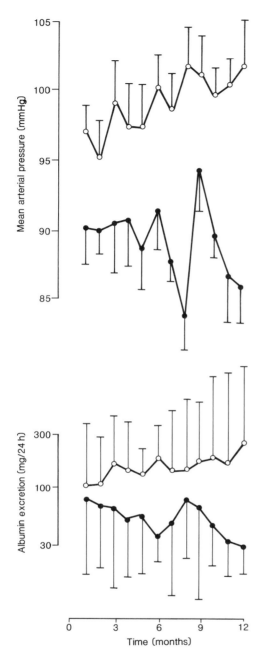

Fig. 17 Changes in blood pressure and albuminuria in 10 enalapril-treated (●) and 10 placebo-treated (○) microalbuminuric insulin-dependent diabetic patients. Adapted from Marre et al. 1988.

on patients with normal albumin excretion rates (Pedersen et al. 1988).

Antihypertensive treatment appears, therefore, to be protective and to delay the progression of established diabetic nephropathy; if started at the stage of microalbuminuria it may even prevent, or at least retard, the onset of clinically overt renal disease. Caution must be applied in the extrapolation of data primarily based on reduction of urinary protein excretion, and long-term trials are needed to establish a definite effect on the rate of deterioration of glomerular function.

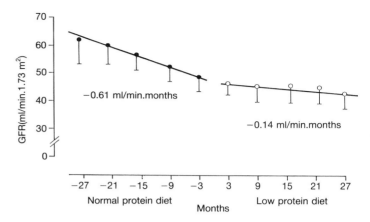

Fig. 18 Rate of loss of ^{51}Cr-EDTA GFR in 19 insulin-dependent diabetic patients with clinical nephropathy during normal protein diet (●) and during low protein diet (○). The reduction in the rate of GFR decline during the low protein diet was highly significant ($p < 0.001$). Adapted from Walker et al. 1989.

Dietary treatment

A reduction of protein intake has long been advocated for the treatment of chronic renal failure (Bergstrom et al. 1984), and diabetic nephropathy is no exception. Early studies by Attman et al. (1983) provided no evidence of a beneficial effect of protein restriction (as low as 0.3 g/kg.day) on the progression of renal disease in insulin-dependent diabetic patients, but this lack of a therapeutic effect may have been due to the advanced stage of the disease, as indicated by serum creatinine levels at entry of about 800 μmol/l. Recent results in diabetic patients with only moderate impairment of renal function seem more encouraging. Barsotti et al. (1983) and Evanoff et al. (1987a) have claimed that a low protein diet retards the progression of diabetic nephropathy in insulin-dependent diabetic patients. These studies have, however, been criticized since renal function was assessed by creatinine clearance or the reciprocal of serum creatinine levels, two indices that are unreliable markers of renal function in the setting of established renal disease and in patients fed a low protein diet (Viberti et al. 1983a; Shemesh et al. 1985; Walser et al. 1988). Other important confounding variables such as blood pressure were also not controlled for.

Two long-term longitudinal studies used reliable markers of glomerular function. In a 5-year self-controlled study of 19 insulin-dependent diabetics with moderate renal failure at entry (mean GFR about 60 ml/min) Walker et al. (1989) showed that the mean rate of decline of GFR, measured as ^{51}Cr-EDTA clearance, was reduced from 0.61 ml/min.month on a 1.13 g/kg body weight/day protein diet to 0.14 ml/min.month on 0.67 g/kg body weight.day (Fig. 18). The individual response to the low protein diet was, however, heterogeneous: four patients failed to respond and seven showed a non-significant reduction in the rate of decline of GFR. The low protein diet reduced albuminuria from 467 to 340 mg/24 h, and was able to check the progressive rise in the fractional clearance of albumin. The effects of the diet seemed to be independent of blood pressure changes, which accounted for only 11 per cent of the decreased rate of loss of GFR.

In a controlled 5-year prospective study of 35 insulin-dependent patients with nephropathy, in whom GFR was assessed by iodoiothalamate clearance, Zeller *et al.* (1990) found that the rate of GFR decline in the group on a protein and phosphorus restricted diet was 0.26 ml/min.month. By contrast, the GFR of a control group on normal protein intake declined at a significantly faster rate of 1.01 ml/min.month. A remarkable finding in this study was that 50 per cent of the patients on dietary therapy showed no decline in GFR during a 31-month follow-up period. Blood pressure was claimed to be well and equally controlled in both groups.

Diets restricted to 0.5 to 0.6 g/kg body weight/day of protein seem to have no long-term detrimental effects on nutritional status. Anthropometric measurements were not affected by such diets, and several authors noted an increase in serum albumin during protein restriction (Barsotti *et al.* 1983; Evanoff *et al.* 1987*a*; Walker *et al.* 1989).

Short-term studies of the physiological mechanisms of action of low protein diets in diabetic patients indicate that this treatment leads to a reduction in the fractional clearance of albumin, IgG, and broad-sized neutral dextrans, and thus to an improvement of glomerular membrane selective permeability, while not affecting GFR or renal plasma flow (Rosenberg *et al.* 1987; Bending *et al.* 1988). Micropuncture studies in the streptozotocin-induced diabetic Munich-Wistar rat suggest that the antiproteinuric effect of a severely protein restricted diet may be modulated by changes in intraglomerular pressure (Zatz *et al.* 1985). However, the mode of action of a low protein diet is complex, unlikely to be mediated simply by haemodynamic changes (Pinto *et al.* 1988; Fogo and Ichikawa 1989), and probably also relates to changes in other nutritional components such as phosphorus and lipids (Walker *et al.* 1989).

Reduction of dietary protein by approximately 50 per cent has also been shown to reduce the fractional clearance of albumin in patients with microalbuminuria (D. Cohen *et al.* 1987) and to lower GFR in patients with hyperfiltration (Wiseman *et al.* 1987*b*), independently of changes in glucose control and blood pressure.

The prognostic significance of these changes remains obscure as no long-term studies have been performed in patients with early renal disease. It also remains to be determined whether good blood pressure control and a low protein diet may have an additive sparing effect on renal function. It is of interest that renal functional changes similar to those produced by a low protein diet can be induced in normal individuals by the administration of a vegetarian diet with normal protein content. The mediators of this effect are not entirely understood but are likely to involve changes in glucagon secretion and the production of renal prostaglandins (Kontessis *et al.* 1990).

Glycaemic control

Blood glucose control appears to have a limited impact on the progression of diabetic renal failure. A positive correlation between levels of HbA_{1c} and rate of decline of GFR was reported (Nyberg *et al.* 1987) but was not subsequently confirmed by others (Viberti *et al.* 1987*a*). Long-term correction of hyperglycaemia for nearly 2 years by continuous subcutaneous insulin infusion, did not affect the rate of fall of GFR or the increasing rates of albumin and IgG fractional clearances in insulin-dependent patients with persistent proteinuria (Viberti *et al.* 1983*b*). Similarly poor results have been obtained in a controlled study of intermittently proteinuric patients given intensified insulin treatment (Bending *et al.* 1986). It would appear that there is a point of no return for the kidney, beyond which the nephropathic process becomes self-perpetuating and independent of the diabetic metabolic abnormality which initiated it.

The effects of reducing elevated blood glucose levels at earlier stages of renal disease are more encouraging. The exaggerated albuminuric response to exercise in insulin-dependent diabetic patients is corrected by a relatively short period of better diabetic control (Viberti *et al.* 1981*a*; Koivisto *et al.* 1981; Vittinghus and Mogensen 1982). A significant correlation between indexes of blood glucose control and albumin excretion rate have been reported by several authors in insulin-dependent patients without persistent proteinuria (Viberti *et al.* 1982*a*; Wiseman *et al.* 1984*b*; Feldt-Rasmussen *et al.* 1986). Strict metabolic control by continuous subcutaneous insulin infusion has been effective in reducing the albumin excretion rate in patients with microalbuminuria (Viberti *et al.* 1979; Kroc Collaborative Study Group 1984; Bending *et al.* 1985) and in preventing the progressive increase in the fractional clearance of albumin in the long term (Feldt-Rasmussen *et al.* 1986). In the Steno study, 28 per cent of conventionally-treated patients progressed to clinical nephropathy over a 2-year period: no progression was seen in the group given intensified insulin treatment (Feldt-Rasmussen *et al.* 1986). Similar reductions in albumin excretion rate can be obtained by multiple injection therapy, provided similar levels of blood glucose control are achieved, suggesting that it is the blood glucose concentration rather than the method of treatment that matters (Dahl-Jørgensen *et al.* 1988). Improvement of blood glucose control by diet or oral therapy also reduced microalbuminuria in non-insulin-dependent patients (Vasquez *et al.* 1984; Schmitz *et al.* 1989*b*).

The results of these studies are consistent with the epidemiological observation of a four-fold increase in the risk of developing nephropathy in insulin-dependent diabetic patients with a record of poor glycaemic control (Krolewski *et al.* 1985, 1988; Warram *et al.* 1990), and suggest a beneficial effect of tight glucose control on progression of microalbuminuria. Whether this effect may be translated to prevention of end-stage renal failure remains to be established.

Correction of hyperglycaemia shortly after diagnosis of insulin-dependent diabetes will induce a decrease in GFR, which in some (Mogensen and Andersen 1975) but not all (Christiansen *et al.* 1982*a*) patients is accompanied by a reduction in the increased kidney size. Recent evidence suggests that the increased kidney size is associated with an exaggerated renal response to amino-acid infusion, and that both can be corrected by 3 weeks of intensified insulin therapy (Tuttle *et al.* 1990). With more long-standing disease, however, near-normoglycaemia lowers GFR while not affecting kidney hypertrophy. Cessation of strict glycaemic control in patients with large kidneys leads to a prompt return of GFR to hyperfiltering levels (Wiseman *et al.* 1985*a*). It has been suggested that the establishment of irreversibility of kidney enlargement signals the onset of a phase of progressive renal disease in which overproduction of intrarenal growth factors perpetuates renal damage (Kleinman and Fine 1988).

Other treatment modalities

A small number of studies on the effect of aldose reductase inhibitors have been reported. Whereas a reduction of GFR (Pedersen *et al.* 1989) and a decrease in albumin excretion rate (Blohmé and Smith 1989) were noted in insulin-dependent patients with either normal albumin excretion rate or incipient nephropathy, a controlled study on non-insulin-dependent subjects with microalbuminuria failed to show any effects on renal function, blood pressure or glucose control after a short period of treatment (Cohen et al. 1989). Longer-term studies will probably be needed to clarify the role of this class of compounds in human disease.

It has been claimed that treatment with aspirin and dipyridamole may delay the progression of chronic renal failure in some patients with diabetic nephropathy (Donadio et al. 1988). This study was however uncontrolled, did not allow for the confounding effect of other variables, and to date has not been confirmed.

Other manifestations of diabetic renal disease

Renal papillary necrosis

Diabetes has long been associated with renal papillary necrosis: its prevalence in autopsy cases averages 4.4 per cent (Mujais 1984), but this figure may be an underestimation since diabetes has been found in up to 50 per cent of patients with renal papillary necrosis (Mandel 1952; Lauler et al. 1960; Eknoyan et al. 1982).

In a recent prospective series, renal papillary necrosis was diagnosed in 24 per cent of 76 consecutive insulin-dependent patients with normal serum creatinine in whom a urography was performed (Groop et al. 1989). It tends to occur in long-standing disease, and affects both kidneys in up to 65 per cent of patients (Mujais 1984). When unilateral, the contralateral kidney becomes involved in the ensuing years. It is more frequent in women, particularly those with recurrent urinary tract infection.

Viewed in the past as an acute devastating condition often leading to sepsis and death, it has become clear that it can also be nearly asymptomatic and follow a more indolent course, with bouts of urinary infection and/or renal colic. Microscopic haematuria has been reported to be more frequent when this condition is present, and pyuria will often be present even in the absence of documented infection, a finding that should alert the clinician to the possibility of underlying silent papillary necrosis (Eknoyan et al. 1982; Groop et al. 1989). Proteinuria is often present but is usually modest (< 2 g/24 h), and patients with persistent proteinuria do not seem to have a higher incidence of renal papillary necrosis than those without (Groop et al. 1989). The urographic appearances of 'moth-eaten' calyxes and the 'ring-shadow' image of the necrotic papilla are highly suggestive of this condition.

The management of acute renal papillary necrosis in the diabetic is compounded by the need to maintain adequate metabolic control. If obstruction is present, its relief is urgent and mandatory, and will thus often determine the success of associated antibiotic therapy. In the chronic, indolent form, the use of non-steroidal anti-inflammatory agents may further compromise medullary circulation, and should therefore be avoided.

Autonomic neuropathy and the bladder

It is difficult to ascertain the true prevalence of autonomic neuropathy of the bladder in diabetes because of the insidious onset of the condition. In the older literature, this uncertainty is reflected by prevalence figures that range from 1 to 26 per cent (Rundles 1945; Martin 1953): more recent studies, where diagnosis was based on urodynamic criteria, report a prevalence of bladder autonomic dysfunction in long-standing diabetes of about 40 per cent (Frimodt-Møller 1976).

The first abnormality, usually detected in asymptomatic patients, is impairment of sensation with decreased awareness of bladder distension, due to involvement of proprioceptive afferent fibres. As a result, micturition occurs at progressively larger bladder volumes, and this, together with progressive damage of the parasympathetic innervation of the detrusor, leads to weaker bladder contraction, incomplete emptying, and increasing residual volume. Involvement of the efferent sympathetic innervation to the trigone may lead to functional incompetence of the vesicoureteral junction and to incomplete relaxation of the internal sphincter during micturition (Mahony et al. 1977; deGroat and Booth 1980).

Patients are often unaware of the extent of their bladder abnormality. Symptoms are few and, at the initial stages, may be confined to disappearance of previous nocturia, with less frequent daytime voiding of large volumes of urine. These changes may go unnoticed for a long time. Later, a weaker stream produced only on straining, terminal dribbling, and involuntary stream interruption or overflow incontinence due to detrusor-urethral sphincter dissynergia, may become apparent. From the early stages, recurrent and/or persistent urinary tract infections may be a consequence of incomplete emptying and/or reflux (Ellenberg 1966; Kahan et al. 1970).

Recognition of this insidious course will assist early diagnosis. Proper consideration of recurring urinary infections in a long-standing diabetic may lead to detection of an unsuspected enlarged bladder by physical examination, or alternatively to disclosure of residual urine by ultrasound scanning. Diabetics may also develop other conditions (i.e., prostatic hypertrophy, asymptomatic stone) that may interfere with bladder function, and overt symptoms may occur due to associated treatment with anticholinergic drugs (Gibberd 1981; Rubinow and Nelson 1982). Uroflowmetry (mictigraphy) (Ewing and Clark 1986), and cystometrogram studies (Bradley 1980) will characterize the bladder dysfunction, and may be invaluable for accurate prognosis and decisions about treatment.

Detection of autonomic bladder dysfunction in a diabetic should lead to adoption of a policy of voluntary, regular voiding by the patient even in the absence of subjective urge. Suprapubic manual pressure will help complete emptying. Intermittent or temporary catheterization and associated parasympathomimetic drug treatment with betanechol chloride may lead to reduction of bladder distension and recovery of detrusor function (Frimodt-Møller and Mortensen 1980). The effect of this drug seems unpredictable, however, and its use may be limited by side-effects (Ellenberg 1980). In the presence of detrusor-urethral dissynergy, centrally acting muscle relaxants or an α-adrenergic blocking agent, depending on whether dysfunction involves the external or the internal sphincter, may help. Bladder neck resection is sometimes very successful, and more complex operations to reduce bladder capacity are sometimes performed. Long term catheterization may be the only solution in severely disabled patients. The most serious complication of neurogenic bladder is, however, an intractable urinary infection that may render the patient unsuitable for renal transplantation.

Pregnancy in diabetic nephropathy

Until recently pregnancy was discouraged in women with diabetic nephropathy, and therapeutic abortion was recommended because of the very poor fetal outcome (Pedersen 1977). During the last decade, more than 100 pregnancies in patients with insulin-dependent diabetes and nephropathy have been reported (Kitzmiller et al. 1981; Jovanovic and Jovanovic 1984; Dicker et al. 1986; Grenfell et al. 1986; Reece et al. 1988). Both the impact of pregnancy on nephropathy and the outcome of gestation appear more encouraging.

The increase in creatinine clearance which occurs in normal gestation is seen in only about 56 per cent of pregnant women with diabetic nephropathy during the first trimester (Jovanovic and Jovanovic 1984; Dicker et al. 1986). In patients with initial creatinine clearances of 50 ml/min or greater, elevated blood pressure ($\geq$140/90 mmHg) is present in approximately one-third during the first trimester, and in just over half at the end of pregnancy. In women with more reduced renal function, these proportions increase to about 80 per cent.

Pre-eclampsia is reported in 26 per cent of all diabetic nephropathic pregnancies: its diagnosis is dependent upon the detection of an acute rise in blood pressure, serum creatinine, and proteinuria. Other signs of deranged hepatic and central nervous system function, and thrombocytopenia with elevated fibrin-split products, may coexist. Changes in serum uric acid have been found to be of no pathognomonic significance in patients with diabetic nephropathy (Kitzmiller et al. 1981).

An increase in proteinuria, sometimes of massive proportions, is common in pregnant women with diabetic nephropathy. Sixteen per cent of patients with creatinine clearance greater than 50 ml/min, excrete more than 3 g/24 h of protein in the first trimester; this rises to 71 per cent in the third trimester. Pooled data suggest that about 50 per cent of patients reach 24-h proteinuria greater than 5 g, and 17 per cent reach greater than 10 g (Kitzmiller et al. 1981; Jovanovic and Jovanovic 1984; Dicker et al. 1986; Grenfell et al. 1986; Reece et al. 1988).

Anaemia was found to relate significantly to the decreased renal function, and haematocrits of 28 per cent or less and/or haemoglobin of 10 g/dl or less were seen in 42 per cent of patients (Kitzmiller et al. 1981; Reece et al. 1988).

The general impression is that the natural history of nephropathy is not adversely affected by pregnancy. A marked reduction in proteinuria is common following delivery, and in the 27 patients in whom proteinuria was measured within 1 year, it had returned to preconception values or below in 67 per cent. The rate of decline of creatinine clearance was calculated in 23 patients followed for 6 to 35 months postdelivery, and was found to be 0.85 ml/min.month (Kitzmiller et al. 1981). A reduction in renal function after pregnancy was more likely in patients with more severe proteinuria or hypertension during the first trimester. Overall, the impact of pregnancy on the kidney seemed not to be different from that observed in other renal diseases (Katz et al. 1980; Hou et al. 1985).

Fetal survival in a pooled series of five studies was 91.3 per cent (Kitzmiller et al. 1981; Jovanovic and Jovanovic 1984; Dicker et al. 1986; Grenfell et al. 1986; Reece et al. 1988). The unacceptably high rates of spontaneous abortions and of congenital malformations seen in women with diabetic nephropathy prior to the 1980s (Pedersen 1977; Hare and White 1977) have been considerably reduced, to about 9.5 per cent. There is ample evidence that achievement of normal levels of HbA_{1c} prior to conception reduces the frequency of malformations in diabetic pregnancies in general (Fuhrman et al. 1983; Buschard et al. 1990; Hanson et al. 1990), and the same may apply to women with diabetic nephropathy (Jovanovic and Jovanovic 1984; Peterson and Jovanovic 1986; Miodovnik et al. 1988). Increases in blood pressure, fetal distress, and premature labour are responsible for a high rate of preterm deliveries (about 55 per cent), which is, however, not different from that reported in women with non-diabetic renal disease and depressed renal function (Hou et al. 1985).

Pregnancy in a woman with diabetic nephropathy should be carefully considered and planned. Its outcome depends to a great extent on the joint efforts of the patient and of a team of experts in a specialized centre. The ultimate decision should consider the long-term prognosis for the patient with diabetic nephropathy and the increased risk to the fetus.

Urinary tract infections

Urinary tract infections have been considered to be more frequent in patients with diabetes, probably due to reports of renal histological evidence of interstitial inflammation and scarring in 10 to 40 per cent of diabetics (Young and Clancy 1955; Gelman et al. 1959; Fabre et al. 1982; Heptinstall 1983; Kasinath et al. 1983; Ditscherlein 1985). These histological findings could arise from other conditions, such as ischaemia, reflux nephropathy, or renal papillary necrosis, and are difficult to distinguish from those of 'chronic pyelonephritis' (Thomsen 1965; Heptinstall 1983; Taft et al. 1990). Several surveys have failed to show an increased incidence of urinary infection in diabetics compared to control populations (Huvos and Rocha 1959; O'Sullivan et al. 1961; Vejlsgaard 1966a). However, some workers have reported that women with diabetes have an incidence of bacteriuria that is approximately double that of non-diabetic controls (Vejlsgaard 1966a; Pometta et al. 1967; Vejlsgaard 1973a; Forland et al. 1977). Diabetic nephropathy seems to be associated with an increased frequency of infection only in pregnancy (Vejlsgaard 1966b, 1973b). A large proportion of urinary tract infections are asymptomatic, and their pathogenic role is not clear. However, in patients with diabetic nephropathy even asymptomatic infection may be associated with worsening renal function, and efforts should be directed to its eradication.

Upper urinary tract infections can lead to severe complications in the diabetic. Perinephric abscesses, which can be bilateral (Bevan et al. 1989), are more frequent in diabetic patients, and may have an insidious onset followed by persistent fever and rigours. Urinary cultures are not infrequently negative, and a tender mass in the flank can sometimes be felt (Thorley et al. 1974); computerized tomography will help establish the diagnosis (Bova et al. 1985). Prolonged antibiotic therapy is usually needed.

Infections of the kidney with anaerobic gas-forming organisms are more common in diabetic patients, and have recently been reviewed by Evanoff et al. (1987b). More than 90 per cent of cases of emphysematous pyelonephritis occur in patients with diabetes. This condition is more frequent in women and is usually caused by *Escherichia coli*, although *Candida* sp. and *Cryptococcus neoformans* have also been reported. Fever, abdominal pain, nausea, and vomiting in an obviously ill patient are common presenting features. Elevated WBC count, creatinine, and grossly elevated blood glucose are common, and frank pyuria is usually present. The infection is bilateral in approximately 10 per cent of cases. Plain abdominal radiographic examination, intravenous urography, or ultrasound scanning are diagnostic in about 85 per cent of cases, showing gas bubbles extending along the renal pyramids and collecting under the perirenal fascia, or, after its rupture, extending into the adjoining retroperitoneal space. Computerized tomography will be needed in the remaining cases, and will define more precisely the location of the gas. Emphysematous pyelonephritis has a guarded prognosis. Medical treatment alone results in a mortality of about 60 per cent when gas is confined to the kidney, and of over 80 per cent when it has extended to the perirenal spaces. Surgical treatment with nephrectomy is a life-saving procedure and should not be delayed, but still carries a mortality rate of about 20 per cent. Emphysematous pyelitis refers to the presence of gas within the collecting system only, and 50 per cent of cases occur in diabetic patients, particularly women. Its clinical presentation is similar to that of emphysematous pyelonephritis, although renal failure may be absent and glucose control may remain adequate. Radiographic examination by plain film, intravenous urography, or ultrasound shows gas outlining the pelvicalyceal system, and, sometimes, the ureters. Obstruction and dilatation are not uncommon associated features needing urgent correction. Medical supportive treatment and antibiotic therapy are usually effective, although the overall mortality is still around 20 per cent, even with adequate treatment.

Acute renal failure in diabetics

Toxicity from contrast media (see also Chapter 8.4)

For 25 years (Bergman *et al.* 1968), diabetic patients have been recognized to be at extra risk from procedures which involve the injection of large amounts of contrast media. This predisposition is strongly related to renal function at the time of the study, being reported in the past as zero in diabetics with normal function, 50 per cent in those with plasma creatinine of 200 μmol/l, and 90 to 100 per cent in those with advanced renal failure (Coggins and Fang 1988). It was also suggested that young insulin-dependent diabetics with nephropathy are particularly at risk (Harkonen and Kjellstrand 1979).

To what extent this vulnerability arises from direct toxicity of the contrast media, and how much from coincident dehydration or desalination, is far from clear. However, acute renal failure has been recorded even in carefully hydrated individuals in whom pulmonary wedge pressures and cardiac output were measured (Fang *et al.* 1980; Coggins and Fang 1988).

Today, although transient increases in plasma creatinine may be seen and indicate temporary renal dysfunction (Schwab *et al.* 1989; Parfrey *et al.* 1989; Brezis and Epstein 1989), actual renal failure is extremely uncommon in patients appropriately prepared for examination. Smaller doses of newer, low osmolar non-ionic contrast material have probably contributed to this welcome decline in incidence, but neither Schwab *et al.* (1989) nor Parfrey *et al.* (1989), could detect a difference between the older and the newer contrast media. As noted in Chapter 1.6, these agents cause loss of renal tubular enzymes into the urine just as the older agents did.

Rhabdomyolysis (see also Chapter 8.2)

Diabetic patients, in common with other patients who may have prolonged coma, are vulnerable to rhabdomyolysis, and even patients in relatively mild confusion or coma may go into acute renal failure in association with muscle necrosis (Grossman *et al.* 1974). However, as noted in the next section, the rarity of this event is striking, under circumstances in which it might be expected more often.

Acute diabetic ketoacidosis

The striking feature of established acute renal failure in acute ketoacidosis is its extreme rarity. In several large series of patients with ketoacidosis, dating from 20 to 30 years ago (Trever and Cluff 1958; Chazan *et al.* 1969; Beigelman and Warner 1973), a total of 1285 episodes of ketoacidosis were described and only eight cases of acute renal failure were noted. In general recent reports are lacking, but Tunbridge (1979) noted two cases of acute renal failure amongst 448 patients dying of diabetes mellitus under the age of 50, and we have seen two additional cases, both in pubertal girls, one of whom sustained a femoral artery thrombosis (Cameron *et al.* 1975), although both recovered. Conversely, of 1347 cases of acute renal failure reported by Turney *et al.* (1990) over a 35-year period, only 30 occurred in diabetics, of which 19 were a result of ketoacidosis. Histological acute tubular necrosis has been reported in renal biopsies in uncomplicated ketoacidosis (Lennhoff and Herrera 1968).

Non-ketotic hyperosmolar coma has also been associated with acute renal failure (Arieff and Carroll 1972). These patients may have extreme volume depletion (up to one-quarter of total body water being lost), but again established acute renal failure is very rare. Previous diminution of renal function was frequent in the few cases of acute renal failure which occurred, that is, they represented a variety of 'acute on chronic' renal failure.

The reasons for the extreme rarity of acute renal failure in comatose, volume depleted, hypokalaemic, acidotic, sometimes infected individuals is not known, but it is of course tempting to suggest that the osmotic diuresis present throughout the episode protects the tubules, as mannitol may (see Chapter 8.4).

Acute papillary necrosis and urinary infection

As discussed above, diabetics may occasionally go into acute renal failure because of acute severe diffuse pyelonephritis, often accompanied by the sloughing of papillae with ureteric blockage. This is a serious state which requires urgent intervention to provide urinary drainage by percutaneous nephrostomy.

Large vessel disease

Diabetic patients are prone to severe atheroma of major vessels, including the aorta and the origins of the real arteries. This is a cause of chronic renal failure, especially in older maturity onset diabetics, but the final occlusion of a solitary functioning kidney may present as acute renal failure. These patients are also vulnerable to the deleterious side-effects of angiotensin converting enzyme inhibitors, whose use must be carefully considered in the older arteriopathic diabetic.

Other glomerulopathies

Both acute crescentic glomerulonephritis and acute endocapillary glomerulonephritis with oligoanuria have been reported in diabetic subjects (Olivero and Suki 1977; Carstens *et al.* 1982). The prognosis and management does not differ from similar nephritides in non-diabetic subjects.

References

Abdullah, M.S. (1978). Diabetic nephropathy in Kenya. *East Africa Medical Journal*, **55**, 513–8.

Adetuyibi, A. (1976). Diabetes in the Nigerian African: 1. Review of long-term complications. *Tropical and Geographical Medicine*, **28**, 155–9.

Allawi, J., and Jarrett, R.J. (1990). Microalbuminuria and cardiovascular risk factors in Type 2 diabetes mellitus. *Diabetic Medicine*, **7**, 115–8.

Allawi, J., et al. (1988). Microalbuminuria in non-insulin-dependent diabetes: its prevalence in Indian compared with Europid patients. *British Medical Journal*, **296**, 462–4.

Amiel, S.A., Sherwin, R.S., Simonson, D.C., Lauritano, A.A., and Tamborlane, W.V. (1986). Impaired insulin action in puberty. A contributing factor to poor glycaemic control in adolescents with diabetes. *New England Journal of Medicine*, **315**, 215–19.

Amoah, E., Glickman, J.L., Malchoff, C.D., Sturgill, B.C., Kaiser, D.L., and Bolton, W.K. (1988). Clinical identification of non-diabetic renal disease in diabetic patients with type 1 and type 2 disease presenting with renal dysfunction. *American Journal of Nephrology*, **8**, 204–11.

Andersen, A.R., Christiansen, J.S., Andersen, J.K., Kreiner, S., and Deckert, T. (1983). Diabetic nephropathy in type 1 (insulin-dependent) diabetes: an epidemiological study. *Diabetologia*, **25**, 496–501.

Anderson, S., Meyer, T.W., Rennke, H.G., and Brenner, B.M. (1985). Control of glomerular hypertension limits glomerular injury in rats with reduced renal mass. *Journal of Clinical Investigation*, **76**, 612–9.

Anderson, S., Rennke, H.G., and Brenner, B.M. (1986). Therapeutic advantages of converting-enzyme inhibitors in arresting progressive renal disease associated with systemic hypertension in the rat. *Journal of Clinical Investigation*, **77**, 1925–30.

Arieff, A.I., and Carroll, H.J. (1972). Non-ketotic hyperosmolar coma with hyperglycaemia: clinical features, pathophysiology, renal function, acid base balance, plasma-cerebrospinal fluid equilibrium and the effects of therapy in 37 cases. *Medicine*, **51**, 73–94.

Atherton, A., Hill, D.W., Keen, H., Young, S., and Edwards, E.J. (1980). The effect of acute hyperglycaemia on the retinal circulation of the normal cat. *Diabetologia*, **18**, 233–7.

Attman, P.O., Bucht, H., Larsson, O., and Uddebom, G. (1983). Protein reduced diet in diabetic renal failure. *Clinical Nephrology*, **19**, 217–20.

Ausiello, D.A., Kreisberg, J.I., Roy, C., and Karnovsky, M.J. (1980). Contraction of cultured rat glomerular cells of apparent mesangial origin after stimulation with angiotensin II and arginine vasopressin. *Journal of Clinical Investigation*, **65**, 754–60.

Axelroth, L. (1975). Response of congestive heart failure to correction of hyperglycaemia in the presence of diabetic nephropathy. *New England Journal of Medicine*, **293**, 1243–5.

Ballard, D.J., et al. (1988). Epidemiology of persistent proteinuria in type 2 diabetes mellitus. Population based study in Rochester, Minnesota. *Diabetes*, **37**, 405–12.

Ballerman, B.J., Skorecki, K.L., and Brenner, B.M. (1984). Reduced glomerular angiotensin II receptor density in early untreated diabetes mellitus in the rat. *American Journal of Physiology*, **247**, F110–F116.

Bank, N., Klose, R., Aynedjian, H.S., Nguyen, D., and Sablay, L.B. (1987). Evidence against increased glomerular pressure initiating diabetic nephropathy. *Kidney International*, **31**, 898–905.

Barsotti, G., Morelli, E., Giannoni, A., Guiducci, A., Lupetti, S., and Giovannetti, S. (1983). Restricted phosphorus and nitrogen intake to slow the progression of chronic renal failure: a controlled trial. *Kidney International*, **24(S16)**, S278–S284.

Becker, D., and Miller, M. (1960). Presence of diabetic glomerulosclerosis in patients with haemochromatosis. *New England Journal of Medicine*, **263**, 367–73.

Beigelman, P.N., and Warner, N.E. (1973). Thirty two fatal cases of severe diabetic ketoacidosis, including a case of mucormycosis. *Diabetes*, **22**, 847–50.

Beisswenger, P.J. (1976). Glomerular basement membrane. Biosynthesis and chemical composition in the streptozotocin diabetic rat. *Journal of Clinical Investigation*, **58**, 844–52.

Beisswenger, P.J., and Spiro, R.G. (1972). Studies on the human glomerular basement membrane: composition, nature of the carbohydrate units and chemical changes in diabetes mellitus. *Diabetes*, **22**, 180–93.

Bell, E.T. (1953). Renal vascular disease in diabetes mellitus. *Diabetes*, **2**, 376–89.

Bending, J.J., Viberti, G.C., Bilous, R.W., and Keen, H. (1985). Eight-month correction of hyperglycaemia in IDDM is associated with a significant and sustained reduction of urinary albumin excretion rates in patients with microalbuminuria. *Diabetes*, **34 (Supp. 3)**, 69–73.

Bennett, D.M. (1989). 'Microalbuminuria' and diabetes: a critique. Assessment of urinary albumin excretion and its role in screening for diabetic nephropathy. *American Journal of Kidney Diseases*, **13**, 29–34.

Bending, J.J., Viberti, G.C., Watkins, P.J., and Keen, H. (1986). Intermittent clinical proteinuria and renal function in diabetes: evolution and the effect of glycaemic control. *British Medical Journal*, **292**, 83–6.

Bending, J.J., Dodds, R.A., Keen, H., and Viberti, G.C. (1988). Renal response to restricted protein intake in diabetic nephropathy. *Diabetes*, **37**, 1641–6.

Bergman, L.A., Ellison, M.R., and Dunea, G. (1968). Acute renal failure after drip-infusion pyelography. *New England Journal of Medicine*, **279**, 1277.

Bergstrom, J. (1984). Discovery and rediscovery of low-protein diet. *Clinical Nephrology*, **21**, 29–35.

Berkman, J., and Rifkin, H. (1973). Unilateral nodular diabetic glomerulosclerosis (Kimmelstiel-Wilson). Report of a case. *Metabolism*, **22**, 715–22.

Béroniade, V.C., Lefèbvre, R., and Falardeau, P. (1987). Unilateral diabetic glomerulosclerosis: recurrence of an experiment of nature. *American Journal of Nephrology*, **7**, 55–9.

Bertani, T., Olesnicky, L., Abu-Regiaba, S., Glasberg, S., and Pirani, C.L. (1983). Concomitant presence of three different glomerular diseases in the same patient. *Nephron*, **34**, 260–6.

Bertolatus, J.A., Abuyousef, M., and Hunsicker, L.G. (1987). Glomerular sieving of high molecular weight proteins in proteinuric rats. *Kidney International*, **31**, 1257–66.

Bevan, J.S., Griffiths, G.J., Williams, J.D., and Gibby, O.M. (1989). Bilateral renal cortical abscesses in a young woman with type 1 diabetes. *Diabetic Medicine*, **6**, 454–7.

Beyer-Mears, A., Cruz, E., Dillon, P., Tanis, D., and Roche, M. (1983). Diabetic renal hypertrophy diminished by aldose reductase inhibition. *Federation Proceedings*, **42**, 505.

Beyer-Mears, A., Ku, L., and Cohen, M.P. (1984). Glomerular polyol accumulation in diabetes and its prevention by oral sorbinil. *Diabetes*, **33**, 604–7.

Beyer-Mears, A., Cruz, E., Edelist, T., and Varagianuis, E. (1986). Diminished proteinuria in diabetes mellitus by sorbinil, an aldose reductase inhibitor. *Pharmacology*, **32**, 52–60.

Bilous, R.W., Viberti, G.C., Christiansen, J.S., Parving, H.-H., and Keen, H. (1985). Dissociation of diabetic complications in insulin-dependent diabetes: a clinical report. *Diabetic Nephropathy*, **4**, 73–6.

Bjørck, S., Nyberg, G., Mulec, H., Granerus, G., Herlitz, H., and Aurell, M. (1986). Beneficial effects of angiotensin converting enzyme inhibition on renal function in patients with diabetic nephropathy. *British Medical Journal*, **293**, 471–4.

Blagg, C.R., Tenckhoff, H., and Scribner, B.H. (1974). Neuropathy in dialysed diabetic patients. *Kidney International*, **6(S1)**, S86–S89.

Blantz, R.C., Peterson, O.W., Gushwa, L., and Tucker, B.J. (1982). Effect of modest hyperglycaemia on tubuloglomerular feedback activity. *Kidney International*, **22(S12)**, S206–S212.

Blohmé, G. and Smith, U. (1989). Aldose reductase inhibition reduces urinary albumin excretion rate in incipient diabetic nephropathy. *Diabetologia*, **32**, 467A.

Bloodworth, J.M.B. (1978). A re-evaluation of diabetic glomerulosclerosis 50 years after the discovery of insulin. *Human Pathology*, **9**, 439–453.

Boeri, D., Almus, F.E., Maiello, M., Cagliero, E., Rao, L.V.M., and Lorenzi, M. (1989). Modification of tissue-factor mRNA and protein response to thombin and interleukin 1 by high glucose in cultured human endothelial cells. *Diabetes*, **38**, 312–18.

Boerwinkle, E., Turner, S.T., Weinshilboum, R., Johnson, M., Richelson, E., and Sing, C.F. (1986). Analysis of the distribution of sodium lithium countertransport in a sample representative of the general population. *Genetic Epidemiology*, **3**, 365–78.

Borch-Johnsen, K., and Kreiner, S. (1987). Proteinuria: value as predictor of cardiovascular mortality in insulin-dependent diabetes. *British Medical Journal*, **294**, 1651–4.

Borch-Johnsen, K., Andersen, P.K., and Deckert, T. (1985a). The effect of proteinuria on relative mortality in type 1 (insulin-dependent) diabetes mellitus. *Diabetologia*, **28**, 590–6.

Borch-Johnsen, K., Nissen, R.N., and Nerup, J. (1985b). Blood pressure after 40 years of diabetes. *Diabetic Nephropathy*, **4**, 11–12.

Bova, J.G., Potter, J.L., Arevalos, E., Hopens, T., Goldstein, H.M., and Radwin, H.M. (1985). Renal and perirenal infection: the role of computerized tomography. *Journal of Urology*, **133**, 539–43.

Bradley, W.E. (1980). Diagnosis of urinary bladder dysfunction in diabetes mellitus. *Annals of Internal Medicine*, **92**, 323–6.

Brenner, B.M. (1985). Nephron adaptation to renal injury or ablation. *American Journal of Physiology*, **249**, F234–F337.

Brenner, B.M., Hostetter, T.H., and Humes, H.D. (1978). Molecular basis of proteinuria of glomerular origin. *New England Journal of Medicine*, **298**, 826–33.

Brezis, M., and Epstein, F.H. (1989). A closer look at radiocontrast-induced nephropathy. *New England Journal of Medicine*, **320**, 179–81.

Bright, R. (1836). Cases and observations illustrative of renal disease accompanied with the secretion of albuminous urine. *Guy's Hospital Reports*, **1**, 338–400.

Brøchner-Mortensen, J. (1973). The glomerular filtration rate during moderate hyperglycaemia in normal man. *Acta Medica Scandinavica*, **194**, 31–7.

Brøchner-Mortensen, J., and Ditzel, J. (1982). Glomerular filtration rate and extracellular fluid volume in insulin-dependent patients with diabetes mellitus. *Kidney International*, **21**, 696–8.

Brown, D.M., Andres, G.A., Hostetter, T.H., Mauer, S.M., Price, R., and Venkatachalam, M.A. (1982). Proceedings of a task force on animals appropriate for studying diabetes mellitus and its complications. Kidney complications. *Diabetes*, **31 (Suppl. 1)**, 71–81.

Brownlee, M., and Spiro, R.G. (1979). Biochemistry of the basement membrane in diabetes mellitus. *Advances in Experimental Medicine and Biology*, **124**, 141–56.

Brownlee, M., Pongor, S., and Cerami, A. (1983a). Covalent attachment of soluble proteins by nonenzymatically glycosylated collagen: role in the in situ formation of immune complexes. *Journal of Experimental Medicine*, **158**, 1739–44.

Brownlee, M., Vlassara, H., and Cerami, A. (1983b). Nonenzymatic glycosylation reduces the susceptibility of fibrin to degradation by plasmin. *Diabetes*, **32**, 680–4.

Brownlee, M., Vlassara, H., and Cerami, A. (1984). Non-enzymatic glycosylation and the pathogenesis of diabetic complications. *Annals of Internal Medicine*, **101**, 527–37.

Brownlee, M., Cerami, A., and Vlassara, H. (1988). Advanced glycosylation end-products in tissue and the biochemical basis of diabetic complications. *New England Journal of Medicine*, **318**, 1315–21.

Bunker, C.H., and Mallinger, A.G. (1985). Sodium-lithium countertransport, obesity, insulin and blood pressure in healthy premenopausal women. *Circulation*, **72(S3)**, III–296.

Burg, M.B. (1988). Role of aldose reductase and sorbitol in maintaining the medullary intracellular milieu. *Kidney International*, **33**, 635–41.

Buschard, K., Hougaard, P., Mølsted-Pedersen, L., and Kühl, C. (1990). Type 1 (insulin-dependent) diabetes mellitus diagnosed during pregnancy: a clinical and prognostic study. *Diabetologia*, **33**, 31–5.

Cagliero, E., Maiello, M., Boeri, D., Roy, S., and Lorenzi, M. (1988). Increased expression of basement membrane components in human endothelial cells cultured in high glucose. *Journal of Clinical Investigation*, **82**, 735–8.

Cambier, P. (1934). Application de la théorie de Rehberg à l'étude clinique des affections rénales et du diabète. *Annales de Médecine*, **35**, 273–99.

Cameron, J.S. (1979). The natural history of glomerulonephritis. In *Progress in glomerulonephritis*, (ed. P. Kincaid-Smith, A.J. d'Apice, and R. C. Atkins). pp. 1–25, Wiley Medical Publisher, New York.

Cameron, J.S., Ireland, J.T., and Watkins, P.J. (1975). The kidney and renal tract. In *Complications of diabetes*, (ed. H. Keen and J. Jarrett). pp. 99–151, Edward Arnold, London.

Campbell, I.W., *et al.* (1977). Gastric emptying in diabetic autonomic neuropathy. *Gut*, **18**, 462–7.

Carstens, S.A., Lee, A.H., Garancis, J.C., Piering, W.F., and Leman, J. (1982). Rapidly progressive glomerulonephritis superimposed on diabetic glomerulosclerosis. *Journal of the American Medical Association*, **247**, 1453–7.

Castellot, J.J., Hoover, R.L., Harper, P.A., and Karnovsky, M.J. (1985). Heparin and glomerular epithelial cell-secreted heparin-like species inhibit mesangial cell proliferation. *American Journal of Pathology*, **120**, 427–35.

Chachati, A., von Frenckell, R., Foidart-Willems, J., Godon, J.P., and LeFèbvre, P.J. (1987). Variability of albumin excretion in insulin-dependent diabetics. *Diabetic Medicine*, **4**, 441–5.

Chavers, B.M., Bilous, R.W., Ellis, E.N., Steffes, M.W., and Mauer, S.M. (1989). Glomerular lesions and urinary albumin excretion in type 1 diabetes without overt proteinuria. *New England Journal of Medicine*, **320**, 966–70.

Chazan, B.I., Rees, S.B., Balodimos, M.C., Younger, D., and Ferguson, B.D. (1969). Dialysis in diabetics. A review of 44 patients. *Journal of the American Medical Association*, **209**, 2026–30.

Chihara, J., Takebayashi, S., Taguchi, T., Yokoyama, K., Harada, T., and Naito, S. (1986). Glomerulonephritis in diabetic patients and its effect on prognosis. *Nephron*, **43**, 45–9.

Christensen, C.K., Christiansen, J.S., Christensen, T., Hermansen, K., and Mogensen, C.E. (1986). The effect of six months continuous subcutaneous insulin infusion on kidney function and size in insulin-dependent diabetics. *Diabetic Medicine*, **3**, 29–32.

Christiansen, J.S. (1984). On the pathogenesis of the increased glomerular filtration rate in short-term insulin-dependent diabetics. Thesis, Copenhagen, Laegeforeningens Forslag.

Christiansen, J.S., Gammelgaard, J., Frandsen, M., and Parving, H.-H. (1981a). Increased kidney size, glomeruar filtration rate and renal plasma flow in short-term insulin-dependent diabetics. *Diabetologia*, **20**, 451–6.

Christiansen, J.S., Frandsen, M., and Parving, H.-H. (1981b). Effect of intravenous glucose infusion on renal function in normal man and in insulin-dependent diabetics. *Diabetologia*, **21**, 368–73.

Christiansen, J.S., Frandsen, M., and Parving, H.-H. (1981c). The effect of intravenous insulin infusion on kidney function in insulin-dependent diabetes mellitus. *Diabetologia*, **20**, 199–204.

Christiansen, J.S., Gammelgaard, J., Ørskov, H., Andersen, A.R., Telmer, S., and Parving, H.-H. (1981d). Kidney function and size in normal subjects before and during growth hormone administration for one week. *European Journal of Clinical Investigation*, **11**, 487–90.

Christiansen, J.S., Gammelgaard, J., Tronier, B., Svendsen, P.A., and Parving, H.-H. (1982a). Kidney function and size in diabetics before and during initial insulin treatment. *Kidney International*, **21**, 683–8.

Christiansen, J.S., Gammelgaard, J., Frandsen, M., Ørskov, H., and Parving, H.-H. (1982b). Kidney function and size in type 1 (insulin-dependent) diabetic patients before and during growth hormone administration for one week. *Diabetologia*, **22**, 333–7.

Christiansen, J.S., Feldt-Rasmussen, B., and Parving, H.-H. (1985). Short-term inhibition of prostaglandin synthesis has no effect on the elevated glomerular filtration rate of early insulin-dependent diabetes. *Diabetic Medicine*, **2**, 17–20.

Close, C.F., on behalf of the MSC group. (1987). Sex, diabetes duration and microalbuminuria in type 1 (insulin-dependent) diabetes mellitus. *Diabetologia*, **30**, 508A.

Coco, M., Aynedjan, H.S., and Bank, N. (1988). Effect of galactose and aldose reductase inhibition on renal haemodynamics. *Clinical Research*, **36**, 626A.

Cogan, D.G. (1984). Aldose reductase and complications of diabetes. *Annals of Internal Medicine*, **101**, 82–91.

Coggins, C.H., and Fang, S.-T. (1988). Acute renal failure associated with antibiotics, anaesthetic agents, and radiographic contrast agents. In *Acute renal failure*, (ed. B. M. Brenner and M. G. Lazarus). pp. 319–33, Churchill Livingstone, New York.

Cohen, A.H., Mampaso, F., and Zamboni, L. (1977). Glomerular podocyte degeneration in human renal disease. An ultrastructural study. *Laboratory Investigation*, **37**, 30–4.

Cohen, D., Dodds, R., and Viberti, G.C. (1987). Effect of protein restriction in insulin-dependent diabetics at risk of nephropathy. *British Medical Journal*, **294**, 795–8.

Cohen, D.L., Close, C.F., and Viberti, G.C. (1987). The variability of overnight urinary albumin excretion in insulin-dependent diabetic and normal subjects. *Diabetic Medicine*, **4**, 437–440.

Cohen, D.L., Allawi, J., Brophy, K., Keen, H., and Viberti, G.C. (1989). Tolerance, safety and effects of Statil, an aldose reductase inhibitor in type 2 (non-insulin-dependent) diabetic patients with microalbuminuria. *Diabetologia*, **32**, 477A.

Cohen, M.P., and Ku, L. (1984). Inhibition of fibronectin binding to matrix components by non-enzymatic glycosylation. *Diabetes*, **33**, 970–4.

Cohen, M.P., and Surma, M.L. (1984). Effect of diabetes on in vivo metabolism of ^{35}S-labeled glomerular basement membrane. *Diabetes*, **33**, 8–12.

Cohen, M.P., Urdanivia, E., Surma, M., and Ciborowski, C. (1981). Non-enzymatic glycosylation of basement membranes in in vitro studies. *Diabetes*, **30**, 367–71.

Cohen, M.P., Saini, R., Klepser, H., and Vasanthi, L.G. (1987). Fibronectin binding to glomerular basement membrane is altered in diabetes. *Diabetes*, **36**, 758–63.

Collins, V.R., Dowse, G.K., Finch, C.F., Zimmet, P.Z., and Linnane, A.W. (1989). Prevalence and risk factors for micro- and macroalbuminuria in diabetic subjects and entire population of Nauru. *Diabetes*, **38**, 602–10.

Cortes, P., Dumler, F., Goldman, J., and Levin, N.W. (1987). Relationship between renal function and metabolic alterations in early streptozotocin-induced diabetes in rats. *Diabetes*, **36**, 80–7.

Corvilain, J., and Abramow, M. (1962). Some effects of human growth hormone on renal haemodynamics and on tubular phosphate transport in man. *Journal of Clinical Investigation*, **41**, 1230–5.

Cotunnius, D. (1764). De ischiade nervosa commentarius. Naples, Simonios.

Cowie, C.C., Port, F.K., Wolfe, R.A., Savage, P.J., Moll, P.P., and Hawthorne, V.M. (1989). Disparities in incidence of end-stage renal disease according to race and type of diabetes. *New England Journal of Medicine*, **321**, 1074–9.

Dachs, S., Churg, J., Mautner, W., and Grishman, E. (1964). Diabetic nephropathy. *American Journal of Pathology*, **44**, 155–68.

Dadone, M.M., Hasstedt, S.J., Hunt, S.C., Smith, J.B., Ash, K.O., and Williams, R.R. (1984). Genetic analysis of sodium-lithium countertransport in ten hypertension-prone kindreds. *American Journal of Medicine*, **17**, 565–77.

Dahl-Jørgensen, K., Hanssen, K.F., Kierulf, P., Bjøro, T., Sandvik, L., and Aagenæs, Ø. (1988). Reduction of urinary albumin excretion after 4 years of continuous subcutaneous insulin infusion in insulin-dependent diabetes mellitus. The Oslo study. *Acta Endocrinologica*, **117**, 19–25.

Damsgaard, E.M., and Mogensen, C.E. (1986). Microalbuminuria in elderly hyperglycaemic patients and controls *Diabetic Medicine*, **3**, 430–5.

Daniels, B.S., and Hostetter, T.H. (1989). Aldose reductase inhibition and glomerular abnormalities in diabetic rats. *Diabetes*, **38**, 981–6.

Danielsen, R., Helgason, T., and Jonasson, F. (1983). Prognostic factors and retinopathy in type 1 diabetics in Iceland. *Acta Medica Scandinavica*, **213**, 323–36.

Deckert, T., and Poulsen, J.E. (1968). Prognosis for juvenile diabetics with late diabetic manifestations. *Acta Medica Scandinavica*, **183**, 351–6.

Deckert, T., Poulsen, J.E., and Larsen, M. (1978). Prognosis of diabetics with diabetes onset before the age of thirty one. *Diabetologia*, **14**, 363–70.

Deckert, T., Feldt-Rasmussen, B., Djurup, R., and Deckert, M. (1988). Glomerular size and charge selectivity in insulin-dependent diabetes mellitus. *Kidney International*, **33**, 100–6.

De Cosmo, S., Ruggenenti, P., Walker, J.D., Remuzzi, G., and Viberti, G.C. (1989). Mechanisms of glucose induced glomerular haemodynamic changes in diabetic nephropathy. *Diabetologia*, **32**, 480A.

Deen, W.M., and Satvat, B. (1981). Determinants of glomerular filtration of proteins. *American Journal of Physiology*, **241**, F162–F170.

DeFronzo, R.A. (1980). Hyperkalaemia and hyporeninaemic hypoaldosteronism. *Kidney International*, **17**, 118–34.

DeFronzo, R.A., Cooke, C.R., Andres, R., Faloona, G.R., and Davis, P.J. (1975). The effect of insulin on renal handling of sodium potassium, calcium and phosphate in man. *Journal of Clinical Investigation*, **55**, 845–55.

deGroat, W.C., and Booth, A.M. (1980). Physiology of the urinary bladder and urethra. *Annals of Internal Medicine*, **92**, 312–15.

Del Castillo, E., Fuenzalida, R., and Uranga, J. (1977). Increased glomerular filtration rate and glomerulopressin activity in diabetic dogs. *Hormone and Metabolism Research*, **9**, 46–53.

Derby, L., Laffel, L.B.M., and Krolewski, A.S. (1988). Risk of diabetic nephropathy declines with age in type 1 (insulin-dependent) diabetes. *Diabetologia*, **31**, 485A.

Dicker, D., Feldberg, D., Peleg, D., Karp, M., and Goldman, J.A. (1986). Pregnancy complicated by diabetic nephropathy. *Journal of Perinatal Medicine*, **14**, 299–307.

Ditscherlein, G. (1985). Renal histopathology in hypertensive diabetic patients. *Hypertension*, **7 (Suppl. 2)**, 29–32.

Ditzel, J., and Brøchner-Mortensen, J. (1983). Tubular reabsorption rates as related to glomerular filtration in diabetic children. *Diabetes*, **32 (Suppl. 2)**, 28–33.

Ditzel, J., and Junker, K. (1972). Abnormal glomerular filtration rate, renal plasma flow, and renal protein excretion in recent and short-term diabetics. *British Medical Journal*, **2**, 13–19.

Ditzel, J., and Schwartz, M. (1967). Abnormal glomerular filtration in short-term insulin-treated diabetic subjects. *Diabetes*, **16**, 264–7.

Ditzel, J., Brøchner-Mortensen, J., and Kawahara, R. (1982). Dysfunction of tubular phosphate reabsorption related to glomerular filtration and blood glucose control in diabetic children. *Diabetologia*, **23**, 406–10.

Doi, T., *et al.* (1988). Progressive glomerular sclerosis develops in transgenic mice chronically expressing growth hormone and growth hormone releasing factor but not in those expressing insulin-like growth factor-1. *American Journal of Pathology*, **131**, 398–403.

Donadio, J.V., Ilstrup, D.M., Holley, K.E., and Romero, J.C. (1988). Platelet-inhibitor treatment of diabetic nephropathy: a 10-year prospective study. *Mayo Clinic Proceedings*, **63**, 3–15.

Drury, P.L. (1983). Diabetes and arterial hypertension. *Diabetologia*, **24**, 1–9.

Dullaart, R.P.F., Dikkeschei, L.D., and Doorenbos, H.H. (1989). Alterations in serum lipids and apolipoproteins in male type 1 (insulin-dependent) diabetic patients with microalbuminuria. *Diabetologia*, **32**, 685–9.

Dumler, F., Kumar, V., Romanski, N.M., Cortes, R., and Levin, N. (1987). Renal involvement in type 2 diabetes mellitus: a clinicopathologic study of the Henry Ford Hospital experience. *Henry Ford Hospital Medical Journal*, **35**, 221–5.

Eggers, P.W. (1988). Effect of transplantation on the Medicare end-stage renal disease program. *New England Journal of Medicine*, **318**, 223–9.

Eknoyan, G., Quinibi, W.Y., Grissom, R.T., Tuma, S.N., and Ayus, J.C. (1982). Renal papillary necrosis: an update. *Medicine*, **61**, 55–73.

Ellenberg, M. (1966). Diabetic neurogenic vesical dysfunction. *Archives of Internal Medicine*, **117**, 348–354.

Ellenberg, M. (1980). Development of urinary bladder dysfunction in diabetes mellitus. *Annals of Internal Medicine*, **92**, 321–3.

Ellis, E.N., Steffes, M.W., Gøetz, F.C., Sutherland, D.E.R., and Mauer, S.M. (1985). Relationship of renal size to nephropathy in type 1 (insulin-dependent) diabetes. *Diabetologia*, **28**, 12–15.

Ellis, E.N., Steffes, M.W., Gøetz, F.C., Sutherland, D.E.R., and Mauer, S.M. (1986). Glomerular filtration surface in type 1 diabetes mellitus. *Kidney International*, **29**, 889–94.

Ellis, E.N., Steffes, M.W., Chavers, B., and Mauer, S.M. (1987). Observations of glomerular epithelial cell structure in patients with type 1 diabetes mellitus. *Kidney International*, **32**, 736–41.

Esmatjes, E., *et al.* (1985). Renal haemodynamic abnormalities in patients with short-term insulin-dependent diabetes mellitus: role of renal prostaglandins. *Journal of Clinical Endocrinology and Metabolism*, **60**, 1231–6.

Evanoff, G.V., Thompson, C.S., Brown, J., and Weinman, E.J. (1987*a*). The effect of dietary protein restriction on the progression of diabetic nephropathy. A 12 month follow-up. *Archives of Internal Medicine*, **147**, 492–5.

Evanoff, G.V., Thompson, C.S., Foley, R., and Weinman, E.J. (1987*b*). Spectrum of gas within the kidney. Emphysematous pyelonephritis and emphysematous pyelitis. *American Journal of Medicine*, **83**, 149–54.

Ewing, D.J. and Clark, F. (1986). Autonomic neuropathy: its diagnosis and prognosis. *Clinics in Endocrinology and Metabolism*, **15**, 855–88.

Fabre, J., Balant, L.P., Dayer, P.G., Fox, H.M., and Vernet, A.T. (1982). The kidney in maturity onset diabetes mellitus: a clinical study of 510 patients. *Kidney International*, **21**, 730–8.

Fahr, T. (1942). Über Glomerulosklerose. *Virchows Archiv für Pathologische Anatomie und Physiologie und für klinische Medizin*, **309**, 16–33.

Falk, R.J., Scheinman, J.I., Mauer, S.M., and Michael, A.F. (1983). Polyantigenic expansion of basement membrane constituents in diabetic nephropathy. *Diabetes*, **32** (**Suppl. 2**), 34–9.

Fang, L.S., Sirota, R.A., Ebert, T.A., and Lichtenstein, N.S. (1980). Low fractional excretion of sodium with contrast media-induced acute renal failure. *Archives of Internal Medicine*, **140**, 531–3.

Farquhar, A., MacDonald, M.K., and Ireland, J.T. (1972). The role of fibrin deposition in diabetic glomerulosclerosis: a light, electron and immunoflourescence microscopy study. *Journal of Clinical Pathology*, **25**, 657–67.

Farquhar, M.G. (1981). The glomerular basement membrane: a selective macromolecular filter. In *Cell biology of extracellular matrix*, (ed. E. D. Hay). pp. 335–378, Plenum, New York.

Feldman, J.N., Hirsch, S.R., Beyer, M.B., James, W.A., L'Esperance, F.A., and Friedman, E.A. (1982). Prevalence of diabetic nephropathy at time of treatment for diabetic retinopathy. In *Diabetic renal-retinal syndrome*, Vol. 2, (ed. E. A. Friedman and F. A. L'Esperance). pp. 9–20, Grunne and Stratton, New York.

Feldt-Rasmussen, B. (1986). Increased transcapillary escape rate of albumin in type 1 (insulin-dependent) diabetic patients with microalbuminuria. *Diabetologia*, **29**, 282–6.

Feldt-Rasmussen, B., and Mathiesen, E.R. (1984). Variability of urinary albumin excretion in incipient diabetic nephropathy. *Diabetic Nephropathy*, **3**, 101–3.

Feldt-Rasmussen, B., Mathiesen, E.R., and Deckert, T. (1986). Effect of two years of strict metabolic control on progression of incipient nephropathy in insulin-dependent diabetes. *Lancet*, **2**, 1300–4.

Feldt-Rasmussen, B., *et al.* (1987). Central role for sodium in the pathogenesis of blood pressure changes independent of angiotensin, aldosterone and cathecolamines in type 1 (insulin-dependent) diabetes mellitus. *Diabetologia*, **30**, 610–7.

Fiaschi, E., Grassi, B., and Andres, G. (1952). La funzione renale nel diabete mellito. *Rassegna Fisiopatologica, Clinica e Terapeutica*, **4**, 373–410.

Finn, R. and Harmer, D. (1979). Etiological implications of sex ratio in glomerulonephritis. *Lancet*, **2**, 1194.

Fioretto, P., *et al.* (1990). Impaired renal response to a meat meal in insulin-dependent diabetes: role of glucagon and prostaglandins. *American Journal of Physiology*, **258**, F675–F683.

First, M.R., Patel, V.B., Pesce, R.J., Bramlage, R.J., and Pollack, V.E. (1978). Albumin excretion by the kidney. The effect of osmotic diuresis. *Nephron*, **20**, 171–5.

Fogo, A. and Ichikawa, I. (1989). Evidence for the central role of glomerular growth promoters in the development of sclerosis. *Seminars in Nephrology*, **9**, 329–42.

Fogo, A., Yoshida, Y., and Ichikawa, I. (1988). Angiotensin converting enzyme inhibitor suppresses accelerated growth of glomerular cells in vivo and in vitro. *Kidney International*, **33**, 296.

Forland, M., Thomas, V., and Shelokov, A. (1977). Urinary tract infections in patients with diabetes mellitus. *Journal of the American Medical Association*, **238**, 1924–6.

Foster, D.W. (1989). Insulin resistance—a secret killer? *New England Journal of Medicine*, **320**, 733–4.

Fox, M., Thier, S., Rosenberg, L., and Segal, S. (1964). Impaired renal tubular function induced by sugar infusion in man. *Journal of Clinical Endocrinology and Metabolism*, **24**, 1318–27.

Frey, J., Zager, P., Jackson, J., Eaton, P., and Scavini, M. (1989). Aldose reductase activity mediates renal prostaglandin production in streptozotocin diabetic rats. *Kidney International*, **35**, 292.

Frimodt-Møller, C. (1976). Diabetic cystopathy. I. A clinical study of the frequency of bladder dysfunction in diabetes. *Dannish Medical Bulletin*, **23**, 267–78.

Frimodt-Møller, C. and Mortensen, S. (1980). Treatment of diabetic cystopathy. *Annals of Internal Medicine*, **92**, 327–8.

Fuhrmann, K., Reiher, H., Semmler, K., Fischer, F., Fischer, M., and Glöckner, E. (1983). Prevention of congenital malformations in infants of insulin-dependent diabetic mothers. *Diabetes Care*, **6**, 219–23.

Gardete, L.M., Silva-Graça, A., Boavida, J.M., Cruz, M., Carreiras, F., and Nunes-Correa, J. (1986). Microalbuminuria—an early marker of developing microangiopathy. *Diabetologia*, **29**, 539A.

Gatling, W., Knight, C., Mullee, M.A., and Hill, R.D. (1988). Microalbuminuria in diabetes: a popultion study of the prevalence and an assessment of three screening tests. *Diabetic Medicine*, **5**, 343–7.

Gellman, D.D., Pirani, C.L., Soothill, J.F., Muehrcke, R.C., and Kark, R.M. (1959). Diabetic nephropathy: a clinical and pathologic study based on renal biopsies. *Medicine (Baltimore)*, **38**, 321–67.

Ghiggeri, G.M., Candiano, G., Delfino, G., and Queirolo, C. (1985). Electrical charge of serum and urinary albumin in normal and diabetic humans. *Kidney International*, **28**, 168–177.

Gibb, D.M., Tomlinson, P.A., Dalton, N.R., Turner, C., Shah, V., and Barratt, T.M. (1989). Renal tubular proteinuria and microalbuminuria in diabetic patients. *Archives of Disease in Childhood*, **64**, 129–34.

Gibberd, F.B. (1981). The neurogenic bladder. *Clinics in Obstetrics and Gynaecology*, **8**, 149–60.

Glick, A.D., Jacobson, H.R., and Haralson, M.A. (1990). Evidence for type I collagen synthesis in diabetic glomerulosclerosis. *Kidney International*, 37, 507.

Goldfarb, S., Simmons, D.A., and Kern, E.F.O. (1986). Amelioration of glomerular hyperfiltration in acute experimental diabetes mellitus by dietary myo-inositol supplementation and aldose reductase inhibition. *Transactions of the Association of American Physicians*, 99, 67–72.

Gøtzsche, O., Gundersen, H.J.G., and Østerby, R. (1981). Irreversibility of glomerular basement membrane accumulation despite reversibility of renal hypertrophy with islet transplantation in early experimental diabetes. *Diabetes*, 30, 481–5.

Grenfell, A., Brudenell, J.M., Doddridge, M.C., and Watkins, P.J. (1986). Pregnancy in diabetic women who have proteinuria. *Quarterly Journal of Medicine*, 59, 379–86.

Grenfell, A., Bewick, M., Parsons, V., Snowden, S., Taube, D., and Watkins, P.J. (1988). Non-insulin-dependent diabetes and renal replacement therapy. *Diabetic Medicine*, 5, 172–6.

Grond, J., Schilthuis, M.S., Koudstaal, J., and Elema, J. (1982). Mesangial function and glomerular sclerosis in rats after unilateral nephrectomy. *Kidney International*, 22, 338–43.

Groop, L., Laasonen, L., and Edgren, J. (1989) Renal papillary necrosis in patients with IDDM. *Diabetes Care*, 12, 198–202.

Grossman, R.A., Hamilton, R.W., Morse, B.M., Penn, A.S., and Goldberg, M. (1974). Non-traumatic rhabdomyolysis and acute renal failure. *New England Journal of Medicine*, 291, 807–11.

Gundersen, H.J.G., and Østerby, R. (1977). Glomerular size and structure in diabetes mellitus. II. Late abnormalities. *Diabetologia*, 13, 43–8.

Guthrow, C.E., Morris, M.A., Day, J.F., Thorp, S.R., and Baynes, J.W. (1979). Enhanced non-enzymatic glycosylation of human serum albumin in diabetes mellitus. *Proceedings of the National Academy of Sciences (USA)*, 76, 4258–61.

Haffner, S.M., *et al.* (1989). Proteinuria in Mexican Americans and Non-Hispanic Whites with NIDD. *Diabetes Care*, 12, 530–6.

Halliburton, I.W. and Thomson, R.Y. (1965). Chemical aspects of compensatory renal hypertrophy. *Cancer Research*, 25, 1882–7.

Hanson, U., Persson, B., and Thunell, S. (1990). Relationship between haemoglobin A_{1c} in early type 1 (insulin-dependent) diabetic pregnancy and the occurrence of spontaneous abortion and fetal malformation in Sweden. *Diabetologia*, 33, 100–4.

Hare, J.W. and White, P. (1977). Pregnancy in diabetes complicated by vascular disease. *Diabetes*, 26, 953–5.

Harkonen, S. and Kjellstrand, C.M. (1977). Exacerbation of diabetic renal failure following intravenous pyelography. *American Journal of Medicine*, 63, 939–46.

Hasslacher, C., Stech, W., Wahl, P., and Ritz, E. (1985). Blood pressure and metabolic control as risk factors for nephropathy in type 1 (insulin-dependent) diabetes. *Diabetologia*, 28, 6–11.

Hasslacher, C., Ritz, E., Tschøpe, W., Gallasch, G., and Mann, J.F.E. (1988). Hypertension in diabetes mellitus. *Kidney International*, 34(S25), S133–S137.

Hasslacher, C., Ritz, E., Wahl, P., and Michael, C. (1989). Similar risks of nephropathy in patients with type 1 or type 2 diabetes mellitus. *Nephrology Dialysis and Transplantation*, 4, 859–63.

Hatch, F.E.Jr. and Parrish, A.E. (1961). Apparent remission of a severe diabetic on developing the Kimmelstiel-Wilson syndrome. *Annals of Internal Medicine*, 54, 544–9.

Hegedüs, L., Christensen, N.J., Mogensen, C.E., and Gundersen, H.J.G. (1980). Oral glucose increases urinary albumin excretion in normal subjects but not in insulin-dependent diabetics. *Scandinavian Journal of Clinical and Laboratory Investigation*, 40, 479–82.

Heptinstall, R.H. (1983). Diabetes mellitus and gout. In *Pathology of the Kidney*, (ed. R.H. Heptinstall). (3rd edn). pp. 1397–1453, Little, Brown and Co., Boston.

Hirose, K., Tsuchida, H., Østerby, R., and Gundersen, H.J.G. (1980). A strong correlation between glomerular filtration rate and filtration surface area in diabetic kidney hyperfunction. *Laboratory Investigation*, 43, 434–7.

Hommel, E., Parving, H.-H., Mathiesen, E., Edsberg, B., Nielsen, M.D., and Giese, J. (1986). Effect of captopril on kidney function in insulin-dependent diabetic patients with nephropathy. *British Medical Journal*, 293, 467–70.

Hommel, E., Carstensen, H., Skøtt, P., Larsen, S., and Parving, H.-H. (1987a). Prevalence and causes of microscopic haematuria in type 1 (insulin-dependent) diabetic patients with persistent proteinuria. *Diabetologia*, 30, 627–30.

Hommel, E., Mathiesen, E., Arnold-Larssen, S., Edsberg, B., Olsen, U.B., and Parving, H.-H. (1987b). Effect of indomethacin on kidney function in type 1 (insulin-dependent) diabetic patients with diabetic nephropathy. *Diabetologia*, 30, 78–81.

Hommel, E., Mathiesen, E.R., Giese, J., Nielsen, M.D., Schutten, H.J., and Parving, H.-H. (1989). On the pathogenesis of arterial pressure elevation early in the course of diabetic nephropathy. *Scandinavian Journal of Clinical and Laboratory Investigation*, 49, 537–44.

Hostetter, T.H., Troy, J.C., and Brenner, B.M. (1981). Glomerular haemodynamics in experimental diabetes mellitus. *Kidney International*, 19, 410–5.

Hostetter, T.H., Rennke, H.G., and Brenner, B.M. (1982). The case for intrarenal hypertension in the initiation and progression of diabetic and other glomerulopathies. *American Journal of Medicine*, 72, 375–80.

Hou, S.H., Grossman, S.D., and Madias, N.E. (1985). Pregnancy in women with renal disease and moderate renal insufficiency. *American Journal of Medicine*, 78, 185–94.

Humphrey, L.L., Ballard, D.J., Frohnert, P.P., Chu, C.P., O'Fallon, M., and Pallumbo, P.J. (1989). Chronic renal failure in non-insulin-dependent diabetes mellitus. *Annals of Internal Medicine*, 111, 788–796.

Huvos, A. and Rocha, H. (1959). Frequency of bacteriuria in patients with diabetes mellitus. A controlled study. *New England Journal of Medicine*, 261, 1213–6.

Ichikawa, I., Yoshida, Y., Fogo, A., Purkerson, M.L., and Klahr, S. (1988). Effect of heparin on the glomerular structure and function of remnant nephrons. *Kidney International*, 34, 638–44.

Ikeda, K., Kida, H., and Oshima, A. (1990). Participation of type VI collagen fibres in formation of diabetic nodular lesions. *Kidney International*, 37, 252.

Inomata, S., Nakamoto, Y., Inoue, M., Itoh, M., Ohsawa, Y., and Masamune, O. (1989). Relationship between urinary albumin excretion and renal histology in non-insulin-dependent diabetes mellitus: with reference to the clinical significance of microalbuminuria. *Journal of Diabetic Complications*, 3, 178–88.

Ireland, J.T. (1970). Diagnostic criteria in the assessment of glomerular capillary basement membrane lesions in newly diagnosed juvenile diabetics. In *Early Diabetes*, (ed. R. A. Cameirini-Davalos and H. S. Cole). pp. 273–5, Academic Press, New York.

Ireland, J.T., Patnaik, B.K., and Duncan, L.J.P. (1967a). Glomerular ultrastructure in secondary diabetics and normal subjects. *Diabetes*, 16, 628–35.

Ireland, J.T., Patnaik, B.K., and Duncan, L.J.P. (1967b). Effect of pituitary ablation on the renal arteriolar and glomerular lesions in diabetes. *Diabetes*, 16, 636–42.

Ireland, J.T., Viberti, G.C., and Watkins, P.J. (1982). The kidney and renal tract. In *Complications of diabetes*, (2nd edn), (ed. H. Keen and J. Jarrett). pp. 137–78, Edward Arnold, London.

Jacobson, M., Sharma, Y.R., Cotlier, E., and Hollander, J.D. (1983) Diabetic complications in lens and nerve and their prevention by sulindac or sorbinil: two novel aldose reductase inhibitors. *Investigative Ophthalmology and Visual Science*, 24, 1426–9.

Jarrett, R.J., Verma, N.P., and Keen, H. (1976). Urinary albumin excretion in normal and diabetic subjects. *Clinica Chimica Acta*, 71, 55–9.

Jarrett, R.J., Keen, H., and Chakrabarthi, R. (1982). Diabetes, hyperglycaemia, and arterial disease. In *Complications of diabetes*, (2nd edn), (ed. H. Keen and J. Jarrett). pp. 179–204, Edward Arnold, London.

Jarrett, R.J., Viberti, G.C., Argyropoulos, A., Hill, R.D., Mahmud, U., and Murrel, T.J. (1984). Microalbuminuria predicts mortality in non-insulin-dependent diabetes. *Diabetic Medicine*, 1, 17–9.

Jensen, P.K., Christiansen, J.S., Steven, K., and Parving, H.-H. (1981). Renal function in streptozocin-diabetic rats. *Diabetologia*, 21, 409–14.

Jensen, P.K., Steven, K., Blæhr, H., Christiansen, J.S., and Parving, H.-H. (1986). Effects of indomethacin on glomerular haemodynamics in experimental diabetes. *Kidney International*, 29, 490–5.

Jensen, T. (1989). Increased plasma level of von Willebrand factor in type 1 (insulin-dependent) diabetic patients with incipient nephropathy. *British Medical Journal*, 298, 27–8.

Jensen, T., Borch-Johnsen, K., Kofoed-Enevoldsen, A., and Deckert, T. (1987a). Coronary heart disease in young type 1 (insulin-dependent) diabetic patients with and without diabetic nephropathy: incidence and risk factors. *Diabetologia*, 30, 144–8.

Jensen, T., Borch-Johnsen, K., and Deckert, T. (1987b). Changes in blood pressure and renal function in patients with type 1 (insulin-dependent) diabetes mellitus prior to clinical diabetic nephropathy. *Diabetes Research*, 4, 159–62.

Jensen, T., Stender, S., and Deckert, T. (1988). Abnormalities in plasma concentrations of lipoproteins and fibrinogen in type 1 (insulin-dependent) diabetic patients with increased albumin excretion. *Diabetologia*, 31, 142–5.

Jensen, T., Feldt-Rasmussen, B., Bjerre-Knudsen, J., and Deckert, T. (1989). Features of endothelial dysfunction in early diabetic nephropathy. *Lancet*, 1, 461–3.

Jerums, J., Cooper, M.E., Seeman, E., Murray, R.M.L., and McNeil, J.J. (1987). Spectrum of proteinuria in type 1 and type 2 diabetes. *Diabetes Care*, 10, 419–27.

Joint Working Party on Diabetic Renal Failure. (1988). Renal failure in diabetics in the UK: deficient provision of care in 1985. *Diabetic Medicine*, 5, 79–84.

Joint Working Party on Diabetic Renal Failure. (1989). Treatment and mortality of diabetic renal failure patients identified in the 1985 UK survey. *British Medical Journal*, 299, 1135–6.

Jones, R.H., Hayakawa, H., Mackay, J.D., Parsons, V., and Watkins, P.J. (1979). Progression of diabetic nephropathy. *Lancet*, 1, 1105–6.

Jones, S.L., Wiseman, M.J., Viberti, G.C., and Keen, H. (1987). Glomerular hyperfiltration and albuminuria—a 5 year prospective study in type 1 (insulin-dependent) diabetes mellitus. *Diabetologia*, 30, 536A.

Jones, S.L., Perico, N., Benigni, A., Remuzzi, G., and Viberti, G.C. (1988). Glomerular filtration rate, extracellular fluid volume and atrial natriuretic factor in insulin-dependent diabetics. *Kidney International*, 33, 268.

Jones, S.L., Close, C.E., Mattock, M.B., Jarrett, R.J., Keen, H., and Viberti, G.C. (1989). Plasma lipid and coagulation factor concentrations in insulin dependent diabetics with microalbuminuria. *British Medical Journal*, 298, 487–90.

Jones, S.L., *et al.* (1990). Sodium-lithium countertransport in microalbuminuric insulin-dependent diabetic patients. *Hypertension*, 15, 570–75.

Jovanovic, R. and Jovanovic, L. (1984). Obstetric management when normoglycaemia is maintained in diabetic pregnant women with vascular compromise. *American Journal of Obstetrics and Gynaecology*, 149, 617–23.

Kador, P.F., Robinson, W.G.Jr., and Kinoshita, J.H. (1985). The pharmacology of aldose reductase inhibitors. *Annual Review of Pharmacology and Toxicology*, 25, 691–714.

Kahan, M., Goldberg, P.D., and Mandell, E.E. (1970). Neurogenic vesical dysfunction and diabetes mellitus. *New York State Journal of Medicine*, 2, 2448–55.

Kahn, C.B., Paman, P.G., and Zic, Z. (1974). Kidney size in diabetes mellitus. *Diabetes*, 23, 788–92.

Kamenetzky, S.A., Bennet, P., Dippe, S.E., Miller, M., and LeCompte, P.M. (1974). A clinical and histologic study of diabetic nephropathy in the Pima Indians. *Diabetes*, 23, 61–8.

Kanwar, Y.S., Rosenzweig, L.J., Linker, A., and Jakubowski, M.L. (1983). Decreased de novo synthesis of glomerular proteoglycans in diabetes: biochemical and autoradiographic evidence. *Proceedings of the National Academy of Sciences (USA)*, 80, 2272–5.

Kasinath, B.S., Musais, S.K., Spargo, B.H., and Katz, A.I. (1983). Non diabetic renal disease in patients with diabetes mellitus. *American Journal of Medicine*, 75, 613–7.

Kasiske, B.L., O'Donnell, M.P., and Keane, W.F. (1985). Glucose induced increases in renal haemodynamic function. Possible modulation by renal prostaglandins. *Diabetes*, 34, 360–4.

Kasiske, B.L., O'Donnell, M.P., Cleary, M.P., and Keane, W.F. (1988). Treatment of hyperlipidaemia reduces glomerular injury in obese Zucker rats. *Kidney International*, 33, 667–72.

Katz, A.I., Davison, J.M., Hayslett, J.P., Singson, E., and Lindheimer, M.D. (1980). Pregnancy in women with kidney disease. *Kidney International*, 18, 192–206.

Keen, H., Chlouverakis, C., Fuller, J.H., and Jarrett, R.J. (1969). The concomitants of raised blood sugar: studies in newly detected hyperglycaemics. II. Urinary albumin excretion, blood pressure and their relation to blood sugar levels, *Guy's Hospital Reports*, 118, 247–52.

Keen, H., Track, N.S., Sowry, G.S.C. (1975). Arterial pressure in clinically apparent diabetics. *Diabetic Metabolisme*, 1, 159–78.

Kefalides, N.A. (1974). Biochemical properties of human glomerular basement membrane in normal and diabetic kidneys. *Journal of Clinical Investigation*, 53, 403–7.

Khalifa, A., and Cohen, M.P. (1975). Glomerular protocollagen lysil hydroxylase activity in streptozotocin diabetes. *Biochimica Biophysica Acta*, 386, 332–9.

Kikkawa, R., Umemura, K., Haneda, M., Arimura, T., Ebata, K., and Shigeta, Y. (1987). Evidence for existence of polyol pathway in cultured rat mesangial cells. *Diabetes*, 36, 240–3.

Kimmelstiel, P., and Wilson, C. (1936). Intercapillary lesions in glomeruli of kidney. *American Journal of Pathology*, 12, 83–97.

Kimmelstiel, P., Kim, O.J., and Beres, J. (1962). Studies on renal biopsies specimens with the aid of the electron microscope. I. Glomeruli in diabetes. *American Journal of Pathology*, 38, 270–7.

Kimmelstiel, P., Osawa, G., and Beres, J. (1966). Glomerular basement membrane in diabetics. *American Journal of Pathology*, 45, 21–31.

Kincaid-Smith, P., and Whitworth, J.A. (1988). Haematuria and diabetic nephropathy. In *The Kidney and hypertension in diabetes mellitus*, (ed. C. E. Mogensen). pp. 81–9, Martinus Nijhoff, Boston.

Kitzmiller, J.L., *et al.* (1981). Diabetic nephropathy and perinatal outcome. *American Journal of Obstetrics and Gynaecology*, 141, 741–5.

Klein, R., Klein, B.E.K., Moss, S.E., Davis, M.D., and DeMets, D.L. (1986). The Wisconsin epidemiology study of diabetic retinopathy: proteinuria and retinopathy in a population of diabetic persons diagnosed prior to 30 years of age. In *Diabetic renal-retinal syndrome*, Vol. 3, (ed. E. A. Friedman and F. A. L'Esperance). pp. 245–64, Grune and Straton, New York.

Klein, R., Klein, B.E.K., Moss, S., and DeMets, D.L. (1988). Proteinuria in diabetes. *Archives of Internal Medicine*, 148, 181–6.

Kleinman, K.S., and Fine, L.G. (1988). Prognostic implications of renal hypertrophy in diabetes mellitus. *Diabetes/Metabolism Reviews*, 4, 179–89.

Knowler, W.C., Bennett, P.H., and Nelson, R.G. (1988). Prediabetic blood pressure predicts albuminuria after development of NIDDM. *Diabetes*, 37 (**Supp. 1**), 120A.

Kobayashi, K., *et al.* (1981). Idiopathic membranous glomerulonephritis associated with diabetes mellitus. *Nephron*, 28, 163–8.

Kofoed–Enevoldsen, A., Borch-Johnsen, K., Kreiner, S., Nerup, J., and Deckert, T. (1987). Declining incidence of persistent proteinuria in type 1 (insulin-dependent) diabetic patients in Denmark. *Diabetes*, **36**, 205–9.

Koivisto, V.A., Huttunen, N.P., and Vierikko, P. (1981). Continuous subcutaneous insulin infusion corrects exercised-induced albuminuria in juvenile diabetes. *British Medical Journal*, **282**, 778–9.

Kokko, J.P. (1973). Proximal tubule potential difference. Dependence on glucose, HCO_3, and amino-acids. *Journal of Clincal Investigation*, **52**, 1362–7.

Kontessis, P.S., Jones, S., Dodds, R.A., Trevisan, R., Nosadini, R., Fioretto, P., Borsato, M., Sacerdoti, D., and Viberti, G.C. (1990). Renal, metabolic and hormonal responses to ingestion of animal and vegetable proteins. *Kidney International*, **38**, 136–44.

Kostraba, J.N., *et al*, (1989). Contribution of diabetes duration before puberty to development of microvascular complications in IDDM subjects. *Diabetes Care*, **12**, 686–93.

Kroc Collaborative Study Group. (1984). Blood glucose control and the evolution of diabetic retinopathy and albuminuria. A preliminary multicentre trial. *New England Journal of Medicine*, **311**, 365–72.

Krolewski, A.S., Warram, J.H., Chriestlieb, A.R., Busick, E.J., and Kahn, C.R. (1985). The changing natural history of nephropathy in type 1 diabetes. *American Journal of Medicine*, **78**, 785–94.

Krolewski, A.S., *et al*. (1987a). Magnitude and determinants of coronary artery disease in juvenile-onset, insulin-dependent diabetes mellitus. *American Journal of Cardiology*, **59**, 750–5.

Krolewski, A.S., Warram, J.H., Rand, L.I., and Kahn, C.R. (1987b). Epidemiologic approach to the etiology of type 1 diabetes mellitus and its complications. *New England Journal of Medicine*, **317**, 1390–8.

Krolewski, A.S., *et al*. (1988). Predisposition to hypertension and susceptibility to renal disease in insulin-dependent diabetes mellitus. *New England Journal of Medicine*, **318**, 140–5.

Kroustrup, J.P., Gundersen, H.J.G., and Østerby, R. (1977). Glomerular size and structure in diabetes mellitus. III. Early enlargement of the capillary surface. *Diabetologia*, **13**, 207–10.

Kunzelman, C.L., Knowler, W.C., Pettitt, D.J., and Bennett, P.H. (1989). Incidence of proteinuria in type 2 diabetes mellitus in the Pima Indians. *Kidney International*, **35**, 681–7.

Laffel, L.M.B., Krolewski, A.S., Rand, L.I., Warram, J.H., Chriestlieb, A.R., and D'Elia, J.A. (1987). The impact of blood pressure on renal function in insulin-dependent diabetes. *Kidney International*, **31**, 207.

Lauler, D.P., Schreiner, G.E., and David, A. (1960). Renal medullary necrosis. *American Journal of Medicine*, **29**, 132–56.

Lee, C.S., Mauer, S.M., Brown, D.M., Sutherland, D.E.R., Michael, A.F., and Najarian, J.S. (1974). Renal transplantation in diabetes mellitus in the rat. *Journal of Experimental Medicine*, **139**, 793–800.

Lennhoff, M., and Herrera, J. (1968). Acute renal failure in diabetic acidosis. *Lancet*, **1**, 758.

Lervang, H.-H., Jensen, S., Brøchner-Mortensen, J., and Ditzel, J. (1988). Early glomerular hyperfiltration and the development of late nephropathy in type 1 (insulin-dependent) diabetes mellitus. *Diabetologia*, **31**, 723–9.

Leung, D.Y.M., Glagov, S., and Mathews, M.B. (1976). Cyclic stretching stimulates synthesis of matrix components by arterial smooth muscle cells in vitro. *Science*, **191**, 475–7.

Leyssac, P.P. (1976). The renin angiotensin system and kidney function. A review of contributions to a new theory. *Acta Physiologica Scandinavica*, **442(S)**, 1–52.

Li, L.K., Trevisan, R., Walker, J.D., and Viberti, G.C. (1990). Overactivity of Na^+/H^+ antiport and enhanced cell growth in fibroblasts of type 1 (insulin-dependent) diabetics with nephropathy. *Kidney International*, **37**, 199.

Loon, N., Nelson, R., and Myers, B.D. (1990). Glomerular barrier abnormality in new onset NIDDM in Pima Indians. *Kidney International*, **37**, 513.

Lorenzi, M., Cagliero, E., and Toledo, S. (1985). Glucose toxicity for human endothelial cells in culture: delayed replication, disturbed cell cycle, and accelerated death. *Diabetes*, **34**, 621–7.

Lorenzi, M., Montisano, D., Toledo, S., and Barrieux, A. (1986). High glucose induces DNA damage in cultured human endothelial cells. *Journal of Clinical Investigation*, **77**, 322–5.

Lorenzi, M., Nordberg, J., and Toledo, S. (1987a). High glucose prolongs cell-cycle traversal of cultured human endothelial cells. *Diabetes*, **36**, 1261–7.

Lorenzi, M., Toledo, S., Boss, G.R., Lane, M.J., and Montisano, D.F. (1987b). The polyol pathway and glucose-6-phosphate in human endothelial cells cultured in high glucose concentrations. *Diabetologia*, **30**, 222–7.

Lorenzi, M., Montisano, D.F., Toledo, S., and Wong, H.C.H. (1987c). Increased single strand breaks in DNA of lymphocytes from diabetic subjects. *Journal of Clinical Investigation*, **79**, 653–6.

Ludvigson, M.A., and Sorenson, R.L. (1980). Immunohistochemical localization of aldose reductase. II. Rat eye and kidney. *Diabetes*, **29**, 450–59.

Lundbaek, K., *et al*. (1970). Diabetic angiopathy and growth hormone. *Lancet*, **2**, 131–3.

Mahnensmith, R.L., and Aronson, P.S. (1985). The plasma membrane sodium-hydrogen exchanger and its role in physiological and pathological processes. *Circulation Research*, **56**, 773–88.

Mahony, D.J., Laferte, R.O., and Blais, D.J. (1977). Integral storage and voiding reflexes. Neurophysiologic concept of continence and micturition. *Urology*, **9**, 95–106.

Malins, J. (1968). Clinical diabetes mellitus. Eyre & Spottiswoode, London, 170 pp.

Mandel, E.E., (1952). Renal medullary necrosis. *American Journal of Medicine*, **13**, 322–7.

Mangili, R., Bending, J.J., Scott, G., Li, L.K., Gupta, A., and Viberti, G.C. (1988). Increased sodium-lithium countertransport activity in red cells of patients with insulin-dependent diabetes and nephropathy. *New England Journal of Medicine*, **318**, 146–50.

Marre, M., Chatellier, G., Leblanc, H., Guyene, T.T., Menard, J., and Passa, P. (1988). Prevention of diabetic nephropathy with enalapril in normotensive diabetics with microalbuminuria. *British Medical Journal*, **297**, 1092–5.

Marshall, S.M., and Alberti, K.G.M.M. (1989). Comparison of the prevalence and associated features of abnormal albumin excretion in insulin-dependent and non-insulin-dependent diabetes. *Quarterly Journal of Medicine*, **70**, 61–71.

Martin, M.M. (1953). Diabetic neuropathy. A clinical study of 125 cases. *Brain*, **76**, 594–624.

Mathiesen, E.R., Øxenboll, B., Johansen, K., Svendsen, P.A., and Deckert, T. (1984). Incipient nephropathy in type 1 (insulin-dependent) diabetes. *Diabetologia*, **26**, 406–10.

Mathiesen, E.R., Saurbrey, N., Hommel, E., and Parving, H.-H. (1986). Prevalence of microalbuminuria in children with type 1 (insulin-dependent) diabetes mellitus. *Diabetologia*, **29**, 640–3.

Mathiesen, E.R., Borch-Johnsen, K., Jense, D.V., and Deckert, T. (1989). Increased survival in patients with diabetic nephropathy. *Diabetologia*, **32**, 884–6.

Mattock, M.B., *et al*. (1988). Coronary heart disease and albumin excretion rate in type 2 (non-insulin-dependent) diabetic patients. *Diabetologia*, **31**, 82–7.

Mattock, M., *et al*. (1990). Microalbuminuria as a predictor of mortality in non-insulin-dependent diabetic patients: results from a three year prospective study. *Diabetologia*, **33**, 49A.

Mauer, S.M., Steffes, M.W., Sutherland, D.E.R., Najarian, J.S., Michael, A.F., and Brown, D.M. (1975). Studies of the rate of regression of the glomerular lesions in diabetic rats treated with pancreatic islet transplantation. *Diabetes*, **24**, 280–5.

Mauer, S.M., *et al*. (1976). Development of diabetic vascular lesions in normal kidneys transplanted into patients with diabetes mellitus. *New England Journal of Medicine*, **295**, 916–20.

Mauer, S.M., Steffes, M.W., Azar, S., Sandberg, S.K., and Brown, D.M. (1978). The effects of Goldblatt hypertension on development of the glomerular lesions of diabetes mellitus in the rat. *Diabetes*, **27**, 738–44.

Mauer, S.M., Steffes, M.W., Chern, M., and Brown, D.M. (1979). Mesangial uptake and processing of macromolecules in rats with diabetes mellitus. *Laboratory Investigation*, **41**, 401–6.

Mauer, S.M., Steffes, M.W., and Brown, D.M. (1981). The kidney in diabetes. *American Journal of Medicine*, **70**, 603–12.

Mauer, S.M., Steffes, M.W., Connet, J., Najarian, J.S., Sutherland, D.E.R., and Barbosa, J. (1983). The development of lesions in the glomerular basement membrane and mesangium after transplantation of normal kidneys into diabetic patients. *Diabetes*, **32**, 948–52.

Mauer, S.M., Steffes, M.W., Ellis, E.N., Sutherland, D.E.R., Brown, D.M., and Gøetz, F.C. (1984). Structural-functional relationships in diabetic nephropathy. *Journal of Clinical Investigation*, **74**, 1143–55.

Mauer, S.M., et al. (1989). Long-term study of normal kidneys transplanted into patients with type 1 diabetes. *Diabetes*, **38**, 516–23.

Melvin, T., Kim, Y., and Michael, A.F. (1984). Selective binding of IgG$_4$ and other negatively charged plasma proteins in normal and diabetic human kidney. *American Journal of Pathology*, **115**, 443–6.

Michels, L.D., Davidman, M., and Keane, W.F. (1981). Determinants of glomerular filtration and plasma flow in experimental diabetic rats. *Journal of Laboratory and Clinical Medicine*, **98**, 869–85.

Miller, K., and Michael, A.F. (1976). Immunopathology of renal extracellular membranes in diabetes mellitus. Specificity of tubular basement membrane immunoflourescence. *Diabetes*, **25**, 701–8.

Miodovnik, M., et al. (1988). Major malformations in infants of IDDM mothers. Vasculopathy and early first-trimester poor glycaemic control. *Diabetes Care*, **11**, 713–8.

Moel, D.I., Safirstein, R.L., McEvoy, R.C., and Hsueh, W. (1987). Effect of aspirin on experimental diabetic nephropathy. *Journal of Laboratory and Clinical Medicine*, **110**, 300–7.

Mogensen, C.E., (1971a). Glomerular filtration rate and renal plasma flow in short-term and long-term juvenile diabetes mellitus. *Scandinavian Journal of Clinical and Laboratory Investigation*, **28**, 91–100.

Mogensen, C.E., (1971b). Kidney function and glomerular permeability to macromolecules in early juvenile diabetes. *Scandinavian Journal of Clinical and Laboratory Investigation*, **28**, 79–90.

Mogensen, C.E. (1971c). Urinary albumin excretion in early and long term juvenile diabetes. *Scandinavian Journal of Clinical and Laboratory Investigation*, **28**, 183–93.

Mogensen, C.E. (1971d). Maximum tubular reabsorption capacity for glucose and renal haemodynamics during rapid hypertonic glucose infusion in normal and diabetic subjects. *Scandinavian Journal of Clinical and Laboratory Investigation*, **28**, 101–9.

Mogensen, C.E. (1976a). Renal function changes in diabetes. *Diabetes*, **25**, 872–9.

Mogensen, C.E. (1976b). Progression of nephropathy in long-term diabetics with proteinuria and effect of initial antihypertensive treatment. *Scandinavian Journal of Clinical and Laboratory Investigation*, **36**, 383–8.

Mogensen, C.E. (1984). Microalbuminura predicts clinical proteinuria and early mortality in maturity onset diabetes. *New England Journal of Medicine*, **310**, 356–60.

Mogensen, C.E. (1986). Early glomerular hyperfiltration in insulin-dependent diabetics and late nephropathy. *Scandinavian Journal of Clinical and Laboratory Investigation*, **46**, 201–6.

Mogensen, C.E., and Andersen, M.J.F. (1973). Increased kidney size and glomerular filtration rate in early juvenile diabetes. *Diabetes*, **22**, 706–12.

Mogensen, C.E., and Andersen, M.J.F. (1975). Increased kidney size and glomerular filtration rate in untreated juvenile diabetics: normalization by insulin treatment. *Diabetologia*, **11**, 221–4.

Mogensen, C.E., and Christensen, C.K. (1984). Predicting diabetic nephropathy in insulin-dependent diabetic patients. *New England Journal of Medicine*, **311**, 89–93.

Mogensen, C.E., and Vittinghus, E. (1975). Urinary albumin excretion during exercise in juvenile diabetes: a provocation test for early abnormalities. *Scandinavian Journal of Clinical and Laboratory Investigation*, **35**, 295–300.

Mogensen, C.E., Christensen, N.J., and Gundersen, H.G.J. (1978). The acute effect of insulin on renal haemodynamics and protein excretion in diabetics. *Diabetologia*, **15**, 153–7.

Mogensen, C.E., Christensen, C.K., and Vittinghus, E. (1983). The stages in diabetic renal disease with emphasis on the stage of incipient diabetic nephropathy. *Diabetes*, **32**(Suppl. 2), 64–78.

Mogensen, C.E., et al. (1986). Microalbuminuria: an early marker of renal involvement in diabetes. *Uraemia Investigation*, **9**, 85–9.

Mohan, P.S., Carter, W.G., and Spiro, R.G. (1990). Occurrence of type VI collagen in extracellular matrix of renal glomeruli and its increase in diabetes. *Diabetes*, **39**, 31–7.

Moorhead, J.F., Chan, M.K., El-Nahas, M., and Varghese, Z. (1982). Lipid nephrotoxicity in chronic progressive glomerular and tubulointerstitial disease. *Lancet*, **2**, 1309–11.

Morelli, E., Loon, N., Meyer, T., Peters, W., and Myers, B.D. (1990). Effects of converting-enzyme inhibition on barrier function in diabetic glomerulopathy. *Diabetes*, **39**, 76–82.

Mower, P., Aynedjian, H., Silverman, S., Wilkes, B., and Bank, N. (1989). Sorbinil prevents glomerular hyperperfusion in diabetic rats. *Kidney International*, **35**, 433.

Mujais, S.K. (1984). Renal papillary necrosis in diabetes mellitus. *Seminars in Nephrology*, **4**, 40–7.

Myers, B.D., Winetz, J.A., Chui, F., and Michaels, A.S. (1982). Mechanisms of proteinuria in diabetic nephropathy: a study of glomerular barrier function. *Kidney International*, **21**, 633–41.

Nakamura, Y., and Myers, B.D. (1988). Charge selectivity of proteinuria in diabetic glomerulopathy. *Diabetes*, **37**, 1202–11.

Nelson, R.G., et al. (1988a). Incidence of end-stage renal disease in type 2 (non-insulin-dependent) diabetes mellitus in Pima Indians. *Diabetologia*, **31**, 730–6.

Nelson, R.G., Pettitt, D.J., Carraher, J.M., Naird, H.R., and Knowler, W.C. (1988b). Effect of proteinuria on mortality in NIDDM. *Diabetes*, **37**, 1499–1504.

Nelson, R.G., Kunzelman, C.L., Pettitt, D.J., Saad, M.F., Bennett, P.H., and Knowler, W.C. (1989). Albuminuria in type 2 (non-insulin-dependent) diabetes mellitus and impaired glucose tolerance in Pima Indians. *Diabetologia*, **32**, 870–6.

Neugarten, J., Feiner, H.D., Schacht, R.G., Gallo, G.R., and Baldwin, D.S. (1982). Aggravation of experimental glomerulonephritis by superimposed clip hypertension. *Kidney International*, **22**, 257–63.

Ng, L.L., Simmons, D., Frigh, V., Garrido, M.C., and Bomford, J. (1990). Effect of protein kinase C modulators on the leucocyte Na^+-H^+ antiport in type 1 (insulin-dependent) diabetic subjects with albuminuria. *Diabetologia*, **33**, 278–284.

Nosadini, R., et al, (1989). Increased Na^+/H^+ countertransport activity is associated with cardiac hypertrophy and insulin resistance in hypertensive type 1 (insulin-dependent) diabetic patients. *Diabetologia*, **32**, 523A.

Nyberg, G., Blohme, G., and Norden, G. (1987). Impact of metabolic control in progression of diabetic nephropathy. *Diabetologia*, **30**, 82–6.

O'Donnell, M.P., Kasiske, B.L., Daniels, F.X., and Keane, W.F. (1986). Effects of nephron loss on glomerular haemodynamics and morphology in diabetic rats. *Diabetes*, **35**, 1011–5.

O'Donnell, M.P., Kasiske, B.L., and Keane, W.F. (1988). Glomerular haemodynamics and structural alterations in experimental diabetes mellitus. *FASEB Journal*, **2**, 2339–47.

Okuda, S., Oh, Y., Onoyama, K., Fujimi, S., and Omae, T. (1984). Autologous immune-complex nephritis in streptozotocin-induced diabetic rats. *Nephron*, **37**, 166–73.

Olivero, J., and Suki, W.N. (1977). Acute glomerulonephritis complicating diabetic nephropathy. *Archives of Internal Medicine*, **137**, 732–4.

Olson, J.L. (1984). Role of heparin as a protective agent following reduction of renal mass. *Kidney International*, **25**, 376–82.

Olson, J.L., Hostetter, T.H., Rennke, H.G., Brenner, B.M., and Venkatachalam, M.A. (1982). Altered glomerular permselectivity and progressive sclerosis following extereme ablation of renal mass. *Kidney International*, **22**, 112–26.

Orloff, M.J., Yamanaka, N., Greenleaf, G.E., Huang, Y.-T., Huang, D.-G., and Leng, X.-S. (1986). Reversal of mesangial enlargement in rats with long-standing diabetes by whole pancreas transplantation. *Diabetes*, **35**, 347–54.

Ortola, F.V., Ballerman, B.J., Anderson, S., Mendez, R.E., and Brenner, B.M. (1987). Elevated plasma atrial natriuretic peptide levels in diabetic rats. Potential mediators of hyperfiltration. *Journal of Clinical Investigation*, **80**, 670–4.

Østerby, R. (1973). A quantitative electron microscopic study of mesangial regions in glomeruli from patients with short-term juvenile diabetes mellitus. *Laboratory Investigation*, **29**, 99–110.

Østerby, R. (1975). Early phases in the development of diabetic glomerulopathy. A quantitative electron microscopy study. *Acta Medica Scandinavica*, **S574**, 1–82.

Østerby, R., and Gundersen, H.J.G. (1975). Glomerular size and structure in diabetes mellitus. I. Early abnormalities. *Diabetologia*, **11**, 225–9.

Østerby, R., and Gundersen, H.J.G. (1980). Fast accumulation of basement membrane material and the rate of morphological changes in acute experimental diabetic glomerular hypertrophy. *Diabetologia*, **18**, 493–500.

Østerby, R., Gundersen, H.J.G., Hørlyck, A., Kroustrup, J.P., Nyberg, G., and Westberg, G. (1983). Diabetic glomerulopathy. Structural characteristics of the early and advanced stages. *Diabetes*, **32(Suppl. 2)**, 79–82.

Østerby, R., Gundersen, H.J.G., Nyberg, G., and Aurell, M. (1987). Advanced diabetic glomerulopathy. Quantitative structural characterization of nonoccluded glomeruli. *Diabetes*, **36**, 612–19.

Østerby, R., et al. (1988). A strong correlation between glomerular filtration rate and filtration surface in diabetic nephropathy. *Diabetologia*, **31**, 265–70.

Østerby-Hansen, R. (1965). A quantitative estimate of the peripheral glomerular basement membrane in recent juvenile diabetes. *Diabetologia*, **1**, 97–100.

O'Sullivan, D.J., Fitzgerald, M.G., Meynell, M.J., and Malins, J.M. (1961). Urinary tract infection. A comparative study in the diabetic and general populations. *British Medical Journal*, **1**, 786–8.

Palmisano, J.J., and Lebowitz, H.E. (1989). Renal function in Black Americans with type 2 diabetes. *Journal of Diabetic Complications*, **3**, 40–4.

Pappenheimer, J.R., Renkin, E.M., and Barrero, L.M. (1951). Filtration diffusion and molecular sieving through peripheral capillary membranes. A contribution to the pore theory of capillary permeability. *American Journal of Physiology*, **167**, 13–46.

Parfrey, P.S., et al. (1989). Contrast induced renal failure in patients with diabetes mellitus, renal insufficiency, or both. *New England Journal of Medicine*, **320**, 143–9.

Parthasarathy, N., and Spiro, R.G. (1982). Effect of diabetes on the glycosaminoglycan component of the human glomerular basement membrane. *Diabetes*, **31**, 738–41.

Parving, H.-H., and Hommel, E. (1989). Prognosis in diabetic nephropathy. *British Medical Journal*, **29**, 230–3.

Parving, H.-H., et al. (1976). The effect of metabolic regulation on microvascular permeability to small and large molecules in short-term juvenile diabetics. *Diabetologia*, **12**, 161–6.

Parving, H.-H., Noer, I., Kehlet, H., Mogensen, C.E., Svendsen, P.A., and Heding, L.G. (1977). The effect of short-term glucagon infusion on kidney function in normal man. *Diabetologia*, **13**, 323–5.

Parving, H.-H., et al. (1979). Effect of metabolic regulation on renal leakiness to dextran molecules in short-term insulin-dependent diabetics. *Diabetologia*, **17**, 157–60.

Parving, H.-H., Christiansen, J.S., Noer, I., Tronier, B., and Mogensen, C.E. (1980). The effect of glucagon infusion on kidney function in short-term type 1 (insulin-dependent) juvenile diabetics. *Diabetologia*, **19**, 350–4.

Parving, H.-H., Smidt, U.M., Friisberg, B., Bonnevie-Nielsen, V., and Andersen, A.R. (1981). A prospective study of glomerular filtration rate and arterial blood pressure in insulin-dependent diabetics with diabetic nephropathy. *Diabetologia*, **20**, 457–61.

Parving, H.-H., Øxenboll, B., Svendsen, P.A., Christiansen, J.S., and Andersen, A.R. (1982). Early detection of patients at risk of developing diabetic nephropathy. A longitudinal study of urinary albumin excretion. *Acta Endocrinologica*, **100**, 550–5.

Parving, H.-H., Viberti, G.C., Keen, H., Christiansen, J.S., and Lassen, N.A. (1983). Haemodynamic factors in the genesis of diabetic microangiopathy. *Metabolism*, **32**, 943–9.

Parving, H.-H., Andersen, A.R., Smidt, U.M., Hommel, E., Mathiesen, E.R., and Svendsen, P.A. (1987). Effect of antihypertensive treatment on kidney function in diabetic nephropathy. *British Medical Journal*, **294**, 1443–7.

Parving, H.-H., et al. (1988a). Prevalence of microalbuminuria, arterial hypertension, retinopathy and neuropathy in patients with insulin-dependent diabetes. *British Medical Journal*, **296**, 156–160.

Parving, H.-H., Hommel, E., and Smidt, U.M. (1988b). Protection of kidney function and decrease in albuminuria by captopril in insulin-dependent diabetics with nephropathy. *British Medical Journal*, **297**, 1086–91.

Parving, H.-H., Hommel, E., Nielsen, M.D., and Giese, J. (1989). Effect of captopril on blood pressure and kidney function in normotensive insulin dependent diabetics with nephropathy. *British Medical Journal*, **299**, 533–6.

Parving, H.-H., Gall, M.A., Skøtt, P., Jørgensen, H.E., Jørgensen, F., and Larsen, S. (1990). Prevalence and causes of albuminuria in non-insulin-dependent diabetic (NIDDM) patients. *Kidney International*, **37**, 243.

Pasternack, A., Kasanen, A., Sourander, L., and Kaarsalo, E. (1985). Prevalence and incidence of moderate and severe clinical renal failure in South Western Finland. *Acta Medica Scandinavica*, **218**, 173–80.

Pedersen, J. (1977). *The pregnant diabetic and her newborn*. (2nd ed.). Munksgaard, Copenhagen.

Pedersen, M.M., Schmitz, A., Pedersen, E.B., Danielsen, H., and Christiansen, J.S. (1988). Acute and long-term renal effects of angiotensin converting-enzyme inhibition in normotensive, normoalbuminuric insulin-dependent diabetic patients. *Diabetic Medicine*, **5**, 562–9.

Pedersen, M.M., Christiansen, J.S., and Mogensen, C.E. (1989). Renal effects of an aldose reductase inhibitor (Statil) during 6 months treatment in type 1 (insulin-dependent) diabetic patients. *Diabetologia*, **32**, 516A.

Peterson, C.M., and Jovanovic, L. (1986). Natural history of the diabetic renal-retinal syndrome during pregnancy. In *Diabetic renal-retinal syndrome*, Vol 3, (ed. E. A. Friedman and F. A. L'Esperance). pp. 471–80, Grunne and Stratton, New York.

Peterson, P.A., Evrin, P.E., and Berggård, I. (1969). Differentiation of glomerular, tubular, and normal proteinuria: determinations of urinary excretion of β_2-microglobulin, albumin, and total protein. *Journal of Clinical Investigation*, **48**, 1189–98.

Pettitt, D.J., and Saad, M.F. (1988). Inheritance of predisposition to renal insufficiency in diabetic men. *Diabetes*, **37(Suppl. 1)**, 51A.

Pillay, V.K.G., Gandhi, V.C., Sharma, B.K., Smith, E.C., and Dunea, G. (1972). Effect of hydration and frusemide given intravenously on proteinuria. *Archives of Internal Medicine*, **130**, 90–2.

Pinto, J.R., Bending, J.J., Dodds, R., and Viberti, G.C. (1988). Failure of low-protein diet to restore the physiological response to protein ingestion in proteinuric type 1 (insulin-dependent) diabetic patients. *Diabetologia*, **31**, 531A.

Pinto, J.R., Walker, J.D., Turner, C.D., Beesley, M., and Viberti, G.C. (1990). Renal response to lowering of arterial pressure by angiotensin converting enzyme inhibitor or diuretic therapy in insulin-dependent diabetic patients with nephropathy. *Kidney International*, **37**, 516.

Pometta, D., Rees, S.B., Younger, D., and Kass, E.H. (1967). Asymptomatic bacteriuria in diabetes mellitus. *New England Journal of Medicine*, **276**, 1118–21.

Pugh, J.A., Stern, M.P., Haffner, S.M., Eifler, C.W., and Zapata, M. (1988). Excess incidence of treatment of end-stage renal disease in Mexican Americans. *American Journal of Epidemiology*, **127**, 135–44.

Pugliese, G., et al. (1990). Modulation of haemodynamics and vascular filtration changes in diabetic rats by dietary myo-inositol. *Diabetes*, **39**, 312–22.

Puig, J.G., et al. (1981). Relation of kidney size to kidney function in early insulin-dependent diabetes. *Diabetologia*, **21**, 363–7.

Purkerson, M.L., Hoffsten, P.E., and Klahr, S. (1976). Pathogenesis of the glomerulopathy associated with renal infarction in rats. *Kidney International*, **9**, 407–17.

Purkerson, M.L., Joist, J.H., Greenberg, J.M., Kay, D., Hoffsten, P.E., and Klahr, S. (1982). Inhibition by anticoagulant drugs of the progressive hypertension and uraemia associated with renal infarction in rats. *Thrombosis Research*, **26**, 227–240.

Purkerson, M.L., Joist, J.H., Yates, J., Valdes, A., Morrison, A., and Klahr, S. (1985). Inhibition of thromboxane synthesis ameliorates the progressive kidney disease of rats with subtotal renal ablation. *Proceedings of the National Academy of Sciences (USA)*, **82**, 193–7.

Purkerson, M.L., Tollefsen, D.M., and Klahr, S. (1988). N-desulfated/acetylated heparin ameliorates the progression of renal disease in rats with subtotal renal ablation. *Journal of Clinical Investigation*, **81**, 69–74.

Rao, K.V., and Crosson, J.T. (1980). Idiopathic membranous glomerulonephritis in diabetic patients. Report of three cases and review of the literature. *Archives of Internal Medicine*, **140**, 624–7.

Rasch, R. (1979a). Prevention of diabetic glomerulopathy in streptozotocin diabetic rats by insulin treatment. Kidney size and glomerular volume. *Diabetologia*, **16**, 125–8.

Rasch, R. (1979b). Prevention of diabetic glomerulopathy in streptozotocin diabetic rats by insulin treatment. The mesangial regions. *Diabetologia*, **17**, 243–8.

Rasch, R. (1980). Prevention of diabetic glomerulopathy in streptozotocin diabetic rats. Albumin excretion. *Diabetologia*, **18**, 413–6.

Rasch, R. (1984). Tubular lesions in streptozotocin-diabetic rats. *Diabetologia*, **27**, 32–7.

Rasch, R., and Holck, P. (1988). Ultrastructure of the macula densa in streptozotocin diabetic rats. *Laboratory Investigation*, **59**, 666–72.

Rasch, R., and Østerby, R. (1989). Lack of influence of aldose reductase inhibitor treatment for 6 months on the glycogen nephrosis in streptozotocin diabetic rats. *Diabetologia*, **32**, 532A.

Rate, R.G., et al. (1983). Diabetes mellitus in Hopi and Navajo Indians. *Diabetes*, **32**, 894–9.

Rave, K., Heineman, L., Sawicki, P., Hohmann, A., and Berger, M. (1987). Increased concentration of atrial natriuretic peptide in type 1 (insulin-dependent) diabetic patients with glomerular hyperfiltration. *Diabetologia*, **30**, 573A.

Rayer, P. (1840). In *Traité des maladies du rein*, Vol. 2, (ed. Baillère, Tindall and Cox). Paris.

Reece, E.A., et al. (1988). Diabetic nephropathy: pregnancy performance and fetomaternal outcome. *American Journal of Obstetrics and Gynaecology*, **159**, 56–66.

Remuzzi, A., Viberti, G.C., Ruggenenti, P., Battaglia, C., Pagai, R., and Remuzzi, G. (1990). Glomerular response to hyperglycaemia in human diabetic nephropathy. *American Journal of Physiology*, **259**, F545–F552.

Rifkin, H., Parker, J.G., Polin, E.B., Berkman, J.I., and Spiro, D. (1948). Diabetic glomerulosclerosis. *Medicine*, **27**, 429–57.

Rohrbach, D.H., Wagner, C.W., Star, V.L., Martin, G.R., and Brown, K.S. (1983). Reduced synthesis of basement membrane heparan sulfate proteoglycan in streptozotocin-induced diabetic mice. *Journal of Biological Chemistry*, **258**, 11676–7.

Rollo, J. (1798). *Cases of the Diabetes Mellitus*, (2nd edn). Dilly, London.

Root, H.F., Mirsky, S., and Ditzel, J. (1959). Proliferative retinopathy in diabetes mellitus; review of eight hundred forty seven cases. *Journal of the American Medical Association*, **169**, 903–9.

Rosenberg, M.E., Swanson, J.E., Thomas, B.L., and Hostetter, T.H. (1987). Glomerular and hormonal responses to dietary protein intake in human renal disease. *American Journal of Physiology*, **253**, F1083–F1090.

Ross, R., and Glomset, J.A. (1976). The pathogenesis of atherosclerosis. *New England Journal of Medicine*, **295**, 369–77.

Rostand, S.G., Kirk, K.A., Rutsky, E.A., and Pate, B.A. (1982). Racial differences in the incidence of treatment for end-stage renal disease. *New England Journal of Medicine*, **306**, 1276–9.

Rowe, D.J.F., Bagga, H., and Betts, P.B. (1985). Normal variation in the rate of albumin excretion and albumin to creatinine ratios in overnight and daytime urine collections in non-diabetic children. *British Medical Journal*, **291**, 693–4.

Rubinow, D.R., and Nelson, J.C. (1982). Tricyclic exacerbation of undiagnosed diabetic uropathy. *Journal of Clinical Psychiatry*, **943**, 210–2.

Ruggenenti, P., Viberti, G.C., Battaglia, C., Perticucci, E., Remuzzi, G., and Remuzzi, A. (1990). Low-dose enalapril and glomerular selective function in insulin-dependent diabetics. *Kidney International*, **37**, 519.

Rundles, R.W. (1945). Diabetic neuropathy: general review with report of 125 cases. *Medicine*, **24**, 111–60.

Saito, Y., et al. (1988). Mesangiolysis in diabetic glomeruli; its role in the formation of nodular lesions. *Kidney International*, **34**, 389–96.

Samanta, A., Burden, A.C., Feehally, J., and Walls, J. (1986). Diabetic renal disease: differences between Asian and White patients. *British Medical Journal*, **293**, 366–7.

Sampson, M.J., Chambers, J., Sprigings, D., and Drury, P.L. (1990). Intraventricular septal hypertrophy in type 1 diabetic patients with microalbuminuria or early proteinuria. *Diabetic Medicine*, **7**, 126–131.

Sasaki, A., Horiuchi, N., Hasegawa, K., and Uehara, M. (1986). Risk factors related to the development of persistent albuminuria among diabetic patients observed in a long-term follow-up. *Journal of the Japanese Diabetes Society*, **29**, 1017–23.

Sawicki, P.T., Heineman, L., Rave, K., Hohmann, A., and Berger, M. (1988). Atrial natriuretic factor in various stages of diabetic nephropathy. *Journal of Diabetic Complications*, **2**, 207–9.

Schambelan, M., Blake, S., Sraer, J., Bens, M., Nivez, M.-P., and Wahbe, F. (1985). Increased prostaglandin production by glomeruli isolated from rats with streptozotocin-induced diabetes mellitus. *Journal of Clinical Investigation*, **75**, 404–12.

Scheinman, J.I., Fish, A.J., and Michael, A.F. (1974). The immunohistopathology of glomerular antigens. The glomerular basement membrane, collagen, and actomyosin antigens in normal and diseased kidneys. *Journal of Clinical Investigation*, **54**, 1144–54.

Schmitz, A., and Vaeth, M. (1988). Microalbuminuria: a major risk factor in non-insulin dependent diabetes. A 10 year follow-up study of 503 patients. *Diabetic Medicine*, **5**, 126–34.

Schmitz, A., Gundersen, H.J.G., and Østerby, R. (1988). Glomerular morphology by light microscopy in non-insulin-dependent diabetes mellitus—lack of glomerular hypertrophy. *Diabetes*, **37**, 38–43.

Schmitz, A., Christensen, T., and Jensen, F.T. (1989a). Glomerular filtration rate and kidney volume in normo-albuminuric non-insulin-dependent diabetics—lack of glomerular hyperfiltration and renal hypertrophy in uncomplicated NIDDM. *Scandinavian Journal of Clinical and Laboratory Investigation*, **49**, 103–8.

Schmitz, A., Hansen, H.H., and Christensen, T. (1989b). Kidney function in newly diagnosed type 2 (non-insulin-dependent) diabetic patients, before and during treatment. *Diabetologia*, **32**, 434–9.

Schober, E., Pollack, A., Coradello, H., and Lubec, G. (1982). Glycosylation of glomerular basement membrane in type 1 (insulin-dependent) diabetic children. *Diabetologia*, **23**, 485–7.

Schwab, S.J., et al. (1989). Contrast nephrotoxicity: A randomized controlled trial of a nonionic and an ionic radiographic contrast agent. *New England Journal of Medicine*, **320**, 149–53.

Seaquist, E.R., Gœtz, F.C., Rich, S., and Barbosa, J. (1989). Familial clustering of diabetic kidney disease. Evidence for genetic susceptibility to diabetic nephropathy. *New England Journal of Medicine*, **320**, 1161–5.

Segel, M.C., Lecky, J.W., and Slasky, B.S. (1984). Diabetes mellitus: the predominant cause of bilateral renal enlargement. *Radiology*, **153**, 341–2.

Seyer-Hansen, K. (1976). Renal hypertrophy in streptozotocin diabetic rats. *Clinical Science and Molecular Medicine*, **51**, 551–5.

Seyer-Hansen, K., Hansen, J., and Gundersen, H.J.G. (1980). Renal hypertrophy in experimental diabetes. A morphometric study. *Diabetologia*, **18**, 501–5.

Seyer-Hansen, K. (1983). Renal hypertrophy in experimental diabetes mellitus. *Kidney International*, **23**, 643–6.

Shaklai, N., Garlick, R.L., and Bunn, H.F. (1984). Nonenzymatic glycosylation of human serum albumin alters its conformation and function. *Journal of Biological Chemistry*, **259**, 3812–7.

Shemesh, O., Golbetz, H., Kriss, J.P., and Myers, B.D. (1985). Limitations of creatinine as a filtration marker in glomerulopathic patients. *Kidney International*, **28**, 830–8.

Shimomura, H., and Spiro, R.G. (1987). Studies on the macromolecular components of human glomerular basement membrane and alterations in diabetes: decreased levels of heparan sulfate proteoglycan and laminin. *Diabetes*, **36**, 374–81.

Silberberg, J.S., Barre, P.E., Prichard, S.S., and Sniderman, A.D. (1989). Impact of left ventricular hypertrophy on survival in end-stage renal disease. *Kidney International*, **36**, 286–90.

Silva, F.G., Pace, E.H., Burns, D.K., and Krous, H. (1983). The spectrum of diabetic nephropathy and membranous glomerulopathy: report of two cases and review of the literature. *Diabetic Nephropathy*, **2**, 28–32.

Siperstein, M.D., Unger, R.H., and Madison, L.L. (1968). Studies of muscle capillary basement membranes in normal subjects, diabetic and pre-diabetic patients. *Journal of Clinical Investigation*, **47**, 1973–99.

Skøtt, P., *et al.* (1989). Effects of insulin on kidney function and sodium excretion in healthy subjects. *Diabetologia*, **32**, 694–9.

Solerte, S.B., Fioravanti, M., Spriano, P., Aprile, C., Patti, A.L., and Ferrari, E. (1987). Plasma atrial natriuretic peptide, renal haemodynamics and microalbuminuria in short-term type 1 (insulin-dependent) diabetic patients with hyperfiltration. *Diabetologia*, **30**, 584A.

Spiro, R.G., and Spiro, M.J. (1971). Effect of diabetes on the biosynthesis of the renal glomerular basement membrane: studies on the glucosyltransferase. *Diabetes*, **20**, 641–8.

Spühler, O., and Zollinger, H.U. (1943). Direfe diabetische Glomerulosklerose. *Deutsche Archive Klinik Medizin*, **190**, 321–79.

Stalder, G., and Schmid, R. (1959). Sever functional disorders of glomerular capillaries and renal haemodynamics in treated diabetes mellitus during childhood. *Annals of Paediatrics*, **193**, 129–38.

Steffes, M.W., Brown, D.M., and Mauer, S.M. (1978). Diabetic glomerulopathy following unilateral nephrectomy in the rat. *Diabetes*, **27**, 35–41.

Steffes, M.W., Brown, D.M., Basgen, J.M., and Mauer, S.M. (1980). Amelioration of mesangial volume and surface alterations following islet transplantation in diabetic rats. *Diabetes*, **29**, 509–15.

Steffes, M.W., Sutherland, D.E.R., Gøetz, F.C., Rich, S.S., and Mauer, S.M. (1985). Studies of kidney and muscle biopsy specimens from identical twins discordant for type 1 diabetes mellitus. *New England Journal of Medicine*, **312**, 1282–7.

Steffes, M.W., Østerby, R., Chavers, B., and Mauer, S.M. (1989). Mesangial expansion as a central mechanism for loss of kidney function in diabetic patients. *Diabetes*, **38**, 1077–81.

Stevens, V.J., Rouzer, C.A., Monnier, V.M., and Cerami, A. (1978). Diabetic cataract formation: potential role of glycosylation of lens crystallins. *Proceedings of the National Academy of Science (USA)*, **75**, 2918–22.

Taft, J.L., Billson, V.R., Nankervis, A., Kincaid-Smith, P., and Martin, F.I.R. (1990). A clinical-histological study of individuals with diabetes mellitus and proteinuria. *Diabetic Medicine*, **7**, 215–21.

Taguma, Y., *et al.* (1985). Effect of captopril on heavy proteinuria in azotemic diabetics. *New England Journal of Medicine*, **313**, 1617–20.

Tchobroutsky, G. (1979). Prevention and treatment of diabetic nephropathy. In *Advances in nephrology*, Vol. 9, (ed. J. Hamburger, J. Crosnier, J.-P. Grunfeld, and M. H. Maxwell). pp. 63–86, Year Book Medical Publishers, Chicago.

Thomsen, A.C. (1965). The kidney in diabetes mellitus. PhD Thesis, Munksgaard, Copenhagen.

Thomsen, O.F., Andersen, A.R., Christiansen, J.S., and Deckert, T. (1984). Renal changes in long-term type 1 (insulin-dependent) diabetic patients with and without clinical nephropathy; a light microscopic, morphometric study of autopsy material. *Diabetologia*, **26**, 361–5.

Thorley, J.D., Jones, S.R., and Sanford, J.P. (1974). Perinephric abscess. *Medicine*, **53**, 441–51.

Tilton, R.G., *et al.* (1989). Prevention of haemodynamic and vascular albumin filtration changes in diabetic rats by aldose reductase inhibitors. *Diabetes*, **38**, 1258–70.

Tomlanovich, S., Deen, W.M., Jones, H.W., Schwartz, H.C., and Myers, B.D. (1987). Functional nature of glomerular injury in progressive diabetic glomerulopathy. *Diabetes*, **36**, 556–65.

Trever, R.W., and Cluff, L.E. (1958). The problem of increasing azotemia during management of diabetic acidosis. *American Journal of Medicine*, **24**, 368–75.

Trevisan, R., *et al.* (1987). Ketone bodies increase glomerular filtration rate in normal man and in patients with type 1 (insulin-dependent) diabetes mellitus. *Diabetologia*, **30**, 214–21.

Tunbridge, W.G.M. (1979). Factors contributing to the deaths of diabetics under the age of 50. *Lancet*, **2**, 569–72.

Turney, J.H., Marshall, D.H., Brownjohn, A.M., Ellis, C.M., and Parsons, F.M. (1990). The evolution of acute renal failure, 1956–1988. *Quarterly Journal of Medicine*, **74**, 83–104.

Tuttle, K., Perusek, M., DeFronzo, R., and Kunau, R. (1990). Increased renal reserve and size regress with strict glycaemic control in insulin-dependent diabetes mellitus. *Kidney International*, **37**, 261.

Ueda, J., Nakanishi, H., Miyazaki, M., and Abe, Y. (1977). Effects of glucagon on the renal haemodynamics of dogs. *European Journal of Pharmacology*, **41**, 209–12.

Uranga, J., Frenzalida, R., Rapoport, A.L., and Del Castillo, E. (1979). Effect of glucagon and glomerulopressin on the renal function of the dog. *Hormone and Metabolism Research*, **11**, 275–9.

Vannini, P., *et al.* (1984). Lipid abnormalities in insulin-dependent diabetic patients with albuminuria. *Diabetes Care*, **7**, 151–4.

Vasquez, B., *et al.* (1984). Sustained reduction of proteinuria in type 2 (non-insulin-dependent) diabetes following diet-induced reduction of hyperglycaemia. *Diabetologia*, **26**, 127–33.

Vejlsgaard, R. (1966a). Studies on urinary infections in diabetics. I. Bacteriuria in patients with diabetes mellitus and in control subjects. *Acta Medica Scandinavica*, **179**, 173–82.

Vejlsgaard, R. (1966b). Studies on urinary infections in diabetics. II. Significant bacteriuria in relation to long-term diabetic manifestations. *Acta Medica Scandinavica*, **179**, 183–8.

Vejlsgaard, R. (1973a). Studies on urinary infections in diabetics. III. Significant bacteriuria in pregnant diabetics and in matched controls. *Acta Medica Scandinavica*, **193**, 337–41.

Vejlsgaard, R. (1973b). Studies on urinary infections in diabetics. IV. Significant bacteriuria in pregnancy in relation to age of onset, duration of diabetes, angiopathy and urologic symptoms. *Acta Medica Scandinavica*, **193**, 343–46.

Velosa, J.A., Glasser, R.J., Nevins, T.E., and Michael, A.F. (1977). Experimental model of focal sclerosis. II. Correlations with immunopathologic changes, macromolecular kinetics, and polyanion loss. *Laboratory Investigation*, **36**, 527–34.

Venkatachalam, M.A., and Rennke, H.G. (1978). The structural and molecular basis of glomerular filtration. *Circulation Research*, **43**, 337–47.

Viberti, G.C., and Keen, H. (1984). The patterns of proteinuria in diabetes mellitus. Relevance to pathogenesis and prevention of diabetic nephropathy. *Diabetes*, **33**, 686–92.

Viberti, G.C., and Wiseman, M.J. (1986). The kidney in diabetes: significance of the early abnormalities. *Clinics in Endocrinology and Metabolism*, **15**, 753–82.

Viberti, G.C., Jarrett, R.J., McCartney, M., and Keen, H. (1978). Increased glomerular permeability to albumin induced by exercise in diabetic subjects. *Diabetologia*, **14**, 293–300.

Viberti, G.C., Pickup, J.C., Jarrett, R.J., and Keen, H. (1979). Effect of control of blood glucose on urinary excretion of albumin and beta-2 microglobulin in insulin-dependent diabetes. *New England Journal of Medicine*, **300**, 638–41.

Viberti, G.C., Haycock, G.B., Pickup, J.C., Jarrett, R.J., and Keen, H. (1980). Early functional and morphologic vascular renal consequences of the diabetic state. *Diabetologia*, **18**, 173–5.

Viberti, G.C., Pickup, J.C., Bilous, R.W., Keen, H., and Mackintosh, D. (1981a). Correction of exercise-induced microalbuminuria in insulin-dependent diabetics after 3 weeks of subcutaneous insulin infusion. *Diabetes*, **30**, 818–23.

Viberti, G.C., Strakosch, C.R., Keen, H., Mackintosh, D., Dalton, N., and Home, P.D. (1981b). The influence of glucose-induced hyperinsulinaemia on renal glomerular function and cirulcating catecholamines in normal man. *Diabetologia*, **21**, 436–9.

Viberti, G.C., Mackintosh, D., Bilous, R.W., Pickup, J.C., and Keen, H. (1982a). Proteinuria in diabetes mellitus: role of spontaneous and experimental variation of glycaemia. *Kidney International*, **21**, 714–20.

Viberti, G.C., Hill, R.D., Jarrett, R.J., Argyropoulos, A., Mahmud, U., and Keen, H. (1982b). Microalbuminuria as a predictor of clinical nephropathy in insulin-dependent diabetes mellitus. *Lancet*, **1**, 1430–2.

Viberti, G.C., Mogensen, C.E., Keen, H., Jacobsen, R.J., Jarrett, R.J., and Christiansen, C.K. (1982c). Urinary excretion of albumin in normal man: the effect of water loading. *Scandinavian Journal of Clinical and Laboratory Investigation*, **42**, 147–51.

Viberti, G.C., Bilous, R.W., Mackintosh, D., and Keen, H. (1983a). Monitoring glomerular function in diabetic nephropathy. *American Journal of Medicine*, **74**, 256–64.

Viberti, G.C., Bilous, R.W., Mackintosh, D., Bending, J.J., and Keen, H. (1983b). Long term correction of hyperglycaemia and progression of renal failure in insulin-dependent diabetes. *British Medical Journal*, **286**, 598–602.

Viberti, G.C., Mackintosh, D., and Keen, H. (1983c). Determinants of the penetration of proteins through the glomerular barrier in insulin-dependent diabetes mellitus. *Diabetes*, **32 (Suppl. 2)**, 92–5.

Viberti, G.C., Keen, H., Dodds, R., and Bending, J.J. (1987a). Metabolic control and progression of diabetic nephropathy. *Diabetologia*, **30**, 481–2.

Viberti, G.C., Keen, H., and Wiseman, M.J. (1987b). Raised arterial pressure in parents of proteinuric insulin-dependent diabetics. *British Medical Journal*, **295**, 515–7.

Viberti, G.C., Benigni, A., Bognetti, E., Remuzzi, G., and Wiseman, M.J. (1989). Glomerular hyperfiltration and urinary prostaglandins in type 1 diabetes mellitus. *Diabetic Medicine*, **6**, 219–23.

Vittinghus, E., and Mogensen, C.E. (1982). Graded exercise and protein excretion in diabetic man and the effect of insulin treatment. *Kidney International*, **21**, 725–9.

Vlassara, H., Brownlee, M., and Cerami, A. (1986). Novel macrophage receptor for glucose-modified proteins is distinct from previously described scavenger receptors. *Journal of Experimental Medicine*, **164**, 1301–9.

Vora, J., Thomas, D.M., Dean, J., Peters, J., Owens, D., and Williams, J.D. (1990). Renal function and albumin excretion rate in 62 newly presenting non-insulin dependent diabetics (NIDDM). *Kidney International*, **37**, 245.

Wahl, P., Deppermann, D., and Hasslacher, C. (1982). Biochemistry of glomerular basement membrane of the normal and diabetic human. *Kidney International*, **21**, 744–9.

Walker, J.D., et al. (1989). Restriction of dietary protein and progression of renal failure in diabetic nephropathy. *Lancet*, **2**, 1411–14.

Walser, M., Drew, H.H., and LaFrance, N.D. (1988). Creatinine measurements often yield false estimates of progression in chronic renal failure. *Kidney International*, **34**, 412–8.

Warram, J., Derby, L., Laffel, L., and Krolewski, A.S. (1990). Role of mean arterial pressure in the development of persistent proteinuria. *Kidney International*, **37**, 404.

Watanabe, Y., Nunoi, K., Maki, Y., Nakamura, Y., and Fujishima, M. (1987). Contribution of glycaemic control to the levels of N-acetyl-β-D-glucosaminidase and serum NAG in type 1 (insulin-dependent) diabetes mellitus without proteinuria. *Clinical Nephrology*, **28**, 227–31.

Watkins, P.J. (1973). Facial sweating after food: a new sign of autonomic diabetic neuropathy. *British Medical Journal*, **1**, 583–7.

Watkins, P.J., et al. (1972). The natural history of diabetic renal disease. A follow-up study of a series of renal biopsies. *Quarterly Journal of Medicine*, **41**, 437–56.

Webb, R.C., and Bohr, D.F. (1981). Recent advances in the pathogenesis of hypertension: consideration of structural, functional, and metabolic vascular abnormalities resulting in elevated arterial resistance. *American Heart Journal*, **102**, 251–64.

West, K.M., Erdreich, L.J., and Stober, J.A. (1980). A detailed study of risk factors for retinopathy and nephropathy in diabetes. *Diabetes*, **29**, 501–8.

Westberg, N.G., and Michael, A.F. (1972) Immunohistopathology of diabetic glomerulosclerosis. *Diabetes*, **21**, 163–74.

Westberg, N.G., and Michael, A.F. (1973). Human glomerular basement membrane: chemical composition in diabetes mellitus. *Acta Medica Scandinavica*, **194**, 39–47.

White, P., and Graham, C.A. (1971). The child with diabetes. In *Joslin's Diabetes Mellitus*, (11th edn), (ed. A. Marble, P. White, R. F. Bradley, and L. P. Krall). pp. 539–60, Lea & Febiger, Philadelphia.

WHO (World Health Organization) Multinational Study of Vascular Disease in Diabetes. (1985). Prevalence of small vessel and large vessel disease in diabetic patients from 14 centres. *Diabetologia*, **28**, 615–40.

Wibell, L. (1974). Studies of β_2-microglobulin in human serum, urine and amniotic fluid. Thesis, Abstracts of Uppsala Dissertations from the Faculty of Medicine, 183.

Williams, S.K., and Siegel, R.K. (1985). Preferential transport of non-enzymatically glycosylated ferritin across the kidney glomerulus. *Kidney International*, **28**, 146–52.

Williams, S.K., Devenny, J.J., and Bitensky, M.W. (1981). Micropinocytic ingestion of glycosylated albumin by isolated microvessels: possible role in pathogenesis of diabetic microangiopathy. *Proceedings of the National Academy of Sciences (USA)*, **78**, 2393–7.

Williamson, J.R., and Kilo, C. (1977). Current status of capillary basement-membrane disease in diabetes mellitus. *Diabetes*, **26**, 65–73.

Williamson, J.R., et al. (1986). Sex steroid dependency of diabetes-induced changes in polyol metabolism, vascular permeability and collagen cross-linking. *Diabetes*, **35**, 20–7.

Winetz, J.A., Golbetz, H.V., Spencer, R.J., Lee, J.A., and Myers, B.D. (1982). Glomerular function in advanced human diabetic nephropathy. *Kidney International*, **21**, 750–6.

Winocour, P.H., Durrington, P.N., Ishola, M., Anderson, D.C., and Cohen, H. (1987). Influence of proteinuria on vascular disease, blood pressure, and lipoproteins in insulin-dependent diabetes mellitus. *British Medical Journal*, **294**, 1648–51.

Wiseman, M., and Viberti, G.C. (1983). Kidney size and GFR in type 1 (insulin-dependent) diabetes mellitus revisited. *Diabetologia*, **25**, 530.

Wiseman, M.J., Viberti, G.C., and Keen, H. (1984a). Threshold effect of plasma glucose in the glomerular hyperfiltration in diabetes. *Nephron*, **38**, 257–60.

Wiseman, M.J., Viberti, G.C., Mackintosh, D., Jarrett, R.J., and Keen, H. (1984b). Glycaemia, arterial pressure and micro-albuminuria in type 1 (insulin-dependent) diabetes mellitus. *Diabetologia*, 2, 401–5.

Wiseman, M.J., Saunders, A.J., Keen, H., and Viberti, G.C. (1985a). Effect of blood glucose on increased glomerular filtration rate and kidney size in insulin-dependent diabetes. *New England Journal of Medicine*, 312, 617–21.

Wiseman, M.J., Redmond, S., House, F., Keen, H., and Viberti, G.C. (1985b). The glomerular hyperfiltration of diabetics is not associated with elevated levels of glucagon and growth hormone. *Diabetologia*, 28, 718–21.

Wiseman, M.J., Mangili, R., Alberetto, M., Keen, H., and Viberti, G.C. (1987a). Glomerular response mechanisms to glycaemic changes in insulin-dependent diabetics. *Kidney International*, 31, 1012–8.

Wiseman, M.J., Bognetti, E., Dodds, R., Keen, H., and Viberti, G.C. (1987b). Changes in renal function in response to protein restricted diet in type 1 (insulin-dependent) diabetic patients. *Diabetologia*, 30, 154–9.

Wu, V.Y., Wilson, B., and Cohen, M.P. (1987). Disturbances in glomerular basement membrane glycosaminoglycans in experimental diabetes. *Diabetes*, 36, 679–83.

Yoshida, Y., Fogo, A., Shiraga, H., Glick, A.D., and Ichikawa, I. (1988). Serial micropuncture analysis of single nephron function in subtotal renal ablation. *Kidney International*, 33, 851–5.

Young, K.R., and Clancy, C.F. (1955). Symposium on diabetes and obesity. Urinary tract infections complicating diabetes mellitus. *Medical Clinics of North America*, 39, 1665–70.

Yudkin, J.S., Forrest, R.D., and Jackson, C.A. (1988). Microalbuminuria as a predictor of vascular disease in non-diabetic subjects. *Lancet*, 2, 530–3.

Zatz, R., Meyer, T.W., Rennke, H.G., and Brenner, B.M. (1985). Predominance of haemodynamic rather than metabolic factors in the pathogenesis of diabetic glomerulopathy. *Proceedings of the National Academy of Sciences (USA)*, 82, 5963–7.

Zatz, R., Dunn, B.R., Meyer, T.W., Anderson, S., Rennke, H.G., and Brenner, B.M. (1986). Prevention of diabetic glomerulopathy by pharmacologic amelioration of glomerular capillary hypertension. *Journal of Clinical Investigation*, 77, 1925–30.

Zavaroni, I., et al. (1989). Risk factors for coronary artery disease in healthy persons with hyperinsulinaemia and normal glucose tolerance. *New England Journal of Medicine*, 320, 702–6.

Zeller, K.R., Jacobson, H., and Raskin, P. (1990). The effect of dietary protein and phosphorus restriction on renal function in diabetic nephropathy—results of a 5 year study. *Kidney International*, 37, 246.

4.2 Amyloidosis

G.K. VAN DER HEM AND M.H. VAN RIJSWIJK

Introduction, historical aspects

Although amyloidosis must have occurred over many centuries, the nature of the disease and its different forms have gradually been unravelled during the past 150 years. The first description of this disease by recognizing a hitherto unknown substance in tissues came from Rudolf Virchow in 1854. He was the first to use the name amyloid, as the waxy substance was thought to be a carbohydrate (starch or cellulose), because of the special blue staining with iodine and a violet staining with sulphuric acid. This theory was proved incorrect a few years later by Friedreich and Kekulé (1859), who showed that the substance was a protein. Nevertheless, the term 'amyloid' has remained in use for this fascinating disease to the present day.

A major advance in diagnosis was introduced by Bennhold in 1922 by staining amyloid with the dye Congo red. A few years later Divry (1927) demonstrated a characteristic double refraction of amyloid after staining with Congo red. This method has been adopted as the standard for demonstrating the presence of amyloid in tissue sections.

Subsequently amyloid tissue was investigated by Cohen and Calkins using electron microscopy (1959). They found a fibrillar composition of amyloid, with fine non-branching fibrils with a diameter of 7 to 9 nm. In 1968 Eanes and Glenner discovered a typical X-ray diffraction pattern of amyloid with a so-called cross-β-pleated sheet formation. The amyloid fibrils appeared to have a structure of polypeptide chains which are arranged perpendicularly to the axis of the fibril. Because of this finding Glenner in 1980 proposed the name of β-fibrilloses as a common denominator of all types of amyloidosis. Although this term is a far better justification of the nature of the disease, the name amyloidosis is still in common use.

During the following years amyloid proteins have been shown to differ with regard to their primary structure. In patients with amyloidosis associated with multiple myeloma, the amyloid was shown in 1970 by Glenner and coworkers to consist of immunoglobulin light chains. The amyloid protein was designated AL (amyloid protein, light chain derived). This AL amyloid was also found in 'primary' amyloidosis, where the origin of the disease was not clear.

In secondary amyloidosis following long-standing inflammatory conditions such as suppurative infections or rheumatoid arthritis, the major component of the amyloid appeared to be a unique protein unrelated to immunoglobulins. The protein found in secondary amyloid was designated AA (amyloid protein A). Later on, this AA amyloid was also demonstrated in tissue specimens of patients with familial Mediterranean fever.

The deposition of AA amyloid was found to be associated with the occurrence of a precursor protein in the blood, designated SAA (serum amyloid A), being an acute-phase reactant. This apolipoprotein is produced by hepatocytes under the influence of an inflammatory process. Subsequently several different amyloid proteins were recognized. A typical example is that of familial amyloidotic polyneuropathy, in which the amyloid fibrils are derived from a variant prealbumin.

In patients who receive regular dialysis for many years a

Table 1 Clinical classification of amyloid syndromes

	'Typical' distribution	'Atypical' distribution
	Nephropathy	Nephropathy Cardiomyopathy Glossopathy Neuropathy
Associated	Systemic amyloidosis associated with chronic inflammatory diseases (protein AA)	Systemic amyloidosis associated with monoclonal gammopathy (protein AL)
Idiopathic	Idiopathic systemic amyloidosis with 'typical' distribution (protein AA)	Idiopathic systemic amyloidosis with 'atypical' distribution (protein AL)
Familial	Familial systemic amyloidosis associated with FMF (protein AA)	Familial amyloidotic polyneuropathy (protein TTR)
Localized	Amyloid tumours (protein AL) Cutaneous amyloid	

special form of amyloid has recently been discovered, derived from β_2-microglobulin (see Chapter 10.3).

Although there have been numerous advances in recognizing the association of the different forms of amyloidosis with different amyloid proteins, there have been few advances in treatment of forms of amyloidosis. As many patients with generalized amyloidosis show renal involvement as a major clinical feature, their treatment by regular haemodialysis and renal transplantation has attracted much attention in the last decade.

Definition of amyloidosis

The generic name of amyloidosis is applied to a family of diseases which have, as their common denominator, a loss of organ function due to the extracellular deposition of a fibrillar protein with a β-pleated sheet formation.

This structure is responsible for the green birefringence with polarized light after staining with Congo red. This is the hallmark of the diagnosis in laboratory investigations of tissue suspected of containing amyloid.

Classification of amyloidosis

In the past decades several schemes for classification of the different forms of amyloidosis have been proposed.

Reimann et al. (1935) suggested a division in primary amyloidosis without a detectable underlying disease, amyloidosis associated with multiple myeloma, secondary amyloidosis following chronic diseases such as rheumatoid arthritis and chronic infections, hereditary amyloidosis, and localized amyloid. Other classification schemes are based on the type of distribution of the amyloid in various organs of the patient. A 'typical' distribution points to amyloid mainly found in kidneys, liver, and spleen, whereas an 'atypical' distribution is associated with amyloid in kidneys, tongue, heart, and nervous system.

Although these classifications appear to differ they merge to a certain extent when the type of amyloid protein is taken into consideration. The so-called secondary amyloidosis, which has generally a 'typical' distribution, is characterized by the protein AA, whereas primary amyloidosis and myeloma-associated amyloidosis, often with an 'atypical' distribution, have been shown to go with protein AL.

During the third International Symposium on Amyloidosis in 1980 the hope was expressed that 'a definitive clinicopathological classification will be possible when the chemical nature (composition) of all the different amyloid proteins has been elucidated'.

In addition to the common types of generalized AA and AL amyloidosis a number of familial types and localized forms of amyloidosis may be recognized (Table 1). In addition, in Table 2 a survey is given of the different types of amyloidosis and their respective amyloid proteins.

Biochemical investigations on the composition of amyloid

Although it had already been proved in the preceding century that amyloid mainly consists of proteins, data on the composition of amyloid fibrils have become available only in the last 20 years. This was possible after Pras and coworkers succeeded in 1968 in obtaining pure amyloid fibrils by using an extraction method with distilled water after previous elimination of saline-soluble proteins. The protein constituents of the amyloid fibrils could be separated by gel filtration and investigated by biochemical procedures such as amino-acid sequence analysis and immunochemical characterization. Subsequently it appeared that amyloid fibrils do not have always the same composition. Indeed, depending on the disease which facilitates the formation of amyloid, fundamental differences were recognized in the primary structure of amyloid proteins. Nevertheless amyloid fibrils found in various organs of the same patient generally show the same primary structure.

AL amyloid

In 1971 Glenner et al. discovered that AL amyloid proteins are homologous with immunoglobulin light chains or N-terminal fragments of these chains. These AL proteins are found in idiopathic (primary) amyloidosis and in amyloidosis associated with plasma cell dyscrasias (myelomatosis, Waldenström's macroglobulinaemia, and benign monoclonal gammopathy). The AL proteins extracted from amyloid tissue of a patient are chemically identical to urinary Bence-Jones protein, if present, of the same patient. Intact light chains as well as different-sized amino-terminal fragments of the same light chains are often present in the same amyloid preparation. This suggests that the fragments are the result of enzymatic cleavage of the intact chains. The AL proteins generally consist of the variable region together with a part of the constant region of the light chain. The preponderance of lambda over kappa chains in AL amyloidosis suggest that lambda chains are more prone to fibril formation than kappa chains. In addition most of the Bence-Jones proteins that have been demonstrated to form amyloid-like fibrils *in vitro* are of the lambda type. There is no light chain protein unique for AL amyloid. The AL amyloid proteins show considerable differences in molecular size, ranging from 5000 to 23 000 Da. A number of different light chain subgroups have been identified by amino-acid sequence analysis of AL proteins from different patients.

AA amyloid

The AA amyloid protein, found in the secondary (or reactive) amyloidosis and in amyloid depositions in patients with familial Mediterranean fever, was first described in 1971 by Benditt *et al*.

This AA protein appeared to have a different primary structure to that of AL proteins. Protein AA appeared to have a unique, hitherto unknown, amino-acid sequence. AA proteins of varying size have been found in different preparations. Molecular weights vary from 4500 to 9200 Da, corresponding to 45 to 83 amino-acid residues. Most of the AA proteins in humans contain 76 amino-acid residues. There is a cross-reactivity between all AA proteins, since antisera raised against a particular AA protein also react with other AA proteins.

AA proteins are probably derived from a larger precursor protein in the circulation, known as serum amyloid A protein (SAA). SAA has a molecular weight of 12 500 Da and consists of 104 amino-acid residues. The precursor-product-concept has been proven by Husebekk and coworkers (1985) by injecting mice with human SAA. Subsequently human AA proteins were isolated from amyloid deposits in the mice. This shows that the mouse is able to convert human SAA to AA, and to incorporate AA in its own amyloid fibrils. AA and SAA have an identical N-terminal amino-acid sequence but differ in molecular weight. AA is thought to be derived from SAA by proteolytic cleavage.

The precursor protein SAA is formed in the liver and is complexed mainly to high density lipoproteins. Under normal

Table 2 Summary of the different amyloid syndromes, classified according to the distribution of the amyloid deposits (systemic or localized), the mode of inheritance and the nature of the amyloid

Amyloid syndrome	Amyloid protein*	Precursor protein
Systemic non-hereditary amyloidosis		
Associated with immunocyte dyscrasia	+/AL	Light chains
Associated with chronic diseases		
Chronic inflammatory conditions	+/AA	SAA
Malignancies	+/AA	SAA
Idiopathic	+/AL/AA	Light chains/SAA
Systemic hereditary amyloidosis		
Neuropathic forms		
Amyloid neuropathy type I	+/AFp	Prealbumin
Amyloid neuropathy type II	+/AFi	Prealbumin
Amyloid neuropathy type III	–	–
Amyloid neuropathy type IV	–	–
Non-neuropathic forms		
Amyloid nephropathy of Ostertag	–	–
Amyloidosis of familial Mediterranean fever	+/AA	SAA
Amyloid nephropathy with deafness and urticaria	–/AA	–
Amyloid cardiomyopathy–Denmark	+/?	Prealbumin
Amyloid cardiomyopathy with persistent atrial standstill	–	–
Localized hereditary amyloidosis		
Hereditary cerebral haemorrhage–Iceland	+/?	Gamma-trace
Lattice dystrophy of the cornea	–	–
Hereditary corneal amyloid	–	–
Papular cutaneous amyloid	–	–
Poikilodermal cutaneous amyloid	–	–
Bullous cutaneous amyloid	–	–
Medullary carcinoma of the thyroid in MEA 2 with amyloid deposits	+/AEt	Calcitonin
Localized non-hereditary amyloidosis		
Tumours of bronchial tree/urinary tract	+/AL	Light chains
Senile cardiac amyloid	+/AScl	Prealbumin
Idiopathic atrial amyloid	–	Atrial natriuretic peptides
Cutaneous amyloid (lichenoid, macular)	–	–
Endocrine organ or tissue-related		
Medullary carcinoma of the thyroid	+/AEt	Calcitonin
Insulinoma	+/AE	Insulin
Tumours of intestine and pancreas	+/AE	Hormone
Specific solid tumours with amyloid	–	–
Alzheimer's disease	+/?	β-protein
Articular amyloid in haemodialysis	+/?	β_2-microglobulin

SAA = serum amyloid protein A.
* Designated according to the guidelines for nomenclature[2]. + Derivation based on immunological or amino-acid sequence data. – Not yet characterized; ? Chemical derivation established, still to be designated.

circumstances it is present in only minute amounts in the circulation, but during inflammation for example, the concentration of SAA is greatly increased, concurrent with an increase in the concentration of other acute-phase reactants, such as C-reactive protein (CRP), which is also produced in the liver.

SAA production is probably induced by a monokine, interleukin 1, released by activated mononuclear phagocytes. It appears that the formation of AA amyloid requires an increased production of SAA. It is still unknown why only a small percentage of patients with elevated SAA levels go on to develop amyloidosis. A retardation of the breakdown of SAA should also be considered as a factor in the pathogenesis of AA amyloidosis.

Amyloid proteins related to prealbumin

In heredofamilial amyloidosis with polyneuropathy, reported from Portugal, Japan, and Sweden, the amyloid fibrils have been shown to consist of a variant of prealbumin. The difference consists of a substitution of valine for methionine in position 30 of the normal plasma prealbumin. In a Jewish form of heredofamilial amyloidosis with polyneuropathy a different substitution was shown. Amyloid fibrils with prealbumin variants have a high affinity for peripheral nerves.

Other prealbumin-like proteins have been found in patients with familial amyloid cardiomyopathy, and in acquired senile cardiac amyloidosis.

Endocrine-related amyloid proteins (AE)

In endocrine tumours, such as medullary carcinoma of the thyroid gland amyloid fibrils have been isolated which contain proteins with a molecular weight of 5700 Da, being chemically related to calcitonin. This protein AE_t is presumably a proform of calcitonin.

Amyloid β-proteins

In amyloid fibril extracts of meningeal vessels of patients with Alzheimer's disease a protein with a molecular weight of 4200 Da (β-protein) has been isolated. In patients with Down's syndrome over 40 years of age the same amyloid protein has been found by Glenner and coworkers.

β-Microglobulin

Recently, in patients treated with regular haemodialysis for many years amyloid formation has been demonstrated, leading to a compression of the median or ulnar nerve in the wrist, and periarticular deposition in bone. Clinically this is recognized by a carpal tunnel syndrome or a spondyloarthropathy (see Chapter 10.3). In the amyloid preparations the fibrils consist of proteins very similar to $β_2$-microglobulin (molecular weight 11 000 Da). This $β_2$-microglobulin amyloid also can be found in other organs, but does not usually involve liver, spleen, or heart. It has been thought that high ambient $β_2$-Microglobulin concentrations are responsible for this amyloid formation and might be the result of insufficient removal through the membranes of the dialyser. $β_2$-Microglobulin is almost the only protein which will form β-pleated sheets spontaneously at high concentrations, without proteolysis.

Amyloid P-component (protein AP)

Protein AP is a glycoprotein which is found in almost all forms of amyloid deposits. It is a non-fibrillar protein and is conspicuous

Table 3 Age at diagnosis and sex of 144 patients with systemic AA and AL amyloidosis

Age (years)	AA		AL	
	Male	Female	Male	Female
< 20	1	1	0	0
20–29	3	3	0	0
30–39	7	4	2	0
40–49	8	5	2	2
50–59	11	11	13	10
60–69	7	15	9	6
70–79	6	9	2	6
≥ 80	0	0	1	0
Median age	53	59	56	58
Range	12–77	18–74	32–80	46–78

by its pentagonal structure. In normal serum a component with antigenic identity with AP is found. This protein is designated as SAP (serum amyloid P-component). Both AP and SAP have a molecular weight of 23 500 Da. It has been shown that SAP binds to amyloid fibrils in a calcium-dependent way *in vitro*, but it is unknown whether this P-component of amyloid contributes to the deposition of amyloid fibrils and if so, in which way.

In addition there is a partial amino-acid sequence homology between SAP and C-reactive protein, the latter being a acute-phase reactant in man. This is not the case for SAP, which shows a fairly constant level even during inflammation.

Clinical manifestations of amyloidosis

As amyloidosis is in many instances a generalized multisystem disease due to deposition of amyloid proteins in various organs, the clinical manifestations may be manifold. Symptoms also depend on the amount of amyloid proteins deposited in the tissues of the various organs.

Although the kidneys are often involved in amyloid deposition it is appropriate to give a survey of the other tissues in the body in which amyloid can be found, either as a part of generalized amyloidosis, or as a single-organ manifestation. These will be discussed first, as they are often accompanied by symptoms which may cause more problems in the patient than the renal involvement and which may contribute to a great extent to the variable clinical picture.

Age and sex incidence

During the last 25 years a large experience with patients suffering from amyloidosis has been amassed in our department. In 1986 a survey of 144 patients with AA and AL amyloidosis was published (Janssen *et al.* 1986). A diagnosis of systemic AA amyloidosis was made in 91 patients, while 53 patients were found to have AL amyloidosis. There were 72 male (43AA and 29AL) and 72 female patients (48AA and 24AL). Amyloid of both types is predominantly a disease of the middle-aged and the elderly. The age distribution at the time of the diagnosis is shown in Table 3.

Associated diseases

These 144 patients with systemic non-hereditary amyloidosis were also divided with regard to the clinical diseases responsible for the development of amyloidosis. A comparison was made with the results of the biochemical and histochemical characterization of the amyloid proteins involved (Table 4).

The introduction of more sensitive techniques such as immunoelectrophoresis and immunofixation of serum and urine, and of immunofluorescence on bone marrow specimens has enabled the existence of a monoclonal gammopathy to be demonstrated in all patients (even in the absence of overt bone marrow abnormalities) with idiopathic ('primary') systemic amyloidosis and atypical distribution. Before the introduction of these techniques this type of amyloidosis was considered to be truly 'idiopathic'. At present the designation 'systemic amyloidosis associated with monoclonal gammopathy' seems to be more appropriate. This is illustrated in Table 4, in which patients indicated as 'multiple myeloma and macroglobulinaemia' did show manifest plasma cell abnormalities on optical microscopy of bone marrow. Patients indicated as 'monoclonal gammopathy' did not show bone marrow plasma cell abnormalities on optical microscopy, but were found to have a clear preponderance of either kappa- or lambda-bearing plasma cells and in addition appeared to excrete a distinct (but usually small) amount of monoclonal light chains on immunofixation of concentrated urine. Those patients in whom no associated disease could be detected ($n=7$) were diagnosed before 1972 when bone marrow immunofluorescence and immunofixation of the urine were not yet included in the regular diagnostic work-up of patients with amyloidosis.

Presenting symptoms

The clinical features which led to the diagnosis of amyloidosis in each of the 144 cases are recorded in Table 5. The most frequent presenting symptom was of renal origin both in the AA (88 per cent) and in the AL group (29 per cent), although the spectrum of initial symptoms was more diverse in the latter and comprised a considerable number (21 per cent) of symptoms related to cardiac dysfunction. The median duration of presenting symptoms until the diagnosis of amyloidosis was made, was 2 months in the AA group (range 0–75) and 3 months (range 0–63) in the AL group (AA versus AL, $p>0.05$).

Organ involvement

Cardiac involvement

Infiltration of amyloid to the heart occurs often (34–80 per cent) in patients with AL amyloidosis and is in these cases frequently the cause of death. In AA amyloidosis cardiac involvement is reported to be uncommon, although perivascular cardiac amyloid deposition is not infrequently present at autopsy in AA patients. Amyloid deposition in the heart muscle tissue or in the cardiac vasculature may lead to congestive heart failure, arrhythmias (including cardiac arrest), and to cardiomegaly. The most characteristic, but not specific electrocardiographic feature in cardiac amyloidosis is a diminished voltage pattern. On echocardiography the ventricles are poorly mobile and may show a 'bright' echogenic appearance throughout.

Gastrointestinal tract

Gastrointestinal involvement in amyloidosis may result in obstruction, malabsorption, ulceration, haemorrhage, diarrhoea, and constipation. AL amyloidosis can be accompanied by infiltration of the tongue causing macroglossia. This symptom has been reported in 10 to 20 per cent of AL patients (Kyle and Greipp 1983); it may interfere with eating, and if severe can cause airway obstruction. In some of our patients macroglossia preceded other symptoms of amyloidosis, whereas in most patients macroglossia develops in the course of the disease. Amyloidosis of the stomach may show resemblance to gastric carcinoma. In the literature prepyloric obstruction has been

Table 4 Clinical and histochemical classification of 144 patients with systemic non-hereditary amyloidosis

Associated disease		No. of patients	M-component	$KMnO_4$ test	Amyloid protein
Rheumatoid arthritis		51	50−, 1+*	Sensitive	AA
Recurrent pulmonary infections		10	−	Sensitive	AA
Crohn's disease		5	−	Sensitive	AA
Ankylosing spondylitis		5	−	Sensitive	AA
Tuberculosis		3	−	Sensitive	AA
Osteomyelitis		2	−	Sensitive	AA
Familial Mediterranean fever		2	−	Sensitive	AA
Hodgkin's disease		2	−	Sensitive	AA
Agammaglobulinaemia		1	−	Sensitive	AA
Psoriatic arthritis		1	−	Sensitive	AA
Infected burns		1	−	Sensitive	AA
Recurrent abscesses		1	−	Sensitive	AA
Paraganglioma		1	−	Sensitive	AA
Carcinoma of the lung		1	−	Sensitive	AA
None		5	−	Sensitive	AA
Total number of patients with AA		91			
Multiple myeloma	(BM+)	24	+	Resistant	AL
Monoclonal gammopathy	(BM−)	20	+	Resistant	AL
Macroglobulinaemia	(BM+)	2	+	Resistant	AL
None	(BM−)	7	−	Resistant	AL
Total number of patients with AL		53			

BM+ = abnormal aspect/number of bone marrow plasma cells; BM− = apparently normal aspect/number of bone marrow plasma cells.
* In one patient with rheumatoid arthritis an M-component was present in the serum; the clinical picture and the result of the $KMnO_4$ method were in accordance with AA amyloid.

Table 5 Presenting features in 144 patients with systemic AA and AL amyloidosis (number of cases and percentage in parentheses)

Presenting symptom	All cases	AA	AL
Proteinuria	73(51)	63(70)	10(19)
Renal insufficiency (creatinine clearance < 10 ml/min)	15(10)	11(13)	4(8)
Renal function loss (creatinine clearance > 10 ml/min)	6(4)	5(5)	1(2)
Heart failure	10(7)	0(0)	10(19)
Paraesthesia	7(5)	0(0)	7(13)
Fatigue	7(5)	2(2)	5(9)
Diarrhoea	6(4)	4(4)	2(4)
Gross bleeding	5(3)	4(4)	1(2)
Intestinal pseudo-obstruction	3(2)	1(1)	2(4)
Carpal tunnel syndrome	2(1)	0(0)	2(4)
Ascites	3(2)	0(0)	3(6)
Dysphagia	2(1)	0(0)	2(4)
Sicca syndrome	1(1)	0(0)	1(2)
Orthostatic hypotension	1(1)	0(0)	1(2)
Muscle weakness	1(1)	0(0)	1(2)
Heart block	1(1)	0(0)	1(2)
Haematuria	1(1)	1(1)	0(0)
Total	144(100)	91(100)	53(100)

reported (Kyle and Greipp 1983; Cooley 1953) which may cause repeated vomiting.

Malabsorption and diarrhoea can be the result of muscle infiltration or autonomic neuropathy, associated with vascular insufficiency and bacterial overgrowth. Stephens (1962) reported a syndrome of internal pseudo-obstruction, defined as the occurrence of clinical signs of mechanical obstruction of the intestine without organic occlusion of the lumen. It is important to recognize this condition in order to avoid surgical treatment, which would be ineffective and is associated with a very high mortality (Legge et al. 1970). Hepatic involvement is very common in both AL and AA amyloidosis, although it is only occasionally accompanied by serious liver function disturbances (Cohen and Skinner 1982). Only hepatomegaly and elevation of serum alkaline phosphatase are often found. In addition, in AL amyloidosis, intrahepatic cholestasis has been reported in a minority of cases; so far this has not been reported in AA amyloidosis. Portal hypertension and formation of ascites may also occur. It must be kept in mind that hepatomegaly in patients with amyloidosis can also be caused by heart failure.

The nervous system

Patients with systemic amyloidosis can develop peripheral neuropathy as a result either of infiltration of the nerve by amyloid, or by compression of the nerve by surrounding deposits of amyloid, as for instance in the carpal tunnel syndrome. The infiltrative type of neuropathy is usually distal, symmetric and progressive, affects mainly the lower limbs and is reported in 8 to 17 per cent of AL patients (Kyle and Greipp 1983; Benson et al. 1975). Nervous involvement can also lead to autonomic dysfunction, which may present as orthostatic hypotension, gastrointestinal disturbances, bladder dysfunction, or impotence. To date no peripheral or autonomic neuropathy has been reported in AA amyloidosis.

Amyloid can occur in the central nervous system in senile plaques and in blood vessels in Alzheimer's disease and Down's syndrome patients over 40 years of age (Glenner and Wong 1984).

Endocrine and exocrine glands

Amyloid can be found in the thyroid gland in cases of medullary carcinoma and may also occur as amyloid goitre in systemic AA or AL amyloidosis. A characteristic feature of amyloid goitre is the rapid growth of the thyroid, often leading to surgical exploration on the suspicion of a neoplasm. The adrenal glands are frequently involved in primary AL amyloidosis, which can lead clinically to Addison's disease.

Dysfunction of exocrine glands has been reported in the literature with the development of Sicca syndrome.

Musculoskeletal

Myopathy can develop in amyloidosis, and is mostly caused by amyloid deposition in the walls of small vessels resulting in ischaemia (Bruni et al. 1977). Infiltration of muscle tissue itself by amyloid has also been described.

As amyloid can be deposited in synovial membranes and synovial fluid or in the articular cartilage, joint symptoms lead to the suspicion of rheumatoid arthritis. Clinical signs in this condition may also include morning stiffness, symmetric joint swelling, and synovial thickening with effusions. A rather typical feature is the development of 'shoulder pads' in AL amyloid arthropathy.

Haematologic abnormalities

Haematologic changes may be found in the form of haemorrhagic diathesis, thought to be related to amyloid infiltration of blood vessels resulting in an increased fragility. Bleeding can occur in the skin, in the gastrointestinal tract, in the bronchi, and after invasive diagnostic procedures. Although clotting tests in general do not show abnormal findings, fibrinogenopenia, increased fibrinolysis and selective deficiencies of clotting factors have been reported, the last for instance in an isolated factor X deficiency by Furie et al. (1977).

Skin

Skin involvement is frequently seen in patients with amyloidosis leading to papule and plaque formation, usually in folds of the skin. Amyloid can be demonstrated by skin biopsy both in patients with AL and AA amyloidosis and this technique is frequently used as a diagnostic procedure.

Renal involvement in amyloidosis

The spectrum of renal symptoms and signs in amyloidosis is wide. Proteinuria is the most frequently noted symptom, ranging from symptomless proteinuria to a severe nephrotic syndrome, the latter sometimes still present in the situation of renal insufficiency. Cohen reported in 1967 that the asymptomatic proteinuria phase may last for years. The time elapsing from slight proteinuria to the development of a full-blown nephrotic syndrome is unknown. In cases of familial Mediterranean fever, it was reported by Sohar et al. (1967) to last for 3 to 5 years. The prevalence of a nephrotic syndrome in AL amyloidosis has been reported as 35 per cent (Kyle and Greipp 1983) and in AA amyloidosis as 50 per cent (Cohen and Calkins 1959).

In a series of 53 patients we found a more severe proteinuria in AL patients (median 8.5 g/24 h; range 5–22) than in AA patients (median 5 g/24 h; range 0–20) ($p<0.01$). Serum albumin concentrations, as expected, showed a negative correlation with severity of proteinuria. Proteinuria in excess of 5 g/24 h was found in 90 per cent of AL patients and in 49 per cent of AA patients. Only two AA patients had proteinuria below 0.5 g/24 h. There was no significant correlation between the degree of urinary protein excretion and the creatinine clearance (Fig. 1).

With regard to selectivity of the urinary protein excretion, 59 per cent of the patients had a selective proteinuria (defined as IgG clearance/albumin clearance × 100) at the time of renal biopsy, whereas no significant correlation was found between selectivity and the amount of protein, excreted in the urine. A negative correlation was found between the selectivity index and the creatinine clearance in this series (Fig. 2).

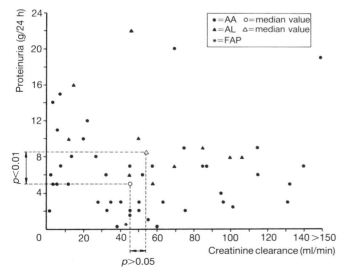

Fig. 1 Creatinine clearance and proteinuria at the time of diagnosis in 54 patients with biopsy-proven renal amyloidosis.

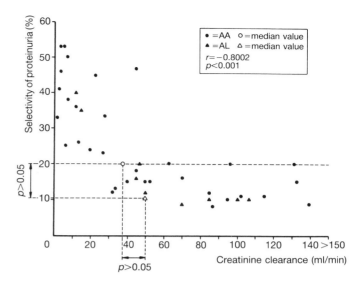

Fig. 2 Creatinine clearance and selectivity of the proteinuria at the time of diagnosis in 41 patients with biopsy-proven renal amyloidosis.

Haematuria is not a conspicuous symptom of amyloidosis, although microscopic haematuria may be present, occasionally accompanied by red blood cell casts (Cohen 1967). We found isolated haematuria in 10 of 53 (20 per cent) patients (9 AA, 1 AL); it constituted the presenting symptom in only one (AA) patient.

Hypertension is, according to the literature, relatively uncommon and was found in 20 per cent of a series of 56 patients and in 35 per cent of a series of 32 patients with more advanced renal failure (Bentwich et al. 1971). Sohar et al. (1967), however, have pointed out that in familial forms of amyloidosis hypertension is noted more frequently. In the study of Janssen (1985) in our series of 53 patients with biopsy-proven amyloidosis (43 AA, 10 AL) hypertension, defined by a diastolic blood pressure over 90 mmHg, was found in 11 patients (8 AA and 3 AL). There was no correlation between the extent of elevation of blood pressure and the degree of renal function loss.

On the other hand long-standing amyloidosis is regularly accompanied by postural hypotension, which can be a disabling symptom (Cohen 1967; Kyle and Greipp 1983). This probably arises from both autonomic neuropathy and adrenal dysfunction. With progression of the renal involvement, there is a decline in renal function, subsequently leading to severe renal insufficiency, necessitating dialysis.

With regard to the size of the kidneys in amyloidosis, radiological studies have revealed that the kidneys are generally of normal length and width, whereas with advancing renal failure the kidneys tend to become smaller than normal (Brandt et al. 1968; Ekelund 1977; Elkin 1980). This is in contrast with the long-held belief that in amyloidosis the kidneys are always enlarged, as was mentioned in the original description by pathologists in the 19th century.

In the presence of thrombosis of the renal vein, which is a well-known complication in amyloidosis, the size of one or both kidneys may be increased. Existence of renal vein thrombosis therefore should be suspected when the size of the kidneys appears to have increased compared to previous radiographs.

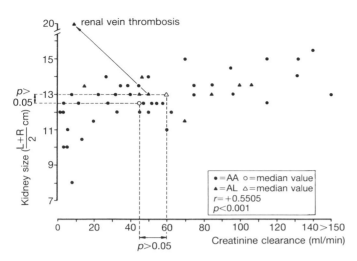

Fig. 3 Creatinine clearance and kidney size at the time of diagnosis in 49 patients with biopsy-proven renal amyloidosis.

Finally, measuring the kidney size in plain films of the abdomen or intravenous urograms revealed that the kidney size in most instances was normal, with a distinct tendency to decrease with declining renal function (Fig. 3).

Pathology of renal amyloidosis

Renal amyloidosis is a major feature both in AA and in AL amyloidosis. In AA amyloidosis the kidneys are almost always affected, whereas in AL amyloidosis this is found less frequently (reportedly in about 35 per cent of cases).

Gross anatomy

Students have always been told that renal amyloidosis is associated with large pale smooth kidneys. This terminology is almost as diagnostic as the butterfly rash in systemic lupus erythematosus. Undoubtedly this may be true, and the weight of the two kidneys in amyloidosis may be double the normal weight, but in our experience in many autopsy reports on patients with amyloidosis the size of the kidneys appeared normal or only slightly increased. This was confirmed in radiographic studies during life, in which kidney size was shown to decrease with declining renal function. Regularly contracted granular kidneys are found in amyloidosis, particularly in cases with long-standing hypertension and/or infection. An exception may be the occurrence of renal vein thrombosis, which is associated with renal enlargement.

Microscopic investigations
Glomeruli

Glomerular involvement is found in all cases of AA amyloidosis, whereas in AL amyloidosis this is present in about 50 per cent of cases studied.

The mesangium is the first part of the glomerulus where amyloid can be demonstrated, in the form of deposits without hypercellularity. This latter phenomenon serves as a diagnostic criterion to differentiate it from early abnormalities found in diabetes mellitus. In later stages, the amyloid acquires nodule-like appearances with compression of mesangial cells, even into the lumen of adjacent capillary loops.

In early stages the capillary walls show no amyloid involvement. In later stages amyloid can be also demonstrated in the walls of the capillaries (see Plate section, Plate 1(a)).

During the evolution of the amyloidotic involvement of the glomeruli amyloid deposits can be shown adjacent to the glomerular basement membrane. Amyloid may be deposited both in a subepithelial and in a subendothelial distribution. The subepithelially situated amyloid deposits sometimes have an appearance of spikes as are usually found in membranous glomerulopathy, but differ in being finer, more numerous, and often larger. Also, in amyloidosis the distribution of these 'spikes' is less diffuse and more patchy (Plate 1(c)). Sometimes amyloid deposits can also be demonstrated in the wall of Bowman's capsule and on the inside of the tubular basement membrane which is not found in membranous glomerulopathy. In a very limited number of cases epithelial cell proliferation with crescent formation is present. In rare cases giant cells containing an amyloid substance staining positively with PAS have been described.

In the full development of amyloidosis the glomerulus will be 'consumed' by amyloid deposits giving a homogeneous appearances without hypercellularity. In these full-blown cases sclerosis will eventually develop and tubular atrophy will follow.

Tubules

Usually, in early stages no amyloid can be demonstrated in tubular tissue. Later on, deposits can be found along the tubular basement membrane and the vasa recta. The deposits along the tubular basement membrane are predominantly demonstrated in the distal tubules and the loops of Henle. Gradually the evolving picture is easier to recognize and this stage will be followed by tubular atrophy. Sometimes the tubules are filled with casts, staining positively for amyloid, especially in patients with AL amyloidosis due to multiple myeloma. Sometimes a large amount of interstitially located amyloid is found in the medulla; this can be so extensive that anoxic necrosis of tubuli can ensue.

Blood vessels

Amyloid can be shown in the wall of blood vessels, comprising the adventitia as well as the intimal layer. Extensive amyloid formation in the wall of blood vessels can cause occlusion of the vessel lumen. Arterioles as well as venules may contain amyloid deposits. There is no clear correlation between the amount of amyloid in the glomeruli and the amount of amyloid found in vessel walls.

A well-known complication of amyloidosis is thrombosis of the renal veins, starting in the small venous vessels in the kidney and subsequently spreading from capillaries and venules to larger vessels and even to the main renal vein and from there eventually to the inferior vena cava (Barclay et al. 1960).

The formation of thrombi in the renal venous circulation is thought to be facilitated by dehydration of the patient. Undoubtedly also a reduced renal blood flow due to progressive renal lesions will also act as a contributive factor in the formation of thrombi. In addition, altered composition of the blood, including a decrease of antithrombin III levels may also play a role (see Chapter 3.4 for a discussion of coagulopathy in nephrotic patients and renal venous thrombosis).

Immunofluorescence studies

Some reports have been published on immunofluorescence studies of glomeruli in renal amyloidosis. The results are not homogeneous; Jerath and coworkers (1980) found in 10 of 27 biopsies no positive immunofluorescence. In 14 biopsies complement C3 and immunoglobulins could be demonstrated, whereas special investigations for the presence of IgE, albumin, and transferrin were negative in all biopsies. In nine of 10 biopsies studied, kappa or lambda chains could be found. Turkish investigators (Müftüoglu et al. 1977) found IgG and C3 complement in all of 15 biopsies, whereas in five of six cases studied IgA and IgM was demonstrated.

If positive, the immunofluroescence studies show a homogeneous, non-granular, non-linear staining, predominantly in the mesangium. Diffuse non-specific binding of immunoglobulins and complement by amyloid is generally assumed to occur.

Electron microscopic studies

Many studies in humans and in experimental animals have shown that amyloid is composed of fine fibrils 8 to 10 nm in diameter with a length up to 1 μm. In the beginning most of these fibrils are found in the mesangium. They have also been found in mesangial cells.

With electron microscopic studies also, fibrils can be shown in the capillary walls, with a possible location on both sides of the basement membrane. In the subepithelial location the fibrils have a perpendicular appearance in relation to the basement membrane which explains the 'spike' or 'brush' formation. Occasionally the amyloid shows signs of digestion and resorption, especially in places where new basement membrane material is laid down on the epithelial side of the deposits.

Histologic staining methods for amyloid

In recent years methods have been developed to distinguish between the nature of amyloid deposits in tissues. It has already been mentioned that Congo red-stained amyloid in polarized light shows a characteristic apple-green birefringence (Plate 1(a), (b)) and moreover, that the type of amyloid can be analysed by amino-acid sequence analysis. An indirect method of differentiating between AA and AL amyloid proteins is the incubation of tissue sections in a 5 per cent potassium permanganate solution (Wright et al. 1977; van Rijswijk et al. 1979). After incubation with the solution, AA amyloid can no longer be stained with Congo red, whereas the staining of AL amyloid with Congo Red is unaffected. This loss of affinity for Congo Red by AA amyloid after incubation in potassium permanganate solution is probably due to interference with the structure of the protein. This method has been shown to be simple and reliable and can be carried out in every laboratory.

In special laboratories for amyloid research, immunochemical methods are used to define the amyloid proteins in tissue by using purified specific antisera against AA and AP proteins (Plate 1(d)). With this immunohistochemical method, using the unlabelled antibody peroxidase–antiperoxidase (PAP) method of Sternberger, a very good correlation was obtained in a study comparing the value of the potassium permanganate method with the PAP method (Janssen et al. 1985). In the immunohistochemical method the anti-AP antiserum served as an indicator for the presence of amyloid, whereas the anti-AA antiserum defined the presence of AA amyloid proteins. By showing such a close correlation, the authors concluded that the potassium permanganate method, by its simplicity, is a very good tool for differentiation between AA and AL amyloid proteins. For the immunohistochemical characterization of AL amyloid, a panel of antisera raised against a number of different AL proteins is needed since AL proteins differ from patient to patient. Only when the constant region of the Ig light chains is part of the AL protein will the amyloid deposits react with regular anti-light chain antisera. Such a panel of anti-AL antisera, reacting with specific light chain subgroups is not yet available for routine use.

Hereditary amyloidosis

In the last 40 years different types of inherited amyloidosis have been described (Pras 1986). Most of these forms of hereditary amyloidosis are marked by systemic deposition of amyloid proteins, while in others there is only a localized deposition of amyloid at specific sites, e.g. in the skin in familial lichen amyloidosis, or in the tumour and its metastases in familial medullary carcinoma of the thyroid gland. The inherited systemic amyloidoses can be divided into three groups.

1. In the nephropathic group, proteinuria is the main feature, progressing to a nephrotic syndrome, followed later by renal insufficiency.
2. The neuropathic group is marked by a progressive symmetrical polyneuropathy in the lower or upper limbs or in the face. In this group prealbumin is the protein subunit of amyloid fibrils.
3. In the cardiopathic group severe heart failure occurs leading to death.

To date no familial amyloidosis has been described based on AL amyloidosis.

The group of familial systemic amyloidosis with nephropathy as the predominant phenomenon includes familial Mediterranean fever, amyloidosis with febrile urticaria and deafness (Muckle-Wells syndrome); scattered reports have also been made on nephropathic amyloidosis in families of different ethnic origin.

Familial Mediterranean fever

Familial Mediterranean fever (FMF) is the most frequent cause of familial systemic amyloidosis, and was first described by Sohar et al. from Israel (1967), where in the meantime over 1500 patients have been treated (Pras et al. 1984). The disease is also found in other countries in the Mediterranean area, and appears to be transmitted as an autosomal recessive trait. FMF is most common in Sephardic Jews, Anatolian Turks, Armenians, and Arabs of the Middle East.

There are two phenotypes of FMF: the most frequently found type I is recognized by repeated febrile attacks of short duration preceding the development of amyloidosis, while in phenotype II the renal symptoms precede the development of amyloidosis without the occurrence of febrile attacks.

The characteristic febrile attacks affect many patients with FMF in the first decade of life; 90 per cent of the patients have manifest symptoms before the age of 20 years. The periods of fever are often accompanied by episodes of peritonitis, pleuritis, synovitis or erythema, also marked by a short duration.

During these febrile episodes acute-phase reactants such as SAA are highly elevated. Remarkably, SAA levels tend to remain elevated in the symptom-free intervals. The amyloid protein in the tissues is of the AA type. The major manifestation of amyloidosis in FMF is located in the kidneys. Clinically the spectrum varies from asymptomatic proteinuria, via a severe nephrotic syndrome, to the development of renal insufficiency at a relatively young age.

At present many of these patients with end-stage renal failure due to amyloidosis are treated by regular dialysis or renal transplantation.

Amyloidosis with febrile urticaria and deafness

This was described by Muckle and Wells (1962) in consecutive generations of a family showing an autosomal dominant trait transmission. This syndrome is characterized by short periods of fever, complicated by an urticarial rash. Progressive perceptive deafness is a second manifestation of this form of familial nephropathic amyloidosis. In latter years this syndrome has also been reported in other families.

Analysis of the proteins extracted from amyloid tissue in these patients has revealed a homology with protein AA.

Miscellaneous

Other observations of nephropathic systemic amyloidosis have been made in different families of varying ethnic origin, e.g. Swedish, Polish, Irish, English, and German. These examples of familial hereditary amyloidosis are different from FMF and the Muckle–Wells syndrome.

Diagnosis of amyloidosis

To make the diagnosis of amyloidosis it is mandatory to obtain a tissue specimen, and to demonstrate the characteristic apple-green birefringence with polarized light after staining with Congo Red. In routine optical microscopy the waxy homogeneous acellular appearance of amyloid may be easily missed.

It is important to consider amyloidosis in the differential diagnosis of patients with particular clinical manifestations. In our opinion, in too many patients the diagnosis of amyloidosis is postponed until the patient is referred to a centre with experience and interest in this disease.

The diagnosis of amyloidosis should be considered in the following categories of patients.

1. Patients with chronic conditions associated with suppurative infections and/or chronic inflammatory disorders (AA amyloidosis). This category includes patients with rheumatoid arthritis (see Chapter 4.8), ankylosing spondylitis, tuberculosis, bronchiectasis, osteomyelitis, cystic fibrosis, paraplegia complicated by infections, leprosy (see Chapter 3.14), Crohn's disease, Behçet's syndrome (Bemelman et al. 1989), Takayasu's arteritis (van der Meulen et al. 1989), and congenital hypogammaglobulinaemia. In the literature also severe drug abuse complicated by multiple skin abscesses has been reported to be associated with amyloidosis (Jacob et al. 1978; Meador et al. 1979; Scholes et al. 1979). When a patient suffering from (or who has suffered from) one of the diseases mentioned develops or is found to have proteinuria, with or without a nephrotic syndrome, amyloidosis should be considered and a biopsy should be taken. If the biopsy is positive, it is most likely that in such cases AA amyloid will be present.

2. AL amyloidosis has to be considered especially in patients with symptoms of nephropathy, cardiomyopathy, neuropathy, or macroglossia, and can be proven by an appropriate biopsy and staining methods. However, it may present as isolated proteinuria or a nephrotic syndrome without any other clinical manifestation. Because of the age distribution demonstrated in Table 3, AL amyloidosis should be considered in any proteinuric patient over the age of 50 years; about one in five nephrotics over 60 years of age has AL amyloidosis. In patients with AL amyloidosis serum protein electrophoresis and investigation of the bone marrow is necessary to establish a diagnosis of either multiple myeloma, macroglobulinaemia, or monoclonal gammopathy as a cause of amyloidosis.

3. In heredofamilial syndromes with symptoms of peripheral neuropathy, nephropathy, and cardiac disease amyloidosis should be suspected.

4. In other patients with symptoms of a diffuse non-inflammatory illness or with intractable congestive heart failure the differential diagnosis should include amyloidosis. In these cases mesenchymal tissues such as blood vessels, heart, or gastrointestinal tract, or parenchymal tissues (kidney, liver, spleen, adrenal glands) can be involved.

The role of tissue biopsy

Biopsies can be taken from many tissues, even from tissues which do not show any sign of involvement in the disease process, e.g. the rectum.

In our experience the rectal biopsy is a simple procedure, often leading to the diagnosis. It is important that the biopsy specimen includes submucosal tissue with vessels in it. Otherwise, it may lead to false-negative results. Other procedures used as a screening test include subcutaneous fat aspiration with a fine needle, and biopsies of gingival tissues. The sensitivity of the latter procedure, however, is inferior compared to the rectal biopsy.

If an appropriate rectal biopsy is negative, a renal biopsy can be performed. This is important in the presence of unexplained proteinuria or a nephrotic syndrome, since in at least 10 per cent of patients the renal biopsy may contain amyloid, even when the rectal biopsy is negative (Cohen 1967). In a comparative investigation of rectal and renal biopsies in a series of 58 patients with systemic amyloidosis in our own department, the renal biopsy showed amyloid deposition in 43 of 46 AA patients, and in 10 of 11 AL patients (Janssen 1985). In the same study the rectal biopsy contained amyloid in 35 of 45 AA patients, and in eight of 10 AL patients (Table 6).

In three patients with rheumatoid arthritis with a positive rectal biopsy, the renal biopsy showed no amyloid deposition, but disclosed local focal glomerulonephritis in two patients and membranous glomerulopathy in a third (see Chapter 4.8).

In conclusion, in a minority of cases of amyloidosis rectal and renal biopsies do not show concordant results, and it may be necessary in certain circumstances to take biopsies from both sites. In our experience no major bleeding problems related to biopsy procedures are observed.

As has been indicated earlier, the classification of amyloid tissue in biopsies can be achieved by the potassium permanga-

Table 6 Results of kidney and rectal biopsies in 58 patients with systemic amyloidosis

Type of amyloid	No. of patients	Kidney + Rectum +	Kidney + Rectum −	Kidney − Rectum +	Kidney + Rectum NP
AA	46	32	10	3	1
AL	11	7	2	1	1
Familial amyloidotic polyneuropathy	1	1	0	0	0
Total	58	40	12	4	2

NP = not performed.

nate ($KMnO_4$) method. AA amyloid is $KMnO_4$-sensitive, as staining of amyloid with Congo red is completely abolished after exposure to $KMnO_4$. On the other hand AL amyloid is $KMnO_4$-resistant, as Congo red affinity is unaffected by prior exposure to $KMnO_4$.

Using an antiamyloid AA-PAP method Janssen et al. (1985) showed that the results of the $KMnO_4$ method are in accordance with the more specific method of immunohistochemical typing of amyloid with polyclonal and monoclonal antibodies to protein AA. If available this should be the method of choice.

Bone marrow aspiration should be performed in the diagnostic work-up of patients with amyloidosis. It is essential to perform immunofluorescence with antibodies directed against immunoglobulin classes and to light chains. This technique enables the detection of monoclonal gammopathy even in the absence of overt plasma cell dyscrasia as is often the case in 'primary' amyloidosis. If a bone-marrow biopsy is performed, Congo red staining should be specifically requested, as amyloid deposits may be found in 70 per cent of bone-marrow biopsies of patients with AL amyloidosis, whereas in our experience amyloid deposits in bone marrow are never encountered in patients with AA amyloidosis.

Laboratory investigation

Renal function

Renal insufficiency in amyloidosis has been attributed primarily to glomerular amyloid deposits. Although glomeruli are almost invariably affected, renal failure does not develop only as a result of glomerular amyloid deposition. Even in severe glomerular amyloidosis glomerular function, as measured by creatinine clearance, may be only slightly disturbed (Muehrcke et al. 1955; Watanabe et al. 1975). Mackensen et al. (1977) found that the degree of interstitial fibrosis is more important for the impairment of renal function than glomerular amyloid deposition. Renal decompensation was explained by a reduction of the cross-sectional area of the postglomerular vessels by interstitial fibrosis, with a consequent decrease of the renal blood flow leading to a diminished clearance by predamaged glomeruli. Others have pointed to the role of amyloid deposition in blood vessels with regard to renal function. Törnroth et al. (1980) concluded that vascular amyloidosis, interstitial fibrosis with concomitant tubular atrophy, and glomerular hyalinization correlated with the degree of renal insufficiency, whereas no clear correlation was found between glomerular amyloid and renal failure. It appears that vascular amyloid deposition is of major importance in the development of renal failure, as it leads to hyalinization of glomeruli and interstitial fibrosis.

Renal tubular function was studied by Janssen et al. (1985) in our department with $^{99}Tc^m$-dimercaptosuccinate (DMSA), which is normally reabsorbed in proximal tubules. Van Luijk et al. demonstrated in 1984, that in proximal tubular dysfunction there is a markedly reduced renal uptake of the radiopharmaceutical by an apparently accelerated renal loss of $^{99}Tc^m$-DMSA due to impaired reabsorption. Further studies have shown that the clearance of $^{99}Tc^m$-DMSA represents proximal tubular function which may be disturbed in interstitial nephritis and fibrosis. In 29 patients with histologically confirmed renal amyloidosis, $^{99}Tc^m$-DMSA clearance was measured concomitantly with GFR and ERPF, using ^{125}I-sodium iothalamate and ^{131}I-sodium hippurate respectively, as described by Donker et al. (1977).

In amyloid nephropathy the relative clearance of $^{99}Tc^m$-DMSA in relation to the ^{125}I-sodium iothalamate clearance appeared to be increased in 86 per cent of patients. In control patients with a decreased GFR due to non-interstitial diseases the relative clearance of $^{99}Tc^m$-DMSA was within the normal range (below 16 per cent of the GFR) irrespective of renal function (Fig. 4).

From this investigation we concluded that proximal tubular dysfunction is a frequent finding in amyloid nephropathy. Moreover, renal tubular dysfunction correlated with the degree of interstitial damage. Therefore it appears that the relative clearance of $^{99}Tc^m$-DMSA is a useful non-invasive test to estimate the degree of interstitial damage in renal amyloidosis.

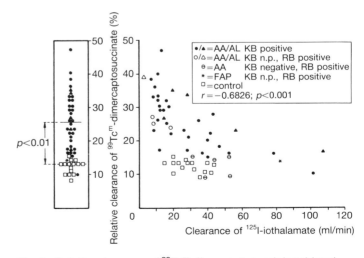

Fig. 4 Relative clearance of $^{99}Tc^m$-dimercaptosuccinic acid and glomerular filtration rate in patients with amyloid nephropathy ($r = -0.6826$) compared with a control group of patients with other renal diseases ($p < 0.01$); KB = rectal biopsy, NP = not performed.

Measurement of SAA levels (serum amyloid A)

As was mentioned earlier, AA amyloid is thought to be derived from the precursor protein serum amyloid A (SAA). There is immunochemical identity as well as N-terminal amino-acid sequence homology between SAA and AA amyloid. Progressive AA amyloidosis is associated with elevated SAA levels. From animal models, there is direct evidence that AA amyloid is derived from SAA.

SAA is elevated in active rheumatoid arthritis, ulcerative colitis, tuberculosis, neoplastic diseases, and familial Mediterranean fever. In immunodeficiency diseases like hypogammaglobulinaemia, SAA levels may be elevated due to infectious complications. Most of these diseases are known to be associated with AA amyloidosis.

SAA appears to be a marker for the acute-phase reaction comparable to C-reactive protein (CRP). Both proteins may be used to assess activity of the infectious and non-infectious inflammatory conditions.

In our laboratory SAA levels are measured by radioimmunoassay, while C-reactive protein levels are measured using either commercially available radial immunodiffusion plates or an automatic rate nephelometer. SAA levels higher than 0.2 μg/ml were only found in the presence of active inflammatory disease or in the presence of neoplastic diseases associated with biochemical signs of inflammation. SAA levels showed a very good correlation with the clinical activity of the inflammatory process. Favourable clinical responses following the institution of appropriate treatment are reflected by a decrease of SAA levels.

In our experience SAA measurements play an important role in the diagnosis of AA amyloidosis and the monitoring of the activity of the process leading to deposition of AA amyloid.

Search for monoclonal gammopathy

In a search for a monoclonal gammopathy as the possible precipitating event it is necessary to perform immunofluorescent or (being the most sensitive technique) immunofixation studies of serum and concentrated urine. Even in the absence of a monoclonal component in the serum, it is generally possible to detect minute amounts of Bence-Jones proteinuria in patients with AL amyloidosis using immunofixation of a concentrated sample of the 24-h urine production. In combination with immunofluorescence of bone marrow plasma cells these techniques provide very sensitive tools for the detection of monoclonal gammopathy.

Therapy of amyloidosis

General remarks

The therapeutic possibilities in the treatment of amyloidosis are restricted. As amyloidosis is caused by extracellular deposition of non-phagocytosable insoluble proteins therapy should be aimed at reducing the production of these amyloid precursor proteins.

In patients with AL amyloidosis therapy with melphalan and prednisone has been in use for many years with the purpose of reducing the number of pathological plasma cells in the bone marrow capable of producing immunoglobulin light chain proteins (see Chapter 4.3). Kyle and Greipp (1983) observed that AL patients had a better prognosis under this combined treatment than the placebo group. Proteinuria in the treatment group decreased, in some cases by more than 50 per cent, and in a minority of cases even disappeared. These authors mention however, that in two of their patients showing a clinical remission of the nephrotic syndrome, amyloid deposition in follow-up biopsies was more severe!

In patients with AA amyloidosis, in general the treatment should be directed at the underlying infection or inflammation. Eradication or suppression of the inflammatory process is required to halt the progression of amyloidosis. In some cases (e.g. osteomyelitis, bronchiectasis) surgery will be necessary. General treatment of amyloidosis is also directed at symptoms originating from the process located in various organs. In cases with renal amyloidosis and the nephrotic syndrome, treatment is based on sodium restriction, protein restriction, and administration of diuretics (see Chapter 3.3).

In patients with severe proteinuria due to amyloidosis in our department, indomethacin treatment has been used to reduce the excretion of protein. In a number of patients it has been possible to reduce the proteinuria to such an extent that the serum albumin increased and oedema disappeared. However, in our experience patients with extensive renal amyloidosis and a nephrotic syndrome in the presence of renal failure do not appear to be suitable candidates for treatment with NSAIDs when creatinine clearance has decreased below 25 ml/min, as in these cases acute renal failure may be superimposed, necessitating haemodialysis.

On the other hand, in refractory cases of renal amyloidosis with massive oedema due to a nephrotic syndrome in the presence of severe renal insufficiency, it may actually be advisable 1979to abolish renal function by administration of NSAIDs in order to stop the loss of protein. In such a situation, it is preferable to treat the patient with regular dialysis and to alleviate the symptoms of the severe nephrotic syndrome plus uraemia. The clinical condition of the patients may improve considerably by this 'medical nephrectomy' procedure.

Specific therapy
Colchicine

Colchicine treatment was described first by Goldfinger in patients with familial Mediterranean fever (1972). Daily use of colchicine in a dosage of 1 to 2 mg results in prevention of reduction of the periodic febrile attacks and abdominal complaints. It appears that SAA levels which are elevated during the attacks are affected favourably by colchicine.

Zemer et al. (1986) described a significantly reduced risk of development of a nephrotic syndrome in FMF patients taking colchicine on a regular daily schedule compared to FMF patients using this drug irregularly. In patients with amyloidosis other than that associated with FMF, colchicine has been used but has no effect.

Dimethyl sulphoxide

Dimethyl sulphoxide (DMSO) was reported by Isobe and Osserman in 1976 to dissolve amyloid fibrils *in vitro*. They also found that administration of DMSO in mice with experimentally induced amyloidosis resulted in a reduction of the amount of amyloid. Later, this was confirmed by Kedar et al. (1977). Ravid et al. (1977) found amyloid-like material in the urine of patients with renal amyloidosis after a single dose of DMSO. In contrast, van Rijswijk et al. (1979a, b) could not obtain convincing evidence for the dissolution of amyloid, either by repeated biopsies, or by demonstration of amyloid-like material in the urine.

Nevertheless in a number of patients with AA amyloidosis due to rheumatoid arthritis, treatment with DMSO resulted in remarkable improvement of renal function. DMSO was administered either orally in a dosage of 5 g dissolved in 25 ml distilled water three times daily, or intravenously (5 g in 50 ml NaCl 0.9 per cent three times daily). These patients also experienced a decrease in pain, and an increase of joint mobility during treatment with DMSO. Another striking observation was the decrease in SAA and C-reactive protein levels, indicating a decrease of inflammatory activity. Not all patients with AA amyloidosis due to rheumatoid arthritis showed improvement. This has, in our department, led to the policy of treating such patients with DMSO for about 3 weeks to establish whether there is any improvement. In the case of a favourable response the regression of renal function may come to a halt.

The side-effects of DMSO may include transient elevation of serum transaminase levels and nausea. A real social problem, however, is the appalling smell produced by the patient consuming DMSO, due to the conversion of a small percentage of this drug into dimethylsulphide. This frequently results in patients declining or stopping the treatment.

At the moment, it seems that the beneficial effect of DMSO, if present, is related to a distinct anti-inflammatory effect on rheumatoid arthritis, with a consequent decrease of SAA. DMSO has been shown to be an effective oxygen radical scavenger (van Rijswijk et al. 1983). As was mentioned earlier, reduction in renal function is influenced by a renal interstitial reaction. It is conceivable also that DMSO may reduce the interstitial inflammatory reaction, thereby improving the renal function. Thus those patients with rapidly declining renal function in particular, presumably with renal interstitial reaction, may be expected to benefit from this treatment and should be offered it.

Corticosteroids and cytostatics

Either elimination of the associated inflammatory disease or non-specific suppression of the inflammatory activity may favourably alter the course of AA amyloidosis. As long ago as 1959, Parkins and Bywaters noted favourable effects of corticosteroids, which are very effective in reducing the magnitude of the acute-phase reaction, including the production of C-reactive protein and SAA.

Ahlmen et al. (1987) and Berglund et al. (1987) reported a beneficial effect of treatment with cytostatic drugs in patients with rheumatic disease complicated by amyloidosis. These findings are in accordance with experience in our department, where patients with rheumatoid arthritis and amyloidosis are treated with prednisolone in combination with azathioprine (van Rijswijk et al. 1983, 1986). At present it seems reasonable to treat patients with systemic AA amyloidosis associated with rheumatoid arthritis with cytostatic drugs such as cyclophosphamide, chlorambucil, or azathioprine, either alone or in combination with corticosteroids. As the effect of treatment with cytostatic drugs may take weeks or even months to appear, it may be necessary to prescribe corticosteroids in addition to ensure an immediate reduction of the acute-phase response, and in particular to block the synthesis of serum amyloid A precursor. An optimal dosage schedule should be achieved by monitoring SAA and C-reactive protein levels. Because of the relatively short half-life of prednisolone this drug should preferably be administered in two or three doses over 24 h, e.g. 2.5 mg three times daily.

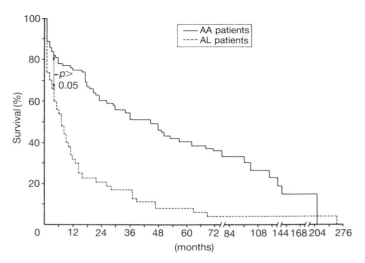

Fig. 5 Survival of 91 patients with systemic AA amyloidosis and 53 patients with systemic AL amyloidosis. Statistically significant differences between AA and AL amyloidosis exist from 5 months on ($p<0.05$). No significant differences were found between patients with myeloma-associated AL amyloidosis and patients with 'primary' amyloidosis.

Administration once a day or on alternate days may induce a rebound acute-phase reaction including elevation of SAA levels.

Prognosis of amyloidosis

The prognosis of a patient in whom the diagnosis of amyloidosis is made, is poor compared with that of most other glomerular or renal diseases. Prognosis depends on (a) the type of amyloid present, (b) the involvement of different organs, especially the kidneys, and (c) the underlying disease. The time-course of the disease is variable and difficult to delineate, as in most cases the point when amyloid formation starts is unknown.

In AL amyloidosis survival is very restricted, as is indicated in the study of Kyle and Greipp (1983), who found a median survival time of only 12 months in 229 patients. The survival time in patients with frank myeloma was even shorter at 5 months (see Chapter 4.3 also), whilst in patients without myeloma (i.e. 'primary' amyloidosis) a mean survival of 13 months was calculated. Other smaller series from renal units show similar poor survival rates.

For AA amyloidosis a more prolonged course can be achieved Median survival times from diagnosis reported in the literature vary from 30 to 60 months (Triger and Joekes 1973; Wright and Calkins 1981; Browning et al. 1985; Wegelius et al. 1980).

In a series of 144 patients with amyloidosis in our own department (91 AA and 53 AL) median survival time after diagnosis was 7 months in the AL and 45 months in the AA patients (Janssen et al. 1986). Thirty-eight per cent of the AA and 6 per cent of the AL patients survived 5 years or more. The maximum survival from the time of diagnosis was 22 years in the AL and 18 years in the AA group. From 4 months after diagnosis prognosis was significantly worse in AL compared with AA amyloidosis (Fig. 5). No significant differences were found between patients with myeloma-associated AL amyloidosis and patients with 'primary' AL amyloidosis.

In 98 per cent of our patients, death was directly related to the

Table 7 Cause of death in 63 patients with systemic AA and 51 patients with systemic AL amyloidosis

Cause of death	Total no. of cases	AA	AL
Cardiac failure	34	2	32
Renal insufficiency	30	22	8
Infection	12	10	2
Bleeding	7	5	2
Perforated diverticula	5	5	0
Myocardial infarction	4	3	1
Hepatic failure	3	1	2
Stroke	2	2	0
Carcinoma	2	2	0
Glossopathy	1	0	0
Tuberculosis	1	1	0
Ischaemic colitis	1	0	1
Unknown	12	10	2

No statistically significant differences were found between patients with myeloma-associated AL amyloidosis and patients with 'primary' AL amyloidosis.

amyloid process (Table 7). Among the AL patients cardiac failure was the most common cause of death and accounted for 63 per cent of deaths, which is more than in other observations in the literature.

Although the presenting symptoms were of renal origin in 88 per cent of our AA patients, renal insufficiency accounted for death in only 35 per cent of these, in part because a growing number of patients is now treated by haemodialysis.

Gastrointestinal involvement (bleeding, perforation) was related to death in 18 per cent of the AA patients (Table 7).

Renal replacement therapy in amyloidosis

Irreversible renal insufficiency due to amyloidosis can now be treated with dialysis and/or renal transplantation. Over several years, reports on renal replacement therapy in amyloidosis have appeared in the literature. Most of these reports, however, comprise only a limited number of observations, and do not provide sufficient insight into the value of these treatment modalities.

Dialysis

Avram (1978) reported maintenance haemodialysis treatment in five patients, of whom only two survived more than 28 months. Jones (1976) collected data from 29 patients with amyloidosis treated with regular haemodialysis in different European centres by means of a questionnaire. The survival rates of these patients were similar to the overall survival rates of 18 750 patients receiving dialysis treatment in the hospital or the home, up to 6 months, but were definitely lower at 12 and 24 months of treatment. Seven patients died during dialysis, and in six of these death was due to a circulatory disorder, including cardiac failure and ventricular dysrhythmia probably due to cardiac amyloidosis. Persistent hypotension in one patient was caused by adrenal insufficiency.

In this questionnaire no question was asked about special difficulties with regard to vascular access because of amyloidosis. Janssen et al. (1986) described their experience with haemodialysis treatment in 11 patients (nine AA, two AL). The median survival after diagnosis was 133 months in the patients treated with haemodialysis, compared with 29 months in the group with biopsy-proven renal amyloidosis who did not receive dialysis treatment. The 5-year survival was 34 per cent for the patients treated with hospital haemodialysis.

Browning et al. (1985) reported successful treatment of five patients with continuous ambulatory peritoneal dialysis for periods up to 30 months (three AA, two AL). The incidence of peritonitis in these patients was less than the overall incidence in their peritoneal dialysis group.

It is now apparent that patients with renal insufficiency due to renal amyloidosis should be treated by regular haemodialysis or by continuous ambulatory peritoneal dialysis, since these treatment modalities not only improve the quality of life, but also undoubtedly prolong it.

It is important to evaluate the cardiac status of the patient before the start of treatment, as cardiac amyloidosis may predispose to hypotension during haemodialysis. It should also be kept in mind that persistent hypotension may be due to amyloid deposition in the adrenal glands. In these cases corticosteroid treatment may be necessary. It can be argued that in patients with hypotension, treatment with continuous ambulatory peritoneal dialysis is preferable.

In 1984 Pras et al. reported that in Israel patients with amyloidosis made up 6 per cent of the dialysis population, while in Europe this was only 0.6 per cent. Recent statistical data obtained from the Registry of the European Dialysis and Transplant Association (Brunner et al. 1989) show that the acceptance rate between 1977 and 1987 rose only slightly from 1.2 per cent to 1.45 per cent for patients with amyloidosis (Fig. 6).

In Fig. 7 it is clearly shown that the acceptance of amyloidosis patients for renal replacement therapy is high in Finland, Luxembourg, Norway, and Sweden, compared with other European countries. The average acceptance rate in countries reporting to the European Dialysis and Transplant Association was 1.6 per cent. At the end of 1987 about 940 patients with amyloidosis were under treatment, either with hospital or home haemodialysis, or with continuous ambulatory peritoneal dialysis. The latter category of patients was small in comparison with the number of patients treated with haemodialysis (Table 8).

Finally, the Registry recorded a 2-year survival of 71 per cent for the group aged 15 to 44 years, and 51 per cent for the age group 45 to 64 years for patients with amyloidosis who commenced renal replacement therapy between 1980 and 1984. Five-

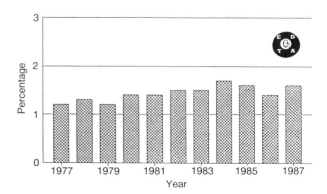

Fig. 6 Proportional acceptance rate of patients with amyloidosis necessitating renal replacement therapy in 1987 for all patients known to the EDTA/ERA registry and selected countries. (Reproduced with kind permission of the Registry of the EDTA/ERA.)

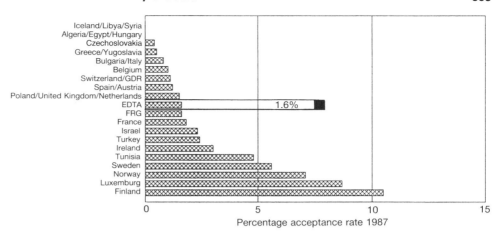

Fig. 7 Proportion of new patients within EDTA/ERA registry for renal replacement therapy with amyloidosis as primary renal disease 1977–1987. (Reproduced with kind permission of the Registry of the EDTA/ERA.)

Table 8 Contribution of modes of treatment in amyloidosis as primary renal disease for renal replacement therapy

Number of patients on renal replacement therapy (31 December 1987)	1150
Female	513
Male	637
Hospital haemodialysis	817(71)
Home haemodialysis	35(3)
Intermittent peritoneal dialysis	15(1)
CAPD	88(8)
Functional graft	195(17)

Percentages in parentheses.

year survival in the two groups was 52 and 25 per cent respectively (Brunner *et al.* 1988).

Renal transplantation

Renal transplantation has been performed for more than 20 years in patients with amyloidosis and end-stage renal failure (Belzer *et al.* 1968; Cohen *et al.* 1971). In subsequent years other observations have been published, but most of them concerned only small numbers of patients.

In 1976 a survey of 10 patients was produced as a result of a questionnaire of renal units reporting to the European Dialysis and Transplant Association (Jones 1976). The patient survival in the amyloidosis group at 6 months after transplantation was lower than that of the total transplanted population. After 1 and 2 years this difference increased considerably. Six patients died, 1 to 48 months after transplantation with a mean survival time of 12.7 months. In five of these six patients death was caused by infection. The results after 6 months however, showed that in the early phase after transplantation amyloidosis was not responsible for special problems with regard to the surgical procedure. In 1980, a leading article in *The Lancet* concluded that patients with amyloidosis should not be excluded from consideration for transplantation.

A large single-centre controlled study on renal transplantation from Finland comprised 45 patients with amyloidosis, treated from 1973 to 1981 (Pasternack *et al.* 1986). All patients received cadaveric grafts. Three patients had 'primary' AL amyloidosis; 33 patients suffered from rheumatoid arthritis or other forms of chronic polyarthritis, and in nine patients amyloidosis was secondary to some other chronic inflammatory disease.

The patient survival in the 45 patients with amyloidosis was, at 3 months postoperatively, already significantly lower compared with a matched control group of 45 grafted patients with end-stage renal failure due to chronic glomerulonephritis. The survival rate of the amyloidotic patients was 63 per cent after 1 year and 51 per cent after 2 years. In the control group the survival rates were 86 per cent and 79 per cent respectively. During the 4-year follow-up period of this study, 23 patients with amyloidosis and 10 patients in the control group died. There was no statistically significant difference in the age at death between the patients in both groups.

The graft survival rates however, were similar in the patients with amyloidosis compared to those with glomerulonephritis (57 per cent after 1 and year and 45 per cent after 2 years). In both groups cardiovascular disease and infection were responsible for deaths, while graft failure in both groups was mainly due to rejection and to a minor extent to operative and vascular complications.

Four patients with amyloidosis died from late complications of the generalized disease. Graft failure in these four in the final stage was due to amyloid deposition in the transplanted kidney, *de novo* membranous glomerulopathy, chronic rejection, and irreversible acute rejection respectively.

Recurrence of amyloid in allografts

Due to the nature of the disease, recurrence of amyloidosis in kidney transplants may be expected. At present it is unknown how frequently this occurs. No exact data on the frequency of this complication have been reported.

In the study of Pasternack *et al.* (1986), in four of the 45 patients with amyloidosis who received a cadaveric graft, amyloid was found in the kidneys at autopsy. These patients had lived with a transplanted kidney for at least 1 year. On the other hand, no amyloid was found in the kidneys of 27 other patients. It was estimated by the authors that recurrence of amyloidosis in a grafted kidney surviving for 1 year or longer may be found in approximately 20 per cent of cases. It is not predictable which patients in particular are at risk.

At the end of 1987, the Registry of the European Dialysis and Transplant Association possessed data concerning 200 patients with amyloidosis, who were alive with a transplanted kidney. In the Registry, however, so far no data are available with regard to the possible recurrence of amyloidosis in the graft.

In conclusion, renal transplantation is a valuable treatment modality for patients with amyloidosis and end-stage renal failure, and should be considered seriously, provided the condition of the patients does not contraindicate surgery.

There is no reason to avoid transplantation in patients with amyloidosis because of the uncertain future with regard to the recurrence of amyloidosis. Even when this does occur, it is still uncertain how long the transplanted kidney will survive. As this may be for a number of years, transplantation in these patients is the preferred treatment, as in most other patients with end-stage renal failure.

Bibliography

Mandema, E., Ruinen, L., Scholten, J.H., and Cohen, A.S., eds. (1968). *Amyloidosis*. Proceedings of the symposium in Groningen, the Netherlands, Excerpta Medica Foundation, Amsterdam.

Wegelius, O. and Pasternack, A., eds. (1976). *Amyloidosis*. Proceedings of the 5th Sigrid Juselius Foundation Symposium. Academic Press, London.

Glenner, G.G., Pinho e Costa, P., and Falcao de Freitas, A., eds. (1980). *Amyloid and Amyloidosis*. Proceedings of the Third International Symposium on Amyloidosis. Portugal, Excerpta Medica, Amsterdam.

Glenner, G.G., Osserman, E.F., Benditt, E.P., Calkins, E., Cohen, A.S., and Zucker-Franklin, D., eds. (1986). *Amyloidosis*. Proceedings of the Fourth International Symposium on Amyloidosis. New York, Plenum Press, New York.

Heptinstall, R.H. (1983). *Pathology of the Kidney*, pp. 1014–36. Little, Brown and Company. Boston.

Marrink, J. and van Rijswijk, M.H., eds. (1986). *Amyloidosis*. Martinus Nijhoff Publishers, Dordrecht.

Isobe, T., Araki, S., Uchino, F., Kito, S., and Tsubura, E., eds. (1988). *Amyloid and Amyloidosis*. Symposium in Hakone, Japan. Plenum Press, New York.

Vaamonde, C.A. and Pardo, V. (1988). Multiple myeloma and amyloidosis. In *Strauss and Welt's Diseases of the Kidney*, 4th edn. (eds. R.W. Schrier and C.W. Gottschalk), pp. 2457–80. Little, Brown and Company, Boston.

References

Ahlmen, M., Ahlmen, J., Svalander, C., and Bucht, H. (1987). Cytotoxic drug treatment of reactive amyloidosis in rheumatoid arthritis with special reference to renal insufficiency. *Clinical Rheumatology*, **6**, 27–38.

Avram, M.M. (1978). Survival in uremia due to systemic diseases. *Kidney International*, suppl. **8**, S55–S60.

Barclay, G.P.T., Cameron, H.M., and Loughridge, L.W. (1960). Amyloid disease of the kidney and renal vein thrombosis. *Quarterly Journal of Medicine*, **29**, 137–51.

Belzer, F.O., Ashby, B.S., Gulyassy, P.F., and Powell, M. (1968). Successful seventeen-hour preservation and transplantation of human cadaver kidney. *New England Journal of Medicine*, **278**, 608–10.

Bemelman, F.J., Krediet, R.T., Schipper, M.E.I., and Arisz, L. (1989). Renal involvement in Behçet's syndrome. *Netherlands Journal of Medicine*, **34**, 148–53.

Benditt, E.P., Eriksen, N., Hermodson, M.A., and Ericsson, L.H. (1971). The major protein of human and monkey amyloid substance: common properties including unusual N-terminal amino acid sequences. *FEBS Letters*, **19**, 169–73.

Bennhold, H. (1922). Eine spezifische Amyloid färbung mit Kongorot. *Münchener Medizinische Wochenschrift*, **69**, 1537.

Benson, M.D., Brandt, K.D., Cohen, A.S., and Cathcart, E.S. (1975). Neuropathy, M-components and amyloid, *Lancet*, **i**, 10–12.

Bentwich, Z., Rosenmann, E., and Eliakim, M. (1971). Prevalence of hypertension in renal amyloidosis: correlation with clinical and histological parameters. *American Journal of Medical Sciences*, **262**, 93–100.

Berglund, K., Keller, C., and Thysell, H. (1987). Alkylating cytostatic treatment in renal amyloidosis secondary to rheumatic disease. *Annals of Rheumatic Diseases*, **46**, 757–62.

Brandt, K., Cathcart, E.S., and Cohen, A.S. (1968). A clinical analysis of the course and prognosis of forty-two patients with amyloidosis. *American Journal of Medicine*, **44**, 955–69.

Browning, M., et al. (1983). Renal involvement in systemic amyloidosis. *Proceedings of the European Dialysis and Transplant Association*, (EDTA), **20**, 595–600.

Browning, M.J., et al. (1984). Continuous ambulatory peritoneal dialysis in systemic amyloidosis and end-stage renal disease. *Journal of the Royal Society of Medicine*, **77**, 189–92.

Browning, M.J., et al. (1985). Ten years' experience of an amyloid clinic–a clinicopathological survey. *Quarterly Journal of Medicine*, **54**, 213–27.

Bruni, J., Bilbao, J.M., and Pritzker, K.P.H. (1977). Myopathy associated with amyloid angiopathy. *Canadian Journal of Neurological Sciences*, **4**, 77–80.

Brunner, F.P., et al. (1988). International review of renal replacement therapy: strategies and results. In *Replacement of Renal Function by dialysis*. 3rd edn. (ed. J.F. Maher), pp. 697–719. Kluwer Academic, Boston.

Brunner, F.P., et al. (1989). Combined report on regular dialysis and transplantation in Europe, XIX, 1988. *Nephrology, Dialysis, Transplantation*, **4**, Suppl. 4.

Calkins, E. and Cohen, A.S., (1960). Diagnosis of amyloidosis. *Bulletin of the Rheumatic Diseases*, **10**, 215–18.

Cohen, A.S. and Calkins, E. (1959). Electron microscopic observations on a fibrous component in amyloid of diverse origins. *Nature (London)*, **183**, 1202–3.

Cohen, A.S. (1967). Amyloidosis. *New England Journal of Medicine*, **277**, 522–30, 574–83, 628–38.

Cohen, A.S., Briccetti, A.B., Harrington, J.T., and Mannich, J.A. (1971). Renal transplantation in two cases of amyloidosis. *Lancet*, **ii**, 513–16.

Cohen A.S. and Skinner, M. (1982). Amyloidosis of the liver. In *Disease of the Liver*, (eds. L.B. Schiff, and E.R. Schiff), pp. 1081–99. J.B. Lippincott and Co., Philadelphia.

Cooley, R.N. (1953). Primary amyloidosis with involvement of the stomach. *American Journal of Roentgenology*, **70**, 428–36.

Divry, P. and Florkin, M. (1927). Sur les propriétés optique de l'amyloide. *Comptes Rendus Societé de Biologie (Paris)*, **97**, 1808–10.

Donker, A.J.M., van der Hem, G.K., Sluiter, W.J., and Beekhuis, H. (1977). A radioisotope method for simultaneous determination of the glomerular filtration rate and the effective renal plasma flow. *Netherlands Journal of Medicine*, **20**, 97–103.

Eanes, E.D. and Glenner, G.G. (1968). X-ray diffraction studies on amyloid filaments. *Journal of Histochemistry and Cytochemistry*, **16**, 673–7.

Ekelund, L. (1977). Radiological findings in renal amyloidosis. *American Journal of Roentgenology*, **129**, 851–3.

Elkin, M., ed. (1980). *Radiology of the urinary system*. vol. 2, pp. 1014–32. Little, Brown and Company, Boston.

Friedreich, N. and Kékulé, A. (1859). Zur Amyloidfrage. *Virchows Archiv für Pathologische Anatomie und Physiologie*, **16**, 50–65.

Furie, B., Green, E., and Furie, B.C. (1977). Syndrome of acquired factor X deficiency and systemic amyloidosis. *In vivo* studies of the metabolic fate of factor X. *New England Journal of Medicine*, **297**, 81–5.

Glenner, G.G., Harbaugh, J., Ohms, J.I., Harada M., and Cuatrecasas, P. (1970). An amyloid protein: the amino-terminal variable fragment of an immunoglobulin light chain. *Biochemical Biophysical Research Communications*, **41**, 1287–9.

Glenner, G.G., Terry, W.D., Harada, M., Isersky, C., and Page, D.L. (1971). Amyloid fibril proteins: proof of homology with immunoglobulin light chains by sequence analysis. *Science*, **172**, 1150–1.

Glenner, G.G. (1980), Amyloid deposits and amyloidosis. The β-fibrilloses. *New England Journal of Medicine*, **302**, 1283–92, 1333–43.

Glenner, G.G. and Wong, C.W. (1984). Alzheimer's disease and Down's syndrome: sharing of a unique cerebrovascular amyloid fibril protein. *Biochemical Biophysical Research Communications*, **122**, 1131–5.

Goldfinger, S. (1972). Colchicine for familial Mediterranean fever. *New England Journal of Medicine*, 287, 1302.

Husebekk, A., Skogen, B., Husby, G., and Marhaug, G. (1985). *In vivo* degradation of protein SAA and incorporation in amyloid fibrils. *Scandinavian Journal of Immunology*, 21, 283–7.

Isobe, T. and Osserman E.F. (1976). Effect of dimethylsulfoxide (DMSO) in the treatment of amyloidosis. In *Amyloidosis*, (eds. O. Wegelius, and A. Pasternack), pp. 553–64. Academic Press, London.

Jacob, H., Charytan, C., Rascoff, J.H., Golden, R., and Janis, R. (1978). Amyloidosis secondary to drug abuse and chronic skin suppuration. *Archives of Internal Medicine*, 138, 1150–1.

Janssen, S. (1985). Clinical and diagnostic features of amyloidosis. Thesis, University of Groningen.

Janssen, S., Elema, J.D., van Rijswijk, M.H., Limburg, P.C., Meijer, S., and Mandema, E. (1985). Classification of amyloidosis: immunohistochemistry versus the potassium permanganate method in differentiating AA from AL amyloidosis. *Applied Pathology*, 3, 29–38.

Janssen, S., van Rijswijk, M.H., Meijer, S., Ruinen, L., and van der Hem, G.K. (1986). Systemic amyloidosis: a clinical survey of 144 cases. *Netherlands Journal of Medicine*, 29, 376–85.

Jerath, R.S., Valenzuela, R., and Guerrero, I. (1980). Immunofluorescence findings in human renal amyloidosis. *American Journal of Clinical Pathology*, 74, 630–5.

Jones, N.F. (1976). Renal amyloidosis: pathogenesis and therapy. *Clinical Nephrology*, 6, 459–64.

Kedar, I., Greenwald, M., and Ravid, M. (1977). Treatment of experimental murine amyloidosis with dimethylsulfoxide. *European Journal of Clinical Investigation*, 7, 149–50.

Kyle, R.A. and Greipp, P.R. (1983). Amyloidosis (AL). Clinical and laboratory features in 229 cases. *Mayo Clinic Proceedings*, 58, 665–83.

Legge, D.A., Wollaeger, E.E., and Carlson, H.C. (1970). Intestinal pseudo-obstruction in systemic amyloidosis. *Gut*, 11, 764–7.

Luijk, W.H.J. van, Ensing, G.J., Meijer, S., Donker, A.J.M., and Piers, D.A. (1984). Is the relative ^{99m}Tc-DMSA clearance a useful marker of proximal tubular dysfunction? *European Journal of Nuclear Medicine*, 9, 439–42.

Mackensen, S., Grund, K.E., Bader, R., and Bohle, A. (1977). The influence of glomerular and interstitial factors on the serum creatinine concentration in renal amyloidosis. *Virchows Archives A. Pathological Anatomy and Histology*, 375, 159–68.

Meador, K.H., Sharon, Z., and Lewis E.J. (1979). Renal amyloidosis and subcutaneous drug abuse. *Annals of Internal Medicine*, 91, 565–667.

Meulen, J. van der, Gupta, R.K., Peregrin, J.H., Al Adnani, M.S., and Johny, K.V. (1989). Takayasu's arteritis and nephrotic syndrome in a patient with crossed renal ectopia. *Netherlands Journal of Medicine*, 34, 142–7.

Muckle, T.J. and Wells, M. (1962). Urticaria, deafness and amyloidosis: a new heredofamilial syndrome. *Quarterly Journal of Medicine*, 31, 235–48.

Muehrcke, R.C., Pirani, C.L., Pollak, V.E., and Kark, R.M. (1955). Primary renal amyloidosis with the nephrotic syndrome studied by serial biopsies of the kidney. *Guy's Hospital Reports*, 104, 295–311.

Müftüoglu, A.U., Erbengi, T., and Harmanci, M. (1977). Renal amyloidosis: immunofluorescence and electron microscopic studies. *Israel Journal of Medical Science*, 13, 1102–8.

Parkins, R.A. and Bywaters, E.G.L. (1959). Regression of amyloidosis secondary to rheumatic arthritis. *British Medical Journal*, 1, 536–40.

Pasternack, A., Ahonen, J., and Kuhlbäck, B. (1986). Renal transplantation in 45 patients with amyloidosis. *Transplantation*, 42, 598–601.

Pras, M., Schubert, M., Zucker-Franklin, D., Rimon, A., and Franklin, E.C. (1968). The characterization of soluble amyloid prepared in water. *Journal of Clinical Investigation*, 47, 924–33.

Pras, M., *et al.* (1984). Recent advances in familial Mediterranean fever. *Advances in Nephrology*, 13, 261.

Pras, M. (1986). The hereditary amyloidoses. In *Amyloidosis*, (eds. J. Marrink and M.H van Rijswijk), pp. 185–93. Martinus Nijhoff Publishers, Dordrecht.

Ravid, M., Kedar, I., and Sohar, E. (1977). Effects of a single dose of DMSO on renal amyloidosis. *Lancet*, i, 730–1.

Reimann, H.A., Koucky, R.F., and Eklund, C.M. (1935). Primary amyloidosis limited to tissue of mesodermal origin. *American Journal of Pathology*, 11, 977–88.

Rijswijk, M.H. van and Heusden, C.W.G.J. van (1979). The potassium permanganate method. A reliable method for differentiating amyloid AA from other forms of amyloid in routine laboratory practice. *American Journal of Pathology*, 97, 43–58.

Rijswijk, M.H. van, Donker, A.J.M., and Ruinen, L. (1979a). Dimethylsulfoxide in amyloidosis. *Lancet*, i, 207–208.

Rijswijk, M.H. van, Donker, A.J.M., and Ruinen, L. (1979b). Treatment of renal amyloidosis with dimethylsulfoxide (DMSO). *Proceedings of the European Dialysis and Transplant Association (EDTA)*, 16, 500–5.

Rijswijk, M.H. van, *et al.* (1980). Successful treatment with dimethylsulfoxide of human amyloidosis secondary to rheumatoid arthritis. In *Amyloid and Amyloidosis*. (eds. G.G. Glenner, P.P. Costa, and A.F. Freitas), pp. 570–7. Excerpta Medica, Amsterdam.

Rijswijk, M.H. van (1981). Amyloidosis. Thesis, University of Groningen.

Rijswijk, M.H. van, Ruinen, L., Donker, A.J.M., Blecourt, J.J. de, and Mandema, E. (1983). Dimethylsulfoxide in the treatment of AA amyloidosis. *Annals of the New York Academy of Sciences*, 411, 67–83.

Rijswijk, M.H. van, Leeuwen, Donker, A.J.M., and Mandema, E. (1986). Treatment of systemic AA amyloidosis. In *Amyloidosis* (eds. G.G. Glenner, E.F. Osserman, E.P. Benditt, E. Calkins, A.S. Cohen, and D. Zucker-Franklin), pp. 571–82. Plenum Press, New York.

Scholes, J., Derosena, R., Appel, G.B., Wellington, J., Boyd, M.T., and Pirani, C.L. (1979). Amyloidosis in chronic heroin addicts with the nephrotic syndrome. *Annals of Internal Medicine*, 91, 26–9.

Sohar, E., Gafni, K., Pras, M., and Heller, H. (1967). Familial Mediterranean fever: a survey of 470 cases and review of the literature. *American Journal of Medicine*, 43, 227–53.

Stephens, F.O. (1962). The syndrome of intestinal pseudo-obstruction. *British Medical Journal*, 1, 1248–50.

Törnroth, T., Falck, H.M., Wafin, F., and Wegelius, O. (1980). Renal amyloidosis in rheumatic disease: a clinicopathological correlative study. In *Amyloid and Amyloidosis*, (eds. G.G. Glenner, P.P. Costa, and A.F. Freitas,) pp. 191–9. Excerpta Medica, Amsterdam.

Triger, D.R. and Joekes, A.M. (1973). Renal amyloidosis. A fourteen-year follow-up. *Quarterly Journal of Medicine*, 42, 15–40.

Virchow, R. (1854). Uber eine in Gehirn und Rückenmark des Menschen aufgefundene Substanz mit der chemischen Reaktion der Cellulose. *Virchows Archiv für Pathologische Anatomie und Physiologie*, 6, 135–8.

Watanabe, T. and Saniter, T. (1975). Morphological and clinical features of renal amyloidosis. *Virchows Archives A. Pathological Anatomy and Histology*, 366, 125–5.

Wegelius, O., Wafin, F., Falck, H.M., and Törnroth, T. (1980). Follow-up study of amyloidosis secondary to rheumatic disease. In *Amyloid and Amyloidosis*, (eds. G.G. Glenner, P.P. Costa, and A.F. Freitas), pp. 183–90. Excerpta Medica, Amsterdam.

Wright, J.R., Calkins, E. and Humphrey, R.L. (1977). Potassium permanganate reaction in amyloidosis. A histologic method to assist in differentiating forms of the disease. *Laboratory Investigation*, 36, 274–81.

Wright, J.R. and Calkins, E. (1981). Clinical-pathologic differentiation of common amyloid syndromes. *Medicine*, 60, 429–49.

Zemer, D., Pras, M., Sohar, E., Modan, M., Cabili, S., and Gafni, J. (1986). Colchicine in the prevention and treatment of amyloidosis of familial Mediterranean fever. *New England Journal of Medicine*, 314, 1001–5.

4.3 Kidney involvement in plasma cell dyscrasias

LUIGI MINETTI

Introduction

The term 'plasma cell dyscrasias' is used to describe the diseases associated with proliferative disorders of plasma cells and their immediate precursors, the activated B cells. It includes syndromes caused by quantitative and qualitative alterations in immunoglobulin synthesis: whole immunoglobulin molecules may be secreted in greatly increased amounts, or an excess of individual heavy or light immunoglobulin chains may be produced. The most common examples are the Bence-Jones protein, composed of free monoclonal light chains, and 'aberrant' forms such as Ig fragments or other atypical immunoglobulin molecules.

The pathological effects of plasma cell dyscrasias arise either as a direct result of infiltration of the bone marrow by abnormal cells, or indirectly through their products. Uncontrolled proliferation of B cells can also have secondary effects on the B-cell maturation pathway, resulting in depressed humoral immunity. Some abnormal B-cell clones synthesize autoantibodies, most commonly with antiglobulin specificity, some of which behave as cryoglobulins and produce immune complex type lesions, with or without complement activation. High concentrations of monoclonal immunoglobulin, especially IgM, can also produce symptoms directly, through their effects on plasma viscosity: such patients present with the hyperviscosity syndrome. Some incomplete or aberrant forms with unusual properties can have specific deleterious effects, such as that of Bence-Jones proteins on the kidney. Interactions of immunoglobulin subunits or fragments with the tissues can result in amyloidosis.

Kidney disease is a predominant feature of multiple myeloma, primary amyloidosis, and light chain deposition disease, the most important plasma cell dyscrasias.

Immunoglobulins and their characterization

Immunoglobulin molecules consist of two pairs of identical polypeptide chains, heavy chains and light chains respectively, held together by disulphide bonds. Both kinds of chain consist of a series of structurally homologous domains, each containing about 110 amino-acid residues. The amino-terminal domains of both heavy and light chains vary in amino-acid sequence, and are referred to as the variable (V) portion of the molecule. The constant (C) domains of each chain are relatively invariant in sequence. Light chains have one variable (V_L) and one constant (C_L) domain; heavy chains have one variable (V_H) and three or four constant (C_H) domains (Fig. 1). Analysis of specific antigenic determinants located on the constant domains allows five classes of heavy chain to be defined; these correspond to the five isotypes of immunoglobulin: IgG, IgA, IgM, IgD, IgE; IgG and IgA can further be divided into subclasses (IgG_1, IgG_2, IgG_3, IgG_4; IgA_1, IgA_2). There are two types of light chain (kappa and lambda). The variable domains of heavy and light chains (V_L and V_H) are aligned so as to be adjacent to one another and constitute the Fab region of the molecule responsible for the specific properties of the antibody. Within these domains there are hypervariable regions, characterized by extreme variability of amino-acid sequence, that constitute the antigen-binding site, which is unique to each molecule and conveys idiotype and antibody specificity. Constant region domains convey isotypes and allotypes and are involved in the biological and metabolic properties of immunoglobulin molecules. The C_H domain at the carboxy-terminal end constitutes the complement binding (Fc) portion of the molecule. The structure of immunoglobulin is described in greater detail in Chapter 4.1.

Isotypes are the same in all normal individuals of the same species while allotypes are small inherited differences between individuals of the same specifies. Idiotypes, the antigenic determinants unique to the molecule produced by a single cell clone, are the only suitable markers of monoclonal proliferations.

Differences in variable region sequence are responsible for the distinctive properties of light chains, including the isoelectric point which determines the electric charge and electrophoretic

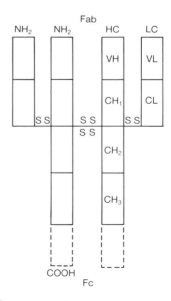

Fig. 1 Schematic structure of the immunoglobulin molecule. For explanations see text.

mobility (Solomon 1976). During electrophoretic analysis immunoglobulins move according to their electric charge and form a broad peak in the gamma region. When a monoclonal immunoglobulin is present at an adequate concentration a sharp spike appears in this region, called an M (monoclonal) component. The use of immunoelectrophoresis allows a monoclonal immunoglobulin to be identified and its intact or incomplete nature to be established.

A small quantity of free light chains of both types is normally secreted by plasma cells; most of these are cleared by the kidney and less than 10 mg/day appears in the urine. A free monoclonal light chain can be produced in large quantities in multiple myeloma, and appears as an M component in the urine, where it is called Bence-Jones protein.

Monoclonal immunoglobulins and renal damage

M components are sometimes detected in patients with other malignancies but only occasionally and at quite low concentrations; however, they are detectable in serum and/or urine of the majority of patients with plasma cell dyscrasias.

All types of immunoglobulin product can cause renal injury: direct or indirect toxic effects can cause glomerular or tubular dysfunction. The most important pathogenic role has been attributed to the free monoclonal light chains and related fragments. These can deposit in intratubular casts (myeloma cast nephropathy), precipitate as crystals in tubular cells (Fanconi syndrome), deposit in tubular and glomerular basement membranes and mesangium (light chain deposition disease), or condense as amyloid fibrils in the walls of blood vessels (AL amyloidosis). Disturbances in electrolyte and water metabolism causing dehydration and hypovolaemia or hypercalcemia may precipitate acute renal failure and contribute to the development of chronic renal failure. Patients with plasma cell dyscrasias are also susceptible to uric acid nephropathy, obstructive nephropathy, neoplastic infiltration of the kidney, acute tubular necrosis induced by radiographic contrast agents, or renal impairment caused by non-steroidal anti-inflammatory drugs.

The frequent involvement of the kidney in plasma cell dyscrasias arises from its central role in the metabolism of light chains. The glomerular filtration of light chains depends on the molecular weight of their circulating forms: monomeric forms (molecular weight 22 000 Da) are filtered freely while the filtration rate of dimers (molecular weight 44 000 Da) is about 8 per cent of inulin clearance. Kappa light chains exist primarily in monomeric forms whereas lambda light chains circulate as dimeric forms. Normally, 99 per cent of filtered light chains are reabsorbed and catabolized by proximal tubule cells. After binding of their cationic zone to the negatively charged brush-border membrane the light chains are pinocytosed and degraded by hydrolytic enzymes to amino acids and peptides in the same way as other proteins of a similar molecular weight. Tubular damage, irrespective of the cause, decreases reabsorption and degradation of light chains and thus increases their excretion. The large amounts of filtered light chains in patients with myeloma exceeds the reabsorptive capacity of the tubule and they appear in high concentrations in the urine. Impaired renal function reduces their filtration and catabolism and leads to higher plasma concentrations. Every part of the kidney (tubular cells and lumina, glomeruli, interstitium and vessels) is, therefore, exposed to the effects of quantitatively or qualitatively altered monoclonal light chains.

The kidney in multiple myeloma

Myeloma kidney (myeloma cast nephropathy)

Approximately 50 per cent of patients with multiple myeloma have renal impairment, and the term myeloma kidney has been widely used to describe the most frequent form of renal injury. The most distinctive features of myeloma kidney are large intratubular casts, which give it the name myeloma cast nephropathy or Bence-Jones cast nephropathy. These intratubular casts cause tubular atrophy and interstitial fibrosis (Rota et al. 1987). A minority of patients with myeloma present with other types of renal injury, including acute tubular necrosis, nodular glomerulosclerosis, proliferative glomerulonephritis, nephrocalcinosis, plasma cell infiltration, or amyloidosis.

Myeloma cast nephropathy develops only when Bence-Jones proteinuria is present but the severity depends on a variety of factors in addition to the quantity of these proteins in the urine. It is relatively more frequent in light chain disease (up to 41 per cent in the series of Rota et al. (1987)) and in patients with IgD myeloma than with the more common IgG, IgM, and IgA myelomas. It is probably more common in patients with Bence-Jones proteins of the lambda type.

Urinalysis is the most important approach to the diagnosis of Bence-Jones proteinuria. Traditionally it is detected by the heat precipitation test but unfortunately it is neither sensitive nor specific. It depends on very high concentrations of Bence-Jones protein (about 200 mg/dl) but even so false-positive results can occur in renal insufficiency because of the excess of polyclonal light chains in the urine (Kyle 1988). Dipsticks for urinary protein are similarly insensitive, and Bence-Jones proteins have to be detected by formal methods for quantifying proteinuria, such as the sulphosalicylic acid method. Immunoelectrophoresis and immunofixation are the definitive tests for Bence-Jones proteins, and these should be performed on adequately concentrated urine (up to 50-fold when the Bence-Jones proteinuria is less than 100 mg/dl) to allow low concentrations of a monoclonal light chain to be identified (Solomon and Weiss 1988). Immunofixation is particularly helpful when there are other proteins in the urine: monoclonal heavy chains or immunoglobulin fragments, polyclonal light chains, or large amounts of albumin and transferrin sometimes obscure the presence of Bence-Jones proteinuria.

In 522 cases studied by Cooper et al. (1984) the most common abnormality in urine other than Bence-Jones proteinuria was a selective tubular proteinuria, i.e. raised concentrations of low molecular weight proteins such as α_1-microglobulin and α_1-acid glycoprotein.

Light-chain nephrotoxicity

The quantity of Bence-Jones proteins excreted in urine depends on the tumour cell mass and on renal function. Any factor adversely affecting renal function reduces light chain catabolism and, paradoxically, increases the excretion of Bence-Jones proteins (Clyne and Pollak 1981). However, a direct relationship between the amount of Bence-Jones protein excreted and functional and morphological signs of nephrotoxicity has never been demonstrated. Chemical and functional properties of the Bence-Jones proteins themselves almost certainly influence renal

Table 1 Nephrotoxicity of Bence-Jones proteins

Bence-Jones protein related factors
  Filtered load and tubular fluid concentration
  Type (kappa versus lambda)
  Structural configuration and molecular weight (monomer/dimer/tetramer; aberrant forms)
  Charge density, isoelectric point (pI)
  Behaving as antibody, tendency to aggregate

Host-related factors
  Glomerular filtration rate
  Tubular transport
  Synthesis and catabolism rates
  Dehydration
  Intratubular pH
  Interaction with Tamm-Horsfall, or other proteins
  Influences of Ca^{2+} and anaemia
  Infection
  Radiocontrast agents, aminoglycosides, NSAIDs, anticancer drugs (BCNU, vincristine)

damage, possibly with a contribution from host factors (Sanders *et al.* 1988*b*; Baylis *et al.* 1988) (see Table 1).

Several experimental models, *in vitro* and *in vivo*, have been used to investigate the nephrotoxic properties of Bence-Jones proteins. Koss *et al.* (1976) reproduced the morphological appearances of myeloma cast nephropathy in mice by a single intraperitoneal injection of a human Bence-Jones protein, but were unable to detect casts with two other human Bence-Jones proteins using the same protocol. Solomon and Weiss (1988), using a similar mouse model, observed the formation of intratubular casts after the injection of a kappa Bence-Jones protein from a patient with multiple myeloma, who had severely impaired renal function and was excreting about 6 g of this protein daily. The Bence-Jones protein from another patient, who had normal renal function despite massive proteinuria (about 40 g a day) produced no renal lesions in mice. Other studies with mice have confirmed this variability but have shown that there is complete concordance between mouse model and the pathological finding in the donor. The extent of cast formation is dose-dependent and reversible with time. It is potentiated by dehydration and hindered by vigorous hydration.

Injection of the $F(ab')_2$ variable portion of nephrotoxic Bence-Jones proteins produced the same pathology as the intact protein, whereas the Fc section was ineffective (Solomon and Weiss 1988). This is not surprising as the variable region of light chains determines their isoelectric point (pI), solubility, and affinity for other proteins.

There are marked differences in the electrophoretic mobilities of Bence-Jones proteins which parallel the relative anionic charge/unit mass of the light chain (Hill *et al.* 1983). Experiments on hydropenic rats suggested that myeloma light chains only cause intratubular casts and renal failure if their pI is greater than 6.2 (Clyne *et al.* 1979), but the data are controversial. More recently Hill *et al.* (1983) reported that light chains with high pI values were more prone to induce cast formation, although these were not typical of the usual myeloma casts. Coward *et al.* (1984) reported that light chain pI was inversely correlated with creatinine clearance in 23 patients with multiple myeloma. Nevertheless, Melcion *et al.* (1984) have shown that severe renal failure in myeloma patients can occur despite a wide range of pI in the light chains; Cooper and MacLennan (1988) observed that light chains exhibit a wide range of pI values from pH 3 to 9, and within a given light chain there is a considerable microheterogeneity of charge producing several bands on isoelectrofocusing. In a study of the pI spectrotypes of 43 light chains Johns *et al.* (1986) found no correlation between the mid-point pI and serum creatinine concentrations or clinical course of renal function.

Low molecular weight proteins, which include plasma proteins, hormones and enzymes, as well as free light chains, are filtered by the glomerulus, and reabsorbed and catabolized by the proximal tubular cells (Maak 1975). Some of these, such as α_1-microglobulin, α_1-acid glycoprotein, β_2-microglobulin and retinol-binding protein, can be conveniently used to quantify tubular function (Kusano *et al.* 1985). In a study of over 500 patients, Cooper *et al.* (1980) found that there is an almost complete correlation between increased urinary excretion of α_1-microglobulin and α_1-acid glycoprotein and Bence-Jones proteinuria which was independent of light chain isotype and electric charge. In a later study of 43 light chains isolated from the urine of patients with myeloma, Johns *et al.* (1986) found that the urinary α_1-microglobulin concentration was greater in patients whose light chains had higher pI values, provided that patients with renal failure were excluded. In other words Bence-Jones proteins with a high pI inhibited the tubular reabsorption of α_1-microglobulin more than those with a more acidic pI, but retinol-binding protein and β_2-microglobulin were less readily affected (Cooper and MacLennan 1988). This is analogous to the effect of substances with a positively charged N-terminal group, such as lysine, which are able to inhibit tubular resorption of low molecular weight proteins. Evidence of tubular damage in the form of increased excretion of N-acetyl-β-D-glucosaminidase can be detected even in patients with very low levels of Bence-Jones proteinuria (Coward *et al.* 1985).

Abnormalities of tubular function in patients with Bence-Jones proteinuria have been reported (Smithline *et al.* 1976). When renal tissue slices are incubated with Bence-Jones protein concentrations similar to those that are supposed to exist in proximal tubular fluid, a number of tubular cell metabolic processes are inhibited, including gluconeogenesis, PAH uptake, and Na/K-ATPase activity (McGeoch *et al.* 1978). In a study on 87 patients with plasma cell dyscrasias, 37 of whom had Bence-Jones proteinuria, Colussi *et al.* (1988) confirmed the significant association between this proteinuria and tubular functional abnormalities in patients with well preserved GFR. They found no correlation between the amount or the type, kappa or lambda, of Bence-Jones proteins in urine and the severity of tubular abnormalities, including reduced reabsorption of β_2-microglobulin and amino acids, and a significantly raised clearance of lithium. Increased free water clearance and lower urine osmolality attained during maximal water diuresis, indicative of increased sodium delivery to postproximal segments of the tubule (Rombolà et al. 1987), suggested that the main effect of Bence-Jones proteins on the tubule was the inhibition of proximal reabsorption of sodium and sodium-cotransported solutes. Distal tubular defects are rare in myeloma patients and invariably associated with global renal damage (Mallick and Williams 1988).

In conclusion, although isolated tubular defects were quite variable and a generalized tubular dysfunction is by no means the rule, abnormal tubular handling of sodium can cause dehydration (Colussi et al. 1988).

Pathogenesis of cast nephropathy

Analysis of the composition of the casts by immunofluorescence, combined with results from experimental models have provided an insight into myeloma cast nephropathy. The casts contain the light chain of the same type as the Bence-Jones protein, together with Tamm-Horsfall protein; albumin and antigenic determinants of other immunoglobulins can often be detected as well. Koss et al. (1976) have defined the relation between light chains and Tamm-Horsfall protein experimentally; injection of light chains into mice was followed immediately by the production of light chain casts but Tamm-Horsfall protein was only found 1 day later.

It is clear that a variety of factors play a role in the pathogenesis of myeloma cast nephropathy. The first problem is why it develops in some patients and not in others. The experimental studies, just cited, have shown that Bence-Jones proteins have similar effects when injected into mice as they did in the patient from which they were derived. The physicochemical properties of individual Bence-Jones proteins are believed to be partly responsible for these differences. It is also possible that host factors are involved. Smolens et al. (1983) have studied a unique strain of rats (LOU/c) which spontaneously develop light chain secreting tumours which can be removed and transplanted into histocompatible rats who do not normally develop the tumours (LOU/m). This allows the effects of Bence-James proteins on the recipients to be analysed under controlled conditions. Possible recipient-related effects of tumour growth were circumvented in another model, in which individual Bence-Jones proteins, with pI values of 4.3, 5.2, 6.7 and 7.6, were collected from tumour bearing animals and given to non-tumour bearing animals, placed on a water-restricted acid diet to maximally acidify their urine (Smolens et al. 1986). The renal changes in LOU/c rats were induced by the administration of the individual Bence-Jones proteins to LOU/m rats for 5 days. Nephrotoxicity was observed only in animals which either produced or received negatively charged Bence-Jones proteins (pI 4.3 and 5.2); i.e. in contrast to other reports, nephrotoxicity was not associated with higher pI. Severe distal nephron and proximal tubule lesions cast nephropathy were observed in rats given the protein with a pI 5.2. The results of the study seemed to confirm that nephrotoxicity may be a unique intrinsic property of Bence-Jones proteins.

Ronco et al. (1988) found lambda Bence-Jones proteins in two-thirds of 34 patients with multiple myeloma; patients with renal failure had Bence-Jones proteins with pI values between 5.2 and 8.9. They also observed severe tubular lesions in mice bearing plasmacytomas producing Bence-Jones protein with low pI. Both sets of data suggest that pI alone does not determine whether Bence-Jones proteins form casts; other as yet unidentified properties or host factors must be involved. Possibilities include less reabsorption and lysosomal degradation by proximal tubular cells, and impaired proximal tubular function, both of which increase delivery of Bence-Jones proteins to the distal nephron. Bence-Jones proteins are thought to coprecipitate with Tamm-Horsfall protein in the distal tubules to form casts. At a low urinary pH, Bence-Jones proteins with a high pI (>5.6) will be positively charged; Tamm-Horsfall protein has pI of 3.2 and negative charge: the resulting electrostatic interaction will facilitate coprecipitation. Precipitation would also be favoured by dehydration, interaction of radiographic contrast agents with Tamm-Horsfall protein, and hypercalcaemia (Smolens and Kreisberg 1985).

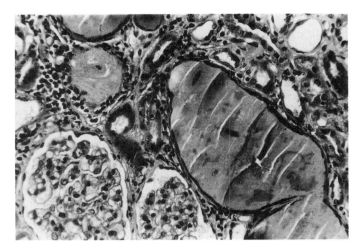

Fig. 2 Myeloma cast nephropathy. Light microscopy. Tubular casts with fractured aspect and polychromatophily. Normal glomeruli. Interstitial peritubular inflammatory reaction (Masson's trichrome staining, original magnification × 125).

Large casts appear to obstruct the nephron. Ronco et al. (1988) demonstrated Tamm-Horsfall protein in Bowman's space in 16 of 18 biopsies performed in patients with myeloma cast nephropathy, suggesting intratubular back-flow of urine, possibly due to the combined effects of increased intratubular pressure and reduced GFR. It is also possible that Bence-Jones proteins are directly toxic to proximal tubular cells and may aggravate the effect of intratubular obstruction (Sanders et al. 1988b). This may be linked to their accumulation in cells and to lysosomal changes observed in some patients, which may be severe enough to cause acute tubular necrosis. Tubular basement membrane rupture may favour the reaction of monocytes/macrophages with cast components, leading to interstitial oedema and inflammation.

Histological features

Myeloma cast nephropathy is characterized by intratubular casts with distinctive features that distinguish them from the intratubular casts found in other nephropathies (Pirani 1988). They are found almost exclusively in the distal convoluted tubules and collecting ducts. Light microscopy reveals large, refractile, bright casts, often multilayered or lamellated (Fig. 2), containing numerous fracture lines, with angulated outlines, and sometimes crystals. The larger crystals are a unique feature of myeloma cast nephropathy. Many casts are surrounded by multinucleated giant cells (Fig. 3) of monocytic origin (Pirani 1988) which seem to be phagocytosing the cast or its fragments. The presence of crystals within the casts, and less commonly within adjacent tubular cells, and the presence of giant cells around the cast are characteristic of myeloma cast nephropathy (Pirani 1988). The casts stain with eosin, PAS, Alcian Blue and show polychromasia with trichrome preparations. Using immunofluorescence, the casts are predominantly stained with antisera to the abnormal light chain, but this is often accompanied by cross-reaction to the other light chain, immunoglobulins, albumin, complement components, and Tamm-Horsfall protein (Cohen 1988).

The electron microscopic features are also characteristic, showing the casts to be composed of large elongated crystals or fragments with a filamentous substructure. By contrast, other casts are finely granular, homogeneously electron dense or coarsely granular with needle-shaped crystals and microspherical particles (Fig. 4). Small crystals can be observed in phagolysosomes of tubular cells (Fig. 5). Both large and small crystals presumably consist of polymerized and perhaps partially degraded light chain proteins (Pirani et al. 1987).

Proximal tubular cells are often atrophic and show other nonspecific changes. Distal tubule dilatations with flattening of the epithelial lining are common; exfoliation of epithelial cells and ruptures of the tubular basement membrane can be occasionally observed, and are associated with interstitial inflammation. The severity of interstitial fibrosis, present in about 25 per cent of cases, corresponds to the degree of tubular atrophy and vessel sclerosis.

Clinical manifestations and course

Multiple myeloma is characterized by bone marrow plasmacytosis (more than 10 per cent) with clustering of atypical plasma cells, a serum peak of M protein (>3 g/dl), M protein in urine, and reduced levels of the uninvolved immunoglobulins. There may be progressive lytic bone lesions. Symptoms are by no

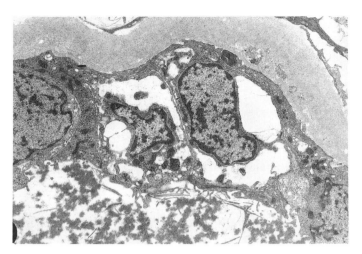

Fig. 4 Myeloma cast nephropathy. Electron microscopy. Coarse granular material and needle-shaped crystals in a tubular lumen. Marked changes of tubular cells (original magnification × 10 000).

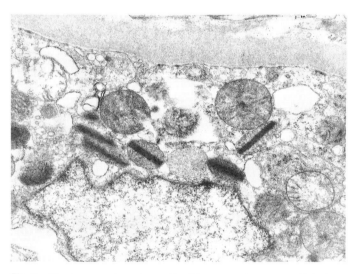

Fig. 5 Myeloma cast nephropathy. Electron microscopy. Tubular epithelial cell containing phagolysosomes with crystalline inclusions (original magnification × 10 000).

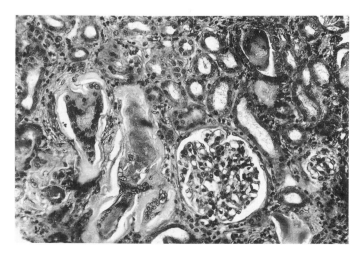

Fig. 3 Myeloma cast nephropathy. Light microscopy. Macrophage reaction (multinucleated giant cells) partially surrounding the cast; in one, note the presence of negative image of cholesterol crystals (AFOG stain, original magnification × 125).

means universal, but are dominated by bone pain and fractures, anaemia, recurrent infections and, less frequently, hypercalcaemia, hyperviscosity, and bleeding tendency. Kidney involvement may present clinically as acute or rapidly progressing renal failure, but slowly progressing chronic renal failure is more usual (Pasquali et al. 1987). Some less severe or early morphological changes can coexist with normal renal function and the severity of cast formation correlates poorly with the amount of Bence-Jones proteinuria and the severity of renal failure (De Fronzo et al. 1987). The severity of tubular and interstitial damage is the only histological feature that correlates with renal outcome. Nephrotic syndrome is unusual in myeloma cast nephropathy and, when present, is always due to associated amyloidosis or light chain deposition disease. Slowly progressing chronic renal failure is sometimes punctuated by episodes of reversible or irreversible worsening due to dehydration, infection, drug nephrotoxicity, hypercalcaemia, and so on. Additional factors which may contribute to progression to end-stage renal disease include plasma cell infiltration or light chain deposition, nephrocalcinosis and renal amyloidosis. Acute renal failure occurs in up to 10 per cent of patients and the major precipitants are dehydration, hypercalcaemia, infection, use of radiocontrast agents, uric acid nephropathy, hyperviscosity, nephrotoxic drugs, and non-steroidal anti-inflammatory drugs. High urinary levels of Bence-Jones proteins and low urinary pH may also be predisposing factors.

Other forms of renal impairment

The Fanconi syndrome

A generalized proximal tubular dysfunction (the so-called Fanconi syndrome) has been described in myeloma patients. It is always associated with Bence-Jones proteinuria and is quite rare (Coward and Mallick 1985).

Three different effects of Bence-Jones proteins on the renal tubule need to be considered. First, that increased excretion of low moleculor weight proteins is caused by direct competition for reabsorption, rather than proximal tubular injury. This has been called selective tubular proteinuria (Cooper et al. 1984) and is found in almost all myeloma patients, very few of whom have other evidence of tubular dysfunction (Scarpioni et al. 1981).

Second, Bence-Jones proteins may inhibit tubular cell function directly (Sanders et al. 1987), and this may be responsible for the abnormal handling of sodium by the proximal tubule which has been reported in some cases (Colussi et al. 1988); it may also cause the Fanconi syndrome.

Lastly, anatomical injury to proximal and distal tubular cells can be a result of cast formation, direct toxic effects on proximal tubular cells, and secondary interstitial inflammation; more rarely from amyloidosis, granular light chain deposition, and hypercalcaemia.

Fanconi syndrome has been described in myeloma patients with normal, or almost normal, findings at renal biopsy (Smithline et al. 1976). Intracellular crystal-like inclusions are found in some cases but are not as specific or as common as previously thought (Pirani 1988). There are reports that the Fanconi syndrome can precede the development of overt myeloma (Maldonado et al. 1975).

According to the literature the GFR is frequently impaired in myeloma patients with the Fanconi syndrome, but the histological pattern of myeloma cast nephropathy does not significantly differ from that of patients without tubular dysfunction (Chan et al. 1987). In short, a histological counterpart of Fanconi syndrome has yet to be identified (Thorner et al. 1983).

The tubular dysfunction usually involves reabsorption of all proximally-handled solutes (Na/fluid, glucose, amino acids, uric acid, phosphorus, bicarbonate, low molecular weight proteins), which distinguishes it from the tubular proteinuria commonly associated with Bence-Jones proteinuria. Close scrutiny of published cases of light chain-related Fanconi syndrome, however, shows that in occasional patients the tubular dysfunction may not be generalized, since reabsorption of individual solutes, such as glucose (Smithline et al. 1976), bicarbonate (Maldonado et al. 1975), uric acid (Rao et al. 1987), and even amino acids (Rao et al. 1987) may be spared in some patients. In a series of 37 patients with Bence-Jones proteinuria, 16 had at least one tubular transport abnormality (Colussi et al. 1988). However an almost continuous spectrum of severity of tubular abnormalities was observed, and only three patients had four or more transport abnormalities, i.e. the classical Fanconi syndrome. Thus it is likely that patients with the complete Fanconi syndrome have the most severe tubular defects. Direct inhibition of tubular cell metabolism and function could be responsible for both the tubular abnormalities and the Fanconi syndrome and the functional and anatomical alterations may be different aspects of light chain nephrotoxicity (Sanders et al. 1988a).

Hypercalcaemia

About 25 per cent of patients with multiple myeloma develop hypercalcaemia at some time (Kyle 1975), and it is not infrequently the presenting manifestation. Hypercalcaemia results from increased bone reabsorption, due to the local activation of osteoclasts by a lymphokine, osteoclast activating factor, and is accompanied by low PTH and 1,25-dihydroxy-vitamin D_3 levels. Osteoclast activating factor is heterogeneous; the large form has a molecular weight of 12 500 to 25 000 and the small form has molecular weight of 1330 to 3500 Da (Mundy and Raisz 1977).

This factor was originally identified in culture medium from normal leucocytes, and has subsequently been detected in supernatants from cultured neoplastic cells from patients with multiple myeloma, Burkitt's lymphoma, and malignant lymphoma. It has not been detected in the blood, and prostaglandins may be required for its production and/or release; its effects are inhibited by corticosteroids. Hypercalcaemia is either absent or mild until impaired renal function limits calcium excretion: acute deterioration of renal function due to dehydration, light chain toxicity or hypercalcaemia itself often results in severe hypercalcaemia, which in turn aggravates dehydration and renal damage.

Mild hypercalcaemia (<12 mg/dl) can often be successfully managed with corticosteroids combined with chemotherapy for the underlying proliferative disorder. Moderate (12–15 mg/dl) and severe (>15 mg/dl) hypercalcaemias must be treated as emergencies: dehydration is always present in these patients and fluid replenishment is the first priority. Saline (1–2 l in the first 2 h), followed by an additional 4 l/day, with the appropriate amounts of potassium and bicarbonate, can be given safely. Frusemide, at doses of 10 to 40 mg intravenously every 6 h to 80 to 100 mg intravenously every 2 to 4 h, has been advocated in order to maximize urinary calcium excretion; however strict monitoring and continous replacement of water and electrolyte losses are mandatory in this treatment. Intravenous administration of salmon calcitonin is usually safe and effective. An initial dose of 4 MRC units kg body weight is followed by 2 MRC units kg every 6 h until plasma calcium levels fall to normal. Since its effect is rapid but short-lived, it must be given together with drugs which have longer lasting effects, such as dichloromethylene diphosphonate, and aminohydroxypropylidene diphosphonate, which inhibit osteoclastic activity without affecting bone mineralization (Raisz 1980) (see Chapter 14.1).

Dichloromethylene diphosphate is infused slowly at a dose of 2.5 mg/kg for the first day, and thereafter at 5 mg/kg.day until plasma calcium reaches normal; oral administration (1600 mg/day), or intermittent intramuscular administration (100 mg once or twice a week) can then be used indefinitely to prevent recurrences. Aminohydroxypropylidene diphosphonate is given at a dose of 10 mg/kg orally for 5 days followed by 5 mg/kg for another 5 days ; its effects appear to be particularly long-lasting. However corticosteroids are most effective in the control of myeloma-associated hypercalcaemia, since they inhibit osteoclast activating factor-induced bone resorption. Prednisone (40–80 mg/day) is usually effective in 5 to 10 days. Definitive control of hypercalcaemia requires effective chemotherapy for the underlying multiple myeloma; chronic prednisolone maintenance therapy (10–20 mg/day) may be helpful in the control of hypercalcaemia in unresponsive patients.

Mithramycin, a very effective drug for controlling hypercalcaemia, has rarely been used since the introduction of the new diphosphonates.

Hyperviscosity syndrome

Plasma-cell dyscrasias also cause the hyperviscosity syndrome, which is defined as a permanent or temporary increase in blood viscosity accompanied by specific signs and symptoms, including neurological manifestations, ocular haemorrhages and exudates, peripheral neuropathies, and bleeding. This syndrome is sometimes associated with impaired urinary concentration capacity disproportionate to the decrease in GFR. Blood viscosity is increased relative to the concentration of circulating M protein, and the syndrome develops as a result of the increased resistance to blood flow when serum viscosity is greater than 4 (normal range 1.4–1.8 (Ostwald viscometer at 37°C: plasma/water ratio)). Hyperviscosity syndrome appears at variable levels of M protein (more often when serum M concentration is greater than 5 g/dl) and depends on the immunoglobulin class and its stereochemical structure. It is more frequent with IgM, and rarer with IgA and IgG, occurring in 33 per cent of patients with Waldenstrom macroglobulinaemia, in 10 to 20 per cent of patients with IgA myeloma, and less commonly still in IgG myeloma.

In addition to chemotherapy, control of hyperviscosity syndrome may require the immediate removal of the M protein (Busnach et al. 1986). Transient inprovements in the bleeding tendency and neurologic manifestations have been reported in more than 75 per cent of episodes treated acutely by plasma exchange (Avnstorp et al. 1985). An intensive course, 6 to 8 treatments over 10 to 14 days, with exchange of at least one plasma volume at each time has been recommended in addition to chemotherapy to overcome the complications of hyperviscosity syndrome (Busnach et al. 1988).

Treatment and outcome

The principal aim of therapy, the improvement of renal function, can be pursued in several ways including: (i) the correction or prevention of conditions that predispose to or precipitate renal failure (dehydration, use of contrast media and non-steroidal anti-inflammatory drugs, etc.); (ii) sustained alkaline diuresis; (iii) dialysis; (iv) reduction of serum and urine light chain levels by chemotherapy and plasmapheresis.

Supportive therapy is the first step, but chemotherapy should be started immediately. A fast acting regimen is preferred in patients presenting acute renal failure with a large tumour mass; a 10 to 15 day regimen of vincristine plus doxorubicin plus prednisone plus alkylating agent (for dosages see Alexanian and Dreicer 1982) is preferred to the slower acting melphalan plus prednisone (Alexanian and Barlogie 1988). Full dosages of all drugs are required to obtain a response. Aggressive chemotherapy can be very dangerous in patients with renal failure, and drug dosage may need to be reduced in some cases. Nevertheless, myeloma must be treated, and renal status does not seem to influence the response. The myeloma must be controlled even when renal failure is irreversible. The need for chemotherapy can be questioned in some patients with advanced renal failure, even though indicated in most patients (Mallick 1988).

The tumour mass does influence renal status and response to chemotherapy. Alexanian and Barlogie (1988) studied 389 patients; seven of 110 patients with a low tumour mass, 13 of 155 with an intermediate tumour mass and 44 of 124 with a high tumour mass had renal failure. About two-thirds of the patients with serum creatinine concentrations above 3 mg/dl had a large tumour mass. The clinical response rate was not significantly different in patients with or without renal failure, and the duration of the remission was roughly similar. The median survival time of patients with mild or severe renal failure was about 9 months shorter than that of patients with normal renal function (22 versus 32). Six per cent of patients with renal failure died during the first 4 weeks of treatment.

Disappearance of Bence-Jones proteinuria, a reduction of more than 50 per cent of the serum M component and normalization of hypercalcaemia are prerequisites, for remission, and have been obtained up to 89 per cent of myeloma patients (Kyle and Greipp 1983).

Improved renal function occurs in more than 50 per cent of patients (Bernstein and Humes 1982), but it may take months to achieve stable renal function independent of dialysis. In the series of Rota et al. (1987) recovery took up to 210 days (median more than 50 days), whereas reversibility was more rapid in the patients of Alexanian and Barlogie (1988).

Monitoring the efficacy of chemotherapy is based primarily on serial analyses of Bence-Jones proteinuria and serum M protein, by measuring serum concentrations of β_2 microglobulin, which is determined both by the renal function and by the residual tumour load, and lastly monitoring of bone-marrow plasmacytosis (Alexanian and Barlogie 1988).

Renal biopsy is indicated in patients presenting acute or rapidly progressing renal failure as it may reveal renal lesions other than myeloma cast nephropathy and give some insights for prognosis. Prognostic criteria have been based both on the degree of kidney impairment and the tumour mass (Durie and Salmon 1975). At present, the median survival is 36 months for all patients with myeloma, calculated on the basis of a recent review of the literature (Rota et al. 1987). Patients who recovered normal renal function may have a disease course and prognosis similar to that of myeloma patients without renal failure (Cosio et al. 1981), but persistent renal failure implies a poorer prognosis. Infection and renal failure dominate early and late mortality.

Treatment of end-stage renal failure by haemodialysis or chronic ambulatory peritoneal dialysis does not seem to prolong survival significantly except in patients who already have a good prognosis (Coward et al. 1983). Renal transplantation has been advocated for patients with the best prognosis, and who have already responded to chemotherapy (Alexanian 1985); in a few cases it has been successful (Cosio et al. 1981; Mallick 1988).

Light chain deposition disease

Light chain deposition disease is a recently identified systemic disease with predominant kidney involvement, and is associated with plasma cell dyscrasias. In 1973 Antonovych et al. reported that the nodules in glomeruli of some myeloma patients contained deposits of a single type of light chain. In 1976 Randall et al. recognized the systemic nature of the disease by demonstrating light chain deposition in extrarenal tissues, sometimes without overt myeloma. Subsequent reports have indicated that nodules are the typical glomerular lesion (Gallo et al. 1980) but that mesangial hypercellularity, non-nodular mesangial enlargement, and even virtually normal glomeruli can also be seen in this disease (Seymour et al. 1980; Ganeval 1988). Roughly 80 per cent of reported patients have had kappa light chain deposits and 20 per cent have lambda light chains (Gallo and Buxbaum 1988).

Typically, light chain deposition disease is seen in patients with myeloma, but it also occurs in other plasma cell dyscrasias and sometimes in patients without detectable monoclonal plasma cell neoplasms or other malignancies. The widespread use of immunofluorescence examination of renal biopsies has increased the number of cases diagnosed and it is thought to be less rare than previously believed; about 75 cases are recorded in the literature. According to Ganeval et al. (1984) it is as common as amyloidosis in patients with multiple myeloma and kidney involvement.

Deposits with a similar appearance but containing both monoclonal light and heavy chain determinants have been reported occasionally, and Gallo and Buxbaum (1988) have used the term of light and heavy chain deposition disease to describe these. The generic term 'monoclonal immunoglobulin deposition disease' has been used to describe both types of deposition disease. Some authors (Tubbs et al. 1981; Cohen 1984; Fang 1985; Gerlag et al. 1986; Bangerter et al. 1987) prefer the term light chain nephropathy to light chain deposition disease; unfortunately it does not emphasize the systemic nature of the disease. On the other hand the older term of glomerular lesions of myeloma does not cover the cases without multiple myeloma.

Pathophysiology

In a few patients the presence of a plasma cell dyscrasia has to be inferred purely from the deposition of monoclonal light chain in tissues (Alpers et al. 1985); in other patients the diagnosis can be confirmed from examination of the bone marrow. Some patients may have also a small amount of monoclonal light chain (the same as that in tissues) in serum and/or urine (Gallo and Buxbaum 1988). Two-thirds of patients have evidence of an underlying plasma cell dyscrasia (multiple myeloma in most cases) to account for the light chain deposition (Troussard et al. 1987). Whether or not the remaining cases represent clinically silent plasma cell disorders is still uncertain. In five patients with light chain deposition diseases and no detectable myeloma at presentation, neither myeloma nor other malignancies were observed after prolonged follow-up, amounting to more than 10 years in one patient (Ganeval 1988). Nevertheless light chain deposition disease is almost certainly linked to the production of a monoclonal light chain, whether or not myeloma is present.

Immunofluorescence examination of the bone marrow and immunoglobulin biosynthesis studies confirm that light chain deposition disease is one of plasma cell dyscrasias, since a monoclonal plasma cell population has been found in every case, even in those without serum and/or urine monoclonal immunoglobulin, and with a normal percentage of plasma cells (Preud'homme 1988). Furthermore experiments using in-vitro incorporation of radioactive amino acids in bone marrow cultures from patients with light chain deposition disease have demonstrated that light chains of smaller or larger size than normal with unusual glycosylation and a strong tendency to polymerize are produced. In a few cases abnormal heavy chains are also produced. In some cases these structurally abnormal

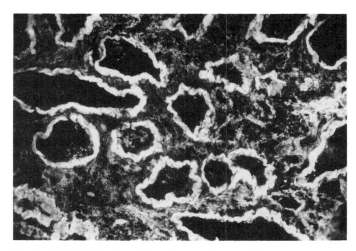

Fig. 6. Light chain deposition disease. Immunofluorescence. Bright ribbon-like deposits of kappa light chains along the tubular basement membranes (original magnification × 250).

immunoglobulins are not detectable in serum and/or urine because their antigenic determinants are altered.

These abnormalities of immunoglobulins, particularly their abnormal glycosylation and their tendency to polymerize, may explain the deposition of light chains in tissues, and may have a direct role in the pathogenesis of the disease (Solling et al. 1984, Palant et al. 1986; Bradley et al. 1987). The correlation between the composition of the deposits and the findings in in-vitro biosynthesis experiments are strongly suggestive (Preud'homme 1988).

Abnormalities of immunoglobulin synthesis and deposition in tissues have also been demonstrated in patients with AL amyloidosis (Ganeval et al 1984); both types of disease have been observed in some patients (Jacquot et al. 1985).

Renal pathology

On immunofluorescence, the pathognomonic finding in light chain deposition disease is bright diffuse linear staining of the basement membranes of tubules (Fig. 6), Bowman's capsule and glomerular capillaries for a single light chain isotype. Occasionally light chains are deposited in the walls of arteries and arterioles. Staining is more intense in glomerular nodules than in capillary walls (Fig. 7) and mesangial staining is less bright when nodular lesions are absent. Sometimes deposits are found only along tubule basement membranes. In some cases heavy chains and various complement components have been found in the same sites as monoclonal light chains.

Mesangial nodules seen on light microscopy were emphasized in the first reports (Fig. 8), but they are not as frequent as previously thought and are absent in about 50 per cent of the patients. Varying degrees of mesangial hypercellularity (McLeish et al. 1985; Venkataseshan et al. 1988) and, in some cases, focal and segmental extracapillary proliferation (Confalonieri et al. 1988), capillary wall thickening and double contours and microaneurysms in dilated capillaries (Sinnah and Cohen 1985) can be observed in both nodular and non-nodular glomeruli. Ten per cent of patients have near-normal glomeruli (Ganeval 1988). The glomerular nodules stain positively with eosin, periodic acid-Schiff and silver impregnation, but are negative for amyloid. Nodules are not typical of the late stages of the disease

Fig. 7. Light chain deposition disease. Immunofluorescence. Intensely positive for kappa light chains on glomerular nodules and tubular basement membranes (original magnification × 250).

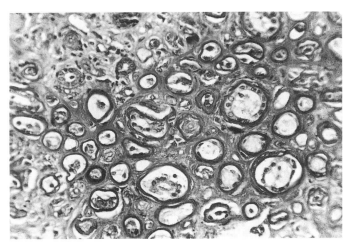

Fig. 9 Light chain deposition disease. Light microscopy. Tubular basement membrane thickening with ribbon-like appearance and occasional splitting (PAS staining, original magnification × 250).

(Ganeval 1988), although progression to nodules has been documented in a few cases (Gallo and Buxbaum 1988).

Tubular involvement is constant and quite striking. The tubule basement membranes are thickened (Fig. 9) and periodic acid-Schiff staining seems to differentiate two layers, an inner one which is intensely positive, and an outer one which is less so. The association of these tubular lesions with the typical cast nephropathy is uncommon in patients with light chain deposition disease and multiple myeloma (Hill *et al.* 1983). Interstitial fibrosis and vessel sclerosis can also be observed.

On electron microscopy, there are electron-dense, fine, uniformly granular deposits that occupy the lamina rara interna in glomerular basement membranes (Fig. 10) and the outer (peritubular) aspect of the tubular basement membranes (Noel *et al.* 1984) (Fig. 11). The electron density of these deposits is quite variable; sometimes they are arranged in uniformly spaced punctate clusters, or they may appear as confluent granular masses (Fig. 12). Only the light chain deposits appear electron dense in mesangial nodules.

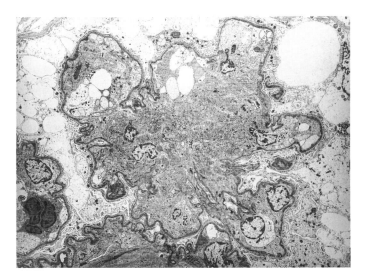

Fig. 10 Light chain deposition disease. Electron microscopy. Mesangial area greatly expanded by a nodule. Capillary walls outlined by electron-dense subendothelial deposits along the glomerular basement membrane (original magnification × 2500).

Clinical features

Overt plasma cell dyscrasia can be found at presentation in about two-thirds of patients with light chain deposition disease. Multiple myeloma is the most common, but Waldenstrom's macroglobulinaemia, chronic lymphocytic leukaemia, and large-cell lymphoma have all occasionally been reported. The clinical presentation points to prominent kidney involvement in almost all patients.

Three clinical patterns have been recognized: nephrotic syndrome, which occurs in less than one-third of patients; rapidly progressive renal failure in another third, and a slowly progressing chronic renal failure, sometimes with proteinuria, sometimes with a pattern suggesting tubular interstitial nephropathy in the remaining cases. Hypertension is quite rare at presentation.

Typical nodular lesions are found in the glomeruli in less than

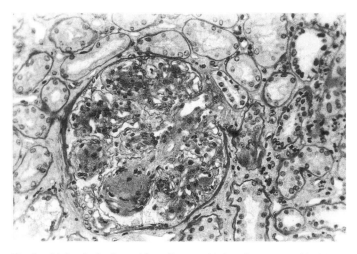

Fig. 8. Light chain deposition disease. Light microscopy. Nodular glomerulosclerosis with moderate hypercellularity (PAS staining, original magnification × 250).

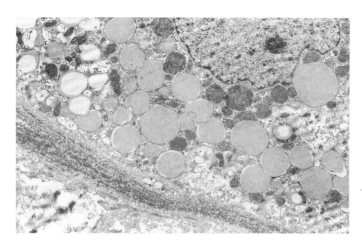

Fig. 11 Light chain deposition disease. Electron microscopy. Multilayer deposition of dense granular material along the tubular basement membrane. Several phagolysosomes in the cytoplasm of the tubular cell (original magnification × 9000).

50 per cent of patients with nephrotic syndrome, and in 25 per cent of the remainder. In the majority of cases, progression of renal failure is relentless and they reach end-stage kidney disease in a few months. Diagnosis requires a high level of suspicion and can be missed without immunofluorescence with antibodies to individual light chains. Subendothelial electron-dense deposits of light chains were demonstrated only by immunoelectron microscopy in three patients with nephrotic syndrome and near normal glomeruli on light microscopy reported by Sanders et al. (1988b).

Therapy and prognosis

The prognosis of light chain deposition disease is influenced by associated diseases, severity of renal lesion and degree of systemic involvement. Many patients die from associated myeloma. The importance of extrarenal visceral lesions is variable, and depends on the extent of the deposits: liver and heart failure are the principal causes of death. In many patients, however, the disease remains asymptomatic and renal function can be stable for a long time.

Since light chain deposition disease is always linked to quantitative and/or qualitative alterations in immunoglobulin production by a plasma cell clone, cytostatic therapy has been considered as treatment. Favourable effects of melphalan and cyclophosphamide given together with prednisone have been reported in a few cases. Although treated patients seem to do better than untreated ones (Bradley et al. 1987; Ganeval 1988) the variable evolution of the disease makes the evaluation of treatment difficult. Cytostatic therapy is indicated only in the presence of important extrarenal manifestations in patients with end-stage renal disease. Some patients survive for many years on regular dialysis and a few patients have received a kidney transplant (Alpers et al. 1985). In some of these, recurrence has been observed (Pirson 1988).

Kidney involvement in primary and myeloma associated amyloidosis

The kidney is commonly involved in amyloidosis secondary to multiple myeloma (see Chapter 4.2). The amyloid proteins are the products of the processing by cells of normally occurring proteins, the amyloid precursors (Glenner 1980; Cohen et al. 1983). The type of amyloid protein and its precursor, identified by means of their biochemical characterization, are the basis of any present classification of amyloidosis (Table 2).

AL is the most commonly encountered amyloid in the kidney. Some cases have mixed amyloid (AL + AA) deposits but often one of the components is barely detectable; for instance, traces of AA in myeloma associated amyloidosis or of AL in AA amyloidosis.

Only primary and myeloma amyloidosis (AL amyloidosis) will be discussed in this chapter as renal involvement in secondary (AA amyloidosis) and other forms of amyloidosis is dealt with in Chapter 4.2. Glenner et al. (1971) were the first to use amino-acid sequence analysis to demonstrate the homology between N-terminal sequence of primary (AL) amyloid proteins and immunoglobulin light chains. As the amyloid proteins were smaller than whole light chain molecules they could have resulted from enzymatic cleavage. Mice, repeatedly injected 1 to 2 months with Bence-Jones proteins obtained from patients with AL amyloidosis, develop AL amyloid demonstrated by deposition of Congo Red positive, birefringent, fibrillar material within the blood vessel walls and interstitial tissue of the kidney (Solomon and Weiss 1988). Only certain light chains are amyloidogenic, which may explain why amyloidosis is observed in only about 10 per cent of patients with multiple myeloma. On the other hand cleavage of the molecule does not seem to be essential for the transformation of the variable segment into an amyloid fibril because other amyloid proteins, such as those of familial amyloidotic polyneuropathy or β_2m-amyloidosis have the same size as the intact precursor molecule (prealbumin and β_2-microglobulin, respectively), and because, in rare cases, AL proteins are composed of the entire light chain molecule (Shirahama et al. 1984). The molecular weight of AL proteins ranges from 5000 to 23 000 Da.

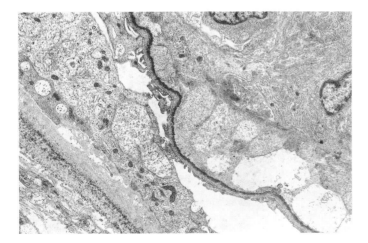

Fig. 12 Light chain deposition disease. Electron microscopy. Continuous layer of granular electron-dense deposits along the lamina rara interna. Multilayered deposition of dense granular material along the basement membrane of the Bowman capsule (original magnification × 7200).

Table 2 Classification of systemic forms of amyloidosis

Clinicopathological forms	Amyloid protein	Precursor
Primary	AL	Light chains (kappa or lambda)
Myeloma-associated		
Secondary	AA	Serum amyloid A
Familial	A prealbumin	Prealbumin
Haemodialysis-associated	A β_2-microglobulin	β_2-microglobulin

Amyloidogenesis

Amyloidogenesis appears to be an intricate process in which three phases can be distinguished: the synthesis and secretion of the precursor protein; enzymatic degradation which transforms the precursor in amyloid protein; and polymerization of amyloid protein into fibrils.

Several factors are involved in this transformation. For simplicity they can be separated into precursor-related factors and environment-related mechanisms. Secretion of free light chains seems to be a constant feature, even in patients with primary amyloidosis with a normal percentage of plasma cells in the bone marrow and in whom monoclonal plasma cells or monoclonal immunoglobulins cannot be detected, since experiments with bone marrow cells have shown that large amounts of free light chains are secreted (Preud'homme 1988). The predominance of lambda type light chains in most patients with myeloma associated and primary amyloidosis, with a reversal of the normal kappa/lambda ratio, suggests that something related to the primary structure of the light chain renders it more susceptible to amyloidogenesis. Normally, lambda light chains have a greater tendency to form dimers than kappa light chains and are relatively difficult to cleave enzymatically. The AL amyloid substance is usually made of light chain fragments containing the entire V region, or the V region plus about 50 residues of the C region. As a result, antisera to whole light chains may react with AL amyloid proteins weakly or not at all.

Light chains are also characterized by multiple V region-related subgroups that have been defined chemically and serologically. Four kappa chain subgroups, and six lambda chain subgroups have been characterized (Solomon and Weiss 1987). Solomon et al. (1982) have found that one particular subgroup, lambda VI, is preferentially associated with AL amyloidosis. This relatively uncommon subgroup was identified in 17 lambda VI Bence-Jones proteins from patients with AL amyloidosis. In kappa chain-related AL amyloidosis a disproportionately high number of the subgroup kappa II was observed.

A feature of light chains which relates to the variable region is their susceptibility to proteolysis. Nine out of 10 immunoglobulin biosynthesis studies in patients with primary amyloidosis produced light chain fragments (Preud'homme 1988) which were extremely unusual in multiple myeloma without amyloidosis. Light chains from AL amyloidosis patients appear to be especially susceptible to enzymatic degradation and amyloid substance can be produced in vitro by proteolysis of free monoclonal light chains. The largest amounts of light chain fragments are found in AL amyloidosis patients without Bence-Jones proteins in urine, probably because of postsecretory degradation (Preud'homme et al. 1980). All of these findings support the hypothesis that light chains are secreted as intact molecules in primary amyloidosis and that certain features of their primary structure determine whether they are amyloidogenic. Whether these factors relate to propensity to enzymatic degradation and to precipitation of the resulting peptides as insoluble tissue deposits is debatable. Other investigators have suggested that defective processing of light chains could lead to peptide fragments which are not easily catabolized and prone to self-assembling and cross-linking (Shirahama 1988). Defective processing by monocytes/macrophages could then be responsible for AL amyloidogenesis, but as yet the evidence is lacking (Zucker-Franklin 1988). Data from experimental models of AA amyloidosis suggest that fibrillogenesis may be facilitated by an enzyme secreted by the stimulated macrophage which favours the polymerization of amyloid peptides in insoluble fibrils.

Macrophages may be involved in the removal of deposited amyloid, and their activity could influence the development of amyloidosis and the size of the deposits. The distribution of amyloid is probably due to the type of precursor, its kinetics and catabolism. It is also conceivable that certain amyloidogenic light chains function as antibodies and bind to specific tissued antigens giving rise to selective organ deposition (Solomon and Weiss 1988).

Clinical manifestations

Renal involvement in primary AL amyloidosis is common. Proteinuria is the main presenting feature in most patients (Ogg et al. 1981); it was reported in 82 per cent of 229 patients with proven primary amyloidosis (Kyle and Greipp 1983) and is equally common in myeloma-associated amyloidosis (Banfi et al. 1988). Primary amyloidosis is predominantly a disease of the elderly: mean ages of 85 patients with primary amyloidosis and 14 with myeloma-associated amyloidosis studied by Banfi et al. (1988) were 57 ± 12 years and 60 ± 7 years respectively. The nephrotic syndrome was the most common clinical manifestation in both groups (85 versus 78 per cent) and was associated with a reduced GFR in many cases. Isolated proteinuria was rarely observed. Cardiac involvement was present in more than 50 per cent, as was hypotension (25 per cent). Hepatomegaly, splenomegaly, and neuropathy were more frequent in patients with myeloma-associated amyloidosis and macroglossia and gastrointestinal tract abnormalities were more common in primary amyloidosis patients. A monoclonal protein in serum and/or urine was found in 45 per cent, and an excess of plasma cells in bone marrow in 23 per cent of primary anyloidosis patients. In the cases reported by Kyle et al. (1986), lambda light chains predominated (83 versus 31 cases) and free monclonal light chains were present in the urine in 75 per cent of patients. Overall an M component was found in serum and/or urine in 89 per cent of patients.

Morphologic features

Two patterns of amyloid deposition in glomeruli are observed in early stages: nodular (mesangial) and diffuse (parietal); the latter is predominant in later stages (Fig. 13). Vascular and interstitial amyloid depositions are also important features, and correlate with glomerular amyloid deposits (Banfi et al. 1988). Drop-like deposits in Bowman's capsule may reflect the deposition of the same sustance as in glomeruli. Glomerular microaneurysms, deposition of amyloid in afferent and efferent arterioles, and accumulation of amyloid along tubular basement membranes are other distinctive features. The presence of argyrophilic spicular structures along glomerular basement membrane, Bowman's capsule, and tubular basement membrane is now considered diagnostic (Fig. 14).

Important findings on electron microscopy are segmental loss or breaks of the glomerular basement membrane, which is replaced by amyloid fibrils (Nakamoto et al. 1984), spicular

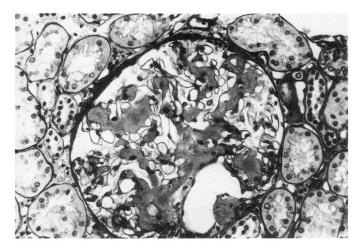

Fig. 13 AL amyloidosis. Light microscopy. Glomerular pattern. The mesangial stalk is widened by deposition of weakly PAS-positive material. Few cells are present in the glomerulus. The glomerular basement membranes appear normal (PAS staining, original magnification × 250).

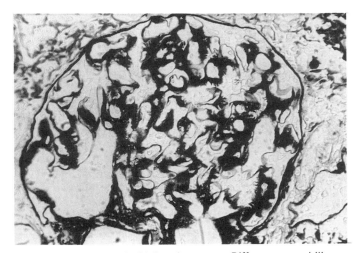

Fig. 14 AL amyloidosis. Light microscopy. Diffuse argyrophilic spike-like structres at the subepithelial side of glomerular basement membranes. (PASM staining, original magnification × 400).

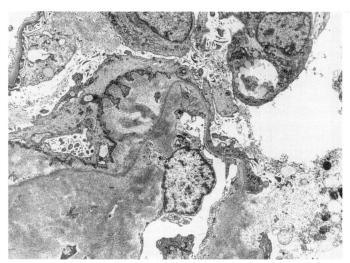

Fig. 15 AL amyloidosis. Electron microscopy. Large amounts of fibrillary material deposited in mesangium and glomerular basement membrane with spike-like subepithelial aspects (original magnification × 5000).

deformation, and newly synthesized subepithelial lamina (Fig. 15).

Outcome and therapy

AL amyloidosis has been reported in 6 to 15 per cent of patients with multiple myeloma. While approximately 60 per cent of patients with multiple myeloma have an initial response to chemotherapy, primary amyloidosis is generally considered to be chemoresistant. In primary amyloidosis nephrotic syndrome may resolve after treatment with melphalan and prednisone (Kyle et al. 1982). However, of 25 patients with AL amyloidosis and overt multiple myeloma only 16 were judged to have amyloidosis responsive to chemotherapy (melphalan plus prednisone or a variety of drugs such as vincristine or cyclophosphamide) (Fielder and Durie 1986). Patients who responded were much less likely to have cardiac amyloidosis, but more likely to have unusually high levels of pretreatment β_2-microglobulin in serum, and a kappa type of monoclonal immunoglobulin in urine (Fielder and Durie 1986). There was no significant difference in survival in a double blind study on 55 patients, who were randomly assigned to receive melphalan + prednisone or placebo (Kyle and Greipp 1978). However the response to chemotherapy is the most important factor in recognizing long-term survivors (Kyle and Greipp 1983).

The presence or absence of myeloma in patients with AL amyloidosis does not influence survival according to Alexanian and Barlogie (1988) (see Chapter 4.2), whereas in the series studied by Banfi et al. (1988) mortality was higher in the former (58 versus 42 per cent). Half of 168 AL amyloidosis patients reported by Kyle et al. (1986) were dead at 1 year; the factors which had a highly significant influence on survival during the first year were congestive heart failure, Bence-Jones proteinuria, hepatomegaly and weight loss.

Twelve patients of the series of Banfi et al. (1988) were treated by regular dialysis, and six were still alive after more than 2 years of treatment.

References

Alexanian, R. (1985). Ten years survival in multiple myeloma. *Archives of Internal Medicine*, **145**, 2073.

Alexanian, R. and Barlogie, B. (1988). Implications of renal failure in multiple myeloma. In *The Kidney in Plasma Cell Dyscrasias* (eds L. Minetti, G. D'Amico, and C. Ponticelli), pp. 259–63. Kluwer, Dordrecht.

Alexanian, R. and Dreicer, R. (1982). Chemotherapy for multiple myeloma. *Cancer*, **53**, 583–8.

Alpers, C.E., Tu, W.H., Hopper, J. Jr., and Biava, C.G. (1985). Single light chain subclass (kappa chain) immunoglobulin deposition in glomerulonephritis. *Human Pathology*, **16**, 294–304.

Antonovych, T., Lin, C., Parrish, E., and Mostofi, K. (1973). *Light chain deposits in multiple myeloma*, (Abstract) 6th Annual Meeting American Society of Nephrology, Washington.

Avnstorp, C., Nielsen, H., Drachmann, O., and Hippe, E. (1985). Plasmapheresis in hyperviscosity syndrome. *Acta Medica Scandinavica*, **217**, 133–7.

Banfi, G., et al. (1988). Renal amyloidosis. Retrospective collaborative study. In *The Kidney in Plasma Cell Dyscrasias* (eds L. Minetti, G. D'Amico, and C. Ponticelli), pp. 211–19. Kluwer, Dordrecht.

Bangerter, A.R. and Murphy, W.M. (1987). Kappa light chain nephropathy. A pathologic study. *Virchows Archives*, **410**, 531–9.

Baylis, C., Falconer-Smith, J., and Ross, B. (1988). Glomerular and tubular handling of differently charged human immunoglobulin light chains by the rat kidney. *Clinical Science*, **74**, 639–44.

Bernstein, S. and Humes, H. (1982). Reversible renal insufficiency in multiple myeloma. *Archives of Internal Medicine*, **142**, 2083–6.

Bradley, J.R., Thiru, S., and Evans, D.B. (1987). Light chains and the kidney. *Journal of Clinical Pathology*, **40**, 53–60.

Busnach, G., Dal Col, A., Brando, B., Perrino, M.L., Brunati, C., and Minetti, L. (1986). Efficacy of a combined treatment with plasma exchange and cytostatics in macroglobulinemia. *International Journal of Artificial Organs*, **9**, 267–70.

Busnach, G., et al. (1988). Prevention and treatment of hyperviscosity syndrome. In *The Kidney in Plasma Cell Dyscrasias* (eds L. Minetti, G. D'Amico, and C. Ponticelli), pp. 247–57. Kluwer, Dordrecht.

Chan, K.W., Ho, F.C., and Chan, M.K. (1987). Adult Fanconi syndrome in kappa light chain myeloma. *Archives of Pathology and Laboratory Medicine*, **111**, 139–42.

Clyne, D.H. and Pollak, V.E. (1981). Renal handling and pathophysiology of Bence Jones proteins. *Contributions to Nephrology*, **24**, 78–87.

Clyne, D.H., Pesce, A.J., and Thompson, R.E. (1979). Nephrotoxicity of Bence-Jones proteins: the importance of isoelectric point. *Kidney International*, **16**, 345–52.

Cohen, A.H. (1984). Pathology of light chain nephropathies. In *Nephrology* Vol. 2, (ed. R.R. Robinson), pp. 898–902. Springer-Verlag, New York.

Cohen, A.H. (1988). The pathogenesis of cast nephropathy. In *The Kidney in Plasma Cell Dyscrasias*, (eds L. Minetti, G. D'Amico, and C. Ponticelli), pp. 123–31. Kluwer, Dordrecht.

Cohen, A.S., Shirahama, T., Sipe, J.D., and Skinner, M. (1983). Amyloid proteins, precursors, mediator, and enhancer. *Laboratory Investigation*, **48**, 1–4.

Colussi, G., et al. (1988). Clinical spectrum of tubular disorders from light chains. In *The Kidney in Plasma Cell Dyscrasias*, (eds L. Minetti, G. D'Amico, and C. Ponticelli), pp. 191–209. Kluwer, Dordrecht.

Confalonieri, R., et al. (1988). Light chain nephropathy: histological and clinical aspects in 15 cases. *Nephrology, Dialysis, Transplantation*, **2**, 150–6.

Cooper, E.H. and MacLennan, I.C.M. (1988). Effect of light chains on proximal tubule function. In *The Kidney in Plasma Cell Dyscrasias*, (eds. L. Minetti, G. D'Amico, and C. Ponticelli), pp. 85–92. Kluwer, Dordrecht.

Cooper, E.H., Forbes, M.A., Crockson, R.A., and MacLennan, I.C.M. (1984). Proximal renal tubular function in myelomatosis: observations in the fourth Medical Research Council trial. *Journal of Clinical Pathology*, **37**, 852–8.

Cosio, F.G., Pence, T.V., Shapiro, F.L., and Kjellstrand, C.M. (1981). Severe renal failure in multiple myeloma. *Clinical Nephrology*, **15**, 206–10.

Coward, R.A., Mallick, N.P., and Delamore, I.W. (1983). Should patients with renal failure associated with myeloma be dialysed? *British Medical Journal*, **287**, 1575–8.

Coward, R.A., Delamore, I.W., Mallick, N.P., and Robinson, E.L. (1984). The importance of urinary immunoglobulin light chain isoelectric point (pI) in nephrotoxicity in multiple myeloma. *Clinical Science*, **66**, 229–32.

Coward, R.A., Mallick, N.P., and Delamore, I.W. (1985). Tubular function in multiple myeloma. *Clinical Nephrology*, **24**, 180–5.

De Fronzo, R.A., Cooke, C.R., Wright, J.R., and Humphrey, R.L. (1987). Renal function in patients with multiple myeloma. *Medicine*, **57**, 151–66.

Durie, B.G.M. and Salmon, S.E. (1975). A clinical staging system for multiple myeloma. *Cancer*, **36**, 842–54.

Fang, L.S.T. (1985). Light chain nephropathy. *Kidney International*, **27**, 582–92.

Fielder, K. and Durie, B.G.M. (1986). Primary amyloidosis associated with multiple myeloma. *American Journal of Medicine*, **80**, 413–8.

Gallo, G. and Buxbaum, J. (1988). Monoclonal immunoglobulin deposition disease: immunopathologic aspects of renal involvement. In *The Kidney in Plasma Cell Dyscrasias* (eds L. Minetti, G. D'Amico, and C. Ponticelli), pp. 171–81. Kluwer, Dordrecht.

Gallo, G.R., et al. (1980). Nodular glomerulopathy associated with non-amyloidotic kappa light chain deposits and excess immunoglobulin light chain synthesis. *American Journal of Pathology*, **99**, 621–44.

Ganeval, D. (1988). Kidney involvement in light chain deposition disease. In *The Kidney in Plasma Cell Dyscrasias* (eds L. Minetti, G. D'Amico, and C. Ponticelli), pp. 221–28. Kluwer, Dordrecht.

Ganeval, D., Noel, L.H., Preud'homme, J.L., Droz, D., and Grunfeld, J.P. (1984). Light-chain deposition disease: its relation with AL-type amyloidosis. *Kidney International*, **26**, 1–9.

Gerlag, P.G.C., Koene, R.A.P., and Berden, J.H.M. (1986). Renal transplantation in light chain nephropathy: case report and review of the literature. *Clinical Nephrology*, **25**, 101–04.

Glenner, G.G. (1980). Amyloid deposits and amyloidosis: the β-fibrillosis. *New England Journal of Medicine*, **302**, 1283–92, 1333–43.

Glenner, G.G., Terry, W., Harada, M., Isersky, C., and Page, D. (1971). Amyloid fibril proteins: proof of homology with immunoglobulin light chains by sequence analysis. *Science*, 172, 1150–1.

Hill, G.S., Morel-Maroger, L., Mery, J.Ph., Brouet, J.Cl., and Mignon, F. (1983). Renal lesions in multiple myeloma: their relationships to associated protein abnormalities. *American Journal of Kidney Diseases*, **2**, 423–38.

Jacquot, C.C., et al. (1985). Association of systemic light chain deposition disease and amyloidosis: a report of three patients with renal involvement. *Clinical Nephrology*, **24**, 93–8.

Johns, E.A., Turner, R., Cooper, E.H., and MacLennan, I.C.M. (1986). Isoelectric points of urinary light chains in myelomatosis: analysis in relation to nephrotoxicity. *Journal of Clinical Pathology*, **36**, 833–7.

Koss, M.N., Pirani, C.L., and Osserman, E.P. (1976). Experimental Bence Jones cast nephropathy. *Laboratory Investigation*, **34**, 579–91.

Kusano, E., Suzuki, M., Asano, Y., Ifoh, Y., Tagagi, K., and Kawai, T. (1985). Human alpha 1 microglobulin and its relationship with renal function. *Nephron*. **41**, 344–50.

Kyle, R.A. (1975). Multiple myeloma; review of 869 cases. *Mayo Clinic Proceedings*, **50**, 29–40.

Kyle, R.A. (1988). Kidney in multiple myeloma and related disorders. In *The Kidney in Plasma Cell Dyscrasias* (eds L. Minetti, C. D'Amico, and C. Ponticelli), pp. 141–51. Kluwer, Dordrecht.

Kyle, R.A. and Greipp, P.R. (1978). Primary systemic amyloidosis: comparison of melphalan and prednisone versus placebo. *Blood*, **52**, 818–27.

Kyle, R.A. and Greipp, P.R. (1983). Amyloidosis (AL). Clinical and laboratory features in 229 cases. *Mayo Clinic Proceedings*, **58**, 665–83.

Kyle, R.A., Wagoner, R.D., and Holly, K.E. (1982). Primary systemic amyloidosis. Resolution of the nephrotic syndrome with melphalan and prednisone. *Archives of Internal Medicine*, **142**, 144–7.

Kyle, R.A., Greipp, P.R., and O'Fallon, W.M. (1986). Primary systemic amyloidosis: multivariate analysis for prognostic factors in 168 cases. *Blood*, **1**, 220–4.

Maak, T. (1975). Renal handling of low molecular weight proteins. *American Journal of Medicine*, **58**, 57–64.

Maldonado, J.E., Velosa, J.A., Kyle, R.A., Wagoner, R.D., Holley, K.E., and Salassa, R.M. (1975). Fanconi syndrome in adults. A manifestation of a latent form of myeloma. *American Journal of Medicine*, **58**, 354–64.

Mallick, N.P. (1988). Uraemia in myeloma: management and prognosis. In *The Kidney in Plasma Cell Dyscrasias* (eds L. Minetti, G. D'Amico, and C. Ponticelli), pp. 265–70. Kluwer, Dordrecht.

Mallick, N.P. and Williams, G. (1988). Glomerular and associated tubular injury by light chains. The spectrum of damage and effect of treatment. In *The Kidney in Plasma Cell Dyscrasias*, (eds L. Minetti, G. D'Amico, and C. Ponticelli), pp. 77–84. Kluwer, Dordrecht.

McGeoch, J., Ledingham, J., Smith, J.F., and Ross, B. (1978). Inhibition of active-transport sodium-potassium ATPase by myeloma protein. *Lancet*, **i**, 17–18.

McLeish, K.R., Gohara, A.F., and Gillespie, C. (1985). Mesangial proliferative glomerulonephritis associated with multiple myeloma. *American Journal of Medical Science*, **290**, 114–17.

Melcion, C., *et al.* (1984). Renal failure in myeloma: relationship with isoelectric point of immunoglobulin light chains. *Clinical Nephrology*, **22**, 138–43.

Mundy, G.R. and Raisz, L.G. (1977). Big and little forms of osteoclast activating factor. *Journal of Clinical Investigation*, **60**, 122–28.

Nakamoto, Y., Hamanaka, S., Akihama, T., Miura, A.B., and Uesaka, Y. (1984). Renal involvement patterns of amyloid nephropathy: a comparison with diabetic nephropathy. *Clinical Nephrology*, **22**, 188–94.

Noel, L.H., Drox, D., Ganeval, D., and Grunfeld, J.P. (1984). Renal granular monoclonal light chain deposits: morphological aspects in 11 cases. *Clinical Nephrology*, **21**, 2163–9.

Ogg, C.S., Cameron, J.S., Williams, D.G., and Turner, D.R. (1981). Presentation and course of primary amyloidosis of the kidney. *Clinical Nephrology*, **15**, 9–13.

Palant, C.E., Bonitati, J., Bartholomew, W.R., Brentjens, J.R., Walshe, J.J., and Bentzel, C.J. (1986). Nodular glomerulosclerosis associated with multiple myeloma. Role of light chain isoelectric point. *American Journal of Medicine*, **80**, 98–102.

Pasquali, S., *et al.* (1987). Renal histological lesions and clinical syndromes in multiple myeloma. *Clinical Nephrology*, **2**, 222–8.

Pirani, C.L. (1988). Histological, histochemical and ultrastructural features of myeloma kidney. In *The Kidney in Plasma Cell Dyscrasias*, (eds L. Minetti, G. D'Amico, and C. Ponticelli), pp. 154–69. Kluwer, Dordrecht.

Pirani, C.L., Silva, F., D'Agati, V., Chander, P., and Striker, L.M. (1987). Renal lesions in plasma cell dyscrasias: ultrastructural observations. *American Journal of Kidney Diseases*, **10**, 208–21.

Pirson, Y. (1988). Sundries on therapy of multiple myeloma and amyloidosis and their renal consequences—A forum. In *The Kidney in Plasma Cell Dyscrasias*, (eds L. Minetti, G. D'Amico, and C. Ponticelli), p. 286. Kluwer, Dordrecht.

Preud'homme, J.L. (1988). Immunoglobulin synthesis in plasma cell dyscrasias with renal lesions. In *The Kidney in Plasma Cell Dyscrasias*, (eds L. Minetti, G. D'Amico, and C. Ponticelli), pp. 31–43. Kluwer, Dordrecht.

Preud'homme, S.J., *et al.* (1980). Synthesis of abnormal immunoglobulins in lymphoplasmacytic disorders with visceral light chain deposition. *American Journal of Medicine*, **69**, 703–10.

Preud'homme, J.L., *et al.* (1982). La maladie des dépôts de chaines légères ou d'immunoglobulines monoclonales: concepts physiopathogéniques. *Nouvelle Presse Medicale*, **11**, 3259–63.

Raisz, L.G. (1980). New diphosphonates to block bone resorption. *New England Journal of Medicine*, **302**, 347–8.

Randall, R.E., Williamson, W.C.Jr., Mullinax, F., Tung, M.Y., and Still, W.J.S. (1976). Manifestations of systemic light chain deposition. *American Journal of Medicine*, **60**, 293–9.

Rao, D.S., Parfitt, A.M., Villanueva, A.R., Dorman, P.J., and Kleerekoper, M. (1987). Hypophosphatemic osteomalacia and adult Fanconi syndrome due to light-chain nephropathy. Another form of oncogenous osteomalacia. *American Journal of Medicine*, **82**, 333–8.

Rombolà, G., Colussi, G., De Ferrari, M.E., Surian, M., Malberti, F., and Minetti, L. (1987). Clinical evaluation of segmental tubular reabsorption of sodium and fluid in man: lithium vs free water clearances. *Nephrology, Dialysis, Transplantation*, **2**, 212–18.

Ronco, P., *et al.* (1988). Pathophysiological aspects of myeloma cast nephropathy. In *The Kidney in Plasma Cell Dyscrasias* (eds L. Minetti, G. D'Amico, and C. Ponticelli), pp. 93–104. Kluwer, Dordecht.

Rota, S., *et al.* (1987). Multiple myeloma and severe renal failure: a clinicopathologic study of outcome and prognosis in 34 patients. *Medicine*, **66**, 126–37.

Sanders, P.W., Herrera, G.A., and Galla, J.H. (1987). Human Bence Jones protein toxicity in rat proximal tubule epithelium *in vivo*. *Kidney International*, **32**, 851–61.

Sanders, P.W., Herrera, G.A., Chen, A., Booker, B.B., and Galla, J.H. (1988*a*). Differential nephrotoxicity of low molecular weight proteins including Bence Jones proteins in the perfused rat nephron *in vivo*. *Journal of Clinical Investigation*, **82**, 2086–96.

Sanders, P.W., Herrera, G.A., and Galla, J.H. (1988*b*). Light chain-related tubulointerstitial nephropathy. In *The Kidney in Plasma Cell Dyscrasias*, (eds L. Minetti, G. D'Amico, and C. Ponticelli), pp. 117–21. Kluwer, Dordrecht.

Scarpioni, L., *et al.* (1981). Glomerular and tubular proteinuria in myeloma. Relationship with Bence Jones proteinuria. *Contributions to Nephrology*, **26**, 89–102.

Seymour, A.E., Thompson, A.J., Smith, P.S., Woodroffe, A.J., and Clarkson, A.R. (1980). Kappa light chain glomerulosclerosis in multiple myeloma. *American Journal of Pathology*, **101**, 557–80.

Shirahama, T. (1988). Amyloidogenesis and light chains. In *The Kidney in Plasma Cell Dyscrasias*, (eds L. Minetti, G. D'Amico, and C. Ponticelli), pp. 57–66. Kluwer, Dordrecht.

Shirahama, T., Cohen, A.S., and Skinner, M. (1984). Immunohistochemistry of amyloid. In *Advances in Immunohistochemistry*, (ed. R.A. De Lellis), pp. 277–302. Masson Publishing, New York.

Sinniah, R. and Cohen, A.H. (1985). Glomerular capillary aneurysms in light-chain nephropathy. An ultrastructural proposal of pathogenesis. *American Journal of Pathology*, **118**, 298–305.

Smithline, N., Kassirer, J.P., and Cohen, J.J. (1976). Light chain nephropathy. Renal tubular dysfunction associated with light chain proteinuria. *New England Journal of Medicine*, **294**, 71–4.

Smolens, P. and Kreisberg, J.R. (1985). Modest hypercalcemia can markedly increase the nephrotoxicity of Bence Jones protein. *Kidney International*, **27**, 238.

Smolens, P., Venkatachalam, M., and Stein, J.H. (1983). Myeloma kidney cast nephropathy in a rat model of multiple myeloma. *Kidney International*, **24**, 192–204.

Smolens, P., Barnes, J.L., and Stein J.H. (1986). Effect of chronic infusion of Bence Jones proteins on rat renal function and histology. *Kidney International*, **30**, 874–82.

Solling, K., Solling, J., and Nielsen, J.L. (1984). Polymeric Bence Jones proteins in serum in myeloma patients with renal insufficiency. *Acta Medica Scandinavica* **216**, 495–502.

Solomon, A. (1976). Bence Jones proteins and light chains of immunoglobulins: Part I and Part II. *New England Journal of Medicine*, **294**, 17–23, 91–98.

Solomon, A. and Weiss, D.T. (1987). Serologically defined V region subgroups of human λ light chains. *Journal of Immunology*, **139**, 824–30.

Solomon, A. and Weiss, D.T. (1988). A perspective of plasma cell dyscrasias: clinical implications of monoclonal light chains in renal disease. In *The Kidney in Plasma Cell Dyscrasias* (eds L, Minetti, G. D'Amico, and C. Ponticelli), pp. 3–18. Kluwer, Dordrecht.

Solomon, A., Frangione, B., and Franklin, E.C. (1982). Bence Jones proteins and light chains of immunoglobulins. Preferential association of the Vλ_{VI} subgroup of human light chains with amyloidosis AL. *Journal of Clinical Investigation*, **70**, 453–60.

Thorner, P.S., Bedard, Y.C., and Fernandes, B.J. (1983). Lambda-light-chain nephropathy with Fanconi's syndrome. *Archives of Pathology and Laboratory Medicine*, **107**, 654–7.

Troussard, X., *et al.* (1987). Polymorphisme de la maladie del dépots de chaines légères. A propos de trois observations. *Révue de Medicine Interne*, **9**, 41–7.

Tubbs, R.R., Gephardt, G.N., McMahon, J.T., Hall, P.M., Valenzuela, R., and Vidt, D.G. (1981). Light chain nephropathy. *American Journal of Medicine*, **71**, 263–9.

Venkataseshan, V.S., Faraggiana, T., Hughson, M.D., Buchwald, D., Olesniky, L., and Goldstein, M.H. (1988). Morphologic variants of light-chain deposition disease in the kidney. *American Journal of Nephrology*, **8**, 272–9.

Zucker-Franklin, D. (1988). Renal amyloidosis; new perspectives. In *The Kidney in Plasma Cell Dyscrasias*, (eds L. Minetti, G. D'Amico, and C. Ponticelli), pp. 45–55. Klower, Dordrecht.

4.4 Sarcoidosis

SABINE KENOUCH AND JEAN-PHILIPPE MÉRY

Sarcoidosis is a multisystem disorder in which the kidney may be involved. The sarcoid granuloma, the histologic hallmark of the disease, was first described by Schaumann (1933) and has since been studied extensively, especially with regard to its structural and functional characteristics (Basset *et al.* 1988). Typically, the florid granuloma is a rounded cellular collection composed of epithelioid and multinucleate giant cells surrounded by a rim of lymphocytes and macrophages in association with various degrees of fibrosis. Giant cells appear to result from the coalescence of epithelioid cells, both cellular types sharing the same ultrastructural characteristics, notably the presence of numerous cytoplasmic vesicles. Giant cells may contain inclusion bodies which are not specific. Lymphocytes are mainly T lymphocytes. While CD4 cells are interspersed with epithelioid cells, CD8 cells predominate in the peripheral rim of the granuloma. Unlike the situation in tuberculous granulomas, caseous necrosis is usually absent or restricted to small central areas of the granuloma.

In recent years, numerous immunological studies have greatly improved understanding of the pathophysiology of the disease. The mechanisms of the granuloma formation are summarized in Fig. 1. It is assumed that lymphocytes and macrophages are locally activated at the sites of active lesions; the stimulus of this activation is still unknown. Activated macrophages co-operate with T lymphocytes to induce activated T-helper lymphocyte proliferation. The last release interleukin-2 which is responsible for both further recruitment of circulating T lymphocytes and their proliferation. Activated T-helper cells also release a chemotactic factor for monocytes and an inhibitor factor of macro-

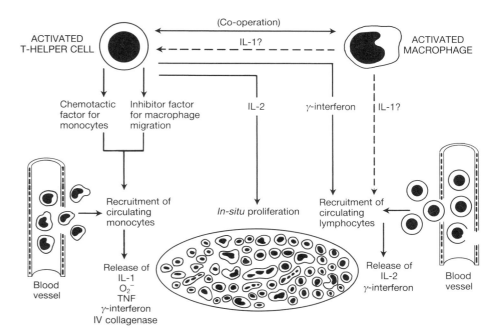

Fig. 1 Mechanisms of granuloma formation (adapted from Semenzato *et al.* 1988). Abbreviations: gamma-IFN=gamma interferon; O_2^- = oxygen peroxide; IL-1=interleukin 1; IL-2=interleukin 2; TNF=tumour necrosis factor.

phage migration which contribute to the recruitment of circulating monocytes and the accumulation of macrophages in the granuloma. Both newly recruited activated monocytes and T-helper lymphocytes release lymphokines and other mediators which in their turn enhance the inflammatory process at the site of granuloma formation. Several additional pathogenetic mechanisms remain unknown or controversial: for example, the exact origin of interleukin-1 (IL-1) which is found in epithelioid cells but not in multinucleated cells and its role in T-cell activation are not clearly established. Other cell populations may also be involved such as endothelial cells which could be another source of IL-1, or mast cells which can release histamine.

Several functional or structural renal disorders may occur in association with sarcoidosis. Granulomatous interstitial infiltration, disorders due to abnormal calcium homeostasis and glomerulopathy have been observed.

Granulomatous interstitial nephritis

The exact prevalence of granulomatous interstitial nephritis (GIN) in sarcoidosis is ill-defined; differing estimations have been reported from necropsy and biopsy series. It has been found in 7 and 13 per cent of cases in two necropsy series respectively (Branson and Park 1954; Longcope and Freiman 1952). On the other hand, recent reviews of biopsy series from the literature showed a mean prevalence of 34 per cent (Romer 1980) with values ranging from 15 to 40 per cent (Muther et al. 1981). These discrepancies may reflect different attitudes among nephrologists to the requirement for renal biopsy in patients with sarcoidosis. Moreover, since granulomatous infiltration of the interstitium may be clinically silent and granulomas may be absent in a small biopsy specimen when they are scarce, the true frequency of granulomatous interstitial nephritis is likely to be underestimated.

Detailed descriptions of the pathology of sarcoid granulomatous interstitial nephritis have been made by Berger and Relman (1955) and by MacDonald Cameron (1956). The number of granulomas varies widely between cases. In some instances, granulomas are rare in the biopsy specimen while in others they are numerous and widespread throughout the interstitium both in the cortex and the medulla. They have the general appearance of sarcoid granulomas as described above (Fig. 2). Some granulomas may have small arteries at the centre (Bottcher 1959; Michielsen et al. 1985; Turner et al. 1977). Various degrees of fibrosis which may result in tubular atrophy and degeneration may be associated with cellular interstitial infiltration. Glomeruli appear normal or show only slight mesangial hypertrophy and thickening of the basement membrane. Electron microscopy has occasionally shown fusion of epithelial foot processes (Falls et al. 1972; Farge et al. 1986). No significant immune deposits either in glomeruli or in tubules have been shown by immunofluorescence microscopy except for the case described by Cuppage et al. (1988) in which IgA and IgG were found within the cytoplasm of some interstitial cells. We have been able to study the renal biopsy of a patient with a fluorescent antiserum to angiotensin-converting enzyme (ACE) (Méry and Kenouch 1988, p. 71). Epithelioid cells stained strongly, giving a bright fluorescence to all granulomas (Fig. 3). We know of no other report of such a fluorescent study in sarcoid granulomatous interstitial nephritis. However, Pertschuk et al. (1981), studying with an anti-ACE serum sarcoid and non-sarcoid granulomas of other tissues than the kidney, found that positive staining was obtained only in sarcoid granulomas. They concluded that this method could help to differentiate sarcoidosis from other granulomatous diseases. However, Grönhagen-Riska et al. (1988) found positive staining with anti-ACE serum in other diseases: both histiocytes of a pleural rheumatoid nodule and macrophages of a renal transplant (during an acute rejection reaction) stained positively.

Cells infiltrating the interstitium were characterized using monoclonal antibodies by Cheng et al. (1989) in two patients with sarcoidosis. They comprised a significant number of B cells,

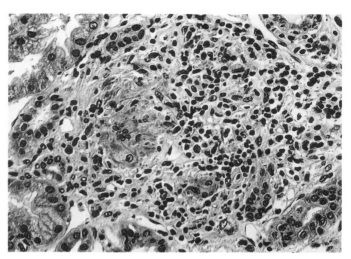

Fig. 2 Interstitial granuloma. Note the presence of lymphocytes, histiocytes, and a few polymorphonuclear leucocytes (PMN), and foci of epithelioid cells (Masson's trichrome × 900). (By courtesy of Dr Béatrice Mougenot.) (From Méry, J.Ph. and Kenouch, S. in Séminaires d'Uro-Néphrologie Pitié-Salpêtrière (1988), with permission from Masson, Paris.)

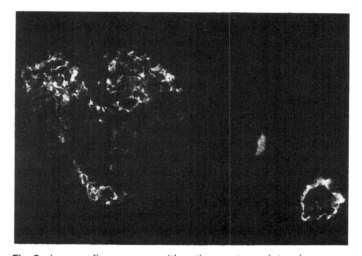

Fig. 3 Immunofluorescence with antiserum to angiotensin-converting enzyme. Top left: bright staining of epithelioid cells of an interstitial granuloma. Bottom right: normal staining of epithelial cells of proximal convoluted tubule. (By courtesy of Dr Béatrice Mougenot. Antiserum was kindly provided by Dr François Alhenc-Gélas.) (From Méry, J.Ph. and Kenouch, S. in Séminaires d'Uro-Néphrologie Pitié-Salpêtrière (1988), with permission from Masson, Paris.)

plasma cells and T cells with a predominance of helper cells. These workers also found an increased tubular display of HLA DR compared with controls, and suggested that this could result in an activation of T-helper lymphocytes and propagation of the immune local reaction.

Sarcoid infiltration of the kidney is responsible for renal failure in most cases. Indeed, in a recent review of 59 cases in the literature, Méry and Kenouch (1988, p. 73) found only three patients whose renal function was normal; among the others, serum creatinine was above 800 μmol/l in 12 (20 per cent). Decline of renal function usually progresses slowly over weeks or months, but some patients may present with acute renal failure (Cuppage et al. 1988; Korzets et al. 1985; Michielsen et al. 1985; Tanneau et al. 1987; Warren et al. 1988). Hypertension which is usually absent early in the course of the nephropathy often occurs later, secondary to progressive sclerosis of the parenchyma, possibly related to long-term corticotherapy. Proteinuria is either absent or mild, with a tubular pattern on electrophoresis. Leucocytes and granular casts are often present in the urine. Tubular abnormalities are frequent and are found in about 50 per cent of cases investigated (Méry and Kenouch 1988, p. 74). Various tubular disorders, such as orthoglycaemic glycosuria, urinary wasting of potassium or sodium, Fanconi syndrome, decreased urinary concentrating ability or overt nephrogenic diabetes insipidus, proximal or distal tubular acidosis have all been described. It is uncertain whether the presence of interstitial lesions is the only factor responsible for these tubular abnormalities, and hypergammaglobulinaemia which is often present in sarcoidosis may also play a pathogenetic role.

Granulomatous infiltration of the kidney often results in an enlarged kidney which may mimic polycystic kidney disease (Coburn et al. 1967; Guédon et al. 1967) or renal carcinoma (Amouroux et al. 1967; Leng-Lévy et al. 1965). Gallium-67 citrate scan may show an uptake of the tracer by the kidneys but it can also be negative in active granulomatous interstitial nephritis (Méry and Kenouch 1988, p. 74). Moreover, it appears of little value in the assessment of evolution under treatment (Pagniez et al. 1987). The same reservations apply for results of serum ACE determination: indeed, serum ACE level may remain normal in cases of active granulomatous interstitial nephritis with severe renal failure (Pagniez et al. 1987; Singer and Evans 1986).

In most cases granulomatous interstitial nephritis is associated with extrarenal manifestations that makes the diagnosis of sarcoidosis easy. In some patients however, renal involvement is isolated, preceding other localizations of the disease for several months or years (Ford et al. 1978; Guédon et al. 1967; Nédelec et al. 1986). It is possible that some cases published as isolated granulomatous interstitial nephritis are in fact localized forms of sarcoidosis (Barbiano di Belgioioso et al. 1980; Bolton et al. 1976; Caruana et al. 1982). Differential diagnosis with the syndrome described in 1975 by Dobrin et al. must be considered; this is characterized by the association of acute interstitial nephritis, anterior uveitis, and epithelioid granulomas in bone marrow and lymph nodes. The renal lesion consists of an interstitial infiltrate mainly composed of mononuclear cells with a few eosinophils. Although no 'mature' interstitial granulomas are found in Dobrin's syndrome, the cellular composition of interstitial infiltrates does not differ from that observed in sarcoidosis and in some cases the cellular arrangement almost makes the outline of a granuloma. It is possible, therefore, that some cases described as Dobrin's syndrome could in fact be atypical forms of sarcoidosis (Kenouch et al. 1989).

It is generally agreed that sarcoid granulomatous interstitial nephritis should be treated with corticosteroids. To be effective, initial treatment usually requires a daily dosage of prednisone or prednisolone of 1 to 1.5 mg/kg body weight. Following initiation of corticosteroid therapy, renal function can improve dramatically. For example, renal replacement therapy could be discontinued after 15 days of corticosteroid in the patient described by Singer and Evans (1986). Renal tubular disorders regress concurrently with improvement of renal function. It is most important to recognize the necessity for prolonged corticosteroid treatment. Indeed, treatment of less than 6 months duration is frequently followed by relapses of the nephropathy (Guénel and Chevet 1988, p. 256). In our opinion, the initial posology should be progressively tapered after two months and then switched to an alternative-day regimen. Treatment should be maintained for at least 1 year. Serial renal biopsies usually show regression of granulomas in parallel with the improvement of renal function (Méry and Kenouch 1988, p. 75); however, such a parallel course is not always seen (Guénel and Chevet 1988, p. 256). Interstitial fibrosis often develops in association with focal glomerulosclerosis and vascular lesions (see references in Méry and Kenouch 1988, p. 76); the latter, being ascribed to long-term corticotherapy, may be responsible for the delayed occurrence of hypertension which in turn may contribute to the progression of renal failure.

Renal consequences of abnormal calcium metabolism

Abnormal calcium metabolism resulting in hypercalcaemia and/or hypercalciuria is often present in sarcoidosis. Though the frequency of these calcium disorders cannot be assessed precisely, it is generally assumed that they account for renal impairment more frequently than does granulomatous interstitial nephritis. It is noteworthy that the two processes, i.e. abnormal calcium homeostasis and sarcoid infiltration of the kidney, may coexist in the same patient. Reports dealing with calcium metabolism in sarcoidosis often focus on hypercalcaemia and do not always provide accurate data on hypercalciuria. In spite of these reservations, it is clear that hypercalciuria is more common than hypercalcaemia. Hypercalciuria has been found with a frequency ranging from 7.5 to 65 per cent of cases in different series while the prevalence of hypercalcaemia was only 2.5 to 17 per cent (Coburn and Barbour 1984, p. 407).

The mechanism of hypercalcaemia in sarcoidosis is now well established. Early observations had supported the hypothesis suggesting an increase of calcium intestinal absorption which was ascribed to an increased sensitivity to vitamin D. Indeed, the occurrence in patients with sarcoidosis of both low-dose vitamin D-induced episodes of hypercalcaemia and exacerbations of hypercalcaemia or hypercalciuria during the summer months suggested either overproduction of active vitamin D or hypersensitivity of target organs to it. A major step in the understanding of the mechanisms of hypercalcaemia was taken when several groups demonstrated an increased concentration of serum 1,25 dihydroxyvitamin D_3 (1,25$(OH)_2D_3$) in patients with active sarcoidosis (Bell et al. 1979; Papapoulos et al. 1979; Zerwekh et al. 1980). Later, the report by Barbour et al. (1981) on an anephric patient with sarcoidosis who had hypercalcaemia and high serum levels of 1,25$(OH)_2D_3$ argued strongly for an extrarenal source of calcitriol. A similar conclusion was reached by Maesaka et al. (1982) who observed the same abnormalities in a patient with

4.4 Sarcoidosis

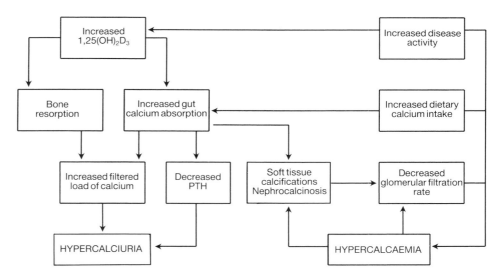

Fig. 4 Mechanisms of altered calcium metabolism in sarcoidosis (adapted from Coburn and Barbour 1984).

end-stage renal disease. Afterwards, Adams et al. (1983) proved the sarcoid granulomas to be the site of calcitriol production, and indeed demonstrated that alveolar macrophages of a patient with sarcoidosis were able to convert in-vitro $25(OH)D_3$ to a $1,25(OH)_2D_3$-like metabolite. Adams and Gacad (1985) ultimately characterized this metabolite as $1,25(OH)_2D_3$ and demonstrated 1α hydroxylation of vitamin D_3 by cultured alveolar macrophages from patients with sarcoidosis (Adams et al. 1985). Mason et al. (1984) also demonstrated that a sarcoid lymph-node homogenate could convert $25(OH)D_3$ to a $1,25(OH)_2D_3$-like metabolite, whereas lymph nodes from normal controls failed to do so. Further investigations showed that 1α hydroxylating activity of sarcoid alveolar macrophages differs from that found in the mammalian kidney in three main ways: (1) macrophages are not able to convert $25(OH)D_3$ to $24,25(OH)_2D_3$; (2) 1α hydroxylation is not inhibited by preincubation with $1,25(OH)_2D_3$; (3) 1α hydroxylation is very susceptible to inhibition by glucocorticoids (Adams and Gacad 1985). It should be emphasized that macrophagic 1α hydroxylating activity is not specific to sarcoid granulomas: hypercalcaemia secondary to an extrarenal production of calcitriol has also been demonstrated in some other granulomatous diseases (Reichel et al. 1989).

The effects of calcitriol release by activated macrophages on calcium homeostasis correlate with the activity of the disease and the extent of the inflammatory process. In active stages of the disease it seems likely that large amounts of calcitriol are released into the circulation with corresponding systemic consequences. A scheme of the presumed physiopathologic events leading to hypercalcaemia and/or hypercalciuria is given in Fig. 4. Due to the particular properties of the macrophage 1α hydroxylase, as discussed above, there is no negative feedback of $1,25(OH)_2D_3$. Increased circulating levels of calcitriol are responsible for increased calcium intestinal absorption and presumably also for increased bone resorption. These two factors lead to an increase of the calcium filtered load which, in association with suppressed PTH activity, results in high urinary calcium excretion. In many patients with sarcoidosis, hypercalciuria may remain the sole abnormality of calcium metabolism for months or years. Hypercalcaemia occurs only if any of the following conditions are also present: large dietary calcium intake, higher amounts of serum $1,25(OH)_2D_3$ reached at times of more active disease, or decrease of glomerular filtration rate. Hypercalcaemia in turn may further reduce glomerular filtration rate through its effects on renal blood flow, thus limiting renal calcium excretion ability. This state of positive calcium balance may result in deposition of calcium in soft tissues and nephrocalcinosis. Ultimately, even in silent phases of the disease when calcitriol is no longer produced in excess, patients may remain hypercalciuric due to mobilization of the body calcium stores, and perhaps also to persistent bone resorption.

Renal consequences of altered calcium metabolism in sarcoidosis consist of both functional and structural disorders.

Reduction of glomerular filtration rate, which often occurs in patients with sarcoidosis, is usually ascribed to the effects of hypercalcaemia on glomerular haemodynamics (see Chapter 14.1). Hypercalcaemia can also result in impairment of urinary concentrating ability (see Chapter 5.5). It is uncertain whether hypercalciuria per se can induce tubular disorders; however, impaired urinary concentrating ability has been observed in some normocalcaemic patients with hypercalciuria (Lebacq et al. 1970).

Radiologically demonstrable nephrocalcinosis is rare in sarcoidosis and is found in less than 5 per cent of cases. Its true frequency is probably higher, especially in patients with renal insufficiency (Coburn and Barbour 1984, p. 419). Nephrocalcinosis seems directly related to long-standing hypercalcaemia.

Nephrolithiasis occurs in about 10 per cent of sarcoid patients. It may be responsible for obstructive nephropathy.

Unlike decrease of glomerular filtration rate and impairment of urinary concentrating ability, both nephrocalcinosis and nephrolithiasis do not regress after normalization of calcium metabolism.

Corticosteroids have long proved efficient in the treatment of abnormalities of calcium metabolism in sarcoidosis. Long before the pathogenesis of hypercalcaemia was clarified, their efficacy was used as an argument supporting the hypothesis of increased intestinal calcium absorption. Their inhibitory effect on the macrophage 1α hydroxylase activity which has been demonstrated in vitro explains their efficacy. Numerous studies using different corticosteroids at various dosages have been reported, all showing an overall efficacy. For example, Barbour et al. (1981) used prednisone at an initial daily dosage of 40 mg which was tapered to 20 mg on alternative days 4 months later. Prednisone was also used successfully by Maesaka et al. (1982) at a lower dosage of 10 mg on alternate days. Nordal et al. (1985) used prednisolone at an initial daily dosage of 35 mg. It appears from all these studies that corticosteroids have a rapid effect, with normalization of both serum calcium levels and renal function within a few days. The effective dosage of corticosteroids was proved to be lower in general than that required for the treatment of active forms of the disease, especially those with pulmonary involvement and/or granulomatous interstitial nephritis. Following corticosteroid therapy, serum calcium rapidly decreases, usually in parallel with a decrease of serum $1,25(OH)_2D_3$ whereas serum $25(OH)D_3$ remains normal; urinary calcium excretion decreases while faecal calcium excretion increases. However, high urinary calcium levels may persist in some patients; this can be explained either by persistent bone resorption on which corticosteroids would be ineffective or by secondary calcium mobilization from body stores. The whole duration of treatment cannot be standardized; the decision for withdrawal of the drug is dependent upon the response of serum and urinary calcium levels following progressive tapering of dosage.

Chloroquine has been proposed as an alternative treatment, especially in cases of intolerance or contraindications to corticosteroids. Hunt and Yendt (1963) had noticed that chloroquine successfully reduced both hypercalcaemia and disease activity in a patient with sarcoidosis. Long-term chloroquine therapy (500 mg/day) was followed by normalization of blood and urinary calcium levels together with a decrease of serum $1,25(OH)_2D_3$ level in two patients with hypercalcaemia and hypercalciuria who were intolerant to corticosteroids. (O'Leary et al. 1986). Barré et al. (1987) have reported on a patient with end-stage renal failure, who had experienced two renal transplant rejections. She had received haemodialysis thereafter for 4 years when hypercalcaemia developed, and sarcoidosis was diagnosed. Because of previous side-effects of corticosteroid therapy, she was given hydroxychloroquine (500 mg/day); normalization of serum calcium and $1,25(OH)_2D_3$ levels was achieved within 24 weeks of treatment. Short-course chloroquine therapy also proved to be efficient and 3 days of treatment were sufficient to decrease serum $1,25(OH)_2D_3$ in a patient reported by Adams et al. (1989). The mechanism by which 4-aminoquinoline derivatives may be efficient in calcium disorders of sarcoidosis is unknown. The finding, as in patients treated by corticosteroids, of a decrease of serum $1,25(OH)_2D_3$ levels while $25(OH)D_3$ remains normal also suggests an inhibition of the conversion of $25(OH)D_3$ to $1,25(OH)_2D_3$ by these drugs. Indeed, Adams et al. (1989) recently showed that chloroquine completely inhibited the $25(OH)D_3$-1-hydroxylation reaction when cultured pulmonary alveolar macrophages from a patient with sarcoidosis were exposed to a chloroquine concentration of 1 μmol/l or greater; the corresponding half-maximal effective concentration (ED_{50}) of the drug seemed to these quthors an order of magnitude lower than the serum concentration achieved in patients receiving usual doses of chloroquine. The mechanism by which chloroquine inhibited macrophage $1,25(OH)_2D_3$ synthesis is not known. Barré et al. (1987) suggested that hydroxychloroquine inhibits the $1,25(OH)_2D_3$-induced coalescence of mononuclear cells which brings them together to form multinucleated giant cells. This could impair granuloma formation and subsequently calcitriol-induced osteoclastic osteolysis, and account for the successful result achieved by chloroquine in the patient of Hunt and Yendt (1963) and in one of the patients of O'Leary et al. (1986) in whom bone resorption was assumed to be responsible for persistent hypercalciuria.

The efficacy of other modes of therapy is not impressive. Restriction of dietary calcium intake does not invariably lead to a decrease of calcium intestinal absorption. The use of various calcium-binding agents (inositol hexophosphate, cellulose phosphate, orthophosphate) has led to conflicting results; moreover, it is frequently followed by serious intestinal side-effects.

Corticosteroids which offer constant and rapid efficacy remain the first-choice drug for calcium disorders of sarcoidosis. Chloroquine or hydroxychloroquine may be an alternative choice in cases of intolerance or contraindications to corticosteroids. They may be particularly efficient in those patients in whom abnormal calcium metabolism is presumed to be mostly the consequence of bone resorption.

Glomerular involvement

Glomerular involvement is rare in sarcoidosis. Various glomerular lesions have been reported: membranous glomerulonephritis, focal and segmental glomerulosclerosis, diffuse endocapillary proliferative glomerulonephritis, and crescentic glomerulonephritis (Jones and Fowler 1989; Molle et al. 1986; T.K. Taylor et al. 1979; R.G. Taylor et al. 1982; Vanhille et al. 1986). It is likely that abnormalities of both humoral and cellular immune system present in patients with sarcoidosis are predisposing factors for the development of immune complex-type glomerulonephritis.

The frequency of membranous glomerulonephritis is slightly greater than that of other types associated with sarcoidosis (Taylor et al. 1982; Vanhille et al. 1986) and among the 39 cases reviewed by Vanhille et al. in 1986, there were 13 cases of membranous glomerulonephritis. The frequency of membranous glomerulonephritis seems lower among cases of primary glomerulonephritis, which makes a chance association between sarcoidosis and membranous glomerulonephritis unlikely. Clinical manifestations of membranous glomerulonephritis in sarcoidosis are similar to those in idiopathic membranous glomerulonephritis. Patients present in most cases with the nephrotic syndrome; mild microscopic haematuria is often present and renal function is occasionally impaired, as in the case described by Taylor et al. (1979) in which membranous glomerulonephritis was associated with epithelial crescents. Pathologic findings are not diagnostic. They consist of thickening of the glomerular basement membrane with subepithelial deposits of IgG and C3. In most cases, membranous glomerulonephritis occurred late in the course of an overt sarcoidosis. Beneficial effects of high-dose prednisone therapy were recently reported in two patients with sarcoidosis-associated membranous glomerulonephritis (Jones and Fowler 1989; Oliver Rotellar et al. 1990).

We know of no reports of minimal change disease in sarcoidosis. However, we have recently observed a 31-year-old black woman with sarcoidosis of 6 years duration who developed massive proteinuria (6 g/day) together with a relapse of the disease. While she was undergoing evaluation, proteinuria disappeared. A renal biopsy showed normal glomeruli by both light and immunofluorescence microscopy; electron microscopy was not performed. Two weeks later, she had a transient recurrence of massive proteinuria (10 g/day) which again spontaneously disappeared after a few days. Corticosteroid therapy was started because of pulmonary involvement. Four months later, at her latest evaluation, proteinuria had not recurred. Such a case could be explained by a functional and transient increase of glomerular permeability to proteins secondary to the release of some lymphokines by activated immune cells.

Rainfray et al. (1988) recently reported a case of renal amyloidosis associated with sarcoidosis and found three other cases in the literature. Though elevated concentrations of serum amyloid associated (SAA) protein in sarcoidosis (Rubinstein *et al.* 1989) may predispose to AA amyloidosis, the nature of the amyloid protein has not been precisely assessed in any of these cases.

Miscellaneous

Retroperitoneal involvement secondary to lymph-node enlargement has been reported by Castrignano *et al.* (1988) in four of 14 patients with long-standing sarcoidosis who had abdominal CT scans. In two patients, involvement of the lymph nodes was responsible for ureteral compression and hydronephrosis. In one of these, acute renal failure was severe enough to require haemodialysis for 12 days while high-dose steroid therapy was initiated. Ureteral compression progressively regressed and renal function returned to normal within a month.

Godin *et al.* (1980) reported on a patient with sarcoidosis in whom hypertension developed. Aortography showed extensive stenosis of the right renal artery. Surgical exploration disclosed extensive periaortic and perirenal fibrosis which caused extrinsic compression of the right renal artery. In addition, pathological examination disclosed an epithelioid infiltration in the adventitia of the renal artery, suggestive of sarcoid angiitis.

Bibliography

Adams, J.S. and Gacad, M.A. (1985). Characterization of 1-α-hydroxylation of vitamin D_3 sterols by cultured alveolar macrophages from patients with sarcoidosis. *Journal of Experimental Medicine*, **161**, 755–65.

Adams, J.S., Sharma, O.P., Gacad, M.A., and Singer, F.R. (1983). Metabolism of 25-hydroxyvitamin D_3 by cultured pulmonary alveolar macrophages in sarcoidosis. *Journal of Clinical Investigation*, **72**, 1856–60.

Adams, J.S., *et al.* (1985). Isolation and structural identification of 1,25-dihydroxy vitamin D_3 produced by cultured alveolar macrophages in sarcoidosis. *Journal of Clinical Endocrinology and Metabolism*, **60**, 960–6.

Barbour, G.L., Coburn, J.W., Slatopolsky, E., Norman, A.W., and Horst, R.L. (1981). Hypercalcaemia in an anephric patient with sarcoidosis: Evidence for extra-renal generation of 1,25-dihydroxyvitamin D. *New England Journal of Medicine*, **305**, 440–3.

Basset, F., Soler, P., and Hance, A.J. (1988). Sarcoidosis—from granuloma formation to fibrosis. In *Sarcoidosis and Other Granulomatous Disorders*, (eds. C. Grassi, G. Rizzato and E. Pozzi), pp. 235–46. Elsevier Science Publishers, Amsterdam.

Bell, N.H., Stern, P.H., Pantzer, E., Sinha, T.K., and De Luca, H.F. (1979). Evidence that increased circulating 1-alpha, 25-dihydroxyvitamin D is the probable cause for abnormal calcium metabolism in sarcoidosis. *Journal of Clinical Investigation*, **64**, 218–25.

Berger, K.W. and Relman, A.S. (1955). Renal impairment due to sarcoid infiltration of the kidney. Report of a case proved by renal biopsies before and after treatment with cortisone. *New England Journal of Medicine*, **252**, 44–9.

Coburn, J.W. and Barbour, G.L. (1984). Vitamin D intoxication and sarcoidosis. In *Hypercalciuric states. Pathogenesis, Consequences and Treatment*, (eds. F.L. Coe), pp. 379–433. Grune and Stratton, Orlando.

Coburn, J.W., Hobbs, C., Johnson, G.S., Richert, J.H., Shinaberger, J.H., and Rosens, S. (1967). Granulomatous sarcoid nephritis. *American Journal of Medicine*, **42**, 273–83.

Grönhagen-Riska, C., Koivisto, V., Riska, H., Von Willebrand, E., and Fyhrquist, F. (1988). ACE in physiologic and pathologic conditions. In *Sarcoidosis and Other Granulomatous Disorders*. (eds. C. Grassi, G. Rizzato and E. Pozzi), pp. 203–11. Elsevier Science Publishers, Amsterdam.

Longcope, W.T. and Freiman, D.G. (1952). A study of sarcoidosis based on combined investigation of 160 cases including 30 autopsies from Johns Hopkins Hospital and Massachusetts General Hospital. *Medicine*, **31**, 132–40.

MacDonald Cameron, H. (1956). Renal sarcoidosis. *Journal of Clinical Pathology*, **9**, 136–41.

Méry, J.Ph. and Kenouch, S. (1988). Les atteintes de l'interstitium rénal au cours des maladies systémiques. In *Séminaires d'Uro-Néphrologie Pitié-Salpêtrière* (eds. C. Chatelain and C. Jacobs), pp. 57–89. Masson, Paris.

Papapoulos, S.E. (1979). 1,25-dihydroxycholecalciferol in the pathogenesis of the hypercalcaemia of sarcoidosis. *Lancet*, **i**, 627–30.

Pertschuk, L.P., Silverstein, E., and Friedland, J. (1981). Immunohistologic diagnosis of sarcoidosis. Detection of angiotensin-converting enzyme in sarcoid granulomas. *American Journal of Clinical Pathology*, **75**, 350–4.

Reichel, H., Koeffler, H.P., and Norman, A.W. (1989). The role of the vitamin D endocrine system in health and disease. *New England Journal of Medicine*, **320**, 980–91.

Schaumann, J. (1933). Etude anatomo-pathologique et histologique sur les localisations viscérales de la lymphogranulomatose bénigne. *Bulletin de la Société Francaise de Dermatologie et de Syphiligraphie*, **40**, 1167–78.

Semenzato, G. (1988). The sarcoid granuloma formation—immunology. In *Sarcoidosis and Other Granulomatous Disorders*, (eds. C. Grassi, G. Rizzato and E. Pozzi), pp. 73–87. Elsevier Science Publishers, Amsterdam.

Singer, F.R. and Adams, J.S. (1986). Abnormal calcium homeostasis in sarcoidosis. *New England Journal of Medicine*, **315**, 755–7.

References

Adams, J.S., Diz, M.M., and Sharma, O.P. (1989). Effective reduction in the serum 1,25-dihydroxyvitamin D and calcium concentration in sarcoidosis-associated hypercalcemia with short-course chloroquine therapy. *Annals of Internal Medicine*, **111**, 437–8.

Amouroux, J., Michon, J., Gurly, R., Olivieri, P., Michaud, B., and Roujeau, J. (1967). Sarcoïdose pseudo-tumorale du rein. *Archives d'Anatomie Pathologique*, **15**, A.301–3.

Barbiano di Belgioioso, G., *et al.* (1980). Granulomatous interstitial nephritis as a possible isolated manifestation of sarcoidosis: report of a case. In *Abstracts of the XVIIth Congress of European Dialysis and Transplantation Association*, p. 27.

Barré, P.E., Gascon-Barré, M., Meakins, J.L., and Goltzman, D. (1987). Hydroxychloroquine treatment of hypercalcemia in a patient with sarcoidosis undergoing haemodialysis. *American Journal of Medicine*, **82**, 1259–62.

Bolton, W.K., Atuk, N.O., Rametta, C., Sturgill, B.C., and Spargo, B.H. (1976). Reversible renal failure from isolated granulomatous renal sarcoidosis. *Clinical Nephrology*, **5**, 88–92.

Bottcher, E. (1959). Disseminated sarcoidosis with a marked granulomatous arteritis. *Archives of Pathology*, **68**, 419–23.

Branson, J.H. and Park, J.H. (1954). Sarcoidosis hepatic involvement: presentation of case with fatal liver involvement, including autopsy findings and review of evidence for sarcoid involvement of liver as found in literature. *Annals of Internal Medicine*, **40**, 111–4.

Caruana, R.J., Carr, A.A., and Rao, R.N. (1982). Idiopathic granulomatous nephritis in a patient with hypertension and an atrophic kidney. *Nephron*, **32**, 83–6.

Castrignano, L., Fasolini, G., Nozza, M., and Schieppati, A. (1988). Retroperitoneal involvement in sarcoidosis. In *Sarcoidosis and Other Granulomatous Disorders*, (eds. C. Grassi, G. Rizzato and E. Pozzi), pp. 383–4. Elsevier Science Publishers, Amsterdam.

Cheng, H.F., Nolasco, F., Cameron, J.S., Hildreth, G., Neild, G., and Hartley, B. (1989). HLA-DR display by renal tubular epithelium and phenotype of infiltrate in interstitial nephritis. *Nephrology Dialysis and Transplantation*, **4**, 205–15.

Cuppage, F.E., Emmott, D.F., and Duncan, K.A. (1988). Renal failure secondary to sarcoidosis. *American Journal of Kidney Diseases*, **11**, 519–20.

Dobrin, R.S., Vernier, R.L., and Fish, A.J. (1975). Acute eosinophilic interstitial nephritis and renal failure with bone marrow-lymph node granulomas and anterior uveitis. A new syndrome. *American Journal of Medicine*, **59**, 325–33.

Falls, W.F., Randall, R.E. Jr., Sommers, S.C., Staey, W.K., Larkin, E.J., and Still, W.J.S. (1972). Nonhypercalcemic sarcoid nephropathy, *Archives of Internal Medicine*, **130**, 285–91.

Farge, D., Lioté, F., Turner, M., Barre, P., and Jothy, S. (1986). Granulomatous nephritis and chronic renal failure in sarcoidosis. Long-term follow-up studies in two patients. *American Journal of Nephrology*, **6**, 22–7.

Ford, M.J., Anderton, J.L., and Mac Lean, N. (1978). Granulomatous sarcoid nephropathy. *Postgraduate Medical Journal*, **54**, 416–7.

Godin, M, Fillastre, J.P., Ducastelle, T., Hemet, J., Morere, P., and Nouvet, G. (1980). Sarcoidosis. Retroperitoneal fibrosis, renal arterial involvement, and unilateral focal glomerulosclerosis. *Archives of Internal Medicine*, **140**, 1240–2.

Guédon, J., Mathieu, F., Chomé, J., Chebat, J., Safar, M., and Küss, R. (1967). Sarcoïdose rénale à forme pseudo-tumorale révélatrice de l'affection. *Presse Médicale*, **75**, 265–8.

Guénel, J., and Chevet, D. (1988). Néphropathies interstitielles de la sarcoidose. Effet de la corticothérapie et évolution à long terme. Etude rétrospective de 22 observations. *Néphrologie*, **9**, 253–7.

Hunt, B.J. and Yendt, E.R. (1963). The response of hypercalcemia in sarcoidosis to chloroquine. *Annals of Internal Medicine*, **59**, 554–65.

Jones, B. and Fowler, J. (1989). Membranous nephropathy associated with sarcoidosis. Response to prednisone. *Nephron*, **52**, 101–2.

Kenouch, S., Belghiti, D., Di Costanzo, P., and Méry, J.Ph. (1989). Un nouveau syndrome: les néphropathies interstitielles aiguës avec uvéite. *Annales de Médecine Interne*, **140**, 169–72.

Korzets, Z., Schneider, M., Taragan, R., Bernheim, J., and Bernheim, J. (1985). Acute renal failure due to sarcoid granulomatous infiltration of the renal parenchyma. *American Journal of Kidney Diseases*, **6**, 250–3.

Lebacq, E., Desmet, V., and Venhaegen, H. (1970). Renal involvement in sarcoidosis. *Postgraduate Medical Journal*, **46**, 526–9.

Leng-Lévy, J., David-Chaussé, J., Martin-Dupont, Cl., and Aparicio, M. (1965). Maladie de Besnier-Boeck-Schaumann à localisation rénale. *Bulletins et Mémoires de la Société Médicale des Hôpitaux de Paris*, **116**, 907–11.

Maesaka, J.K., Bautman, V., Pablo, N.C., and Shakamuri, S. (1982). Elevated 1,25-dihydroxyvitamin D levels in a patient with sarcoidosis and end-stage renal disease. *Archives of Internal Medicine*, **142**, 1206–7.

Mason, R.S., Frankel, T., Chan, Y.-L., Lissner, D., and Posen, S. (1984). Vitamin D conversion by sarcoid lymph node homogenate. *Annals of Internal Medicine*, **100**, 59–61.

Michielsen, P., Van Damme, B., Hauglustaine, D., Waer, M., Verberckmoes, R., and Vanrenterghem, Y. (1985). Insuffisance rénale aiguë: première manifestation d'une sarcoïdose familiale. *Néphrologie*, **6**, 151–2.

Molle, D., Baumelou, A., Beaufils, H., Vannier, R., and Legrain, M. (1986). Membranoproliferative glomerulonephritis associated with pulmonary sarcoidosis. *American Journal of Nephrology*, **6**, 386–7.

Muther, R.S., McCarron, D.A., and Bennett, W.M. (1981). Renal manifestations of sarcoidosis. *Archives of Internal Medicine*, **141**, 643–5.

Nédelec, G., Didelot, F., Giudicelli, C.P., and Doco, J.B. (1986). Néphropathie interstitielle et insuffisance rénale dans la sarcoïdose (letter to the Editor). *Presse Médicale*, **15**, 623.

Nordal, K.P., Dahl, E., Halse, J., Aksnes, L., and Aarskog, D. (1985). Rapid effect of prednisolone on serum 1,25-Dihydroxycholecalciferol in hypercalcemic sarcoidosis. *Acta Medica Scandinavica*, **218**, 519–23.

O'Leary, T.J., Jones, G., Yip, A., Lohnes, D., Cohanim, M., and Yendt, E.R. (1986). The effects of chloroquine on serum 1,25-dihydroxyvitamin D and calcium metabolism in sarcoidosis. *New England Journal of Medicine*, **315**, 727–30.

Oliver Rotellar, J.A., Garcia Ruiz, C., and Martinez Vea, A. (1990). Response to prednisone in membranous nephropathy associated with sarcoidosis. *Nephron*, **54**, 195.

Pagniez, D.C., Macnamara, E., Beuscart, R., Wambergue, F., Dequiedt, P., and Taquet, A. (1987). Gallium scan in the follow-up of sarcoid granulomatous nephritis. *American Journal of Nephrology*, **7**, 326–7.

Rainfray, M., Meyrier, A., Valeyre, D., Tazi, A., and Battesti, J.P. (1988). Renal amyloidosis complicating sarcoidosis. *Thorax*, **43**, 422–3.

Romer, F.K. (1980). Renal manifestations and abnormal calcium metabolism in sarcoidosis. *Quarterly Journal of Medicine*, **49**, 233–47.

Rubinstein, I., Knecht, A., De Beer, F.C., Baum, G.L., and Pras, M. (1989). Serum amyloid-A protein concentrations in sarcoidosis. *Israel Journal of Medical Sciences*, **25**, 461–2.

Singer, D.R.J. and Evans, D.J. (1986). Renal impairment in sarcoidosis: granulomatous nephritis as an isolated cause (two case reports and review of the literature). *Clinical Nephrology*, **26**, 250–6.

Tanneau, R., Hervé, J.P., Guiserix, J., Gentric, A., Leroy, J.P., and Clèdes, J. (1987). Insuffisance rénale aiguë anurique par infiltration interstitielle granulomateuse révélatrice d'une sarcoïdose (letter to the Editor). *Annales de Médecine Interne*, **138**, 147.

Taylor, T.K., Senekjian, H.O., Knight, T.F., Györkey, F., and Weinman, E.J. (1979). Membranous nephropathy with epithelial crescents in a patient with pulmonary sarcoidosis. *Archives of Internal Medicine*, **139**, 1183–5.

Taylor, R.G., Fisher, C., and Hoffbrand, B.I. (1982). Sarcoidosis and membranous glomerulonephritis: a significant association. *British Medical Journal*, **284**, 1297–8.

Turner, M.C., Shin, M.L., and Ruley, E.J. (1977). Renal failure as a presenting sign of diffuse sarcoidosis in an adolescent girl. *American Journal of Diseases of Children*, **131**, 997–1000.

Vanhille, Ph., et al. (1986). Glomérulonéphrite rapidement progressive à dépôts mésangiaux d'IgA au cours d'une sarcoïdose. *Néphrologie*, **5**, 207–9.

Warren, G.V., Sprague, S.M., and Corwin, H.L. (1988). Sarcoidosis presenting as acute renal failure during pregnancy. *American Journal of Kidney Diseases*, **12**, 161–3.

Zerwekh, J.E., Pak, C.Y.C., and Kaplan, R.A. (1980). Pathogenetic role of 1-alpha, 25-dihydroxyvitamin D in sarcoidosis and absorptive hypercalciuria: Different response to prednisone therapy. *Journal of Clinical Endocrinology and Metabolism*, **51**, 381–6.

4.5.1 Pathogenesis of angiitis

F.J. VAN DER WOUDE AND L.A. VAN ES

Pathogenesis of angiitis

Introduction

Angiitis or vasculitis refers to an inflammatory reaction in the wall of any blood vessels. The clinical presentation is very diverse; it may present as a primary disease or be associated with other diseases. In some patients, blood vessels of a specific size are involved, in others blood vessels of various sizes are affected. In clinical practice the most common type of systemic vasculitis affects small blood vessels, usually postcapillary venules. In other forms of vasculitis, however, the diameter of the affected blood vessels and their location varies as does the histology. Morphologically, arteritis is characterized by proliferation and swelling of endothelial cells, invasion with neutrophils, fibrinoid necrosis, and disruption of the internal elastic lamina giving neutrophils access to the media and the adventitia. The wall of capillaries and postcapillary venules consists only of a layer of endothelial cells and the basement membrane. In leucocytoclastic vasculitis these blood vessels are surrounded by neutrophils and nuclear debris together with fibrinoid necrosis and haemorrhage. The acute phase of leucocytoclastic vasculitis can be followed by a chronic phase characterized by a mononuclear infiltrate. Other types of vasculitides start primarily with vascular infiltration by T lymphocytes and macrophages; there are experimental models analogous to these vasculitides, which may not have the same aetiology as their human counterparts, but have certainly been instructive with respect to pathogenesis. These models will be discussed here as far as they are relevant to the understanding of human vasculitis. Unfortunately, animal models of granulomatous vasculitides are not very suitable and have not been studied extensively. At the end of this chapter the pathogenesis of granulomatous vasculitides will be discussed on the basis of the immunological and histological abnormalities found in patients.

Pathogenesis of vasculitis in animals

Traditional views on the pathogenesis of angiitis or vasculitis are mainly based in histopathological and immunological studies in laboratory animals (Cohrane and Koffler 1973). The animal model that has provided most of our current understanding of the pathogenesis of vasculitis is serum sickness in rabbits.

Serum sickness as a model for necrotizing vasculitis

In the acute model, rabbits given a single intravenous dose of bovine serum albumin (Dixon 1978) develop necrotizing arteritis in their coronary arteries, arthritis, and acute glomerulonephritis about 1 week later. At the time when the vascular lesions develop complexes of bovine serum albumin and its antibodies can be demonstrated in the circulation and in the wall of arteries and glomerular capillaries. In the chronic model, bovine serum albumin is given daily. Rabbits with strong antibody responses start with manifestations resembling acute serum sickness. Most develop glomerulonephritis without arteritis, but some animals receiving multiple high doses of bovine serum albumin produce high levels of antibodies and develop small vessel vasculitis, resembling human leucocytoclastic vasculitis (Brentjens et al. 1975).

Formation of circulating antigen–antibody complexes

The composition of antigen–antibody complexes formed in the circulation depends on the absolute and relative amounts of antigens and antibodies present, the valency and the size of the antigen, the affinity and isotype of the antibodies, and the participation of complement. Rabbits developing acute serum sickness have circulating soluble complexes containing moderate antigen excess and with a molecular weight of 10^6 Da. The antibodies are usually of the IgG isotype, although IgM antibodies can also be detected; their role in the development of immune complex disease is less well studied. Systemic IgA responses have been studied in other animal models after rechallenge with antigens that were previously encountered on mucosal surfaces. Experimental studies have demonstrated that circulating IgA antibodies induced by oral or intrabronchial immunization can form complexes after an intravenous rechallenge. These IgA complexes can become deposited in the glomeruli and cause glomerulonephritis.

Elimination of circulating antigen–antibody complexes

Acute serum sickness does not develop in all rabbits receiving intravenous bovine serum albumin: some animals develop high titres of high affinity antibodies. The resulting complexes formed in antibody excess are efficiently removed by Kupffer cells in the liver whose Fc receptors interact with the Fc portions of the antibodies. The in-vivo role of C3 receptors in the elimination of immune complexes is still unclear because complement-depleted rabbits clear preformed immune complexes as efficiently as animals with an intact complement system (Mannik and Arend 1971). In vitro, Kupffer cells and macrophages do not differ in their capacity to endocytose immune complexes; in vivo, however, circulating immune complexes are mainly taken up by Kupffer cells in direct contact with blood flowing through the hepatic sinusoids. Macrophages located in the interstitium of the spleen on the other hand are surrounded by a less free flowing part of the circulation. The capacity of the liver to remove circulating immune complexes is determined by the number of Kupffer cells and their degree of activation (Dieselhoff-Den Dulk et al. 1979), the density of Fc and C3 receptors (Nishi et al. 1981), and the rate of the hepatic blood flow.

Formation of deposits in blood vessels

Circulating immune complexes may be deposited in the vessel wall. Local immune complex formation may also occur due to circulating antibodies which react with antigen trapped in the vessel wall. In the experimental Arthus reaction, active or passive immunization results in circulating antibodies able to react with antigen deposited in vessel walls following earlier subcutaneous injection. The subsequent sequence of events is similar to those which occur after deposition of circulating immune complexes, with local activation of complement and influx of neutrophils. Theoretically it is also possible for antibodies directed against intrinsic vascular structures such as endothelial antigens to induce local formation of immune deposits.

The role of vascular permeability

In rabbits with acute serum sickness arteritis is mainly found in areas of turbulent blood flow with high shear stress at the vessel wall, as can be expected at the aortic outflow tract, bifurcations, and at ostia of aortic branches. The arteritis can also be produced locally at the site of an experimentally induced coarctation (Kniker and Cochrane 1968). It is clinically relevant that bovine serum albumin can be detected by immunofluorescence in vessel walls for only a few days. Ultrastructurally, immune deposits can be found between the endothelial cells and the internal elastic membrane and it is assumed that these represent circulating immune complexes that have gained access to the subendothelial space via transient opening of interendothelial clefts (Majno and Pallade 1961). In acute serum sickness in rabbits, the increased permeability is probably mediated by antigen-specific IgE antibodies bound to mast cells, basophils and possibly macrophages. Antigen causes clustering of these IgE antibodies, which in turn leads to transduction of an activation signal, resulting in release of stored mediators such as histamine, heparin, and proteolytic enzymes and the synthesis of mediators such as platelet activating factor and leukotrienes such as LTB4, LTC4, and LTD4. In rabbits, platelet activating factor induces aggregation of platelets with the release of their histamine. Acute serum sickness does not develop in rabbits which fail to develop this IgE-mediated histamine release; serum sickness can also be prevented by treatment with antihistamines (Cochrane 1963). These observations strongly suggest that histamine is the vasoactive substance mediating the deposition of circulating immune complexes in rabbits. In man, however, platelets contain more serotonin and less histamine.

In addition to vascular permeability, the size of circulating immune complexes determines whether deposits are formed in vessel walls; complexes prepared *in vitro* which have a molecular weight of 1000 kDa do not localize in vessel walls when administered to guinea-pigs together with histamine (Cochrane and Hawkins 1968). However, complexes with a molecular weight greater than 1000 kDa localize subendothelially in blood vessels when administered with histamine. The sieving properties of the internal elastic lamina and the basement membrane of smaller blood vessels presumably determine this selective deposition; complexes smaller than 1000 kDa are thought to diffuse through these membranes whereas larger complexes are retained. In active models of immune complexes disease it is technically impossible to determine which complexes localize preferentially in vessel walls. In acute serum sickness less than 0.005 per cent of the injected antigen can be detected in glomerular deposits (Wilson and Dixon 1970) and more than 99 per cent is eliminated by the mononuclear phagocyte system, mainly the Kupffer cells in the liver. The proportion of complexes that will localize preferentially in vessel walls is so small that they cannot be identified by following the change of the ultracentrifugation profile of circulating complexes with time.

The mediators which participate in other experimental models of vasculitis remain unknown. Other vasoactive mediators such as leukotrienes, bradykinin, serotonin, angiotensin II, and platelet activating factor may play a role. The leukotrienes LTB4, LTC4, and LTD4 have been shown to influence the permeability of postcapillary venules (Dahlén *et al.* 1981). Leukotriene B4 also induces adhesion of leucocytes to endothelium of these venules *in vivo*. It should be noted here that the development of increased vascular permeability and the diapedesis of leucocytes do not necessarily represent two separate steps in the pathogenesis of vasculitis. Experimental studies have shown that the trans-endothelial migration of leucocytes itself induces a significant increase in vascular permeability (Issekutz 1981). Furthermore, activation of inflammatory cells such as neutrophils, basophils, monocytes, and macrophages can lead to the release of mediators such as platelet activating factor which, in turn, may influence vascular permeability both by a direct effect on endothelium and by inducing the release of other mediators.

The inflammatory reaction

Complement plays an important role in the development of the inflammatory process which follows the deposition of immune complexes in vessel walls. Rabbits that have been treated with cobra venom in order to deplete their complement system do not develop arteritis, although they still develop acute glomerulonephritis (Henson and Cochrane 1971). Such rabbits show deposition of bovine serum albumin and IgG but not of C3, and the characteristic neutrophil infiltration and necrosis of media and adventitia are absent. In rabbits with an intact complement system neutrophils are attracted to the subendothelial deposits by the chemotactic effects of complement fragments such as C5a. Neutropenia induced by cytostatic agents also prevents the development of arteritis in acute serum sickness.

Animal models for mononuclear vasculitis

Glomerulonephritis and vasculitis develop spontaneously in 10 per cent of NZB mice and in 20 per cent of NZB/W hybrids. Fibrinoid necrosis of the wall of small and medium-sized arteries, associated with circumferential lymphocytes and plasma cells, occurs in various organs, including the heart, spleen, thymus, lymph nodes, pancreas, and gastrointestinal tract. Certain strains of mice suffering from acute or chronic infections with the lymphocytic choriomeningitis virus also develop vasculitis. Mice that survive the acute infection, or which are infected neonatally, become chronic carriers of the virus and develop glomerulonephritis and arteritis. Viral antigens are present in circulating immune complexes in these animals and antibodies to the virus can be eluted from the glomeruli. Susceptibility to this immune complex disease is linked to the major histocompatibility (H–2) complex of the mouse. Glomerulonephritis with acute or chronic arteritis occurs in mink infected with Aleutian disease virus which affects small and medium-sized muscular arteries, especially those of the heart, the brain, and kidney. Animals with the most severe disease have acute arteritis with infiltrating neutrophils, fibrinoid necrosis, and disruption of the internal elastic membrane, but subacute lesions, consisting of proliferation of endothelial cells and a dense perivascular infiltration with lymphocytes and plasma cells, are more common. This illustrates that classic necrotizing neutrophilic vasculitis and mononuclear vasculitis can be encountered as an expression of the same disease.

Little is known about the role of cell-mediated immunity in the pathogenesis of systemic vasculitis. T lymphocytes from BALB/c mice which have been sensitized by culture with endothelial cells isolated from brain capillaries of outbred mice cause vasculitis when injected into syngeneic recipients (Hart et al. 1983). This affects not only the capillaries of the brain, but also the veins, capillaries, and arteries of other organs. This reaction shows many similarities with the T lymphocytes-directed rejection process produced against intrinsic transplantation antigens of the vascular endothelium.

T lymphocytes can also be sensitized to vascular smooth muscle cells. In lupus-prone MRL/1pr mice, which develop spontaneous mononuclear vasculitis, collections of Thy 1^+, Ly1^+, L3T$_4^-$ T cells are found around small to medium-sized muscular arteries at the age of 8 weeks. By 12 weeks mononuclear inflammatory cells are seen next to hypertrophied vascular smooth muscle cells. T lymphoblasts subsequently infiltrate the media, resulting in lysis of smooth muscle cells (Moyer et al. 1987). T lymphocytes from MRL/1pr mice are capable of attacking syngeneic smooth muscle cells but not those from H–2K compatible control mice (Moyer et al. 1984); this has been attributed to the expression of MHC class II antigens and the release of interleukin 1 by the smooth muscle cells of MRL/1pr, but not control, mice.

Cell-mediated immune reactions can also be directed against extrinsic, and possibly exogenous, antigens, for example those of viral or bacterial origin. This is the delayed type hypersensitivity reaction which is initiated by presentation of antigen to CD4$^+$ T helper lymphocytes. Activated T helper cells proliferate and stimulate other lymphocytes and macrophages to localize in the vessel wall, where they are activated in turn and release lymphokines. These cells in turn stimulate circulating monocytes and resident macrophages to accumulate locally and become activated, producing a mononuclear cell infiltrate which is transformed into a granulomatous reaction as the phagocytes differentiate into giant cells.

Experimental granuloma formation

Granulomas can be produced experimentally by the induction of insoluble immune complexes and by cell-mediated immunity. When large doses of insoluble complexes are injected subcutaneously in rats, they are taken up by infiltrating macrophages and after about 10 days lymphocytes, epithelioid cells, and giant cells appear in the lesion. Granulomas of small pulmonary venules and veins can also be produced by regular intravenous injection of high doses of bovine serum albumin into rabbits which produce excessively high levels of antibodies (Germuth and Pollack 1967).

Granulomatous vasculitis can also be induced by T lymphocytes. Mice, sensitized to a hapten conjugate and injected with hapten conjugated polyacrylamide beads, develop typical delayed type hypersensitivity reactions after 24 to 48 h, with the development of epithelioid granulomas after 72 to 96 h (Ginsburg et al. 1982). Similar granulomas can also be induced in unprimed mice by transfer of T lymphocytes. There is no evidence that vasculitis with granuloma formation is always a cell-mediated process, but T lymphocytes are abundant in the vascular lesions, which suggests their involvement in pathogenesis.

Pathogenesis of vasculitides in man

The aetiology and pathogenesis are unknown in the vast majority of patients with vasculitis, but the associations with a wide variety of different diseases clearly suggest considerable diversity. The wide clinical and histological heterogeneity makes it highly unlikely that the pathogenesis will be the same in all cases, but attempts to clarify this confusion have led to a multitude of classifications for these diseases which reflect our ignorance. A widely used classification is based on the types of blood vessels involved (Table 1).

Small vessel vasculitis is the most common form of the disease. In some patients, it occurs as a primary or idiopathic disorder; in others it is associated with other conditions such as connective tissue diseases, or complicates an infection. It is uncertain whether small vessel vasculitides are basically different from polyarteries, although they certainly differ in their clinical detectability. The widespread use of angiography and muscle biopsy detects vasculitides affecting small vessels as well as arteries. It has been suggested that polyarteritis and small vessel vasculities are not separate disease entities but are rather an expression of more severe disease. Polyarteritis affecting not only aortic branches and middle-sized arteries, but also small vessels and veins, occur as 'polyangiitis overlap' syndrome, which has characteristics of two types of vasculitis, for example, polyarteritis and Churg–Strauss syndrome, giant cell arteritis and polyarteritis, or giant cell arteritis and Wegener's granulomatosis.

In the last decade, immunological abnormalities in patients with vasculitis have been studied intensively, and vasculitides with similar immunological abnormalities and similar histopathology have been classified into four groups:

1. leucocytoclastic (small vessel) vasculitis
2. mononuclear vasculitis
3. necrotizing arteritis
4. granulomatous vasculitis

Although it is still not possible to delineate distinct pathogenetic mechanisms for each group, it is reasonable to assume that the histological difference reflects different mechanisms. The histopathology of vasculitides will be discussed here in relation to possible immune mechanisms.

Leucocytoclastic vasculitides

Leucocytoclastic vasculitis affects mainly the postcapillary venules and is most easily detected in the skin, where it usually presents as indurated purpura, but may also appear as urticaria, macules, nodules, vesicles, or skin ulcers. Histologically there is perivascular infiltration by polymorphonuclear leucocytes, together with nuclear debris, fibrinoid necrosis, and extravasation of erythrocytes.

Since rabbits receiving daily multiple and high doses of bovine serum albumin develop leucocytoclastic vasculitis (Brentjes et al. 1975), leucocytoclastic vasculitis in man was assumed to be due to deposition of circulating immune complexes. The evidence for an immune complex-mediated pathogenesis in human leucocytoclastic vasculitis will therefore be discussed first.

Table 1 Classification of vasculitis on the basis of blood vessels involved

	Aorta	Aortic branches	Middle-sized arteries	Small arteries	Arterioles	Capillaries	Venules	Veins
Small vessel vasculitis								
Primary								
Hypersensitivity angiitis	−	−	−	−	−	+	++	−
Anaphylactoid purpura	−	−	−	−	−	+	++	−
Urticarial vasculitis	−	−	−	−	−	+	++	−
Erythema multiforme	−	−	−	−	−	+	++	−
Secondary								
Postinfectious vasculitis	−	−	−	+	+	+	++	−
Mixed cryoglobulinaemia	−	−	−	+	+	+	++	−
Rheumatoid vasculitis	−	−	−	+	+	+	++	+
Vasculitis in SLE	−	−	−	+	+	+	++	+
Sjögren's syndrome	−	−	+	+	+	+	++	+
Polyarteritis								
Wegener's granulomatosis	−	−	+	+	+	+	++	+
Churg-Strauss syndrome	−	−	+	+	+	+	++	+
Relapsing polychondritis	−	+	+	+	+	+	+	+
Lymphomatoid granulomatosis	−	−	+	+	−	−	−	+
Behçet's syndrome	−	+	+	+	+	+	++	+
Giant cell arteritis	+	+	+	+	+	−	−	−
Polyarteritis nodosa	−	+−	++	++	+	−	−	−
Kawasaki's disease	−	++	+	−	−	−	−	−
Takayasu's disease	++	++	+	−	−	−	−	−

Circulating antigen–antibody complexes

Although circulating immune complexes are found in a variety of diseases, the fact that very few patients develop vasculitis has raised doubts about their pathogenicity.

In order to prove an immune complex-mediated mechanism in a patient with vasculitis the following requirements should be fulfilled: firstly antigen–antibody complexes should be demonstrated in the circulation, and their concentration should correlate clinically and histologically with disease activity, and the complexes should disappear when the disease remits. Secondly, the complement activating capacity of these complexes should be proven since, in laboratory animals at least, complement activation is a requirement for the development of an inflammatory response. Thirdly, deposits of antigen, antibodies, and complement should be detectable in the vessel walls when immune complexes are present in the circulation;. Since circulating antigens and antibodies are in a dynamic equilibrium with the complexes in the circulation; however the possibility that antigens localize first in the vessel wall to react at a later stage with circulating antibodies cannot be excluded.

The finding of circulating immune complexes has very little diagnostic value, but their presence in vasculitis may indicate the pathogenesis of the disease.

Since the nature of the antigens involved in the pathogenesis of systemic vasculitis is not usually known, circulating immune complexes are detected in clinical practice by antigen-independent methods, based on the properties which they acquire as several Fc parts of antibodies combine (Theofilopoulos and Dixon 1979; McDougal and McDuffie 1985). These include the ability to react with C1q, the first component of the classical pathway of complement, adherence to Fc receptors of phagocytes or B lymphocytes, and increased binding to rheumatoid factors or experimentally induced anti-immunoglobulins. When complement is activated C3b and C3d are covalently linked to the complex, allowing it to interact with CR-1 or CR-3 receptors on phagocytes and cultured cells such as the Raji cell (Theofilopoulos and Dixon 1979).

Not all patients with circulating immune complexes develop vasculitis or glomerulonephritis, as has been shown in studies of hepatitis B virus infections. In a hyperendemic area in Alaska, 4600 of every 100 000 Indians were positive for hepatitis B surface antigen (HBs Ag), and 25 per cent carried anti-HBs antibodies (McMahon *et al.* 1989). Different studies have found a prevalence of circulating immune complexes in HBsAg-carriers between 5 and 10 per cent: extrapolating these figures to the Alaskin Indians, 230 to 460 individuals per 100 000 inhabitants would be HBsAb-positive and have circulating immune complexes, while the incidence of systemic vasculitis was 13 cases over 12 years or 7.7 cases per 100 000 inhabitants (McMahon *et al.* 1989). The risk of developing leucocytoclastic vasculitis or systemic vasculitis in patients with hepatitis B-antigenaemia is relatively low. This also applies to patients with other infections, with circulating immune complexes, and tumours (Theofilopoulos and Dixon 1979). The factors which influence the development of immune complex-mediated vasculitis are shown in Table 2.

The effect of immune complex concentration

The level of circulating immune complexes is expressed in microequivalents of IgG or IgA aggregates with comparable reactivity. There is no absolute correlation between the level of circulating immune complexes and the development of vasculitis in various diseases; although patients with rheumatoid arthritis may have high levels of circulating immune complexes, systemic vasculitis occurs in only a minority. However, patients with rheumatoid vasculitis usually have higher immune complex levels than patients without.

4.5.1 Pathogenesis of angiitis

The effect of size and composition of immune complexes

The size of immune complexes is determined by their composition. Various clinical studies which have attempted to correlate the size of complexes with clinical activity and with manifestations of disease have largely been unsuccessful.

The extremely low incidence of vascular complications in patients with cancer has been attributed to the small size of the complexes. A study of patients with systemic lupus erythematosus showed a correlation between the presence of IgG-containing complexes and diffuse proliferative glomerulonephritis (Wener et al. 1987). Patients with acute hypersensitivity angiitis have predominantly IgA-containing complexes (Table 3) whereas immune complexes seen in patients with chronic hypersensitivity angiitis contain IgA as well as IgG (Kauffmann et al. 1980). Patients with Henoch-Schönlein purpura have large IgA-containing complexes in the circulation, whereas smaller IgA-immune complexes are found in patients with IgA nephropathy (Van Es et al. 1987).

The vasculitides in group A of Tables 3 and 4 are all associated with the presence of circulating immune complexes and vascular deposits of IgM, IgG, IgA, and C3 in the active phase of the disease, suggesting a role for the complexes in the pathogenesis of vasculitis. IgG-containing complexes are present in most types of leucocytoclastic vasculitis, whereas IgA-containing immune complexes are present in acute phases of vasculitis and in hypersensitivity angiitis (Kauffmann et al. 1980), Henoch-Schönlein purpura (Levinsky and Barratt 1979; Kauffmann et al. 1980) and acute rheumatoid vasculitis (Westedt et al. 1986). The development of leucocytoclastic vasculitis in patients with Henoch-Schönlein purpura coincides with the presence of large IgA complexes in the circulation and IgA- and C3-containing deposits in the walls of cutaneous venules. These complexes disappear as the patient recovers from the disease (Kauffmann et al. 1980).

IgG-containing immune complexes are generally detected in chronic forms of leucocytoclastic vasculitis, such as those associated with hypocomplementaemic urticaria, infections, rheumatic diseases, systemic lupus erythematosus, Sjögren's syndrome, and mixed cryoglobulinaemia (Table 3). Mixed complexes containing IgM, IgG, and IgA are frequently found in mixed cryoglobulinaemia, rheumatoid vasculitis, infectious mononucleosis, and cytomegalovirus infections. Although antigen is unknown in the majority of patients, HBsAg and virus particles have been detected in patients with essential mixed cryoglobulinaemia (Levo et al. 1977). HBsAg has also been found in dialysis patients with polyarteritis nodosa. In a study by Drüeke et al. (1980), three of 266 HBsAg carriers and none of the seronegative patients developed polyarteritis. HBsAg has also been demonstrated in cryoprecipitate from a patient with leucocytoclastic vasculitis (Gower et al. 1978) and in glomerular deposits of children with membranous or membranoproliferative glomerulonephritis (Kleinknecht et al. 1979); this result is limited to areas where chronic hepatitis B is endemic. The diverse manifestations of the vascular complications with hepatitis B virus illustrate the multiple manifestations of a single disease. Antigens have been demonstrated in complexes and deposits from isolated patients with bacterial infections, but this has not been pursued systematically. Autologous IgG is also probably involved in complex formation; exogenous antigens have not been demonstrated in circulating immune complexes from patients with rheumatoid arthritis. IgG is likely to be the major autoantigen; rheumatoid factor can form complexes in synovial fluid by self-association. Idiotypic determinants of immugolobulins may also act as autoantigens: antibodies directed against IgM idiotypes have been demonstrated in cryoglobulins (Geltner et al. 1980). Antigens which have been demonstrated in glomerular deposits include those associated with DNA, HBsAg, HBcAg, cytomegalovirus and Epstein-Barr virus, Streptococcus, Staphylococcus, spirochaetes, and trypanosomes. When these patients exhibit simultaneous systemic vasculitis, it can be assumed that this has been caused by the same antigen. Such studies have been performed on individual cases in technically advanced research laboratories; in clinical practice it is not feasible to search for such antigens.

The capacity of complexes to activate complement

Complement plays a protective and an inflammatory role in the pathogenesis of vascultitis. The covalent binding of C3b to circulating immune complexes promotes the binding and degradation of these complexes by mononuclear phagocytes. CR1 receptors on erythrocytes may bind to immune complexes and transport them to Kupffer cells in the liver (Schifferli et al. 1986). Immune complexes are also solubilized by complement activation (Schifferli et al. 1981), a mechanism that protects patients against the inflammatory effects of circulating immune complexes. The high incidence of glomerulonephritis and vasculitis in C3-deficient patients (Roord et al. 1983) is due not only to the lack of these mechanisms, but also to a higher antigenic load resulting from the high incidence of bacterial infections in these patients.

Serum complement levels are normal in hypersensitivity angiitis and anaphylactoid purpura or Henoch-Schönlein syndrome, (Table 3). The acute phase of anaphylactoid purpura is associated with elevated C3d levels, reflecting C3 activation (Van Es et al. 1987). Although hypocomplementaemia is common in the other forms of leucocytoclastic vasculitis represented in Table 3, the mechanisms responsible are very different. In so-called hypocomplementaemic urticarial vasculitis, serum C1q levels are disproportionately low compared to C4 and C3 levels; autoantibodies directed against the collagen portion of C1q cause precipitation of this component (Marder et al. 1984), and similar autoantibodies are also found in patients with systemic lupus erythematosus (Uwatoko and Mannik 1988).

C1q, C4, and C3 levels are frequently low in postinfectious vasculitis (Table 3). Activation of classical and alternative pathways has been described in patients with bacterial endocarditis associated with vasculitis (Kauffmann et al. 1980). Activation of

Table 2 Factors influencing the deposition of circulating immune complexes and the development of vasculitis

1. Properties of immune complexes
   The concentration of circulating immune complexes
   The size and composition of the immune complexes
   The capacity to activate complement
   The properties of the antigen involved
   The charge of the immune complex

2. Most factors
   The capacity of the mononuclear phagocyte system (MPS) to eliminate circulating immune complexes efficiently
   The activation of mediator systems that increase vascular permeability
   The existence of haemodynamic conditions that favour deposition of circulating immune complexes

Table 3 Immunological abnormalities in the circulation of patients with small vessel vasculitis

	Immune complexes			Antibodies					Complement			Eosinophilia
	IgG-Cx	IgA-Cx	Cryo	RF	C1q	Endoth	c-ANCA	p-ANCA	C1q	C4	C3	
Leucocytoclastic vasculitis												
Hypersensitivity angiitis	+	+	−	−	NR	NR	NR	NR	N	N	N	+
Anaphylactoid purpura	−	+	−	−	NR	NR	−	NR	N	N	N	−
Urticarial vasculitis	+	−	−	−	+	NR	NR	NR	↓	↓	↓	−
Postinfectious vasculitis	+	−	+	+	NR	NR	NR	NR	↓	↓	↓	−
Rheumatoid vasculitis	+	+	+	+	NR	NR	−	+	N	↓	N	−
Vasculitis in SLE	+	−	+	+	+	+	−	+	↓	↓	↓	−
Sjögren's syndrome	+	−	+	+	NR	NR	−	NR	N	↓	↓	−
Mixed cryoglobulinaemia	+	−	+	+	NR	NR	−	NR	↓	↓	N	−
Mononuclear vasculitis												
Urticarial vasculitis	−	NR	−	−	−	NR	NR	NR	N	N	N	−
Erythema multiforme	−	NR	−	−	NR	NR	NR	NR	N	N	N	−
Behçet's syndrome	+	NR	−	−	NR	NR	−	NR	N	N	N	−

IgG- and IgA-Cx = IgG- and IgA-containing immune complexes; cryo = cryoglobulins; RF = rheumatoid factor; p-ANCA = perinuclear antineutrophil cytoplasmic antibodies; c-ANCA = cytoplasmic ANCA. NR = not reported; N = normal; ↓ = lowered.

the alternative pathway is frequently associated with thrombocytopenia and disseminated intravascular coagulation. Hypocomplementaemia is also common in patients with vasculitis associated with rheumatoid arthritis or systemic lupus erythematosus, but not in such patients without vasculitis. This has been attributed to the relatively small complexes found in patients with rheumatoid arthritis. In patients who develop so-called rheumatoid vasculitis, C4 levels can become severely depressed, and levels correlate with disease activity (Scott et al. 1981). Patients with systemic lupus erythematosus may show a decrease in C3 levels in association with an exacerbation, but the predictive value of this finding is low. The complement profile in patients with vasculitis associated with Sjögren's syndrome has not been studied systematically, but patients with vasculitis and central nervous system disease have been reported to show activation of the terminal complement sequence (Alexander et al. 1988). Levels of complement components C1, C4, and C2 are frequently low in mixed cryoglobulinaemia, but C3 levels are usually normal (Table 3). These levels do not correlate with the presence of purpura, arthralgia, fever, or abdominal pain (Tarantino et al. 1978). The lack of C3 consumption has been attributed to the effects of C4b binding protein (Haydey et al. 1980).

Immune elimination

The low incidence of vasculitis in patients suffering from various conditions associated with circulating immune complexes has been attributed to the efficient removal of pathogenic complexes system. The function of this system can now be measured in terms of removal of immune complexes by the spleen. This technique depends upon the sensitization of ^{51}Cr-labelled autologous erythrocytes with IgG anti-Rhesus antibodies at a density at which complement fixation does not occur (Frank et al. 1979). In initial studies, in patients with systemic lupus erythematosus, this clearance rate showed an inverse correlation with the level of C1q-binding immune complexes and disease activity (Frank et al. 1979). Later studies confirmed the correlation with disease activity, but not with immune complex levels (Parris et al. 1982; Van der Woude et al. 1984). Dambyant et al. (1982) found an

Table 4 Deposit formation and histological abnormalities in blood vessels of patients with small vessel vasculitis

	Deposit formation				Vascular infiltrate						Necrosis
	IgM	IgG	IgA	C3	PMN	Eosinophils	Mast cells	T cells	Macrophages	Giant cells	
Leucocytoclastic vasculitis											
Hypersensitivity angiitis	−	−	+	+	+	+	+	−	−	−	+
Anaphylactoid purpura	+	−	+	+	+	−	−	−	−	−	+
Hypocomplementaemic urticarial vasculitis	+	+	−	+	+	+	−	+	−	−	−
Postinfectious vasculitis	+	+	+	+	+	−	−	−	−	−	+
Rheumatoid vasculitis	+	+	+	+	+	−	−	−	−	−	+
Vasculitis in SLE	+	+	+	+	+	−	−	+	+	−	+
Sjögren's syndrome	+	+	+	+	+	−	−	+	+	−	+
Mixed cryoglobulinaemia	+	+	+	+	+	−	−	−	+	−	+
Mononuclear vasculitis											
Normocomplementaemic urticarial vasculitis	−	−	−	−	+	−	−	+	−	−	−
Erythema multiforme	+	−	−	+	−	−	−	+	+	−	−
Behçet's syndrome	−	−	−	−	−	−	−	+	+	+	−

PMN = polymorphonuclear leucocytes.

accelerated clearance in patients with necrotizing vasculitis. Similar studies have been performed using soluble radiolabelled IgG aggregates (Lobatto et al. 1988): although patients with systemic lupus erythematosus had a prolonged clearance of these aggregates, the $T_{\frac{1}{2}}$ did not correlate with immune complex levels or with disease activity. Thus there is little evidence to suggest that immune complex diseases in patients develop due to saturation of the mononuclear phagocyte system, as has frequently been proposed. Interestingly, erythrocytes from patients with systemic lupus erythematosus bind signficantly fewer aggregates than erythrocytes from control individuals (Lobatto et al. 1988). A lower density of CR1 receptors in these patients has been reported (Miyakawa et al. 1981), but it is still unclear whether this is a primary defect or acquired as a result of masking by cell-bound immune complexes.

The role of vascular permeability in deposition of complexes

Several investigators have shown that immune complexes circulating in patients become localized in vessel walls following local injection of histamine (Braverman and Yen 1975; Gower et al. 1977; Jorizzo et al. 1983) or by exposure of the skin to cold (Eady et al. 1981).

The sequence of events leading to the development of leucocytoclastic vasculitis has been studied extensively in a patient with cutaneous vasculitis provoked by cold or by stroking (dermographism) (Soter et al. 1978). Five minutes after provocation hypogranulation of mast cells was seen, with extrusion of granules. At 30 min a perivascular infiltrate of neutrophils could be observed together with deposition of fibrin and endothelial cell swelling and necrosis. At 1 h eosinophils appeared in the infiltrate, and the number of neutrophils and the amount of fibrin increased. In this patient mast cells seemed to play a primary role in the development of the vasculitis.

Patients with leucocytoclastic vasculitis are frequently hypocomplementaemic and positive for antibodies and antibodies to Ro-antigen; these serological abnormalities are not present in patients with mononuclear vasculitis. In Sjögren's syndrome leucocytoclastic vasculitis in cutaneous venules can occur adjacent to mononuclear vasculitis (Molina et al. 1985). In essential mixed cryoglobulinaemia leucocytoclastic vasculitis of small vessels and mononuclear vasculitis of small and medium-sized vessels can occur in the same patient (Vital et al. 1988) This overlap between leucocytoclastic and mononuclear vasculitis demonstrates that two histological types may occur as the expression of the same disease. As in experimental serum sickness, the vasculitis frequently starts with an acute infiltrate of polymorphonuclear leucocytes followed by infiltration with mononuclear leucocytes.

Haemodynamic factors

Leucocytoclastic vasculitis frequently develops in areas with a high hydrostatic pressure in capillaries and venules. It frequently affects dependent parts of the body such as the lower limbs, or areas that experience pressure such as the buttocks. Local vasculitis has also been observed exclusively in the hand of a dialysis patient at the side of the arteriovenous fistula (unpublished observation). It is not unusual to see new vasculitic lesions develop after a sphygmomanometer cuff has been applied. These clinical observations suggest that haemodynamic factors play a role in the pathogenesis of vasculitis, possibly by creating favourable conditions for the localization of circulating immune complexes in vessel walls.

The inflammatory reaction

The inflammatory reaction in leucocytoclastic vasculitis is generally assumed to depend on complement activation, and this is certainly the case in acute serum sickness. However, the situation in man may be more complicated. Patients with rheumatoid arthritis may have deposits of immunoglobulins and C3 in cutaneous vessel walls, without signs of vasculitis (Conn et al. 1976). A striking observation was made in a Dutch family with inherited C3 deficiency, two members of which developed classic leucocytoclastic vasculitis. In the acute phase IgM- and IgA-containing complexes could be demonstrated in the circulation together with vascular deposits lacking C3 (Roord et al. 1983). Factors other than C5a may therefore play a role in chemotaxis and in the inflammatory reaction of leucocytoclastic vasculitis (see above).

Antibody-mediated vasculitis

Autoantibodies directed against endothelial antigens have recently been detected in sera from patients with systemic lupus erythematosus (see Table 7) (Clines et al. 1984).

Serum IgG from 88 per cent of patients with active lupus bound to venous endothelium cultured from umbilical cords (Cines et al. 1984). This binding also occurred with isolates 7 S. IgG was first isolated lacking the Fc-part of the antibody, suggesting true antiendothelial activity of a 7 S IgG antibody. However, the sera also contained high molecular weight IgG fractions and IgG aggregates that can bind to endothelial cells, particularly in the presence of complement (Cines et al. 1984), leading to the binding of C3 to endothelial cells. Binding of IgG aggregates stimulates the release of the prostaglandins 6-keto $F_{1\alpha}$ and prostacyclin. Platelets did not adhere to these endothelial cell cultures unless they had been preincubated with IgG aggregates. Surface molecules of endothelial cells have received recent attention, and in addition to transplantation antigens such as ABO, HLA, and EM alloantigens, they may also express Fc, C3, and C1q receptors (Ryan et al., 1981). Theoretically, binding to these receptors could be the first step in the sequence by which IgG aggregates and IgM- and IgG-containing immune complexes cause injury to the vessel wall.

It has been known for many years that sera from patients with hypocomplementaemic urticarial vasculitis contain C1q-precipitins with a sedimentation coefficient of 7 S (Marder et al. 1984). These precipitins do not bind C1q by their Fc region but by the F(ab)$_2$ part of the molecule, and are probably antibodies directed against the collagen-like region of C1q (see Table 7). Anti-C1q antibodies have also been found in patients with systemic lupus erythematosus (Uwatoko and Mannik 1988) and have been correlated with nephritis, dermatitis, hypocomplementaemia, anti-ds DNA antibodies, and the presence of immune complexes detected by the solid-phase C1q binding assay. These observations suggest that the results of C1q-binding assays will have to be reconsidered as they can no longer be interpreted as evidence of circulating immune complexes. Further studies are required to distinguish between the pathogenetic role of C1q-antibodies and that of C1q-binding immune complexes (Hack and Belmer 1986).

Other antibodies have also been reported to correlate with the development of vasculitis. There is, however, no evidence to suggest that these antibodies are directed at endothelial antigens or can bind to endothelium, either directly or indirectly. Such antibodies include anti-Sm antibodies in systemic lupus erythe-

Table 5 Immunological abnormalities in the circulation of patients with polyarteritis or granulomatous vasculitis

	Immune complexes			Antibodies					Complement			Eosinophilia
	IgG-Cx	IgA-Cx	Cryo	RF	C1q	Endoth	c-ANCA	p-ANCA	C1q	C4	C3	
Necrotizing arteritis												
Polyarteritis nodosa	+	−	+	−	NR	−	+	+	N	N	N	+
Kawasaki's disease	+	+	+	−	NR	+	NR	NR	N	N	N	−
Granulomatous vasculitis												
Wegener's granulomatosis	−	−	−	+	NR	−	+	+	N	N	N	N
Churg-Strauss syndrome	−	−	−	+	NR	−	+	NR	N	N	N	+
Relapsing polychondritis	−	NR	−	+	NR	−	NR	NR	N	N	N	N
Lymphomatoid granulomatosis	−	−	−	−	NR	−	−	−	N	N	N	−
Giant cell arteritis	−	−	−	−	NR	−	NR	NR	N	N	N	−
Takayasu's disease	−	−	−	−	NR	−	NR	NR	N	N	N	−

IgG- and IgA-Cx = IgG- and IgA-containing immune complexes; cryo = cryoglobulins; RF = rheumatoid factor; p-ANCA = perinuclear antineutrophil cytoplasmic antibodies; c-ANCA = cytoplasmic ANCA. NR = not reported; N = normal.

matosus patients (Beaufils *et al.* 1983), antibodies to the Ro- and SSA-antigen in Sjögren's syndrome (Reichlin and Wasicek 1983), and antilaminin antibodies in patients with leucocytoclastic vasculitis accompanied by hepatitis, leucopenia, and circulating anticardiolipin antibodies (Lassoned *et al.* 1988). Circulating antibodies as well as immune complexes may therefore play a role in the pathogenesis of systemic vasculitis.

Mononuclear vasculitides

Mononuclear vasculitis also occurs in normocomplementaemic urticarial vasculitis, erythema multiforme, and Behçet's syndrome (Tables 3 and 4). C1q-binding immune complexes and hypocomplementaemia are uncommon in erythema multiforme (Table 3) and IgG or IgA deposits are not found (Table 4). In Behçet's syndrome the common underlying lesion is vasculitis, with lymphocytes, monocytes, and some mast cells present around venules and arterioles. Vascular changes are characterized by a perivascular infiltration with mononuclear cells without necrosis (Table 4). In larger vessels the infiltrate usually starts in the adventitia and consists mainly of lymphocytes. In later stages macrophages accumulate and infiltration of the tunica media occurs.

It is unclear whether mononuclear vasculitides should be considered as a variant of antibody- or immune complex-mediated leucocytoclastic vasculitis, or as an expression of cell-mediated immunity. The occasional finding of deposits of immunoglobulins and C3 in erythema multiforme and Behçet's syndrome might suggest a humoral immune mechanism. Serum immune complexes are absent or present in very low levels, while complement levels are normal (Table 3). The presence of T lymphocytes in the perivascular infiltrate supports the idea that these lesions are caused by T cell-mediated cellular immunity, but direct evidence is lacking.

Necrotizing polyarteritides

The pathogenesis of necrotizing polyarteritides is still controversial. Circulating immune complexes are found less frequently and less consistently than in leucocytoclastic vasculitis (Table 5). The immune complexes may contain HBsAg, particularly in intravenous drug abusers, but this antigen does not seem to be important in other patients. Serum C3 levels can be low, but are frequently normal; peripheral blood eosinophils are frequently increased. Cytoplasmic and perinuclear antineutrophil cytoplasmic antibodies have been found in patients with microscopic polyarteritis and idiopathic rapidly progressive glomerulonephritis (Falk and Jenette 1988); some, but not all, of the latter are directed against myeloperoxidase.

Histologically, the lesions are characterized by infiltration of the intima and media with polymorphonuclear leucocytes, with fragmentation of the internal elastic membrane and fibrinoid necrosis (Table 6). Lymphocytes, macrophages, and giant cells are not present and immunofluorescence studies are contradictory; some investigators have shown vascular deposits of immunoglobulins and C3, whereas others have not.

Polyarteritis nodosa is a necrotizing polyarteritis which occurs predominantly in medium-sized muscular arteries (see Chapter 4.5.3). The lesions are segmental and in different stages of development. The predilection of these lesions for the ostia of branches and bifurcations has been explained by the turbulence and shear stress at those sites, which create favourable conditions for the deposition of circulating immune complexes. The possible role of HBsAg in the pathogenesis of polyarteritis (Shusterman and London 1984) has already been discussed.

Kawasaki's disease (also known as mucocutaneous lymph node syndrome) is another vasculitic disease which is interesting from a pathogenetic standpoint. It is an acute febrile illness of unknown aetiology which mainly affects infants and is characterized clinically by diffuse mucosal inflammation, indurative oedema of hands and feet, a rash, and lymphadenopathy. The coronary arteries are the predominant sites of the disease in about 15 to 20 per cent of children, although other arteries may be affected. Numerous immunological abnormalities have been reported in this disease, including high levels of circulating immune complexes, increased numbers of CD_4^+ helper cells, decreased number of CD_8^+ cytotoxic/suppressor cells, polyclonal B-cell activation, impaired granulocyte chemotaxis, and production of a factor that causes platelet aggregation. IgM antibodies which react against venous endothelium (Table 7) have been reported (Leung *et al.* 1986). The exact nature of the target antigen is unknown, but it cannot be induced on autologous fibroblasts.

In the acute state of the disease, polymorphonuclear leucocytes and plasma cells form an intense perivascular infiltrate, and there is inflammation of the intima of coronary arteries and the

4.5.1 Pathogenesis of angiitis

Table 6 Vascular abnormalities in patients with polyarteritis or granulomatous vasculitis

	PMN	Eosinophils	Mast cells	Plasma cells	T cells	Macrophages	Giant cells	Granuloma	Necrosis
Necrotizing arteritis									
Polyarteritis nodosa	+	+	−	−	−	−	−	−	+
Kawasaki's disease	+	−	−	+	+	+	−	−	+
Granulomatous vasculitis									
Wegener's granulomatosis	+	+	−	−	+	+	+	+	+
Churg-Strauss syndrome	−	++	−	+	+	+	+	+	+
Relapsing polychondritis	−	−	−	−	+	+	+	+	+
Lymphomatoid granulomatosis	−	−	−	−	++	+	−	−	−
Giant cell arteritis	−	−	−	−	+	+	+	+	−
Takayasu's disease	−	−	−	+	+	+	+	+	+

PMN = polymorphonuclear leucocytes.

vasa vasorum. After about 10 days the acute infiltrate is gradually replaced by plasma cells and lymphocytes (Table 6). Panvasculitis and the coronary aneurysms develop 12 to 18 days after the onset of disease. There are relatively low levels of circulating immune complexes in Kawasaki's disease, and their presence does not correlate with the development of coronary complications (Melish 1987), and it is doubtful whether they have any pathogenetic significance. Complement levels can be elevated but are rarely depressed.

Granulomatous vasculitides

The granulomatous vasculitides can be distinguished from the previous two groups by the absence of circulating immune complexes and hypocomplementaemia. Vascular deposits of immunoglobulins and complement are not usually present, and the vascular infiltrate consists of T lymphocytes, macrophages, and giant cells. Polymorphonuclear leucocytes are not prominent, although leucocytoclastic vasculitis can occur in patients with Wegener's granulomatosis.

The vascular infiltrate in Wegener's granulomatosis is predominantly composed of T lymphocytes and macrophages or histiocytes. Arterial and venous lesions include mural microabscesses composed of aggregates of neutrophils surrounded by activated lymphocytes and palisading histiocytes.

Table 7 Antibodies associated with systemic vasculitis

1. Antibodies directed against endothelial antigens
   (a) Complement-fixing IgM antibodies directed against endothelial antigens induced by γ-interferon, in Kawasaki's disease
   (b) Complement-fixing IgG antibodies directed against endothelial antigens in SLE
2. Antibodies correlating with manifestations of vasculitis
   (a) Anti-C1q antibodies in patients with SLE and hypocomplementaemic urticarial vasculitis
   (b) Anti-Sm antibodies in SLE
   (c) Antibodies to Ro (SSA) in Sjögren's syndrome
   (d) Autolaminin antibodies in patients with SLE-like syndrome
   (e) IgG antineutrophil cytoplasmic antibodies (ANCA) in Wegener's granulomatosis and systemic vasculitis
   (f) IgA anticytoplasmic antibodies in Henoch-Schönlein purpura

The CD_4^+ helper/inducer lymphocytes are present in larger numbers than the CD_8^+ cytotoxic/suppressor T lymphocytes. The extravascular lesions also consist of lymphocytes and epithelioid histiocytes including multinucleated giant cells (Mark et al. 1988). Reactions to tests with Tuberculin and Candida antigens, and the macrophage migration inhibition assay are frequently normal in these patients.

Relapsing polychondritis resembles Wegener's granulomatosis (McAdam et al. 1976), but may also show overlap with systemic lupus erythematosus, rheumatoid arthritis, Sjögren's syndrome, and temporal arteritis (Chang-Miller et al. 1987). The vascular lesions are characterized by granulomatous vasculitis, with infiltration by lymphocytes, macrophages, plasma cells, and giant cells. Sera from these patients contain autoantibodies directed against type II collagen, and antibody titres correlate with disease activity (Foidart et al. 1978). The disease has been reported to be transiently induced in neonates by placental transfer of these antibodies (Arundell and Haserick 1960), suggesting it is mediated by an IgG antibody.

The vascular lesions in lymphomatoid granulomatosis are different from those in the forms of granulomatous vasculitides previously described. The vascular infiltrate consists predominantly of lymphoreticular cells with features of immunoblasts, lymphocytes, and plasma cells (Foley et al. 1987; Gaulard et al. 1988). Macrophages are sparse, and epithelioid cells and giant cells are uncommon. Necrosis is not present. About 12 per cent of these patients will develop a malignant lymphoma. These data taken together suggest that lymphomatoid granulomatosis is mediated by T lymphocytes, which are either malignant from the start or are easily transformed into a malignant clone.

The infiltrating cells in giant cell arteritis consist of lymphocytes, macrophages, giant cells, and plasma cells (Ettlinger et al. 1978). These cells may concentrate in a perivascular location, but more often infiltrate all layers of the affected large or medium sized artery. The majority of lymphocytes carry the cell surface marker of T-helper cells (Banks et al. 1983), and most express HLA-DR, suggesting that they are activated T cells (Andersson et al. 1988). The giant cells may contain ingested fibres of the vascular elastic membrane. There are no observations that provide clues to the pathogenesis of giant cell arteritis. No cellular immune responses directed against normal or diseased arterial wall antigens have been detected. The smooth muscle cells do not express HLA-DR antigens. The presence of circulating immune complexes is controversial, and immunoglobulin deposits are only seen in a small number of patients. The other serological and histological abnormalities do not support an immune complex-mediated pathogenesis and it is more likely that cell-mediated immune mechanisms play a role.

Takayasu's disease is characterized by proliferative or granulomatous lesions in the aortic and pulmonary branches, including the subclavian, the carotid, the mesenteric, and the renal arteries (Lupi-Herrera et al. 1977). The proliferative lesions consist of lymphocytes and plasma cells, whereas the granulomatous lesion consists of coagulation necrosis surrounded by epithelioid histiocytes and Langhans-type giant cells. These changes have a close relationship to the vasa vasorum and spread from the adventitia over the entire media (Nasu 1962; Hall et al. 1985). Sera from these patients do not contain circulating immune complexes, rheumatoid factor, or antinuclear antibodies (Shelhamer et al. 1985). Individuals with HLA-DR4 and the MB3 determinant of the HLA-DQ locus seem to be genetically susceptible to this disease. More intensive immunological studies will have to be performed in order to be able to define the pathogenesis.

Autoantibodies and the pathogenesis of angiitis

Sera from about 50 per cent of patients with Wegener's granulomatosis contain rheumatoid factor activity, whereas cryoglobulins and circulating immune complexes are uncommon and serum complement concentrations are normal (Table 5). Recently, the presence of antibodies directed against components of azurophilic granules present in granulocytes (antineutrophil cytoplasmic antibodies) have been studied in patients with vasculitis and glomerulonephritis (Van der Woude et al. 1985). The importance of these antibodies in vasculitis was originally established in studies with sera from patients with Wegener's granulomatosis (see also Chapter 4.5.3). Three major types of autoantibodies that react with cytoplasmic constituents of neutrophils have been recognized. Cytoplasmic antineutrophil cytoplasmic antibody is seen as a diffuse, granular staining with central accentuation of the cytoplasm of neutrophils and some monocytes, but not of lymphocytes, on standard indirect immunofluorescence ethanol-fixed leucocytes. Perinuclear antineutrophil cytoplasm antibody produces a perinuclear staining pattern on immunofluoresence. Antimyeloperoxidase antibodies (Falk and Jenette 1988) are usually of the perinuclear type, but may sometimes be cytoplasmic. IgA antineutrophil cytoplasmic antibodies which react with an acidic extract of neutrophils were discovered in Henoch-Schönlein patients (Van den Wall Bake et al. 1987). This antibody cannot be detected using sera in the standard immunofluorescence assay. However, these sera do produce a positive cytoplasmic staining. Immunofluorescence assays of cytoplasmic antineutrophil cytoplasmic antibody have clinical value in the diagnosis of Wegener's granulomatosis. Its positive predictive value for Wegener's granulomatosis is around 90 per cent. The value of this assay in monitoring disease activity seems to be limited; relapses usually occur when titres are high but increasing titres are not necessarily followed by relapses. The occurrence of cytoplasmic antineutrophil cytoplasmic antibody in patients with other forms of vasculitis and idiopathic necrotizing and crescentic glomerulonephritis has cast some doubt on the specificity of these antibodies for Wegener's disease. Cytoplasmic antineutrophil cytoplasmic antibodies can also be found in patients with microscopic polyarteritis (see Chapter 4.5.3), but there has been some discussion about the distinction between this form of vasculitis and Wegener's disease, since microscopic polyarteritis is not a term commonly used outside the United Kingdom.

Wegener's disease may initially present as apparently isolated glomerulonephritis, and other manifestations of the disease may not become apparent for several years. Most studies agree that more than 90 per cent of patients with Wegener's granulomatosis are positive for cytoplasmic antineutrophil cytoplasmic antibodies, while perinuclear antibodies are present in only about 10 per cent. Conversely, most patients with idiopathic necrotizing and crescentic glomerulonephritis are positive for the perinuclear antibodies. Cytoplasmic and perinuclear antibodies are present in roughly equal numbers of patients with periarteritis nodosa. It has therefore been proposed (Falk and Jenette 1989) that these patients form a pathological continuum ranging from purely renal to widespread systemic vascular injury.

There are no animal models of Wegener's disease and speculation about its pathogenesis and aetiology can only be based on clinicopathological and laboratory findings in patients. Friedrich Wegener himself postulated that a microbial infection or hyperreaction to an extrinsic agent was the cause of the disease (Wegener 1939). Most relapses in Wegener's disease are associated with infection by common pathogens, and several groups claim that antibiotic therapy with sulphomethoxazole–trimethoprim has a beneficial effect on the course of the disease. Infection of the respiratory tract may trigger relapses by releasing myeloid lysozomal enzymes that elicit a humoral and cellular autoimmune response. Antineutrophil antibodies could theoretically play a direct role in pathogenesis by causing granulocyte lysis and subsequent tissue necrosis. Since cytoplasmic antineutrophil cytoplasmic antibodies may bind to both a 29 kDa serine protease within the neutrophil and to a much larger antigen present in neutrophil supernatant and sputum (Daha et al. 1989) it is attractive to speculate that the target antigen may be both the proteinase and the complex formed by the proteinase and its inhibitor. Proteinase-antibody complexes may be protected against inactivation by inhibitors and transported through the body to areas such as the kidneys. If the proteinases in these complexes retain their enzymatic activity, they might cause local damage. It has recently been shown (Daha et al. 1989) that T lymphocytes from several patients with Wegener's granulomatosis are stimulated to proliferate by this antigen, in contrast to lymphocytes from healthy controls. In the light of these findings, it is attractive to speculate that T cells directed against this antigen might invade lesions which contain proteinase-antibody complexes. The presence of tissue necrosis and such proteinase-antibody complexes might then induce granuloma formation.

Bibliography

Cochrane, C.G. and Koffler, D. (1973). Immune complex disease in experimental animals and man. *Advances in Immunology*, **16**, 185–264.

Dixon, F.J. (1978). Experimental serum sickness. *Immunological Diseases* (Ed. M. Samter), pg 253, Little, Brown and Co. Boston.

McDougal, J.S. and MacDuffie, F.C. (1985). Immune complexes in man: detection and clinical significance. *Advances in Clinical Chemistry*, **24**, 1–60.

Theofilopoulos, A.N. and Dixon, F.J. (1979). The biology and detection of immune complexes. *Advances in Immunology*, **28**, 89–220.

References

Alexander, E.L., Provoost, T.T., Sanders, M.E., Frank, M.M., and Joiner, K.A. (1988). Serum complement activation in central nervous system disease of Sjögren's syndrome. *American Journal of Medicine*, **85**, 513–8.

Andersson, R., Hansson, G.K., Söderström, T., Jonsson, R., Beugtsson, B.Å., and Nordborg, E. (1988). HLA-DR expression in the vascular lesion and circulating T-lymphocytes of patients with giant cell arteritis. *Clinical and Experimental Immunology*, **73**, 82–7.

Arundell, F.W. and Haserick, J.R. (1960). Familial chronic atrophic polychondritis. *Archives of Dermatology*, **82**, 439–41.

Banks, P.M., Cohen, M.D., Ginsburg, W.W., and Hunder, G.G. (1983). Immunohistologic and cytochemical studies of temporal arteritis. *Arthritis and Rheumatism*, **26**, 1201–7.

Beaufils, M., Kouki, F., Mignon, F., Camus, J-P., Morel-Maroyer, L., and Richet, G. (1983). Clinical significance of anti-Sm antibodies in systemic lupus erythematosis. *American Journal of Medicine*, **74**, 201–5.

Braverman, I.M. and Yen, A. (1975). Demonstration of immune complexes in spontaneous and histamine-induced lesions and in normal skin of patients with leukocytoclastic angiitis. *Journal of Investigative Dermatology*, **64**, 231–9.

Brentjes, J.R., et al. (1975). Experimental chronic serum sickness in rabbits that received daily multiple and high doses of antigen: a systemic disease. *Annals of the New York Academy of Sciences*, **254**, 603–13.

Chang-Miller, A., et al. (1987). Renal involvement in relapsing poychondritis. *Medicine*, **66**, 202–17.

Churg, A. (1983). Pulmonary angiitis and granulomas revisited. *Human Pathology*, **14**, 868–83.

Cines, D.B., Lyss, A.P., Reeber, M., Bina, M., and DeHoratius, J. (1984). Presence of complement-fixing anti-endothelial cell antibodies in systemic lupus erythematosus. *Journal of Clinical Investigation*, **73**, 611–25.

Cochrane, C.G. (1963). Studies on the localization of circulating antigen-antibody complexes and other macromolecules in vessels. II. Pathogenic and pharmacodynamic studies. *Journal of Experimental Medicine*, **118**, 503–13.

Cochrane, C.G. and Hawkins, D. (1968). Studies on circulating immune complexes. III. Factors governing the ability of circulating complexes to localize in blood vessels. *Journal of Experimental Medicine*, **127**, 137–54.

Conn, D.L., Schroeter, A.L., and McDuffie, F.C. (1976). Cutaneous vessel immune deposits in rheumatoid arthritis. *Arthritis and Rheumatism*, **19**, 15–9.

Daha, M.R., Kramps, J.A., Schrama, E., Van Es, L.A., and Van der Woude, F.J. (1989). Isolation from purulent sputum of an antigen reactive with antibodies in Wegeners serum. In Proceedings of the 2nd International ANCA workshop (ed. F. J. van der Woude). *Netherlands Journal of Medicine*, in press.

Dahlén, S-E., et al. (1981). Leukotrienes promote plasma leakage and leukocyte adhesion in postcapillary venules: *In vivo* effects with relevance to the acute inflammatory response. *Proceedings of the National Academy of Sciences USA*, **78**, 3887–91.

Dambyant, C., Thivolet, J., Viala, J.J., Ville, D., and Boyer, J. (1982). Clearance mediated by splenic macrophage membrane receptors for immune complexes in cutaneous vasculitis. *Journal of Investigative Dermatology*, **78**, 194–9.

Diesselhoff-den Dulk, M.M.C., Croffon, R.W., and Van Furth, R. (1979). Origin and kinetics of Kupffer cells during an acute inflammatory response. *Immunology*, **37**, 7–14.

Drüeke, T., et al. (1980). Hepatitis B antigen-associated periarteritis nodosa in patients undergoing long-term hemodialysis. *American Journal of Medicine*, **68**, 86–90.

Eady, R.A.J., Keahey, T.M., Sibbold, R.G., and Black, A.K. (1981). Cold urticaria with vasculitis: report of a case with light and electron microscopic, immunofluoroscene and pharmacological studies. *Clinical and Experimental Dermatology*, **6**, 355–66.

Ettlinger, R.E., Hunder, G.G., and Ward, L.E. (1978). Polymyalgia rheumatica and giant cell arteritis. *Annual Review of Medicine*, **29**, 15–22.

Falk, R.J. and Jennette, J.C. (1988). Anti-neutrophil cytoplasmic autoantibodies with specificity for myeloperoxidase in patients with systemic vasculitis and idiopathic necrotizing and crescentic glomerulonephritis. *New England Journal of Medicine*, **318**, 1651–7.

Foidart, J.-M., et al. (1978). Antibodies to type II collagen in relapsing polychondritis. *New England Journal of Medicine*, **299**, 1203–6.

Foley, J.F., Linder, J., Koh, J., Severson, G., and Purtilo (1987). Cutaneous necrotizing granulomatous vasculitis with evolution to T cell lymphoma. *American Journal of Medicine*, **82**, 839–44.

Frank, M.M., Hamburger, M.I., Lawley, T.J., Kimberley, R.P., and Platz, P.H. (1979). Defective reticuloendothelial system Fc-receptor function in systemic lupus erythematosus. *New England Journal of Medicine*, **300**, 518–23.

Gaulard, Ph., et al. (1988). Lethal midline granuloma (polymorphic reticulosis) and lymphoid granulomatosis. *Cancer*, **62**, 705–10.

Geltner, D., Franklin, E.C., and Frangione, B. (1980). Antiidiotypic activity in the IgM fractions of mixed cryoglobulins. *Journal of Immunology*, **125**, 1530–5.

Gephardt, G.N., Ahmad, M., and Tubbs, R.R. (1983). Pulmonary vasculitis (Wegener's granulomatosis). Immunohistochemical study of T and P cell markers. *American Journal of Medicine*, **74**, 100–4.

Germuth, F.G. and Pollack, A.D. (1967). Immune complex diseases. III. The granulomatous manifestations. *Johns Hopkins Medical Journal*, **12**, 254–62.

Ginsburg, C.H., et al. (1982). Antigen- and receptor-driven regulatory mechanisms. X. The induction and suppression of hapten-specific granuloma. *American Journal of Pathology*, **106**, 421–31.

Gower, R.G., Sams, W., Thorne, E.G., Kohler, P.F., and Claman, H.N. (1977). Leukocytoclastic vasculitis: sequential appearance of immunoreactants and cellular changes in serial biopsies. *Journal of Investigative Dermatology*, **69**, 477–84.

Gower, R.G., Sausker, W.F., Kohler, P.F., Thorne, G.E., and McIntosh, R.M. (1978). Small vessel vasculitis caused by hepatitis B virus immune complexes. *Journal of Allergy and Clinical Immunology*, **62**, 222–8.

Hack, C.E. and Belmer, A.J.M. (1986). The IgG detected in the C1q solid-phase immune complex assay is not always of immune complex nature. *Clinical Immunology and Immunopathology*, **38**, 120–8.

Hall, S., Barr, W., Lie, J.T., Stanson, A.W., Kazmier, F.J., and Hunder, G.G. (1985). Takayasu arteritis. A study of 32 North American patients. *Medicine*, **64**, 89–99.

Hart, M.N., Sodewasser, K.L., Cancilla, P.A., and DeBault, L.E. (1983). Experimental autoimmune type of vasculitis resulting from activation of mouse lymphocytes to cultured endothelium. *Laboratory Investigation*, **48**, 419–27.

Haydey, R.P., Patarroyo de Rojas, M., and Gigli, I. (1980). A newly described control mechanism of complement activation in patients with mixed cryogloblinemia (cryoglobulins and complement). *Journal of Investigative Dermatology*, **74**, 328–32.

Henson, P.M. and Cochrane, C.G. (1971). Acute immune complex disease in rabbits. The role of complement and of a leucocyte-dependent release of vasoactive amines from platelets. *Journal of Experimental Medicine*, **133**, 554–71.

Issekutz, A. (1981). Vascular response during acute neutrophilic inflammation. Their relationship to *in vivo* neutrophil emigration. *Laboratory Investigation* **45**, 435–41.

Kniker, W.T. and Cochrane, C.G. (1968). The localization of circulating immune complexes in experimental serum sickness. *Journal of Experimental Medicine*, **127**, 199–135.

Jorizzo, J.L., Daniels, J.C., Apisarnthanara, P., Gonzalez, E.B., and Cavallo, T. (1983). Histamine-triggered localized vasculitis in patients with seropositive rheumatoid arthritis. *Journal of the American Academy of Dermatology*, **9**, 845–51.

Kauffmann, R.H., et al. (1980). Circulating and tissue-bound immune complexes in allergic vasculitis: relationship between immunoglobulin class and clinical features. *Clinical and Experimental Immunology*, **41**, 459–70.

Kleinknecht, C., Levy, M., Peix, A., Broyer, M., and Courtecuisse, V. (1979). Membranous glomerulonephritis and hepatitis B surface antigen in children. *Journal of Pediatrics*, **95**, 946–52.
Lassoned, K., et al. (1988). Antinuclear autoantibodies specific for laminins. *Annals of Internal Medicine*, **108**, 829–33.
Leung, D.Y.M., Collins, T., Lapierre, L.A., Gaha, R.S., and Pober, J.S. (1986). Immunoglobulin M antibodies present in the acute phase of Kawasaki syndrome lyse cultured vascular endothelial cells stimulated by gamma interferon. *Journal of Clinical Investigation*, **77**, 1428–35.
Levinsky, R.J. and Barratt, T.M. (1979). IgA immune complexes in Henoch-Schöenlein purpura. *Lancet*, **ii**, 110–1103.
Levo, Y., Gorevic, P.D., Kassab, H.J., Zucker-Franklin, D., and Franklin, E.C. (1977). Association between hepatitis B virus and essential mixed cryoglobulinemia. *New England of Medicine*, **296**, 1501–4.
Lobatto, S., et al. (1988). Abnormal clearance of soluble aggregates of human immunoglobulin G in patients with systemic lupus erythematosus. *Clinical and Experimental Immunology*, **72**, 55–9.
Lupi-Herrera, E., Sanchez-Torres, G., Marcushamer, J., Mispireta, J., Horwitz, S., and Vela, J.E. (1979). Takayasu's arteritis. Clinical study of 107 cases. *American Heart Journal*, **93**, 94–103.
Majno, G. and Palade, G.E. (1961). Studies on inflammation. The effect of histamine and serotonin on vascular permeability: an electron microscopic study. *Journal of Biophysical and Biochemical Cytology*, **11**, 571–605.
Mannik, M. and Arend, W.P. (1971). Fate of preformed immune complexes in rabbits and rhesus monkeys. *Journal of Experimental Science*, **134**, 14S–31S.
Marder, R.J., Potempa, L.A., Verrier Jones, J., Toriumi, D., Schmid, F., and Gewurz, H. (1984). Assay, purification and further characterization of 7S C1q-precipitins (C1q-p) in hypocomplementemic vasculitis urticaria syndrome and systemic lupus erythematosus. *Acta Pathologica Microbiologica Scandinavica Section C Supplement*, **284**, 25–34.
Mark, E.J., Matsubara, O., Tan-Liu, M.S., and Fienberg, R. (1988). The pulmonary biopsy in the early diagnosis of Wegener's (pathergic) granulomatosis. *Human Pathology*, **19**, 1065–72.
McAdam, L.P., O'Haulen, M.A., Bleustone, R., and Pearson, C.M. (1976). Relapsing polychondritis: prospective study of 23 patients and a review of the literature. *Medicine*, **55**, 193–215.
McMahon, B.J., Heyward, W.L., Templin, D.W., Clement, D., and Lanier, A.P. (1989). Hepatitis B-associated polyarteritis nodosa in Alaskan Eskimos: clinical and epidemiologic features and long-term follow-up. *Hepatology*, **9**, 97–101.
Melish, M.E. (1987). Kawasaki sydrome: a 1986 perspective. *Rheumatic Disease Clinics of North America*, **13**, 7–17.
Miyakawa, Y., Yameda, A., and Kosaka, K. (1981). Defective immune adherence (C3b) receptor on red cells from patients with systemic lupus erythematosus. *Lancet*, **ii**, 493–7.
Molina, R., Provost, T.T., and Alexander, E.L. (1985). Two types of inflammatory vascular disease in Sjögren's syndrome. *Arthritis and Rheumatism*, **28**, 1251–8.
Moyer, C.F. and Reinisch, C.L. (1984). The role of vascular smooth muscle cells in experimental autoimmune vasculitis. *American Journal of Pathology*, **117**, 380–90.
Moyer, C.F., Strandberg, J.D., and Reinisch, C.L. (1987). Systemic mononuclear-cell vasculitis in MRL/Mp-1r/1pr mice. *American Journal of Pathology*, **127**, 229–42.
Nasu, T. (1962). Pathology of pulseless disease. *Angiology*, **14**, 225–42.
Nishi, T., Bhan, A.K., Collins, A.B., and McCluskey, R.T. (1981). Effect of circulating immune complexes on Fc and C3 receptors of Kupffer cells *in vivo*. *Laboratory Investigation*, **44**, 442–8.
Parris, T.M., Kimberly, R.P., Inman, R.D., McDougal, J.S., Gibofsky, A., and Christian, C.L. (1982). Defective Fc-receptor-mediated function of the mononulcear phagocyte system in lupus nephritis. *Annals of Internal Medicine*, **97**, 526–32.
Reichlin, M. and Wasicek, C.A. (1983). Clinical and biologic significance of antibodies to Ro/SSA. *Human Pathology*, **14**, 401–5.
Roord, J.J., et al. (1983). Inherited deficiency of the third component of complement associated with recurrent pyogenic infections, circulating immune complexes and vasculitis in a Dutch family. *Pediatrics*, **71**, 81–7.
Ryan, U.S., Schultz, D.R., and Ryan, J.W. (1981). Fc and C3 receptors on pulmonary endothelial cells: induction by injury. *Science*, **214**, 557–8.
Schifferli, J.A., Morris, S.M., Dash, A., and Peters, D.K. (1981). Complement-mediated solubilisation in patients with SLE nephritis or vasculitis. *Clinical and Experimental Immunology*, **46**, 557–64.
Schifferli, J.A., Ng, Y.C., and Peters, D.K. (1986). The role of complement and its receptors in the elimination of immune complexes. *New England Journal of Medicine*, **315**, 488–95.
Scott, D.G.I., Bacon, P.A., Allen, C., Elson, C.J., and Wallington, T. (1981). IgG rheumatoid factor, complement and immune complexes in rheumatoid synovitis and vasculitis: comparative and serial studies during cytotoxic therapy. *Clinical and Experimental Immunology*, **43**, 54–63.
Shelhamer, J.H., Volkman, D.J., Parrillo, J.E., Lawley, T.J., Johnston, M.R., and Fauci, A.S. (1985). Takayasu's arteritis and its therapy. *Annals of Internal Medicine*, **103**, 121–6.
Shusterman, N. and London, W.T. (1984). Hepatitis B and immune complex disease. *New England Journal of Medicine*, **310**, 43–6.
Soter, N.A., Mihon, M.C., Dvorak, H.F., and Austen, K.F. (1978). Cutaneous necrotizing venulitis: a sequential analysis of the morphological alterations occurring after mast cell degranulation in a patient with a unique syndrome. *Clinical and Experimental Immunology*, **32**, 46–58.
Tarantino, A., Anelli, A., Costantino, A., De Vecchi, A., Monti, G., and Massaro, L. (1978). Serum complement pattern in essential mixed cryoglobulinemia. *Clinical and Experimental Immunology*, **32**, 77–85.
Uwatoko, S. and Mannik, M. (1988). Low-molecular weight C1q-binding immunoglobulin G in patients with systemic lupus erythematosus consists of autoantibodies to the collagen-like region of C1q. *Journal of Clinical Investigation*, **82**, 816–4.
Van der Woude, F.J., et al. (1984). Reticuloendothelial Fc-receptor function in SLE patients. *Clinical and Experimental Immunology*, **55**, 473–80.
Van der Woude, F.J., et al. (1985). Auto-antibodies to neutrophils and monocytes: A new tool for diagnosis and a marker of disease activity in Wegener's granulomatosis. *Lancet*, **i**, 425–9.
Van Es, L.A., Kauffmann, R.H., and Valentijn, R.M. (1987). Renal manifestations of systemic disease. *Pediatric Nephrology* (2nd edn.) (ed. M. A. Holliday, T. M. Barrat, and R. C. Vernier), p. 492–498, Williams and Wilkins, Baltimore.
Vital, C., et al. (1988). Peripheral neuropathy with essential mixed cryoglobulinemia: biopsies from 5 cases. *Acta Neuropathologica*, **75**, 605–10.
Wegener, F. (1939). Uber eine eigenartige rhinogene Granulomatose mit besonderer Beteiligung des Arteriensystem und der Nieren. *Beiträge zur Pathologischen Anatomie und Allgemeinen Pathologie*, **102**, 37–68.
Wener, M.H., Mannik, M., Schwartz, M.M., and Lewis, E.J. (1987). Relationship between renal pathology and the size of circulating immune complexes in patients with systemic lupus erythematosus. *Medicine*, **66**, 85–97.
Westedt, M.L., Daha, M.R., Baldwin, W.M. III, Stijnen, Th., and Cats, A. (1986). Serum immune complexes containing IgA appear to predict erosive arthritis in a longitudinal study in rheumatoid arthritis. *Annals of Rheumatic Diseases*, **45**, 809–15.
Wilson, C.B. and Dixon, F.J. (1970). Antigen quantitation in experimental immune complex glomerulonephritis I. Acute serum sickness. *Journal of Immunology*, **105**, 279–290.

4.5.2 The nephritis of Henoch-Schönlein purpura

GEORGE B. HAYCOCK

Henoch-Schönlein purpura is one of a number of synonyms applied to the clinical syndrome comprising a characteristic skin rash, abdominal colic, joint pain and glomerulonephritis. Other names include anaphylactoid purpura, purpura rheumatica, and peliosis rheumatica: all of these incorporate the skin lesion in the name of the condition, and by conventional definition the rash is the *sine qua non* of the diagnosis. In a detailed account of 12 personal cases and 46 others collected from the literature, Gairdner (1948) argued that the condition should be known as the Schönlein-Henoch syndrome: the name order from historical precedence, and syndrome rather than purpura because the cutaneous lesion is not a 'pure' purpura, and indeed does not always show purpuric features. These arguments notwithstanding, Henoch-Schönlein purpura is by far the most commonly used term today: for this reason and, arguably, in the interest of euphony, I shall use it in this chapter.

Historical background

Isolated case reports of patients suffering from what was probably Henoch-Schönlein purpura appeared at the beginning of the 19th century (Willan 1808; Rook 1958) but the association between an erythematous or purpuric rash and joint pains was first clearly described by Schönlein (1837). Schönlein's former pupil, Henoch (1874), described four children with the combination of rash, colic, bloody diarrhoea, and joint pains and in a later report (1899) added haemorrhagic nephritis to the list of components of the syndrome, thus completing the modern definition of the disease. A little later Macalister (1906), in a review of 127 children with purpura seen at Guy's Hospital and affiliated institutions, identified 15 cases of 'Henoch's purpura'. A modern review of this report suggests that as many as 24 of Macalister's cases may have had Henoch-Schönlein purpura: he emphasized the frequency with which urinary abnormalities persisted after the acute illness had subsided, and indeed gave it as his opinion that: 'Henoch's purpura may really be considered as a variety of Bright's disease, and, therefore, the ultimate prognosis is usually unfavourable'. Around the turn of the last century, Osler published a series of papers on the condition, culminating in a review of 29 cases (1914) in which he emphasized the similarity between Henoch-Schönlein purpura and serum sickness. This was the first published suggestion that allergic or immunological mechanisms might be involved in the aetiology of the disease, although Frank (1915) independently introduced the term anaphylactoid into the terminology of the condition. The concept that Henoch-Schönlein purpura might be an allergic or idiosyncratic response to infection was first promoted by Glanzmann (1916).

The paper by Gairdner (1948), previously referred to, included a detailed and comprehensive description of the signs and symptoms of Henoch-Schönlein purpura which has never been bettered as a clinical account of the major features of the condition. Another excellent review of the disease was contributed by Ackroyd (1953), who distinguished it from other forms of allergic purpura due to foods, drugs, and infections. Allen *et al.* (1960) of the Boston Children's Hospital reported what is still the largest single group of patients with Henoch-Schönlein purpura to be described in detail, comprising 131 children aged between 6 months and 16 years (only the renal features of the 141 cases of Koskimies *et al.* (1981) were published). The size of the database allowed these authors to perform a detailed analysis of various clinical and other factors and this study, combined with that of Gairdner, provides a definitive overview of the clinical variations of Henoch-Schönlein purpura and the relative frequency of its various manifestations and presentations. The introduction in the 1960s of percutaneous renal biopsy as a routine investigative procedure in children paved the way for the independent and simultaneous publication of two large series (Meadow *et al.* 1972; Habib and Levy 1972) describing the histopathological features of the nephritis of Henoch-Schönlein purpura, and demonstrating the value, and indeed the indispensability, of histology as a predictor of outcome. Current research interest is focused mainly on the nature and pathophysiology of the underlying disease process and on treatment of the renal disease: progress has been somewhat disappointing in both of these areas.

Clinical features

Although Henoch-Schönlein purpura occurs at all ages and in both sexes, it is mainly a disease of early childhood, with most cases presenting under 10 years of age, and there is a male:female predominance of about 2:1. A recent history of an intercurrent infection, usually respiratory, is common (see below). Although any of the four major components of the syndrome (rash, joint pain, abdominal symptoms, and renal disease) may present in advance of the others, it is rare for the renal disease to do so. The most common combination is rash, arthralgia, and colic appearing simultaneously or within a day or two of each other, the nephritis either appearing later or being present at onset only in the form of urinary abnormalities. Knowledge that abdominal pain can precede other manifestations should lead the clinician to look carefully for other features of the syndrome, particularly the rash, in any child presenting with acute abdominal pain or bleeding. The disease is rare in adults: in a report from the United Kingdom Medical Research Council Glomerulonephritis Registry, Knight and Cameron (personal communication) discovered only 372 patients in the world literature, to which they added 47 (a further 16 patients, all with nephritis, have since been reported by Fogazzi et al. (1989)). Based on the United Kingdom figures, they estimate the incidence in adults over the age of 20 years to be only 1:10 000 000—admittedly a minimum estimate, since not all cases will have been reported, but clearly far lower than that in children. There are no consistent differences between the clinical characteristics of the disease in children and adults.

The disease is more common in the winter months, with a peak between January and March in the northern hemisphere (Allen et al. 1960; Cream et al. 1970; Levy et al. 1976; Neilsen 1988). The racial incidence is difficult to establish accurately, since many published reports either omit reference to it or are based on ethnically relatively uniform populations (Koskimies et al. 1974; Yoshikawa et al. 1987). However, in North America white subjects are much more commonly affected than black subjects: only one of the 131 cases of Allen et al. (1960) was black, and anecdotal reports from the United Kingdom support this impression. It is common in Japan: Yoshikawa et al. (1987) biopsied 128 children for Henoch-Schönlein purpura during a 10-year period in two children's departments in the Japanese cities of Kobe and Tokyo, which suggests an incidence at least as high as that seen in white Europeans and Americans. Figures for southern Asia (India, Pakistan, Bangladesh and Sri Lanka) are lacking, but the disease certainly occurs in children of this ethnic group resident in England, without obvious under- or overrepresentation as compared with the white population. Familial cases have occasionally been described. It is often stated that Henoch-Schönlein purpura is more common in the temperate zones of the world than in the tropics, although hard data on this are scarce.

The skin

The rash of Henoch-Schönlein purpura is distinctive, both in its distribution and in the nature of the lesions. It is initially urticarial rather than purpuric in character with small, wheal-like spots appearing on the extensor surfaces of the arms and legs, especially round the ankles and over the buttocks and elbows. The abdomen and chest are spared, as is the back above a horizontal line drawn between the iliac crests. Over the next day or two the lesions become more sharply defined, pinkish maculopapules which darken due to extravasation of blood within and around them: at this stage they no longer fade on pressure and the rash is correctly described as purpuric. The individual lesions are mostly less than 1 cm in diameter, but may coalesce to form larger discoloured patches; ecchymoses occur occasionally, sometimes associated with dermal necrosis and scarring (purpura necrotica), and haemorrhagic bullae are rare. At its height, the eruption is pleomorphic and consists of flattish pink or red papules with frankly purpuric macules within and between them. The lesions regress gradually, fading from purple to brown and disappearing over about 2 weeks. The distribution is generally symmetrical. The face and scalp are usually spared but are occasionally involved, especially in younger children. Lesions can be induced by mild trauma, such as measuring the blood pressure, during the acute phase of the disease, but an interval of about a day typically elapses between the provoking stimulus and the appearance of the eruption. This contrasts with the immediate purpura seen following the application of a sphygmomanometer cuff in a patient with thrombocytopenic purpura (positive Hess's test) on the one hand, and the acute 'wheal and flare' reaction of immediate hypersensitivity on the other. There may be a single episode of the rash with subsequent rapid resolution, but recurrent crops of lesions occur in other patients. In the large series of Allen et al. (1960) the average duration of the illness was 3.9 weeks with a range of 3 days to 2 years. About one-third of patients had symptoms for less than 2 weeks, another one-third from 2 to 4 weeks and the remaining third for more than 4 weeks. The skin rash preceded other manifestations of the disease in more than half of all cases.

Joint manifestations

Joint pain occurs in two-thirds of all cases, and is the presenting symptom in one-quarter. The large joints, especially the ankles and knees, are principally affected. Several joints are usually involved either simultaneously or sequentially, but sometimes only one is affected. The arthralgia varies from mild to moderately severe, rarely if ever mimicking the exquisite pain and tenderness of acute rheumatic fever. Swelling usually, but not always, accompanies the pain: clinical and radiological examination confirms that this is due to periarticular oedema without effusion or enlargement of the joint space. Arthralgia usually lasts for a shorter time than the skin rash but, like the rash, may recur with exacerbations of the disease. The dramatic response to salicylates characteristic of the true rheumatic joint is not seen. Affected joints are never permanently damaged.

Oedema

Generalized hypoproteinaemic oedema may be seen in Henoch-Schönlein purpura as a feature of the nephrotic syndrome or protein-losing enteropathy (see below). In addition, localized oedema occurs in patients with a normal plasma albumin concentration. The periarticular swelling referred to in the previous paragraph is an example of this, but it also occurs periorbitally and in the scalp: a dramatic example is illustrated in the paper by Allen et al. (1960). Diffuse, bruised looking swelling of the dorsum of the hand and foot, merging into periarticular oedema of the corresponding wrist or ankle joint, is often observed. The mechanism underlying this type of oedema is not well understood, but it is presumably a manifestation of altered small vessel permeability due to vasculitis.

Abdominal symptoms

Abdominal symptoms occur in the majority of patients: 78 per cent of those reported in the series of Allen et al. (1960) and Meadow et al. (1972) which between them comprise 219 cases, had abdominal pain, ileus, melaena, haematemesis, or a combination of these. The abdominal pain is colicky, frequently severe, and may mimic an abdominal emergency, especially when it is the first manifestation of the syndrome. Ten of the 219 children previously referred to had a laparotomy before Henoch-Schönlein purpura had been diagnosed: one of these had an intussusception, but no surgical condition was present in the other nine. Intestinal bleeding of some degree is present in up to 80 per cent of all cases, varying from a positive stool test for occult blood in an asymptomatic patient to life-threatening haemorrhage requiring blood transfusion. Intussusception in Henoch-Schönlein purpura was first described by Vierhuff (1893). A series of 20 cases was reviewed by Wolfsohn in 1947, who observed that the interval between the onset of Henoch-Schönlein purpura and the development of intussusception varied from 3 days to 5 years; he also emphasized the difficulty of

4.5.2 The nephritis of Henoch-Schönlein purpura

distinguishing between intussusception and intestinal colic on clinical grounds alone in these patients. Intussusception occurs frequently enough (eight of 219 in the series of Allen *et al.* and Meadow *et al.*) to warrant a high index of suspicion in all patients with persistent or unusually severe abdominal pain or bleeding; however, the use of modern imaging techniques including ultrasonography should prevent children without this complication being subjected to unnecessary laparotomy.

Chronic intestinal obstruction due to ileal stricture has occasionally been seen as a late complication of Henoch-Schönlein purpura (Young 1964; Lombard *et al.* 1986). The mechanism is not clear, but the narrowing is presumably the result of inflammatory or ischaemic damage to the bowel with subsequent fibrosis. Other rare abdominal manifestations of the disease include pancreatitis (Garner 1977; Puppala *et al.* 1978), intestinal perforation (Smith and Krupski 1980), and massive gastric haemorrhage (Weber *et al.* 1983). The scrotum and its contents are not uncommonly involved in boys, the symptoms and signs resembling those of testicular torsion (O'Regan and Robitaille 1981). This is usually due to oedema of the scrotum itself but in at least one surgically explored case the testis, epididymis, and cord were found to be swollen and inflamed (Fitzsimmons 1968). The diagnosis is particularly difficult if scrotal pain and swelling are the first symptoms of the disease: the results of nuclear and ultrasonographic scrotal imaging studies may permit a child to be spared an unnecessary surgical exploration (Clark and Kramer 1986).

Table 1 Data from 23 studies comprising 2032 unselected patients with Henoch-Schönlein purpura to show the incidence of renal involvement

Reference	Numbers studied	Percentage with renal disease
Gairdner (1948)	12	92
Philpott (1952)	40	48
Oliver and Barnett (1955)	26	42
Wedgwood and Klaus (1955)	36	38
Rubin (1955)	65	20
Derham and Rogerson (1956)	94	49
Pratesi and Rizzuto (1956)	57	18
Tveterås (1956)	38	24
Bywaters *et al.* (1957)	52	38
Allen *et al.* (1960)	131	41
Burke *et al.* (1960)	88	44
Sterky and Thilen (1960)	224	22
Vernier *et al.* (1961)	45	62
Roberts *et al.* (1962)	298	12
Kobayashi *et al.* (1965)	82	63
Lelong and Deroubaix (1968)	36	64
Cream *et al.* (1970)	77	49
Striker *et al.* (1973)	59	59
Haahr *et al.* (1974)	203	30
Koskimies *et al.* (1974)	91	41
Mota-Hernandez *et al.* (1975)	96	57
Chiappo *et al.* (1978)	41	44
Koskimies *et al.* (1981)	141	28
		43 +/− 18

Urinary abnormalities +/− other features.

Renal disease

The renal manifestations of Henoch-Schönlein purpura are those of glomerulonephritis: haematuria, proteinuria, cylindruria, oliguria with fluid retention, oedema and hypertension, and impaired glomerular filtration rate: haematuria is usually the earliest sign that the kidney is involved. The nephrotic syndrome is seen in some. Renal involvement is the most important aspect of the disease in affected patients, since it may cause permanent functional impairment and occasionally leads to end-stage renal failure. Glomerulonephritis is rarely, if ever, the presenting feature of Henoch-Schönlein purpura: in the series reported by Allen *et al.* (1960) the onset of haematuria was preceded by at least two of the other major manifestations of the disease in all of the 53 children with evidence of renal disease. However, the likelihood and severity of renal involvement cannot be predicted from the severity of the non-renal symptoms and signs.

The reported prevalence of renal disease in Henoch-Schönlein purpura varies according to the criteria used to define it. If haematuria and/or proteinuria at some stage of the illness is taken as the criterion, rates varying from 12 to 92 per cent have been reported in the literature, although most were between 30 and 60 per cent (Table 1). Eleven of Gairdner's (1948) cases had haematuria: this was gross in four. One of these died following what would now be called rapidly progressive glomerulonephritis and in three others urinary abnormalities persisted after the acute phase of the illness. In the series of Allen *et al.* (1960) 53 of 131 patients developed haematuria: this was macroscopic on at least one occasion in 40 per cent and microscopic only in the other 60 per cent. Eighteen patients had impaired renal function (blood non-protein nitrogen concentration >40 mg/dl, equivalent to plasma urea >14 mmol/l); 12 of these were otherwise asymptomatic apart from raised blood pressure, while six had clinical features of acute glomerulonephritis including oliguria and oedema. The 12 asymptomatic patients recovered completely, as did one of the six with clinical nephritis. Two of the latter, however, died of rapidly progressive glomerulonephritis, two survived with impaired renal function, and one was lost to follow up. The findings of Koskimies *et al.* (1981), in a group of 141 unselected cases collected over 13 years, were broadly similar: 39 patients (28 per cent) had urinary abnormalities persisting for more than 4 weeks from the clinical onset of the disease, and were followed for a mean period of 7 years. One patient deteriorated to end-stage renal failure during this period, and two others had evidence of chronic glomerulonephritis. Thus, urinary abnormalities occur in 25 to 50 per cent of patients not previously selected on the basis of having known renal disease, but serious or progressive glomerulonephritis is seen in less than 5 per cent and probably not more than 1 to 2 per cent of the total. The pathology, prognosis, and management of the renal manifestations of Henoch-Schönlein purpura are discussed below.

Nervous system

Neurological symptoms and signs occur in a significant proportion of children with Henoch-Schönlein purpura. The most common of these are headache, behavioural changes, and seizures, although numerous other manifestations have been described including aphasia, hemi- and quadriplegia, chorea, ataxia, and peripheral nerve lesions. In an exhaustive literature review, Belman *et al.* (1985) reported three cases of their own and cited more than 30 references describing nervous system involvement: the reader is referred to that paper for further details. The nervous system manifestations are usually transient, permanent sequelae being fortunately rare. On the published evidence it is difficult or impossible to say whether these features represent vasculitis of the nervous system or the consequences of hypertension and metabolic disturbances. There is no doubt that hypertensive encephalopathy accounts for some, perhaps the majority, although some of the published cases were normotensive at the time of presentation.

Other organs and systems

Cardiac and pulmonary involvement have been described in a few patients. Imai and Matsumoto (1970) reported a 14-year-old boy with conduction abnormalities and congestive cardiac failure; however, the antistreptolysin O titre was greatly elevated and it is possible that the patient had coincident rheumatic fever. A patient of Abdel-Hadi et al. (1981) suffered a myocardial infarction during the course of Henoch-Schönlein purpura. Fatal pulmonary haemorrhage has been described (Jacome 1967), although whether this case was really one of Wegener's granulomatosis or Goodpasture's syndrome accompanied by a purpuric rash cannot be determined since diagnostic tests for these conditions were not available at the time of this report. However, the diagnosis of Henoch-Schönlein purpura appears secure in a more recently described case (Shichiri et al. (1987), which responded to oral corticosteroid therapy. Another rare complication is ureteritis, which may be complicated by stenosis or perforation: 10 cases were reviewed by Kher et al. (1983).

Aetiology

Many patients with Henoch-Schönlein purpura report a history of an intercurrent infection, usually respiratory, at or shortly before the onset of the illness. Gairdner (1948) cultured β-haemolytic streptococci from the throats of five of his original 11 patients and all of six subsequent cases—11 positive cultures from 17 patients in all. This contrasted with only one positive culture from 10 control children in the same wards suffering from different (unspecified) conditions. He therefore concluded that: 'infections, particularly streptococcal respiratory infections, play a dominant aetiological role in many cases of the syndrome'. However, subsequent careful studies (Bywaters et al. 1957; Allen et al. 1960; Ayoub and Hoyer 1969) failed to find convincing evidence for a higher incidence of streptococcal infection in patients with Henoch-Schönlein purpura than in controls, and infection with this organism cannot be incriminated as the cause of the disease in most patients, if indeed in any.

There are sporadic reports of Henoch-Schönlein purpura following infection with a wide variety of other micro-organisms, including Gram-positive and -negative bacteria, mycobacteria and viruses of several different groups (Table 2). Most of these reports simply describe a temporal association between the infection and the onset of Henoch-Schönlein purpura, although in some a rise in specific antibody titre coinciding with the appearance of the syndrome adds some weight to the inference that exposure to the organism precipitated the disease. At best, the evidence is indirect and circumstantial and no particular micro-organism had been proved to be specifically involved in the aetiology of the syndrome. A recent study by Neilsen (1988) of the epidemiology of 1222 cases of Henoch-Schönlein purpura, recorded in the Danish National Patient Register between 1977 and 1984, used a mathematical model to detect evidence of clustering in space and time, as would be expected if the disease occurred as a complication of a communicable infection. No such evidence was found: in fact, the observed pattern of scatter of the disease followed closely that predicted by the null hypothesis that each case occurred at random and independent of all other cases. In contrast, Farley et al. (1989) found evidence of clustering of cases within an outbreak of Henoch-Schönlein purpura in Hartford County, Connecticut, and interpreted their findings as meaning that case-to-case transmission of the disease had occurred. The different geographic and temporal scale of the two studies may reconcile these apparently contradictory conclusions: a localized epidemic of any infectious agent capable of precipitating the disease would be expected to produce a local cluster of cases, but this effect might be obscured by the larger number of cases presenting at the national level as a result of other pathogens at large in the community during the same period. Taken together with the number and variety of infections which have been implicated as precipitants of Henoch-Schönlein purpura, these important observations suggest strongly that, rather than being an idiosyncratic response to infection with a particular, specific micro-organism (as in the case of acute rheumatic fever), the putative series of immune events leading to the clinical expression of Henoch-Schönlein purpura may be stimulated by any one of many infectious agents in susceptible individuals.

Allergy to drugs, food, and various exogenous factors other than micro-organisms has frequently been postulated, but seldom proved, in the aetiology of Henoch-Schönlein purpura. Winkelmann and Ditto (1964) noted a history of drug exposure in 20 of 36 patients with allergic vasculitis: the individual drugs, as well as others described more recently, are listed in Table 2. Ackroyd (1953) culled from the literature nine reports describing 23 patients in whom food allergy appeared to be responsible for Henoch-Schönlein purpura according to reasonably rigorous criteria: exposure to the suspected food repeatedly and consistently caused episodes of the disease, while its withdrawal was equally consistently followed by remission. The foods identified are given in Table 2: egg, milk, wheat, chocolate, and beans were the most frequently reported. Repeated attacks of Henoch-Schönlein purpura following alcohol ingestion have been described in two adult patients (Knight and Cameron, personal communication).

Henoch-Schönlein purpura has occasionally been reported in association with a miscellany of other events and diseases, including insect bites (Burke and Jellinek 1954), cancer (Cairns et al. 1978; Kauffmann et al. 1980), exposure to cold (Rogers et al. 1971), and blunt trauma (Talbot et al. 1988). One case occurred in a patient with monoclonal IgA gammopathy (Dosa et al. 1980), and cytotoxic treatment of the underlying condition was associated with remission of Henoch-Schönlein purpura, suggesting a causal, rather than a casual, relationship between the two. The fact that the abnormal paraprotein in this case was of the IgA class is of particular interest in view of the possible role of IgA in the pathophysiology of the disease (see below). This case apart, the variety of drugs, chemicals, foods, and diseases proposed as putative causes of Henoch-Schönlein purpura is so great that they must be regarded as non-specific triggers of the syndrome, as with micro-organisms and pathogens.

4.5.2 The nephritis of Henoch-Schönlein purpura

Table 2 Environmental agents which have been implicated in the causation of Henoch-Schönlein purpura

Agent	Reference
Micro-organisms	
β-Haemolytic Streptococcus	Gairdner (1948)
Mycobacterium tuberculosis	Dalgleish and Ansell (1950)
Varicella zoster	Pedersen and Petersen (1953)
Vaccinia	Jiminez and Darrington (1968)
Haemophilus parainfluenzae	Cream *et al.* (1970)
Streptococcus pneumoniae	Cream *et al.* (1970)
Rubella	Meadow *et al.* (1972)
Measles	Meadow *et al.* (1972)
Mycoplasma pneumoniae	Liew and Kessel (1974); Steare (1988)
Yersinia enterocolitica	Rasmussen (1982)
Human parvovirus	Lefrère *et al.* (1986)
Human immunodeficiency virus	Velji (1986)
Staphylococcus sp.	Montoliu (1987)
Legionella sp.	Bull *et al.* (1987)
Influenza vaccine	Patel *et al.* (1988)
Drugs	
Aspirin	Winkelmann and Ditto (1964)
Erythromycin	Winkelmann and Ditto (1964)
Griseofulvin	Winkelmann and Ditto (1964)
Iodide	Winkelmann and Ditto (1964)
Penicillin	Winkelmann and Ditto (1964)
Phenacetin	Winkelmann and Ditto (1964)
Phenothiazines	Winkelmann and Ditto (1964)
Quinidine	Winkelmann and Ditto (1964)
Sulphonamide	Winkelmann and Ditto (1964)
Tetracycline	Winkelmann and Ditto (1964)
Thiazide diuretics	Cream *et al.* (1970)
Chlorpromazine	Aram (1987)
Paracetamol dihydrocodeine	Richards and Lindley (1987)
Thiram	Duell and Morton (1987)
Foods	
Nuts	Ackroyd (1953)
Blackberries	Ackroyd (1953)
Egg	Ackroyd (1953)
Milk	Ackroyd (1953)
Potato	Ackroyd (1953)
Wheat	Ackroyd (1953)
Meat (various)	Ackroyd (1953)
Fish	Ackroyd (1953)
Chocolate	Ackroyd (1953)
Chicken	Ackroyd (1953)
Tomato	Ackroyd (1953)
Alcohol	Knight and Cameron (unpublished)

The papers by Winkelmann and Ditto (1964) and Ackroyd (1953) are reviews: the reader is referred to them for the source references.

Pathophysiology

Several clinical and laboratory features of Henoch-Schönlein purpura suggest that the syndrome has an immunological basis. Firstly, in most cases the onset of the disease is apparently triggered by exposure to a foreign agent or substance, usually a pathogen, a drug, or a food, and the chronology of events is consistent with the development of an immune response. Secondly, immunoglobulins and components of the complement system are deposited in the glomeruli and skin of affected patients. Thirdly, changes in the concentration of circulating IgA and the presence in serum of IgA-containing immune complexes and rheumatoid factors have been observed in a high proportion of cases. Nevertheless, although it is widely accepted that these changes are probably causally related to the disease, the precise mechanisms involved remain to be fully elucidated.

Alterations in the IgA immune system

The serum concentration of IgA is increased in about half the patients in the first 3 months of the disease (Immonen and Kouvulainen 1967; Trygstad and Stiehm 1971). Similä *et al.* (1977) found that IgM was also increased, but the increase over control values was only significant in patients with evidence of persisting renal disease after the acute phase of the illness. Several studies indicate that serum IgA in Henoch-Schönlein purpura is polymeric (Egido *et al.* 1980; Danielsen *et al.* 1984; Kauffmann *et al.* 1981), in contrast to that in normal subjects which is mainly (>98 per cent) monomeric (Conley and

Delacroix 1987). Egido et al. (1980) found that IgA oligomers (molecular weight <10^6 Da) in serum were increased. However, the study of Kauffmann et al. (1981), using precipitation with 3 per cent polyethylene glycol to separate macromolecules of different sizes, showed that much of the circulating IgA in Henoch-Schönlein purpura is in the molecular weight range above 10^6 Da, and suggested that this probably represents immune complexes. Cryoglobulins containing IgA and properdin have been described in children with Henoch-Schönlein purpura (Garcia-Fuentes et al. 1977).

Other workers have detected IgA-containing immune complexes by a variety of different assay techniques. Levinsky and Barratt (1979), using inhibition of agglutination of coated latex particles, found such complexes in 13 of 18 children with Henoch-Schönlein purpura; this finding did not discriminate between those with and those without glomerulonephritis. However, they also found immune complexes containing IgG in eight of nine with, but none of nine without, nephritis. Taking this finding in conjunction with the observation that IgA can inhibit the clearance by phagocytes of IgG-containing complexes from the circulation (Wilton 1978), Levinsky and Barratt (1979) postulated that IgG-containing immune complexes may be the direct cause of tissue damage in Henoch-Schönlein purpura, with IgA-containing immune complexes playing an indirect role. This hypothesis receives further support from the more recent observation that clearance of immune complexes is impaired in active disease (Davin et al. 1985). IgA-containing immune complexes have also been demonstrated by an anti-IgA inhibition assay (Kauffmann et al. 1980), by IgA conglutinin binding (Doi et al. 1984; Coppo et al. 1984) and by the Raji-cell assay (Hall et al. 1980; Wenner and Safai 1983) in 46 to 73 per cent of patients.

Additional evidence of a role for the IgA component of the immune system in the pathogenesis of Henoch-Schönlein purpura derives from studies of the cellular production and control of immunoglobulin synthesis. In-vitro production of both IgA and IgG by B cells in increased in patients compared to controls (Bannister et al. 1983), a finding more marked in active than quiescent disease (Casanueva et al. 1983; Williams et al. 1987). The study of Beale et al. (1982) suggests that enhanced immunoglobulin production may result from the impaired ability of T cells to suppress B-cell function. Alterations in immunocyte function are not confined to the blood: Bene et al. (1986) showed that the plasma cell IgA:IgG ratio was increased in tonsillar lymphoid tissue in patients with classical Henoch-Schönlein purpura although, interestingly, it was normal in two patients with the clinical syndrome of Henoch-Schönlein purpura but no tissue deposits of IgA. This last observation, and the description by Martini et al. (1985) of typical disease in a child with complete selective deficiency of serum and secretory IgA and absence of IgA deposits in skin and kidney biopsies, inevitably call into question the central role of IgA-related mechanisms in the pathophysiology of the disease. It must be conceded, at the present stage of knowledge, that the serum, secretory, and tissue abnormalities of IgA so consistently found in Henoch-Schönlein purpura might be an epiphenomenon secondary to some other, as yet unknown, pathophysiological process.

Putative antigens in Henoch-Schönlein purpura

Although the term immune complex is widely interpreted as meaning a macromolecular assembly containing antibody and antigen, no specific antigen has been consistently isolated from immune complexes in Henoch-Schönlein purpura, and the immunological specificity of circulating IgA has been poorly documented. Davin et al. (1987) found IgA and IgG antibodies to anti-α-galactosyl residues, a common component of the cell wall of mucosal pathogens, in the serum of six of eight children with Henoch-Schönlein purpura and haematuria, but neither of two children without haematuria at the time of study. Furthermore, six of the eight haematuric patients had raised levels of polyethylene glycol-precipitable IgA-containing immune complexes in the serum, with discordance between the presence of antibody to α-galactosyl and IgA-containing immune complexes in only one case. Incubation of sera with α-galactosyl glucoside, but not with D-glucose, reduced the IgA content of the immune complexes, suggesting that they contained α-galactosyl antibody of the IgA class. These interesting results may go some way towards explaining the well known clinical association between respiratory infections and exacerbations of haematuria in patients with Henoch-Schönlein purpura, but the fact that similar antibodies were found in eight of 10 subjects with proven mucosal infections with *Mycobacterium pneumoniae* or *Escherichia coli* but without evidence of Henoch-Schönlein purpura, renal disease, or haematuria means that the question as to why certain individuals develop Henoch-Schönlein purpura in response to such infections, while the majority do not, remains unanswered.

Other recent studies suggest that the role of IgA in the syndrome may be less straightforward than the traditional idea that IgA-antigen complexes are deposited in and around blood vessels of affected tissues, leading to complement activation, inflammation, and tissue damage. Saulsbury (1986) found IgA rheumatoid factor (an IgA antibody with specificity for antigenic determinants located in the Fc portion of IgG) in sera of patients with Henoch-Schönlein purpura, and in a subsequent study (Saulsbury et al. 1987) showed that this rheumatoid factor, mostly containing IgA of the IgA_1 subclass, was a constituent of polyethylene glycol precipitable immune complexes in the disease. He also found a strong correlation ($r = 0.72$, $p<0.001$) between the amount of IgA and IgG in such precipitates from patients with Henoch-Schönlein purpura but not from controls. Saulsbury's hypothesis (1987), that deposition of IgA rheumatoid factor immune complexes in small vessels might be the cause of the leucocytoclastic vasculitis characteristic of the disease, is interesting, particularly in view of the earlier observations of Levinsky and Barratt (1979) concerning an apparent role for IgG-containing immune complexes in the nephritis of Henoch-Schönlein purpura. The presence of IgA rheumatoid factor in the serum of some patients with active disease, but none with quiescent disease, was confirmed by Knight et al. (1988). Another intriguing observation is that of van der Wall Bake et al. (1987), who found IgA antineutrophil cytoplasmic antibody in both monomeric and polymeric fractions of serum IgA. However, Knight et al. (1988) failed to find these antibodies, or anti-DNA or anticardiolipin antibodies, in their patients, even during the active phase of the disease: further studies are indicated to resolve these discordant findings. If the presence of IgA antineutrophil cytoplasmic antibody is confirmed, its possible value as a diagnostic test, as in the case of IgG antibodies in Wegener's granulomatosis (van der Woude et al. 1985), awaits further evaluation.

Role of the complement system

Most studies of the complement system in Henoch-Schönlein purpura have found normal levels of the major serum factors and the CH_{50} (Evans et al. 1973; Kauffmann et al. 1980). However, Kauffmann et al. (1980) also found raised serum C3d levels, interpreted as evidence of increased C3 turnover, in children with active disease. Garcia-Fuentes et al. (1978) observed a reduced CH_{50} in nine of 23 children with active disease and low properdin levels in five of 17. This finding, taken with the observation by the same group (Garcia-Fuentes et al. 1977) that circulating immune complexes in Henoch-Schönlein purpura contain properdin, and with the consistently normal serum concentrations of C1q and C4, indicates activation of the alternative complement pathway. This conclusion is further supported by the histopathological evidence (see below). A few patients with C2 deficiency and relapsing Henoch-Schönlein purpura have been described (Sussman et al. 1973; Gelfand et al. 1975). The nature and significance of this association are obscure: most subjects with C2 deficiency appear completely healthy.

Similarities and differences between Henoch-Schönlein purpura and idiopathic IgA nephropathy

The renal histology of Henoch-Schönlein purpura is indistinguishable from that of idiopathic IgA nephropathy (Berger's disease). This has led many observers to speculate that the two conditions, as conventionally defined, may in reality be different manifestations of the same disease process. In other words, IgA nephropathy is Henoch-Schönlein purpura without involvement of other systems or, conversely, Henoch-Schönlein purpura is IgA nephropathy complicated by systemic vasculitis. Since the definitions of the two diseases are both purely descriptive, and the cause of neither is understood at the basic pathophysiological level, the question is not currently answerable in meaningful terms: indeed, one might legitimately ask whether the question has any real logical content. Be that as it may, the renal disease of Henoch-Schönlein purpura has many similarities with, and some differences from, that of IgA nephropathy.

Similarities

The glomerular deposits are predominantly IgA_1 in both diseases: J chain is also present (Conley et al. 1987; Lomax-Smith et al. 1983). Serum total and macromolecular (polymeric) IgA levels are increased in the acute phase of IgA nephropathy and in Henoch-Schönlein purpura (Lopez-Trascasa et al. 1980), as is the IgA concentration of pharyngeal washings (Tomino et al. 1983). Circulating IgA-containing immune complexes, detected by a wide variety of methods, are present in a similar proportion of patients with the two diseases: the relevant literature is reviewed in detail by Feehally (1988). The ratio of IgA- to IgG-producing plasma cells in tonsillar, but not jejunal, lymphoid tissue is increased in both diseases (Bene et al. 1983; 1986; Hene et al. 1987). The anti-α-galactosyl antibodies and immune complexes referred to above are present at similar titres in both (Davin et al. 1987), and spontaneous in-vitro IgA production by peripheral blood mononuclear cells is increased in the active phase of both diseases compared with controls (Williams et al. 1987). Immunogenetic studies give similar, if inconsistent, results: the prevalence of the HLA class 1 antigen BW35 has been reported to be increased in both IgA nephropathy (Berthoux et al. 1978) and Henoch-Schönlein purpura (Nyulassy et al. 1977), but later reports failed to confirm the association in either (Le Petit et al. 1982; Sengar et al. 1984). An increased frequency of homozygous null C4 phenotypes has been described in both (McLean et al. 1984). A case has been reported of IgA nephropathy and Henoch-Schönlein purpura occurring in the same child at an interval of 2 years (Hughes et al. 1988). Meadow and Scott (1985) described the simultaneous appearance of Henoch-Schönlein purpura in one of a pair of monozygotic twins and IgA nephropathy in the other, following the same, proven, adenovirus infection. This last report, in particular, makes it difficult to account for the differences between IgA nephropathy and Henoch-Schönlein purpura on the basis of genetically determined differences in immune function, since the two patients were almost certainly genetically identical.

Differences

The clinical features of IgA nephropathy and Henoch-Schönlein purpura differ considerably. Although the latter is seen at all ages, it is predominantly a childhood disease, while most cases of IgA nephropathy present in adult life with a minority occurring in children. The renal disease in both may be located at any point on the clinical spectrum of glomerulonephritis from minor urinary abnormalities to a full blown acute nephritic syndrome: the nephrotic syndrome and/or 'rapidly progressive glomerulonephritis' may be seen in either. However, overt 'nephritic' and nephrotic features are more commonly seen in the early phase of Henoch-Schönlein nephritis than in IgA nephropathy, while the recurrent haematuria syndrome and asymptomatic proteinuria are more typically observed in the latter. More importantly, the natural history of the two conditions seems to be different: the late outcome of Henoch-Schönlein purpura is strongly predicted by the initial presentation and renal histological findings, whereas Yoshikawa et al. (1987) found no correlation between clinical or histological features at diagnosis and progression to renal failure in 206 children with IgA nephropathy. Similarly, studies of adult patients with IgA nephropathy indicate that a high proportion develop progressive deterioration of renal function despite the absence of identifiable early predictors of poor outcome (Nicholls et al. 1984; D'Amico et al. 1985). Recently Demaine et al. (1988) found abnormal polymorphisms in the heavy chain switch region of chromosome 14 in IgA nephropathy but not in Henoch-Schönlein purpura: this might indicate subtle differences in immune function between the two conditions. In a study of the cell types and numbers involved in the mesangial proliferative response in various forms of nephritis, using monoclonal antibodies, Nolasco et al. (1987) found increases in total leucocytes, monocytes/macrophages, and both the CD4 and CD8 T-cell subsets in Henoch-Schönlein purpura but not IgA nephropathy. At first sight this seems to contradict the findings of Yoshioka et al. (1989), who found comparable increases in glomerular monocytes/macrophages in both conditions, as well as intraglomerular deposition of cross-linked fibrin in a degree proportional to the numbers of mononuclear cells. Other differences between the two diseases were found in the latter study, however. In IgA nephropathy, but not in Henoch-Schönlein purpura, there was a significant positive correlation between the degree of monocyte infiltration and the amount of proteinuria. In addition, monocyte infiltration declined with time elapsed from the clinical onset of disease in Henoch-Schönlein purpura but remained constant in IgA nephropathy. These findings are consistent with the conclusion based on previously cited clinical and morphological studies (Nicholls et al. 1984; D'Amico et al. 1985; Yoshikawa et al. 1987): Henoch-Schönlein purpura is usually an acute, self-limiting illness in which the renal prognosis depends mainly on the severity of the original insult, while idiopathic IgA nephropathy is frequently a chronic, progressive glomerular disease. Whether this difference is due to host factors connected with innate differences in the function of the IgA-related immune system, or to exogenous factors such as the identity, and perhaps the degree of persistence, of the putative antigen or antigens involved, cannot be determined on the basis of current knowledge.

Histopathology

The skin

Histologically, Henoch-Schönlein purpura is a leucocytoclastic vasculitis, with perivascular accumulation of inflammatory cells, mostly polymorphonuclear leucocytes and mononuclear cells with occasional eosinophils, surrounding the capillaries and post-capillary venules of the corium. The presence of fragmented nuclear material in the infiltrate attests to the leucocytoclastic nature of the process. Extravasation of red blood cells occurs, as would be expected in a purpuric lesion; fibrinoid necrosis, and thrombi containing leucocytes and platelets, are seen in the more aggressive lesions but only in vessels surrounded by the leucocytic infiltrate (Vernier et al. 1961). Immunofluorescent staining reveals the presence of IgA, C3 and fibrin/fibrinogen in vessels and connective tissue of clinically involved skin, without C1q or C4 (Baart de la Faille-Kuyper et al. 1976; Giangiacomo and Tsai 1977), adding further support to the hypothesis of a role for activation of the alternative complement pathway. In clinically uninvolved skin, positive staining for IgA and C3 is seen in some, but not all, cases, and then only in capillary walls (Baart de la Faille-Kuyper et al. 1976; Zuckner et al. 1977). IgG and IgM are found in small quantities in a minority of patients. No staining for IgA was seen in patients with systemic or discoid lupus erythematosus, lipoid nephrosis, or normal controls (Zuckner et al. 1977).

Other organs

Anecdotal reports have appeared of the histology of various other organs, including the lungs, various segments of the gastrointestinal tract and the adrenal cortex (Levitt and Burbank 1953; Norkin and Weiner 1960). The lesion is essentially identical to that in the skin.

The kidney

The renal histopathology of Henoch-Schönlein purpura was comprehensively described by Heptinstall (1983), with an exhaustive bibliography, and also by Habib (1983); the reader is referred to these works for a detailed list of original sources.

Light microscopy

The characteristic glomerular lesion of Henoch-Schönlein purpura is a focal and segmental proliferative glomerulonephritis: focal, in that not all glomeruli are involved, and segmental, in that some lobules are spared even in affected glomeruli. The proliferative process mainly affects mesangial cells, but in the more severe cases infiltration with polymorphonuclear leucocytes may be seen, and variable epithelial (extracapillary) proliferation occurs, leading to crescent formation. Nolasco et al. (1987) used monoclonal antibodies to type the mononuclear cells in the glomerular mesangium: total leucocytes, monocytes/macrophages, and both CD4 and CD8 T-cell subsets were increased. The findings were similar to those in type 1 mesangiocapillary glomerulonephritis and diffuse proliferative lupus nephritis, while patients with microscopic polyarteritis and with anti-glomerular nephritis basement membrane antibodies showed more marked increases in the numbers of the same cell types. Perhaps rather unexpectedly, idiopathic IgA nephropathy did not show a significant increase in the numbers of any of the cell types studied. In a somewhat similar study Yoshioka et al. (1989) found monocytes/macrophages to be the predominant cell type in the mesangium in both Henoch-Schönlein purpura and IgA nephropathy; CD4 and CD8 cells were also increased compared with controls, but it is not clear from the paper whether these increases were statistically significant. Yoshioka et al. (1989) found cross-linked fibrin in the glomeruli in amounts proportional to the intensity of the monocytic infiltrate, and found that the magnitude of proteinuria was correlated with the number of monocytes detected in the glomeruli.

The spectrum of severity of renal involvement is highly variable, from minimal changes (no abnormalities visible on light microscopy) to a fulminant, necrotizing glomerulonephritis with large crescents surrounding 100 per cent of glomeruli (Fig. 1). Towards the more severe end of the spectrum, the essentially focal and segmental distribution of the lesion may be obscured, with all glomeruli involved to a variable extent, but the process rarely or never shows the diffuse uniformity seen in classical postinfectious glomerulonephritis. As emphasized by Counahan and Cameron (1977), the crescent formation of Henoch-Schönlein purpura differs from that seen in many forms of 'rapidly progressive nephritis' in that segmental extracapillary proliferation is often seen overlying areas of segmental mesangial and endocapillary proliferation, forming small crescents encircling less than 50 per cent of the circumference of the glomerulus. These 'non-circumferential crescents' may not necessarily carry the same grim prognostic significance as those seen, for example, in idiopathic rapidly progressive glomerulonephritis, Goodpasture's syndrome, and polyarteritis/Wegener's granulomatosis (see below). Focal areas of glomerulosclerosis may be seen in biopsies performed late in the disease, usually adherent to Bowman's capsule (Vernier et al. 1961): these are the result of healing or scarring of the original proliferative lesion, rather than a primary sclerosing process (Heptinstall 1983).

A number of attempts have been made to construct a classification of the renal lesion in Henoch-Schönlein purpura according to the severity of the changes, mainly with a view to predicting the clinical outcome. All depend on measurement of two variables: mesangial proliferation and extracapillary proliferation (crescent formation). The schema of Meadow et al. (1972) has been widely employed and, with minor adaptations, has been adopted by the International Study of Kidney Disease in Childhood (ISKDC). The ISKDC classification is given in Table 3, slightly modified after Heaton et al. (1977) (in the original version, group II was not subdivided into focal and diffuse). The lesion classified as group VI is morphologically indistinguishable from membranoproliferative glomerulonephritis type 1, with exaggerated lobulation of the tuft, marked mesangial proliferation, and subendothelial interposition of mesangial cells with 'double contouring' of the glomerular basement membrane.

4.5.2 The nephritis of Henoch-Schönlein purpura

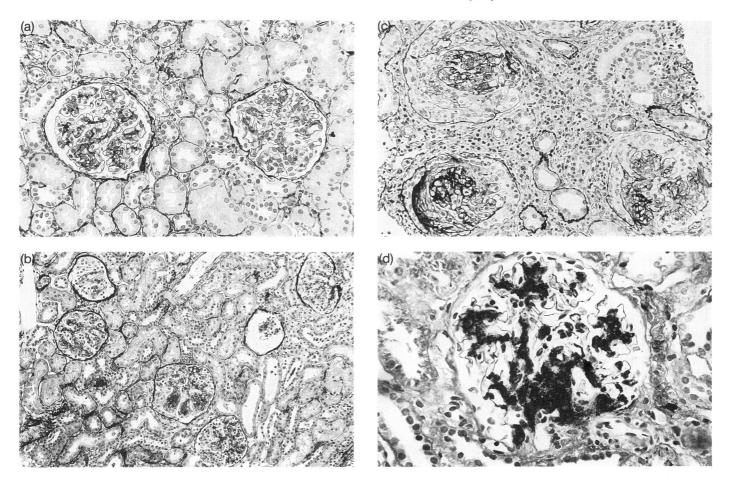

Fig. 1 (a) Henoch-Schönlein purpura nephritis grade IIa. The glomerulus on the right of the figure is within normal limits. That on the left shows mild segmental mesangial hypercellularity and increase in mesangial matrix. (Silver-methenamine). (b) Henoch-Schönlein purpura nephritis grade IIIa. The glomeruli show variable focal mesangial proliferation and sclerosis, ranging from near normal (top right) to extensively involved (lower centre). Small cellular crescents are seen in a minority of glomeruli (silver-methenamine). (c) Henoch-Schönlein purpura nephritis grade Vb. Crescents are present in the majority of glomeruli (100 per cent in this example). The tufts are contracted and show diffuse mesangial hypercellularity and sclerosis. The interstitium is extensively infiltrated by mononuclear cells and marked tubular atrophy is present (silver-methenamine). (d) Glomerulus from a child with Henoch-Schönlein purpura nephritis, grade IIb, showing heavy deposition of IgA in a mesangial distribution (immunoperoxidase stain). (Photomicrographs by courtesy of Dr. R. B. Hartley.)

Immunofluorescent and immunoperoxidase microscopy

The most consistent renal finding in the nephritis of Henoch-Schönlein purpura is immunoglobulin deposition, mainly IgA, in the glomerular mesangium and to a lesser extent in the capillary walls (Urizar et al. 1968; Berger 1969) (Fig. 1(d)). These and subsequent studies (reviewed by Counahan and Cameron 1977) demonstrated the presence of IgA, IgG, and IgM in glomeruli of 90 per cent, 69 per cent, and 30 per cent, respectively, of patients studied. The corresponding percentages for C3, C4, C1q, and properdin were 80 per cent, 2 per cent, and 69 per cent; fibrin was found in 74 per cent. In the majority of cases the immunoglobulins were present in a predominantly mesangial distribution, but in those with histological changes of grades V and VI it was mainly pericapillary with very little mesangial deposition (Heaton et al. 1977). The mesangial IgA is mostly of the IgA_1 subclass, with small amounts of IgA_2 in the mesangium and capillary walls (Conley et al. 1980). As with the immunological findings in the blood, discussed above, the consistent finding of properdin and the relative absence of C4 and C1q suggests activation of the alternative, rather than the classical, complement pathway (Giangacomo and Tsai 1977).

Electron microscopy

Examination by electron microscopy reveals, in addition to the changes visible by optical examination, the presence of electron-dense deposits in the mesangium, the matrix of which is typically somewhat expanded, and also in a subendothelial position. Small subepithelial deposits are seen much less often, and are also visible on light microscopy using oil immersion of 1 μm sections of Araldite-embedded tissue stained with toluidine blue. Using electron microscopy, Heaton et al. (1977) found mesangial deposits in the majority of cases, subendothelial deposits in about half and subepithelial deposits only rarely. However, the toluidine blue method revealed mesangial deposits in all, and deposits on both sides of the glomerular basement membrane rather more frequently (Table 4); the latter were more prominent in biopsies with more severe changes on conventional optical microscopy (grades III–VI), and in three cases large,

Table 3 Classification of Henoch-Schönlein purpura glomerulonephritis recommended by pathologists of the International Study of Kidney Disease in Childhood (ISKDC), as modified by Heaton et al. (1977)

I	Minimal changes
II	Pure mesangial (a) Focal (b) Diffuse
III	Mesangial proliferative glomerulonephritis with less than 50 per cent crescents: (a) Focal (b) Diffuse
IV	Mesangial proliferative glomerulonephritis with 50–75 per cent crescents: (a) Focal (b) Diffuse
V	Mesangial proliferative glomerulonephritis with more than 75 per cent crescents: (a) Focal (b) Diffuse
VI	Membranoproliferative (mesangiocapillary) glomerulonephritis

Table 4 Comparison of conventional electron microscopy with optical microscopy of 1 μm sections stained with toluidine blue as a means of localizing 'immune deposits' in glomeruli in children with Henoch-Schönlein purpura (Heaton et al. 1977). The latter method is more sensitive

Method	n	Mesangial	Subendothelial	Subepithelial
Electron microscopy	19	13 (68%)	9 (47%)	3 (16%)
Toluidine blue	25	24 (96%)	12 (48%)	6 (24%)

subepithelial 'humps' were present. The greater sensitivity of the toluidine blue method over electron microscopy may reflect the fact that a larger sample of tissue can be examined in a given time by the former, due to the lower magnification.

A recent study using the advanced technique of immuno-electron microscopy (Yoshiara et al. 1987) showed that the distribution of immunospecific IgA reaction product closely followed that of the 'deposits' seen on conventional electron microscopy: this product was identified in the mesangium and subepithelial region in 100 per cent of 14 biopsies from children with Henoch-Schönlein purpura, and subepithelially in about half of the cases. C3 reaction product was seen in the same distribution, but the deposits were smaller and less extensive than those of the IgA reaction product: IgG reaction product was seen rarely and IgM reaction product not at all. This study provides the best evidence to date that the electron-dense deposits seen on conventional electron microscopy are indeed composed of, or at least contain, constituents of the immune system especially IgA, and that they are the same material that is demonstrated by immunofluorescent and immunoperoxidase staining under light microscopy. The probable pathogenic role of these deposits, particularly those in a subepithelial position, is supported by the findings of another electron microscopy study (Yoshikawa et al. 1986) in which lytic lesions of the glomerular basement membrane were seen in close association with deposits in 36 of 72 biopsies from children with Henoch-Schönlein purpura.

Prediction of renal outcome

All published experience indicates that the great majority of patients, both children and adults, recover completely with no measurable impairment of renal function. In large studies of unselected children (Allen et al. 1960; Koskimies et al. 1974; Kobayashi et al. 1977) death or end-stage renal disease ensued in seven of 245 patients and persisting urinary abnormalities were present in 5 to 15 per cent after several years' follow up. Because of selection bias, patients referred to renal units have an apparently worse outcome (Meadow et al. 1972; Counahan et al. 1977; Yoshikawa et al. 1981, 1987; Habib 1983). Studies of unselected adults are few: in four reports (Cream et al. 1970; Debray et al. 1971; Bar-On and Rosenmann 1972; Tay and da Costa 1976) nine of 227 patients died or developed end-stage renal disease: this is not significantly different from the paediatric experience ($\chi^2 = 0.44$; $p = NS$). As in the paediatric literature, reports describing patients selected for having nephritis inevitably give a worse impression (Lee et al. 1986; Fogazzi et al. 1989).

Since the long-term prognosis of Henoch-Schönlein purpura is almost entirely determined by the behaviour of the nephritis, it would be useful to identify a subgroup or groups of patients at increased risk of progressive renal disease, who it would be appropriate to investigate carefully and perhaps to offer treatment. Several groups have examined this question in terms of both clinical and histological criteria.

Clinical correlations

The major clinical predictors of outcome were identified by Meadow et al. (1972): subsequent studies, in the main, have confirmed their findings but added little that is new (Koskimies et al. 1974; Counahan et al. 1977; Yoshikawa et al. 1981, 1987; Habib 1983). Meadow et al. (1972) also introduced a system of clinical grading at follow up that has been widely adopted in more recent publications (Table 5).

As might be expected, the more severe the renal presentation the greater the likelihood of both the worse grades of biopsy change on light microscopy (grades IV to VI) and a clinical outcome in category C or D. The findings in different studies are fairly consistent. Patients with haematuria alone, or haematuria

Table 5 Classification of clinical status at follow-up of patients with Henoch-Schönlein purpura nephritis (Meadow et al. 1972)

Outcome group		
A		*Normal* Normal physical examination; no urinary abnormality; normal renal function
B		*Minor urinary abnormality* Normal physical examination; haematuria (microscopic ± intermittent macroscopic) and/or proteinuria < 1 g/24 h; normal renal function
C		*Active renal disease* Proteinuria > 1 g/24 h ± hypertension; normal renal function
D		*Renal insufficiency* GFR < 60 ml/min.1.73 m^2; actual or renal death (dialysis or transplantation)

4.5.2 The nephritis of Henoch-Schönlein purpura

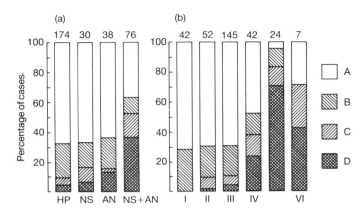

Fig. 2 Comparison of (a) clinical and (b) histological features at presentation with eventual outcome. HP=haematuria and/or proteinuria; NS=nephrotic syndrome; AN=acute nephritic syndrome. Histological categories I to VI as defined in Table 3; outcome categories A to D as defined in Table 5 (based on pooled data from Koskimies *et al.* 1974; Counahan *et al.* 1977; Yoshikawa *et al.* 1981, 1987).

and proteinuria of less than nephrotic proportions, rarely have any residual loss of renal function. Patients with the nephrotic syndrome but no 'nephritic' features (azotaemia, hypertension, oliguria) also have a favourable outcome with relatively few exceptions. On the other hand, children with the acute nephritic syndrome have a 10 to 20 per cent risk of a category 4 outcome, while those with mixed 'nephritic-nephrotic' features at presentation have the worst prognosis, with half in category C or D and as many as a third in category D. Figure 2(a), a compilation from four studies (Koskimies *et al.* 1974; Counahan *et al.* 1977; Yoshikawa *et al.* 1981, 1987) comprising 318 children, illustrates the relationship of clinical presentation to outcome.

The paper by Meadow *et al.* (1972) was not included in the above analysis because the report by Counahan *et al.* (1977) describes the same patients at a later period in the evolution of the disease. It is noteworthy that some patients in all clinical categories (other than D) deteriorated between 2 years and >6.5 years after onset (Counahan *et al.* 1977). In particular, six of 11 patients in category C at 2 years had progressed to D in the interval, while of 25 in category B at 2 years, four had progressed to C and two to D (although 14 had improved to A status). The study by Bunchman *et al.* (1988) also sounds a cautionary note concerning the confidence with which long-term predictions can be made from clinical features at presentation. Sixteen patients who had progressed to end-stage renal disease due to Henoch-Schönlein nephritis were compared with 16 who had recovered completely after an average follow-up period of 14 years. Creatinine clearance and major clinical features were the same in both groups at onset of the illness, except that the group with end-stage renal disease were more likely to have had gross haematuria. The best prediction of final outcome was obtained by creatinine clearance at 3 and 5 years. No patient with a 3-year creatinine clearance above 110 ml/min/1.73 m^2 progressed to end-stage renal disease, while no patient with a 3-year creatinine clearance below 80 ml/min.1.73 m^2 escaped this outcome. The range 80 to 110 ml/min.1.73 m^2 included some who eventually progressed and some who recovered; at 5 years, however, the creatinine clearance in all those who recovered had improved to above 95 ml/min.1.73 m^2, while in all of those who eventually developed end-stage renal disease it was below this value. Although this paper is a valuable reminder that deterioration of renal function may occur late in the course of the disease, and that predictions from the clinical features at presentation are not foolproof, it should be emphasized that the design of the study involved selecting two groups of children, after the event, according to their eventual outcome, and deliberately matching them as closely as possible for presenting features. This design inevitably minimizes any differences that might have been present between patients who recovered and those who did not; a study of the whole group of 100 biopsied children from whom the study groups were selected might well have provided a different emphasis.

The study of Levy *et al.* (1976) of 100 patients of whom 81 were followed for at least a year, republished by Habib (1983), was also omitted from Fig. 2 because the systems of classification used were slightly different from those in Tables 3 and 5: however, the findings were essentially the same as those in the other studies reported.

Figure 2(a) does not distinguish between patients with haematuria alone and those with haematuria and mild to moderate proteinuria, because in neither of the studies of Yoshikawa *et al.* (1981, 1987), which between them accounted for more than half of the patients, was this distinction made. The papers that do differentiate these categories, however (Koskimies *et al.* 1974; Counahan *et al.* 1977; Levy *et al.* 1976) report no group D outcome in any child with isolated haematuria, which therefore appears to be the most benign presentation of all, as might be expected. It should be remembered that patients with a history of Henoch-Schönlein purpura may subsequently develop haematuria for other reasons, and unrelated causes should be excluded by appropriate investigations. In summary, patients with haematuria but without proteinuria have an almost completely benign prognosis, while ascending amounts of proteinuria (<1 g/24 h: >1 g/24 h: nephrotic syndrome) are associated with progressively worse outcome. Patients with the acute nephritic syndrome, and especially those with a mixed nephritic-nephrotic presentation, have the worst prognosis of all and are at high risk of chronic renal failure.

Against this background, it is our policy at Guy's Hospital to perform renal biopsy in patients with Henoch-Schönlein purpura who have the following characteristics: (i) nephritic and/or nephrotic presentation; (ii) persistent heavy proteinuria (+ + or more on dipstick testing or >1 g/day); (iii) urinary abnormalities with hypertension and/or reduced renal function; (iv) persisting urinary abnormalities with evidence of continuing or relapsing extrarenal disease activity.

Clinicopathological correlations

The findings on renal biopsy offer the best available prediction of renal outcome. Figure 2(b) is complied from the same sources as Fig. 2(a). Biopsy was technically unsatisfactory in six of the 318 children whose clinical data are shown in Fig. 2(a), so that the total analysed in Fig. 2(b) is 312. As the figure shows, the proportion of cases progressing to category D strongly correlates with the histological grade, 70.8 per cent of patients with grade V biopsy changes and 23.8 per cent of those in grade IV having this outcome. Grade VI (membranoproliferative) changes are also associated with a poor outlook, although the small number of cases in this group means that the exact proportions shown in Fig. 2(b) should not be given too much weight. The study by Bunchman *et al.* (1988) showed a similar outcome in the worse biopsy grades: end-stage renal disease ensued in three of seven patients in group IV (43 per cent), four of five in group V (80 per cent), and three of six in group VI (50 per cent). The apparently very high incidence of adverse outcome in the less severe biopsy categories—one of two in group I (50 per cent) and two of six in group III (33 per cent)—is almost certainly an artefact of the selection process used in this study, as discussed in the previous section.

The series reported by the Paris group (Levy *et al.* 1976; Habib 1983) used a slightly different classification in which biopsies were divided into those with (proliferative) and without (non-proliferative) mesangial cell proliferation: each of these groups was then subdivided by the percentage of glomeruli showing crescent formation. In both proliferative and non-proliferative groups a bad prognosis was associated with numerous crescents, especially if more than 75 per cent of glomeruli were involved, and the proliferative group children generally had a worse outcome than non-proliferative children. However, the excess of group D children in the group with proliferation is entirely accounted for by the fact that no non-proliferative biopsy showed more than 70 per cent crescents. As in the other studies reviewed here, and which form the basis of Fig. 2(b), the number of crescents appears to be a specific predictor of poor outlook and an independent effect of other histological variables is not convincingly shown.

The relationship between the presence and distribution of electron-dense 'deposits' in the glomeruli and clinical course has been examined by several groups. Heaton *et al.* (1977) observed that patients with subepithelial deposits, and particularly those with deposits present at all three glomerular sites (mesangial, subendothelial, and subepithelial) were more likely to have a poor outcome than those with mesangial deposits alone, or mesangial and subendothelial but not subepithelial deposits. The numbers were small, however, and there were exceptions in both directions. Yoshikawa *et al.* (1981) reported similar findings: in their series, no child without subepithelial deposits had a grade D outcome, while only those with deposits in all three locations had grade V or VI histology. These authors also observed a correlation between the presence of electron-dense, 'lead shot' microparticles in mesangial and subendothelial locations and poor outcome: nine of 16 patients with this finding were in category C or D after a mean follow up period of 5.8 years while only one of 19 of those without it was in category C and none was in category D. In a more recent study (Yoshikawa *et al.* 1986), the presence of lytic lesions in the glomerular basement membrane, apparently spatially associated with subepithelial dense deposits and neutrophil polymorphonuclear leucocytes, was also associated with other predictors of a deteriorating clinical course: they were seen in 89 per cent of nine patients with nephrotic syndrome, 70 per cent of 37 with heavy proteinuria but only 35 per cent of 51 with slight or no proteinuria.

Further study of the clinical associations with glomerular ultrastructural changes in Henoch-Schönlein purpura is indicated to confirm and amplify these interesting observations. In the meantime, the presence and number of enveloping crescents on optical microscopy remains the most reliable predictor of the likely clinical course in individual cases.

Treatment

General measures

No specific treatment has ever been shown, in a controlled fashion, to be of any value in the treatment of any aspect of Henoch-Schönlein purpura. The majority of cases are mild and clearly need no treatment, or only supportive measures. If the onset of the disease is associated with a bacterial infection, it is reasonable to administer an appropriate antibiotic. Fluid and electrolyte balance should be monitored in any child with significant renal or gastrointestinal disease, particularly in those with generalized or dependent oedema. If the oedema is associated with hypoalbuminaemia, the cause may be renal, gastrointestinal, or, possibly, transudation of albumin into the interstitial space due to increased capillary permeability: the nephrotic syndrome should not be diagnosed without documentation of urinary protein loss of commensurate magnitude. Intravenous infusion of albumin or plasma is helpful in selected cases, but should be reserved for those with clinical evidence of circulatory volume contraction: low central venous pressure, reduced peripheral perfusion (increased central-to-peripheral temperature gap) and low urinary sodium concentration (<10 mmol/l). Many patients with oedema, including some of those with hypoalbuminaemia, have a predominantly 'nephritic' illness with normal or increased circulating volume and the administration of plasma expanders in such cases is dangerous.

Extrarenal manifestations

Analgesics should be given as necessary: both joint and abdominal pain may be severe, and the potency of analgesia that is appropriate is commonly underestimated, particularly in younger children who are unable to articulate their distress. Blood transfusion may be necessary for patients with gastrointestinal bleeding. Suspicion of intussusception or other major abdominal events (see above) should be investigated urgently in consultation with a surgeon familiar with the disease and its complications. Abdominal ultrasonography has greatly facilitated the investigation of the acute abdomen in patients with Henoch-Schönlein purpura (Clark and Kramer 1986), but contrast enema examination may occasionally be necessary. 'Exploratory laparotomy' without proper preoperative evaluation is inappropriate in the modern era, and should be relegated to history, since unnecessary abdominal surgery in a sick patient with Henoch-Schönlein purpura is poorly tolerated and may adversely affect the outcome.

Allen *et al.* (1960) reviewed the literature and added their own experience of the effects of corticosteroids on the abdominal and systemic features of Henoch-Schönlein purpura. They concluded that in about one-third of patients, particularly those with oedema and gastrointestinal signs and symptoms, steroids produced immediate and dramatic benefit. In their own series of 131 patients about half received steroids: none of those treated developed an intussusception, while four of the untreated patients did so. Although these observations were not prospectively controlled, they have strongly influenced subsequent practice. Most paediatricians have seen instances of striking improvement, especially in abdominal pain, within hours of starting steroids, often enough to be convinced that a trial of oral prednisolone or a comparable drug should be instituted in patients with severe intestinal symptoms in whom a surgical complication has been ruled out. No other drug treatment has ever shown to be of any benefit.

Renal disease

It must be emphasized that most patients with Henoch-Schönlein purpura have either no clinically apparent renal disease or only transient urinary abnormalities. No treatment is needed in the mildly affected majority. Even among those with more established renal disease, manifested as nephrotic syndrome or acute nephritic syndrome, less than half have a poor long-term outcome (Fig. 2(a)), and chronic renal failure is seen frequently only among biopsied patients in histological grades IV to VI, especially V (Fig. 2(b)), who are quite rare. The results of treatment of this very small high-risk group are therefore very different to assess because of small numbers and the consequent lack of controlled data.

The one aspect of the nephritis of Henoch-Schönlein purpura in which treatment is of definite benefit, as in other chronic renal diseases, is control of blood pressure. This applies during the acute phase of the illness, but more importantly in the long term in those with evidence of incomplete resolution of the active disease (persistent urinary abnormalities and/or reduced GFR). Blood pressure should therefore be monitored carefully during and after the clinical illness and treated if found to be raised. Surveillance should be continued while urinary abnormalities persist and probably for some years thereafter, since cases of hypertension have been observed several years after apparent normalization of renal function and the urinary sediment. It is our practice at Guy's to continue hospital follow up to 2 years after disappearance of haematuria and proteinuria, and to recommend biannual blood pressure checks by the general practitioner thereafter.

Steroids given by mouth in 'conventional' doses (prednisolone 1–2 mg/kg.day or 30–60 mg/m^2.day) have never been shown to be of benefit in nephritis associated with Henoch-Schönlein purpura. Counahan et al. (1977), in a follow-up study of the patients first reported from Guy's Hospital by meadow et al. (1972), found that treatment with steroids or 'immunosuppressives' was associated with a poorer outcome than no treatment. Not surprisingly, this could be explained by the patients with the most severe disease being more likely to have received treatment than those less severely affected: when this was taken into account the apparent difference disappeared. The results in the other large published series cited earlier in this chapter are similar: there are no scientific grounds for using oral steroids in this dosage in these patients. In a recent letter from Czechoslovakia, Buchanec et al. (1987) claimed that the routine administration of steroids to children with Henoch-Schönlein purpura at the clinical onset of the disease, before evidence of renal involvement was apparent, reduced the risk of nephropathy developing subsequently compared with children not so treated. This conclusion was based on retrospective, uncontrolled data, and would require confirmation in a prospective, randomized study before being accepted as valid.

Various combinations of steroids with other immunosuppressive drugs (usually azathioprine or an alkylating agent), anticoagulants, and antiplatelet drugs have been tried in Henoch-Schönlein purpura patients with severe nephritis as in other forms of rapidly progressive glomerulonephritis, but most series include one or two cases of Henoch-Schönlein purpura in a larger heterogeneous group and it is not possible to separate them from the others. For patients with rapidly progressive glomerulonephritis as a whole, taking extensive crescent formation as the definition, this therapy induced improvement in about 55 per cent of patients compared with only about 11 per cent of untreated (retrospective, pooled) controls (Haycock 1988). The patients with Henoch-Schönlein purpura seem to have responded in a manner indistinguishable from those with rapidly progressive glomerulonephritis due to other causes, but it should be emphasized that the numbers are so small that only a tentative conclusion can be drawn as to the benefit of this type of treatment in the nephritis of Henoch-Schönlien purpura.

Kim et al. (1988) compared treatment with rifampicin, which is known to have immunosuppressive properties (group A), with prednisolone and either azathioprine or cyclophosphamide (group B) in children with Henoch-Schönlein purpura and the nephrotic syndrome. They claimed a more favourable outcome in group A, in that all of seven were protein-free 6 months after treatment while only two of five of group B were protein free. In addition, group A patients were more likely than group B to show a reduction in intensity of immunofluorescent staining between biopsies performed before and after treatment. However, at final follow up only four of seven in group A, and two of five in group B were in clinical category A, the remainder in both groups being in category B ($\chi^2=0.34$, $p=$NS). Two group B patients improved their histological grade from IIIb to II, while no such improvement was seen in any group A patient. The authors claim that rifampicin improves the renal prognosis in Henoch-Schönlein purpura nephritis is not supported by the published data, particularly since all but one patient had biopsy changes in grades I–III, and would therefore have been expected to do well without treatment of any kind.

In recent years, two other treatments have been advocated for rapidly progressive glomerulonephritis, including the form associated with Henoch-Schönlein purpura. These are very high dose intravenous methylprednisolone ('pulse therapy', typically 600 mg/m^2.dose), originally introduced by Cole et al. (1976), and plasma exchange, first used in presumed immune complex glomerulonephritis in 1977 (Lockwood et al. 1977; Becker et al. 1977). As with the combined immunosuppressive regimens referred to above, isolated cases of nephritis with Henoch-Schönlein nephritis are included in numerous reports of rapidly progressive glomerulonephropathies of various aetiologies and it is impossible to distinguish them from the rest. For patients with rapidly progressive glomerulonephritis as a whole, pulse therapy and plasma exchange seem to be about equally effective, with about 75 per cent of patients showing improvement in renal function (Haycock 1988). One study (Robson et al. 1981) included details of seven patients treated with pulse therapy: initial GFR was 34 ± 11 ml/min.1.73 m^2, and after a follow-up interval of 34 ± 6 months, six of these had a GFR above 80 ml/min.1.73 m^2. Two patients with more than 65 per cent crescent formation, treated similarly by Ferraris et al. (1983), also did well after 24 and 38 months respectively.

At Guy's Hospital, which is a regional referral centre for children with renal disease, 31 children with Henoch-Schönlein purpura fulfilled our criteria for renal biopsy during the 10-year period 1979–89. The fact that only two of these had biopsy appearances in grade V and one in grade IV confirms the rarity of histological rapidly progressive glomerulonephritis in the setting of Henoch-Schönlein purpura. These three, and five others with grade III biopsies and deteriorating renal function, were treated with pulse therapy. The three patients with the most severe biopsy abnormalities, one of whom was dialysed for oliguric renal failure, were also given plasma exchange. After a mean follow-up period of 33 months (range 12–72), three are in category A, one is in category B, and four in category C with GFR of 67, 75, 88, and 145 ml/min. 1.73 m^2, respectively (the boy who was dialysed had the lowest GFR). Despite the lack of control data, these results and those mentioned above differ more from the untreated outcome than would be expected by chance or sampling error. We therefore recommend that children with Henoch-Schönlein purpura and nephritis and either grade IV or worse histology, or deteriorating renal function, should be treated with pulse therapy initially, and that if no convincing improvement is seen after three or four doses plasma exchange should be added. The occasional child with late deterioration of renal function (>3–6 months after onset of renal disease) is probably best managed by good supportive care, particularly as regards blood pressure control. There is even less evidence of benefit from immunosuppressive therapy in this group than in those treated early in the disease.

Henoch-Schönlein purpura and transplantation

In 1972 Bar-On and Rosenmann described recurrence of both renal and extrarenal features of Henoch-Schönlein purpura in a 26-year-old man who had received a kidney transplant from a live related donor—his twin brother. Similar, sporadic reports subsequently appeared of children (Baliah et al. 1974; Sakai et al. 1975) and another adult (Nast et al. 1987), all of whom received transplants from living related donors all of whom suffered eventual graft loss. Two more extensive series have also been published, one from Paris describing the clinical and histological features of 11 cadaveric transplants placed in children with a primary diagnosis of Henoch-Schönlein purpura (Habib 1983), and another from Tokyo describing the results of 17 grafts, 12 from living related donors and five cadaveric, placed in 15 children (Hasegawa et al. 1989). These two reports differ substantially in their findings.

Habib et al. (1983) found recurrence of clinical disease in none of the children, but mesangial IgA deposits were present in seven of nine grafts biopsied. The immunoglobulin deposition was not accompanied by evidence of glomerulonephritis visible on light microscopy, apart from some 'mesangial hypertrophy'. In contrast, Hasegawa et al. (1989) found both clinical and histological recurrence in five of their 12 living related grafts, histological recurrence alone in four and no recurrence in three. Clinical recurrence occurred in none of the five cadaveric grafts in this series; five of these were biopsied, three three times and one twice, and IgA deposits were not seen on any occasion. The results of retransplantation in the Japanese series are particularly noteworthy. One child whose first graft (from his mother) was lost to recurrent disease received a second cadaveric graft which showed no clinical recurrence at 22 months and no evidence of IgA in biopsies at 1 h, 2 months, and 1 year post-transplant. Another had lost her first living related graft to recurrence of Henoch-Schönlein purpura while still in the care of another centre: this is the same child previously reported by Sakai (1975). She then received a second, cadaveric, graft which was lost to primary non-function without having been biopsied. A third cadaveric graft functioned and was free of recurrence when biopsied for the third time 1 year after transplantation. In this context it may be noted that the adult reported by Nast et al. (1987), who lost his first living related graft to recurrent disease showed no evidence of recurrence following a second, cadaveric, graft up to the time of reporting (9 months post-transplantation).

The published evidence, then, strongly suggests that patients with Henoch-Schönlein purpura nephritis do well with cadaveric transplants despite a high incidence of mesangial IgA deposition in the grafts, while patients receiving kidneys from living related donors are subject to a high rate of clinical disease recurrence. This may affect both the kidney and other systems, and leads to graft loss in many cases. Contrary to folklore, waiting a long time after the original disease presentation before performing the transplant affords little or no protection against recurrence in grafts from living related donors: the five patients with clinical recurrence in the report of Hasegawa et al. (1989) received their grafts after intervals of 27 to 101 months (mean 46 months). In the current state of knowledge it would seem prudent to recommend that patients with renal failure due to Henoch-Schönlein purpura should be offered cadaveric transplants, but organs from living related donors should probably not be used because the rate of graft loss due to recurrent disease is unacceptably high.

Bibliography

Ackroyd, J.F. (1953). Allergic purpura, including purpura due to foods, drugs and infections. *American Journal of Medicine*, 14, 605–32.

Allen, D.M., Diamond, L.K., and Howell, D.A. (1960). Anaphylactoid purpura in children (Schönlein-Henoch syndrome). *American Journal of Diseases of Children*, 99, 833–54.

Counahan, R., and Cameron, J.S. (1977). Henoch-Schönlein nephritis. *Contributions to Nephrology*, 7, 143–65.

Cream, J.J., Gumpel, J.M., and Peachey, R.D.G. (1970). Schönlein-Henoch purpura in the adult. A study of 77 adults with anaphylactoid or Schönlein-Henoch purpura. *Quarterly Journal of Medicine*, 39, 461–84.

Gairdner, D. (1948). The Schönlein-Henoch syndrome (anaphylactoid purpura). *Quarterly Journal of Medicine*, 17, 95–122.

Habib, R. (1983). Purpura rheumatoïde. In *Néphrologie Pédiatrique*, (3rd edn). (eds. P. Royer, R. Habib, H. Mathieu, and M. Broyer), pp. 342–50. Flammarion, Paris.

Heptinstall, R.H. (1983). Schönlein-Henoch syndrome; lung hemorrhage and glomerulonephritis, or Goodpasture's syndrome. In *Pathology of the Kidney*, (3rd edn). (ed. R. H. Heptinstall), pp. 741–91. Little, Brown, Boston and Toronto.

Koskimies, O., Rapola, J., Savilahti, E., and Vilska, J. (1974). Renal involvement in Schönlein-Henoch purpura. *Acta Paediatrica Scandinavica*, 63, 357–63.

Meadow, S.R., Glasgow, E.F., White, R.H.R., Moncrieff, M.W., Cameron, J.S., and Ogg, C.S. (1972). Schönlein-Henoch nephritis. *Quarterly Journal of Medicine*, 41, 241–58.

Urizar, R.E., Michael, A.E., Sisson, S., and Vernier, R.L. (1968). Anaphylactoid purpura. II. Immunofluorescent and electron microscopic studies of the glomerular lesions. *Laboratory Investigation*, 19, 437–50.

Vernier, R.L., Worthen, H.G., Peterson, R.D., Colle, R.D., and Good, R.A. (1961). Anaphylactoid purpura. I. Pathology of the skin and kidney and frequency of streptococcal infection. *Pediatrics*, 27, 181–93.

References

Abdel-Hadi, O., Hartley, R.B., Greenstone, M.A. (1981). Myocardial infarction—a rare complication of Henoch-Schönlein purpura. *Postgraduate Medical Journal*, 57, 390–2.

Aram, H. (1987). Henoch-Schönlein purpura induced by chlorpromazine. *Journal of the American Academy of Dermatology*, 17, 139–40.

Ayoub, E.M. and Hoyer J. (1969). Anaphylactoid purpura: streptococcal antibody titers and beta-1 globulin levels. *Journal of Pediatrics*, **75**, 193–201.

Baart de la Faille-Kuyper, E.H., *et al.* (1976). Occurrence of vascular IgA deposits in clinically normal skin of patients with renal disease. *Kidney International*, **9**, 424–9.

Baliah, T., Kim, K.H., Anthone, S., Anthone, R., Montes, M., and Andres, G.A. (1974). Recurrence of Henoch-Schönlein purpura glomerulonephritis in transplanted kidneys. *Transplantation*, **18**, 343–6.

Bannister, K.M., Drew, P.A., Clarkson, A.R., and Woodroffe, A.J. (1983). Immunoregulation in glomerulonephritis, Henoch-Schönlein purpura and lupus nephritis. *Clinical and Experimental Immunology*, **53**, 384–90.

Bar-On, H. and Rosenmann, E. (1972). Schönlein-Henoch syndrome in adults. A clinical and histological study of renal involvement. *Israeli Journal of Medical Sciences*, **8**, 1702–15.

Beale, M.G., Nash, G.S., Bertovich, M.J., and MacDermott, R.P. (1982). Similar disturbances in B cell activity and regulatory T cell function in Henoch-Schönlein purpura and systemic lupus erythematosus. *Journal of Immunology*, **128**, 486–91.

Belman, A.L., Leicher, C.R., Moshé, S.L., and Mezey, A.P. (1985). Neurologic manifestations of Schönlein-Henoch purpura: report of three cases and review of the literature. *Pediatrics*, **75**, 687–92.

Becker, G.J., d'Apice, A.J.F., Walker, R.G., and Kincaid-Smith, P. (1977). Plasmapheresis in the treatment of glomerulonephritis. *Medical Journal of Australia*, **2**, 693–6.

Bene, M.C., Faure, G., Hurault de Ligny, B., Kessler, M., and Duheille, J. (1983). Quantitative immunohistomorphometry of the tonsillar plasma cells evidences an inversion of IgA versus IgG secreting cell balance. *Journal of Clinical Investigation*, **71**, 1342–7.

Bene, M.C., Hurault de Ligny, B., Faure, G., Kessler, M., and Duheille, J. (1986). Histoimmunological discrepancies in primary IgA nephropathy and anaphylactoid purpura sustain relationships between mucosa and kidney. *Nephron*, **43**, 214–16.

Berger, J. (1969). IgA glomerular deposition in renal disease. *Transplantation Proceedings*, **1**, 939–44.

Berthoux, F.C., *et al.* (1978). HLA BW35 and mesangial IgA glomerulonephritis. *New England Journal of Medicine*, **298**, 1034–5.

Bull, P.W., Scott, P.W., and Breathnach, S.M. (1987). Henoch-Schönlein purpura associated with Legionnaire's disease. *British Medical Journal*, **294**, 220.

Buchanec, J., Galanda, V., Minarik, M., and Zibolen, M. (1987). Can the nephropathy of Schönlein-Henoch purpura (syndrome) be prevented by the early administration of steroids? *Clinical Nephrology*, **28**, 156.

Bunchman, T.E., Mauer, S.M., Sibley, R.K., and Vernier, R.L. (1988). Anaphylactoid purpura: characteristics of 16 patients who progressed to renal failure. *Pediatric Nephrology*, **2**, 393–7.

Burke, D.M. and Jellinek, J.I. (1954). Nearly fatal case of Schönlein-Henoch syndrome following insect bite. *American Journal of Diseases of Children*, **88**, 772–4.

Burke, E.C., Mills, S.D., and Stickler, G.B. (1960). Nephritis associated with anaphylactoid purpura in childhood: clinical observations and prognosis. *Proceedings of the Mayo Clinic*, **35**, 641–8.

Bywaters, E.G.L., Isdale, I., and Kempton, J.J. (1957). Schönlein-Henoch syndrome—evidence for a group A beta-haemolytic streptococcal aetiology. *Quarterly Journal of Medicine*, **26**, 161–77.

Cairns, S.A., Mallick, N.P., Lawler, W., and Williams, G. (1978). Squamous cell carcinoma of bronchus presenting with Henoch-Schönlein purpura. *British Medical Journal*, **2**, 474–5.

Casanueva, B., Rodriguez-Valverde, V., Merino, J., Arias, M., and Garcia-Fuentes, M. (1983). Increased IgA-producing cells in the blood of patients with active Henoch-Schönlein purpura. *Arthritis and Rheumatism*, **26**, 854–60.

Chiappo, G.F., Mignone, F., Oggero, R., Licata, D., and Cordero, A. (1978). Complicanze renale nella sindrome di Schönlein-Henoch in eta pediatrica. *Minerva Pediatrica*, **30**, 1933–41.

Clark, W.R., and Kramer, S.A. (1986). Henoch-Schönlein purpura and the acute scrotum. *Journal of Pediatric Surgery*, **21**, 991–2.

Cole, B.R., Brocklebank, J.T., Kienstra, R.A., Kissane, J.M., and Robson, A.M. (1976). 'Pulse' methylprednisolone therapy in the treatment of severe glomerulonephritis. *Journal of Pediatrics*, **88**, 307–14.

Conley, M.E., Cooper, M.D., and Michael, A.F. (1980). Selective deposition of immunoglobulin A in IgA nephropathy, anaphylactoid purpura nephritis and systemic lupus erythematosus. *Journal of Clinical Investigation*, **66**, 1432–6.

Conley, M.E. and Delacroix, D.L. (1987). Intravascular and mucosal immunoglobulin A: two separate but related systems of immune defence. *Annals of Internal Medicine*, **106**, 892–9.

Coppo, R., *et al.* (1984). IgA1 and IgA2 immune complexes in primary IgA nephropathy and Henoch-Schönlein purpura. *Clinical and Experimental Immunology*, **57**, 583–90.

Counahan, R., *et al.* Prognosis of Henoch-Schönlein nephritis in children. *British Medical Journal*, **2**, 11–14.

Dalgleish, P.G., and Ansell, B.M. (1950). Anaphylactoid purpura in pulmonary tuberculosis. *British Medical Journal*, **1**, 225–7.

D'Amico, G., *et al.* (1985). Idiopathic IgA mesangial nephropathy. Clinical and histological study of 374 patients. *Medicine*, **64**, 49–60.

Danielsen, H., Eriksen, E.F., Johansen, A., and Solling, J. (1984). Serum immunoglobulin sedimentation patterns and circulating immune complexes in IgA nephropathy and Schönlein-Henoch nephritis. *Act Medica Scandinavica*, **215**, 435–41.

Davin, J.C., Vandenbroeck, M.C., Foidart, J.B., and Mahieu, P.R. (1985). Sequential measurements of the reticulo-endothelial system function in Henoch-Schönlein disease of childhood. Correlations with various immunological parameters. *Acta Paediatrica Scandinavica*, **74**, 201–6.

Davin, J.C., Malaise, M., Foidart, J., and Mahieu, P. (1987). Anti-alpha-galactosyl antibodies and immune complexes in children with Henoch-Schönlein purpura or IgA nephropathy. *Kidney International*, **31**, 1132–9.

Debray, J., Krulik, M., and Giorgi, H. (1971). Le purpura rheumatoide (syndrome de Schönlein-Henoch) de l'adulte. A propos de 22 observations. *Semaine des Hôpitaux de Paris*, **28**, 1805–19.

Demaine, A.G., Rambausek, M., Knight, J.F., Williams, D.G., Welsh, K.I., and Ritz, E. (1988). Relation of mesangial IgA glomerulonephritis to polymorphism of immunoglobulin heavy chain switch region. *Journal of Clinical Investigation*, **81**, 611–14.

Derham, R.J. and Rogerson, M.M. (1956). The Schönlein-Henoch syndrome with particular reference to renal sequelae. *Archives of Disease in Childhood*, **31**, 364–8.

Doi, T., Kanatsu, K., Sekita, K., Yoshida, H., Nagai, H., and Hamashima, Y. (1984). Detection of IgA class circulating immune complexes bound to anti-C3D antibody in patients with IgA nephropathy. *Journal of Immunological Methods*. **69**, 95–104.

Dosa, S., Cairns, S.A., Mallick, M.P., Lawler, W., and Williams, G. (1980). Relapsing Henoch-Schönlein syndrome with renal involvement in a patient with an IgA monoclonal gammopathy. *Nephron*, **26**, 145–8.

Duell, P.B. and Morton, W.E. (1987). Henoch-Schönlein purpura following thiram exposure. *Archives of Internal Medicine*, **147**, 778–9.

Egido, J., *et al.* (1980). A possible common pathogenesis of the mesangial IgA glomerulonephritis in patients with Berger's disease and Schönlein-Henoch syndrome. *Proceedings of the European Dialysis and Transplant Association*. **17**, 660–6.

Evans, D.J., *et al.* (1973). Glomerular deposition of properdin in Henoch-Schönlein syndrome and idiopathic focal nephritis. *British Medical Journal*, **3**, 326–8.

Farley, T.A., Gillespie, S., Rasoulpour, M., Tolentino, N., Hadler, J.L., and Hurwitz, E. (1989). Epidemiology of a cluster of Henoch-Schönlein purpura. *American Journal of Diseases of Children*, **143**, 798–803.

Feehally, J. (1988). Immune mechanisms in glomerular IgA deposition. *Nephrology Dialysis and Transplantation*, **3**, 361–78.

Ferraris, J.R., Gallo, G.E., Ramirez, J., Iotti, R., and Gianantonio, C. (1983). 'Pulse' methylprednisolone therapy in the treatment of acute crescentic glomerulonephritis. *Nephron*, 34, 207–8.

Fitzsimmons, J.S. (1968). Uncommon complications of anaphylactoid purpura. *British Medical Journal*, 4, 431–2.

Fogazzi, G.B., et al. (1989). Long term outcome of Schönlein-Henoch nephritis in the adult. *Clinical Nephrology*, 31, 60–6.

Frank, E. (1915). Die essentielle Thrombopenie (Konstitutionelle Purpura: Pseudo-Hämophilie). *Berliner Klinische Wochenschrift*, 52, 4548.

Garcia-Fuentes, M., Chantler, C., and Williams, D.G. (1977). Cryoglobulins in Henoch-Schönlein purpura. *British Medical Journal*, 2, 163–5.

Garcia-Fuentes, M., Martin, A., Chantler, C., and Williams, D.G. (1978). Serum complement components in Henoch-Schönlein purpura. *Archives of Disease in Childhood*, 53, 417–19.

Garner, J.A. McV. (1977). Acute pancreatitis as a complication of anaphylactoid (Henoch-Schönlein) purpura. *Archives of Disease in Childhood*, 52, 971–2.

Gelfand, E.W., Clarkson, J.E., and Minta, J.O. (1975). Selective deficiency of the second component of complement in a patient with anaphylactoid purpura. *Clinical Immunology and Immunopathology*, 4, 269–76.

Giangacomo, J. and Tsai, C.C. (1977). Dermal and glomerular deposition of IgA in anaphylactoid purpura. *American Journal of Diseases of Children*, 131, 981–3.

Glanzmann, E. (1916). *Jahrbuch für Kinderheilkunde*, 83, 271.

Goldbloom, R.B. and Drummond, K.N. (1968). Anaphylactoid purpura with massive gastrointestinal hemorrhage and glomerulonephritis. *American Journal of Diseases of Children*, 116, 97–102.

Haahr, J., Thomsen, K., and Sparrevohn, S. (1974). Renal involvement in Henoch-Schönlein purpura. *British Medical Journal*, 4, 405–6.

Habib, R. and Levy, M. (1972). Les nephropathies du purpura rheumatoide chez l'enfant. *Archives Françaises de Pédiatrie*, 29, 305–24

Hall, R.P., Lawley, T.J., Heck, J.A. and Katz, S.I. (1980). IgA-containing circulating immune complexes in dermatitis herpetiformis, Henoch-Schönlein purpura, systemic lupus erythematosus and other diseases. *Clinical and Experimental Immunology*, 40, 431–7.

Hasegawa, A., et al. (1989). Fate of renal grafts with recurrent Henoch-Schönlein purpura nephritis in children. *Transplantation Proceedings*, 21, 2130–3.

Haycock, G.B. (1988). The treatment of glomerulonephritis in children. *Pediatric Nephrology*, 2, 247–55.

Heaton, J.M., Turner, D.R., and Cameron, J.S. (1977). Localization of glomerular 'deposits' in Henoch-Schönlein nephritis. *Histopathology*, 1, 93–104.

Hene, R.J., Schuurman, H.J., and Kater, L. (1987). IgA subclass-containing plasma cells in the jejunum in primary IgA nephropathy and Henoch-Schönlein purpura. *Abstracts of the Xth International Congress of Nephrology*, p. 380.

Henoch, E. (1874). Über eine eigenthümliche Form von Purpura. *Berliner klinische Wochenschrift*, 11, 641.

Henoch, E. (1899). *Vorlesungen über Kinderkrankheiten*, Vol. 10, p. 839. A. Hirschwald, Berlin.

Hughes, F.J., Wolfish, N.M., and McLaine, P.N. (1988). Henoch-Schönlein syndrome and IgA nephropathy: a case report suggesting a common pathogenesis. *Pediatric Nephrology*, 2, 389–92.

Imai, T. and Matsumoto, S. (1970). Anaphylactoid purpura with cardiac involvement. *Archives of Disease in Childhood*, 45, 727–9.

Immonen, P. and Kouvulainen, K. (1967). Immunoglobulin levels in anaphylactoid purpura. *Scandinavian Journal of Clinical and Laboratory Investigations*, 95, 115.

Jacome, A.F. (1967). Pulmonary haemorrhage and death complicating anaphylactoid purpura. *Southern Medical Journal*, 60, 1003–4.

Jiménez, E.L. and Darrington, H.J. (1968). Vaccination and Henoch-Schönlein purpura. *New England Journal of Medicine*, 279, 1171.

Kalowski, S. and Kincaid-Smith, P. (1973). Glomerulonephritis in Henoch-Schönlein syndrome. In *Glomerulonephritis: morphology, natural history and treatment* (eds. P. Kincaid-Smith, T. H. Mathew, and E. L. Becker), pp. 1123–1132. Wiley, New York.

Kauffmann, R.H., Herrmann, W., Meyer, C.J.L.M., Daha, M.R., and van Es, L.A. (1980). Circulating IgA-immune complexes in Henoch-Schönlein purpura. A longitudinal study of their relationship to disease activity and vascular deposition of IgA. *American Journal of Medicine*, 69, 859–66.

Kauffmann, R.H., van Es, L.A., and Daha, M.R. (1981). The specific detection of IgA immune complexes. *Journal of Immunological Methods*, 40, 117–29.

Kher, K.K., Sheth, K.J., and Makker, S.P. (1983). Stenosing ureteritis in Henoch-Schönlein purpura. *Journal of Urology*, 129, 1040–2.

Kim, P.K., Kim, K.S., Lee, J.K., Lee, J.S., Jeong, H.J., and Choi, I.J. (1988). Rifampicin therapy in Henoch-Schönlein purpura nephritis accompanied by nephrotic syndrome. *Child Nephrology and Urology*, 9, 50–6.

Knight, J.F., et al. (1988). IgA rheumatoid factor and other autoantibodies in acute Henoch-Schönlein purpura. *Contributions to Nephrology*, 67, 117–20.

Kobayashi, O., Wada, H., Kanasawa, M., and Kamiyama, T. (1965). The anaphylactoid purpura-nephritis in childhood. *Acta Medica et Biologica (Niigata)*, 13, 181–97.

Kobayashi, O., Wada, H., Okawa, K., and Takeyama, I. (1977). Shonlein-Henoch's syndrome in children. In *Contributions to Nephrology*, Vol. 4 (eds. G. M. Berlyne, and S. S. Giovanetti), p. 48, Karger, Basel.

Koskimies, O., Mir, S., Rapola, J., and Vilska, J. (1981). Henoch-Schönlein nephritis: long term prognosis of unselected patients. *Archives of Disease in Childhood*, 56, 482–4.

Lee, H.S., Koh, H.I., Kim, M.J., and Rha, H.Y. (1986). Henoch-Schönlein nephritis in adults: a clinical and morphological study. *Clinical Nephrology*, 26, 125–30.

Lefrère, J.J., et al. (1986). Henoch-Schönlein purpura and human parvovirus infection. *Pediatrics*, 78, 183–4.

Lelong, M., and Deroubaix, P. (1968). Le purpura rheumatoide de Schönlein-Henoch: étude de 36 observations. *Lille Medicine*, 13, 307–12.

Le Petit, J.C., et al. (1982). Genetic investigations in mesangial IgA nephropathy. *Tissue Antigens*, 19, 108–14.

Levinsky, R.J. and Barratt, T.M. (1979). Immune complexes in Henoch-Schönlein purpura. *Lancet*, ii, 1100–3.

Levitt, L.M. and Burbank, B. (1953). Glomerulonephritis as a complication of Schönlein-Henoch syndrome. *New England Journal of Medicine*, 248, 530–6.

Levy, M., Broyer, M., Arson, A., Levy-Bentolila, A., and Habib, R. (1976). Anaphylactoid purpura nephritis in childhood: natural history and immunopathology. In *Advances in Nephrology*, Vol. 6. (eds. J. Hamburger, J. Crosnier and M. H. Maxwell), p. 183, Year Book, Chicago.

Liew, S.W. and Kessel, I. (1974). *Mycoplasma* pneumonia preceding Henoch-Schönlein purpura. *Archives of Disease in Childhood*, 49, 912–3.

Lockwood, C.M., et al. (1977). Plasma exchange and immunosuppression in the treatment of fulminating immune complex glomerulonephritis. *Lancet*, i, 63–7.

Lomax-Smith, J.D., Zabrowarny, L.A., Howarth, G.S., Seymour, A.E., and Woodroffe, A.J. (1983). The immunochemical characterization of mesangial IgA deposits. *American Journal of Pathology*, 113, 359–64.

Lombard, K.A., Shah, P.C., Thrasher, T.V., and Grill, B.B. (1986). Ileal stricture as a late complication of Henoch-Schönlein purpura. *Pediatrics*, 77, 396–8.

Lopez-Trascasa, M., Egido, J., Sanchez, J., and Hernando, L. (1980). Iga glomerulonephritis (Berger's disease): evidence of high serum levels of polymeric IgA. *Clinical and Experimental Immunology*, 42, 247–54.

Macalister, G.H.K. (1906). The prognostic significance of purpura in children (second of two parts). *Guy's Hospital Gazette*, 20, 196–204.

Martini, A., Ravelli, A., Notarangelo, L.D., Burgio, V.L., and Plebani, A. (1985). Henoch-Schönlein syndrome and selective IgA deficiency. *Archives of Disease in Childhood*, **60**, 160–1.

McLean, R.H., Wyatt, R.J., and Julian, B.A. (1984). Complement phenotypes in glomerulonephritis: increased frequency of homozygous null C4 phenotypes in IgA nephropathy. *Kidney International*, **26**, 855–60.

Meadow, S.R. and Scott, D.G. (1985). Berger disease: Henoch-Schönlein syndrome without the rash. *Journal of Pediatrics*, **106**, 27–32.

Montoliu, J. (1987). Henoch-Schönlein purpura complicating staphylococcal endocarditis in a heroin addict. *American Journal of Nephrology*, **7**, 137–9.

Mota-Hernandez, F., Valbuena-Paz, R., and Gordillo-Paniagua, G. (1975). Long term prognosis of anaphylactoid purpura nephropathy. *Paediatrician*, **4**, 52–9.

Nast, C.C., Ward, H.J., Koyle, M.A., and Cohen, A.H. (1987). Recurrent Henoch-Schönlein purpura following renal transplantation. *American Journal of Kidney Diseases*, **9**, 39–43.

Neilsen, H.E. (1988). Epidemiology of Schönlein-Henoch purpura. *Acta Paediatrica Scandinavica*, **77**, 125–31.

Nicholls, K.M., Fairley, K.F., Dowling, J.P., and Kincaid-Smith, P.S. (1984). The clinical course of mesangial IgA associated nephropathy in adults. *Quarterly Journal of Medicine*, **53**, 227–50.

Nolasco, F.E.B., Cameron, J.S., Hartley, B., Coelho, A., Hildreth, G., and Reuben, R. (1987). Intraglomerular T cells and monocytes in nephritis: study with monoclonal antibodies. *Kidney International*, **31**, 1160–6.

Norkin, S. and Wiener, J. (1960). Involvement of other organs in Henoch-Schönlein syndrome. *American Journal of Clinical Pathology*, **33**, 55–65.

Nyulassy, S., et al. (1977). The HLA system in glomerulonephritis. *Clinical Immunology and Immunopathology*, **7**, 319–23.

Oliver, T.K.R. and Barnett, H.L. (1955). The incidence and prognosis of nephritis associated with anaphylactoid (Schönlein-Henoch) purpura in children. *American Journal of Diseases of Childhood*, **90**, 544–7.

O'Regan, S. and Robitaille, P. (1981). Orchitis mimicking testicular torsion in Henoch-Schönlein purpura. *Journal of Urology*, **126**, 834–5.

Osler W. (1914). Visceral lesions of purpura and allied conditions. *British Medical Journal*, **1**, 517–25.

Patel, U., Bradley, J.R., and Hamilton, D.V. (1988). Henoch-Schönlein purpura after influenza vaccination. *British Medical Journal*, **296**, 1800.

Pedersen, F.K. and Petersen, E.A. (1953). Varicella followed by glomerulonephritis. *Acta Paediatrica Scandinavica*, **64**, 886–90.

Philpott, M.G. (1952). The Schönlein-Henoch syndrome in childhood with particular reference to the occurrence of nephritis. *Archives of Disease in Childhood*, **27**, 480–1.

Pratesi, G. and Rizzuto, A. (1956). La glomerulonefrite nel corso della sindrome di Schoenlein-Henoch. *Rassegna di Fisiopatologia Clinica e Terapeutica*, **28**, 443–50.

Puppala, A.R., Cheng, J.C., and Steinhever, F.U. (1978). Pancreatitis—a rare complication of Schönlein-Henoch purpura. *American Journal of Gastroenterology*, **69**, 101–4.

Rasmussen, N.H. (1982). Henoch-Schönlein purpura after yersiniosis. *Archives of Disease in Childhood*, **57**, 322–3.

Richards, A.J. and Lindley, D.C. (1987). Henoch-Schönlein purpura associated with co-dydramol. *British Journal of Rheumatology*, **65**, 65.

Roberts, F.B., Slater, R.J., and Laski, B. (1962). The prognosis of Henoch-Schönlein nephritis. *Canadian Medical Association Journal*, **87**, 49–51.

Robson, A.M., Rose, G.M., Cole, B.R., and Ingelfinger, J.R. (1981). The treatment of severe glomerulonephritis in children with intravenous methylprednisolone 'pulses'. *Proceedings of the 8th International Congress on Nephrology, Athens, S-6.* pp. 305–311. Karger, Basel.

Rogers, P.W., Bunn, S.M., Kurtyman, M.C., and White, M.C. (1971). Schönlein-Henoch syndrome associated with exposure to cold. *Archives of Internal Medicine*, **128**, 782–6.

Rook, A. (1958). William Heberden's cases of anaphylactoid purpura. *Archives of Disease in Childhood*, **33**, 271.

Rubin, M. (1955). In discussion following paper of Oliver and Barnett cited above. *American Journal of Diseases of Childhood*, **90**, 545–6.

Sakai, T., Tanaka, T., Kasai, N., Shinagawa, I., and Endo, T. (1975). Recurrence of Henoch-Schönlein purpura (HSP) glomerulonephritis (GN) in transplanted kidney. *VIth International Congress of Nephrology: Abstracts of Free Communications*, abstract 1023.

Saulsbury, F.T. (1986). IgA rheumatoid factor in Henoch-Schönlein purpura. *Journal of Pediatrics*, **108**, 71–6.

Saulsbury, F.T. (1987). The role of IgA rheumatoid factor in the formation of IgA-containing immune complexes in Henoch-Schönlein purpura. *Journal of Clinical and Experimental Immunology*, **23**, 123–7.

Shichiri, M., Tsutsumi, K, Yamamoto, I., Ida, T., and Iwamoto, H. (1987). Diffuse intrapulmonary hemorrhage and renal failure in adult Henoch-Schönlein purpura. *American Journal of Nephrology*, **7**, 140–2.

Schönlein, J.L. (1837). *Allgemeine und specielle Pathologie und Therapie*, Vol. 1. (3rd edn), p. 48.

Sengar, D.P., Achyara, C.D., and Wolfish, N.M. (1984). HLA specificities, lymphocyte subsets, and mitogenic response in Henoch-Schönlein purpura nephritis. *International Journal of Pediatric Nephrology*, **5**, 197–200.

Similä, S., Kouvalainen, K., and Lanning, M. (1977). Serum immunoglobulin levels in the course of anaphylactoid purpura in children. *Acta Paediatrica Scandinavica*, **66**, 537–40.

Smith, H.J. and Krupski, W.C. (1980). Spontaneous intestinal perforation in Schönlein-Henoch purpura. *Southern Medical Journal*, **73**, 603–10.

Steare, S.E. (1988). *Mycoplasma pneumoniae* infection associated with Henoch-Schönlein purpura. *Journal of Infection*, **16**, 305–7.

Sterky, G. and Thilen, A. (1960). A study on the onset and prognosis of acute vascular purpura (the Schönlein-Henoch syndrome) in children. *Act Paediatrica (Uppsala)*, **49**, 217–29.

Striker, G.E., Quadracci, L.J., Larter, W., Hickman, R.O., Kelly, M.R., and Schaller, J. (1973). The nephritis of Henoch-Schönlein purpura. In *Glomerulo-nephritis: morphology, natural history and treatment* (eds. P. Kincaid-Smith, T. H. Mathew, and E. L. Becker), pp. 1105–21. Wiley, New York.

Sussman, M., Jones, J.H., Almeida, J.D., and Lachmann, P.J. (1973). Deficiency of the second component of complement associated with anaphylactoid purpura and presence of mycoplasma in the serum. *Clinical and Experimental Immunology*, **14**, 531–9.

Talbot, D., Craig, R., Falconer, S., Tomson, D., and Milne, D.D. (1988). Henoch-Schönlein purpura secondary to trauma. *Archives of Disease in Childhood*, **63**, 1114–15.

Tay, C.H. and da Costa, J.L. (1976). Henoch-Schönlein purpura in adult Singaporeans—a ten year study of 107 patients. *Annals of the Academy of Medicine (Singapore)*, **5**, 11–20.

Tomino, Y., Endoch, M., Miura, M., Nomoto, Y., and Sakai, H. (1983). Immunopathological similarities between IgA nephropathy and Henoch-Schönlein purpura (HSP) nephritis. *Acta Pathologica Japonica (Tokyo)*, **33**, 113–22.

Trygstad, C.W. and Stiehm, E.R. (1971). Elevated serum IgA levles in anaphylactoid purpura. *Pediatrics*, **47**, 1023–8.

Tveterås, E. (1956). Anaphylactoid purpura (Schönlein-Henoch syndrome) complicated by nephritis. *International Archives of Allergy and Applied Immunology (Basel)*, **9**, 274–80.

van der Wall Bake, A.W.L., Lobatto, S., Jonges, L., Daha, M.R., and van Es, L.A. (1987). IgA antibodies directed against cytoplasmic antibodies of polymorphonuclear leukocytes in patients with

Henoch-Schönlein purpura. *Advances in Experimental Medicine and Biology*, **216**, 1593–8.
van der Woude, F.J., *et al.* (1985). Autoantibodies against neutrophils and monocytes: tool for diagnosis and marker of disease activity in Wegener's granulomatosis. *Lancet*, **i**, 425–9.
Velji, A.M. (1986). Leukocytoclastic vasculitis associated with positive HTLV-III serological findings. *Journal of the American Medical Association*, **256**, 2196–7.
Vierhuff, J. (1893). Zur Casuistic der Darminvagination. *St. Petersburg Medizinische Wochenschrift*, **10**, 369.
Weber, T.R., Grosfield, J.L., Bergstein J., Fitzgerald, J. (1983). Massive gastric haemorrhage: an unusual complication of Henoch-Schönlein purpura. *Journal of Pediatric Surgery*, **18**, 576–8.
Wedgwood, R.J.P. and Klaus, M.H. (1955). Anaphylactoid purpura (Schönlein-Henoch syndrome); long-term follow-up study with special reference to renal involvement. *Pediatrics*, **16**, 196–205.
Wenner, N.P. and Safai, B. (1983). Circulating immune complexes in Henoch-Schönlein purpura. *International Journal of Dermatology*, **22**, 383–5.
Willan, R. (1808). *On cutaneous diseases*. Johnson, London.
Williams, D.G., Perl, S.J., Knight, J.F., and Harada, T. (1987). Immunoglobulin production *in vitro* in IgA nephropathy and Henoch-Schönlein purpura. *Seminars in Nephrology*, **7**, 322–4.
Wilton, J.M.A. (1978). Suppression by IgA of IgG-mediated phagocytosis by human polymorphonuculear phagocytes. *Clinical and Experimental Immunology*, **34**, 423–8.
Winkelmann, R.J.P. and Ditto, W.B. (1964). Cutaneous and visceral syndromes of necrotizing or "allergic" angiitis: a study of 38 cases. *Medicine*, **43**, 59–89.
Wolfsohn, H. (1947). Purpura and intussusception. *Archives of Disease in Childhood*, **22**, 242–7.
Yoshiara, S., Yoshikawa, N., and Matsuo, T. (1987). Immunoelectron microscopic study of childhood IgA nephropathy and Henoch-Schönlein nephritis. *Virchows Archiv*, **412**, 95–102.
Yoshikawa, N., White, R.H.R., and Cameron, A.H. (1981). Prognostic significance of the glomerular changes in Henoch-Schönlein nephritis. *Clinical Nephrology*, **16**, 223–9.
Yoshikawa, N., Yoshiara, S., Yoshiya, K., and Matsuo, T. (1986). Lysis of the glomerular basement membrane in children with IgA nephropathy and Henoch-Schönlein nephritis. *Journal of Pathology*, **150**, 119–26.
Yoshikawa, N., *et al.* (1987). Henoch-Schoenlein nephritis and IgA nephropathy in children; a comparison of clinical course. *Clinical Nephrology*, **27**, 233–7.
Yoshioka, K., Takemura, T, Aya, N., Akano, N., Miyamoto, H., and Maki, S. (1989). Monocyte infiltration and cross-linked fibrin deposition in IgA nephritis and Henoch-Schönlein purpura nephritis. *Clinical Nephrology*, **32**, 107–12.
Young, D.G. (1964). Chronic intestinal obstruction following Henoch-Schönlein disease. *Clinical Pediatrics*, **3**, 737–40.
Zuckner, J., Tsai, C., Giangiacomo, J., Baldassare, A.R., and Auclair, R. (1977). IgA deposition in normal and purpuric skin of patients with Henoch-Schönlein purpura. *Arthritis and Rheumatism*, **20** (Suppl. 2), 395–7.

4.5.3 Systemic vasculitis

GILLIAN GASKIN AND CHARLES D. PUSEY

Historical aspects and classification

The term 'systemic vasculitis' encompasses a variety of conditions which may affect the kidney by damaging its blood supply. Virtually any size or type of vessel may be involved within the spectrum of vasculitis; involvement of glomerular capillaries leads to focal necrotizing glomerulonephritis, while disease affecting larger arteries causes renal infarction and ischaemia.

The entity of vasculitis was first appreciated by Kussmaul and Maier (1866), who described a condition characterized clinically by fever with muscle, gastrointestinal, and renal disease, and pathologically by nodular swellings along the course of medium-sized arteries with histological evidence of inflammatory change within the vessel wall. They named the condition 'periarteritis nodosa'. It was later appreciated that the pathological process arose from within the arterial wall rather than engulfing the vessel from without, and the term 'polyarteritis' rather than 'periarteritis', as suggested by Ferrari (1903), came into general use. In the early part of the 20th century, with the increasing awareness of systemic necrotizing vasculitis as a distinct disease, the term became increasingly used to describe a more heterogeneous group of vasculitic illnesses, within which distinctive clinical and histological patterns gradually emerged, leading to the classifications which we use today.

The first description of the clinical syndrome which we now know as Wegener's granulomatosis came from Klinger (1931), who described a patient with nephritis and uraemia, together with destructive sinusitis, systemic arteritis, and splenic granulomata. Friedrich Wegener (1936), a German pathologist, reported clinical and pathological findings in three patients with the same pattern of disease. He went on to describe this illness in detail, distinguishing it from periarteritis nodosa, and referring to it as 'a peculiar rhinogenic granuloma with particular involvement of the arterial system and the kidney' (Wegener 1939; Socias *et al.* 1987). As further case reports appeared during the 1940s, the disease took his name and became known as 'Wegener's granulomatosis'.

Microscopic polyarteritis was first defined by Davson *et al.* in 1948. On the basis of postmortem histology and the clinical details, they divided 14 patients with a diagnosis of polyarteritis nodosa into two groups: those with extensive glomerular changes and those without. The former group of nine patients had a febrile illness with a combination of respiratory, abdominal, and rheumatic symptoms; four died in uraemia with a normal blood pressure. All but one showed patchy fibrinoid necrosis of glomerular tufts and there were varying degrees of crescent formation. This group was defined as showing a microscopic form of periarteritis nodosa, distinct from other types of glomerulonephritis recognized at that time. Of the five cases in their study without extensive glomerular lesions, three had severe hypertension and four had arterial lesions typical of periarteritis nodosa. The pathological findings in a further six patients were reported by Wainwright and Davson (1950). In describing the form of disease typically affecting small vessels it is logical to drop the 'nodosa' suffix as aneurysms and subcutaneous nodules are not seen. The name 'microscopic polyarteritis' is not universally accepted, with some authors using the term 'hypersensitivity vasculitis' instead (Zeek 1952; Fauci et al. 1978), and others reporting 'necrotizing glomerulonephritis with vasculitis' (Furlong et al. 1987). Confusion arises in the use of the former: some authors use it to mean a disorder with predominant cutaneous involvement rather than a potentially fatal disseminated vasculitis. Microscopic polyarteritis is therefore to be preferred as a succinct descriptive term which does not imply a pathogenesis peculiar to this form of vasculitis (Heptinstall 1983).

Churg and Strauss (1951) described the pathological features of another variant of polyarteritis nodosa, which had previously been recognized clinically by Rackemann and Greene (1939) and Harkavy (1941). They studied 13 cases who came to autopsy after an illness characterized by asthma, fever, hypereosinophilia, cardiac failure, peripheral neuropathy, and gastrointestinal and renal disease. Vascular abnormalities similar to those of periarteritis nodosa were seen in most of the patients, with nodular swellings in small arteries of many organs. Microscopically, the lesions consisted of segmental fibrinoid necrosis in the vessel wall, often with aneurysm formation, with an inflammatory response containing many eosinophils in and around the vessel. Inflammatory extravascular lesions were noted, eosinophils being the predominant early cell type. Giant cells were prominent in older lesions and granulomatous nodules were identified, with necrosis and eosinophils at the core. Renal findings commonly included vasculitis in arcuate arteries, afferent arterioles and glomerular capillaries, often with intense eosinophil infiltration. Focal segmental glomerular lesions were present in most cases but rarely involved the majority of glomeruli. Interstitial nephritis with a heavy eosinophilic infiltrate was also described. They concluded that this represented a discrete syndrome, separate from classical polyarteritis nodosa without asthma, and named it allergic granulomatosis and angiitis.

In 1957 Rose and Spencer studied a group of 111 patients seen at nine British teaching hospitals with a diagnosis of polyarteritis nodosa and were struck by the differences between those patients with lung disease, who also had a tendency to form granulomata, and those without. Some, with marked eosinophilia and asthma, corresponded to patients with the allergic granulomatosis and angiitis of Churg and Strauss (1951) while others, with granulomatous lesions in the lungs and upper airways, could be considered to have Wegener's granulomatosis (Wegener 1936; Godman and Churg 1954). The term polyarteritis nodosa could then be reserved for cases without granulomata, characterized by a necrotizing vasculitis involving medium-sized muscular arteries and leading to aneurysm formation.

These disorders, sometimes grouped together as the polyarteritis group of vasculitides or the necrotizing vasculitides, are summarized in Table 1. This classification is based broadly on the size of vessel affected, but the distinctions between groups are not absolute. Overlaps between small and medium vessel vasculitides are well recognized (Fauci et al. 1978b), as are patients with both necrotizing glomerulonephritis on renal biopsy and aneurysms visible on arteriography. Churg–Strauss syndrome often involves both medium and small vessels. Occasionally

Table 1 Primary systemic vasculitis

Vessel size	With granulomata	Without granulomata
Small	Wegener's granulomatosis	Microscopic polyarteritis
Medium	Churg-Strauss syndrome	Polyarteritis nodosa
Large	Giant cell arteritis*	
	Takayasu's arteritis*	

* Granulomatous arteritis may sometimes be seen in these diseases.

giant cell arteritis and Takayasu's arteritis, which both typically affect large arteries, may involve small vessels in the kidney. In addition, there are numerous other conditions in which vasculitis and renal disease occur together including Henoch–Schönlein purpura (Chapter 4.5.2), systemic lupus erythematosus (Chapter 4.6.2), rheumatoid disease (Chapter 4.8), and mixed essential cryoglobulinaemia (Chapter 4.9) but these will not be discussed further.

Of greatest importance to the nephrologist are those conditions which involve small vessels and lead to a necrotizing glomerulonephritis. If unrecognized this may progress rapidly to renal failure, but if treated adequately with immunosuppressive agents it often responds dramatically. Wegener's granulomatosis and microscopic polyarteritis are the diseases most commonly concerned and this chapter will focus on these disorders.

Wegener's granulomatosis

Definitions

Following Wegener's original descriptions of the pathological features of this disease, Fahey et al. (1954) published a series of seven further cases, and the pathological findings were analysed and reviewed by Godman and Churg (1954). They defined the illness by the presence of necrotizing granulomatous lesions in the respiratory tract, a generalized vasculitis involving both arteries and veins, and a focal necrotizing glomerulonephritis with evolution of a granulomatous pattern. Pathological confirmation of all three features was required to make a firm diagnosis. These criteria were extremely strict and only applicable because, in the absence of effective treatment, most cases came to autopsy, permitting a full anatomical diagnosis.

Fauci et al. (1983), assessed a large series of patients, and adopted a more flexible definition that allowed living patients to be classified. A definite diagnosis of Wegener's granulomatosis required a patient to have clinical evidence of disease in at least two of the following three areas: upper airways, lungs (both in the form of a granulomatous vasculitis) and kidneys (in the form of a glomerulonephritis). Histological confirmation was required from at least one, and preferably two, of these sites. This definition is widely used now, but in practice, the criterion of demonstrating granulomata can be difficult to satisfy. Biopsies from the upper respiratory tract frequently demonstrate non-specific inflammatory changes or may be too necrotic to interpret. Proof of granulomatous histology of lung lesions often requires open lung biopsy, and renal biopsies will demonstrate granulomata in less than one-third of patients (Ronco et al. 1983). We therefore diagnose Wegener's granulomatosis in patients with a multisystem vasculitis affecting small vessels, with prominent involvement of the upper and lower respiratory tracts and kidneys. Our usual source of histological confirmation is a renal biopsy which shows focal necrotizing glomerulonephritis.

Although the above descriptions define Wegener's granulomatosis as a severe and generalized illness, in some cases with typical histology the disease appears to be localized to one or two organ systems. Carrington and Liebow (1966) designated this 'limited' disease.

Which features serve to distinguish Wegener's granulomatosis from the other vasculitic illnesses discussed elsewhere in this chapter (Fauci *et al.* 1978)? Respiratory tract involvement in small vessel vasculitis favours the diagnosis of Wegener's granulomatosis rather than microscopic polyarteritis, although alveolar haemorrhage may occur in both. However, granulomata are not seen in microscopic polyarteritis. Churg–Strauss syndrome typically involves the upper airways and lung, and histology may reveal granulomata, small vessel vasculitis and focal necrotizing glomerulonephritis. However, it may be differentiated from Wegener's granulomatosis by the presence of asthma, eosinophilic tissue infiltrates and peripheral blood eosinophilia. Polyarteritis nodosa can be distinguished by the involvement of medium-sized arteries, and the lack of respiratory disease or glomerulonephritis.

Epidemiology

While Wegener's granulomatosis is clearly an uncommon disease, its true incidence and prevalence are hard to establish. The number of cases seen at any one centre reflects not only the prevalence of the disease but also the pattern of referral and the diagnostic criteria used. The size of published series may, however, give a rough guide. The National Institutes of Health described the treatment of 85 cases over a period of 21 years (Fauci *et al.* 1983); the Mayo Clinic (DeRemee 1988) refers to a total experience of 151 patients; while Hammersmith Hospital has treated 79 new patients (unpublished observations) between 1975 and 1989. Most other series are smaller, describing 10 to 20 patients collected over a number of years. An increasing incidence has been suggested (Andrassy *et al.* 1988), on the basis that some centres have treated more patients in recent years. Whether this is due to increasing incidence, changing referral patterns, or increasing recognition of the disease is uncertain.

Data on the age and sex incidence are more readily available. There is a male predominance with a male:female ratio of roughly 1.3:1 (Fauci *et al.* 1983; Gaskin *et al.* 1989). The disease affects patients of all ages but presents most commonly in middle life, with a mean age at presentation in the late forties. It is a rare disease in childhood, and presentation in children may be atypical (Hall *et al.* 1985). The age at presentation in our own experience is shown in Fig. 1, which illustrates a skewed distribution favouring older age groups.

The majority of patients reported in the literature are Caucasoid, although the disease has been described in Negroid, Asian, and Hispanic individuals in series from Europe and North America. Bambery *et al.* (1987) have reported 11 cases from North India. A true predilection for particular racial types has not been distinguished from the structure of the populations from which the patients are drawn.

Several pieces of evidence suggest a genetic predisposition to Wegener's granulomatosis. Affected siblings have been described by both Muniain *et al.* (1986) and Knudsen *et al.* (1988). In the latter report the illness developed in sisters who had lived apart for 25 years, making a predisposing genetic factor more likely than a common environmental stimulus. We know of two sisters with the disease (D. Rainford, personal communication)

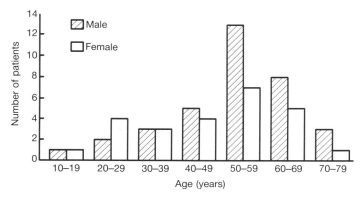

Fig. 1 Distribution of sex and age at presentation in 60 patients with Wegener's granulomatosis treated at Hammersmith Hospital.

but have also studied twins discordant for the disease. Serological tissue typing in 17 patients (Elkon *et al.* 1983) revealed an increased incidence of HLA DR2: 65 per cent of Wegener's patients carried DR2 compared to 21 per cent of a control population. An increase in HLA B8 which did not reach statistical significance was also noted. Katz *et al.* (1979) had earlier reported an increased frequency of HLA B8 in a group of 31 patients when compared to a control population. Our recent studies using molecular biological methods have shown a marked under-representation of HLA DR3 (Spencer *et al.* unpublished observations). The role of the HLA system in susceptibility to Wegener's granulomatosis is not yet clearly defined.

Environmental factors predisposing to Wegener's granulomatosis have not in general been identified, although one author noted that many patients with the disease have a long history of chronic respiratory infection (Pinching *et al.* 1983).

Clinical features

The spectrum of disease

Although the upper and lower respiratory tracts and kidneys are, by definition, the characteristic sites of disease in Wegener's granulomatosis, the overall clinical picture varies considerably from patient to patient and may also change in any one patient as the illness progresses. The pattern of disease reported in different series may reflect the special interest of the reporting unit; nephrologists will inevitably see more patients with severe renal disease than other specialists. The clinical manifestations within different organ systems will be considered in turn, before looking at their representation in the major series.

Non-specific features

Constitutional features are very common, with many patients reporting malaise and weight loss at presentation. Fever may be recorded in up to 94 per cent of patients at diagnosis (Pinching *et al.* 1983).

Upper respiratory tract

Upper airways disease occurs in over 90 per cent of patients, and is characterized histologically by necrotizing granulomata and clinically by a wide spectrum of symptoms and signs. Disease affecting the nose may present with symptoms due to inflammation and ulceration, such as epistaxis, rhinorrhoea, and nasal discomfort, or with symptoms resulting from obstruction of drainage of the paranasal sinuses or nasolacrimal duct. Destruction of cartilage may follow soft tissue necrosis and lead to the characteristic saddle-nose deformity (Drake-Lee et al. 1988), shown in Fig. 2. Gross destruction of bone or of the overlying skin is unusual and should raise suspicions of lethal midline granuloma or other underlying neoplasm.

There is commonly disease within the sinuses causing facial pain and symptoms of purulent sinusitis, and this may be confirmed by plain radiography or CT scanning. It should be remembered that sinusitis and soft tissue thickening on radiographs are common in the 'healthy' population.

Otitis media and conductive hearing loss are frequently observed and may be caused by middle ear involvement or granulomatous tissue in the nasopharynx causing eustachian tube blockage. Mastoid and tympanic granulomata have also been recognized.

Inflammation may involve the larynx (Waxman and Bose 1986), trachea (Hellman et al. 1987b), and large airways (Cohen et al. 1984). While this can lead to haemoptysis or hoarseness, it may only become apparent when the airways are visualized at bronchoscopy, or may reveal itself later when critical subglottic stenosis occurs. Pulmonary function tests, with analysis of the I_{50} from the flow-volume loop, may be needed for diagnosis of extrathoracic airway narrowing (Ross et al. 1990).

Pulmonary disease

Various forms of disease of the lower respiratory tract are apparent at presentation in around 90 per cent of cases. Symptoms are not specific and include cough, dyspnoea, haemoptysis, and chest pain, with or without pleuritic characteristics. Lesions may occur in the lung parenchyma, typically in the form of granulomata seen as rounded, and frequently cavitating, lesions on radiography (Fig. 3); histology may be required to distinguish these from neoplasms if diagnostic features in other organs are

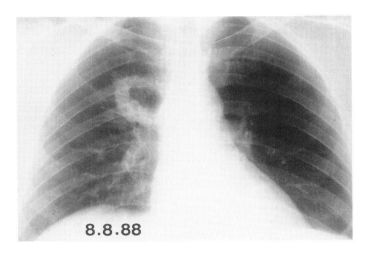
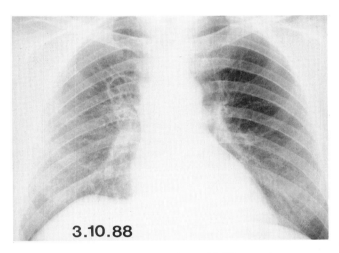

Fig. 3 Chest radiographs in a patient with Wegener's granulomatosis demonstrating resolution of a cavitating granuloma with immunosuppressive therapy.

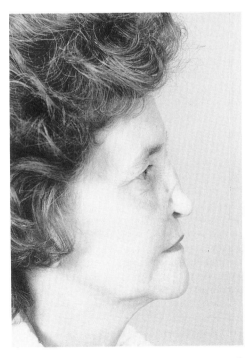

Fig. 2 Characteristic saddle-nose deformity in a patient with Wegener's granulomatosis.

absent. Microscopically these lesions contain areas of necrosis with granuloma formation (Godman and Churg 1954; Mark et al. 1988).

Granulomata can also be found in the bronchi. These contribute to haemoptysis, or may be an additional finding at bronchoscopy. One particularly ominous, though not specific, manifestation of disease in the lungs is alveolar haemorrhage, which has been associated with a high early mortality (Haworth et al. 1985). While there is a clear association between cigarette smoking and lung haemorrhage in antiglomerular basement membrane antibody-mediated disease, this is not so in Wegener's granulomatosis. Intercurrent infection and fluid overload may however lead to an exacerbation. Alveolar haemorrhage may be revealed by haemoptysis and dyspnoea, together with anaemia, hypoxia, and diffuse alveolar shadowing on radiography (indistinguishable from the appearance of lung haemorrhage in microscopic polyarteritis as shown in Fig. 10); increased transfer factor for carbon monoxide provides confirmation (Ewan et al. 1976). Using a combination of these criteria for diagnosis, pulmonary haemorrhage was diagnosed in 42 per cent

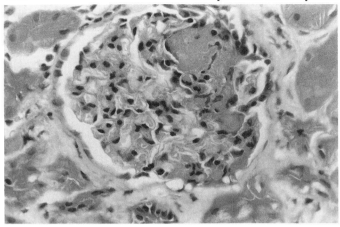

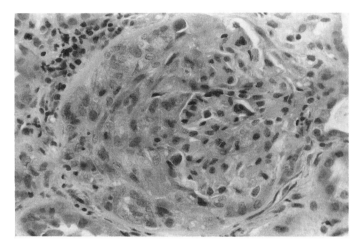

Fig. 4 Light microscopy of a renal biopsy from a patient with Wegener's granulomatosis demonstrating (a) segmental necrosis of glomerular tuft and (b) cellular crescent formation with compression of glomerular tuft.

of our patients with generalized disease (Gaskin et al. 1989). Many patients simply have ill-defined infiltrates apparent on chest radiography (Rijckaert et al. 1983) and some have pleural effusions.

Renal disease
This is common but rarely leads to the presenting symptoms, and may not be apparent at the time of presentation. Suspicion of early renal lesions should arise in patients with abnormal urinary sediment, microscopic haematuria, proteinuria, and granular and red cell casts. More advanced disease will lead to impairment of renal function; this may progress rapidly over days or weeks to oliguria and dialysis-dependent renal failure. Macroscopic haematuria, hypertension (except in the presence of advanced renal failure), and frank nephrotic syndrome are uncommon. Renal biopsy generally shows focal segmental necrotizing glomerulonephritis, which may become diffuse in advanced disease. Extracapillary proliferation and crescent formation are seen in cases with a clinically rapidly progressive nephritis. Glomerular lesions of varying ages may coexist (Fig. 4). Vasculitis may be found within the kidney, typically affecting interlobular arteries and arterioles (Novak et al. 1982; Yoshikawa and Watanabe 1984). Although granulomata have been described within the interstitium and around glomeruli, and have been held to be a specific and diagnostic finding, they can be demonstrated unequivocally in only a minority of biopsies (Ronco et al. 1983). Some authors consider that granulomata form only around glomeruli severely damaged by necrosis (Yoshikawa and Watanabe 1984) or where there is crescent formation associated with exudation of immunoglobulins and complement components into Bowman's space (Bhathena et al. 1987). An interstitial infiltrate of inflammatory cells is common, and Ten Berge et al. (1985) report that this comprises mainly T lymphocytes, predominantly of the helper (CD4) phenotype.

Immunofluorescence findings are variable; American authors often consider nephritis in this condition to be 'pauci-immune', yet immunoglobulins and complement are demonstrable in a large proportion of patients. However, these deposits are seen particularly in necrotic or sclerotic lesions, and tend to be scanty. Electron microscopy only rarely shows immune deposits (Balow et al. 1978; Ronco et al. 1983). The histology of the renal lesion appears similar whether or not immune deposits are detected.

Clinical features of small vessel systemic vasculitis
Eye disease occurs in over 50 per cent of patients with Wegener's granulomatosis (Fauci et al. 1983; Gaskin et al. 1989) and may be the first clinical feature. The spectrum of disease includes conjunctivitis, episcleritis, corneal ulceration, uveitis, retinal vasculitis, optic neuropathy, orbital masses and cellulitis, and obstruction of the nasolacrimal duct. Significant ocular morbidity may occur, and Bullen et al. (1983) reported the need for enucleation of three eyes in a series of 40 patients with ophthalmic disease.

Neurological disease may affect the brain, cranial, or peripheral nerves. Cerebral disease may be due to vasculitis, locally arising granulomata or occasionally spread of granulomatous disease from the nasopharynx (Drachmann 1962). There may be major neurological deficits due to the involvement of large arteries (Satoh et al. 1988), as well as more subtle physical signs or alteration in conscious level. Seizures may occur, although metabolic factors are difficult to exclude in patients with severe renal failure. Peripheral nerve lesions are more common, typically those of mononeuritis multiplex.

Vasculitic skin rashes, isolated necrotic lesions, and evidence of vasculitis in the form of nailbed infarcts or splinter haemorrhages occur frequently. Soft tissue infarction of the extremities, although more often associated with large vessel vasculitides, has also been observed.

Gastrointestinal disease is well recognized, and may present with abdominal pain, diarrhoea, or gastrointestinal bleeding (Camilleri et al. 1983). Inspection of the intestinal mucosa, either at endoscopy or surgery, reveals scattered purpuric lesions with ulceration. Serious complications of major bleeding (Haworth and Pusey 1984) or bowel perforation (Geraghty et al. 1986) may occur, although remission of gastrointestinal disease in parallel with recovery from the systemic illness is usual. Severe ulceration of the oral mucosa occurs in a number of patients.

Cardiac involvement is rarely reported, although any part of the heart may be affected. Valvular lesions may mimic infective endocarditis (Gerbracht et al. 1987), and dysrhythmias (Allen et al. 1984) and dilated cardiomyopathy (Fauci et al. 1983) have been reported. Pericarditis is also described, but in some patients uraemia could be responsible.

Musculoskeletal symptoms are often among the first reported. Both arthralgia and arthritis (Noritake et al. 1987), occasionally

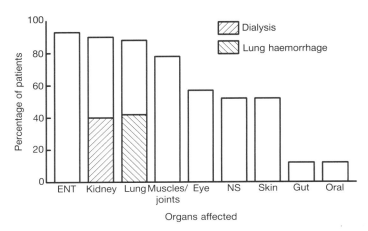

Fig. 5 Clinical features at presentation in 60 patients with Wegener's granulomatosis treated at Hammersmith Hospital.

with erosive change (Jacobs et al. 1987), are observed and myalgia is common.

Other unusual lesions may occur, more often recognized at autopsy than in life. These include granulomata in the prostate, parotid glands, and cervical vertebrae, and vasculitis in the spleen, adrenal glands, ovaries, testes, pancreas, oesophagus, and gallbladder. Inflammatory change has also been described in the liver and thyroid gland. Anterior and posterior pituitary disease with diabetes insipidus, have both been reported (Haynes and Fauci 1978; Lohr et al. 1988; Hurst et al. 1983.)

Severity of disease

Although Wegener's granulomatosis may present with any of the diverse manifestations described above, not all series describe the same spectrum or severity of disease. The early reports of Wegener (1936) and Fahey et al. (1954) were of disseminated disease progressing relentlessly to death from renal and respiratory failure; lack of effective treatment presumably allowed study of the full natural history. More recently, Fauci's large series (Fauci et al. 1983) detailed a wide variety of clinical features, but included few patients who were critically ill at presentation; the majority of patients had evidence of renal involvement but only 11 per cent had functional renal impairment. The experience of nephrologists contrasts with that of Fauci, since the referral practice attracts patients with severe renal and extrarenal disease. The extent of disease, and the severity of disturbance of organ function in patients seen at the Hammersmith Hospital is outlined in Fig. 5. It is worth noting that 40 per cent had dialysis-dependent renal failure and 42 per cent had lung haemorrhage.

A further contrast is the pattern of disease in 'limited Wegener's granulomatosis', which would not satisfy the strict diagnostic criteria set out by Fahey et al. (1954) or, in some cases, the criteria of Fauci et al. (1983). Limited Wegener's granulomatosis was first documented by Carrington and Liebow (1966), who reported 16 patients with predominantly pulmonary disease. All had abnormal chest radiographs and 14 of 16 had constitutional symptoms, but other extrapulmonary disease was limited. None of the patients had unequivocal evidence of a glomerulonephritis, but five had granulomata in the kidney at autopsy. Although the disease in this series was anatomically limited, it was not benign, since five patients died as a direct result of severe pulmonary lesions. D'Cruz et al. (1989) recently emphasized that some patients at presentation have disease predominantly involving the upper respiratory tract without accompanying renal disease. Gross (1989) perceives limited Wegener's granulomatosis as a clearly defined early and benign stage of the disease, which switches at a later date to become generalized and life-threatening. This cannot be applied to every case and some patients present a with short history of generalized disease without evidence of previous respiratory involvement. Indeed, in others, renal disease may be the first manifestation of Wegener's granulomatosis, with lung lesions following several years later (Woodworth et al. 1987).

Diagnosis

While history and examination may suggest the diagnosis of Wegener's granulomatosis, firm clinicopathological diagnosis depends on establishing the presence of granulomatous necrotizing vasculitis with prominent involvement of the respiratory tracts and kidneys. Ultimately, this requires proof by histology, but many other investigations can yield supporting evidence.

Haematology and biochemistry

Routine blood count frequently demonstrates anaemia, typically of a normocytic normochromic pattern but occasionally with iron deficiency due to lung or gastrointestinal haemorrhage. A neutrophil leucocytosis is common, but low levels of eosinophilia have also been reported (Ronco et al. 1983). Thrombocythaemia is frequently seen and the mean platelet count in one series (Pinching et al. 1983) was 612×10^9/l. Erythrocyte sedimentation rate will invariably be elevated (Fauci et al. 1983) as will C-reactive protein, and the latter is more specific in renal failure; C-reactive protein concentrations change more rapidly and more closely reflect disease activity (Hind et al. 1984b). As part of the same acute phase response, albumin may be low and alkaline phosphatase elevated. There may be hyperglobulinaemia, especially affecting IgA (Fauci et al. 1971; Ten Berge et al. 1985).

Serum urea and creatinine concentrations may be increased, commensurate with the degree of renal damage, but values in the normal range do not exclude the presence of nephritis. Proteinuria indicates the presence of glomerular disease, but is only rarely in the nephrotic range. The urine sediment will usually show red cells and granular and red cell casts, but in some cases shows less abnormality than expected from the histological severity of nephritis (Fauci and Wolff 1973). These and other non-specific findings are summarized in Table 2.

Immunology

Immunological abnormalities, including rheumatoid factors and immune complexes, are common (Pinching et al. 1983; Ronco et al. 1983; Littlejohn et al. 1985). The sensitivity of the assays used influences the proportion of patients found to have circulating immune complexes, but overall between one-third and two-thirds of patients are positive. Antibodies binding to cultured endothelial cells have recently been identified in the sera of patients with systemic vasculitis, including Wegener's granulomatosis (Brasile et al. 1989; Frampton et al. 1989; Ferraro et al. 1990), but these are not invariably present and have also been detected in other rheumatological disorders. More specific is the finding of antibodies against neutrophil cytoplasmic antigens (ANCA) in almost all patients with active and generalized

Table 2 Non-specific laboratory findings in vasculitis

Haematology	Anaemia
	Leucocytosis
	Thrombocytosis
	Elevated ESR
Biochemistry	Evidence of renal impairment
	Elevated alkaline phosphatase
	Hypoalbuminaemia
	Elevated C-reactive protein
Urine findings	Proteinuria
	Microscopic haematuria
	Granular and red cell casts
Immunology	Hyperglobulinaemia
	Immune complexes
	Rheumatoid factors

disease (Van der Woude et al. 1989). Such antibodies were first described in illnesses compatible with systemic vasculitis, thought to be associated with viral infection, by Davies et al. (1982). However, Van der Woude et al. (1985) were the first to appreciate an association with Wegener's granulomatosis. They detected ANCA by indirect immunofluorescence in 25 of 27 patients with active Wegener's granulomatosis. Other workers have confirmed a close association of these antibodies with Wegener's granulomatosis and describe a correlation with disease activity (Lüdemann and Gross 1987; Savage et al. 1987; Cohen Tervaert et al. 1989; Nölle et al. 1989). They appear to occur less frequently in patients with more limited disease. The estimated frequency in such patients is around 60 per cent. Recent studies distinguish ANCA found in Wegener's granulomatosis from those found in microscopic polyarteritis, polyarteritis nodosa, Churg–Strauss syndrome and idiopathic rapidly progressive glomerulonephritis. In Wegener's granulomatosis, immunofluorescence staining of alcohol-fixed normal human neutrophils is diffusely distributed throughout the cytoplasm and is granular in appearance. This pattern is termed cytoplasmic or 'C-ANCA'. Recent work suggests that the target antigen is a 29 kDa serine proteinase found in neutrophil primary granules (Goldschmeding et al. 1989; Niles et al. 1989; Lüdemann et al. 1990a).

In the other group of vasculitic syndromes, the immunofluorescence pattern is more commonly perinuclear, and is termed 'P-ANCA'. This pattern correlates in most cases with antibody binding to the neutrophil enzyme myeloperoxidase (Falk and Jennette 1988). This distinction is not absolute, and antibodies to other neutrophil enzymes, including elastase (Goldschmeding et al. 1989) and lactoferrin (Lesavre et al. 1990), have been described. However, all authors agree on the correlation between C-ANCA and Wegener's granulomatosis. When the diagnosis is suspected, therefore, blood should be analysed for the presence of ANCA, either by indirect immunofluorescence or by one of the numerous solid phase immunoassays which are becoming available (Lockwood et al. 1987; Lüdemann et al. 1988; Cohen Tervaert et al. 1990; Rasmussen et al. 1990; reviewed by Rasmussen and Daha 1990).

Radiology

The chest radiograph may show rounded opacities corresponding to granulomata, or diffuse bilateral shadowing due to alveolar haemorrhage. Sometimes a less specific pattern of infiltration or pleural effusion is seen. Computed tomography may confirm the presence of lesions poorly visualized on plain radiography (Cordier et al. 1990). Radiological sinus abnormalities, including fluid levels and soft tissue thickening, are very common, but bone destruction is uncommon. Requests for other radiological examinations should be guided by the clinical presentation, e.g. laryngeal tomography in the presence of stridor. While imaging may point to a suitable site from which to seek histology, it will not in itself provide a specific diagnosis.

Histopathology

Firm pathological diagnosis requires confirmation of the presence of necrotizing granulomata. Tissue is perhaps most easily obtained from the upper airways; unfortunately biopsy specimens are frequently necrotic or show non-specific inflammation (Littlejohn et al. 1985). Lung histology obtained at open-lung biopsy is more reliable, but this is invasive. For this reason transbronchial biopsy has been advocated, but the specimens are often too small and fragmented to be of value. By comparison, renal biopsy is more likely to give an adequate amount of tissue, and will usually show a focal necrotizing glomerulonephritis; however, granulomata and vascular lesions are much less commonly seen. Our practice is to accept these renal biopsy findings, in the presence of clinical and radiological evidence of granulomatous disease in the upper and lower respiratory tracts, as sufficient evidence for the diagnosis of Wegener's granulomatosis. Histological examination of nasal lesions is routinely attempted, but open lung biopsy is performed only if there is no other route to diagnosis.

Investigations and differential diagnosis

Other investigations will often be needed to exclude other possible diagnoses. Lung haemorrhage and rapidly progressive nephritis should provoke a search for antiglomerular basement membrane (anti-GBM) antibodies (see Chapter 3.11), though the clinical picture of widespread systemic disease favours Wegener's granulomatosis. Rarely there is evidence of both diseases. Anti-GBM antibodies have been detected in the serum of some patients with systemic vasculitis, and there may be linear deposition of immunoglobulin on the glomerular basement membrane in patients with biopsy proven granulomatous disease (Jayne et al. 1990). Wahls et al. (1987) report a patient who presented with biopsy proven pulmonary lesions and developed a typical picture of anti-GBM disease 6 months later. Conversely, O'Donoghue et al. (1989) report the development of systemic vasculitis associated with ANCA appearing days or weeks after the serologically confirmed diagnosis of anti-GBM disease. Other multisystem diseases with a predilection for lungs and kidneys (Leatherman et al. 1984; Leavitt and Fauci 1986), in particular systemic lupus erythematosus, can be excluded by appropriate investigations.

Differentiation of Wegener's granulomatosis from other primary vasculitic illnesses may be more difficult. Henochowitz et al. (1986) and Yousem and Lombard (1988) reported cases of small vessel vasculitis where marked tissue eosinophilia was seen in the absence of peripheral blood eosinophilia or asthma, suggesting an overlap with Churg–Strauss syndrome. Some patients with Wegener's granulomatosis have involvement of medium sized arteries, demonstrable by the presence of aneurysms on renal arteriography similar to those seen in polyarteritis nodosa (Moutsopoulos et al. 1983), but this is rare.

Treatment

Historical aspects

An early report by Fahey et al. (1954) described the benefit of cortisone and adrenocorticotrophic hormone treatment in one patient and commented on 'good evidence that these drugs controlled the intensity of the inflammatory process'; but this verdict was not universally held. Walton (1958) later described 10 patients with Wegener's granulomatosis, two of whom were treated with cortisone and had temporary remission of symptoms; they survived for 12 and 21 months respectively, in contrast to the survival of 1 to 10 months in his untreated patients. In a review of all previously reported cases, most of which were untreated, he calculated a mean survival of 5 months and noted that 81 per cent had died within 1 year. By 1967, 26 steroid treated patients had been reported in the literature and mean survival was 12.5 months (Hollander and Manning 1967). Despite the use of high doses of prednisone (a mean of 44 mg daily) the disease eventually became refractory. Thus, while corticosteroids represented the first step in development of therapy, their use alone was inadequate.

Fahey et al. (1954) had used intravenous nitrogen mustard in one patient who became resistant to steroid therapy, and reported a temporary improvement in symptoms. Further use of alkylating agents followed, in individual cases, in the 1960s. One such patient treated with chlorambucil, in whom prolonged remission from generalized disease was achieved, was reported by Hollander and Manning (1967). Novack and Pearson (1971) reported the successful treatment of four patients with generalized disease with cyclophosphamide. Fauci et al. (1971), encouraged by Novack's results, studied nine patients, seven of whom had generalized disease, treated with oral cyclophosphamide alone at an initial dose of 100 to 125 mg daily; all showed clinical improvement within 3 weeks of commencing therapy. In 1973 the same group reported the use of cytotoxic agents in greater detail (Fauci and Wolff 1973). Fourteen of 18 patients received an adequate trial of cyclophosphamide, and one received azathioprine instead. The starting oral dose of 1 to 2 mg/kg daily was maintained for 10 to 14 days and increased by 25 mg increments after similar intervals until clinical response was seen, or until serious toxicity required a reduction in dose. Cyclophosphamide was given intravenously at the higher dose of 2 to 4 mg/kg for the first few days in three patients with fulminant disease. Short courses of corticosteroids were used in addition, in all but four patients, when constitutional symptoms or inflammatory vasculitic lesions were severe. On this regimen, two patients died of active disease, but the remaining 13 went into remission. At the time of publication, one of these had died of coronary artery disease, one had relapsed after 21 months and 11 had stayed in remission; six were off all therapy. Reza et al. (1975) reported similar benefit from the use of cyclophosphamide in 10 patients followed for up to 10 years. These preliminary series paved the way for development of a regimen of cyclophosphamide and corticosteroids, and by 1983 Fauci et al. were able to report use of these drugs in 85 patients with follow-up of up to 21 years.

The NIH regimen

Initial therapy comprised daily oral cyclophosphamide at a dose of 2 mg/kg together with oral prednisolone at 1 mg/kg daily. The daily steroids were continued for 1 to 2 weeks, and then converted to an alternate day regimen over a period of 1 to 2 months. Typically, a dose of 20 mg on alternate days was achieved by 6 to 12 months. The dose was then tapered until the patient was receiving cyclophosphamide alone, provided there was no evidence of disease activity. The cyclophosphamide was continued at the starting dose, toxicity permitting, until the patient had been in complete clinical remission for at least 1 year. A tapering schedule was then begun, in which the dose of cyclophosphamide was lowered by 25 mg decrements every 2 to 3 months. The drug was either eventually discontinued, or maintained at a dosage below which the patient had evidence of disease activity. The guiding principle to minimize the risk of infection was never to allow the total white cell count to fall below 3000 to 3500/mm^3, modifying the dose as necessary. Patients who could only tolerate low doses of cyclophosphamide were given additional prednisolone to support the white cell count. Patients with fulminant disease received more aggressive induction therapy, with higher initial doses of both prednisolone (2 mg/kg for a few days) and cyclophosphamide (4–5 mg/kg), adjusted subsequently according to the leucocyte count. The regimen is summarized in Table 3. No other group has reported such a large number of successfully treated patients and Fauci's treatment has become the 'gold standard' with which to compare other treatment strategies.

More recent studies

Development of treatment in recent years has concentrated on improvements in several areas: management of fulminant disease; regimens with lower toxicity than long-term cyclophosphamide for maintenance therapy and for treatment of limited disease; and the use of better indices of disease activity or markers for risk of relapse, to allow individual tailoring of therapy.

Table 3 Treatment of Wegener's granulomatosis

	NIH	Hammersmith
Induction	Prednisolone 1 mg/kg daily Cyclophosphamide 2 mg/kg daily	Prednisolone 60 mg daily for most adults Cyclophosphamide 3 mg/kg daily (2 mg/kg if > 55)
Fulminant disease	Prednisolone 2 mg/kg daily Cyclophosphamide 4–5 mg/kg daily	Standard induction drug therapy Plasma exchange 4 l daily for albumin (5–10 days) or methylprednisolone 500 mg IV × 3 if plasma exchange not possible (rarely used)
Maintenance	Prednisolone tapering alternate day dose Cyclophosphamide starting dose maintained until remission > 1 year	Prednisolone tapering daily dose Azathioprine 1–2 mg/kg daily until remission > 1 year (unless relapse on treatment, when cyclophosphamide is used)

Treatment of severe disease

Two other forms of treatment may have a special place in the treatment of fulminant disease, in addition to standard drug therapy. The use of plasma exchange to treat crescentic nephritis due to Wegener's granulomatosis was proposed in the 1970s, with the rationale that humoral factors involved in pathogenesis might be removed (Lockwood et al. 1977). The combination of prednisolone, cyclophosphamide, and plasma exchange was associated with rapid renal recovery in five of nine patients with severe renal failure, although the exact contribution of plasma exchange to this could not be determined. A controlled trial of plasma exchange in focal necrotizing glomerulonephritis, which included 23 patients with Wegener's granulomatosis, was then performed and demonstrated a significant benefit of plasma exchange over the use of immunosuppressive drugs alone only in patients with dialysis dependent renal failure (C.D. Pusey et al. unpublished observations). A smaller multicentre controlled trial using a less intensive regimen of plasma exchange and containing only two patients with Wegener's granulomatosis did not show additional benefit (Glöckner et al. 1988). Plasma exchange has been used successfully by other groups in uncontrolled series (Thysell et al. 1982; Fuiano et al. 1988), but its contribution in these patients is difficult to assess. Our experience suggests that it may also be useful for severe non-renal disease, such as pulmonary haemorrhage and neurological involvement.

Bolus doses of intravenous methylprednisolone have also been used as an adjunct to conventional immunosuppression in patients with oliguric crescentic nephritis (Bolton and Couser 1979; Stevens et al. 1983; Bolton and Sturgill 1989) and in those with fulminant systemic vasculitis (Fuiano et al. 1988). However, a controlled comparison of plasma exchange and methylprednisolone has not been made. Both treatments have disadvantages; plasma exchange is costly, mainly due to the need for albumin replacement solutions, carries the risks of allergic reactions, transfer of infection, bleeding due to anticoagulation and depletion of clotting factors, and requires vascular access procedures. Methylprednisolone is cheaper, but may be associated with an increased susceptibility to infection and with the increased likelihood of long-term corticosteroid toxicity, including avascular necrosis of bone.

Reducing drug toxicity and treatment of limited disease

Cyclophosphamide has proved the most effective drug in the treatment of Wegener's granulomatosis but is not without significant toxicity, chiefly to the bone marrow, bladder, and reproductive organs (Schein and Winokur 1975). Serious infections may also occur, particularly if insufficient regard is paid to reduction or cessation of therapy when the leucocyte count begins to fall.

Bladder toxicity associated with cyclophosphamide is mediated by the effect of a toxic metabolite, acrolein. Fauci et al. (1983) reported cystitis in 27 of 85 patients, which was severe in three and required cessation of the drug in nine. Stilwell et al. (1988), described 17 cases of haemorrhagic cystitis in 111 patients with Wegener's granulomatosis treated with oral cyclophosphamide, three of whom developed bladder cancer. The patients with haemorrhagic cystitis had a longer duration of treatment and a larger total dose of cyclophosphamide than those without. In our series of 60 patients, in whom long-term cyclophosphamide has been avoided where possible, two have suffered haemorrhagic cystitis, and one has developed bladder malignancy.

Cyclophosphamide frequently causes reversible leucopenia, with the risk of opportunist infection. In our experience, the elderly are especially prone to this problem, and in patients over 55 years we limit the starting dose of cyclophosphamide to 2 mg/kg. In addition to short-term marrow problems, cyclophosphamide has been associated with later development of myelodysplastic syndromes and haematological malignancies, including Hodgkin's (Colburn et al. 1985) and non-Hodgkin's lymphomas (Ambrus and Fauci 1984), and acute erythroid (Ten Berge et al. 1985), myeloid, and myelomonoblastic leukaemias (Ohyashiki et al. 1986).

Gonadal dysfunction is a serious and inevitable complication of long-term cyclophosphamide. Fauci et al. (1983) reported invariable oligo- or amenorrhoea in women and severe oligospermia in two men. Hair loss has been variably described as 'troublesome' and 'relatively insignificant' in other series.

The risks of cyclophosphamide are associated with the duration and total dose of treatment and it is logical to use it for the shortest possible time, and in the minimum effective dose. Since Wegener's granulomatosis generally follows a relapsing course, it is rarely possible to discontinue therapy at an early stage. Attention has therefore been directed at safer alternatives for maintenance therapy. One approach, explored by Gross (1989) is to move from oral cyclophosphamide to monthly intravenous pulses in order to reduce the total dose administered. Similar regimens have been used in the treatment of lupus nephritis (Austin et al. 1986). The change to intravenous dosage is made in patients with leucopenia or grumbling disease, or may be used from the outset, but there is no firm evidence that it is more effective (Steppat and Gross 1989). Intravenous pulses of cyclophosphamide have been used by others as part of initial therapy in specific cases (Fuiano et al. 1988), or as part of a protocol of treatment of limited disease (D'Cruz et al. 1989), but as yet there is no clear evidence of their value in generalized disease.

Others have looked to alternatives to cyclophosphamide for maintenance therapy. Our own practice, which has also been reported by others (Fuiano et al. 1988), is to use azathioprine as the maintenance cytotoxic agent once remission has been achieved (Table 3). This satisfactorily maintains remission in many patients and carries fewer risks, but some patients relapse and require reversion to cyclophosphamide; however, such relapse is rarely severe and permanent renal damage is unusual if diagnosed promptly. We view the increased frequency of relapse as compensated by the preservation of reproductive capacity and reduced risk of haematological and urothelial malignancy. Azathioprine has also been used successfully as initial therapy in conjunction with steroids in limited disease (Harrison 1989).

Cotrimoxazole has been proposed as an alternative agent in the treatment of Wegener's granulomatosis. DeRemee et al. (1985) reported 12 patients treated with antimicrobial agents, and further reports of successful treatment followed (Axelson et al. 1987; West et al. 1987; Israel 1988; Yuasa et al. 1988). These patients form a heterogeneous group in which cotrimoxazole was used in different ways: as sole initial treatment; as sole treatment for relapse; as a maintenance agent in remission; and as additional therapy in patients on standard treatment. The majority of patients were said to respond. The spectrum of active disease at the time of treatment varied, but in most cases was limited to the respiratory tract. Only one patient with active renal disease achieved a lasting remission on cotrimoxazole alone. One difficulty in ascribing a direct beneficial effect to cotrimoxazole is that of distinguishing active disease from superimposed infection. Both may cause upper and lower respiratory tract symptoms, fever, and elevation of the ESR, and in several cases described as improving on cotrimoxazole therapy reported data are inadequate confidently to exclude infection.

Several reports describe prolonged maintenance of remission on cotrimoxazole. This could be explained by a reduction in the frequency of infection, which is known to provoke relapse in some patients (Pinching *et al.* 1980). There is presently no good evidence to support its use as a first line agent in severely ill patients and its proponents advocate limiting its use to disease showing a slow or indolent course, using a regimen of one double-strength tablet two to three times daily, initially for a trial period of 6 to 8 weeks. Recently, encouraged by DeRemee's results, Steppat and Gross (1989) have used this protocol as a maintenance therapy in eight patients who entered remission on conventional immunosuppression. Three experienced severe relapse, two with rapidly progressive nephritis, after several months. They also used cotrimoxazole alone to treat three patients with limited disease. In two, there was no evidence of disease progression during follow-up (periods of 13 and 27 months respectively) but a third required cyclophosphamide after 3 months. The debate about the usefulness of this treatment continues (DeRemee 1988; Leavitt *et al.* 1988).

Another agent proposed for the treatment of Wegener's granulomatosis is cyclosporin A. Gremmell *et al.* (1988) described two cases with respiratory and renal disease in whom the drug was used. In the first, cyclosporin 5 mg/kg was added when the patient had failed to respond to 16 days of treatment with high-dose prednisolone and 2 mg/kg cyclophosphamide; subsequent improvement was rapid but a delayed effect of cyclophosphamide in inducing remission cannot be excluded. In the second case cyclosporin was used as the sole therapy for a dialysis dependent patient with glomerulonephritis on biopsy, who subsequently recovered renal function. Borleffs *et al.* (1987) reported a patient in whom progression of Wegener's granulomatosis may have been prevented by cyclosporin. Additional data come from transplanted patients who have been treated with cyclosporin as monotherapy to prevent graft rejection. We have one patient who has remained in remission for 5 years on this therapy and another who has experienced a non-renal relapse (unpublished data).

Treatments which are no longer commonly used include cytotoxic agents other than cyclophosphamide and azathioprine, such as chlorambucil (Israel and Patchefsky 1975) and actinomycin D which was associated with a high incidence of opportunist infection (Ten Berge *et al.* 1985). Neither has supplanted cyclophosphamide as the mainstay of therapy. Local radiotherapy was in the past applied to upper respiratory tract lesions, with some success in achieving local control of disease, but it made no impact on the progression to systemic disease. We have used total lymphoid irradiation in desperate circumstances in one patient, without success.

In limited disease, when there are no immediately life-threatening features, the risks of drug toxicity must be carefully balanced against the benefits of treatment. Regimens avoiding cyclophosphamide have been advocated, but the possibility of transition to more aggressive disease must be considered.

Monitoring of treatment

Institution of therapy at diagnosis is the first step in a scheme of management which may need to continue for decades. Wegener's granulomatosis has a continuing tendency to relapse, yet no two patients will behave in exactly the same way, so treatment must be tailored to the needs of the individual. It is important to be able to diagnose active disease, and to identify patients at risk from relapse. Assessment of disease activity used to depend on the patient's symptoms, together with investigations such as the haemoglobin, leucocyte and platelet counts, renal function, urinary sediment, chest radiograph, and ESR. However the ESR is slow to reflect changes in disease activity, and estimations of C-reactive protein are more reliable (Hind *et al.* 1984*b*). Nonetheless, these markers are not specific, and sometimes reflect the presence of infection rather than active vasculitis. Serial assays for ANCA, as a non-invasive marker of disease activity are an advance; titres correlate in general with disease activity (Van der Woude *et al.* 1985; Nölle *et al.* 1989; Cohen Tervaert *et al.* 1989), and relapse occurs only in the presence of detectable ANCA (Van Es and Wiik 1990), although active disease is not invariably present when the antibody is detectable. Clinical evidence of relapse may be preceded by a rise in the titre of ANCA (Heaton *et al.* 1990). The hope is that sensitive and specific solid-phase assays for this antibody will not only permit identification of active disease, but will also allow prediction of relapse, so that potentially toxic therapy can be reserved for such patients.

Outcome

Untreated disease

The outcome of untreated disease was poor, with a mean survival of only 5 months (Walton 1958). A few patients survived for up to 4 years, but this was exceptional. Overall, around 80 per cent of patients died within 1 year and 90 per cent died within 2 years. The presence of progressive renal failure was a particularly ominous sign before dialysis was generally available, although death from respiratory failure was also common. The introduction of corticosteroids extended mean survival to 12.5 months, but Wegener's granulomatosis remained a serious and usually fatal illness until the introduction of cytotoxic therapy, and notably of cyclophosphamide. The outcome using such regimens, based particularly on the results published by Fauci *et al.*, and on our own experience of patients with severe vasculitis, will be discussed here. Therapy for Wegener's granulomatosis may be considered in two phases: first, aggressive immunosuppression to halt progression of the disease and to induce remission; and second, maintenance therapy to prevent relapse. The outcome may similarly be divided into the response to initial therapy and the long-term prognosis.

Response to initial therapy

Fauci *et al.* (1983) reported induction of complete remission—that is to say the absence of evidence of active disease—in 79 of 85 patients (93 per cent). The remaining six patients died as a result of active disease.

In our series of 60 patients, renal disease was present in all but six and was often severe. Twenty-four patients required dialysis within 24 h of admission, and mean serum creatinine concentration in the 30 non-dialysis dependent patients with renal disease was 341 μmol/l. This contrasts with the incidence of 'renal disease' in Fauci's series of 85 per cent but 'functional renal impairment' in only 11 per cent (Fauci *et al.* 1983). After initial therapy, at 2 months, 49 of 60 patients were alive and only four required regular dialysis. Overall, 14 of 24 patients initially

Table 4 Outcome of Wegener's granulomatosis in series describing severe disease

Author	No. of patients	Index of severity	Survival
Brandwein et al. (1983)	13	13/13 renal disease 4/13 'fulminant vasculitis' 2/13 lung haemorrhage and oliguric renal failure	4/13 alive at 1 year
Littlejohn et al. (1985)	17	12/17 renal disease 5/17 creatinine > 350 μmol/l	13/16 alive at 1 year
Ten Berge et al. (1985)	12	12/12 renal disease 7/12 oliguric	9/12 alive at 6 months
Gaskin et al. (1989)	60	24/60 dialysis-dependent 25/60 lung haemorrhage	43/60 alive at 1 year

needing dialysis had recovered independent renal function and were alive at 2 months with a mean serum creatinine of 225 μmol/l. Of the 30 non-dialysis dependent patients with renal disease, 26 were alive at 2 months with independent renal function (mean serum creatinine 221 μmol/l). There were 11 early deaths, four within 2 weeks before an adequate trial of therapy had been given. Eight deaths were due to massive pulmonary haemorrhage, infection being an additional factor in four. Seven deaths were in patients initially requiring dialysis, although three of these had a partial response to treatment in that they had recovered renal function before death from non-renal causes.

Long-term outcome

Patient survival

Fauci's 85 patients were followed for a mean of 51 months. At the end of this period, 75 of the 79 achieving a remission with initial therapy were still alive and four had died from causes unrelated to their disease. Fifty-three patients were still receiving treatment.

In our group of more severely affected patients, actuarial survivals were 70 per cent at 1 year and 56 per cent at 5 years. The mean period of follow-up was 97 months, and at the end of this time 31 patients were still alive. Opportunist infection caused three deaths, all within the first year, while fatal relapse occurred in two patients, in the fifth and sixth years respectively. Three deaths were related to dialysis dependent renal failure and the remaining 10 deaths were due to causes unrelated to the disease or its treatment. Whilst these results are clearly worse than Fauci's, they are similar to others involving patients with comparably severe disease, in particular advanced renal failure (see Table 4). Ten Berge et al. (1985) reported 12 patients with severe renal disease, seven of whom were oliguric at presentation; three died during the first 6 months, and a further six died during follow-up. Similar results were described by Brandwein et al. (1983) and Littlejohn et al. (1985). Severe systemic vasculitis, renal disease, and the presence of lung haemorrhage were highlighted as particularly ominous. Serra et al. (1984), from Guy's Hospital, drew attention to the poor outcome of a mixed group of patients with renal vasculitis, even without life-threatening extrarenal manifestations such as lung haemorrhage. In this group, actuarial survivals were 54 per cent and 38 per cent at 1 and 5 years, respectively. Oligoanuria and extensive crescent formation on biopsy were adverse prognostic features. One note of optimism is the trend towards earlier referral noted by Fuiano et al. (1988), reporting 4 years later from the same unit. Actuarial survival was 77 per cent at 5 years in patients presenting with renal vasculitis, and the improvement was attributed in part to seeing patients earlier in the course of disease, reflected in the smaller numbers with advanced renal failure at presentation.

Renal function

Fauci reported a 30 per cent risk of progression to end-stage renal failure in those patients with glomerulonephritis and 'azotaemia' before starting immunosuppressive therapy.

Many patients who are dialysis dependent on admission, with severe glomerulonephritis on biopsy, recover renal function with intensive treatment (Gaskin et al. 1989). In this respect, the rapidly progressive nephritis of Wegener's granulomatosis differs from that of anti-GBM disease (see Chapter 3.11). The relationship between severity of renal disease and outcome at 1 year in 29 patients treated at Hammersmith Hospital is summarized in Fig. 6: this illustrates that while most of the deaths are associated with advanced renal failure at presentation, patients with severe renal vasculitis may recover renal function. There are no patients in this group on dialysis at 1 year. However, a proportion of the patients with independent renal function after initial therapy later progress to end-stage renal failure. Ten of 45 such patients in our series (Gaskin et al. 1989) later required dialysis. Most had a long history of untreated or partially-treated disease and had extensive glomerular sclerosis and interstitial damage on initial renal biopsy. These patients made an incomplete recovery with initial treatment, and renal function declined inexorably, presumably due to progressive scarring. Three patients were preci-

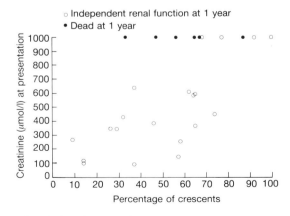

Fig. 6 Graph showing the relationship between serum creatinine at presentation and proportion of glomeruli with crescents. Only biopsies containing 10 or more glomeruli were analysed. Outcome at 1 year is also indicated.

pitated on to dialysis by renal relapse of Wegener's granulomatosis. Fortunately, prognosis on renal replacement therapy is fair, and patients in several series, including our own, have had successful renal transplants (Fauci et al. 1976; Brandwein et al. 1983; Littlejohn et al. 1985). However, relapse may occur after commencement of dialysis and both extrarenal and renal relapses have been described in transplant patients not receiving cyclophosphamide (Curtis et al. 1983, our unpublished observations).

Relapse

Relapsing disease is a major problem in Wegener's granulomatosis. Twenty-five of 83 patients in Fauci's series and 26 of 60 in our own had episodes of relapse after initial therapy. These were especially likely to occur on tapering therapy (Fauci et al. 1983) or after infection (Pinching et al. 1980). Relapses may occur several years after initial presentation, and multiple episodes of relapse may be seen in the same patient. Recurrent disease may also be seen in patients on long-term dialysis and after transplantation. Since cyclophosphamide appears to be the most effective drug for maintenance as well as for induction of remission, patients with a tendency to recurrent relapse may require its long-term administration, with the attendant risks. As discussed previously, assays for ANCA may prove to be useful in identifying relapse early and as a marker for risk of relapse. Use of this investigation should allow a more rational approach to maintenance immunotherapy.

Microscopic polyarteritis

Definitions

Davson et al. (1948) first distinguished this disease from classical polyarteritis nodosa by the presence of extensive glomerular disease. Later investigators have supported this division into microscopic and classical forms, noting differences in the long-term outcome in the two groups (Heptinstall 1983; Balow 1985). Microscopic polyarteritis differs from Wegener's granulomatosis in the lack of granuloma formation and rarity of upper airways involvement, but has many features in common due to necrotizing inflammation in small vessels. We define this condition as a small vessel systemic vasculitis with focal segmental necrotizing glomerulonephritis, but without granulomatous disease of the respiratory tract (Savage et al. 1985). Many authors (Serra et al. 1984; Parfrey et al. 1985; Croker et al. 1987; Couser 1988), but not all (Furlong et al. 1987), consider that idiopathic focal necrotizing and crescentic glomerulonephritis without anti-GBM antibodies is a renal-limited form of this vasculitis (assuming that it is possible to exclude more generalized involvement in life). This is supported by the similar age at presentation, the presence of non-specific constitutional symptoms and laboratory findings similar to those of microscopic polyarteritis, and the identical histological changes on renal biopsy. Furthermore, some patients presenting with isolated rapidly progressive glomerulonephritis go on to develop evidence of extrarenal vasculitis, both conditions respond to steroids and immunosuppressive therapy, and ANCA have been detected in both (Falk and Jennette 1988; Nässberger et al. 1989; Cohen Tervaert et al. 1990).

Epidemiology

Like Wegener's granulomatosis, microscopic polyarteritis is an uncommon disease, and accurate incidence and prevalence

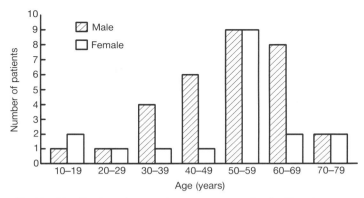

Fig. 7. Distribution of sex and age at presentation in 49 patients with microscopic polyarteritis treated at Hammersmith Hospital.

figures are not available. The incidence of the two diseases is probably roughly similar. Scott et al. (1982) estimated an incidence of 4.6 new cases per year of systemic vasculitis of the polyarteritis type per million of population, based on the number of patients treated in a district hospital over a period of 8 years.

The disease shows a male predominance and presents most commonly in middle age (Savage et al. 1985; Coward et al. 1986; Adu et al. 1987). The age and sex distribution in 49 patients treated at Hammersmith Hospital between 1974 and 1988 is shown in Fig. 7. Most reported cases are European Caucasoids, although the disease may sometimes be seen in Asian and Negroid individuals. The experience of the Guy's Hospital group suggests an under-representation of Negroid patients, with only one in a series of 120 cases of microscopic polyarteritis and Wegener's granulomatosis; this is significantly fewer than expected from the ethnic mix of the catchment population (J.S. Cameron, personal communication).

Studies of associations with major histocompatibility complex antigens have been performed by Elkon et al. (1983) and Müller et al. (1984). The former study did not show a significant association with any particular HLA type, but included both classic and microscopic forms of polyarteritis in a total of only 15 patients. Müller's study of serological HLA class II types in 24 patients with idiopathic rapidly progressive glomerulonephritis demonstrated a statistically significant increase in HLA DR2. It also demonstrated a significant association with the properdin factor B phenotype BfF. Our recent studies using molecular biological techniques have shown an under-representation of HLA DR3, as in Wegener's granulomatosis (Spencer et al. unpublished observations).

Vasculitic illnesses with the clinical features of microscopic polyarteritis have been associated with D-penicillamine treatment (Sternlieb et al. 1975; Banfi et al. 1983, and others, reviewed by Devogelaer et al. 1987). We have studied a patient who developed lung haemorrhage and rapidly progressive nephritis following penicillamine therapy for rheumatoid arthritis. Anti-neutrophil cytoplasmic antibodies and anti-myeloperoxidase antibodies were detected in her blood, suggesting that the penicillamine-induced disease shares serological as well as clinical features of microscopic polyarteritis. The antihypertensive agent hydralazine has been associated with rapidly progressive nephritis (Björk et al. 1985), sometimes with systemic symptoms (Mason and Lockwood 1986), and separately with cutaneous vasculitis (Peacock and Weatherall 1981; Finlay et al. 1981). The

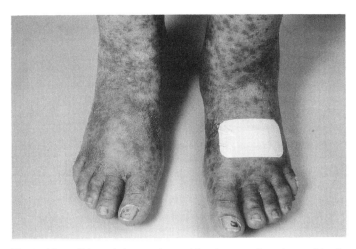

Fig. 8 Vasculitic rash in a patient with microscopic polyarteritis. A plaster covers the site of skin biopsy.

ability of this drug to cause a lupus-like illness is well known and, although these patients did not meet the American Rheumatology Association criteria for a diagnosis of systemic lupus erythematosus, all had antinuclear factors detectable in the serum. A patient with silicosis has developed systemic vasculitis (Arnalich et al. 1989). Two reports suggest a possible link with infection: Davies et al. (1982) reported similar illnesses, associated with ANCA, in patients with serological evidence of infection with an arbovirus, Ross River virus; while Falk et al. (1990) noted a seasonal variation in presentation of necrotizing and crescentic glomerulonephritis associated with ANCA. Recently there have been reports of necrotizing vasculitis associated with infection by the human immunodeficiency virus (Calabrese et al. 1989). The association with hepatitis B virus infection reported in polyarteritis nodosa (Gocke et al. 1970; Duffy et al. 1976; Sergent et al. 1976; Ronco et al. 1983; Guillevin et al. 1988) has not been reproduced in series concentrating on microscopic vasculitis. However, there are no environmental factors consistently present in patients with microscopic polyarteritis.

Clinical features

Clinical findings are the product of a widespread small-vessel vasculitis. By definition, the kidneys are invariably affected by a focal necrotizing glomerulonephritis, with crescent formation in many cases. This results in proteinuria, haematuria, urinary casts, and functional renal impairment. Hypertension may be seen at presentation in a proportion of patients (21 per cent (Adu et al. 1987), 29 per cent (Savage et al. 1985), 41 per cent (Coward et al. 1986)) but fluid overload and the presence of long-standing renal disease may be important contributory factors. However, hypertension is rarely as prominent a feature as it is in polyarteritis nodosa.

As in Wegener's granulomatosis, constitutional features and joint and muscle symptoms are common, and a similar vasculitic disease may be seen in the eyes, nervous system, mouth, skin, and gastrointestinal tract (Moore and Fauci 1981; Camilleri et al. 1983; Savage et al. 1985; Coward et al. 1986). Purpuric rashes (Fig. 8) and intestinal vasculitis are said to occur with greater frequency in microscopic polyarteritis. The lung may be involved, often leading to alveolar haemorrhage, presenting as dyspnoea, haemoptysis, and anaemia (Haworth et al. 1985; Adu et al. 1987). Diagnosis of pulmonary haemorrhage may be made in the same way as for Wegener's granulomatosis, and carries a similarly high early mortality. Necrotizing granulomata in the upper respiratory tract are not seen, but minor symptoms relating to the ear, nose, and throat are often reported, including a history of sinusitis or deafness (Savage et al. 1985). However these are common in the general population and it is not certain that they represent part of the vasculitic illness. Figure 9 illustrates the spectrum of organ involvement in our experience of microscopic polyarteritis.

Diagnosis

Diagnosis of microscopic polyarteritis depends on the demonstration of small vessel vasculitis affecting the kidneys and at least one other organ system, in the absence of typical clinical, radiological or histological evidence of necrotizing granulomata. History and examination may suggest the diagnosis and the results of non-specific blood tests will be supportive. These apply equally to Wegener's granulomatosis, as described above, and are summarized in Table 2.

Antineutrophil cytoplasmic antibodies have been described in microscopic polyarteritis (Savage et al. 1987), although the findings are more heterogeneous than in Wegener's granulomatosis. In some cases, indirect immunofluorescence on alcohol-fixed normal human neutrophils may show a granular cytoplasmic pattern, but more often there is staining over the cell nucleus, confined to the edge of the nucleus or in the immediately surrounding cytoplasm. Combinations of these patterns are also seen. The nuclear staining is specific to granulocytes (granulocyte-specific antinuclear antibodies) and was first observed in patients with rheumatoid disease (Wiik et al. 1974). These patterns are grouped as 'perinuclear' or 'P-ANCA' and are often associated with antibodies directed against the neutrophil granule enzyme myeloperoxidase (Falk and Jennette 1988; Lee et al. 1990; Cohen Tervaert et al. 1990). Antibodies against another enzyme, elastase, are reported less frequently but may also give a perinuclear fluorescence pattern (Goldschmeding et al. 1989; Lesavre et al. 1990). Patients with 'idiopathic' necrotizing glomerulonephritis may show the same patterns of antibody specificity, lending support to their inclusion as renal-limited microscopic vasculitis (Falk and Jennette 1988; Cohen Tervaert et al. 1990; Daha and Falk 1990). Jayne et al. (1989) described a

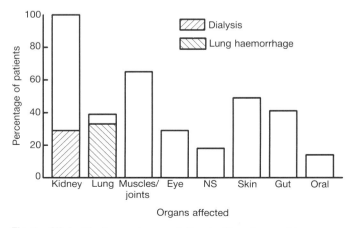

Fig. 9 Clinical features at presentation in 49 patients with microscopic polyarteritis treated at Hammersmith Hospital.

4.5.3 Systemic vasculitis

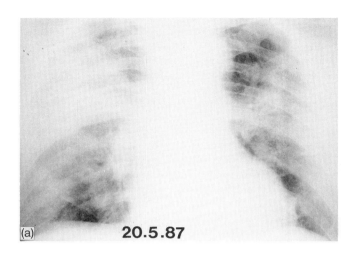

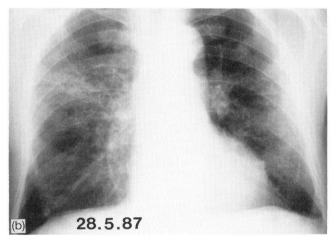

Fig. 10 Chest radiographs in a patient with microscopic polyarteritis demonstrating rapid resolution of pulmonary haemorrhage with immunosuppressive therapy including plasma exchange.

subgroup of patients with nephritis and severe lung haemorrhage who had antineutrophil cytoplasmic antibodies only of the IgM class, but Cohen Tervaert et al. (1990) reported that similar patients may have IgG antibodies against myeloperoxidase. Assays for ANCA have become an important non-invasive aid to diagnosis and may point to the non-Wegener's group of vasculitides when a perinuclear pattern or antimyeloperoxidase activity is seen.

Chest radiography may reveal evidence of alveolar haemorrhage and a typical pattern is illustrated in Fig. 10, which shows the appearance before treatment and a week afterwards. There is rapid resolution with immunosuppressive therapy. Pleural effusions are occasionally seen. No other radiographic investigations are routinely required and requests should be guided by the clinical picture. Angiography does not demonstrate aneurysms in this condition unless there is an overlap with larger vessel disease.

Definitive diagnosis requires histological confirmation, almost invariably from renal biopsy showing focal necrotizing glomerulonephritis with crescent formation. Vascular inflammation and necrosis, other than within the glomerular capillaries, is seen in only a minority of biopsies (Savage et al. 1985). If present, it may be found in the smaller interlobular arteries and arterioles and rarely in arteries of arcuate size (Heptinstall 1983). Müller et al. (1988) used monoclonal antibodies to examine cell types in renal biopsies showing crescent formation. T lymphocytes, predominantly of the $CD4^+$ phenotype, were identified within interstitial infiltrates, as in the study of Wegener's granulomatosis by Ten Berge et al. (1985), while lymphocytes of the $CD8^+$ phenotype were also common within the glomeruli. Monocyte/macrophages were present in Bowman's space. HLA Class II antigen expression by glomerular cells was found to be reduced commensurate with the degree of glomerular damage, although recent studies (F. Coelho, B. Hartley, and J.S. Cameron, unpublished observations) have not confirmed this. Expression by tubular cells was increased. Many other investigators have looked at the cellular composition of glomerular crescents and this is discussed in detail in Chapter 3.10. In microscopic polyarteritis, as in Wegener's granulomatosis, immunofluorescence shows scanty or absent immunoglobulin deposition. Biopsy from other organs, such as skin or intestine, may confirm the presence of a vasculitis.

Treatment

Much is written on the treatment of 'polyarteritis nodosa', although some of the patients included had small vessel disease rather than, or as well as, classical polyarteritis nodosa. This section will therefore cover protocols used for the polyarteritis group of illnesses as a whole, and those specifically employed in microscopic polyarteritis.

Induction therapy

Oral corticosteroids

The earliest attempts to treat systemic vasculitis used oral corticosteroids alone. Microscopic and classical polyarteritis were not separated in reports which suggested an improvement in overall survival from 13 per cent at 5 years in untreated patients to around 50 per cent in steroid-treated patients (Frohnert and Sheps 1967; Sack et al. 1975). However, deaths due to active vasculitis and renal failure were still frequent and the use of steroids alone was clearly inadequate.

Cytotoxic agents

Cyclophosphamide had proved very effective in inducing remission and controlling disease in Wegener's granulomatosis, and interest grew in the use of cytotoxic agents in polyarteritis. Fauci reported success in steroid-resistant disease, initially in two patients with classic polyarteritis nodosa (Fauci et al. 1978a) and later in a mixed group of patients with systemic necrotizing vasculitis, using oral cyclophosphamide at a dose of 2 mg/kg daily in conjunction with oral steroids (Fauci et al. 1979). Leib et al. (1979) described treatment in 64 patients with polyarteritis. It is not clear how many had predominantly small vessel disease and only 22 had abnormal renal function, but the response to a combination of steroid and cytotoxic therapy (using azathioprine in most cases) was dramatic. Untreated patients had a 5-year survival of 12 per cent; in patients treated with steroids alone this was 53 per cent, and in patients treated with steroids and cytotoxic agents it rose to 80 per cent. Cohen et al. (1980) described retrospectively a similar group of patients and found less convincing benefit, with only 55 per cent survival at 5 years. This was no better than the outcome of patients treated with steroids alone, but it seems likely that the patients selected to receive additional treatment were those with the most severe disease at the outset. Recent reports of treatment in microscopic polyarteritis (Savage *et al*. 1985; Coward *et al*. 1986; Adu *et al*. 1987), have described the use of a combination of steroids and cytotoxic agents in most patients, as have reports of patients with focal segmental necrotizing glomerulonephritis with a variety of other clinical features (Serra *et al*. 1984; Fuiano *et al*. 1988). Whether the use of cyclophosphamide confers any benefit compared with azathioprine is less clear. Good results have been achieved with azathioprine (Leib *et al*. 1979) and the toxicity of cyclophosphamide is well documented. However, there are some patients who achieve an inadequate response to azathioprine (Fauci *et al*. 1979) and cyclophosphamide is widely regarded as more effective in Wegener's granulomatosis. Both Balow (1985) and Couser (1988) advocate the use of cyclophosphamide in microscopic polyarteritis, and our practice is to use the same treatment regimen as in Wegener's granulomatosis (Table 3).

Cyclophosphamide is generally administered orally in an initial daily dose of 2 to 3 mg/kg. Some authors have suggested the use of intermittent intravenous cyclophosphamide, with the theoretical advantage of lower long-term toxicity. Fort and Abruzzo (1988) reported a case of polyarteritis, with both small and medium sized vessel involvement, who deteriorated after initial improvement on oral steroids and cyclophosphamide, but in whom intravenous cyclophosphamide and methylprednisolone induced remission. In contrast, Adu *et al*. (1987) reported a higher mortality (although not significantly so) in patients aggressively treated in this way. Others have used intravenous cyclophosphamide as part of an uncontrolled protocol (Fuiano *et al*. 1988) in which it is impossible to discern its value. A preliminary report of a prospective study in ANCA-positive focal necrotizing glomerulonephritis has not shown any difference between outcome on oral and intravenous cyclophosphamide therapy (Falk *et al*. 1990).

Methylprednisolone

Intravenous pulses of methylprednisolone, followed by oral prednisolone, have been shown to be effective treatment for idiopathic crescentic nephritis in numerous uncontrolled studies. A typical regimen involves administration of 30 mg/kg as an infusion over a period of 20 min, repeated on alternate days to a total of three doses. The patient should be volume replete and have normal electrolytes at the start of therapy to minimize the risk of inducing arrhythmias. This is followed by oral therapy, with prednisolone dose tapering from 1 mg/kg.day over a period of months (Couser 1988; Bolton and Sturgill 1989). Short-term adverse effects include risk of infection, and in the longer term cataracts may develop in up to 25 per cent of patients when high-dose steroids are continued.

In patients with obvious extrarenal vasculitis, intravenous methylprednisolone has been used in conjunction with cytotoxic agents. This therapy was effective in achieving remission in many of the patients described by Coward *et al*. (1986). Sustained improvement in renal function was reported by McLelland *et al*. (1989).

Plasma exchange

Numerous studies have set out to examine whether plasma exchange has a place in routine treatment of systemic vasculitis, or whether it might be indicated in particular circumstances. These have already been discussed with respect to the treatment of Wegener's granulomatosis. To recapitulate briefly, early uncontrolled data from Lockwood *et al*. (1977) and Hind *et al*. (1983) suggested that patients with severe disease could respond to an immunosuppressive regimen including plasma exchange. In a subsequent controlled trial designed to analyse the effect of plasma exchange over and above that of prednisolone and cyclophosphamide in focal necrotizing glomerulonephritis, additional benefit was demonstrated only in those patients who were dialysis dependent at presentation (C.D. Pusey *et al*. unpublished observations). Thysell (1982), in an uncontrolled study in a heterogeneous group of patients, and Burran (1986), in a study of plasma exchange in addition to intravenous methylprednisolone in idiopathic rapidly progressive glomerulonephritis, reported responses to treatment even in dialysis dependent patients, but it is difficult to be certain that plasma exchange was responsible. A controlled trial by Glöckner *et al*. (1988), which included patients with a variety of disorders, failed to demonstrate a definite advantage of plasma exchange, but use of the treatment in non-responding 'control' patients may have confused the results.

McLelland *et al*. (1989) and Bruns *et al*. (1989) found plasma exchange and intravenous methylprednisolone to be equally effective in uncontrolled series. Bruns *et al*. cautioned against the risks of infection in both treatments, and a review by Couser (1982) found these to be similar. Both Adu (1987) and Rondeau (1989) noted the risk of sepsis in patients in whom multiple immunosuppressive strategies are combined, although our experience suggests that the dose of prednisolone is more important than plasma exchange (Cohen *et al*. 1982). Our own regimen is identical to that for treatment of Wegener's granulomatosis (Table 3). Plasma exchange is added in patients with fulminant disease in the form of dialysis dependent renal failure, severe lung haemorrhage, or other life-threatening complications.

Anticoagulation and antiplatelet therapy

Experimental data suggesting a benefit of anticoagulation in crescentic nephritis led to clinical trials of heparin followed by warfarin or antiplatelet agents in the 1970s. Some success was achieved in improving renal function, but the risk of bleeding was significant. Moreover, these results were usually achieved in conjunction with prednisolone and cytotoxic agents and the benefit directly attributable to the anticoagulation was uncertain. Therapeutic regimens rarely employ these agents today.

Maintenance therapy

The need for maintenance therapy is highlighted by the number of patients reported in different series experiencing relapse after initial remission is achieved. Our policy is to use prednisolone and azathioprine, gradually tapering to discontinuation or until a point is reached where relapse tends to occur. Most patients continue on therapy for several years; those with relapsing disease stay on long-term treatment and often need cyclophosphamide. No generalization can be made about the duration of treatment needed and adjustments are made to suit the individual patient. Laboratory indices of disease activity are invaluable in making this assessment, and, as in Wegener's granulomatosis, serial measurements of C-reactive protein (Hind *et al*. 1984*a*) and ANCA (Heaton *et al*. 1990) may be particularly useful.

Outcome

The analysis of outcome of microscopic polyarteritis is fraught with the same problems of diagnostic criteria and labelling. Much of this section will therefore be based on our own experience of microscopic polyarteritis and that of other nephrologists in the United Kingdom who define the condition similarly. There are many reports of outcome in idiopathic crescentic nephritis, and as there is reason to believe that this represents a limited form of vasculitis, these results will also be summarized.

Response to initial therapy

As in Wegener's granulomatosis, aggressive initial immunosuppression is effective in inducing remission, and allowing recovery of independent renal function in many dialysis dependent patients.

In our own series of 49 patients treated since 1974, 14 (29 per cent) required dialysis within 24 h of admission and mean creatinine in the remainder was 456 μmol/l. At 2 months, 38 of 49 were alive with independent renal function and three of 49 were on regular dialysis. Six of 14 dialysis dependent patients had recovered renal function, six had died, three first having recovered some renal function, and two were on dialysis.

Adu et al. (1987) examined outcome at 3 months in 43 patients, of whom 20 of 43 (47 per cent) required dialysis soon after admission. Twenty-nine of 43 were alive, 26 with independent renal function. Patients with a creatinine greater than 500 μmol/l had a higher mortality; at 3 months eight of 24 were alive with independent renal function, 13 had died and three were on regular dialysis.

Coward et al. (1986) studied 27 patients with microscopic polyarteritis, of whom 11 (41 per cent) required dialysis in their initial illness. Twenty-four were alive at 3 months and recovery of renal function occurred in eight of 11 patients requiring dialysis during their initial illness; in two patients this occurred after 4 months of dialysis.

Early deaths in these three series were primarily from respiratory complications (16 of 28 deaths), pneumonia and lung haemorrhage being most common. The risks of immunosuppression with death from sepsis, or undertreatment with death from active vasculitis are finely balanced. Death due to myocardial infarction and subarachnoid haemorrhage occurred in six cases and two cases, respectively. Whether these reflected active vasculitis is not known.

Long-term outcome

Patient survival

The same three series give information on long-term patient survival. Actuarial survivals in Coward's series show a 1-year survival of 81 per cent and a 5-year survival of 64 per cent (Coward et al. 1986). Actuarial survival at 1 year in Adu's series was 62 per cent (Adu et al. 1987). In our series, where 39 patients were followed for 5 years, actuarial survivals were 71 per cent at 1 year, 64 per cent at 5 years. Late deaths in these series were due to a variety of causes including vascular disease and infections. Although our series shows a high early mortality, with 15 of 20 deaths occurring in the first year (six due to lung haemorrhage), it demonstrates that long-term survival is possible, even after a severe initial illness.

Survival in recent studies of mixed groups of patients with renal vasculitis (Fuiano et al. 1988) is similar and is outlined in Table 5.

Renal function

The initial improvement in renal function following treatment is often maintained. Adu et al. (1987) found that renal function remained stable in the year after initial therapy, and the same was true in all but two of Coward's patients (Coward et al. 1986). In our study, six patients with independent renal function at 2 months were accepted on to chronic dialysis programmes at between 14 months and 12 years after the initial illness. None of these had been dialysis dependent initially. Overall actuarial renal survival from the time of diagnosis was 65 per cent at 1 year and 50 per cent at 5 years.

Prognosis on renal replacement therapy is satisfactory, and Coward et al. (1986) noted no special problems related to either haemodialysis or chronic ambulatory peritoneal dialysis. A review of outcome of end-stage renal disease due to rare causes by Nissenson and Port (1990) showed a survival in the polyarteritis group of illnesses comparable to that in control populations: 62 per cent at 3 years. Six of our patients have undergone renal transplantation, with a successful outcome in four; one patient had a poorly functioning graft after an initial rejection episode, and finally lost the graft with focal necrotizing glomerulonephritis, and another patient died of opportunist infection during treatment for a late rejection episode.

Relapse

Relapse of microscopic polyarteritis may occur months or years after the initial illness and is especially likely on reducing or stopping therapy (Coward et al. 1986). Relapse may have serious consequences including death or dialysis dependent renal failure, and withdrawal of treatment should be gradual and only performed when there is no evidence of disease activity. As in Wegener's granulomatosis, serial monitoring of ANCA may be valuable in predicting those patients at risk of relapse (Heaton et al. 1990; G. Gaskin et al., unpublished observations).

Idiopathic rapidly progressive glomerulonephritis

Many reports of outcome in patients treated for focal necrotizing glomerulonephritis include both patients with systemic vasculitis and those with isolated, or 'idiopathic', rapidly progressive glomerulonephritis. They document a wide range of responses to therapy, which is not surprising in view of the heterogeneity in diagnosis, severity of disease, and treatment. However, it is clear that intravenous methylprednisolone followed by oral prednisolone can be effective in restoring renal function in many patients. Since there is reason to believe that isolated focal necrotizing glomerulonephritis is a renal-limited form of small vessel vasculitis, we adopt the same treatment regimen and use a combination of steroids and cytotoxic agents.

The subject of treatment of focal necrotizing glomerulonephritis in general, and whether or not idiopathic rapidly progressive glomerulonephritis merits separate consideration, is discussed further in Chapter 3.10.

Table 5 Outcome of microscopic polyarteritis

Author	No. of patients	Survival
Adu et al. (1987)	43	62% at 1 year
Coward et al. (1986)	27	81% at 1 year 64% at 5 years
Fuiano et al. (1988)	20	77% at 2 years 77% at 5 years
Hammersmith Hospital series	49	71% at 1 year 64% at 5 years

Risk factors for poor prognosis in microscopic polyarteritis and idiopathic rapidly progressive nephritis

Oliguria and advanced renal failure at presentation have been highlighted as risk factors for failure of renal recovery and high mortality in many series. Older patients are at greater risk in mixed vasculitis series (Sack *et al.* 1975), microscopic polyarteritis (Adu *et al.* 1987), and in focal necrotizing glomerulonephritis in general (Fuiano *et al.* 1988; Rondeau *et al.* 1989; Wilkowksi *et al.* 1989). Whether the presence of systemic disease associated with renal vasculitis alters the prognosis is open to debate. Weiss and Crissman (1985) found no difference in outcome, while Croker *et al.* (1987) found a higher morbidity in patients without systemic disease. They attributed the difference to later presentation with more advanced renal damage where medical help was not sought for systemic symptoms. Velosa (1987), in an editorial comment based on data reported by Heilman *et al.* (1987), suggested that the presence of vasculitis might imply a better prognosis independently of other variables, but the conclusions were drawn from a study in which presenting renal function and treatment administered were quite different in the two groups. Indeed, in general, cytotoxic drugs have been used less frequently in isolated rapidly progressive glomerulonephritis than in systemic vasculitis. One systemic manifestation clearly associated with a high early mortality is the presence of lung haemorrhage.

Polyarteritis nodosa

Definition

Polyarteritis nodosa, as first described by Kussmaul and Maier (1866), is characterized clinically by fever, muscular, gastrointestinal, and renal disease, and pathologically by nodular swellings along the course of medium-sized arteries with histological evidence of inflammation in the outer part of the vessel wall. Thus it is a necrotizing vasculitis involving medium-sized muscular arteries and leading to aneurysm formation.

Clinical features

Polyarteritis nodosa is typically a disease of men in middle life, with a male:female ratio of 2:1 and presentation in the fifth or sixth decades (Frohnert and Sheps 1967; Sack *et al.* 1975; Guillevin *et al.* 1988). An association with hepatitis B virus infection is recognized (Gocke *et al.* 1970; Duffy *et al.* 1976; Sergent *et al.* 1976) and has been reported in up to 36 per cent of cases from France (Ronco *et al.* 1983; Guillevin *et al.* 1988) although the incidence is much less in the United Kingdom (Scott *et al.* 1982). A similar clinical picture has been described in drug abusers (Citron *et al.* 1970) and in those with human immunodeficiency virus infection (Calabrese *et al.* 1989).

The clinical features are those of tissue infarction, haemorrhage, and organ dysfunction, typically affecting the gastrointestinal tract, nervous system, muscles, and soft tissues. Constitutional features are common, in the form of fever, malaise, and weight loss. Renal involvement may present with loin pain and haematuria due to renal infarction, or renal impairment due to ischaemia. Glomerulonephritis is reported, presumably due to an overlap with microvascular disease. Hypertension is common, although the reported frequency varies from 25 (Sack *et al.* 1975) to 71 per cent (Travers *et al.* 1979). Systemic features of accelerated phase hypertension are sometimes seen (Guillevin *et al.* 1988). The pathognomonic subcutaneous nodules overlying arteries are not invariably seen in life (Sack *et al.* 1975, Travers *et al.* 1979).

Diagnosis

Non-specific indices of inflammation are usually seen (see Table 2), and both C-ANCA (Lüdemann *et al.* 1990*b*; Nässberger *et al.* 1989) and P-ANCA (G. Gaskin and C.D. Pusey, unpublished observations) have been identified. These antibodies have also been detected in hepatitis B-associated disease, typically with a perinuclear immunofluorescence pattern (Daha and Falk 1990). The use of differing diagnostic criteria in the reported studies has made it difficult to address the question of ANCA in polyarteritis nodosa satisfactorily. Confirmation of diagnosis requires the demonstration of medium-sized artery involvement. Typically this leads to formation of multiple aneurysms with vessel narrowing and irregularity, which may be seen on arteriography. These findings may often be demonstrable in the renal circulation, although studies of the hepatic circulation are said to give the best yield of positive findings (Travers *et al.* 1979). The same study demonstrated aneurysms in 10 of 17 patients with suspected polyarteritis nodosa; five patients in the non-aneurysm group underwent renal biopsy which revealed a focal necrotizing glomerulonephritis, suggesting that they had microscopic polyarteritis rather than larger vessel disease.

The diagnosis may be confirmed morphologically, for which muscle biopsy has been widely used. Renal biopsy is logical in the presence of clinical evidence of renal involvement and is an effective means of diagnosis (Ronco *et al.* 1983), but carries a higher than usual risk of bleeding or formation of an arteriovenous fistula (Curran *et al.* 1967). The histological picture is one of fibrinoid necrosis of arcuate or occasionally of interlobular arteries, with a marked inflammatory response within and surrounding the vessel (Heptinstall 1983). Destruction of the internal elastic lamina may be seen, even in healed or healing lesions and aneurysms may be observed. Hypertensive changes may be seen in smaller arteries and arterioles, but are clearly not specific. There may be evidence of ischaemic damage, with tubular atrophy and periglomerular fibrosis, and occasionally areas of infarction. Focal necrotizing glomerulonephritis reflects smaller vessel involvement and is seen only in presumed 'overlap' cases.

Treatment

Treatment conventionally involves corticosteroids, but many authors would advocate using an additional cytotoxic agent (Leib *et al.* 1979; Fauci *et al.* 1979). Guillevin *et al.* (1988) used a combination of steroids and plasma exchange with or without cyclophosphamide. We use the same protocol as for small vessel vasculitis.

Outcome

Frohnert and Sheps (1967) reviewed 130 patients examined at the Mayo Clinic between 1945 and 1962, of whom 30 per cent had a raised blood urea at presentation. They reported a 5-year survival from diagnosis of 13 per cent in untreated patients, compared to 48 per cent in patients treated with corticosteroids or adrenocorticotrophic hormone. A lower survival was noted in those with more impaired renal function, and indeed the chief cause of early death was renal failure. However, no renal histology was given and it is unclear how many of these patients had small vessel disease with a glomerulonephritis. Sack *et al.* (1975) reported a 5-year survival of 57 per cent in a group of 40 patients, including several with features of Churg–Strauss syndrome and relatively few (29 per cent) with renal disease. Most patients had been treated with corticosteroids and older patients had a poorer outcome. Fauci reported success in achieving remissions in steroid-resistant disease using a combination of prednisolone and cyclophosphamide (Fauci et al. 1978, 1979) while Leib et al. (1979), reported a 5-year survival of 80 per cent using a combination of steroids and a cytotoxic agent—in most cases azathioprine. Guillevin et al. (1988) described outcome in 165 patients with polyarteritis nodosa, including patients with Churg–Strauss syndrome (49 patients had asthma) and probably a small number with microscopic arteritis (16 patients had glomerular lesions). Patients treated before 1973 received corticosteroids alone; those treated subsequently received steroids and plasma exchange and were randomized for the addition of cyclophosphamide. Five-year survival was 63 per cent in the group as a whole, 59 per cent in the steroids-only group, 77 per cent in the steroids and plasma exchange group, and 82 per cent in the group receiving triple therapy. These differences did not reach statistical significance. In general, death is most likely to be due to extrarenal vasculitis in classic polyarteritis nodosa, in contrast to microscopic polyarteritis (reviewed by Balow 1985). Complications of intestinal vasculitis were a major cause of death in Guillevin's series (1988).

Churg–Strauss syndrome

Definitions

This disease was clearly defined by Churg and Strauss (1951) as a variant of polyarteritis nodosa, and patients with this syndrome have been described subsequently in numerous series, including reports concentrating on polyarteritis nodosa (Rose and Spencer 1957; Sack et al. 1975; Guillevin et al. 1988) and reports limited to what is known as Churg–Strauss syndrome (Chumbley et al. 1977; Lanham et al. 1984). Just as Godman and Churg (1954) set out strict pathological criteria for the diagnosis of Wegener's granulomatosis, Churg and Stauss devised diagnostic criteria from postmortem studies, including demonstration of necrotizing vasculitis, eosinophilic tissue infiltration, and extravascular granulomata. Lanham et al. (1984) suggested that the pattern of disease was sufficently distinctive to allow a more flexible clinical definition. This demanded the coexistence of systemic vasculitis involving two or more extrapulmonary organs with asthma and peripheral blood eosinophilia (in excess of $1.5 \times 10^9/l$.) In particular they felt that it was impractical to insist on the demonstration of granulomata to confirm the diagnosis.

Clinical features

Extrarenal disease

Lanham et al. (1984) documented the clinical features in 16 patients and reviewed the literature for details in a further 138 cases. Respiratory disease is characterized by asthma, usually of relatively late onset. Infiltrates on the chest radiograph are common, although often transient, and pleural effusions, which may contain large numbers of eosinophils, are described. Disease of the upper respiratory tract is common in the form of allergic rhinitis and nasal polyps. The symptoms of respiratory involvement often antedate the appearance of vasculitis by several years. Eosinophilia exceeding $1.5 \times 10^9/l$ occurred in all Lanham's patients with a mean peak eosinophil count of $8.4 \times 10^9/l$. This is considerably higher than the small elevations in eosinophil count sometimes seen in polyarteritis nodosa and Wegener's granulomatosis. Eosinophilic tissue infiltrates may occur in the absence of peripheral blood eosinophilia. Treated patients may not show a characteristic eosinophilia, since this is rapidly suppressed by corticosteroids. Vasculitic manifestations of the disease affect particularly the heart, skin, bowel, and the musculoskeletal and nervous systems. Mononeuritis multiplex is described in two-thirds of cases. As in other forms of small vessel vasculitis, life-threatening alveolar haemorrhage may occur (Clutterbuck and Pusey 1987) although this is rare. Involvement of medium sized arteries can be confirmed at arteriography.

Renal disease

In general, renal disease is not a major feature although histological changes in the kidney were clearly described by Churg and Strauss. In an unselected series of 30 patients reported by Chumbley et al. (1977), only one patient had 'renal failure', six had microscopic haematuria, and three had slight elevations of urea and creatinine. The Hammersmith experience differs, but referral bias may have led to an over-representation of severe vasculitis and renal disease. In Lanham's series (1984), most of the patients had some degree of renal involvement, and Clutterbuck et al. (1990) described renal disease in 16 of 19 patients; nephrotic syndrome occurred in three and serum creatinine concentration was greater than 500 μmol/l in four patients. The urine sediment was abnormal in all but one patient with renal disease, generally with microscopic haematuria and granular casts, and proteinuria was detected in 12 of the 16. The dominant histological pattern was a focal segmental glomerulonephritis. In general, vascular lesions and eosinophilic infiltration were less common, although in one biopsy an intense interstitial infiltrate was seen.

Antineutrophil cytoplasmic antibodies in Churg–Strauss syndrome

Antineutrophil cytoplasmic antibodies have recently been reported in patients with Churg–Strauss syndrome (Wathen and Harrison 1987; Cohen Tervaert et al. 1990). We have confirmed this observation in our patients, detecting both P-ANCA and C-ANCA in different individuals, and have also demonstrated anti-myeloperoxidase activity in all patients positive by immunofluorescence (unpublished observations). These antibodies are therefore useful for diagnosis of this syndrome, and their presence illustrates the immunological relationship with other forms of systemic vasculitis. As in other forms of vasculitis there is no direct evidence for pathogenicity of ANCA, and their relationship to the characteristic eosinophil infiltration (which may be directly pathogenic) is unknown. (Tai et al. 1984).

Treatment

Corticosteroids have been the mainstay of therapy in most reported series (Chumbley et al. 1977; Lanham et al. 1984). This approach proved adequate for the majority of patients, but doses were sometimes high and side-effects troublesome, and cytotoxic agents were added in a minority of cases. In Clutterbuck's series (1990), which had a greater proportion of renal disease, additional immunosuppression was used in 11 of 19, with plasma exchange in five. The consensus view from these series is that steroids alone may be insufficient to control severe disease, and that cytotoxic agents should then be added. Our preference would be to use cyclophosphamide, as for other forms of systemic vasculitis. Single case reports describe the use of intravenous methylprednisolone (MacFadyen et al. 1987) and intravenous cyclophosphamide (Chow et al. 1989), for severe respiratory disease.

The vasculitic illness is usually of limited duration, and gradual reduction or withdrawal of treatment is therefore possible. Occasionally, disease activity continues and maintenance therapy with a combination of prednisolone and azathioprine is suggested (Lanham et al. 1984).

Outcome

Initial therapy is effective in most patients and long-term survival is usual. In Chumbley's series 5-year survival was 62 per cent and median patient survival was over 9 years. The same is true even in cases with more severe renal disease; 10 of 19 patients in Clutterbuck's series had a normal serum creatinine at 2 to 10 years after diagnosis. Cardiac disease is the single most common cause of death in reported series. Late relapse may sometimes occur, but is less frequent than in Wegener's granulomatosis.

Renal disease in other vasculitic illnesses

Giant cell arteritis

Giant cell arteritis is a disease of the elderly, usually characterized clinically by temporal arteritis or polymyalgia rheumatica. The major complications are ophthalmic and overt renal disease is rare, although microscopic haematuria and low levels of proteinuria with normal renal function may occur at presentation (Sonnenblick et al. 1989). More noteworthy are occasional reports of widely disseminated visceral giant cell arteritis, affecting both large arteries in various organs and smaller vessels within the kidney (Lie 1978). Elling and Kristiensen (1980) reported a patient with polymyalgia rheumatica who developed fatal renal failure associated with florid intrarenal vasculitis affecting all the intrarenal arteries and arterioles. Others (Droz et al. 1979; O'Neill et al. 1976) have reported the coexistence of typical giant cell arteritis affecting the temporal arteries and focal necrotizing glomerulonephritis, leading to marked renal impairment. One case of temporal arteritis with membranous nephropathy has been described (Truong et al. 1985).

Takayasu's arteritis

Takayasu's arteritis is typically a disease of young women which causes inflammation in large arteries, particularly the aorta and its main branches. Stenoses and occlusions, and also dilatations and aneurysms result. The main renal arteries are frequently involved (Hall et al. 1985) but glomerular disease is rare. Minor mesangial proliferative changes have been reported by Takagi et al. (1984) and others. Focal segmental necrotizing glomerulonephritis with crescent formation has been described in one case, which responded to treatment with pulse methylprednisolone (Hellman et al. 1987a).

Behçet's syndrome

This disease is characterized clinically by ocular inflammation, oral and genital ulceration, and variable thrombotic, neurological, pulmonary, and rheumatic features, and pathologically by a leucocytoclastic vasculitis chiefly involving veins, venules, and capillaries. Renal involvement is rare, and one difficulty in associating patterns of nephritis with Behçet's syndrome is the difficulty in making a certain diagnosis. Kansu et al. (1977) reported a focal necrotizing glomerulonephritis in a patient with oral ulceration, polyarthritis, myocarditis, and vasculitic scrotal ulceration, where the unifying diagnosis was Behçet's syndrome. A report by Landwehr et al. (1980) suggested crescentic nephritis in this disorder, although the underlying illness was suspicious of Wegener's granulomatosis. Herreman et al. (1982) studied 10 patients with a diagnosis of Behçet's syndrome, and found one case with a focal segmental mesangial proliferation with necrosis and crescent formation (although the extent of the latter is not clear). Recently, Donnelly et al. (1989) reported two cases of crescentic glomerulonephritis in Behçet's syndrome. The clinical features in both included oral and genital ulceration; the first patient had a cutaneous vasculitis and the second, episcleritis. Both were treated with prednisolone and cyclophosphamide. The first patient, who also received intravenous methylprednisolone, responded well to therapy; the second, in whom compliance was poor, showed progression of renal damage.

Relapsing polychondritis

This is another rare inflammatory condition in which renal disease has been recognized. It is characterized by episodic inflammation and destruction of cartilaginous structures, with diverse clinical manifestations including features of systemic vasculitis in some patients. Chang-Miller et al. (1987) have summarized a variety of case reports showing renal involvement, typically in the form of a segmental necrotizing glomerulonephritis, with a rapidly progressive course with crescent formation in some. They also reviewed 129 patients with relapsing polychondritis seen at the Mayo Clinic over a period of 40 years, and estimated an incidence of renal involvement of 22 per cent. Renal biopsies were performed in 11 cases; mesangial proliferation was seen in all, and segmental necrotizing glomerulonephritis in eight. Renal disease was diagnosed more commonly in older patients, and evidence of systemic vasculitis was more likely in those patients with renal disease. Actuarial survival was significantly worse in the presence of renal involvement, and cytotoxic therapy in addition to corticosteroids was advocated for most patients with glomerular disease.

Kawasaki disease

Kawasaki disease, or mucocutaneous lymph node syndrome, is a febrile illness of childhood characterized by a multisystem disease with systemic vasculitis, often with life-threatening coronary involvement. Lytic antiendothelial antibodies have been identified in this condition (Leung et al. 1986) and antineutrophil cytoplasmic antibodies may also be detected (Savage et al. 1989). Renal disease is not usually a feature. Renal histology was reported in one case with microscopic haematuria and proteinuria to show mesangial hypercellularity and interstitial infiltrates, without renal vasculitis (Salcedo et al. 1988).

Conclusion

In the absence of a clear understanding of their pathogenesis, division of the illnesses which are grouped together as systemic vasculitis is problematic and has caused much debate. In the classification outlined here, the different syndromes are defined by varying clinical and pathological characteristics, and show different patterns of renal disease. The small vessel diseases, Wegener's granulomatosis and microscopic polyarteritis, are commonly associated with focal necrotizing glomerulonephritis,

4.5.3 Systemic vasculitis

and the syndrome of idiopathic rapidly progressive glomerulonephritis may represent a limited form of microscopic polyarteritis. Several of these conditions share an association with antibodies directed against neutrophil cytoplasmic antigens, although the fine specificity appears to vary with clinical diagnosis. Treatment has been developed along broadly similar lines in each, with corticosteroids and cytotoxic drugs the mainstay of therapy, but response to treatment and long-term prognosis may differ. However, it is clear that in all these illnesses early recognition and prompt institution of therapy are crucial in achieving the best possible outcome.

References

Adu, D., Howie, A.J., Scott, D.G.I., Bacon, P.A., McGonigle, R.J.S., and Michael, J. (1987). Polyarteritis and the kidney. *Quarterly Journal of Medicine*, 62, 221–37.

Allen, D.C., Doherty, C.C., and O'Reilly, D.P.J. (1984). Pathology of the heart and the cardiac conduction system in Wegener's granulomatosis. *British Heart Journal*, 52, 674–8.

Ambrus, J.L.Jr. and Fauci A.S. (1984). Diffuse histiocytic lymphoma in a patient treated with cyclophosphamide for Wegener's granulomatosis. *American Journal of Medicine*, 76, 745–7.

Andrassy, K., Koderisch, J., Waldherr, R., and Rufer, M. (1988). Diagnostic significance of anticytoplasmic antibodies (ACPA/ANCA) in detection of Wegener's granulomatosis and other forms of vasculitis. *Nephron*, 49, 257–8.

Arnalich, F., et al. (1989). Polyarteritis nodosa and necrotising glomerulonephritis associated with long-standing silicosis. *Nephron*, 51, 544–7.

Austin, H.A., et al. (1986). Therapy of lupus nephritis. Controlled trial of prednisone and cytotoxic drugs. *New England Journal of Medicine*, 514, 614–9.

Axelson, J.A., Clark, R.H., and Ancerewicz, S. (1987). Wegener granulomatosis and trimethoprim-sulphamethoxazole (letter). *Annals of Internal Medicine* 107, 600.

Balow, J.E. (1985). Renal Vasculitis. *Kidney International*, 27, 954–64.

Balow, J.E., Antonovych, T., Fauci, A.S., and Wilson, C.B. (1978). The nephritis of Wegener's granulomatosis (abstract). *Kidney International*, 14, 706.

Bambery, P., et al. (1987). Wegener's granulomatosis in North India. An analysis of eleven patients. *Rheumatology International*, 7, 243–7.

Banfi, G., Imbasciati, E., Guerra, L., Mihatsch, M.J., and Ponticelli, C. (1983). Extracapillary glomerulonephritis with necrotising vasculitis in D-penicillamine treated rheumatoid arthritis. *Nephron*, 33, 56–60.

Bhathena, D.B., Migdal, S.D., Julian, B.A., McMorrow, R.G., and Baehler, R.W. (1987). Morphologic and immunohistochemical observations in granulomatous glomerulonephritis. *American Journal of Pathology*, 126, 581–91.

Björk, S., Svalander, C., and Westberg, G. (1985). Hydralazine-associated glomerulonephritis. *Acta Medica Scandnavica*, 218, 261–9.

Bolton, W.K. and Couser, W.G. (1979). Intravenous pulse methylprednisolone therapy of acute crescentic rapidly progressive glomerulonephritis. *American Journal of Medicine*, 66, 495–502.

Bolton, W.K. and Sturgill, B.C. (1989). Methylprednisolone therapy for acute crescentic rapidly progressive glomerulonephritis. *American Journal of Nephrology*, 9, 368–75.

Borleffs, J.C.C., Derksen, R.H.W.M., and Hené, R.J. (1987). Treatment of Wegener's granulomatosis with cyclosporin (letter). *Annals of the Rheumatic Diseases*, 46, 175.

Brandwein, S., Esdaile, J., Danoff, D., and Tannenbaum, H. (1983). Wegener's granulomatosis. Clinical features and outcome in 13 patients. *Archives of Internal Medicine*, 143, 476–9.

Brasile, L., Kremer, J.M., Clarke, J.L., and Cerilli, J. (1989). Identification of an autoantibody to vascular endothelial cell-specific antigens in patients with systemic vasculitis. *American Journal of Medicine*, 87, 74–86.

Bruns, F.J., Adler, S., Fraley, D.S., and Segel, D.P. (1989). Long-term follow-up of aggressively treated idiopathic rapidly progressive glomerulonephritis. *American Journal of Medicine*, 86, 400–6.

Bullen, C.L., Liesegang, T.J., McDonald, T.J., and DeRemee, R.A. (1983). Ocular complications of Wegener's granulomatosis. *Ophthalmology*, 90, 279–90.

Burran, W.P., Avasthi, P., Smith, K.J., and Simon, T.L. (1986). Efficacy of plasma exchange in severe idiopathic rapidly progressive glomerulonephritis. A report of ten cases. *Transfusion*, 26, 382–7.

Calabrese, L.H., et al. (1989). Systemic vasculitis in association with human immunodeficiency virus infection. *Arthritis and Rheumatism*, 32, 569–76.

Camilleri, M., Pusey, C.D., Chadwick, V.S., and Rees, A.J. (1983). Gastrointestinal manifestations of systemic vasculitis. *Quarterly Journal of Medicine*, 52, 141–9.

Carrington, C.B. and Liebow, A.A. (1966). Limited forms of angiitis and granulomatosis of Wegener's type. *American Journal of Medicine*, 41, 497–527.

Chang-Miller, A., et al. (1987). Renal involvement in relapsing polychondritis. *Medicine*, 66, 202–17.

Chow, C.C., Li, E.K.M., and Lai, F.M.M. (1989). Allergic granulomatosis and angiitis (Churg–Strauss syndrome): response to 'pulse' intravenous cyclophosphamide. *Annals of the Rheumatic Diseases*, 48, 605–8.

Chumbley, L.C., Harrison, E.G., and DeRemee, R.A. (1977). Allergic granulomatosis and angiitis (Churg–Strauss syndrome). Report and analysis of 30 cases. *Mayo Clinic Proceedings*, 52, 477–84.

Churg, J. and Strauss, L. (1951). Allergic granulomatosis, allergic angiitis and periarteritis nodosa. *American Journal of Pathology*, 27, 277–301.

Citron, B.P., et al. (1970). Necrotising angiitis associated with drug abuse. *New England Journal of Medicine*, 283, 1003–11.

Clutterbuck, E.J. and Pusey, C.D. (1987). Severe alveolar heamorrhage in Churg–Strauss syndrome. *European Journal of Respiratory Diseases*, 71, 158–63.

Clutterbuck, E.J., Evans, D.J., and Pusey, C.D. (1990). Renal involvement in Churg–Strauss syndrome. *Nephrology Dialysis Transplantation*, 5, 1–7.

Cohen, J., Pinching, A.J., Rees, A.J., and Peters, D.K. (1982). Infection and immunosuppression. A study of the infective complications of 75 patients with immunologically-mediated disease. *Quarterly Journal of Medicine*, 51, 1–15.

Cohen, M.I., Gore, R.M., August, C.Z., and Ossoff, R.H. (1984). Tracheal and bronchial stenosis associated with mediastinal adenopathy in Wegener granulomatosis: CT findings. *Journal of Computer Assisted Tomography*, 8, 327–9.

Cohen, R.D., Conn, D.L., and Ilstrup, D.M. (1980). Clinical features, prognosis and response to treatment in polyarteritis. *Mayo Clinic Proceedings*, 55, 146–55.

Cohen Tervaert, J.W., et al. (1989). Association between active Wegener's granulomatosis and anticytoplasmic antibodies. *Archives of Internal Medicine*, 149, 2461–5.

Cohen Tervaert, J.W., et al. (1990). Autoantibodies against myeloid lysosomal enzymes in crescentic glomerulonephritis. *Kidney International*, 37, 799–806.

Colburn, K.K., Cao, J.D., Krick, E.H., Mortensen, S.E., and Wong, L.G. (1985). Hodgkin's lymphoma in a patient treated for Wegener's granulomatosis with cyclophosphamide and azathioprine. *Journal of Rheumatology*, 12, 599–602.

Cordier, J.F., Valeyre, D., Guillevin, L., Loire, R., and Brechot, J.M. (1990). Pulmonary Wegener's granulomatosis. A clinical and imaging study of 77 cases. *Chest*, 97, 906–12.

Couser, W.G. (1982). Idiopathic rapidly progressive glomerulonephritis. *American Journal of Nephrology*, 2, 57–69.

Couser, W.G. (1988). Rapidly progressive glomerulonephritis: classification, pathogenetic mechanisms, and therapy. *American Journal of Kidney Diseases*, 11, 449–64.

Coward, R.A., Hamdy, N.A.T., Shortland, J.S., and Brown, C.B. (1986). Renal micropolyarteritis: a treatable condition. *Nephrology Dialysis Transplantation*, **1**, 31–7.

Croker, P.B., Lee, T., and Gunnells, J.C. (1987). Clinical and pathological features of polyarteritis nodosa and its renal-limited variant: primary crescentic and necrotising glomerulonephritis. *Human Pathology*, **18**, 38–44.

Curtis, J.J., Diethelm, A.G., Herrera, G.A., Crowell, W.T., and Whelchel, J.D. (1983). Recurrence of Wegener's Granulomatosis in a cadaver renal allograft (letter). *Transplantion*, **36**, 452–4.

Curran, R.E., Steinberg, I., and Hagstrom, J.W.C. (1967). Arteriovenous fistula complicating percutaneous renal biopsy in polyarteritis nodosa. *American Journal of Medicine*, **43**, 465–70.

Daha, M.R. and Falk, R.J. (1990). Anti-myeloperoxidase antibodies and clinical associations. (Proceedings of the second international workshop on anti-neutrophil cytoplasmic antibodies.) *Netherlands Journal of Medicine*, **36**, 152–3.

Davies, D.J., Moran, J.E., Niall, J.F., and Ryan, G.B. (1982). Segmental necrotising glomerulonephritis with antineutrophil antibody: possible arbovirus aetiology? *British Medical Journal*, **285**, 606.

Davson, J., Ball, J., and Platt, R. (1948). The kidney in periarteritis nodosa. *Quarterly Journal of Medicine*, **17**, 175–202.

D'Cruz, D.P., Baguley, E., Asherson, R.A., and Hughes, G.R.V. (1989). Ear, nose, and throat symptoms in subacute Wegener's granulomatosis. *British Medical Journal*, **299**, 419–22.

DeRemee, R.A. (1988). The treatment of Wegener's granulomatosis with trimethoprim/sulphamethoxazole: illusion or vision? *Arthritis and Rheumatism*, **31**, 1068–72.

DeRemee, R.A., McDonald, T.J., and Weiland, L.H. (1985). Wegener's granulomatosis: observations on treatment with antimicrobial agents. *Mayo Clinic Proceedings*, **60**, 27–32.

Devogelaer, J.P., Pirson, Y., Vandenbroucke, J.M., Cosyns, J.P., Brichard, S., and Nagent de Deuxchaisnes, C. (1987). D-penicillamine induced crescentic glomerulonephritis: report and review of the literature. *Journal of Rheumatology*, **14**, 1036–41.

Donnelly, S., Jothy, S., and Barre, P. (1989). Crescentic glomerulonephritis in Behçet's syndrome—results of therapy and review of the literature. *Clinical Nephrology*, **31**, 213–8.

Drachman, D.A. (1962). Neurological complications of Wegener's granulomatosis. *Archives of Neurology*, **8**, 145–55.

Drake-Lee, A.B., Bickerton, R.C., and Milford, C. (1988). Wegener's granulomatosis and nasal deformity. *British Journal of Clinical Practice*, **42**, 348–50.

Droz, D., Noel, L.H., Leibowitch, M., and Barbanel, C. (1979). Glomerulonephritis and necrotising angiitis. In *Advances in Nephrology*, (ed. J. Hamburger, J. Crosnier, J. P. Grunfeld and M.H. Maxwell.) Vol. 8. pp. 343–63. Year Book Medical Publishers, Chicago.

Duffy, J., *et al.* (1976). Polyarthritis, polyarteritis and hepatitis B. *Medicine*, **55**, 19–37.

Elkon, K.B., Sutherland, D.C., Rees, A.J., Hughes, G.R.V., and Batchelor, J.R. (1983). HLA antigen frequencies in systemic vasculitis: increase in HLA-DR2 in Wegener's granulomatosis. *Arthritis and Rheumatism*, **26**, 102–5.

Elling, H. and Kristensen, I.B. (1980). Fatal renal failure in polymyalgia rheumatica caused by disseminated giant cell arteritis. *Scandinavian Journal of Rheumatology*, **9**, 206–8.

Ewan, P.W., Jones, H.A., Rhodes, C.G., and Hughes, J.M.B. (1976). Detection of intrapulmonary haemorrhage with carbon monoxide uptake. Application in Goodpasture's syndrome. *New England Journal of Medicine*, **295**, 1391–6.

Fahey, J.L., Leonard, E., Churg, J., and Godman, G. (1954). Wegener's granulomatosis. *American Journal of Medicine*, **17**, 168–79.

Falk, R.J. and Jennette, J.C. (1988). Anti-neutrophil cytoplasmic autoantibodies with specificity for myeloperoxidase in patients with systemic vasculitis and idiopathic necrotising and crescentic glomerulonephritis. *New England Journal of Medicine*, **318**, 1651–7.

Falk, R.J., Hogan, S.L., and Jennette, J.C. (1990). A prospective cohort study of 70 patients with anti-neutrophil cytoplasmic autoantibody-associated glomerulonephritis (abstract). *Kidney International*, **37**, 256.

Fauci, A.S. and Wolff, S.M. (1973). Wegener's granulomatosis: studies in eighteen patients and a review of the literature. *Medicine*, **52**, 535–61.

Fauci, A.S., Wolff, S.M., and Johnson, J.S. (1971). Effect of cyclophosphamide upon the immune response in Wegener's granulomatosis. *New England Journal of Medicine*, **285**, 1493–6.

Fauci, A.S., *et al.* (1976). Successful renal transplantation in Wegener's granulomatosis. *American Journal of Medicine*, **60**, 437–40.

Fauci, A.S., Doppman, J.L., and Wolff, S.M. (1978a). Cyclophosphamide-induced remissions in advanced polyarteritis nodosa. *American Journal of Medicine*, **64**, 890–4.

Fauci, A.S., Haynes, B.F., and Katz, P. (1978b). The spectrum of vasculitis. Clinical, pathologic, immunologic, and therapeutic considerations. *Annals of Internal Medicine*, **89**, 660–76.

Fauci, A.S., Katz, P., Haynes, B.F., and Wolff, S.M. (1979). Cyclophosphamide therapy of severe systemic necrotising vasculitis. *New England Journal of Medicine*, **301**, 235–8.

Fauci, A.S., Haynes, B.F., Katz, P., and Wolff, S.M. (1983). Wegener's granulomatosis: prospective clinical and therapeutic experience with 85 patients for 21 years. *Annals of Internal Medicine*, **98**, 76–85.

Ferrari, E. (1903). Ueber Polyarteriitis acuta nodosa (sogenannte Periarteriitis nodosa) und ihre Beziehungen zur Polymyositis und Polyneuritis acuta. *Beitrage der Pathologische Anatomie*, **34**, 350.

Ferraro, G., *et al.* (1990). Anti-endothelial cell antibodies in patients with Wegener's granulomatosis and micropolyarteritis. *Clinical and Experimental Immunology*, **79**, 47–53.

Finlay, A.Y., Statham, B., and Knight, A.G. (1981). Hydralazine induced necrotising vasculitis. *British Medical Journal*, **282**, 1703–4.

Fort, J.G. and Abruzzo, J.L. (1988). Reversal of progressive necrotising vasculitis with intravenous pulse cyclophosphamide and methylprednisolone *Arthritis and Rheumatism*, **31**, 1194–8.

Frampton, G., Lockwood, C.M., and Cameron, J.S. (1989). Antibodies to endothelial cells and neutrophil cytoplasm in Wegener's granuloma and microscopic polyarteritis (abstract). *Nephrology Dialysis Transplantation*, **4**, 426.

Frohnert, P.P. and Sheps, S.G. (1967). Long-term follow-up study of periarteritis nodosa. *American Journal of Medicine*, **43**, 8–14.

Fuiano, G., Cameron, J.S., Raftery, M., Hartley, B.H., Williams, D.G., and Ogg, C.S. (1988). Improved prognosis of renal microscopic polyarteritis in recent years. *Nephrology Dialysis Transplantation*, **3**, 383–91.

Furlong, T.J., Ibels, L.S., and Eckstein, R.P. (1987). The clinical spectrum of necrotising glomerulonephritis. *Medicine*, **66**, 192–201.

Gaskin, G., Bateson, K., Evans, D.J., Rees, A.J., and Pusey, C.D. (1989). Response to treatment and long-term follow-up in 60 patients with Wegener's granulomatosis (abstract). *Nephrology Dialysis Transplantation*, **4**, 832–3.

Geraghty, J., Mackay, I.R., and Smith, D.C. (1986). Intestinal perforation in Wegener's granulomatosis. *Gut*, **27**, 450–1.

Gerbracht, D.D., Savage, R.W., and Scharff, N. (1987). Reversible valvulitis in Wegener's granulomatosis. *Chest*, **92**, 182–3.

Glöckner, W.M. *et al.* (1988). Plasma exchange and immunosuppression in rapidly progressive glomerulonephritis: a controlled, multi-center study. *Clinical Nephrology*, **29**, 1–8.

Gocke, D.J., Hsu, K., Morgan, C., Bombardieri, S., Lockshin, M., and Christian, C.L. (1970). Association between polyarteritis and Australia antigen. *Lancet*, **ii**, 1149–53.

Godman, G.C. and Churg, J.C. (1954). Wegener's granulomatosis. Pathology and review of the literature. *Archives of Pathology*, **58**, 533–53.

Goldschmeding, R., *et al.* (1989). Wegener's granulomatosis autoantibodies identify a novel diisopropylfluorophosphate-binding protein in the lysosomes of normal human neutrophils. *Journal of Clinical Investigation*, **84**, 1577–87.

Gremmel, F., Druml, W., Schmidt, P., and Graninger, W. (1988). Cyclosporin in Wegener granulomatosis (letter). *Annals of Internal Medicine*, **108**, 491.

Gross, W.L. (1989). Wegener's granulomatosis. New aspects of the disease course, immunodiagnostic procedures, and stage-adapted treatment. *Sarcoidosis*, **6**, 15–29.

Guillevin, L., Le The Huong Du, Godeau, P., Jais, P., and Wechsler, B. (1988). Clinical findings and prognosis of polyarteritis nodosa and Churg–Strauss angiitis: a study in 165 patients. *British Journal of Rheumatology*, **27**, 258–64.

Hall, S., Barr, W., Lie, J.T., Stanson, A.W., Kazmier, F.J., and Hunder, G.G. (1985a). Takayasu arteritis. A study of 32 North American patients. *Medicine*, **64**, 89–99.

Hall, S.L., Miller, L.C., Duggan, E., Mauer, S.M., Beatty, E.C., Hellerstein, S. (1985b). Wegener granulomatosis in pediatric patients. *Journal of Pediatrics*, **106**, 739–44.

Harrison, D.F.N. (1989). Ear, nose, and throat symptoms in subacute Wegener's granulomatosis (letter). *British Medical Journal*, **299**, 791.

Harkavy, J. (1941). Vascular allergy. Pathogenesis of bronchial asthma with recurrent pulmonary infiltrations and eosinophilic polyserositis. *Archives of Internal Medicine*, **67**, 709.

Haworth, S.J. and Pusey, C.D. (1984). Severe intestinal involvement in Wegener's granulomatosis. *Gut*, **25**, 1296–1300.

Haworth, S.J., Savage, C.O.S., Carr, D., Hughes, J.M.B., and Rees, A.J. (1985). Pulmonary haemorrhage complicating Wegener's granulomatosis and microscopic polyarteritis. *British Medical Journal*, **290**, 1775–8.

Haynes, B.F. and Fauci, A.S. (1978). Diabetes insipidus associated with Wegener's granulomatosis successfully treated with cyclophosphamide. *New England Journal of Medicine*, **299**, 764.

Heaton, A., Jayne, D., Brownlee, A., Evans, D.B., and Lockwood, C.M. (1990). Sequential antineutrophil cytoplasm antibody titres in the management of systemic vasculitis (abstract). *Kidney International*, **37**, 440.

Heilman, R.L., Offord, K.P., Holley, K.E., and Velosa, J.A. (1987). Analysis of risk factors for patient and renal survival in crescentic glomerulonephritis. *American Journal of Kidney Diseases*, **9**, 98–107.

Hellman, D.B., Hardy, K., Lindenfeld, S., and Ring, E. (1987a). Takayasu's arteritis associated with crescentic glomerulonephritis. *Arthritis and Rheumatism*, **30**, 451–4.

Hellman, D., Laing, T., Petri, M., Jacobs, D., Crumley, R., and Stulbarg, M. (1987b). Wegener's granulomatosis: isolated involvement of the trachea and larynx. *Annals of Rheumatic Diseases*, **46**, 628–31.

Henochowicz, S., Eggensperger, D., Pierce, L., and Barth, W.F. (1986). Necrotising systemic vasculitis with features of both Wegener's granulomatosis and Churg–Strauss vasculitis. *Arthritis and Rheumatism*, **29**, 565–9.

Heptinstall, R.H. (1983). Polyarteritis (periarteritis) nodosa, other forms of vasculitis, and rheumatoid arthritis. In *The Pathology of the kidney*. (ed. R. H. Heptinstall) Vol.2 (3rd edn) pp. 793–807. Little Brown, Boston.

Herreman, G., et al. (1982). Behçet's syndrome and renal involvement: a histological and immunofluorescent study of eleven renal biopsies. *American Journal of Medical Science*, **284**, 10–17.

Hind, C.R.K., Paraskevakou, H., Lockwood, C.M., Evans, D.J., Peters, D.K., and Rees, A.J. (1983). Prognosis after immunosuppression of patients with cresentic nephritis requiring dialysis. *Lancet*, **i**, 263–5.

Hind, C.R.K., Savage, C.O., Winearls, C.G., and Pepys, M.B. (1984a). Objective monitoring of disease activity in polyarteritis by measurement of serum C reactive protein concentration. *British Medical Journal*, **288**, 1027–30.

Hind, C.R.K., Winearls, C.G., Lockwood, C.M., Rees, A.J., and Pepys, M.B. (1984b). Objective monitoring of activity in Wegener's granulomatosis by measurement of serum C-reactive protein concentration. *Clinical Nephrology*, **21**, 341–5.

Hollander, D. and Manning, R.T. (1967). The use of alkylating agents in the treatment of Wegener's granulomatosis. *Annals of Internal Medicine*, **67**, 393–8.

Hurst, N.P., Dunn, N.A., and Chalmers, T.M. (1983). Wegener's granulomatosis complicated by diabetes insipidus. *Annals of Rheumatic Diseases*, **42**, 600–1.

Israel, H.L. (1988). Sulphamethoxazole-trimethoprim therapy for Wegener's granulomatosis. *Archives of Internal Medicine*, **148**, 2293–5.

Israel, H.L. and Patchefsky, A.S. (1975). Treatment of Wegener's granulomatosis of the lung. *American Journal of Medicine*, **58**, 671–3.

Jacobs, R.P., Moore, M., and Brower, A. (1987). Wegener's granulomatosis presenting with erosive arthritis. *Arthritis and Rheumatism*, **30**, 943–6.

Jayne, D.R.W., Jones, S.J., Severn, A., Shaunak, S., Murphy, J., and Lockwood, C.M. (1989). Severe pulmonary haemorrhage and systemic vasculitis in association with circulating anti-neutrophil cytoplasm antibodies of IgM class only. *Clinical Nephrology*, **32**, 101–6.

Jayne, D.R.W., Marshall, P.D., Jones, S.J., and Lockwood, C.M.L. (1990). Autoantibodies to GBM and neutrophil cytoplasm in rapidly progressive glomerulonephritis. *Kidney International*, **37**, 965–70.

Kansu, E., Deglin, S., Cantor, R.I., Burke, J.F., Cho, S.Y., and Cathart, R.T. (1977). The expanding spectrum of Behçet syndrome. A case with renal involvement. *Journal of the American Medical Association*, **237**, 1855–6.

Katz, P., Alling, D.W., Haynes, B.F., and Fauci, A.S. (1979). Association of Wegener's granulomatosis with HLA-B8. *Clinical Immunology and Immunopathology*, **14**, 268–70.

Klinger, H. (1931). Grenzformen der Periarteritis Nodosa. *Frankfurter Zeitschrift für Pathologie*, **42**, 455–80.

Knudsen, B.B., Joergensen, T., and Munch-Jensen, B. (1988). Wegener's granulomatosis in a family. *Scandinavian Journal of Rheumatology*, **17**, 225–7.

Kussmaul, A. and Maier, R. (1866). Über einer bischer nicht beschriebere eigenthumliche Arterienerkrankung (Periarteritis Nodosa), die mit Morbus Brightii und rapid fortschreitender allemeiner Muskellähmung einhergeht. *Deutsche Archiv für Klinische Medizin*, **1**, 484–516.

Landwehr, D.M., Cooke, C.L., and Rodriguez, G.E. (1980). Rapidly progressive glomerulonephritis in Behçet's syndrome. *Journal of the American Medical Association*, **244**, 1709–11.

Lanham, J.G., Elkon, K.B., Pusey, C.D., and Hughes, G.R. (1984). Systemic vasculitis with asthma and eosinophilia: a clinical approach to the Churg–Strauss syndrome. *Medicine*, **63**, 65–81.

Leatherman, J.W., Davies, S.F., and Hoidal, J.R. (1984). Alveolar haemorrhage syndromes: diffuse microvascular lung haemorrhage in immune and idiopathic disorders. *Medicine*, **63**, 343–61.

Leavitt, R.Y. and Fauci, A.S. (1986). Pulmonary Vasculitis. *American Review of Respiratory Diseases*, **134**, 149–166.

Leavitt, R.Y., Hoffman, G.S., and Fauci, A.S. (1988). The role of trimethoprim/sulphamethoxazole in the treatment of Wegener's granulomatosis. *Arthritis and Rheumatism*, **31**, 1073–4.

Lee, S.S., Adu, D., and Thompson, R.A. (1990). Anti-myeloperoxidase antibodies in systemic vasculitis. *Clinical and Experimental Immunology*, **79**, 41–46.

Leib, E.S., Restivo, C., and Paulus, H.E. (1979). Immunosuppressive and corticosteroid therapy of polyarteritis nodosa. *American Journal of Medicine*, **67**, 941–7.

Lesavre, P., Chen, N., Nusbaum, P., Mecarelli, L., and Noël, L-H. (1990). Antineutrophil cytoplasm antibodies (ANCA) with antilactoferrin activity in vasculitis (abstract). *Kidney International*, **37**, 442.

Leung, D.Y.M., Collins, T., Lapierre, L.A., Geha, R.S., and Prober, J.S. (1986). Immunoglobulin M antibodies present in the acute phase of Kawasaki syndrome lyse cultured vascular endothelial cells stimulated by gamma interferon. *Journal of Clinical Investigation*, **77**, 1428–35.

Lie, J.T. (1978). Disseminated visceral giant cell arteritis. Histopathological description and differentiation from other granulomatous vasculitides. *American Journal of Clinical Pathology*, **69**, 299–305.

Littlejohn, G.O., Ryan, P.J., and Holdsworth, S.R. (1985). Wegener's granulomatosis: clinical features and outcome in seventeen patients. *Australian and New Zealand Journal of Medicine*, **15**, 241–5.

Lockwood, C.M., et al. (1977). Plasma-exchange and immunosuppression in the treatment of fulminating immune-complex crescentic nephritis. *Lancet*, **i**, 63–7.

Lockwood, C.M., Bakes, D., Jones, S., Whitaker, K.B., Moss, D.W., Savage, C.O.S. (1987). Association of alkaline phosphatase with an autoantigen recognised by circulating antineutrophil antibodies in systemic vasculitis. *Lancet*, **i**, 716–21.

Lohr, K.M., Ryan, L.M., Toohill, R.J., and Anderson, T. (1988). Anterior pituitary involvement in Wegener's granulomatosis. *Journal of Rheumatology*, **15**, 855–7.

Lüdemann, G. and Gross, W.L. (1987). Autoantibodies against cytoplasmic structures of neutrophil granulocytes in Wegener's granulomatosis. *Clinical and Experimental Immunology*, **69**, 350–7.

Lüdemann, J., Utecht, B., and Gross, W.L. (1988). Detection and quantification of antineutrophil cytoplasmic antibodies in Wegener's granulomatosis by ELISA using affinity-purified antigen. *Journal of Immunological Methods*, **114**, 167–74.

Lüdemann, J., Utecht, B., and Gross, W.L. (1990a). Anti-neutrophil cytoplasm antibodies in Wegener's granulomatosis recognize an elastinolytic enzyme. *Journal of Experimental Medicine*, **171**, 357–362.

Lüdemann, J., et al. (1990b). Anti-neutrophil cytoplasm antibodies in Wegener's granulomatosis: immunodiagnostic value, monoclonal antibodies and characterisation of the target antigen. *Netherlands Journal of Medicine*, **36**, 157–62.

MacFadyen, R., Tron, V., Keshmiri, M., and Road, J.D. (1987). Allergic angiitis of Churg–Strauss syndrome. Response to pulse methylprednisolone. *Chest*, **91**, 629–31.

Mark, E.J., Matsubara, O., Tan-Liu, N.S., and Fienberg, R. (1988). The pulmonary biopsy in the early diagnosis of Wegener's (pathergic) granulomatosis. A study based on 35 open lung biopsies. *Human Pathology*, **19**, 1065–71.

Mason, P.D. and Lockwood, C.M. (1986). Rapidly progressive nephritis in patients taking hydralazine. *Clinical and Laboratory Immunology*, **20**, 151–3.

McLelland, P., Williams, P.S., Stevens, M.E., and Bone, J.M. (1989). Letter to *Nephrology Dialysis Transplantation*, **4**, 917.

Moore, P.M. and Fauci, A.S. (1981). Neurologic manifestations of systemic vasculitis. A retrospective and prospective study of the clinicopathological features and responses to therapy in 25 patients. *American Journal of Medicine*, **71**, 517–24.

Moutsopoulos, H.M., Avgerinos, P.C., Tsampoulas, C.G., and Katsiotis, P.A. (1983). Selective renal angiography in Wegener's granulomatosis. *Annals of the Rheumatic Diseases*, **42**, 192–5.

Müller, G.A., Gebhardt, M., Kömpf, J., Baldwin, W.M., Ziegenhagen, D., and Bohle, A. (1984). Association between rapidly progressive glomerulonephritis and the properdin factor BfF and different HLA-D region products. *Kidney International*, **25**, 115–8.

Müller, G.A., Müller, C.A., Markovic-Lipkovski, J., Kilper, R.B., and Risler, T. (1988). Renal major histocompatibility complex antigens and cellular components in rapidly progressive glomerulonephritis identified by monoclonal antibodies. *Nephron*, **49**, 132–9.

Muniain, M.A., Moreno, J.C., and Gonzalez-Campora, R. (1986). Wegener's granulomatosis in two sisters. *Annals of the Rheumatic Diseases*, **45**, 417–21.

Nässberger, L., Sjöholm, A.G., Bygren, P., Thysell, H., Højer-Madsen, M., and Rasmussen, N. (1989). Circulating anti-neutrophil cytoplasm antibodies in patients with rapidly progressive glomerulonephritis and extracapillary proliferation. *Journal of Internal Medicine*, **225**, 191–6.

Niles, J.L., McCluskey, R.T., Ahmad, M.F., and Arnaout, M.A. (1989). Wegener's granulomatosis autoantigen is a novel neutrophil serine proteinase. *Blood*, **74**, 1888–93.

Nissenen, A.R. and Port, F.K. (1990). Outcome of end-stage renal disease in patients with rare causes of renal failure. III: systemic/vascular disorders. *Quarterly Journal of Medicine*, **74**, 63–74.

Nölle, B., Specks, U., Lüdemann, J., Rohrbach, M.S., DeRemee, R.A., and Gross, W.L. (1989). Anticytoplasmic autoantibodies: their immunodiagnostic value in Wegener granulomatosis. *Annals of Internal Medicine*, **111**, 28–40.

Noritake, D.T., Weiner, S.R., Bassett, L.W., Paulus, H.E., and Weisbart, R. (1987). Rheumatic manifestations of Wegener's granulomatosis. *Journal of Rheumatology*, **14**, 949–51.

Novack, S.N. and Pearson, C.M. (1971). Cyclophosphamide therapy in Wegener's granulomatosis. *New England Journal of Medicine*, **284**, 938–42.

Novak, R.F., Christiansen, R.G., and Sorensen, E.T. (1982). The acute vasculitis of Wegener's granulomatosis in renal biopsies. *American Journal of Clinical Pathology*, **78**, 367–71.

O'Donaghue, D.J., Short, C.D., Brenchley, P.E.C., Lawler, W., and Ballardie, F.W. (1989). Sequential development of systemic vasculitis with anti-neutrophil cytoplasmic antibodies complicating antiglomerular basement membrane disease. *Clinical Nephrology*, **32**, 251–5.

Ohyashiki, K., Kocova, M., Ryan, D.H., Rowe, J.M., and Sandberg, A.A. (1986). Secondary acute myeloblastic leukaemia with a Ph translocation in a treated Wegener's granulomatosis. *Cancer Genetics Cytogenetics*, **19**, 331–3.

O'Neill, W.M., Hammar, S.P., and Bloomer, H.A. (1976). Giant cell arteritis with visceral angiitis *Archives of Internal Medicine*, **136**, 1157–60.

Parfrey, P.S., et al. (1985). The spectrum of diseases associated with necrotising glomerulonephritis and its prognosis. *American Journal of Kidney Diseases*, **6**, 387–96.

Peacock, A. and Weatherall, D. (1981). Hydralazine-induced necrotising vasculitis. *British Medical Journal*, **282**, 1120–1.

Pinching, A.J., Rees, A.J., Pussell, B.A., Lockwood, C.M., Mitchison, R.S., and Peters, D.K. (1980). Relapses in Wegener's granulomatosis: the role of infection. *British Medical Journal*, **281**, 836–8.

Pinching, A.J., et al. (1983). Wegener's granulomatosis: observations on 18 patients with severe renal disease. *Quarterly Journal of Medicine*, **52**, 435–60.

Rackemann, F.M. and Greene, J.E. (1939). Periarteritis nodosa and asthma. *Transactions of the Association of American Physicians*, **54**, 112–8.

Rasmussen, N. and Daha. M.R. (1990). Concluding remarks on solid phase assays. (Proceedings of the second international workshop on antineutrophil cytoplasmic antibodies.) *Netherlands Journal of Medicine*, **36**, 137–42.

Rasmussen, N., Sjölin, C., Isaksson, B., Bygren, P., and Wieslander, J. (1990). An ELISA for the detection of anti-neutrophil cytoplasmic antibodies (ANCA). *Journal of Immunological Methods*, **127**, 139–145.

Reza, M.J., Dornfeld, L., Goldberg, L.S., Bluestone, R., and Pearson, C.M. (1975). Wegener's granulomatosis. Long-term followup of patients treated with cyclophosphamide. *Arthritis and Rheumatism*, **18**, 501–6.

Rijckaert, R., Lamont, H., Verhoye, P., Tasson, J., Pauwels, R., and van der Straeten, M. (1983). Wegener's granulomatosis: radiographic and clinical pattern. Report of six cases. *European Journal of Respiratory Diseases*, **67**, 216–9.

Ronco, P., et al. (1983). Immunopathological studies of polyarteritis nodosa and Wegener's granulomatosis: a report of 43 patients with 51 renal biopsies. *Quarterly Journal of Medicine*, **52**, 212–23.

Rondeau, E., et al. (1989). Plasma exchange and immunosuppression for rapidly progressive glomerulonephritis: prognosis and complications. *Nephrology, Dialysis, Transplantation*, **4**, 196–200.

Rose, G.A. and Spencer, H. (1957). Polyarteritis nodosa. *Quarterly Journal of Medicine*, **26**, 43–81.

Ross, C.N., Tam, F.W.K., Winter, R.J.D., Pusey, C.D., and Rees, A.J. (1990). Anti-neutrophil cytoplasmic antibodies and subglottic stenosis (letter). *Lancet*, **335**, 1231–2.

Sack, M., Cassidy, J.T., and Bole, G.G. (1975). Prognostic factors in polyarteritis. *Journal of Rheumatology*, **2**, 411–20.

Salcedo, J.R., Greenberg, L., and Kapur, S. (1988). Renal histology of mucocutaneous lymph node syndrome (Kawasaki disease). *Clinical Nephrology*, **29**, 47–51.

Satoh, J., et al. (1988). Extensive cerebral infarction due to involvement of both anterior cerebral arteries by Wegener's granulomatosis. *Annals of the Rheumatic Diseases*, **47**, 606–11.

Savage, C.O.S., Winearls, C.G., Evans, D.J., Rees, A.J., and Lockwood, C.M. (1985). Microscopic polyarteritis: presentation, pathology and prognosis. *Quarterly Journal of Medicine*, **56**, 467–83.

Savage, C.O.S., Winearls, C.G., Jones, S., Marshall, P.D., and Lockwood, C.M. (1987). Prospective study of radioimmunoassay for antibodies against neutrophil cytoplasm in diagnosis of systemic vasculitis. *Lancet*, **i**, 1389–93.

Savage, C.O.S., Tizard, J., Jayne, D.R.W., Lockwood, C.M., and Dillon, M.J. (1989). Antineutrophil cytoplasmic antibodies in Kawasaki syndrome. *Archives of Diseases of Childhood*, **64**, 462.

Schein, P.S. and Winokur, S.H. (1975). Immunosuppressive and cytotoxic chemotherapy: long-term complications. *Annals of Internal Medicine*, **82**, 84–95.

Scott, D.G.I., Bacon, P.A., Elliott, P.J., Tribe, C.R., and Wallington, T.B. (1982). Systemic vasculitis in a district general hospital 1972–1980: clinical and laboratory features, classification and prognosis of 80 cases. *Quarterly Journal of Medicine*, **51**, 292–311.

Sergent, J.S., Lockshin, M.D., Christian, C.L., and Gocke, D.J. (1976). Vasculitis with hepatitis B antigenaemia: long-term observations in nine patients. *Medicine*, **55**, 1–18.

Serra, A., et al. (1984). Vasculitis affecting the kidney: presentation, histopathology and long-term outcome. *Quarterly Journal of Medicine*, **53**, 181–207.

Socias, R., Geraint James, D., and Pozniak, A. (1987). Wegener and Wegener's granulomatosis. *Thorax*, **42**, 920–1.

Sonnenblick, M., Nesher, G., and Rosin, A. (1989). Nonclassical organ involvement in temporal arteritis. *Seminars in Arthritis and Rheumatism*, **19**, 183–90.

Specks, U., Wheatley, C.L., McDonald, T.J., Rohrbach, M.S., and DeRemee, R.A. (1989). Anticytoplasmic autoantibodies in the diagnosis and follow-up of Wegener's granulomatosis. *Mayo Clinic Proceedings*, **64**, 28–36.

Steinman, T.I., Jaffe, B.F., Monaco, A.P., Wolff, S.M., and Fauci, A.S. (1980). Recurrence of Wegener's granulomatosis after kidney transplantation. Successful re-induction of remission with cylophosphamide. *American Journal of Medicine*, **68**, 458–60.

Steppat, D. and Gross, W.L. (1989). Stage-adapted treatment of Wegener's granulomatosis. *Klinische Wochenschrift*, **67**, 666–71.

Sternlieb, I., Bennett, B., and Scheinberg, I.H. (1975). D-Penicillamine-induced Goodpasture's syndrome in Wilson's disease. *Annals of Internal Medicine*, **82**, 673–6.

Stevens, M.E., McConnell, M., and Bone, J.M. (1983). Aggressive treatment with pulse methyl prednisolone or plasma exchange is justified in rapidly progressive glomerulonephritis. *Proceedings of the European Dialysis and Transplant Association/European Renal Association*, **19**, 724.

Stilwell, T.J., Benson, R.C.Jr., DeRemee, R.A., McDonald, T.J., and Weiland, L.H. (1988). Cyclophosphamide-induced bladder toxicity in Wegener's granulomatosis. *Arthritis and Rheumatism*, **31**, 465–70.

Takagi, M., et al. (1984). Renal histological studies in patients with Takayasu's arteritis. Report of 3 cases. *Nephron*, **36**, 68–73.

Tai, P.C., Holt, M.E., Denny, P., Gibbs, A.R., Williams, B.D., and Spry, C.J. (1984). Deposition of eosinophil cationic protein in allergic granulomas and vasculitis: the Churg–Strauss syndrome. *British Medical Journal*, **289**, 400–2.

Ten Berge, I.J.M., et al. (1985). Clinical and immunological follow-up of patients with severe renal disease in Wegener's granulomatosis. *American Journal of Nephrology*, **5**, 21–9.

Thysell, H., et al. (1982). Immunosuppression and the additive effect of plasma exchange in treatment of rapidly progressive glomerulonephritis. *Acta Medica Scandinavica*, **212**, 107–14.

Travers, R.L., Allison, D.J., Brettle, R.P., and Hughes, G.R.V. (1979). Polyarteritis nodosa: a clinical and angiographic analysis of 17 cases. *Seminars in Arthritis and Rheumatism*, **8**, 184–99.

Truong, L., Kopelman, R.G., Williams, G.S., and Pirani, C.L. (1985). Temporal arteritis and renal disease. Case report and review of the literature. *American Journal of Medicine*, **78**, 171–5.

Van der Woude, F.J., et al. (1985). Autoantibodies against neutrophils and monocytes: tool for diagnosis and marker of disease activity in Wegener's granulomatosis. *Lancet*, **i**, 425–9.

Van der Woude, F.J., Daha, M.R., and Van Es, L.A. (1989). The current status of neutrophil cytoplasmic antibodies. *Clinical and Experimental Immunology*, **78**, 143–8.

Van Es, L.A. and Wiik, A. (1990). ANCA: clinical association and use in disease monitoring. (Proceedings of the second workshop on antineutrophil cytoplasmic antibodies). *Netherlands Journal of Medicine*, **36**, 146–51.

Velosa, J.A. (1987). Idiopathic crescentic glomerulonephritis a systemic vasculitis? (Editorial). *Mayo Clinic Proceedings*, **62**, 145–7.

Wahls, T.L., Bonsib, S.M., and Schuster, V.L. (1987). Coexistent Wegener's granulomatosis and anti-glomerular basement membrane disease. *Human Pathology*, **18**, 202–5.

Wainwright, J. and Davson, J. (1950). The renal appearances in the microscopic form of periarteritis nodosa. *Journal of Pathology and Bacteriology*, **62**, 189–96.

Walton, E.W. (1958). Giant-cell granuloma of the respiratory tract (Wegener's granulomatosis). *British Medical Journal*, **2**, 265–70.

Wathen, C.W. and Harrison, D.J. (1987). Circulating anti-neutrophil antibodies in systemic vasculitis. *Lancet*, **i**, 1037.

Waxman, J. and Bose, W.J. (1986). Laryngeal manifestations of Wegener's granulomatosis: case reports and review of the literature. *Journal of Rheumatology*, **13**, 408–11.

Wegener, F. (1936). Über generalisierte, septische Gefässerkrankungen. *Verhandlungen der Deutschen Pathologischen Gesellschaft*, **29**, 202–10. Translation (1987): On generalised septic vessel diseases. *Thorax*, **42**, 918–9.

Wegener, F. (1939). Über eine eigenartige rhinogene Granulomatose mit besonderer Beteiligung des Arteriensystems und der Nieren. *Beitrage der Pathologische Anatomie*, **102**, 36–68.

Weiss, M.A. and Crissman, J.D. (1985). Segmental necrotising glomerulonephritis: diagnostic, prognostic, and therapeutic significance. *American Journal of Kidney Diseases*, **6**, 199–211.

West, B.C., Todd, J.R., and King, J.W. (1987). Wegener granulomatosis and trimethoprimsulphamethoxazole. Complete remission after a twenty-year course. *Annals of Internal Medicine*, **106**, 840–2.

Wheeler, G.E. (1981). Cyclophosphamide-associated leukaemia in Wegener's granulomatosis. *Annals of Internal Medicine*, **94**, 361–3.

Wiik, A., Jensen, E., and Friis, J. (1974). Granulocyte-specific antinuclear factors in synovial fluids and sera from patients with rheumatoid arthritis. *Annals of Rheumatic Diseases*, **33**, 515–22.

Wilkowski, M.J., et al. (1989). Risk factors in idiopathic renal vasculitis and glomerulonephritis. *Kidney International*, **36**, 1133–41.

Wolff, S.M., Fauci, A.S., Horn, R.G., and Dale, D.C. (1974). Wegener's granulomatosis. *Annals of Internal Medicine*, **81**, 513–25.

Woodworth, T.G., Abuelo, J.G., Austin, H.A., and Esparza, A. (1987). Severe glomerulonephritis with late emergence of classic Wegener's granulomatosis. *Medicine*, **66**, 181–91.

Yoshikawa, Y. and Watanabe, T. (1984). Granulomatous glomerulonephritis in Wegener's granulomatosis. *Virchows Archiv.*, **402**, 361–72.

Yousem, S.A. and Lombard, C.M. (1988). The eosinophilic variant of Wegener's granulomatosis. *Human Pathology*, **19**, 682–8.

Yuasa, K., Tokitsu, M., Goto, H., Kato, H., and Shimada, K. (1988). Wegener's granulomatosis: diagnosis by transbronchial lung biopsy, evaluation by gallium scintigraphy and treatment with sulphamethoxazole/trimethoprim (letter). *American Journal of Medicine*, **84**, 371.

Zeek, P.M. (1952). Periarteritis nodosa: a critical review. *American Journal of Clinical Pathology*, **22**, 777–90.

4.6.1 Recent advances in the pathogenesis of systemic lupus erythematosus

LAURENT JACOB, JEAN-PAUL VIARD AND JEAN-FRANÇOIS BACH

Introduction

Systemic lupus erythematosus (SLE) is an autoimmune disease characterized by an immune dysregulation leading to the production of a number of autoantibodies, prominent amongst which are antinuclear antibodies. Its aetiology is still poorly understood. More is known of its multifactorial pathogenesis.

The possible role of environmental factors, in particular viral infection, has often been discussed. Several strains of lupic mice strongly express the gp70 retroviral glycoprotein and gp70-anti-gp70 immune complexes have been found in their serum (Theofilopoulos and Dixon 1985). The production of anti-gp70 antibodies is not specific for lupic strains (Dixon 1985) but the anti-gp70 response nevertheless appears genetically linked to the anti-DNA antibody response (Shirai *et al.* 1986).

The influence of hormonal factors in the development of SLE is demonstrated by the high frequency of the disease in the lupic female, both in the human and in (NZB×NZW)F1 (B/W) mice, as well as by the frequent occurrence of a clinically active state during pregnancy. In addition, elevated levels of oestrogen metabolites and reduced levels of plasma androgens have been demonstrated in female lupic patients (Jungers *et al.* 1982).

Different arguments for a primary abnormality of T and B cells have been put forth, on the basis of studies made in various strains of lupic mice (Theofilopoulos and Dixon 1985). Recent studies would seem to indicate the common presence of T-cell abnormalities in the various lupic mouse strains, which could be responsible for the production of pathogenic antinuclear antibodies, such as cationic anti-double-stranded (ds) DNA antibodies (Datta *et al.* 1987). By another method, a good model for SLE can be seen in the murine graft-versus-host disease responsible for glomerulonephritis (Gleichmann *et al.* 1972) and antinuclear antibodies (Van Elven *et al.* 1981). The relative role played by humoral and cellular immunological disorders in SLE is still debated. Thus, polyclonal B cell activation is considered as the immunological cornerstone of the disease (Izui *et al.* 1978). T-cell dysregulation could be responsible for uncontrolled B-cell activity (Krakauer *et al.* 1976), but anti-T-cell autoantibodies could in turn contribute to the T-cell subpopulation abnormalities (Shirai *et al.* 1978). However, it has been shown by studies in SLE patients and murine models of SLE that the disease does not necessarily rely on the same immunological abnormalities in all individual cases, and pathogenic considerations should take this into account.

Anti-DNA antibodies

Anti-DNA antibody specificities

Most studies on anti-DNA antibodies present in the serum of SLE patients and (NZB×NZW)F1 (B/W) mice have shown a great diversity in their specificities. Some antibodies react with double-stranded DNA and are directed against conformational antigenic determinants of the phosphate deoxyribose backbone. These antibodies do not recognize purine and pyrimidine bases. Furthermore, it has been shown that these antibodies do not constitute a homogeneous family of antibodies. In particular, it has been demonstrated that they can recognize different sizes of double-stranded DNA fragments. Other anti-DNA antibodies are specific for single-stranded DNA, polynucleotides, purine and pyrimidine bases, or sequences of bases. Yet other anti-DNA antibodies react with both single-stranded and double-stranded DNA.

However, two major considerations have limited the study of the fine specificities of DNA-binding antibodies in SLE sera. First, it is difficult to characterize the DNA used as antigen in terms of single-stranded versus double-stranded state: an 0.1 per cent denaturation of a native DNA preparation makes it reactive with anti-single-stranded DNA antibodies. Conversely, single-stranded DNA may, in the physicochemical conditions of the antigen-antibody reaction, exhibit secondary structures and consequently react with anti-double-stranded DNA antibodies. Second, almost every immunologic system is complex in that the antibodies produced against a given immunogen are quite heterogeneous. In murine and human SLE sera, the situation is further complicated by the presence of many groups of antibodies capable of reacting with distinct nucleic acids.

Therefore, the use of a non-homogeneous antigen and of sera containing a complex mixture of antibodies makes it impossible to distinguish between homologous and heterologous reactions and renders the fine specificity studies of anti-DNA antibodies difficult to perform in competition experiments.

Hybridoma technology overcomes most of the problems encountered with SLE sera by allowing the isolation of pure monoclonal antibodies. So far, several groups have obtained monoclonal anti-DNA antibodies following fusion between myeloma lines and spleen cells from unimmunized, spontaneously autoimmune mice. The main pitfall of these studies on monoclonal antibodies remains, however, that the antibodies studied may not be representative of the most common or most pathogenic antibodies present in patients' sera.

Tron reported the production of monoclonal anti-DNA antibodies from B/W spleen cells, the antigenic specificities of which were studied in competition experiments (Tron et al. 1980). The antigenic specificities were demonstrated to be identical and directed against the conformational determinant of the B-helical form of double-stranded DNA. Conversely, Schwartz's group in Boston reported the production of anti-DNA autoantibodies by cloned hybridoma from MRL/l mice that had the capacity to bind single-stranded DNA and several different nucleic acid antigens (Andrzejewski et al. 1980). Moreover, each of these monoclonal antibodies appeared to recognize the same epitope on different polynucleotides. These findings suggested the polyspecificity of these monoclonal anti-single-stranded DNA.

Cross-reactions of anti-DNA antibodies

Monoclonal anti-DNA antibodies obtained from lupic patients or mice can also react with phospholipid (Lafer et al. 1981). This observation could explain false positive serological tests for syphilis or anticoagulant activity observed in SLE. Other cross-reactions have been described against bacterial antigens (Shoenfeld et al. 1986), proteoglycans (Faaber et al., 1984), cytoskeletal proteins (André-Schwartz et al. 1984), other nuclear antigens (Migliorini et al. 1987), or cell surface proteins (Jacob et al. 1984). Experiments described below urge caution in the interpretation of these 'cross-reactions'. The role of DNA-anti DNA immune complexes present in the hybridoma supernatants should be taken into consideration. Some of these complexes could bind to the so-called 'cross reactive antigens'.

Structures recognized at the cell surface by autoantibodies present in SLE

Histones and nucleosomes

Rekvig and Hannestad showed that autoantibodies from sera of SLE patients react with the cell surface of leucocytes (Rekvig and Hannestad 1979). These autoantibodies called 'X-ANA' have been partially purified from SLE sera by absorption on cytoplasmic membranes of viable human leucocytes. They react with histones and especially with octamers composed of H_2A, H_2B, H_3 and H_4 histones. These data suggest that histones are present at the surface of cells. It has been suggested that the epitope recognized by these autoantibodies could be located in the NH_2-terminus of H_2A or H_2B histone (Rekvig and Hannestad 1980). The same group also showed that autoantibodies react with DNA of 140 bp coiled around a histone octamer. Histones or nucleosomes could therefore be present at the surface of cells. These data were confirmed in 1985 by Holers and Kotzin who detected the presence of DNA and histones at the surface of human monocytes (Holers and Kotzin 1985). Lastly, monoclonal anti-DNA and antihistone antibodies bind to viable cells.

All these data, however, are still debated, especially since the presence of a few lysed cells leads to non-specific binding of DNA and histones at the cell surface.

DNA receptor

Bennett recently showed that DNA binds to the surface of monocytes, T and B cells, platelets, and polymorphonuclear neutrophils (Bennett et al. 1985). This binding could correspond to a ligand-receptor interaction, leading to the concept of a DNA receptor. The binding of radioactive DNA to cells is a saturable phenomenon, inhibited by non-radioactive DNA in excess, but not by RNA. The binding of exogenous DNA to its receptor could be responsible for the internalization and the degradation of DNA into oligonucledotides. The molecular weight of this receptor is 30 kDa. A functional defect of this receptor causing an alteration of exogenous DNA binding to the surface of mononuclear cells, could exist in SLE or other connective tissue diseases. Finally, autoantibodies against this 30 kDA DNA-binding protein have been found in one SLE patient.

Binding of a monoclonal anti-dsDNA autoantibody to a 94 kDa cell-surface protein via nucleosomes or a DNA-histone complex

It has recently been shown, using indirect immunofluorescence, that a monoclonal anti-ds DNA antibody purified on a DNA cellulose column, binds to the cell surface. Immunoblot analysis has shown that the antibody binds to cell-surface protein with a molecular weight of 94 kDa. The binding only occurs when nucleosomes or a DNA-histone complex are added (Jacob et al. 1989). It is assumed that the purified monoclonal anti-DNA antibody has a greater affinity for the DNA-histone complex than for DNA alone. This anti-DNA antibody acts like an antinucleosome antibody (nucleosomes are DNA–histone complexes).

Anti-94 kDa cell-surface protein antibodies have been observed in the serum of several SLE patients in the active state. These antibodies probably correspond to antibodies complexed to nucleosomes. The sera of these human SLE patients have been passed through a DNA cellulose column. The purified antibodies reacted with the 94 kDa cell-surface protein only when the DNA–histone complex was added (Jacob et al. 1989).

In fact, two populations of anti-DNA antibodies could exist in SLE sera, free antibodies and antibodies bound to DNA–histone complexes (nucleosomes). The proportion of bound antibodies could be increased by cell lysis. These data could be useful for understanding the mechanism of the pathogenic action of anti-DNA autoantibodies. The complexed antibodies bind to the 94 kDa cell-surface protein which could play the role of target molecule in SLE.

Pathogenicity of anti-DNA antibodies

Koffler et al. (1967) and Krishnan and Kaplan (1967) in the human, and Lambert and Dixon (1968) and Izui et al. (1977) in B/W mice showed the presence of anti-DNA antibodies, DNA and complement in kidney eluates, suggesting the pathogenic role of DNA-anti DNA immune complexes. Moreover, even if major individual exception exists, most workers have observed a significant overall correlation (before treatment) between high anti-DNA serum titres and the severity of the disease, particularly with glomerulonephritis (Pincus et al. 1969; Tron and Bach 1977) as well as with the amount and the diffuse character of immunoglobulin glomerular deposits (Hill et al. 1978). There is a correlation between the Ig chain of the anti-DNA antibodies and the severity of the disease. The presence of anti-DNA antibodies of IgG classes is more frequently observed in severe disease with kidney involvement (Rothfield and Stollar 1967). Similarly, anti-DNA antibodies of the IgG class are observed in kidney eluates of SLE patients (Winfield et al. 1977). In murine lupus, Talal et al. have shown that the IgM/IgG switch of anti-DNA antibodies occurs earlier in females than in males, who develop the disease later and inconstantly (Talal et al. 1981). Anti-DNA antibodies of IgG classes that fix complement are correlated with the presence of kidney involvement.

The pathogenic role of anti-DNA antibody subpopulations is further suggested by the demonstration that anti-DNA antibodies present in glomerular eluates from MRL/1 and B/W mice and in cryoglobulins from SLE patients, have a more restricted heterogeneity than serum antibodies in isoelectric focusing studies (Dang and Harbeck 1979; Ebling and Hahn 1980). Hahn showed that cationic anti-DNA antibodies with high isoelectric points were more frequently nephritogenic than anionic antibodies (Hahn 1982). These DNA–anti-DNA immune complexes could be formed in the circulation (Bruneau and Benveniste 1979; Tron et al. 1982) or in situ (Izui et al. 1977).

Recent studies showed that the size of DNA present in the circulation and within DNA–anti-DNA immune complexes of lupic patients is small (200 bp)(Raptis and Menard 1980; Sano and Morimoto 1981). Finally, some authors have suggested that histones could bind to renal glomeruli and trigger the local formation of immune complexes (Schmiedeke et al. 1989). These concepts suggest that nucleosomes could be the antigen present within immune complexes. Antibodies complexed to nucleosomes could correspond to the pathogenic antibodies involved in SLE. These complexed antibodies could bind to a cell-surface protein which could play the role of target molecule.

Idiotypic network and systemic lupus erythematosus

Autoimmunity and idiotypic network

Antibodies bear their own antigenic determinants, characteristic for each molecule and its specificity, idiotypic determinants or idiotopes, able to induce the production of anti-idiotypic antibodies. Jerne suggested the existence of a functional network based on interactions between idiotypes and anti-idiotypes (Jerne 1974). Such a network could play a fundamental role in the expression and modulation of the immune repertoire, especially in the suppression of immune responses against certain autoantigens. The study of idiotypic interactions during induced or spontaneous autoimmune processes is an interesting approach to the regulation (activation or suppression) of autoantibody production. The nature and distribution of idiotypic determinants borne by different autoantibodies of the same or different specificities is useful for the understanding of genetic mechanisms controlling the synthesis of antibodies directed against self components.

Numerous authors have suggested several mechanisms for the production of pathogenic autoantibodies in the context of idiotypic networks (Zanetti 1985): (1) autoreactive B cells could be directly activated by autoanti-idiotypic antibodies specific for their own idiotypes; (2) anti-idiotypic antibodies could mimic the structure of autoantigens (internal image phenomenon, demonstrated particularly in the case of antireceptor and antiligand immunity), and stimulate autoreactive B cells; (3) antibodies directed against external antigens could fortuitously share idiotypic determinants with autoantibodies. This idiotypic cross-reaction could explain the autoimmune reaction through the action of anti-idiotypic antibodies reacting with cross-idiotypes.

SLE and recurrent idiotopes

Recurrent idiotypic determinants (idiotopes) borne by different antibodies of the same specificity produced by different individuals, sometimes even of different specificities, have been demonstrated for autoantibodies in SLE, especially for anti-DNA antibodies. This observation raises the question of the molecular genetic mechanisms responsible for the synthesis of these autoantibodies (Horsfall and Isenberg 1988). For example, the 16/6 idiotype borne by an IgM human monoclonal anti-single-stranded DNA antibody (Schoenfeld et al. 1983) is found in 50 per cent of SLE patients. Moreover, its presence is better correlated with progression of the disease than anti-DNA antibodies (Shoenfeld et al. 1983). In the same vein, the 3I idiotype (Diamond and Salomon 1983) is found in 90 per cent of patients. Similar results have been found for several human and mice monoclonal antibodies and polyclonal sera. Recurrent idiotypic determinants have been detected within anti-DNA antibodies from different animal species (Eilat et al. 1985). Nevertheless, private idiotopes specific for each individual have also been found in lupic humans (Diamond and Salomon 1983) and mice (Jacob and Tron 1984; Tron et al. 1983).

Some recurrent idiotopes are common to anti-DNA autoantibodies and antisoluble nuclear antigen autoantibodies observed in SLE. Certain idiotopes borne by monoclonal anti-DNA antibodies (16.6 in man, H130 in mice) are shared by human and murine anti-Sm, anti-RNP, and anti-SSA antibodies (Migliorini et al. 1987; Pisetsky et al. 1985).

SLE and idiotypes: role of regulatory idiotypes

The presence of recurrent idiotopes within anti-DNA antibodies, the other antibodies involved in SLE, and in other autoimmune systems, would tend to support the preferential use of certain genes coding for variable regions of autoantibodies. More generally, the cellular regulation of autoantibody and idiotypic expression is still to be elucidated. In a small number of patients, a converse expression between idiotypic-bearing anti-DNA antibodies and the corresponding autologous anti-idiotype during the course of the disease has been demonstrated (Abdou et al. 1981). This result agrees with the data obtained in vitro, where anti-idiotype antibodies were shown to inhibit the production of various autoantibodies (anti-La, Sm/RNP, DNA) by patients' circulating lymphocytes or by hybridoma cells (Abdou et al. 1981; Epstein et al. 1987).

Idiotypic immunointervention in SLE

All the data presented here suggest that idiotypic interactions could have a major functional role in the pathogenesis of SLE. Various attempts at idiotype-based experimental immunomanipulation have been performed in lupic mice. The injection of monoclonal anti-idiotypic antibodies directed against major recurrent idiotypes of anti-DNA antibodies has been shown to improve the disease and prolong survival of B/W mice (Hahn and Ebling 1983, 1984). However, the effect is transient, a consequence of the appearance of anti-DNA antibodies bearing other idiotypes. The suppression of anti-DNA antibody production has also been observed in MRL/l mice (Mahana et al. 1987a). Conversely, lupus disease has been induced by the injection of human monoclonal anti-DNA antibodies bearing the 16/6 recurrent idiotype (Mendlovic *et al.* 1988). Following anti-16/6 immunization, murine anti-16/6 antibodies and anti-anti-16/6 antibodies with anti-DNA activity, as well as anti-RNP, anti-Sm, anti-SS-A, anti-SS-B, and anticardiolipin antibodies have been observed. Increased erythrocyte sedimentation rate, leucopenia, proteinuria, mesangial deposition of immune complexes bearing the 16/6 idiotype, and glomerular sclerosis have also been found.

Lupus and idiotype: the future

At present it has appeared easier to stimulate rather than to depress the production of autoantibodies by idiotypic manipulation. However, experimental manipulation could lead to new perspectives in both pathogenesis and therapeutics of human disease. The use of anti-idiotypic antibodies linked to toxins could allow suppression of cellular clones involved in the production of idiotypes associated to pathogenic antibodies.

Nucleotide sequence of genes coding for natural pathogenic monoclonal anti-DNA antibodies

As discussed above, two populations of monoclonal anti-DNA antibodies have been described: monoclonal anti-DNA antibodies of the IgG class, specific for double-stranded DNA, and monoclonal anti-DNA antibodies of the IgM class, specific for single-stranded DNA, similar to polyspecific natural antibodies described by Avraméas (Mahana *et al.* 1987b). Schwartz *et al.* analysed the sequence of genes coding for monoclonal anti-DNA antibodies of the IgM class produced from lupic patients. The nucleotidic sequence of genes coding for these autoantibodies is very similar from one immunoglobulin to another, different hybridomas expressing the same VH gene (Dersimonian *et al.* 1987). These authors concluded that monoclonal anti-DNA antibodies are encoded by non-mutated genes in germ-line configuration.

Conversely, Theofilopoulos *et al.* have recently analysed the nucleotidic sequence of genes coding for monoclonal anti-DNA antibodies produced in MRL/lpr/lpr lupic mice (Koffler *et al.* 1988). Six monoclonal anti-DNA antibodies appeared to be coded by different VH genes. Sequences correspond to mutated genes. These antibodies are similar to those produced in response to extrinsic antigens in normal mice.

A monoclonal anti-DNA antibody developed in our laboratory by Tron *et al.* (1980), has been sequenced by D. Eilat. Its nucleotidic sequence also appeared to correspond to a mutated gene.

Finally, it is difficult to present a unified concept of the mutated or conserved nature of the nucleotidic sequence of immunoglobulins coding for anti-DNA antibodies. We suggest, however, that IgM monoclonal anti-DNA antibodies specific for single-stranded DNA could be in general coded for by germ-line genes, in contrast to IgG monoclonal anti-DNA antibodies with higher affinities, specific for double-stranded DNA, that are encoded by mutated genes as if their production was antigen driven. The nature of the driving antigen (DNA or a cross-reactive antigen) remains, however, to be determined.

Mediation of injury in lupus nephritis

The formation of antigen—antibody complexes within the glomerulus does not directly induce tissue injury. Rather, damage occurs as a consequence of activation of other cellular and humoral mediator systems. Because of the difficulty in producing experimental glomerular injury by deposition of preformed immune complexes, most studies of the immune renal injury have been carried out in models of *in-situ* immune complex formation. Five mechanisms have been established by which glomerular immune-complex deposits can initiate tissue injury (Couser 1985).

Direct effect of antibody deposition alone The reaction of IgG antibody with glomerular basement membrane (GBM) antigens can markedly increase glomerular permeability to protein independently of complement or inflammatory cells. Anti-GBM antibodies may be found in SLE using current radioimmunoassays but their prevalence and rates are not particularly high. However, the role of autoantibodies directed against other antigenic structures than GBM should not be excluded.

Direct effect of complement Studies of passive Heymann nephritis in rats made C6-deficient demonstrate a requirement for the C5b–9, or membrane attack complex portion of the complement system for full development of proteinuria. The presence of neoantigen of the membrane attack complex in glomerular immune deposits in human renal diseases, including lupus nephritis and idiopathic membranous nephropathy suggests that a similar mechanism may be operative in humans.

Complement neutrophil-mediated glomerular injury The major mechanism of complement-mediated immune complex-induced tissue injury has long been thought to be the generation of chemotactic peptides, primarily C5a, resulting in neutrophil attraction to the site of immune deposits and, as a consequence, tissue injury by release of toxic products of neutrophil activation adjacent to the glomerular basement membrane (GBM). The mechanism of neutrophil-mediated injury has been presumed to involve proteolytic digestion of GBM by enzymes locally by invading neutrophils. However, this mechanism has only been demonstrated in certain models of anti-GBM disease, where several observations document the importance of neutrophils.

Complement-independent, cell-mediated glomerular injury Again neutrophils appear to be able to mediate glomerular injury independently of complement when Fc portions of immune deposits are accessible for initiating immune adherence. Macrophages also participate in immune-complex disease through complement-independent mechanisms. Such mechanisms are prominent in glomeruli in several models of immune-complex nephritis as well as in human disease.

Specifically-sensitized cells The role of antibody-independent cellular hypersensitivity to fixed or planted antigenic components of glomerular immune complexes in human disease has long been postulated but has usually been dismissed. However, experimental support for this mechanism is now accumulating. Thus, the cellular arm of the immune response to antigens that induce immune-complex nephritis may play a previously unappreciated role in producing these lesions. This mechanism could explain the favourable effect of cyclosporin noted in the lupus nephritis of both mice (Mountz *et al.* 1987) and some humans (Feutren *et al.* 1986)

Immune dysregulation in SLE

The immunological studies in SLE patients and of the four main murine lupus models (NZB, B/W, BXSB and MRL/l mice) have brought considerable information on the immune dysregulation associated with SLE. The cellular abnormalities found in these different models are very heterogeneous, supporting the view that SLE is more a syndrome than a single disease. The various clinical syndromes could possibly result from distinct types of immunological dysfunction.

B-cell hyperactivity

B cell hyper-responsiveness and 'spontaneous' polyclonal activation are constant features of SLE patients and animal models of SLE (Bach 1982). Using very low density B-cell cultures, it has been shown that B cells from young BXSB and B/W mice, but not from young MRL/l mice, are more responsive than normal B cells to B-cell stimulating signals (Theofilopoulos and Dixon 1985).

It is worth noting that B cells from SLE patients with very active disease generally do not respond to polyclonal B cell activation *in vitro*, very probably because of previous polyclonal activation *in vivo*, which generates differentiated autoantibody-secreting cells.

Genetic studies indicate the existence of an intrinsic B-cell abnormality in most lupus-prone mouse strains, which seems necessary, if not sufficient, for the development of autoimmunity, and which is linked to polyclonal B cell activation.

Interestingly, the introduction of the *Xid* gene (an X-linked gene coding for a defect in a subset of mature B cells (Lyb3+, Lyb5+) in lupus mice significantly retards or stops their autoimmune manifestations.

Many reports have been published recently on the possible role of a subpopulation of B cells, bearing the CD5 antigen (Leu 1 in man, Ly1 in the mouse), in the production of autoantibodies. NZB mice present with high levels of Ly1+ B cells. Data are less clear cut in human SLE. It has also been shown that *S. aureus*-stimulated Leu1+ B cells produce rheumatoid factors and that EBV-transformed Leu1+ B cells secrete rheumatoid factor and anti-single-stranded DNA antibodies (Casali *et al.* 1987) or even anti-double-stranded DNA (Sthoeger *et al.* 1989). However, Leu1+ B cells do not only produce autoantibodies, since they can also produce antibodies against exogenous antigens (such as tetanus toxoid and β-galactosidase), and patients with elevated levels of Leu1+B cells (such as rheumatoid arthritis patients) also show increased numbers of Leu1−B cells. This makes it more likely that the Leu1 antigen is only one of the B cell markers associated with autoantibody production. Another possibility is that autoantibody-associated V genes rearrangements might occur preferentially in CD5+ B cells (Zouali *et al.* 1988).

Disturbed T-cell function
Murine SLE
Suppressor T-cell dysfunction: the NZB mouse paradigm

T-suppressor cell function has been found to be defective in NZB mice, but studies in other lupus-prone strains brought conflicting results to the attractive hypothesis that murine SLE is due to a lack of suppressive activity. Additionally, the nature (and even, for some authors) the existence of T-cell mediated suppressor mechanisms has been questioned (Möller *et al.* 1980).

Suppressor T-cell depletion could occur in SLE due to anti-lymphocyte antibodies as found in NZB mice (natural thymocytotoxic autoantibodies), but such autoantibodies are not found at a high level in MRL or BXSB mice, despite fulminant SLE-like disease. Furthermore, studies of recombinant inbred NZB mice show the absence of correlation between natural thymocytic autoantibodies and other autoantibodies (Theofilopoulos and Dixon 1985). Finally, the concept of a suppressor T-cell deficit remains attractive but needs to be documented. In any case, it is probably not the exclusive aetiologic factor but could contribute to onset and/or persistence of the disease. Its antigen specificity and cellular mechanisms need to be established.

Helper T-cell dysfunction: the MRL/l mouse paradigm

The key role of T help in the development of murine SLE has been emphasized by recent studies in MLR/l lupic mice. These mice are characterized by lupus-like clinical and immunological disorders and massive lymphadenopathy induced by the lymphoproliferation *lpr* gene. These proliferating T-lymphoid cells bear an unusual phenotype: Thy 1.2+, L3T4−, Lyt2−, Ly5+, B220+. In these mice, neonatal thymectomy as well as depletion of Thy1.2+ and L3T4+ cells (Wofsy *et al.* 1984) and cyclosporin A treatment (Mountz *et al.* 1987) improve the disease.

The study of three non-*lpr* strains, NSF1, B/W, and MRL+/+ mice, has shown the existence of a common cellular abnormality specifying the production of supposedly pathogenic (i.e. cationic) anti-DNA IgG antibodies, at the age these mice begin to develop glomerular disease (Datta *et al.* 1987). Spleen cells from these mice contain two populations of helper T cells that are responsible for inducing B cells to produce cationic IgG autoantibodies to both single-stranded and double-stranded DNA: one population bears the classical L3T4+, Lyt2− phenotype, whereas the second population is L3T4−, Lyt2−, which is similar to the phenotype of proliferating T cells of MRL/l mice.

Also, T cells from MRL/l mice spontaneously produce factors that induce B cell differentiation and Ig secretion. Conversely, it appears paradoxical that, despite evidence for increased T-helper activity for B cells, T-helper activity for T-cell responses (interleukin-2 production, interleukin-2 responsiveness and *in vitro* cytotoxic T-lymphocyte generation) was shown to be reduced.

Surprisingly, it has recently been shown that MRL/l mice and not MRL+/+ mice have an age-dependent loss of T-helper function involving MHC-self-restricted L3T4+ T-helper responses and not Lyt2+ T-helper responses. This defect is associated with the appearance of a suppressor cell activity that selectively inhibits L3T4+ MHC-self-restricted responses. Similar suppressor cell activity, due to L3T4+ cells, has also been reported in murine graft versus host models of autoimmunity (also characterized by excessive T-helper activity for B cells), suggesting that these suppressor cells represent a common immunoregulatory mechanism in the setting of autoimmunity (Via and Shearer, 1988). The pathogenic role of this late onset suppression is open to speculation.

Lastly the abnormal expression of oncogenes in MRL/l mice should be noted. A regulatory mechanism, rather than a structural alteration of some oncogenes could be responsible for the elevated expression of c-myb in MRL/l L3T4−, Lyt2− T cells (Evans *et al.* 1987).

Murine graft-versus-host disease: a model of SLE

Murine graft versus host reactions that occur in (C57BL/10× DBA/2)F1 mice receiving DBA/2 cells are considered to be good models of human SLE. Immune complex glomerulonephritis (Gleichmann *et al.* 1972) as well as the production of antinuclear antibodies and especially anti-dsDNA (Van Elven *et al.* 1981), anti-ssDNA and antihistone antibodies (Portanova *et al.* 1985) have been described in such mice. The cellular mechanism responsible for this graft-versus-host-induced autoimmunity has been studied by Gleichmann's group who postulated that it is due to the interaction of donor-derived alloreactive helper T cells carrying the Lyt-1+2− surface phenotype and recipient-derived self-reactive B cells (Gleichmann *et al.* 1984).

Human SLE

Among the wide spectrum of autoantibodies produced by SLE patients, antilymphocyte antibodies have long been known to show a high prevalence in the disease; however, they are not specific for the disease since they are also found in other autoimmune diseases and in infectious states. As SLE patients present with lymphopenia and impaired T-cell function, with decreased delayed hypersensitivity, autologous mixed lymphocyte reaction, and T-cell-mediated responses, the relationship between anti-lymphocyte antibodies and abnormal T-cell function has been intensively studied (Morimoto and Schlossman 1987).

Conflicting results have been published concerning suppressor T-cell function in human SLE. Several studies showed a defect in concanavalin A-induced suppressive activity, and attempts have been made to correlate this abnormality with the specific reactivities of antilymphocyte antibodies equivalent to natural thymocytic autoantibodies of NZB mice. Heterogeneity of SLE patients in this respect, as well as in the relative numbers of CD4+ and CD8+ T cells, have made it difficult to consider SLE as a disease associated with precise cellular abnormalities.

Abnormalities in CD4+ subpopulations of the peripheral blood have more recently been demonstrated in SLE patients. Among CD4+ cells, two subpopulations have been distinguished: CD4+4B4+ cells respond to soluble antigen and induce B cells to secrete Ig; CD4+2H4+ cells induce CD8+ pre-suppressor cells to become CD8+ suppressor effector cells (Morimoto *et al.* 1985). Lupus patients show a markedly decreased percentage of 2H4+ cells in their peripheral blood lymphocytes, compared with normal controls. This decrease is associated with decreased suppression of pokeweed mitogen-induced IgG synthesis by CD8+ cells. Anti-2H4 antibody blocks the suppressor-inducer function of CD4+ 2H4+ cells activated by autologous mixed lymphocyte reaction (Morimoto and Schlossman 1987).

Furthermore, SLE patients with severe renal disease and/or thrombocytopenia, without systemic manifestations, have a decrease in CD4+ cells and a relative increase in CD8+ cells, resulting (but rather inconsistently) in a low CD4:CD8 ratio. Anti-T cell antibodies in these patients, reactive either with CD4 suppressor/inducer cells or with both CD4+ suppressor/inducer cells and CD8+ suppressor effector cells have been observed in a significant number of patients. The demonstration that CD4+ cells are reduced in SLE patients in the presence of cold-reactive antilymphocyte antibodies preferentially cytotoxic for this subpopulation, provides further support for the view that reduction of circulation CD4+ cells is a consequence of anti-T cell autoantibody production, but that does not exclude the possibility that *in vivo* CD4+ cell depletion could play a role in the abnormal T-cell function characteristically found in SLE patients with active disease (Winfield *et al.* 1987).

Role of the thymus
Murine SLE

Thymectomy in newborns has a protective effect against the immunological disease of MRL/l mice, but has an accelerating effect on the disease of NZB and B/W mice (Theofilopoulos and Dixon 1985).

Grafting a thymus (from MRL/l or MRL+/+ mice) in thymectomized MRL/l mice is followed by the reappearance of the disease, showing that the thymus is necessary for the expression of a defect borne by pre-T cells. This emphasizes the importance of the immunological function of the thymic epithelium in this model of autoimmunity. It remains to be determined, however, whether the thymus intervenes by selecting positively the autoreactive clones or by non-antigen-specific mechanisms, notably via the secretion of thymic hormones. The putative role of thymic hormones is also suggested by the premature cessation of thymic hormone secretion noted in all strains of lupus mice (Bach *et al.* 1973) as assessed both by determination of seric levels and number of thymic hormone producing cells. Note, however, that this thymic hormone abnormality could be secondary to a defect in the stem cells that colonize the thymus since H-2 compatible bone marrow grafting restores thymic hormone secretion in NZB mice (Dardenne *et al.* 1988). It should also be noted that the treatment of B/W mice with thymic hormone may slow down or accelerate the time course of the autoimmune disease according to the parameters considered (Israel-Biet *et al.* 1983). The abnormal presence of crystals in the thymic epithelium of lupus mice (Dardenne *et al.* 1989) is intriguing but remains unexplained.

Other experiments confirm that a primitive abnormality is borne by bone-marrow stem cells. The disease can be transferred into irradiated recipients by the bone marrow cells of SLE-prone mice, and the resulting disease bears the characteristic features of the disease of the donor (Bach 1982).

Human SLE

Very few data support the view that the thymus plays a role in the development of human SLE. It has been reported from necropsy studies that SLE thymus contain numerous germinal centres with plasmocytes, but this abnormality could be secondary to the disease (Bach 1982). Thymic hormone levels are decreased but, as discussed above, this could also be secondary. However, it should be noted that thymectomy in patients with myasthenia gravis has been followed in a few cases (in genetically predisposed persons) by the appearance of autoimmune syndromes resembling SLE (Laukaitis and Borenstein 1989).

Complement system dysfunction in SLE
Physiological aspects: immune complexes and the complement system

Much attention has been paid to complement abnormalities in SLE. Complement plays a key role in the clearance of immune complexes. Physiologically, complement-activating immune complexes formed in the bloodstream rapidly adhere, via C3b, to the C3b receptor (called CR1) of erythrocytes. CR1 and factor I degrade C3b to C3bi and C3b d,g. These fragments can bind to other specific complement receptors, notably on tissue monocytes. Red cells carry immune complexes to the liver and spleen where they are transferred to the mononuclear phagocytes.

Defects in this system can lead to inappropriate immune

complex deposition. This can occur in cases of excessive immune complex formation, deficient antibody response, deficiency in a component of the classical complement pathway up to C3, or in quantitative or qualitative abnormalities of complement and immunoglobulin receptors (Walport and Lachmann 1988).

Complement deficiency and immune complex-related disease

Inherited complement deficiencies have been shown to be associated with the development of autoimmune diseases. SLE or SLE-like syndrome occur in more than 80 per cent of persons with complete deficiency of C1q, C1r/C1s, C4, or C3. Similarly, more than 60 per cent of the C2-deficient patients reported in the literature have SLE or SLE-related syndromes. Such patients are characterized by a defect in the inhibition of immune complex deposition. Inherited complete deficiency states are present in roughly 1 per cent of SLE patients.

Partial deficiency of C4 is far more frequent, involving 10 to 15 per cent of SLE patients. C4 is present in two highly homologous forms, coded for by two genes termed C4A and C4B. Experimental studies suggest that C4A is more efficient than C4B in preventing immune complex precipitation (Fielder *et al*. 1983; Walport and Lachmann 1988, Atkinson 1988).

Complement receptor dysfunction in SLE

The C3b/C4b receptor, known as CR1 has been much studied in SLE, because of its important role in immune-complex clearance. Erythrocytes of SLE patients have a reduction of approximately 50 per cent in the number of their C3b receptors, compared with normal controls. Restriction fragment length polymorphism (RFLP) analysis has shown that two codominantly inherited alleles correlate with high and low numbers of red cell CR1, but there does not seem to be an increased frequency of the gene coding for low levels of CR1 in family studies of SLE patients.

CR1 levels also vary with disease activity, and are lowered in conditions associated with complement activation. The suggestion that this abnormality is acquired is also reinforced by the disappearance of CR1 from red cells transfused into SLE patients. Note also that anti-CR1 antibodies have been found in a few patients with very low levels of CR1.

A low molecular weight form of CR1 has been identified with higher frequency in SLE patients. Finally, it is likely that both inherited and acquired abnormalities influence the state of C3b receptors in SLE patients (Walport and Lachmann 1988; Atkinson 1988) but acquired changes are probably predominant.

Genetic control of SLE

Murine SLE

The different murine strains predisposed to SLE are genetically heterogeneous: they differ in their origins, MHC haplotypes, Ig allotypes, and other immunologic characteristics. Such heterogeneity strongly indicates that murine SLE is not linked to simple genetic markers (Theofilopoulos and Dixon 1985).

Additionally, molecular studies of the immunoglobulin heavy chain variable region (Igh-V) gene complex of lupus mice, and the sequencing of MRL/l rheumatoid factors have revealed that the autoantibody response in lupus mice involves a variety of VH and VL genes as well as D and JH genes similar to those used in antibodies to exogenous antigens. These autoantibodies therefore derive from the same germ line repertoire as antibodies to external antigens in normal mice (Dixon 1985).

Genetic studies have been useful in defining the respective influence of the genetic background and of genetic accelerating factors in the development of lethal SLE in lupus-prone mice.

The NZB and (NZB× NZW)F1 mice

NZB mice (H-2^d MHC haplotype) develop autoimmune manifestations by the age of 6 months. Their disease seems to be dependent on the presence of at least six different genes. SLE features are more prominent in the (NZB×MZW)F1 mouse than in the NZB parental strain. The NZW parent (H-2z MHC haplotype) does not present with SLE, but its genetic contribution to the very severe autoimmune disease of (NZB×NZW)F1 hybrids seems to be of major importance. At least three gene loci of NZW mice could contribute to the exacerbation of the relatively mild autoimmune disorders encountered in NZB mice. NZB mice usually produce IgM antinuclear autoantibodies, but do not produce high levels of IgG anti-DNA antibodies. As CD4+ (L3T4+) cells are required for IgG anti-DNA antibody production in F1 animals, several studies have looked for a genetic NZW influence of T-cell function. It has been found that the NZW T-cell receptor exhibits an unusual structure, with deletion of the C-β-1, D-β-2 and J-β-2 gene segments. Nevertheless, recent studies of F1×NZB backcrosses have shown that the only NZW genetic contribution to the F1 disease seems to be linked to, or even probably located within, the NZW MHC (Kotzin and Palmer 1988). Additionally, segregation studies have shown that a number of non-MHC genes control the individual autoimmune response against defined autoantigens (DNA, red cells, lymphocytes).

MRL mice

MRL+/+ (MRL/n) mice (H-2^k MHC haplotype) develop a relatively mild autoimmune syndrome, which is under the influence of multiple genes. MRL mice have a genetic background which derives from L6 (75 per cent), AKR (13 per cent), C3H (12 per cent) and C57BL/6 (0.3 per cent) mice. The course of the disease is dramatically accelerated in MRLlpr/lpr mice (MRL/l mice) which are homozygous for the lymphoproliferation gene *lpr*. In MRL/l mice, the massive lymphoproliferation of T cells with an unusual immature phenotype that nevertheless express T-cell receptor proteins, is probably pivotal in the excessive T-cell help to normal B cells which results in the severe SLE observed in this strain.

BXSB mice

The SLE-like disease of BXSB mice (H-2b MHC haplotype) is dependent on multiple genes. The genetic accelerating factor of disease in these mice is Y-linked, transmitted by and expressed only in males.

Human SLE

Genetic factors are certainly involved in human SLE, since family studies of patients indicate that more than 10 per cent of SLE patients have at least one relative affected by the disease. Relatives of SLE patients may show only mild manifestations of autoimmunity, such as hypergammaglobulinaemia (5 to 20 per cent of cases, compared with 3 to 4 per cent of normal subjects), antinuclear antibodies (7 to 20 per cent, versus less than 2 per cent in normal subjects), rheumatoid factor (10 to 15 per cent versus less than 6 per cent in normal subjects), antilymphocyte or antiphospholipid antibodies. Other autoimmune diseases, such as rheumatoid arthritis can also be encountered in the families of SLE patients (Bach 1982). In addition, the study of monozygotic twins shows a concordance rate of more than 40 per cent for the disease; the influence of ethnic origin is established by the high frequency of SLE in the North American black population and in Asiatics.

Associations of SLE with MHC class II and class III genes has been observed in a variety of ethnic groups. These associations are potentially relevant to the aetiopathogenesis of SLE, since MHC class II molecules present antigens to T-helper cells, and MHC class III gene products (complement components) play a major role in immune-complex clearance.

HLA-DR2, DR3, and DQw1 antigens and deficiency in C2 or C4 ('C4A or C4B null') are significantly increased in SLE patients, compared with normal controls. The wide spectrum of MHC genes reported to be associated with SLE, and the relative weakness of these associations, underline the polygenic nature and the heterogeneity of human SLE (Hochberg et al. 1985; Fronek et al. 1988).

Interestingly, clinical subsets of the disease, e.g. renal SLE, seem to be more homogeneous genetically than the whole population of SLE patients. In addition HLA alleles associate strongly with autoantibody specificity against Ro and La antigens, regardless of race and even disease (SLE or Sjögren's syndrome): anti-Ro- and anti-La-positive patients are very often DR3- or DR2-positive, and the DQw1/DQw2 heterozygotes have the highest autoantibody levels (Arnett et al. 1988).

Nevertheless, the lack of absolute concordance for the disease in monozygotic twins indicates that environmental events, T-cell receptor gene rearrangements or somatic mutations might be involved in the appearance of SLE.

General conclusion

Although important progress has been made in understanding the pathogenesis of SLE much remains to be elucidated. The understanding of effector mechanisms is still based on the role of anti-DNA antibodies, but possibly with a specific target involving membrane phenomena. Several arguments suggest that DNA is complexed to histones within nucleosomes. Autoantibodies complexed to nucleosomes could bind to a cell-surface protein which could play the role of target molecule in SLE.

At the immunogenetic level, the analysis of nucleotide sequences of monoclonal anti-DNA antibodies is useful in understanding the origin of the immune response against self components: whether these authoantibodies are selected by autoantigen or rather, represent the exacerbation of anti-self specificity secondary to polyclonal B cell activation is unclear, and may vary from patient to patient and antibody to antibody. There are only fragmentary answers to this basic question and studies that are under way should add to our knowledge of this phenomenon. Finally, these new pathogenetic considerations could lead to new therapeutic approach, notably based on idiotypic manipulation.

References

Abdou, N., Wall, H., Lindsley, H.B., Halsey, J.F., and Susuki, T. (1981). Network theory in autoimmunity. *In vitro* suppression of serum anti-DNA antibody binding to DNA by anti-idiotypic antibody in systemic lupus erythematosus. *Journal of Clinical Investigation*, **67**, 1297.

Andrzejewski, D., Stollar, B.D., Latot, T.M., and Schwartz, R.S. (1980). Hybridoma autoantibodies to DNA. *Journal of Immunology*, **124**, 1499–1502.

André-Schwartz, J., Datta, S.K., Shoenfeld, Y., Isenberg, D.A., Stollar, B.D., and Schwartz, M.S. (1984) Binding of cytoskeletal proteins by monoclonal anti-DNA lupus autoantibodies. *Clinical Immunology and Immunopathology*, **31**, 261–29.

Arnett, F.C., Goldstein, R., Duvic, M., and Reveille, J.D. (1988). Major histocompatibility complex genes in systemic lupus erythematosus, Sjögren's syndrome and polymyositis. *American Journal of Medicine*. **85**, (Suppl. 6A), 38–41.

Atkinson, J.P. (1988) Complement deficiency. Predisposing factor to autoimmune syndromes. *American Journal of Medicine*, **85** (Suppl. 6A), 45–47.

Bach, J.F. (1982), Pathology of immune complexes. Systemic lupus erythematosus. In *Immunology* (Ed. J.F. Bach), Wiley Medical Publications, New York.

Bach, J.F., Dardenne, M., and Salomon, J.C. (1973) Studies on thymus products. IV. Absence of serum 'thymus activity' in adult NZB and (NZB × NZW)F$_1$ mice. *Clinical and Experimental Immunology*, **14**, 247–56.

Bennett, R.M., Gabor, G.T., and Merritt, M.M (1985). DNA binding to human leukocytes. Evidence for a receptor-mediated association, internalization and degradation. *Journal of Clinical Investigation* **76**, 2182–90.

Bruneau, C. and Benveniste, J. (1979). Circulating DNA: anti-DNA complexes in systemic lupus erythematosus. Detection and characterization by ultracentrifugation. *Journal of Clinical Investigation*, **64**, 191–8.

Casali, P., Burastero, S.E., Nakamura, M., Inghirami, G., and Notkins, A.L. (1987). Human lymphocyte making rheumatoid factor and antibody to ssDNA belong to Leu-1+B-cell subset. *Science*, **236**, 77–83.

Couser, W.G. (1985). Mechanisms of glomerular injury in immune complex disease. *Kidney International*, **28**, 569–83.

Dang, H. and Harbeck, R.A. (1979). A comparison of anti-DNA antibodies from serum and kidney eluates of NZB× NZW F1 (NZB/W) mice (Abstract). *Arthritis and Rheumatism*, **22**, 603.

Dardenne, M., Savino, W., and Bach, J.F. (1988). Reconstitution of thymic endocrine function in autoimmune mice by bone marrow transplant (Abstract). *Immunointervention in Autoimmune Diseases*, Paris, France, 15–17 June, 1988.

Dardenne, M., Savino, W., Nabarra, B., and Bach, J.F. (1989). Male BXSB mice develop a thymic hormonal dysfunction with presence of intraepithelial crystalline inclusion. *Clinical Immunology and Immunopathology*, **52**, 392–405.

Datta, S.K., Patel, H., and Berry, D. (1987). Induction of a cationic shift in IgG anti-DNA autoantibodies. Role of helper cells with classical and novel phenotype in three models of lupus nephritis. *Journal of Experimental Medicine*, **165**, 1252–68.

Dersimonian, H., Schwartz, R.S., Barrett, K.J., and Stollar, B.D. (1987). Relationship of human variable region heavy chain germline genes to genes encoding anti-DNA autoantibodies. *Journal of Immunology*, **139**, 2496–501.

Diamond, B. and Solomon G. (1983). A monoclonal antibody that recognizes anti-DNA antibodies in patients with systemic lupus erythematosus. *Annals of the New York Academy of Sciences*, **418**, 379–85.

Dixon, F.J. (1985). Murine lupus. A model for human autoimmunity. *Arthritis and Rheumatism*, **28**, 1081–8.

Ebling, F. and Hahn B.H. (1980). Restricted subpopulations of DNA antibodies in kidneys of mice with systemic lupus: comparison of antibodies in serum and renal eluates. *Arthritis and Rheumatism*, **23**, 392–403.

Eilat, D., Fischel, R., and Zlotnick, A. (1985). A central anti-DNA idiotype in human and murine systemic lupus erythematosus. *European Journal of Immunology*, **15**, 368–75.

Epstein, A., Greenberg, M., Diamond, B., and Grayzel, A.I. (1987). Suppression of anti-DNA antibody synthesis in vitro by a cross-reactive anti-idiotypic antibody. *Journal of Clinical Investigation*, **79**, 997–1000.

Evans, J.L., Boyle, W.J., and Ting, J.P.Y. (1987). Molecular basis of elevated C-myb expression in the abnormal L3T4-, Lyt-2- T lymphocytes of autoimmune mice. *Journal of Immunology*, **139**, 3497–505.

Faaber, P., Capel, M.J.A., Rijke, G.P.M., Vierwinder, G., Van de Putte, C.B.A., and Koene, R.A.P. (1984). Cross reactivity of anti-DNA antibodies with proteoglycans. *Clinical and Experimental Immunology*, **55**, 502–8.

Feutren, G., Querin, S., Tron, F., Noël, L.H., Chatenoud, L., Lesavre, P., and Bach, J.F. (1986). The effect of cycolosporine in patients with systemic lupus. *Transplantation Proceedings*, **18**, 643–4.

Fielder, A.H., et al. (1983). Family study of the major histocompatibility complex in patients with systemic lupus erythematosus: importance of null alleles of C4A and C4B in determining disease susceptibility. *British Medical Journal*, **286**, 425–8.

Fronek, Z., et al. (1988). Major histocompatibility complex associations with systemic lupus erythematous. *American Journal of Medicine*, **85** (suppl. 6), 42–4.

Gleichmann, H., Gleichmann, E., Andre-Schwartz, J., and Schwartz, R.J. (1972), Chronic allogeneic disease. III. Genetic requirements for the induction of glomerulonephritis. *Journal of Experimental Medicine*, **135**, 516–18.

Gleichmann, E., Pals, S.T., Rolink, A.G., Radaskiewicz, T., and Gleichmann, H. (1984). Graft-versus-host reactions: clues to the etiopathology of a spectrum of immunological diseases. *Immunology Today*, **5**, 324–32.

Hahn, B.H. (1982). Characteristics of pathogenic subpopulation of antibodies to DNA. *Arthritis and Rheumatism*, **25**, 747–52.

Hahn B.H. and Ebling, F.M. (1983). Suppression of murine lupus nephritis by administration of an anti-idiotypic antibody to anti-DNA. *Journal of Immunology*, **71**, 1728–36.

Hahn B.H. and Ebling F.M. (1984). Suppression of murine lupus nephritis by administration of an anti-idiotypic antibody to anti-DNA. *Journal of Immunology*, **132**, 187–90.

Hill, G.S., Hinglais, N., Tron, F., and Bach, J.F. (1978). Systemic lupus erythematosus. Morphological correlations with immunologic and clinical date at the time of biopsy. *American Journal of Medicine*, **64**, 61–8.

Hochberg, M.C., et al. (1985). Systemic lupus erythematosus: a review of clinico-laboratory features and immunogenetic markers in 150 patients with emphasis on demographic subsets. *Medicine*, **64**, 285–95.

Holers, V.M. and Kotzin, B.L. (1985). Human peripheral blood monocytes display surface antigens recognized by monoclonal antinuclear antibodies. *Journal of Clinical Investigation*, **76**, 991–8.

Horsfall, A.C. and Isenberg, D.A. (1988). Idiotypes and autoimmunity: a review of their role in human disease. *Journal of Autoimmunity*, **1**, 7–30.

Israel-Biet, D., Noël, L.H., Bach, M.A., Dardenne, M., and Bach, J.F. (1983). Marked reduction of DNA antibody production and glomerulopathy in thymulin (FTS–Zn) or cyclosporin A treated (NZB×NZW)F$_1$ mice. *Clinical and Experimental Immunology* **54**, 359–65.

Izui, S., Mc Conahey, P.J., and Dixon F.J (1978). Increased spontaneous polyclonal activation of B lymphocytes in mice with spontaneous autoimmune disease. *Journal of Immunology*, **121**, 2213–19.

Izui, S., Lambert, P.H., and Miescher, P.A. (1977). In vitro demonstration of a particular affinity of glomerular basement membrane and collagen for DNA: a possible basis for a local formation of DNA anti-DNA complexes in systemic lupus erythematosus. *Journal of Experimental Medicine*, **144**, 428–43.

Izui, S., Lambert, P.H., Türler, H., and Miescher, P.A. (1977). Features of systemic lupus erythematosus in mice injected with lipopolysaccharides. Identification of circulating DNA and renal localization of DNA: anti-DNA complexes. *Journal of Experimental Medicine*, **145**, 1115–30.

Jacob, L. and Tron, F. (1984). Induction of anti-DNA autoanti-idiotypic antibodies in (NZB × MZW)F1 mice: possible role for specific immune suppression. *Clinical Experimental Immunology*, **58**, 293–9.

Jacob, L., Tron, F., Bach, J.F., and Louvard D. (1984). A monoclonal anti-DNA antibody also binds to cell-surface protein (s). *Proceedings of the National Academy of Sciences of the USA*, **81**, 3843–5.

Jacob, L., et al. (1989). A monoclonal anti-double stranded DNA autoantibody binds to a 94 kDA cell-surface protein on various cell types via nucleosomes or a DNA-histone complex. *Proceedings of the National Academy of Sciences of the USA*, **86**, 4669–73.

Jerne, N.K. (1974). Towards a network theory of the immune system. *Annales d'Immunologie de l'Institut Pasteur (Paris)*, **125C**, 373–89.

Jungers, P., Nahoul, K., Pelissier, C., Dougados, M., Tron, F., and Bach, J.F. (1982). Low plasma androgens in women with active or quiescent systemic lupus erythematosus. *Arthritis and Rheumatism*, **25**, 454–7.

Koffler, D., Shur, P.H., and Kunkel, H.G. (1967). Immunologic studies concerning the nephritis of SLE. *Journal of Experimental Medicine*, **126**, 607–23.

Koffler, R., et al. (1988) Immunoglobulin K light chain variable region gene complex organization and immunoglobulin genes encoding anti-DNA autoantibodies in lupus mice. *Journal of Clinical Investigation*, **82**, 852–60.

Kotzin, B.L. and Palmer, E. 1988. Genetic contributions to lupus-like disease in NZB × NZW mice. *American Journal of Medicine*, **85**, (Suppl. 6A), 29–31.

Krakauer, R.S., Waldmann, T.A., and Strober, W. (1976). Loss of suppressor T cells in adult NZB/NZW mice. *Journal of Experimental Medicine*, **144**, 662–73.

Krishnan C.S. and Kaplan, M.H. (1967). Immunopathologic studies of systemic lupus erythematosus. II. Antinuclear reaction of globulin eluted from homogenates and isolated glomeruli of kidneys from patients with lupus nephritis. *Journal of Clinical Investigation*, **46**, 569–79.

Lafer, E.M., et al. (1981). Polyspecific monoclonal lupus autoantibodies reactive with both polynucleotides and phospholipids. *Journal of Experimental Medicine*, **153**, 897–909.

Lambert, P.H. and Dixon F.J. (1968). Pathogenesis of glomerulonephritis of NZB/W mice. *Journal of Experimental Medicine*, **127**, 507–22.

Laukaitis, J.P. and Borenstein, D.G. (1989). Multiple autoimmune diseases in predisposed patients. *Arthritis and Rheumatism*, **32**, 119–20.

Mahana, W., Guilbert, B., and Avrameas, S. (1987a). Suppression of anti-DNA antibody production in MRL mice by treatment with anti-idiotypic antibodies. *Clinical and Experimental Immunology*, **70**, 538–45.

Mahana, W., Matsiota, P., and Avrameas, S. (1987b). Both natural and induced murine anti-TNP antibodies possess anti-DNA activity. *Annales de l'Institut Pasteur*, **138**, 805–14.

Mendlovic S., et al. (1988). Induction of a systemic lupus erythematosus-like disease in mice by a common human anti-DNA idiotype. *Proceedings of the National Academy of Sciences of the USA*, **85**, 2260–5.

Migliorini, P., Ardman, B., Kaburaki, J., and Schwartz, R.S. (1987). Parallel sets of autoantibodies in MRL-1pr/1pr mice. An anti-DNA, anti-Sm RNP, anti-gp 70 network. *Journal of Experimental Medicine*, **105**, 485–99.

Möller, E., Ström, H., and Al-Balaghi, S. (1980). Role of polyclonal activation in specific immune responses. *Scandinavian Journal of Immunology*, **12**, 177–82.

Morimoto, C., Letvin, N.L., Distaso, J.A., Aldrich, W.R., and Schlossman, S.F. (1985). The isolation and characterization of the human suppressor inducer T cell subset. *Journal of Immunology*, **134**, 1508–15.

Morimoto, C. and Schlossman, S.F. (1987). Anti-lymphocyte antibodies and systemic lupus erythematosus. *Arthritis Rheumatism*, **30**, 225–8.

Mountz, J.D., Smith, H.R., Wilder, R.L., Reeves, J.P., and Steinberg, A.D. (1987). CsA therapy in MRL/lpr/lpr mice: amelioration of immunopathology despite autoantibody production. *Journal of Immunology*, **138**, 157–63.

Pincus, T., Schur, P.H., Rose, J.A., Decker, J.L., and Talal, N. (1969). Measurement of serum DNA-binding activity in systemic lupus erythematosus. *New England Journal of Medicine*, **281**, 701–5.

Pisetsky, D.S., Hoch, S.O., Klatt, C.L., O'Donnel, M.A., and Keene, J.D. (1985). Specificity and idiotypic analysis of a monoclonal anti-Sm antibody with anti-DNA activity. *Journal of Immunology*, **135**, 4080–5.

Portanova, J.P., Claman, H.N., and Kotzin, B.L. (1985). Autoimmunization in murine graft-versus-host disease. I. Selective production of antibodies to histones and DNA. *Journal of Immunology*, **135**, 3850–4.

Raptis, L. and Menard, H.. (1980). Quantification and characterization of plasma DNA in normals and patients with systemic lupus erythematosus. *Journal of Clinical Investigation*, **66**, 1391–9.

Rekvig, O.P. and Hannestad, K. (1979). The specificity of human autoantibodies that react with both cell nuclei and plasma membranes: the nuclear antigen is present on mononucleosomes. *Journal of Immunology*, **13**, 2673–81.

Rekvig, O.P. and Hannestad, K. (1980). Human autoantibodies that react with both cell nuclei and plasma membranes display specificity for the octamer of histones H_2A, H_2B, H_3 and H_4. *Journal of Experimental Medicine*, **152**, 1720–33.

Rothfield, N.F. and Stollar, B.D. (1967). The relation of immunoglobulin class, pattern of antinuclear antibody, and complement fixing antibodies to DNA in sera from patients with systemic lupus erythematosus. *Journal of Clinical Investigation*, **46**, 1785–94.

Sano, H. and Morimoto, C. (1981). Isolation of DNA from DNA/anti-DNA antibody immune complexes in systemic lupus erythematosus. *Journal of Clinical Investigation*, **126**, 538.

Schmiedeke, T.M.J., Stöckl, F.W., Weber, R., Sugisaki, Y., Batsford, S., and Vogt, A. (1989). Histones have high affinity for the glomerular basement membrane. Relevance for immune complex formation in lupus nephritis. *Journal of Experimental Medicine*, **169**, 1879–94.

Shirai, T., et al. (1986). Naturally occurring antibody response to DNA is associated with the response to retroviral gp70 in autoimmune New Zealand mice. *Arthritis and Rheumatism*, **29**, 242–50.

Shirai, T., Hayakawa, K., Okumura K., and Tada T. (1978). Differential cytotoxic effect of natural thymocytotoxic autoantibodies of NZB mice on functional subsets of T cells. *Journal of Immunology*, **120**, 1924–9.

Shoenfeld, Y., Isenberg, D.A., Rauch, J., Madaio, M.P., Stollar, B.D., and Schwartz, R.S. (1983). Idiotypic cross reactions of monoclonal human lupus autoantibodies. *Journal of Experimental Medicine*, **158**, 718–30.

Shoenfeld, Y., et al. (1986). Monoclonal anti-tuberculosis antibodies react with DNA and monoclonal anti-DNA autoantibodies react with *Mycobacterium tuberculosis*. *Clinical and Experimental Immunology*, **66**, 255–61.

Sthoeger, Z.M., et al. (1989). Production of autoantibodies by CD5-expressing B lymphocytes from patients with chronic lymphocytic leukemia. *Journal of Experimental Medicine*, **169**, 255–68.

Talal, N., Roubinian, J.F., Shear, H., Hom, J.T., and Miyasaica, N. (1981). Progress in the mechanisms of autoimmune disease. In *Progress in Immunology*. (Eds. M. Fougereau. and J. Dausset.) p. 889. Academic Press, New York.

Theofilopoulos, A.N. and Dixon, F.J. (1985), Murine model of systemic lupus erythematosus. *Advances in Immunology*, **37**, 269–390.

Tron, F. and Bach, J.F. (1977). Relationships between antibodies to native DNA and glomerulonephritis in systemic lupus erythematosus. *Clinical and Experimental Immunology*, **28**, 426–32.

Tron, F., Charron, D., Bach, J.F., and Talal, N. (1980). Establishment and characterization of a permanent murine hybridoma secreting monoclonal anti-DNA antibody. *Journal of Immunology*, **125**, 1805–9.

Tron, F., Jacob, L., and Bach, J.F. (1983). Murine monoclonal anti-DNA antibodies with an absolute specificity for DNA have a large amount of idiotypic diversity. *Proceedings of the National Academy of Sciences of the USA*. **80**, 6024–7.

Tron, F., Letarte, J., Roque-Antunes Barreira, M.C., and Lesavre, P. (1982). Specific detection of DNA: anti-DNA immune complexes in human SLE sera using murine monoclonal antibody. *Clinical and Experimental Immunology*, **49**, 481–7.

Van Elven, E.H., Van der Veen, F.M., Rolink, A.G., Issa, P., Duin, T.M., and Gleichmann, E. (1981). Diseases caused by reactions of T lymphocytes to incompatible structures of the major histocompatibility complex. V. High titres of IgG autoantibodies to double-stranded DNA. *Journal of Immunology*, **127**, 2435–8.

Via, C.S. and Shearer, G.M. (1988). Functional heterogeneity of L3T4+T cells in MRL/lpr/lpr mice. L3T4+T cells suppress major histocompatibility complex-self-restricted L3T4+ T helper cell function in association with autoimmunity. *Journal of Experimental Medicine* **168**, 2165–81.

Walport, M.J. and Lachman, P.J. (1988). Erythrocyte complement receptor type I, immune complexes, and the rheumatic diseases. *Arthritis and Rheumatism*, **31**, 153–8.

Winfield, J.B., Faiferman, I., and Koffler, C. (1977). Avidity of anti-DNA antibodies in serum in IgG glomerular eluates from patients with systemic lupus erythematosus. *Journal of Clinical Investigation*, **59**, 90–6.

Winfield, J.B., Shaw, M., Yamada, A., and Minota, S. (1987). Subset specificity of antilymphocyte antibodies in systemic lupus erythematosus. II. Preferential reactivity with T4+ cells is associated with relative depletion of autologous T4+ cells. *Arthritis and Rheumatism*, **30**, 162–8.

Wofsy, D., Hardy, R.R., and Seaman, W.E. (1984). The proliferating cells in autoimmune MRL/lpr mice lack L3T4, an antigen on helper T cells that is involved in the response to class II major histocompatibility antigens. *Journal of Immunology*, **132**, 2686–9.

Zanetti, M. (1985). The idiotype network in autoimmune processes. *Immunology Today*, **6**, 299–302.

Zouali, M., Stollar, D.B., and Schwartz, R.S. 1988. Origin and diversification of anti-DNA antibodies. *Immunological Reviews*, **105**, 137–59.

4.6.2 Systemic lupus erythematosus (clinical)

CLAUDIO PONTICELLI AND GIOVANNI BANFI

Definition

The term lupus has been used for several centuries to indicate a variety of facial skin lesions. The definition of systemic lupus erythematosus (SLE) was first applied in 1872 by Kaposi, who described two types of disease, one characterized by a cutaneous eruption associated with lymphadenopathy and often preceded by arthritis, the other characterized by diffuse rash, pleurisy, pneumonia, and sometimes mental disturbances. The significance of the systemic manifestations of the disease was further stressed by Osler (1904) who suggested a vascular basis for SLE. The most full clinical delineation of SLE was probably made in 1935 by Baehr. Since then the major advance came with the discovery of the LE cell phenomenon (Hargraves et al. 1948) which not only represented a tool to distinguish SLE from other rheumatic diseases but rapidly led to demonstration that the disease has an autoimmune pathogenesis.

Today the term lupus erythematosus refers to an inflammatory disorder characterized by autoantibody overproduction. In its benign form, lupus presents as coin or disc-shaped cutaneous lesions without clinical signs or symptoms of systemic involvement (discoid lupus). However lupus erythematosus can exist as a potentially fatal systemic disease with widespread organ involvement, variable and protean clinical manifestations, and serologic evidence of systemic disease.

Epidemiology

SLE is predominantly a disease of women, with a female: male ratio of about 9:1 (Estes and Christian 1971; Wallace et al. 1982; Hochberg et al. 1985a, b). Although SLE usually develops during the child-bearing years, it spares neither the neonate (Watson et al. 1984) nor persons in advanced age (Catoggio et al. 1984). The prevalence of SLE ranges around 50 per 100 000 in white races (Fessel 1974). Epidemiologic surveys based on hospitalized patients reported a considerably higher incidence in black and in Asian populations (Serdula and Rhoads 1979; Hochberg 1985a). However the actual difference in racial prevalence remains difficult to estimate as long as studies do not include both in- and outpatients.

The familial occurrence of SLE as well as the development of other immunologically mediated diseases and serological abnormalities in relatives of patients with SLE is well established (Arnett 1987).

Kidney involvement is frequent in SLE. Overt clinical evidence of renal disease varies from 35 to 75 per cent of patients with well-documented SLE (Dubois and Tuffanelli 1964; Estes and Christian 1971; Wallace et al. 1982; Hochberg et al. 1985b). However, clinical data probably underestimate the prevalence of lupus nephritis. Studies with renal biopsy showed in fact that SLE involves the kidney in almost all the cases in which adequate tissue analysis is available. Even in the absence of proteinuria or abnormal urinary sediment, the renal biopsy is rarely free of detectable abnormalities, especially if evaluated by immunofluorescence or electron microscopy (Cavallo et al. 1977; Mahajan et al. 1977; Font et al. 1987).

Diagnosis of systemic lupus erythematosus

Clinical criteria of disease

In 1971 a subcommittee of the American Rheumatism Association proposed a classification of SLE which was based mainly upon clinical criteria (Cohen et al. 1971). Since the data were collected in the early 1960s these criteria have been revised (Tan et al. 1982). A new classification has been proposed which takes into account 11 criteria, including some serologic tests widely used in clinical practice (Table 1). According to this classification a diagnosis of SLE should be made if any four or more of the 11 criteria are present serially or simultaneously during any interval of observation. When compared with the 1971 criteria, the revised criteria showed gains in sensitivity (96 vs. 88 per cent) and in specificity (96 vs. 95 per cent). The most sensitive criteria were the positivity of antinuclear antibodies, arthritis, and immunologic disorders. The most specific criteria were discoid rash, neurologic disorders, malar rash, photosensitivity, oral ulceration, renal abnormalities, and immunologic disorders.

These criteria are now well accepted and generally used. However, while useful for the diagnosis of SLE they are of little help in assessing the activity of the disease. In fact a patient might accumulate four or more criteria over the years but could have no criteria present at a single time-point. To assess the clinical activity better, Urowitz et al. (1984) analysed several variables associated with activity in patients with SLE. The following variables, in order of discriminant importance, were found to be associated with activity.

1. arthritis;
2. abnormal laboratory tests (leucopenia, hypocomplementaemia, anti-DNA antibodies);
3. rash, mucous membrane ulcers, alopecia;
4. pleurisy—pericarditis;
5. seizures, psychosis, brain syndrome, lupus headache;
6. vasculitis—skin or digital ulceration;
7. haematuria.

The presence of two or more of these variables predicted active SLE in 100 per cent of cases.

4.6.2 Systemic lupus erythematosus (clinical)

Table 1 Sensitivity and specificity of the revised criteria for classification of systemic lupus erythematosus (adapted from Tan et al. 1982)

Criterion	Sensitivity (%)	Specificity (%)
Malar rash	57	96
Discoid rash	18	99
Photosensitivity	43	96
Oral ulcers	27	96
Arthritis (two or more joints)	86	37
Pleurisy–pericarditis	56	86
Renal abnormalities (persistent proteinuria > 0.5 g daily; cellular casts)	51	94
Neurologic disorders (seizures, psychosis)	20	98
Haematological disorders (haemolytic-anaemia, leucopenia, lymphocytopenia, thrombocytopenia)	59	89
Immunologic disorders (anti-DNA, anti-Sm antibodies, false positive tests for syphilis, positive LE cell preparation)	85	93
Antinuclear antibody	99	49

Serological and immunological tests

Antibodies

The typical serological abnormality of SLE consists of an overproduction of autoantibodies directed against numerous autoantigens (see Chapter 4.6.1). These antigens can be well-defined macromolecules such as DNA, RNA, and ribosomal constituents or antigens defined by precipitin reactions in agar gels which have been generally identified by the first two letters of the patient from whom the serum originated. This group includes antigens Sm, Mo, Ro, La, Ma, and proliferating cell nuclear antigen. Antibodies can also form against phospholipid antigens and cell membrane components (Table 2).

Table 2 Principal autoantibodies in SLE

Antinuclear antibodies
  Native DNA
  Denaturated DNA
  Single-stranded RNA
  Double-stranded RNA
  Cell nuclear antigens
  MA

Against extractable nuclear antigens
  Histones
  Sm
  Ro (SSA)
  La (SSB)
  Ribonucleoprotein

Against components of cytoplasm
  Ro (SSA)
  Ribosomes
  Lysosomes

Antiphospholipid antibodies
  Cardiolipin
  Lupus anticoagulant
  Brain tissue

Against cell membrane determinants
  Red cell
  White cell
  Platelets

Antibodies to DNA can be divided into two categories corresponding to the two macromolecular forms of DNA. These are native (n or double-stranded) and denaturated (ss or single-stranded) DNA. Anti-nDNA antibodies have the highest specificity for the diagnosis of SLE. Demonstration of these antibodies requires impeccably native DNA in the assay procedure. A test which has gained increasing favour is the immunofluorescent test utilizing *Crithidia luciliae* as substrate. This trypanosome contains within it kinetoplast circular DNA which is double-stranded by all criteria. Anti-ssDNA antibodies can be found also in patients with other connective tissue diseases or active hepatitis. In some 5 per cent of patients with SLE antinuclear antibodies cannot be detected. These patients generally develop anti-Ro antibodies and do not present abnormalities of complement system. In these patients clinical renal involvement is rare.

Patients with SLE also develop antibodies to non-DNA-containing soluble antigens. The most common antigens are histones, Sm, ribonucleoprotein (nRNP), Ro, and La (Table 3). Histone antibodies are considered to be highly specific for SLE. The histones consist of five discrete polypeptides, classified in two groups, one called core histone and the other, more heterogenous, which includes several subgroups of histones (Hess 1988). Histone antibodies are frequently found in drug-induced lupus while their incidence is lower in idiopathic SLE. Antibodies to nRNP antigen are more frequent than antibodies to Sm. However while anti-Sm antibodies have been found only in SLE, the highest titres of anti-nRNP antibodies have been seen also in mixed connective tissue disease suggesting a lower specificity for this marker. Nephritis is generally mild and follows a benign course in patients with antibodies to Sm. A low incidence of renal disease has been reported also in patients with anti-nRNP antibodies, unless anti-DNA antibodies are also present. Ro and La are two antigens that evoke autoantibody formation in SLE. There is evidence today that Ro and La are immunologically identical respectively to the antigen SSA and SSB, which had been first detected in Sjögren's syndrome. The exact cellular localization of the Ro (SSA) and La (SSB) is a moot point. La (SSB) is probably located in both the nucleus and cytoplasm while Ro (SSA) is mainly localized in the cytoplasm but is present also in the cell nucleus. Anti-Ro (SSA) antibodies are

Table 3 Specificity of antinuclear antibodies

Antibodies anti-	Associated diseases
ds DNA	SLE (60%), mixed connective tissue disease (10–20%)
ss DNA	SLE (60%) rheumatoid arthritis (40%), Sjögren's syndrome (25%), mixed connective tissue disease (25%), progressive systemic sclerosis (10%), drug-induced lupus (10%)
Histones	Drug-induced lupus (95%) other ANAs absent, SLE (30%) other ANAs present, rheumatoid arthritis (25%)
n RNP	Mixed connective tissue disease specific if other ANAs absent (95–100%), SLE (30%), progressive systemic sclerosis (5%), Sjögren's syndrome, rheumatoid arthritis (2–4%)
Sm	SLE (30–40%) specific
Ro (SSA)	Sjögren's syndrome (70%), SLE (30%) 'ANA-negative lupus', rheumatoid arthritis, neonatal lupus
La (SSB)	Sjögren's syndrome (60%), SLE (15%) neonatal lupus
Ma	SLE (20%) severe course

unaffected by age, duration of disease, or clinical activity. Although these antibodies are not of major pathogenetic significance in SLE, they may have considerable diagnostic importance, especially in patients with a negative antinuclear antibody test by immunofluorescence. Patients with both anti-Ro (SSA) and anti-La (SSB) have a lower incidence of nephritis and milder disease than those with anti-Ro (SSA) alone (Reichlin 1981). Two other antibodies may occur in a small group of SLE patients. One is directed against a proliferating cell nuclear antigen, and the other against an acidic nuclear protein called Ma antigen. This antibody occurs in a subset of patients with particularly severe SLE.

Another group of antibodies is directed against phospholipid antigens. These antibodies are present in about 5 to 10 per cent of patients with SLE. Two of these antibodies, anticardiolipin and the so-called 'lupus anticoagulant', are frequently associated with thrombosis, thrombocytopenia, abortion, and cerebral disease in patients with SLE who may or may not have antinuclear antibody (Hughes 1983). The lupus anticoagulant is an acquired immunoglobulin which inhibits the phospholipid portion of the prothrombin activator. The reasons for the thrombotic tendency are still unclear. However, there is some experimental and clinical evidence that antiphospholipid antibodies affect platelet membranes (Hughes 1988). Antiphospholipid antibodies and widespread small vessel thrombosis have been observed also in clinical conditions other than SLE, such as idiopathic recurrent deep-vein thrombosis, arterial gangrene, early coronary disease, chorea, livedo reticularis, and multi-infarct dementia.

Hypocomplementaemia

Many patients with SLE have an activation of the classic component cascade with consumption of the early components C1q, C4, and C3. Serum C4 levels generally decrease earlier and to a greater extent than C3 in the exacerbations of the disease. Alternative pathway activation of the complement can also occur in SLE. This activation may account for the low C3 levels found in patients with congenital deficiency of C2, C1r, or C1s which are essential components of the classic pathway. The specificity of hypocomplementaemia for the diagnosis of SLE is not agreed upon. Low serum C3 levels are sometimes present only during periods of disease activity. Moreover hypocomplementaemia has been reported in some patients with rheumatoid arthritis and in other pathological conditions that may resemble SLE, such as bacterial endocarditis and essential cryoglobulinaemia. However the presence of both low C3 levels and high anti-DNA antibody titre is 100 per cent correct in predicting the diagnosis of SLE when applied to a patient population in which that diagnosis is considered (Weinstein et al. 1983).

Immune complex assays

Circulating immune complexes can be detected and quantified by a variety of assays. The correlation between the assays is limited however, suggesting that the heterogeneity of complexes can account for their variable detection by the different methods. Intermediate or small-sized complexes are probably more nephritogenic than others. However, it should be pointed out that complexes which deposit in tissue escape measurement despite their potential for damage. Hence the clinical utility of immune-complex determinations remains low.

Lupus band test

Deposits of immunoglobulins and/or complement at the dermal–epidermal junction of lesional or normal skin can be found in about 70 per cent of patients with SLE. However, these deposits can be observed also in several patients with rheumatoid arthritis and in about 50 per cent of patients with mixed connective tissue disease. The specificity of the lupus band test increases with the number of proteins detected at the dermal–epidermal junction. The predictive value is low with IgM, which is the single most common protein found either in SLE or in other diseases. On the other hand, C4, properdin, and IgA have the greatest predictive value for the diagnosis of SLE but their sensitivity is quite low (Smith et al. 1984). Therefore this test may only be of value in establishing a diagnosis of SLE when rigorous criteria are adopted.

Clinical features of lupus nephritis

Renal syndromes at presentation

Only one-quarter of patients with renal lupus present with renal disease as the first manifestation. Arthritis and/or facial rash usually precede nephritis (Cameron 1989). The kidney involvement in SLE may manifest clinically with virtually any possible form of presentation of a renal disease. Patients with lupus nephritis may have initially only mild urinary abnormalities such as microhaematuria, red cells casts, or asymptomatic proteinuria. These abnormalities may be intermittent so that urine can appear 'normal' when examined sporadically.

Some 40 per cent of patients show at onset a frank nephrotic syndrome which does not differ from that seen in primary glomerulonephritis. Others present with a nephrotic syndrome usually accompanied by various degrees of impairment of renal function. Sometimes the patient is referred to the nephrologist because of an already established chronic renal failure. These renal disorders are generally associated with some other signs or symptoms of SLE and with the typical serologic abnormalities.

In about 5 per cent of cases, however, glomerulonephritis is the only initial manifestation of SLE and several years may elapse before extrarenal features develop or antinuclear antibodies become detectable in the blood. This unusual presentation is more frequent in males and in patients over 40 years. A picture of membranous nephropathy is often found at renal biopsy in these cases (Adu *et al.* 1983).

Acute renal failure can exceptionally represent the first manifestation of the disease. The onset of oligoanuria is generally accompanied by severe extrarenal manifestations of SLE and by strong serologic activity. Heavy glomerular involvement with widespread capillary thrombi can be found in these patients (Ponticelli *et al.* 1974) but in some cases tubulointerstitial lesions are predominant with only mild glomerular abnormalities (Tron *et al.* 1978). More rarely extensive crescents are seen at renal biopsy in anuric patients. In some cases postpartum or postabortum acute renal failure may be the first manifestation of unrecognized SLE (Imbasciati *et al.* 1984).

In other cases the first renal symptoms are related to an incomplete or complete renal tubular acidosis, which has been ascribed to mononuclear cell infiltration, deposition of tubular antigen–antibody complexes or formation of antibodies directed against the tubular basement membranes. Patients with nephrocalcinosis, nephrolithiasis, hypokalaemia, or renal magnesium wasting consequent to distal renal tubular acidosis should be evaluated for the possible presence of SLE (Carvana *et al.* 1985).

Finally, some patients with well-established SLE present no urinary or renal function abnormalities but may show various findings of renal disease at kidney biopsy. In some of these patients with clinically silent renal disease a diffuse proliferative glomerulonephritis has even been found (Leehey *et al.* 1982). The prevalence of diffuse nephritis ranges between zero (Cavallo *et al.* 1977) and 45 per cent (Mahajan *et al.* 1977). As patients with clinically occult diffuse lupus nephritis are younger than patients with more benign glomerular lesions an age-dependent factor may account for the discrepancies of different reports. According to Eiser *et al.* (1979) it is possible that diffuse proliferative glomerulonephritis occurring in young patients is less likely to be clinically manifest. There are several unsolved problems for the clinician. Should a renal biopsy be performed in a young patient without urinary abnormalities but with active SLE? If biopsy shows the presence of a diffuse nephritis even in the absence of clinical signs or symptoms should treatment be given? Should therapy be aggressive in view of the potential of diffuse lupus nephritis or careful in view of the clinical absence of renal damage?

The clinical course of lupus nephritis

The clinical outcome of lupus nephritis is highly variable. Moreover most patients with SLE and renal involvement are given corticosteroid and/or cytotoxic agents which can modify the course of the disease. It is therefore difficult to assess the natural history for these patients. Bearing in mind that many exceptions can occur and that transitions between groups may be frequent, it is possible to attempt to categorize the clinical course of lupus nephritis in the following groups.

Some patients maintain minimal proteinuria and microhaematuria over a long period without developing nephrotic syndrome, hypertension, or renal insufficiency. In this group, the course is characterized by intermittent remission and recurrence of the mild renal manifestations usually associated with the fluctuations of activity of the systemic disease. In these patients the final outcome is predominantly determined by the severity of SLE in other systems than by renal disease. However it is possible that patients with clinically mild or silent renal disease may eventually develop severe kidney involvement.

In some 40 to 60 per cent of patients the renal involvement manifests with a picture of nephrotic syndrome accompanied by microhaematuria and 'telescoped' urinary sediment with or without hypertension. The natural clinical course may be slowly progressive with about 50 per cent of patients developing hypertension and uraemia within 10 years. The extrarenal and serologic activities of SLE are usually mild to moderate in these patients.

In a third group of patients nephrotic syndrome, haematuria, hypertension, and renal insufficiency generally accompanied by other signs and symptoms of SLE activity may be present from the onset or can develop early after the diagnosis of nephritis. Remission of the nephrotic syndrome and of renal insufficiency can occur after institution of an adequate therapy. The course of these patients may be punctuated by renal and/or extrarenal flares alternated with periods of clinical quiescence. When untreated most, if not all, patients with such clinical features progress to end-stage renal failure or die from other complications of SLE within 2 years from clinical onset (Pollak *et al.* 1964).

A few patients can manifest a tumultuous progressive course with death or irreversible renal failure in spite of therapy. These patients show marked hypertension, papilloedema, heart failure, encephalopathy and subacute renal failure. This clinical picture strongly suggests a severe underlying vascular involvement (Baldwin *et al.* 1970). Patients with SLE may also develop acute anuric renal failure, either related to an exacerbation of the disease or to the use of anti-inflammatory non-steroidal drugs. A syndrome resembling thrombotic thrombocytopenic purpura with thrombocytopenia, microangiopathic haemolytic anaemia, seizures, and renal dysfunction may also complicate the course of SLE in exceptional cases. Although this syndrome more often occurs in pregnant women with lupus anticoagulant (Kincaid-Smith *et al.* 1988) it has also been observed outside pregnancy either in patients with clinically and serologic quiescent SLE or during a flare-up of the disease (Gelfand *et al.* 1985).

Clinical aspects of SLE in childhood

The prevalence of SLE in children ranges around 0.6 per 100 000 (Fessel 1974). Onset of the disease before the age of 5 years is possible but extremely rare and most patients are diagnosed in the adolescent age group with a peak between 12 and 14 years. Under 10 years the sex distribution is nearly equal. It is only after puberty that the striking female preponderance can be seen.

Children who develop SLE are often very ill at the time of presentation with severe multisystem involvement (Emery 1986; Cameron 1989). Weight loss, anorexia, malaise, and fatigue occur in 40 to 50 per cent of patients. Fever is common but must be distinguished from a superimposed infection, favoured by leucopenia and decreased immune status. Pneumococcal infections and varicella pose the most serious problems. Mucocutaneous findings are particularly frequent and severe. About 75 per cent of children with SLE have arthritis or arthralgias which can sometimes be confused with a juvenile chronic arthritis. One of the most common presentations of childhood lupus is that of haematological abnormalities. About 50 per cent of patients present anaemia with a positive Coombs' test and leucopenia, sometimes severe. Thrombocytopenia occurs in about 25 per cent of children. Pleuritis is the most common expression of pulmonary involvement. Pericarditis and/or myocarditis occur in about 25 per cent of children and can be a presenting finding. Fatal cardiac infarction has also been reported in young patients, some with inactive SLE. Neurologic manifestations occur in 15 to 30 per cent of patients. Grand mal seizures are the most frequent manifestations. Headache of clinical significance is also frequent. Severe psychoneurosis (phobias and depression) can occur in 25 to 30 per cent of children. Minor psychiatric abnormalities and steroid-related psychosis are even more frequent.

Renal involvement is present in almost 80 per cent of children with SLE. The presentation of renal disease is variable. Some 60 per cent of patients have non-nephrotic proteinuria and 40 per cent have a nephrotic syndrome. Microhaematuria is almost constant. About 50 per cent of patients have a deterioration of renal function and 17 per cent present with acute renal failure. Hypertension is common; at onset 30 to 40 per cent of patients are hypertensive (Cameron 1989). The incidence and the type of renal disease at biopsy are poorly correlated with clinical signs or urine abnormalities. A high frequency of pathologic transition may occur (Platt et al. 1982). The course is variable. Many children continue to show active disease with multisystem involvement which is sometimes refractory to treatment and leads to early death. In other patients the treatment of the first flare-up may be followed either by a prolonged quiescence or by sudden relapses, sometimes preceded by solar exposure. Drug administration or infection, growth retardation, and delayed sexual maturation, especially in boys, are very serious problems in children treated with steroids. The prognosis, which was almost uniformly fatal until a few decades ago, has improved. From a review of recent paediatric literature Cameron (1989) found a 10-year survival rate ranging between 48 and 80 per cent. Infections and uncontrollable active lupus account for most deaths. End-stage renal failure in children is unusual, being often deferred until the late teens. Other causes of death, such as cerebritis, cardiac failure, or pulmonary haemorrhage, played a minor role.

In summary, both the clinical presentation and the outcome of SLE in children are characterized by widespread multisystem involvement and by hectic clinical activity. Therefore the disease in childhood needs careful evaluation of the patient, adequate treatment, and close monitoring to prevent complications either by the active disease or by its overtreatment.

Clinical aspects of SLE in the elderly

Lupus nephritis is an uncommon problem in the elderly, since only 4 per cent of cases of SLE present after the age of 60 (Murray and Raij 1987). The presentation is generally insidious. In some cases the initial manifestations are polymyalgia rheumatica syndrome or rheumatoid-like arthritis, but in most patients the first symptoms are interstitial lung disease or serositis. Baker et al. (1979) reported a low prevalence of cutaneous manifes-

Table 4 Association between drugs and lupus syndrome

Well-defined association with lupus
   Hydralazine
   Procainamide
   Isoniazid
   Methyldopa
   Chlorpromazine
   Quinidine
Probable association with lupus
   Anticonvulsants
   Antithyroid drugs
   Penicillamine
   Captopril
   Sulphasalazine
   β-Blockers
   Lithium
Possible association with lupus
   p-Aminosalicylic acid
   Oestrogens
   Gold salts
   Penicillin
   Griseofulvin
   Reserpine
   Tetracycline

tations, neuropsychiatric disorders and especially lymphadenopathy with relatively common subclinical liver involvement. On the contrary, Ballou et al. (1982) found that renal disease, central nervous system disorders, cutaneous manifestations and haematologic abnormalities occurred with similar frequency in older and in younger patients. Several older patients have overt clinical features of Sjögren's syndrome, making a differential diagnosis difficult (Jonsson et al. 1988). Only a few patients have antibodies to DNA but most have antibodies to Ro (SSA) and to La (SSB). Hypocomplementaemia is less frequent than in younger patients (Catoggio et al. 1984). Only some 10 to 15 per cent of older patients with SLE have overt nephritis (Wilson et al. 1981). Renal disease seems to run a more slowly progressive course than in younger patients, but surprisingly there are very few reports on the outcome of older patients with lupus nephritis.

Drug-related lupus

Many drugs may induce antinuclear antibody formation with or without clinical symptoms of lupus (Table 4). Some of these drugs contain hydrazino or amino groups (hydralazine, procainamide, isoniazid) others contain sulphydryl groups (penicillamine, captopril, propylthiouracil). The drugs implicated may combine with nuclear histones evoking the production of autoantibodies directed against histones rather than against DNA, as in idiopathic SLE. Drugs inducing lupus might also interact with lymphocytes and promote the production of autoreactive lymphocytes and of antilymphocyte antibodies which alter the immunoregulation. Some common clinical features of drug-related lupus include fever, myalgias, arthralgias, pleurisy, and pericarditis which may have an abrupt onset. Malar rash, alopecia, discoid lesions, hypocomplementaemia, severe anaemia, or leucopenia are unusual. Renal and central nervous system involvement are very rare.

The two drugs that most commonly induce lupus are hydralazine and procainamide. Hydralazine was the first drug shown to induce SLE. A daily dose of hydralazine greater than 400 mg or a total dose greater than 100 g considerably increases the risk of developing lupus. Patients who develop clinical lupus are often (but not always) slow acetylators, whereas asymptomatic patients are predominantly normal acetylators. The clinical course is indistinguishable from that of idiopathic SLE, except for the lower incidence of visceral involvement. There are only a few cases with nephropathy, generally recognized to be nephrosclerosis or pyelonephritis, but focal proliferative lupus nephritis has also been observed (Neparstek et al. 1984).

Procainamide is the most frequent cause of drug-induced lupus. Some 50 to 100 per cent of patients taking the drug for 1 year or more have positive antinuclear antibodies and 5 to 10 per cent of them present symptoms of lupus. The mean age in procainamide-induced lupus is 57.5 years and the prevalence is almost as frequent in men as in women. Adenopathy, gastric symptoms, and central nervous system changes are almost completely absent. Musculoskeletal changes, serositis, and pulmonary infiltrates (30 per cent) are the most common features (Dubois and Wallace 1987). Renal involvement is extremely rare. The few reported cases showed a focal or membranous picture at renal biopsy (Foucar et al. 1979). The treatment of drug-related lupus is drug withdrawal. Following cessation of the causative agent, symptoms subside within a few days to some weeks without additional therapy. Corticosteroids are only rarely needed.

A number of drugs (sulphonamides, diuretics, oral antidiabetics, some antibiotics, non-steroidal anti-inflammatory agents, danazol, cimetidine) may provoke a cutaneous flare in patients with pre-existing SLE. These drugs may either act as photosensitizing effectors or promote a hypersensitivity reaction (Dubois and Wallace 1987).

Pregnancy and SLE

Since many women with SLE are of child-bearing age and have normal fertility, the clinician has often to face the problem of pregnancy in patients with lupus nephritis. In this condition the influence of lupus nephritis on fetal outcome, the obstetrical complications, and the influence of pregnancy on SLE should be taken into account.

Influence of lupus nephritis on fetal outcome

A high incidence of abortion, perinatal death, and prematurity has been reported in women with SLE. Even in pregnancies completed before clinical onset of SLE several fetal losses, generally spontaneous abortions, occur. Fetal wastage is strikingly higher in patients in whom onset of SLE occurs during pregnancy or in the immediate postpartum period (Imbasciati et al. 1984; Bobrie et al. 1987). In pregnancies initiated after the diagnosis of SLE has been made fetal prognosis depends not only on SLE activity but also on the presence of impaired renal function, nephrotic syndrome, or hypertension (Cameron 1984). It has been reported that fetal loss is particularly frequent in women with circulating lupus anticoagulant (Hughes 1983). This antibody might favour placental thrombosis resulting in variable placental ischaemia, impaired placental development and function with either subsequent impairment of fetal growth or death. Prednisone (40–60 mg per day) and aspirin (75 mg per day) have been successfully used to suppress lupus anticoagulant in women with a prior history of repeated abortions (Lubbe et al. 1983). However the role of this therapy in patients with subclinical SLE and idiopathic habitual abortion has been challenged by others. Some investigators reported uncomplicated pregnancies with a live birth at term in patients with SLE and untreated lupus anticoagulant and concluded against an extensive use of prednisone and aspirin treatment in these patients (Petri et al. 1987; Stafford-Brady et al. 1988). On the other hand Lockshin et al. (1987) actually observed a fetal loss higher than 50 per cent in patients with lupus anticoagulant and/or anticardiolipin antibodies. Yet prednisone and aspirin did not influence fetal mortality in these patients.

Prematurity is a constant hazard in lupus pregnancy. About one-third of liveborn infants are delivered before 36 weeks and suffer from the complications of prematurity. However, in experienced hands, perinatal and postpartum deaths are rare.

Neonatal lupus

This uncommon syndrome may develop in some newborns generated by mothers with SLE. The syndrome is characterized by a transient dermatitis, haematological abnormalities, systemic features, and isolated congenital heart block. Serologic investigations have demonstrated the almost universal presence of La (SSB) and/or Ro (SSA) antibodies in the sera of infants with neonatal lupus and their mothers (Watson et al. 1984). This supports the hypothesis that the neonatal syndrome is caused by a transplacental passage of these antibodies. The cutaneous lesions spontaneously disappear with minimal scarring, but healing may be delayed for many months in occasional cases. A more important but rare complication of neonatal lupus involves impairment of the fetal cardiac condition system, with congenital atrial ventricular block. This disorder is caused by an extensive fibrosis that may replace the septal musculature in the area of the atrial ventricular node. It seems more associated with La (SSB) than with Ro (SSA) antigen. Normal children following the birth of a child with neonatal lupus have not as yet been reported.

Obstetrical complications

Women with SLE are at high risk of developing pregnancy-induced hypertension. About one-third of completed pregnancies may be complicated by the development of hypertension and/or proteinuria. This complication seems to be independent of the clinical activity of lupus nephritis when the patient entered pregnancy. It is difficult, however, to attribute hypertension and proteinuria either to an obstetrical complication or to an exacerbation of lupus nephritis. As complement may be activated even in normal pregnancies and hypocomplementaemia is common in pregnancy-induced hypertension, the measurement of complement or its components is of little help for a differential diagnosis (Lockshin et al. 1987).

Influence of pregnancy on lupus nephritis

There are different views about the course of lupus nephritis in pregnancy. Many clinicians think that the the outcome of the disease may be benign if there are no clinical signs of SLE activity before conception. However in some 20 per cent of patients without previous clinical evidence of SLE, lupus nephritis first manifests during pregnancy or in the immediate postpartum or postabortum period. In a few patients renal disease manifests only with proteinuria and/or hypertension. However in many cases lupus nephritis bursts out abruptly together with other signs and symptoms of SLE. Nephrotic syndrome and increased serum creatinine are frequent and a diffuse proliferative glomerulonephritis is usually seen at kidney biopsy (Bobrie et al. 1987).

In some cases acute renal failure develops with clinical and histological features similar to those observed in postpartum renal failure (Imbasciati et al. 1984). When pregnancy starts after the onset of SLE, recurrence or exacerbation of renal disease may occur during pregnancy or after delivery. Although exacerbations occur more frequently in patients whose SLE was active at conception, severe renal flares can be observed even in patients who were in complete remission at conception (Imbasciati et al. 1984; Bobrie et al. 1987). Therapy may fail to reverse renal failure in some of these patients. Recently, Kincaid-Smith et al. (1988) reported that pregnant women with circulating lupus anticoagulant are particularly prone to develop moderate to severe renal insufficiency, caused by a thrombotic microangiopathy.

In summary, SLE may adversely affect the fetal outcome while pregnancy can worsen the course of lupus nephritis. These risks are particularly elevated in patients with active lupus. The chances of favourable fetal and renal outcomes are better in women with stable and prolonged remission, normal renal function, and blood pressure at conception. The fact that the most severe renal complications seem to occur at the interruption of pregnancy and in women who had no clinical evidence of SLE and who therefore were not adequately treated, seems to suggest that corticotherapy may have a protective effect. Thus, although controlled studies are lacking, it seems reasonable to give a course of prednisone in the last few weeks of pregnancy and for at least 2 months after delivery in order to prevent renal flares. High-dose corticosteroids should be given after abortion or after delivery in untreated patients with evidence of SLE activity. Careful monitoring of blood pressure and an appropriate treatment of hypertension are of great importance. While waiting for the results of further studies the early administration of low-dose aspirin may be recommended in patients with lupus anticoagulant.

Correlations between clinical activity and serology

The diagnosis of SLE has been greatly facilitated by the use of serological tests, but whether these tests may be helpful in assessing the clinical activity of SLE is still controversial. Elevated concentrations of antinuclear antibodies have been found to correspond to clinical activity in some studies (Epstein 1977; Scopelitis et al. 1980) but other authors have reported high levels of anti-DNA antibodies during quiescent disease (Gladman et al. 1979). Neither CH50 nor other complement components proved to be good indicators of renal disease in patients taking immunosuppressive agents (Cameron et al. 1976; Morrow et al. 1982). Using different assay systems circulating immune complexes have been found to be accurate markers of disease activity in patients with lupus nephritis (Levinsky et al. 1977; Cairns et al. 1980). A C1q solid-phase assay seems to be a better marker of clinical activity than the assays using a fluid phase C1q binding (Abrass et al. 1980) but false results can be observed even with this technique (Morrow et al. 1982). Among the other tests proposed for monitoring the clinical status of lupus nephritis, the lupus band test (Morris et al. 1979) and erythrocyte sedimentation rate are of little if any help while lymphocyte count is often affected by the concomitant use of corticosteroids.

In summary, no single test can be considered as a reliable indicator of lupus nephritis activity. It has been reported that even the use of a panel of immunological and haematological tests can classify less than 50 per cent of patients into their correct clinical grades (Morrow et al. 1982). Waiting for a test or a combination of tests which can better assess the clinical activity of the disease, the nephrologist should carefully look at the 'renal' signs of activity rather than at serological tests. Study of urinary sediment, quantization of urine protein excretion, and measurement of glomerular filtration rate can be considered as reliable indices of the activity of the disease. Of course the presence of extrarenal signs and symptoms of SLE and the variations of simple serological tests, such as serum C3 and C4 levels, DNA-binding activity, and platelet count may be helpful in confirming the activity of the disease in doubtful cases. Continuous normalization of complement and of DNA-binding capacity may suggest a benign course (Appel et al. 1978).

Pathological features of lupus nephritis

Basic lesions

Hypercellularity of the glomerular tuft is frequent. It can be segmental or global, focal or diffuse but is almost never uniform. Hypercellularity may be due to proliferation of mesangial, endothelial, epithelial cells and to some extent to monocytes and polymorphonuclear cell exudation. There are often areas of necrosis of the tuft, usually confined to a lobule. They are constituted of granular and ill-defined material weakly eosinophilic or staining for fibrin ('fibrinoid necrosis'). Nuclear fragments and polymorphs intermingle with this material. Proliferating epithelial cells may encircle the necrotic area and adhere to the Bowman capsule-forming crescent. Haematoxylin bodies can be found in less than 10 per cent of cases in the areas of necrosis or inflammation, at times in the capillary lumen. Haematoxylin bodies, the only pathognomonic lesion in SLE nephritis, are mainly observed in florid cases and rarely if ever in doubtful ones when they might be of use. They appear on haematoxylin–eosin staining as round or oval, violet-pink, amorphous structures without a limiting membrane (Fig. 1). On ultrastructure examination they are electron-dense nuclear masses with light areas in the centre sometimes surrounded by degenerated cytoplasma.

The presence of immune deposits in the glomeruli is the hallmark of lupus nephritis. The amount and location of deposits are closely related to the severity and pattern of glomerular changes (Dujovne et al. 1972). Mesangial deposits are usually present even in patients without overt renal disease. Subepithelial and/or subendothelial deposits may be associated with these in mesangial region. They are often irregularly distributed and when massive give to the thickened capillary wall the rigid and refractile appearance of the classic 'wire loops' (Fig. 1). Subepithelial deposits are usually uneven in size and distribution, in contrast to idiopathic membranous nephritis. The amount of subepithelial deposits is often inversely correlated with that of subendothelial deposits. Subendothelial deposits may be so prominent as to bulge out into the lumen forming the so-called 'hyaline thrombi' (Fig. 1). True fibrin thrombi, although rare, may be observed at the same time as severe glomerular lesions.

Immunofluorescence invariably reveals the presence in glomerular deposits of strong staining for IgG, followed in intensity by IgM and IgA, although sometimes IgA may be dominant. All complement components may be found, the most usual being C3, C1q, and C4. A 'full house' of these immunoreactants is found in some 25 per cent of patients and is highly characteristic of lupus nephritis (Cameron 1989).

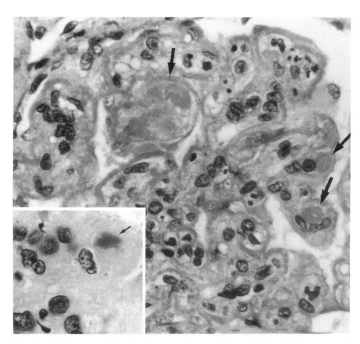

Fig. 1 Diffuse proliferative lupus glomerulonephritis. Diffuse wiry appearance of the thickened capillary walls. Some loops are stuffed with inflammatory cells while others are plugged by so-called hyaline thrombi (arrows) (Masson × 480). Insert: haematoxylin body (arrow) (HE × 1200).

On electron microscopy, electron-dense deposits may appear more granular than those in other types of glomerulonephritis. In about 6 per cent of cases a substructure with a fingerprint-like pattern can be seen (Fig. 2(a)). This finding has been regarded as a sensible marker of concomitant or subsequent development of overt lupus nephritis (Alpers et al. 1984). It is much more frequent in SLE than in other disease but is not specific for SLE. Other findings which are frequent in lupus while exceptional in other diseases are tubuloreticular inclusions ('virus-like' particles) resembling myxoviruses (Fig. 2(b)). They are mostly located within the cytoplasm of endothelial cell of glomeruli, interstitial capillary, and arterioles. Originally thought to represent viral material they seem more probably derived from degenerative cellular changes.

Classification of glomerular lesions. The WHO classification

Although not universally accepted, the WHO classification remains the most commonly used in clinical nephrology and has represented the basis for more recent classification (McCluskey 1975) (Table 5 and Fig. 3).

Class I. Normal kidney

Completely normal glomeruli not only at light microscopy but also on immunofluorescence and electron microscopy may be seen in SLE patients (Cavallo et al. 1977; Mahajan et al. 1977; Font et al. 1987). This condition is, however, quite rare in those with renal manifestations.

Class II. Mesangial glomerulonephritis

This is observed in 10 to 30 per cent of patients depending on the criteria used for indication to renal biopsy. The changes are characterized by the presence of IgG, C3, and sometimes IgM and IgA deposits confined to the mesangium. Tubulointerstitial and vascular changes are very rare. This class has been subdivided into two subgroups: IIA, characterized by pure mesangial deposits on immunofluorescence without glomerular changes and IIB in which some degree of mesangial hypercellularity is also present. Most class IIA patients do not have detectable urinary abnormalities and present normal DNA-binding capacity and normal complement profile. Conversely the majority of class IIB patients have some abnormalities of the urinary sediment, mild to moderate proteinuria, and positive serology. Renal function is usually preserved.

Class III. Focal proliferative glomerulonephritis

Some 10 to 25 per cent of patients present this form of nephritis. Glomerular changes involving not only the mesangium but also the periphery of the tuft are limited to 50 per cent or less of the glomeruli. At light microscopy they are characterized by sharply delineated areas of segmental proliferation. In these areas subendothelial immune deposits are irregularly distributed along the capillary walls. Neutrophils and mononuclear cell infiltration, foci of fibrinoid necrosis with karyorrhexis and crescent formation may be associated with proliferative lesions. Other glomeruli may show healing stages with scars, capsular adhesions, and/or global sclerosis. Mild diffuse mesangial enlargement may accompany segmental lesions but the capillary walls of uninvolved segments are normal. At immunofluorescence all the glomeruli appear to be involved. There are mesangial deposits of IgG, C3, and frequently also of IgM, IgA, and C1q. On electron microscopy there are subendothelial and more rarely subepithelial deposits, mainly confined to segmental proliferative lesions. With regard to the extension of the parietal deposits there is some disagreement in admitting (Hill 1983) or excluding (Grishman and Churg 1982) their presence outside the segmental lesions in class III.

These patients generally have proteinuria and erythrocyturia. Nephrotic syndrome is present in about one-third of patients. Renal insufficiency is rare. Anti-DNA antibodies are elevated in almost all patients and in 50 per cent of the cases C3 and C4 are lowered.

Class IV. Diffuse proliferative glomerulonephritis

This form accounts for 20 to 60 per cent of the cases in series coming from renal units. The lesions are quite similar to those of class III but differ in extension and severity. The proliferative lesions involve more than 50 per cent of glomeruli. They are diffuse but still with irregular distribution from segment to segment. The capillary walls are generally thickened with 'wire-loop' appearance caused by extensive subendothelial deposits. Segmental necrosis with leucocyte infiltration is frequent. Crescent formation may also occur. This finding is usually associated with severe renal disease. Haematoxylin bodies may occasionally be observed in these areas. A variant of class IV is the membranoproliferative form in which circumferential subendothelial extension of mesangial cells is a prominent feature giving rise to the lobular aspect and to the double-contour appearance of the capillary loops. Necrotizing changes and crescent formation are less frequent.

Immunofluorescence studies show diffuse mesangial, subendothelial, and endomembranous staining in class IV disease. Subepithelial deposits are less abundant and are irregularly distributed. IgG and C3 are constantly found while IgM and IgA

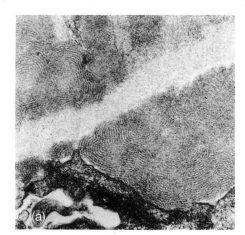
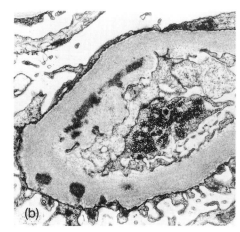

Fig. 2(a) Fingerprint-like appearance of electron-dense subendothelial and subepithelial deposits (× 85 000). **(b)** 'Virus-like' particles in the cytoplasm of an endothelial cell (× 20 000).

are often present but fainter. C1q is almost universally present, brighter at times than C3. Ultrastructure confirms numerous and often confluent mesangial and parietal deposits which in severe cases permeate the capillary wall from side to side.

Patients with class IV disease have usually heavy proteinuria, haematuria, and 'telescoped' sediment. Many of them are hypertensive and present variable degree of renal insufficiency. Most patients have positive anti-DNA antibodies and hypocomplementaemia.

Class V. Membranous glomerulonephritis

The membranous form is observed in 10 to 20 per cent of patients with lupus nephritis. It shares the main histologic features and staging criteria of the idiopathic form. A characteristic feature is the rather uniform capillary wall thickening due to the presence of numerous subepithelial and intramembranous immune deposits. Immunofluorescent microscopy shows diffuse granular staining with IgG, IgM, C3, C1q. Instead IgA and C4 are more variable and fainter. Subepithelial deposits are usually more irregular in shape and in size than those observed in the idiopathic form. Slight and diffuse mesangial expansion and proliferation are often seen and mesangial deposits are present, which differs from the idiopathic form. When present in a membranous nephropathy these features together with the occasional finding of 'finger-print' aspect of deposits should raise the suspicion of lupus.

Moderate to severe proteinuria is the dominant urine abnormality. About two-thirds of the patients become nephrotic. Haematuria is found in about 50 per cent of the patients while moderate renal insufficiency and hypertension are noted in some 25 per cent of the cases at presentation. DNA binding is moderately elevated whereas antinuclear antibodies may be negative. Serum complement and C4 may be normal or slightly reduced. In some patients membranous nephritis may occur months or even years before systemic manifestations and clear-cut immunologic features of SLE develop (Adu *et al.* 1983).

Unusual forms of lupus nephritis
Crescentic glomerulonephritis

Crescents are frequently noted in classes III and IV where they are usually segmental and focal in distribution. In rare cases epithelial proliferation is so prominent and diffuse as to give rise to the picture of severe crescentic nephritis with a rapid downhill course.

Interstitial nephritis

About 60 to 70 per cent of patients with lupus nephritis show some degree of interstitial inflammation, tubular damage and interstitial fibrosis. Together with inflammatory lesions some 30 to 50 per cent of cases present immune deposits (mostly IgG and C3) along tubular basement membranes, into interstitium and along peritubular capillary and arteriolar walls (Brentjens *et al.* 1975; Park *et al.* 1986). Usually the degree of interstitial lesions correlates with the severity of glomerular changes. On rare occasions, however, severe and predominant interstitial nephritis without glomerular involvement may cause clinical manifestations of renal disease such as tubular acidosis, salt-losing nephritis, Fanconi syndrome (Carvana *et al.* 1985), and sometimes acute renal failure (Tron *et al.* 1978). More frequently there is a progressive decline of renal function. The presence of immune deposits in tubulointerstitial and vascular structures has suggested that tubulointerstitial damage may be immune complex mediated (Brentjens *et al.* 1975).

Table 5 Patterns of histologic changes of glomerulonephritis (GN) in systemic lupus erythematosus (WHO classification)

WHO Class	Light microscopy		Immunofluorescence (Ig/C deposits)		Electron microscopy (Electron-dense deposits)		
	Mes	PCW	Mes	PCW	Mes	SEn	SEp
I Normal	0	0	0	0	0	0	0
IIA Mesangial deposits	0	0	+	0	+	0	0
IIB Mesangial hypercellularity	+	0	+	0	+	0	0
III Focal, segmental proliferative glomerulonephritis	+	+	++	+	++	+	±
IV Diffuse proliferative glomerulonephritis	++	++	++	++	++	++	±
V Membranous glomerulonephritis	+	++	++	++	+	±	++

Mes = mesangial, PCW = peripheral capillary wall, SEn = subendothelial, SEp = subepithelial, Ig = immunoglobulin, C = complement.

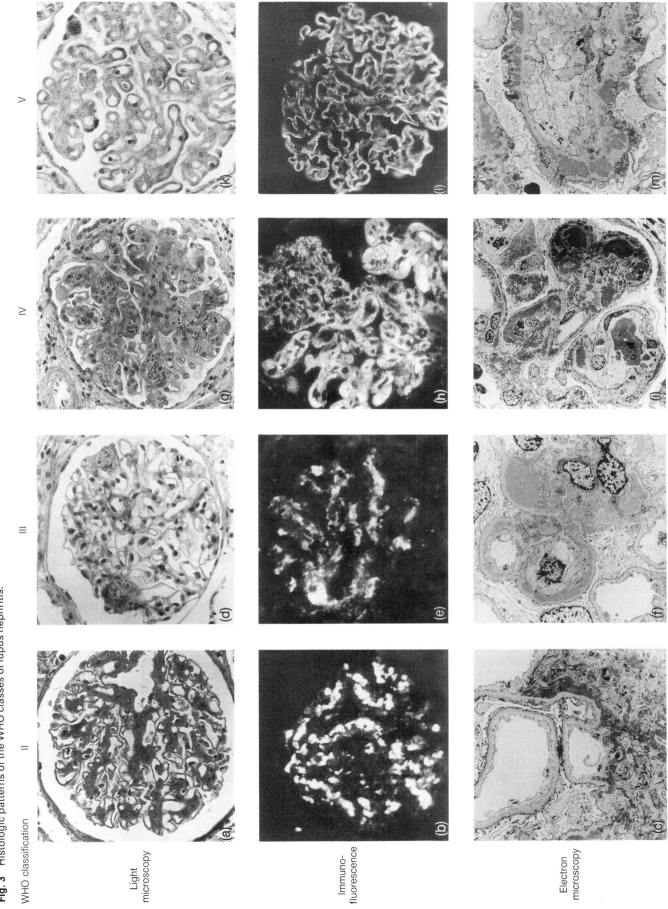

Fig. 3 Histologic patterns of the WHO classes of lupus nephritis.

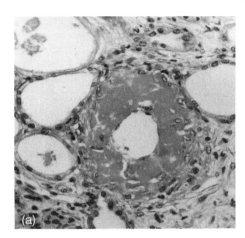 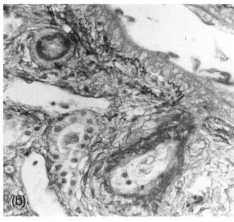

Fig. 4 Different aspects of lupus vasculitis. Left: an interlobular artery shows circumferential lumpy and massive proteinaceous deposits permeating the intima and partially extending to the media. Note the absence of cellular inflammatory reaction. Right: a small arteriole (upper left) is occluded by fibrinoid material which extends to the intima; another segment of interlobular artery (lower right) shows massive infiltration of the intima with fibrinoid material focally invading the muscular layer. Fluffy material also accumulates beneath the swollen endothelial cells (Masson × 250).

Interstitial cellular infiltration consisting mainly of lymphocytes can be found with equal frequency even in the absence of tubulointerstitial immune deposits (Park *et al.* 1986). This suggests that cellular immunity too may have a role in the pathogenesis of interstitial nephritis (Cameron 1988).

'Lupus vasculitis'

A non-inflammatory renal microangiopathy can be observed in about 10 per cent of patients mostly in those with diffuse proliferative nephritis. Although similar to microangiopathic lesions seen in malignant hypertension or in thrombotic thrombocytopenic purpura, this vasculopathy has some distinctive features. Lesions involve arterioles and interlobular arteries. They are characterized by destruction of the endothelial layer and massive precipitation in the intima of IgG, C3, and fibrin deposits, only rarely with mural extension and myocyte necrosis. Absence of inflammatory reaction is typical and makes the term 'vasculitis' inappropriate. Occlusion of the lumen with the same protein components, or predominantly with fibrin, frequently accompanies mural lesions (Fig. 4). The prominence of immunoreactants in the arteriolar precipitates strongly suggests that these lesions may result from deposition of these components as the primary event and thrombosis triggered as secondary phenomenon (Bhathena *et al.* 1981). Patients who present such lesions are likely to develop renal failure in a short time.

Thrombotic thrombocytopenic purpura

The clinicopathologic features of thrombotic thrombocytopenic purpura may occur exceptionally in SLE patients. This complication usually carries a bad prognosis for renal function, and for the patient overall.

Glomerular thrombosis

Recently some investigators have reported an unexpected high frequency of intraglomerular thrombotic lesions associated with proliferative but also with membranous lupus nephritis (Kant *et al.* 1981). Only a minority of patients had associated arteriolar thrombosis. Though less rare than previously believed these glomerular alterations seem to be fairly unusual and in most cases associated with the presence of circulating anticoagulants.

Amyloidosis

Sporadic patients with lupus develop renal amyloidosis. The majority of the cases reported, however, showed the coexistence of chronic infections or other rheumatic diseases or lymphoproliferative malignancies. The rarity of this complication may be explained by the usually normal plasma level of the amyloid A protein precursor (SSA) in spite of the chronic inflammatory nature of SLE.

Transformations

The increasing use of histologic examination in lupus nephritis has shown that during the course of the disease, transformations among forms can occur either spontaneously or as a result of treatment. The exact rate of transformation is difficult to assess for several reasons; these include the different criteria used for classifications, the changing therapeutic attitude in the last years, and the small number of patients undergoing second biopsies. Transformation from mild to severe forms has been reported more frequently than the converse. This may reflect the rarity of histologic re-evaluation in patients with a favourable course. Classes II may transform into class III or IV in some 15 to 20 per cent of cases. Baldwin (1982) has regarded this event as the real occurrence of renal disease rather than as a transformation. Class III seems the most unstable form with a rate of transformation (mainly to class IV) of between 20 and 40 per cent. Transition from class V to either class III or class IV is rather unusual, being reported in some 7 per cent of cases. Classes II and IV appear to transform to class V at an even lower rate (2.5 per cent). The rarity of transformation between membranous and proliferative glomerulonephritis could suggest the existence of different pathologic mechanisms in these types of nephritis. There are no clinical or pathologic features which might predict transformation. Transition to a more severe form is usually heralded by a sudden increase of proteinuria or by worsening of renal function. Viewed retrospectively, some patients who eventually transformed had more severe lesions at presentation than the others of the same class.

Clinicopathologic correlations

The histological form of nephritis at initial renal biopsy may help to predict the outcome. Although few patients may show a late transformation into severe forms and may eventually progress to uraemia, most patients of class II usually have an excellent prognosis even in the long term.

Much controversy exists on the validity of distinction between class III and class IV. Patients of both categories may show prominent signs of extrarenal involvement as well as marked serological alterations. Today most investigators agree that an involvement of less than 50 per cent of glomeruli (class III) identifies patients with distinctly better kidney survival than those with more extensive lesions (class IV). In the long term, however, at least in some series, class III patients may tend to have a survival rate similar to that of those in class IV (Appel *et al.* 1987). This form has been considered to have an ominous prognosis, and in many series still has the worst outcome when compared to other forms. Nevertheless some recent papers have reported a 10-year survival rate of over 80 per cent in diffuse proliferative lupus nephritis, with most patients maintaining prolonged stabilization of renal function (Austin *et al.* 1986; Ponticelli *et al.* 1987).

Membranous lupus nephritis has a relatively indolent course. Patient survival at 5 years lies between 80 and 90 per cent but in some series it abates to roughly 50 per cent, at 10 years (Appel *et al.* 1987). Also for this class the different survival rates reported by different investigators may depend partly upon the histologic criteria adopted for patient selection. For example class V patients with superimposed glomerular proliferation and extensive subendothelial deposits have a worse prognosis than patients with pure membranous pattern (Banfi *et al.* 1985). When inflammatory lesions are included, the outcome is similar to that of diffuse proliferative lupus nephritis (Schwartz *et al.* 1984).

Although there is a recognized trend for distinctive clinical features and long-term outcome in the different histologic categories they do not represent absolute and distinct clinicopathologic entities. As already discussed, some patients without clinical signs of renal involvement may present significant glomerular lesions including those of classes III, V, and exceptionally IV (Cavallo *et al.* 1977; Mahajan *et al.* 1977). It is possible, however, that repeated and precise laboratory tests would have disclosed renal abnormalities after a reasonable period of time in many of the reported cases. Though in the majority of cases renal lesions in individual patients fall in one of the major patterns, in about 10 to 20 per cent mixed forms are present which cannot be easily classified. Cases which develop transformation during the course of the disease are not rare and frequently present mixed patterns. For these reasons a recent modified version of the WHO classification has been proposed which includes subclasses with a mixed pattern (Churg and Sobin 1982) (Table 6). Moreover, some authorities are inclined to regard the histological picture of any individual patient as one point in a dynamic process rather than as a fixed pathologic entity (Kashgarian *et al.* 1982; Hill 1983). This assumption may be valid mostly for the proliferative forms.

The 'activity' and 'chronicity' indices

The predictive power of renal biopsy may be improved by the so-called 'chronicity' and 'activity' indices obtained by summing the semiquantitative scores of certain histological features whose prognostic significance was already emphasized by Pollak and Pirani (1969) several years ago (Table 7). Carette *et al.* (1983) and Austin *et al.* (1986) found that both indices could predict renal failure in patients with diffuse lupus nephritis. The effect of the chronicity index was particularly strong. Patients with a high chronicity index had a much poorer preservation of renal function at 5 years than those with a low index. Conversely, most patients with very low chronicity index did well without needing immunosuppressive therapy (Fig. 5). It has been pointed out, however, that the chronicity index applies only to diffuse proliferative glomerulonephritis and not to other lupus-related renal diseases (Lewis *et al.* 1987). Other investigators have reported that chronicity index determined on entry biopsy was not as significant a predictor of outcome as it was on second biopsies (Magil *et al.* 1988). The value of the activity index has long been recognized. The amount of acute inflammatory change in the glomerulus is a good indicator of prognostic potential, and is particularly useful in deciding whether a treatment should be aggressive.

Other prognostic clues

Several investigators have reported that the amount and the distribution of immune deposits within the glomeruli, studied by immunofluorescence and electron microscopy, correlate with clinical features and course of lupus nephritis better than histological changes (Grishman *et al.* 1973; Hill 1983). Since the location of deposits correlates with WHO classes to a large extent, the

Table 6 Classification of lupus nephritis from the International Study of Kidney Disease in Children (Churg and Sobin 1982)

I. *Normal glomeruli*
  A. Nil
  B. Normal by light microscopy but deposits on immunofluorescence and electron microscopy
II. *Pure mesangiopathy*
  A. Mild hypercellularity (+)
  B. Moderate hypercellularity (++)
III. *Focal and segmental glomerulonephritis*
  A. Active necrotizing lesions
  B. Active and sclerosing lesions
  C. Sclerosing lesions
IV. *Diffuse glomerulonephritis*
  A. Without segmental lesions
  B. With active necrotizing lesions
  C. With active and sclerosing lesions
  D. With sclerosing lesions
V. *Diffuse membranous glomerulonephritis*
  A. Pure membranous glomerulonephritis
  B. Associated with lesions of Class II (A or B)
  C. Associated with lesions of Class III (A, B or C)
  D. Associated with lesions of Class IV (A, B, C or D)
VI. *Advanced sclerosing glomerulonephritis*

Table 7 Renal pathology scoring system in lupus nephritis*

Activity index	Chronicity index
Glomerular abnormalities	
1. Cellular proliferation	1. Glomerular sclerosis
2. Fibrinoid necrosis, karyorrhexis	2. Fibrous crescents
3. Cellular crescents	
4. Hyaline thrombi, wire loops	
5. Leucocyte infiltration	
Tubulointerstitial abnormalities	
1. Mononuclear-cell infiltration	1. Interstitial fibrosis
	2. Tubular atrophy

Each factor is scored from 0 to 3.
* Fibrinoid necrosis and cellular crescents are weighted by a factor of 2. Maximum score of the activity index is 24, and of the chronicity index is 12. (From Austin, H.A. *et al.* (1983). *American Journal of Medicine*, **75**, 382–91, with permission.)

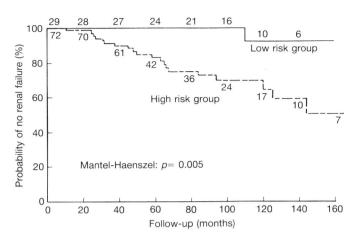

Fig. 5 Probability of maintaining life-supporting renal function in patients with active lupus nephritis identified as being at high or low risk, according to the presence of chronic histologic changes. (From Austin, H.A., III, et al. (1986). *New England Journal of Medicine*, **314**, 617, with permission.)

information based on distribution of deposits parallels that based on classification (Banfi *et al.* 1985).

Recent attention has been focused on tubulointerstitial and vascular lesions. These changes have been disregarded in previous classifications but although unusual may have great prognostic impact. Interstitial nephritis, with or without immune deposits, in rare cases may dominate the pathologic picture and produce severe renal failure (Tron *et al.* 1978; Park *et al.* 1986). Independently of the severity of glomerular changes, extraglomerular deposits have been shown to be a significant prognostic factor in diffuse proliferative lupus nephritis (Magil *et al.* 1988). Necrotizing non-inflammatory arteritis and glomerular capillary thrombosis, more often associated with severe glomerular changes, are forerunners of an extremely poor prognosis (Bhathena *et al.* 1981; Kant *et al.* 1981).

Recent researches on the identification with monoclonal antibodies of the different subsets of inflammatory cells infiltrating the interstitium as well as the glomeruli have yielded conflicting results (Cameron 1988). T lymphocytes seem to predominate in the interstitium as well as in glomeruli. The extent of infiltration appears to correlate with the disease activity with a preponderance of CD4 subset in the more active forms (D'Agati *et al.* 1986; Castiglione *et al.* 1988). These studies which explore the possible role of cellular immunity in the pathogenesis of lupus nephritis and other studies on the role of immunoglobulin subclasses have stimulated new approaches to the clinicopathologic correlations of renal disease in SLE.

The role of renal biopsy in lupus nephritis

There is a wide divergence of opinion concerning the value of renal biopsy in SLE (McCluskey 1982). There are few doubts, however, that this procedure is often irreplaceable in assessing diagnosis and prognosis. Renal biopsy is the only tool which permits correct classification of lupus nephritis. Moreover, in several patients lupus is first recognized by renal histologic examination.

Much has been debated about the prognostic value of renal biopsy. Fries *et al.* (1978) concluded that histological examination does not provide better information than that obtained by clinical and serological data. However, a careful analysis of active and chronic changes at renal biopsy can actually increase the possibility of assessing the prognosis (Whiting-O'Keefe *et al.* 1982; Austin *et al.* 1984; Banfi *et al.* 1985), although the spontaneous or the drug-induced transformation of the histologic picture may dissipate the long-term prognostic value of these indices in individual patients.

In evaluating the prognostic significance of histologic examination the point in the course of the disease when biopsy is performed should also be taken into account. In untreated patients with recent onset of renal disease, biopsy provides information often imprecisely predicted by clinical observation alone. The assessment of the severity of the lesions together with the overall pathologic picture can help in forecasting prognosis and in deciding the initial therapeutic strategy. On the other hand, in a few cases renal biopsy may reveal a pathologic process other than that related to SLE and may dictate different forms of therapy.

Histologic examination may also be indicated during the course and the treatment of renal disease, either when there is a sudden worsening of renal function, or increasing or relapsing proteinuria or persistence of important renal abnormalities in spite of treatment. In these circumstances biopsy may pose some problems in the interpretation. Careful evaluation of potentially reversible versus irreversible lesions may guide clinical management. Treatment decisions are sometimes difficult in patients with lupus nephritis because of the chronic and protean nature of the disease and because of concern over the efficacy and toxicity of current medications. In selected patients renal biopsy, in conjunction with the entire clinical picture, may assist in identifying those who may or may not benefit from prolonged and hazardous treatments.

Prognosis

Patient survival

A progressive improvement in the survival rate of patients with lupus nephritis has been noted in recent years. While the life expectancy at 5 years was only 25 to 40 per cent in the decade 1960 to 1970 (Pollak *et al.* 1964; Baldwin *et al.* 1970) it has risen to 75 to 85 per cent in the period 1970 to 1980 (Donadio *et al.* 1978; Cameron *et al.* 1979) and ranges now between 85 and 95 per cent (Appel *et al.* 1987; Ponticelli *et al.* 1987). Earlier diagnosis and therapy as well as advances in specific and symptomatic treatment probably account for this better survival.

Several factors may influence the survival of patients with SLE. Epidemiologic studies identified black race (Kaslow and Masi 1978; Gordon *et al.* 1981), male sex (Wallace *et al.* 1981), and poor socioeconomic status (Studenski *et al.* 1987) as variables predictive of a worse outcome. There is more uncertainty about age. Some authors have reported a bad prognosis for children (Cassidy *et al.* 1977) but others did not (Caeiro *et al.* 1981). A benign course has been predicted for older patients (Baker *et al.* 1979) but other studies did not show evidence of diminished lupus-related mortality in the elderly (Ballou *et al.* 1982; Studenski *et al.* 1987). The severity of underlying renal disease may also influence survival (Appel *et al.* 1987).

Infections and lupus-related organ involvement, mainly active nephritis, are the most frequent causes of death in the first few years after diagnosis (Rosner *et al.* 1982; Rubin *et al.* 1985). In the long term, active SLE may persist or reappear in the course of the disease accounting for some deaths. However late mortality is mainly due to atherosclerotic heart disease. It has been estimated that the proportionate mortality from myocardial infarction is 10 times greater in SLE patients than in an age- and sex-matched population (Rosner et al. 1982). Corticosteroid therapy, hypertension, nephrotic syndrome hyperlipidaemia, and renal failure may all contribute to the development of an accelerated atherosclerosis. Hyperaggregable platelets, deficient fibrinolysis, and circulating antiphospholipid antibodies, which are frequent findings in patients with SLE (Cameron 1989), can expose to an increased risk of thrombosis. It has also been speculated that abundant circulating immune complexes may interact with other risk factors for cardiovascular disease (Correia et al. 1985). A less frequent cause of late mortality is represented by malignancy. An increased incidence of cancer has been reported in patients with SLE (Lewis et al. 1976). Prolonged immunosuppressive therapy can further expose these patients to the risk of malignancy (Penn 1981).

Kidney survival

Not only the patient survival rate but also the renal prognosis of patients with lupus nephritis has considerably improved over the years. While in the past most patients with kidney involvement died or developed end-stage renal failure within a few years (Pollak et al. 1964), a recent multicentre survey of 536 Italian patients with lupus nephropathy reported that about 80 per cent of patients still retained adequate renal function 10 years after diagnosis (Gruppo Italiano Studio Nefrite Lupica, unpublished data).

Some investigators found that neither hypertension nor renal dysfunction at presentation influenced the long-term outcome, while the presence of the nephrotic syndrome at initial renal biopsy was associated with an increased probability of renal failure developing (Appel et al. 1987). However, others reported that nephrotic syndrome, hypertension, or serum creatinine were significant predictors of outcome while urinary protein was not (Magil et al. 1988). The prognostic significance of histological lesions has already been discussed.

Therapy

Mild renal lesions

Patients with minor renal abnormalities (proteinuria less than 1 g per day, inactive urine sediment, normal renal function and blood pressure) and with minimal or mesangial lesions at kidney biopsy do not require specific therapy. However, periodical clinical surveillance of these patients is mandatory to allow early diagnosis and treatment of possible transformations. While renal disease remains clinically silent, therapeutic measures should be aimed at controlling the extrarenal activity of SLE. Exposure to ultraviolet light should be avoided to prevent skin lesions and reactivation of the disease. Arthralgias, arthritis, myalgias, and fever may respond to rest and salicylates in about 50 per cent of cases. Patients with SLE are, however, at risk of hepatitis and diminished renal function from salicylates which therefore need to be used with caution (Kimberly and Plotz 1977). Chloroquine (250–500 mg daily) or hydroxychloroquine (200–400 mg daily) have been used successfully for the same symptoms as well as for control of cutaneous manifestations (Rothfield 1988). These agents may produce dermatitis, nausea, or leucopenia in 5 to 10 per cent of patients. A rare but severe side-effect is retinopathy, initially characterized by small paracentral relative scotomas which can progress to a bilateral, permanent visual field abnormality (Easterbrook 1988). Periodic ophthalmic evaluation is therefore needed in patients taking these drugs. Non-steroidal anti-inflammatory drugs can be helpful in controlling arthralgias, myalgias, and fever. However, these agents should be handled with caution, as they may produce renal function deterioration through their interference with vasodilating prostaglandin production (Patrono and Pierucci 1986). When these measures are insufficient or contraindicated, small doses of prednisone (0.2–0.5 mg/kg body weight per day) may be used. Whenever possible corticosteroids should be given in a single dose, between 7 and 9 a.m., in order to reduce the side effects. Large doses of corticosteroids are recommended if severe extrarenal manifestations of SLE occur, namely cerebritis, carditis, pleuritis, haemolytic anaemia, leucopenia, and thrombocytopenia.

Focal proliferative glomerulonephritis

Whether to treat patients with focal lesions affecting only a minority of glomeruli is still controversial. If there are neither clinical nor severe histological features of disease activity, many clinicians prefer to spare powerful but toxic drugs and go on with symptomatic treatment alone. Others suggest the use of small doses of corticosteroid and/or cytotoxic agents to inhibit the immunological activity and to prevent potential transformation into more severe forms of lupus nephritis (Lee et al. 1984). Patients with diffuse segmental lesions, important proteinuria, elevated serum creatinine, and nephritic sediment should be treated similarly to those with diffuse proliferative lupus glomerulonephritis.

Membranous glomerulonephritis

There is general agreement that patients with membranous lupus nephritis, symptomless proteinuria, and stable renal function should be given only symptomatic treatment in order to control extrarenal disease. The management of patients with nephrotic syndrome is still controversial. Corticosteroids are usually ineffective in reducing proteinuria, and there is as yet little information about the usefulness of cytotoxic agents in reversing nephrotic syndrome. There is today controlled evidence that a 6-month regimen consisting of corticosteroids (1 g of methylprednisolone intravenously for 3 days followed by 0.5 mg/kg body weight per day of prednisone in months 1, 3, 5) and chlorambucil (0.2 mg/kg per day in months 2, 4, 6) can favour remission of the nephrotic syndrome and prevent renal dysfunction in patients with idiopathic membranous nephropathy (Ponticelli et al. 1989). In view of the strict similarities between the idiopathic and lupus forms of membranous glomerulonephritis a trial with such a schedule might be offered to patients with severe nephrotic syndrome. In patients with rapid deterioration of renal function a new biopsy should be carried out to detect conversion to diffuse proliferative glomerulonephritis or the development of concomitant interstitial nephritis or vasculitis. An aggressive treatment with high-dose intravenous methylprednisolone pulses and/or cytotoxic agents is warranted in the few cases which show a rapidly progressive impairment of renal function.

Diffuse proliferative and severe focal proliferative glomerulonephritis

These are the most serious forms of renal disease in SLE. Vigorous therapy is recommended when nephritic sediment, renal insufficiency, and/or nephrotic syndrome are present. There is more uncertainty for the cases of clinically-silent or mild lupus nephritis. Whether to treat these patients is still controversial. Probably low-dose corticosteroid and/or cytotoxic therapy are sufficient to prevent possible impairment of renal function without exposing an asymptomatic patient to the iatrogenic risks of a vigorous treatment. Aggressive approaches should be reserved for treating severe extrarenal flares or clinically active renal disease.

Several different therapeutical agents and schedules have been suggested for the management of severe lupus nephritis.

Corticosteroids

Even today corticosteroids represent the cornerstone for the treatment of SLE, but how to give these agents in severe lupus nephritis is a matter of debate. In the early 1960s, Pollak *et al.* (1964) reported that low-dose prednisone was unable to interfere with the downhill course of diffuse proliferative lupus glomerulonephritis, while high-dose prednisone could improve the outcome, at least in some patients. Since then, many clinicians start treatment with prednisone at doses of at least 1 mg/kg body weight per day whenever a diagnosis of diffuse lupus nephritis is made. The doses are tapered only when renal disease and serological parameters improve. This may happen only after several months in some patients, and others may continue to manifest active and progressive renal disease which is often refractory to an increased dosage of prednisone. On the other hand, patients with milder forms of lupus nephritis who could be treated with lower doses are equally exposed to the iatrogenic toxicity of vigorous corticosteroid therapy. As a consequence, many patients given high-dose prednisone suffer from devastating steroid-related morbidity—which includes hypertension, infection, accelerated atherosclerosis, obesity, diabetes, aseptic bone necrosis, cataracts, and myopathy—and some 20 to 30 per cent of patients die within 5 years either from active SLE or from infection (Rubin *et al.* 1985). It is now clear that protracted high-dose prednisone should be avoided, at least in patients with less severe disease. In these patients initial prednisone dosage should not exceed 1 mg/kg body weight per day and should be tapered after 4 to 8 weeks. Patients with good clinical response should be allocated to low-dose maintenance therapy and switched to an alternate-day regimen when possible. For patients resistant to this schedule alternative approaches can be tried.

Today, a number of units treat diffuse lupus nephritis and renal flares with high-dose steroid 'pulses' (Kimberly 1982). Intravenous pulse methylprednisolone in doses of up to 1 g per day is given for 3 or more consecutive days, followed by oral prednisone in decreasing doses. Pulse therapy may obtain rapid resolution of clinical extrarenal symptoms and a slower improvement in serologic activity. Serum creatinine may rapidly decrease, particularly in patients who had rapid deterioration of their renal function but in other cases renal function tends to ameliorate slowly. Reduction of urinary protein excretion is also variable but usually occurs more slowly than the improvement in renal function, often after several weeks or some months (Ponticelli *et al.* 1982). There is a general impression that once remission has been achieved with methylprednisolone pulse therapy it can be maintained by low-dose prednisone, preventing most steroid-related side-effects.

Exceptional cases of sudden death, arrhythmias and anaphylaxis have been reported in severely ill organ transplant recipients after rapid injections of large doses of methylprednisolone. The relationship between these complications and steroid administration is not well documented. However, it is advisable to infuse the amount of methylprednisolone over at least 30 min and into a peripheral rather than a central vein to minimize these possible side-effects. Hyperglycaemia and a transient hypercoagulable state may also occur after methylprednisolone pulse therapy. On the other hand, steroid-related side-effects in the long term are considerably less numerous and less severe with a strategy based on methylprednisolone pulse therapy and low-dose maintenance prednisone than with prolonged high-dose oral prednisone (Kimberly 1982; Ponticelli *et al.* 1987). The mechanisms responsible for the enhanced affectiveness of pulse therapy are incompletely understood. The excess steroid may intercalate into the lipid bilayer of target cell membranes. The ensuing decrease in membrane fluidity blocks some enzymatic activities causing a marked reduction in inflammatory cell function (Jacob 1985). Moreover, at very high concentrations methylprednisolone can exert some anti-inflammatory and immunosuppressive effects not observed with standard doses (Kimberly 1982) and can inhibit complement activation (Weiler and Packard 1982).

In summary, in view of the need for protracted corticotherapy in most SLE patients schedules which maximize the therapeutic index of steroids should be preferred. These could be based upon short courses of intravenous high-dose methylprednisolone whenever renal and extrarenal flares occur. The maintenance treatment with oral prednisone should be kept at the lowest possible dosage: it is worth pointing out that many patients with lupus nephritis can maintain stable remission with 0.2 to 0.5 mg/kg body weight per day oral prednisone. An alternate-day regimen may further protect from steroid toxicity (Axelrod 1976). On the other hand, in patients with persisting activity of SLE combination with a cytotoxic agent and/or further course of steroid pulse therapy may be considered.

Cytotoxic agents

Several cytotoxic drugs which can interfere with the immune and the inflammatory response have been tried in SLE. However, data on some agents, such as methotrexate and chlorambucil, are scanty and still inconclusive, and most studies have concentrated on azathioprine and cyclophosphamide. Conflicting results have been reported from randomized trials comparing the effects of prednisone alone versus prednisone plus azathioprine or cyclophosphamide. Many of these studies, however, involved too small a number of patients to reach adequate statistical power. In order to achieve a sufficient number of patients for analysis, Felson and Anderson (1984) pooled the data of eight controlled trials in which patients had been randomly assigned to receive either prednisone alone or prednisone plus azathioprine or cyclophosphamide. They observed that patients who had been given cytotoxic therapy had significantly less deterioration of renal function and were less likely to have end-stage renal disease or to die from renal failure than patients given steroid alone. However, the cumulative mortality from non-renal causes was similar in the two groups, possibly as the result of more severe toxicity of the combined therapy. It is interesting that this pooled analysis showed that azathioprine preserved renal function more effectively than cyclophosphamide and was also associated with a lower rate of non-renal deaths.

The role of cytotoxic agents has been carefully studied in a prospective, randomized fashion by the group of the National Institute of Health at Bethesda (Austin *et al.* 1986). A cohort of

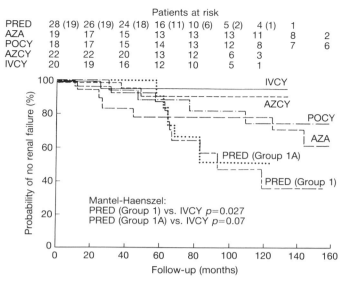

Fig. 6 Probability of maintaining life-supporting renal function in 107 patients with active lupus nephritis, according to treatment group. PRED denotes prednisone, AZA azathioprine, POCY oral cyclophosphamide, AZCY combined oral azathioprine and cyclophosphamide, and IVCY intravenous cyclophosphamide. Under the heading 'Patients at risk', the number in parentheses refers to Group 1A. The 95 per cent confidence limits at 7 years (median follow-up) for Groups 1, 1A, and 2 to 5 were 0.83 to 0.35, 0.84 to 0.26, 0.98 to 0.58, 1.00 to 0.62, 1.00 to 0.76, and 1.00 to 0.84, respectively. (From Austin, H.A., III, et al. (1986). *New England Journal of Medicine*, **314**, 616, with permission.)

111 patients with lupus nephritis was randomly allocated to receive one of the following treatments.

1. Prednisone, at the initial dose of 1 mg/kg body weight per day up to 8 weeks, followed by tapering to alternate-day doses. Flares of SLE were treated with additional cycles of high-dose prednisone.
2. Azathioprine up to 4 mg/kg body weight per day.
3. Oral cyclophosphamide up to 4 mg/kg body weight per day.
4. Combined oral azathioprine and cyclophosphamide up to 1 mg of each per kg body weight per day.
5. Intravenous infusion over 30 to 60 min of cyclophosphamide every 3 months, at initial doses of 0.75 g/m^2.

If the leucocyte nadir did not fall below 4000/mm^3 the subsequent doses were raised to a maximum of 1 g/m^2. If the nadir was lower than 2000 the subsequent dose of cyclophosphamide was reduced to 0.5 g/m^2. In all the groups treated with cytotoxic agents moderate doses of prednisone, 0.5 mg/kg body weight per day, were also given. The conclusions of this important study were:

1. The risk of developing renal failure increased in patients treated with prednisone alone after approximately 5 years of follow-up (Fig. 6).
2. Although more favourable outcomes were observed in patients assigned to any of the cytotoxic-drug regimens, only the comparison of patients treated with intravenous pulse cyclophosphamide versus prednisone reached statistical significance.
3. Five cases of cancer occurred in patients treated with prolonged oral cytotoxic therapy while no case of malignancy developed in patients treated with intermittent pulses of cyclophosphamide.
4. Progressive bone marrow depression, gonadal toxicity, and haemorrhagic cystitis occurred only in patients treated with oral cytotoxic drugs.
5. The risk of infections was greater for patients given prednisone alone.

A substantial amelioration of severe SLE with monthly intravenous administration of cyclophosphamide for 6 months has been observed in another study (McCune et al. 1988). Unfortunately, follow-up was too short to allow any evaluation of this therapy in the long term. Treatment was accompanied by a progressive decline in CD3, CD4, CD8, and B cells. After discontinuation of therapy, B cells tended to return to baseline values while T-cell subsets showed persistent decreases. T-cell proliferative responses, however, were similar to entry values.

It is still unclear whether the effects of intravenous cyclophosphamide on lymphocyte markers and functions may predispose patients to severe infection or neoplasia in the long term. In spite of the demonstrated efficacy of intermittent pulse cyclophosphamide, many clinicians still prefer to use oral cytotoxic agents combined with low doses of prednisone. Excellent long-term results have been reported with the protracted use of azathioprine limiting the administration of high-dose steroids to clinical flares (Cameron et al. 1979; Ponticelli et al. 1987).

In summary, it seems from the studies available that immunosuppressive drugs and steroids together are more effective than prednisone alone, and may permit lower doses of prednisone to be used in the long term. According to Felson and Anderson (1984), trials which failed to reach this conclusion were probably biased by inadequate sample sizes. Among oral cytotoxic agents azathioprine is safer and may be more effective than cyclophosphamide, and should be preferred for long-term treatment, being probably less oncogenic. When given intermittently by intravenous 'pulses' cyclophosphamide is only marginally more effective than oral cytotoxic drugs, but seems better tolerated in the long term. On the other hand, nausea and vomiting are common within the first days after pulse cyclophosphamide, and a variable alopecia is extremely frequent. An increased risk of herpes zoster infection may also be predicted for patients given this therapy. Large intravenous hydration is needed during the first 24 h to assure polyuria and prevent haemorrhagic cystitis.

Plasmapheresis

In theory, repeat plasmapheresis could remove pathogenic immune complexes, inflammatory mediators, and toxic antibodies. Several uncontrolled studies reported that plasmapheresis may benefit some patients with severe lupus nephritis. It must be emphasized, however, that in most studies patients were also treated aggressively with steroid and cytotoxic agents. It is therefore difficult to separate the effects of pharmacological treatments from those of plasmapheresis. Two controlled trials failed to show any favourable effect of plasma exchange in either patients with mild SLE (Wei et al. 1983) or in patients with severe diffuse lupus nephritis (Lewis and Lachin 1986). Although plasmapheresis might be useful in controlling severe cerebritis or devastating cutaneous disorders of SLE, its indications in lupus nephritis should be reappraised. However, this therapy might have some role in the rare cases of haemolytic uraemic syndrome complicating SLE (Gelfand et al. 1985; Kincaid-Smith et al. 1988).

Cyclosporin

In autoimmune (NZB x NZW) F1 mice cyclosporin prevents the deposition of immune complexes in the kidneys and the subsequent development of glomerulonephritis and proteinuria (Gunn and Ryffel 1986). Experience with this drug in human SLE is still preliminary. Miescher et al. (1987) gave cyclosporin (at a dose of 5 mg/kg per day) together with steroids to 14 SLE patients in whom the disease could not be controlled with a combined steroid–antimetabolite treatment. With the exception of two patients who did not respond, the combined regimen was followed by diminution of the activity scores, improved kidney function, and reduced proteinuria. No unequivocal signs of renal toxicity could be detected after a mean follow-up of 22 months. Favourable clinical results have also been obtained in another study, in which 13 patients were treated for 9 to 18 months with cyclosporin, 5 mg/kg.day. However, the clinical improvement was not associated with any serologic change. Six patients showed reversible signs of nephrotoxicity, and eight developed hypertension. Interruption of cyclosporin was followed by relapse or worsening of SLE in five patients (Feutren et al. 1988). Although these results must be confirmed by larger controlled studies, with longer follow-up the use of cyclosporin as steroid-sparing agent might offer new alternatives to the current therapeutic strategies.

Selective thromboxane antagonists

Vasoactive eicosanoids have a role in modulating renal haemodynamics in patients with lupus nephritis. In particular, thromboxane A_2 could exert vasoconstrictive and contractile mesangial action (Mené et al. 1988). The infusion of a selective thromboxane antagonist in patients with lupus nephritis increases both glomerular filtration rate and renal blood flow by about 25 per cent. However, low-dose aspirin, which inhibits thromboxane production from platelets, does not affect renal function. This suggests that cells other than platelets are the source of thromboxane A_2 which impairs renal function (Pierucci et al. 1989). Safe and long-lasting receptor antagonists or syntethase inhibitors are now becoming available and may have some therapeutic applications in lupus nephritis.

Total lymphoid irradiation

Total lymphoid irradiation is a potent immunosuppressive regimen which can induce long-term remission in several experimental autoimmune diseases, including lupus-like syndrome in (NZB x NZW) F1 mice (Kotzin and Strober 1979). In a recent study, 15 patients with diffuse proliferative glomerulonephritis, moderate or high chronicity index at renal biopsy, and nephrotic syndrome were treated with total lymphoid irradiation, total dose 20 Gy during a 4 to 6 week interval (Strober et al. 1987). One patient committed suicide and another died from progressive renal failure. The other 13 patients were alive with stable renal function. Six of them completely stopped steroid therapy 18 to 65 months after irradiation but in two it was necessary to reinstitute prednisone because of reactivation of renal disease. Localized infections and irregular menstruation were the most frequent side effects. The oncogenicity of this regime in lupus over 10 to 20 years is as yet unknown.

Anticoagulants

Platelet involvement and microthrombosis have been reported in some cases of severe lupus nephritis. Because of these findings heparin and antiplatelet agents have been used in patients with acute renal function deterioration (Ponticelli et al. 1974; Cameron et al. 1979). Beneficial results have been reported but controlled evidence supporting favourable effects of this treatment is lacking.

Pollak and coworkers have suggested the use of ancrod in SLE with thrombosis. Ancrod is a thrombin-like enzyme from Malayan pit viper venom which splits only fibrinopeptide A from the fibrinogen molecule and does not activate factor XIII. In many patients with SLE and glomerular and vascular microthrombi, intravenous infusion of ancrod resulted in prompt normalization of the impaired fibrinolytic system. In these patients serial histologic studies showed a striking decrease or disappearance of microvascular thrombosis while thrombosis persisted in the fibrinolysis non-responders (Kim et al. 1988).

Androgens

Steroid sex hormones greatly influence normal immune mechanisms and autoimmune disease, with androgen suppressing and oestrogens augmenting. Androgens can exert a therapeutic effect in some murine models of lupus (Talal 1986). In SLE patients there is a deviation in the normal metabolism of oestradiol with retained oestrogenic activity. There is also evidence for decreased total androgens. The possible utilization of androgens in the treatment of human autoimmune diseases is currently under study.

Practical approach to therapy of diffuse proliferative lupus nephritis

In choosing the therapeutic strategy for patients with diffuse proliferative lupus nephritis, the clinician should remember that an aggressive treatment is needed in patients with high clinical or histological activity in order to prevent irreversible deterioration of renal function, and that in most patients therapy should be continued for years or even decades in order to control the activity of SLE. Some physicians face these problems by following fixed therapeutic regimens drawn from protocols originally studied for controlled trials. This choice has the advantage of using an extensively investigated therapy, of which the benefits and the side-effects are well recognized, at least in the short term. The potential disadvantage of fixed schedules is unnecessary overtreatment in patients with milder disease. Others prefer a more flexible strategy based on short courses of vigorous therapy during flares of the disease and on low-dose treatment during quiescent phases. This choice may be biased by the subjectivity of the clinician but can allow better modulation of therapy accordingly to the activity of SLE.

Among the fixed schedules that of the every-3-months intravenous pulse cyclophosphamide type seems to be safer and more effective than other protocols. It has not been established how long this therapy should be given. Balow (1986) suggested that cyclophosphamide pulses should be continued for approximately 5 years or until a remission of lupus nephritis has been sustained for 2 years. The long-term risks of such prolonged treatment with a cytotoxic agent are still to be assessed. The group of the National Institute of Health is comparing the results of monthly intravenous pulses of cyclophosphamide for 6 months to those with the same therapy followed by a maintenance regimen of pulse cyclophosphamide given every 3 months for an additional 2

Table 8 A suggested flexible strategy for treatment of lupus nephritis

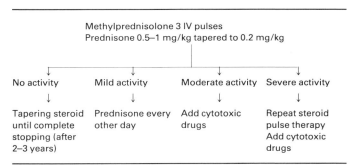

years. No data from this ongoing trial are yet available; it would offer the advantage of a shorter (and therefore less dangerous) period of administration of cyclophosphamide.

Among the flexible strategies we prefer the administration of three consecutive intravenous methylprednisolone pulses of 1 g each at the diagnosis of diffuse lupus nephritis (Table 8). The course is repeated whenever a renal flare-up (arbitrarily defined as a rapid increase in serum creatinine of at least 30 per cent over the basal values, and/or an increase in proteinuria of at least 2 g per day or a doubling in the case of nephrotic proteinuria) or an extrarenal exacerbation (fever with arthralgias and serositis, cerebritis) occurs. According to the clinical activity of the disease after pulse therapy, oral prednisone is given in a single morning dose of 0.5 to 1 mg/kg per day and progressively tapered to the lowest dose possible. Cyclophosphamide, 1 mg/kg per day, if therapy is scheduled for a few weeks or azathioprine, 1 mg/kg per day, if more prolonged treatment is foreseen are added in patients with more active disease. It is interesting that when two large series in which patients with diffuse proliferative lupus nephritis were treated according to either of these different therapeutic philosophies were compared, no difference could be seen between the two studies in the 10-year pure kidney survival rates which ranged around 90 per cent (Austin et al. 1986; Ponticelli et al. 1987). In the rare patients who do not respond to these therapies, or have contraindications to corticosteroid and/or cytotoxic agents, a cautious trial with cyclosporin is justified.

Since protracted therapy with corticosteroids or cytotoxic agents may expose the patient to the risk of invalidating or even life-threatening complications, it is important to know whether and when therapy can eventually be stopped in patients with diffuse proliferative lupus nephritis. Trials of abrupt cessation of treatment resulted in severe 'flares', and deterioration of kidney function (Aptekar et al. 1972). However in patients with stable clinical remission a careful and gradual decrease of drug dosage until complete withdrawal has been successful (Ponticelli et al. 1988). After interruption of therapy 13 of 14 patients showed stable renal function, stable or improved proteinuria, and no clinical flares after a mean follow-up of 3 years. In the remaining patient a nephrotic syndrome reappeared after more than 2 years but completely reversed after reinstitution of therapy. In eight patients low levels of serum C3 or elevated DNA-binding activity persisted or reappeared after stopping therapy.

End-stage renal disease
Dialysis
Even during dialysis treatment patients with SLE maintain the characteristics of a variable and erratic disease. Some patients, usually those who had a slowly progressive renal dysfunction and inactive disease before dialysis, are easy to manage, may tolerate decrease of immunosuppression to minimal levels (Correia et al. 1984) and have life-expectancy comparable to that of dialysis patients with disorders other than SLE (Coplon et al. 1983). Another subset of patients, more often those who reached dialysis after a rapidly progressive course, maintain clinical and serologic activity (Correia et al. 1984). Many of them die shortly after starting dialysis but a few patients may recover renal function and discontinue dialysis (Kimberly et al. 1983). The choice of therapy in these instances is critical. An aggressive treatment might favour a recovery of renal function in a consistent percentage of patients and might prevent deaths from active lupus. On the other hand, uraemia increases the risks of iatrogenic complications, particularly infection and bone marrow suppression. There are no rules that may substitute for clinical practice in these difficult instances.

After the first months, patients generally do well whether they are treated with haemodialysis or peritoneal dialysis (Cantaluppi 1988). Clinical and serologic manifestations of SLE improve, probably as a result of the reduction in immune responsiveness induced by uraemia, and corticosteroid, as well as cytotoxic agents, can be completely stopped. Rehabilitation is excellent and many patients return to normal physical activity. However some patients may present clinical flares, including fever, rash, myalgias, serositis, cerebritis, and haematological abnormalities usually associated with serologic activity. These cases seem to be more frequent in patients undergoing peritoneal dialysis, perhaps as an effect of better immune reactivity (Rodby et al. 1987), but they can also occur in haemodialysis patients (Sires et al. 1989). These patients require maintenance prednisone therapy which exposes them to increased risks of infection and atherogenesis. Exceptionally the clinical activity is particularly severe and refractory to corticosteroids. No specific dialysis-related problems have been reported in SLE patients with the possible exceptions of an increased risk of fistula thrombosis in some patients. More frequent peritonitis in heavily immunosuppressed peritoneal dialysis patients may also be expected.

Renal transplantation
The results of cadaveric renal transplantation in SLE patients are comparable to those obtained in patients with other diseases. Renal transplantation may therefore be recommended in SLE (Roth et al. 1987). It may be advisable, however, to postpone renal transplantation by at least 1 year after entry to dialysis both to allow a 'wash-out' of previous corticosteroid and cytotoxic therapy in a patient who has been, and will be, immunosuppressed for years and to wait for a potential reversibility of renal failure. Lupus remains inactive in the majority of renal transplant recipients (Correia et al. 1984). Very few cases of recurrent lupus nephritis have been reported. They generally occur in patients with persistent serologic and clinical activity and do not affect renal allograft function (Kumano et al. 1987).

A low graft survival rate has been reported in SLE patients who received transplanted kidneys from their parents (Cats et al. 1984). This has been related to increased organ vulnerability, perhaps related to disease susceptibility. This study was based on only 20 patients which may have biased the results. However a possible lower rate of success should be discussed with patient and donor in cases of parental donor grafts.

References

Abrass, C.K., Nies, K.M., Louie, J.S., Border, W.A., and Glassock, R.J. (1980). Correlation and predictive accuracy of circulating immune complexes with disease activity in patients with systemic lupus erythematosus. *Arthritis and Rheumatism*, 23, 273–82.

Adu, D., et al. (1983). Late onset systemic lupus erythematosus and lupus-like disease in patients with apparent idiopathic glomerulonephritis. *Quarterly Journal of Medicine*, 52, 471–87.

Alpers, C.E., Hopper, J.Jr., Bernstein, M.J., and Biava, C.G. (1984). Late development of systemic lupus erythematosus in patients with glomerular 'finger-print' deposits. *Annals of Internal Medicine*, 100, 66–8.

Arnett, F.C. (1987). Familial SLE, the HLA system and the genetics of lupus erythematosus. In *Dubois Lupus Erythematosus* (3rd. edn) (eds. D.J. Wallace and E.L. Dubois), pp. 161–84. Lea & Febiger, Philadelphia.

Appel, G.B., Cohen, D.J., Pirani, C.L., Meltzer J., and Estes, D. (1987). Long-term follow-up of patients with lupus nephritis. *American Journal of Medicine*, 83, 877–85.

Appel, A.E., Sablay, L.B., Golden, R.A., Barland, P., Grayzel, A.I., and Bank, N. (1978). The effect of normalization of serum complement and anti-DNA antibody on the course of lupus nephritis. *American Journal of Medicine*, 64, 274–83.

Aptekar, R.G., Decker, J.L., and Steinberg, A.D. (1972). Exacerbation of SLE nephritis after cyclophosphamide withdrawal. *New England Journal of Medicine*, 286, 1159–60.

Austin, H.A. III, et al. (1983). Prognostic factors in lupus nephritis. *American Journal of Medicine*, 75, 382–91.

Austin, H.A. III, et al. (1986). Therapy of lupus nephritis: controlled trial of prednisone and cytotoxic drugs. *New England Journal of Medicine*, 314, 614–19.

Austin, H.A. III, Muenz, L.R., Joyce, K.M., Antonovych, T.T., and Balow, J.E. (1984). Diffuse proliferative lupus nephritis: identification of specific pathologic features affecting renal outcome. *Kidney International*, 25, 689–95.

Axelrod, L. (1976). Glucocorticoid therapy. *Medicine*, 55, 39–65.

Baehr, G., Klemperer, P., and Schiffrin, A. (1935). A diffuse disease of the peripheral circulation usually associated with lupus erythematosus and endocarditis. *Transactions of the Association of American Physicians*, 50, 139–42.

Baker, S.B., Rovira, J.R., Campion, E.W., and Mills, J.A. Late onset of systemic lupus erythematosus (1979). *American Journal of Medicine*, 66, 727–32.

Baldwin, D.S. (1982). Clinical usefulness of the morphology classification of lupus nephropathy. *American Journal of Kidney Diseases*, 2 (suppl.), 142–9.

Baldwin, D.S., Lowenstein, J., Rothfield, N.F., Gallo, G., and McCluskey, R.T. (1970). The clinical course of the proliferative and membranous forms of lupus nephritis. *Annals of Internal Medicine*, 73, 929–42.

Ballou, S.P., Khan, M.A., and Kushner, I. (1982). Clinical features of systemic lupus erythematosus. *Arthritis and Rheumatism*, 25, 55–60.

Balow, J.E. (1986). Lupus nephritis: natural history, prognosis and treatment. *Clinics in Immunology and Allergy*, 6, 353–66.

Banfi, G., et al. (1985). Morphologic parameters in lupus nephritis: their relevance for classification and relationship with clinical and histological findings and outcome. *Quarterly Journal of Medicine*, 55, 153–8.

Bhathena, D.B., Sobel, B.J., and Migdal, S.D. (1981). Noninflammatory renal microangiopathy of systemic lupus erythematosus. (Lupus vasculitis). *American Journal of Nephrology*, 1, 144–59.

Bobrie, G., Liote, F., Houillier, P., Grünfeld, J.P., and Jungers, P. (1987). Pregnancy in lupus nephritis and related disorders. *American Journal of Kidney Diseases*, 9, 339–43.

Brentjens, J.R., et al. (1975). Interstitial immune complex nephritis in patients with systemic lupus erythematosus. *Kidney International*, 7, 342–50.

Caeiro, F., Michielson, F.M.C., Bernstein, R., Hughes, G.R.V., and Ansell, B.M. (1981). Systemic lupus erythematosus in childhood. *Annals of the Rheumatic Diseases*, 40, 325–31.

Cairns, S.A., London, A., and Mallick, N.P. (1980). The value of three immune complex assays in the management of systemic lupus erythematosus: an assessment of immune complex levels, size and immunochemical properties in relation to disease activity and manifestations. *Clinical and Experimental Immunology*, 40, 273–82.

Cameron, J.S. (1984). Nuovi aspetti del lupus eritematoso sistemico. *Giornale Italiano di Nefrologia*, 1, 63–72.

Cameron, J.S. (1988). Aspects immunologiques des néphrites tubulo-interstitielles primitives et secondaires. In *Actualités Néphrologiques de l'Hôpital Necker* (eds. J. Crosnier, J.L. Funck Brentano, J.F. Bach, and J.P. Grünfeld), pp. 223–61. Flammarion, Paris.

Cameron, J.S. (1989). The treatment of lupus nephritis. *Pediatric Nephrology*, 3, 350–62.

Cameron, J.S., Lessof, M.H., Ogg, C.S., Williams, B.D., and Williams, D.G. (1976). Disease activity in the nephritis of systemic lupus erythematosus in relation to serum complement concentrations. DNA-binding capacity and precipitating anti-DNA antibody. *Clinical and Experimental Immunology*, 25, 418–27.

Cameron, J.S., et al. (1979). Systemic lupus with nephritis: a long-term study. *Quarterly Journal of Medicine*, 48, 1–24.

Cantaluppi, A. (1988). CAPD and systemic diseases. *Clinical Nephrology*, 30(S.1), S8–S12.

Carette, S., et al. (1983). Controlled studies of oral immunosuppressive drugs in lupus nephritis. *Annals of Internal Medicine*, 99, 1–8.

Caruana, R.J., Barish, C.F., and Buckalew, V.M. (1985). Complete distal renal tubular acidosis in systemic lupus: clinical and laboratory findings. *American Journal of Kidney Diseases*, 6, 59–63.

Cassidy, J.T., Sullivan, D.B., Petty, R.E., and Ragsdale, C. (1977). Lupus nephritis and encephalopathy. Prognosis in 58 children. *Arthritis and Rheumatism*, 20, 315–21.

Castiglione, A., Bucci, A., Fellin, G., D'Amico, G., and Atkins, R.C. (1988). The relationship of infiltrating renal leucocytes to disease activity in lupus or cryoglobulinemic glomerulonephritis. *Nephron*, 50, 14–23.

Catoggio, L.J., Skinner, R.P., Smith, G., and Maddison, P.J. (1984). Systemic lupus erythematosus in the elderly: clinical and serologic characteristics. *Journal of Rheumatology*, 11, 175–81.

Cats, S., Terasaki, P.I., Perdue, S., and Mickey, M.R. (1984). Increased vulnerability of the donor organ in related kidney transplants for certain diseases. *Transplantation*, 37, 575–9.

Cavallo, T., Cameron, W.R., and Lapenas, D. (1977). Immunopathology of early and clinically silent lupus nephropathy. *American Journal of Pathology*, 87, 1–15.

Churg, J. and Sobin, L.H. (1982). Lupus nephritis. In *Renal Disease* (eds. J. Churg and L.H. Sobin), pp. 127–31. Igaku-Shoin, Tokyo.

Cohen, A.S., et al. (1971). Preliminary criteria for the classification of systemic lupus erythematosus. *Bulletin of Rheumatic Diseases*, 21, 643–8.

Coplon, N.S., Diskin, C.J., Petersen, J., and Swenson, R.S. (1983). The long-term clinical course of systemic lupus erythematosus in end-stage renal disease. *New England Journal of Medicine*, 308, 186–90.

Correia, P., Cameron, J.S., Ogg, C.S., Williams, D.G., Bewick, M., and Hicks, J.A. (1984). End-stage renal failure in systemic lupus erythematosus with nephritis. *Clinical Nephrology*, 22, 293–302.

Correia, P., et al. (1985). Why do patients with lupus nephritis die? *British Medical Journal*, 1, 126–31.

D'Agati, V., Appel, G.B., Estes, D., Knowles II, D.M., and Pirani, C.L. (1986). Monoclonal antibody identification of infiltrating mononuclear leukocytes in lupus nephritis. *Kidney International*, 30, 573–81.

Donadio, J.V., Jr., Holley, K.E., Ferguson, R.H., and Ilstrup, D.M. (1978). Treatment of lupus nephritis with prednisone and combined prednisone and cyclophosphamide. *New England Journal of Medicine*, 299, 1151–5.

4.6.2 Systemic lupus erythematosus (clinical)

Dubois, E.L. and Wallace, D.J. (1987). Drugs that exacerbate and induce systemic lupus erythematosus. In *Dubois' Lupus Erythematosus* (eds. D.J. Wallace and E.L. Dubois), pp. 450–69. Lea & Febiger, Philadelphia.

Dubois, E.L. and Tuffanelli, D.L. (1964). Clinical manifestations of systemic lupus erythematosus. Complete analysis of 520 cases. *Journal of the American Medical Association*, **190**, 104–9.

Dujovne, I., Pollak, V.E., Pirani, C.L., and Dillan, M.C. (1972). The distribution and character of glomerular deposits in systemic lupus erythematosus. *Kidney International*, **2**, 33–50.

Easterbrook, M. (1988). Ocular effects and safety of antimalarial agents. *American Journal of Medicine*, **85(4A)**, 23–9.

Eiser, A., Katz, S., and Swartz, C. (1979). Clinically occult diffuse proliferative lupus nephritis. An age-related phenomenon. *Archives of Internal Medicine*, **139**, 1022–5.

Emery, H. (1986). Clinical aspects of systemic lupus erythematosus in childhood. *Pediatric Clinics of North America*, **33**, 1177–90.

Epstein, W.V. (1977). Laboratory tests in rheumatic disease. *Medical Clinics of North America*, **61**, 377–87.

Estes, D. and Christian, C.L. (1971). SLE: a prospective analysis. *Medicine*, **50**, 85–95.

Felson, D.T. and Anderson, J. (1984). Evidence for the superiority of immunosuppressive drugs and prednisone over prednisone alone in lupus nephritis. *New England Journal of Medicine*, **311**, 1528–33.

Fessel, W.J. (1974). Systemic lupus erythematosus in the community. *Archives of Internal Medicine*, **134**, 1027–31.

Feutren, G., *et al.* (1987). Effects of cyclosporine in severe systemic lupus erythematosus. *Journal of Pediatrics*, **111**, 1063–8.

Font, J., Torras, A., Cervera, R., Darnell, A., Revert, L., and Ingelmo, M. (1987). Silent renal disease in systemic lupus erythematosus. *Clinical Nephrology*, **27**, 283–8.

Foucar, E., Erickson, D.G., and Tung, S.K. (1979). Glomerulonephritis in procainamide-induced lupus erythematosus. Report of a case and review of the literature. *Journal of Clinical and Laboratory Immunology*, **2**, 79–84.

Fries, T.F., Pova, J., and Liang, M.H. (1978). Marginal benefit of renal biopsy in systemic lupus erythematosus. *Archives of Internal Medicine*, **138**, 1386–9.

Gelfand, J., Truong, L., Stern, L., Pirani, C.L., and Appel, G.B. (1985). Thrombotic thrombocytopenic purpura syndrome in systemic lupus erythematosus: treatment with plasma infusion. *American Journal of Kidney Diseases*, **6**, 154–60.

Gladman, D.D., Urowitz, M.B., and Keystone, E.C. (1979). Serologically active clinically quiescent systemic lupus erythematosus. *American Journal of Medicine*, **66**, 210–15.

Gordon, M.F., Stolley, P.D., and Schinnar, R. (1981). Trends in recent systemic lupus erythematosus mortality rates. *Arthritis and Rheumatism*, **24**, 762–9.

Grishman, E. and Churg, J. (1982). Focal segmental lupus nephritis. *Clinical Nephrology*, **17**, 5–13.

Grishman, E., Porush, J.G., Lee, S.L., and Churg, J. (1973). Renal biopsies in lupus nephritis: correlation of electron microscopic findings with clinical course. *Nephron*, **10**, 25–36.

Gunn, H.C. and Ryffel, B. (1986). Successful treatment of autoimmunity in (NZB x NZW) F1 mice with cyclosporin and (Nva)-cyclosporin: II Reduction of glomerulonephritis. *Clinical and Experimental Immunology*, **64**, 234–42.

Hargraves, M.M., Richmond, H., and Morton, R. (1948). Presentation of two bone marrow elements: the 'tart' cell and the 'LE' cell. *Proceedings of the Staff Meetings of Mayo Clinic*, **23**, 25–33.

Hess, E. (1988). Drug-related lupus. *New England Journal of Medicine*, **318**, 1460–2.

Hill, G. (1983). Systemic lupus erythematosus and mixed connective tissue disease. In *Pathology of the Kidney* (ed. R.H. Heptinstall), pp. 839–906. Little Brown, Boston.

Hochberg, M.C., *et al.* (1985a). The incidence of systemic lupus erythematosus in Baltimore, Maryland 1970–1977. *Arthritis and Rheumatism*, **28**, 80–6.

Hochberg, M.C., *et al.* (1985b). Systemic lupus erythematosus. A review of clinico-laboratory findings and immunogenetic markers in 150 patients with emphasis on demographic subsets. *Medicine*, **64**, 285–93.

Hughes, G.R.V. (1983). Thrombosis, abortion, cerebral disease and the lupus anticoagulant. *British Medical Journal*, **27**, 1088–9.

Hughes, G.R.V. (1988). An immune mechanism in thrombosis. *Quarterly Journal of Medicine*, **69**, 753–4.

Imbasciati, E., *et al.* (1984). Lupus nephropathy and pregnancy. A study of 26 pregnancies in patients with systemic lupus erythematosus and nephritis. *Nephron*, **36**, 46–51.

Jacob, H.S. (1985). Pulse steroid in hematologic diseases. *Hospital Practice*, **8**, 87–94.

Jonsson, H., *et al.* (1988). The effect of age on clinical and serologic manifestations in unselected patients with systemic lupus erythematosus. *Journal of Rheumatology*, **15**, 505–9.

Kant, K.S., Pollak, V.E., Weiss, M.A., Glueck, H.I., Miller, M.A., and Hess, E.V. (1981). Glomerular thrombosis in systemic lupus erythematosus: prevalence and significance. *Medicine*, **60**, 71–86.

Kaposi, M.K. (1872). Neue Beitrage zur Kenntniss des Lupus Erythematosus. *Archiv von Dermatologie und Syphilogie*, **4**, 36–47.

Kashgarian, M. (1982). New approaches to clinical pathologic correlation in lupus nephritis. *American Journal of Kidney Diseases*, **1 (suppl.)**, 164–9.

Kaslow, R.A. and Masi, A.T. (1978). Age, sex, and race effects on mortality from systemic lupus erythematosus in the United States. *Arthritis and Rheumatism*, **21**, 473–9.

Kim, S., *et al.* (1988). Fibrinolysis in glomerulonephritis treated with ancrod: renal functional, immunologic and histopathologic effects. *Quarterly Journal of Medicine*, **69**, 879–95.

Kimberly, R.P. (1982). Pulse methylprednisolone in SLE. *Clinics in Rheumatic Diseases*, **8**, 261–78.

Kimberly, R.P. and Plotz, P. (1977). Aspirin-induced depression of renal function. *New England Journal of Medicine*, **296**, 418–24.

Kimberly, R.P., Lockshin, M.D., Sherman, R.L., Mouradian, J., and Saal, S. (1983). Reversible 'end-stage' lupus nephritis. *American Journal of Medicine*, **74**, 361–8.

Kincaid-Smith, P., Fairley, K.F., and Kloss, M. (1988). Lupus anticoagulant associated with renal thrombotic microangiopathy and pregnancy-related renal failure. *Quarterly Journal of Medicine*, **69**, 795–815.

Kotzin, B.L. and Straber, S. (1979). Reversal of NZB/NZW disease with total lymphoid irradiation. *Journal of Experimental Medicine*, **150**, 371–8.

Kumano, K., *et al.* (1987). A case of recurrent lupus nephritis after renal transplantation. *Clinical Nephrology*, **27**, 94–8.

Lee, H.S., Majjais, S.K., Kasinath, B.S., Spargo, B.H., and Katz, A.I. (1984). Course of renal pathology in patients with systemic lupus erythematosus. *American Journal of Medicine*, **77**, 612–20.

Leehey, D.J., Katz, A.I., Azaran, A.H., Aronson, A.J., and Spargo, B.H. (1982). Silent diffuse lupus nephritis: long-term follow-up. *American Journal of Kidney Diseases*, **2(S)**, 188–96.

Levinsky, R.J., Cameron, J.S., and Soothill, J.F. (1977). Serum immune complexes and disease activity in lupus nephritis. *Lancet*, **i**, 564–7.

Lewis, E. and Lachin, J. (1986). Primary outcomes in the controlled trial of plasmapheresis therapy (PPT) in severe lupus nephritis. *American Society of Nephrology*, **51A** (abstract). Published in *Kidney International*, 1987, **3**.

Lewis, R.B., Castor, C.W., Knisley, R.E., and Bole, G.G. (1976). Frequency of neoplasia in systemic lupus erythematosus and rheumatoid arthritis. *Arthritis and Rheumatism*, **19**, 1256–60.

Lewis, E.J., Kawala, K., and Schwartz, M.M. (1987). Histologic features that correlate with the prognosis of patients with lupus nephritis. *American Journal of Kidney Diseases*, **10**, 192–7.

Lockshin, M.D., Qamar, T., and Druzin, M.L. (1987). Hazards of lupus pregnancy. *Journal of Rheumatology*, **13S**, 214–17.

Lubbe, W.F., Palmer, S.J., Butter, W.S., and Liggins, G.C. (1983). Fetal survival after prednisone suppression of maternal lupus anticoagulant. *Lancet*, **i**, 1361–3.

Magil, A.B., *et al.* (1988). Prognostic factors in diffuse proliferative lupus glomerulonephritis. *Kidney International*, **34**, 511–17.

Mahajan, S.K., Ordonez, N.G., Feitelson, P.J., Lim, V.S., Spargo, B.H., and Katz, A.I. (1977). Lupus nephropathy without clinical renal involvement. *Medicine*, **54**, 493–501.

McCluskey, R.T. (1975). Lupus nephritis I. In *Kidney Pathology Decennal 1966–1975* (ed. S.C. Sommers), pp. 435–60. Appleton Century Crofts, New York.

McCluskey, R.T. (1982). The value of the renal biopsy in lupus nephritis. *Arthritis and Rheumatism*, **25**, 867–75.

McCune, W.J., Golbus, J., Zeldes, W., Bohlke, P., Dunne, R., and Fox, D.A. (1988). Clinical and immunologic effects of monthly administration of intravenous cyclophosphamide in severe systemic lupus erythematosus. *New England Journal of Medicine*, **318**, 1423–31.

Mené, P., Dubyak, G.R., Abboud, H.E., Scarpa, A., and Dunn, M.J. (1988). Phospholipase C activation by prostaglandins and thromboxane A in cultured mesangial cells. *American Journal of Physiology*, **255**, F1059–F1069.

Miescher, P.A., Favre, H., Chatelanat, F., and Mihatsch, M.J. (1987). Combined steroid–cyclosporin treatment of chronic autoimmune diseases. *Klinische Wochenschrift*, **65**, 727–36.

Morel-Maroger, L., *et al.* (1976). The cause of lupus nephritis: contribution of serial renal biopsies. *Advances in Nephrology*, **6**, 79–118.

Morris, R.J., Guggenheim, S.J., McIntosh, R.M., Rubin, R.L., and Kohler, P.F. (1979). Simultaneous immunologic studies of skin and kidney in systemic lupus erythematosus. *Arthritis and Rheumatism*, **22**, 864–70.

Morrow, W.J.W., Isenberg, D.A., Todd-Pokropek, A., Parry, H.F., and Snaith, M.L. (1982). Useful laboratory measurements in the management of systemic lupus erythematosus. *Quarterly Journal of Medicine*, **51**, 125–38.

Murray, B.M. and Raij, L. (1987). Glomerular disease in the aged. In *Renal Function and Disease in the Elderly* (eds. J.F. Macias Nunez and J.S. Cameron), pp. 298–320. Butterworths, London.

Neparstek, Y., Kupolovic, J., Tur-Kaspa, R., and Rubingev, D. (1984). Focal glomerulonephritis in the course of hydralazine-induced lupus syndrome. *Arthritis and Rheumatism*, **27**, 822–5.

Osler, W. (1904). On the visceral manifestation of the erythema group of skin diseases. *American Journal of Medical Sciences*, **127**, 1–12.

Park, M.H., D'Agati, V., Appel, G.B., and Pirani, C.L. (1986). Tubulointerstitial disease in lupus nephritis: relationship to immune deposits, interstitial inflammation, glomerular changes, renal function and prognosis. *Nephron*, **44**, 309–19.

Patrono, C. and Pierucci, A. (1986). Renal effects of nonsteroidal anti-inflammatory drugs in chronic glomerular disease. *American Journal of Medicine*, **81S**, 71–83.

Penn, I. (1981). Depressed immunity and the development of cancer. *Clinical and Experimental Immunology*, **46**, 459–74.

Petri, M., Golbus, M., Anderson, R., Whiting-O'Keefe, Q., Corash, L., and Hellmann, D. (1987). Antinuclear antibody, lupus anticoagulant, and anticardiolipin antibody in women with idiopathic habitual abortion. *Arthritis and Rheumatism*, **30**, 601–6.

Pierucci, A., *et al.* (1989). Improvement of renal function with selective thromboxane antagonism in lupus nephritis. *New England Journal of Medicine*, **320**, 421–5.

Platt, J.L., Burke, B.A., Fish, A.J., Kim, Y., and Michael, A.F. (1982). Systemic lupus erythematosus in the first two decades of life. *American Journal of Kidney Diseases*, **1(S)**, 212–22.

Pollak, V.E., Pirani, C.L., and Schwartz, F.D. (1964). The natural history of the renal manifestations of systemic lupus erythematosus. *Journal of Laboratory and Clinical Medicine*, **63**, 537–50.

Pollak, V.E. and Pirani, C.L. (1969). Renal histologic findings in systemic lupus erythematosus. *Mayo Clinic Proceedings*, **44**, 630–44.

Ponticelli, C., Imbasciati, E., Brancaccio, D., Tarantino, A., and Rivolta, E. (1974). Acute renal failure in systemic lupus erythematosus. *British Medical Journal*, **3**, 716–19.

Ponticelli, C., *et al.* (1982). Treatment of diffuse proliferative lupus nephritis by intravenous high-dose methylprednisolone. *Quarterly Journal of Medicine*, **31**, 16–24.

Ponticelli, C., Zucchelli, P., Moroni, G., Cagnoli, L., Banfi, G., and Pasquali, S. (1987). Long-term prognosis of diffuse lupus nephritis. *Clinical Nephrology*, **28**, 263–71.

Ponticelli, C., Moroni, G., and Banfi, G. (1988). Discontinuation of therapy in diffuse proliferative lupus nephritis. *American Journal of Medicine*, **85**, 275.

Ponticelli, C., *et al.* (1989). A randomized trial of methylprednisolone and chlorambucil in idiopathic membranous nephropathy. *New England Journal of Medicine*, **320**, 8–13.

Reichlin, M., (1981). Current perspectives on serological reactions in SLE patients. *Clinical and Experimental Immunology*, **44**, 1–10.

Rodby, R.A., Korbet, S.M., and Lewis, E.J. (1987). Persistence of clinical and serologic activity in patients with systemic lupus erythematosus undergoing peritoneal dialysis. *American Journal of Medicine*, **83**, 613–18.

Rosner, S., *et al.* (1982). A multicenter study of outcome in systemic lupus erythematosus II. Causes of death. *Arthritis and Rheumatism*, **25**, 612–17.

Roth, D., Milgrom, M., Esquenazi, V., Strauss, J., Zilleruelo, G., and Miller, J. (1987). Renal transplantation in systemic lupus erythematosus: one center's experience. *American Journal of Nephrology*, **7**, 367–74.

Rothfield, N. (1988). Efficacy of antimalarials in systemic lupus erythematosus. *American Journal of Medicine*, **85(4A)**, 53–6.

Rubin, L.A., Urowitz, M.B., and Gladman, D.D. (1985). Mortality in systemic lupus erythematosus: the bimodal pattern revisited. *Quarterly Journal of Medicine*, **216**, 87–98.

Schwartz, M.M., Kawala, K., Roberts, J.L., Humes, C., and Lewis, E.J. (1984). Clinical and pathological features in membranous glomerulonephritis of systemic lupus erythematosus. *American Journal of Nephrology*, **4**, 301–11.

Scopelitis, E., Biundo, J.J., and Alspaugh, M.A. (1980). Anti-SS-A antibody and other anti-nuclear antibodies in systemic lupus erythematosus. *Arthritis and Rheumatism*, **23**, 287–93.

Serdula, M.K. and Rhoads, G.G. (1979). Frequency of systemic lupus in different ethnic groups in Hawaii. *Arthritis and Rheumatism*, **22**, 328–33.

Sires, R.L., Adler, S.G., Louie, J.S., and Cohen, A.H. (1989). Poor prognosis in end-stage lupus nephritis due to nonautologous vascular access site-associated septicemia and lupus flares. *American Journal of Nephrology*, **9**, 279–84.

Smith, C.D., Marino, C., and Rothfield, N.F. (1984). The clinical utility of the lupus band test. *Arthritis and Rheumatism*, **27**, 382–7.

Stafford-Brady, F.J., Gladman, D.D., and Urowitz, M.B. (1988). Successful pregnancy in systemic lupus erythematosus with an untreated lupus anticoagulant. *Archives of Internal Medicine*, **148**, 1647–8.

Strober, S., *et al.* (1987). Lupus nephritis after total lymphoid irradiation: persistent improvement and reduction of steroid therapy. *Annals of Internal Medicine*, **107**, 689–90.

Studenski, S., Allen, N.B., Caldwell, D.S., Rice, J.R., and Polisson, R.P. (1987). Survival in systemic lupus erythematosus. *Arthritis and Rheumatism*, **30**, 1326–32.

Talal, N. (1986). New therapeutic approaches to autoimmune disease. *Springer Seminars in Immunopathology*, **9**, 105–16.

Tan, E.M., *et al.* (1982). The 1982 revised criteria for the classification of systemic lupus erythematosus. *Arthritis and Rheumatism*, **25**, 1271–7.

Tron, F., Ganeval, D., and Droz, D. (1978). Immunologically-mediated acute renal failure of nonglomerular origin in the course of systemic

lupus erythematosus (SLE). *American Journal of Medicine*, **67**, 529–32.
Urowitz, M.B., Gladman, D.D., Tozman, C.S., and Goldsmith, C.H. (1984). The lupus activity criteria count. *Journal of Rheumatology*, **11**, 783–7.
Wallace, D.J. Podell, T., Weiner, J., Klinenberg, J.R., Forouzesh, S., and Dubois, E.L. (1981). Systemic lupus erythematosus: survival patterns. *Journal of the American Medical Association*, **245**, 934–8.
Wallace, D.J., *et al.* (1982). Lupus nephritis. Experience with 230 patients in a private practice from 1950 to 1980. *American Journal of Medicine*, **72**, 209–20.
Watson, R.M., Lane, A.T., Barnett, N.K., Bias, W.B., Arnett, F.C., and Prevost, T.T. (1984). Neonatal lupus erythematosus. *Medicine*, **63**, 362–78.
Wei, N., *et al.* (1983). Randomized trial of plasma exchange in mild systemic lupus erythematosus. *Lancet*, **i**, 17–21.
Weiler, J.M. and Packard, B.D. (1982). Methylprednisolone inhibits the alternative and amplification pathways of complement. *Infection and Immunity*, **38**, 122–6.
Weinstein, A., Bordwell, B., Stone, B., Tibbetts, C., and Rothfield, N.F. (1983). Antibodies to native DNA and serum complement (C3) levels. *American Journal of Medicine*, **74**, 206–16.
Whiting-O'Keefe, Q., *et al.* (1982). The information content from renal biopsy in systemic lupus erythematosus. *Annals of Internal Medicine*, **96**, 718–23.
Wilson, H.A., *et al.* (1981). Age influences the clinical and serologic expression of systemic lupus erythematosus. *Arthritis and Rheumatism*, **24**, 1230–5.

4.7 Scleroderma—systemic sclerosis

CAROL M. BLACK

Introduction

Uncontrolled and irreversible proliferation of apparently normal connective tissue (LeRoy 1981; Bailey and Black 1988) along with striking vascular changes (Campbell and LeRoy 1975) are the major hallmarks of systemic sclerosis. These processes considerably impair the normal physiological functions of many organs and result in increased morbidity and mortality. In this multisystem multistage disease, each target organ progresses through stages of inflammation, fibrosis, and atrophy, not necessarily at the same time or the same speed. The kidney is no exception, although the vascular aspects are often more pronounced. Renal disease is the most feared complication, because of its typically rapid onset and, until recently, its almost inevitable progression to renal failure.

Systemic sclerosis is uncommon but not rare; clinical descriptions over the past 100 years have provided a spectrum of disease that varies widely in severity and extent from single patches of hard skin (morphoea) of no more than cosmetic importance, to life-threatening systemic illness (LeRoy 1985; Bailey and Black 1988). It has also come to include a presclerotic vasospastic disease and sclerotic syndromes that mimic systemic sclerosis and include some induced by occupational, environmental, or metabolic stimuli (Table 1). These syndromes must be distinguished from localized and generalized scleroderma.

In its localized form, although the skin shows a marked inflammatory and fibrotic reaction in general, there are no systemic features (Jablonska and Lovell 1988). Generalized scleroderma can be separated into three distinct subsets: diffuse cutaneous systemic sclerosis, limited cutaneous systemic sclerosis or scleroderma sine scleroderma (Table 2). Inflammatory, autoimmune, vascular and fibrotic features occur in many organs and are often irreversible. Although less prevalent than rheumatoid arthritis or systemic lupus erythematosus, systemic sclerosis is more difficult to treat: it is often resistant to therapy and has a high morbidity and considerable mortality.

Epidemiology and genetic studies

Scleroderma occurs throughout the world and in all races. In the United States the incidence is approximately 10 new cases per million population annually (Medsger and Masi 1979) and the average annual mortality has been reported to be between 2.1 and 2.8 per million population; the average age of onset is between 30 and 50 years. There is a marked female preponderance, particularly in the reproductive years, with the ratio of female to male cases varying from 15:1 to 3:1, depending on study and country of origin. Both the incidence and sex ratio of systemic sclerosis change notably with age of onset, and possibly with intervening pregnancies, which suggests that hormonal factors may play an important role in initiating and perpetuating the condition.

Inheritance is also important in the development of systemic sclerosis. The disease is associated with HLA Class II antigens DR5 and DR3 and with the complement C4 null alleles, which are also encoded within the major histocompatibility complex. There is an increased prevalence of antinuclear antibodies in family members of scleroderma probands and also a link between the presence of antinuclear antibodies and HLA type (Briggs *et al.* 1986; Welsh and Black 1988).

Diagnosis

The diagnosis of early scleroderma, or better still the recognition of prescleroderma, is very difficult. However, capillary change (dilatation with or without drop out) plus positive serology (ANA^+, $Scl\,70^+$, ACA^+) (Table 3) can help identify 'connective tissue disease prone patients. A high level of suspicion in the early oedematous phase of the disease is also helpful.

The subsequent development of firm, taut, hidebound skin proximal to the metacarpophalangeal joints permits a definite diagnosis of systemic sclerosis in over 90 per cent of patients (Masi *et al.* 1980). Changes distal to these joints are called

Table 1 Spectrum of scleroderma and scleroderma-like syndromes

Raynaud's phenomenon
  Idiopathic Raynaud's phenomenon (Raynaud's disease)
  Secondary Raynaud's phenomenon (Raynaud's syndrome)
Scleroderma
  Localized forms
    Morphoea
      Plaques
      Guttate
      Generalized
    Linear scleroderma
    Encoup de sabre (with or without facial atrophy)
  Systemic sclerosis
    Limited cutaneous systemic sclerosis
    Diffuse cutaneous systemic sclerosis
Occupational and environmental agents
  Vinyl chloride
  Silica dust
  Toxic oil
  Organic solvents
  Aromatic hydrocarbons, e.g. toluene, benzene aliphatic hydrocarbons
    Chlorinated, e.g. trichlorethylene, perchloroethylene
    Non-chlorinated, e.g. naphtha-n-hexane hexachloroethane
  Epoxy resins
  Drugs
    Bleomycin
    Carbidopa and L-5-hydroxytryptophan
    Diethylpropion hydrochloride and mazindol
    Pentazocine
Breast augmentation
  Paraffin, silicon
Metabolic-genetic (pseudosclerodermas)
  Acromegaly
  Amyloidosis
  Carcinoid syndrome
  Heritable premature ageing syndromes (progeria, Rothmund's Werner's)
  Insulin-dependent diabetes mellitus (digital sclerosis)
  Lichen sclerosus et atrophicus
  Phenylketonuria
  Porphyrias
  Scleroderma (with or without paraproteinuria)
  Scleromyxoedema (with or without paraproteinuria)
Inflammatory–immunological
  Chronic graft-versus-host disease
  Eosinophil fasciitis
  Overlap syndromes (systemic sclerosis with rheumatoid arthritis, dermatomyositis/polymyositis, SLE, Sjogren's syndrome)
  Undifferentiated connective tissue syndromes
Localized systemic sclerosis and visceral disease
  Amyloidosis
  Oesophageal hypomobility (diabetes mellitus, ageing)
  Idiopathic pulmonary fibrosis
  Infiltrative cardiomyopathies
  Intestinal hypomobility syndromes
  Malignant hypertension (hyper-reninaemic, accelerated)
  Occupational, environmental and drug-associated interstitial disease (see vascular changes above)
  Sarcoidosis

sclerodactyly and are not necessarily diagnostic of systemic sclerosis, since they may occur in other conditions. The extent of skin involvement also distinguishes the three prognostically different subsets (Fig. 1). Skin biopsy is usually no more diagnostically sensitive than the experienced touch. Overlap syndromes occur and there are also patients who do not fit tidily into any category; terms such as undifferentiated connective tissue syndromes (LeRoy 1987) and mixed connective tissue disease have emerged to describe these patients: the former seems a better and more accurate term than the latter, which was introduced by Sharp *et al.* (1972) to describe patients with features of myositis, arthritis, lupus and scleroderma. These patients had high titres of circulating antibodies to ribonucleoprotein. Unfortunately, the clinical syndrome, the autoantibodies, and the response to therapy have not proved to be specific.

The connective tissue and vasculature

Some knowledge of the content and regulation of the connective tissue and the structure of blood vessels is essential in order to understand the disease and its particular renal manifestations.

Connective tissue

Connective tissue has many components (Table 4), most of which are large multidomain proteins that interact with one another and with the cells embedded in their matrix. The tissue has many functions which are required for normal development, repair and physiological activity. Connective tissue plays an important supportive role in the morphological arrangement of cells and organs and also interacts with the cytoskeleton to determine the shape of the cell. Besides these mechanical properties, connective tissue has vital biological functions, modulating cell growth, migration, and differentiation (Caplan 1986; Krieg *et al.* 1988). Collagen is the most abundant protein in the connective tissue and its excessive and uncontrolled synthesis is the hallmark of systemic sclerosis. Collagen will therefore be discussed in considerable detail and the other structural components to a much lesser degree.

Collagen

Mammals synthesize at least 12 different types of collagen which are, for the main part, tissue specific (Kuhn 1986). Table 5 shows the molecular and macromolecular structure of these collagen types, as well as their composition and tissue distribution.

Broadly, the collagens can be classified into three groups, interstitial or fibre-forming collagens (Types I, II, III), basement membrane collagen (IV), and minor collagens (V, VI, VII, VIII, IX, X, XI, XII) which are present in tiny amounts in many tissues. Collagens Type I and III represent 80 to 90 per cent of the total collagens in skin, ligaments and blood vessels whereas Type IV is the main component of basement membranes.

All collagen molecules have a similar structure: three helical polypeptide chains possessing the repeated sequence $(Gly-X-Y)_n$ wound into a stable triple helix to produce a rod-like molecule, where Gly is glycine, X is often proline and Y is hydroxyproline. The stability of the triple helix is determined by the proportion of hydroxyproline. Other subtle variations in the primary sequence result in differences in the flexibility of the molecule. In addition, extensive post-translational modification of the molecule provides additional properties to accommodate the multitude of interactions with itself and other extracellular

4.7 Scleroderma—systemic sclerosis

Table 2 Subsets of systemic sclerosis

Diffuse cutaneous systemic sclerosis
  Onset of Raynaud's phenomenon within 1 year of onset of skin changes (puffy or hidebound)
  Truncal and acral skin involvement
  Presence of tendon friction rubs
  Early and significant incidence of interstitial lung disease, oliguric renal failure, diffuse gastrointestinal disease, and myocardial involvement
  Absence of anticentromere antibodies
Limited cutaneous systemic sclerosis*
  Isolated Raynaud's phenomenon for years (occasionally decades)
  Skin involvement limited to hands, face, feet (acral)
  A significant late incidence of pulmonary hypertension, trigeminal neuralgia, skin calcifications, telangiectasia
  A high incidence of anticentromere antibodies (70–80%)
  Dilated nailfold capillary loops without capillary dropout
Scleroderma *sine* scleroderma
  Visceral disease without cutaneous involvement, e.g.
    Oesophageal hypomotility, duodenal dilatation with malabsorption, wide-mouthed colonic sacculations
    Raynaud's phenomenon, dilated nailford capillary loops, oesophageal hypomotility, oliguric renal failure
    Raynaud's phenomenon, dilated nailfold capillary loops, oesophageal hypomotility, pulmonary hypertension and/or interstitial lung disease

* Also termed CREST syndrome (calcinosis, Raynaud's phenomenon, oesophageal hypomotility, sclerodactyly and telangectasia).

Table 3 Autoantibodies showing specificity for systemic sclerosis

Autoantibody specificity	Approximate frequency (%)	Disease subset
ACA	30	CREST variant (60–98%). Not in childhood systemic sclerosis
Scl–70	20	Diffuse systemic sclerosis (30–60%); severe interstitial lung disease
Antinucleolar (high titre)	40	No apparent association with clinical expression
Uracil-specific anti-RNA	76	Diffuse systemic sclerosis; highest titres in active disease
Ku	4	Systemic sclerosis–polymyositis overlap
PM-Scl	4	Systemic sclerosis–polymyositis overlap
Scl-4	3	CREST variant
Scl-6	5	CREST variant

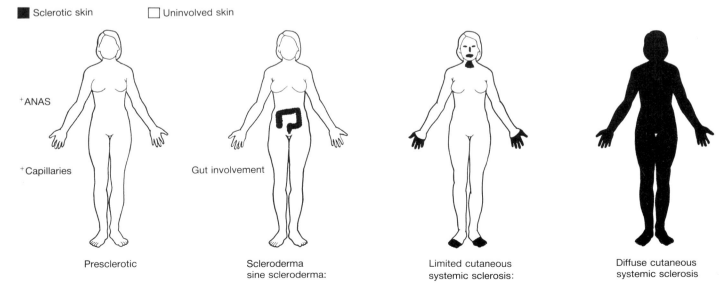

Fig. 1 Scleroderma patients: pictorial representation of suggested subgroups.

Table 4 The structural components of the connective tissue

Collagens I–XII
Proteoglycans
Fibronectin
Elastin
Laminin
Fibrillin
Osteonectin/SPARC/BM-40
Entactin/nidogen
Osteopontin

matrix components. These modifications result in the well-known diverse functional properties of collagen.

The interstitial collagens

These are the most common collagens and include Types I and III, which are frequently codistributed in the extracellular matrix. These collagens are synthesized in a precursor form with large globular peptides located at both ends. The peptides are then cleaved in the extracellular space and the triple helical molecules with short non-helical sequences at the amino and carboxy-terminal ends associate laterally in a quarter stagger manner (Fig. 1).

Type I collagen is the major collagen of bone, tendon, and skin, but is also found in internal organs. It is almost certainly required for the mechanical strength of the tissue.

The highest concentration of Type III collagen occurs in elastic tissues such as major blood vessels, lung, uterus, and the gastrointestinal tract. It is present to a significant extent as fine fibre and is thought to represent the reticulin fibres described by histochemists. It is present in the more distensible types of connective tissue and may be responsible for their flexibility. Following self assembly there is extensive cross linking of individual fibres after collagen molecules have condensed (Bailey et al. 1974).

The basement membrane collagen type IV

Type IV collagen is the major structural component of the basement membrane of all tissues, where it is closely associated with other basement membrane specific macromolecules such as laminin, entactin, and heparan sulphate proteoglycan (Farquhar 1981). These extracellular structures act as an underlying support for epithelial cells and provide continuous sheaths surrounding muscle, fat, nerve, and smooth muscle cells. The Type IV molecule contains similar structural features to the fibre-forming collagens but, unlike interstitial collagens, these aggregate via their terminal regions rather than by lateral association. Further, the retention of the large N- and C-terminal peptides and the flexibility of the normally rigid triple helical domain due to the presence of short non-helical regions along its length also restrict most lateral associations, though some may occur. These restrictions result in the formation of an open non-fibrillar network. The polymer structure is stabilized by the same cross linking mechanism that operates in the fibrillar collagens, despite the different organization of the molecules (Fig. 2).

The minor collagens

Collagen Types V–XII have not been fully defined at a molecular level, and their functions are not completely understood. They are present in tissues in only small amounts, but have a major influence on architecture and function. Type V may play a regulatory role in limiting the diameter of the major collagens; Type VI may provide linking fibres between the major collagens; and Type VII is a major component of anchoring fibrils.

Metabolism of collagen

Collagen synthesis takes place via a series of exquisitely controlled modifications, and a complex series of enzymatic reactions which occur during its degradation. The collagen genes are very complex and, like other eukaryotic genes, are characterized by frequent interruptions of the coding sequence (exons—about 50 in number) with a series of large non-coding sequences (introns) (de Crombrugghe and Pastan 1982).

The unprocessed RNA which is transcribed from the gene is a direct copy and therefore contains both exons and introns. The pro α chains are synthesized by translation of the mRNA on the ribosomes; this involves around 50 excision and splicing steps. The collagens are first synthesized as pre-pro-collagens with remarkably large signal sequences of 100 amino acids. The nascent chains then pass into the cisterna of the rough endoplasmic reticulum where a complex series of post-translational enzymic modifications occur both intracellularly and extracellularly (Fig. 3). The signal peptides are removed from the amino-terminus of all three chains, lysyl and prolyl residues are hydroxylated, and certain hydroxylysyl residues are then glycosylated. Cofactors such as ascorbic acid, Fe^{2+} and α-ketoglutarate are required by these enzymes whose activity is terminated with triple helix formation.

Following secretion of the procollagen molecules, specific proteases cleave the N- and C-terminal peptides. The molecules then immediately arrange themselves side by side in a quarter staggered fashion to form fibrils which are stabilized by the cross linking described previously.

The mature collagen fibrils are degraded in the extracellular space by a collagenase which cleaves the molecules at a specific point within the triple helix. The mammalian collagenase is synthesized by fibroblasts and mainly cleaves collagens Types I and III but also has some activity for Type II. Other collagens have highly specific collagenases. The resulting three-quarter and one-quarter fragments can be degraded by a variety of other proteases.

Collagenase is synthesized in an inactive precursor form and the presence of several inhibitors in the connective tissue and

Table 5 Molecular composition and location of known genetically distinct collagens

Type	Molecular composition	Tissue location
I	$[\alpha 1(I)]_2, \alpha 2(I)$	Skin, tendon, bone, dentine
II	$[\alpha 1(II)]_3$	Cartilage, disc, vitreous
III	$[\alpha 1(III)]_3$	Vascular system, skin, gut
IV	$[\alpha 1(IV)]_2, \alpha 2(IV) + \alpha 3(IV)$	Basement membrane
V	$[\alpha 1(V)]_2, \alpha 2(V) + [\alpha 1(V), \alpha 2(V), \alpha 3(V)]$	Cell-associated skin, vascular tissue
VI	$[\alpha 1(VI), \alpha 2(VI), \alpha 3(VI)]$	Vascular system
VII	Unknown	Skin, amniotic membrane
VIII	$\alpha 1(VIII)$	Endothelial cells
IX	$[\alpha 1(IX), \alpha 2(IX), \alpha 3(IX)]$	Cartilage, disc
X	$[\alpha 1(X)]$	Mineralizing cartilage
XI	$(1\alpha, 2\alpha, 3\alpha)$	Cartilage
XII	Similar to Type IX	Cartilage

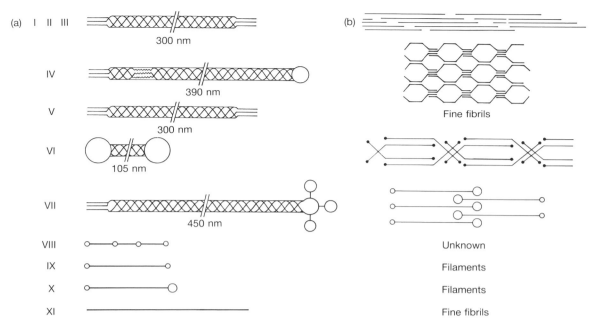

Fig. 2 (a) Molecular structure of the various collagen types. The triple helices are of different lengths and there are globular domains at the end. (b) Schematic diagrams of the supramacromolecular organizations developed by different collagen types.

body fluids implies complex control mechanisms regulating the turnover of collagen (Etherington 1980).

Regulation

The regulatory mechanisms which determine the type, amount, and spatial deposition of the different collagens are clearly critical to the maintenance of appropriate levels of tissue collagens. Such regulation could be transcriptional, translational, or post-translational but the available evidence indicates that it is predominantly transcriptional. Levels of the procollagen mRNA usually correspond with the rate of synthesis of the corresponding collagen propeptides, and this also appears to be the case in systemic sclerosis (Kahari 1987).

Since little is known about nuclear events that control collagen synthesis, attempts to inhibit excess collagen synthesis have concentrated on the post-transitional stage in collagen assembly, both within the cell and extracellularly. This is the rationale underlying attempts to modify fibrotic disorders with drugs such as D-penicillamine and colchicine (Uitto et al. 1984).

Proteoglycans

Proteoglycans are macromolecules that have a protein core to which glycosaminoglycan chains are covalently bound. There are four classes of glycosaminoglycans: keratan sulphates, chondroitin and dermatan sulphates, heparan sulphate/heparins, and hyaluronic acid. All are polyanionic macromolecules containing a repeating disaccharide of a hexosamine and, generally, hexuronic acid (Hardingham 1981). The biosynthesis of proteoglycans is complex. Different types of proteoglycan have different core proteins, implying the existence of many different genes.

The proteoglycans have a wide variety of functions, many of which arise by their interactions with other molecules. Heparan sulphate proteoglycans are associated with cell membrane receptors and may be involved in cell-matrix interactions and cell adhesion. Heparan sulphate may also be involved in binding fibronectin and laminin. These proteoglycans are also important regulators of the normal filtration process in the glomerular basement membrane. Dermatan sulphate proteoglycans in tendons interact with collagen fibrils and may limit radial growth of the fibril.

Glycoproteins

The extracellular matrix contains many glycoproteins which seem to be important for its integrity. Little is known about these glycoproteins in systemic sclerosis and only two, fibronectin and laminin, have been investigated at all in this disease.

The blood vessel

Vascular injury is crucial to the pathogenesis of systemic sclerosis and a knowledge of the structure and functional interactions of the vessel wall is important in trying to understand the possible mechanisms which initiate and perpetuate the disease (Stemerman 1974; Pratt et al. 1985). Many extracellular matrix proteins are found in the vessel wall, including collagens Types I, III, IV, V, VI, and VIII, proteoglycans, entactin, and laminin. Most of the injury in systemic sclerosis involves the intimal regions of the blood vessels, which represent about one-quarter of the vessel wall thickness and are composed of endothelial cells and a basement membrane (basal lamina) which anchors the endothelial cells and contributes to their selective permeability properties.

Endothelial cells divide infrequently but synthesize and secrete many kinds of molecules in addition to providing non-thrombotic surfaces. They play a key role in directing neutrophils, lymphocytes, and monocytes into the interstitial space. Endothelial cells are held together by occluding tight junctions and communicate by gap junctions. They also make contact with smooth muscle cells by penetrating through the elastic lamellae.

The basement membrane is composed of a lamina densa, which is mainly Type IV collagen but also contains Types V and VI surrounded on either side by a lamina rara composed mainly of proteoglycans, which serve as ionic barriers. This extracellular

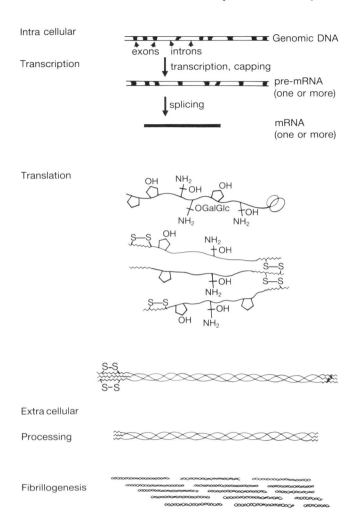

Fig. 3 Schematic representation of the various stages in the biosynthesis of collagen from intracellular transcription and translation of the gene to extracellular processing.

matrix is absent in the microvessels where there is a discrete pericapillary space.

The media of arteries are largely unaffected in systemic sclerosis but consist of a meshwork of elastic fibrils and Type I collagen which, working together with smooth muscle cells, probably prevent excessive distension. Type III collagen is also found in the 'elastin containing' blood vessels. The fibrils are small and laid down in continuous sheaths, providing a less bulky structural support at sites where diffusion and metabolic interactions are more important than tensile strength.

Pathogenesis and pathology

Our knowledge of the pathogenesis of systemic sclerosis is still rudimentary, but it is clear that several complicated events occur, simultaneously or in sequence. These almost certainly include inflammation, autoimmune attack, vascular damage, fibroblast activation by cytokines and growth factors and interactions between different components of the extracellular matrix (Campbell and LeRoy 1975; Korn 1980; Bailey and Black 1988; Krieg and Meurer 1988).

It is uncertain what initiates the process, but the interstitial fibroblast and the endothelial cell are strong candidates. Endothelial cell damage, perivascular mononuclear cell infiltrates, platelet activation, and vasomotor and permeability changes are present in target tissues before fibrosis is prominent and the eventual fibrotic lesion often appears to be adjacent to small blood vessels. However, if blood vessels are the initial target, additional factors are needed to explain why dense fibrosis occurs instead of the usual process of repair.

It is possible that subsets of fibroblasts, activated by unknown factors, are responsible for the connective tissue disease. Understanding the role of cytokines and collagen gene regulation may be a key to the understanding not only of systemic sclerosis but also to the unregulated fibrosis seen in liver cirrhosis, primary pulmonary fibrosis, and atherosclerosis.

Clinical manifestations

The clinical features in each affected organ result from various combinations of inflammation, fibrosis, and atrophy. Different organs may be at different stages of the process, and this may sometimes remit spontaneously or even regress. The clinical features are also influenced by the age of onset. Juvenile scleroderma is a special category, and is predominantly localized: in the uncommon event of a child developing diffuse disease with renal involvement, the expression and prognosis are the same as in the adult.

The first symptom in most patients with limited cutaneous systemic sclerosis is Raynaud's phenomenon, followed by limited skin changes, calcinosis and telangiectasia. In contrast, patients with diffuse systemic sclerosis most frequently have oedematous skin which thickens, with or without arthritis, tendon rubs and involvement of other organs. It is very rare for renal scleroderma to be the first manifestation of the disease. Detailed descriptions of the involvement of organs other than the kidney can be found in other texts and reviews.

Renal disease

Renal disease has been recognized as the most common cause of death in scleroderma in virtually all series reported over the last 30 years. Modern therapy is fortunately changing this situation, prognosis can be improved by the earliest possible recognition of renal involvement, or better still by identification of the population at risk.

The earliest reports did not suggest a direct relationship between skin disease and the kidney. In describing the first reported case Auspitz (1863) concluded that 'there is no evidence to support the causal relationship between the kidney disease and skin disease as such'. Eighty years later clear descriptions of the histological changes emerged. Masugi and Ya-Shu (1938) and Talbott et al. (1939) described intimal hyperplasia and fibrinoid degeneration in the interlobular renal arteries, and Goetz (1945) described the widespread pathological changes in the small blood vessels. However, it was not until the early 1950s that the direct relationship between systemic sclerosis and acute renal failure was accepted, after Moore and Sheehan (1952) described the clinical and histological abnormalities in the kidneys of three patients with scleroderma who were dying of uraemia.

The overall significance of renal disease in systemic sclerosis before the introduction of angiotensin converting inhibitors in

Table 6 Criteria for renal involvement in systemic sclerosis*

Renal vascular pathologic changes	60–80%
Decreased renal blood flow or GFR	50%
Clinical 'markers' of renal disease	45%
Proteinuria (1+ or greater)	36%
Hypertension (BP > 140/90 mmHg)	24%
Azotaemia (BUN > 25 mg/dl)	19%
Rapidly progressive acute renal insufficiency associated with malignant hypertension (scleroderma renal crisis)	5–15%

* Compiled from Cannon et al. 1974; D'Angelo et al. 1969; Kovalchik et al. 1978; Medsger et al. 1971; Steen et al. 1984.

the mid 1970s is clearly illustrated in the survival figures reported by Medsger et al. (1971). In their initial study, all 16 patients who developed renal scleroderma were dead within 1 year. In a later study all 17 patients with kidney involvement died within 10 months of onset; and the overall cumulative survival rate was 35 per cent after 7 years of follow up (Medsger and Masi 1973). The factors indicating a worse prognosis included older age, male sex, and black race. Renal disease occurred relatively early in the course of the disease, almost invariably within the first 5 years and especially in winter. However, the most important risk factor for renal disease is diffuse skin involvement. Obviously the 5 per cent of patients who have scleroderma sine scleroderma are much more difficult to identify. Similar results have been reported by Cannon et al. (1974).

The reported frequency and type of renal involvement in systemic sclerosis varies, depending on the clinical and pathological criteria used, and how they were defined. Criteria for renal involvement in systemic sclerosis, compiled by Shapiro and Medsger (1988) are shown in Table 6, and the frequency of renal involvement in the various subsets of scleroderma is shown in Table 7. These data are taken from the University of Pittsburgh for the period 1972 to 1984.

Spectrum of renal disease

At one end of the spectrum is a slowly-progressive form of chronic renal disease in which the relationship to systemic sclerosis is tenuous; at the other end are patients with acute scleroderma renal crisis, which is sudden and rapidly progressive. These two clinical situations cannot always be clearly separated. The criteria for diagnosing scleroderma renal crisis are well defined, and the frequency with which it occurs is consistent at between 5 and 15 per cent of reported patients (Medsger et al. 1971; Eason et al. 1981; Traub et al. 1983). However, the reported frequency of the various types of renal involvement are influenced by referral bias, the type of scleroderma (diffuse or limited), and disease duration at the time of assessment.

The vascular abnormalities of intimal hyperplasia, fibrinoid necrosis, and diminished renal blood flow are the hallmark of renal involvement but are not completely specific for systemic sclerosis (Cannon et al. 1974). Proteinuria, hypertension, and renal failure—clinical markers of renal disease—are frequent: Cannon et al. (1974) identified at least one of these abnormalities in 45 per cent of 210 patients with systemic sclerosis, and Tuffanelli and Winkelman (1961) reported 109 instances of proteinuria in a series of 271 patients, although only 19 had severe renal failure. Mild proteinuria may, of course, be coincidental and may lead to an over estimate of the frequency of renal scleroderma. Hypertension may also be coincidental, and other causes of renal disease are not always easy to exclude.

Scleroderma renal crisis

Scleroderma renal crisis is a distinctive medical emergency which requires prompt intervention. It can occur at any time during the disease but is most common during the first 5 years, often during the winter months. The best available predictor of renal crisis is rapid progression of skin disease in the preceding few months. All patients with diffuse disease should have 4-monthly urine collections for creatinine clearance and proteinuria measurements and should be carefully observed in the first 5 years of the disease. A GFR of less than 60 ml/min, protein excretion of more than 500 mg/24 h, or a sudden change in either, should prompt the measurement of resting levels of renin; if these are increased, treatment should be initiated.

Symptoms of the crisis usually present acutely. The pulse rate is increased and patients develop headaches, visual phenomena and convulsions due to accelerated hypertension. Symptoms and signs of left ventricular failure may follow rapidly. The glomerular filtration rate and renal blood flow are decreased, and serum creatinine increases. Oliguria and anuria may follow and death from renal failure can occur within 7 to 10 days in untreated patients. Proteinuria is almost universal; although it may present long before the renal crisis develops, it often increases with the crisis, though not to nephrotic levels. Microscopic haematuria and granular casts may be present, as in other forms of accelerated hypertension; this situation may be complicated by microangiopathic haemolytic anaemia.

Hypertension is suggestive of renal involvement in scleroderma but is common in the general population. About one-third of the population of the United Kingdom have a diastolic blood pressure of 90 mmHg or more on a single reading, but the frequency drops to about 5 per cent if six readings are taken (Sleight 1988). In the United States, the prevalence of hypertension in the general adult population has been estimated to be 20 per cent (Roberts and Maurer 1977); it is not surprising that hypertension does not always presage scleroderma renal crisis, but may be present for several years, sometimes long before the onset of the connective tissue disease itself (Shapiro and Medsger 1988). Black patients with systemic sclerosis appear to have a particularly high incidence of renal crisis, with a prevalence of 21 per cent compared with 7 per cent for the white population (Traub et al. 1983); this may be related to the twofold increase in the prevalence of essential hypertension in the black population in the United States and its tendency to follow a more severe course. Elevated blood pressure as a sign of scleroderma renal crisis is almost always accompanied both by the fundal changes of malignant hypertension and a very high plasma renin activity.

The rapid development of renal failure is the other hallmark of renal crisis and alone is sufficient to make the diagnosis, especially as there are several reports of normotensive renal failure in systemic sclerosis (Moore and Sheehan 1952; Dichoso 1976; Helfrich et al. 1989). This suggests that a large change in blood pressure may not be required to precipitate vascular damage. Additional mechanisms which may be operative in scleroderma renal crisis in some patients include hyper-reactivity of blood vessels or blood pressure to a number of mediators such as catacholamines—these mechanisms may be enhanced by a drop in temperature (Rodnan et al. 1964; Winkelman et al. 1977).

Table 7 Frequency of renal involvement in systemic sclerosis subsets (University of Pittsburgh 1972–1984)

	Diffuse scleroderma (n = 351) (%)	CREST syndrome (n = 346) (%)	Overlap syndrome (n = 49) (%)
Proteinuria (2+ by dipstick or > 0.5 g/24 h)	22	16	12
Hypertension (>140/90 mmHg)	31	26	25
Azotaemia (serum creatinine > 1.3 mg/dl)	25	13	10
Scleroderma renal crisis	19	1	6

The renin-angiotensin system in scleroderma renal crisis

Plasma renin activity increases markedly with the onset of renal injury and may rise acutely to extreme levels, but there is no convincing evidence that a rise precedes the onset of scleroderma renal crisis. Traub et al. (1983) reported 13 cases in which plasma renin activity was coincidentally measured from 1 week to 1 year prior to the development of the renal crisis: none of the patients had significantly increased levels. Similar results have been reported by others (Fleischmajer and Gould 1975; Fiocco et al. 1978). A functional Raynaud's-like renal vasoconstriction may be superimposed on more chronic structural changes; this hypothesis is consistent with the data provided by Kovalchik et al. (1978) who performed renal biopsies on nine patients with scleroderma and normal renal function. Plasma renin activity was slightly elevated in three of four patients with microscopic vascular damage and normal in the other, but it rose significantly in all four after cold pressor testing. The resting and cold pressor levels remained normal in patients without microscopic changes. Xenon clearance studies provide additional support for the importance of functional vascular narrowing (Urai et al. 1958; Cannon et al. 1974). In these studies renal cortical perfusion was found to be decreased in systemic sclerosis patients, and much more so at times of renal crisis. Decreased perfusion is likely to cause ischaemia of the juxtaglomerular apparatus, increased renin secretion and the formation of large amounts of angiotensin II which, in turn, would cause both renal and generalized vasoconstriction. Thus, although increased renin secretion is probably a secondary phenomenon, it certainly adds to the vicious circle of vascular constriction and damage.

Indolent renal disease

Proteinuria may be the only clinical manifestation in this much more chronic form of renal disease which is non-specific, and often non-progressive.

Histopathology of the kidney

There are histological changes in the renal blood vessels and the interstitium as well as changes in the amount of basement membrane and other matrix collagens. The distribution of the collagens I, III, IV and V in normal kidneys has been described by Roll et al. 1980 and by Black et al. 1983 (Fig. 4(a–d)). Immunofluorescent studies with type specific anticollagen antibodies have revealed a marked increase in Type III collagen around large blood vessels and throughout the kidney, but particularly in the interstitium (Fig. 4(b,f)), and a similar but less obvious increase in Type I collagen (Fig. 4(a,e)). The tubular and glomerular basement membranes also appear thickened and show an increased intensity of staining with antibody to Type IV collagen (Fig. 4(c,g)). There is increased Type V collagen in the glomerular mesangial matrix as well as the interstitium (Fig. 4(d,h)). These changes are found even in patients without clinical evidence of renal disease. Biochemical quantification of Type I and III collagens in normal and affected kidney shows a

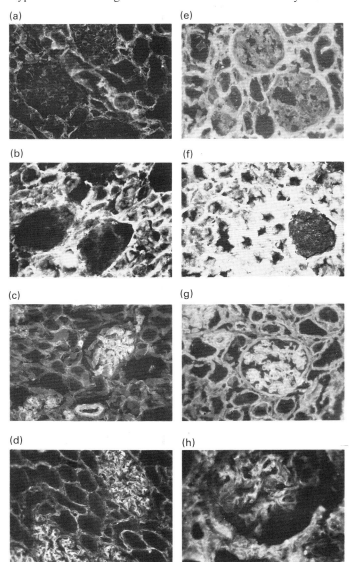

Fig. 4 Immunofluorescent staining of normal (a–d) and systemic sclerosis kidney (e–h) stained with antibodies to Type I (a–e), Type III (b and f), Type IV (c and g) and Type V (d and h) collagen ×100. An increase in staining is evident for all types of collagen but is particularly marked for Type III.

threefold increase in Type III compared with Type I, consistent with immunological data. This altered ratio persists long after the early phase of the disease.

Increased collagen synthesis must, if progressive, interfere with renal function. It has been suggested that the primary injury in systemic sclerosis is to the blood vessel and that this is caused by a circulating factor. The initial response may be a thickening of capillary basement membrane which restricts local nutrition and causes an increase in the number or activity of fibroblasts synthesizing collagen. Excess collagen then restricts nutrient supply still further, exacerbating organ dysfunction.

Arterial lesions

Microscopically, there are two types of arterial lesion in the acute renal disease of scleroderma. The first, affecting small interlobular and arcuate arteries, is characteristic of scleroderma. The second, involving the small arteries and arterioles supplying the glomeruli and glomerular capillaries, is indistinguishable from changes seen in malignant hypertension. Larger arteries (outer diameters > 500 μm) are normal or show atherosclerotic changes.

The earliest change in acute renal disease is oedema, followed by marked proliferation of intimal cells and accumulation of mucoid ground substance composed of glycoproteins and mucopolysaccharides (Fig. 5). This proliferation may damage the internal elastic lamina allowing smooth muscle-like cells, capable of synthesizing collagen and elastin, to migrate into the intima from the media, which is often stretched and thinned. Fibrous thickening in the adventitia and periadventitial tissue is another feature of systemic sclerosis, seldom found in other causes of malignant hypertension. Gross narrowing of the blood vessels results in vascular insufficiency with atrophy of perfused tissue; thrombosis can also lead to necrosis and infarction of glomeruli and tubules.

Subintimal or intramural fibrinoid necrosis of the small arteries and arterioles supplying glomeruli and glomerular capillaries is the second typical lesion of scleroderma. These changes are indistinguishable from those found in malignant hypertension

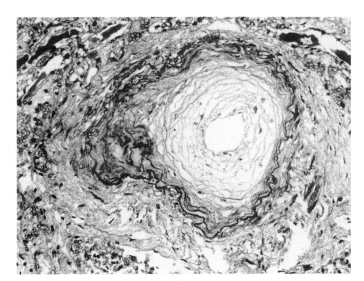

Fig. 5 Interlobular artery showing intimal thickening and accumulation of mucin (haematoxylin and eosin, × 250).

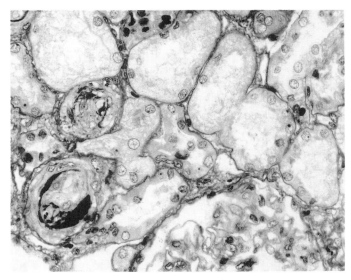

Fig. 6 Renal cortex with two arterioles showing fibrinoid necrosis and thrombus formation (haematoxylin and eosin, × 500).

and can occur in patients with systemic sclerosis but without previously elevated blood pressure or clinical renal disease (D'Angelo et al. 1969; Fig. 6).

Glomerular changes which have been reported include focal or diffuse basement membrane thickening, with progressive glomerulosclerosis. There may also be non-specific hyperplasia of the juxtaglomerular apparatus. The tubules appear to be secondarily affected by the vascular insufficiency and the most prominent changes are flattening of the epithelial cells with hyaline droplet degeneration. The interstitial stroma may show a marked increase in fibrous tissue and small collections of lymphocytes and/or plasma cells.

Immunohistological studies are inconclusive and non-specific. Deposits of immunoglobulins, most notably IgM, complement components (including C1q, C3 and C4), and fibrinogen have occasionally been reported (Lapernas et al. 1978). However, C3 deposition has also been reported in patients without clinical renal disease (Kovalchik et al. 1978). In a few cases antinuclear antibodies have been eluted from the kidney, raising the possibility that immune complexes are involved in the pathogenesis. However, electron-dense material has not been found on electron microscopy. The identification of factor VIII in vessel walls may be due to activation of the coagulation cascade (Kahaleh et al. 1981) or may merely reflect vascular injury.

Chronic renal disease

Patients who survive scleroderma renal crisis may develop similar but less florid proliferative changes in the interlobular and arcuate arteries. Even those who have never had a renal crisis may show reduplication of elastic fibres, sclerosed glomeruli, tubular atrophy, and interstitial fibrosis, presumably reflecting the chronic changes of scleroderma.

Pathogenesis of renal involvement

The vasculopathy of systemic sclerosis proceeds at different rates in different organs and may be present in many patients who do not develop scleroderma renal crisis. Renal lesions have been found in postmortem studies of patients suffering from systemic sclerosis but who died of non-renal complications (D'Angelo *et al*. 1969), and physiological studies in the 1950s showed that, although less than 10 per cent of the patients have a reduced GFR, 80 per cent had a reduction in renal plasma flow. Presumably the renal crisis is precipitated by some new and sudden event which changes the course of the disease. Various factors such as pregnancy, hormones, corticosteroid therapy, accelerated fibroblast activity reflected in an increasing skin score, and cold-induced vasospasm have been considered, but their roles are poorly defined.

Management of scleroderma renal crisis

Survival after renal crisis ranged from 1 to 3 months before the introduction of the angiotensin converting enzyme inhibitors (Medsger *et al*. 1971; Traub *et al*. 1983), but these have transformed the prognosis. The first report of the reversal of scleroderma renal crisis appeared in 1979 (Lopez-Ovejero 1979) and an increasing number of similar cases have been described since then (Traub *et al*. 1983). The inhibition of angiotensin II production rapidly lowers blood pressure and stops renal deterioration; this may even be reversed if treatment is started early.

Most studies of angiotensin converting enzyme inhibitors have used captopril, which has a number of side-effects that may be related to the sulphydryl group on the molecule. These include rashes, leukopenia, and proteinuria and are discussed in more detail elsewhere. D-Penicillamine has similar sulphydryl groups and so the two drugs should not normally be used together. Following an initial dose of 12.5 mg of captopril there should be careful monitoring of the blood pressure: patients who are already fluid depleted can be very sensitive to the drug. The dose should be increased by 12.5 to 25-mg increments every 6 to 8 h to maintain the blood pressure below 140/90 mmHg. Control is usually obtained using 150 mg/day, although up to 300 mg may be required.

Other angiotensin converting enzyme inhibitors, such as enalapril, lack the sulphydryl group. These are under consideration and preliminary data (T. Medsger Jr., personal communication) indicate that enalapril is as effective as captopril and can be given safely to patients who have had captopril-related toxicity. Additional hypotensive drugs, including β-blockers or even potent vasodilators such as minoxidil may be required; in these cases a diuretic is usually needed as well. Nitroprusside can be used if the patient is unable to tolerate oral medication. Dialysis may be needed if renal function deteriorates sufficiently; some patients regain renal function after many months of dialysis and treatment with angiotensin converting enzyme inhibitors. Plasma renin activity tends to remain high in patients on regular dialysis and may even increase, making blood pressure control more difficult. Under these circumstances bilateral nephrectomy may be life-saving: this drastic step should be considered only when all other measures have failed. Nephrectomy may not solve the problem, and recurrence of hypertension after bilateral nephrectomy has been reported (Traub *et al*. 1983). Kidney transplantation has resolved the clinical features of scleroderma renal crisis in a number of instances, but recurrence of renal scleroderma has been reported occasionally (Traub *et al*. 1983).

A number of therapies no longer have a place in the management of renal scleroderma, including anticoagulants and antithrombotic drugs. Steroid therapy should also be avoided since it may precipitate or even accelerate the renal crisis.

In summary, scleroderma renal crisis is a medical emergency in which the renin-angiotensin system has a particular role. Administration of angiotensin converting enzyme inhibitors appears to be the most promising form of therapy to date; these should be given rapidly and aggressively in order to prevent renal deterioration and death.

Conclusion

Therapy for systemic sclerosis is difficult. No therapy has emerged which, when tested in a controlled manner, can halt either the excessive fibrosis or vascular damage in the skin or internal organs. Current therapy is, therefore, directed towards reducing the clinical effects of the disease (Jayson 1988) and providing a general understanding of the process which helps both the patient and family cope with the various problems associated with a chronic disease. Several patient support groups are available and useful, and the individual patient must be taught a programme of self-care to fit his or her own particular needs. This may include physical exercises, joint protection including splinting, and skin care with emphasis on protection against the cold.

The long-term hope is that a better understanding of connective tissue metabolism and the complex interactions between different cellular systems in systemic sclerosis will allow for a more rational therapy of the disease and a better quality of life for the patient.

References

Auspitz, H. (1863). Ein Beitrag zur Lehre vom Haut-Sklerem der Erwachsenen. *Wiener Medizinische Wochenschrift*, **13**, 739, 755, 772, 788.

Bailey, A.J. and Black, C.M. (1988). The role of the connective tissue in the pathogenesis of scleroderma. In *Systemic Sclerosis—Scleroderma* (eds. M.I.V. Jayson and C.M. Black), pp. 75–105. Wiley, Chichester.

Bailey, A.J., Robins, S.P., and Balian, G. (1974). Biological significance of the intermolecular crosslinks of collagen. *Nature*, **251**, 105–9.

Black, C.M., Duance, V.C., Sims, T.J. and Light, N.D. (1983). An investigation of the biochemical and histological changes in the collagen of the kidney and skeletal muscle in systemic sclerosis. *Collagen Related Research*, **3**, 231–44.

Briggs, D.C., Laurent, R., Black, C.M., and Welsh, K.I. (1986). A strong association between null alleles at the C4A locus in the major histocompatibility complex and systemic sclerosis. *Arthritis and Rheumatism*, **29**, 1274–7.

Campbell, P.M. and LeRoy, E.C. (1975). Pathogenesis of systemic sclerosis: a vascular hypothesis. *Seminars in Arthritis and Rheumatism*, **4**, 351–68.

Cannon, P.J., Hassar, M., Case, D.B., Casarella, W.J., Sommers, S.C., and LeRoy, E.C. (1974). The relationship of hypertension and renal failure in scleroderma (progressive systemic sclerosis) to structural and functional abnormalities of the renal cortical circulation. *Medicine*, **53**, 1–46.

Caplan, A.I. (1986). The extracellular matrix is instructive. *Bioessays*, **5**, 129–32.

D'Angelo, W.A., *et al.* (1969). Pathologic observations in systemic sclerosis (scleroderma). A study of 58 autopsy cases and 58 matched controls. *American Journal of Medicine*, **46**, 428–40.

de Crombrugghe, B., and Pastan. I. (1982). Structure and regulation of a collagen gene. *Trends in Biochemistry*, **7**, 11–13.

Dichoso, C.C. (1976). The kidney in progressive systemic sclerosis (scleroderma). In *The Kidney in Systemic Disease—Perspectives in Nephrology and Hypertension, Vol 3.* (eds. W.N. Suki and G. Eknoyan), pp. 57–74. Wiley, New York.

Eason, R.J., Tan, P.L., and Gow, P.J. (1981). Progressive Systemic Sclerosis in Auckland: A ten-year review with emphasis on prognostic features. *Australian and New Zealand Journal of Medicine*, **11**, 657–62.

Etherington, D.J. (1980). Proteinases in connective tissue breakdown. In *Protein Degradation in Health and Disease*, Ciba Foundation Symposium, No. 75, pp. 87–103. Excerpta Medica, Amsterdam.

Farquhar, M.G. (1981). The glomerular basement membrane: a selective macromolecular filter. In *Cell Biology of the Extracellular Matrix*, (ed. E.D. Hay)., pp. 335–78. Plenum, New York.

Fiocco, U., et al. (1978). Plasma renin activity in progressive systemic sclerosis. *Bollettino della Societa Italiana di Biologia Sperimentale*, **54**, 2507–10.

Fleischmajer, R. and Gould, A.P. (1975). Serum renin and renin substrate levels in scleroderma. *Proceedings of the Society for Experimental Biology and Medicine*, **150**, 374–9.

Goetz, R.H. (1945). The pathology of progressive systemic sclerosis (generalised scleroderma) with special reference to changes in viscera. *Clinical Proceedings*, **4**, 337–92.

Hardingham, T. (1981). Proteoglycans: their structure, interactions and molecular organisation in cartilage. *Biochemical Society Transactions*, **9**, 489–97.

Helfrich, D.J., Banner, B., Steen, V.D., and Medsger, T.A. (1989). Normotensive renal failure in systemic sclerosis. *Arthritis and Rheumatism*, **32**, 1128–34

Jablonska, S. and Lovell, C.R. (1988). Localized scleroderma. In *Systemic Sclerosis—Scleroderma* (ed. M.I.V. Jayson and C.M. Black), pp. 303–18. Wiley, Chichester.

Jayson, M.I.V. (1988). Treatment of Systemic Sclerosis. In *Systemic Sclerosis–Scleroderma* (eds. M.I.V. Jayson and C.M. Black, pp. 289–301. Wiley, Chichester.

Kahaleh, M.B., Osborn, I., and LeRoy, E.C. (1981). Increased factor VIII/von Willebrand factor antigen and von Willebrand factor activity in scleroderma and in Raynaud's phenomenon. *Annals of Internal Medicine*, **94**, 482–4.

Kahari, V.M., Multimaki, P., and Vuorio, E. (1987). Elevated proα2[I] collagen mRNA levels in cultured scleroderma fibroblasts results from an increased transcription rate of the corresponding gene. *FEBS Letters*, **215**, 331–4.

Korn, J.H. (1980). Interaction of immune and connective tissue cells. *International Journal of Dermatology*, **19**, 487–95.

Kovalchik, M.T., et al. (1978). The kidney in progressive systemic sclerosis. A prospective study. *Annals of Internal Medicine*, **89**, 881–7.

Krieg, T. and Meurer, M. (1988). Systemic scleroderma: clinical and pathophysiological aspects. *Journal of the American Academy of Dermatology*, **18**, 457–84.

Krieg, T.H., Hein, R., Hatamochi, A., and Aumailley, M. (1988). Molecular and clinical aspects of connective tissue. *European Journal of Clinical Investigation*, **18**, 105–23.

Kuhn, K. (1986). The collagen family, variations in the molecular and supermolecular structure. *Rheumatology, Connective tissue: Biological and Clinical Aspects*, Vol. 10, (eds. K. Kuhn and T. Krieg), pp. 29–69. Karger, Basel.

Lapernas, D., Rodnan, G.P., and Cavallo, T. (1978). Immunopathology of the renal vascular lesion of progressive systemic sclerosis (scleroderma). *American Journal of Pathology*, **91**, 243–56.

LeRoy, E.C. (1981). Scleroderma [systemic sclerosis]. In *Textbook of Rheumatology* (eds. W.N. Kelly, E.D. Harris, Jr., S. Ruddy, and C.B. Sledge), pp. 1211–30. Saunders, Philadelphia.

LeRoy, E.C. (1985). Scleroderma (systemic sclerosis). In *Textbook of Rheumatology*, 2nd ed. (eds. W.N. Kelly, E.D. Harris, Jr., S. Ruddy, and C.B. Sledge, pp. 1183–205. Saunders, Philadelphia.

LeRoy, E.C. (1987). Systemic sclerosis. In *Cecil Textbook of Medicine* (eds. J.B. Wyngaarden and L.H. Smith, Jr.) pp. 2018–24. Saunders, Philadelphia.

Lopez-Ovejero, J.A., et al. (1979). Reversal of vascular and renal crises of scleroderma by oral angiotensin-converting enzyme blockade. *New England Journal of Medicine*, **300**, 1417–9.

Masi, A.T., et al. [1980]. Preliminary criteria for the classification of systemic sclerosis (scleroderma). *Arthritis and Rheumatism*, **23**, 581–90.

Masugi, M. and Ya-Shu (1938). Die diffuse Sklerodermia und ihre Gefasveranderung. *Virchows Archiv* [Pathologie Anatomie], **302**, 39–62.

Medsger, T.A., Jr., Masi, A.T., Rodnan, G.P., Benedik, T.G., and Robinson, H. (1971). Survival with systemic sclerosis (scleroderma). *Annals of Internal Medicine*, **75**, 369–76.

Medsger, T.A., Jr. and Masi, A.T. (1973). Survival with scleroderma–II. A life-table analysis of clinical and demographic factors in 358 US veteran patients. *Journal of Chronic Diseases*, **26**, 647–60.

Medsger, T.A., Jr. and Masi, A.T. (1979). Epidemiology of progressive systemic sclerosis. *Clinics in Rheumatic Diseases*, **5**, 15–25.

Moore, H.C., and Sheehan, H.L. (1952). The kidney of scleroderma. *Lancet*, **i**, 68–70.

Pratt, B.M., Form, D., and Madri, J.A. (1985). Endothelial cell–extracellular matrix interactions. *Annals of the New York Academy of Science*, **460**, 274–88.

Roberts, J. and Maurer, K. (1977). Blood pressure levels of persons 6–74 years, United States, 1971–1974. In *Vital and Health Statistics* **11**, p. 203. DHEW Publication, Washington.

Rodnan, G.P., Shapiro, A.P., and Krifcher, E. (1964). The occurrence of malignant hypertension and renal insufficiency in progressive systemic sclerosis (diffuse scleroderma). (Abstract). *Annals of Internal Medicine*, **60**, 737.

Roll, F.J., Madri, J.A., Albert, J., and Furthmayr, H. (1980). Codistribution of collagen types IV and AB_2 in basement membranes and mesangium of the kidney. *Journal of Cell Biology*, **85**, 597–616.

Shapiro, A.P. and Medsger, T.A., Jr. (1988). Renal involvement in systemic sclerosis. In *Diseases of Kidney* 4th edn. (eds. R. Schreiner and C. Gottschalk), pp. 2272–83. Little, Brown & Co, USA.

Sharp, G.C., et al. (1972). Mixed connective tissue disease. An apparently distinct rheumatic disease syndrome associated with a specific antibody to an extractable nuclear antigen (ENA). *American Journal of Medicine*, **52**, 148–59.

Sleight, P. (1988). Essential hypertension. In *Oxford Textbook of Medicine*, 2nd edn. (eds D.J. Weatherall, J.G.G. Ledingham, and D.A. Warrell), pp. 360–82. Oxford Medical Publications, Oxford.

Steen, V.D., et al. (1984). Factors predicting the development of renal involvement in progressive systemic sclerosis. *American Journal of Medicine*, **76**, 779–86.

Stemerman, M.B. (1974). Vascular intimal components: precursors of thrombosis. *Progress on Haemostasis and Thrombosis*, **2**, 1–47.

Talbott, J.H., et al. (1939). Dermatomyositis with scleroderma, calcinosis and renal endarteritis associated with focal cortical necrosis. *Archives of Internal Medicine*, **63**, 476–96.

Traub, X.M. (1983). Hypertension and renal failure (scleroderma renal crisis) in progressive systemic sclerosis. Review of a 25-year experience with 68 cases. *Medicine*, **62**, 335–52.

Tuffanelli, D.L., and Winkelman, R.K. (1961). Systemic scleroderma: a clinical study of 727 cases. *Archives of Dermatology*, **84**, 359–71.

Uitto, J., et al. (1984). Pharmacological inhibition of excessive collagen deposition in fibrotic diseases. *Federation Proceedings*, **43**, 2815–820.

Urai, L., Nagy, Z., Szinay, G., and Waltner, W. (1958). Renal function in scleroderma. *British Medical Journal*, **2**, 1264–6.

Welsh, K.I. and Black, C.M. (1988). Environmental and genetic factors in scleroderma. In *Systemic Sclerosis—Scleroderma* (eds. M.I.V. Jayson and C.M. Black), pp. 33–47. Wiley, Chichester.

Winkelman, R.K., Goldyne, M.E., and Linscheid, R.L. (1977). Influence of cold on catecholamine response of vascular smooth muscle strips from resistance vessels of scleroderma skin. *Angiology*, **28**, 330–9.

4.8 Rheumatoid arthritis, mixed connective tissue disease, and polymyositis

PAUL EMERY AND D. ADU

Rheumatoid arthritis

Introduction

Renal disease is a well-recongnized cause of ill health and death in patients with rheumatoid arthritis (RA). Three broad categories of renal disease occur in patients with RA. The first and by far the most common is due to nephrotoxicity from the drugs used in the treatment of RA (see also Chapter 2.1.). Treatment with gold and penicillamine leads to proteinuria and a glomerulonephritis in between 10 and 30 per cent of patients and this is often severe enough to cause a nephrotic syndrome. Non-steroidal anti-inflammatory drugs are widely used for pain relief and are associated with the development of a variety of renal syndromes ranging from a reversible reduction in glomerular filtration rate to acute renal failure either due to an acute tubular necrosis or an acute interstitial nephritis. The latter may be complicated by nephrotic range proteinuria (see Chapter 6.3).

A second major cause of renal disease in RA is renal amyloid found in between 7 and 17 per cent of patients at autopsy. The clinical presentation is with proteinuria, a nephrotic syndrome, or renal impairment and progression to renal failure is inevitable. Finally, after years of debate, there is now good evidence that RA may be associated with the development of a glomerulonephritis. The main types of glomerulonephritis described are a mesangial proliferative glomerulonephritis with or without IgA deposits, a membranous nephropathy, and a focal segmental necrotizing glomerulonephritis of the vasculitic type.

Prevalence and patterns of renal disease in RA

The prevalence of renal disease in RA has been examined using two types of study, based either on death certificates or on autopsy. Each has its limitations, and there is still no agreement on the incidence of renal disease in patients with RA.

Death certificates

Studies of mortality in patients with RA based on death certificates show that there is an excess of deaths from renal failure accounting for between 3 and 12 per cent of deaths. Renal amyloid was reported as the cause of death in 1.5 to 9 per cent of cases(Table 1). Interpretation of these studies is complicated by differences in the definitions of RA, the duration of disease at death, the treatment received, the accuracy of death certification and the method of statistical analysis.

Autopsy studies (Table 2)

In these studies the proportion of renal failure ranged from 9 to 27 per cent and of renal amyloid from 8 to 17 per cent. In the study of Mutru *et al.* (1976) 27 per cent of patients with RA died from renal failure compared with 3 per cent of age- and sex-matched controls. In the patients with RA 17 per cent had renal amyloidosis compared with 1 per cent of controls. This high prevalence of renal failure in the autopsy studies is at variance with clinical experience.

Autopsy renal pathology

In addition to renal amyloid, autopsy studies in the 1940s showed glomerular endothelial proliferation in between 13 per cent

Table 1 Incidence of renal disease in rheumatoid arthritis: death certificate studies

Study	Country	Number of patients	Amyloid (%)	Renal failure (%)
Cobb *et al.* 1953	USA	130	3.1	10
Rasker and Cosh 1981	UK	43	7	11.6
Prior *et al.* 1984	UK	199	1.5	3.0
Laasko *et al.* 1986	Finland	356	8.7	11.8

Table 2 Autopsy studies in rheumatoid arthritis

Study	Country	Number of patients	Amyloid (%)	Renal failure (%)
Missen and Taylor 1956	UK	47	17	
Mutru *et al.* 1976	Finland	41	17	27
Ramirez *et al.* 1981	USA	76	8	9
Boers *et al.* 1987	Holland	132	11	23

Table 3 Renal biopsy studies in rheumatoid arthritis

Study	No. of patients	Clinical status	Normal	Mesangial proliferative glomerulonephritis	Membranous glomerulonephritis	Amyloid	Tubulo-interstitial nephritis	Other
Pollak et al. 1962	41	Normal, proteinuria	21	0	0	4	0	16
Brun et al. 1965	32	Normal, proteinuria, renal impairment	11	0	1	4	9	7
Salomon et al. 1974	18	Normal, proteinuria, microscopic haematuria	11	7	0	0	0	0
Orjavik et al. 1981	14	Proteinuria, nephrotic syndrome	0	5	0	7	0	2
Sellars et al. 1983	30	Proteinuria, microscopic haematuria, nephrotic syndrome	0	13	9	1	4	3
Horden et al. 1984	21	Microscopic haematuria	1	15	1	0	1	3
Helin et al. 1986	39	Proteinuria, nephrotic syndrome, renal impairment	3	11	9	16	0	0
Adu et al. (unpublished work)	47	Proteinuria, nephrotic syndrome, renal impairment	0	10	13	6	10	8

Table 4 Amyloid protein

Primary amyloid	AL	Immunoglobulin light chain
Myeloma-associated	AL	Immunoglobulin light chain
Secondary amyloid	AA	Amyloid protein A
Chronic haemodialysis-associated	AH	β_2-microglobulin
Heredofamilial amyloidosis	AP	Prealbumin
Local amyloid	AD/AE	Keratin/precalitain or calitain gene-related protein
Amyloid associated with ageing	AS	β-protein

(Fingerman and Andres 1943) and 63 per cent (Baggenstoss and Rosen 1943) of cases. However, many of these patients had infections and it is probable that their glomerulonephritis was caused by this. Subsequently autopsy studies have differed in the pattern of renal disease seen (Table 2). It is likely that some of these differences are due to selection bias and referral patterns. Thus some studies have shown a high incidence of glomerulonephritis and renal vasculitis (Ramirez et al. 1981; Boers et al. 1987) with little in the way of renal papillary necrosis. In contrast in the study of Nanra and Kincaid-Smith (1975), 30 per cent of patients had a renal papillary necrosis and no mention was made of either glomerulonephritis or amyloid.

Renal biopsy studies in RA

Clinical studies in patients with RA have not provided a consistent pattern of renal disease in this disorder. The first renal biopsy studies in patients with RA were performed to ascertain whether or not there was a specific rheumatoid glomerulonephritis. Most patients had minimal or minor urinary abnormalities and these studies (Pollak et al. 1962; Brun et al. 1965) were characterized by a high proportion of normal biopsies, and the almost complete absence of glomerulonephritis (Table 3). These careful studies laid the foundation for the belief that patients with RA did not develop a glomerulonephritis other than as a consequence of treatment with gold or penicillamine.

Studies in the 1970s and 1980s investigated patients with RA and significant renal disease; these studies have provided good evidence of glomerulonephritis in those patients who had not been treated with gold or penicillamine. None of these studies (Table 3) has compared the prevalence of glomerulonephritis in patients with RA with that in healthy controls. Ideally such a study should be performed to establish beyond all argument that glomerulonephritis may develop as a consequence of RA. This seems likely and is important for two reasons. The first is because the development of proteinuria and renal impairment in a patient with RA does not always indicate renal amyloid with its inexorable progression to renal failure, and the second is because some types of renal disease such as vasculitis are amenable to treatment with steroids and cyclophosphamide (see below).

Amyloidosis

Amyloidosis is the extracellular deposition of the fibrillar protein amyloid at one or more sites in the body. The types of amyloid protein are summarized in Table 4 and a full discussion of the various forms of amyloidosis and its biochemical structure is given in Chapter 4.2. There are two aspects of relevance to rheumatic diseases. The first is the clinical situation where amyloidosis mimics arthritis, and second is the much greater problem of amyloidosis occurring in patients with rheumatoid arthritis (RA).

Rheumatic symptoms in amyloidosis

Rheumatic symptoms occur in the following contexts.

1. Haemodialysis arthropathy where raised β_2-microglobulin levels result in deposition of fibrils, producing the characteristic clinical picture of an entrapment neuropathy (especially carpal tunnel syndrome), bone cysts, and destructive arthropathy with a predilection for the shoulder and wrist.
2. Amyloidosis which involves articular structures and mimicking other rheumatic complaints.

Only in the amyloidosis associated with multiple myeloma do

joint symptoms occur frequently. Here a clinical picture of symmetrical small joint involvement with subcutaneous nodules can lead to the erroneous diagnosis of RA. Helpful distinguishing features are that in myeloma the stiffness is short-lived, the joints are rarely very tender, erosions are uncommon, and (most useful), despite the presence of nodules the serum does not contain rheumatoid factor. A further clue is that there is often excessive soft tissue swelling for the degree of inflammation and in the shoulder this produces a characteristic 'shoulder pad' appearance.

Secondary amyloidosis (Cohen 1989)

Secondary amyloidosis results from deposition of fibrils containing AA (amyloid A) protein. AA is antigenically related to SAA (serum amyloid A) with which it has NH_2-terminal homology. It is thought that SAA is the precursor of AA protein, being converted to the latter, probably by COOH-terminal proteolysis. As an acute-phase reactant SAA is increased (up to 1000-fold) by inflammatory/infective stimuli. Although a raised SAA level is essential for the development of secondary amyloidosis it is not a sufficient cause on its own. There is evidence, for juvenile RA at least, that genetic factors are important in determining the development of amyloid. Furthermore there appear to be three isotypes of SAA possibly with propensities to produce AA.

Frequency and clinical presentation

RA is the commonest disease producing secondary amyloidosis in developed countries (see Chapter 4.2.). There are no precise figures on the incidence of the disorder. At autopsy, prevalence rates of 8 to 17 per cent are found (Mutru *et al.* 1976), whilst data from biopsy show a lower prevalence of around 5 to 10 per cent (Tribe 1966). There is some evidence for a decline in prevalence over the last 20 years perhaps due to better therapy leading to suppression of the acute-phase response. Males are relatively more prone to amyloidosis.

Renal problems are the chief clinical manifestation of secondary amyloidosis and the major cause of death; in one study only 3 per cent of RA patients with amyloidosis did not have pathological changes in the kidney (Wegelius 1980). Proteinuria is the most common presenting sign; others include acute or chronic renal failure and renal vein thrombosis. Occasionally there may be gastric bleeding or malabsorption. The diagnosis is based on histological examination of either rectal biopsy or abdominal fat aspiration. If these are negative in the presence of renal signs then renal biopsy is indicated. Histological Congo red staining (potassium permanganate sensitive in reactive amyloidosis and resistant for primary amyloidosis) which is birefringent in polarized light is characteristic of amyloid. Monoclonal and polyclonal antibodies that specifically bind AA are now available and these are of use for histological diagnosis.

Survival after diagnosis of amyloid is for a mean of 4 to 5 years in patients with RA (Wegelius 1980), and it is a major problem in juvenile RA patients, accounting for more than 40 per cent of deaths, the majority due to renal failure.

Treatment (see Chapter 4.2)

There is no specific therapy for AA, the general principle being suppression of the underlying chronic antigenic stimulation. Uncontrolled evidence suggests that aggressive treatment of RA may be effective in delaying renal function deterioration. Treatment of renal failure from amyloid with CAPD, haemodialysis and renal transplantation is becoming increasingly frequent (see Chapter 4.2).

Other renal histology

Proliferative glomerulonephritis

This was reported mostly in the early autopsy studies, in which it was referred to as a 'glomerulitis' and characterized by endothelial proliferation (Baggenstoss and Rosenberg 1943; Fingerman and Andrus 1943; Pollak *et al.* 1962). More recently in the autopsy study of Boers *et al.* (1987) five out of 132 cases had a proliferative glomerulonephritis (diffuse in one and focal in four). However, recent biopsy studies in patients with RA have reported little in the way of endothelial proliferative glomerulonephritis.

Mesangial proliferative glomerulonephritis

Recently several studies in patients with RA and microscopic haematuria have reported a mild mesangial proliferative glomerulonephritis, mostly without immune deposits, in patients with RA and microscopic haematuria. Sellars *et al.* (1983) described a mesangial proliferative glomerulonephritis in 13 out of 30 patients. The biopsies of nine were examined by immunofluorescent microscopy and this showed mesangial IgA deposits in two. Helin *et al.* (1986) reported a mild mesangial glomerulonephritis in 11 of 19 patients with RA and haematuria or proteinuria. Three of the 11 had glomerular mesangial IgA deposits and a further seven had deposits of IgM, IgG and C3. Details of these and other studies (Salomon *et al.* 1974; Orjavik 1981; Horden *et al.* 1984) are summarized in Table 3. The majority of these patients were taking, or had recently been taking gold or penicillamine, suggesting that their mild mesangial proliferative glomerulonephritis was drug-related.

Mesangial IgA glomerulonephritis

In the studies of Sellars *et al.* (1983) and Helin *et al.* (1986) several patients were described with a mesangial proliferative glomerulonephritis and mesangial IgA deposits. These studies raised the possibility of an association between RA and mesangial IgA glomerulonepritis. In most patients mesangial IgA nephropathy is idiopathic (Berger and Hinglais 1968) but it has also been reported in patients with cirrhosis of the liver, dermatitis herpetiformis, mycosis fungoides, arthritis after yersinia infection and seronegative spondyloarthritis. Recently we reported four patients with RA, microscopic haematuria, and proteinuria in whom renal biopsy showed a mesangial IgA glomerulonephritis (Fig. 1); we also reviewed the literature on this association (Beaman *et al.* 1987). Three of these patients had never received gold or penicillamine and in the one patient who had received gold this had been discontinued at least 10 years before presentation. These data provide circumstantial evidence of an association between RA and mesangial IgA nephropathy. As in idiopathic mesangial IgA nephropathy the pathogenetic mechanisms of this disorder in patients with rheumatoid arthritis are unclear. It may be relevant that raised serum levels of IgA and also of IgA rheumatoid factors are found in some patients with RA. There are no long-term studies of renal function in patients with rheumatoid arthritis and mesangial IgA glomerulonephritis.

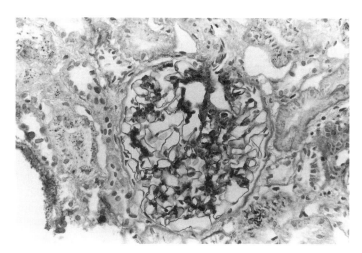

Fig. 1 Glomerulus from a patient with rheumatoid arthritis showing mesangial proliferation and mesangial deposits of IgA (immunoperoxidase staining for IgA × 100). (Courtesy of Dr A. J. Howie.)

Membranous nephropathy

The most common cause of membranous nephropathy in patients with RA is gold or penicillamine therapy. However, there are now many reports of patients with RA who had not been on gold or penicillamine who have a membranous nephropathy (Fig. 2). The numbers of reported cases make it unlikely that this association is coincidental. Recently Honkanen et al. (1987) reported four patients with RA and membranous nephropathy and reviewed the literature. Only one of their four patients had received gold and that was 16 years before the renal biopsy. We have recently observed six patients with RA and a membranous nephropathy (Berisa et al. in press). None had been treated with penicillamine and two had received gold which was discontinued 17 and 13 years before renal biopsy. None of the patients in our study or in the previous reports (reviewed by Honkanen 1987) had clinical or serological evidence of systemic lupus erythematosus, which may lead to a membranous nephropathy.

Renal vasculitis

The clinical spectrum of RA includes a systemic necrotizing vasculitis with involvement of blood vessels ranging in size from capillaries to small- and medium-sized arteries. The clinical presentation includes nail-fold infarcts, a leucocytoclastic vasculitis, a peripheral neuropathy, pericarditis, gastrointestinal infarcts, and renal vasculitis (Scott et al. 1981). Although renal abnormalities are found in about 25 per cent of patients with rheumatoid vasculitis (Scott et al. 1981), there are few reports of renal histology in these patients. Recently Kuznetsky et al. (1986) reported four patients with RA and a segmental necrotizing glomerulonephritis of the vasculitic type. All had immunoglobulin deposits which were focal and segmental or in one case characteristic of membranous nephropathy. Renal function was significantly impaired in all four, two of whom were taking gold or penicillamine. Three patients were treated with prednisolone and cyclophosphamide; renal function improved in two and deteriorated in one.

In the autopsy study of Boers et al. (1987), 18 of the 132 patients had a systemic vasculitis. Of 18 cases, eight had a large vessel renal arteritis and of these four had a proliferative glomerulonephritis. Of the 10 cases without a renal arteritis, two had a severe proliferative glomerulonephritis, two a membranous nephropathy and one amyloid. We have recently observed a further five patients with RA and a segmental necrotizing glomerulonephritis (one also with membranous nephropathy). Two of our patients had been on penicillamine and we could not exclude the possibility that the necrotizing glomerulonephritis was due to this drug (Falck et al. 1979). Four of our five patients had clinical evidence of extrarenal vasculitis and all patients had significant renal impairment which in two was severe enough to require dialysis. All five patients were treated with prednisolone and cyclophosphamide with improvement in their renal function.

These studies show that in addition to a renal arteritis, patients with RA may develop a segmental necrotizing glomerulonephritis shown by light microscopy to be similar to that seen in microscopic polyarteritis and Wegener's granulomatosis. It differs from these disorders by having more prominent deposition of immunoglobulins within glomeruli but responds to treatment with prednisolone and cyclophosphamide in a similar manner.

Renal disease in juvenile rheumatoid arthritis

Renal failure is the cause of death in approximately 38 per cent of patients with juvenile rheumatoid arthritis (JRA). The spectrum of renal involvement in JRA has been reviewed by Anttila (1972). Proteinuria is found in between 3 and 12 per cent and microscopic haematuria in 3 to 8 per cent of these patients. Nephrotic range proteinuria is commonly due to renal amyloid, found in between 1.2 and 6.7 per cent of patients with JRA, whilst haematuria and proteinuria may be due to amyloid or to gold treatment.

In one study (Anttila and Laaksonen 1969), 47 of 638 patients (7.4 per cent) with JRA had renal disease. In a study of renal histology (57 renal biopsy, three autopsy) of a relatively unselected group of patients with JRA (Anttila 1972) renal amyloid was detected in two (21.7 per cent) and an interstitial nephritis in

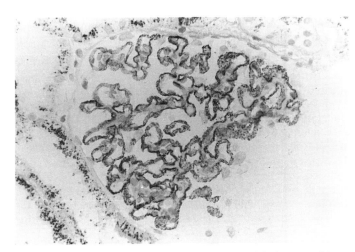

Fig. 2 Membranous nephropathy in a patient with rheumatoid arthritis who had not been treated with gold or penicillamine. Note deposits of IgG on outer surface of the glomerular basement membrane (immunoperoxidase staining of IgG × 125). (Courtesy of Dr A. J. Howie.)

eight patients (13.3 per cent). It is possible that the interstitial nephritis in these patients was due to treatment with phenylbutazone or salicylates.

Tubulointerstitial nephritis

Renal biopsy studies in patients with rheumatoid arthritis and juvenile rheumatoid arthritis have reported that up to 28 per cent of patients with renal abnormalities have a tubulointerstitial nephritis (Anttila 1972; Brun et al. 1965; Sellars et al. 1983). This is often attributed to the renal toxicity of analgesic agents, including NSAIDs. It is, however, probable that these disorders may, in themselves, be associated with the development of a tubulointerstitial nephritis (see Chapter 8.10).

Gold and penicillamine nephropathy

There is no cure for patients with RA. The aim of disease-modifying antirheumatic drugs (DMARDs), therefore, is to suppress disease activity. Gold and D-penicillamine are the prototypes of this group of drugs and despite the introduction of alternatives remain the mainstay of therapy. Of the newer drugs the renal toxicity of cyclosporin A is discussed later. However we will focus mainly on the first two drugs.

Despite intense investigation, the mode of action of these drugs is unknown although the immunological consequences of an improvement in disease activity are well documented. They have the highest prevalence of side-affects produced by any drugs outside those used in malignancy. The incidence of toxicity, up to 70 per cent, is more frequent than lack of efficacy; withdrawal of the drugs as a result of these adverse reactions occurs in around 40 per cent of patients (Huskisson et al. 1974). Therefore the avoidance or reduction of toxicity would greatly improve patient care.

Clinical features of drug-induced disease

In general gold and D-penicillamine produce very similar problems (Huskisson et al. 1974) and where not specifically mentioned, the following discussion applies to both drugs. The side-effects of these drugs can be divided temporally into two groups. The first group consists of those side-effects occurring early in therapy. These are believed to be due to a direct toxic effect of the drug and examples include anorexia and certain rashes. The second group consists of those reactions which occur after 3 months and maximally around 6 months. These are believed to have an immunological basis and renal toxicity comes within this category.

The immunological reactions most frequently observed are glomerulonephritis and thrombocytopenia. Drug-induced autoimmune disorders such as systemic lupus erythematosus (D-penicillamine) occur less frequently (Emery and Panayi 1989). The prevalence of side-effects increases with the dose of D-penicillamine (Williams et al. 1983). This has not been shown for gold probably because it is given intermittently according to a fixed regimen.

The most frequent presenting feature is proteinuria; haematuria is relatively rare (and seen more frequently with penicillamine). Proteinuria occurs in approximately 10 per cent of patients receiving gold and up to 30 per cent of patients taking penicillamine. This progresses to the nephrotic syndrome in 30 and 16 per cent respectively. Haematuria occurring in the context of therapy with these drugs still requires the exclusion of other causes.

Renal function

In general these drugs do not produce a significant drop in GFR although a wide range of renal function has been documented. A secondary fall in GFR may occur secondary to hypovolaemia and renal damage in severe cases of the nephrotic syndrome. Furthermore, there is no significant deterioration in renal function with time provided the drug is stopped.

Pathology
Membranous glomerulonephritis

About 80 per cent of patients who present with D-penicillamine or gold-induced proteinuria will have membranous glomerulonephritis. Epimembranous spikes and a mild increase in mesangial cells are usually seen, and the diagnosis can be confirmed with immunofluorescence/immunoperoxidase microscopy which shows granular subepithelial deposits of predominantly IgG. On electron microscopy electron-dense deposits are seen (see Chapter 3.7).

Mesangial glomerulonephritis

Mesangial glomerulonephritis is found in patients with RA irrespective of therapy. The prevalence is probably increased after treatment with disease-modifying antirheumatic drugs and for gold at least there is an increased association between this histological appearance and haematuria. Immunofluorescence may reveal either granular deposits of immunoglobulin (predominantly IgG) and complement or may be negative particularly in the case of D-penicillamine-induced disease.

Minimal change nephropathy

This can occur in association with the use of disease-modifying antirheumatic drugs (Lee et al. 1965). Electron microscopy shows fusion of epithelial cell foot processes.

Other renal lesions

Tubulointerstitial disease   This is found in up to 10 per cent of patients in association with mild low molecular weight proteinuria. The outlook is good with rapid resolution on withdrawal of the drug. A proliferative glomerulonephritis with crescents has also been reported.

Autoantibody production   D-Penicillamine therapy has been associated with a wide range of autoantibodies, and these may lead directly to renal problems. In particular the development of antiglomerular basement membrane antibody and the clinical picture of Goodpasture's syndrome (Sternlieb et al. 1975) (which has also been demonstrated in the absence of this antibody) is recognized, and antibodies against DNA may lead to a drug-induced syndrome of systemic lupus erythematosus. Whilst most drug-induced lupus syndromes have antibodies directed against histones, occasionally patients with D-penicillamine-induced disease may have antibodies against double stranded DNA and may develop significant renal disease (Chalmers et al. 1982).

Management

An increasing experience of patients with renal toxicity to D-penicillamine and gold has revealed a benign long-term outcome (Collins 1987, Hall 1988). The management of renal toxicity

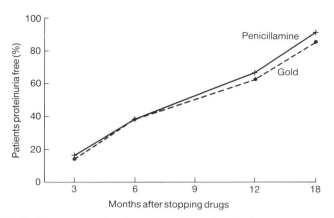

Fig. 3 Percentage of patients free of proteinuria after discontinuing gold or penicillamine (Hall *et al.* 1987, 1988).

occurring *de novo* on therapy is now determined by practical issues. These include the response of the patient, the level of proteinuria, and the presence of any deterioration in renal function. In general a falling albumin, proteinuria greater than 2 g/24 h, or a reduced GFR are considered indications for stopping treatment. No specific immunosuppression is required for the drug-induced disorder, although supportive measures are given as indicated.

In those patients without renal abnormalities before treatment renal biopsy should be confined to those who have deteriorating renal function, or who fail to improve after withdrawal of the drug. In practice, therefore, patients who develop proteinuria are often managed empirically by a reduction in either the frequency or the size of the dose of the drug, with withdrawal of the drug only if any of the adverse factors are present. Regular monitoring of proteinuria and GFR are mandatory. Haematuria, unless it resolves when the drug is stopped requires investigation in the usual manner.

Long-term outcome

After cessation of the drug, proteinuria peaks at around a month then gradually disappears; the majority of patients will have clear urine by 1 year and almost all will achieve this by 2 years (Fig. 3) (Hall 1988; Hall *et al.* 1988). Renal function does not deteriorate in uncomplicated cases. Given the genetic basis (the same for both drugs, see below) of most renal toxicity it is not surprising that rechallenge with the same drug at the same dose usually leads to a recurrence of the renal problem, although a lower dose may be tolerated. The dilemma of whether to restart the drug is now less of a problem because of the increasing number of alternative therapies.

Predicting toxicity

HLA and drug toxicity

Soon after the demonstration of an increased prevalence of the Class II MHC antigen HLA DR4 in patients with RA it was shown that the risk of developing certain immunological adverse reactions to gold was increased in patients with HLA DR3 (Panayi *et al.* 1978; Wooley *et al.* 1980). Several studies have now confirmed this observation and it has been extended to D-penicillamine. Seventy-five per cent of patients with idiopathic membranous glomerulonephritis are HLA DR3 positive (see Chapter 3.7) and HLA DR3 confers a relative risk of 14.0 to 32.0 for gold-induced nephropathy and 3.2 to 10.0 for D-penicillamine-induced nephropathy (Klouda *et al.* 1979).

Metabolic factors which determine disease-modifying drug-induced toxicity

It is known that polymorphisms exist for certain metabolic pathways. These are genetically determined and are investigated using probe drugs. Adverse reactions to drugs metabolized by these pathways occur only or much more severely in the individuals who metabolize poorly. The ability for sulphoxidation was originally assessed in patients with toxicity due to D-penicillamine because of the similarity between the structure of the drug and the probe-drug carbocysteine. The relative risk of toxicity was found to be 7.5 in those patients with poor sulphoxidation. HLA DR3 status was also assessed and although the two risk factors were not additive the possession of either DR3 or poor sulphoxidation produced a relative risk of 25.0 (Emery *et al.* 1984). Subsequently, poor sulphoxidation ability has been examined in patients treated with gold and has been shown to be a significant risk factor for gold toxicity also (Madhok *et al.* 1987). The mechanism for this association between sulphoxidation status and toxicity is unknown. As the only common structural similarity between gold and penicillamine is a thiol group (gold is usually given as aurothiomalate), the metabolism of this may be the critical determinant of toxicity. However, it is known that poor sulphoxidation is part of a number of other genetically determined 'linked' enzyme systems (e.g. glucuronidation and S-methylation) and it is possible that one of these is important in inducing the immunological abnormalities.

Specific pathogenic mechanisms

Heymann and Lund (1951) showed that injection of autologous renal cortex in complete Freund's adjuvant induced specific autoantibodies against the renal tubular epithelium, and that this was associated with the development of a membranous nephropathy. Gold and penicillamine can cause toxic damage to the tubular epithelium, and gold deposits can be demonstrated in renal tubules many years after use, but whether this relates to their ability to cause a membranous glomerulonephritis is unknown.

The mechanisms involved in the production of other autoantibodies by D-penicillamine and to a lesser extent gold have been the subject of speculation. Theories include the suggestion that the drug acts as a hapten by binding to tissue components and thus directly stimulates autoantibody production; that an immunomodulatory role of the drug allows release of previously suppressed B cell clones; and that the drug interacts with pre-existing autoantibodies altering their affinity and possibly their valency which (by altering the size of the antibody antigen complex) could lead to immune complex formation (Gleichmann and Gleichmann 1981).

Cyclosporin A nephrotoxicity

The use of cyclosporin A in patients with rheumatoid arthritis is increasing and preliminary data suggest it is capable of disease modification. Early studies were characterized by a high dropout rate due to renal toxicity. Almost all patients showed a rise in creatinine and fall in GFR, and it was felt that concomitant administration of non-steroidal anti-inflammatory drugs with a consequent loss of renal prostaglandins was a major factor in their renal impairment. Recent evidence from randomized studies (Canadian Clinical Epidemiology Rheumatology Research Group, 1989) suggests that low dose (less than 5 mg/kg adjusted for any rise in creatinine) can greatly reduce the deterioration in renal function and lead to high patient acceptability. The deterioration in renal function has been shown to be reversible for patients treated for 6 months; data on patients on long-term therapy are awaited (see also Chapter 11.5).

Analgesic nephropathy

Patients with RA often takes analgesics over many years. Until the 1960s analgesia was obtained with aspirin or less commonly with compound analgesic drugs. More recently the newer non-steroidal anti-inflammatory drugs such as indomethacin have been widely used. Renal papillary necrosis and chronic tubulointerstitial nephritis leading to renal failure was a well-recognized complication of compound analgesics, in particular of those that contained phenacetin (Cove-Smith and Knapp 1978). These renal complications also occurred in patients with RA (Cove-Smith and Knapp 1978). Most patients with RA were however treated with aspirin as the sole analgesic agent and the occasional observation at autopsy of renal papillary necrosis in these patients led to suggestions that aspirin on its own was nephrotoxic (Lawson and Maclean 1966; Nanra and Kincaid-Smith 1975). The New Zealand Rheumatism Association (1974) study of 763 patients with RA and 145 patients with osteoarthritis, however, found no association between aspirin intake and renal dysfunction and concluded that the risk of nephrotoxicity from this drug was of a low order.

Phenacetin was withdrawn by the 1970s and at this time aspirin was supplanted by the newer non-steroidal anti-inflammatory drugs (NSAIDs) and paracetamol. NSAIDs are potentially nephrotoxic (Dunn 1984) and their renal side-effects are described in Chapter 6.2. In patients with RA these drugs may lead to a reversible reduction in glomerular filtration rate, acute tubular necrosis, an acute interstitial nephritis often with heavy proteinuria, renal papillary necrosis, and chronic tubulointerstitial nephritis (Adams et al. 1986; Unsworth et al. 1987). Paracetamol is also widely used and there are suggestions that it may be nephrotoxic (Sandler et al. 1989).

Mixed connective tissue disease

Some patients with a connective tissue disorder do not fit easily into the accepted definitions of a single disease. In 1972 Sharp et al. reported 25 patients in whom there was an overlap of the clinical features of systemic lupus erythematosus, systemic sclerosis and polymyositis. Sera from all of these patients contained antibodies to a RNAase sensitive extractable nuclear antigen subsequently shown to be ribonucleoprotein. These patients lacked antibody to Sm, had infrequent and low titres of antibody to double-stranded DNA and normal or raised serum levels of complement. They termed this disorder mixed connective tissue disease (MCTD). Subsequent studies have included patients with features at onset similar to rheumatoid arthritis and emphasized the asynchronous development of overlapping clinical features of different connective tissues in this syndrome (Bennett and O'Connell 1980).

In the original report of Sharp et al. (1972), of 25 patients with MCTD, only one developed evidence of renal disease. A follow-up study 4 to 5 years later showed that renal disease remained an infrequent complication of MCTD (Nimelstein et al. 1980). Subsequent studies in patients with MCTD have, however, shown that the prevalence of renal involvement was high ranging from 10 to 40 per cent in adults (reviewed by Kitridou et al. 1986) and up to 47 per cent in children (Singsen et al. 1980).

Clinical presentation

The clinical presentation of renal disease in patients with MCTD usually takes the form of symptomless proteinuria or haematuria often with mild renal impairment. Some patients with MCTD however develop a nephrotic syndrome and rarely there is progression to renal failure (Kitridou et al. 1986).

Histological changes

In 1986 Kitridou et al. reported the renal histological changes in 12 patients with MCTD and reviewed previous reports in 64 patients. Membranous nephropathy and a mesangial proliferative glomerulonephritis were the most common histological changes found in 34 and 30 per cent of cases respectively. A focal or diffuse proliferative glomerulonephritis was found in 17 per cent, a mixed lesion with membranous nephropathy in 5 per cent and in 7 per cent renal histology was normal. Seventeen per cent of patients had evidence of vascular sclerosis. Immunofluorescent microscopy of glomeruli in patients with MCTD have shown immunoglobulin and complement deposits and on electron microscopy dense deposits are found consistent with an immune-complex-mediated glomerulonephritis (Kitridou et al. 1986). In children widespread vascular intimal proliferation has been reported (Singsen 1980).

Treatment

Treatment of renal disease in MCTD is with steroids initially in high doses, tapered to a maintenance low dose over weeks (Glassock et al. 1982; Kitridou et al. 1986). In the study of Kitridou et al. treatment with high-dose steroids of patients with a nephrotic syndrome was associated with a significant reduction of proteinuria in 62 per cent of episodes. Whether patients with renal disease resistant to steroids would benefit from the addition of immunosuppressant drugs is not known. Some 14 per cent of patients with MCTD and renal disease reviewed by Kitridou et al. (1986) developed chronic renal failure. The long-term mortality of MCTD has varied from 7 per cent over a mean follow up of 7 years (Sharp 1981) to 30 per cent over a period of 3 to 25 years after onset of disease (Nimelstein et al. 1980).

Polymyositis

Inflammatory diseases of skeletal muscle may be idiopathic or secondary to a wide variety of disorders (see Bradley 1985, for a review). Primary adult idiopathic polymyositis usually presents with insidious proximal muscle weakness and muscle pain often without previous illness although some patients have a preceding febrile illness, Raynaud's phenomenon, or arthralgia. Rarely severe muscle weakness develops abruptly and this may be associated with myoglobinuria and diagnosed by raised serum levels of creatine kinase.

Acute renal failure has been reported infrequently in patients with polymyositis and has been attributed to rhabdomyolysis and myoglobinuria (Sloan et al. 1978). The clinical evolution of the acute renal failure is similar to that seen with other causes of non-traumatic rhabdomyolysis. However, treatment with high-dose steroids is necessary in patients with polymyositis. There are occasional reports of a mesangial proliferative glomerulonephritis with mesangial deposits of immunoglobulin and complement in patients with polymyositis (Dyck *et al.* 1979).

Bibliography

Anttila, R. and Laaksonen, A.L. (1969). Renal disease in juvenile rheumatoid arthritis. *Acta Rheumatologica Scandinavica*, **15**, 99–111.

Bradley, W.G. (1985). Inflammatory diseases of muscle. In *Textbook of Rheumatology* 2nd edn (ed. W.N. Kelly, E.D. Harris, S. Ruddy, and C. B. Sledge), **Vol. 2**, pp. 1225–45, Saunders, Philadelphia.

Cohen, A.S. (1989). In *Amyloidosis in Arthritis and Allied Conditions*. (ed. D.J. McCarty), pp. 1273–1293. Lea and Febiger, Philadelphia.

Dunn, M.J. (1984). Non-steroidal anti-inflammatory drugs and renal function. *Annual Review of Medicine*, **35**, 411–28.

Hall, C.L. (1988). Gold nephropathy. *Nephron*, **50**, 265–72.

Kitridou, R.C., Akmal, M., Turkel, S.B., Ehresmann, G.R., Quismorio, F.P., and Massry, S.G. (1986). Renal involvement in mixed connective tissue disease; a longitudinal clinicopathologic study. *Seminars in Arthritis and Rheumatism*, **16**, 135–45.

Laasko, M., Mutru, O., Osomaki, H., and Koota, K. (1986). Mortality from amyloidosis and renal diseases in patients with rheumatoid arthritis. *Annals of Rheumatic Diseases* **45**, 663–7.

References

Adams, D.H., Michael, J., Bacon, P.A., Howie, A.J., McConkey, B., and Adu, D. (1986). Non-steroidal anti-inflammatory drugs and renal failure. *Lancet*, **i**, 57–60.

Anttila, R. (1972). Renal involvement in juvenile rheumatoid arthritis. A clinical and histopathological study. *Acta Paediatrica Scandinavica*, Supplement, **227**, 1–73.

Anttila, R. and Laaksonen, A.L. (1969). Renal disease in juvenile rheumatoid arthritis. *Acta Rheumatologica Scandinavica*, **15**, 99–111.

Baggenstoss, A.H. and Rosenberg, E.F. (1943). Visceral lesions associated with chronic infectious (rheumatoid) arthritis. *Archives of Pathology*, **35**, 503–16.

Beaman, M., Adu, D., Howie, A.J., McConkey, B., Michael, J., and Popert, A.J. (1987). Rheumatoid arthritis and IgA nephropathy. *British Journal of Rheumatology*, **26**, 299–302.

Bennett, R.M. and O'Connell, D.J. (1980). Mixed connective tissue disease: a clinicopathologic study of 20 cases. *Seminars in Arthritis and Rheumatism*, **10**, 25–51.

Berger, J. and Hinglais, N. (1968). Les dépôts intercapillaries d'IgA-IgG. *Journal d'Urologie et de Nephrologie*, **74**, 694–5.

Boers, M., *et al.* (1987). Renal findings in rheumatoid arthritis: clinical aspects of 132 necropsies. *Annals of the Rheumatic Diseases*, **46**, 658–63.

Bradley, W.G. (1985). Inflammatory diseases of muscle. In *Textbook of Rheumatology*, 2nd edn. (ed. W.N. Kelly, E.D. Harris, S. Ruddy and C.B. Sledge) **Vol. 2**, pp. 1225–1245, Saunders, Philadelphia.

Brun, C., Olsen, T.S., Raaschou, F., and Sorensen, A.W.S. (1965). Renal biopsy in rheumatoid arthritis. *Nephron*, **2**, 65–81.

Canadian Clinical Epidemiology Rheumatology Research Group (1989). Randomized trial of low dose Cyclosporin A in severe RA. *Arthritis and Rheumatism*, **supplement 32**, 517.

Chalmers, A., *et al.* (1982). Systemic lupus erythematosus during penicillamine therapy for rheumatoid arthritis. *Annals of Internal Medicine*, **97**, 659–63.

Cobb, S., Anderson, F., and Bauer, W. (1953). Length of life and cause of death in rheumatoid arthritis. *New England Journal of Medicine*, **249**, 553–6.

Cohen, A.S. (1989). In *Amyloidosis in Arthritis and Allied Conditions*. (ed. D.J. McCarty), pp. 1273–1293. Lea and Febiger, Philadelphia.

Collins, A.J. (1987). Gold treatment for rheumatoid arthritis reassurance on proteinuria. *British Medical Journal*, **295**, 739–40.

Cove-Smith, J.R. and Knapp, M.S. (1978). Analgesic nephropathy: an important cause of chronic renal failure. *Quarterly Journal of Medicine*, **47**, 49–69.

Dunn, M.J. (1984). Non-steroidal anti-inflammatory drugs and renal function. *Annual Review of Medicine*, **35**, 411–28.

Dyck, R.F., *et al.* (1979). Glomerulonephritis associated with polymyositis. *Journal of Rheumatology*, **6**, 336–44.

Emery, P. and Panayi, G.S. (1989). Autoimmune reactions to D-penicillamine. In *Autoimmunity and Toxicology*, (ed.M.E. Kammuller, N. Bloksma, W. Seinan), pp. 167–182, Elsevier, Amsterdam.

Emery, P., *et al.* (1984). D-Penicillamine induced toxicity in rheumatoid arthritis. The role of sulphoxidation status and HLA DR3. *Journal of Rheumatology*, **11**, 626–32.

Falck, H.M., Tornroth, T., Kock, B., and Wegelius, O. (1979). Fatal renal vasculitis and minimal change glomerulonephritis complicating treatment with penicillamine. *Acta Medica Scandinavica*, **205**, 133–8.

Fingerman, D.L. and Andrus, F.C. (1943). Visceral lesions associated with rheumatoid arthritis. *Annals of the Rheumatic Diseases*, **3**, 168–81.

Glassock, R.J., Goldstein, D., Akmal, M., Kitridou, R., and Koss, M. (1982). Recurrent acute renal failure in a patient with mixed connective tissue diseases. *American Journal of Nephrology*, **5**, 276–90.

Gleichmann, E. and Gleichmann, H. (1981). Spectrum of disease caused by alloreactive T cells. In *Immunoregulation in Autoimmunity*. (ed. R.S. Krakamer), **Vol. 13**, pp. 73–84, Elsevier, North Holland.

Hall, C.L. (1988). Gold nephropathy. *Nephron*, **50**, 265–72.

Hall, C.L., Fothergill, N.J., Blackwell, M.M., Harrison, P.R., Mackenzie, J.C., and MacIver, A.G. (1987). The natural course of gold nephropathy: long term study of 21 patients. *British Medical Journal*, **295**, 745–84.

Hall, C.L., *et al.* (1988). Natural course of penicillamine nephropathy: a long term study of 33 patients. *British Medical Journal*, **296**, 1083–6.

Helin, H., Korpela, M., Mustonen, J., and Pasternack, A. (1986). Mild mesangial glomerulopathy—a frequent finding in rheumatoid arthritis patients with haematuria or proteinuria. *Nephron*, **42**, 224–30.

Heymann, W. and Lund, H.Z. (1951). Nephrotic syndrome in rats. *Pediatrics*, **7**, 691–706.

Honkanen, E., Tornroth, T., Patterson, E., and Skrifvars, B. (1987). Membranous glomerulonephritis in rheumatoid arthritis not related to gold or D-penicillamine therapy: a report of four cases and review of the literature. *Clinical Nephrology*, **27**, 87–93.

Horden, L.D., Sellars, L., Morley, A.R., Wilkinson, R., Thompson, M., and Griffiths, I.D. (1984). Haematuria in rheumatoid arthritis: an association with mesangial glomerulonephritis. *Annals of the Rheumatic Diseases*, **43**, 440–3.

Huskisson, E.C. *et al.* (1974). Trial comparing D-penicillamine and gold in rheumatoid arthritis. *Annals of the Rheumatic Diseases*, **33**, 532–5.

Kitridou, R.C., Akmal, M., Turkel, S.B., Ehresmann, G.R., Quismoria, F.P., and Massry, S.G. (1986). Renal involvement in mixed connective tissue disease: a longitudinal clinicopathologic study. *Seminars in Arthritis and Rheumatism*, **16**, 135–45.

Klouda, P.T., *et al.* (1979). Strong association between idiopathic membranous nephropathy and HLA DR3. *Lancet*, **2**, 770–2.

Kuznetsky, K.A., Schwartz, M.M., Lohmann, L.A., and Lewis E.J. (1986) Necrotizing glomerulonephritis in rheumatoid arthritis. *Clinical Nephrology*, **26**, 257–64.

Laasko, M., Mutru, O., Osomaki, H., and Koota, K. (1986). Mortality from amyloidosis and renal diseases in patients with rheumatoid arthritis. *Annals of the Rheumatic Diseases*, **45**, 663–7.

Lawson, A.A.H. and Maclean, N. (1966). Renal disease and drug therapy in rheumatoid arthritis. *Annals of the Rheumatic Diseases*, **25**, 441-9.

Lee, J.C., *et al.* (1965). Renal lesions associated with gold therapy: light and electron microscopic studies. *Arthritis and Rheumatism*, **8**, 1–13.

Madhok, R., Capell, H., and Waring, R.H. (1987). Does sulphoxidation state predict gold toxicity in rheumatoid arthritis? *British Medical Journal*, **294**, 483.

Missen, G.A.K. and Taylor, J.D. (1956). Amyloidosis in rheumatoid arthritis. *Journal of Pathology and Bacteriology*, **71**, 179–92.

Mutru, O., Koota, K., and Isomaki, H. (1976). Causes of death in autopsied RA patients. *Scandinavian Journal of Rheumatology*, **5**, 239–40.

Nanra, R.S. and Kincaid-Smith, P. (1975). Renal papillary necrosis in rheumatoid arthritis. *Medical Journal of Australia*, **1**, 194–7.

New Zealand Rheumatism Association Study (1974). *British Medical Journal*, **1**, 593–6.

Nimelstein, S.H., Brody, S., McShane, D., and Holman, H.R. (1980). Mixed connective tissue disease: a subsequent evaluation of the original 25 patients. *Medicine*, **59**, 239–48.

Orjavik, O., Brodwll, E.K. Oystese, B., Natvig, J.B., and Mellbye, O.J. (1981). A renal biopsy study with light and immunofluorescent microscopy in rheumatoid arthritis. *Acta Medica Scandinavica*, **Supplement 645**, 9–14.

Panayi, G.S., Wooley, P., and Batchelor, J.R. (1978). Genetic basis of rheumatoid disease: HLA antigens, disease manifestations and toxic reactions to drugs. *British Medical Journal*, **2**, 1326–8.

Pollak, V.E., Pirani, C.L., Steck, I.E., and Kark, R.M. (1962). The kidney in rheumatoid arthritis: studies by renal biopsy. *Arthritis and Rheumatism*, **5**, 1–8.

Prior, P., Symmons, D.P.M., Scott, D.L., Brown, R., and Hawkins, C.F. (1984). Cause of death in rheumatoid arthritis. *British Journal of Rheumatology*, **23**, 92–9.

Ramirez, G., Lambert, R., and Bloomer, H.A. (1981). Renal pathology in patients with rheumatoid arthritis. *Nephron*, **29**, 124–6.

Rasker, J.J. and Cosh, J.A. (1981). Cause and age of death in a prospective study of 100 patients with rheumatoid arthritis. *Annals of the Rheumatic Diseases*, **40**, 115–20.

Salomon, M.I., Gallo, G., Poon, T.P., Goldblatt, M.V., and Tchertkoff, K. (1974). The kidney in rheumatoid arthritis. A study based on renal biopsies. *Nephron*, **12**, 297–310.

Sandler, D.P., *et al.* (1989). Analgesic use and chronic renal disease. *New England Journal of Medicine*, **320**, 1238–43.

Scott, D.G.I., Bacon, P.A., and Tribe, C.R. (1981). Systemic rheumatoid vasculitis: a clinical and laboratory study of 50 cases. *Medicine*, **60**, 288–97.

Sellars, L., Siamopoulos, K., Wilkinson, R., Leokapand, T., and Morley, A.R. (1983). Renal biopsy appearances in rheumatoid disease. *Clinical Nephrology*, **20**, 114–20.

Sharp, G.C. (1981). Mixed connective tissue disease and overlap syndromes. In *Textbook of Rheumatology*, 1st edn (eds. W.N. Kelley, E.D. Harris, S. Ruddy, and C.B. Sledge), vol. 2, pp. 1151–61, Saunders, Philadelphia.

Sharp, G.C., Irvin, W.S., Tan, E.M., Gould, G.R., and Holman, H.S. (1972). Mixed connective tissue disease—an apparently distinct rheumatic disease syndrome associated with a specific antibody to an extractable nuclear antigen. *American Journal of Medicine*, **52**, 148–59.

Singsen, B.H., Swanson, V.L., Bernstein, B.H., Henser, E.T., Hanson, V., and Landing, B.H. (1980). A histological evaluation of mixed connective tissue disease in childhood. *American Journal of Medicine*, **68**, 710–17.

Sloan, M.F., Franks, A.J., Exley, K.A., and Davison, A.M. (1978). Acute renal failure due to polymyositis. *British Medical Journal*, **2**, 1457.

Sternlieb, I., Bennett, B., and Scheinburg, I.H., (1975). D-Penicillamine induced Goodpasture's disease in Wilson's disease. *Annals of Internal Medicine*, **82**, 673–6.

Tribe, C.R. (1966). Amyloidosis in rheumatoid arthritis. In *Modern Trends in Rheumatology*, (ed. A.G.S. Hill), pp. 121–38, Butterworth, London.

Unsworth, J., Sturman, S., Lunec, J., and Blake, D.R. (1987). Renal impairment associated with non-steroidal anti-inflammatory drugs. *Annals of the Rheumatic Diseases*, **46**, 233–6.

Wegelius, O., Wafin, F., Falck, M. N., and Tornroth, T. (1980). Follow up of amyloidosis secondary to rheumatic disease. In *Amyloid and Amyloidosis*, (eds. G.G. Glenner, P.P. de Costa, and A.F. de Freitas) pp. 337–42, Excerpta Medica, Amsterdam.

Williams, N.J., Ward, J.R., and Reading, J.C. (1983). Low dose D-penicillamine therapy in rheumatoid arthritis. *Arthritis and Rheumatism*, **26**, 581–92.

Wooley, P.H., *et al.* (1980). HLA-DR antigens and toxic reactions to sodium aurothiomalate and D-penicillamine in patients with rheumatoid arthritis. *New England Journal of Medicine*, **303**, 300–2.

4.9 Essential mixed cryoglobulinaemia

GIUSEPPE D'AMICO

Definition and classification of cryoglobulinaemias

Cryoglobulinaemia is a pathologic condition in which the blood contains immunoglobulins that have the property of reversible precipitation in the cold. According to the classification based on the chemistry of the cryoglobulins involved, proposed by Brouet *et al.* (1974) and universally accepted, there are three types of cryoglobulinaemia (Table 1). In type I cryoglobulinaemia, the cryoprecipitable immunoglobulin is a single monoclonal type, and is found in patients with multiple myeloma, Waldenström macroglobulinaemia or idiopathic monoclonal gammopathy. Type II and III cryoglobulinaemias are mixed types, with at least two immunoglobulins. In both a polyclonal IgG is bound to another immunoglobulin, which is an antiglobulin, i.e., acts as an anti-IgG rheumatoid factor. The important difference between these two types of mixed cryoglobulinaemias is that in type II the antiglobulin component, which is usually of the IgM class, is monoclonal, while in type III it is polyclonal.

Roughly 60 to 75 per cent of all cryoglobulinaemias are found in patients with other identifiable illnesses such as connective tissue diseases, infectious or lymphoproliferative disorders, hepatobiliary diseases, and immunologically-mediated glomerular diseases, and can therefore be considered 'secondary mixed cryoglobulinaemias'. However, in approximately 30 per cent of all mixed cryoglobulinaemias, no underlying or associated dis-

4.9 Essential mixed cryoglobulinaemia

Table 1 Classification of cryoglobulinaemias and associated diseases. Only essential type II mixed cryoglobulinaemia produces the type of cryoglobulinaemic glomerulonephritis described in this chapter

Type I Single monoclonal IgA, or IgG, or IgM	Type II Polyclonal IgG bound to monoclonal anti-IgG rheumatoid factor (usually IgM)	Type III Polyclonal IgG bound to polyclonal anti-IgG rheumatoid factor (usually IgM)
1. Multiple myeloma	1. B-lympocytic neoplasm	1. Autoimmune diseases (SLE, polyarteritis nodosa; rheumatoid arthritis; scleroderma; Sjögren's syndrome; Henoch-Schönlein purpura)
2. Waldenström's macroglobulinaemia	2. Diffuse lymphoma	2. Infectious diseases (mononucleosis; cytomegalovirus; hepatitis B; subacute bacterial endocarditis; leprosy; malaria; schistosomiasis; toxoplasmosis; AIDS)
3. Chronic lymphocytic leukaemia	3. Chronic lymphocytic leukaemia	3. Miscellaneous (primary proliferative glomerulonephritis; lymphoma; chronic hepatitis; biliary cirrhosis)
4. Idiopathic monoclonal gammopathy	4. Sjögren's syndrome	4. Essential
	5. Essential	

ease is found and the cryoglobulinaemia is referred to as 'essential'. The clinical syndrome of essential mixed cryoglobulinaemia (EMC) was first described by Meltzer et al. (1966) as characterized by purpura, weakness, arthralgia, and, in some of the patients, by glomerular lesions. This description has been extended in many later reports. These have confirmed that the syndrome can be associated with both type II and type III cryoglobulins. Type III outnumbers type II EMC in patients referred to rheumatologists, whereas surveys based on descriptions of renal involvement have shown a large prevalence of type II EMC, almost always with a monoclonal IgM kappa component. While in the few cases of type III EMC with renal involvement the glomerular lesions were variable and non-specific, in type II EMC, in which IgMk was the monoclonal component, a specific well-characterized pattern of glomerular disease has been described in recent years, to which we will refer in this chapter.

Pathogenesis of type II essential mixed cryoglobulinaemia

While the polyclonal anti-IgG immunoglobulins involved in type III EMC derive from perturbation and magnification of the physiologic mechanism of production of antiglobulin concerned with immunoregulation and are probably antigen-driven, the monoclonal anti-IgG immunoglobulins of type II EMC may derive from the abnormal proliferation of a special clone of B lymphocytes and are probably the consequence of a lymphoproliferative disorder, even though patients with type II EMC show (with rare exceptions of late progression to lymphoma) very little evidence of lymphoid malignancy, and their dominant manifestation results from the tissue deposition of IgM/IgG immune complexes.

The location and size of the clone responsible for this rheumatoid factor production is being investigated, using a panel of mouse monoclonal rheumatoid factors. Increased percentages of idiotype-positive ($\mu+$) B cells have been found in both the peripheral blood and in the morphologically normal bone marrow of all tested patients (Winearls et al. 1986).

Preliminary data demonstrating immunoglobulin-gene rearrangements in the DNA extracted from peripheral blood cells (and lymph node cells) further strengthen this concept (Winearls and Sissons 1988). We have also found that peripheral B lymphocytes from EMC patients produce larger quantities of IgM-rheumatoid factor than normal mononuclear cells, both spontaneously and after in-vitro stimulation with pokeweed mitogen (Meroni et al. 1986).

It is possible that the polyclonal IgG to which monoclonal IgM binds is already bound to an antigen, to form an immune complex, so that monoclonal IgM acts as an anti-immune complex antibody. However, such antigen(s) have not yet been identified in the majority of cases. Hepatitis B surface antigen has been considered to be a possible antigen in a minority of patients (Levo et al. 1977). Recently we have demonstrated the presence of IgM antiviral capsid antigen of Epstein-Barr virus in a great number of patients with type II EMC, and have detected by DNA studies that Epstein-Barr virus genome was incorporated in the bone marrow cells of all individuals tested with type II, but not into bone marrow cells of those with type III cryoglobulinaemia. This suggests an association between type II-EMC and persistent Epstein-Barr virus infection (Fiorini et al. 1986, 1988). This is tantalizing because of the fact that infection of peripheral blood lymphocytes with EBV 'in vitro' is known to induce the synthesis of IgM anti-IgG antibody and the development of permanent B-cell lines (Slaughter et al. 1978).

Whatever the mechanism of binding, it is still unclear why the immunoglobulins involved in type II EMC precipitate in the cold. The intrinsic physicochemical properties of the individual components are not sufficient to explain the phenomenon because both immunoglobulins must be present for cryoprecipitation to occur. Factors such as antigen–antibody ratio, Fc–Fc interactions, reduced solubilization due to hypocomplementaemia, binding with fibronectin, and hydrophobic properties of proteins have been considered (reviewed by Cordonnier et al. 1987).

Renal pathology

Glomeruli

The majority of patients have a peculiar type of membranoproliferative exudative glomerulonephritis (Fig. 1) which has been

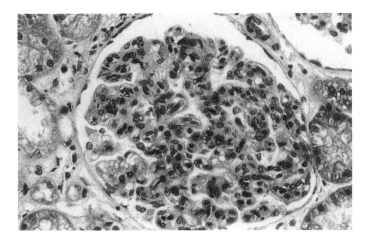

Fig. 1 EMC-associated membranoproliferative exudative glomerulonephritis. The picture shows marked intracapillary hypercellularity, diffuse thickening of the glomerular basement membrane with double-contoured appearance (Masson trichrome, × 250).

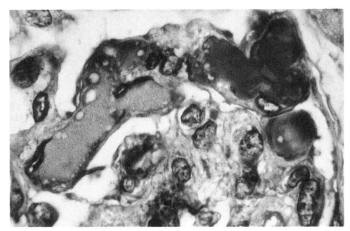

Fig. 3 Mixed cryoglobulinaemic glomerulonephritis. At higher magnification double contours and enormous endoluminal 'thrombi' are particularly evident (Silver stain, × 1000).

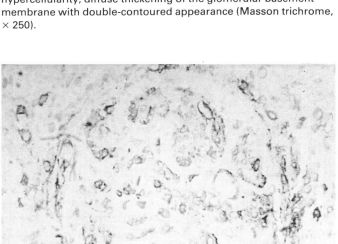

Fig. 2 Immunoperoxidase staining of a cryostat section from a patient with EMC-glomerulonephritis. There is prominent intraglomerular leucocyte infiltration which is positively labelled with the monoclonal antibody PHM1.

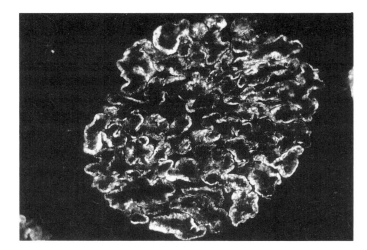

Fig. 4 Granular deposits of IgM along the capillary walls in a subendothelial pattern (× 400).

defined by Mazzucco et al. (1986) and called 'cryoglobulinaemic glomerulonephritis'. The pattern of glomerular involvement differs from that in the idiopathic type of membranoproliferative glomerulonephritis as well as from the diffuse proliferative glomerulonephritis of systemic lupus in the following characteristic features.

1. Endocapillary proliferation is caused by an infiltration of leucocytes, mainly monocytes, (Fig. 2) which is often massive (Monga et al. 1979 and 1986; Ferrario et al. 1985 and 1986). Four times more cells per glomerulus can be counted in this disease than in severe diffuse proliferative lupus nephritis, and the degree of monocyte infiltration is inconsistent in idiopathic membranoproliferative glomerulonephritis. Some T lymphocytes, with a prevalence of OKT_8 cytotoxic suppressor lymphocytes, are also present (Castiglione et al. 1988).

2. There are large amorphous, eosinophilic, PAS-positive, Congo red-negative intraluminal deposits of variable size and diffusion, lying against the inner side of the glomerular capillary wall or completely filling the capillary lumen (the so-called 'intraluminal thrombi'). These are present in more than one-third of all patients, especially in those with more acute renal disease and more massive proliferation and exudation (Fig. 3). By immunofluorescence the composition of these deposits is identical to that of circulating cryoglobulin (IgG and IgM), suggesting that they are locally trapped or precipitated cryoglobulins. When subendothelial deposits prevail, the immunofluorescence pattern is similar to that found in idiopathic membranoproliferative glomerulonephritis (diffuse intense granular staining of peripheral loops), usually accompanied by C3 together with IgG and IgM (Fig. 4). When intraluminal thrombi are also present, intense massive intraluminal staining with IgM and IgG filling the capillary lumen coexist (Fig. 5). The identity of the deposits with circulating cryoglobulins is confirmed by the antiglobulin

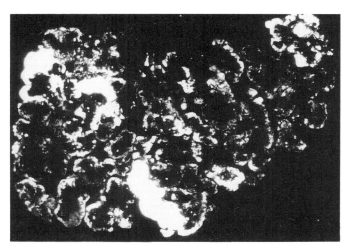

Fig. 5 Intense staining of intraluminal 'thrombi' filling the capillary lumina, associated with irregular segmental staining along the capillary walls (IgG, × 400).

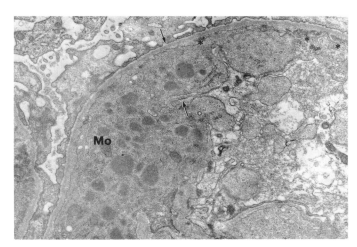

Fig. 7 A monocyte (Mo) in close contact with subendothelial deposits (asterisks) and duplication of basement membrane (arrows) (× 8000). (By courtesy of the Department of Pathology, S. Carlo Hospital, Milan.)

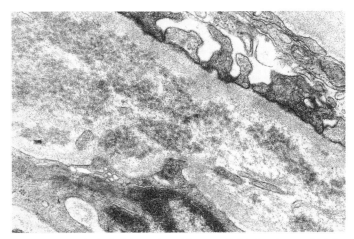

Fig. 6 Subendothelial deposits with crystalloid structure, which in cross-section appear to be made up of tubular units (magnification × 22 000). (By courtesy of the Department of Pathology, S. Carlo Hospital, Milan.)

activity of the eluted IgM (Maggiore et al. 1982) and by the demonstration that the same idiotype of the circulating monoclonal IgM rheumatoid factor is present in the intraluminal deposits (Sinico et al. 1988). By electron microscopy, the deposits are frequently amorphous immune complex-like deposits, but often they show a peculiar fibrillar or crystalloid structure (Cordonnier et al. 1975; Feiner and Gallo 1977), which is identical to that seen in the *in-vitro* cryoprecipitate of the same patient: cylindrical bodies of 100 nm to 1 μm long, cross sections of which appear as annular bodies, with a light centre, a dense ring and a lighter peripheral protein coat (Fig. 6).

3. Thickening of the glomerular basement membrane with double-contoured appearance is more diffuse and obvious than in lupus nephritis and in idiopathic membranoproliferative glomerulonephritis. As can be seen more clearly by electron microscopy the double contour is mainly due to the peripheral interposition of monocytes (Fig. 7), while mesangial matrix and cell interposition is less prominent than in lupus nephritis and in idiopathic membranoproliferative glomerulonephritis. Monocytes are very often in close contact with the subendothelial amorphous or crystalloid deposits and appear to be involved in their degradation.

In some patients the monocyte infiltration is less consistent, there are usually no intraluminal thrombi, and the picture of 'cryoglobulinaemic glomerulonephritis' we have described becomes less typical. A picture of mild segmental mesangial proliferation, without significant monocyte infiltration and alteration of capillary walls is found in about 10 per cent of our patients. Subendothelial faint irregular segmental parietal staining of some peripheral loops is demonstrated in these cases by immunofluorescence. A picture of lobular glomerulonephritis, with evident mesangial cell proliferation (but also with monocyte infiltration), mesangial matrix expansion, and areas of centrolobular sclerosis is found sometimes with a clinical picture of intense proteinuria, often in the nephrotic range (another 10 per cent of our biopsied patients). Only in this last small group of patients is mesangial sclerosis a consistent phenomenon. Glomerular segmental and global sclerosis are rather mild and inconstant in the majority of patients with cryoglobulinaemic glomerulonephritis, even many years after the onset of the renal disease, as has been shown by repeated biopsies. In spite of the intense intracapillary proliferation and exudation, extracapillary proliferation is an uncommon finding, even in the more acute stages. When present, it is always focal and segmental, never circumferential.

Interstitium

Infiltration of mononuclear leucocytes (monocytes, dendritic cells, and T lymphocytes, mainly OKT_8 + cytotoxic-suppressor lymphocytes) is usually found in the acute stages of the renal disease, and interstitial fibrosis is not a prominent finding, even in late biopsies.

Vessels

Vasculitis of small and medium-sized arteries is found in more than one-third of patients biopsied because of renal disease and in a larger percentage of postmortem specimens. It is seen most frequently in patients with an acute clinical presentation and intense membranoproliferative exudative glomerulonephritis, with or without massive intraluminal thrombi. It can also be found in the absence of obvious glomerular damage and is sometimes associated with other signs of systemic vasculitis. It is characterized by fibrinoid necrosis of the arteriolar wall, with infiltration of monocytes around the wall, which is granuloma-like in some cases. In biopsies taken at later stages, perivascular infiltration appears to be replaced by perivascular fibrosis. Intimal fibrosis of interlobular arteries is often found in patients with EMC, even in the absence of vasculitic lesions.

Clinical features

Age, sex and geographical distribution

Patients with type II EMC usually present in the fourth to fifth decades of life, although first presentation in younger patients has been reported. Women outnumber men in unselected populations, and to a lesser extent in patients with glomerulonephritis. The incidence varies in different geographical areas; the majority of cases have been reported in the Mediterranean countries, Italy, France, Spain, and Israel.

Systemic and renal symptoms at presentation

Renal symptoms are usually a late manifestation of type II EMC. On the average they appear some years after the onset of the extrarenal signs; 4 years according to Gorevic *et al.* (1980) and 6 years according to Tarantino *et al.* (1981). Systemic signs are mainly purpura, arthralgias, leg ulcers, systemic vasculitis, Raynaud's phenomenon, hepatic involvement, and peripheral neuropathy. However, the concomitant appearance of renal and systemic signs of the disease is rather frequent. In some patients, the renal involvement may be the main presenting feature of the disease, before the appearance of the more diagnostic purpura. The frequency of renal involvement has varied from 20 to 50 per cent in different series reported by rheumatologists. Incidence of the most typical extrarenal signs was the same for populations of patients reported by nephrologists and for unselected populations with EMC.

An acute nephritic syndrome characterized by haematuria, usually macroscopic, severe proteinuria, hypertension, and sudden rise of BUN, is present at onset of renal disease in approximately 25 per cent of cases and is complicated by acute oliguric renal failure in no more than 5 per cent of them. In the acute stage, biopsy shows the typical cryoglobulinaemic glomerulonephritis, i.e., a severe membranoproliferative exudative glomerulonephritis with intense monocyte infiltration and massive intraluminal deposits which obliterate many capillary lumina.

However, isolated proteinuria with microscopic haematuria is the most frequent presenting renal syndrome and is sometimes associated with signs of moderate chronic renal insufficiency or, less frequently, proteinuria in the nephrotic range. The corresponding histological picture is of a membranoproliferative glomerulonephritis with less conspicuous intraluminal deposits and monocyte infiltration, sometimes with a lobular pattern, while immunofluorescence shows subendothelial deposits of IgM, IgG, and C3, sometimes segmental. A minority of patients present with isolated urinary abnormalities and may have a rather non-specific picture of mild segmental mesangial proliferation, without significant monocyte infiltration and capillary wall alterations.

Arterial hypertension is frequently found at the time of the apparent onset of renal disease, even in patients without nephritic syndrome.

Natural history and prognosis of the renal disease

The course of the renal disease is variable. In nearly one-third of patients, even in those who presented an acute nephritic syndrome or severe nephrotic syndrome there is a remission of renal symptoms. Remission after an acute nephritic syndrome may occur even before treatment is started, and is associated with disappearance of the massive intraluminal deposits that are a characteristic feature of this presentation. In one-third of patients, the renal disease has a rather indolent course and, in spite of the persistence of urinary abnormalities, does not progress to renal failure for several years. In up to 20 per cent of patients, reversible clinical exacerbations such as nephritic syndrome or nephrotic syndrome occur during the course of the disease, sometimes associated with flare-up of systemic signs. New episodes of nephritic syndrome are frequently associated with the reappearance of massive intraluminal thrombi and intense monocyte infiltration; individual patients may have more than one exacerbation.

A moderate degree of renal insufficiency, if not present already at clinical onset, is frequently found in later stages of the disease. However, progression to end-stage renal failure is less common than was believed in the past, even in patients with multiple relapses. Chronic uraemia developed in only 10 per cent of patients reported in the literature, usually several years after the onset of renal symptoms (Tarantino *et al.* 1981; Ferrario *et al.* 1986). Only six of 108 patients with renal involvement studied in Milan required regular dialysis treatment after a mean follow up of more than 10 years, whereas 27 had died from extrarenal complications (Tarantino *et al.* 1986). The most frequent causes of death in essential mixed cryoglobulinaemia are systemic vasculitis, infections, and cardiovascular and cerebrovascular accidents. The last two are certainly favoured by the arterial hypertension, which is a very common early complication, frequently severe and difficult to control. It may have an accelerated course in some patients.

It has been stated that the outcome of patients with essential mixed cryoglobulinaemia is worse if there is renal involvement. This is no longer so, probably because treatment of the disease and its complications is now better. In our series of more than 100 patients with EMC and renal involvement, the mortality rate was 30 per cent 10 years after the beginning of the disease, similar to that reported by Popp *et al.* (1980) for patients without renal disease.

The extrarenal features of EMC, i.e. purpura, arthralgia, peripheral neuropathy, and hepatic involvement, complicate the course in patients with renal involvement as often as in those without evidence of renal disease. The presence of widespread extrarenal involvement, and in particular of systemic vasculitis, is associated with a more severe renal disease (Barbiano di Belgioioso *et al.* 1986) and a worse prognosis (Tarantino *et al.* 1986).

Laboratory findings

The amount of circulating cryoglobulins may vary between patients, and in the same patient at different times. Cryocrit value can range between 2 and 70 per cent, with large variations during the course of the disease. Serum titres of IgM rheumatoid factor are also frequently increased. There is no correlation between these laboratory parameters and the degree of activity of the disease, but in our experience a high cryoglobulin concentration appears to be correlated with a bad prognosis (Tarantino et al. 1986). The serum complement results are highly suggestive of EMC nephritis. Early components (C1q, C4) and CH50 are usually very low, C3 is slightly but significantly reduced and later components (C5 and C9), C3 PA, and C1 INH tend to be higher than in normal controls. The C3 breakdown product C3d is sometimes increased. The complement pattern changes little with variations of disease activity. In fact, early complement components tend to remain low, whatever the degree of activity of the systemic and renal disease. Markers of type B viral hepatitis infection, and in particular persistent HBs antigenaemia, have been reported with variable frequency by different investigators. It has been reported that the laboratory signs indicating hepatitis B infection were so frequent that they were considered a possible cause of the systemic disease (Levo et al. 1977). Others have been unable to confirm this association even in the same geographical areas (Popp et al. 1980; D'Amico et al. 1984).

Pathophysiology of the renal damage in type II EMC

Cryoglobulinaemic glomerulonephritis is an immune-complex glomerulonephritis. The finding of the cryoprecipitating immunoglobulins in the capillary lumen by electron microscopy and the demonstration by immunofluorescence of subendothelial deposits of the same immunoglobulin classes as those in the circulating cryoglobulins, together with complement, suggest that cryoglobulins act at the glomerular level as trapped immune complexes which bind complement. However, local precipitation for purely physicochemical mechanisms is an alternative explanation, for example as a consequence of higher endocapillary cryoglobulin concentrations induced by the filtration process in the glomerulus (Lockwood 1979). This seems to be the likely explanation in cases in whom massive intraluminal accumulation of cryoglobulins is found histologically, together with an immunofluorescent pattern of almost exclusively intraluminal staining of thrombi. Two main clinical-histological syndromes correspond to these two mechanisms of glomerular deposition of cryoglobulins.

1. Time-limited, sometimes recurrent bouts of intense cryoglobulin precipitation in the glomerulus, which give rise to reversible acute nephritic syndromes and to a histological picture of membranoproliferative exudative glomerulonephritis with thrombi, which is also reversible when the accumulated monocytes and precipitated intraluminal cryoglobulins have been disposed of.

2. A milder clinical syndrome, characterized by urinary abnormalities, sometimes with proteinuria in the nephrotic range, associated with the deposition of cryoglobulins (and possibly of other circulating immune complexes) on the subendothelial aspects of the glomerular basement membrane and a histological picture of membranoproliferative exudative glomerulonephritis, sometimes of the lobular type, without thrombi and with variable degrees of monocytic infiltration. In the latter syndrome activation and peripheral circumferential expansion of mesangial cells and matrix is less marked and more slowly progressing than in idiopathic membranoproliferative glomerulonephritis, but the reasons for this are uncertain. Therefore, mesangial sclerosis and progressive irreversible impairment of renal function are indolent, inconstant phenomena even in untreated patients with cryoglobulinaemic glomerulonephritis.

Irrespective of why they were deposited in the glomerulus, locally trapped cryoglobulins trigger endocapillary accumulation of monocytes from the bloodstream. A large number of monocytes in the capillary lumen and in the double contoured capillary walls of the glomeruli are hallmarks of the disease. These cells accumulate in the more acute clinical-histological syndrome, when there can be up to 80 cells per glomerulus; they are also present in the more chronic phases of the renal disease. Monocytes act as scavenger cells in renal tissue as evidenced by the presence of large phagolysosomes in their cytoplasm and by their close contact with the deposits of cryoglobulins. But it is highly probable that they also behave as mediators of the local damage through liberation of lytic enzymes and other humoral factors able to induce proliferation of resident cells (Nathan et al. 1980). The participation of cellular immunity in this type of renal damage must be considerate since T lymphocytes, mainly of cytotoxic-suppressor phenotype, are found together with monocytes in both the glomerulus and the interstitium.

Treatment

The evaluation of the efficacy of therapy for type II essential mixed cryoglobulinaemia is difficult for the following reasons: (1) the rarity of the disease; (2) the variable correlation of renal and extrarenal clinical manifestations in individual as well as in different patients; and (3) the occurrence of spontaneous recovery of the acute exacerbations of the renal disease.

Oral steroids and/or cyclophosphamide have been used for renal exacerbations, but with conflicting results. However, we believe that this traditional immunosuppressive therapy is justified, especially to control the renal and extrarenal signs and to avoid further exacerbations in patients with less acute stages of the disease. The cautious use of maintenance doses of cyclophosphamide seems most useful for these.

When there is a nephritic syndrome with rapid deterioration of renal function, especially when signs of systemic vasculitis are evident, we use either short courses of intravenous methylprednisolone pulses, as proposed by De Vecchi et al. (1983), to obtain the best cytotoxic and anti-inflammatory effect with the minimum of side-effects, and/or intensive plasma exchange or cryofiltration (at least three times a week for a few weeks) to remove circulating cryoglobulins, to prevent their intravascular precipitation and to restore the functional capacity of the overloaded reticuloendothelial system (Pusey et al. 1981; Berkman and Orlin 1980; Bombardieri et al. 1983; D'Amico et al. 1984).

In our hands, the combined use of intravenous steroid pulses (three pulses of 1 g of methylprednisolone on consecutive days) followed by oral doses of 0.5 to 0.7 mg/kg body weight of prednisone (tapered in a few weeks to small maintenance doses), plasma exchange, and cyclophosphamide (2–3 mg/kg body weight daily for at least 6 months), provided good control of the episodes of renal exacerbation for more than 40 patients.

α-Interferon, an inhibitor of B-cell proliferation, has been recently suggested for the treatment of cryoglobulinaemic patients (Bonomo et al. 1987), but its role needs further investigation.

In conclusion

In essential mixed cryoglobulinaemias in which anti-IgG rheumatoid factor is a monoclonal IgM (Type II-EMC), cryoglobulins can fix in the glomerulus in a subendothelial position, to give a peculiar type to immune complex membranoproliferative glomerulonephritis and/or can precipitate by purely physicochemical mechanisms in the capillary lumen to give huge deposits which tend to obliterate acutely the capillary space.

An acute reversible syndrome is usually associated with the massive precipitation of cryoglobulins in the capillary lumina, while a more chronic and discrete clinical syndrome, characterized by urinary abnormalities, with a proteinuria which is sometimes in the nephrotic range, is associated with the prevalent deposition of cryoglobulins on the subendothelial aspect of the glomerular basement membrane.

A peculiar characteristic of both types of renal involvement is the fact that trapped cryoglobulins trigger intracapillary accumulation of monocytes from the bloodstream. The presence of a large number of monocytes in the capillary lumina and in the double contoured capillary walls is the hallmark of the disease. Another peculiar aspect of EMC glomerulonephritis is that, in spite of the intense monocyte recruitment, activation and expansion of mesangial cells and matrix is less marked and more slowly progressing than in idiopathic membranoproliferative glomerulonephritis, and mesangial sclerosis with progressive irreversible impairment of renal function is an indolent, inconstant phenomenon even in untreated patients.

Bibliography

Brouet, J.C., Clauvel, J.P., Danon, F., Klein, M., and Seligman, M. (1974). Biological and clinical significance of cryoglobulins. A report of 86 cases. *American Journal of Medicine*, 67, 775–8.

Cordonnier, D., et al. (1982). Lésions Rénales chez 10 malades porteurs de cryoglobulines mixtes IgM-IgG de Type II. In *Actualités Nephrologiques de l'Hôpital Necker*. (eds. J. Hamburger, J. Crosnier, and J.L. Funk-Brentano), pp. 319–41, Flammarion, Paris.

Cordonnier, D.J., Renversez, J.C., Vialtel, P., and Dechelette, E. (1987). The kidney in mixed cryoglobulinemias. *Springer Seminars in Immunopathology*, 9, 395–415.

D'Amico, G., Ferrario, F., Colasanti, G., and Bucci, A. (1984). Glomerulonephritis in essential mixed cryoglobulinemia (EMC). In *Proceedings of the XXI Congress of the European Dialysis and Transplant Association*, (eds. A.M. Davison, and P.J. Guillou), pp. 527–48, Pitman, London.

D'Amico, G., Colasanti, G., Ferrario, F., Sinico, R.A., Bucci, A., and Fornasieri, A. (1987). L'atteinte rénale dans la cryoglobulinémie mixte essentielle: une type particuliere de néphropathie à médiation immunologique. In *Actualités Nephrologiques de l'Hôpital Necker*. (eds. J. Crosnier, J.L. Funk-Brentano, J.F. Bach, and J.P. Grünfeld), pp. 201–19, Flammarion, Paris.

D'Amico, G., Colasanti, G., Ferrario, F., and Sinico, R.A. (1989). Renal involvement in essential mixed cryoglobulinemia. *Kidney International*, 35, 1004–14.

Ferrario, F., et al. (1986). Histological and immunohistological features in essential mixed cryoglobulinemia glomerulonephritis. In *Antiglobulins, Cryoglobulins and Glomerulonephritis*. (eds. C. Ponticelli, L. Minetti, and G. D'Amico), pp. 193–202, M. Nijhoff, Dordrecht.

Gorevic, P.D., et al. (1980). Mixed cryoglobulinemia: clinical aspects and long-term follow-up of 40 patients. *American Journal of Medicine*, 69, 287–308.

Grey, H.H. and Kohler, P.F. (1973). Cryoglobulins. *Seminars in Hematology*, 10, 87–112.

Invernizzi, F., et al. (1983). Secondary and essential cryoglobulinemias. Frequency, nosological classification and long-term follow-up. *Acta Haematologica*, 70, 73–82.

Lockwood, C.M. (1979). Lymphoma, cryoglobulinemia and renal disease. *Kidney International*, 16, 522–30.

Meltzer, M., Franklin, E.C., Elias, K., McCluskey, R.T., and Cooper, N. (1966). Cryoglobulinemia – a clinical and laboratory study. II. cryoglobulins with rheumatoid factor activity. *American Journal of Medicine*, 40, 837–56.

Monga, G., Mazzucco, G., Coppo, R., Piccoli, G., and Coda, R. (1986). Glomerular findings in mixed IgG-IgM cryoglobulinemia. Light, electron microscopy, immunofluorescence and histochemical correlations. *Virchows Archivs* (*Zell Pathologie*), 20, 185–96.

Morel-Maroger, L. and Mery, J.P. (1972). Renal lesions in mixed IgG–IgM essential cryoglobulinemia. In *Proceedings of the 5th Congress of Nephrology*. (ed. H. Villarreal), pp. 173–8, Karger, Basel.

Tarantino, A., et al. (1981). Renal disease in essential mixed cryoglobulinemia. Long-term follow-up of 44 patients. *Quarterly Journal of Medicine*, 50, 1–30.

Wang, A. and Wang, I.Y. (1986). Intrinsic properties inducing precipitation of cryoglobulins. In *Antiglobulins, Cryoglobulins and Glomerulonephritis*. (eds. C. Ponticelli, L. Minetti, and G. D'Amico), pp. 101–12, M. Nijhoff, Dordrecht.

References

Barbiano di Belgioioso, G., et al. (1986). Clinical and histological correlations in essential mixed cryoglobulinemia (EMC) glomerulonephritis. In *Antiglobulins, Cryoglobulins and Glomerulonephritis*. (eds. C. Ponticelli, L. Minetti, and G. D'Amico), pp. 203–10, M. Nijhoff, Dordrecht.

Berkman, E.M. and Orlin, J.B. (1980). Use of plasmapheresis and partial plasma exchange in the management of patients with cryoglobulinemia. *Transfusion*, 20, 171–8.

Bombardieri, S., et al. (1983). Prolonged plasma exchange in the treatment of renal involvement in essential mixed cryoglobulinemia. *International Journal of Artificial Organs*, 6, 47–50.

Bonomo, L., Casato, M., Afeltra, A., and Caccamo D. (1987). Treatment of idiopathic mixed cryoglobulinemia with alpha interferon. *American Journal of Medicine*, 83, 726–30.

Castiglione, A., Bucci, A., Fellin, G., D'Amico, G., and Atkins, R.C. (1988). The relationship of infiltrating renal leucocytes to disease activity in lupus and cryoglobulinemic glomerulonephritis. *Nephron*, 50, 14–23.

Cordonnier, D.D., Martin, H., Groslambert, P., Micouin, C., Chenais, F., and Stoebner, P. (1975). Mixed IgG–IgM cryoglobulinemia with glomerulonephritis. Immunochemical fluorescent and ultrastructural study of kidney and *in vitro* cryoprecipitate. *American Journal of Medicine*, 59, 867–72.

De Vecchi, A., Montagnino, G., Pozzi, C., Tarantino, A., Locatelli, F., and Ponticelli, C. (1983). Intravenous methylprednisolone pulse therapy in essential mixed cryoglobulinemia nephropathy. *Clinical Nephrology*, 19, 221–7.

Feiner, H. and Gallo, G. (1977). Ultrastructure in glomerulonephritis associated with cryoglobulinemia. *American Journal of Pathology*, 88, 145–62.

Ferrario, F., Castiglione, A., Colasanti, G., Barbiano di Belgioioso, G., Bertoli, S., and D'Amico, G. (1985). The detection of monocytes in human glomerulonephritis. *Kidney International*, 28, 513–19.

Fiorini, G., Bernasconi, P., Sinico, A.R., Chianese, R., Pozzi, F., and

D'Amico, G. (1986). Increased frequency of antibodies to ubiquitous viruses in essential mixed cryoglobulinemia. *Clinical Experimental Immunology*, **64**, 65–70.
Fiorini, G., Sinico, R.A., Winearls, C., Custode, P., De Giuli-Morghen, C., and D'Amico, G. (1988). Persistent Epstein-Barr virus infection in patients with type II essential mixed cryoglobulinemia. *Clinical Immunology and Immunopathology*, **47**, 262–9.
Levo, Y., Gorevic, P.D., Kassab, H.J., Tobias, H., and Franklin, E.C. (1977). Association between hepatitis B virus and essential mixed cryoglobulinemia. *New England Journal of Medicine*, **296**, 1501–4.
Maggiore, Q., et al. (1982). Glomerular localization of circulating antiglobulin activity in essential mixed cryoglobulinemia. *Kidney International*, **21**, 387–94.
Mazzucco, G., Monga, G., Casanova, S., and Cagnoli, L. (1986). Cell interposition in glomerular capillary walls in cryoglobulinemic glomerulonephritis. Ultrastructural investigation of 23 cases. *Ultrastructural Pathology*, **10**, 355–61.
Meroni, P.L., et al. (1986). *In vitro* synthesis of IgM rheumatoid factor by lymphocytes from patients with essential mixed cryoglobulinemia. *Clinical Experimental Immunology*, **65**, 303–10.
Monga, G., Mazzucco, G., Barbiano di Belgioioso, G., and Busnach, G. (1979). The presence and possible role of monocyte infiltration in human chronic proliferative glomerulonephritis. *American Journal of Pathology*, **94**, 271–84.
Nathan, C.F., Murray, H.W., and Cohen, Z.A. (1980). Current concepts: the macrophage as an effector cell. *New England Journal of Medicine*, **303**, 622–6.
Popp, J.W., Ienstag, J.L., Wands, J.R., and Bloch, K.J. (1980). Essential cryoglobulinemia without evidence for hematitis B virus infection. *Annals of Internal Medicine*, **92**, 379–83.
Pusey, C.D., Schifferli, J.A., Lockwood, C.M., and Peters, D.K. (1981). Use of plasma exchange in the management of mixed essential cryoglobulinemia. *Artificial Organs*, (**suppl. 5**), 183–9.
Sinico, A.R., Winearls, C.G., Sabadini, E., Fornasieri, A., Castiglione, A., and D'Amico, G. (1988). Identification of glomerular immune deposits in cryoglobulinemia glomerulonephritis. *Kidney International*, **34**, 1–8.
Slaughter, L., Carson, D.A., Jensen, F.B., Holbrook, and Vaughan, J.H. (1978). *In vitro* effects of Epstein-Barr virus on peripheral mononuclear cells from patients with rheumatoid arthritis. *Journal of Experimental of Medicine*, **148**, 1429–34.
Tarantino, A., et al. (1986). Prognostic factors in essential mixed cryoglobulinemia nephropathy. In *Antiglobulins, Cryoglobulins and Glomerulonephritis*. (eds. C. Ponticelli, L. Minetti, and G. D'Amico), pp. 219–25, M. Nijhoff, Dordrecht.
Winearls, C.G., et al. (1986). Identification of circulating MRF-producing cells in essential mixed cryoglobulinemia. *Nephrology Dialysis Transplantation*, **1**, 78 (Abstract).
Winearls, C.G. and Sissons, J.P. (1988). Use of idiotype markers for cellular detection of monoclonal rheumatoid factors. *Springer Seminars in Immunopathology*, **10**, 67–74.

4.10 Sjögren's syndrome and overlap syndromes

PATRICK J.W. VENABLES

Introduction

Sjögren's syndrome and the overlap syndromes, including mixed connective tissue disease, represent a substantial part of the clinical load of the rheumatologist. In these conditions overt renal disease is rare, occurring in about 10 per cent of patients. The most common route of presentation to a nephrologist is therefore via a rheumatological referral. Occasionally the patients present with renal disease, in which case the underlying connective tissue disease is usually obvious. However, Sjögren's syndrome may not be recognized, as the dominant symptoms and signs of the disease, xerostomia and xerophthalmia are easily missed.

Sjögren's syndrome

Classification

Sjögren's syndrome represents a group of diseases characterized by a common pathological feature, namely inflammation and destruction of exocrine glands. The salivary and lachrymal glands are principally involved giving rise to dry eyes and mouth. Other exocrine glands including those of the pancreas, sweat glands, and mucus-secreting glands of the bowel, bronchial tree, and vagina may be affected. It was originally described as the triad of dry eyes, dry mouth, and rheumatoid arthritis (Sjögren 1933). It is now classified as (i) primary Sjögren's syndrome where the disease exists on its own and (ii) secondary Sjögren's syndrome, where it is associated with another autoimmune rheumatic disease.

Primary Sjögren's syndrome

Primary Sjögren's syndrome occurs when the disease predominantly affects exocrine glands. Most of the patients have systemic or extraglandular features which, by definition, are not numerous enough to fulfil criteria for other connective tissue diseases.

Secondary Sjögren's syndrome

Secondary Sjögren's syndrome occurs when sicca symptoms are associated with another connective tissue disease, particularly rheumatoid arthritis (RA) when it occurs in approximately one-third of patients. The ocular and oral symptoms of Sjögren's syndrome (SS) with RA are similar to those of primary SS, although they usually occur after the onset of the arthritis and tend to be relatively mild. Secondary SS also occurs in the autoimmune connective diseases, SLE, scleroderma, polymyositis, and primary biliary cirrhosis.

Clinical features
Exocrinopathy
Sjögren's syndrome is the second most common autoimmune rheumatic disease (Shearn 1977). The exact prevalence is not known since there are thought to be many sufferers whose symptoms are so mild that they do not seek medical advice. It is nine times more common in women than men and its onset is at any age from 15 to 65 years. The patients rarely complain of dry eyes, but rather a gritty sensation, soreness, photosensitivity, or intolerance of contact lenses. In early disease, patients occasionally complain of excessive watering or deposits of dried mucus in the corner of the eye and recurrent attacks of conjunctivitis. The dry mouth is often manifest as the 'cream cracker' sign, inability to swallow dry food without fluid, or the need to wake up in the night to take sips of water. About half of the patients complain of recurrent parotid swelling, sometimes misdiagnosed as recurrent mumps. When the swelling is excessively painful it is often due to secondary bacterial infection.

On examination, xerostomia can be detected as a diminished salivary pool, a dried fissured tongue, often complicated by angular stomatitis and chronic oral candidiasis. The eyes may be reddened and roughened due to shallow erosions in the conjunctiva. Occasionally the front of the eye is eroded to reveal strands of underlying collagen leading to the appearance of filamentary keratitis.

Exocrine glands other than the salivary and lachrymal glands may be affected. Dry skin and dry hair are symptoms frequently elicited on direct questioning. About 30 per cent of patients have diminished vaginal secretions and may present with dyspareunia. Involvement of the gastrointestinal tract leads to reflux oesophagitis or gastritis due to lack of protective mucus secretion, and some patients complain of constipation which may be attributed to defective mucus in the colon and rectum. Pancreatic failure leading to malabsorption syndromes occurs rarely.

Extraglandular features
Sjögren's syndrome is a systemic disease. Almost all patients complain of fatigue and depression. Occasionally weight loss and fever, mimicking an occult malignancy, may be the presenting symptoms. Other features are those attributable to circulating immune complexes, particularly in those patients with antibodies to Ro (SS-A) and La (SS-B). Prominent amongst these are nonerosive arthritis, Raynaud's phenomenon, and a purpuric vasculitis on the lower legs associated with ankle oedema. A high incidence of abnormalities of pulmonary function has been described though these are rarely clinically significant. Patients may present with a syndrome indistinguishable from polymyalgia rheumatica and it is well recognized that polymyositis may be associated with SS. Alexander et al. (1986) have noted a wide range of neurological diseases including central nervous system disorders resembling multiple sclerosis. However, some selection bias may account for the unusually high frequency of such complications in their patients. Clinically significant renal disease occurs in about 10 per cent of patients (Pease et al. 1989).

Sjögren's lymphoma
About 5 per cent of patients with Sjögren's syndrome eventually develop lymphomas. These may present as progressive (as

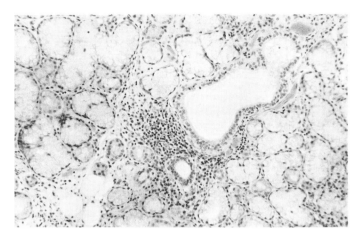

Fig. 1 Biopsy of minor salivary gland from a patient with Sjögren's syndrome. There is an inflammatory infiltrate surrounding the ducts with the acini being relatively spared.

opposed to intermittent) massive enlargement of the salivary glands, with diffuse lymphadenopathy or with skin deposits. This malignant change, which is thought to arise from chronic B cell stimulation, may be much commoner than previously thought since a substantial proportion of patients with extraglandular disease show evidence of oligoclonal B cell hyperactivity.

Diagnostic tests
Tests for keratoconjunctivitis sicca and xerostomia
Keratoconjunctivitis sicca can be detected by Schirmer's test, tear break-up time, and Rose Bengal staining (Kincaid 1987), and xerostomia by a reduced parotid salivary flow rate and by reduced uptake and clearance on isotope scans (Daniels 1987). It is important to remember that both salivary and lachrymal function decline with age and may be impaired in conditions other than SS (Morrow and Isenberg 1987). One cause of diagnostic confusion arises from treatment with drugs with anticholinergic side-effects, the most frequent being the tricyclic antidepressants.

Labial gland biopsies
Biopsy and histology of the labial glands from behind the lower lip provides the most definitive diagnostic test. The area is anaesthetized with lignocaine containing adrenalin and an incision 1.5 cm long allows access to between five and 10 glands 2 to 4 mm in diameter which are removed by simple blunt dissection. A diagnosis of Sjögren's syndrome depends on finding foci of periductular infiltrates of at least 50 lymphocytes and/or plasma cells at a density of more than one focus per 4 mm^2 (Fig. 1).

Haematology and chemical pathology
The majority of patients with SS have a raised ESR and a mild normocytic anaemia. Leucopenia and thrombocytopenia occur in about 50 per cent of patients, and a referral from the haematology department with leucothrombocytopenia is an increasingly common route of presentation. One of the most remarkable features of primary SS is a polyclonal hypergammaglobulinaemia which can cause levels of IgG of up to 50 g/l. Complement levels are usually normal, even in the presence of immune complex dis-

ease, although C4 levels can sometimes be reduced, because of the link between SS and the C4A null gene.

Antibodies to antigens Ro (SS-A) and La (SS-B)

Antibodies to the ribonucleoprotein antigens Ro (SS-A) and La (SS-B) are found in about 65 and 50 per cent of patients with primary SS respectively (Alspaugh et al. 1976; Tan 1982; Venables et al. 1989). Anti-La (SS-B) is virtually restricted to patients with primary SS or SS/SLE whereas anti-Ro (SS-A) occurs with other diseases, particularly SLE, and occasionally polymyositis, rheumatoid arthritis and apparently healthy subjects.

Other autoantibodies

Rheumatoid factors, as measured by routine assays, occur in all forms of SS. They do not distinguish SS/RA from other forms of Sjögren's syndrome, and their detection in primary SS is a common reason for misdiagnosing such patients as having RA (Venables 1989). Similarly antinuclear antibodies detected by immunofluorescence, often in the absence of anti-Ro (SS-A) or anti-La (SS-B), can occur. Both rheumatoid factors and immunofluorescence, although not diagnostically specific, can help in distinguishing Sjögren's syndrome from non-autoimmune causes of sicca symptoms (Fox et al. 1986). Antisalivary gland antibodies occur mainly in patients with SS/RA. This antibody assay is not widely available and has not established itself as a useful diagnostic test.

Immunogenetics

Primary Sjögren's syndrome is strongly associated with HLA DR3, and the linked genes B8, and DQ2, and C4A null gene (Harley et al. 1986). Within the primary SS group, those with anti-La represent a subset which show an even more striking association with HLA DR3, 90 to 100 per cent of patients with this antibody being HLA-DR3 positive. This suggests that the anti-La positive patients with SS may be the most homogeneous subgroup both clinically and immunogenetically. Anti-Ro is associated with DR2 and the linked DQ1 gene, as well as DR3, and this may reflect the wider diagnostic associations of this antibody. Rheumatoid arthritis with secondary SS, is associated with DR4, not DR3 (Venables 1989). There are no clear tissue types related to scleroderma with SS.

Diagnostic criteria

The application of diagnostic criteria, although not important in clinical practice, is essential for the standardization of any research involving patient groups, particularly with a disease, or group of diseases, as heterogeneous as SS. Criteria used currently include those of Daniels et al. (1975), Manthorpe et al. (1981) and Fox et al. (1986), all of which depend on the demonstration of keratoconjunctivitis sicca, xerostomia, and a positive labial gland biopsy. Those of Fox et al. are the most practical and are the first to recognize the importance of immunopathology in the disease by including autoantibodies. Four criteria were proposed: (i) dry eyes (by Schirmer's test and by Rose Bengal or fluorescein staining); (ii) dry mouth (symptoms and decreased salivary flow rate); (iii) lymphocytic infiltrates on lip biopsy; (iv) demonstration of serum autoantibodies (rheumatoid factors, antinuclear antibodies, Ro or La antibodies). Four criteria represent 'definite' and three 'possible' SS.

A new classification of Sjögren's syndrome

The relationship of Sjögren's syndrome to other rheumatic diseases has been clarified by associations between clinical subsets, tissue types, and autoantibody specificities, suggesting that three subsets of the disease are now distinguishable (Table 1).

1. Primary Sjögren's syndrome
2. Sjögren's associated overlap syndromes
3. Secondary Sjögren's syndrome with rheumatoid arthritis.

The justification for the separation of SS/RA from other types of secondary Sjögren's syndrome is based on the findings that the autoantibodies (antisalivary gland) and immunogenetics (DR4) are quite different from SS/SLE (anti-Ro/La and DR3) (Maini 1987). Because of the clinical, serological, and immunogenetic similarities between SLE and primary SS, it was proposed that SS/SLE should be regarded as an overlap syndrome rather than a form of secondary SS (Venables 1989).

Aetiopathogenesis of Sjögren's syndrome

Viruses

Obvious aetiological candidates for triggering autoimmunity in SS are viruses which infect the salivary gland. Sialotropic viruses such as Epstein-Barr virus (EBV) and cytomegalovirus have been examined with conflicting reports of abnormal responses to inflection (Shillitoe et al. 1982; Venables et al. 1985). Using DNA hybridization techniques, EBV has been detected in

Table 1 Clinical, serological and immunogenetic features of primary SS, SS/SLE overlap and secondary SS with RA

	Primary SS	SS/SLE	SS/RA
Arthritis	Non-erosive	Non-erosive	Erosive
Raynaud's phenomenon	++	+++	±
Purpura	++	++	±
Digital infarcts	−	±	++
Subcutaneous ulcers	−	±	++
Leucopenia	++	+++	− (except Felty's)
Rheumatoid factors	++	+	++
Anti-Ro* (%)	65	90	1
Anti-La* (%)	50	60	0
DR3* (%)	63	60	30
DR4* (%)	24	20	68

* Based on data from Pease et al. (1989).

parotid gland (Wolf et al. 1984) and in labial biopsies (Fox et al. 1986; Venables et al. 1989) although the evidence to date suggests that the virus is simply using the glands as a site of persistence rather than being responsible for the inflammation in SS. Nevertheless, because of case reports of patients with infectious mononucleosis developing Sjögren's syndrome (Whittingham et al. 1985), EBV is still considered a major candidate for involvement in its pathogenesis.

Immunohistology

The most remarkable finding in inflamed salivary glands is dense epithelial class II (and to a lesser extent class I) MHC antigen expression (Fox et al. 1985; Lindahl et al. 1985) on ducts and acini. The majority of the lymphocytes infiltrating the salivary gland are of the CD4 helper/inducer phenotype with a relative paucity of CD8 (suppressor/cytotoxic) cells (Adamson et al. 1983). About 10 per cent of the lymphocytes are B cells. There are also plentiful plasma cells whose cytoplasm contains IgA. Horsfall et al. (1988) have demonstrated anti-La (SS-B) idiotypes within these plasma cells as well as specific concentration of IgA anti-La (SS-B) and IgA rheumatoid factors in saliva (Horsfall et al. 1989) suggesting local production of autoantibody within the gland.

All these findings suggest that the disease starts with a persistent virus infection of salivary epithelium which, mediated by interferon γ (Fox et al. 1985) induces class II together with the expression of viral and/or host antigens. The cellular response, predominantly of the CD4 phenotype and unchecked by the deficient suppressor cells, leads to an immune attack on salivary epithelium. The production of autoantibodies results in the formation of immune complexes which in turn leads to the extra-glandular features of the disease.

Like patients with RA, patients with primary Sjögren's syndrome have an increased number of CD5 positive B cells in their peripheral blood (Plater-Zyberk et al. 1985). The CD5 antigen, normally present on T cells, is also found on B cells from patients with chronic lymphocytic leukaemia and in fetal blood, suggesting that the cells are of primitive ontogeny and showing an intriguing link with B cell malignancy. In mice the equivalent cell secretes autoantibodies. If a similar function is defined in man, it is possible that the CD5 positive B cell may play a role, not only in the autoimmunity of SS, but also in the lymphomas that occasionally complicate it.

Renal disease in Sjögren's syndrome

Clinically significant renal disease is rare in SS, and most descriptions in the literature are based on accumulated single case reports. In a recent study of 89 SS patients in our unit (Pease et al. 1989) clinically significant renal disease, manifest as proteinuria, hyperchloraemia, or acidosis, was detected in only four patients. Three of 42 patients with primary SS had renal tubular acidosis, and one patient with SS/SLE overlap had glomerulonephritis.

The two major immunopathological lesions of Sjögren's syndrome, the exocrinopathy and the circulating immune complex disease, reflect the major types of lesion seen in the kidney. The first of these is the interstitial nephritis in which plasma cells and lymphocytes surround the tubules, a pathological change which is strikingly reminiscent of the changes around the ducts in the inflamed salivary gland. The second is immune complex glomerulonephritis, often present in the SS/SLE overlap group, and perhaps representing a feature of SLE rather than the Sjögren's lesion itself.

Interstitial nephritis

Interstitial glomerulonephritis in SS results predominantly in tubular dysfunction. This was initially reported as nephrogenic diabetes insipidus in 10 of 62 patients by Bloch et al. (1965). Kassan and Talal (1987) suggested that this lesion may have been secondary to hypokalaemia associated with renal tubular acidosis rather than a direct effect of the inflammation itself. However, in a more recent study (Shiozawa et al. 1987), defective urinary concentration ability, as measured by the Fishberg test, was detected in nine of 11 patients with primary SS, and there was no relationship between this defect and renal tubular acidosis. This provided evidence that the failure to concentrate urine was not secondary to renal tubular acidosis, and may have been a direct result, or even a very sensitive indicator, of subclinical interstitial nephritis.

The classical lesion of SS is Type 1 renal tubular acidosis (affecting the distal convoluted tubule) which results in defective hydrogen ion excretion, leading to hyperchloraemic acidosis with secondary hypokalaemia. Type 2 renal tubular acidosis, involving the proximal tubule and leading to impaired bicarbonate reabsorption is said to be less frequent (Kassan and Talal 1987), although it has been described in association with Fanconi's syndrome (Shearn and Tu 1968). In these patients severe pathological changes, such as tubular atrophy and extensive fibrosis were present. More subtle changes in proximal tubular function, detected as reduced phosphate reabsorption in association with normal parathormone levels, were found in six of 17 patients with subclinical renal disease (Shiozawa et al. 1987).

At least half the patients presenting with renal tubular acidosis have Sjögren's syndrome (Talal 1971) and, conversely, clinically important renal tubular acidosis occurs in about 10 per cent of patients with SS. Latent disease, detected only by means of an acid-load test in patients with normal bicarbonate or chloride levels, occurs in about one-third of patients with primary SS or SS/SLE overlap syndromes (Shearn and Tu 1968; Siamopoulos et al. 1986; Shiozawa et al. 1987) and rarely in SS/RA. The study by Shiowaza et al. (1987) suggested that the frequency of mild interstitial nephritis may be even higher than this. Among seven renal biopsies from SS patients, five showed interstitial inflammation, though only two of the patients had evidence of renal tubular acidosis.

Presenting features

Renal tubular acidosis almost always presents as a biochemical defect rather than with any clinical symptoms or signs. In one of the few studies reporting clinical features, Shioji et al. (1970) described muscle weakness in three of four patients with SS and renal tubular acidosis. The weakness was reversed by treatment with alkalis in all three cases suggesting it was due to the renal tubular acidosis rather than the SS. In other studies (Shearn and Tu 1968; Siamopoulos et al. 1986; Shiozawa et al. 1987), common presenting features were hypokalaemia and hyperchloraemia. The plasma bicarbonate level seemed to be a surprisingly insensitive screening test. Urinary findings have also been unimpressive (Kassan and Talal 1987; Shiozawa et al. 1987) with mild proteinuria and occasional hyaline casts or white cells. Osteomalacia and/or nephrocalcinosis are rare sequelae of renal tubular acidosis in SS.

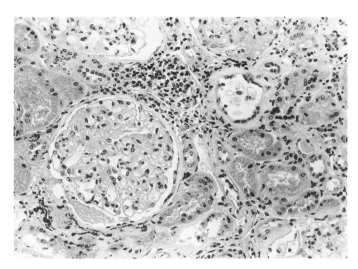

Fig. 2 Renal biopsy from a patient with renal tubular acidosis. There is an infiltrate of lymphocytes and plasma cells around the tables with complete sparing of the glomeruli.

Histology
As in the salivary gland, the target cells for the inflammatory response in the kidney appear to be the tubular epithelial cells. These are surrounded by lymphocytes and plasma cells (Fig. 2). In three reported cases where both renal and salivary gland biopsies were available in the same patients, the density of inflammation was much less in the kidney than in the salivary gland (Shioji et al. 1970). In the salivary gland, aggregates amounting to ectopic lymphoid follicles were seen, whereas in the corresponding renal tissue the infiltrates were much more scattered. Supporting Talal's prediction that similar pathological processes were occurring in the two tissues (Talal 1971), is the finding that, as in the salivary gland, most of the infiltrating T cells in the kidney are CD4 positive (Takaya et al. 1985; Rosenberg et al. 1988). In the majority of cases the tubular epithelium appears normal (Shioji et al. 1970). In severe disease there may be tubular atrophy or even necrosis of renal tubules which can rarely lead to glomerular hyalinization and renal failure (Gerhardt et al. 1978). Interstitial fibrosis is a common finding which appears to occur roughly in proportion to disease duration.

Immunopathogenesis
In early studies, the pathogenesis of the lesion was thought to be related to the hypergammaglobulinaemia characteristic of SS (McCurdy et al. 1967). Two possible mechanisms have been proposed: (1) that filtered immunoglobulin damages the renal tubules and; (2) that hypergammaglobulinaemia leads to increased plasma viscosity which in turn causes disturbances in the tubular microcirculation. In favour of such a link has been the known association between renal tubular acidosis and hyperglobulinaemia, both in idiopathic renal tubular acidosis and other diseases in which renal tubular acidosis is found such as SLE, primary biliary cirrhosis and other connective tissue diseases, all of which may be characterized by very high IgG levels. On the other hand Shioji et al. (1970) found no difference between the serum IgG levels in SS patients with or without renal tubular acidosis. They also described a patient with severe interstitial nephritis in whom immunoglobulin levels were normal. In the same study a careful search revealed no significant vascular abnormalities in renal biopsies from four patients with renal tubular acidosis.

It is now generally thought that it is SS itself rather than the associated hypergammaglobulinaemia which is causally linked with interstitial nephritis. Kassan and Talal (1987) argued that most of the connective tissue diseases associated with hypergammaglobulinaemia and tenal tubular acidosis are also associated with SS. It was also suggested that many patients with apparently idiopathic renal tubular acidosis have SS if it is looked for (Kassan and Talal 1987). Talal (1971), noting the analogy between the peritubular infiltrates in the kidney with the periductular lesion within the salivary gland, speculated that viruses within epithelial glands could be responsible for the local accumulation of inflammatory cells. It is tempting to speculate that the high frequency of occult inflammatory liver disease (Trevino et al. 1987) in primary SS may be due to similar mechanisms.

Diagnosis and treatment
Renal tubular acidosis only requires treatment if there are overt clinical or biochemical abnormalities such as hyperchloraemia, hypokalaemia, or evidence of acidosis, or if there are complications such as osteomalacia or nephrocalcinosis. The investigations require nothing more than routine examination of the blood electrolytes, including chloride and bicarbonate, and the urine. An acid-load test (see Chapters 1.3 and 1.5) may be performed to confirm the diagnosis. A renal biopsy is rather academic unless glomerulonephritis needs to be excluded. Rather less invasive and often more informative is a labial biopsy for the diagnosis of the associated Sjögren's syndrome. The biopsy is considerably more reliable than the greatly overused Schirmer test which often misses early disease (Shioji et al. 1970; Venables et al. 1989). Antibodies to Ro and La are useful serological markers.

Renal tubular acidosis is best treated with sodium bicarbonate in sufficient dose to normalize bicarbonate and chloride levels. Secondary features such as hypokalaemia and osteomalacia will be corrected and even mild azotaemia and proteinuria disappear (Kassan and Talal 1987). Renal tubular acidosis in itself is not usually an indication for treatment of the underlying Sjögren's syndrome although there is good evidence that it responds to either steroids or cyclophosphamide (100 mg/day) (Talal and Moutsopoulos 1987)

Immune complex glomerulonephritis
Prevalence
Glomerulonephritis is very rare in SS. Even in patients with SS and fully-developed SLE the frequency of nephritis is only 10 per cent, considerably less than the frequency in SLE without Sjögren's syndrome (Venables 1986; Pease et al. 1989). The observation by Wasiceck and Reichlin (1982) that SLE patients with anti-La (SS-B) had a relatively low frequency of renal disease would suggest that the immunopathology of SS may, in some way, be protective against nephritis.

Immunohistology
Membranous nephritis is the most common reported glomerular lesion in SS. Moutsopoulos et al. (1978) described three cases, two with some proliferative features. All three had primary Sjögren's syndrome although extraglandular features were prominent, suggesting that they were tending towards the SS/SLE

overlap part of the spectrum. More recently two out of 36 SS patients in Greece were described with membranous or membranoproliferative lesions (Siamopoulos *et al.* 1986). Occasional cases of proliferative nephritis have also been described. These are associated with cryoglobulinaemia, hypocomplementaemia, and evidence of widespread immune complex disease (Kassan and Talal 1987).

Mixed connective tissue disease

Definition

Mixed connective tissue disease (MCTD) is an overlap syndrome combining features of SLE, scleroderma, and polymyositis. It was originally defined by the finding of an antibody to an RNAse sensitive component of a tissue extract called extractable nuclear antigen (Sharp *et al.* 1972). The major antigenic component of this mixture subsequently turned out to be the polypeptides of a nuclear ribonucleoprotein termed U1 RNP (Tan 1982). The clinical features included a high frequency of Raynaud's phenomenon, arthritis, polymyositis, and fibrosing alveolitis and a low frequency of renal disease. The importance of the recognition of this syndrome was based on the claim that it had a good prognosis and response to steroid therapy.

Mixed connective tissue disease as a distinct entity

The association between anti-nRNP and the clinical features described by Sharp *et al.* have been confirmed by a number of workers. Whether this constitutes a distinct entity has been the subject of continuous and often fruitless arguments over the last 15 years. Our own studies of patients presenting with polymyositis (Venables 1986*a*) or fibrosing alveolitis (Chapman *et al.* 1984) suggested that patients with anti-nRNP seemed to merge into the spectrum of autoimmune rheumatic diseases rather than appearing particularly distinctive.

Prognosis and response to treatment

The original claims that patients with the disease had a good prognosis or had a particularly good response to steroids have not stood the test of time (Bennett 1985; Venables 1986*b*). Adult patients with MCTD have subsequently developed typical scleroderma, and children with features of the disease have evolved into typical SLE. The mortality of MCTD (about 5 per cent in 5 years) is similar to SLE. There is no evidence that differentiating MCTD significantly alters treatment. Patients with MCTD require the same treatment for their dominant clinical problem, whether or not they have anti-nRNP antibody.

Renal disease in mixed connective tissue disease

Membranous nephritis

When originally described, MCTD was associated with a low frequency of renal disease (Sharp *et al.* 1972). Bennett claimed that kidney involvement in MCTD was as high as 25 per cent (Bennett 1985) and that membranous nephritis was the most common lesion. Most other studies, perhaps including patients with milder disease, would suggest a frequency of about 10 per cent. These considerations are of limited significance to renal physicians as most of these patients would be diagnosed as SLE and treated as such. Such treatment would be entirely appropriate, as over 60 per cent of MCTD patients also fulfil criteria for SLE (Venables 1986). A diagnosis of MCTD or SLE represents an arbitrary choice of nomenclature. The link between anti-U1 RNP and membranous nephritis may reflect the high frequency of antibodies to ribonucleoproteins in membranous nephritis in SLE (Venables *et al.* 1983; Field *et al.* 1988). As in SLE, membranous nephritis in MCTD is rather slow to evolve and responds relatively poorly to steroids (Bennett 1985; Venables 1986*b*).

Other SLE-like renal lesions

Because MCTD is closely related to SLE, any lupus-associated renal lesion can be found. This is particularly important in children, as MCTD can rapidly differentiate into full-blown lupus with severe proliferative glomerulonephritis.

Scleroderma kidney

A few cases of vascular intimal hyperplasia affecting renal vessels, indistinguishable from scleroderma kidney, have also been described in association with MCTD (Bennett 1985). Such patients have developed accelerated hypertension and renal failure, as in idiopathic scleroderma.

References

Adamson, T.C., Fox, R.I., Frisman, D.M., and Howell, F.V. (1983). Immunohistologic analysis of lymphoid infiltrates in primary Sjögren's syndrome using monoclonal antibodies. *Journal of Immunology*, **130**:203–8.

Alexander, E.L., Malinow, K., Lijewski, J.E., Jerdan, M.S., Provost, T.T., and Alexander, G.E. (1986). Primary Sjögren's syndrome with central nervous system dysfunction mimicking multiple sclerosis. *Annals of Internal Medicine*, **104**:323–30.

Alspaugh, M.A., Talal, N., and Tan, E.M. (1976). Differentiation and characterization of autoantibodies and their antigens in Sjögren's syndrome. *Arthritis and Rheumatism*, **19**:216–22.

Bennett, R.M. (1985). Mixed connective tissue disease and overlap syndromes. In *Textbook of Rheumatology*, 2nd edn, (eds. W.D. Kelly, E.D. Harris, S. Ruddy, and C.B. Sledge), pp. 1115–33, W.B. Saunders, Philadephia.

Bloch, K.H., Buchanan, W.W., Wohl, M.J., and Bunim, J.J. (1965). Sjögren's syndrome: a clinical, pathological and serological study of sixty-two cases. *Medicine* (Baltimore), **44**:187–231.

Chapman, J.R., *et al.* (1984). Definition and clinical relevance of antibodies to nuclear ribonucleoprotein and other nuclear antigens in patients with cryptogenic fibrosing alveolitis. *American Review of Respiratory Disease*, **130**:439–43.

Daniels, T.E., Silverman, S., Michalski, J.P., Greenspan, J.S., Sylvester, R.A., and Talal, N. (1975). The oral component of Sjögren's syndrome. *Oral Surgery*, **39**:875–85.

Daniels, T.E. (1987). Oral manifestations of Sjögren's syndrome. In *Sjögren's syndrome: clinical and immunological aspects*. (eds N. Talal, H.M. Moutsopoulos, and S.S. Kassan), pp. 15–24, Springer-Verlag, Berlin.

Field, M., Williams, D.G., Charles, P., and Maini, R.N. (1988). Specificity of anti-Sm antibodies by ELISA for systemic lupus erythematosus: increased sensitivity for detection using purified peptide antigens. *Annals of the Rheumatic Diseases*, **47**:820–5.

Fox, R.I., Bumol, T., Fantozzi, R., Bone, R., and Schreiber, R. (1985). Expression of histocompatibility antigen HLA DR by salivary epithelial cells in Sjögren's syndrome. *Arthritis and Rheumatism*, **29**:1105–11.

Fox, R.I., Robinson, C., Kozin, F., and Fowell, F.V. (1986). Sjögren's syndrome: proposed criteria for classification. *Arthritis and Rheumatism*, **29**:577–85.

Fox, R.I., Pearson, G., and Vaughan, J.H. (1986). Detection of Epstein-Barr virus associated antigens and DNA in salivary gland biopsies from patients with Sjögren's syndrome. *Journal of Immunology*, **137**:3162–8.

Gerhardt, R.E., Loebl, D.H., and Rao, R.N. (1978). Interstitial immunofluorescence in nephritis of Sjögren's syndrome. *Clinical Nephrology*, **10**:201–7.

Harley, J.B., Reichlin, M., Arnett, F.C., Alexander, E.L., Bias, W.B., and Provost, T.T. (1986). Gene interaction at HLA-DQ enhances autoantibody production in primary Sjögren's syndrome. *Science*, **232**:1145–7.

Horsfall, A.C., Venables, P.J.W., Mumford, P.A., Allard, S.A., and Maini, R.N. (1988). Distribution of idiotypes in human autoimmune disease. *Biochemical Society Transactions*, **16**:341.

Horsfall, A.C., Venables, P.J.W., Allard, S.A., and Maini, R.N. (1989). Coexistent anti-La antibodies and rheumatoid factors bear distinct idiotypic markers. *Scandanavian Journal of Rheumatology*, (Supplement) **75**:84–8.

Kassan, S.S. and Talal, N. (1987). Renal disease in Sjögren's syndrome. In *Sjögren's syndrome: clinical and immunological aspects*. (eds. N. Talal, H.M. Moutsopoulos, and S.S. Kassan), pp. 96–101, Springer-Verlag, Berlin.

Kincaid, M.C. (1987). The eye in Sjögren's syndrome. In *Sjögren's Syndrome: Clinical and Immunological Aspects*. (eds. N. Talal, H.M. Moutsopoulos, and S.S. Kassan), pp. 25–33, Springer-Verlag, Berlin.

Lindahl, G., Hedfors, E., Klareskog, L., and Forsum, U. (1985). Epithelial HLA-DR expression and T lymphocyte subsets in salivary glands in Sjögren's syndrome. *Clinical and Experimental Immunology*, **61**:475–82.

Maini, R.N. (1987). The relationship of Sjögren'syndrome to rheumatoid arthritis. In *Sjögren's Syndrome: Clinical and Immunological Aspects*. (eds. N. Talal, H.M. Moutsopoulos, and S.S. Kassan), pp. 15–24, Springer-Verlag, Berlin.

Manthorpe, R., Frost-Larsen, K., Isager, H., and Prause, J.U. (1981). Sjögren's syndrome: a review with emphasis on immunological features. *Allergy*, **36**:139–53.

McCurdy, D.K., Cornwell, G.G., and DePratti, V.J. (1967). Hyperglobulinemic renal tubular acidosis. Report of two cases. *Annals of Internal Medicine*, **67**:110–17.

Morrow, J. and Isenberg, D. (1987). Sjögren's syndrome. In *Autoimmune Rheumatic Diseases*, pp. 203–33, Blackwell Scientific Publications, London.

Moutsopoulos, H.M., Balow, J.E., Cawley, J.T., Stahl, N.I., Autonovych, T.T., and Chused, T.M. (1978). Immune complex glomerulonephritis in sicca syndrome. *American Journal of Medicine*, **64**:955–60.

Pease, C.T., Shattles, W., Charles, P.J., Venables, P.J.W., and Maini, R.N. (1989). Clinical, serological and HLA pheonotypes subsets in Sjögren's syndrome. *Clinical and Experimental Rheumatology*, **7**, 181–4.

Plater-Zyberk, C., Maini, R.N., Lam, K., Kennedy, T.D., and Janossy, G. (1985). A rheumatoid arthritis B cell subset expresses a phenotype similar to that in chronic lymphatic leukemia. *Arthritis and Rheumatism*, **28**:971–6.

Rosenberg, M.E., Schendel, P.B., McCurdy, F.A., and Platt, J.L. (1988). Characterization of immune cells in kidneys from patients with Sjögren's syndrome. *American Journal of Kidney Disease*, **11**:20–2.

Sharp, G.C., Irwin, W.S., Tan, E.M., Gould, R.G., and Holman, H.R. (1972). Mixed connective tissue disease – an apparently distinct rheumatic disease syndrome associated with antibody to extractable nuclear antigen. *American Journal of Medicine*, **52**:148–59.

Shearn, M.A. (1977). Sjögren's syndrome. *Medical Clinics of North America*, **61**:271–82.

Shearn, M.A. and Tu, W.H. (1968). Nephrogenic diabetes insipidus and other defects of renal tubular dysfunction in Sjögren's syndrome. *American Journal of Medicine*, **39**:312–18.

Shillitoe, E.J., Daniels, T.E., Whitcher, J.P., Strand, C.V., Talal, N. and Greenspan, J.S. (1982). Antibody to cytomegalovirus in patients with Sjögren's syndrome as detected by enzyme-linked immunosorbent assay. *Arthritis and Rheumatism*, **25**:260–5.

Shioji, R., Furuyama, T., Onodera, S., Saito, H., Ito, H., and Sasaki, Y. (1970). Sjögren's syndrome and renal tubular acidosis. *American Journal of Medicine*, **48**:456–63.

Shiozawa, S., Shiozawa, K., Shimizu, S., Nakada, M., Isobe, T., and Fugita, T. (1987). Clinical studies of renal disease in Sjögren's syndrome. *Annals of the Rheumatic Diseases*, **46**:768–72.

Siamopoulos, K.C., Mavridis, A.K., Elisaf, M., Drosos, A.A., and Moutsopoulos, H.M. (1986). Kidney involvement in primary Sjögren's syndrome. *Scandinavian Journal of Rheumatology*, (supplement) **61**:156–60.

Sjögren, H. (1933). Zur Kenntnis der Keratoconjunctivitis sicca (keratitis filiformis bei hypofunktion der Tranendrusen). *Acta Ophthalmologica*, **11**:1–15.

Takaya, M., *et al.* (1985). T lymphocyte subsets of the infiltrating cells in the salivary gland and kidney of a patient with Sjögren's syndrome associated with nephritis. *Clinical and Experimental Rheumatology*, **3**:259–63.

Talal, N. (1971). Sjögren's syndrome, lymphoproliferation and renal tubular acidosis. *Annals of Internal Medicine*, **74**:633–4.

Talal, N. and Moutsopoulos, H.M. (1987). Treatment of Sjögren's syndrome. In *Sjögren's syndrome: Clinical and Immunological Aspects*, (eds. N. Talal, H.M. Moutsopoulos, and S.S. Kassan), pp. 291–5, Springer-Verlag, Berlin.

Tan, E.M. (1982). Autoantibodies to nuclear antigens (ANA): their immunobiology and medicine. *Advances in Immunology*, **33**:167–240.

Trevino, H., Tsianos, E.B., and Schenker, S. (1987). Gastrointestinal and hepatobiliary features in Sjögren's syndrome. In *Sjögren's syndrome: Clinical and Immunological Aspects*, (eds. N. Talal, H.M. Moutsopoulos, and S.S. Kassan), pp. 96–101, Springer-Verlag, Berlin.

Venables, P.J.W., Tung, Y., Woodrow, D.F., Moss, J., and Maini, R.N. (1983). Relationship of antibodies to soluble cellular antigens with histological manifestations of renal disease in SLE. *Annals of Rheumatic Diseases*, **42**:17–22.

Venables, P.J.W., Ross, M.G.R., Charles, P.J., Melsom, R.D., Griffiths, P.D., and Maini, R.N. (1985). A seroepidemiological study of cytomegalovirus and Epstein-Barr virus in rheumatoid arthritis and sicca syndrome. *Annals of the Rheumatic Diseases*, **44**:742–6.

Venables, P.J.W. (1986a). Antibodies to nucleic acid binding proteins: their clinical and aetiological significance. MD Thesis, University of Cambridge.

Venables, P.J.W. (1986b). Overlap syndromes and mixed connective tissue disease. In *Copeman's Textbook of the Rheumatic Diseases*, 6th edn, (ed. J.T. Scott), pp. 1269–77, Churchill Livingstone, Edinburgh.

Venables, P.J.W., Tio, C.G., Baboonian, C., Hughes, R.A., Griffin, B.E., and Maini, R.N. (1989). Persistence of Epstein-Barr Virus in salivary gland biopsies from patients with Sjögren's syndrome and healthy controls. *Clinical and Experimental Immunology*, **75**, 359–64.

Venables, P.J.W., Shattles, W., Pease, C.T., Ellis, J.E., Charles, P.J., and Maini, R.N. (1989). Anti-La (SS-B): a diagnostic criterion for Sjögren's syndrome? *Clinical and Experimental Rheumatology*, **7**:175–80.

Wasicek, C.A. and Reichlin, M. (1982). Clinical and serological differences between systemic lupus erythematosus patients with antibodies to Ro versus patients with antibodies to Ro and La. *Journal of Clinical Investigation*, **69**:835–43.

Whittingham, S., McNeilage, J., and McKay, I.R. (1985). Primary Sjögren's syndrome after infectious mononucleosis. *Annals of Internal Medicine*, **102**:490–7.

Wolf, H., Haus, M., and Wilmes, E. (1984). Persistence of Epstein-Barr virus in the parotid gland. *Journal of Virology*, **51**:795–8.

4.11 Sickle-cell disease and the kidney

L.W. STATIUS VAN EPS

Historical introduction

The discovery of human sickle cells and of sickle-cell anaemia was announced in the form of a case report presented at the 25th annual meeting of the Association of American Physicians at the New Willard Hotel in Washington, D.C. on the morning of 5 May 1910. All eminent American internists of that era were present, but there was no discussion of Herrick's paper, the discoverer of this new abnormality, and this new disease.

It was published that same year (Herrick 1910) and gives an accurate account of many of the now well-known features of the disease, 'the peculiar elongated and sickle-shaped red blood corpuscles', also the most important feature of sickle-cell nephropathy, an inability of the kidney to concentrate urine normally. All these observations were made by the study of one patient during a period of 6 years, a young black student from Granada, West Indies, staying in Chicago.

Between 1946 and 1949 Pauling and coworkers made brilliant discoveries identifying this familiar autosomal codominant disorder as an abnormality of the haemoglobin molecule introducing sickle-cell anaemia as 'a molecular disease' (Pauling et al. 1949).

Ingram (1956) was the first to use electrophoresis in combination with chromatography of the tryptic peptides of normal and sickle-cell haemoglobins, thus producing the so-called 'fingerprints'. He discovered that there was only a very small difference between the two globins: the exact nature of the defect was the substitution of valine for glutamic acid at the sixth residue of the β-chain (146 amino acids, α-chain 141 amino acids), causing a slight change in the three-dimensional spatial configuration of the haemoglobin molecule.

Glutamic acid is a charged amino acid and is therefore very soluble in water; valine is uncharged and poorly soluble in water. The loss of charge explains the slower migration in electrophoresis. Furthermore it has been confirmed that the uncharged, poorly soluble valine residues create a 'sticky spot' which causes the haemoglobin molecules to adhere to one another and form elongated structures that distort the red cells into their characteristic sickle shape (Edelstein 1986).

Definitions

The sickling phenomenon (Figs. 1, 2 and 3)

The one-point mutation at the sixth residue of the β-chain is the cause of intracellular polymerization of haemoglobin resulting in all the clinical manifestations of sickle-cell disease. When deoxygenation or acidosis or hyperosmolality leading to an increase in intracellular haemoglobin concentration occurs a special arrangement of the individual molecules is achieved with formation of long chains and subsequent alignment of haemoglobin elements. A sickle cell is formed. This is a rigid structure causing increased viscosity of whole blood, stagnation in the microcirculation, formation of microthrombi (aggregation of thrombocytes) and microinfarctions. The mechanical fragility of sickled cells is significantly increased causing haemolysis. These phenomena are, as a rule, reversible, when the physicochemical

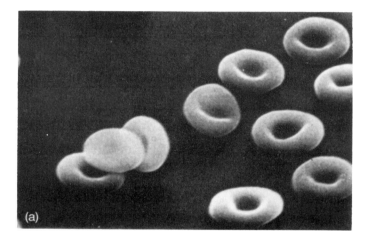

Fig. 1 (a) Normal red blood cells as seen by scanning electron microscopy. (b) Scanning electron microscopy on sickle cells. Intracellular polymerization has changed the form and consistency of the red blood cells, which have been transformed into bizarre forms and have become rigid. The cause is the formation of intracellular elongated fibres as a result of polymerization of haemoglobin molecules. There is increased viscosity of the microcirculation and the formation of microthrombi. (Reproduced from Barnhart, M.D., Henry, R.L. and Lusher, J.M. (1981). *Sickle Cell*. 3rd edn. Upjohn, Kalamazoo, by permission.)

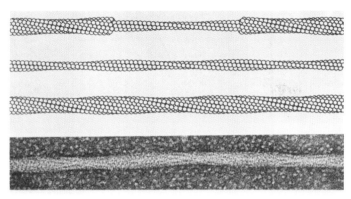

Fig. 2 Lower part: electron microscopy of a polymerized fibre formed by haemoglobin S molecules. The three upper parts show the three-dimensional reconstruction of such a fibre. The lowest of the three is a complete fibre, which consists of 14 helical structures arranged together. In the uppermost part a combination of inner and outer filaments is shown. (Reproduced from Edelstein, S.J., *Structures of the Fibers of Hemoglobin S. Human Hemoglobin and Hemoglobinopathies. A review to 1981.* (1981). University of Texas, Galveston, with permission.)

changes causing polymerization return to normal. A small percentage become irreversible sickle cells. The sickling cycle damages the cell membrane. Loss of fragments of membrane during the sickle–unsickle process has been observed (Jensen and Lessin 1970). Membrane proteins are destroyed by the polymerization process (Franck *et al.* 1985). There is evidence that the interaction of membrane proteins and calcium may play a role in fixing the sickle cell in the irreversible form (Palek 1977).

Sickle-cell anaemia (HbSS) is characterized by severe haemolytic anaemia (haemoglobin 5.9–9.6 g/dl), 80 per cent or more haemoglobin S (Hb-S), while the remainder is fetal haemoglobin (Hb-F). These patients suffer periodic painful crises involving the joints, bones, muscles, and abdomen; they are as a rule icteric and have a high percentage ($\pm$ 20 per cent) of reticulocytes. Their mean red-cell survival is 30 days ($N = 120$ days). These sickle-cell crises are caused by microvascular occlusions (thrombo-occlusive crises) (Chen 1984). Sometimes only a haemolytic crisis occurs, without pain but characterized by an acute decrease of haemoglobin and increase in serum bilirubin. The number of circulating irreversibly-sickled cells seems to determine the haemolytic component of the disease (Serjeant *et al.* 1969), which is significantly increased during the so-called haemolytic crisis. Very seldom an 'aplastic crisis' can be observed in sickle-cell anaemia with a rapid decrease in haemoglobin content and disappearance of reticulocytes (Van der Sar 1967). It is probably caused by an infection with parvovirus B 19. A sequestration crisis—an important cause of death in Jamaican children with sickle-cell disease—is characterized by acute enlargement of the spleen trapping a significant proportion of the red-cell mass, causing a precipitate fall in haemoglobin level and the risk of death from circulatory failure (Serjeant 1985). Recurrences may be prevented by chronic transfusion programmes or splenectomy.

In the heterozygote form, the sickle-cell trait (Hb-AS), only 30 to 45 per cent of the haemoglobin is HbS. This can be considered a benign condition. Under extreme physical stress and exhaustion, however, sickling and even crises can occur. Renal abnormalities such as hyposthenuria, increasing with age, do occur: transient attacks of haematuria and papillary necrosis with renal colic are occasionally observed. Haemoglobin content is normal or slightly decreased.

The hybrid sickling disorders

Haemoglobin S in combination with several other abnormal haemoglobins or with α- or β-thalassaemia have been described. These double heterozygotes manifest themselves in typical but variable clinical manifestations. Haemoglobin C with a substitution of lysine for glutamic acid at the sixth position of the β-chain causes a mild haemolytic anaemia, splenomegaly, and target cells. The homozygote form has completely normal renal functions (Statius van Eps *et al.* 1970*a*). In the double heterozygote form with 50 per cent HbS, the sickle-cell haemoglobin C disease (HbSC), the clinical manifestations are almost as severe as in HbSS, with crises especially during pregnancy and with signifi-

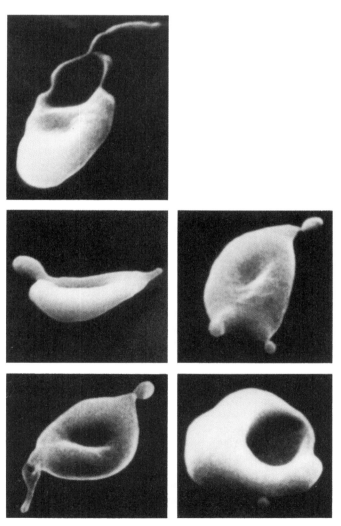

Fig. 3 Scanning electron microscopy of sickle erythrocytes. The effect of intracellular polymerization is demonstrated by the abnormal structures and the formation of spicules and vesicles. (Reproduced from Barnhart, M.D., Henry, R.L., and Lusher, J.M. (1981). *Sickle Cell*. 3rd edn. Upjohn, Kalamazoo, with permission.)

cant microthrombi in the retinal vessels. Impaired renal concentrating capacity and other renal abnormalities as in HbSS, are present (River *et al.* 1961; Statius van Eps *et al.* 1970a).

Haemoglobin E, also a point mutation in the β-chain, is primarily found in South-east Asia with many homozygotes in Thailand. The combination of haemoglobin E and S is uncommon (HbSE). It was first described in the Eti-Turks, a small Arabic speaking group in southern Turkey, who possess the highest incidence of HbS (16.8 per cent) in the white race. Clinical manifestations are limited to a mild hypochromic anaemia.

Haemoglobin D Punjab (HbD Punjab) is found in India and Pakistan, and also in the American black population. In combination with HbS (SD haemoglobinopathy) the clinical state varies from mild to quite severe with arthralgia, abdominal pain, splenic enlargement and anaemia. Some patients suffer from periodic painful crises. A renal concentration defect has been described (Cawein 1966).

Haemoglobin O Arab is a less common β-chain variant in which lysine is substituted for glutamic acid at position 121. Its origin is found in African people living in Arab territories. It is now also found in the American black population.

As the interaction of HbO Arab and HbS enhances gelling, the double heterozygote SO-disease has all the clinical features of sickle-cell anaemia.

Unlike sickle-cell–α-thalassaemia with variable clinical manifestations, the double heterozygote sickle-cell–β-thalassaemia (HbS–β-thal) has all the severe clinical abnormalities of HbSS with a higher haemoglobin content (50–70 per cent of normal).

Haemoglobin C Harlem has two substitutes on the β-chain and has been observed only in heterozygotes. Sickling and hyposthenuria are the characteristic abnormalities (Bookchin *et al.* 1968).

Haemoglobin C Georgetown, a variant of haemoglobin C, resembles HbS in its tendency to sickle. The same is the case with sickle-cell haemoglobin D Iran, where the substitution of glutamine for glutamic acid is at position 22 of the β-chain.

HbD Punjab has the same substitution, but at position β_{121}. Hb S-D Iran has been identified in a Jamaican family of West African ancestry and is a benign sickle-cell syndrome (Serjeant *et al.* 1982).

Mild sickle-cell anaemia has recently been associated with the combination β^s and an α-globin mutant α Montgomery (Prchal *et al.* 1989). It implies that the interaction between Hb Montgomery and HbS renders the haemolysate less likely to polymerize, consistent with the mild clinical course.

Our description of most of the sickle-cell variants has been extensive in order to stress the variability of these conditions. It should be realized that besides the original homozygote sickle-cell anaemia and the sickle-cell trait the many double heterozygotes have created a large group of haemoglobinopathies characterized by the phenomenon of sickling. This signifies that under conditions of hypoxaemia, hyperosmolality, and acidosis, intracellular polymerization can occur, although in variable degrees. For the nephrologist it is of utmost importance to realize that conditions are optimal for sickling to occur, especially in the vasa recta of the inner renal medulla. This is because of the existing hyperosmolality as a result of countercurrent multiplication, and because the inner medulla is oxygen-poor. Stagnation of the circulation through the vasa recta and ultimately obliteration of these vessels by microthrombi are the most important events in the creation of those renal abnormalities which characterize sickle-cell nephropathy.

Any sickling disorder can cause these abnormalities, but in a most variable way. There are also important factors counteracting intracellular polymerization and this causes variability of the renal abnormalities.

Factors affecting gelation

Allison (1957) and Singer and Singer (1953), measured the minimal concentration of haemoglobin required for gelation (polymerization) and found that the type and quantity of non-S haemoglobin had a marked influence. HbC was more effective than HbA in reducing minimum gelling concentration, while HbF was almost inert. This explains the clinical severity of patients with HbSC, the absence of clinical manifestations in HbAS and the protection afforded against sickling by the presence and quantity of HbF.

The following observations are interesting examples. Both HbD Punjab and Hb O Arab have amino acids that replace the amino-acid residue in position 121 of the β-chain and both interact strongly with HbS. Patients with HbS-D Punjab or HbS-O Arab disease can have severe clinical manifestations (Conley 1980).

Haplotypes of the β^s globin gene cluster

In the search for intrauterine detection of the sickle cell gene with the employment of restriction enzymes it was noticed, when using the restriction enzyme HpaI, that there were conflicting results. In most cases the β-globin gene was incorporated in large restriction fragments of 7.0 or 7.6 kb. However, in many individuals who were known to possess the sickle mutation, the β-globin gene was found on a 13-kb fragment, the presence of such a fragment being a marker for the sickle mutation. The appearance of the sickle mutation in β-globin genes on both 7.6 kb and 13 kb restriction fragments has intriguing implications for the origin of sickling (Edelstein 1986). The sickle mutation on a larger restriction fragment indicates that there had been two mutations of chromosome 11; the sickle mutation at (1) the position of the sixth amino acid of the β-chains; and (2) in the region flanking the β-globin gene altering one base responsible for producing the 7.6 kb fragment. As this region changed the larger 13-kb fragment was produced. As individuals possess the sickle mutation, but not the flanking mutation, this suggests that the sickle mutation arose independently on more than one occasion.

This suggestion that there were multiple origins of the sickle cell mutation was made by Kan and Dozy (1980), who studied the occurrence of 7.6-kb and 13-kb fragments in various populations. In the Niger region and the surrounding areas of West Africa the 13-kb fragment is found, whereas the population nearer the Zaire river has only the 7.6-kb fragment.

Pagnier *et al.* (1984) have discovered a third independent sickle-cell mutation in Africa, localized in Senegal. This pattern of sickle-cell anaemia may be less severe because individuals with this mutation produce higher levels of haemoglobin F (Nagel *et al.* 1985).

A particular mild form of sickle-cell anaemia found in Saudi Arabia appears to be due to elevated levels of fetal haemoglobin. This means that at present we can determine four β^s haplotypes with the help of restriction enzymes. The Benin type (most common in Benin and Algeria), the Central African Republic type, the Senegal type, clinically benign because of the production of high levels of glycoyl-γ-fetal haemoglobin ($^G\gamma$-HbF) and finally the Saudi Arabian type with the highest level of $^G\gamma$-HbF and therefore the mildest clinical manifestations.

The influence of fetal haemoglobin

Intracellular polymerization is much influenced by the concentration of HbS within the red cell, by intracellular oxygen tension, and pH. The quantity and type of other haemoglobins in the red cell are also of great importance. Fetal haemoglobin protects against sickling, but is not uniformly distributed among the red cells of HbSS patients. The higher the HbF content the longer such a cell will survive.

Inheritance of the genes for HbS and for the hereditary persistence of fetal haemoglobin results in the syndrome sickle-cell hereditary persistence of fetal haemoglobin (S-HPFH). It is characterized by an even intracellular distribution of high levels of HbF, which inhibit polymerization of HbS and hence intravascular sickling (Murray et al. 1988). These patients have a completely normal red blood cell count, a HbF percentage of 25 to 30 and no clinical manifestations of sickle cell disease. Unfortunately concentrating capacity and other renal functions have not been investigated in S-HPFH patients.

The percentage of HbF in HbSS patients without HPFH is never higher than 20 to 25. A high HbF level has been considered as a factor contributing to the mildness of sickle-cell disease and to long survivals, although not all investigators agree (Aluoch 1985). It has recently been reported that increased HbF in SS disease is determined by a factor (Gγ) linked to the β^s gene from one parent (Milner et al. 1984). The level of HbF modulates HbSS disease especially in certain populations (Pembrey et al. 1978) and individual patients. It now seems that it is not so much the HbF level but rather the type of HbF (Gγ and Aγ) as determined by the β^s-related haplotype, which controls the Gγ ratio of the HbF (Gilman and Huisman 1984).

Geography and epidemiology

The world distribution of haemoglobin S, C, and the thalassaemia genes are shown in Fig. 4. The S gene is most concentrated in two areas of West Central Africa: Nigeria and Ghana on the one hand and Gabon and Zaire on the other. Sickle-cell heterozygotes (HbAS) are found in over one-quarter of the newborn population. The gene is also concentrated in two other areas: the north-eastern corner of Saudi Arabia and in East Central India (Bunn and Forget 1986). The clinical manifestations in homozygotes from Africa are much more severe than from the other two locations, although the gene in Saudi Arabia and India results from migration out of East Africa (Gelpi 1973). Transport of slaves from East Africa to the Persian Gulf flourished from 200 to 1500 AD.

The slave trade from West Africa to the New World (from the 17th to the 19th century) carried the sickle-cell gene and the disease to the Americas and the Caribbean (Serjeant 1981).

Epidemiological evidence has indicated that sickle-cell heterozygotes (HbAS) are protected against malaria (the severe form caused by *Plasmodium falciparum*) which explains the very high incidence of the sickle-cell gene in malaria regions in Africa and other regions.

Eradication of malaria in North America resulted in a fall in trait-rate from 20 per cent to the current frequency rate of 8 per cent. Improved medical treatment of the homozygotes (protection against infections, blood transfusions) resulted in survival of one HbSS for every 40 HbAS. The present trait incidence in West Africa is much higher: the average frequency is 20 per cent (from 12.4 per cent in Ghana up to 27 per cent in Sierra Leone). In some regions an incidence of up to 50 per cent has been observed.

At present the multiple mutation theory receives most support, especially as a result of restriction endonuclease analysis. The 13-kb mutation arose in a small geographical area in what is now Upper Volta and Ghana. It is found in 70 per cent of the American black population who have HbS. Because the HbS and HbC genes protect against malaria the frequency of these genes increased and with it the linked 13-kb mutation (Kan and Dozy 1980).

Although HbAS individuals are protected against malaria, young HbSS individuals have a significant decreased resistance to bacterial infections, with the result that most African HbSS

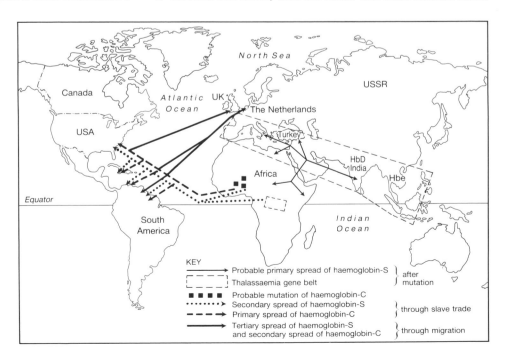

Fig. 4 World distribution of haemoglobins S,C,D,E and of the thalassaemia gene. (Reproduced from Aluoch, J.R. (1985). *Sickle Cell Disease in the Netherlands*. Thesis, University of Amsterdam, with permission.)

children die in their first 5 years of life without treatment. The survival rate is now much better with better medical supervision.

The cause of these infections arises mainly from loss of the function of the spleen: a progressive diminution in size occurs as a result of repeated infarctions. By the age of 7 SS children are generally functionally asplenic (Edelstein 1986). Pearson *et al.* (1969), observing Howell Jolly bodies in the blood of children with sickle-cell anaemia suggested the paradox of splenic hypofunction despite clinical splenomegaly. No uptake of radioactive technetium was seen in these enlarged spleens.

Asplenic patients (after splenectomy or in sickle-cell anaemia) are most susceptible to fulminant bacterial infections, and in particular to pneumococcal infection and sepsis, meningococcus, streptococcus, and *H. influenzae*. The incidence of bacterial meningitis is 25 times higher in HbSS children. The incidence of salmonellosis was found to be 300 times higher. The incidence of silent bacteriuria and urinary tract infections is much increased in sickle-cell disease.

Najjar (1975) discovered a tetrapeptide, tuftsin, originating from the spleen, which stimulate phagocytosis by increasing opsonization. Tuftsin has been synthesized and a radioimmunoassay developed to measure its serum concentration. Spirer *et al.* (1977, 1980) found a serum tuftsin concentration of 255.71 ± 10.27 ng/ml in 35 normal subjects. After elective splenectomy in 20 patients or staging in Hodgkin's disease (38 patients), serum tuftsin concentration had decreased to 118. ± 7.89 and 126.05 ± 10.27 ng/ml respectively.

Spirer *et al.* (1980) measured serum tuftsin in 21 patients with sickle-cell disease. In 14 patients with homozygote sickle-cell anaemia serum tuftsin concentration was significantly decreased (154.3 ± 35.1 ng/ml; $p < 0.01$). Comparable results have been observed by Winkelstein and Drachman (1968). They studied serum opsonizing activity for the pneumococcus with a phagocytic test using normal peripheral leucocytes incubated in normal serum and in serum from patients with sickle-cell disease. Heat-labile serum opsonizing activity was markedly deficient in serum of patients, compared with normal children.

Wilson *et al.* (1976) measured factors B and D as well as the total activity of the alternative pathway of complement activation in sera from 29 patients with sickle-cell anaemia and 18 normal controls. Total alternative pathway activity was reduced in the patients. Factor D levels were normal, whereas factor B levels were significantly depressed to about 50 per cent of normal.

This deficiency may be an important limiting factor in opsonification and may have an important bearing on the susceptibility to infection of patients with sickle-cell anaemia.

This susceptibility should not be underestimated as bacterial infection accounts for 30 to 40 per cent of the mortality in patients with sickle-cell anaemia. The very dangerous infections with pneumococci and meningococci are probably caused by the asplenism, responsible for a deficiency in tuftsin and of factor B. These factors are most important in stimulating opsonization of these bacteria by granulocytes.

Other mechanisms may be important in the frequent occurrence of osteomyelitis due to salmonella organisms. Kaye and Hook (1963) have suggested an impairment in the ability of the reticuloendothelial system to destroy phagocytized salmonella in the presence of haemolytic anaemia; such an impairment was observed in mice. Furthermore thrombosis, infarction, and necrosis of bones in sickle-cell disease may increase susceptibility to salmonella or other infections. Also microinfarctions and necrosis in the kidney may predispose to pyelonephritis.

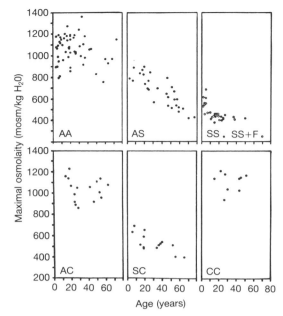

Fig. 5 Relationship between renal concentrating capacity (maximal osmolality) and age in normal subjects (HbAA), sickle-cell trait (HbAS), homozygote sickle-cell anaemia (HbSS, HbSS+F), haemoglobin C trait (HbAC), and sickle-cell haemoglobin C disease (HbSC). In homozygote sickle-cell anaemia concentrating capacity decreases significantly in young patients. It stabilizes in those older than 10 years at a mean of 434 ± 21 (SD) mosmol. In HbAS and HbSC a more gradual decrease with age is observed. (Reproduced from Statius van Eps, L.W., *et al.* (1970). *Clinica Chimica Acta*, **27**, 501–11, with permission.)

Sickle-cell disease is becoming a worldwide problem. After the Second World War it has become important in Western Europe, since a considerable number of the people among whom HbS occurs have migrated there, especially from the former colonies in the Caribbean to Great Britain, France, and The Netherlands (Aluoch 1985).

Renal manifestations of sickle-cell disease

Impaired concentrating capacity

Much attention has been given by many investigators (Keitel *et al.* 1956; Schlitt and Keitel 1960; Levitt *et al.* 1960; Hatch *et al.* 1967; Statius van Eps *et al.* 1967; Rubin 1968; Statius van Eps *et al.* 1970*a*; de Jong *et al.* 1982*a*; Statius van Eps and de Jong 1988; Statius van Eps and Leeksma 1989) to the remarkable paradox that in homozygote sickle-cell disease (HbSS) and other sickle-cell variants patients have a normal or even supernormal glomerular filtration rate (GFR)) and effective renal plasma flow (ERPF), while their concentrating capacity can be considerably disturbed.

The results of our studies of maximal concentrating capacity in a large number of patients with different haemoglobinopathies are reproduced in Fig. 5 (Statius van Eps *et al.* 1970*a*).

Analysing these results a definite relationship can be observed between concentrating capacity and age for subjects with sickle-cell trait (HbAS), homozygote sickle-cell anaemia (HbSS and

Fig. 6 The effects of multiple transfusions of normal blood given to a 4-year-old boy with homozygote sickle-cell anaemia (HbSS). There is a significant improvement in concentrating capacity, only to return to pretransfusion levels after discontinuation of the transfusions and disappearance of HbA from the blood. (Reproduced from Statius van Eps, L.W., et al. (1967). *Clinica Chimica Acta*, **17**, 449–61, with permission.)

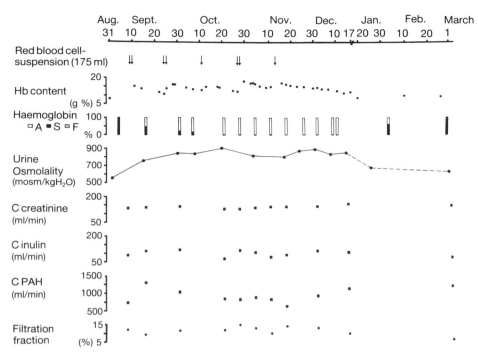

HbSS-F), and sickle-cell haemoglobin C disease (HbSC). In HbAS there is a gradual decline in maximal urinary osmolality from normal up to the age of 20 and declining to about 400 mosm/kg at 70 to 80 years. In HbSC patients the progressive gradual decrease in concentrating capacity is of a more severe nature although in the older patients it does not decrease below 400 mosm/kg. The patients with homozygote sickle-cell disease show a different pattern. There is a rapid decrease in children below the age of 10 to a maximal concentrating capacity of 400-450 mosm/kg. No further decrease in older subjects can, however, be observed. All these patients had normal or even supernormal GFR and effective renal plasma flow.

A second important observation in sickle-cell nephropathy has been the discovery that in young children renal concentrating capacity can be improved by multiple blood transfusions with normal (HbAA) red cells (Figs. 6 and 8). In older adults this improvement was not observed (Fig. 7).

Both Keitel *et al*. (1956) and Statius van Eps *et al*. (1967) in prolonged observations in 14 subjects of different ages (2–40 years) have shown that the capacity to improve renal concentrating capacity with multiple transfusions of HbAA blood is progressively lost with age (Fig. 8). No consistent changes in renal haemodynamics were observed during these experiments.

The results, summarized in Fig. 8, show that in young children under 10 years of age repeated transfusions are able to normalize concentrating capacity completely. This suggests that the concentrating defect is reversible in children. The basic phenomenon could be increased blood viscosity creating defective circulation through the vasa recta. The increased blood viscosity is the result of red cells undergoing sickling in the hyperosmotic inner medulla. The defective circulation through the vasa recta will disturb normal countercurrent multiplication and exchange making normal renal concentration impossible. Multiple transfusions of normal blood will normalize viscosity in the vasa recta and restore a normal circulation through these vessels.

Perillie and Epstein (1963) observed *in vitro* that increases in the osmolality of sickle-cell suspensions above 600 mosm cause sickling together with increased viscosity. Normalizing osmolality caused the cells to unsickle rapidly.

The increase in viscosity is a logical consequence of intracellular polymerization of HbS. Normal red blood cells are characterized by extreme pliability. A sickled erythrocyte lacks this pliability and passes through small vessels with great difficulty. Impairment in countercurrent multiplication and exchange will result in inefficient removal of reabsorbed fluid from collecting ducts into the inner medulla. It will prevent both the active and

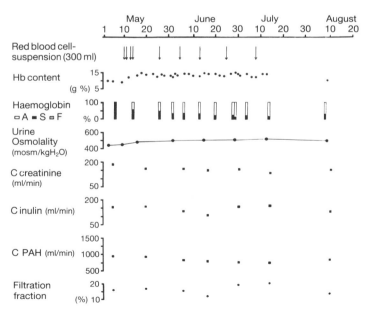

Fig. 7 The effects of multiple transfusions of normal blood in a 40-year-old woman with sickle-cell anaemia (HbSS). At this age no improvement in concentrating capacity is observed. (Reproduced from Statius van Eps, L.W., et al. (1967). *Clinica Chimica Acta*, **17**, 449–61, with permission.)

4 The patient with systemic disease affecting the kidney

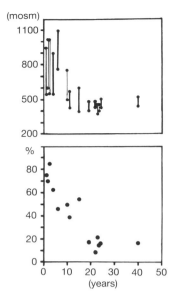

Fig. 8 Relationship between age and the ability to reverse the defect in renal concentrating capacity by multiple transfusions of normal blood in patients with sickle-cell anaemia. The upper part represents the maximal urinary osmolality achieved before transfusion (lower points of each vertical line) and after multiple transfusion with normal blood (upper point of each vertical line) in patients with sickle-cell anaemia ranging in age from 2 to 40 years. (Eight patients, thick lines: from Statius van Eps, L.W., et al. (1967). *Clinica Chimica Acta*, **17**, 449–61; and six patients, thin lines: from Keitel, H.L. (1956). *Journal of Clinical Investigation*, **35**, 998–1007 with permission.) The lower part of the figure indicates the percentage increase in maximal urinary osmolality resulting from multiple transfusions with normal blood. (Reproduced from Statius van Eps, L.W., et al. (1967). *Clinica Chimica Acta*, **17**, 449–61, with permission.)

passive accumulation of solute in the interstitium of the inner medulla necessary to achieve a maximal concentrated urine during antidiuresis. Dilution of the hyperosmotic inner medulla will occur.

In HbSS subjects above the age of 15 years concentrating capacity is fixed at 400 to 450 mosmol and does not improve with multiple blood transfusions with HbA erythrocytes or any other measure (Fig. 7). It is logical to assume that the original functional disturbances have developed into irreversible structural abnormalities (Statius van Eps et al. 1970a). Increased viscosity and repeated vasa recta occlusion by microthrombi produced by sickled cells causes obliteration of these vessels, ischaemia and multiple infarctions of the inner medulla and papillae. The result will be either scar formation or necrosis, which can even cause papillary necrosis with complete destruction of the papilla.

The process of slow and gradual obliteration of arterioles, compared with the attacks of severe painful crises in most bones and muscles, is not only observed in the inner renal medulla, although circumstances in that part of the kidney are most favourable. The hyperosmotic surroundings decrease the mean corpuscular volume of the red blood cells causing an increase in intracellular haemoglobin concentration, which favours polymerization.

The phenomenon of slow obliteration of the microcirculation, without the acute crises, is also observed in bones, causing for instance aseptic necrosis of the humoral or femoral head. In the skin it leads to leg ulcers, in the retina to vaso-occlusion, and occurs even in the lungs, causing pulmonary hypertension in the third or fourth decade of life.

Obliteration of the vasa recta as proposed above results in a permanent loss of countercurrent multiplication and countercurrent exchange in the long loops of Henle, but not in the short loops descending from the cortical nephrons. The result will be an irreversibly reduced concentrating capacity to values of 400 to 450 mosm/kg, which exists at the boundary between the outer and the inner medulla (Fig. 9), this being the boundary between the short loops and the long loops of Henle. It should be realized that in the normal human kidney approximately 85 per cent of the nephrons are located in the cortex and have these short loops. Their capacity for increasing interstitial osmolality is restricted to values of 400 to 450 mosm. The remaining 15 per cent are juxtamedullary nephrons with long loops of Henle, which descend into the inner medulla and papillae and are accompanied by parallel running hairpin vasa recta. They are able to increase interstitial osmolality to a maximum of about 1.200 mosm at the tip of the papilla. During antidiuresis the kidney is for this reason able to achieve a maximal urine osmolality of more than 1000 to 1200 mosm (Jamison 1973). In sickle-cell disease this highly specialized spatial configuration of parallel running long loops of Henle and vasa recta, called zonation (Fig. 9) (Plakke and Pfeiffer 1964; Schmidt Nielsen and O'Dell 1961), is lost. As a result the sickle-cell kidney has lost its ability to create a high osmolality in the inner medulla through normal countercurrent multiplication and exchange and normal 'trapping' of solute.

Maximally concentrated urine is produced by extraction of water from collecting duct fluid as it passes through the hypertonic medullary interstitium in the presence of antidiuretic hormone. Sickle-cell anaemia patients are not able to increase urinary osmolality above 400 to 500 mosm in spite of excessive stimulation with ADH (Fig. 10).

Microradioangiographic studies (Fig. 11)

Microradioangiographic studies of normal and sickle-cell kidneys (HbAA, HbSS, and HbSC patients) (Statius van Eps et al. 1970b) offer the visible proof of the abnormalities in the renal vasculature, in particular of the vasa recta in sickle-cell nephropathy. They show that with age vasa recta disappear and are completely absent in older HbSS patients.

By losing the concentrating function of the juxtamedullary nephrons with their long loops of Henle the sickle-cell kidney resembles the beaver kidney. This animal has only short loops of Henle, no inner medulla, and no vasa recta (Schmidt-Nielsen and O'Dell 1961). Like the sickle-cell kidney the beaver cannot concentrate its urine above 500 mosm. Both are characterized by renal homogeneity: they have only functional cortical nephrons.

Indomethacin and concentrating capacity

Indomethacin improves renal concentrating capacity in normal subjects (Fig. 12) (de Jong et al. 1982a; Allon et al. 1988), most probably because it promotes sodium reabsorption in the medulla with a consequent rise in papillary sodium and chloride concentration. However it does not increase urinary osmolality in HbSS patients deprived of water (Fig. 12) (de Jong et al. 1982a). This suggests a defect in the capacity to trap solute in the medulla in sickle-cell anaemia. The same may occur for reabsorbed urea.

Indomethacin administration to sickle-cell anaemia (HbSS)

4.11 Sickle-cell disease and the kidney

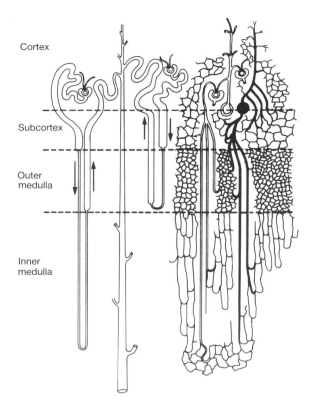

Fig. 9 On the left-hand side a juxtamedullary nephron with its long loops of Henle is illustrated; on its right is a cortical nephron with short loops. Their relationship to the cortex, the outer and the inner medulla of the kidney can be observed. On the right-hand side the vasculature consisting of vasa recta and collaterals are drawn separately. In reality the vasa recta and the long loops of Henle run parallel to each other, making countercurrent multiplication and countercurrent exchange possible. The result is the trapping of solute in the renal medulla creating the high osmolality necessary to produce a concentrated urine. (Reproduced from Fourman, J. and Moffat, D.B. (1971). *The Blood Vessels of the Kidney*. Blackwell, Oxford, with permission.)

patients caused a marked fall in fractional urea excretion and a rise in serum urea (de Jong *et al*. 1980). As urea is not trapped sufficiently, serum urea concentration increases, but there is no increase in medullary osmolality. As a result there is no effect on urinary osmolality. This agrees with the observation (Rubin 1968) that the ability to concentrate urine was not increased in HbSS children placed on a high-protein diet, as observed in normal subjects.

Generation of negative free water and the diluting capacity

In Fig. 12(b) the effect of indomethacin on urinary osmolality in the water-loaded state in normal subjects and sickle-cell patients is reproduced (de Jong *et al*. 1982*a*). Patients with sickle-cell anaemia are capable of diluting their urine normally (Hatch *et al*. 1967; Keitel *et al*. 1956; Schlitt and Keitel 1960).

Under conditions of water diuresis the decrease in urinary osmolality is identical in normal subjects and in HbSS patients. This was also the case for the percentage of filtered water excreted (C_{H_2O}/GFR). The capacity to reabsorb solute in the thick portion of the medullary ascending limb of Henle's loop is apparently normal.

This combination of a defect in renal concentrating capacity with a normal diluting capacity is characteristic of sickle-cell anaemia. During indomethacin administration, however, the HbSS patient is not able to excrete water normally. There was a greater fall in C_{H_2O}/GFR in sickle-cell subjects compared to normal controls and urinary osmolality rose from 42 to 125 mosm (de Jong *et al*. 1982*a*) (Fig. 12(b)). Identical results have recently been obtained by Allon *et al*. (1988). The reason for the rise in urine osmolality is not readily apparent, but obviously reflects the net result of prostaglandin inhibition on solute and water handling in the HbSS kidney. These results suggest that renal prostaglandins are apparently of importance in the normal diluting capacity in sickle-cell anaemia.

Allon *et al*. (1988) suggest that the increase in urine osmolality in sickle-cell patients following indomethacin has two possible explanations. One is an increase in the reabsorption of solute-free water in the collecting duct. As vasopressin is completely suppressed in both the patient and the control group, this is unlikely. The second explanation is a decrease in solute reabsorption in the diluting segment of the nephron. It was indeed observed that the fractional reabsorption of distally delivered solute decreased from 86 to 69 per cent. This signifies that the fraction of solute delivered to the diluting segment which remained in the final urine increased from 14 to 31 per cent, an increase of 120 per cent. This is contrary to the antinatriuresis expected; this occurs when blocking prostaglandins with non-steroidal anti-flammatory drugs such as indomethacin. Both de Jong *et al*. (1982*a*) and Allon *et al*. (1988) observed a pronounced fall in FE_{Na} of about 40 per cent in sickle-cell patients following indomethacin. The overall effect of prostaglandin inhibition is antinatriuretic with the exception of the medullary thick ascending limb of Henle, which has a natriuretic effect (Stokes 1979), which is enhanced in HbSS.

The capacity to generate negative free water (Tc_{H_2O}) has been studied in HbSS by several investigators (Whitten and Younes 1960; Levitt *et al*. 1960; Hatch *et al*. 1967).

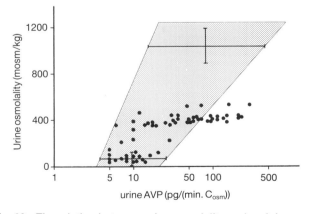

Fig. 10 The relation between urine osmolality and arginine vasopressin (AVP) excretion. The mean ±2 SD for AVP excretion and urine osmolality in control subjects, both during water deprivation and water loading, is given. The individual values for HbSS patients are shown. (Reproduced from Statius van Eps, L.W. and de Jong, P.E. (1988). The kidney in sickle cell disease. In *Diseases of the Kidney*, (eds. R.W. Schrier and C.W. Gottschalk), 4th edn. pp. 2561–83. Little, Brown, Boston with permission.)

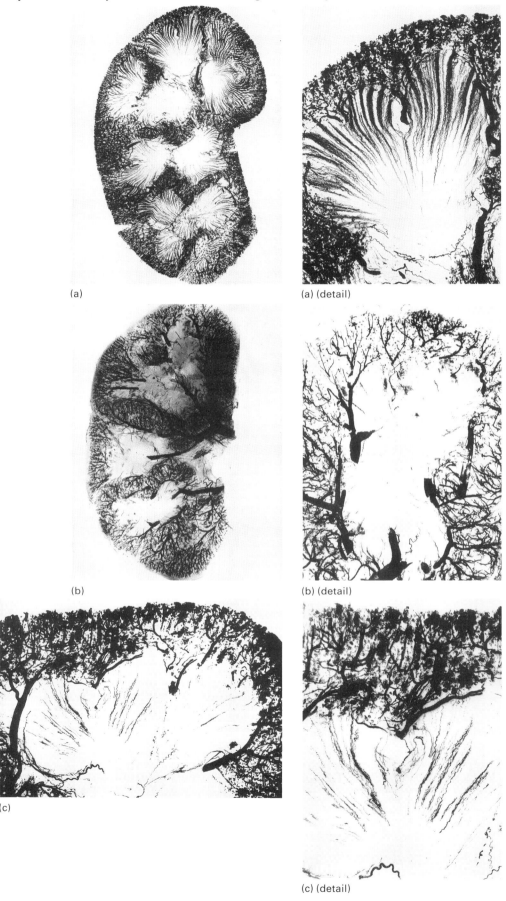

Fig. 11 Microradioangiographs from (a) a normal man with normal haemoglobin (HbAA); (b) a patient with homozygote sickle-cell anaemia (HbSS); and (c) a patient with sickle-cell–haemoglobin C disease (HbSC). (Reproduced from L.W. Statius van Eps et al. (1970). Lancet, **i**, 450–2, with permission.)

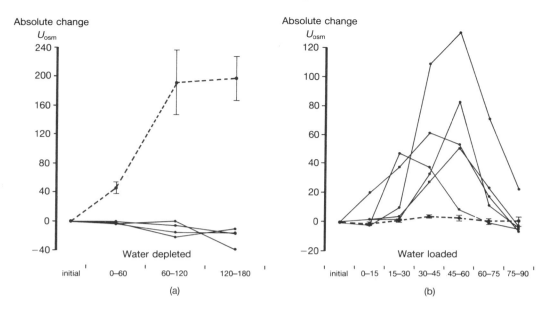

Fig. 12 The effect of indomethacin (75 mg as a suppository) in the water-depleted state (a) and the effect of indomethacin (0.25 mg/kg body weight IV) in the water-loaded state (b). The broken lines represent the means ± standard error of the mean in control subjects, and the continuous lines represent the individual data in patients with sickle-cell anaemia. (From P.E. de Jong et al. (1982). The influence of indomethacin on renal concentrating and diluting capacity in sickle-cell nephropathy. *Clinical Science*, **63**, 53–8, with permission.)

Normal Tc_{H_2O} levels were found in HbSS children, but in most cases Tc_{H_2O} after mannitol loading was lower in sickle-cell anaemia. After saline infusion Hatch et al. (1967) and Forrester and Alleyne (1977) agreed that Tc_{H_2O} was impaired in HbSS patients. From these studies it has been concluded that whereas a normal free water clearance indicates a normal transport mechanism in the thick portion of the ascending limb (outer medulla) only, a normal Tc_{H_2O} is dependent on adequate function of the part of Henle's loop in the inner medulla. This signifies that patients with sickle-cell anaemia are only able to increase osmolality to the level that can be generated in the outer medulla (about 450 mosmol). The conclusion is that in sickle-cell nephropathy the outer medullary circulation is adequate and the vascular pathology is confined to the inner zone (Statius van Eps and de Jong 1988).

Urinary acidification defect

Sickle-cell nephropathy is characterized by an incomplete form of renal tubular acidosis (Goossens et al. 1972; Ho Ping Kong and Alleyne 1971; Oster et al. 1976a; de Jong et al. 1983). Fortunately this defect does not cause a systemic acidosis, which has the potential danger of increasing sickling. On the contrary a respiratory alkalosis is encountered in the majority of these patients (Ho Ping Kong and Alleyne 1969). For this reason, routine treatment with alkali is not indicated. During a sickle-cell crisis lactic acidosis can occur. It is advisable, in severe crisis, to estimate arterial pH and blood gases and institute the appropriate therapeutic measures.

When given an acid load (Wrong and Davies 1959) the majority of patients with sickle-cell anaemia were unable to decrease urinary pH below 5.3. In normal subjects values below 5.0 are observed. Titratable acid and total hydrogen ion excretion are lower in HbSS and HbSC, but ammonia excretion is normal.

De Jong et al. (1983) demonstrated that the increased ammonium excretion induced by acid loading was reduced by indomethacin. This suggests that the assumed enhanced prostaglandin synthesis in sickle-cell disease (de Jong et al. 1984) may be important in maintaining a normal ammoniagenesis in sickle-cell disease.

Giving HbSS patients a maximal acidifying stimulus such as an infusion of sodium sulphate urinary pH can be lowered to that in normal subjects. These patients apparently require a greater than normal stimulus to excrete the expected quantity of hydrogen ion. In patients with sickle-cell trait, renal acidification is normal (Oster et al. 1976b).

None of the patients studied were acidaemic or hyperchloraemic before acid loading and there were no proximal tubular reabsorptive defects. The acidification defect should therefore be classified as the incomplete syndrome of distal renal tubular acidosis (Wrong and Davies 1959; Buckalew et al. 1968; Goossens et al. 1972). The microcirculatory disturbances in the inner medulla and papillae may well interfere with the ability of the collecting ducts to maintain the necessary hydrogen ion and electrochemical gradients.

Haematuria

Abel and Brown (1948) were the first to discover a relationship between sickle-cell disease and haematuria. The gross haematuria, which can occur both in homozygotes (HbSS) and in heterozygotes (HbAS, HbSC) can be a dramatic clinical event; the loss of blood, sometimes accompanied by severe renal colic, can be massive. Most frustrating for the clinician and for the patients are those cases which are resistant to therapy and in which the urinary bleeding continues for weeks and even for months. Not surprisingly, in many cases a nephrectomy was performed, although this should be considered an unnecessary and inadvisable therapeutic intervention. The clinical problem can be complicated because blood clots can imitate neoplasms on intravenous pyelography.

Many investigations have been published on this major pathologic manifestation of sickle-cell nephropathy (Goodwin et al. 1950; Chapman et al. 1955; Mostofi et al. 1957; Knochel 1969; Statius van Eps and Leeksma 1989). Gross haematuria may occur at any age and is more common in males than in females. The cause of the haematuria is probably related to the pathological events in the inner medulla and renal papillae of sickle-cell patients. Disturbances of the microcirculation as a result of increased sickling in the hyperosmotic surroundings of that part of the kidney causes microthrombi, infarctions, and obliteration of vasa recta (Statius van Eps et al. 1970b). In 21 sickle-cell kidneys removed because of massive and protracted haematuria Mostofi et al. (1957) were impressed by the absence of significant gross alterations. The lesions were inconspicuous and could easily be missed. They found severe stasis in peritubular capillaries, especially in the medulla. There was extravasation of blood, mainly in the collecting tubules. These observations agree with our present knowledge of the pathologic processes in the microcirculation of the inner medulla in sickle-cell nephropathy. Extravasation of blood as a result of disruption and interruption of arteriolae rectae circulation and vessel wall injury could well explain the bleeding. The presence in urine of urokinase, a fibrinolytic enzyme that converts plasminogen to plasmin which in turn lyses fibrin clots, undoubtedly contributes to the prolonged and persistent bleeding in some cases. However, in the majority of cases bleeding eventually subsides spontaneously, although, as mentioned before, it may last for weeks or even months. In some cases the haematuria follows trauma to the renal area. The left kidney is affected four times more frequently than the right kidney; in 10 per cent of cases the bleeding is bilateral.

For the clinician it is important to remember that most patients with the sickle-cell gene have microscopic haematuria. When this symptom is found on routine examination of the urine, it is advisable to perform a sickle-cell test before more extensive investigations are performed. It should now be considered also in people of white and other non-negroid races, because of the present-day worldwide distribution of the HbS gene.

Gross haematuria has been observed in patients with sickle-cell trait and von Willebrand's disease in the absence of extrarenal bleeding (Brody et al. 1977). These patients have been treated successfully with cryoprecipitate. It is therefore advisable to rule out other haemorrhagic disorders in sickle-cell patients with haematuria.

Urinary tract infections

The decreased resistance against infection in sickle-cell anaemia as a result of functional splenectomy has been discussed. In women with the sickle-cell trait, notwithstanding a normal spleen, the incidence of asymptomatic bacteriuria during pregnancy and the puerperium is distinctly higher than in non-pregnant HbS women and in normal subjects (Whalley et al. 1964; Pathak et al. 1969). The observation has been related to the increased incidence of 'pyelonephritis' observed by Amin and Ragbeer (1972) at autopsy in sickle-cell patients in Jamaica. Since the abnormalities in the renal medulla as a result of vascular obliteration, necrosis, and fibrosis could well be the result of the sickling process itself, it is necessary to view their conclusion cautiously. Silent bacteriuria should be considered in relation to urinary tract infections. Pyelonephritis or urosepsis may precipitate a crisis as may other infections. For these reasons all sickle-cell subjects in crisis should have a search performed for foci of infection, including any in the urinary tract. Urine should be examined microscopically and bacteriologically. In children 'mixed' bacteriuria is common. Organisms usually isolated include *E. coli*, enterobacteriaceae, and Klebsiella.

Papillary necrosis

Renal papillary necrosis is a frequent occurrence in sickle-cell disease, in homozygotes as well as in those with sickle-cell trait and in double heterozygotes. It has been extensively reviewed by Vaamonde (1984). Harrow et al. (1963) were the first to stress the importance of this complication when they reported their findings in five patients with sickle-cell trait and described the characteristic radiographic abnormalities. The distinctive abnormalities of the renal medulla and papillae in sickle-cell nephropathy are obliteration of vasa recta, and medullary and papillary necrosis and fibrosis.

In 45 patients gross painless haematuria was the most common symptom reported. Renal colic caused by the passage of blood clots or ruptured particles of necrotic papillae is less frequent.

At present, papillary necrosis is most frequently diagnosed as an incidental finding during renal radiography in asymptomatic sickle-cell patients (Vaamonde 1984). For this reason many cases are overlooked and remain undiagnosed. Microscopic haematuria is commonly found and a number of patients have renal abnormalities suggestive of a urinary tract infection. The urine should be tested for the presence of pieces of necrotic papillary tissue. As a rule sequestration of necrotic papillae is not accompanied by a decrease in renal function. The mild medullary form involving only the tip of the papilla, and the papillary form or total papillary necrosis, can be distinguished (Günther 1937). In the sickle-cell nephropathy there is a slow progression in time with the quantity of papillae involved, both kidneys demonstrating the lesions.

Intravenous pyelography is the method of choice to diagnose papillary necrosis. The observed abnormalities result from penetration of the radiocontrast material into the spaces resulting from loss of parenchyma or caused by necrosis. The necrotic papilla detaches with the formation of a cavity or remains attached with the possibility of calcium deposits, which explains the many and variable radiographic signs described by different investigators (Lindvall 1978). In sickle-cell nephropathy the partial papillary necrosis is most frequently encountered with a sinus or cavity formation within one or more papillae (McCall et al. 1978).

The frequency of sickle-cell nephropathy as a cause of renal papillary necrosis is unknown. Eknoyan et al. (1982) report an incidence of 15 per cent and Flaster et al. (1975) a much higher one of 36 per cent. The other causes of papillary necrosis are obstructive uropathy, pyelonephritis, analgesic nephropathy, and diabetes mellitus. Recent studies with intravenous pyelography show an incidence of this complication in sickle-cell disease of 40 per cent (range 23–67) (Eckert et al. 1974; McCall et al. 1978).

Vaamonde (1984) discusses data from 131 cases of renal papillary necrosis in 334 patients with HbS. There were more cases of HbSS disease (248) than cases with sickle-cell trait (50) although the frequency of HbaS is 40 times greater than HbAS. Perhaps the fact that the vascular abnormalities of the renal papilla develop much more slowly in patients with sickle-cell trait than in homozygotes explains its much lower incidence in the former (HbAS). In 168 published cases of renal papillary necrosis 51 per cent occurred in patients with HbSS, 34 per cent in HbAS, 12 per cent in HbSC, and 3 per cent in HbS Thal.

Papillary necrosis in patients with sickle-cell disease was found in relatively young individuals compared to those with other causes. Most subjects were younger than 40 years (87 per cent). The mean age in 124 patients summarized from the literature was 29 years with a range of 4 to 68 years (Vaamonde 1984).

It is important to realize that in the majority of patients with sickle-cell disease the prognosis of papillary necrosis itself is good. The lesions are not progressive as in other causes of papillary necrosis.

Renal potassium metabolism

In sickle-cell nephropathy there is a defect both in hydrogen-ion excretion and in the excretion of potassium (De Fronzo et al. 1979; Batlle et al. 1982). Potassium excretion primarily reflects distal nephron secretion. Not surprisingly surgical papillectomy significantly suppresses the increase in potassium excretion following KCl loading (Finkelstein and Hayslett 1974). De Fronzo et al. (1979) observed that after an intravenously administered KCl load patients with sickle-cell anaemia excreted significantly less potassium than control subjects. The defect could not be explained by hypoaldosteronism. Plasma renin activity and plasma aldosterone concentration were normal in these patients both under standard conditions and after volume contraction. The defect could not be explained by abnormalities in baseline insulin levels, nor by decreased delivery of sodium to the distal exchange site, nor by enhanced faecal losses of potassium. Also, after administration of sodium sulphate and frusemide, renal potassium excretion in HbSS patients was lower than in normal subjects.

As with the defect in the excretion of water and of hydrogen ions, the abnormality in potassium metabolism could be the result of ischaemic damage of that part of the distal nephron responsible for potassium excretion. The exact nature of the impaired potassium excretion in sickle-cell nephropathy needs further investigation and clarification.

Potassium excretion has been found normal in the sickle-cell trait. The impairment in renal potassium excretion in sickle-cell anaemia is unaccompanied by elevated serum potassium, even during acute KCl loading.

Batlle et al. (1982) observed hyperkalaemic hyperchloraemic metabolic acidosis in sickle-cell nephropathy (three with HbSS, two with HbAS, and one with HbSC). All had impaired potassium excretion. Five of these patients had slight to significant decrease of GFR and they all had a metabolic acidosis. Three of these patients had a selective aldosterone deficiency, two with normal and one with low plasma renin activity.

A hyporeninaemic hypoaldosteronism in sickle-cell disease has also been observed (De Fronzo 1980; Yoshino et al. 1982). It has been suggested that a disturbance in renin production by the juxtaglomerular apparatus is the primary cause, leading to a decreased aldosterone production and an impaired ability to excrete potassium loads. The hyperkalaemia responded favourably to treatment with mineralocorticosteroids.

Acute glomerular disease, proteinuria, and the nephrotic syndrome

Acute glomerulonephritis in HbSS patients has been reported by several investigators (Strauss et al. 1975; Nicholson 1977). Post-streptococcal aetiology was less frequent than expected, although its occurrence in three patients older than 35 years could have a relationship with the altered immune status or the presence of leg ulceration (Nicholson 1977).

Important findings in sickle-cell patients were made by Pardo et al. (1975) and Strauss et al. (1975), who demonstrated the deposition in the glomerulus of immune complexes derived from renal tubular epithelial antigen. They presumed that renal tubular antigen was released by ischaemic damage in the medulla stimulating antibody production, with deposition of immune-complexes in the glomerular capillary wall. The result is an immune-complex nephritis with a better prognosis than the rare occurrence of tubular antigen glomerulonephritis in non-sickle-cell patients.

Another cause of glomerular damage in sickle-cell patients is the formation of aggregates of sickle cells in glomerular capillaries, and the afferent and efferent arterioles, resulting in glomerular fibrosis. Occasionally embolism of glomeruli with necrotic bone marrow has been observed (Serjeant 1985).

Proteinuria and the nephrotic syndrome

All these glomerular pathological changes cause proteinuria and can develop into the nephrotic syndrome. In general appreciable proteinuria is a common finding in sickle-cell disease, occurring in about 30 per cent of patients. Massive proteinuria (> 3 g/24h) is less common, but is frequent enough to suggest a relationship between sickle-cell disease and the nephrotic syndrome (Berman and Tublin 1959; Sweeney et al. 1962).

Tejani et al. (1985) observed the frequent occurrence of glomerulosclerosis in children with sickle-cell disease. It can be assumed (de Jong and Statius van Eps 1985) that the hyperfiltration in sickle-cell anaemia patients makes them more prone to develop glomerulosclerosis, this being another cause of proteinuria. Elfenbein et al. (1974) observed the nephrotic syndrome in sickle-cell patients as a result of membranous glomerulopathy.

As sickle-cell disease is characterized by a predisposition to venous thrombosis, renal vein thrombosis should be ruled out in all sickle-cell patients in whom massive proteinuria develops.

Renal insufficiency

As a rule renal function deteriorates with age in HbSS patients. In the majority of cases this is as a result of parenchymal damage due to sickling and progressive occlusion of the renal microcirculation. In the medulla this happens in young subjects, but in patients 4 years and older definite vascular abnormalities are also observed in the cortex. In these patients chronic renal failure is an important cause of morbidity and mortality (Morgan and Serjeant 1981; Thomas et al. 1982). In the younger age group haemodynamics the cause is usually the glomerulopathies discussed above.

Serjeant (1985) suggests that renal failure in HbSS disease is multifactorial with contributions from progressive glomerular sclerosis, tubular damage, and cortical and medullary infarction and infection.

Hyperfiltration and glomerulosclerosis

It has been postulated that the medullary tissues in sickle-cell disease secrete increased amounts of prostaglandins in response to ischaemia (de Jong and Statius van Eps 1985) and that this may contribute to the supernormal renal plasma flow and GFR seen in most young patients with sickle-cell anaemia (Etteldorf et al. 1955) as well as in several older patients in the age group 37 to 58 years (Alleyne et al. 1975).

In support of this hypothesis is the demonstration by de Jong et al. (1980) that after treatment with the non-steroidal anti-inflammatory drug indomethacin, which inhibits cyclo-oxygenase, a reduction in renal haemodynamics (effective renal plasma flow and GFR) in young patients with sickle-cell anaemia is observed. These observations are supported by recent work (Allon et al. 1988).

Focal segmental glomerulosclerosis is seen in most sickle-cell patients (Tejani et al. 1985). Glomerular hyperfiltration has been attributed a pathogenetic role in this glomerulopathy (Brenner *et al* 1982) (see Chapter 9.3). These pathogenetic mechanisms may well contribute to the high incidence of end-stage renal disease in sickle-cell nephropathy (Thomas *et al*. 1982).

A characteristic finding in the kidneys of sickle-cell patients is glomerular enlargement and congestion, most frequent in children beyond the age of 2-years and most marked in juxtamedullary glomeruli (Bernstein and Whitten 1960; Buckalew and Someren 1974). There is a distinct difference between children with sickle-cell disease and normal children. In adult patients with HbSS and normal GFR glomerular size was larger than in HbAS and normal subjects. As the HbSS patients age progressive ischaemia and fibrosis with obliteration of the glomeruli is observed (Bernstein and Whitten 1960; Walker *et al*. 1971).

As our knowledge of the relation between glomerular hypercirculation and hyperfiltration and the development of glomerulosclerosis increases, more attention has been given to the occurrence of glomerulosclerosis in sickle-cell anaemia. In 13 HbSS children who had renal biopsy for persistent proteinuria or the nephrotic syndrome eight had glomerulosclerosis and five mesangial proliferation (Tejani *et al*. 1985). Those with glomerulosclerosis were older at onset of the nephropathy. There are more reports of the nephrotic syndrome with focal glomerulosclerosis in HbSS in the literature (Statius van Eps and de Jong 1988).

Hypertension and sickle-cell disease

The incidence of hypertension in the black population in the United States between the ages of 18 and 74 is 28.2 per cent (Health Statistics Series II No. 203, 1977). About 10 years ago reports were published surveying blood pressure in sickle-cell disease (de Jong *et al*. 1982b; Grell *et al*. 1981; Johnson and Giorgo 1981). Diastolic pressures greater than 90 mmHg were observed in only 2 per cent of patients with this disease. In a sex- and age-matched comparison, blood pressure levels were lower in HbSS patients (de Jong *et al*. 1982b). As the inner renal medulla and the juxtamedullary glomeruli play an important role in the conservation of water and sodium, the disturbed function of this part of the kidney causing renal loss of salt, hyponatraemia, and hyperaldosteronism (Radel *et al*. 1976; Carpentieri *et al*. 1977) has been related to the protection against hypertension in sickle-cell disease. However, other investigators found normal salt conservation by the sickle-cell kidney (Hatch *et al*. 1967; de Jong *et al*. 1978a; Hatch *et al*. 1989). Although in the experiments of Hatch *et al*. (1989), urinary sodium conservation in HbSS patients could only be achieved owing to an impressive increase in plasma renin activity, plasma renin activity in sickle-cell patients was four times as high as in control subjects on a 10 mosmol Na/24 h diet to achieve the same quantity of 24-h excretion of urinary sodium. This is a highly significant difference ($p < 0.01$). The increase in plasma renin activity was not accompanied by elevation of blood pressure.

As renal prostaglandins play an important role in regulating intrarenal blood flow, especially the intrarenal flow distribution between cortical and juxtamedullary nephrons, a relationship between the antihypertensive effect in HbSS and abnormalities in renal prostaglandin production has also been suggested, especially since vasoconstriction and ischaemia of the renal medulla stimulate prostaglandin production in interstitial cells of the medulla.

De Jong *et al*. (1984) found a normal excretion of prostaglandin E_2 in sickle-cell patients both in the dehydrated and in the water loaded state. Excretion of prostaglandin $F_{2\alpha}$, however, was decreased resulting in an increased prostaglandin $E_2/F_{2\alpha}$ ratio. As prostaglandin E_2 and prostaglandin $F_{2\alpha}$ have different and sometimes opposite effects on renal haemodynamics, renin release and sodium and water excretion these results could explain the protective action against hypertension.

In a recent report (Hatch *et al*. 1989) found a prevalence of 4 per cent of hypertension in 371 sickle-cell patients from their centre in Memphis, United States. A study of 1349 adult sickle-cell patients from 27 centres by these investigators revealed a similar low prevalence of hypertension.

Hatch *et al*. (1989) investigated 10 sickle-cell patients (five HbSS and five HbSC) and 11 age-matched black controls. Plasma renin activity was significantly greater in the sickle-cell patients than in the control subjects at all levels of sodium intake. The difference in plasma renin activity between the two groups was accentuated with progressive sodium restriction. The sickle-cell patients had also significantly greater forearm blood flow, than in normal subjects or in non-sickle chronic anaemic controls.

Forearm vascular resistance was significantly lower in the sickle-cell patients, than in either control group. The pressor dose of angiotensin II required to effect a 20 mmHg rise in diastolic blood pressure was greatly increased, compared with the control subjects.

The findings of hyper-reninaemia and normotension suggest the possibility of altered vascular responsiveness to endogenous angiotensin II in sickle-cell patients; the experiments discussed above support this suggestion.

The combination of hyper-reninism, increased production of renal prostaglandins, and normotension of sickle-cell nephropathy is also encountered in Bartter's syndrome (see Chapter 5.4). In this syndrome, the pathology of the renal medulla, consisting of fibrosis and increase and hyperplasia of the interstitial cells (Erkelens and Statius van Eps 1973) is very similar to that found in sickle-cell kidneys. Clinical, pathological, and hormonal similarities between Bartter's syndrome and sickle-cell nephropathy may help to understand intrarenal mechanisms and the contribution of the renal medulla in the regulation of intrarenal circulation and hormone production.

Macroscopic and microscopic pathology

Sydenstricker *et al*. (1923) were the first to describe postmortem studies of kidneys from patients with sickle-cell disease. After the discovery of electrophoretic and other techniques for differentiating between the disease (HbSS), the trait (HbAS), and other sickling haemoglobinopathies more sophisticated functional and morphologic studies could be carried out.

Gross anatomy

Kidneys of patients with sickle-cell disease are usually of near normal size and most do not show gross alterations (Alleyne *et al*. 1975). In the past kidneys have frequently been removed because of severe and prolonged haematuria (see above). They may demonstrate submucosal haemorrhages in the pelvis, medulla, and cortex (Mostofi *et al*. 1957). In many cases, the origin of the bleeding remains undetected.

Autopsy studies on patients with calicectasis show lesions consistent with acute and chronic pyelonephritis. However these lesions could also be caused by ischaemia, and necrosis as the result of the sickling process itself. The occurrence and pathogenesis of papillary necrosis have already been discussed (see above).

In sickle-cell nephropathy the lesions are primarily located in the medulla. The continuously recurring process of intravascular sickling, increased viscosity, the formation of microthrombi, ischaemia, and necrosis causes severe alterations in the microcirculation, loss of the highly specialized structure of parallel running loops of Henle and vasa recta making countercurrent multiplication and exchange impossible. Necrosis, ischaemia, and fibrosis can be encountered, especially in the inner medulla and the renal papillae.

In contrast, abnormalities in the cortex such as cortical infarctions are seldom reported. In patients over 40 years of age, 20 per cent exhibited irregularity of the renal outline on intravenous urography (Morgan and Serjeant 1981).

Microradioangiographic studies of the microcirculation in sickle-cell nephropathy have already been discussed. The changes observed are the result of occlusion of vasa recta and offer visible proof of the structural basis for the functional changes observed (Editorial 1970).

Microscopy

Sydenstricker et al. (1923) observed prominent glomeruli distended with blood and necrosis and pigmentation of tubular cells. The glomerular enlargement and congestion are more frequent in children older than 2 years and are most marked in juxtamedullary glomeruli (Bernstein and Whitten 1960; Buckalew and Someren 1974). As discussed before, glomerular size has been found significantly greater in HbSS patients than in normal subjects. Elfenbein et al. (1974) observed that glomeruli of patients with sickle-cell disease were 87 and 74 per cent larger, respectively than controls and patients with sickle-cell trait. Iron deposition, primarily in the proximal tubule, is a common finding, similar to other forms of haemolytic anaemia.

Ultrastructural pathology

In adults with sickle-cell anaemia normal epithelial foot processes and normal glomerular basement membranes are found. No specific changes in sickle-cell nephropathy are encountered. Possibly the electron-dense material with a homogeneous and sometimes granular or lamellated structure in the cytoplasm of mesangial cells may be the earliest effects of sickle-cell disease on the kidney (Pitcock et al. 1970). In patients with sickle-cell anaemia and the nephrotic syndrome much more striking electron microscopic changes have been reported (Elfenbein et al. 1974).

Four sickle-cell patients, three with the nephrotic syndrome and one with acute poststreptococcal proliferative glomerulonephritis, were examined by electron microscopy. In three patients reduplication of the GBM was more extensive than was appreciated by light microscopy, and mesangial cell cytoplasm was interposed between the outer original GBM and the inner basement-membrane-like material. No electron-dense deposits were seen. In the reduplicated GBM most capillary loops were more or less affected and the outer membranes were often thicker than usual. Scattered unbound aggregates of electron-dense granules, about 25 nm in diameter, were seen in endothelial and mesangial cells. Siderosomes in the glomeruli, small, membrane-bound, and spare were smaller and less granular than in tubular epithelium. It has been suggested that intracapillary fragmentation and phagocytosis of sickled red cells may play a role in the pathogenesis of the nephrotic syndrome in sickle-cell anaemia (Antonovych 1971).

Pitcock et al. (1970) showed that the early renal changes in sickle-cell anaemia without significant proteinuria included mild mesangial proliferation and the increase in GFR and effective renal plasma flow should therefore be regarded as an intrinsic characteristic of sickle-cell nephropathy most probably related to the medullary abnormalities.

De Jong et al. (1980) have suggested that the ischaemic damage to the renal medulla would stimulate prostaglandin synthesis leading to hyperfiltration. In their experiments indomethacin, a powerful inhibitor of prostaglandin synthesis, caused a significant fall in GFR and effective renal plasma flow, when given to HbSS patients, but not in control subjects. Similar observations have recently been reported by Allon et al. (1988). It is suggested that prostaglandins may play an important role in maintaining a normal or even supernormal GFR and effective renal plasma flow in sickle-cell anaemia. There may also be a relationship between the supernormal haemodynamics and the observed increased reabsorption capacity of the proximal renal tubule in HbSS (de Jong et al. 1978a).

The decline in renal function with age could be the result of the sickling phenomenon itself, causing progressive obliteration of the renal microvasculature, in particular of the vasa recta in the medulla. Alternatively, it could be the result of progressive glomerulosclerosis because of the hypercirculation and hyperfiltration.

Proximal tubular function

Proximal tubular activity, both secretory and reabsorptive, appears to be supranormal.

Tubular transport maximum of p-aminohippurate is elevated in sickle-cell anaemia, in particular in children (Etteldorf et al. 1952).

Fractional creatinine excretion is increased. Creatinine secretion displays a 20 to 29 per cent rise in fractional excretion in HbSS patients compared with normal subjects (de Jong et al. 1980; Allon et al. 1988). This signifies that creatinine clearance overestimates GFR considerably in sickle-cell anaemia. Inulin or iodothalamate clearances are preferred when estimating GFR in sickle-cell patients.

Other evidence of an increased proximal tubular secretory capacity has been obtained with data on uric acid excretion. Although an overproduction of uric acid in sickle-cell patients can be expected, serum concentration was not found to be elevated. Urate clearance, however, was significantly increased (Walker and Alexander 1971; Diamond et al. 1975). The increased urate clearance could be suppressed by pyrazinamide offering proof of the existence of an increased tubular secretion of urate. As urate clearance and renal function decreases with age in sickle-cell patients, the incidence of hyperuricaemia increases and also the possibility of attacks of gout (Ball and Sorensen 1970; Rotschild et al. 1980). Sometimes attacks of gout are difficult to differentiate from acute vaso-occlusive crisis or sickle-cell arthropathy. Serum uric acid estimations can be helpful in diagnosis.

Besides the increased secretory capacity of the proximal

tubule, sickle-cell patients also demonstrate an elevated proximal tubular reabsorption capacity. Maximal tubular reabsorption of phosphate per litre of glomerular filtrate is significantly increased, causing an elevation of serum phosphate (de Jong et al. 1978b). This leads to the conclusion that the high phosphate reabsorption reflects an increased reabsorptive activity of the proximal tubule (de Jong et al. 1978b). Sodium reabsorption in the proximal tubule parallels phosphate reabsorption. This suggests that the increased maximal tubular reabsorption of phosphate is accompanied by increased sodium reabsorption. This mechanism could explain the increased plasma volume in steady-state HbSS patients reported by several investigators (Barreras et al. 1966; Wilson and Alleyne 1976; Hatch et al. 1989). Alternatively, this increased proximal tubular reabsorption of sodium could be an important correcting homeostatic mechanism for defects in water and sodium conservation by the disturbed inner medullary functions.

Allon et al. (1988) also suggest 'that the enhanced natriuretic effect of prostaglandins in the ascending limb of Henle in sickle-cell patients in the face of normal salt delivery to the diluting segment in these patients under baseline conditions implies an enhanced solute reabsorption by the proximal tubule'.

Further support for an increased activity of the proximal tubule in sickle-cell anaemia is the increased tubular uptake of β_2-microglobulin in this condition (de Jong et al. 1981) and the positive correlation between the reabsorption of phosphate and β_2-microglobulin.

It is an attractive hypothesis that HbSS patients have a defect in renal medullary function with a tendency to lose water and sodium and to increase prostaglandin production. The latter will increase effective renal plasma flow, GFR, and proximal tubular activity. The ultimate result of this compensation will be normal homeostasis of fluid and electrolytes.

Intracellular mechanisms and the sickling phenomenon

Factors influencing intracellular polymerization of haemoglobin S have been discussed before. Recently they have become the subject of intensive investigation because many aspects of the polymerization process are still a matter of speculation. Hofrichter et al. (1974) have formulated the kinetic hypothesis of sickle-cell disease, schematically represented in Fig. 13 (Eaton and Hofrichter 1987; Mozzarelli et al. 1987).

Deoxyhaemoglobin S concentration, osmolality, pH, and temperature influence the so-called 'delay time', the time necessary for the formation of nuclei, composed of haemoglobin molecules, large enough to initiate the polymerization process. Delay time can range from milliseconds to days.

Deoxygenation occurs in the capillaries. Transit time is the time necessary to pass through the capillary system. According to the kinetic hypothesis sickling occurs normally in the venous system and normalization of the red blood cell occurs after reoxygenation in the lungs. Under conditions of stress, dehydration or acidosis, deoxygenation occurs more rapidly and polymerization occurs in the capillaries producing sickling, stasis, and tissue damage. Delay time and transit time, according to this hypothesis are the important factors influencing sickling. Nucleation will progress to chain formation, growth of the fibre, and alignment of the fibres.

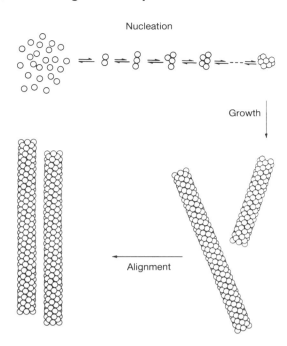

Fig. 13 Schematic reproduction of the sequence of events according to the 'kinetic hypothesis in sickle-cell disease' proposed by Eaton and Hofrichter. The process of polymerization starts with nucleation, followed by growth of the fibre and alignment of the fibres. According to this hypothesis sickling occurs normally in the venous system, but under conditions of dehydration, acidosis or hypoxaemia deoxygenation occurs more rapidly and polymerization occurs in the capillaries producing sickling, stasis and tissue damage. (Reproduced from Hofrichter, J. et al. (1976).) In Proceedings of the Symposium on Molecular and Cellular Aspects of Sickle cell disease, Dallas, Texas, December 1975. (eds. J.I. Hercules, et al.), DHEW Publication 76–1007, NIH, Bethesda.

Renal hormones

Erythropoietin (EPO) levels in sickle-cell anaemia have been found to be elevated (Alexanian 1973; Haddy et al. 1979; de Klerk et al. 1981), as could be expected as EPO titres tend to increase at decreasing haemoglobin concentrations. However, there are great differences in EPO titres at comparable haemoglobin concentrations and between the different causes of anaemia. EPO titres in patients with pure red cell aplasia, for instance, were 10-fold higher than in patients with homozygous sickle-cell anaemia (de Klerk et al. 1981). Alexanian (1973) observed in four of his eight patients with sickle-cell anaemia rather low EPO levels for the measured haematocrit values, although still significantly elevated. Although an increased erythropoietin level in sickle-cell anaemia is of great importance in maintaining a high reticulocyte production (about 20 per cent), the relatively low EPO levels in relation to the significantly decreased haemoglobin concentration has led to some speculation. It is possible that the structural and functional abnormalities in sickle-cell nephropathy are of influence in the production of EPO, causing a relatively lower production of this hormone in the sickle-cell kidney, when compared with other forms of anaemia. Morgan and Serjeant (1981), in a study of HbSS patients over 40 years, observed a decrease in GFR and age and a decrease in GFR and haemoglobin concentration. They suggested a reduced EPO production.

Plasma renin activity and aldosterone concentration are found to be normal in HbSS patients in steady-state conditions (De Fronzo *et al.* 1979; de Jong *et al.* 1980). However, Hatch *et al.* (1989) observed an impressive increase in plasma renin activity in sickle-cell patients on a salt-poor diet: the increase was four times as high as in control subjects (see section on hypertension in sickle-cell disease). Some patients with hyporeninaemic hypoaldosteronism and hyperkalaemia have been observed in sickle-cell disease (see section on renal potassium metabolism).

High plasma renin activity has been described in a HbSS patient with intermittent hypertension occurring during a painful crisis. Blood pressure became normal after the crisis had subsided (Sellers 1978).

Renal prostaglandins have an important role in the regulations of intrarenal blood flow. They are probably most important in intrarenal distribution of flow between cortical and juxtamedullary nephrons. The main production sites of these hormones are the interstitial medullary cells and the collecting duct cells of the kidney (Nissen and Anderson 1968).

The production of renal prostaglandins is promoted by vasoconstrictor stimuli (Zins 1950), as occurs as a result of hypocirculation or ischaemia of the renal medulla. The interstitial cells of the medulla in patients with sickle-cell anaemia contain aggregates of granular electron-dense material (Pitcock *et al.* 1970), a finding compatible with increased prostaglandin synthesis in these cells.

Renal prostaglandins can either be studied by direct measurement in the urine, or indirectly by inhibiting prostaglandin synthesis by a non-steroidal anti-inflammatory drug such as indomethacin. As already discussed indomethacin does not change GFR and effective renal plasma flow in control subjects, but produces a significant fall in these renal functions in HbSS patients (de Jong *et al.* 1980; Allon *et al.* 1988). This means that prostaglandins play an important role in maintaining a normal or even a supranormal GFR and effective renal plasma flow in sickle-cell anaemia and are most probably responsible for the glomerular hyperfiltration observed in sickle-cell children and young adults. This elevation in renal blood flow is most likely localized in the juxtamedullary nephrons, causing an increase in their GFR.

During indomethacin administration the HbSS patient is not able to excrete water normally: there was a greater fall in the percentage of filtered water excreted and urinary osmolality rose from 42 to 125 mosmol (Fig. 12(b)). This phenomenon has been discussed in the paragraph on diluting capacity.

In HbSS patients a similar sodium retention to that in normal subjects was observed after indomethacin administration. Unlike normal subjects, who showed both water and sodium retention and a rise in body weight, the HbSS patients had sodium retention only and a rise in serum osmolality, but no increase in body weight (de Jong and Statius van Eps 1985).

As discussed, the assumed enhanced prostaglandin synthesis in sickle-cell disease may be important in maintaining the increased ammonia production induced by acid loading.

De Jong *et al.* (1984) observed a normal prostaglandin E_2 excretion in patients with sickle-cell disease, both in the dehydrated and in the water-loaded state. Prostaglandin $F_{2\alpha}$ excretion, however, was decreased, resulting in a higher prostaglandin $E_2/F_{2\alpha}$ ratio than in normal subjects. They attributed these findings to a defect in 9-ketoreductase activity. They also postulate that, since prostaglandin E_2 and prostaglandin $F_{2\alpha}$ have different and sometimes opposite effects on renal haemodynamics and sodium and water excretion (Levenson *et al.* 1982), the abnormal balance between these two prostaglandins may cause some of the abnormalities of sickle-cell nephropathy.

Blood coagulation and vasa recta occlusion in sickle-cell disease

The presence of intravascular coagulation in sickle-cell disease, in particular during a vaso-occlusive crisis, is suggested by the following events. Platelet counts are decreased and platelet turnover increased; hyporeactive 'exhausted' platelets, elevated fibrinopeptide A and fragment D-dimer levels are all observed in patients suffering from a vaso-occlusive crisis. Renal vasa recta occlusion may reflect activation of coagulation (Statius van Eps and Leeksma 1989). Local, as well as systemic, factors may promote the formation of microthrombi, leading to eventual vasa recta obliteration.

The sickling of erythrocytes in the inner medulla leads to increased blood viscosity. Sickle erythrocytes and sickle-cell spicules and vesicles have been shown to accelerate blood clotting *in vitro*. Their procoagulant activity has been attributed to the presence of phosphatidylserine in the outer leaflet of their membrane (Chiu *et al.* 1981).

The increased adherence of sickle erythrocytes to monocytes and endothelial cells have been extensively investigated and recently reviewed (Statius van Eps and Leeksma 1989). Hypertonicity, as occurs in the vasa recta, may facilitate adherence (Hebbel *et al.* 1981).

Diminished prostacyclin-regenerating ability and increased prostacyclin inactivation in sickle-cell anaemia may explain the increased platelet aggregation in sickle-cell disease. This agrees with our observations that intravenous prostacyclin can have a beneficial effect on sickle-cell crisis provided therapy is started within 2 h after the onset of symptoms (L.W. Statius van Eps, unpublished observations). There is an increased platelet turnover and thrombocytopenia during vaso-occlusive crises, followed by a rebound thrombocytosis after clinical recovery, the latter stimulating *in-vivo* platelet aggregate formation. The functional asplenia characteristic of sickle-cell anaemia leads to thrombocytosis and the release of hyper-reactive megathrombocytes (Kenny *et al.* 1980).

Treatment

General measures

Knowing the abnormalities of sickle-cell disease and the sickle-cell kidney in particular, general measures should aim at the prevention or rapid correction of further decrease in the already low haemoglobin concentration, prevention of crises and prevention or rapid and adequate treatment of infections.

It is important that the patient is advised to avoid altitudes above 2500 m, either in unpressurized aeroplanes or on mountains above this level, without oxygen. In general, heavy exercise should be avoided as lactic acidosis or relative oxygen deficiency can cause a crisis. Transfusions with filtered normal red blood cells should always be given before operations, both major and minor. In critical medical situations such as infections or sepsis, shock as a result of Gram-negative bacteria, or severe blood loss, rapid blood transfusions are mandatory. Adequate antibiotic therapy, when necessary should start as soon as possible as the granulocytes of sickle-cell patients have a decreased opsonization capacity.

A routine 5 mg of folic acid daily has been shown to be an important and necessary dietary supplementation in sickle-cell anaemia.

Fluid intake

As discussed in the section on impaired concentrating capacity loss of the ability to concentrate urine to more than 450 mosm is the most important and consistent abnormality in sickle-cell nephropathy. Under normal circumstances this renal abnormality has few clinical consequences. As a rule subjects drink more than the minimum amount of water necessary to excrete the average solute load of 600 to 800 mosm per day.

The obligatory urine volume to excrete such a load in adult sickle-cell patients amounts to about 2000 ml per day, which is not an unusual volume for a normal person. In children the obligatory urine volume is less, and will vary with surface area. Fluid deprivation, inability to ingest fluid, or excessive fluid loss, however, will lead more rapidly to dehydration in HbSS patients than in normal individuals. The clinician should be very much aware of this danger, especially as dehydration increases the danger of sickling and crisis. Therefore, the fluid intake of patients with sickle cell disease should never be limited. On the contrary, larger quantities of intravenous fluid are very important in their treatment. On the other hand, care should be taken to avoid overloading the patient with intravenous fluid, especially during a sickle-cell crisis; this might tip the patient into circulatory overload and cause backward failure. A close watch on fluid status is necessary in all cases. Monitoring fluid intake and output carefully, and when necessary, making corrections by increasing the input or by giving frusemide in cases of circulatory overload, are advisable.

Repeated measurements of plasma electrolytes, pH, and blood gases and their correction where necessary are more than usually important in the care of critically-ill sickle-cell patients. Metabolic acidosis can then be corrected; this is important in prevention of the intracellular polymerization of haemoglobin S, which can lead to more severe crises. To correct acidosis sodium bicarbonate can be added to the intravenous infusion in appropriate quantities. Decreased arterial oxygen concentrations should also be corrected by giving oxygen, another factor important in the sickling process.

In hot climates and in clinical syndromes attended by vomiting, diarrhoea, infection, sepsis, and fever, the sickle-cell patient is more vulnerable to rapid dehydration or desalination than non-sicklers (Statius van Eps and de Jong 1988). Fluid balance can best be controlled by measuring urine production, which should be kept above 2000 ml/24 h in adults. In children the target will be less, in proportion to size.

In critically-ill sickle-cell patients, multiple transfusions with normal blood, erythrocyte transfusions or, if volume is a problem, exchange transfusions may be life-saving. Preoperative transfusions of normal blood to increase the haemoglobin concentration to 15 g/dl and decrease the HbS concentration below 30 per cent in patients with sickle-cell anaemia is important in their management.

Treatment of haematuria

Gross haematuria is a major clinical complication of sickle-cell nephropathy, and can be massive and life threatening. Nevertheless, conservative measures are the treatment of choice. In the majority of cases bleeding eventually subsides spontaneously, although it may last for weeks or even months. Therapeutic measures include complete bedrest, alkalinization of the urine, maintenance of a high rate of urine flow by excessive fluid administration, infusion of mannitol or other diuretic agents such as frusemide or even infusion of distilled water (500 ml in 15 min) (Knochel 1969; Marynick et al. 1977).

We have observed good results with repeated transfusions of normal haemoglobin blood or red blood cell suspensions. By these measures blood loss is corrected and the percentage of normal haemoglobin A is increased. Hyperbaric oxygenation has also been employed. Nephrectomy should only be considered when the haemorrhage is life-threatening and multiple transfusions are not possible.

In 1965 Immergut and Stevenson showed the favourable therapeutic effect of ε-aminocaproic acid (EACA) in the control of haematuria in sickle-cell patients. This drug is a potent inhibitor of urokinase, a fibrinolytic enzyme in the urine, which is capable of destroying fibrin and thus dissolving clots in the urinary tract. Complete inhibition of fibrinolytic activity occurs with 8 g EACA daily in divided doses. A high urinary concentration is obtained after oral or parenteral ingestion. The results obtained with EACA in the treatment of sickle-cell haematuria have been impressive, although the therapeutic regimens have varied widely (Statius van Eps and de Jong 1988). Since excessive dosages (12 g/day) can cause thrombotic complications, it is advisable to give no more than 2 to 3 g/day in divided dosages. As urinary levels of EACA are 50 to 100 times those of plasma, repeated oral dosages will guarantee adequate sustained urinary levels. If necessary EACA can be increased to 3 to 4 g four times daily. EACA should be added to the therapeutic measures, when bedrest, blood transfusions, urinary alkalinization (sodium bicarbonate, several grams four times daily) and diuresis with high volumes of fluid, preferably hypotonic solutions and frusemide (40 mg twice daily) have not been successful (Statius van Eps and de Jong 1988).

The newer antifibrinolytic drug, tranexamic acid, is a promising agent. It is seven to 10 times more potent than EACA. It will be worthwhile to run clinical trials in sickle-cell haematuria.

Treatment of urinary tract infection

As discussed above, urinary tract infections are a major problem in sickle-cell disease, because of the increased incidence of silent bacteriuria and the role of pyelonephritis or urosepsis in the cause or precipitation of a crisis. For this reason these infections should be looked for and treated properly, especially in young children who cannot complain of urinary symptoms.

When laboratory facilities make it possible, tests can be performed to distinguish upper from lower urinary tract infections, such as the estimation of urinary β_2-microglobulin (Schardijn et al. 1984). A pyelonephritis demands more intensive treatment and frequent checks.

Treatment of end-stage renal disease

The clinician has to decide which type of renal replacement therapy is the best. Both haemodialysis (Friedman et al. 1974) and transplantation (Spector et al. 1978; Chatterjee 1980; Gonzalez-Carillo et al. 1982), have been performed successfully, and the results do not differ from those in non-sickle-cell individuals. An interesting observation is the increased haematocrit values after transplantation as compared with HbSS patients with normal renal function.

De Jong (unpublished observation) observed in one HbSS patient, who had a renal transplantation, a gradual decrease in renal concentrating capacity reaching a maximum osmolality of 400 mosm within 12 months after transplantation.

Miner *et al.* (1987) have described the case of a 34-year-old black male, who underwent a successful renal transplantation. He tolerated the procedure well and was discharged with a serum creatinine level of 0.9 mg/dl (80 μmol/1). Some 3.5 years after transplantation he experienced a permanent decline in renal function. Biopsy revealed multiple features supporting recurrence of sickle-cell nephropathy as the cause of the deterioration. There were no pathologic abnormalities as seen in acute cellular or humoral rejection, but the abnormalities of sickle-cell nephropathy were conspicuous.

As for all types of surgical procedures, a patient with sickle-cell anaemia should be prepared properly for transplantation by multiple transfusions of normal blood to bring his haemoglobin content above 8 μmol/l and the HbS percentage below 30 per cent. Otherwise he runs extra risks such as severe crisis and massive sickling.

References

Abel, M.S. and Brown, C.R (1948). Sickle cell disease with severe hematuria simulating renal neoplasm. *Journal of the American Medical Association*, **136**, 624–5.

Alexanian, R. (1973). Erythropoietin excretion in bone marrow failure and hemolytic anemia. *Journal of Laboratory and Clinical Medicine*, **82**, 438–45.

Alleyne, G.A.O., Statius van Eps, L.W., Addae, S.K., Nicholson, G.D., and Schouten, H. (1975). The kidney in sickle cell anemia. Editorial Review. *Kidney International*, **7**, 371–9

Allison, A.C. (1957). Properties of sickle cell haemoglobin. *Biochemical Journal*, **65**, 212–9.

Allon, M., Lawson, L., Eckman, J.R., Delaney, V., and Bourke, E. (1988). Effects of nonsteroidal antiinflammatory drugs on renal function in sickle cell anemia. *Kidney International*, **34**, 500–6.

Aluoch, J.R. (1985). *Sickle cell disease in the Netherlands*. Thesis, University of Amsterdam. Foris Publications.

Amin, U.F. and Ragbeer, M.M.S. (1972). The prevalence of pyelonephritis among sicklers and non-sicklers in an autopsy population. *West Indian Medical Journal*, **21**, 166.

Antonovych, T.T. (1971). Ultrastructural changes in glomeruli of patients with sickle cell disease and nephrotic syndrome. *American Society of Nephrology* (Abstr.).

Ball, G.V. and Sørenson, L.B. (1970). The pathogenesis of hyper-uricemia and gout in sickle cell anemia. *Arthritis and Rheumatism*, **13**, 846–8.

Batlle, D., Itsarayoung Yen, K., Arruda, J.A.L., and Kurtzman, N.A. (1982). Hyperkalemic hyperchloremic metabolic acidosis in sickle cell hemoglobinopathies. *American Journal of Medicine*, **72**, 188–92.

Barreras, L., Diggs, L.W., and Lipscomb, A. (1966). Plasma volume in sickle cell disease. *Southern Medical Journal*, **59**, 456–61.

Berman, L.B. and Tublin, I. (1959). The nephropathies of sickle-cell disease. *Archives of Internal Medicine*, **103**, 602–6.

Bernstein, J. and Whitten, C.F. (1960). Histological appraisal of the kidney in sickle cell anemia. *Archives of Pathology*, **70**, 407–18.

Bookchin, R.M., Davis, R.P., and Ranney, H.M. (1968). Clinical features of hemoglobin C$_{\text{Harlem}}$, a new sickling hemoglobin variant. *Annals of Internal Medicine*, **68**, 8–18.

Brenner, B.M., Meyer, T.W., and Hostetter, T.H. (1982). Dietary protein intake and the progressive nature of kidney disease: The role of hemodynamically mediated glomerular injury in the pathogenesis of progressive glomerular sclerosis in aging, renal ablation, and intrinsic renal disease. *New England Journal of Medicine*, **307**, 652–9.

Brody, J.I., Levison, S.P., and Chung Ja Jung (1977). Sickle cell trait associated with von Willebrand syndromes. *Annals of Internal Medicine*, **86**, 529–33.

Buckalew, V.M. and Someren, A. (1974). Renal manifestations of sickle cell disease. *Archives of Internal Medicine*, **133**, 660–9.

Buckalew, V.M., McCurdy, D.K., Ludwig, G.D., Chaykin, L.B., and Elkington, J.R. (1968). Incomplete renal tubular acidosis: physiologic studies in three patients with a defect in lowering urine pH. *American Journal of Medicine*, **45**, 32–42.

Bunn, H.F. and Forget, B.G. (1986). *Haemoglobin: Molecular, Genetic and Clinical Aspects*. W.B. Saunders Company, Philadelphia.

Carpentieri, U., Matustik, M.C., and Meyer, W.J.III. (1977). Hyperaldosteronism and urinary sodium loss in sickle cell anaemia. *Blood*, **50** (supply. 1), 102.

Cawein, M.J., Lappat, E.J., Brangle, R.W., and Farley, C.H. (1966). Haemoglobin S-D disease. *Annals of Internal Medicine*, **64**, 62–70.

Chapman, A.Z., Reeder, P.S., Friedman, I.A., and Baker, L.A. (1955). Gross hematuria in sickle cell trait and sickle cell hemoglobin-C disease. *American Journal of Medicine*, **19**, 773–81.

Chatterjee, S.N. (1980). National study of natural history of renal allografts in sickle cell disease or trait. *Nephron*, **25**, 199–201.

Chen, S. (1984). Rheology of sickle cells and the microcirculation. *New England Journal of Medicine*, **311**, 1567–9.

Chiu, D., Lubin, B., Roelofsen, B., and van Deenen, L.L.M. (1981). Sickled erythrocytes accelerate clotting *in vitro*: an effect of abnormal membrane lipid asymmetry. *Blood*, **58**, 398–401.

Conley, C.L. (1980). Sickle cell anemia the first molecular disease. In *Blood, Pure and Eloquent*. (Ed. M.M. Wintrobe) 318–71. McGraw-Hill, New York.

De Fronzo, R.A., Taufield, P.A., Black, H., McPhedran, P., and Cooke, C.R. (1979). Impaired renal tubular potassium secretion in sickle cell disease. *Annals of Internal Medicine*, **90**, 310–6.

DeFronzo, R.A. (1980). Hyperkalemia and hyporeninemic hypoaldosteronism. *Kidney International*, **17**, 118–34.

Diamond, H.G., Meisel, A., Sharon, E., Holden, D., and Cacatian, A. (1975). Hyperuricosuria and increased tubular secretion of urate in sickle cell anemia. (1975). *American Journal of Medicine*, **59**, 796–802.

Eaton, W.A. and Hofrichter, J. (1987). Hemoglobin S and sickle cell disease. *Blood*, **70**, 124–66.

Eckert, D.E., Jonutis, A.G., and Davidson, A.J. (1974). The incidence and manifestations of urographic papillary abnormalities in patients with S hemoglobinopathies. *Radiology*, **113**, 59–63.

Edelstein, S.J. (1986). *The sickled cell. From myths to molecules*. Harvard University Press, Cambridge, Massachusetts.

Editorial (1970). Visible proof. *Lancet* i: 456–7.

Eknoyan, G., Quinibi, W.Y., and Grisson, R.T. (1982). Renal papillary necrosis: An update. *Medicine (Baltimore)* **61**, 55–73.

Elfenbein, I.B., Patchefsky, A., Schwartz, W., and Weinstein, A.G. (1974). Pathology of the glomerulus in sickle cell anemia with and without nephrotic syndrome. *American Journal of Pathology*, **77**, 357–76.

Erkelens, D.W. and Statius van Eps, L.W. (1973). Bartter's syndrome and erythrocytosis. *American Journal of Medicine*, **55**, 711–19.

Etteldorf, J.N., Tuttle, A.H., and Clayton, G.W. (1952). Renal function studies in pediatrics I. Renal hemodynamics in children with sickle cell anemia. *American Journal of Diseases of Children*, **83**, 185–91.

Etteldorf, J.N., Smith, J.D., Tuttle, A.H., and Diggs, L.W. (1955). Renal hemodynamic studies in adults with sickle cell anemia. *American Journal of Medicine*, **18**, 243–8.

Finkelstein, F.O. and Hayslett, J.P. (1974). Role of medullary structures in the functional adaptation of renal insufficiency. *Kidney International*, **6**, 419–25.

Flaster, S., Lome, L.G., and Presman, D. (1975). Urologic complications of renal papillary necrosis. *Urology*, **5**, 331–6.

Forrester, T.E. and Alleyne, G.A.O. (1977). Excretion of salt and water by patients with sickle cell anemia: effect of a diuretic and solute diuresis. *Clinical Science*, **55**, 523–7.

Franck, P.F.H., et al. (1985). Uncoupling of the membrane skeleton from the lipid bilayer. The cause of accelerated phospholipid flip-flop leading to an enhanced procoagulant activity of sickled cells. *Journal of Clinical Investigation*, **75**, 183–91.

Friedman, E.A., et al. (1974). Uremia in sickle cell anemia treated by maintenance hemodialysis. *New England Journal of Medicine*, **291**, 431–5.

Gelpi, A.P. (1973). Migrant populations and the diffusion of the sickle-cell gene. *Annals of Internal Medicine*, **79**, 258–64.

Gilman, J.G. and Huisman, T.H.J. (1984). Two independent genetic factors in the β-globin gene cluster are associated with high G levels in the HbF of SS patients. *Blood*, **64**, 452–7.

Gonzalez-Carillo, M., Rudge, C.J., Parsons, V., Bewick, M., and White, J.M. (1982). Renal transplantation in sickle cell disease. *Clinical Nephrology*, **18**, 209–10.

Goodwin, W.E., Alston, E.F., and Semans, J.H. (1950). Hematuria and sickle cell disease: unexplained gross unilateral renal hematuria in Negroes, coincident with the blood sickling trait. *Journal of Urology*, **63**, 79–96.

Goossens, J.R., Statius van Eps, L.W., Schouten, H., and Giterson, A.L. (1972). Incomplete renal tubular acidosis in sickle cell disease. *Clinica Chimica Acta*, **41**, 149–56.

Grell, G.A.C., Alleyne, G.A.O., and Serjeant, G.R. (1981). Blood pressure with homozygous sickle cell disease. *Lancet* **ii**, 1166.

Günther, G.W. (1937). Die Papillennekrosen der Niere bei Diabetes. *Münchener Medizinischen Wochenschrift*, **84**, 1695–9.

Haddy, T.B., Lusher, J.M., Hendricks, S., and Trosko, B.K. (1979). Erythropoiesis in sickle cell anaemia during acute infection and crisis. *Scandinavian Journal of Haematology*, **22**, 289–95.

Harrow, B.R., Sloane, J.A., and Liebman, N.C. (1963). Roentgenologic demonstration of renal papillary necrosis in sickle-cell trait. *New England Journal of Medicine*, **268**, 969–76.

Hatch, F.E., Culbertson, J.W., and Diggs, L.W. (1967). Nature of the renal concentrating defect in sickle cell disease. *Journal Clinical Investigation*, **46**, 336–45.

Hatch, F.E., Crowe, L.R., Miles, D.E., Young, J.P., and Portner, M.E. (1989). Altered vascular reactivity in sickle hemoglobinopathy. A possible protective factor from hypertension. *American Journal of Hypertension*, **2**, 2–8.

Hebbel, R.P., Moldow, C.F., and Steinberg, M.H. (1981). Modulation of erythrocyte endothelial interactions and the vaso-occlusive severity of sickling disorders. *Blood*, **58**, 47–52.

Herrick, J.B. (1910). Peculiar elongated and sickle shaped red blood corpuscles in a case of severe anemia. *Archives of Internal Medicine*. **6**, 517–21.

Ho Ping Kong, H. and Alleyne, G.A.O. (1969). Acid-base status of adults with sickle-cell anemia. *British Medical Journal*, **3**, 271–3.

Ho Ping Kong, H. and Alleyne, G.A.O. (1971). Studies on acid excretion in adults with sickle cell anemia. *Clinical Science*, **41**, 505–18.

Hofrichter, J., Ross, P.D., and Eaton, W.A. (1974). Kinetics and mechanism of deoxyhemoglobin S gelation: a new approach to understanding sickle cell disease. *Proceedings of the National Academy of Science of the USA*, **71**, 4864–8.

Immergut, M.A. and Stevenson, T. (1965). The use of epsilon amino caproic acid in the control of hematuria associated with hemoglobinopathies. *Journal of Urology*, **93**, 110–11.

Ingram, V.M. (1956). A specific chemical difference between the globins of normal human and sickle-cell anaemia haemoglobin. *Nature*, **178**, 792–4.

Jamison, R.L. (1973). Intrarenal heterogeneity. The case for two functionally dissimilar populations of nephrons in the mammalian kidney. *American Journal of Medicine*, **54**, 281–9.

Jensen, W.N. and Lessin, L.S. (1970). Membrane alterations associated with hemoglobinopathies. *Seminars in Hematology*, **7**, 409–26.

Johnson, C.S. and Giorgio, A.J. (1981). Arterial blood pressure in adults with sickle cell disease. *Archives of Internal Medicine*, **141**, 891–3.

De Jong, P.E., de Jong-van den Berg, L.T.W., Donker, A.J.M., and Statius van Eps, L.W. (1978a). The role of prostaglandins and renin in sickle-cell nephropathy. *Netherlands Journal of Medicine*, **21**, 67–72.

De Jong, P.E., de Jong-van de Berg, L.T.W., and Statius van Eps, L.W. (1978b). The tubular reabsorption of phosphate in sickle cell nephropathy. *Clinical Science and Molecular Medicine*, **55**, 429–34.

De Jong, P.E., de Jong-van den Berg, L.T.W., Sewrajsing, G.S., Schouten, H., Donker, A.J.M., and Statius van Eps, L.W. (1980). The influence of indomethacin on renal haemodynamics in sickle cell anemia. *Clinical Science*, **59**, 245–50.

De Jong, P.E., de Jong-van den Berg, L.T.W., Sewrajsingh, G.S., Schouten, H., Donker, A.J.M., and Statius van Eps, L.W. (1981). Beta-2-microglobulin in sickle cell anemia—evidence of increased tubular reabsorption. *Nephron*, **29**, 138–41.

De Jong, P.E., de Jong-van den Berg, L.T.W., de Zeeuw, D., Donker, A.J.M., Schouten, H., and Statius van Eps, L.W. (1982a). The influence of indomethacin on renal concentrating and diluting capacity in sickle cell nephropathy. *Clinical Science*, **63**, 53–8.

De Jong, P.E., Landman, H., and Statius van Eps, L.W. (1982b). Bloodpressure in sickle cell disease. *Archives of Internal Medicine*, **142**, 1239–40.

De Jong, P.E., de Jong-van den Berg, L.T.W., Schouten, H., Donker, A.J.M., and Statius van Eps, L.W. (1983). The influence of indomethacin on renal acidification in normal subjects and in patients with sickle cell anemia. *Clinical Nephrology*, **19**, 259–64.

De Jong, P.E., et al. (1984). Urinary prostaglandins in sickle cell nephropathy: a defect in 9-ketoreductase activity. *Clinical Nephrology*, **22**, 212–3.

De Jong, P.E. and Statius van Eps, L.W. (1985). Sickle cell nephropathy. New insights into its pathophysiology. *Kidney International*, **27**, 711–7.

Kan, Y.W. and Dozy, A.M. (1980). Evolution of the hemoglobin S and C genes in world population. *Science*, **209**, 388–91.

Kaye, D. and Hook, E.W. (1963). Influence of hemolysis on susceptibility to Salmonella infection: additional observations. *Journal of Immunology*, **91**, 518–27.

Keitel, H.G., Thompson, D., and Itano, H.A. (1956). Hyposthenuria in sickle cell anemia: A reversible renal defect. *Journal Clinical Investigation*, **35**, 998–1007.

Kenny, M.W., George, A.J., and Stuart, J. (1980). Platelet hyperactivity in sickle cell disease: a consequence of hyposplenism. *Journal of Clinical Pathology*, **33**, 622–5.

de Klerk, G., Rosengarten, P.C.J., Vet, R.J.W.M., and Goudsmit, R. (1981). Serum erythropoietin (ESF) titers in anemia. *Blood*, **58**, 1164–80.

Knochel, J.P. (1969). Hematuria in sickle cell trait. *Archives of Internal Medicine*, **123**, 160–5.

Levenson, D.J., Simmons, C.J., and Brenner, B.M. (1982). Arachidonic acid metabolism, protaglandins and the kidney. *American Journal of Medicine*, **72**, 345–74.

Levitt, M.F., Hauser, A.D., Levy, M.S., and Polimeros, D. (1960). The renal concentrating defect in sickle cell disease. *American Journal of Medicine*, **29**, 611–22.

Lindvall, N. (1978). Radiological changes in renal papillary necrosis. *Kidney International*, **13**, 93–106.

Marynick, S.P., Ramsey, E.J., and Knochel, J.P. (1977). The effect of bicarbonate and distilled water on sickle cell trait hematuria and *in-vitro* studies on the interaction of osmolality and pH on erythrocyte sickling in sickle cell trait. *Journal of Urology*, **118**, 793–6.

McCall, I.W., et al. (1978). Urographic findings in homozygous sickle cell disease. *Radiology*, **126**, 99–104.

Milner, P.F., Leibfarth, J.D., Ford, J., Barton, B.P., Grenett, H.E., and Garver, F.A. (1984). Increased HbF in sickle cell anemia is determined by a factor linked to the β^s gene from one parent. *Blood*, **63**, 64–72.

Miner, D.J., Jorkasky, D.K., Perloff, L.J., Grossman, R.A., and Tomaszewski, J.E. (1987). Recurrent sickle cell nephropathy in a transplanted kidney. *American Journal of Kidney Diseases*, **10**, 306–13.

Morgan, A.G. and Serjeant, G.R. (1981). Renal function in patients over 40 with homozygous sickle-cell disease. *British Medical Journal*, **282**, 1181–3.

Mostofi, F.K., Vorder Bruegge, C.F., and Diggs, L.W. (1957). Lesions in kidneys removed for unilateral hematuria in sickle cell disease. *Archives of Pathology*, **63**, 336–51.

Mozzarelli, A., Hofrichter, J., and Eaton, W.A. (1987). Delay time of hemoglobin S polymerization prevents most cells from sickling *in vivo*. *Science*, **237**, 500–6.

Murray, N., Serjeant, B.E., and Serjeant, G.R. (1988). Sickle cell—hereditary persistence of fetal haemoglobin and its differentiation from other sickle cell syndromes. *British Journal of Haematology*, **69**, 89–92.

Nagel, R.L., *et al.* (1985). Two hematologically and genetically distinct forms of sickle cell anemia: the Senegal type and the Benin type. *New England Journal of Medicine*, **312**, 880–4.

Najjar, V.A. (1975). Defective phagocytosis due to deficiencies involving the tetrapeptide tuftsin. *Journal of Pediatrics*, **87**, 1121–4.

National Health survey for years 1971–74. *Vital and Health Statistics Series II*. No. 203. U.S. Dept. of HHS, Public Health Service, 1977.

Nicholson, G.D. (1977). Post-streptococcal glomerulonephritis in adult Jamaicans with or without sickle cell anaemia. *West Indian Medical Journal*, **26**, 78–84.

Nissen, H.M. and Anderson, H. (1968). On the localization of prostaglandin dehydrogenase activity in the kidney. *Histochemie*, **14**, 189.

Oster, J.R., Lespier, L.E., Lee, S.M., Pellegrini, E.L., and Vaamonde, C.A. (1976*a*). Renal acidification in sickle cell disease. *Journal of Laboratory and Clinical Medicine*, **88**, 389–401.

Oster, J.R., Lee, S.M., Lespier, L.E., Pellegrini, E.L., and Vaamonde, C.A. (1976*b*). Renal acidification in sickle cell trait. *Archives of Internal Medicine*, **136**, 30–5.

Pagnier, J., *et al.* (1984). Evidence for the multi-centric origin of the sickle cell hemoglobin gene in Africa. *Proceedings of the National Academy of Science of the USA*, **81**, 1771–3.

Palek, J. (1977). Red cell membrane injury in sickle cell anaemia. *British Journal of Haematology*, **35**, 1–9.

Pardo, V., Strauss, J., Kramer, H., Ozawa, T., and McIntosh, R.M. (1975). Nephropathy associated with sickle cell anemia: an autologous immune complex nephritis II. Clinicopathologic study of seven patients. *American Journal of Medicine*, **59**, 650–9.

Pathak, U.N., *et al.* (1969). Bacteriuria of pregnancy: Results of treatment. *Journal of Infectious Disease*, **120**, 91–103.

Pauling, L., Itano, H.A., Singer, S.J., and Wells, J.C. (1949). Sickle cell anemia, molecular disease. *Science*, **110**, 543–8.

Pearson, H.A., Spencer, R.P., and Cornelius, E.A. (1969). Functional asplenia in sickle cell anemia. *New England Journal of Medicine*, **281**, 923–6.

Pembrey, M.E., Wood, W.G., Weatherall, D.J., and Perrine, R.P. (1978). Fetal hemoglobin production and the sickle cell gene in the oases of Eastern Saudi Arabia. *British Journal of Haematology*, **40**, 415–29.

Perillie, P.E. and Epstein, F.H. (1963). Sickling phenomenon produced by hypertonic solutions: A possible explanation for the hyposthenuria of sicklemia. *Journal of Clinical Investigations*, **42**, 570–80.

Pitcock, J.A., Muirhead, E.E., Hatch, F.E., Johnson, J.G., and Kelly, B.J. (1970). Early renal changes in sickle cell anemia. *Archives of Pathology*, **90**, 403–10.

Plakke, R.K. and Pfeiffer, E.W. (1964). Blood vessels of the mammalian renal medulla. *Science*, **146**, 1683–5.

Prchal, J.T., *et al.* (1989). Mild sickle cell anemia associated with alpha globin mutant alpha Montogomery. *The American Journal of Medicine*, **86**, 232–6.

Radel, E.G., Kochen, J.A., and Finberg, L. (1976). Hyponatremia in sickle cell disease: a renal salt-losing state. *Journal of Prediatrics*, **88**, 800–5.

River, G.L., Robbins, A.B., and Schwartz, S.O. (1961). S-C hemoglobin: A clinical study. *Blood*, **18**, 385–416.

Rotschild, B.M., Sienknecht, C.W., Kaplan, S.B., and Spindler, J.S. (1980). Sickle cell disease associated with uric acid deposition disease. *Annals of Rheumatic Disease*, **39**, 392–5.

Rubin, M.I. (1968). Effect of dietary protein on the renal concentrating process in sickle cell anemia. *American Journal Diseases of Children*, **115**, 262–66.

Van der Sar, A. (1967). Aplastic sickle cell crisis. A report on four cases. *Tropical and Geographical Medicine*, **19**, 273–85.

Schardijn, G.H.C., Statius van Eps., L.W., Pauw, W., Hoefnagel, C.A., and Nooijen, W.J. (1984). Comparison of reliability of tests to distinguish upper from lower urinary tract infections. *British Medical Journal*, **289**, 282–7.

Schlitt, L. and Keitel, H.G. (1960). Pathogenesis of hyposthenuria in persons with sickle cell anemia or the sickle cell trait. *Pediatrics*, **26**, 249–54.

Schmidt-Nielsen, B. and O'Dell, R. (1961). Structure and concentrating mechanism in the mammalian kidney. *American Journal of Physiology*, **200**, 1119–24.

Sellers, B.B. (1978). Intermittent hypertension during sickle cell crisis. *Journal of Pediatrics*, **92**, 941–3.

Serjeant, G.F., Serjeant, B.E., and Milner, P.F. (1969). The irreversibly sickled cell: a determinant of haemolysis in sickle cell anaemia. *British Journal of Haematology*, **17**, 527–33.

Serjeant, G.R. (1981). Observation on the epidemiology of sickle cell disease. *Transactions of the Royal Society for Tropical Medicine and Hygiene*, **75**, 228.

Serjeant, B., Myerscough, E., Serjeant, G.R., Higgs, D.R., and Moo-Penn, W.F. (1982). Sickle cell-hemoglobin D Iran: a benign sickle cell syndrome. *Hemoglobin*, **6** (1), 57–9.

Serjeant, G.R. (1985). *Sickle Cell Disease*. Oxford University Press, Oxford.

Singer, K. and Singer, L. (1953). Studies on abnormal haemoglobins VIII. The gelling phenomenon of sickle cell hemoglobin: its biologic and diagnostic significance. *Blood*, **8**, 1008–23.

Spector, D., *et al.* (1978). Painful crises following renal transplantation in sickle cell anemia. *American Journal of Medicine*, **64**, 835–9.

Spirer, Z., Weisman, Y., Zakuth, V., Fridkin, M., and Bogair, N. (1980). Decreased serum tuftsin concentrations in sickle cell disease. *Archives of Diseases in Childhood*, **55**, 566–7.

Spirer, Z., *et al.* (1977). Decreased tuftsin concentrations in patients, who have undergone splenectomy. *British Medical Journal*, **2**, 1574–6.

Statius van Eps, L.W. and de Jong, P.E. (1988). The kidney in sickle cell disease. In *Diseases of the Kidney*, 4th edn., (eds. R.W. Schrier and C.W. Gottschalk), Vol. III, pp. 2561–83. Little, Brown, Boston.

Statius van Eps, L.W. and Leeksma, O.C. (1989). Sickle cell nephropathy and haemostasis. In *Haemostasis and the Kidney*. (eds. G. Remuzzi and E.C. Rossi), pp. 169–90. Butterworths, London.

Statius van Eps, L.W., Schouten, H., la Porte-Wijsman, L.W., and Struyker Boudier, A.M. (1967). The influence of red blood cell transfusions on the hyposthenuria and renal hemodynamics of sickle cell anemia. *Clinica Chimica Acta*, **17**, 449–61.

Statius van Eps, L.W., Schouten, H., ter Haar Romeny-Wachter, C.C., and la Porte-Wijsman, L.W. (1970*a*). The relation between age and renal concentrating capacity in sickle cell disease and hemoglobin C disease. *Clinica Chimica Acta*, **27**, 501–11.

Statius van Eps, L.W., Pinedo-Veels, C., de Vries, C.H., and de Koning, J. (1970*b*). Nature of concentrating defect in sickle cell nephropathy; microradioangiographic studies. *Lancet* **i**, 450–2.

Stokes, J.B. (1979). Effect of prostaglandin E_2 on chloride transport across the rabbit thick ascending limb of Henle: selective inhibition of the medullary portion. *Journal of Clinical Investigation*, **64**, 495–502.

Strauss, J., Pardo, V., Koss, M.N., Griswold, W., and McIntosh, R.M. (1975). Nephropathy associated with sickle cell anemia: An autologous immune complex nephritis. *American Journal of Medicine*, **58**, 382–7.

Sydenstricker, V.P., Mulherin, W.A., and Houseal, R.W. (1923). Sickle cell anemia, report of two cases in children with necropsy in one case. *American Journal of Diseases of Children*, **26**, 132–54.

Sweeney, M.J., Dobbins, W.T., and Etteldorf, N.J. (1962). Renal disease with elements of the nephrotic syndrome associated with sickle cell anemia. *Journal of Pediatrics*, **60**, 42–51.

Tejani, A., Nicastri, A., Chen, C.K., Sen, D., Phadke, K., and Adamson, O. (1985). Renal lesions in sickle cell nephropathy in children. *Nephron*, **39**, 352–5.

Thomas, A.N., Pattison, C., and Serjeant, G.R. (1982). Causes of death in sickle cell disease in Jamaica. *British Medical Journal*, **285**, 633–5.

Vaamonde, C.A. (1984). Renal papillary necrosis in sickle cell hemoglobinopathies. *Seminars in Nephrology*, **4**, 48–64.

Walker, B.R., Alexander, F., Birdsall, T.R., and Warren, R.L. (1971). Glomerular lesions in sickle cell nephropathy. *Journal of the American Medical Association*, **215**, 437–40.

Walker, B.R. and Alexander, F. (1971). Uric acid excretion in sickle cell anemia. *Journal of the American Medical Association*, **215**, 255–8.

Whalley, P.J., Martin, F.G., and Pritchard, J.A. (1964). Sickle cell trait and urinary tract infection during pregnancy. *Journal of the American Medical Association*, **189**, 903–6

Whitten, C.F. and Younes, A.A. (1960). A comparative study of renal concentrating ability in children with sickle cell anemia and in normal children. *Journal of Laboratory and Clinical Medicine*, **55**, 400–15.

Wilson, W.A. and Alleyne, G.A.O. (1976). Total body water, extra cellular and plasma volume compartments in sickle cell anemia. *West Indian Medical Journal*, **25**, 241–50.

Wilson, W.A., Hughes, G.R.V., and Lachmann, P.J. (1976). Deficiency of factor B of the complement system in sickle cell anaemia. *British Medical Journal*, **1**, 367–9.

Winkelstein, J.A. and Drachman, R.N. (1968). Deficiency of pneumococcal serum opsonizing activity in sickle-cell disease. *New England Journal of Medicine*, **279**, 459–66.

Wrong, O. and Davies, H.E.F. (1959). The excretion of acid in renal disease. *Quarterly Journal of Medicine*, **28**, 259–313.

Yoshino, M., Amerian, R., and Broutbar, N. (1982). Hyporeninemic hypoaldosteronism in sickle cell disease. *Nephron*, **31**, 242–4.

Zins, G.R. (1975). Renal prostaglandins. *American Journal of Medicine*, **58**, 14–24.

SECTION 5
The patient with tubular disease

5.1 Fanconi syndrome
5.2 Isolated defects of tubular function
5.3 Renal tubular acidosis
5.4 Bartter's syndrome
5.5 Nephrogenic diabetes insipidus

5.1 Fanconi syndrome

J. BRODEHL

Definition

The renal Fanconi syndrome (de Toni-Debré-Fanconi syndrome) is characterized by generalized disturbance of tubular function, while glomerular filtration is not primarily affected. It leads to excessive losses of amino acids, glucose, phosphate, bicarbonate, and other organic and inorganic substrates handled by the proximal and distal tubules. The metabolic consequences are acidosis, hypophosphataemia, hypocalcaemia, dehydration, rickets, osteoporosis, and growth retardation. This syndrome may either be congenital or acquired, primary or secondary, complete or incomplete. (For recent reviews see Brewer 1985; Brodehl 1978(a); Chesney 1980; Roth *et al.* 1981; Foreman and Segal 1987.)

Pathophysiology

The function of the renal tubules is to reclaim the bulk of solutes and water filtered by the glomerulus. This is achieved by a series of steps, which involve the transport systems of the brush border membrane at the luminal site, the energy producing and transferring processes within the cells, and the Na^+K^+-ATPase at the basolateral membrane, which actively exchanges intracellular sodium for extracellular potassium. The main energy expenditure is concerned with the movement of sodium from the luminal to the peritubular fluid. The activity of Na^+K^+-ATPase at the basolateral membrane lowers intracellular sodium concentration and generates an electrochemical gradient across the luminal membrane, which is used for the movement of sodium from the lumen into the cell (see Chapter 2.3). The transport of other substrates as amino acids, glucose, and phosphate is coupled with the transport of sodium and uses the energy inherent in the electrochemical gradient for sodium. The carrier systems of the brush border are thus able to move solutes from the luminal into the intracellular site even against a chemical gradient which exists, for instance, for amino acids. The movement from intracellular site to peritubular fluid across the basolateral membrane occurs for most solutes by facilitated diffusion, which is carrier mediated but not energy dependent.

Besides the transcellular transport there exists another pathway for solute movement by the paracellular route. The intercellular junctional complexes are tight in the distal, but 'leaky' in the proximal tubule. It is claimed that one-third to one-half of sodium and most of the water is reabsorbed via the paracellular route through these proximal junctional complexes. Movement of salt and water along this pathway is affected by the peritubular oncotic pressure and the state of extracellular fluid volume, but net movements of other solutes, i.e. glucose and amino acids are not.

The disturbances leading to the Fanconi syndrome could be caused by defects at various steps of the transporting systems (Gonick and Kramer 1985; Foreman and Segal 1987) (Fig. 1). Thus, the pathomechanism of the Fanconi syndrome is heterogenous and currently only incompletely understood. There are at least six possibilities:

1. a defective influx
2. a back leakage from the intracellular fluid into the lumen
3. a decreased cellular efflux across the basolateral membrane which leads to an intracellular accumulation of the substrate
4. an increased cellular influx from the blood
5. a defect in energy generation or transduction to the plasma membrane
6. an increase in backflux across the tight intercellular junctions.

There is evidence that some of these mechanisms are operative both in patients and in experimental models. A defective influx (1) was found by Manz *et al.* (1984) in a patient lacking proximal tubular brush borders. The increased back leakage from the cell into the lumen (2) may be the cause of the Fanconi syndrome induced experimentally by maleic acid (Bergeron *et al.* 1976). A decreased cellular efflux across the basolateral membrane (3) has been incriminated by Manz *et al.* (1987) in the Fanconi syndrome associated with glycogenosis (Fanconi-Bickel syndrome). A

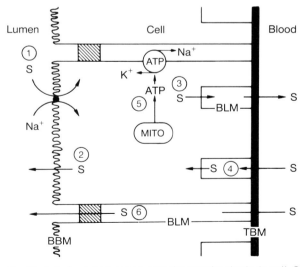

Fig. 1 Model of solute transport by the proximal tubule cell. S = solutes, BBM = brush-border membrane, TBM = tubular basement membrane, BLM = basolateral membrane. For further explanation see text. (Taken from Foreman, J.W., and Segal, S. (1987). Fanconi syndrome. In *Paediatric Nephrology*, (eds. M.A. Holliday, T.M. Barratt, and R.L. Vernier), pp. 547–65, Williams and Wilkins, Baltimore, with permission.)

defect in energy generation (5) has been postulated as the underlying cause of the Fanconi syndrome in metabolic diseases such as hereditary fructose intolerance, galactosaemia and cytochrome-c-oxidase deficiency. The heterogenous pathophysiology of the Fanconi syndrome explains the variety of facets observed in the different aetiological types of the syndrome.

Phenomenologically the Fanconi syndrome is characterized by multiple transport defects. This is in contrast to the specific tubulopathies, in which the inactivity or absence of a specific carrier or receptor induce only singular defects, while all the other tubular functions remain undisturbed. In the Fanconi syndrome although multiple functions of the proximal and/or distal tubules are involved, the degree of the disturbance and its extent along the nephron vary according to the underlying disease. The cardinal symptoms of the Fanconi syndrome are hyperaminoaciduria, glucosuria, and phosphate wasting. Other symptoms may be associated such as defects in bicarbonate reabsorption, renal acidification, urate reabsorption, urinary concentration, potassium conservation, reabsorption of sodium, calcium and proteins, and secretion of p-aminohippuric acid. The metabolic consequences of urinary substrate losses are found as long as the glomerular filtration is sufficiently high to provide enough substrate filtration. This means that in severe renal insufficiency, although often accompanied by multiple tubular dysfunctions, the Fanconi syndrome becomes clinically irrelevant.

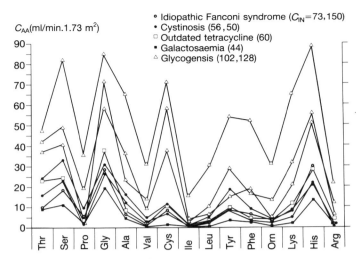

Fig. 2 Clearance rates of free amino acids in children with the Fanconi syndrome due to various aetiologies.

Hyperaminoaciduria

Amino acids are filtered freely through the glomerular basement membrane: their concentration in the ultrafiltrate equals that in plasma water (Eisenbach et al. 1975). They are subsequently reabsorbed by active processes. Tubular reabsorption is so effective that 98 to 99 per cent of the filtered amino acids are retrieved with an eventual urinary loss of only 1 to 2 per cent, which is called the physiologic aminoaciduria. Any aminoaciduria larger than this amount is labelled hyperaminoaciduria (Brodehl 1978b).

Every amino acid detected in the blood is found in the urine; however, the pattern of amino acids is quite different in the two fluids (Brodehl and Bickel 1973). This is due to the fact that individual amino acids are handled specifically by the tubular cells, independent of their plasma concentration. The degree of reabsorption ranges from 92.8 ± 1.8 (for histidine) to 99.8 ± 0.1 per cent (for valine). For most amino acids there is more than one type of active tubular transport system.

The pattern of physiologic aminoaciduria is quite constant in healthy children and adults. In early infancy the rate of urinary excretion of amino acids is higher than in later life (infantile hyperaminoaciduria) (Brodehl and Gellissen 1968).

In the Fanconi syndrome all amino acids are hyperexcreted, but to various extents. In the common type of Fanconi syndrome produced by inborn errors of metabolism, toxic intermediates alter energy generation within the tubular cells. The clearance rates of free amino acids in eight children with the common Fanconi syndrome caused by various metabolic disturbances are depicted in Fig. 2. The patterns are the same in all cases, i.e. high excretion rates of those amino acids, which exhibit higher excretion rates physiologically (serine, glycine, cystine, lysine, histidine), and lower excretion rates of those with lower rates physiologically (proline, valine, isoleucine, ornithine, and arginine). Mean values of percentage tubular reabsorption are shown in Fig. 3 and compared with normal values: the pattern of hyperaminoaciduria in the Fanconi syndrome appears similar to the pattern of normal aminoaciduria, but on an elevated level. Thus, the hyperaminoaciduria of the common type of Fanconi syndrome is an exaggeration of normal aminoaciduria and is thus called 'generalized' hyperaminoaciduria. It is quite different from specific hyperaminoacidurias, in which singular transport carriers are disturbed.

A different pattern of hyperaminoaciduria was described in a child, who was completely unable to reabsorb amino acids. Clearance rates were the same for all amino acids and approached glomerular filtration rate (Manz et al. 1984). This Fanconi syndrome was caused by a complete absence of brush borders in the proximal tubules as demonstrated by electron

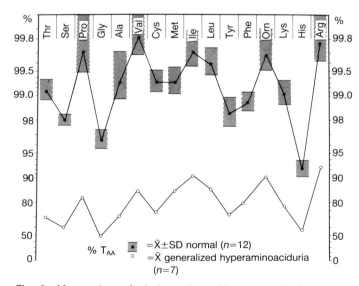

Fig. 3 Mean values of tubular amino acid reabsorption in percentage of filtered amino acids (%TAA) in normal children (black dots with shaded area) in comparison with mean values of seven children with the renal Fanconi syndrome due to various aetiologies (open dots).

5.1 Fanconi syndrome

microscopy in the renal biopsy. Hyperaminoaciduria was severe with an extremely flat pattern, depicted in Fig. 4.

The pathomechanism by which generalized hyperaminoaciduria of the common type is produced is not thoroughly understood. Estimation of tubular net reabsorption indicates that hyperaminoaciduria is not due to the saturation of tubular reabsorptive capacity (Brodehl 1978a). The cellular disorder seems related either to an alteration of energy generation or transfer system within the cells or to a defect of the luminal membranes, leading to increased leakage of amino acids from the intracellular pool.

From the clinical point of view it is noteworthy that amino acid losses do not lead to deficiencies, and, therefore, there is no need for supplements of amino acids or proteins.

Glucosuria

Glucosuria is the second cardinal symptom of the Fanconi syndrome. It is of renal origin and is therefore usually accompanied by normal blood glucose levels. In the Fanconi syndrome, the renal threshold for glucose is reduced or completely abolished, as shown in Fig. 5 (personal observations in five children with Fanconi syndrome). When the tubules are loaded with more glucose there is a gradual increase in the rate of net reabsorption. Extra glucose loading, however, can be dangerous in patients with Fanconi syndrome, especially in those with cystinosis, due to the potassium lowering effect (Cherry and Surawicz 1962). After glucose loading with blood glucose levels usually enough to saturate the maximal capacity the maximal reabsorptive capacity (Tm_G) is significantly lower than normal (Fig. 5). In the case of very large glucose loading, however, it is within the normal range. Occasionally a depression in Tm_G is observed after prolonged glucose loading (Reem et al. 1967; personal observation). Thus the glucosuria of the common type of Fanconi syndrome is neither described by type A renal glucosuria according to Reubi (1963), i.e. low threshold and low maximal reabsorption, nor by type B, i.e. low threshold but normal maximal reabsorption. It is characterized by a low threshold, a low maximal glucose reabsorption at saturation glucose blood levels, but normal values of Tm_G during excessive glucose loading. In the case of Fanconi syndrome with absent proximal tubular brush borders (Manz et al. 1984), there was virtually no net glucose reabsorption, i.e. glucose clearance approached glomerular filtration rate. In this respect it was comparable to the type O of renal glucosuria, recently described by Oemar et al. (1987).

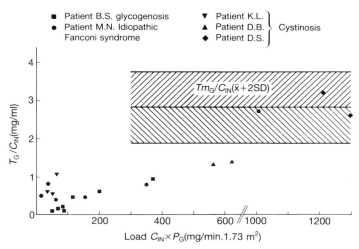

Fig. 5. Tubular glucose reabsorption (T_G/C_{IN}) in relation to filtered glucose ($C_{IN} \times P_G$) in children with the Fanconi syndrome.

Glucosuria is variable, usually below 10 g per day. The only exception is the glycogenosis with Fanconi syndrome (Fanconi-Bickel syndrome) in which glucosuria is massive, reaching up to 200 g per day. It has been suggested (Manz et al. 1987) that this type of Fanconi syndrome results from a defective glucose efflux through the basolateral membrane, which leads to a glucose cellular overload. Clinical consequences are usually mild or absent except from the glucose induced osmotic diuresis. Hypoglycaemia and ketonaemia are encountered only in patients with severe glucosuria, especially those with glycogenosis. Both respond to dietary treatment.

The pathomechanism of glucosuria in the Fanconi syndrome is still incompletely understood. It is best explained by an inability of the tubular cell to sustain a glucose gradient across the luminal membrane, an anomaly similar to that postulated for hyperaminoaciduria.

Phosphate wasting

The impairment in renal handling of inorganic phosphate is the third cardinal symptom of the Fanconi syndrome. It is referred to as phosphate diabetes or hyperphosphaturia, although the actual amounts of phosphate excreted in the urine are not higher than in normal subjects. Urinary phosphate reflects phosphate intake, i.e. is a measure of phosphate turnover. Phosphate excretion is exaggerated only in relation to the serum phoshate level.

In the Fanconi syndrome the tubular reabsorption of phosphate is lowered together with plasma phosphate, as shown in Fig. 6. The close correlation between renal phosphate-threshold and plasma phosphate, shown to exist in normal subjects

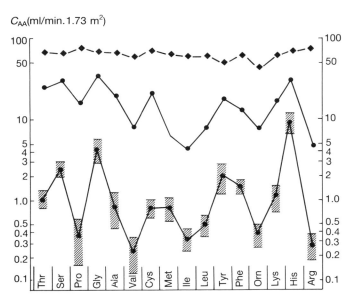

Fig. 4 Pattern of clearance rates of free amino acids in the Fanconi syndrome with absence of tubular brush borders (upper curve), in the common Fanconi syndrome due to metabolic disorders (middle curve), and in normal children (lower curve). (Taken from Manz, F. et al. (1984). Clinical Nephrology, 24, 149–57, with permission.)

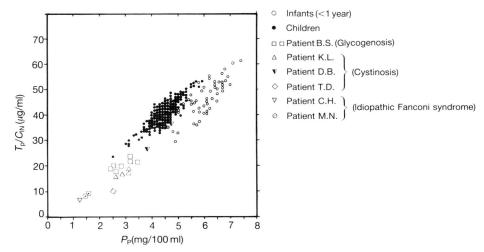

Fig. 6. The correlation between fractional phosphate reabsorption (T_P/C_{IN}) and plasma phosphate concentration (P_P) in infants and children (white and black dots) compared with children with the Fanconi syndrome (normal values from Brodehl et al. 1982).

(Brodehl et al. 1982), is preserved in the Fanconi syndrome. Hypophosphataemia is thus caused by the inability of the kidney to reabsorb phosphate in adequate amounts. Loading with phosphate does not increase phosphate reabsorption (Brodehl 1978a). Results of calcium loading on phosphate reabsorption are conflicting; some observations describe no or subnormal effects (Wilson and Yendt 1963), and others demonstrate a definite increase of phosphate reabsorption (Callis et al. 1970). Intestinal absorption of phosphate is normal in the Fanconi syndrome (Wilson and Yendt 1963).

In long-term studies of the Fanconi syndrome the impaired phosphate reabsorption remains fairly constant, although it can be modified by extrarenal factors such as changes in extracellular fluid volume, vitamin D intake, acidosis, and hormonal levels, especially of parathormone. In contrast to earlier reports, patients with the Fanconi syndrome and normal GFR have normal serum levels of immunoreactive parathormone (Fanconi et al. 1974), which was confirmed in bone biopsy specimens (Brodehl 1976). The conversion of 25-hydroxy-vitamin D to 1,25-dihydroxy-vitamin D is impaired in the experimental Fanconi syndrome (Brewer et al. 1977), a finding accounting for the low levels of 1,25-dihydroxy-vitamin D observed in patients despite the profound hypophosphataemia.

The pathophysiology of impaired phosphate handling is not well understood. It is not clear whether it is primarily related to an alteration of the luminal brush border, or to an increased leakage of intracellular phosphate in the more distal parts of the nephron (Bergeron et al. 1976), or both (Brewer et al. 1983). In some cases with incipient incomplete Fanconi syndrome the disturbance in phosphate conservation seems to follow the other cardinal symptoms, i.e. it would be secondary.

Hypophosphataemia and phosphate depletion are the major factors contributing to the alterations in the skeletal system in the Fanconi syndrome (Brodehl 1976; Wilson and Yendt 1963). The bones show a wide spectrum of disorders depending on the age of onset of the Fanconi syndrome. They include defective mineralization and demineralization (rarefication), rickets, osteomalacia, and osteoporosis. Radiographic findings are not different from other types of defective bone mineralization. Only patients with glomerular insufficiency show in addition signs of azotaemic osteodystrophy with secondary hyperparathyroidism. The changes are accompanied by growth retardation, fractures and pseudofractures, and crippling deformities of the bone. Most can be prevented by early and continuous treatment.

Other tubular symptoms
Acidosis

Acidosis is a frequent finding in the Fanconi syndrome, caused mainly by defective bicarbonate reabsorption in the proximal tubule, called type 2 renal tubular acidosis (see Chapters 5.3 and 7.3). The reduced renal threshold for bicarbonate results in a hyperchloraemic metabolic acidosis (Morris and Sebastian 1983). At plasma bicarbonate levels below the renal threshold, the urinary pH can be lowered to 5.4 or less, demonstrating the ability of the distal tubule to establish a normal hydrogen ion gradient (Kamm and Fischer 1972). In some cases, however, distal acidification is also disturbed, as shown by an inability to lower urinary pH even in the presence of severe metabolic acidosis (Lee et al. 1972). Tubular ammoniagenesis is unimpaired in the absence of renal failure (Houston et al. 1968). Sustained metabolic acidosis contributes to hypercalciuria and skeletal changes. Large amounts of alkali are required to correct the acidosis.

Uric acid wasting

Hypouricaemia is a common finding in the Fanconi syndrome, mainly in adults (Wallis et al. 1957; Reem et al. 1967; Newcombe 1975). In children, it is rarely reported (Houston et al. 1968). It is due to a high urate clearance rate. Since the exact mechanism of renal urate handling is unclear in humans, it is not known whether decreased proximal resorption or increased distal secretion is involved in the hyperuricosuria. The clinical significance of lower serum urate is unknown. There have been no reports on urate lithiasis in the Fanconi syndrome.

Hypokalaemia

Hypokalaemia is a serious threat to patients with Fanconi syndrome. It is mainly due to renal loss of potassium (Sebastian et al. 1971). The clearance rates for potassium can be elevated to more than twice the glomerular filtration rate, demonstrating thereby net tubular secretion (Brodehl et al. 1965). Although acidosis may increase the potassium loss, the correction of acidosis does not necessarily re-establish renal potassium conservation. Since the proximal tubular reabsorption of sodium is disturbed, the distal tubules become overloaded with sodium, leading to an increased resorption of sodium in exchange to potassium.

The potassium loss may be aggravated by impaired hydrogen ion secretion and by increased plasma renin activity and hyperaldosteronism. Potassium depletion may ensue, leading to constipation, muscle weakness, growth retardation, polyuria, and even sudden death. Potassium supplements should be given regularly in adequate amounts; conditions, tests, and drugs likely to lower serum potassium should be avoided.

Sodium loss

Sodium wasting is another defect present in the Fanconi syndrome. In some cases the renal salt wasting is so severe that hyponatraemia with metabolic alkalosis develops simulating Bartter's syndrome in spite of the lowered bicarbonate threshold (Houston et al. 1968). Supplementation with sodium chloride in these patients increases weight, extracellular volume and GFR, with an eventual reappearance of acidosis.

The pathomechanism of disturbed sodium reabsorption is intimately related to the basic metabolic defect of the Fanconi syndrome, as the transtubular transport of amino acids, glucose, and phosphate is sodium dependent. The clinical symptoms depend to a large extent on the degree of involvement of the distal parts of the nephron. Patients with severe renal salt loss complain of orthostatic hypotension and exhibit signs of salt craving.

Hypercalciuria

Hypercalciuria is frequently found in patients with the Fanconi syndrome (Wilson and Yendt 1963; Kruse and Barthels 1979), while serum calcium is normal or slightly reduced, depending on the degree of acidosis, the rate of GFR, the state of bone mineralization, and vitamin D supply. There is evidence that the tubular calcium transport may be disturbed secondarily to the defect in sodium reabsorption.

Vitamin D and its metabolites have been found to be low (Chesney et al. 1980) or normal (Fanconi et al. 1974, Kruse and Barthels 1979). Correspondingly the excretion of cAMP is in the normal range, and responds adequately to loading with calcium, phosphate, or exogenous parathormone. The last finding, however, has not been confirmed in the maleic rate model (Wald et al. 1984), in which defective hydroxylation of 25-hydroxyvitamin D_3 is also observed (Brewer et al. 1977). The intestinal absorption of calcium was originally thought to be impaired but later investigations with oral calcium loading did not confirm the defective transport (Wilson and Yendt 1963). An increase in plasma phosphate is followed by an increase in intestinal calcium absorption (Nagant de Deuxchaisnes and Krane 1967; Lee et al. 1972). Secondary complications of hypercalciuria, i.e. nephrocalcinosis or nephrolithiasis, are rare, possibly because the kidney is protected by a high urinary output and an increased excretion of organic substances such as citrate. Calcium excretion rates should be monitored during treatment with vitamin D or its metabolites.

Polyuria

Polyuria, polydipsia, and episodes of dehydration are prominent clinical features of the Fanconi syndrome. Data on maximal urinary concentrating ability in patients with the Fanconi syndrome are uncommon (Fanconi and Bickel 1949; Lee et al. 1972; Katzir et al. 1988), since water deprivation can be hazardous.

Polyuria is due both to the osmotic diuresis of glucose, and to a concentrating defect of the distal tubule and collecting ducts, possibly related to hypokalaemia. Fanconi syndromes due to light-chain proteinuria and multiple myeloma are especially prone to distal concentrating defects. As a consequence of dehydration, episodes of 'unexplained' fever and vomiting occur which are often the first clinical signs of Fanconi syndrome in infants and young children, especially those with cystinosis.

Proteinuria

Almost all patients with the Fanconi syndrome exhibit abnormal proteinuria, as already indicated by the early name for the syndrome (Fanconi 1936) — nephrotic glucosuric dwarfism. Sophisticated methods of urinary protein determination have disclosed three types of proteinuria: (1) a prerenal proteinuria, as in paraproteinaemias and multiple myeloma, (2) a glomerular proteinuria, as in late stages of cystinosis or intoxications, and (3) a tubular proteinuria, which derives either from tubular cells or from impaired reabsorption of filtered low molecular weight proteins. In many cases there are mixed types of proteinuria, due to the cause and stage of the Fanconi syndrome. The amount of proteins excreted is mostly low to moderate and is rarely in the nephrotic range. There is a preponderance of tubular proteins (Barratt and Crawford 1970), typically of low molecular weight (10 000–50 000 Da) including enzymes, immunoglobulin light chains, and small protein hormones. The clinical consequences of tubular proteinuria are still uncertain.

Miscellaneous

Low plasma carnitine levels (β-hydroxy-γ-trimethylaminobutanoic acid) have been reported in children with the Fanconi syndrome (Bernardini et al. 1985; Steinmann et al. 1987). The urinary excretion and the fractional excretion rates of carnitine and acylcarnitine are considerably increased compared with controls. Carnitine excretion is closely correlated with total α-amino-acid excretion. Failure to reabsorb free and acylcarnitine results in plasma and muscle carnitine deficiencies, which could contribute to the muscular hypotonicity often encountered in patients with the Fanconi syndrome. However, oral substitution with carnitine increases plasma carnitine level, but not the low muscle carnitine concentration (Gahl et al. 1988). It is thus questionable whether carnitine substitution can alter the muscular symptoms of the Fanconi syndrome.

Urinary excretion of peptides has been investigated in patients with Wilson's disease (Asatoor et al. 1976). Hydroxyproline-containing peptides from collagen degradation are excreted in significant higher amounts than in controls. In rats treated with maleic acid, increased peptiduria results at least partially from a diminished tubular reabsorption (Asatoor et al. 1979).

Increased lactate excretion has been observed in a family with idiopathic Fanconi syndrome and is considered as one of the earliest signs of incipient tubulopathy (Brenton et al. 1981). Hyperlacticaciduria has been confirmed by other investigators (Jonas et al. 1989), who also found significantly increased levels of glyceraldehyde in the urine of 16 children with the Fanconi syndrome. Levels of both substances are unrelated to either increased serum levels, administration of electrolyte solutions, or incidental formation during derivatization procedures. They appear related directly to renal proximal tubular dysfunction. Their clinical relevance remains to be elucidated.

Tubular secretion of p-aminohippurate (PAH) may be impaired in some categories of Fanconi syndrome out of proportion to the degree of glomerular filtration rate. This has been noted especially in cystinosis (Brodehl and Bickel 1973), in multiple myeloma (Sirota and Hamerman 1954), and in intoxication with degraded tetracycline (Brodehl et al. 1968). In other patients with comparably severe tubular symptoms, as in glycogenosis (Brodehl et al. 1969), there is no such disproportionate impairment of PAH secretion, indicating differences in the pathomechanisms of these entities. The clearance and maximal tubular secretion of PAH (C_{PAH}, Tm_{PAH}) are affected similarly, producing high values for the filtration fraction (C_{IN}/C_{PAH}) (where C_{IN} is inulin clearance) and extremely low fractional maximal PAH secretion (Tm_{PAH}/C_{IN}), respectively. Since renal extraction of PAH is extremely low in these cases (Fellers et al. 1960), it cannot be used in the estimation of renal plasma flow.

Assessment of tubular dysfunction

The first step in the evaluation of the renal handling of amino acids is the measurement of free amino acids by screening or semiquantitative methods such as thin-layer chromatography (Bremer et al. 1981). If hyperaminoaciduria is suspected a timed urine collection should be examined by more accurate methods such as column chromatography or gas chromatography. The most sensitive parameter of tubular amino-acid handling is the percentage tubular amino-acid reabsorption, which can easily be calculated by the equation

$$\%T_{AA} = \left(1 - \left(\frac{U_{AA} \times P_{CR}}{P_{AA} \times U_{CR}}\right)\right) \times 100$$

where U_{AA} and P_{AA} are amino-acid concentrations in urine and plasma, and U_{CR} and P_{CR} are creatinine concentrations in urine and plasma. The $\%T_{AA}$ is independent of amino-acid loading, since there is almost no limit in the maximal capacity of amino-acid reabsorption. Therefore, it can be measured at any time during the day in non-fasted conditions, and obviates the need for accurately timed urine collection. The blood for amino-acid and creatinine measurement has to be drawn at the time of urinary collection. Normal values for $\%T_{AA}$ are depicted in Fig. 3.

Glucose in the urine should be measured by specific enzymatic methods, since the common reduction method is too insensitive and the reagents react with other carbohydrates. Glucose content of a timed urine collection should be correlated with the blood glucose level. The renal glucose threshold is expressed by T_G/GFR, where T_G is calculated from the expression

$$GFR \times P_G - U_G V$$

measured by clearance procedure. The maximal glucose reabsorption (Tm_G) is determined after glucose loading; its determination, however, is not necessary for clinical diagnosis.

The best parameter to assess the tubular handling of phosphate is the maximal phosphate reabsorption corrected by glomerular filtration rate (Tm_P/GFR). As shown recently (Brodehl et al. 1988) this can easily be calculated by the equation

$$Tm_P/GFR = P_P - \left(\frac{U_P \times P_{CR}}{U_{CR}}\right)$$

where P_P and U_P are phosphate concentration in plasma and urine, and P_{CR} and U_{CR} are creatinine concentrations in plasma and urine. Since fasting values of phosphate reabsorption are already 'maximal', no exogenous phosphate loading is necessary (Stark et al. 1986; Brodehl et al. 1988). Therefore, Tm_P/GFR can be assessed from a spontaneously voided urine specimen and a simultaneously drawn blood specimen on a fasting ambulatory basis. The nomogram of Walton and Bijvoet (1975) to calculate Tm_P/GFR overestimates the values in cases of low phosphate loading, and therefore has no advantage over values obtained using the equation (Brodehl et al. 1988).

Table 1 Aetiological classification of Fanconi syndrome

1	*Primary (idiopathic) Fanconi syndrome*
1.1	Common type
1.1.1	Hereditary (autosomal dominant, autosomal recessive(?), X-linked recessive(?))
1.1.2	Sporadic
1.2	Brush-border type
2	*Secondary (symptomatic) Fanconi syndrome*
2.1	In inborn errors of metabolism
2.1.1	Of amino acids
2.1.1.1	Cystinosis (autosomal recessive)
2.1.1.2	Tyrosinaemia type I (autosomal recessive)
2.1.2	Of carbohydrates
2.1.2.1	Glycogenosis (autosomal recessive)
2.1.2.2	Galactosaemia (uridyl transferase deficiency) (autosomal recessive)
2.1.2.3	Hereditary fructose intolerance (autosomal recessive)
2.1.3	Of other or unknown causes
2.1.3.1	Lowe's syndrome (X-linked recessive)
2.1.3.2	Wilson's disease (autosomal recessive)
2.1.3.3	Cytochrome-c-oxidase deficiency (autosomal recessive)
2.2	In acquired diseases
2.2.1	Multiple myeloma
2.2.2	Nephrotic syndrome
2.2.3	Transplanted kidney
2.2.4	Tumour
2.3	In intoxications
2.3.1	Heavy metals (mercury, uranium, lead, cadmium)
2.3.2	Maleic acid, lysol, toluene
2.3.3	Outdated tetracycline, methyl-3-chromone
2.3.4	Cisplatin, ifosfamide

Other calculated indexes such as urinary phosphate concentration, phosphaturia per minute, phosphate clearance ($C_P = U_P V/P_P$), fractional clearance (C_P/GFR) or percentage tubular phosphate reabsorption ($\%TRP = 100 - C_P/GFR$) depend heavily on actual phosphate intake and are thus not useful for diagnostic purposes (Brodehl et al. 1988).

Disturbances of electrolyte reabsorption, renal acidification and concentration, PAH secretion and urinary protein excretion are evaluated by standard procedures as described in the relevant chapters of this volume.

Clinical features of Fanconi syndrome

The Fanconi syndrome represents a rather uniform response of the renal tubules to various exogenous and/or endogenous insults, sparing the glomeruli. From the aetiological point of view two main types should be considered (Table 1): the primary and the secondary Fanconi syndrome. In the primary type no underlying cause or systemic disease has so far been recognized. This classification is preliminary as some aetiological factors may yet be identified. This group is thus heterogenous.

The secondary type is caused by metabolites derived from inborn errors of metabolism, acquired diseases, or intoxications. In some diseases both kidney and other organs (liver, intestine) have the defective enzyme, while in others these are located in extrarenal tissues only.

In addition there are incomplete forms of the Fanconi syndrome, in which at least one of the cardinal features (hyperaminoaciduria, glucosuria, phosphate wasting) is missing. Such

5.1 Fanconi syndrome

cases fall into two categories. Some are an incomplete expression of the Fanconi syndrome, in which the tubular disturbance is not fully developed. This could be due to a transitional stage of the disease, to a low toxicity of the damaging metabolite, or a higher resistance of the intracellular mechanism to the toxin. These *formes frustes* of the Fanconi syndrome are encountered in intoxications leading to glucohyperaminoaciduria, or in idiopathic types which later progress to a complete Fanconi syndrome (Patrick *et al.* 1981). The second category of so-called incomplete Fanconi syndromes encompasses various defects of multiple carrier systems, such as glucoglycinuria (Käser *et al.* 1962), glucophosphaturia (Dent 1952), and glucoglycinephosphaturia (Scriver *et al.* 1964, Brodehl 1978b). These specific tubular defects should not be listed under the term of Fanconi syndrome.

The clinical features depend on the type of Fanconi syndrome and the underlying disease. These are described in the following chapters. The physician dealing with patients with the Fanconi syndrome should first determine quantitatively the extent of the tubular defect and, second, try to identify the cause of the disease by clinical methods, familial history, and biochemical analyses. The exact diagnosis is most important, since specific or symptomatic treatment is available for several types. A differential diagnosis for the most common types of Fanconi syndrome is outlined in Table 2.

Primary Fanconi syndrome

The primary form of Fanconi syndrome is diagnosed by exclusion of all known causes leading to the complex tubular defect. From a pathophysiological point of view two categories have to be distinguished, one with generalized disturbance of tubular cell function, probably produced by defective intracellular energy production or transfer (common type), and the second with absence of proximal tubular brush borders, possibly caused by a congenital defect in the biochemical synthesis of brush-border constituents (brush-border type).

The primary Fanconi syndrome of the common type, initially considered to occur only in adults, may develop at any age and last over a long life-span. Some cases are familial with an autosomal dominant trait (Hunt *et al.* 1966; Brenton *et al.* 1981; Patrick *et al.* 1981), an autosomal recessive trait (Illig and Prader 1961), and an X-linked recessive trait (Neimann *et al.* 1968), but most occur sporadically without any evidence of inheritance (Houston *et al.* 1968; Briggs *et al.* 1972; Lee *et al.* 1972). The primary Fanconi syndrome, therefore, is a heterogenous group and will remain so as long as the tubular enzymatic defects are not identified.

Reports on renal morphology in the common type are scarce. By light microscopy no specific alterations were observed (Houston *et al.*1968). In three members of a family with idiopathic Fanconi syndrome only non-specific signs of interstitial nephritis were detected by light or electron microscopy (Neimann *et al.* 1968). In an autopsy specimen, marked changes in the tubular epithelium and interstitium were found (Hunt *et al.* 1966). A renal biopsy of one case showed dilatation of the tubules on light microscopy (Brodehl 1976). The diameter of the tubules measured up to 300 μm, which means that some were as wide as Bowman's capsules. Parts of the proximal tubular epithelium were swollen and contained a fine granular material, while other parts were completely normal. On electron microscopy grossly enlarged mitochondria were found with laterally or marginally dislocated cristae and occasional electron-dense material.

The clinical symptoms are characterized in children by failure to thrive, anorexia, polydipsia and polyuria, episodes of vomiting, dehydration, unexplained fever, rickets despite adequate supplementation with vitamin D, and growth retardation. In adults the symptoms are mostly related to defective bone mineralization (osteomalacia, bone pain, fractures), muscular weakness, and fatigue. Moderate to severe hyperchloraemic metabolic acidosis with hypokalaemia, and hypophosphataemia can be detected, together with a generalized hyperaminoaciduria and glucosuria. The glomerular filtration rate is usually normal, but may decline later in the course. Tubular PAH secretion is not severely impaired. The prognosis is variable but may be uncertain in regard to kidney survival. Some patients develop chronic renal insufficiency over a period of 10 to 30 years (Hunt *et al.* 1966; Lee *et al.* 1972; Brenton *et al* 1981), which then requires renal transplantation (Briggs *et al.* 1972; Patrick *et al.*

Table 2 Features of diseases associated with the renal Fanconi syndrome

Disease	Inheritance	Enzyme defect	Renal signs					Extrarenal signs	Treatment
			Hyperaminoaciduria	Glucosuria	Phosphate wasting	Tubular PAH transport	Glomerular filtration rate		
Primary Fanconi syndrome	AD, AR, XLR	Unknown	++	++	++	N	Slow ↓	Occasionally cataract, hearing loss	Symptomatic
Cystinosis	AR	Unknown	++	++	++	+++	Progressive ↓	Corneal crystals, short stature	Symptomatic, cysteamine
Tyrosinaemia type I	AR	Fumarylacetoacetate hydrolase	++	(+)	++	N	N	Hepatopathy	Low phenylalanine and tyrosine diet
Glycogenosis	AR	Unknown	+++	+++	+++	N	N	Hepatomegaly, short stature	Symptomatic
Galactosaemia	AR	Galactose 1-phosphate-uridyl transferase	+	+	+	+	Slight ↓	Cataract, hepatopathy	Galactose free
Fructose intolerance	AR	Fructose 1-phosphate aldolase B	++	(+)	(+)	N	Slight ↓	Hepatopathy, hypoglycaemia	Fructose free
Lowe's syndrome	XLR	Unknown	+	(+)	+	N	Slow ↓	Cataract, buphthalmus, mental retardation	Symptomatic
Wilson's disease	AR	Unknown	++	(+)	+	+	N	Hepatopathy, corneal ring, encephalopathy	Penicillamine
Cytochrome-c-oxidase deficiency	?	Cytochrome-c-oxidase	++	++	+	?	N	Severe myopathy, lactic acidosis	Symptomatic

+ = Slightly impaired; ++ = Moderately impaired; +++ = Severely impaired; N = normal.

1981). In one case with maternal graft a recurrence of Fanconi syndrome has been reported (Briggs et al. 1972).

A single case of the brush-border type of idiopathic Fanconi syndrome has been described in an infant with severe tubular dysfunction from the first day of life (Manz et al. 1984). At the age of 30 months and 4 years GFR was reduced to 60 ml/min.1.73 m^2, and tubular reabsorption of glucose, phosphate and amino acids (see Fig. 4) was virtually absent. The distal tubule compensated for the proximal loss of electrolytes and seemed to be undisturbed. A renal biopsy specimen showed proximal tubular cells with various heights and a widespread absence of brush borders, while the distal tubules showed adaptive changes with prominent apical vacuolization and basolateral interdigitations. The child survived the first years of life, but suffered from severe somatic and psychomotor retardation.

Cystinosis

For a complete discussion on cystinosis see Chapter 18.5.

Tyrosinaemia type I

Tyrosinaemia type I is caused by a deficiency of fumarylacetoacetate hydrolase (FAH), catalysing the last step of tyrosine degradation (Goldsmith 1983). It leads to an accumulation of maleyl- and fumarylacetoacetate and succinylacetone which have toxic effects on renal tubular function, hepatic enzymes of tyrosine catabolism, and porphobilinogen synthase. It is still undecided whether tyrosinaemia type I represents a nosologic entity or whether it is a syndrome caused by a variety of disturbances. In many cases it is transmitted as an autosomal recessive trait. It has to be differentiated from transient neonatal hypertyrosinaemia and from benign tyrosinaemia type II (Richner-Hanhart syndrome).

Tyrosinaemia type I is clinically expressed in an acute and a chronic form. Symptoms in the acute form start very early after birth with severe metabolic disturbances, including hepatomegaly with jaundice, coagulation defects, melaena, ascites, oedema, and sepsis-like fever. Death occurs rapidly in hepatic failure in most untreated infants (Halvorsen et al. 1966). The chronic form is characterized by the appearance during childhood of a failure to thrive, vitamin D-resistant rickets, growth retardation, nodular liver cirrhosis, hypoglycaemic attacks associated with hypertrophied islets of Langerhans, and, in some cases, mild to moderate mental retardation (Gentz et al. 1965). Death occurs mostly from acute metabolic exacerbations, liver cirrhosis, or hepatoma (Dehner et al. 1989), but not from renal insufficiency. Biochemically the disease is characterized by high plasma levels of tyrosine, often accompanied by methionine, biological evidence of liver impairment including high alpha-fetoprotein serum levels, and by urinary excretion of large amounts of tyrosyl compounds (mainly p-hydroxyphenyllactic acid, p-hydroxyphenylpyruvic acid, and p-hydroxyphenylacetic acid). The enzymatic diagnosis is possible in lymphocytes, fibroblast and liver specimens (Berger et al. 1981; Kvittingen et al. 1983) and, for prenatal diagnosis, in chorionic villi material (Kvittingen et al. 1986a).

The renal symptoms are usually severe, especially in the later stage, and present the fully developed Fanconi syndrome, while glomerular filtration remains normal. Hyperaminoaciduria in the acute stage reflects both the tubular defect and the increased plasma levels of amino acids due to liver damage (Gentz et al. 1965; Brodehl 1978a). As a consequence the typical pattern of generalized hyperaminoaciduria may be obscured. Glucosuria is mild and sometimes intermittent, and is often associated with low blood glucose levels. The tubular reabsorption of phosphate is severely impaired leading to low plasma phosphate levels and vitamin D-resistant rickets (Aronsson et al. 1968). Acidification defects have not been intensively looked for, but their presence can be deduced from the severe metabolic acidosis encountered in this disease. Proteinuria is of tubular origin and mostly mild.

A marked increase in urinary excretion of succinylacetone and succinylacetoacetate excretion (1–2 mmol/24 h) is considered pathognomonic for the disease (Grenier et al. 1982; Lindbladt and Steen 1982). Succinylacetone is probably responsible for the tubular dysfunction, since it inhibits the tubular function at both the mitochondrial and membrane levels (Spencer et al. 1988).

Treatment relies on a low phenylalanine and tyrosine diet supplemented with cysteine hydrochloride. In the acute neonatal form it is rarely effective, since the progressive liver damage has already occurred before birth (Hostetter et al. 1983). In the chronic form dramatic improvement has been observed: plasma tyrosine drops to normal levels, tyrosyluria stops and all tubular symptoms subside (Aronsson et al. 1968; Fairney et al. 1968). The final outcome remains doubtful since cirrhosis has developed in all cases so far reported. Liver transplantation has been successfully performed in a few cases (Kvittengen et al. 1986b; Dehner et al. 1989). After liver transplantation most biochemical abnormalities disappear, but low-grade excretion of succinylacetone persists, since the enzymatic defect is still present in the kidney.

Glycogenosis with Fanconi syndrome

The glycogenosis with Fanconi syndrome is a rare autosomal recessive disease, of which 19 cases have been described in the literature (Manz et al. 1987). It does not belong to any one of the 10 or more known types of glycogenosis, especially not to type I (hepatic glucose 6-phosphatase deficiency). The enzymatic defect has not been identified. Both liver and kidneys are affected. It has been suggested that the defect is located in the basolateral membrane transport systems for glucose and galactose in both hepatocytes and proximal tubular cells (Manz et al. 1987).

The disease is characterized by the combination of hepatic accumulation of glycogen (more than 8 g/100g wet tissue) and a severe Fanconi syndrome. Most cases are diagnosed in infancy and early childhood. The Fanconi syndrome is pronounced and includes all the characteristic laboratory findings. Generalized hyperaminoaciduria is massive (Fig. 2), phosphate reabsorption is severely impaired (Fig. 6) resulting in rickets, osteoporosis, and growth retardation, acidosis is moderate to severe, and the excretion of organic non-aminated acids greatly increased. However, the tubular secretion of PAH is not affected (Brodehl et al. 1969).

The most disturbed function is tubular glucose reabsorption. The rates of glucose excretion in several of these patients are listed in Table 3. They are certainly higher than in other types of the Fanconi syndrome, and occur while the patients are normoglycaemic or hypoglycaemic. The ratio of endogenous glucose clearance to inulin clearance (C_G/C_{IN}) in one case stayed constantly at 0.7 and did not change under glucose loads of between 50 and 350 mg/min.1.73m^2 (Brodehl et al. 1969). These values indicate that only 30 per cent of filtered glucose is reabsorbed, perhaps through non-active mechanisms. Fellers et al. (1967)

5.1 Fanconi syndrome

Table 3 Maximal urinary glucose excretion in glycogenosis with Fanconi syndrome (renal glucose losing syndrome)

Authors		Age (years)	Patient's maximum		
			U_G (g/100 ml)	$U_G V$	
				(g/24 h)	(g/24 h.1.73 m^2)
Fanconi and Bickel	(1949)	4	6.0	64	240
Rotthauwe et al.	(1963)	4	5.8	55	206
Odievre	(1966)				
Patient R.L.		2	?	50	210
Patient B.B.		6	?	21	74 (average)
Lampert and Mayer	(1967)	4	8.0	73	230
Bauer	(1968)	5	?	53	178
Brodehl et al.	(1969)	1	5.4	43	200
Garty et al.	(1974)	3	?	35–50	173
Brivet et al.	(1983)	5	?	43–94	143–325

U_G = urinary glucose concentration; $U_G V$ = rate of excretion of glucose.

compared the severe glucosuria in these patients with the effect of phlorizin, and called it the pseudophlorizin diabetes syndrome. We prefer the term renal-glucose-losing syndrome (Brodehl et al. 1969). It has recently been called the Fanconi-Bickel syndrome because of the first description by Fanconi and Bickel (1949).

The prognosis of this type of Fanconi syndrome is fairly good as long as acute dehydration, hypoglycaemia, and acidosis during infections or surgical procedures is prevented by intravenous administration of fluids and alkali. GFR is not altered and the tubular defects tend to improve with age. Treatment is only symptomatic but should be meticulously followed. Our patient (Fig. 7) has been under observation for 24 years and is doing well.

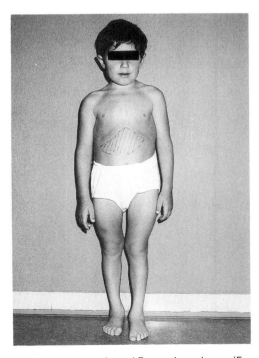

Fig. 7 Boy with glycogenosis and Fanconi syndrome (Fanconi-Bickel syndrome) at the age of 10 years (authors observations; some of his renal functions are given in Figs. 5 and 6).

Galactosaemia

Galactosaemia is an inherited disease of galactose metabolism transmitted as an autosomal recessive trait. It is caused by the deficient activity of galactose 1-phosphate uridyl transferase, which catalyses the reaction of galactose 1-phosphate with uridine-diphosphate-glucose (UDP glucose) to give UDP galactose and glucose 1-phosphate (Segal 1983). Deficiency of this enzyme leads to accumulation of galactose 1-phosphate intracellularly with eventual damage to the liver, brain, kidney, and lens. The clinical signs in untreated patients are hepatomegaly with jaundice, mental retardation, renal Fanconi syndrome, and bilateral cataract. Galactosaemia due to deficient galactokinase, which forms galactose 1-phosphate from galactose, is only associated with cataracts, but not with renal, liver or brain impairment. Early detection of galactosaemia is possible in newborns by screening methods: the incidence ranges from 1:18 000 to 1:110 000.

The renal symptoms start early after exposure to lactose by milk feeding. They are characterized by hyperaminoaciduria, impaired phosphate reabsorption, bicarbonate wasting, tubular proteinuria, and slight to moderate reduction in GFR (Brodehl 1978a). Hyperaminoaciduria is generalized, and is usually not as severe as that seen in cystinosis or glycogen storage disease (Fig. 2). The tubular reabsorption of phosphate, expressed as T_P/GFR, is significantly reduced, while the reabsorption of glucose is not, or is only minimally, impaired, Therefore, mellituria in galactosaemic infants is due to galactosuria, not to glycosuria, which means that only non-specific reducing tests (Benedict's test, Clinitest tablets) are positive, but not specific glucose tests (Clinistix, Combistix).

Proteinuria is a constant finding in untreated patients, and is of the tubular type. Glomerular filtration rate and PAH clearance are significantly reduced in the acute stage, and the filtration fraction is elevated.

Treatment relies on the elimination of galactose from the diet by the exclusion of milk and milk products. This is relatively simple, since galactose is not an essential nutrient. After introduction of a galactose-free diet, symptoms of galactose intoxication disappear, normal growth and development resume and all renal symptoms improve rapidly (Brodehl 1978a). It takes about 1 to 2 weeks for the hyperaminoaciduria to develop after exposing the child to galactose (Komrower et al. 1956) or to remit after discontinuing galactose intake (Hsia et al. 1954). This time course is longer than in hereditary fructose intolerance. Recently, however, late sequelae of galactosaemia have been reported, despite a continuous galactose-free diet: ovary dysfunction with hypergonadotropic hypogonadism (Kaufman et al. 1981), and progressive cerebral dysfunction (Böhles et al. 1986).

Hereditary fructose intolerance

Hereditary fructose intolerance is caused by a deficiency of fructose 1-phosphate aldolase B (Gitzelmann et al. 1983). The enzyme cleaves fructose 1-phosphate to D-glyceraldehyde and phosphodihydroxyacetone for further conversion into glucose or CO_2 and H_2O. Fructose 1-phosphate cannot be metabolized and accumulates in those tissues that normally extract fructose rapidly and convert it to glucose. Only the proximal tubules, the liver and the small intestine contain the whole set of enzymes for fructose elimination, i.e. fructokinase, fructose 1-phosphate aldolase, and triokinase, while renal medulla lacks these enzymatic systems (Kranhold et al. 1969). Therefore the toxic effect of fructose 1-phosphate is limited to the small intestine, liver, and kidney.

Hereditary fructose intolerance is transmitted as an autosomal recessive trait and has a reported incidence of 1:20 000. It must be differentiated from essential fructosuria (fructose kinase deficiency), galactose–glucose malabsorption, and hereditary fructose 1,6-diphosphatase deficiency. With the exception of one report (Bakker et al. 1974) the last syndrome has not been associated with renal dysfunction.

In patients with hereditary fructose intolerance ingestion of fructose is followed by severe, life-threatening bouts of vomiting, diarrhoea, hypoglycaemia, acidosis, and coma. Chronic fructose ingestion leads to failure to thrive, hepatomegaly with jaundice progressing to liver cirrhosis, and renal Fanconi syndrome. If unrecognized, the condition may end in cachexia and death. Patients beyond infancy seem to increase their tolerance for fructose; they usually develop a strong aversion to fruits and sweets, thereby avoiding intoxication. There is unfortunately no screening method allowing an early diagnosis. Diagnosis relies on a nutritional history, and clinical and laboratory investigation. It may be confirmed by a fructose tolerance test (cave hypoglycaemia) and an aldolase assay on liver biopsy or intestinal mucosa (Steinmann and Gitzelmann 1981).

The renal disturbance is characterized by a severe dysfunction of the proximal tubule, while the distal tubule is preserved and compensates for some proximal functional defects. The Fanconi syndrome in the hereditary fructose intolerance is characterized by hyperaminoaciduria, proximal tubular acidosis and proteinuria, but only mild or absent glucosuria and phosphate wasting. The hyperaminoaciduria is generalized (Levin et al. 1968) and usually accompanied by normal plasma amino acids. However, in severe hepatic involvement, plasma amino acids may be elevated, especially tyrosine and methionine, leading to a distorted urinary pattern, which could be misinterpreted as acute tyrosinaemia. The onset of hyperaminoaciduria after exposure to fructose is almost instantaneous (Levin et al. 1968; Morris 1968); return to normal levels after fructose-free diet occurs within 2 weeks.

The bicarbonate reabsorption is rapidly disturbed after oral fructose loading (Morris 1968). At normal plasma bicarbonate levels it is reduced by 20 to 30 per cent, but in moderately severe acidosis, bicarbonate disappears from the urine and a minimal pH can be achieved. These findings are typical of proximal tubular acidosis (type II renal acidosis). In infants, severe metabolic acidosis with hyperlactic acidaemia is observed. In adults, chronic tubular acidosis may lead to nephrocalcinosis and urolithiasis (Higgins and Varney 1966; Mass et al. 1966).

Glucosuria is usually absent or mild. Fructose is excreted only when the patient is exposed to it. Phosphate reabsorption may be disturbed, as suggested by hypophosphataemia with chronic sucrose ingestion. Acute fructose loading is followed by sharp decline in plasma phosphate, produced by a rapid phosphate uptake by those tissues that accumulate fructose 1-phosphate and are unable to metabolize it further. Phosphate excretion is not increased after fructose loading (Morris 1968). Rickets have not been reported in this syndrome. Proteinuria is related to the acute intoxicating effect of fructose.

Treatment of hereditary fructose intolerance is simple as soon as the diagnosis is established and the acute phase of intoxication is overcome. All fructose- and sucrose-containing foods and sweets should be avoided; most patients beyond infancy develop a strong aversion to foods containing sucrose. Affected patients should never be exposed to infusions containing fructose, sorbitol, or xylitol. Since it must be assumed that many carriers of the enzyme defect remain undetected, it has been proposed that infusion fluids should never contain fructose, sorbitol, or xylitol, which could be deleterious for unrecognized patients (Lameire et al. 1978; Müller-Wiefel et al. 1983; Galaske et al. 1986).

Oculocerebrorenal dystrophy (Lowe syndrome)

The oculocerebrorenal syndrome was first described by Lowe and coworkers in 1952; more than 100 cases have since been reported. It is characterized by congenital eye abnormalities (cataract, glaucoma, buphthalmos), cerebral dysfunction leading to severe mental retardation, muscular hypotonia with diminished or absent reflexes, arthropathy resembling rheumatoid arthritis, and both functional and morphological renal abnormalities. The renal findings are initially those of the Fanconi syndrome, and later of slowly progressing glomerular insufficiency. The underlying metabolic defect is unknown, but may involve the metabolism of glycoaminoglycans (Fukui et al. 1981; Wisniewski et al. 1984).

The condition is transmitted as an X-linked recessive trait and appears in males with a few exceptions (Sagel et al. 1970). A higher incidence of cataracts and corneal opacities has been reported in the mothers of affected boys, a finding that has not been confirmed (Holmes et al. 1969).

The Fanconi syndrome is usually mild. Hyperaminoaciduria appears after the first few months of life (Vis et al. 1972), tends to decrease after the sixth year (Hambraeus et al. 1970), and may be absent in adults (Phelip et al. 1973). The generalized pattern of aminoaciduria is characterized by a more pronounced involvement of the dibasic amino acids and cystine (Manz et al. 1978). Whether this is of any significance with regard to the underlying metabolic defect remains doubtful. A postulated defect in ornithine metabolism could not be confirmed (Chutorian and Rowland 1966). The intestinal absorption of dibasic amino acids seems to be also impaired (Colombo 1971).

Glucosuria is mild and sometimes intermittent. The defect on phosphate reabsorption is variable, but can be severe (Texier et al. 1971). The renal acidosis is due to proximal bicarbonate wasting responsive to changes of extracellular fluid volume (Oetliker and Rossi 1969). Proteinuria is often present from early infancy and may approach the nephrotic range. Glomerular filtration decreases with age and finally chronic renal insufficiency can emerge.

The morphology of the kidney in the early stage is normal on light microscopy (Witzleben et al. 1968). Electron microscopy, however, reveals very early distinct changes: in the glomeruli there is slight swelling of the endothelial cells with vacuolization, thickening, and splitting in the capillary basement membranes and fusion of the foot processes (Van Acker et al. 1967); in the proximal tubules there is shortening of the brush borders, enlargement of the mitochondria, and distortion or complete disappearance of mitochondrial cristae (Ores 1970; Sagel et al. 1970). In later stages advanced glomerular changes with membrane thickening, increased glomerular cellularity, and glomerular fibrosis are observed; there is also scattered thickening of renal tubular basement membranes, multifocal tubular atrophy with dilatations, and interstitial fibrosis (Habib et al. 1962).

The treatment of Lowe syndrome is symptomatic, in the absence of specific therapy. It follows the natural course of the disease, characterized by three stages. During the early infantile stage the main emphasis lies on the management of ocular and central nervous symptoms; during the second stage from late infancy to middle childhood it is focused on the correction of the metabolic abnormalities due to the Fanconi syndrome, and during the last stage it is concerned with advancing renal insufficiency and progressive joint involvement (Athreya et al. 1983). No attempts at renal replacement therapy have yet been reported, probably because of severe mental retardation.

Wilson's disease

Wilson's disease (hepatolenticular degeneration) is a familial disease of copper metabolism (Danks 1983). The primary enzyme defect is still unknown. The main abnormality is a severely impaired biliary copper excretion and incorporation of copper into caeruloplasmin, a plasma α_2-globulin. These defects lead to an excessive intracellular accumulation of copper in the liver with subsequent overflow into other tissues, such as brain, cornea, and renal tubules. The disease is transmitted as an autosomal recessive trait. Its incidence is estimated to be between 1 and 2 per 100 000.

Patients present signs of liver disease in 42 per cent of the cases, neurological symptoms in 34 per cent, psychiatric and behavioural abnormalities in 10 per cent, haematological symptoms in 12 per cent, and renal symptoms in 1 per cent (Sternlieb and Scheinberg 1979). In children symptoms start only occasionally before the age of 6 years and are mostly related to active hepatitis or cirrhosis. The full spectrum of the disease includes cirrhosis of the liver, a haemolytic anaemia, osteoarthropathy, tubulopathy, bizarre neurological manifestations, psychiatric abnormalties, and greenish-brown rings at the limbus of the cornea (Kayser-Fleischer rings). Biochemically there is an excessive accumulation of copper in many tissues, especially in the liver, in most instances a reduction of serum caeruloplasmin levels, and an increased urinary copper excretion. The diagnosis is confirmed by the radiocopper test (Wesch et al. 1980).

The renal symptoms can start at any stage of Wilson's disease (Bearn et al. 1957; Leu et al. 1970; Wilson and Goldstein 1973), but are rare in children. The GFR decreases as the disease progresses but does not lead to chronic renal insufficiency as patients die earlier from extrarenal complications. PAH clearance and maximal tubular PAH secretion are even more impaired than GFR, which indicates the preponderance of tubular disturbance. Hyperaminoaciduria has a generalized pattern of variable severity (Schønheider et al. 1971). Plasma amino acids are not elevated unless there is a severe hepatic failure. Glucose reabsorption is mildy impaired: at normal blood glucose levels glucosuria is not always present, but Tm_G is reduced. Phosphate reabsorption is also impaired leading to hypophosphataemia and osteomalacia in some patients. Urate excretion is increased, with eventual hypouricaemia. The ability to acidify the urine is reduced. Proteinuria is usually slight.

Morphologic studies in the early stage of Wilson's disease reveal either no significant alteration by light microscopy (Gilsanz et al. 1960), or only a slightly flattened, faintly staining tubular epithelium without recognizable brush borders (Reynolds et al. 1966). On electron microscopy the brush borders are sparse or absent, the cellular network is destroyed, and electron-dense bodies are present in the subapical areas of the tubular cytoplasm (Elsas et al. 1971); they probably represent metalloproteins. The mitochondria show increased density, and their cristae are not usually well preserved. In postmortem specimens the tubular epithelium is damaged, and staining with rubeamic acid reveals intracytoplasmatic copper granules (Wolff 1964). The copper content of kidney tissue is greatly increased.

The deranged copper metabolism can be treated successfully with D-penicillamine or other chelating agents (Walshe 1975, 1982). D-penicillamine produces a striking increase in copper excretion associated with marked clinical improvement, particularly when treatment is started early (Schønheider et al. 1970; Elsas et al. 1971). If they are not too far advanced, all types of renal dysfunction subside. There are, however, occasionally serious side-effects of D-penicillamine. In these cases oral treatment with zinc sulphate appears to be an effective alternative (Hoogenraad et al. 1984). In cases with advanced liver disease, however, liver transplantation has been successfully performed (Sternlieb 1984) with an attendant cure of the disturbed copper metabolism.

Cytochrome-c-oxidase deficiency

The Fanconi syndrome due to cytochrome-c-oxidase deficiency has been described by Van Biervliet et al. (1977). The cytochromes are haem-containing enzymes of the electron transporting chain within the mitochondria. Transport of electrons is coupled to the production of energy. The mitochondria use this energy for synthesis of ATP or other energy consuming reactions. Van Biervliet observed three infants of a family, with generalized muscular weakness since birth, a mitochondrial myopathy associated with lactic acidosis, and a Fanconi syndrome. The severe myopathy led to lethal asphyxia at the age of 11 to 13 weeks. Four other infants with the same clinical picture and the Fanconi syndrome have been described (Sengers et al. 1984). All died at the age of 7 to 15 weeks. There are, however, other cases with cytochrome-c-oxidase deficiency without Fanconi syndrome.

The Fanconi syndrome is characterized by generalized hyperaminoaciduria, renal glucosuria, decreased phosphate reabsorption, proteinuria, and hypercalciuria. Blood creatinine and urea are normal. Persistent metabolic acidosis is mainly due to severe hyperlacticacidaemia. In the muscles the cytochrome-c-oxidase is decreased to between 10 and 17 per cent of normal. Decreased activity was also noted in the kidney (38 per cent of normal) in one infant (DiMauro et al. 1980). Impaired electron transport and energy generation in the tubular cells may be the cause of the Fanconi syndrome.

Fanconi syndrome with abnormal proteinuria

Most acquired Fanconi syndromes in adults are related to abnormal proteinurias. The proteinuria can be caused either by extrarenal production of abnormal proteins of low molecular weight, which are filtered by the glomeruli and then reabsorbed, or by increased basement membrane permeability to proteins in glomerular diseases. Synthesis of abnormal proteins inducing the Fanconi syndrome is observed in multiple myeloma, hypergammaglobulinaemia, immunoglobulin light chain disease, Sjögren's syndrome, and amyloidosis. These diseases are extremely rare in infancy and childhood. They are not accompanied usually by the complete Fanconi syndrome, but rather by partial defects such as renal tubular acidosis and nephrogenic diabetes insipidus or by acute or chronic renal failure.

Multiple myeloma was first recognized to be associated with the Fanconi syndrome by Sirota and Hamerman (1954). Since then a number of cases have been described (Maldonado et al. 1975; Walb et al. 1980). The onset of the Fanconi syndrome may precede the diagnosis of myeloma by many years (up to 13 (Maldonado et al. 1975), or 16.5 (Sewell and Dorreen 1984)). The renal symptoms of the Fanconi syndrome are often severe, and tubular PAH secretion is grossly impaired. Chronic renal failure is common in the late stages. Successful treatment of the myeloma may correct the Fanconi syndrome and the other renal disorders (Scheele 1976).

Histologic examination discloses degenerative changes of the tubular cells, particularly of the proximal nephron (Engle and Wallis 1957; Lee et al. 1972; Maldonado et al. 1975). Crystalline inclusion bodies can be visualized by electron microscopy. In experimental studies, Bence-Jones proteins inhibit the uptake of PAH (Preuss et al. 1967), whereas proteins derived from nephrotic patients fail to do so (Preuss et al. 1974). $\varkappa$-Type Bence-Jones protein injected *in vivo* induce proximal tubular lesions in rats (Clyne et al. 1974).

Proteinurias of immunoglobulin light chains (Thorner et al. 1983), hypergammaglobulinaemia (Kamm and Fisher 1972), Sjögren's syndrome (Walker et al. 1971), and amyloidosis (Rochman et al. 1980) have also been associated with the Fanconi syndrome. The pathomechanism is similar to that in multiple myeloma.

Tubular disturbances leading to the Fanconi syndrome are also observed in acute or chronic interstitial nephritis (Bergstein and Litman 1975; Levy et al. 1978; Burghard et al. 1984). Interstitial nephritis may be produced by viral or bacterial infection, or by reactions to drugs, or may occur spontaneously. It can be associated with uveitis and usually is accompanied by moderate or severe impairment of glomerular filtration. In the acute type generalized hyperaminoaciduria, moderate glucosuria phosphate wasting, acidosis, tubular proteinuria, and polyuria are the most prominent features. Renal biopsy reveals diffuse interstitial mononuclear infiltration and interstitial oedema, and the tubules are deformed, with flattened epithelium. Immunofluorescence microscopy shows linear staining of thickened tubular basement membranes. In the chronic type, diffuse interstitial fibrosis with scattered mononuclear cells prevails. Treatment of the acute interstitial nephritis with steroids may lead to complete recovery of renal function. In the chronic type treatment is rarely effective.

The occurrence of the Fanconi syndrome in patients with the nephrotic syndrome is less frequently reported (Shwayder et al. 1976). In most patients the nephrotic syndrome starts in early childhood and proves resistant to steroid therapy (Hooft and Vermassen 1960). The tubular symptoms appear 1 to 4 years after the onset of the nephrotic syndrome, and are characterized by the Fanconi syndrome, polyuria, renal acidosis, hypocalcaemia, tetany, rickets, and growth retardation. The course is generally unaffected by therapy and renal death usually occurs within 2 to 8 years.

Renal biopsy in an early stage reveals minimal glomerular lesions, associated with dilatation, and atrophy of the proximal tubules, thickening of the basement membranes, and interstitial fibrosis (Royer et al. 1963; Shwayder et al. 1976). In the late stage, marked tubulointerstitial lesions and advanced glomerular hyalinizations are observed. In distal tubules birefringent crystals may be detected by light microscopy (Hooft and Vermassen 1960).

The pathogenesis of the combination of glomerular and tubular disease is still an open question. It is undecided whether the glomerular or the tubular lesion is the primary event, or whether both are produced simultaneously by a systemic immunological injury. All three pathogenic ways seem possible, as discussed by Shwayder et al. (1976) and Levy et al. (1978).

Other causes of Fanconi syndrome

The appearance of the Fanconi syndrome in the transplanted kidney may be caused either by the existence of an unknown systemic metabolic disturbance leading to an 'idiopathic' Fanconi syndrome in the original kidney and, thus, reappearance of the disturbance in the graft (Briggs et al. 1972), or by ischaemia, rejection, hyperparathyroidism, or immunosuppressive therapy (Vertuno et al. 1974; Better 1980; Brodehl et al. 1984). These insults to the graft may also produce less complex tubular dysfunctions as tubular acidosis, phosphate leak, glucosuria, and polyuria (Györy et al. 1969; Moorhead et al. 1974; Hoyer et al. 1988).

A tumour-related Fanconi syndrome has been described by Leehey et al. (1985). The complete Fanconi syndrome, associated with a non-ossifying fibroma of the left tibia, resolved promptly after removal of the tumour.

Exogenous intoxications

Various toxic substances and drugs produce disturbances in tubular function without primarily affecting GFR. The degree of tubular dysfunction varies according to the type, the duration, and the severity of exposure to the toxic substance. The defects are mostly reversible after exposure to the toxin is stopped. A very careful history is thus mandatory to exclude the possibility of exogenous intoxications in all patients with tubular dysfunction.

Three categories of substances may induce tubular intoxication: heavy metals, organic compounds, and drugs. In humans heavy metals such as lead, cadmium, uranium, and mercury have been incriminated in acute and chronic tubular disturbances (Goyer et al. 1972; Lee et al 1972; Iesato et al. 1977). A mild Fanconi syndrome has been reported in a 18-month-old infant following long-term application of a mercury-containing solution for difficult teething (Rützler 1973). Most metals induce an experimental tubulopathy utilized to elucidate the pathomechanism of the disturbance (Foulkes 1983; Mori et al. 1984). In these models the Fanconi syndrome appears related to disturbances in the energy-producing or energy-transporting mechanisms of the cell, reflected by morphologic changes in the mitochondria, and alterations of sodium transport and intracellular homeostasis. A direct toxic effect on the brush-border membrane has also been suggested (Foulkes 1983).

Although many organic compounds are nephrotoxic, reports on tubulopathies in humans related to accidental exposure to those substances are rare. Lysol intoxication associated with anuria may be followed by hyperaminoaciduria and glucosuria in the polyuric phase (Spencer and Franglen 1952), a finding that should be considered as non-specific, however, since it is commonly observed after acute tubular necrosis (Emslie-Smith et al. 1956). Toluene exposure by glue-sniffing has also been reported to be accompanied by the Fanconi syndrome (Moss et al. 1980).

Adverse reactions to drugs are a more common cause of Fanconi syndrome. Outdated or degraded tetracyclines cause a reversible Fanconi syndrome in both children and adults, even when taken in recommended dosages (Frimpter et al. 1963; Mavromatis 1965; Brodehl et al. 1968; Montoliu et al. 1981). The clinical picture is characterized by polyuria, dizziness, muscular weakness, acidosis, and sometimes severe neurological symptoms. The Fanconi syndrome is associated with impaired GFR and PAH clearance and, in two cases, with the excretion of Bence-Jones proteins (Frimpter et al. 1963; Mavromatis 1965). Renal biopsy reveals striking vacuolization of the distal tubular epithelium, hypercellularity of the glomerular tufts, prominent capillary loops and basement membrane thickening. Most changes prove reversible. The intoxicating substance is anhydro-4-epitetracycline formed from tetracycline under the influence of heat, moisture, and low pH. This substance produces similar symptoms in dogs and rats (Lindquist and Fellers 1966). The accidental ingestion of a structurally related compound, methyl-3-chromone (diachrome), has been followed by a Fanconi syndrome (Otten and Vis 1968).

Other drugs associated with the Fanconi syndrome include 6-mercaptopurine (Butler et al. 1965), gentamicin, and cephalotin (Schwartz and Schein 1978; Russo and Adelman 1980), cisplatin (Dentino et al. 1978; Davis et al. 1980) and, recently, ifosfamide (Smeitink et al. 1988; Moncrieff and Foot 1989). The toxic effect of cisplatin is immediately manifest whereas that of ifosfamide is delayed for several months, which would indicate a different pathomechanism.

The substance most widely used for experimental induction of the Fanconi syndrome is maleic acid, the *cis* isomer of fumaric acid. Berliner and coworkers (1950) showed that the intravenous injection of maleate into acidotic dogs interfered with urinary acid excretion and was followed by an increased output of sodium, phosphate, and bicarbonate. Harrison and Harrison (1954) injected neutralized maleic acid intraperitoneally in rats kept on a low phosphorus diet and observed marked phosphaturia, hyperaminoaciduria and glucosuria. Many amino acids are involved (Günther et al. 1979), with an associated increase in the excretion of peptides (Asatoor et al. 1979), lysozymes (Fujita and Itakura 1978), and proteins (Christensen and Maunsbach 1980).

In vitro, maleic acid impairs the uptake of amino acids by renal cortical slices, both by retarding influx and by accelerating efflux from the cells (Rosenberg and Segal 1964). The accumulation of radioactive amino acids in the kidney is reduced after injection of maleate (Bergeron and Vadeboncoer 1971). Microinjection studies in kidneys treated with maleate demonstrate that the luminal uptake of labelled leucine is not reduced, but that the transcellular transport from the peritubular site to the tubular lumen is enhanced, indicating a decreased ability of the tubular cell to retain absorbed amino acids. It is thus suggested that the increased membrane permeability causes the transport defect in the Fanconi syndrome (Bergeron et al. 1976). This hypothesis is supported by the demonstration of an increased tubular leakage for lissamine green, and a decreased net transport of sodium and phosphate (Maesaka and McCaffery 1980).

The ATP content of tubular cells treated with maleate is reduced and tubular Na^+-K^+-ATPase is inhibited. The transport defect of urate, phosphate, calcium, magnesium, and potassium produced by maleate may be secondary to the inhibition of sodium transport (Gonick and Kramer 1985). Isolated rat kidney brush-border membrane vesicles are not affected by maleic acid (Reynolds et al. 1978). In free-flow micropuncture studies, maleic acid reduces the reabsorption of sodium and phosphate both in the proximal and the distal tubule (Brewer et al. 1983). The renal tubules are refractory to parathormone in the maleic acid rate model, despite an apparently normal adenylate cyclase activation (Wald and Popovtzer 1984). Finally, maleic acid impairs the conversion of 25-hydroxy-vitamin D to 1,25-dihydroxy-vitamin D (Brewer et al. 1977).

Spontaneous Fanconi syndrome in dogs

Spontaneous Fanconi syndrome in dogs offers a useful model for the study of renal tubular transport defects (Bovee et al. 1979). Fractional glucose reabsorption is reduced between 31 and 82 per cent. During glucose loading tubular reabsorption falls to less than 10 per cent. The hyperaminoaciduria is due to a decreased tubular reabsorption. Its pattern is generalized, but not quite comparable with that of the human Fanconi syndrome. Fractional phosphate reabsorption ranges from 47 to 79 per cent. *In vitro*, uptake of lysine, glycine and α-methyl-D-glucoside by renal cortical slices is impaired. Histopathology is normal except for enlarged nuclei in some tubular cells.

Principles of treatment

Treatment of the Fanconi syndrome can either be specific or symptomatic. Specific treatment is possible in secondary types where the underlying metabolic defect is recognized. The accumulation of toxic metabolites can be reduced or avoided, mostly by dietary means as in transferase deficiency galactosaemia, hereditary fructose intolerance, tyrosinaemia type I, or by enhanced excretion as in Wilson's disease, or heavy metal intoxication. In drug-induced Fanconi syndrome, withdrawal of the drug may correct the defect. If the specific treatment is carefully continued, a complete return to normal can usually be obtained.

The symptomatic treatment for all other types of the Fanconi syndrome is equally important as it may almost completely compensate for the deranged tubular function. It may not only prolong survival, but also provide in most cases full rehabilitation and a normal life with full activity and well-being. In children it heals or prevents rickets and skeletal deformities, and thus avoids disabilities and severe growth retardation. However a normal growth rate cannot be achieved in all cases.

Symptomatic treatment is directed toward compensation for the consequences of excessive urinary losses, i.e. acidosis, hypokalaemia, hypophosphataemia, polyuria, and vitamin D resistance. Hyperaminoaciduria, glucosuria, proteinuria, and hypouricaemia are not associated with overt clinical symptoms and, therefore, do not require special treatment.

Acidosis, due to bicarbonate losses, requires large amounts of alkali. Depending on the degree of tubular impairment, doses of 2 to 10 mmol/kg.day or more are required in the form of sodium bicarbonate, citrate, or lactate. Alkali should be offered in divided doses every 4 to 5 h during the daytime, the amount needed being adjusted to the serum bicarbonate level. Most patients are potassium depleted and should receive a potassium supplement as bicarbonate, citrate, lactate, or phosphate depending on the levels of these anions in the plasma. The amount of potassium needed is adapted to the level in serum; it ranges from 2 to 4 mmol/kg.day.

In some cases the amount of alkali required to compensate for the severe bicarbonate loss is more than the patient can tolerate. In such patients, hydrochlorothiazide (2–3 mg/kg.day) has been successful (Callis et al. 1970) as extracellular volume depletion raises the bicarbonate threshold and improves the acidosis. Indomethacin, also reported to exert a beneficial effect on the tubulopathy (Haycock et al. 1982), should be given with caution to avoid further impairment of GFR.

Hypophosphataemia, due to decreased tubular phosphate reabsorption, can be compensated partly by oral supplementation of 1 to 3 g neutral phosphate per day (Phosphate Sandoz) given every 4 to 5 h throughout the daytime. Intestinal discomfort and diarrhoea may ensue but may improve after reduction of the dosage. In some cases phosphate supplementation may produce or aggravate hypocalcaemia and stimulate hyperparathyroidism. Administration of additional oral calcium and vitamin D may prevent hypocalcaemia.

Treatment with vitamin D is indicated in rickets or osteomalacia, although the nature of the vitamin D resistance is not well understood. The starting dose of vitamin D should be 5000 units per day or 0.25 to 0.5 μg/day of 1,25-dihydroxy-vitamin D, to be increased gradually. Careful monitoring of vitamin D treatment is necessary to avoid toxic side-effects, especially nephrocalcinosis. The best parameter is urinary calcium excretion, which should not exceed 6 mg/kg.day or 0.25 mg Ca/mg creatinine. It is still unclear whether ordinary vitamin D or hydroxylate compounds are better in regard to efficacy and side-effects.

Polyuria is frequent in the Fanconi syndrome. Sufficient fluid should be provided to avoid dehydration. In practice, we prepare a fixed amount of fluid for the whole day (1–2 l) in which the required salts were dissolved. This fluid is taken at regular intervals in divided doses. Additional fluids may be ingested *ad libitum*.

Finally, in the few patients who progess to renal insufficiency dialysis and transplantation may be indicated (see cystinosis).

Bibliography

Brewer, E.D. (1985). The Fanconi syndrome: clinical disorders. In *Renal Tubular Disorders* (eds. H.C. Gonick, and V.M. Buckalew), pp. 475–544, Marcel Dekker, New York.

Brodehl, J. (1978a). The Fanconi syndrome. In *Pediatric Kidney Disease* (ed. C.M. Edelmann), pp. 955–87, Little, Brown, Boston.

Chesney, R.W. (1980). Etiology and pathogenesis of the Fanconi syndrome. *Mineral Electrolyte Metabolism*, 4, 303–16.

Foreman, J.W. and Segal, S. (1987). Fanconi syndrome. In *Paediatric Nephrology* (eds. M.A. Holliday, T.M. Barratt, and R.L. Vernier), pp. 547–65, Williams and Wilkins, Baltimore.

Roth, K.S., Foreman, J.W., and Segal, S. (1981). The Fanconi syndrome and mechanisms of tubular transport dysfunction. *Kidney International*, 20, 705–16.

References

Aronsson, S., Engleson, G., Jagenburg, R., and Palmgren, B. (1968). Long-term dietary treatment of tyrosinosis. *Journal of Pediatrics*, 72, 620–7.

Asatoor, A.M., Milne, M.D., and Walshe, J.M. (1976). Urinary excretion of peptides and of hydroxyproline in Wilson's disease. *Clinical Science and Molecular Medicine*, 51, 369–78.

Asatoor, A.M., Bending, M.R., and Milne, M.D. (1979). Peptiduria in experimental Fanconi syndrome in rats. *Clinical Science and Molecular Medicine*, 57, 277–9.

Athreya, B.H., Schumacher, R., Getz, H.D., Norman M.E., Borden, S., and Witzleben, C.L. (1983). Arthropathy of Lowe's (oculo-cerebro-renal) syndrome. *Arthritis and Rheumatism*, 26, 728–35.

Bakker, H.D., De Bree, P.K., Ketting, F., Van Sprang, F.J., and Wadman, S.K. (1974). Fructose-1, 6-diphosphatase deficiency: another enzyme defect which can present itself with the clinical features of 'tyrosinosis'. *Clinica Chimica Acta*, 55, 41–7.

Barratt, T.M. and R. Crawford. (1970). Lysozyme excretion as a measure of renal tubular dysfunction in children. *Clinical Science and Molecular Medicine*, 39, 457–65.

Bauer, B. (1968). Debre-DeToni-Fanconi Syndrom mit Glykogenose, der Leber. *Klinische Wochenschrift*, 46, 317–28.

Bearn, A.G., Yü, T.T., and Gutman, A.B. (1957). Renal function in Wilson's disease. *Journal of Clinical Investigation*, 36, 1107–14.

Berger, R., et al. (1981). Deficiency of fumarylacetoacetase in a patient with hereditary tyrosinemia. *Clinica Chimica Acta*, 114, 37–4.

Bergeron, M. and Vadeboncoer. (1971). Microinjections of L-leucine into tubules and peritubular capillaries of the rat: II. The maleic acid model. *Nephron*, 8, 367–74.

Bergeron, M., Dubord, L., and Hausser, C. (1976). Membrane permeability as a cause of transport defects in experimental Fanconi syndrome. A new hypothesis. *Journal of Clinical Investigation*, 57, 1181–9.

Bergstein, J. and Litman, N. (1975). Interstitial nephritis with anti-tubular-basement-membrane antibody. *New England Journal of Medicine*, 292, 875–8.

Berliner, R.W., Kennedy, T.J., and Hilton, J.G. (1950). The effect of maleic acid on renal function. *Proceedings of the Society for Experimental Biology and Medicine*, 75, 791–4.

Bernardini, I., Rizzo, W.B., Dalakas, M., Bernar, J., and Gahl, W.A. (1985). Plasma and muscle free carnitine deficiency due to renal Fanconi syndrome. *Journal of Clinical Investigation*, 75, 1124–30.

Better, O.S. (1980). Tubular dysfunction following kidney transplantation. *Nephron*, 25, 209–13.

Böhles, H., Wenzel, D., and Shin Y.S. (1986). Progressive cerebellar and extrapyramidal motor disturbances in galactosaemic twins. *European Journal of Pediatrics*, 145, 413–7.

Bovee, K.C., Joyce, T., Blazer-Yost, B., Goldschmidt, M.S., and Segal, S. (1979). Characterization of renal defects in dogs with a syndrome similar to the Fanconi syndrome in man. *Journal of the American Veterinary Medical Association*, 174, 1094–9.

Bremer, H.J., Duran, M., Kamerling, J.P., Przyrembel, H., and Wadman, S.K. (1981). *Disturbances of Amino Acid Metabolism: Clinical Chemistry and Diagnosis*. Urban & Schwarzenberg, München.

Brenton, D.P., Isenberg, D.A., Cusworth, D.C., Garrod, P., Krywawych, S., and Stamp, T.C.P. (1981). The adult presenting idiopathic Fanconi syndrome. *Journal of Inherited Metabolic Diseases*, 4, 211–5.

Brewer, E.D., Tsai, H.C., Szeto, K.S., and Morris, R.C. Jr. (1977). Maleic acid-induced impaired conversion of 25 (OH) D3 to 1,25 (OH)2 D3: implications for Fanconi's syndrome. *Kidney International*, 12, 244–52.

Brewer, E.D., Senekjian, H.O., Ince, A., and Weiman, E.J. (1983). Maleic acid-induced reabsorptive dysfunction in the proximal and distal nephron. *American Journal of Physiology*, 245, 339–44.

5.1 Fanconi syndrome

Briggs, W.A., Kominami, N., Wilson, R.E., and Merrill, J.P. (1972). Kidney transplantation in Fanconi syndrome. *New England Journal of Medicine*, **286**, 25.

Brivet, M., Moatti, N., Corriat, A., Lemmonier, A., and Odievre, M. (1983). Defective galactose oxidation in a patient with glycogen storage disease and Fanconi syndrome. *Pediatric Research*, **17**, 157–61.

Brodehl, J. (1976). Tubular Fanconi syndrome with bone involvement. In *Inborn Errors of Calcium and Bone Metabolism* (eds. H. Bickel, and J. Stern), pp. 191–213, MTP Press, Lancaster.

Brodehl, J. (1978b). Renal hyperaminoacidurias. In *Pediatric Kidney Disease* (ed. C.M. Edelmann), pp. 1047–79, Little, Brown, Boston.

Brodehl, J. and Gellissen, K. (1968). Endogenous renal transport of free amino acids in infancy and childhood. *Pediatrics*, **42**, 395–404.

Brodehl, J. and Bickel, H. (1973). Aminoaciduria and hyperaminoaciduria in childhood. *Clinical Nephrology*, **1**, 149–68.

Brodehl, J., Hagge, W., and Gellissen, K. (1965). Die Veränderungen der Nierenfunktion bei der Cystinose. Teil I: Die Inulin-, PAH- und Elektrolyt-Clearance in verschiedenen Stadien der Erkrankung. *Annales Paediatrici (Basel)*, **205**, 131–54.

Brodehl, J., Gellissen, K., Hagge, W., and Schumacher, H. (1968). Reversibles renales Fanconi-Syndrom durch toxisches Abbauprodukt des Tetrazyklins. *Helvetica Paediatrica Acta*, **23**, 373–83.

Brodehl, J., Gellissen, K., and Hagge, W. (1969). The Fanconi syndrome in hepato-renal glycogen storage disease. In *Progress in Nephrology* (eds. G. Peters, and F. Roch-Ramel), pp. 241–3, Springer, Berlin.

Brodehl, J., Gellissen, K., and Weber, H.P. (1982). Postnatal development of tubular phosphate reabsorption. *Clinical Nephrology*, **17**, 163–7.

Brodehl, J., Ehrich, J.H.H., Krohn, H.P., Offner, G., and Byrd, D. (1984). Kidney transplantation in nephropathic cystinosis. In *Paediatric Nephrology* (ed. J. Brodehl, and J.H.H. Ehrich), pp. 172–8, Springer, Berlin.

Brodehl, J., Krause, A., and Hoyer, P.F. (1988). Assessment of maximal tubular phosphate reabsorption: comparison of direct measurement with the nomogram of Bijvoet. *Pediatric Nephrology*, **2**, 183–9.

Burghard, R., Brandis, M., Hoyer, P.F., Ehrich, J.H., Galaske, R.G., and Brodehl J. (1984). Acute interstitial nephritis in childhood. *European Journal of Pediatrics*, **142**, 103–10.

Butler, H.E., Morgan, J.M., and Smythe, C.M. (1965). Mercaptopurine and acquired tubular dysfunction in adult nephrosis. *Archives of Internal Medicine*, **116**, 853–6.

Callis, L., Castello, F., Fortuny, G., Vallo, A., and Ballabriga, A. (1970). Effect of hydrochlorothiazide on rickets and on tubular acidosis in two patients with cystinosis. *Helvetica Paediatrica Acta*, **25**, 602–19.

Cherry, J.D. and Surawicz, B. (1962). Unusual effects of potassium deficiency on the heart of a child with cystinosis. *Pediatrics*, **30**, 414–24.

Chesney, R.W., Rosen, J.F., Hamstra, A.J., and De Luca, H.F. (1980). Serum 1,25 dihydroxyvitamin D levels in normal children and in vitamin D disorders. *American Journal of Diseases of Children*, **134**, 135–9.

Christensen, E.I. and Maunsbach, A.B. (1980). Proteinuria induced by sodium maleate in rats: effects on ultrastructure and protein handling in renal proximal tubule. *Kidney International*, **17**, 771–87.

Chutorian, A. and Rowland, L.P. (1966). Lowe's syndrome. *Neurology*, **16**, 115–22.

Clyne, D.H., Brendstrup, L., and First, M.R. (1974). Renal effects of intraperitoneal kappa chain injection: induction of crystals in renal tubular cells. *Laboratory Investigations*, **31**, 131–42.

Colombo, J.P. (1971). Hinweis auf eine tubuläre und intestinale Malabsorption der Aminosäuren beim Lowe-Syndrom. *Schweizerische Medizinische Wochenschrift*, **101**, 968–72.

Danks, D.M. (1983). Hereditary disorders of copper metabolism in Wilson's disease and Menke's disease. In *The Metabolic Basis of Inherited Disease*, (ed. H.B. Stanbury, J.B. Wyngaarden, D.S. Fredrickson, J.L. Goldstein, and M.S. Brown), pp. 1251–68, McGraw-Hill, New York.

Davis, S., Kessler, W., Haddad, B.M., and Maesaka, J.K. (1980). Acute renal tubular dysfunction following cis-dichlordiammine platinum therapy. *Journal of Medicine; Clinical Experimental and Theoretical*, **11**, 133–41.

Dehner, L.P., Snover, D.C., Sharp, H.L., Ascher, N., Nakhleh, R., and Day, D.L. (1989). Hereditary tyrosinemia type I (chronic form). Pathologic findings in the liver. *Human Pathology*, **20**, 149–58.

Dent, C.E. (1952). Rickets and osteomalacia from renal tubule defects. *Journal of Bone and Joint Surgery, British Volume*, **34**, 266–75.

Dentino, M., Luft, F.G., Yum, M.V., and Einhorn, L.H. (1978). Long term effects of cis-diammine dichloride platinum (CDDP) on renal function and structure in man. *Cancer*, **41**, 1274–81.

DiMauro, S., et al. (1980). Fatal infantile mitochondrial myopathy and renal dysfunction due to cytochrome-c-oxidase deficiency. *Neurology*, **30**, 795–804.

Eisenbach, G.M., Weise, M., and Stolte, H. (1975). Amino acid reabsorption in the rat nephron. Free flow micropuncture study. *Pflügers Archiv: European Journal of Physiology*, **357**, 63–76.

Elsas, L.J., Hayslett, J.P., Spargo, B.H., Durant, J.L., and Rosenberg, L.E. (1971). Wilson's disease with reversible renal tubular dysfunction: correlation with proximal tubular ultrastructure. *Annals of Internal Medicine*, **75**, 427–33.

Emslie-Smith, D., Johnstone, J.H., Thomson, M.B., and Lowe, K.G. (1956). Aminoaciduria in acute tubular necrosis. *Clinical Science*, **15**, 272–6.

Engle, R.L. and Wallis, L.A. (1957). Multiple myeloma and the adult Fanconi syndrome: I. Report of a case with crystal-like deposits in the tumor cells and in the epithelial cells of the kidney. *American Journal of Medicine*, **22**, 5–12.

Fairney, A., Francis, D., Ersser, R.S., Seakins, J.W.T., and Cottom, D. (1968). Diagnosis and treatment of tyrosinosis. *Archives of Disease in Childhood*, **43**, 540–7.

Fanconi, G. (1936). Die nicht diabetischen Gykosurien und Hyperglykämien des älteren Kindes. *Jahrbuch Kinderheilkunde*, **147**, 299–338.

Fanconi, G. and Bickel, H. (1949). Die chronische Aminoacidurie (Aminosäurendiabetes) oder nephrotisch-glukosurischer Zwergwuchs bei der Glykogenose und der Cystinkrankheit. *Helvetica Paediatrica Acta*, **4**, 359–96.

Fanconi, A., Fischer, J.A., and Prader, A. (1974). Serum parathyroid hormone concentration in hypophosphatemic vitamin D resistant rickets. *Helvetica Paediatrica Acta*, **29**, 187–94.

Fellers, F.X., Ko, K.W., and Nicolaidou, M. (1960). A defect of tubular secretory function in the de Toni-Fanconi syndrome. *American Journal of Diseases of Children*, **100**, 588.

Fellers, F.X., Piedrahita, V., and Galan, E.M. (1967). Pseudophlorhizin diabetes. *Pediatric Research*, **1**, 304.

Foulkes, C.F. (1983). Tubular sites of action of heavy metals, and the nature of their inhibition of amino acid reabsorption. *Federation Proceedings; Federation of American Societies for Experimental Biology*, **42**, 2965–68.

Frimpter, G.W., Tinpanelli, A.E., Eisenmenger, W.J., Stein, H.S., and Ehrlich, L.I. (1963). Reversible Fanconi syndrome caused by degraded tetracycline. *Journal of the American Medical Association*, **184**, 111–3.

Fujita, T. and Itakura. (1978). Renal handling of lysozyme in experimental Fanconi syndrome. *Journal of Laboratory and Clinical Medicine*, **92**, 135–40.

Fukui, S., Yoshida, H., Tanaka, T., Sakano, T., Usui, T., and Yamashina, I. (1981). Glycoaminoglycan synthesis by cultured skin fibroblasts from a patient with Lowe's syndrome. *Journal of Biological Chemistry*, **256**, 10313–18.

Gahl, W.A., et al. (1988). Oral carnitine therapy in children with cystinosis and renal Fanconi syndrome. *Journal of Clinical Investigation*, **81**, 549–60.

Galaske, R.G., Burdelski, M., and Brodehl, J. (1986). Primär polyurisches Nierenversagen und akute gelbe Leberdystrophie nach Infusion von Zuckeraustauschstoffen im Kindesalter. *Deutsche Medizinische Wochenschrift*, **111**, 978–83.

Garty, R., Cooper, M., and Tabachnik, E. (1974). The Fanconi syndrome associated with hepatic glycogenosis and abnormal metabolism of galactose. *Journal of Pediatrics*, **85**, 821–3.

Gentz, J., Jagenburg, R., and Zettererström, R. (1965). Tyrosinemia: an inborn error of tyrosine metabolism with cirrhosis of the liver and multiple renal tubular defects (de Toni-Debre-Fanconi-syndrome). *Journal of Pediatrics*, **66**, 670–96.

Gilsanz, V., Barrera, A., and Anaya, A. (1960). The renal biopsy in Wilson's disease. *Archives of Internal Medicine*, **105**, 758–61.

Gitzelmann, R., Steinmann, B., and Van den Berghe, G. (1983). Essential fructosuria, hereditary fructose intolerance, and fructose-1, 6-diphosphatase deficiency. In *The Metabolic Basis of Inherited Disease* (eds. J.B. Stanbury, J.B. Wyngaarden, D.S. Fredrickson, J.L. Goldstein, and M.S. Brown), pp. 118–40, McGraw-Hill, New York.

Goldsmith, L.A. (1983). Tyrosinemia and related disorders. In *The Metabolic Basis of Inherited Disease* (eds. J.B. Stanbury, J.B. Wyngaarden, D.S. Fredrickson, J.L. Goldstein, and M.S. Brown), pp 287–99, McGraw-Hill, New York.

Gonick, H.C. and Kramer, H.J. (1985). Pathogenesis of the Fanconi Syndrome. In *Renal tubular disorders* (eds. H.C. Gonick, and V.M. Buckalew), pp. 545–607, Marcel Dekker, New York.

Goyer, R.A., Tsuchiya, K., Leonard, D.L., and Kahyo. (1972) Aminoaciduria in Japanese workers in the lead and cadmium industries. *American Journal of Clinical Pathology*, **57**, 635–42.

Grenier, A., Lescault, A., Laberge, C., Gagne, R., and Mamer, O. (1982). Detection of succinylacetone and the use of its measurement in mass screening for hereditary tyrosinemia. *Clinica Chimica Acta*, **123**, 933–9.

Günther, R., Silbernagl, S., and Deetjen, P. (1979). Maleic acid induced aminoaciduria, studied by free flow micropuncture and continuous microperfusion. *Pflügers Archiv: European Journal of Physiology*, **383**, 109–14.

Györy, A.Z., Stewart, J.H., George C.R.P., Tiller, D.J., and Edwards, K.D.G. (1969). Renal tubular acidosis, acidosis due to hyperkalaemia, hypercalcaemia, disordered citrate metabolism and other tubular dysfunction following human renal transplantation. *Journal of Medicine, Clinical, Experimental and Theoretical*, **38**, 231–54.

Habib, R., Bargeton, E., Brissaud, H.E., Raynaud, J., and Le Ball, J.-C. (1962). Constatations anatomiques chez un enfant atteint d'un syndrome de Lowe. *Archives Francaises de Pediatrie*, **19**, 945–60.

Halvorsen, S., Pande, H., Loken, A.C., and Gjessing, L.R. (1966). Tyrosinosis: a study of 6 cases. *Archives of Diseases in Childhood*, **41**, 238–49.

Hambraeus, L., Pallisgaard, G., and Kildeberg, P. (1970). The Lowe syndrome: observations on the amino acid metabolism in a 2-year-old affected boy. *Acta Paediatrica Scandinavica*, **59**, 631–6.

Harrison, H.E. and Harrison, H.C. (1954). Experimental production of renal glucosuria, phosphaturia and aminoaciduria by injection of maleic acid. *Science*, **120**, 606–8.

Haycock, G.B., Al-Dahhan, J., Mak, R.H.K., and Chantler, C. (1982). Effect of indomethacin on clinical progress and renal function in cystinosis. *Archives of Disease of Childhood*, **57**, 934–9.

Higgins, R.B. and Varney, J.K. (1966). Dissolution of renal calculi in a case of hereditary fructose intolerance and renal tubular acidosis. *Journal of Urology*, **95**, 291–6.

Holmes, L.B., McGowan, B.L., and Efron, M.L. (1969). Lowe's syndrome: a search for the carrier state. *Pediatrics*, **44**, 358–64.

Hoogenraad, T.U., Van Den Hamer, C.J.A., and Vanhattum, J. (1984). Effective treatment of Wilson's disease with oral zinc sulphate: two case reports. *British Medical Journal*, **289**, 273–6.

Hooft, C. and Vermassen, A. (1960). De Toni-Debre-Fanconi syndrome in nephrotic children. A review. *Annales Paediatrici Basel*, **194**, 193–216.

Hostetter, M.K., Levy, H.L., Winter, H.S., Knight, G.J., and Haddow, J.W. (1983). Evidence for liver disease preceding amino acid abnormalities in hereditary tyrosinemia. *New England Journal of Medicine*, **308**, 1265–7.

Houston, I.B., Boichis, H., and Edelmann, C.M., Jr. (1968). Fanconi syndrome with renal sodium wasting and metabolic alkalosis. *American Journal of Medicine*, **44**, 638–46.

Hoyer, P.F., Brodehl, J., Byrd, D.J., Krohn, H.P., Oemar, B.S., and Offner, G. (1988). Effect of cyclosporine on the renal tubular amino acid handling after kidney transplantation. *Transplantation*. **46**, 73–8.

Hsia, D.Y.Y., Hsia, H.H., Green, S., Kay, M., and Gellis, S.S. (1954). Amino aciduria in galactosemia. *American Journal of Diseases in Childhood*, **88**, 458–65.

Hunt, D.D., Stearns, G., McKinley, J.B., Frowing, E., Hicks, P., and Bonfiglio, M. (1966). Long-term study of family with Fanconi syndrome without cystinosis (de Toni-Debre-Fanconi syndrome). *American Journal of Medicine*, **40**, 492–510.

Iesato, K., Wakashin, M., Wakashin, Y., and Tojo, S. (1977). Renal tubular dysfunction in Minamata disease. Detection of renal tubular antigen and beta-2-microglobulin in the urine. *Annals of Internal Medicine*, **86**, 731–7.

Illig, R. and Prader, A. (1961). Primäre Tubulopathien: II. Ein Fall von idiopathischem Gluko-Amino-Phosphat-Diabetes (De Toni-Debre-Fanconi syndrome). *Helvetica Paediatrica Acta*, **16**, 622–46.

Jonas, A.J., Lin, S.-N., Conley, S.B., Schneider, J.A., Williams, J.C., and Caprioli, R.C. (1989). Urine glyceraldehyde excretion is elevated in the renal Fanconi syndrome. *Kidney International*, **35**, 99–104.

Kamm, D.E. and Fischer, M.S. (1972). Proximal renal tubular acidosis and the Fanconi syndrome in a patient with hypergammaglobulinemia. *Nephron*, **9**, 208–19.

Käser, H., Cottier, P., and Antener, I. (1962). Gluco-glycinuria, a new familial syndrome. *Journal of Pediatrics*, **61**, 386–94.

Katzir, Z., Shvil, Y., Landau, H., Kidrony, G., and Popovtzer, M.M. (1988). Nephrogenic diabetes insipidus, cystinosis and vitamin D. *Archives of Disease in Childhood*, **63**, 548–50.

Kaufman, F.R., Kogut, M.D., Donnell, G.N., Goebelsmann, U., March, C., and Koch, R. (1981). Hypogonadotropic hypogonadism in female patients with galactosemia. *New England Journal of Medicine*, **304**, 994–8.

Komrower, G.M., Schwarz, V., Holzel, A., and Goldberg, L. (1956). A clinical and biochemical study of galactosemia: a possible explanation of the nature of the biochemical lesion. *Archives of Disease in Childhood*, **31**, 254–64.

Kranhold, J.F., Loh, D., and Morris, R.C. (1969). Renal fructose-metabolising enzymes: significance in hereditary fructose intolerance. *Science*, **165**, 402–3.

Kruse, K. and Barthels, H. (1979). Hypercalciuria in idopathic Fanconi syndrome. *European Journal of Pediatrics*, **131**, 247–54.

Kvittingen, E.A., Halvorsen, S., and Jellum, E. (1983). Deficient fumarylacetoacetate fumarylhydrolase activity in lymphocytes and fibroblasts from patients with hereditary tyrosinemia. *Pediatric Research*, **14**, 541–4.

Kvittingen, E.A., et al. (1986a). Prenatal diagnosis of hereditary tyrosinaemia type I by determination of fumarylacetoacetase in chorionic villus material. *European Journal of Pediatrics*, **144**, 597–8.

Kvittingen, E.A., et al. (1986b). Liver transplantation in a 23-year old tyrosinaemia patients: effect on the renal tubular dysfunction. *Journal of Inherited Metabolic Diseases*. **9**, 216–24.

Lameire, N., Mussche, M., Baele, G., Klint, J., and Ringoir, S., (1978). Hereditary fructose intolerance: a difficult diagnosis in the adult. *American Journal of Medicine*, **65**, 416–23.

Lampert, F. and Mayer, H. (1967). Glykogenose der Leber mit Galaktoseverwertungsstörung und schwerem Fanconi-Syndrom. *Zeitschrift für Kinderheilkunde*, **98**, 133–45.

Lee, D.B.N., Drinkhard, J.P., Rosen, V.J., and Gonick, H.C. (1972). The adult Fanconi syndrome. *Medicine*, **51**, 107–38.

Leehey, D.J., Ing, T.S., and Davgirdas, J.T. (1985). Fanconi syndrome associated with a non-ossifying fibroma of bone. *American Journal of Medicine*, **78**, 708–10.

Leu, M.L., Strickland, G.T., and Gutman, R.A. (1970). Renal function in Wilson's disease: response to penicillamine therapy. *American Journal of the Medical Sciences*, **260** 381–98.

Levin, B., Snodgrass, G.J.A.I., Oberholzer, V.G., Burgess, E.A., and Dobbs, R.H. (1968). Fructosemia: observation of seven cases. *American Journal of Medicine*, **45**, 826–38.

Levy, M., Gagnadoux, M.-F., Beziau, A., and Habib, R. (1978). Membranous glomerulonephritis associated with antitubular and antialveolar basement membrane antibodies. *Clinical Nephrology*, **10**, 158–65.

Lindblad, B. and Steen, G. (1982). Identification of 4, 6-dioxohepatonic acid (succinylacetone), 3,5-dioxooctanedioic acid (succinylacetoacetate) and 4-oxo-6-hydroxy-hepatonic acid in the urine from patients with hereditary tyrosinemia. *Biomedical Mass Spectrometry*, **9**, 419–24.

Lindquist, R.R. and Fellers, F.X. (1966). Degraded tetracycline nephropathy: functions, morphologic and histochemical observations. *Laboratory Investigations*, **15**, 864–76.

Lowe, C.U., Terrey, M., and McLachlan, E.A. (1952). Organic-aciduria, decreased renal ammonia production, hydrophthalmus, and mental retardation. *American Journal of Diseases of Childhood*, **83**, 164–84.

Maesaka, J.K. and McCaffery, M. (1980). Evidence for renal tubular leakage in maleic acid-induced Fanconi syndrome. *American Journal of Physiology*, **239**, 507–13.

Maldonado, J.E., Velosa, J.A., Kyle, R.A., Wagoner, R.D., Holley, K.E., and Salassa, R.M. (1975). Fanconi syndrome in adults. A manifestation of a latent form of myeloma. *American Journal of Medicine*, **58**, 354–64.

Manz, F., Bremer, H.J., and Brodehl, J. (1978). Renal transport of amino acids in children with oculocerebrorenal syndrome. *Helvetica Paediatrica Acta*, **33**, 37–44.

Manz, F., et al. (1984). Idiopathic de Toni-Debre-Fanconi syndrome with absence of proximal tubular brush border. *Clinical Nephrology*, **24**, 149–57.

Manz, F., et al. (1987). Fanconi-Bickel syndrome. *Pediatric Nephrology*, **1**, 509–18.

Mass, R.E., Smith, W.R., and Walsh, J.R. (1966). The association of hereditary fructose intolerance and renal tubular acidosis. *American Journal of the Medical Sciences*, **252**, 516–23.

Mavromatis, F. (1965). Tetracycline nephropathy. *Journal of the American Medical Association*, **193**, 191–4.

Moncrieff, M. and Foot, A. (1989). Fanconi syndrome after ifosfamide. *Cancer Chemotherapy Reports*, **23**, 121–2.

Montoliu, J., Carrera, M., Darnell, A., and Revert, L. (1981). Lactic acidosis and Fanconi's syndrome due to degraded tetracycline. *British Medical Journal*, **283**, 1576–7.

Moorhead, J.F., et al. (1974). Hypophosphatemic osteomalacia after cadaveric renal transplantation. *Lancet*, **i**, 694–7.

Mori, Y., et al. (1984). Renal tubular disturbance induced by tributyl-tin oxide in guinea pigs: a secondary Fanconi syndrome. *Clinical Nephrology*, **21**, 118–25.

Morris, R.C. (1968). An experimental renal acidification defect in patients with hereditary fructose intolerance: II. Its distinctions from classic renal tubular acidosis; its resemblance to the renal acidification defect associated with the Fanconi syndrome of children with cystinosis. *Journal of Clinical Investigation*, **47**, 1648–63.

Morris, R.C. and Sebastian A. (1983). Renal acidosis and the Fanconi syndrome. In *The Metabolic Basis of Inherited Disease* (eds. J.B. Stanbury, J.B. Wyngaarden, D.S. Fredrickson, J.L. Goldstein, and M.S. Browon), pp. 1808–43, McGraw-Hill, New York.

Moss, A.N., Gabow, P.A., Kaehny, W.D., Goodman, S.I., and Haut, L.L. (1980). Fanconi syndrome and distal tubular acidosis after glue sniffing. *Annals of Internal Medicine* **92**, 69–70.

Müller-Wiefel, D.E., Steinmann, B., Holm-Hadulla, M., Wille, L., Schärer, K., and Gitzelmann R. (1983). Infusionsbedingtes Nieren-u. Leberversagen bei undiagnostizierter hereditärer Fructose-Intoleranz. *Deutsche Medizinische Wochenschrift*, **108**, 985–9.

Nagant de Deuxchaisnes, C. and Krane, M. (1967). The treatment of adult phosphate diabetes and Fanconi syndrome with sodium phosphate. *American Journal of Medicine*, **43**, 508–43.

Neimann, N., Pierson, M., Marchal, C., Rauber, G., and Grignon, G. (1968). Nephropathie familial glomerulotubulaire avec syndrome de Toni-Debre-Fanconi. *Archives Francaises de Pediatrie*, **25**, 43–69.

Newcombe, D.S. (1975). Fanconi syndrome. In *Inherited Biochemical Disorders and Uric Metabolism* (ed. D.S. Newcombe), pp. 209–42, HM + Publishers, Aylesbury.

Odievre, M. (1966). Glycogenose hepato-renale avec tubulopathie complex. *Revue Internationale d'Hepatologie*, **26**, 1–70.

Oemar, B.S., Byrd, D.J., and Brodehl, J. (1987). Complete absence of tubular glucose reabsorption: a new type of renal glucosuria (type O). *Clinical Nephrology*, **27**, 156–60.

Oetliker, O. and Rossi, E. (1969). The influence of extracellular fluid volume on the renal bicarbonate threshold: a study of two children with Lowe's syndrome. *Pediatric Research*, **3**, 140–8.

Ores, R.O. (1970). Renal changes in oculo-cerebro-renal syndrome of Lowe. *Archives of Pathology*, **89**, 221–5.

Otten, J. and Vis, H.L. (1968). Acute reversible renal tubular dysfunction following intoxications with methyl-3-chrome. *Journal of Pediatrics*, **73**, 422–5.

Patrick, A., Cameron, J.S., and Ogg, C.S. (1981). A family with a dominant form of idiopathic Fanconi syndrome leading to renal failure in adult life. *Clinical Nephrology*, **16**, 289–92.

Phelip, X., Bocquet, B., Gras, J.R., Bouvier, M., Cabanel, G., and Le Jeune, E. (1973). Fibrose palmo-plantaire extensive au cours d'un syndrome de Lowe. *Revue du Rheumatisme*, **40**, 597–601.

Preuss, H.G., Hammack, W.J., and Murdaugh, H.V. (1967). The effect of Bence Jones protein on the *in vitro* function of rabbit renal cortex. *Nephron*, **5**, 210–6.

Preuss, H.G., Weiss, F.R., Iammarino, R.M., Hammack, W.J., and Murdaugh, H.V. (1974). Effects on rat kidney slice function *in vitro* of proteins from the urines of patients with myelomatosis and nephrosis. *Clinical Science and Molecular Medicine*, **46**, 283–94.

Reem, H.G., Isaacs, M., and Vanamee, P. (1967). Renal transport of urate, phoshate and glucose in the Fanconi syndrome. *Journal of Clinical Endocrinology and Metabolism*, **127**, 1141–1150.

Reubi, F.C. (1963). *Clearance Tests in Clinical Medicine*, Thomas, Springfield, Illinois.

Reynolds, R., McNamara, P.D., and Segal, S. (1978). On the maleic acid induced Fanconi syndrome: effects on transport by isolated rat kidney brushborder membrane vesicles. *Life Sciences*, **22**, 39–44.

Reynolds, E.S., Tannen, R.L., and Tyler, H.R. (1966). The renal lesions in Wilson's disease. *American Journal of Medicine*, **40**, 518–27.

Rochman, J., Lichtig, C., Osterweill, D., Tatarsky, I., and Eidelman, S. (1980). Adult Fanconi syndrome with renal tubular acidosis in association with renal amyloidosis. *Archives of Internal Medicine*, **140**, 1361–3.

Rosenberg, L.E. and Segal, S. (1964). Maleic acid-induced inhibition of amino acid transport in rat kidney. *Biochemical Journal*, **92**, 345–52.

Rotthauwe, H.W., et al. (1963). Glykogenose der Leber mit Aminoacidurie und Glukosurie. *Klinische Wochenscrhrift*, **41**, 818–25.

Royer, P., Delaitre, R., Mathieu, H., Gerbeaux, S., Habib, R., and Koegel, R. (1963). Le syndrome nephrotique avec insuffisance tubulaire globale et tetanie recidivante. *Annales de Pediatrie*, **10**, 583–596.

Russo, J.C. and Adelman, R.D. (1980). Gentamycin induced Fanconi syndrome. *Journal of Pediatrics*, **96**, 151–3.

Rützler, L. (1973). Passageres tubuläres Syndrom durch Vergiftung mit einer organischen Quecksilberverbindung (Glyceromerfen) bei einem 1 1/2 jährigen Mädchen. *Schweizerische Medizinische Wochenschrift*, **103**, 678–81.

Sagel, I., Ores, R.O., and Yuceoglu, A.M. (1970). Renal function and morphology in a girl with oculo-cerebro-renal syndrome. *Journal of Pediatrics*, **77**, 124–7.

Scheele, C. von (1976). Light chain myeloma with features of the adult Fanconi syndrome: six years remission following one course of Melphalan. *Acta Medica Scandinavica*, **199**, 533–7.

Schønheider, F., Gregerson, G., Hansen, H.E., and Skov, P.E. (1971). Renal clearances of different amino acids in Wilson's disease before and after treatment with penicillamine. *Acta Medica Scandinavica*, **190**, 395–9.

Schwartz, J.H. and Schein, P. (1978). Fanconi syndrome associated with cephalothin and gentamycin therapy. *Cancer*, **41**, 769–72.

Scriver, C.R., Goldbloom, R.B., and Roy, C. (1964). Hypophosphatemic rickets with renal hyperglycinuria, renal glucosuria and glycylprolinuria. *Journal of Pediatrics*, **34**, 357–71.

Sebastian, A., McSherry, E., and Morris, R.C., Jr. (1971). Renal potassium wasting in renal tubular acidosis (RTA). *Journal of Clinical Investigation*, **50**, 231–43.

Segal, S. (1983). Disorders of galactose metabolism. In *The Metabolic Basis of Inherited Disease*, (eds. J.B. Stanbury, J.B. Wyngaarden, D.S. Fredrickson, J.L. Goldstein, and M.S. Brown), pp. 167–91, McGraw-Hill, New York.

Sengers, R.C.A., Stadhouders, A.M., and Trijbels, J.M.F. (1984). Mitochondrial myopathies. Clinical, morphological and biochemical aspects. *European Journal of Pediatrics*, **141**, 192–207.

Sewell, R.L. and Dorreen, M.S. (1984). Adult Fanconi syndrome progressing to multiple myeloma. *Journal of Clinical Pathology*, **37**, 1256–8.

Shwayder, M., Ozawa, T., Boedecker, E., Guggenheimn, S., and McIntosh, R.M. (1976). Nephrotic syndrome associated with Fanconi syndrome. *Annals of Internal Medicine*, **84**, 433–7.

Sirota, J.H. and Hamerman, D. (1954). Renal function studies in an adult subject with Fanconi syndrome. *American Journal of Medicine*, **16**, 138–52.

Smeitink, J., Verreussel, M., Schröder, C., and Lippens, R. (1988). Nephrotoxicity associated with ifosfamide. *European Journal of Pediatrics*, **148**, 164–6.

Spencer, A.G. and Franglen, G.T. (1952). Gross aminoaciduria following a lysol burn. *Lancet*, **i**, 190–2.

Spencer, P.D., Meddow, M.S., Moses, L.C., and Roth, K.S. (1988). Effects of succinylacetone on the uptake of sugars and aminoacids by brush border vesicles. *Kidney International*, **34**, 671–7.

Stark, H., Eisenstein, B., Tieder, M., Rachmel, A., and Alpert, G. (1986). Direct measurement of T_P/GFR: a simple and reliable parameter of renal phosphate handling. *Nephron*, **44**, 125–8.

Steinmann, B., Bachmann, C., Colombo, J.-P., and Gitzelmann, R. (1987). The renal handling of carnitine in patients with selective tubulopathy and with Fanconi syndrome. *Pediatric Research*, **21**, 201–4.

Steinmann, B. and Gitzelmann, R. (1981). The diagnosis of hereditary fructose intolerance. *Helvetica Paediatrica Acta*, **36**, 297–316.

Sternlieb, I., (1984). Wilson's disease: indications for liver transplants. *Hepatology*, **4**, 159–79.

Sternlieb, I. and Scheinberg, I.H. (1979). Wilson's disease. In *Liver and Biliary Disease: Pathophysiology, Diagnosis, Management*. (eds. R. Wright, K.G.M. Alberti, S. Karran, and G.H. Millward-Sadler), pp. 774–804, W.B. Saunders, London.

Texier, J.L., Jully, G., and Bach, C. (1971). Syndrome de Lowe. *Annales de Pediatrie*. **18**, 825–30.

Thorner, P.S., Bedard, Y.C., and Fernandes, B.J. (1983). λ-Lightchain nephropathy with Fanconi syndrom. *Archives of Pathology*, **107**, 654–7.

Van Acker, K.J., Roels, H., Beelaerts, W., Pasternack, A., and Valke, R. (1967). The histologic lesions of the kidney in the oculo-cerebro-renal syndrome of Lowe. *Nephron*, **4**, 193–214.

Van Biervliet, J.P.G.M., et al. (1977). Hereditary mitochondrial myopathy with lactic acidemia, a De Toni-Fanconi-Debre syndrome, and a defective respiratory chain in voluntary striated muscles. *Pediatric Research*, **11**, 1088–93.

Vertuno, L.L., Preuss, H.G., Argy, W.P., and Schreiner, G.E. (1974). Fanconi syndrome following homotransplantation. *Archives of Internal Medicine*, **133**, 302–5.

Vis, H.L., Vainsel, M., Willaert, H., and Meur, G. (1972). Lowe's syndrome. *Pathologia Europaea (Bruxelles)*, **7**, 307–19.

Walb, D., Wohlenberg, H., Rumpelt, H.J., Schmidt, H., and Thomas, L. (1980). Fanconi Syndrom des Erwachsenen bei Frühmyelom mit monoklonaler Gammapathie IgG, Typ Kappa. *Deutsche Medizinische Wochenschrift*, **105**, 1355–9.

Wald, H. and Popovtzer, M.M. (1984). Renal refractoriness to PTH in experimental Fanconi syndrome: evidence for intact adenylate cyclase activation. *Pflügers Archiv: European Journal of Physiology*, **403**, 116–9.

Walker, B.R., Alexander, F., and Tannenbaum, P.J. (1971). Fanconi syndrome with renal tubular acidosis and light chain proteinuria. *Nephron*, **8**, 103–7.

Wallis, L.A. and Engle, R.L., Jr. (1957). The adult Fanconi syndrome: II. Review of eighteen cases. *American Journal of Medicine*, **22**, 12–23.

Walshe, J.M. (1975). Wilson's Disease (hepatolenticular degeneration). In *The Treatment of Inherited Metabolic Disease* (ed. D.N.Raine), pp. 171–90, Medical and Technical Publishing Co., Lancaster.

Walshe, J.M. (1982). Treatment of Wilson's disease with trientine (triethylene tetramine) dihydrochloride. *Lancet*, **i**, 643–7.

Walton, R.J. and Bijvoet, O.L.M. (1975). Nomogram for derivation of renal threshold phosphate concentration. *Lancet*, **ii**, 309–10.

Wesch, H., Przuntek, H., and Feist, D. (1980). Morbus Wilson. Rasche Diagnose und Differenzierung heterozygoter und homozygoter Anlageträger mit 64 $CuCl_2$. *Deutsche Medizinische Wochenschrift*, **105**, 483–8.

Wilson, D.R. and Goldstein, N.P. (1973). Renal urate excretion in patients with Wilson's disease. *Kidney International*, **4**, 331–6.

Wilson, D.R. and Yendt, E.R. (1963). Treatment of adult Fanconi syndrome with oral phosphate supplement and alkali. *American Journal of Medicine*, **35**, 487–511.

Wisniewski, K.E., Kieras, F.J., French, J.H., Houck, G.E., and Ramos, P.L. (1984). Ultrastructural, neurological, and glycosaminoglycan abnormalities in Lowe's syndrome. *Annals of Neurology*, **16**, 40–9.

Witzleben, C.L., Schoen, E.J., Tu, W.H., and McDonald, L.W. (1968). Progressive morphologic renal changes in the oculo-cerebro-renal syndrome of Lowe. *American Journal of Medicine*, **44**, 319–24.

Wolff, S. (1964). Renal lesions in Wilson's disease. *Lancet*, **i**, 843–845.

5.2 Isolated defects of tubular function

GEORGE B. HAYCOCK

In addition to reclaiming two-thirds of the volume of filtered salt and water, the proximal tubule is responsible for the complete, or nearly complete, reabsorption of the filtered load of numerous important extracellular fluid solutes, including bicarbonate, uric acid, glucose, phosphate, amino acids and certain organic anions, by means of a series of specific, mostly sodium-coupled, epithelial transport systems. Proximal tubular handling of sodium, chloride, potassium and bicarbonate are discussed elsewhere (Chapter 5.3), as is the renal Fanconi syndrome (Chapter 5.1). This chapter will describe general and specific aspects of defects of proximal tubular handling of glucose, amino acids, phosphate, and uric acid.

Transport mechanisms in the proximal tubule—general aspects

Although transepithelial solute transport is usually spoken of as a single process, the reabsorption of any substance from the fluid perfusing the proximal tubule involves at least two separate steps: movement from lumen into the tubular epithelial cell (the entry step), and from the cell interior into the peritubular interstitial fluid (the exit step). A complete description should include a third step: reabsorption of transported fluid from the peritubular interstitium into the plasma perfusing the peritubular capillary plexus. The latter is governed by Starling forces and is undoubtedly extremely important in the overall regulation of volume reabsorption. In so far as anything that affects the rate of sodium reabsorption will have a secondary effect on the transport of other substances, the reabsorption of those solutes that are the subject of this chapter must be influenced by these haemodynamic factors, but in a non-specific way and this aspect of the process will not be further considered here.

The entry step

Glucose, amino acids, phosphate, and organic anions, including citrate and lactate, enter the proximal tubular cell across the apical (luminal, brush border) membrane by a series of cotransport systems, each of which binds one or more sodium ions and its specific substrate and carries them across the cell membrane. The presence of sodium stimulates transport of the cotransportate and *vice versa* (Kinne *et al.* 1975; Hoffmann *et al.* 1976; Wright *et al.* 1980; Silbernagl 1988; Cheng and Sacktor 1981; Frömter 1982). The energy for sodium transport is provided by the lumen-to-cell interior concentration gradient, maintained in turn by removal of sodium from the cell across the basolateral cell membrane (Fig. 1). The energy for the transport of charged systems is provided by the membrane voltage. Thus the cotransportate (glucose, phosphate etc.) is transported against its own concentration gradient by a process of secondary active transport which may be electrogenic or electroneutral depending on the stoichiometry of the transport system and the charge on the cotransportate: in other words, there may or may not be net transfer of a (positive) charge from lumen to cell interior.

The exit step

Sodium ions entering the cell with the cotransportate are pumped out by primary, energy consuming, active transport across the basolateral membrane by Na^+, K^+-ATPase which provides energy for the entry of cotransportate against a concentration gradient on the luminal side by secondary active transport. The various cotransportates leave the cell across the basolateral membrane by transport systems that are generally different from those involved in the entry step. These are substrate-specific, mostly sodium-independent and are active processes in some cases (Na^+, Ca^{2+}) and facilitated diffusion pathways in others. Further details are given below in relation to the individual transported substances.

Although the sodium exit step provides the energy for transepithelial transport of solute across the proximal tubule, in the case of phosphate at least the entry step (luminal sodium-phosphate cotransport) is thought to be rate limiting, and is the major site of action of physiological factors which modulate phosphate reabsorption, such as parathyroid hormone (PTH) and 1,25-dihydroxyvitamin D (Stoll *et al.* 1979; Hruska and Hammerman 1981; Lang *et al.* 1986; Murer 1988). Although direct evidence is lacking, it seems likely that the same is true of the other substances under consideration, in view of the specificity of the systems involved and the lack of evidence of a general-

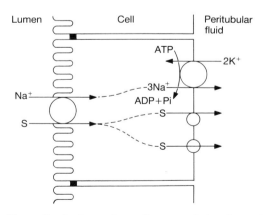

Fig. 1 Generalized scheme for sodium-coupled reabsorption of a solute, S, in the proximal tubule. The energy-consuming step is the extrusion of sodium across the basolateral cell membrane by sodium, potassium-ATPase (upper right in diagram). This maintains a gradient favouring sodium entry across the luminal membrane via a sodium/S cotransporter (left): S is thus carried against its concentration gradient by secondary active transport.

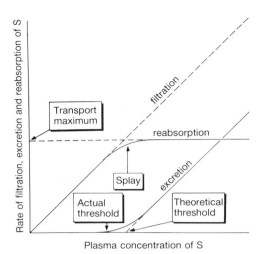

Fig. 2 'Family' of titration curves describing the reabsorption of a solute, S, by saturable active transport in the proximal renal tubule. The curve for excretion of S is plotted from direct measurements, that for filtration of S is calculated by multiplying plasma S by glomerular filtration rate, and that for reabsorption is obtained by subtracting the former from the latter. The terms transport maximum, splay, and threshold are explained in the text.

ized abnormality of proximal tubular sodium transport in the specific defects considered below. Conversely, a defect of Na^+, K^+-ATPase-mediated sodium extrusion from the cell would be predicted to have major effects on all the sodium-coupled entry systems, and this appears to be the case in at least one experimental model of the generalized Fanconi syndrome (Hong Que et al. 1982).

Renal clearance studies

The intact kidney is capable of reabsorbing virtually 100 per cent of filtered glucose, phosphate and amino acids when these are present in low concentration. If the plasma concentration of such a solute (S) is manipulated by infusing it under controlled conditions it is possible to plot the excretion of S (E_S) against its plasma concentration. If inulin or an equivalent marker of glomerular filtration rate (GFR) is infused at the same time, the rate of filtration of S (F_S) can be calculated; by subtracting E_S from F_S the rate of tubular reabsorption (R_S) may be derived. The resulting family of titration curves is shown in Fig. 2. The form of the curve for R_S indicates that above a certain value no more solute can be reabsorbed, despite a further increase in the filtered load. This limiting value for reabsorption is known as the tubular maximal reabsorptive capacity for the solute (T_mS). If the point of inflexion of the curve of R_S (the point at which the ascending part meets the horizontal part) were a sharp angle, T_mS would be reached at precisely the plasma concentration of the solute above which it begins to appear in the urine—the renal threshold for the given solute. In actual titration studies this angle is bridged by a curved splay, the degree of which varies with the solute, being large for glucose and small for some amino acids such as lysine. Where splay is significant, two threshold values can be quoted: an actual threshold, which will be somewhat lower than the plasma concentration at which T_mS is reached, and a theoretical threshold, which is the projection on the abscissa of the theoretical point of inflexion of the R_S curve if there were no splay. The theoretical threshold is obtained by dividing T_mS by GFR, and is commonly abbreviated to T_mS/GFR. It is convenient to factor individual values of E_S, F_S, and R_S by GFR before plotting the curves such that, for example, a given point would be expressed in mg (or mmol) of solute per 100 ml GFR; in this case, T_mS/GFR can be read off directly as the projection on the ordinate of the horizontal portion of the R_S curve. The alternative method, of plotting filtered load (GFR × plasma S) on the abscissa against E_S, F_S and R_S on the ordinate is mathematically equivalent: glomerulotubular balance dictates that T_mS will change proportionately with changes in GFR, and either method of correcting for GFR will compensate for this effect. It is important to note that T_mS and T_mS/GFR, when experimentally derived, are in part determined by the conditions under which the measurements are made: in particular, changes in extracellular fluid volume alter the threshold value for all proximally reabsorbed solutes. If titration experiments are performed by a method which causes progressive volume expansion as a result of the infusion of the solute the threshold will progressively rise and it will prove impossible to demonstrate a T_mS.

Significance of T_m

If the reabsorption curve for a given solute exhibits a T_m, (i.e. eventually becomes parallel to the abscissa) the transporting mechanism is saturable. In molecular terms, this means that all receptor sites for the solute on the luminal side of the carrier protein are occupied and the rate of transport is therefore limited by the next step in the process: the rate at which receptor sites can be vacated on the cytoplasmic side of the membrane and presented again to the luminal fluid for reuse. A T_m can be demonstrated for glucose, phosphate, and most amino acids, but it has not been possible to demonstrate a T_m for other substances, histidine for example, probably because it is so high that it is not experimentally attainable at non-toxic plasma concentrations (Wright et al. 1947). The kinetics of saturable proximal tubular reabsorption are described by the Michaelis–Menten equation:

$$T = \frac{Tm \times S}{Km + S}$$

where T = rate of transport, T_m is the transport maximum, S is the substrate concentration in glomerular filtrate and K_m is the Michaelis–Menten constant for the transport system.

Significance of threshold

By definition, below the renal threshold value for a particular solute S, the urine contains no S or only trace amounts. This can only occur if the solute is reabsorbed by active transport (either primary or secondary) since absorption of the last vestiges of the solute from the tubular fluid into the cell must take place against a concentration gradient unless its intracellular concentration is zero, which is not the case for any of the solutes under consideration in this chapter.

Significance of splay

The existence of splay can be explained in at least two ways, which are not mutually exclusive. The first explanation rests on the concept of nephron heterogeneity: if individual nephrons had slightly different thresholds for S, the urine would not be free of S unless its plasma concentration was below the lowest single nephron threshold value. Similarly, the whole kidney T_m would not be reached until the plasma concentration exceeded the highest single nephron threshold value. In this model the

5.2 Isolated defects of tubular function

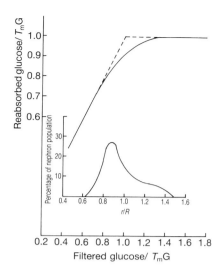

Fig. 3 Theoretical explanation of splay in the normal glucose reabsorption curve according to the nephron heterogeneity hypothesis. The term r/R is the ratio of GFR to glucose reabsorptive capacity for each individual nephron (r) factored by the ratio of whole kidney GFR to whole kidney glucose reabsorptive capacity (R). The extended right hand tail of the frequency distribution curve is due to the presence of a minority of nephrons with functional glomerular preponderance (glomerulotubular imbalance) (modified from Smith et al. 1943).

curve of the splay joins these two extreme threshold values, and its exact shape is a function of the distribution of single nephron threshold values in the total nephron population (Fig. 3). The second explanation concerns the affinity of the carrier protein receptor for the solute: the combination of a solute with its receptor B is reversible and obeys the law of mass action:

$$K = \frac{[S][B]}{[SB]}$$

where K is the dissociation constant and the terms in brackets the activities of S, B and the SB complex respectively. The higher the value of K, the lower the affinity of B for S and the broader the splay; in the limiting case where $K=0$ there would be no splay. The significance of splay is discussed further in the section on glycosuria.

Glycosurias

Normal glucose reabsorption

Glucose titration curves were first plotted in dogs by Shannon and Fisher (1938) and in humans by Smith et al. (1943). A typical study in a normal human subject is depicted in Fig. 4: a conspicuous feature of the curve is the considerable splay, such that actual or minimum threshold for glucose is 8 to 11 mmol/l (145–200 mg/dl). The T_m for glucose (T_mG) corrected for surface area (mmol or mg/min.1.73 m^2) is similar in children and adults but is significantly lower in infants: for example, Smith et al. (1943) obtained a value (mean ± SD) of 1.96 ± 0.45 mmol/min.1.73 m^2 in adults, while Brodehl et al. (1972) found corresponding values of 2.01 ± 0.53 in children but only 1.18 ± 0.39 in infants. Although this has been interpreted by some as evidence of tubular immaturity (Tudvad 1949) the discrepancy is entirely explicable as a consequence of the low GFR which is characteristic of the first few months of life. Thus values for T_mG/GFR (mmol/l) are 15.1 ± 2.5 in adults (Smith et al. 1943), 15.7 ±2.6 in children and 16.3 ± 4.1 in infants (Brodehl et al. 1972): there is no evidence of glomerulotubular imbalance for glucose even in premature neonates (Arant et al. 1974).

Causes of glycosuria

Small amounts of glucose are present in the urine of normal individuals: the definition of 'abnormal' glycosuria is therefore somewhat arbitrary. According to the data of Elsas and Rosenberg (1969), glucose excretion rates above about 2.75 mmol (500 mg)/day.1.73 m^2 are unequivocally abnormal, and this seems a reasonable criterion by which to define significant glycosuria. Glycosuria may be due to hyperglycaemia in the presence of normal renal glucose handling (overload or hyperglycaemic glycosuria), or to abnormalities of tubular glucose transport at normal blood glucose concentrations (renal glycosuria). A list of causes of hyperglycaemic and renal glycosurias is given in Table 1. Hyperglycaemic glycosuria will not be further discussed here; the Fanconi syndrome is described in Chapter 5.1. The remainder of this section will deal with renal glycosurias.

Primary (isolated) renal glycosurias
Definition and incidence

Primary renal glycosuria is defined as abnormal glucose excretion at normal plasma glucose concentrations, in the absence of other abnormalities of renal function. Its incidence varies markedly depending on the criteria used to define it. Transient glycosuria occurs during the course of a glucose tolerance test in up to 65 per cent of subjects investigated for glycosuria (Lawrence 1947); this seems an excessively lenient standard by which to define a 'disease' (high sensitivity, low specificity). Conversely, Marble (1932) required glycosuria to be present in all urine samples tested, including those obtained after an overnight fast: such a stringent test will be highly specific but relatively insensitive: many subjects with T_mG/GFR or a renal glucose threshold outside the normal range conventionally defined as mean −2 SD will be missed. The application of these criteria resulted in isolated glycosuria being attributed to renal causes in less than 1 in 500 of a large series of cases (Marble 1939).

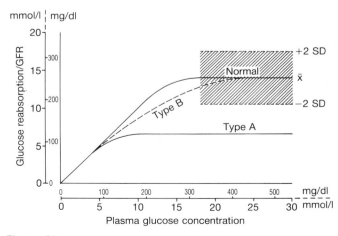

Fig. 4 Glucose reabsorption curves for normal subjects and patients with the two types of renal glycosuria (see text for details).

Table 1 Causes of glycosuria

Hyperglycaemic (overload) glycosuria
  Diabetes mellitus
    Type 1
    Type 2
  Iatrogenic
    Drugs
      Glucocorticoids
      Catecholamines
      Glucagon
    Infusion fluids
      Dextrose solutions
      Total parenteral nutrition
Renal glycosuria
  Type A (reduced T_mG/GFR)
  Type B (increased splay)
  Glycosuria of pregnancy
  Glucose-galactose malabsorption
  Fanconi syndrome
  Drug-induced
Unclassified
  Raised intracranial pressure
  Catabolic stress (burns, major trauma)
  Endocrine causes
  Infection
  Malignancy

Pathophysiology

Clearance studies suggest that there are two types of renal glycosuria which show different abnormalities of the glucose titration curve (Reubi 1954). Type A renal glycosuria is characterized by an abnormally low T_mG/GFR, although the general conformation of the curve is normal; in type B, T_mG and T_mG/GFR are normal but the splay is exaggerated, leading to a lowering of the renal glucose threshold (Fig. 3). The physiological interpretation of type A glycosuria is relatively straightforward: there is presumably a defect in the carrier mechanism, common to all nephrons, such that single nephron T_mG is uniformly subnormal. The position with respect to type B glycosuria is less satisfactory; just as the existence of the normal splay can be explained in one of two ways (see above), so can its exaggeration in this condition. The hypothesis that it is due to nephron heterogeneity with variable degrees of internephron glomerulotubular imbalance failed to find morphological support in the study of Monasterio *et al.* (1964), in which the glomerular and tubular dimensions of 50 microdissected nephrons from two subjects with renal glycosuria were no more variable than those from normal individuals. The kinetic analysis of Woolf *et al.* (1966) is more satisfactory. According to conventional Michaelis–Menten kinetics (see above) both types of renal glycosuria can be explained by abnormalities of the carrier protein: if the number of carrier sites were reduced, but the affinity (K_m) of the carrier molecules for glucose was normal, a type A glucose reabsorption curve would be predicted. Conversely, if K_m were reduced but the number of carrier sites was normal, splay would be increased with a lowered actual glucose threshold but normal T_mG/GFR, although T_mG would only be achieved at a higher plasma glucose (and therefore filtered load) than normal: this is precisely the situation in type B renal glycosuria. An additional confounding factor is introduced by the work of Turner and Moran (1982) who found evidence that glucose carriers in brush border membrane vesicles had different stoichiometries and affinities in different segments (S1 and S3) of the proximal tubule: it is not clear how this finding relates to the results of clearance studies in whole subjects and animals.

The matter is complicated further by the finding of cases of both type A and type B glycosuria in the same family (Elsas *et al.* 1971). Furthermore, reported values for T_mG in patients with renal glycosuria appear to form a single distribution with no clear separation into two subgroups, as would be expected if types A and B really represented the consequences of two distinct abnormalities of glucose transport (Krane 1978). An alternative explanation might be that the glucose titration curve observed in any individual, whether 'normal' or exhibiting renal glycosuria, results from the interaction of several genetically determined characteristics, one (or more) determining T_mG and one (or more) determining splay. Thus subjects classified as having type A renal glycosuria (low T_mG/GFR) may independently have normal or increased splay. Whether isolated renal glycosuria truly represents variable aspects of a single inherited abnormality, or the manifestation of two or more independently inherited characteristics is at present unclear, and clarification will probably only come with a more detailed understanding of the molecular basis of tubular glucose transport and the identification of markers for the responsible genes.

Clinical features and diagnosis

Primary (isolated) renal glycosuria is a benign condition whose major clinical significance is its distinction from other, more serious causes of glycosuria. It is usually detected accidentally, when a urine specimen is tested routinely (as for insurance purposes) or during investigation of an unrelated disorder. The amount of glycosuria varies from a minimum (by definition) of 500 mg/day to 100 g/day or more, with most subjects excreting 1 to 30 g/day. Even at the higher recorded levels of glucose excretion symptoms such as polyuria, polydipsia, and dehydration are rare. No treatment is indicated, which is fortunate since none is available.

The diagnosis is established by exclusion of other causes of glycosuria, the most important of which is diabetes mellitus. The finding of normal random or fasting blood glucose concentrations in the presence of glycosuria virtually excludes diabetes, but it is wise to perform a glucose tolerance test to detect the occasional patient with mild or early diabetes, whose blood glucose is unstable and only intermittently raised. The absence of aminoaciduria, phosphaturia, hypokalaemia, and the proximal type of renal tubular acidosis excludes the Fanconi syndrome. Patients with glycosuria due to glucose-galactose malabsorption present acutely in infancy with intestinal symptoms; this condition does not arise in the differential diagnosis of the asymptomatic patient with glycosuria. The finding of glycosuria in asymptomatic first degree relatives of the patient supports the diagnosis but is not essential.

Genetics

Primary renal glycosuria is commonly familial and presumably has an inherited basis. As discussed in the preceding section, the distinction between types A and B glycosuria is insecure and it is unhelpful to attempt to distinguish between them from the genetic point of view with our current knowledge. Hjarne (1927) found vertical transmission of glycosuria through several generations in a large Swedish pedigree, and concluded that the disorder was inherited in an autosomal dominant fashion.

However, in this study the magnitude of the glycosuria was not taken into account. Elsas *et al.* (1971) found that heavy, continuous glycosuria behaved as an autosomal recessive trait, in that heavy glycosuria was seen in some siblings of index cases but mild or no glycosuria was found in the parents. Given that mild glycosuria is found in some obligate heterozygotes it would be more satisfactory to classify familial renal glycosuria as partially recessive, in that some abnormality can be demonstrated in heterozygotes with much more marked abnormality in homozygotes. By analogy with sickle-cell disease, it seems logical to follow the example of Khachadurian and Khachadurian (1964) in labelling homozygotes as having renal glycosuria while heterozygotes have 'renal glycosuria trait'.

Other (secondary) renal glycosuria

Glucose-galactose malabsorption

This rare disease, first described by Lindquist and Meeuwisse (1962), presents with severe fermentative diarrhoea in infancy, as soon as the child is given a feed containing glucose, galactose or any polysaccharide containing either or both of these hexoses—normally on the first day of life. It is recessively inherited, and is due to absence of the small intestinal glucose-galactose transporter (Elsas *et al.* 1970). The fact that patients with this disease also have renal glycosuria, apparently of type B (Meeuwisse 1970), is of no clinical importance but indicates that the intestine and the renal tubule have at least one glucose transporter in common. Conversely, the fact that the great majority of subjects with renal glycosuria have no abnormality of intestinal sugar absorption suggests that there may be more than one glucose transport system in the nephron, only one of which is inherited with the intestinal glucose-galactose transporter. Independent evidence for this is provided by the findings of Turner and Moran (1982), referred to in the previous section.

Glucoglycinuria

Käser *et al.* (1962) described a boy with cystic fibrosis who also had type B renal glycosuria and glycinuria without generalized aminoaciduria. Thirteen of 45 relatives were found to have both renal glycosuria and glycinuria, indicating an abnormality of a common transport mechanism or very close genetic linkage of two transport systems. The familial pattern suggested autosomal dominant inheritance. No further patients with this disorder have yet been described.

Glycosuria with hyperphosphaturia and glycinuria

The combination of severe hyperphosphaturic rickets, type A glycosuria with severely depressed T_mG and increased urinary excretion of glycine and glycyl-proline was described by Scriver *et al.* (1964b). Absence of generalized aminoaciduria or other features of the Fanconi syndrome suggest that this is a distinct genetic entity, but the nature of the underlying transport defects and the manner in which they are connected is unknown.

Glycosuria of pregnancy

As with other glycosurias, the apparent incidence of glycosuria of pregnancy varies as a function of the diagnostic criteria used. Probably the most clinically relevant definition is a positive dipstick test in the setting of a routine antenatal clinic: on this basis Chen *et al.* (1976) reported an incidence of 1.7 per cent. Clearance studies have thrown considerable light on the mechanism of this disorder. Christensen (1958) found a reduced actual glucose threshold with normal T_mG; however, the importance of relating T_mG to GFR is illustrated by the studies of Welsh and Sims (1960), who clearly showed that T_mG/GFR is lower in glycosuric pregnant women than in non-glycosuric pregnant women at all levels of GFR. They also showed that T_mG/GFR was lower even in non-glycosuric pregnant subjects that in non-pregnant controls. The findings of Davison and Hytten (1975) were similar, but they showed in addition that although glucose reabsorption had improved when their subjects were restudied 8 to 10 weeks after delivery, it remained less efficient in those who had been glycosuric when pregnant than in those who had not.

Two points emerge clearly from these results. Firstly, T_mG/GFR falls during pregnancy in all women, due to the well known increase in GFR not being accompanied by a parallel increase in T_mG. One of the major circulatory changes in pregnancy is volume expansion, which is known to inhibit proximal tubular sodium reabsorption: as suggested in the introductory part of this chapter, the coupling that exists between reabsorption of sodium and its cotransportates (including glucose) in the proximal tubule might enforce the observed parallel reduction of fractional reabsorption of glucose. This interpretation is supported by the fact that urinary excretion of amino acids (Hytten and Cheyne 1972) and uric acid (Semple *et al.* 1974) is also increased during pregnancy. Secondly, there is considerable variation in renal glucose handling among the general population, with those heterozygous for primary renal glycosuria ('renal glycosuria trait': see above) having a T_mG/GFR lower than controls, although not always being clinically glycosuric. It may well be that women glycosuric in pregnancy are simply manifesting the physiological effect of pregnancy on a lower than average T_mG/GFR and that some of them represent mild (perhaps heterozygous) cases of primary renal glycosuria. The main importance of glycosuria of pregnancy is its distinction from diabetes mellitus by means of a glucose tolerance test. If this is normal the condition should be regarded as benign: no dietary or other treatment is indicated.

Aminoacidurias

Normal amino-acid reabsorption (see also Chapter 5.1)

With the exception of tryptophan, which is 60 to 90 per cent protein bound, amino acids are present in plasma in free solution and their concentrations in glomerular filtrate are the same as those in plasma water. In health, reabsorption is nearly complete, ranging from 97 to 99.9 per cent of the filtered load in adults and older children, with the exception of histidine which is only 90 to 95 per cent reabsorbed (Brodehl and Gellison 1968; Yü *et al.* 1969; Silbernagl 1988). In infants aged less than 4 months the fractional reabsorption of all amino acids is slightly lower than in children and adults, but is still >97 per cent except for histidine ($\bar{x} \pm SD$, 86.6 ± 5.1 per cent), glycine (87.6 ± 6.4 per cent), and serine (93.4 ± 2.7 per cent) (Brodehl and Gellison 1968). Premature infants have a generalized aminoaciduria with fractional reabsorption of many amino acids of below 90 per cent, presumably reflecting functional immaturity of the transport mechanisms (O'Brien and Butterfield 1963). The resulting pattern of plasma concentrations and urinary excretion rates for the main amino acids in normal children is shown in Fig. 5. The pattern in healthy adults is essentially identical: infants excrete significantly more threonine, asparagine, serine, proline, glycine, and alanine than children or adults.

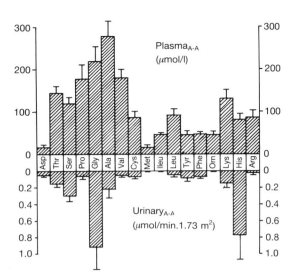

Fig. 5 Plasma concentrations (above) and urinary excretion rates (below) of free amino acids (A-A) measured in 12 normal children aged 2 to 13 years: the bars represent one standard deviation. Values in adults are essentially the same (data from Brodehl and Gellissen 1968).

The multiplicity and chemical diversity of amino acids are matched by their epithelial transport systems in the proximal tubule (Silbernagl 1988). Most, and probably all, amino acids are reabsorbed by more than one transporter: one which is specific to each amino acid and characterized by high specificity and affinity (K_m) and low maximal transport capacity (V_{max}), and one shared with other members of a group of chemically related amino acids, characterized by low K_m and high V_{max}. Membrane vesicle studies have identified five carrier systems in the latter, group-specific category (Sacktor 1978): these transport neutral and cyclic amino acids, glycine and imino acids, cystine and dibasic amino acids, dicarboxylic amino acids, and β-amino acids, respectively. The existence of transporters specific to single amino acids is revealed by the inborn errors of metabolism in which there is excessive loss of an individual amino acid without increased excretion of the other amino acids normally reabsorbed by the corresponding group carrier. The best documented example is isolated hypercystinuria without increased excretion of dibasic amino acids (Brodehl et al. 1967): experimental evidence of a specific cystine transporter has been found in isolated rat proximal tubules (Foreman et al. 1980). It may be that the different types of entry step transporter are located in different segments of the proximal tubule: this appears to be the case for histidine (Lingard et al. 1973) and possibly proline (Völkl et al. 1979). Much work remains to be done in characterizing the various transport systems for amino acids in the proximal tubule, and in determining their relative importance and their inter-relationships.

As far as is known, both monospecific and group-specific amino-acid transport systems conform to the general model described earlier in this chapter: a sodium coupled entry step, which is probably rate limiting, and a sodium independent exit step. However, sodium dependent uptake of amino acids has also been demonstrated across the basolateral membrane of the pars recta, but not the pars convoluta, of the proximal tubule (Barfuss et al. 1980). It has also been suggested that this mechanism is necessary to furnish the cells of the more 'downstream' segments of the tubule with amino acids for their own internal needs, since the tubular fluid at the end of the proximal tubule contains very low concentrations of amino acids (Kinne 1985). Presumably a proportion of the amino acids absorbed from the lumen in more proximal segments would be used for this purpose.

Titration studies for individual amino acids show typical T_m-limited tubular reabsorption much as for glucose, but the degree of splay varies greatly: for example, lysine has almost no splay (Wright et al. 1947) while for glycine the splay extends over at least a six-fold range of filtered load (Pitts 1943) (Fig. 6).

Mechanisms of aminoaciduria

Theoretically, impaired tubular reabsorption of amino acids could result from abnormalities of any of the successive steps in the transport process. Possible defects include: (i) the transporter for the entry step (either the group transporter or the specific transporter) might be absent or defective; (ii) the luminal membrane might be excessively leaky to the amino acids in question, leading to secondary back leak into the lumen; (iii) the exit step might be abnormal; (iv) some metabolic process linking the entry step to the exit step might be deficient, leading to accumulation of substrate within the cell and the generation of a lumen-to-cell gradient too steep for the entry process to surmount. The exact mechanism is not known for any of the renal aminoacidurias, but the absence of any evidence of an associated abnormality of sodium transport and the existence of disorders in which there is increased excretion of whole groups of amino acids favours the entry step as the site of the abnormality in most, if not all, of the disorders to be discussed below.

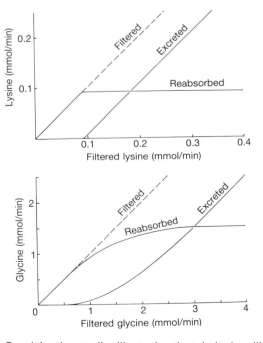

Fig. 6 Renal titration studies illustrating the tubular handling of amino acids with varying degrees of splay: lysine exhibits a sharp inflexion and almost no splay, while the curves for glycine are markedly splayed (data from Pitts 1943 and Wright et al. 1947).

5.2 Isolated defects of tubular function

Table 2 Causes of non-renal (overflow) aminoaciduria

Argininaemia
Citrullinaemia
Cystathioninaemia
Glycinaemia
Histidinaemia
Homocystinuria
Hydroxylysinaemia
Hydroxyprolinuria
Isoleucine-leucinaemia
Lysinaemia
Branded chain aminoaciduria
  (maple syrup urine disease)
Ornithinaemia
Prolinaemia
Sarcosinaemia
Threoninaemia
Tryptophanaemia
Tyrosinaemia
Valinaemia

Non-renal (overflow) aminoacidurias

As in the case of glucose, if the plasma concentration of an amino acid exceeds the renal threshold, urinary excretion of that amino acid will be increased, leading to overflow or non-renal aminoaciduria. Numerous abnormalities have been described in this category (Table 2). In most cases, deficiency of an enzyme necessary for further metabolism of the amino acid leads to its accumulation in the blood: clinical disease may be caused either by toxicity of the amino acid or an alternative metabolite, or by deficiency of a 'downstream' metabolite. Since, by definition, renal function is normal in these conditions they are not described further here. The interested reader is referred to standard textbooks of paediatrics (Forfar and Arneil 1983; Behrman and Vaughan 1987) and metabolic medicine (Scriver et al. 1989).

Renal aminoacidurias

By definition, renal aminoacidurias are those in which one or more amino acids are excreted in excess of normal due to a defect in renal tubular reabsorption. In accordance with the concept of group-specific and individual amino acid specific transporters discussed above, some diseases involve abnormal excretion of whole groups of amino acids while others are characterized by loss of an individual amino acid. Table 3 lists the major renal aminoacidurias classified according to which amino acid(s) are excreted in excess.

Classical cystinuria

The renal defect

Cystine stones were first identified in the urinary tract in the early 19th century (Wollaston 1810). Although Garrod proposed the then revolutionary hypothesis that cystinuria was caused by an inborn error of metabolism as early as 1908, it was not until 1951 that Dent and Rose showed that the disease was due to a defect of renal tubular reabsorption of cystine as well as the dibasic amino acids arginine, lysine, and ornithine. Part and parcel of this observation was the recognition that cystine and the dibasic amino acids share a common transport system. This insight was based firstly on the consistent finding of increased excretion of all four acids in classical cystinuria and, secondly, on the structural similarities between them. It was reinforced by the finding of Robson and Rose (1957) that the urinary excretion of cystine, arginine and ornithine was increased in normal humans by lysine infusion, suggesting competition for tubular reabsorption, but not in patients with cystinuria, in whom the transporter is presumed to be lacking.

An additional observation, not explained by defective tubular reabsorption alone, is that cystine excretion may exceed glomerular filtration rate, indicating net secretion. This finding has been noted in cystinuric patients both spontaneously (Crawhall et al. 1967) and during lysine infusion (Lester and Cusworth 1973) as well as in dogs (Webber et al. 1961). The most likely explanation for this is uptake of cystine into tubular cells across the peritubular (basolateral) membrane with secondary, possibly passive, diffusion into the lumen. The existence of such a mechanism can be explained teleologically thus: normally, the luminal fluid in the more distal segments of the proximal tubule (and presumably the distal tubule also) contains virtually no amino acids. The metabolic needs of the tubule cells themselves for amino acids must be met from another source, which can only be from the peritubular capillary blood via the basolateral membrane. In normal subjects, any cystine leaking into the lumen will probably be reclaimed by the entry transporter but in cystinuric patients, in whom the transporter is lacking, it is added to that filtered contributing to a total renal clearance greater than glomerular filtration rate.

The intestinal defect

The polyamines cadaverine and putrescine are present in greatly increased amounts in the urine of cystinuric patients (von Udransky and Baumann 1889). These compounds are decarboxylation products of lysine and arginine, respectively: Milne and coworkers reasoned that their presence could be explained by bacterial degradation of unabsorbed lysine and arginine in the gut, and confirmed this interpretation by noting increased urinary polyamine excretion following feeding of these amino acids (Milne et al. 1961; Asatoor et al. 1962). The presence of a defect in intestinal absorption of cystine and the dibasic amino acids is also supported by the finding that urinary cystine

Table 3 Classification of renal aminoacidurias

A. Cystine and dibasic amino acids
   Classical cystinuria*
   Isolated hypercystinuria
   Dibasic aminoaciduria
     Type 1
     Type 2
   Lysinuria

B. Neutral amino acids
   Hartnup disease*
   Methionine malabsorption
   Histidinuria

C. Glycine and imino acids
   Iminoglycinuria*
   Glycinuria

D. Dicarboxylic amino acids
   Dicarboxylic aminoaciduria*

* These disorders are thought to represent defects of group-specific amino-acid transport systems.

excretion is not increased by cystine feeding (Brand et al. 1935), and has been confirmed by direct measurement of uptake in jejunal biopsy material (Thier et al. 1964). The intestinal defect is of little or no clinical consequence, however. Sufficient cystine and dibasic amino acids for nutritional needs are probably absorbed as dipeptides (Hellier et al. 1970): this is supported by the previously cited studies of Brand and associates (1935) in which protein feeding, but not cystine feeding, was found to be followed by increased urinary cystine excretion.

Genetics

As defined clinically, cystinuria is inherited as an autosomal recessive gene, but at the molecular level at least three subtypes can be distinguished on the basis of intestinal transport characteristics and whether or not obligate heterozygotes have measurably increased urinary excretion of cystine and dibasic amino acids (Harris et al. 1955; Rosenberg et al. 1966). As classified by Rosenberg et al. (1966), type 1 (by far the most common) shows absence of in-vitro intestinal uptake of cystine and dibasic amino acids and no rise in plasma cystine following oral administration in homozygotes, with no abnormal aminoaciduria in heterozygotes. Types 2 and 3 differ from type 1 in that heterozygotes have increased urinary excretion of cystine and dibasic amino acids, although these are well below homozygote levels. Type 2 is associated with intestinal transport of cystine (less than normal but more than type 1) but not of dibasic amino acids, while in type 3 intestinal transport of cystine and dibasic amino acids is present and an oral cystine load is followed by a rise in its plasma concentration.

The three types are indistinguishable clinically, and the offspring of parents with different subtypes behave as stone forming homozygotes. This suggests that different abnormal genes are present at the same locus. It also means that renal tubular and intestinal transport of cystine and dibasic amino acids must involve multiple mechanisms with incomplete genetic linkage between them: further study of the genetics of cystinuria using the methods of modern molecular biology promises to illuminate further the transport processes involved.

Clinical features

The true incidence of cystinuria is difficult to determine. The number of asymptomatic homozygotes is unknown, and some heterozygotes of types 2 and 3 may be diagnosed as homozygotes unless quantitative estimation of cystine excretion is performed. Estimates of incidence based on population studies range from 1:100 000 (Bostram and Hambraeus 1964) to as high as 1:2500 in Sephardic Jews of Libyan origin (Weinberger et al. 1974). An even higher incidence of 1:1100 was claimed by Thalhammer (1975) among 70 000 newborns screened in Vienna, although the same author reported figures between 1:15 000 and 1:25 000 for newborns in Boston and Prague, similar to that found by Smith et al. (1979) in Sydney.

The symptomatology of cystinuria is entirely related to stone formation. Cystine stones may manifest themselves in any of the ways common to stones in the urinary tract: renal colic, haematuria, infections, or the urethral passage of stones. Urethral obstruction occurs occasionally, usually in males. If stones form bilaterally and are not adequately treated, chronic renal failure may ensue. Cystine lithiasis may present at any age, but most cases come to light in childhood or early adult life. Radiologically the stones are radio-opaque, but typically less so than calcium stones, and often have a uniform 'ground-glass' appearance. They may become very large and can form complete 'staghorn' casts of the pelvicalyceal system.

Although one or two reports have suggested that the incidence of mental handicap and psychiatric illness is higher in patients with cystinosis than in the general population (Berry 1962; Scriver et al. 1970), psychometric assessment failed to reveal any differences between a group of homozygous cystinurics and controls (Gold et al. 1977). It may be that the apparent high incidence in inmates of mental institutions is mainly a reflection of the fact that such patients are often subjected to repeated metabolic investigation, coupled with the fact that the baseline incidence of cystinuria in most populations is not known and probably underestimated.

Diagnosis

The diagnosis of cystinuria rests on the demonstration of increased urinary excretion of cystine. Cystine crystals may be seen on microscopy and are diagnostic. A simple, colorimetric screening test is available (the cyanide-nitroprusside test), and if positive should be confirmed by quantitation of the excretion rate of cystine and the dibasic amino acids. This is necessary since homocysteine gives a false positive cyanide-nitroprusside test and because some heterozygotes for types 2 and 3 cystinuria may also give a positive reaction. Normal individuals excrete less than 10 nmol cystine/mol creatinine (often expressed as <20 nmol 1/2-cystine, i.e. cysteine) while patients homozygous for cystinuria exceed this value at least 10-fold. Cystine excretion in heterozygotes is intermediate in value between that in normals and that in homozygotes.

Treatment

The objective of medical treatment is to prevent supersaturation of the urine with dissolved cystine, thus preventing crystallization and stone formation. Two strategies are employed: to reduce the urinary concentration of cystine, and to increase its solubility. At physiological urine pH (4.5–7.5) the upper limit of solubility is 1.0 to 1.25 mmol/l (240–300 mg/l); solubility increases progressively at pH above 7.5. Alkalinization of the urine is therefore desirable, and alkali (sodium bicarbonate or equivalent) should be given in several divided doses to maintain pH above 7.5. This is difficult to achieve continuously, and a high water intake should also be encouraged. Quantitation of the cystine excretion rate in the individual patient provides a guideline for the minimum recommended water intake: 1 l of water should be ingested daily for every mmol cystine (2 mmol 1/2-cystine). This also is difficult to achieve, particularly at night since it requires an intake of 4 to 10 l daily. A combination of alkali and high water intake is effective and safe, however, and is still the initial treatment of choice.

Patients who are unable to comply with the above regimen may be treated with D-penicillamine by mouth in a dose of 30 mg/kg.day to a maximum of 2 g/day in adults. D-Penicillamine reacts with cystine to produce a mixed penicillamine–cysteine disulphide which is much more soluble than cystine (Crawhall et al. 1963). The drug is highly effective but, unfortunately, causes many side-effects, including proteinuria with membranous glomerulonephritis and the nephrotic syndrome (Rosenberg and Hayslett 1967). Other toxic effects of penicillamine include rashes and arthralgia, bone marrow depression, epidermolysis, and loss of taste sensation. Pyridoxine supplements (10 mg/day) should be prescribed for patients on penicillamine, since the action of pyridoxine is antagonized by the drug (Jaffe et al.

1964). Other sulphydryl compounds have a similar effect. Mercaptopropionyl glycine was initially thought to be less toxic than penicillamine (Hautmann et al. 1977) but subsequent experience suggests that it offers little, if any, advantage.

Recent observations on the relationship between dietary sodium intake and urinary cystine excretion suggest a third possible approach to treatment. In a study primarily designed to investigate the effect of glutamine on urinary cystine excretion, Jaeger et al. (1986) noted that glutamine reduced cystine excretion in cystinuric patients on a high sodium intake (300 mmol/day) but not on a lower sodium intake (150 mmol/day); however, basal cystine excretion was lower on the lower sodium diet than on the higher, and lower still when sodium intake was reduced to 50 mmol/day. The mechanism of this relationship is not known, but the effect may well be due to changes in extracellular fluid volume, known to have an important influence on proximal tubular reabsorption of sodium and solutes reabsorbed by sodium coupled systems (see above). The therapeutic possibilities of this observation are obvious but have not yet been assessed clinically. If a low sodium diet is prescribed, any accompanying alkali therapy would presumably have to be given in a sodium free form, such as potassium citrate.

Once stones have formed in the kidney they should be treated according to their merits by surgery, lithotripsy, or percutaneous nephrolithotomy by a urological team experienced in the management of urolithiasis. Stone formation during medical treatment indicates either inadequate treatment or (commonly) non-compliance.

Other disorders of cystine and dibasic amino-acid excretion

Isolated hypercystinuria
Two children, a brother and sister, have been described with excessive urinary cystine excretion without dibasic aminoaciduria (Brodehl et al. 1967). Neither had stone formation, and urine amino acid excretion was normal in both parents and another sibling. The existence of this condition is evidence for a tubular reabsorptive pathway for cystine distinct from that which is shares with the dibasic amino acids.

Dibasic aminoacidurias
Several reports have described individuals with increased urinary excretion of arginine, lysine, and ornithine without cystinuria. There appear to be two main types. The first, also called lysinuric protein intolerance (Perheentupa and Visakorpi 1965), is characterized by dibasic aminoaciduria and intestinal malabsorption of dibasic amino acids, presumably due to a defect in a transport system common to the renal tubule and intestine. This is a severe disease, inherited as an autosomal recessive, in which protein loading produces hyperammonaemia with vomiting, weakness, and neurological symptoms including coma. The mechanism is thought to be intracellular depletion of arginine and ornithine, which are important components of the urea cycle. Intravenous infusion of arginine prevents the acute adverse response to protein loading (Simell et al. 1975). Oral supplementation with dibasic amino acids appears to have helped some patients but not others: presumably the underlying intestinal defect limits its effectiveness (Awrich et al. 1975). More recently, citrulline supplementation has been used and appears promising, since it is absorbed by a pathway independent of that for the dibasic amino acids but can be converted to arginine and ornithine in the liver (Rajantie et al. 1980). A combination of oral supplementation with citrulline and lysine (which cannot be made from citrulline) and moderate protein restriction is the current treatment of choice.

A second type of dibasic aminoaciduria was described by Whelan and Scriver (1968a) in numerous individuals from two generations of family. They showed increased excretion of dibasic amino acids without cystinuria: hyperaminaemia was not seen, and affected individuals appeared clinically normal. Sporadic reports have appeared of other, single cases which may represent this second type or, possibly, newly described entities. They are extremely rare and the nature of the biochemical transport abnormality has not been elucidated.

Lysinuria
One severely mentally retarded child who excreted excessive amounts of lysine but not arginine, ornithine, or cystine, and who had an intestinal defect of absorption also limited to lysine has been identified (Omura et al. 1976). This case attests yet further to the complexity and multiplicity of transport systems for dibasic amino acids in both intestine and renal tubule.

Hartnup disease
Hartnup disease takes its name from the English family in which it was first described.

The renal defect
There is incomplete tubular reabsorption of the neutral (monoamino, monocarboxylic) amino acids: alanine, asparagine, glutamine, histidine, isoleucine, leucine, methionine, phenylalanine, serine, threonine, tryptophan, tyrosine, and valine. Only about 50 per cent of the filtered load of most of these is reabsorbed, apart from histidine of which virtually 100 per cent of the filtered load is excreted. Absorption of other amino acids is normal. The renal tubular defect is presumably due to deficiency of the neutral amino acid group transporter.

The intestinal defect
The presence of an abnormality of intestinal amino acid absorption was suspected by Baron et al. (1956) because indoles and indicans, products of the bacterial breakdown of tryptophan, were present in increased amounts in urine and faeces of affected patients. This suspicion was confirmed by the finding of Scriver and Shaw (1962) of unabsorbed neutral amino acids in the stool, the pattern being similar to that in urine, and by the detailed studies of Milne et al. (1960), who demonstrated several abnormalities following the oral administration of a tryptophan load. These included: (i) a much smaller rise in plasma tryptophan than was observed in normal subjects; (ii) an increase in faecal tryptophan; (iii) absent conversion of tryptophan to indolyl-3-pyruvic acid; and (iv) a further increase in the already high urinary and faecal excretion rate of indoles and indicans. *In-vitro* confirmation of the transport defect was provided by Shih et al. (1971), who demonstrated impaired uptake of tryptophan and methionine in mucosal biopsy specimens. The metabolic consequences of this absorptive defect are presumably ameliorated by the fact that dipeptides containing tryptophan, histidine, and phenylalanine are absorbed normally by patients with Hartnup disease (Navab and Asatoor 1970; Asatoor et al. 1970).

Genetics
Hartnup disease is inherited as an autosomal recessive trait. Its incidence has been estimated at about 1:20 000 (Levy 1973).

Clinical features

The clinical features of Hartnup disease are due to deficiency of nicotinamide, which is partly derived from dietary tryptophan: the intestinal malabsorption of this amino acid, probably exacerbated by increased urinary losses, leads to a rate of nicotinamide production that varies from borderline to inadequate, according to variations in dietary intake and, perhaps, metabolic need. The major signs and symptoms of the disease are a scaly, photosensitive erythematous rash that is clinically identical to pellagra (niacin deficiency), cerebellar ataxia, psychiatric symptoms including emotional liability and confusional states, and occasional mental retardation. All these are intermittent and variable in severity except the mental retardation which, however, is found in only a minority of patients. Many affected patients remain completely asymptomatic, which may be a consequence of good dietary intake. In symptomatic patients there tends to be improvement with age.

Diagnosis

Diagnosis is straightforward and rests on demonstrating increased excretion of all of the neutral amino acids but not the dibasic, dicarboxylic, or imino acids or glycine. This investigation should be performed in all patients presenting with a pellagra-like rash, otherwise unexplained cerebellar ataxia or intermittent 'psychiatric' symptoms, and in relatives of known cases. Some affected infants will be detected as a result of routine urine screening for phenylketonuria and other inborn errors of metabolism. Full chromatographic analysis of urinary amino acids should be performed in all cases to avoid confusion with the renal Fanconi syndrome and the (even rarer) abnormalities of tubular reabsorption of individual neutral amino acids.

Treatment

Nicotinamide supplementation reverses the cutaneous and neuropsychiatric features of the disease (Halvorsen and Halvorsen 1963): it is wise to prescribe this for all affected subjects, even the asymptomatic, since the treatment is harmless and the possible effects of subclinical nicotinamide deficiency are unknown.

Other disorders of renal transport of neutral amino acids

Methionine malabsorption (oasthouse syndrome)

Two children with increased urinary excretion of methionine have been described (Smith and Strang 1958; Hooft *et al.* 1965). Both also had large amounts of α-hydroxybutyric acid in the urine, and in the patient described by Hooft *et al.* (1965) methionine was also present in the faeces, suggesting a specific defect of methionine transport in both intestinal and renal tubule. Since α-hydroxybutyric acid is a product of methionine breakdown, it is presumed to be formed in the colon by bacterial degradation, whence it is absorbed into the bloodstream and excreted in the urine. Clinical features were similar in the two cases: white hair, oedema, mental retardation, and seizures. The urine had an offensive odour, said to be similar to that of an oasthouse (a building in which hops are dried for use in the brewing of beer); hence the alternative name for the condition.

It is postulated that the presence of α-hydroxybutyric acid, rather than methionine deficiency, is responsible for the clinical features of the condition since a low methionine diet resulted in disappearance of α-hydroxybutyric acid from the urine and a striking clinical improvement.

Histidinuria

Sabater (1976) described two mentally retarded brothers with isolated histidinuria, accompanied by intestinal malabsorption of the amino acid. In accordance with the assumption that the disease is recessively inherited, both parents exhibited a mild intestinal absorptive defect for histidine but neither had histidinuria. These unique cases provide direct evidence for a specific histidine transporter common to intestine and renal tubule.

Disorders of renal glycine and imino-acid transport

Available evidence suggests that the renal tubular reabsorption of glycine and the imino acids proline and hydroxyproline is somewhat analogous to that of cystine and the dibasic amino acids: a group-specific carrier exists that transports both glycine and the imino acids, but specific systems for each are also present (Silbernagl 1988). This view is supported by *in-vitro* experiments. Indirect evidence for the group specific carrier was provided by Scriver *et al.* (1964), who found that the (overflow) increase in urinary proline excretion observed in familial hyperprolinaemia was accompanied by increased excretion of hydroxyproline and glycine. They interpreted this as resulting from competitive inhibition of reabsorption of hydroxyproline and glycine by the increased filtered load of proline. In contrast, developmental studies in rats (Webber 1967), dogs (Blazer-Yost and Jazyk 1979), and in man (Brodehl 1976) show that tubular reabsorption of imino acids matures more rapidly than that of glycine, suggesting that separate carrier systems are involved. The results of *in-vitro* studies were reviewed by Foreman and Segal (1985), who concluded that there are probably three carriers: one for imino acids, one for glycine, and a third, probably of lower affinity, that transports both.

Iminoglycinuria

The first description of a patient with increased urinary excretion of proline, hydroxyproline, and glycine but no generalized aminoaciduria was in a French child with early onset seizures (Joseph *et al.* 1958). Later reports suggest that iminoglycinuria is probably a benign condition (Procopis and Turner 1971), the apparent association with neurodevelopmental disease suggested by early reports being due to selection bias. It is inherited as an autosomal recessive trait, with an estimated incidence of 1:20 000 (Levy 1973), and, as with classical cystinuria, several allelic variants probably occur. In some (Goodman *et al.* 1967), but not all (Tancred *et al.* 1970), there is an associated defect of intestinal absorption of glycine and imino acids; furthermore, heterozygotes in some pedigrees excrete increased amounts of glycine (Whelan and Scriver 1968b) while those in other affected families do not (Bank *et al.* 1972). Although of no clinical importance, this genetic heterogeneity indicates that the inheritance and molecular biology of glycine and imino acid transport may be even more complex than the 'three transporter' model described above.

Isolated glycinuria

Several patients in whom hyperglycinuria without iminoaciduria occurred at normal plasma glycine concentrations have been described; this is apparently inherited as a dominant trait. In the first pedigree described (de Vries *et al.* 1957) the abnormality was confined to females, and was associated with calcium oxalate stone formation in three of four affected individuals. The relationship between the glycinuria and the urolithiasis is obscure, and had not been found in other affected families (Greene *et al.* 1973).

However, as pointed out in the previous paragraph, some obligate heterozygotes for iminoglycinuria exhibit isolated glycinuria, and it may be that those subjects described as having isolated, dominantly inherited glycinuria are in fact carriers of iminoglycinuria in families without a homozygous member: the heterozygous state would, of course, mimic a dominant mode of inheritance. This question cannot be resolved by the available evidence.

Glucoglycinuria
Glucoglycinuria is discussed in the section on glycosuria (above).

Disorders of renal dicarboxylic amino acid transport
Two patients with heavy dicarboxylic amino acid (aspartic and glutamic acid) excretion have been described: one was mentally retarded and hypothyroid (Teijama *et al.* 1974) but the other was clinically normal (Melancon *et al.* 1977). Obviously, no conclusions can be drawn as to any possible causative link between the aminoaciduria and the other abnormalities in the first case. In both patients the renal clearance of aspartic and glutamic acid was greater than the glomerular filtration rate, indicating net secretion. The mechanism of this secretion is totally obscure: it might be accounted for by the same putative explanation that has been proposed for the apparent secretion of cystine in some cases of cystinuria (see above), but this is no more than speculation. The apparent normality of both sets of parents, and the fact that one patient was male and other female, suggests autosomal recessive inheritance.

Phosphaturias (see also Chapters 14.1 and 14.2)

Normal renal phosphate excretion
Tubular phosphate reabsorption
Like glucose and amino acids, inorganic phosphate (P_i) is filtered at the glomerulus and reabsorbed in the tubule, the rate of excretion being the difference between the two. Since the filtration rate of inorganic phosphate is the product of glomerular filtration rate and plasma concentration, it is apparent that the several-fold changes in inorganic phosphate excretion that occur with little or no measurable change in plasma concentration must be largely mediated by changes in the tubular reabsorption rate (tubular secretion of inorganic phosphate has never been demonstrated in man or other mammals).

Clearance studies generate titration curves of the form discussed at the beginning of this chapter, indicating T_m-limited active transport (Fig. 7). The most useful index of tubular handling of inorganic phosphate is the theoretical threshold, T_mP/GFR. It is not practical to measure this directly in the clinical setting but it can be derived indirectly from the TRP (fractional reabsorption of filtered inorganic phosphate, calculated as

$$1 - (U : P_{\text{inorganic phosphate}} / U : P_{\text{creatinine}})$$

and plasma concentration of inorganic phosphate. The nomogram constructed by Walton and Bijvoet (1975) is reproduced in Fig. 8.

One important difference between renal excretion of inorganic phosphate and that of glucose and amino acids is that, under normal conditions, 10 to 15 per cent of filtered inorganic

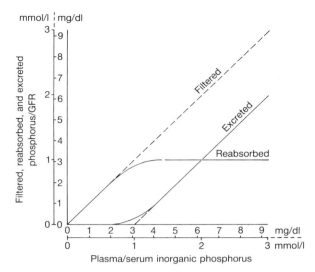

Fig. 7 Normal renal tubular reabsorption of inorganic phosphate.

phosphate is excreted in contrast to only trace amounts of the other substances: in other words, the renal threshold is slightly but significantly below the normal plasma concentration. This is to be expected, since inorganic phosphate must be excreted in an amount equal to dietary intake (with due allowance for growth or markedly anabolic or catabolic states) if external balance is to be maintained. The same logic predicts that excretion will vary directly with dietary intake, and this is indeed the case: loading with inorganic phosphate leads to progressive phosphaturia, while depletion leads to the virtually complete disappearance of inorganic phosphate from the urine.

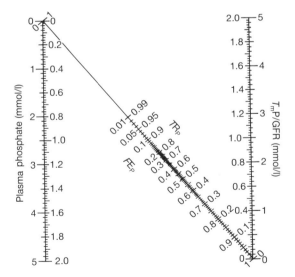

Fig. 8 Nomogram for estimation of the theoretical renal threshold for inorganic phosphate (T_mP/GFR) from the plasma phosphate concentration and the fractional excretion (or reabsorption) of phosphate. FE_P is the fractional excretion of filtered phosphate (calculated as the urine-to-plasma phosphate concentration ratio divided by the urine-to-plasma creatinine ratio). TR_P is the fractional reabsorption of filtered phosphate ($1-FE_P$) (redrawn from Walton and Bijvoet 1975).

The major site of inorganic phosphate reabsorption is the proximal tubule. It has been claimed that a significant proportion is reabsorbed in the distal tubule (Poujeol et al. 1977; Amiel 1980) and that this distal transport is PTH-independent (Pastoriza-Munoz et al. 1978). Other careful studies have failed to find evidence of distal inorganic phosphate transport, although as much as 20 per cent of the filtered load may be absorbed in the pars recta (S3 segment) of the proximal tubule, at which site it would be identified as 'postproximal' by conventional proximal tubular micropuncture methodology (Greger et al. 1977a; 1977b); this may explain the apparently discrepant results mentioned above.

Mechanism of reabsorption

Inorganic phosphate enters the epithelial cell from the tubular lumen by sodium coupled, secondary active transport, as do glucose and amino acids. The carrier probably transports two sodium ions with each phosphate ion: it accepts the monovalent ($H_2PO_4^-$) and divalent (HPO_4^{2-}) forms with approximately equal affinities. A consequence of this is that divalent inorganic phosphate transport is electroneutral, whereas that of monovalent inorganic phosphate is electrogenic, one positive charge being transported with each ion (Hoffman et al. 1976). Recent, detailed reviews of the reabsorptive mechanism are available (Lang et al. 1986; Murer 1988). The process is fuelled by active removal of sodium across the basolateral membrane by Na^+,K^+-ATPase mediated transport, thus maintaining a lumen-to-cell gradient favouring the entry of sodium and its cotransportate (in this case phosphate).

Factors influencing reabsorption of inorganic phosphate

Parathyroid hormone (PTH)

PTH inhibits inorganic phosphate reabsorption at all tubule sites known to be capable of such transport (Lau et al. 1980). Experimental evidence indicates that, in the proximal tubule at least, PTH binds to a receptor on the basolateral membrane causing activation of adenylate cyclase and an increase in intracellular cyclic adenosine 3',5'-monophosphate (cAMP), leading in turn to reduced permeability of the luminal (brush border) membrane to inorganic phosphate (Evers et al. 1978). More recently, other intracellular events initiated by PTH have been identified, including activation of phospholipase C with the production of 1,2-diacylglycerol and inositol 1,4,5-triphosphate. 1,2-Diacylglycerol, in turn, activates protein kinase C while inositol 1,4,5-triphosphate leads to an increase in intracellular $[Ca^{2+}]$. The studies of Quamme et al. (1989a; 1989b) suggest that activation of protein kinase C is of physiological importance in the phosphaturic action of PTH while elevation of intracellular $[Ca^{2+}]$ is not. Interactions between the cAMP and protein kinase C-dependent actions of PTH remain to be elucidated.

PTH has two additional actions that would be expected to reduce absorption of inorganic phosphate: inhibition of proximal tubular sodium reabsorption (Agus et al. 1971) and alkalinization of proximal tubular fluid (Arruda et al. 1977). The pK of the $HPO_4^{2-}/H_2PO_4^-$ buffer system is 6.8: at alkaline pH most of the urinary inorganic phosphate will be in the divalent form (HPO_4^{2-}), while at pH 6.8 it will mostly be in the monovalent form. Since the intracellular pH is thought to average about 7.2 to 7.3, inorganic phosphate will be converted to the divalent form on entering the cell, thus maintaining a gradient for entry of the monovalent but not the divalent ion. Whether the actions of PTH on sodium absorption and hydrogen ion secretion are of physiological significance in the control of inorganic phosphate excretion is not known.

PTH release is stimulated by a fall in plasma levels of ionized calcium and is inhibited by 1,25-dihydroxycholecalciferol. Interactions between PTH, vitamin D metabolites, and circulating calcium and inorganic phosphate concentrations are complex and incompletely understood.

Vitamin D

1,25-Dihydroxycholecalciferol, the active metabolite or hormone synthesized from vitamin D, affects renal excretion of inorganic phosphate by at least two mechanisms. First, the direct action of 1,25-dihydroxycholecalciferol in promoting intestinal calcium absorption leads to an increase in plasma $[Ca^{2+}]$ and hence to suppression of PTH release. Second, 1,25-dihydroxycholecalciferol has a direct inhibitory action on the parathyroid glands which is independent of its calcaemic effect. Thus, administration of 1,25-dihydroxycholecalciferol or its analogues to intact subjects is antiphosphaturic due to suppression of PTH. Whether vitamin D metabolites have a direct action on tubular transport of inorganic phosphate, distinct from their indirect effects via changes in PTH, plasma inorganic phosphate, and $[Ca^{2+}]$ remains controversial: the question is reviewed in detail by Avioli (1980).

Phosphate

Changes in dietary intake of inorganic phosphate induce parallel changes in urinary excretion. These effects occur in the absence of PTH (Steele and de Luca 1976; Trohler et al. 1976): furthermore, the phosphaturic response to PTH is modulated in some way by phosphate intake, being blunted by depletion and enhanced by repletion (Bonjour et al. 1978). The effect appears to be mediated by up-regulation of the brush border transporter, since prior depletion enhances uptake by rat proximal tubule brush border vesicles (Stoll et al. 1979). It is likely that this change is induced by reduction in the intracellular inorganic phosphate content, since it is independent of plasma levels.

Other factors

Numerous hormones, ions and drugs, including calcitonin, growth hormone, glucocorticoids, thyroxine, calcium, magnesium, lithium, bicarbonate, glucose, and amino acids, as well as diuretic drugs, have been found to affect renal excretion of inorganic phosphate. Glucose and amino acids both cause phosphaturia, probably by competing for a common energy source (Na^+, K^+-ATPase-dependent sodium extrusion across the basolateral membrane). The effects of the other substances listed are of unknown physiological significance, if any: they are discussed in more detail by Chesney (1985).

Disorders of renal tubular phosphate transport

Familial X-linked hypophosphataemic rickets

Familial X-linked hypophosphataemic rickets is the most common inherited disorder of phosphate metabolism, and has been claimed to be the leading cause of rickets in developed countries (Fraser and Scriver 1976). It is known by several synonyms, of which the most widely used is vitamin D-resistant rickets (not to be confused with vitamin D-dependency rickets, see below). Familial X-linked hypophosphataemic rickets is unusual, if not unique, among inborn errors of human metabolism in that there exists a mutant strain of mice (the *Hyp* strain) which exhibits an apparently identical transport defect, also X-linked. Investigation of this mouse has thrown much light on the human disease: the description of the pathophysiology of the condition that follows is a composite, based on clinical studies in the human and experimental studies in the *Hyp* mouse. While it is legitimate to question the validity of extrapolating directly from mouse to human, no significant differences have yet been demonstrated between the human disease and the mouse 'model' and it seems reasonable to accept data based on the *Hyp* mouse as relevant to the human disease in the absence of direct experimental data on the latter.

The renal defect
The characteristic abnormality of renal function in familial X-linked hypophosphataemic rickets is hypophosphataemia with urinary inorganic phosphate wasting, in the absence of glycosuria, aminoaciduria, or other abnormalities of tubular reabsorption. Plasma inorganic phosphate is typically below 0.8 mmol/l (2.5 mg/dl) while T_mP/GFR is <0.56 mmol/l (1.8 mg/dl). Both plasma levels of inorganic phosphate and T_mP/GFR are somewhat higher in affected children than in adults, as is the case in normal subjects (Stickler *et al.* 1970).

Although there is still controversy concerning the mechanism of impaired tubular inorganic phosphate reabsorption (Chesney 1985), studies in the *Hyp* mouse point clearly to an intrinsic defect in the sodium phosphate entry transporter in the proximal tubular brush border. *In vitro* uptake of inorganic phosphate by isolated brush border vesicles is markedly impaired, while uptake of glucose and amino acids is normal (Tenenhouse and Scriver 1978; Kiebzak and Dousa 1985; Lyon *et al.* 1986). In contrast, inorganic phosphate uptake by renal cortical slices from *Hyp* mice is normal, indicating normal transport in the basolateral membrane (Tenenhouse *et al.* 1978). Dietary inorganic phosphate restriction causes uptake by brush border membrane vesicles to increase, but to a level well below that seen in similar vesicles from normal mice (Tenenhouse and Scriver 1979). Total parathyroidectomy does not improve phosphaturia in the *Hyp* mouse (Cowgill *et al.* 1979), nor does subtotal parathyroidectomy in human patients (Tarwalker *et al.* 1974). These *in-vivo* and *in-vitro* observations strongly suggest that the transport defect in both human familial X-linked hypophosphataemic rickets and the *Hyp* mouse is intrinsic to the tubule. However, one recent piece of evidence is discordant with this view: in an experiment in which *Hyp* mice were cross-circulated with normal litter mates, in such a way that the normal mice were perfused with blood from the *Hyp* mice, the normal animals developed phosphaturia (Meyer *et al.* 1987). If this observation is confirmed the above conclusion may need to be revised, but at present the bulk of clinical and experimental data favours a primary tubular defect.

The intestinal defect
In their original description of familial X-linked hypophosphataemic rickets, Albright *et al.* (1937) postulated a defect of calcium absorption as the primary abnormality in this disorder, leading to hypocalcaemia and secondary hyperparathyroidism. This hypothesis is no longer tenable for several reasons: plasma total and ionised calcium levels are normal, and plasma PTH is also normal or only slightly increased. Nevertheless, there is evidence for a defect of small intestinal absorption of calcium and inorganic phosphate in both familial X-linked hypophosphataemic rickets and the *Hyp* mouse. Oral calcium loading produces only a minimal rise in plasma calcium concentration and urine calcium excretion in affected patients (Albright *et al.* 1937; Saville *et al.* 1955; Stickler 1963); both balance studies (Meyer *et al.* 1984) and *in-vitro* studies of isolated intestinal segments (Meyer *et al.* 1986) demonstrate calcium and inorganic phosphate malabsorption in juvenile, but not in adult, *Hyp* mice. The mechanism of this malabsorption, and how it is linked to the renal phosphate losing abnormality, is not yet clear, but it is probably a consequence of the abnormality of vitamin D metabolism that is found in both the human disease and the *Hyp* mouse (see below).

Parathyroid hormone in familial X-linked hypophosphataemic rickets
Immunoreactive parathyroid hormone (iPTH) levels are normal in most patients with familial X-linked hypophosphataemic rickets (Arnaud *et al.* 1971; Roof *et al.* 1972), although some workers have reported slightly raised levels (Hahn *et al.* 1975). Even these levels are much lower than those seen in dietary (vitamin D-deficiency) rickets, untreated renal osteodystrophy and vitamin D-dependency rickets, however, and since iPTH levels were suppressed by calcium infusion without correcting the hyperphosphaturia or hyperphosphataemia (Hahn *et al.* 1975) it is unlikely that PTH has an important primary role in familial X-linked hypophosphataemic rickets. As previously mentioned, the fact that parathyroidectomy fails to return T_mP/GFR in *Hyp* mice to normal (Cowgill *et al.* 1979) leads to the same conclusion for the animal model. It is possible that the modestly elevated iPTH levels reported by some, if validated, may be secondary to the abnormality of vitamin D metabolism discussed below.

Vitamin D metabolism in familial X-linked hypophosphataemic rickets
Plasma concentrations of 1,25-dihydroxycholecalciferol are normal to low in patients with familial X-linked hypophosphataemic rickets (Scriver *et al.* 1978; Lyles *et al.* 1982), although since hypophosphataemia is normally a stimulus to renal 1-α hydroxylase activity they would be expected to be in the high normal range or above. The finding of Insogna *et al.* (1983) that phosphate deprivation actually causes plasma levels of 1,25-dihydroxycholecalciferol to fall, as opposed to the rise seen in normal subjects, adds further support to the contention that vitamin D metabolism is altered in familial X-linked hypophosphataemic rickets, and that 1,25-dihydroxycholecalciferol levels are lower than would be expected in a hypophosphataemic normal subject. It has been shown both in patients (Lyles *et al.* 1982; Insogna *et al.* 1983) and in *Hyp* mice (Tenenhouse 1984; Nesbitt *et al.* 1986) that the 1α-hydroxylase response to PTH and calcium restriction is subnormal, suggesting a down-regulation of the response of this enzyme system to physiological stimuli. In addition, *in-vitro* inactivation of 1,25-dihydroxycholecalciferol and other vitamin D metabolites by 24-hydroxylation is increased by a factor of 2 in mitochondria from renal tubular cells of *Hyp* mice compared with controls (Tenenhouse *et al.* 1988), suggesting that 1,25-dihydroxycholecalciferol levels may be functionally low as a result of both decreased production and increased elimination. An accelerated plasma disappearance rate for 1,25-dihydroxycholecalciferol has indeed been described in familial X-linked hypophosphataemic rickets (Seino *et al.* 1984). These abnormalities of vitamin D metabolism cannot explain the phosphaturia of this disease, but may be an important determinant of the variable severity of bone disease, which is not correlated with the degree of phosphaturia or the plasma

inorganic phosphate concentration and therefore demands another explanation.

Genetics

It is now clear that familial X-linked hypophosphataemic rickets is inherited as an X-linked dominant condition with variable expression, especially in females, provided that hypophosphataemia rather than bone disease is taken as the diagnostic criterion (Rasmussen and Anast 1983; Rasmussen and Tenenhouse 1989). If clinical bone disease is required for the diagnosis, about half of all affected females will be missed: this caused some confusion in earlier reports. The gene has now been mapped to the distal part of the short arm of the X chromosome. A second mutant strain of hypophosphataemic mice, the *Gy* strain, has recently been described (Lyon *et al.* 1986), in which the abnormal gene is also situated on the short arm of the X chromosome but at a different locus from the *Hyp* mutation. *Gy* mice display a pattern of hypophosphataemic, phosphaturic rickets apparently identical to that in *Hyp* mice, but in addition have neurological abnormalities and cochlear deafness. An audiometric study of 22 patients with familial X-linked hypophosphataemic rickets (Boneh *et al.* 1987) revealed sensorineural deafness, which apparently bred true, in five: families were internally consistent with respect to the presence or absence of deafness. This raises the intriguing possibility that two genetically distinct forms of the human disease exist, analogous to the *Hyp* and *Gy* mouse models, respectively. Genes on the X chromosome are highly conserved in evolutionary terms, and it is at least possible that the human and murine diseases are caused by genes common to the two species.

Clinical features

The major clinical features of familial X-linked hypophosphataemic rickets are growth retardation and rickets, the latter mainly affecting the legs. Affected infants appear normal at birth, but statural growth delay is usually evident by the age of 6 months to 2 years. The rickets becomes apparent after weight bearing, usually with smooth bowing of femur and tibia leading to marked genu valgum. It has been claimed that short stature is due entirely to restricted growth of the lower limbs (McNair and Stickler 1969), but careful measurement shows that the arms are also short, although proportionately less so than the legs (Steendijk and Herweijer 1984). Craniosynostosis affects some patients and dentition is abnormal in all, although the abnormality is different from the enamel hypoplasia characteristic of vitamin D-deficiency rickets, consisting of defective dentine maturation with enlarged pulp chambers (Rakocz *et al.* 1982); recurrent dental abscesses are common (Tulloch and Andrews 1983). Radiologically, the appearances are those of rickets: widening and fraying of the epiphyseal plates associated with coarse trabeculation and osteomalacia of the shafts of the long bones. Features of hyperparathyroidism (subperiosteal bone resorption, osteitis fibrosa cystica) are usually minimal or absent as might be expected. The histomorphometric features of familial X-linked hypophosphataemic rickets are those of osteomalacia in both cortical and trabecular bone, decreased calcification rate and increased mineralization lag time, and reduction in the occurrence of bone remodelling units (Marie and Glorieux 1981).

After growth is complete, disabling bone pain and arthritis affecting the large limb joints are common. Ectopic ossification occurs at points of muscle attachment to the skeleton and in periarticular regions (Polisson *et al.* 1985). Considerable disability may result from these late manifestations (Moser and Fessel 1974).

There is great variability of expression of the skeletal features of the disease, especially in females. An analysis by Rasmussen and Tenenhouse (1989) of 143 patients, all (by definition) with hypophosphataemia, revealed 47 females but only two males without bone disease. This 'sexual dimorphism' is seen in other X-linked dominant disorders, for example Alport's syndrome (Flinter and Bobrow 1988): it is probably due to 'lyonization' of the X chromosome, in which one X chromosome in each female nucleus is permanently inactivated at an early stage of embryogenesis. During subsequent cell division, the progeny of each cell in which this has occurred express the same X chromosome as the parent cell. Assuming the inactivation process to have been randomly determined, either the normal or the abnormal X chromosome may be the predominantly active one in the individual as a whole, accounting for the wide range of clinical severity observed. The alternative theory, that the female behaves as a heterozygote while males are hemizygous, would account for females being less severely affected than males but not for the great range of variability. Both *Hyp* and *Gy* mutant mice show the same sex difference in the expression of the gene as affected humans. The effect of the disease on final height is also highly variable. In each of three series the mean adult height for both sexes lay between -2 and -3 population standard deviation scores (SDS), with a standard deviation of about 1.5 (Steendijk and Latham 1971; Stickler and Morgenstern 1989; Reusz *et al.* 1990). This distribution implies that about 40 per cent of patients will have a final height within the lower half of the normal range, but the most severely affected 40 per cent will lie between -4 and -6 SDS, equivalent to 136 to 149 cm for men and only 129 to 140 cm for women.

Diagnosis

The characteristic diagnostic findings in familial X-linked hypophosphataemic rickets are phosphaturia in the presence of hypophosphataemia (low T_mP/GFR), normocalcaemia and low urinary calcium excretion, normal plasma iPTH and urinary cAMP, and normal or slightly reduced plasma 1,25-dihydroxycholecalciferol. The diagnosis of vitamin D-deficiency rickets (nutritional rickets) is usually apparent from the history of dietary vitamin D deficiency and lack of exposure to sunlight. Plasma total and ionized calcium concentration is reduced, iPTH levels are raised, and, by definition, plasma 1,25-dihydroxycholecalciferol level is very low. Prompt response to physiological doses of vitamin D confirms the diagnosis. Vitamin D-dependency rickets (see below) mimics the biochemical changes of nutritional rickets closely but is unresponsive to physiological doses of vitamin D and a history of dietary deficiency and absence of exposure to sunlight is lacking. The rare disease hereditary hypercalciuric, hypophosphataemic rickets has many similarities with familial X-linked hypophosphataemic rickets but urinary calcium excretion is high: so far, it has only been clearly described in one pedigree (see below). Oncogenic osteomalacia with phosphaturia is biochemically indistinguishable from familial X-linked hypophosphataemic rickets, but is an acquired disease of adults: the absence of an affected parent and the presence of the causative tumour serve to differentiate the two.

There are no other abnormalities of tubular transport in either familial X-linked hypophosphataemic rickets or in oncogenic osteomalacia with phosphaturia. Patients with both vitamin D-deficiency and vitamin D-dependency rickets have generalized aminoaciduria, secondary to hyperparathyroidism (Fraser et al. 1967): this resolves with treatment. Children with the Fanconi syndrome due to cystinosis may present in infancy with hypophosphataemic rickets but other components of the syndrome (glycosuria, aminoaciduria, proximal renal tubular acidosis, and hypokalaemia) are present by the second year of life: corneal and leucocyte accumulation of cystine is diagnostic. Plasma and urinary levels of inorganic phosphate may be normal in the first few months of life in patients with familial X-linked hypophosphataemic rickets; a child known to be genetically at risk for the disease must therefore be observed for at least a year from birth before reassurance can be given that he or she is not affected. Asymptomatic females may be difficult to diagnose: however, low fasting plasma inorganic phosphate in the presence of phosphaturia is invariably present in asymptomatic mothers of affected children, confirming the presence of abnormal renal phosphate handling even in the absence of bone disease.

Treatment

The therapy of familial X-linked hypophosphataemic rickets is unsatisfactory and difficult. The first treatment for which success was claimed was vitamin D_3 (cholecalciferol) in supraphysiological doses (10 000–300 000 U/day) (Winters et al. 1958; Harrison et al. 1966; Stickler et al. 1970). Any benefit was bought at a high price, however, since hypercalcaemia occurred frequently, with the attendant risks of hypercalciuria, nephrocalcinosis, and renal damage. This is not surprising given the primacy of renal phosphate wasting in the pathogenesis of the bone disease, and from the early 1960s oral phosphate supplements were introduced (Frame et al. 1963; West et al. 1964). Although a net positive inorganic phosphate balance could be achieved by administration of a combination of oral inorganic phosphate and vitamin D, large doses (2–4 g oral phosphate/day) and frequent administration (4–6 doses/day) are necessary, making compliance a considerable problem. Diarrhoea is a frequent complication of oral high dose phosphate therapy, although this can be overcome to some extent by starting with a small dose which is then gradually increased.

Recognition that 1,25-dihydroxycholecalciferol levels are inappropriately low relative to the low plasma inorganic phosphate, combined with the availability of synthetic 1α-hydroxylated vitamin D derivatives (1α-hydroxycholecalciferol and 1,25-dihydroxycholecalciferol) led to numerous therapeutic trials, both long term and short term, of these agents, alone and in combination with phosphate supplements: comprehensive references to these are given in reviews by Chesney (1985) and Rasmussen and Tenenhouse (1989). Most workers have concluded that both 1α-hydroxycholecalciferol and 1,25-dihydroxycholecalciferol improve intestinal absorption of both calcium and inorganic phosphate, heal rachitic bone lesions, and improve linear growth. Dosage is empirical and should be adjusted to the highest level attainable without hypercalcaemia and hypercalciuria: this is usually 25 to 100 ng/kg.day. This treatment does not correct tubular inorganic phosphate wasting, however, and phosphate supplements should be given to maintain the plasma concentration within the normal range for age (Greenberg et al. 1960).

Although sustained treatment with oral inorganic phosphate and 1α-hydroxycholecalciferol or 1,25-dihydroxycholecalciferol is now accepted treatment for familial X-linked hypophosphataemic rickets, a cautionary note was recently sounded by Stickler and Morgenstern (1989). In a retrospective review of 52 patients treated at the Mayo Clinic since 1934, and who had reached the age of 18 or over in 1989, they found no difference in final height between patients treated with vitamin D_2 (calciferol), 1,25-dihydroxycholecalciferol and phosphate in various combinations and patients who had never been treated at all, and no difference between groups in bone, joint, and muscle symptom scores. They also pointed out that previous studies were all uncontrolled, and often relied on growth velocity analysis for their claims of improved growth rates: reports describing an improved final height SDS are lacking. Most importantly, three treated patients developed end-stage renal failure in their twenties, almost certainly as a result of calcific renal damage secondary to treatment: evidence of medullary nephrocalcinosis had previously been noted in treated children by Goodyer et al. (1987). A rebuttal from Reusz et al. (1990), showing modest but statistically significant improvement in SDS (approximately -2.8 to -2.2) in 12 patients treated for at least 2 years with 1,25-dihydroxycholecalciferol and phosphate supplements is inconclusive since the comparable growth pattern for untreated patients is not known: the same reservation applies to the similar experience of Stamp and Goldstein (1990) and Balsan and Tieder (1990). The suggestion by Stickler and Morgenstern (1989) that prospective, controlled trials of treatment are needed to resolve this problem is attractive, but the disease is sufficiently rare that this could probably only be achieved on a multicentre basis. In the meantime, it is imperative that patients treated with vitamin D analogues and phosphate supplements are closely monitored with regards to plasma and urine calcium levels and glomerular filtration rate.

The use of diuretic (thiazide plus amiloride) adjuvant therapy has been proposed by Alon and Chan (1985) as a means of reducing urinary phosphate excretion, presumably as a result of extracellular fluid volume contraction. Experience with this approach is very limited, and it would be premature to recommend it for general use, particularly in view of recent anxieties about possible adverse affects of long-term diuretic drug treatment. In the light of the experience of Stickler and Morgenstern (1989), the well-known effect of thiazides in elevating the plasma calcium concentration is an additional reason for caution in their use in familial X-linked hypophosphataemic rickets.

Vitamin D-dependency rickets type I

This rare disorder is inherited as an autosomal recessive trait which presents as severe rickets of early onset (3 months of age or less), hypocalcaemia with tetany and convulsions, hyperphosphaturia and hypophosphataemia, and dental enamel hypoplasia (Fraser and Salter 1958; Prader et al. 1961). Aminoaciduria is present, probably due to the associated severe hyperparathyroidism. In contrast to familial X-linked hypophosphataemic rickets, vitamin D in large doses (100 000 U/day) heals the rickets, restores plasma calcium and inorganic phosphate levels to normal and reverses the hyperparathyroidism and its consequences (hyperphosphaturia, aminoaciduria). Physiological doses of 1α-hydroxycholecalciferol (Reade et al. 1975) and 1,25-dihydroxycholecalciferol (Delvin et al. 1981) are curative, and it is clear that the disease is the consequence of defective conversion of 25-hydroxycholecalciferol to 1,25-dihydroxycholecalciferol (Fraser et al. 1973). As would be predicted, circulating endogenous 1,25-dihydroxycholecalciferol levels are very low (Scriver et al. 1978) while those of 25-hydroxycholecalciferol, its immediate precursor, are normal in untreated patients and high in those treated with vitamin D (Reade et al. 1975).

Vitamin D-dependency rickets type II

This is clinically and biochemically similar to type I disease, with the important differences that endogenous levels of 25-hydroxycholecalciferol and 1,25-dihydroxycholecalciferol are high, and the disease is resistant to doses of 1α-hydroxycholecalciferol and 1,25-dihydroxycholecalciferol that are curative in type I disease (Marx et al. 1978; Liberman et al. 1980). It is probable that more than one mechanism exists for end-organ resistance to the action of 1,25-dihydroxycholecalciferol. Absence or deficiency of both nuclear (Eil et al. 1981) and cytoplasmic (Feldman et al. 1982) receptors for 1,25-dihydroxycholecalciferol has been demonstrated in cultured fibroblasts from affected patients. However, an example has been reported of resistance to 1,25-dihydroxycholecalciferol with normal binding of the hormone to fibroblasts: a postreceptor abnormality has been postulated in this instance (Griffen et al. 1982). Indirect evidence for end-organ resistance is provided by the finding that intestinal calcium absorption is subnormal despite high levels of 1,25-dihydroxycholecalciferol (Marx et al. 1978; Tsuchuya et al. 1980). Vitamin D dependency rickets type II is inherited in autosomal recessive fashion: affected members of some, but not all, pedigrees have alopecia and extreme growth retardation in addition to the other features of the disease. It must be presumed that the molecular basis for end-organ resistance to 1,25-dihydroxycholecalciferol differs in those with and those without these phenotypic features.

Treatment with large doses of 1α-hydroxycholecalciferol (3–6 μg/kg.day) and calcium supplements is effective in reversing the biochemical and radiological abnormalities in type II vitamin D-dependency rickets (Takeda et al. 1987). Alopecia is not improved, and the effect on growth rate has not yet been evaluated.

Hereditary hypophosphataemic rickets with hypercalcaemia

This rare, familial disease has some similarities to familial X-linked hypophosphataemic rickets but is clearly distinct from it. The clearest account is to be found in the description of a large, consanguineous Arab pedigree in which nine of 59 inter-related family members investigated were found to have hereditary hypophosphataemic rickets with hypercalcaemia and a further 21 had lesser degrees of biochemical abnormality but were clinically normal (Tieder et al. 1985; 1987). The unfathomable complexity of Bedouin tribal genealogy makes the genetics of the disease difficult to ascertain. The existence of two apparent degrees of severity, the milder approximately twice as common as the more severe, within a multiply inbred family tree is suggestive of autosomal recessive inheritance with partial expression in the heterozygote, but the reported pattern is also compatible with a variably expressed dominant gene.

The clinical features are those of rickets and osteomalacia, short, stature, bone deformity and pain with muscle weakness. Biochemical findings include phosphaturia with hypophosphataemia and reduced T_mP/GFR, normal plasma calcium and PTH concentrations, and increased plasma 1,25-dihydroxycholecalciferol and urinary calcium excretion. Those with the milder, subclinical form of the disease (?heterozygotes) are clinically normal but have plasma phosphate and 1,25-dihydroxycholecalciferol levels and urinary calcium excretion intermediate between those of fully affected patients and controls. Oral loading tests show increased intestinal absorption of calcium and phosphate, while oral phosphate administration also leads to endogenous PTH release and increased phosphaturia (lowered T_mP/GFR). Infusion of PTH causes an increase in phosphaturia and further depression of T_mP/GFR. Taken together, these findings indicate that the primary abnormality in hereditary hypophosphataemic rickets with hypercalcaemia is a PTH-independent renal tubular phosphate leak, while the increased intestinal absorption of calcium and phosphate and resulting hypercalciuria are secondary to increased production of 1,25-dihydroxycholecalciferol appropriate to the degree of phosphate depletion. If this interpretation is correct, the key difference between hereditary hypophosphataemic rickets with hypercalcaemia and familial X-linked hypophosphataemic rickets lies in the subnormal response of renal 1α-hydroxylase to phosphate depletion in the latter. Further studies into the cellular defect in hereditary hypophosphataemic rickets with hypercalcaemia will be important not only in the elucidation of the disease itself but also for the light it may shed on the mechanism of the much more common familial X-linked hypophosphataemic rickets.

The only available treatment is phosphate supplementation: oral administration of 1 to 2 g neutral phosphate in divided daily doses has been reported to improve the clinical symptoms and signs of the disease, including increased growth rate, as well as the radiological and histological features of rickets and osteomalacia.

Hypophosphataemic rickets with hypercalciuria and microglobulinuria

Dent and Friedman (1964) described two children with a form of hypophosphataemic rickets accompanied by hypercalciuria with low to normal plasma calcium, moderate aminoaciduria, and 'tubular' proteinuria that appears to be distinct from both familial X-linked hypophosphataemic rickets and hereditary hypophosphataemic rickets with hypercalcaemia: an apparently identical case was investigated by Carey and Hopfer (1987). Unlike hereditary hypophosphataemic rickets with hypercalcaemia, the hypercalciuria appears to be due to a primary defect in tubular calcium reabsorption, not to excessive intestinal absorption. This view is supported by the finding that long-term hydrochlorothiazide administration reduced urinary calcium excretion by 60 per cent, with eventual healing of the rickets. The cause of this rarely recognized disease is unknown: no evidence exists as to whether it is inherited or acquired.

Oncogenic hypophosphataemic osteomalacia

Oncogenic hypophosphataemic osteomalacia, also known as oncogenous rickets with phosphaturia, is a paraneoplastic syndrome with biochemical features similar or identical to those of familial X-linked hypophosphataemic rickets, the main differences being that it is an acquired disease and that cure is effected by removal of the causative tumour. The tumour is usually of connective tissue or bone, and may be benign or malignant: a list of more than a dozen tumour types associated with this syndrome was compiled by Chesney (1985) from 22 reports. It has also been reported in syndromes of multiple epidermal naevi (Aschinberg et al. 1977; Carey et al. 1986).

The disease may present at any age from childhood to old age, with insidious and progressive weakness, bone pain and, in children, growth failure and bony deformities. The radiological appearances are those of rickets and/or osteomalacia depending on the age of onset. Biochemically there is hyperphosphaturia with hypophosphataemia, in the presence of normal plasma calcium concentration and normal or only slightly abnormal plasma level of iPTH and 1,25-dihydroxycholecalciferol.

Since removal of the tumour is followed by clinical and biochemical cure in most cases, it is difficult to escape the conclusion that the tumour produces a humoral factor (not PTH) that causes the syndrome. No such factor has yet been identified: not only is it immunologically different from PTH (since it does not react in PTH-specific radioimmunoassays) but it must also be functionally different, since hypercalcaemia is not a feature of the condition. The similarities between oncogenic hypophosphataemic rickets and familial X-linked hypophosphataemic rickets have led to speculation that a similar humoral factor might be the cause of the latter. However, apart from the single, so far unconfirmed, observation that cross-circulation between a *Hyp* mouse and a normal litter mate can induce phosphaturia in the control animal (Meyer *et al.* 1987), no direct evidence has been adduced in favour of a circulating phosphaturic factor in familial X-linked hypophosphataemic rickets. Most of the clinical and experimental evidence (see above) implicates an intrinsic defect of proximal tubular cells as the cause of the latter, and the similarities between it and oncogenic hypophosphataemic osteomalacia may be either coincidental or due to activation of a final common pathway affecting tubular phosphate reabsorption and vitamin D metabolism due to different proximate causes.

Treatment of oncogenic hypophosphataemic osteomalacia is by identification and excision of the underlying tumour. If this is impossible, either because the tumour cannot be found or is inoperable, phosphate supplementation and administration of 1α-hydroxycholecalciferol or 1,25-dihydroxycholecalciferol may provide some symptomatic relief, but the results are often disappointing. Patients with the clinical and biochemical abnormalities of oncogenic hypophosphataemic osteomalacia, but in whom an exhaustive search fails to find a tumour, are sometimes said to have 'sporadic hyophosphataemic osteomalacia'. The relationship between these patients and those with typical tumours of mesenchyme is obscure, and in the absence of a satisfactory explanation of the mechanism of the abnormalities of mineral metabolism it is impossible, and probably meaningless, to say whether or not these groups of patients suffer from the same disease.

Isolated defects in renal tubular urate handling

Hypouricaemia

Excess loss of urate occurs as part of the Fanconi syndrome (see Chapter 5.1), but an isolated defect of tubular handling of urate with otherwise normal proximal tubular function, was first reported in 1950 by Praetorius and Kirk, and a number of other patients have been reported since. Some patients have no renal symptoms, being discovered as a result of an incidental finding of hypouricaemia, but a number present with renal colic, haematuria, and/or urinary tract stones of urate or calcium oxalate, presumably the result of the hyperuricosuria. Renal hypouricaemia needs to be distinguished from that arising from xanthine oxidase deficiency, which again may present with stones, on this occasion formed from xanthine and not urate, or from ingestion of the xanthine oxidase inhibitor allopurinol. Physiological hypouricaemia occurs during pregnancy in response to volume expansion (see Chapter 15.1). Approximately 0.5 per cent of the population have hypouricaemia from these various causes (Hisatome *et al.* 1989).

Some patients with congenital renal hypouricaemia have a family history, while others have no affected relatives: the mode of inheritance is not clear, but is most consistent with an autosomal recessive pattern. A few families have associated hypercalciuria and osteoporosis (de Vries and Sperling 1979). It is noticeable that the majority of patients described to date have been Israeli or Japanese, which might suggest that the genetic defect or defects responsible is more common in these races than others.

All other renal functions, including glomerular filtration, appear normal and remain so, the only abnormality being a fractional excretion of urate (FE_{ur}; C_{ur}/GFR) of more than the normal 9 to 14 per cent; in most patients this has been calculated to be between 30 and 50 per cent, but in one patient the FE_{ur} actually exceeded the creatinine clearance, at 105 per cent (Akaoka *et al.* 1977), even without pharmacological manipulation. Two other patients in the same series had an FE_{ur} of 95 per cent and 88 per cent respectively, and another was reported at 91 per cent (Matsuda *et al.* 1982).

Studies using uricosuric agents with different actions on the complex bidirectional handling of urate in the renal tubule (see Chapter 7.4), such as pyrazinamide, probenecid, and benzbromarone (Garcia Puig *et al.* 1983; Colussi *et al.* 1987; Diamond 1989), suggest considerable heterogeneity in the defect of urate transport in these patients. The supposed action of the drugs suggests that defects in presecretory reabsorption occur in most patients (Greene *et al.* 1973), but others appear to show defective postsecretory reabsorption (Barrientos *et al.* 1979), while in a few both defects (Sperling *et al.* 1974) or even enhanced tubular secretion (Schichiri *et al.* 1987) have been suggested. In one patient, a two to three-fold increase in the T_{max} for *p*-aminohippurate was also documented (Matsuda *et al.* 1982).

Isolated increases in renal excretion of urate have also been described in patients with some malignant tumours, including carcinomas, lymphomas and plasma cell disorders (Mitnick and Beck 1979; Mir and Delamore 1974; Tykarski *et al.* 1988). Not all of the tubular defect is explicable on the basis of a lowered net reabsorption of urate because of volume expansion following inappropriate secretion of vasopressin or vasopressin-like peptides (see Chapter 6.4). Again, heterogeneity of responses to pyrazinamide, benzbromarone, and probenecid have been reported; in some cases the defect reversed following successful treatment of the tumour, and returned with relapse (Bennett *et al.* 1972; Kay and Gottlieb 1973).

Hyperuricaemia

Patients with an isolated reduction in FE_{ur} have been described, most recently by Calabrese *et al.* (1990). These patients present with precocious gout and/or renal failure, and the condition has been termed familial juvenile gouty nephropathy by McCusick. The condition, which is further discussed in Chapter 21.2, is characterized by dominant inheritance of a lowered FE_{ur}, ranging from 3 to 8 per cent. Both sexes are affected, and there may be an excess of females. No studies of drugs affecting urate handling have been described in these patients, and the nature of the defect remains obscure, although clearly a defect of tubular secretion of urate could explain the findings.

Bibliography

Avioli, L. (1980). Effects of vitamin D and its metabolites on renal handling of phosphate. In *The renal handling of phosphate* (ed. S. G. Massry, and H. Fleisch) pp. 197–208. Plenum Press, New York.

Behrman, R.E. and Vaughan, V.C. (eds.) (1987). *Nelson Texbook of Pediatrics*. 13th ed. Philadelphia, Saunders.

Berry, H.K. (1962). Detection of metabolic disorders among mentally retarded children by means of paper spot tests. *American Journal of Mental Deficiency*, **66**, 555–60.

Brodehl, J. (1976). Postnatal development of tubular amino acid reabsorption. In *Amino acid transport and uric acid transport*, (ed. S. Silbernagl, F. Lang, and R. Greger), 128–35. Georg Thieme, Stuttgart.

Chesney, R.W. (1985) Phosphaturic syndromes. In *Renal tubular disorders: pathophysiology, diagnosis and management* (ed. H. C. Gonick and V. M. Buckalew), 201–38. Marcel Dekker, New York and Basel.

Foreman, J.W. and Segal, S. (1985). Aminoacidurias. In *renal tubular disorders: pathophysiology, diagnosis and management* (ed. H. C. Gonick, and V. M. Buckalew), 148–50. Marcel Dekker, New York and Basel.

Forfar, J.O. and Arneil, G.C. (eds.) (1983). *Textbook of Paediatrics*. 3rd ed. Edingburgh, Churchill Livingstone.

Fraser, D. and Salter, R.B. (1958). The diagnosis and management of the various types of rickets. *Pediatric Clinics of North America*, vol. 5, 417–441.

Kinne, R. (1985). Biochemical aspects of tubular transport. In *Renal tubular disorders: pathophysiology, diagnosis and management* (ed. H. C. Gonick and V. M. Buckalew), pp. 1–45. Marcel Dekker, New York.

Krane, S.N. (1978). Renal glycosuria. In *The metabolic basis of inherited disease* (ed. J. B. Stanbury, J. B. Wyngarten, and D. S. Fredrickson), 4th ed., pp. 1607–17. McGraw-Hill, New York.

Levy, H.L. (1973). Genetic screening. In *Progress in human genetics* (ed. H. Herns, and K. Hirschorn), **4**, pp. 1–104. Plenum Press, New York.

Rasmussen, H. and Anast, C. (1983). Familial hypophosphatemic rickets and vitamin-D-dependent rickets. In *The metabolic basis of inherited disease* (ed. J. B. Stanbury, J. B. Wyngaarden, D. S. Frederickson, J. L. Goldstein, and M. S. Brown), 5th ed. pp. 1743–73. McGraw-Hill, New York.

Rasmussen, H. and Tenenhouse, H.S. (1989). Hypophosphatemias. In *The metabolic basis of inherited disease* (ed. C. R. Scriver, A. L. Beaudet, W. S. Sly., and D. Valle), 6th ed. pp. 2581–604. McGraw-Hill, New York.

Scriver, C.R., Beaudet, A.L., Sly, W.L., and Valle, D. (eds.) (1989). *The Metabolic basis of inherited disease*. 6th ed. New York, McGraw-Hill.

Silbernagl, S. (1988). The renal handling of amino acids and oligopeptides. *Physiological Reviews*, **68**, 911–1007.

Thalhammer, O. (1975). Frequency of inborn errors of metabolism, especially PKU, in some representative screening centers around the world: a collaborative study. *Humangenetik*, **30**, 273–86.

References

Agus, Z.S., Puschett, J.B., Senesky, D., and Goldberg, M. (1971). Mode of action of parathyroid hormone and cyclic adenosine 3', 5'-monophosphate on renal tubular phosphate reabsorption in the dog. *Journal of Clinical Investigation*, **50**, 617–26.

Akaoka, I., Nishizawa, T., Yano, E., Kamatani, N., Nishida, Y., and Sasaki, S. (1977). Renal urate secretion in five cases of hypouricemia with isolated defect of urate transport. *Journal of Rheumatology*, **4**, 86–94.

Albright, F., Butler, A.M., and Bloomberg, E. (1937). Rickets resistant to vitamin D therapy. *American Journal of Diseases of Children*, **54**, 529–47.

Alon, U. and Chan, J.C.M. (1985). Effect of hydrochlorothiazide and amiloride in renal hypophosphatemic rickets. *Pediatrics*, **75**, 754–63.

Amiel, C. (1980). Sites of renal tubular reabsorption of phosphate. In *The renal handling of phosphate*, (ed. S. G. Massry and H. Fleisch), pp. 39–55. Plenum Press, New York.

Arant, B.S., Jr., Edelmann, C.M., Jr., and Nash, M.A. (1974) The renal reabsorption of glucose in the developing canine kidney. A study of glomerulotubular balance. *Pediatric Research*, **8**, 638–46.

Arnaud, C., Glorieux, F., and Scriver, C.R. (1971). Serum parathyroid hormone in X-linked hypophosphatemia. *Science*, **173**, 845–7.

Arruda, J.A.L., Nascimento, L., Westenfelder, C., and Kurtzman, N.A. (1977). Effect of parathyroid hormone on urinary acidification. *American Journal of Physiology*, **232**, F423–9.

Asatoor, A.M., et al. (1970). Intestinal absorption of two dipeptides in Hartnup disease. *Gut*, **11**, 380–7.

Asatoor, A.M., Lacey, B.W., London, B.R., and Milne, M.D. (1962). Amino acid metabolism in cystinuria. *Clinical Sciences*, **23**, 285–304.

Aschinberg, L.C., Solomon, L.M., Zeis, P.M., Justice, P.M., and Rosenthal, I.M. (1977). Vitamin D-resistant rickets associated with the epidermal nervus syndrome: demonstration of a phosphaturic substance in the dermal lesions. *Journal of Pediatrics*, **91**, 56–60.

Awrich, A.E., Stackhouse, W.J., Cantrell, J.E., patterson, J.J., and Rudman, D. (1975). Hyperdibasic-aminocaciduria, hyperammonaemia, and growth retardation. Treatment with arginine, lysine and citrulline. *Journal of Pediatrics*, **87**, 731–8.

Balsan, S. and Tieder, M. (1990). Hypophosphataemic rickets and final height. *Lancet*, **335**, 536–7.

Bank, H., Crispin, M., Ehrlich, D., and Szeinberg, A. (1972). Iminoglycinuria: a defect of renal tubular transport. *Israeli Journal of Medical Science*, **8**, 606–12.

Barfuss, D.W., Mays, J.M., and Schafer, J.A. (1980). Peritubular uptake and transepithelial transport of glycine in isolated proximal tubules. *American Journal of Physiology*, **238**, F234–333.

Baron, D.N., Dent, C.E., Harris, H., Hart, E.W., and Jepson, J.B. (1956). Hereditary pellagra-like skin rash with temporary cerebellar ataxia, constant renal amino-aciduria and other bizarre biochemical features. *Lancet*, **1**, 421–8.

Barrientos, A., Perez Diaz, V., Diaz Gonzalez, R., and Rodocio, J.L. (1979). Hypouricemia by defect in the tubular reabsorption. *Archives of Internal Medicine*, **139**, 787–9.

Bennett, J., Bond, J., Singer, I., and Gottlieb, A. (1972). Hypouricemia in Hodgkin's disease. *Annals of Internal Medicine*, **76**, 751–6.

Blazer-Yost, B. and Jezyk, P.F. (1979). Free amino acids in the plasma and urine of dogs from birth to senescence. *American Journal of Veterinary Research*, **40**, 832–8.

Boneh, A., Reade, T.M., Scriver, C.R., and Rishikof, E. (1987). Audiometric evidence for two forms of X-linked hypophosphatemia in humans, apparent counterparts of *Hyp* and *Gy* mutations in mouse. *American Journal of Human Genetics*, **27**, 997–1003.

Bonjour, J.P., Trohler, U., Preston, C., and Fleisch, H. (1978). Parathyroid hormone and renal handling of Pi: effect of dietary Pi and diphosphonates. *American Journal of Physiology*, **234**, D497–F505.

Bostrom, H. and Hambraeus, L. (1964). Cystinuria in Sweden. VII. Clinical, histopathological and medico-sociological aspects of the disease. *Acta Medica Scandinavica*, **175 (Suppl. 411)**, 1–128.

Brand, E., Cahill, G.H., and Harris, M.M. (1935). Cystinuria. II. The metabolism of cysteine, methionine and glutathione. *Journal of Biological Chemistry*, **109**, 69–83.

Brodehl, J. and Gellison, K. (1968). Endogenous renal transport of free amino acids in infancy and childhood. *Pediatrics*, **42**, 395–404.

Brodehl, J., Gellison, K., and Kowalewski, S. (1967). Isolated cystinuria (without lysine-ornithine-argininuria) in a family with hypocalcemic tetany. *Klinische Wochenschrift*, **45**, 38–40.

Brodehl, J., Franken, A., and Gellison, K. (1972). Maximal tubular reabsorption of glucose in infants and children. *Acta Paediatrica Scandinavica*, **61**, 413–20.

Calabrese, G., Simmonds., H.A., Cameron, J.S., and Davies, P.M. (1990). Precocious familial gout with reduced fractional urate clearance and normal purine enzymes. *Quarterly Journal of Medicine*, **75**, 441–50.

Carey, D.E. and Hopfer, S.M. (1987). Hypophosphatemic rickets with hypercalciuria and microglobulinuria. *Journal of Pediatrics*, **111**, 860–3.

Carey, D.E., et al. (1986). Hypophosphatemic rickets/osteomalacia in linear sebaceous nevus syndrome: a variant of tumour-induced osteomalacia. *Journal of Pediatrics*, **109**, 994–1000.

Chen, W.W., Sese, L., Tankatsen, P., and Tricomi, V. (1976). Pregnancy associated with renal glycosuria. *Obstetrics and Gynecology*, **47**, 37–40.

Cheng, L. and Sacktor, B. (1981). Sodium gradient-dependent phosphate transport in renal brush border membrane vesicles. *Journal of Biological Chemistry*, **256**, 1556–64.

Christensen, P. (1958). Tubular reabsorption of glucose during pregnancy. *Scandinavian Journal of Clinical and Laboratory Investigation*, **10**, 364–73.

Colussi, G., *et al.* (1987). Pharmacological evaluation of urate renal handling in humans: Pyrazinamide test vs combined pyrizinamide and probenecid administration. *Nephrology Dialysis Transplantation*, **2**, 10–16.

Cowgill, L.D., Goldfarb, S., Lau, K., Slatopolsky, E., and Agus, Z.S. (1979). Evidence for an intrinsic renal tubular defect in mice with genetic hypophosphatemic rickets. *Journal of Clinical Investigation*, **63**, 1203–10.

Crawhall, J.C., Scowen, E.F., Thompson, C.J., and Watts, R.W.E. (1967). The renal clearance of amino acids in cystinuria. *Journal of Clinical Investigation*, **46**, 1162–70.

Crawhall, J.C., Scowen, E.F., and Watts, R.W.E. (1963). Effects of penicillamine on cystinuria. *British Medical Journal*, **1**, 585–90.

Davison, J.M. and Hytten, F.E. (1975). The effects of pregnancy on the renal handling of glucose. *British Journal of Obstetrics and Gynaecology*, **82**, 374–81.

de Vries, A., Kochwa, S., Lazebnik, J., Frank, M., and Djaldetti, M. (1957). Glycinuria, a hereditary disorder associated with nephrolithiasis. *American Journal of Medicine*, **23**, 408–15.

de Vries, A. and Sperling, O. (1979). Inborn hypouricemia due to isolated renal tubular defect. *Biomedicine*, **30**, 75–80.

Delvin, E.E., Glorieux, F.H. Marie, P.J., and Pettifor, J.M. (1981). Vitamin D dependency: replacement therapy with calcitriol. *Journal of Pediatrics*, **99**, 26–34.

Dent, C.E. and Friedman, M. (1964). Hypercalciuric rickets associated with renal tubular damage. *Archives of Disease in Childhood*, **39**, 240–9.

Dent, C.E. and Rose, G.A. (1951). Amino acid metabolism in cystinuria. *Quarterly Journal of Medicine*, **20**, 205–18.

Diamond, H.S. (1989). Interpretation of pharmacologic manipulation of urate transport in man. *Nephron*, **51**, 1–5.

Eil, C., Liberman, V. A., Rosen, J.F., and Marx, S.J. (1981). A cellular defect in hereditary vitamin D dependent rickets type II: Defective nuclear uptake of 1,25-dihydroxyvitamin D in cultured skin fibroblasts. *New England Journal of Medicine*, **304**, 1588–91.

Elsas, L.J. and Rosenberg, L.E. (1969). Familial renal glycosuria: a genetic reappraisal of hexose transport by kidney and intestine. *Journal of Clinical Investigation*, **48**, 1845–54.

Elsas, L.J., Hillman, R.E., Patterson, J., and Rosenberg, L.E. (1970). Renal and intestinal hexose transport in familial glucose-galactose malabsorption. *Journal of Clinical Investigation*, **49**, 576–85.

Elsas, L.J., Busse, D., and Rosenberg, L.E. (1971). Autosomal recessive inheritance of renal glycosuria. *Metabolsim*, **20**, 968–75.

Evers, C., Murer, H., and Kinne, R. (1978). Effect of parathyrin on the transport properties of isolated renal brush-border vesicles. *Biochemical Journal*, **172**, 49–56.

Feldman, D., *et al.* (1982). Vitamin D resistant rickets with alopecia: cultured skin fibroblasts exhibit defective cytoplasmic receptors and unresponsiveness to 1,25 $(OH)_2D_3$. *Journal of Clinical Endocrinology and Metabolism*, **55**, 1020–2.

Flinter, F.A. and Bobrow, M. (1988). The application of molecular biology to the prenatal diagnosis of renal disease. *Pediatric Nephrology*, **2**, 343–50.

Foreman, J.W., Hwang, S.M., and Segal, S. (1980). Transport interactions of cystine and dibasic amino acids in isolated rat renal cortical tubules. *Metabolism*, **29**, 53–61.

Frame, B., Smith, R.W., Jr., Fleming, J.L., and Hanson, G. (1963). Oral phosphates in vitamin D-refractory rickets and osteomalacia. *American Journal of Diseases of Children*, **106**, 147–53.

Fraser, D. and Scriver, C.R. (1976). Familial forms of vitamin D-resistant rickets revisited: X-linked hypophosphatemia and autosomal recessive vitamin D dependency. *American Journal of Clinical Nutrition*, **29**, 1315–29.

Fraser, D., Kooh, S.W., and Scriver, C.R. (1967). Hyperparathyroidism as the cause of hyperaminoaciduria and phosphaturia in human vitamin D deficiency. *Pediatric Research*, **7**, 425–35.

Fraser, D., Kooh, S.W., Kind, H.P., Holick, M.F., Tanaka, Y., and DeLuca, H.F. (1973). Pathogenesis of hereditary vitamin D-dependent rickets: an inborn error of vitamin D metabolism involving defective conversion of 25-hydroxyvitamin D to 1-alpha, 25-dihydroxyvitamin D. *New England Journal of Medicine*, **289**, 817–22.

Frömter, E. (1982) Electrophysiological analysis of rat renal sugar and amino acid transport. I. Basic phenomena. *Pflügers Archiv* **393**, 179–89.

Garcia Puig, J., *et al.* (1983). Renal handling of uric acid in normal subjects by means of the pyrizinamide and probenecid tests. *Nephron*, **35**, 183–6.

Garrod, A.E. (1908). Inborn errors of metabolism. *Lancet*, **2**, pp 1, 73, 142 and 214.

Gold, R.J.M., Dobrinski, M.J., and Gold, D.P. (1977). Cystinuria and mental deficiency. *Clinical Genetics*, **12**, 329–32.

Goodman, S.I., McIntyre, C.A., and O'Brien, D. (1967). Impaired intestinal transport of proline in a patient with familial iminoaciduria. *Journal of Pediatrics*, **71**, 246–9.

Goodyer, P.R., Kronick, J.B., Jequier, S., Reade, T.M., and Scriver, C.R. (1987). Nephrocalcinosis and its relationship to treatment of hereditary rickets. *Journal of Pediatrics*, **111**, 700–4.

Greenberg, B.G., Winters, R.W., and Graham, J.B. (1960). The normal range of serum inorganic phosphorus and its utility as a discriminant in the diagnosis of congenital hypophosphatemia. *Journal of Clinical Endocrinology*, **20**, 364–79.

Greene, M.L., Marcus, R., Aurbach, G.D., Kazam, E.S., and Seegmiller, J.E. (1972). Hypouricemia due to isolated renal tubular defect. Dalmatian dog mutation in man. *American Journal of Medicine*, **53**, 361–7.

Greene, M.L., Lietman, P.S., Rosenberg, L.E., and Seegmiller, J.E. (1973). Familial hyperglycinemia: new defect in renal tubular transport of glycine and imino acids. *American Journal of Medicine*, **54**, 265–71.

Greger, R., *et al.* (1977a) Site of renal phosphate reabsorption. Micropuncture and microperfusion study. *Pflügers Archive*, **369**, 111–8.

Greger, R., *et al.* (1977b) The postproximal site of phosphate reabsorption in presence and absence of parathyroid hormone. *Advances in Experimental Medicine and Biology*, **81**, 149–51.

Griffin, J.E., Chandler, J.S., Haussler, M.R., and Zerwekh, J.E. (1982). Receptor-positive resistance to 1,25-dihydroxyvitamin D: a new cause of osteomalacia associated with impaired induction of 24-hydroxylase in fibroblasts. *Clinical Research*, **30**, 524A.

Hahn, C.J., Scharp, C.R., Halstead, L.R., Haddad, J.G., Karl, D.M., and Avioli, L.V. (1975). Parathyroid hormone status and renal responsiveness in familial hypophosphatemic rickets. *Journal of Clinical Endocrinology and Metabolism*, **41**, 936–7.

Halvorsen, K. and Halvorsen, S. (1963). Hartnup disease. *Pediatrics*, **31**, 29–38.

Harris, H., Mittwoch, U., Robson, E.B., and Warren, F.L. (1955). Phenotypes and genotypes in cystinuria. *Annals of Human Genetics*, **20**, 57–91.

Harrison, H.E., Harrison, H.C., Lifschitz, F., and Johnson, A.D. (1966). Growth disturbance in hereditary hypophosphatemia. *American Journal of Diseases of Children*, **112**, 290–7.

Hautmann, R., Terhorst, B., Stuhlsatz, H.W., and Lutzeyer, W. (1977). Mercaptopropionylglycine: a progress in cystine stone therapy. *Journal of Urology*, **117**, 628–30.

Hellier, M.D., Perrett, D., and Holdsworth, L.E. (1970). Dipeptide absorption in cystinuria. *British Medical Journal*, **4**, 782–3.

Hisatome, I., et al. (1989). Cause of persistent hypouricemia in outpatients. *Nephron*, **51**, 13–6.

Hjärne, U. (1927). A study of orthoglycaemic glycosuria with particular reference to its heritability. *Acta Medica Scandinavica*, **67**, 422–571.

Hoffmann, N., Thees, M., and Kinne, R. (1976). Phosphate transport by isolated renal brush border vesicles. *Pflügers Archiv*, **362**, 147–56.

Hooft, C., Timmermans, J., Snoeck, J., Antener, I., Oyaert, W., and van den Hende C. (1965). Methionine malabsorption syndrome. *Annals of Pediatrics*, **205**, 73–84.

Hong Que, N.T., Gmaj, P., and Angielski, S. (1982) Uncoupling of Na^+-dependent solute transport in renal brush border membranes of maleate-treated rats. *Acta Biochimica Polonica*, **29**, 275–87.

Hruska, K.A. and Hammerman, M.R. (1981) Parathyroid hormone inhibition of phosphate transport in renal brush border vesicles from phosphate-depleted dogs. *Biochimica et Biophysica Acta*, **645**, 351–6.

Hytten, F.E. and Cheyne, G.A. (1972). The aminoaciduria of pregnancy. *Journal of Obstetrics and Gynaecology of the British Commonwealth*, **79**, 424–32.

Insogna, K.L., Broadus, A.E., and Gertner, J.M. (1983). Impaired phosphorus conservation and 1,25 dihydroxyvitamin D generation during phosphorus deprivation in familial hypophosphatemic rickets. *Journal of Clinical Investigation*, **71**, 1562–9.

Jaeger, P., Portmann, L., Saunders, A., Rosenberg, L.E., and Thier, S.O. (1986). Anticystinuric effects of glutamine and of dietary sodium restriction. *New England Journal of Medicine*, **315**, 1120–3.

Jaffe, I.A., Altmann, K., and Merryman, P. (1964). The antipyridoxine effect of penicillamine in man. *Journal of Clinical Investigation*, **43**, 1869–73.

Joseph, R., Ribierre, M., Job, J.-C., and Girault, M. (1958). Maladie, familiale asociante des convulsions à début très précoce, une albuminorachie et une hyperaminoacidurie. *Archives Françaises de Pédiatrie*, **15**, 374–87.

Käser, H., Cottier, P., and Antener, I. (1962). Glucoglycinuria, a new familial syndrome. *Journal of Pediatrics*, **61**, 386–94.

Kay, M.E. and Gottlieb, A.J. (1973). Hypouricemia in Hodgkin's disease. *Cancer, New York*, **32**, 1508–11.

Khachadurian, A.K. and Khachadurian, L.A. (1964). The inheritance of renal glycosuria. *American Journal of Human Genetics*, **165**, 189–94.

Kiebzak, G.M. and Dousa, T.P. (1985). Thyroid hormones increase renal brush border membrane transport of phosphate in X-linked hypophosphatemic (*Hyp*) mice. *Endocrinology*, **117**, 613–9.

Kinne, R., Murer, H., and Kinne-Saffran, E. (1975). Sugar transport by renal plasma membrane vesicles. Characterisation of the systems in the brush border microvilli and basal-lateral membranes. *Journal of Membrane Biology*, **21**, 375–95.

Lang, F., Greger, R., Knox, F.G., and Oberleithner, H. (1981) Factors modulating the renal handling of phosphate. *Renal Physiology*, **4**, 1–16.

Lang, F., Meissner, G., and Rehwald, W. (1986). Electrophysiology of sodium-coupled transport in proximal renal tubules. *American Journal of Physiology*, **250**, F953–F962.

Lau, K., Goldfarb, S., and Goldberg. M. (1980). The effects of parathyroid hormone on renal phosphate handling. In *The renal handling of phosphate*, (ed. S. G. Massry, and H. Fleisch), pp. 115–135. Plenum Press, New York.

Lawrence, R.D. (1947). Symptomless glycosurias: differentiation by sugar tolerance test. *Medical Clinics of North America*, **31**, 289–97.

Lester, F.T. and Cusworth, D.C. (1973). Lysine infusion in cystinuria: theoretical renal thresholds for lysine. *Clinical Science*, **44**, 99–111.

Liberman, U.A., et al. (1980). End-organ resistance to 1,25-dihydroxycholecalciferol. *Lancet*, **1**, 504–7.

Lindquist, B. and Meeuwisse, G.W. (1962). Chronic diarrhoea caused by monosaccharide malabsorption. *Acta Paediatrica Scandinavica*, **51**, 674–85.

Lingard, J., Rumrich, G., and Young, J.A. (1973). Kinetics of L-histidine transport in the proximal convolution of the rat nephron studies using the stationary microperfusion technique. *Pflügers Archiv*, **342**, 13–28.

Lyles, K.W., Clark, A.G., and Drezner, M.K. (1982). Serum 1, 25-dihydroxyvitamin D levels in subjects with X-linked hypophosphatemic rickets and osteomalacia. *Calcif Tissue International*, **34**, 125–30.

Lyon, M.F., Scriver, C.R., Baker, L.R.I., Tenenhouse, H.S., Kronick, J., and Mandla, S. (1986). The *Gy* mutation: another cause of X-linked hypophosphatemia in mouse. *Proceedings of the National Academy of Sciences of the USA*, **83**, 4899–903.

Machler, M., et al. (1986). X-linked dominant hypophosphatemia is closely linked to DNA markers DXS41 and DCS43 at Xp22. *Human Genetics*, **73**, 271–5.

Marble, A., Joslin, E.P., Dublin, L.I., and Marks, H.H. (1939). Studies in diabetes mellitus. VII. Non-diabetic glycosuria. *American Journal of Medical Sciences*, **197**, 533–56.

Marie, P.J. and Glorieux, F.H. (1981). Histomorphometric study of bone remodeling in hypophosphatemic vitamin D-resistant rickets. *Metabolic Bone Disease and Related Research*, **3**, 31–8.

Marx, S.J., et al. (1978). A familial syndrome of decrease in sensitivity to 1,25-dihydroxyvitamin D. *Journal of Clinical Endocrinology and Metabolism*, **47**, 1303–10.

Matsuda, O., Shiigai, T., Ito, Y., Aonuma, K., and Takeuchi, J. (1982). A case of familial renal hypouricemia associated with increased secretion of para-aminohippurate and idiopathic edema. *Nephron*, **30**, 178–86.

McNair, S.L. and Stickler, G.B. (1969). Growth in familial hypophosphatemic vitamin D-resistant rickets. *New England Journal of Medicine*, **281**, 511–16.

Meeuwisse, G.W. (1970). Glucose-galactose malabsorption: studies on renal glycosuria. *Helvetica Paediatrica Acta*, **25**, 13–24.

Melancon, S.B., Dallaire, L., Lemieux, B., Robitaille, P., and Potier, M. (1977). Dicarboxylic aminoaciduria: an inborn error of amino acid conservation. *Journal of Pediatrics*, **91**, 422–7.

Meyer, M.H., Meyer, R.A., Jr., and Iorio, R.J. (1984). A role for the intestine in the bone disease of juvenile X-linked hypophosphatemic mice: malabsorption of calcium and reduced skeletal mineralization. *Endocrinology*, **115**, 1464–70.

Meyer, R.A., Jr., Meyer, M.H., Erickson, P.R., and Korkor, A.B. (1986). Reduced absorption of ^{45}calcium from isolated duodenal segments 'in vivo' in juvenile but not adult X-linked hypophosphatemic mice. *Calcified Tissue International*, **38**, 95–102.

Meyer, R.A., Jr., Meyer, M.H., and Gray, R.W. (1987). Humoral origin of X-linked hypophosphatemia in mice suggested by parabiosis. *Federation Proceedings*, **46**, 1393.

Milne, M.D., Crawford, M.A., Girao, C.B., and Loughbridge, L. (1960). The metabolic disorder in Hartnup disease. *Quarterly Journal of Medicine*, **29**, 407–21.

Milne, M.D., Asatoor, A.M., Edwards, K.D.G., and Loughridge, L.W. (1961). The intestinal absorption defect in cystinuria. *Gut*, **2**, 323–37.

Mir, M.A. and Delamore, I.W. (1974). Hypouricemia and proximal renal tubular dysfunction in acute myeloid leukaemia. *British Medical Journal*, **3**, 775–777.

Mitnick, P.D. and Beck, L.H. (1979). Hypouricemia and malignant neoplasm. A new case of xanthinuria. *Archives of Internal Medicine*, **139**, 186–7.

Monasterio, G., Oliver, J., Muiesan, G., Pardelli, G., Marinozzi, V., and MacDowell, M. (1964). Renal diabetes as a congenital tubular dysplasia. *American Journal of Medicine*, **37**, 44–61.

Moser, C.R. and Fessel, W.J. (1974). Rheumatic manifestations of hypophosphatemia. *Archives of Internal Medicine*, **134**, 674–8.

Murer, H. (1988). Transport of sulfate and phosphate in small intestine and renal proximal tubule: methods and basic properties. *Comparative Biochemistry and Physiology. A: Comparative Physiology (Oxford)* **90**, 749–55.

Navab, F. and Asatoor, A.M. (1970). Studies on intestinal absorption of amino acids and a dipeptide in a case of Hartnup diease. *Gut*, **11**, 373–9.

Nesbitt, T., Drezner, M.K., and Lobaugh, B. (1986). Abnormal parathyroid hormone stimulation of 25-hydroxyvitamin D-1 alpha-hydroxylase activity in the hypophosphatemic mouse. Evidence for a generalized defect of vitamin D metabolism. *Journal of Clinical Investigation*, **77**, 181–7.

O'Brien, D. and Butterfield, L.J. (1963). Further studies on renal tubular conservation of free amino acids in early infancy. *Archives of Disease in Childhood*, **38**, 437–42.

Omura, K., et al. (1976). Lysine malabsorption syndrome: a new type transport defect. *Pediatrics*, **57**, 102–5.

Pastoriza-Munoz, E, Colindres, R., Lassiter, W., and Lechene, C. (1978). Effect of parathyroid hormone on phosphate reabsorption in rat distal convolution. *American Journal of Physiology*, **235**, F321-30.

Perheentupa, J. and Visakorpi, J.K. (1965). Protein intolerance with deficiency transport of basic amino acids: another inborn error of metabolism. *Lancet*, **2**, 813–16.

Pitts, R.F. (1943). A renal reabsorptive mechanism in the dog common to glycine and creatinine. *Americal Journal of Physiology*, **140**, 156–67.

Polisson, P.P., et al. (1985). Calcification of entheses associated with X-linked hypophosphatemic osteomalacia. *New England Journal of Medicine*, **313**, 1–6.

Poujeol, P., Corman, B., Touvay, C., and de Rouffignac, C. (1977). Phosphate reabsorption in rat terminal nephron segments: intrarenal heterogeneity and strain differences. *Pflügers Archiv*, **371**, 39–44.

Prader, A., Illig, R., and Heierli, E. (1961). Eine besondere Form der primären Vitamin D resistenten Rachitis mit Hypocalcaemie und autosomal dominanten Erbgang: die hereditäre Pseudo-mangelrachitis. *Helvetica Paediatrica Acta*, **16**, 452–68.

Praetorius, E. and Kirk, J.E. (1950). Hypouricemia with evidence for tubular elimination of uric acid. *Journal of Laboratory and Clinical Medicine*, **35**, 865–8.

Procopis, P.G. and Turner, B. (1971). Iminoaciduria: a benign renal tubular defect. *Journal of Pediatrics*, **79**, 419–22.

Puschett, J.B. (1982). Are all of the renal tubular actions of parathyroid hormone mediated by the adenylate cyclase system? *Mineral and Electrolyte Metabolsim*, **7**, 281–4.

Quamme, G., Pfeilschifter, J., and Murer, H. (1989a). Parathyroid hormone inhibition of Na^+/phosphate cotransport in OK cells: requirement of protein kinase C-dependent pathway. *Biochimica et Biophysica Acta*, **1013**, 159–65.

Quamme, G., Pfeilschifter, J., and Murer, H. (1989b). Parathyroid hormone inhibition of Na^+/phosphate cotransport in OK cells: intracellular $[Ca^{2+}]$ as a second messenger. *Biochimica et Biophysica Acta*, **1013**, 166–72.

Rajantie, J., Simell, O., Rapola, J., and Perheentupa, J. (1980). Lysinuric protein intolerance: a two year trial of dietary supplementation therapy with citrulline and lysine. *Journal of Pediatrics*, **97**, 927–32.

Rakocz, M., Keating, J., III., and Johnson, R. (1982). Management of the primary dentition in vitamin D-resistant rickets. *Oral Surgery, Oral Medicine, Oral Pathology*, **54**, 166–71.

Reade, T.M., et al. (1975). Response to crystalline 1-alpha-hydroxyvitamin D_3 in vitamin D dependency. *Pediatric Research*, **9**, 593–9.

Reubi, F.C. (1954). Glucose titration in renal glycosuria. In *Ciba Foundation Symposium on the Kidney* (ed. A.A.G. Lewis and G.E.W. Wolstenholme), pp. 96–106. J. and A. Churchill, London.

Reusz, G.S., Brodehl, J., Krohn, H-P., and Ehrich, J.H.H. (1990). Hypophosphataemic rickets. *Lancet*, **335**, 178.

Robson, E.B. and Rose, G.A. (1957). The effects of intravenous lysine on the renal clearances of cystine, arginine and ornithine in normal subjects, in patients with cystinuria and their relatives. *Clinical Science*, **16**, 75–91.

Roof, B.S., Piel, C.F., and Gordan, G.S. (1972). Nature of defect for familial vitamin D-resistant rickets (VDRR) based on radio-immunoassay for parathyroid hormone (PTH). *Transactions of the Association of American Physicians*, **85**, 172–80.

Rosenberg, L.E. and Hayslett, J.P. (1967). Nephrotoxic effects of penicillamine in cystinuria. *Journal of the American Medical Association*, **201**, 698–9.

Rosenberg, L.E., Downing, S., Durant, J.L., and Segal, S. (1966). Cystinuria: biochemical evidence for three genetically distinct diseases. *Journal of Clinical Investigation*, **45**, 365–71.

Sabater, J., Ferre, C., Pulioli, M., and Maya, A. (1976). Histidinuria: a renal and intestinal histidine transport deficiency found in two mentally retarded children. *Clinical Genetics*, **9**, 117–24.

Sacktor, B. (1978). Mechanisms and specifications of amino acid transport in proximal tubule luminal membrane vesicles. In *Renal function* (ed. G. H. Giebisch, and E. F. Purcell), pp. 221–29. Josiah Macy Jr. Foundation, New York.

Saville, P.D., Nassim, J.R., Stevenson, F.H., Mulligan, L., and Carey, M. (1955). The effect of A.T. 10 on calcium and phosphorus metabolism in resistant rickets. *Clinical Science*, **14**, 489–99.

Schichiri, M., Iwamoto, H., and Shiigai, T. (1987). Hypouricemia due to increased tubular urate secretion. *Nephron*, **45**, 31–4.

Scriver, C.R. and Shaw, K.N. (1962). Hartnup disease: an example of genetically determined defective cellular amino acid transport. *Canadian Medical Association Journal*, **86**, 232.

Scriver, C.R., Efron, M.L., and Schafer, I.A. (1964a). Renal tubular transport of proline, hydroxyproline and glycine in health and in familial hyperprolinema. *Journal of Clinical Investigation*, **43**, 374–85.

Scriver, C.R., Goldbloom, R.B., and Roy, C. (1964b). Hypophosphatemic rickets with renal hyperglycinuria, renal glycosuria and glycylprolinuria. *Pediatrics*, **34**, 357–71.

Scriver, C.R., Whelan, D.T., Clow, C.L., and Dallaire, L. (1970). Cystinuria: increased prevalence in patients with mental disease. *New England Journal of Medicine*, **283**, 783–6.

Scriver, C.R., Reade, T.M., deLuca, H.F., and Hamstra, A.J. (1978). Serum 1,25-dihydroxyvitamin D levels in normal subjects and patients with hereditary rickets or bone disease. *New England Journal of Medicine*, **299**, 976–80.

Seino, Y., et al. (1984). Activity of renal 25-hydroxyvitamin D_3-1 alpha hydroxylase in a case of X-linked hypophosphatemic rickets. *European Journal of Pediatrics*, **142**, 219–22.

Semple, P.F., Carswell, W., and Boyle, J.A. (1974). Serial studies of the renal clearances of urate and inulin during pregnancy and after the puerperium in normal women. *Clinical Science and Molecular Medicine*, **47**, 559–65.

Shannon, J.A. and Fisher, S. (1938). The renal tubular reabsorption of glucose in the normal dog. *American Journal of Physiology*, **122**, 765–74.

Shih, V.E., Bixby, E.M., Alpers, D.H., Bartosocas, C.S., and Thier, S.O. (1971). Studies of intestinal transport defect in Hartnup disease. *Gastroenterology*, **61**, 445–53.

Simell, O., Perheentupa, J., Rapola, J., Visakorpi, J., and Eskelin, L.-E. (1975). Lysinuric protein intolerance. *American Journal of Medicine*, **59**, 229–40.

Smith, A.J. and Strang, L.B. (1958). An inborn error of metabolism with the urinary excretion of alpha-hydroxybutyric acid and phenylpyruvic acid. *Archives of Disease in Childhood*, **33**, 109–13.

Smith, A., Yu, J.S., and Brown, D.A. (1979). Childhood cystinuria in New South Wales: results in children who were followed up after being detected by urinary screening in infancy. *Archives of Disease in Childhood*, **54**, 676–81.

Smith, H.W., Goldring, W., Chasis, H., Ranges, H.A., and Bradley, S.E. (1943). The application of saturation methods to the study of glomerular and tubular function in the human kidney. *Journal of the Mount Sinai Hospital*, **10**, 59–108.

Sperling, O., Boler, P., Weinberger, A., and de Vries, A. (1974). Hypouricemia, hypercalciuria and decreased bone density: a hereditary syndrome. *Annals of Internal Medicine*, **80**, 482–7.

Stamp, T. and Goldstein, A.J. (1990). Hypophosphataemic rickets and final height. *Lancet*, **335**, 536.

Steele, T.H. and de Luca, H.F. (1976). Influence of dietary phosphorus on renal phosphate reabsorption in the parathyroidectomized rat. *Journal of Clinical Investigation*, **57**, 867–74.

Steendijk, R. and Herweijer, T.J. (1984). Height, sitting height and leg length in patients with hypophosphatemic rickets. *Acta Paediatrica Scandinavica*, **73**, 181–4.

Steendijk, R. and Latham, S.C. (1971). Hypophosphatemic vitamin D-resistant rickets; an observation on height and serum inorganic phosphate in untreated cases. *Helvetica Paediatrica Acta*, **26**, 179–84.

Stickler, G.B. (1963). External calcium and phosphorus metabolism in resistant rickets. *Journal of Pediatrics*, **63**, 942–8.

Stickler, G.B., Beabout, J.W., and Riggs, B.L. (1970). Vitamin D-resistant rickets: clinical experience with 41 typical familial hypophosphatemic patients and 2 atypical non-familial cases. *Mayo Clinic Proceedings*, **45**, 197–218.

Stickler, G.B. and Morgenstern, B.Z. (1989). Hypophosphataemic rickets: final height and clinical symptoms in adults, **2**, 902–5.

Stoll, R., Kinne, R., Murer, H., Fleisch, H., and Bonjour, J.P. (1979). Phosphate transport by rat renal brush border membrane vesicles: influence of dietary phosphate, thyroparathyroidectomy and 1,25-dihydroxyvitamin D_3. *Pflügers Archiv*, **380**, 47–52.

Takeda, E., *et al.* (1987). 1 alpha-hydroxyvitamin D_3 treatment of three patients with 1,25 dihydroxyvitamin D-receptor-defect rickets and alopecia. *Pediatrics*, **80**, 97–101.

Tancredi, G., Guazzi, G., and Auricchio, S. (1970). Renal iminoglycinuria without intestinal malabsorption of glycine and imino acids. *Journal of Pediatrics*, **76**, 386–92.

Tarwalkar, Y.B., Musgrave, J.E., Buist, N.R.M., Campbell, R.A., and Campbell, J.R. (1974). Vitamin D-resistant rickets and parathyroid adenomas: renal transport of phosphate. *American Journal of Diseases of Children*, **128**, 704–8.

Teijema, H.L., van Gelderen, H.H., Geisberts, M.A.H. and Laurent de Angulo, M.S.L. (1974). Dicarboxylic aminoaciduria: and inborn error of glutamate and aspartate transport with metabolic implications, in combination with a hyperprolinemia. *Metabolism*, **23**, 115–23.

Tenenhouse, H.S. and Scriver, C.R. (1978). The defect in transcellular transport in the nephron is located in brush-border membranes in X-linked hypophosphatemia (*Hyp* mouse model). *Canadian Journal of Biochemistry*, **56**, 640–6.

Tenenhouse, H.S. and Scriver, C.R. (1979). Renal adaptation to phosphate deprivation in the *Hyp* mouse with X-linked hypophosphatemia. *Canadian Journal of Biochemistry*, **57**, 938–44.

Tenenhouse, H.S., Scriver, C.R., McInnes, R.R., and Glorieux, F.H. (1978). Renal handling of phosphate *in vivo* and *in vitro* by the X-linked hypophosphatemic male mouse: evidence for a defect in the brush border membrane. *Kidney International*, **14**, 236–44.

Tenenhouse, H.S., Yip, A., and Jones, G. (1988). Increased renal catabolism of 1,25-dihydroxyvitamin D_3 in murine X-linked hypophosphatemic rickets. *Journal of Clinical Investigation*, **81**, 461–5.

Thakker, R.V., *et al.* (1987). Bridging markers defining the map position of X-linked hypophosphatemic rickets. *Journal of Medical Genetics*, **24**, 756–60.

Thier, S., Fox, M., Segal, S., and Rosenberg, L.E. (1964). Cystinuria: in vitro demonstration of an intestinal transport defect. *Science*, **143**, 482–4.

Tieder, M., *et al.* (1985). Hereditary hypophosphatemic rickets with hypercalciuria. *New England Journal of Medicine*, **312**, 611–7.

Tieder, M. (1987). 'Idiopathic' hypercalciuria and hereditary hypophosphatemic rickets. *New England Journal of Medicine*, **316**, 125–9.

Trohler, U., Bonjour, J.P., and Fleisch, H. (1976). Inorganic phosphate homeostasis: renal adaptation to the dietary intake in intact and thyroparathyroidectomized rats. *Journal of Clinical Investigation*, **57**, 264–73.

Tsuchuya, T., *et al.* (1980). An unusual form of vitamin D-dependent rickets in a child: alopecia and marked end-organ insensitivity to biologically active vitamin D. *Journal of Clinical Endocrinology and Metabolism*, **51**, 685–90.

Tudvad, F. (1949). Sugar reabsorption in prematures and full term babies. *Scandinavian Journal of Clinical Investigation*, **1**, 218–83.

Tulloch, E.N. and Andrews, F.F.H. (1983). The association of dental abscesses with vitamin D-resistant rickets. *British Dental Journal*, **154**, 136–8.

Turner, R.J. and Moran, A. (1982). Heterogeneity of sodium-dependent D-glucose transport sited along the proximal tubule: evidence from vesicle studies. *American Journal of Physiology*, **242**, F406–14.

Tykarski, A. (1988). Mechanism of hypouricemia in Hodgkin's disease. Isolated defect in post-secretory reabsorption of uric acid. *Nephron*, **50**, 217–19.

Völkl, H., Silbernagl, S., and Deetjen, P. (1979). Kinetics of L-proline reabsorption in rat kidney by continuous microperfusion. *Pflügers Archiv*, **382**, 115–21.

von Udransky, L. and Baumann, E. (1889). über das Vorkommen von Diaminen, sogennaten Ptomainen, bei Cystinurie. *Zeitschrift für Physiologische Chemie*, **13**, 562–94.

Walton, R.J. and Bijvoet, O.L.M. (1975). Nomogram for the derivation of renal tubular threshold phosphate concentration. *Lancet*, **2**, 309–10.

Webber, W.A. (1967). Amino acid excretion patterns in developing rats. *Canadian Journal of Physiology and Pharmacology*, **45**, 867–72.

Webber, W.A., Brown, J.L., and Pitts, R.F. (1961). Interactions of amino acids in renal tubular transport. *American Journal of Physiology*, **200**, 380–6.

Weinberger, A., Sperling, O., Rabinovitz, M., Brosh, S., Adam, A., and de Vries, A. (1974). High frequency of cystinuria among Jews of Libyan origin. *Human Heredity*, **24**, 568–72.

Welsh, G.W. and Sims, E.A.H. (1960). The mechanism of renal glycosuria in pregnancy. *Diabetes*, **9**, 363–9. West, C.D., Blanton, J.C., Siverman, F.N., and Holland, N.H. (1964). Use of phosphate salts as an adjunct to vitamin D in the treatment of hypophosphatemic vitamin D-refractory rickets. *Journal of Pediatrics*, **64**, 469–77.

Whelan, D.T. and Scriver, C.R. (1968a). Hyperdibasic-aminoaciduria: an inherited disorder of amino acid transport. *Pediatric Research*, **2**, 525–39.

Whelan, D.T. and Scriver, C.R. (1968b). Cystathioninuria and iminoglycinuria in a pedigree: a perspective on counseling. *New England Journal of Medicine*, **278**, 924–7.

Winters, R.W., Graham, J.B., Williams, T.F., McFalls, V.W., and Burnett, C.H. (1958). A genetic study of familial hypophosphatemia and vitamin D-resistant rickets with a review of the literature. *Medicine*, **37**, 97–142.

Wollaston, W.H. (1810). On cystic oxide: a new species of urinary calculus. *Transactions of the Royal Society of London*, **100**, 223–30.

Woolf, L.I., Goodwin, B.L., and Renold, A.E. (1966). Tm-limited reabsorption and the genetics of renal glycosuria. *Journal of Theoretical Biology*, **11**, 10–21.

Wright, H.R., Russo, H.F., Skeggs, H.R., Patch, E.K., and Beyer, K.H. (1947). The renal clearance of essential amino acids: arginine, histidine, lysine and methionine. *American Journal of Physiology*, **149**, 130–4.

Wright, S.H., Kippen, I., Klinenberg, J.R., and Wright, E.M. (1980). Specificity of the transport system for tricarboxylic acid cycle intemediates in renal brush borders. *Journal of Membrane Biology*, **57**, 73–82.

Yü. T.-F., Adler, M., Bobrow, E., and Gutman, A.B. (1969). Plasma and urinary amino acids in primary gout, with special references to glutamine. *Journal of Clinical Investigation*, **48**, 885–94.

5.3 Renal tubular acidosis

J. RODRIGUEZ-SORIANO

Definitions and classification

The term renal tubular acidosis is applied to a group of transport defects in the reabsorption of filtered bicarbonate, the excretion of hydrogen ion, or both. The tubular acidosis syndromes commonly occur in patients with relatively normal glomerular function. They are associated with hyperchloraemia and a normal plasma anion gap, that is, sodium − (chloride + bicarbonate) = 8–16 mmol/l (Emmett and Narins 1977). In contrast, the term uraemic acidosis is applied to those with advanced renal insufficiency in whom metabolic acidosis is accompanied by normo- or hypochloraemia and an increase in the plasma anion gap (Warnock 1988).

On clinical and pathophysiological grounds, tubular acidosis can be separated into several types (McSherry 1981; Batlle 1983) (Table 1) as follows.

Proximal tubular acidosis (type 2)

Under ordinary circumstances, virtually all filtered bicarbonate is reabsorbed along the nephron. If the plasma bicarbonate exceeds the renal excretion threshold (by exogenous administration of alkali, for example), bicarbonate reabsorption is incomplete and the increased urinary excretion gradually lowers the plasma concentration to a level below the threshold. Bicarbonate excretion then ceases and new steady state is reached. The renal bicarbonate threshold varies with age: it is 25 to 26 mmol/l in adults, 23 to 24 mmol/l in children and 22 mmol/l in infants (Edelmann *et al.* 1967; Broyer *et al.* 1969).

Patients with proximal acidosis have a diminished renal bicarbonate threshold; bicarbonate is excreted by them at lower plasma levels than in normal individuals of similar age. For them, a steady state is maintained when the plasma bicarbonate is in the acidaemic range. A characteristic feature is their intact ability to lower urinary pH and to excrete adequate amounts of titrable acid and ammonium when their plasma bicarbonate lies below their individual renal threshold. Conversely, when their plasma bicarbonate is normalized by the administration of alkali, urinary bicarbonate wastage becomes apparent; the urine is highly alkaline and contains a large amount of filtered bicarbonate (more than 10 per cent).

Proximal tubular acidosis may be primary or isolated but more often occurs as part of Fanconi's syndrome.

Distal tubular acidosis (type 1)

This type is characterized by the inability to lower the urinary pH maximally (below 5.5) under the stimulus of systemic acidaemia. The impaired excretion of titrable acid and ammonium is secondary to this defect. In general, bicarbonate reabsorption is quantitatively normal, but, because of the elevated pH, there will be a degree of bicarbonaturia (less than 5 per cent of the filtered load).

Distal acidosis can develop when there is a true failure of the distal nephron to secrete hydrogen ion ('secretory' defect) or when this capacity is intrinsically intact but secondarily impaired. Such non-secretory defects are caused either by the inability of the distal nephron to generate and maintain a negative potential difference across the luminal epithelium ('voltage-dependent' defect) or by its inability to create a steep hydrogen ion gradient between lumen and tubular cells ('gradient' defect). Patients with distal tubular acidosis usually have normo- or hypokalaemia, but when there is a severe 'voltage-dependent' defect the associated impairment in potassium secretion will lead to hyperkalaemia (Batlle 1981; Batlle *et al.* 1981; Kurtzman 1983).

The concept of distal tubular acidosis has now been expanded to include patients whose urinary pH is appropriately low during systemic acidaemia but who have a low P_{CO_2} in alkaline urine, indicating a defect in the 'rate' of hydrogen ion secretion (Batlle *et al.* 1982a). It is not yet clear if such patients have a new and

Table 1 Pathophysiological classification of renal tubular acidosis (RTA)

Type	Pathophysiology
Proximal RTA (type 2)	Impaired proximal HCO_3^- reabsorption
Isolated	Isolated defect in proximal HCO_3^- reabsorption
Associated with other tubular defects	Multiple defect in proximal tubular reabsorption (Fanconi's syndrome)
Distal RTA (type 1)	Impaired distal H^+ secretion
Secretory defect ('classic' distal RTA)	H^+-pump failure
Gradient defect	Increased back-leak of secreted H^+
Voltage-dependent defect	Reduced luminal electronegativity
Rate-dependent defect	Intact distal acidification but reduced ability to increase urine P_{CO_2}
Combined proximal and distal RTA	Impaired distal H^+ secretion associated with impaired HCO_3^- reabsorption
Distal RTA with HCO_3^- wasting in infants and young children	Permanent H^+-pump failure associated with a transient defect in proximal HCO_3^- reabsorption
Hyperkalaemic RTA (type 4)	Intact distal acidification but impaired ammoniagenesis

different type of distal acidosis or simply a similar but milder defect.

In children, distal tubular acidosis usually appears as a primary, genetic disorder, whereas in adults it is most often secondary and acquired.

Combined proximal and distal tubular acidosis (type 3)

In some patients it is difficult to distinguish between proximal and distal acidosis because they have features of both. There is a striking reduction in their tubular reabsorption of bicarbonate but, in contrast to pure proximal acidosis, they also cannot maximally acidify their urine despite severe systemic acidaemia. With a normal or reduced plasma bicarbonate, the urinary loss of base may exceed the net excretion of acid and thus contribute significantly to the development of acidosis (Morris et al. 1972; Rodriguez-Soriano et al. 1975).

This pattern of distal tubular acidosis with associated bicarbonate wasting (once termed type 3 renal tubular acidosis) is usually found in infants and young children who apparently have a primary distal acidosis; it is not a separate form.

Hyperkalaemic tubular acidosis (type 4)

This type accompanies a large number of hyperkalaemic states and is probably not a single pathophysiological entity. It is characterized by a normal ability to acidify the urine after an acid load and to increase its P_{CO_2} after a bicarbonate load, provided that glomerular function is not greatly impaired. Renal bicarbonate reabsorption is reduced at normal plasma concentrations of bicarbonate but this reduction (less than 10 per cent of the filtered load) is not enough to implicate an associated proximal defect. The tubular abnormality concerns the ability to excrete sufficient ammonium and is a direct consequence of the increased cellular stores of potassium. Aldosterone deficiency plays an important contributory role (Batlle 1981).

This type of tubular acidosis occurs most frequently in adults with low aldosterone and generally it appears with an underlying renal parenchymal disease.

The physiology of renal acidification

Mechanisms of proximal bicarbonate reabsorption

Filtered bicarbonate must be completely reabsorbed to allow for the full development of distal mechanisms of urinary acidification. The proximal convoluted tubule normally reabsorbs 80 to 90 per cent of filtered bicarbonate; reabsorption continues along the proximal straight tubule but little or no bicarbonate is reabsorbed in Henle's loop, so that less than 5 to 10 per cent of filtered bicarbonate is offered for reabsorption in the distal segments (Cogan et al. 1979). The reabsorption of bicarbonate is mainly accomplished by active cellular transport, which includes a luminal Na^+/H^+ antiporter, working in parallel with a basolateral membrane Na^+–HCO_3^- symport (Aronson 1983) (Fig. 1).

In the proximal tubules, carbonic acid (H_2CO_3) is formed within the cell by the hydration of CO_2, a reaction catalysed by a soluble cytoplasmic carbonic anhydrase C (or II). This enzyme is in functional contact with the luminal fluid. The H_2CO_3 ionizes and the hydrogen ion is secreted in exchange for luminal sodium (Na^+/H^+ antiporter). This mechanism is electroneutral, driven

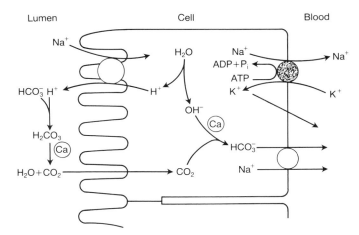

Fig. 1 Cellular mechanism of bicarbonate reabsorption in the proximal convoluted tubule.

by a lumen-to-cell sodium gradient, and stimulated by intracellular acidosis and inhibited by high concentrations of amiloride. In addition to Na^+/H^+ exchange, active hydrogen ion transport, mediated by either a H^+-ATPase or a redox-driven H^+-pump and located at the brush border, may also contribute to reabsorbing part of luminal bicarbonate. Bicarbonate generated within the cell leaves it across the basolateral membrane by passive 3 HCO_3^-–1 Na^+ cotransportation. The secreted hydrogen ion reacts with filtered bicarbonate to form luminal carbonic acid, which quickly dissociates into CO_2 and water by the luminal action of membrane-bound carbonic anhydrase. Luminal CO_2 can freely diffuse back into the cell to complete the reabsorptive cycle.

The proximal tubule is not completely impermeable to bicarbonate and this anion can back-diffuse passively along the paracellular pathways. As much as 20 per cent of filtered bicarbonate may be reabsorbed by this carbonic anhydrase-independent process.

Regulation of proximal bicarbonate reabsorption

Numerous microperfusion experiments and in-vivo studies have examined the role of various physiological determinants of proximal bicarbonate reabsorption (Cogan and Alpern 1984; Maddox et al. 1987).

Luminal bicarbonate concentration

Active bicarbonate reabsorption displays saturation kinetics with respect to the luminal concentration of bicarbonate. Studies in the rat have shown that when luminal bicarbonate concentration exceeds 45 mmol/l, transcellular proton secretion reaches a maximal rate, although total bicarbonate reabsorption still increases due to passive back-diffusion of non-dissociated bicarbonate.

Luminal flow rate

Bicarbonate absorption in the rat proximal convoluted tubule increases with flow rate, so it has become customary to express absolute bicarbonate reabsorption per unit of GFR. The effect of flow rate is mainly due to changes in the luminal bicarbonate concentration.

Peritubular bicarbonate concentration and P_{CO_2}

Alterations in peritubular bicarbonate concentration have the opposite effect on proximal acidification to that of changes in luminal bicarbonate concentration. Metabolic acidosis stimulates bicarbonate reabsorption and, conversely, metabolic alkalosis suppresses it. Changes in peritubular bicarbonate concentration predominantly regulate the rate of proton secretion, since cell pH is modified by changes in the driving force for bicarbonate leaving across the basolateral membrane.

It is also well established that a decrease in plasma P_{CO_2} inhibits proximal bicarbonate reabsorption and that, conversely, a chronic elevation in plasma P_{CO_2} stimulates it. However, acute increases in P_{CO_2} lead to very little or no change in proximal bicarbonate reabsorption *in vivo*. The influence of plasma P_{CO_2} on proximal acidification is also mediated by changes in cellular pH.

Extracellular fluid volume

Fractional reabsorption of bicarbonate, as with that of sodium chloride, is suppressed by expansion and enhanced by contraction of the extracellular fluid volume. That there is no apparent Tm for bicarbonate reabsorption when this is measured in volume-contracted subjects indicates that the long-standing observation of a maximal rate of bicarbonate reabsorption (at 2.8 mmol/100 ml glomerular filtration) was indeed the experimental consequence of volume expansion itself (Kurtzman 1970). The rate of bicarbonate to sodium reabsorption is not influenced by changes of effective arterial blood volume and allows the assessment of other factors acting on bicarbonate reabsorption.

The mechanism whereby changes in the volume of the extracellular fluid influence bicarbonate reabsorption is only partly known. It may be a change in luminal permeability to bicarbonate that alters the amount passively back-diffused or also a secondary change in proton secretion caused by altered absorption of sodium chloride.

Chloride

Metabolic alkalosis is invariably present in states of chloride depletion. As chloride depletion is almost always associated with contraction of the extracellular fluid volume, it was earlier concluded that proximal bicarbonate absorption was enhanced through changes in volume homeostasis and that volume expansion was essential for correction of chloride-depletion alkalosis (Jacobson and Seldin 1983). It now appears that bicarbonate reabsorption is directly enhanced by chloride deficiency and that alkalosis may be corrected by provision of chloride without volume expansion or changes in the intranephronal distribution of fluid reabsorption (Rosen *et al.* 1988). Chloride does not, however, play a direct regulatory role in bicarbonate reabsorption. The effect of chloride depletion on the maintenance of chloride-depletion alkalosis may be due either to a decrease in GFR through tubuloglomerular feedback or to a stimulation of distal acidification through linked proton and chloride secretion, independent of sodium.

Potassium

Although microperfusion experiments in rats have clearly shown that bicarbonate reabsorption is markedly increased where there is potassium depletion, there is no convincing proof that the metabolic alkalosis in potassium-depleted subjects is related to potassium depletion *per se* rather than to accompanying factors, such as chloride deficiency or mineralocorticoid excess (Jacobson and Seldin 1983). Conversely, the metabolic acidosis frequently encountered in hyperkalaemia may depend on mineralocorticoid deficiency or impaired renal ammoniagenesis rather than on a reduced proximal reabsorption of bicarbonate.

Parathyroid hormone, calcium, and phosphate

Acute administration of parathyroid hormone suppresses proximal reabsorption of filtered bicarbonate, through a change in active proton secretion mediated by an inhibition of Na^+/H^+ antiporter activity (Bichara *et al.* 1986). The clinical relevance of such acute experiments is a subject of discussion: although some cases of primary hyperparathyroidism have features of metabolic acidosis, there is also evidence that in man chronic administration of parathyroid hormone or $1,25\text{-}(OH)_2$ vitamin D results in sustained metabolic alkalosis. This is due to an increase in the set point at which plasma bicarbonate concentration is regulated by the kidney (Hulter 1985). Parathyroid hormone, by mobilizing skeletal alkali, is a major determinant of extrarenal buffering capacity and this effect on bone tends also to increase the plasma bicarbonate. The net effect of the hormone on plasma bicarbonate would be the result not only of its circulating level but also of the state of renal function, the capacity of the bone to buffer acid loads, the level of cytosolic calcium and the state of the body stores of phosphate (Arruda and Kurtzman 1980).

The effect of extracellular calcium on the renal handling of bicarbonate is unclear because plasma calcium and parathyroid hormone are intimately connected and because acute variations in calcaemia also induce profound changes in haemodynamics that in turn may affect urinary acidification. There is evidence, however, that acute hypocalcaemia enhances renal absorption of bicarbonate and that acute hypercalcaemia depresses it. The specific effect of chronic hypercalcaemia on renal bicarbonate reabsorption (independent of variations in parathyroid hormone or $1,25\text{-}(OH)_2$ vitamin D_3) is not completely elucidated because the possible renal effect is largely masked by the extrarenal effects. When chronic hypercalcaemia is induced by sustained infusion of calcium salts, a metabolic acidosis ensues due to release of hydrogen ion from bone tissue; when chronic hypercalcaemia is induced by prolonged administration of parathyroid hormone or $1,25\text{-}(OH)_2$ vitamin D_3, the base released from bone leads to metabolic alkalosis (Hulter 1985).

In the dog, phosphate depletion causes a decrease in proximal bicarbonate reabsorption, but this effect is not observed in the rat or in man. As pointed out by Arruda and Kurtzman (1980), the site of the nephron involved and the type of defect of urinary acidification observed may vary between species and in relation to the degree of phosphate depletion. Moreover, phosphate depletion enhances extrarenal buffering capacity in the rat, an effect that counteracts the tendency to cause acidaemia.

Neurohormonal factors

In the rat, experiments have shown that a change in renal nerve activity may affect sodium bicarbonate and sodium chloride reabsorption in the superficial proximal tubules: an increase in α-adrenergic tone increases bicarbonate reabsorption while denervation depresses it (Cogan 1986). Some hormones, such as angiotensin II, may also stimulate bicarbonate reabsorption in the rat proximal tubule (Liu and Cogan 1987).

Mechanisms of distal urinary acidification

The distal nephron performs three functions in urinary acidification (Simpson 1971): (1) it reclaims the small fraction of bicarbonate (less than 10 per cent) that escapes proximal reabsorption; (2) it titrates divalent basic phosphate (HPO_4^{2-}), which is converted to the monovalent acid form ($H_2PO_4^-$) or titratable acid; and (3) it accumulates ammonia (NH_3) intraluminally, which buffers hydrogen ion to form non-diffusible ammonium (NH_4^+). Hydrogen ion secretion and titration of non-bicarbonate urinary buffers lowers the urinary pH to as little as 4.5 under conditions of acid loading. Distal secretion of hydrogen ion generates an equimolar amount of bicarbonate in the renal tubular cells, which is delivered to the blood and body fluids. To maintain acid–base equilibrium, the amount of bicarbonate generated (once filtered bicarbonate is completely reabsorbed) should be equivalent to the amount of hydrogen ion delivered to the body fluids through catabolism of foodstuffs, approximately 1 mmol/kg.day. In infancy and childhood, endogenous hydrogen ion input is greater (about 2 mmol/kg.day) because skeletal growth releases hydrogen ions from bone.

Secretion of hydrogen ion

Most knowledge of the mechanisms of distal urinary acidification comes from *in-vitro* studies of the toad and turtle urinary bladders, which resemble the mammalian collecting duct in their capacity to develop large pH gradients between serosal and mucosal membranes (Al-Awqati 1978; Steinmetz 1986). In the isolated bladder, hydrogen ion secretion results from a specific hydrogen ion pump, that is, an electrogenic mechanism not directly linked to sodium reabsorption and mainly present in the α-type of carbonic anhydrase-rich cells. The pump is an H^+-translocating ATPase (Stone and Xie 1988), and the movement of protons against a gradient consumes energy, which comes from the hydrolysis of ATP. The rate of hydrogen ion secretion is highly dependent on the electrochemical gradient for protons across the mucosal membrane. Although hydrogen ion secretion is sodium-independent, sodium transport influences the rate of proton secretion by generating a lumen-negative potential difference across the epithelium. The amphibian bladders can also secrete bicarbonate. This secretion takes place in the β-type of carbonic anhydrase-rich cells, which have reversed polarity of the membrane elements to that of the α-type.

The function of the proton pump is also influenced by the availability of cellular hydrogen ion, which depends on the rapid exit of formed bicarbonate from the cell. Formation of bicarbonate is catalysed by intracellular carbonic anhydrase; its removal is facilitated by an electroneutral mechanism involving 1:1 chloride/bicarbonate exchange. The protein transporting the anions through the serosal membrane (band 3 protein) is probably identical to that present in erythrocyte membranes. Inhibiting carbonic anhydrase by acetazolamide or blocking bicarbonate exit by disulphonic stilbenes will have a direct effect of hydrogen ion secretion due to the increase in intracellular pH.

The application of knowledge obtained from the voltage-influenced, amphibian urinary bladders to interpretation of the mammalian collecting duct suggests that there may be two functionally distinct segments in the cortex and medulla (Madsen and Tisher 1986). In the cortical collecting tubule, the intercalated cells are involved in both hydrogen ion and bicarbonate secretion, whilst the principal cells appear to be in charge of sodium reabsorption and potassium secretion. The concept is that there

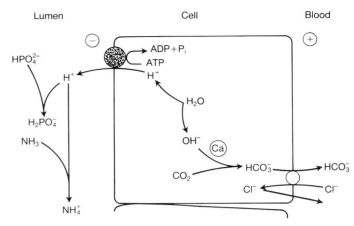

Fig. 2 Mechanism of hydrogen ion secretion in the A-intercalated cell of the cortical collecting tubule.

are two populations of intercalated cells which differ both functionally and morphologically: the A cell is similar to the α-type of carbonic anhydrase-rich cell of the amphibian bladder and is responsible for hydrogen ion secretion; the B cell is similar to the β-type cell and is responsible for bicarbonate secretion. Band 3 protein is present only in the basolateral membranes of A cells, thus supporting the notion that B cells are functionally and structurally different. The cortical collecting tubule is lumen-negative: hydrogen ion secretion, although dependent on the proton pump, is also much influenced by active sodium transport; agents that decrease sodium reabsorption and diminish the lumen-negative potential difference across the epithelium alter the rate of hydrogen ion secretion. Bicarbonate secretion is electroneutral and occurs in exchange with luminal chloride; it seems to operate in systemic metabolic alkalosis but its role in overall acid–base homeostasis is not completely known (McKinney and Burg 1977) (Fig. 2).

The outer medullary collecting tubule is capable of very high rates of hydrogen ion transport and plays a key role in the formation of an acid urine and the trapping of ammonia (Stone *et al.* 1983). It appears to be a unique segment that does not reabsorb sodium or secrete potassium and whose only function is hydrogen ion secretion, which takes place in cells morphologically identical to intercalated cells. The medullary collecting tubule is lumen-positive and hydrogen ion must be actively secreted against the electric gradient via an electrogenic, sodium-independent process modulated by a powerful H^+-ATPase. Hydrogen ion secretion is not inhibited by agents that block sodium transport but is influenced by aldosterone through a mechanism which is independent of sodium delivery or reabsorption.

The inner medullary collecting tubule is also involved in urinary acidification, at least in the rat, but its physiology is still poorly understood. A specific feature of the rat inner medullary collecting tubule is that cells lining it, although resembling principal cells rather than intercalated cells, are actively involved in hydrogen ion secretion (Selvaggio *et al.* 1988).

Formation of urine P_{CO_2}

The increase in urinary P_{CO_2} that occurs after maximal alkalinization may be the direct consequence of distal hydrogen ion secretion (DuBose 1982). The secreted hydrogen ion reacts with bicarbonate present in the lumen of collecting tubules to form carbonic acid, which slowly dissociates into CO_2 and water because luminal carbonic anhydrase is not present in the distal nephron. This delay in H_2CO_3 dehydration allows for the development of a high urinary $P\text{co}_2$ as the tubular fluid reaches the most distal parts of the nephron, where an unfavourable back-diffusion of CO_2 is present. Micropuncture studies (DuBose 1982; DuBose and Caflish 1985) have shown that the rise in $P\text{co}_2$ occurs along the collecting tubule, with trapping of CO_2 in the medullary countercurrent system. Dehydration of H_2CO_3 is also slow in the renal pelvis and bladder due to the unfavourable relationship between surface area and volume.

Urinary $P\text{co}_2$ also increases markedly after the administration of neutral phosphate, provided that the pH is close to the pK of the phosphate buffer system (i.e., 6.8). In a highly alkaline urine, phosphate concentration does not influence the generation of $P\text{co}_2$ because almost all phosphate is in the dibasic form (HPO_4^{2-}) and is not capable of donating hydrogen ion to titrate bicarbonate. At a urinary pH of about 6.8, however, half of the phosphate is in the monobasic form ($H_2PO_4^-$) and is able to titrate bicarbonate and give rise to the $P\text{co}_2$ (Stinebaugh et al. 1977).

Ammoniagenesis

Trapping of secreted hydrogen ion by ammonia to form ammonium is one of the main mechanisms in maintaining normal acid–base homeostasis. Under normal circumstances, ammonium excretion is only slightly greater than titratable acid excretion, but during chronic metabolic acidosis ammonium excretion increases markedly and far exceeds the excretion rate of titrable acid. Ammonia is synthetized principally in the proximal tubule cells and is transferred into the final urine by a series of specialized transport processes that involve secretion into proximal tubules, absorption by the loop of Henle, accumulation in the renal medullary interstitium, and secretion into the collecting tubules (Good and Knepper 1985). It was long held that ammonia, a non-ionic base, diffused passively into the tubular lumen whereas the intraluminally formed ammonium could not retro-diffuse because the luminal membrane was impermeable to it. This 'diffusion trapping' does probably occur in parts of the nephron, such as the proximal tubule or collecting tubule, but in other segments, such as the ascending loop of Henle, there is a direct transport of ammonium from lumen to interstitium. The accumulation of ammonia in the medullary interstitium increases the driving force for diffusional entry of ammonia into the collecting tubules, a process facilitated by the high acidity of the tubular fluid at this level (Lee and Simon 1987).

Ammonia is mainly formed in the mitochondria of proximal tubule cells by deamination of glutamine by the enzyme phosphate-dependent glutaminase. Other mitochondrial enzymes, such as glutamate dehydrogenase, may also contribute to the formation of ammonia by deamination of glutamate. Cytosolic pathways of glutamine metabolism, involving glutamic synthetase, glutaminase II, phosphate-independent glutaminase, glutamyl transferase and the purine-nucleotide cycle, probably play a minor role in the total production of ammonia by the kidney (Tannen 1977; Lemieux et al. 1982). Stimulation of ammonia production by chronic metabolic acidosis depends on enhanced glutamine entry into the mitochondria and/or activation of phosphate-dependent glutaminase. This adaptive process is modulated by enhanced renal gluconeogenesis. What intimal mechanism leads to enhanced glutaminase activity during metabolic acidosis is not precisely known, but low intracellular pH and other factors, such as diminution of intracellular potassium and activation of glucocorticoid hormones, may play an important role (Tannen 1977; Lemieux et al. 1982).

Regulation of distal urinary acidification
Blood pH

Distal hydrogen ion secretion is increased in acute or chronic metabolic acidosis. Schwartz and Cohen (1978) have proposed that distal acidification is regulated by variations in the rate of delivery or reabsorption of the different anions presented to the acidification sites, independent of changes in systemic acid–base equilibrium. There is, however, experimental evidence that distal acidification is also directly influenced by blood and/or intracellular pH (Gougoux et al. 1980). Probably, acidaemia is necessary to augment the renal acid excretion but, on the other hand, the anion profile contributes to modulation of the steady state of plasma bicarbonate during chronic metabolic acidosis (Madias and Zelman 1986).

Blood $P\text{co}_2$

Acute and chronic variations in ventilation may also induce a secondary change in distal urinary acidification: in the dog and the rat, hypocapnia results in a significant diminution of hydrogen ion secretion, while hypercapnia directly stimulates it.

Sodium transport and transepithelial potential difference

In the cortical collecting tubule the rate of active hydrogen ion secretion is indirectly influenced by sodium transport and is exquisitely sensitive to alterations in transepithelial electrical gradient. This segment is lumen-negative due to active sodium transport and thus the secretion of positively charged ions, such as hydrogen ion or potassium, is strongly favoured. The secretion of hydrogen ion is especially influenced by the type and quantity of non-bicarbonate sodium salts delivered and by the avidity of this segment for sodium reabsorption, which is mainly determined by the action of aldosterone. An increase in the distal delivery of the sodium salts of poorly reabsorbable anions (e.g., by infusion of sodium sulphate or administration of frusemide) will increase the lumen-negative potential difference across the epithelium and enhance the secretion of hydrogen ion and potassium. Conversely, blocking distal sodium reabsorption (e.g., by amiloride) will inhibit the secretion of hydrogen ion and potassium (Batlle 1986a)

Aldosterone

Aldosterone influences distal acidification through several mechanisms (Marver and Kokko 1983); (1) it enhances sodium transport in late distal and cortical collecting tubules and thereby increases the lumen-negative potential difference across epithelium, so favouring both hydrogen ion and potassium secretion; (2) it also enhances the conductance for protons of the luminal membrane, by a mechanism independent of its action on sodium transport; (3) it has a direct effect on ammonium excretion by increasing ammonia synthesis, and this effect appears to be independent of simultaneous changes in cellular potassium; and (4) aldosterone regulates potassium homeostasis and accordingly has also an indirect effect on ammonia production by the kidney.

Potassium

Alterations in potassium homeostasis may influence distal acidification by modulating the rate of ammonia synthesis and by influencing aldosterone production and release. Potassium itself has a major effect on ammonia synthesis: potassium depletion stimulates it, whereas potassium excess inhibits it (Sastrasinh and Tannen 1983). This is reflected in changes in urinary pH. In hypokalaemia, increased renal production of ammonia increases the available urinary buffer and mimics a defect of acidification. Conversely, in hyperkalaemia, total hydrogen ion excretion may be impaired, despite a low urinary pH, by virtue of a reduction in buffer for titration.

Potassium may also have a direct effect on distal urinary acidification, independent of its effects on ammonia production or on aldosterone secretion. In the dog (Pichette et al. 1982) and rat (Kornandakieti and Tannen 1984), potassium depletion is a primary stimulus of distal acidification, but potassium loading inhibits it.

Proximal renal tubular acidosis

Pathogenesis

As already mentioned, the renal defect in proximal tubular acidosis is in the tubular reabsorption of bicarbonate. The magnitude of the bicarbonate loss in most patients with proximal acidosis (more than 15 per cent of the filtered load) indicates that the abnormality is in the proximal tubule (Morris et al. 1972). Moreover, the unimpaired ability to acidify the urine when the plasma bicarbonate is below the individual's renal bicarbonate threshold supports the conclusion that distal acidification mechanisms are intact (Rodriguez-Soriano et al. 1967a; Rodriguez-Soriano and Edelmann 1969).

The nature of the proximal tubular defect leading to impaired bicarbonate reabsorption is not known. In the dog and rat, intravenously administered maleic acid induces a complex dysfunction of the proximal tubule with impaired reabsorption of filtered bicarbonate (Al-Bander et al. 1982). The mechanism by which maleic acid causes experimental Fanconi's syndrome is also unknown, but the fact that prior administration of phosphate attenuates the severity of the tubular dysfunction indicates that the nephrotoxicity of that salt is related to impaired oxidative metabolism in mitochondria (Al-Bander et al. 1985).

Giving the dog several lysine derivatives (L-lysine monochloride, isoelectric L-lysine, sodium lysinate) also inhibits bicarbonate reabsorption without affecting other functions of the proximal tubules (Gougoux et al. 1980). Lysine, when given in a form that can accept a proton, accumulates inside proximal tubular cells and by titrating available hydrogen ion prevents the intraluminal titration of filtered bicarbonate.

The pathogenetic factors involved in human proximal tubular acidosis are probably multiple and of varying significance. Changes in either intrinsic or extrinsic renal factors that participate in the tubular reclamation of filtered bicarbonate may be implicated, depending on the specific aetiology. In most cases, proximal acidosis occurs in clinical Fanconi's syndrome and is associated with a defect in proximal sodium transport (Rodriguez-Soriano et al. 1980). A metabolic impairment of the proximal tubule that restricts energy-requiring, sodium dependent reabsorption in this segment may account for this. The finding of increased urinary excretion of glyceraldehyde in children with Fanconi's syndrome suggests that the metabolic impairment could be located in the glycolytic metabolism of the proximal tubule (Jonas et al. 1989). In Fanconi's syndrome, other circumstances leading to reduced proximal reabsorption of bicarbonate (chronic hypocapnia, extracellular expansion of fluid volume, hyperkalaemia, hyperparathyroidism, intracellular phosphate depletion) should also be considered as possible contributory factors (Morris 1983).

The pathogenesis of isolated proximal tubular acidosis is also unknown. These patients have many of the functional characteristics of those receiving carbonic anhydrase inhibitors, but there is no convincing proof that a deficiency in carbonic anhydrase underlies the proximal defect (Edelmann 1985). It is of interest that patients with proven genetic absence of renal carbonic anhydrase II develop a distal rather than a proximal defect of acidification (Sly et al. 1985a). In such cases, the positive renal response to intravenous acetazolamide suggests that luminal carbonic anhydrase is the product of a different gene and that it is unaffected in this particular type of carbonic anhydrase II mutation (Sly et al. 1985b).

Clinical features

Proximal tubular acidosis is most commonly diagnosed in children, either as a primary and isolated entity or accompanied by other proximal tubular dysfunctions, such as glycosuria, hyperaminoaciduria, hyperphosphaturia and hyperuricosuria (Fanconi's syndrome). In adults, it occurs most often in acquired proximal tubulopathy (Table 2).

Stunted growth is a cardinal feature of the disease in children. Rickets or osteomalacia are never seen unless there is hypophosphataemia, as occurs in Fanconi's syndrome (Brenner et al. 1982). Nephrocalcinosis and urolithiasis are also never seen in proximal tubular acidosis, even when hypercalciuria is present (Brenner et al. 1982). Hypokalaemia and related signs are restricted to Fanconi's syndrome.

Primary proximal tubular acidosis

A non-familial and transient type of proximal acidosis has been described in nine infants with an isolated defect in bicarbonate reabsorption that was without identifiable cause or evidence of other abnormality (Rodriguez-Soriano et al. 1967b; Nash et al. 1972). The presenting symptoms were growth failure and persistent vomiting in early infancy. Alkali therapy induced a rapid increase in growth rate and could be discontinued after several years without the reappearance of symptoms. The self-limiting course of this infantile proximal tubular acidosis contrasts with that of primary distal acidosis, which appears to be permanent even when onset is in infancy.

The familial occurrence of isolated proximal tubular acidosis has also been reported and the course is persistent. Brenes et al. (1977) describe a family with nine affected members aged 2 to 27 years. The presence of the disorder in several generations suggested autosomal dominant transmission. Winsnes et al. (1979) describe two brothers with proximal acidosis and other associated anomalies (mental retardation, corneal opacities, cataract, glaucoma, enamel defects), indicating that an autosomal recessive type may also occur.

Secondary proximal tubular acidosis

As mentioned above, proximal acidosis is almost always part of Fanconi's syndrome, either idiopathic or caused by cystinosis, Lowe's syndrome, Wilson's disease, hereditary fructose intolerance, hereditary tyrosinaemia or glycogenosis. It may also be

Table 2 Causes of proximal renal tubular acidosis

I. *Primary or isolated*
1. Sporadic (transient in infancy)
2. Genetically transmitted (persistent)
   Autosomal dominant
   Autosomal recessive

II. *Secondary to inherited or acquired diseases*
1. Associated with other inherited dysfunctions of the proximal tubule
   Idiopathic Fanconi syndrome
   Cystinosis
   Lowe's syndrome
   Hereditary fructose intolerance
   Tyrosinaemia
   Galactosaemia
   Glycogen storage disease
   Metachromatic leucodystrophy
   Mitochondrial myopathies
2. Drugs and toxins
   Carbonic anhydrase inhibitors (acetazolamide)
   Outdated tetracycline
   Methyl-3-chromone
   Streptozotocin
   Valproic acid
   Heavy metals (cadmium, lead, copper, mercury)
3. Miscellaneous
   Vitamin D deficiency
   Subacute necrotizing encephalopathy (Leigh's syndrome)
   Tetralogy of Fallot
   Intestinal malabsorption
   Hyperparathyroidism
   Renal cystic disease
   Hereditary nephritis
   Renal transplantation
   Multiple myeloma
   Sjögren's syndrome
   Amyloidosis
   Chronic active hepatitis
   Recurrent renal stone disease

induced by drugs and toxins such as degraded tetracycline, methyl-3-chromone, streptozotocin, valproic acid and heavy metals (cadmium, copper, lead, mercury). Proximal tubular acidosis occurs in a variety of other conditions including vitamin D deficiency, nephrotic syndrome, Leigh's syndrome, and Fallot's tetralogy in children, and renal transplantation, hyperparathyroidism, multiple myeloma, Sjögren's syndrome, amyloidosis, and chronic active hepatitis in adults (Edelmann 1985; Morris 1983). In intestinal malabsorption, the associated presence of hyperparathyroidism, hypocalcaemia, vitamin D deficiency and hyperphosphaturia appears necessary for the clinical expression of proximal tubular acidosis (Muldowney *et al.* 1970). It is of interest that 15 per cent of those who form recurrent kidney stones have a proximal rather than a distal type of tubular acidosis (Backman *et al.* 1980). This proximal dysfunction may be a consequence not a cause of the nephrolithiasis.

Pathophysiology of symptoms

Many of the clinical peculiarities of proximal tubular acidosis depend upon the fact that it is not accompanied by hydrogen ion retention, increased bone buffering and hypercalciuria. Increased distal delivery of bicarbonate also enhances distal tubular reabsorption of calcium, thus contributing to the absence of hypercalciuria. Even when urinary calcium excretion is increased because of Fanconi's syndrome, the normal or elevated excretion of citrate and the highly acidic urine lead to adequate solubilization of urinary calcium and no nephrocalcinosis or lithiasis. The increase in urinary citrate, despite the presence of systemic acidaemia, may be due to inhibition of proximal citrate reabsorption caused by the elevation of both luminal (Brennan *et al.* 1988) and intracellular pH (Halperin *et al.* 1989).

The absence of radiographic evidence of skeletal lesions where there is no hypophosphataemia indicates that acidosis *per se* does not necessarily cause bone disease. Even though there is experimental evidence that acute metabolic acidosis impairs the conversion of 25-(OH) vitamin D to 1,25-$(OH)_2$ vitamin D_3, this inhibition is not apparent in chronic metabolic acidosis in man (Cunningham *et al.* 1984). That vitamin D metabolism is disturbed in maleic acid-induced Fanconi's syndrome has led to the suggestion that reduced levels of vitamin D metabolites could play a role in the development of the skeletal lesions seen in man. However, several studies have shown that, although plasma levels of 1,25-$(OH)_2$ vitamin D_3 in patients with the syndrome may occasionally be low (Baran and Marcy 1984), there are generally no abnormalities in vitamin D metabolism in those with bone disease (Chesney *et al.* 1984).

The renin–aldosterone axis is hyperactive in some patients with proximal tubular acidosis, especially in Fanconi's syndrome. The most probable cause is chronic contraction of fluid volume due to continuous renal loss of sodium and bicarbonate. Increased distal delivery of sodium promotes potassium excretion, an effect further enhanced by the state of hyperaldosteronism. Potassium wasting may become even more severe when large doses of sodium bicarbonate are given therapeutically (Sebastian *et al.* 1971).

An unanswered question is why children with isolated proximal tubular acidosis have retarded growth. The assumption is that acidosis *per se*, even in absence of disturbed mineral metabolism or impaired activation of vitamin D, can lead to failure of growth by interfering with growth hormone secretion or response (McSherry 1981).

Diagnosis

Proximal tubular acidosis should be suspected in patients who have a normal plasma anion gap and hyperchloraemic metabolic acidosis, especially when associated with other features such as hypokalaemia, renal glycosuria, hyperaminoaciduria, hyperphosphaturia with hypophosphataemia, and hyperuricosuria. Only exceptionally will metabolic acidosis appear as an isolated feature (Table 3).

If the spontaneous metabolic acidosis is severe enough (plasma bicarbonate of less than 15 to 18 mmol/l), pH testing of early morning urine is very useful. A pH of 5.5 or less, accompanied by an abundant excretion of ammonium (greater than 40 μmol/min. 1.73 m^2), confirms the diagnosis of proximal acidosis if gastrointestinal loss of bicarbonate is excluded. If a low urinary pH is not found, despite some degree of acidaemia, an acute test for ammonium chloride loading should be done to exclude definitively the diagnosis of distal tubular acidosis.

The diagnosis of proximal acidosis can be confirmed by bicarbonate titration. The finding of a low renal threshold for bicarbonate and an elevated excretion of bicarbonate (more than 10 to 15 per cent of the amount filtered) with a normal plasma

Table 3 Differential features of various types of renal tubular acidosis (RTA)

	Proximal RTA	Distal RTA			Hyperkalaemic RTA
		'Classic'	With HCO_3^- wasting	Hyperkalaemic	
During metabolic acidosis (spontaneous or after acid load)					
P_K	Normal or decreased	Normal or decreased	Normal or decreased	Increased	Increased
Urine anion gap	Negative	Positive	Positive	Positive	Positive
Minimal urinary pH	< 5.5	> 5.5	> 5.5	> 5.5	< 5.5
$U_{NH_4^+}V$	Normal	Decreased	Decreased	Decreased	Decreased
FE_K	Normal or increased	Increased	Increased	Decreased	Decreased
$U_{Ca}V$	Normal	Increased	Increased	Increased	Normal or decreased
$U_{Cit}V$	Normal or increased	Decreased	Decreased	Decreased	Normal or decreased
During normal $P_{HCO_3^-}$ concentration (after an alkaline load)					
FE_{bicarb}	> 10–15%	< 5%	> 5–15%	< 5%	> 5–10%
Urine–blood P_{CO_2} gradient	> 20 mmHg	< 20 mmHg	< 20 mmHg	< 20 mmHg	≤ 20 mmHg (depends on GFR)
Presence of other tubular defects	Common	Rare	Rare	Rare	Rare
Presence of nephrocalcinosis or lithiasis	Rare	Common	Common	Possible	Absent
Response to standard doses of alkali therapy (2–3 mmol/kg.day)	Refractory	Good	Refractory	Good	Variable

bicarbonate (more than 20 to 22 mmol/l) is characteristic (Fig. 3). Fractional excretion (FE) of bicarbonate can be calculated by dividing the clearance rate of bicarbonate by the clearance rate of creatinine:

$$FE_{Bicarb} = U_{Bicarb} \times P_{Cr}/U_{Cr} \times P_{Bicarb}$$

Although it is said that fractional excretion of bicarbonate should exceed 15 per cent to confirm the existence of proximal tubular acidosis, experiments have shown that this approach underestimates the proximal defect because, even if such an amount is rejected proximally, bicarbonate reabsorption may proceed in the functionally intact distal tubule (Batlle and Chan 1989). In my opinion, a defect in proximal bicarbonate reabsorption is better uncovered by examining bicarbonate excretion when the plasma bicarbonate is below normal.

Diagnosis can also be established by plotting urinary pH as a function of plasma bicarbonate: the points are displaced to the left of the normal relationship, and the patient's renal bicarbonate threshold can be estimated by observing the concentration of plasma bicarbonate at which there is an abrupt fall in urinary pH to about 6.0 or 6.5 (Fig. 3).

Bicarbonate titration is a difficult test that requires intravenous infusion of sodium bicarbonate. In clinical practice it can be replaced by oral administration of sodium bicarbonate over several days, aiming to bring the plasma bicarbonate up to the normal range. At that moment, both fractional excretion of bicarbonate and the urinary P_{CO_2} should be determined. A high FE_{Bicarb} (more than 10 to 15 per cent) and a normal urine minus blood P_{CO_2} gradient (more than 20 mmHg) are diagnostic of proximal tubular acidosis if hyperkalaemia is absent. Urinary P_{CO_2} is only diagnostic if the pH is greater than 7.8 for at least 30 min before collecting the urine sample. The bicarbonate threshold may also be estimated by this method. Treatment is discontinued to allow for a progressive and gradual decrease in plasma bicarbonate; paired samples of blood and urine are

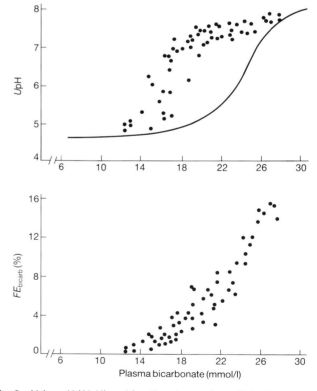

Fig. 3 Urine pH (UpH) and fractional excretion of bicarbonate (FE_{Bicarb}) as a function of plasma bicarbonate concentration in two children with proximal renal tubular acidosis. Curve indicates the relationship present in normal children.

examined once or twice daily to identify the plasma bicarbonate at which urinary pH falls abruptly to between 6.0 and 6.5.

Treatment

In proximal acidosis, treatment must compensate for the urinary losses of bicarbonate and for the endogenous production of acid. Because patients with this disorder usually excrete 10 per cent or more of the filtered amount of bicarbonate when the plasma bicarbonate is brought within the normal range, the minimum starting dose of alkali is between 6 and 10 mmol/kg.day, but larger doses are often needed. Citrate excretion is not decreased so there is no special need to give the anion in the form of citrate, and sodium bicarbonate can be given with excellent results. When proximal tubular acidosis occurs in Fanconi's syndrome, at least half of the alkali usually necessary to sustain correction of the acidosis should be given as a potassium salt. Children with isolated proximal acidosis do not specifically require potassium supplementation. Treatment should aim to maintain as normal a plasma bicarbonate as possible day and night, so doses should be frequent, with the last one as late as possible at night.

In very severe forms of proximal tubular acidosis, alkali alone may be ineffective because of gastrointestinal intolerance to the enormous doses needed, or because of rapid loss of the base into the urine. In these circumstances, hydrochlorothiazide may be useful because it improves bicarbonate reabsorption and allows the dose of bicarbonate to be reduced. The initial dose is about 1.5 to 2 mg/kg.day, but after correction of the acidosis a smaller dose may be sufficient (Rampini *et al.* 1968). An additional benefit of hydrochlorothiazide in patients with Fanconi's syndrome is the diminution of urinary calcium and phosphate excretion with subsequent improvement of osteomalacia (Callis *et al.* 1970). When hydrochlorothiazide is used, potassium supplements are always necessary to prevent hypokalaemia.

It must be recognized that alkali alone will not heal the coexisting bone lesions in Fanconi's syndrome. Hypophosphataemia and osteomalacia are best treated with phosphate supplements and 1,25-$(OH)_2$ vitamin D_3.

Distal renal tubular acidosis

Pathogenesis

Distal tubular acidosis is characterized functionally by the inability of the distal nephron to decrease the urinary pH below 5.5 and increase the urinary excretion of ammonium, regardless of the extent of systemic acidosis. The exact nature of the acidification defect is unknown. Based on numerous observations in man and in experimental animals, several distinctly different pathogenic types of distal acidosis are recognized. These subtypes have been distinguished by differences in the response of the urinary pH to sulphate infusion or frusemide administration, and of urinary P_{CO_2} to sodium bicarbonate or sodium phosphate loading (Batlle 1981; Batlle 1983; Kurtzman 1983; Batlle 1986*a*).

1. Failure of the distal nephron to secrete hydrogen ion because of a reduced number, or through impaired function of the hydrogen ion secreting pump ('secretory' defect).
2. Inability of the distal nephron to generate or maintain a lumen-negative potential difference across the epithelium due to a defect in distal sodium transport ('voltage-dependent' defect). The abolishment of this potential difference in the cortical collecting tubule will reduce both hydrogen ion and potassium secretion and may lead to hyperkalaemia.
3. Inability to create a steep hydrogen ion gradient across the distal nephron due to back-leak of secreted hydrogen ion or to increased luminal entry of cellular bicarbonate ('gradient' defect). This type could result from increased permeability of the luminal membrane to hydrogen ion, carbonic acid or bicarbonate.

Other possible mechanisms, such as reduced availability of cellular hydrogen ion (due to defective hydroxyl disposal or impaired exit of cellular bicarbonate to the peritubular fluid) or increased distal secretion of bicarbonate, are less well documented.

That there are different functional types of distal tubular acidosis has gained strong support from the micropunctures experiments of DuBose and Caflish (1985), showing that rats receiving amiloride or lithium, or after ureteral obstruction, have a defect in hydrogen ion secretion secondary to abolishment of the lumen-negative potential difference. Amphotericin B, however, only impaired the ability to maintain a pH gradient in the collecting tubule despite an intact ability to secrete hydrogen ion.

That these subtypes occur in man is also supported by many clinical studies. The view that prevailed for many years was that primary distal tubular acidosis, either hereditary or sporadic, resulted from the inability of the distal nephron to produce a significant gradient for hydrogen ions between blood and tubular fluid, regardless of the degree of systemic acidosis. This view was first challenged by Halperin *et al.* (1974) because patients with distal acidosis could not increase urinary P_{CO_2} when a favorable gradient (urine pH higher than blood pH) was imposed by the administration of sodium bicarbonate. Their conclusion that a defect in hydrogen ion secretion was the most probable cause was confirmed by our studies in children (Rodriguez-Soriano *et al.* 1985) (Fig. 4). We have shown that the high urinary pH cannot be modified by sulphate infusion or frusemide and that the low urinary P_{CO_2} found after bicarbonate loading is not increased by phosphate loading.

The exact nature of the cellular defect that causes the failure to secrete hydrogen ion is unknown. It is probably located in the luminal membrane, and hypothetical mechanisms include impaired function or diminished number of proton secretory pumps. Reduced availability of cellular hydrogen ion may be involved in that small group of patients with distal tubular acidosis, recessive osteopetrosis, cerebral calcification, and deficiency of carbonic anhydrase II (Sly *et al.* 1985*a*).

Batlle (1981, 1986*a*) and Kurtzman (1983) have made extensive functional studies in adult patients with secondary, acquired distal acidosis. A true failure of hydrogen ion secretion has been suspected in normo- or hypokalaemic patients with renal transplant rejection or chronic interstitial nephritis from various causes. In other circumstances, hyperkalaemia coexists with the distal acidifying defect (hyperkalaemic distal tubular acidosis). These patients probably have a 'voltage-dependent' defect which precludes both hydrogen ion and potassium transport. The abnormality in sodium transport may be apparent on a normal diet or only become evident when the subject is maintained on a low sodium diet. Plasma aldosterone concentration may be normal or increased, but in some patients there is aldosterone deficiency. Clinical examples include obstructive uropathy, interstitial nephritis associated with sickle-cell disease, and renal transplant rejection. In children, this type of defect may be observed in salt-losing congenital adrenal hyperplasia (Rodriguez-Soriano *et al.* 1986).

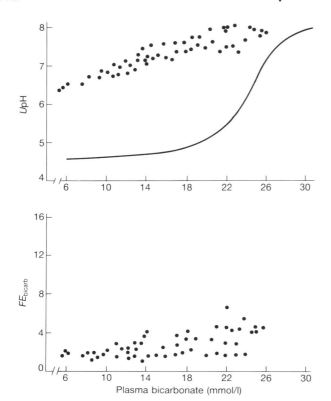

Fig. 4 Urine pH (UpH) and fractional excretion of bicarbonate (FE$_{\text{Bicarb}}$) as a function of plasma bicarbonate concentration in three children with primary distal renal tubular acidosis. The curve indicates the relationship present in normal children.

Urinary acidification may also be affected by decreased distal delivery of sodium, as observed in oedema-forming states such as hepatic cirrhosis or the nephrotic syndrome. Sodium sulphate or frusemide increase the arrival of sodium at the collecting duct and completely correct the acidification defect (Rodriguez-Soriano et al. 1982a).

A clinical counterpart to the experimental defects of acidification induced by lithium or amiloride is not always found. Patients treated with lithium for affective disorders more often have a diminished capacity to concentrate the urine; the defect in distal acidification is generally not manifested by systemic acidosis (incomplete distal tubular acidosis). Only bicarbonate loading will reveal the existence of a 'rate-dependent' defect of urinary acidification by demonstrating a failure to increase appropriately urine P_{CO_2} under such circumstances (Batlle et al. 1982a). The only clinical state that can be attributed with some certainty to a 'gradient' defect is the form of distal acidosis found in patients treated with amphotericin B (McCurdy and Elkinton 1968). That a hypokalaemic type of distal tubular acidosis develops in individuals who sniff toluene has been seen as an example of a 'gradient' defect. However, experiments with the turtle bladder have shown that toluene, unlike amphotericin B, does not cause hydrogen ion back-diffusion; the defect is thus best explained by decreased conductance of protons through the transport pathway (Batlle et al. 1988a).

Clinical features

Distal tubular acidosis is almost always a primary condition in children, but in adults it is generally acquired and often found in the context of an immune-mediated disease. Prominent clinical features include impairment of growth, polyuria, hypercalciuria, nephrocalcinosis and/or urolithiasis, and potassium depletion (Wrong and Feest 1980; Santos and Chan 1986; Caruana and Buckalew 1988). Renal failure is a threatening complication caused by progressive nephrocalcinosis and recurrent nephrolithiasis.

Primary distal tubular acidosis

Primary or idiopathic distal acidosis is the most frequent expression of the disease in children, although it may also occur in adults. Most cases are sporadic but in some instances the disease is inherited as an autosomal dominant trait. The defect varies considerably in severity in different families, and some patients do not have systemic acidosis despite the presence of the characteristic acidification defect (incomplete distal tubular acidosis). Another autosomal dominant type is related to idiopathic hypercalciuria, which appears to be the inherited defect because some members of the family have a raised urinary calcium excretion unaccompanied by a defect in acidification (Buckalew et al. 1974).

Primary distal tubular acidosis occurs in both sexes, with a slight predominance of females. Usually it is not diagnosed before 2 years of age and occasionally not before adulthood. There is no question, however, that it is congenital, and diagnosis in early infancy is possible if attention is paid to non-specific symptoms such as vomiting, constipation, polyuria, dehydration, irritability, hypotonia, and failure to thrive. Growth retardation is most evident beyond infancy and may be the only clinical abnormality (Nash et al. 1972).

Rickets or osteomalacia are rarely encountered. In a systematic review of 44 treated patients with distal tubular acidosis (17 children and 27 adults), skeletal abnormalities were only evident radiographically in one adult patient with azotaemia (Brenner et al. 1982). We have observed advanced rickets in just one 8-year-old child with untreated disease. Thus, bone lesions develop only after a clinical course of several years.

Nephrocalcinosis is an early complication and may be visualized by ultrasonography or radiography even during the first month of life. Only when treatment is delayed does this progress to significant calcification of the renal medulla. This progression can be halted if adequate treatment is instituted in infancy (McSherry 1981; Santos and Chan 1986).

Urolithiasis is more common in adults than in children; it may occur together with nephrocalcinosis or appear as an isolated complication. Nearly half of the adults with idiopathic or acquired distal tubular acidosis may present clinically with kidney stone (Harrington et al. 1983), whereas a defect in distal urinary acidification may be found in about 20 per cent of adults who have recurrent stone formation (Backman et al. 1980). The calculus is calcium phosphate (brushite), although calcium oxalate and struvite stones may also be found.

Polyuria, due to a renal concentrating defect, is almost always present, even in cases diagnosed very early in life. In our experience, this abnormality persists unmodified despite adequate correction of the other symptoms, probably as a consequence of early nephrocalcinosis.

Renal loss of potassium is increased and may result in severe hypokalaemia, or even cause periodic paralysis, circulatory collapse and cardiac arrhythmia. Vague musculoskeletal symptoms (arthralgia, myalgia, low back pain) may be the presenting symptoms in as many as 52 per cent of affected adults but are rare in children (Harrington et al. 1983).

Blood chemistry demonstrates hyperchloraemia, metabolic acidosis and normo- or hypokalaemia. Hypokalaemia was present in 28 per cent of 58 cases (16 children and 42 adults) studied by Caruana and Buckalew (1988). Exceptionally, hypokalaemic metabolic alkalosis may be the initial feature in infancy. Hyperkalaemia, instead of hypokalaemia, may also be the presenting abnormality; in a group of children, plasma potassium was transiently elevated in just over half (Santos and Chan 1986). Blood phosphate may be low, with normal or elevated levels of calcium. Plasma magnesium is generally normal or increased. The finding of profound hypomagnesaemia should suggest the diagnosis of a different and unrelated syndrome of renal magnesium wasting, hypercalciuria, nephrocalcinosis, and secondary distal tubular acidosis (Rodriguez-Soriano et al. 1987). Glomerular function is normal in the young child but a progressive decrease may take place over the years as consequence of advancing parenchymal damage.

The urinary pH is usually between 6.0 and 7.0, with low rates of titrable acid and ammonium excretion. The finding of abundant sodium excretion and a low ammonium excretion are pivotal for the diagnosis of distal tubular acidosis because the pH may be as high as 6.0 in states of chronic metabolic acidosis of extrarenal origin if there is sodium depletion or increased urinary availability of ammonia (Batlle et al. 1987; Halperin et al. 1988). A small degree of proteinuria of 'tubular' type may be found. Leucocyturia is frequent, even in absence of urinary tract infection. Glycosuria and hyperaminoaciduria are absent. Other urinary findings are increased excretion of phosphate, calcium and potassium, and reduced excretion of citrate. Hypercalciuria is an outstanding feature and, although not uniformly present in some series this could be due to the effect of alkali treatment or to the coexistence of a low GFR (Caruana and Buckalew 1988; Santos and Chan 1986). In our experience, urinary calcium excretion is inversely related to the plasma bicarbonate, and normal or almost normal levels are only achieved after adequate and sustained correction of the acidosis (Rodriguez-Soriano et al. 1982b).

The histological appearance of the kidney is normal in the early stages of the disease but nephrocalcinosis may be seen as early as in the first month of life. When calcium is deposited in the kidney, it is generally accompanied by chronic interstitial nephritis of variable degree, with cellular infiltration, tubular atrophy, and glomerular sclerosis. Immunofluorescence findings are constantly negative.

The prognosis for primary distal tubular acidosis is relatively good, provided that diagnosis is established early enough to prevent advanced nephrocalcinosis. The defect in acidification is permanent and any attempt to withdraw treatment is followed by the reappearance of metabolic acidosis and related symptoms. With adequate treatment, however, the growth rate accelerates, calcium excretion reverts to normal, and further deposition of calcium in the kidneys is arrested (McSherry and Morris 1978; Rodriguez-Soriano et al. 1982b).

Clinical variants of primary distal tubular acidosis

Combined proximal and distal renal tubular acidosis This is a constant expression of the disease during the first years of life (McSherry et al. 1972; Rodriguez-Soriano et al. 1975). The excessive loss of bicarbonate results from a defect in bicarbonate reabsorption which progressively disappears with advancing age. By the school years, urinary bicarbonate excretion is similar to or only slightly greater than that characteristically observed in adults with distal tubular acidosis, that is, about 1 to 3 mmol/kg.day.

Incomplete distal renal tubular acidosis Wrong and Davies (1959) found this in three patients with nephrocalcinosis who did not have metabolic acidosis. Although these patients could not acidify urine, a high rate of ammonium excretion compensated for their limited excretion of titrable acid. More cases have been reported, mostly as a result of screening the families of a propositus with complete distal tubular acidosis (Norman et al. 1978).

Associated nerve deafness An association of primary distal tubular acidosis with deafness is not exceptional. The observation of consanguinity in some families and of two or more sibs affected in some others suggest autosomal recessive inheritance. There is great variation in the presentation of deafness, which may become manifest from birth to late childhood (Donckerwolcke et al. 1976).

Transient acidosis A transient tubular acidosis was frequently reported in Great Britain from 1945 to 1955 (Lightwood et al. 1953). This condition, once termed Lightwood's syndrome, rapidly declined in frequency in the following years, suggesting that it was the result of some unrecognized environmental factor, such as vitamin D, sulphonamide, or mercury (Lightwood and Butler 1963).

Secondary distal tubular acidosis

Distal acidosis may occur as a secondary part of a number of systemic or renal disorders, and in adults is often acquired (Table 4). Most frequently, distal acidosis is related to an autoimmune disease and occurs in women who have raised plasma levels of different tissue antibodies and features of one or more disorders such as Sjögren's syndrome, rheumatoid arthritis, lupus erythematosus, and hepatic or thyroid disease (Wrong and Feest 1980; Morris 1981). The exact nature of the renal tubular damage in these cases of immune-mediated disease is unknown. There is no evidence that the tubular injury is caused by autoantibodies to components of the distal tubule or to Tamm–Horsfall glycoprotein.

Pathophysiology of symptoms

All symptoms of distal tubular acidosis are attributable to the primary defect in acidification of the urine and the metabolic abnormalities that follow. The balance of hydrogen ion in these patients is constantly positive because the endogenously produced acid cannot be excreted and accumulates in the body. Buffering of the excess hydrogen ion by bone salts leads to hypercalciuria through dissolution of bone calcium carbonate (Bushinsky and Lechleider 1987). Chronic metabolic acidosis has been directly implicated in the pathogenesis of osteomalacia and osteopenia because these bone lesions improve after prolonged alkali therapy without the need to increase the administration of vitamin D. Experimental chronic metabolic acidosis in the rat is followed by histological evidence of osteopenia but not of osteomalacia (Kraut et al. 1986), so alterations in the secretion of parathyroid hormone or in the metabolism of vitamin D may be implicated as contributory factors. However, as already mentioned, chronic metabolic acidosis in the human does not affect the conversion of 25-(OH) vitamin D_3 to 1,25-(OH)$_2$ vitamin D_3

Table 4 Causes of distal renal tubular acidosis

I. *Primary or idiopathic*
1. Permanent
   Adult form or 'classic'
   With bicarbonate wasting in infancy and early childhood
   Incomplete
   With nerve deafness
2. Transient (in infancy?)

II. *Secondary*
1. Disorders of mineral metabolism
   Primary hyperparathyroidism
   Hypercalcaemic hyperthyroidism
   Vitamin D intoxication
   Familial hypophosphataemia with profound phosphate depletion
   Familial idiopathic hypercalciuria with nephrocalcinosis
   Familial hypomagnesaemia–hypercalciuria with nephrocalcinosis
2. Immune-mediated disorders
3. Hyponatriuretic states
   Hepatic cirrhosis
   Nephrotic syndrome
4. Drugs and toxins
   Amphotericin B
   Lithium
   Toluene
5. Renal diseases
   Renal transplantation
   Medullary sponge kidney
   Obstructive uropathy
6. Genetic disorders
   Syndrome of recessive osteopetrosis (carbonic anhydrase II deficiency)
   Hereditary fructose intolerance with nephrocalcinosis
   Ehlers–Danlos syndrome
   Hereditary elliptocytosis
   Sickle-cell disease
   Wilson's disease
   Type 1 primary hyperoxaluria
7. Endocrine disorders
   Hypothyroidism
   Salt-losing congenital adrenal hyperplasia

(Cunningham *et al.* 1984), and plasma levels of the latter metabolite are not significantly different in patients with distal tubular acidosis from those in control subjects (Chesney *et al.* 1984; Preminger *et al.* 1987). Patients with distal tubular acidosis frequently have inadequate intestinal absorption of calcium, which is not dependent on low vitamin D levels and which improves with alkali therapy (Preminger *et al.* 1987). The association of low intestinal absorption of calcium and high urinary excretion could lead to negative calcium balance, secondary hyperparathyroidism, and bone disease.

Studies of chronic acid loading in normal individuals indicate that hypercalciuria results from release of skeletal calcium. An associated factor in the induction of hypercalciuria is a reduction in the distal tubular reabsorption of calcium related to decreased delivery of bicarbonate, an effect that is independent of vitamin D or parathyroid hormone (Adams *et al.* 1979). Parathyroid hormone, however, may exert a direct calciuric effect during metabolic acidosis (Batlle *et al.* 1982*b*) and thus contribute to the increased urinary calcium excretion if hyperparathyroidism is present (Coe and Firpo 1975).

Nephrocalcinosis and lithiasis have been attributed to the combined effects of hypercalciuria, elevated urinary pH, and hypocitraturia. Citrate is a chelator of calcium and an inhibitor of calcium phosphate and oxalate crystallization. The reduced excretion of citrate is probably due to the intracellular acidosis and potassium depletion. Cytoplasmic acidosis induces citrate entry into the mitochondria through stimulation of the citrate carrier in the inner membrane; this leads to a fall in the concentration of cytoplasmic citrate and secondary stimulation of citrate absorption from the luminal and peritubular sides of the proximal tubular cell (Simpson 1983). Correction of hypocitraturia by potassium bicarbonate is often incomplete and, in this regard, potassium citrate should be preferred.

Renal wasting of sodium and potassium are common features in distal tubular acidosis. Hyperkaliuria resulting in hypokalaemia and intracellular potassium depletion may be explained as a consequence of the small but constant urinary loss of bicarbonate, with contraction of the extracellular fluid volume, secondary hyper-reninism and hyperaldosteronism, and increased potassium secretion by the distal nephron. In contrast to the effect in patients with proximal tubular acidosis, correction of the acidosis usually leads to improvement or potassium wasting (Sebastian *et al.* 1971). Renal conservation of sodium and chloride often remains impaired, despite correction of the acidosis, because there is a permanent defect in distal reabsorption of sodium chloride (Rodriguez-Soriano *et al.* 1980).

Diagnosis

The diagnosis of distal tubular acidosis is easily accomplished (see Table 3). Patients have a normal plasma anion gap, hyperchloraemic metabolic acidosis and a urinary pH above 5.5. It should be borne in mind that the pH only indicates the activity of free hydrogen ion, which is less than 1 per cent of the total amount of protons excreted by the distal nephron. An individual excreting large amounts of ammonium may have a urinary pH as high as 6.0 without this implying that there is defective distal acidification. Urinary pH must always be evaluated in conjunction with ammonium excretion in order properly to assess acidification (Halperin *et al.* 1988). The urinary sodium level should also be known before the diagnosis of distal tubular acidosis is considered, because an elevated urinary pH may be associated with low sodium excretion if there is sodium depletion (Batlle *et al.* 1987).

When systemic acidosis is absent or mild, it is more difficult to establish the diagnosis of incomplete distal tubular acidosis and, in this situation, an acute ammonium chloride loading test may become necessary. This consists of the oral administration of 75 to 100 mmol/m^2 NH$_4$Cl, usually in the form of crushed granules mixed with lemon juice or a soft drink and given over the course of 1 h. Urine must be collected for at least the following 6 h. A practical protocol is to administer the load at home, at midnight, and to study the patient when coming to the hospital at 8.00 am. on the following day. Normal individuals submitted to this test should decrease their urinary pH to below 5.5. and increase their excretion rates of titrable acid and ammonium to values greater than 35 and 45 μmol/min. 1.73 m^2, respectively. In general, urinary pH and acid excretion should be evaluated at plasma bicarbonate concentrations of less than 18 mmol/l in children and of 20 mmol/l in adults, that is, a few millimoles below the renal bicarbonate threshold. In cases of hepatic or gastric intolerance to ammonium chloride, oral administration of calcium chloride

5.3 Renal tubular acidosis

Table 5 Pathogenic diagnosis of distal renal tubular acidosis

Diagnostic feature	'Secretory' defect	'Voltage-dependent' defect		'Gradient' defect
		Mild	Severe	
Urine pH during acidaemia	High	High	High	High
Urine pH during sodium sulphate infusion	High	Low	High	Low
Urine pH after frusemide	High	Low	High	Low
Urine–blood P_{CO_2} gradient in alkaline urine	Low	Low	Low	Normal
Urine–blood P_{CO_2} gradient after phosphate administration	Low	Normal	Low	?
Plasma potassium	Normal/low	Normal/high	High	Normal/low
Urinary potassium excretion after sulphate or frusemide	Normal/high	Normal	Low	Normal/high

(2 mmol/kg) or an intravenous administration of arginine-HCl (200–250 ml/m² of 10 per cent solution, over 2–3 h) may be used, respectively. Even though these short tests do not allow for a maximal increase in ammonium excretion, they are completely reliable for clinical use.

An alternative to the acid-loading test is to stimulate distal urinary acidification by sodium sulphate infusion or acute frusemide administration (Batlle 1986b; Rastogi et al. 1985). Both substances enhance hydrogen ion and potassium secretion in the collecting cortical tubule by increasing distal sodium delivery and generating a high lumen-negative potential difference across the epithelium. The avidity to reabsorb sodium distally may be increased by giving 9α-fludrocortisone, 12 h before the beginning of the test, but this is not usually necessary.

Acute frusemide administration produces results similar to those obtained after sodium sulphate infusion, but the test is more easily performed. In adults, frusemide may be given orally (80 mg) but in children we prefer to administer it as an intravenous bolus, at the dose of 1 mg/kg. Urine should be collected for at least 3 h because, although maximal natriuresis and kaliuresis are immediate, the maximal acidification is observed 120 to 180 min after giving the drug. Simultaneous determination of plasma renin activity and aldosterone concentration, before and 3 h after drug administration, also allows assessment of the renin-aldosterone axis. Normal subjects should attain a urinary pH of less than 5.5. The test is very useful in clinical practice but it should not be a substitute for ammonium chloride loading, because a failure to acidify the urine after frusemide does not necessarily imply the presence of an irreversible defect of distal acidification (Rodriguez-Soriano and Vallo 1988).

The finding by Batlle et al. (1982a) that some patients with incomplete distal tubular acidosis could lower the urinary pH during acidosis but have a low urinary P_{CO_2} during bicarbonate loading means that this test is required in selected circumstances. For assessment of urinary P_{CO_2} the pH and bicarbonate concentration should increase up to 7.8 and 80 mmol/l, respectively. Sodium bicarbonate may be given intravenously (3 ml/min.1.73 m² of 1 M solution) or orally (4 g/1.73 m², as a single dose). Normal individuals should be able to increase the urine to blood P_{CO_2} gradient to values higher than 20 mmHg, the mean response being situated between 35 and 45 mmHg.

The increase in urinary P_{CO_2} may also be evaluated after giving neutral phosphate. We have found in children that an oral phosphate load is as effective as a phosphate infusion in inducing a significant increase in P_{CO_2}. A total dose of 54 mg/kg of elemental phosphorus is divided into three doses, which are given at 8-h intervals. If the urine pH is less than 6.8 at the time of giving the third dose, oral sodium bicarbonate (2 g/1.73 m²) is also given. Normal children were all able to increase the urine to blood P_{CO_2} to values higher than 40 mmHg (Vallo and Rodriguez-Soriano 1984). Similar reference data are not available in adults.

The assessment of plasma potassium, urinary pH, and urinary P_{CO_2} under different functional conditions will allow for the differentiation of the various pathogenic types of distal tubular acidosis (Table 5). In presence of a 'secretory' defect ('classic' distal tubular acidosis), plasma potassium is usually normal or low, urinary pH fails to fall below 5.5 during acidosis or after sodium sulphate or frusemide, and the urinary P_{CO_2} remains low after either sodium bicarbonate or neutral phosphate loading. However, potassium excretion is normal or increased, and rises further after sodium sulphate or frusemide. In cases secondary to a 'voltage-dependent' defect (hyperkalaemic distal tubular acidosis), urinary pH may fall and the P_{CO_2} may increase after frusemide or phosphate, respectively, but if the defect is severe enough these responses will be absent. In this situation, the only differential sign of a 'secretory' defect will be the presence of hyperkalaemia and the failure to increase potassium excretion after sodium sulphate or frusemide. A 'gradient' defect may be differentiated from a 'secretory' defect by the normal urine to blood P_{CO_2} gradient after sodium bicarbonate loading and by the normal drop in urinary pH that follows sulphate or frusemide.

Treatment

Treatment of distal tubular acidosis consists of giving enough alkali to maintain correction of the acidosis and normalization of the urinary excretion of calcium and citrate. Potassium is also needed, regardless of the plasma potassium, and in cases of severe hypokalaemia, it should be given before correcting the acidosis. Although the effects of sodium bicarbonate on acid–base homeostasis are indistinguishable from those of sodium citrate, the latter should be clearly preferred because it is more effective in increasing urinary excretion of citrate. Often, the amount of sodium or potassium bicarbonate needed to correct the acidosis and hypercalciuria is not enough to raise the low citrate excretion. Treatment is best accomplished by a solution of sodium and potassium citrate: 100 g of each salt in 1 l of distilled water provide approximately 2 mmol of citrate per ml. The total dose of citrate is mainly determined by the concomitant urinary excretion of bicarbonate. The amount of alkali needed for treatment can be estimated to be equal to the urinary bicarbonate excretion plus about 2 mmol/kg.day in children and 1 mmol/kg.day in adults, to compensate for endogenous acid production. Therapeutic requirements may be as high as 10 to 14 mmol/kg.day during the first year of life, falling to about 3 mmol/kg.day in older children. In adults, the dose required varies between 0.5 and 3 mmol/kg.day.

Treatment should aim not only to correct the acidosis but also to normalize calcium excretion and thus prevent further nephrocalcinosis and nephrolithiasis. Unfortunately, parenchymal calcification already present is not reversible. The efficacy of treatment started early in infancy is demonstrated by normal growth and complete arrest of further nephrocalcinosis (Rodriguez-Soriano et al. 1982b). Also, the rate of stone formation decreases markedly in satisfactorily treated adults (Preminger et al. 1985).

Other therapeutic measures are of little or doubtful use. Vitamin D is not generally needed; hydrochlorothiazide is not only of limited benefit but is also potentially dangerous, because it may aggravate the hypokalaemia. It must be recognized that treated patients with distal tubular acidosis are at risk of pseudoephedrine intoxication when this drug is used because its renal tubular reabsorption is increased in alkalinized urine (McSherry 1981).

Hyperkalaemic renal tubular acidosis

Pathogenesis

Hyperkalaemic (type 4) tubular acidosis is found in subjects with sustained hyperkalaemia and is demonstrated by a normal ability to acidify the urine (pH less than 5.5), with an ammonium excretion that is decreased out of proportion to any possible simultaneous reduction in GFR. Bicarbonate wasting (less than 10 per cent of filtered bicarbonate load) may also be present at normal plasma bicarbonate concentrations.

The pathogenesis is complex, although most patients have low circulating levels of aldosterone or a decrease in the renal tubular response to this hormone. The importance of this finding is emphasized by those who equate the terms type 4 tubular acidosis and aldosterone deficiency (Kurtzman 1983). As mentioned above, aldosterone plays an important role in the control of distal hydrogen ion secretion. Mineralocorticoid deficiency in dogs, induced by adrenalectomy and selective replacement of glucocorticoids, results in hyperkalaemic metabolic acidosis (Hulter et al. 1977), and the occurrence of metabolic acidosis is not obviated by preventing the appearance of hyperkalaemia (Hulter et al. 1979). However, metabolic acidosis also appears in chronic potassium loading of adrenalectomized dogs maintained on complete steroid replacement and reduced phosphate intake (Hulter et al. 1983). In a model of selective aldosterone deficiency in the rat, micropuncture studies by DuBose and Caflish (1988) have shown that the acidification defect may be the result of a decrease in ammonia production and delivery to the loop of Henle, an impaired transfer from loop to collecting duct, and a reduction in the rate of hydrogen ion secretion by the collecting duct.

In man, hyperkalaemic tubular acidosis may be found in pure aldosterone deficiency but is more often associated with other contributing factors (renal insufficiency, volume contraction, and administration of potassium-retaining drugs). Many patients have the syndrome of hyporeninaemic hypoaldsteronism, a condition particularly found in adults with mild to moderate renal insufficiency due to diabetic nephropathy or other chronic interstitial nephritis (DeFronzo 1980; Schambelan et al. 1980). The pathogenesis of the metabolic acidosis in these cases is not only attributable to hypoaldosteronism but also to the hyperkalaemia itself. The role of aldosterone deficiency is possibly demonstrated by the beneficial effect of long-term fludrocortisone, which results in large enough increase in ammonium excretion to ameliorate the metabolic acidosis. However, the increase in ammonium excretion may also be due to correction of the hyperkalaemia rather than to a direct effect of mineralocorticoid therapy on ammoniagenesis. Conversely, return of the plasma potassium to normal by dietary potassium restriction or administration of potassium-binding resins is followed by correction of the acidosis, but such an effect could be due in part to extrarenal mechanisms. The improvement in the acidosis that follows prolonged frusemide administration should be attributed to the known effects of this diuretic on renal acid and potassium excretion. The exact parts played by aldosterone deficiency and hyperkalaemia in the development of metabolic acidosis in patients with the syndrome of hyporeninaemic hypoaldosteronism remain to be determined.

Two other important associated factors in the development of metabolic acidosis in patients with aldosterone deficiency are volume contraction and the administration of potassium-retaining drugs. When a patient with moderate renal failure and hyporeninaemic hypoaldosteronism is put on a salt-restricted diet he or she will continue to lose sodium, and the ensuing volume depletion will lead to enhanced proximal reabsorption of sodium chloride and a relative dilution of plasma bicarbonate. The resulting decrease in distal delivery of sodium chloride will also preclude an appropriate level of hydrogen ion secretion. A number of drugs (potassium-sparing diuretics, heparin, indomethacin, captopril, cyclosporin) may aggravate the hyperkalaemia and precipitate metabolic acidosis in a predisposed patient.

Tubular unresponsiveness to aldosterone may also be found in chronic renal insufficiency (Arruda et al. 1981). These patients have a normal or high plasma aldosterone, do not lose sodium under standard dietary conditions, and required supraphysiological doses of fludrocortisone to benefit the hyperkalaemia and metabolic acidosis. The characteristics of the acidification defect are identical to those described in aldosterone deficiency. Those with aldosterone deficiency and/or aldosterone resistance may occasionally present with features of hyperkalaemic distal tubular acidosis. Conceivably, structural damage of the renal parenchyma may impair both distal mechanisms of hydrogen ion and potassium secretion and result in a 'voltage-dependent' defect of acidification.

Hyperkalaemic tubular acidosis may also occur where an expanded plasma volume results in inhibition of renin and aldosterone secretion and hyperkalaemia. This may take place during the acute phase of acute glomerulonephritis or in the context of so-called 'chloride-shunt' syndrome (Gordon's syndrome), in which there is a genetically determined, tubular hyper-reabsorption of sodium chloride (Gordon 1986). The pathogenesis of the acidosis is directly related to the hyperkalaemia.

Clinical features and diagnosis

The causes of hyperkalaemic tubular acidosis are shown in Table 6. In general, the clinical features depend on the underlying disease rather than on metabolic acidosis itself. Nephrocalcinosis and lithiasis are absent, and bone lesions are only apparent in uraemic subjects.

Table 6 Causes of hyperkalaemic renal tubular acidosis

I. *Primary or idiopathic*
   'Early childhood' hyperkalaemia (transient)
II. *Secondary*
1. Aldosterone deficiency without intrinsic renal disease
   Salt-losing congenital adrenal hyperplasia
   Isolated hypoaldosteronism
   Addison's disease
2. Hyporeninaemic hypoaldosteronism in patients with chronic renal disease
   Diabetic nephropathy
   Lupus nephritis
   Sickle-cell nephropathy
   Nephrosclerosis
   Other interstitial nephritis
3. Hyporeninaemic hypoaldosteronism in patients with acute glomerulonephritis
4. Renal tubular disorders
   Primary pseudohypoaldosteronism of infancy
   Secondary pseudohypoaldosteronism
     Obstructive uropathy in infancy
     Renal vein thrombosis in infancy
   'Chloride-shunt' syndrome (Gordon's syndrome)
5. Drugs and toxins (contributory factors)
   Potassium chloride supplements
   Potassium-sparing diuretics
   Heparin
   Indomethacin and other prostaglandin inhibitors
   Captopril
   Cyclosporin

Table 7 Causes of metabolic acidosis

I. *With normal plasma anion gap*
1. Renal loss of bicarbonate
   Administration of carbonic anhydrase inhibitors (acetazolamide)
   Renal tubular acidosis
   Post-hypocapnic acidosis
2. Gastrointestinal loss of bicarbonate
   Diarrhoea
   Ileostomy
   Fistulas
   Ureterosigmoidostomy
   Ileal bladder and conduit
3. Miscellaneous
   Administration of HCl, NH_4Cl, arginine-HCl, lysine-HCl
   Parenteral nutrition
   'Dilution' acidosis
II. *With elevated plasma anion gap*
1. Increased production of acid
   Ketoacids (starvation, diabetes mellitus, ketogenic diets, ethanol-associated)
   Lactic acidosis and other organic acidosis
   Toxins (e.g., salicylate, methanol, ethylglycol, paraldehyde)
2. Decreased excretion of acid
   Acute renal failure
   Chronic renal failure

Hyperkalaemic tubular acidosis should be differentiated from hyperkalaemic distal tubular acidosis (see Table 3). The first diagnostic step in a case with hyperkalaemia and acidosis is to determine GFR and urine pH. The finding of a low pH (less than 5.5) during spontaneous or induced acidaemia will establish the diagnosis of hyperkalaemic tubular acidosis. Also, the urine to blood $P{CO_2}$ gradient in alkaline urine will be greater than 20 mmHg in this type of acidosis, provided that GFR is not much diminished (higher than 40 ml/min) and there is no marked defect in urinary concentrating ability. The presence of some degree of bicarbonaturia when the plasma bicarbonate is normal may bring about some confusion with proximal tubular acidosis, but differential diagnosis is easy if attention is paid to the associated impairment in the excretion rate of ammonium in acidaemia.

The best evaluation of these patients is by an acute frusemide test. This will allow assessment of both the distal acidifying capacity and the response of the renin–aldosterone axis to acute volume depletion. Patients with hyporeninaemic hypoaldosteronism will show a characteristic response: the urine will be strongly acid with low ammonium and potassium, and the plasma renin activity and aldosterone concentration will remain subnormal. Patients with expanded plasma volume and secondary inhibition of renin and aldosterone secretion will have, in contrast, a response indistinguishable from that of controls.

Treatment

The cause of the hyperkalaemic tubular acidosis should be found for adequate management. Potassium-retaining drugs should be discontinued at once. If there is volume contraction, it should be corrected; this is critical in those with salt-losing aldosterone deficiency. In those with chronic renal disease and hyporeninaemic hypoaldosteronism, treatment decisions should be made from the symptoms and the severity of the hyperkalaemia. Acidosis and hyperkalaemia may be corrected by high doses of fludrocortisone (Sebastian *et al*. 1977). Prolonged treatment, however, may be complicated by excessive salt retention with increased risk of cardiovascular compromise or arterial hypertension. Restriction of potassium intake, potassium-binding resins or, preferably, periodic administration of frusemide will allow the dose of fludrocortisone to be reduced and so prevent such complications (Sebastian *et al*. 1984). In many cases, sodium bicarbonate at a dose of 1.5 to 2 mmol/kg.day will also be necessary.

Patients with primary or secondary pseudohypoaldosteronism do not respond to exogenous mineralocorticoids and must be treated with salt supplements. Sustained correction of volume contraction is followed in most cases by correction of the hyperkalaemia and metabolic acidosis, without need of additional alkali. Mineralocorticoids are also ineffective in Gordon's syndrome: in this disease, hyperkalaemic metabolic acidosis is rapidly improved by dietary salt restriction or diuretics. Frusemide has been recommended but it aggravates the hypercalciuria frequently present in this syndrome. The ideal treatment now is the administration of hydrochlorothiazide, which is as effective as frusemide in the reversal of hyperkalaemia and metabolic acidosis but also corrects the hypercalciuria (Rodriguez-Soriano *et al*. 1989).

Diagnostic approach to metabolic acidosis

In seeking the cause of metabolic acidosis one should consider all circumstances that may lead to gain of acid or loss of base from body fluids (Table 7). In general, tubular acidosis should be

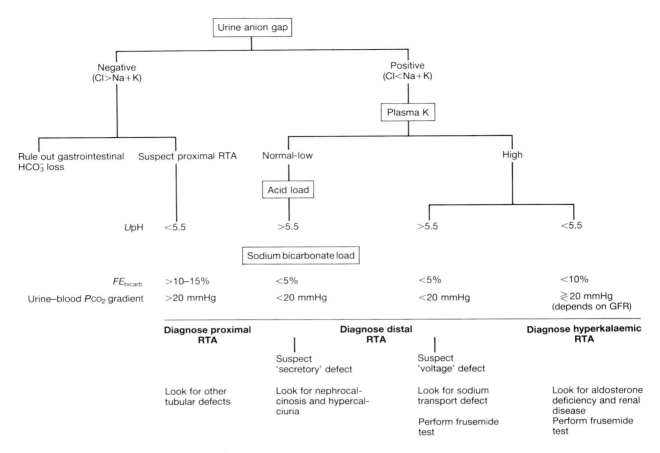

Fig. 5 Algorithm for the differential diagnosis of hyperchloraemic metabolic acidosis.

suspected when metabolic acidosis is accompanied by hyperchloraemia and a normal plasma anion gap. A normal gap reflects loss of bicarbonate from the extracellular fluid via the gastrointestinal tract or the kidney, dilution of extracellular buffer, or administration of hydrochloric acid or its precursors. The distinction of tubular acidosis from gastrointestinal loss of bicarbonate may be greatly facilitated by study of the urinary anion gap. This may provide an indirect index of urinary ammonium excretion in patients with sustained metabolic acidosis (Goldstein et al. 1986; Batlle et al. 1988b). The anion gap (also called urinary net charge) is calculated from the measured concentration of electrolytes in the urine: (sodium + potassium) − chloride. On a regular diet, the amounts of calcium and magnesium are small and the excretion of phosphate, sulphate, and organic acids varies very little. The difference in excretion of these anions and divalent cations is in the order of 80 mmol of anions per day. The urinary anion gap reflects, therefore, the excretion of ammonium, which is the only monovalent cation that is not actually measured.

When a urine sample from a patient with metabolic acidosis has a negative anion gap, the ammonium excretion must exceed 80 mmol/day in adults (Goldstein et al. 1986) and 50 mmol/day. 1.73 m^2 in children (Rodriguez-Soriano and Vallo 1990), suggesting the presence of gastrointestinal or renal loss of bicarbonate or prior administration of an acid load. When the urine has a positive anion gap, the rate of ammonium excretion must be lower than 80 mmol/day in adults or 50 mmol/day. 1.73 m^2 in children, suggesting the presence of a distal acidification defect.

Cautious use of the anion gap should be made where the urine contains large amounts of unusual anions such as ketoacids, salicylate, or penicillin. In these instances, the above formula for calculating the anion gap will underestimate urinary ammonium excretion (Halperin et al. 1988).

In patients with a negative anion gap, differential diagnosis should be established between proximal tubular acidosis and gastrointestinal loss of bicarbonate. Diagnosis of proximal acidosis is rapidly established by the demonstration of a high renal bicarbonate excretion at normal plasma bicarbonate concentration (Fig. 5). In those with a positive anion gap or with low urinary ammonium excretion, the immediate step in the diagnostic work-up is to measure the plasma potassium. If this is normal or low, the finding of an inability to lower the urinary pH below 5.5 will confirm the diagnosis of distal tubular acidosis. This diagnosis is further supported by the finding of an incapacity to increase urinary $P{\rm co}_2$ after loading with sodium bicarbonate. In isolated distal tubular acidosis, the fractional excretion of bicarbonate should not exceed 5 per cent; more than this bicarbonate should suggest the presence of a combined proximal and distal tubular acidosis.

In hyperkalaemic patients, the first diagnostic step is also to determine the urinary pH. Finding a high urine pH during spontaneous or induced metabolic acidosis will allow the separation of a small group of patients with hyperkalaemic distal tubular acidosis, caused by a severe 'voltage-dependent' defect. When there is a normal ability to acidify the urine and ammonium excretion is particularly low, the diagnosis of hyperkalaemic (type 4) tubular acidosis is established and diagnostic work-up should investigate the underlying cause.

References

Adams, N.D., Gray, R.W., and Lemann, J.Jr. (1979). The calciuria of increased fixed acid production: evidence against a role for parathyroid hormone and 1,25-(OH)$_2$ vitamin D. *Calcified Tissue International*, **27**, 233–9.

Al-Awqati, Q. (1978). H$^+$ transport in urinary epithelia. *American Journal of Physiology*, **235**, F77–88.

Al-Bander, H.A., Weiss, M.H., Humphreys, M.H., and Morris, R.C.Jr. (1982). Dysfunction of the proximal tubule underlies maleic acid-induced type II renal tubular acidosis. *American Journal of Physiology*, **243**, F604–11.

Al-Bander, H., Etheredge, S.B., Pankert, T., Humphreys, M.H., and Morris, R.C.Jr. (1985). Phosphate-loading attenuates renal tubular dysfunction induced by maleic acid in the dog. *American Journal of Physiology*, **248**, F513–21.

Aronson, P.S. (1983). Mechanisms of active H$^+$ secretion in the proximal tubule. *American Journal of Physiology*, **245**, F647–59.

Arruda, J.A.L. and Kurtzman, N.A. (1980). Hyperparathyroidism and metabolic acidosis: a complex interaction of multiple factors. *Nephron*, **26**, 1–6.

Arruda, J.A.L., Batlle, D.C., Sehy, T., Roseman, M.K., Boronowski, R.L., and Kurtzman, N.A. (1981). Hyperkalemia and renal insufficiency: role of aldosterone deficiency and tubular unresponsiveness to aldosterone. *American Journal of Nephrology*, **1**, 160–7.

Backman, U., Danielson, B.G., Johansson, G., Ljunghall, S., and Wikström, B. (1980). Incidence and clinical importance of renal tubular defects in recurrent renal stone formers. *Nephron*, **25**, 96–101.

Baran, D.T. and Marcy, T.W. (1984). Evidence for a defect in vitamin D metabolism in a patient with incomplete Fanconi syndrome. *Journal of Clinical Endocrinology and Metabolism*, **59**, 998–1001.

Batlle, D.C. (1981). Hyperkalemic hyperchloremic metabolic acidosis associated with selective aldosterone deficiency and distal renal tubular acidosis. *Seminars in Nephrology*, **1**, 260–74.

Batlle, D.C. (1983). Renal tubular acidosis. *Medical Clinics of North America*, **67**, 859–78.

Batlle, D.C. (1986*a*). Segmental characterization of defects in collecting tubule acidification. *Kidney International*, **30**, 546–54.

Batlle, D.C. (1986*b*). Sodium-dependent urinary acidification in patients with aldosterone deficiency and in adrenalectomized rats: effect of furosemide. *Metabolism*, **35**, 862–60.

Batlle, D.C. and Chan, Y.L. (1989). Effect of L-arginine on renal tubular bicarbonate reabsorption in the rat kidney. *Mineral and Electrolyte Metabolism*, **15**, 187–94.

Batlle, D.C., Sehy, J.T., Roseman, M.K., Arruda, J.A.L., and Kurtzman, N.A. (1981). Clinical and pathophysiological spectrum of acquired distal renal tubular acidosis. *Kidney International*, **20**, 389–96.

Batlle, D.C., Grupp, M., Gaviria, M., and Kurtzman, N.A. (1982*a*). Distal renal tubular acidosis with intact capacity to lower urinary pH. *American Journal of Medicine*, **72**, 751–8.

Batlle, D.C., Itsarayoungyen, K., Hays, S., Arruda, J.A.L., and Kurtzman, N.A. (1982*b*). Parathyroid hormone is not anticalciuric during chronic metabolic acidosis. *Kidney International*, **22**, 264–71.

Batlle, D.C., Van Riotte, A., and Schlueter, W. (1987). Urinary sodium in the evaluation of hyperchloremic metabolic acidosis. *New England Journal of Medicine*, **316**, 140–4.

Batlle, D.C., Sabatini, S., and Kurtzman, N.A. (1988*a*). On the mechanism of toluene-induced renal tubular acidosis. *Nephron*, **49**, 210–8.

Batlle, D.C., Hizon, M., Cohen, E., Gutterman, C., and Grupta, R. (1988*b*). The use of the urinary anion gap in the diagnosis of hyperchloremic metabolic acidosis. *New England Journal of Medicine*, **318**, 594–9.

Bichara, M., Mercier, O., Paillard, M., and Leviel, F. (1986). Effects of parathyroid hormone on urinary acidification. *American Journal of Physiology*, **251**, F444–53.

Brenes, L.G., Brenes, J.M., and Hernãndez, M.M. (1977). Familial proximal renal tubular acidosis. A distinct clinical entity. *American Journal of Medicine*, **63**, 244–52.

Brennan, S., Hering-Smith, K., and Hamm, L.L. (1988). Effect of pH on citrate reabsorption in the proximal convoluted tubule. *American Journal of Physiology*, **255**, F301–6.

Brenner, R.J., *et al.* (1982). Incidence of radiographically evident bone disease, nephrocalcinosis and nephrolithiasis in various types of renal tubular acidosis. *New England Journal of Medicine*, **307**, 217–21.

Broyer, M., Proesmans, W., and Royer, P. (1969). La titration des bicarbonates chez l'enfant normal et au cours des diverses néphropathies. *Revue Française d'Etudes Cliniques et Biologiques*, **6**, 556–67.

Buckalew, V.M.Jr., Purvis, M.L., Shulman, M.G., Herndon, C.N., and Rudman, D. (1974). Hereditary renal tubular acidosis. Report of a 64 member kindred with variable clinical expression including idiopathic hypercalciuria. *Medicine*, **53**, 229–54.

Bushinsky, D.A. and Lechleider, R.J. (1987). Mechanism of proton-induced bone calcium release: calcium carbonate dissolution. *American Journal of Physiology*, **253**, F998–1005.

Callis, L., Castello, F., Fortuny, G., Vallo, A., and Ballabriga, A. (1970). Effect of hydrochlorothiazide on rickets and on renal tubular acidosis in two patients with cystinosis. *Helvetica Paediatrica Acta*, **25**, 602–19.

Caruana, R.J. and Buckalew, V.M.Jr. (1988). The syndrome of distal (type 1) renal tubular acidosis. Clinical and laboratory findings in 58 cases. *Medicine*, **67**, 84–99.

Chesney, R.W., Kaplan, B.S., Phelps, M., and DeLuca, H.F. (1984). Renal tubular acidosis does not alter circulating levels of calcitriol. *Journal of Pediatrics*, **104**, 51–5.

Coe, F.L. and Firpo, J.L. (1975). Evidence for mild reversible hyperparathyroidism in distal renal tubular acidosis. *Archives of Internal Medicine*, **135**, 1485–9.

Cogan, M.G. (1986). Neurogenic regulation of proximal bicarbonate reabsorption and chloride reabsorption. *American Journal of Physiology*, **250**, F22–6.

Cogan, M.G. and Alpern, R.J. (1984). Regulation of proximal bicarbonate reabsorption. *American Journal of Physiology*, **247**, F387–95.

Cogan, M.G., Maddox, D.A., Lucci, M.S., and Rector, F.C.Jr. (1979). Control of proximal bicarbonate reabsorption in normal and acidotic rats. *Journal of Clinical Investigation*, **64**, 1168–80.

Cunningham, J., Bikle, D.D., and Avioli, L.V. (1984). Acute, but not chronic, metabolic acidosis disturbs 25-hydroxyvitamin D$_3$ metabolism. *Kidney International*, **25**, 47–52.

DeFronzo, R.A. (1980). Hyperkalemic and hyporeninemic hypoaldosteronism. *Kidney International*, **17**, 118–34.

Donckerwolcke, R.A., Van Biervliet, J.P., Koorevaar, G., Kuitjen, R.H., and Van Stekelenburg, G.J. (1976). The syndrome of renal tubular acidosis with nerve deafness. *Helvetica Paediatrica Acta*, **65**, 100–4.

DuBose, T.D.Jr. (1982). Hydrogen ion secretion by the collecting duct is a determinant of the urine to blood P_{CO_2} gradient in alkaline urine. *Journal of Clinical Investigation*, **69**, 145–56.

DuBose, T.D.Jr. and Caflish, C.R. (1985). Validation of the difference in urine to blood carbon dioxide tension during experimental models of distal renal tubular acidosis. *Journal of Clinical Investigation*, **75**, 1116–23.

DuBose, T.D.Jr. and Caflish, C.R. (1988). Effect of selective aldosterone deficiency on acidification in nephron segments of the rat inner medulla. *Journal of Clinical Investigation*, **82**, 1624–32.

Edelmann, C.M.Jr. (1985). Isolated proximal (type 2) renal tubular acidosis. In *Renal tubular disorders* (ed. H.C. Gonick and V.M. Buckalew Jr.), pp. 261–79. New York, Marcel Drekker.

Edelmann, C.M.Jr., Rodriguez-Soriano, J., Boichis, H., Gruskin, A.B., and Acosta, M.I. (1967). Renal bicarbonate reabsorption and

hydrogen ion excretion in normal infants. *Journal of Clinical Investigation*, **46**, 1309–17.
Emmett, M. and Narins, R.G. (1977). Clinical use of the anion gap. *Medicine*, **56**, 38–54.
Goldstein, M.B., Bear, R., Richardson, R.M.A., Mardsen, P.A., and Halperin, M.L. (1986). The urine anion gap: A clinically useful index of acid excretion. *American Journal of Medical Sciences*, **292**, 198–202.
Good, D.W. and Knepper, M.A. (1985). Ammonia transport in the mamalian kidney. *American Journal of Physiology*, **248**, F459–71.
Gordon, R.D. (1986). Syndrome of hypertension and hyperkalaemia with normal glomerular filtration rate. *Hypertension*, **8**, 93–102.
Gougoux, A., et al. (1980). Effect of blood pH on distal hydrogen ion secretion. *Kidney International*, **17**, 615–21.
Halperin, M.L., Goldstein, M.S., Haig, A., Johnson, M.D., and Stinebaugh, B.J. (1974). Studies on the pathogenesis of type I (distal) renal tubular acidosis as reveled by the urinary P_{CO_2} tensions. *Journal of Clinical Investigation*, **53**, 669–77.
Halperin, M.L., Richardson, R.M.A., Bear, R.A., Magner, P.O., Kamel, K., and Ethier, J.H. (1988). Urine ammonium: the key to the diagnosis of distal renal tubular acidosis. *Nephron*, **50**, 1–4.
Halperin, M.L., Kamel, K.S., Ethier, J.H., and Magner, P.O. (1989). What is the underlying defect in patients with isolated, proximal renal tubular acidosis? *American Journal of Nephrology*, **9**, 265–8.
Harrington, T.M., Bunch, T.W., and Van den berg, C.J. (1983). Renal tubular acidosis. A new look at treatment of musculoskeletal and renal disease. *Mayo Clinic Proceedings*, **58**, 354–60.
Hulter, H.N. (1985). Effects and interrelationships of PTH, Ca^{2+}, vitamin D and Pi in acid–base homeostasis. *American Journal of Physiology*, **248**, F739–52.
Hulter, H.N., Zenicki, L.P., Harbottle, J.A., and Sebastian, A. (1977). Impaired renal H+ secretion and NH_3 production in mineralocorticoid-deficient glucocorticoid-replete dogs. *American Journal of Physiology*, **232**, F136–46.
Hulter, N.H., Licht, J.H., Glynn, R.D., and Sebastian, A. (1979). Renal acidosis in mineralocorticoid deficiency is not dependent on NaCl depletion or hyperkalemia. *American Journal of Physiology*, **236**, F283–94.
Hulter, H.N., Toto, R.D., Ilnicki, L.P., and Sebastian, A. (1983). Chronic hyperkalemic renal tubular acidosis induced by KCl loading. *American Journal of Physiology*, **244**, F255–64.
Jacobson, H.R. and Seldin, D.W. (1983). On the generation, maintenance, and correction of metabolic alkalosis. *American Journal of Physiology*, **245**, FF425–32.
Jonas, A.J., Lin, S.W., Conley, S.B., Schneider, J.A., Williams, J.C., and Caprioli, R.C. (1989). Urine glyceraldehyde excretion is elevated in the renal Fanconi syndrome. *Kidney International*, **35**, 99–104.
Kornandakieti, C. and Tannen, R.L. (1984). Hydrogen ion secretion by the distal nephron in the rat: Effect of potassium. *Journal of Laboratory and Clinical Medicine*, **104**, 293–303.
Kraut, J.A., Mishler, D.R., Singer, F.R., and Goodman, W.G. (1986). The effects of metabolic acidosis on bone formation and bone resorption in the rat. *Kidney International*, **30**, 694–700.
Kurtzman, N.A. (1970). Regulation of renal bicarbonate reabsorption by extracellular volume. *Journal of Clinical Investigation*, **49**, 586–95.
Kurtzman, N.A. (1983). Acquired distal renal tubular acidosis. *Kidney International*, **24**, 807–19.
Lee, H.L. and Simon, E.E. (1987). Roles and mechanisms of urinary buffer excretion. *American Journal of Physiology*, **253**, F595–605.
Lemieux, G., Vinay, P., and Gougoux, A. (1982). Renal ammoniagenesis. *Advances in Nephrology*, **11**, 371–97.
Lightwood, R. and Butler, N. (1963). Decline in primary infantile renal acidosis. Aetiological implications. *British Medical Journal*, **1**, 855–7.
Lightwood, R., Payne, W.W., and Black, J.A. (1953). Infantile renal acidosis. *Pediatrics*, **12**, 628–44.
Liu, F-Y. and Cogan, M.G. (1987). Angiotension II: a potent regulator of acidification in the rat early proximal convoluted tubule. *Journal of Clinical Investigation*, **80**, 272–5.
Maddox, D.A., Deen, W.M., and Gennari, F.J. (1987). Control of bicarbonate reabsorption in the proximal convoluted tubule. *Seminars in Nephrology*, **29**, 667–74.
Madias, N.E. and Zelman, S.J. (1986). The renal response to chronic mineral acid feeding: A re-examination of the role of systemic pH. *Kidney International*, **29**, 667–74.
Madsen, K.M. and Tisher, C.C. (1986). Structural-functional relationships along the distal nephron. *American Journal of Physiology*, **250**, F1–15.
Marver, D. and Kokko, J.P. (1983). Renal target sites and the mechanism of action of aldosterone. *Mineral and Electrolyte Metabolism*, **9**, 1–18.
McCurdy, D.K. and Elkinton, J.R. (1968). Renal tubular acidosis due to amphotericin B. *New England Journal of Medicine*, **276**, 124–30.
McKinney, T.D. and Burg, M.B. (1977). Bicarbonate transport by rabbit cortical collecting tubules. Effect of acid and alkaline loads in vivo on transport in vitro. *Journal of Clinical Investigation*, **60**, 766–8.
McSherry, E. (1981). Renal tubular acidosis in childhood. *Kidney International*, **20**, 799–809.
McSherry, E. and Morris, R.C.Jr. (1978). Attainment and maintenance of normal stature with alkali therapy in infants and children with classic renal tubular acidosis. *Journal of Clinical Investigation*, **61**, 509–27.
McSherry, E., Sebastian, A., and Morris, R.C.Jr. (1972). Renal tubular acidosis in infants: the several kinds, including bicarbonate-wasting renal tubular acidosis. *Journal of Clinical Investigation*, **51**, 499–514.
Morris, R.C.Jr. (1981). Renal tubular acidosis. *New England Journal of Medicine*, **304**, 418–20.
Morris, R.C.Jr. (1983). Renal tubular acidosis and Fanconi syndrome. In *The metabolic basis of inherited disease* (ed. J.B. Stanbury, J.B. Wyngaarden, D.S. Fredrickson, J.C. Goldstein, and M.S. Brown), pp. 1808–43. New York, McGraw-Hill.
Morris, R.C.Jr., Sebastian, A., and McSherry, E. (1972). Renal acidosis. *Kidney International*, **1**, 322–40.
Muldowney, F.P., Donohoe, J.F., Freaney, R., Kampff, C., and Swan, M. (1970). Parathormone-induced renal bicarbonate wastage in intestinal malabsorption and in chronic renal failure. *Irish Journal of Medical Science*, **3**, 221–31.
Nash, M.A., Torrado, A., Greifer, I., Spitzer, A., and Edelmann, C.M.Jr. (1972). Renal tubular acidosis in infants and children. Clinical course, response to treatment and prognosis. *Journal of Pediatris*, **80**, 738–48.
Norman, M.E., Feldman, N.I, Cohn, R.M., Roth, K.S., and McCurdy, D.K. (1978). Urinary citrate excretion in the diagnosis of renal tubular acidosis. *Journal of Pediatrics*, **92**, 394–400.
Pichette, C., Tam, S.C., Chen, C.B., Goldstein, M., Stinebaugh, B., and Halperin, M. (1982). Effect of potassium on distal hydrogen ion secretion in the dog. *Journal of Laboratory and Clinical Medicine*, **100**, 374–84.
Preminger, G.M., Sakhale, K., Skurda, C., and Pak, C.Y.C. (1985). Prevention of recurrent calcium stone formation with potassium citrate therapy in patients with distal renal tubular acidosis. *Journal of Urology*, **134**, 20–3.
Preminger, G.M., Sakhale, K., and Pak, C.Y.C. (1987). Hypercalciuria and altered intestinal calcium absorption occurring independently of vitamin D in incomplete distal renal tubular acidosis. *Metabolism*, **36**, 176–9.
Rampini, S., Fanconi, A., Illig, R., and Prader, A. (1968). Effect of hydrochlorothiazide on proximal renal tubular acidosis in a patient with idiopathic 'DeToni–Debré–Fanconi syndrome'. *Helvetica Paediatrica Acta*, **23**, 13–21.
Rastogi, S., Bayliss, J.M., Nascimento, L., and Arruda, J.A.L. (1985). Hyperkalaemic renal tubular acidosis: Effect of furosemide in humans and rats. *Kidney International*, **28**, 801–7.

Rodriguez-Soriano, J. and Edelmann, C.M.Jr. (1969). Renal tubular acidosis. *Annual Review of Medicine*, **20**, 363–82.

Rodriguez-Soriano, J. and Vallo, A. (1988). Renal tubular hyperkalaemia in childhood. *Pediatric Nephrology*, **2**, 489–509.

Rodriguez-Soriano, J. and Vallo, A. (1990). Renal tubular acidosis. *Pediatric Nephrology*, **4**, 268–75.

Rodriguez-Soriano, J., Boichis, H., Stark, H., and Edelmann, C.M.Jr. (1967a). Proximal renal tubular acidosis. A defect in bicarbonate reabsorption with normal urinary acidification. *Pediatric Research*, **1**, 81–98.

Rodriguez-Soriano, J., Boichis, H., and Edelmann, C.M.Jr. (1967b). Bicarbonate reabsorption and hydrogen ion excretion in children with renal tubular acidosis. *Journal of Pediatrics*, **71**, 802–13.

Rodriguez-Soriano, J., Vallo, A., and Garcia-Fuentes, M. (1975). Distal renal tubular acidosis in infancy: A bicarbonate-wasting state. *Journal of Pediatrics*, **86**, 524–32.

Rodriguez-Soriano, J., Vallo, A., Castillo, G., and Oliveros, R. (1980). Renal handling of water and sodium in children with proximal and distal renal tubular acidosis. *Nephron*, **25**, 193–8.

Rodriguez-Soriano, J., Vallo, A., Castillo, G., and Oliveros, R. (1982a). Defect in urinary acidification in nephrotic syndrome and its correction by furosemide. *Nephron*, **32**, 308–13.

Rodriguez-Soriano, J., Vallo, A., Castillo, G., and Oliveros, R. (1982b). Natural history of primary distal renal tubular acidosis treated since infancy. *Journal of Pediatrics*, **101**, 669–76.

Rodriguez-Soriano, J., Vallo, A., Castillo, G., and Oliveros, R. (1985). Pathophysiology of primary distal renal tubular acidosis. *International Journal of Pediatric Nephrology*, **6**, 71–8.

Rodriguez-Soriano, J., Vallo, A., Castillo, G., Oliveros, R., and Fernandez-Garnica, M. (1986). Hyperkalemic distal renal tubular acidosis in salt-losing congenital adrenal hyperplasia. *Acta Paediatrica Scandinavica*, **75**, 425–32.

Rodriguez-Soriano, J., Vallo, A., and Garcia-Fuentes, M. (1987). Hypomagnesaemia of hereditary renal origin. *Pediatric Nephrology*, **1**, 465–72.

Rodriguez-Soriano, J., Vallo, A., and Dominguez, M.J. (1989). 'Chloride-shunt' syndrome: An overlooked cause of renal hypercalciuria. *Pediatric Nephrology*, **3**, 113–21.

Rosen, R.A., Julian, B.A., Dubovsky, E.V., Galla, J.H., and Luke, R.G. (1988). On the mechanism by which chloride corrects metabolic alkalosis in man. *American Journal of Medicine*, **84**, 449–58.

Santos, F. and Chan, J.C. (1986). Renal tubular acidosis in children. Diagnosis, treatment and prognosis. *American Journal of Nephrology*, **6**, 289–95.

Sastrasinh, S. and Tannen, R.L. (1983). Effect of potassium on renal NH_3 production. *American Journal of Physiology*, **244**, F383–91.

Schambelan, M., Sebastian, A., and Biglieri, E.G. (1980). Prevalence, pathogenesis, and functional significance of aldosterone deficiency in hyperkalemic patients with chronic renal insufficiency. *Kidney International*, **17**, 89–101.

Schwartz, W.B. and Cohen, J.J. (1978). The nature of the renal response to chronic disorders of acid-base equilibrium. *American Journal of Medicine*, **64**, 417–28.

Sebastian, A., McSherry, E., and Morris, R.C. Jr. (1971). Renal potassium wasting in renal tubular acidosis (RTA): its occurrence in types 1 and 2 RTA despite sustained correction of systemic acidosis. *Journal of Clinical Investigation*, **50**, 667–78.

Sebastian, A., Schambelan, M., Lindefeld, S., and Morris, R.C.Jr. (1977). Amelioration of metabolic acidosis with fludrocortisone therapy in hyporeninemic-hypoaldosteronism. *New England Journal of Medicine*, **297**, 576–83.

Sebastian, A., Schambelan, M., and Sutton, J.M. (1984). Amelioration of hyperchloremic acidosis with furosemide therapy in patients with chronic renal insufficiency and type 4 renal tubular acidosis. *American Journal of Nephrology*, **4**, 287–300.

Selvaggio, A.M., Schwartz, J.H., Bengele, H.H., Gordon, F.D., and Alexander, E.A. (1988). Mechanisms of H^+ secretion by inner medullary collecting duct. *American Journal of Physiology*, **254**, F391–400.

Simpson, D.P. (1971). Control of hydrogen ion homeostasis and renal acidosis. *Medicine*, **50**, 503–41.

Simpson, D.P. (1983). Control of hydrogen ion homeostasis and renal acidosis. *Medicine*, 50, 503–41.

Simpson, D.P. (1983). Citrate excretion: a window on renal metabolism. *American Journal of Physiology*, **244**, F223–34.

Sly, W.S., *et al*. (1985a). Carbonic anhydrase II deficiency in 12 families with the autosomal recessive syndrome of osteopetrosis with renal tubular acidosis and cerebral calcification. *New England Journal of Medicine*, **313**, 139–45.

Sly, W.S., Whyte, M.P., Krupin, T., and Sundaram, V. (1985b). Positive renal response to intravenous acetazolamide in patients with carbonic anhydrase II deficiency. *Pediatric Research*, **19**, 1033–6.

Steinmetz, P.R. (1986). Cellular organization of urinary acidification. *American Journal of Physiology*, **251**, F173–87.

Stinebaugh, B.J., Schloeder, F.X., Gharafry, E., Suki, W.N., Goldstein, M.B., and Halperin, M.L. (1977). Mechanism by which neutral phosphate infusion elevates urine P_{CO_2}. *Journal of Laboratory and Clinical Medicine*, **89**, 946–58.

Stone, D.K. and Xie, X.S. (1988). Proton translocating ATPases: Issues in structure and function. *Kidney International*, **33**, 767–74.

Stone, D.K., Seldin, D.W., and Jacobsen, H.R. (1983). Mineralocorticoid modulation of rabbit medullary collecting duct acidification. A sodium-independent effect. *Journal of Clinical Investigation*, **72**, 77–83.

Tannen, R.L. (1977). Ammonia metabolism. *American Journal of Physiology*, **235**, F265–77.

Vallo, A. and Rodriguez-Soriano, J. (1984). Oral phosphate-loading test to the assessment of distal urinary acidification in children. *Mineral and Electrolyte Metabolism*, **10**, 387–90.

Warnock, D.G. (1988). Uremic acidosis. *Kidney International*, **34**, 278–87.

Winsnes, A., Monn, E., Stokke, O., and Feyling, T. (1979). Congenital, persistent proximal type of renal tubular acidosis in two brothers. *Acta Paediatrica Scandinavica*, **60**, 861–8.

Wrong, O. and Davies, H.E.F. (1959). The excretion of acid in renal disease. *Quarterly Journal of Medicine*, **28**, 259–313.

Wrong, O.M. and Feest, T.G. (1980). The natural history of distal renal tubular acidosis. *Contributions in Nephrology*, **21**, 137–44.

5.4 Bartter's syndrome

R.O.B. GANS AND S.J. HOORNTJE

Introduction

After a preliminary report in 1960, Bartter and his coworkers described in 1962 two patients with growth retardation, aldosteronism with hypokalaemic alkalosis, hyper-reninism, a normal blood pressure, relative resistance to the pressor effects of exogenously administered angiotensin-II, and hypertrophy and hyperplasia of the juxtaglomerular apparatus of many glomeruli. Though its prevalence is not exactly known, Bartter's syndrome does very occasionally appear as a cause of hypokalaemia. Nevertheless, the syndrome has aroused great interest in many clinical investigators because it may provide new insights in renal electrolyte metabolism and the pathophysiology of hypertension.

It is questionable whether there is a unifying hypothesis which explains all the key features of the syndrome. Even after exclusion of those conditions that may mimic Bartter's syndrome, many patients who otherwise meet all the criteria of this disease show differences in results obtained from renal tubular function studies. Apparently, Bartter's syndrome is essentially heterogenous and, therefore, a syndrome in the etymological sense of the word.

Clinical features

With only approximately 200 cases described over the last 30 years, Bartter's syndrome is relatively rare. However, its prevalence may well be higher than was previously assumed. In a large retrospective study of 28 patients from Sweden, the figure of 19 per million population was estimated (Rudin 1988). It is remarkable that a substantial number of these patients were detected by chance in a routine setting.

Patients with Bartter's syndrome come from many geographic areas and include all races although there is a preponderance of black people. Females are slightly more often affected than men. The syndrome has been diagnosed as early as in the 20th week of pregnancy, and as late as in people in their fifties (Proesmans *et al.* 1985; Rudin 1988).

The familial incidence of Bartter's syndrome is marked but vertical inheritance has only once been described (Gill 1980). The occurrence in siblings and in the offspring of consanguineous matings suggests autosomal recessive inheritance (Bartter 1962). The description of four patients in one generation of two related families enforces this assumption (Hogewind *et al.* 1981). No HLA linkage has been established (Delaney *et al.* 1984). The detection of asymptomatic carriers is difficult. Heterozygotes with the responsible genes have no clear-cut symptoms or biochemical abnormalities characteristic for the syndrome. Adrenaline and ADP-induced platelet aggregatory defects have been put forward as consistent markers of the carrier state (Pereira and van Wersch 1983), but these defects vary with sodium intake and can be manipulated therapeutically (Stoff *et al.* 1980), making them far from perfect markers of the syndrome.

The clinical symptoms of Bartter's syndrome (Table 1) are dominated by hypokalaemia. Muscular weakness, polyuria associated with enuresis, and recurrent tetany with positive Chvostek and Trousseau signs are predominant symptoms. Proximal muscle weakness may cause incapacity and force families to seek medical help. Flaccid tetraparesis has been described but its occurrence is rare despite the often severe hypokalaemia. Gastrointestinal symptoms include anorexia and constipation due to renal water loss and hypokalaemic ileus. A craving for salt, vinegar and sour pickles is occasionally mentioned. Rickets, nephrocalcinosis, chondrocalcinosis, gout, impaired glucose tolerance and erythrocytosis have also, though rarely, been reported. Despite defects in platelet aggregation, a bleeding diathesis is absent.

Clinically, Bartter's syndrome can be divided into at least two groups: one group with an early (infancy), and the other with a late onset of symptoms. Neonatal Bartter's syndrome is characterized by the intrauterine onset of polyuria, leading to polyhydramnios between the 22nd and 24th weeks of gestation. Repeated drainage by amniocentesis is necessary to prevent preterm labour and birth, although this generally occurs anyway (Proesmans *et al.* 1985). Failure to thrive, as well as tetany, polyuria and polydipsia, and muscle weakness are leading symptoms in infants (Chan 1980). In adults, fatigue, proximal muscle weakness, and tetany are the most common presenting features (White 1972). Dehydration and sodium wasting are infrequent, although more prominent at an early age (Robson *et al.* 1979). This is probably partly due to the increased susceptibility to water and sodium loss at this age, especially during concomitant childhood illnesses, e.g. gastroenteritis. Finally, it should be noted that an increasing number of asymptomatic patients will probably be identified in routine laboratory measurements.

Laboratory features

Hypokalaemia is one of the hallmarks of the syndrome with levels sometimes as low as 1.5 mmol/l. Most of the patients hitherto described had levels below 2.5 mmol/l. Hypochloraemia is normally less obvious; values regularly observed range from 62 to 93 mmol/l. Increased plasma bicarbonate levels (28–45 mmol/l) are a common feature. The resulting pH may be mitigated, however, by alveolar hypoventilation (as a compensatory mechanism or due to hypokalaemia *per se*) or after the development of renal insufficiency. Both hypochloraemia and metabolic alkalosis are more severe in infants and young children (Robson 1979). Despite hypokalaemia, urinary potassium wasting is quite substantial, as is chloride loss. The latter accounts for the fact that the metabolic alkalosis is 'saline (chloride)-resistant' despite

Table 1 Clinical manifestations of Bartter's syndrome

	Percentage of patients positive during course of disease	
	Children ($n = 39$)	Adults ($n = 61$)
Muscle weakness	41	40
Paresis	3	8
Fatigue	3	21
Tetany	26	21
Paraesthesia	3	2
Polyuria	28	10
Polydipsia	26	5
Nocturia	3	5
Enuresis	12	0
Constipation	8	3
Salt craving	12	5
Dehydration	15	0
Orthostatic hypotension	–	6
Failure to thrive	31	–
Delayed growth	51	–
Short stature	–	11
Hypogonadism	–	2
Mental retardation	8	0
Gout	0	2
Chondrocalcinosis	0	5
Asymptomatic	10	37

Data collected from 100 consecutive cases reported in the English literature.

the often contracted extracellular volume. The urinary concentrating ability is frequently impaired and may range from normal to about 300 mosmol/kg. Hyponatraemia is sometimes recorded. Blood urea nitrogen, serum creatinine, and creatinine clearance are normal, though progression to renal failure may occur. Meyer et al. (1975) noted a high frequency of hyperuricaemia with a possible role for alkalosis, dehydration, and/or angiotensin-II in the decreased clearance of uric acid. Hypomagnesaemia with renal magnesium wasting is often but not always present (Gill 1980; Evans et al. 1981). Although hypercalciuria has been occasionally reported, most patients do have hypocalciuria (Rudin et al. 1988). This dissociation of renal divalent cation handling points to an intrinsic tubular defect similar to that observed in chronic cisplatinum nephropathy, a toxic nephropathy known to mimic Bartter's syndrome. Plasma renin level is increased. The high concentration of active renin is associated with an absence of prorenin (Chan et al. 1981). Plasma aldosterone is usually increased also. Its level results from two opposite forces: stimulation of aldosterone production due to increased angiotensin-II generation on the one hand and inhibition of aldosterone production due to hypokalaemia on the other. Vascular resistance to exogenous angiotensin-II, but also noradrenaline, has been repeatedly demonstrated.

Urinary prostaglandin excretion measured by both radioimmunoassay and chemical methods is increased (Verberckmoes et al. 1976; Gill et al. 1976). Increased renal synthesis of prostaglandin E_2 is the most consistent, (although not invariable) attendant abnormality found. Urinary excretion of prostaglandin $F_{2\alpha}$ is usually increased while the excretion of metabolites of prostacyclin is either normal (2,3-dinor-6-keto-prostaglandin $E_{1\alpha}$) or increased (6-keto-prostaglandin $F_{1\alpha}$) (Dunn 1981). There is only circumstantial evidence to support non-renal or systemic overproduction of prostaglandins.

The third renal vasoactive system, the kallikrein-kinin system, is also stimulated (Bourke and Delaney 1981). Plasma bradykinin and urinary kallikrein are increased; however, levels of urinary kinins are below normal.

Finally, both high and normal plasma levels of the atrial natriuretic peptide have been reported (Gordon et al. 1986; Yamada et al. 1986; Doorenbos et al. 1988).

Renal histopathology

Little is known about renal histology in Bartter's syndrome. Renal biopsies are mentioned in about 50 cases of the 200 patients so far described; descriptions usually run to only one or two sentences. This is the more remarkable as Bartter's syndrome is a true syndrome both for its aetiology and its pathophysiology. This is also reflected by the variety of renal histopathological diagnoses such as membranoproliferative glomerulonephritis, interstitial nephritis, nephrocalcinosis, and sponge kidney, which have all been reported in association with the syndrome.

Hyperplasia and hypertrophy of the juxtaglomerular apparatus are the central pathological abnormalities (Bartter 1962; McLaren and MacDonald 1982; Taugner et al. 1988). In most cases all three parts of the apparatus are increased—the juxtaglomerular and the non-granular epithelial cells in the wall of the afferent arteriole, the macula densa of the distal convoluted tubule, and the polkissen, a group of extraglomerular mesangial cells in the triangular region formed by the afferent and efferent arteriole and the distal tubule (Bloom and Fawcett 1975). The predominant feature is hyperplasia of the cells in the wall of the afferent arteriole from the point of contact between the latter and the macula densa up to the vascular branching at the hilum of the glomerulus. In these cells all the signs of an increased renin synthesis can be observed. Ultrastructural studies with electron microscopy have shown 'hypertrophy' of the rough endoplasmoreticulum and of the Golgi complexes, with paracrystalline deposits (possibly prorenin) associated with an increased synthesis of renin. Renin has been identified by immunocytochemistry in both mature secretory granules and protogranules. However, some cases of Bartter's syndrome have a paucity of secretory granules in the hyperplastic juxtaglomerular apparatus, a unique and puzzling finding (Wald et al. 1971). There are also some sporadic descriptions of atrophy, and considerable flattening and necrotic degeneration of the macula densa cells (Cannon 1968). It has been suggested that such a deranged sensor structure could contribute to the inappropriate and excessive renin secretion because of failure of the feedback control.

In addition to juxtaglomerular hyperplasia, light microscopy may show a variety of other more or less obvious abnormalities such as enlarged hypercellular glomeruli, membranous thickening of glomerular capillary walls, mesangial hypercellularity, crescent formation, and periglomerular fibrosis (Gill 1980). The changes observed in small arteries and arterioles deserve special attention. Smooth muscle cells may be replaced by cells of a juxtaglomerular type. Thickening and sclerosis of these smaller vessels also occur. These renal arteriolar lesions presumably lead to a decreased perfusion through afferent arterioles setting renin release at high(er) levels which, in turn, may cause vasculotoxic effects. Remarkably, atrophic glomeruli are often associated with hyperplastic juxtaglomerular apparatus and progression towards renal insufficiency have been mentioned.

Despite the increased production of renal prostaglandins described in most, if not all patients with Bartter's syndrome, hyperplasia of renomedullary interstitial cells has been described in only a few patients (Verberckmoes et al. 1976). Allowing for sampling errors, the rarity of this finding is intriguing. Hyperplasia of renomedullary interstitial cells could be a rare primary abnormality. Alternatively, the rapid disappearance, within a few days, of intracytoplasmic granules in, and shrinkage of, renomedullary cells has been described in a hypokalaemic and hyperprostaglandinuric patient after potassium repletion (Toyoshima and Watanabe 1988). Therefore, previous potassium repletion could have obscured the finding of renal medullary cell hyperplasia on renal biopsy in an unknown number of cases.

Pathophysiology

The primary aetiology of Bartter's syndrome is still unknown (Table 2). There are several reasons for this. First, the study of the syndrome is impeded by its low incidence. All data are derived from reports of a single or only a few cases. Patients studied so far also constitute a heterogeneous group. Several (extra) renal disorders can present with all the clinical features of Bartter's syndrome and its renal histopathologic changes, which suggests that entities of different aetiology can result in a common pathophysiological sequence. Study of the syndrome is also limited by the fact that all consistently reported abnormalities are inter-related and each can affect the others by positive or negative feedback control. Bartter's syndrome can, therefore, be considered medicine's most remarkable case of 'which came first', or 'the chicken and the egg' enigma (Bourke and Delaney 1981).

Most authors adduce a defect in sodium chloride reabsorption

Table 2 Postulated causes of Bartter's syndrome

Sodium-chloride wasting tubular defect
   Defect of the proximal tubule
   Defect of the distal tubule
   Combined proximal and distal tubular defects
   Defect of the thick ascending limb of Henle's loop
   As part of a generalized membrane defect
Potassium-wasting renal tubular defect
Renal overproduction of prostaglandins
Decreased vascular responsiveness to angiotensin-II
Primary juxtaglomerular hyperplasia
Primary abnormality of the kallikrein–kinin system
Primary elevation of atrial natriuretic peptide

as the proximate defect in Bartter's syndrome (Stein 1985). Patients can usually, but not always, conserve sodium normally when dietary salt is restricted, albeit with signs of extracellular volume depletion. Robson et al. (1979) showed that the ability to conserve salt in Bartter's syndrome may be age-related, with younger patients demonstrating greater salt wasting than patients over 3 years of age. The site of decreased sodium reabsorption within the nephron has been the subject of much debate (Stein 1985). Some workers reported a defective proximal renal tubular sodium reabsorption. Additional abnormalities of the proximal tubule function, e.g. aminoaciduria and bicarbonaturia have been recorded only once. A decreased sodium reabsorption in the distal tubule has been noted by others, while combined proximal and distal defects have also been found. In these situations salt wasting is prominent and hypokalaemia occurs as a consequence of increased sodium-for-potassium 'exchange' in the distal tubule, brought about by increased sodium and fluid delivery along this segment.

A defect in sodium, potassium and chloride ($Na^+2Cl^-K^+$) cotransport in the thick ascending limb of Henle is now considered to be the most likely cause of Bartter's syndrome (Gill and Bartter 1978). This hypothesis is supported by the finding of an impaired renal diluting ability in several patients with the syndrome. During water loading, when release of ADH is suppressed, the renal clearance of solute-free water (C_{H_2O}) provides a reliable estimate of the rate of sodium chloride reabsorption in the thick ascending limb. In several patients with Bartter's syndrome the C_{H_2O} has been found to be below normal in proportion to the amount of solute delivered to the diluting segment. This alteration in the cotransport will also lead to a reduced rate of potassium reabsorption, since as much as 30 to 40 per cent of the filtered load of potassium may be reabsorbed in the ascending limb (Greger 1988). The maintenance of hypertonicity in the medullary interstitium is also impaired, accounting for the renal concentration defect found in the syndrome. Other factors such as prostaglandin excess and hypokalaemia may also be involved in this concentration defect. The excessive urinary excretion of magnesium found in about 20 per cent of patients can also be attributed to a defect in this segment of the tubule. It follows that abuse of diuretic agents with an inhibitory effect in this part of the tubule, such as frusemide, may lead to a clinical and biological picture resembling Bartter's syndrome. The hypothesis of a defect in the sodium, potassium, and chloride ($Na^+2Cl^-K^+$) cotransport in the loop of Henle is attractive as it unifies all the characteristics of the syndrome, but it should be borne in mind that the absence of inappropriate water reabsorption in the collecting duct is a prerequisite for a correct interpretation of the renal clearance studies. ADH levels were measured only once during water loading and were not detectable. This does not exclude, however, inappropriate water reabsorption independent of ADH, or an exaggerated renal responsiveness to ADH conditioned by as yet unidentified factors. Moreover, the validity of tests of renal diluting ability has recently been challenged (Kurtz et al. 1983). Finally, it should be pointed out that normal free water clearances have been described in patients with Bartter's syndrome (Kurtz et al. 1981).

A potassium-wasting renal tubular defect is another possible primary cause of Bartter's syndrome (Bourke and Delaney 1981; Stein 1985). It could account for the hyper-reninaemia, the angiotensin-II resistance, low urinary kinins, and increased urinary prostaglandin E_2. Potassium wasting could also cause the renal chloride reabsorption defect. However, several observations strongly argue against a primary role of a potassium-wasting renal tubular defect. First, normokalaemia may fail to reverse the hyper-reninaemia or the angiotensin-II resistance (Norby et al. 1976). However, in the absence of balance data, a lasting intracellular potassium depletion cannot be ruled out in this study. Paradoxically, another study reports a rise in plasma renin after potassium supplementation. Second, data on the effect of potassium on the regulation of the prostaglandin metabolism are conflicting. Neither acute potassium depletion induced by 3 days of frusemide, nor chronic potassium depletion accompanying primary aldosteronism or DOCA administration increase prostaglandin E excretion. Third, the avid sodium retention seen in experimental human potassium depletion is not a feature of the syndrome, although there are clinical states attended by hypokalaemia where this is also absent. Fourth, magnesium depletion caused by renal magnesium wasting may play a critical role in the pathogenesis of renal potassium wasting in some patients with Bartter's syndrome. Experimentally induced magnesium depletion in normal subjects leads to reversible renal potassium loss while long-term magnesium administration can induce renal conservation of potassium in some patients with Bartter's syndrome (Evans et al. 1981). Be this as it may, some patients exhibiting all the features of Bartter's syndrome have excessive urinary potassium loss as the only renal tubular defect (Stein 1985).

Garrick et al. (1985) suggest that a generalized membrane defect could underlie the various renal transport abnormalities described above. In fact, an increased intracellular sodium concentration in erythrocytes has been reported, although there is little agreement concerning the various transport mechanisms involved. They postulate an increase in cell membrane sodium permeability with secondary stimulation of (Na^+, K^+)-ATPase. This disorder would enhance potassium secretion in the renal distal and collecting tubules. The ensuing potassium depletion with its concomitant stimulation of the renin–angiotensin-II–aldosterone axis, prostaglandin synthesis, and the kinin system would result in the clinical picture of Bartter's syndrome. Since a variety of cells are supposed to be affected, the reduced vascular responsiveness to angiotensin-II could be a consequence of reduced cytosolic calcium activity due to the increased cell–sodium entry. Although resting calcium concentration in neutrophils did not differ between patients with Bartter's syndrome and normal controls, the intracellular mobilization of calcium is lower. The fact that hypokalaemia from any cause can raise the intracellular sodium content of erythrocytes tends to invalidate the hypothesis which suggests a generalized membrane defect. Also, a return to normal of the intracellular sodium content and rate constants for the sodium efflux variables studied has been reported after correction of hypokalaemia; the latter suggests that the intrinsic sodium transport processes in erythrocytes are normal in the syndrome (Korff et al. 1984).

Renal overproduction of prostaglandins has been mentioned as the primary underlying abnormality (Dunn 1981). By direct action and through stimulation of natriuresis these compounds stimulate renin secretion thereby promoting angiotensin-II generation, aldosterone release, and potassium wasting. They also have a direct effect on aldosterone biosynthesis, act locally to reduce vascular tone, and modulate the vascular response to vasoactive substances including angiotensin-II. Although natriuresis induced by prostaglandins is probably due to a non-specific haemodynamic effect similar to that produced by other vasodilators, prostaglandin E_2 has been shown to decrease sodium reabsorption in isolated rabbit collecting ducts. In addition, it has been reported to decrease chloride reabsorption in isolated thick medullary ascending limbs; it may, therefore, account for the chloride wasting. The increased delivery of sodium to the distal tubules would account for the increased potassium loss. Several arguments militate against a primary role for prostaglandins in the pathogenesis of the syndrome. The same prostaglandin profile can be found in conditions that mimic Bartter's syndrome, e.g. chronic vomiting or diuretic abuse (Dunn 1981). Also, some patients do not have increased prostaglandin production and excretion. Furthermore, most patients have experienced only partial improvement with inhibitors of prostaglandin synthesis. Although the potassium balance is improved, non-steroidal anti-inflammatory drugs generally fail to restore normokalaemia, even when urinary excretion of prostaglandin E_2 becomes completely normal. Neither the concentration defect nor chloride wasting are reversed. Secondary overproduction of prostaglandins is, therefore, a more probable explanation. Angiotensin-II, kallikrein and kinin, as well as vasopressin are potent stimulators of phospholipase A and could be responsible for the increased renal production of prostaglandins. There is ample proof that hypokalaemia itself has resulted in increased urinary excretion of prostaglandin E_2. According to more recent studies, both the failure of potassium loading to reduce excretion of prostaglandin E_2 and its normalization on water restriction suggest that its increased excretion in Bartter's syndrome may be the result of the polyuria *per se* (Bourke and Delaney 1981). Since all abnormal features investigated were sometimes completely restored to normal by indomethacin treatment (Verberckmoes et al. 1976; Donker et al. 1977), it is possible that there is a subset of patients with Bartter's syndrome in whom primary overproduction of prostaglandins occurs.

Originally, a decreased vascular responsiveness to the pressor effects of angiotensin-II was suggested as the proximate cause of the syndrome (Bartter 1962). This was supposed to lead to increased renin production, and therefore juxtaglomerular hyperplasia. The increased angiotensin-II could, in turn, account for the increased levels of aldosterone, kallikrein, and prostaglandins (Bourke and Delaney 1981). It could also contribute to the potassium and sodium loss in this condition. Again, several facts argue against a primary role for the decreased vascular responsiveness. It has since been shown that a variety of extrarenal disorders which cause hypokalaemia are associated with a low pressor responsiveness (Gill 1980). In addition, administration of saralasin (a competitive inhibitor of angiotensin-II) results in a significant fall in blood pressure in patients with

Bartter's syndrome (Bourke and Delaney 1981). These observations suggest that the increases in plasma renin activity and subsequently in angiotensin-II are, at least in part, a homeostatic response to maintain blood pressure. However, the low extracellular volume found in most patients (Boer et al. 1983) cannot be the sole explanation for the activated renin–angiotensin-II (and adrenergic) system and the vascular hyporesponsiveness. From several studies it appears that volume expansion does not always reduce the overproduction of renin and/or aldosterone to normal levels, nor did it always improve pressor responsiveness. Higher than normal plasma bradykinin as well as enhanced production of vasodilatory prostaglandins probably also contribute to the hyporesponsive state. However, a primary angiotensin-II resistance cannot be completely ruled out since some workers did observe an unaltered angiotensin-II resistance even when extracellular volume was expanded and potassium and plasma renin levels were restored to normal.

A primary role for juxtaglomerular hyperplasia in the pathogenesis of the syndrome was suggested by Brackett et al. (1968). They postulated that the arteriolar medial hypertrophy found on renal biopsy results in a lowered afferent arteriolar pressure with consequent hypersecretion of renin. Most, but not all studies seem to support this view in that infusion of angiotensin-II fails to lower plasma renin. The fact that the renin–angiotensin system is still responsive to physiological stimuli such as changes of posture and expansion of the extracellular volume, does not alter this assumption since a similar lack of autonomy can be found in primary adenomas of the pituitary or parathyroid glands. The absence of hypertension, a prominent feature of renin-secreting tumours, is more difficult to reconcile with this hypothesis. Patients with Bartter's syndrome are generally normotensive, although hypertension may ensue with the development of renal insufficiency. Angiotensin-II tachyphylaxis, which has been demonstrated in normal man and in other high renin, angiotensin-II resistant states, must therefore be postulated. Hypokalaemia is another feature hard to explain. In primary renin-secreting conditions hypokalaemia is thought to be secondary to the ensuing hyperaldosteronism. However, both bilateral adrenalectomy and aldosterone antagonists, e.g. spironolactone or aminoglutethimide, have failed to correct the severe hypokalaemia in this condition (Bourke and Delaney 1981). In addition, aldosterone levels may be normal or even reduced in some patients. It has also been demonstrated that the same histologic lesions of juxtaglomerular hyperplasia can be found in hypokalaemia due to other causes such as chronic vomiting, chronic laxative abuse, familial chloride diarrhoea, and cystinosis, which suggests that these histological findings are a secondary phenomenon (Gill 1980).

Finally, abnormalities of the kallikrein–kinin system and elevated levels of atrial natriuretic peptide have been implicated in the pathogenesis of Bartter's syndrome. Consistent data, however, are scarce and sometimes contradictory (Bourke and Delaney 1981; Gordon et al. 1986; Sasaki et al. 1987; Doorenbos et al. 1988).

Differential diagnosis

Bartter's syndrome may be confused with other conditions leading to hypokalaemic alkalosis (Fig. 1). Surreptitious vomiting and abuse of laxatives or diuretics are the most common causes of unexplained hypokalaemia (Veldhuis et al. 1979; Fleischer et al. 1969; Jamison et al. 1982). Any attempt to distinguish between these conditions and Bartter's syndrome by utilizing expensive hormone assays is unrewarding. In patients with covert vomiting and surreptitious abuse of laxatives a low urinary potassium may point to the gastrointestinal tract as the most likely source of the potassium loss (Fig. 1). However, in these patients kaliuresis may be inappropriately sustained as discussed elsewhere (Chapter 7.3). Urinary chloride determination is a most useful diagnostic test to avoid this pitfall (Veldhuis 1979). If chloride excretion is low during a chloride intake of approximately 100 mmol/day and associated with a much higher sodium and potassium excretion, Bartter's syndrome can be ruled out and an abnormality of the gastrointestinal tract is likely. It should not be forgotten, however, that severe potassium depletion per se (serum potassium below 2.0 mmol/l) may impair chloride reabsorption (see Chapter 7.3). Surreptitious diuretic abuse is more difficult to ascertain. True detective work and (multiple) screens of the urine for the presence of diuretics may be necessary to confirm a suspicion of hidden diuretic abuse or to establish a diagnosis of Bartter's syndrome.

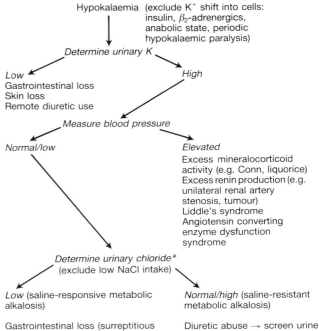

Fig. 1 Algorithm for the diagnosis of Bartter's syndrome.

Other, less common, conditions may mimic Bartter's syndrome (Fig. 1). Dietary chloride deficiency, as observed in infants ingesting almost exclusively chloride-deficient feeding formulas can lead to severe hypokalaemic alkalosis (Grossman et al. 1980). These patients can also readily be distinguished by a low urinary chloride excretion. Congenital or acquired magnesium-losing nephropathy, chemotherapy, familial chloride diarrhoea, and mucoviscidosis can all produce hypokalaemic alkalosis and should be duly considered in the differential diagnosis (Evans et al. 1981; Gill 1980; Lieber et al. 1984). Normotensive primary aldosteronism can be discerned by a low plasma renin activity and the presence of an aldosterone-producing adenoma (Shiroto et al. 1980). A recently reported congenital hypokalaemic tubular disorder resembling Bartter's syndrome in other aspects, can be distinguished by a marked increase of not only renal, but also systemic prostaglandin E_2 activity (Seyberth et al. 1985). Liddle's syndrome, another disorder with hypokalaemia through unknown causes, can be readily differentiated by the mere presence of hypertension. The same holds true for another recently reported syndrome named 'angiotensin-II converting enzyme dysfunction syndrome', which is characterized by hypokalaemic alkalosis, hyper-reninism, increased angiotensin-I concentration, but normal levels of angiotensin-II and aldosterone (Umeki et al. 1988).

When confronted with a patient with hypokalaemic alkalosis and a normal blood pressure and none of the above conditions, a renal biopsy may be considered as a next step. The renal histopathological changes in 'primary', 'true', or 'idiopathic' Bartter's syndrome are first and foremost juxtaglomerular hypertrophy and hyperplasia, which may be present at birth or even antenatally (Proesmans et al. 1985). Bartter's syndrome, however, has been described as a presenting syndrome in patients with a renal biopsy compatible with proliferative glomerulonephritis, interstitial fibrosis, cystinosis, and sponge kidney (Arant et al. 1970; Gill 1980). Recently, there have been a few reports without juxtaglomerular hyperplasia in multiple renal biopsies, which makes it unlikely that the last finding is due to a sampling error. The clinical features of this syndrome of 'congenital renal alkalosis' would have been indistinguishable from Bartter's syndrome (Calgagno 1979; Barakat et al. 1986). The same holds true for a syndrome described in two sibships with proximal renal tubulopathy. The tubular basement membrane was thickened with intensive staining of the tubular cells (Gullner et al. 1983; Aalbers and Leembuis 1987). Clearly, a renal biopsy is indicated in Bartter's syndrome when study of the pathophysiological origins and hereditary traits is contemplated.

Treatment

The main objective of treatment in Bartter's syndrome is the correction of potassium depletion and its related symptoms. Potassium supplementation has been widely used, often in conjunction with potassium-sparing diuretics (Cannon et al. 1968; Rudin et al. 1988). Quite frequently large dosages of oral potassium chloride are needed, exceeding 160 mmol per day (sometimes up to 500 mmol). Unfortunately, such amounts are not feasible because of gastric intolerance or diarrhoea. Spironolactone (10–15 mg/kg.day), triamterene (10 mg/kg.day) and amiloride (10–40 mg/day) are often only transiently effective (Simatupang et al. 1978; Chan 1980; Griffing et al. 1982). Even the combination of oral potassium and potassium-sparing diuretics restores serum potassium levels to normal in only a minority of patients. Inhibitors of prostaglandin synthesis can correct hyper-reninism and hyperaldosteronism, but again hypokalaemia (with a few exceptions) is only partially corrected (Gill 1976; Donker 1977; Dunn 1981). A combination of small dosages of prostaglandin synthethase inhibitors with supplemental oral potassium and potassium-sparing diuretics may allow a greater degree of normalization of the abnormalities than has been possible so far with either agent alone (Cunningham et al. 1979). Propranolol has been tried, but its efficacy has not been proved (Modlinger et al. 1973; Solomon and Brown 1975). Hypomagnesaemia, if present, should be treated with magnesium chloride. This treatment often improves the symptoms of tetany and muscle cramps.

Finally, long-term treatment of Bartter's syndrome with angiotensin-II converting enzyme (ACE) inhibitors has been shown to result in a rise in serum (and total body) potassium in some, but not all, patients (Aurell and Rudin 1983; Van de Stolpe et al. 1987). As these drugs may initially cause a substantial fall in blood pressure and occasionally result in a (reversible) renal function loss, they should be used with caution (Hené et al. 1987). Nevertheless, ACE inhibitors may be of some value especially in symptomatic and therapy-resistant cases, alone or in combination with potassium sparing diuretics (Clementsen et al. 1989).

Treatment with one of these drugs alone usually fails to correct hypokalaemia. Combination therapy is usually necessary. Unfortunately, the efficacy and tolerance are difficult to evaluate in the absence of adequate follow-up data.

Prognosis

Since patients with Bartter's syndrome constitute a heterogeneous group, and adequate cohort studies are sadly lacking, statements regarding the prognosis are hard to give. It is estimated that up to one-third of patients with an onset of symptoms in infancy suffer mental retardation (Simopoulos 1979). The rarity of the latter in adults, however, is a reassuring observation (Table 1). In early childhood severe growth retardation can also be found. Although Simopoulos (1979) reported a delayed adolescent growth spurt which does nevertheless finally result in a normal stature, the data from Rudin et al. (1988) suggest that this is certainly not always the case. The prognosis with regard to renal function is also difficult, with renal insufficiency occurring in some patients. Ironically, the first patient described by Bartter finally presented with uraemia, hypertension, and normokalaemia, and was also the first patient on maintenance haemodialysis.

Bibliography

Bartter, F.C., Ponove, P., Gill, J.R., and MacCardle, R.C. (1962). Hyperplasia of the juxtaglomerular complex with hyperaldosteronism and hypokalemic alkalosis. *American Journal of Medicine*, **33**, 811–28.

Bourke, E. and Delaney, V. (1981). Bartter's syndrome—a dilemma of cause and effect. *Nephron*, **27**, 177–86.

Chan, J.C.M. (1980). Bartter's syndrome. *Nephron*, **26**, 155–62.

Dunn, M.J. (1981). Prostaglandins and Bartter's syndrome. *Kidney International*, **19**, 86–102.

Gill, J.R. (1980). Bartter's syndrome. *Annual Reviews of Medicine*, **31**, 405–19.

Stein, J.H. (1985). The pathogenetic spectrum of Bartter's syndrome. *Kidney International*, **28**, 85–93.

References

Aalbers, R. and Leemhuis, M.P. (1987). An uncommon variant of Bartter's syndrome. *Netherlands Journal of Medicine*, **30**, 295–300.

Arant, B.S., Brackett, N.C., Young, R.B., and Still, W.J.S. (1970). Case studies of siblings with juxtaglomerular hyperplasia and secondary aldosteronism associated with severe azotemia and renal rickets—Bartter's syndrome or disease? *Pediatrics*, **46**, 344–61.

Aurell M. and Rudin, A. (1983). Effect of captopril on blood pressure, renal function, the electrolyte balance and the renin–angiotensin system in Bartter's syndrome. *Nephron*, **33**, 274–8.

Barakat, A.Y., Francis, Y.K., and Mufarrij, A.A. (1986). Hypokalemic alkalosis, hyperreninemia, aldosteronism, normal blood pressure and normal juxtaglomerular apparatus – a new syndrome of renal alkalosis. *International Journal of Pediatric Nephrology*, **7**, 99–100.

Bloom, W. and Fawcett, D.W. (eds.) (1975). The urinary system. In *A Textbook of Histology*, 10th edn., pp 782–4. W. Saunders Co., London.

Boer, P., Hené, R.J., Koomans, H.A., Nieuwenhuis, M.G., Geyskes, G.G., and Dorhout Mees, E.J. (1983). Blood and extracellular fluid volume in patients with Bartter's syndrome. *Archives of Internal Medicine*, **143**, 1902–5.

Brackett, N.C., Koppel, M., Randall, R.E., and Nixon, W.P. (1968). Hyperplasia of the juxta-glomerular complex with secondary aldosteronism without hypertension (Bartter's syndrome). *American Journal of Medicine*, **44**, 803–19.

Calcagno, P.L. (1979). Congenital renal alkalosis. *Pediatrics Research*, **13**, 1379–81.

Cannon, P.J., Leeming, J.M., Sommers, S.C., Winters, R.W., and Laragh, J.H. (1968). Juxtaglomerular cell hyperplasia and secondary hyperaldosteronism (Bartter's syndrome): a reevaluation of the pathophysiology. *Medicine*, **47**, 107–31.

Chan, L.L., Osmond, D.H., Balfe, J.W., and Halperin, M.L. (1981). Plasma 'prorenin'-renin in Bartter's syndrome, cystic fibrosis, and chloride deficiency, and the effect of prostaglandin synthetase inhibition. *Journal of Laboratory and Clinical Medicine*, **97**, 785–90.

Clementsen, P., Hoegholm, A., Laerkholm Hansen, C., Damkjaer, M., Christensen, P., and Giese, J. (1989). Bartter's syndrome – treatment with potassium, spironolactone and ACE-inhibitor. *Journal of Internal Medicine*, **225**, 107–10.

Cunningham, R.J., Brouhard, B.H., Berger, M., Petrusick, T., and Travis, L.B. (1979). Long-term use of propranolol, ibuprofen, and spironolactone in the management of Bartter's syndrome. *Pediatrics*, **63**, 754–6.

Delaney, V., Watson, A.J., Pollack, M., Dupont, B., and Bourke, E. (1984). HLA Typing in Bartter syndrome. *American Journal of Medical Genetics*, **19**, 779–82.

Donker, A.J.M., De Jong, P.E., Statius van Eps, L.W., Brentjens, J.R.H., Bakker, K., and Doorenbos, H. (1977). Indomethacin in Bartter's syndrome. Does the syndrome represent a state of hyperprostaglandism? *Nephron*, **19**, 200–13.

Doorenbos, C.J., Daha, M.R., Buehler, F.R., and van Brummelen, P. (1988). Effects of posture and saline infusion on atrial natriuretic peptide and haemodynamics in patients with Bartter's syndrome and healthy controls. *European Journal of Clinical Investigation*, **18**, 369–74.

Evans, R.A., Carter, J.N., George, C.R.P., Walls, R.S., Newland, R.C., McDonnell, G.D., and Lawrence, J.R. (1981). The congenital 'magnesium-losing kidney'. *Quarterly Journal of Medicine*, **197**, 39–52.

Fleischer, N., *et al.* (1969). Chronic laxative-induced hyperaldosteronism and hypokalemia simulating Bartter's syndrome. *Annals of Internal Medicine*, **70**, 791.

Garrick, R., Ziyadeh, F.N., Jorkasky, D., and Goldfarb, S. (1985). Bartter's syndrome: a unifying hypothesis. *American Journal of Nephrology*, **5**, 379–84.

Gill, J.R., *et al.* (1976). Bartter's syndrome: a disorder characterized by high urinary prostaglandins and a dependence of hyperreninemia on prostaglandin synthesis. *American Journal of Medicine*, **61**, 43–51.

Gill, J.R. and Bartter, F.C. (1978). Evidence for a prostaglandin-independent defect in chloride reabsorption in the loop of Henle as a proximal cause of Bartter's syndrome. *American Journal of Medicine*, **65**, 765–72.

Griffing, G.T., Komanicky, P., Aurecchia, S.A., Sindler, B.H., and Melby, J.C. (1982). Amiloride in Bartter's syndrome. *Clinical Pharmacology and Therapeutics*, **31(6)**, 713–18.

Greger, R. (1988). Chloride transport in thick ascending limb, distal convolution, and collecting duct. *Annual Reviews of Physiology*, **50**, 111–22.

Gordon, R.D., Tunny, T.J., Klemm, S.A., and Hamlet, S.M. (1986). Elevated levels of plasma atrial natriuretic peptide in Bartter's syndrome fall to normal with indomethacin: implications for atrial natriuretic peptide regulation in man. *Journal of Hypertension*, **4**, S555–S558.

Grossman, H., Duggan, E., McCamman, S., Welchert, E., and Hellerstein, S. (1980). The dietary chloride deficiency syndrome. *Pediatrics*, **66**, 366–74.

Gullner, H.G., Bartter, F.C., Gill, J.R., Dickman, P.S., Wilson, C.B., and Tiwari, J.L. (1983). A sibship with hypokalemic alkalosis and renal proximal tubulopathy. *Archives of Internal Medicine*, **143**, 1534–40.

Hené, R.J., Koomans, E.J., Dorhout Mees, E.J., v.d. Stolpe, A., Verhoef, G.E.G., and Boer, P. (1987). Correction of hypokalemia in Bartter's syndrome by enalapril. *American Journal of Kidney Diseases*, **9**, 200–5.

Hogewind, B.L., van Brummelen, P., and Veltkamp, J.J. (1981). Bartter's syndrome: an autosomal recessive disorder? *Acta Medica Scandinavica*, **209**, 463–7.

Jamison, R.L., *et al.* (1982). Surreptitious diuretic ingestion and pseudo-Bartter's syndrome. *American Journal of Medicine*, **73**, 142.

Korff, J.M., Siebens, A.W., and Gill, J.R. (1984). Correction of hypokalemia corrects the abnormalities in erythrocyte sodium transport in Bartter's syndrome. *Journal of Clinical Investigation*, **74**, 1724–9.

Kurtz, I., *et al.* (1981). Familial chloride-resistant renal alkalosis and hypokalemia with fasting hypercalciuria and medullary nephrocalcinosis: a unique variant of Bartter's syndrome without impaired renal diluting ability. *Clinical Research*, **29**, 555.

Kurtz, I., *et al.* (1983). The results of tests of renal diluting ability do not support the hypothesis that NaCl transport in the loop of Henle is impaired in Bartter's syndrome. *American Society of Nephrology*, 33A.

Lieber, I.H., Stoneburner, S.D., Floyd, M., and McGuffin, W.L. (1984). Potassium-wasting nephropathy secondary to chemotherapy simulating Bartter's syndrome. *Cancer*, **54**, 808–10.

McLaren, K.M. and MacDonald, M.K. (1982). Histological and ultrastructural studies of the human juxtaglomerular apparatus in Bartter's syndrome and renal artery stenosis. *Journal of Pathology*, **136**, 181–97.

Meyer, W.J., Gill, J.R., and Bartter, F.C. (1975). Gout as a complication of Bartter's syndrome. *Annals of Internal Medicine*, **83**, 56–9.

Modlinger, R.S., Nicolis, G.L., Krakoff, L.R., and Gabrilove, J.L. (1973). Some observations on the pathogenesis of Bartter's syndrome. *New England Journal of Medicine*, **289**, 1020.

Norby, L., Mark, A.L., and Kaloyanides, G.J. (1976). On the pathogenesis of Bartter's syndrome: report of studies in a patient with this disorder. *Clinical Nephrology*, **6**, 404–13.

Ohlsson, A., Sieck, U., Cumming, W., Akhtar, M., and Serenius, F. (1984). A variant of Bartter's syndrome. *Acta Pediatrica Scandinavica*, **73**, 868–74.

Pereira, R.R. and van Wersch, J. (1983). Inheritance of Bartter's syndrome. *American Journal of Medical Genetics*, **15**, 79–84.

Proesmans, W., et al. (1985). Bartter syndrome in two siblings—antenatal and neonatal observations. *International Journal of Paediatrics and Nephrology*, **6**, 63–70.

Robson, W.L., Arbus, G.S., and Balfe, J.W. (1979). Bartter's syndrome. *American Journal of Diseases of Children*, **133**, 636–9.

Rudin, A. (1988). Bartter's syndrome—a review of 28 patients followed for 10 years. *Acta Medica Scandinavica*, **224**, 165–71.

Rudin, A., Aurell, M., and Wilske, J. (1988). Low urinary calcium excretion in Bartter's syndrome. *Scandinavian Journal of Urology and Nephrology*, **22**, 35–9.

Sasaki, H., Okumura, M., Kawasaki, T., Kangawa, K., and Matsuo, H. (1987). Indomethacin and atrial natriuretic peptide in pseudo-Bartter's syndrome. *New England Journal of Medicine*, **316**, 167. .

Seyberth, H.W., Rascher, W., Schweer, H., Kuehl, P.G., Mehls, O., and Schaerer, K. (1985). Congenital hypokalemia with hypercalciuria in preterm infants: A hyperprostaglandinuric tubular syndrome different from Bartter's syndrome. *Journal of Pediatrics*, **107**, 694–701.

Shiroto, H., et al. (1980). Normotensive primary aldosteronism. *American Journal of Medicine*, **69**, 603–6.

Simatupang, T.A., Rado, J.P., Boer, P., Geyskes, G.G., Vos, J., and Dorhout Mees, E.J. (1978). Pharmacologic studies in Bartter's syndrome: opposite effects of treatment with anti-kaliuretic and anti-prostaglandin drugs. *International Journal of Clinical Pharmacology and Biopharmacy*, **16(1)**, 14–18.

Simopoulos, A.P. (1979). Growth characteristics in patients with Bartter's syndrome. *Nephron*, **23**, 130–5.

Solomon, R. and Brown, R. (1975). Bartter's syndrome. New insights into the pathogenesis and treatment. *American Journal of Medicine*, **59**, 575–83.

Stoff, J.S., Stemerman, M., Steer, M., Salzman, E., and Brown, R.S. (1980). A defect in platelet aggregation in Bartter's syndrome. *American Journal of Medicine*, **68**, 171–80.

Taugner, R., Waldherr, R., Seyberth, H.W., Erdos, E.G., Menard, J., and Schneider, D. (1988). The juxtaglomerular apparatus in Bartter's syndrome and related tubulopathies. *Virchows Archiv A: Pathology, Anatomy, Histopathology*, **412**, 459–70.

Toyoshima, H. and Watanabe, T. (1988). Rapid regression of renal medullary granular change during reversal of potassium depletion nephropathy. *Nephron*, **48**, 47–53.

Umeki, S., Akira Terao, F.I.C.A., and Sawayama, T. (1988). A new syndrome: angiotensin-converting enzyme dysfunction syndrome: differential diagnosis and pathogenesis—case reports. *Journal of Vascular Diseases*, 58–67.

Van de Stolpe, A., Verhoef, G.E.G., Hené, R.J., Koomans, H.A., and van der Vijver, J.C.M. (1987). Total body potassium in Bartter's syndrome before and during treatment with enalapril. *Nephron*, **45**, 122–5.

Veldhuis, J.D., Bardin, C.W., and Demers, L.M. (1979). Metabolic mimicry of Bartter's syndrome by covert vomiting. *American Journal of Medicine*, **66**, 361–3.

Verberckmoes, R., van Damme, B., Clement, J., Amery, A., and Michielsen, P. (1976). Bartter's syndrome with hyperplasia of renomedullary cells: successful treatment with indomethacin. *Kidney International*, **9**, 302–7.

Wald, M.K., Perrin, E.V., and Bolande, R.P. (1971). Bartter's syndrome in early infancy. *Pediatrics*, **47**, 254–63.

White, M.G. (1972). Bartter's syndrome. *Archives of Internal Medicine*, **129**, 41–7.

Yamada, K., et al. (1986). Atrial natriuretic peptide in Bartter's syndrome. *Lancet*, **i**, 273.

5.5 Nephrogenic diabetes insipidus

DANIEL G. BICHET

A nephrogenic failure to concentrate urine maximally may be due either to a defect in vasopressin-induced water permeability of the distal tubules and collecting ducts, or to insufficient build-up of the corticopapillary interstitial osmotic gradient, or to a combination of these two factors (Valtin et al. 1985). Thus the broadest definition of the term 'nephrogenic diabetes insipidus' comprises then any antidiuretic hormone-resistant urinary concentrating defect, including medullary disease with low interstitial osmolality, renal failure, and osmotic diuresis. In its narrower sense, it describes only those conditions in which antidiuretic hormone release fails to induce the expected rise in water permeability of the cortical and medullary collecting duct (Table 1) (Magner and Halperim 1987).

These various conditions included in Table 1 are described in other Chapters of this book, as indicated in the Table, and the reader is referred to them for further clinical discussion. This Chapter concentrates upon the inherited forms of diabetes insipidus, but the discussion of the countercurrent system and the elaboration of concentrated urine described here are relevant to all these different clinical manifestations of diabetes insipidus in its widest sense.

Urine concentration and the countercurrent system

Urine is not concentrated by an active water transport from tubule fluid to blood, since such a system would require a tremendous expenditure of metabolic energy. It has been estimated that more than 300 times the energy needed by an active salt transport and passive water equilibration system would be required since, while salt concentrations are about 0.15 mmol/l, water concentrations are about 55 mmol/l. Instead, the urine is concentrated with relatively little expenditure of metabolic energy by a complex interaction between the loops of Henle, the medullary interstitium, the medullary blood vessels or vasa recta, and the collecting tubules. This mechanism of urine concentration is called the countercurrent mechanism because of the anatomic arrangement of the tubules and vascular elements. Tubular fluids move from the cortex toward the papillary tip of the medulla via the proximal straight tubule and the thin descending limbs. The tubules then loop back toward the cortex so that the direction of the fluid movement is reversed in the ascending limbs. Similarly the vasa recta descend to the tip of the

Table 1 Causes of nephrogenic diabetes insipidus

Nephrogenic diabetes insipidus (restrictive definition): water permeability not increased by AVP
  Idiopathic (congenital)
  Hypercalcaemia (see Chapter 14.1)
  Hypokalaemia (see Chapter 7.2)
  Drugs (see Chapter 2.1)
    Lithium (see also Chapter 6.7)
    Demeclocycline
    Amphotericin B
    Methoxyflurane
    Diphenylhydantoin
    Nicotine
    Alcohol
Defective medullary countercurrent function
  Renal failure, acute and chronic (especially interstitial nephritis) (see Chapter 9.1)
  Obstructive uropathy (see Chapter 16.2)
  Medullary damage
    Sickle-cell anaemia and trait (see Chapter 4.11)
    Amyloidosis (see Chapter 4.2)
    Sjögren's syndrome (see Chapter 4.10)
    Sarcoidosis (see Chapter 4.4)
    Hypercalcaemia (see Chapter 14.1)
    Hypokalaemia (see Chapter 7.2)
    Protein malnutrition
    Cystinosis (see Chapters 5.1, 18.5.1)

Reproduced from Magner and Halperin (1987). *Medicine North America*, **15**, 2987–97, with permission.

papilla and then loop back toward the cortex. This arrangement of tubule segments and vasa recta allows the two fundamental processes of the countercurrent mechanism to take place, namely, countercurrent multiplication and countercurrent exchange (Valtin 1983; Jamison and Oliver 1982).

Antidiuretic hormone control of water transport

The antidiuretic hormone in humans is arginine-vasopressin (AVP), a cyclic nonapeptide. In the presence of this hormone, the entire collecting tubule system becomes permeable to water and the kidney is able to take advantage of the osmotic pressure gradient between the concentrated medullary interstitium and the dilute collecting tubule fluid. In the human kidney, the maximum achievable osmolality is around 1200 mmol/kg (Fig. 1) and, since the obligatory waste product excretion of substances like urea, sulphate, and phosphate amounts to about 600 mmol/day, at least 0.5 l of water per day must be excreted by the kidney.

Osmotic and non-osmotic stimulation

The regulation of AVP release from the posterior pituitary is dependent primarily on two mechanisms, involving the osmotic and non-osmotic pathways (Robertson 1987a).

The osmotic regulation of AVP is dependent on osmoreceptor cells in the anterior hypothalamus, which probably recognize changes in extracellular fluid osmolality by alterations in their volume. Cell volume is decreased most readily by substances that are restricted to the extracellular fluid, such as hypertonic saline or hypertonic mannitol, which not only enhance osmotic water movement from the cells but also very effectively stimulate AVP release. In contrast, hypertonic urea, which moves rapidly into the cells neither readily alters cell volume nor effectively stimulates AVP release (Zerbe and Robertson 1983). The osmoreceptor cells are very sensitive to changes in the osmolality of extracellular fluid: under conditions of fluid deprivation, 1 per cent increase in osmolality stimulates AVP release. During water ingestion a 1 per cent decrease in osmolality suppresses such release.

AVP release can also be induced by large decrements in blood volume or blood pressure (>10 per cent), which are sensed by stretch and pressoreceptors in the central venous and arterial system.

The secretion of AVP is regulated by changes in blood osmolality, volume and pressure (Robertson 1987a), detected by specialized hypothalamic and cardiovascular sensors, known as osmo- or baroreceptors, which act on the neurohypophysis via convergent neural pathways (Robertson *et al.* 1985). A variety of hypothalamic neurotransmitters, including monoamines and neuropeptides, are involved in the control of AVP release (Leibowitz 1988). Noradrenaline in the supraoptic nuclei, as well as in the paraventricular nuclei, has a primary excitatory effect on release, probably mediated through α_1-adrenergic receptors (Randle *et al.* 1986). Angiotensin II is also a potent stimulant of AVP release (Sklar and Schrier 1983), while β-adrenergic receptors (Leibowitz 1988) and opioid receptors (Robertson *et al.* 1985b) may be involved in inhibition of its release.

The osmotic stimulation of AVP release by dehydration or

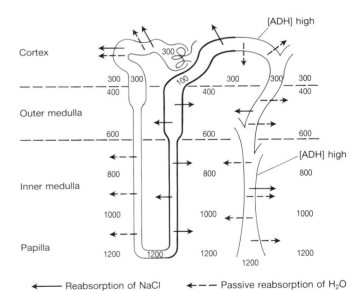

Fig. 1 The countercurrent multiplication system. During antidiuresis, heavy boundaries indicate very low permeability to water. The numbers refer to the osmolality (mosmol/kg H$_2$O) of either intratubular or interstitial fluid. Solid arrows denote reabsorption of NaCl, which is active except in the thin ascending limbs of Henle, where it may be largely passive; arrows with dashed line denote passive reabsorption of water. The number of arrows in each part of the nephron signifies semiquantitatively the amounts of solute transported relative to water. For example, in ascending limbs of Henle, solute is reabsorbed to the virtual exclusion of water (but not complete exclusion, since renal membranes are not wholly impervious to water). (From H. Valtin, *Renal Function. Mechanisms Preserving Fluid and Solute Balance in Health*, 2nd edn., Little Brown, Boston. p. 162, with permission.)

5.5 Nephrogenic diabetes insipidus

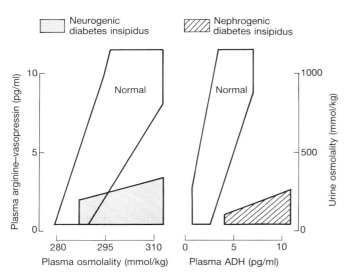

Fig. 2 Plasma arginine–vasopressin, plasma osmolality and urine osmolality. (a) Left part of figure: relationship between plasma AVP and plasma osmolality during hypertonic saline infusion. Patients with primary polydipsia and nephrogenic diabetes insipidus have values within the normal range (open area) in contrast to patients with neurogenic diabetes insipidus who show subnormal plasma AVP responses (stippled area). (b) Right part of figure: relationship between urine osmolality and plasma AVP during dehydration and water loading. Patients with neurogenic diabetes insipidus and primary polydipsia have values within the normal range (open area) in contrast to patients with nephrogenic diabetes insipidus who have hypotonic urine despite high plasma AVP (cross-hatched area). (Reproduced from Zerbe and Robertson 1984, with permission.)

hypertonic saline infusion or both is regularly used to test the vasopressin secretory capacity of the posterior pituitary. This secretory capacity can be assessed directly by comparing plasma AVP levels measured sequentially during the dehydration procedure with the normal values and then correlating these plasma levels with simultaneously obtained urinary osmolality measurements (Fig. 2) (Zerbe and Robertson 1981).

Release of AVP can also be assessed indirectly by measuring plasma and urine osmolalities at regular intervals during the dehydration test (Miller et al. 1970). The maximal urinary osmolality obtained during dehydration is compared with the maximal urinary osmolality obtained after the administration of either pitressin or 1-desamino(8-d-arginine)vasopressin, as described in Section 1.

The non-osmotic stimulation of AVP release is used to assess the vasopressin secretory capacity of the posterior pituitary of a rare group of patients with the 'essential hypernatraemia and hypodypsia' syndrome (Bichet et al. 1985). Although some of these patients may have partial central (neurogenic) diabetes insipidus, they respond normally to non-osmolar AVP release signals such as hypotension, emesis, and hypoglycaemia (Bichet et al. 1985). In all other cases of suspected central (neurogenic) diabetes insipidus these non-osmotic stimulation tests will not give additional clinical information (Baylis et al. 1981).

Cellular actions of vasopressin

The antidiuretic hormone interacts with two types of receptors: V_1 mediate the effects of AVP on vascular smooth muscle, and V_2 mediate its effects on renal tubules. V_1 receptors are linked to the phosphoinositol second messenger system, whereas V_2 receptors activate adenylate cyclase (Jard 1985). Once released from the posterior pituitary, AVP exerts its biological action on water excretion by binding to V_2 receptors on the basolateral membrane of the collecting duct. This binding causes an increase in adenylate cyclase activity which in turn catalyses the formation of cyclic adenosine 3',5'-monophosphate (cyclic AMP) from adenosine triphosphate (ATP). The receptor–adenylate cyclase complex appears to consist of three membrane proteins: (1) the receptor that recognizes the hormone, (2) a regulatory subunit, and (3) the catalytic subunit. Considerable evidence indicates that the hormonal activation of adenylate cyclase requires the presence of guanine nucleotides and that the regulatory subunit is a guanine nucleotide binding protein. The AVP-induced increase in cyclic AMP formation ultimately increases the permeability of the luminal cell membrane to water by the insertion of a preformed 'water channel' (Brenner et al. 1987).

Unfortunately, although this knowledge would suggest that the measurement of urinary cyclic AMP should help to distinguish between central (neurogenic) and nephrogenic diabetes insipidus, this has not been the case (Uttley et al. 1978; Bichet et al. 1989).

Quantitating renal water excretion

Osmotic and non-osmotic polyuric states

Diabetes insipidus is characterized by the excretion of abnormally large volumes (more than 30 ml/kg body weight/day) of hypo-osmotic urine (<250 mmol/kg). Polyuria occurs either with normal solute excretion and a depressed water retaining ability of the tubules or with high solute excretion and a normal ability to retrieve filtered water. The former is called diabetes insipidus, the latter osmotic diuresis. Osmotic diuresis occurs when excess solute is excreted, as with glucose in the polyuria of diabetes mellitus. Other agents that produce osmotic diuresis are mannitol, urea, glycerol, contrast media, and loop diuretics. Osmotic diuresis should be considered when solute excretion exceeds 60 mmol/h.

Osmolar clearance and free-water clearance

The quantitation of water excretion has been facilitated by the concept that urine flow can be divided into two components (Berl and Schrier 1986). One component is the urine volume needed to excrete solutes at the concentration of the solutes in the plasma. This isotonic component has been termed 'osmolar clearance' (C_{osm}). The other component is called 'free-water clearance' (C_{H_2O}) and is the theoretical volume of solute-free water that has to be added to (positive C_{H_2O}) or removed from (negative C_{H_2O} or $T^c_{H_2O}$) the isotonic portion of the urine (C_{osm}) to create either hypotonic or hypertonic urine, respectively. These calculations can be found in Berl and Schrier (1986).

Clinical and biochemical characteristics of nephrogenic diabetes insipidus

Congenital
Genetics

Congenital nephrogenic diabetes insipidus is a rare X-linked disorder associated with tubular resistance to AVP. In affected families, 50 per cent of the males are symptomatic and completely unresponsive to vasopressin; female patients show variable degrees of polyuria and polydipsia secondary to X-chromosome inactivation.

Four 'classical' families with congenital nephrogenic diabetes insipidus have been described (Forssman 1942, 1975; Williams and Henry 1947; Cannon 1955; Bode and Crawford 1969). Bode and Crawford (1969) have made the interesting suggestion that most of the cases of congenital nephrogenic diabetes insipidus in North America can be traced to descendants of an Ulster Scotsman who arrived in Nova Scotia in 1761 on the ship *Hopewell*.

The two largest groups of male patients with congenital nephrogenic diabetes insipidus in North America (one in Nova Scotia (Bode and Crawford 1969; Crawford and Bode 1975), the second in Utah (Cannon 1955)) are probably descendants of these early immigrants (Bode and Crawford 1969). However, this disease has also been found in French-Canadians with no apparent relationship with 'Ulster's Scotsmen' (Bichet *et al.* 1988, 1989), in North American black subjects (Feigin *et al.* 1970; Bianchine *et al.* 1971; Libber *et al.* 1986), in an Australian aboriginal kindred (Schultz and Lines 1975), and in a Samoan pedigree (Nakano 1969). Furthermore, numerous descriptions of affected patients have been published in Europe (Forssman 1942, 1975; Wattiez *et al.* 1957; Niaudet *et al.* 1985; Kirman *et al.* 1956; Lewis and Thomas 1986) and Japan (Chung and Mantell 1952; Ohzeki *et al.* 1984; Ohzeki 1985). Most of the cases described suggest an X-linked inheritance; in all of the seven pedigrees studied by Crawford and Bode (1975), involving over 600 individuals and 74 affected males, are consistent with an X-linked inheritance. An autosomal dominant type of inheritance has been mentioned (Cannon 1955; Ohzeki *et al.* 1984; Niaudet *et al.* 1985). However, the description of these cases is poorly documented since the clinical features of the male patients (polyuria, polydipsia, episodes of dehydration during infancy with mental retardation) are absent. Male-to-male transmission is a characteristic of dominant inheritance and rules out an X-linked inheritance except when an affected male patient fathers an affected boy with a female carrier as the mother.

Clinical characteristics

The clinical characteristics (Forssman 1942; Waring *et al.* 1945; Williams and Henry 1947; Crawford and Bode 1975; Niaudet *et al.* 1985) include hypernatraemia, hyperthermia, mental retardation and repeated episodes of dehydration in early infancy. Mental retardation, a probable consequence of repeated episodes of dehydration, was prevalent in the Crawford and Bode (1975) study, in which only nine of 82 patients (11 per cent) had normal intelligence. Of the 16 affected males that I examined in Montreal and Halifax 11 (69 per cent) had some degree of mental retardation. Early recognition and treatment of congenital nephrogenic diabetes insipidus with an abundant intake of water allows a normal life-span with a normal physical and mental development (Niaudet *et al.* 1985; D.G. Bichet, personal observations). Two characteristics suggestive of congenital nephrogenic diabetes insipidus are the familial occurrence and the confinement of mental retardation to male patients. It is then tempting to assume that the family described in 1892 by McIlraith and discussed by Reeves and Andreoli (1989) was affected by congenital nephrogenic diabetes insipidus. Lacombe (1841) and Weil (1884) described an autosomal dominant type of transmission without any associated mental retardation. The descendants of the family originally described by Weil were later found to have autosomal dominant central (neurogenic) diabetes insipidus (Weil 1908; Camerer 1935; Dölle 1951) a now well-characterized entity (Pedersen *et al.* 1985; Repaske *et al.* 1990). Patients with autosomal dominant central (neurogenic) diabetes insipidus retain some limited capacity to secrete arginine-vasopressin during severe dehydration and the polyuro-polydipsic symptoms usually appear after the first year of life (Pedersen *et al.* 1985; Os *et al.* 1985), when the infant's demand for water is more likely to be understood by adults.

The early symptomatology of the nephrogenic disorder and its severity in infancy is clearly described by Crawford and Bode (1975). Although the polydipsia and polyuria are often overlooked during infancy, the concentration defect is demonstrable within 6 days of birth, and parents aware of the disorder in the family have reported that their affected infants show a distinct preference for water over milk as early as 3 weeks of age. Pregnancies leading to the birth of affected infants are usually complicated by polyhydramnios. The infants are irritable, cry almost constantly and, although eager to suck, will vomit milk soon after ingestion unless prefed with water. The history given by mothers often includes persistent constipation, erratic, unexplained fever, and failure to gain weight. Even though the patients characteristically show no visible evidence of perspiration, increased water loss during fever or in warm weather exaggerates the symptoms. Unless the condition is recognized early, children will experience frequent bouts of hypertonic dehydration, sometimes complicated by convulsions or death; mental retardation is a frequent consequence of these episodes. The intake of large quantities of water, combined with the patient's voluntary restriction of dietary salt and protein intake lead to hypocaloric dwarfism beginning in infancy. Affected children frequently develop lower urinary tract dilatation and obstruction, probably secondary to the large volume of urine produced (Streitz and Streitz 1988). Dilatation of the lower urinary tract is also seen in primary polydipsic patients and in patients with central (neurogenic) diabetes insipidus (Boyd *et al.* 1980; Gautier *et al.* 1981). Chronic renal insufficiency may occur by the end of the first decade of life and could be the result of episodes of dehydration with thrombosis of the glomerular tufts (Crawford and Bode 1975).

Clinical and biochemical diagnosis

The family history will usually provide evidence for a presumptive diagnosis in an affected infant. However several generations may have passed since the birth of the last affected male and a 'sporadic' case may then be diagnosed. The diagnosis of congenital nephrogenic diabetes insipidus could be rapidly confirmed by a short dehydration test followed by vasopressin administration (see investigation of a patient with polyuria) (Fig. 3). Plasma sodium, and plasma and urine osmolality measurements should be immediately available at intervals during dehydration procedures.

In our clinical research unit, plasma sodium and plasma and urinary osmolalities are measured at the beginning of each dehydration procedure and at regular intervals (usually hourly) thereafter, depending on the severity of the polyuric syndrome.

For example, an 8-year-old patient (31 kg body weight) with a clinical diagnosis of congenital nephrogenic diabetes insipidus continued to excrete large volumes of urine (300 ml/h during a short 4-h dehydration test. During this time, the patient was suffering from severe thirst, his plasma sodium was 155 mmol/l, his plasma osmolality was 310 mmol/kg and his urinary osmolality

5.5 Nephrogenic diabetes insipidus

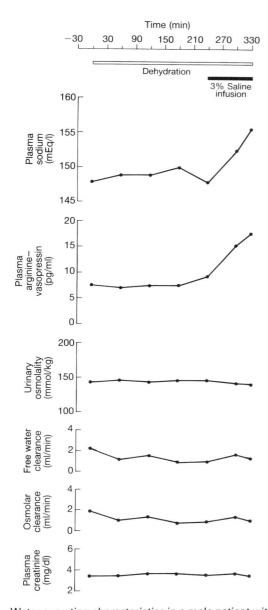

Fig. 3 Water excretion characteristics in a male patient with congenital nephrogenic diabetes insipidus. The patient (age 45, weight 50 kg, height 1.6 m) had a personal and family history pathognomonic of hereditary nephrogenic diabetes insipidus. His genealogical tree was published by Huard and Picard (1960). During the first years of his life he suffered repeated episodes of dehydration and is now mentally retarded. He has spent the past 20 years without medication but is fitted with a permanent urinary catheter in a chronic care institution. During the past 5 years, he has had multiple episodes of urinary tract infection and dehydration. At the time of this study, his serum creatinine was 3.5 mg/dl (309 μmol/l) with a creatinine clearance of 20 ml/min. 1.73 m². This has been attributed to chronic pyelonephritis. The patient was dehydrated for a total of 330 min and received a 0.1 ml/kg.min 3 per cent saline infusion during the last 120 min of the test. Plasma sodium increased from 147 to 155 mEq/l and very high concentrations of plasma vasopressin were measured (from 7 to 15 pg/ml). However, urinary osmolality remained below 150 mmol/kg and free water clearance, osmolar clearance and plasma creatinine were unchanged during the procedure.

was 85 mmol/kg. The patient received 1 μg of DDAVP intravenously and was allowed to drink water. Repeated urinary osmolality measurements demonstrated a complete urinary resistance to this agent.

It would have been dangerous and unnecessary to prolong the dehydration further in this young patient. Thus, the usual overnight dehydration should not be used in patients, especially children, with severe polyuria and polydipsia (more than 30 ml/kg. day). Great care should be taken to avoid any severe hypertonic state, arbitrarily defined as a plasma sodium >155 mEq/l.

Abnormal renal and extrarenal V_2 receptor responses in male patients with congenital nephrogenic diabetes insipidus

In male patients with congenital nephrogenic diabetes insipidus, urinary osmolality does not change after administration of AVP or DDAVP, and maximal urinary osmolality remains below 250 mmol/kg (Fig. 4). The plasma endogenous vasopressin concentrations also remain within the normal range and increase normally during dehydration or hypertonic saline infusion (Fig. 5). These results indicate that the renal V_2 receptor responses are abnormal.

Two extrarenal actions of DDAVP have been described: it stimulates release of two coagulation factors—factor VIIIc and von Willebrand factor (Mannucci et al. 1975, 1981; Richardson and Robinson 1985). It also decreases blood pressure and peripheral resistance with a concomitant increase in plasma renin activity. The latter effect can also be induced by DDAVP or (4-valine, 8-D-arginine) vasopressin (VDAVP), another V_2-selective agonist (Williams et al. 1986; Schwartz et al. 1985). The haemodynamic

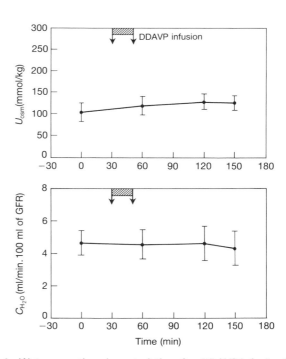

Fig. 4 Water excretion characteristics after DDAVP infusion in six male patients with congenital nephrogenic diabetes insipidus. DDAVP (0.3 μg/kg up to a maximum of 24 μg) was infused from 0 to 20 min. Urinary osmolality (U_{osm}), free water clearance (C_{H_2O}) and osmolar clearance (not shown) remained unchanged following the infusion. Mean ±SEM are represented.

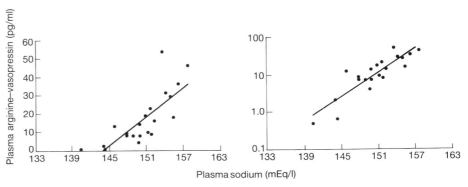

Fig. 5 Osmotic stimulation of arginine–vasopressin in five male patients with congenital nephrogenic diabetes insipidus. Plasma arginine–vasopressin was measured after various lengths of dehydration or during an hypertonic saline infusion. Linear (left) and log-linear (right) relationships are represented.

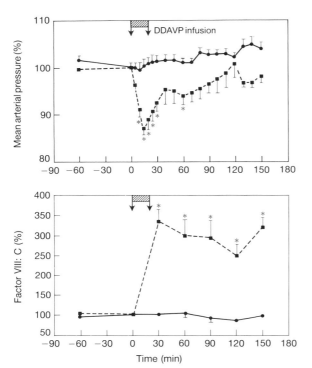

Fig. 6 Mean arterial blood pressure and Factor VIIIc responses to DDAVP infusion in normal subjects (■) (eight studies) and in 14 male patients (●) with congenital nephrogenic diabetes insipidus. In normal subjects, DDAVP decreased mean arterial blood pressure (by 10 to 15 per cent) and increased Factor VIIIc (three-fold). Mean arterial blood pressure and Factor VIIIc did not change in patients with congenital nephrogenic diabetes insipidus. Asterisks indicate significant differences from baseline (values at 0 and 30 min). See Bichet et al. 1988, 1989 for further details.

responses to VDAVP have been observed in anephric dogs (Liard 1988), and an increase in coagulation factors has been observed in anephric patients (Mannucci et al. 1975). Thus, these haemodynamic and coagulation responses are probably mediated by extrarenal vasopressin V_2 receptors.

Bichet et al. (1988, 1989) found that blood pressure did not decrease and that plasma renin activity and the release of coagulation factors did not increase in response to the administration of DDAVP in 14 male patients with congenital nephrogenic diabetes insipidus (Fig. 6). Similar results have been obtained by other investigators (Derkx et al. 1987; Knoers et al. 1990). These results would suggest an abnormal extrarenal V_2 receptor response in patients with congenital nephrogenic diabetes insipidus.

In a recent series of experiments (Bichet et al. 1989), we observed that administration of DDAVP increased plasma cyclic adenosine cyclic AMP levels in normal subjects but had no effect in 14 male patients with congenital nephrogenic diabetes insipidus (Fig. 7). Intermediate responses were observed in obligatory carriers of the disease possibly corresponding to half of the normal receptor responses (carriers have one normal and one defective gene). These results would suggest an altered precyclic AMP stimulation mechanism in patients with congenital nephrogenic diabetes insipidus (Bichet et al. 1989).

Normal V_1 receptor responses in male patients with congenital nephrogenic diabetes insipidus

These abnormal V_2 receptor responses are in contrast to the apparently normal V_1-mediated vascular or pressor responses to AVP in patients with congenital nephrogenic diabetes insipidus, in whom administration of the hormone induces skin blanching, abdominal cramps, increased arterial blood pressure, and prostaglandin E_2 excretion (Bichet et al. 1988). These patients also have normal platelet V_1 receptors since their platelet binding (of tritiated AVP) and functional platelet characteristics are normal (Lonergan et al. 1990).

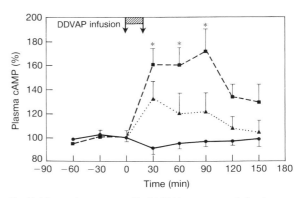

Fig. 7 Evidence for a precyclic AMP V_2 receptor defective mechanism in patients with congenital nephrogenic diabetes insipidus. Plasma cyclic AMP responses to DDAVP infusion (0.3/kg from 0 to 20 min) into 14 male patients with congenital nephrogenic diabetes insipidus (●), 11 obligatory carriers of the nephrogenic diabetes insipidus gene (▲) and five normal subjects (■). Asterisks indicate significant difference from baseline values (at −60, −30 and 0 min) (from Bichet et al. 1989).

Is there a heterogeneity of physiopathological mechanisms in patients with hereditary nephrogenic diabetes insipidus?

Most of the cases of congenital nephrogenic diabetes insipidus, which I have studied in Montreal and Halifax over the past few years are males with indisputable X-linked inheritance. In these families, women might express polyuro-polydipsic symptoms, but they are never afflicted with severe dehydration episodes or the consequent mental retardation. This phenotypic homogeneity of our patients was reflected in the homogeneity of the responses observed after administration of DDAVP: in the male patients urinary osmolality and plasma concentrations of the coagulation factors and cyclic AMP did not increase.

Moses et al. (1988), however, suggested that two distinct pathophysiological mechanisms could be involved in X-linked congenital nephrogenic diabetes insipidus, since the concentration of the coagulation factors increased normally in the absence of an antidiuretic effect in one female patient after the administration of DDAVP. However, this result could be secondary to random X-chromosome inactivation. Early in the development of the female embryo, one of the two X chromosomes in every cell becomes permanently inactivated in a random process known as lyonization (Lyon 1966). All mature cell lineages in normal women are thus composed of a mixture of cells in which one or other of the X chromosomes is active. Thus, the gene for congenital nephrogenic diabetes insipidus could be inactivated at the extrarenal level but expressed to only a lesser degree or not at all at the renal level. Thus, the observations of Moses et al. do not suggest two different pathophysiological mechanisms.

Different results were obtained by Brenner et al. (1988), who infused DDAVP into four patients (two males) with congenital nephrogenic diabetes insipidus. In these patients, the plasma concentrations of factor VIII and the von Willebrand factor rose normally, two- to three-fold, whereas the urinary osmolality remained low. The phenotypic characteristics of these patients were unusual, since both the males and the females were severely affected by repeated episodes of dehydration in infancy and suffered mental retardation as a result (Hochberg 1977). These latter cases have not been studied with X-specific molecular probes. The possible correspondence of this phenotypic heterogeneity to a genotypic heterogeneity would make an interesting study.

Genetic screening for congenital nephrogenic diabetes insipidus with polymorphic desoxyribonucleic acid (DNA) probes

Mothers and daughters of affected males are 'obligatory' carriers of the nephrogenic diabetes insipidus gene (by definition according to the law of X-linked inheritance): mothers transmit the 'defective' gene to their affected sons, whereas affected males transmit their only (defective) X chromosome to their daughters.

Non-obligatory female carriers (sisters of affected male patients, or sisters of obligatory carriers) are very difficult, if not impossible, to detect by biochemical testing because of the X-chromosome inactivation mechanism ('lyonization'). For example, obligatory carriers are rarely polyuro-polydipsic and their coagulation factors show a continuum of responses (normal to absent) to DDAVP administration (Bichet et al. 1988, 1989). At best, carrier detection based on pedigree analysis and on coagulation and haemodynamic responses to DDAVP administration could predict the genotype of 60 to 70 per cent of obligatory carriers. It has recently become possible to detect obligatory carriers using polymorphic DNA probes since a close linkage has been found between the disease locus and markers of the subtelomeric region of the long arm of the X chromosome (Kambouris et al. 1988; Knoers et al. 1988a,b, 1989; Bichet et al. 1990b) (Fig. 8).

In conclusion, in congenital nephrogenic diabetes insipidus the renal and extrarenal V_2 receptor responses are abnormal, whereas the V_1 receptor responses are normal with the defective gene coding for the defects localized on the X chromosome at position q28.

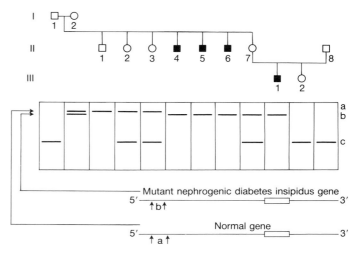

Fig. 8 Restriction length fragment polymorphism study in a family with congenital nephrogenic diabetes insipidus. Schematic representation of an autoradiogram obtained after DNA cleavage with the enzyme Taq I and hybridization with the molecular probe St_{14-1}. In subjects II-1, II-2 and II-3, allele a is associated with the normal gene (coding for normal V_2 responses) and allele b with the mutant nephrogenic diabetes insipidus gene (in subjects I-2, II-4, II-5, II-6, II-7 and III-1). The normal gene (in subjects I-2, II-8 and III-2), which is not shown at the bottom of the figure is also associated with polymorphic allele c. Subjects II-2, II-3 and III-2 have inherited allele a (and c) and are predicted non-carriers. The following gene order has been proposed by Knoers et al. (1989): centromere-F9-DXS98-F8/CBD, CBP-DXS52/NDI-DXS134-telomere (F9 = Factor IX; F8 = Factor VIII; CBD = colour blindness, partial, deutan series; CBP = colour blindness, partial, protan series; NDI = nephrogenic diabetes insipidus).

Treatment

An abundant unrestricted water intake should always be provided and affected male patients should be carefully followed during their first years of life (Niaudet et al. 1984, 1985). Water should be offered every 2 h day and night, and temperature, appetite and growth should be monitored (Behrman et al. 1987). The parents of these children easily accept setting their alarm clock every 2 h during the night! Hospital admission may be necessary to allow continuous gastric feeding. A low osmolar and sodium diet, hydrochlorothiazide (1–2 mg/kg.day) and indomethacin (1.5–3.0 mg/kg) substantially reduce water excretion (Blalock et al. 1977; Alon and Chan 1985; Libber et al. 1986) and are helpful in the treatment of children. Many adult patients receive no treatment. The mechanism by which prostaglandin synthesis inhibitors cause a reduction in urine volume in patients with congenital nephrogenic diabetes insipidus is not clear. They might increase water reabsorption in collecting duct cells by increasing the response to AVP (Bonvalet et al. 1987). However, the presence of this hormone is not a prerequisite for the antidiuretic effect of prostaglandin synthesis inhibitors since these compounds also reduce water excretion in the Brattleboro rat (Stoff et al. 1981), an animal model of central (neurogenic) diabetes insipidus. Non-steroidal anti-inflammatory drugs also reduce water excretion in humans with low concentrations of AVP resulting from water loading (Schwertschlag et al. 1986). The decreased urinary volume and free water clearance observed in this latter study were attributed to a decrease in delivery of water and solute to the diluting segments of the nephron. Thus, the site at which prostaglandin synthesis inhibitors decreased solute reabsorption was probably proximal to the ascending limb of Henle's loop (Schwertschlag et al. 1986). Similar results were obtained in a patient with congenital nephrogenic diabetes insipidus (Usberti et al. 1985): indomethacin decreased the delivery of solute to the diluting segment (expressed as $C_{H_2O} + C_{Na} + C_K$) and did not affect the reabsorption in the diluting segment (expressed as $C_{H_2O} / [C_{H_2O} + C_{Na} + C_K]$). These studies are not definitive answers since the calculations of clearances are only rough estimates of solutes delivered to specific segments of nephrons (for critics of clearance methodology, see Schuster and Seldin 1985).

Acquired

The acquired form of nephrogenic diabetes insipidus is much more common than the congenital form, but it is rarely severe. The ability to elaborate a hypertonic urine is usually preserved, in spite of the impairment of the maximal concentrating ability of the nephrons. Polyuria and polydipsia are therefore moderate (3–4 l/day).

The more common causes of acquired nephrogenic diabetes insipidus are listed in Table 1. Lithium administration has become the most frequent cause: a recent review (Boton et al. 1987) reported this abnormality in at least 54 per cent of 1105 unselected patients on chronic lithium therapy. Nineteen per cent of these patients had polyuria, as defined by a 24-h urine output exceeding 3 l. The mechanism whereby lithium causes polyuria has been extensively studied. Lithium inhibits adenylate cyclase in a number of cell types, including the renal epithelium (Christensen et al. 1985; Cogan et al. 1987), and the lithium concentration in the urine of patients on well-controlled lithium therapy (i.e. 10–40 mmol/l) is sufficient to exert this effect. Recently, measurements of adenylate cyclase activity in membranes isolated from a cultured pig kidney cell line (LLC-PK$_1$) revealed that lithium concentrations in the range of 10 mmol/l interfered with the hormone stimulated guanyl nucleotide regulatory unit (G_s) (Goldberg et al. 1988). In patients on long-term lithium therapy, amiloride has been proposed to prevent the uptake of lithium in the collecting ducts, thus preventing the inhibitory effect of intracellular lithium on water transport (Batlle et al. 1985).

Investigation of a patient with polyuria

Plasma sodium and osmolality are maintained with normal limits (136–143 mmol/l for plasma sodium, 275–290 mmol/kg for plasma osmolality) by a thirst–AVP-renal axis. Thirst and AVP release, both stimulated by increased osmolality, have been termed a 'double negative' feed-back system (Leaf 1979): even when the AVP-limb of this system is lost, the thirst mechanism still preserves the plasma sodium and osmolality within the normal range, but at the expense of pronounced polydipsia and polyuria. Thus, the plasma sodium concentration or osmolality of an untreated patient with diabetes insipidus may be slightly higher than the mean normal value, but the values usually remain within the normal range, and these small increases have no diagnostic significance.

Theoretically, it should be relatively easy to differentiate between central (neurogenic) diabetes insipidus, nephrogenic diabetes insipidus, and primary polydipsia. A comparison of the osmolality of urine obtained during dehydration from patients with central (neurogenic) diabetes insipidus or nephrogenic diabetes insipidus with that of urine obtained after the administration of AVP should reveal a rapid increase in osmolality only in the central (neurogenic) diabetes insipidus patients. Urinary osmolality should increase normally in response to moderate dehydration in primary polydipsia patients.

However, these distinctions may not be as clear as one might expect (Robertson 1985). Firstly, chronic polyuria of any aetiology interferes with the maintenance of the medullary concentration gradient and this 'washout' effect diminishes the maximum concentrating ability of the kidney. The extent of the blunting varies in direct proportion to the severity of the polyuria and is independent of its cause. Hence, for any given level of basal urine output, the maximum urine osmolality achieved in the presence of saturating concentrations of AVP is depressed to the same extent in patients with primary polydipsia, central (neurogenic) diabetes insipidus and nephrogenic diabetes insipidus (Fig. 9). Secondly, most patients with central (neurogenic) diabetes insipidus maintain a small, but detectable, capacity to secrete AVP during severe dehydration, and urinary osmolality may then rise above plasma osmolality. Thirdly, many acquired nephrogenic diabetes insipidus patients have an incomplete deficit response to AVP and concentrated urine could again be obtained during dehydration testing. Finally, all polyuric states (whether central (neurogenic), nephrogenic, or psychogenic) can induce large dilatations of the urinary tract and bladder (Boyd et al. 1980; Gautier et al. 1981). As a consequence, the urinary bladder of these patients has an increased residual capacity and changes in urinary osmolalities induced by diagnostic manoeuvres might be difficult to demonstrate.

Indirect test

The measurements of urinary osmolality after dehydration or vasopressin administration is usually referred to as 'indirect testing', since vasopressin secretion is indirectly assessed through changes in urinary osmolalities (Miller et al. 1970). During this procedure the patient is maintained on a complete fluid restriction regimen until urinary osmolality reaches a plateau, as indicated by an hourly increase of less than 30 mmol/kg for at least three successive hours. After measuring the plasma osmolality, five units of aqueous vasopressin are administered subcutaneously. Urinary osmolality is measured 30 and 60 min later. The last urinary osmolality value obtained before the vasopressin injection and the highest value obtained after the injection are compared. The patients are then separated into five categories according to previously published criteria (Miller et al. 1970; Table 2).

5.5 Nephrogenic diabetes insipidus

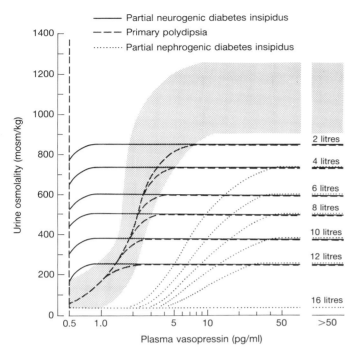

Fig. 9 The relationship between urine osmolality and plasma vasopressin in patients with polyuria of diverse aetiology and severity. Note that for each of the three categories of polyuria neurogenic diabetes insipidus, nephrogenic diabetes insipidus, and primary polydipsia, the relationship is described by a family of sigmoid curves that differ in height. These differences in height reflect differences in maximum concentrating capacity due to 'washout' of the medullary concentration gradient. They are proportional to the severity of the underlying polyuria (indicated in l/day at the right end of each plateau) and are largely independent of the aetiology. Thus, the three categories of diabetes insipidus differ principally in the submaximal or ascending portion of the dose-response curve. In patients with partial neurogenic diabetes insipidus, this part of the curve lies to the left of normal, reflecting increased sensitivity to the antidiuretic effects of very low concentrations of plasma AVP. In contrast, in patients with partial nephrogenic diabetes insipidus, this part of the curve lies to the right of normal, reflecting decreased sensitivity to the antidiuretic effects of normal concentrations of plasma AVP. In primary polydipsia, this relationship is relatively normal. (Reproduced from Robertson 1985, with permission.)

Direct test

The two approaches of Zerbe and Robertson are used (1981).

First approach

During the dehydration test, plasma is collected hourly during dehydration, and assayed for vasopressin. The results are plotted on a nomogram depicting the normal relationship between plasma sodium or osmolality and plasma AVP in normal subjects (Fig. 2). If the relationship between plasma vasopressin and osmolality falls below the normal range, the disorder is diagnosed as central (neurogenic) diabetes insipidus.

Second approach

Partial nephrogenic diabetes insipidus and primary polydipsia can be differentiated using the second approach, which analyses the relationship between plasma AVP and urinary osmolality at the end of the dehydration period (Fig. 9). However, a definitive differentiation between these two disorders might be impossible since a normal or even supranormal AVP response to increased plasma osmolality occurs in polydipsic patients. None of the patients with psychogenic or other forms of severe polydipsia studied by Robertson has ever shown any evidence of pituitary suppression (Robertson 1985).

Benefits of direct testing

Zerbe and Robertson (1981) found that all seven of the cases of severe neurogenic diabetes insipidus diagnosed by the standard indirect test were confirmed by the plasma vasopressin assay. However, two of six patients diagnosed as having partial neurogenic diabetes insipidus by the indirect test had normal vasopressin secretion as measured by the direct assay; one was found to have primary polydipsia and the other nephrogenic diabetes insipidus. Moreover, three of ten patients diagnosed as having primary polydipsia by the indirect test had clear evidence of partial vasopressin deficiency by the direct assay (Zerbe and Robertson 1981). These patients were thus wrongly diagnosed as primary polydipsic. The limitations of the indirect test are described in Fig. 10. The benefits of combined direct and indirect testing of the arginine–vasopressin function have been discussed by Stern and Valtin (1981). The diagnosis of primary polydipsia remains one of exclusion and its causes could be psychogenic (Barlow and de Wardener 1959) or inappropriate thirst (Robertson 1987b). Psychiatric patients with polydipsia and hyponatraemia have unexplained defects in urinary dilution, the osmoregulation of water intake and the secretion of vasopressin (Goldman et al. 1988).

Table 2 Recognition of partial defects in antidiuretic hormone secretion. Urinary responses to fluid deprivation and exogenous vasopressin

	No. cases	Maximum U_{osm} with dehydration	U_{osm} after vasopressin	Percentage change (U_{osm})	U_{osm} increase after vasopressin (%)
Normal subjects	9	1068 ± 69	979 ± 79	−9 ± 3	< 9
Complete central diabetes insipidus	18	168 ± 13	445 ± 52	183 ± 41	> 50
Partial central diabetes insipidus	11	438 ± 34	549 ± 28	28 ± 5	> 9, < 50
Nephrogenic diabetes insipidus	2	123.5	174.5	42	< 50
Compulsive water-drinking	7	738 ± 53	780 ± 73	5.0 ± 2.2	< 9

Urinary osmolality (U_{osm}) in mmol/kg. Reproduced from Miller et al. (1970), Annals of Internal Medicine, **73**, 721–9, with permission.

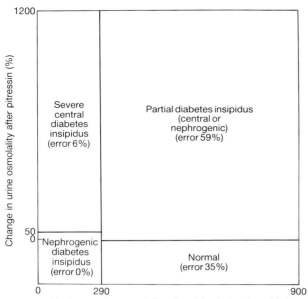

Fig. 10 Direct and indirect tests of AVP function in patients with polyuria of diverse aetiology. The indirect tests were done and interpreted as described by Miller *et al.* (1970). The direct tests, which are based on an immunoassay of plasma AVP were done and interpreted as described by Zerbe and Robertson (1981). The error rates indicate the percentage of patients in each of the four categories that were misdiagnosed by the indirect tests. (Reproduced from Robertson 1985, with permission.)

Table 3 Differential diagnosis of diabetes insipidus

1. Measure plasma osmolality and/or sodium concentration under conditions of *ad libitum* fluid intake. If they are above 295 mmol/kg and 143 mmol/l, the diagnosis of primary polydipsia is excluded and the workup should proceed directly to step 5 and/or 6 to distinguish between neurogenic and nephrogenic diabetes insipidus. Otherwise:

2. Perform a dehydration test. If urinary concentration does not occur before plasma osmolality and/or sodium reach 295 mmol/kg or 143 mmol/l, the diagnosis of primary polydipsia is again excluded and the workup should proceed to step 5 and/or 6. Otherwise:

3. Determine the ratio of urine to plasma osmolality at the end of the dehydration test. If it is less than 1.5, the diagnosis of primary polydipsia is again excluded and the workup should proceed to step 5 and/or 6. Otherwise:

4. Perform a hypertonic saline infusion with measurements of plasma vasopressin and osmolality at intervals during the procedure. If the relationship between these two variables is subnormal, the diagnosis of diabetes insipidus is established. Otherwise:

5. Perform a vasopressin infusion test. If urine osmolality rises by more than 150 mosmol/kg above the value obtained at the end of the dehydration test, nephrogenic diabetes insipidus is excluded. Alternatively:

6. Measure urine osmolality and plasma vasopressin at the end of the dehydration test. If the relationship is normal, the diagnosis of nephrogenic diabetes insipidus is excluded

Data taken from Robertson (1987*a*).

Therapeutic trial

In selected patients with an uncertain diagnosis, a closely monitored trial of DDAVP (10 μg intranasally twice a day) may be used to distinguish partial nephrogenic diabetes insipidus from partial neurogenic diabetes insipidus and primary polydipsia. If this dosage of DDAVP causes a significant antidiuretic effect, nephrogenic diabetes insipidus is effectively excluded. If polydipsia and polyuria are both abolished and plasma sodium does not fall below the normal range, the patient probably has central (neurogenic) diabetes insipidus. Conversely if there is a reduction in urine output without a reduction in water intake and hyponatraemia appears, the patient probably has primary polydipsia. Since fatal water intoxication is a remote possibility, the trial should be carried out with close monitoring.

Recommendations

For a full account of the recommendations in the differential diagnosis of diabetes insipidus see Table 3.

Bibliography

Bichet, D.G. (1990). Diabetes insipidus and vasopressin. In *Diagnostic Endocrinology*, (eds. W.T. Moore and R. Eastman), pp. 111–26. B.C. Decker Inc., Toronto.

Crawford, J.D. and Bode, H.H. (1975). Disorders of the posterior pituitary in children. In *Endocrine and Genetic Diseases of Childhood and Adolescence*, (2nd edn.), (ed. L.I. Gardner), pp. 126–58. W.B. Saunders, Philadelphia.

Niaudet, P., Dechaux, M., Leroy, D., and Broyer, M. (1985). Nephrogenic diabetes insipidus in children. In *Frontiers of Hormone Research*, Diabetes insipidus in man, Vol. 13, (eds. P. Czernichow and A.G. Robinson). pp. 224–31. S. Karger, Basel.

Reeves, W.B. and Andreoli, T.E. (1989). Nephrogenic diabetes insipidus. In *The metabolic basis of inherited disease*, (eds. C.R. Scriver, A.L. Beaudet, W.S. Sly and D. Valle). pp. 1985–2011. McGraw-Hill, New York.

Robertson, G.L. (1987*a*). Posterior pituitary. In *Endocrinology and Metabolism*, (2nd edn.), (eds. P. Felig, J.D. Baxter, A.E. Broadus, and L.A. Frohman), pp. 338–85. Mc Graw-Hill, New York.

Robertson, G.L. (1988). Differential diagnosis of polyuria. *Annual Review of Medicine*, **39**, 425–442.

Valtin, H. (ed.) (1983). Concentration and dilution of urine: H_2O balance. In *Renal Function. Mechanisms Preserving Fluid and Solute Balance in Health*, (2nd edn.), pp. 161–194. Little Brown and Company, Boston.

References

Alon, U. and Chan, J.C. (1985). Hydrochlorothiazide-amiloride in the treatment of congenital nephrogenic diabetes insipidus. *American Journal of Nephrology*, **5**, 9–13.

Barlow, E.D. and de Wardener, H.E. (1959). Compulsive water drinking. *Quarterly Journal of Medicine*, **28**, 235–58.

Batlle, D.C., von Riotte, A.B., Gaviria, M., and Grupp, M. (1985). Amelioration of polyuria by amiloride in patients receiving long-term lithium therapy. *New England Journal of Medicine*, **312**, 408–14.

Baylis, P.H., Gaskill, M.G., and Robertson, G.L. (1981). Vasopressin secretion in primary polydipsia and cranial diabetes insipidus. *Quarterly Journal of Medicine*, **199**, 345–58.

Behrman, R.E., Vaughan, V.C., and Nelson, W.E. (eds.) (1987). 17.30 Nephrogenic diabetes insipidus. In *Nelson Textbook of Pediatrics*, pp. 1136–7. W.B. Saunders, Philadelphia.

Berl, T. and Schrier, R.W. (1986). Disorders of water metabolism. In *Renal and Electrolytic Disorders*, (ed. R.W. Schrier), pp. 1–77. Little Brown, Boston.

5.5 Nephrogenic diabetes insipidus

Bianchine, J.W., Stambler, A.A., and Harrison, H.E. (1971). Nephrogenic (vasopressin-resistant) diabetes insipidus with the usual X-linked inheritance. *Birth Defects*, 7, 280–1.

Bichet, D.G., Levi, M., and Schrier, R.W. (1985). Polyuria, dehydration, overhydration. In *The Kidney: Physiology and Pathophysiology*, (eds. D.W. Seldin and G. Giebisch), pp. 951–84. Raven Press, New York.

Bichet, D.G., *et al.* (1988). Hemodynamic and coagulation responses to 1-desamino [8-D-arginine] vasopressin in patients with congenital nephrogenic diabetes insipidus. *New England Journal of Medicine*, 318, 881–7.

Bichet, D.G., *et al.* (1989). Epinephrine and dDAVP administration in patients with congenital nephrogenic diabetes insipidus. Evidence for a pre-cylic AMP V_2 receptor defective mechanism. *Kidney International*, 36, 859–66.

Bichet, D.G., Hendy, G.N., Hirsch, D.J., Lonergan, M., Arthus, M.F., and Mandel, J.L. (1990). Congenital nephrogenic diabetes insipidus: from the ship Hopewell to restriction fragment length polymorphism studies. *Kidney International*, 37, 246A.

Blalock, T.J., Gerron, G., Quiter, E., and Rudman, D. (1977). Role of diet in the management of vasopressin-responsive and -resistant diabetes insipidus. *American Journal of Clinical Nutrition*, 30, 1070–6.

Bode, H.H. and Crawford, J.D. (1969). Nephrogenic diabetes insipidus in North America. The Hopewell hypothesis. *New England Journal of Medicine*, 280, 750–4.

Bonvalet, J.P., Pradelles, P., and Farman, N. (1987). Segmental synthesis and actions of prostaglandins along the nephron. *American Journal of Physiology*, 253, F377–87.

Boton, R., Gaviria, M., and Batlle, D.C. (1987). Prevalence, pathogenesis, and treatment of renal dysfunction associated with chronic lithium therapy. *American Journal of Kidney Diseases*, 10, 329–45.

Boyd, S.D., Raz, D., and Ehrlich, R.M. (1980). Diabetes insipidus and nonobstructive dilation of urinary tract. *Urology*, 26, 266–9.

Brenner, B., Coe, F.L., and Rector, F.C. (eds.) (1987). Transport functions of the renal tubules. In *Renal physiology in health and disease*, pp. 27–56. W.B. Saunders Company, Philadephia.

Brenner, B., Seligsohn, U., and Hochberg, Z. (1988). Normal response of factor VIII and von Willebrand factor to 1-deamino-8D-arginine vasopressin in nephrogenic diabetes insipidus. *Journal of Clinical Endocrinology and Metabolism*, 67, 191–3.

Camerer, J.W. (1935). Eine ergänzung des Weilschen diabetes-insipidus-stammbaumes. *Archiv für Rassen- und Gesellschafts-hygiene Biologie*, 28, 382–3.

Cannon, J.F. (1955). Diabetes insipidus clinical and experimental studies with consideration of genetic relationships. *Archives of Internal Medicine*, 96, 215–72.

Christensen, S., Kusano, E., Yusufi, N.K., Murayama, N., and Dousa, T.P. (1985). Pathogenesis of nephrogenic diabetes insipidus due to chronic administration of lithium in rats. *Journal of Clinical Investigation*, 75, 1869–79.

Chung, R.C. and Mantell, L.K. (1952). Urographic changes in diabetes insipidus. *Journal of the American Medical Association*, 150, 1307–8.

Cogan, E., Svoboda, M., and Abramow, M. (1987). Mechanisms of lithium-vasopressin interaction in rabbit cortical collecting tubule. *American Journal of Physiology*, 252, F1080–7.

Derkx, F.H.M., Brink, H.S., Merkus, P., Smiths, J., Brommeer, E.J.R., and Schalekamp, M.A.D.H. (1987). Vasopressin V_2-receptor-mediated hypotensive response in man. *Journal of Hypertension*, 5, S107–9.

Dölle, W. (1951). Eine weitere erganzung des Weilschen diabetes-insipidus-stammbaumes. *Zeitschrift für Menschliche Vererbungs- und Konstitutionslehre*, 30, 372–4.

Feigin, R.D., Rimoin, D.L., and Kaufman, R.L. (1970). Nephrogenic diabetes insipidus in a Negro kindred. *American Journal of Diseases of Children*, 120, 64–8.

Forssman, H. (1942). On the mode of hereditary transmission in diabetes insipidus. *Nordisk Medicin*, 16, 3211–3.

Forssman, H. (1975). Modern medical history. The recognition of nephrogenic diabetes insipidus. *Acta Medica Scandinavica*, 197, 1–6.

Gautier, B., Thieblot, P., and Steg, A. (1981). Mégauretère, mégavessie et diabète insipide familial. *Semaine des Hôpitaux de Paris*, 57, 60–1.

Goldberg, H., Clayman, P., and Skorecki, K.L. (1988). Mechanism of Li inhibition of vasopressin-sensitive adenylate cyclase in cultured renal epithelial cells. *American Journal of Physiology*, 255, F995–1001.

Goldman, M.B., Luchins, D.J., and Robertson, G.L. (1988). Mechanisms of altered water metabolism in psychotic patients with polydipsia and hyponatremia. *New England Journal of Medicine*, 318, 397–403.

Hochberg, Z. (1977). Nephrogenic diabetes insipidus in five siblings. *Harefuah*, 92, 118–20.

Huard, G. and Picard, J.L. (1960). Insuffisance congénitale du tubule rénal. *Union Médicale du Canada*, 89, 711–18.

Jamison, R.L. and Oliver, R.E. (1982). Disorders of urinary concentration and dilution. *American Journal of Medicine*, 72, 308–22.

Jard, S. (1985). Vasopressin receptors. In *Frontiers of Hormone Research*, Diabetes insipidus in man, Vol. 13, (eds. P. Czernichow and A.G. Robinson), pp. 89–104. S. Karger, Basel.

Kambouris, M., Dlouhy, S.R., Trofatter, J.A., Conneally, P.M., and Hodes, M.E. (1988). Localization of the gene for X-linked nephrogenic diabetes insipidus to Xq28. *American Journal of Medical Genetics*, 29, 239–46.

Kirman, B.H., Black, J.A., Wilkinson, R.H., and Evans, P.R. (1956). Familial pitressin-resistant diabetes insipidus with mental defect. *Archives of Disease in Childhood*, 31, 59–66.

Knoers, N., v.d. Heyden, H., v. Oost, B.A., Ropers, H.H., Monnens, L., and Willems, J. (1988*a*). Nephrogenic diabetes insipidus: close linkage with markers from the distal long arm of the human X chromosome. *Human Genetics*, 80, 31–8.

Knoers, N., v.d. Heyden, H., v. Oost, B.A., Monnens, L., Willems, J., and Ropers, H.H. (1988*b*). Linkage of X-linked nephrogenic diabetes insipidus with DXS52, a polymorphic DNA marker. *Nephron*, 50, 187–98.

Knoers, N., v.d. Heyden, H., v. Oost, B.A., Monnens, L., Willems, J., and Ropers, H.H. (1989). Three-point linkage analysis using multiple DNA polymorphic markers in families with X-linked nephrogenic diabetes insipidus. *Genomics*, 4, 434–7.

Knoers, N., Brommer, E.J.P., Willems, H., v. Oost, B.A. and Monnens, L.A.H. (1990). Fibrinolytic responses to 1-desamino-8-D-arginine-vasopressin in patients with congenital nephrogenic diabetes insipidus. *Nephron*, 54, 322–6.

Lacombe, U.L. (1841). *De la polydipsie*. Thesis of Medicine. No. 99. pp. 1–87. Imprimerie et Fonderie de Rignoux, Paris.

Leaf, A. (1979). Neurogenic diabetes insipidus. *Kidney International*, 15, 572–80.

Leibowitz, S.F. (1988). Impact of brain monoamines and neuropeptides on vasopressin release. In *Vasopressin. Cellular and Integrative Functions*, (eds. A.W. Cowley Jr, J.F. Liard and D.A. Ausiello), pp. 379–88. Raven Press, New York.

Lewis, D.J. and Thomas, J.P. (1986). Treatment of nephrogenic diabetes insipidus. *New England Journal of Medicine*, 315, 1292–3.

Liard, J.F. (1988). Effects of a specific antidiuretic agonist on cardiac output and its distribution in intact and anephric dogs. *Clinical Sciences*, 74, 293–9.

Libber, S., Harrison, H., and Spector, D. (1986). Treatment of nephrogenic diabetes insipidus with prostaglandin synthesis inhibitors. *Journal of Pediatrics*, 108, 305–11.

Lonergan, M., Arthus, M.F., and Bichet, D.G. (1990). Platelet vasopressin receptors are normal in patients with congenital nephrogenic diabetes insipidus. *Kidney International*, 37, 250A.

Lyon, M.F. (1966). X chromosome inactivation in mammals. *Advances in Teratology*, 1, 25–54.

Magner, P.O. and Halperin, M.L. (1987). Polyuria—a pathophysiological approach. *Medicine North America*, **15**, 2987–97.

Mannucci, P.M., Aberg, M., Nilsson, I.M., and Robertson, B. (1975). Mechanism of plasminogen activator and factor VIII increase after vasocative drugs. *British Journal of Haematology*, **30**, 81–93.

Mannucci, P.M., Canciani, M.T., Rota, L., and Donovan, B.S. (1981). Response of factor VIII/von Willebrand factor to dDAVP in healthy subjects and patients with haemophilia A and von Willebrand's disease. *British Journal of Haematology*, **47**, 283–93.

McIlraith, C.H. (1892). Notes on some cases of diabetes insipidus with marked family and hereditary tendencies. *Lancet*, **2**, 767–8.

Miller, M., Dalakos, T., Moses, A.M., Fellerman, H., and Streeten, D.H.P. (1970). Recognition of partial defects in antidiuretic hormone secretion. *Annals of Internal Medicine*, **73**, 721–9.

Moses, A.M., Miller, J.L., and Levine, M.A. (1988). Two distinct pathophysiological mechanisms in congenital nephrogenic diabetes insipidus. *Journal of Clinical Endocrinology and Metabolism*, **66**, 1259–64.

Nakano, K.K. (1969). Familial nephrogenic diabetes insipidus. *Hawaii Medical Journal*, **28**, 205–8.

Niaudet, P., Dechaux, M., Trivin, C., Loirat, C., and Broyer, M. (1984). Nephrogenic diabetes insipidus: Clinical and pathophysiological aspects. In *Advances in Nephrology*, (eds, J.P. Grünfeld and M.H. Maxwell), Vol. 13, pp. 247–60. Year Book Medical Publishers, Chicago.

Ohzeki, T. (1985). Urinary adenosine 3',5'-monophosphate (cAMP) response to antidiuretic hormone in diabetes insipidus (DI): comparison between congenital nephrogenic DI type 1 and 2, and vasopressin sensitive DI. *Acta Endocrinologica*, **108**, 485–90.

Ohzeki, T., Igarashi, T., and Okamoto, A. (1984). Familial cases of congenital nephrogenic diabetes insipidus type II: Remarkable increment of urinary adenosine 3', 5'-monophosphate in response to antidiuretic hormone. *Journal of Pediatrics*, **104**, 593–5.

Os, I., Aakesson, I., and Enger, E. (1985). Plasma vasopressin in hereditary cranial diabetes insipidus. *Acta Medica Scandinavica*, **217**, 429–34.

Pedersen, E.B., et al. (1985). Familial cranial diabetes insipidus: a report of five families: genetic, diagnostic and therapeutic aspects. *Quarterly Journal of Medicine*, **57**, 883–96.

Randle, J.C.R., Mazurek, M., Kneifel, D., Dufresne, J., and Renaud, L.P. (1986). α_1-Adrenergic receptor activation releases vasopressin and oxytocin from perfused rat hypothalamic explants. *Neuroscience Letters*, **65**, 219–23.

Repaske, D.R., Phillips, J.A., Kirby, L.T., Tze, W.J., D'Ercole, A.J., and Battey, J. (1990). Molecular analysis of autosomal dominant neurophyphyseal diabetes insipidus. *Journal of Clinical Endocrinology and Metabolism*, **70**, 752–7.

Richardson, D.W. and Robinson, A.G. (1985). Desmopressin. *Annals of Internal Medicine*, **103**, 228–39.

Robertson, G.L. (1985a). Diagnosis of diabetes insipidus. In *Frontiers of Hormone Research*, Diabetes insipidus in man, Vol. 13, (eds. P. Czernichow and A.G. Robinson). pp. 176–89, S. Karger, Basel.

Robertson, G.L., Oiso, Y., Vokes, T.P., and Gaskill, M.B. (1985b). Diprenorphine inhibits selectively the vasopressin response to hypovolemic stimuli. *Transactions of the Association of American Physicians*, **98**, 322–33.

Robertson, G.L. (1987a). Posterior pituitary. In *Endocrinology and Metabolism*, (eds. P. Felig, J.D. Baxter, A.E. Broadus, and L.A. Frohman). pp. 338–85, McGraw-Hill, New York.

Robertson, G.L. (1987b). Dipsogenic diabetes insipidus: A newly recognized syndrome caused by a selective defect in the osmoregulation of thirst. *Transactions of the Association of American Physicians*, **100**, 241–9.

Schultz, P. and Lines, D.R. (1975). Nephrogenic diabetes insipidus in an Australian aboriginal kindred. *Humangenetik*, **26**, 79–85.

Schuster, V.L. and Seldin, D.W. (1985). Renal clearance. In *The Kidney Physiology and Pathophysiology*, Vol. 1, (eds. D.W. Seldin and G. Giebisch). pp. 365–95, Raven Press, New York.

Schwartz, J., Laird, J.F., Ott, C., and Cowley, A.W. Jr. (1985). Hemodynamic effects of neurohypophyseal peptides with antidiuretic activity in dogs. *American Journal of Physiology*, **249**, H1001–8.

Schwertschlag, U., Gerber, J.G., Barnes, J.S., and Nies, A.S. (1986). A dual effect of nonsteroidal anti-inflammatory drugs on renal water handling in humans. *Journal of Pharmacology and Experimental Therapeutics*, **238**, 653–8.

Sklar, A.H. and Schrier, R.W. (1983). Central nervous system mediatros of vasopressin release. *Pharmacological Reviews*, **63**, 1243–79.

Stern, P. and Valtin, H. (1981). Verney was right, but . . . *New England Journal of Medicine*, **305**, 1581–2.

Stoff, J.S., Rosa, R.M., Silva, P., and Epstein, F.H. (1981). Indomethacin impairs water diuresis in the DI rat: role of prostaglandins independent of ADH. *American Journal of Physiology*, **241**, F2231–7.

Streitz, J.M.Jr. and Streitz, J.M. (1988). Polyuric urinary tract dilatation with renal damage. *The Journal of Urology*, **139**, 784–5.

Usberti, M., et al. (1985). Mechanism of action of indomethacin in tubular defects. *Pediatrics*, **75**, 501–7.

Uttley, W.S., Atkinson, B., and Adams, A. (1978). Cyclic adenosine monophosphate excretion in urine of patients and carriers of congenital nephrogenic diabetes insipidus. *Journal of Inherited Metabolic Disease*, **1**, 75–7.

Valtin, H., North, W.G., Edwards, B.R., and Gellai, M. (1985). Animal models of diabetes insipidus. In *Frontiers of Hormone Research*, Diabetes insipidus in man, Vol. 13 (eds. P. Czernichow and A.G. Robinson), pp. 105–26. S. Karger, Basel.

Waring, A.J., Kajdi, L., and Tappan, V. (1945). A congenital defect of water metabolism. *American Journal of Diseases of Children*, **69**, 323–4.

Wattiez, R., Loeb, H., Bellens, R., and Van Geffel, R. (1957). Diabète insipide pitressino-résistant. *Helvetica Paediatrica Acta*, **6**, 643–62.

Weil, A. (1884). Ueber die hereditare form des diabetes insipidus. *Archives fur Pathologische Anatomie und Physiologie and fur Klinische Medicine (Virchow's Archives)*, **95**, 70–95.

Weil, A. (1908). Ueber die hereditare form des diabetes insipidus. *Deutches Archiv fur Klinische Medizin*, **93**, 180–290.

Williams, R.M. and Henry, C. (1947). Nephrogenic diabetes insipidus transmitted by females and appearing during infancy in males. *Annals of Internal Medicine*, **27**, 84–95.

Williams, T.D.M., Lightman, S.L., and Leadbeater, M.J. (1986). Hormonal and cardiovascular responses to dDAVP in man. *Clinical Endocrinology (Oxford)*, **24**, 89–96.

Zerbe, R.L. and Robertson, G.L. (1981). A comparison of plasma vasopressin measurements with a standard indirect test in the differential diagnosis of polyuria. *New England Journal of Medicine*, **305**, 1539–46.

Zerbe, R.L. and Robertson, G.L. (1983). Osmoregulation of thirst and vasopressin secretion in man: The effect of various solutes. *American Journal of Physiology*, **244**, E607–14.

Zerbe, R.L. and Robertson, G.L. (1984). Disorders of ADH. *Medicine North America*, **13**, 1570–1574.

INDEX

Page numbers for key sections are given in **bold type**, and for tables in *italic type*. '*vs*' denotes differential diagnosis.

AA amyloid, *see* amyloid
abdominal pain,
　haemodialysis 1454–5
　in Henoch–Schönlein purpura
　　595, 596, 606
　　treatment 606–7
　see also acute abdomen
abdominal wall, prune belly
　syndrome 2034
ABO autoimmune haemolytic
　anaemia, group A
　transplantation recipients 2345
abortion, septic abortion 1932–3
abortion-related acute renal failure
　1056–7
abscess,
　haemodialysis, long-term 1452–3
　renal 93, 103, 105, 143, 1691
　renal congenital 2027–8
　renal cortical abscess 1695–6,
　　1705–6
　renal perinephric 1696, 1706
　in diabetes mellitus 531
　see also kidney(s), abscess; renal
　　abscess
abscess, visceral,
　crescentic glomerulonephritis 421
　infection-associated
　　glomerulonephritis 470
acanthosis nigricans, renal
　transplantation 1584
accelerated hypertension **2124–36**
　see also hypertension
acebutolol, properties, in relation to
　use in impaired renal function
　2053
acetaminophen, *see* paracetamol
acetaminophenone
　gluthathione depletion 169
　mixed-function oxidase system
　　169
acetanilide, first use 803
acetate,
　dialysate, disequilibrium
　　syndrome 1400–1
　diet, acid and base production
　　932–3, *933*
　peritoneal dialysis solution 1481
acetazolamide,
　avoiding in kidney disease 191
　erythropoietin production effects
　　1350
　glaucoma 206
　hyperphosphataemia 206
　nephrocalcinosis and 1893–4
　nephrotic syndrome, diuretics
　　potentiation 270–1
　renal tubular acidosis 1861
　restrictions in use 213
　structural formula *199*
　toxicity 213
　urinary alkalinization 206
　see also carbonic anhydrase
　　inhibitors
acetoacetate,
　instability 928
　plasma concentration 929
　urate excretion decrease 829

acetoheximide, creatinine clearance
　and measurement, interfering
　substance *32*
acetohydroxamic acid, urease
　inhibitor 1666
acetophenetidine, *see* phenacetin
N-acetyl-β-D-glucosaminidase 165
　A and B deficiency
　　Sandhoff disease 2206
　　Tay–Sachs disease 2206
　excretion 564
　glomerular capillaries 133
　proximal tubular function,
　　lithium 854
　in urine
　　marker of hypertension 18
　　value of assay 19
　see also aminoglycosides
acetylcholine, release,
　hypomagnesaemia and 1813
acetylcysteine, for overdose of
　paracetamol 2315
acetylsalicylic acid, *see* aspirin;
　salicylates
acetyltransferase deficiency, liver
　glomerulonephritis 360
acid excretion, *see* renal tubular
　acidosis; tubular function
α-1 acid glycoprotein,
　decreased drug protein binding
　　1240
　protein binding, competition for
　　sites 176
acid, titrable,
　excretion, daily urinary *923*
　see also metabolic acidosis
acidaemia,
　acid load, buffering 921
　defined 919
　K$^+$ regulation
　　acid–base status 899
　　metabolic and respiratory
　　　acidosis 896
　metabolic acidosis, K$^+$ regulation
　　896
　nomogram 919
　P_{CO_2}, treatment 960
　profound, mixed acid–base
　　disorders 958
　respiratory acidosis, K$^+$
　　regulation 896
　respiratory response 921–2
　systemic, distal renal tubular
　　acidosis 763
　treatment, early,
　　contraindications 934
　see also metabolic acidosis; renal
　　tubular acidosis; respiratory
　　acidosis
acidification defects, *see* renal
　acidification; renal tubular
　acidosis
acidosis, *see* metabolic acidosis;
　acid–base disorders; renal
　tubular acidosis; respiratory
　acidosis
aciduria, and renal damage from
　incompatible blood transfusion
　995

acid–base disorders **917–60**
　acute renal failure 1019–20
　blood transfusion 959
　dialysis in 959–60
　hypothermia 960
　mixed **956–60**
　　associated clinical conditions *957*
　　nomogram 919
　　respiratory and metabolic
　　　disorders 957–60
　　triple disorders 959–60
　normal values, extracellular
　　fluids *917*
　parenteral nutrition 960
　primary, secondary disorders *920*
　respiratory origin 950–2
　terminology and definitions 919
acid–base ratio,
　pregnancy 1916–17
　renal regulation 1916–17
acid–base status,
　hypercalcaemia causing 1885
　potassium homeostasis 896
　reabsorption, phosphate 1787
Acinetobacter calcoaceticus,
　urinary tract infection 1661
acne, following renal
　transplantation 1583
acral necrosis, diabetics with
　chronic renal failure 1641
acrocephalo–polydactylous
　dysplasia, features and
　associated abnormalities *2038*
acrolein 620
acromegaly, elevation of GFR 29
acroparasthaesiae, Anderson–Fabry
　disease 2207
ACTH, *see* adrenocorticotrophic
　hormone
actin,
　Bowman's capsule parietal layer
　　137
　mesangial cells 136
　podocytes 133
actinic keratosis,
　azathioprine-associated 1584
　chronic renal failure 1393
α-actinin, podocytes 133
actinomycin D,
　radiation injury potentiation
　　849
　in Wegener's granulomatosis
　　621
　Wilms' tumour 2260–1
active transport,
　bicarbonate 764
　proximal tubule 741–2
activity/chronicity indices,
　glomerular abnormalities 145
　tubulointerstitial abnormalities
　　145
actomyosin, in glomerular
　basement membrane 524
acute abdomen,
　causes 1284
　chronic renal failure, differential
　　diagnosis 1284
　haemodialysis patients 2311
　see also abdominal pain

acute on chronic renal failure
　1158–61
　causes 1158–61, *1159*
　contrast media 1186
　in diabetes mellitus 532
　drugs 1159
　interstitial nephritis, acute 1161
acute cortical necrosis,
　nephrocalcinosis 1888
　renal failure, acute, pregnancy
　　1933
acute endocapillary
　glomerulonephritis, *see*
　　endocapillary
　　glomerulonephritis, acute
acute fatty liver of pregnancy 1057,
　1933–4
acute intermittent porphyria 2285
acute interstitial nephrosis, chronic
　renal failure, assessment 1161
acute intravascular haemolysis,
　paroxysmal nocturnal
　haemoglobinuria 2280
acute nephritic syndrome, *see*
　nephritic syndrome, acute
acute renal failure **969–1146**
　acute insults 993
　acute postinfectious
　　glomerulonephritis 1070–1
　amino acids, insulin effects on
　　uptake 1259
　anaemia 973–4
　animal models 78
　anuria, radiological tests 73
　assessment 969–70
　bacterial infections 1021
　oliguria **971–7**
　radiological assessment **974–7**
　urinary biochemical indices 988
bacterial infections
　assessment 1021
　management 1021
　sepsis **978–80**
　tropics **1133–4**
　urinary tract infection 1692
bleeding diathesis in 2344
causes
　in children *1122*
　glomerulonephritis and
　　vasculitis **1060–71**
　intrinsic renal pathology *989*,
　　989–90
　prerenal/renal/postrenal 969–70
　tubulointerstitial disease 989–90
cell injury
　basement membrane damage
　　997
　biochemistry **996–7**
　calcium overload 997
　cell swelling 997
　pathogenesis **990–6**
　see also acute renal failure, renal
　　damage
children 1036–7, **1122–3**
chronic glomerulonephritis
　with crescentic change 1070
　without crescentic change 1071
classical, tropics 1129
clinical examination 970–1

Index

acute renal failure (*cont.*):
  clinical presentation, cholesterol embolism 1078–80
  complications
    investigations 978–80
    radiological assessment 979
  continuous renal replacement therapy **1026–33**
  in diabetics **532**
  diuretics 204–5
    clinical use 204
  drug-induced, incidence 168
  drug-induced glomerulonephritis 159–63
  haemodialysis, intermittent *vs* continuous 1027–33
  haemofiltration **1026–33**
    patient supervision **1031–3**
  haemorrhagic syndrome, *see* Hantavirus
  hepatorenal syndrome, liver cirrhosis 2310
  histopathology, minimal change nephrotic syndrome 301
  hospital-acquired renal failure 971
  hypercalcaemia 1773
  hyperkalaemia 908–9
  immunological markers 974
  immunological mediation 974
  interstitial nephritis **1084–94**
  investigations 969–80
    biochemical disturbances 973
    biopsy, renal 977–8
      indications 144
    clinical assessment 969–71
    clinical examination 970–1
    community-acquired renal failure 969–70
    complications 978–80
    computed tomography scanning 106
    contrast media injection 78
    diagnosis, establishing 969–71
    haematological disturbances 973–4
    indications for renal biopsy 143–4
    non-oliguric form 977
    nuclear imaging 124–5
    oliguria, assessment **971–7**
    radiological assessment **974–7**
    recovery phase 980
    ultrasound, with radiography 72–3
  ischaemic insult as cause 1035–8
    case histories *1036*
  ischaemic renal disease **1077–83**
  leptospirosis **1132**
    occupational risks 9
  lithium-induced 67
  in liver disease **1098–107**
  malaria **1130–2**
  malignant aetiology 476
  management **1015–23**
    acute respiratory failure 1022
    bleeding defects 1022
    dietary management 1020–1
    drugs and GFR 1022
    gastrointestinal disorders 1021–2
    infections 1021
    patient survival factors *1022*
    prerenal failure 1015–19
    prevention 1019
    reoperation 1021
    symptomatic treatment 1019–22
    water and electrolytes 1019–20

acute renal failure (*cont.*):
  mechanisms of excretory failure **997–1001**
    resistance and susceptibility to further damage 1000
  melioidosis **1132–3**
  minimal change nephrotic syndrome and 301
  in multiple myeloma, treatment 568
  in myeloma cast nephropathy 567
  neonates **1110–20**, 1605
  nephrotic syndrome complicated by ARF 290–1, **1034–9**
  neurological involvement 2325–6
    clinical features 2326
    treatment 2326
  no pre-existing disease, acute insults 993
  non-oliguric 977
    tropics 1129
  nutrition, total parenteral nutrition *1021*
  obstetric, *see* obstetric acute renal failure
  obstructive/non-obstructive, CT scanning 105–6
  oliguria, assessment, differentiation of prerenal oliguria from ARF 971, 971–7
  pathogenesis, shock 1044
  pathophysiology **982–1001**
    intrinsic renal disease 989–90
    mechanisms causing excretory failure 997–1001
    minimal change nephrotic syndrome 301
    normal renal circulation 982–6
    pathogenesis of renal damage 990–7
    prerenal failure 986–8
    urinary obstruction 988–9
    see also acute renal failure, renal damage
  peripartum, differential diagnosis *1057*
  in polymyositis 684
  pre-existing, predisposing factor for contrast-associated nephropathy 78–9
  pregnancy **1056–7**
  prerenal failure 969, **986–8**
    causes 969, *1015*
    defined 1015
    differential diagnosis *1016*
    drug-induced 987
    management 1016
  prevention **1017–19**
    amino acids 1018
    animal models 1018
    ATP–magnesium chloride 1018
    calcium channel blockers 1018
    diuretics 1017–18
    dopamine 1018
    other drugs 1018
    thromboxane synthetase inhibitor 1018
  primary glomerulonephritis **1070–2**
  prognosis 976–7, 1022–3
  protein synthesis 1259
  radiocontrast media-induced 996, 1009–10, 1185, 2334
  radiological tests 72–3
  recovery phase, investigations 980

acute renal failure (*cont.*):
  renal damage 990–5
    haemodynamically-mediated 990
    haemoglobin and 995
    incompatible blood transfusion 995
    malaria and 995
    myoglobin and 995
    non-homogeneous renal oxygenation 990–1
    renal cortical necrosis 993
    renal ischaemia 992–3
    renal perfusion reduction 991
    sepsis 991
    see also acute renal failure, cell injury
  renal replacement therapy **1026–33**
  rhabdomyolysis 2323
  sepsis **978–80**
  snake bite **1134–5**
  sodium excretion, FE_{Na} 972
  symptoms and signs *970*
  in systemic lupus erythematosus (SLE) 649, 1069
  in pregnancy 652
  treatment, *see* acute renal failure, management
  tropics **1124–38**
    obstetrics 1138
    surgery 1138
  tubular injury **1006–13**
    anaesthetic agents 1009
    antibiotics 994, 1010–11
    chemotherapy 1011
    cyclosporin 1012–13
    glycols 1009
    hypercalcaemia 1012
    hyperoxaluria 1012
    metals causing nephrotoxicity 993, 1008–9
    neoplastic disease 1012
    organic solvents 993, 1009
    pigment nephropathy 1006–8
    poisoning 1011
    radiographic contrast media 996, 1009–10
    see also acute renal failure, renal damage; acute tubular necrosis; cell injury
  typhoid fever 460
  unilateral 70–2, 73
  urinary tract infection 1692
  'vasomotor' type complicating nephrotic syndrome **1034–9**
  viral infections **1134**
  see also acute tubular necrosis
acute tubular necrosis,
  acute renal failure in liver disease 1101
  animal models 990
  causes 989–90
    heavy metal intoxications 1106–7
  children 1123
  chronic renal failure 1158
  differential diagnosis 972, *1016*
  DTPA scan 125
  early onset phase 125
  end-stage renal failure 1234
    proportional distribution by age at commencement *1229*
  heavy metals exposure 838
  hyperemesis gravidarum complicating 1933

acute tubular necrosis (*cont.*):
  long-term recovery 1001
  mechanisms of excretory failure **997–1001**
    animal models 990, 997
    haemodynamic factors 999
    resistance and susceptibility to further damage 1000
    tubular factors 997–9
    tubuloglomerular feedback 999–1000
    urinary volume 1000–1
  mercuric chloride nephrotoxicity 844
  nephrotic syndrome complicated by ARF 1039
  non-steroidal anti-inflammatory drugs 821
  pathogenesis of renal damage
    haemodynamically-mediated 990
    non-homogeneous renal oxygenation 990–1
    renal perfusion reduction 991
    sepsis 991
  patients at risk, recommendations 1018–19
  prerenal failure, radionuclide studies 125
  prognosis 1022–3
  renal biopsy contraindications 144
  role 1462
  snake bite **1134–5**
  in transplantation 97
  treatment, prostacyclin infusion 1055
  ultrasonography 977
  uranium, animal model 845
  vs diffuse parenchymal renal disease, ultrasonography 93
  see also acute renal failure
acute tumour lysis syndrome 2319
acute-phase reactant, serum amyloid A (SAA) as 548, 556
acyclovir,
  creatinine increase 2306
  cytomegalovirus infections, prophylaxis in renal transplantation 1533
  handling in renal impairment 184
  herpes simplex and zoster 2306
Addison's disease, hypoaldosteronism 909
adenine arabinose monophosphate, hepatitis B 1288
adenine nucleotides, alcohol-stimulated increased turnover and 829
adenine phosphoribosyl transferase, deficiency 1865
adenoma sebaceum, tuberous sclerosis 2048
adenomata,
  aldosterone-producing 787
  computed tomography scanning 103
  oncocytoma 103
adenosine,
  arteriolar myocytes of afferent arterioles, effects 26, 27
  mediation, A_1 and A_2 receptors 984
adenosine deaminase binding protein, assay 19

Index

adenylate cyclase,
  inhibition by lithium, diabetes insipidus, central 796
  V_2 renal receptors and 791
ADH, human, *see* arginine vasopressin
adhesion proteins,
  endothelial binding 1052
  receptors ELAM-1, ICAM-1 1052
adipsia 887
ADPKD, *see* polycystic disease, autosomal dominant
adrenal glands,
  in amyloidosis 550
  chronic renal failure, function tests 1325
  function tests 1325
adrenal insufficiency,
  causes 1325
  hyponatraemia 876
adrenaline,
  arteriolar myocytes of afferent arterioles, effects 26
  exercise and, hypertensive subjects 1461
  hypokalaemia 902
α-adrenoceptor agonists, potassium homeostasis 896, 899
α-1-adrenoceptor agonists,
  heart mass, effect of pharmacological interventions on nephrectomized animal models *1266*
  renin secretion and *2071*
α-2-adrenoceptor agonists, hypertension of end-stage renal failure 2121
β-2-adrenoceptor agonists,
  hyperkalaemia, therapy 910
  potassium, transcellular shifts 902
  potassium homeostasis 896, 899
  see also adrenaline; fenoterol; salbutamol; terbutaline
α-1 and α-2-adrenoceptor antagonists, renal hypertension, treatment 2055–6
β-2-adrenoceptor antagonists, *see* beta-adrenoceptor blockers
adrenocorticotrophic hormone,
  chronic renal failure 1325
  ectopic production, potassium depletion 906
  function tests, CRF 1325
  neonates, role 53
  no suppression in uraemia by dexamethasone 1325
  rapidly progressive osteoporosis, nephrocalcinosis 1891
  resistance, urinary dilution tests 42
  secretion, frequency and 2
adrenogenital syndrome,
  C-21 hydroxylase deficiency 909
  3-β-OH-dehydrogenase deficiency 909
  hyperkalaemia, hypoaldosteronism 909
adrenoleuco-dystrophy, neonatally lethal, features and associated abnormalities *2038*
adriamycin,
  drug-induced nephropathy 162, 163
  metastatic renal carcinoma 2248
  radiation injury potentiation 849
  renal pelvis tumour 2269

adriamycin (*cont.*):
  tubular injury in acute renal failure 1011
  Wilms' tumour 2260–1
adriamycin nephrosis,
  animal models
    dietary protein restriction *1210*
    experimental, functional and structural parameters *1194*, *1195*
    glomerular scarring, progressive chronic renal failure 1197
    glomerular scarring and hypertension 1196
    haemodynamic and structural glomerular parameters *1214*
  nephropathy *1201*
adynamic bone disease, aluminium-related osteopathy 1369, 1376
aerobactin production, *Escherichia coli* 1666
afferent arterioles, *see* renal arteries
aflatoxins, and carcinoma 2238
African patients, skin-lightening cream 161
AGCP (1-alkyl-2-acetyl-*sn*-glycero-3-phosphorylcholine), *see* platelet activating factor
ageing kidney, *see* elderly patients
agenesis, *see* kidney(s), absence
AGR, *see* aniridia–genitourinary–retardation triad
agranulocytosis, analgesic nephropathy and 813
AIDS,
  complications, nephrotoxicity of treatment 1010
  PNP, nephrotoxicity of treatment 1011
AIDS-associated nephropathy 464–6
  electrolyte disorders 466
  glomerulonephritis 464–6
  oropharynx involvement 1279
air embolism, haemodialysis 1429
ajmaline, cholestasis 2313
AL amyloid, *see* amyloid
Alagille syndrome, features and associated abnormalities *2036*, 2216, 2311
alanine:glyoxalate aminotransferase, absence, primary hyperoxaluria 2226
alanine, excretion, pregnancy 1916
alanine aminopeptidase 163, 165
albumin,
  albumin:creatinine ratios, in diabetes mellitus 510
  albumin:globulin ratio, pregnancy-induced changes, rat models 1912
  cationic isoalbumins, minimal change nephrotic syndrome 309
  deposition, on glomerular basement membrane in diabetes 517
  distribution, Starling forces 267–8
  excretion 514
    in diabetic nephropathy 510
    diurnal variation 509
    elderly patients 176
    enalapril treatment, reducing 527–8
    increased levels, *see* albuminuria
    normal 509

albumin (*cont.*):
  excretion (*cont.*):
    predictive value in diabetes *510*
    restricted protein intake, reducing 528
    urine flow association 515
  infusion 271
    in Henoch–Schönlein purpura 606
    natriuresis promotion 269
    nephrotic syndrome 271
  nephrotic syndrome
    levels and hyperlipidaemia 286–7
    metabolism, summary 291
    platelet function 283
  serum
    at equilibrium 238
    calculation 238
    protein binding sites decrease 176
  size of molecule, and filtration barrier of glomerulus 136
albuminuria,
  in diabetes mellitus
    cardiovascular death risk 508
    glomerular hyperfiltration correlation 509
    increased, in early phase 509
    prediction from blood pressure 507
  exercise-induced, diabetes control relation 510, 529
  nephrotic syndrome, mercury occupational exposure 844
  unreliability of serum albumin for calculation 238
  see also albumin, excretion; microalbuminuria
albumin–zinc, loss in nephrotic syndrome 290
albuterol, hyperkalaemia, therapy 910
alcaptonuria, nephrocalcinosis *1897*
alcohol,
  and essential hypertension 8
  and liver glomerulonephritis 360
  metabolic pathway *931*
alcohol abuse,
  adenine nucleotides, increased turnover 829
  hyperuricaemia and 829
  'moonshine'
    arsenic toxicity 846
    lead toxicity 838
  renal papillary necrosis, differential diagnosis 815
alcoholism,
  central pontine myelinosis 881
  glomerulonephritis 2311
  in hypomagnesaemia 1811
  hypophosphataemia in 1792
  ketoacidosis 930
    acid–base disorders 959
  renal transplantation, contraindications 1534
alcuronium, avoiding in kidney disease 186
aldose reductase 518
  inhibitors, in diabetes mellitus 518, 529–30
aldosterone
  antagonists 200
    NSAID interactions 823
  assay 2070
  blood and urinary concentrations, young and elderly patients 63

aldosterone (*cont.*):
  deficiency
    distal renal tubular acidosis 771–2
    hyperkalaemic renal tubular acidosis and 776
    hyporeninaemic hypoaldosteronism 207
    renal tubular acidosis 43, *938*
  endocapillary glomerulonephritis, acute 407
  high/low aldosterone, mineralocorticoid excess states 905
  hydrogen ion secretion 766
  hyperkalaemia and 787
  increase, Bartter's syndrome 783
  low sodium diet, young and elderly subjects *63*
  and magnesium 1808
  metabolic acidosis 940
  mineralocorticoid action, 'escape' 264
  nephrotic syndrome 267
  potassium homeostasis 896, 898–9
  pregnancy-induced increase 1919, 1962
    animals as gestational models 1920
  primary hyperaldosteronism, hypernatraemia 883
  renal response
    neonates 51, 52
    preterm neonates 52–3
  secretion, factors affecting *2071*
  in sickle-cell disease 715
  sodium reabsorption 264
  urinary acidification 767
  see also hyperaldosteronism; mineralocorticoids; renin–angiotensin–aldosterone
aldosteronism, primary/secondary, hypertension, metabolic alkalosis and 948
Aleutian disease, infection-associated glomerulonephritis 462
Aleutian disease virus 584
alfacalcidol, action, chronic renal failure 1770
alfentanyl, handling in kidney disease 185
alginates, handling in kidney disease 189
alkalaemia,
  defined 919
  following treatment for metabolic acidosis 955
  metabolic acidosis, K^+ regulation 896
  potassium, transcellular shifts 902
  respiratory acidosis, K^+ regulation 896
  see also metabolic alkalosis; respiratory alkalosis
alkali gain,
  exogenous alkali gain, metabolic alkalosis 943–4
  see also bicarbonate
alkali load, elimination, metabolic alkalosis 943
alkali replacement therapy, hypokalaemia 903

Index

alkaline phosphatase,
  aluminium-related osteopathy 1370
  bone isoenzyme, marker of hyperparathyroid bone disease 1370, 1384
  brush border membrane 1784
  calcitonin effects on calcium *1768*
  elevated, intestinal isoenzyme 1282
  gamma-glutamyl transferase levels 1370
  increased, marker of renal cell carcinoma 2242
  normal range, hypercalcaemia risk 1384
  osteitis fibrosa 1370, *1774*
  osteitis fibrosa cystica 1854
  osteomalacia *1774*
  transplantation, normo/hypercalcaemic recipients *1775*
  ureteric obstruction, animal models 2013
  in urine 18
alkalosis, *see* metabolic alkalosis; respiratory alkalosis
1-alkyl-2-acetyl-*sn*-glycero-3-phosphorylcholine, *see* platelet activating factor
alkylating agents,
  minimal change nephrotic syndrome 316–18
  adults 328
  neoplasia 2320
  see also chlorambucil; cyclophosphamide; mechlorethamine
allergy,
  in Henoch–Schönlein purpura aetiology 598
  minimal change nephrotic syndrome, aetiology *299*
allogeneic reactions, autoimmune glomerulonephritis 249
allografts,
  mouse models 1558
  privileged sites and tissues 1545
  rejection, *see* rejection of allograft
allopurinol,
  benefits *vs* risks 834
  in chronic renal failure 834
    gout familial syndrome, treatment 833
  dosage 1185
    reduction with creatinine concentrations 832
  drug-induced interstitial nephritis 1090
  drug-induced lithiasis 167
  gouty nephropathy 2294
  hypersensitivity reactions 1090
  interactions with diuretics 208
  prophylaxis
    acute uric acid nephropathy 832
    recurrent gouty arthritis 835
  renal stone disease 1853
  side-effects, caused by oxpurinol reabsorption in low protein diet 178
  toxicity 834
  treated/non-treated patients, progression to renal failure *835*
  urate concentration changes, fractional urate clearance 828
  xanthine oxidase inhibition 834
xanthinuria 1866

allotypes, defined 562
alpha-1-adrenoceptor agonists, *see* (α-1-)adrenoceptor agonists
alpha-1-antitrypsin deficiency, *see* (α-1-)antitrypsin deficiency
alphafetoprotein, congenital nephrotic syndrome of Finnish type 2219
Alport's syndrome,
  assessment 1155
  classification 2160, 2203–4
  clinical features
    hearing defect 2197
    ocular defects 2197
    other abnormalities 2197–8
  crescentic glomerulonephritis 422
  diagnosis 1155
  differential diagnosis
    benign familial haematuria 2203
    non-progressive hereditary nephritis 2202–3
    progressive hereditary nephritis without deafness 2202–3
    thin-basement membrane nephropathy 2203
  end-stage renal failure 1233
    proportional distribution by age at commencement *1229*
  genetics 2197–8, 2201–2
    counselling 2202
    molecular genetics 2160–1
    X-chromosome markers, DXS3 locus 2160
    X-linkage 2201–2
  glomerular changes, schematic representation 151
  Goodpasture antigen 2200–1
  hereditary nephritis, and Goodpasture's disease 439, 452
  juvenile *vs* adult type 2198
  pathology and pathogenesis 2198–201
  renal biopsy
    EM findings 150
    indications 145
  renal hypertension and *2048*
  renal transplantation 1515
    and replacement therapy 2201, 2204
Alport's syndrome-like hereditary nephritis 2160
alprenolol, properties, in relation to use in impaired renal function *2053*
altitude acclimatization, erythropoietin and *1347*
aluminium compounds **1446–8**
  absorption, citrate ingestion 1378
  antacids, with exchange resins 944–5
  in dialysate, quality standard 1422, 1446–8
  dietary intake, drinking water 1368–9
  dihydroxyaluminium sulphate, effect on renal scarring *1205*
  haemodialysis
    anaemia in 1351
    liver changes 1282
  parathormone effects 1760
  plasma binding 1369
  toxicity, *see* aluminium intoxication

aluminium hydroxide,
  calcium carbonate *vs*, phosphate binding 1305
  compliance aspects 1471
  dose
    maximum 1383
    and side-effects *1382*
  drug-induced lithiasis 168
  effect on renal scarring *1205*
  handling in kidney disease 189
  phosphate binding, drug binding 176
  replacement by phosphate antacid 1022
aluminium intoxication **1368–78**
  Al levels, risk of bone disease 1371
  aluminium-induced bone disease **1446, 1447**
  assessment 1370–1
  bone disease, CAPD 1483–4
  bone histology 1369
  causes, citrate ingestion 1378
  children 1612
  clinical features 1377–8
  desferrioxamine
    infusion test 1371
    removal, parathormone effects 1760
    treatment 1774
  dialysate 1401
  dialysis encephalopathy 2329
    reasons for diagnosis 1401
  dialysis-induced dementia 1623
  dietary intake 1368–9
  differential diagnosis 1774
  encephalopathy, children 1613
  encephalopathy and 1400
  haemodialysis patients 1446
  hypercalcaemia and 1369, 1774
  neurological threat in infancy 1614
  parathormone resistance 1369
  parathyroid tissue, electron microscopic changes 1370
  signs 1446
  skeletal effects, direct 1369
  sources 1446–7
  treatment 1447–8
  uraemic syndrome 1244
  see also aluminium overload; aluminium-related osteopathy
aluminium overload,
  aluminium-related osteopathy, direct measurement 1376
  animal models 1369
  bone histology 1374–5
    Aluminon stain 1374–5, 1376
    cortical *vs* trabecular changes 1369
    electron microscopic changes 1369
  children 1612
  drinking water 1368–9
  haemodialysis patients 1351
  hypercalcaemia in dialysis patients 1773
  hyperparathyroidism 1368–9
  left ventricular function 1266
  phosphate binders 1368
aluminium-containing compounds, handling in kidney disease 189
aluminium-induced bezoars and faecalomas 1285

aluminium-related osteopathy **1446, 1447**
  adynamic bone disease 1369, 1376
  alkaline phosphatase 1370
  bone histology 1379
  clinical features 1377–8
  diagnostic procedures 1370–6
  differential diagnosis 1370, *1380*
  following renal transplantation 1585
  laboratory tests *1371*
  management 1385
  radiography 1371–3
  serum biochemistry *1371*
  see also aluminium intoxication; osteitis fibrosa; osteomalacia
alveolar haemorrhage,
  in microscopic polyarteritis 624, 625
  in Wegener's granulomatosis 615
alveolar ventilation,
  response to metabolic acidosis 921–2
  response to metabolic alkalosis 942
Alzheimer's disease, amyloid deposits 548, 550
Amanita, poisoning
  acute renal failure and 1011, 1136
  renal and hepatic failure 2310
ambulatory continuous ECG, *see* ECG, Holter monitoring
amelogenesis imperfecta 1892
amikacin,
  CAPD peritonitis regimens *1489*
  doses and therapeutic plasma concentrations of aminoglycosides, handling in kidney disease *182*
  nephrotoxicity 1010
amiloride,
  Bartter's syndrome 787
  contraindications in hypoaldosteronism 910
  creatinine clearance, interference 32
  lithium
    clearance, sensitivity 41
    polyuria 856
    therapy, prevention of tubular uptake 796
  magnesium excretion 1808
  nephrotic syndrome 270
  potassium maintenance therapy 904
  renal tubular acidosis *938*
  sodium reabsorption blocking 200
  structural formula *199*
  urinary acidification 772
amines, heterocyclic, decreased drug protein binding 1245
amino acids,
  absorption, uraemic animal models 1281
  acute renal failure, prevention of 1018
  chronic renal failure
    insulin resistance 1259
    levels in serum and tissue 1258–9
    patterns (plasma and intracellular) *1258*
  clearance rates
    Fanconi syndrome *724*

Index

amino acids (*cont.*):
  clearance rates (*cont.*):
    normal children *vs* Fanconi syndrome 725
  dibasic
    Lowe syndrome 732
    urinary excretion 749
  dicarboxylic, transport disorders 751
  diet, acid and base production 932–3, *933*
  dietary intake, acute renal failure *1020*
  excretion, pregnancy and 1916
  insulin-stimulated exchange, uraemia 1253
  loss in haemodialysis 1260
  metabolism, uraemic liver 1283
  neutral, transport disorders 750–1
  peritoneal dialysis solution 1481
  transport, entry and exit steps 746
  transporters 746
  tubular reabsorption, normal 745–6
    children 745–6
    mean values, children with Fanconi syndrome *724*
    neonates 745
aminoacidurias **745–51**
  causes
    non-renal (overflow) *747*
    renal *747*
  classification *747*
  dibasic 749
  Fanconi syndrome 724, 724–5
  hyperaminoaciduria defined 724
  hypercalcaemia causing 1885
  mechanisms 746
  renal and non-renal (overflow) 747
  see also hyperaminoaciduria;
ε-aminocaproic acid (EACA),
  antifibrinolysis, following renal biopsy 142
  in sickle-cell disease 716
aminoglycosides,
  N-acetyl-β-D-glucosaminidase excretion 18
  acute on chronic renal failure 1186
  animal models of nephrotoxicity 994
  CAPD peritonitis regimens *1489*
  chronic renal failure, accumulation 1185
  cyclosporin therapy
    interactions *1563*
    nephrotoxicity 1573
  drug-induced acute toxic tubular nephropathy 163
  elderly patients, inappropriate use 67
  handling in kidney disease 182
  interactions with diuretics 208
  lysosomal permeability 169
  magnesium wasting and 1810
  membrane permeability 168
  nephrosis, experimental, proteinuria 269
  nephrotoxicity 994, 1010, 1573
    in preterm neonates 54
    neurological side-effects 2331
  non-depolarizing neuromuscular blocking agents 186
  potassium depletion 906
  renal complications of 2343

aminoglycosides (*cont.*):
  see also *N*-acetyl-β-D-glucosaminidase; amikacin; gentamicin; kanamycin; neomycin; netilmycin; streptomycin; tobramycin
aminohydroxypropylidene diphosphonate, in hypercalcaemia of multiple myeloma 567
δ-aminolaevulinate dehydratase activity, Balkan nephropathy 860
amiodarone,
  antiarrhythmic drugs, ventricular arrhythmias 187
  atrial fibrillation 187
  pulmonary complications of 2339
amitryptiline, prostatitis 1690
ammonia (ammonium),
  aldosterone effects 767
  excretion
    daily urinary *923*
    hyperkalaemic renal tubular acidosis 776
    pH and *923*
    renal tubular acidosis 937
    urinary pH and 774
  formation
    drug malabsorption 176
    hydrogen ion secretion 767
  and renal potassium depletion 901
  homeostasis 899
  tubular reabsorption, elderly patients 60–1
ammoniagenesis, tubular, adaptive 1205
ammonium chloride acidification test 774, 1851
  chronic renal failure 44
  renal tubular acidosis 43
ammonium exchange resins, urinary appearance 22
ammonium hydrogen urate stone disease, infected, lower tract 1830, 1831–2
amoxycillin,
  drug-induced interstitial nephritis 1088
  urinary tract infection 1686, 1687
amoxycillin + clavulanic acid, dose adjustment *181*
amphetamines,
  drug-induced renal vasculitis 167
  renal excretion, pK_A and pK_B 178
amphotericin,
  bladder irrigation 1748
  CAPD peritonitis regimens *1489*
  chronic renal failure, accumulation 1185
  fungal infections 1748
  haemofiltration/haemodialysis 180
  handling in renal impairment 184
  magnesium wasting and 1810
  membrane permeability 168
  nephrotoxicity 994, 1010, 1748
  permeability defect, renal tubular acidosis 938
  renal complications of 2343
ampicillin,
  CAPD peritonitis regimens *1489*
  dose adjustment *181*

ampicillin (*cont.*):
  drug-induced interstitial nephritis 1088
  drug-induced nephropathy 163
  hypokalaemia and metabolic alkalosis 947
  Listeria infections, pregnancy and 1339
  pregnancy 1929
  urinary appearance 22
  urinary tract infection 1686, 1687, 1693
amylase,
  chronic renal failure changes, pancreatic isoenzyme 1283
  properties, effects on kidney, significance of elevation *229*
amyloid,
  AA (amyloid protein A) 545
    AL amyloid differentiation 553
    amyloidogenesis 572
    composition/structure 547–8, 680
    precursor, *see* amyloid, SAA (serum amyloid A)
  AL (amyloid protein, light chain derived) 545, 571
    composition/structure 546–7, 571
    lambda chain 547, 572
    molecular weights 547, 571
    proteolysis 572
  biochemical investigations **546–8**
  composition/structure 545, **546–8**, 571
  endocrine-related proteins (AE) 548
  fibrils in urine, EM microscopy 19
  β-microglobulin amyloid 546, 548, 571
  P-component (protein AP) 548, 1153
  β-proteins 548
  proteins related to prealbumin 548
  radiological tests 72
  SAA (serum amyloid A) 545
    conditions with elevation 556
    in familial Mediterranean fever 554
    measurement 556
    production 547–8
    in sarcoidosis 581
    structure 547–8, 680
  SAP (serum amyloid P-component) 548
  staining, thioflavin T 391
  subepithelial deposits ('spikes') 552, 553, 573
amyloid nephropathy, familial, features and associated abnormalities *2036*
amyloid-associated arthropathy, complications of renal disease 2297–8
amyloidosis **545–61**
  AA 545, 546, 547
    cancer and 2317
    clinical features 549, 680
    Crohn's disease 2310
    diagnosis 554, 680
    prognosis 557, 680
    treatment 556, 680
  age/sex incidence 548, *548*

amyloidosis (*cont.*):
  AL **571–3**, 1152
    amyloidogenesis, phases 572
    cancer and 2317
    clinical features 549, 550, 572
    diagnosis 554, 556
    immunoglobulin abnormalities 569, 571
    morphological/pathological features 552, 572–3
    prognosis 557–8, 573
    treatment and outcome 556, 573
  amyloid hand, haemodialysis, long-term patients 1442
  cancer, secondary to 2317
  cardiac 549, 558
  cause of death *558*, 680
  chronic renal failure in elderly patients 1623
  classification **546**, *546*, *547*, 572
    typical/atypical distributions *546*
  clinical manifestations **548–51**, 572, 680
    organ involvement 549–51
    presenting symptoms 549, *550*
    rheumatic symptoms 550, 679–80
  see also amyloidosis, renal
  definition **546**
  diagnosis **554–5**, 680, 1152–3, 1380
  differential diagnosis, mesangiocapillary glomerulonephritis 392
  diseases associated 548–9, *549*, 571
  end-stage renal failure, proportional distribution by age at commencement *1229*
  ethnic factors 9
  familial Mediterranean fever 1153, 2300
  Fanconi syndrome 734
  febrile urticaria and deafness with (Muckle-Wells syndrome) 553, 554
  gastrointestinal bleeding, haemodialysis patients 1284
  glomerulopathies
    acute renal failure in liver disease 1104
    in the tropics 488, 491–2
  haemodialysis-associated (β₂-microglobulin) 548, 1380
  hereditary 548, **553–4**, 572
    diagnosis 554
    groups 553
  histologic staining 146
  historical aspects **545–6**
  idiopathic ('primary') 545, 549
  see also amyloidosis, AL
  laboratory investigations **555–6**
    renal function 555
  leprosy aetiology 491
  localized hereditary/non-hereditary 547
  lymphoma 482
  malignancy-associated 481
  β-2-microglobulin deposition long-term haemodialyis 1282
  see also β-2-microglobulin amyloidosis
  myeloma-associated 549, 571–3, 572, 679
  see also amyloidosis, AL
  paraplegia 2276

amyloidosis (cont.):
  predictive and prognostic value 145
  primary, see amyloidosis, AL
  prognosis **557–8**, 573, 680
  renal
    complications of 2294
    electron microscopy 553, 573
    gross anatomy 552
    histological staining methods 545, 553
    immunofluorescence 553, 555
    leprosy 461
    malignancy-associated glomerulonephritis 478
    microscopy 552–3, 572
    nodules, diabetic nephropathy vs. 524
    pathology **552–3**, 572–3
    sarcoidosis associated 580–1
    signs/symptoms **551–2**
  renal biopsy contraindication 143
  renal cell carcinoma 478, 2242
  renal replacement therapy **558–60**, 573
    acceptance rate 558
    dialysis 558–9
    recurrence in allografts 559–60
    survival after 558, 559
    renal transplantation 559–60, 1515
    recurrence following 1574
  schistosomiasis 1734, 1737, 1739
  secondary 545, 546, 547, 571, 680, 1153
    rheumatoid arthritis and, see rheumatoid arthritis
    see also amyloidosis, AA
  skin lesions 2301
  in SLE 656
  spondyloarthropathy 2297
  systemic
    hereditary/non-hereditary 547
    monoclonal gammopathy associated 549
  tissue biopsy role 554, 555
  treatment **556–7**, 573, 680, 2294
  tuberculosis aetiology 488
  see also β-2-microglobulin amyloidosis
amyloidosis–fibrillosis, renal biopsy, EM findings 150
AN69S haemodiafiltration, comparisons 1247
AN69S haemodialysis, comparisons 1247
anaemia,
  acute renal failure 973–4, 1041
  antiglomerular basement membrane disease 440–1
  CAPD 1482
  children
    chronic renal failure 1609–10
    treatment with erythropoietin 1610, 1611
  chronic
    acute uric acid nephropathy 832
    see also anaemia, chronic renal failure (below)
  chronic haemorrhagic, analgesic nephropathy 813
  haemodialysis patients, pregnancy 1948
  haemolytic anaemia, acute renal failure 2341

anaemia (cont.):
  management, haemodialysis 1353
  microangiopathic, acute renal failure **1041–57**, 2341
  normochromic–normocytic type, renal carcinoma 2241
  in pregnancy 531
  renal failure, erythropoietin, recombinant human 1518
  transplantation patients, causes 2345
anaemia, chronic renal failure **1344–56**
  animal models, treatment with erythropoietin 1217
  assessment 1156–7, 1167
  bleeding diathesis 1352
  cardiovascular effects 1353
  characteristics 1348–9
  children 1356
  clinical aspects 1351–6
  clinical consequences 1352
  erythropoietin increase 1347
  ferrokinetic studies, normal vs haemodialysis 1346
  haemodialysis
    management 1353
    pregnancy and 1338
  haemoglobin, dialysis patients 1352
  haemolytic, ferrokinetic studies 1346
  left ventricular hypertrophy and, end-stage renal disease 1265
  management 1353–6
    erythropoietin replacement therapy 1353–6
    steroids 1353
  pathogenesis 1348–50
  prevalence 1351
  prevention 1353
  uraemic syndrome 1238
anaesthetic agents,
  handling in kidney disease 185
  tubular injury, acute renal failure 1009, 1010
analgesic nephropathy **803–16**
  analgesic-associated diseases, relative risks 808
  animal models 809–10
  assessment 1150
  carcinoma and haematuria 1454
  carpal tunnel syndrome 813
  clinical course and prognosis 815
  clinical features 812–14
    extrarenal manifestations 813–14
    psychosomatic aspects 814
    renal manifestations 813
  complications
    cardiovascular 813
    gastrointestinal 813
    haematological 813–14
    skeletal 814
  defined 804
  dehydration 811
  diagnosis, EDTA registry 1231–3
  diagnosis/differential diagnosis 814–15
  epidemiology 804–7
  European vs English-speaking drug preferences 803–4
  genetic factors 811
  gout, urate clearance reduced 831
  historical summary 803–4

analgesic nephropathy (cont.):
  neoplastic complications of 2320
  nephrocalcinosis 1895
  pathology 811–12
    capillary sclerosis 811–12
    cortical parenchymal damage 812
    phases 1, 2 and 3 811
    renal papillary necrosis 812
  pathophysiology 809–11
  prevalence
    autopsy studies 807
    contries with renal replacement therapy 806–5
    various countries 805
  prevention 816
  radiological signs 814
  renal failure, allopurinol/non-allopurinol treated patients 835
  renal papillary necrosis, differential diagnosis 815
  in rheumatoid arthritis **684**
  sex of patient 811, 812
  smoking 2267
  therapy 815–16
  tumour type 813, 815, 2267
  ulceration and 2310
  ultrasonography 815
analgesics,
  analgesic-associated disease, relative risks 8
  and carcinoma 2238, 2267
  chronic consumption, defined 804
  drug-related interstitial nephritis 1089
  European vs English-speaking subjects, drug preferences 803–4
  kidney disease, see analgesic nephropathy
  prostaglandin hydroperoxidase-mediated co-oxidative action 810
  prostaglandin synthesis, inhibition 811
  see also analgesic nephropathy
anaphylactoid purpura, see Henoch–Schönlein purpura
anaphylotoxin C5a, haemodialysis increase 1432
anastomotic site, see arteriovenous fistula
ANCA, see anti-neutrophil cytoplasmic antibodies
ancrod,
  in lupus nephritis 662
  nephrotoxic nephritis, defibrination 257
Anderson–Fabry disease (α-galactosidase A deficiency) 2210, 2300
  assessment 1155
  carrier state detection 2208
  clinical features 2206–7
  diagnosis 2208
  enzyme replacement 2209
  management 2209
  metabolic defect 2207
  molecular genetics 2208–9
  prenatal exclusion, male fetus 2208
  presentation 2208
  renal involvement 2207–8
  renal transplantation 1513, 2209–10

Andrade syndrome, features and associated abnormalities 2036
androgen receptors, hormone therapy, renal cell carcinoma 2248
androgens,
  carcinogenesis in uraemic anaemia 2320
  erythropoietin production 1353
  in lupus nephritis 662
  women, chronic renal failure 1615
aneurysms,
  aortic, renal effects 2335
  cirsoid aneurysm, renal arteriography 113
  following renal biopsy 143
  intracranial aneurysm 2177–8
  in polyarteritis nodosa 628
  renal artery stenosis 2097–8
  vascular access 1414–15
angiitis,
  hypersensitivity, see hypersensitivity vasculitis
  see also vasculitis
angina, elderly patients on haemodialysis 1625
angiodysplasia, gastrointestinal bleeding, incidence 1283–4
angiography, see renal angiography
angiokeratoma corporis diffusum, Fabry's disease 2300
angiokeratomata,
  Anderson–Fabry disease (α-galactosidase A deficiency) 2207
  differential diagnosis 2207
angiolymphoid hyperplasia, and eosinophilia (Kimura's disease) 2281
angioma, renal arteriography 113
angiomyolipoma,
  computed tomography scanning 103
  histological appearance 2243
  pathology 2243
  treatment 2243
  tuberous sclerosis 2181
angioplasty, see percutaneous transluminal renal angioplasty
angiotensin converting-enzyme, acute renal failure in diabetics 1638
antisera, in sarcoid granulomatous interstitial nephritis 577
  assay 2070
  in granulomatous interstitial nephritis 578
angiotensin converting-enzyme inhibitors,
  action, renal function in diabetic/nondiabetic nephropathy 1214
  acute renal failure 1016
    cause of 1078–9
    ischaemic renal disease 1078
    animal models 1214
    haemodynamic and structural glomerular parameters 1214
  Bartter's syndrome 787
  blood pressure
    pregnancy 1920
    volume expansion 272
  cardiovascular disease 1177–8
  caution in use 2054
  children, dosage 2145, 2148

Index

angiotensin converting-enzyme inhibitors (*cont.*):
chronic renal failure, toxicity 1185
compared with other drugs 1176
congestive heart failure 272
in diabetes mellitus 520, 527
dialysis patients 1273
dose, new maximum 2054
erythropoietin production effects 1350
glomerular filtration
 in the ischaemic kidney 2106
 reduction 180
glomerular scarring, prevention 1195
hyperkalaemia-inducing 909
hypertension 1177–8
 accelerated 2135
 cough in 2339
 in diabetes with chronic renal failure 1646
 end-stage renal failure 2121
 handling in kidney disease 188
 hyponatraemia 879
 insulin response, enhanced 1648
 interactions with diuretics 208
 left ventricular hypertrophy, regression 1273
 liver effects 2315
 neonates 1116
 nephrotic syndrome, no ARF 1039
 pregnancy
  blood pressure 1920
  contraindications 1962
  pregnancy-induced hypertension 1963
 prerenal failure **986–7**
 progressive chronic renal failure 1214
 proteinuria 1214
  nephrotic syndrome 271
  renovascular hypertension 321
 renal artery stenosis, prerenal failure **986–7**
 renal failure 159
 renal hypertension, treatment 2053–5
 renal transplantation 1531
  hypertension treatment 1536, 1598
  renal artery stenosis 1527–8
  vascular complications 1526
 renin secretion and *2071*
 renin-dependent hypertension 1273
 renin–angiotensin system and 2064
 in scleroderma renal crisis 676
 side-effects 2054–5
 thiazide diuretics and 212
angiotensin II,
 arteriolar myocytes of afferent arterioles
  effects 26
  interactions 25, 27
 bicarbonate reabsorption 765
 diabetes, pressor dose 1646
 exogenous, pregnancy-induced hypertension 1965
 glomerular filtration in the ischaemic kidney 2106
 hyponatraemic hypertensive syndrome 2131
 inhibition, in scleroderma renal crisis 676

angiotensin II (*cont.*):
 inhibition of prostaglandin synthesis 987
 mesangial cells 136
 pressor effects, postulation in Bartter's syndrome 785
 pressor response
  angiotensin sensitivity test 1963
  pregnancy-induced hypertension 1963
 receptor agonist, saralasin test 2106
 receptor density, reduced in diabetes mellitus 513
 tachyphylaxis, postulation in Bartter's syndrome 787
 see also renin–angiotensin–aldosterone system
angiotensinogen, assay 2070
angiotensins,
 angiotensin II (octapeptide) 984
 angiotensin III (heptapeptide) 984
 antidiuretic hormone and thirst stimulation 262
 I, II and III, assay 2070
 reduced renal blood flow, NSAIDs 821
 sensitivity test 1963
 thirst sensation, ageing changes 65
 thirst stimulation 262
 urate clearance reduced in normal subjects 831
aniline, poisoning 1011
animal models,
 acute uric acid nephropathy 832
 anaemia, left ventricular function and diameter 1265
 analgesics *810*
 Bence–Jones protein nephrotoxicity 564
 diabetic nephropathy
  early, glomerular hyperfiltration 512, 513
  hyperglycaemic role in pathogenesis 517
  renal hypertrophy 514
 exercise training, progressive chronic renal failure 1217
 fractional urate clearance, extracellular volume expansion effects 828
 Goldblatt rat models of hypertension 1595, 2059–65, 2075–9, 2101
 heart weight, pharmacological interventions *1266*
 hyperuricaemia 825
 interstitial myocardial fibrosis, end-stage renal disease 1266–7
 left ventricle, micromorphometric analysis *1266*
 murine lupus nephritis 251
 myeloma cast nephropathy pathogenesis 565
 progressive chronic renal failure **1192–217**
  age-related factors 1208
  dietary factors 1209–12
  genetic factors 1206–7
  glomerular scarring 1192–203
  immune factors 1208
  metabolic and hormonal factors 1208–9

animal models (*cont.*):
 progressive chronic renal failure (*cont.*):
  pharmacological interventions 1212–17
  sex-related factors 1207–8
  tubulointerstitial scarring 1203–6
 SLE 639–40, 642
 tubular reabsorption, urate/uric acid 827
 vasculitis pathogenesis **583–5**
  granulomatous 585
  mononuclear 584–5
  see also serum sickness
 see also individual animal models
animal toxins,
 nephrotoxicity 1125
 see also specific animals
anion gap,
 differential diagnosis 778
 venous blood, normal values *917*
anion gap in metabolic acidosis,
 anion gap increased 926–34
 anion gap normal 934–42
 classification *925*
 genesis and diagnostic implications 925–6
 ionic composition of plasma *925*
 see also lactic acidosis
anion gap in metabolic alkalosis, increase 943
anion gap in renal tubular acidosis, diagnosis 778
anions, metabolic alkalosis, poorly reabsorbable, administration 947
aniridia,
 candidate gene 2255
 Wilms' tumour 2254
aniridia–genitourinary–retardation triad, Wilms' tumour risk 2254–5
ankylosing spondylitis, IgA nephropathy 361
antacids, aluminium, with exchange resins 944–5
anti-neutrophil cytoplasm antibodies, Wegener's granulomatosis 2338
antiarrhythmic drugs,
 cardiac drugs 187–8
 haemodialysis, hypokalaemia 1625
 ventricular arrhythmias 187
antibasement membrane antibodies,
 pathogenic role 169–70
 see also antiglomerular basement membrane antibodies
antibiotics,
 CAPD peritonitis regimens *1489*
 drug-induced acute toxic tubular nephropathy 166
 lipophilic, intracystic penetration 2173
 nephrotoxicity 994, **1010–11**
 neurological side-effects 2331
 renal transplantation 1517–18
 Staphylococcus spp, haemodialysis patients 1453
 urinary tract infection 1686–7, 1693–4
  children 1713–15
antibodies,
 deposition, glomerular injury 639

antibodies (*cont.*):
 in mononuclear vasculitis pathogenesis *590*
 in necrotizing polyarteritides pathogenesis 590, *591*
 and rejection of allograft, *see* rejection of allograft, acute/chronic rejection
 schistosomiasis 1732
 see also immunoglobulins; *individual antibodies*
antibody test, hepatitis C 1168
antibody-coated bacteria test,
 bacteriuria 1664
 children 1664
 urinary tract infection 1682
antibody-dependent cell-mediated cytotoxicity, antiglomerular basement membrane-mediated glomerulonephritis 256
antibody-mediated glomerulonephritis **251–4**
 circulating immune complexes 251–2
 free circulating antibodies 253–4
 planted glomerular antigens 254
 structural glomerular basement membrane antigens 253
 structural glomerular cellular antigens 253–4
 subepithelial deposits *254*
 immunization with glomerular basement membrane structural antigens 253
 progression of chronic renal failure 1208
 see also IgA nephropathy
antibody-mediated vasculitis 589–90
anticancer drugs,
 drug-induced acute toxic tubular nephropathy 165–6
 drug-induced nephropathy 162
anticardiolipin antibodies 648
 thrombophlebitis following contrast media injection 77
anti-Clq antibodies 589
anticoagulants,
 animal models 1215–16
 defibrotide 1055
 haemodialysis 2344–5
 extracorporeal blood circuit and clotting system 1427
 haemolytic–uraemic syndrome 1055
 handling in kidney disease 190
 heparin 1427–8
 high risk patients 1428
 in lupus nephritis 662
 mesangiocapillary glomerulonephritis 401
 in microscopic polyarteritis 626
 renal effects 2335
 renal transplantation 1518
anticonvulsants,
 cytochrome P-450 187
 eclampsia effects 1973–4
 handling in kidney disease 186
 neurological side-effects 2331
 vitamin D effects 1763
 vitamin D metabolism changes 1282
antidepressants,
 contraindicated, elderly patients 65

Index

antidepressants (*cont.*):
  elderly patients, inappropriate use 67
  handling in kidney disease 186
  volume of distribution 180
antidiarrhoeal agents, loperamide 189
antidiuretic hormone, *see* arginine vasopressin
anti-DNA antibodies,
  graft versus host disease 249
  in SLE/SLE models **636–8**, 647
    binding to histones/nucleosomes 637
    clinical activity correlation 652
    cross-reactions 637
    DNA receptor defect 637
    DNA size 638
    double-stranded DNA 636, 647
      binding via nucleosomes 637
    in focal proliferative glomerulonephritis 653
    genes coding, sequences 639
    IgG 637–8, 639, 642
    IgM 638, 639
    immunofluorescent assay 647
    monoclonal 637, 639
    pathogenicity 637–8
    populations (free and DNA–histone complexed) 637
    recurrent idiotypes 638
    single-stranded DNA 637, 638, 639, 647
    specificities 636, 647, *648*
  see also antinuclear antibodies
antiemetics, metoclopramide 189
antifibrinolysis, aminocaproic acid 142
antifibrinolytic agents, gastrointestinal bleeding 1284
antifungal agents,
  CAPD peritonitis regimens *1489*
  handling in renal impairment 184
  therapy 1748–9
antigen presentation,
  and antigen processing 240–1
  endothelial cells of glomerular capillaries, MHC II 132
  immune response to allograft 1544–5
  see also major histocompatibility complex (MHC)
antigens,
  autoantibody specificity in SLE 647
  in immune complexes in leucocytoclastic vasculitis 587
antigen–antibody complexes, *see* immune complexes
antiglomerular basement membrane,
  antibodies 442, 446, 618
  cell-mediated glomerulonephritis 255
  collagen IV 239
  diagnosis of Goodpasture's disease 442–3
  immunologically-mediated mercury-induced 844–5
  pathogenic role 169–70, 446

antiglomerular basement membrane-mediated glomerulonephritis 247, 1514
antibody-dependent cell-mediated cytotoxicity 256
in renal transplantation 1514, 1531
see also autoimmune glomerulonephritis
antihelminthics, handling in kidney disease 185
antihistamines, handling in kidney disease 187
antihypertensive drugs,
  accelerated hypertension 2129
  action on renin–angiotensin–aldosterone system 2071–2
  adverse effects 1178
  animal models 1212–14
    effect on glomerular scarring *1213*
    renal ablation, effects *1195*
  children 2147–8
    dosage *2148*
  classes 1176
  diabetes, treatment 527, 1647–8
  diuretics and, consequences for therapy 202
  haemodialysis patients, recommendations *1463*
  intrarenal effects 1176
  lipoprotein changes and 211
  liver effects 2315
  non-steroidal anti-inflammatory drugs 823
  platelet effects and therapy 1968
  pregnancy-induced hypertension therapy 1969
  progressive renal disease and 2085–6
  renin secretion and *2071*
  side-effects, neurological involvement 2331
  sodium retention 273
  see also specific substances
antihypertensive lipids, interstitial cells, antihypertensive polar/neutral renomedullary lipid 2064–5
antiidiotype antibodies, mercury-induced autoimmune glomerulonephritis 250
anti-inflammatory agents,
  action 190
  progressive chronic renal failure 1217
anti-La antibodies, *see* La antigen (SSB)
antiliver/kidney microsome antibodies, cytochrome P-450 isoenzyme 2315
antilymphocyte antibodies 641
antilymphocyte preparations, rejection of allograft 1571
antimalarial therapy, prophylaxis, neglect 10
antimicrobials, handling in kidney disease *181*
antimitotic immunosuppressive drugs, mesangiocapillary glomerulonephritis 401
antimyeloperoxidase antibodies 592, 618, 625

antineoplastic treatment, renal complications of 2318–19
antineutrophil cytoplasmic antibodies (ANCA) 239–40, 1150
  antigenic targets 1066
  in Churg–Strauss syndrome 629
  cytoplasmic (C-ANCA) 592, 618
  diagnosis of Goodpasture's disease 443
  in microscopic polyarteritis 624
  perinuclear (P-ANCA) 592, 618, 624
  in polyarteritis nodosa 628
  in Wegener's granulomatosis 592, 600, 617–18
  therapy monitoring 621
antinuclear antibodies,
  autoimmunity and 478
  in scleroderma 675
  in Sjögren's syndrome 695
  in SLE 647
  see also anti-DNA antibodies
antiphospholipid antibodies 648
  fetal loss 5
α-2-antiplasmin, nephrotic syndrome *282*
antiplatelet therapy,
  haemolytic–uraemic syndrome 1055
  IgA nephropathy, primary *358*
  in lupus nephritis 662
  mesangiocapillary glomerulonephritis 401
  in microscopic polyarteritis 626
  pregnancy-induced hypertension 1966–8
antipneumococcal vaccines, nephrotic syndrome 278
antiprotozoal agents, handling in kidney disease 185
antipyrine, analgesic-associated kidney disease 803
anti-Ro antibodies, *see* Ro antigen (SSA)
antisalivary gland antibodies, in Sjögren's syndrome 695
antithrombin III, nephrotic syndrome *281*, 282
antithymin-mediated glomerulonephritis 257
antithymocyte globulin,
  rejection of allograft 1571
  treatment period 1556–7
α-1-antitrypsin, nephrotic syndrome *282*
α-1-antitrypsin deficiency,
  crescentic glomerulonephritis 422
  membranoproliferative glomerulonephritis 2216
antitubular basement membrane disease,
  pathogenesis 1092–3
  target antigen 3M–1 1092
  see also interstitial nephritis
antitumour antibodies, glomerular deposits 476
antitumour drugs, *see* anticancer drugs
antiulcer drugs,
  omeprazole 189
  proton pump inhibitors 189
antiurease drugs, infected stone disease 1864
antiviral agents, handling in renal impairment 184

anuria, phenylbutazone, non-steroidal anti-inflammatory drugs 823
anxiety–hyperventilation syndrome 955
aorta,
  coarctation, children 2143
  fibrous plaque, *see* retroperitoneal fibrosis
  retroperitoneal fibrosis, computed tomography scanning 106
  ultrasonography
    para-aortic lymphadenopathy 92
    retroperitoneal fibrosis 92
aortic aneurysm, renal effects 2335
aortic valves, *see* heart valves
aortorenal bypass,
  intravenous digital angiography 113
  renal artery stenosis 2108
APD, new therapy, hyperparathyroidism 1856
apical (luminal, brush border) membrane, *see* brush border, luminal
Apodemus, Hantavirus disease carrier 1143
apolipoproteins, *see* lipoproteins
arabinose, excretion, pregnancy 1916
arachidonic acid, derivatives 256–7
arachidonic acid pathway, prostaglandin synthesis 820
arcuate arteries,
  arteriography 112
  lesions in scleroderma 675
Argentinian haemorrhagic fever *1045*
arginine,
  diet, acid and base production 932–3, *933*
  excretion, urate excretion 830
arginine HCl,
  hyperkalaemia 907
  loading test 775
arginine vasopressin,
  antidiuretic hormone control of water transport 791
  arteriolar myocytes of afferent arterioles
    effects *26*
    interactions 25, 27
  AVP-dependent water channels 40
  blood osmolality and 870
  cellular actions 791
  DDAVP treatment 886
  diabetes insipidus, nephrogenic, lithium toxicity 853
  differential diagnosis of polyuria 796
    direct / indirect tests *798*
    direct test of urine osmolality 797
    indirect test of urine osmolality 796
  extrarenal actions 793–4
    cAMP levels 794
    factor VIIIc responses 793–4
    von Willebrand factor 793
  failure of central release 41
  immunoreactive, concentration *vs* osmolality 43

arginine vasopressin (*cont.*):
  inappropriate secretion 262
    criteria *878*
    diuretics 207, 879
    glucocorticoid deficiency 879
    hypothyroidism 879
    hypouricaemia 830, 832
    postoperative state 878–9
    potassium deficiency 879
    tumours 878
  kidney(s) and 870–2
  malignant hypertension and 2132
  medullary perfusion pressure 26
  mesangial cells 136
  nephrotic syndrome 267
  nicotine-mediated release 874
  non-osmotic regulation 870
  non-osmotic release 41
  normal function 790–1
  normal V_1 receptor responses 794
  osmotic and nonosmotic
    stimulation 790–1
  osmotic regulation 869–70
  osmotic thresholds in pregnancy
    1917
  perfusion pressure 26
  plasma osmolality and *791*, 885
  potassium homeostasis 899
  pregnancy
    clearance rate 1918
    renal vasodilatation 1915
  prohormone 869
  reduced renal blood flow,
    NSAIDs 821
  renal vascular control, mediation
    984–5
  resistant urinary concentrating
    defects, *see* diabetes insipidus,
    nephrogenic
  secretion defects, arginine
    vasopressin responses *798*
  sodium delivery, distal nephron
    function 41
  trial, differential diagnosis of
    partial diabetes insipidus 797
  urine osmolality and *791*, 886,
    1898
  urine osmolality relation, in sickle-
    cell disease 706, 707
  V_1 and V_2 renal receptors 791
  vasoconstriction of afferent
    arterioles 26
  vasopressinase 1918
  volume depletion response, young
    and elderly subjects 65
  volume homeostasis, animal
    gestational models 1921
  volume status and release 1918
  water deprivation tests in young
    and elderly subjects 65
aristolochic acid, Balkan
  nephropathy 860
Armanni–Ebstein lesion 523–4
arrhythmias,
  24-hour ECG recording, uraemic
    patient with cardiac problems
    1271–2
  acute, hyperkalaemia 187
  acute myocardial infarction,
    cardiac drugs 187
  diuretic-induced hypokalaemia
    209
  elderly patients on haemodialysis
    1625
  Holter monitoring, assessment
    1271–2

arrhythmias (*cont.*):
  hypomagnesaemia and 1813
  interstitial myocardial fibrosis
    and 1267
  magnesium therapy 1813
  uraemic patient 1270
  ventricular, cardiac drugs 187
arsenic, toxicity 845–6, 1008
arsine,
  inhalation effects, acute tubular
    necrosis 845
  nephrotoxicity 1008
arterial hypertension, diuretics,
  clinical use 201–2
arterial occlusive disease, end-stage
  renal failure 1268
arterial superficialization,
  aneurysm 1415
  vascular access 1409–10
arterial thrombosis, nephrotic
  syndrome **278–84**
arteries,
  endothelium, integration of signals
    in vessel wall 985
  lesions, in scleroderma 675
  vessel wall
    effects of sodium 2119
    integration of signals 985
    media composition 672
    vascular tone 985
  see also renal arteries; *specific
    vessels*
arteriography,
  endovascular interventional
    radiology 113–14
  indications 112–13
  normal arteriogram 112
  percutaneous transluminal
    angioplasty 114
  radiographic contrast-induced
    acute renal failure 996
  rapid, indications 113
  technical considerations 111–12
  see also renal angiography
arteriohepatic dysplasia, features
  and associated abnormalities
  2036
arterioles,
  in amyloidosis 552
  arteriolar fibroid necrosis,
    accelerated hypertension 2131,
    2134
  cyclosporin, nephrotoxicity,
    histological changes 1562
  hyalinization, in diabetes 524
  immune deposits, in SLE 656
  lesions, in diabetic nephropathy
    523
  obliteration, in sickle-cell anaemia
    (HbSS) 706
arteriolosclerosis,
  liver glomerulonephritis *360*
  schematic representation 152
arteriosclerosis,
  ageing changes 58
  arterial fibrosis, schematic
    representation 152
  definition disagreements 58
  hyperuricaemia as risk factor 835
  parenchymal consequences,
    computed tomography
    scanning 105
  schematic representation 152
  see also atheroma; ischaemia;
    renal ischaemia

arteriovenous fistula,
  access, haemofiltration 1426–7
  alternatives 1405
  anastomotic site 1406–7
    aneurysm 1414
  Cimino shunt 265
  creation and selection of type
    1405–6
  daily care 1411
  following renal biopsy 143
  insufficient blood flow 1407, 1411
  left ventricular hypertrophy, end-
    stage renal disease 1265
  patency factors 1408
  patency rates 1411
  renal arteriography 113
  in renal transplantation 97
  side-to-side 1407
  sore-thumb syndrome 1414
  stenosis, imaging 1440–1
  surgery 1407–9
  types of anastomosis 1407
arteriovenous shunt,
  chronic renal failure 1188
  non-homogeneous renal
    oxygenation 990–1
  thrombosis in 1363
arteriovenous shunt dermatitis,
  chronic renal failure 1392
arteritis 583
  end-stage renal failure 1233
  giant cell, *see* giant cell arteritis
  necrotizing, *see* necrotizing
    polyarteritides
  renal, in rheumatoid arthritis 681
  Takayasu's arteritis 2324
  see also vasculitis
arthralgia,
  in Henoch–Schönlein purpura 596
  renal disease in 2293
arthritis,
  gouty, urate/uric acid
    concentrations 835
  of hyperuricaemia, urate
    arthropathy contraindicated
    834
  septic, complication of renal
    disease 2296
arthropathy,
  amyloid-associated arthropathy
    2297
  haemodialysis, long-term
    patients 1442
arylsulphatase A deficiency,
  metachromic leukodystrophy
  2206
asbestos, and carcinoma 2238
Ascaris lumbricoides,
  mebendazole 185
ascites,
  central blood volume reduction
    203
  in dialysis patients **1449–50**
    causes *1449*
  diuretics, clinical use 202–3
  nephrogenic, diagnosis 1284
  specific complications of therapy
    203
ascorbic acid,
  excretion, pregnancy 1916
  hyperoxaluria 1857
  oxalate, plasma and 2229–30
  total parenteral nutrition in ARF
    1021

ash leaf spots, tuberosclerosis 2300
Ask–Upmark kidney (segmental
  hypoplasia) 2088–9
  children 2142
  end-stage renal failure 1231
aspartate, diet, acid and base
  production 932–3, *933*
aspergillosis 1746–7
Aspergillus infection,
  opportunistic infections 1746
  renal transplantation 1579
asphyxiating thoracic dystrophy,
  features and associated
  abnormalities *2038*
aspiration, interventional
  ultrasonography 97
aspiration pneumonitis, triple
  metabolic and respiratory
  disorders *957*
aspirin,
  N-acetyl-β-D-glucosaminidase
    excretion 18
  action 190
  antiplatelet therapy, pregnancy-
    induced hypertension 1966–7
  in diabetic nephropathy
    treatment 530
  glomerular filtration rate
    reduction, in diabetes 513
  handling in kidney disease 186
  nephrotic syndrome 284
  pregnancy
    CLASP study 1968
    neonatal effects 1967
    pregnancy-induced hypertension
      prophylaxis, low-dose 1967–8
    prostacyclin inhibition 1966–7
  renal papillary necrosis
    association 684, 822
  thromboxane A_2 inhibition 1967
  see also salicylates
aspirin–dipyridamole, animal renal
  ablation models, effects *1195,
  1216*
aspirin–paracetamol–caffeine
  (APC) medication,
  addition of centrally-acting agents,
    analgesic nephropathy 816
  defined 809
asplenism, in sickle-cell anaemia
  704
assessment of renal function, *see*
  clinical assessment; *specific
  procedures*
asterixis,
  acute renal failure 2326
  uraemic encephalopathy 1398
asthma,
  β-agonists 189
  in Churg–Strauss syndrome 613,
    629
  diuretics in 207
Aström syndrome 2231
atenolol,
  animal models, effect on
    glomerular scarring *1213*
  eliminated by kidney 187
  haemodialysis patients,
    recommendations *1463*
  pregnancy-induced hypertension
    therapy 1970, 1971
  properties, in relation to use in
    impaired renal function *2053*

Index

atheroembolic disease of the kidney,
  acute on chronic renal failure 1160
  assessment, chronic renal failure 1160
  chronic renal failure
    assessment 1160
    prevalence 1268
  clinical features and diagnosis 2283–4
  epidemiology and aetiology 2283
  management 2284
  parenchymal consequences, computed tomography scanning disease 105
  renal effects 2334–5
  see also arteriosclerosis; renal artery stenosis; vascular diseases
atherogenesis, uraemic patients 1257
  ischaemic heart disease **1267–70**
atherosclerosis,
  accelerated 1462
  cholesterol embolism 1079
  foot lesions 1642
  ischaemic heart disease, diabetes and chronic renal failure 1642
  renal artery, differential diagnosis 2098
  in SLE 659
  see also atherogenesis; coronary atherosclerosis; renal arterial disease; renal artery stenosis
atherosis, acute, uteroplacental bed 1960
athletes, hypomagnesaemia 1811
ATN, see acute tubular necrosis
ATP, renal blood flow, net constriction 26
ATPase pump, salt-wasting, ageing changes 63
ATP–magnesium chloride, acute renal failure, prevention of 1018
atracurium, handling in kidney disease 186
atrial fibrillation,
  cardiac drugs 187
  digoxin 188
atrial natriuretic peptide,
  acute renal failure, prevention of 1018
  animal gestational models 1922
  arteriolar myocytes of afferent arterioles, effects 26
  endocapillary glomerulonephritis, acute 407
  glomerular hyperfiltration association 513
  IgA nephropathy 342
  liver cirrhosis 272
  mesangial cells 136
  neonates, role 53
  nephrotic syndrome 267
  obstruction 2014
    animal models 2012
  plasma concentration, hydration state marker 263
  pregnancy, increase 1919, 1962
  renal vascular control, mediation 984
  renal vasodilatation
    gestational animal models 1915

atrial natriuretic peptide (cont.):
  renal vasodilatation (cont.):
    human pregnancy 1915
  renin–angiotensin–aldosterone system interactions 53
  sodium retention 264
  stimulation, Bartter's syndrome 783
  stretch receptors, release 263
  uraemia 1326
  uraemic syndrome 1243
atriopeptin III, acute renal failure, prevention of 1018
atrioventricular shunts, infected, crescentic glomerulonephritis 421
augmentin, urinary tract infection 1686
auranofin,
  efficacy 160
  nephrotoxicity 160
aurothiomalate, see gold nephropathy
Australia and New Zealand, renal transplantation, skin neoplasia 1582
autoantibodies,
  inhibition by anti-idiotype antibodies 638
  mechanisms for production 638
  in mixed connective tissue disease (MCTD) 684, 698
  penicillamine therapy association 682
  pathogenesis 683
  in scleroderma 669
  in Sjögren's syndrome 694, 695
  in SLE, see systemic lupus erythematosus (SLE)
  in vasculitis pathogenesis 589, 590, 591, 592
  see also individual antibodies
autoimmune diseases,
  classification 246, 247
  see also individual diseases
autoimmune glomerulonephritis **246–51**
  animal models
    Brown–Norway rats 247
    bursectomized chickens 247
    antiglomerular basement membrane-mediated glomerulonephritis 247
  classification 246, 247
  drug-induced 249
    other than mercury 251
  H-2 susceptible mice 249
  Heymann's nephritis 247–8
  mercury-induced
    in Brown-Norway rats 249–51
    in other species 251
  murine lupus nephritis 251
  polyclonal activation context 248–51
    allogeneic reactions 249
    bacterial product-induced 248
    graft-versus-host disease 249
    parasite-induced 248–9
    transplantation tolerance model 249
  see also antiglomerular basement membrane-mediated glomerulonephritis
autoimmune response,
  defined 245–6
  idiotypic network and 638

autoimmune response (cont.):
  induction, leading to glomerulopathy **246–51**
  self-antigen reactions 246
autonomic nervous system, dysfunction 1401
  haemodialysis, disorders associated with left ventricular hypertrophy 1266
  in uraemia 1461
autonomic neuropathy,
  Anderson–Fabry disease 2207
  chronic renal failure, impotence, erectile 1331
  in diabetes mellitus 512, 530
    chronic renal failure 1641–2, 1643, 1647
  uraemic polyneuropathy 2328
autosomal dominant polycystic disease, see polycystic disease
avascular necrosis, bone 2298
avascular osteonecrosis, renal transplantation, complication 1537
AVP, see arginine vasopressin
azabenzenarsenate-keyhole limpet haemocyanin, crescentic glomerulonephritis 255
5-azacytidine,
  antineoplastic treatment, renal complications 2319
  drug-induced acute toxic tubular nephropathy 166
azathioprine,
  activity increased with allopurinol treatment 834
  animal models, pregnancy effects 1339
  breast feeding 1339
  cancer risk 2320
  hepatic veno-occlusive disease 2315
  IgA nephropathy 358
  interactions, cyclophosphamide 1282
  leucopenia, transplantation patients 2345, 2346
  liver effects 2315
  in lupus nephritis 660–1
    recommended strategy 663
  membranous nephropathy 383–4
  in microscopic polyarteritis 626
  minimal change nephrotic syndrome
    steroid-resistant 320
    steroid-responsive 318
  myelodysplasia 2345
  pregnancy effects 1339
  rejection of allograft
    action 1554–5
    acute 1571
    change to cyclosporin 1572
    deleterious effects 1555
  renal transplantation
    breast feeding 1951
    discontinuation in cancer treatment 1540
    leucocyte abnormalities 1950
    liver toxicity in pregnancy 1950
    malignancy following 1582–3
    mode of action and side-effects 1555
    monitoring dosage 1950
    multiple drug protocols 1556
    optimal regimen 1590

azathioprine (cont.):
  renal transplantation (cont.):
    preconception counselling 1949
    pregnancy 1950
    septicaemia and bone marrow hypoplasia 1534–5
    skin disorder 1583
    skin lesions 1584
    withdrawal effects 1538
  steroid-resistant nephrotic syndrome 320
  steroid-responsive nephrotic syndrome 318
  triple therapy, dose 2345
  venous thrombosis, incidence compared with cyclosporin therapy 1562
  in Wegener's granulomatosis 620
azlocillin, CAPD peritonitis regimens 1489
azosemide, action 199
azotaemia,
  causes 1099
  defining 1112
  neonates, see neonates, acute renal failure
  progressive, ischaemic renal disease 1078
  sodium loss 1884
  unexplained 1078, 1079–80
aztreonam,
  CAPD peritonitis regimens 1489
  handling in kidney disease 183

B cells,
  abnormalities in SLE 6, 636
  activation, antiself Ia autoreactive T cells 845
  anti-DNA antibodies 249
  CD5 positive 640, 696
  clonal anergy 246
  deficiency in uraemic syndrome 1239
  in essential mixed cryoglobulinaemia pathogenesis 687
  IgA-secreting 252
  mitogens, immunodeposition 248
  monoclonal light chains, glomerulopathy/myeloma 229, 239
  polyclonal activation, bacterial lipopolysaccharide 248
  proliferative disorders 562
  responses, uraemic patients 1315
  in sarcoid granulomatous interstitial nephritis 577
  self-reactive 246
  in Sjögren's syndrome 696
  spontaneous polyclonal activation 640
  tolerance, autoimmunity 246
β-2-microglobulin, see (β-2-)microglobulin (amyloidosis)
β-adrenoceptor blockers, see beta-adrenoceptor blockers
bacitracin, nephrotoxicity 1010
back pain,
  chronic renal failure 1166
  and loin pain
    chronic renal failure 1166–7
    vesicoureteric reflux 1995
  peritoneal dialysis 1495
    elderly patients 1626

Index

bacterial endocarditis, *see* endocarditis
bacterial endotoxins,
  haemodialysis, monocyte activation 1316
  pathogenesis of renal damage 991–2
bacterial infections,
  acute renal failure 1021
  glomerulopathies, acute renal failure in liver disease 1104
  haemodialysis 1439, 1452–3
  infection-associated glomerulonephritis 458–62
  liver disease, associated with acute renal failure 1106
  nephrotic syndrome and 276–8
  pathogenesis, activation of coagulation system 1045
  in sickle-cell anaemia 703, 704
  susceptibility in sickle-cell anaemia 703, 704
  tropics, acute renal failure **1133–4**
bacterial interstitial nephritis, focal or multifocal, urinary tract infection 1691
bacterial lipopolysaccharide, polyclonal activation, B cells 248
bacteriuria,
  antibody-coated bacteria test 1664
  asymptomatic **1683–5**
    children 1700–1, 1991
    complicated 1684
    pregnancy 1684, 1928–9
      in sickle-cell disease 710
    prevalence 1683–4
    uncomplicated 1684
  contamination, distinguished from bladder infection 1677
  criteria 5
  defined 1660–1, 1676–7
  estimation of bacterial count 1662–3
  incidence in diabetes 531
  males, numbers in bladder urine 1677
  mixed cultures, numbers in bladder urine 1677
  pad culture method 1663
  serum antibody titre 1664
  tetrazolium reduction test 1663
  theoretical distribution of viable bacterial count *1660*
  treatment 1684
    failure, defined 1678
  see also urinary tract infection
Bacteroides, periurethral flora 1660
BAL (British antilewisite), mercury chelation 845
BALB/c mice 585
Balkan nephropathy **857–863**
  aetiological factors 858–60
    cadmium 858
    lead 858
    silica 858
    trace elements 858–9
    bacteria/viruses 859
  clinical features 861
  diagnostic imaging 862–3
  end-stage renal failure 1234
    proportional distribution by age at commencement *1229*
  epidemiological characteristics 858

Balkan nephropathy (*cont.*):
  functional changes 861–2
  fungal and plant toxins 860
    aristolochic acid 860
    ochratoxin A 860
  genetic factors 859, 860
  geographical distribution 857–8
  immune mechanisms 860
  morphological investigations *862*, 862–3
  prevention 863
  treatment 863
  upper tract *vs* bladder tumours 2266–7
Bancroftian filariasis, *Wuchereria bancrofti* 495
barbiturates,
  accumulation 2332
  antagonism of nephrotoxic potential of phenacetin and acetylsalicylic acid 810
  cyclosporin therapy, interactions *1563*
  hepatic metabolism of cyclosporin 2332
  vitamin D effects 1763
Bardet–Biedl syndrome,
  clinical features 2231
  features and associated abnormalities *2037*
  glomerulosclerosis 2216
  nephronophthisis, autosomal recessive 2193
  see also Laurence–Moon–Bardet–Biedl syndrome
barium salts, hypokalaemia 902
baroreceptors, arginine vasopressin release, ageing changes 65
Barraquer–Simon's disease, partial lipodystrophy 2301
Bartter's syndrome 712, **782–7**
  aetiology 949
  characteristics 904
  clinical features 782, 1894
  defined 782, 1894
  diagnosis, algorithm 786
  differential diagnosis 786–7, 1894, 2310
  diuretic abuse 905
  Fanconi syndrome, differential diagnosis 727
  hydramnios in 1894, *1897*
  incidence 782
  laboratory features 782–3
  neonates 782
  nephrocalcinosis 1894
  pathogenesis 949
  pathophysiology 784–6, 904–5
  potassium depletion 904–5
  pregnancy, hypokalaemia 1917
  prostaglandins, overproduction 905
  renal histopathology 783–4
  renin inhibitory effects, NSAIDs 820
  sulindac ineffectiveness 823
  treatment and prognosis 787
basal cell carcinoma,
  chronic renal failure 1393
  glomerulonephritis 2305
  immunosuppressed patients 2307
  risk following renal transplantation 1539, 1582

basal metabolic rate, and body surface and weight 1296–7
basement membrane,
  Alport's syndrome 2198–201
  anti-GBM antibodies 446
  collagen Type IV in 670, 671
  damage, biochemistry 997
  filtration properties 446
  laminin 446
  pathology, crescentic glomerulonephritis 425–7
  see also antiglomerular basement membrane disease; *cell types specific conditions*; glomerular basement membrane; tubular basement membrane
basement membrane lesions,
  Alport's syndrome, lens 2201
  anti-glomerular antibodies, immunologically-mediated mercury-induced 844–5
  nail–patella syndrome 2214–15
  nephronophthisis 2189–90, 2195
  thin-basement membrane nephropathy in Alport's syndrome 2203
basement membrane nephritis, antiglomerular, renal transplantation 1514
BCG, immunosuppressive therapy, metastatic renal carcinoma 2248–9
BCNU, drug-induced acute toxic tubular nephropathy 166
beaver kidneys, structure 706
Beckwith–Wiedemann syndrome, chromosome 11 mutation 2256
  Wilms' tumour 2254
bee stings, *see* insect stings
Behçet's disease, neurological involvement 2324
Behçet's syndrome,
  mononuclear vasculitis in 590
  nephritis 2295
  rapidly progressive glomerulonephritis 1070
  renal disease in 630
  skin lesions 2303
Bence–Jones cast nephropathy 563
  see also multiple myeloma, renal involvement
Bence–Jones protein 546, 547, 556, 563
  diagnosis/detection 563
  Fanconi syndrome 734
  heat test 1152
  isoelectric point (pI) 564, 565
  low molecular weight protein excretion, correlation 564
  nephrotoxicity 563–5
    experimental models 564, 565
  quantity excreted 563
  renal tubular dysfunction/injury 567
  tubular function abnormalities associated 564
bendroflumethiazide, carbohydrate intolerance 210
benign familial haematuria, differential diagnosis, Alport's syndrome 2203
benign nodular cortical nephrocalcinosis 1889

benzafibrate, nephrotic hyperlipidaemia, effects 288
benzbromarone,
  fractional urate clearance 827
  urate excretion increase *829*
  uricosuric effect, inhibition of postsecretory reabsorption of urate 826
benzodiazepines,
  handling in kidney disease 186
  neurological side-effects 2331
benzylpenicillin, dose adjustment *181*
Berger's disease, *see* IgA nephropathy, primary
Bertin, column of,
  arteriography 112
  hypertrophy, and analgesic syndrome 814
  pseudotumours 90
Bertin, septa of (cortical islands), CT 75
beryllium,
  interstitial nephritis 845
  non-renal effects 845
  urate excretion decrease *829*
beta lactamines, drug-induced interstitial nephritis 1088
beta-adrenoceptor blockers,
  acute supraventricular tachycardia, cardiac drugs 187
  angina, prescribing 187–8
  calcium channel blockers and, contraindications 1273
  children 2147
    dosage *2148*
  chronic renal failure
    half-lives 1273
  left ventricular volume increase 1274
  diabetes
    intermittent claudication 1647
    sexual dysfunction 1647
  frusemide and, effect on nephrectomized animal models *1266*
  heart mass, effect of pharmacological interventions on nephrectomized animal models *1266*
  hyperkalaemia and 907
  hypertension 1178
    end-stage renal failure 2121
    handling in kidney disease 188
  hypoglycaemia risk 1646
  liver effects 2315
  NSAID interactions 823
  pregnancy-induced hypertension therapy 1970–1
  renal hypertension, treatment 2052–3
  renal transplantation
    hypertension treatment 1598
    portal hypertension 1540
  renin secretion and *2071*
beta-agonists, asthma 189
beta nechol chloride 530
bezafibrate,
  handling in kidney disease 191
  triglyceride lowering 1258
bezoars, bowel obstruction/perforation, complications of transplantation 1285

Vol. 1 – Sections 1–5.   Vol. 2 – Sections 6–11.   Vol. 3 – Sections 12–21.

Index

BFU-E, *see* burst-forming unit-erythroid
bicarbonate,
  active transport 764
  adaptive tubular ammoniagenesis 1205
  administration
    overproduction acidosis 944
    and precursors, metabolic alkalosis 944
  alcoholic ketoacidosis 930
  'apparent volume of distribution' *footnote* 920
  bicarbonaturia
    in distal renal tubular acidosis 777
    respiratory acidosis) 953
  blood transfusion data 960
  buffering 920–1
    respiratory acidosis 951
    respiratory alkalosis 954
    response to metabolic alkalosis 942, *943*
  chloride depletion 765
  contraindications to administration
    diabetic ketoacidosis 929
    lactic acidosis 928
  diabetic ketoacidosis 928–9
  diarrhoea, alkali loss from intestine 934
  disequilibrium bicarbonaturia 945
  excretion
    fractional, formula 770
    proximal renal tubular acidosis 763
  extracellular fluids
    normal values *917*
    volume 765
  fractional reabsorption 765
  and [H^+] concentration *952, 956*
  Henderson equation 918
  homeostasis 920
  hydrogen ion generation tests
    early morning urine pH 41
    loading and threshhold 44
    urine:plasma P_{CO2}/ gradient 44
  hyperkalaemia, therapy *910*
  hypertonic, inadvertent administration causing hypernatraemia 883
  increase (primary hypercapnia: respiratory acidosis) 951
  loading test
    differential diagnosis 778
    renal tubular acidosis 775
  loss
    Fanconi syndrome 736
    hyperkaliuria 774
    proximal renal tubular acidosis 768
    tubular reabsorption defect in children 773
  luminal concentration 764
  luminal flow rate 764
  measurement 1855
  neurohormonal factors 765
  normal respiratory metabolism 950–1
  parathyroid hormone, calcium and phosphate 765
  P_{CO_2}, and [H], nomogram *919*
  peritubular concentration and P_{CO_2} 765

bicarbonate (*cont.*):
  potassium depletion 765
  rapid infusion, avoidance in neonates 54
  reabsorption
    from intestine, Na/H exchanger 934
    and hydrogen excretion, neonates 52
  regulation, renal acidification 764–5
  renal generation, excessive, metabolic alkalosis 947–8
  secretion by pancreas, chronic renal failure changes 1283
  spontaneous metabolic acidosis 769
  titration, fractional excretion 770
  tubular reabsorption 951–2
    fructose loading 732
    inhibition 198
    isolated defect 768
    mechanisms 764
    regulation 764
    renal acidification 764
  urine pH, infants 43
  see also CO_2, P_{CO2}; metabolic acidosis; metabolic alkalosis
bicarbonate-binding drugs, *see* calcium chloride; cholestyramine
Bickel Fanconi syndrome, glycogenosis 2311
biguanides,
  contraindications in diabetic nephropathy 1645
  handling in kidney disease 189
  see also metformin; phenformin
bile-acid binding proteins, nephrotic hyperlipidaemia, effects *288*
bilharzia, praziquantel 185
  see also schistosomiasis
biliary dilatation, polycystic disease, autosomal dominant 2177
biliary dysgenesis, polycystic disease, autosomal recessive 2169
biliary fibroadenomatosis, polycystic disease, autosomal dominant 2177
Binswanger's encephalopathy, elderly patients on haemodialysis 1623
biopsy **141–53**
  bone marrow 555
  labial gland 694
  lung 1063
  rectal, *see* rectal biopsy
  skin 1063
biopsy, renal,
  acute renal failure 977–8
    preparations 978
  in amyloidosis 554, *555*
  analgesic nephropathy 814
  children 1711
  chronic renal disease, pregnancy 1939–40
  chronic renal failure 1163–4
  complications *142*
    aneurysms 143
    arteriovenous fistula 143
    dysfunction 143
    haematuria 142
    peritoneal haemorrhage 142–3

biopsy, renal (*cont.*):
  contraindications
    coagulation disorders 143
    hypertension 143
    other 143, 1941
    solitary kidney 143
    urinary tract infections 143
  cyclosporin, nephrotoxicity 1565
    vascular injury 1562
  diagnosis, most frequent *2079*
  eclampsia and pre-eclampsia, contraindications 1941
  end-stage renal failure 1233
  endocapillary glomerulonephritis, acute, indications 408
  evaluation
    analytic approach 147–50
    analytical items *148*
    guidelines 147
    morphometry 153
    semi-quantitative approach 150–3
    stepwise evaluation *147*
  Goodpasture's disease 443, 443–4
  in Henoch–Schönlein purpura 605–6, 607
  IgA nephropathy, primary **345–9**
    predictive value 350–1
  indications
    asymptomatic urinary findings 145
    chronic insufficiency 145
    failure, acute 144
    idiopathic nephrotic ssyndrome 143–4
    systemic diseases 145
    transplantation 145–6
  lithium effects 855
  liver glomerulonephritis 359
  macrohaematuria, IgA nephropathy and 340–1
  minimal change nephrotic syndrome 303
  in myeloma cast nephropathy 568
  NSAID-induced acute tubulointerstitial nephritis 822
  NSAID-induced chronic interstitial nephritis 822
  predictive and prognostic value 145
  pregnancy, chronic renal disease 1939–40
  preoperative and postoperative care 142
  procedure, percutaneous/open 141
  rapidly progressive glomerulonephritis 1063
  rejection of allograft, assessment 1529–30
  in rheumatoid arthritis *679*, 679
  role, in lupus nephritis 658
  specimen processing
    electron microscopy 146–7
    immunohistology 146
    light microscopy 146
  in systemic lupus erythematosus (SLE) 649
biotin,
  deficiency, uraemic peripheral polyneuropathy 1398
  total parenteral nutrition in ARF *1021*
biphosphonates,
  hyperparathyroidism 1856

Birmingham Reflux Study Group, controlled trials of surgery 1995–6
bismuth, nephrotoxicity 1008
bisoprolol, properties, in relation to use in impaired renal function *2053*
Biuret method, detection of urine protein 227, 228
black races, nephropathy susceptibility 1207
blackwater fever, haemolysis with haemoglobinuria, and acute renal failure 995
bladder, *see* urinary bladder
Blastomyces dermatitidis, natural history 1742
blastomycosis 1742–3
bleeding,
  haemodialysis 1428
  see also haemorrhage
bleeding tendency, in hyperviscosity syndrome 568
bleeding time, shortened/lengthened 142
bleomycin,
  handling in kidney disease 191
  metastatic renal carcinoma 2248
  radiation injury potentiation 849
bleomycin, vinblastin and cisplatin, haemolytic uraemic syndrome 849
  thrombotic microangiopathy resembling radiation nephritis 849
blood biochemistry, renal stone disease 1850–1
blood extravasation, in sickle-cell disease 710
blood flow, haemodialysis, elderly patients *1625*
blood glucose, *see* glucose; glycaemic control
blood groups,
  Lewis antigens 1703
  p1 antigen 1703
blood loss, haemolytic–uraemic syndrome 1041
blood P_{CO_2}, urinary acidification 767
blood pH, urinary acidification 767
blood pressure,
  accelerated hypertension 2130
    timespan and thresholds 2130
  children 2139–40
  chronic renal failure 1166
    progression 1175
  control
    in diabetics, nephropathy treatment **526–8**
    in Henoch–Schönlein purpura 607
    return of renal function 1151
  control in haemodialysis **1458–64**
  control in renal disease **2058–66**
    animal models 2058–61
    renin–angiotensin system 2061–4
    renomedullary antihypertensive function 2064–5
    sodium and water excretion 2058–60
    sympathetic nervous system 2065–6
    transplantation studies 2060–1
  defined, children 2141

Index

blood pressure (*cont.*):
  determinants, children 2140–1
  diabetes 1648–9, *1649*
    antihypertensive drugs *1649*
    in diabetic nephropathy 507, 511, 527
    in non-insulin-dependent diabetes 507
  diastolic, diabetic nephropathy progression and 510
  hyperkalaemia, changes 907
  increase, accelerated hypertension 2131
  Korotkoff sounds 11
  lead 841
  measurement
    ambulatory 11
    at clinical examination 11
    at home 11
    difficulties 2047
    in hypertensive children 2139–40
  microalbuminuria association *515*, 515, 525
  neonates 1116
  normotension, defined 1176
  percutaneous transluminal renal angioplasty effects *2108*
  potassium depletion effects 902
  pregnancy
    mild renal disease *1938*, 1938, *1939*
    normal changes 1962
    significant threshold values 1957
    renal artery reconstruction effects *2111*
    renal transplantation, recipient-dependent factors 1595
  sodium reabsorption and 264
  standards, children 2140
  target, essential hypertension 2056
  see also hypertension; pregnancy-induced hypertension
blood pump, haemodialysis 1419
blood tests, rapidly progressive glomerulonephritis *1062*
blood transfusion,
  acid–base disorders 959
  citrate in 960
  complications of
    cytotoxic antibody formation 1353
    iron overload 1353
    viral disease 1353
  erythropoietin and 1353
  exchange, tropics 1130
  heart mass, effect on nephrectomized animal models *1266*
  hepatitis non-A, non-B transmission 1288, 1290
  incompatible, and acute renal failure 995
  metabolic alkalosis and 944
  pre renal transplantation 1510, 1516
  renal transplantation 1589–90
    haemosiderosis following 1584
  in sickle-cell disease 716, 717
  renal concentrating ability 705
blood urea, interpretaion 238
blood vessels,
  in amyloidosis 552
  in essential mixed cryoglobulinaemia 690

blood vessels (*cont.*):
  injury, in scleroderma pathogenesis 672, 675
  in scleroderma pathogenesis 671–2
  wall structure 671
  see also specific blood vessels
blood viscosity, increased,
  in plasma cell dyscrasias 568
  in sickle-cell disease 700, 705, 715
blood volume,
  blood volume preventing factors 265
  expansion, Starling forces 264–5
  nephrotic syndrome 267–8
  normal values *1344*
  see also fluid; plasma volume
Bloom's syndrome, skin lesions 2300
blue diaper syndrome 1886
BN52021, platelet activation factor antagonist, cyclosporin, nephrotoxicity, and 1561
body mass,
  blood pressure and, recipient-dependent factors 1595
  diet, glomerular disease 238
Boerhaave syndrome 2312
Bohr effect, metabolic acidosis 924
Bolivian haemorrhagic fever *1045*
bone,
  calcium, total *1753*
  calcium absorption 1764–5
    in uraemia 1764–5
  complications following renal transplantation 1585
    aluminium associated 1585
  development, and growth in healthy children 1296
  growth zone, rickets 1612
  histology 1374–6
    aluminium burden 1374–5
    Masson–Goldner stain 1374
    normal mineralization rate 1174, 1376
    parathormone activity increased 1374
    von Kossa stain 1374
    Yamshidi needle 1374
  lead storage measurement and release 838, 839–40
  mass, methods 1374
  mineralization process **1789–90**
    hormones 1789, *1790*
    resorption 1789
  minerals, metabolic acidosis and 924
  phosphate distribution *1789*, 1789
  remodelling events
    chronic renal failure 1765
    hormonal modulation *1790*
    normal bone 1765
  resorption
    calcitonin inhibition 1757
    increased in malignancy 1776
    in multiple myeloma 567
    in sarcoidosis 579
  turnover, biochemical indices 1775
bone disease,
  adynamic, aluminium-related osteopathy 1369, 1376
  aseptic necrosis, following renal transplantation 1585
  avascular necrosis 2298

bone disease (*cont.*):
  bone cysts, haemodialysis, long-term patients 1442
  bone GLA protein, marker of hyperparathyroid bone disease 1370, 1384
  bone-metastasizing renal tumour of childhood 2258
  chronic renal failure 1157, 1166, 1186, **1365–85**
    clinical syndromes 1376–81
    diagnostic procedures 1370–6
    differential diagnosis *234*
    management 1381–5
    pathogenesis of disturbed calcium and phosphate metabolism 1365–70
    rare causes *234*
    vitamin D 1186
    see also aluminium-related osteodystrophy; osteitis fibrosa; osteomalacia; osteonecrosis
  Fanconi syndrome 726
  hypophosphataemic non-rachitic, nephrocalcinosis *1897*
  nephronophthisis, autosomal recessive 2193
  sepsis, amyloidosis 2294
bone marrow,
  aspiration, amyloidosis diagnosis 555
  biopsy, amyloidosis diagnosis 555
  depression, gout treatment 2296
  growth factor production 1345
  intrathymic differentiation *242*
  uraemic inhibitors of erythropoiesis, postulated substances 1350
bone marrow transplantation,
  acute renal failure following 2343
  haemolytic–uraemic syndrome 2343
  radiation nephritis-resembling syndrome 851
boric acid, inhibition of bacterial growth 1662
Bourneville's disease,
  tuberous sclerosis complex children 2169
  computed tomography scanning 103
bovine serum albumin, cationic, glomerular basement membrane 252
  planted glomerular antigens 254
Bowen's disease, skin lesions 2307
Bowman's capsule **129–39**
  in amyloidosis 552, 573
  basement membrane 129, 131, 132–4
  capsular drop 523, 525, 573
  longitudinal section 130
  pathology, crescentic glomerulonephritis 423–5
  podocytes 131, 132–4
Brachmann–de Lange syndrome, features and associated abnormalities *2036*
brain,
  pathology, pregnancy-induced hypertension 1962
  see also central nervous system; cerebral; neurological abnormalities

branchio-oto-renal syndrome, features and associated abnormalities *2036*
Branhamella catarrhalis, urinary tract infection 1661
Brattleboro rat model,
  diabetes insipidus, central 796
  compensatory hypertrophy in progressive chronic renal failure 1204
breast cancer,
  osteosclerotic metastases 1776
  parathormone-related protein 1776
breast milk,
  phosphate content 1617
  sodium content 52
breast-feeding,
  azathioprine therapy in renal transplantation 1950
  cyclosporin therapy in renal transplantation 1951
  drugs contraindicated 1946
  renal transplant mothers 1951
Bright's disease 505
British antilewisite, mercury chelation 845
bromobenzine, nephrotoxicity 169
bromocryptine,
  treatment of impotence 1335
  see also dopaminergic agonists
bromoethylamine hydrobromide, renal medullary destruction 2065
bronchial carcinoma, metastases, computed tomography scanning 103
Brown-Norway rats, antitubular basement membrane disease 1092
Brucella abortus, liver disease associated with acute renal failure 1106
Brucella melitensis, infection-associated glomerulonephritis 461
Bruch's membrane, anti-GBM antibodies 442
brush border,
  antigens
    BB-50 assay 19
    urine 18–19
  luminal
    bicarbonate reabsorption 764
    disturbance in Fanconi syndrome 723–4, 726
    idiopathic type 730
    proximal tubule 741–2
  proximal cells, sodium–phosphate cotransport 1783–4
brushite, calcium mineralization 1789
bullous disorders, chronic renal failure 1392, 1393
bumetanide,
  action 199
  chronic renal failure 204, 272
  nephrotic syndrome 270
buprenorphine, handling in kidney disease 186
burst-forming unit-erythroid, in erythropoiesis evaluation 1344
BXSB mice 642

C3b-receptor, membrane proteins of podocytes 133
C5b-9 membrane attack complex, glomerular damage 229
  membranous nephropathy 376–7
C-peptide, renal catabolism *1318*
C-reactive protein,
  serum amyloid P-component homology 548, 556
  in Wegener's granulomatosis 617
    monitoring of therapy 621
Cacchi and Ricci disease, children 2169
cachectin, *see* tumour necrosis factor
cadmium **841–4**
  Balkan nephropathy 858
  blood
    β-2-microglobulin 842, 843
    serum creatinine and 843
  blood pressure effects 843
  metabolism 842
    urinary excretion and renal cadmium content 842
  nephrotoxicity 1008–9
  occupational exposure 841–2
    permissible level 843
  renal cell carcinoma 2238
  total body stores, non-occupational exposed adults 842
cadmium nephropathy,
  chronic interstitial nephritis 843
  clinical manifestations 842
  diagnosis 843–4
  lead nephropathy compared 842
  management 844
  β-2-microglobulin 843
  osteomalacia 843
  proximal tubule dysfunction 842
caeruloplasmin–copper,
  loss in nephrotic syndrome 289–90
  Wilson's disease 733
Caesarean section, renal transplant mothers 1951
café au lait spots, neurofibromatosis 2300
caffeine,
  antagonism of nephrotoxic potential of phenacetin and acetylsalicylic acid 810
  relative risks of analgesic-associated disease, end-stage renal failure 808
calcidiol (25,hydroxyvitamin D$_3$), loss, CAPD 1368
calcifediol,
  action, chronic renal failure 1771
  conversion, sarcoidosis 1773
  vitamin D metabolic transformations 1762
calciferol, *see* vitamin D
calcification(s),
  children 1611
  cutaneous and subcutaneous, chronic renal failure 1393
  ectopic distribution 2298
  of necrotic cells, cyclosporin toxicity 149
  preterm neonates, longterm frusemide administration 54
  soft tissue 2298
  tubulointerstitial scarring and 1205

calcification(s) (*cont.*):
  valvular heart disease, dialysis patients 1275
  see also nephrocalcinosis; renal stones
calcinosis, tumoural 1794
calciphylaxis,
  acute, hyperphosphataemia 1793
  in osteitis fibrosa 1377
calcitonin **1761–2**
  bone remodelling *1790*
  bone resorption, inhibition 1757
  degradation 1761
  hypercalcaemia, inhibition of bone resorption 1777
  magnesium and 1807
  osteoclast inhibition 1777
  phosphate transport and 1787
  salmon, in hypercalcaemia of multiple myeloma 567
  secretion
    chronic renal failure 1761, *1762*
    endogenous, normal subjects *1762*
  structure 1761
calcitriol,
  action, chronic renal failure 1770
  biosynthesis 1367
    disturbances in early renal failure *1366*
    extrarenal synthesis 1368
  clearance rate, altered 1366–7
  deficiency, secondary hyperparathyroidism 1366
  haemodialysis
    hypercalcaemia *1759*
    treatment effects 1367
  hypercalciuria of pregnancy 1917
  levels
    following transplantation 1378
    T-cell lymphoma 1777
  osteitis fibrosa
    prophylaxis 1381
    safety 1381
    selection of patients 1381–2
    treatment 1384
  parathormone in 1365
  parathyroid glands, effects *1365*
  putative bone resorption factor 1776
  therapy
    efficacy estimation 1384
    haemodialysis 1367
    long-term 1381
    pulse therapy 1384
    renal growth 1381
    risk of extraskeletal calcifications 1384
    and tolerance 1384
  thiazide diuretics, interactions 1381
  in transplantation, resumption of synthesis 1777
  see also 1,25-dihydroxyvitamin D$_3$; vitamin D
calcium,
  abnormal metabolism in sarcoidosis **578–80**
    calcitriol production 579
  absorption, transplantation, normo/hypercalcaemic recipients *1775*
  albumin-adjusted values *1754*
  biologically active/ionized 1754, *1755*

calcium (*cont.*):
  calcidiol treatment, parathormone effects *1760*
  calcium-regulating hormones **1758–63**
  chronic renal failure 1186
    metabolism, pathogenesis of disturbed function **1365–70**
  concentration in dialysate 1368, 1383–4
  deficiency, lead intoxication, hypertension 834
  dialysate, CAPD 1756
  dietary intake
    animal models of transport 1788
    calcium stone disease 1839–40
    chronic renal failure 1367
    foods high in *1850*
    increase vitamin D intoxication 1773
    phosphate reabsorption and 1788
    RDA for children with chronic renal failure 1307
    total parenteral nutrition in ARF *1021*
  dietary restriction, animal models of progressive glomerular scarring 1212
  dietary supplementation, children with chronic renal failure 1307–8
  end-stage renal failure, calcium responses 1757
  excretion
    hypercalcaemia causing 1885
    in pregnancy 1917
    thiazide diuretics 1852
  extracellular transport **1764–8**
    bone 1764–5
    integrated effects 1767
    intestine 1764
    kidney 1765–7
    transient and steady states 1768
    uraemia of chronic renal failure 1764–8
  homeostasis **1753–8**
    chronic renal failure **1753–8**
    error correction 1757
    hypomagnesaemia and 1812
    index of bone resorption 1757
    set-point 1755–7
  hormone regulation **1758–63**
  hypercalcaemia *1754*
  hypocalcaemia *1754*
  increase
    chronic renal failure 1886
    differential diagnosis 973
  intestinal absorption, increased in sarcoidosis 579
  intoxication, calcium carbonate 1891
  intracellular increase 168
  ionized in plasma, assay 1851
  measurement 1854
  metabolism
    acidosis 924–5
    disorders, renal transplantation 1537
    disturbance in Fanconi syndrome 727
    elderly patients 61
    excretion rate—normal values, neonates 51

calcium (*cont.*):
  metabolism (*cont.*):
    and phosphorus transport, neonates 52
    see also bone complications
  overload, irreversibility of ischaemic injury 997
  parathyroid hormone and 765
  phosphate transport 1787
    animal models 1787
  plasma **1753–5**
    daily turnover 1757
    distribution *1754*
    error correction 1757
    measurement 1854
    metabolism, nephrotic syndrome, vitamin D-binding protein loss 290
    minimal change nephrotic syndrome 302
    movements 1755
    raised/normal *1861*
    set-point 1755–7
  renal transplantation
    metabolic problems 1378–80
    normo/hypercalcaemic recipients *1775*
  total body distribution *1753*
  tubular reabsorption, thiazide diuretics 207
  tubulointerstitial scarring, effects on 1205
  see also hypercalcaemia; hypercalciuria; hypocalcaemia; vitamin D
calcium acetate,
  dose 1383
  phosphate binder, dose and side-effects *1382*
calcium alginate, phosphate binder, dose and side-effects *1382*
calcium antagonists, *see* calcium channel blockers
calcium carbonate,
  administration in children, safety 1305
  calcium intoxication 1891
  contraindications 189
  dose 1383
  infected stone disease 1831–2
  phosphate binder 1186
    advantages 1382–3
    dose and side-effects *1382*
  vs aluminium, phosphate binding 1305
calcium channel blockers,
  acute renal failure, prevention of 1018
  animal models of chronic renal failure 1215
  effect on glomerular scarring *1213*
  β-blockers and, contraindications 1273
  children 2147–8
    dosage *2148*
  chronic renal failure, left ventricular volume increase 1274
  cyclosporin therapy effects 1561
  cyclosporin-induced renal vasoconstriction 1598
  erythropoietin production effects 1350
  hypertension 188, 1178
    accelerated 2135

Index

calcium channel blockers (*cont.*):
  hypertension (*cont.*):
    chronic renal failure 1273
    diabetes 1647
    end-stage renal failure 2121
    hypokalaemia 902
    platelet effects and therapy 1968
    pregnancy-induced hypertension therapy 1972
    protective effect 171
    renal effects 1176
    renal hypertension, treatment 2052–3
    renal transplantation, hypertension treatment 1598
    uteroplacental blood flow 1972
  see also diltiazem; felodipine; nifedipine; verapamil
calcium chloride,
  bicarbonate-binding, alkali loss 934–5
  loading test 775
calcium citrate,
  dose 1383
  phosphate binder, dose and side-effects *1382*
calcium edetate, lead mobilization test 834, 839
calcium gluconate,
  administration, acute hypercalcaemia 1887
  antidote to magnesium sulphate 1974
  hyperkalaemia, therapy *910*
calcium ipodate, acute uric acid nephropathy 832
calcium oxalate,
  arthritis-associated 2296
  inhibitors of crystallization in urine *1828*
  uraemic toxin 2229–30
  see also oxalate
calcium oxalate stone disease, lower tract 1830
calcium oxalate stones, formation 751
calcium phosphate,
  chronic dialysis patients 834
  in hyperphosphataemia 1793
  infected stone disease 1831–2
  inhibitors of crystallization in urine *1827*
calcium pump, hypomagnesaemia and 1813
calcium salts, hyperkalaemia, therapy 910
calcium stone disease **1832–42**
  aetiological risk factors 1835
  crystallization modifiers 1834–5
  epidemiological factors **1836–42**
    demographic factors 1836–7
    diet 1838–42
    environmental, climatic and seasonal factors 1837
    fluid intake 1837
    occupation 1838
    social class 1838
    stress 1837
    summary *1837*
  geographical factors *1822*, 1832, 1835
  hyperoxaluria 1832, 1835
  hypercalciuria 1833, 1834
  hyperoxaluria 1833
  hyperuricosuria 1834
  idiopathic 1832–5

calcium stone disease (*cont.*):
  immobilization 1836
  magnesium 1835
  milk–alkali syndrome 1836
  multifactorial causes 1832, 1835
  paraplegia 1836
  pregnancy 1917
  primary 1832–5
  risk factors 1832, 1835
  sarcoidosis 1836
  secondary 1835–6
    enteric hyperoxaluria 1835
    hereditary hyperoxaluria 1835
    medullary sponge kidney 1836
    primary hyperparathyroidism 1835
    renal tubular acidosis 1835
  tubular ectasia 1836
  urinary pH 1834
  urinary risk factors *1831*
  urine volume 1833
  vitamin D intoxication 1836
calcium urolithiasis, *see* calcium stone disease
calcium-regulating hormones **1758–63**
caliectasis, ageing changes 57
Callilepis poisoning, acute renal failure and 1136
calomel (mercurous sulphate), toxicity 844
calyces,
  forming a track 1870
  Hodson's lines 82–3
  IVU imaging, delayed opacification 83
  stones, removal 1872–3
  see also renal calyces
cAMP levels,
  DDAVP-stimulated release 794
  phosphate regulation 1786
$CaNa_2EDTA$ lead mobilization test, *see* edetate
cancer, chemotherapy *see* cytotoxic drugs
cancomycin, handling in kidney disease *182*
Candida, other opportunistic species 1743–6
Candida albicans,
  infection, renal transplantation 1580
  infection-associated glomerulonephritis 466
  opportunistic infections 1743–6
  sexually transmitted infections 1686
candidosis **1743–6**
  acute disseminated 1745
  candiduria 1744–5
  fungal ball of the bladder 1745, 1746
  pyelonephritis 1743–4
  sources of infection 1745–6
  treatment 1746
  urethritis 1745
candiduria 1744–5
canrenoate, gynaecomastia 213
CAPD, *see* peritoneal dialysis
CAPD peritonitis 1450, **1486–92**
  see also peritonitis, CAPD
capillaries,
  capillary leak 265
  colloids, capillary retention coefficient 265

capillaries (*cont.*):
  haemoglobin polymerization in 714
  pressure
    interstitial fluid pressure and volume 265
    lymph flow and 265
    Starling forces 265
  see also glomerular capillaries
capillary sclerosis,
  analgesic-associated, prevalence in autopsy studies 807
  basement membrane thickening 811
  phenacetin abuse 811–12
  renal pelvis, unidentified cause 811
capreomycin,
  drug dosage, renal impairment *184*
  renal tuberculosis *1727*
captopril,
  children, treatment of hypertension 2148
  digital subtraction angiography *2147*
  dose, new maximum 2054
  drug-induced interstitial nephritis 1090
  drug-induced nephropathy 162
  erythropoietin 1350
    production effects 1350
  haemodialysis patients, recommendations *1463*
  hypertension
    accelerated 2135
    responsiveness, renal transplantation 1595
  immune-mediated acute interstitial nephritis 2335
  liver effects 2315
  membranous glomerulonephritis 162
  membranous nephropathy, induction 2295
  neonates, treatment of hypertension 2148
  nephrotic range proteinuria 321
  proteinuria due to renovascular hypertension 321
  renin–angiotensin–aldosterone system blockade 2072
  in scleroderma 676
  see also angiotensin converting-enzyme inhibitors
captopril scintigraphy test,
  nuclear imaging 122
  plasma renin activity 2104
  renal artery stenosis 2102
  renovascular hypertension 122, 2103
carbamazepine,
  cyclosporin therapy, interactions *1563*
  hepatic metabolism of cyclosporin 2332
  neurological side-effects 2331–2
carbenicillin, hypokalaemia and metabolic alkalosis 947
carbenoxolone, mineralocorticoid activity, hypokalaemia 905

carbohydrates,
  dietary intake
    high level, age-related glomerular scarring in rats 1212
    RDA for children with chronic renal failure *1307*
    refined 1841
  dietary restriction, triglyceride lowering 1257–8
  intolerance
    diuretic-induced 210–11
    predisposing factors 210
    therapeutic consequences 211
  metabolism
    disorders of **1251–4**
    growth hormone–somatomedin axis 1324
    nephrotic syndrome 289
    uraemic syndrome 1239
  nephrocalcinosis, associated substances 1887
  refined, dietary intake 1841
  starvation acidosis 930
carbon dioxide,
  daily production 950
  P_{CO_2}
    above/below 5.3 kPa 958
    bicarbonate and in acidaemia 960
    peritubular concentration 765
    extracellular fluids, normal values *917*
    Henderson equation 918
    increase (primary hypercapnia: respiratory acidosis) 951
    nomogram 919
    urinary, distal tubule acidification 766–7
  venous blood, normal values *917*
carbon monoxide, erythropoietin increase 1347
carbon tetrachloride,
  drug-induced liver and kidney necrosis 1107
  nephrotoxicity 993, 1009
carbonate apatite, Randall's plaques 1896
carbonic anhydrase, activity 950–1
carbonic anhydrase inhibitors,
  aspirin toxicity 206
  drug-induced lithiasis 167
  glaucoma 206
  mountain sickness 206
  proximal diuretics 197–9
  renal tubular acidosis 936
  toxicity 213
  urinary alkalinization 206
  see also acetazolamide
3-carboxy-4-methyl-5-propyl-2-furanpropanoic acid, protein binding, competition for sites 176
carbutamide, drug-induced nephropathy 163
carcinoembryonic antigen, glomerular deposits 477
carcinoma,
  glomerulonephritis in 234–5
  incidence in patients with glomerulopathy 479
  renal involvement *476*
  direct/indirect effects 476–7
  renal tubular epithelial antigen 478

carcinoma (*cont.*):
see also renal cell carcinoma; *specific names, organs and regions*; tumours
cardiac, see also heart
cardiac amyloidosis 549, 558
cardiac drugs, arrhythmias in acute myocardial infarction 187
cardiac failure,
  chronic renal failure 1159
  diuretics 188
  renal prostaglandins 821
  NSAID-induced water retention 823
  severe chronic, diuretics and 202
  vascular access complications 1412
  see also congestive heart failure
cardiac glycosides, interactions with diuretics 208
cardiac grafts, models, experimental solutions 1558
cardiopulmonary bypass,
  acute renal failure, complications of 2334
  elderly patients on haemodialysis 1624
cardiovascular collapse, contrast media injection 77
cardiovascular disease,
  accelerated hypertension, *see* hypertension, accelerated
  action of prostaglandins, NSAIDs toxicity 820
  Anderson–Fabry disease 2207
  cardiac thrombus, acute renal failure and 1079
  cardiogenic shock, monitoring 1032
  cardiorespiratory arrest, mixed acid–base disorders 957, 958
  cardiorespiratory disorders, renal syndromes **2334–6**
  children
    chronic renal failure 1610
    mortality causes *1610*
  complicating renal transplantation 1512–13
  complications of analgesic nephropathy 813
  deaths from
    in diabetic nephropathy 506, 508
    in SLE 659
  diabetic nephropathy and chronic renal failure 1642–3
  elderly patients with chronic renal failure 1624
  encephalopathy and 2324
  following renal transplantation
    causes 1512–13, 1540–1, 1575–7
    mortality 1575–6
    patients on dialysis 1576
    prevention 1577
  in Henoch–Schönlein purpura 598
  hyperuricaemia 835 813
  insulin resistance association 211, 521
  metabolic alkalosis 943
  nephrotic syndrome, risk profiles 286
  renal syndromes **2334–6**
  renal transplantation and 1577
  respiratory acidosis 953
  respiratory alkalosis 955
  uraemic patients **1264–75**

cardiovascular disease (*cont.*):
  uraemic syndrome, clinical features 1237–8
  valvular abnormalities, polycystic disease, autosomal dominant 2178
  in Wegener's granulomatosis 616
  see also coronary artery disease
cardiovascular system,
  changes, accelerated hypertension 2130
  examination in acute renal failure 970
  haemodialysis 1438–9
  haemofiltration, monitoring 1032
  hypomagnesaemia and 1813
  impairment in neonatal lupus 651
  instability in elderly patients on haemodialysis 1625
  monitoring in shock 1032
  pregnancy **1956–7**
  pregnancy-induced hypertension 1962–3
  renal transplantation, hypertension, effect on 1597
  thoracic pain, haemodialysis patients 1452
carmustine, antineoplastic treatment, renal complications 2319
carnitine,
  deficiency
    left ventricular hypertrophy and, end-stage renal disease 1266
    uraemic peripheral polyneuropathy 1398
  excretion, Fanconi syndrome 727
Caroli's disease, *see* polycystic disease, autosomal recessive
carp bile, acute renal failure and 1012, **1135**
carpal tunnel syndrome 548, 550, 1400
  analgesic nephropathy 813
  haemodialysis, long-term 1441–2
caseation, granulomas 1720
casts,
  cast nephropathy, *see* multiple myeloma, renal involvement intratubular
    Bence–Jones protein nephrotoxicity and 564
    in multiple myeloma 563, 564
    histology 565–6
  in urinary sediment 20–1
  see also red cell casts
catecholamines,
  catecholamine-secreting tumours, problems 2256
  diabetes, pressor dose 1646
  diuretics and 211
  increased, uraemia 1461
  potassium homeostasis 896
  transcellular shifts 902
cathartics, hypertonic dehydration 883
catheterization,
  catheter samples, urinary tract infection 1680–1
  percutaneous transluminal renal angioplasty 2106–7
  retrograde femoral, intra-arterial injection 111
  techniques 2011

catheterization (*cont.*):
  urinary tract infection, susceptibility factors 1670
catheters,
  access methods 1405
  CAPD peritonitis 1490
  exit-site/tunnel infection 1493–4
  implantation 1493
  peritoneal dialysis **1492–4**
  Tenckhoff 1492
  TWH 1493
  vascular access 1410
CAVH, *see* haemofiltration, continuous arteriovenous
CAVHD, *see* haemodialysis, continuous arteriovenous
CCNU,
  chemotherapy, metastatic renal carcinoma 2248
  drug-induced acute toxic tubular nephropathy 166
cell-mediated glomerulonephritis, proteinuria present/absent 255
cell-mediated immunity,
  glomerular injury mediated by, in SLE 639
  in mononuclear vasculitis 590
  in pathogenesis of vasculitis in animals 585
  see also T cells
cell-mediated tubulointerstitial disease, pathogenesis 1093–4
cellulitis, nephrotic syndrome 277
cellulose phosphate, magnesium effects 1852
centipede bite, acute renal failure and 1136
central nervous system,
  calcium, total *1753*
  central pontine myelinosis, hyponatraemia 880–1
central nervous system disease,
  haemolytic–uraemic syndrome *1047*, 1047
  infection(s), following renal transplantation 1536, 1579
  renal disease and **2323–32**
cephaloridine,
  nephrotoxicity 994, 1010
  onycholysis 1394
cephalosporins,
  acute on chronic renal failure 1186
  CAPD peritonitis 182
  CAPD peritonitis regimens *1489*
  creatinine clearance and measurement, interfering substance 32
  dose adjustment *182*
  drug-induced acute toxic tubular nephropathy 164
  handling in kidney disease *181, 182*
  interactions with diuretics 208
  nephrotoxicity 994, 1010
  pregnancy 1929
  renal abscess 1696
  urinary tract infection, upper tract 1693
cephalotin, Fanconi syndrome, caused by 735
ceramidase deficiency, Farber disease 2206

cerebral circulation, autoregulation in hypertensive encephalopathy 2133, 2134
cerebral function,
  accelerated hypertension 2134
  cystinosis 2222
cerebral oedema,
  bicarbonate administration 929–30
  see also oedema
cerebrospinal fluid, pleocytosis, Churg–Strauss vasculitis 2324
cerebrovascular disease,
  accelerated hypertension 2134
  Anderson–Fabry disease 2207
  cerebral aneurysm, polycystic disease 2324
  cerebral arterial spasm, nimodipine 2178
  cerebrovascular accident, haemodialysis 1439, 1448
  nephronophthisis 2193
  and renal disease 2324
cerebro–hepato-renal syndrome, features and associated abnormalities *2038*
cervical carcinoma, renal transplantation, incidence 1540, 1583
cGMP, *see* (cyclic) guanosine 3,5,monophosphate (cGMP)
Chagas' disease,
  benzidazole 185
  ketoconazole 185
Charcot joint, diabetes 1640
Charcot–Marie–Tooth syndrome 2216
CHARGE association, features and associated abnormalities *2039*
chelation therapy 1665–6
  lead intoxication 838
  see also edetate
chemical medullectomy model of hypertension 2065
chemotherapy,
  metastatic renal carcinoma 2248
  combinations 2248
  in multiple myeloma 568
  see also cytostatic therapy; cytotoxic agents; *specific drugs*;
children,
  acute renal failure **1122–3**
    causes *1122*
    ischaemic insult as cause *1036*, 1037
    outcome 1123
  AIDS-related nephropathy 466
  aluminium intoxication 1612
  amino acids, normal tubular reabsorption 745, *746*
  anaemia in chronic renal failure 1356, 1609–10
    erythropoietin replacement therapy, r-HuEPO 1356
    treatment with erythropoietin 1609–10
  angiotensin converting-enzyme inhibitors 2148
    dosage *2148*
  antibody-coated bacteria test 1664
  antihypertensives 2147–8
    dosage *2148*
  Ask–Upmark kidney 2142
  asymptomatic bacteriuria 1991

Index

children (cont.):
  Bartter's syndrome, clinical features 783
  β-adrenoceptor blockers 2147
    dosage 2148
  bladder stones 1861
  Bloom's syndrome 2300
  calcium channel blockers 2147–8
    dosage 2148
  CAPD 1473
  captopril, treatment of hypertension 2148
  chronic renal failure 1155–7, **1605–18**
    anaemia 1356, 1609–10
    assessment **1155–7**
    bone disease, growth and development 1157
    cardiovascular disease 1610, 1610–11
    causes 1155
    congenital renal disorders 1605–9
    developmental problems **1613–16**
    dialysis and growth rates 1616–17
    dialysis and transplantation 1617–18
    dietary intervention 1616–17
    Fanconi syndrome **1608–9**
    growth 1157, 1608, 1616
      management 1301–9
    hyperparathyroidism, management 1305
    hypertension 1610, 2144–5, 2145
    incidence 1605
    left ventricular hypertrophy 1611
    management 1616–17
    neurological problems **1613–14**, 1613–14
    psychological problems 1614
    pubertal problems 1614–16
    renal osteodystrophy 1611–13
    rickets 1157
    special clinical features 1609–16
    underlying disorders 1606
    undernutrition 1613
  clonidine
    emergencies 2149
    treatment of hypertension 2148
  combined proximal and distal tubule acidosis 764
  congenital anomalies of the urinary tract **2023–41**
    megaureter 2003
  congenital mesoblastic nephroma 2258
  cortical abscess 1705–6
  crescentic glomerulonephritis 422
  cyst(s), renal **1607–8**
    Cacchi and Ricci disease 2169
    congenital nephrotic syndrome of Finnish type 2169
    cortical vs medullary cysts 2169
    cystic dysplasia 2169, 2180
    glomerulocystic kidneys 2169
    medullary cystic disease 2169
    medullary sponge kidney 2169, 2180
    microcysts, with diffuse chronic tubulointerstitial pyelonephritis 2169
    multicystic dysplasia 2169, 2180

children (cont.):
  cyst(s), renal (cont.):
    multilocular cyst 2180
    multiple malformation syndrome, associated with 2169
    pluricystic hypoplasia 2169
    simple cyst 2180
  cystinosis 2219–24
  diarrhoea, bloody, E. coli, verotoxin-producing strain 1046, 1048–9
  diazoxide, emergencies 2149
  diet
    FAO/WHO recommendations 1616
    soybean diet, chloride depletion 950
  dihydroxyadenine stones **1865**
  disequilibrium syndrome 1431
  distal renal tubular acidosis 763–4, 772–3
  diuretics 2147
    dosage 2148
  nephrotic syndrome complicated by ARF 1039
  Drash syndrome, diffuse mesangial sclerosis 479
  enalapril, advantages 2148
  end-stage renal disease 1610
  endocrine changes and growth 1297–8
  epididymo-orchitis 1706
  epiphyseal slipping 1611–12
  Fabry's disease, α-galactosidase A deficiency 2300
  Fanconi syndrome **1608–9**
  flow diagram for investigations 74
  genetic diagnosis, data bank 1155
  GFR increase, early life 1297
  gout, familial juvenile 2294
  growth in healthy children **1294–301**
    endocrine control 1297–8
    head circumference 1613
    height
      and blood pressure 2140
      British Standards tables 1299
      mean, and SDS 1299
      mean growth velocity 1300
      measurement 1298–300
      prediction 1295–6
      skeletal age 1300, 1301
      see also growth
    sexual maturation 1298
    stature and skeletal changes 1294–6
    weight, body composition and renal function 1296–7
  growth in osteitis fibrosa 1376
  growth in uraemia of chronic renal failure 1295, 1301–9, 1302, 1608, 1616
    corticosteroid drugs 324–5
    intellect 1303
    management 1305–9
    pathogenesis 1303–5
    sexual maturation 1302–3
    see also children, growth
  haemodialysis 1431
    access 1605
    achieving normal growth 1308
    continuous arteriovenous 1123
    growth rates 1616–17
    mortality causes 1610
    patient survival 1618

children (cont.):
  haemodialysis (cont.):
    psychological aspects 1472–4
    vascular access creation 1406
  haemolytic–uraemic syndrome 1047, 2341
    familial 1049–50
    gastrointestinal disorders differential diagnosis 1047
    outcome 1056
    platelet aggregating factor 1053
    vitamin E 1055
  height, see children, growth
  Henoch–Schönlein purpura 2303
  hepatitis B nephropathy 463
  hydralazine
    emergencies 2149
    treatment of hypertension 2148
  hypernatraemia 887
  hypertension **2139–49**
    blood pressure measurements 2139–40
    causes 2141–5
    chronic, causes 2141, 2142
    chronic renal failure 1610, 2144
    clinical presentation 2145
    diagnostic approach 2145–6
    emergencies, drug treatment 2149
    investigations, primary and supplementary 2145, 2146
    neurofibromatosis 2300
    prevalence 2141
    renal parenchymal disease 2142–3
    renovascular hypertension 2143–4
    transient 2141–2
    transplantation and 2144–5
    treatment 2146–9
      minoxidil 2148
  hypomagnesaemia 1810, 1813
    treatment 1814
  hypomagnesaemia–hypercalciuria syndrome 1894
  hyponatraemia 879
  indomethacin, steroid-resistant nephrotic syndrome 1039
  infections
    nephrotic syndrome 276–8
    streptococcal infections, endocapillary glomerulonephritis, acute 406
    see also children, urinary tract infections
  intrarenal reflux 1983–4
  lead mobilization test 839
  lead nephropathy, acute 838
  lipoid nephrosis, corticosteroid therapy 143
  Lowe's syndrome 1796, 1797
  magnesium-losing nephropathy 1894
  malakoplakia 1706
  membranous glomerulonephritis, hepatitis B nephropathy 463
  membranous nephropathy 370
  minimal change nephrotic syndrome 298–327, **316–23**
    incidence of minimal change 298
  Munchausen syndrome, 'by proxy' 9
  nephronophthisis, autosomal recessive 2169, 2189–95

children (cont.):
  nephrotic syndrome
    decreased permeability for large macromolecules 268
    infections 276–8
    pneumococcal peritonitis 277
    prevalence 487
    pulmonary artery thrombosis 280
    renal vein thrombosis 281
    steroid-resistance, indomethacin 1039
    urinary tract infections 277
  neurofibromatosis, hypertension 2300
  nifedipine, emergencies 2149
  nutrition
    early life 1297
    FAO/WHO recommendations 1616
  osteological changes 1611–13
  pericarditis 1610
  perinephric abscess 1706
  pneumococcal peritonitis 277
  polycystic disease, autosomal dominant **2167–70**
  polycystic disease, autosomal recessive **2163–7**
  poststreptococcal glomerulonephritis, prognosis 459
  Potter syndrome 1613
  primary hypomagnesaemia 1810
  pubertal problems
    chronic renal failure 1614–16
    see also children, growth
  pulmonary artery thrombosis 280
  pyelonephritis, xanthogranulomatous 1706
  reflux nephropathy, rarity in normal kidney 1991
  renal carbuncle 1706
  renal cell carcinoma 2258
  renal replacement therapy, mortality causes 1610
  renal stones 1847, 1861
    radiolucency 1865
  renal transplantation 1511–12, 1591
    cystinosis 1513
  renal tubular acidosis, management 936
  renal vein thrombosis 281
  sexual abuse 1716
  SLE in
    clinical features 649–50
    prognosis 650, 658
  sodium nitroprusside, emergencies 2149
  steroid-resistant nephrotic syndrome, indomethacin 1039
  Takayasu's arteritis, ethnic factors 2144
  urinary tract infection **1699–716**
    abnormal tract 1705
    acute effects 1706–7
    advice to parents and children 1716
    antibiotics 1713–15
    asymptomatic 1715
    asymptomatic bacteriuria 1700–1
    clinical effects 1706
    clinical presentation 1704–7
    diagnosis 1707–9, 1991
    differential diagnosis 1711

children (*cont.*):
  urinary tract infection (*cont.*):
    epidemiology 1699–701
    grossly abnormal tract 1715
    host defence mechanisms *1701*
    host predisposing factors 1701–3
    investigations 1709–12
    long-term effects 1707
    management 1712–16, 1998
    morbidity and mortality 1700
    organisms *1701–3*
    pathogenesis 1701–4
    prophylaxis 1715
    radiological tests 73, 74
    recurrence 1700
    sexual abuse 1716
    single dose and short course treatment 1715
    vesicoureteric reflux **1991–2**, *1992*
  urinary tract obstruction 1122–3
  urine
    collection 1707–8
    colony counts 2208–1709
    microscopy 1708
  vesicoureteric reflux **1983–98**
    controlled trials of surgery 1995–6
    natural history 1990
  Wilms' tumour **2253–66**
    hypertension and 2093
  xanthogranulomatous pyelonephritis 1706
  see also neonates
Chlamydia trachomatis, sexually transmitted infections 1686, 1688
chlorambucil,
  IgA nephropathy, primary *358*
  in lupus nephritis 659
  membranous nephropathy, controlled studies 383–4
  minimal change nephrotic syndrome 317
    steroid-resistant 320
  neoplastic complications 2320
  toxicity 322
  in Wegener's granulomatosis 619, 621
  see also alkylating drugs
chloramine,
  anaemia and 1351
  haemolysis and 1351
chlordiazepoxide, handling in kidney disease 186
chloride,
  administration, metabolic alkalosis 945–6
  deficiency, dietary 787
  depletion
    causes 904
    chloride diarrhoea 947
    metabolic alkalosis 945–6
    potassium depletion 904
    renal acidification 765
    treatment 904
    vomiting, metabolic alkalosis 945
  excretion, daily urinary *923*
  measurement 1855
  plasma, measurement 1855
  reabsorption from intestine, Na/H exchanger 934
  total parenteral nutrition in ARF *1021*

chloride (*cont.*):
  transport, Bartter's syndrome and 784
  transport in nephron 198
  urinary, diagnosis of Bartter's syndrome 786
chloride shunt disease, *see* drug-induced pseudohypoaldosteronism type II
chlormethiazole, eclampsia effects 1974
chloroquine,
  in calcium abnormalities in sarcoidosis 580
  handling in kidney disease 185
  in lupus nephritis 659
chlorothiazide, uricosuric effect, inhibition of postsecretory reabsorption of urate 826
chlorpromazine, cholestasis 2313
chlorpropamide,
  drug-induced nephropathy 163
  elderly patients, inappropriate use 65, 67
  elimination in chronic renal failure 1645
  hypoglycaemia 189
  see also hypoglycaemic agents, oral; sulphonylurea compounds
chlorthalidone,
  diuretic-induced hypokalaemia 209
  nephrotic syndrome 270
cholangiocarcinoma, tropics, acute renal failure and 1138
cholecalciferol,
  metabolic transformations to vitamin D 1762
  osteitis fibrosa, treatment 1384
  supraphysiological dose 755
  see also vitamin D
cholecystitis, acute acalculous 979
cholecystokinin, elevated, chronic renal failure 1280
cholera, acute renal failure and 1134
cholestasis,
  anicteric, causes *2313*
  chronic, Alagille syndrome 2216
  drug-induced *2314*
  transplantation 2313
cholesterol,
  chronic renal failure 1254–5
  dietary and pharmacological lipid manipulations *1202*
  metabolic transformations to vitamin D 1762
  minimal change nephrotic syndrome 302
  and triglycerides, nephrotic syndrome 285
cholesterol embolism, azotaemia, acute renal failure *1078*, 1079–80
  ischaemic renal disease 1079
cholesterol ester, 'Maltese crosses' in urinary sediment 20
cholestyramine,
  animal models, dietary and pharmacological lipid manipulations *1202*, 1216
  bicarbonate-binding, alkali loss 934–5
  in chronic renal failure 1181, 1186

cholestyramine (*cont.*):
  nephrotic syndrome, treatment 287–8
  pruritus, treatment 1391
choline deficiency, liver glomerulonephritis *360*
chondrocalcinosis, spondyloarthropathy in 2297
chondrodysplasia calcificans punctata, rhizomelic type, features and associated abnormalities *2038*
chondroitin-4-sulphate,
  inhibitor of calcium oxalate crystallization in urine *1828*
  inhibitor of calcium phosphate crystallization in urine *1827*
chorionic gonadotrophin, *see* CG
chorionic villus sampling, polycystic disease, autosomal dominant 2182
Christmas disease, renal complications of 2343
chromium, toxicity 845–6
chromosome 1, translocations, AGR patient 2255
chromosome 3,
  loci, small cell carcinoma 2162
  marker for Balkan nephropathy 860
chromosome 6,
  IgA nephropathy, primary, HLA antigens 352, *353*
  see also MHC system
chromosome 11 deletions, aniridia–genitourinary–retardation patients 2255
  inactivation of homologous alleles 2256
  Wilms' tumour 2256
chromosome 14, IgA nephropathy, primary, immunoglobulins organization and expression 352–3
chromosome 16,
  dihydroxyadenine stones 1865
  loci
    ADPKD1, closely linked markers 2160
    genetic linkage map 2159
    HBA, ADPKD alleles 2157–8
    translocations, AGR patient 2255
chromosome 19, IgA nephropathy, primary, C3 frequencies 352, *353*
chromosome duplications, features and associated abnormalities *2040*
chromosome loci, X chromosome, location of common disease loci 2209
chronic ambulatory peritoneal dialysis, *see* peritoneal dialysis
chronic non-obstructive atrophic pyelonephritis, *see* reflux nephropathy
chronic obstructive airways disease, acetazolamide 206
chronic renal failure, ischaemic heart disease, assessment 1161
  ischaemic heart disease assessment 1161
  chronic renal failure 1161
chronic pyelonephritis, nephrocalcinosis 1889

chronic renal failure **1149–402**, **1605–50**
  acidosis balance 1183
  acute on chronic, *see* acute on chronic renal failure
  acute interstitial nephrosis, assessment 1161
  acute tubular necrosis 1158
  aetiology uncertain
    EDTA classification 1228–30
    proportional distribution by age at commencement *1229*
    proportional distribution by age at commencement of treatment *1230*
  age-related factors 1208, 1210
  Alport's syndromes, assessment 1155
  amino acids
    levels in serum and tissue 1258–9
    patterns (plasma and intracellular) *1258*
  amyloidosis, assessment 1152–3
  anaemia, *see* anaemia, chronic renal failure
  analgesic nephropathy, assessment 1150
  animal models 1196–7
    exercise training 1217
  assessment **1149–70**
    follow-up assessment 1164–7
    follow-up list *1165*
    initial assessment 1149–64
    investigations *1162*, 1162–4
    predialysis assessment 1167–70
    radiological investigations *1163*
    recommendations 1187–8
    renal function 1174–5
  atheroembolic disease
    assessment 1160
    prevalence 1268
  backache and loin pain 1166–7
  biopsy, renal 1163–4
  bleeding diathesis in 2344
  blood pressure 1166, 1175
  bone disease 1166, 1186–7, **1365–85**
  calcium homeostasis **1753–8**
  cardiovascular-risk factors 1177, 1185–6
  children **1155–7**, **1605–18**
  chronicity **1155–8**
  clinical features 2326–7
  COAD, ischaemic heart disease, assessment 1161
  coagulation abnormalities **1361–3**
  congenital causes 1155
  congestive heart failure 1186
  course 1175
  defined 1173
  dehydration in 1158, 1182
  diabetes 188
    age and 5-year survival *1637*
    assessment 1151, 1161
    cardiovascular mortality 1268
    see also diabetic nephropathy
  diabetic nephropathy **1635–50**
  diagnosis **1149–55**
  dietary manipulations *1209*, 1209–10
    age-related factors 1210
    animal models 1209–10
    calcium and sodium 1212
    carbohydrates 1212

Index

chronic renal failure (*cont.*):
 dietary manipulations (*cont.*):
  lipids 1212
  monitoring compliance 1165–6
  phosphorus 1211–12
  protein 1210–11
  protein degradation 1259
  protein restriction 1178–81
  protein synthesis 1259
 diuretics, clinical use 204
 drugs, *see* chronic renal failure, pharmacological interventions
 elderly patients
  bone disease in 1623
  cardiovascular disease 1624
  causes 1622–3
  and expected survival of elderly normal subjects *1622*
  haemodialysis-related problems 1624–5
  malnutrition 1623
  neurological problems 1623
  peritoneal dialysis-related problems 1625–6
  psychosocial problems 1624
  rheumatological problems 1624
  treatment **1621–32**
   age of commencement *1621*
   haemodialysis 1624–5
   peritoneal dialysis 1625–6
   suitability for treatment 1622
   transplantation 1627–32
   withdrawal of treatment 1627
 encephalopathy **1398–400**
 end-stage renal failure **1227–35**
 endocrine disorders **1317–26**
 erythropoietin, deficiency 1349
 exercise training, animal models 1217
 Fabry's disease, assessment 1155
 gastrointestinal effects **1278–92**
  bleeding 1284
  symptoms pre dialysis 1280
  upper tract bleeding 1284
 glomerular filtration 271–2
 glomerulonephritis, drug-induced **159–63**
 glomerulonephritis, rapidly progressive
  assessment 1150, 1154
  crescentic conversion 1160
 gout, assessment 1151
 growth in children, *see* growth in uraemia of chronic renal failure
 haemorrhagic diathesis 1187, 1361–3
 heart changes 271–2
 hepatitis B, vaccination 1188
 hypercalcaemia 1185
 hypercorticism 1209
 hyperkalaemia and GFR 908–9
 hyperlipidaemia 1185, 1186
  assessment 1167
  dietary and pharmacological lipid manipulations *1202*
  in progressive CRF 1201–2
 hyperphosphataemia 1793, 1794
 hypertension 271–2, 1176, **2117–22**
  essential hypertension 1151
  malignant hypertension 1184
  risk factor in end-stage renal failure 1268
 hyperuricaemia 1185
 hypotension 1158

chronic renal failure (*cont.*):
 immune response in **1312–16**
  immunotactoid glomerulopathy 2281–3
 incidence, in non-insulin-dependent diabetics 508
 infections 1159–60, 1184
 investigations 2327
  ammonium excretion tests 44
  computed tomography scanning 106–7
  nuclear imaging 125–6
  phosphate excretion tests 44
  ultrasound, with radiography 72–3
 iron, kinetic studies 1349
 ischaemic heart disease, assessment 1161
 in light chain deposition disease 570
 liver changes **1282–3**
 loop diuretics, resistance 204
 management
  initial **1173–88**
  monitoring progression 1165
  volume and electroylytes 1182–4
 medullary cystic disease, assessment 1155
 metabolic acidosis 1183
 metabolic alkalosis 1183
 metabolic derangements, glomerular scarring 1208–9
 metabolic disorders **1251–60**
 in myeloma cast nephropathy 566
 nail–patella syndrome, assessment 1155
 nephron reduction **1191–217**
 nephrotoxins 2285–6
 neurological involvement **2326–7**
  mental state 2326
  neuropsychiatric alterations **1396–402**
  pathogenesis 2326
  skin manifestations, assessment 1157–8
 obstructive/non-obstructive, computed tomography scanning 105–6
 organic, drug-induced nephropathy 159–68
 overfilling 271–2
 oxalosis, assessment 1153–4
 papillary necrosis 1160
 parathyroidectomy, indications *1384*
 pharmacological interventions **1212–17**
  angiotensin converting-enzyme inhibitors 1214–15
  anticoagulants 1215–16
  antihypertensives 1212–14
  antiplatelet agents 1215–16
  assessment 1159
  calcium channel blockers 1215
  dopaminergic agents 1215
  drug-induced glomerulonephritis **159–63**
  effects on drug metabolism 177
  erythropoietin 1217
  immunosuppressive and antiinflammatory agents 1217
  lipid-lowering agents 1216–17
  prostaglandins 1217
 phosphorus, restriction 1181
 polycystic disease, autosomal dominant, assessment 1154–5

chronic renal failure (*cont.*):
 potassium balance, adaptation 908–9
 potassium depletion 1183
 pregnancy 1187
 progressive **1191–217**
  drugs and toxins 1185
  see also progressive renal failure
 radiocontrast media 1185
 radiography **1371–3**
 rate of progression 6
 reflux nephropathy 1993–5
  assessment 1154
 renal adaptation 271
 renal artery stenosis, assessment 1151
 renal function
  assessment 1174–5
  dysfunction, superimposed 1184–5
  retarding decline 1175–81
 renal replacement therapy, USA total statistical data *1628*
 renal transplantation
  diagnosis of recurring diseases 1153–5
  rejection of allograft 1538
  see also renal transplantation
 renal tuberculosis 1728
 renal vein thrombosis, assessment 1160
 schistosomiasis, assessment 1152
 sex-related factors 1207–8
 sexual dysfunction 1157, **1329–39**
  in sickle-cell disease 711
 skin manifestations **1390–4**
  assessment 1157
 SLE, assessment 1151
 sodium depletion 1182
 sodium retention 1182
 superimposed renal dysfunction 1184–5
  factors *1184*
 systemic infections 1160
 terminal, prediction from GFR 37
 treatment 272
  potassium-sparing diuretics 272
 tuberculosis, assessment 1152
 uraemia, monitoring 1167
 uraemic syndrome **1236–47**
 urinary tract infections 1159–60, 1184
 urinary tract obstruction 1185
  assessment 1149–50
 vascular disorders with skin pathology 1393
 vitamin D metabolism 177
 volume depletion 1184
 water clearance 1182
 water intoxication 1182
Churg–Strauss syndrome **629–30**
 antineutrophil cytoplasmic antibodies in 629
 clinical features 629
 definitions 629
 diagnosis 1066
 differential diagnosis 614, 618, 2284
 historical aspects and terminology 613
 skin lesions 2303
 treatment and outcome 629–30
 Wegener's granulomatosis overlap 613, 618

Churg–Strauss vasculitis, polyarteritis nodosa 2324
chylomicrons, defined 1255
cilistatin, CAPD peritonitis regimens *1489*
cilofungin, fungal infections 1748
cimetidine,
 cisplatin, antineoplastic treatment, renal complications 2319
 drug-induced interstitial nephritis 1090
 handling in kidney disease 189
 hyperparathyroid bone disease *1760*
 interference with creatinine clearance 30
 percentage protein binding according to renal function *1240*
Cimino shunt, arteriovenous fistula 265
ciprofloxacin,
 CAPD peritonitis regimens *1489*
 drug-induced acute toxic tubular nephropathy 164
 drug-related interstitial nephritis 1089
 handling in kidney disease 183
 intracystic penetration 2173
circadian rhythms, urate/uric acid, renal handling 828
circulating immune complexes, *see* immune complex(es)
circulatory shock, paradoxical susceptibility to ischaemia 990
circumoval precipitin test, schistosomiasis 1731
cirrhosis,
 haemodialysis patients 1289, 1292
 see also liver disease, cirrhosis
cirrhotic glomerulonephritis, acute renal failure in liver disease 1105
cirsoid aneurysm, renal arteriography 113
cisplatin,
 N-acetyl-β-D-glucosaminidase excretion 18
 acute renal failure, prevention **1019**
 biotransformation 165
 cimetidine, antineoplastic treatment, renal complications 2319
 drug-induced acute toxic tubular nephropathy 165
 Fanconi syndrome, caused by 735
 handling in kidney disease 191
 hydration 165
 methylprednisone, antineoplastic treatment, renal complications 2319
 nephrotoxicity 2318–19
 phosphomycin, antineoplastic treatment, renal complications 2319
 systemic hypertension 165
 tubular dysfunction 165
 tubular injury in acute renal failure 1010, 1011
 vascular effects 165
 verapamil, antineoplastic treatment, renal complications 2319

cisplatinum, magnesium wasting and 1810
citrate,
　diet, acid and base production 932–3, *933*
　distal renal tubular acidosis, treatment 775
　excretion
　　case study 1860
　　dietary factors 1834
　　oestrogens 1834
　inhibitor of calcium oxalate crystallization in urine *1828*
　inhibitor of calcium phosphate crystallization in urine *1827*
　oral administration, calcium stone disease 1834
　phosphate binder 1447
　　dose and side-effects *1382*
　prevention of phosphate precipitation from urine *1893*
　reduced, acetazolamide and 1861
CLASP study, aspirin 1968
clathrin-coated pits,
　gp 330 antigen 247
　podocytes 253
clavulanic acid, amoxycillin and, dose adjustment *181*
cleaning fluids, nephrotoxicity 1009
clear cell carcinoma, see renal cell carcinoma
clear cell hypernephroma, von Hippel–Lindau disease 2181
clearance tests **29–40**
Clethrionomys, Hantavirus disease carrier 1143
climate, tropics, diseases of 1124
clindamycin,
　CAPD peritonitis regimens *1489*
　nephrotoxicity 1010
　renal transplantation, immediate peri-operative administration 1522
clinical examination **11–15**
　adults 11–12
　children 12–13
　elderly patients 13
　examination of kidneys and urinary tract 11
　general examination 11–12
　neonates 12
　patient follow-up 14–15
　see also blood pressure, measurement
clinical history of presenting complaint **6–10**
　dietary history 8–9
　drug history 7–8
　ethnic/geographical factors 9–10
　factitious history 9
　family history 10
　gynaecological/obstetric history 7
　occupational history 9
　past history 7
　social history 9
　systematic enquiry 10
clinical presentation of renal disease **3–6**
　asymptomatic/symptomatic 3–6
　micturition disorders 3–4
　oedema 5
　pain 5

clodronate,
　hypercalcaemia, management 1777
　hyperparathyroidism 1856
clofibrate,
　acute on chronic renal failure 1159
　animal renal ablation models, effects *1195*
　dietary and pharmacological lipid manipulations *1202*, 1216
　handling in kidney disease 191
　nephrotic hyperlipidaemia, effects 288
　toxicity 1186
　triglyceride lowering 1258
clomiphene, sexual function effects 1335
clonidine,
　children
　　emergencies *2149*
　　treatment of hypertension 2148
　haemodialysis patients, recommendations *1463*
　percentage protein binding according to renal function *1240*
　renal hypertension 2055–6
　renal impairment 188
Clostridium,
　haemorrhagic fever *1045*
　phospholipase C infusion, animal models 1054
clotting factors, deficiencies, in amyloidosis 550
cloxacillin, onycholysis 1394
COAD, see chronic obstructive airways disease
coagulation abnormalities,
　chronic renal failure **1361–3**
　contraindication to renal biopsy 143
　disseminated intravascular coagulation, pneumococcus pneumonia 460
　haemolytic–uraemic syndrome 1054
　neurological involvement 2324–5
　obstetric acute renal failure 1932
　platelets and, pregnancy-induced hypertension 1964–5
　in sickle-cell disease 709, **715**
　thromboembolic complications, nephrotic syndrome 282–4
　upper gastrointestinal bleeding, chronic renal failure 1284
　uraemic syndrome 1238
coagulation factors,
　activation
　　by factor XII following contrast media 77–8
　　infections 1045–6
　　mechanisms *1044*
　endothelial cells of glomerular capillaries 132
　glomerulonephritis, pathogenesis 257
　list and molecular weights *281*
　nephrotic syndrome
　　changes to concentrations *281*
　　summary of alterations 282
cobalt, administration, erythropoietin increase 1347

cobra venom factor, decomplementation 256
Coccidioides immitis, natural history 1743
coccidioidomycosis 1743
Cockayne syndrome,
　features and associated abnormalities *2037*
　skin lesions 2300
codeine,
　antagonism of nephrotoxic potential of phenacetin and acetylsalicylic acid 810
　handling in kidney disease 186
coffee drinking 2238
colchicine,
　action 190
　in acute renal failure 2296
　in amyloidosis 556
colestipol, nephrotic syndrome, treatment 287–8
colic,
　acute renal, IVU imaging 83
　haemodialysis patients 1454–5
colistin, nephrotoxicity 1010
collagen 668–71
　basement membrane, see collagen, Type IV
　in blood vessel walls 671
　classification/types 668
　composition and location *670*
　deposits, renal biopsy, diagnostic relevance 150
　distribution
　　normal 674
　　in scleroderma kidneys 674
　genes 670
　interstitial 670
　mineralization process *1789*
　minor types 670
　post-translational modification 668, 670, 671
　signal peptides 670
　structure 668, 671
　synthesis 670–1, 672
　　increased in diabetic nephropathy 518
　　increased in scleroderma 674–5
　　inhibition 671
　　regulation 671
　type I 670, 672
　type III 670, 672
　　increase in scleroderma 674
　type IV (basement membrane) 670, 671
　　collagen 135
　　cross-linking, in glomerular basement membrane 517
　　in diabetic nephropathy 517, 524
　　Goodpasture antigen, Alport's syndrome 2199–201
　　in scleroderma 674
　type V 670
　　increase, in scleroderma 674
　type VI, in diabetic nephropathy 524
　see also glomerular collagen
collagenase 670–1
　synthesis 670–1
collagen–vascular diseases, neurological involvement 2324
collecting system,
　arginine vasopressin insensitivity in elderly patients 65

collecting system (cont.):
　filling defects 83–4
　gas vs staghorn calculi, ultrasonography 95
　non-obstructive dilatation 2017
　see also renal sinus
collecting tubule,
　A and B cells 766
　hydrogen ion transport 766
　transport of sodium, chloride, and water 198
colon,
　colonic diverticulitis, haemodialysis patients 1454–5
　colonic necrosis, sorbitol 911
　diverticula, polycystic disease, autosomal dominant 2179
　dysfunction following renal transplantation 1582
　microbial changes, in uraemia 1282
　obstruction/perforation, management 1285
　perforation, following transplantation 1281
　phosphate transport 1788
　potassium
　　excretion 897
　　homeostasis, long-term haemodialyis 1282
　　secretion, renal failure, chronic 908
　structural changes, chronic renal failure 1281–2
colour flow mapping, Doppler echography 86–7
coma,
　diabetic ketoacidosis, hypokalaemia 903
　hypernatraemia 888
　hyponatraemia 880–1
　metabolic alkalosis 943
　non-ketotic hyperosmolar 532
　uraemic encephalopathy 1399
combined proximal and distal renal tubular acidosis 764, 773
complement,
　activation 255
　alternative pathway, in Henoch–Schönlein purpura 601, 602, 603
　chronic renal failure 1315
　dinitrophenol-conjugated antigens 357
　by Factor XII following contrast media 77–8
　in leucocytoclastic vasculitis 587–8, 589, 602
　stimulation of dialyser membranes by inflammatory mediators 1432–3
C2 deficiency, in Henoch–Schönlein purpura 601
C3 deficiency
　in SLE 648
　in vasculitis 587, 589
C3 nephritic factor 239, 393–4, 411
C3 receptors, immune complex elimination in animals 583
C3 turnover, increased in Henoch–Schönlein purpura 601

Index

complement (*cont.*):
C3b/C4b receptor (CR1), reduced/defective in SLE 642
C4 deficiency
 in SLE 642, 648
 in vasculitis 587
C5b-9 229
 membranous nephropathy 376–7
cell-mediated action 255–6
 direct effect 256
C1q antibodies 589
 assays in SLE 652
 in diffuse proliferative glomerulonephritis, in SLE 654
 in hypocomplementaemic urticarial vasculitis 589
 in leucocytoclastic vasculitis 587
consumption, detection 239
deficiency
 in SLE 642, 648
 see also hypocomplementaemia
deposits
 in Henoch–Schönlein purpura 603, 604
 in scleroderma 675
 in SLE 652, 653
 in essential mixed cryoglobulinaemia 588, 691
 in experimental vasculitis 584
 in Henoch–Schönlein purpura 601, 602, 603, 604
 immune complex clearance by 641
 independent glomerular injury, in SLE 639
 in leucocytoclastic vasculitis 587, 588, 601
 liver glomerulonephritis 359
membrane attach complex, *see* membrane attach complex
nephritogenicity, terminal components 411
nephrotic syndrome, abnormalities 278
neutrophil-mediated glomerular injury mediated by 639
poststreptococcal endocapillary glomerulonephritis 408, 411
and proteinuria, glomerulonephritis–type diseases 408
rapidly progressive glomerulonephritis, tests *1062*
in SLE
 deposits localization 639, 652, 653
 dysfunction 641–2, 648
 mediation of renal injury 639
complement cascade,
C3a and C5a activation during haemodialysis 1315
C5-deficient/sufficient mice 1208
schema 255
schistosomiasis 1732
complement markers,
assay 239
C3 nephritic factor 239, 393–4, 411
membrane attack complex 239
mesangiocapillary glomerulonephritis 393–5
computed tomography **99–107**
Bertin, septa of (cortical islands) 75

computed tomography (*cont.*):
chronic urinary tract obstruction 2015
contrast media 99
contrast-associated nephropathy 78
cysts 75
doubtful renal mass 75
flow diagram for investigations 73
genitourinary tuberculosis 1724
indications 71
renal infarction 76
renal mass 2246
upper urinary tract obstruction 2008
concanavalin A, planted glomerular antigens 254
concentrating capacity, *see* renal concentrating capacity
congenital disorders of the urinary tract **1605–9, 2035–41**
 autosomal dominant *2036*
 autosomal recessive *2037–8*
 chromosomal abnormalities *2040–1*
 mesoblastic nephroma 2258
 reference lists 2035
 renal dysplasia, end-stage renal failure 1231
 sporadic mutations *2039*
 teratogenic abnormalities *2040*
 X-linked recessive *2038*
 X-ray exposure *in utero* 81–2
 see also cyst(s); familial juvenile nephronophthisis; Fanconi syndrome
congenital nephrotic syndrome, neonates 2169, 2264–5
congestive heart failure,
 chronic renal failure 1186
 compensatory hyperaldosteronism 2335
 diabetes and chronic renal failure 1643
 diuretics and 202
 GFR in 2335
 hypokalaemia 2335
 hyponatraemia 877
 management 2335
 in renal disease, clinical features 2337
 renal effects 2334
 salt and water retention 877
 secondary fluid retention 272
 sodium retention and oedema 2335
 see also cardiac failure
Congo red 545, 553, 554
connective tissue **668–72**
 components *670*
 see also collagen; proteoglycans
Conn's syndrome, pregnancy, hypokalaemia 1917
constipation,
 dialysis patients 1281
 ispaghula husk 189
 lactulose 189
 laxatives 189
 senna 189
 sodium picosulphate 189
contact-system activators,
 activation by factor XII following contrast media 77–8
continuous ambulatory peritoneal dialysis, *see* peritoneal dialysis

contraception,
 dialysis patients 1948
 oral, *see* oral contraceptive
 renal transplantation patients 1952
 IUCD contraindications 1952
contraction alkalosis 949
contrast media, *see* radiocontrast media
convulsions,
 uraemic encephalopathy 1399
 see also coma; seizures
Coomassie method, detection of urine protein 227, 228
copper,
 anaemia and 1351
 haemolysis and 1351
 loss in nephrotic syndrome 289–90
 metabolism, disorders, *see* Wilson's disease
 see also cuprophan membranes
copper sulphate, nephrotoxicity 1125, **1137**
cor pulmonale, *Schistosoma mansoni* 1737, 1739
cornea verticillata 2207
corneal dystrophy,
 Anderson–Fabry disease (α-galactosidase A deficiency) 2207
 hypercalcaemia, signs 1891
 see also eye disorders
coronary angiography,
 renal patients, indications 1268
 uraemic patient with cardiac problems 1272
coronary angioplasty, percutaneous transluminal, dialysis patients 1274
coronary artery bypass,
 chronic renal failure, timing 1161
 diabetes and chronic renal failure 1643
 end-stage renal failure, dialysis patients 1274
 renal patients, results 1268
coronary artery disease,
 chronic renal failure
  calcification without stenosis 1269
  left ventricular hypertrophy causing clinical features 1269
  prevalence 1268–9
  risk factors 1267–8, 1268
  treatment 1274
  with/without ischaemic heart disease 1269–70
 coronary artery spasm 1269
 in diabetes mellitus 512
 diagnosis, accuracy 1269
 hypertriglyceridaemia, risk factor in end-stage renal failure 1268
 ischaemic heart disease, diabetes and chronic renal failure 1642
 Kawasaki's disease 590
 percutaneous transluminal coronary angioplasty, dialysis patients 1274
 in renal disease
  diagnosis 2336–7
  treatment 2337
 risk, persistent proteinuria in diabetes mellitus 506
 significant, defined 1269

coronary artery disease (*cont.*):
 treatment, chronic renal failure 1274
 uraemic patients **1267–70**
  risk factors 1267–8
 see also cardiovascular disease; coronary angiography
coronavirus disease, Balkan nephropathy 860
cortical abscess 1695–6
 children 1705–6
 clinical features 1695
 differential diagnosis 1695
 pathology 1695
 treatment 1695–6
cortical cystic disease, children 2169
cortical granulomas 1720
cortical necrosis,
 end-stage renal failure 1234
 see also acute cortical necrosis
corticomedullary nephrosis 1886
corticosteroids,
 in amyloidosis 557, 558
 blocking activation of IL-1, IL-6 genes 1553
 contrast media injection, before 78
 cyclosporin and
  minimal change nephrotic syndrome *328*, 328
  renal transplantation, hypertension effects 1597
 effects on growth, children 1308, 1309
 growth curves 324–5
 growth effects, children with uraemia 1305
 handling in kidney disease 190
 in Henoch–Schönlein purpura 606–7
 IgA nephropathy, primary *358*
 immunosuppressive therapy in renal transplantation
  action *1554*
  hypertension-causing debate 1596
  multiple drug protocols 1556
 in lupus nephritis 659, 660
  cytotoxic agent comparison 660–1
 membranous nephropathy, controlled studies *382*, 382–3
 in microscopic polyarteritis 625, 626
 minimal change nephrotic syndrome
  adults 327–9
  children, complications 322
  children, resistance 95–321
  children, responsive 314–16, 321
  dosage 314
  remission 314
 steroid-resistant, recommendations 321
 steroid-responsive, recommendations 321
neurological side-effects 2332
NSAID-induced acute tubulointerstitial nephritis 822
NSAID-induced vasculitis 822
in polyarteritis nodosa 628
pregnancy
 animal models 1339
 fetal effects 1339

corticosteroids (*cont.*):
  in sarcoidosis
    calcium abnormalities 579–80
    granulomatous interstitial nephritis 578
  side-effects 322
    skin 2306
  in SLE, in pregnancy 652
  statural growth 324–5
  therapy for lipoid nephrosis 143
  thrombosis, nephrotic syndrome 283–4
  toxicity, Cushingoid syndrome 322
  in Wegener's granulomatosis 619
  see also other specific substances; prednisolone; prednisone; steroids
corticotrophin-releasing hormone, hypothalamic, ACTH function tests 1325
cortisol,
  abnormal metabolism, hypertension, metabolic alkalosis and 948
  chronic renal failure, accumulation 1325
  Cushing's syndrome, diagnosis 1325
  increase, animal models of growth 1304
  pregnancy, increase 14
cortisol-binding protein, loss in nephrotic syndrome 290
cortisone, radiation injury potentiation 849
Corynebacterium, infected stone disease 1863
co-trimoxazole,
  with cyclosporin, nephrotoxicity 1573
  drug-related interstitial nephritis 1088–9
  handling in kidney disease 183
  intracystic penetration 2173
  prostatitis 1689
  renal excretion, pK_A and pK_B 178
  renal transplantation
    contraindications 1579
    Pneumocystis carinii infection 1579
    prevention of infections 1578
    urinary tract infections, prophylaxis 1533
  urinary tract infection 1686, 1687
  in Wegener's granulomatosis 620–1
countercurrent mechanism, defective medullary function 789–90
  urinary concentration 789–90
countercurrent multiplication 707
  loss, in sickle-cell anaemia 705, 706
cows' milk, *see* milk
creatine, conversion to creatinine, 'goulash effect' 34
creatinine,
  acute tubular necrosis *1016*
  age effects 35
  age and sex differences 59
  albumin ratio, in diabetes mellitus 510
  allopurinol/non-allopurinol treated patients 835

creatinine (*cont.*):
  anion concentration and total CO_2 *933*
  assessment 1174
    neonates 50
  catabolism 238
  contrast-associated nephropathy 78
  creatinine clearance and *1158, 1175*
  cyclosporin, nephrotoxicity 1563
  cytomegalovirus infection, renal transplantation 1527
  in diabetic nephropathy 510–11
  endogenous, clearance 1174
  excretion
    coefficient of variation 33, 34–5
    equilibration after change in muscle mass 34
  generation 31
  glomerular filtration rate *1174*
    calculation 238
  glomerular filtration rate measurement 30, 178
  glomerular filtration rate prediction 511
    age effects 35
    calculation of GFR 32–6
    children 35
    as indicator of GFR 34–5
    obese patients 35
    simple estimation 59–60
    unreliability 36
  hypertension with renal failure, lead-associated 840
  increase, acyclovir treatment 2306
  interfering substances *32*
  lead-induced focal interstitial nephritis marker 839
  levels
    in Churg–Strauss syndrome 629
    commencing osteitis fibrosa prophylaxis 1381
  measurement 31–2
    reciprocal and logarithm 36–7
    substances interfering *32*
  neonates 1112–13
  Pneumocystis carinii infection, renal transplantation 1527
  prerenal failure *1016*
  progression of renal failure 1174–5
  reabsorption 32
  rejection of graft 1538
  renal function decline, pregnancy 1937
  renal handling 32–3
  transplantation, normo/hypercalcaemic recipients *1775*
  uraemic syndrome 1243
  urea and, pregnancy changes 1921
  urine/plasma creatinine, oliguria *vs* prerenal failure *988*
creatinine clearance 31–7
  accelerated hypertension 2134
  children and young adults *33*
  in clinical practice 33–4
    pregnancy changes 1921
  congenital cyanotic heart disease 2277
  extrarenal clearance in advanced renal failure 31
  glomerular filtration rate creatinine concentration *1937*
  measurement 30, 178

creatinine clearance (*cont.*):
  in Henoch–Schönlein purpura outcome prediction 605
  IgA nephropathy, primary 347
  interfering substances *32*
  inulin clearance ratio 30
  Jaffé reaction 31
  Janovsky complex with picrate 31
  measurement 31–7
    glomerular filtration rate 30, 178
    substances interfering *32*
  in oliguria 972
  paroxysmal nocturnal haemoglobinuria 2280
  pregnancy 531, 1910–11
    mild chronic disease *1938*
  pregnancy-induced hypertension 1961
  in renal amyloidosis 551
  renal transplantation, normo/hypercalcaemic recipients *1775*
  in sickle-cell disease 713
  time *vs* 1176
  uraemic retention marker 1245–6
  uraemic syndrome 1238
  urate excretion reduced and, chronic renal failure 831
  urate ratio 826
  various ages 59
cremophor, vehicle for cyclosporin 1560
crescentic glomerulonephritis **418–32**
  acute, in diabetics 532
  acute renal failure and 1070
  α-1-antitrypsin deficiency 2216
  azabenzenearsenate-keyhole limpet haemocyanin 255
  characteristic immunohistologic patterns 147, 149
  children 422
  clinical features 423
  creatinine levels 450
  defined 418–20, 1061
  differential diagnosis 2284
  epidemiology 423
  histology 391
  malignancy-associated 480
  pathogenesis 425–7
  pathology 423–5
    electron microscopy 425
    histopathology 423–5
    immunohistology 425
  poststreptococcal 459
  prognosis 431
  renal transplantation, recurrence following 1574
  silicoproteinosis 845
  in SLE 654
  treatment 427–31
  treatment and outcome *1067*
  varieties
    antiglomerular basement membrane disease 420
    chronic glomerulonephritis 422
    drug hypersensitivity 422
    with granular immune deposits 421
    malignancy-associated 422
    mesangiocapillary glomerulonephritis 391
    renal microscopic vasculitis 420
    systemic immune complex disease 421–2
    systemic infections 421

crescentic glomerulonephritis (*cont.*):
  see also IgA nephropathy, primary
crescents,
  cellular, prognostic indications 145
  formation in Henoch–Schönlein purpura 602, 603, *604*, 606
Crithidia luciliae 647
Crohn's disease, amyloidosis AA 2310
Crow–Fukase syndrome, POEMS syndrome 2285
cryoglobulin(s) 686, 687
  deposition 688–, 691
  in Henoch–Schönlein purpura 600
  see also essential mixed cryoglobulinaemia (ECM)
cryoglobulinaemia,
  definition and classification **686–7**, *687*
  essential mixed, *see* essential mixed cryoglobulinaemia (ECM)
  renal biopsy, EM findings 150
  type I 686, *687*
  type II 686, *687*
    see also essential mixed cryoglobulinaemia (ECM)
  type III 686, *687*
  types I, II and III, renal aspects 2343
cryoprecipitate,
  alternatives 1363
  blood-borne disease 1363
  uraemic haemorrhage 1362–3
cryptococcosis 1746
Cryptococcus neoformans infection 1746
  following renal transplantation, management 1536
crystallization modifiers, calcium stone disease 1834–5
crystalluria,
  causes **1825–1829**
  direct nephrotoxicity 994
  in urinary sediment 21, 22
cuprophan membranes *1247*, 1315, 1316
  haemodialysis 1315
    comparisons *1247*
    monocyte activation 1316
Cushingoid syndrome, steroid toxicity 322
Cushing's disease, calcium stone disease 1836
Cushing's syndrome,
  hypercalciuria 1861
  hypertension, metabolic alkalosis and 948
  potassium depletion 906
cutaneous and subcutaneous disease, *see* skin changes
CV3988, competitive inhibition, antihypertensive polar renomedullary lipid 2065
cyanide–nitroprusside test, cystine 748
cystine stones 1865
cyanotic congenital heart disease 2277, 2334
  endocarditis and 2334
cyclo-oxygenase inhibitors, erythropoietin and 1350
  erythropoietin production effects 1350

Index

cyclo-oxygenase inhibitors (*cont.*):
  further impairment in cyclosporin nephrotoxicity 1561
  GFR and, rat models 1914
cyclophosphamide,
  antiglomerular basement membrane disease 1065
  bladder cancer risk 2320
  crescentic glomerulonephritis 428
  drug-induced acute toxic tubular nephropathy 166
  handling in kidney disease 191
  IgA nephropathy *358*
  interactions, azathioprine 1282
  leucopenia and 2346
  in lupus nephritis 660–1
    intravenous 661
    recommended strategy 663
    schedules and advantages 662–3
  membranous nephropathy, controlled studies 383–4
  metastatic renal carcinoma 2248
  microscopic polyarteritis 625, 626, 1066
    intravenous 626
  minimal change nephrotic syndrome 316–17
    steroid-resistant 319–20
  neoplastic complications 2320
  radiation injury potentiation 849
  renal complications of 2343
  renal transplantation, replacing cyclosporin in glomerulonephritis 1515
  risks 431
  side-effects 322
  toxicity 322
  toxicity and side-effects 620
  in Wegener's granulomatosis 619, 620
    intravenous pulses 620
  Wilms' tumour 2260–1
  see also alkylating drugs
cycloserine, renal tuberculosis *1727*
cyclosporin,
  breast feeding 1339
  cancer risk 2321
  children, renal transplantation 1512
  corticosteroids and, minimal change nephrotic syndrome *328*, 328
  dose dependent reactions 1012–13
  dose regimens 1565
  enzyme induction 187
  haemolytic–uraemic syndrome 1048, 1050, 2346
  handling in kidney disease 191
  hepatic metabolism of 2332
  hepatotoxicity 2315
    liver transplantation 1564
  high dose, acute renal failure 1012–13
  hyperkalaemia-inducing 909
  hypertension-associated, prevalence *1565*, 1596, *1597*
  IgA nephropathy, primary *358*
  liver effects 2315
  lung infections 2338
  in lupus nephritis 662
  magnesium wasting and 1810
  metabolic pathway 2315
  metabolism in liver, monitoring 1563

cyclosporin (*cont.*):
  minimal change nephrotic syndrome
    adults *328*, 328
    children 318
    steroid-resistant 320
  nephrotoxicity 323, **683–4**, 1538–9, 1555–6, **1560–6, 1572–3**
    acute **1560–2, 1572–3**
      pathophysiology 1560
    chronic **1562–6, 1573**
      hyperkalaemia 1564
      hyperuricaemia 1564
      management 1564–5, 1573
      monitoring 1563, 1573
      nitrogen, salt and water retention 1563–4
      pathophysiology 1562, 1573
    clinical features 1563–4, 1572–3
    diagnosis 1564–57
      blood values 1565
    drug interactions *1563*, 1573
    experimental vascular injury 1562
    gingival hypertrophy 1566, 1584
    hepatotoxicity 1566
    histological changes 1562, 1572
    hypertension 1565–58, 1573
      prevalence *1565*
    hypertrichosis 1566
    induction, criteria 1563
    malignancy 1566
    management 1564–5, 1573
    mechanism 1572
    multiple renal 168
    neurotoxicity 1566
    non-Hodgkin's lymphoma 1566, 1582
    prevention 1564
    prognosis 1565
    rejection *vs* drug nephrotoxicity 1527
    reversibility of acute nephrotoxicity 1560
    time taken after commencement 1563
    vascular changes in kidney 1573
  neurological side-effects 2332
  non-steroidal anti-inflammatory drugs and 823
  pharmacokinetics 1555–6
  pregnancy effects 1339
  primary effect (vasoconstriction) 1560
  rejection of allograft action 1555–6
  methodology for blood concentration determination 1529
  rejection *vs* drug nephrotoxicity 1527
  renal transplantation
    avoidance in immediate post-transplantation period 1565
    blood transfusion effect 1590
    children 1308, 1512
    cyclophosphamide replacing in glomerulonephritis 1515
    diabetes 1650
    dosage 1573
    gum hypertrophy 1566, 1584
    hydration protocol *1518*, 1519
    hypertensive effects 1565, 1596–7, *1597*

cyclosporin (*cont.*):
  renal transplantation (*cont.*):
    immediate post-operative administration 1522–3
    liver dysfunction 1584
    multiple drug protocols 1556
    optimal regimen 1590
    oxalosis 1513
    pancreatitis 1582
    preconception counselling 1949
    pregnancy and 1950
    short-term 1598
    skin lesions 1584
    skin tumours 1582
    treatment commencement 1518
    withdrawal effects 1538, 1572
  side-effects, skin 2306
  steroid-resistant nephrotic syndrome 320
  steroid-responsive nephrotic syndrome 318
  toxicity
    tubular lesions 149
    vascular basis 987–8
  in Wegener's granulomatosis 621
  withdrawal, and kidney function 1538, 1572
  see also FK-506
cyscain, and carcinoma 2238
cyst(s), renal,
  acquired renal cystic disease 2180
  animal models 2183
  aspiration using ultrasonography 90
  children **1607–8**
  classification *2181*
  computed tomography scanning 99–100
  CT 75
  cystic dilatation, intratubular oxalate 1205
  decompression 2172, 2175
  differential diagnosis
    carcinoma 88–90
    solid renal mass, MRI 108
    tumour, computed tomography scanning 100
  DMSA scans 126
  end-stage renal failure, proportional distribution by age at commencement *1229*
  induction by chemicals 2183
  infections 1696
  intracystic haemorrhage, computed tomography scanning 100
  lithium toxicity 855–6
  medullary sponge kidney 2180, 2286, 2287
  multilocular cyst 2180
  nephronophthisis 2189–90
  parapelvic cysts, computed tomography scanning 106
  pathophysiology 2183
  radionuclide scanning 2245
  renography 126
  simple, hypertension and 2090–1
  tuberous sclerosis 2181
  ultrasonography 88–90, 2245
  unilateral multicystic kidney, hypertension and 2090–1
  see also multicystic kidney disease; polycystic disease, autosomal dominant & autosomal recessive

cysteamine,
  cystinosis, renal transplantation 1513, 1574
  cystinosis treatment 1609
  mobilization of cystine 2224
cysteine, diet, acid and base production 932–3, *933*
cystic dysplasia, children 2169, 2180
cystic fibrosis, glucoglycinuria and 745
cystic nephroma, multilocular 103
cystine,
  demonstration,
    cyano–nitroprusside test 748
  excretion
    glutamine effects 749
    pregnancy and 1916
  involvement, Lowe syndrome 732
  mobilization 2224
  normal excretion 748
  transporter, evidence 746
  see also cystinuria
cystine lithiasis, *see* cystine stone disease
cystine stone disease **1829–30**
  clinical features 748
  geographical factors *1822*
  urethral obstruction 748
  urinary risk factors *1831*
cystine stones **1864–5**
  lithotripsy 1875
cystine storage disease, *see* cystinosis
cystinosis,
  children
    survival *1609*
    treatment 1608–9
  classification of types 2220–1
  diagnosis 2222–3
  end-stage renal failure 1233
    proportional distribution by age at commencement *1229*
  extra-renal signs and symptoms 2221–2
  Fanconi syndrome 1608
  incidence 2220
  metabolic defect 2220–1
  PAH excretion 727
  renal signs and symptoms 2221
  renal transplantation 1513
    recurrence following 1574
  treatment
    renal replacement therapy 2223–4
    specific treatment 2224
    symptomatic 2223
  see also cystinuria
cystinuria **1864–5**
  classical 747–9
  clinical features 748
  defined 1864
  diagnosis 748, 1864–5
  differentiation from cystinosis 2220
  epidemiology and aetiology 1864
  genetics 748
  inheritance 1864
  intestinal defect 747–8
  pregnancy 1942
  renal defect 747
  treatment 748–9, 1865
  types 1, 2 and 3 748
cystine disorders, other 749–50
  see also cystinosis

Index

cystitis, *see* urinary tract infection, cystitis and urethritis
cystography,
　retrograde 81
　retrograde cystourethrogram, genitourinary tuberculosis 1722
cystoscopy, asymmetric trigone 71
cystotubular ectasia, polycystic kidney disease, infantile 1607
cysts, cystic disease, *see* polycystic disease; cyst(s), renal
cytochrome P-450,
　activation, toxicity of drugs 810
　cyclosporin A metabolic pathway 2315
　vitamin D metabolite transformations 1384
cytochrome P-450 isoenzyme, antiliver/kidney microsome antibodies 2315
cytochrome P-450 system, anticonvulsants 187
　drug effects *191*
cytochrome-*c*-oxidase deficiency, Fanconi syndrome *729*, 733–4
cytokines,
　biological activities *1313*
　endothelial effects 1052
　endotoxin-mediated effects 1052
　haemodialysis, induction of acute phase inflammatory response 1432–3
　mediation of vascular damage in accelerated hypertension 2132
　mercury-induced autoimmune glomerulonephritis 250–1
　OKT3 antibody, production 1527
　vascular injury 257
cytomegalovirus infection,
　hepatitis
　　in chronic renal failure 1289, 1290
　　prevention, control and prophylaxis 1290
　infection-associated glomerulonephritis 464
　renal dysfunction following renal transplantation 1532
　　acyclovir prophylaxis 1533
　renal transplantation 1579, 1580
　　acyclovir prophylaxis 1533
　　antibody status 1510
　　diagnosis 1534
　　gancyclovir 1534
　　granulocyte-associated immunoglobulins 1534
　　pneumonia management 1536
cytosine arabinoside, drug-induced acute toxic tubular nephropathy 166
cytostatic therapy,
　in amyloidosis 557
　in light chain deposition disease 571
cytotoxic agents,
　clearance of cancer chemotherapy agents *191*
　lung infections 2338
　in lupus nephritis 660–1
　in microscopic polyarteritis 625–6
　in polyarteritis nodosa 628
　renal complications of 2343
　tubular injury in acute renal failure 1011

cytotoxic agents (*cont.*):
　in Wegener's granulomatosis 619, 620, 621
　see also chemotherapy; cytostatic therapy; *individual drugs*
cytotoxic antibodies, HLA antigens 1170

Daentl syndrome, features and associated abnormalities *2038*
Dahl salt-sensitive rats,
　glomerular reduction 1196, 1197
　renal influence on blood pressure control 2060
　sodium balance, renin and sodium depletion *2060*
　sodium and water excretion 2059–60
Dalmation dog mutation, renal tubular defect 830
danazol, IgA nephropathy, primary *358*
Dane particles (hepatitis B virus), structure *1286*
dapsone:antidapsone antibodies, leprosy 492
dapsone,
　drug-induced nephropathy 163
　IgA nephropathy, primary *358*
daunomycin, and carcinoma 2238
DDAVP, *see* arginine vasopressin; desamino-*cis*-1,8-D-arginine vasopressin
de Toni–Debre–Fanconi syndrome, *see* Fanconi syndrome
deafness, *see* hearing defects
Debaryomyces, infection 1748
deep vein thrombosis, *see* thromboembolism
defibrotide, haemolytic–uraemic syndrome 1055
dehydration,
　analgesic-associated renal papillary necrosis 815
　chronic **1853**
　chronic renal failure 1158, 1182
　in hypercalcaemia of multiple myeloma 567
　hypertonic dehydration 883
　in sickle-cell disease 716
　see also water balance
deletions, features and associated abnormalities *2040*
delivery, *see* fetal surveillance, and timing of delivery
demeclocycline,
　arginine vasopressin antagonist, prostaglandin biosynthesis effect 179
　inappropriate secretion of ADH syndrome 207
　nephrogenic diabetes insipidus, causing 887
dementia, in dialysis patients 1448–9
dendritic cells, immune response to allograft, MHC antigens 1545
dengue haemorrhagic fever 1045
dense intramembranous deposit disease,
　in renal transplantation 1531
　see also mesangiocapillary glomerulonephritis, type II

dental enamel, absent, hypocalciuria and 1892, *1897*
dental and paranasal screening, transplantation, prior to 1509
Dent's disease 1891, *1897*
　hypercalciuria 1893, 1896
　prognosis 1898
deoxycoformycin, haemolytic–uraemic syndrome 1048
deoxycorticosteroid-induced potassium excretion, abolished in pregnancy by progesterone 1917
deoxycorticosteroid-induced sodium excretion, pregnancy 1919–20
deoxycorticosterone,
　dialysis, K$^+$ load 896
　mineralocorticoid action, 'escape' 264
　pregnancy and 1919–20
deoxycorticosterone–salt model of hypertension 2059, *2060*
　animal models
　　dietary protein restriction *1210*
　　haemodynamic and structural glomerular parameters *1214*
　　nifedipine 1213
　　renal ablation studies 1196
　　in C5-deficient mice 1208
　　sodium balance, renin and sodium depletion *2060*
dermatitis,
　arteriovenous shunt dermatitis, chronic renal failure 1392
　drug-induced bullous dermatitis, chronic renal failure 1393
dermatomyositis, skin lesions 2302
dermatophytes, immunosuppression, asymptomatic carriage 2307
desamino-*cis*-1,8-D-arginine vasopressin,
　exogenous administration 42
　shortened bleeding time 142, 143
　see also arginine vasopressin
desferrioxamine,
　aluminium assessment 1383
　aluminium reduction 1448, 2331
　aluminium-related bone disease, treatment 1385
　desferrioxamine-associated mucormycosis 2338
　dialysate, aluminium intoxication 1401
　infusion test 1371
　before parathyroidectomy, Al toxicity exclusion 1385
　side-effects 2331
　therapy, longterm in osteitis fibrosa 1385
　thrombocytopenia 2346
desmopressin,
　alternatives 1363
　bleeding time 1451
　uraemic haemorrhage 1363
detrusor sphincter dyssynergia 2275
　bladder pressure 1985
　in diabetes mellitus 530
developmental problems, children, *see* growth
dexamethasone,
　adrenal insufficiency 1325
　phosphate transport and 1787

dextran,
　acute renal failure in diabetics 1638
　glomerular damage 229
　inducing IgA nephropathy 357
　minimal change nephrotic syndrome 308
　neutral, clearance 514, 527, 529
　peritoneal dialysis solution 1481
dextropropoxyphene, handling in kidney disease 186
dextrose, peritoneal dialysis solution 1481
diabetes (mellitus) **505–45**, **1635–50**
　acute renal failure in 532
　antihypertensive treatment 2086
　assessment 1151
　chronic renal failure 1161
　autonomic neuropathy in 512, 530
　bacterial focal or multifocal 1691
　cardiac functional disorders 1266
　chronic renal failure 188
　　assessment 1161
　　autonomic neuropathy 1641–2, 1643, 1647
　　cardiovascular mortality 1268
　　early haemodialysis, need for 1648
　　see also diabetic nephropathy
　continuous ambulatory peritoneal dialysis **1649–50**
　cardiovascular instability 1649
　coronary artery disease risk 1269–70
　diabetic amyotrophy 1641
　and diabetic nephropathy 1640
　diabetic embryopathy, features and associated abnormalities *2040*
　diabetic foot, chronic renal failure 1641–2
　elderly patients
　　chlorpropamide contraindicated 67
　　glycaemia and the ageing kidney 61
　end-stage renal failure, IDD/NIDD *1229*
　ethnic factors 10
　glomerulopathies, associated, renal biopsy indications 145
　haemodialysis
　　bone disease 1649
　　cardiovascular accidents 1649
　　insulin resistance, partial reversal 1648
　　malnutrition in 1648
　　parathyroid hormone levels 1649
　　psychological aspects 1471–2
　　vascular access 1649
　　vision in 1649
　hypophosphataemia in 1792
　infant of diabetic mother, hypomagnesaemia 1810–11
　juvenile-onset diabetes mellitus
　　renal transplantation and pregnancy 1950
　urate clearance increased 831
　ketoacidosis 928–9
　　potassium depletion 903–4
　membranous nephropathy association 526
　non-diabetic (nephropathy) renal disease 525–6
　non-renal complications 511

Index

diabetes (mellitus) (*cont.*):
  pancreas transplantation 1512, 1650
  prescribing 188
  renal artery stenosis 1638
  renal disease manifestations 530–2, 1638
    diffuse acute bacterial nephritis, computed tomography 105
    end-stage renal failure, IDD/NIDD *1229*
    glomerular diseases 526
    mesangiocapillary glomerulonephritis, differential diagnosis 392
    nephropathy, *see* diabetic nephropathy
    renal papillary necrosis 530, 1160
      differential diagnosis 815
    urinary tract infections 530, 531–2
  see also diabetic nephropathy
  renal transplantation in 1584–5
    peripheral vascular disease 1650
  renin–angiotensin–aldosterone system 1646
  rubiosis 2301
  skin lesions 2301
  streptozotocin-induced, animal models of adaptive haemodynamic change in renal ablation 1193–4
  type I, minimal change nephrotic syndrome 300
  type II
    elderly patients
      hyporeninaemic hypoaldosteronism 1646
      renovascular hypertension 1646
      renal failure 1151
    types I and II, proteinuria and chronic renal failure *1636*
    unrelated renal failure, factors *1154*
  untreated early, elevation of GFR 29
  urinary tract infection, susceptibility factors 1670
  see also diabetic nephropathy
diabetes insipidus, central (neurogenic) 885–6
  Brattleboro rat model 796
  causes 884, 885
  severe, differential diagnosis 798
  treatment 207, 886, 887
  vs nephrogenic *884*
  water deprivation, responses *885*
diabetes insipidus, nephrogenic **789–98**
  acquired 41, 886
  causes *790*
  clinical features 791–6
  arginine vasopressin,
    exogenous administration 42
    plasma osmolality and urine osmolality *791*
  congenital form 886
    abnormal renal and extrarenal V$_2$ receptor responses 793–4
    biochemical diagnosis 792–3
    clinical features 791–6
    DNA probes 795
    genetic screening 795

diabetes insipidus, nephrogenic (*cont.*):
  congenital form (*cont.*):
    genetics 791–2
    normal renal V$_1$ receptor responses 794
    physiopathological mechanisms 795
    RLFP studies *795*
    treatment 795–6
  contrast-associated nephropathy 78
  differential diagnosis *798*
  lithium toxicity 853–4
  in Sjögren's syndrome 696
  smoking as risk factor 9
  water excretion characteristics *793*
  see also hypercalcaemia; hypokalaemia; polyuria
diabetes insipidus, partial central (neurogenic) 791
  differential diagnosis *798*
diabetes insipidus, pregnancy-related 886
diabetic nephropathy **505–45**
  albuminuria in, *see* albuminuria; microalbuminuria
  animal models, functional and structural parameters *1194*
  cardiovascular mortality 506, 508
  chronic renal failure **1635–50**
    angiotensin converting-enzyme inhibitors 1644
    assessment 1636–8, 1643
    cardiac problems 1642–3
    check list *1644*
    clinical consequences *1641*
    continuous ambulatory peritoneal dialysis 1649–50
    dialysed patients, causes of death *1637*
    dietary protein 1644
    dyslipidaemia 1648
    evaluation of patient 1637–8
    foot in diabetes 1641–2
    glycaemic control 1644–5
    haemodialysis 1648–9
    hypertension management 1645–8
    magnitude of problem 1635–6
    management of patient *1643*, 1643–50
    monitoring 1643–4
    pressor responsiveness 1645
    prognosis 1636
    renal transplantation 1512, 1515, 1574, 1650
      patient and graft survival *1637*
    risk factors 1643–4, 1644
    vs other causes 1637–8
  clinical course **508–12**
    early phase 508–10
    late phase (clinical nephropathy) 510–11
    see also end-stage renal failure
  definition 505, 1637
  dietary protein restriction, animal models *1210*
  differential diagnosis 524
  end-stage renal failure 1233
  epidemiology **505–8**
    age at diagnosis 506
    cumulative incidence 505, 507
    insulin-dependent 505–6
    non-insulin-dependent 507–8
    risk factors 506

diabetic nephropathy (*cont.*):
  haemodialysis, survival rates 1437
  insulin requirements, predialytic phase 1645
  natural history **508–12**
  pathogenesis **517–21**
    familial/genetic pathways 520–1
    haemodynamic/hypertrophic pathways 519–20
    metabolic pathways 517–19
    possible sequence of events 521, 522
  pathology **521–6**
    electron microscopy 524–5
    immunopathology 524
    light microscopy 521–4
    structure–function 525
  pathophysiology **512–17**
    early functional changes 512–15
    glomerular filtration rate decline 516
    glomerular hyperfiltration 512–14
    hyperlipidaemia 516–17
    late functional changes 516–17
    microalbuminuria 514–15
    proteinuria 516
    renal hypertrophy 514
    tubular function 515
    see also glomerular filtration rate; microalbuminuria; proteinuria
  pregnancy in 530–1, *1939*, 1942
  progression of chronic renal failure 1175
    antihypertensive therapy 1213
  progressive chronic renal failure 1175, 1213
  renal biopsy indications 525–6
  renal transplantation 1512, 1515, 1650
    patient and graft survival *1637*
    recurrence following 1574
  survival and long-term outcome 506, 508
  susceptibility to 520
  treatment **526–30**
    aldose reductase inhibitors 529–30
    blood pressure control 526–8
    dietary 528–9
    glycaemic control 529
  see also diabetes mellitus
diabetic neuropathy, chronic renal failure 1640–1
diabetic retinopathy,
  chronic renal failure 1639–40
  common features *1640*
  prorenin 1646
diachrome, Fanconi syndrome 734
diagnostic tests in haemodialysis 1436
dialysate,
  acetate-containing
    disequilibrium syndrome 1400–1
    hypoxaemia 2339
  aluminium intoxication 1401
  bacterial endotoxin in 1316
  bicarbonate *vs* acetate 1273
  buffering 959
  calcium deficient 1771, *1772*
  calcium levels 1368, 1383–4, 1756
  composition 1423
  and extracellular fluid, comparison *1423*

dialysate (*cont.*):
  microbiological quality standard 1422
  peritoneal dialysis **1480–2**
    characteristics *1481*
    osmotic agents *1481*
  quality standards 1446
  REDY system 1423–4
  regeneration 1423–4
  single pass system 1421–2
  tap water preparation 1422–3
  ultrafiltration control 1424–5
dialysers,
  clearance 1421
  high-flux large-surface 1437
  hollow fibre 1420
  parallel plate 1420
  reuse rates 1437
  ultrafiltration control 1441
dialysis, *see* haemodialysis; haemofiltration; peritoneal dialysis; *type of dialysis*
dialysis dementia 1448–9
dialysis disequilibrium syndrome 2329
dialysis membranes 1420–1
  complement activation 1315
  cuprophan membranes *1247*, 1315, 1316
  ethylene–vinyl copolymer, patients at risk of haemorrhage in 1362
  inflammatory mediators, stimulation of 1432–3
  non-specific immune response 1315
  particles 1445–6
  polyacrylonitrile 1315
dialysis tubing 1419
  phthalate-containing plasticizers 1332
  plasticizers 1432
  polyvinyl chloride, photosensitizer causing bullous disorders 1392
  spallation of particles 1432, 1445
dialysis-associated encephalopathy 1401
  aluminium intoxication 2329
    clinical features and management 2329
dialysis-associated headache 2330
dialysis-refractory hypertension, *see* hypertension, dialysis-refractory
diaminodiphenylsulphone, *see* dapsone
diarrhoea,
  alkaline secretion loss from intestine 934
  in amyloidosis 550
  bloody
    differential diagnosis 1047
    E. coli, verotoxin-producing strain 1046, 1048–9
  chloride diarrhoea 947
  chronic, nephrolithiasis 2310
  diet precipitating hypernatraemia 882–3
  haemolytic–uraemic syndrome 2310
  hypernatraemia resulting 882–3
  hyponatraemia 879
  laxative abuse, anion gap measurement 934
  potassium depletion 903

Vol. 1 – Sections 1–5. Vol. 2 – Sections 6–11. Vol. 3 – Sections 12–21.

diarrhoea (cont.):
  tropics, acute renal failure and 1134
  see also intestinal fluids, loss
diatrizoate, secretion by renal tubule 38
diazepam,
  eclampsia, management 1973
  handling in kidney disease 186
  percentage protein binding according to renal function *1240*
diazotization reaction, Griess test for nitrite 1663
diazoxide,
  children, emergencies *2149*
  contraindications 2056
  neonates 1116
  pregnancy-induced hypertension therapy 1971
    contraindications 1972
  side-effects 2056
dibromochloropropane, nephrotoxicity 169
DIC, see disseminated intravascular coagulation
dichloroacetate, administration in lactic acidosis 928
dichloromethylene diphosphate, in hypercalcaemia of multiple myeloma 567
diesel oil, nephrotoxicity 1009
diet,
  acute renal failure, requirements of patients *1020*, 1020–1
  animal proteins
    calcium stone disease **1840–1**
    purine consumption 1842
  body mass, glomerular damage 238
  calcium, foods high in *1850*
  calcium stone disease 1837, 1839–40
    animal proteins 1840–1
    calcium 1839–40
    controlled studies 1838–9, *1839*
    fibre 1841
    oxalate 1841
    refined carbohydrates 1841
  carbohydrates, refined, calcium stone disease 1841
  children, FAO/WHO recommendations *1616*
  dietary history, presenting complaint and **6–9**
  dietary therapy, diabetic nephropathy 528–9
  energy intake, FAO/WHO recommendations *1616*
  fibre, calcium stone disease 1841
  glomerulopathy, body mass 238
  IgA nephropathy, primary 358
  normal, acid and base production 932–3, *933*
  oligoantigenic diet, minimal change nephrotic syndrome 312
  oxalate 1856–7
    calcium stone disease 1841
    foods high in *1850*
  phosphate restriction 753
  potassium depletion 899–900, 903
  potassium in 897, 899
  renal stone disease 1849, *1850*, 1852

diet (cont.):
  vegetarian diets 29, 529
  see also nutrition; *specific substances*
diethylcarbamazine,
  filariasis 185
  Wuchereria bancrofti 185
Tc-99m-diethylene triamine penta-acetic acid, renal imaging,
  in chronic obstruction 71, 123–4
  chronic renal failure 126
  clearance studies 1174
  contrast media in MRI 108
  hydronephrosis 74, *75*
  indications 74
  ischaemic renal disease *1080*, 1080
  measurement of glomerular filtration 116, 1174
  use of DTPA Sn 38–9
  properties 117
  transplantation, monitoring graft function 126–8
  see also dimercaptosuccinate, technetium-labelled
diffuse acute bacterial nephritis, computed tomography scanning 104–5
diffuse chronic tubulointerstitial nephritis, with microcysts, children 2169
diffuse glomerulonephritis, glomerular scarring 1192
diffuse mesangial proliferation, histology 305–7
  see also IgA nephropathy, primary
diffuse mesangial sclerosis 2219
  differential diagnosis 307
  Drash syndrome, children 479
  histology 308
diffuse parenchymal renal disease, Doppler findings 94
  type 1 renal disease 93–4
  ultrasonography 93–4
diffuse proliferative glomerulonephritis,
  in SLE 649, 653–4, 655
  treatment in SLE 659–63
digital subtraction angiography, see renal angiography
digitalis,
  in haemodialysis 1270
  intoxication
    hyperkalaemia 907
    hypokalaemia precipitating 901
    hypomagnesaemia and 1813
    treatment by haemoperfusion 1273
    side-effects, heart failure 1273
digitoxin, in uraemia 1273
digoxin,
  antiarrhythmics, cardiac drugs 187
  atrial fibrillation 188
    prescribing 187
  clearance, haemofiltration 1032
  elderly patients, inappropriate use 66
  hypomagnesaemia and 1813
  inactivation 1273
  inappropriate use, elderly patients 66
  protein binding, renal impairment 176
  toxicity, atrial fibrillation 187
  in uraemia 1273

digoxin (cont.):
  volume of distribution 180
dihydropropoxymethylguanine, see ganciclovir
dihydrotachysterol, action, chronic renal failure 1770
1,24-dihydroxy vitamin D, putative bone resorption factor 1776
1,25-dihydroxy vitamin D_3 61, 577–80
  phosphaturias 752–7
  in sarcoidosis
    chloroquine reducing 580
    increased levels 577, 579
    macrophages producing 579, 580
    physiopathologic events due to 579
  see also calcitriol; vitamin D
24,25-dihydroxy vitamin D_3 61
dihydroxyadenine stone disease 1842
dihydroxyadenine stones **1865**
diltiazem,
  angina, calcium-channel blockers 188
  cyclosporin therapy in renal transplantation
    effects 1561
    interactions *1563*
  cyclosporin-induced renal vasoconstriction 1598
  haemodialysis patients, recommendations *1463*
  hypertension, chronic renal failure 1273
  prevention of post transplant acute renal failure 1018
  see also calcium channel antagonists
dilution tests 42
dilutional acidosis 935
2,3-dimercaptopropanol, British antilewisite, mercury chelation 845
Tc-99m-dimercaptosuccinate, aluminium-related osteopathy 1372
  in amyloidosis 555
  captopril scintigraphy, renovascular hypertension 2103
  children 1711, 1712
  chronic renal failure 126
  indications 71, 72, 73
  β-2-microglobulin amyloidosis 1375
  neonates 2026
  osteitis fibrosa 1372
  osteomalacia 1375
  osteonecrosis 1373
  properties 118
  reflux/duplex kidney 119
  renal mass lesions 126
  vesicoureteric reflux *2146*
  Wilms' tumour 2256
  see also scintigraphy
dimethyl sulphoxide (DMSO),
  in amyloidosis 556–7
  side-effects 557
dinitrophenol-bovine serum albumin complex,
  IgA nephropathy 252, 255
  experimental models 356–7
dinitrophenol-conjugated antigens, complement activation 357

Diodrast, clearance, ageing changes 58
dip-slide method for bacterial count 1662, 1663
Dipetalonema viteae 495
diphenhydramine, in polyclonal immune globulins therapy 1557
diphenylhydantoin,
  drug-induced interstitial nephritis 1090
  dysregulation of the immune system 170
diphenylthiazole, induction of cystic disease, rats 2183
diphosphonate,
  effect on renal scarring *1205*
  effects, hyperparathyroidism *1776*
  hypercalcaemia, management 1777
diphtheria, acute interstitial nephritis 1091
dipyridamole,
  in diabetic nephropathy treatment 530
  focal segmental glomerulosclerosis 321
  membranous nephropathy 383–4
  nephrotic syndrome 284
dipyrone, relative risks of analgesic-associated disease, end-stage renal failure 808
diquat, intoxication 1011
Dirofilaria immitis 495
disequilibrium syndrome 1400–1
  children 1431
  haemodialysis 1428, 2329
disopyramide, indications, GFR 187
disseminated intravascular coagulation,
  acute renal failure 973–4, *974*
  associated clinical conditions *1043*
  clinical features 1042–4
  clotting factors and 1042
  defined 1042
  differential diagnosis 1045
  haematological abnormalities *1044*
  haemolytic–uraemic syndrome 2341
  pathogenesis *1044*, 1044
  pathological activation *1044*
  pneumococcus pneumonia 460
  pregnancy-induced hypertension 1965
distal renal tubular acidosis, see renal tubular acidosis, distal
distal renal tubule, see renal tubules, distal
diuresis,
  action of prostaglandins, NSAIDs and 820
  extracellular volume expansion effects, fractional urate clearance 828
  post obstruction 2012–13
  potassium depletion 906
  prolonged, 'washout period' 42
diuretics **197–213**
  abuse
    Bartter's syndrome 905
    diagnosis 786
    idiopathic oedema 273

Index

diuretics (cont.):
  acute renal failure, prevention of 1017–18
  adverse reactions **208–13**
    diuretic-induced changes in renal function 212
    diuretic-induced impotence 212
    diuretic-induced ototoxicity 212, 2331
    diuretic-induced skin reactions 213
    drugs interfering with secretion 201
    hepatotoxicity 2315
    hypophosphataemia 1792
    intoxications 208
    liver effects 2315
    metabolic disturbances 208–12
    urinary voiding difficulty 212
  antihypertensives
    combinations 201
    importance for therapy 202
    morbidity and mortality 201–2
  asthma 207
  basis for use, pelvic transit time 117
  calcium balance, effects 210
  carbohydrate intolerance 210–11
  carbonic anhydrase inhibitors 197–9
  cardiac failure 188
  cellular mechanisms of action 197–200
  children 2147
    dosage *2148*
    nephrotic syndrome complicated by ARF 1039
  chronic administration
    urate retention 829
    uricosuria 829
  clinical use **201–8**
    acute renal failure 204–6
    arterial hypertension 201–2
    chronic renal failure 204
    drug interactions 208
    liver cirrhosis and ascites 202–3
    nephrotic syndrome 204
    other indications 206–8
  combinations, antihypertensive action 201
  congestive heart failure, and pulmonary oedema 202
  direct effects 202
  diuretic renography, nuclear imaging 123
  efficacy, nephropathic states 176
  hyperkalaemia and 210
  hypernatraemia, elderly patients 883
  hypertension 1177
    end-stage renal failure 2120–1
    prescribing in kidney disease 188
  hypokalaemia and 209–10
  hyponatraemia and 208–9
  inappropriate secretion of ADH syndrome 207, 879
  interactions with other drugs
    allopurinol 208
    angiotensin converting-enzyme inhibitors 208
    antibiotics, aminoglycosides 208
    cardiac glycosides 208
    lithium 208
    NSAIDs 208, 823
  magnesium depletion and 210

diuretics (cont.):
  magnesium excretion 1808
  metabolic alkalosis 210, 946–7
  mixed metabolic disorders *957*
  nephropathic states, protein binding 176
  non-steroidal anti-inflammatory drugs 823
  salt and water retention, hyponatraemia 823
  oedema treatment 270–1
  organ specificity 200
  potassium
    depletion 209–10, 904
    excess 210
    see also hyperkalaemia; hypokalaemia
  prostaglandins, action, and NSAIDs toxicity 820
  proximal tubule diuretics 197–9
  renal hypertension, treatment 2052
  renin secretion and *2071*
  resistance, shunt and complications 203
  sodium channel blockers 200
  sodium–potassium cotransporter inhibitors 199
  thrombosis, nephrotic syndrome 283–4
  trial of, nephrotic syndrome complicated by ARF 1039
  triple metabolic and respiratory disorders *957*
  volume homeostasis control, pregnancy and 1922
  see also specific substances
diverticulitis,
  acute abdomen in 1284
  management 1285
diverticulosis, peritoneal dialysis, elderly patients 1626
djenkol bean poisoning, acute renal failure and 1136
DMSA, see dimercaptosuccinate
DNA, antibodies to, see anti-DNA antibodies
DNA probes, diabetes insipidus, nephrogenic, congenital form 795
DNA receptor 637
  defect in SLE 637
DNA-cytophotometry, staging of renal cell carcinoma 2239
Dobrin's syndrome 578
DOCA–salt hypertension, see deoxycorticosterone–salt model of hypertension
Döhle bodies, hereditary macrothrombocytopenia, nephritis and deafness 2199
dopamine,
  acute renal failure
    diuretic non-response 206
    prevention of 1018
  and exercise, hypertensive subjects 1461
  with frusemide, in respiratory distress syndrome 54
  pregnancy, renal vasodilatation 1915
  preterm neonates, prevention of renal side-effects of indomethacin 54
  renal effects 1178, 1179

dopaminergic agonists,
  animal models of chronic renal failure 1215
  treatment of impotence 1335
Doppler echography **86–97**
  Doppler effect 86–7
  duplex 975
  pulsatility index 88
  see also ultrasonography
Down syndrome,
  amyloid deposits 548, 550
  features and associated abnormalities *2040*
doxazosin,
  hypertension 188
  see also α_1 adrenoceptor agonists
doxorubicin, see adriamycin
doxycycline, handling in kidney disease 185
Drash syndrome,
  diffuse mesangial sclerosis and 308, 479
  features and associated abnormalities *2037*
  nephropathy with pseudohermaphroditism 2219
  Wilms' tumour 2254
  see also Wilms' tumour
drug-induced and drug-related disease,
  acute renal failure in nephrotic syndrome 1038–9
  acute toxic tubular nephropathy 163
  autoimmune glomerulonephritis
    mercury 249–51
    other than mercury 251
  bullous dermatitis, chronic renal failure 1393
  cholestasis *2314*
  hepatotoxicity and nephrotoxicity 1106–7
  interstitial nephritis 166–7, **1087–90**
    clinical features 1087–8
    diagnosis *1087*
    diuretics 1039, 1089–90
    drug types *1087*, 1088–90
  lithiasis 167
  liver disease, cholestasis in 2313
  nephritis, eosinophils, urinary sediment changes 20
  nephropathy **159–71**
    immunological mechanism 169
    mechanisms involved in renal lesions 168–70
    prevention 170–1
    toxic mechanisms 168–9
  pseudohypoaldosteronism type II 776, 777
  renal disease, rheumatology drugs 2295
  renal vasculitis 167
drugs,
  absorption and distribution, elderly patients 66
  abuse, amyloidosis associated with 554
  acute/chronic renal failure, for specific substances, see chronic renal failure, pharmacological interventions

drugs (cont.):
  chronic renal failure
    acute on chronic renal failure 1159
    interventions **1212–17**
    toxicity 1185
  clearance 178–9, *180*
    factors *180*
    haemofiltration and haemodialysis 1032
    sieving coefficient 180
  covert use, 'idiopathic oedema syndrome' and potassium depletion 8
  dialysis
    dialysed/non-dialysed *180*
    first-order kinetics 180
  dose/unit time 179
  elderly patients
    absorption and distribution 66
    elimination 66
    'unofficial use' 7
  elimination rate fraction Q_r 179
  experimental models of glomerulonephritis 169–70
  first-order kinetics 178
  haemofiltration, dosage 1032
  half-life 178
  handling in kidney disease **175–92**, *181*
  in Henoch–Schönlein purpura aetiology 598, *599*
  hepatotoxicity **1282–3**
  interactions with diuretics 208
  nephrotoxicity
    interactions with cyclosporin *1563*
    renal transplantation 1527–8
    see also specific substances
  parent drugs, metabolites and possible adverse effects *177*
  pharmacokinetic changes in renal impairment **175–92**
    dialysis and haemofiltration 180–1
    distribution 176–7
    gastrointestinal absorption 176
    metabolism 177
    peritoneal absorption 176
    renal excretion 177–9
    renal haemodynamics 179–80
    see also specific substances
  prescribing in kidney disease **181–92**
    major errors 175
    see also specific substances
  protein binding, uraemic syndrome *1240*
  tubular secretion 177
  uraemic patient with cardiac problems 1273–4
  urinary sediment changes 22
  volume of distribution 178–9
  see also drug-induced and drug-related disease; *individual drugs, drug groups*
DSA, see renal angiography, digital subtraction angiography
DTPA, see diethylene triamine penta-acetic acid, $^{99}Tc^m$
ductus arteriosus,
  failure to close, frusemide administration 54
  patent, NSAIDs and 821

Index

ductus arteriosus (*cont.*):
  preterm neonates, prevention of renal side-effects of indomethacin 54
  treatment with indomethacin in neonates 54
duodenum,
  chronic renal failure, microbial changes 1281
  chronic renal failure effects 1280
  dysfunction following renal transplantation 1582
duplex kidney(s), *see* kidney(s), duplex
dwarfism, nephrotic glucosuric, *see* Fanconi syndrome
dysautonomia, familial 2216
dyslipidaemia, diabetic nephropathy 1645–8
dyslipoproteinaemia, uraemic **1255–8**
  atherogenesis 1257
  dialysis in 1257
  pathogenesis 1256
  and progression of disease 1257
  treatment 1257–8
dyspepsia, renal transplantation, management 1582
dysplagia hypospadias syndrome, features and associated abnormalities 2036
dysproteinaemias 2301
  neurological involvement 2325
dysuria, causes 1685–6

EACA, *see* aminocaproic acid (EACA)
ear lesions, relapsing polychondritis 2303
Ebola virus, haemorrhagic fever *1045*
echocardiography, uraemic patient with cardiac problems 1271
eclampsia and pre-eclampsia,
  acute fatty liver of pregnancy and 1057, 1933
  acute tubular necrosis 1933
  chlormethiazole 1974
  defined 1956, 1958
  diazepam 1973
  differential diagnosis 1973
  pregnancy-induced disease, hypertension 1956
  fulminating, management 1972–3
  GFR reduction 1933
  hypertension, essential, differential diagnosis 1940–1
  incidence 1959, 1973
  management of eclampsia **1973–4**
  MHC HLA DR4 1959
  mortality 1959
  phenytoin 1974
  pregnancy-induced disease, hypertension 1972–3
  prevention, diuretics 1923
  proteinuria 1940
    significant threshold value 1957
ectopic kidney(s), *see* kidney(s), ectopic
ectopic pregnancy, renal transplantation 1949

ectrodactyly–ectodermal dysplasia–cleft syndrome, features and associated abnormalities 2036
Cr-51-edetate,
  assessment of glomerular filtration 38
  CaNa$_2$EDTA lead mobilization test 839
  clearance, GFR estimation 1174
  elderly patients 60
  lead mobilization test 839
  management of lead intoxication 840–1
  measurement of GFR 116
EDRF, *see* endothelial derived relaxing factor
EDTA, *see* (Cr-51-)edetate; European Dialysis and Transplant Association
efferent arterioles, capillaries of the glomerulus 129, 130, 131
eicosapentaenoic acid,
  animal models of progressive glomerular scarring *1209*
  dietary and pharmacological lipid manipulations *1202*
  dietary supplementation, uraemia 1212
  IgA nephropathy, primary *358*
  triglyceride lowering 1258
Einstein–Stokes radii, measurement of molecules 29, 30
elastase, antibodies to 618, 624
elderly normal subjects, expected survival *1622*
elderly patients **56–67**
  ageing processes 1623
  cadaver donation 1631–2
  chronic renal failure, treatment **1621–32**
  clinical examination 13
  clinical importance of age-related changes in renal function
    cross-sectional studies 57
    renal pharmacokinetics 66–7
    summary 67
  continuous ambulatory peritoneal dialysis 1626
  drugs (NSAIDs), 'unofficial consumption' 7
  electrolytes, psychiatric disturbance and 63–6
  end-stage renal disease, treatment **1621–32**
  glomerular filtration rate
    creatinine clearance 59–60
    β-2-microglobulin 60
    radioisotopic methods 60
  haemodialysis **1473–4, 1625, 1627–32**
  heavy metals exposure causing interstitial nephritis, misdiagnosis 838
  hyponatraemia, predisposing factor 879, *880*
  incidence of main renal disorders *13*
  inulin clearance, glomerular filtration rate 59
  long-term recovery from acute tubular necrosis 1001
  muscle mass changes, potassium depletion 64

elderly patients (*cont.*):
  nocturia 62, 65, 1156
  plasma volume 65
  prostaglandin metabolism, NSAIDs and 821
  psychiatric disturbance, electrolytes 65–6
  radioisotopic methods, glomerular filtration rate 60
  renal transplantation, *see* renal transplantation, elderly patients
  SLE in, clinical features 650
  sodium balance
    distal and proximal nephrons 62
    salt wasting - causes 63–4
    site of sodium loss within nephron 62–3
    sodium loading 61–2
    tubular handling 62
  tubular reabsorption
    acid excretion 61
    bicarbonate reabsorption 61
    calcium and phosphate 61
    glucose 61
    titrable acid 61
    urinary pH 61
  unexplained azotaemia *1078*, 1079–80
  urinary pH 61
electrocardiogram,
  24-hour ECG recording, uraemic patient with cardiac problems 1271–2
  Brody effect 1271
  chronic renal failure 1271
  Holter monitoring, arrhythmias, assessment 1271–2
  uraemic patient with cardiac problems 1271
electrolyte disorders, neurological effects 2332
electrolytes,
  balance
    and fluid balance, neonates 1115–20
    haemofiltration 1031
  depletion, acute renal failure, recovery 980
  haemofiltration, balance 1031
  total parenteral nutrition *1021*
electron microscopy,
  glomerular structure 129–39
  renal biopsy
    diagnostic relevance 150
    specimen processing 146–7
Elejalde syndrome, features and associated abnormalities *2038*
ELISA,
  antiglomerular basement membrane antibodies 239
  schistosomiasis 1731
elliptocytosis, renal tubular acidosis 1893
Ellis van Creveld's syndrome, nephronophthisis, autosomal recessive 2193
embolization,
  renal cell carcinoma 2247–8
  contraindications 2246
  see also renal embolization
embryoma, *see* Wilms' tumour
emergencies,
  accelerated hypertension 2124–36
  hypertensive child 2147–8
  drug treatment *2149*

emergencies (*cont.*):
  main renal disorders *13*
  paraquat poisoning 1137
  sodium nitroprusside administration 2135
Emopremarin, shortened bleeding time, uraemic patients 142
emphysematous pyelitis 532
emphysematous pyelonephritis 531–2
  computed tomography scanning 105
enalapril,
  albumin excretion reduction in diabetes 527–8
  animal models
    effect on glomerular scarring *1213*
  renal ablation, effects *1195*
  children, advantages 2148
  crescentic glomerulonephritis 422
  haemodialysis patients, recommendations *1463*
  hypertension, accelerated 2135
  liver effects 2315
  in scleroderma 676
encephalopathy,
  aluminium
    children 1613
    see also aluminium intoxication
  Binswanger's encephalopathy, elderly patients on haemodialysis 1623
  carbonic anhydrase inhibitors 213
  chronic renal failure 1158, **1398–1400**
    clinical features 1398–9
    differential diagnosis 1399–400
    electrophysiology 1399
    pathogenesis 1399
  cystinosis 1609
  dementia in dialysis patients 1449
  dialysis encephalopathy 1401
  history 2325
  lead intoxication-associated, blood concentrations 838
end-stage renal failure,
  anaemia, left ventricular hypertrophy and 1265
  causes **1227–35**
    geographical areas *1230*
    percentage causes by groups of diseases *1228*
    proportional distribution by age at commencement *1229*
    sex-related factors *1230*
    women at commencement of renal replacement therapy *1232*
  children 1610
  in diabetes mellitus 505, 511–12
    incidence 508
    insulin-dependent 506
    non-insulin-dependent 507–8
    proteinuria onset time and 506, 507–8, 511
  diagnosis 1235
    schema *1235*
  elderly patients, treatment **1621–32**
  erythropoietin, production and deficiency 1349
  gout as cause 833

Index

end-stage renal failure (*cont.*):
  hypertension 1458–9, **2117–22**
  and left ventricular hypertrophy 1265
  hypothyroidism in 1319–20
  left ventricular hypertrophy **1264–7**
  left ventricular systolic/diastolic malfunction **1264–7**
  in multiple myeloma 568
  parathormone, and bone resorption *1761*
  percutaneous transluminal coronary angioplasty, dialysis patients 1274
  pubertal problems 1615
  reflux nephropathy 1993–5
    in children 119–20
  in SLE 659
  ultrasonography 93–4
  vascular disease, proportional distribution by age at commencement *1229*
  see also chronic renal failure
endocapillary glomerulonephritis, acute,
  clinical features 406–8
  complement depletion (C3) 411
  defined **405–15**
  diagnosis 408
  epidemiology 405
  genetic aspects 411–12
  immune deposits and infiltrating cells 414
  infectious agents associated 405
  pathology 412
  poststreptococcal 409–11
  prognosis 414–15
  renal fuctional reserve 415
  treatment 412–14
endocapillary proliferation, in essential mixed cryoglobulinaemia 688
endocarditis,
  cyanotic congenital heart disease 2334
  dialysis patients 1275
  following urinary tract infection 1693
  in renal disease 2337–8
  see also infective endocarditis
endocrine conditions, prescribing 188–9
endocrine disorders,
  chronic renal failure **1317–26**
    growth hormone–somatomedin axis 1322–4
    mechanisms 1317–19
    pathogenesis *1318*
    pituitary–adrenal axis 1324–5
    thyroid disorders 1319–22
  prohormones, disturbed activation 1319
  target tissues, disturbed responses 1319
endocrine glands, in amyloidosis 550
endorphins, uraemic syndrome 1243–4
endoscopy **1877–9**
  combined techniques 1880–1
  complications 1879
  indications 1879
  instrumentation 1877–8
  integrated stone management 1879–80

endoscopy (*cont.*):
  results 1879
  techniques 1878
endostreptosin, streptococcal antigens 409
endothelial antigens,
  autoantibodies to 589, 590, *591*, 617
endothelial leucocyte adhesion molecule (ELAM 1) 257
endothelin,
  arteriolar myocytes of afferent arterioles 25, *26*
  kidney vasoconstriction 2132
  mediation in cyclosporin-induced endothelial injury 1560–1
  vascular tone, integration of signals in vessel wall 985
endothelium,
  cyclosporin-induced endothelial injury 1560
  functions 671
  glomerular capillaries **129–36**, 131–2, 133–6
  glucotoxicity 519
  haemolytic–uraemic syndrome, prostacyclin deficiency 1054
  injury 1052
    von Willebrand factor release 1053
  ionizing radiation effects 849
  platelet-derived growth factor effects 1052
  proliferation, in glomerulonephritis of rheumatoid arthritis 680
  in scleroderma 672
  sickle erythrocytes, adherence to 715
  vascular proliferation 680, 1052
endothelium-derived relaxing factor,
  cyclosporin-induced decrease in activity 1560
  endothelial dysfunction 2132
  nitric oxide 25, *26*, 27
  vascular tone 985
endothelium-derived vasoconstrictor peptide, see endothelin
endotoxins,
  activity, lipopolysaccharide 1665
  pathogenesis of renal damage 991–2
  renal cortical necrosis 993
  Schwartzmann reaction 993
endovascular interventional radiology **113–14**
energy intake,
  children with chronic renal failure 1306, *1307*
  FAO/WHO recommendations *1616*
energy metabolism, children with chronic renal failure 1303
enterobacteria, perinephric abscess 1696, 1706
enterocystoplasty, renal tuberculosis 1727, 1728
entry step, proximal renal tubule 741
enuresis, covert urinary tract infection 1705
enzyme defects, inherited, ketoacidosis 930

enzymuria 18–19
eosinophil chemotactic factor, schistosomiasis 1731
eosinophilia,
  acute renal failure 973–4
  angiolymphoid hyperplasia 300, 2281
  in Churg–Strauss syndrome 629
  interstitial nephritis 972, 1084
  lymphoid granuloma, in Orientals (Kimura's disease) 300, 2281
eosinophils,
  drug-induced nephritis, urinary sediment changes 20
eosinophiluria, renal biopsy indications 144
epa-1 antigen, immune response 1546
epidermal growth factor,
  glomerular scarring, progressive 1192
  glomerular sclerosis, role 1203
  proximal tubule cells, effect on 1204
epididymo–orchitis,
  in children 1706
  tuberculosis 1720–1
  urinary tract obstruction 2010
epidural anaesthesia, urinary retention 2010
epiphyseal slipping, children 1612
epispadias 2035
epithelial benign tumours,
  multilocular cystic nephroma 103
  oncocytoma 103
epithelial cell carcinoma, analgesic nephropathy and 813
epithelial cells 1668
  apical (luminal, brush border) membrane 741
  cell shedding 1671
  foot processes
    in diabetes mellitus 524
    in sarcoid granulomatous interstitial nephritis 577
  potassium transport 897–8
  renal tubular epithelial antigen, carcinoma and 478
epithelial transport mechanisms, amino acids 746
  entry step 741
  exit step 741–2
  proximal tubule 741–3
epithelioid cells, in sarcoid granuloma 576, 577
epoietin, see erythropoietin, human recombinant
Epstein–Barr virus (EBV),
  essential mixed cryoglobulinaemia association 687
  hepatitis 1289
  IgM anti-IgG antibody induction 687
  infection-associated glomerulonephritis 466
  liver disease associated with acute renal failure 1106
  lymphoproliferative disorders, transplantation patients 2346
  renal transplantation 1581
  Sjögren's syndrome aetiopathogenesis 695–6
ergotamine, retroperitoneal fibrosis and 104

ERPF (effective renal plasma flow) 1910, 1919
erythema multiforme, mononuclear vasculitis in 590
erythrocyte sedimentation factor, renal cell carcinoma 2242
erythrocytes,
  cell casts
    in diabetes mellitus 511
    non-diabetic (nephropathy) renal disease 526
  cell membrane damage in sickling cycle 701
  changes, haemolytic–uraemic syndrome 1054
  development, see erythropoiesis
  in dialysate, detection 1422
  distortion, assessment 231
  erythropoietin replacement therapy, r-HuEPO 1353–6
  isomorphic/dysmorphic, in urinary sediment 19–20
  malonyldialdehyde, haemolytic–uraemic syndrome 1054
  mass 1344
  normal values *1344*
  permeability, spurious hyperkalaemia 907
  phosphate in *1789*
  potassium, elderly patients 64
  precursors 1344
  progenitors 1344–5
  receptors 1344
  in sickle-cell anaemia 700–1
  see also haemoglobin; sickle-cell disease
  sodium content, ageing changes 63
  superoxide dismutase, deficiency in haemolytic–uraemic syndrome 1054
  6-thioguanine nucleotides, measurement, thiopurine methyltransferase metabolism 1950
  uraemic haemorrhage, erythropoietin and 1362
erythrocytosis,
  minimal change nephrotic syndrome 302
  polycystic disease, autosomal dominant 2176
  renal cell carcinoma 2242
erythromycin,
  CAPD peritonitis regimens *1489*
  cyclosporin therapy in renal transplantation interactions *1563*
  nephrotoxicity 1573
  drug-related interstitial nephritis 1089
  haemofiltration/haemodialysis 180
  handling in kidney disease 183
erythropoiesis,
  chronic renal failure 1282
  normal
    factors required 1345
    red cell progenitors, action upon 1344–5
  uraemic inhibitors 1350
    animal models 1350
    postulated substances 1350

Index

erythropoietin,
  anaemia, children 1610, 1611
  androgens and 1353
  bone marrow production 1345
  calcium channel blockers and 1350
  deficiency
    anaemia in 2344
    chronic renal failure 1349
    uraemic syndrome 1245
  drug-induced effects 1350
  elevated, renal cell carcinoma 2242
  end-stage renal failure, deficiency 1349
  foot in diabetes 1642
  gene 1346
  haemodialysis, sexual function effects 1335
  homology
    other proteins 1346
    other species 1346
    recombinant 1346
  human recombinant 1462
    adverse effects 1355–6
    anaemia in acute renal failure 1518, 2344
      treatment 1187
    benefits 1355
    iron deficiency and 2344
    neurological effects 2332
    pharmacokinetics 1355
    r-HuEPO availabilty 1353–4
  hypertensive encephalopathy and 2133–4
  immunoreactive, estimates, anaemia and normal blood *1349*
  left ventricular function and diameter 1265
  levels
    changes *1347*
    states causing changes *1347*
  obstruction and 2013
  physiology
    factors required 1346–7
    mode and site of production 1346
    oxygen sensory control 1347
  platelet count and 1356
  production
    drugs and 1350
    end-stage renal failure 1349
    polycystic disease 1349
    renal artery stenosis 1350
    renal tumours 1350
    transplantation 1349–50
  receptors 1344–5
    animal models 1344
    expression 1345
  recombinant human, *see* erythropoietin, human recombinant
  red blood cells and 1362
  renal production, progressive chronic renal failure 1318
  in sickle-cell disease 714–15
  specific binding sites 1345
  steroids and 1353
  synthesis, feedback *1347*
Escherichia coli,
  α haemolysin and β haemolysin 1666
  abscess 470
  adherence properties 1666–9

Escherichia coli (*cont.*):
  aerobactin production 1666
  bacteraemia, haematogenous infection 1678
  capsular (K) antigens 1665
  cystitis, recurrent, pregnancy and 1928
  enterochelin (enterobactin) 1666
  faecal, pathogenicity 1678
  fimbriae, haemagglutination properties 1666–9
  haemolysin production, and pathogenicity 1666
  incidence in urinary tract infection 1685
  infected stone disease 1831, 1863
  infection-associated glomerulonephritis 461–2
  iron acquisition and haemolysin production 1665–6
  mannose and mannosides 1666
  mannose-resistant fimbriae 1667
  microbial virulence factors 1665–9
  neonates 1699, 1701
  O-serotypes
    serum antibody titre 1664
    upper tract infection 1712
  p-fimbriae 1704
  percentage of strains and markers *1668*
  periurethral flora 1660
  plasmid pColV 1666
  resistance to serum bactericidal effect 1665
  schistosomiasis 1736
  somatic (O) antigenic lipopolysaccharides (endotoxins) 1665
  upper urinary tract infection 1681
  urinary tract infection 1661
    spermicides 1685
  uropathogenic virulence factors 1665
  verotoxin-producing strain
    animal models 1048–9
    haemolytic–uraemic syndrome 1046, 1048
    infection outbreaks *1049*
  virulence factors *1703*, 1703–4
    adaptations of organism 1928
essential mixed cryoglobulinaemia (ECM) **687–92**
  acute reversible syndrome 691, 692
  age, sex and geographical distribution 690
  antigens in 587, 687
  B cell clone producing monoclonal anti-IgG 687
  causes of death 690
  clinical features **690–1**
    extrarenal 690
    renal 690
  clinico-histological syndromes 691, 692
  complement levels 588, 691
  Epstein–Barr virus association 687
  hepatitis B virus association 687, 691
  IgM-rheumatoid factor in 686, 687, 689, 691
  immune complexes
    deposition 688–9, 691, 692
    size/composition 587, 687

essential mixed cryoglobulinaemia (ECM) (*cont.*):
  laboratory findings 691
  leucocytoclastic/mononuclear vasculitis overlap 589, 690
  natural history and prognosis 690–1
  pathogenesis **687**
  renal damage, pathophysiology **691**
  renal disease in 1069
  renal pathology **687–90**
    glomeruli 687–9
    interstitium 689
    membranoproliferative glomerulonephritis 687–8, 691, 692
    vessels 589, 690
  treatment **691–2**
    evaluation of efficacy 691
ethacrynic acid,
  nephrotic syndrome 270
  urinary appearance 22
ethambutol,
  drug dosage, renal impairment *184*
  drug-related interstitial nephritis 1089
  renal tuberculosis *1727*
  urate excretion decrease *829*
ethanol, *see* alcohol
ethiofos, *see* WR-2721
ethionamide, renal tuberculosis *1727*
ethnic factors,
  hypertension, and progressive renal failure 2126
  malignant hypertension risk 2128
  plant toxin remedies, use 1125
  presenting complaint 9–10
  skin-lightening cream, use 161
  Takayasu's arteritis, children 2144
ethylene dichloride, nephrotoxicity 993
ethylene glycol,
  intoxications *931*, 931–2
  diagnosis 972
  metabolic pathway *931*
  nephrotoxicity 1009
  oxalate ingestion 1895
ethylene oxide, first use syndrome 1429
etidronate, hypercalcaemia, management 1777
Europe,
  analgesic nephropathy, north–south gradient 804
  European vs English-speaking analgesic preferences 803–4
European Dialysis and Transplant Association,
  causes of end-stage renal failure, classification **1227–35**
  growth in chronic renal failure, data 1301
  renal replacement therapy in amyloidosis 558–9
European Dialysis and Transplant Association Registry 804
  hypertension incidence 2128
European Dialysis and Transplant Association–European Renal Association registry,
  paediatrics 1605
  renal replacement therapy *1624*

excretory urography, renal mass 2244
exercise,
  albuminuria and 510, 529
  marathon-running, hyperkalaemia 908
  Na-K ATPase 897
  potassium depletion 903
  potassium homeostasis 897
  urinary sediment changes 19
exercise testing, uraemic patient with cardiac problems 1271
exercise training,
  animal models, progressive chronic renal failure 1217
  triglyceride lowering 1258
exit step, proximal tubule 741–2
exocrine glands, in Sjögren's syndrome 693, 694
experimental models, *see* animal models
exstrophic anomalies 2035
extracellular fluid,
  calcium transport **1764–8**
  compartments, Starling forces 264–5
  and dialysate, comparison *1423*
  Henderson equation 918
  normal values *917*
  phosphate distribution *1789*, 1789
  tissue perfusion fluid, formula 264–5
  ultrafiltration, formula 264–5
  volume
    glomerular hyperfiltration and 513
    renal acidification 765
  volume expansion
    reabsorption, phosphate 1787
    urate/uric acid, increased clearance causing hypouricaemia 830
  see also fluid homeostasis; oedema
extracellular matrix, abnormalities in diabetic nephropathy 518–19
extracorporeal immunoadsorption, HLA antibodies in transplantation 1516–17
extraction efficiency E, measurement by nuclear imaging 116–17
  see also renal plasma flow
eye disorders,
  Alport's syndrome 2197
  cystinosis 2221–2
  nail–patella syndrome 2215
  nephronophthisis, autosomal recessive 2193
  in Wegener's granulomatosis 616
eyes, clinical examination 12

Fab region of immunoglobulin 562
Fabry's disease, *see* Anderson–Fabry disease
facio–auriculo–vertebral syndrome, features and associated abnormalities *2037*, *2039*
factitious history, presenting complaint 9
factor D, reduced in sickle-cell anaemia 704

Index

factor V, nephrotic syndrome 281, 282
factor VIII:von Willebrand factor, index of injury, renal circulation 1053
factor VIII,
  congenital nephrogenic diabetes insipidus 886
  cryoprecipitate, shortened bleeding time 142
  renal circulation, thrombocytopenia 1053
  in scleroderma 675
  SLE-like inhibitor, gastrointestinal bleeding in chronic renal failure 1284
factor VIII coagulant,
  DDAVP-stimulated release 793–4
  pregnancy-induced hypertension 1963
factor X 281, 282
  pregnancy-induced hypertension 1964
factor XI, pregnancy-induced hypertension 1964
factor XII 281, 282
  activation by contrast media 77–8
  pregnancy-induced hypertension 1964
failure to thrive, neonates 1605–6
familial amyloidotic polyneuropathy 545
familial dysautonomia 2216
familial hyperphosphataemia 1798
familial juvenile nephronophthisis 1608
  ophthalmological disease 1607
familial Mediterranean fever 553–4
  amyloidosis 1153, 2300
  treatment 556
family history,
  presenting complaint 9
  see also congenital disease; inherited disorders
Fanconi syndrome 567, **723–36**
  acquired 734
  acute plumbism 839
  assessment of tubular dysfunction 728
  associated diseases, list 729
  causes 1608
  children **1608–9**
  classification, aetiological 728
  clinical features 728–35
  common features 1796–7
  in cystinosis 2219–24
  cytochrome-c-oxidase deficiency 729, 733–4
  defined 723
  differential diagnosis, Hartnup disease 750
  dogs, spontaneous 735
  familial syndromes, classification 1796
  fructose intolerance, hereditary 729, 732
  galactosaemia 729, 731
  glycogenosis 729, 730–1
  immunoglobulin light chain proteinuria 996
  incomplete syndrome 728–9
  lead intoxication and 838, 839
  liver cirrhosis 2311
  Lowe syndrome 729, 732–3

Fanconi syndrome (cont.):
  morphology of the kidney
    Lowe syndrome 733
    Wilson's disease 733
  myeloma association 567
  nephrocalcinosis 1893–4
  oculocerebrorenal syndrome 729, 732–3
  pathophysiology **723–8**
    acidosis 726
    carnitine excretion 727
    glycosuria 725
    hyperaminoaciduria 724–5
    hypercalciuria 727
    hypokalaemia 726–7
    hyponatraemia 727
    lactate excretion 727
    p-aminohippurate, tubular secretion 727–8
    peptides excretion 727
    phosphate wasting 725–6
    polyuria 727
    proteinuria 727
    uric acid wasting 726
  physiological findings 1797
  primary 729, 729–30
  proximal tubular dysfunction 567
  renal morphology 729
  renal transplantation 734
  renal tubular acidosis 44
    type II 207
  in Sjögren's syndrome 696
  treatment 735–6
  tumour-related 734
  tyrosinaemia type I 729, 730
  urate/uric acid, increased clearance 830
  variants, tubular reabsorption defect of sodium pump 742
  Wilson's disease 729, 733
Fanconi's familial nephronophthisis, see nephronophthisis
Farber disease, ceramidase deficiency 2206
fat,
  mass and body weight, by age 1296
  skinfold measurement 1301
fats, saturated,
  age-related glomerular scarring 1212
  animal models of progressive glomerular scarring 1209
fatty acids,
  heparin and 1322
  omega-3 polyunsaturated, chronic renal failure effects 1181
Fawn–Hooded rats, accelerated glomerular lesions 1207
Fc region of immunoglobulin 562
felodipine, animal models, effect on glomerular scarring 1213
Felson's silhouette sign 79
femoral artery, superficialization 1410
fenoprofen,
  acute interstitial nephritis and renal failure 67
  interstitial nephritis 823
  renal papillary necrosis 822
fenoterol, hypokalaemia 902
fentanyl, handling in kidney disease 185

fertility, chronic renal failure 1337–8
fetal alcohol syndrome, features and associated abnormalities *2040*
fetal antigens, re-expressed 477
fetal surveillance, and timing of delivery 1938–9
  nipple stimulation test 1939
  oxytocin challenge test 1939
  renal transplantation 1951
fetal survival,
  in diabetic nephropathy 531
  in lupus nephritis 651
fetoplacental unit, renal dilatation and 1914
fetus,
  abnormalities
    antenatal diagnosis 2025
    obstetric ultrasound 2025
    Bartter's syndrome 1894
  development, urinary system 2023–4
  embryology 2023–4
  nephrocalcinosis 1894
  recurrent loss, antiphospholipid antibodies 7
  ultrasonography of the urinary tract 1995
  urinary system, development 2023–4
  X-ray exposure *in utero* 81–2
fever of unknown origin, haemodialysis 1452–3
fibre, dietary intake 1841
fibric acid derivatives, nephrotic hyperlipidaemia, effects 288
'fibrillary' nephritis 2281–3
β-fibrilloses, see amyloidosis
fibrin, deposition in animal models of glomerulonephritis 257
fibrin cap, in diabetic nephropathy 523, 525
fibrinogen,
  in glomerular basement membrane 524
  nephrotic syndrome 281, 282
fibrinoid deposits, systemic lupus, indications 145
fibrinoid necrosis,
  animal models 584
  in Churg–Strauss syndrome 613
  in polyarteritis nodosa 628
  in scleroderma 673, 675
  vascular diseases with renal involvement, schematic representation 152
fibrinolysis,
  failure in haemolytic–uraemic syndrome 1054
  inhibition, sickle-cell disease treatment 716
  platelets and, pregnancy-induced hypertension 1964–5
  pregnancy-induced hypertension 1963
fibrinolytic proteins,
  list and molecular weights 281
  nephrotic syndrome 281, 282
fibrinopeptide β1–42, pregnancy-induced hypertension 1964
fibroblast growth factor, role in glomerular sclerosis 1203

fibroid necrosis, renal arterioles 2131, 2134
fibromuscular dysplasia,
  differential diagnosis 2098
  renal arterial disease 1078, 2097–8
  see also renal artery stenosis
fibronectin, mesangial cells 137
fibro-osteoclasia 1376–7
  see also osteitis fibrosa
fibular nerve, conduction velocities 1397
filariasis,
  diethylcarbamazine 185
  glomerulopathy **495–6**
  infection-associated glomerulonephritis 468
filling defects, collecting system 85
filtration fraction,
  defined 1910
  pregnancy, decrease 1910, 1912
filtration in the glomerulus, see glomerular filtration
filtration pressure equilibrium, defined 1911
  pregnancy changes 1912
fimbriae, uropathogenic bacteria 1666–9
fimbrial receptors, oral administration 1672
fine-needle aspiration biopsy, indications in transplantation 145
  see also biopsy, renal
Finnish congenital nephrotic syndrome 2169, 2218–65
first use syndrome, haemodialysis 1429
fistula,
  leading to bicarbonate loss 935
  potassium depletion 935
fixation,
  electron microscopy 146–7
  light microscopy 146
FK–506,
  rejection of allograft, action 1555–6
  see also cyclosporin
flank pain syndrome, non-steroidal anti-inflammatory drugs 823
flecainide, antiarrhythmics, cardiac drugs 187
fleroxacin, CAPD peritonitis regimens *1489*
flucloxacillin, dose adjustment *181*
fluconazole,
  dosage nomogram *1749*
  fungal infections 1748
  handling in kidney disease 185
flucytosine,
  CAPD peritonitis regimens *1489*
  dosage nomogram *1748*
  fungal infections 1748
  handling in renal impairment 184
  see also fluorocytosine
fludrocortisone,
  hyperkalaemic renal tubular acidosis 776, 777
  mineralocorticoid replacement 909
fluid depletion,
  chronic renal failure 1184
  hyponatraemia 875–6
  stimulation of renin–aldosterone axis 935–6
  see also hypovolaemia

Index

fluid deprivation test 42
  diabetes insipidus 884
fluid homeostasis,
  and electrolyte balance, neonates 1115–19
  pregnancy 1918–1920
    diuretics and 1922
    sodium excretion, increase/decrease 1919–1920
    sodium handling 1918–1919
  pregnancy, animals as gestational models 1921
  volume receptors 262–3
  volume status
    assessment 874–5
    hypernatraemia 882
fluid intake,
  calcium stone disease 1837
  in sickle-cell disease 716
fluid replacement, renal transplantation, stable vs metabolizable molecules 1521
fluid restriction, and fluid balance 1115–16
fluid retention **262–74**
  acute nephritic syndrome, defined 266
  anaemia in chronic renal failure 1349
  circulatory assessment, uraemic patient with cardiac problems 1271
  haemodialysis, long-term 1441
  idiopathic oedema 273
  IgA nephropathy 342
  left ventricular hypertrophy and, end-stage renal disease 1265
  liver cirrhosis 272
  nephrotic syndrome, defined 266–7
  in renal failure in diabetics 511
  secondary
    congestive heart failure 272
    idiopathic oedema 272
    liver cirrhosis 272
    pregnancy 272–3
  uraemia, left ventricular hypertrophy 1265
  see also ascites; extracellular fluid; oedema; water balance
fluid volume contraction, obstetric acute renal failure 1932
fluid volume expansion,
  neonates, contraindications 1115
  pregnancy-induced hypertension, SG catheter 1973
  renal transplantation 1518–19
fluorescein-conjugated antihuman immunoglobulin 1664
fluoride,
  dietary uptake, fluorosis 1380
  inhibitor of calcium phosphate crystallization in urine *1827*
5-fluoro-2-deoxyuridine, infusional chronotherapy 2248
fluorocytosine,
  candidosis 1746
  creatinine clearance and measurement, interfering substance *32*
  see also flucytosine
fluorosis, nephrocalcinosis 1896
5-fluorouracil, haemolytic–uraemic syndrome 1048

foam cells,
  Alport's syndrome 2199
  atheroma, hyperlipidaemia in nephrotic syndrome 291
  lecithin:cholesterol acyltransferase deficiency, familial 2211
focal acute bacterial nephritis, computed tomography scanning 105
focal glomerular sclerosis,
  interstitial fibrosis with 578
  in renal transplantation 1531
  see also nephrotic syndrome with focal glomerular sclerosis
focal necrotizing glomerulonephritis,
  crescent formation with 624
  cytotoxic therapy 626
  idiopathic 623, 624, 627
  in microscopic polyarteritis 623, 624
  plasma exchange in 620
  treatment and outcome *1067*
  vasculitis and 1065–8
  in Wegener's granulomatosis 613, 616
focal proliferative glomerulonephritis,
  in SLE 653, 655, 659
  severe, treatment 659–63
focal segmental fibrosis, histology 304
focal segmental glomerulonephritis,
  in Churg–Strauss syndrome 629
focal segmental glomerulosclerosis,
  characteristic immunohistologic patterns 147, 149
  dipyridamole 321
  electron microscopy 305
  familial steroid-resistant nephrotic syndrome 2219
  glomerular disease in the tropics 487
  glomerulonephritis 1737
  glycogenosis, type I 2311
  histology 304–5
    electron microscopy 305
  hypertension in 1198
  immunofluorescence studies 306
  minimal change disease vs. nephrotic syndrome 291
  minimal change nephrotic syndrome, single disease concept 298
  nephrotic syndrome 291
  pathogenesis, mesangial cells 313
  renal transplantation, recurrence following 1573
  in sickle-cell disease 711–12
  see also minimal change nephrotic syndrome
focal and segmental proliferative glomerulonephritis, in Henoch–Schönlein purpura 602, *604*
folate,
  excretion, pregnancy and 1916
  RDA 1351
  supplement, in sickle-cell disease 716
  total parenteral nutrition in ARF *1021*

folate deficiency,
  children with chronic renal failure 1308
  haemodialysis patients 1351
  iron and 1351
Folin/Lowry method, detection of urine protein 227, 228
follicle stimulating hormone, effects 1298
  uraemia
    females 1337
    males 1333
food allergy, in Henoch–Schönlein purpura aetiology 598, *599*
foot, atherosclerosis lesions 1642
foot in diabetes,
  chronic renal failure 1641–2
  erythropoietin treatment 1642
  neuropathic foot 1641–2
Forbes' disease, enzyme defect *2210*
formalin,
  anaemia and 1351
  haemolysis and 1351
free radicals, mediation of injury, renal circulation 1052
free water clearance (C_{H_2O}) 871–2
frost, uraemic, terminal renal failure 1390
fructose 1-phosphate, accumulation 732
fructose 1-phosphate aldolase B, deficiency 732
fructose,
  chronic administration, acute uric acid nephropathy 832
  excretion, pregnancy and 1916
  hyperuricaemia and 830
  peritoneal dialysis solution 1481
fructose intolerance,
  chromosomal locus *1796*
  fructose administration, proximal tubule dysfunction 830
  hereditary *729*, 732
    clinical features 732
    differential diagnosis 732
fructose kinase, deficiency, differential diagnosis 732
fructosuria, essential, differential diagnosis 732
frusemide,
  action on ascending limb 124
  acute administration, renal tubular acidosis test 775, 776
  acute renal failure
    animal models 205
    prevention 171
  adverse effects 206
  basis for use, pelvic transit time 117
  chronic renal failure 204, 272, 1647
  diuretic renography 124
  dosage 204
    advanced renal insufficiency 1273
    high dose 204
  drug-induced interstitial nephritis 1089–90
  drug-induced lithiasis 168
  in hypercalcaemia of multiple myeloma 567
  hyponatraemia 881
  hypoproteinaemia 271

frusemide (cont.):
  inappropriate secretion of ADH syndrome 207
  intravenous, ototoxicity 270, 271
  magnesium excretion 1808
  neonates 1115, 1116
    preterm neonates with respiratory distress syndrome 54
    nephrocalcinosis following 1892
    side-effects 54
  nephrotic syndrome 270
  nephrotoxicity, synergism with antibiotics, aminoglycosides 1010
  NSAID interactions 823
  prevention of acute renal failure **1017–18**
  renal transplantation, immediate post-operative administration 1522
  resistance, nephrotic syndrome 203
  structural formula *199*
  uraemia, extrarenal clearance 272
  uraemic patient with cardiac problems 1272
  urinary appearance 22
  urinary pH *footnote* 937
  see also diuretics; loop diuretics
Fryns syndrome, features and associated abnormalities *2037*
fucose, excretion, pregnancy and 1916
fumarylacetoacetate, accumulation 730
fumarylacetoacetate hydrolase, deficiency 730
fungal infections,
  aetiological diagnosis and treatment 1748–9
  aspergillosis 1746–7
  blastomycosis 1742–3
  candidosis **1743–6**
  coccidioidomycosis 1743
  cryptococcosis 1746
  histoplasmosis 1742
  infection-associated glomerulonephritis 466
  mucormycosis 1747
  paracoccidioidomycosis 1743
  skin 2307
  unusual infections 1747
fungi,
  poisoning
    acute renal failure and 1136
    renal and hepatic failure 2310
furanpropanoic acid derivatives, decreased drug protein binding 1240, 1245

G hypospadias syndrome, features and associated abnormalities *2036*
G protein, membrane-associated, integration of signals in vessel wall 985
GABA, hypomagnesaemia and 1813
galactocerebroside-β-galactosidase deficiency, Krabbe disease 2206
galactokinase, deficiency, galactosaemia 731

galactorrhoea,
  haemodialysis, treatment 1338
  in haemodialysis patients 1338
galactosaemia,
  Fanconi syndrome 729, 731
  galactokinase deficiency 731
galactose 1-phosphate,
  accumulation 731
galactose 1-phosphate uridyl
    transferase, deficiency 731
galactose–glucose malabsorption
    745
α-galactosidase A deficiency, see
    Anderson–Fabry disease
α-galactosyl residues, antibodies in
    Henoch–Schönlein purpura 600,
    601
Galerina poisoning, acute renal
    failure and 1136
gallamine, avoiding in kidney
    disease 186
gallbladder, radiology, acute
    acalculous cholecystitis 979
gallium scan,
  children 1712
  differential diagnosis 976
gallium-67 citrate scan, in sarcoid
    granulomatous interstitial
    nephritis 578
gamma camera studies,
  iodine-123, advantages 117
  iodine-131, contraindications 117
  nuclear imaging using
    radiopharmaceuticals 117
gamma-glutamyl transferase,
  HELLP syndrome, pregnancy-
    induced hypertension 1961
ganciclovir,
  cytomegalovirus infection,
    creatinine elevation 1527
  dose in renal impairment 2343
  handling in renal impairment 184
  rapidly progressive
    glomerulonephritis 1064
gangrene,
  diabetic foot 1642
  peripheral, CRF 1393
Gardnerella vaginalis, urinary tract
    infection 1661
Gardner–Diamond syndrome,
  membranous nephropathy 375
gastric acid hypersecretion,
  chronic renal failure 1280
  dialysis patients 1280
  gastrin, elevated 1280
gastric acid hyposecretion, gastritis
    and 1280
gastric alkalosis,
  H_2 receptor antagonists 946
  Zollinger–Ellison syndrome 946
gastric inhibitory polypeptide,
    elevated, chronic renal failure
    1280
gastric mucosa,
  hypertrophy, chronic renal
    failure 1279
  permeability changes pre dialysis
    1279
gastric secretions,
  loss
    consequences, schema 946
    metabolic alkalosis 946
gastric surgery, renal
    transplantation, prior to 1517

gastrin, hypochlorhydria and 1280
gastritis, erosive, analgesic
    nephropathy and 813
gastrointestinal bleeding,
  angiodysplasia 1283–4
  causes 2311
  chronic renal failure 1283–4
  drug therapy 1284
  following transplantation 1285
  haemodialysis patients 1283
  in Henoch–Schönlein purpura 596
  hypercatabolism 978
  β-2-microglobulin amyloidosis
    2311
  ulceration 2311
  visceral amyloidosis 1284
gastrointestinal complications,
  of analgesic nephropathy 813
  following renal transplantation
    1582
gastrointestinal disorders,
  in acute nephritic syndrome 2311
  amyloidosis AA 2310
  CAPD 2312
  carcinoma, complication of
    glomerulonephritis 2310
  haemodialysis 2311–12
  nephrotic syndrome 2311
  renal cell carcinoma 2242
  with renal problems 2310–12
  in transplantation 2312
gastrointestinal drugs, see H_2-
    receptor antagonist, prescribing
gastrointestinal fluids, loss,
    various causes 934
  see also diarrhoea
gastrointestinal infections,
  following renal transplantation
    1580
gastrointestinal obstruction,
  chronic, in Henoch–Schönlein
    purpura 597
gastrointestinal pseudo-obstruction,
  in amyloidosis 550
gastrointestinal tract,
  in amyloidosis 549
  chronic renal failure effects
    1278–92
  following transplantation 1285
  hepatitis 1285–92
  management of common clinical
    problems 1283–5
  on structure and function
    1279–83
  upper tract changes 1279–80
  drug malabsorption 176
  examination in acute renal failure
    970
  in Henoch–Schönlein purpura 596
  transport, potassium homeostasis
    897
  in Wegener's granulomatosis 616
gastrointestinal ulceration, see
    peptic ulcer
gastroparesis, in diabetics with
    chronic renal failure 1641
Gaucher disease,
  glucocerebroside–β-galactosidase
    deficiency 2206, 2212
GBM, see basement membrane;
    glomerular basement membrane
gel chromatography,
  disadvantages 1241

gel chromatography (cont.):
  uraemic syndrome, identification
    of metabolites 1241–2
gelatin, peritoneal dialysis solution
    1481
gemfibrozil,
  chronic renal failure 1181, 1186
  hyperlipidaemia 1216
  handling in kidney disease 191
  nephrotic hyperlipidaemia,
    effects 288
  triglyceride lowering 1258
genetic engineering, rejection of
    allografts, future prospects 1558
genetics of renal disorders
    2155–233
  Alport's syndrome 2160–1
  polycystic disease, autosomal
    dominant 2157–60
  predisposition
    family history 10
    main groups of inherited
      diseases 10
  renal cell carcinoma 2161–2
  see also chromosome loci; X-
    linkage
genital oedema, peritoneal dialysis
    1494–5
genitourinary abnormalities,
    incidence, Wilms' tumour 2254
genitourinary tuberculosis 1720–1
  IVU 1721–3
gentamicin,
  animal models of nephrotoxicity
    994
  CAPD peritonitis regimens 1489
  cyclosporin therapy in renal
    transplantation, interactions
    1563
  doses and therapeutic plasma
    concentrations of
    aminoglycosides, handling in
    kidney disease 182
  Fanconi syndrome, caused by 735
  magnesium wasting and 1810
  nephrotoxicity 994, 1010
  low in preterm neonates 54
  prostatitis 1689
  renal failure 169
  upper tract infection 1693, 1696
geophagia, potassium depletion
    903
Geotrichum, infection 1748
Gerota's fascia, thickening
  polycystic disease 101
  renal abscess 105
giant cell arteritis 613, 630
  giant cells 591
  treatment 1069
  vascular infiltrate 591
giant cells,
  interstitial granuloma 1085
  in myeloma cast nephropathy 566
  in sarcoid granuloma 576
gingival hypertrophy 1566, 1584
  cyclosporin, nephrotoxicity 1566
glafenine,
  drug-induced lithiasis 167
  drug-related interstitial nephritis
    1089
glaucoma,
  carbonic anhydrase inhibitors 206
  secondary, diabetic retinopathy
    1640

glibenclamide,
  elimination in chronic renal
    failure 1645
  hypoglycaemia 189
  see also sulphonylurea compounds
glicazide, hypoglycaemia 189
α-globin,
  hypervariable region (HRV) 2157
  polymorphism mechanism 2157
β-globin, see haemoglobin, β-globin
glomerular, see also glomerulo-;
    glomerulus(i)
glomerular basement membrane
    131, 134–7
  accumulation, in diabetic
    nephropathy 518–19
  ageing changes 57
  Alport's syndrome 2199
  in amyloidosis 552, 573
  autoantibodies, autoimmune
    glomerulonephritis 239, 247,
    618
  biochemical composition 135
  carbohydrate abnormalities, in
    diabetic nephropathy 519
  cationic bovine serum albumin
    252
  connections to mesangial cells 137
  electronegative charge 135
  EM microscopy, morphometry
    153
  filtration mechanism 229–30
  focal segmental glomerular
    sclerosis 313
  glomerulonephritis, autoimmune
    antibodies 239, 247, 618
  heparan sulphate synthesis,
    reduced 516, 519
  immunohistology,
    mesangiocapillary
    glomerulonephritis 389, 391
  injury and mesangial
    proliferation 313
  laminae rara interna/externa and
    densa 134
  in light chain deposition disease
    570
  liver glomerulonephritis 359–60
  lytic lesions, in Henoch–Schönlein
    purpura 606
  membranous nephropathy, spike
    deposits 372–3
  minimal change nephrotic
    syndrome
    generalized disorder of negative
      charges 310
    sialic acid content 308–9
  negative charge loss,
    permselectivity 268
  new formation, renal biopsy,
    diagnostic relevance 150
  non-enzymatic glycosylation
    affecting 517
  proteoglycans 135
  reduplication, in sickle-cell
    disease 713
  spike deposits, membranous
    nephropathy 372–4
  structural antigens 253
  thickening
    in diabetes mellitus 518–19, 522,
      523, 524, 525
    in essential mixed
      cryoglobulinaemia 689
    pathology 524, 525

Index

glomerular basement membrane (*cont.*):
  type IV collagen 135
  width 135
  see also glomerular capillary wall; glomerulus
glomerular basement membrane disease,
  diagnosis 1150
glomerular sclerosis in progressive chronic renal failure 1203
nephritis, renal transplantation 1514
glomerular capillaries,
  afferent/efferent arterioles 129, 130, 131
  endothelial structure 131–2
    antigen system 132
    fenestrations, number/area 132
    luminal membrane 132
    selective permeability 135–6
    ultrastructure 133–4
  infiltration, in essential mixed cryoglobulinaemia 688, 691, 692
  intraluminal thrombi, in essential mixed cryoglobulinaemia 688–9
  length, increase in diabetes mellitus 514
  surface area, reduction in diabetes mellitus 512, 516, 525
  ultrafiltration coefficient K_f
    defined 1911
    pregnancy-induced changes, rat models 1912, 1913
    and SNGFR, pregnancy 1912
glomerular capillary wall,
  in amyloidosis 552
  as blood–urine barrier 514
  in diabetes mellitus, glycaemic control effect 529
  electrical charge 514
  glycosylated protein transport 515
  loss of negative charge and microalbuminuria 514–15, 518
  shunt pathway in diabetic nephropathy 516
  sialic acid component 516
  size selectivity defect in diabetic nephropathy 516
  subendothelial deposits, *see* subendothelial deposits
  subepithelial deposits, *see* subepithelial deposits
  thickened
    in diabetic nephropathy 518–19, 522, 523, 524
    in lupus nephritis 652, 653
    wire loop appearance 652, 653
  see also endothelial cells; glomerular basement membrane
glomerular filtration rate **26–39, 135–6**
  action of prostaglandins, NSAIDs and 820, *821*
  age and sex differences 59
  anaemia and 1156
  angiotensin converting-enzyme inhibitors 180
  area of available surface 26
    transmembrane pressure *vs* osmotic pressure 27–8
  attenuation of decline in renal function 1180

glomerular filtration rate (*cont.*):
  calculation, creatinine, plasma 238
  children 1297
  chronic renal failure 272
  contrast media, acute on chronic renal failure 1186
  creatinine clearance
    creatinine concentration, plasma and 1937
    elderly patients 59–60
  determinants, uncomplicated nephrotic syndrome 1034–5
  in diabetes mellitus
    blood pressure control affecting 527
    early increase, *see* glomerular hyperfiltration
    pathophysiology of decline 516
    progressive decline 510, 516
    restricted protein intake improving 528
  drugs
    diuretic-induced changes 212
    filtration and tubular secretion 178
    renal clearance, equation 177
  elderly patients **59–60**
    creatinine clearance 59–60
    inulin clearance 59
    β-2-microglobulin 60
    radioisotopic methods 60
  gestational increase, rat models 1914
  in gold/penicillamine nephropathy 682
  growth hormone–somatomedin axis and 1324
  hyperfiltration
    protein-induced hypertrophy 1210
    in sickle-cell disease 711–12, 715
  hyperfiltration in diabetes mellitus 509, 512–14
    hyperglycaemia mediating 512–13, 516
    metabolic/hormonal mediators 512–14
    pathophysiology 512–14, 521
  impairment
    defined, chronic renal failure 1301
    in Fanconi syndrome 567
    (up to 75%), freedom from symptoms 1937
  increase
    cyclosporin therapy in renal transplantation 1560
    in pregnancy, longterm effects 1922
    proximal tubular reabsorption 263
  infantile cystinosis 1608
  inulin clearance 1174
  elderly patients 59
  lithium toxicity 854
  maintenance, in renal disease 2055
  measurement
    children and young adults 33
    creatinine clearance 31–7
    diurnal variation 29
    drug pharmacology 178
    DTPA 1174
    EDTA 1174

glomerular filtration rate (*cont.*):
  measurement (*cont.*):
    filtration fraction (FF) 116
    filtration pressure equilibrium, pregnancy-induced changes 1911
    fractional urate clearance C_{ur}, effects of various disorders *827, 828*
    I-125-iothalamate 38
    inulin clearance 29–30, 1174
    nuclear imaging using radiopharmaceuticals **116–28**
    plasma β2-microglobulin 37, 60
    pregnancy 29
    problems 30
    radioisotopic methods 60
    reference ranges for GFR 30–1
    renal clearance formula 29
    'renal reserve' 29
    simple estimation [130 − age in years] 59–60
    size of subject 30
  mesangial cells, reduction of K_f 1561
  β-2-microglobulin 37
    elderly patients 60
  minimal change nephrotic syndrome 302
  neonates 50, **51–2**, 1297
    at birth 1699
  nephrotic syndrome, determinants in uncomplicated nephrotic syndrome 1034–5
  normal physiology 982–3
    charge-selective and size-selective barrier 28
    determinants 27
    formula for renal clearance 29
    microphysiology 26–7
    regulation of GFR 27–8
    single nephron GFR 982–3
    variations in GFR 28–9
  NSAIDs 179–80
  obstruction effects 2011–12
  oxalate, plasma and 2229
  pregnancy-induced changes
    animal models 1911–13
    hypertension 1961
    increase 1910–11, 1922
  proteins, restriction 1180
  reduction
    acute tubular necrosis 1933
    in hypercalcaemia of sarcoidosis 579
    impaired renal haemodynamics 179
    potassium excretion 897
    thyroid metabolism and 1320
  renal autoregulation 1015
  renal clearance studies 742, 742–3
    proximal renal tubule 742–3
    solute titration rate curves
      splay 742–3
      threshold 742
      transport maximum, T_m 742
  renal transplantation, normo/hypercalcaemic recipients *1775*
  in sickle-cell disease 711, 713, 714, 715
  single nephron GFR (SNGFR)
    animal gestational models
      plasma volume expansion 1915
      prolactin and 1914
      tubuloglomerular feedback system 1922

glomerular filtration rate (*cont.*):
  single nephron GFR (SNGFR) (*cont.*):
    pregnancy-induced changes 1911, 1912
    sodium excretion and 264
    tubuloglomerular feedback system, animal gestational models 1922
    ultrafiltration coefficient, in diabetes mellitus 512
glomerular scarring, *see* glomerular sclerosis/scarring
glomerular sclerosis/scarring,
  adaptation to nephron loss 1192–7
  animal models 1192
  adaptive haemodynamic changes 1193–6
  capillary lumen obliteration, schematic representation of changes 151
  determinants of progression
    growth factors and 1203
    lipids and 1201–2
    platelets and monocytes 1202–3
    stages I, II and III 1199–201, 1203
  stages and similarity to atherosclerosis 1201
  glomerular adaptive changes 1198
  hypertension, experimental nephropathy 1196–7
  hypertension and 1198
  indications 145
  modulating factors
    age-related factors 1208
    anticoagulants 1215–16
    antiplatelet agents 1215
    dietary factors 1209
    genetic factors 1206–7
    immune factors 1208
    metabolic and hormonal factors 1208–9
    sex-related factors 1207–8
  nephrectomy and, age-related factors 1208
  proteinuria, role 1198–9
  sexual dysfunction, glomerular sclerosis/scarring 1208
  stages
    cellular injury (stage I) 1199–201
    atherosclerosis (stage II) 1201–3
    glomerular sclerosis and fibrosis (stage III) 1203
  therapeutic implications
    dietary intervention 1210–12
    pharmacological intervention 1212–17
glomerulis, *see* glomerulonephritis, proliferative
glomerulocystic kidneys,
  children 2169
  familial 2182
glomerulonephritis,
  N-acetyl-β-D-glucosaminidase excretion 18
  acute
    pregnancy and 1941
    renal biopsy indications 144
    in sickle-cell disease 711
  acute endocapillary 532
  acute renal failure **1060–71**
  acute postinfectious 1070–1
  alcoholism 2311
  antibody-mediated, animal models 1208

Index

glomerulonephritis (*cont.*):
 a–GBM, characteristic immunohistologic patterns 147, 149
 cancer, secondary to 2316–17
 characteristic immunohistologic patterns 147, 149
 chronic
  children 2142
   incidence 2143
  without crescentic change 1071
 crescentic, *see* crescentic glomerulonephritis
 crescentic change
  acute on chronic renal failure 1070, 1160–1
  in Henoch–Schönlein purpura 602, 603, *604*, 606
 Crohn's disease and 2310
 cryoglobulinaemic 688, 691, 692
 diagnosis
  EDTA registry 1230
  geographical distribution at commencement of treatment *1231*
  proportional distribution by age at commencement of treatment *1231*
 drug-induced renal failure 159–63
 end-stage renal failure, proportional distribution by age at commencement *1229*
 endocapillary, *see* endocapillary glomerulonephritis, acute
 experimental models 169–70
 extracapillary, renal transplantation 1514
 gold/penicillamine associated 682
 granulomatous, in Wegener's granulomatosis 616
 haemolytic–uraemic syndrome superimposed on 1050
 heavy metals exposure, differential diagnosis 838
 in Henoch–Schönlein purpura 597, 602
  classification *604*
 hepatitis B-associated 496–7
 immunologically-mediated mercury-induced 844–5
 infection-associated **456–70**
 liver, *see* liver glomerulonephritis
 membranoproliferative, *see* membranoproliferative glomerulonephritis
 membranous, *see* membranous glomerulonephritis
 mercury salts 169–70
 mesangial, *see* mesangial glomerulonephritis
 neoplastic disease 234
 non-steroidal anti-inflammatory drugs 822
 in NZB mice 584
 in polyarteritis nodosa 628
 pregnancy
  animal studies 1946–8
  herpesvirus varicellae 2304–5
 primary
  acute interstitial infiltration 149
  acute renal failure in **1070–2**
  renal transplantation 1513–14
 protocol for study 237–8
 radiological tests 72

glomerulonephritis (*cont.*):
 rapidly progressive, *see* rapidly progressive glomerulonephritis
 recurrence in graft 1154
 recurrent, in transplantation 1531–2
 renal biopsy, EM findings 150
 in renal transplantation, first appearance 1532
 renal transplantation
  recurrence following 1573
  recurrence rates *1574*
 in rheumatoid arthritis 678, 680–1
 schistosoma-specific 1736–7
 secondary, renal transplantation 1514–15
 sex-related factors 1207–8
 in SLE 2293
 streptococcal impetigo, following 2304
 in Wegener's granulomatosis 592
 Yersinia infection 2310
 see also lupus nephritis; nephritic syndrome; *specific forms of glomerulonephritis*
glomerulopathy,
 diet, body mass 238
 IgA deposits 252
 see also immunoglobulin A nephropathy
 immune mechanisms **240–58**
 immune reactants
  antibody-mediated glomerulonephritis 251
  circulating immune complexes 251–2
  IgA nephropathy 251
  localization mechanism 251–5, *252*
  serum sickness 251–2
 immunotactoid 2281–3
 incidence in neoplasia 479
 induction of autoimmune response **246–51**
 liver disease, associated with acute renal failure 1104–5
 malignancy-associated glomerulonephritis
  aetiology 477
  incidence 479
  'presentation pyramid' 235
  rheumatoid 2295
  silicon-associated, without silicosis 845
  Sjögren's syndrome 2295
  see also nephrotic syndrome
glomerulopressin 513
glomerulosclerosis,
 Bardet–Biedl syndrome 2216, 2226
 in diabetic nephropathy
  functional significance 525
  haemodynamic changes causing, evidence for/against 519, 520
  hypertrophic changes causing 520, 521
 focal, renal transplantation 1513–14
 focal segmental 2219
 in Henoch–Schönlein purpura 602
 nodular, in light chain deposition disease 570
 in sickle-cell disease 711, 711–12

glomerulus-specific autoimmunity, antiglomerular basement membrane-mediated glomerulonephritis 247
 Heymann's nephritis 247–8
glomerulus(i) **129–39**
 abnormalities, activity/chronicity indices 145
 accelerated hypertension 2128
 adaptive hyperperfusion 1195
 afferent tone, animal models 1197
 AL amyloid deposition 552, 572, 573
 anatomy, ageing changes 57
 blood supply 985
 charge- and size-selective filter 28
 collagen
  tumour necrosis factor 1203
  type II to type IV changes, glomerular sclerosis in progressive chronic renal failure 1203
 cryoglobulin precipitation 688–9, 691, 692
 cysts, nephronophthisis 2190
 disease **227–388**
  in diabetics 526
  proforma 236–7
 disease in the tropics **486–98**
  filarial glomerulopathy 495–6
  primary
   focal segmental glomerulosclerosis 487
   IgA nephropathy 488
   membranous nephropathy 488
   minimal change nephropathy 486–7
   poststreptococcal nephritis 488
  primary/secondary, prevalence *487*
  secondary
   amyloidosis 488, 491–2
   SLE 488
  see also specific infections
 drugs, filtration and 177
 endothelial cells
  antigen system 132
  thrombomodulin secretion 257
  von Willebrand factor VIII secretion 257
  see also glomerular capillaries
 endotheliosis, pregnancy-induced hypertension 1960
 enlargement
  in Bartter's syndrome 784
  in sickle-cell disease 712, 713
 epithelial cells, monoclonal antibodies, recognition 253
 fibrosis, in sickle-cell disease 711
 filtration barrier, characteristics 135–6
 filtration mechanism 229–30
  Pappenheimer model 229–30
  see also glomerular filtration rate
 foamy changes, hyperlipidaemia in progressive chronic renal failure 1202
 glomerular capillaries, *see* glomerular capillaries
 haemodynamics
  pregnancy, animal studies 1946–8
  pregnant animal models 1912
 haemolytic–uraemic syndrome 1051

glomerulus(i) (*cont.*):
 hyalinization, in amyloidosis 555
 hyalinosis 1693
 hyperperfusion, adaptive 1195
 hypertension in 1192–3
 hypertrophy
  adaptive, role 1196
  adaptive changes 1198
  animal models 1193
  compensatory changes in nephron loss 1192
  experimental rat models 1193
  hypertension susceptibility 1198
  triggering factors 1192
 immune deposits, in lupus nephritis 652
 injury mediated by immune complexes 639
 intraglomerular pressure, elevated
  angiotensin converting enzyme inhibitor effect 527
  in early renal disease in diabetics 512, 520, 521, 527
 ischaemia, interstitial nephritis 1086
 juxtamedullary glomeruli
  ageing changes 58
  loss 57
 leucocyte infiltration, schematic representation 151
 longitudinal section 130
 mesangial cells 28, 130, 136–7
  extraglomerular 130, 136–7, 138
  haemodynamics, regulation 136–7
  proliferation, schematic representation 151
 mesangial closure device 138
 microaneurysms, renal ablation 1199, 1200
 nephritic, schema *1194*
 nodules, *see* nodules
 'non-selective pore' 268
 numbers 129
  ageing changes 57
 parietal epithelium 130, 137
 permeability for large macromolecules, children 268
 permselectivity 268
 proteinuria, defined 17, 18
 radiation nephritis appearance 850
 in renal amyloidosis 552, 572, 573
 renal biopsy
  cell changes 148, 151
  deposits 148, 151
  evaluation *149*
  extracellular structures 148, 151
  schematic representation of changes 151
 in sarcoid granulomatous interstitial nephritis 577
 in scleroderma 675
 sclerosis
  ageing changes 57
  and vascular changes 57
 size increase, oligomeganephric hypoplasia 1606–7
 subepithelial deposits, schematic representation 151
 surface capillaries, Munich-Wistar rat, pregnancy 1912
 'tactoids' 2281–3
 thrombosis
  animal models 1052

Index

glomerulus(i) (*cont.*):
 thrombosis (*cont.*):
  in SLE 656
 transcapillary hydraulic pressure, elevated in early diabetes 512
 tubular glomerular feedback mechanism 139
 tuft, hypercellularity, in lupus nephritis 652
 ultrastructure 129–39
 vascular changes, and sclerosis 57
 visceral epithelium (podocytes) 130, 131, 132–4
 volume, increased in diabetes mellitus 525
 see also glomerular basement membrane; *specific diseases of the glomerulus*
glucagon,
 chronic renal failure
  hyperglucagonaemia 1254
  role 1254
  serum levels 1251, 1280
 glomerular filtration rate increase 513
 hyperkalaemia, enhancement of secretion 907
 magnesium and, animal models 1807
 phosphate transport and 1787, 1791
 protein-loading, and elevation of GFR 29
 renal haemodynamics affected by 513
 somatostatin blocking 26
 transcellular K^+ distribution 896
glucocerebroside–β-galactosidase deficiency, Gaucher disease 2206
glucocorticoids,
 bone remodelling *1790*
 deficiency, inappropriate secretion of ADH 878
 excess, potassium depletion 906
 growth action 1297
 magnesium and 1807
 phosphate transport and 1787, 1791
 potassium homeostasis 899
 renal ablation, animal models *1209*
glucoglycinephosphaturia, incomplete Fanconi syndrome 729
glucoglycinuria 745
 incomplete Fanconi syndrome 729
glucohyperaminoaciduria, incomplete Fanconi syndrome 729
glucophosphaturia, incomplete Fanconi syndrome 729
glucose 6-phosphatase deficiency, *see* von Gierke's disease
glucose 6-phosphate dehydrogenase,
 deficiency, haemolysis and 1042
 ureteric obstruction, animal models 2013
glucose,
 blood
  glomerular filtration rate relationship 513
  see also glycaemia, control; hyperglycaemia

glucose (*cont.*):
 dietary restriction
  animal models 1212
  animal models of progressive glomerular scarring *1209*
 diuresis induced by 515
 excretion, pregnancy changes 1922
 insulin and, hyperkalaemia, therapy 910
 intolerance
  hypokalaemia 211
  thiazide diuretics 210
 intolerance in chronic renal failure 1251–4
  clamp technique 1251–2
  glucagon, role of 1254
  growth hormone role 1324
  insulin resistance 1251–3
 measurement, specific enzymatic methods 728
 metabolism, inhibition of insulin-mediated glucose uptake in CRF 1303
 osmotic diuresis
  hypernatraemia 883
  pseudo-hyponatraemia masking hypernatraemia 883
 polymers, peritoneal dialysis solution 1482
 pregnancy
  animals as gestational models 1920
  gestational hormones and 1916
  glycosuria increase 1915–16
  maximal transfer concept 1915–16
 reabsorption in proximal tubule
  children with Fanconi syndrome 725
  glycogenosis with Fanconi syndrome 730, *731*
  glycosurias 743–5
  normal 743
  normal/disordered curves *743*
 renal glucose-losing syndrome 730, *731*
 renal handling, ageing kidney 61
 total parenteral nutrition in ARF *1021*
 toxic effect on endothelial cells 519
 uraemic syndrome 1239
glucose oxidase paper test strips, lactose and 1916
glucose oxidase test 1681
glucose–galactose malabsorption 745
glucose–galactose transporter 745
α-glucosidase, in urine 18
glucosuria, *see* glycosuria
glutamate 700
 diet, acid and base production 932–3, *933*
 excretion, pregnancy and 1916
 glutamate deaminase, ammonia formation 767
glutamine,
 cystine excretion, effects 749
 metabolism 923
α-glutamyltransferase, in urine 18
γ-glutamyltransferase, alkaline phosphatase levels 1370
glutaric acidaemia type II, features and associated abnormalities *2038*

glutathione,
 depletion, acetaminophenone 169
 neutrophil and free radical-mediated injury, renal circulation 1052
glutathione reductase, uric acid stones *1861*
glycaemia,
 control
  chronic renal failure, haemoglobin A1c 1645
  in diabetic nephropathy 529
  kidney size relationship 509, 529
 diabetes contraindicated, ageing kidney 61
 diabetic nephropathy, control arguments 1644–5
glyceraldehyde, increased excretion, Fanconi syndrome 768
glycerol,
 high-salt diet 171
 peritoneal dialysis solution 1481
glyceryl guaiacolate, urate excretion increase *829*
glycine,
 excretion, pregnancy and 1916
 normal tubular reabsorption 745
 transport disorders 750
 urate excretion increase *829*, 830
glycinurias, isolated 750–1
glycirrhizinic acid,
 inhibition of 11-β-hydroxysteroid dehydrogenase 948
 see also liquorice
glycogen storage diseases,
 enzyme defects *2210*
 metabolic pathways *2212*
 type 1, uric acid stones *1861*
glycogenosis,
 Bickel Fanconi syndrome 2311
 Fanconi syndrome *729*, 730–1
 type I, focal segmental glomerulosclerosis 2311
glycollate, excess, hyperoxaluria 1857, 1858
glycols, nephrotoxicity 1009
β-1-glycoprotein,
 pregnancy test 1337
 see also cytokines
glycoproteins 671
glycosaminoglycans 671
 Fanconi syndrome 732
 in glomerular basement membrane 519
 heparin-like
  inhibitor of calcium oxalate crystallization in urine *1828*
  inhibitor of calcium phosphate crystallization in urine *1827*
glycosphingolipid metabolism, disorder, *see* Anderson–Fabry's disease
glycosurias 743–5
 causes 743, *744*
 defined 743, 1885
 Fanconi syndrome 725, 730–1
  types *729*
 glycinuria and 745
 hereditary fructose intolerance 732
 hyperphosphaturia and 745
 Lowe syndrome 732
 massive 725, 730–1

glycosurias (*cont.*):
 pregnancy 745
 primary (isolated) renal glycosurias 743–5
  clinical features and diagnosis 744
  defined 743
  genetics 744–5
  incidence 743
  pathophysiology 744
 secondary glycosurias 745
 types A and B 744
glycosylation of proteins, non-enzymatic 515
 diabetic nephropathy pathogenesis 517–18
glycosyltransferase 519
glycoyl-γ-fetal haemoglobin 703
glycyrrhizic acid, liquorice, sodium retention 273
glyoxalate metabolism,
 disorders 2226
 schema *2227*
Goeminne syndrome, features and associated abnormalities *2038*
gold nephropathy 160–1, **682–3**
 antibodies to tubular basement membrane 170
 autoimmune glomerulonephritis 249
 clinical features 682
 incidence 682
 long-term outcome 683
 management 682–3
 membranous nephropathy 370, 375, 380, 2295
 pathology 682
  membranous glomerulonephritis 681, 682
 predictive factors 683
Goldblatt models of hypertension 519, 520, 2059–65
 renal artery stenosis 2101
 renal structure effects 2075–9
 renin elevation and inhibition of renin–angiotensin system 2062
 single artery 1595
 sodium balance, renin and sodium depletion 2060
Goldenhar complex, *see* facio–auriculo–vertebral syndrome
Goldston syndrome, features and associated abnormalities *2037*
gonadal dysfunction,
 children 1614–15
 cyclophosphamide causing 620
gonadal toxicity,
 cyclophosphamide, minimal change nephrotic syndrome 322
gonadotrophic hormones, ectopic production, renal cell carcinoma 2244
gonadotrophic releasing-hormone, dysfunction 1615
 secretion 1298
gonadotrophins,
 gonadal resistance to, haemodialysis 1336
 haemodialysis 1336–7
Goodpasture antigen,
 Alport's syndrome 2199–201
 distribution 439
 glomerular basement membrane 446

Goodpasture antigen (*cont.*):
  renal transplantation, Alport's syndrome 1515
Goodpasture's syndrome **438–52, 1064–5**
  aetiology 439, 445
  antiglomerular basement membrane disease **438–52**, 1064–5
  cell-mediated immunity 446–7
  clinical features 440–2
  crescentic glomerulonephritis 420
    treatment 429–30
  crescents and creatinine levels 450
  defined 438
  diagnosis 442–5
  disease associations 440
  end-stage renal failure, proportional distribution by age at commencement *1229*
  environmental influences 445–6
  epidemiology 438–40
  incidence 438–40
  occupational risks 9
  outcome 1065
  pathogenesis 446–8
  penicillamine inducing 161, 682
  relapse and recurrence 451
  renal transplantation following 452
  smoking as risk factor 9, 1064
  treatment 448–51, 1065
  see also antiglomerular basement membrane
Gordon's syndrome 776, 777
Gottron's papules, dermatomyositis 2302
'goulash effect', meat creatinine 34
gout **830–9**
  assessment 1151–2
  complication of renal disease 2296
  end-stage renal failure 1234
    proportional distribution by age at commencement *1229*
  familial syndrome 2294
    with renal failure 833
  HGPRT deficiency 2294
  lead intoxication and 2294
  lead mobilization test, undiagnosed lead nephropathy 839
  lead nephropathy, hypertension and 838–9
  Lesch–Nyhan syndrome, HGPRT deficiency 2294
  primary
    renal failure in 833
    'underexcretors' *vs* 'overproducers' 830–1
    urate excretion 830–1
  saturnine
    lead intoxication and 833–4
    urate excretion 830–1
  secondary, chronic renal failure and 834
  in sickle-cell disease 713
  skin lesions 2301
  treatment, depression of bone marrow 2296
  urate assessment 2295
  uric acid stones *1861*, 1862
gouty arthritis, hyperuricaemia, diuretic-induced 212
gouty nephropathy 833
  allopurinol in, renal function 2294

gp70 retroviral glycoprotein 636
graft artery stenosis, *see* renal artery stenosis, renal transplantation
graft versus host disease, animal models, idiopathic nephrotic syndrome 311
  autoimmune glomerulonephritis 249
  mechanism 250
  mercury-induced autoimmune glomerulonephritis, similarity 250
  murine model 636, 640–1
Gram-negative organisms, *Limulus polyphemus* (horseshoe crab) amoebocyte lysates 1663
  sepsis, pregnancy 1930
α-granule serine proteinase, antineutrophil cytoplasm antibodies 239
granulocyte-associated immunoglobulins, in cytomegalovirus infection, test 1534
granulocyte-specific antinuclear antibodies 624
granulomas,
  bladder 1735
  caseation 1720
  cavitating, in Wegener's granulomatosis 615
  cortical 1720
  experimental, formation 585
  glomerular, in Wegener's granulomatosis 616
  interstitial, in sarcoidosis 577
  interstitial granuloma 1085
  medullary 1720
  papillary 1720
  parenchymal 1723
  sarcoid, *see* sarcoid granuloma
  schistosomiasis 1733, 1735
  tubercular 1720
granulomatous disease,
  chronic urinary tract obstruction 2020
  glomerulonephritis, in Wegener's granulomatosis 616
  interstitial nephritis, sarcoid, *see* sarcoidosis
  periaortitis of the ureters and 2020
  vasculitis 2324
    skin lesions 2303
  see also vasculitis
Griess test for nitrite,
  diazotization reaction 1663
  Stix test 1681
griseofulvin, handling in kidney disease 185
growth,
  nephrotic syndrome, persistent 324–6
  statural, corticosteroid drugs 324–5
growth factors, glomerulonephritis, pathogenesis 257
growth in healthy children **1294–301**
  bone development 1296
  pubertal 1615–16
  race differences 1296
  seasonal differences 1296

growth hormone,
  action 1297–8
    general metabolic effects 1297
  animal models 1304
    growth hormone-deficient rats 1208–9
  bone remodelling *1790*
  chronic renal failure 1239, 1322–3
  enzyme immunoassay evaluation 1322
  glomerular filtration rate increase 513
  increase, glomerular hypertrophy 1208–9
  kinetics, CRF 1322–3
  in malnutrition 1322
  phosphate transport and 1787
  plasma binding, and peripheral action 1323
  plasma levels, CRF 1322
  receptor 1297
  release 1297
  somatotroph function tests 1323
  therapeutic use
    availability and cost 1308
    children with chronic renal failure 1304
    following transplantation and corticosteroid therapy 1309
    leukaemia and 1304
  thyrotrophin-releasing hormone stimulation, increase in uraemic subjects 1322
  uraemic syndrome 1239
growth hormone–somatomedin axis,
  carbohydrate metabolism 1324
  chronic renal failure disorders 1322–4
  growth in uraemic children 1324
  kidney function 1324
growth retardation **1301–9**
  intellect 1303
  management 1305–9
  pathogenesis 1303–5
  proximal renal tubular acidosis 768, 769
  rickets and, cystinosis 2221–2
  sexual maturation 1302–3
  stature and skeletal changes 1301–2
  see also rickets
growth in uraemic children **1301–9**
  chronic renal failure 1616–17
  growth hormone treatment
    IGF:binding protein 3 1324
    prognosis 1324
    somatomedin blockade in 1324
  growth hormone–somatomedin axis 1324
guanidines,
  guanido compounds, neurotoxins 1397
  NK responses to interleukin-2 1243
  toxic effects 1243
cyclic guanosine 3,5,monophosphate (cGMP), action 1915
renal vasodilatation
  gestational animal models 1915
  human pregnancy 1915
guanyl nucleotide regulatory unit 796

gynaecological and obstetric disorders,
  clinical history 7
  renal transplantation 1952
gynaecomastia,
  canrenoate 213
  in haemodialysis patients 1336
  overlap syndrome 1336
  spironolactone long-term therapy 213

H_2 receptor agonists, gastrointestinal bleeding 1021
H_2 receptor antagonists, gastric alkalosis 946
  upper gastrointestinal bleeding, chronic renal failure 1284
  vomiting in haemodialysis 1283
H-Y antigen, immune response 1546
haemangiopericytoma, hypertension and 2092–3
haemarthroses, complication of renal disease 2297
haematocrit,
  nephrotic syndrome 267
  normal values *1344*
haematological abnormalities,
  in amyloidosis 550
  in SLE in children 650
  in Wegener's granulomatosis 617
haematological disorders,
  accelerated hypertension 2131
  analgesic nephropathy 813–14
  cryoglobulinaemia 2343
  following renal transplantation 1585–6
  haemolytic disease 2341
  haemophilia and Christmas disease 2343
  renal disorders **2340–6**
  sickle-cell disease 2340–1
haematological malignancies,
  hypercalcaemia, corticosteroids 1777
  leukaemia 2342
  lymphoma 2342
  metastatic infiltration 2318
  myeloma 2341–2
  renal cell carcinoma 2241–2
  see also individual malignancies
haematological values,
  acute renal failure 973–4
  haemolytic–uraemic syndrome 1054
  normal, adults *1344*
  'pregnancy norms' 1931
haematoma,
  post cannula insertion 1406
  renal biopsy leading to 142
  subcapsular, computed tomography scanning 107
haematoxylin bodies 652, 653
haematuria **230–6**
  algorithm for screening 232
  Alport's syndrome 2197–204
  in amyloidosis (renal) 551
  analgesic nephropathy 813
  benign familial haematuria 2203
  detection 230–1
  in diabetes mellitus, non-diabetic (nephropathy) renal disease 526

haematuria (cont.):
  drug-induced glomerulopathy 234
  glomerular causes
    clinical approach **230–2**
    clinical syndrome in relation to history of glomerular disease 235
    dysmorphic erythrocytes 19–20
    follow-up 235
    infection-associated 234
    origin 231–2
    paraprotein-associated 234
  glomerulonephritis, rapidly progressive 233
  gross, IVU imaging 84
  in Henoch–Schönlein purpura 597, 605
  IgA nephropathy, primary 349–50
  isolated, renal biopsy indications 145
  isolated macroscopic 233
  in lupus nephritis 654
  macrohaematuria
    clinical features 340
    precipitating events 340
  microscopic 6
    IgA nephropathy indicator 341
    isolated 234
    with proteinuria 233
  minimal change nephrotic syndrome 302
  painless 75
  polycystic disease 75, 2172
  pregnancy 1930–1
  radiological tests 75–6
  recurrent 6
  renal biopsy leading to 142
  renal cell carcinoma 2240
  in renal papillary necrosis 530
  screening algorithm 232
  in sickle-cell disease 709–10
    treatment 716
  site of bleeding and 2–5
  tubular casts 231
  tumours 75
  unexplained, renal arteriography 113
  vs other causes of red/brown coloration 4
  see also microhaematuria
haemochromatosis,
  glomerulopathies, acute renal failure in liver disease 1104
  haemodialysis, liver changes 1282
haemodiafiltration, continuous, techniques 1029
haemodialysis **1405–505**
  abdominal pain 1454–5
  acetate
    effects 1273
    high efficiency, metabolic acidosis 934
  acid–base disorders 959–60
  acute abdomen 1284
  acute renal failure **1026–33**
  adaptation, see haemodialysis, psychological aspects
  adequacy **1430–1**
    defined 1430
  adrenal function in 1325
  air embolism 1429
  alternatives 1273
  aluminium overload 1351
  amino acids, loss 1260
  in amyloidosis 558–9

haemodialysis (cont.):
  anaemia
    ferrokinetic studies *1346*
    prevalence 1351
  analgesic nephropathy
    availability 816
    prognosis 815
  antiarrhythmics 1270
    hypokalaemia 1625
  anticoagulants 2344–5
  arthropathy 679
  ascites, renal failure, chronic 272
  assessment of suitability 1169–70
    checklist *1169–70*
  β2-microglobulin removal 1431
  bioincompatibility **1445–6**
  bleeding time 1451
  blood flow
    arteriovenous fistula 1411
    measurement, vascular access *1413*
  blood loss 1351
  blood pressure control **1458–64**
  blood pump 1419
  bone erosion 1374, 1377
  calcitriol treatment, parathormone secretion 1367
  calcium *1754*
    absorption efficiency *1763*
    homeostasis 1771
    pathogenesis of disturbed function 1368
  cancer risk 2320
  cardiovascular disorders
    arrhythmias 1270
    digitalis 1270
    endocarditis 1275
    ischaemic heart disease 1270
    left ventricular hypertrophy 1265
  carpal tunnel syndrome 1400
  cerebrovascular accident during, mortality 1576
  children **1431**
    access 1605
    achieving normal growth 1308
    growth rates 1616–17
    mortality causes *1610*
    patient survival 1618
  chronic haemodialysis
    calcium phosphate deposition 834
    computed tomography scanning 100
  chronic renal failure, patient survival *1627*
  Cimino–Brescia fistula 1625
  cirrhosis 1289, 1292
  coagulation abnormalities 1361–3
  colonic diverticulitis 1454–5
  commencement criteria 1427
  compliance, see haemodialysis, psychological aspects
  complications 1417–33, **1438–41**
    acute 1428–30
    air embolism 1429–30
    bleeding 1428
    cardiovascular system 1438–9
    cerebrovascular accident 1439, 1448
    disequilibrium syndrome 1428
    first use syndrome 1429
    hypercalcaemia 1429
    hyperchloraemic acidosis 1429
    hypotension 1429

haemodialysis (cont.):
  complications (cont.):
    inappropriate use of equipment 1429–30
    infections 1439
    long-term **1436–55**
    matrix stones 1445
    renal cysts 1443–4
    technical faults 1429
    thrombotic 1268, 1363
  continuous arteriovenous **1026–32**
    children 1123
  contraception in haemodialysis patients 1948
  death rates, see haemodialysis, mortality
  diabetic nephropathy 1648–9
  diagnostic tests 1436
  dialysis-linked syndrome, immune response 1315
  disequilibrium syndrome 1400–1, 2329
  DOCA, K+ load 896
  efficiency
    HPLC comparisons *1247*
    kinetic data for urea NCDS *1246*
  elderly patients 1473–4, **1627–32**
    angina 1625
    angiodysplasia in 1281
    arrhythmias 1625
    blood flow, compared with younger patients *1625*
    cardiovascular instability 1625
    heparin 1623
    interdialytic complications 1625
    intradialytic complications 1625
    vascular access 1625
  encephalopathy in 1401
  end-stage renal failure
    uraemic pericarditis and pericardial effusion 1274–5
    valvular heart disease 1275
  erythropoietin in, sexual function effects 1335
  extracorporeal blood circuit 1418
    biocompatibility 1432
    and coagulation system 1428
  fever of unknown origin 1452–3
  fistulae 1268
  folate deficiency 1351
  galactorrhoea 1338
  gastrointestinal bleeding 2311
  gonadotrophins, gonadal resistance to 1336
  gynaecomastia 1336
  'haemodialysis bone disease', aluminium overload 1369
  haemofiltration vs haemodialysis 180
  haemoglobin levels 1352
  haemolysis in 1351
  haemorrhage in, platelet defects 1361
  headache 2330
  heparin, regional heparinization 1362
  heparin administration, guidelines 1433
  heparin pump 1419
  heparin-free 1362
  heparinization, regional 1029, 1362, 2345
  hepatitis A, management and prophylaxis 1290

haemodialysis (cont.):
  hepatitis B 2313–15
    exclusion of carriers from renal units 1287
    management and prophylaxis 1289, 1290
    screening 1168–9
  high efficiency, mixed metabolic and respiratory disorders *957*
  history 1418
  HIV testing 1169
  hypercalcaemia *1455*
  hyperphosphataemia, resistance to dietary control 1794
  hypotension 1429, 2338
  hypoxaemia in 2339
  ileostomy, electrolyte management 1283
  ileus 1454
  intermittent vs continuous, advantages/disadvantages 1027–33
  iron deficiency 1351
  iron overload 1282
  KT/V assessment 1430–1
  liver changes 1282
  long-term **1436–55**
    β-2-microglobulin deposition and amyloidosis 1282
    thoracic pain 1452
  in lupus nephritis (SLE) 663
  malnutrition **1439**
  medical complications **1436–55**
  membranes 1420-1421; dialysis membranes
  β-2-microglobulin amyloidosis 548, *572*, 2330
  modalities
    silicone particles in liver from pump 1282
    uraemic patient with cardiac problems 1272–3
  mortality
    causes of death 1437–9, *1438*
    survival rates 1437
  in multiple myeloma 568
  need for, symptoms 1167
  needles and tubing 1410–11, 1419
  neonates 1117–19
  nephrogenic ascites, diagnosis 1284
  neurological involvement 2329–30
  non-compliance, see haemodialysis, psychological aspects
  nutrition in 1021, 1439
  nutritional problems 2330
  obstetric acute renal failure 1932
  overdialysis 1440
  oxypurinol reduction 832
  peritoneal dialysis vs. elderly patients 1626
  phosphate, pathogenesis of disturbed function 1368
  polycystic disease, autosomal dominant 2184–5
  pre renal transplantation 1518
  predialysis assessment 1168, *1169*
  pregnancy 1337–8, **1948–9**
    nutrition in 1948
    obstetric acute renal failure 1932
  pressure monitoring 1419
  priapism 1335–6
  principle 1417–18
  prolactin levels 1333
  psychiatric problems 1400

haemodialysis (cont.):
  psychological aspects **1465–75**
    adaption 1469–70
    children 1472–4
    compliance 1470–1
    denial 1468
    depression 1468
    diabetes 1471–2
    stress 1468–9
  Quinton catheter 1625
  rapidly progressive glomerulonephritis 1063–4
  regional heparinization 1029, 2345
  renal replacement therapy in patients with 2321–2
  schistosomiasis 1739
  in scleroderma 676
  sexual dysfunction **1329–39**
  sexual function
    menstruation 1336
    ovulation 1336
    sex hormone levels 1336
  short-term 1430
  in sickle-cell disease 716
  silicone uptake 1368
  social assessment 1168
  spironolactone, K^+ load 896
  subdural haematoma 2329
  surgery, perioperative treatment 1451–2
  techniques
    arteriovenous connection 1028
    balancing 1029–30
    continuous arteriovenous haemofiltration 1028–9
    continuous dialysis 1029
    continuous haemodiafiltration 1029
    continuous venovenous haemofiltration 1029
    heparinization 1029
    venovenous connection 1028
  thrombotic complications 1363
  thrombotic complications of fistulae, salicylates 1268
  thyroid hormone-binding globin levels 1320
  thyroid hormones, heparin and 1322
  tropics 1130
  ulceration 2311
  underdialysis **1439–41**
  uraemic polyneuropathy 2328
  uraemic syndrome, side-effects mimicking uraemic toxicity 1240
  urate/uric acid removal, vs peritoneal dialysis 832
  vascular access **1405–16**
  vascular access site 2330
  vomiting **1453–4**
    management 1283
    non-occlusive bowel infarction 1284
  see also dialysate; dialysis; dialysis membranes; dialysis tubing
haemodynamic changes of pregnancy,
  animal gestational model 1911–13
  renal vasculature changes 1913–15
haemodynamics, renal,
  diabetic nephropathy
    pathogenesis 519–20

haemodynamics, renal (cont.):
  glucagon effects 513
  in sickle-cell disease 711, 713, 714, 715
haemofilters *1117*
haemofiltration,
  acute renal failure **1026–33**
  advantages 1273, 1426
  anticoagulation during 1427
  antihypertensive effects 1464
  arteriovenous fistula access 1426–7
  clearance of drugs 1032
  clinical results 1032–3
  continuous arteriovenous **1026–33**, *1117*
    children 1123
    neonates 1117–19
  continuous vs intermittent treatment, development of lethality 1032
  contraindications 1033
  drugs, dosage 1032
  indications 1033
  numbers of patients 1426
  organization 1426
  patient selection 1426
  patient supervision 1031–2
  principles 1425
  techniques
    balancing 1029–30
    commercial solutions 1030–1
    composition of solutions *1031*
    continuous venovenous haemodiafiltration 1029
    haemofilters 1030
    heparinization 1029
  vs haemodialysis 180
haemoglobin,
  acute renal failure and 1006
  carbamylation by isocyanate, uraemic syndrome 1242–3
  chronic renal failure 1187
  β-globin
    Benin type 702
    gene 702
    haplotypes 702–3
    mutations, in sickle-cell disease 700, 702–3
  levels in dialysis patients 1352
  normal haematological values *1344*
  physiology of iron 1345
  polymerization
    blood viscosity increase 705
    factors affecting 702, 714
    kinetic hypothesis 714
    in sickle-cell disease 700, 701, 702, 714
  properties, effects on kidney, significance of elevation *229*
  rhabdomyolysis, and ARF 995
haemoglobin A1c,
  glycaemic control in chronic renal failure 1645
  overconcentration, uraemic syndrome 1242
haemoglobin C 701
  haemoglobin polymerization affected by 702
haemoglobin C Georgetown 702
haemoglobin C Harlem 702
haemoglobin D Punjab 702
haemoglobin E 702

haemoglobin F (fetal, Hb-F) 701, 702
  protection against sickling **703**
  in sickle-cell anaemia 703
haemoglobin Montgomery 702
haemoglobin O Arab 702
haemoglobin S 701
haemoglobinopathies 702
  see also sickle-cell disease
haemolymphatic state 185
haemolysin,
  α and β forms, *Escherichia coli* 1666
  production 1665
haemolysis,
  acute renal failure, cause 974
  acute renal failure and **1041–57**
  anaemia in chronic renal failure 1351
  causes and mechanisms *1043*
  diagnosis and treatment 1007–8
  haemolytic–uraemic syndrome 1054
  HELLP syndrome, pregnancy-induced hypertension 1961
  rhabdomyolysis, and ARF 995
haemolytic anaemia,
  investigations *1043*
  microangiopathic, acute renal failure 2341
  renal complications of 2341
haemolytic–uraemic syndrome **1041–57**
  acute renal failure 974
  haematological changes 1041–2
  aetiology 1048
  bleomycin, vinblastin and cisplatin induced, resembling radiation nephritis 849
  cancer and 2317
  children 1122, 2341
    hypertension 2142–3
    incidence 2143
  clinical features 1047
  CNS signs and symptoms *1047*
  cyclosporin in 2346
  cyclosporin nephrotoxicity 1560, 1562
  diarrhoea and 2310
  end-stage renal failure 1233
    proportional distribution by age at commencement *1229*
  endothelial injury and 1052–3
  epidemiology 1049
  geographical factors 1047
  haematological values 1054
  infection-associated glomerulonephritis 462, 466
  investigations
    biochemical changes 1055
    haematological changes 1054
  microangiopathic haemolytic anaemia 2341
  mitomycin 162, 163, 1048, 1050
  mitomycin-induced, resembling radiation nephritis 849
  nephrocalcinosis in 1888, 1890
  pathogenesis 1051–4
  pathology
    arterial 1051
    glomerular 1051
    tubular 1051
  postpartum 1057
  pregnancy and 1934

haemolytic–uraemic syndrome (cont.):
  radiation nephritis, similarities 850
  renal biopsy, EM findings 150, 151
  renal complications of 2341
  renal transplantation 1515
  recurrence following 1574
  shock and disseminated intravascular coagulation 1042–6
  superimposed on glomerulonephritis 1050
  thrombotic thrombocytopenic purpura 1046–56
    aetiology 1048–9
    epidemiology 1049
    investigations 1054–6
    pathogenesis 1047, 1050–4
  tropics, acute renal failure and 1138
  vascular diseases with renal involvement, schematic representation 152
  vitamin E deficiency 1052
haemophilia, renal complications of 2343
Haemophilus influenzae, infection-associated glomerulonephritis 466, 470
Haemophilus spp., urinary tract infection 1661
haemorrhage,
  in amyloidosis 550
  chronic renal failure
    management 1361–3
    pathogenesis 1361
    thrombotic complications 1363
  gastrointestinal,
    haemolytic–uraemic syndrome 1041
  haemodialysis in patients at risk 1428–9
  pulmonary–renal syndrome 1061
  skin, causes in ARF *1042*
haemorrhagic diathesis,
  chronic renal failure 1187
  renal biopsy contraindication 143
haemorrhagic fever,
  experimental 1045
  renal symptoms 1091
  RNA viruses *1045*
haemorrhagic fever with renal syndrome, see Hantavirus disease
haemosiderosis, following blood transfusion in renal transplantation 1584
halides, intoxication, loop diuretics 208
halofenate,
  animal models, dietary and pharmacological lipid manipulations 1216
  dietary and pharmacological lipid manipulations *1202*
Hansel's stain, leucocytes 20
Hansenula, infection 1748
Hantavirus disease **1143–5**
  in Asia 1145
  differential diagnosis 1144
  in Europe 1143–5
  haemorrhagic fever 1091
  pathogenesis, activation of coagulation system 1045

haptoglobin, increased, marker of renal cell carcinoma 2241, 2242
Hartnup disease,
  clinical features 750
  diagnosis 750
  genetics 749
  intestinal defect 749
  renal defect 749
  treatment 750
'hatter's shakes', mercury nephrotoxicity 844
hCG,
  elevated, symptoms, men/women 2242
  osmoregulation in pregnancy and 1918
headache,
  accelerated hypertension 2130
  analgesic nephropathy, discontinued 815
hearing defects,
  Alport's syndrome 2198
  distal renal tubular acidosis and 773
  nephrocalcinosis and *1897*
  renal disease in association *2202*
  renal tubular acidosis, distal (type 1) 773
  thiazide diuretics 2331
  and X-linked hypophosphataemic rickets 754
heart,
  changes, accelerated hypertension 2130
  hypokalaemia-induced alterations in function 900–1
  mass, effect of pharmacological interventions on animal models *1266*
  stretch receptors, atrial natriuretic peptide release 263
  see also cardiac; cardiovascular disease; congestive heart failure
heart disease, congenital,
  cyanotic 2277
  in neonates, acute renal failure 1119–20
  urate, renal handling 831
heart failure, *see* cardiac failure; cardiovascular disease and cv. system; congestive heart failure
heart sounds, clinical examination 12
heart valves,
  aortic abnormalities, polycystic disease 2336
  aortic calcifications 2336
  calcification, dialysis patients 1275
  mitral valves
    abnormalities in polycystic disease 2336
    calcifications 2336
  prosthetic, haemolytic anaemia in 2341
heat stroke **1137–8**
heat test, Bence–Jones proteinuria 1152
heavy metals **837–846**
  N-acetyl-β-D-glucosaminidase excretion 18
  intoxications, Fanconi syndrome 734–5
  nephrotoxicity 993
  occupational exposure 837

heavy metals (*cont.*):
  see also specific metals
height, *see* children, growth; growth in healthy children; growth in uraemic children
HELLP syndrome, liver, pregnancy-induced hypertension 1961
helminth infestation, handling in kidney disease 185
Hemasite 1406, 1407
heminevrin, handling in kidney disease 186
Henderson equation, extracellular fluids 918
Henle's loop,
  countercurrent multiplier effect 871–2
  maximal urine osmolality and 706, 707
  potassium transport 897–8
  sodium chloride reabsorption defect in ascending limb, ageing kidney 64, 65
  sodium reabsorption, load dependence 264
  TAL, potassium secretion 897–8
  thick ascending limb
    compensatory hypertrophy in progressive chronic renal failure 1204
    dietary-protein induced hypertrophy 1210
    function 871–2
    reabsorption of sodium 263
    transport of sodium, chloride, and water 198
  vasa recta 871–2
Henoch–Schönlein purpura **595–612**
  aetiology **598**
    allergies 598
    environmental agents *599*
    infections 595, 598
  autoantibodies in 592, 600
  clinical features **595–8**
    abdominal 596–7
    joint pain/swelling 596
    nervous system 597–8
    skin 595, 596
  complement activation 601, 602, 603
  crescentic glomerulonephritis 421
    treatment 205
  differential diagnosis, mesangiocapillary glomerulonephritis 392
  end-stage renal failure, proportional distribution by age at commencement *1229*
  epidemiological factors 596
  histopathology **602–4**
  historical background **595**
  idiopathic IgA nephropathy comparison 601–2
  IgA nephropathy and 339
  leucocytoclastic vasculitis in 587, 600, 602
  mesangioproliferative glomerulonephritis 2317
  pathophysiology **599–602**
    complement role 587, 601, 602
    IgA alterations 587, 599–600
    immune complexes 587, 600, 601
    putative antigens 600

Henoch–Schönlein purpura (*cont.*):
  prediction of outcome 601, **604–6**
    clinical *604–605*
    clinicopathological 605–6
  rapidly progressive glomerulonephritis 1070
  renal involvement 597
    classification *604*
    electron microscopy 603–4
    immune deposits 587, 603, *604*
    immunofluorescence/ immunoperoxidase findings 603
    incidence *597*
    late deterioration 605
    light microscopy 602–3
  renal transplantation **608**, 1514
  skin lesions 2303
  terminology 595
  treatment **606–8**
    extrarenal features 606–7
    renal disease 607–8
heparan sulphate,
  basement membrane 446
  decrease, congenital nephrotic syndrome of Finnish type 307
  glomerular basement membrane 135
  proteoglycans 671
  reduced synthesis in glomerular basement membrane 516, 519
heparin,
  administration in haemodialysis 1433
  alternatives 1362
  animal renal ablation models, effects *1195*, *1216*
  anticoagulation 1427–8
  elderly patients on haemodialysis 1623
  haemodialysis 1362
  hypoaldosteronism 909–10
  N-desulphated-acylated, experimental nephropathies 1216
  nephrotic syndrome 284
  pruritus treatment 1392
  pump, haemodialysis 1419
  rebound 1362
  regional heparinization 1362
  haemodialysis 1029, 2345
hepatic cysts, in polycystic disease 100–1
hepatic disease, *see* liver disease; *specific diseases*
hepatic fibrosis,
  infantile polycystic kidney disease 1607
  schistosomiasis and 1737
hepatic fibrosis, congenital, polycystic disease, autosomal dominant 2177
hepatic siderosis, iron overload, uroporphyrin decarboxylase inhibition 1282
hepatic triglyceride lipase, action in chronic renal failure 1255–6
hepatitis,
  drug-induced 2315–16
  gastrointestinal effects **1285–92**
  renal cell carcinoma 2242
hepatitis A,
  in chronic renal failure 1285–6, 1290
  postexposure prophylaxis 1290

hepatitis A (*cont.*):
  tropics, acute renal failure and 1134
hepatitis B,
  active immunization
    categories in low prevalence areas 1290
    reduction of new cases *1286*
  carriers
    exclusion from renal units 1287
    prevalence 1288
  children 462–3
  chronic active
    diagnosis 2314
    liver changes 2314
    spontaneous seroconversion 2314
  in chronic renal failure 1286–91
    management 1287
    vaccination under-responsiveness 1312
  clinical features, chronology *1287*
  consequences of infection *1286*
  detection in liver biopsy 2314
  essential mixed cryoglobulinaemia association 687, 691
  glomerulopathies, acute renal failure in liver disease 1104
  haemodialysis policy 1168–9, 1286, 2313–15
    exclusion of carriers from renal units 1287
    postexposure prophylaxis 1290–1
  HBsAg
    antigenaemia 496
    appearance during incubation 1286
    assessment 1163, 1168
    carriers, immune complexes in 586
    in polyarteritis nodosa 587, 590, 628
    prevalence world-wide *1288*
    virus surface antigen proteins 1163, 1168, 2314–15
  infection-associated glomerulonephritis 462–3, 496–7
    polyarteritis nodosa 463–4
  markers, new 2314
  occupationally-acquired infection 1291
  polyarteritis nodosa and 1068
  postexposure prophylaxis 1290
  prevention, control and prophylaxis 1289
  renal transplantation 1581, 2313–15
    antibody status 1510
  schistosomiasis and 1730
  tropics, acute renal failure and 1134
  vaccination
    chronic renal failure 1188
    in haemodialysis, potential graft recipients 1533
  uraemic patients 2314
  vaccines 1188, 1291
  vascular disease pathogenesis 1107
  virus structure *1286*
hepatitis C,
  antibodies
    antibody test 1168

Index

hepatitis C (*cont.*):
  antibodies (*cont.*):
    prevalence 1288
    isolation and immunoassay 1288
    markers, new 2314
    renal transplantation 1534, 1581
hepatitis in chronic renal failure **1285–92**
  cytomegalovirus hepatitis 1289, 1290
  Epstein–Barr virus 1289
  following renal transplantation 1291–2
  hepatitis A 1285–6, 1290
  hepatitis B 1286–91
  hepatitis delta 1289, 1290
  hepatitis non-A, non-B 1287–9, 1290
hepatitis delta,
  in chronic renal failure 1289, 1290
  prevention, control and prophylaxis 1290
hepatitis non-A, non-B,
  in chronic renal failure 1287–90
  therapy 1289
  incidence after blood transfusion 1288
  postexposure prophylaxis 1290
  prevention, control and prophylaxis 1290
hepatointestinal clinical syndromes 1734–5
hepatolenticular degeneration, *see* Wilson's disease
hepatorenal syndrome **1102–4**
  acute fatty liver of pregnancy and 1934
  acute renal failure 2310
  mixed metabolic and respiratory disorders *957*
  prostaglandin metabolic abnormalities, NSAIDs and 821
hepatosplenic schistosomiasis 1737
hepatotoxicity, drug-induced, with nephrotoxicity 1106–7
heptapeptides, *see* angiotensin III
herbicides, intoxication 1011
hereditary haemorrhagic telangiectasia, urinary tract lesions 2300
hereditary macrothrombocytopenia, nephritis and deafness 2199
hereditary nephritis,
  end-stage renal failure, proportional distribution by age at commencement *1229*
  progressive, classification *2203*
  renal transplantation, IgG linear fixation 2201
hernias,
  peritoneal dialysis 1494–5
  elderly patients 1626
herpes simplex,
  renal transplantation 1581, 1583, 1584
  cervical carcinoma 1583
  skin lesions 1584
herpes zoster,
  causing renal problems 2305
  renal transplantation, skin lesions 1584
herpesvirus, varicellae, skin lesions 2304–5
Hers' disease, enzyme defect *2210*

hexamine, contraindications 1687
Heymann antigen, membrane proteins of podocytes 133
Heymann nephritis 161
  active 376
  animal models
    functional and structural parameters *1194*
    LEW or PVG/c rats 247–8
    various rat strains *1201*
  basement membrane structural antigens, experiments 253
  glomerular scarring, experimental nephropathy and hypertension 1196
  membranous nephropathy 374, 376
  model 639
  passive 376
HGPRT deficiency, *see* Lesch–Nyhan syndrome
5-HIAA, *see* 5-hydroxyindoleacetic acid
I-123-hippuran, chronic renal failure 126
I-131-hippuran,
  ERPF (effective RPF) 117
  gamma-camera imaging techniques 40
hippuric acid, decreased drug protein binding 1240, 1243
histamine,
  pruritus and 1391
  release, in serum sickness model 584
  vasodilatation of afferent arterioles 26, 27
histidine,
  diet, acid and base production 932–3, *933*
  excretion, pregnancy and 1916
  histidinuria 750
  normal tubular reabsorption 745
histones, autoantibodies in SLE 647, 650
  binding 637
Histoplasma capsulatum,
  infection-associated glomerulonephritis 466
  natural history 1742
histoplasmosis 1742
history, *see* clinical history of presenting complaint
HIV,
  focal segmental glomerulosclerosis 462
  renal transplantation and 1510, 1581
  management 1536
  testing, dialysis 1169
HLA antigens,
  cytotoxic antibodies 1170, 1516–17
  DR3-B8 haplotype, autoimmune diseases and membranous nephropathy 376, *379*
  DR3-DR7, minimal change nephrotic syndrome and 312
  gold/penicillamine nephrotoxicity prediction 683
  Goodpasture's disease 445
  Henoch–Schönlein purpura association 601

HLA antigens (*cont.*):
  HLA Class I A, B, C antigens
    crossmatching in renal transplantation 1570, 1587–9
    HLA B18, marker for Balkan nephropathy 860
    reactivity against peripheral blood lymphocytes (PBL) 1549
  HLA Class II DP, DR, DQ antigens
    renal transplantation 1548–9
    cadaver/host outcome 1548
    crossmatching 1570, 1587–9
    relatedness of donor/host 1548
  HLA DR1, poststreptococcal endocapillary glomerulonephritis 411
  HLA DR2, Goodpasture's disease 445
  HLA DR3
    gold/penicillamine nephrotoxicity association 683
    Sjögren's syndrome association 695, 696
  HLA DR4
    endocapillary glomerulonephritis, acute 411
    pregnancy-induced hypertension 1959
  HLA DR
    and HLA DQ expression, elderly and young subjects *1630*
    tubular epithelium, and interstitial cell infiltrates 1086
  IgA nephropathy association 601
  major/minor
    genetic control 1548
    transplantation 1544
  mesangiocapillary glomerulonephritis 395
  microscopic polyarteritis association 623
  minimal change nephrotic syndrome and 312
  renal transplantation, cytotoxic antibodies
    pretransplant transfusion 1516
    removal by plasma exchange/ extracorporeal immunoadsorption 1516–17
  scleroderma association 667
  Sjögren's syndrome association 695
  SLE association 643
  Wegener's granulomatosis association 614
  see also major histocompatibility complex (MHC)
HLA-identical sibs, bone marrow transplantation, graft-*vs*-host disease 1544
HMG–CoA reductase inhibitors,
  chronic renal failure 1181, 1186
  hypertension 2051
  nephrotic hyperlipidaemia, effects 288
  triglyceride lowering 1258
  see also lovastatin
Hodgkin's lymphoma,
  minimal change disease 2342
  minimal change nephrotic syndrome 299
  nephropathy and 482–3

Hodgkin's lymphoma (*cont.*):
  renal complications of 2317
  tumour-related virus 482
  urate clearance 830
Hodson's lines, calyces 82–3
Holtzman rat, renal hypertensive studies 1196
hormone therapy, androgen receptors, renal cell carcinoma 2248
hormones, disorders, *see* endocrine disorders; *specific substances*
hospital-acquired acute renal failure 971
Howell–Jolly bodies 704
HPLC chromatography, uraemic syndrome, identification of metabolites 1241–2
HTLV-1-associated myelitis, retroviral infections 2323
Hunner's ulcer 1686
hyaline droplet nephropathy, nephrocalcinosis 1887
hyaline thrombi 652, 653
hyalinization,
  arteriolar 2216
  in diabetes mellitus 524
  vascular sclerosis, progressive chronic renal failure 1206
  glomerular, in amyloidosis 555
hybridoma technology, anti-DNA antibodies 637
hydantoin derivatives, drug-induced nephropathy 163
hydatid disease, mebendazole 185
hydralazine,
  action and side-effects 1971
  animal models
    effect on glomerular scarring *1213*
    renal ablation effects *1195*
  children
    emergencies 2149
    treatment of hypertension 2148
  crescentic glomerulonephritis 422
  haemodialysis patients, recommendations *1463*
  hepatocellular hepatitis 2315
  hypertension 188
    end-stage renal failure 2121
  liver effects 2315
  lupus induced by 651
  microscopic polyarteritis association 623
  neonates 1116
  treatment of hypertension 2148
  platelet effects and therapy 1968
  pregnancy-induced hypertension therapy 1971–2
  rapidly progressive glomerulonephritis 1070
  see also vasodilators
hydration protocol, renal transplantation 1518–19
hydrocephalus syndrome, features and associated abnormalities *2037*
shunt nephritis 469–70
hydrochlorothiazide,
  animal models, effect on glomerular scarring *1213*
  bicarbonate, tubular reabsorption 771
  children, diabetes insipidus, nephrogenic 795

hydrochlorothiazide (*cont.*):
  contraindications 776
  metabolic acidosis in Fanconi
      syndrome 736
  nephrotic syndrome 270
  structural formula *199*
hydrogen ion secretion,
  aldosterone 766
  ammoniagenesis 767
  ATPase, active hydrogen
      transport 764, 766
  blood P_{CO_2} 767
  blood pH 767
  distal tubule acidification 763, 766
  hypercalcaemia causing 1885
  mechanisms 766
  renal regulation in pregnancy
      1916–17
  RTA patients, P_{CO_2} values 1893
  secretory defect, distal renal
      tubular acidosis 763, 766
  tests of generation
      acid-loading tests 43
      chronic renal failure and
          ammonium excretion 44
      plasma bicarbonate and early
          morning urine pH 43
      urine:plasma P_{CO_2}/ gradient,
          bicarbonate loading and
          threshhold 44
hydronephrosis,
  differential diagnosis 2267
  familial, features and associated
      abnormalities *2036*
  gross, ultrasonography 91
  hypertension and 2089–90
  intermittent 81
  mild, ultrasonography 91–2
  pregnancy *1931*, 1931
  radiological tests 74
  see also urinary tract, dilated,
      upper
hydrothorax, peritoneal dialysis
      1495
hydroureter,
  differential diagnosis 2267
  pregnancy *1931*, 1931
3-hydroxy-3-methylglutaryl
      coenzyme A reductase
      inhibitors, *see* HMG coA
      reductase inhibitors
β-hydroxy-γ-
      trimethylaminobutanoic acid, *see*
      carnitine
hydroxyapatite, inflammatory
      deposits 2298
2-hydroxybenzoylglycine, protein
      binding, competition for sites
      176
β-hydroxybutyrate, urate excretion
      decrease *829*
hydroxybutyrate dehydrogenase,
      haemolytic–uraemic syndrome
      1055
α-hydroxybutyric acid, methionine
      malabsorption 750
hydroxychloroquine, in lupus
      nephritis 659
1α-hydroxycholecalciferol,
  hypophosphataemic rickets 755
  vitamin D-dependency rickets
      type II 756
hydroxyethyl starch,
  accumulation, haemodialysis
      patients 1282

hydroxyethyl starch (*cont.*):
  acute renal failure in diabetics
      1638
2-hydroxyhippuric acid, protein
      binding, competition for sites
      176
o-hydroxyhippuric acid, uraemic
      syndrome 1243
p-hydroxyhippuric acid, uraemic
      syndrome 1243
5-hydroxyindoleacetic acid
      clearance, chromatography 40
hydroxyl radicals, superoxide ion
      169
1α hydroxylase,
  activity in macrophage, in
      sarcoidosis 579
  corticosteroids inhibiting 579
  deficiency, clinical features 1798
  phosphate, suppression 1382
  response, parathyroid hormone
      753
11 or 17-hydroxylase deficiency,
      hypertension, metabolic alkalosis
      and 948
hydroxylysine, increased in
      glomerular basement membrane
      in diabetes 519
hydroxyproline,
  Fanconi syndrome 727
  transplantation, normo/
      hypercalcaemic recipients *1775*
hydroxypyruvate, disorders of
      metabolism 2227
11-β-hydroxysteroid
      dehydrogenase,
  deficiency 1895
  abnormal peripheral cortisol
      metabolism 948
  inhibition by glycirrhizinic acid
      948
  nephrocalcinosis 1895
  liquorice intoxication 8
  mineralocorticoid 'apparent excess
      syndrome' 905
5-hydroxytryptamine, pregnancy-
      induced hypertension 1964
hydroxyurea,
  infected stone disease 1864
  urease inhibitor 1666
hyperaldosteronism,
  accelerated hypertension 2131
  compensatory, hypokalaemia
      2335
  liver glomerulonephritis *360*
  magnesium wasting and 1810
  primary
      hypernatraemia 883
      hypokalaemia 905
hyperaminoacidurias **730–3**
  in galactosaemia 731
  glycogenosis with Fanconi
      syndrome 730
  hereditary fructose intolerance
      732
  Lowe syndrome 732
  measurement 728
  Wilson's disease 733
  see also aminoacidurias
hypercalcaemia,
  acute
      acute renal failure in 1012
      treatment 1012, 1777
  acute renal failure 1773
  recovery 980

hypercalcaemia (*cont.*):
  aluminium intoxication 1369,
      1774
  animal models 1883
  bicarbonate depletion 765
  biochemical measurements, *vs*
      normocalcaemic recipients
      1775
  bone problems, differential
      diagnosis 1381
  calcium channel blockers and
      1885
  causes *1773*
  chronic renal failure 1185
  dialysis-independence 1772–3
  clinical features *1769*, 1896–7
  dialysis patients 1773
  differential diagnosis 1774
  end-stage renal failure,
      proportional distribution by age
      at commencement *1229*
  familial
      benign nephrocalcinosis 1896
      complications of 1896, *1897*
  following transplantation,
      transient *vs* static 1379
  glomerular filtration rate reduced
      by 579
  haematological malignancies 1777
  haemodialysis patients *1455*
  hyperparathyroidism 1774, 1854
      metabolic alkalosis and 949
  hypertension, transient 1884
  hypertensive subjects 1461
  infantile 1891
  irrelevant, causes *1455*
  left ventricular function, end-stage
      renal disease 1266
  magnesium handling 1808
  malignant aetiology 476, 1776–7
  management 1012, 1777, 1898
      transplantation patient 1777
  in multiple myeloma 567–8
  natural history 1898
  neonates 1116
  nephrocalcinosis causation 1891
  nephropathy, treatment 1885
  pathology 1883
      tubular damage 1883
  polyuric syndromes, nephrogenic
      diabetes insipidus 886–7
  renal cell carcinoma 2242
  renal failure, transient 1883
  renal function effects 1777
  renal transplantation **1775–7**
      following 1585
      management 1777
  renal vessels, vasoconstriction
      1884
  reversal following cessation of
      vitamin D *1771*
  reversible hypercalcaemic
      nephropathy 1883, 1891
  in sarcoidosis 578, 579
      mechanism 578–9
  symptoms *1854*
  transient and steady state 1768
  tubular defects
      acid-base state 1885
      aminoaciduria 1885
      calcium excretion 1885
      hydrogen ion secretion 1885
      magnesium excretion 1885, 1887
      metabolic alkalosis 1885
      phosphate excretion 1885

hypercalcaemia (*cont.*):
  tubular defects (*cont.*):
      potassium excretion 1884
      renal glycosuria 1885
      sodium loss 1884
      water handling 1884
  tumours, complications of 2317
  uraemic encephalopathy 1399
  vitamin D intoxication 1767
      milk–alkali syndrome 1773
      vitamin D treatment 1773
hypercalciuria **1851–3**
  absence in RTA 1893
  absorptive 1833
  calcium stone disease 1833, 1834
  causes, by plasma calcium 1861
  Cushing's syndrome 1861
  Dent's disease 1893, 1896
  dietary intake of calcium 1834
  Fanconi syndrome 727
  idiopathic *1897*
      defined 1851–2
      nephrocalcinosis 1891
      treatment 1852–3
  pregnancy, mechanism 1917
  renal 1834
  renal tubular acidosis-associated
      774, 1835
  in sarcoidosis 578, 579
      mechanism 579
  skeletal breakdown 1834
  sponge kidney and 1895–6
  thiazide diuretics 1852
  vitamin D and 1833
hypercapnia,
  chronic, respiratory acidosis and
      metabolic alkalosis following
      treatment 958
  post-hypercapnic acidosis 935
  primary
      respiratory acidosis 951
          causes *953*
          treatment 953
      see also respiratory acidosis
hypercatabolism,
  causes 978
  defined 978
  inadequate nutrition 978
hyperchloraemia,
  renal tubular acidosis 777–8
  in Sjögren's syndrome 696
hyperchloraemic metabolic
      acidosis,
  differential diagnosis, algorithm
      778
  distal renal tubular acidosis
      diagnosis 774
hyperchlorhydria 1280
hypercholesterolaemia,
  diabetic nephropathy 1648
  familial *1897*
hypercorticism, chronic renal
      failure 1209
hypercystinuria, isolated 749
hyperemesis gravidarum,
  acute tubular necrosis,
      complicating 1933
  obstetric acute renal failure *1933*
hypergammaglobulinaemia,
  Fanconi syndrome 734
  renal tubular acidosis link 697
  in Sjögren's syndrome 694–5, 697
hyperglycaemia,
  in diabetic nephropathy
      pathogenesis 517–18

Index

hyperglycaemia (cont.):
 following renal transplantation 1576
 glomerular filtration rate elevation 512–13, 516
 glucagon, role 1254
 hyperkalaemia in diabetics 907
 proteinuria risk in diabetes mellitus 506
 pseudo-hyponatraemia 873
 serum sodium effects 873–4
hyperglycinuria, isolated 750–1
hyperkalaemia **906–11**
 acetazolamide 206
 acute arrhythmias 187
 acute treatment 1182–3
 aetiology 907–10, *908*
  excessive potassium retention 908
  hypoaldosteronism 909–10
  renal failure 908–9
  spurious hyperkalaemia 907
  transcellular shifts 907–8
 aldosterone deficiency/resistance and 41
 children with chronic renal failure 1306
 clinical sequelae 906–7, *907*
  cardiac complications 906
  endocrine effects 907
  fluid and electrolytes 907
  neurological complications 907
  systemic haemodynamics 907
 cyclosporin therapy in renal transplantation 1564
 drug-induced 907
 glucagon secretion, enhancement 907
 hypertension and, renin–angiotensin–aldosterone system blockade 2073
 hyporeninaemic hypoaldosteronism 823
 insulin secretion, enhancement 907
 management *910*, 910–11
  tropics 1129
 neonates 1116
 oliguric patients 1019
 pacemaker conduction abnormalities 906
 in renal failure in diabetics 512
 renal transplantation, cyclosporin therapy 1564
 renal tubular acidosis and, *see* renal tubular acidosis, hyperkalaemic
 rhabdomyolysis, acute renal failure and 1008
 sodium chloride wasting, with hyporeninaemic hypoaldosteronism 1305
hyperkalaemic periodic paralysis, characteristics 908
hyperkalaemic tubular acidosis, *see* renal tubular acidosis, hyperkalaemic (type 4)
hyperlacticaciduria, Fanconi syndrome 727
hyperlipidaemia,
 animal models, dietary and pharmacological lipid manipulations *1202*, 1216
 chronic renal failure 1185, 1186
 progressive 1201–2

hyperlipidaemia (cont.):
 in diabetic nephropathy 516–17, 521
 diuretic-induced
  pathogenesis 211
  therapeutic consequences 211
 following renal transplantation 1540, 1576
 HAPPHY study 211
 incidence, diuretic therapy 211
 lipid lowering agents 191
 management 2051
 MRFIT study 211
 nephrotic syndrome **284–9**
  focal segmental glomerulosclerosis and 291
  pseudo-hyponatraemia 873
hypermagnesaemia **1815–16**
 aetiology 1815
 clinical features 1815
 emergencies 1816
 general aspects of magnesium metabolism **1802–9**
 management 1816
 pathogenesis 1815
hypernatraemia **882–9**
 assessment of volume status 882
 causes *882–3*
 children 887
 defined 882
 elderly patients 887–8
 essential 887
 experimental models 888
 neonates 53
 outcome 888–9
 pseudo-hyponatraemia masking hypernatraemia 883
 signs and symptoms 888
 total body sodium
  decreased 882–3
  increased 883
  normal 883–7
 treatment 889
hypernephroma, *see* Wilms' tumour
hyperoestrogenaemia, liver glomerulonephritis *360*
hyperoxaluria **1856–8**
 ascorbate 1857
 calcium stone disease 1833
  geographical factors 1832, 1835
 causes *1856*
 dietary intake 1856–7
 enteric, calcium stone disease 1836
 glyceric/glycollic, *see* hyperoxaluria, primary
 hereditary, calcium stone disease 1836
 nephrocalcinosis 1895
 primary *2226*, 2226–30
  acute renal failure 1012
  assessment 2228
  clinical features 2227–8
  glycolic *1897*
  management 2228–9
  metabolic defects 2226–7
  natural history 4
  pathogenesis 1895
  prognosis 1898
  type 1 glyceric 1857, 1895, *1897*, 2226–7
  type 2 glycollic 1858, 1895, *1897*, 2226–7
 secondary *2226*
 steatorrhoea 1857

hyperparathyroid, *see* hyperparathyroid bone disease; hyperparathyroidism
hyperparathyroid bone disease,
 in advanced renal failure 1367–8
 cimetidine *1760*
 diagnostic procedures 1370–6
 differential diagnosis 1370
 in early renal failure 1365–7
 parathormone in 1370
 radiography
  recommended *1371*
  rugger jersey spine 1373
  signs *1372*
 serum biochemistry *1371*
 treatment 1384–5
 vitamin D treatment, calcium set-point *1760*
hyperparathyroidism,
 biphosphonates 1856
 bone disease in, *see* hyperparathyroid bone disease
 chronic renal failure
  in advanced renal failure, maintaining factors 1367
  aluminium overload 1368–9
  animal models, calcium levels 1366
  calcitriol role in initiation 1366
  children, management 1305, 1307
  dialysis procedures 1368
  elderly patients, suppression 1623
  iatrogenic perturbations 1368–9
  lead mobilization test and 834
  left ventricular hypertrophy 1265
  phosphate levels in 1365
  platelet defects, uraemic haemorrhage 1361
  pre/post renal transplantation 1510
  renal transplantation, reversibility 1378
  terminal renal failure 1367
  tertiary 1767
  testosterone levels in 1332
  uraemic syndrome 1244
 discriminant index 1855
 epiphyseal slipping, children 1612
 fifth gland 1855
 hypercalcaemia 1774, 1854
  metabolic alkalosis and 949
 nephrocalcinosis 1891
 oestrogens 1856
 preterm neonates, longterm frusemide administration 54
 primary **1853–6**
  biochemistry 1854–5
  calcium stone disease 1835
  defined 1853
  follow-up 1856
  medical treatment 1856
  presentation 1853–4
  surgical treatment 1855–6
  tumour localization 1855
  types and frequency 1853
 renal transplantation, complication 1537
 renal tubular acidosis-associated 774, 1855
 rickets and 1612

hyperparathyroidism (cont.):
 secondary, *see* hyperparathyroidism, chronic renal failure
 tertiary
  aluminium intoxication and 1774
  chronic renal failure 1767
  differential diagnosis 1774
  tumours
   CT scan 1855
   parathormone concentrations 1855
   radiolabelling 1855
   surgical treatment 1855
   types I and II 1853
  uraemic peripheral polyneuropathy 1398
hyperphosphataemia **1793–5, 1798–9**
 calciphylaxis, acute 1793
 carbonic anhydrase inhibitors 206
 causes *1794*
 chronic renal failure and 1793, 1794
 clinical features 1793
 control 1794
 familial
  classification *1796*
  pseudohypoparathyroidism 1798
 haemodialysis, resistance to dietary control 1794
 hypocalcaemia in 1772
 inheritance, mechanisms and treatment *1799*
 nephropathy 2319
 physiological findings *1797*
 prevention 1382–3
 treatment 1382–3
hyperphosphaturia, Fanconi syndrome 725
hyperpigmentation, and analgesic syndrome 814
hyperpipecolic acidaemia, features and associated abnormalities *2038*
hyperprolactinaemia,
 in chronic renal failure 1337
 haemodialysis, treatment 1338
hyperreninaemia, loop diuretics 199
hypersensitivity reactions, direct, interstitial nephritis 1094
hypersensitivity vasculitis 587, 613
 see also microscopic polyarteritis
hypertension **2047–152**
 accelerated **2124–36**
  aetiology 2127
  arteriolar fibroid necrosis 2131, 2134
  blood pressure reduction 2135
  clinical presentation 2130–2
  controlled trials 2124
  defined 2124
  haematology 2131
  haemolytic anaemia in 2341
  hypertensive encephalopathy 2132–4
  incidence and prognosis 2124–7
  investigations 2131
  management 2134–6
  natural history 2125
  renal function and renal survival 2126–7, 2134
  renal pathology **2127–30**
   afferent arterioles 2128

hypertension (*cont.*):
  accelerated (*cont.*):
    renal pathology (*cont.*):
      antihypertensive treatment effects 2129
      glomeruli 2128
      interlobular arteries 2127–8
      pathology in other organs 2130
      and renal dysfunction 2129
      and secondary malignant hypertension 2129
      tubules and interstitium 2128–9
      specific disease 2127
      see also hypertension, malignant
  N-acetyl-β-D-glucosaminidase in urine 18
  angiotensin, chronic infusion causing 2061–2
  angiotensin converting-enzyme inhibitors, handling in kidney disease 188
  animal models 1196–7
    glomerular scarring 1196–7
    sodium balance, renin and sodium depletion *2060*
    sodium excess 2118–19
    see also blood pressure control
  benign hypertension, radiation nephritis 851
  β-adrenoceptor blockers, handling in kidney disease 188
  calcium deficiency, lead intoxication and 834
  CAPD 1486
  chemical medullectomy model 2065
  children 2139–49
    chronic renal failure 1610
    defined 1610
  chronic renal failure **2117–22**
    classification 1176
    control in progression 1213–14
    focal glomerulosclerosis 1198
    IgA nephropathy 1198
    membranoproliferative glomerulonephritis 1198
    non-pharmacological treatment 1177
    paraplegia 2276
    pharmacological treatment 1177–8
    polycystic disease 1198
    potassium depletion 1177
    practical approach 1176–7
    progression 1176
    renal artery stenosis 1151
    risk factor in end-stage renal failure, prevalence 1268
    sodium restriction 1177
  clinical approach **2047–56**
  corticosteroids, toxicity 322
  cyclosporin-related
    clinical features 1565
    management 1565–6, 1599–600
    pathogenesis 1565
    prevalence *1565*
  defined 1176, 2047
    children 2141
  detection 1177
  diabetes and chronic renal failure
    antihypertensive treatment 1647–8
    common problems *1646*
    determinant of coronary disease 1643
    differential diagnosis *1646*

hypertension (*cont.*):
  diabetes and chronic renal failure (*cont.*):
    nephropathy, management 1645–8
  dialysis-refractory *1462*
  diuretics, handling in kidney disease 188
  drug-induced 8
  encephalopathy and 1400, **2132–4**
    cerebral autoregulation and 2133
    erythropoietin 2133–4
    nephritic syndrome, acute, treatment 412
  end-stage renal failure 1233, 1458–9, 2117
    clinical pictures 2120
    left ventricular hypertrophy 1265
    pathogenesis 2117–20
    renin-dependent 1459
    sodium-dependent 1459–60
    therapy 2120–2
  essential
    animal models of progressive glomerular scarring 1212–13
    assessment 1151
    criterion for treatment 2047
    development of renal impairment 2126
    incidence 2047
    pregnancy 1956
    progressive renal disease and **2083–6**
    renal diseases and pre-eclampsia, differential diagnosis 1940–1
    target blood pressure 2056
    vs renal causes 2047–8
    see also hypertension, malignant
  in essential mixed cryoglobulinaemia 690
  Goldblatt models of hypertension 2059–65, 2075–9, 2101
  haemodialysis
    decrease 1460
    long-term patients 1330, 1438
    treatment **1463–4**
  HAPPHY trial 202
  hyperkalaemia and, renin–angiotensin–aldosterone system blockade 2073
  hypertensive crisis 1464
  hyperuricaemia in 831
    renal handling of urate/uric acid 831
    secondary to associated factors 835
  investigations
    essential hypertension vs renal causes *2049*
    indications for search for renal causes *2049*
  IPPSH trial 202
  kidney donors, unilateral nephrectomy 2093–4
  kidney structure effects **2075–81**
  lead-associated 839–40
  malignant
    aetiology 2125
    chronic renal failure 1184
    diabetes, renal artery stenosis 1646
    end-stage renal failure 1233
    haemodialysis patients 1462

hypertension (*cont.*):
  malignant (*cont.*):
    haemolytic–uraemic syndrome and 1050
    IgA nephropathy 2338
    incidence and prognosis 2124–7, 2125
    outcome 2125
    pathogenesis, vasopressin 2132
    radiation nephritis 851
    in scleroderma 675
    secondary, and accelerated hypertension 2129
    vascular diseases with renal involvement, schematic representation 152
    see also hypertension, accelerated
  metabolic alkalosis and
    aldosteronism, primary 948
    aldosteronism, secondary 948
    cortisol, abnormal metabolism 948
    Cushing's syndrome 948
    11 or 17-hydroxylase deficiency 948
    Liddle's syndrome 948–9
    mineralocorticoids, exogenous 948
  in microscopic polyarteritis 624
  MRFIT trial, morbidity and mortality 201
  neonates 1116
  neurofibromatosis, children 2300
  nuclear imaging
    captopril test 122
    uptake function and transit times 121–2
  paradoxical, haemodialysis patients 1462–3
  in polyarteritis nodosa 628
  pregnancy, see pregnancy-induced hypertension
  progressive renal disease and **2083–6**
    evidence of deleterious effect **2083–5**
  renal, renomedullary antihypertensive function **2065–6**
  renal biopsy, contraindications 143
  renal disease and **2047–56**
    amyloidosis 551
    analgesic nephropathy 813
    blood pressure control **2058–66**
    chemical medullectomy model 2065
    clinical history 2048
    effector pathway 2063–4
    glomerular scarring, experimental nephropathy 1196–7
    IgA nephropathy 342, 343
    IgA nephropathy, primary 350
    investigations 2049–51
    membranous glomerulonephritis 1198
    nephrocalcinosis 1896
    nephrosclerosis 2079 and **2079–81**
    pathogenesis 2051
    physical examination 2048–9
    primary renal disease 2047
    reflux nephropathy 1992–3
    renal artery stenosis **2096–112**
    renal cell carcinoma 2242

hypertension (*cont.*):
  renal disease and (*cont.*):
    renovascular hypertension 119–22, **2096–112**
    reversal of renovascular hypertension 2062–3
    surgical therapy 2056
    treatment 1177–8, 2051–6
    unilateral parenchymal disease **2088–94**
  vs essential hypertension 2047–8
  renal transplantation, following **1536–7, 1594–600**, 2094
    causes 1595–7
    diagnostic strategy 1598–9
    effect on graft and patient outcome 1597
    incidence 1576
    management 1599–600
    prevalence 1536, 1576, *1594*
    severity *1594*
    therapy and results 1597–8
  renal vasculature effects **2075–81**
  renin–angiotensin–aldosterone system and 2014, **2069–74**
  renovascular 119–22, **2096–112**
    branch artery stenosis 122
    clinical history 6, 7
    large-vessel vs small-vessel 121–2
    parenchymal transit times 121
    PTA 114
    radionuclide studies 119, 121–2
    smoking as risk factor 9
    see also hypertension, renal disease and renal artery stenosis; renal occlusive disease
  in scleroderma 673, 675
  sickle-cell disease and 712
  in SLE 649
  in children 650
  systemic, see hypertension, essential
  treatment **2051–6**
    angiotensin converting-enzyme inhibitors 1177–8
    β-adrenoceptor blockers 1178
    calcium entry blockers 1178
    diabetic nephropathy 1645–8
    diuretics 1177
    erythropoietin replacement therapy, r-HuEPO 1355–6
    non-pharmacological treatment 1178
    vasodilators 1178
  tuberculosis and 1721
  unilateral renal parenchymal disease and **2088–94**
    aetiology *2088*
    diagnosis *2088*
    see also blood pressure
hypertensive rat, spontaneous, haemodynamic and structural glomerular parameters *1214*
'hypertensive stress test', pregnancy 1959
hyperthermia, malignant, potassium loss from muscle 908
hyperthyroidectomy, pruritus and 1391, 1392
hyperthyroidism, nephrocalcinosis 1896
  pruritus 1391
hypertrichosis, cyclosporin, nephrotoxicity 1566

Index

hypertriglyceridaemia,
  coronary artery disease, risk factor in end-stage renal failure 1268
  diabetic nephropathy 1648
  following renal transplantation 1540
  uraemic, liver changes 1283
hypertyrosinaemia, transient neonatal 730
hyperuricaemia,
  alcohol abuse and 829
  allopurinol 189–90
  animal model 825
  cardiovascular damage and 835
  chronic renal failure 1185
  cyclosporin therapy in renal transplantation 1564
  in diabetes mellitus type II 831
  diuretic-induced, gouty arthritis and 212
  endothelial damage and 835
  fructose and 830
  in hyperlipidaemia 831
  in hyperlipoproteinaemia type IV 831
  hypertension in 831
    renal handling of urate/uric acid 831
    secondary to associated factors 835
  isolated 757
  ketoacidosis-related 832
  nephroangiosclerosis, early indicator 831
  pregnancy 1958
  toxaemia of pregnancy 832
hyperuricosuria, see urate/uric acid, acute uric acid nephropathy
hyperviscosity syndrome 562, 568
  M protein levels 568
  see also blood viscosity
hypoalbuminaemia,
  congenital, filtration fraction 268
  in Henoch–Schönlein purpura 606
hypoaldosteronism,
  C-21 hydroxylase deficiency 909
  3-β-OH-dhydrogenase deficiency 909
  drug-induced 909
  hyperkalaemia 909–10
  hyporeninaemic
    hyperkalaemia 909
    lead intoxication 839
    non-steroidal anti-inflammatory drugs 823
    in sickle-cell disease 711
  therapy 909
  treatment 940
  potassium excretion 909
  pseudohypoaldosteronism type I 940
  pseudohypoaldosteronism type II 910
    distal renal tubular acidosis 910, 938, 939
hypocalcaemia,
  bicarbonate reabsorption 765
  calcium in 1754
  causes 1772
  chronic renal failure
    CAPD 1771
    haemodialysis 1771
    mechanisms 1769–71
  clinical features 1769

hypocalcaemia (cont.):
  hyperphosphataemia and, chronic renal failure 1367
  management 1770
  parathormone effects 1761
  parathyroidectomy 1772
  pathogenesis 1761
  rhabdomyolysis, acute renal failure and 1008
  transient and steady state 1768
  tumours, complications of 2317
hypocalciuria, Bartter's syndrome 783
hypocapnia, metabolic acidosis, respiratory response 921
hypocapnia, primary,
  pregnancy 955
  respiratory alkalosis
    causes 955
    treatment 955
  see also respiratory alkalosis
hypocapnia, secondary, metabolic acidosis 955
hypochloraemia, Bartter's syndrome 782
hypochlorhydria 1280
  gastric acid hypersecretion, chronic renal failure 1280
hypocomplementaemia,
  atheroembolic disease of the kidney 2284
  in glomerulonephritis 393–6
    activators of complement 393
    chronic or persistent, see mesangiocapillary glomerulonephritis
    correlation of abnormalities with disease 394
    evolution with time 394
    inherited deficiency 394, 396
    partial lipodystrophy and 394–5
    types I - III, MCGN 394
  in leucocytoclastic vasculitis 587, 588
  in SLE 642, 648
  see also complement
hypodipsia 887
hypoglycaemia,
  chlorpropamide 189
  chronic renal failure 1254
  glibenclamide 189
  glicazide 189
  hypoglycaemic agents, oral elimination in chronic renal failure 1645
  see also chlorpropamide; glibenclamide
  stress reaction 1325
    suppression in uraemia 1325
  sulphonylurea compounds and 1645
  tolbutamide 189
hypokalaemia 900–6
  acute renal failure, recovery 980
  aetiology 902–6
    extrarenal potassium depletion 903
    potassium depletion 902–3, 903
    redistribution 902
    renal potassium depletion 903–6
    transcellular shifts 902
  aldosterone and 787
  Bartter's syndrome 782–7
    NSAIDs and 820
  cardiac glycosides and 208
  pathophysiology 209

hypokalaemia (cont.):
  clinical sequelae 900–2
    cardiac complications 900–1
    endocrine effects 901–2
    fluid and electrolyte disorders 901
    neuromuscular function 901
    renal effects 901
    systemic haemodynamics 902
  compensatory hyperaldosteronism 2335
  differential diagnosis 2310
  digoxin toxicity, elderly patients 67
  diuretic-induced
    clinical consequences 209
    prevalence 209
  Fanconi syndrome 726–7
  glucose intolerance 211
  haemodialysis, antiarrhythmics 1625
  hypomagnesaemia 1811–12
  juxtaglomerular apparatus hyperplasia 787
  Liddle's syndrome 787
  malignant aetiology 476
  renal tubular acidosis, disease processes 939, 1859
  in Sjögren's syndrome 696
  spontaneous, chronic renal failure 1183
  treatment 208–10, 906
  tumours, complications of 2318
  unexplained, covert vomiting, laxative abuse 787
  see also potassium depletion
hypokalaemic distal renal tubular acidosis 937–9
hypokalaemic metabolic alkalosis, differential diagnosis 787
  infancy 773
hypokalaemic periodic paralysis, acquired, thyrotoxicosis 902
  inherited, hypokalaemia 902
hypomagnesaemia 1809–14
  aetiology 1772, 1809–11
  athletes 1811
  Bartter's syndrome 783
  children
    cardiovascular features 1813
    primary hypomagnesaemia 1810
  clinical features 1811–13
    hypokalaemia 1811–12, 1812
    neuromuscular and CNS features 1812–13
  congenital conditions 1811
  diagnosis 1813–14
  following blood transfusion 1811
  general aspects of magnesium metabolism 1802–9
  hypocalcaemia in 1772
  lactation and 1811
  management 1814
  neonates 1116
  pathogenesis 1809–11
    alcoholism 1811
    increased excretion 1810–11
    prophylaxis 1814
    reduced absorption 1809–10
    reduced intake 1809
  treatment 1814
  prevalence 1809
  stress-induced 1811

hypomagnesaemia–hypercalciuria syndrome, nephrocalcinosis 1894
hyponatraemia **872–82**
  acute, treatment 881
  adrenal insufficiency 876
  AIDS-related complex 466
  angiotensin converting-enzyme inhibitors 879
  assessment of volume status 874–5
  causes 476, 873
  children 879
  chronic asymptomatic 881
  clinical approach 872
  coma 880–1
  congestive heart failure 877
  defined 872
  diarrhoea 879
  diuretic-induced, pathogenetic factors 208–9
  diuretics and 879
  elderly patients 879–80
  Fanconi syndrome 727
  fluid intake 209
  glucocorticoid deficiency 879
  high risk patients 881
  hypothyroidism 879
  inappropriate secretion of ADH syndrome 207
  laboratory variables 875
  liver cirrhosis 877
  magnesium 210
  malignant aetiology 476
  mechanisms 875
  neonates 53
  nephrotic syndrome 877
  neurological abnormalities 880–1
  NSAID-induced water retention 823
  overcorrection 881
  postoperative state 878
  potassium deficiency 879
  prevention 881
  pseudo-hyponatraemia 872–4
    causes 873
  psychogenic polydipsia 874
  renal failure 878
  severe chronic 881
  signs, symptoms and outcome 880–1
  smoking and 874
  total body sodium
    decreased 875–6, 943
    inappropriate secretion of ADH 878–9
    increased 876–8
    normal 878–9
  treatment 881–2
  tumours 878
    complications of 2318
hyponatraemic hypertensive syndrome, accelerated hypertension 2131
hypophosphataemia **1790–3, 1795–8**
  in alcoholism 1792
  causes **1790–2**
    decreased intake 1792
    diuretics 1792
    gastrointestinal loss 1792
    multiple causes 1792
    PTH secretion 1790–1
    transplantation 1791
    tubular function impaired 1790
    tumour-associated 1791–2

hypophosphataemia (cont.):
  clinical features **1790–3**
    following transplantation 1791
    signs and symptoms *1791*
  defined 1790
  in diabetes 1792
  familial
    classification *1796*
    other genetic defects 1797–8
    other phosphate transport defects 1796–7
    primary phosphate transport defects 1795–6
    X-linked rickets 1795
  Fanconi syndrome 726
  following transplantation 1378
  inheritance, mechanisms and treatment *1798*
  oral phosphate, metabolic acidosis in Fanconi syndrome 736
  pathogenesis 1790
  treatment 1792–3
  tumours, complications of 2317
  in vitamin D-resistant rickets 753, 754
hypophosphataemic rickets, cystinosis 755
  familial X-linked hypophosphataemic rickets 753–5
  hereditary hypophosphataemic rickets with hypercalcaemia 756
  with hypercalciuria and microglobulinuria 756
  oncogenic osteomalacia 756–7
hypophosphatasia, nephrocalcinosis *1897*
hypophysectomy, animal models, resistance to glomerular scarring 1208–9
hypoplasia, oligomeganephric 1606–7
hypoproteinaemia,
  causes 5
  drug binding, reduction 8
  nephrotic syndrome 1035
  proteinuria, and oedema 5
hyporeninaemic hypoaldosteronism 776, 777, 909
  aldosterone deficiency 207
  angiotensin converting-enzyme inhibitors and 1647
  diabetes, type II, elderly patients 1646
  with hyperkalaemia and sodium chloride wasting 1305
  lead exposure 839
  non-steroidal anti-inflammatory drugs 823
hyposthenuria, children with chronic renal failure 1306
hypotension,
  in cardiac amyloidosis 558
  chronic renal failure 1158
  haemodialysis 1429, 2338
  nephrotic syndrome complicated by ARF 1039
  obstetric acute renal failure 1932
  peritoneal dialysis, elderly patients 1626
  postural
    in diabetes mellitus 512
    in renal amyloidosis 551

hypothalamo–pituitary–thyroid axis, chronic renal failure 1320–1
hypothalamus,
  function tests 1325
  hypoglycaemic stress reaction, suppression in uraemia 1325
  pulse generator, chronic renal failure 1616
hypothermia, acid–base disorders 960
hypothyroidism,
  congenital
    infantile nephrocalcinosis 1894
    nephrocalcinosis 1894–5
  end-stage chronic renal failure 1319–20
  gynaecomastia in 1336
  haemodialysis, treatment 1338
  inappropriate secretion of ADH 878
  uraemia 1320
hypouricaemia,
  familial, uric acid stones *1861*
  Fanconi syndrome 726
  inappropriate secretion of ADH 832
  isolated 757
  pregnancy 832
  syndromes causing 830
hypovolaemia,
  action of prostaglandins, NSAIDs toxicity 820
  arginine vasopressin release and 1918
  causes *1015*, 1015–16
  concept, nephrotic syndrome 269
  in nephrotic syndrome 267
  nephrotic syndrome 290–1
  overtreatment 273–4
  prerenal failure 1015
  thirst sensation 65
hypoxaemia, haemodialysis-related 2339
hypoxanthine, toxicity, uraemic syndrome 1244
hypoxanthine-guanine phosphoribosyltransferase deficit,
  acute uric acid nephropathy *832*
  allopurinol treatment 833
  gout 2294
hypoxia, *see* lactic acidosis
hypo–hyperkalaemia **895–911**
hypo–hypernatraemia **869–89**
hysterectomy, septic abortion 1933

ibopamine, chronic tubulointerstitial nephropathy 1215
ichthyosiform dermatosis 1390–1
idiopathic IgA nephropathy, *see* immunoglobulin A (IgA) nephropathy
idiopathic membranous nephropathy, treatment 659
idiopathic multicentric osteolysis 2216
idiopathic primary nephrotic syndrome, *see* minimal change nephrotic syndrome
idiotopes 638
  recurrent 638

idiotypes 562
  anti-idiotypes 245
  defined 245
idiotype network theory, immune response 245
idiotypic network **638–9**
  16/6 idiotype 638
  autoimmunity and 638
  future prospects in SLE 639
  immunointervention in SLE 638–9
  recurrent idiotopes 638
  regulatory idiotypes role 638
ifosfamide,
  drug-induced acute toxic tubular nephropathy 166
  Fanconi syndrome, caused by 735
  renal complications of 2343
ileostomy,
  renal stone formation 1831
  uric acid stones and 1862
ileus,
  haemodialysis patients 1454
  hypokalaemia precipitating 901
iliac horns, nail–patella syndrome 2214–15
iliac sign, pregnancy 1909
imidazoles,
  derivatives, decreased drug protein binding 1245
  fungal infections 1748
  handling in kidney disease 184
imino acids,
  transport disorders 750
  see also hydroxyproline; proline
iminoglycinuria 750
imipenem, CAPD peritonitis regimens 1489
imipenem/cilastin, handling in kidney disease 183
imipramine,
  percentage protein binding according to renal function *1240*
  prostatic enlargement in elderly patients 67
immersion, 'head out of water', natriuresis 269
immobilization, calcium stone disease 1836
immune complex(es),
  circulating
    antibody-mediated glomerulonephritis 251–4
    liver glomerulonephritis *360*
    mesangiocapillary glomerulonephritis 395
    complement role in clearance of 641
  deposition
    complement deficiency and 642, 648
    in essential mixed cryoglobulinaemia 688–9, 691, 692
    in lupus nephritis 652
    polyreactive autoantibodies 254
    in SLE, *see* systemic lupus erythematosus (SLE)
    see also subendothelial deposits; subepithelial deposits
  deposition sites 410
  detection 586
  DNA-anti-DNA, pathogenicity 637–8

immune complex(es) (*cont.*):
  glomerular injury mediated by 639
  in Henoch–Schönlein purpura pathophysiology 600, 601
  hepatitis B nephropathy 463
  IgA, *see* immunoglobulin A nephropathy, primary
  IgG, IgA nephropathy, primary 355–6
  liver glomerulonephritis 359
  minimal change nephrotic syndrome 311
  in necrotizing arteritis 590
  size/composition
    in essential mixed cryoglobulinaemia 587, 687
    in leucocytoclastic vasculitis 587, 600
    in SLE 587, 638
  in vasculitis pathogenesis
    in animals, *see* serum sickness
    leucocytoclastic, *see* leucocytoclastic vasculitis
    mononuclear *590*, 590
    necrotizing *590*, 590
  see also immunoglobulin A; immunoglobulin G
immune complex glomerulonephritis,
  in essential mixed cryoglobulinaemia 687–8, 691, 692
  in sickle-cell disease 711
  in Sjögren's syndrome 697–8
  see also glomerulonephritis
immune complex mediated tubulointerstitial disease, pathogenesis 1093
  see also interstitial nephritis
immune complex nephritis
  autoheterologous, antihypertensive treatment 2086
  hypertension effects 2084
  lymphocytes 410–11
immune deficiency, uraemic syndrome 1238
immune deposits, crescentic glomerulonephritis 421, 425
immune mechanisms of glomerular damage **240–58**
immune response **240–6**
  to allograft **1543–54**
  see also renal transplantation, immunology
  chronic renal failure effects **1312–16**
    B cell responses 1315
    complement activation 1315–16
    mediators 1312–14
    monocyte activation 1316
    renal replacement therapy 1315
    specific immune responses in uraemia 1314–15
  dialysis-linked syndrome 1315
  idiotype network theory 245
  mucosal, provoking IgA nephropathy 357
  simplified schema 1543–4
  site, poststreptococcal glomerulonephritis 410
  soluble immune response suppressor 311–12
immune system **240–6**
  and antigen presentation 240–1

Index

immune system (cont.):
  mucosal, IgA nephropathy 357
immuno-electron microscopy, in Henoch–Schönlein purpura 604
immunofluorescence,
  anti-DNA antibodies detection 647
  in Henoch–Schönlein purpura 603
  interstitial nephritis 1086
  linear staining
    causes 443
    Goodpasture's disease 443, 444, 446
    minimal change nephrotic syndrome 306–7
  in renal amyloidosis 553, 555
immunoglobulin A,
  antineutrohil cytoplasmic antibodies 592, 600
  deposits, immunofluorescence studies, minimal change nephrotic syndrome 307
  dinitrophenol-bovine serum albumin complex 252, 255
  in Henoch–Schönlein purpura 598, 599–600
  deficiency 600
  IgA$_1$ in mesangium 603
  in immune complexes 587, 600, 601
  in-vitro production increased 600
  increased serum levels 599
  mesangial deposits 603, 608
  polymeric nature 599–600
  recurrence in graft 608
  rheumatoid factor 600
  site of deposits 603
  IgA:IgG ratio, increased in Henoch–Schönlein purpura 600, 601
  immune complexes 587, 600, 601
    deposition in serum sickness model 583
    impaired clearance 600
  liver glomerulonephritis 359–60
  polymeric immune complexes, nephritogenesis 357
  properties and biological functions 244
  in rheumatoid arthritis
    mesangial deposits 680
    serum levels 680
  rheumatoid factors 600, 696
immunoglobulin A glomerulonephritis,
  coeliac disease and 2310
  schistosoma-specific 1737
  see also immunoglobulin A nephropathy
immunoglobulin A glomerulopathy, acute renal failure in liver disease 1104–5
immunoglobulin A nephropathy,
  accelerated hypertension risk 2127
  antibody-mediated glomerulonephritis 251
  associated diseases 360–1
  Berger's disease 601
    Henoch–Schönlein purpura comparison 601–2
    monocyte infiltration 601, 602
    rheumatoid arthritis association 680

immunoglobulin A nephropathy (cont.):
  biopsy, renal, benign nephrosclerosis 2079
  crescentic glomerulonephritis 422
  ethnic factors 9
  glomerular disease in the tropics 488
  hepatitis B
    HBs antigenaemia and 496
    nephropathy 463
  hypertension in 1198
  immunoglobulin A serum, levels 344
  laboratory findings 344–5
  malignancy-associated 480
  pathogenesis 354–8
    bone marrow role 356
    experimental models 356–8
    immunologically cellular abnormalities 354–5
    immunologically serum abnormalities 355
    mucosal factors 356
    other immunological abnormalities 355–6
  poststreptococcal 459
  pregnancy and 1941
  primary **339–59**
    age, predictive value 351
    atrial natriuretic peptide and 342
    biopsy, renal, predictive value 350–1
    characteristics 361–2
    children, matrix predominance 347
    clinical features 340–2
    clinical outcome 342–3
    creatinine, clearance 347
    defined, various nephropathies 339
    epidemiology and immunogenetics 351–4
    experimental models 252, 356–8
    extrarenal manifestations 342
    haematuria 349–50
    histology
      electron microscopy 349
      grading 345
      grading: morphological lesions *346*
      immunofluorescent deposits 347–8
    hypertension 342, 343, 350
    immunofluorescent deposits, predictive value 351
    immunoglobulin A serum, antibodies 344
    pregnancy 343
    prognostic markers 349–51
    proteinuria 344–5, 350
    renal biopsy 345–9
    renal insufficiency *342*, 350
    renal survival 343–4
    serum IgA, predictive value 351
    sex, predictive value 351
    therapy *358*, 358–9
    see also mesangioproliferative glomerulonephritis
  Reiter's syndrome 2295
  renal insufficiency 342, 343, 350
  renal transplantation 1514
    recurrence following 1573
  tubular atrophy 149
  see also immunoglobulin A glomerulonephritis

immunoglobulin D,
  myeloma 563
  properties and biological functions 244
immunoglobulin E,
  properties and biological functions 244
  vascular permeability mediated by 584
immunoglobulin F-II, chromosome 11, Wilms' tumour 2256
immunoglobulin G,
  anti-DNA antibodies 637–8, 639, 642
  antibodies to (anti-IgG), see rheumatoid factors
  autologous sialic acid-depleted, animal models 409
  basement membrane deposition, antiglomerular basement membrane antibodies 239
  in cryoglobulinaemia 686
  cytomegalovirus-specific, renal transplantation 1580
  deposition, on glomerular basement membrane in diabetes 517, 524
  deposits, immunofluorescence studies, mesangial immunoglobulin G deposits 306
  Gm allotypes, Goodpasture's disease 445
  Goodpasture's disease 444
  idiopathic nephrotic syndrome 311
  IgG glomerulonephritis, schistosoma-specific 1737
  immune complexes
    in Henoch–Schönlein purpura 587, 600
    in leucocytoclastic vasculitis 587
  nephrotic syndrome 277
  production, B cells from haemodialysed patients 1315
  properties and biological functions 243–4
  in Sjögren's syndrome 694–5
  staphylococcal protein A, adsorption of immunoglobulin G antibodies 450
  subclasses G1 to G4 244
immunoglobulin M,
  anti-DNA antibodies 638, 639
  cytomegalovirus-specific, renal transplantation 1580
  idiopathic nephrotic syndrome 311
  IgM mesangial nephropathy, characteristic immunohistologic patterns 149, 306
  in Kawasaki's disease 590
  myeloma 568
  properties and biological functions 244
  rheumatoid factors, see rheumatoid factors
immunoglobulins **243–6**
  abnormal, in light chain deposition disease 569
  allotypes 243
  antiphosphorylcholine antibodies 248
  biological functions 243–4

immunoglobulins (cont.):
  characterization 562–3
  charge 410
  cold precipitation, see cryoglobulin(s)
  deposits
    in lupus nephritis 652, 653
    in scleroderma 675
    see also immune complex(es), deposition
  generation of antibody diversity 245
  granulocyte-associated, in cytomegalovirus infection 1534
  heavy chains 244, 562
    deposition 569
    gene rearrangement *245*
    polymorphisms of switch region 601
  idiotypes
    defined 245
    regulation of immune response 245
  and immune complexes 410
  isotypes 243
  in leucocytoclastic vasculitis 587, *588*
  light chains 244–5, 562
    AL amyloid proteins homology 546, 571
    deposition, see light chain deposition disease
    as dimers/monomers 563, 572
    Fanconi syndrome 734
    free 563
    gene rearrangement *245*
    kappa 562, 563, 572
    lambda 547, 562, 563, 572
    metabolism/excretion 563
    nephrotoxicity 563–5
    proteinuria, renal damage and 996
    proteolysis 572
    subgroups 572
  molecular biology and genetics 244–5
  monoclonal, see monoclonal antibodies
  papain, pepsin cleavage sites 243
  polyclonal, see polyclonal immune globulins
  rapidly progressive glomerulonephritis, tests *1062*
  recombination and generation of diversity 244
  rheumatoid factors 248
  self-recognition, tolerance and autoimmunity 245–6
  structure **562–3**
    primary/secondary/tertiary 243
  switching 245
  T cell receptor, structural similarities 241
  see also individual immunoglobulin classes
immunohistology,
  immunofixation, qualitative analysis of proteinuria 18
  immunofluorescence, immune sera 146
  immunoperoxidase, immune sera 146
  primary glomerulonephritis, renal biopsy 147, 149

immunohistology (*cont.*):
  renal biopsy specimen processing 146
immunological nephropathies, animal models 169–70
immunoperoxidase microscopy, in Henoch–Schönlein purpura 603
immunosuppressive agents,
  autoimmune renal disease 2320
  cancer risk 2321
  combined, minimal change nephrotic syndrome, steroid-resistant 320
  crescentic glomerulonephritis 428
  haemolytic–uraemic syndrome and 1048
  handling in kidney disease 190
  in Henoch–Schönlein purpura 607
  IgA nephropathy *358*
  liver effects 2315
  mesangiocapillary glomerulonephritis 401
  metastatic renal carcinoma 2248–9
  minimal change nephrotic syndrome
    adults *328*, 328
    children **316–323**
    steroid-resistance 319–20
  neurological side-effects 2332
  progressive chronic renal failure 1217
  see also cytotoxic agents
immunosuppressive agents in renal transplantation,
  cancer risk 1540
  cancer treatment, discontinuation 1540
  elderly patients 1630
  hypertension 1536
  optimal, discussion 1590–1
  polyclonal immune globulins *1556*
  prior to transplantation 1518
  rejection of allograft 1530, **1554–8**
  steroids, side-effects *1554*
  see also azathioprine; corticosteroids; cyclosporin; OKT3
immunotactoid glomerulopathy 2281–3
immunothrombocytopenia, IgA nephropathy 361
IMN (idiopathic membranous nephropathy), *see* membranous nephropathy
impetigo 406
  streptococcal, glomerulonephritis following 2304
impila (*Callilepis*) poisoning, acute renal failure and 1136
impotence, erectile,
  in haemodialysis patients **1329–34**
  arterial factors 1329–30
  diagnosis 1334
    SKATT test 1330
  diuretic-induced 212
  drug-induced 1332
  hormonal abnormalities 1332–4
  neurogenic and psychogenic 1331–2
  treatment
    bromocryptine and dopaminergic agonists 1335
    hormonal therapy 1334–5
    SKATT therapy 1334

impotence, erectile (*cont.*):
  treatment (*cont.*):
    surgery 1334
    zinc therapy 1335
    venous leakage 1330–1
inappropriate secretion of ADH syndrome, *see under* arginine vasopressin
incipient nephropathy 509
incontinence, urge incontinence, urinary tract infection 1685
indacrinone, urate excretion increase *829*
indirect haemagglutination tests, schistosomiasis 1731
indium-111-labelled leucocyte scintigraphy, renal cysts 2173
indoles,
  indicans, Hartnup disease 749
  neurotoxins 1397
  toxicity, uraemic syndrome 1244
indomethacin,
  in amyloidosis 556
  children
    diabetes insipidus, nephrogenic 795
    steroid-resistant nephrotic syndrome 1039
  diuretics, interactions 208
  erythropoietin production effects 1350
  GFR, reduction, in diabetes 513
  metabolic acidosis in Fanconi syndrome 736
  neonates, preterm, prevention of renal side-effects 54
  nephrotoxicity 1010
  prostaglandin inhibition 1962
  renal papillary necrosis 822
  renal tubular absorption enhancement, growth effects 1305
  in sickle-cell disease
    renal concentrating capacity 706–7, 715
    renal haemodynamics reduced 711, 713, 715
    urinary osmolality changes 707, 709, 715
  sodium reabsorption, effects on 63, 706
  steroid-resistant nephrotic syndrome
    children 1039
    risks 323
  water excretion, reduction, diabetes insipidus, nephrogenic 795, 796
  see also prostaglandin synthetase inhibitors
indoramin, *see* α_1 adrenoceptor agonists
indoxyl sulphate, protein binding, competition for sites 176, 1240, 1243
Industrial Injuries Act, occupationally-acquired infection 1291
infantile hypercalcaemia 1891
infantile nephrocalcinosis, congenital, hypothyroidism 1894
infantile polycystic kidney disease 1607
infants, preterm, *see* neonates, preterm; children

infected stone disease **1831–2**, 1862–4
  aetiology 1862
  composition of stones 1831–2
  geographical factors *1822*
  infecting organisms 1863
  organisms 1831
  stone composition 1863
  struvite 1666
  symptoms 1863
  treatment 1863–4
  urinary risk factors *1831*
infection-associated glomerulonephritis **456–70**
  bacterial infections 458–62
  fungal infections 466
  infective endocarditis 468–9
  introduction 456–8
  parasitic infections 466–8
  renal consequences *457*
  rickettsial infections 462
  viral infections 462–6
  visceral abscess 470
  Yersinia 462, 2310
infection-associated nephropathy 456–8
infections,
  see pyelonephritis; bacterial, fungal, protozoal and viral infections; *specific causative organisms and specific conditions*; urinary tract infections
infectious mononucleosis, acute interstitial nephritis 1091
infective endocarditis,
  crescentic glomerulonephritis 421
  treatment 431
  infection-associated glomerulonephritis 468–9
  organisms associated *468*
  see also endocarditis
inferior vena cava,
  carcinoma invasion, ultrasonography 89
  clear cell carcinoma, regional extension 102
  iliocavogram to ascertain patency 115
inflammation, chronic renal failure, erythropoietin in 1351
inflammatory bowel disease, chronic urinary tract obstruction 2020
  periaortitis of the ureters and 2020
inflammatory disorders, amyloidosis associated with *549*, 554
inflammatory mediators, stimulation of dialyser membranes 1432–3
influenza A virus infection, glomerulonephritis 466
  haemolytic uraemic syndrome 466
infusion fluids, fructose intolerance, precautions 732
inherited diseases, main groups 10
inositol phospholipids, determination of progressive glomerular scarring 1193
insect stings,
  acute renal failure and **1135**
  poisoning 1011

insecticides, decreased drug protein binding 1245
insulin,
  binding, skeletal muscle 1252
  bone remodelling *1790*
  chronic renal failure
    actions on non-glucose metabolism 1253
    extrarenal catabolism, reduction 1318
    mediation of glucose uptake 1303
    protein synthesis defect 1304
    resistance, *see* insulin, resistance in chronic renal failure
    secretion 1253–4
    serum levels 1251
  continuous ambulatory peritoneal dialysis 188
    dose required 1649
  diabetic ketoacidosis 928–9
  diabetic nephropathy, pregnancy 1942
  growth factor action 1297
  handling in kidney disease 188
  hyperkalaemia
    enhancement of secretion 907
    therapy 910
  infusion, glomerular filtration rate reduction 513
  membrane receptors 1252
  phosphate transport and 1787
  potassium
    depletion 901–2
    transcellular shifts 902
  release, uraemic syndrome 1239
  renal catabolism *1318*
  resistance
    cardiovascular disease 211
    cardiovascular events association 521
    diuretic-induced 210
    sodium–lithium countertransport association 521
    uraemic syndrome 1239
  resistance in chronic renal failure
    amino acids 1259
    cellular mechanisms 1252–3
    clamp studies 1251
    site of resistance 1251–2
    uraemic milieu 1253
  sodium reabsorption increased by, in diabetes 515
  transcellular K^+ distribution, Na-K ATPase pump 896
  urinary albumin excretion increased 515
insulin-like growth factor(s),
  binding proteins, elevation in uraemic children 1319
  children with chronic renal failure 1304
  determination of progressive glomerular scarring 1192
  growth hormone and 1304
  radioimmunoassays 1323
  see also somatomedins
insulin-like growth factor I,
  action 1297
  animal models, growth hormone–somatomedin axis and 1324
  bioassay 1297
  bone remodelling *1790*

Index

insulin-like growth factor I (*cont.*):
  increase, independence to growth hormone 1204
  pro-hormone, elevation in uraemia 1319
  renal plasma flow increase, eicosanoid-dependent mechanism 1324
  in uraemia, decreased 1323
insulin-like growth factor II, uraemia, elevated 1323
intellectual growth, retardation in chronic renal failure 1303
intensive care, drug therapy 190
interferon,
  acute renal failure 2306
  bone remodelling *1790*
  drug-induced nephropathy 163
  in essential mixed cryoglobulinaemia 692
  focal sclerosis 2306
  HBV infection 497
  metastatic renal carcinoma *2249*, 2249
interleukin-1,
  bone remodelling *1790*
  C3a and C5a stimulation during haemodialysis 1316
  chronic renal failure, levels 1368
  function 1314
  IL-α and IL-β 1314
  interleukin-6 genes, corticosteroids blocking activation 1553
  putative bone resorption factor 1776
  release from monocytes, bacterial endotoxin in dialysate 1316
  in sarcoid granuloma formation 577
  transplantation tolerance model 249
interleukin-1 receptor, T cell activation 1314
interleukin-2,
  function 1314
  helper T cells, cytoxic capacity of cytotoxic T cells 1554
  interaction with antigen-activated T cells 1553
  measurement 1571
  production in uraemia 1314–15
  in sarcoid granuloma formation 576
  T cell activation 1553
interleukin-2 receptor,
  activation antigens, immunology of transplantation 1543–4, 1546
  immune response 1313
  soluble, production in uraemia 1315
interleukin-4, immune response to antigen 1543–4
interlobular arteries,
  lesions in scleroderma 675
  see also renal arterial supply
intermittent claudication, diabetes, β-adrenoceptor blockers 1647
International Society of Paediatric Oncology, Wilms' tumour studies *2261*
International Study of Kidney Disease in Childhood, classification 602, *604*, 657

interstitial cells,
  antihypertensive lipids 2064–5
  see also renomedullary antihypertensive function
interstitial disease,
  hyperplasia in Bartter's syndrome 784
  infiltrates, membranous nephropathy 374
  nephrogenic diabetes insipidus 887
interstitial fibrosis,
  in amyloidosis 555
  causes 149–50
interstitial granuloma 1085
interstitial lung disease, renal effects 2336
interstitial myocardial fibrosis, diffuse, and collagen deposition in uraemia 1266
  left ventricular hypertrophy and, end-stage renal disease 1266–7
interstitial nephritis **803–66**
  acute
    acute on chronic renal failure 1161
    antitubular basement membrane disease 1092–3
    causes 149
    cell-mediated tubulointerstitial disease 1093–4
    direct hypersensitivity reactions 1094
    diuretic-induced 1039
    drug-related 1087–90
    eosinophilia 972
    general features 1084–7
    idiopathic 1091
    immune-complex mediated tubulointerstitial disease 1093
    incidence 1084
    infections and 1090–1
    malignancy and 1091
    management 1039
    pathogenesis 1092–4
    in systemic disease 1091
  acute renal failure **1084–94**
  analgesic nephropathy and, consequence of renal papillary necrosis? 811
  analgesic-associated, prevalence in autopsy studies 807
  bacterial focal or multifocal 1691
  beryllium-induced 845
  causes, patients with liver disease *1106*
  children 1122
  chronic
    cadmium nephropathy 843
    NSAID-induced, renal biopsy 822
  drug-induced 166–7
    diffuse infiltration of leucocytes 149
  drug-related 1087–90
  Fanconi syndrome, clinical features 734
  fenoprofen-associated 823
  heavy metals exposure, differential diagnosis 838
  lead intoxication-associated 838–9
  pathogenesis, animal models 1092–4
  renal biopsy indications 144

interstitial nephritis (*cont.*):
  sarcoid granulomatous, *see* sarcoidosis
  in Sjögren's syndrome 696–7
  in SLE 654, 658
  of unknown aetiology 804
  see also drug-related interstitial nephritis
interstitial nephropathy,
  chronic focal, lithium toxicity 855
  renal papillary necrosis, increase with use of NSAIDs 804
interstitial pressure, renal perfusion 983
interstitial scarring, *see* tubulointerstitial scarring
interstitial sclerosis, indications 145
interstitium,
  ageing changes 58
  collagen increase, in scleroderma 674
  in essential mixed cryoglobulinaemia 689
  inflammatory cellular infiltrates
    acute rejection
      assessment 1530
      histology 1570–1
      treatment 1530
    peritoneal cavity 1478
  see also tubulointerstitial disorders
intima, proliferation, in scleroderma 675
intoxications 931–3, *933*
  diuretics 208
  ethanol 930, *931*
  ethylene glycol *931*, 931–2
  Fanconi syndrome 734–5
  metabolic acidosis 931–3
  methanol *931*, 931–2
  mixed metabolic and respiratory disorders *957*
  nitroprusside reaction 932
  paraldehyde, metabolic acidosis 932
  salicylates, metabolic acidosis 932
  treatment 932
  see also other specific substances
intra-arterial injection, retrograde femoral catheterization 111
intracellular pumps, Na-K ATPase, K^+ regulation 895
intracranial aneurysm, polycystic disease, autosomal dominant 2177–8
intranuclear inclusion bodies,
  lead marker 839
  lead sequestration 838
intrarenal reflux, demonstration 1983
intrarenal vasculature, arteriography 112
intratubular negative potential, reduced, renal tubular acidosis *938*
intravenous coagulation,
  renal biopsy, EM findings 150
  see also disseminated intravascular coagulation
intravenous pyelography,
  defined 99
  in sickle-cell disease 710
intravenous urography **76–85**
  antegrade/retrograde pyelography 72–3

intravenous urography (*cont.*):
  chronic urinary tract obstruction 2014–15
  contraindications in pregnancy 81
  contrast-associated nephropathy 78
  ectopic kidney(s) 71
  essential hypertension *vs* renal causes 2050
  genitourinary tuberculosis 1721–3
  hydronephrosis, hypertension and 2090
  indications 71, 72
    urinary tract infection *1683*
  neonatal kidneys 1113–15, 2027
  normal clinical picture 82–3
  plain films 79–80
  procedures 80–3
  reflux nephropathy 1711, 1988
  rejection of allograft, contraindications in 1529
  renal papillary necrosis, analgesic nephropathy 814
  renal pelvis tumours 2267–8
  renovascular hypertension 2103
  retrograde cystography 81
  retrograde urethrocystography 81
  retrograde/anterograde ureteropyelography 81
  small *vs* absent kidneys 71
  suprapubic urethrocystography 81
  urinary tract, dilated, upper 74, 75
  urinary tract infection 1682–3
  urinary tract obstruction 2006–8
  X-ray exposure 81–2
intravenous volume expansion, neonates, contraindications 1115
intrinsic pathway, *see* coagulation factors
intussusception, in Henoch–Schönlein purpura 596–7
inulin clearance 1174
  assessment, neonates 50
  children and young adults *33*
  diseased kidney(s) 30
  elderly patients, glomerular filtration rate 59
  glomerular filtration rate, elderly patients 59
  'gold standard' 30
  increase following meat meal, pregnancy 1914
  measurement of renal clearance 29–30
  pregnancy, increase 1910, 1914
iodide, thyroid metabolism, animal models 1320
iodinated contrast media, *see* radiocontrast media; *specific substances*
I-125-*o*-iodohippurate, clearance 39–40
I-123-*o*-iodohippurate,
  effective renal plasma flow 116–17
  extraction efficiency E 116
  measurement of renal plasma flow 116
  red cell uptake 117
  in renal imaging, properties 117
I-131-*o*-iodohippurate, in renal imaging, properties 117
iohexol 77

[50] Index

ionics, see radiocontrast media, hyperosmolar
ionizing radiation,
 animal models of nephritis 849
 lupus nephritis 662
 radiation nephritis prevention 851
 total-lymphoid irradiation, immunosuppressive therapy prior to renal transplantation 1518
 see also radiation nephritis
ionophore A, phosphate transport 1787
iopamidol 77, 79
iopanoic acid, acute uric acid nephropathy 832
I-125-iothalamate,
 assessment of glomerular filtration 38
 clearance, GFR estimation 1174
 obstruction studies 123–4
 transplantation, monitoring graft function 126–8
ioversol 77
ioxaglate, ionic contrast medium 77
ipophosphamide, antineoplastic treatment, renal complications 2319
iritis, acute interstitial nephritis 1091
iron,
 as bacterial nutrient 1665
 blood transfusion in renal transplantation, hamosiderosis 1584
 chronic renal failure, kinetic studies 1349
 dietary intake 1345
 exchange, erythropoiesis evaluation 1345
 ferrokinetic studies 1349
 folate deficiency and 1351
 neutrophil and free radical-mediated injury 1052
 normal urinary excretion 1665
 total parenteral nutrition in ARF *1021*
 transferrin loss in nephrotic syndrome 289
 tubulointerstitial scarring, effects on 1206
iron deficiency,
 erythropoietin replacement therapy 1356
 haemodialysis patients 1351
 loss, causes 1345
iron overload,
 blood transfusion 1353
 calciphylaxis 1377
 cutaneous porphyria tarda 1282
irradiation, see ionizing radiation; radiation nephritis
ischaemia,
 acute, computed tomography scanning 105
 chronic, parenchymal atrophy 105
 hyperlipidaemia, nephrotic syndrome 284
 local causing hypertension, measurement of divided renal functioon 119
 'rim sign', computed tomography scanning 105

ischaemia (cont.):
 see also arteriosclerosis; ischaemic heart disease; renal ischaemia
ischaemic heart disease,
 assessment, chronic renal failure 1161
 chronic renal failure, assessment 1161
 defined 1269
 diabetes, chronic renal failure 1642
 differential diagnosis 2334
 end-stage renal failure
  non-coronary problems *1269*
  potential causes 1269
 silent, 3 types 1270
ISKDC (International Study of Kidney Disease in Childhood) classification,
 Henoch–Schönlein purpura 602, *604*
 lupus nephritis 657
isoniazid,
 drug dosage, renal impairment *184*
 drug-related interstitial nephritis 1089
 renal transplantation, tuberculosis treatment 1518
 renal tuberculosis *1727*
isopropyl alcohol, metabolic pathway *931*
isotope renogram, essential hypertension vs renal causes 2050
isotypes 562
isovaleric acidaemia, ketoacidosis 930
ispaghula husk, constipation 189
'itai-itai byo' (ouch-ouch disease),
 cadmium ingestion 858
 cadmium nephropathy 843
IVU, see intravenous urography

Jaffé reaction, creatinine clearance 31
Janovsky complex, creatinine with picrate 31
jejunum, chronic renal failure, microbial changes 1281
jelly-fish stings, acute renal failure and 1136
Jeune's syndrome (asphyxiating thoracic dystrophy),
 cystic disease 2169
 nephronophthisis, children 2193
Jewitt system for classification of renal tumours 2268
Johanson–Blizzard syndrome, features and associated abnormalities *2037*
joints,
 in amyloidosis 550
 manifestations, of Henoch–Schönlein purpura 596
 pain, renal transplantation, management 1537
jump grafts 1405
juvenile polycystic kidney disease, end-stage renal failure 1233

juvenile rheumatoid arthritis (JRA),
 amyloidosis associated 680
 renal disease in **681–2**
juxtaglomerular apparatus,
 afferent arterioles, tubulo–glomerular feedback 25
 arterial pressure sensor 263
 components 138–9
 hyperplasia
  accelerated hypertension 2131
  and hypertrophy, Bartter's syndrome 783–4, 786, 1894
  hypokalaemia and 787
 longitudinal section 130, 131, 138–9
 pathology, renin, decreased secretion 909
 regulatory function 139
 renin release, pregnancy-induced hypertension 1963
 transplantation studies 2065
 tubuloglomerular feedback 983–4
 volume receptor 263
juxtaglomerular cell tumour, hypertension and 2092–3

kala–azar, glomerulonephritis 468
kallikrein, nephritic syndrome, acute 407–8
kallikrein–kinin system,
 activation by Factor XII following contrast media 77–8
 end-stage renal failure, hypertension of 2120
 stimulation, Bartter's syndrome 783
Kallmann syndrome, features and associated abnormalities *2038*
kanamycin, doses and therapeutic plasma concentrations of aminoglycosides, handling in kidney disease *182*
Kaposi's sarcoma,
 cyclosporin, nephrotoxicity 1566, 1583
 pseudo-Kaposi's sarcoma in CRF 1393
 renal transplantation 2320–1
 diagnosis 1539–40
Kawasaki disease 590, 590–1, 630
 mucocutaneous lymph-node syndrome 2303
Kayser–Fleischer rings 733
kdkd mice, tubulointerstitial nephritis model 1093
keratinization abnormalities, chronic renal failure 1390–1
keratoconjunctivitis sicca 694
keratotic papules 1391
ketanserin, renin secretion and *2071*
α-keto acids, phosphate binder, dose and side-effects *1382*, 1383
ketoacidosis,
 alcoholism 930
 defined 928
 diabetes 928–9
  acute renal failure in 532
 following refeeding post starvation 832
 inherited enzyme defects 930
 late phase 935

ketoacidosis (cont.):
 repair 928
ketoconazole,
 CAPD peritonitis regimens *1489*
 cyclosporin therapy in renal transplantation 1573
 interactions *1563*
 fungal infections 1748
 handling in kidney disease 184
ketonaemia, nitroprusside reaction 932
ketone bodies,
 defined 928
 glomerular filtration rate increase 513
ketoprofen, interaction with methotrexate 166
kidney(s),
 abscess
  cortical 1695
  diagnosis 1691, 1695–7
  incidence 1691
  perinephric 1696
  see also abscess, renal; renal abscess
 absence
  congenital 2027–8
  defined *2029*
  diagnosis 2028–9
  computed tomography scanning 106
  indications 71
  vs multicystic dysplastic disease 91
 ageing kidney, anatomy 57–8
 alkaline phosphatase activity, intestinal isoenzyme 1282
 anatomy
  elderly patients 57–8
  ultrasonography 87
 antihypertensive renal system 2119–20
 Ask–Upmark kidney 1231
 autoregulation 983–4
 bilateral palpable *2048*
 blood pressure control and **2058–66**
  animal models 2059–61
  Goldblatt hypertension models 2059
  homeostasis mechanisms 2117–19
  sodium and water excretion 2058–60
 calcium, see headings under calcium; hyper- and hypocalcaemia
 cancer **2237–66**
  diagnostic work-up 2244–6
  molecular genetics **2155–61**
  prognosis 2249–51
  treatment 2246–9
 'cement' 1722
 clearance 29–39
  see also specific substances
 compensatory hypertrophy 1989
 congenital abnormalities
  antenatal diagnosis 2025
  diagnostic errors 2025
  investigations 2026–7
  presentation 2025–6
  ureteral valves 2089
  urinary tract **2023–41**
 cysts, see cyst, renal; polycystic disease

Vol. 1 – Sections 1–5.   Vol. 2 – Sections 6–11.   Vol. 3 – Sections 12–21.

Index

kidney(s) (*cont.*):
  'dead', renal transplantation 1523
  development
    characteristics 1110
    imaging studies 1113–15
    maldevelopment 2027–30
    see also kidney(s), embryology; kidney(s), malformations; kidney(s), neonates
  diluting capacity, in sickle-cell disease 707–8
  dimensions, see kidney(s), size
  drug-associated disorders *2308*
    relating to drug therapy for the skin 2305–6
  duplex
    congenital, nuclear imaging 120
    reflux nephropathy, measurement of divided renal functioon 119
  ectopic 2029
    crossed, radionuclide studies 120
    hypertension and 2089
    IVU 71
    nuclear imaging 120
    pregnancy and 1942–3
  embryology 2023–4
    renal function 2024
    structure 2024
    tubular function 2024
    urine production 2024–5
  endstage obstruction 83
  enlargement in pregnancy 1909–10
  erythropoietin
    oxygen sensory function 1347
    physiology 1346–7
  extraction efficiency, defined 116
  fetal, see kidney(s), development
  function
    absolute measurement of each kidney, nuclear imaging 120–1
    defined 24
    see also renal function
  gamma-camera imaging 39–40
  haemodynamic changes, pregnancy 1911–13
  histology, lithium effects 855–6
  horseshoe, congenital, nuclear imaging 120
  hydrogen ion generation tests 42–4
  hypoplastic kidney, hypertension and 2089
  imaging, see renography
  impairment and renal infection 1692
  inflammation, focal 1691
  innervation 985
  interstitium, ageing changes 58
  Jewitt system for classification of tumours 2268
  large
    aetiology *84*
    radiological tests *72*
  localization by nuclear imaging 119
  loss, end-stage renal failure *1229*
  macroscopic appearance, ageing changes 57
  malformations **2035–41**
    extent and time of manifestation *2029*
    terminology *2029*

kidney(s) (*cont.*):
  mass, children 1710–11
  'mastic' 1722, 1723
  native
    renal transplantation
      embolization 1598
      nephrectomy 1598
  neonates 1110, **2024–5**
    ultrasound *in utero* studies 1114–15
    see also kidney(s), development
  nerves, efferent 985
  neural reflexes, role 2120
  neural system, afferent innervation 2120
  obstruction, see urinary tract obstruction
  oxygen consumption, see oxygen
  oxygen sensory function 1347
  physiology
    free water clearance (CH_2O) 871–2
    haemodynamics 263–4
    hypokalaemia-induced alterations in function 901
    osmolar clearance (C_{osm}) 871–2
    performance limits 264
    water balance regulation 870–2
    water volume and osmolar clearance 871–2
  pregnancy, animals as gestational models 1911–12
  ptopic, hypertension and 2089
  radiological tests **70–6**
    angiography **111–15**
    CT **99–107**
    IVU **76–85**
      normal IVU imaging 82–3
    MRI **107–8**
    ultrasonography **86–97**
  remnant
    antihypertensive treatment *2086*
    hypertension effects *2084*
  remnant kidney model, antihypertensive therapy, role in glomerular sclerosis 1212–13
  removal, see nephrectomy
  renal architecture 870–2
  renal nerves 2066
  renal plasma flow, elderly patients 58–9
  rupture
    following renal transplantation 1575
    non-traumatic 1931
    Wunderlich triad 2243
  scrum kidney 405
  size
    ageing changes 57
    at birth 1699
    children 1710–11
    compensatory hypertrophy 1989
    depth, measurement 121
    in diabetes mellitus 508–9
      early phase 508
      glomerular hyperfiltration pathophysiology 512
      glycaemic control relationship 509, 529
      increase 514
    granulomatous interstitial nephritis 578
    large kidney *72*, 84, 1989
    normal size 974
    radiology 974–5

kidney(s) (*cont.*):
  size (*cont.*):
    in renal amyloidosis 551, 552
    small
      aetiology *84*
      parenchymal transit time 121–2
      radiological tests *72*
      vs absent, ultrasonography 71
    solitary
      pregnancy and 1923, 1942–3
      renal biopsy contraindication 143
    small unilateral, hypertension and **2088–9**
    transplantation donor 1511
    tumours, options 2246–7
    unilateral, hypertension and 2089
  sponge 1895–6
  stenotic, blood pressure and 2129
  structure, hypertension effects **2075–81**
  trauma
    assessment, ultrasonography 96
    classification of blunt renal injuries 2092
    hypercatabolism and 978
    hypertension and 2091–2
    neonates, assessment with ultrasonography 96–7
    perirenal effusions, computed tomography scanning 106
    renal arteriography, indications 113
    retroperitoneal fluid collections, computed tomography scanning 107
  tubuloglomerular feedback 983–4
  unilateral
    compensatory glomerular hypertrophy 1193
    failure 70–2, *73*
    multicystic kidney, hypertension and 2090–1
    see also kidney(s), solitary
  urine concentration, failure 41
  whole kidney transit time index (WKTTI) 117
kidney donors,
  cadaver grafts 1548–9, 1586–7
  elderly patients 1631–2
  living related donors 1628–9
  hypertensive donors 1523, 1536, 1596
  living donors 1511, 1517, 1586
  unilateral nephrectomy, hypertension and 2093–4
  see also renal transplantation
kidney stones, see renal stones
Kimura's disease 2281
  minimal change nephrotic syndrome 300
Kjeldahl method, detection of urine protein 227, 228
Klebsiella,
  infected stone disease 1831
  mannose-resistant fimbriae 1667
  urinary tract infection 1661
Klebsiella pneumoniae, infection-associated glomerulonephritis 462
Klinefelter syndromes, features and associated abnormalities *2041*
Klippel–Trenaunay syndrome, children 2144

Klippel–Trenaunay–Weber dysplasia, features and associated abnormalities *2039*
Kluyveromyces, infection 1748
knee pathology, pseudogout 1376
Koch, Robert, tuberculosis 1719
Koletsky rats, accelerated glomerular lesions 1207
Korean haemorrhagic syndrome, see Hantavirus
Korotkoff sounds, blood pressure 11
Kr-85, gamma-camera imaging 40
Krabbe disease, galactocerebroside–β-galactosidase deficiency 2206
Kupffer cells, immune complex elimination 583
Kussmaul breathing 2339
  alcoholic ketoacidosis 930
Kyanasur Forest disease *1045*
Kyrle's disease, keratotic papules 1391

La antigen (SSB), antibodies to, HLA associations 695
  in Sjögren's syndrome 694, 695, 696
  in SLE 647–8, 651, 697
labetalol,
  haemodialysis patients, recommendations *1463*
  hypertension, accelerated 2135
  platelet effects and therapy 1968
  pregnancy-induced hypertension therapy 1970–1
  properties, in relation to use in impaired renal function *2053*
labial gland, biopsy 694
lachrymal glands, in Sjögren's syndrome 693, 694
β-lactamines, drug-induced interstitial nephritis 1088
lactate,
  diet, acid and base production 932–3, *933*
  excretion, Fanconi syndrome 727
D-lactate, lactic acidosis 927
lactate, urate excretion decrease 829
lactation,
  and hypomagnesaemia 1811
  see also breast-feeding
lactic acidosis,
  bicarbonate administration 944
  causes 926, *927*, 927
  classification *927*
  D-lactate 927
  lactate metabolism 926
  treatment 928
lactic dehydrogenase, increased, marker of renal cell carcinoma 2242
Lactobacillus, periurethral flora 1660
lactose, lactosuria of pregnancy 1916
lactulose, hypertonic dehydration 883
laminae rara interna/externa and densa,
  basement membrane of glomerular apparatus 134

laminae rara interna/externa and densa (cont.):
  renal biopsy, diagnostic relevance 150
laminin, basement membrane 446
Laplace's law 1197
large bowel, see colon
large kidney(s), see kidney(s), size
large vessel disease, in acute renal failure in diabetes 532
Laurence–Moon–Bardet–Biedl syndrome,
  children 2169
  clinical features 2231
  see also Bardet–Biedl syndrome
laxatives,
  abuse
    diarrhoea, anion gap measurement 934
    potassium deficiency 167
    potassium depletion 903
    unexplained hypokalaemia 787
  constipation 189
  ispaghula husk 189
  lactulose 189
  senna 189
  sodium picosulphate 189
LCAT, see lecithin:cholesterol acyltransferase; lecithin-cholesterol acyltransferase
LE cell phenomenon 646
lead 837–41
  blood concentrations, exposed/unexposed subjects 838
  bone storage and release 838, 839–40
  mobilization test, calcium edetate 834, 839
  occupational exposure
    exposure levels 841
    and hypertension 840
  renal cell carcinoma and 2238
  urate reabsorption increase, reduced clearance 831
lead intoxication 833–8, 1008
  acute, lead colic 839
  bone storage and release 838, 839–40
  calcium deficiency, hypertension and 834
  chelation therapy 838
  chronic renal failure and 834
  cumulative lifelong exposure 838
  gout 833–4, 2294
  occult, unknown origin 834
  saturnine gout and 833–4
  treatment 1008
  urate, reduced clearance 831
  see also lead nephropathy
lead nephropathy,
  acute 838
  Balkan nephropathy 858
  chronic 838–40
    management 840–1
  chronic renal failure, assessment 1162
  hypertension and, low level exposure 840
  occupational risks 9
  secondary gout and 834
  urate excretion decrease 829
Leber's amaurosis,
  nephronophthisis 2192, 2193
lecithin:cholesterol acyltransferase, deficiency 287
  chronic renal failure 1202

lecithin:cholesterol acyltransferase (cont.):
  deficiency (cont.):
    nephrotic syndrome 287
    renal biopsy, EM findings 150
    uraemia 1256
  deficiency, familial
    Balkan nephropathy, deficit and 860
    clinical features 2211
    genetics 2211–12
    metabolic defect 2211–12
    management 2212
    renal involvement 2211
  HDL 238
liver glomerulonephritis 360
left heart failure, malignant hypertension and 2130
left ventricle, uraemia, micromorphometric analysis in animal models 1266
left ventricular hypertrophy,
  anaemia and, animal models 1265
  children, chronic renal failure 1611
  diabetes and chronic renal failure 1643
  dialysis 1265
    elderly patients 1624
    prevalence in 1269
  end-stage renal disease 1264–7
    anaemia 1265
    arteriovenous fistula 1265
    autonomic nervous system 1266
    carnitine deficiency 1266
    diabetes 1266
    fluid retention 1265
    hypertension 1265
    influence of treatment modality 1267
    interstitial myocardial fibrosis 1266–7
    β-2 microglobulin amyloidosis 1266
    secondary hyperparathyroidism 1265–6
    systolic/diastolic malfunction 1264–7
    thiamine deficiency 1266
  left heart failure, hypertension and 2130
  renal transplantation 1577
  uraemia, functions affecting function and mass 1265
legionnaire's disease,
  renal transplantation 1579
  tubulointerstitial nephritis 462
leiomyomatosis, diffuse, renal disease and nerve deafness 2199
leiomyosarcoma,
  pathology 2243–4
  treatment 2244
Leishmania donovani 468
leishmaniasis,
  infection-associated glomerulonephritis 468
  sodium stibogluconate 185
leprosy,
  amyloidosis and 491
  autoantibodies 492
  glomerular disease, clinical features 491
  glomerulonephritis 461, 492–3
  handling in renal impairment 184
Leptospira, liver disease associated with acute renal failure 1106

leptospirosis,
  acute interstitial nephritis 1091
  acute renal failure **1132**
  occupational risks 9
  infection-associated glomerulonephritis 461
Lesch–Nyhan syndrome,
  acute uric acid nephropathy 832
  HGPRT deficiency 2294
  uric acid stones and 1861
Leu-I antigen 640
leucine, excretion, pregnancy and 1916
leucine aminopeptidase 165
  in urine 18
leucocyte esterase test 1680
leucocytes,
  abnormalities
    cyclosporin therapy in renal transplantation 1950
    in renal impairment 2344
  diffuse infiltration
    drug-induced interstitial nephritis 149
    schematic representation 151
  haemolytic–uraemic syndrome 1054
  infiltrating polymorphonuclear, reactive oxygen species 256
  renal biopsy, diagnostic relevance 150
  spurious hyperkalaemia 907
  in urinary sediment 20
  urine 1662
leucocytoclastic vasculitis,
  pathogenesis 583, **585–90**
  antibody-mediated 589–90
  criteria to prove mechanism 586
  in Henoch–Schönlein purpura 587, 600, 602
  immune complexes 586–9, 600
    antigens 587, 600
    complement activation 587–8, 589, 601
    concentration 586
    detection 586
    elimination 588–9
    haemodynamic factors in deposition 587, 589
    immunoglobulin classes 587, 588, 600
    size/composition 587, 600
    vascular permeability role in deposition 589
  inflammatory reaction 589
  mononuclear vasculitis overlap 589
leucopenia,
  cyclophosphamide causing 620
  in SLE in children 650
  transplantation patients, azathioprine 2345
leukaemia,
  acute, potassium depletion 906
  acute renal failure in 2342
  metastatic infiltration 2318
  potassium determinations, spurious hyperkalaemia 907
  renal involvement 482, 482–3
leukaemia virus infection, Gross, infection-associated glomerulonephritis 462
leukaemoid reaction, renal cell carcinoma 2242

leukotrienes,
  arteriolar myocytes of afferent arterioles
    effects 26
    interactions 27
  pregnancy-induced hypertension 1965–6
  prostaglandin synthesis and 820
  synthesis 256
  in vasculitis pathogenesis in animals 584
levamisole,
  drug-induced nephropathy 163
  minimal change nephrotic syndrome, relapsing 319
  toxicity 323
Lewis rats, progressive age-related glomerular lesions 1206–11551207
LFA-1, adhesion protein, mediation of neutrophil and free radical-mediated injury 1052
lichen simplex 1391
Liddle's syndrome 1895
  hypertension, metabolic alkalosis and 948–9
  mineralocorticoid excess 905–6
  nephrocalcinosis 1895
lidocaine, pruritus treatment 1392
light chain deposition disease 563, **568–71**
  byproduct of tumours 1012
  clinical features 570–1
  differential diagnosis, mesangiocapillary glomerulonephritis 392
  nodules 569–70
  diabetic nephropathy vs. 524
  pathophysiology 569
  renal pathology 569–70
  therapy and prognosis 571
light chains, see Bence–Jones protein; immunoglobulins
light and heavy chain deposition disease 569
light microscopy **19–22**
  renal biopsy specimen processing 146
Lightwood's syndrome 773
lignocaine,
  antiarrhythmics, cardiac drugs 187
  arrhythmias in acute myocardial infarction, cardiac drugs 187
Limulus polyphemus (horseshoe crab) amoebocyte lysates, Gram-negative organisms 1663
linoleic acid,
  animal models
    progressive glomerular scarring 1209
  renal ablation, effects 1195
  dietary and pharmacological lipid manipulations 1202
lipid-lowering agents,
  animal models 1216
  renal ablation, effects 1195
  chronic renal failure 1216
  hyperlipidaemia 191
  hypertension 2051
  see also specific substances
lipids,
  abnormalities, in diabetic nephropathy 516–17, 521

Index

lipids (cont.):
  antihypertensive polar/neutral renomedullary lipid 2064–5
  blood, hydrochlorothiazide therapy 211
  chronic renal failure 1181
  dietary restriction, effect on renal scarring 1212
  metabolism
    disorders of **1254–8**
    in dementia 1448
    nephrotic syndrome **284–9**
    uraemia, abnormalities *1256*
  renomedullary lipids, hypertension of end-stage renal failure 2120
  total parenteral nutrition in ARF *1021*
  in urinary sediment 20
lipiduria, nephrotic syndrome 286
lipodystrophy,
  Barraquer–Simon's disease 2301
  glomerulonephritis in 1154
  partial, in glomerulonephritis, and mesangiocapillary glomerulonephritis 394–5
  renal hypertension, and *2048*
lipoid nephrosis, corticosteroid therapy 143
lipomatosis of renal sinus, computed tomography scanning 106
lipopolysaccharide,
  endotoxin activity 1665
  exposure, animal models 248
lipoprotein lipase,
  action in chronic renal failure 1255–6
  animal model *1256*
  lipid transformations, nephrotic syndrome 287
lipoproteins,
  albumin synthesis and 287
  apolipoprotein C II
    nephrotic syndrome 286
    VLDL, nephrotic syndrome 286
  apolipoproteins
    risk factor in end-stage renal failure 1268
    uraemia 1255
  β-VLDL, in uraemia 1255
  HDL, nephrotic syndrome 238, 285–6
  HDL precursors 1255
  hepatic lipoproteins, synthesis in nephrotic syndrome 287
  IDL, nephrotic syndrome 285–6
  LDL
    apoB/E receptor in liver 1255
    nephrotic syndrome 285–6
    metabolism
      proteinuria 238
      scheme *1255*
    nephrotic syndrome 285–6
  sodium–lithium countertransport association 520–1
  uraemia, degradation 1255–6
  VLDL, nephrotic syndrome 285–6
liposarcoma,
  diagnosis, computed tomography scanning 103
  pathology 2244
  treatment 2244

liquorice,
  intoxication, 11-hydroxysteroid dehydrogenase 8
  mineralocorticoid activity, hypokalaemia 905
  sodium retention 273
lissencephaly type II, features and associated abnormalities *2037*
Listeria monocytogenes infection, following renal transplantation
  CNS infections 1579–80
  management 1536
  pregnancy following transplantation, ampicillin 1339
lisuride hydrogen maleate, inhibition of prolactin, sexual function effects 1335
lithiasis, see renal stone disease
lithium **853–856**
  amiloride and, diabetes insipidus, central 796
  clearance 853
    amiloride sensitivity 41
    distal nephron function 41
    index of proximal tubule reabsorption, pregnancy 1918
    nephrotic syndrome 269
  diabetes insipidus, central 796
  elderly patients, inappropriate use 67
  handling in kidney disease 186
  inappropriate secretion of ADH syndrome 207
  inappropriate use, elderly patients 67
  interactions with diuretics 208
  nephrogenic diabetes insipidus, causing 887
  nephrotoxicity 1009
  renal function effects
    distal tubule acidification impairment 854
    duration of treatment 854
    glomerular filtration rate 854
    polyuria, polydipsia, nephrogenic diabetes insipidus 853–4
    proximal tubular function 854
    site of action 854
    urinary concentrating ability impairment 854
  renal histology effects
    acute changes 855
    chronic focal interstitial nephropathy 855
    cystic lesions 855–6
    glomerular lesions 856
    therapeutic options 856
  toxicity, increased risks 856
  transport 520–1
  urinary acidification 772
lithotomy, see percutaneous renal surgery
lithotripsy **1873–81**
  combined techniques 1880–1
  complications 1875–7
  history 1873–4
  indications and contra-indications 1875
  integrated stone management 1879–80
  principles 1874–5
lithotripters, types 1873–5, 1877
livedo reticularis, oxalosis 1154

liver,
  biopsy, haemodialysis patients 2313–14
  calcium, total *1753*
  chronic renal failure **1282–3**
    drugs and 1282–3
    erythropoiesis 1282
    hepatitis **1285–92**
    metabolic disorders 1283
    morphology 1282
  drug effects 2315–16
  dysfunction
    following renal transplantation 1584
    renal cell carcinoma 2242
  fibrosis
    congenital, polycystic disease, autosomal dominant 1607, 2177
    schistosomiasis and 1737
  haematoma, spontaneous subcapsular 1361
  HELLP syndrome, pregnancy-induced hypertension 1961
  physiological relationship with kidneys 1098–9
  pregnancy, acute fatty liver of pregnancy 1933–4
  siderosis, uroporphyrin decarboxylase inhibition 1282
  transplantation, cyclosporin, hepatotoxicity 1564
liver cyst(s),
  polycystic disease, autosomal dominant 2176–7
  von Meyenberg's complexes 2177
liver disease **2312–16**
  acute fatty liver of pregnancy 1057
  acute renal failure in **1098–107**
    acute tubular necrosis 1101
    classification *1099*
    diagnostic approach 1098–9
    haemodynamic and biochemical consequences 1099–101
    hepatorenal syndrome **1102–4**
    nephropathies associated 1104–7
    postrenal acute renal failure 1101
  chronic, renal syndromes 2310–11
  cirrhosis
    acute renal failure of hepatorenal syndrome 2310
    atrial natriuretic peptide 272
    diuretics
      clinical use 202–3
      specific complications 203
    Fanconi syndrome 2311
    hyponatraemia 877
    renal handling of urate/uric acid 832
    renal prostaglandins 820–1
    NSAIDs toxicity 820
    salt and water retention 877
    secondary fluid retention 272
    Wilson's disease 2311
  complicating renal transplantation 1533–4
  cyclosporin, nephrotoxicity 1566
  end-stage liver disease 1540
  failure, mixed metabolic and respiratory disorders 957
  glomerulonephritis
    clinical features 359

liver disease (cont.):
  glomerulonephritis (cont.):
    laboratory findings 359
    pathogenesis 360
    renal biopsy 359–60
  nephronophthisis, autosomal recessive 2194
  polycystic disease, autosomal dominant 2176–7
  schistosomiasis 1734–5
  see also hepatic; liver cirrhosis
lobar nephronia 1692
lobular glomerulonephritis, see mesangiocapillary glomerulonephritis
loiasis, infection-associated glomerulonephritis 468, 495
loin pain,
  chronic renal failure 1166
  vesicoureteric reflux 1995
loin pain–haematuria syndrome 2277–9
Loken–Senior syndrome 2231
  see also Senior syndrome
lomustine, antineoplastic treatment, renal complications 2319
loop diuretics 199
  acute renal failure
    clinical studies 205–6
    pathogenesis 205
    animal models 205
  cardiac failure 188
  chronic renal failure 204
  congestive heart failure 202
  high dose 204
  hypercalcaemia 207
  hypertension of end-stage renal failure 2120–1
  hypokalaemia and 209
  magnesium excretion 1808
  mechanism of action 199
  metabolic alkalosis, oedematous disorders 947
  natriuresis 269
  renal hypertension, treatment 2052
  resistance 203
Looser zones 1612
  aluminium-related osteopathy 1372, 1374
  fractures, aluminium-related osteopathy 1377
loperamide, antidiarrhoeal agents 189
LOU/c and LOU/m rats 565
lovastatin,
  animal renal ablation models, effects *1195*
  in chronic renal failure 1181, 1186
  in diabetics receiving ACE inhibitors, myositis cause 8
  dietary and pharmacological lipid manipulations *1202*, 1216
  nephrotic hyperlipidaemia, effects 288
  see also HMG–CoA reductase inhibitors
Lowe syndrome,
  children 1795, 1796, 1797
  chromosomal locus *1796*
  Fanconi syndrome 729, 732–3
  features and associated abnormalities *2038*
  nephrocalcinosis 1891, *1897*

Index

lower urinary tract obstruction, *see* urinary tract obstruction, acute lower *and* chronic lower
lpr gene 640, 642
lung disease,
　cancer, renal effects 2336
　interstitial
　　sarcoidosis 2336
　　systemic sclerosis 2336
　see also pulmonary; respiratory
lupus,
　definition 646
　discoid 646
　drug-related *650*, 650–1
　　penicillamine 682
　systemic, *see* systemic lupus erythematosus (SLE)
lupus antibodies, thrombophlebitis following contrast media injection 77
lupus anticoagulant 648, 649, 651
lupus band test 648
lupus nephritis,
　clinical course 649
　clinical features **648–52**
　　at presentation 648–9
　　in children 650
　clinicopathologic correlations 656–8
　course assessment, biopsy role 658
　flares 649, 652, 663
　mediation of injury 639
　murine 251
　neonatal 651
　outcome prediction 656–7, 657–8
　pathology **652–8**
　　activity/chronicity indices 657
　　basic lesions 652–3
　　class II (mesangial glomerulonephritis) 653, 655
　　class III (focal proliferative glomerulonephritis) 653, 655
　　class IV (diffuse proliferative glomerulonephritis) 653–4, 655
　　class V (membranous glomerulonephritis) 654, 655
　　ISKDC classification *657*
　　mixed connective tissue disease relationship 698
　　outcome prediction 656–7, 657–8
　　scoring system 145, *657*
　　transformations 656, 657
　　unusual forms 654–6
　　vascular lesions 656, 658
　　WHO classification 653–4, *654*, 655
　pregnancy in 651–2
　　effect on fetal outcome 651
　　obstetrical complications 651
　prevalence 646
　prognosis 656, **658–9**
　　in children 650, 658
　　factors influencing 658
　　kidney survival 659
　　pathological features 656–7, 657–8
　　patient survival 658–9
　recurrence in grafts 663
　remissions and recurrences 649
　renal biopsy,
　　EM findings 150

lupus nephritis (*cont.*):
　renal biopsy (*cont.*):
　　role 658
　renal replacement therapy 663
　　dialysis 663
　　transplantation 663
　therapy **659–63**
　　cessation of 663
　　corticosteroids 659, 660
　　cytotoxic agents 660–1
　　diffuse proliferative glomerulonephritis 659–63
　　fixed schedules, disadvantages 662
　　flexible strategy *663*
　　focal proliferative glomerulonephritis 659
　　maintenance 660
　　membranous glomerulonephritis 659
　　mild lesions 659
　　practical approaches 662–3
　　severe focal proliferative glomerulonephritis 659–63
　tubular atrophy 149
　see also systemic lupus erythematosus (SLE)
lupus vasculitis 656
lupus-like syndrome, D-penicillamine 161
luteinizing hormone,
　haemodialysis 1333
　　female patients 1337
　renal catabolism, CRF 1318
　secretion 1298
　testicular resistance, haemodialysis patients 1333
luteinizing hormone releasing hormone 1333
luteotrophic hormones, gestation, renal haemodynamics and 1914
lymph nodes,
　enlargement in sarcoidosis 581
　tumours, computed tomography scanning 102
lymphadenectomy, regional, in treatment of renal cell carcinoma 2247
lymphadenopathy,
　computed tomography scanning 104
　vs MRI 108
lymphatic system, peritoneal cavity 1478
lymphocele, renal transplantation 1528, 1575
lymphocytes,
　immune complex nephritis 410–11
　renal transplantation 1522
　in sarcoid granuloma 576
　see also B cells; T cells;
lymphocytic choriomeningitis virus 584
lymphokines, soluble immune response suppressor, steroid-responsive nephrotic syndrome 311
lymphoma,
　acute renal failure in 2342
　amyloidosis 482
　glomerulonephritis in 234–5
　minimal change disease 2342
　renal involvement *482*, 482–3
　renal transplantation
　　incidence and type 1540
　　risk 1539

lymphoma (*cont.*):
　Sjögren's 694
　ultrasonography 90
　see also Hodgkin's lymphoma
lymphomatous disorders,
　lymphomatoid granulomatosis 591
　minimal change nephrotic syndrome 299
lymphoproliferative disorders,
　acute uric acid nephropathy 832
　Epstein–Barr virus infection, transplantation patients 2346
lymphotoxin, *see* tumour necrosis factor-β
lysine,
　diet, acid and base production 932–3, *933*
　excretion, pregnancy and 1916
　lysinuria 749
　protein intolerance 749
lysol, intoxication, Fanconi syndrome 735
lysosomes,
　N-acetyl-β-D-glucosaminidase, in urine 18
　cadmium accumulation 842
　permeability, aminoglycosides 169
lysozyme,
　properties, effects on kidney, significance of elevation 229
　urinary, potassium depletion 906
lysyl-hydroxylase 518

M component (monoclonal) 563, 566
　AL (primary) amyloidosis 572
　see also monoclonal antibodies
Ma antigen 648
McArdle's disease,
　enzyme defect *2210*
　non-traumatic rhabdomyolysis, and ARF 995
McGivvon–Lubinsky syndrome, nephrocalcinosis 1892
macroangiopathy,
　diabetics, extrarenal problems 1638–43
　uraemic diabetics, extrarenal problems 1638–43
α-1-macroglobulin, nephrotic syndrome 282
macroglossia 549
macrophages 256
　activation
　　alloantigen-specific T cells 1552
　　pro-inflammatory cytokines 1552
　in amyloidosis development, role 572
　calcitriol production in sarcoidosis 579
　corticosteroid-treated, IL-1, IL-6 (and IL-2) blocking 1554
　1α hydroxylating activity 579
　immune complex elimination 583
　marker or rejection 1571
　receptor, for glycosylated proteins 518
　in sarcoid granuloma formation 576

macrophages (*cont.*):
　see also monocyte/macrophage lineage
macrothrombocytopenia, nephritis and deafness, hereditary 2199
macula densa,
　glomerular apparatus 130, 131, 138–9
　granular cells 139
　hyperkalaemia, renin secretion 907
　see also renal tubules, distal; macula densa
　structure 138–9
Tc-99m-MAG3, *see* mercaptoacetyltriglycine
magnesium **1802–16**
　aldosterone and 1808
　biological importance 1802
　body content and distribution 1802–3
　calcium stone disease 1835
　in chronic renal failure 1368
　deficiency
　　cisplatin 165
　　deficit estimation 1814
　　haemodialysis depletion, parathormone effects 1759
　　magnesium-losing nephropathy children 1894
　　familial nephrocalcinosis 1894
　depletion
　　metabolic alkalosis 949
　　potassium, depletion 906
　　renal potassium wasting 906
　dietary intake 1803, 1809
　　alcoholism 1811
　　malabsorption 1809
　　total parenteral nutrition in ARF *1021*
　dose, in chronic renal failure 1772
　emergencies 1814
　excretion
　　acidosis and alkalosis 1808
　　diuretics 1808
　　hypercalcaemia 1808
　　increased 1810–11
　　load test 1814
　　measurement 1814
　　phosphate depletion 1808
　homeostasis
　　hypercalcaemia causing excretion 1885, 1887
　　intestinal handling 1804
　　nutritional aspects 1803
　　renal handling 1805
　　turnover 1803–4
　hyponatraemia and 210
　increased, differential diagnosis 973
　inhibitor of calcium oxalate crystallization in urine *1828*
　inhibitor of calcium phosphate crystallization in urine *1827*
　parathormone effects, uraemia 1759
　phosphate binding 1774
　pruritus and 1392
　reabsorption defects 1894
　transport
　　animal models 1805–6
　　calcitonin and 1807
　　filtered load 1806
　　in nephron 1806–7
　　parathormone and 1807

Vol. 1 – Sections 1–5.　Vol. 2 – Sections 6–11.　Vol. 3 – Sections 12–21.

Index

magnesium (*cont.*):
 transport (*cont.*):
  segmental transport 1805–6
 see also hypermagnesaemia;
  hypomagnesaemia
magnesium ammonium phosphate,
 infected stone disease 1831–2
magnesium carbonate, phosphate
 binder, dose and side-effects
 1382, 1383
magnesium chloride and ATP,
 acute renal failure, prevention
 of 1018
magnesium hydroxide,
 antacids, with exchange resins
  944–5
 drug-induced lithiasis 168
 phosphate binder, dose and side-
  effects *1382*, 1383
magnesium sulphate, eclampsia
 effects 1974
magnesium trisilicate,
 mixture, handling in kidney
  disease 189
 silica stones 1866
magnesium-losing nephropathy,
 hypokalaemic alkalosis 787
magnetic resonance imaging **107–8**
 intracranial aneurysm, in
  polycystic disease, autosomal
  dominant 2178
 renal mass 2246
 T1 and T2 contrast scans 2246
major histocompatibility complex
 (MHC),
 antigen presentation and
  processing 240–1
 class II antigens, in Sjögren's
  syndrome 696
 class II/III genes, in SLE 643
 glomerulonephritis, prognosis
  1207
 and immune response 1629
 induction of immune response to
  allograft 1544
 membranous nephropathy and
  371–2
 minimal change nephrotic
  syndrome, HLA and 312
 polymorphism 1544
 see also HLA antigens
malabsorption, in amyloidosis 550
malabsorption syndrome,
 β-2-microglobulin amyloidosis
  2311–12
 pyridoxine deficiency 1857
malakoplakia,
 children 1706
 diagnosis 1693–5
 Michaelis-Gutman body 1693–5
 treatment 1693–5
malaria,
 acute renal failure and 1045–6,
  1130–2
 falciparum malaria, and acute
  renal failure 995
 glomerular disease **493–6**
  falciparum malaria 493
  incidence 493
  quartan malarial nephropathy
   494–5
 infection-associated
  glomerulonephritis 466–7
 sickle-cell trait and 703
 treatment 185

malate, diet, acid and base
 production 932–3, *933*
maleic acid,
 action on kidney 735
 animal model, proximal renal
  tubular acidosis 768
 Fanconi syndrome
  induction 735
 vitamin D metabolism 769
malformation syndromes **2035–41**
 reference lists 2035
 see also kidney(s), malformations
malignancy,
 hypercalcaemia 476, 1776–7
 minimal change nephrotic
  syndrome 299
 schistosomiasis 1734, 1735
 *see also specific names, organs and
  regions*; tumours
malignancy-associated glomerular
 disease,
 autoimmunity and 478
 clinical features 478
 glomerular appearances 480–1
 glomerulopathy pathogenesis
  477–8
 incidence of glomerulopathy in
  neoplasia 479
 nephropathy, causal relationship
  475–6
 prognosis 481–2
 therapy effects 477
 types involved 479–80, *480*
 viral causes 477–8
malignant hypertension, *see*
 hypertension, malignant
malignant melanoma, *see*
 melanoma
malignant nephrosclerosis 2081
malignant obstruction 2018
 see also urinary tract obstruction
Mallory–Weiss syndrome,
 gastrointestinal bleeding 2311,
  2312
malnutrition,
 children with chronic renal
  failure 1303
 elderly patients with chronic renal
  failure 1623
 haemodialysis 1439
 insulin-like growth factor(s),
  children with chronic renal
  failure 1304
 see also nutrition
Maltese crosses, cholesterol ester,
 urinary sediment 20
malylacetoacetate, accumulation
 730
mannitol,
 action 204
 cellular mechanisms 197
 acute renal failure 204, 204–5, 996
 adverse effects 206
 basis for use, pelvic transit time
  117
 clinical studies 205
 elderly patients 206
 intravenous volume expansion,
  neonates 1115
 magnesium excretion 1808
 prevention of acute renal failure
  171, 205, **1017**
 urate/uric acid, fractional urate
  clearance C_{ur} 830

mannose and mannosides,
 Escherichia coli 1666
mannose-resistant fimbriae,
 Klebsiella 1667
 Proteus 1667
Mantoux test, tuberculosis 1152
MAP (magnesium ammonium
 phosphate) stones 1831–2
maple-syrup urine disease,
 ketoacidosis 930
Marburg virus, haemorrhagic fever
 1045
marking-nut poisoning, acute renal
 failure and 1136
Masson–Goldner stain, bone
 histology 1374
matrix stones, haemodialysis
 patients 1444–5
matrix substance A,
 inhibitor of calcium oxalate
  crystallization in urine *1828*
 inhibitor of calcium phosphate
  crystallization in urine *1827*
May–Grünwald–Giemsa stain, fine-
 needle aspiration biopsy 145
May–Hegglin anomaly, hereditary
 macrothrombocytopenia,
 nephritis and deafness 2199
mdr-1 (multidrug resistance-1-
 gene) 2248
mean parenchymal transit time
 (MPTT), defined 116–17
measles,
 infection-associated
  glomerulonephritis 466
 remission in proteinuria 457
meat creatinine, 'goulash effect' 34
meat-eating,
 calcium stone disease 1834
 see also diet; proteins
mebendazole, handling in kidney
 disease 185
mechlorethamine,
 minimal change nephrotic
  syndrome 317–18
 steroid-resistant 320
 toxicity 322–3
 see also alkylating drugs
Meckel syndrome,
 children 2169
 features and associated
  abnormalities *2037*
media, of arteries 672
medroxyprogesterone therapy,
 androgen receptors, metastatic
  renal carcinoma 2248
medulla, renal, *see* renal medulla
medullary cystic disease,
 assessment 1155
 children 2169
 end-stage renal failure 1233
 proportional distribution by age
  at commencement *1229*
 see also nephronophthisis,
  autosomal dominant
medullary sponge kidney **1858–9**,
 2286–90
 associated inherited conditions
  2287
 cause of renal stones 1836, 1858
 children 2169
 clinical features 2287
 defined 1858, 2286
 diagnosis 1858

medullary sponge kidney (*cont.*):
 nephrocalcinosis 1895–6, 1897
 pathogenesis 2288
 radiological features 2287
 treatment 1858–9, 2288
Mee's lines, nails 1393
mefanamic acid, renal papillary
 necrosis 822
megaureter, congenital 2003
meglumine iothalamate, *see*
 iothalamate
melanoma,
 antigen and antibody, glomerular
  deposits 477
 melanuria, paraffin oxidation
  products 2305
 renal transplantation 1582
melanosis cutis, chronic renal
 failure, vitamin D deficiency
 1370
melioidosis, acute renal failure
 1132–3
melphalan, handling in kidney
 disease 191
membrane attack complex,
 antigen deposition in capillary
  walls, human
  glomerulonephritis 256
 C5b-9
  glomerular damage 229
  membranous nephropathy 376–7
 garland pattern 411, *412*
 poststreptococcal endocapillary
  glomerulonephritis 411
membranes,
 mass transfer area coefficient
  1479
 stability 1479
 ultrafiltration 1479–80
 see also basement membrane
membranoproliferative
 glomerulonephritis,
 8 weeks/16 weeks prednisone
  therapy 144
 acute renal failure in liver disease
  1105
 α-1-antitrypsin deficiency 2216
 characteristic immunohistologic
  patterns 147, 149
 crescentic glomerulonephritis 422
 exudative, essential mixed
  cryoglobulinaemia 687–8, 690,
  691, 692
 Henoch–Schönlein purpura *604*
 hypertension in 1198
 idiopathic, differential diagnosis
  308
 membrane attack complex
  antigens 256
 renal biopsy 144
  EM findings 150
 renal transplantation 1531–2
  with isolated C3 deposits 1532
 schistosomiasis 488
 Sjögren's syndrome 698
 SLE 653, 655
 steroid therapy with longterm
  platelet inhibitors 144
 types I/II, schematic
  representation of glomerular
  changes 151
 see also mesangiocapillary
  glomerulonephritis
membranous glomerulonephritis,
 cancer, secondary to 2316–17

membranous glomerulonephritis (cont.):
  captopril and 162
  D-penicillamine 161
  drug-induced 159–63
  gold/penicillamine causing 160, 681, 682
  hepatitis B nephropathy 463
  hypertension in 1198
  malignancy-associated 480
  nephrotic syndrome, tumour removal 479
  sarcoidosis 580
  Sjögren's syndrome 697–8
  SLE 654, 655, 657
    treatment 659
  with subendothelial deposits, in renal transplantation 1531
membranous glomerulopathy, experimental 253
  mechanisms 253
membranous nephritis, mixed connective tissue disease 684, 698
membranous nephropathy 370–85
  aetiology 375–6
    drug-induced 375
    infections 375
    malignancy-associated 375
  characteristic immunohistologic patterns 147, 149
  children 370
  clinical features 370–1
    and symptoms, profiles 371
  complications 371
  crescentic glomerulonephritis 422
  de novo renal transplantation 375
  defined 370
  diabetes mellitus 526
  epidemiology 370
  glomerular disease in the tropics 488
  gold/penicillamine causing 681, 682
  HBV and 496–7
  idiopathic, treatment 659
  immunogenetic abnormalities 371–2
  induction
    aurothiomalate 2295
    captopril 2295
    mercaptopropionylglycine 2295
    penicillamine 2295
  mixed connective tissue disease 684
  natural history 377–8
  nodules, diabetic nephropathy vs. 524
  outcome
    complete remission/renal failure 381
    controlled studies 382, 384
    uncontrolled studies
      treated patients 381
      untreated patients 378
  pathogenesis 376–7
  pathology 372–5
  prognostic factors
    age 378
    B18-BfF1-DR3 antigens 372
    class I and II antigens, phenotype frequency 379
    DR3 and B8 antigens 372, 379
    immunogenetic factors 379–80
    nephrotic syndrome 379
    sex 378–9

membranous nephropathy (cont.):
  prognostic factors (cont.):
    stage of glomerular lesion 379
    renal transplantation 1514
    recurrence following 1574
    renal vein thrombosis, frequency and prevalence 280
    rheumatoid arthritis 681
  secondary 370
    idiopathic vs 371
    incidence 375
  staging 373
  treatment
    controlled studies 382–3
    delayed 383–4
    side-effects 384
    uncontrolled studies 380–1
  tubular atrophy 149
MEN, see multiple endocrine neoplasia
menarche, see menstruation; puberty
meningitis, sickle-cell anaemia 704
meningococcal meningitis, infection-associated glomerulonephritis 462
meningomyelocele, neurogenic bladder 2033
  renal consequences 2275–6
menstruation,
  dysfunction in haemodialysis 1336, 1337
  amenorrhoea 1336, 1337
  arrest 1337
  hypermenorrhoea 1337
  menstrual irregularities, chronic renal failure 1337
Tc-99m-mercaptoacetyltriglycine, obstruction studies 123–4
  in renal imaging, properties 117–18
  renal plasma flow, measurement 116
mercaptopropionylglycine, cystinuria 749
  induction of membranous nephropathy 2295
6-mercaptopurine,
  activity increased with allopurinol treatment 834
  Fanconi syndrome 735
  thiopurine methyltransferase metabolism 1950
mercury 844–5
  elemental, no nephrotoxicity 844
  forms, nephrotoxicity 844
  indicator of exposure 844
mercury compounds,
  chelation, British antilewisite 845
  drug-induced nephropathy 161
  glomerulonephritis 169–70
    immunologically mediated 844–5
  high-salt diet 171
  intoxication 1008
    Fanconi syndrome 734–5
  membranous glomerulonephritis 161
  metabolism 844
  nephrotoxicity 844–5, 993
    diagnosis 845
    iatrogenic 1008
    management 845
  occupational exposure 844

mercury compounds (cont.):
  organomercurials, nephrotoxicity 844
    chlormerodrin 844
    mercuhydrin 844
    p-chloromercuribenzoate 844
  organomercurials, neurotoxicity
    methylmercury 844
    phenylmercury 844
  phospholipid degradation 168, 169
  polyclonal activator 170
mercury-induced autoimmune glomerulonephritis,
  in Brown-Norway rats 249–51
  cytokines 250–1
  NZ rabbit models 251
  in other species 251
mesangial cells 130–1, 136–7
  basement membrane connections 137
  contractility 983
  cyclosporin therapy in renal transplantation effects 1561
  extraglomerular mesangial cells 130, 136–7, 138
  fibronectin 137
  functions 136
    mesangial matrix 136–7
  glomerular characterization 28
  glomerular scarring 1199
  growth factors, role in glomerular sclerosis 1203
  hepatitis B nephropathy 463
  interposition, renal biopsy, diagnostic relevance 150
  K_f, cyclosporin therapy in renal transplantation effects 1561
  mesangial angles 131, 134
  proliferation, glomerular changes, schematic representation 151
  see also mesangial matrix
mesangial glomerulonephritis,
  gold salts 160
  gold/penicillamine-induced 682
  SLE 653
mesangial hypercellularity, lupus nephritis 653
mesangial IgA, see immunoglobulin A nephropathy
mesangial IgA glomerulonephritis, rheumatoid arthritis association 680, 681
mesangial IgM, IgG, IgA deposits, immunofluorescence studies, minimal change nephrotic syndrome 306–7
mesangial interposition, schematic representation of glomerular changes 151
mesangial matrix,
  amyloid deposits 552
  contractile anomalies diabetes 524
  expansion in diabetes mellitus
    excess glycated proteins role 518
    functional significance 525
    haemodynamic disturbances role 519
    pathogenesis 518–19
    pathology 522, 523
    protein deposition role 520
    volume 524
  expansion in light chain deposition disease 569

mesangial matrix (cont.):
  expansion in lupus nephritis 653
  Henoch–Schönlein purpura 602, 603, 608
  immune deposits
    Henoch–Schönlein purpura 603, 608
    lupus nephritis 652, 653, 654
    rheumatoid arthritis 680
  nodules 569, 570
  schematic representation of glomerular changes 151
  see also mesangial cells
mesangial nephropathy, characteristic immunohistologic patterns 147, 149
mesangial overload, glomerular scarring and, experimental nephropathy in rats 1201
mesangial sclerosis, essential mixed cryoglobulinaemia 689, 691
mesangiocapillary glomerulonephritis 388–402
  aetiology and pathogenesis
    animal models 395
    associated disorders 395–6
    hypocomplementaemias 396
    recuurence after transplantation 396–7
  characteristics 388
  clinical features 397–8
  D-penicillamine 161
  defined 388
  diagnosis 398
  differential diagnosis
    amyloidosis 392
    diabetes mellitus 392
    Henoch–Schönlein purpura 392
    light chain nephropathy 392
    postinfectious nephritis 392–3
    SLE 392
  glomerulonephritis 1737
  hepatitis B nephropathy 463
  histology and pathology 389–93
    clinical interpretation of histology 391–3
    crescentic MCGN 391
  immunofluorescent deposits 347, 348
  immunohistology, diagnosis of types I and II 389, 391
  malignancy-associated 480
  nodules, diabetic nephropathy vs. 524
  nomenclature 388
  outcome and prognostic features 398–400
    age at presentation 399
    geographical variations 399
    histological types 399
  pathology
    complement system 393–5
    family studies and genetics 395
    immune complexes 395
    platelets 395
  recurrence in renal transplantation 396–7
  poststreptococcal 459
  renal transplantation, recurrence following 1573
  schistosomiasis 488, 490
  treatment 400–2
    anticoagulants, antiplatelet drugs 401
    antimitotic immunosuppressive drugs 401

Index

mesangiocapillary
  glomerulonephritis (*cont.*):
  treatment (*cont.*):
    drug trials, controlled and
      uncontrolled *400*, *401*
    plasma exchange 402
    steroids 400–1
  type I (subendothelial; classic)
    association with infections 397
    clinical features 398
    histology and pathology 389–90
  type II (dense deposit disease)
    clinical features 398
    histology and pathology 390–1,
      396
    recurrence after transplantation
      397
  type III (mixed membranous and
    proliferative)
    clinical features 398
    histology and pathology 391
  type IV 391
  types I, II and III
    characteristics 388–9
    clinical associations *392*
  for specific detail see other
    references on pp 389–402
  see also immunoglobulin A
    nephropathy, primary
mesangiocapillary glomerulopathy,
  complement markers 239
mesangioproliferation,
  IgM deposits, minimal change
    nephrotic syndrome 306
  schematic representation
    of glomerular changes 151
mesangioproliferative
  glomerulonephritis,
  acute renal failure in liver disease
    1104–5
  chronic, *see* mesangiocapillary
    glomerulonephritis
  diffuse, histology 305–6
  ethnic factors 9
  Henoch–Schönlein purpura 602,
    604, 2317
    crescents 602, 603, *604*, 606
  mixed connective tissue disease
    684
  renal transplantation 1514
  rheumatoid arthritis 680
  see also rapidly progressive
    glomerulonephritis
mesenchymal benign tumours,
  sarcoma, computed tomography
    scanning 103
mesenchymal tumours 2243–4
mesothelium, peritoneal cavity
  1477–8
metabolic acidosis,
  acute renal failure 1019–20
  carbonic anhydrase inhibitors 213
  causes 777
  causes, anion gap, classification
    925
  causes, anion gap increased
    926–34
    alcoholic ketoacidosis 930, *931*
    diabetic ketoacidosis 928–30
    haemodialysis, high efficiency
      acetate 934
    intoxications, various 931–3
    ketoacidosis due to inherited
      enzyme defects 930
    lactic acidosis 926–8
    starvation ketoacidosis 930

metabolic acidosis (*cont.*):
  causes, anion gap normal 934–42
    acid administration 940
    alkali loss, various causes 934–6
    renal acid excretion,
      decreased:distal renal tubular
      acidosis 936–40
  children with growth retardation
    1305
  chronic, osteomalacia,
    osteopenia 773
  chronic renal failure 1183
  clinical approach and evaluation
    940–2
    plasma anion gap and potassium
      concentration 941
    renal response, assessment 941
    respiratory response 941
  defined 919
  diagnostic approach in renal
    tubular acidosis 777–9
  dilutional acidosis 935
  Fanconi syndrome 726, 736
  hyperchloraemic
    hyperkalaemic-induced
      suppression of renal
      ammonia 909
    hyperkalaemic 907
  hyperkalaemic hyperchloraemic,
    sickle-cell disease 711
  hypocapnia, secondary 955
  integrated response
    buffering 920–1
    elimination of acid load 922–4
    respiratory response 921–2
  magnesium excretion 1808
    animal models 1808
  mechanism 919
  metabolic alkalosis, mixed 956–7
  mineral acid-induced, potassium
    redistribution 907
  neonates 52, 53
    'late metabolic acidosis' 52, 54
    preterm neonates 53–4
  overproduction acidosis 940, *957*
  pathogenesis 920
  pathophysiology 920–6
    anion gap 925–6
    clinical consequences 924–5
  post-hypocapnic acidosis 935
  proximal (type II; bicarbonate-
    wasting) renal tubular acidosis
    935–6
  respiratory acidosis, mixed 958
  respiratory alkalosis, mixed 957–9
  respiratory response 921–2, *922*
    appropriateness 941
  responses and secondary changes
    920
  sickle-cell disease, management
    716
  spontaneous 769
  type IV renal tubular acidosis,
    hyperkalaemia 911
  uraemia 932–3
    animal model, increased
      catabolism 1305
    calcium effects 1367
metabolic alkalosis **942–51**
  accelerated hypertension 2131
  acid losses and chloride depletion
    945–7
    chloride diarrhoea 947
    diuretics 946–7

metabolic alkalosis (*cont.*):
  acid losses and chloride depletion
    (*cont.*):
    gastric secretion loss 946
    post-hypercapnic alkalosis 947
    villous adenoma 947
  alkalaemia in 943
  anions, poorly reabsorbable,
    administration 947
  bicarbonate, renal generation,
    excessive 947–8
    with hypertension 948–9
    without hypertension 949
  carbonic anhydrase inhibitors 206
  causes **943–50**
    miscellaneous 949–50
  chronic metabolic, potassium
    depletion 899, 903
  chronic renal failure 1183
  classification *944*, 948
  clinical approach 950
  clinical consequences 943
    metabolic effects 943
    neurological abnormalities 943
  defined 919, 950
  exogenous alkali gain 943–5
    bicarbonate (and precursors)
      administration 944
    milk–alkali syndrome 945
    non-absorbable antacids 944–5
    transfusions, plasmapheresis
      944–5
  hypercalcaemia causing 1885
  hypertension and
    primary aldosteronism 948
    secondary aldosteronism 948
  hypokalaemia in 943
  integrated response
    alkali load, elimination 943
    buffering 942
    respiratory response 942–3
  magnesium excretion, animal
    models 1808
  metabolic acidosis, mixed 956–7
  pathogenesis 942–3
  potassium depletion 901
  refeeding alkalosis 949
  respiratory acidosis, mixed 958–9
  respiratory alkalosis, mixed 957–8
  responses and secondary changes
    920
  treatment, saline responsiveness
    950
  see also Bartter's syndrome
metabolic disorders,
  chronic renal failure **1251–60**
  hypercatabolism 978
metabolic and respiratory acid–base
  disorders 2339
  mixed **957–60**
  neurological effects 2332
metachromic leukodystrophy,
  arylsulphatase A deficiency 2206
metal-binding proteins,
  loss in nephrotic syndrome 289
metallothionein, cadmium
  transport 842
metaraminol, priapism 1336
metastatic infiltration 2318
  haematological malignancies 2318
metformin,
  contraindications in diabetic
    nephropathy 1645
  handling in kidney disease 189
  induced lactic acidosis 189
  see also biguanides

methaemoglobin, acute renal
  failure and 1007
methanol,
  intoxications *931*, 931–2
  metabolic pathway *931*
  metabolites, creatinine clearance
    and measurement, interfering
    substance *32*
methenamide, contraindications
  1687
methicillin,
  drug-induced acute toxic tubular
    nephropathy 166
  drug-induced interstitial nephritis
    1088
  immunological origin 170
methimazole, drug-induced
  nephropathy 163
methionine,
  diet, acid and base production
    932–3, *933*
  excretion, pregnancy and 1916
  malabsorption 750
methotrexate,
  antineoplastic treatment, renal
    complications 2319
  crystalluria, nephrotoxicity 994
  drug-induced acute toxic tubular
    nephropathy 166
  handling in kidney disease 191
  interaction with ketoprofen 166
  metastatic renal carcinoma 2248
  percentage protein binding
    according to renal function
    1240
  renal complications of 2343
  tubular injury in acute renal
    failure 1011
methoxyflurane,
  crystalluria, nephrotoxicity 994
  oxalate ingestion 1895
methyl CCNU, drug-induced acute
  toxic tubular nephropathy 166
2-methylacetoacetyl-CoA thiolase
  deficiency, ketoacidosis 930
methyl-3-chromone, Fanconi
  syndrome 734
methyldopa,
  haemodialysis patients,
    recommendations *1463*
  hepatotoxicity 2315
  hydralazine and 1971
  liver effects 2315
  pregnancy-induced hypertension
    therapy 1969–70, 1971
  renal hypertension 2056
  renal impairment 188
methylene blue, administration in
  lactic acidosis 928
methylene chloride,
  nephrotoxicity 1009
methylglucamine salts of contrast
  media, *see* radiocontrast media,
  hyperosmolar
methylguanidine, toxic effects 1243
methylmalonic acidaemia,
  ketoacidosis 930
methylpolysiloxane, silica stones
  1866
methylprednisolone,
  Goodpasture's disease 449
  pulsed
    essential mixed
      cryoglobulinaemia 691–2

methylprednisolone (cont.):
  pulsed (cont.):
    Henoch–Schönlein purpura 607, 608
    lupus nephritis 660
      recommended strategy 663
      mechanisms of enhanced effectiveness 660
    microscopic polyarteritis 626
    plasma exchange vs 1068
    sudden death and side-effects 660
    Wegener's granulomatosis 620
  rapidly progressive glomerulonephritis, adverse effects *1063*
  see also prednisolone
methylprednisone, cisplatin, antineoplastic treatment 2319
methysergide, followed by retroperitoneal fibrosis 106
metoclopramide, antiemetics 189
metolazone,
  cardiac failure 188
  GFR and 204
  nephrotic syndrome 270
  clinical use 203
metoprolol,
  elimination by kidney 187
  haemodialysis patients, recommendations *1463*
  pregnancy-induced hypertension therapy 1970
  properties, in relation to use in impaired renal function *2053*
metrifonate, schistosomiasis 1738
metronidazole,
  CAPD peritonitis regimens *1489*
  handling in kidney disease 183
mexilitene, antiarrhythmics, cardiac drugs 187
mezlocillin, CAPD peritonitis regimens *1489*
mice,
  Hantavirus disease carrier 1143
  see also animal models
Michaelis-Gutman body, malakoplakia 1693–5
Michaelis–Menten constant, renal clearance studies 742
miconazole,
  CAPD peritonitis regimens *1489*
  fungal infections 1748
  handling in kidney disease 184
microalbuminuria,
  definition 509
  diabetes mellitus
    arterial pressure association *515*, 515, 525
    concomitants of 515
    functional significance 525
    glomerular origin 514
    mechanisms/pathophysiology 514–15
    prevalence 509, 510
    prognostic significance 510
    as sign of early renal disease 510
    sodium–lithium countertransport rates 520
  diagnosis 509
  see also albumin, excretion; albuminuria
microangiopathic haemolytic anaemia 2341
  acute renal failure **1041–57**, 2341

microangiopathy,
  chronic renal failure 1393
  diabetics, extrarenal problems 1638–43
microcephaly 1613–14
  minimal change nephrotic syndrome and 300
microfibrils, renal biopsy, diagnostic relevance 150
α-1-microglobulin,
  placental barricade, clinical potential 1115
  urinary excretion 564
β-2-microglobulin 514
  accumulation, chronic renal failure 1244
  arthropathy, following renal transplantation 1585
  assessment of glomerular filtration 37
  β-pleated sheet formation 548
  cadmium nephropathy 842, 843
  chromium toxicity 846
  chronic renal failure, levels 1368
  deposition and amyloidosis, long-term haemodialyis 1282
  and dialyser pore size 1244
  excretion rate, diabetes mellitus 514, 516
  glomerular filtration rate, elderly patients 60
  marker of tubular damage 165, 228
  osteoarthropathy, radiography *1372*
  placental barricade, clinical potential 1115
  plasma, and inulin clearance 37
  proximal tubular function, lithium 854
  radio-iodinated, imaging procedures 1372–3
  reabsorption, neonates 52
  tubular uptake, sickle-cell disease 714
  urinary, cadmium absorption and 844
  various ages *60*
β-2-microglobulin amyloidosis 546, 548
  bone scintigrams 1375
  CAPD 1244
  carpal tunnel syndrome 1400
  complication of renal disease 2297–8
  dialysis-associated, diagnosis 1380
  gastrointestinal bleeding 1284, 2311
  haemodialysis 548, *572*, 2330
    long-term 1441, 1443
  haemodialysis retention product 1244
  imaging procedures 1372–3
  laboratory tests *1371*
  left ventricular hypertrophy, end-stage renal disease 1266
  malabsorption syndrome 2311–12
  marker of tubular damage 228
  neurological involvement 2325
  skeletal changes 2297
  treatment, renal complications 2297–8
microglobulins, see α-1-microglobulin; β-2-microglobulin

microhaematuria,
  diabetes mellitus 511
  essential mixed cryoglobulinaemia 690
  juvenile rheumatoid arthritis 681
  sickle-cell disease 710
  SLE 649
    children 650
  see also haematuria
Microplasma hyminis, urinary pathogen 1686, 1688
microscopic polyarteritis 587, 592, **623–8**
  anti-GBM antibodies 443
  clinical features 624
  cyclophosphamide 1066
  definition 623, 1066
  diagnosis 624–5
  epidemiology 623–4
  historical aspects 612
  outcome 625, 626–8
    long-term 627
    renal function 627
    response to initial therapy 627
    relapses 627
  renal-limited form 623, 624
  renal transplantation 1514
  risk factors for poor prognosis 628
  treatment 625–6
    maintenance 626
    renal replacement therapy 627
  vascular infiltrate 625
  Wegener's granulomatosis vs. 623
microscopic polyarteritis infections 624
microscopy,
  light **19–22**
  renal biopsy specimen processing 146
microscopy of urine, see urine, microscopy
Microtus, Hantavirus disease carrier 1143
micturating cystourethrography,
  need for alternative methods 122–3
  neonates 2026
  normal IVU 80
  vesicoureteral reflux, nuclear imaging 122–3
  vesicoureteric reflux 1986
    grading system 1986
micturition, control, children 1699
micturition disorders,
  dysuria 4
  frequency 3–4
  nocturia 3–4
  prostatic enlargement 4
  see also urine, frequency
midazolam, handling in kidney disease 186
Milan hypertensive rats,
  glomerular lesions, protection 1207
  renal hypertensive studies 1196
  renal influence on blood pressure control 2060
milk,
  cows' milk, breast milk compared 883
  dietary exclusion 731
  dietary intake, chronic glomerulonephritis 1888
  osmotic load 883

milk formula,
  aluminium 1614
  phosphate content 1617
milk–alkali syndrome 945
  calcium stone disease 1836
  hypercalcaemia, vitamin D intoxication 1773
  metabolic alkalosis 945
  nephrocalcinosis 1891
  peptic ulceration 1836
mineralocorticoids,
  'apparent excess syndrome' 905
  excess
    high/low aldosterone *905*
    potassium depletion 905–6
  exogenous
    hypertension, metabolic alkalosis and 948
    inhibition of 11-β-hydroxysteroid dehydrogenase 948
  11-β-hydroxysteroid dehydrogenase inhibition 905
  magnesium and 1807
  pregnancy, potassium excretion/retention 1917
  replacement, fludrocortisone 909
  salt-retaining stimulus, 'escape' 264
  see also aldosterone
minimal change glomerulonephritis, renal biopsy 144
  EM findings 150
  steroid therapy with longterm platelet inhibitors 144
minimal change glomerulopathy, drugs and 162
  glomerular damage 230
  gold salts 160
  nephrotic syndrome complicated by ARF 1039
minimal change nephropathy,
  cancer, secondary to 2317
  glomerular disease in the tropics 486–7
  gold/penicillamine associated 682
  malignancy-associated 481
minimal change nephrotic syndrome,
  adults, treatment and outcome **327–9**
  aetiology and inheritance 299–300
  age at onset 300
  children 268, **298–327**
    aetiology 299
  clinical features 300–1
  defined 298
  differential diagnosis
    congenital nephrotic syndrome 307
    diffuse mesangial sclerosis 307
    idiopathic membranoproliferative glomerulonephritis 308
    membranous nephropathy 308
    quartan malaria nephropathy 308
  epidemiology 298–9
  focal segmental glomerulosclerosis, single disease concept 298
  GFR in 268
  histology 269–70, 303–7
    diffuse mesangial proliferation 305–7
    electron microscopy 303

Vol. 1 – Sections 1–5.   Vol. 2 – Sections 6–11.   Vol. 3 – Sections 12–21.

Index

minimal change nephrotic syndrome (cont.):
  histology (cont.):
    lesions vs minimal change disease 269, 270
    relationship between histological patterns 307
  history 298
  immune complexes 311
  immunofluorescence studies 306
  incidence 298
  infections, occult 314
  inheritance 299–300
  laboratory abnormalities
    blood chemistry 302
    haematology 302
    renal function 302
    urinalysis 301–2
  MHC system
    DR3-DR7 312
    HLA and 312
  minimal change disease, defined 276
  paraneoplastic 300
  pathogenesis
    allergy, IgE 312–13
    glomerular polyanion, loss 308
    oligoantigenic diet 312
    possible 283
  pathophysiology
    immune system 310–13
    proteinuria mechanisms 308–10
  predisolone-induced remission 270
  relapses 315–16
    pregnancy 315
    treatment 315–16
  renal failure, acute 301
  renal transplantation, recurrence of nephrotic syndrome 327
  steroid-resistant
    outcome 325
    symptoms and signs 319
  steroid-responsive
    soluble immune response suppressor 311–12
    symptoms and signs 319
  treatment and outcome, adults 327–9
  treatment and outcome, children 313–22
    alkylating agents 316–18
    alkylating agents, complications 322
    complications related to treatment 322–3
    corticosteroid complications 322
    corticosteroid-resistant nephrosis 319–21
    corticosteroid-responsive nephrosis 315–19, 321
    corticosteroids 314–15
    initial treatment 314–15
    levamisole 319
    levamisole, complications 323
    long-term outcome 323–7
    NSAIDs, complications 319, 323
    overview 313–14
    practical recommendations 321–2
  vascular permeability factor 309–10
  see also focal segmental glomerulosclerosis; nephrotic syndrome

minor histocompatibility antigens, immune response to allograft 1544
minoxidil,
  children, treatment of hypertension 2148
  haemodialysis patients, recommendations 1463
  hypertension
    accelerated 2135
    end-stage renal failure 2121
    renal hypertension 2056
  side-effects 2056
Miranda syndrome, features and associated abnormalities 2037
misoprostol, cyclosporin therapy in renal transplantation 1561
mithramycin,
  antineoplastic treatment, renal complications 2319
  drug-induced acute toxic tubular nephropathy 166
  hypercalcaemia of multiple myeloma 568
  tubular injury in acute renal failure 1011
mitochondria,
  disordered permeability 168–9
  giant, cyclosporin toxicity 149
mitogens, PHA lymphocyte response 372
mitomycin,
  antineoplastic treatment, renal complications 2319
  drug-induced nephropathy 162–3
  glomerular thrombosis 1052
    animal models 1052–3
  haemolytic uraemic syndrome 849
  haemolytic–uraemic syndrome 162, 163, 1048, 1050
  handling in kidney disease 191
  renal complications of 2343
  renal pelvis tumour 2269
  thrombotic microangiopathy resembling radiation nephritis 849
  tubular injury in acute renal failure 1011
mitral valves, see heart valves
mixed connective tissue disease (MCTD) 684, 698
  autoantibodies 698
  as distinct entity 698
  renal disease 698
    presentation/histology 684, 698
    treatment/prognosis 684, 698
  scleroderma kidney 698
  skin lesions 2302
mixed cryoglobulinaemia, IgA nephropathy 361
mixed-function oxidase system, acetaminophenone 169
molecular genetics of renal disorders 2155–62
  forward vs reverse genetics 2155
  genetic linkage map 2159
  reverse genetics 2155–7
molybdenum, xanthine oxidase formation 1866
Monckeberg's sclerosis 512
monoclonal antibodies 563
  future prospects, immunosuppressive therapy in renal transplantation 1558

monoclonal antibodies (cont.):
  interstitial infiltrate characteristics 1085–6
  light chain deposition disease 569
  OKY–1581, animal renal ablation models 1195, 1215, 1216
  renal damage due to 563
  urinary examination 973
  see also Bence–Jones protein; OKT3 monoclonal antibodies
monoclonal gammopathy,
  amyloidosis associated with 549, 555, 571–3
  crescentic glomerulonephritis 421
  detection/search amyloidosis investigations 555, 556
  IgA, Henoch–Schönlein purpura aetiology 598
  neurological involvement 2325
  see also amyloidosis, AL
monoclonal light chains,
  B cells, glomerulopathy/myeloma 229, 239
  properties, effects on kidney, significance of elevation 229
monocyte/macrophage lineage 256
monocytes,
  granulomatous response 1720
  infiltration
    essential mixed cryoglobulinaemia 688, 689, 691, 692
    Henoch–Schönlein purpura 601, 602
    IgA nephropathy 601, 602
    poststreptococcal endocapillary glomerulonephritis 411
  role in glomerular sclerosis 1202–3
  sickle erythrocytes adherence to 715
mononeuritis multiplex wasting, diabetic neuropathy 1641
mononeuropathy, uraemic polyneuropathy 2328
mononuclear infiltration, inflammatory, in endocapillary glomerulonephritis, acute 410–11
morphine, renal colic 2010
morphometry,
  renal biopsy
    light/EM microscopy 153
    specimen evaluation 153
Moschkowitz syndrome, see haemolytic–uraemic syndrome
motor and sensory neuropathy, uraemic polyneuropathy 2327–9
mountain sickness, acute, carbonic anhydrase inhibitors 206
MPTT (mean parenchymal transit time) 116–17
MR fimbriae, receptors and potential receptors 1668
MRL/l mice 585, 640
  anti-DNA antibodies 637
  genes controlling autoimmune features 642
  helper T-cell dysfunction 640
  non-lpr strains 640
Muckle-Wells syndrome 553, 554
mucocutaneous lymph node syndrome, see Kawasaki's disease

mucormycosis 1747
  desferrioxamine-associated 2338
mucosal immune system, IgA nephropathy, role in provoking 357
Muercke's striae, nails 1393
multicystic dysplasia, children 2169, 2180
multicystic kidney disease, in childhood 1607
multidrug resistance-1-gene, resistance to chemotherapy 2248
multilocular cyst, children 2180
multiple bilateral parapelvic cysts 2180
multiple endocrine neoplasia, type 1/2, nephrocalcinosis in 1897
multiple malformation syndrome, associated with renal cyst(s) 2169
multiple myeloma,
  cast nephropathy, vs contrast-associated nephropathy 78
  Fanconi syndrome
    clinical features 734
    PAH excretion 727
  immunoglobulin light chain proteinuria 996
  neurological involvement 2325
  paraproteinaemia, pseudo-hyponatraemia 873
  radiographic contrast-induced acute renal failure 996
  renal biopsy,
    casts 148
    indications 144, 145
  renal involvement 563–7
    amyloidosis associated, see amyloidosis
    cast nephropathy 563–7
    clinical course/manifestations 566–7
    histology 565–6
    light-chain nephrotoxicity 563–5
    pathogenesis 565
    staining 566
    Fanconi syndrome 567
    hypercalcaemia 567–8
    prognostic criteria 568
    remission 568
    treatment and outcome 568
  tubular lesions, renal biopsy 148
multiple sclerosis, bladder symptoms 2323
Munchausen syndrome, 'by proxy', in children 9
Munich–Wistar rats,
  glomerular hypertrophy 1193
  progressive age-related glomerular lesions 1206–3
  renal ablation studies 1193
  streptozotocin-induced diabetes 1193–4
  sex-related factors 1208
  see also Wistar Kyoto rats
MURCS association, features and associated abnormalities 2039
murexide test, xanthine stones 1866
muromonab, see OKT3
muscle damage,
  acute renal failure and 1008
  rhabdomyolysis, diagnosis and treatment 1007–8

Index

muscle mass,
  and body weight, by age *1296*
  changes in elderly patients 64
muscle weakness, Sjögren's syndrome 696
musculoskeletal system, amyloidosis 550
mushroom poisoning, acute renal failure and 1011, 1136
muzolimine, adverse effects 204
mycobacterial infections, renal impairment 184
  see also leprosy; tuberculosis
Mycobacterium, liver disease associated with acute renal failure 1106
Mycobacterium leprae, antigens 492
Mycobacterium smegmatis 1719, 1724
Mycobacterium tuberculosis 1719
  identification 1724
  other pathogenic spp 1719
  three populations 1724–5
Mycoplasma pneumoniae, infection-associated glomerulonephritis 462
mycosis fungoides, glomerulonephritis 2305
mycotoxin(s), Balkan nephropathy and 860
myelodysplasia, azathioprine 2345
myeloma,
  acute renal failure in 2342
  amyloidosis associated, *see* amyloidosis
  Bence-Jones proteinuria and hypercalcaemia 1777
  bone resorption and 1776
  hypercalcaemia of malignancy 1776
  IgD 563
  IgM 568
  proteinuria, monoclonal light chains, B cells *229*, 239
  renal complications of 2341
  see also multiple myeloma
myeloma kidney, *see* multiple myeloma
myelomatosis,
  end-stage renal failure, proportional distribution by age at commencement *1229*
  hypercalcaemia in 1777
myeloperoxidase, antibodies to 592, 618, 625
myeloproliferative disease,
  acute uric acid nephropathy 832
  uric acid stones *1861*
myocardial infarction, haemodialysis, long-term patients 1438
myocardial ischaemia, *see* coronary artery disease; ischaemic heart disease
myoclonus, acute renal failure 2326
myogenic hypothesis, renal perfusion 983
myoglobin,
  acute renal failure and 1007
  properties, effects on kidney, significance of elevation *229*
  rhabdomyolysis, and ARF 995

myoglobinuria, causes *1007*
myoinositol,
  neurotoxins 1397
  supplementation, glomerular filtration rate reduced 518
  uraemic syndrome 1243
myosin,
  mesangial cells 136
  podocytes 133
myotonic dystrophy, features and associated abnormalities *2036*

$Na^+2Cl^-K^+$ cotransporter,
  (inhibitors, *see* loop diuretics)
  locations 200
  thiazide diuretics 200
Na^+-H^+ exchanger, ion reabsorption from intestine 934
Na^+-K^+-ATPase pump,
  dysfunction, renal tubular acidosis *938*
  exercise 897
  inhibitors
    antiulcer drugs 189
    increased peripheral resistance 2119
  intracellular K^+ regulation 895
  mechanisms 766
  ouabain-like inhibitors, hyperkalaemia 909
  salt-wasting, ageing changes 63
NAD,
  phosphate reabsorption, regulation 1786
  phosphaturic response to PTH 1785
nadolol,
  haemodialysis patients, recommendations *1463*
  properties, in relation to use in impaired renal function *2053*
nails,
  brown nail areas 1157, 1158
  clinical examination 12
  dyschromia, chronic renal failure 1393–4
  fungal infections 2307
  half-and-half nail 1394
  nail-fold telangiectasia, mixed connective tissue disease 2302
  onycho-osteodysplasia *2036*
  onycholysis, cephaloridine 1394
nail–patella syndrome 2299
  assessment 1155
  eye disorders 2215
  features and associated abnormalities *2036*
  genetics 2215–16
  renal biopsy, EM findings 150
  renal involvement 2214–15
nalidixic acid,
  drug-induced bullous disorders 1393
  renal transplantation, immediate peri-operative administration 1522
β-naphthylamine, urinary bladder tumours 2266
natriuresis,
  action of prostaglandins, NSAIDs and 820
  loop diuretics 269
  paradoxical 262
  plasma renin 269

necrobiosis lipoidica, diabetes 2301
necrotizing polyarteritides **590–1**
  immunological abnormalities 590
  Kawasaki's disease *590*, 590–1, 630
  pathogenesis **590–1**
  serum sickness as model, *see* serum sickness
  vascular abnormalities *591*
  see also polyarteritis nodosa
necrotizing vasculitis, *see* necrotizing polyarteritides; vasculitis
needles,
  haemodialysis 1410–11, 1419
  for puncture, vascular access 1410–11
  and tubing
    haemodialysis 1410–11, 1419
    see also dialysis tubing
Neisseria gonorrhoeae, sexually transmitted infections 1686
neocarzinostatin, chemotherapy, metastatic renal carcinoma 2248
neonatal lupus 651
neonates **50–4, 2035–41**
  abnormalities
    clinical presentation 2025–6
    dimercaptosuccinate scan 2026
    malformations of the kidney **2035–41**
    micturating cystourethrography 2026
    ultrasound 2026
  acute renal failure **1110–20**, 1605
    aetiology 1111–12
    diagnosis 1112–15
    incidence 1110–11
    new therapies 1119
    outcome 1119–20
    subclinical form 1120
    treatment 1115–19
  adult polycystic kidney disease in childhood 1607
  amino acids, normal tubular reabsorption 745
  anatomy 1699
  assessment of renal function 50–4
  Bartter's syndrome 782
  captopril, treatment of hypertension 2148
  clinical examination 12
  Cockayne's syndrome 2300
  congenital chronic renal failure 1606
  congenital nephrotic syndrome 2169, 2264–5
  diazepam effects 1973
  electrolyte disturbances
    hyper/hyponatraemia 53
    metabolic acidosis 53–4
  Escherichia coli
    infection 1699, 1701
    septicaemia 1704
  exstrophic anomalies 2035
  failure to thrive 1605–6
  fluid and electrolyte balance 1115–19
  fractional urate clearance 828
  frusemide, side-effects 54
  glomerular filtration rate 51–2
    normal 1297
  growth patterns 1294–5

neonates (*cont.*):
  hormonal regulation of renal function
    ADH 53
    atrial natriuretic peptide 53
    renin–angiotensin–aldosterone system 52–3
  hydralazine, treatment of hypertension 2148
  hypertyrosinaemia 730
  hyponatraemia 879
  immunological defence mechanisms 1699, 1702–3
  infantile polycystic kidney disease 1607
  intravenous urography 1113–15, 2027
  investigations 2026–7
  kidneys, characteristics 1110, 2024–5
  microcephaly 1613–14
  multicystic dysplastic disease 91, 1607
  multiple malformation syndromes 2169
  obstructive uropathy 1606
  pelviureteric junction obstruction 2030, 2031–2
  polycystic disease, autosomal recessive 2163–7
  ultrasonography 91
  posterior urethral valves 2032–3
  prerenal failure 54
  prerenal lesions 1120
  pressure flow studies 2027
  preterm
    aldosterone excretion 52–3
    creatinine clearance 50
    dopamine administration 54
    fractional urate clearance 828
    frusemide administration 1892
    'late metabolic acidosis' 52
    β-2-microglobulin reabsorption 52
  nephrocalcinosis following RDS 1892
  patent ductus arteriosus, NSAIDs and 821
  respiratory distress syndrome 54
  sodium excretion 51, 52
  sodium losing state 52
  urinary K:Na ratio 51
  prostacyclin stimulating factor 1053
  prune belly syndrome 2034
  radionuclide nephography 2026
  RDS, nephrocalcinosis following 1892
  reflux nephropathy 1606
  renal agenesis, *vs* multicystic dysplastic disease 91
  renal parenchymal lesions 1121
  renal transplant mothers, neonatal problems 1339, *1949*, 1950
  renovascular hypertension, causes 2144
  SLE
    heart block and other defects 1945
    neonatal lupus syndrome 1945
  taurine deficiency 888
  technetium scan 2026
  tethered cord syndrome 2033
  trauma, assessment with ultrasonography 96–7

Index

neonates (cont.):
  tubular transporting capacity
    bicarbonate reabsorption and hydrogen excretion 52
    calcium and phosphorus transport 52
    peptide transport 52
    sodium transport 52
    urinary concentrating capacity 52
  ultrasonography, anatomy of kidney 87
  urinary tract infection 1699–700
  urinary transport disorders 2030–4
  vesicoureteric reflux, natural history 1990
neoplasia,
  incidence in glomerulopathy 479
  minimal change nephrotic syndrome 299
  proliferative glomerulonephritis, IgA nephropathy 360
  sclerosing glomerulonephritis, IgA nephropathy 360–1
  see also malignancy-associated glomerular disease; *specific conditions; tumours*
nephrectomy,
  animal models 1198
    outcome 1193
  bilateral
    and hypertension 1595
    role 1464
    scleroderma 676
  complications of, adrenal insufficiency 1325
  contraindications 2279
  glomerular scarring and, age-related factors 1208
  hypertrophy of contralateral organ 1193
  indications 2108
  'medical' 113–14
  metastatic disease, rationale 2247
  native kidneys
    renal transplantation 1598
    renin secretion present/absent 1599
  polycystic disease, autosomal dominant, prior to transplantation 2184
  renal carcinoma 2246–7
    bilateral tumours 2246
    caval involvement 2247
    prognostic factors 2249, *2250*
    solitary kidney 2247
    solitary metastases 2247
  renal transplantation, prior to 1515–16, 2184
  renal tuberculosis 1726–8
  ureteral stump cancer following 2269
  vs simple stone removal, renography 119
nephritic syndrome,
  acute
    children 406, 1122
    endocapillary glomerulonephritis, acute 406
    essential mixed cryoglobulinaemia 690
    Henoch–Schönlein purpura 601, 605

nephritic syndrome (cont.):
  acute (cont.):
    and oedema, differential diagnosis 408
    renal fluid retention mechanisms 266
    treatment 412–14
    urinary symptoms 6
  development, settings 233
  pathophysiology 266
  presentation 235
  see also glomerulonephritis
nephritis,
  chronic, crescentic glomerulonephritis, treatment 431
  complications of, rheumatological conditions 2295
  diffuse acute bacterial nephritis 105
  'fibrillary' nephritis 2281–3
  focal acute bacterial nephritis 105
  of Henoch–Schönlein purpura, see Henoch–Schönlein purpura
  hereditary
    Goodpasture's disease 452
    pregnancy and 1941
  of immune aetiology
    pathology 413
    treatment 412–14
  interstitial, see interstitial nephritis
  isolated, differential diagnosis 408
  nephrotoxic, functional and structural parameters *1194*
  postinfectious, differential diagnosis, mesangiocapillary glomerulonephritis 392
  systemic lupus erythematosus (SLE), see lupus nephritis
  see also glomerulonephritis; pyelonephritis; *specific nephritides*
nephroangiosclerosis,
  hyperuricaemia as indicator 831
nephroblastoma,
  diagnosis, computed tomography scanning 103
  see also rhabdomyosarcoma; Wilms' tumour
nephroblastomatosis,
  classification 2258
  defined 2257
  intralobar, differential diagnosis 2258
  precursor of Wilms' tumour 2258
nephrocalcin, inhibitor of calcium oxalate crystallization in urine 1828
nephrocalcinosis **1882–99**
  acute cortical necrosis 1888
  analgesic nephropathy 1895
  autosomal dominant *1897*
  autosomal recessive *1897*
  Bartter's syndrome 1894
  benign nodular cortical 1889
  chemical 1883–5
    clinical features 1883
    pathogenesis 1883
    treatment 1885
    tubular defects 1884–5
  chronic glomerulonephritis 1888
  chronic pyelonephritis 1889
  definitions 1882
  distal renal tubular acidosis and 773, 1892–4
  drug-induced 1893–4

nephrocalcinosis (cont.):
  end-stage renal failure, proportional distribution by age at commencement *1229*
  familial, list *1897*
  familial benign hypercalcaemia 1896
  familial magnesium-losing nephropathy 1894
  Fanconi syndrome 1893–4
  fetus 1894
  fluorosis and 1896
  11β-hydroxysteroid dehydrogenase deficiency 1895
  hypercalcaemia causation 1891
  hypertension 1896
  hyperthyroidism 1896
  hypomagnesaemia–hypercalciuria syndrome 1894
  hypothyroidism, congenital 1894–5
  idiopathic hypercalciuria 1891
  infantile form 1891, 1898
  Liddle's syndrome 1895
  McGivvon–Lubinsky syndrome 1892
  macroscopic **1888–98**
    cortical 1888–9
    medullary 1889–98
  medullary sponge kidney 1895–6, 1897
  microscopic
    animal models 1886–8
    corticomedullary 1886
    hyaline droplet nephropathy 1887
    management 1885–6
    pelvic 1887
  neonates 1892
  oxalosis 1889, 1895
  papillary calcification 1895
  persistence 1898
  proximal tubule dysfunction 1896
  radiological, causes *1889*, 1896
  Randall's plaques 1896
  rapidly progressive osteoporosis 1891
  renal papillary necrosis 1895
  renal tubular acidosis *938*
  sarcoidosis 579
  secondary, urinary pH in 1896, 1897
  symptomless 1896
  thyroid protective action 1896
  tram-line clarification 1888
  transplantation 1889
  ultrasonography 95
  vs papillary necrosis, ultrasonography 95
  X-linked *1897*
  see also calcifications; renal calculous disease
nephrography, see arteriography; renal angiography
nephrolithiasis, see renal stone disease
nephrolithotomy, see percutaneous renal surgery
nephromegaly, see renal hypertrophy
nephronophthisis,
  end-stage renal failure 1233
  familial juvenile nephronophthisis 1607
  historical notes 2188–9

nephronophthisis, autosomal dominant 2194–5
nephronophthisis, autosomal recessive,
  age 2194
  associated disorders
    CNS dysfunction 2193
    eye, liver, brain, bone involvement *2193*
    liver involvement 2194
    skeletal involvement 2193–4
    tapetorenal degeneration 2192–3
  children 2169, 2194
  clinical features 2191
  differential diagnosis 2195
  heredity 2192
  imaging 2191–2
  incidence 2190–1
  non-recurrence 2192
  pathology 2189–90
nephronophthisis–medullary cystic disease, autosomal dominant 2194–5
  differential diagnosis 2195
nephrons,
  distal convoluted tubule
    potassium handling 897
    potassium secretion *898*
  function 870–2
  general aspects, see Bowman's capsule; collecting system; glomerulus(i); Henle's loop; renal tubules
  loss in renal failure
    age-related factors 1208
    compensatory changes 1192
    dietary factors 1209
    dietary intervention 1210–12
    genetic factors 1206–7
    glomerular scarring 1192–7
    immune factors 1208
    metabolic and hormonal factors 1208–9
    modulating factors 1206–10
    pharmacological intervention 1212–17
    sex-related factors 1207–8
    therapeutic implications 1210–17
    tubulointerstitial scarring 1203–6
    vascular sclerosis 1206
  numbers 129
nephropathy,
  animal model
    adaptive haemodynamic change in renal ablation 1193–4
    outcome of renal ablation *1194*
  cast, see multiple myeloma, renal involvement
  contrast-associated nephropathy 78–9
  end-stage renal failure, proportional distribution by age at commencement *1229*
  and infections 456–8
  liver disease, associated with acute renal failure 1104–7
  pathophysiology 78–9
  predisposing factors 78
  vasomotor nephropathy 999
  see also glomerulonephritis; immunoglobulin A nephropathy; *specific nephropathies*

nephropathy-associated tumours,
  benign tumours 480
  malignancies *480*
nephrosclerosis **2079–81**
  benign 2079
    frequency *2079*
  decompensated benign 2080
  focal glomerulosclerosis and 2080
  malignant 1050, 2081
  primary/secondary 1050
  structural changes 2079–81
nephrosialidosis 2212
nephrostomy, percutaneous
  fluid collection 79
  ultrasonography 92–3
nephrotic syndrome,
  acute
    and oedema, differential
      diagnosis 408
    schistosomiasis, salmonella
      bacteraemia 460
  acute renal failure 290–1
    complicated by urinary sodium
      1039
    complicating **1034–9**
    management 1039
    pathophysiology and clinical
      features 1035–9
  AL (primary) amyloidosis 572
  albumin distribution, Starling
    forces 267–8
  aldosterone 267
  antidiuretic hormone 267
  atrial natriuretic peptide 267
  carbohydrate metabolism 289
  children, incidence of minimal
    change (idopathic) nephrotic
    syndrome 298
  clinical consequences **276–91**
  clinical/physiological findings 267
    blood pressure 267
    blood volume 267
  congenital
    Alport's syndrome 2198
    diffuse mesangial sclerosis 2219
    neonates 2169, 2264–5
  congenital, of Finnish type 2169,
    2264–5
    classification 2218
    differential diagnosis 307
    glomerular damage 230
    pathology 2218
    prevention 2219
    treatment 2219
  congenital, focal segmental
    glomerulosclerosis 2219
  corticosteroid therapy prior to
    renal biopsy
    adults 144
    children 143–4
  defined 266–7, 276
  diabetes mellitus 511
  diuretics
    clinical use 203–4
    resistance 203
  drug binding 290
  extracellular fluid volume, blood
    volume and bp 266
  Fanconi syndrome 734
  focal glomerular sclerosis
    Bardet–Biedl syndrome 2231
    in transplantation 1531
  gastrointestinal disorders in 2311
  glomerular damage 229–30

nephrotic syndrome (*cont.*):
  glomerular filtration rate
    determinants 1034–5
  gold-associated 160
  haematocrit 267
  Henoch–Schönlein purpura 601,
    605
  histological types, prevalence,
    children *487*
  histology
    diffuse mesangial proliferation
      305–7
    focal segmental
      glomerulosclerosis 304–5
    minimal change disease 303–4
    relationship between histological
      patterns 307
  history 266–7
  hyperlipidaemia 283, 284–9
    albumin levels 286–7
    cardiovascular disease risk 286
    causes of lipid alterations 286
    cholesterol and triglycerides 285
    hepatic lipoprotein synthesis 287
    ischaemic vascular disease 284
    lecithin:cholesterol
      acyltransferase 287
    lipiduria 286
    lipoproteins 285–6
    LPL activity reduction 287
    nicotinic acid 288
    treatment, resins,
      cholestyramine and
      colestipol 287–8
  hypertension and 267
  hyponatraemia 877
  hypovolaemia concept 267, 269,
    290–1
  contraindications 269
  idiopathic
    indications for renal biopsy
      143–4
    renal biopsy 143–4
  IgA nephropathy
    geographical distribution 341
    spontaneous remission 342
  immunotactoid glomerulopathy
    2281–3
  incidence, worldwide hospital
    admissions *486*
  induced and spontaneous change,
    effects 269
  initiation and recovery 269
  infections
    causes 277–8
    cellulitis 277
    clinical features 276
    incidence 276
    peritonitis, primary 276–7
    treatment and prophylaxis 278
    urinary tract infections 277
    viral infections 277
  investigations, clearance methods
    269
  ischaemic vascular disease risk
    284
  light chain deposition disease 570
  lipid metabolism **284–9**
  lung infections 2338
  lupus nephritis 648–9, 653
  malignancy-associated
    glomerulonephritis 478
    membranous glomerulonephritis,
      tumour removal 479

nephrotic syndrome (*cont.*):
  membranous nephropathy 371,
    379
  mercury, occupational exposure
    844
  minimal change, 8 weeks/16 weeks
    prednisone therapy 144
  minimal lesion, lithium toxicity
    856
  myeloma cast nephropathy 566
  neonates 2169, 2264–5
  neoplasia
    children 479
    incidence 479
  noradrenaline 267
  oedema in **262–74**
    treatment 270–1
  plasma renin activity 267
  pregnancy 1946
  prevalence, histological types *487*
  protein malnutrition 291
  proteinuria 2280
    massive obesity 2280
  renal amyloidosis 551
  renal biopsy indications 144
  renal function 268
  renal haemodynamics 268–9
  renal vein thrombosis, frequency
    and prevalence *280*
  salt and water retention 877
  sickle-cell disease 711, 713
  SLE 659
  sodium and water excretion 269
  thrombosis
    complications **278–84**
    corticosteroids 283
    diuretics 283–4
    treatment 283–4
  transport proteins 289–90
  tubular dysfunction 291
  ultrafiltration coefficient,
    decrease 269
  urinary acidification 772
  urinary sediment, 'oval fat
    bodies' 20
  see also minimal change nephrotic
    syndrome
nephrotomography, renal mass
  2244
nephrotoxic nephritis,
  ancrod defibrination 257
  animal models 1035, 1208, *1209*
    parathyroidectomy 1208–9
  anti-GBM F(ab436)$_2$ fragment
    255
  antihypertensive treatment *2086*
  catalase administration 256
  dietary protein restriction, animal
    models *66*
  functional and structural
    parameters *1194*
  glomerular basement membrane,
    addition of antibody-containing
    serum 255
  glomerular scarring, experimental
    nephropathy and hypertension
    1196
  hypertension effects *2084*
nephrotoxins 2285–6
  contrast media, *see* radiocontrast
    media
  direct nephrotoxicity 993–4
  drug-induced, with
    hepatotoxicity 1106–7
  plant toxins of tropics 1125

nephrotoxins (*cont.*):
  see also cyclosporin; *other specific
    substances*
nephroureterectomy, ureteral
  stump cancer following
  incomplete 2269
netilmicin,
  advantages, elderly patients 67
  CAPD peritonitis regimens *1489*
  doses and therapeutic plasma
    concentrations of
    aminoglycosides, handling in
    kidney disease *182*
  nephrotoxicity 1010
neural conduction,
  changes, peripheral
    polyneuropathy *1397*
  velocities, uraemic peripheral
    polyneuropathy 1397
neuroblastoma, vanilmandelic acid
  2256
neurofibromatosis,
  café au lait spots 2300
  children, differential diagnosis
    2144
neurogenic bladder 1709–10, 2033
neurological abnormalities,
  amyloidosis 550
  associated with divided renal
    function 120
  Henoch–Schönlein purpura 597–8
  hyperkalaemia 907
  hypernatraemia 888
  hyponatraemia 880–1
    central pontine myelinosis 880–1
  respiratory acidosis 953
  respiratory alkalosis 955
  SLE children 650
neurological disease,
  renal disease and **2323–32**
  uraemic syndrome, clinical
    features 1237–8
  Wegener's granulomatosis 616
neurological problems,
  children **1613–14**
  elderly patients with chronic renal
    failure 1623
neuromuscular blocking agents,
  depolarizing 186
  narcotic analgesics, handling in
    kidney disease 186
  non-depolarizing, handling in
    kidney disease 186
  non-narcotic analgesics, handling
    in kidney disease 186
neuromuscular function,
  examination in acute renal failure
    970
  hypokalaemia-induced
    alterations 901
neuromyopathy, renal cell
  carcinoma 2242
neuropathy of renal failure, *see*
  diabetic neuropathy
neuropsychiatric alterations in
  uraemia **1396–402**
neurotoxins, peripheral
  polyneuropathy 1397
neutrophil chemotactic factor,
  ionizing radiation causing
  release 849
neutrophils,
  activation, pregnancy-induced
    hypertension and 1965–6

Index

neutrophils (cont.):
  endothelial binding, LFA-1 mediation 1052
  free radical-mediated injury and, renal circulation 1052
  glomerular injury mediated by 639
  haemolytic–uraemic syndrome 1054
  infiltration of glomeruli 255–6
  leucocytoclastic vasculitis 583
  pregnancy-induced hypertension, activation and 1965–6
  vascular damage, pathophysiology 1965–6
nicotinamide, deficiency, Hartnup disease 750
nicotinate,
  excretion, pregnancy and 1916
  nephrotic hyperlipidaemia, effects *287*
  nephrotic syndrome, treatment 288
  urate excretion decrease *829*
  urate reabsorption 829
nicotinic receptor antagonists, handling in kidney disease 186
Niemann–Pick disease, sphingomyelin deficiency 2206
nifedipine,
  animal models
    effect on glomerular scarring *1213*
    prevention of acute renal failure 1018
  children, emergencies *2149*
  contrast media toxicity 1185
  cyclosporin-induced renal vasoconstriction 1598
  haemodialysis patients, recommendations *1463*
  hypertension
    accelerated 2135
    chronic renal failure 1273
    hypertensive crisis 1464
  platelet effects and therapy 1968
  pregnancy-induced hypertension, therapy 1972
nimodipine, cerebral arterial spasm 2178
niridazole, schistosomiasis 1738, 1739
nitrates,
  anaemia and 1351
  angina, prescribing 187–8
  haemolysis and 1351
  handling in kidney disease 188
nitrendipine,
  contrast media toxicity 1185
  hypertension, chronic renal failure 1273
  pregnancy-induced hypertension, therapy 1972
nitrofurantoin,
  children 1715
  hepatotoxicity 2315
  pregnancy 1684, 1929
  pulmonary complications of 2339
  urinary tract infection 1686, 1687
  pregnancy 1684
nitrogen, salt and water retention, cyclosporin therapy in renal transplantation 1563
nitroprusside,
  reaction, ketonaemia 932
  scleroderma 676

nitrosamines,
  and carcinoma 2238
  in tobacco smoke 2238
nitrosureas,
  antineoplastic treatment, renal complications 2319
  drug-induced acute toxic tubular nephropathy 166
  tubular injury in acute renal failure 1011
Nocardia asteroides infection,
  following renal transplantation 1579
  management 1535, 1536
nocturia,
  assessment, chronic renal failure 1156
  elderly patients 62, 65, 1156
  pathogenesis 1156
nocturnal enuresis, causes 1995
nodules,
  diabetic nephropathy 521–2, 524
    location and incidence 522
  glomerular, light chain deposition disease 568, 569, 570
  granulomatous, Churg–Strauss syndrome 613
  mesangial, light chain deposition disease 569, 570
non-Hodgkin's lymphoma,
  cyclosporin, nephrotoxicity 1566, 1582
  renal transplantation 2320–1
non-steroidal anti-inflammatory drugs (NSAIDs) **819–23**
  action 190, 819
  acute renal failure 67, 1016
    in nephrotic syndrome 1038–9
  acute tubular necrosis 821
  acute tubulointerstitial nephritis 821–2
  amyloidosis 556
  analgesic nephropathy and 810–11, 816
  antihypertensives and 823
  chronic renal failure 822
    toxicity 1185
  classification *820*
  clinical syndromes associated 821–3
  cyclosporin and 823
  interactions 823, 1563
    nephrotoxicity 1573
    enhancement 1561
  diuretics and 821, 823
  drug-induced nephropathy 162
    acute toxic tubular 164, 166
  drug-related interstitial nephritis 1087, *1088*, 1089
  elderly patients, inappropriate use 67
  fenoprofen-induced tubulointerstitial nephritis 821
  flank pain syndrome 823
  GFR, prostaglandin biosynthesis effect 179–80
  glomerulonephritis 822
  IgA nephropathy, primary *358*
  inappropriate use, elderly patients 67
  infusion 271
  inhibition of prostaglandin synthesis 987

non-steroidal anti-inflammatory drugs (NSAIDs) (cont.):
  interactions with diuretics 208
  interstitial nephritis 162
  lupus nephritis 659
  minimal change nephrotic syndrome
    steroid-resistant 320
    steroid-responsive 320
  nephrotic syndrome 162
  nephrotoxicity 678, 684
    multiple renal 168
  prescription data 819
  prostaglandins
    biochemistry and 819–20
    renal, and systemic disease 820–1
    synthesis, inhibition 811
  reduction of water excretion 796
  renal failure 159
  renal side-effects
    incidence 819
    specific drugs 823
  salt and water retention 823
  sodium retention 273
  steroid-resistant nephrotic syndrome 320
    risks 323
  steroid-responsive nephrotic syndrome 319
  vasculitis and 822
non-obstructive dilatation of collecting system 2017
noradrenaline,
  and exercise, hypertensive subjects 1461
  nephrotic syndrome 267
  pregnancy, renal vasodilatation 1915
  urate clearance reduced in normal subjects 831
  vascular reactivity and 1461
NSAIDs, *see* non-steroidal anti-inflammatory drugs
nuclear imaging **116–28**
  clinical applications 118–28
    measurement of individual kidney function 118–21
    outflow obstruction 123–4
    renal failure 124–6
    renal mass lesions 126–8
    renovascular hypertension 121–2
    vesicoureteral reflux 122–3
  radiopharmaceuticals 116–18
nucleosomes,
  DNA-anti-DNA complexes 638
  SLE autoantibodies binding to 637
nutrition,
  acute renal failure
    requirements of patients *1020*, 1020–1
    total parenteral nutrition *1021*
  CAPD 1484–5
  children with chronic renal failure 1306
  children on haemodialysis, achieving normal growth 1308
  haemodialysis 1021, 1439, 2330
    pregnancy and 1948
  inadequate, hypercatabolism and 978
  infants with chronic renal failure, neurological problems 1614

nutrition (cont.):
  malnutrition
    children with chronic renal failure 1303
    insulin-like growth factor 1304
  parenteral nutrition
    acid–base disorders 960
    acute renal failure *1021*
    urate/uric acid increased clearance causing hypouricaemia 830
  protein:energy ratio, chronic renal failure 1616–17
  tropics, diseases of 1124
  see also diet
nystatin, toxicity 1748
NZB mice,
  anti-DNA antibodies 636
  genes controlling autoimmune features 642
  mononuclear vasculitis 584
  suppressor T-cell dysfunction 640

oasthouse syndrome 750
obesity, proteinuria of nephrotic syndrome 2280
obstetric acute renal failure **1931–2**
  acute cortical necrosis and 1933
  acute fatty liver of pregnancy 1933–4
  acute tubular necrosis and 1933
  causes *1932*
  dialysis, role 1932
  haemolytic–uraemic syndrome 1934
  miscellaneous causes 1934
  septic shock 1932–3
  summary 1934–5
  tropics 1138
obstructive jaundice, tropics, acute renal failure and 1138
obstructive lung disease, triple metabolic and respiratory disorders *957*
obstructive pyelonephrosis, perinephric abscess 1697
obstructive uropathy,
  diagnosis 1606–7
  multiple sclerosis 2323
  schistosomiasis 1735–6
  upper tract infections 1693–5
  see also urinary tract obstruction
occupational history, presenting complaint 9
occupational renal disease,
  calcium stone disease 1837
  heavy metals exposure 837–8
  see also specific substances
  renal stone disease 1849
Ochoa syndrome, features and associated abnormalities *2036*
ochratoxin A,
  Balkan nephropathy and 860
  carcinoma 2238
octapeptides, *see* angiotensin II
oculo–cerebro–renal syndrome, *see* Lowe syndrome
oedema **262–74**
  clinical history 5
  congestive heart failure 272
    sodium retention and 2335
  endocapillary glomerulonephritis, acute 406
  heart disease 272

Vol. 1 – Sections 1–5. Vol. 2 – Sections 6–11. Vol. 3 – Sections 12–21.

oedema (*cont.*):
  Henoch–Schönlein purpura 596, 606
  idiopathic
    diuretic abuse 273
    potassium depletion 8
    secondary fluid retention 273
  localized 265
  minimal change nephrotic syndrome, steroid-resistant 321
  nephrotic syndrome, treatment 270–1
  NSAID-induced salt and water retention 823
  oedema preventing factors 265
  pathophysiology of formation
    peripheral capillaries 265
    primary/secondary oedema 265–6
    Starling forces 264–5
  periarticular 596
  periorbital, cause 266
  pregnancy, causes 1957
  Quinke's, contrast media injection 77
  renal failure, diabetics 511
  scoring system 237
  treatment
    diuretics 270–1
    sodium restriction 270
    ultrafiltration, mechanical 271
  see also ascites; cerebral oedema; extracellular fluid, volume expansion; fluid retention
oesophagus,
  CAPD effects 1279
  chronic renal failure effects 1279
oestrogens,
  animal gestational models 1921
  bone remodelling *1790*
  citrate excretion 1834
  conjugated, uraemic haemorrhage 1363
  hyperparathyroidism 1856
  induction of carcinoma
    inhibition 2238
    models 2238
oestradiol, chronic renal failure 1615
  pregnancy-associated increase 1920
  sodium retention 273
ofloxacin, CAPD peritonitis regimens *1489*
OIH, *see* I-123/131-*o*-iodohippurate
Okamoto rat, renal hypertensive studies 1196
OKT3 monoclonal antibodies,
  lymphoproliferative disorders, transplantation 2346
  neurological effects 2332
  rejection of allograft 1571
    adverse effects 1558
    reversal 1557–8
    side-effects 1571
    treatment 1530–1
  renal transplantation
    cytokine production 1527
    perioperative use, trials 1558
    renal failure, induction 1527
  see also monoclonal antibodies
OKY–1581,
  animal renal ablation models, effects *1195*, 1215, *1216*

OKY–1581 (*cont.*):
  see also thromboxane synthetase inhibitors
oligomeganephric hypoplasia 1606–7
oligomeganephronia, compensatory glomerular hypertrophy 1193
oliguria,
  acute
    urinary deposit, MA to nucleated cells 973
    urinary examination 973
  acute tubular necrosis *vs* prerenal failure, biochemical indices 988
  assessment in acute renal failure **971–7**, *971*
  biochemical indices, acute renal failure *vs* acute tubular necrosis 1016
  causes 4
  defined 971
  differentiation of prerenal oliguria from ARF *971*, 971–7
  laboratory assessment 971–7
  neonates, *see* neonates, acute renal failure
  post-renal causes 970
  pre-renal causes 969
    differential diagnosis 972
  prognostic indicator, acute renal failure in neonates 1119–20
  renal causes 970
omeprazole, antiulcer drugs 189
Omsk haemorrhagic fever *1045*
Onchocerca volvulus,
  filarial glomerulopathy 495–6
  infection-associated glomerulonephritis 468
onchocerciasis, river blindness 468, 495–6
oncocytoma, computed tomography scanning 103
oncogenes, abnormal expression, MRL/l mice 640
oncogenic hypophosphataemic osteomalacia 756–7
oncology and the kidney **2316–22**
onion skin lesions 2127
  vascular diseases with renal involvement, schematic representation 152
onycho-osteodysplasia, hereditary, features and associated abnormalities *2036*
onycholysis, cephaloridine 1394
ophthalmological disorders,
  familial juvenile nephronophthisis 1607
  renal–ocular syndrome 1091–2
  see also eye disorders
opiates, in kidney disease 186
Opitz–Frias syndrome, features and associated abnormalities *2036*
opportunistic infections,
  AIDS-associated 1010–11
  Aspergillus 1579, 1746
  Candida, other opportunistic species 1743–6
  Candida albicans 466, 1743–6
  fungal infections 1743–9
  Pneumocystis carinii infection, *see Pneumocystis carinii* infection
opsonizing ability, reduced sickle-cell anaemia 704

optic fundi, changes, accelerated hypertension 2130
oral contraceptive,
  cyclosporin therapy in renal transplantation, interactions *1563*
  hypertension and 2048
  renal transplantation 1952
oral mucosal sensitization, Ni-alloy crowns, IgA nephropathy 360
organic anions, preformed, diet, acid and base production 932–3, *933*
organic solvents, nephrotoxicity 993, 1009
ornithine, excretion, pregnancy and 1916
oropharynx,
  AIDS nephropathy 1279
  chronic renal failure effects 1279
oro–facial–digital syndrome, features and associated abnormalities *2037*, *2038*
  polycystic disease 2301
  type I
    polycystic disease, autosomal dominant 2181–2
    X-linkage 2181–2
orthostatic (postural) hypotension, *see* hypotension, postural
Osler's manoeuvre,
  pseudohypertension, measurement of BP 11
Osler's sign, defined 1646
osmolar clearance C_{osm} 871–2
  water excretion, defined 791
osmoreceptors, thirst sensation, elderly subjects 65
osmoregulation 262–3
  oliguria *vs* prerenal failure 988
  pregnancy 1917–18
osmotic agents, peritoneal dialysis solution 1481–2
osmotic diuresis,
  calculation, confirmation 884
  glucose
    hypernatraemia 883
    pseudo-hyponatraemia masking hypernatraemia 883
  see also water diuresis
osmotic diuretics 197
  action 200
  magnesium excretion 1808
  see also mannitol
osmotic nephrosis, contrast-associated nephropathy 78
osteitis fibrosa,
  alkaline phosphatase activity *1774*
  aluminium-related osteopathy 1369
  bone histology 1374, 1377
  bone scintigrams 1375
  clinical features 1376–7
  diagnostic procedures 1370–6
  hypercalcaemia in 1370
  laboratory tests *1371*
  management
    conservative treatment 1384
    prophylaxis 1381–4
    surgery 1384–5
    parathyroid glands in 1384–5
  pathogenesis 1770
  radiography 1372
  serum biochemistry *1371*

osteitis fibrosa cystica 1853–4
  alkaline phosphatase 1854
  radiology 1855
osteo-onychodysplasia, *see* nail–patella syndrome
osteoarthritis, erosive, haemodialysis 2297
osteocalcin (bone GLA protein),
  calcitriol therapy, efficacy marker 1370, 1384
  marker of hyperparathyroid bone disease 1370, 1385
osteochondrodysplasia, features and associated abnormalities *2038*
osteoclast activating factor 567
osteolysis, idiopathic multicentric 2216
osteomalacia,
  alkaline phosphatase activity *1774*
  aluminium-related osteopathy 1369
  bone histology 1378
    aluminium-related pathology 1378
  scintigrams 1375
  cadmium nephropathy 843
  chronic metabolic acidosis 773
  diagnostic procedures 1370–6
  pathogenesis 1770
  vitamin D treatment, calcium set-point 1760
  see also aluminium, bone disease
osteomyelitis,
  sickle-cell anaemia 704
  spinal, following urinary tract infection 37
osteonecrosis,
  clinical features 1379–80
  steroid-induced, following transplantation 1379–80
  technetium scans 1373
osteopenia, following transplantation 1380
osteoporosis,
  chronic renal failure 1765
  corticosteroids, toxicity 322
  rapidly progressive, nephrocalcinosis 1891
  steroid-induced 1891
  thiazide diuretics and 207
osteosclerosis, ground glass appearance 1372, 1373
ototoxicity,
  diuretic-induced 200, 213
  interactions with aminoglycosides 208
  pathogenesis 212
  prevention 212
ouabain, binding sites, hypokalaemia and 208
oval fat bodies, urinary sediment, nephrotic syndrome 20
ovarian cysts, menstruation and, in haemodialysis 1337
Ovary test, vascular permeability factor 309
overdistension syndrome, *see* hydroureter; hydronephrosis
overlap syndrome, gynaecomastia 1336
overload proteinuria, defined 17
ovulation, haemodialysis and 1336, 1337

Index

oxalate,
  crystalluria, nephrotoxicity 994
  dietary intake 1833, 1841
    foods high in *1850*
  ethylene glycol intoxication 972
  excess production 1895
  excretion, thiazide diuretics 207
  intratubular, tubulointerstitial scarring and 1205
  metabolism 1833
  poisoning 1895
  production, metabolic pathways 2226
  uraemic toxin 2229–30
oxalosis,
  acquired, causes 1233
  chronic renal failure 1380
    assessment 1153–4
  end-stage renal failure 1233
    proportional distribution by age at commencement *1229*
  nephrocalcinosis 1889, 1895
  recurrence in graft 1153
  renal transplantation 1513
    recurrence following 1574
oxamniquine, schistosomiasis 185, 1738
oxazacort, minimal change nephrotic syndrome 314
oxazolidinedione derivatives, drug-induced nephropathy 163
3-oxoacid-CoA transferase deficiency, ketoacidosis 930
oxprenolol,
  elimination by kidney 187
  pregnancy-induced hypertension therapy 1970, 1971
  properties, in relation to use in impaired renal function *2053*
oxygen,
  administration, sickle-cell disease 716
  consumption in kidneys, and energy dependence of transport 986
  countercurrent arrangement of kidney vessels, and ischaemia 986
  reactive oxygen species 256
    formation, schema 256
    glomerulonephritis and 256
    metabolism 256
    toxic effects 256
oxygen transport, metabolic acidosis, Bohr effect 924
oxygen free radicals,
  radiation-induced 849
  tubulointerstitial scarring, effects on 1206
oxygen sensor, erythropoietin production 1347
oxygenation,
  non-homogeneity 990–1
  renal cortical hypoperfusion in vivo, P_{O_2} variation 990–1
oxypurinol, clearance and creatinine clearance 835
  see also allopurinol
oxytocin challenge test 1939

P-fimbriae 1667, 1668

pacemaker conduction abnormalities, hyperkalaemia 906
pad culture method, bacteriuria 1663
Paget–von Schrötter syndrome, haemodialysis 1440
PAH, *see* p-aminohippurate
pain,
  bladder obstruction 5
  chronic pain states and analgesic abuse 814
  clinical history 6
  perirenal abscess 5
  pyelonephritis 5
pain relief, epidural block, continuous segmental 1941
paired filtration dialysis, comparisons *1247*
palpation, clinical examination 11
pamidronate, hypercalcaemia, management 1777
p-amino-salicylic acid, renal tuberculosis *1727*
p-aminohippurate,
  clearance
    cardiopulmonary bypass patients 39
    in glycosuria 39
    increase following meat meal, pregnancy 1914
    measurement of renal blood flow 58
    measurement of renal plasma flow 116
    normal adults and children 39
    Wilson's disease 733
  congenital cyanotic heart disease 2277
  effective renal plasma flow 116–17
  extraction efficiency E 116
  renal secretion, cystinosis 2221
  tubular secretion, Fanconi syndrome 727–8
pancreas transplantation,
  diabetic patients 1512, 1650
  pregnancy and 1950
  with renal transplantation 1650
pancreatic cyst(s), polycystic disease, autosomal dominant 2179
pancreatic disorders,
  chronic renal failure changes 1283
  cystinosis 2222
  following renal transplantation 1582
pancreatic hormones, potassium homeostasis 896
pancreatitis,
  acute, complication of transplantation 1285
  cyclosporin-associated, renal transplantation 1582
  proliferative endarteritis 2130
pancuronium, avoiding in kidney disease 186
pantothenic acid, total parenteral nutrition in ARF *1021*
PAP (peroxidase–antiperoxidase) method, in amyloidosis 553, 555
papaveretum, avoiding in renal impairment 186
papaverine, SKATT test and therapy 1330, 1334

papilla, papillary, *see* renal papilla
papillary necrosis, *see* renal papillary necrosis
papilloma, haemodialysis patients 1454
papillomavirus infection,
  cervical carcinoma 1583
  renal transplantation 1581
Papillon–League syndrome, oral facial digital syndrome 2301
papillotoxins 810
papovavirus infection, renal transplantation 1581
Pappenheimer model, glomerular filtration mechanism 229–30
para-aortic lymphadenopathy, diagnosis, ultrasonography 92
paracetamol,
  chronic intake, low renal risk 809
  drug-induced liver and kidney necrosis 1107
  drug-related interstitial nephritis 1089
  glutathione depletion 169
  handling in kidney disease 186
  mixed-function oxidase system 169
  overdose
    acetylcysteine 2315–16
    hepatotoxicity 2315
  phenacetin replacement 804
  in polyclonal immune globulins therapy 1557
  prostaglandin synthesis, minimal inhibition 811
  relative risks of analgesic-associated disease
    end-stage renal failure 808
    renal insufficiency 808
    renal papillary necrosis 808
    total exposure 809
  as single substance, and analgesic nephropathy 808, 816
  synergism with other analgesics 810
Paracoccidioides brasiliensis, natural history 1743
paracoccidioidomycosis 1743
paraldehyde, intoxications, metabolic acidosis 932
parapelvic cysts, computed tomography scanning 106
paraplegia,
  calcium stone disease 1836
  chronic renal failure 2275–6
  renal consequences 2275–6
paraproteinaemia, pseudo-hyponatraemia 873
paraquat,
  intoxication 1011
  nephrotoxicity 169, 1125, **1137**
parasite-induced autoimmune glomerulonephritis 248–9
parasitic infections,
  glomerulopathies, acute renal failure in liver disease 1104–5
  infection-associated glomerulonephritis 466–8
parathormone **1758–61**
  actions **1760–1**, 1786
    bone effects 1760
    negative feedback 1759
    target organs 1760
    uraemia 1760

parathormone (*cont.*):
  ageing changes 61
  assays *1758*, 1851
  bicarbonate
    calcium and 765
    phosphate depletion 765
  bone histology 1374
  bone remodelling *1790*
  calcium homeostasis
    calcidiol treatment *1760*
    calcium deposition 1208–9
    calcium set-point regulation 1756
    chronic renal failure 1755
  calcium-mobilization, skeletal resistance 1367
  calciuric effect 774
  cardiac effects 1265
  chronic renal failure
    bone resorption *1761*
    set-point disturbance 1759
  compensatory renal growth 1212
  degradation and excretion, chronic renal failure 1367
  diabetes, haemodialysis 1649
  elevated, low GFR 1365
  glucose intolerance in chronic renal failure and 1253
  1α-hydroxylase response 753
  hyperparathyroid bone disease 1370
  hypomagnesaemia and 1812
  intestinal phosphate transport 1788
  magnesium handling 1807
    animal models 1807
    plasma levels, following renal transplantation 1378
    measurement 1854
  modulation by vitamin D 1786
  multiple actions, and toxicity 1377
  neurotoxicity 1398
  phosphate homeostasis 765, 1786, 1788
  phosphate reabsorption 752, 1786
  phosphaturic response 1785, 1786
  PI breakdown 1786
  plasma levels *1758*
    immunoreactive 1758
    measurement 1854
  pre-pro-parathormone mRNA 1365
  protein degradation and 1259–60
  renal transplantation
    following 1378
    normo/hypercalcaemic recipients *1775*
  rickets, familial X-linked hypophosphataemic rickets 753
  secretion **1759–60**
    WR-2721 1777
  sodium transport, exit step 741
  steroid-induced secretion 1380
  structure 1758
  synthetic human (1-38)
    parathyroid hormone, early renal failure 1366
  uraemia, β-cell function 1253
  uraemic syndrome, oversecretion 1244
parathormone receptor,
  abnormal coupling 1798
  decreased renal effect 1798–9

Vol. 1 – Sections 1–5.   Vol. 2 – Sections 6–11.   Vol. 3 – Sections 12–21.

Index

parathormone-related peptide,
  activity and structure 1776
  putative bone resorption factor 1776
parathyroid glands,
  adenomata, in patients with nephrocalcinosis, ultrasonography 95–6
  bone disease
    advanced renal failure 1367–8
    early renal failure 1365–7
    calcitriol effects *1365*
  carcinoma, treatment 1385
  coagulation 1385
  fifth gland 1855
  hyperplasia
    apoptosis following transplantation 1379
    hypercalcaemia in 1777
    reduction by gallium nitrate 1385
    surgery 1384–5
    WR-2721 (ethiofos) administration 1385
    see also parathyroidectomy
  imaging procedures 1373–4
  involution following transplantation 1379
parathyroidectomy,
  animal models of nephrotoxic nephritis 1208–9
  cardiac function effects 1266
  chronic renal failure, indications 1384
  renal transplantation, prior to 1517
  see also parathyroid glands, hyperplasia
paratopes, see idiotopes
parenteral nutrition, see nutrition
paroxysmal nocturnal haemoglobinuria,
  acute renal failure 2280
  aetiology 2279–80
  morphological and functional changes 2280
passenger leucocytes, immune response to allograft 1545
patellar hypoplasia, nail–patella syndrome 2214–15
peliosis hepatis 2313
  associated conditions *2313*
pellagra, differential diagnosis, Hartnup disease 750
pelvic irradiation, safeguards 81, 82
pelvic malignancy, ureteric obstruction 2003–4
pelvic nephrocalcinosis 1887
pelvis of the kidney, see renal pelvis
pelviureteric junction obstruction 2030, 2031–2
  management 2017, 2018
D-penicillamine,
  autoimmune glomerulonephritis 249
  crescentic nephritis 2295
  cystine stones 1865
  in cystinuria 748
  drug-induced nephropathy 161
  genetic predisposition to nephrotoxicity 162
  induction of membranous nephropathy 2295

D-penicillamine (*cont.*):
  microscopic polyarteritis association 623
  nephropathy cause 8
  nephrotoxicity 422
    autoantibody production 682, 683
    clinical features 682
    incidence 682
    long-term outcome 683
    management 682–3
    membranous nephropathy 681, 682
    pathology 682
    predictive factors 683
  rapidly progressive glomerulonephritis 1070
  in scleroderma 676
  Wilson's disease 733
penicillin,
  CAPD peritonitis regimens *1489*
  chronic renal failure, toxicity 1185
  clearance, haemofiltration 1032
  dose adjustment *181*
  drug-induced disease
    acute toxic tubular nephropathy 167
    hypokalaemia and metabolic alkalosis 906, 947
    interstitial nephritis 1088
    renal vasculitis 167
  handling in kidney disease *181*
  neurological side-effects 2331
penile function, see impotence, erectile
pentamidine,
  handling in kidney disease 183
  Pneumocystis carinii infection and 183
  Pneumocystis pneumonia 1011
  trypanosomiasis 185
peptic ulcer, see ulcer(s)
peptides,
  transport, neonates 52
  unidentified
    inhibition of insulin-mediated glucose uptake 1303
    insulin-dependent 1253
  uraemic syndrome 1243–4
percutaneous transluminal coronary angioplasty, end-stage renal failure, dialysis patients 1274
percutaneous transluminal renal angioplasty **1870–3**, **2106–7**
  blood pressure effects *2108*
  complications 114, *2106*
  contraindications 114
  indications 114
  ischaemic renal disease 1081–2, *1082*
  results 114, 1873, 2107
  technique 114, 2107
percutaneous transluminal renal surgery,
  history 1870
  indications/contraindications 1872
  results 1873
  techniques 1870–2
perfusion pressure flow studies, neonates 2027

periaortitis of the ureters 2018–20
  in absence of ureteric obstruction
    granulomatous disease 2020
    inflammatory bowel disease 2020
  clinical features 2019
  diagnosis 2019–20
  management 2020
  pathogenesis 2019
periarteritis nodosa, see polyarteritis nodosa
pericarditis,
  children 1610
  chronic renal failure 1186
  and pericardial effusion, in uraemia 1274–5
  in uraemic patients on haemodialysis 1428
perinephric abscess 1696
  children 1706
  clinical features 1696
  in diabetes mellitus 531
  diagnosis 1696
  treatment 1696
perineum, hygiene, children and infants 1714–15
perinuclear antineutrophil cytoplasm antibodies (P-ANCA) 592, 618, 624
peripartum renal failure, differential diagnosis *1057*
  vascular catastrophe 1057
peripheral circulation, coagulation disorders 1044
peripheral polyneuropathy, amyloidosis 550
  children 1613
  chronic renal failure 1157, **1396–8**
  clinical features 1396
  diabetes mellitus 512
  differential diagnosis 1398
  electrophysiology 1396–7
  hypomagnesaemia and 1813
  pathogenesis 1397–8
  treatment 1398
  uraemia, impotence, erectile 1331
peripheral vascular disease,
  continuous ambulatory peritoneal dialysis, elderly patients 1626
  diabetes, post transplantation 1650
  in renal disease 2338
  renal effects 2335
perirenal effusions,
  shifting urinary tract components, computed tomography scanning 107
  trauma, computed tomography scanning 106
peritoneal cavity,
  effects of peritoneal dialysis 1478
  fluid transport to/from 1480
  morphology 1477–8
  physiology 1478–80
  Staverman reflection coefficient 1479
peritoneal dialysis **1477–500**
  acid–base disorders 959–60
  adrenal function 1325
  anaemia, prevalence 1351
  animal models, sodium and water excretion 2059
  assessment of suitability 1169–70
  β-2-microglobulin amyloidosis 1443
  buffering of dialysate 959

peritoneal dialysis (*cont.*):
  calcium
    absorption efficiency *1763*
    pathogenesis of disturbed function 1368
  calcium homeostasis and 1771
  cancer risk 2320
  carbohydrate metabolism 1254
  catheters **1492–4**
  children 1123, 1473
    chronic renal failure 1617–18
    side-effects 1617–18
  clinical results 1482–92
    aluminium bone disease 1483–4
    anaemia 1482
    hypertension 1486
    nutritional aspects 1484–5
    peritonitis **1486–92**
    renal osteodystrophy 1482–4
    uraemia control 1482
  continuous cyclic (CCPD) **1498–9**
  cost and cost-effectiveness 1497–8
  diabetic nephropathy 1649–50
  dialysate, see dialysate
  dyslipoproteinaemia 1257
  efficiency
    HPLC comparisons *1247*
    kinetic data for urea NCDS *1246*
  elderly patients 1626
    back pain 1626
    dialysate leakage 1626
    diverticulosis 1626
    haemodialysis vs peritoneal dialysis 1626
    hernias 1626
    hypotension 1626
    outcome 1626
    peritonitis 1626
  erythropoietin replacement therapy, r-HuEPO 1354
  gastrointestinal disorders 1284, 2312
  genital oedema 1494–5
  glucose, continuous absorption 1257
  haemodialysis vs 1626
  haemoglobin levels 1352
  hepatitis B screening 1168–9
  hernias 1494–5
  insulin 188
  intermittent **1499**
  ischaemic heart disease, lack of clinicopathological correlation 1270
  left ventricular hypertrophy and 1267
  left ventricular mass decrease 1267
  in lupus nephritis (SLE) 663
  neonates 1117–19
  nightly **1499**
  nutrition in 1021
  obstetric acute renal failure 1932
  outcomes **1495–8**
    EDTA 1496
    quality of life 1497
    UK 1496
    United States 1496
  peritonitis
    differential diagnosis 1284
    see CAPD peritonitis
  phosphate, pathogenesis of disturbed function 1368
  physiology 959–60
  pregnancy 1338, **1948–9**

Vol. 1 – Sections 1–5.   Vol. 2 – Sections 6–11.   Vol. 3 – Sections 12–21.

Index

peritoneal dialysis (*cont.*):
  principles 1480
  protein peritoneal loss 1257
  rapidly progressive
    glomerulonephritis 1063–4
  renal transplantation and 1511
  sexual dysfunction **1329–39**
  social assessment 1168
  techniques 1480
  thyroid hormones 1321
  thyroxine-binding proteins, loss 1320
  tidal **1499**
  uraemic polyneuropathy 2329
  vitamin D loss 1763
  Y system 1480
peritoneovenous shunt, diuretics resistance 203
peritoneum, drug malabsorption 176
peritonitis,
  CAPD 1450, **1486–92**
    aetiology and diagnosis 1487–8
    cephalosporins 182
    defined 1486
    differential diagnosis 1284
    host-defence 1490–1
    incidence 1490
    management 1488
    outcome 1490
    pathogenesis 1490–2
    prevention 1490
    relapsing 1488
    sclerosing 1492
    signs and symptoms 1486–7
  infectious, in dialysis patients 1450
  minimal change nephrotic syndrome and 300
  peritoneal dialysis
    elderly patients 1626
    ultrafiltration, loss 1491–2
  primary, nephrotic syndrome 276–7
  renal transplantation 1582
  ultrastructural changes 1478
peritubular capillary plexus, proximal tubule 741
peroxidase–antiperoxidase (PAP) method, in amyloidosis 553, 555
peroxisomal alanine:glyoxalate aminotransferase deficiency 1857, 1858
peroxisomal disorders, features and associated abnormalities *2038*
pethidine, handling in kidney disease 186
Petriellidium, infection 1748
Peutz–Jegher's syndrome, features and associated abnormalities *2036*
pH,
  extracellular fluids, normal values *917*
  and [H$^+$], conversion table *918*
  urine, see urine
phaeochromocytoma, children, renal artery compression 2144
phagocytosis, reduced in sickle-cell anaemia 704
phenacetin **803–16**
  nephritis, see analgesic nephropathy
  nephrocalcinosis and 1893–4
  nephrotoxicity 684

phenacetin (*cont.*):
  paracetamol *vs* 809
  prostaglandin synthesis, minimal inhibition 811
  relative risks of analgesic-associated disease
    creatinine elevation 808
    renal insufficiency 808
    renal papillary necrosis 808
    synergism with other analgesics 810
  renal papillary necrosis 822
  as single substance
    and analgesic nephropathy 808, 816
    availability 809
  withdrawal/restriction from market 816
phenazone, relative risks of analgesic-associated disease, end-stage renal failure 808
phenazopyridine, drug-induced lithiasis 168
phenformin,
  contraindications in diabetic nephropathy 1645
  see also biguanides
phenindione,
  drug-induced interstitial nephritis 1090
  drug-induced nephropathy 163
phenols,
  neurotoxins 1397
  poisoning 1011
  toxicity, uraemic syndrome 1244
phentolamine, SKATT test and therapy 1330, 1334
phenylacetylurea, creatinine clearance and measurement, interfering substance 32
phenylalanine, excretion, pregnancy and 1916
phenylbutazone,
  anuria, non-steroidal anti-inflammatory drugs 823
  crescentic glomerulonephritis 422
  renal papillary necrosis 822
  urate excretion increase 829
phenylhydroxylamine, poisoning 1011
phenytoin,
  binding 2331
  cyclosporin therapy in renal transplantation, interactions 1563
  eclampsia effects 1974
  handling in kidney disease 186
  hepatic metabolism of cyclosporin 2332
  IgA nephropathy, primary *358*
  neurological side-effects 2332
  percentage protein binding according to renal function *1240*
  pregnancy, enhanced effects 1974
  protein binding, reduction of unbound drug 176
phlebotomy, porphyrea cutanea tarda 1392–3
phlegmatia alba dolens, see thrombophlebitis
phlorizin, renal glucose-losing syndrome 731
phosphate binders 1186
  aluminium intoxication 1774

phosphate binders (*cont.*):
  calcium carbonate *vs* aluminium 1305
  compliance aspects 1471
  indications 1382–3
  types, dose and side-effects *1382*
  see also aluminium
phosphate stones, removal, endoscopy 1872, 1873
phosphates **1782–90**
  balance **1782–90**
  breast milk 1617
  calcium effects 1770
  calcium transport, neonates 52
  chronic renal failure 1186, **1365–70**
  dietary deficiency
    gastrointestinal loss *1792*
    homeostasis 1788
  dietary excess, animal models 1788
  dietary increase, renal stone disease 1853
  dietary intake
    animal models 1785, 1788
    and calcium stone disease 1834–5
    changes 752
    and excretion 1785
    intestinal reabsorption 1788
    normal/low 1382
    RDA for children with chronic renal failure 1307
    total parenteral nutrition in ARF *1021*
  dietary restriction
    animal models of progressive glomerular scarring 1209
    chronic renal failure 1181, 1186
    effect on renal scarring 1205
  disorders of tubular reabsorption 752–7
  distribution
    body stores *1789*
    bone 1789–90
    in cells 1789
    extracellular fluid *1789*
  excretion
    hypercalcaemia causing 1885
    index (IPE) 1782, 1789–90
  excretion tests
    chronic renal failure 44
    normal values, neonates 51
  extracellular fluid *1789*, 1789
    to cell shift, causes *1792*
  glomerular filtration 1782
    Donnan correction 1782
  1α hydroxlase suppression 1382
  i.v. administration 1777
  increased
    differential diagnosis 973
    renal calcification 1886
  index of excretion (IPE) 1782, 1789–90
  inhibitor of calcium oxalate crystallization in urine *1828*
  inhibitor of calcium phosphate crystallization in urine *1827*
  intestinal handling 1788
  loading test, renal tubular acidosis 775
  magnesium excretion 1808
  magnesium wasting and 1810
  malabsorption, rickets 1788

phosphates (*cont.*):
  management of hypercalcaemia 1769
  measurement 728, 1854
  metabolism
    chronic renal failure, pathogenesis of disturbed function **1365–70**
    disorders, renal transplantation 1537
  elderly patients 61
  milk formula 1617
  normal tubular reabsorption 751–2, *752*
    parathormone 752
    proximal tubule 752
    vitamin D 752
  oral, metabolic acidosis in Fanconi syndrome 736
  organic, uraemic syndrome 1244
  and parathormone 1786, 1788
  pathogenesis of renal bone disease 1770
  phosphate-wasting Fanconi syndrome 725–6
  physiology of phosphate balance **1782–90**
  plasma, measurement 1854
  precipitation from urine 1893
  reabsorption effects, vitamin D 1786–7
  renal calcifications 1886
  renal clearance, reduced by insulin in diabetes 515
  renal handling **1782–5**
    post obstruction 2013
  renal threshold
    nomogram *1783*
    nomogram T_mP/GFR 752
  renal transplantation, normo/hypercalcaemic recipients *1775*
  retention, hypocalcaemia 1769
  transport **1783–5**
    apical entry 1783–4
    basolateral 1784–5
    cellular mechanisms 1783–5
    ECF to cell shift *1792*
    index 1782
    nomogram *1783*
    sodium–phosphate cotransport 1784
    tubular localization 1783
  tubular reabsorption 1783, **1785–8**
    acid-base status 1787
    adaptation in low intake 1785
    animal models 1783
    extracellular volume expansion 1787
    intestinal 1788
    non-hormonal factors 1787–8
    normal children *vs* Fanconi syndrome *726*
    parathormone 1786
    plasma calcium concentration 1787
    post obstruction 2013
    reduced in insulin-dependent diabetes 515
    requirement for phosphates 1784–5
    rickets in Fanconi syndrome 730
    in sickle-cell disease 714
    vitamin D 1786–7
  tubulointerstitial scarring, effects on 1205

Vol. 1 – Sections 1–5.   Vol. 2 – Sections 6–11.   Vol. 3 – Sections 12–21.

phosphates (*cont.*):
 urinary excretion, *see* phosphates, excretion
 urinary pH and 1834
 vitamin D, inappropriate production 1318
phosphatidylcholine, peritoneal cavity 1478
phosphaturias **751–7**
 see also rickets
phosphocitrate,
 inhibitor of calcium oxalate crystallization in urine *1828*
 inhibitor of calcium phosphate crystallization in urine *1827*
3-phosphocitric acid, effect on renal scarring *1205*
phosphofructokinase, glucagon/insulin ratio 1253
 uraemic muscle 1253
phosphohexose isomerase, increased, marker of renal cell carcinoma 2242
phosphoinositol second messenger system, V_1 renal receptors and 791
phospholipase A_2, arachidonic acid pathway and 820
phospholipids, antibodies to, in SLE 648
phosphomycin, cisplatin, antineoplastic treatment, renal complications 2319
phosphonocarboxylic acids, inhibition of sodium–phosphate cotransport 1784
phosphoribosyl pyrophosphate synthetase, uric acid stones and *1861*
phosphorin, brush border membrane 1784
phosphoserine, diet, acid and base production 932–3, *933*
photochemotherapy, psoriasis 2305
photodermatitis, Cockayne's syndrome 2300
photophobia, cystinosis 1608, 1609
photosensitivity,
 Rothman–Thomson syndrome 2300
 thiazide diuretics and 213
photosensitizers, drug-induced bullous disorders 1392, 1393
phototherapy, pruritus treatment 1392
phthalate-containing plasticizers, haemodialysis tubing 1332
phycomycosis, defined 1747
Physalia, jelly-fish stings, acute renal failure and 1136
phytanic acid, excess, Refsum's disease 2212
phytohaemagglutinin, inhibition, nephrotic syndrome 311
 response, elderly patients *1629*
pica, lead nephropathy, acute, in children 838
Pichia, infection 1748
picrate, Jaffé reaction, Janovsky complex with creatinine 31
pIgAN, *see* immunoglobulin A nephropathy, primary
pigeons, cryptococcosis 1746

pigment nephropathy **1006–8**
 causes *1007*
 diagnosis and treatment 1007–8
 mechanisms 1007
pigmentation abnormalities, chronic renal failure 1390
Pima Indians 507, 508, 520
pindolol,
 elimination by kidney 187
 platelet effects and therapy 1968
 properties, in relation to use in impaired renal function *2053*
piperacillin,
 CAPD peritonitis regimens *1489*
 dose adjustment 181
 drug-induced interstitial nephritis 1088
pipercuronium, avoiding in kidney disease 186
piretanide,
 action 199
 chronic renal failure 204
piridoxilate, drug-induced lithiasis 168
piromidic acid, drug-related interstitial nephritis 1089
piroxicam, NSAID-induced vasculitis 822
pitressin, antidiuretic hormone control of water transport 791
pituitary, anterior, allografting, rat models 1914
pituitary function tests 1325
pituitary hormones,
 chronic renal failure
  function tests 1325
  plasma levels 1325
 increase, glomerular hypertrophy 1208–9
 see also ACTH; cortisol
pituitary–adrenal axis, chronic renal failure 1324–5
pityriasis versicolor 2307
placenta,
 ischaemia 1960
 pregnancy-induced hypertension 1963
 physiology, pregnancy-induced hypertension 1960
 renin production 1962
placental trophoblastic tumour, glomerular DIC appearances 481
plant toxins, tropics, acute renal failure and 1125, 1136
planted glomerular antigens, pathogenesis of glomerulonephritis 254
plaque, fibrous, *see* retroperitoneal disease, fibrosis
plasma,
 acetoacetate
  concentration 929
  instability 928
 composition, cations/anions *925*
 electrolyte composition
  respiratory acidosis 953
  respiratory alkalosis 955
 fresh frozen, dosage 1055
 ionic composition *925*
 phosphate in *1789*
plasma cell dyscrasias **562–76**
 AL amyloidosis in, *see* amyloidosis
 hypercalcaemia in 567–8

plasma cell dyscrasias (*cont.*):
 hyperviscosity syndrome in 568
 renal impairment in 567–8
 see also light chain deposition disease; multiple myeloma
plasma exchange,
 adverse effects, rapidly progressive glomerulonephritis *1063*
 antiglomerular basement membrane disease 449–50, 1065
 crescentic glomerulonephritis 429
 haemolytic–uraemic syndrome 1055
 Henoch–Schönlein purpura 607
 HLA antibodies in transplantation 1516–17
 hyperviscosity syndrome 568
 IgA nephropathy, primary *358*
 lupus nephritis 661–2
 metabolic alkalosis and 944
 methylprednisolone *vs* 1068
 microscopic polyarteritis 626
 polyarteritis nodosa 629
 and proteinuria, contraindications 1531
 Wegener's granulomatosis 620
plasma renin, *see* renin, plasma renin activity
plasma skimming, non-homogeneous renal oxygenation 990–1
plasma volume,
 decrease, pregnancy-induced hypertension 1961
 disease and, elderly patients 65
 regulation 264–5
 young and elderly subjects *66*
 see also blood volume; fluid volume; renal plasma flow
plasmapheresis, *see* plasma exchange
plasminogen activator,
 ionizing radiation causing reduction 849
 role in glomerular sclerosis 1202–3
Plasmodium,
 B cell mitogen 249
 prescribing 185
Plasmodium berghei 493, 495
Plasmodium cynomolgi 495
Plasmodium falciparum **493–6**
 glomerulonephritis 467
Plasmodium malariae **493–6**
 infection-associated glomerulonephritis 466–7
 membranous nephropathy, tropical areas 375
Plasmodium vivax,
 glomerulonephritis 467
plasticizers, phthalate-containing, haemodialysis tubing 1332
platelet activating factor 584
 antihypertensive polar renomedullary lipid and 2064–5
 arteriolar myocytes of afferent arterioles, effects 26
 cyclosporin, nephrotoxicity 1561
 neutrophil and free radical-mediated injury 1052
platelet aggregating factor,
 E. coli verotoxin and 1053

platelet aggregating factor (*cont.*):
 endothelial injury and 1053
 renal circulation, thrombocytopenia 1053
platelet-derived growth factor,
 glomerular sclerosis, role 1202–3
 glomerulonephritis-mediating 357
 progressive glomerular scarring 1192
platelets,
 abnormalities, management 1022
 activation, mesangiocapillary glomerulonephritis 395
 defects
  haemodialysis and 1361
  uraemic haemorrhage 1361
 endothelial injury and **1053–5**
 fibrinolysis, pregnancy-induced hypertension 1964–5
 haemolytic–uraemic syndrome 1054
 hyperkalaemia, spurious 907
 increased aggregation, in sickle-cell disease 715
 nephrotic syndrome, alterations in function 282–3
 normal values *1344*
 parathormone and 1361
 pregnancy, consumption 1958
 pregnancy-induced hypertension 1964–5
 role in glomerular sclerosis 1202–3
 transfusions 1362
 uraemic abnormalities, bleeding diathesis in 2344
pleuropericardial disease 2339
plumbism, *see* lead intoxication
pluricystic hypoplasia, children 2169
pneumococcal infections, haemolytic–uraemic syndrome 1049
pneumococcus pneumonia 460
 in sickle-cell anaemia 704
Pneumocystis carinii infection,
 co-trimoxazole, prophylaxis in renal transplantation 1533
 following renal transplantation 1579
 management 1535, 1536
 trimethoprim, nephrotoxicity 1527
pneumonia,
 following renal transplantation, management 1535–6
 see also Pneumocystis carinii infection
podocalyxin, luminal membrane, capillaries of the glomerulus 132, 133
podocytes,
 cytoskeleton 133
 EM detail 131, 132–4
 filtration slits on endothelial cells of glomerular capillaries 132
 foot process effacement, diagnostic relevance 150
 fusion, glomerular damage 229
 injury and mesangial proliferation 313
 luminal membrane, glycocalyx 132–3
 membranous nephropathy 372–4

Index

POEMS syndrome, Crow–Fukase syndrome 2285
poisoning, see intoxications
polkissen, defined 783
polyacrylonitrile membranes, haemodialysis 1315
polyamines,
  markers of renal cell carcinoma 2242
  toxicity, uraemic syndrome 1244
polyangiitis overlap syndrome 585
polyanions and cations, peritoneal dialysis solution 1481
polyarteritis 585, *586*, 590, 613
  microscopic, see microscopic polyarteritis
  serum C reactive protein 1167
  see also necrotizing polyarteritides; vasculitis, granulomatous
polyarteritis nodosa **628–9**
  Churg–Strauss vasculitis 2324
  clinical features 628, 1068
  definition 628
  diagnosis 628
  HBsAg in 587, 590
  hepatitis B infection 462, 463
    glomerulonephritis 463–4
  historical aspects and terminology 612, 613
  immunological abnormalities *590*
  outcome 628–9
  pregnancy 1943
  treatment 625–6, 628, 1069
polyarthropathy, chronic, complication of renal disease 2297–8
polychondritis, relapsing, see relapsing polychondritis
polyclonal activation, see autoimmune glomerulonephritis
polyclonal immune globulins, action 1556–8
  immunosuppressive therapy in renal transplantation *1556*
  over-immunosuppression 1557
  prophylaxis of renal transplantation 1557
  unpredictability 1557
polycystic disease,
  defined *2029*
  features and associated abnormalities *2036*
  gout and, urate clearance reduced 831
  haematuria 75
  hepatic cysts, computed tomography scanning 101
  hypertension in 1198
  infected renal cyst 1696
  multicystic dysplastic disease 91
  pregnancy *1939*, 1942
  radiological tests *72*
  renal biopsy contraindication 143
  renal failure, allopurinol/non-allopurinol treated patients 835
  renal transplantation, nephrectomy and 1516
  tuberosclerosis 2300
  ultrasonography 75, 90–1
polycystic disease, adult, in childhood 1607–8

polycystic disease, adult dominant, computed tomography scanning 99–100
  early features 90–1
polycystic disease, autosomal dominant **2171–85**
  animal models 2183
  antenatal counselling 2182
  assessment 1154–5
  basement membrane gene expression 2183–4
  cancer in 2174
  cardiac valvular abnormalities 2178
  cerebral aneurism 2324
  children **2167–70**
    antenatal diagnosis 2168
    clinical features 2167–8
    defined 2167
    differential diagnosis 2168–9
    genetics 2168
    incidence 2143
    prognosis 2168
  chronic renal failure 1154–5
  classification *2181*
  defined 2167
  diagnosis 2179–80
    genetic linkage markers 2159
    mutations 2159–60
  differential diagnosis, autosomal recessive polycystic disease 2181
  differential diagnosis in adults
    genetic cystic disease of the kidney 2180–1
    non-genetic cystic disease of the kidney 2180
  end-stage renal failure 1233
    proportional distribution by age at commencement *1229*
  erythrocytosis 2176
  erythropoietin, deficiency 1349
  erythropoietin production 1349
  extra-renal manifestations 2176–9
  fever, causes *2172*
  genetic counselling 2182
  genetic linkage studies 2179–80
    HVR marker 2179–80
  haemodialysis/CAPD 2184
  hypertension in 2174–5
    renal failure 2176
  hyperuricaemia and gout 2176
  imaging 2179
  infections 2172–3
  intracranial aneurysm 2177–8
  liver complications 2185
  liver involvement 2176–7
  molecular genetics
    genetic heterogeneity 2158
    genetic linkage markers in diagnosis 2159
    mutations, diagnosis 2159–60
    reverse genetics 2157–60
  nephrogenic diabetes insipidus 887
  non-renal complications 2185
  orofacialdigital syndrome 2301 type I 2181–2
  pathophysiology 2182–4
  pre-calyceal tubular ectasia, and renal calculous disease 2173
  prevalence 2171
  renal complications **2171–6**, 2184
  renal failure
    chronic 2175–6

polycystic disease, autosomal dominant (*cont.*):
  renal failure (*cont.*):
    hypertension effects 2176
    male sex 2176
    pregnancy 2176
    progression, determinants and pattern 2175–6
    urinary tract infections 2176
  renal replacement therapy 2184–5
  screening 2176
  tuberous sclerosis 2181
  see also cyst(s), renal
polycystic disease, autosomal recessive,
  antenatal diagnosis 2167
  children **2163–7**
    clinical features 2163–4
    defined 2163
    liver involvement 2164–6
    prognosis 2166
  differential diagnosis, polycystic disease, autosomal dominant 2181
  genetics 2166–7
  pathology 2166
polycystic disease, infantile 1607
  cystotubular ectasia 1607
  prenatal diagnosis 91
polycythaemia,
  congenital cyanotic heart disease 2277
  in renal impairment 2344
  transplantation patients 2345
polydipsia, see polyuria, polydipsia and
polyhydramnion, Bartter's syndrome 782
polymorphonuclear leucocytes, bladder 1671
polymyalgia rheumatica, features in SLE 650
  renal failure in 630
polymyositis **684–5**
  differential diagnosis 2283
  primary adult idiopathic 684
  Sjögren's syndrome association 694
polymyxins, nephrotoxicity 164, 994, 1010
polyneuropathy,
  familial amyloidotic *2325*
  uraemic 2327–9
    haemodialysis 2328
polyol pathway, in diabetic nephropathy pathogenesis 518
polyomavirus infection, renal transplantation 1581
polypeptide hormones, renal catabolism 1317–18
polysulphone haemodialysis, comparisons *1247*
polyuria **796–7**
  acute renal failure established and recovering 1000–1
  tropics 1129
  arginine vasopressin function *798*
  assessment 883–4
  Bartter's syndrome 782–7
  causes 4
  children with uraemia, indomethacin 1305
  clinical approach 884
  differential diagnosis 796, *884*, 884

polyuria (*cont.*):
  Fanconi syndrome 727
  hypokalaemia-induced 901
  investigations
    direct test 797
    indirect test 796
    therapeutic trial 797
  osmotic and nonosmotic states 791
  polydipsia and **796–797**
    children with uraemia, indomethacin 1305
    differential diagnosis 856
    hypokalaemia-induced 901
    lithium toxicity nephrogenic diabetes insipidus 853–4
    primary, differential diagnosis 884
    psychogenic 874
    differential diagnosis 856
  urine osmolality, and plasma vasopressin 797
  washout effect 796
  water deprivation, response *885*
  see also diabetes insipidus, nephrogenic
polyuronic acid, phosphate binding 1448
Pompe's disease, enzyme defect *2210*
Ponceau S method, detection of urine protein 227, 228
porphyria, acute intermittent porphyria 2285
porphyria cutanea tarda, iron overload 1282
  phlebotomy 1392–3
  photosensitizers 1392
porphyrins, levels, bullous disorders 1392, 1393
portal hypertension, beta blockade 1540
positron emission tomography, renal blood flow assessment 2106
postcapillary venules, vasculitis 583, 585
  wall 583
posterior urethral valves, urinary tract obstruction 2032–3
post-hypercapnic alkalosis, metabolic alkalosis 947
post-hypocapnic acidosis 935
postoperative state, inappropriate secretion of ADH 878
post-partum haemolytic syndrome 1934
postrenal acute renal failure 1101
  neonates 1121
poststreptococcal endocapillary glomerulonephritis **405–11**
  characteristic immunohistologic patterns 147, 149
  clinical features 458, 458–9
  children and elderly patients *407*
  crescentic glomerulonephritis 421
    treatment 430–1
  epidemic 458
  garland pattern 411, 412
  glomerular disease in the tropics 488
  immune deposits *414*
  management 459
  nephritic syndrome 266
  nephritis in 459
  pathology 411, 412, 413

poststreptococcal endocapillary
    glomerulonephritis (cont.):
  prognosis 459
  renal fuctional reserve *415*
  starry sky immune deposition 412
  treatment 412–14
  see also endocapillary
    glomerulonephritis, acute
postural hypotension, *see*
  hypotension, postural
postural proteinuria 5
posture, sodium handling in
  pregnancy and 1920
potassium,
  administration, children with
    chronic renal failure 1306
  age and body K^+ *895*
  clearance rates, Fanconi
    syndrome 726
  defective tubular secretion 910
  deficiency
    hypercalcaemia 1884–5
    inappropriate secretion of
      ADH 879
    laxative abuse 167
  depletion
    acute renal failure, recovery 980
    Bartter's syndrome 787
    chronic renal failure 1183
    clinical sequelae *900*
    diarrhoeal loss 934
    distal renal tubular acidosis
      772–3
    elderly patients 64
    Fanconi syndrome 726–7
    hypertension 1177
    'idiopathic oedema syndrome' 6
    renal acidification 765
    sodium retention 273
    ureteral–gastrointestinal
      linkage 935
    see also hypokalaemia;
      potassium, deficiency (above)
  depletion, extrarenal
    diarrhoea 903
    dietary 899–900, 903
    exercise 903
    geophagia 903
    laxative abuse 903
    metabolic alkalosis 901, 949
    villous adenoma of rectum 903
  depletion, renal aetiology 903–6
    acute leukaemia 906
    aminoglycoside-induced 906
    Bartter's syndrome 904–5
    chloride depletion 904
    diabetic ketoacidosis 903–4
    glucocorticoid excess 906
    magnesium, depletion 906
    mineralocorticoid excess 905–6
    penicillin-induced 906
    renal tubular acidosis 903
  depletion, renal effects 901
  dietary intake
    elderly patients 64
    oliguric patients 1019
    total parenteral nutrition in
      ARF *1021*
  elderly patients 64
  excessive retention 908
  excretion
    defective in sickle-cell anaemia
      711
    hypercalcaemia causing 1885
    normal in sickle-cell trait 711
    in pregnancy 1917

potassium (cont.):
  handling, nephron 897
  homeostasis **895–900**
    aldosterone effects 767–8
    body K^+ distribution 895
    gastrointestinal transport 897
    hypomagnesaemia and 1811–12
    intracellular K^+ regulation 895
    transcellular K^+ distribution
      895–7
    acid–base status 896
    α-adrenergic agonists 896
    β-2-adrenergic agonists 896
    aldosterone 896
  inability to conserve 64
  large bowel secretion, renal
    insufficiency 1282
  maintenance therapy, diuretic-
    treated patients 904
potassium-sparing drugs
  Bartter's syndrome 787
  hyperkalaemic renal tubular
    acidosis 776
  potassium-wasting renal defect
    785
  renal tubular acidosis 774
  pregnancy, animals as gestational
    models 1920–1
  red cells, elderly patients 64
  renal handling, young and elderly
    subjects *64*
  replacement
    chloride, bicarbonate
      (precursors), phosphates 906
    rate 906
  transcellular K^+ distribution
    catecholamines 896
    cell K^+ content 897
    exercise 897
    factors *895*
    osmolality 896–7
    pancreatic hormones 896
    plasma 898
  transport 897–900
    α-adrenergic agonists 899
    β-2-adrenergic agonists 899
    aldosterone 898–9
    ammonia 899
    anions 899
    arginine vasopressin 899
    Bartter's syndrome and 784
    dietary K^+ 899–900
    flow rate 898
    glucocorticoids 899
    plasma K^+ concentration 898
    sodium delivery 898
    systemic pH 899
    transepithelial potential
      difference 899
  uptake, insulin-stimulated 1253
  urinary acidification 768
  see also hyperkalaemia;
    hypokalaemia
potassium bicarbonate (precursors),
  replacement therapy 906
potassium canrenoate,
  gynaecomastia 213
potassium chloride,
  hypokalaemia treatment 210
  replacement therapy 906
potassium citrate,
  children 1714–15
  oral administration 1834

potassium perchlorate, drug-
  induced nephropathy 163
potassium permanganate, in
  amyloidosis 553, 554–5
potassium phosphate, replacement
  therapy 906
potassium-losing diuretics,
  combined therapy 201
potassium-sparing diuretics 199,
  200
  ascites 203
  combined therapy 201
  contraindications in diabetes
    1647
  hyperkalaemia 210
    hypoaldosteronism 910
  hypertension of end-stage renal
    failure 2121
  interactions, angiotensin
    converting-enzyme inhibitors
    208
  interactions with NSAIDs 208
  metabolic alkalosis 947
  renal hypertension, treatment
    2052
Potter syndrome,
  children 1613
  facies, in bilateral renal agenesis
    2028
PRA (plasma renin activity), *see*
  renin
pravastatin, nephrotic
  hyperlipidaemia, effects 288
praziquantel,
  bilharzia 185
  schistosomiasis 185, 1738
prazosin,
  haemodialysis patients,
    recommendations *1463*
  hypertension 188
  percentage protein binding
    according to renal function
    1240
  prostatitis 1690
  see also $α_1$ adrenoceptor agonists
precancerous and cancerous lesions,
  chronic renal failure 1393
prednisolone,
  advantages, minimal change
    nephrotic syndrome 314
  antiglomerular basement
    membrane disease 1065
  crescentic glomerulonephritis 429
  cyclosporin therapy in renal
    transplantation, interactions
    1563
  handling in kidney disease 190
  in Henoch–Schönlein purpura 607
  membranous nephropathy 383–5
  mesangiocapillary
    glomerulonephritis 400–1
  nephrotic syndrome, remission of
    minimal change nephrotic
    syndrome 270
  rejection of allograft
    with OKT3 m a treatment
      1530–1
    'pulse' technique, treatment
      1530
  renal transplantation, optimal
    regimen 1590
  in sarcoidosis 579
  in Wegener's granulomatosis 619
  see also methylprednisolone

prednisone,
  advantages, minimal change
    nephrotic syndrome 314
  alternate-day course in nephrotic
    syndrome 144
  handling in kidney disease 190
  in lupus nephritis 659, 660
    avoidance of high-dose 660
    cytotoxic agents comparison
      660–1
    in pregnancy 651
    recommended strategy 663
  membranous nephropathy,
    controlled studies *382*, 382–4
  renal failure risk 661
  renal transplantation,
    preconception counselling 1949
  in sarcoidosis 579
pre-eclampsia, *see* eclampsia and
  pre-eclampsia
pregnancy **1909–80**
  acute fatty liver 1057, **1933–4**
  acute renal failure **1056–7**
  animal models
    normal renal haemodynamic
      changes 1911–13
    normal renal physiological
      changes 1911–15, 1920–2
    normal renal tubular function
      1920–1
    normal renal vasculature
      function 1913–15
    volume homeostasis 1921–2
  assessment of renal function
    1921–2
  asymptomatic bacteriuria 1684
    prevalence 1993
  in sickle-cell disease 710
  bed rest, studies 1969
  chronic renal disease **1936–48**
    antenatal strategy and decision-
      making 1938
    blood pressure 1938, *1939*
    fetal surveillance and timing of
      delivery 1938–9
    renal biopsy, role 1939–40
    summary *1939*
  chronic renal failure 1187, 1337–8
    prospects *1937*
  diabetes insipidus 886
  diabetic nephropathy 530–1
  gestational hormones, renal
    dilatation and 1914
  GFR, normal elevation 29
  glomerulonephritis, herpesvirus
    varicellae 2304–5
  glycosuria 745
  haematuria 1930–1
  haemodialysis 1337–8
  hydramnios 1338
  haemolytic–uraemic syndrome,
    postpartum 1057
  hypertension 651
    hypertensive stress test 1959
    and pre-eclampsia, renal
      disorders, differential
      diagnosis 1940–1
  hypocapnia, chronic 955
  hypouricaemia 832
  IgA nephropathy 343
  iliac sign 1909
  IVU
    contraindications 81
    indications 1683

Index

pregnancy (*cont.*):
  longterm effects **1946**
    animal studies **1946–8**
  mixed metabolic and respiratory disorders *957*
  nephrotic syndrome 1946
  normal osmoregulation 1917–18
  normal renal haemodynamic changes **1910–13**
    animal models 1911–13
    human pregnancy 1910–11
  normal renal physiological changes **1909–23**
    clinical relevance 1922–3
  normal renal tubular function **1915–17, 1920–1**
    animal models 1920–1
    human pregnancy 1915–17
  normal renal vasculature function, animal models 1913–15
  normal structural changes, urinary tract 1909–10
  obstruction of ureter(s) 2004
  peritoneal dialysis 1338, **1948–9**
  polycystic disease, autosomal dominant 2176
  pre-eclampsia 531, 1993
    and transplant rejection episodes 1338–9
  pregnancy test, β-1-glycoprotein 1337
  prepregnancy assessment of renal function *1937*
  reflux nephropathy 1993
    management 1998
  renal autoregulation 1913
  renal blood flow 272
  renal complications **1928–35**
    urinary tract infections 1928–30
  renal function, prepregnancy assessment and *1937*
  renal function impairment **1937–48**
    mild/moderate 1937
    severe 1938–40
  renal function tests 1922
  renal transplantation and 1338–9, 1591
    infections 1339
    post-partum transplant rejection 1339
    pre-eclampsia 1338
    time interval 1338
  renin activity 2071
  secondary fluid retention 272–3
  SLE and, *see* lupus nephritis
  toxaemia
    hyperuricaemia 832
    nephrocalcinosis in 1888
    secondary fluid retention 273
  tropics, acute renal failure and 1138
  ultrasound studies **1114–15**
  underlying renal disease **1936–52**
    advisibility of conception 1936, 1952
    CAPD patients 1949
    chronic renal disease 1936–48
    haemodialysis patients 1948–9
    renal transplantation 1949–52
  urinary tract infection, susceptibility factors 1670
  volume homeostasis **1918–20, 1921–2**
    animal models 1921–2

pregnancy (*cont.*):
  volume homeostasis (*cont.*):
    human pregnancy 1918–20
    X-ray exposure of embryo *in utero* 81–2
  see also fetoplacental unit; fetus; pregnancy-induced hypertension; *specific substances*
pregnancy-induced hypertension 651, **1956–74**
  chronic hypertension *vs* 1958
  classification 1956
  clinical features 1958
  defined 1956
  diagnostic criteria 1956–8
  and differential diagnosis 1940–1
  differentiation from pre-eclampsia 1956
  eclampsia *vs* 1958
  epidemiology
    genetics 1959
    outcome for fetus 1960
    outcome for mother 1959
    risk factors 1958–9
  haemodialysis 1948
  haemodynamic status 1973
  'hypertensive stress test' 1959
  incidence 1959
  management **1966–72**
    anti-platelet therapy 1966–8
    first, second and third line therapy 1969
  maternal age 1959
  MHC HLA DR4 1959
  mild renal disease *1938*, 1938, *1939*
  pathology **1960–6**
    brain pathology 1962
    cardiovascular and renin–angiotensin systems 1962–3
    coagulation, fibrinolysis, and platelets 1964–5
    liver and HELLP syndrome 1961
    neutrophils 1965–6
    prostacyclin and vascular endothelium 1965
    renal pathology 1960–1
    uteroplacental bed 1960
  pre-eclampsia, fulminating 1972–65
  prophylaxis
    aspirin low-dose 1967–8
    roll-over test 1967
  risks, animal models 1913
  significant threshold values of BP 1957
  see also eclampsia
prenatal diagnosis, polycystic disease 91
pre-pro-collagen 670
prerenal failure, defined 1099
  see also acute renal failure, prerenal failure
prerenal lesions, neonates 1120
prescleroderma 667, 669
pressure flow studies, neonates 2027
preterm infants, *see* neonates, preterm
priapism,
  in haemodialysis patients 1335–6
  SKATT therapy 1335

primodone, handling in kidney disease 186–7
primrose oil, dietary and pharmacological lipid manipulations *1202*
probenecid,
  action 190
  drug-induced nephropathy 163
  fractional urate clearance 827
  urate excretion increase 829
    pyrazinamide suppression test 826
probucol,
  in chronic renal failure 1181
  nephrotic hyperlipidaemia, effects 288
  triglyceride lowering 1258
procainamide,
  lupus induced by 651
  ventricular arrhythmias 187
prochlorperazine, handling in kidney disease 187
procollagen molecules 670
proforma, *appendix*, glomerular disease 236–7
progesterone,
  animals as gestational models 1921
  levels in uraemia 1333
  pregnancy
    desoxy-corticosteroid-induced potassium excretion abolished 1917
    sodium excretion and 1919
    renal haemodynamics, rat models 1914
progressive, creatinine clearance 531, 1910–11, *1938*, 1961
progressive chronic renal failure **1191–217**
  determinants of progressive renal scarring 1192–206
    glomerular scarring 1192–203
    tubulointerstitial scarring 1203–6
    vascular sclerosis 1206
  hypertension and **2083–6**
  modulating factors of progressive renal scarring 1206–9
  proteinuria 1192
  therapeutic implications 1210–17
progressive glomerulonephritis, *see* rapidly progressive glomerulonephritis
progressive renal failure, early, hyperuricaemia and 834–5
  renal handling of urate/uric acid 831
  urate clearance reduced 831
  urate excretion per nephron enhanced 831
proguanil, handling in kidney disease 185
prolactin,
  animals as gestational models
    chronic hyperprolactinaemia 1914
    effects 1921
    pseudopregnancy, ovine model 1914
    renal haemodynamics, rat models 1914
  circadian rhythm 1333
  effects, sexual dysfunction 1333

prolactin (*cont.*):
  haemodialysis, female/male patients 1333, 1337
  inhibition, bromocriptine 1335
  uraemic patients, sexual dysfunction 1333
  see also hyperprolactinaemia
proliferative endarteritis, untreated hypertension 2129
proliferative glomerulonephritis, characteristic immunohistologic patterns 147, 149
  in rheumatoid arthritis 680
  see also glomerulonephritis; rapidly progressive glomerulonephritis
proliferative retinopathy, *see* diabetic retinopathy
propantheline, contraindications 2008
propionic acidaemia, ketoacidosis 930
propionylhydroxamic acid, infected stone disease 1863
propofol, handling in kidney disease 185
propranolol,
  elimination by kidney 187
  haemodialysis patients, recommendations *1463*
  haemofiltration/haemodialysis 180
  neonates 1116
  percentage protein binding according to renal function *1240*
  platelet effects and therapy 1968
  properties, in relation to use in impaired renal function *2053*
prostacyclin,
  animal models of prevention of acute renal failure 1018
  antiplatelet therapy 1966
  antithrombosis 1428
  decreased secretion, renin deficiency 909
  deficiency
    in pregnancy-induced hypertension 1964–5
    antiplatelet therapy 1966
  endothelial effects 1965
  infusion, treatment of acute tubular necrosis 1055
  pregnancy-induced hypertension 1964–5
  in sickle-cell disease 715
  vascular endothelium and, pregnancy-induced hypertension 1965
  see also prostaglandin I_2
prostacyclin stimulating factor, defined 1053
  failure in haemolytic–uraemic syndrome 1053
  neonates 1053
prostaglandin E_1, cyclosporin therapy in renal transplantation 1561
prostaglandin E_2,
  arteriolar myocytes of afferent arterioles, effects 26
  chronic renal failure and 1217
  compensatory hypertrophy in progressive chronic renal failure 1204

Index

prostaglandin E_2 (*cont.*):
  endocapillary glomerulonephritis, acute 407–8
  excretion, in sickle-cell disease 712, 715
  increased excretion, Bartter's syndrome 783, 785
  increased systemic activity hypokalaemia 787
  marker of renal cell carcinoma 2242
  pregnancy, increase 1920
  renal effects 820, *821*
  synthesis 820
  urinary, cyclosporin therapy in renal transplantation 1561
prostaglandin $F_{1\alpha}$,
  (6-keto) 513
  cyclosporin therapy in renal transplantation 1561
prostaglandin $F_{2\alpha}$,
  excretion, in sickle-cell disease 712, 715
  increased excretion, Bartter's syndrome 783
  potassium intake 907
prostaglandin F_2,
  endocapillary glomerulonephritis, acute 407
  renal effects 820, *821*
  synthesis 820
prostaglandin G_2,
  renal effects 820, *821*
  synthesis 820
prostaglandin I_2,
  arteriolar myocytes of afferent arterioles, effects *26*
  bone remodelling *1790*
  cyclosporin therapy in renal transplantation 1561
  pregnancy-induced hypertension 1963
  renal effects 820, *821*
  see also prostacyclin
prostaglandin I_2 synthetase 820
prostaglandin synthetase inhibitors,
  prerenal failure 987
  renal stone disease 1853
  renal tubular absorption enhancement, growth effects 1305
  susceptibility in patients receiving cyclosporin 988
  see also indomethacin
prostaglandins,
  animal models 1217
  Bartter's syndrome 905
  biochemistry 819–20
  chronic renal failure 821, 1217
  cirrhosis 820–1
  end-stage renal failure, hypertension of 2120
  excess, postulation in Bartter's syndrome 785
  heart failure 821
  increased excretion, Bartter's syndrome 783
  inhibition
    antinatriuretic effects 707, 714
    glomerular filtration rate decrease 513, 713, 715
    see also indomethacin
  interactions with diuretics 208
  mesangial cells 136

prostaglandins (*cont.*):
  metabolic abnormalities, hepatorenal syndrome 821
  pregnancy 1962
    renal dilatation 1914
  prostaglandin hydroperoxidase-mediated co-oxidative action of analgesics 810
  renal
    potassium depletion 901–2
    production stimuli 715
    renal vascular control, mediation 984
  renal actions 820, *821*
  renal depressor system, role 1462
  renin/angiotensin system, pregnancy and 1920
  in sickle-cell disease 715
    diluting capacity and 707
    increased synthesis 707, 709, 711, 713, 715
    renal blood flow and 711, 712, 713, 715
  sodium depletion 821
  sodium reabsorption 63
  synthesis 256–7
    in glucagon effect on renal haemodynamics 513
    inhibition and prerenal failure 987
    opposing afferent constriction 263
    in sickle-cell disease 707, 709, 711, 713, 715
  vasodilatation 159
  cyclosporin-induced decrease in activity 1561
prostate carcinoma,
  chromosomal rearrangements 2161
  osteosclerotic metastases 1776
prostatic enlargement 1690
  benign, ultrasonography 92
  drug-induced 67
  elderly patients, renal replacement therapy 1622–3
  frequency and 2
  micturition disorders 4
  symptoms 1149
  urinary obstruction, benign cause 476
  see also urinary tract obstruction, lower tract
prostatitis,
  acute bacterial 1689
  chronic bacterial 1689
  defined 1688
  diagnosis 1688–9
  non-bacterial 1689
  pathogenesis 1688
  split urine sample, collection 1688
  treatment 1689–90
    surgery 1690
prostatodynia 1690
protein C,
  deficiency
    calciphylaxis 1377
    skin disorders 1393
  nephrotic syndrome 282
  pregnancy-induced hypertension 1964
protein kinase C, proximal tubule cells, effect on 1204
protein S, nephrotic syndrome 282

proteinase, in Wegener's granulomatosis pathogenesis 592
proteins,
  binding
    competition for sites 1240, 1243
    renal impairment, drug malabsorption *176*
  catabolism, interpretaion 238
  children
    protein:energy ratio, CRF 1616–17
    requirements and safe levels *1307*, 1607, *1616*
  degradation
    acute renal failure 1020
    following nephrectomy 1259
    parathyroid hormone 1259–60
  dietary increase, animal models 1210–11
  dietary intake
    acute renal failure *1020*
    calcium stone disease 1840–1
    children's requirements *1307*, 1607, *1616*
    chronic renal failure *1307*
    creatinine 1175
    creatinine clearance 29
    estimation 1181
    excessive intake and stone formation 8
    GFR elevation 29
    malnutrition, nephrotic syndrome 291
    RDA *1307*, 1607
    renal blood flow 26
    see also diet, animal proteins
  dietary restriction **1178–81**
    adverse effects 1181
    attenuation of progression of symptoms 1179
    background 1178
    clinical approach 1180
    diabetic nephropathy treatment 520, 528
    dietician 1181
    extent 1180
    follow up 1181
    institution 1181
    low intake regimes, diabetic nephropathy 1644
    mechanisms of action in diabetics 529
    protective effects 1210–11
    type of protein 1210
    underlying renal disease 1179
    uraemic symptoms, relief 1178
    wasting syndrome 1284
  dietary restriction, animal models
    of adaptive haemodynamic change in chronic renal ablation 1193
    chronic renal failure 1210–11
    enzymes in uraemic models 1281
    progressive glomerular scarring *1209*
    renal ablation models *1195*
    various nephropathies *1210*
  dysproteinaemias 2301
  elderly patients on haemodialysis, rda 1623
  glycosylation 515, 517–18
  isoelectric points, renal damage and 996

proteins (*cont.*):
  low in diet, see proteins, dietary restriction
  malnutrition, nephrotic syndrome 291
  metabolism
    chronic renal failure 1303
    disorders of **1258–60**
    disturbance in uraemia 1259–60
    uraemic liver 1283
  molecular size, and GFR 28
  plasma, minimal change nephrotic syndrome 302
  sialoproteins, podocalyxin 132, 133
  synthesis, uraemia 1259
  transport proteins 289–90
  tuberoglomerular feedback 29
  type, chronic renal failure 1180
  in urine
    tests available *228*
    see also proteinuria
  see also diet; *specific nutrients* and *trace elements*
proteinuria **227–36**
  24-h collection, normal limits 227
  accelerated hypertension 2131
  albumin, and other plasma proteins 238
  algorithm for screening 232
  in amyloidosis 551, 556
    AL (primary) 551, 572
    in rheumatoid arthritis 678, 680
  analgesic nephropathy 813
  angiotensin converting-enzyme inhibitors 1214
  animal models
    GFR and increased glomerular permeability 839
    glomerular scarring 1197–8
  antiglomerular basement membrane antibodies 255
  asymptomatic 5
    renal biopsy indications 145
  Bence–Jones, see Bence–Jones protein
  benign, defined 17
  cadmium nephropathy 841
  cell-mediated glomerulonephritis 255
  classification 17
  congenital cyanotic heart disease 2277
  creatinine clearance and 30
  detection 227, *228*
  diabetes types I and II, and chronic renal failure *1636*
  in diabetic nephropathy 505
    in definition of late phase 505, 510
    end-stage renal disease frequency and 506, 507, 508
    end-stage renal failure onset time 511
    insulin-dependent 506
    non-insulin-dependent 507, 508
    pathophysiology 516
    prevalence 505, 507, 526
    raised blood pressure relationship 511
    size selectivity defect 516
    survival 506
  in diabetic vs. non-diabetic renal disease 526
  drug-induced glomerulopathy 234

Vol. 1 – Sections 1–5. Vol. 2 – Sections 6–11. Vol. 3 – Sections 12–21.

Index

proteinuria (*cont.*):
 eclampsia and pre-eclampsia 1940
  differentiation from pregnancy-induced hypertension 1956
  significant threshold value 1957
 in essential mixed cryoglobulinaemia 690
 experimental aminoglycoside nephrosis 269
 Fanconi syndrome 727, 734
 fenoprofen-induced tubulointerstitial nephritis 821
 glomerular, defined 17, 18
 glomerular causes **227–36**
  clinical syndrome in relation to history of glomerular disease 235
  detection and analysis 227
  Fanconi syndrome 727
  follow-up 235
  haematuria and *233*, 233–4
  infection-associated 234
  neoplastic disease 234–5
  oedema and *233*
  orthostatic 227–8
  overflow 228, *229*
  paraprotein-associated 234
  pathological
   damaged glomerular filter 229–30
   defined 228
   various diseases 230
  postural 232
  source 227
  symptomless isolated 232–3
  tubular dysfunction 228–9
 glomerular sclerosis, initiation 1197–8
 in gold and penicillamine nephropathy 682, 683
 haematuria and 233
 health, variations in loss 227
 in Henoch–Schönlein purpura 605
 IgA nephropathy 344–5
  primary 350
 intermittent, abandonment of concept 510
 isolated symptomless 232
 in juvenile rheumatoid arthritis 681
 lipoprotein metabolism 238
 low molecular weight, cadmium nephropathy 842
 in lupus nephritis 654
 malignancy-associated glomerulonephritis 478
 markers of tubular damage 227, 228–9
 massive
  amyloidosis 2294
  in obesity 2280
  renal transplantation, recurrent nephrotic syndrome 326
 massive transient, in sarcoidosis 580
 mechanisms 308
 microscopic haematuria and, IgA nephropathy 341
 minimal change nephrotic syndrome 301–2
  proteins contained 302
  monomeric light chains, tubular catabolism *229*, 238–9
 nephrotic range and mean value 301

proteinuria (*cont.*):
 with oedema 233
  and visible haematuria 233
 origins 1197
 orthostatic (postural) proteinuria 5, 227, 232
 overload, defined 17
 and plasma exchange 1531
 pregnancy 531
  changes 1922
  false-negative/false-positive results 1957
  massive 1937
  significant threshold value 1957
 prerenal, Fanconi syndrome 727
 progressive chronic renal failure 1192
 protein/creatinine ratio 18
 qualitative analysis 18
 recurrent focal glomerular sclerosis, in renal transplantation 1531
 reflux nephropathy, prognosis 1993
 remission in measles 457
 in renal papillary necrosis 530
 renal pathology 5
 renal transplantation, nephrotic syndrome 326
 in scleroderma 673
 screening algorithm 232
 selective tubular 18, 567
  in amyloidosis 551
  in multiple myeloma 563
 and serum complement, glomerulonephritis-type diseases 408
 in sickle-cell disease 711
 stix test 5
 sulphosalicylic acid test 5
 symptomless isolated 232
 tests available *228*
 tubular, defined 17, 18
 tubular damage, causes *229*
 tubulointerstitial scarring, effects on 1206
 variations in loss in health 227
proteoglycans 671
 biosynthesis and functions 671
 decreased synthesis, polycystic disease, autosomal dominant 2183
 glomerular basement membrane 135
Proteus,
 infected stone disease 1831, 1863
 mannose-resistant fimbriae 1667
 perinephric abscess 1696
 renal calculus disease, urinary tract infection 1692
 and urease 1666
 urinary tract infection 1661
  children 1701
 xanthogranulomatous pyelonephritis 1694
Proteus mirabilis,
 urease 1666
 urinary tract infection 1661
Proteus vulgaris,
 urease 1666
 urinary tract infection 1661
proto-oncogenes *c-myc* and *c-fos*, adaptive renal hyperplasia 1193
proton pumps, *see* Na^+-K^+-ATPase pump

protons, permeability defect leak-back, renal tubular acidosis *938*
proximal diuretics 197–9
 carbonic anhydrase inhibitors 197–9
proximal renal tubular acidosis, *see* renal tubular acidosis, proximal
proximal renal tubule, *see* renal tubule, proximal
prune belly syndrome 2034, *2039*
prurigo nodularis 1391
pruritus,
 chronic renal failure 1391–2
 uraemia 1157, 1158
 vitamin A involvement 1391
Pseudoallescheria, infection 1748
pseudoaneurysms, angiomyolipoma 2243
pseudocavities, renal tuberculosis 1728
pseudogout,
 complication of renal disease 2296
 in osteitis fibrosa 1376
pseudohermaphroditism, with nephropathy, Drash syndrome 2219
pseudohypertension,
 defined, Osler's sign 1646
 Osler's manoeuvre 11
pseudohypoaldosteronism, types I & II 207
pseudohypoparathyroidism,
 type 1A 1798
 type 1B 1799
pseudo-Kaposi's sarcoma in CRF 1393
Pseudomonas aerogenosa,
 ciprofloxacin 183
 infected stone disease 1831
 urinary tract infection 1661
Pseudomonas pseudomallei,
 melioidosis, acute renal failure **1132–3**
pseudophlorizin diabetes syndrome 731
pseudopolycythaemia, transplantation patients 2345
pseudoporphyria cutanea, chronic renal failure 1392–3
pseudopregnancy,
 animal gestational model
  GFR increase, initiation 1914
  renal dilatation 1914
pseudotumours,
 DMSA scans 126
 ultrasonography 90
pseudoxanthoma elasticum, renal artery abnormalities 2301
psoralen, drug-induced nephropathy 163
psoriasis,
 acute uric acid nephropathy *832*
 photochemotherapy 2305
psoriatic arthritis, IgA nephropathy 361
psychiatric disturbance,
 elderly patients
  electrolyte changes 65–6
  electrolyte levels and 63–6
psychogenic polydipsia 874
 differential diagnosis 856
psychological aspects
 of haemodialysis **1465–75**
 of uraemia 1400

psychosocial problems, elderly patients with chronic renal failure 1624
PTFE graft,
 seroma 1415
 vascular access method 1409
ptopic kidney(s), hypertension and 2089
PTRA, *see* percutaneous transluminal renal angioplasty
PTTI, *see* renal parenchymal transit time index
pubertal and developmental problems, children **1613–16**
puberty,
 delayed
  chronic renal failure 1302
  growth measurement 1299–300
 endocrine changes and growth 1297–8
 normal growth during 1295–6, 1298–300
 sexual maturation 1298, 1301
 stages *1301*
pulmonary artery thrombosis, nephrotic syndrome 280
pulmonary calcifications, in renal disease 2338
pulmonary congestion, left ventricular hypertrophy, uraemic patient 1272
pulmonary disease, in Wegener's granulomatosis 615–16
pulmonary embolism,
 mixed metabolic and respiratory disorders *957*
 prevalence in nephrotics 278
pulmonary haemorrhage,
 Goodpasture's disease 439, 441–2
 pathogenesis 447–8
 in Henoch–Schönlein purpura 598
 in microscopic polyarteritis 624
 pulmonary–renal syndrome 1061
 in Wegener's granulomatosis 615
pulmonary infection,
 renal transplantation 1535–6, 1578–9
 see also Pneumocystis carinii infection
pulmonary oedema,
 diuretics and 202
 mixed metabolic and respiratory disorders *957*
 nephritic syndrome 266
 pregnancy, causes 1957
 pulmonary infection following renal transplantation, management 1535–6
 renal failure, chronic 272
pulmonary–renal syndrome 439
 differential diagnosis *2336*
 pulmonary haemorrhage 1061
 in renal disease 2338
purines,
 dietary, uric acid stones and 1862
 nucleotides, dephosphorylation 829, 830
 toxicity, uraemic syndrome 1244
puromycin nephropathy,
 animal models 1195
 glomerular damage 229, 1197
 functional and structural parameters *1194*

puromycin nephrosis,
  animal models
    dietary protein chronic
      restriction *1210*
    haemodynamic and structural
      glomerular parameters *1214*
  glomerular scarring, progressive
    chronic renal failure 229, 1197
purpura,
  chronic relapsing, and polyclonal
    gammaglobulinaemia 2301
  hyperglobulinaemia 2301
putative virulence factors, produced
  by *Escherichia coli 1669*
putrescine, renal cell carcinoma
  2242
PVG/c rats,
  adaptive glomerular hypertrophy
    1196
  progressive age-related glomerular
    lesions 1206–7
  renal ablation studies 1193
pyelitis, emphysematous 532
pyelocalyceal urothelial tumours,
  computed tomography scanning
  106
pyelography,
  antegrade, indications 2268
  retrograde 1726
  filling defects 2268
  translumbar 1725
  upper urinary tract obstruction
    2008
pyelonephritis,
  acute
    children 1705, 1706
    clinical features 1691
    histopathology 1691
    pregnancy and 1930, 1933
    respiratory distress, with/
      without, pregnancy *1930*
    septicaemia 1090
    urinary tract infection 1690–1
  acute bacterial, adenosine
    deaminase binding protein
    assay 19
  candidosis 1743–4
  chronic
    children 2142
      incidence 2143
    end-stage renal failure and
      scarring 119–20
    hypertension and 2089
    pathogenesis 1990
    urinary tract infection 1691–2
  chronic renal failure and *1159*
  computed tomography scanning
    104–5
  cryptococcal 1746
  diagnosis, EDTA registry 1230–1
  emphysematous 531–2
  end-stage renal failure,
    proportional distribution by age
      at commencement *1229*
  experimental models 1669–70
  pregnancy *1939*, 1941
  radiological tests 71
  in sickle-cell disease 710
  suppurative, in diabetes 71
  ultrasonography 93
  xanthogranulomatous 103, 105,
    1694
    children 1706
    clinical features 1694

pyelonephritis (*cont.*):
  xanthogranulomatous (*cont.*):
    diagnosis 1694
    pathology 1694
    treatment 1694
    ultrasonography 92
  see also pyonephrosis
pyeloureteric junction, stricture in
  renal tuberculosis 1725, 1727
pyonephrosis,
  diagnosis, ultrasonography 91–2
  see also pyelonephritis
pyrazinamide,
  drug dosage, renal impairment
    184
  inhibition of tubular secretion of
    urate/uric acid 826
  renal tuberculosis *1727*
  suppression test 826, 827
    advanced renal failure 831
    fractional urate clearance *827*
    urate reabsorption *829*, 829
pyrazolone,
  kidney damage 810
  as single substance, and analgesic
    nephropathy 808, 816
pyridoxine,
  deficiency
    children with chronic renal
      failure 1308
    malabsorption syndrome 1857
    nephrocalcinosis animal models
      1887
  handling in renal impairment 184
  penicillamine and, in cystinuria
    748
  renal transplantation, tuberculosis
    treatment 1518
  total parenteral nutrition in ARF
    1021
pyrithioxine, rheumatoid arthritis,
  and drug-induced nephropathy
  162
pyrophosphate,
  excretion, and phosphate 1834–5
  inhibitor of calcium oxalate
    crystallization in urine *1828*
  inhibitor of calcium phosphate
    crystallization in urine *1827*
pyruvate, uptake, uraemic muscle
  1253
pyruvate dehydrogenase,
  dichloroacetate, stimulation 928
pyuria,
  defined 1677
  demonstration 1680
  diagnoses 1693
  microscopy 1680
  sterile 5

Quality Adjusted Life Years, renal
  transplantation *vs*
  haemodialysis 1591
Queensland nephritis, lead
  intoxication-associated 838, 839
quinidine,
  drug-induced nephropathy 163
  ventricular arrhythmias 187
Quinke's oedema, contrast media
  injection 77
quinolones,
  CAPD peritonitis regimens *1489*
  drug-induced lithiasis 168

quinolones (*cont.*):
  drug-related interstitial nephritis
    1089
  schistosomiasis 1738

r-HuEPO, *see* erythropoietin,
  human recombinant
race-related factors, nephropathy
  susceptibility 1207
radiation nephritis 849–51
  acute, chronic *vs* 851
  benign hypertension in 851
  classification 850–1
  clinical manifestations 850–1
  criteria 2319
  hypertension and 2089
  malignant hypertension in 851
  management 851
  pathogenesis 849
  pathology 849–50
  prevention 851
radiation nephropathy 477
radical surgery, renal mass 2246–7
radiocontrast media 76–85
  acute on chronic renal failure 1186
  acute renal failure 996, 1009–10,
    1185, 2334
    case histories 1038
    diabetics 1638
    prevention *1019*
    risk factors *1010*
  acute uric acid nephropathy 832
  allergic accidents 77
  amount, proportionate to renal
    function 1638
  anti/procoagulant properties 77
  antibodies 79
  availability 77
  chronic renal failure 1185
  contrast nephropathy, computed
    tomography scanning 106
  drug-induced acute toxic tubular
    nephropathy 164
  handling in kidney disease 191
  historical note 76
  hyperosmolar/low-osmolality 77
  intra-arterial injection 111
  intravenous injection 111–12
  ionic/non-ionic 77
  nephrotoxicity 78–9
    cardiovascular complications 77
    lethal accidents 77
    minor complications 77
    pathophysiology 77–8
    prevention using saline solution
      79
    recommendations for
      prevention 79
  new agents 79
  Thorotrast, drug-induced acute
    toxic tubular nephropathy 164
  time intervals between injections
    78
  toxicity, in diabetes 532
  urate excretion increase 829
radioimmunoassays, adenosine
  deaminase binding protein 19
radioisotopic methods,
  elderly patients, glomerular
    filtration rate 60
  glomerular filtration rate 37–40
  elderly patients 60
  nuclear imaging **116–18**
  uptake and output components
    118

radioisotopic methods (*cont.*):
  see also scintigraphy
radiological tests,
  conventional **76–85**
  haematuria 75–6
  hydronephrosis 74
  rationale **70–6**
  renal failure 72–3
  renal insufficiency 1163
  unilateral kidney failure 70–2, *73*
  urinary tract infections 73–4
radiolucent stones, *see* uric acid
  stones; cystine stones;
  dihydroxyadenine stones;
  xanthine stones
radionuclide angiography, uraemic
  patient with cardiac problems
  1272
radionuclide micturating
  cystourethrography 1987
radionuclide nephography,
  diuresis-enhanced, possible
  responses 2027
  neonates 2026
radionuclides,
  radiopharmaceuticals, *see*
  radioisotopic methods
  renal mass 2245
radiotherapy, renal mass 2247
Randall's plaques,
  carbonate apatite 1896
  nephrocalcinosis 1896
ranitidine, handling in kidney
  disease 189
rapidly progressive
  glomerulonephritis,
  acute renal failure **1060–4**
  assessment 1150, 1154
  blood tests *1062*
  causes 439
  CMV infection, gancyclovir 1064
  defined 438, 1060–1
  diagnosis 1061–3
  differential diagnosis 1061
  disorders causing *1061*, 1061
  haematuria 233
  in Henoch–Schönlein purpura
    597, 607
  idiopathic
    diagnosis 1066
    risk factors for poor prognosis
      628
    treatment 627
    treatment and outcome *1067*
  management 1063–4
    approach *1064*
  plasma exchange effects *1063*
  other diseases and 1070
  outcome *1068*
  secondary 1069–70
  systemic vasculitis-associated
    1065–8
  treatment, plasma exchange *1068*
  see also crescentic
    glomerulonephritis;
    Goodpasture's syndrome;
    progressive glomerulonephritis
rapidly progressive osteoporosis,
  nephrocalcinosis 1891
rash, in Henoch–Schönlein
  purpura 595, 596
rats,
  Hantavirus disease carriers 1143
  laboratory diet, mineral content
    1887

Index

rats (*cont.*):
  nephrocalcinosis, comparisons with man 1887
  see also animal models; *specific genetic strains*
RCC, *see* renal cell carcinoma
receptors and potential receptors, MR fimbriae 1668
rectal biopsy, in amyloidosis 554, 555
red eye,
  uraemia 1793
  see also calcifications, ectopic distribution
red eye syndrome, children 1611
refeeding alkalosis, metabolic alkalosis 949
reflux nephropathy **1983–99**
  animal models 1990
  assessment 1154
  children
    chronic pyelonephritis and end-stage renal failure 119–20
    hypertension in 2142
    incidence 2143
  classification 1989, 1990
  clinical presentations *1991*, 1991–6
  congenital, nuclear imaging 120
  DMSA scan 119
  hypertension 1992–3, 2089
    management 1998
  imaging
    computed tomography scanning 1989
    intravenous urography 1988
    radionuclide scanning 1989
    ultrasound 1989
  measurement of divided renal function 119
  neonates 1606
  parenchymal scarring, grading 1988
  pathogenesis 1990
  pregnancy *1939*, 1941
    management 1998
  progression to end-stage renal failure 1984
  renal papillary necrosis, differential diagnosis 815
  urinary calculi 1995
  urinary tract infection 1692–3
  see also vesicoureteric reflux
Refsum's disease, renal tubular changes 2212
regulator proteins,
  nephrotic syndrome *281*, 282
  list and molecular weights *281*
  see also coagulation factors
Reiter's syndrome, IgA nephropathy 2295
rejection of allograft,
  accelerated rejection 1570
    defined 1528
    differential diagnosis 1571
  acute interstitial infiltration 149
  acute rejection **1528–31, 1538, 1570–1**
    clinical features 1570–1
    defined 1528
    diagnosis **1528–30**
    management 1571
    treatment **1530–1, 1554–6**
    see also rejection of allograft, hyperacute rejection

rejection of allograft (*cont.*):
  acute vascular 97
  adenosine deaminase binding protein assay 19
  aetiology of rejection 1538
    drug toxicity 1538–9
    fever of unknown origin 1453
    see also rejection of allograft, immunology; *specific toxic substances*
  blocking T cells, experimental solutions 1558
  chronic rejection
    antibodies 1547
    creatinine criteria 1538
    differential diagnosis 1572
    hypertension-related causes 1536, 1596
    hypertensive vs normotensive recipients 1596
    management 1538, 1572
    pathology 1538, 1571–2
    starting point 1538
  chronic renal failure 1538
  classification 1528
  cytology/pathology
    immunofluorescence studies 1530
    interstitial lesions 1530
  cytomegalovirus-induced renal dysfunction 1532
  diagnosis **1528–30**
    colour Doppler ultrasound 1529
    cyclosporin concentration 1529
    cytology/pathology 1529–30
    early 1571
    immunological monitoring 1529
    radiology 1529
    radionuclides 127–8
    renal biopsy 1571
    renal function 1529
  differential diagnosis 1571, 1572
  effector phase of immune response 1545–6
  encephalopathy 2330
  failure rates 1509
  histological signs 1538, 1570–2
  hyperacute rejection 1524, 1547, 1570
    defined 1528
    preformed antibodies 1547
  immunology **1543–54**, 1570
    T cells activation 1551–4
  immunosuppressive therapy **1554–8**
  mechanisms **1551–4**
  non-renal complications 1539–41
  OKT3 monoclonal antibodies 2332
  pregnancy and 1950
  specificity, non-specific effector cells 1546, 1547
  treatment **1530–1**
    effects on colon 1281
    multiple drug protocols 1556
  ultrasonography 97
relapsing polychondritis 591, 630
  ear lesions 2303
  rapidly progressive glomerulonephritis and 1070
  renal effects 2335
remnant kidney model,
  antihypertensive therapy, role in glomerular sclerosis 1212–13
  see also kidney(s), remnant

renal ablation, animal models,
  adaptive haemodynamic changes 1193–4
  anticoagulants, effects *1216*
  compensatory hypertrophy 1193–4
  dietary and pharmacological intervention *1195*
  dietary protein restriction *1210*
  functional and structural parameters *1194*
  haemodynamic and structural glomerular parameters *1214*
  hormone manipulations 1208, *1209*
renal abscess 1691
  acute parenchymal, vs renal cell carcinoma 103
  computed tomography scanning 105
  cortical 1695–6, 1705–6
  perinephric 531, 1696, 1706
  renal biopsy contraindication 143
  ultrasonography 93
  vs renal cyst, ultrasonography 93
  see also abscess, renal
renal acidification,
  bicarbonate reabsorption 764
  bicarbonate regulation 764–5
  chloride depletion 765
  extracellular fluid volume 765
  physiology 764–8
  potassium depletion 765
renal acidosis, *see* metabolic acidosis; renal tubular acidosis
renal adenocarcinoma, *see* renal cell carcinoma
renal adenoma,
  defined 2239
  in dialysis patients 1450–1
renal agenesis, *see* kidney(s), absence; kidney(s), congenital abnormalities
renal angiography **111–15**, 2050
  arteriography, renovascular hypertension 2102–3
  contrast-associated nephropathy 78
  digital subtraction angiography 2245–6
  assessment 975
  captopril and *2147*
  essential hypertension vs renal causes 2050
  image collecting 112
  intravenous 112
    essential hypertension vs renal causes 2050
    ischaemic renal disease *1080*, 1080–1
    renal mass 2245–6
  renovascular hypertension 2103
  selective renal angiography 2245
  vascular grafts 1411
  see also arteriography
renal aplasia, defined *2029*
renal arterial disease,
  chronic renal failure in elderly patients 1622
  contrast-associated nephropathy 78
  embolism
    acute, and thrombosis 2110

renal arterial disease (*cont.*):
  embolism (*cont.*):
    acute loss of perfusion, radionuclide studies 125
    fibromuscular dysplasia 1078
    haemolytic–uraemic syndrome 1051
  ischaemic **1077–83**
    animal models 992
    circulatory shock, paradoxical susceptibility 990
    countercurrent arrangement of vessels 986
    distribution of cell injury, selective cellular susceptibility 992–3
    non-ischaemic injury and, differentiation 990
    pathogenesis of renal damage 990–1
    selective relief 992–3
    see also renal arterial disease; renal artery stenosis; renal circulation
  ischaemic disease **1077–83**
    angiography 1080–1
    clinical presentation 1078–80
    diagnostic procedures 1080–1
    epidemiology and aetiology 1077–8
    management 1081–3
  loin pain–haematuria syndrome 2277
  occlusive disease
    acute 2110–12
    acute dissection 2112
    chronic 2110
    indications for operative management *2109*
    problems of invasive treatment 2110–12
    symptoms, diagnosis and management *2111*
  pathology in accelerated hypertension 2127–8
renal arteries,
  abnormalities, pseudoxanthoma elasticum and 2301
  afferent arterioles
    accelerated hypertension 2127
    ageing changes 58
    arteriolar fibroid necrosis of accelerated hypertension 2131, 2132
    arteriolar medial hypertrophy 2128
    capillaries of the glomerulus 129, 130, 131
    resistance and vascular tone 25
    vascular sclerosis, progressive chronic renal failure 1206
    vasoactive substances, effects 26
  ageing changes 58
  aneurysm, renal biopsy contraindication 143
  arteriography **111–14**
  endoprosthesis 115
  endovascular interventional radiology 113–14
  fibromuscular dysplasia 1078, 2097–8
  interlobular arteries, accelerated hypertension 2127
  intra-arterial injection 111
  intravenous digital angiography 112

Index

renal arteries (*cont.*):
  intravenous injection 111–12
  IVU *vs* Doppler sonography 96
  percutaneous transluminal
      angioplasty 114, 2106–7
  pulsatility index
    defined 88
    Doppler findings in diffuse
        parenchymal renal disease 94
    in renal vein thrombosis 96
  renal resistance 985
  retrograde femoral
      catheterization 111
  right, stenosis in sarcoidosis 581
  selective catheterization
    PTA 114
    renal embolization 113
  stenosis, *see* renal artery stenosis
  supplying tumour,
      ultrasonography 90
  wall
    integration of signals 985
    vascular tone 985
  see also renal artery stenosis ;
      renal blood flow; renal
      circulation
renal artery, *see* renal arteries
renal artery occlusive disease, *see
  also* renal arterial disease; renal
  artery sclerosis
renal artery sclerosis, ischaemic
  disease **1077–83**
renal artery stenosis **2096–112**
  analgesic nephropathy 813
  angioplasty, predicting outcome
      122
  assessment 1151
  atherosclerosis in 2097–8
  children 2143
    differential diagnosis 2144
  clinical features *2101*
  defined 2096
  diabetes, malignant hypertension
      1646
  in diabetics 1638
  diagnosis 2101–4
    diagnostic approaches 2105
    differential diagnosis 2098
  Doppler sonography 96
  epidemiology and aetiology
      2096–7
  erythropoietin
    animal models 1350
    deficiency 1350
    production 1350
  natural history 2098–9
    atherosclerosis 2098–9
    fibromuscular dysplasia 2099
    progression *2100*
  pathogenesis 2097–8
  pathophysiology 2099–101
  presentation 2102
  prevalence 2096
  PTA 114
  radiology 2098–9
  radionuclide studies 121
  renal haemodynamic adaptation
      2054
  renal transplantation 97
    angiotensin-converting enzyme
        inhibitors 1527–8
    diagnosis 1598–600
    hypertension mechanism 1595
    prevalence 1595

renal artery stenosis (*cont.*):
  renal transplantation (*cont.*):
    stenosis of transplant kidney
        1536, 1575
    treatment 2105–12
      angiotensin converting-enzyme
          inhibitors, effects **986–7**
      nephrectomy indications 2108
      percutaneous transluminal renal
          angioplasty 2106–7
      problems of invasive treatment
          2110–12
      reconstruction 2108–10, *2111*
    urate clearance reduced 831
    without associated hypertension
        58
  see also hypertension,
      renovascular; renal arterial
      disease, occlusive disease
renal artery thrombosis,
  following renal transplantation
      1526–7, 1574–5
  treatment and prophylaxis 284
renal artery to renal vein transit
    time, renal radionuclide studies
    118
renal autoregulation 1015
  pregnancy, animal models 1913
renal blood flow **24–6, 39–40**
  action of prostaglandins, NSAIDs
      and 820, *821*
  acute tubular necrosis, vasomotor
      nephropathy 999
  calculated filtration fraction,
      increase with age 58
  diminished, in scleroderma 673
  distribution 24–5
  impairment, congestive heart
      failure 272
  intrapelvic pressure and 1985
  PAH clearance 39
  pregnancy 272
  prostaglandin synthesis inhibition,
      mechanism of papillary damage
      by analgesics/NSAIDs 811
  reduced, NSAIDs and 821
  regulation of vascular tone 25–6
  renal artery to renal vein transit
      time 118
  resistance within kidney 24
  vascular resistance, increase with
      age 58
  other influences 26
renal bone disease, *see* bone
    disease, chronic renal failure;
    specific disorders
renal calculous disease, *see* renal
    stones; renal stone disease
renal calyces,
  'bunch of grapes' appearance,
      medullary sponge kidney 2287
  dilatation, increase in pregnancy
      1909–10
  megacalyx 2017
  see also calyces
renal carbuncle 1695–6
  children 1706
renal catabolism, polypeptide
    hormones 1317–18
renal cell carcinoma **2237–51**
  aetiology 2237–9
  amyloidosis 478
  biological markers 2242–3
  children 2258
  chromosomal rearrangements
      2161

renal cell carcinoma (*cont.*):
  chromosome abnormalities,
      associated syndromes *2041*
  clear cell type 2239
  clinical features 2240–2, *2292*
    amyloidosis 2242
    erythrocytosis 2242
    haematological syndromes
        2241–2
    haematuria 2240
    hepatic dysfunction 2242
    hypercalcaemia 2242
    hypertension 2242
    metabolic syndromes 2241
    non-urological symptoms 2241–2
    pain 2240–1
    palpable mass 2241
  computed tomography scanning
      100–2
  demographic factors 2237
  diagnostic work-up 2244–6
  in dialysis patients 1450–1, 1454
  differential diagnosis
    angiomyolipoma 2243
    leiomyosarcoma 2243–4
    renal cyst 88–90
    renal oncocytoma 2243
  environmental factors 2237
  epidemiology and aetiology
      2237–9
    aetiological factors 2237–8
    chemical agents 2238
    hormones 2238
    physical agents 2238
    viruses 2238
  erythropoietin, deficiency 1350
  ESR 2242
  granular cell type 2239
  growth patterns 102
  and haematuria, analgesic abuse
      1454
  hormone-dependence 2238
  hypertension and 2093
  hypervascularization 101
  hypovascularization 101–2
  incidence and prevalence 2237
  local and regional extension 102
  male:female ratio 2237
  management 101–2
  metastases
    chemotherapy 2248
    dissemination routes 2239–40
    embolization 2247–8
    hormone therapy 2248
    immunotherapy 2248–9
    interferon therapy 2249
    surgical approach 2247
  molecular genetics 2161–2, 2239
  MRI 107–8
  pathology 2239–40, *2240*
  prevalence 2237
  prognosis 2249–51
  renal transplantation-transmitted
      1582
  renal tubular epithelial antigen
      478
  sarcomatoid type 2239
  small cell type 2239
  staging *102*, 2240
    prognostic factors 2249, *2250*
  treatment 2246–9
  ultrasonography 88–90
  varicocele of acute onset 2241
  viruses 2238–9

renal cell carcinoma (*cont.*):
  in von Hippel–Lindau disease
      2161–2, 2300
renal circulation,
  collateral circulation 1077
  cytokine-mediated injury 1052
  endothelial injury, animal models
      1052–3
  hypertension effects **2075–81**
  injury **1051–6**
    factor VIII:von Willebrand
        factor index 1053
    investigations 1054–5
    pathogenesis 1051–4
    treatment and outcome 1055–6
  see also haemolytic–uraemic
      syndrome
  ischaemic disease **1077–83**
    clinical presentation 1078–80
    diagnostic procedures 1080–1
    epidemiology and aetiology
        1077–8
    management 1081–3
  microscopic vasculitis, crescentic
      glomerulonephritis 420–1, 429
  neutrophil and free radical-
      mediated injury 1052
  physiology
    glomerular filtration 982–3
    glomerular and tubular blood
        supply 985–6
    renal autoregulation 983
    renal vascular control,
        mediation 984–5
    tubuloglomerular feedback
        983–4
    vessel wall signals, integration
        985
  renal artery to renal vein transit
      time 118
  responsivity, pregnant animal
      models 1913–14
  thrombocytopenia 1053
  vascular resistance 985–6
    increase, cyclosporin therapy
        1560
  vasodilatation, responsivity,
      pregnant animal models
      1913–14
  see also renal arteries; renal blood
      flow; renal perfusion; *specific
      conditions*
renal clearance, *see* glomerular
  filtration; *specific substances*
renal concentrating capacity,
    impaired
  sickle-cell disease 704–7, 716
    indomethacin action 706–7
    microradioangiography 706,
        708
    transfusions effect on 705
  Sjögren's syndrome 696
  mechanism 706
  see also urine, osmolality
renal corpuscle, *see* glomerulus
renal cortex,
  analgesic nephropathy, changes
      with progressive exposure 812
  cortical abscess 1695–6, 1705–6
  cortical necrosis, pathogenesis
      993
  cortical perfusion, decreased, in
      scleroderma 674
renal cyst, *see* cyst(s), renal;
    polycystic disease

Index

renal depressor system, role 1462
renal dilatation, pseudopregnancy, rat models 1914
renal disorders,
  cardiorespiratory syndromes **2336–9**
  cardiovascular disease **2334–6**
  clinical presentation, asymptomatic/symptomatic **3–6**
  dysplasia
    clinical features 2029–30
    defined *2029*
  haematological aspects **2344–6**
  haematological complications of **2340–6**
  molecular genetics **2155–62**
  thrombotic complications of 2345
renal embolization,
  aims 113–14
  'medical nephrectomy' 113
  risks 113–14
renal enlargement, *see* kidney(s), size; renal hypertrophy
renal failure,
  early, urate clearance reduced 831
  hyponatraemia 878
  in sarcoid granulomatous interstitial nephritis 578
  in scleroderma 673
  see also acute on chronic renal failure; acute renal failure; chronic renal failure; end-stage renal failure
  progressive renal failure
renal function,
  function testing **24–44**
  functional reserve, loss in diabetes mellitus 516
  haemodynamics 263–4
  see also renal tubules, function
renal glomerulus, *see* glomerulus
renal glycosuria, *see* glycosurias
renal hamartoma, hypertension and 2092–3
renal histology, *see* kidney(s), histology
renal hypertension 119–22, **2096–112**
  see also hypertension, renal; renal arterial disease; renal artery stenosis;
renal hypertrophy,
  in diabetes mellitus 509
    in glomerulosclerosis
      pathogenesis 520
      pathology 523
      pathophysiology 514
      sodium–hydrogen antiport controlling 521
    granulomatous interstitial nephritis 578
  see also kidney(s), size
renal hypoplasia,
  clinical features 2029
  defined *2029*
  diagnosis 1606
  management 1606–7
renal infarction,
  computed tomography scanning 76, 104–5
  CT 76
  DMSA scans 126
  radiological tests *72*

renal infarction (*cont.*):
  retroperitoneal fluid collections, computed tomography scanning 107
renal infections, *see* pyelonephritis; *specific causative organisms*; urinary tract infections
renal insufficiency,
  in amyloidosis 555, 558
  analgesic nephropathy 813
  chronic, indications for renal biopsy 145
  in sickle-cell disease 711
  in SLE 649
renal interstitium, accelerated hypertension 2127
renal malformations **2035–41**
renal mass,
  children 1710–11
  erythropoietin, production and deficiency 1350
  parenchymatous, computed tomography scanning 99–103
renal mass, diagnosis,
  algorithm for diagnosis 2244
  arteriography 113
  computed tomography 2246
  diagnostic work-up 2244–6
    plain X-ray 2244
  excretory urography 2244
  IVU imaging 83
  magnetic resonance imaging 2246
  mimicking renal cell carcinoma 2242
  nephrotomography 2244
  nuclear imaging 126–8
  radical surgery 2246–7
  radionuclide scanning 2245
  radiotherapy 2247
  single photon renal scintigraphy 2245
  ultrasonography 88–90, 2245
  see also cysts, renal; renal infarctions; *specific lesions*; tumours; tumours, *specific names, organs and regions*
renal medulla,
  free radicals, oxidative action of analgesics 810
  glutathione content 810
  intracellular scavenger mechanism 810
  prostaglandin hydroperoxidase-mediated co-oxidative action of analgesics 810
  sickle-cell disease
    inner medullary defects 705, 706, 707
    ischaemic damage 706, 713
    pathology 713
renal microscopic vasculitis,
  crescentic glomerulonephritis 420–1
  treatment 429
renal multicystic dysplasia, defined *2029*
renal nerves, functions 2066
renal obstruction,
  obstructive uropathy, renal papillary necrosis in analgesic nephropathy 815, 816
  ultrasonography 975–6
  see also urinary tract obstruction
renal oncocytoma,
  arteriographic signs 2243
  metastasis 2243

renal osteodystrophy,
  analgesic nephropathy 813
  CAPD 1482–4
  children 1611–13
    treatment 1304–5
  lead mobilization test and 834
  prevention and treatment 1613
  surgery, contraindications 1613
  see also aluminium-related osteopathy; hyperparathyroid bone disease; osteitis fibrosa
renal papilla,
  calcifications 1890
  pathogenesis of damage 810–11
    genetic predisposition 811
    haemodynamic factors 811
renal papillary blush, medullary sponge kidney 2286, 2287
renal papillary necrosis,
  acute, diabetes mellitus 530, 532
  acute on chronic renal failure 1160
  analgesic nephropathy
    aspirin 684
    changes with progressive exposure 812
    prevalence in autopsy studies 807
    radiological signs 814
  calcifications 1895
    analgesic nephropathy 813
    calcified sloughed papilla 2003, 2010
  chronic renal failure 1160
  diabetes 1160
  diagnosis 1160
  non-steroidal anti-inflammatory drugs 822
  paroxysmal nocturnal haemoglobinuria 2280
  pathogenesis
    analgesic synergism 810
    combination toxicity 810
    radiological signs, analgesic nephropathy 814
  sickle-cell disease 706, 710, 1160
    frequency 710
  vs nephrocalcinosis, ultrasonography 95
renal parenchymal disease,
  abscess 103
  clear cell carcinoma 101–2
  computed tomography scanning 106
  diffuse, type 1 renal disease, ultrasonography 93–4
  graft disease, hypertension, diagnosis 1595, *1599*
  infections, computed tomography scanning 104–5
  lesions, neonates 1113, 1121
  other tumours 102–3
  scarring, detection of focal abnormalities, radionuclide studies 119–20
  unilateral
    aetiology *2088*
    diagnosis *2088*
    hypertension and **2088–94**
renal parenchymal impulse retention fraction, defined 117
renal parenchymal transit time index, defined 116

renal pelvis,
  capillary sclerosis, unidentified cause 811
  dilatation, increase in pregnancy 1909–10
  fibrotic thickening 1723
  laminar flow, residence time 117
  pelvic transit time (PVTT)
    defined 117
    diuretic renography 117
  stones, removal 1872–3
renal pelvis and ureter tumours **2266–71**
  aetiology 2266
  histological grading *2267*
  incidence 2266
  investigations 2267–8
  Jewitt system for classification of tumours 2268
  staging 2267
  treatment 2269
    summary of data on surgery *2270*
  surgery 2269
renal perfusion,
  assessment 975
  autoregulation 983
  countercurrent arrangement of vessels, and ischaemia 986
  first-pass, radionuclide studies 125
  interstitial pressure 983
  investigations 975
  myogenic hypothesis 983
  renal resistance 985
  sepsis 980
  see also renal blood flow; renal plasma flow; renal vascular supply
renal perinephric abscess 1696
renal plasma flow,
  congenital cyanotic heart disease 2277
  in early diabetes mellitus 508, 512
  effective renal plasma flow, pregnancy-induced increase 1919
  increase, rat models 1913–14
  measurement by nuclear imaging 116–17
  plasma volume expansion, gestational animal models 1915
  pregnancy
    effective renal plasma flow 1910
    GFR and 1911
    increase/decrease during 1910
    rat models 1913
    renal autoregulation 1913
  in sickle-cell disease 711, 713, 714, 715
  see also renal blood flow
renal pyramids, gas bubbles along 531
renal radionuclide study, *see* radionuclides; renography; scintigraphy
renal replacement therapy,
  acute renal failure **1026–33**
  children, mortality causes *1610*
  chronic renal failure, *see* children; chronic renal failure; diabetes; elderly patients
  EDTA–European Renal Association registry *1624*
  classification 1228–30
  proportional distribution by age at commencement *1229*

renal replacement therapy (*cont.*):
  elderly patients **1627–32**
  end-stage nephropathy, prevalence *1995*
  immune response in 1315
  in patients with cancer 2322
  proportional distribution of primary renal diseases, by age at commencement of treatment *1232*
  side-effects mimicking uraemic toxicity 1240
  uraemic syndrome 1240
  women, causes of end-stage renal failure *1232*
  see also haemodialysis; peritoneal dialysis; renal transplantation
renal scarring, children 2142
renal sinus,
  calculi, ultrasonography 94
  gas *vs* staghorn calculi, ultrasonography 95
  lipomatosis, computed tomography scanning 106
  transitional cell carcinoma *vs* calculi 94
  ultrasonography 87, 90, 94
  see also collecting system
renal sodium depletion, *see* sodium, depletion
renal stone disease,
  allopurinol 1853
  analgesic nephropathy 813
  biochemical investigations 1849–51
  biochemical risk factors 1851
  cadmium nephropathy 843
  children 1847
  chronic diarrhoea and 2310
  diet 1849
  differential diagnosis
    transitional cell carcinoma, ultrasonography 89–90
    tumour 94
  distal renal tubular acidosis 772
  drug-induced lithiasis 167–8
  epidemiology 1847
  history-taking 1848
  hyperuricosuria 832
  male/female incidence 2003
  management 2009–10
    medical management **1847–66**
    surgical management **1869–81**
  measurement of divided renal function 119
  medullary sponge kidney 2287
  obstruction, emergencies 2010
  occupation 1849
  paraplegia 2276
  polycystic disease, autosomal dominant 2173
  prostaglandin I_2 synthetase inhibitors 1853
  proximal renal tubular acidosis 769
  reflux nephropathy 1995
  renal transplantation, following 1575
  sarcoidosis 579
  schistosomiasis 1736
  travel 1848
  triamterene-induced 213
  types of disease **1829–38**
    historical factors 1822
    percentages *1849*
    world-wide examples *1822*

renal stone disease (*cont.*):
  ultrasonography 89–90, 94
  water intake 1849
renal stones,
  aetiological factors **1822–43**
  chemical analysis 1848, 1849
  children 1861
  composition 1848
  crystallization **1825–9**
    crystalluria, causes 1825–9
    increased supersaturation 1826
    inhibitors of calcium oxalate *1828*
    inhibitors of calcium phosphate *1827*
    inhibitors, decreased excretion 1826–9
    promotors, increased excretion 1828
  formation **1822–9**
    composite theories 1829
    crystal growth factor 1829
    crystallization 1825–9
    formation product ratio–activity product ratio score 1829
    saturation-inhibitor index 1829
    theories 1823–5
    urinary tract infection, organisms 1661
  geographical factors *1822*
  historical aspects 1864
  idiopathic hypercalciuria 1891
  impaction 1875
    sites 2003
  infrared spectroscopy 1848
  lithotripsy
    complications of 1875–6
    infections 1876–9
    size 1875
  obstruction, drainage of system 2010
  prevention of regrowth 1666
  radiology 1848
  removal **1869–81**
    complications of 1872–3
    contraindications 1872
    disintegration 1871–2
    endoscopy 1871–2
    radiography 1872
    ultrasound breaking 1872
    see also endoscopy
  removal *vs* nephrectomy, renography 119
  spontaneously radio-opaque, plain films 79
  staghorn calculi
    radiological tests 72
    ultrasonography 94–5
  types, percentages *1849*
  ureteric, radiological tests 73
  X-ray crystallography 1848
  see also renal stone disease; *specific stone types*
renal surgery, percutaneous, *see* percutaneous transluminal renal surgery
renal transplantation **1509–601**
  ABO incompatible kidneys, living donors 1517
  acute interstitial nephritis 1091
  acute pancreatitis 1285
  acute tubular necrosis 1523–4
  acute rejection 149

renal transplantation (*cont.*):
  adrenal insufficiency 1325
  allograft rejection, *see* rejection of allograft
  allograft survival, and patient survival *1637*
  Alport's syndrome 2201, 2204
    autoimmune disease 2201
  amyloidosis 559–60
    graft survival rates 559
    recurrence in allografts 559–60
  anaemia, causes 2345
  analgesic nephropathy, prognosis 815, 816
  angiodysplasia in 1281
  antibiotics 1517–18
    immediate peri-operative administration 1522
  anticoagulation 1518
  assessment of complications
    indications for renal biopsy 145–6, *146*
    radionuclide studies 126–8
    surgical 97
    ultrasonography 96–7
    vascular 97
  biochemical measurements, normo/hypercalcaemic recipients 1775
  blood group A recipients, ABO autoimmune haemolytic anaemia 2345
  blood pressure control 2060–1
  'blood pressure follows kidney' concept 2065
  blood transfusion 1589–90
    donor-specific 1590
    pre-transplant 1589
    timing 1590
  bone complications 1585
    avascular osteonecrosis 1537
    hyperparathyroidism 1537
  bowel obstruction/perforation 1285
  cadaver grafts 1548–9, 1586–7
  cancer, *see* renal transplantation, malignancy
  cardiac function 1577
  children 1511–12, 1591
    access 1605
    catch-up growth 1302
    corticosteroid therapy effects on growth 1308
    cystinosis 1513
    graft survival 1618
    hypertension and 2144–5, *2145*
    patient survival 1618
  cholestasis 2313
  chronic renal failure, criteria 1188
  CMV infection 464
  complications following
    bone problems **1378–80**
    calcium metabolism 1378–80
    cardiac changes 1267
    hepatitis 1291–2
    hypercalcaemia, renal transient *vs* static 1379
    mineral metabolism 1169–379
    osteopenia 1380
    paraplegia 2276
    sexual function 1336
    uraemia-induced cardiac changes, reversal 1267

renal transplantation (*cont.*):
  complications following (*cont.*):
    urological problems 1512, 1525–6, 1575, 1579, 2018
  contraception 1952
  crossmatching serum/donor lymphocytes 1570
    B and T cell positive crossmatch 1549
    outcome 1549–50
    successful transplantation 1549
    see also renal transplantation, hyperacute rejection
  cystinosis 2223
  de novo membranous nephropathy 375–6
  diabetes **1650**
    chronic renal failure 1650
    diabetic nephropathy 1650
    with pancreas transplantation 1650
    peripheral vascular disease 1649
    steroids, effects 1650
  diabetic nephropathy 1512, 1515
  diagnosis of recurring diseases 1153–5
  dialysis
    cerebrovascular accident during, mortality 1576
    peritoneal dialysis patients 1495, 1511
    post surgery 1522
    pre surgery 1518
    see also types of dialysis
  donated kidney
    ischaemic storage 1524
    see also kidney donors
  drug nephrotoxicity
    summary 1527–8
    see also specific substances
  drug non-compliance, withdrawal effects 1538
  elderly patients 1512, **1627–32**
    as cadaver donors 1631–2
    donor age and time to kidney function *1632*
    longterm graft survival *1632*
    results *1631*
    failure, causes and recipient age *1630*
    five-year percentage survival *1630*
    immunological system and immunosuppression
      macrophage function 1629
      MHC antigens and immune response 1629–30
      T cell function 1629
    living related donors 1628–9
    one-year graft survival *1632*
    outcome 1630–1
    selection of patients 1628
      Europe and USA 1628
    summary and conclusions 1631–2
  erythropoietin
    deficiency 1349
    production 1349–50
    recombinant human, anaemia of renal failure 1518
  failure, vesicoureteric junction antireflux procedure 1984
  Fanconi syndrome 734
  fluid replacement 1521
  gastrointestinal effects 1285, 1582
    bleeding 1281, 1285

Index

renal transplantation (*cont.*):
  gastrointestinal effects (*cont.*):
    infections 1580
  gastrointestinal surgery, prior to transplantation 1517
  Goodpasture's disease following 452
  gout 2296
  graft revascularization techniques
    percutaneous transluminal angioplasty 1597–8
    saphenous bypass 1597–8
  gynaecological problems 1952
  haemolytic–uraemic syndrome 1050
  Henoch–Schönlein purpura **608**
    cadaveric 608
  hepatitis B 2313–15
  hereditary nephritis, IgG linear fixation 2201
  HLA and matching 1516–17, 1544, 1548–9, 1587–9
  hydration protocol 1518–19
  hyperacute rejection 1524
    see also rejection of allograft; renal transplantation, crossmatching
  hypercalcaemia 1768, **1775–7**
    acute management 1777
    incidence 1775
    vs normocalcaemia *1775*
  hypertension following **1536–7, 1594–600**, 2094
    causes 1595–7
    cyclosporin immunosuppressive therapy 1565–6
    diagnostic clues *1599*
    diagnostic strategy 1598–600
    effect on graft and patient outcome 1597
    management 1536–7, 1598–600
    native kidney hypertension 1595, 1599
    prevalence and severity 1594–5
    revascularization procedure 1599
    therapy and results 1597–8
  hypertensive donors 1523
    transmission of hypertension tendency 1536, 1596
  hypophosphataemia 1378
  immune response to allograft **1544–8**
    antibody 1547
    antigen presentation 1544–5
    effector phase 1545–6
    genetic control 1545
    immunogenicity 1544
    major histocompatibility antigens 1544
    minor histocompatibility antigens 1544
    non-specific effector cells 1547
    primary target 1547
    privileged sites and tissues 1545
    sensitization site 1545
    suppressor cells 1547
    T cell activation **1551–4**
    T cell role 1543–4, 1546
  immunological causes of renal failure **1528–32**
  immunology **1543–50**
  immunosuppression 1518, 1532–3, 1590–1
    pregnancy and 1950

renal transplantation (*cont.*):
  infants of mothers with allografts 1339
  infections 1532–6, **1577–82**
    CNS infections 1579
    cytomegalovirus infection 1532, 1536, 1579, 1580
      clinical features 1534
      diagnosis 1580
      live attenuated vaccination of potential recipients 1533
      prevention and treatment 1580–1
    Epstein–Barr virus infection 1581
    hepatitis B 1510, 1533, 1581
    hepatitis C 1534, 1581
    herpes simplex infection 1581
    HIV infection 1510, 1536, 1581
    isolated fever 1534–5
    late 1539
    presentation 1578
    prevention pre-, peri- and postoperatively 1533–4, 1577–8
    pulmonary infection 1535–6, 1578–9
      causes *1578*
      see also Pneumocystis carinii infection
    viral infections 1533–4
    wound infections 1580
    see also renal transplantation, long-term complications; renal transplantation, *specific infections*
  ischaemic storage of allograft 1524
  joint pain, management 1537
  in juvenile-onset diabetes mellitus, pregnancy and 1950
  left ventricular hypertrophy 1577
  leucopenia in 2345
  living donors 1586
    ABO incompatible kidneys 1517
    age of donor, minimum 1511
    legislation 1586
    monozygotic twins 1586
    related 1511
    risk to donor 1586
    siblings 1586
    spouse donation 1586
    unrelated donors 1586
  long-term complications **1570–91**
    bone complications 1585
    cardiovascular disease following 1575–7
    diabetes mellitus 1584–5
    gastrointestinal complications 1582
    haematological complications 1585–6
    infections 1577–82
    liver dysfunction 1584
    malignancy 1582–3
    recurrence of disease 1573–4
    skin lesions 1583–4
    vascular complications 1574–5
    see also rejection of allograft; renal transplantation, non-renal complications
  lymphoma 2330
  lymphoproliferative disorders 2345–6

renal transplantation (*cont.*):
  malignancy
    following transplantation 1582–3
    previous to transplantation 1513
    recurrence or persistence 2322
    risk 1539, 1582–3, 2320–1
  management of the transplant recipient **1520–41**
    clinical care 1521–2
    early management 1524–37
    immediate management 1521
    laboratory data 1522–15191523
    longterm management 1537–41
    perioperative management 1521–3
    steroid therapy effects 2330–1
    see also renal transplantation, preparation of the recipient
  models, experimental solutions 1558
  mouth ulcers 1285
  multiple myeloma 568
  nephrocalcinosis 1889
  nephrotic syndrome, recurrence 311, 326
  neurological involvement 1401, **2330–1**
    infections 2330
    lymphoma 2330
    rejection encephalopathy 2330
    steroid therapy effects 2330–1
    stroke 2331
    TIAs 2331
  non-renal complications
    cancer 1539–40, 1582–3
    cardiovascular complications 1540–1, 1575–7
    cardiovascular disease pre-existing 1512–13
    early 1532–7
    end-stage liver disease 1540
    Fabry's disease 1513
    gastrointestinal complications 1582
    hepatitis B infection 1510, 1533, 1581
    hepatitis C infection 1534, 1581
    left ventricular hypertrophy 1577
    liver disorder 1584
    neoplasia 1582–3
    neurological symptoms 1539
    oxalosis 1513
    papovavirus infection 1581
    peptic ulcers 1509
    skin disorder 1541, 1583–4
    vascular disorders 1526–7
    see also renal transplantation, long-term complications
  nuclear imaging 126–8
  obstruction, diagnosis 97
  oxalate, primary hyperoxaluria 2229
  parenchymal graft disease 1595
  post-operative care
    laboratory data 1522
    lymphocyte counts 1522
    triple therapy 1522
  predialysis, assessment of suitability 1170
  pregnancy and 1338–1339, 1591, 1949–52
    advisability 1952
    breast feeding 1951
    ectopic pregnancy 1949

renal transplantation (*cont.*):
  pregnancy (*cont.*):
    fetal surveillance and delivery 1951
    follow-up of patients 1951
    immunosuppressive therapy 1950
    juvenile-onset diabetes mellitus and 1950
    neonatal problems *1949*
    preconception counselling 1949
    transplant rejection 1950
  preparation of the recipient **1509–19**
    assessment 1509–11
    blood transfusion 1516
    gastric surgery 1517
    high risk patients 1512–13
    highly sensitized patients 1516–17
    immediate pre-operative preparation 1517–19
    immunological examination 1511
    nephrectomy 1515
    parathyroidectomy 1517
    radiological examination 1511
    sensitized patient **1549–50**
    special measures 1515–17
    splenectomy 1517
    urological problems 1512
  proteinuria, massive, recurrent nephrotic syndrome 326
  proximal tubule dysfunction 1791
  psychological aspects 1474–6
  pulmonary artery pressure 1521, 1523
  quality of life 1591
  recurrent disease, risk 1512–15, 1573–4
    Alport's syndrome 1515
    amyloidosis 1515, 1574
    cardiovascular disease 1512
    cystinosis 1574
    diabetic nephropathy 1512, 1515, 1574
    Fabry's disease 1513
    glomerulonephritis 1573–4, *1574*
      mesangiocapillary 396, 1514, 1574
      primary 1513–14
      secondary 1514–15
    haemolytic–uraemic syndrome 1515
    malignancy 1513
    oxalosis 1574
    SLE 1515, 1574
    systemic sclerosis 1515
    see also renal transplantation, malignancy
  regrafts 1587
  rejection, *see* rejection of allograft
  remaining kidney, prognosis 1511
  renal failure, delayed
    diagnosis 1524–5
    immunological causes 1528–32
    immunology, *see also* rejection of allograft
    non-immunological causes 1525–8
  renal failure, immediate
    acute tubular necrosis (ATN) 1523–4
    dead kidney 1523
    hyperacute rejection 1524
    hypovolaemia 1523

renal transplantation (*cont.*):
  renal failure, immediate (*cont.*):
    vascular occlusion 1524
  results
    cadaver grafts 1548–9, 1586–7
    living donors 1586
    regrafts 1587
  revascularization techniques
    1597–8
  rhinocerebral mucormycosis 1747
  schistosomiasis 1739
  scleroderma 676
  sensitized patients 1549–50
    HLA antibodies 1516
  septicaemia in 1534–5
  sexual function 1591
  sickle-cell disease 716–17
  SLE 663
  strongyloidiasis, prophylactic
    thiabendazole 10
  survival
    live *vs* cadaver donors 1951
    rates 1520–1
  temporary tolerance of graft, non-
    specific immunosuppressive
    therapy 1532
  therapy, diabetes 1650
  thrombotic complications of
    2345–6
  tolerance model of autoimmune
    glomerulonephritis 249
  transplant glomerulonephropathy,
    renal biopsy, EM findings 150
  tuberculosis 1533
  tubular lesions, cyclosporin
    toxicity 149
  uraemic polyneuropathy 2329
  ureters, anastomosis 1525
  urinary tract
    fistula 1526
    infections 1579
    obstruction 1525–6, 2018
  urological problems 1512, 1575
  varicella zoster infection 1581
  vascular lesions 1526–7
    cyclosporin, nephrotoxicity 1562
    renal arteriography 113
  white cell count 2345
  see also allografts; rejection of
    allograft
renal trauma, *see* kidney(s), trauma
renal tubular acidosis **763–79**,
  1859–61
  aetiology 1859–60
  aldosterone deficiency 43
  allied conditions 1860
  analgesic nephropathy 813
  autosomal recessive,
    nephrocalcinosis 1893
  bicarbonate-wasting, *see* renal
    tubular acidosis, proximal
  classification 763–4
    algorithm 778
    differential diagnosis 770
    types 1–4 763–764, *770*
  clinical chemistry 1860
  combined, in Sjögren's syndrome
    696
  combined (type 3) 764, 773
  defined 763, 771, 1859
  diagnostic approach to metabolic
    acidosis 777–9
  distal (type 1) **763–4, 771–6,**
    936–40
    Atlanta and Oklahoma
      syndromes 1797–8

renal tubular acidosis (*cont.*):
  distal (type 1) (*cont.*):
    causes *774*
    children 772–3
    clinical features 772–3
    defects
      hypotheses 771–2
      types 771
    diagnosis 774–5
      algorithm 778
    differential diagnosis 1894
    incomplete 772, 773
    mechanisms *938*, 1859
    nephrocalcinosis 1892–4
    nerve deafness 773
    osteopetrosis and 1893
    pathogenesis 771–2, *938*
    pathophysiology of symptoms
      773–4
    physiological findings *1796*
    primary 772–3
    San Francisco and Philadelphia
      syndromes 1797
    secondary 773–6
    transient 773
    treatment 697, **775–6**, 939, 1860
  distal *vs* proximal, plasma
    bicarbonate and urine pH 44
  elliptocytosis and 1893
  familial, *see* renal tubular acidosis,
    distal (type 1)
  hydrogen ion generation tests
    42–4
    acid-loading tests 43
    chronic renal failure and
      ammonium excretion and 42
    plasma bicarbonate and early
      morning urine pH 43
    urine:plasma P_{CO_2} gradient and
      42
  hypercalciuria 774, 1835
  hyperchloraemic acidosis 1893
  hyperkalaemic (type 4) 207, 764,
    776–7, 939–40
    causes 777
    classification 763, *938*
    algorithm 778
    clinical features **776–7**, 938
    diagnosis, algorithm 778
    hyperkalaemia-associated
      disease processes *940*
    pathogenesis 776
    rate defects 939, *940*
    treatment 777
    voltage defects *940*
  hyperparathyroidism 1855
  incomplete sponge kidney 1895–6
  incomplete syndrome 1893
    in sickle-cell disease 709
  lithium-induced 67
  loop diuretics and 207
  medullary sponge kidney 1836
    obstruction and 2012
  pathophysiology *763*
  physiology 764–8
  potassium depletion 903
  presenting features 696
  primary, nephrocalcinosis 1893
  proximal (type 2) **763**, **768–71**,
    935–6
    bicarbonate reabsorption,
      fructose loading 732
    causes 769, *936*
    children 936
    uraemic growth retardation
      1305

renal tubular acidosis (*cont.*):
  proximal (type 2) (*cont.*):
    clinical features 768–9
    combined and distal renal
      tubular acidosis 764, 773
    de Toni–Debre–Fanconi
      syndrome 207
    diagnosis 769–71
      algorithm 778
    familial 768
    isolated 768
    maleic acid, animal model 768
    pathogenesis 768, 935–6
    pathophysiology of symptoms
      769
    primary/secondary 768
    treatment 771
    in SLE 649, 771
    urinary and arterial P_{CO_2} values
      1893
  renal tubular atrophy, cyclosporin,
    nephrotoxicity 1562
  renal tubular disease and
    dysfunction **723–800**
    Anderson–Fabry disease 2208
    Bartter's syndrome **782–7**
    diabetes insipidus, nephrogenic
      789–98
    Fanconi syndrome **723–36**
    isolated defects of tubular
      function **741–57**
    see also renal tubular acidosis;
      tubulointerstitial disease
  renal tubular reabsorption **741–3**
    ammonium ion 60–1
      elderly patients 60–1
    cyclosporin therapy effects 1561
    dysfunction in nephrotic
      syndrome 291
    elevated capacity in sickle-cell
      disease 714
    endocapillary glomerulonephritis,
      acute 407
    GFR increase 263
    minimal change nephrotic
      syndrome 302
    normal renal handling 264
    potassium 897–8
    renal clearance studies
      Michaelis–Menten constant 742
      splay 742–3, *743*
      threshold 742
      transport maximum, T_m 742
    sodium 870–2
    sodium pump *741*
    solute titration curves *742*
    urate/uric acid, animal model 827
  renal tubular secretion,
    cyclosporin therapy effects 1561
    urate, pyrazinamide suppression
      test 826
  renal tubules,
    in amyloidosis 552
    atrophy, and thickening,
      nephronophthisis 2189–90
    basement membrane
      anti-TBM antibodies,
        membranous nephropathy
        374
      granular deposits, animal
        models 1093
      immune deposits, in SLE 654
      light chain deposition 569, 570,
        571
    casts, haematuria 231
    cells, in urinary sediment 20, 21

renal tubules (*cont.*):
  damage
    acute renal failure **1006–13**
    markers 227, 228–9
    nephrotoxins *1006*
    proximal tubule, causes *229*
    see also interstitial nephritis
  defects, Dalmation dog mutation
    830
  defects of tubular function
    acute uric acid nephropathy *832*
    all tubular transport,
      pyrazinamide effect nil 830
    aminoacidurias **745–51**
    combined urate reabsorption,
      pyrazinamide effect 830
    glycosurias **743–5**
    hypercalcaemia and 1883–5
    hyperuricaemia **757**
    hypouricaemia **757**, 830
    inherited hypouricaemic
      syndromes, acute uric acid
      nephropathy *832*
    nephrocalcinosis, chemical
      1884–5
    phosphaturias **751–7**
    postsecretory urate reabsorption,
      pyrazinamide effect
      increased 830
    presecretory urate reabsorption,
      pyrazinamide effect
      decreased 830
    urate hypersecretion,
      pyrazinamide effect
      enhanced 830
  distal
    function
      arginine vasopressin
        exogenous 42
        plasma 42
      concentrating ability 41
      lithium clearance 41
      phosphate reabsorption 1783
      potassium handling 897
      potassium secretion 898
      sodium transport 40, 198
      delivery in absence of AVP
        41
      urinary dilution capacity 42
      urine osmolality 41–2
        early morning, and fluid
          deprivation tests 42
    gradient defect 771, 775
    hypercalcaemia effects 1884–5
    tubular damage 1883
    increased length in diabetes
      mellitus 514
    macula densa, hyperplasia and
      hypertrophy in Bartter's
      syndrome 783
    obstruction effects 2012
    secretory defect 771, 775
    urinary acidification
      mechanisms 766–7
      regulation 767–8
      secretion of hydrogen ion 766
    voltage-dependent defect 771,
      775
  ectasia, calcium stone disease
    1836
  epithelial antigen
    deposition, in sickle-cell disease
      711
    renal cell carcinoma 478
  epithelial cells, in Sjögren's
    syndrome 697

renal tubules (*cont.*):
  fibrosis, and tubulointerstitial infiltrate 1206
  function
    abnormalities
      in amyloidosis 555
      in multiple myeloma 564
    acid excretion, ageing kidney 60–1
    ageing changes 57–8
    changes in diabetes mellitus 515
    cyclosporin therapy effects 1561
    elderly patients **60–1**
    inhibited by Bence–Jones protein 567
    mean proximal tubules, volume 58
    neonates **52–3**
    normal 723
    post obstruction 2012–13
    pregnancy 1915–17
      animals as gestational models 1920–1
    tests **40–4**
      sodium handling in the nephron 40–4
      upper tract infection 1712
      see also renal tubules, distal; renal tubules, proximal
  tubular atrophy
    GFR impairment and 147
    indications 145
  tubules, atrophy, chronic disease 149
  glomerular feedback mechanism 139
  injury, *see* renal tubules, damage; renal tubules, lesions
  and interstitium, accelerated hypertension 2128–9
  lesions
    in diabetic nephropathy 523–4
    renal biopsy, evaluation *148*
  necrosis, *see* acute tubular necrosis
  nephropathy, drug-induced acute toxic 163
  proteinuria, defined 17, 18
  proximal
    active transport 741–2
    acute tubular necrosis, mechanisms of excretory failure 999–1000
    bicarbonate reabsorption, dog and other animals 765
    diuretics 197–9
    dysfunction
      in amyloidosis 555
      cadmium nephropathy 842
      in Fanconi syndrome 567
      fructose administration 830
      nephrocalcinosis 1896
    epithelial transport mechanisms 741–3
    Fanconi syndrome 723–4
    function
      sodium reabsorption 40
      see also aminoaciduria; Fanconi syndrome; glycosuria; phosphaturia; renal tubules, function
    growth factors, effects on 1204
    hypercalcaemia, tubular damage 1883

renal tubules (*cont.*):
  proximal (*cont.*):
    hypermetabolism, tubulointerstitial scarring 1204–5
    increased length in diabetes mellitus 514
    mean proximal tubular volume 58
    model of solute transport 723
    obstruction, intratubular oxalate 1205
    pars convoluta 746
    pars recta, sodium-dependent uptake of amino acids 746
    phosphate reabsorption 752, 1783
    reabsorption, lithium clearance as index, pregnancy 1918
    renal clearance studies 742–3
    sickle-cell disease **713–14**
    sodium transport 741–2
    solute titration curves *742*
    succinate uptake and 118
    transport mechanisms
      entry step 741
      exit step 741–2
      general aspects 741–3
      sodium, chloride, and water 198
    tubular obstruction 998–9
    vacuolarization in hypokalaemia 901
  transporting defects, Fanconi syndrome 723–4
  see also renal tubular acidosis; renal tubular reabsorption
renal tumours, *see* renal mass; renal pelvis and ureter tumours; *specific names, organs and regions*
renal vascular system, *see* renal arteries; renal circulation; renal veins
renal vein stenosis, in transplantation 97
renal vein thrombosis,
  acute on chronic renal failure 1160
  in amyloidosis 551, 552
  assessment 975
  chronic renal failure 1160
  detection, Doppler sonography 96
  following renal transplantation 1575
  membranous nephropathy and 371
  nephrotic syndrome
    diagnosis and evaluation 281
    frequency and prevalence *280*
    treatment 281–2
  parenchymal consequences, computed tomography scanning 105
  selective catheterization 115
  see also venous thrombosis
renal veins,
  ageing changes 58
  arteriography 112
  carcinoma invasion, ultrasonography 89
  clear cell carcinoma, regional extension 102
  left, retro-aortic course, computed tomography scanning 106

renal veins (*cont.*):
  loin pain–haematuria syndrome 2277
  occlusion, radiological tests *72*
  selective catheterization 115
  tumour extension, MRI 108
  venography 115
renal–ocular syndrome 1091–2
renin,
  assay 2069
    immunoloradiometric assay 2069
  hypertension and
    chronic renal failure 1610
    role 2064
  increased, renal cell carcinoma 2242
  inhibitors, renin secretion and *2071*
  juxtaglomerular apparatus, decreased secretion 909
  low sodium diet, young and elderly subjects *63*
  in lymph, periarteriolar space 984
  menstrual cycle 2071
  plasma renin activity
    correlation with active renin *2070*
    cyclosporin therapy 1561
    increased
      in scleroderma 673, 674, 676
      in sickle-cell disease 712, 715
    measurement 2104
    indications *2072*
  natriuresis 269
  nephrotic syndrome 267
  normal, in sickle-cell disease 715
  pregnancy 2071
  reduced, in diabetes mellitus 513
  renal cell carcinoma 2242
  renovascular hypertension 2062
  unilateral multicystic kidney and 2090–1
  polycystic disease, autosomal dominant 2175
  potassium depletion 901–2
  pregnancy
    activity 2071
    increase 1962
  production disturbance, in sickle-cell disease 711
  prorenin in diabetic retinopathy 1646
  release
    action of prostaglandins, NSAIDs and 820, *821*
    factors affecting *2071*
    in pregnancy-induced hypertension 1963
  renal activity, and plasma renin, correlation *2070*
  renal vein renin, measurements 2104–5
  renin-dependent hypertension 1459
  renin-producing tumours, hypertension and 2092–3
  secretion, cystic expansion 2175
  sodium and 2119
  tests (ratios), value in ischaemic renal disease *1080*, 1081
reninoma, hypertension and 2092–3

renin–angiotensin–aldosterone system **2064**, **2069–74**, 2101
  accelerated hypertension 2131, 2132
  acute blockade 2073
  acute manipulations, detection of hypertension 2072–4
  angiotensin converting-enzyme inhibitors 2064
  animal gestational model 1914–15
  antihypertensives, action 2071–2
  assay of components
    aldosterone 2070
    angiotensin converting-enzyme 2070
    angiotensinogen 2070
    angiotensins 2070
    renin 2069–70
  blood pressure control and 2061–4
  cyclosporin therapy 1561
  diabetes 1646
  end-stage renal failure 2119
  endocapillary glomerulonephritis, acute 407
  hyperactivity, role in hypertension 1460–1
  hyperkalaemic tubular acidosis 777
  hypertension and 2014
  hyponatraemic hypertensive syndrome 2132
  mediation of vascular damage 2132
  native kidney-induced hypertension 1595
  neonates 52–3
  paracrine and autocrine mechanisms 2132
  physical parameters 2070–1
  placenta 1962
  pregnancy
    pregnancy-induced hypertension 1962–3
    and prostaglandins 1920
    renal dilatation 1914–15
    proximal renal tubular acidosis 769
    renal vascular control 1914–15
    mediation 984
  in scleroderma renal crisis 674
  vascular control 25, *26*, 27
  volume depletion-induced stimulation 935–6
renography,
  correction for depth of kidneys 118
  defined 118
  errors, sources 118–19
  hypertension due to local ischaemia 119
  nephrolithiasis 119
  nuclear imaging using radiopharmaceuticals **116–28**
  probe *vs* gamma camera studies 117
  pyelonephritic scarring 119–20
  reflux nephropathy 119
  renal localization 119
  renal radionuclide study, activity–time curve 118
  supply curve (blood clearance curve $B(t)$) 118
  see also specific procedures
renomedullary lipids,
  antihypertensive function
    end-stage renal failure 2120

renomedullary lipids (*cont.*):
 antihypertensive function (*cont.*):
  sympathetic activity **2064–5**
renovascular disease, children, causes and incidence *2143*, 2143
renovascular hypertension 119–22, **2096–112**
 see also hypertension, renal; renal arterial disease; renal artery stenosis
resins,
 cation-exchange therapy, hyperkalaemia 911
 nephrotic syndrome, treatment 287–8
respiratory acidosis **951–4**
 causes *953*
 causes and treatment 953–4
 clinical approach 953–4
 clinical features and pathologic consequences 953–4
  cardiovascular dysfunction 953
  CNS dysfunction 953
  plasma electrolytes, composition 953
 defined 919
 integrated response
  acid load elimination 952–3
  buffering 951
  renal response 951–2
 K^+ regulation 896
 metabolic acidosis, mixed 958
 metabolic acidosis and metabolic alkalosis 959
 metabolic alkalosis, mixed 958–9
 metabolic response, nomogram *952*
 pathogenesis, integrated response 951–3
 pathophysiology 951–3
 responses and secondary changes *920*
respiratory acid–base disorders, renal effects 2336
respiratory alkalosis **954–6**
 chronic, post-hypocapnic acidosis unveiling metabolic acidosis 935
 clinical approach 956
 clinical features and pathologic consequences
  cardiovascular effects 955
  CNS dysfunction 955
  plasma electrolyte composition 955
 defined 919
 integrated response
  alkali load elimination 954–5
  buffering 954
  renal response 954–5
 metabolic acidosis, mixed 957–9
 metabolic acidosis and metabolic alkalosis 959
 metabolic alkalosis, mixed 957–8
 metabolic response, nomogram *954*
 pathophysiology 954–5
 responses and secondary changes *920*
 in sickle-cell disease 709
respiratory distress syndrome,
 following renal transplantation 1535–6
 mixed metabolic and respiratory disorders *957*

respiratory distress syndrome (*cont.*):
 preterm neonates, dopamine with frusemide 54
respiratory and metabolic disorders, mixed **957–60**
respiratory tract,
 in Churg–Strauss syndrome 629
 examination in acute renal failure 970
 in Wegener's granulomatosis 613, 614–15
restless leg syndrome, uraemic peripheral polyneuropathy 1396–8
restriction endonucleases 2155–7
restriction fragment length polymorphisms, number in human genome 2155–7
reticulocytes, normal values *1344*
retinoids, safety 2305–6
retinol, uraemic xerosis pathology 1390–1
retinol-binding protein, marker of tubular damage 228
retinopathy,
 chloroquine/hydroxychloroquine side-effect 659
 diabetic 511
  absence in non-diabetic renal disease 526
 Goodpasture's disease, anti-GBM antibodies 442–3
 haemolytic–uraemic syndrome 1047
retrograde cystography, *see* cystography
retroperitoneal disease,
 computed tomography scanning 106–7
 fibrosis 1150
  computed tomography scanning 104
 diagnosis, ultrasonography 92
 idiopathic 2018–20
  diagnosis 106
 secondary, diagnosis 106–7
 fluid collections, computed tomography scanning 107
 retroperitoneal tumour, computed tomography scanning 104
 ureteric obstruction 2004
retroviral infections,
 acute interstitial nephritis 1091
 HTLV-1-associated myelitis 2323
reverse genetics, defined 2155–7
Reye's syndrome, acute fatty liver of pregnancy and 1933
rhabdoid tumour, Wilms' tumour 2258
rhabdomyolysis,
 acute renal failure 2323
 acute uric acid nephropathy 832
 in diabetes mellitus 532
 diagnosis and treatment 1007–8
 hypokalaemia precipitating 901
 mannitol 205
 myoglobin, and ARF 995
 in polymyositis, renal failure due to 684
 volume replacement and diuresis 206
rhabdomyosarcoma,
 histogenesis 2244
 pathology 2244

rheumatoid arthritis (RA) **678–84**
 amyloidosis associated 554, 556, 678, **679–81**
 clinical presentation 678, 680
 diagnosis 680
 frequency 680
 treatment/prognosis 680
 analgesic nephropathy in **684**
 antirheumatic drugs (DMARDs) 682
 cyclosporin A nephrotoxicity in **683–4**
 D-penicillamine 161
 glomerulonephritis 678, 679
  membranous 681, 682
  mesangial IgA 680
  mesangial proliferative 680
  proliferative 680
 gold nephropathy
  nephrotic syndrome 160
  see also gold nephropathy
 immune complexes in 587
 juvenile, IgA nephropathy 361
 penicillamine nephropathy, *see* D-penicillamine
 rapidly progressive glomerulonephritis 1070
 renal disease
  incidence *678–9*
  patterns and prevalence 678–*679*
 Sjögren's syndrome association 693, *695*, 695
 vasculitis asssociated, *see* rheumatoid vasculitis
 see also juvenile rheumatoid arthritis (JRA)
rheumatoid factors,
 IgA
  in Henoch–Schönlein purpura 600
  in rheumatoid arthritis 680
  in Sjögren's syndrome 696
 IgM
  Epstein–Barr virus inducing 687
  in essential mixed cryoglobulinaemia 686, 687, 689, 691
  immune complex deposition 689, 691
 in Sjögren's syndrome 695, 696
rheumatological conditions **2293–8**
 complications of, nephritis 2295
 drugs, drug-induced renal disease 2295
 elderly patients with chronic renal failure 1624
 rheumatoid arthritis, skin lesions 2302
 vasculitis 681
  complement in 588
Rhodotorula, fungaemia 1747
ribonucleoprotein (nRNP), antibodies 647, 684, 698
rickets,
 familial X-linked hypophosphataemic rickets
  clinical features 754
  diagnosis 754–5
  genetics 754
  intestinal defect 753
  parathormone in 753
  renal defect 753
  treatment 755
  vitamin D metabolism 753

rickets (*cont.*):
 growth retardation, cystinosis 2221–2
 hereditary hypophosphataemic rickets with hypercalcaemia 756
 hyperparathyroidism and 1612
 hypophosphataemic rickets with hypercalciuria and microglobulinuria 756
 oncogenic hypophosphataemic osteomalacia 756–7
 phosphate malabsorption 1788
 skeletal deformities, children 1611–13
 vitamin D-dependent rickets
  classification *1796*
  nephrocalcinosis *1897*
  physiological findings *1796*
  types I and II 755, 756
 vitamin D-resistant rickets **1795–8**
  classification *1796*
  physiological findings *1797*
  X-linked hypophosphataemic rickets
   clinical features 1795
   nephrocalcinosis *1897*
   pathogenesis 1795, *1796*
   treatment 1795
Rickettsia, acute renal failure **1134**
rickettsial infections, infection-associated glomerulonephritis 462
rifampicin,
 CAPD peritonitis regimens *1489*
 crescentic glomerulonephritis 421, 422
 cyclosporin therapy, interactions *1563*
 drug dosage, renal impairment *184*
 drug-induced nephropathy 163
  acute toxic tubular nephropathy 167
  interstitial nephritis 1089
 in Henoch–Schönlein purpura 607
 rapidly progressive glomerulonephritis 1070
 renal tuberculosis *1727*
right atrium, stretch receptors, atrial natriuretic peptide release 263
'rim sign', computed tomography scanning 105
ristocetin, nephrotic syndrome, hyperaggregability 283
river blindness, onchocerciasis 468, 495
Ro antigen (SSA), antibodies to,
 HLA associations 643, 695
 in Sjögren's syndrome 694, 695
 in SLE 643, 647, 648, 650, 651
Roberts syndrome, features and associated abnormalities *2037*
Rocky Mountain spotted fever, infection-associated glomerulonephritis 462
roll-over test, pregnancy-induced hypertension 1967
Rothman–Thomson syndrome, skin lesions 2300
RPF, *see* renal plasma flow
RPGN, *see* rapidly progressive glomerulonephritis

Index

Rubinstein–Taybi syndrome, features and associated abnormalities 2039
rugger jersey spine, hyperparathyroid bone disease 1373
Russell–Silver syndrome, features and associated abnormalities 2039

Saccharomyces, infection 1748
saddle-nose deformity 615
safflower oil, dietary and pharmacological lipid manipulations 1202
salbutamol, hypokalaemia 902
salicylates,
 anaemia and 813
 high-dose, urate excretion increase 829
 low-dose, urate excretion decrease 829
 lupus nephritis, cautions on use 659
 prostaglandin synthesis inhibition, mechanism of papillary damage 811
 renal excretion, pK_A and pK_B 178
 sulphosalicylic acid test for proteinuria 5
 thrombotic complications, dialysis fistulae 1268
 toxicity
  analgesic-associated kidney disease 803
  cadmium nephropathy 843
  carbonic anhydrase inhibitors 206
  end-stage renal failure 808
  kidney damage 810
  metabolic acidosis 932
  peptic ulcer 813
  single substance, and analgesic nephropathy 816
  synergism with other analgesics 810
 see also aspirin
salivary glands, in Sjögren's syndrome 693, 694, 696
 histology 697
Salmonella infections, glomerulonephritis 460
 schistosomiasis and 1730, 1737
 tropics **1133**
salt depletion, *see* sodium, depletion
salt-sensitive rat, *see* Dahl rat
Samoyed hereditary glomerulopathy 2201
Sanarelli–Schwatzman reaction, pregnancy, *see* Schwatzman reaction
Sandhoff disease, N-acetyl-β-glucosaminidase A and B deficiency 2206
SAP, *see* amyloidosis, serum
saperoconazole, fungal infections 1748
saphenous vein graft, *see* vascular grafts
saralasin,
 antihypertensive action 1460, 1461
 renal transplantation 1595

saralasin (*cont.*):
 Bartter's syndrome, blood pressure fall 785–6
 saralasin test 2106
sarcoid granuloma, formation mechanisms 576–7
sarcoidosis **576–82**
 acute interstitial nephritis 1091
 acute renal failure 1106
 calcium abnormalities in, *see* calcium; hypercalcaemia; hypercalciuria
 calcium stone disease 1836
 causes 1773
 diagnosis 1381
 differential diagnosis 577
 glomerular involvement **580–1**
 granulomatous interstitial nephritis (GIN) **577–8**
  differential diagnosis 578
  pathology 577–8
  prevalence 577
  therapy 578
 hypervitaminosis D 1883, 1891
 hypercalcaemia in 1773
 interstitial nephritis 2293–4
 liver disease associated with acute renal failure 1106
 magnesium wasting and 1810
 renal amyloidosis associated 580–1
 renal disease in 2293, 2336
 renal failure, isolated 144
 retroperitoneal involvement 581
sarcoma, mesenchymal origin, computed tomography scanning 103
scabiosis, and streptococcal pharyngitis, diagnosis and treatment 414
scarlatina, *see* poststreptococcal glomerulonephritis; streptococcus
scarlet fever, acute interstitial nephritis 1091
SCH 39304, fungal infections 1748
Schirmer test 697
schisis association, features and associated abnormalities 2039
Schistosoma bovis 1729
Schistosoma haematobium **488–90, 1729–39**
 antigens 1731
 glomerulonephritis 467
 life cycle 1730
 praziquantel 185, 1738
 in urine 21
Schistosoma intercalatum 1729
Schistosoma japonicum **488–90**, 1729, 1731, 1733, 1736
 drugs against 1738
 glomerulonephritis 467
Schistosoma mansoni **488–90**
 antigens 1731
 cor pulmonale 1737, 1739
 distribution 1729
 drugs against 1738
 glomerulonephritis 467
 overt glomerulopathy 1736
 oxamniquine 185, 1738
Schistosoma mattheei 1729
Schistosoma mekongi 1729
schistosomiasis,
 amyloidosis 1734, 1737, 1739
 assessment 1152

schistosomiasis (*cont.*):
 clinical syndromes 1734–8
  extraurinary 1737–8
  hepatointestinal 1734–5
  urinary 1734
  urinary tract 1735–7
 epidemiology 1729–30
 glomerulopathy **488–90**
  acute, salmonella bacteraemia 460
  animal models 249
  clinical features 467, 489
  incidence 489
  infection-associated glomerulonephritis 467
  overt 1736
  pathogenesis 490
  renal histology 489–90
  treatment 185, 490, 1738
 granulomas 1733, 1735
 immunology 1730–3
 malignancy 1734, 1735
 obstructive uropathy 1735–6
 pathology 1733–4
 treatment 185, 490, 1738–9
 vasculitis 1733
 vesicoureteric reflux 1736
Schönlein–Henoch syndrome, *see* Henoch–Schönlein purpura
Schwartzman reaction,
 endotoxins 993
 glomerular thrombosis 1052
 pregnancy 1933, 1934
scintigraphy,
 chronic urinary tract obstruction 2015–16
 isotope renogram, essential hypertension vs renal causes 2050
 radioisotope renogram, renovascular hypertension 2103
 static 2050
 see also single photon renal scintigraphy
sclerodactyly 667–8
scleroderma **667–77**
 autoantibodies in 669
 clinical manifestations **672–6**
 diagnosis **667–8**
 end-stage renal failure, proportional distribution by age at commencement 1229
 epidemiology and genetic studies 667
 generalized 667, 669
  diffuse 669, 669
  limited cutaneous 669, 669
 haemolytic–uraemic syndrome 1050
 juvenile 672
 localized 667
 pathogenesis/pathology **672**
  vascular injury 671–2, 675
 pregnancy and 1943
 prognostic factors and survival 673
 renal biopsy, EM findings 150
 renal crisis 673, 676
  management 676
 renal disease 672–3
  chronic 675
  criteria 673
  frequency 673, 674

scleroderma (*cont.*):
 renal disease (*cont.*):
  histopathology 674–5
  indolent 674
  mixed connective tissue disease, similarity 698
  pathogenesis 675–6
  renin-angiotensin system in 674
  spectrum of 673
 sine scleroderma 669, 669
 skin lesions 2302
 vascular diseases with renal involvement 2294
  schematic representation 152
scleroderma-like syndromes 668
scorpion sting, acute renal failure and 1136
scrotum, Henoch–Schönlein purpura 597
scrub typhus, acute renal failure and 1134
scrum kidney 405
sea water, osmotic load, hypernatraemia 883
segmental glomerulonephritis, rheumatoid arthritis associated 681
vasculitis and 1065–8
segmental hypoplasia, *see* Ask–Upmark kidney
seizures,
 acetazolamide and 206
 acute renal failure 2326
 control in eclampsia 1973–4
selective renal angiography, renal mass 2245
selenium, Balkan nephropathy and 859
self-antigen reactions, autoimmune response 246
Semecarpus poisoning, acute renal failure and 1136
semustine, antineoplastic treatment, renal complications 2319
senile dementia, in dialysis patients 1448–9
Senior's syndrome, nephronophthisis 2192, 2193
sepsis,
 acute renal failure **978–80**
 nephrotic syndrome 1038
 animal models, prevention of renal failure 1018
 nephrotic syndrome 1038
 treatment 278
 pathogenesis of renal damage 991–2
septic abortion, renal failure, acute and 1932–3
septic arthritis, complication of renal disease 2296
septic shock,
 action of prostaglandins, NSAIDs toxicity 820
 following renal transplantation 1534–5
 mixed metabolic and respiratory disorders 957
 triple metabolic and respiratory disorders 957
septicaemia,
 acute pyelonephritis 1090
 haemodialysis 1439, 1452–3
 interstitial nephritis 1090
 renal transplantation 1579
 treatment 1534–5

Index

sequential ultrafiltration dialysis, alternative to haemodialysis 1273
serine,
  excretion, pregnancy and 1916
  normal tubular reabsorption 745
seroma, vascular graft 1415–16
serotonin,
  5-hydroxyindoleacetic acid clearance 40
  hypomagnesaemia and 1813
Serratia, urinary tract infection 1661
serum C reactive protein, polyarteritis 1167
serum complement, and proteinuria, glomerulonephritis–type diseases 408
serum sickness,
  antibody-mediated glomerulonephritis 251–2
  glomerular thrombosis, animal models 1053
  inflammatory reaction 584
  as necrotizing vasculitis model 583–4
  immune complexes, deposition in vessels 583–4
sex chromosome abnormalities, *see* X-linked abnormalities
sex differences,
  amyloidosis incidence 548, 548
  diabetic nephropathy risk 506
  microalbuminuria in diabetics 509
  Wegener's granulomatosis incidence 614
sexual abuse, children 1716
sexual dysfunction,
  chronic renal failure 1207–8, **1329–39**
    assessment 1157
    females 1336–9
    males 1329–36
  haemodialysis patients **1329–39**
  impaired fertility
    females 1337–8
    males 1335
  impaired spermatogenesis 1335
  impotence 1329–35
  priapism 1335–6
sexual maturation, growth retardation in chronic renal failure 1302–3
sexually transmitted infections, vaginitis 1686
Shigella dysenteriae,
  haemolytic–uraemic syndrome 1049
  infections, tropics, acute renal failure and **1133–4**
short-rib polydactyly syndrome, features and associated abnormalities
  Majewski type *2038*
  Saldino–Noonan type *2038*
shunt nephritis,
  hydrocephalus 469–70
  organisms associated *469*
shunts,
  external, construction 1406
  Hemasite 1406, 1407
  staphylococcal colonization 460

shunts (*cont.*):
  types 1405
  see also arteriovenous fistula
SIADH, *see* arginine vasopressin, inappropriate secretion
sialic acid,
  in glomerular basement membrane 516
  removal from basement membrane, streptococcal neuraminidase 410
  sialic acid-depleted IgG 409
sialoglycoprotein,
  in pathogenesis of diabetic nephropathy 519
  podocalyxin 132, 133
sickle cells 700
  irreversible 701
sickle-cell anaemia (HbSS) 701
  aplastic crises 701
  arteriole obliteration 706
  bacterial infections susceptibility 703, 704
  fetal haemoglobin levels 703
  haemolytic crises 701
  hypouricaemia 830
  papillary necrosis 1160
  phagocytosis/opsonizing ability reduced 704
  renal concentrating capacity 705–6
    age affecting 706
  renal impairment in 2341
  sequestration crises 701
  sickle-cell crises 701
    fluid management 716
  skin lesions 2304
  tuftsin levels in 704
sickle-cell disease **700–20**
  blood coagulation in 709, **715**
  clinical features 2340–1
  definitions **700–3**
  dialysis 2340–1
  epidemiology 703–4
  geographical distribution 703
  historical aspects 700
  homozygous, *see* sickle-cell anaemia (HbSS)
  hypertension and 712
  kinetic hypothesis 714
  multiple mutation theory 702, 703
  pathology **712–13**
    gross anatomy 712–13
    microscopy 713
    ultrastructural 713
  prostaglandin synthesis 707, 709, 712, 715
  renal manifestations **704–12**
    acute glomerulonephritis 711
    diluting capacity 707, 709, 715
    glomerular hyperfiltration in 711–12
    glomerulosclerosis 711, 711–12
    haematuria 709–10
    impaired concentrating capacity 704–7, 715, 716
    negative free water generation 707, 709
    nephrotic syndrome 711, 713
    papillary necrosis 710
    potassium metabolism 711
    proteinuria 711
    proximal tubular function **713–14**
    renal hormones in **714–15**

sickle-cell disease (*cont.*):
  renal manifestations (*cont.*):
    renal insufficiency 711
    urinary acidification defect 709
    urinary tract infections 710
    vasa recta occlusion 702, 705, 706, 709, 713, **715**
  skin lesions 2304
  treatment **715–17**
    of end-stage renal disease 716–17
    fluid intake 716
    general measures 716
    of haematuria 716
    of urinary tract infections 716
sickle-cell haemoglobin C disease (HbSC) 701–2
  renal concentrating capacity 705
  see also sickle-cell disease
sickle-cell haemoglobin D Iran 702
sickle-cell hereditary persistence of fetal haemoglobin (S-HPFH) 703
sickle-cell nephropathy, renal papillary necrosis, differential diagnosis 815
sickle-cell trait, heterozygous (HbAS) 701
sickle-cell trait (HbAS) 701
  see also sickle-cell disease
sickle-cell–α-thalassaemia 702
sickle-cell–β-thalassaemia 702
sickling disorders, hybrid 701–2
sickling phenomenon 700–1
  intracellular mechanisms and **714**
  see also haemoglobin, polymerization
siderophores, chelating agents 1665
siderosis, uroporphyrin decarboxylase inhibition, iron overload 1282
siderosomes 713
silica,
  Balkan nephropathy and 858
  occupational exposure, renal disease 845
silica stone disease 1842
silica stones **1866**
silicoproteinosis, crescentic glomerulonephritis 845
silicosis 845
  crescentic glomerulonephritis 422
silicates,
  drug-induced lithiasis 168
  interstitial nephritis, animal model 845
silicone uptake, spallation of particles during haemodialysis 1282, 1368, 1432, 1445
silver, drug-induced nephropathy 161
simvastatin,
  handling in kidney disease 191
  nephrotic hyperlipidaemia, effects 288
single nephron GFR (SNGFR), *see* glomerular filtration rate, single nephron GFR (SNGFR)
single photon renal scintigraphy, detection of focal abnormalities 120
  prediction of coronary atherosclerosis 1643
  renal mass 2245
  see also scintigraphy

sinusitis, in Wegener's granulomatosis 615
Sjögren's syndrome **693–700**
  aetiopathogenesis 695–6
  Epstein–Barr virus 695–6
  autoantibodies in 694, 695
  classification **693–4**
    new *695*, 695
  clinical features 694–5
    exocrinopathy 694
    extraglandular 694
  complement in 588, 694
  diagnostic criteria 695
  diagnostic tests 694–5
  Fanconi syndrome 734
  features in SLE, in elderly 650
  glomerulopathy 2295
  IgA nephropathy 361
  immunogenetics 695
  immunohistology 696
  leucocytoclastic/mononuclear vasculitis overlap 589
  lymphoma in 694
  primary 693, *695*
  renal disease in **696–8**
    diagnosis/treatment 697
    immune complex glomerulonephritis 697–8
    immunopathogenesis 697
    interstitial nephritis 696–7
    presenting features 696
    renal tubular acidosis in 696–7
    hyperglobulinaemia link 697
  rheumatoid arthritis association 693, *695*, 695
  secondary 693–4, *695*
  SLE association *695*
SKATT test, impotence, erectile 1330
SKATT therapy, impotence, erectile 1334
skeletal age,
  changes, in chronic renal failure 1302
  methods of assessment 1300–1
  sexual maturation and 1302
skeletal calcium, total *1753*
skeletal complications of analgesic nephropathy 814
skeletal deformities, children 1611–13
skeletal muscle, paralysis, hypokalaemia precipitating 901
skin,
  allografts, survival, HLA-identical sibs 1548
  calcium, total *1753*
  examination in acute renal failure 970
skin diseases/disorders **2299–308**
  acquired conditions 2301–4
  in amyloidosis 551
  bullous disorders 1392, 1393
  cancer, risk following renal transplantation 1539
  chronic renal failure **1390–4**
    arteriovenous shunt dermatitis 1392
    assessment 1156
    cutaneous and subcutaneous calcifications 1393
    drug-induced bullous dermatitis 1393

Index

skin diseases/disorders (*cont.*):
  chronic renal failure (*cont.*):
    keratinization abnormalities 1390–1
    nail dyschromia 1393–4
    pigmentation abnormalities 1157, 1390
    precancerous and cancerous lesions 1393
    pruritus 1391–2
    pseudoporphyria cutanea 1392–3
    vascular disorders with skin pathology 1393
  clinical examination 12
  diuretic-induced 213
  haemorrhage, causes in ARF *1042*
  in Henoch–Schönlein purpura 596, 602
  immunosuppression, complications of 2306–7
  infections
    bacterial 2306
    fungal 2306–7
    viral 2307
  inherited conditions
    autosomal dominant 2298–301
    autosomal recessive 2299
    X-linked 2299
  photosensitivity, thiazide diuretics and 213
  pigmentation abnormalities, chronic renal failure 1157, 1390
  rash of secondary syphilis, mimicking property 2303–4
  renal problems caused by 2298–308, *2304*, 2304–5
  renal transplantation complicating 1541
    following 1583–4
  skin necrosis, chronic renal failure 1393
  tumours, Australia, following renal transplantation 1582
skin-lightening cream, African use 161
skinfold measurement 1301
sleeping sickness 185
slow virus disease, Balkan nephropathy and 860
Sm antigen, antibodies to 647
small bowel,
  acute obstruction 1284
    management 1285
  calcium absorption 1764
    in uraemia 1764
  enzymes in uraemic animal models 1281
  non-occlusive bowel infarction 1284
  obstruction, haemodialysis 1454
  phosphate reabsorption 1788
  phosphate transport 1788
  structural changes, chronic renal failure 1281
small bowel disease, enteric hyperoxaluria 1835
small intestinal glucose–galactose transporter 745
small kidneys, *see* kidney(s), size
Smith–Lemili–Opitz syndrome, features and associated abnormalities *2037*
smoking,
  analgesic nephropathy 2267

smoking (*cont.*):
  Goodpasture's disease 9, 441, 445, 1064
    pulmonary haemorrhage 448
  hyponatraemia 874
  nitrosamines in tobacco smoke 2238
  renal cell carcinoma 2238
  renal transplantation 1577
  risk factor for renal disorders 9
SN-GFR, *see* glomerular filtration rate, single nephron-GFR
snake bite,
  acute renal failure **1134–5**
  poisoning 1011
  venoms and their actions **1046**
nephrocalcinosis in 1888
social class, calcium stone disease 1837
social history, presenting complaint and 7
sodium,
  aged kidney, *see* elderly patients
  in breast milk 52
  conservation, in sickle-cell disease 712
  cotransport, sodium–phosphate 1783–7
  cotransporter, Na$^+$2Cl$^-$K$^+$ 200
  cotransporter inhibitors, *see* loop diuretics
  countertransport, sodium–lithium 520–1, 2051
  delivery
    in absence of AVP, distal nephron function 41
    potassium homeostasis 898
  depletion
    acute tubular necrosis, oedema and 273–4
    causes in aged kidney 63
    chronic renal failure 1182
    Fanconi syndrome 726–7
    hypercalcaemia causing 1885
    overtreatment 273–4
    rarity 274
    renal prostaglandins 821
  dietary deprivation, renin effects *2071*
  dietary intake
    children with chronic renal failure 1306
    excesssive, *see* sodium, excessive intake
    low sodium diet 62
    oliguric patients 1019
    total parenteral nutrition in ARF *1021*
  dietary restriction
    animal models of progressive glomerular scarring *1209*, 1212
    ascites 203
    hypertension 1177
    hypertension of end-stage renal failure 2120
    hypertensive diabetes 1647
    low sodium diet 62
    nephrotic oedema 203
    renin effects *2071*, 2119
  elderly patients
    salt wasting - causes 63–4
    site of sodium loss within nephron 62–3
    sodium loading 61–2
    tubular handling 62

sodium (*cont.*):
  excessive intake 8
    animal models 2118
    hypertension, chronic renal failure 1610
    hypertension and 2118, *2119*
    peripheral resistance 2119
    renal perfusion and 988
    vessel wall 2119
  excretion
    blood pressure control and 2058–60
    daily urinary *923*
    effector pathways 2061
    FE$_{Na}$, differential diagnosis 972
    fractional
      oliguria *vs* prerenal failure *988*
      tubular injury 988
    low sodium diet 62
    neonates 1113
    nephrotic syndrome 269
    normal values, neonates 51
    normal *vs* impaired *1182*
    perceived signal 2059–60
    pre-renal causes of uraemia 972
    type of disease and 988
    in volume homeostasis, increase/decrease 1919–20
  homeostasis
    end-stage renal failure 2118–19
    haemodialysis long-term patients 1441
    low sodium diet 62
    normal renal handling 263–4
    regulation 262–3
    tubular handling, distal and proximal nephrons 62
  hyperglycaemia, serum sodium effects 873–4
  normal renal handling 263–4
  osmolality, standard equation 873
  'overflow' *vs* 'underfilling' theories 202
  permeability defect, postulation in Bartter's syndrome 785
  pregnancy
    animals as gestational models 1921
    excretion, increase/decrease *1919*, 1919–20
    mean values through gestation 1923
  pump
    hypertension predisposition 2061
    hypomagnesaemia and 1813
    ion reabsorption from intestine 934
    Na$^+$-K$^+$-ATPase, *see* Na$^+$-K$^+$-ATPase pump
    pregnancy changes 1919
    sodium–hydrogen antiport 521
    tubular reabsorption *741*
    tubular reabsorption defect, Fanconi syndrome variant 742
  pump inhibitors, increase, in uraemia 1461–2
  reabsorption
    Bartter's syndrome and 784
    blood pressure 264
    from intestine, Na/H exchanger 934

sodium (*cont.*):
  reabsorption (*cont.*):
    glucose-coupled, glomerular filtration rate and 513
    increased in diabetes mellitus 513, 515, 521
    indomethacin promoting 706
    inhibition by Bence–Jones protein 565
    inhibitors 200
    insulin effect in diabetes mellitus 515
    post obstruction 2013
    in sickle-cell disease 714
    theories 202–3
  reabsorption inhibitors, *see* thiazide diuretics
  renal perfusion and 988
  renin and *2071*, 2119
  retention
    chronic renal failure 1182
    minimal change nephrotic syndrome 302
    non-steroidal anti-inflammatory drugs 823
    renal effects 2334
  and water retention
    cirrhosis 877
    congestive heart failure 877
    mechanism 876–7
    nephrotic syndrome 877
    renal failure 878
  salt-losing congenital adrenal hypoplasia 771
  salt-losing nephropathy, renal cell carcinoma 2242
  serum, postoperative monitoring 878, 881–2
  sodium and water excretion, nephrotic syndrome 269
  sodium-dependent hypertension 1459–60
  transport
    Bartter's syndrome and 784
    distal nephron function 40
    in nephron 198
    proximal tubule 741–2
      entry step 741
      exit step 741–2
    tubular reabsorption, neonates 52
    urinary acidification 767
  volume status and 1459–60
  wasting
    causes in aged kidney 63
    renal tubular acidosis 774
  see also hypernatraemia; hyponatraemia; Na$^+$2Cl$^-$K$^+$ cotransporter; Na$^+$-K$^+$-ATPase inhibitors
sodium aurothiomalate, *see* aurothiomalate
sodium bicarbonate,
  handling in kidney disease 189
  in renal tubular acidosis, in Sjögren's syndrome 697
  see also bicarbonate
sodium bromate, poisoning 1011
sodium channel blockers,
  potassium-sparing diuretics 200
sodium channels, locations 200
sodium chlorate, poisoning 1011
sodium chloride,
  administration in children 1305, 1306

sodium chloride (*cont.*):
   deficiency, chronic renal failure 1305
   wasting, infants 1305
sodium etidronate, hyperparathyroidism 1856
sodium iothalamate clearance (^{125}I-labelled), in amyloidosis 555
sodium nitroprusside,
   children, emergencies *2149*
   hypertension, accelerated 2135
   test, tests (ratios) 1081
sodium polystyrene sulphonate, hyperkalaemia, therapy *910*, 911
sodium stibogluconate, leishmaniasis 185
sodium sulphate, infusion, distal renal tubular acidosis diagnosis 775
sodium valproate,
   binding 2331
   cyclosporin therapy, interactions *1563*
sodium–hydrogen antiport 521
sodium–lithium countertransport, contraindications 2051
sodium–lithium countertransport rates,
   in diabetic nephropathy 520
   insulin resistance association 521
   lipid abnormalities and 520–1
sodium–phosphate cotransport,
   adaptation in low intake 1785
   allosteric regulation 1784
   cell culture (LLCPK1) 1785
   pH and 1784, 1787
   proximal cells, brush border 1783–4
   transport of phosphate 1784
solar keratosis 2307
soluble immune response suppressor, *see* lymphokines
solute titration curves,
   splay 742–3, *743*
   threshold 742
   transport maximum, T_m 742
   tubular reabsorption *742*
solvent abuse,
   Fanconi syndrome 735
   Goodpasture's disease 445–6
   nephrotoxicity 1009
solvents, organic, *N*-acetyl-β-D-glucosaminidase excretion 18
somatomedins,
   assessment 1323
   binding proteins 1-3 in children 1324
   bioassays 1323
   growth hormone axis
     and body growth 1324
     carbohydrate metabolism 1324
     kidney function 1324
   increase, growth hormone administration 1304
   inhibitors
     binding proteins 1-3 1324
     circulating 1323
     inactive IGF-I precursor competing at receptor 1323–4
   see also insulin-like growth factors
somatostatin, blocking by glucagon 26
somatotroph function tests, uraemic patients 1323
sorbinil 518

sorbitol,
   colonic necrosis 911
   hypertonic dehydration 883
   increased in diabetes mellitus 518
   peritoneal dialysis solution 1481
sore-thumb syndrome, venous hypertension 1414
sotolol, properties, in relation to use in impaired renal function *2053*
soybean diet, chloride depletion, metabolic alkalosis 950
specific gravity of urine, use of urinometer 16
spermatogenesis, haemodialysis 1335
spermicides, urinary tract infection and 1685
spermine, spermidine,
   renal cell carcinoma 2242
   toxicity, uraemic syndrome 1244
   uraemic syndrome, suppression of haemopoiesis 1238
sphingolipidoses, major *2206*
sphingomyelin deficiency, Niemann–Pick disease 2206
spider bite, acute renal failure and 1136
spinal cord,
   evaluation in neonates 2033
   tethered cord syndrome 2033
spindle-shaped cells, leiomyosarcoma 2243
spirochaete infections, *see* leptospirosis
spironolactone,
   ascites 203
   Bartter's syndrome 787
   contraindications in hypoaldosteronism 910
   dialysis, K$^+$ load 896
   interference with creatinine clearance 32
   long-term therapy 213
   potassium canrenoate metabolites 213
   toxicity 213
   urate/uric acid effects 829
splay, solute titration curves, defined 742, *743*
spleen, cyst(s), polycystic disease, autosomal dominant 2179
splenectomy, renal transplantation, prior to 1517
spondylitis, imaging procedures 1373, 1375
spondyloarthropathy,
   in chondrocalcinosis 2297
   chronic, complication of renal disease 2297–8
   haemodialysis, long-term patients 1442
spontaneously hypertensive rat, sodium balance, renin and sodium depletion *2060*
Sporothrix, infection 1748
Sprague–Dawley rats,
   phosphate-restricted diet 1211–12
   progressive age-related glomerular lesions 1206–7
squamous cell carcinoma,
   chronic renal failure 1393
   glomerulonephritis 2305
   immunosuppressed patients 2307

squamous cell carcinoma (*cont.*):
   management 2269
   risk following renal transplantation 1539, 1582
squamous epithelial cells, in urinary sediment 20, 21
SSA antigen, *see* Ro antigen
SSB antigen, *see* La antigen
staghorn calculi 72, 94–5
   primary hyperoxaluria 2227
   removal 1872–3
Staphylococcus aureus,
   abscess 470
   bacteraemia, haematogenous infection 1678
   cortical abscess 1695, 1696
   endocarditis in dialysis patients 1275
   haemodialysis patients 1453
   neonates 1701–2
   perinephric abscess 1696
   septic shock, renal transplantation following 1534–5
Staphylococcus epidermidis,
   endocarditis in dialysis patients 1275
   haemodialysis patients 1453
   periurethral flora 1660
   shunt nephritis 466
Staphylococcus saprophyticus 1666
   adherence to uroepithelial cells 1667
   incidence in urinary tract infection 1685
   numbers in bladder urine 1677
   urinary tract infection 1661
Staphylococcus spp.,
   colonization, shunt(s) 460
   infected stone disease 1831
   infection-associated glomerulonephritis 459–60
   staphylococcal protein A, IgG antibodies, adsorption 450
   staphylococcal septicaemia 460
Starling forces,
   distribution of body fluids 264
   disturbance 264–5
   localized oedema 265
   peripheral capillaries 265
   pregnancy toxaemia 273
starvation, ketoacidosis 930
statil 518
Staverman reflection coefficient, peritoneal cavity 1479
steal syndrome, vascular access 1413–14
steatorrhoea, hyperoxaluria 1857
stem cells, intrathymic differentiation *242*
steroid-induced osteonecrosis, following transplantation 1379–80
steroid-resistant nephrotic syndrome,
   8 weeks/16 weeks prednisone therapy 144
   focal segmental glomerosclerosis 2219
steroids,
   erythropoietin and 1353
   in Henoch–Schönlein purpura 606, 607
   11β-hydroxysteroid dehydrogenase deficiency, nephrocalcinosis 1895

steroids (*cont.*):
   immunosuppressive therapy *1554*
   withdrawal in transplantation 1591
   lung infections 2338
   mesangiocapillary glomerulonephritis 400
   in mixed connective tissue disease 684, 698
   potassium-losing effects, pregnancy 1917
   progressive nephropathy 1217
   renal transplantation, withdrawal effects 1538
   skin changes, following renal transplantation 1541, 1583
   toxicity, Cushingoid syndrome 322
   see also corticosteroids; *specific substances*
sting procedure 1714
Stix test, Griess test 1681
stocking glove syndrome, uraemic peripheral polyneuropathy 1396–8
stomach,
   chronic renal failure effects 1280
   dysfunction following renal transplantation 1582
   see also gastric
stones, *see* renal stones
Streptococcus,
   infected stone disease 1831
   see also poststreptococcal glomerulonephritis
Streptococcus agalactiae, urinary tract infection 1661
Streptococcus pneumoniae,
   nephrotic syndrome
     antipneumococcal vaccines 278
     complement deficiency 278
     primary peritonitis 276–7
Streptococcus pyogenes,
   streptococcal impetigo, glomerulonephritis 2304
Streptococcus spp.,
   β-haemolytic, primary peritonitis in nephrotic syndrome 276–7
streptococcal antigens, endostreptosin 409
streptococcal infections
   haemolytic group A, poststreptococcal endocapillary glomerulonephritis 406
   haemolytic group A receptor, Fc fragment of IgG 409–10
   impetigo, glomerulonephritis 2304
   infection-associated glomerulonephritis 458–9
   see also poststreptococcal glomerulonephritis
streptococcal neuraminidase disease marker 409
   IgG deposition 459
   pneumococcus pneumonia 460
streptococcal pharyngitis, diagnosis and treatment 414
streptococcal proteinase, disease marker 409
streptococcus pneumonia, clinical features 460
streptokinase, renal effects 2335

Index

streptomycin,
  drug dosage, renal impairment 184
  renal tuberculosis 1727
streptozotocin,
  antineoplastic treatment, renal complications 2319
  and carcinoma 2238
  tubular injury in acute renal failure 1011
  urate excretion increase 829
streptozotocin diabetic rats 513, 518, 529
streptozotocin-induced diabetes,
  animal models of adaptive haemodynamic change in renal ablation 1193–4
  hypertension effects *2084*
stress,
  calcium stone disease 1837
  in dialysis patients 1468–9
stroke,
  hyperkalaemia, protective effects 907
  transplantation, following 2331
  young persons, imaging 2324
Strongyloides stercoralis, handling in kidney disease 185
strongyloidiasis, in transplantation patients, developing years after contact 10
struvite,
  infection stones 1666
  see also magnesium ammonium phosphate
subdural haematoma, haemodialysis 2329
subendothelial deposits,
  in essential mixed cryoglobulinaemia 688–, 691
  fingerprint-like appearance 653, 654
  in Henoch–Schönlein purpura 603, 604
  in lupus nephritis 652
  space widening, radiation nephritis 849, 850
subepithelial deposits,
  amyloid (spikes) 552, 553, 573
  fingerprint-like appearance 653, 654
  in Henoch–Schönlein purpura 603, 604
  in lupus nephritis 652, 653, 654
succinylacetone, accumulation 730
succinylcholine, potassium permeability 907
sucralphate, gastrointestinal bleeding 1021
Sudan III and IV, lipid staining 20
sugars, pregnancy, excretion increase and 1916
sulbactam, CAPD peritonitis regimens *1489*
sulindac 518
  drug-induced hypoaldosteronism, therapy 909
  interactions with diuretics 208
  minimal change nephrotic syndrome 1089
  prerenal failure 987
  renal sparing effect 823
  uraemia, conversion to active metabolite 177

sulphamethoxazole,
  CAPD peritonitis regimens *1489*
  trimethoprim and, *see* co-trimoxazole
sulphate,
  inhibitor of calcium oxalate crystallization in urine *1828*
  inhibitor of calcium phosphate crystallization in urine *1827*
sulphate salts, sodium tubular reabsorption increase 947
sulphinpyrazone,
  fractional urate clearance 827
  renal failure, urate clearance 826
  urate excretion increase 829
sulphonamides,
  crystalluria, nephrotoxicity 994
  drug-induced interstitial nephritis 1088–9
  drug-induced lithiasis 167
  drug-induced renal vasculitis 167
  handling in kidney disease 183
  hepatotoxicity 2315
  nephrotoxicity 1010
  pregnancy 1929
  urinary appearance 22
  urinary tract infection 1686
  pregnancy 1684
sulphonylurea compounds,
  elimination in chronic renal failure 1645
  hypoglycaemia and 1645
  see also chlorpropamide; glibenclamide
sulphosalicylic acid test, proteinuria 5
sulphoxidation, gold/penicillamine nephrotoxicity prediction 683
sunlight, exposure, cancer risk following renal transplantation 1539, 1582–3
superoxide dismutase, cytoplasmicturition-mitochondrial 256
superoxide ion, hydroxyl radicals 169
suprapubic aspiration of urine, children 1708
  urinary tract infection 1680
suprapubic urethrocystography 81
suprofen, flank pain syndrome 823
sural nerve, conduction velocities 1397
suramin, trypanosomiasis 185
surface area/weight calculations 30
surgical procedures,
  haemodialysis patients medication 1451
  perioperative treatment 1451–2
  to renal vascular supply
    ischaemic renal disease 1081–2, *1082*
    operative risk vs discriminant analysis *1082*
    see also angioplasty; *specific percutaneous procedures*
  tropics, acute renal failure 1138
  see also nephrectomy; renal transplantation
sweating, hypernatraemia 887
swimmer's itch 1738
Swyer syndrome (46,XY gonadal dysgenesis), features and associated abnormalities *2038*

sympathetic nervous system, blood pressure control in renal disease **2065–6**
syphilis,
  acquired, nephropathy 461
  congenital, nephropathy 461
  infection-associated glomerulonephritis 461
  secondary, skin rash 2303–4
systemic lupus erythematosus **646–67**
  acute interstitial nephritis 1091
  acute renal failure 1069
  assessment 1151
  autoantibodies 589, 641, *647*, 647–8
    anti-La in pregnancy 1944
    anti-Ro (SS-A) in pregnancy 1944, 1946
    anticardiolipin, in pregnancy 1944, 1945, 1946
    antiphospholipid, in pregnancy 1944, 1945
    binding to histones/nucleosomes 637, 647
    pregnancy 1944–6
    to DNA, *see* anti-DNA antibodies
    to extractable nuclear antigens 647
    to phospholipid antigens *647*, 648
  classification criteria 646, *647*
  clinical activity
    assessment 646, *647*
    serology correlation 652
  clinical features **648–52**
    in children 649–50
    drug-related 650–1
    in elderly 650
    see also lupus nephritis
  crescentic glomerulonephritis 421
  cutaneous flares, drug-induced 651
  definition **646**
  diagnosis **646–8**
    clinical criteria 646–*647*
    lupus band test 648
    serological/immunological 647–8
  differential diagnosis, mesangiocapillary glomerulonephritis 392
  DNA receptor defect 637
  encephalopathy and 1400
  end-stage renal failure, proportional distribution by age at commencement *1229*
  epidemiology **646**
  ethnic factors 9
  experimental idiotypic immunointervention 638–9
  general notes 2293
  genetic control **642–3**
  glomerular disease in the tropics 488
  glomerular involvement, isolated glomerulopathy 234
  glomerulopathies, acute renal failure in liver disease 1104
  hypertension and renal failure 1944–5
  IgA nephropathy and 339–40
  immune complexes 589, 637
  assays 648

systemic lupus erythematosus (*cont.*):
  immune complexes (*cont.*):
    DNA-anti-DNA 637–8
    elimination 588–9
    size 587, 638
  lupus coagulant, pregnancy 1944
  malignancy risk 659
  membranous nephropathy 374
  microangiopathic haemolytic anaemia 2341
  models 639–40, 642
  mortality, causes 658–9
  neonates, heart block and other defects 1945
  neoplastic complications of 2320
  pathogenesis **636–45**
    anti-DNA antibodies, *see* anti-DNA antibodies
    B-cell abnormalities 636, 640
    complement dysfunction 641–2
    of glomerular injury **639**
    hormonal factors 636
    idiotopes and, *see* idiotypic network
    immune dysregulation **639–42**
    T-cell abnormalities, *see* T cells
    thymus role 641
  pregnancy **1943–6**
    autoantibodies 1944–6
    drugs 1945–6
    effect on systemic lupus erythematosus 1943–4
    monitoring 1945
    see also lupus nephritis
  prognosis and therapy, *see under* lupus nephritis
  prophylaxis, aspirin low-dose 1967
  renal involvement
    classification 145
    see also lupus nephritis
  renal transplantation 1515
    recurrence following 1574
  Sjögren's syndrome association 695
  skin lesions 2302
  thrombotic complications 1050
systemic sclerosis,
  pregnancy 1943
  renal effects 2336
  renal transplantation 1515
  see also scleroderma

T antigen (Thomsen–Friedenreich antigen), haemolytic–uraemic syndrome 1049
T cell(s) **240–3**
  accessory molecules 242–3
  activation
    cascade of activation-associated events 1552
    initiation 1552
    proliferation signals 1553
    rejection of allograft **1551–4**
    T-cell growth factor secretion 1553
  antibodies 1312, *1313*
  antigen presentation and processing 240–1
  antigen recognition complex
    constitution 1557
    stucture 1552
  antigen-receptor–CD3 complex, monoclonal antibodies 1557

T cell(s) (*cont.*):
  antiself Ia autoreactive T cells, B-cell activation, immunologically-mediated mercury-induced 845
  autoantibodies to 636
  autoreactive potential, elimination 240–2
  blocking, experimental solutions 1558
  CD antigens 1312, *1313*
    CD4+ infiltration, poststreptococcal endocapillary glomerulonephritis 411
  CD4, see T cells, helper
  CD4:CD8 ratio, in SLE 641
  CD25, see interleukin-2 receptor
  cytoplasmic second messengers 1552
  cytotoxic, see T cell(s), suppressor/cytotoxic
  depletion, Heymann's nephritis 247–8
  differentiation clusters 1312, *1313*
  epithelial ICAM-1 adhesion molecules 1547
  function, elderly patients 1629
  glomerular damage, essential mixed cryoglobulinaemia 688, 691
  glomerular localization 255
  growth factor secretion
    growth factor β, proteoglycan production 257
    T cell activation 1553
    see also interleukin-2
  helper T cells
    deficiency in uraemic syndrome 1239
    dysfunction in SLE/SLE models 640, 641
    in giant cell arteritis 591
    immunology, role 1543–4, 1546
    in lupus nephritis 658
    phenotype (CD4+), immune response to allograft 1545–6
    in sarcoid granuloma formation 576–7
    in sarcoid granulomatous interstitial nephritis 578
    in Sjögren's syndrome 696, 697
    in vasculitis pathogenesis in animals 585
    in Wegener's granulomatosis 591
  idiopathic nephrotic syndrome, impaired responses to polyclonal activators 310
  immune response **1312–14**
    specific response in uraemia **1314–16**
  urinary tract infection 1671–2
  immunology, role 1543–4, 1546
  immunosuppressive therapy, multiple drug protocols 1556
  interstitial infiltration
    in essential mixed cryoglobulinaemia 689
    in lupus nephritis 658
  membrane markers 1312, *1313*
  in microscopic polyarteritis 625
  nephrotic syndrome, impaired function 278
  NK cells, elderly patients 1629

T cell(s) (*cont.*):
  NSAID-induced acute tubulointerstitial nephritis 822
  ontogeny and thymic selection 241
  over-immunosuppression, polyclonal immune globulins 1557
  pre-T cells, intrathymic differentiation *242*
  proliferation in response to proteinase 592
  self-reactive 246
  in Sjögren's syndrome 696
  in SLE, abnormalities 636, 640–1
    human 641
    murine 640–1
  subsets and functions 242
  suppressor/cytotoxic
    CD4+, CD8+ cells, rejection of allograft 1547
    cytotoxic T phenotype (CD8+), immune response to allograft 1545
    dysfunction in SLE/SLE models 640, 641
    idiopathic nephrotic syndrome 310, 311
    L3T4+ cell inhibition 640
    paucity in Sjögren's syndrome 696
    soluble immune response suppressor 311–12
  T3+ cells, rejection of allograft, indications 1529
  T cell markers 241
  T cell receptor 241, 642
    αβ, γδ *242*
    characteristics *242*
    genes *241*
    multiple forms 1558
  thymic selection theory 242
  thymus gland, thymic selection 241
  tolerance at B cell level 241–2
  type I allergic reactions 1094
T cell leukaemia/lymphoma, calcitriol levels 1777
  epidermotrophic cutaneous 2305
  glomerulonephritis 2305
tabaciere, alternatives to A-V fistula 1405
'tactoids', glomerulopathy, immunotactoid 2281–3
Takayasu's arteritis 592, 613, 630
  children, ethnic factors 2144
  neurological involvement 2324
  treatment 1069
Tamm–Horsfall protein,
  anti-Tamm–Horsfall glycoprotein 1664
  binding of *Escherichia coli* fimbriae 1671
  haem protein aggregation 995
  inhibitor of calcium oxalate crystallization in urine *1828*
  inhibitor of calcium phosphate crystallization in urine *1827*
  in myeloma cast nephropathy 565
  origin 227
  pI and precipitation 565
  pregnancy, increased excretion 1917
  protein casts, nephronophthisis 2190

Tamm–Horsfall protein (*cont.*):
  protein co-precipitation 996
  response to 1985
  synthesis 1664
  tubulointerstitial disease model 1093
  urinary casts 20, 21
tap water, see water, haemodialysis, preparation
tapetorenal degeneration, nephronophthisis, autosomal recessive 2192–3
TAR syndrome, features and associated abnormalities *2037*
Tauri disease, enzyme defect 2210
taurine,
  deficiency, neonates 888
  excretion, pregnancy and 1916
  hypomagnesaemia and 1813
Tay–Sachs disease, *N*-acetyl-β-glucosaminidase deficiency 2206
TBM, see tubular basement membrane
technetium scan, see (Tc-99m-)dimercaptosuccinate, scintigraphy
teeth,
  deficient enamel 1892
  phosphate in *1789*
Teflon, suburothelial injection 1714
teicoplanin,
  CAPD peritonitis regimens *1489*
  handling in kidney disease *182*
temporal arteritis, see giant cell arteritis
tendons, rupture, following transplantation 2298
terazosin, hypertension 188
terbutaline, hypokalaemia 902
terfenadine, handling in kidney disease 187
terminal renal failure,
  calcification in 1373
  uraemic syndrome 1239
terodiline hydrochloride, prostatitis 1690
testis,
  function, chronic renal failure 1615
  testicular growth, chronic renal failure 1614
  undescended, prune belly syndrome 2034
testosterone,
  effects, sexual function 1334–5
  impotence, erectile, and 1332
  luteinizing hormone effects 1298
  in uraemia 1332
tetany, acute renal failure 2326
tethered cord syndrome 2033
tetrachlorethane, nephrotoxicity 1009
tetracycline,
  action 185
  acute on chronic renal failure 1159
  Fanconi syndrome 735
  handling in kidney disease 183
  hepatotoxicity 1106–7
  hypercatabolism and 978
  nephrotoxicity 994, 1106–7
  PAH excretion 727
  pregnancy 1929
  renal clearance, osmotic diuresis and urea increase 179

tetrazolium reduction test, bacteriuria 1663
thalamus, injury, hypodipsia and adipsia 887
thalidomide embryopathy, features and associated abnormalities *2040*
thallium, intoxication, loop diuretics 208
thallium-201 perfusion imaging, uraemic patient with cardiac problems 1272
theophylline,
  hypokalaemia 902
  percentage protein binding according to renal function *1240*
thiamine deficiency, left ventricular hypertrophy and, end-stage renal disease 1266
thiazide diuretics **200–2**
  angiotensin converting-enzyme inhibitors and 212
  bone mineralization and 207
  calcitriol and, interactions 1381
  calcium excretion 1852
  calcium nephrolithiasis 207
  carbohydrate intolerance 210
  carbonic anhydrase inhibitors, photosensitivity 213
  chronic renal failure, toxicity 1185
  congestive heart failure 202
  deafness 2331
  diabetes insipidus 207
  drug-induced interstitial nephritis 1089–90
  familial X-linked hypophosphataemic rickets 755
  hypercalciuria 1852
  hypertension
    antihypertensives 201
    diabetes 1647
    end-stage renal failure 2120–1
    renal, treatment 2052
  hyponatraemia 879
  inhibitors of early distal NaCl tubular reabsorption 200
  kaliuresis in hypoaldosteronism 910
  lithium-induced polyuria 856
  magnesium excretion 1808
  mechanism of action 200
  metabolic alkalosis, oedematous disorders 947
  NSAID interactions 823
  with oral loop diuretics 176
  potentiation of frusemide 1273
  'reversed autoregulatory response' 201
thick ascending limb, see Henle's loop, thick ascending limb
thin-basement membrane nephropathy, Alport's syndrome 2203
thiobendazole, handling in kidney disease 185
thioflavin T, amyloid staining 391
6-thioguanine nucleotides, measurement, thiopurine methyltransferase metabolism 1950
thiopentone, handling in kidney disease 185

thiopronine, rheumatoid arthritis, and drug-induced nephropathy 162
thiopurine methyltransferase, renal transplantation, 6-mercaptopurine metabolism 1950
5-thiopyridoxine, rheumatoid arthritis, and drug-induced nephropathy 162
thiotepa, renal pelvis and ureter tumour 2269
thiouracils, thyroid disease 189
thirst,
  cerebral dysfunction and 63
  compulsive 42
  hypodipsia, adipsia 887
  non-osmotic regulation 869
  osmotic regulation 869
  psychogenic polydipsia 874
  reduced sensation in elderly subjects 65
Thomsen–Friedenreich antigen (T antigen), haemolytic–uraemic syndrome 1049
thoracic pain, long-term haemodialysis 1452
Thorotrast,
  carcinogenicity 2238
  drug-induced acute toxic tubular nephropathy 164
  see also radiocontrast media
threadworms, mebendazole 185
threonine, excretion, pregnancy and 1916
thrombocytopenia,
  mechanisms 1043
  platelet changes in acute renal failure 1042
  renal circulation 1053
    Factor VIII 1053
    platelet aggregation 1053
thrombocytopenic purpura 596
thrombocytosis, minimal change nephrotic syndrome 302
thromboembolic lung disease 2338
thromboembolism,
  coagulopathy, nephrotic syndrome complications 282–4
  nephrotic syndrome **278–84**
    coagulopathies in 282–4
    complications 278–82
    treatment and prophylaxis 284
  see also arterial thrombosis; renal vein thrombosis; venous thrombosis
thrombophlebitis, contrast media injection 77
thromboplastin, partial thromboplastin time, target in haemodialysis 1433
thrombosis,
  complication of haemodialysis 1363
  endothelial pathology, and clinical symptoms 1269
  following renal transplantation 1576
    cyclosporin vs azathioprine 1576
  glomerular, in SLE 656
  intraluminal, in essential mixed cryoglobulinaemia 688
  renal vein, see renal vein thrombosis

thrombosis (cont.):
  risk, in SLE 659
thrombotic microangiopathy, bleomycin, vinblastin and cisplatin induced, resembling radiation nephritis 849
  radiation nephritis, similarities 850
  vascular diseases with renal involvement, schematic representation 152
thrombotic thrombocytopenic purpura **1046–56**
  defined 1055
  idiopathic 2341
  plasma factor causing platelet aggregation 1053
  prostacyclin deficiency 1053
  radiation nephritis, similarities 850
  in SLE 656
  treatment 1055–6, 1056
thromboxane A_2,
  arteriolar myocytes of afferent arterioles, effects 26
  cyclosporin, nephrotoxicity 1561
  in lupus nephritis 662
  pregnancy-induced hypertension 1964
  prostacyclin effects 1965
  renal effects 820, 821
  role in glomerular sclerosis 1202–3
  synthesis 257, 820
thromboxane antagonists, in lupus nephritis 662
thromboxane B_2,
  glomerulonephritis-type diseases 408
  urinary, cyclosporin, nephrotoxicity 1561
thromboxane synthetase, inhibitors, animal renal ablation models 1195, 1215, 1216
  cyclosporin therapy 1561
β-thrombulin, role in glomerular sclerosis 1202–3
thymus,
  cell preparations, immunosuppressive therapy 1556
  in SLE pathogenesis 641
  thymic hormones
    in SLE pathogenesis 641
    zinc loss, lymphocyte function in nephrotics 278
  thymic selection theory, T cells 241, 242
thyroid gland/hormones,
  action
    CRF 1321
    uraemia 1321
  in amyloidosis 548, 550
  bone remodelling 1790
  CAPD 1321–2
  chronic renal failure disorders 1319–22
  deficiency, chronic renal failure 1304
  end-stage chronic renal failure, primary hypothyroidism and 'sick euthyroid' syndrome 1321
  end-stage chronic renal failure levels compared 1321
  function, magnesium and 1807

thyroid gland/hormones (cont.):
  growth action 1297
  kinetics, CRF 1320
  levels, uraemic syndrome 1239
  metabolism, CRF 1320
  protective action, nephrocalcinosis 1896
  thyroid hormone-binding globin levels in CAPD 1320
  levels in haemodialysis 1320
  nephrotic syndrome 290
  thyroid-stimulating hormone nephrotic syndrome 290
  response to TRH, CRF 1320
thyroidectomy, animal models of nephrotoxic nephritis 1208–9
thyrotoxicosis, hypokalaemic periodic paralysis, acquired 903
thyrotrophin-releasing hormone delayed rise in chronic renal failure 1320
  growth hormone, increase in uraemic subjects 1322
thyroxine
  phosphate transport and 1787
  thyroid disease 189
ticarcillin, CAPD peritonitis regimens 1489
ticlopidine, treatment of progressive glomerular sclerosis 1215
ticrynafen, urate excretion increase 829
tienilic acid 829
  hepatotoxicity 2315
timolol, properties, in relation to use in impaired renal function 2053
tissue perfusion fluid, volume 264–5
toadstool poisoning, acute renal failure and 1011, 1136
tobramycin,
  CAPD peritonitis regimens 1489
  doses and therapeutic plasma concentrations of aminoglycosides, handling in kidney disease 182
  magnesium wasting and 1810
  nephrotoxicity 1010
tolbutamide,
  drug-induced nephropathy 163
  hypoglycaemia 189
tolmetin, minimal change nephrotic syndrome 1089
toluene,
  distal (type I; classic) 938
  hypokalaemia 902
  intoxication, Fanconi syndrome 735
  nephrotoxicity 1009
tonsillectomy, IgA nephropathy, primary 358
torasemide,
  action 199
  chronic renal failure 204
total lymphoid irradiation, immunosuppressive therapy prior to renal transplantation 1518
  in lupus nephritis 662
total parenteral nutrition in acute renal failure 1021
tourniquet application, hyperkalaemia, spurious 907

Townes–Brocks syndrome, features and associated abnormalities 2036
toxic nephropathy, diagnosis, EDTA registry 1231–3
Toxoplasma gondii infection, following renal transplantation, management 1535, 1536
  glomerulonephritis 467–8
trace elements,
  Balkan nephropathy 858–9
  dialysate 1446
trace metals,
  inhibitor of calcium oxalate crystallization in urine 1828
  inhibitor of calcium phosphate crystallization in urine 1827
toxicity, uraemic syndrome 1244
tranexamic acid,
  gastrointestinal bleeding 1284
  in sickle-cell disease 716
tranquillizers, handling in kidney disease 186
transaminases,
  hypertransaminasaemia, renal transplantation 1534
  papaverine induction of 1334
transferrin,
  anaemia, uptake 1349
  erythropoietin replacement therapy, r-HuEPO 1353–6
  loss in nephrotic syndrome 289
  lymphocyte function in nephrotics 278
  transferrin–zinc, loss in nephrotic syndrome 290
transforming growth factor-α, putative bone resorption factor 1776
transforming growth factor-β, bone remodelling 1790
  determination of progressive glomerular scarring 1192
  glomerular collagen 1203
  proximal tubule cells, effect on 1204
transfusions, see blood transfusions
transglomerular pressure gradient, elevated in diabetes mellitus 514
transient neonatal hypertyrosinaemia 730
transitional cell carcinoma, incidence 2266
  renal pelvis and ureter tumours 2266
  ultrasonography 89–90, 92
  differentiation of stages 1/2 92
  differentiation of stages 3/4 93
  ureteral tumours 2266
  vs renal calculi, ultrasonography 89–90
transitional epithelial cells, in urinary sediment 20, 21
transketolase, dysfunction 1397
transport maximum, T_m, solute titration curves 742
transport proteins, nephrotic syndrome, losses 289–90
trauma, see kidney(s), trauma
travel, renal stone disease 1848
trehalase, in urine 18
Treponema,
  infection-associated glomerulonephritis 461

Index

Treponema (*cont.*):
  serology, ethnic factors 10
TRH, *see* thyrotrophin-releasing hormone
triamterene,
  ascites 203
  Bartter's syndrome 787
  contraindications in hypoaldosteronism 910
  creatinine clearance, interference 32
  drug-induced interstitial nephritis 1089–90
  drug-induced lithiasis 167
  magnesium excretion 1808
  nephrotic syndrome 270
  and NSAIDs, acute renal failure and 1016
  potassium maintenance therapy 904
  renal tubular acidosis 938
  sodium reabsorption blocking 200
  toxicity 213
  urinary appearance 22
Trichomonas vaginalis, in urine 21
trichosis trichuria, mebendazole 185
Trichosporon beigelii, fungaemia 1747–8
triglycerides,
  delivery as chylomicrons 1255
  see also lipids
tri-iodide contrast media,
  drug-induced acute toxic tubular nephropathy 164
  see also radiocontrast media
trimethoprim,
  CAPD peritonitis regimens 1489
  children 1715
  cyclosporin therapy, interactions 1563, 1573
  handling in kidney disease 183
  renal transplantation contraindications 1579
  nephrotoxicity 1527
  urinary tract infection 1686–7
  pregnancy 1684
trinitrophenol-bovine serum albumin complex, IgA nephropathy 252, 255
triploidy, features and associated abnormalities 2040
trisomies, features and associated abnormalities 2040
tropics, diseases of,
  acute renal failure 1124–38
  acute tubular necrosis in 1125–6
  clinical manifestations 1129
  epidemiology and aetiology incidence 1125
  management 1129–38
  pathogenesis 1126–9
  pathology 1125–6
  climate 1124
  glomerular disease, *see* glomerular disease in the tropics
  glucose 6-phosphate dehydrogenase deficiency 1124–5
  Hantavirus disease 1143–5
  nutrition 1124
  Plasmodium malariae-associated membranous nephropathy 375
  socioeconomic factors 1124

Trypanosoma cruzi, *see* Chagas' disease
trypanosomiasis,
  glomerulonephritis, animal models 249
  pentamadine 185
  suramin 185
tryptophan, Hartnup disease 749
TSH, *see* thyroid stimulating hormone
TTC, *see* tetrazolium reduction test
tuberculosis,
  amyloidosis, secondary 488
  chronic renal failure assessment 1152
  diagnosis 1152
  drug dosage, renal impairment 184
  end-stage renal failure 1234
  proportional distribution by age at commencement 1229
  ethnic factors 10
  hypertension and 2091
  infection-associated glomerulonephritis 460–1
  IVU imaging 85
  lung infections in 2338
  renal transplantation 1533
tuberculosis, renal 1719–28
  aetiology 1719
  clinical features 1720–1
  closed cavity 1728
  drugs 1727
  incidence 1719
  laboratory data 1724, 1726
  management 1724–7
  non-functioning kidney 1728
  pathogenesis 1719–20
  radiology 1721–4
  specific cases 1727–8
  ureteritis 1728
tuberous sclerosis,
  angiomyolipoma 2181
  ash leaf spots 2300
  features and associated abnormalities 2036
  neurological involvement 2325
  polycystic disease, autosomal dominant 2181
  renal hypertension and 2048
tuberous sclerosis complex, angiomyolipoma 2243
  Bourneville's disease 103
tubocurarine, avoiding in kidney disease 186
tubules, *see* renal tubular –; renal tubules
tubulin, podocytes 133
tubuloglomerular feedback 983–4
tubulointerstitial disease,
  activity/chronicity indices 145
  cell-mediated, animal models 1093
  gold and penicillamine association 682
  immune-complex mediated, animal models 1093
  infiltrate, and fibrosis 1206
  liver disease, associated with acute renal failure 1105–7
  progressive chronic renal failure, compensatory hypertrophy 1203–6

tubulointerstitial disease (*cont.*):
  progressive reduction in nephrons 1203–6
  see also interstitial nephritis
tubulointerstitial nephritis,
  acute
    non-steroidal anti-inflammatory drugs 821–2
    NSAID-induced, renal biopsy 822
  acute renal failure in liver disease 1105–7
  animal model 170
  cadmium nephropathy 842–3
  chronic, NSAID-associated 823
  edetate therapy 841
  fenoprofen-induced 821
  heavy metals exposure, differential diagnosis 838
  hyporeninaemic hypoaldosteronism 909
  immunological origin 170
  in juvenile rheumatoid arthritis 682
  legionnaire's disease 462
  and uveitis (TINU), renal–ocular syndrome 1091–2
  see also analgesic nephropathy; interstitial nephritis
tuftsin,
  normal concentrations 704
  in sickle-cell disease 704
tumour lysis syndrome,
  acute 2319
  acute renal failure in 1012, 2343
  acute urate nephropathy 482, 832
  clinical features 1795
tumour necrosis factor,
  acute phase protein sysnthesis 1314
  anti-GBM antibody 447
  β, osteoclast activating factor 1776
  endothelial effects 448
  glomerular collagen 1203
  release from monocytes, bacterial endotoxin in dialysate 1316
  synergism with IL-1 1314
tumour suppressor genes,
  retinoblastoma, RB1 2161
  sporadic and hereditary tumours 2161
tumoural calcinosis 1794
tumours,
  and the kidney 2316–22
  analgesic abuse, risk 815
  analgesic nephropathy and 813
  antineoplastic treatment, complications 2318–19
  associated glomerulonephritis, *see* malignancy-associated glomerular disease
  autoimmune renal disease 2320
  byproducts and acute renal failure, light chain nephropathy 1012
  complications of renal disease 2316–18, 2320–1
  antineoplastic treatment 2318–19
  cyclosporin, nephrotoxicity 1566
  in dialysis patients 1450–1
  differential diagnosis 1723
  renal stone disease 94
  embolization, preoperative 113

tumours (*cont.*):
  end-stage renal failure, proportional distribution by age at commencement 1229
  factor release, causing hypercalcaemia 1776
  Fanconi syndrome 734
  haematuria and 75
  haemolytic anaemia, microangiopathic 1050
  haemolytic–uraemic syndrome 1050
  hypercalcaemia of malignancy 1776
  hypophosphataemic osteomalacia 756–7
  hypouricaemia and 757
  imaging
    computed tomography scanning 101–3, 106
    dilated/undilated urinary tracts, radiological tests 73
    DMSA scans 126
    IVU imaging 83
    ultrasonography 75, 88–90
    vs cyst, computed tomography scanning 100
    vs renal calculous disease, ultrasonography 94
  inappropriate secretion of ADH 878
  molecular genetics 2155–61
  non-renal, in dialysis patients 1451
  oncology and the kidney 2316–22
  parenchymal origin 101–3
  polycystic disease, autosomal dominant 2174
  PTH-related protein 1786
  renal replacement therapy in patients with 2321–2
  renal transplantation, risk 1539, 1582–3, 2322
  renin-producing, hypertension and 2092–3
  staging, MRI 107–8
  symptoms, classical triad 71
  treatment
    discontinuation 1540
    renal complications of 2318–19
  tumour suppressor genes, sporadic and hereditary tumours 2161
  two-mutation hypothesis 2255
  see also malignancy; renal mass; renal pelvis and ureter tumours; *specific names, organs and regions*
typhoid, renal involvement 460
tyrosinaemia type I, Fanconi syndrome 729, 730
tyrosine, excretion, pregnancy and 1916

U1 RNP, antibodies to 698
ulcer(s),
  acute stress, dialysed patients 1279–80
  analgesic nephropathy and 2310
  colonic, chronic renal failure 1281
  haemodialysis 2311
  peptic ulcer
    analgesic nephropathy and 813, 815

Index

ulcer(s) (cont.):
 peptic ulcer (cont.):
  in chronic renal failure 1279–80
  dialysed patients 1279–80
  gastrointestinal bleeding 2311
  incidence, dialysed patients 1281
  milk–alkali syndrome 1836
  renal transplantation 1509
   management 1582
 renal transplantation
  large bowel 1281
  oral 1285
Ullrich–Turner syndrome, features and associated abnormalities 2041
ulnar deviation, renal osteodystrophy 1611
ultrafiltration, extracellular fluid, formula 264–5
ultrafiltration coefficient 269
ultrasonography **86–97**
 acute tubular necrosis 977
 analgesic nephropathy 815
 anatomy of kidney, adult/neonate 87
 basic principles 86–7
 duplex Doppler ultrasound 975
 essential hypertension *vs* renal causes 2050
 flow diagram for investigations 73
 in utero studies **1114–15**
 indications 71
 interventional 97
  applications 97
 lithotripsy 1875
 neonates 2026
 renal mass 2245
 renal obstruction 975–6
 screening at the renal unit 979
 small *vs* absent kidneys 71
 technique 87
 upper tract obstruction 2008
 vesicoureteric reflux 1987
upper urinary tract obstruction, *see* urinary tract obstruction, acute upper *and* chronic upper
uraemia,
 acute, *see* acute renal failure
 acute on chronic renal failure, causes 1158–61
 anaemia in, *see* anaemia, chronic renal failure
 animal models, interstitial myocardial fibrosis 1266
 arrhythmia 1270
 atherogenesis 1257
  pro/anti-atherogenic factors 1268
 autonomic nervous system 1461
 bile acid distribution, disordered 1281
 bleeding diathesis in 2344
 bleeding time, shortened 142
 cancer in 1450–1
 carbohydrate metabolism 1251–4
 cardiological disorders in uraemic patients **1264–75**
 children **1155–7, 1605–18**
  causes *1122*
 chronic, in essential mixed cryoglobulinaemia 690
 chronic renal failure, *see* chronic renal failure
 creatinine and 1167
 in diabetes mellitus 511

uraemia (cont.):
 dialysis
  CAPD 1482
  long-term haemodialysis **1436–55**
  need for, symptoms 1167
 diet, acid and base production 932–3
 dietary control 1179
 drugs
  non-renal clearance 177
  sulindac conversion 177
 growth in children, *see* growth in uraemic children
 hypoglycaemic stress reaction, suppression 1325
 lipid metabolism 1254–8
  abnormalities *1256*
  lipid synthesis 1255
  lipoprotein degradation 1255–6
 liver changes 1282
 metabolic acidosis 932–3
 pericarditis and pericardial effusion 1274–5
 pruritus 1157, 1158
 serum repressors, IL-2 bioavailability 1314
 sexual dysfunction 1157
 T cell responses 1314–15
 see also uraemic syndrome
uraemic cardiomyopathy, haemodialysis, long-term patients 1438
uraemic cystic disease of the kidney, Doppler findings 94
uraemic encephalopathy, differential diagnosis *1399*
uraemic frost, terminal renal failure 1390
uraemic hypertriglyceridaemia, liver changes 1283
uraemic lung 272
uraemic patient with cardiac problems **1270–4**
 evaluation 1270–2
  24-hour ECG 1271–2
  circulatory assessment 1271
  coronary angiography 1272
  ECG 1271
  echocardiography 1271
  exercise testing 1271
  physical examination 1270
  radionuclide angiography 1272
  thallium-201 perfusion imaging 1272
 treatment 1272–4
  dialysis modalities 1272–3
  drug therapy 1273–4
  fluid overload 1272
uraemic pericarditis and pericardial effusion 1274–5
uraemic peripheral polyneuropathy 1396–8
 autonomic neuropathy 2328
 differential diagnosis *1398*
 mononeuropathy 2328
 motor and sensory neuropathy 2327–8
uraemic syndrome **1236–47**
 biochemical alterations
  drug protein binding 1240
  enzymatic processes 1240
 clinical characteristics 1237–40
 deficiencies 1246
 differential diagnosis 1239
 endocrinology 1240

uraemic syndrome (cont.):
 haemodialysis 1430–1
 neuropsychiatric alterations **1396–402**
 pathophysiology *1237*
 renal replacement therapy, interfering factors 1240
 terminal renal failure in 1239
 uraemic solute retention 1240–7
  elimination patterns 1246–7
  general classification of solutes 1240–1
  major retention products 1242–5
  markers 1245–6
  plasma concentration, factors influencing 1245
  separation and identification methods 1241–2
 see also uraemia
uraemic xerosis, chronic renal failure 1390–1
uranium, acute tubular necrosis, animal model 845
uranyl nitrate, high-salt diet 171
urate arthropathy, contraindicated in arthritis of hyperuricaemia 834
urate/uric acid **825–35**
 acute uric acid nephropathy
  acute renal failure complicating 832
  animal model 832
  predisposing factors *832*
  prevention 832
  tumour lysis syndrome 482
 assessment, gout 2295
 biochemistry 1861
 chronic uric acid nephropathy, causes *1830, 1831*
 circadian rhythms 828
 crystalline deposits
  endothelial damage 833
  interstitial tissue 833
 crystalluria
  nephrotoxicity 994
  uric acid stones and *1862*
 daily production 825
 elevated plasma levels, endogenous production in various disorders 829
 endothelial damage 835
 excretion, in sickle-cell disease 713
 fractional urate clearance *827*
 effects of various disorders *828*
 gouty arthritis, hyperuricaemia 212
 haemodialysis *vs* peritoneal dialysis 832
 nephropathy, chronic, tumours and 2318
 plasma, new values 1861
 pregnancy
  animals as gestational models 1920
  plasma 1921–2
  uric acid clearance/GFR changes 1915
 production
  gout 2295
  increased, differential diagnosis 973
 renal handling
  chronic renal disease 831
  diabetes 831

urate/uric acid (cont.):
 renal handling (cont.):
  drugs affecting *829*
  evaluation 827
  factors affecting 828–30
  glomerular filtration 825
  gout, primary and saturnine 830–1
  hypertension 831
  hypouricaemic syndromes 830
  inappropriate secretion of ADH 832
  liver cirrhosis 832
  new values 1861
  pathophysiological effects 825–7
  postsecretory reabsorption 826–7
  pregnancy 832
  reabsorption 826
  tubular secretion 826
 retention, diuretics, chronic administration 829
 serum, arthritis risk and 835
 solubility 825
 total body/plasma 825
 transport model 826
 tubular reabsorption model 827
 tubular secretion
  cyclosporin therapy effects 1561
  pyrazinamide suppression test 826
 uric acid clearance/GFR changes, pregnancy 1915
 wasting, Fanconi syndrome 726
 water diuresis effects 828
 see also uric acid stone(s) (disease)
urea,
 clearance, time *vs*, protein restriction *1179, 1180*
 criterion for pre-renal oliguria 969
 excretion
  estimation of protein intake 1181
  rate, assessment in children 1306
 generation rates, correlation regression analysis *1246*
 kinetic data *1246*
 magnesium excretion 1808
 plasma, interpretaion 238
 pregnancy, mean values through gestation 1923
 renal function decline, pregnancy and 1937
 serum, indomethacin, in sickle-cell disease 707
 time averaged concentration (TAC) 1430
 uraemic syndrome, toxic effects 1242–3
 urine/plasma urea, oliguria *vs* prerenal failure *988*
Ureaplasma urealyticum,
 infected stone disease 1831, 1863
 urinary tract infection 1661, 1686, 1688
urease inhibitor,
 acetohydroxamic acid 1666
 hydroxyurea 1666
urease in *Proteus* 1666
ureter(s),
 anastomosis, renal transplantation 1525
 capillary sclerosis resulting from analgesic nephropathy 811
 catheterization 1681
 children 1711–12

Vol. 1 – Sections 1–5.   Vol. 2 – Sections 6–11.   Vol. 3 – Sections 12–21.

ureter(s) (cont.):
  catheterization (cont.):
    Fairley technique 1712
    localization of urinary tract infection 1664, 1681
  colonization, neonates 1702
  connective tissue, hyperplasia in pregnancy 1909
  dilatation 1876
    increase in pregnancy 1909–10
  ectopic 2034–5
  fibrotic thickening 1723
  normal IVU imaging 83
  obstruction
    carcinoma 476
    see also urinary tract obstruction
  orifice to bladder, competency grading *1984*
  periaortitis 2018–20
  peristalsis
    animal models 2004, 2005
    failure 2003
    failure due to fibrosis 2004
    normal function 2006
    pressures 2006
  retroperitoneal fibrosis 2018–20
  smooth muscle, hypertrophy in pregnancy 1909
  standing, IVU imaging 83
  stenosis 1695
    following renal transplantation 1575
  stones, removal 1872–3, 1878–9
  stricture 1879
  tone, increase in pregnancy 1909
  tumours, see renal pelvis and ureter tumours
ureterectomy, ureteral stump cancer following 2269
ureteroceles
  embryological aspects 2034
  ultrasonography 92
  see also urinary tract
ureteral valves 2089
  hypertension and 2089
ureteral–gastrointestinal linkage, leading to bicarbonate loss 935
ureteric stones, removal 1872–3, 1878–9
ureteritis, obstructive 1728
ureteritis cystica 1707
ureterocystostomy, urinary tract fistula, renal transplantation 1526
ureteroperitoneal diversion, experimental nephropathy *1194*
  renal ablation models *1195*
ureteropyelography, retrograde and anterograde 81
ureteroscopy **1877–0**
  results 1879
  rigid/flexible ureteroscopes 1877–8
  techniques 1878
ureterosigmoid anastomosis, pregnancy and 1942
ureterovesical junction obstruction 2032
urethra,
  bacterial passage 1679
  catheterization 1679
    techniques 2011
  normal IVU imaging 83
  strictures, causes 2005
urethral obstruction, cystine lithiasis 748

urethral syndrome,
  defined 1685
  urinary tract infection 1685–6
urethral valves,
  posterior
    neonates 1115
    urinary tract obstruction 2032–3
    and primary hypoplasia 1606
urethritis, see urinary tract infection, cystitis and urethritis
urethrocystography,
  retrograde 81
  suprapubic 81
  voiding 80, 81
urethrosigmoidostomy, renal tubular acidosis 1860–1
uric acid stone disease **1830–1**
  epidemiology 1861
  geographical factors *1822*
  treatment 1862
  urinary risk factors *1831*
uric acid stones **1861–2**
  biochemistry 1861
  causes 1861–2
  diagnosis 1862
  differential diagnosis 1862
  stone formation in polycystic disease, autosomal dominant 2173
uricosuric drugs,
  fractional urate clearance 827
  handling in kidney disease 189
  inhibition of urate secretion 826
  pyrazinamide suppression test 826, 827
  see also allopurinol
uridine diphosphate glucose (UDP), galactosaemia 731
urinary bladder,
  antibacterial system 1671
  autonomic neuropathy in diabetes 530
  bacteria, establishment 1679–80
  carcinoma
    analgesic nephropathy and 813
    cyclophosphamide and 2320
  catheterization 1681
  children
    abnormalities 1702
    host factors predisposing to infection 1702–3
    neurogenic bladder 1702
    stones 1861
  cyclophosphamide toxicity 620
  detrusor hypertrophy 2004–5
  detrusor instability, vesicoureteric reflux and 1990
  distension, in diabetes 530
  dyssynergy 2275
  egg shell calcification 1152–3
  elderly patients, diuretics and 212
  epispadias 2035
  epithelial cells 1671
  exstrophy 2035
  Fairley technique 1712
  fibrosis 1720, 1722
  fungal ball of the bladder, candidosis 1745, 1746
  granulomas 1735
  ground-glass mucosa 1735
  neurogenic bladder 1709–10, 2033
  normal IVU imaging 83
  polymorphonuclear leucocytes 1671

urinary bladder (cont.):
  pressure, detrusor sphincter dyssynergia 1985
  residual capacity, differential diagnosis of polyuria 796
  sandy patches 1735
  sensation impairment 530
  spastic 2275
  stones, see renal stones; renal stone disease
  symptoms, multiple sclerosis 2323
  tumours
    aetiology 2266
    chromosomal rearrangements 2161
    incidence, subsequent to renal pelvis tumours 2266
    β-naphthylamine 2266
    transitional cell carcinoma recurrence 2269
  ulcers, schistosomiasis 1735
  ureteric orifice to urinary bladder, competency grading *1984*
  urinary tract infection, non-specific resistance mechanisms 1671
  voiding 1679–80
  washout 1681
    urinary tract infection 1681
    method 1664
  see also urinary tract infections
urinary bladder obstruction,
  causes 2004–5
  diagnosis, ultrasonography 91–3
  high-pressure, leading to renal failure 74
  nuclear imaging
    diuretic renography 123
    effect on renogram 123–4
  pathology, causative, ultrasonography 92
  pelviureteric junction, ultrasonography 92
  pressure/flow study, Whitaker test 71, 72
  pyelonephrosis, computed tomography scanning 105
  radiological tests 71
  radionuclide studies 125
  ultrasonography 91
  urinary bladder obstructive syndrome 83
  vs upper urinary tract dilatation, radiological tests 74
  see also urinary tract obstruction
urinary diversion,
  leading to bicarbonate loss 935
  permanent, pregnancy and 1942
  renal tubular acidosis 1860–1
  urinary tract obstruction 2018
urinary stones,
  excessive intake of animal protein 8
  pregnancy *1939*, 1941
urinary tract,
  colonization 1660
    introital, periurethral, and preputial colonization 1660
    origin of uropathogenic bacteria 1660
  congenital abnormalities **2023–41**
    investigations 2026–7
  defence mechanisms *1671*

urinary tract (cont.):
  dilatation of lower tract
    diabetes insipidus, nephrogenic 791
    pregnancy *1931*
  dilatation of upper tract
    radiological tests 74, *75*
    see also hydronephrosis
  duplication **2034–5**
  ecological system 1659–60
    anatomical aspects 1659
    composition of urine 1659–60
    functional aspects 1659
  embryology 2023–4
  fetus, ultrasonography 1995
  fistula, renal transplantation 1526
  lower, imaging of dysfunction 1990
  obstruction, see urinary tract obstruction
  rupture, pregnancy *1931*
  ureteral valves 2089
  see also ureter(s)
urinary tract infections **1659–750**
  abnormal tract 1705
  N-acetyl-β-D-glucosaminidase excretion 18
  acute renal failure, recovery period 980
  analgesic nephropathy
    causing interstitial nephritis 811
    late complications 813
  antimicrobials, handling in renal impairment 184
  children **1699–716**
    chronic pyelonephritis 1661
    diagnosis 1991
    management 1998
    radiological tests 73, 74
    vesicoureteric reflux 1991–2
    vesicoureteric reflux and 1991
  chronic renal failure 1159–60
    treatment 1184
  complicated/uncomplicated 1677
    classifications *1677*
  contraindications to renal biopsy 143
  cystitis and urethritis **1685–8**
    acute, pregnancy 1929
    candidosis 1745
    clinical features 1685
    cystitis cystica and glandularis, schistosomiasis 1735
    haemorrhagic 620
    histopathology 1685
    incidence 1685
    low-dose prophylaxis 1687
    in men, radiological tests 73–4
    microbiology 1659–72
    natural history 1685
    pathogens 1685
    postcoital treatment 1687–8
    recurrent, pregnancy 1929–30
    recurrent infections 1687
    single infections 1685, 1686
    supportive measures 1687
    treatment 1686–7
    urethral syndrome 1685–6
    vaginitis and 1686
    in women, radiological tests 73
  defences of urinary tract 1670–2
  in diabetes mellitus 530, 531–2
    acute renal failure 532
  diagnosis **1680–3**
    antibody-coated bacteria test 1682

Index

urinary tract infections (*cont.*):
  diagnosis (*cont.*):
    biochemical screening tests 1681
    bladder washout 1681
    catheter samples 1680–1
    clean-catch midstream samples 1680
    cytoscopy 1683
    gamma-camera localization 1682
    Gram-stained films 1681
    IVU 1682
    lower tract, defined 1681
    serum antibody response 1681
    single-dose infection 1682
    special investigations 1682–3
    suprapubic aspiration 1680
    ultrasound 1683
    upper tract, defined 1681
    ureteral catheterization 1681
  experimental models 1669–70
    ascending models 1669–70
    haematogenous models 1669
  haematogenous route of infection 1659
  hospital-acquired 1661
  host susceptibility factors *1670*
  host-microbe interaction 1665–9
  immune response 1671
    cellular responses 1671
    humoral responses 1671
  infected stone disease 1831
  laboratory diagnostic methods 1661–4
  localization 1664
    invasive methods 1664
    non-invasive methods 1664–5
  management 1998
    failure, defined 1678
    pregnancy 1929
  metastatic infection 1693
  microbiology and defences **1659–72**
  natural history 1661
  nephrotic syndrome 277
  non-specific resistance mechanisms 1671
    bladder 1671
    normal flora 1671
    polymorphonuclear leucocytes 1671
  paraplegia 2275–6
  pathogenesis **1678–80**
    establishment within bladder 1679–80
    periurethral colonization 1679
    route of infection 1678
    source of bacteria 1678–9
  polycystic disease, autosomal dominant 2172–3
  potential for kidney damage 1677
  pregnancy 1670, 1928–30
    management 1929
    nitrofurantoin 1684
    postpartum evaluation 1929
    relapses and reinfection 1929
    sulphonamides 1684
    trimethoprim 1684
  prevention 1672
  radiological tests **73–4**, 73–4
  recurrent
    defined 1677
    hypertension and 1693
    reinfection 1677–8
      differential diagnosis *1678*
    relapsing
      defined 1677

urinary tract infections (*cont.*):
  relapsing (*cont.*):
    persistence 1694
    relationship between adhesion and infection 1668
  renal transplantation 1579
    contraindication 1509
    prophylaxis 1533
  in sickle-cell disease 710
    treatment 716
  susceptibility factors *1670*
    adherence 1670
    catheterization 1670
    diabetes mellitus 1670
    obstruction and vesicoureteric reflux 1670
    pregnancy 1670
  trimethoprim, nephrotoxicity 1527
  upper tract **1690–4**
    acute renal failure 1692
    bacteraemia and metastatic infection 1693
    bacterial interstitial nephritis, focal or multifocal 1691
    calculi 1690
    diagnosis 1693
    hypertension 1693
    incidence 1690
    medullary susceptibility 1690
    obstruction 1690
    pathogenesis 1690
    pyelonephritis, acute 1690–1
    pyelonephritis, chronic 1691–2
    reflux nephropathy in 1692–3
    renal impairment and 1692
    treatment 1693–4
    urodynamic factors 1690
    vesicoureteric reflux 1690
    virulence factors 1690
  Ureaplasma urealyticum 1686, 1688
  women
    cystitis, recurrent 1686–8
    radiology 73
    reflux nephropathy in 1991
  see also bacterial infections; bacteriuria; prostatitis; *specific organisms*
urinary tract obstruction **2002–21**
  acute lower tract obstruction 2010–11
    clinical features 2010
    investigations 2011
    management 2011
    pathogenesis 2010
  acute upper tract obstruction 2005–10
    animal models 2005
    causes, other 2010
    clinical features 2006
    effects on renal blood flow 2005–6
    histopathological changes 2006
    imaging 2006–9
    lesions within lumen 2003
    lesions within wall 2003
    management 2009–10
    normal ureteric peristalsis 2005
    renal response 2005
  animal models 988–9
  antenatal diagnosis, postnatal evaluation 2026
  assessment 1149–50
  carcinoma 476
  causes *2003*, 2003–5

urinary tract obstruction (*cont.*):
  children 1122–3
  chronic 83
  chronic lower tract obstruction 2020–1
    clinical features 2021
    investigations 2021
    lesions within wall 2004–5
    management 2017–18, 2021
    pathogenesis 2020
    pressure from outside 2005
  chronic renal failure 1185
  chronic upper tract obstruction 2011–16
    clinical features 2014
    differential diagnosis 2017
    effects on kidney 2011
    effects on renal function **2011–14**
    investigations 2014–16
    management 2017–18
    pressure flow studies 2016
    pressure from outside 2003–4
    significant incomplete 2016
    strictures 2003
  computed tomography scanning 103–4
  defined 2031
  diagnosis 2031
  duration, recovery and 2012
  incidence 2002
  malignant obstruction 2018
  obstetric acute renal failure 1932
  patients with urinary diversion 2018
  pelvic malignancy 2003–4
  pelviureteric junction obstruction 2030, 2031–2
    imaging 2007
    management 2018
  posterior urethral valves 2032–3
  pregnancy 2004
  prostatic enlargement 476
  radiological tests *72*
  relief, and renal function 2031
  renal transplantation 1525–6
  silent 1150
  sodium and water conservation 989
  susceptibility factors 1670
  ureteroceles 2034
  ureterovesical junction obstruction 2032
  see also urinary transport disorders
urinary tract stones, *see* renal stones
urinary transport disorders **2030–4**
  see also urinary tract obstruction
urinary volume, *see* urine, volume
urinary, *see also* urine
urine,
  24-hour collection 1162–3, 1849–50
    assessment in children 1306
  acid excretion, daily *923*
  acidification 766–8, 1687
    aldosterone 767
    bicarbonate loading 772
    blood P_{CO_2} 767
    blood pH 767
    children 1714
    defect in sickle-cell disease 709
    distal renal tubular acidosis 773–4
    neonates 51
    nephrotic syndrome 772

urine (*cont.*):
  acidification (*cont.*):
    sodium transport, and transepithelial potential difference 767
  acidification test, medullary sponge kidney 2287
  acute retention, drug-induced 67
  alkalinization 766–7
    carbonic anhydrase inhibitors 206
  ammmonium excretion, daily *923*
  analysis **16–22**
    anion gap as an approximation of ammonium ion generation 44
    brush border antigens 18–19
    calcium:creatinine ratio, neonates 51
    cellular constituents
      casts 20–1
      crystals 21, 22
      erythrocytes 19–20
      leucocytes 20
      lipids 20
      main renal diseases 22
      'Maltese crosses' (cholesterol) 20
      organisms 21
      renal tubular cells 20
      squamous epithelial cells 20
      transitional epithelial cells 20
    chemical examination 16–19
    colour, odour and appearance *4*, 16
    composition, alteration 4–5
    enzymuria 18–19
    glucose in 17
    haemoglobin
      false-negatives and false-positives 17
      presence 17
    leucocytes 1662
    microscopy **19–22**
    P_{CO_2} 44
    pH changes
      elderly patients 61
      measurement 16–17
      neonates 51
      regulation 17
    physical examination 16
    potassium:sodium ratio, neonates 51
    procedures *1162*
    protein
      detection 17
      false-negatives and false-positives 17, *18*
      normal amounts 17
      protein/creatinine ratio 18
      in proteinuric diabetics 511
    red/brown coloration, causes *4*
    sediment
      and drugs 22
      main renal diseases 22
      microsopy 19–22, *22*
    specific gravity 16
    see also urine, microscopy
  bacterial count
    bacteriological methods 1663
    biochemical methods 1663
    chemical methods 1663
    children 1708–9
    quantitative methods 1662–3
    semiquantitative methods 1663
  biochemical indices
    acute renal failure *1016*

urine (cont.):
  biochemical indices (cont.):
    oliguria vs prerenal failure 972, *988*
  catheter samples 1680
  chloride excretion, daily *923*
  concentration
    ageing changes 65
    capacity in neonates 52
    circadian rhythms 65
    failure 41
    GFR and, ageing changes 65
    impairment *1156*
    water deprivation tests in young and elderly subjects 65
  conventional urine culture 1663
  creatinine clearance, *see* creatinine
  crystalluria
    pH 1834
    stone identification 1848, 1850
  dilution
    ageing kidney 65
    plasma volume regulation and 63
  examination
    monoclonal antibodies to nucleated cells *973*
    oliguria, prerenal *vs* ARF *972*
  flow, albumin excretion association 515
  frequency
    ADH secretion and 2
    bacteriuria 1680
    causes 1686
    prostatic enlargement and 2
    see also micturition disorders
  inhibitor of calcium oxalate crystallization in urine *1828*
  inhibitor of calcium phosphate crystallization in urine *1827*
  leak following renal transplantation 1575
  microscopy 1662
    children 1708
    formed elements of urinary sediment **19–22**
    immunofluorescence 19
    phase-contrast 19
    procedures 1162
  osmolality
    arginine vasopressin 1898
    differential diagnosis of polyuria 796
    distal nephron function 41–2
    early morning, and fluid deprivation tests 42
    indomethacin effect 707, 709
    maximal 706
    neonates 51
    oliguria *vs* prerenal failure *988*
    in sickle-cell disease 705–6
    indomethacin effect 707, 709
    see also renal concentrating capacity
  output
    neonates 1112, 1113
    see also oliguria; polyuria
  P_{CO_2} 766–7
    alkaline diuresis *footnote* 937
    test, contraindications 1893
  pH
    crystalluria 1834
    distal renal tubular acidosis and 773
    fractional excretion of bicarbonate 770

urine (cont.):
  pH
    indications 1850
    morning 1850
    secondary nephrocalcinosis 1896, 1897
    urate excretion 828
  red cell casts, *see* erythrocytes, cell casts
  retention
    acute 2010
    elderly patients, drug-induced 67
    see also urinary tract obstruction
  sediment
    oliguria, prerenal *vs* ARF *972*
    in SLE 649, 653
    telescoped 649, 654
  sex-associated differences 1660
  sodium
    excretion, daily *923*
    oliguria, prerenal *vs* ARF *972*, *988*
  specimen collection 30, 33
    24-hour 1162–3, 1306, 1849–50
    catheter specimens 1662
    catheterisation 30
    children 1662, 1707
    errors in pregnancy 1910
    infants 1662
    midstream specimens 1662, *1680*
    storage and transport 1662
    split urine samples, collection *1688*
  stone formation, risk factors *1830*
  suprapubic needle aspiration 1662
  testing, children 1708
  urate/uric acid oversaturation 825
  urine:plasma creatinine, oliguria *vs* prerenal failure *988*
  urine:plasma osmolality ratio 972
  urine:plasma urea
    creatinine and urate ratios 972
    oliguria *vs* prerenal failure *988*
  voiding, difficulty, diuretic-induced 212
  volume
    calcium stone disease 1833
    differentiation of prerenal oliguria from ARF *971*, 971–2
    disorders 4
    in established and recovering acute renal failure 1000–1
    in sickle-cell disease 716
  see also specific substances; urinary –
urinoma, renal transplantation 1526
urinometer, specific gravity of urine 16
urofuranic acids, decreased drug protein binding 1245
urokinase,
  IgA nephropathy, primary *358*
  in sickle-cell disease 710, 716
uroporphyrin decarboxylase inhibition, hepatic siderosis, iron overload 1282
uroporphyrinogen decarboxylase deficiency 1392
urosepsis 1661
urticarial vasculitis,
  hypocomplementaemic 587, 589
  normocomplementaemic 590

uteroplacental bed, pathology, pregnancy-induced hypertension 1960
uterus, dialysis, cancer risk 2320
UV light, *see* phototherapy
uveitis, acute interstitial nephritis 1091

V_1 renal receptors 791
V_2 renal receptors 791
vagina, normal flora 1671
vaginal carcinoma, following renal transplantation 1582
vaginitis, sexually transmitted infections 1686
valine 700
  excretion
    pregnancy and 1916
    urate excretion and 830
4-valine,8-arginine vasopressin (VDAVP), renin stimulation 793–4
valproic acid, handling in kidney disease 186
valvular heart disease, dialysis patients 1275
Van Allen syndrome with neuropathy, features and associated abnormalities *2036*
vancomycin,
  CAPD peritonitis regimens *1489*
  chronic renal failure, accumulation 1185
  drug-induced acute toxic tubular nephropathy 164
  drug-related interstitial nephritis 1089
  haemofiltration/haemodialysis 180
  neurological side-effects 2331
  perinephric abscess 1696
vanilmandelic acid, neuroblastoma 2256
varicella zoster infection, renal transplantation 1581
varicocele of acute onset, renal cell carcinoma 2241
vasa recta,
  Henle's loop 871–2
  in maximal urine osmolality development 706, 707
  in sickle-cell disease
    impaired circulation 702, 705, 713
    obliteration of 706, 713, **715**
vascular access **1405–16**
  aneurysm 1414–15
  arterial superficialization 1409–10
  arteriovenous fistula 1406–9
  blood flow measurement *1413*
  catheters 1410
  classification *1405*
  complications 1412–15
  construction 1406
  evaluation and examination 1411
  infections 1412–13
  maintenance long-term 1415–16
  names of accesses in practice 1405
  needles for puncture 1410–11
  results and outcome 1411–12
  selection of type *1406*
  steal syndrome 1413–14

vascular access (cont.):
  surgery procedures, perioperative treatment 1451–2
  timing of formation 1405–6
  types *1405*
  vascular graft 1409
vascular diseases,
  causing acute renal failure in patients with liver disease 1107
  following renal transplantation, causes *1576*, 1576–7
  necrotizing vasculitis, schematic representation 152
  renal biopsy indications 144
  renal involvement
    computed tomography scanning 105
    schematic representation 152
  with skin pathology, chronic renal failure 1393
  see also atheromatous disease; renal arterial disease; renal artery sclerosis; *specific conditions*
vascular endothelium,
  pregnancy-induced hypertension effects 1965–6
  prostacyclin and 1965
  see also endothelium
vascular grafts,
  aneurysm 1415
  complications 1412–15
  implantation 1409
  indications 1409
  patency rates 1411
  saphenous bypass, renal transplantation 1597–8
  seroma 1415–16
  varieties and characteristics 1409
  vascular access 1409
  see also arteriovenous fistula
vascular injury,
  cyclosporin, nephrotoxicity
    experimental 1562
    in man 1562
vascular permeability,
  in immune complex deposition, in leucocytoclastic vasculitis 589
  in serum sickness model 584
vascular permeability factor,
  minimal change nephrotic syndrome 309–10
  renal transplantation, nephrotic syndrome 311
vascular resistance, renal, reduced in early diabetes mellitus 512
vascular sclerosis,
  progressive reduction in nephrons 1206
  progressive renal failure 1206
vascular smooth muscle,
  cyclosporin therapy effects 1561
vascular system, peritoneal cavity 1478
vasculitis **583–636**
  acute renal failure **1060–71**
  anti-GBM antibodies 443
  antineutrophil cytoplasmic antibodies 1150
  classification 585, *586*, 613
  differential diagnosis 440, 2284
  drug-induced interstitial nephritis 1086
  drug-induced renal 167

Index

vasculitis (cont.):
  in essential mixed
    cryoglobulinaemia 690
  granulomatous *591–2*, *613*, 2324
    animal models 585
    autoantibodies in pathogenesis
      of 592
    classification *590*, *613*
    immunological abnormalities
      590
    skin lesions 2303
    vascular abnormalities *591*
    see also Churg–Strauss
      syndrome; Wegener's
      granulomatosis
  hypersensitivity, see microscopic
    polyarteritis
  leucocytoclastic, see
    leucocytoclastic vasculitis
  medium-sized vessels *613*
  mononuclear *590*
    animal models 584–5
    leucocytoclastic vasculitis
      overlap 589
  necrotizing
    in malignant hypertension 2129
    serum sickness model 583–4
    see also necrotizing
      polyarteritides
  non-steroidal anti-inflammatory
    drugs 822
  pathogenesis *583–94*
    in animals *583–5*
    autoantibodies in 592
    in man *585*
  renal disease in 2293
  renal effects 2335, 2336
  renal microscopic
    crescentic glomerulonephritis
      420–1
    treatment 428, 429
    isolated nephritis 428
  rheumatoid, see rheumatoid
    vasculitis
  schistosomiasis 1733
  small vessel 585
    deposit formation and histology
      588
    immunological abnormalities
      588
    in microscopic polyarteritis 623,
      624, 625
    Wegener's granulomatosis, see
      Wegener's granulomatosis
    see also leucocytoclastic
      vasculitis; vasculitis,
      mononuclear
  systemic *612–36*
    antibodies associated *590*, *591*
    biopsy, renal 1066
    classification *590*, **612–13**, *613*
    in essential mixed
      cryoglobulinaemia 690
    granulomatous, see vasculitis,
      granulomatous
    larger vessels 1068–9
    rapidly progressive
      glomerulonephritis-
      associated **1065–8**
    in rheumatoid arthritis 681
    skin lesions 2303
    see also necrotizing
      polyarteritides
  urticarial, see urticarial vasculitis

vasculitis (cont.):
  see also arteritis; granulomatous
    disease, vasculitis; polyarteritis;
    specific syndromes
vasoactive hormones,
  lead interactions 840
  see also arginine vasopressin;
    atrial natriuretic peptide;
    cGMP; dopamine;
    noradrenaline; prostaglandins
vasoactive intestinal peptide,
  elevated, chronic renal failure
    1280
vasoactive substances,
  effect on arteriolar myocytes of
    afferent arterioles 26
  in serum sickness model 584
vasoconstriction,
  experimental induction 159
  potassium depletion effects 902
  renal, in scleroderma 674
  thrombosis and 1044
vasodilators,
  hypertension 188, 1178
  hypertension of end-stage renal
    failure 2121
  renin secretion and *2071*
  see also hydralazine; minoxidil
vasopressin, see arginine
  vasopressin
vasopressinase,
  excessive, pregnancy-related
    diabetes insipidus 886
  pregnancy and 1917–18
VATER association, features and
  associated abnormalities *2039*
vecuronium, avoiding in kidney
  disease 186
vegetarian diets,
  in diabetic nephropathy 529
  lower creatinine clearance 29
venography, indications 115
venous hypertension, sore-thumb
  syndrome 1414
venous thrombosis,
  cyclosporin therapy, reduced
    incidence compared with
    azathioprine 1562
  deep vein thrombosis, prevalence
    in nephrotics 278
  nephrotic syndrome **278–84**
  see also renal vein thrombosis
venous tumour, see renal vein,
  tumour
ventilatory response, acidaemia
  921
verapamil,
  animal models
    effect on glomerular scarring
      1213
    prevention of acute renal failure
      1018
    renal ablation effects *1195*
  antiarrhythmics, cardiac drugs
    187
  cisplatin, antineoplastic
    treatment 2319
  cyclosporin-induced renal
    vasoconstriction 1598
  erythropoietin and 1350
  haemodialysis patients,
    recommendations *1463*
  hypertension, chronic renal
    failure 1273
  liver effects 2315

verotoxin, see Escherichia coli
vesicoureteric junction,
  defects 2003
  inflammation leading to
    vesicoureteric reflux 1984
vesicoureteric reflux **1983–99**
  animal models 1985
  antibiotics 1713
  children 1702, 1707
    hypertension in 2142
    transient 1707
  clinical presentations *1991*
  controlled trials of surgery 1995–6
    Birmingham Reflux Study
      Group 1995–6
    International Reflux Study
      Committee 1996
  defect, defined 1983
  demonstration and imaging
    1986–90
  diagnosis 1987
    associated abnormalities 1995
    grading system 1986
    loin pain in 1995
  intrarenal reflux 1983
  management
    hypertension 1998
    surgical correction 1996–7
    urinary calculi 1998
    urinary tract infections 1998
  micturating cystogram, and
    nuclear imaging 122–3
  micturating cystourethrography
    1986
  natural history 1990
  neonates
    crossed ectopia 2030
    detection 1710
    management plan *53*
  pathogenesis 1984–6
  prevalence 1983
  prevention 1714
  primary familial 1995, *1996*
  radiological tests 74
  radionuclide micturating
    cystourethrography 1985
  retrograde (direct) radionuclide
    studies 123
  schistosomiasis 1736
  secondary causes 1984
  surgery
    controlled trials 1995–6
    indications 1995
    management 1714
    techniques 1996
  technetium scan *2146*
  ultrasound 1987
  urinary tract infection 1690
    susceptibility factors 1670
  see also reflux nephropathy
vessels, see blood vessels; *specific
  vessels*
vigabatrin, accumulation 2332
villous adenoma,
  intestinal fluid loss 934
  metabolic alkalosis 947
  rectum, potassium depletion 903
  treatment 947
vimentin, podocytes 133
vincristine,
  metastatic renal carcinoma 2248
  Wilms' tumour 2260
viper, see snake bites

viral hepatitis, see hepatitis B; *etc*;
  hepatitis in chronic renal failure
viral infections,
  acute interstitial nephritis 1091
  glomerulopathies, acute renal
    failure in liver disease 1104
  haemodialysis 1439
  IgA nephritis, animal models 358
  infection-associated
    glomerulonephritis 462–6
  nephrotic syndrome 277
  pathogenesis, activation of
    coagulation system 1045
  skin diseases 2307
  uraemic syndrome 1049
virulence factors, putative,
  produced by Escherichia coli
    1669
viruses, renal cell carcinoma and
  2238–9
visceral abscess,
  crescentic glomerulonephritis 421
  infection-associated
    glomerulonephritis 470
visceral epithelium (podocytes),
  glomerular apparatus 130, 131,
  132–4
vitamin A, uraemic xerosis
  pathology 1390–1
vitamin C, megadose, urate
  excretion increase *829*, 830
vitamin D **1762–3**
  $1\alpha,25$-dihydroxyvitamin D_3, see
    $1,25$-dihydroxy vitamin D_3
  $24,25$-dihydroxyvitamin D_3 61
  actions 1762–3
  analogues, advantages 1384
  calcium transport, in uraemia
    1764
  cholecalciferol
    compliance measurement 1370
    $1,25$-dihydroxycholecalciferol,
      phosphaturias 752–7
    see also cholecalciferol
  chronic renal failure 1762–3
    inappropriate production 1318
  deficiency
    aluminium-related osteopathy
      1370
    hyperparathyroid bone disease
      1370
    renal osteodystrophy 1611
    uraemic syndrome 1245
  effects
    anticonvulsants 1763
    barbiturates 1763
  familial X-linked
    hypophosphataemic rickets 753
  hypercalciuria 1833
  hypervitaminosis D 1883, 1891
  interactions with PTH 1786
  intestinal phosphate transport
    1788
  intoxication
    calcium stone disease 1836
    clinical features 1794–5
    differential diagnosis 1381
    magnesium wasting and 1810
    risk and treatment 1794–5
  metabolic transformations 1762
  metabolism
    abnormalities, early renal
      failure 1366
    chronic renal failure 177

Index

vitamin D (*cont.*):
  metabolism (*cont.*):
    disturbance in Fanconi syndrome 726, 727
    familial X-linked hypophosphataemic rickets 753
    maleic acid-induced Fanconi syndrome 769
    urinary obstruction 2014
  metabolites
    calcitriol levels, following transplantation 1378
    hypercalcaemia in dialysis patients 1775
    phosphate reabsorption effects 1786
    sarcoidosis 1891
    treatment of rickets 1795
    use 1775
    see also individual metabolites
  oat cell carcinoma 1776
  osteitis fibrosa
    prophylaxis 1381–2
    selection of patients 1381–2
  phosphate and 1786, 1788
    reabsorption effects 752, 1786
  plasma levels 1833
  renal bone disease 1186
  resistance, end-stage chronic renal failure *1763*
  sodium transport, exit step 741
  stone formers 1833
  supplementation, hyperparathyroidism in children 1307
  transport 1763
  treatment
    hypercalcaemia 1773
    risk and 1771
  X-linked hypophosphataemic rickets 752–5
  see also calcitriol; cholecalciferol; rickets
vitamin D-binding protein, loss in nephrotic syndrome, calcium metabolism and 290
vitamin D-resistant rickets 752–3, 1795
  mouse *Hyp* model 752–3
vitamin E,
  deficiency in animals 1052
  haemolytic–uraemic syndrome 1052
    children 1055
  prostacyclin stimulating factor and 1053
vitamins, total parenteral nutrition in ARF *1021*
volume, see also fluid volume; plasma volume
volume of distribution,
  drugs 178–9
  renal failure, chronic 176–7
volume overload, see fluid overload
vomiting,
  chloride depletion
    metabolic alkalosis 945
    and proton depletion 904
  haemodialysis patients **1453–4**
  mixed metabolic disorders *957*
  unexplained hypokalaemia 787
von Gierke's disease 930
  clinical features 2210
  management 2210–11

von Gierke's disease (*cont.*):
  renal involvement 2210
von Hippel–Lindau disease, children 2169
  clear cell hypernephroma 2181
  cysts and carcinomas 102
  features and associated abnormalities *2036*
  multiple renal cysts 2181
  renal cell carcinoma in 2161–2, 2174, 2300
  skin lesions 2300
von Kossa stain, bone histology 1374
von Meyenberg's complexes, liver cyst(s) 2177
von Willebrand factor,
  congenital nephrogenic diabetes insipidus 886
  DDAVP-stimulated release 793
  nephrotic syndrome *281*, 282
  pregnancy-induced hypertension 1963–4
  ristocetin hyperaggregability 283
  storage sites, desmopressin and 1363
  see also Factor VIII
von Willebrand's disease 710
  cryoprecipitate 1362
vulval carcinoma, following renal transplantation 1582

Waldenstrom's macroglobulinaemia,
  acute renal failure 2343
  hyperviscosity syndrome associated 568
Waldeyer's sheath, hypertrophy in pregnancy 1909
warfarin,
  animal renal ablation models, effects *1195*, *1216*
  embryopathy, features and associated abnormalities *2040*
  handling in kidney disease 190
  nephrotic syndrome 284
  renal effects 2335
warts, skin changes 1541
wasp stings, see insect stings
wasting syndrome, management 1284
water,
  haemodialysis, for
    preparation 1422–3
    quality standards 1422
  negative free, generation, in sickle-cell disease 707–8
  total body, proportion in elderly patients 65
water balance,
  acute renal failure 1019
  ADH 869–72
  ageing changes 64–5
  antidiuretic hormone 869–72
  extra-intracellular destinations of water 262
  free water clearance C_{H_2O} 871–2
  water excretion, defined 791
  handling, assessment 874
  psychogenic polydipsia 874
  regulation 869–72
  renal response to load 874
  thirst
    disorders 874, 887

water balance (*cont.*):
  thirst (*cont.*):
    regulation, osmotic/non-osmotic 869
    thirst sensation, ageing changes 65
  see also dehydration; fluid retention
water clearance, chronic renal failure 1182
water depletion,
  acute renal failure, recovery 980
  indomethacin effect on urinary osmolality 707, 709
water deprivation test 65
  children, contraindications 885
  diabetes insipidus 885
water diuresis,
  calculation, confirmation 884
  differential diagnosis 884
  extracellular volume expansion effects, fractional urate clearance 828
  see also osmotic diuresis
water drinking,
  cerebral dysfunction and 63
  compulsive 42
  hypodipsia, adipsia 887
  osmotic/non-osmotic regulation 869
  psychogenic polydipsia 874
  thirst sensation, ageing changes 65
water excretion,
  blood pressure control and 2058–60
  diabetes insipidus, nephrogenic, DDAVP infusion *793*
  nephrotic syndrome 269
  normal vs impaired *1182*
  osmolar clearance 791
  osmotic and nonosmotic states 791
  reduction, drugs 796
water handling,
  hypercalcaemia 1885
  neonates 1113
water hardness, calcium stone disease 1837
water intake, renal stone disease 1849
water intoxication, chronic renal failure 1182
water loading, indomethacin effect on urinary osmolality 707, 709
water retention, non-steroidal anti-inflammatory drugs 823
water therapy, renal stone disease 1852
water transport,
  antidiuretic hormone control, normal function 790–1
  in nephron 198
Wegener's granulomatosis **613–23**
  antineutrophil cytoplasmic antibodies (ANCA) 239–40, 592, 617–18, 621
  biopsy, lung 1063
  Churg–Strauss syndrome overlap 613, 618
  clinical features 592, **614–17**
    disease severity 617
    presenting symptoms 592
    small vessel vasculitis 616–17
  crescentic glomerulonephritis 429
  cyclophosphamide 1066

Wegener's granulomatosis (*cont.*):
  definition **613**
  diagnosis 592, **617–18**, 1062, 1066, 1150
    haematology/biochemistry 617
    histopathology 618
    immunological 592, 617–18
    radiology 615, 618
  differential diagnosis 614, 618, 623
  end-stage renal failure 1233
    proportional distribution by age at commencement *1229*
  epidemiology **614**
  historical aspects 612
  limited 617
  neurological involvement 2324
  outcome **621–3**, *622*
    long-term 622–3
    predictive factors 622
    renal function 622–3
    response to initial therapy 621–2
    untreated disease 621
  pathogenesis 592
  pregnancy 1943
  relapse 623
  renal disease 616
    glomerulonephritis in 592, 613, 616
    vascular infiltrate 591, 616
    vasculitis 616, 622
  renal transplantation, recurrence following 1574
  respiratory tract involvement 613, 614–15
  rheumatoid factor activity 592, 617
  treatment **619–21**, *619*
    historical aspects 619
    of limited disease 620
    maintenance *619*, 620, 621
    monitoring 621
    NIH regimen 619
    reducing drug toxicity 620–1
    of severe disease *619*, 620
weight, body composition, and renal function, children 1296–7
Weil's disease, infection-associated glomerulonephritis 461
Wernicke's syndrome, elderly patients on haemodialysis 1623
Wernicke–Korsakoff's syndrome, hypomagnesaemia and 1812
West Nile virus, Balkan nephropathy and 859
wheal and flare reaction 596
Whitaker test,
  pressure flow studies 2027
  pressure/flow study, for obstruction 71, 72
white cells, normal values *1344*
  see also leucocytes
white piedra, fungal infections 1747
whole blood activated partial thromboplastin time, target in haemodialysis 1433
whole kidney transit time index (WKTTI) 117
Wiedemann–Beckwith syndrome, features and associated abnormalities *2039*
Williams' syndrome 1891, *1897*, 1897
  features and associated abnormalities *2039*

Index

Wilms' tumour **2253–64**
 3-year survival *2260*
 adults 2257
 age of diagnosis *2254*
 associated conditions 2254–5
 bilateral tumours 2257, 2261–2
 bone-metastasizing renal tumour of childhood 2258
 chromosomal rearrangements 2161
 clinical features 2256–7
 defined 2253
 diagnosis, computed tomography scanning 103
 differential diagnosis
  intralobar nephroblastomatosis 2258
  teratoma 2258
 embryology 2253–4
 epidemiology 2254
 follow-up and screening 2262–3, *2263*
 genetics 2254, 2255–6
 histology 2258
 hypertension and 2093
 International Society of Paediatric Oncology studies, summary *2261*
 investigations 2257
 malignant rhabdoid tumour 2258
 North American studies, summaries *2261*, *2262*
 pathology 2257
 pre-existing disease 2262
 prognostic features 2257–9, 2263
 relapse, patterns and treatment 2263
 single kidneys 2257, 2262
 staging *2259*, 2259
 'teratoid' 2258
 treatment
  chemotherapy 2260–1
  radiotherapy 2261
  relapse 2263
  surgery 2259–60
 variants and other tumours 2258
 see also Drash syndrome

Wilson's disease 1897
 chromosomal locus *1796*
 Fanconi syndrome 729, 733
 liver cirrhosis 2311
 nephrocalcinosis *1897*
 urate/uric acid, increased clearance 830
Winter syndrome, features and associated abnormalities *2037*
Wistar Kyoto rats,
 models of hypertension 2060
 progressive age-related glomerular lesions 1206–7
 renal ablation studies 1193
 see also Munich–Wistar rats
World Health Organization (WHO), classification of lupus nephritis 653–4, *654*, 655
WR-2721 (ethiofos),
 effect on renal scarring *1205*
 hyperparathyroid hyperplasia, treatment 1385
 parathormone secretion 1777
Wright's stain, leucocytes 20
Wuchereria bancroftia, filarial glomerulopathy 495–6
Wunderlich triad, tumour rupture 2243

X chromosome,
 location of common disease loci 2209
 lyonization 754
 markers, DXS loci, Alport's syndrome 2160–1
X-ANA autoantibodies 637
X-linked disorders *2038*
 Alport's syndrome 2201–2
 congenital nephrogenic diabetes insipidus 886
 diabetes insipidus, nephrogenic 791–6
 familial X-linked hypophosphataemic rickets 753–5

X-linked disorders (*cont.*):
 glycosuria of Fanconi syndrome 729
 hypophosphataemic rickets *1897*
 Lowe syndrome 732
 mesangiocapillary glomerulonephritis 395
 nephrocalcinosis and 1891, *1897*
 skin and kidney 2299
 X-linked hypophosphataemic rickets, and deafness 754
X-ray exposure in radiology 81–2
X-specific fimbriae 1669
xanthine,
 nephropathy 2319
 toxicity, uraemic syndrome 1244
 xanthinuria, incidence of xanthine stones 1866
xanthine oxidase, allopurinol inhibition of 834
xanthine oxidase deficiency 1865–6
xanthine stone disease 1842
xanthine stones **1865–6**
xanthogranulomatous pyelonephritis 1694
 children 1706
 computed tomography scanning 103, 105
 diagnosis 2244
 hepatic dyfunction, reversible 2242
 pathology 2244
xerosis, chronic renal failure 1390–1
xerostomia, in Sjögren's syndrome 694
Xid gene 640
xylitol,
 nephrotoxicity 1009
 peritoneal dialysis solution 1481
xylose, excretion, pregnancy and 1916
D-xylose, renal failure, chronic, drug malabsorption 176

yellow fever *1045*
Yersinia arthritis, IgA nephropathy 361
Yersinia infection,
 acute interstitial nephritis 1091
 glomerulonephritis and 462, 2310
yohimbine, impotence, erectile, therapy 1335

Zellweger's cerebrohepatorenal syndrome,
 children 2169
 features and associated abnormalities *2038*
zidovudine, handling in renal impairment 184
zinc,
 anaemia and 1351
 Balkan nephropathy and 858
 deficiency
  testosterone and 1333
  uraemic peripheral polyneuropathy 1398
  uraemic syndrome 1245
 impotence, erectile, therapy 1335
 lymphocyte function in nephrotics 278
zinc sulphate, Wilson's disease 733
Zollinger–Ellison syndrome, gastric alkalosis 946
zomepirac, minimal change nephrotic syndrome 1089
Zucker rats 520
 accelerated glomerular lesions 1207
 hyperlipidaemia in progressive chronic renal failure 1201–2
zygomycosis, defined 1747
zymogens,
 nephrotic syndrome 281–2
 see also coagulation factors